Encyclopedia of REPRODUCTION

Volume 1   A–En

Editorial Board

Editors-in-Chief

Ernst Knobil

Jimmy D. Neill

Associate Editors

Eli Y. Adashi
Department of Obstetrics and Gynecology
University of Utah
Salt Lake City, Utah

Fuller W. Bazer
Albert B. Alkek Institute of Biosciences and Technology
Texas A&M University
Houston & College Station, Texas

Ian P. Callard
Department of Biology
Boston University
Boston, Massachusetts

Kenneth G. Davey
Department of Biology
York University
North York, Ontario, Canada

Claude Desjardins
College of Medicine
University of Illinois at Chicago
Chicago, Illinois

Marc E. Freeman
Department of Biological Science
Florida State University
Tallahassee, Florida

Michael J. K. Harper
Conrad Program
Arlington, Virginia

Paul Licht
College of Letters and Science
University of California, Berkeley
Berkeley, California

John S. Pearse
Institute of Marine Sciences
University of California, Santa Cruz
Santa Cruz, California

Donald W. Pfaff
Rockefeller University
New York, New York

Mary Lake Polan
Department of Gynecology and Obstetrics
Stanford University School of Medicine
Stanford, California

Jerome F. Strauss, III
Department of Obstetrics and Gynecology
University of Pennsylvania
Philadelphia, Pennsylvania

Peter Thomas
University of Texas Marine Science Institute
Port Aransas, Texas

H. Allen Tucker
Department of Animal Science
Michigan State University
East Lansing, Michigan

John C. Wingfield
Department of Zoology
University of Washington
Seattle, Washington

Encyclopedia of REPRODUCTION

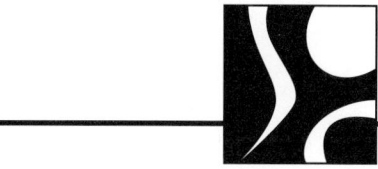

Volume 1   A–En

Editors-in-Chief

Ernst Knobil

The H. Wayne Hightower Professor in the Medical Sciences
and
Ashbel Smith Professor,
University of Texas Health Sciences Center
Houston, Texas

Jimmy D. Neill

Distinguished Professor
University of Alabama at Birmingham

ACADEMIC PRESS

San Diego   London   Boston   New York   Sydney   Tokyo   Toronto

This book is printed on acid-free paper.

Copyright © 1998 by ACADEMIC PRESS

All Rights Reserved.
No part of this publication may be reproduced or transmitted in any form or by any means, electronic or mechanical, including photocopy, recording, or any information storage and retrieval system, without permission in writing from the publisher.

Academic Press
a division of Harcourt Brace & Company
525 B Street, Suite 1900, San Diego, California 92101-4495, USA
http://www.apnet.com

Academic Press Limited
24-28 Oval Road, London NW1 7DX, UK
http://www.hbuk.co.uk/ap/

Library of Congress Catalog Card Number: 98-84463

International Standard Book Number: 0-12-227020-7 (set)
International Standard Book Number: 0-12-227021-5 (vol. 1)
International Standard Book Number: 0-12-227022-3 (vol. 2)
International Standard Book Number: 0-12-227023-1 (vol. 3)
International Standard Book Number: 0-12-227024-X (vol. 4)

PRINTED IN THE UNITED STATES OF AMERICA
98 99 00 01 02 03 MM 9 8 7 6 5 4 3 2 1

Contents

Contents by Subject Area	xxiii
Preface	xxxiii
Guide to the Encyclopedia	xxxv

A

Abortion 1
　Steven J. Sondheimer

Acanthocephala 6
　D. W. T. Crompton

Acrosome Reaction 17
　Gregory S. Kopf

Activin and Activin Receptors 26
　Ralph H. Schwall

Adrenal Androgens 35
　Collin B. Smikle

Adrenal Hyperplasia, Congenital Virilizing 43
　Peter A. Lee, Selma F. Witchel

Adrenarche 51
　Frank Gonzalez

Aedes aegypti 61
　Alexander S. Raikhel, Thomas W. Sappington

Aggressive Behavior 77
　David A. Edwards

Agnatha 83
　Stacia A. Sower, Aubrey Gorbman

Allantochorion (Chorioallantois) 91
　Fuller W. Bazer

Allantoic Fluid 93
　Fuller W. Bazer

Allantois 95
　Fuller W. Bazer

Allatostatins 97
　Stephen S. Tobe

Altitude, Effects on Humans 107
　Lorna G. Moore

Altricial and Precocial Development in Birds 113
　F. M. Anne McNabb

Altruism in Insect Reproduction 118
　Laurence Packer

Altruistic Behavior, Vertebrates 129
　R. Haven Wiley

Amenorrhea 135
　Sarah Berga

Amniocentesis 144
　Nancy C. Rose, Tzazil Ayala

Amniotic Fluid 149
　Robert A. Brace

Amphibian Ovarian Cycles 154
　Alberta M. Polzonetti-Magni

Amphibian Reproduction, Overview 161
　Marvalee H. Wake

Androgen Inhibitors/Antiandrogens 166
　Vivian L. Fuh, Elizabeth Stoner

Androgen Insensitivity Syndromes 174
　James Aiman

Androgens Bernard Robaire	180	Avian Reproduction, Overview Tony D. Williams	325
Androgens, Effects in Birds Cheryl F. Harding	188	Avian Reproductive System, Developmental Endocrinology James E. Woods, Robert C. Thommes	337
Androgens, Effects in Mammals Shalender Bhasin	197		
Androgens, Subavian Species Kyle W. Selcer, Jeffrey W. Clemens	207	**B**	
Andrology: Origins and Scope Philip Troen	214	Benign Prostatic Hyperplasia (BPH) Kevin T. McVary, John T. Grayhack	349
Annelida Damhnait McHugh	219	Birds, Diversity of Scott V. Edwards	358
Anterior Pituitary Robert B. Page	224	Blastocyst Richard M. Schultz	370
Antiestrogens Donald P. McDonnell	229	Blood–Testis Barrier Brian P. Setchell	375
Antiprogestins Irving M. Spitz	238	Brachiopoda Stephen A. Stricker	382
Apgar Score Joseph Dancis, Karen D. Hendricks-Muñoz	248	Breast Cancer Laura Esserman, Hope Wallace	389
Aphids Jim Hardie	251	Breast Disorders Stefanie S. Jeffrey, Diana O. Cua	402
Aplysia Stephen Arch	255	Breastfeeding Carla D. Harris, Eloise D. Clawson	411
Apoptosis (Cell Death) Paul F. Terranova, Christopher C. Taylor	261	Breeding Strategies for Domestic Animals George R. Foxcroft	419
Armadillo Richard D. Peppler	274	Brood Parasitism in Birds Alfred M. Dufty, Jr.	425
Aromatization Alan J. Conley, Karen W. Walters	280	Bruce Effect Peter Brennan	433
Artificial Insemination, in Animals William L. Flowers	291	Bryozoa (Ectoprocta) Robert Woollacott	439
Artificial Insemination, in Humans Douglas T. Carrell, Deborah Cartmill	302	**C**	
Asexual Reproduction Kerstin Wasson	311	*Caenorhabditis elegans* Dave Pilgrim	449
Autonomic Nervous System and Reproduction Karen J. Berkley	320	Captive Breeding of Wildlife Barbara S. Durrant	458

Cardiovascular Adaptation to Pregnancy Margaret K. McLaughlin, Robin F. Gandley	467	Circadian Rhythms Fred W. Turek	614
Castration, Effects in Humans (Male) Harald H. J. Hoekstra, Mels F. van Driel, Pax H. B. Willemse	475	Circannual Rhythms Irving Zucker, Brian J. Prendergast	620
		Circumcision M. Sean Esplin	627
Castration, Effects in Nonhuman Mammals (Female) Sandra J. Legan	480	Clitoris Ursula Kuhnle	631
		Cloning Mammals by Nuclear Transfer Randall S. Prather	638
Castration, Effects in Nonhuman Mammals (Male) Graeme B. Martin, David R. Lindsay	486	Cnidaria Daphne Gail Fautin	645
Cats David E. Wildt, Janine L. Brown, William F. Swanson	497	Conjugation in Ciliates Peter J. Bruns	653
Cattle Roy Fogwell	510	Contraceptive Methods and Devices, Female Malcolm Potts, Diana S. Wolfe	661
CBG (Corticosteroid-Binding Globulin) Geoffrey L. Hammond	526	Contraceptive Methods and Devices, Male William J. Bremner, M. Cristina Meriggiola	667
Cephalochordata M. Dale Stokes	529		
Cervical Cancer James A. Roberts, B. Hannah Ortiz	536	Copulation, Mammals Robert L. Meisel	675
Cervix Kamran S. Moghissi	546	Corpora Lutea of Nonmammalian Species Giovanni Chieffi, Gabriella Chieffi Baccari	680
Cesarean Delivery Linda J. Heffner	553		
Chaetognatha George L. Shinn	559	Corpus Allatum Stephen S. Tobe	688
Chelicerate Arthropods W. Reuben Kaufman	564	Corpus Cardiacum, Insects Barry G. Loughton	698
Chickens, Control of Reproduction in Peter J. Sharp	572	Corpus Luteum (CL) Gordon D. Niswender, Jennifer J. Juengel, Eric W. McIntush	703
Choriocarcinoma Anthony C. Evans, Jr.	581		
Chorionic Gonadotropin, Human Robert E. Canfield, Stephen Birken, John O'Connor, Leslie Lobel	587	Corpus Luteum of Pregnancy Richard L. Stouffer	709
		Corpus Luteum Peptides O. David Sherwood, Phillip A. Fields	718
Chorionic Gonadotropins, Nonhuman Mammals Marylynn Barkley, Mary B. Zelinski-Wooten	601	Cotyledonary Placenta Stephen P. Ford	730

Cricetidae (Hamsters and Lemmings)	739
Bruce D. Goldman, David A. Freeman	
Critical Period, Estrous Cycle	748
Lewis C. Krey	
Crustacea	755
Hans Laufer, Matthew Landau	
Cryopreservation of Embryos	765
C. Matthew Peterson, Akiyasu Mizukami, Douglas T. Carrell	
Cryopreservation of Sperm	773
Rupert P. Amann	
Cryptorchidism	784
George W. Kaplan, Irene M. McAleer	
Ctenophora	792
George I. Matsumoto	
Cycliophora	800
Peter Funch, Reinhardt Møbjerg Kristensen	
Cytokines	809
Sarah A. Robertson	

D

Decidua	823
Linda C. Giudice, Juan C. Irwin	
Deciduoma	836
Geula Gibori, Yan Gu	
Deer	842
Edward D. Plotka	
DHEA (Dehydroepiandrosterone)	858
John E. Nestler, Nancy Pahle	
Diapause	863
David L. Denlinger, Seiji Tanaka	
Dihydrotestosterone	872
Shutsung Liao, Richard A. Hiipakka	
Diploptera punctata	880
Barbara Stay	
Discoidal Placenta	890
John J. Rasweiler, IV, Nilima K. Badwaik	
Dogs	902
Cheryl S. Asa	
Dorsal Bodies in Mollusca	910
A. Saber M. Saleuddin	
Drosophila	917
Marla B. Sokolowski, John Ewer	

E

Ecdysiotropins and Ecdysiostatins	929
Henry H. Hagedorn	
Ecdysteroids	933
Henry H. Hagedorn	
Echinodermata	940
Maria Byrne	
Ectopic Pregnancy	955
Christos Coutifaris	
Egg, Avian	963
Carol M. Vleck	
Egg Coverings, Insects	971
Michael P. Kambysellis, Lukas Margaritis, Elysse M. Craddock	
Eicosanoids	991
William J. Silvia	
Ejaculation	1002
Kevin E. McKenna	
Elasmobranch Reproduction	1009
Thomas J. Koob	
Elephants	1018
Keith Hodges, Cheryl Niemuller, Janine Brown	
Embryogenesis, Mammalian	1029
Carol A. Burdsal	
Embryo Transfer	1037
George E. Seidel, Jr.	
Endogenous Opioids	1043
John A. Russell, C. H. Brown, R. W. Caron	

Endometriosis 1061
Camran Nezhat, Farr Nezhat, Ceana Nezhat

Endometrium 1067
Linda C. Giudice

Endotheliochorial Placentation 1078
Vibeke Dantzer

Energetics of Reproduction 1085
Cynthia Carey

Energy Balance, Effects on Reproduction 1091
George N. Wade

Environmental Estrogens 1100
Stephen H. Safe

Contents of Other Volumes

VOLUME 2

Contents by Subject Area
Preface
Guide to the Encyclopedia

E

Epididymis
Trevor G. Cooper

Epitheliochorial Placentation
Vibeke Dantzer

Equine Chorionic Gonadotropin
Janet F. Roser

Erection
George J. Christ

Erythroblastosis Fetalis
Donald J. Dudley, G. Marc Jackson

Estrogen Action, Behavior
Lynda Uphouse, Sharmin Maswood

Estrogen Action, Bone
Robert Marcus

Estrogen Action, Breast
Serdar E. Bulun, Evan R. Simpson

Estrogen Action on the Female Reproductive Tract
Kenneth S. Korach, Jonathan Lindzey

Estrogen Effects and Receptors, Subavian Species
Marina Paolucci, Noemi Custodia, Ian P. Callard

Estrogen, Effects in Birds
Barney A. Schlinger, Colin J. Saldanha

Estrogen Replacement Therapy
Rogerio A. Lobo

Estrogen Secretion, Regulation of
Koji Yoshinaga

Estrogens, Overview
Carolyn L. Smith

Estrous Cycle
Neena B. Schwartz, Signe M. Kilen

Estrus
Janice E. Thornton, Patricia D. Finn

Eunuchoidism
Victor Y. Fujimoto, Michael R. Soules

F

Fallopian Tube
Michael P. Diamond, Diaa M. El-Mowafi

Family Planning
Allan Rosenfield, Victoria L. Dunning

Female Reproductive Disorders, Overview
Robert W. Rebar

Female Reproductive System, Amphibians
Rakesh K. Rastogi, Luisa Iela

Female Reproductive System, Birds
Patricia Johnson

Female Reproductive System, Fish
Martin A. Connaughton, Katsumi Aida

Female Reproductive System, Humans
Bruce R. Carr

Female Reproductive System, Insects
Erwin Huebner

Female Reproductive System, Nonhuman Mammals
Robert A. Dailey

Female Reproductive System, Reptiles
Valentine A. Lance

Female Sterilization
Richard M. Soderstrom

Fertility and Fecundity
Philip J. Dziuk

Fertilization
Gerald Schatten

Fetal Adrenals
Robert B. Jaffe, Samuel Mesiano

Fetal Alcohol Syndrome
Nancy C. Rose, David M. Stamilio

Fetal Anomalies
E. Albert Reece, Arnon Wiznitzer

Fetal Diagnosis, Invasive
Carl P. Weiner, Celeste Sheppard

Fetal Growth and Development
William W. Hay, Jr.

Fetal Hormones
Theresa M. Siler-Khodr

Fetal Lung Development
Philip L. Ballard, Susan Guttentag

Fetal Membranes
Jerome F. Strauss, III, Erdal Budak

Fetal Monitoring and Testing
Iraj Forouzan

Fetal–Placental Unit
Bruce R. Carr

Fetal Surgery
N. Scott Adzick, Theresa M. Quinn

α-Fetoprotein and Triple Screening
Jacob A. Canick

Fetus, Overview
Timothy A. Cudd

Fish, Modes of Reproduction in
Rudolf Reinboth

Follicular Atresia
J. Yeh, G. D. Chen, R. H. Oliver

Follicular Development
Bradley J. Van Voorhis

Follicular Steroidogenesis
Bradley J. Van Voorhis

Follistatin
David M. Robertson, Christopher Gilfillan

Freemartin
Sherrill E. Echternkamp

FSH (Follicle-Stimulating Hormone)
Leo E. Reichert, Jr.

G

Galactorrhea
Howard A. Zacur

Gametes, Overview
James M. Robl, Rafael A. Fissore

Gene Transfer, Sperm-Mediated
Jorge A. Piedrahita, Jagdeece J. Ramsoondar

Genitalia
Mary Min-chin Lee

Germ Layers
Jonathan J. H. Pearce

Global Zones and Reproduction
Franklin Bronson

Gnathostomulida
Wolfgang Sterrer

GnRH (Gonadotropin-Releasing Hormone)
P. Michael Conn, L. Jennes, J. A. Janovick

GnRH Pulse Generator
Jon E. Levine

Gonadogenesis, Female
Michael K. Skinner, Jeffrey A. Parrott

Gonadogenesis, Male
Mary Min-chin Lee, Jose Teixeira

Gonadotropes
Gwen V. Childs

Gonadotropin Biosynthesis
Raymond Counis

Gonadotropin Receptors
David Puett

Gonadotropin Secretion, Control of
Charles A. Blake

Gonadotropins, Overview
M. Ram Sairam

Graafian Follicle
J. Yeh, G. D. Chen, R. H. Oliver

Granulosa Cells
Kenneth H. H. Wong, Eli Y. Adashi

Growth Factors
Asgerally T. Fazleabas, J. Julie Kim

Guinea Pig, Female
Reinhold J. Hutz, Amanda L. Trewin

Gynecomastia
Samuel Smith

H

Hemichordata
Gary M. King

Hemochorial Placentation
Jerome F. Strauss, III, Erdal Budak

Hemocoelic Insemination
Kenneth G. Davey

Hermaphroditism
Kerstin Wasson

Hirsutism
William J. Butler

HIV Infection and AIDS
Penelope J. Hitchcock,
Kearston Schmidt

Homosexuality
Vivienne Cass

Hormonal Contraception
Malcolm Potts, Claire Norris

Hormonal Control of the Reproductive Tract, Subavian Vertebrates
Ian P. Callard, Vicki Abrams-Motz,
Georgia Giannoukos, Lisa A. Sorbera

Hormone Receptors, Overview
David Puett, Adviye Ergul

Hormones and Reproductive Behaviors, Fish
Harold H. Zakon, Kent D. Dunlap

Hormones of Pregnancy
Glen E. Hofmann

Horses
Dan C. Sharp

Human Placental Lactogen (Human Chorionic Somatomammotropin)
Michael Freemark

Hybridization
Michael L. Arnold

Hydra
Vicki J. Martin

Hyenas
Stephen E. Glickman,
Christine M. Drea,
Elizabeth M. Coscia

Hyperprolactinemia
Howard A. Zacur

Hypogonadism
Andrew J. Friedman

Hypophysectomy
Donald C. Johnson

Hypopituitarism
William W. Hurd

Hypospadias
Steven G. Docimo, Ranjiv Mathews

Hypothalamic–Hypophysial Complex (Pituitary Portal System)
Robert B. Page

Hypoxia, Effect on Reproduction
Charles A. Ducsay

Hysterectomy
Howard T. Sharp

I

IGF (Insulin-like Growth Factor)
James M. Hammond

Immunocytochemistry
William C. Okulicz

Immunology of Reproduction
Joan S. Hunt, Peter M. Johnson

Implantation
Daniel D. Carson

Impotence
Irwin Goldstein, Lawrence S. Hakim,
Ajay Nehra

Infections in Pregnancy
Jack Ludmir

Infertility
Lawrence C. Udoff, Eli Y. Adashi

Inhibin
Ralph H. Schwall

Insect Accessory Glands
Cedric Gillott

Insect Reproduction, Overview
Kenneth G. Davey

Interferons
Troy L. Ott

Internal Fertilization in Birds and Mammals
Sally D. Perreault, John D. Kirby

Interrenal Gland, Stress Response and Reproduction
Wilfrid Hanke

Intersexuality in Mammals
R. H. F. Hunter

Intrauterine Growth Restriction and Mechanisms of Fetal Growth
Victor K. M. Han

Intrauterine Position Phenomenon
Frederick S. vom Saal,
Mertice M. Clark,
Bennett G. Galef, Jr.,
Lee C. Drickamer,
John G. Vandenbergh

In Vitro **Fertilization**
Lewis C. Krey, Alan S. Berkeley

J

Juvenile Hormone
Gerard R. Wyatt

K

Kallmann's Syndrome
Lisa M. Halvorson

Kamptozoa (Entoprocta)
Kerstin Wasson

Kinorhyncha
Birger Neuhaus

Klinefelter's Syndrome
Fady I. Sharara

L

Labor and Delivery, Human
Peter W. Nathanielsz,
Gordon C. S. Smith

Lactational Amenorrhea
Amy Banulis, William D. Schlaff

Lactational Anestrus
Jeffrey S. Stevenson

Lactation, Human
Margaret C. Neville

Lactation, Nonhuman
Robert J. Collier

Lactogenesis
R. Michael Akers

Lactotrophs
Tom E. Porter

Leiomyoma
Linda C. Giudice, Salli Tazuke

Leukemia Inhibitory Factor
Colin L. Stewart

Leydig and Sertoli Cells, Nonmammalian
Jeffrey Pudney

Leydig Cells
Matthew P. Hardy,
Benson T. Akingbemi, Ren-Shan Ge

LH (Luteinizing Hormone)
George R. Bousfield

Local Control Systems in Reproduction
John A. McCracken

Locusts
Cedric Gillott

Lordosis
Donald W. Pfaff

Luteinization
Anthony J. Zeleznik

Luteolysis
John A. McCracken

Luteotropic Hormones
Gilbert S. Greenwald

Lymphokines
Wenbin Tuo, Fuller W. Bazer,
Wendy C. Brown

VOLUME 3

Contents by Subject Area
Preface
Guide to the Encyclopedia

Magnocellular System
R. John Bicknell

Male Reproductive Disorders
Ivan Damjanov

Male Reproductive System, Amphibians
Riccardo Pierantoni

Male Reproductive System, Birds
Timothy R. Birkhead

Male Reproductive System, Fish
Florence Le Gac, Maurice Loir

Male Reproductive System, Human
John F. Redman

Male Reproductive System, Insects
Cedric Gillott

Male Reproductive System, Nonhuman Mammals
Larry Johnson, Gabrielle U. Falk,
Genevieve E. Spoede

Male Reproductive System, Reptiles
Daniel H. Gist

Mammary Gland Development
Jose Russo, Irma H. Russo

Mammary Gland, Overview
Isabel A. Forsyth

Marine Invertebrate Larvae
Craig M. Young

Marine Invertebrates, Modes of Reproduction in
Jan A. Pechenik

Marsupials
Marilyn B. Renfree, Geoffrey Shaw

Mate Choice, Overview
Patricia Adair Gowaty

Mating Behaviors, Insects
William H. Cade

Mating Behaviors, Mammals
Michael J. Baum

Mating Behaviors, Invertebrates Other Than Insects
Janet Leonard

Median Eminence
Ann-Judith Silverman

Meiosis
V. Polanski, Jacek Z. Kubiak

Meiotic Cell Cycle, Oocytes
Nava Dekel

Menarche
Harry Hatasaka

Menopause
Lawrence C. Udoff, Eli Y. Adashi

Menstrual Cycle
Sarah L. Berga

Menstrual Disorders
Isaac Schiff, Shimon Segal

Menstruation
Linda R. Nelson

Metamorphosis, Insects
Fred Nijhout

Microtinae (Voles)
Theresa M. Lee

Migration, Amphibians
Raymond D. Semlitsch,
Travis J. Ryan

Migration, Birds
Peter Berthold

Migration, Fish
Graham Young

Migration, Insects
Kenneth Wilson

Migration, Reptiles
David W. Owens

Milk, Composition and Synthesis
Harold M. Farrell, Jr.

Milk Ejection
Jonathan B. Wakerley

Mollusca
A. Saber M. Saleuddin

Molt and Nuptial Color
Christopher W. Thompson

Monotremes
Mervyn Griffiths

**Morning Sickness and
Hyperemesis Gravidarum**
Marc Jackson, Marcelo F. Noguera

Myriapoda
Richard L. Hoffman

N

Naked Mole-Rats
Christopher G. Faulkes

Nematodes and Related Phyla
Denis J. Wright

Nematomorpha
Andreas Schmidt-Rhaesa

Nemertea
James M. Turbeville

Nervus Terminalis
Marlene Schwanzel-Fukuda

Nesting, Birds
Joanna Burger

Neuroendocrine Systems
George Fink

Neurohypophysial Hormones
Hans H. Zingg

Neuropeptides
Andrés F. Negro-Vilar, Brian J. Arey,
Francisco J. López

Neurosecretion
Harold Gainer

Neurotransmitters
William F. Ganong

Nutritional Factors and Reproduction
Gary L. Williams

Nutritional Factors and Lactation
Michael J. VandeHaar

O

Obstetric Anesthesia
Brett B. Gutsche, David L. Hepner

Olfaction and Reproduction
Dietrich L. Meyer, Rakesh K. Rastogi

Onychophora
Michael T. Ghiselin

Oocyte and Embryo Transport
Horacio B. Croxatto,
Manuel Villalón, Luis Velasquez

Oocyte, Mammalian
Anne Byskov, Maria Strömstedt

**Oocyte Maturation and Spawning
in Starfish**
Takeo Kishimoto

Oogenesis, in Mammals
Roger G. Gosden, Helen M. Picton

Oogenesis, in Nonmammalian Vertebrates
Charles A. Lessman

Oostatins, Folliculostatins, and Antigonadotropins, Insects
Terry S. Adams

Opossums
John D. Harder, Leslie M. Jackson

Orchitis
Wolfgang Weidner, W. Krause

Orgasm
Kevin E. McKenna

Orthonectida
John S. Pearse

Osteoporosis
Claude D. Arnaud, E. Bruce Roe

Ovarian Cancer
Mark K. Dodson, Jason L. Johnson

Ovarian Cycle, Mammals
Michel Ferin

Ovarian Cycle, Teleost Fish
Izhar A. Khan, Peter Thomas

Ovarian Cycles and Follicle Development in Birds
Alan L. Johnson

Ovarian Function in the Perimenopause
Elizabeth B. Connell

Ovarian Hormones, Overview
Shao-Yao Ying, Zhong Zhang

Ovarian Innervation
Sergio Ojeda, Gregory A. Dissen

Ovary, Overview
Janice Bahr, Humphrey H. C. Yao

Oviposition in Molluscs
Jeffrey L. Ram

Ovulation
Lawrence L. Espey

Ovulation and Oviposition, Insects
Marc J. Klowden

Oxytocics
Laird Wilson, Jr., Ramkrishna Mehendale

Oxytocin
Carol Sue Carter, A. Courtney DeVries

P

Pampiniform Plexus
Brian P. Setchell

Parasites and Reproduction
Jack J. O'Brien

Parasitoids
Nancy E. Beckage

Paraspermatozoa
Alan N. Hodgson

Parental Behavior, Arthropods
Gary A. Polis, Joseph D. Barnes, Andrew N. Beld, C. Todd Jackson

Parental Behavior, Birds
John D. Buntin

Parental Behavior, Mammals
Michael Numan

Parthenogenesis and Natural Clones
Robert C. Vrijenhoek

Parturition, Nonhuman Mammals
Michael Fields, Anna-Riitta Fuchs

Pelvic Inflammatory Disease (PID)
Thomas E. Snyder

Pelvic Nerve
Karen J. Berkley

Pelvimetry
Samuel Parry, Mark A. Morgan

Penis
Paul F. Schelhammer, Gerald H. Jordan

Pheromones, Fish
Norman Stacey, Peter Sorensen

Pheromones, Insects
Jeremy N. McNeil, Johanne DeLisle, Claude Everaerts

Pheromones, Mammals
John G. Vandenbergh

Phoronida
Russel L. Zimmer

Photoperiodism, Vertebrates
Randy J. Nelson

Pigeons
Richard F. Johnston

Pigs
Rodney D. Geisert

Pineal Gland, Melatonin Biosynthesis and Secretion
Stuart E. Dryer

Pineal Gland, Regulatory Function
Fred W. Turek

Pituitary Gland, in Fish
Martin P. Schreibman,
Lucia Magliulo-Cepriano

Pituitary Gland, Overview
Béla Halász

Placenta and Placental Analogs in Elasmobranchs
William C. Hamlett

Placenta and Placental Analogs in Reptiles and Amphibians
Daniel G. Blackburn

Placenta: Implantation and Development
Kurt Benirschke

Placental and Decidual Protein Hormones, Human
Stuart Handwerger

Placental Gas Exchange
Lawrence D. Longo

Placental Lactogens
Daniel I. H. Linzer

Placental Nutrient Transport
Colin P. Sibley

Placental Steroidogenesis in Primate Pregnancy
Eugene D. Albrecht, Gerald J. Pepe

Placozoa
Vicki Buchsbaum Pearse

Platyhelminthes
Seth Tyler

PMS (Premenstrual Syndrome)
Ellen W. Freeman

Poecilogony
Glenys D. Gibson

Polycystic Ovary Syndrome
Richard S. Legro

Polyspermy
R. H. F. Hunter

Porifera
Paul E. Fell

Postdate (Postterm) Pregnancy
Brian J. Koos, Jennifer Claman

Postpartum Depression
Joseph F. Mortola

Prader-Willi Syndrome
Shahab S. Minassian

Preeclampsia/Eclampsia
James N. Martin, Everett F. Magann

Pregnancy in Dogs and Cats
Patrick W. Concannon,
John Verstegen

Pregnancy in Farm Animals
Troy L. Ott

Pregnancy in Humans, Overview
Carmen L. Regan

Pregnancy in Other Mammals
Lloyd L. Anderson

Pregnancy, Maintenance of
Fuller W. Bazer

Pregnancy, Maternal Recognition of
Thomas E. Spencer

Pregnancy, Metabolic Changes in
William W. Hay, Jr.

Prenatal Genetic Screening
Deborah A. Driscoll

Preterm Labor and Delivery
Steve N. Caritis,
Douglas A. Woelkers

Priapulida
Christian Lemburg,
Andreas Schmidt-Rhaesa

Primates, Nonhuman
Bill Lasley, Susan Shideler

Primordial Germ Cells
Peter J. Donovan

VOLUME 4

Contents by Subject Area
Preface
Guide to the Encyclopedia

Progesterone Actions on Behavior
Anne M. Etgen

Progesterone Actions on Reproductive Tract
Francesco J. DeMayo, Cindee R. Funk

Progesterone Effects and Receptors, Subavian Species
Marina Paolucci, Noemi Custodia, Ian P. Callard

Progestins
Thomas P. Burris

Prolactin, Overview
Nadine Binart, Vincent Goffin, Christopher J. Ormandy, Paul A. Kelly

Prolactin, Actions of
James A. Rillema

Prolactin Inhibitory Factors
Michael Selmanoff

Prolactin, in Nonmammalian Vertebrates
E. Gordon Grau, Gregory M. Weber

Prolactin Secretion, Regulation of
György Nagy, Pal Gööz, Katalin M. Horváth, Béla E. Tóth

Prostate Cancer
James M. Kozlowski

Prostate Gland
Chung Lee, Lynn Janulis

Prostate-Specific Antigen
Joseph E. Oesterling, Ricardo Beduschi

Protein Hormones of Primate Pregnancy
Gerald J. Pepe, Eugene D. Albrecht

Protozoa
O. Roger Anderson

Pseudocyesis
Joseph F. Mortola

Pseudopregnancy
Mary S. Erskine

Puberty Acceleration
John G. Vandenbergh

Puberty, in Humans
Thomas A. Klein

Puberty, in Nonhuman Primates
Tony M. Plant

Puberty, in Nonprimate Mammals
Douglas L. Foster, Francis J. P. Ebling

Puberty, Precocious
Leo Plouffe, Jr.

Puerperal Infections
John W. Riggs, Jorge D. Blanco

Puerperium
Harish M. Sehdev

Rabbits
Josephine B. Miller

Radioimmunoassay
Terry M. Nett, Jennifer M. Malvey

Receptors for Hormones, Overview
Kevin J. Catt

Reflex (Induced) Ovulation
Arnold L. Goodman

Regulation of Sertoli Cells
Michael D. Griswold

Relaxin, Mammalian
Russell V. Anthony

Relaxin, Nonmammalian
Thomas J. Koob

Reproductive Senescence, Human
Charles V. Mobbs

Reproductive Senescence, Nonhuman Mammals
Anne N. Hirshfield, Jodi A. Flaws

Reproductive Technologies, Overview
Alan O. Trounson

Reproductive Toxicology
Robert E. Chapin

Reptilian Reproduction, Overview
David Crews

Reptilian Reproductive Cycles
Valentine A. Lance

Respiratory Distress Syndrome
Rebecca A. Simmons

Rhodnius prolixus
Kenneth G. Davey

Rhombozoa
John S. Pearse

Rhythms, Lunar and Tidal
Peter P. Fong

Rodentia
Franklin H. Bronson

Rotifera
Robert Lee Wallace

Ruminants
William W. Thatcher

S

Seals
Daniel P. Costa, Daniel E. Crocker

Seasonal Reproduction, Birds
Alistair Dawson

Seasonal Reproduction, Fish
Jon P. Nash

Seasonal Reproduction, Mammals
Robert L. Goodman

Seasonal Reproduction, Marine Invertebrates
John S. Pearse

Sea Urchins
John S. Pearse

Semen
Gail S. Prins

Seminal Vesicles
Lawrence S. Ross

Sertoli Cells, Function
Michael D. Griswold, Lonnie D. Russell

Sertoli Cells, Overview
Lonnie D. Russell

Sex Chromosomes
Baccio Baccetti, Giulia Collodel

Sex Determination, Environmental
Reynaldo Patiño, Carlos A. Strüssman

Sex Determination, Genetic
Józefa Styrna

Sex Differentiation in Amphibians, Reptiles, and Birds, Hormonal Regulation
Tyrone B. Hayes

Sex Differentiation, Psychological
Nancy G. Forger

Sex Ratios
Ronald J. Ericsson, Scott A. Ericsson

Sex Skin
Fred B. Bercovitch

Sexual Attractants
Lee C. Drickamer

Sexual Dysfunction
Steven M. Petak

Sexual Imprinting
David B. Miller

Sexually Transmitted Diseases
Paul Summers

Sexual Selection
Anders P. Møller

SHBG (Sex Hormone-Binding Globulin)
William Rosner

Sheehan's Syndrome
Peter J. Snyder

Sheep and Goats
Duane H. Keisler

Sipuncula
Mary E. Rice

Social Insects, Overview
Wolf Engels, Klaus Hartfelder

Song in Arthropods
Glenn K. Morris

Songbirds and Singing
Eliot A. Brenowitz

Spawning, Marine Invertebrates
John H. Himmelman

Sperm Activation, Arthropods
Julian Shepherd

Spermatogenesis, Overview
Rex A. Hess

Spermatogenesis, Disorders of
Claude Desjardins, Thorsten Diemer

Spermatogenesis, Hormonal Control of
Barry R. Zirkin

Spermatogenesis, in Nonmammals
Gloria V. Callard, Ian P. Callard

Spermatogenetic Cycle in Fish
Takeshi Miura

Spermatophores in the Arthropods
Heather C. Proctor

Spermatozoa
Clarke F. Millette

Sperm Capacitation
J. Michael Bedford, Nicholas L. Cross

Spermiogenesis
Richard Oko, Yves Clermont

Sperm Transport
James W. Overstreet, Mary A. Scott

Sperm Transport, Arthropods
Julian Shepherd

SRY Gene
Grace Lee, Mert Bahtiyar

Sterility
Bradley S. Hurst

Steroid Hormones, Overview
Terry R. Brown

Steroidogenesis, Overview
Margaret M. Hinshelwood

Steroid Hormone Receptors
Nancy H. Ing

Stress and Reproduction
Thomas H. Welsh, Jr.,
Nann Kemper-Green,
Kimberly N. Livingston

Substance Abuse and Pregnancy
Mark A. Morgan, Sara J. Marder

Suckling Behavior
Edward O. Price

Suprachiasmatic Nucleus
Robert Y. Moore

Surfactant
Aron B. Fisher

Symbiosis
Mary Beth Saffo

Tardigrada
Roberto Bertolani, Lorena Rebecchi

Teleosts, Viviparity
John P. Wourms

Temperature, Effects on Testicular Function
Jeffrey B. Kerr

Teratogens
Robert L. Brent, David A. Beckman

Territorial Behavior, Overview
Judy Stamps

Testicular Cancer
Gary D. Steinberg

Testicular Developmental Anomalies
Jay Radhakrishnan

Testis, Overview
Larry Johnson, Tobin A. McGowen,
Genevieve E. Keillor

Testosterone Biosynthesis
Douglas M. Stocco

Theca Cells
Paul F. Terranova, Katherine F. Roby

Theca Cell Tumors
Richard Leach, Nilsa Ramirez

Thyroid Hormones, in Subavian Vertebrates
David O. Norris

Tocolytic Agents
George A. Macones, Martha E. Rode

Transgenic Animals
Jorge A. Piedrahita, Karen Moore

Trophoblast to Human Placenta
Harvey J. Kliman

Tsetse Flies
R. H. Gooding

Tubal Surgery
Alan H. DeCherney, Mikio A. Nihira

Tumors of the Female Reproductive System
Basil C. Tarlatzis, Th. Agorastos

Tunicata (Urochordata)
Andrew Todd Newberry

Turner's Syndrome
David H. Barad

Twinning
Kurt Benirschke

U

Ullrich Syndrome
James Aiman

Ultradian Hormone Rhythms
Johannes D. Veldhuis

Ultrasound
Frank A. Chervenak, Edith D. Gurewitsch

Umbilical Cord
Harvey Kliman

Uterine Anomalies
John A. Rock, Bradley S. Hurst

Uterine Contraction
Robert E. Garfield, Venu Jain, George R. Saade

Uterus, Human
David A. Grainger

Uterus, Nonhuman
Frank F. Bartol

V

Vagina
Raymond E. Papka, Sonya J. Williams

Varicocele
Terry T. Turner

Vasectomy
Sherman J. Silber

Vitellogenins and Vitellogenesis
Gary J. LaFleur, Jr.

Viviparity and Oviparity: Evolution and Reproductive Strategies
Daniel G. Blackburn

Vomeronasal Organ
Michael Meredith

W

Whales and Porpoises
Daniel P. Costa, Daniel E. Crocker

Whitten Effect
John G. Vandenbergh

Wolffian Ducts
Terry W. Hensle, Harry Fisch

Y

Yolk Proteins, Invertebrates
G. R. Wyatt

Yolk Sac
Robert W. McGaughey

Z

Zygotic Genomic Activation
Carol Warner, Ginger Exley

Contributors
Glossary of Key Terms
Subject Index

Contents by Subject Area

GAMETES, FERTILIZATION, AND EARLY EMBRYOGENESIS

Acrosome Reaction
Adrenarche
Allantochorion (Chorioallantois)
Amniotic Fluid
Apoptosis (Cell Death)
Blastocyst
Choriocarcinoma
Chorionic Gonadotropin, Human
Cloning Mammals by Nuclear Transfer
Cotyledonary Placenta
Cryopreservation of Embryos
Cytokines
DHEA (Dehydroepiandrosterone)
Eicosanoids
Embryogenesis, Mammalian
Embryo Transfer
Endogenous Opioids
Endotheliochorial Placentation
Epitheliochorial Placentation
Fallopian Tube
Fertilization
Fetal Adrenals
Follistatin
Gametes, Overview
Germ Layers
Gonadogenesis, Female
Gonadogenesis, Male
Gonadotropes
Granulosa Cells
Hemochorial Placentation
Hormone Receptors, Overview
IGF (Insulin-like Growth Factor)
Implantation
In Vitro Fertilization
Lactotrophs
Leydig Cells
Meiosis
Meiotic Cell Cycle, Oocytes
Neuropeptides
Neurosecretion
Neurotransmitters
Oocyte and Embryo Transport
Oocyte, Mammalian
Oogenesis, in Mammals
Pineal Gland, Regulatory Function
Polyspermy
Primordial Germ Cells
Reproductive Technologies, Overview
Reproductive Toxicology
Sex Chromosomes
Sex Determination, Environmental
Sex Determination, Genetic
Sex Skin
SRY Gene
Testicular Developmental Anomalies
Wolffian Ducts
Yolk Sac
Zygotic Genomic Activation

REPRODUCTION IN HUMANS AND EXPERIMENTAL PRIMATES

Abortion
Adrenal Hyperplasia, Congenital Virilizing
Adrenarche
Altitude, Effects on Humans

Anterior Pituitary
Apgar Score
Artificial Insemination, in Humans
Autonomic Nervous System and Reproduction
Chorionic Gonadotropin, Human
Circumcision
Contraceptive Methods and Devices, Female
Contraceptive Methods and Devices, Male
Cryopreservation of Embryos
Embryo Transfer
Eunuchoidism
Family Planning
Female Reproductive System, Humans
Fetal Alcohol Syndrome
FSH (Follicle-Stimulating Hormone)
Genitalia
Gonadotropin Biosynthesis
Gonadotropin Receptors
Gonadotropin Secretion, Control of
Gonadotropins, Overview
Gynecomastia
Hirsutism
HIV Infection and AIDS
Homosexuality
Hormonal Contraception
Human Placental Lactogen (Human Chorionic Somatomammotropin)
Hyperprolactinemia
Hypogonadism
Hypophysectomy
Hypopituitarism
Hypospadias
Hypothalamic-Hypophysial Complex (Pituitary Portal System)
Hypoxia, Effect on Reproduction
Immunocytochemistry
Immunology of Reproduction
Infertility
In Vitro Fertilization
Kallmann's Syndrome
Klinefelter's Syndrome
Leiomyoma
LH (Luteinizing Hormone)
Lymphokines
Male Reproductive System, Human
Median Eminence
Nervus Terminalis
Neuroendocrine Systems

Orgasm
Osteoporosis
Pituitary Gland, Overview
Prader-Willi Syndrome
Pregnancy in Humans, Overview
Prolactin, Overview
Prolactin, Actions of
Prolactin Inhibitory Factors
Prolactin Secretion, Regulation of
Puberty, in Humans
Puberty, Precocious
Radioimmunoassay
Reproductive Senescence, Human
Reproductive Technologies, Overview
Reproductive Toxicology
Respiratory Distress Syndrome
Sex Differentiation, Psychological
Sex Ratios
Sexual Dysfunction
Sexually Transmitted Diseases
SHBG (Sex Hormone-Binding Globulin)
Sheehan's Syndrome
Sterility
Steroid Hormones, Overview
Steroidogenesis, Overview
Steroid Hormone Receptors
Stress and Reproduction
Suprachiasmatic Nucleus
Turner's Syndrome
Twinning
Ullrich Syndrome

DOMESTIC ANIMALS

Artificial Insemination, in Animals
Breeding Strategies for Domestic Animals
Castration, Effects in Nonhuman Mammals (Female)
Castration, Effects in Nonhumans Mammals (Male)
Cats
Cattle
Chickens, Control of Reproduction in
Cloning Mammals by Nuclear Transfer
Cryopreservation of Sperm
Dogs
Embryo Transfer

Equine Chorionic Gonadotropin
Estrus
Fertility and Fecundity
Freemartin
Guinea Pig, Female
Horses
Pigs
Pregnancy in Dogs and Cats
Pregnancy in Farm Animals
Pregnancy, Maintenance of
Reproductive Technologies, Overview
Ruminants
Sheep and Goats
Transgenic Animals

MAMMALIAN REPRODUCTION

Aggressive Behavior
Allantochorion (Chorioallantois)
Allantois
Androgens, Effects in Mammals
Armadillo
Aromatization
Bruce Effect
Chorionic Gonadotropins, Nonhuman Mammals
Circadian Rhythms
Circannual Rhythms
Copulation, Mammals
Cricetidae (Hamsters and Lemmings)
Deer
Elephants
Embryogenesis, Mammalian
Equine Chorionic Gonadotropin (ECG)
Estrous Cycle
Estrus
Female Reproductive System, Nonhuman Mammals
Freemartin
Global Zones and Reproduction
Hyenas
Internal Fertilization, in Birds and Mammals
Intersexuality in Mammals
Lactation, Nonhuman
Magnocellular System
Male Reproductive System, Nonhuman Mammals
Mammary Gland, Overview
Marsupials
Mating Behaviors, Mammals
Microtinae (Voles)
Monotremes
Naked Mole-Rats
Opossums
Ovarian Cycle, Mammals
Parental Behavior, Mammals
Parturition, Nonhuman Mammals
Pheromones, Mammals
Placental Steroidogenesis in Primate Pregnancy
Pregnancy in Other Mammals
Pregnancy, Maintenance of
Pregnancy, Maternal Recognition of
Primates, Nonhuman
Progesterone Actions on Behavior
Protein Hormones of Primate Pregnancy
Puberty Acceleration
Puberty, in Nonhuman Primates
Puberty, in Nonprimate Mammals
Rabbits
Reflex (Induced) Ovulation
Relaxin, Mammalian
Reproductive Senescence, Nonhuman Mammals
Rodentia
Seals
Seasonal Reproduction, Mammals
Sex Ratios
Territorial Behavior, Overview
Uterus, Nonhuman
Whales and Porpoises
Whitten Effect

AVIAN REPRODUCTION

Altricial and Precocial Development in Birds
Altruistic Behavior, Vertebrates
Androgens, Effects in Birds
Avian Reproduction, Overview
Avian Reproductive System, Developmental Endocrinology
Birds, Diversity of
Brood Parasitism in Birds
Chickens, Control of Reproduction in
Corpora Lutea of Nonmammalian Species
Egg, Avian
Energetics of Reproduction
Estrogen, Effects in Birds

Female Reproductive System, Birds
Internal Fertilization, Birds and Mammals
Interrenal Gland, Stress Response and
  Reproduction
Lactotrophs
Male Reproductive System, Birds
Mate Choice, Overview
Migration, Birds
Molt and Nuptial Color
Nesting, Birds
Ovarian Cycles and Follicle Development in Birds
Parental Behavior, Birds
Photoperiodism, Vertebrates
Pigeons
Pineal Gland, Melatonin Biosynthesis and
  Secretion
Prolactin, Actions of
Prolactin, in Nonmammalian Vertebrates
Relaxin, Nonmammalian
Seasonal Reproduction, Birds
Sex Differentiation in Amphibians, Reptiles, and
  Birds, Hormonal Regulation
Sex Ratios
Sexual Imprinting
SHBG (Sex Hormone-Binding Globulin)
Songbirds and Singing
Territorial Behavior, Overview
Viviparity and Oviparity: Evolution and
  Reproductive Strategies

REPTILES AND AMPHIBIA

Amphibian Ovarian Cycles
Amphibian Reproduction, Overview
Androgens, Subavian Species
Corpora Lutea of Nonmammalian Species
Energetics of Reproduction
Estrogen Effects and Receptors, Subavian Species
Female Reproductive System, Amphibians
Female Reproductive System, Reptiles
Hormonal Control of the Reproductive Tract,
  Subavian Vertebrates
Interrenal Gland, Stress Response and
  Reproduction
Leydig and Sertoli Cells, Nonmammalian
Male Reproductive System, Amphibians
Male Reproductive System, Reptiles

Mate Choice, Overview
Migration, Amphibians
Migration, Reptiles
Oogenesis, in Nonmammalian Vertebrates
Photoperiodism, Vertebrates
Placenta and Placental Analogs in Reptiles and
  Amphibians
Progesterone Effects and Receptors, Subavian
  Species
Prolactin, in Nonmammalian Vertebrates
Reptilian Reproduction, Overview
Reptilian Reproductive Cycles
Sex Differentiation in Amphibians, Reptiles, and
  Birds, Hormonal Regulation
Spermatogenesis, in Nonmammals
Territorial Behavior, Overview
Thyroid Hormones, in Subavian Vertebrates
Vitellogenins and Vitellogenesis
Viviparity and Oviparity: Evolution and
  Reproductive Strategies
Vomeronasal Organ

FISH, ELASMOBRANCHII, AND CYCLOSTOMES

Agnatha
Androgens, Subavian Species
Cephalochordata
Corpora Lutea of Nonmammalian Species
Elasmobranch Reproduction
Energetics of Reproduction
Estrogen Effects and Receptors, Subavian Species
Female Reproductive System, Fish
Fish, Modes of Reproduction
Hormones and Reproductive Behaviors, Fish
Interrenal Gland, Stress Response and
  Reproduction
Leydig and Sertoli Cells, Nonmammalian
Male Reproductive System, Fish
Mate Choice, Overview
Migration, Fish
Oogenesis, in Nonmammalian Vertebrates
Ovarian Cycle, Teleost Fish
Pheromones, Fish
Photoperiodism, Vertebrates
Pituitary Gland, in Fish
Placenta and Placental Analogs in Elasmobranchs

Pituitary Gland, in Fish
Placenta and Placental Analogs in Elasmobranchs
Progesterone Effects and Receptors, Subavian Species
Prolactin, in Nonmammals
Relaxin, Nonmammalian
Rhythms, Lunar and Tidal
Seasonal Reproduction, Fish
Sex Determination, Environmental
Spermatogenetic Cycle in Fish
Teleosts, Viviparity
Thyroid Hormones, in Subavian Vertebrates
Vitellogenins and Vitellogenesis
Viviparity and Oviparity: Evolution and Reproductive Strategies

INVERTEBRATES

Acanthocephala
Aedes aegypti
Allatostatins
Altruism in Insect Reproduction
Annelida
Aphids
Aplysia
Asexual Reproduction
Brachiopoda
Bryozoa (Ectoprocta)
Caenorhabditis elegans
Chaetognatha
Chelicerate Arthropods
Cnidaria
Conjugation in Ciliates
Corpus Allatum
Corpus Cardiacum, Insects
Crustacea
Ctenophora
Cycliophora
Diapause
Diploptera punctata
Dorsal Bodies in Mollusca
Drosophila
Ecdysiotropins and Ecdysiostatins
Ecdysteroids
Echinodermata
Egg Coverings, Insects
Female Reproductive System, Insects
Gnathostomulida
Hemichordata
Hemocoelic Insemination
Hermaphroditism
Hybridization
Hydra
Insect Accessory Glands
Insect Reproduction, Overview
Juvenile Hormone
Kamptozoa (Entoprocta)
Kinorhyncha
Locusts
Male Reproductive System, Insects
Marine Invertebrate Larvae
Marine Invertebrates, Modes of Reproduction in
Mating Behaviors, Insects
Mating Behaviors, Invertebrates Other Than Insects
Metamorphosis, Insects
Migration, Insects
Mollusca
Myriapoda
Nematodes and Related Phyla
Nematomorpha
Nemertea
Onychophora
Oocyte Maturation and Spawning in Starfish
Oostatins, Folliculostatins, and Antigonadotropins, Insects
Orthonectida
Oviposition in Molluscs
Ovulation and Oviposition, Insects
Parasites and Reproduction
Parasitoids
Paraspermatozoa
Parental Behavior, Arthropods
Parthenogenesis and Natural Clones
Pheromones, Insects
Phoronida
Placozoa
Platyhelminthes
Poecilogony
Porifera
Priapulida
Protozoa
Rhodnius prolixus
Rhombozoa
Rhythms, Lunar and Tidal

Sea Urchins
Sipuncula
Social Insects, Overview
Song in Arthropods
Spawning, Marine Invertebrates
Sperm Activation, Arthropods
Spermatophores in the Arthropods
Sperm Transport, Arthropods
Symbiosis
Tardigrada
Tsetse Flies
Tunicata (Urochordata)
Yolk Proteins, Invertebrates

REPRODUCTIVE BEHAVIOR

Aggressive Behavior
Altruism in Insect Reproduction
Altruistic Behavior, Vertebrates
Androgens, Effects in Birds
Androgens, Effects in Mammals
Autonomic Nervous System and Reproduction
Brood Parasitism in Birds
Captive Breeding of Wildlife
Circadian Rhythms
Circannual Rhythms
Copulation, Mammals
Critical Period, Estrous Cycle
Endogenous Opioids
Energy Balance, Effects on Reproduction
Estrogen Action, Behavior
Estrogen, Effects in Birds
Estrous Cycle
Estrus
Fertility and Fecundity
Global Zones and Reproduction
Homosexuality
Hormones and Reproductive Behaviors, Fish
Intersexuality in Mammals
Lordosis
Mate Choice, Overview
Mating Behaviors, Insects
Mating Behaviors, Invertebrates Other Than Insects
Mating Behaviors, Mammals
Migration, Amphibians

Migration, Birds
Migration, Fish
Migration, Insects
Migration, Reptiles
Nesting, Birds
Nutritional Factors and Reproduction
Olfaction and Reproduction
Orgasm
Parasitoids
Parental Behavior, Arthropods
Parental Behavior, Birds
Parental Behavior, Mammals
Pheromones, Fish
Pheromones, Insects
Pheromones, Mammals
Photoperiodism, Vertebrates
Pineal Gland, Melatonin Biosynthesis and Secretion
Progesterone Actions on Behavior
Reproductive Senescence, Human
Reproductive Senescence, Nonhuman
Reptilian Reproductive Cycles
Rhythms, Lunar and Tidal
Seasonal Reproduction, Birds
Seasonal Reproduction, Fish
Seasonal Reproduction, Mammals
Seasonal Reproduction, Marine Invertebrates
Sex Differentiation, Psychological
Sexual Attractants
Sexual Imprinting
Sexual Selection
Song in Arthropods
Songbirds and Singing
Spawning, Marine Invertebrates
Stress and Reproduction
Suckling Behavior
Territorial Behavior, Overview
Ultradian Hormone Rhythms
Vomeronasal Organ

FEMALE REPRODUCTIVE SYSTEMS

Activin and Activin Receptors
Amenorrhea
Antiestrogens
Antiprogestins

Breast Cancer
Breast Disorders
Cervical Cancer
Cervix
Circumcision
Clitoris
Contraceptive Methods and Devices, Female
Corpus Luteum (CL)
Corpus Luteum Peptides
Endometriosis
Endometrium
Environmental Estrogens
Estrogen Action, Behavior
Estrogen Action, Bone
Estrogen Action, Breast
Estrogen Action on the Female Reproductive Tract
Estrogen Replacement Therapy
Estrogen Secretion, Regulation of
Estrogens, Overview
Eunuchoidism
Fallopian Tube
Female Reproductive Disorders, Overview
Female Reproductive System, Amphibians
Female Reproductive System, Birds
Female Reproductive System, Fish
Female Reproductive System, Humans
Female Reproductive System, Insects
Female Reproductive System, Nonhuman Mammals
Female Reproductive System, Reptiles
Female Sterilization
Follicular Atresia
Follicular Development
Follicular Steroidogenesis
Genitalia
GnRH (Gonadotropin-Releasing Hormone)
GnRH Pulse Generator
Gonadogenesis, Female
Graafian Follicle
Hormonal Contraception
Hysterectomy
IGF (Insulin-like Growth Factor)
Infertility
Inhibin
Interferons
Leukemia Inhibitory Factor
Local Control Systems in Reproduction
Luteinization
Luteolysis
Luteotropic Hormones
Menarche
Menopause
Menstrual Cycle
Menstrual Disorders
Menstruation
Neurohypophysial Hormones
Osteoporosis
Ovarian Cancer
Ovarian Cycle, Mammals
Ovarian Function in the Perimenopause
Ovarian Hormones, Overview
Ovarian Innervation
Ovary, Overview
Ovulation
Oxytocics
Pelvic Inflammatory Disease (PID)
PMS (Premenstrual Syndrome)
Polycystic Ovary Syndrome
Progesterone Actions on Behavior
Progesterone Actions on the Reproductive Tract
Progestins
Receptors for Hormones, Overview
Relaxin, Mammalian
Theca Cells
Theca Cell Tumors
Trophoblast to Human Placenta
Tubal Surgery
Tumors of the Female Reproductive System
Turner's Syndrome
Ullrich Syndrome
Uterine Anomalies
Uterus, Human
Uterus, Nonhuman
Vagina

MALE REPRODUCTIVE SYSTEMS

Adrenal Androgens
Androgen Inhibitors/Antiandrogens
Androgen Insensitivity Syndromes
Androgens
Andrology: Origins and Scope
Aromatization

Benign Prostatic Hyperplasia (BPH)
Blood-Testis Barrier
Castration, Effects in Humans (Male)
Castration, Effects in Nonhuman Mammals (Male)
Circumcision
Contraceptive Methods and Devices, Male
Cryopreservation of Sperm
Cryptorchidism
Dihydrotestosterone
Ejaculation
Epididymis
Erection
Eunuchoidism
Gene Transfer, Sperm-Mediated
Genitalia
Gonadogenesis, Male
Gynecomastia
Hypospadias
Impotence
Infertility
Leydig Cells
Male Reproductive Disorders
Male Reproductive System, Amphibians
Male Reproductive System, Birds
Male Reproductive System, Fish
Male Reproductive System, Human
Male Reproductive System, Insects
Male Reproductive System, Nonhuman Mammals
Male Reproductive System, Reptiles
Orchitis
Pampiniform Plexus
Pelvic Nerve
Penis
Prostate Cancer
Prostate Gland
Prostate-Specific Antigen
Regulation of Sertoli Cells
Semen
Seminal Vesicles
Sertoli Cells, Function
Sertoli Cells, Overview
Spermatogenesis, Overview
Spermatogenesis, Disorders of
Spermatogenesis, Hormonal Control of
Spermatozoa
Sperm Capacitation
Spermiogenesis

Sperm Transport
Temperature, Effects on Testicular Function
Testicular Cancer
Testicular Developmental Anomalies
Testis, Overview
Testosterone Biosynthesis
Varicocele
Vasectomy

PREGNANCY

Abortion
Allantoic Fluid
Amniocentesis
Cardiovascular Adaptation to Pregnancy
CBG (Corticosteroid-Binding Globulin)
Cesarean Delivery
Choriocarcinoma
Corpus Luteum of Pregnancy
Cotyledonary Placenta
Cryopreservation of Embryos
Decidua
Deciduoma
Discoidal Placenta
Ectopic Pregnancy
Endotheliochorial Placentation
Epitheliochorial Placentation
Erythroblastosis Fetalis
Family Planning
Fetal Adrenals
Fetal Alcohol Syndrome
Fetal Anomalies
Fetal Diagnosis, Invasive
Fetal Growth and Development
Fetal Hormones
Fetal Lung Development
Fetal Membranes
Fetal Monitoring and Testing
Fetal–Placental Unit
Fetal Surgery
α-Fetoprotein and Triple Screening
Fetus, Overview
Growth Factors
Hemochorial Placentation
Hormones of Pregnancy
Infections in Pregnancy

Intrauterine Growth Restriction and Mechanisms
   of Fetal Growth
In Vitro Fertilization
Labor and Delivery, Human
Morning Sickness and Hyperemesis Gravidarum
Obstetric Anesthesia
Parturition, Nonhuman Mammals
Pelvimetry
Placenta: Implantation and Development
Placental and Decidual Protein Hormones, Human
Placental Gas Exchange
Placental Lactogens
Placental Nutrient Transport
Placental Steroidogenesis in Primate Pregnancy
Postdate (Postterm) Pregnancy
Postpartum Depression
Preeclampsia/Eclampsia
Pregnancy in Dogs and Cats
Pregnancy in Farm Animals
Pregnancy in Humans, Overview
Pregnancy in Other Mammals
Pregnancy, Maintenance of
Pregnancy, Maternal Recognition of
Pregnancy, Metabolic Changes in
Prenatal Genetic Screening
Preterm Labor and Delivery
Protein Hormones of Primate Pregnancy
Pseudocyesis
Pseudopregnancy
Puerperal Infections
Puerperium
Substance Abuse and Pregnancy
Surfactant
Teratogens
Tocolytic Agents
Twinning
Ultrasound
Umbilical Cord
Uterine Contraction

LACTATION

Breast Disorders
Breastfeeding
Cattle
Galactorrhea
Human Placental Lactogen (Human Chorionic
   Somatomammotropin)
Lactational Amenorrhea
Lactational Anestrus
Lactation, Human
Lactation, Nonhuman
Lactogenesis
Mammary Gland Development
Mammary Gland, Overview
Milk, Composition and Synthesis
Milk Ejection
Nutritional Factors and Lactation
Oxytocin
Placental Lactogens
Pregnancy in Humans, Overview
Prolactin, Overview
Prolactin, Actions of
Prolactin Secretion, Regulation of
Suckling Behavior

Preface

The publication of the *Encyclopedia of Reproduction* comes at a most opportune time. Hardly a day goes by when the news media do not report some new dimension in the treatment of infertility or, conversely, controversies associated with the control of fertility and the ethical issues raised by both. Organismal cloning is a matter of constant debate, and the pharmacological correction of erectile dysfunction has become a preoccupation of international dimensions. Procreation remains a subject of universal interest to every segment of society, from scientists to students, from science reporters to the proverbial person on the street.

The present work should serve as a convenient and comprehensive source of information encompassing all aspects of the subject of reproduction as it relates to the entire animal kingdom. It should be as useful to the expert exploring reproductive phenomena outside his or her own field as it is to students and to the educated public at large. Topics for inclusion were initially generated by forming a matrix of systems (gametes, fertilization, and early embryogenesis; reproductive behavior; female reproductive systems; male reproductive systems; pregnancy; and lactation) and of groups of animals (humans and experimental primates; domestic animals; mammals; birds; reptiles and amphibia; fish, elasmobranchii, and cyclostomes; and invertebrates).

A group of outstanding Associate Editors having expertise in one of more of these areas was then recruited. The preliminary list of entries prepared by the Editors was refined and expanded at a meeting with these Associate Editors, who then identified the appropriate authors. Manuscripts were critically reviewed by the Associate Editors and finally scrutinized by us and the editorial staff at Academic Press.

In a work of this kind, errors of omission and of commission are inevitable and we should appreciate having them called to our attention for correction in possible future editions. The 542 entries constituting the work each contain a glossary of terms, a summary introduction, cross-references to related articles, and a reading list. A standard subject index and an index of reproductive systems and zoological groupings are also provided.

Each entry was written to be self-contained, inevitably leading to some overlap of content. We do not view this as a weakness, but instead believe that it will facilitate a reader's search for information by reducing the number of entries that have to be consulted.

The completion of this project demanded the best efforts of a large number of participants. Chief among them are the 700 authors, especially those who wrote articles on short notice so that the publication deadline could be met. The stellar group of 15 Associate Editors, each of whom possesses great breadth of knowledge and who, as a group, span the spectrum of expertise from zoology to animal husbandry to obstetrics and gynecology, also rendered exceptional service.

Finally, we acknowledge the indispensable contributions of the staff at Academic Press: Jasna Markovac, Editor-in-Chief for Biomedical Science, who originally conceived of the Encyclopedia; and Chris Morris, Gail Rice, and Erika Conner, Major Reference Works editors who provided ongoing management of the project.

Ernst Knobil
Jimmy D. Neill

Guide to the Encyclopedia

ORGANIZATION

The *Encyclopedia of Reproduction* is organized to provide the maximum ease of use for its readers. All of the articles are arranged in a single alphabetical sequence by title. Articles whose titles begin with the letters A to En are in Volume 1, articles with titles from Ep through L are in Volume 2, then M through Pri in Volume 3, and Pro to Z in Volume 4.

Volume 4 also includes a complete subject index for the entire work, an alphabetical list of the contributors to the encyclopedia, and a glossary of key terms used in the articles.

Article titles generally begin with the key noun or noun phrase indicating the topic, with any descriptive terms following. For example, "Uterus, Human" is the article title rather than "Human Uterus," and "Migration, Birds" is the title rather than "Bird Migration." This is done so that the same phenomenon or feature can be studied across various groups. For example, all the articles on female reproductive systems in humans, other mammals, birds, etc., appear in one sequence in the Fe- section of the encyclopedia.

INDEX

The Subject Index in Volume 4 contains more than 20,000 entries. The subjects are listed alphabetically and indicate the volume and page number where information on this topic can be found. In addition, the Table of Contents by Subject Area also functions as an index, since it lists all the topics covered in a given area; e.g., the encyclopedia includes 90 different articles dealing with reproduction in invertebrates.

OUTLINE

Each entry in the Encyclopedia begins with a topical outline that indicates the general content of the article. This outline serves two functions. First, it provides a brief preview of the article, so that the reader can get a sense of what is contained there without having to leaf through the pages. Second, it highlights important subtopics that will be discussed within the article. For example, the article "Fallopian Tube" includes subtopics such as "Tubal Disorders," "Tubal Sterilization," and "Assisted Reproductive Techniques Involving the Tube."

The outline is intended as an overview and thus it lists only the major headings of the article. In addition, extensive second-level and third-level headings will be found within the article.

GLOSSARY

The Glossary section contains terms that are important to an understanding of the article and that may be unfamiliar to the reader. Each term is defined in the context of the article in which it is used. Thus the same term may appear as a glossary entry in two or more articles, with the details of the definition varying slightly from one article to another. The encyclopedia has approximately 4,250 glossary entries.

In addition, Volume 4 provides a comprehensive glossary that collects all the core vocabulary of reproductive biology in one A-Z list. This section can be consulted for definitions of terms not found in the individual glossary for a given article.

DEFINING STATEMENT

The text of each article in the encyclopedia begins with an introductory paragraph that defines the topic under discussion and summarizes the content of the article. For example, the article "Energetics in Reproduction" begins with the following statement:

Energetics of reproduction is defined as the amount of energy that an animal expends to reproduce. The energetics of reproduction can include the costs of gamete manufacture, synthesis of secondary sexual characteristics and sex-attractant chemicals (pheromones), and reproductive behavior including territorial defense, nest building, courtship rituals, and parental care. . . .

CROSS-REFERENCES

Almost all articles in the Encyclopedia have cross-references to other articles. These cross-references appear at the conclusion of the article text. They indicate articles that can be consulted for further information on the same topic or for other information on a related topic. For example, the article "Osteoporosis" contains references to the articles "Estrogen Replacement Therapy" and "Menopause."

BIBLIOGRAPHY

The Bibliography section appears as the last element in an article. The reference sources listed there are the authors' recommendations of the most appropriate materials for further research on the given topic. The bibliography entries are for the benefit of the reader and thus they do not represent a complete listing of all the materials consulted by the author in preparing the article.

COMPANION WORKS

The *Encyclopedia of Reproduction* is one of a series of multivolume reference works in the life sciences published by Academic Press. Other such titles include the *Encyclopedia of Human Biology, Encyclopedia of Cancer, Encyclopedia of Toxicology, Encyclopedia of Immunology*, and *Encyclopedia of Microbiology*.

Abortion

Steven J. Sondheimer

University of Pennsylvania

I. Spontaneous Abortion
II. Habitual Abortion
III. Immunologic Causes of Pregnancy Loss
IV. Induced Abortion
V. Procedures Used to Terminate Pregnancy

GLOSSARY

dilatation and curettage The mechanical stretching of the cervical os, allowing access to the uterine cavity for scraping of the wall to remove the contents and lining.

human chorionic gonadotropin A hormone unique to pregnancy produced by the placenta and secreted into the maternal circulation after implantation.

karyotype The genetic material; in humans, 46 chromosomes, including 2 sex chromosomes.

Müllerian anomaly Abnormalities of the developing female reproductive organs including the uterus, upper vagina, cervix, and Fallopian tubes.

ultrasound A radiological technique using sound waves for visualizing the uterus, ovaries, and pregnancy.

An abortion is an early pregnancy loss. For statistical purposes, if a pregnancy either spontaneously or by induced techniques terminates before 20 weeks of gestation (calculated by ultrasound or from history of the first day of the last period) or when the fetus weighs under 500 g, an abortion has occurred. The performance of serum or urinary assays for human chorionic gonadotropins (hCGs), ultrasound, or pathologic identification of placental or fetal tissue establishes that a pregnancy was present. An abortion does not occur unless there has been implantation, which can be detected by identification of hCG in the maternal blood. Sensitive assays for hCG indicate that approximately 50% of pregnancies are lost after implantation, but only 20% of these are clinically diagnosed. If all the tissue has been expelled, this is a complete abortion; if tissue remains in the uterus, this is an incomplete abortion.

I. SPONTANEOUS ABORTION

The term threatened abortion means bleeding in a first-trimester pregnancy with a closed cervix and unknown viability. When the cervix is open, then an inevitable abortion is present, and if tissue has begun to be expelled, an incomplete abortion is present. With modern ultrasound technology and quantitative human chorionic gonadotropin (hCG) assays, first-trimester abortions are being diagnosed before there is clinical evidence for pregnancy wastage and these are termed "missed abortions." In a normal early pregnancy, by transvaginal ultrasound the gestational sac should be found 3 weeks after fertilization (the fifth menstrual week) or when the quantitative hCG in the serum is approximately 2000–3000 mIU/ml. By the sixth menstrual week a fetal pole is usually seen by ultrasound, and by 7 weeks a fetal heartbeat can be identified. Ultrasound evidence of missed abortion includes failure to identify a fetal pole or, if it is initially seen, its subsequent disappearance; the absence of fetal heart activity; or failure to meet the appropriate milestones of first-trimester ultrasound. Failure of hCG levels to rise normally in the early first trimester (approximate doubling every 48–72 hr), when hCG doubling time is normal but no evidence of fetal cardiac activity is detected in the presence of a gestational sac of 17 mm in diameter, or an hCG level of >30,000 mIU/ml indicate a nonviable pregnancy. Usually, this will lead to a spontaneous abortion, either complete or incomplete. The diagnosis of nonviable pregnancies (missed abortions) in which the cervical os is still closed is a modern technological advance.

Sixty percent of first-trimester spontaneous abortions are associated with an abnormal karyotype of the fetal tissue. In contrast, only about 30% of pregnancies which abort in the second trimester spontaneously have abnormal chromosomes and 3% of stillbirths have abnormal chromosomes. The later an abortion occurs, the more likely it is that the chromosomes are normal. Common chromosomal abnormalities are autosomal trisomys, such as trisomies of 13, 16, 18, 21, and 22. The most common single chromosomal abnormality in spontaneous abortions is 45,X, which is the genotype for Turner's syndrome. Most of these chromosomal abnormalities end in spontaneous pregnancy losses. Other common abnormalities are polyploidies.

Aging of the oocytes associated with older chronological age or elevated Day 3 follicle-stimulating hormone (FSH) levels are associated with increased risk of chromosomally abnormal losses. In patients with three or more consecutive losses, approximately 8% of couples will be found to have an abnormal karyotype, most frequently a balanced translocation or occasionally mosiacism, deletion, or inversion.

Systemic illnesses have not been clearly associated with an increased risk of abortion. Diabetes mellitus, hyperthyroidism, and some other chronic illnesses are associated with an increased incidence of placental inadequacy manifested by intrauterine growth retardation or fetal death. The chemotherapeutic drugs methotrexate and aminopterin in therapeutic doses will often cause abortion, especially early in gestation. Anesthetic agents probably do not cause abortion. Acute infection in pregnancy with toxoplasmosis gondii and viruses such as rubella, herpes simplex, cytomegalovirus, measles, and coxsackievirus may provoke pregnancy loss. Epidemiologically, certain environmental factors are associated with a higher risk of abortion. These include cigarette smoking, alcohol, and heavy coffee consumption. Power lines, air travel, video display terminals, exercise, coitus, or heavy activity are not causes of abortion. Endometriosis is probably not a cause of spontaneous pregnancy loss. Women with polycystic ovary syndrome and women who become pregnant following ovulation induction are at a higher risk of abortion, but this also may be related to older age of the patients or closer monitoring. Maternal Müllerian abnormalities, uterine synechaie, or submucosal fibroids are cause of pregnancy loss, but these more commonly occur in the second trimester.

Surgical removal of the corpus luteum prior to 8–10 weeks, when the placenta becomes the major source of progesterone, or destruction of the corpus luteum by total body irradiation >3000 rads have been associated with pregnancy loss.

The possibility that endocrine factors such as inadequate luteal phase or so-called luteal phase defect are causes of pregnancy loss is a subject of debate. There is no evidence that evaluating a patient after a single or even two losses of an inadequate luteal phase increases the chances of a subsequent successful pregnancy. It has not been shown convincingly that progesterone or progestational agents will decrease the incidence of abortion. Clomiphene citrate therapy and human menopausal gonadotropins are associated with occasional inadequate luteal phases and it is common practice in assisted reproductive technologies to supplement the luteal phase with progesterone. Theoretically, this may be more appropriate if a gonadotropin-releasing hormone analog has been used to downregulate the first part of the cycle. By no means, however, has the necessity of progesterone supplementation been clearly documented.

II. HABITUAL ABORTION

Habitual abortion or repetitive pregnancy wastage is defined as three consecutive pregnancy losses before 20 weeks of gestation. Since the risk of one clinically evident spontaneous abortion is 20%, the risk of two consecutive losses is 20% × 20% or approximately 4%. Three consecutive losses will occur in approximately 1% of all women. The greater the number of abortions, especially if these include a second-trimester loss, the greater the likelihood of finding an etiology. After three losses the chances of finding an explanation are increased to an extent that evaluation is probably appropriate.

A luteal phase defect is diagnosed if two midluteal phase endometrial biopsies display a histological pattern that is more than 2 days out of phase based on a luteinizing hormone (LH) surge or the onset of the next period. A basal body temperature chart documenting a luteal phase of <11 days has been used to diagnose luteal phase defect, but this method is probably unreliable. An inadequate luteal phase can be treated with progesterone supplements or clomiphene citrate in the early follicular phase to enhance follicular development. If a specific endocrinopathy can be identified, such as an elevated prolactin level or hypothyroidism, the disorder should be treated.

Uterine anomalies (Müllerian anomalies) are more likely to be associated with abortion and can be diagnosed by a number of radiologic techniques. The hysterosalpingogram is usually the standard approach to diagnosing abnormalities in the structure of the uterus, along with hysteroscopy and endovaginal ultrasound. Abnormal findings should be confirmed by sonohyterograms or magnetic resonance imaging. A uterine septum, partial or complete, is the most well-defined Müllerian anomaly associated with pregnancy loss. A bicornuate uterus, in which there is a single cervix but varying degrees of separation in the midline, may be associated with first- or second-trimester abortion. Possibly the blood supply to the septum is inadequate for appropriate implantation or placentation. Uterine duplication and a uterine septum are associated with preterm labor and breech presentation. Hysteroscopic resection of uterine septa leads to successful pregnancy in approximately 75% of cases. Unfortunately, there have been no randomized prospective studies to convincingly establish the value of surgical repair of uterine abnormalities.

In utero exposure to diethylstilbesterol can lead to an abnormally formed T-shaped uterus which is associated with both first- and second-trimester pregnancy losses and an increased risk of ectopic pregnancy, incompetent cervix, and premature delivery.

Intrauterine synechaie (Asherman's syndrome) usually caused by postabortal or postpartum infection requiring dilation and curettage or by pelvic tuberculosis is an uncommon cause of recurrent abortion but can often be surgically corrected by hysteroscopy. Submucosal fibroids and even intramural fibroids when they are over 3 cm in diameter and close to the cavity are associated with abortion. Myomectomy, either hysteroscopically if feasible or by laparatomy, improves outcome.

No infectious etiology has clearly been demonstrated for recurrent spontaneous abortions. There is evidence that bacterial vaginosis is associated with premature delivery and premature rupture of the membranes but it is not clearly related to first- or second-trimester pregnancy loss. One study found a higher rate of colonization with ureaplasma urealytitcum in women with recurrent abortion.

III. IMMUNOLOGIC CAUSES OF PREGNANCY LOSS

There is controversy regarding immunologic or autoimmune causes of habitual abortion. Placental insufficiency, preeclampsia, and second-trimester pregnancy loss or thrombotic events are associated with antiphospholipid antibody syndrome, a disease in which antibodies to phospholipids are present in maternal serum. Serum lupus anticoagulant or an anticardiolipin antibody or antiphospholipid antibodies increase the risk of placental thrombosis, leading to spontaneous abortion, intrauterine growth retardation, or fetal death. Endothelial cell wall damage and thrombosis or possibly interference with the formation of the syncytiotrophoblast from cytotrophoblasts may be the mechanism of pregnancy loss. This syndrome can be successfully treated with low-dose aspirin and heparin.

Immunologic similarities or differences between individuals as a cause of otherwise unexplained recurrent pregnancy loss have not yet been clearly demonstrated and various suggested therapies such as immunization of the male with his female partner's white cells are still experimental.

Currently, there is no standard approach to the evaluation of the couple with recurrent pregnancy loss. History should include details on each loss, including the weeks gestation of the fetus and serum, ultrasound and pathologic confirmation, and, if possible, chromosomal results. Evaluation for uterine and cervical anomalies, a history of DES exposure, or risk factors for uterine synechaie such as postabortal or postpartum infection with instrumentation or tuberculosis should be performed. Medical conditions such as diabetes, collagen vascular diseases such as SLE, and rheumatoid arthritis, thrombosis, and any autoimmune endocrine or systemic illness should be excluded. Endocrine evaluation should include fasting glucose, prolactin, thyroid function tests, particularly a thyroid-stimulating hormone (TSH) test, and identification of any other endocrine abnormality. Luteal phase evaluation should include a mid-luteal phase endometrial biopsy. Cervical cultures for chlamydia, gonorrhea, mycoplasma, and ureaplasma uriliticum should be considered. Serum tests for antinuclear antibody, anticardiolipin antibody, lupus anticoagulant, and activated partial thromboplastin should be timed to rule out immune factors and parental karyotypes.

Hysterosalpingography, hysteroscopy, sonohysteroscopy, and magnetic resonance imaging can be used to rule out uterine defects, Asherman's syndrome, or submucus fibroids.

IV. INDUCED ABORTION

In the United States, most of Western Europe, Eastern Europe, Russia, China, and Japan, induced abortion is legally available. Data from the Centers for Disease Control and Prevention indicate that 1.33 million legal abortions were performed in the United States in 1993. There are 62 million women in the United States aged 13–44: Half are believed to be fertile and having heterosexual intercourse. Of American women ages 15–44, 2.3% had an abortion in 1993. The number of abortions per 1000 women aged 15–44 (abortion rate) was 22 in 1978, 28 in 1985, and 26 in 1995. Surveys of women in the United States indicate that 50% of pregnancies are unintended; half of these result in elective induced abortion, meaning that one-fourth of all pregnancies are terminated. Forty-seven percent of unplanned pregnancies occur among couples using some method of contraception.

Eighty-nine percent of induced abortions in the United States occur in the first trimester. Overall, 52% of abortions take place 8 weeks or less from the last menstrual period (LMP); 6% of induced abortions take place at 13–15 weeks and 4% at 16–20 weeks LMP.

V. PROCEDURES USED TO TERMINATE PREGNANCY

Early induced abortion before 7 weeks from the LMP can be safely performed by both surgical and medical means. A manual vacuum aspirator does not require an electrical suction machine and is particularly useful in early abortions or for early incomplete abortions. Methotrexate and misoprostol or mifepristone (RU 486) and misoprostol are two medical methods for early pregnancy termination.

Both oral and intramuscular methotrexate protocols have been evaluated for medical abortion. Both require the use of a prostaglandin agent such as misoprostol. Misoprostol is a synthetic prostaglandin E_1 (PgE_1) analog developed to prevent ulcerative effects of chronic use of antiinflammatory agents. This PgE_1 analog produces less severe gastrointestinal symptoms and overall has been safer and less expensive than other prostaglandins.

In failed medical procedures, suction curettage or manual vacuum aspiration can be used to complete the procedure. Both early medical abortion and early surgical abortion are optimally performed with the availability of transvaginal ultrasound and quantitative β hCG levels to confirm complete evacuation and to diagnose ectopic pregnancy. The incidence of continuing pregnancy after early suction curettage with a hand-held syringe in one series was very low, with 3 continuing pregnancies of 2398 abortion procedures performed at <6 weeks of gestation. Medically induced first-trimester abortions are successful in approximately 90% of patients <42 days from the LMP as confirmed by vaginal ultrasound.

First-trimester abortions are usually not performed with preoperative cervical dilation, but both oral and vaginal misoprostol effectively produce cervical dilation prior to a first-trimester procedure. Hydrophilic tents such as laminaria can also be used for this purpose.

Vacuum aspiration (suction curettage) revolutionized the performance of induced abortion and the completion of incomplete abortions up to 14 weeks gestation. After dilation of the cervix to approximately 3 weeks gestations in millimeter by graduated dilators, a suction cannula is inserted in the uterine cavity. Vacuum aspiration has replaced sharp curettage for first-trimester-induced abortions.

Dilation and evacuation (D&E) is commonly used for second-trimester abortions up to 18 weeks and, in a few experienced hands, even further. Because fetal parts are larger, gradual dilation of the cervix with multiple hydrophilic dilators such as multiple laminaria tents is necessary. In the hands of experienced surgeons, D&E is associated with a lower complication rate than techniques which induce uterine contractions and expulsion.

Induction of labor in the second trimester with the intraamniotic instillation of 20% sodium chloride (hypertonic saline), intraamniotic hyperosmolar urea usually supplemented with intravenous oxytoxin, or intraamniotic prostaglandins or vaginal prostaglandins are also commonly used to induce second-trimester abortions.

Second-trimester abortions have a higher complication rate than first-trimester pregnancy terminations. Complications include retained placenta that often can be removed under sedation and paracervical block. This occurs less commonly with D&E procedures. Perforation is more frequent with D&E procedures, but overall in experienced hands the incidence is low. Hemorrhage occasionally complicated by coagulopathy is more common in the second trimester. Cervical laceration and other complications related to dilation are less common when the cervix is softened and dilated with hydrophilic dilators. Hypernatremia with hypertonic sodium chloride has been reported. The stress of labor is an emotional side effect which is not seen with dilation and evacuation procedures. Gastrointestinal side effect and delivery of a fetus with a heartbeat and fever are complications more frequently associated with prostaglandin-induced pregnancy termination. The preponderance of evidence concerning safe legal abortion indicates minimal to no long-term sequalae. Particularly after first-trimester abortions by suction curettage, there is no evidence of increased risk of subsequent prematurity, infertility, or ectopic pregnancy. There may be some increase in midtrimester pregnancy loss in women who have undergone two or more induced abortions. Even in this situation, with modern techniques of gentle and slow dilation the risks appear to be quite small. D&E now can be used to successfully manage fetal death *in utero*.

Transabdominal injection of potassium chloride into the heart or umbilical vein can be used to terminate a pregnancy with an anomalous fetus or reduce the number of fetuses in a multifetal pregnancy.

Human Rho(D) immune globulin should be administered to all patients undergoing a pregnancy termination who are Rh negative. The routine use of postoperative antibiotics may decrease the risks of endometritis. The patient should be promptly seen for any signs or symptoms of retained products or pelvic infection such as prolonged heavy bleeding or worsening pain or fever. The overall risk of death for the mother from childbirth is about 10 times higher than that for abortion in the United States; an abortion is safer than delivery in older women and those with medical problems complicating pregnancy. The risk of death associated with induced abortion increases with gestational age: 1 death for every 600,000 abortions at 8 weeks or less, and 1 per 17,000 at 16–20 weeks. From 1973 to 1987 abortion death rates decrease from 3.4 to 0.4 deaths per 100,000 procedures. Death rates are lower in countries in which abortions are legal and accessible.

Major complications associated with abortion, such as serious pelvic infection, hemorrhage requiring blood transfusion, or unintended major surgery, occur in <1% of all induced abortion procedures in the United States.

The availability of sensitive qualitative and quantitative serum hCG assays permits the accurate diagnosis of pregnancy, even at a very early stage. Abdominal pelvic ultrasound and transvaginal ultrasound facilitate the early diagnosis of intrauterine pregnancy, missed abortion, and ectopic pregnancy. Induced abortion is safer because of advances in the ability to accurately date the gestational age and, therefore, guide the choice of the correct procedure and tools to terminate the pregnancy.

See Also the Following Articles

CHORIONIC GONADOTROPIN, HUMAN; FETAL ANOMALIES; TURNER'S SYNDROME; ULTRASOUND; UTERINE ANOMALIES

Bibliography

Alan Guttmacher Institute (1995). Abortion in the United States, Facts in Brief. Alan Guttmacher Institute.

American College of Obstetrics and Gynecology (ACOG) (1995, September). Early pregnancy loss, Technical Bulletin No. 212. ACOG.

Berkowitz, R. L., Stone, J. L., and Eddleman, K. (1997). One hundred consecutive cases of selective termination of an abnormal fetus in a multifetal gestation. *Obstet. Gynecol.* 90, 606–610.

Crenin, M. D., Vittinghoff, E., Schlaff, E., Klaisle, C., Darney, P., and Dean, C. (1997). Medical abortion with oral methotrexate and vaginal misoprostol. *Obstet. Gynecol.* 90, 611–616.

Dizon-Townson, D., and Branch, D. W. (1997). The antiphosholipid antibody syndrome. *Primary Care Update Ob/Gyn* 4, 92–96.

Edwards, J. (1996). Surgical abortion at less than six weeks gestation. Reproductive Health 1996, Association of Reproductive Health Prof. meeting, Nashville, TN.

Grimes, D. A. (1997). Emergency contraception—Expanding opportunities and primary prevention. *N. Engl. J. Med.* 337, 1078–1079.

Hurd, W. W., Whitefield, R. R., Randolph, J. F., Jr., and Kercher, M. L. (1997). Expectant management versus elective curettage in the treatment of spontaneous abortion. *Fertil. Steril.* 68, 601–606.

Jain, J. K., and Mishell, D. R., Jr. (1997). How clinical studies rate abortion induction with misoprosol. *Contemp. OB/GYN* 42, 157–165.

Pernoll, M. L., and Taylor, C. M. (1994). In *Current Obstetrics and Gynecology, Diagnosis and Treatment* (A. Decherney and M. Pernoll, Eds.), 8th ed.. Appelton & Lange, Norwalk, CT.

Precis, V. (1994). An update in obstetrics and gynecology: Pregnancy termination, pp. 290–297. American College of Obstetrics and Gynecology.

Precis, V. (1994). An update in obstetrics and gynecology: Recurrent spontaneous abortion, pp. 433–438. American College of Obstetrics and Gynecology.

Wilcox, A. J., Weinberg, C. R., O'Connor, J. F., Baird, D. D., Schlatterer, J. P., Canfield, R. E., *et al.* (1988). Incidence of early pregnancy loss. *N. Engl. J. Med.* 319, 189–194.

Acanthocephala

D. W. T. Crompton
University of Glasgow

I. Introduction
II. Sex Determination and Sex Ratio
III. Functional Morphology of the Reproductive System
IV. Gonads and Gametogenesis
V. Sexual Congress, Mating, and Fertilization
VI. Prepatency, Patency, and Fecundity
VII. Embryology and Development
VIII. Research Themes

GLOSSARY

Acanthella (plural, *Acanthellae*) An acanthocephalan development stage in the intermediate host leading to the formation of the cystacanth.

acanthor The stage in an acanthocephalan life history that results from embryogenesis in the eggshells within the female worm, adapted for the infection of the intermediate host.

chromosome A rod-like structure containing DNA in the nucleus of an eukaryotic organism.

cystacanth The end product of acanthocephalan development in an intermediate host, adapted for transmission to a definitive host.

definitive host The host in which a parasite having an indirect life history pattern attains maturity and participates in sexual reproduction.

dioecious Having the two sexes in separate individuals.

gamete A specialized product from the gonads (testis or ovary) of an organism designed to transfer genetic material while participating in fertilization.

intermediate host The host in which a parasite having an indirect life history pattern undergoes development but would not normally attain sexual maturity.

monorchism The condition of having one testis.

paratenic host A host which may be essential for the transport of a parasite between intermediate and definitive hosts but which is not thought to be necessary for development.

polygamy The habit of mating with more than one partner of the opposite sex.

polyspermy The entry of more than one spermatozoon into the oocyte at the time of fertilization.

syncytium A mass of cytoplasm containing several nuclei, not divided by membranes into separate cells.

zygote The product of the fusion of a male (spermatozoon) and female (oocyte) gamete; referred to in metazoan animals as a fertilized egg.

Until 1948, the acanthocephalan worms were included as a group in the Aschelminthes, but then Van Cleave proposed that they merited the status of a phylum in their own right. Recently, results from a limited number of investigations of acanthocephalan spermatozoa and mitochondrial nucleic acid sequences have indicated that acanthocephalans may be closely related to rotifers; in fact, some authorities have questioned their qualifications for phylum status and suggest that acanthocephalans, rotifers, and *Seison* comprise the monophylum Syndermata. However, acanthocephalan reproductive biology has many unique features and further study of these may make a significant contribution to our understanding of the origins and phylogenetic relations of this interesting and highly specialized group of endoparasitic invertebrates.

I. INTRODUCTION

A. Acanthocephalan Features

Acanthocephalan worms are readily recognized anatomically by the possession of (i) a retractile, hook-bearing proboscis; (ii) two lemnisci; (iii) a complex body wall which includes typical nuclei and a lacunar system; (iv) a pseudocoelomic body cavity; (v) cement glands and a copulatory bursa in male worms; and (vi) a uterine bell in female worms (Fig. 1). Much more morphological detail is given by Meyer (1932/1933), Bullock (1969) and Miller and Dunagan (1985). Acanthocephalans do not possess an alimentary tract at any stage of their development or life history; feeding is based on the absorption of nutrients across the tegumentary surface. As far as is known, all species of Acanthocephala are dioecious, sexual maturity is attained in the alimentary tract, usually the small intestine, of vertebrates, and there is no evidence of asexual reproduction. Most acanthocephalans are quite small, ranging in length from about 5 to 30 mm and over 1000 species have been described.

B. Life History Patterns

All known acanthocephalans are endoparasites and all species studied have been found to have an indirect life history pattern involving an invertebrate intermediate host and a vertebrate definitive host. Acanthocephalans which reproduce in top predators invariably rely for transmission on a paratenic host that forms an important prey item in the diet of the top predator. Postcyclic transmission is also known to occur in which subpopulations of mature worms are successfully transferred through predation or cannibalism between definitive hosts. Species of arthropod serve as intermediate hosts; usually crustaceans for acanthocephalans that infect aquatic vertebrate hosts and insects for those that infect terrestrial vertebrate hosts. Generally, specificity for the intermediate host is observed to be high, whereas that for the definitive host appears to be low. The biological significance of what may obviously become a com-

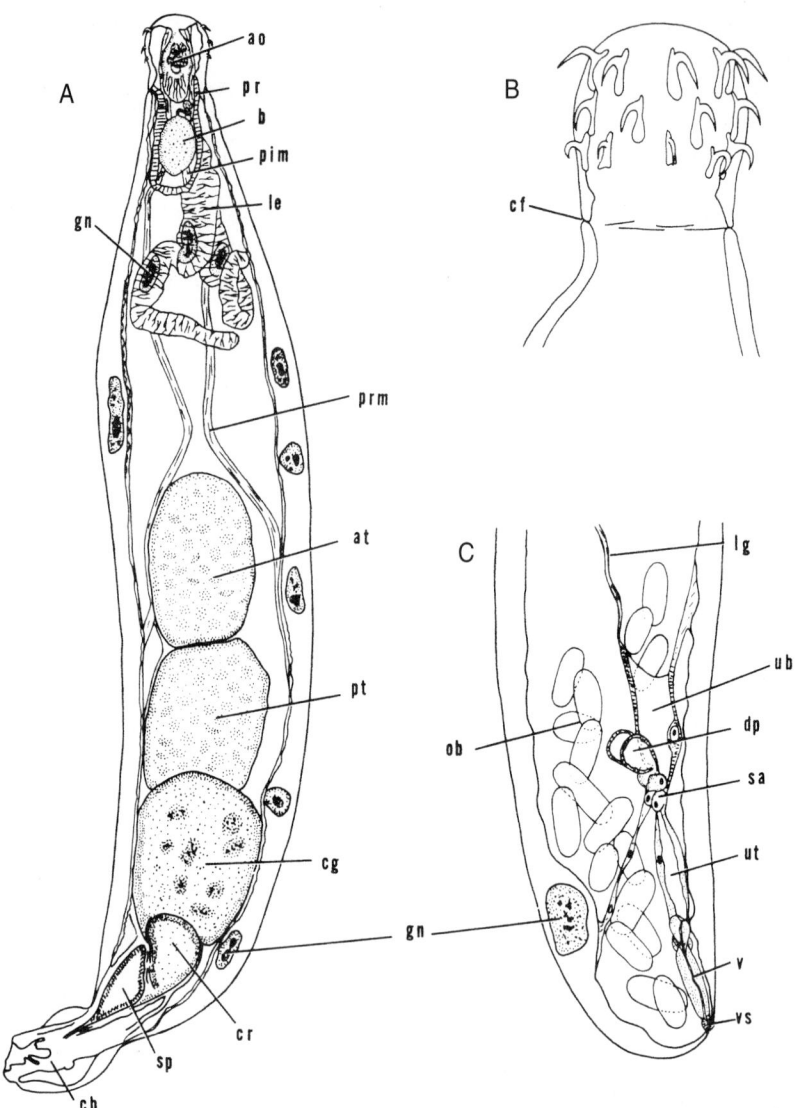

FIGURE 1 General morphology and body plan of adult acanthocephalans based on *Octospiniferoides chandleri*. (A) Adult male. (B) Proboscis and neck region. (C) Posterior end of an adult female. ao, apical organ; at, anterior testis; b, proboscis ganglion; cb, copulatory bursa; cf, cuticular fold; cg, cement gland; cr, cement reservoir; dp, dorsal pouch; gn, giant nucleus; le, lemniscus; lg, ligament; ob, ovarian ball; pim, proboscis invertor muscle; prm, proboscis retractor muscle; pr, proboscis receptacle; pt, posterior testis; sa, selective apparatus; sp, seminal vesicle; ub, uterine bell; ut, uterus; v, vagina; vs, vaginal sphincter. In practice, there is no agreed terminology for acanthocephalan anatomy and the variety of terms used by Crompton and Bullock (1985) may be consulted for reference (reproduced with permission from Bullock, 1969).

plex life history is to enable each individual acanthocephalan worm of a particular species to pass on its genes to the next generation. All its adaptations to parasitism and all its reproductive activity in the context of parasitism are directed to that end.

C. Objectives and Sources of Information

The objective of this article is to offer a concise summary of basic knowledge of the reproductive biology of the Acanthocephala as a foundation to stimulate interest and further research. Introduction to the literature and sources of information are to be found in Meyer (1932/1933), Parshad and Crompton (1981), Crompton and Nickol (1985), and the volumes on the reproductive biology of invertebrates compiled and edited by Adiyodi and Adiyodi (1983–1992).

II. SEX DETERMINATION AND SEX RATIO

The sex of acanthocephalans is established during fertilization by a process involving sex chromosomes. Males are the heterogametic sex (XO or XY) and females the homogametic (XX). The mechanism involving XO is more common; of seven species studied, five have been found to be XO. For example, in young adult *Moniliformis moniliformis* four homologous pairs of metacentric chromosomes (2n = 8) are present of which one pair is much smaller than the others (4 μm compared with 7 μm). During spermatogonial metaphase, seven chromosomes can be observed—three pairs and a solitary small chromosome—indicating that males are determined by an XO mechanism.

When sex determination depends on one homogametic and one heterogametic parent, a sex ratio of 1:1 is to be expected. This ratio is regularly found in what are known to be populations of young worms. There is plenty of experimental evidence indicating that females live longer than males. This differing longevity can be explained as an adaptation favoring reproductive success. If females have a finite number of oocytes (see Section IV,A), once sufficient spermatozoa have been delivered to achieve maximum fertilization events, loss of males will release resources (food and space in the restricted environment of the small intestine) to the females for the provision and development of the shelled acanthors (see Section VI).

Marked sexual dimorphism is observed in some species of acanthocephalans, including the size and shape of the body (females are often larger than males), the distribution of body spines, the size and shape of proboscis hooks, and the position of genital openings. The significance of these differences is not understood: do they promote mate choice or do they correlate with reproductive success or some form of sexual selection? The behavioral ecology of parasites is a topic awaiting much more investigation.

III. FUNCTIONAL MORPHOLOGY OF THE REPRODUCTIVE SYSTEM

The main organs of the reproductive system are associated with the ligament sac or sacs or their derivatives. In some species, the ligament sac breaks down as the female worms mature, and in others the sacs remain as a loose lining in the pseudocoelom. The sacs tend to remain intact in male worms regardless of the species or taxon.

A. Female Tract and Organs

The female reproductive system is contained in the body cavity and consists of ovarian balls (= free-floating ovaries; see Section IV,A), the uterine bell, the uterus, vagina, and gonopore (= efferent duct system). A vaginal ganglion has been found in female *Echinopardalis atrata*. The uterine bell is a particularly interesting structure which not only allows spermatozoa to gain access to the ovaries in the body cavity but also sorts the eggs, discharging those that are fully developed (i.e., contain acanthors; see Section VII) via the efferent duct system and returning to the body cavity those that are not. Details of the structure and probable functioning of the uterine bell have been worked out for *Polymorphus minutus* by Whitfield and reviewed for various species by Miller and Dunagan (Fig. 2). Evidence that the uterine bell has this remarkable egg-sorting function is secured by the fact that eggs in all stages of development are seen in the female worm's body cavity, but only fully developed eggs are normally detected in the feces of an infected host.

B. Male Tract and Organs

The male reproductive tract is much more complex than the representation shown in Fig. 1. In addition to two testes and related ducts, there are the cement

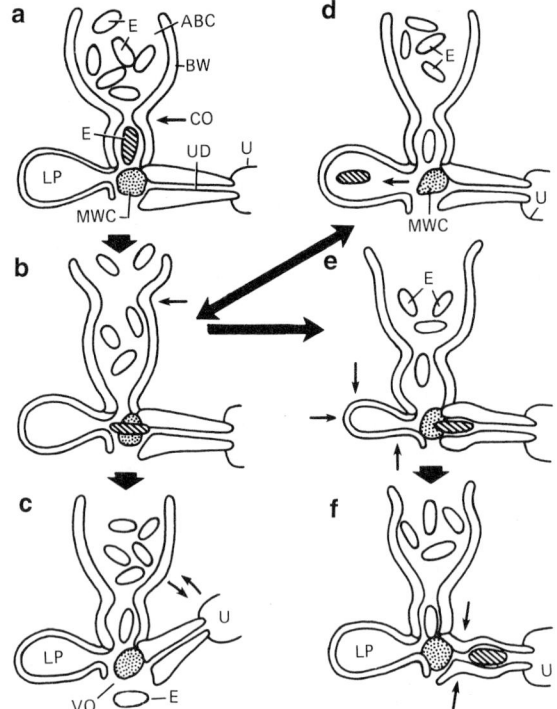

FIGURE 2 The egg-sorting function of the uterine bell. Underdeveloped eggs are assumed to follow sequence (a–c) into the ligament sac or body cavity. Developed eggs (= shelled acanthors; shaded areas) follow sequence (a, b, e, and f) into the uterus. Apparently, eggs cannot move from b to a. ABC, anterior bell chamber; BW, bell wall; CO, common oviduct; E, egg; LP, lateral pocket; MWC, median wall cell; U, uterus; UD, uterine duct; VO, ventral opening (reproduced with permission of Cambridge University Press from Miller and Dunagan, 1985).

glands, other glands, Saefftigen's pouch, the copulatory bursa, and the penis or cirrus. There is also a posterior genital ganglion and a bursal ganglion, neither of which are found in females and which may be assumed to be concerned with copulatory activity. A full description is given by Miller and Dunagan and it should be noted that the bursal ganglion been found in few species.

The spherical or ovoid testes are located in the ligament sac in the anterior, middle, or posterior part of the body; this location is useful for taxonomic work. Monorchic males are frequently found and those of *M. moniliformis* have been shown to be capable of successful insemination. Three general forms of cement gland are recognized and their organization in the body is useful in taxonomy. The secretions of the cement glands, perhaps modified by secretions from the other glands, provide the material used to form the copulatory cap (see Section V). The caps appear to be quite hard and secretions from various glands could lead to some form of irreversible polymerization.

The copulatory bursa, like the proboscis, is eversible and is shaped to envelope the posterior end of the female body where the vagina is situated. The bursa contains diverse musculative and is in some way linked to Saefftigen's pouch. The precise function of this pouch is not yet known, but it is likely to be involved in the movements of the bursa.

IV. GONADS AND GAMETOGENESIS

The gonadial organ primordia arise from the central nuclear mass or entoblast in the middle of the developing acanthor. Gonadial development then continues in close association with the ligament; usually the testes are formed in advance of the ovaries and male worms mature before females.

A. Ovaries, Oogenesis, and Oocytes

As a female acanthocephalan matures, the ovary divides and the division products further divide so that the body cavity or ligament sacs within the body cavity become endowed with numerous free-floating ovaries. These ovarian structures are commonly known as ovarian balls and the process of their formation is referred to as ovarian fragmentation. Under laboratory conditions, female *M. moniliformis* contain on average about 8 ovarian balls at the end of the first week of the course of infection. By 3 weeks postinfection, when insemination and fertilization are beginning in *M. moniliformis*, approximately 2000 ovarian balls are observed per female worm and the number gradually rises to a maximum of about 6000. Similar events occur for other acanthocephalan species, although the numbers produced and the timing of the process vary between species.

In the case of *M. moniliformis*, the mature ovarian balls are prolate spheroids of maximum length 300 μm and each has three components. There are two multinucleate syncytia, named the oogonial syncytium and the supporting syncytium, and a cellular zone. These components have been recognized in acanthocephalan species from the three main taxa of the phylum. The cells are observed to be germline cells in varying stages of development that are budded off from the oogonial syncytium. Those that are observed at the peripheral part of the ovarian ball are mature oocytes, zygotes, or oocytes undergoing atresia. All these cells are embedded in the support syncytium, which forms the interface between the germline cells, spermatozoa, and nutrients in the body fluid of the parent female worm. Full descriptions of ovarian structure and function have been given for *M. moniliformis* (Archiacanthocephala), *P. minutus* (Palaeacanthocephala), and *Pallisentis golvani* (Eoacanthocephala) by Crompton and Whitfield (1974), Atkinson and Byram (1976), and Marchand and Mattei (1976). A diagrammatic interpretation of acanthocephalan ovarian processes is given in Fig. 3. Details of the cytological events leading to the development of the mature ovarian ball have been described for *M. moniliformis* by Asaolu *et al.* (1981).

B. Testes, Spermatogenesis, and Spermatozoa

The sex of male acanthocephalans can be determined approximately from the middle onwards of the period of development in the intermediate host when the rudimentary testes can be seen as thickenings associated with the developing ligament (Table 1). In the case of *P. minutus*, Whitfield found two functional cell types in the testis; small cells measuring about 5–7 μm in diameter with relatively large nuclei and basophilic cytoplasm and larger cells of 10–15 μm in diameter. The small cells are believed to be primordial germ cells or gonocytes and the larger cells are supporting cells. Processes from the supporting cells envelope the gonocytes or extend between them; in later stages of development the supporting cells appear to form a syncytium. Similar observations were made by Asaolu studying *M. moniliformis,* and to avoid confusion with one of the syncytia in the ovarian ball, Asaolu named this the sustaining syncytium in the testis.

The mature testis of *M. moniliformis* consists of the various types of cells involved in spermatogenesis (spermatogonia, two stages of spermatocytes, spermatids, and spermatozoa) arranged in clusters or morulae loosely enveloped by the sustaining syncytium. The coat or outer layer of the testis appears to be derived from the sustaining syncytium.

Spermatogenesis in the Acanthocephala appears to follow the same sequence of stages as have been described in detail for other phyla. First, spermatocytogenesis produces spermatogonia, which increase in number by mitosis to form spermatocytes; then meiosis occurs to form spermatids; and finally spermiogenesis takes place, leading to the production of spermatozoa. These events occur once the male worms are established in an appropriate species of

TABLE 1
Events in the Ontogeny and Reproduction of Various Species of Acanthocephala

	Species		
Event	Macracanthorhynchus hirudinaceus	Moniliformis moniliformis	Polymorphus minutus
Development in IH		49 d (26°C)	60 d (17°C)
Beginning of ovarian fragmentation		IH	IH
Testes primordia in IH		38 d (26°C)	35 (17°C)
Prepatent period	70 d	38 d	22 d
Patent period	~300 d	106 ± 16 d	25 d
Egg output per female worm	24 million	600,000	13,000
Daily egg output per day	260,000	5,500	1,700 (peak)

Note. Abbreviations used: IH, intermediate host; d, days.

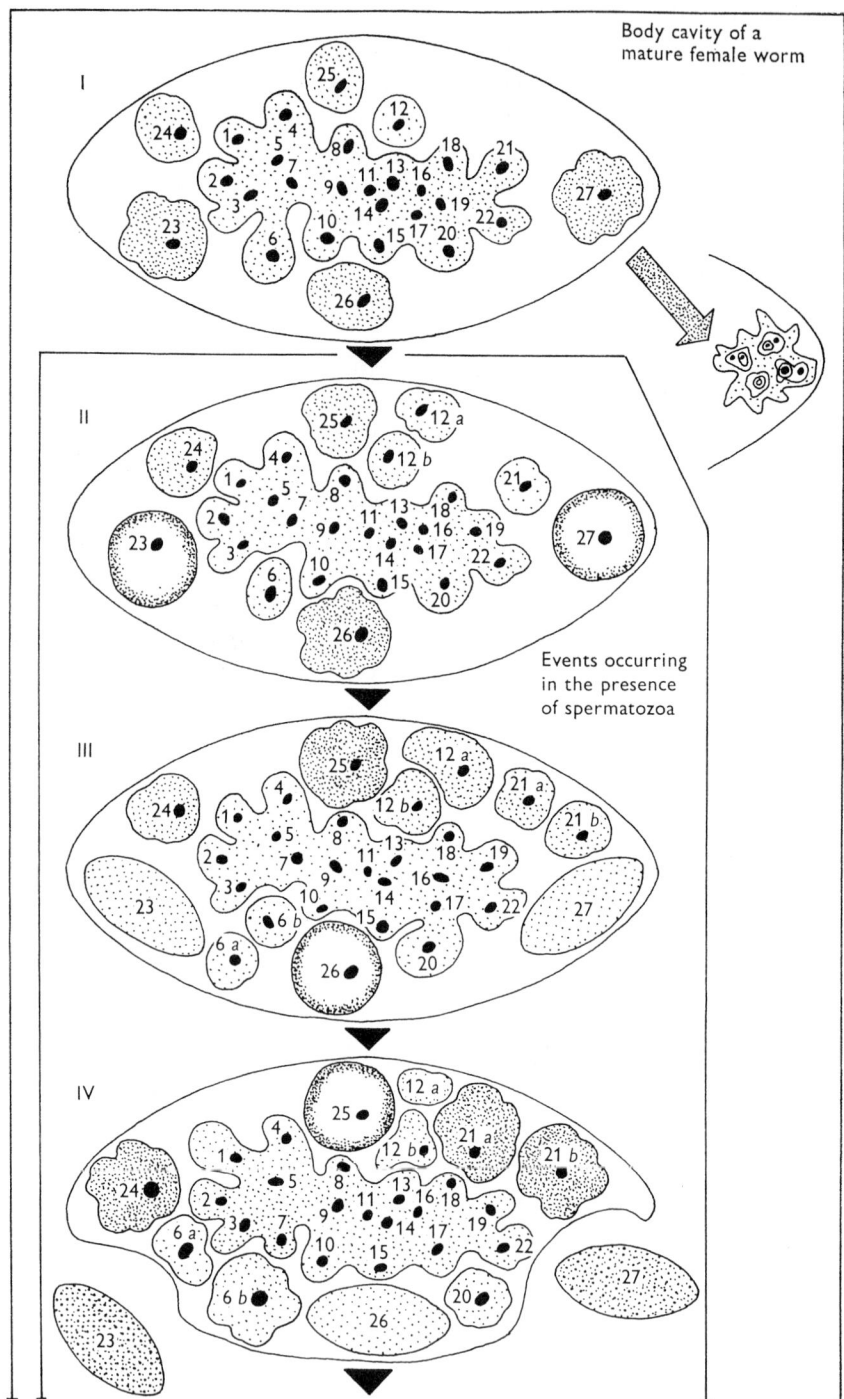

FIGURE 3 Diagrammatic interpretation of cytological processes and oogenesis in an ovarian ball from the body cavity of a mature female acanthocephalan worm. Stippling represents the oogonial syncytium and its cellular products. The unstippled part of the ovarian ball represents the supporting syncytium. The numbers identify nuclei in or emerging from the oogonial syncytium (reproduced with permission of Cambridge University Press from Crompton and Nickol, 1985).

axoneme invariably reveals the 9+2 arrangement of peripheral doublets and central singlets, respectively. In some species, however, other arrangements may occur—for example, 9+0 in *Illiosentis furcatus*.

V. SEXUAL CONGRESS, MATING, AND FERTILIZATION

On the basis of experimental studies of primary infections of *M. moniliformis* in laboratory rats, acanthocephalans may be assumed to be polygamous, with males being capable of inseminating several females. However, each female in a subpopulation of worms may not achieve maximum egg production unless several males are present; the optimum operational sex ratio remains to be determined. Males and females of different ages can mate successfully, leading to the production of infective shelled acanthors. The fact that acanthocephalans locate in fairly precise sites in the small intestines of their definitive hosts favors sexual contact and, in the case of *M. moniliformis*, there appears to be a time corresponding with the onset of insemination when an individual worm is having most physical contact with others. This latter observation may be an experimental artifact resulting from the fact that all the worms had arrived in the rat at the same time.

Spermatozoa are transferred from the male to the female by a copulatory act in which the male bursa is everted and placed over the posterior end of the female. A plug of cement is deposited in the entrance to the vagina and more cement extends over the female's posterior end. These copulatory caps serve to prevent loss of a male's spermatozoa and also delay the introduction of spermatozoa from rival males. In due course, perhaps after a few days, the copulatory caps are lost and shelled acanthors can begin to be released; more spermatozoa can also be acquired.

There has been considerable interest in the finding of copulatory caps on the posterior ends of male worms. This observation is interpreted by some workers as a form of sexual selection in which a dominant male is seen to be temporarily delaying the reproductive activity of other males. Other workers believe that the caps on males are in some way a response from the deteriorating environmental con-

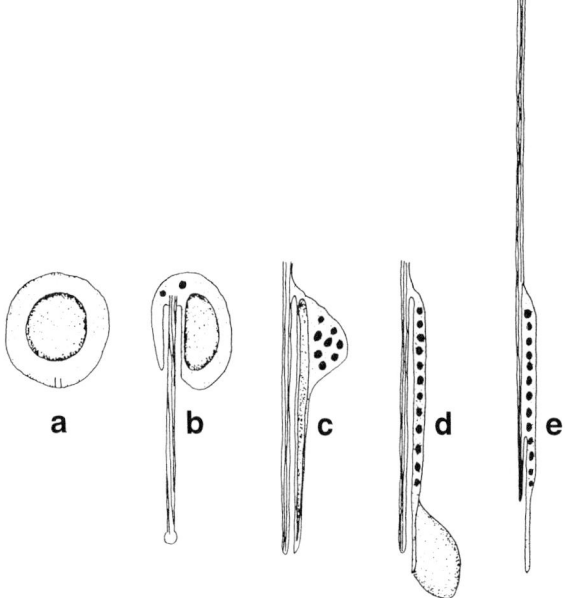

FIGURE 4 Spermiogenesis in *Neoechinorhynchus agilis* (Eoacanthocephala). (a) Young spermatid, (b) first growth of the flagellum, (c) older spermatid with elongated nucleus and extending flagellum, (d) formation of the cytoplasmic droplet prior to its discard, and (e) fully formed spermatozoon in which the main part of the flagellum is seen to be at the anterior part while the initial extension of the flagellum has fused with the spermatozoon body. (Based on studies by Marchand and Mattei and reproduced from Adiyodi and Adiyodi, Vol. 2, 1983. © John Wiley and Sons Limited.).

definitive host. Most is know about spermiogenesis due mainly to the work of Marchand and Mattei, who studied 11 species from the three main taxa of the phylum. The centriole migrates to the anterior part of the spermatid to become located beyond the anterior end of the nucleus. The flagellum begins to develop, the nuclear envelope disintegrates, and a residual portion of cytoplasm, containing the mitochondrion, is discarded as a droplet. These events are summarized in Fig. 4, which is based on investigations by Marchand and Mattei.

Mature spermatozoa are motile, filariform cells measuring from about 20 to 70 μm in length, depending on the species, and consisting of a flagellum and a body or nucleocytoplasmic derivative which contains DNA and glycogen. There is no evidence of either an acrosome or a mitochondrion. The flagellar ditions in dead hosts. This view is difficult to sustain because the caps can also be observed on vigorously moving males recently retrieved from an experimental host. One difficulty that must be explained is the fact that shelled acanthors are discharged through the same opening as spermatozoa enter. The release of shelled acanthors could be disrupted if copulatory caps are always produced following insemination.

Sexual congress and mating behavior among acanthocephalans is clearly more complex than might have been initially imagined. Genes can only be passed to the next generation by copulation between a male and a female worm. There is no evidence of asexual propagation, parthenogenesis, hermaphroditism, or other specialized reproductive strategy. Lawlor *et al.* discussed mate choice in relation to *M. moniliformis* and devised a system for exposing males to excess numbers of females and then comparing the features of the inseminated females. Interestingly, among the females recovered at the end of the experiment, those that had been inseminated were located most anteriorly and were the largest. This pattern of mating adopted by the males is consistent with the maximization of reproductive success; the male genes were being deposited in females with the best access to exogenous resources and so are likely to have the best endogenous resources. Many workers have observed that male acanthocephalans are often less firmly attached than females to the host's intestinal wall, another feature suggesting that active mate choice could be component of male mating behavior.

Once spermatozoa have entered the body cavity via the uterine bell, they are observed to swim actively in the body fluid and to become attached to the surface of the ovarian balls. It is not known how long they can survive in the body cavity of the recipient female worm. The mature oocytes are located near the surface of the ovarian ball just beneath the cortical layer of the supporting syncytium. In the case of *P. minutus* and *P. golvani*, electron microscope observations reveal that the entire spermatozoon enters the mature oocyte as the first stage of fertilization. Once the spermatozoon has entered the oocyte, the flagellum and nucleocytoplasmic derivative separate and the flagellum appears to disintegrate. The fusion of the male and female nuclear elements has still to be observed.

Electron microscope studies have also shown that the cytoplasm of mature oocytes contains numerous electron-dense granules dispersed fairly evenly across the cell. Once the spermatozoon has entered the oocyte, the granules are found placed just beneath the cell surface, perhaps acting as a block to polyspermy but also forming the fertilization membrane which becomes one of the egg-shells.

VI. PREPATENCY, PATENCY, AND FECUNDITY

Prepatency and patency are rather artificial concepts. The prepatent period is the time that passes from the beginning of an infection to the first detection of shelled acanthors in the host's feces. The patent period is the time during which shelled acanthors are produced. Fecundity is defined as the number of shelled acanthors produced per female worm and ultimately depends on the number of nuclei generated by the oogonial tissue in the ovarian balls. Relevant information obtained from acanthocephalans studied under experimental conditions is summarized in Table 1. Experiments have also shown that acanthocephalan reproduction is sensitive to density-dependent constraints and that the quality and quantity of the host's diet is accompanied by effects on sexual development and fecundity.

VII. EMBRYOLOGY AND DEVELOPMENT

For arbitrary convenience, embryological events are taken to be the transformation of the zygote to the shelled acanthor. These processes are crucial to our understanding of acanthocephalan reproduction because they involve sex determination, gonad determination, and fecundity. The production of shelled acanthors occurs inside the adult female worm at the deep body temperature of the definitive host. The acanthors have been shown to be infective to intermediate hosts as soon as they are fully formed; perhaps the relatively low fecundity of acanthocephalans (Table 1) is offset by the equivalent of parental care during the production of shelled acanthors.

Albrecht *et al.* investigated the fine structure of developing acanthors of *P. minutus*, *Neoechinorhynchus rutili*, and *M. moniliformis*, representative species of the three main taxa. The zygote is a single cell and the acanthor is observed to be composed of three syncytia. There are differences regarding the number of divisions that occur to give rise to the syncytia, but in these species and all others studied the central syncytium is found to contain many condensed nuclei (about 400 in an acanthor of *P. minutus*). Albrecht *et al.* conclude that all organs, including gonads, arise from the central syncytium of the fully formed acanthor. How the gonads arise would be the subject of a most intriguing investigation.

Development is taken to be the transformation of the acanthor, after release from its eggshells, through the acanthella stages to the cystacanth stage. These events occur in the intermediate host, in which rate of development is sensitive to ambient temperature. Interestingly, some acanthocephalans using poikilothermic definitive hosts will largely be subject to ambient temperatures throughout their life histories. For example, *N. rutili* can complete its life cycle in three-spined sticklebacks and ostracods in the same freshwater pond. *Moniliformis moniliformis*, in contrast, relies on rat deep body temperature to support embryogenesis and tolerates the vagaries of the temperature in the cockroach's environment while accomplishing development. Even more intriguing is the observation that the spermatozoa of a rat do not develop at rat deep body temperature (hence the descent of the testicles), whereas those of *M. moniliformis* are adapted for development at the higher temperature.

VIII. RESEARCH THEMES

In their review of knowledge of acanthocephalan reproduction, Parshad and Crompton compiled a list of 60 topics requiring further work. Some of these topics still await attention, but future approaches should take advantage of molecular biology techniques, be comparative between species of Acanthocephala to expand knowledge of the group, be comparative with other groups to contribute to our understanding of phylogenetic relations, and be more aware of advances in behavioral ecology. The following topics are suggested as being fruitful themes for research and discovery:

Fecundity: Given that over 1000 species of Acanthocephala have been described, there is a scarcity of information.

Fertilization: How do spermatozoa pass into the ovarian ball and then into the mature oocyte and how do the male and female nuclei fuse?

Mating behavior: Can evidence be obtained to recognize mate choice, sexual selection, and intraspecific arms races?

Gonad development: What are the roles of the nuclei and the cytoplasm in the central syncytium of the acanthor in forming the gonads and germline cells?

Acknowledgments

I thank Professor P. J. Whitfield and Dr. B. B. Nickol for their helpful comments and Mrs. Desho Sahonta for her skilled preparation of the manuscript.

See Also the Following Articles

PARASITES AND REPRODUCTION; ROTIFERA

Bibliography

Adiyodi, K. G., and Adiyodi, R. G. (1983–1992). *Reproductive Biology of Invertebrates*, Vols. 1–5. Chichester, Wiley.

Asaolu, S. O., Whitfield, P. J., Crompton, D. W. T., and Maxwell, L. (1981). Observations on the development of ovarian balls of *Moniliformis* (Acanthocephala). *Parasitology* 83, 23–32.

Albrecht, H., Ehlers, U., and Taraschewski, H. (1997). Syncytial organisation of acanthors of *Polymorphus minutus* (Palaeacanthocephala), *Neoechinorhynchus rutili* (Eoacanthocephala), and *Moniliformis moniliformis* (Archiacanthocephala) (Acanthocephala). *Parasitol. Res.* 83, 326–338.

Atkinson, K. H., and Byram, J. E. (1976). The structure of the ovarian ball and oogenesis in *Moniliformis dubius* (Acanthocephala). *J. Morphol.* 148, 391–426.

Bullock, W. L. (1969). Morphological features as tools and as pitfalls in Acanthocephalan systematics. In *Problems in Systematics of Parasites* (G. D. Schmidt, Ed.), pp. 9–45. University Park Press, Baltimore.

Crompton, D. W. T., and Nickol, B. B. (Eds.) (1985). *Biology of the Acanthocephala*. Cambridge Univ. Press, Cambridge, UK.

Crompton D. W. T., and Whitfield, P. J. (1974). Observations on the functional organisation of the ovarian balls of *Moniliformis* and *Polymorphus* (Acanthocephala). *Parasitology* 69, 429–443.

Lawlor, B. J., Read, A. F., Keymer, A. E., Parveen, G., and Crompton, D. W. T. (1990). Non-random mating in a parasitic worm: Mate choice by males? *Anim. Behav.* 40, 870–876.

Marchand, B., and Mattei, X. (1976). Präsence de flaggelles spermatiques dans les sphÅres ovariennes des äoacanthocephales. *J. Ultrastruct. Res.* 56, 331–338.

Marchand, B., and Mattei, X. (1978). La spermatogenèse des Acanthocéphales J. Flagellogenèse chez un Eoacanthocephala: mise en place et désorganisation de l'axonème spermatique. *J. Ultrastruct. Res.* 63, 41–50.

Meyer, A. (1932/1933). Acanthocephala. In *Bronn's Klassen und Ordnungen des Tierreichs 4*, pp. 1–332 (1932) and 333–582 (1933). Akademische Verlagsgesellschaft, Leipzig.

Miller, D. M., and Dunagan, T. T. (1985). Functional morphology. In *Biology of the Acanthocephala* (D. W. T. Crompton and B. B. Nickol, Eds.), pp. 73–123. Cambridge Univ. Press, Cambridge, UK.

Parshad, V. R., and Crompton, D. W. T. (1981). Aspects of acanthocephalan reproduction. *Adv. Parasitol.* 19, 73–138.

Schmidt, G. D. (1985). Development and life cycles. In *Biology of the Acanthocephala* (D. W. T. Crompton and B. B. Nickol, Eds.), pp. 273–305. Cambridge Univ. Press, Cambridge, UK.

Whitfield, P. J. (1970). The egg sorting function of the uterine bell of *Polymorphus minutus* (Acanthocephala). *Parasitology* 61, 111–126.

Acari

see Chelicerate Arthropods

Acrosome

see Spermatozoa

Acrosome Reaction

Gregory S. Kopf

University of Pennsylvania

I. Introduction
II. Acrosome Biogenesis and Maturation
III. Biological Significance and Site of the Acrosome Reaction
IV. Nature of the Biological Trigger of the Acrosome Reaction
V. Biochemistry of the Acrosome Reaction

GLOSSARY

acrosome A secretory vesicle overlying the nucleus of mammalian sperm, which contains a variety of nonenzymatic and enzymatic proteins; the contents of the acrosome are released following the acrosome reaction.

acrosome reaction An exocytotic event that is an absolute prerequisite to successful fertilization and occurs following sperm binding to the zona pellucida of the egg; exocytosis involves the fusion and vesiculation of the plasma membrane overlying the acrosome and the outer acrosomal membrane.

zona pellucida An oocyte-specific extracellular matrix that surrounds the egg and is responsible for species-specific binding of sperm and induction of the acrosome reaction of bound sperm; the components of the zona pellucida are glycoproteins.

The acrosome is a secretory vesicle overlying the nucleus of the mature spermatozoon, is a product of the Golgi complex, and is synthesized and assembled during spermiogenesis. The contents of the acrosome include structural and nonstructural and nonenzymatic and enzymatic components, and this secretory vesicle is delimited by both the inner and outer acrosomal membranes. The acrosome reaction in mammalian sperm, as in the sperm of other lower species, is a regulated exocytotic event that is an absolute prerequisite to successful fertilization under physiological conditions.

I. INTRODUCTION

In mammalian sperm the acrosome reaction involves the fusion and vesiculation of the plasma membrane overlying the acrosome with the outer acrosomal membrane, thus creating hybrid membrane vesicles. The molecular mechanisms involved in this fusion and vesiculation process are not known. Although this exocytotic event can be induced by both physiological stimuli and pharmacological agents, the molecular mechanisms by which these different stimuli and agents function to induce exocytosis may be dramatically different. The resultant fusion of these membranes leads to the subsequent exposure of the acrosomal contents to the extracellular environment. Both the soluble and insoluble components of the acrosome appear to play important roles in the binding of the acrosome-reacted sperm to the zona pellucida as well as the subsequent penetration of the acrosome-reacted sperm through the zona pellucida.

II. ACROSOME BIOGENESIS AND MATURATION

Although the acrosome is present following spermiogenesis, there remain several questions pertaining to the formation and maturation of this organelle. For example, although prominent biogenesis of the acrosome occurs during the Golgi and cap phases of spermiogenesis, it is not clear when during this developmental process that this organelle actually starts to develop. Furthermore, the acrosome is composed of multiple component proteins, but little is known regarding whether the synthesis of all these components occurs at the same time or whether synthesis is ordered and coordinate. Experimental evidence to date suggests the latter mechanism. The mechanism by which these acrosomal components are targeted to this organelle during biogenesis is also not known. Although spermatogenic cells possess functional mannose-6-phosphate/insulin-like growth factor-II receptors, it is not clear whether these receptors play a role in the transport of glycoproteins to the acrosome or whether targeting occurs primarily through the "default" pathway seen in the transport of proteins in other secretory systems. Finally, once these components are packaged into the acrosome, the functional significance of additional processing of these components (i.e., posttranslational modifications and movement within the organelle) during sperm residence in the testis and/or during residence in the extratesticular male reproductive organs (i.e., epididymis and vas deferens) is not clear. In some species (e.g., guinea pig), the formation of specific domains within the acrosome has been clearly demonstrated, but the mechanism by which this compartmentalization is established is poorly understood and an understanding of the biological role of this compartmentalization is only beginning to be realized. Answers to all these questions will no doubt become apparent when a systematic evaluation of many of the proteins comprising the acrosome is undertaken with respect to transcription, translation, and posttranslational modifications. An understanding of these processes may greatly increase our knowledge of the role of the acrosome in fertilization since it is becoming apparent that this secretory vesicle may have multiple functions. In any event, studies focused on the synthesis and processing of acrosomal components should be considered in the context of the acrosome functioning as a secretory granule and not a modified lysosome, as has been historically suggested.

III. BIOLOGICAL SIGNIFICANCE AND SITE OF THE ACROSOME REACTION

Migration of sperm through the different regions of the female reproductive tract (vagina, cervix, uterus, and oviduct), their interactions with the epithelial cells in these regions, and their interactions with the cumulus oophorus-enclosed metaphase II-arrested egg prior to sperm–egg fusion and fertilization provides a very efficient means to select for the most viable, functionally active sperm. In fact, *in vivo* studies suggest that the number of sperm arriving at the site of fertilization at any particular time is extremely low—on the order of 1–100. Although these specific interactions may vary from species to species, such cellular selection is critical for the maintenance and perpetuation of all species. Since the sperm composition of the ejaculate is quite heterogeneous with respect to cellular age, morphology, motility characteristics, and the ability to undergo capacitation, the acrosome reaction itself may also be considered a selection process designed to ensure the delivery of sperm with optimal fertilization potential to the initial site of sperm–egg interaction proper, i.e., the zona pellucida (ZP), which is a unique extracellular matrix that is a product of the growing oocyte. In many species, sperm that undergo the acrosome reaction prematurely generally display reduced fertilizability due to a failure to penetrate the cumulus oophorus, an increased propensity for binding to the cells comprising the cumulus oophorus thus excluding them from interaction with the ZP, and an inability to establish binding to the ZP. Although there are a few exceptions, it is generally accepted that only acrosome-intact sperm can establish high-avidity binding to the ZP. It should be emphasized, however, that recent data pertaining to the identity, localization, and function of sperm surface-associated ZP-binding proteins/receptors may prompt us to rethink the paradigm of acrosomal status in sperm binding to the ZP. The primary function of the acrosome reaction following ZP binding is to permit the penetration of this extracellular matrix so that these cells can gain access to the perivitelline space. A consequence of ZP penetration is the development of a sharply defined, thin penetration slit in this extracellular matrix which is presumed to result from the limited digestion of the matrix by acrosomal-associated proteases. The relative contribution of mechanical forces generated by sperm motility and the aforementioned enzymatic digestion of the ZP by acrosomal proteases in mediating sperm passage through the ZP, however, is not known. As described later, recent data from a number of different laboratories are beginning to define new potential and more active roles for the components of the acrosomal matrix in numerous aspects of the fertilization process.

Sperm, having undergone the acrosome reaction in response to the ZP and having penetrated the ZP, have the ability to interact and fuse with the plasma membrane of the egg once it arrives in the perivitelline space. Since acrosome-intact sperm do not have the ability to bind and fuse with the egg plasma membrane, the acrosome reaction serves an additional role in fertilization, i.e., the exposure of sperm-associated domains involved in egg plasma membrane binding and fusion. The nature of the sperm- and egg-associated molecules involved in these membrane-binding and fusion events is the subject of great interest and intensive scrutiny.

The acrosome reaction, therefore, is biologically significant from two standpoints. First, in most species studied, it serves as a selection process to ensure the delivery of a subpopulation of viable, acrosome-intact sperm to the ZP, the initial site of sperm–egg interaction proper. Second, the ZP-induced acrosome reaction ensures that acrosome-reacted sperm, which are capable of fusion with the egg plasma membrane, are delivered to the perivitelline space, thus enhancing the probability for successful fertilization.

IV. NATURE OF THE BIOLOGICAL TRIGGER OF THE ACROSOME REACTION

Although there is little doubt regarding the importance of the acrosome reaction in the fertilization process, there is some debate regarding the physiologically relevant trigger of the acrosome reaction *in vivo*, i.e., the acrosome reaction required for successful penetration of the ZP. The origins of this debate likely stem from the fact that many conclusions drawn about the *in vivo* situation are based on results of experiments *in vitro*, and this extrapolation could be misleading. For example, a variety of biological agents, when added to sperm, can induce acrosomal exocytosis. Some of these agents have been demonstrated to function through the activation of specific signal transduction cascades in the sperm, whereas the mechanism of action of others is not known. To date, work in a variety of species has demonstrated that bioactive components of serum (e.g., steroid hormones, neurotransmitters, and products of fatty acid metabolism), glycosaminoglycans present within the follicular fluid and cumulus oophorus matrix complex, and the ZP can function to induce the acrosome reaction. Two of these components, progesterone and the ZP, will be considered in detail since the greatest amount of information is known about their biology and mechanism of action on sperm.

Progesterone was identified as the active component of serum or follicular fluid responsible for the induction of acrosomal exocytosis in human sperm. Progesterone has subsequently been demonstrated to induce acrosomal exocytosis in the sperm of several other mammalian species. Several lines of evidence suggest that this effect is mediated by a sperm surface receptor through a nongenomic pathway, although this receptor has still not been unequivocally identified. The identity of this receptor has, to date, been indirectly inferred by studies in other somatic cells, as well as by pharmacological characterization, but will remain controversial until it has been purified or cloned. The mechanism of action by which this steroid induces the acrosome reaction appears to be through an effect to elevate intracellular Ca^{2+}; however, the signal transduction cascades activated by this steroid differ from those induced by the ZP and have not been characterized in detail. The following are significant questions that must be asked regarding the biological relevance of progesterone as a trigger for the acrosome reaction: (i) Where does this steroid exert its effect on the acrosome reaction in the female reproductive tract? (ii) Do the concentrations of this steroid reach levels high enough to induce the acrosome reaction? and (iii) If this steroid functions to induce the acrosome reaction prior to the interaction of the sperm with the ZP, what is its biological significance? To date, estimates have been made that suggest that the progesterone levels in the oviduct might, under certain conditions, be high enough to induce the acrosome reaction. This, however, creates a problem with regard to the biology of fertilization since it has been demonstrated in many species that acrosome-reacted sperm cannot establish binding to the ZP and thus would not fertilize eggs. One, however, could possibly resolve this conundrum by posing two models of action of this steroid that could be physiologically relevant. First, as stated previously, biological selection of the most functionally competent sperm to fertilize eggs most likely occurs at different regions of the female reproductive tract and might seem to be a reasonable mechanism by which a heterogeneous population of sperm gives rise to the generation of a vanguard population of fertilizing sperm. As stated earlier, the acrosome reaction may represent one mechanism of such a selection process. Progesterone in the female tract may, therefore, induce acrosomal exocytosis of those sperm in the population that are not optimal for fertilization. This hypothesis has not been tested. Second, it is possible that progesterone could function in coordination with the ZP to "prime" the sperm so that they undergo efficient acrosomal exocytosis following binding to the ZP. Although this hypothesis has been proposed using *in vitro* assays, its proof will only be approached experimentally once the identity of the sperm surface-binding proteins/receptors for progesterone and the ZP are identified.

The biologically relevant trigger of the acrosome reaction that has received the greatest attention from an experimental standpoint has been the ZP. In a classical series of experiments, it was demonstrated that free-swimming acrosome-intact sperm establish tight binding to the ZP and then, while bound, undergo the acrosome reaction. These initial experiments were carried out in the mouse but were then performed by many investigators in a variety of species, including human. To date, the preponderance of information regarding the molecular basis of sperm–ZP interaction has been gleaned from studies in the mouse. However, it is becoming apparent that results of studies in this species have translated to similar results in many other species, thereby allowing us to conclude that the ZP is the universal biological trigger of the acrosome reaction in mammals.

In the mouse, the ZP is synthesized and assembled by the growing oocyte during its growth phase and is composed of three major glycoproteins, designated ZP1, ZP2, and ZP3. The genes encoding each of these proteins are unique and are under the control of ZP-specific promotors that ensure their temporal and tissue-specific expression; the ZP is synthesized only by the oocyte and not by somatic cells. ZP1, ZP2, and ZP3 are all highly glycosylated, and this glycosylation is extremely important for conferring specific biological functions to the ZP. It should be noted that different designations have been given to the components of the ZP in other species and this has led to some confusion in nomenclature; a uniform nomenclature should be carefully considered so as to avoid future confusion. Recent results suggest that the coordinate expression of these individual ZP components is essential for secretion and subsequent assembly of this unique egg-associated extracellular matrix. Following assembly, the ZP is composed of ZP2/ZP3 heterodimers that are cross-linked in an organized fashion by ZP1 monomers, giving rise to a three-dimensional, relatively insoluble structure that functions biologically as a matrix. It should be noted that the genes encoding the different ZP components in the mouse have been cloned in several other species and the deduced primary polypeptide structures from these other species (including the human) bear remarkable similarity to one another. These data suggest that the primary protein structures of the ZP components of mammalian eggs are similar to one another and that differences observed in biochemical and biological heterogeneity between species may be influenced by the carbohydrate domains. It is notable in this regard that recent studies have demonstrated that the primary structure of the egg vitelline envelopes of certain fish, the extracellular coats of these oldest of vertebrates, bear a resemblance to the structure of ZP2 and ZP3. These data suggest that specific domains of egg extracellular coats which are involved in sperm recognition and binding are highly conserved in animal evolution.

Free-swimming acrosome-intact sperm establish binding to the ZP in a relatively species-specific fashion and, once bound, undergo the acrosome reaction. As stated previously, acrosome-reacted sperm will not initiate binding to the ZP. In the mouse, the sperm-binding and acrosome reaction-inducing activities of the intact ZP are conferred by ZP3, and these biological activities are confined to both the carbohydrate and the protein regions of ZP3. Moreover, ZP3 binds only to the plasma membrane overlying the acrosome in acrosome-intact sperm, suggesting that this region of the sperm plasma membrane possesses specific ZP3-binding proteins and/or receptors, the identity of which is a subject of much controversy and will be addressed later. Currently, most is known about the functional domains involved in sperm binding; these domains are encoded by the O-linked oligosaccharide chains of ZP3. These functional domains are currently being further refined and will ultimately be valuable for probing the molecular mechanisms underlying these cell–matrix interactions. Work in other species has demonstrated that the respective ZP3 equivalent also possesses functional equivalence, although detailed functional analysis has yet to be carried out. Although the role of ZP2 has not been directly tested, experiments indicate that this ZP component is involved in anchoring the acrosome-reacted sperm on the ZP. In fact, it has been demonstrated that the inner acrosomal membrane, which is retained following the acrosome reaction, is the site for binding of ZP2, suggesting that this particular membrane domain possesses specific ZP2-binding proteins/receptors. It is likely, therefore, that the ZP2/ZP3 heterodimers function together in an integrated manner to regulate sperm–ZP binding and ZP penetration. To date, ZP1 is thought to play solely a structural role in the formation and maintenance of the ZP.

V. BIOCHEMISTRY OF THE ACROSOME REACTION

Given the fact that both progesterone and the ZP interact with the sperm surface to modulate function (i.e., to mediate sperm binding and to initiate acrosomal exocytosis) it is reasonable to postulate that the regulation of these functions may occur through classical signaling cascades known to regulate cellular function in somatic cells. Efforts to understand the biochemistry of the acrosome reaction have focused on three major areas, namely, the mechanism by which the biological trigger of the acrosome reaction is initiated, the intracellular signaling cascades that are activated, and the nature of how membrane fusion leading to exocytosis occurs. Since most of the information regarding the biochemistry of the acrosome reaction in response to physiological-relevant ligands has been gleaned from studies of the ZP-induced acrosome reaction, this article focuses primarily on this ligand.

The ZP-induced acrosome reaction has several hallmarks of a classical ligand-stimulated secretory event. First, a specific ligand associated with the ZP, namely, ZP3, possesses quantitatively all of the sperm-binding and acrosome reaction-inducing activity of the intact ZP; the unique structure and oocyte-specific expression of ZP3 confirms that it is a ligand designed for a very specific function. Second, ZP3 interacts with the plasma membrane of acrosome intact but not acrosome-reacted sperm, and its binding is restricted to the plasma membrane overlying the acrosome, suggesting that specific binding proteins and/or receptors are present and specifically localized to domains on the sperm surface that participate in ZP interaction. The effects of ZP3 on sperm binding and the acrosome reaction are concentration-dependent and saturable, further indicating the existence of specific binding sites. In addition, ZP3 following fertilization is modified (presumably by the action of egg cortical granule-associated glycosidases and proteases) to a form that has lost its ability to bind sperm, suggesting that sperm-associated-binding proteins/receptors bind only biologically active ligand.

A. Zona Pellucida-Binding Proteins/Receptors

The identity and function of sperm-associated ZP3-binding proteins/receptors has been quite controversial in recent years and is still yet to be resolved.

Identification of ZP (or ZP3)-binding proteins/receptors has been hampered by the lack of information pertaining to the precise identity of the active sperm adhesion and acrosome reaction-inducing moieties of the ZP (or ZP3) in any species as well as the seemingly complex nature of the interactions of these moieties with the sperm surface. These complex interactions, which have been examined experimentally by several investigators, do not preclude the possibility that sperm–ZP interactions likely involve multiple interactions between the ZP and the sperm surface, and that sperm–ZP binding and the acrosome reaction result from the formation of functional complexes containing multiple ZP-associated active domains as well as specific sperm-associated ZP-binding proteins and signal-transducing receptors. The formation of these functional complexes, when examined at the biochemical level, may manifest as both high- and low-affinity binding interactions. At this juncture, therefore, it is premature to conclude that there is a single component on the sperm surface that mediates all these biological events and that this single component must be a receptor; it is this very perception that has hindered the advancement of this field. For example, although several receptor candidates have been proposed, no one candidate has satisfied all the properties that one would expect for a specific receptor. Specific criteria to be kept in mind when analyzing potential receptor candidates include (i) the presence in the appropriate region of the sperm cell involved in gamete adhesion and acrosomal exocytosis, (ii) specificity and kinetics of ligand binding, (iii) the presence of receptor candidates in numbers on the sperm surface that are consistent with the ligand-binding kinetics and biology of the effect, (iv) ability to couple to specific signal-transduction systems, and (v) appropriate cell and tissue expression. Many candidates have been proposed as binding proteins/receptors, and it is beyond the scope of this article to discuss all of them in detail; the reader is encouraged to examine the numerous review articles dealing with this topic. For the most part, these candidates have the ability to bind to carbohydrates and, in some cases, may possess enzymatic activities that may or may not be involved in their putative ZP-binding/receptor activity (see Table 2 in Tulsiani et al., 1997). For the purposes of general discussion, only a few of the putative candidates will be discussed briefly, given the fact that they have received the most attention; this does not, in any way, diminish the importance of the other candidates.

Perhaps the most extensively characterized mouse sperm protein that possesses specific binding activity for ZP3 is a protein known as "sp56." This protein has lectin-like properties, has properties of a peripheral membrane protein that is associated with the plasma membrane overlying the sperm head, displays appropriate tissue expression, and bears homology to the family of complement component 4-binding proteins. The domains on sp56 involved in ZP3 binding and the mechanism by which signal transduction leading to acrosomal exocytosis is integrated with ZP3 binding to this protein are not known. Recently, the guinea pig sperm ortholog of sp56 (AM67) was cloned and shown to be present within the acrosomal matrix of the guinea pig sperm. Moreover, reevaluation of the localization of sp56 in mouse sperm demonstrated that it, likewise, was associated with the acrosomal matrix. This apparent discrepancy in localization of mouse sperm sp56 by two different groups (i.e., localization solely to the plasma membrane versus localization to the acrosomal matrix) is of interest and must be resolved by additional experiments. If it is conclusively established that sp56 is a ZP3-binding protein present within the acrosomal matrix, does this rule out its function in mediating sperm binding to the ZP? It is distinctly possible that "acrosome-intact" sperm that establish binding to the ZP via ZP3 (or its functional equivalent in other species) may, in fact, be sperm in which the plasma membrane and outer acrosomal membranes have "docked" with one another and have formed intermediate membrane complexes comprising the vesicular face of the outer acrosomal membrane and the extracellular face of the plasma membrane; such membrane docking and formation of intermediate membrane complexes characterize the events of regulated secretion in other secretory model systems. It has been proposed that small fusion pores that open and close in a dynamic fashion ("flickering pores") may form subsequent to the formation of such intermediate membrane complexes and prior to overt exocytosis. It is entirely possible that stable interactions between an acrosomal matrix-associated sp56 and ZP3 are established during the formation of such a metastable state in which flickering pores precede overt acrosomal exocytosis. The presence of a metastable state following sperm–ZP binding and prior to exocytosis has been documented but has not been characterized in great detail. This new model for sperm–ZP interaction should be examined in greater detail because it could explain apparent "conflicting" data seen by others using different species regarding the acrosomal status of sperm bound to the ZP. Moreover, such a model might focus attention on a new role for the acrosomal matrix in sperm–ZP interaction and ZP penetration.

A mouse sperm protein designated "p95" was proposed as a ZP3 receptor with properties of a receptor tyrosine kinase. This protein has some characteristics of a membrane protein and it was postulated that p95 possessed intrinsic tyrosine kinase activity that was modulated by ZP3. Subsequent purification of this protein demonstrated it to be a germ cell-specific hexokinase (type I) with some unique properties, although its role as a specific ZP3 receptor is unclear. The reported cloning of the human homolog of mouse p95, which was designated as *Hu9*, has revealed that it is *c-mer*, a protooncogene of uncertain function and a member of the *axl* family transforming receptor tyrosine kinases. The function of *Hu9* as a receptor tyrosine kinase for ZP3 in human sperm is unclear.

A specific form of β-galactosyltransferase (GalTase) has also been implicated as a receptor for ZP3 in mouse sperm. This protein has been postulated to mediate sperm–ZP binding by interacting with oligosaccharide residues specifically on ZP3 and to induce acrosomal exocytosis through GalTase aggregation on the cell surface and activation of heterotrimeric Gi proteins. Targeted overexpression of this form of the enzyme in sperm, predicted in theory to yield sperm that have an enhanced ability to interact with the ZP, yields sperm that, in fact, display a reduced ability to bind to the ZP. This is apparently due to their hypersensitivity to ZP3, such that they undergo acrosome reactions precociously/spontaneously and, therefore, have a reduced avidity of binding. These experiments were interpreted as demonstrating that successful fertilization requires an optimal, rather than a maximal, concentration of GalTase moieties on the sperm surface. In contrast, targeted mutation of the GalTase gene yields null males that are fertile, yet their sperm bind less ZP3 than the sperm from wild-type animals and are unable to undergo a ZP3-induced acrosome reaction. Lu and Shur conclude that although ZP3 binding and induction of the acrosome reaction are dispensable for fertilization, these properties impart a physiological advantage to sperm for fertilization. One can conclude that although the sperm-associated GalTase may be important for certain aspects of sperm–ZP interaction, it is not absolutely critical for fertility. One may argue that the observed effects on sperm–ZP interaction could occur as a consequence of secondary effects that result from defective galactosylation of other proteins during spermatogenesis; this has not been tested in a rigorous manner. These data suggest that GalTase is probably not of primary importance in mediating sperm–ZP3 binding leading to fertilization.

In conclusion, although numerous laboratories have identified a number of candidate sperm proteins thought to be involved in sperm–ZP binding and the induction of the acrosome reaction, it is clear that this is a very complex process and that we should rethink our current models for how these interactions occur. It is distinctly possible that some of these aforementioned candidates (as well as others not discussed; see Table 2 in Tulsiani *et al.*, 1997) may represent only a subset of the complex nature of interactions between the sperm surface and the ZP.

B. Zona Pellucida-Mediated Signal Transduction

Although there are still many questions to be answered regarding the identity of the proteins on the sperm surface that mediate ZP binding and acrosomal exocytosis, experiments focusing on mechanisms of transmembrane signal transduction and activation of intracellular effectors have evolved from several laboratories. If one proposes that ZP (or ZP3) effects on sperm are mediated via specific receptors or binding proteins that are coopted into a functional signaling complex, one could argue that transmembrane signal transduction may have similarities to that seen in somatic cells. Since heterotrimeric G proteins play a key role in signal transduction in response to the activation of several classes of receptors, experiments from several laboratories have focused on the role of these GTP-binding proteins on ZP and ZP3-mediated sperm binding and the acrosome reaction. Sperm-associated heterotrimeric G proteins of the G_i class appear to play a key role in regulating ZP or ZP3-mediated signal transduction events leading to the acrosome reaction in mouse, bovine, and human sperm. This has been demonstrated by a variety of experimental approaches. First, G_i proteins are present in the membranes overlying the acrosome, that region of the sperm that interacts with ZP3. Second, functional inactivation of G_i proteins in sperm by pertussis toxin treatment does not block the ability of the sperm to establish binding to the ZP but inhibits the ZP (or ZP3)-induced acrosome reaction. Finally, membranes isolated from sperm display G_i protein activation *in vitro* following the addition of ZP3 but not ZP1 or ZP2. Thus, one of the earliest events following ZP or ZP3 interaction with the sperm surface is transmembrane signaling via this particular class of G proteins. Whether other signal transduction events are entrained is unclear. The activation of sperm G proteins, however, appears to be a universal ZP-mediated event.

C. Intracellular Effectors

As previously summarized, there is substantial evidence that acrosomal exocytosis is a highly regulated process initiated by the ZP and likely mediated by cell surface-binding proteins/receptors and signal-transducing G proteins (at least in mammals). It is likely, therefore, that intracellular regulation of acrosomal exocytosis is similar to other receptor-mediated exocytotic events. Such intracellular signals include changes in ionic conductance, changes in cyclic nucleotide metabolism, and changes in phospholipid metabolism. It is also anticipated that these intracellular effector systems would be modulated in a ZP-binding protein/receptor-dependent fashion. Although there have been numerous reports describing ionic and/or second messenger systems purported to play a role in the induction of the acrosome reaction, there is little information linking such effectors to ligands in a receptor-dependent fashion due to the paucity of knowledge of the receptors and the signal transducers. Furthermore, a description of effector systems correlated with the induction of acrosomal exocytosis cannot necessarily be equated with cause and effect. The current state of knowledge regarding the nature of sperm intracellular signaling during the acrosome reaction in response to the ZP will be addressed.

Studies in both mouse and bull sperm have revealed that elevations in intracellular Ca^{2+}, as well as intracellular pH, represent some of the earliest responses of sperm incubated with ZP or ZP3. Many of these studies have been performed with Ca^{2+} and pH indicator dyes and have reported localized changes to the acrosomal region. In bovine sperm, the ZP-induced Ca^{2+} entry, as well as sperm membrane potential and the acrosome reaction, is dependent on membrane depolarization, and the Ca^{2+} uptake and acrosome reaction are inhibited by antagonists of voltage-dependent Ca^{2+} channels. Pertussis toxin inhibits the ZP (and ZP3)-induced pH changes in mouse sperm, as well as the ZP-induced pH and Ca^{2+} changes in bull sperm, indicating that sperm G_i proteins may regulate such ionic changes. Incubation of bovine sperm under depolarizing conditions, which would activate such voltage-dependent Ca^{2+} channels, bypasses the inhibitory effects of pertussis toxin on the acrosome reaction, suggesting that G_i proteins might regulate such Ca^{2+} channels indirectly. Based on these and other data, a signaling pathway has emerged whereby ZP3 depolarizes the sperm membrane by activating a pertussis toxin-insensitive pathway with characteristics of a poorly selective cation channel. ZP3 also activates a pertussis toxin-sensitive pathway that produces a transient rise in intracellular pH. Together, the membrane depolarization and rise in intracellular pH open voltage-sensitive Ca^{2+} channels, leading to the activation of downstream effectors of the acrosome reaction. These Ca^{2+} channels have characteristics of T-type channels.

Additional studies have suggested that alterations in phospholipid metabolism and/or cyclic nucleotide metabolism may play important intermediary roles in the sperm acrosome reaction. It has been demonstrated that biologically active phorbol diesters and diacylglycerols alter the kinetics of the ZP-mediated acrosome reaction in mouse sperm, thus suggesting that this exocytotic event could be regulated in some manner by protein kinase C. This is further supported by the observation that diacylglycerol is formed in mouse sperm in response to the ZP. The products of phospholipase C turnover (e.g., IP$_3$ and sn-1,2 diacylglycerol), as well as the role of other phospholipases (A$_2$ or D), have not yet been examined in sperm challenged with ZP or ZP3. Recently, it has been demonstrated that the mouse ZP stimulates mouse sperm adenylyl cyclase, suggesting that the ZP can alter sperm cAMP metabolism. This is consistent with a previous observation that solubilized ZP from mouse eggs can cause transient elevations in sperm cyclic AMP concentrations that are dependent on the presence of extracellular Ca^{2+}. These cyclic AMP elevations precede and are correlated with the induction of the acrosome reaction by the ZP, suggesting that cyclic AMP may be a potential participant in the signaling pathway leading to acrosomal exocytosis. It will be of interest to determine whether such intracellular signaling systems are coupled to sperm G$_i$ proteins since these second messenger systems are coupled in a receptor-mediated fashion to G proteins in other cell types.

D. Mechanics of Membrane Fusion

As previously stated, the acrosome reaction has characteristics of a regulated secretory event. However, literally nothing is known regarding the mechanisms governing the fusion of the outer acrosomal and plasma membranes that give rise to the formation of hybrid membrane vesicles and the release of the acrosomal contents. Again, it is reasonable to assume that elements of this event may have properties in common with some of the other well-studied models of regulated secretion. Sperm do contain monomeric G proteins that have been implicated in the secretory process in other systems, but little is known about the function of these proteins. In this regard, it is interesting to note that acrosomal exocytosis in ram sperm induced by A23187 is enhanced by a synthetic peptide corresponding to the effector domain of rab3 (a monomeric G protein), suggesting a role for this protein in exocytosis. In addition, recent work in sea urchin sperm has demonstrated the existence of two homologs of the secretory vesicle cycle, namely, syntaxin and VAMP (synaptobrevin), that are shed during the acrosome reaction; their role in this secretory event is not known. However, major questions to be addressed are whether similar proteins exist in mammalian sperm, whether they localize to the appropriate region of the acrosome, and whether they play a role in ZP-mediated acrosomal exocytosis.

See Also the Following Articles

Fertilization; Oocyte, Overview; Progesterone Actions on Reproductive Tract; Spermiogenesis; Sperm Transport

Bibliography

Arnoult, C., Zeng, Y., and Florman, H. M. (1996). ZP3-dependent activation of sperm cation channels regulates acrosomal secretion during mammalian fertilization. *J. Cell Biol.* **134**, 637–645.

Bookbinder, L. H., Cheng, A., and Bleil, J. D. (1995). Tissue- and species-specific expression of sp56, a mouse sperm fertilization protein. *Science* **269**, 86–89.

Bork, P. (1996). Sperm–egg binding protein or proto-oncogene? *Science* **271**, 1431–1432.

Burks, D. J., Carballada, R., Moore, H. D. M., and Saling, P. M. (1995). Interaction of a tyrosine kinase from human sperm with the zona pellucida at fertilization. *Science* **269**, 83–86.

Dubois, D. H., and Shur, B. D. (1995). Cell surface β1,4-galactosyltransferase—A signal transducing receptor? *Adv. Exp. Med. Biol.* **376**, 105–114.

Foster, J. A., Friday, B. B., Maulit, M. T., Blobel, C., Winfrey, V. P., Olson, G. E., Kim, K. S., and Gerton, G. L. (1997). AM67, a secretory component of the guinea pig sperm acrosomal matrix, is related to mouse sperm protein sp56 and the complement component 4-binding proteins. *J. Biol. Chem.* **272**, 12714–12722.

Garde, J., and Roldan, E. R. (1996). rab 3-peptide stimulates exocytosis of the ram sperm acrosome via interaction with cyclic AMP and phospholipase A2 metabolites. *FEBS Lett.* **391**, 263–268.

Knobil, E., and Neill, J. D. (1994). *The Physiology of Reproduction*. Raven Press, New York.

Kopf, G. S., and Gerton, G. L. (1991). The mammalian sperm acrosome and the acrosome reaction. In *Elements of Mammalian Fertilization, I. Basic Concepts* (P. M. Wassarman, Ed.), CRC Uniscience Series, pp. 153–203. CRC Press, Boca Raton, FL.

Lu, Q., and Shur, B. D. (1997). Sperm from β1,4-galactosyl-transferase-null mice are refractory to ZP3-induced acrosome reactions and penetrate the zona pellucida poorly. *Development* **124**, 4121–4131.

Revelli, A., Massobrio, M., and Tesarik, J. (1998). Nongenomic actions of steroid hormones in reproductive tissues. *Endocr. Rev.* **19**, 3–17.

Schulz, J. R., Wessel, G. M., and Vacquier, V. D. (1997). The exocytosis regulatory proteins syntaxin and VAMP are shed from sea urchin sperm during the acrosome reaction. *Dev. Biol.* **191**, 80–87.

Schultz, R. M., and Kopf, G. S. (1995). Molecular basis of mammalian egg activation. *Curr. Topics Dev. Biol.* **30**, 21–62.

Tulsiani, D. R. P., Yoshida-Komiya, H., and Araki, Y. (1997). Mammalian fertilization: A carbohydrate-mediated event. *Biol. Reprod.* **57**, 487–494.

Ward, C. R., and Kopf, G. S. (1993). Molecular events mediating sperm activation. *Dev. Biol.* **158**, 9–34.

Wassarman, P. (1988). Zona pellucida glycoproteins. *Annu. Rev. Biochem.* **57**, 415–442.

Wassarman, P. (1995). Towards molecular mechanisms for gamete adhesion and fusion during mammalian fertilization. *Curr. Biol.* **7**, 658–664.

Activin and Activin Receptors

Ralph H. Schwall
Genentech, Incorporated

I. Activin
II. Activin Receptors

GLOSSARY

in situ ligand binding A procedure for localization of receptor binding sites by placing a drop of labeled ligand onto a histological section of tissue and then microscopically determining the distribution of the label.

knockout mice A family of mice created from embryonic stem cells in which a specific gene has been deleted by homologous recombination.

serine/threonine kinase An enzyme that transfers the terminal phosphate from ATP to a serine or threonine residue in a substrate protein.

I. ACTIVIN

Activin was first identified as a stimulator of follicle-stimulating hormone (FSH) secretion when this activity was quite unexpectedly noted in side fractions during the purification of inhibin from follicular fluid. The second surprise came when sequence analysis showed that activin is composed of the β subunit of inhibin. The third surprise came shortly thereafter, when an activity known as erythroid differentiation factor was purified to homogeneity and found to have the same sequence as activin. Over the past 10 years, activin has been shown to influence cellular behavior in a number of organ systems, both within the reproductive system and in nonreproductive tissues. Delineating the physiological functions of this complex system is a stimulating challenge for research, a task that is further complicated by the fact that there are several different forms of activin. Toward that end, significant progress is being made in understanding the mechanisms of activin signal transduction. The activin receptor is now known to consist of two membrane-spanning polypeptides; one binds activin and the other is required for intracellular signaling.

Also, like the ligand, there are several different forms of each of the receptor subunits. The intracellular mediators are also beginning to be defined. Mice with deletions in genes for activin and its receptors

TABLE 1
Members of the Activin Superfamily

Activin-A, -B, -AB, others (see text)
TGF-β1–5
Inhibin-A, -B
Müllerian inhibiting substance
Bone morphogenetic protein-2–7
Decapentaplegic (drosophila)
Vg-1 (*Xenopus*)
Nodal

have yielded some unexpected results; however, interpretation of these data is complicated by the potential redundancy among the activins and the receptors. The objective of this article is to provide a broad, general overview of this field.

A. Structure

The name activin refers to a family of proteins that belongs to a much larger superfamily that includes the tumor growth factors-β (TGF-β) and a number of potent differentiation factors (Table 1). Activin-A, activin-B, and activin-AB are 25-kDa, disulfide-linked homo- and heterodimers of the β_A and β_B subunits of inhibin. Recently, three other related genes, named β_C, β_D, and β_E, have been identified (Table 2). It is presumed that each would form homodimers but there is no information available about what types of heterodimers may be formed among themselves or with β_A and β_B.

Within each subunit there are nine cysteines, the placement of which is highly conserved (Fig. 1). In TGF-β2, whose crystal structure has been solved, eight of the cysteines form four intrachain disulfide bonds, whereas one forms an interchain disulfide with the same cysteine in the other subunit of the dimer. The subunit structure resembles a slightly curved left hand, with an α-helix forming the heel, β strands and loops forming the fingers, and the amino terminus protruding as the thumb. In the dimer, the subunits are paired, palms together, with the fingers pointing in opposite directions. The palms form an extensive area of hydrophobic interaction, with the interchain disulfide bond in the center. Based on amino acid sequence conservation, activin is predicted to have a similar structure. Indeed, mutational analysis has confirmed that the analogous cysteine in activin β_A, at position 80, is involved in holding together the activin-A dimer.

B. Biosynthesis and Secretion

Five different activin subunit genes have been identified, with independent but overlapping patterns of gene expression (Table 2). Upregulation of subunit gene expression can be stimulated by gonadotropins in the ovary and the testis through a cAMP-dependent mechanism. Expression of subunit genes

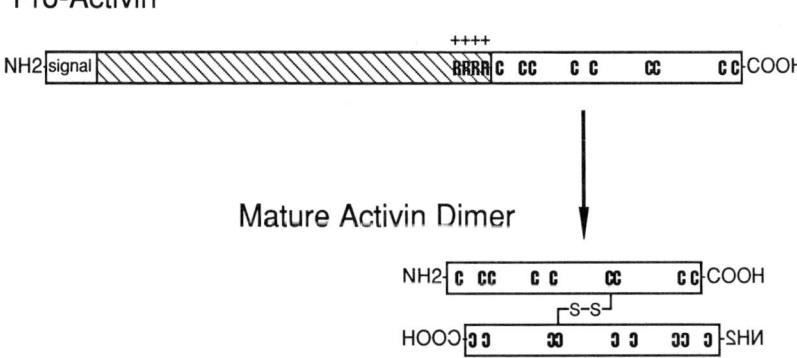

FIGURE 1 Activin structure. Activin is synthesized as a large precursor containing a signal sequence on the amino terminus and the mature β subunit in its carboxy-terminal third. The coding region for the mature subunit is preceded by a short stretch of basic amino acids. Generation of the active protein involves proteolytic processing of the precursor and formation of an antiparallel dimer that is covalently linked by a single disulfide bond between cysteine 80 in each subunit.

TABLE 2
Activin Genes and Their Expression Patterns

Activin subunit gene	Gene expression pattern[a]	Proteins identified
β_A	Placenta, ovary, testis, bone marrow, CNS, liver, adrenal cortex, pituitary, pancreas, vascular smooth muscle	β_A-β_A, β_A-β_B
β_B	Ovary, pituitary, placenta, testis, CNS, pancreas	β_B-β_B, β_A-β_B
β_C	Liver	None
β_D	Xenopus	None
β_E	Liver	None

[a] Tissues are listed in descending order of expression level.

can also be stimulated by phorbol esters. A binding site for the transcription factor AP-2 is found in the 5' region of several of the genes.

The mRNA encodes a much larger precursor protein that contains the mature subunit in its carboxy terminus (Fig. 1). The coding region for the mature subunit is preceded by a short string of basic amino acids, which are a site for proteolytic cleavage. The N-terminal portion of the precursor molecule is required for dimerization of the mature subunit, but they need not be contiguous in one polypeptide, which suggests that dimerization may occur after proteolytic release of the mature subunit. However, little is known about how subunit assembly is regulated. The mature, dimeric protein is secreted rapidly after synthesis, rather than being stored in secretory granules, because it is very difficult to detect activin in cells and in tissue extracts.

Measuring activin secretion has been hampered by the difficulty in developing immunoassays, a task that is complicated by the existence of multiple activin subunits and the presence of common subunits in activin and inhibin. In addition, the extremely high degree of amino acid sequence conservation across species limits immune responses, which has made it hard to elicit antibodies. Monoclonal antibodies that are specific for activin-A and activin-B, and show little cross-reactivity with the appropriate inhibins, have been produced using hypogonadal mice. These antibodies are beginning to be used to develop dimer-specific ELISAs. More widespread use of these highly specific assays will be important to understanding how activin secretion is regulated.

C. Biological Actions

Activin has a number of biological activities in a number of different organ systems. To generalize, most involve stimulation of differentiated function.

1. In the Pituitary

The name activin was proposed by investigators in laboratories headed by Wylie Vale, Nick Ling, and Roger Guillemin at the Salk Institute because this protein has an activity that is opposite to inhibin. That is, it stimulates FSH secretion. Activin increases steady-state FSH-β mRNA and enhances FSH secretion through a constitutive pathway rather than through secretory granules. FSH-β mRNA is stabilized in activin-stimulated cells, but there is also evidence for increased transcription of the FSH-β gene. Compared to gonadotropin-releasing hormone (GnRH), the effects of activin are slow in onset and are selective for FSH without affecting luteinizing hormone (LH). This reflects differences in their mechanisms of action; GnRH stimulates rapid exocytosis of FSH and LH from secretory granules. Indeed, downregulation of gonadotropes to GnRH has no effect on their responsiveness to activin, further supporting distinct mechanisms of action. However, there is also cooperation between these two mechanisms as activin enhances pituitary FSH levels and stimulates GnRH receptor expression, both of which contribute to enhanced GnRH-induced FSH release.

The β_A and β_B subunits are expressed within the pituitary, and activin is secreted by pituitary cells *in vitro*, suggesting an autocrine or paracrine effect on FSH secretion. Further supporting that concept, a neutralizing antibody to activin-B inhibits basal FSH secretion. Such a mechanism also explains why follistatin, an activin-binding protein, inhibits basal FSH secretion in the absence of exogenous activin.

Activin also inhibits the secretion of growth hormone. Both basal and TRH-induced growth hormone (GH) release are affected. The transcription factor *Pit-1*, which is important in regulating GH gene expression, becomes phosphorylated and its binding to the 5' region of the GH gene decreases. These effects have been observed in tumor cell lines, indicating a direct effect on the somatotroph. Whether such actions are important in the changes in growth associated with puberty is unknown.

2. In the Ovary

Activin is a normal component of follicular fluid, which was the source from which activin-A, activin-AB, and activin-B were first isolated. Both the β_A and β_B subunits are expressed in granulosa cells of antral follicles. In primates, and only in primates, small antral follicles express very high levels of β_B but very little β_A, and the expression of β_B diminishes as the follicle matures to the preovulatory stage. It is interesting to speculate that activin-B may be involved in follicle selection, possibly through an autocrine mechanism.

In isolated granulosa cells, activin has a number of activities that would promote follicular development. It stimulates DNA synthesis, increases FSH receptor number, and enhances FSH induction of LH receptors and aromatase activity. Activin also induces germinal vesicle breakdown in isolated oocytes. However, it decreases androgen production by theca cells and progesterone secretion by granulosa luteal cells. The latter observation has been interpreted to suggest that activin may help prevent premature luteinization.

Consistent with the follicle-promoting activities described previously, activin induces growth of isolated preantral follicles and, strikingly, development of an antrum *in vitro*. Systemic injections of activin enhance the ovarian response to PMSG. This occurs even in hypopohysectomized rats, which suggests a direct effect on the ovary. However, the effects of activin may be influenced by the microenvironment of the follicle. When injected intrabursally, it induces atresia. Consistent with that data, it also stimulates expression of IGF-I-BP-5 by granulosa cells *in vitro*, which could contribute to atresia by sequestering IGF-I. Further research is needed to define activin's role in the complex process of follicular selection and development.

3. In the Testis

Activin subunit mRNAs are expressed in Leydig and Sertoli cells, and the protein is secreted by both of these cell types as well as by peritubular myoid cells. The expression varies with the stage of the cycle of the seminiferous tubule, being highest in association with the transition of B spermatagonia to preleptoene spermatocytes. Whether it influences meiosis in these germ cells is unknown. By *in situ* ligand binding, activin is seen to bind to cells in the basal compartment of the seminiferous tubule independent of the stage of the cycle and to early spermatids in stages VII–VIII. Activin stimulates spermatagonial proliferation *in vitro*, with effects on preleptotene spermatocytes and intermediate spermatagonia. Activin also has a profound morphological effect in germ cell–Sertoli cell cocultures, causing, over several days, aggregation into tubular structures. During this reorganization, the germ cells proliferate but remain attached to each other in small aggregates, and cytoplasmic bridges between the germ cells have been observed.

The activin receptor is also expressed in Sertoli cells, and activin inhibits FSH-induced aromatase activity and androgen receptor mRNA. Activin also stimulates inhibin and transferrin secretion by Sertoli cells. Exogenous activin inhibits DNA synthesis in Sertoli cells in the fetal testis but stimulates their proliferation in the early neonatal period.

4. In the Placenta

High levels of activin subunit are expressed in the placenta. Activin stimulates human chorionic gonadotropin secretion by isolated placental cells. The development of specific and sensitive immunoassays has revealed that activin-A levels increase throughout pregnancy and rise sharply even further during labor. Just as rapidly, levels return to normal after parturition. Activin receptors are expressed on trophoblast and fetal membranes. The source of the activin during labor is unclear, as is the physiological role of activin in labor, but this is an interesting avenue for future research.

5. In Nonreproductive Tissues

i. Erythroid Differentiation The first nonreproductive activity identified for activin was its ability to stimulate hemoglobin expression in erythroleukemia cells and in erythroid progenitors. This erythroid differentiating activity is also manifested as enhanced responses to erythropoietin in bone marrow progenitor assays.

ii. Mesoderm Perhaps the most widely studied nonreproductive activity is the induction of mesoderm in the early *Xenopus* embryo. However, there has been controversy in this area because although exogenous activin has profound effects on development, it has been difficult to detect expression of endogenous activin at the appropriate time to implicate it in this function in normal embryos. This issue may be resolved when expression of the newly isolated *Xenopus* β_D subunit is examined in detail. In addition to its effects on mesoderm, activin can also disrupt branching morphogenesis during development of ductile epithelial tissues, such as the salivary gland, pancreas, and kidney.

iii. Liver The activin receptor is expressed in the liver. In isolated hepatocytes activin stimulates glycogenolysis and glucose secretion, inhibits growth factor-induced DNA synthesis, and can induce apoptosis. Infusion of activin in rodents causes apoptosis of hepatocytes surrounding the central vein. The β_A subunit is expressed in the normal liver and is upregulated during liver regeneration, suggesting that activin may have a physiological role in hepatic homeostasis. This concept is further corroborated by the cloning of two new activin subunit genes, β_C and β_E, from liver cDNA libraries.

iv. Cachexia An effect on the liver may be involved in the cachexia observed in inhibin-α knockout mice. These mice develop gonadal tumors after puberty, with a very high prevalence in both males and females, indicating that inhibin-α is a tumor-suppressor gene. As the tumors grow, the mice develop cachexia and die within a few weeks. Testicular βA mRNA levels are elevated 200-fold in these mice, although hypothalamic βA levels are normal, and their circulating activin-A levels are 10-fold greater than normal. It is unclear whether the high levels of activin-A cause tumor formation or are simply associated with it, but there is good evidence that activin is directly involved in the cachexia. The livers from cachectic mice contain the same lesions around the central vein observed in animals infused with activin. In addition, there are lesions in the glandular stomach. When these mice are crossed with activin receptor II knockouts, gonadal tumors develop and serum activin levels remain high, but neither cachexia nor the liver lesions develop. These data convincingly demonstrate that activin is the cause of the cachexia and also implicate the liver lesions as being important in cachexia in this model. It will be of interest to measure activin levels in patients with cachexia to determine whether this observation could have some broader clinical application.

Interestingly, if the inhibin-α knockout mice are gonadectomized before puberty, nearly all will subsequently develop adrenal sex steroid tumors, and the appearance of these tumors is also associated with cachexia. Activin levels in such mice have not yet been published but the same mechanism is likely given that activin subunits are expressed in the adrenal cortex.

v. Pancreas Activin converts an acinar cell line into an insulin-producing cell line. In addition, activin stimulates insulin secretion even in the absence of glucose. The β_A and β_B subunits are expressed in the fetal and adult pancreas. Following treatment with streptozotocin, β_A and β_B immunoreactivities disappear from pancreatic β cells much more rapidly than insulin. Combined with the ability of activin to induce insulin expression, this observation suggests that the diabetes induced by streptozotocin may be secondary to its effects on pancreatic activin.

vi. Adrenal Gland Both the β_A and β_B subunits are expressed in fetal and adult adrenal cortex, and expression of β_A is stimulated by adrenocorticotropin hormone (ACTH). Activin inhibits ACTH-induced steroidogenesis. In the fetal adrenal, it suppresses proliferation of fetal zone cells and enhances the ACTH-induced shift in the cortisol:DHEA sulfate ratio.

vii. Blood Vessels Activin is expressed by smooth muscle cells and inhibits proliferation of endothelial cells. Expression is upregulated in atherosclerotic lesions, localized to the neointima of the diseased artery. Mutations in one of the activin type I receptor-related genes, *ALK1*, have been associated with hereditary hemorrhagic telangiectasia, also known as Osler–Rendu–Weber syndrome, which is an autosomal dominant vascular dysplasia.

viii. Bone Activin promotes chondrogenesis and has a TGF-β-like effect to induce bone formation when applied to the periosteum. Activin subunits are expressed in the dental sockets. βA knockout mice lack incisors and also have prominent defects in craniofacial development.

ix. Nervous System Activin subunits are expressed in the central nervous system, and exogenous activin modulates oxytocin release and feeding behavior. In the eye, an extension of the central nervous system, activin may have an autocrine/paracrine effect in the retina. It is expressed in the pigmented epithelial layer and inhibits serum-induced growth of these cells. Activin is also expressed by choroid cells in the eye, where it induces expression of somatostatin in neurons that innervate this structure from the ciliary ganglion.

D. Regulation of Bioavailability

The ability of activin to induce biological responses in target cells is modulated by two binding proteins, follistatin and α_2-macroglobulin. Follistatin binds activin with high affinity and thereby blocks its interaction with cellular activin receptors (Fig. 2). Alternative splicing of the follistatin mRNA gives rise to forms that are 288, 303, and 318 amino acids in length. Each form can be glycosylated to varying extents, giving rise to a heterogeneous mixture. Glycosylation is not required for activin binding, and all isoforms, regardless of length or glycosylation state, bind activin with similar affinity. The short form also binds to cell surface heparan sulfate proteoglycans, which may act to tether activin on the cell surface. α_2-Macroglobulin is a large protein that binds a wide variety of substances and that is present in high abundance in serum.

E. Activin Knockout Mice

Proteins can have many activities in *in vitro* systems. One of the most powerful methods for elucidating which activities are physiologically relevant is to make mice in which the gene is homozygously deleted. Initially, while in the laboratory of Allan Brad-

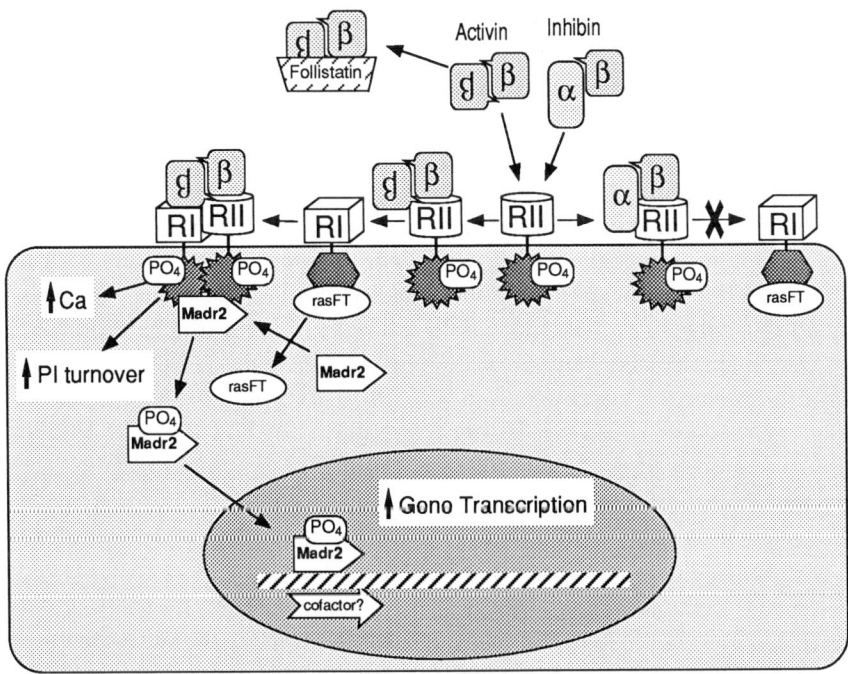

FIGURE 2 Model for activin signal transduction (see text for details). RI, activin type I receptor; RII, activin type II receptor; I-R, inhibin receptor; rasFT, *ras* farnesyl transferase; MAD, mothers against dpp; FAST, forkhead activin signal transducer; PI, phosphoinositide; Ca, calcium.

ley, Martin Matzuk made such mice in which the β_A gene, the β_B gene, or both β_A and β_B genes were deleted. This group has also characterized mice in which follistatin gene and the activin receptor gene (see Section II,C) have been deleted.

The β_A knockouts develop to term and are born but they do not suckle and die shortly thereafter. Many have craniofacial defects in the hard and soft palate, which is probably why they fail to suckle successfully. They lack whiskers and lower incisors. In view of the extensive literature on activin in early embryonic development, it was surprising that the mice develop beyond the first few days after fertilization. Either basic mechanisms of embryonic development in the mouse are different from those of the frog, in which much of the work on activin in mesoderm induction has been done, or other activin subunits can compensate for loss of β_A in this function.

The phenotype of the β_B knockout mice is much more subtle. A small proportion are lost during gestation to unknown causes, but most develop to term and are born normally. The proportion that survive to term varies in different background strains. The mechanism for the incomplete penetrance of the embryonic lethality is unknown. Approximately one-third of the pups are born with their eyes open due to failure of the eyelids to fuse, but again the penetrance varies in different background strains. Of the mice that are born, all appear to develop normally postnatally. Males are fertile. Females are also fertile and have normal pregnancy rates and litter sizes, but problems become apparent near the end of pregnancy. Delivery can be delayed 1 or 2 days, and sometimes the pups die *in utero*. The pups appear normal at birth but all die shortly after with empty stomachs. The mother's mammary glands contain milk but the milk is not let down. That this is the cause of neonatal death is supported by the fact that the pups can be rescued by fostering to a wild-type mother. Whether the defect in milk let-down reflects a function for activin in this pathway in the adult or a failure of these neural pathways to develop properly during embryogenesis is unclear. When the previously discussed mice are crossed to generate mice lacking both β_A and β_B genes, no new defects are observed.

The phenotype of the follistatin knockout mice is quite interesting. As noted previously, follistatin binds activin with high affinity and neutralizes activin's biological activity. No cellular receptors for follistatin have been identified and follistatin has no known biological activity of its own, other than its ability to block activin. In some systems in which follistatin has a direct effect, such as pituitary FSH secretion, there is good evidence that this is secondary to neutralization of endogenous activin. Thus, one would predict that a follistatin knockout mouse would resemble an activin-overexpressing transgenic because the activin would go unchecked. Surprisingly, the follistatin knockouts resemble the activin knockouts in several ways. They develop to term but die shortly after birth. They exhibit some of the defects in craniofacial development, whiskers, and teeth seen the β_A knockouts but also have defects in the ribs and have shiny, taut skin. The cause of neonatal death is failure to breath, associated with atrophic development of the diaphragm and the intercostal muscles. These mice show that there is much to learn about follistatin.

II. ACTIVIN RECEPTORS

A. Structure

When radiolabeled activin is cross-linked to responsive cells and the products are analyzed by polyacrylamide gel electrophoresis, two protein products are identified—one with a molecular weight of ~65 kDa and the other ~80 kDa. By analogy to the TGF-β receptors, these have been designated type I and type II receptors, respectively. Cloning of the activin type II receptor was first achieved by Lawrence Mathews while he was in the laboratory of Wylie Vale. Since then a number of laboratories have made significant contributions to elucidating this complex receptor system, which is shown schematically in Fig. 2.

The type II receptor has an extracellular domain, a single transmembrane domain, and an intracellular domain that is a serine/threonine kinase. It binds activin with high affinity but cannot transmit an intracellular signal without interacting with the type I receptor, also a serine/threonine kinase. The intracellular domains of the type I and type II receptors show strong similarities in their sequences, but their extracellular domains are quite divergent. The type I receptor is incapable of binding activin by itself; however, it binds activin when the type II receptor is also present. Based on cross-linking results, the stoichiometry of activin subunit binding to each of the receptor types appears to be 1:1. In addition, activin subunit monomers, which are produced by mutation of the cysteine residue that forms the interchain disulfide bond, have very low biological activity. (The formation of noncovalent dimers through their hydrophobic palms might account for this mutant having some biological activity.) These types of data suggest that the complex consists of one subunit of the activin dimer interacting with a type II receptor, while the other interacts with a type I receptor. What takes place in the complex to allow the type I receptor to bind activin is unclear. There may be conformational changes in activin secondary to binding by type II receptor or in the type I receptor secondary to its phosphorylation by the type II receptor–activin complex, or the type I receptor may recognize a conformational feature that is uniquely created by the activin/type II receptor complex.

Two different type I and type II receptors have been identified. Moreover, the genes for one of the type II receptors, the ActRIIB receptor, can generate four alternatively spliced products by inclusion or exclusion of two introns on either side of the transmembrane domain. Whether these splice variants have altered function is unresolved. To complicate matters further, because a number of groups have independently been working on some of these genes, a number of names have appeared in the literature and are listed in Table 3.

B. Signaling Mechanisms

A model for activin signal transduction, which is based in part on data from the TGF-β receptor, is shown in Fig. 2. Although activin has major effects on FSH secretion, FSH is not specifically mentioned in Fig. 2 because this pathway has been delineated in other systems, primarily in embryonic development. However, similar mechanisms are presumed to function in the gonadotrope.

TABLE 3
Activin Receptors[a]

Type I receptors
   ActRI (SKRI, Tsk7L, R1, ALK2, ActX1R, XAR3)
   ActRIB (SKRI-1, R2, ALK4)
Type II receptors
   Act RII
   Act RIIB
      Alternative splic variants = ActRIIB1, ActRIIB2, ActRIIB3, ActRIIB4

[a] There are two type I and two type II activin receptors. Some of these genes have also been designated by an assortment of other names, which are shown in parentheses.

The serine–threonine kinase of the type II receptor is constitutively activated and the receptor is phosphorylated even under basal conditions, probably by autophosphorylation. Binding of activin to the type II receptor causes complex formation with the type I receptor, and the type II receptor kinase phosphorylates the intracellular domain of the type I receptor. Critically important for signal transduction is the phosphorylation of threonine at position 206 in the type I receptor; mutation of this residue to a nonphosphorylatable alanine disrupts activin signal transduction, whereas substitution with aspartic acid, which mimics phosphorylated threonine in charge and size, leads to constitutive signaling even in the absence of activin.

Ras farnesyl transferase, an enzyme that attaches a farnesyl lipid group to *$p21^{ras}$*, is associated with the type I receptor in resting cells and dissociates rapidly on ligand stimulation. Farnesylation of *ras* provides a greasy foot that inserts into the plasma membrane, conferring a membrane localization to *$p21^{ras}$* that is required for *ras*'s transforming activity. Although it is not yet known whether binding to the type I receptor modulates the activity of *rasFT* and consequently *ras* itself; it is tempting to speculate that alterations in *ras* farnesylation may play a role in the ability of activin to inhibit proliferation.

The transcription of a number of genes, such as *Mix.2* and *junB*, is upregulated by activin through an activin-responsive element (ARE) in their 5′ region. These transcriptional effects are mediated by *MAD3*, a member of the MAD (*m*others *a*gainst *d*pp) gene family, the first of which was identified by drosophila geneticists based on its effects on the phenotype of flies with mutations in the decapentaplegic gene. (Notice that dpp is in the activin superfamily; see Table I). Other MAD homologs mediate the actions of TGF-β. Although all the details have not yet been delineated for the activin receptor, the results obtained to date, combined with results from the TGF-β receptor, lead to the following scheme. In response to activin, *MAD3* associates with the activin receptor complex and undergoes phosphorylation. Phosphorylated *MAD3* then translocates to the nucleus, where it forms a complex with a protein known as FAST (*f*orkhead *a*ctivin *s*ignal *t*ransducer). This complex binds to the ARE and activates transcription. Whether other components are involved in the transcription complex is currently an area of intense research.

Activin also affects GH secretion (see Section I,C,1), but the effect is inhibitory. The transcription factor *Pit-1* becomes phosphorylated in response to activin, and its binding to regulatory elements in the GH gene decreases. This pathway is not shown in Fig. 2 because the details are poorly understood.

Other intracellular responses to activin include increases in intracellular calcium and phosphatidyl inositide metabolism in pituitary and gonadal cells. In addition, transient hypophosphorylation of the retinoblastoma protein occurs in association with growth inhibition in erythroleukemia cells. The pathways by which these responses are mediated intracellularly are unknown.

Inhibin, which antagonizes activin in most systems, binds the type II receptor through its β subunit (Fig. 2). The affinity of this interaction is approximately 10-fold less than that for activin. Most significantly, however, the inhibin/type II receptor is incapable of forming a complex with the type I receptor and therefore fails to signal. However, by occupying the type II receptor, activin is excluded. This mechanism can account for the competitive antagonism between activin and inhibin. However, recent data indicate that inhibin also interacts with an inhibin-specific binding component (designated I-R in Fig. 2) that is required for its inhibitory actions.

C. Receptor Knockout Mice

Most mice in which the activin RII receptor gene has been homozygously deleted develop normally and thrive after birth but have defects in their reproductive system. FSH levels are suppressed in both sexes. Although males are fertile, the size of the testes is reduced and the age at which they become fertile is markedly delayed. Females are sterile, with small ovaries in which corpora lutea are rarely observed.

The RII knockout mice that do not develop normally have skeletal and facial abnormalities that resemble a congenital condition in humans known as Pierre–Robin syndrome and die shortly after birth. The reason for the low penetrance of the neonatal death phenotype is not understood. The RII knockout mice also tend to develop defects in the mandible that are not observed in the β_A or β_B knockouts. When these data were published, Martin Matzuk noted the "striking lack of overlap" in the phenotypes of the receptor compared to the ligand knockout mice and suggested that there may be other ligands. This idea would be consistent with the recent identification of additional β subunit-related genes.

See Also the Following Articles

Inhibins; Transgenic Animals

Bibliography

Attisano, L., Wrana, J. L., Montalvo, E., and Massague, J. (1996). Activation of signaling by the activin receptor complex. *Mol. Cell Biol.* **16**, 1066–1073.

Baker, J. C., and Harland, R. M. (1996). A novel mesoderm inducer, Madr2, functions in the activin signal transduction pathway. *Genes. Dev.* **10**, 1880–1889.

Chen, X., Rubock, M. J., and Whitman, M. (1996). A transcriptional partner for MAD proteins in TGF-β signalling. *Nature* **383**, 691–696.

Daopin, S., Piez, K. A., Ogawa, Y., and Davies, D. R. (1992). Crystal structure of transforming growth factor-β2: An unusual fold for the superfamily. *Science* **257**, 369–373.

Hasegawa, Y., Eto, Y., Ibuki, Y., and Sugino, H. (1994). Activin as autocrine and paracrine factor in the ovary. *Horm. Res.* **41**(Suppl. 1), 55–62.

Lebrun, J. J., and Vale, W. W. (1997). Activin and inhibin have antagonistic effects on ligand-dependent heteromerization of the type I and type II activin receptors and human erythroid differentiation. *Mol. Cell Biol.* **17**, 1682–1691.

Mathews, L. S. (1994). Activin receptors and cellular signaling by the receptor serine kinase family. *Endocr. Rev.* **15**, 310–325.

Matzuk, M. M., Kumar, T. R., Shou, W., Coerver, K. A., Lau, A. L., Behringer, R. R., and Finegold, M. J. (1996). Transgenic models to study the roles of inhibins and activins in reproduction, oncogenesis, and development. *Recent. Prog. Horm. Res.* **51**, 123–154.

Moore, A., Krummen, L. A., and Mather, J. P. (1994). Inhibins, activins, their binding proteins and receptors: Interactions underlying paracrine activity in the testis. *Mol. Cell Endocrinol.* **100**, 81–86.

Xu, J., McKeehan, K., Matsuzaki, K., and McKeehan, W. L. (1995). Inhibin antagonizes inhibition of liver cell growth by activin by a dominant-negative mechanism. *J. Biol. Chem.* **270**, 6308–6313.

Adrenal Androgens

Collin B. Smikle
University of California, San Francisco

I. Physiology
II. Pathophysiology
III. Treatment
IV. Summary

GLOSSARY

adrenal androgens Steroid hormones secreted primarily by the adrenal gland; include dehydroepiandrostenedione, dehydroepiandrostenedione sulfate, 11 hydroxy-androstenedione, and androstenedione.

antiandrogen A substance capable of antagonizing and reducing the biological effects of androgenic hormones.

congenital adrenal hyperplasia A syndrome resulting primarily from a defect in one of the enzymes involved in steroid biosynthesis but may result from androgen ingestion or androgen producing tumor of the ovary or adrenal gland.

polycystic ovarian disease A heterogeneous group of disorders that may result in oligomenorrhea, anovulation, obesity, bilateral ovarian enlargement, and hirsuitism.

Androgens, produced primarily by the adrenal gland and the gonad, are important during fetal life and after birth. Adrenal androgens maintain pregnancy homeostasis, influence fetal organ maturation, and may play a role in the timing of parturition. After birth, adrenal production diminishes until puberty, rises until the fourth decade of life, then declines thereafter.

I. PHYSIOLOGY

Adrenal androgens play an important role in human development from early embryonic life through adulthood. After 16 weeks gestation, the fetal adrenal gland produces dehydroxyepiandrosterone (DHEA), the main C_{19} steroid precursor for placental estrogen production. The placenta lacks 17α-hydroxylase and can neither synthesize estrogens *de novo* nor convert progesterone or pregnenolone to C_{19} steroids (Fig. 1). After birth, dehydroxyepiandrosterone sulfate (DHEAS) and androstenedione levels drop dramatically, with the nadir at approximately 3 months. The levels of DHEA and androstenedione remain low until adrenarche when the adrenal gland becomes responsive to adrenocortico-

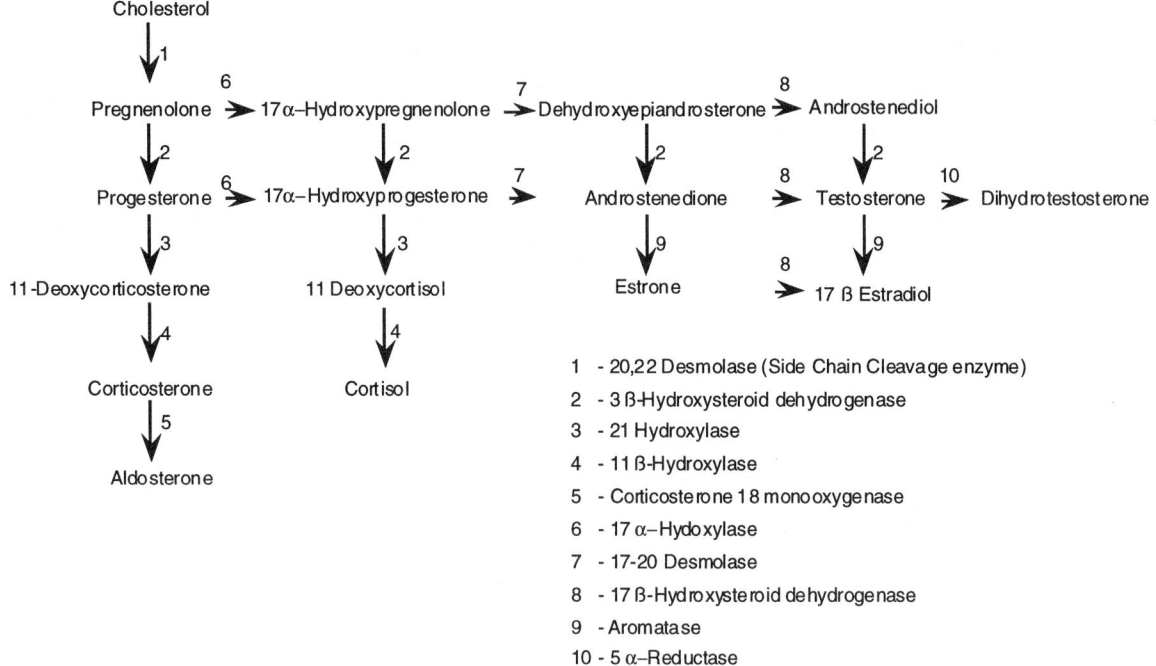

FIGURE 1 Steroidogenic pathway for the formation of androgens, estrogen, and testosterone.

tropin (ACTH) stimulation and androgen production resumes. Adrenal androgen production increases after adrenarche, is maximal during the third and fourth decades of life, And decreasing steadily thereafter. This midlife change in androgen production occurs without any significant change in ACTH or glucocorticoid secretion, suggesting that the adrenal gland becomes less responsive to ACTH stimulation.

A. Androgen Production

Although androgens are produced primarily in the adrenal gland, both the ovaries and peripheral adipose tissues synthesize androgen. After puberty, testosterone (T) is produced not only by the adrenal gland (25%) and the ovary (25%) but also by conversion of circulating DHEA, androstenedione, and estrogen in peripheral tissues (50%). Both androgen and estrogen production by the ovary decrease after menopause. Because estradiol (E_2) decreases by 95% and T and androstenedione decrease by 50 and 20%, respectively, virilization occurs. Increased serum luteinizing hormone (LH) after menopause stimulates the ovarian stroma and increases androgen production.

B. Androgen Action

Male secondary sex characteristics are caused by androgens, primarily testosterone. In the pilosebaceous unit of the skin, testosterone and its active metabolite, dihydrotestosterone, increase sebum production and transform the thin, wispy, vellous hair into darker, thicker terminal hair. In the skin, muscles, and bone, testosterone has an anabolic effect and increases protein production resulting in thicker skin, bones, and muscles. Testosterone causes hypertrophy of the laryngeal mucosa causing deepening of the voice. Excess androgens in men are of little phenotypic consequence but may influence hypertension and cardiovascular disease. However, in women excess androgens produced by the adrenal gland or the ovary may have significant consequences. Hyperandrogenemia in female fetuses may cause ambiguous genitalia or female pseudohermaphroditism, whereas in adult females a variety of pathologic consequences may occur.

C. Androgen Biosynthesis

The first step in androgen biosynthesis is binding of low-density lipoprotein (LDL) cholesterol to LDL receptors on cell membranes within the adrenal glands and the gonad (Fig. 2). The receptor/LDL cholesterol complex is internalized in clathrin-coated pits, aggregated with lysosomes, and hydrolyzed into cholesterol. Cholesterol is then transported to the mitochondrial membrane and converted to pregnenolone by the side chain cleavage (P450$_{scc}$) enzyme. This critical conversion of cholesterol to pregnenolone is the rate-limiting step for all steroid biosynthesis. Pregnenolone crosses from the mitochondria to the smooth endoplasmic reticulum where it is metabolized to DHEA, 11-deoxycortisol, and androstenedione. These intermediate androgens can be released into the circulation, converted to progesterone and estradiol, or transported back to the mitochondria for processing to cortisol or 11-β-hydroxyandrostenedione.

In the adrenal gland, ACTH regulates the production of most of the enzymes involved in steroid biosynthesis, including P450$_{scc}$, P450$_{11\beta}$, P450$_{c21}$, and P450$_{17\alpha}$. In the ovaries and testes, production of these enzymes is also regulated by LH and follicle-stimulating hormone (FSH). All circulating steroid hormones

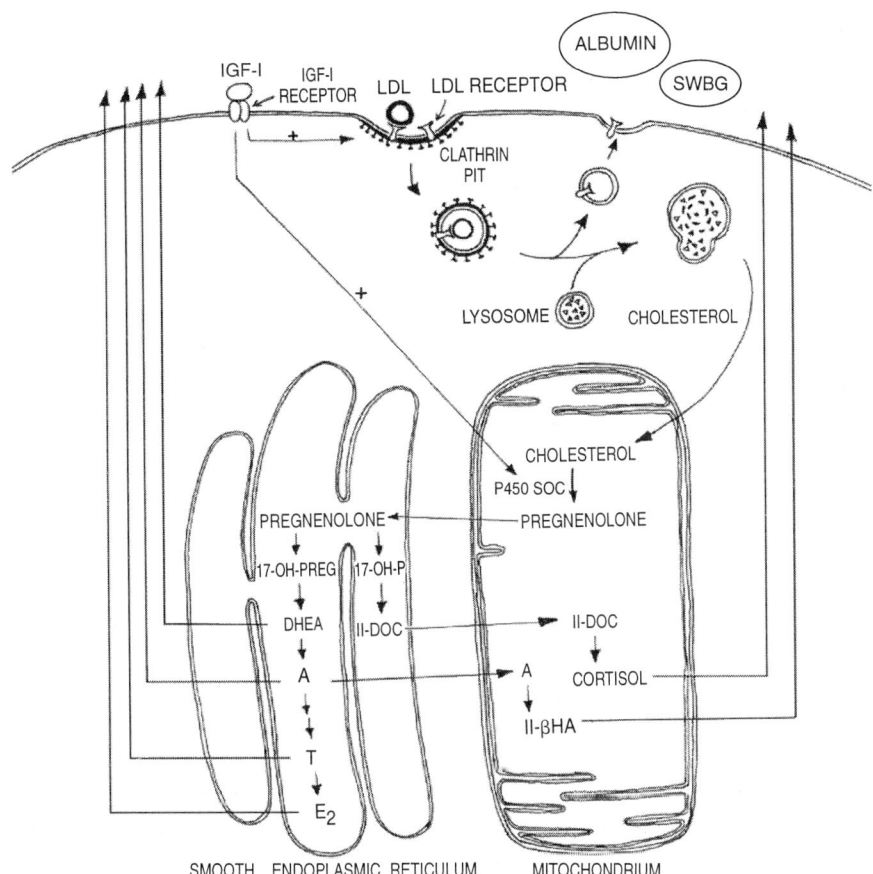

FIGURE 2 Schematic of intracellular steroid biosynthesis from circulating LDL particles. IGF-I, insulin-like growth factor-I; IGFr, IGF-I receptor; LDL, low-density lipoprotein; SHBG, steroid hormone-binding globulin; 17-OHP, 17 hydroxyprogesterone; 17-OH-preg, 17 hydroxypregnenolone; DHEA, dehydroxyepiandreosterone; A, androstenedione; 11-DOC, 11-deoxycorticosterone; 11-βHA, 11β-hydroxyandrosterone; T, testosterone; E$_2$, estradiol.

are transported by albumin and sex hormone-binding globulin (SHBG) produced by the liver, with a small fraction of hormone unbound or "free" in the circulation.

II. PATHOPHYSIOLOGY

Pathological conditions resulting in increased androgen production can begin during fetal life (Table 1). Congenital adrenal hyperplasia (CAH) results from exposure to androgens *in utero* during genitourinary system development. The severity of the exposure may cause female pseudohermaphroditsm or ambiguous genitalia in female fetuses. If CAH is incomplete, affected children may present with precocious puberty and short stature due to premature closure of the epiphyseal plates. In boys, this disorder may go undetected unless there is marked sexual precocity. After puberty, hyperandrogenism may result from adrenal or ovarian tumors, hyperprolactinemia, iatrogenic causes, and polycystic ovarian syndrome (PCOD).

A. Congenital Adrenal Hyperplasia

1. 21-Hydroxylase Deficiency

A defect in the gene encoding the 21-hydroxylase enzyme ($P450_{c21}$ enzyme) causes the majority of all cases of CAH. This gene is located in the human leukocyte antigen complex on the short arm of chromosome 6. Complete or partial enzyme defects will limit conversion of 11-deoxycorticosterone (DOC) and 17-hydroxyprogesterone to 11-deoxycortisol and lead to accumulation of progesterone, 17-hydroxyprogesterone, androstenedione, and testosterone.

In female fetuses with complete enzyme deficiency, external genitalia are virilized *in utero,* whereas male fetuses appear normal. In affected infants marked cortisol deficiency and aldosterone excess result in salt wasting, and these infants present with failure to thrive, hypotonia, and Addisonian crisis. Patients with incomplete enzyme defects have simple virilizing CAH without aldosterone deficiency or salt wasting. Patients with partial $P450_{c21}$ enzyme deficiency may present later in childhood with late onset (non-

TABLE 1
Diagnoses in Patients with Hyperandrogenism

Congenital adrenal, hyperplasia
  21-Hydroxylase deficiency
  11β-Hydroxylase deficiency
  3β-Hydroxysteroid dehydrogenase deficiency
  17α-Hydroxylase deficiency
Cushing's syndrome
  ACTH dependent   ACTH independent
  Ectopic ACTH-secreting   Adrenal adenomas
    tumor   Adrenal carcinoma
  Exogenous ACTH   Adrenal nodular hyperplasia
  Excess pituitary ACTH
Ovarian causes
  Neoplasms
    Sex cord tumors
    Hilar cell tumors
    Adrenal rest tumors
    Germ cell tumors
    Mixed gonadal tumors
  Polycystic ovarian disease
Iatrogenic
  Phentoin
  Diazoxide
  Anabolic steroid
  Corticosteroids
  19-Norsteroid derivatives
Hyerprolactinemia
Acromegaly
Porphyria

classical) CAH. Because of only mild deficiencies in cortisol and aldosterone synthesis, serum electrolytes are normal. Without excess prenatal androgens, virilization of the female fetus does not occur. However, these patients may have precocious puberty, hirsuitism, clitoral enlargement, development of male body habitus, and infertility.

CAH is inherited in an autosomal recessive pattern. One percent of Caucasian women are carriers for the gene responsible for CAH, resulting in a defect in the synthesis of the $P450_{c21}$ enzyme. Three to 10% of women with hyperandrogenism have incomplete 21-hydroxylase deficiency. There are two $P450_{c21}$ genes (CYP21B gene and CYP21P pseudogene) located on chromosome 6 within human leukocyte antigen region. The CYP21P gene is nonfunctional due to insertions, deletions, and point mutations within the gene. These defects within the CYP21B gene cause salt-wasting CAH.

2. Diagnosis

Fetuses affected with defects of the $P450_{c21}$ enzyme can be diagnosed either by genotyping fetal cells (amniocentesis of chorionic villus sampling) or by measuring amniotic fluid 17-OHP and androstenedione concentration. Because the genitourinary tract has already formed at 12 weeks, diagnosis after this time will not prevent virilization of female fetuses. Therefore, affected families should be treated presumptively with dexamethasone (10 {{mu}}g/kg) when pregnancy is diagnosed, and chorionic villus sampling should be offered at 8–10 weeks. Dexamethasone is then continued throughout pregnancy for all female fetuses.

Children and adults with late-onset CAH can be diagnosed by measuring morning 17-OHP or by ACTH stimulation test. Following intravenous cosyntropin ($ACTH_{1-24}$), 17-OHP levels of ≥ 10 ng/ml indicate $P450_{c21}$ enzyme deficiency (Table 2).

B. 3β-Hydroxysteroid Dehydrogenase Deficiency

3β-Hydroxysteroid dehydrogenase deficiency (3β-HSD) is a rare cause of hyperandrogenism. Affected individuals with complete 3β-HSD have absent sex steroids and hypogonadism. Partial enzyme deficiencies cause menstrual irregularities and hirsutism. After diagnostic ACTH challenge, 17-hydroxypregnenolone (17-OH-preg) and DHEA are elevated (Table 2). These patients also have high 17-OH-preg to 17-OHP and DHEA to androstenedione ratios. Treatment with dexamethasone may normalize menstruation and improve hirsuitism and acne in up to 50% of affected patients.

C. 11β-Hydroxylase

The incidence of 11β-hydroxylase ($P450_{c11}$ enzyme) deficiency in patients with hyperandrogenism is 1–5%. The gene for the $P450_{c11}$ enzyme is located on human chromosome 8. Patients with $P450_{c11}$ enzyme deficiencies cannot convert 11-deoxycortisol and 11-deoxycorticosterone to cortisol and corticosterone, respectively (Fig. 1), resulting in elevated plasma aldosterone and hypertension. An exaggerated response in 11-deoxycortisol and DOC after ACTH stimulation confirms the diagnosis. Treatment with glucocorticoids reduces endogenous ACTH and reduces blood pressure to within normal limits.

D. Other Causes of CAH without Hyperandrogenism

Other causes of CAH include congenital lipoid hyperplasia, 17-hydroxylase deficiency, adrenoleukodystrophy, and primary familial xanthomatosis. Both congenital lipoid hyperplasia and 17-hydroxy-

TABLE 2
Laboratory Evaluation of Patients with Hyperandrogenism

	Polycystic ovarian disease	Ovarian tumor	Congenital adrenal hyperplasia	Adrenal neoplasm	Cushing's syndrome	Idiopathic
DHEAS	Nl, or ↑	N	N or ↑	↑↑↑	Nl or ↑	N
Testosterone	Nl to ↑	↑↑	Nl to ↑	Nl to ↑	Nl to ↑	N
Serum LH	Nl to ↑	N	N	N	N	N
ACTH stimulation	N	N	↑↑	Nl to ↑	N	N
Dex suppression	N	N	N	↑	↑	N

Note. Nl, normal; ↑, increased.

lase deficiency are cause by enzymatic defects that limit androgen production. Adrenoleukodystrophy is an X-linked recessive disorder that results in accumulation of esterified cholesterol in all tissues and causes diffuse sclerosis of the brain and adrenals. Primary familial xanthomatosis is caused by a defect in a lysosomal enzyme that prevents cholesterol breakdown. Although these conditions may be lethal, serum androgen concentration is absent or low and affected patients are not virilized.

E. Adrenal Diseases

Some adrenal disorders are associated with hyperandrogenism, including Cushing's syndrome (cortisol excess) or hyperaldosteronism (Table 1). Most patients with adrenal hyperstimulation have overlapping features including obesity, hirsuitism, virilization, and hypertension. Cushing's syndrome is characterized by bilateral adrenal hyperplasia due to hypersecretion of ACTH by a pituitary adenoma or ectopic neuroendocrine tumor. Primary adrenal neoplasms cause Cushing's syndrome in 20–25% of cases. Primary adrenal tumors are usually unilateral, and approximately 50% are malignant. Patients with Cushing's syndrome may have elevated plasma and urinary cortisol, generalized osteoporosis, and, in severe cases, hypokalemia. Patients with corticosterone-secreting tumors usually present with hypertension and hyperaldosteronism.

Micronodular adrenal dysplasia rarely causes hyperandrogenism and occurs more commonly in children than adults. Patients with adrenal neoplasms or micronodular adrenal hyperplasia have low or undetectable plasma ACTH concentrations. Many malignant adrenal tumors have reduced 11β-hydroxylation and increasing serum DHEA, DHEAS, and androsterone. Adrenal tumors may be detected by magnetic resonance imaging (MRI) or computerized tomography (CT) of the adrenal glands and surgically removed. ACTH-secreting pituitary adenomas are also diagnosed by MRI or CT scan. These tumors can be removed via transspenoidal resection of the adenoma.

Cushing's syndrome is diagnosed by a prolonged dexamethasone suppression test. However, ectopic ACTH-secreting tumors may be difficult to detect. These tumors usually arise from bronchogenic carcinomas, medullary carcinoma of the thyroid, tumors of the ovary, pancreas, or thymus, or from bronchial adenomas. Treatment of ectopic ACTH-secreting tumors may be limited to medical or surgical adrenalectomies. Aminoglutethimide, o,p'-DDD, and metyrapone can be used alone or in combination with surgery to treat Cushing's syndrome.

F. Polycystic Ovarian Disease

One of the most common causes of hyperandrogenism in women is PCOD. PCOD may be the end result of one of several underlying disease processes including polycystic ovarian syndrome and hyperandrogenic chronic anovulation. Patients with PCOD often display hirsuitism, anovulation, and oligomenorrhea but rarely have signs of virilization. On physical examination, patients with PCOD may be obese with bilateral ovarian enlargement. Sonographic or laparoscopic evaluation of the ovaries may show multiple small (10-mm) cysts within the ovary and thickened ovarian stroma. Ovarian stromal androgen production is stimulated by LH and hCG, and serum LH levels may be elevated (Table 2). Moreover, the normal ratio of androstenedione to testosterone may be decreased.

In a subset of patients with PCOD, there is a defect in the intracellular signaling of the insulin receptor. This defect results in hyperinsulinemia and adult onset diabetes mellitus. Insulin, which also binds to insulin-like growth factor-I (IGF-I) receptor, affects follicular development in the ovary and steroidogenesis and the adrenal gland. Therefore, hyperinsulinemia increases ovarian and adrenal androgen biosynthesis.

III. TREATMENT

A. Surgical Therapy

Corrective surgery may be used to reduce clitoromegaly, labial fusion, and masculinization effects of adrenal steroid excess in females born with CAH or adrenogenital syndrome. The amount of surgery depends on the extent of virilization and is usually done in early childhood after female gender assignment. Vaginal reconstructive surgery has improved outcomes if delayed until after age 16. Since the upper reproductive tract is intact, normal reproduction is feasible after lower tract abnormalities are corrected. Because untreated CAH may result in heterosexual precocious pseudopuberty with early closure of epiphyseal plates and short stature, medical therapy with glucocorticoids is recommended.

B. Weight Loss

Weight reduction may be useful in treating hyperandrogenism associated with PCOD. Patients who lose weight successfully have lower plasma androstenedione and testosterone levels. Weight loss can also reduce serum insulin, testosterone, progesterone, and LH concentrations, causing many oligomenorrheic women to resume menses and become fertile.

C. Glucocorticoids

Although glucocorticoid treatment can improve hyperandrogenism in patients with enzyme deficiencies, large amounts of potent corticosteroids may be necessary to suppress ACTH successfully. Side effects of glucocorticoid therapy include iatrogenic Cushing's syndrome, adrenal suppression, inability to respond to stress, glucose intolerance, and osteoporosis and aseptic necrosis of the femoral head.

D. Antiandrogens

Patients with hyperandrogenism can also be treated with antiandrogens. Antiandrogens, which bind to androgen receptors and inhibit androgen biosynthesis, are successful in treating peripheral manifestations of hyperandrogenism. Hirsuitism may be successfully treated with spironolactone, cyproterone acetate, and flutamide. Cyproterone acetate acts both as an antiandrogen and as an antiprogestin and is often used in combination with estradiol to treat hirsuitism. Although treatment with hydrocortisone may result in greater reduction in plasma androstenedione and testosterone, patients treated with cyproterone acetate and estradiol may report greater reduction in Ferriman–Gallwey score for hirsuitism.

IV. SUMMARY

Androgens are produced primarily by the adrenal gland and the gonad and may be converted by peripheral adipose tissues to form potent androgens such as dihydrotestosterone. During fetal life, adrenal androgens maintain pregnancy homeostasis, influence fetal organ maturation, and may play a role in the timing of parturition. After birth, adrenal production diminishes until puberty, rises dramatically until the fourth decade of life, and declines slowly thereafter. Pathologic conditions due to abnormalities in the genes encoding the steroidogenic enzymes, the androgen receptor, or the insulin receptor may result in increased androgen production.

The pathologic consequences of hyperandrogenism in women include mild (acne and menstrual irregularities), moderate (infertility, short stature, anovulation, and endometrial hyperplasia), and severe (life-threatening salt wasting, hypertension, Addison's disease, ambiguous genitalia, or female pseudohermaphroditism). The ACTH stimulation test can identify patients with enzymatic defects when baseline laboratory values are abnormal. Treatment for each disease process depends on the underlying etiology and may even begin *in utero* when enzymatic defects are suspected.

See Also the Following Articles

ADRENAL HYPERPLASIA, CONGENITAL VIRILIZING; DHEA (DEHYDROEPIANDROSTERONE); POLYCYSTIC OVARY SYNDROME

Bibliography

Azziz, R. (1996). The adrenal connection. In *Reproductive Endocrinology, Surgery, and Technology* (E. Adashi, J. Rock, and Rosenwaks, Z., Eds.), Vol. 1, pp. 1162–1180. Lippincott-Raven, Philadelphia

Barbieri, R. (1995). Hyperandrogensim. In *Reproductive Medicine and Surgery* (E. Wallach and H. Zucar, Eds.), pp. 209–229. Mosby-Year Book, St. Louis, MO.

Casson, P. R., Hornsby, P. J., and Buster, J. E. (1996). Adrenal androgens, insulin resistance, and cardiovascular disease. *Sem. Reprod. Endocrinol.* **14**, 29–34.

Clark, P. M., Hindmarsh, P. C., Shiell, A. W., *et al.* (1996). Size at birth and adrenocortical function in childhood. *Clin. Endocrinol. Oxford* **45**, 721–726.

Gonzalez, F. (1997). Adrenal involvement in polycystic ovary syndrome. *Sem. Reprod. Endocrinol.* 15, 137–157.

Gonzalez, F., Chang, L., Horab, T., and Lobo, R. A. (1996). Evidence for heterogeneous etiologies of adrenal dysfunction in polycystic ovary syndrome. *Fertil. Steril.* 66, 354–361.

Hsueh, A. (1989). Ovarian hormone synthesis, circulation, and mechanism of action. In *Endocrinology* (L. DeGroot, Ed.), Vol. 3, pp. 1829–1839. Saunders, Philadelphia.

Kornely, E., and Schlaghecke, R. (1994). Complete remission of metastasized adrenocortical carcinoma under o,p'-DDD. *Exp. Clin. Endocrinol.* 102, 50-53.

Kuil, C. W., and Brinkmann, A. O. (1996). Androgens, antiandrogens and androgen receptor abnormalities. *Eur. Urol.* 29(Suppl. 2), 78–82.

McKenna, T. J., and Cunningham, S. K. (1995). Adrenal androgen production in polycystic ovary syndrome. *Eur. J. Endocrinol.* 133, 383–389.

O'Brien, R. C., Cooper, M. E., Murray, *et al.* (1991). Comparison of sequential cyproterone acetate/estrogen versus spironolactone/oral contraceptive in the treatment of hirsutism. *J. Clin. Endocrinol. Metab.* 72, 1008–1013.

Pasquali, R., Antenucci, D., Casimirri, F., *et al.* (1989). Clinical and hormonal characteristics of obese amenorrheic hyperandrogenic women before and after weight loss. *J. Clin. Endocrinol. Metab.* 68, 173–179.

Polderman, K. H., Gooren, L. J., and Heine, R. J. (1996). Effects of physiological and supraphysiological doses of insulin on adrenal androgen levels. *Horm. Metab. Res.* 28, 152–155.

Rodrigues, N. R., Dunham, I., Yu, Y. C., *et al.* (1987). Molecular characterization of the HLA-linked steroid 21-hydroxylase B gene from an individual with congenital adrenal hyperplasia. *EMBO J.* 6, 1653–1661.

Simpson, E., and Waterman, M. (1989). Steroid hormone biosynthesis in the adrenal cortex and its regulation by adrenocorticotropin. In *Endocrinology* (L. DeGroot, Ed.), Vol. 2, pp. 1543–1556. Saunders, Philadelphia.

Speiser, P. W., Laforgia, N., Kato, (1990). First trimester prenatal treatment and molecular genetic diagnosis of congenital adrenal hyperplasia 21-hydroxylase deficiency. *J. Clin. Endocrinol Metab.* 70, 838–848.

Speiser, P. W., Serrat, J., New, M. L., and Gertner, J. M. (1992). Insulin insensitivity in adrenal hyperplasia due to nonclassical steroid 21-hydroxylase deficiency. *J. Clin. Endocrinol. Metab.* 75, 1421–1424.

Spritzer, P., Billaud, L., Thalabard, J. C., Birman, *et al.* (1990). Cyproterone acetate versus hydrocortisone treatment in late-onset adrenal hyperplasia. *J. Clin. Endocrinol. Metab.* 70, 642–646.

van Heerden, J. A., Young, W. F., Jr., Grant, C. S., and Carpenter, P. C. (1995). Adrenal surgery for hypercortisolism—Surgical aspects. *Surgery* 117, 466–472.

White, P. C., and New, M. I. (1992). Genetic basis of endocrine disease 2: Congenital adrenal hyperplasia due to 21-hydroxylase deficiency. *J. Clin. Endocrinol. Metab.* 74, 6–11.

Yen, S. (1991). Chronic anovulation caused by peripheral endocrine disorders. In *Reproductive Endocrinology*: Physiology, Pathophysiology and Clinical Management (S. Yen and R. Jaffe, Eds.), 3rd ed. Saunders, Philadelphia.

Adrenal Hyperplasia, Congenital Virilizing

Peter A. Lee and Selma F. Witchel

University of Pittsburgh School of Medicine and Children's Hospital of Pittsburgh

I. Introduction
II. Female Patients with Congenital Virilizing Adrenal Hyperplasia
III. Male Patients with Congenital Virilizing Adrenal Hyperplasia
IV. Late-Onset Congenital Adrenal Hyperplasia and Manifesting Heterozygotes
V. Summary

GLOSSARY

ambiguous genitalia The physical development resulting from virilization of external genitalia in females or inadequate virilization of genitalia in males. In the case of the virilized female infant, the clitoris is enlarged with varying degrees of labial fusion which may resemble a scrotum (testes are not palpable).

CYP11B1 gene The gene which codes for 11β-hydroxylase, the enzyme which converts 11-deoxycortisol to cortisol.

CYP21 gene The gene which codes for 21-hydroxylase. This enzyme is expressed in the adrenal cortex where it converts 17-hydroxyprogesterone to 11-deoxycortisol and progesterone to deoxycorticosterone. This gene is located on the short arm of chromosome 6 in the class III HLA region. A highly homologous nonfunctional pseudogene (CYP21P) maps in close proximity to CYP21. Most CYP21 mutations associated with 21-hydroxylase deficiency are gene conversion events in which the functional gene has acquired deleterious nucleotide sequences from the pseudogene.

3β-HSD2 gene The gene which codes for the isozyme of 3β-hydroxysteroid dehydrogenase, which is predominantly adrenal and gonad; this enzyme converts Δ^5 steroids, i.e., pregnenolone, to Δ^4 steroids, i.e., progesterone.

newborn screening programs Programs, legally required in some regions, to screen newborn infants for diseases, such as congenital hypothyroidism, phenylketonuria, and galactosemia, using dried whole blood samples collected on filter paper. Newborn screening programs to identify infants affected with 21-hydroxylase deficiency determine whole blood 17-hydroxyprogesterone. For 21-hydroxylase deficiency, screening programs help identify affected males prior to symptoms of adrenal insufficiency and affected females who have been misassigned or assigned to male sex of rearing because of marked virilization of the external genitalia.

virilization In females, clinical features indicative of excessive androgen secretion: If prenatal, ambiguous genitalia including clitoromegaly results; if postnatal, clitoromegaly, premature sexual hair, or hirsutism may occur.

The congenital virilizing adrenal hyperplasias are autosomal recessive disorders characterized by impaired cortisol production and excessive adrenal androgen production. The inadequate cortisol secretion leads to diminished negative feedback inhibition, leading to excessive ACTH secretion, resulting in increased adrenal androgen secretion. Over 95% of patients with congenital virilizing adrenal hyperplasia have 21-hydroxylase deficiency due to mutations in the 21-hydroxylase gene. The other 5% of patients have 11β-hydroxylase deficiency due to mutations in the 11β-hydroxylase gene or 3β-hydroxysteroid dehydrogenase deficiency secondary to mutations in the 3β-hydroxysteroid dehydrogenase type II gene. The specific findings, e.g., hypertension, and pattern of steroid hormones differ for the particular enzyme deficiency. In 21 hydroxylase deficiency, elevated random or stimulated 17-hydroxyprogesterone concentrations are observed.

I. INTRODUCTION

Prior to the availability of glucocorticoid therapy, infertility among male and female patients with congenital adrenal hyperplasia was considered to be inevitable because of the increased adrenal androgen secretion. In affected females, the consequences of excessive adrenal androgen secretion, such as virilization and interference with hypothalamic–pituitary–ovarian axis function, precluded normal feminization and induced chronic anovulation. Among affected males, the extratesticular source of androgen was presumed to suppress the hypothalamic–pituitary–testicular axis and impair spermatogenesis.

Most information in the medical literature regarding fertility in congenital adrenal hyperplasia pertains to 21-hydroxylase deficiency. Interpretation of prior studies is often confounded by the broad spectrum of clinical severity, diverse treatment regimens, age at onset of therapy, compliance with therapy, and adequacy of adrenal androgen suppression. The most worthwhile data involve patients diagnosed promptly after onset of symptoms and treated optimally throughout life. Consequently, because of delay in diagnosis, suboptimal prescribed therapy, and poor compliance with daily medical regimens, many questions remain regarding the potential for fertility among affected individuals, despite the use of glucocorticoid therapy to prevent hyperandrogenism since 1950.

II. FEMALE PATIENTS WITH CONGENITAL VIRILIZING ADRENAL HYPERPLASIA

A. Background

1. Primary Factors

i. Prenatal Virilization Among females, hyperandrogenism secondary to congenital adrenal hyperplasia at any point during life can impair or preclude fertility. During fetal life, the developing brain and ovaries are exposed to excessive adrenal androgens which also virilize the external genitalia. The magnitude of prenatal virilization ranges from mild clitoromegaly to full masculine differentiation. While the ovaries and Müllerian-duct derived uterus, Fallopian tubes, and upper vagina developed normally in most affected females, a few adult patients have been found to have hypoplastic uteri even after adequate estrogen exposure. The pelvis in the newborn female with congenital virilizing adrenal hyperplasia (CVAH) is narrowed in comparison with other females and more like the male pelvis. The degree of male differentiation may be so extensive that affected females are considered to be males with undescended testes. If affected females are raised as males, the potentially fertile female is, of course, unequivocally infertile. Newborn screening programs and greater physician awareness have allowed correction of incorrect sex assignment after missed diagnosis of virilized females with congenital adrenal hyperplasia. Affected females with late-onset CVAH are usually not identified in infancy because their external genitalia are not virilized and their 17-hydroxyprogesterone concentrations are generally below the neonatal screening programs cutoff concentrations. Most such patients present with findings in the peripubertal period which should lead to diagnosis and institution of appropriate therapy.

The virilized external genitalia of the females presenting in the neonatal period require surgical correction. Relevant education and counseling of the parents, and in an age-appropriate fashion, of the patient herself are important. Initial surgical repair involving clitoroplasty, which may include clitoral recession, not clitoridectomy, should begin during the first year of life. Vaginoplasty is recommended during adolescence after counseling sessions with the patient to decide the most appropriate timing of the operation. Vaginal dilatation may be necessary after vaginoplasty; therefore, it is ideal if such surgery can be timed to briefly precede the onset of regular sexual intercourse. With adequate glucocorticoid replacement, suppression of adrenal androgens, and surgical treatment, normal adult sexual function and fertility would be anticipated.

ii. Hyperandrogenism during Childhood and Adolescence It is unclear how excessive androgen secretion during infancy and childhood impacts on ovarian function. During adolescence and young adulthood, excessive androgen concentrations influence gonadotropin secretion and impair follicular maturation. Available data suggest that patients who have suppressed androgen concentrations through most of childhood, adolescence, and adulthood are most likely to have regular menses. Indeed, regular menses in a girl with congenital adrenal hyperplasia indicates appropriate glucocorticoid therapy and generally predicts good potential for fertility.

Excessive adrenal androgen secretion during the peripubertal and postpubertal years results in dysfunction of gonadotropin secretion leading to chronic anovulation due to failure to select a dominant follicle. The phenotype of patients with congenital adrenal hyperplasia, especially untreated mild CVAH, resembles that of polycystic ovary syndrome (PCOS). Features common to congenital adrenal hyperplasia and PCOS include hirsutism, severe acne, oligo/amenorrhea, infertility, increased luteinizing hormone:follicle-stimulating hormone (LH:FSH) ratio, and polycystic ovaries on ultrasound. Some women with congenital adrenal hyperplasia and secondary PCOS appear to require ovarian suppression concomitant with adrenal suppression to prevent symptoms of hyperandrogenism. The finding of exaggerated 17-hydroxyprogesterone responses to gonadotropin-releasing hormone (GnRH) analog provides additional evidence that secondary ovarian hyperandrogenism can develop in some affected women. The precise mechanism through which adrenal hyperandrogenism induces ovarian hyperandrogenism remains to be elucidated.

iii. Pubertal Development With appropriate suppression and good compliance, female pubertal changes progress normally. If the physique at diagnosis and initiation of therapy does not include evidence of chronic hyperandrogenism, such as excessive skeletal and muscular development, normal feminization can be anticipated. Adrenarche, the pubertal physiologic increase in adrenal androgen secretion usually precedes gonadarche; such an increase has not been detectable among patients treated with suppressive glucocorticoid doses. However, given the underlying pathophysiology, it may be most difficult to ascertain adrenarche. Nevertheless, when judging the adequacy of suppression using the endpoints of circulating adrenal androgen levels or urinary excretion of 17-ketosteroids, levels appropriate for age and stage of development are used. Age at menarche varies considerably. Most patients experience menarche slightly later than their peers. Unless menarche is extremely early due to secondary gonadotropin precocious puberty or extremely late due to hyperandrogenism, the age at menarche does not predict menstrual regularity or fertility.

If persistent hyperandrogenism occurs during adolescence, a masculinized physique can be expected with narrow hips, broad shoulders, and male-type muscular configuration. Prolonged hyperandrogenism can lead to male patterns of facial and body hair, temporal hair loss (Fig. 1), clitoromegaly, and deepening of the voice. Hence, inadequately treated patients, whether due to insufficient glucocorticoid therapy or poor compliance, are at increased risk for poor feminization and progressive virilization during puberty which may lead to not only infertility but also poor body image and self-esteem.

2. Confounding Factors

i. Sexual Orientation/Gender Identity The medical literature concerning the psychosocial development of women with congenital adrenal hyperplasia is controversial. While cross-gender role behavior and less sense of femininity exist during childhood, the frequency of homosexuality or homosexual behavior is unclear. The most meaningful data show that there are fewer casual and less intimate social contacts with men. Furthermore, there are fewer sex-

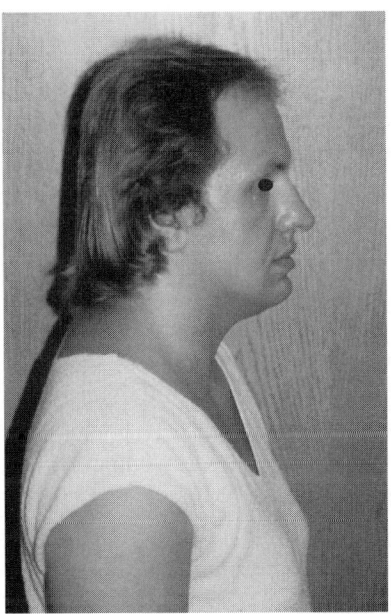

FIGURE 1 Profile of an adult female with congenital adrenal hyperplasia (CYP21) who has been chronically inadequately treated. She has a masculinized body habitus, receding temporal hairline, facial hair, and overall virilized appearance.

ual experiences with men, but there are not more sexual experiences with women than those of control groups. Women with congenital adrenal hyperplasia who have atypical gender identity usually have had a history of delayed or lack of surgical feminization; they also have a history of progressive virilization during childhood and adolescence. It is speculated that prenatal androgen influences gender and sexual orientation. However, physical appearance, social factors, and poor self-perception psychologically may interfere with opportunities to develop heterosexual relationships and make these patients more open to homosexual interactions and the development of homosexual preferences.

ii. Socialization Few affected women have short-term heterosexual relationships. Long-term relationships are rare. Overall, sexual activity is low. Preference for hobbies, sports, activities, and interests typically associated with males may develop during childhood and persist into adolescence. Poor self-esteem related to having a chronic genetic disorder, especially if virilization has occurred, may turn social contacts into major obstacles during adolescence and young adulthood. Masturbation occurs as frequently, if not more so, compared to age-matched females. The capacity for "falling in love" and experiencing erotic feelings may be diminished.

iii. Adequacy of Genital Repair Many patients who have undergone a vaginoplasty feel that they still do not have a functional vagina. Furthermore, among those women who indicate a functionally satisfactory vagina, gynecologic examination often reveals that they do not have a functional vagina. Anxiety and fear of vaginal inadequacy can impede progression toward sexual activity in a heterosexual relationship. Attempts at intercourse, especially when it is impossible because of physical constraints related to genital anatomy, can embarrass and humiliate. A sense of failure regarding sexuality can impact negatively on self-esteem, leading to further social isolation and withdrawal.

B. Fertility

The potential for fertility is good among patients who have regular menses with evidence of ovulation and have regular sexual intercourse with a male sexual partner. Fertility is rare among inadequately treated or poorly compliant patients. To date, there are no controlled studies of women with congenital adrenal hyperplasia desiring and achieving pregnancy. The available data, primarily from the Johns Hopkins Hospital, included those who were or had been sexually active but were not limited to married or cohabiting women desiring children. A recent series included 40 from a total of 80 women who reported heterosexual activity; 25 had simple virilizing SV-CAH and 15 had salt-wasting SW-CAH (more severe). Pregnancies were reported for 15 of the 25 with SV-CAH but only for a single women with SW-CAH. However, data regarding the regularity of unprotected intercourse were not reported. Of note, whereas 20 of 40 (50%) with SV-CAH were married, only 5 of 40 (12.5%) of the women with SW-CAH were married. An earlier report from this same population compared pregnancy rates with age at initiation of glucocorticoid therapy. None of the group members who were untreated until after age 20 years became pregnant; all were anovulatory. Seven of 11 in whom glucocorticoid therapy was initiated between ages 6 and 20 years who were sexually active reported a total of 12 pregnancies.

Among 26 postpubertal females followed at the Children's Hospital of Pittsburgh, 18 were menstruating regularly, 3 irregularly, 3 had oligomenorrhea, and 2 had amenorrhea (Table 1). Menstrual patterns showed no relationship with severity of adrenal hyperplasia based on need for mineralocorticoid replacement (salt-losing or simple virilizing). Two pa-

TABLE 1
Menstrual History of 26 Postmenopausal Females with CVAH Due to 21CYP Gene Mutations

Menstrual history	SV	SW
Regular menses	9	9
Irregular menses	3	0
Oligomenorrhea	1	2
Amenorrhea	1	1

Note. Data show no relationship of menstrual regularity and severity of adrenal hyperplasia as indicated by simple virilizing (SV) or salt wasting (SW).

tients were married and had children. Previous assessment of a portion of this population suggested that good androgen suppression from infancy correlated with regular menses. Good suppression during infancy and puberty but only fair suppression thereafter tended to correlate with regular menses just after menarche with subsequent menses becoming irregular. Not surprisingly, poor suppression tended to predict irregular or absent menses.

From this limited information, it appears that pregnancy can be anticipated among women with CVAH who are menstruating regularly with evidence of ovulation, who have a vagina adequate for intercourse, and who have regular intercourse with a fertile male partner. Data also support the hypothesis that better treatment of adrenal hyperplasia to suppress adrenal androgen secretion during the peripubertal and postpubertal years is associated with more regular and ovulatory menstrual cycles. Significant factors related to apparent infertility are low marriage rates and decreased heterosexual activity. Both of these factors appear to be related to delayed or failure to achieve psychosexual milestones, more homosexual fantasies and experience, and social withdrawal and isolation. A vaginal introitus insufficient for sexual intercourse intensifies the fears of inadequacy regarding sexual function.

C. Pregnancy and Delivery

Once pregnancy has been achieved, adrenal androgen suppression by glucocorticoid therapy should be continued. This will preclude virilization of the unaffected female fetus. Glucocorticoid dosage generally does not need to be increased during pregnancy. Many, but not all, affected women require cesarean sections for delivery due to cephalopelvic disproportion because of an android pelvis.

Genetic counseling regarding risk for bearing affected children should be addressed prior to conception. All forms of congenital adrenal hyperplasia are autosomal recessive. For example, the female patient with 21-hydroxylase deficiency carries two deleterious CYP21 alleles. Genotyping the male partner of an affected female may be beneficial to enable accurate genetic counseling because of the high frequency of heterozygotic carriers for 21-hydroxylase deficiency. If the male partner shows two wild-type alleles, all the offspring will inherit one of their mother's two mutant alleles and will be carriers for 21-hydroxylase deficiency. If the male partner is a carrier for 21-hydroxylase deficiency, there is a 50% chance for each pregnancy to result in an affected child, with each child receiving one mutant maternal allele and either the mutant or wild-type paternal allele.

III. MALE PATIENTS WITH CONGENITAL VIRILIZING ADRENAL HYPERPLASIA

A. Background

1. Potentially Significant Factors

i. Precocious Puberty In the past, males with salt-losing CVAH commonly presented in adrenal crisis during the neonatal period or died without being diagnosed. Males with milder forms typically present with excessive virilization (peripheral precocious puberty or precocious pseudopuberty), accelerated growth rates, and advanced skeletal maturation. In the untreated or inadequately treated male, male physique and genital development are advanced for chronologic age. Typically, testicular volumes are less than anticipated for pubertal development, indicative of lack of gonadotropin stimulation consistent with an extratesticular (adrenal) source of androgen.

However, GnRH-dependent precocious puberty may occur as a consequence of the chronic androgen exposure. This, by definition, is the onset of pubertal gonadotropin secretion leading to testicular testosterone secretion and stimulation of spermatogenesis. Adult short stature is a potential outcome of CVAH and secondary precocious puberty because of premature epiphyseal fusion.

ii. Adequate Adrenal Androgen Suppression With timely use of glucocorticoid therapy to suppress the excessive adrenal androgen secretion, growth rates, skeletal maturity, and pubertal development including usual virilization can be expected to occur normally. If secondary GnRH-dependent precocious puberty has occurred, GnRH analog therapy may be appropriate to help suppress pubertal and skeletal maturation.

B. Fertility

Limited data exist regarding fertility in males with CVAH. Published data from the Johns Hopkins Hospital (JHH) Pediatric Endocrine clinic, observations by the authors in male patients at the Children's Hospital of Pittsburgh (CHP), and a publication of case reports of three men are summarized below. Findings from the JHH report include men whose CVAH was never treated, men who had discontinued treatment, and men on adequate glucocorticoid suppression. Paternity was reported for 10 of 13 married men in the JHH series with conceptions ending in abortion for two additional men.

Among the CHP patients (Table 2), 27 of 30 men had normal adult testicular volume; 2 of the men with decreased size did have normal adult testicular volume previously at the completion of puberty but subsequently became noncompliant with treatment. All 3 of the men in the case reports of infertility had diminished testicular size, subnormal or low normal gonadotropin concentrations, and decreased sperm counts. Decreased numbers of Leydig cell were observed on testicular biopsies performed in 2 men prior to glucocorticoid treatment. Following initiation of glucocorticoid therapy, paternity was achieved in 2 of these 3 men.

A summary of the hormonal data from these three sources shows that testosterone concentrations were in the normal range for all tested (JHH, $n = 18$; CHP, $n = 17$; and 3 in the case report). Serum LH concentrations were normal for 14 men from CHP, normal but below the mean for adult males for the JHH men, and low normal or subnormal for the 3 men in the case reports. The normal testosterone and relatively low LH concentrations suggests suppression of the hypothalamic/pituitary–LH/Leydig cell axis. Presumably, extratesticular steroid hormones derived from peripheral conversion of adrenal androstenedione suppress this axis in untreated or inadequately treated men with CVAH.

Serum FSH levels were normal for 31 of 38 men in the three series. Three men showed elevated FSH concentrations and 4 men had decreased FSH concentrations. Possible mechanisms responsible for abnormal FSH levels could be either suppression of FSH secretion by chronic excessive androgen feedback or excessive secretion of FSH resulting from inadequate negative feedback from the seminiferous tubules, suggesting primary testicular damage.

An important predictor for fertility is spermatogenesis as indicated by sperm analyses. While some adult males with adrenal hyperplasia who have never been treated have been reported to have normal LH, FSH, and testosterone levels, normal sperm counts, and paternity, others have been reported to have azoospermia or oligospermia. Among this latter group are those with improvement of sperm production with adrenal suppression. Reviewing the male CVAH patients described previously, 9 of 11 men in the JHH study and 2 of 3 in the CHP study showed normal sperm counts. A third CHP patient had an abnormal sperm count and markedly elevated FSH levels, suggesting testicular damage.

Thus, fertility can be expected to accompany optimal therapy for adrenal hyperplasia among men. However, the data do suggest that excessive adrenal androgen production may reversibly impair the hypothalamic–pituitary–testicular axis. One caveat is that whereas decreased testicular volume suggests diminished sperm production, normal volume does not imply normal sperm counts.

C. Indications for Treatment of Adult Males with CVAH

While fertility is possible among some untreated males with CVAH, other males who are inadequately treatment may be infertile. Also, at least some of these men have the potential for fertility following the institution of appropriate glucocorticoid therapy. Therefore, it is recommended that adult males continue glucocorticoid therapy to maintain adequate

TABLE 2
Characteristics Relating to Fertility of 30 Men with Adrenal Hyperplasis

Characteristic	Range		
	Normal	Abnormal	Not available
Testicular volume	27	3	0
Serum testosterone	18	0	12
Serum LH	13	1	16
Serum FSH	12	2	16
Semen analysis	2	1	27

suppression of adrenal androgens. Patients with or without replacement therapy are at risk for adrenal crises with excessive stress unless increased glucocorticoids are given.

Another rationale for continued therapy is to decrease the risk of development of adrenal rest tumors secondary to cortisol deficiency provoking increased ACTH stimulation. Untreated or inadequately treated patients are at risk for hyperplasia of adrenal rest tissue within or adjacent to the testes or the abdomen. Hyperplasia of pituitary ACTH-secreting cells, adrenal tumors, and adrenal carcinomas may also be complications of chronic cortisol deficiency and increased ACTH secretion.

IV. LATE-ONSET CONGENITAL ADRENAL HYPERPLASIA AND MANIFESTING HETEROZYGOTES

Late-onset congenital adrenal hyperplasia, also known as nonclassical congenital adrenal hyperplasia, is an allelic variant of the salt-wasting and simple virilizing forms of CVAH. Based on HLA haplotype and steroid hormone concentrations, disease frequency for late-onset CAH has been estimated to be 1/1000 for Caucasians. In certain ethnic groups, disease frequency is higher: 1/27 for Ashkenazi Jews, 1/53 for Hispanics, and 1/333 for Italians. The symptoms of late-onset congenital adrenal hyperplasia, hirsutism (Fig. 2), oligo/amenorrhea, infertility, and acne, lead to an ascertainment bias favoring diagnosis of affected women. Polycystic ovaries may be observed on ultrasound examination. To distinguish women with late-onset CAH from those with ovarian hyperandrogenism, ACTH stimulation tests may be helpful. Typically, the ACTH-stimulated 17-hydroxyprogesterone responses are >1200 ng/dl. Affected males are usually recognized only through family studies because most are asymptomatic. Among patients with late-onset adrenal hyperplasia, chronic hyperandrogenism among females may lead to anovulation, whereas there are no data to indicate infertility among males.

There is also a question of whether a single mutation, if it is severe, may be expressed with relatively mild evidence of hyperandrogenism, perhaps presenting at the age of puberty in females, with prema-

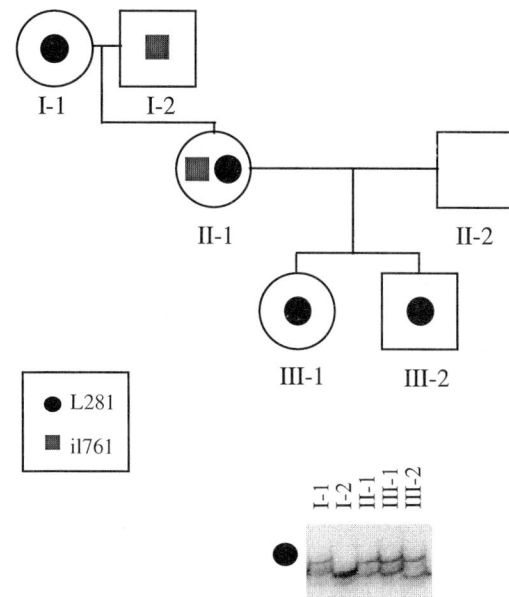

FIGURE 2 Late-onset 21-hydroxylase deficiency. The family pedigree is indicated on the upper portion of the figure. Individual II-1 sought medical attention for hirsutism at age 38. Her ACTH stimulation test showed a stimulated 17-hydroxyprogesterone response of 2600 ng/dl, which confirmed the diagnosis of late-onset 21-hydroxylase deficiency. Subsequently, the 10-year-old daughter was evaluated for premature puberty. The ACTH-stimulated 17-hydroxyprogesterone response obtained from the daughter is consistent with heterozygosity for 21-hydroxylase deficiency (571 ng/dl). Molecular diagnosis was performed using single-strand conformational polymorphism (SSCP) analysis and is illustrated in the autoradiograph in the lower portion of the figure. For this technique, PCR primers specific to the 21-hydroxylase gene amplify an 823-bp PCR product spanning exons 7 and 8. The PCR product is digested with BglI and electrophoresed on a nondenaturing SSCP gel. The mutant Leu281 conformer is the uppermost band on the gel (indicated by the arrow to the left of the autoradiograph). The lower conformer, easily seen for individual I-2, is the normal Val281 allele (indicated by the arrow to the right of the autoradiograph). The maternal grandfather, I-2, and the mother were found to carry the T insertion mutation in exon 7 using allele-specific oligonucleotide hybridization. Thus, the maternal grandfather is a heterozygous carrier of the T insertion mutation and the maternal grandmother is a heterozygous carrier of the Val281 → Leu mutation. Individual II-1 is a compound heterozygote and carries the T insertion mutation on one allele and the Val281 → Leu mutation on her other allele. The children of II-1 inherited their mother's maternal allele, which carries the Val281 → Leu mutation, and are both heterozygous carriers of 21-hydroxylase deficiency. No mutations were detected during molecular genetic analysis of a DNA sample obtained from II-2 (data not shown).

ture pubic hair and a 17-OH progesterone response to ACTH in the heterozygote range. Such a situation could be called manifesting heterozygotes and may be exemplified by individual III-1 in Fig. 2.

V. SUMMARY

Available information indicates that, with appropriate therapy, both males and females can be expected to have normal fertility. Prolonged exposure to excessive androgens among females during and after puberty is associated with chronic anovulation, a secondary form of PCOS. Once this clinical picture has developed, appropriate suppression of adrenal androgens is unlikely to restore ovulation and fertility. Such patients probably require exogenous hormone regimens to induce ovulation. Most important, the stigma of the consequences of the virilization among females influences psychological development and social interactions leading to decreased opportunities to be fertile. The majority of men with CVAH, whether treated with glucocorticoids or not, appear to be fertile. In contrast to affected women, untreated infertile men generally show normalization of spermatogenesis with initiation of glucocorticoid therapy.

See Also the Following Articles

ADRENAL ANDROGENS; ADRENARCHE; SEX DIFFERENTIATION, PSYCHOLOGICAL

Bibliography

Azziz, R., Dewailly, D., and Owerbach, D. (1994). Nonclassic adrenal hyperplasia: Current concepts. *J. Clin. Endocrinol. Metab.* 78, 810–815.

Bonaccorsi, A. C., Adler, I., and Figueiredo, J. G. (1987). Male infertility due to congenital adrenal hyperplasia: Testicular biopsy findings, hormonal evaluation, and therapeutic results in three patients. *Fertil. Steril.* 47, 664–670.

Brunelli, V. L., Chiumello, G., David, M., and Forest, M. G. (1995). Adrenarche does not occur in treated patients with congenital adrenal hyperplasia resulting from 21-hydroxylase deficiency. *Clin. Endocrinol.* 42, 461–466.

Carmina, E., and Lobo, R. A. (1994). Ovarian suppression reduces clinical and endocrine expression of late-onset congenital adrenal hyperplasia due to 21-hydroxylase deficiency. *Fertil. Steril.* 62, 738–743.

Dittmann, R. W., Kappes, M. E., and Kappes, M. H. (1992). Sexual behavior in adolescent and adult females with congenital adrenal hyperplasia. *Psychoneuroendocrinology* 17, 153–170.

Feldman, S., Billaud, L., Thalabard, J.-C., Raux-Demay, M.-C., Mowszowicz, I., Kuttenn, F., and Mauvais-Jarvis, P. (1992). Fertility in women with late-onset adrenal hyperplasia due to 21-hydroxylase deficiency. *J. Clin. Endocrinol. Metab.* 74, 635–639.

Ghizzoni, L., Virdis, R., Vottero, A., Cappa, M., Street, M. E., Zampolli, M., Ibañez, L., and Bernasconi, S. (1996). Pituitary–ovarian responses to leuprolide acetate testing in patients with congenital adrenal hyperplasia due to 21-hydroxylase deficiency. *J. Clin. Endocrinol. Metab.* 81, 601–606.

Gschwend, S., and Lee, P. A. (1993). Adult height and menstrual history among patients with classical 21-hydroxylase deficiency adrenal hyperplasia: Correlation with growth and therapy during development. *Adolescent Pediatr. Gynecol.* 6, 209–213.

Hague, W. M., Adams, J., Rodda, C., Brook, C. G. D., DeBruyn, R., Grant, D. B., and Jacobs, H. S. (1990). The prevalence of polycystic ovaries in patients with congenital adrenal hyperplasia and their close relatives. *Clin. Endocrinol.* 33, 501–510.

Klingensmith, G. J., Garcia, S. C., Jones, H. W., Jr., Migeon, C. J., and Blizzard, R. M. (1977). Glucocorticoid treatment of girls with congenital adrenal hyperplasia: Effects on height, sexual maturation, and fertility. *J. Pediatr.* 90, 996–1004.

Miller, W. L. (1994). Genetics, diagnosis, and management of 21-hydroxylase deficiency. *J. Clin. Endocrinol. Metab.* 78, 241–246.

Mulaikal, R. M., Migeon, C. J., and Rock, J. A. (1987). Fertility rates in female patients with congenital adrenal hyperplasia due to 21-hydroxylase deficiency. *N. Engl. J. Med.* 316, 178–182.

Slijper, F. M. E., van der Kamp, H. J., Brandenburg, H., de Muinck Keizer-Schrama, M. P. F., Drop, S. L. S., and Molenaar, J. C. (1992). Evaluation of psychosexual development of young women with congenital adrenal hyperplasia: A pilot study. *J. Sex Educ. Ther.* 18, 200–207.

Speiser, P. W., Dupont, B., Rubinstein, P., Piazza, A., Kastelan, A., and New, M. I. (1985). High frequency of nonclassical steroid 21-hydroxylase deficiency. *Am. J. Hum. Genet.* 37, 650–667.

Urban, M. D., Lee, P. A., and Migeon, C. J. (1978). Adult height and fertility in men with congenital virilizing adrenal hyperplasia. *N. Engl. J. Med.* 299, 1392–1396.

Adrenarche

Frank Gonzalez
State University of New York at Buffalo

I. Introduction
II. Adrenal Morphology
III. Adrenocortical Steroidogenesis
IV. Control of Adrenocortical Function
V. Summary

GLOSSARY

ACTH Adrenocorticotropin hormone, the anterior pituitary hormone that promotes adrenocortical growth and controls corticosteroid biosynthesis and secretion.

adrenal cortex The outermost portion of the adrenal constituting 90% of the gland and composed of the zone glomerulosa, the zona fasciculata, and the zona reticularis.

CASH Cortical androgen-stimulating hormone, a human pituitary factor proposed as the primary regulator of adrenal androgens due to its ability to stimulate adrenal androgen secretion *in vitro*.

centripetal circulation The adrenal's unique blood supply characterized by a continuous intraglandular capillary and venous network extending from the cortex to the centrally located medulla and ultimately coalescing into a large central vein.

inward "escalator" migration The en masse movement of adrenocortical cells centrally toward the medulla by cell division in the outer cortex, balanced by cell death in the inner cortex.

steroid gradient The progressive increase in corticosteroid concentration from the outer cortex to the inner cortex due to steroid production from concentric adrenocortical layers and inward secretion into the adrenal's centripetal vasculature.

Adrenarche is defined as the progressive rise in circulating adrenal androgens from late childhood to the middle of the second decade associated with the formation of the zona reticularis within the adrenal cortex. This rise in adrenal androgens is considered to be the impetus for the development of secondary sexual hair. In the normal circumstance, initiation of adrenarche is preceded by and independent of gonadal maturation (gonadarche).

I. INTRODUCTION

From fetal life until adolescence, the concentration of circulating adrenal androgens is proportional to adrenal gland size. During fetal life, adrenal androgens produced in the fetal zone of an enlarged adrenal are secreted into the circulation at high levels. Within 1 year after birth, circulating adrenal androgens decline to low levels coinciding with an 80% diminution in the weight of the adrenal and either fetal zone involution or its remodeling into the zona fasciculata. It is not until about age 6 or 7 with the onset of adrenarche that serum adrenal androgens begin to rise again as adrenocortical mass progressively increases and the zona reticularis is formed. These observations suggest a relationship between adrenocortical mass and alterations in adrenal morphology culminating in adrenarche.

Despite the temporal overlap observed with adrenarche and gonadarche, these events are independent of one another such that either can take place without the other. In cases of premature adrenarche, gonadal dysgenesis, and hypogonadal hypogonadism, adrenarche occurs in the absence of puberty. Conversely, in the instance of precocious puberty, gonadarche occurs before the onset of adrenarche. If adrenarche occurs prematurely or is completely absent such as in Addison's disease, the normal timing of gonadarche is unaffected. Thus, adrenal androgens or gonadal estrogens do not play a major role in the timing of puberty.

The pattern of adrenal androgen synthesis and adrenal morphologic alteration that occurs in most animal species during sexual maturation is not comparable to what is observed in humans during adrenarche. In nonprimates, levels of adrenal androgens are generally low. This is apparent in dogs despite zona reticularis development in the canine adrenal. In primates, adrenal androgen concentrations are similar to those in humans. However, rhesus and cynomegalus monkeys and baboons exhibit adrenal androgen levels throughout development that are no different than those during adulthood. Moreover, only chimpanzees exhibit the progressive rise in circulatory adrenal androgens that resembles the pattern in human adrenarche. Therefore, it is probable that the variability in adrenal androgen secretion observed among animal species reflects a fundamental difference in adrenal androgen regulation from one species to another.

To date, the mechanism for the induction of adrenarche remains unknown. Two theories have been proposed to explain this phenomenon. One states that adrenal androgen synthesis is under the control of a specific pituitary factor independent of adrenocorticotropin (ACTH), whereas the other states that an increase in adrenal androgen synthesis is an inevitable result of the changing intraadrenal steroid microenvironment dictated by ACTH and the adrenal's unique centripetal circulation. Although this dilemma has yet to be resolved, a description of adrenal morphology, adrenal steroidogenesis, and the factors believed to control adrenal function in general will provide the necessary background for assigning the appropriate degree of validity to each of these theories.

II. ADRENAL MORPHOLOGY

A. Anatomic Description

The adrenal glands are paired organs located medial to the upper pole of each kidney. The right adrenal is pyramidal shaped, whereas the left adrenal is more flattened and crescent shaped. Each adrenal gland weighs approximately 4 or 5 g in the normal unstressed adult and is composed of two developmentally unrelated tissues surrounded by a thin capsule. The outermost cortex is of ectodermal origin and constitutes 90% of the gland, whereas the more centrally located medulla is of mesodermal origin and makes up only 10% of the gland.

The adrenal cortex is classically divided into three zones known as the zona glomerulosa, the zona fasciculata, and the zona reticularis (Fig. 1). The zona glomerulosa is located immediately beneath the capsule and accounts for only 5% of cortical thickness. Glomerulosa cells are arranged in clumps that may lack continuity with each other. The zona fasciculata is located just below the zona glomerulosa and accounts for 70% of cortical thickness. Fasciculata cells are larger, contain lipid vacuoles for cholesterol storage, and are arranged in a narrow cord-like columnar configuration. Superficially located fasciculata cells possess the largest and most numerous lipid vacuoles, whereas the more deeply lying fasciculata cells have a high content of rough endoplasmic reticulum and mitochondria. The zona reticularis is the deepest zone located closest to the central medulla and ac-

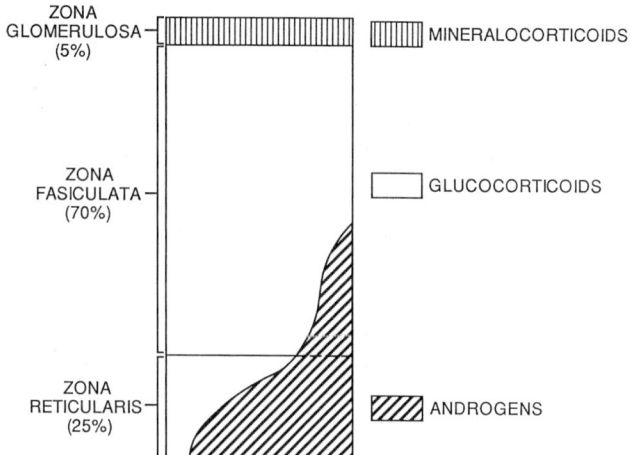

FIGURE 1 Diagrammatic representation of the adrenal cortex. The zonae glomerulosa, fasciculata, and reticularis occupy 5, 70, and 25% of cortical thickness, respectively. Mineralocorticoids are produced exclusively in the zona glomerulosa. However, the zona fasciculata and the zona reticularis produce glucocorticoids and androgens. While the majority of glucocorticoids are produced in the zona fasciculata, most adrenal androgens are produced in the zona reticularis (reproduced with permission from Gonzalez, 1997. © Thieme Medical Publishers).

counts for the remaining 25% of cortical thickness. Reticularis cells have a size in between those of the glomerulosa and the fasciculata, possess a high content of smooth endoplasmic reticulum and mitochondria, and are arranged in an anastomotic network. Therefore, each zone has a characteristic appearance even though boundaries between zones are not always demarcated.

The medulla is a direct extension of the sympathetic nervous system. Medullary cells arranged in clusters synthesize the catecholamines, epinephrine, and norepinephrine that are then stored in secretory granules. Functioning as postganglionic neurons, medullary cells are innervated by preganglionic fibers of the splanchnic nerve, the celiac ganglion, and the subsidiary plexus. Neuronal stimulation leads to release of catecholamines from the secretory granules into the medullary capillaries by exocytosis. Therefore, the medulla is responsible for instantaneous secretion of presynthesized neurohormones upon command.

B. Adrenal Vasculature

The adrenal glands receive their blood supply from three arteries. The inferior phrenic artery, the abdominal aorta, and the renal artery supply the superior, middle, and inferior adrenal arteries, respectively, which coalesce to form a plexus over the capsule of each gland. The capsular plexus gives rise to approximately 40–60 vessels that penetrate the substance of the gland. A few of these penetrating arterioles pass directly through the cortex to supply the medullary capillaries. However, the vast majority of the vessels enter a cortical plexus that is continuous throughout all three zones and assumes the configuration of the respective zonal cellular arrangements. The cortical plexus empties into a large number of venules that either drain in medullary veins or continue into the capillary plexus of the medulla. Thus, the medullary plexus blood supply receives direct but minor input from capsular arterioles that bypass the cortex and major contribution from the cortical plexus venules. The medullary plexus assumes the clustered configuration of medullary cells and ultimately drains into medullary veins, which coalesce into the large central vein that emerges from each adrenal glad. The right

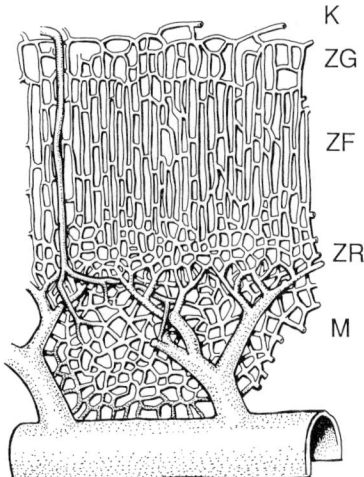

FIGURE 2 Diagrammatic representation of the centripetal vasculature of the adrenal gland. The capsular, cortical, and medullary plexi provide a continuous capillary network spanning the thickness of the gland and ultimately draining into venous tributaries of progressively larger caliber located deep within the medulla where the central vein exists. Approximately 10% of the medullary plexus blood supply arises from capsular arterioles that bypass the cortex. K, capsule; ZG, zona glomerulosa; ZF, zona fasciculata; ZR, zona reticularis; M, medulla (reproduced from Gonzalez *et al.*, *Obstet. Gynecol. Surv.* 45(8), 491–508, 1990, with permission from Williams & Wilkins and adapted from *Johns Hopkins Hospital Reports* 9, 153–289, 1900).

central vein is short (1-cm average) and enters directly into the inferior vena cava. The left central vein is somewhat longer (2-cm average) and enters into the left renal vein. Therefore, the adrenal vasculature exhibits continuity throughout the gland, permitting centripetal blood flow through an intraglandular capillary network that extends from the cortex to the medulla (Fig. 2).

The adrenal gland receives the greatest amount of blood per gram of tissue than any other body organ except for possibly the thyroid gland. The intraglandular capillary network is composed of fenestrated endothelial cells that facilitate rapid passage of substances in or out of the bloodstream. Adrenal blood flow is primarily controlled at the level of the medullary veins and the central vein, both of which possess walls with prominent longitudinal smooth muscle bundles arranged in eccentric fascicles. Postganglionic sympathetic innervation controls the contractile state of these fascicles thereby regulating the proximal intradrenal blood flow. Therefore, the fenestrated capillary network and constriction of venous outflow permits adequate intracellular exposure to the centripetal flow of intrinsic and extrinsic substances, which ultimately affect the intraadrenal microenvironment.

III. ADRENOCORTICAL STEROIDOGENESIS

A. Corticosteroid Descriptions

The adrenal cortex secretes five types of steroid hormones: mineralocorticoids, glucocorticoids, and androgens to a greater extent and estrogens and progestogens to a lesser extent. The principal mineralocorticoid, aldosterone, is produced exclusively by the zona glomerulosa. Aldosterone plays a major role in electrolyte homeostasis and is primarily under the control of the renin–angiotensin system and the plasma concentration of potassium. The anterior pituitary hormone, ACTH, plays only a secondary role in the control of aldosterone that is limited to pathological situations associated with salt imbalance.

Glucocorticoids and androgens are produced in both the zona fasciculata and the zona reticularis. However, the majority of glucocorticoid production occurs in the zona fasciculata ($\approx$93%), whereas the majority of androgen production occurs in the zona reticularis ($\approx$66%). The glucocorticoids, primarily cortisol and to a lesser extent corticosterone, are involved in the overall regulation of carbohydrate metabolism and influence the synthesis of catecholamines in the medulla. In addition, glucocorticoids along with catecholamines aid in coordinating the body's metabolic response to stress.

The principal adrenal androgens are dehydroepiandrosterone (DHEA), dehydroepiandrosterone sulfate (DHEAS), 11β-hydroxyandrostenedione (11βA), and andostenedione. The adrenal accounts for >90% of DHEA, DHEAS, and 11βA as well as slightly less than 50% of androstenedione found in the peripheral circulation. Intracortical conversion of androstenedione also leads to secretion of testosterone of adrenal origin that contributes roughly 25% to circulating levels. As previously stated, adrenal androgens play an important role in sexual maturation during adrenarche after a series of morphological events within the adrenal. In addition, peripheral conversion of adrenal androgens in muscle and adipose tissue provides the bulk of nongonadally derived estrogen.

The ovary is the major source of the body's estrogen and progestogens in women. Secretion of estrogen by the adrenal cortex is quantitatively insignificant except in certain rare cortical neoplasms. Progestogens of adrenal origin are generally precursors of the major corticosteroids and are not secreted in large amounts unless they are substrates of a congenitally deficient adrenal steroidogenic enzyme.

B. Biosynthetic Mechanisms

1. Cholesterol Substrate

The building block for corticosteroid biosynthesis is cholesterol. The major source of adrenal cholesterol is low-density lipoprotein (LDL) present in the circulation. Entry of LDL into cortical cells is accomplished by binding to a high-affinity receptor and subsequent internalization of the LDL–receptor complex. Upon arrival within the cell, the protein and cholesterol ester components of LDL undergo hydrolysis in lysosomes causing release of free cholesterol. While some of the free cholesterol is used for steroid synthesis, the remainder is reesterified and stored in lipid vacuoles. When rapid hormone production is required, the pool of free cholesterol is expanded immediately by the mobilization and hydrolysis of stored cholesterol esters and, to a lesser extent, by *de novo* synthesis from acetate. However, uptake of LDL from the circulation is accelerated simultaneously and, thus, remains the prime source for hormone production.

2. Molecular Enzyme Reactions

In general, the corticosteroid biosynthetic pathway from cholesterol involves the following steps: (i) cholesterol side chain cleavage to form a C_{21} steroid; (ii) hydroxylation of the steroid skeleton at a variety of positions; (iii) oxidation of the 3β-hydroxyl group to a ketone with a subsequent double bond shift from the 5,6 position to the 4,5 position; and (iv) cleavage

ADRENOCORTICAL STEROIDOGENIC ENZYMES	INTRACELLULAR LOCATION	ADRENOCORTICAL CELL TYPE	STEROIDOGENIC FUNCTION
P450$_{SCC}$	MITOCHONDRIA	GLOMERULOSA FASICULATA RETICULARIS	CHOLESTEROL SIDECHAIN CLEAVAGE
P450 21	ENDOPLASMIC RETICULUM	GLOMERULOSA FASICULATA RETICULARIS	21α HYDROXYLATION
P450 11	MITOCHONDRIA	GLOMERULOSA FASICULATA RETICULARIS	11β HYDROXYLATION 18α HYDROXYLATION (GLOMERULOSA ONLY)
3βHSD	ENDOPLASMIC RETICULUM	GLOMERULOSA FASICULATA RETICULARIS	3β HYDROXYL OXIDATION 5,6 TO 4,5 DOUBLE BOND SHIFT
P450 17	ENDOPLASMIC RETICULUM	FASICULATA RETICULARIS	17α HYDROXYLATION 17, 20 BOND CISSION
SULFOTRANSFERASE	CYTOPLASM	RETICULARIS	DHEA SULFATION

FIGURE 3 Adrenocortical steroidogenic enzymes are confined to specific locations within certain adrenocortical cell types and have specific functions culminating in synthesis of mineralocorticoids, glucocorticoids, and androgens (reproduced with permission from Gonzalez, 1997. © Thieme Medical Publishers).

of the bond between carbons 17 and 20 to form a C_{19} steroid in some instances. The majority of these reactions are catalyzed by members of the cytochrome P450 group of enzymes (Fig. 3).

Conversion of cholesterol to pregnenolone is the rate-limiting step in steroidogenesis. The several reactions involved are mediated by the side chain cleavage cytochrome P450 enzyme (P450scc) encoded by a single gene and located within the mitochondria of adrenocortical cells. The carbons in position 20 and 22 are each hydroxylated to form 20α,20ε-dehydrocholesterol with subsequent bond cleavage between carbons 20 and 22 to form pregnenolone and isocaproic aldehyde.

Hydroxylation of a steroid at a number of positions and in a variety of stereoorientations is performed by three additional cytochrome P450 enzymes confined to specific organelles within the cells of certain adrenocortical cells. Throughout the cortex, glucocorticoids or mineralocorticoid precursors are hydroxylated at carbon 21 in the α orientation by P450 21 present in the endoplasmic reticulum and at carbon 11 in the β orientation by P450 11 present in the mitochondria. However, aldosterone synthesis requires an additional hydroxylation in the α orientation as well as methyl oxidation at the 18 position, which are also catalyzed by P450 11 but limited to the mitochondria of glomerulosa cells. On the other hand, synthesis of androgens and of the major glucocorticoid cortisol requires 17α hydroxylation through the action of P450 17 limited to the endoplasmic reticulum of fasciculata and reticularis cells. Therefore, steroid substrates must undergo intracellular transport between organelles possibly with the aid of sterol carrier proteins similar to the hepatic sterol carrier protein. In addition, some cytochrome P450 enzymes can metabolize multiple substrates mediating more than one reaction but often in a zone-specific manner that dictates zonal steroidogenic patterns.

3β-Hydroxol oxidation and the 5,6 to 4,5 double-bond shift are sequentially accomplished by the action of the only noncytochrome P450 enzyme, 3β-hydroxysteroid dehydrogenase-Δ^5 3 ketosteroid isomerase (3βHSD), which is encoded by a single gene and is microsomal in origin. While the oxidation step requires NADP as a hydrogen acceptor, no cofactor is required for the isomerase step.

Adrenal androgen synthesis is further mediated by P450 17 and sulfotransferase. Besides its 17α-hydroxylase activity, P450 17 possesses 17,20 lyase activity which is encoded by the same gene and transforms 17 hydroxylated C_{21} precursors to C_{19} androgens by bond scission between carbons 17 and 20. Sulfotransferase attaches a sulfate group to DHEA to yield DHEA-S. DHEA and its sulfate do not require the action of 3βHSD for their synthesis. Therefore, androgen production is predominant in an adrenocortical microenvironment in which P450 17 and sulfotransferase activities are high and 3βHSD activity is low. Determinants of this microenvironment within the zona reticularis will be discussed later.

IV. CONTROL OF ADRENOCORTICAL FUNCTION

A. The Modulating Role of ACTH

ACTH, a peptide of anterior pituitary origin, plays a dual role in the modulation of adrenocortical function. ACTH stimulates growth of the adrenal cortex during prenatal and postnatal life and exerts primary control over the biosynthesis and secretion of glucocorticoids and most likely that of androgens. The effects of ACTH can be generalized or specific, direct or indirect, and acute or chronic (Fig. 4). Therefore, the mechanisms by which ACTH performs its dual role are variable. Many of these mechanisms still require greater elucidation.

1. Effects on Adrenocortical Growth

The regulation of adrenocortical growth is achieved by actions that lead to cellular hypertrophy and tissue hyperplasia. ACTH promotes cellular hypertrophy by directly inhibiting the synthesis of DNA required for cell replication. On the other hand, ACTH indirectly promotes tissue hyperplasia by rapidly increasing adrenal blood flow and by inducing neovascularization of cortical tissue over time. These actions serve to increase exposure of the adrenal cortex to the nutrients and oxygen necessary for cell proliferation. Therefore, adrenocortical growth, in terms of hypertrophy versus hyperplasia, is the result of the delicate balance between the direct inhibitory and indirect stimulatory effects of ACTH.

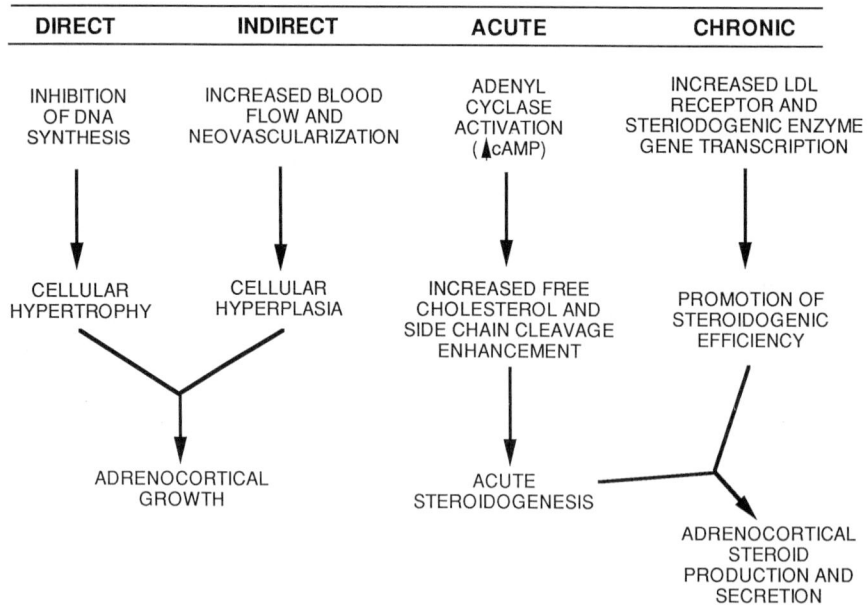

FIGURE 4 The variable actions of ACTH are suitable for its dual roles of inducing adrenocortical growth and stimulating adrenocortical steroidogenesis (reproduced with permission from Gonzalez, 1997. © Thieme Medical Publishers).

2. Effects on Corticosteroid Biosynthesis and Secretion

The immediate and delayed effects of ACTH provide significant impact on the biosynthesis of glucocorticoids and androgens. The acute action of ACTH is mediated by cyclic AMP, which activates a protein kinase that phosphorylates cholesterol ester hydrolase to its active form. In the process, free cholesterol is liberated from cytoplasmic lipid droplets and made available to P450-scc. Other cyclic AMP-mediated phosphoproteins have been postulated to increase cholesterol binding to P450-scc. The chronic action of ACTH efficiently maintains a high rate of steroidogenesis by induction of protein synthesis within cortical cells. The induction process is partially mediated by free cholesterol made available by the acute action of ACTH to simulate synthesis of P450-scc. Once the rate-limiting step has been surmounted, the cascade of substrates made available may induce other enzymes within the steroidogenic pathway. ACTH also increases the number of LDL receptors to promote uptake of LDL from the circulation. Induction of protein synthesis is slow because proteins such as adrenocortical enzymes have long half lives of 3 or 4 days. Therefore, the acute and chronic actions of ACTH intertwine to make cholesterol substrate available to the steroidogenic pathway and to maintain a sufficient level of enzymatic activity for steroidogenic demands.

The factors governing the secretion and regulation of ACTH have direct relevance to the control of glucocorticoid and androgen secretion by ACTH. Corticotropin-releasing hormone (CRH) from the hypothalamus stimulates the release of ACTH from the anterior pituitary. A long negative feedback loop exists by which glucocorticoids inhibit the secretion of ACTH and CRH. Under physiological conditions, ACTH release occurs episodically in a diurnal 24-hr cycle. This pattern of ACTH release is intrinsic to the hypothalamic control system and is independent of feedback control. Secreted ACTH rapidly arrives in the adrenal cortex so that secretion of cortisol and adrenal androgens shows an episodic and circadian variation parallel to that of ACTH with certain exceptions. Testosterone does not demonstrate significant variation in women because of its derivation from several sources, namely, the ovary, the adrenal, and peripheral conversion of androstenedione. However, the minor adrenal component does exhibit a circadian rhythm. DHEA-S also shows negligible variation because it is present in plasma in large quantities and is metabolized slowly.

B. Determinants of Zonal Steroidogenic Patterns

1. Maintenance of Adrenocortical Mass

The acquisition and maintenance of adrenocortical mass throughout prenatal and postnatal life is determined by the relation between the rate of cell division and the rate of cell death (Fig. 5). Cell division is indirectly influenced by ACTH as described previously and is restricted to the outer regions of the cortex, namely, the zona glomerulosa and the outer zona fasciculata. Considerable mitotic activity occurs in the outer cortex when compared to the inner regions and is probably a result of greater nutrient availability because the outer cortex is closer to the arterial blood supply. Cell death is age related and limited to the inner cortex as demonstrated by the increased presence of age pigment (lipofuscin) in the zona reticularis. These observations clearly support the concept that adrenocortical cells are pushed en masse toward the medulla by cell division in the outer cortex, balanced by cell death in the inner cortex. Therefore, glomerulosa cells are pushed inward to become fasciculata cells and eventually end their life span as reticularis cells, indicating that a specific cell will secrete different types of corticosteroids over time through a process of interconversion.

The concept of inward "escalator" migration suggests that cells from all three zones of the adrenal cortex are of the same basic cell types. Other findings also support this contention. Synthetically active adrenocortical cells from any zone exhibit a uniform characteristic during hormone production and secretion upon stimulation. Moreover, release of products into the circulation occurs after a short delay required for the synthesis of steroids from stored lipid precursors. Furthermore, the zona fasciculata and the zona reticularis are probably one functional unit. Clear cells of the outer fasciculata may be relatively

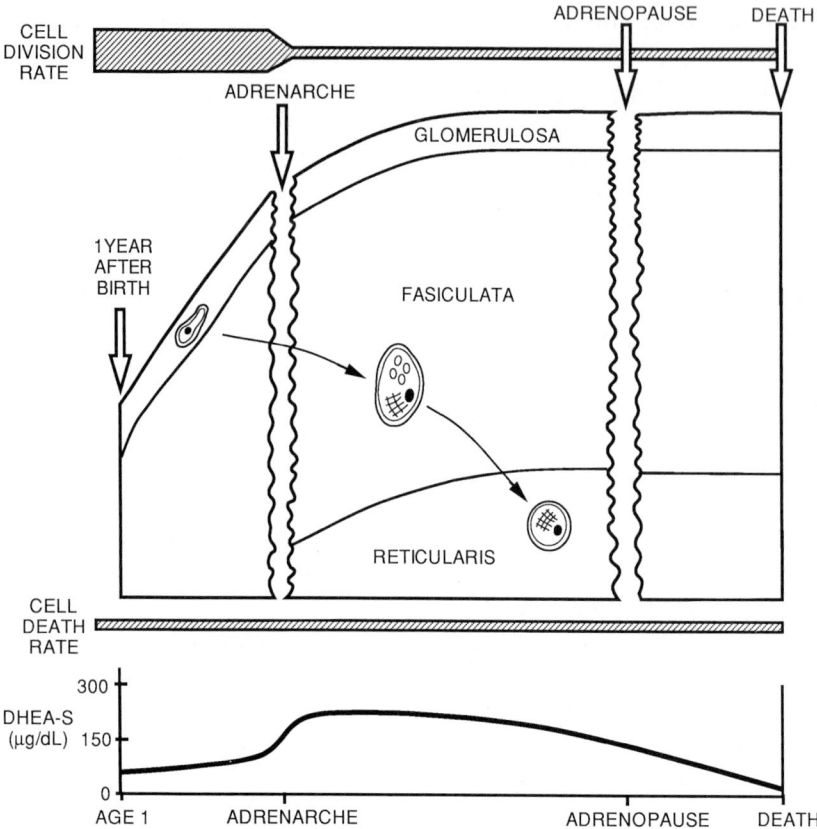

FIGURE 5 Adrenocortical thickness is believed to be related to the ratio of new cell formation in the outer cortex to that of old cell death in the inner cortex. This ratio is increased before adrenarche and proportional after adrenarche. In the process, it is proposed that a given adrenocortical cell will undergo inward "escalator" migration resulting in interconversion to different cell types during its life span. The persistence of the zona reticularis, despite the decline in its production of androgens during the so-called adrenopause, is an unexplained phenomenon (reproduced with permission from Gonzalez, 1997. © Thieme Medical Publishers).

inactive storage cells of surplus precursors, whereas compact cells of the inner fasciculata and reticularis are most likely involved in active synthesis of glucocorticoids and androgens because they are abundant in endoplasmic reticulum and mitochondria. Acute ACTH stimulation causes a change in cellular configuration from clear to compact, causing an apparent increase in the size of the zona reticularis at the expense of the zona fasciculata. Sustained ACTH stimulation converts the remaining clear cells to compact cells, causing extreme precursor depletion and maximal steroidogenic output. Final support is provided by observation of adrenocortical cells placed in culture. Marked differences in corticosteroids produced by cells isolated from different zones disappear over time to yield identical patterns of steroidogenesis. Therefore, cellular function within the adrenal cortex is determined by external stimulatory influences such as ACTH or the renin–angiotensin system and by the specific location of a given adrenocortical cell at the time of stimulation.

2. Establishment of Functional Zonation

i. Reassessment of the Gradient Hypothesis
The mechanism for functional zonation of adrenocortical cells remains speculative. In the past, the gradient hypothesis provided a plausible explanation inclusive of proposed intrinsic factors responsible for zona reticularis formation and primary adrenal androgen secretion during adrenarche. The steroid gradient

across the adrenal cortex created by the adrenal's centripetal blood flow formed the basis of the gradient hypothesis. Central to this hypothesis was the concept that specific corticosteroids termed gradient substances achieved the critical concentrations required to directly inhibit key adrenal enzymes as the flow of blood carried progressively higher amounts of these secreted steroids deeper within the cortex. Thus, it had been proposed that these enzyme inhibitions culminated in local shifts in steroidogenic patterns to establish boundaries between zones.

Until recently, the data accumulated were consistent with the gradient hypothesis. Quantitation of the steroid concentrations within normal adrenal tissue *in vitro* confirmed the presence of an intraadrenal steroid gradient by demonstrating progressive increases in the steroid concentrations from the outer layers to the inner layers of the adrenal cortex. Kinetic studies performed on purified adrenal enzyme extracts indicated efficacious inhibition by specific corticosteroids at concentrations equal to those found within the appropriate cortical layers where zonal boundaries are located. As a consequence, the glomerulosa/fasciculata boundary was thought to be established when gradient substance candidates such as corticosterone or cortisol achieved the critical concentration required to directly inhibit P450 11 18α hydroxylation, resulting in suspension of aldosterone production. In addition, the fasciculata/reticularis boundary was thought to be established with the attainment of sufficient cortical thickness by late childhood through the direct and indirect actions of ACTH to permit corticosterone or another gradient substance candidate, androstenedione, to achieve the critical concentration required for direct inhibition of 3βHSD activity. This latter phenomenon, along with a concomitant increase in the activity of highly ACTH-dependent P450 17, would culminate in development of the zona reticularis and initiation of an adult pattern of adrenal androgen secretion.

The latest investigations are inconsistent with direct enzyme inhibition as the mechanism for decreased 3βHSD activity during adrenarche. Moreover, lower levels of 3βHSD messenger RNA (mRNA) and enzyme are present in the zona reticularis compared to the zona fasciculata in the postadrenarchal adrenal cortex. This suggests that the decrease in 3βHSD activity observed during adrenarche is the result of lowered 3βHSD gene expression. It is possible that the intraadrenal steroid gradient plays a role in lowering 3βHSD gene expression or increasing P450 17 gene expression. In this instance, the appropriate gradient substance could impose its effects through a receptor-mediated mechanism upon attaining the necessary critical concentrations. However, support for this postulate remains to be demonstrated. On the other hand, increased P450 17 activity can result from the chronic action of ACTH which may be mediated by insulin-like growth factors (IGF-I and IGF-II) locally produced within the adrenal. IGF-I receptors have been identified on the surface of adrenocortical cells primarily within the zona reticularis. IGF-I and IGF-II in RNAs and the respective peptides are present in adrenocortical cells and increase in response to ACTH. In addition, adrenal steroidogenesis is promoted by IGF-II through interaction with the IGF-I receptor. Thus, it is possible that either of these IGFs amplifies the ACTH signal to promote steroidogenic efficiency within adrenocortical cells responsible for adrenal androgen synthesis.

ii. A Proposed Cortical Androgen-Stimulating Hormone Contribution to the induction of adrenarche by a hormonal factor distinct from ACTH has been postulated. There are a number of situations to suggest that another factor participates in the regulation of adrenal androgen synthesis. For instance, the stress of anorexia nervosa, other chronic illnesses, and brain trauma promote ACTH hypersecretion which causes serum cortisol levels to rise while DHEA and DHEA-S concentrations fall. With advancing age, during the so-called adrenopause, ACTH and cortisol secretion is minimally altered, whereas adrenal androgen secretion declines without a reduction in adrenal gland weight or a corresponding involution of the zona reticularis. Thus, there is a divergence in the secretion of cortisol and the adrenal androgens despite similar ACTH exposure, unaltered adrenal morphology, or both.

To date, none of the hormones implicated in the induction of adrenarche, such as estrogen, prolactin, or the gonadotropins, exhibit plasma levels that change in parallel with the rises in serum adrenal

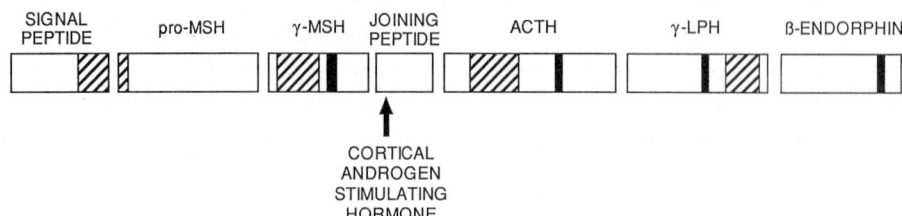

- 18AA, mol. wt.1758: N-Glu-Asp-Val-Ser-Ala-Gly-Glu-Asp-Cys-Gly-Pro-Leu-Pro-Glu-Gly-Gly-Pro-Glu-C
- 18AA-18AA dimer (disulfide linked)
- Naturally occurring peptide and synthetic peptide active *in vitro* in adrenal cells from dog, human, and human fetal adrenal

FIGURE 6 The location within the proopiomelanocortin precursor molecule as well as the amino acid sequence and characteristics of the peptide proposed as the cortical androgen-stimulating hormone (CASH). The putative role of this peptide *in vivo* still remains controversial (adapted from *Polycystic Ovary Syndrome*, p. 178, 1992, with permission from Blackwell Sciences and based on results from Parker *et al.*, *Endocr. J.* 1, 441–445, 1993).

androgens during late childhood. In the case of estrogen, adrenarche occurs before gonadarche indicating that any possible modulating role played by estrogen during gonadarche would be complimentary rather than primary. A human pituitary factor localized to the 18-amino acid joining peptide fragment of pro-opiomelamocortin (POMC) has been called the cortical androgen-stimulating hormone (CASH) due to its ability to bind adrenal cells and stimulate DHEA secretion synergistically with ACTH *in vitro* (Fig. 6). Because the activity of this substance has been shown to be small or undetectable, its contribution to adrenal androgen regulation *in vivo* remains controversial. However, this area of research is still in its infancy such that an increase in the potency of this POMC fragment by dimerization, amidation, or variable POMC processing may still be possible.

V. SUMMARY

ACTH promotes developmental growth and overall steroidogenic efficiency within the adrenal cortex. Another recently isolated human pituitary factor known as CASH has been proposed for separate regulation of adrenal androgen synthesis. However, CASH has low biologic activity and factors that could increase its potency have not yet been characterized. Thus, the role of CASH in the induction of adrenarche remains controversial. Alteration in adrenal morphology may well serve this purpose. The unique centripetal circulation within the adrenal creates a steroid gradient across the cortex. Sufficient adrenocortical mass and cortical thickness are achieved by adolescence under the influence of ACTH to allow maximum increases in corticosteroid concentrations within the deepest layer of the cortex. Whether or not it is the achievement of a high enough concentration of a particular corticosteroid that triggers a receptor-mediated mechanism to lower 3βHSD expression, increase P450 17 gene expression, or both remains to be established. The mediating role of IGFs in promoting ACTH-induced rises in P450 17 activity also requires further elucidation. Once the promotion of adrenal androgen production is accomplished by these enzyme alterations, ACTH continues to initiate and maintain the steroidogenic process.

See Also the Following Articles

Adrenal Androgens; Fetal Adrenals; Steroidogenesis, Overview

Bibliography

Albertson, B. D., Cutler, G. B., and Loriaux, D. L. (1992). Adrenarche: The maturing of the adrenal cortex. In *Polycystic Ovary Syndrome: Current Issues in Endocrinology and Metabolism* (A. Dunairf, J. R. Givens, F. P. Haseltine, and G. Merriam, Eds.), pp. 163–182. Blackwell Scientific, Cambridge, UK.

Anderson, D. C., and Winter, J. S. D. (Eds.) (1985). *Adrenal Cortex*. Butterworth, London.

Dickerman, Z., Grant, D. R., Faiman, C., *et al.* (1984). Intra-adrenal steroid concentrations in man: Zonal differences and developmental changes. *J. Clin. Endocrinol. Metab.* 59, 1031–1036.

Gonzalez, F. (1997). Adrenal involvement in polycystic ovary syndrome. *Sem. Reprod. Endocrinol.* 15(2), 137–157.

James, V. H. T. (1994). Adrenal cortex physiology. In *Clinical Endocrinology* (G. M. Besser and M. O. Thorner, Eds.), 2nd ed., pp. 7.2–7.12. Mosby-Wolfe, London.

Orth, D. N., Kovacs, W. J., and DeBald, C. R. (1992). The adrenal cortex. In *Williams Textbook of Endocrinology* (J. D. Wilson and D. W. Foster, Eds.), 8th ed., pp. 816–890. Saunders, Philadelphia.

Parker, L. (1995). Control of DHEAS secretion. *Sem. Reprod. Endocrinol.* 13, 275–281.

Yeasting, R. A. (1986). Selected morphological aspects of human suprarenal glands. In *The Adrenal Gland* (P. J. Mulrow, Ed.). Elsevier, New York.

Aedes aegypti

Thomas W. Sappington and Alexander S. Raikhel
Michigan State University

I. Introduction
II. Mating Behavior and Insemination
III. Previtellogenesis
IV. Vitellogenesis
V. Postvitellogenesis
VI. The Oviposited Egg

GLOSSARY

chorion The proteinaceous insect egg shell, secreted by the follicular epithelium as two distinct layers, the endochorion (closest to the oocyte plasma membrane) and the exochorion.

clathrin triskelion A three-armed protein complex, with each arm consisting of one each clathrin heavy chain and light chain. Clathrin triskelions attach to the cytoplasmic tails of plasma membrane-bound receptors and self-assemble to form clathrin-coated pits.

follicular epithelium A monolayer of somatic cells surrounding the oocyte and its nurse cells. One of its major functions is to secrete the chorion around the oocyte.

germinal nucleus A small sphere containing the oocyte's chromosomes, separated by a membrane from the nucleoplasm of the surrounding vegetative nucleus.

gonotrophic cycle The series of events involved in the synchronous maturation of a single batch of eggs.

meroistic ovary A type of ovary found in many insects, including *Aedes aegypti*, in which nutritive germ cells supply the developing oocyte with RNA and organelles through cytoplasmic bridges.

nurse cells Nutritive germ cells (seven in *A. aegypti*) that are connected to the oocyte via cytoplasmic bridges, located within the follicle at the anterior pole of the oocyte. Nurse cells supply the developing oocyte with RNA and organelles.

perioocytic space A space between the oocyte and follicle cell surfaces that is created at the beginning of vitellogenesis to allow hemolymph proteins to come in contact with the oocyte plasma membrane.

receptor-mediated endocytosis A specific cellular mechanism for internalizing macromolecules in which a membrane-bound receptor binds a ligand, collects with other receptors in a clathrin-coated pit, and is internalized in a clathrin-coated vesicle.

state of arrest An indefinite pause in the physiological development of the mosquito fat body and ovaries beginning after the completion of the preparatory previtellogenic phase and terminated at the initiation of vitellogenesis by a blood meal.

vegetative nucleus The main nucleus of the vitellogenic oocyte encompassing, but not including, the germinal nucleus which contains the chromosomes.

vitellin The crystallized, delipidated egg-storage form of vitellogenin, the major yolk protein precursor.

The reproductive physiology of the female yellow fever mosquito, *Aedes aegypti*, has been the focus of intensive research for several decades. A number of factors have contributed to its becoming the model insect of choice in this regard. Consequently, most research into the female reproductive physiology of an insect, and most studies of insect gene regulation, involve comparisons to and extrapolations from what is known of *A. aegypti* biology.

I. INTRODUCTION

Aedes aegypti is an important vector of several human and livestock tropical diseases worldwide, so its physiology is of direct interest to medical entomologists. It is a floodwater mosquito, so its eggs are adapted to withstand desiccation for long periods. The ability to store eggs and synchronize hatching makes it much easier to rear and manipulate in the laboratory than other disease vectors, such as malaria mosquitoes (*Anopheles*), so it serves nicely as a model for medically important hematophagous insects. In addition, the blood meal triggers events leading to synchronous development of *A. aegypti* eggs, greatly facilitating experimental dissection of the complex physiological events associated with insect reproduction. This synchronicity also involves the tissue-, sex-, and stage-specific activation of a number of genes, making it an excellent model for studying the molecular mechanisms of gene regulation.

II. MATING BEHAVIOR AND INSEMINATION

Mating behavior is chiefly expressed by the male, with the female's role limited to passively allowing the male to position himself for genital contact or to physically rejecting the male's attempts by kicking with the hindlegs and jerking her abdomen away from his genitalia. Females will mate with males but will not accept sperm until about 2 days after emergence when juvenile hormone III (JH) titers peak and trigger female receptivity. Long-range mate location is accomplished indirectly. After a 1- or 2-day maturation period postemergence, the female seeks a vertebrate (usually human) source of blood, a behavior also probably induced by rising JH titers. Males, which feed only on nectar, also seek out vertebrate hosts as a way of gaining proximity to potential mates. Females locate hosts by orienting to a complex mixture of skin and breath odors detected by various olfactory sensilla on the antennae as well as to exhaled CO_2 detected by capitate pegs on the maxillary palps (components of the mouthparts). Males also possess capitate pegs and are therefore thought to orient to a host, at least partly, via CO_2 detection. At short range, males locate females by auditory stimuli. The wingbeat frequency of a mature *A. aegypti* female, averaging 458 Hz, sets the male's numerous antennal fibrillae vibrating in resonance, and this vibration is transmitted down the antennal shaft to the Johnston's organ, where it is translated into an electrical signal.

Auditory stimulation triggers a progression of stereotyped behaviors in the male: It orients toward the flying female, seizes her from above with its front and middle legs, then maneuvers underneath her body before she lands to assume a venter-to-venter, head-to-head position. The male clamps down on the female's legs with specialized grooved tarsi equipped with greatly elongated tarsal claws, holding her hindlegs back with his middle legs to prevent kicking and using his hindlegs to position her abdomen before thrusting the tip of his abdomen forward to engage the female's genitalia. Copulation is normally completed within about 15 sec. Olfaction is not directly involved in mate location, but there is some evidence that a female contact pheromone must be perceived by the male after seizure before he will attempt to mate.

During copulation, the male's claspers push the female's cerci and postgenital plate dorsally, allowing the teeth on the male's aedeagus to make contact and mesh with the teeth on the female's dorsal vaginal valve (Fig. 1). The latter is consequently pulled posteriorly, having the effect of blocking off the openings to the spermathecal ducts while widening the opening to the bursa copulatrix, into which the male directly ejaculates (Fig. 1). Within 5 min of insemination, the spermatozoa relocate to the central spermathecum and to one of two smaller lateral spermathecae, partly under their own motive power and partly through the aid of muscles of the spermathecal duct. Secretions from glandular cells associated with the spermathecae probably provide nourishment to the sperm, which can be stored viably for more than 60 days. Sperm is released from the spermathecae to fertilize each egg as it is oviposited.

Mating elicits a number of profound changes in female behavior and physiology, mediated by substances introduced by the male's semen. In addition to spermatozoa, the ejaculate contains a large amount of male accessory gland secretion, which remains behind in the bursa copulatrix after the spermatozoa translocate to the spermathecae. The secretory mate-

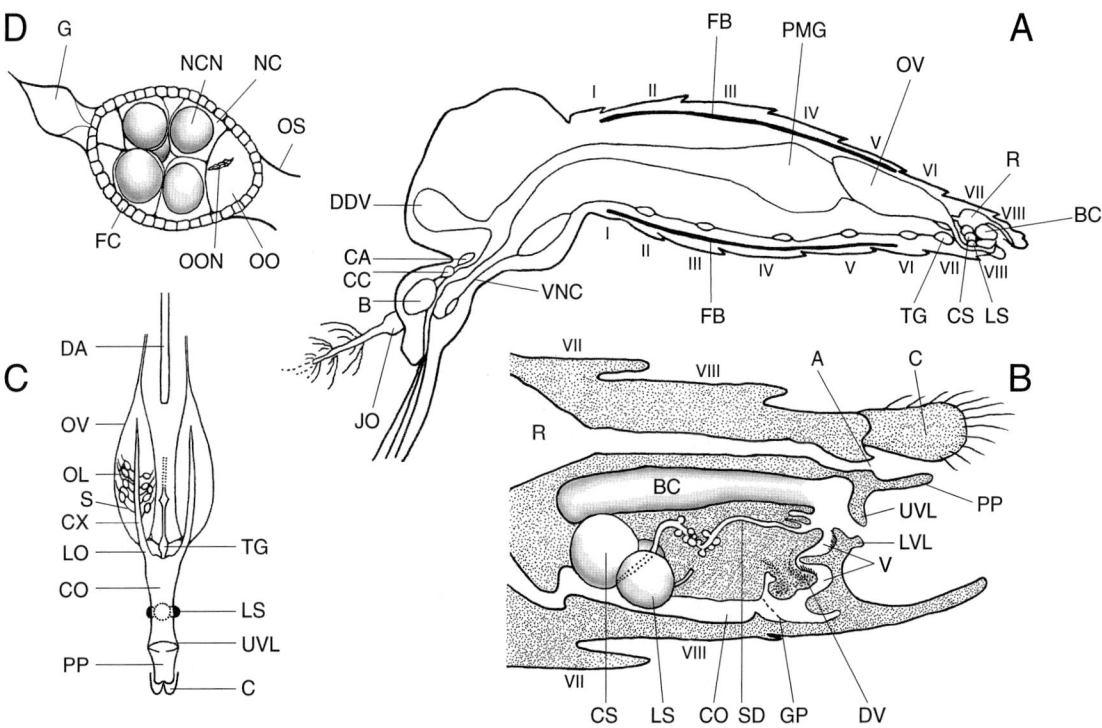

FIGURE 1 Schematic morphology of female *Aedes aegypti* structures important in reproduction. (A) Sagittal section through whole body. (B) Close-up sagittal view of abdominal terminus (modified from J. C. Jones and R. E. Wheeler, *J. Morphol.* 117, 401–424, 1965). (C) Ventral view of reproductive tract (modified from T. J. Curtin and J. C. Jones, *Ann. Entomol. Soc. Am.* 34, 298–313, 1961). (D) Previtellogenic ovariole in state of arrest. Roman numerals indicate abdominal body segments. A, anus; B, brain; BC, bursa copulatrix; C, cercus; CA, corpora allatum; CC, corpora cardiacum; CO, common oviduct; CS, central spermathecum; CX, calyx; DA, dorsal aorta; DDV, dorsal diverticulum; DV, dorsal valve; FB, fat body; FC, follicle cell; G, germarium; GP, gonopore; JO, Johnston's organ; LO, lateral oviduct; LS, lateral spermathecum; LVL, lower vaginal lip; NC, nurse cell; NCN, nurse cell nucleus; OL, ovariole; OO, oocyte; OON, oocyte nucleus; OS, ovarian sheath; OV, ovary; PMG, posterior midgut; PP, postgenital plate; R, rectum; S, ovariolar stalk; SD, spermathecal duct; TG, terminal abdominal ganglion; UVL, upper vaginal lip; V, vagina; VNC, ventral nerve cord.

rial is absorbed into the hemolymph through the single-cell-layer wall of the bursa within 2 or 3 days of mating. One of the accessory gland substances absorbed in this way renders females refractory to insemination for life so that females are in effect monogamous. Females may mate before or after a blood meal, but unmated blood-fed females continue to be attracted to a host, a place where males aggregate. Termination of host-seeking behavior requires a combination of a neuropeptide (Aea-HP-I) released from the female head after a blood meal and a male accessory gland factor. Unmated nutritionally stressed females that have taken a blood meal often do not begin oogenesis, but after mating a male accessory gland substance stimulates a metabolic shift in resource allocation away from female sustenance to egg development. A factor in the male accessory gland secretion also promotes oviposition behavior which is inhibited in unmated gravid females. None of these factors have been isolated, but it appears likely that more than one is involved.

III. PREVITELLOGENESIS

Aedes aegypti mosquitoes use the resources from digested vertebrate blood to synthesize yolk protein precursors (YPP) for provisioning the egg. Egg development (oogenesis) after a blood meal is synchronous and rapid, requiring the precise coordination of a number of complex physiological activities across different tissues, a process referred to *in toto* as vitellogenesis. A newly emerged female needs about 3 days to become competent to engage in the intense physiological demands of vitellogenesis. During this previtellogenic period, the fat body prepares to synthesize and secrete enormous amounts of YPPs, while the oocytes are readied for their uptake and storage. After 3 days, the fat body and ovary enter a state of developmental arrest which persists until a blood meal triggers the cascade of events associated with vitellogenesis.

A. Fat Body

The fat body is the major metabolic organ of insects, analogous to the vertebrate liver. It is a diffuse organ, occurring as sheets or lobes of tissue lining the body cavity, primarily in the abdomen. Its hormonally regulated functions are diverse, changing to meet the specific requirements of different phases of the life cycle. When the adult emerges from the pupal stage, the paired corpora allata (CA) (Fig. 1A) begin to secrete JH, which signals the fat body to prepare for vitellogenesis. JH titers peak at 2 days postemergence (Fig. 2). JH production is halted by a CA inhibitory factor released from arrested stage ovaries and, together with degradation of existing JH by a specific esterase, results in a gradual decline in titers throughout the period of arrest. Fat body trophocytes (the major fat body cell type) respond to JH by increasing ploidy level from $2n$ to $4n$, whereas another 20% become octoploid by the third day postemergence. This increase in DNA provides more template for the rapid transcription of selected genes during vitellogenesis. In addition, ribosomes begin to accumulate. Total RNA increases by about 50% over the first 3 days postemergence, with the highest rate of transcription occurring within the first 2 hr. In addition to rRNA production, mRNA increases twofold over the first 3 days. One species of mRNA, that encoding ribosomal protein L8, is neither immediately translated nor degraded; rather, it accumulates and is translated only after a blood meal, during a second cycle of ribosome accumulation. It is possible that the mRNAs of other housekeeping genes are likewise generated and amassed in preparation for the intense protein synthesis of the vitellogenic phase.

Many of the physiological activities associated with vitellogenesis are regulated by 20-hydoxyecdysone (20E), not JH, but exposure of the previtellogenic fat body to JH is necessary to make it responsive to 20E. The rise in JH titer after adult emergence corresponds temporally to transcription of the mosquito ecdysone receptor (EcR) gene (Fig. 3). EcR mRNA levels peak at 2 days postemergence then drop rapidly. To bind both 20E and DNA response elements, the EcR must heterodimerize with the product of the ultraspiracle gene (USP) (Fig. 3).

B. Ovary

In the mosquito, each of the two ovaries (Fig. 1C) contains about 75 ovarioles of the meroistic poly-

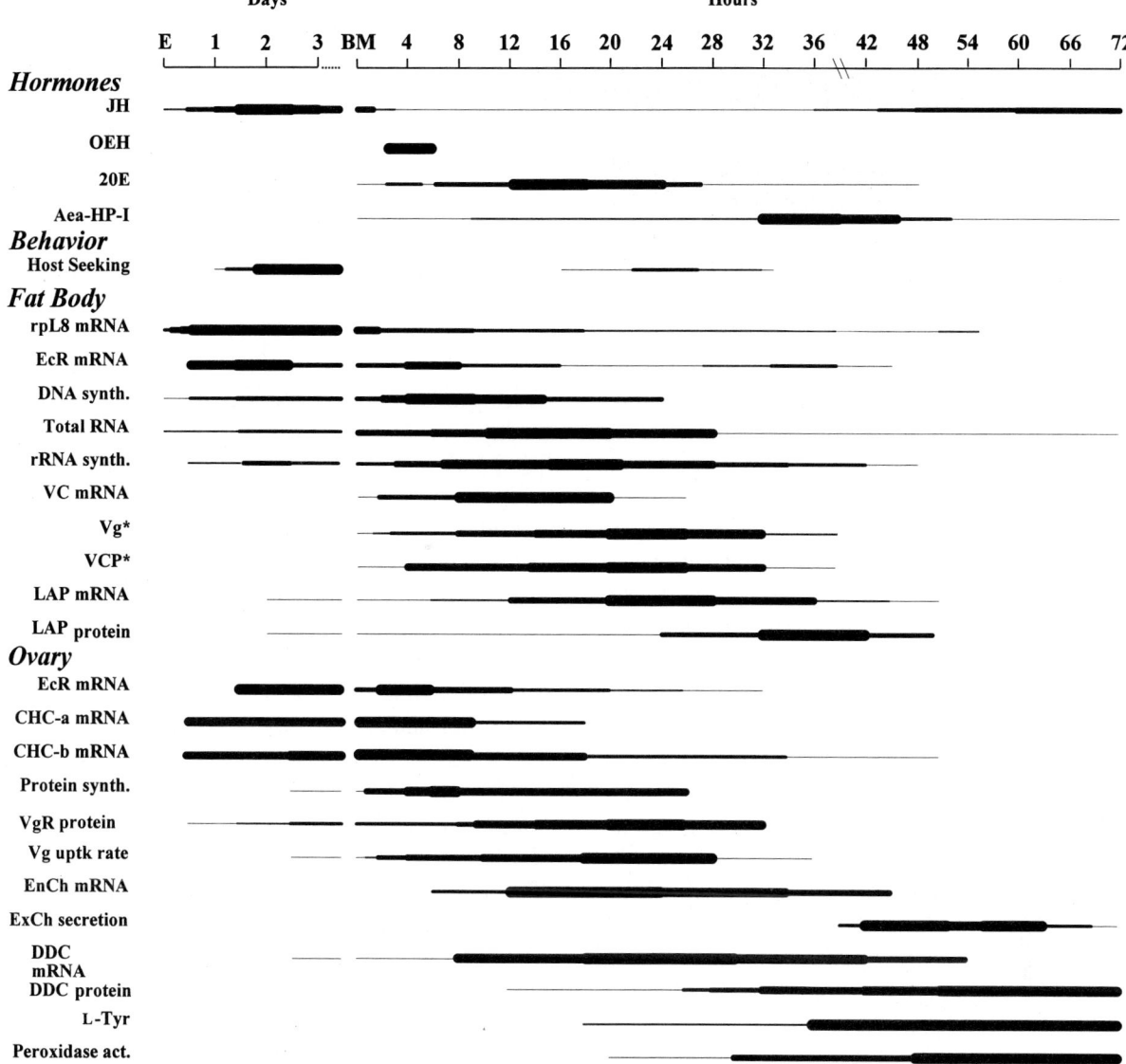

FIGURE 2 Temporal profiles of relative concentrations, rates of synthesis (synth), and other activities with respect to relative hormone titers in *Aedes aegypti* after emergence (E) and after a replete blood meal (BM). JH, Juvenile hormone III; OEH, ovarian ecdysteroidogenic hormone I; 20E, 20-hydroxyecdysone; Aea-HP-I, *A. aegypti* head peptide-I; rpL8, ribosomal protein L8; EcR, ecdysteroid receptor; VC, vitellogenin convertase; Vg*, vitellogenin protein and mRNA (profiles are similar); VCP*, vitellogenic carboxypeptidase protein and mRNA (profiles are similar); LAP, lysosomal aspartic protease; CHC, clathrin heavy chain, isoforms a or b; VgR, vitellogenin receptor, Vg uptk rate, rate of Vg uptake by oocyte; EnCh, endochorion gene 15-a; ExCh, exochorion; DDC, dopa decarboxylase; L-Tyr, L-tyrosine, act, enzymatic activity; ......, state of arrest of indefinite duration. \\, Scale of time intervals changes from 4 hr to 6 hr per tick mark.

trophic type. At emergence, each ovariole is composed of an apical germarium and a single primary follicle (or egg chamber). The follicle consists of one oocyte and seven nurse cells, the products of three cystocyte divisions in the germarium, which are interconnected by cytoplasmic bridges. These germ cells are surrounded by a monolayer of somatic follicular epithelial cells, which are also interconnected by cytoplasmic bridges, whose role is to secrete the components of the egg shell. A follicular stalk con-

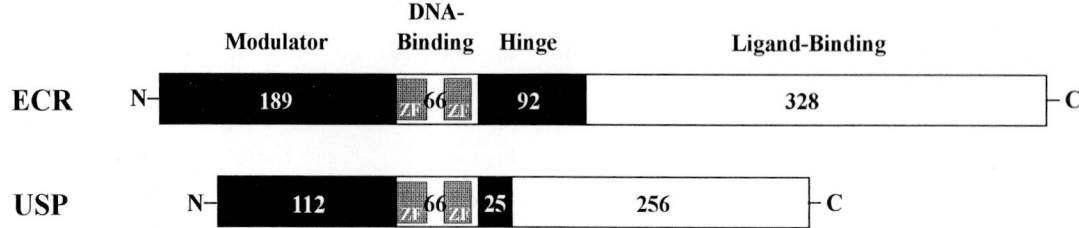

FIGURE 3 Domain structure of the *Aedes aegypti* ecdysteroid receptor (EcR) and ultraspiracle (USP). Both proteins are organized as typical steroid receptors. EcR must heterodimerize with USP before it can bind its ligand, 20-hydroxyecdysone. Numerals, number of residues per domain; ZF, zinc finger involved in DNA binding.

nects the follicle to the calyx, a central tube running the length of the ovary axis, which is continuous at the base of the ovary with the lateral oviduct through which mature eggs pass at oviposition (Fig. 1C). Before the first blood meal, both the calyx and the lateral oviduct are flattened tubes without well-developed lumens. The follicle is surrounded by a several-layered proteinaceous sheath, the basal lamina, which consists of a regular meshwork of square-shaped pores of about 14 nm on a side. The entire ovariole is surrounded by an ovariolar sheath which is connected to the calyx basally and as a filament to the ovarian sheath apically. The mosquito may undergo several cycles of egg production, or gonotrophic cycles, but only one follicle in each ovariole matures per cycle. Secondary follicles separate from the germarium during maturation of primary follicles, but their development is arrested at an early stage and does not resume until after oviposition of the fully developed eggs and a second blood meal.

The main task of the oocyte during vitellogenesis is to internalize massive amounts of YPPs, then route and process them for storage in yolk bodies. Internalization is via receptor-mediated endocytosis: Membrane-bound receptors specifically bind YPPs, then accumulate in clathrin-coated pits which pinch off into the cytoplasm as clathrin-coated vesicles. At emergence, however, the oocyte cortex (the region of the cytoplasm just inside the plasma membrane) is undifferentiated and incapable of YPP internalization. Under the influence of rising JH titers, clathrin-coated pits and vesicles, microvilli, and endosomes appear at the oocyte surface, making it competent to internalize YPPs by 2 or 3 days postemergence.

During this period, mRNAs encoding both clathrin and the receptor for the major YPP, vitellogenin, are produced in the nurse cells and transferred to the oocyte cytoplasm where they accumulate. Vitellogenin receptor mRNA is translated into protein only in the oocyte. Receptor protein increases threefold between the first and third day postemergence and amasses at the oocyte surface. In addition, the follicle doubles in size, from about 50 to 100 μm in length. The oocyte nucleus begins meiosis just before emergence but halts at the diplotene stage of prophase I when the oocyte enters the state of arrest.

Like the oocyte, the follicular epithelial cells are undifferentiated at the time of adult emergence. Under the influence of rising JH, however, they almost double in size, from about 3.0 to 5.5 μm in cross section, and they become well differentiated. Follicle cells undergo several mitotic divisions, increasing in number from about 20 cells per follicle at emergence to about 250 cells 2 days postemergence when mitosis halts. The number of RER cisternae, mitochondria, and Golgi complexes increase substantially by 2 days postemergence, and both the nucleus and nucleolus enlarge.

The source of the prohormone ecdysone (which is activated by conversion to 20E by several tissues) during vitellogenesis is the ovary, and its production is induced by ovarian ecdysteroidogenic hormone I (OEH), as will be discussed later. However, ovarian responsiveness to OEH requires exposure to JH during the previtellogenic period. As in the case of the fat body, some of the events in the ovaries during vitellogenesis are governed by 20E, and it is possible that JH renders previtellogenic ovaries responsive to 20E. Expression of EcR transcripts in the ovary increases after emergence to a peak 2 or 3 days later, corresponding to the peak in JH titer.

IV. VITELLOGENESIS

During the 2- or 3-day preparatory phase of the previtellogenic period, emergence from the pupa stimulates a rise in JH, which in turn sets in motion events that ready the mosquito for vitellogenesis: The fat body is remodeled into an efficient protein factory and becomes responsive to 20E, the ovary is tooled with the necessary machinery for protein uptake and becomes responsive to OEH, and the female becomes receptive to mating and responsive to host stimuli. When the mosquito takes a blood meal, however, JH titers plunge due in part to a doubling of JH esterase activity, and further events are modulated by other hormones, most notably 20E. Vitellogenesis is biphasic in that some vitellogenic events are initiated soon after the blood meal by unknown head factors but are greatly accelerated or intensified several hours later when the ovaries start to produce 20E in response to a pulse of OEH.

A. Initiation Phase

A number of processes are activated in the female mosquito during or shortly after blood feeding. If the blood meal is large enough (~5 μl), stretch receptors in the posterior midgut inhibit host-seeking behavior. Smaller blood meals may not cause enough distension to trigger this inhibition, and host seeking will continue until a sufficient volume of blood is ingested to activate the stretch receptors. A low rate of vitellogenin (Vg) transcription and synthesis begins in the fat body at least as early as 30 min postblood meal (PBM), and by 3 hr PBM, secretory granules carrying Vg and other YPPs appear in the trophocytes. The trophocyte nucleolus becomes more multilobed, and ribosome proliferation resumes. A second cycle of EcR transcription begins, peaking about 6 hr PBM (Fig. 2).

In the ovary, the follicle cells pull away from the surface of the oocyte creating a periooocytic space, remaining in contact with the oocyte only via the latter's microvilli, which begin to increase in number at this time. Mitosis resumes in the follicular epithelium so that cell number approximately doubles by 10 hr PBM. Narrow channels ~20 nm wide, along with a few intercellular spaces, appear between follicle cells, allowing hemolymph to penetrate to the periooocytic space and come in contact with the oocyte surface. The oocyte nucleolus starts to enlarge and synthesize rRNA by 4 hr PBM. Two splice variants of clathrin heavy chain (CHC) (the major protein component of the clathrin triskelion) are transcribed in the ovary during the previtellogenic period, and mRNA levels are highest at 6 hr PBM. One variant is ovary specific and probably encodes a CHC protein used exclusively in YPP internalization. The other is similar to a somatic-specific form and is thought to encode clathrin that is stored for use by the future embryo. In addition, because of alternative polyadenylation, both mRNA variants have shorter 3′ untranslated regions than a third, soma-specific variant coded by the same gene. Fewer instability signals in this 3′ tail presumably renders the ovarian transcripts more stable, permitting accumulation.

A pulsed factor from the head released during the first 5 min of blood meal initiation stimulates a threefold increase in ovarian protein synthesis by 30 min PBM and a threefold increase in the rate of Vg endocytosis. A second phase of more rapid protein synthesis and endocytosis begins by 2.5 hr PBM and is dependent on a factor released from the head from 1 hr through at least 12 hr PBM. Both phases appear to be activated through a cAMP signaling pathway. The identity of the head factor or factors responsible for these early events is unknown. However, it is possible that an insulin-like neurohormone is involved in some of these early activation processes in the ovary. An insulin receptor is expressed primarily in the ovary, and porcine insulin can stimulate a two- or three-fold increase in protein synthesis in previtellogenic mosquito ovaries in vitro. Insulin also stimulates the ovaries to produce ecdysone in vitro. This is significant because 20E activates the second rapid phase of endocytosis, which commences several hours before the major rise in 20E titers begins and which corresponds to a minor peak of ecdysteroid titers observed by several investigators at ~4 hr PBM. Ecdysone receptor transcription is stimulated before 6 hr PBM, by which time EcR mRNA levels are much higher than those in arrested-stage previtellogenic ovaries.

B. Trophic Phase

The trophic phase of vitellogenesis is characterized by rapid synthesis of YPPs in the fat body and rapid growth of the oocyte. It begins when OEH is released from the head and activates ovarian ecdysone synthesis, about 4–6 hr PBM (Fig. 2). OEH (referred to in the earlier literature as egg development neurosecretory hormone) is produced in the medial neurosecretory cells of the brain during the previtellogenic period and then is transported through a pair of nerves to the corpora cardiaca (CC) where it is stored. The release of OEH from the CC requires a combination of several signals: neural stimuli from a distended midgut, increased levels of amino acids in the hemolymph as the blood meal is digested, and an unidentified hormone secreted by the ovaries (the OEH-releasing factor). After OEH release, ecdysteroid titers increase sharply to a peak at 18–24 hr PBM, followed by a rapid decline between 24 and 30 hr PBM. 20E titers regulate the transcription of various genes in both the fat body and the ovary (Fig. 2).

1. Fat Body

Sometime between 4 and 8 hr PBM, rRNA production begins to accelerate, peaking at a three- or fourfold higher rate 12–18 hr PBM, then decreasing to very low rates by 48 hr PBM. After 12 hr PBM the highly lobed nucleoli begin to reverse transform to a compact core, a process which takes about 24 hr. During the first 12 hr PBM, lipid and glycogen stores decrease in the trophocytes, but after this time they increase again as the blood meal is digested and stores are replenished. By 12 hr PBM, RER occupy most of the cytoplasm. After 3 hr PBM, Golgi complexes begin to increase in number, peaking about 18–24 hr PBM, and by this time they contain cisternae two or three times wider than those during the initiation phase of vitellogenesis.

This proliferation of biosynthetic machinery in the fat body is necessary for the intensified production and secretion of YPPs triggered by the rise in 20E titers beginning ~6 hr PBM. At least three species of YPPs are synthesized and secreted by the vitellogenic A. aegypti fat body. By far the most abundant is Vg, a large protein that serves as the major source of amino acids for the future embryo. It is homologous to the vitellogenins of many other oviparous animals including most insects, nematodes, and vertebrates. The other YPPs are the 53-kDa vitellogenic carboxypeptidase (VCP) and a 44-kDa protein (44KP), which is presumably a thiol protease precursor. Both VCP and 44KP are proenzymes internalized by the oocyte and activated at the onset of embryogenesis, probably to digest vitellin (the crystalized storage form of Vg). All three YPPs follow the same kinetics, with expression peaking 24–28 hr PBM. The transcription of at least the Vg and VCP genes occurs in response to 20E. However, 20E does not directly activate these genes but rather represents the apex of a regulatory cascade. It spawns at least one intermediate gene product that activates transcription of Vg and VCP, which lie somewhere downstream in the 20E-response hierarchy.

The processing of the Vg precursor in the fat body has been studied in detail (Fig. 4A). The 6.5-kb Vg mRNA transcript is translated in the RER as a 224-kDa pre-pro-Vg. However, the pre-pro-Vg is cotranslationally glycosylated and posttranslationally phosphorylated to produce a 250-kDa pro-Vg intermediate. The pro-Vg is rapidly cleaved in the RER into two fragments of 190 and 62 kDa by vitellogenin convertase (VC), a fat body-specific 140-kDa member of the subtilisin-like proprotein convertase family of enzymes. VC mRNA transcripts begin to accumulate in the fat body by 3 hr PBM, peaking at 12–18 hr and decreasing to low levels by 24 hr PBM. The VC proenzyme is presumably autoactivated by the cleaving away of an N-terminal propeptide domain, but its activity is relatively low through the first 3 hr PBM. It increases later, but this increase is not under the control of 20E. The pro-Vg is cleaved immediately C-terminal to a dibasic amino acid motif, RYRR, preceded by a predicted β-turn, matching the consensus recognition requirements of other convertase family members (i.e., a RXR/KR motif near a β-turn) (Fig. 4B). Both fragments enter the Golgi complex, where they undergo tyrosine sulfation, in-

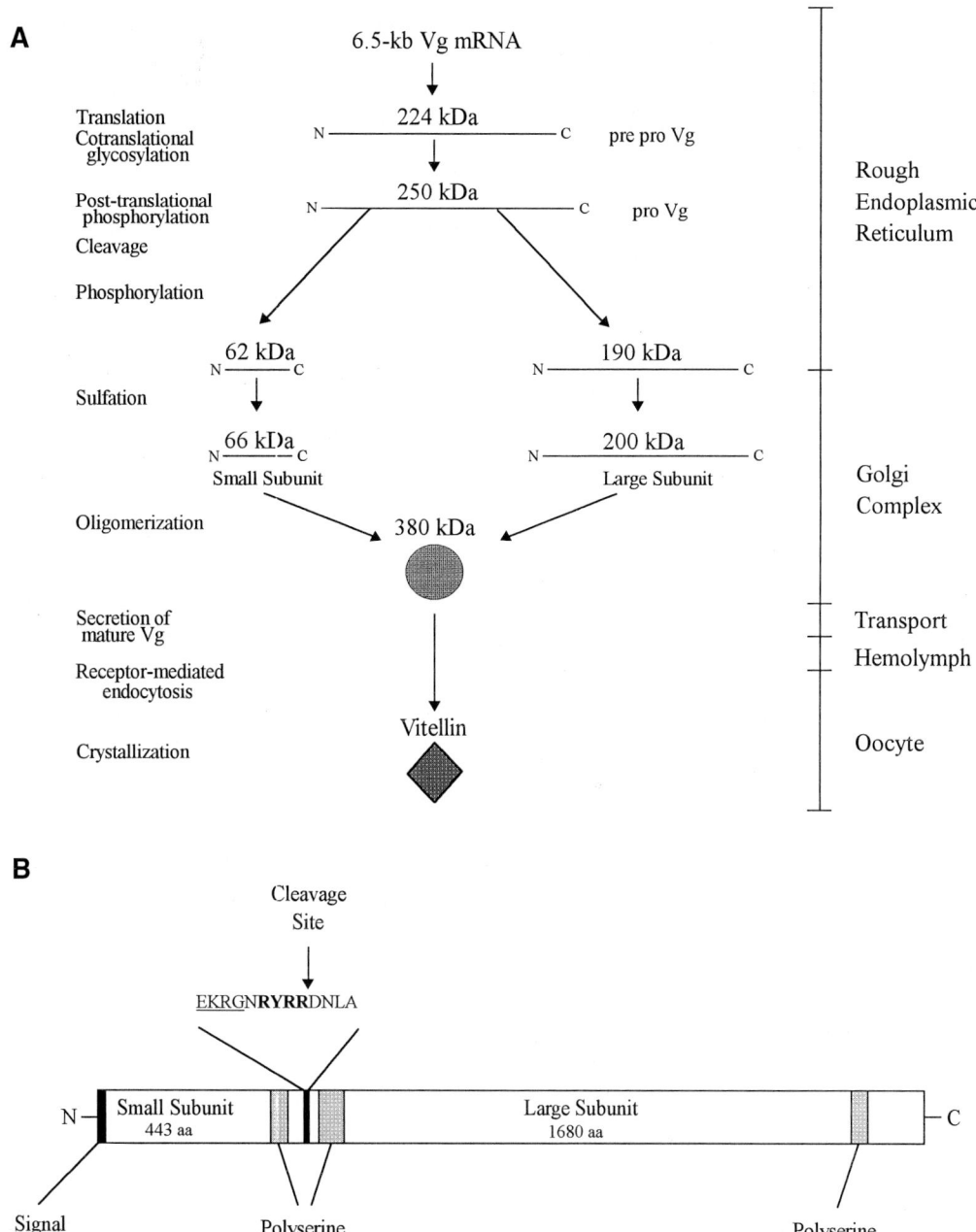

FIGURE 4 Co- and posttranslational processing (A) of the *Aedes aegypti* pre-pro-vitellogenin (B) from synthesis in the fat body to crystallization as vitellin in the oocyte. The cleavage site is indicated by an arrow in B and the amino acid sequence (bold) in the context of a β-turn (underlined), which is recognized by the cleavage enzyme, vitellogenin convertase, is presented at the base of the arrow.

creasing in size to that of the mature large (200 kDa) and small (66 kDa) subunits of Vg. The subunits oligomerize within the Golgi complex to form the mature 380-kDa Vg, which is packaged into condensing vacuoles. The latter develop into large secretory granules, which empty their contents into the hemolymph almost immediately after leaving the Golgi complex. The details of VCP and 44KP processing have not been studied, but they both follow the same pathway through the RER and Golgi complex as Vg, being packaged in and released from the same secretory granules.

2. Ovary

Vitellogenin is soluble in the hemolymph, and Vg titers rise sharply between 6 and 8 hr PBM to a broad

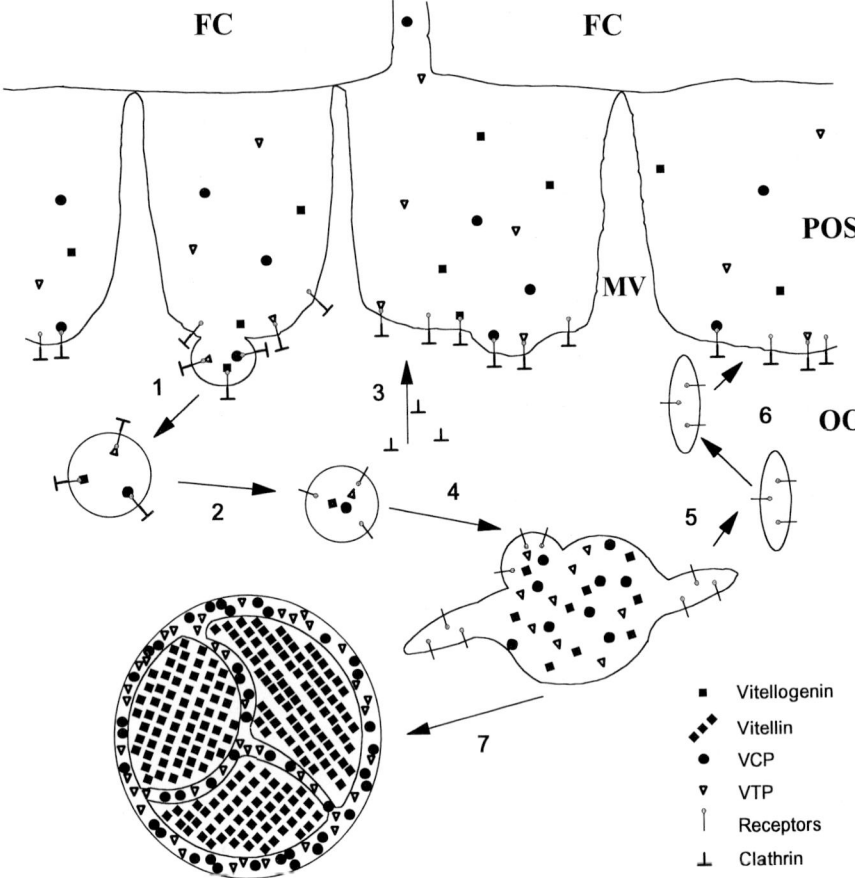

FIGURE 5 Receptor-mediated endocytosis and postendocytotic routing of the *Aedes aegypti* yolk protein precursors, vitellogenin (Vg), vitellogenic carboxypeptidase (VCP), and 44KP. (1) Receptor/ligand complexes accumulate in clathrin-coated pits which pinch off to form coated vesicles. Putative receptors for VCP and 44KP have not yet been isolated or characterized. (2) Clathrin is released from the receptors and the vesicle becomes an early endosome. Acidification of the endosome causes release of ligands from their receptors. (3) The released clathrin triskelions are presumably recycled to the cell surface. (4) The endosome fuses with other endosomes to create a transitional yolk body. Within this structure, ligands are segregated from receptors which accumulate in peripheral arms. (5) The peripheral arms containing the receptors pinch off as tubular compartments and recycle to the surface (6). (7) Delipidated Vg molecules begin to crystallize as vitellin (Vn), and at some point the transitional yolk body no longer fuses with endosomes. In the mature yolk body, crystalline Vn occupies the bulk of the organelle's interior, whereas VCP and 44KP are confined to the peripheral spaces between Vn crystalloids and the organelle membrane [modified from T. W. Sappington and A. S. Raikhel, In *Recent Advances in Insect Biochemistry and Molecular Biology* (E. Ohnishi, H. Sonobe, and S. Y. Takahashi, Eds.), pp. 235–257, Univ. of Nagoya Press, Nagoya, Japan, 1995; and incorporating data from E. S. Snigirevskaya, T. W. Sappington, and A. S. Raikhel, *Cell Tissue Res.* **290**, 175–183, 1997].

peak of 6.0–6.5 μg/μl (~16 or 17 μM) from 16 to 32 hr, after which it drops rapidly. Vg, VCP, and 44KP pass through the basal lamina surrounding the follicle, then through the narrow interfollicular channels, concentrating in the periooctic space. All three YPPs bind to the oocyte surface in the subdomains located between and at the base of microvilli, an area characterized by a dense extracellular glycocalyx and numerous clathrin-coated pits in which they accumulate. Clathrin triskelions attach to the cytoplasmic tail of endocytotic receptors via an adaptor protein, then attach to one another to form a polyhedral lattice that deforms the plasma membrane into a depression. This pit grows as more receptor/clathrin complexes are recruited to it through the mediation of the self-assembling clathrin triskelions. Eventually, it is sealed off by a GTPase, creating an intracellular clathrin-coated vesicle. Whatever ligands are bound to the receptors trapped in the pit are also internalized. This process, referred to as receptor-mediated endocytosis (Fig. 5), was first observed in *A. aegypti* oocytes and is now recognized as the universal mechanism employed by eukaryotic cells for internalizing specific macromolecules.

The mosquito receptor for Vg (VgR) has been extensively characterized biochemically and molecularly. It is a large 205-kDa protein which apparently homodimerizes. The VgR gene is transcribed in the nurse cells, and the 7.3-kb mRNA transcript is translated in the oocyte. Transcription begins by 6 hr PBM, but mRNA quantities do not start to increase rapidly until 6–12 hr PBM, peaking at 24 hr. VgR protein accumulates rapidly from 8 to 24 hr PBM, then declines. The linear increase in VgR parallels a linear increase in rate of Vg uptake from 6 to 24 hr PBM.

Like other endocytotic receptors, the mosquito VgR is anchored in the plasma membrane by a single-pass hydrophobic transmembrane helix. The cytoplasmic tail is relatively short and contains a leucine–isoleucine internalization signal that presumably binds the clathrin adaptor protein. Over 90% of the molecule is extracellular, and its domain structure clearly identifies it as a member of the low-density lipoprotein receptor (LDLR) family, consisting of discreet clusters of repetitive elements arranged in a characteristic manner (Fig. 6). However, the mosquito VgR is about twice as large as vertebrate VgRs and LDLRs, harboring two clusters of cysteine-rich class A repeats instead of one. Class A domains are implicated in the binding of numerous ligands in several other LDLR family members. Little is known about the specific interaction between Vg and VgR at the molecular level, except that it involves both Vg subunits and that phosphorylation of Vg is required. There are probably multiple points of contact between positive residues on the surface of Vg and negative residues in one or both of the VgR class A domains. Binding affinity is very high, with an estimated dissociation constant (K_D) of 15 nM in enriched solutions of solubilized VgR. The receptors for VCP and 44KP have not been identified, but multiple ligands are characteristic of LDLR family receptors, and it is possible that they bind the same receptor as Vg.

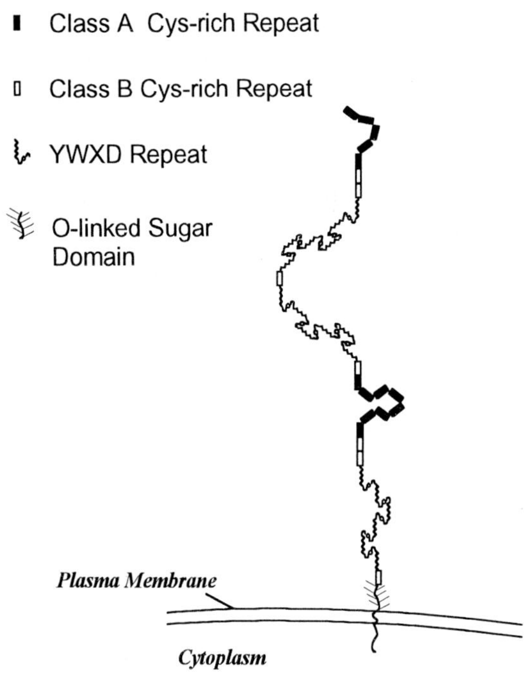

Soon after internalization, the coated vesicle loses its clathrin mantle, becoming an early endosome (Fig. 5). The receptors remain anchored in the vesicle

FIGURE 6 Schematic representation of the *Aedes aegypti* vitellogenin receptor. Like other members of the low-density lipoprotein receptor family, it is a protein characterized by a highly conserved arrangement of modular elements homologous to domains in functionally unrelated proteins.

membrane as the YPPs are released, presumably through acidification of the endosome lumen. In the absence of Vg, endosomes containing nonspecific proteins that were internalized by the oocyte are routed directly to lysosomes for degradation, but the presence of Vg somehow signals the endosomes to fuse with one another, creating transitional yolk bodies (TYB). Released YPPs concentrate in the lumen of the TYB while membranous tubular compartments bud off to recycle the VgR, and probably other receptors, back to the oocyte surface. At some point, TYBs cease fusing with endosomes and begin the transformation to a mature yolk body (MYB). Vg condenses in the core of the yolk body as its load of lipids is removed and eventually crystallizes as vitellin (Vn). Within the MYB, the Vn occurs as two to four separate crystalloid bodies, each surrounded by and interspersed with a noncrystalline matrix. VCP and 44KP, which are intermixed with Vg in endosomes and TYBs, are segregated from the Vn as it crystallizes, collecting on one side of the MYB as a cap. By 48 hr PBM, VCP and 44KP are evenly distributed in the matrix surrounding the crystalline Vn (Fig. 7). Any nonspecific proteins that were routed to the MYB along with the YPPs are segregated into vesicular extensions which bud off into the cytoplasm and fuse with lysosomes.

The length of the oocyte increases eightfold, and the volume several hundred fold, from 6 to 36 hr PBM as YPPs are internalized at ever-increasing rates through 30 hr PBM. In addition to the necessary synthesis of large quantities of the endocytotic machinery required to sustain high rates of YPP internalization, the oocyte must stockpile a sizable supply of histones and ribosomes for use by the future embryo. These activities place a tremendous demand on the synthetic capacity of the follicle. In most insects with meroistic ovaries, the nurse cells manufacture the necessary rRNA which is transferred to the oocyte via cytoplasmic bridges. However, mosquitoes are unusual in that both the nurse cells and the oocyte synthesize rRNA. It is presumed that *A. aegypti* nurse cells increase their capacity for rRNA production by becoming polytene through endoreduplication of

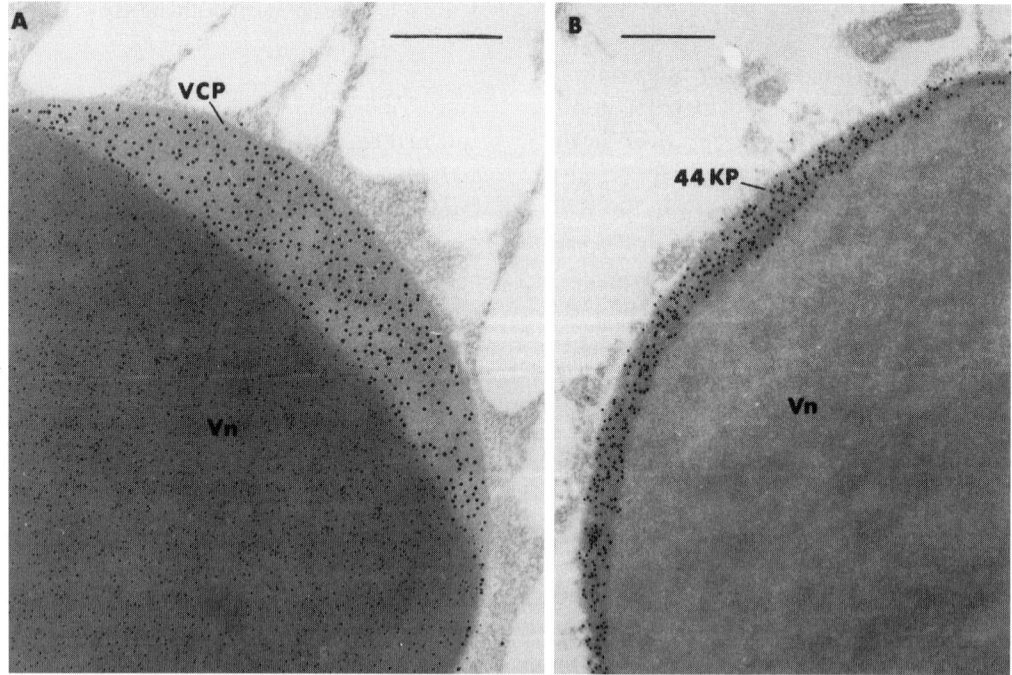

FIGURE 7 Immunolabeling of VCP, 44KP, and Vn in the mature yolk bodies (MYB) of *Aedes aegypti* oocytes. (A) Double labeling of Vn (10-nm gold) and VCP (15-nm gold) in a MYB from an oocyte 24 hr PBM. (B) Single labeling of 44KP (15-nm gold); note that 44KP is distributed in the MYB around the crystalline Vn (unlabeled) as an envelope during the late stages of oocyte development (48 h PBM) (from Snigirevskaya *et al.*, 1997).

chromosomes, as do other mosquitoes, but the stage at which this occurs is not known. The rate of RNA production in *A. aegypti* nurse cells is very high 24–36 hr PBM. The oocyte, on the other hand, amplifies its rRNA production capacity by selectively replicating rDNA. The oocyte nucleolus increases greatly in size from 13 to 20 hr PBM. Sometime during this period, the nucleolar organizer separates from the chromosomes, its DNA is replicated, then it breaks up into numerous rDNA fragments that disperse throughout the nucleoplasm. rRNA synthesis is continuous in the oocyte from 4 to 60 hr PBM. After nucleolar–organizer separation, the chromosomes condense and are enclosed by a membrane, creating a small karyosphere or germinal nucleus, which remains distinct from the ever-growing vegetative nucleus, a condition unique to mosquitoes. By 30 hr PBM, the vegetative nucleus has more than doubled in diameter, with lobes extending out into the cytoplasm.

By 24 hr PBM, the calyx and the lateral oviduct (Fig. 1C) have begun preparing to receive mature eggs for oviposition, both developing a distinct lumen. Cells of both layers increase greatly in size, and by 72 hr PBM, at the completion of development, deep crypts have formed between the epithelial cells, whereas extensive infoldings occur between the muscle cells. These structural modifications allow for stretching during the passage of eggs at ovulation. During the first 24 hr PBM, the inner layer of epithelial cells of the calyx elongate, and by 72 hr the calyx has become a highly distended sac.

The major rise in 20E that begins ~6 hr PBM stimulates the follicle cells to synthesize the endochorion (or vitelline envelope), which is the innermost of the two principal layers of the egg shell. Several proteins of about 20 kDa, one of 64 kDa, and one of 90 kDa are secreted, none of which are glycosylated. 20E induces transcription of at least two vitelline genes, 15a-1 and 15a-2, in the follicle cells beginning ~5 hr PBM, with a peak in transcript abundance at ~20 hr PBM. Granules of endochorionic material appear in the Golgi complexes of the follicle cells by 6–8 hr PBM and in the periooocytic space by 8–10 hr PBM. These granules coalesce to form endochorionic plaques distributed uniformly between oocyte microvilli. Proliferation of synthetic machinery, i.e., RER and Golgi complexes, is evident in follicle cells by 12 hr PBM and continues through ~30 hr PBM. Endochorionic plaques rapidly increase in size from 18 to 30 hr PBM, and follicle cells become columnar in shape during this same period. The continuous presence of 20E is required to maintain endochorion secretion. However, a signal from the oocyte, presumably passed to the follicle cells through gap junctions in the tips of the oocyte's microvilli, is also required because in those follicle cells surrounding the nurse cells, endochorion genes are only minimally expressed and endochorion precursors are not secreted. At 32–36 hr PBM, the endochorionic plaques fuse, secretory activity in the follicle cells declines sharply while they prepare for exochorionic protein production, and the microvilli lose contact with the follicle cells as the former retreat to the oocyte surface.

As will be described later, deposition of the outer layer of the egg, or exochorion, does not begin until about 42 hr PBM, and tanning (hardening) of the eggshell does not begin until the eggs have been oviposited. However, transcription of one of the enzymes involved in *A. aegypti* chorionic tanning, dopa decarboxylase (DDC), begins in the ovary in response to rising 20E titers, with a peak in transcript abundance at 24 hr, followed by a gradual decline through at least 48 hr PBM when transcript level is still relatively high. Translation of the DDC mRNA into protein, though, does not start until 24 hr PBM, when 20E concentrations are falling, suggesting that 20E may act as a translation inhibitor of DDC mRNA. The initial precursor in the dopa pathway of chorion tanning, the amino acid L-tyrosine, is detectable in the ovary by 24 hr, increases five- or sixfold from 24 to 48 hr PBM, then continues to increase at a slower rate through 96 hr, but its source is unknown. Similarly, the peroxidase involved in a second pathway of tanning, involving tyrosine cross-linking, is detectable in the ovary by 24 hr PBM and increases threefold by 48 hr to a level which is maintained through at least 96 hr.

All the primary follicles in the mosquito ovary develop synchronously from emergence through oviposition. However, this first gonotrophic cycle can be followed by three or more synchronous cycles of egg maturation depending on the longevity and nutritional status of the female. High 20E titer induces the separation of a secondary follicle from the germarium at ~20 hr PBM. Some development follows, but these follicles do not advance to the previtellogenic state of arrest until JH titers start to rise again ~36–48 hr PBM.

V. POSTVITELLOGENESIS

The vitellogenic phase ends abruptly between 27 and 30 hr PBM. During this time, termination of both YPP synthesis in the fat body and YPP uptake by the oocyte is initiated. The fat body's protein synthetic machinery is degraded, and the trophocytes are remodeled in preparation for the next gonotrophic cycle. The oocyte ceases to internalize YPPs when it loses contact with the hemolymph through the closing of interfollicular channels and the fusion of the endochorionic plaques. While the fat body undergoes its ultrastructural reorganization, the outer layer of the eggshell is secreted and the egg is prepared for oviposition.

A. Fat Body

The amount of Vg transcript and protein in the fat body peaks between 24 and 30 hr PBM, then declines rapidly, with Vg synthesis halting altogether by 36 hr. The mechanism for terminating Vg gene transcription is not yet clear. It may occur in response to decreasing 20E titers or it may involve feedback inhibition from hemolymph Vg concentrations that increase when oocytes cease Vg uptake. However, the interruption of Vg production in the fat body does not depend simply on turning off YPP gene transcription, but rather involves the active degradation of synthetic and secretory organelles in the trophocytes by the lysosomal system.

At the end of vitellogenesis, beginning between 30 and 36 hr PBM, the biosynthetic machinery in the fat body is broken down by lysosomal enzymes including lysosomal aspartic protease (LAP). The number of lysosomes in the trophocyte cytoplasm increases ~12-fold from 24 to 36 hr PBM, and the activities of several lysosomal enzymes increase dramatically between 24 and 27 hr PBM, remaining high through at least 48 hr PBM. Enhanced fat body-specific transcription of the LAP gene, probably in response to rising 20E titers, begins ~6 hr PBM, with mRNA levels peaking at 24–30 hr and declining rapidly after 36 hr PBM. However, it is likely that 20E negatively regulates LAP translation, so the LAP protein does not begin to accumulate until 20E titers start to fall after 24 hr PBM.

Sometime between 36 and 48 hr PBM, JH titers begin to rise again, at least partly as a result of a sharp drop in JH esterase activity between 42 and 48 hr PBM. The remodeled fat body is incapable of responding to 20E until about 70 hr PBM, when trophocyte ultrastructure resembles that at the previtellogenic state of arrest. As in the case of newly eclosed mosquitoes, competency of the remodeled fat body to respond to 20E requires exposure to JH.

B. Ovary

Rate of Vg uptake by oocytes declines precipitously from 30 to 36 hr PBM because access of hemolymph, and therefore of YPPs, to the surface of the oocyte is restricted by two concurrent events. Permeability of the follicular cell layer is eliminated by sealing junctions that form between follicle cells beginning at ~32 hr PBM. In addition, the endochorionic plaques fuse into an impermeable envelope around the oocyte. During fusion of the endochorion, the follicle elongates, and the nurse cells degenerate between 36 and 48 hr PBM. As the oocyte lengthens, its vegetative nucleus flattens into a long, thin sheet that sends branches throughout the cytoplasm, a phenomenon apparently unique to mosquitoes. Nucleolar fragments extend throughout the nucleoplasm and continue to synthesize rRNA until the nuclear envelope disintegrates soon after the oocyte reaches its maximum length. The small germinal nucleus containing the chromosomes remains at the anterior pole of the oocyte near the nurse cells. As the vegetative nuclear envelope collapses, the chromosomes condense and meiosis resumes. Meiosis proceeds to metaphase I, where it is arrested until oviposition.

Secretory activity of the follicle cells decreases greatly between 30 and 36 hr PBM as synthesis of endochorion proteins is completed. Rapid secretion of exochorion proteins by the follicle cells begins

~38–42 hr PBM and is essentially completed by 60 hr PBM. Unlike the endochorion proteins, most exochorion proteins are glycosylated and range in size from 25 to 50 kDa. The exochorion comprises ~80% of the chorion by weight and has a highly sculpted appearance arising from uneven deposition. After secretion of the chorion is completed, the follicular epithelium separates from the chorion, creating an egg sac in which the egg rests until ovulation.

As JH titers rise between 48 and 72 hr PBM, the secondary follicles (which form at ~20 hr PBM in response to high 20E titer) increase in length and develop to the previtellogenic state of arrest. A female that cannot oviposit immediately after egg maturation (~72 hr PBM) may take a second blood meal, which is digested normally, in anticipation of the next gonotrophic cycle. However, an unidentified oostatic hormone produced by mature primary follicles prevents the secondary follicles from breaking the state of arrest. Its target is unknown, but it probably has the effect of preventing initiation of vitellogenesis in the fat body. Although relatively little research has been done in this area, it appears that the same hormonal events are required to initiate vitellogenesis after a second blood meal as those after the first, i.e., the secondary follicles secrete OEH-releasing factor, the corpora cardiaca release OEH, the ovaries synthesize and secrete 20E, inducing the transcription of YPP genes in the fat body. Presumably, the oostatic hormone blocks one of these events.

C. Host-Seeking Behavior

As the blood meal is digested, distention of the posterior midgut decreases, eventually releasing the nervous inhibition on host-seeking behavior mediated by stretch receptors. However, by ~30 hr PBM, repression of host-seeking behavior is taken over by a small head peptide, Aea-HP-I, released from neurosecretory cells in the brain and possibly from endocrine cells in the midgut. It presumably acts by desensitizing the antennal receptor for lactic acid, one of the major attractants present in human odors. Aea-HP-I titers remain high until ~48–56 hr PBM, after which host seeking and blood feeding resume,
even if oviposition has not yet occurred. The release of Aea-HP-I requires the presence of the postvitellogenic fat body, which has received an unknown signal from the ovaries by 6–8 hr PBM. For Aea-HP-I to be fully effective, however, requires the presence of a protein from the male accessory gland transferred to the female during mating.

D. Preoviposition Behavior

By 48–72 hr PBM, the gravid female becomes responsive to airborne oviposition-site stimuli, i.e., bacterial metabolites including ethyl propionate. Sensitivity at this time requires that the female be mated; a male accessory gland protein removes a physiological block that prevents unmated females from responding to oviposition-site stimuli. A circadian rhythm dictates that preoviposition behavior occurs during late afternoon and early evening, at which time gravid females initiate upwind flight. Females oviposit at multiple sites (~12–120 per gonotrophic cycle) over several days, and the distance they disperse during this time depends on the number of suitable oviposition sites encountered and can exceed at least 400 m. Aedes aegypti is a floodwater mosquito, so it oviposits on surfaces just above the waterline. After landing at a potential oviposition site, the female tests the salinity of the water with tarsal sensillae and then tests the oviposition substrate with tactile setae near the tip of the abdomen. It prefers moist, rough surfaces of low reflectance. There is some evidence that females lay fewer eggs at sites which already contain a high density of eggs.

E. Oviposition

The central nervous system in the head and thorax initiates ovulation and oviposition once tarsal and abdominal sensilla detect a suitable oviposition surface. At ovulation, the follicular epithelium and basal lamina rupture at the base of the egg sac. The egg enters the calyx lumen by breaking through an opening in a group of cells called the basal body, to which the follicular stalk is attached. Not all eggs ovulate at the same time, but those which do are apparently oviposited immediately. The nature of the mechanism controlling selective ovulation at a given oviposition site is unknown. An egg in the base of the calyx lumen is drawn into and through the lateral oviduct by peristaltic contractions of the latter. The normally retracted seventh and eighth abdominal segments telescope outward, straightening the common oviduct through which the egg passes. The vagina contracts, widening the opening of the spermathecal ducts. The spermathecae are repeatedly thrust up and back and release sperm into the vagina as the egg passes. Sperm gain access to the oocyte's plasma membrane through the micropyle, a small pore in the chorion at the anterior tip of the egg. Meiosis in the egg resumes and is completed within 30 min of oviposition. Several eggs may be laid together in a small chain, or they may be scattered individually as the female walks around the site. A sticky substance is associated with egg's surface, which effectively glues it to the substrate.

VI. THE OVIPOSITED EGG

At oviposition, the egg chorion is white, soft, and permeable to water. The egg more than doubles its weight during the first few hours after oviposition through the uptake of water from the moist substrate. Exposure to the atmosphere activates phenol oxidase (PO) in the chorion, an enzyme that hydroxylates free tyrosine residues to form dopa. Some dopa is further converted by dopa decarboxylase to dopamine. Both dopa and dopamine are converted by PO into quinones, which either enter melanization pathways or participate directly in tanning reactions in which they interact with nucleophilic groups on structural proteins to effect cross-linking. In another tanning pathway, the peroxidase that was stockpiled in the ovary during oogenesis is activated at oviposition and catalyzes the cross-linking of tyrosines on structural proteins. By 1 hr after oviposition, the eggs begin to turn brown, and by 4 hr they have become black and relatively hard. Within the first day after oviposition, the serosa secretes a thin, apparently chitonous, serosal cuticle suffused with wax, which renders the egg resistant to water loss. The embryo requires ~4 days to develop to the pharate first instar, before which the egg will not hatch. Eggs remain viable for several months. Hatching is stimulated by the drop in oxygen concentration that occurs when the eggs are flooded by water.

Bibliography

Brogdon. W. G. (1994). Measurement of flight tone differences between female. *Aedes aegypti* and *A. albopictus* (Diptera: Culicidae). *J. Med. Entomol.* **31**, 700–703.

Brown, M. R., Graf, R., and Swiderek, K. (1995). Structure and function of mosquito gonadotropins. In *Molecular Mechanisms of Insect Metamorphosis and Diapause*, pp. 229–238. Industrial Publishing and Consulting, Tokyo.

Chen, J.-S., and Raikhel, A. S. (1996). Subunit cleavage of mosquito pro-vitellogenin by a subtilisin-like convertase. *Proc. Natl. Acad. Sci. USA* **93**, 6186–6190.

Cho, W.-L., and Raikhel, A. S. (1992). Cloning of cDNA for mosquito lysosomal aspartic protease: Sequence analysis of an insect lysosomal enzyme similar to cathepsins D and E. *J. Biol. Chem.* **267**, 21823–21829.

Cho, W.-L., Kapitskaya, M. Z., and Raikhel, A. S. (1995). Mosquito ecdysteroid receptor: Analysis of the cDNA and expression during vitellogenesis. *Insect Biochem. Mol. Biol.* **25**, 19–27.

Christophers, S. R. (1960). *Aëdes aegypti* (L.): *The Yellow Fever Mosquito. Its Life History, Bionomics, and Structure.* Cambridge Univ. Press, Cambridge, UK.

Ciba Foundation (1996). *Olfaction in Mosquito–Host Interactions. Symposium 200.* Wiley, New York.

Clements, A. N. (1992). *The Biology of Mosquitoes, Volume 1. Development, Nutrition, and Reproduction.* Chapman & Hall, New York.

Deitsch, K. W., Chen, J.-S., and Raikhel, A. S. (1995). Indirect control of yolk protein genes by 20-hydroxyecdysone in the fat body of the mosquito, *Aedes aegypti. Insect Biochem. Mol. Biol.* **25**, 449–454.

Dhadialla, T. S., and Raikhel, A. S. (1994). Endocrinology of mosquito vitellogenesis. In *Perspectives in Comparative Endocrinology*, pp. 275–281. National Research Council of Canada.

Dittmer, N. T., and Raikhel, A. S. (1997). Analysis of the mosquito lysosomal aspartic protease gene: An insect housekeeping gene with fat body-enhanced expression. *Insect Biochem. Mol. Biol.* **27**, 323–335.

Ferdig, M. T., Li, J., Severson, D. W., and Christensen, B. M. (1996). Mosquito dopa decarboxylase cDNA characterization and blood-meal-induced ovarian expression. *Insect Mol. Biol.* **5**, 119–126.

Graf, R., Neuenschwander, S., Brown, M. R., and Ackermann, U. (1997). Insulin-mediated secretion of ecdysteroids from mosquito ovaries and molecular cloning of the insulin receptor homologue from ovaries of bloodfed *Aedes aegypti*. *Insect Mol. Biol.* **6**, 151–163.

Hancock, R. G., and Foster, W. A. (1993). Effects of prebloodmeal sugar on sugar seeking and upwind flight by gravid and parous *Aedes aegypti* (Diptera: Culicidae). *J. Med. Entomol.* **30**, 353–359.

Hoc, T. Q. (1996). Quiescent ovarioles in the mosquito *Aedes aegypti*. *Ann. Trop. Med. Parasitol.* **90**, 71–78.

Kapitskaya, M., Wang, S., Cress, D. E., Dhadialla, T. S., and Raikhel, A. S. (1996). The mosquito *ultraspiracle* homologue, a partner of ecdysteroid receptor heterodimer: cloning and characterization of isoforms expressed during vitellogenesis. *Mol. Cell Endocrinol.* **121**, 119–132.

Klowden, M. J. (1997). Endocrine aspects of mosquito reproduction. *Arch. Insect Biochem. Physiol.* **35**, 491–512.

Kokoza, V. A., and Raikhel, A. S. (1997). Ovarian- and somatic-specific transcripts of the mosquito clathrin heavy chain gene generated by alternative 5′-exon splicing and polyadenylation. *J. Biol. Chem.* **272**, 1164–1170.

Li, J. (1994). Egg chorion tanning in *Aedes aegypti* mosquito. *Comp. Biochem. Physiol.* **109A**, 835–843.

Li, J., Hodgeman, B. A., and Christensen, B. M. (1996). Involvement of peroxidase in chorion hardening in *Aedes aegypti*. *Insect Biochem. Mol. Biol.* **26**, 309–317.

Mazzacano, C. A., and Fallon, A. M. (1996). Changes in ribosomal protein rpL8 mRNA during the reproductive cycle of the mosquito, *Aedes aegypti*. *Insect Biochem. Mol. Biol.* **6**, 563–570.

Raikhel, A. S. (1992). Vitellogenesis in mosquitoes. *Adv. Dis. Vector Res.* **9**, 1–39.

Raikhel, A. S., and Dhadialla, T. S. (1992). Accumulation of yolk proteins in insect oocytes. *Annu. Rev. Entomol.* **37**, 217–251.

Raikhel, A. S., and Snigirevskaya, E. S. (1998). Vitellogenesis. In *Microscopic Anatomy of Invertebrates, Vol. 11c, Insecta* (M. Locke and F. Harrison, Eds.), pp. 933–955. Wiley-Liss, New York.

Reiter, P., Amador, M. A., Anderson, R. A., and Clark, G. G. (1995). Short report: Dispersal of *Aedes aegypti* in an urban area after blood feeding as demonstrated by rubidium-marked eggs. *Am. J. Trop. Med. Hyg.* **52**, 177–179.

Sappington, T. W., and Raikhel, A. S. (1998). Molecular characteristics of insect vitellogenins and vitellogenin receptors. *Insect Biochem. Mol. Biol.*, in press.

Snigirevskaya, E. S., Hays, A. R., and Raikhel, A. S. (1997). Secretory and internalization pathways of mosquito yolk protein precursors. *Cell Tissue Res.* **290**, 129–142.

Sutcliffe, J. F. (1994). Sensory bases of attractancy: Morphology of mosquito olfactory sensilla—A review. *J. Am. Mosquito Control Assoc.* **10**, 309–315.

Aggressive Behavior

David A. Edwards
Emory University

I. Introduction
II. Gonadal Hormones and Intermale and Maternal Aggression in Mice and Rats
III. Situational, Experiential, and Sensory Determinants of Aggression in Mice and Rats
IV. Organizing Effects of Testicular Hormones on Aggressive Behavior
V. The Hormonal Activation and Organization of Aggression in Humans

GLOSSARY

agonistic behavior Behaviors shown when animals are in conflict. Common forms include threatening gestures and vocalizations, physical attack, and defensive reactions such as counterattack, submission, and escape.

androgen Any substance that promotes male characteristics. The androgen testosterone is the main hormonal product of the mammalian testis.

congenital adrenal hyperplasia A genetic disorder associated with an enzymatic defect such that, beginning before birth, the adrenal glands of affected individuals produce high levels of androgens.

pheromone Any externally voided substance that provokes or contributes to a specific physiologic and/or behavioral reaction when received, as by taste or smell, by another individual of the same species.

progestin A substance having progestational (like progesterone) activity. Some synthetic progestins also have androgenic properties; in some cases, treatment of pregnant women with androgenic progestins can masculinize the external genitalia and behavior of female offspring resulting from the pregnancy.

thelectomy Surgical removal of the nipples.

Aggressive behavior (aggression) is a physical attack against another animal, although the term is commonly used to include threatening gestures (Fig. 1) and vocalizations. As studied experimentally in nonhuman species, aggression has been categorized according to its style (offensive or defensive), apparent function (predatory or territorial), its protagonists (intermale or maternal), and the factors that seem to control the behavior (pain elicited or hormone dependent). Where aggression serves to gain access to a mate, defend a territory for mating, or protect offspring against predation by intruders, aggressive behaviors have an obvious importance for reproduction.

I. INTRODUCTION

In nonhuman animals, attack and other agonistic behaviors are stereotyped and vary according to species. For example, male mice are highly aggressive toward other males and vigorous fighting and wounding are common. Attacks are directed against an opponent's flanks and rear, and particularly combative males may engage in vigorous bouts of wrestling and biting (Fig. 2), each lasting as long as several seconds. A defeated male signals his submission by assuming an upright sitting posture facing his opponent, and in this posture he is rarely attacked by his opponent.

FIGURE 1 An adult male Western lowland gorilla shows the chest-beat display typical of agonistic threatening in male gorillas. Photograph by Ronald Nadler.

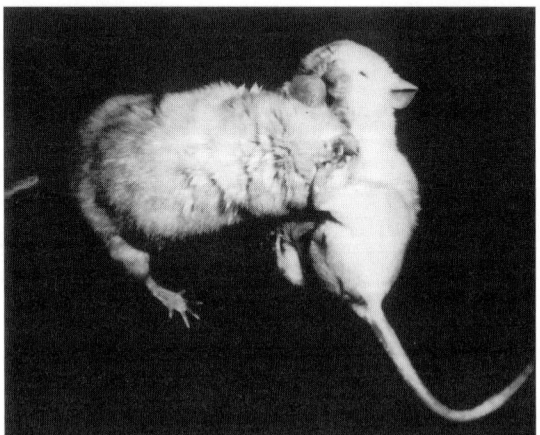

FIGURE 2 Two male mice of the Swiss–Webster strain in combat. Fresh wounds are apparent on the back and flank of the male on the left. The photograph was taken by James Gaddy and originally appeared in *Hormonal Correlates of Behavior, Volume 1* (B. E. Eleftheriou and R. L. Sprott, Eds.), Plenum, New York, 1975.

An attacking male rat, back arched and hair erect, will approach another male with his body parallel to that of his opponent to push at the target male with his hindleg. Lunging attacks are more direct, with the attacking male jumping directly at and either biting or pushing at the flank of his opponent.

Female mice and rats are generally peaceful with each other and with males, but as a pregnant female establishes her nest site during the last half of gestation she typically becomes more aggressive and will begin to lunge threateningly at male or female intruders. Aggressiveness peaks within 2 days after parturition and mothers defending young will direct fierce, rapid, biting attacks against an intruder. With the growth of the young, maternal aggression in mice and rats declines during the third week after parturition and is only infrequently elicited by the end of the week when the young are weaned.

It has not been possible to fruitfully generalize across types of aggression and species to infer common underlying mechanisms. One problem has to do with definition. The word aggression connotes categorically different forms of behavior which, by the very nature of things, are unlikely to have a common basis. Cross-species comparisons are made problematic because it is difficult to specify truly comparable contexts for comparing aggression in different species. Comparisons are also constrained by the fact that an extensive experimental literature exists for only very few species.

Aggression has been most systematically studied in rodents, particularly rats and mice. In these two species, aggressive behavior is influenced in important ways by gonadal hormones interacting with social experience, context, and sensory abilities. A few examples will illustrate some general principles.

II. GONADAL HORMONES AND INTERMALE AND MATERNAL AGGRESSION IN MICE AND RATS

From antiquity it has been generally appreciated that castration of domestic animals results in a gentling of responsiveness to conspecifics and humans. Rodent species, however, were among the first in which the connection between testicular hormones and intermale aggression was made experimentally explicit. In male mice and rats, an increase in aggressiveness coincides with the increase in androgen output by the testes during puberty. Castration before puberty prevents the development of intermale aggression, and administration of testosterone to castrated males increases intermale fighting. Similar results were obtained for other rodent species, and testicular androgens also appear to promote intermale aggression in ungulates and birds. The evidence for the same contribution in nonhuman primates is more inconsistent, perhaps owing to the relatively greater complexity of primate social interactions.

Maternal aggression in mice and rats resembles intermale aggression in form, but the hormones responsible for its expression are secreted by the ovary. The ovarian hormones estrogen and progesterone play an important role in the natural development of maternal care in mice and rats, and hormone injections that are effective in stimulating infant care also increase the aggressiveness of females toward intruders.

Neurons with receptors specific for different gonadal hormones are widely distributed in the central nervous system and gonadal hormones almost certainly act on some of these to increase aggressiveness. However, as of this writing, there is no certain knowledge of where gonadal hormones work to activate either intermale or maternal aggression in any species.

III. SITUATIONAL, EXPERIENTIAL, AND SENSORY DETERMINANTS OF AGGRESSION IN MICE AND RATS

Although gonadal hormones contribute to the arousal of both intermale and maternal aggression, each of these forms of aggression may be modified by situational, experiential, and sensory factors. The examples that follow give some sense of the variety of variables that can influence "hormone-dependent" aggression.

When male mice are housed individually they become more combative toward other males. Although testicular hormones are normally required for the display of intermale aggression, once individually housed males have begun to fight they will continue to fight for many weeks after castration. Although an aggressively naive castrated male will be nonaggressive toward a male of the same age, he will vigorously attack a much younger and smaller male. Opponent size also influences the aggression of female mice: Individually housed females usually behave amicably toward another adult but readily attack a much smaller male or female intruder.

Sense of smell is acutely developed in rodents and the identification of conspecifics as strange or familiar probably depends in large part on olfactory cues. In male mice testicular hormones stimulate the production of attack eliciting pheromones, which explains why males show higher levels of aggression against gonadally intact males than castrates. That females are usually not attacked by males probably owes to the feminine production of pheromones that inhibit aggression.

Deafferentation of the olfactory bulbs renders animals anosmic, profoundly decreasing intermale aggression. Olfactory bulbectomized mice are routinely used as opponents in aggression research with mice because they readily elicit aggressive responses but do not themselves react aggressively toward their attackers. Deafferentation of the olfactory bulbs of rat mothers or isolating mother rats from the olfactory cues of their pups virtually eliminate attacks against intruders, and the ability to detect contrasts between pup and intruder odors appears essential for the display of maternal aggression.

Ovarian hormones, presumably by acting on the central nervous system, promote maternal care and aggressive threatening. Ovarian hormones also work peripherally to cause the growth of the nipples required for pup attachment and suckling. Female mice whose nipples have been surgically removed before parturition show adequate maternal behaviors directed at foster young but do not attack intruders. This is so because the initial expression of the ferocious aggression characteristic of the early postpartum period depends on suckling stimulation provided by pups. Postpartum aggression in mice does not require the physical presence of the pups, but aggressiveness toward an intruder wanes within a few hours of separation. Aggressiveness is reinstated when the pups are returned and allowed to suckle. However, once maternal aggression is firmly established from the combination of ovarian hormonal stimulation and suckling, females will remain aggressive even after thelectomy—provided maternal behavior is maintained by continuous exposure to infant pups.

IV. ORGANIZING EFFECTS OF TESTICULAR HORMONES ON AGGRESSIVE BEHAVIOR

In mammalian sexual differentiation the presence of a single Y chromosome causes the indifferent gonads of a fetal male to develop as testes. Then, testicular androgens—testosterone and some of its metabolites—direct the differentiation of most of the body in the masculine direction. Giving testosterone to a female at the right time in development causes the differentiation of a penis and a completely male-like appearance. Conversely, if the testes are removed or the action of androgens is pharmacologically blocked at the right time of development, a male will develop a completely feminine exterior.

In their classic study having to do with sex differences in mating in guinea pigs, Phoenix *et al.* (1959) proposed that gonadal hormones could affect behavior in at least two fundamentally different ways. First, hormones could work to bring expression to or activate established patterns of behavior. Second, hormones could work developmentally to promote the development of or *organize* neural mechanisms for behavior whose operation would then be reflected in the behavior of the adult animal. The process of neurobehavioral differentiation was thought to parallel the hormone-induced differentiation of the body which had been so clearly demonstrated much earlier by experimental embryologists.

The idea that hormones could organize behavior was particularly important because it suggested that sex differences in behavior are due, at least in part, to the influence of testicular hormones acting during critical periods during development to masculinize neural and psychosexual development. There is now an extensive literature of experimentation on the subject, and it is clear that testicular androgens are a major force in the anatomical, neural, and behavioral masculinization of males in many, if not all, mammalian species. Ovarian secretions may assist in the process of feminization, but they are not necessary for the development of a female. Sex differences in the central nervous system that were predicted by the organization hypothesis have now been documented in a number of mammalian species, including mouse, rat, guinea pig, gerbil, and human. At least for rodent species, some of these have been proven to result from the perinatally occurring action of testicular hormones.

Around Day 13 of gestation, the testes of male mice and rats begin to secrete the androgens that cause the masculine differentiation of the Wolffian ducts and external genitalia as part of a process that continues through the early neonatal period. After birth the testes deliver testosterone into the bloodstream in a surge that peaks about 2 hr after birth. After this peak, serum testosterone falls so that by about 6 hr after birth its level is only slightly higher than that for perinatal females. About a month later the testes reawaken to produce increasing amounts of androgen as the male advances to puberty.

Male mice are highly aggressive, whereas females show little fighting even when injected with testosterone as adults. Research appearing in the late 1960s made it clear that this sex difference is established early in life as the result of stimulation by testicular androgens in the male and the absence of these secretions in the female. The conclusion is based on two facts. First, female mice given testosterone on the day of birth are just as aggressive as normal males when injected with testosterone as adults. Second, male mice castrated on the day of birth are less aggressive than males castrated 10 days after birth when both are given androgens as adults and tested for aggression. In a process that probably begins prenatally and appears to be completed around 6 days after birth, endogenous testicular androgens promote the development of an androgen-sensitive propensity for aggression.

In polytocous (multiple-birth) species, the positioning of fetuses during intrauterine development can be an important source of variation with respect to prenatal exposure to gonadal hormones. By Day 17 of gestation, male mouse fetuses have three times as much circulating testosterone as do females. The fetuses are packed in each uterine horn so that transport of hormones occurs across the fetal membranes by diffusion, and an individual fetus will receive supplements of testosterone from neighboring males. Thus, a female located between two males (a 2M female) will be exposed to more testosterone than a female situated between two females (0M female). As adults 2M females, while not as aggressive as males, are more aggressive than 0M females when the two are tested in pairs, show higher levels of postpartum aggression, and are more likely to fight against male and female intruders when administered testosterone during adulthood. This is an example of a phenomenon known as the intrauterine position effect which makes reference to the capacity of fetuses to have their development modified by exposure to steroids secreted by contiguous fetuses. With respect to aggression, the phenomenon clearly demonstrates that among normal females variation in several kinds of aggression is caused by differences in prenatal exposure to circulating testosterone.

Play fighting, sometimes called rough and tumble play, often resembles intraspecific aggression among adults but does not include serious biting or other behaviors that might be injurious to the participants. Juvenile males in many rodent, ungulate, pinniped, and primate species engage in play fighting more frequently than females. Play fighting in male rats, rhesus monkeys (Fig. 3), and probably most other

FIGURE 3 Two juvenile rhesus monkey males engaged in rough and tumble play. Photograph by Kim Wallen.

species does not require the activating influence of testicular hormones, and the sex difference in this activity clearly results from the organizing effect of testosterone or its metabolite dihydrotestosterone. In rats this organization occurs during the first 10 days after birth; in rhesus monkeys the process is completed prenatally.

V. THE HORMONAL ACTIVATION AND ORGANIZATION OF AGGRESSION IN HUMANS

While there have been many attempts to relate serum androgens and aggression in human males, there is as yet no convincing documentation of a direct link between serum testosterone and aggressive behavior. Frustration, fear, and grief are emotions that predispose to anger in humans, but there is little or no credible evidence of a direct connection between androgens and any of these emotions. To study hormones and aggression in humans it would help to be able to identify individuals particularly predisposed to act aggressively. However, a predisposition to behave aggressively may never be identified as such when it is channeled into socially accepted activities including, but not limited to, art, sport, and business. Moreover, in human societies there are both social and legal prohibitions against many forms of aggression, and even a strong propensity to do violence may be manifest only in the most extreme circumstances. The search for a hormonal basis for human aggression is made even more problematic by the fact that human violence may serve a range of motivations whose only connection is that actions stemming from them are hurtful to others.

Almost three decades have passed since the introduction of the idea that gonadal hormones acting during development could organize neural substrates for adult behavior. As applied to sex differentiation, the hypothesis predicts that male/female differences in fetal and/or neonatal exposure to testicular androgens will be reflected in sex differences in juvenile and adult behavior. The experimental literature is now sufficient to recognize this as a guiding principle in the development of sex differences in the behavior of nonhuman primates, rodents, and other vertebrates. A second principle logically follows: The same hormones that cause the development of between-sex differences in behavior also contribute to within-sex differences in the same behaviors. Behavioral studies of the intrauterine position effect provide an elegant illustration of this principle at work in rodents. In human studies, girls with congenital adrenal hyperplasia exposed to high levels of androgen prenatally and early in the postnatal period show a masculine-like preference for sex-typed toys that appears to reflect some degree of psychological masculinization. The recent discovery that individual differences in gendered behavior in adult women are weakly correlated with variation in the second-semester blood testosterone levels of their mothers provides additional support. Also, women born to mothers given pharmacological doses of synthetic progestins as treatment for at-risk pregnancy are, to some extent, behaviorally masculinized and this includes a propensity for rough and tumble preadolescent play and an enhanced tendency to imagine oneself responding aggressively in hypothetical conflict situations. In every known human society and at all ages studied, human males display more aggressive behavior than females. Within each sex, there is enormous variation in the predisposition to think and act aggressively. It is not yet known whether these behavioral differences are correlated with differences in the organization of the brain, but the prospect is easily imagined given recent discoveries relating human brain structure and sexual orientation, sexual identity, and sex differences in cognitive function. Whether sex differences in the human nervous system result from early hormone stimulation, differences in early experience, or some combination of the two remains to be determined.

See Also the Following Articles

ADRENAL HYPERPLASIA, CONGENITAL VIRILIZING; MATING BEHAVIORS, MAMMALS; PHEROMONES; PROGESTERONE ACTIONS ON BEHAVIOR; SEX DIFFERENTIATION, PSYCHOLOGICAL

Bibliography

Albert, D. J., Walsh, M. L., and Jonik, R. H. (1993). Aggression in humans: What is its biological foundation? *Neurosci. Biobehav. Rev.* 17, 405–425.

Beatty, W. W. (1984). Hormonal organization of sex differences in play fighting and spatial behavior. In *Progress in Brain Research* (G. J. De Vries *et al.*, Eds.), Vol. 61, pp. 315–329. Elsevier, Amsterdam.

Breedlove, S. M. (1992). Sexual dimorphism in the vertebrate nervous system. *J. Neurosci.* **12**, 4133–4142.

Breedlove, S. M. (1994). Sexual differentiation of the human nervous system. *Annu. Rev. Psychol.* **45**, 389–418.

Collaer, M. L., and Hines, M. (1995). Human behavioral sex differences—A role for gonadal hormones during early development? *Psychol. Bull.* **118**, 55–107.

Phoenix, C. H., Goy, R. W., Gerall, A. A., and Young, W. C. (1959). Organizing action of prenatally administered testosterone propionate on the tissues mediating mating behavior in the female guinea pig. *Endocrinology* **65**, 369–382).

Reinisch, J. M., Ziemba-Davis, M., and Sanders, S. A. (1991). Hormonal contributions to sexually dimorphic behavioral development in humans. *Psychoneuroendocrinology* **16**, 213–278.

Svare, B. B. (1990). Maternal aggression. In *Mammalian Parenting* (N. A. Krasnegor and R. S. Bridges, Eds.), pp. 118–132. Oxford Univ. Press, New York.

Udry, J. R. (1994). The nature of gender. *Demography* **31**, 561–573.

vom Saal, F. S. (1983). Models of early hormonal effects on intrasex aggression in mice. In *Hormones and Aggressive Behavior* (B. B. Svare, Ed.), pp. 197–222. Plenum, New York.

Aging and Reproduction

see Reproductive Senesence

Agnatha

Stacia A. Sower and Aubrey Gorbman
University of New Hampshire

I. Introduction
II. Hagfish
III. Lampreys

GLOSSARY

adenohypophysis The anterior lobe of the pituitary gland.
ammocoete A term used for larval lampreys.
ampulla A general term to designate a flask-like dilation of a tubular structure.
anadromy A type of life cycle in which fish migrate from fresh water to seawater and then return to fresh water to spawn.
atresia Degeneration of egg follicles.
gastrulation The process by which the young embryo acquires its three germ layers.
gonadotropin A hormone that is released from the pituitary and travels by general circulation to the gonads to stimulate gametogenesis and steroidogenesis.
gonadotropin-releasing hormone A neurohormone produced by the hypothalamus that stimulates the pituitary gland.
hypothalamus The portion of the diencephalon in the brain

which forms the floor and part of the lateral wall of the third ventricle. It is composed of various nuclei that synthesize and secrete neurohormones.
lobule A small lobe.
metamorphosis A transition from one developmental stage to another, such as that from larval lampreys to young adult lampreys.
oocyte A developing egg in one of two stages.
progyny A form of sex differentiation in which all developing animals are at first female.
steroid hormone A lipid hormone.
vitellogenesis Production of yolk.

I. INTRODUCTION

The relatedness of the hagfish and lampreys, the two most primitive vertebrate taxa, can be judged from certain key anatomical features that they share, but they differ profoundly in other ways. Both groups lack jaws, paired fins, a third semicircular canal, and neural arches over the spinal cord. Both have a single median gonad and a distinctively thin, flat adenohypophysis that is broken up into lobules. They differ in the basic structure of the intestine, in the germ layer origins of the olfactory organ and adenohypophysis, and in the numbers of eggs that they produce. In terms of their relatedness to higher vertebrate taxa, the lampreys share features, or resemble, these taxa much more than do the modern hagfishes.

Hagfishes mostly inhabit a deep marine environment that is relatively free of circadian, or even seasonal changes. Lampreys, on the other hand, either are totally freshwater dwellers or they at least reproduce in fresh water after periods in the sea, migrating between the two. How much the modern hagfishes and lampreys represent or reflect the ancient forms from which they evolved 500 million years ago is a question that still concerns evolutionary biologists and biochemists.

II. HAGFISH

Reproductive patterns in the two agnathan groups, as far as they are known, are clearly divergent. Lamprey reproduction is a visible and thoroughly described event. It follows ritual nest building and mating behavioral phenomena that can be witnessed in freshwater streams and lakes. Because of their inaccessibility, hagfish reproduction and early development have escaped observation and can only be inferred. Lampreys die after fecundation of their eggs. It is possible that hagfish also have only a single reproductive cycle but, again, this feature of their reproduction is not known.

A. The Hagfish Gonad

The gonads of both hagfish sexes are single median structures suspended from the ventral side of the gut, which in turn is suspended from the dorsal body wall (Fig. 1). The gonadal elements are contained in a membrane that extends the full length of the body cavity. The testis occupies the extreme posterior end of the membrane, and it is relatively small. The ovary comprises most of the length of the membrane. In rare bisexual animals both male and female components occupy the same membrane, the ovary anterior and the testis at the extreme posterior (Fig. 1).

1. Sex Differentiation

All posthatching smaller hagfish are female. In the most thoroughly studied species, *Eptatretus stouti*, the gonad contains only small nonvitellogenic oocytes until the animal attains a length of about 20 cm. Thus, sex differentiation is progynous. In male differentiation, in animals larger than 20 cm, the oocytes degenerate and the posterior end of the gonadal membrane develops spermatic follicles (Fig. 1). In female differentiation, the posterior part of the gonad involutes and the long anterior section develops vitellogenic oocytes and, finally, mature eggs.

It is of interest that many populations of hagfish have been reported to consist of variable ratios of males to females, in which the females are more numerous. Thus, it is possible that environmental factors influence sex differentiation.

i. The Ovary The germinative tissue of the adult ovary is restricted to the free edge of the membranous gonad. Numerous small oocytes move upward in the membrane as they grow. It is uncertain whether these growing rounded oocytes actually move or whether their apparent movement is due to growth in depth of the membrane. As the oocytes grow they gradually assume an oval shape and finally a spindle shape. In *E. stouti*, when the oocytes reach a length of about 4.5 mm, many become atretic, and only about 25–30 continue growing as active vitellogenesis begins. It would appear that even after selection of 25–30 eggs for the completion of maturation has begun, production of smaller oocytes continues, but all of these are destined for atresia.

When the selected eggs have reached a length of more than 25 mm their follicular epithelium (granulosa) thickens and begins to secrete a shell, which, of course, precludes further growth of the egg. The follicular epithelium at either end of the egg become thrown into a complexly folded pattern in order to secrete the two crowns of anchor-shaped hooks. A hole in the shell remains (the micropyle) to allow sperm entrance for fertilization. The remarkably complex morphogenetic properties of the hagfish granulosal epithelium deserve more study.

ii. The Testis The gonadal membrane in the posterior few abdominal body segments is much shorter than it is in the ovarian area (Fig. 1). One lateral surface of the membrane in the testicular area becomes thickened and folded and this epithelial area is the source of the spermatic follicles that comprise the testis. Since the testis is so small compared to the ovary, previous authors have pointed out that the volume of spermatozoa produced must be too small to permit a system of external fertilization in the open seawater. Fertilization must be in an enclosed space to achieve a high enough concentration of spermatozoa to ensure success. This is especially so because of limited access to the egg through the micropyle and the usually small number (<30) of eggs.

B. Secondary Sexual Characters

Hagfish have no gonoducts and they present no externally differentiated sexual features. Tsuneki has

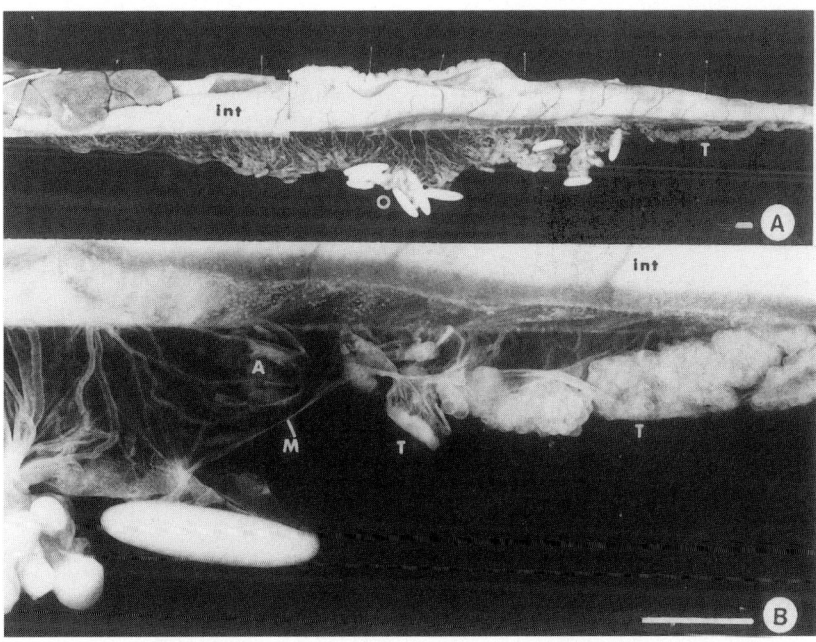

FIGURE 1 Gonad of an adult hermaphroditic hagfish, *Eptatretus stouti*, with the sexual elements in their normal positions, a rare condition. (A) General view. The testis (T) occupies the posterior end of the gonadal membrane. The ovary (O) contains large vitellogenic eggs which normally occupy the full remaining length of the gonadal membrane below the intestine (int). In this specimen the large eggs are normal in shape but reduced in number. (B) A higher magnified view of the junction area between ovary and testis. There is no overlap. The mesovarium (M) contains an atretic egg (A) dorsal to the large egg. There are many more atretic eggs in the mesovarium anterior to the large eggs shown in A. Scale bars at lower right of each photo are equivalent to 5 mm of actual size (reproduced with permission from Gorbman, 1997).

pointed out that in male *E. burgeri* a mucus gland near the cloacal pore is larger in males than in females. It has also been claimed that the thread cells in the slime glands of the skin could be considered secondary sex features because the threads that they form (in both sexes) may be used to enclose and protect the developing eggs on the sea floor.

C. Hypothalamus–Pituitary Axis

1. Neurohypophysis and Adenohypophysis

The hagfish neurohypophysis is a flattened sac-like structure that contains nerve endings of neurons that originate in centers of the hypothalamus. These neurons have been shown to contain two forms of immunoreactive gonadotropin-releasing hormone (GnRH). The adenohypophysis is a thin layer of follicular groups of cells that is coextensive with the neurohypophysis (Fig. 2). This arrangement of the hypothalamo–hypophysial structures resembles that of lampreys, even in lacking any direct nervous or vascular connection between the two. It is a structure that would seem to favor regulation of adenohypophysis secretion by neurohormones diffusing from the neurohypophysis. Unfortunately, the adenohypophysis of hagfish appears to be minimally, if at all, functional. Its cells contain very few granules and attempts to show whether the usual pituitary tropic hormones can be extracted or immunochemically strained have given equivocal results. Furthermore, surgical hypophysectomy has no observable effect on gonadal structure. It is as though the system is there anatomically but in a nonfunctional state.

This may reflect the fact that the deep sea habitat is so lacking in photic or temperature seasonal cyclic clues that there is no longer any need for a functional hypothalamo–hypophysial system.

D. Reproductive Hormones

Just as there seems to be no or poor regulation of hagfish reproduction by hypothalamo–hypophysial peptides, the presence and possible function of sex steroid hormones has poor or inconsistent experimental support. The obviously yolky large eggs of hagfish would seem to require active synthesis of vitelloproteins by the liver. In other vertebrates hepatic vitelloprotein synthesis is stimulated by estrogen. In female hagfish plasma levels of estrogen are low regardless of whether or not their ovaries are vitellogenic. On the other hand, estrogen injection into females does induce a low level of synthesis of protein (including vitelloprotein) in the liver. Furthermore, studies of nuclear binding of estrogen by hepatic cell nuclei have demonstrated the presence of high-affinity, low-capacity binding sites. Such binding sites are absent or diminished in nonvitellogenic females.

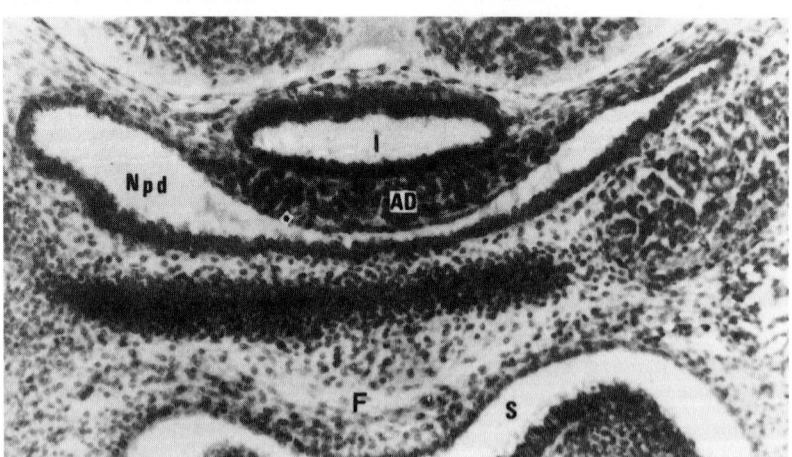

FIGURE 2 Cross section of the head region of an older hagfish embryo. At this level the infundibulum of the diencephalon (I) is cut off from the brain and has an oval shape. Just beneath it is the adenohypophysis (AD), a flattened mass of follicles between the infundibulum and the broad nasopharyngeal cavity (Npd). Separating the Npd from the stomodeum (S) below is a broad shelf of tissue (F) containing skeletal elements. Note that the dorsal epithelium of Npd (*) is very thin in the area where the AD cells have budded off (reproduced with permission from Gorbman, 1997).

III. LAMPREYS

A. Gonad

There are approximately 40 species of lampreys that are classified as parasitic or nonparasitic. Lampreys spawn only once in their lifetime, after which they die. The parasitic lampreys are generally anadromous. All larval lampreys, called ammocoetes, live in fresh water as burrowing organisms in the bottoms of streams or lakes. In the parasitic sea lamprey, sexual maturation is a seasonal, synchronized process. The sea lampreys begin their lives as freshwater ammocoetes, which are blind, filter-feeding larvae. After approximately 5–7 years in freshwater streams, metamorphosis occurs and the ammocoetes become free-swimming, sexually immature lampreys which migrate to the sea or lakes. During the approximately 15-month-long parasitic sea phase, gametogenesis progresses. In males, spermatogonia proliferate and develop into primary and secondary spermatocytes, and in females vitellogenesis occurs. After approximately 15 months at sea, lampreys return to freshwater streams and undergo the final maturational processes resulting in mature eggs and sperm and finally spawning. In nonparasitic species, gonadal differentiation and metamorphosis occur together. Shortly after the end of the maturation of parasitic lamprey species, spawning occurs and the lampreys die.

The gonad in both sexes is unpaired and median and is suspended from the dorsal wall of the body cavity by means of a mesentery containing connective tissue. Lampreys are among the few vertebrates, including teleost fish, that have no intraperitoneal genital ducts. Hardisty has summarized the most important stages in the development of the larval gonads (Fig. 3) as follows:

1. The primordial germ cells appear during gastrulation and then migrate to the gonadal site below the aorta.
2. A median genital ridge develops extending caudally from the pronephric region. At this stage, the gonadal rudiment contains a relatively small number of germ cells and is covered by a peritoneal epithelium.
3. After hatching, for periods varying from 6 months to more than 2 years, the undifferentiated

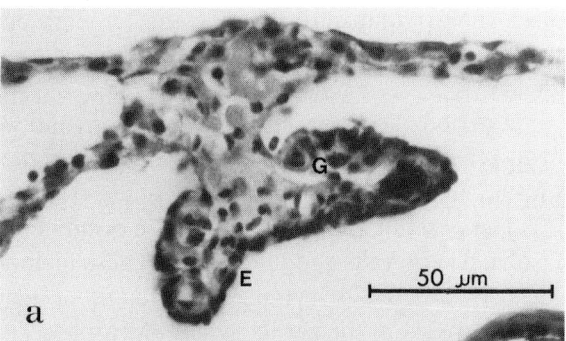
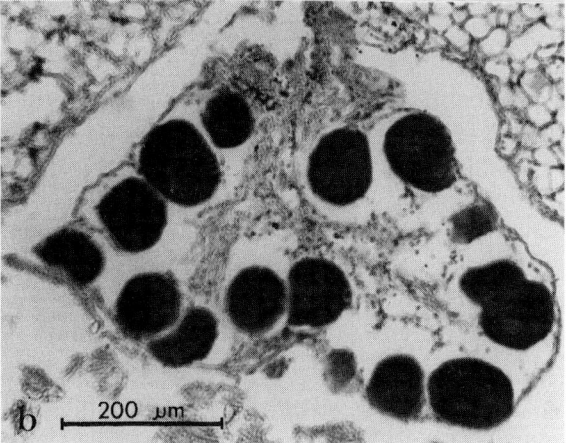
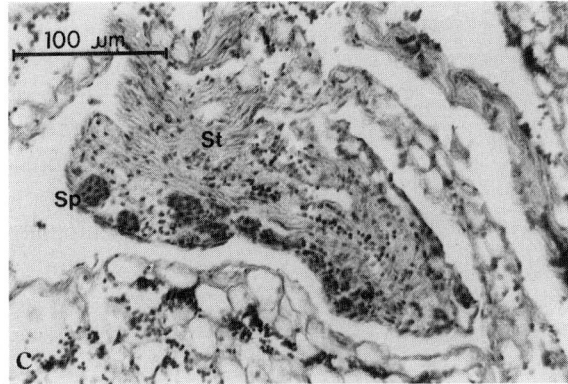

FIGURE 3 (a) Undifferentiated gonad of a 51-mm least brook lamprey larva containing germ cells (G) enclosed by a peritoneal epithelium (E). (b) Differentiated ovary from a 118-mm larva. (c) Testis from a 104-mm larva, illustrating small but well-defined cysts of presumptive spermatogonia (Sp) and extensive stromal tissue (St) (reproduced from Docker and Beamish, 1994, with kind permission from Kluwer Academic Publishers).

gonad shows comparatively little further development, and throughout this stage the germ cells divide slowly.

4. A period of more rapid mitotic division follows, during which the germ cells tend to remain together to form cell nests or cysts.

5. The onset of the meiotic prophase occurs both in isolated germ cells and in the cysts. These meiotic cells may proceed to cytoplasmic growth or may undergo atresia in the earlier stages of prophase.

6. Oogenesis occurs to a variable extent in the great majority, if not all, of ammocoete gonads. However, in the future ovaries it tends to be more synchronous, leading eventually to a gonad containing no germ cells other than oocytes in the cytoplasmic growth phase.

7. In the differentiation of the male gonads, a large proportion of the germ cells undergo degeneration in the earlier stages of the meiotic prophase. Oocytes that survive to the cytoplasmic growth phase are eventually eliminated by atresia. Because of this extensive degeneration of germ cells, the testis is reduced in size and contains only small numbers of residual germ cells. From these undifferentiated elements, nests of primary spermatogonia are developed by renewed mitotic divisions either shortly before or during metamorphosis.

Following these stages, either during or after metamorphosis, gametogenesis continues. In the male parasitic lampreys returning to the rivers in early autumn or late spring, only primary spermatogonia are present in the testes. The testes grow by mitotic division of spermatogonia. At the end of this growth phase, spermatogonia are transformed into spermatocytes. This stage appears to last for several months in river lampreys and for a few weeks in sea lampreys (Fig. 4). The final stages of spermatogenesis occur during the last few weeks before spawning and the testicular ampullae become filled with mature spermatozoa (Fig. 5). Prior to spawning the ampullae break down and the spermatozoa are liberated into the body cavity.

During the final maturational stages in the females, there is an intense rapid gonadal growth period a few weeks before spawning. As sea lamprey enter fresh water, the oocytes are ready for spawning, with most of the cytoplasm filled with yolk platelets. The

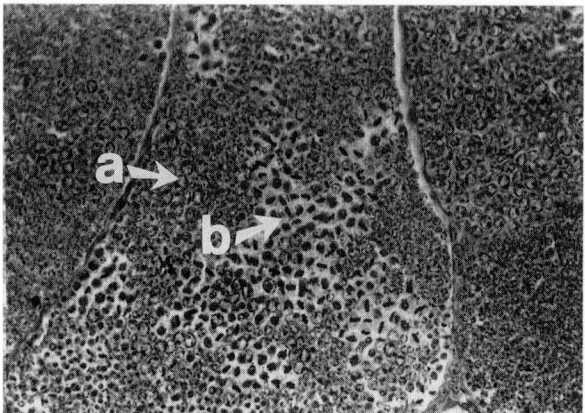

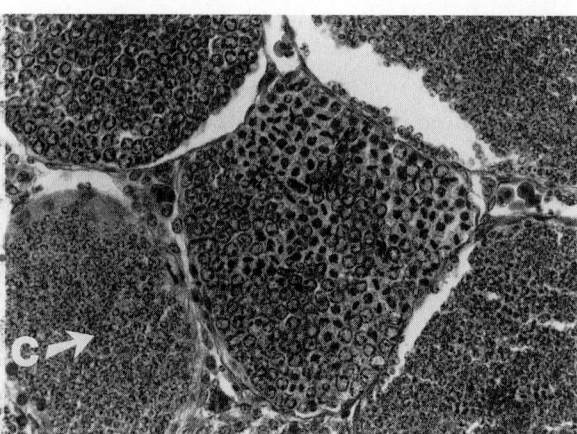

FIGURE 4 Testis of an adult sea lamprey. (Top) a, primary spermatocytes; b, dividing primary spermatocytes. (Bottom) c, spermatids. Magnification, ×400 (reproduced with permission from Fahien and Sower, 1990).

germinal vesicle of the egg is located peripherally, the vitelline membrane is double, and the theca has begun to separate from the vitelline membrane. At ovulation, the thecal follicle cells have separated from the oocytes. Some time before ovulation actually occurs, the granulosa cells are distended with secretory materials (gelatinous material) and show a maximum degree of separation from one another. This gelatinous material, containing granules of acid mucopolysaccharides, is not liberated until ovulation, when the granulosa cells disintegrate leaving remnants attached to the basal pole of the oocyte, forming the adhesive layer which causes the eggs to stick to the substrate when they are finally shed.

The numbers of eggs produced by each lamprey vary with species. The least number of eggs are produced by *Mordacia praecox*, which matures about

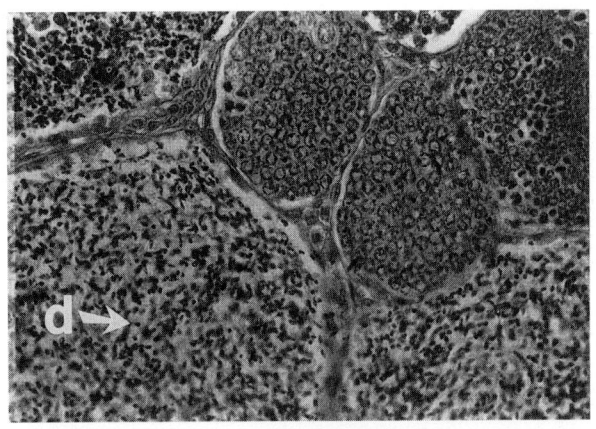

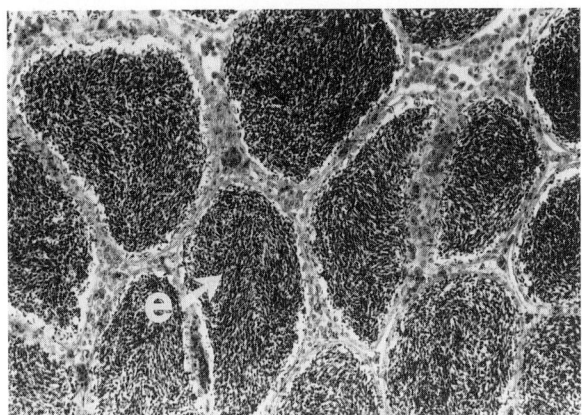

FIGURE 5 Testis of an adult sea lamprey. (Top) d, immature sperm (magnification, ×400). (Bottom) e, mature sperm (magnification, ×200) (reproduced with permission from Fahien and Sower, 1990).

474 eggs at the time of spawning. The largest amount of eggs, estimated to be from 124,000 to 260,000, are produced by the largest lamprey species, *Petromyzon marinus*.

B. Secondary Sexual Characters

In lampreys, both sexes develop secondary sexual characters during the final weeks of reproduction and spawning activity. In both sexes, the fins enlarge at the time of prespawning sexual maturity and the two dorsal fins become continuous. In addition, there are swollen cloacal labia and curvature of the tail, and an erectile "penis" develops in the cloacal region of the male. The tail of the female curves sharply upwards at sexual maturity and in the male there is a slight downward curvature.

C. Hypothalamus–Pituitary Axis

A key neuroendocrine function of the hypothalamus in the control of reproduction is the timed release of the decapeptide, GnRH, which acts on the pituitary to regulate the pituitary–gonadal axis for all vertebrates. Gonadotropins, secreted in response to GnRH, are released from the pituitary gland and are the major hormones influencing steroidogenesis and gametogenesis. Two molecular forms of GnRH have been identified in the sea lamprey, lamprey GnRH-I and lamprey GnRH-III. Gonadotropins have not been identified from either the lampreys or hagfish. Lampreys are the most primitive vertebrates for which there are demonstrated functional roles for multiple GnRH neurohormones that are involved in pituitary reproductive activity. Both lamprey GnRH-I and -III have been shown to induce steroidogenesis and spermiation and/or ovulation in adult sea lampreys. In lampreys undergoing metamorphosis, there is an increase of brain lamprey GnRH-I and -III which coincides with the acceleration of gonadal maturation. In immunocytochemical studies, both lamprey GnRH-I and -III immunoreactions can be found in the cell bodies in the rostral hypothalamus and preoptic area in larval and adult sea lamprey. It is suggested that in the larval stage, most of the irGnRH is lamprey GnRH-III. Thus, GnRH-III may be the more active form during gonadal maturation. Such information suggests that the structure and function of the GnRHs in vertebrates are highly conserved throughout vertebrate evolution.

D. Reproductive Hormones

The physiological roles of gonadal sex steroids and the identity of other potentially important sex hormones need to be clarified in the lamprey. Plasma estradiol and progesterone have been measured as indicators of gonadal activity in the sea lampreys in various physiological studies. Estradiol and progesterone are two steroids that have been demonstrated to be associated with reproductive activity in both female and male lampreys. During the 2 or 3 months prior to ovulation, plasma levels of estradiol fluctuate and then decrease at ovulation in female sea lampreys and the Japanese river lamprey, *Lampetra japonica*. During the 2 or 3 months prior to spermiation, primary and secondary spermatocytes develop into mature sperm. During this time, plasma levels of estradiol fluctuate and increase significantly at spermiation in the sea lamprey and the Japanese river lamprey. In addition, the presence of an estrogen receptor has been demonstrated in the testis of the adult sea lamprey. This supports the notion that estrogen is physiologically significant in testicular function. Testosterone has not been demonstrated to be associated with reproduction in adult lampreys, i.e., the levels are either very low and show no change with spermiation or the levels are undetectable. In an earlier study, plasma estradiol but not testosterone was found elevated in response to administered mammalian GnRH analog in male and female sea lampreys. These studies and the demonstrated absence of androgen receptors in the lamprey testis suggest that testosterone may not have a role during the final spermatogenic phases in adult male lampreys. However, 15-hydroxylated compounds are produced in the gonads of river and sea lampreys. Based on effects of partial hypophysectomy and gonadectomy on plasma levels of various steroids in river lampreys, it has been suggested that the true sex hormones responsible for development of secondary sex characteristics may be 15-hydroxylated derivatives of estradiol and testosterone. However, both the ovary in brook lamprey and testis in sea lamprey have been demonstrated to be capable of synthesizing estradiol. In addition, estradiol and progesterone have been extracted from the ovary. Estradiol from these studies appears to be associated with reproductive activity; however, the nature of sex steroids in lampreys remains unresolved.

See Also the Following Article

FISH, MODES OF REPRODUCTION

Bibliography

Brodal, A., and Fange, R. (1963). *The Biology of Myxine*. Universiteltsforlaget, Oslo.

Docker, M. F., and Beamish, F. W. H. (1994). Age, growth and sex ratio among populations of larval least brook lamprey, *Lampetra aepyptera*: An argument for environmental sex determination. *Environ. Biol. Fish.* **41**, 191–205.

Dodd, J. M., and Dodd, M. H. I. (1985). Evolutionary aspects of reproduction in cyclostomes and cartilaginous fishes. In *Evolutionary Biology of Primitive Fishes* (R. E. Foreman, A. Gorbman, J. M. Dodd, and R. Olsson, Eds.), pp. 295–319. Plenum Press, New York.

Fahien, C. M., and Sower, S. A. (1990). Relationship between brain gonadotropin-releasing hormone and final reproductive period of the adult male sea lamprey, *Petromyzon marinus*. *Gen. Comp. Endocrinol.* **80**, 427–437.

Fernholm, B. (1975). Ovulation and eggs of hagfish *Eptatretus burgeri*. *Acta Zool.* **56**, 199–204.

Gorbman, A. (1983). Reproduction in cyclostomes and its regulation. In *Fish Physiology* (W. S. Hoar, D. J. Randall, and E. M. Donaldson, Eds.), Vol. IX, pp. 1–30. Academic Press, New York.

Gorbman, A. (1990). Sex differentiation in the hagfish *Eptatretus stouti*. *Gen. Comp. Endocrinol.* **77**, 309–323.

Gorbman, A. (1997). Hagfish development. *Zool. Sci.* **14**, 375–390.

Hardisty, M. W. (1971). Gonadogenesis, sex-differentiation and gametogenesis. In *Biology of Lampreys* (M. W. Hardisty and I. C. Potter, Eds.), Vol. 11, pp. 295–360. Academic Press, New York.

Hardisty, M. W. (1979). *Biology of Cyclostomes*. Chapman & Hall, London.

Jesperson, A. (1975). Fine structure of spermiogenesis in Eastern Pacific species of hagfish (Myxinidae). *Acta Zool.* **56**, 189–198.

Koch, E. A., Spitzer, R. H., Pithawalla, R. B., Castillas, F. A., and Wilson, L. J. (1993). The hagfish oocyte at late stages of oogenesis: Structural and metabolic events at the micropylar region. *Tissue Cell* **25**, 259–273.

Lewis, J. C., and McMillan, D. B. (1965). The development of the ovary of the sea lamprey (*Petromyzon marinus* L.). *J. Morphol.* **117**, 425–466.

Sower, S. A. (1990a). Gonadotropin-releasing hormone in primitive fishes. *Prog. Clin. Biol. Res.* **342**, 73–78.

Sower, S. A. (1990b). Neuroendocrine control of reproduction in lampreys. *Fish Physiol. Biochem.* **8**, 365–374.

Sower, S. A. (1998). Brain and pituitary hormones of lampreys, recent findings and their evolutionary significance. *Am. Zool.*, **38**, 15–38.

Von Kupffer, C. (1899). Zur Kopfentwicklung von Bdellostoma. *Sitzungsber. Gesellsch. Morph. Physiol. Munchen* **15**, 21–35.

Allantochorion (Chorioallantois)

Fuller W. Bazer
Texas A&M University

I. Introduction
II. Specialized Structures of the Chorioallantois
III. Hormones of the Chorion
IV. Immunological Protection of the Fetal–Placental Semiallograft
V. Human Chorioamnion

GLOSSARY

allantoic membrane A membrane that develops as an evagination of the hindgut to form a sac filled with water, urine, and nutrients.

chorion The trophoblast and mesoderm derived from the somatopleure to form the outermost membrane of the placenta.

conceptus The embryo/fetus and its associated placental membranes.

placenta The amnion, allantois, and chorion which are membranes that form the unit for exchange of nutrients and gases between the mother and fetus.

trophoblast A layer composed of trophectoderm derived from ectoderm and extraembryonic endoderm that is the outermost placental membrane.

umbilicus The stalk connecting the fetus to the placenta; contains major blood vessels and the urachus, which connects the fetal bladder and allantoic sac.

urachus A duct that transports urine (water and nutrients cleared by the fetal kidney into the bladder) to the allantoic sac.

Allantochorion refers to the fused allantoic and chorionic membranes which form a functional vascularized placenta in most species of mammals. The term trophoblast is derived from a Greek root word referring to feeding or nourishing layer. The trophoblast is composed of trophectoderm derived from ectoderm and extraembryonic endoderm that is the outermost placental membrane.

I. INTRODUCTION

The chorion fuses with the mesoderm of the yolk sac (splanchnopleure) to form a choriovitelline placenta which is short-lived. The allantoic membrane subsequently forms as an evagination of the hindgut (endoderm) as a diverticulum covered by mesoderm (splanchnopleure). The allantoic membrane dilates and extends outside the embryo/fetus except for attachment by the umbilical stalk which contains the urachus for communication between the allantoic sac and fetal bladder. The allantoic membrane acquires an extensive blood supply from caudal branches of the aorta which break up into an extensive vascular net. The allantois and chorion fuse early in gestation so that the vascular system of the allantoic membrane supplies the entire chorioallantoic placenta (Fig. 1).

The chorion eventually fuses with the allantois (splanchnopleure) to form the allantochorion or chorioallantoic placenta that is functional for the remainder of gestation. General functions of the allantochorion are to serve as the interface for absorption of nutrients from maternal uterine secretions, exchange micronutrients (e.g., amino acids, water, electrolytes, and glucose) and gases from the maternal system, secrete both steroid and protein hormones, secrete growth factors necessary for growth and development of the uterus and fetal–placental units, secrete extracellular matrix molecules that allow placental attachment to the uterine wall, and participate in protection of the conceptus from rejection by the maternal immune system.

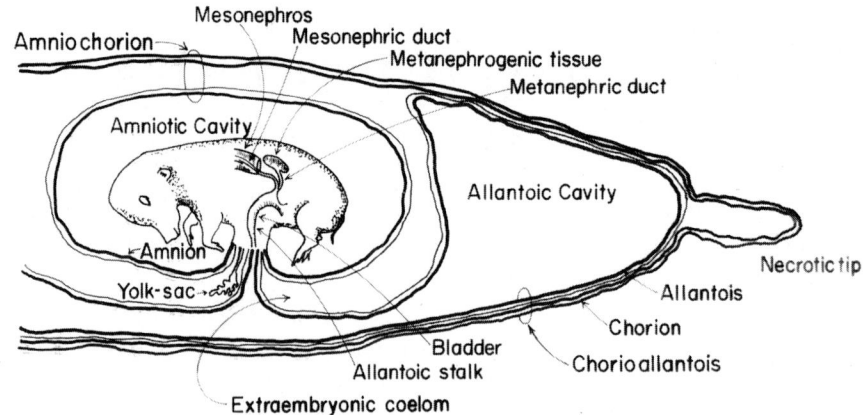

FIGURE 1 Diagram of the chorioallantoic placenta of the pig with various membranes labeled.

II. SPECIALIZED STRUCTURES OF THE CHORIOALLANTOIS

The chorion may differentiate into specialized structures that include the following:

1. Chorionic villi which interdigitate with corresponding villi on endometrial epithelium to increase surface area for exchange of nutrients and gases between the maternal and fetal–placental tissues.

2. Areolae that are specialized areas of tall columnar epithelial cells that form only where the chorion is in direct apposition to the opening of the "mouth" of a uterine gland. The areolae, which number 2500–3000 per placenta in pigs, absorb secretions of the uterine glands and transport them into the fetal circulation by fluid phase pinocytosis.

3. Cotyledons are complex rosettes of villi that form on the chorion that is in direct contact with the endometrial caruncles. These number 70–90 in sheep and around 120 in cattle. They serve to greatly increase surface area for exchange of micronutrients (sugars and amino acids) and gases between the maternal and fetal circulations. The unit formed by interdigitation of a cotyledon and caruncle is the placentome. Approximately 85% of maternal uterine blood flow is directed through the placentome and about 90% of placental blood flow of the fetus is directed to the placentomes.

4. Hemophagus organs/zones are specialized areas into which maternal blood is extravasated and erythrocytes phagocytized to obtain iron for delivery to the fetus.

5. Petrifactions are areas of the chorioallantois into which significant amounts of calcium have been deposited.

6. Hippomanes is the name given to circular to oval-shaped light to dark brown masses of cellular debris and uterine secretions which become lodged in the allantochorion or float free in the allantoic fluid of horse, pig, cow, and sheep. The hippomanes can form evaginations toward the allantoic sac to form peduncles which can eventually pinch off to float free in the allantoic fluid. The hippomanes appear to be concentrated granular material full of cell debris, globules of fat, degenerating erythrocytes, and uterine secretions. Their function, if any, is unknown.

III. HORMONES OF THE CHORION

The chorion secretes steroid hormones (estrogen and/or progesterone), protein hormones (placental lactogen and equine chorionic gonadotropin), growth factors (fibroblast growth factor, transforming growth factor betas, etc.). The chorioallantois also expresses many extracellular matrix molecules associated with implantation.

IV. IMMUNOLOGICAL PROTECTION OF THE FETAL–PLACENTAL SEMIALLOGRAFT

The conceptus is a semiallograft that one would expect to be rejected by the maternal system. However, the chorion appears to prevent this by expressing histocompatibility antigens that are monomorphic, molecules such as Fas ligand that prevent immune cell-mediated killing of the placenta, and/or secretion of high-molecular-weight molecules that may provide a barrier to immune cells. There may be additional "immunoprotective" mechanisms elicited by the chorioallantois to protect the conceptus from the maternal immune system to allow these two systems to coexist.

V. HUMAN CHORIOAMNION

In humans and other primates the allantois does not form; therefore, the placental type is an amniochorion. The chorion is then bathed directly in maternal blood in the uterine decidua for absorption of nutrients and gases to support fetal–placental development. These species have hemochorial-, endotheliochorial-, or hemoendothelial-type placenta. In species with these types of placenta, the chorion performs many of the same roles as in species with epitheliochorial placentae. Also, the amniotic fluid is the supporting fluid for the fetus to allow symmetrical development and to prevent fetal tissues from adhering to fetal membranes.

See Also the Following Articles

ALLANTOIC FLUID; ALLANTOIS

Bibliography

Perry, J. S. (1981). The mammalian fetal membranes. *J. Reprod. Fertil.* **62**, 321–335.

Steven, D. H. (Ed.). *Comparative Placentation*. Academic Press, New York.

Allantoic Fluid

Fuller W. Bazer
Texas A&M University

I. Introduction
II. Functions of Allantoic Fluid
III. Abnormal Allantoic Fluid Accumulation

GLOSSARY

allantoic membrane A membrane that develops as an evagination of the hindgut to form a sac filled with water, urine, and nutrients.

conceptus The embryo/fetus and its associated placental membranes.

placenta The amnion, allantois, and chorion which are membranes that form the unit for exchange of nutrients and gases between the mother and fetus.

umbilicus The stalk connecting the fetus to the placenta; contains major blood vessels and the urachus which connects the fetal bladder and allantoic sac.

urachus A duct that transports urine (water and nutrients cleared by the fetal kidney into the bladder) to the allantoic sac.

Allantoic fluid is a major fluid pool in the conceptus that serves to expand the chorioallantois and as a site for storage of soluble nutrients. The allantoic fluid derives from the transport of water from the maternal system to the conceptus.

I. INTRODUCTION

The allantoic membrane forms as an evagination of the hindgut (endoderm) as a diverticulum covered by mesoderm (splanchnopleure) that dilates and extends outside the embryo/fetus except for attachment by the umbilical stalk. The umbilical stalk contains the urachus for transport of urine from the bladder and for accumulation of water and nutrients from the maternal system. The driving force for expansion of the allantoic sac and formation of a chorioallantoic placenta is the accumulation of water which occurs rapidly in early gestation. For example, allantoic fluid volume increases from 1 ml on Day 18 to about 250 ml on Day 30 of gestation, then decreases to about 50 ml by Day 45 before increasing again to about 500 ml on Day 55 of pregnancy in pigs (Fig. 1). After Days 55–60 of gestation, allantoic fluid volume in pigs decreases to term (115 days).

The early literature implies that allantoic fluid derives from the mesonephros; however, the kidney redistributes water but does not produce water. Recently, it was demonstrated that the allantoic fluid derives from the maternal system and that it is actively transported across the chorioallantois and into the fetal circulation and/or directly into the allantoic sac.

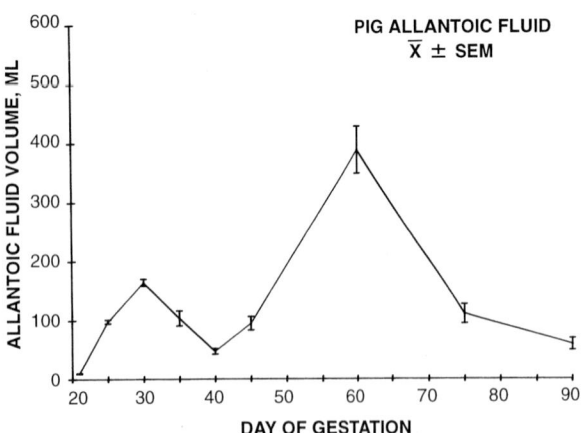

FIGURE 1 Changes in allantoic fluid volume in the fetal pig during gestation.

II. FUNCTIONS OF ALLANTOIC FLUID

Allantoic fluid serves a number of functions critical to pregnancy. First, it serves to expand the chorioallantoic placenta to ensure that it establishes adequate placental surface area to support development of the embryo/fetus. Second, it is rich in electrolytes, sugars, proteins, etc. and, therefore, serves as a nutrient reservoir for the developing embryo/fetus. Allantoic fluid contains high levels of glucose; however, fructose is the primary hexose sugar in species such as ruminants, pigs, and marine mammals, in which glycogen is not stored in the placenta. The allantoic fluid is also enriched in amino acids which can give rise to nitric oxide (arginine) and give rise to polyamines (citrulline) essential for growth and development of tissues. Water is a key nutrient for fetal–placental growth and development since every gram of tissue deposited contains 0.9–0.95 g of water. Vitamins, minerals, growth factors, protein hormones, and steroids are other molecules present in allantoic fluid.

The rate of accumulation of water and other nutrients into the allantoic sac is affected by active transport processes involving sodium and potassium ions. This process can be stimulated by proteins hormones from both the posterior pituitary gland (e.g., arginine vasopressin) and anterior pituitary gland (e.g., prolactin), placenta (e.g., placental lactogens), and sex steroids from the ovaries and placenta. There is also evidence that adrenal corticosteroids can influence the volume of allantoic fluid and its rate of accumulation.

III. ABNORMAL ALLANTOIC FLUID ACCUMULATION

Hydroallantois is a condition characterized by excessively large volumes of allantoic fluid which have especially high concentrations of sodium but low concentrations of potassium. In cattle, allantoic fluid has been reported to weigh about 200 kg in fetuses with hydroallantois compared to 80 kg in normal fetuses. The cause of this condition is not known, but researchers suspect that hormonal imbalances, placental defects, and/or genetic defects may be responsible.

See Also the Following Articles

ALLANTOCHORION; ALLANTOIS

Bibliography

Bazer, F. W. (1989). Allantoic fluid: Regulation of volume and composition. In *Fetal and Neonatal Body Fluids: The Scientific Basis for Clinical Practice* (R. A. Brace, M. G. Ross, and J. E. Robillard, Eds.), pp. 135–154. Perinatology, Ithaca, NY.

Perry, J. S. (1981). The mammalian fetal membranes. *J. Reprod. Fertil.* **62**, 321–335.

Steven, D. H. (Ed.). *Comparative Placentation.* Academic Press, New York.

Wintour, E. M. (1989). Maternal influences on fetal fluid and electrolyte balance. In *Fetal and Neonatal Body Fluids: The Scientific Basis for Clinical Practice* (R. A. Brace, M. G. Ross, and J. E. Robillard, Eds.), pp. 272–287. Perinatology, Ithaca, NY.

Allantois

Fuller W. Bazer

Texas A&M University

I. Introduction
II. Functions of the Allantoic Membrane
III. Humans Lack an Allantoic Membrane

GLOSSARY

allantoic membrane A membrane that develops as an evagination of the hindgut to form a sac filled with water, urine, and nutrients.

conceptus The embryo/fetus and its associated placental membranes.

placenta The amnion, allantois, and chorion which are membranes that form the unit for exchange of nutrients and gases between the mother and fetus.

umbilicus The stalk connecting the fetus to the placenta; contains major blood vessels and the urachus which connects the fetal bladder and allantoic sac.

urachus A duct that transports urine (water and nutrients cleared by the fetal kidney into the bladder) to the allantoic sac.

The allantoic membrane forms as an evagination of the hindgut (endoderm) as a diverticulum covered by mesoderm (splanchnopleure). The allantoic membrane dilates and extends outside the embryo/fetus except for attachment by the umbilical stalk, which contains the urachus for communication between the allantoic sac and fetal bladder. The allantoic membrane acquires an extensive blood supply from caudal branches of the aorta which break up into an extensive vascular net. The allantois and chorion fuse early in gestation so that the vascular system of the allantoic membrane supplies the entire chorioallantoic placenta.

I. INTRODUCTION

The placental membranes of most subprimate mammals are of the chorioallantoic type and composed of the following membranes: yolk sac, amnion, allantois, and chorion. The yolk sac develops from about Day 14, is maximum in size by about Day 18, and becomes very inconspicuous after Day 20 of pregnancy. The allantoic membrane forms as an evagination of the hindgut (endoderm) as a diverticulum covered by mesoderm (splanchnopleure) (Fig.

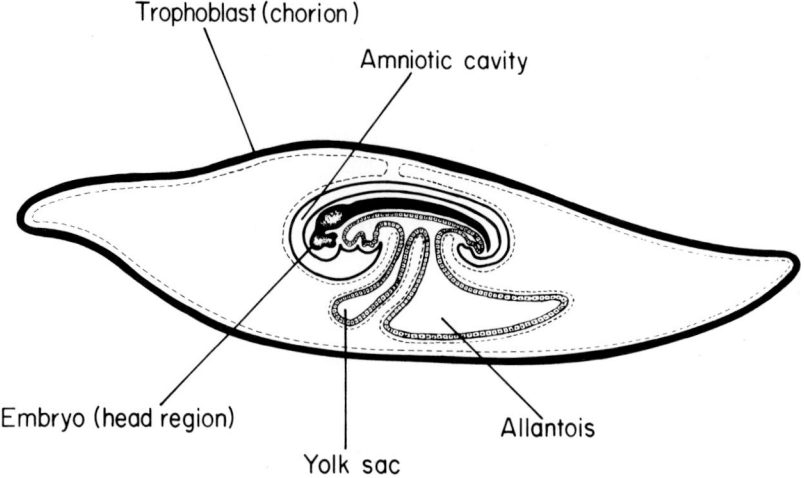

FIGURE 1 Diagram of the fetal membranes to depict outgrowth of the allantoic membrane.

1). The allantoic membrane dilates and extends outside the embryo/fetus except for attachment by the umbilical stalk, which contains the urachus for communication between the allantoic sac and fetal bladder.

II. FUNCTIONS OF THE ALLANTOIC MEMBRANE

The allantoic membrane acquires an extensive blood supply from caudal branches of the aorta which break up into an extensive vascular net. The allantois and chorion fuse early in gestation so that the vascular system of the allantoic membrane supplies the entire chorioallantoic placenta. In the pig the allantoic membrane does not expand to fill the entire chorionic vesicle; therefore, the extreme ends of the chorion become necrotic due to lack of a vascular supply from the allantoic membrane. The allantoic membrane forms a sac and accumulates allantoic fluid which is of both maternal and fetal (urine) origin. The allantoic fluid contains nutrients, hormones, growth factors, etc. which accumulate for either immediate recycling into the fetal circulation (e.g., fetal transferrin) or for future use (e.g., fructose) by the conceptus (embryo/fetus and its associated membranes). The allantoic membrane is derived from splanchnopleure (mesoderm and endoderm) and can transport nutrients into the fetal circulation.

III. HUMANS LACK AN ALLANTOIC MEMBRANE

In humans, the allantoic membranes does not form a sac filled with allantoic fluid. Rather, the allantoic membrane is represented by an endodermal outgrowth from the hindgut that fuses with the chorion and develops blood vessels, and this participates in formation of the umbilical cord.

See Also the Following Articles

ALLANTOCHORION; ALLANTOIC FLUID

Bibliography

Perry, J. S. (1981). The mammalian fetal membranes. *J. Reprod. Fertil.* 62, 321–335.
Steven, D. H. *Comparative Placentation.* Academic Press, New York.

Allatostatins

Stephen S. Tobe
University of Toronto

I. Introduction
II. Chemical Structure of Allatostatins and the Allatostatin Gene
III. Distribution and Expression of Allatostatins
IV. Roles of Allatostatins
V. Mode of Action of Allatostatins
VI. Degradation of Allatostatins

GLOSSARY

corpora allata Insect endocrine glands associated with and receiving innervation from the central nervous system; the site of production of juvenile hormone.

juvenile hormone Sesquiterpenoid compounds, derived from farnesyl pyrophosphate, which regulate metamorphosis and reproduction in many insect species.

neurosecretory cells Neuroendocrine cells, which possess characteristics of both endocrine cells and neurons, located in or associated with the central nervous system, containing peptides or amines which are released upon depolarization (e.g., following an action potential).

open reading frame The frame, one of three possible, in which mRNA is potentially translated into protein; the series of triplet codons encoding amino acids are uninterrupted by a translation termination codon.

preprohormone The entire polypeptide encoded by a mRNA for a peptide hormone(s) before processing by proteases to remove signal sequences and other nonfunctional regions.

Allatostatins are a family of peptides defined originally on the basis of their ability to inhibit juvenile hormone biosynthesis in insects. To date, three different families have been described: one characterized by Tyr/Phe-Xaa-Phe-Gly-Leu-amide carboxyl terminus and found in cockroaches, locusts, crickets, moths, and flies; one characterized by Xaa-Trp-Xaa-Asp-Leu-Xaa-Gly-Gly-Trp-amide and found only in crickets; and a 16-mer nonamidated peptide (pGlu-Val-Arg-Phe/Tyr-Arg-Gln-Cys-Tyr-Phe-Asn-Pro-Ile-Ser-Cys-Phe-OH) found in moths and flies and not yet identified in other orders. The ability to regulate juvenile hormone biosynthesis precisely, through inhibitory pathways, appears to be an important phenomenon in insects in view of the role of juvenile hormone in both metamorphosis and reproduction in most species. This function may have appeared on more than one occasion in various insect species.

I. INTRODUCTION

The corpora allata of insects are the defined sites of biosynthesis and release of the insect juvenile hormone. This hormone is responsible for the timely metamorphosis of larval insects through to the adult stage. Juvenile hormone is also a principal regulator of female reproductive function, including vitellogenesis, in many species. The production of juvenile hormone is known to be precisely regulated and neuropeptides are now known to be major factors involved in this regulation.

Changes in the production of juvenile hormone can be associated with metamorphic events in larvae and with oocyte growth and maturation in adult females. The presence of juvenile hormone at inappropriate times during metamorphosis can result in production of "superlarvae," retention of larval characteristics in the adult stage, or sterility, whereas its absence can result in premature metamorphosis and sterility. In the adult female, the presence of juvenile hormone at inappropriate times can also result in sterility, as a consequence of inhibition of oogenesis or embryogenesis, or oocyte resorption, whereas in its absence, oocyte growth and vitellogenesis cease. These consequences indicate the importance of the precise regulation of juvenile hormone production in insects.

Several modes of regulation of juvenile hormone production can be envisaged. For example, at appropriate times, hormone production could be stimulated by tropic or stimulatory factors (allatotropins); removal of the stimulatory signal would result in a decline in hormone production to basal levels. Alternatively, juvenile hormone production could be inhibited or reduced below a basal level by negative or static factors (allatostatins). Removal of the inhibitory signals would result in an increase in juvenile hormone biosynthesis. Stimulation or inhibition of basal levels of juvenile hormone production (modulation) clearly operate in opposite fashions, and a combination of these two signaling mechanisms could permit a complex, high level of control over hormone production. Cyclic changes in juvenile hormone biosynthesis during egg production, as has been observed during vitellogenesis, could thus occur as a consequence of both allatotropic and allatostatic signals. For example, in Fig. 1, the dramatic increase in juvenile hormone biosynthesis associated with the onset of vitellogenesis could occur as a result of an allatotropic signal or the removal of an allatostatic signal. Similarly, the decline in juvenile hormone biosynthesis at the completion of vitellogenesis could be attributable to removal of the allatotropic signal or to the reimposition of an allatostatic signal. However, there is little evidence for a dual regulatory mechanism in most systems studied to date.

II. CHEMICAL STRUCTURE OF ALLATOSTATINS AND THE ALLATOSTATIN GENE

The allatostatins are families of peptides originally defined on the basis of their abilities to inhibit the biosynthesis of juvenile hormone by corpora allata. Over 50 years ago, Scharrer and colleagues hypothesized the existence of factors originating in the brain able to inhibit the corpora allata, although their occurrence had been inferred considerably earlier from experiments involving nerve severance, brain ablation, or cautery of regions of the brain. These original experiments used cockroaches as model insects because of the relative ease in performing simple nerve transection surgery and in assessing the effects of such interventions on the corpora allata (as determined by volume/size of these glands).

The first family of allatostatins was discovered in cockroaches (*Diploptera punctata*, the Pacific beetle cockroach) in 1989. They are characterized by a common carboxyl terminus, Tyr/Phe-Xaa-Phe-Gly-Leu-amide, where Xaa = Gly, Ala, Ser, or Asn. Subsequently, additional allatostatins have been identified in this and other cockroach species and virtually all possess the common carboxyl terminus, with the exception of the terminal Leu, which has been replaced by the conservative substitution, Ile, in rare instances (see Fig. 2, allatostatin 13 and 14). Additional allatostatins of this family have now been identified in crickets, locusts, and flies, again with the characteristic pentapeptide carboxyl terminus (Table 1). In the case of cockroaches and locusts, on the basis of the deduced coding regions for the allatostatin preprohormone derived from cDNA libraries of brains from each species (by PCR), it is now

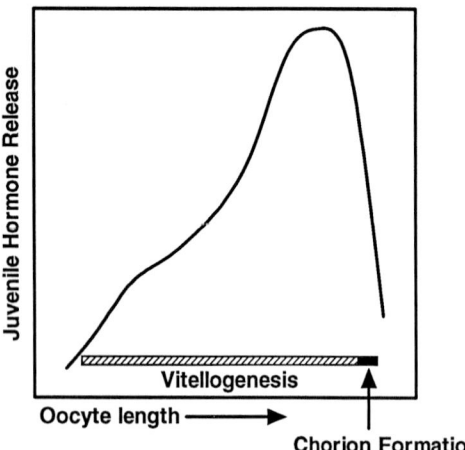

FIGURE 1 Relationship between basal oocyte length and juvenile hormone release in the cockroach *Diploptera punctata*. As oocytes grow, juvenile hormone production increases during the period of vitellogenesis. At the completion of vitellogenesis and the onset of chorion formation, juvenile hormone production decreases dramatically.

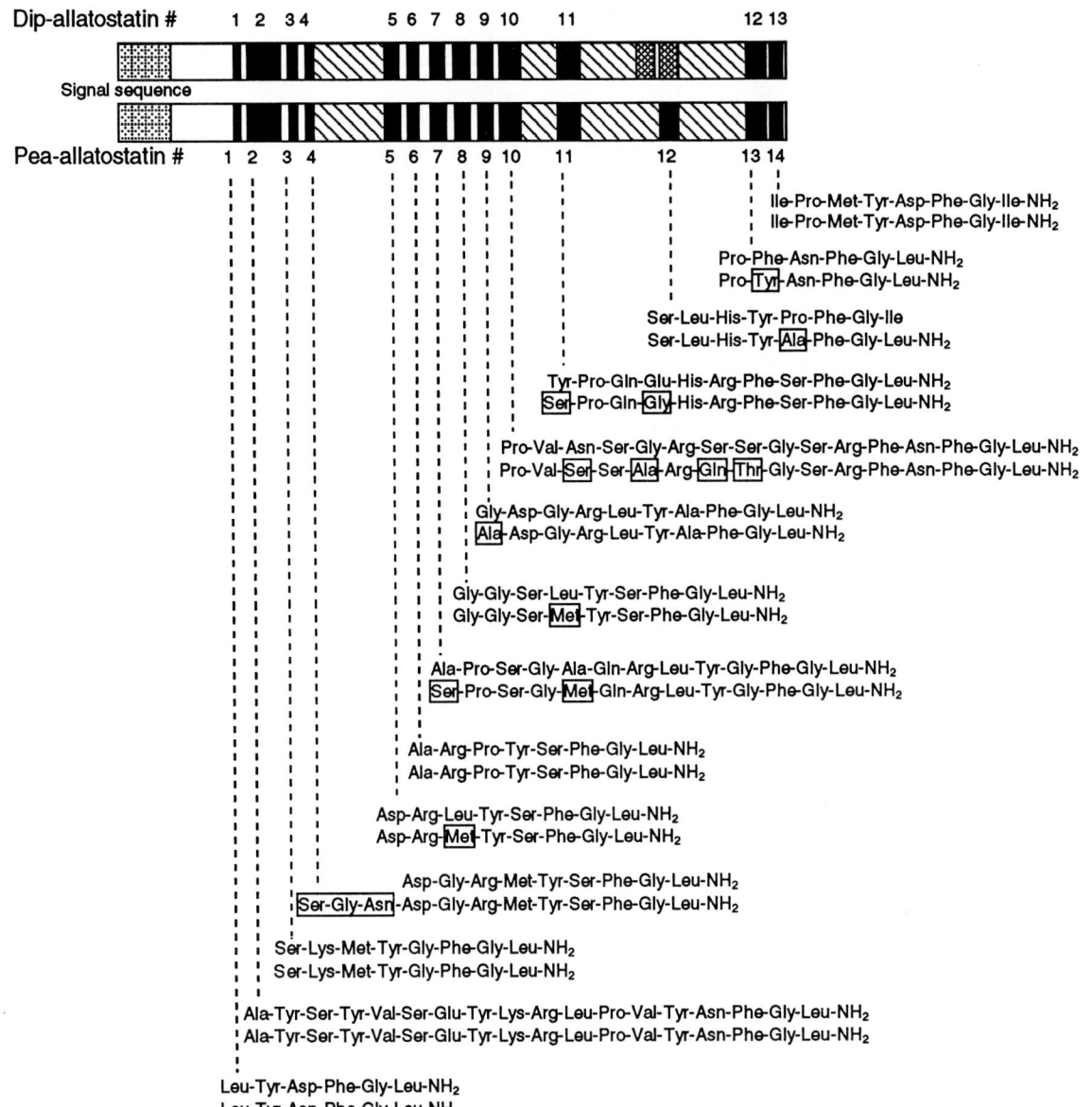

FIGURE 2 Schematic representation of the allatostatin precursor polypeptides in the cockroaches *Diploptera* and *Periplaneta*. The hydrophobic leader sequence (signal sequence) precedes an untranslated region (clear), which is then followed by the allatostatin peptides. Black boxes represent individual allatostatins which are numbered relative to their position to the amino terminus, and the amino acid sequences of each are shown below. Differences between the two species are shown in boxes. Acidic regions are represented as diagonal areas. Within the third acid region of *Diploptera* are two peptide sequences which do not have the α-amidation signal, whereas in *Periplaneta*, one of these does (adapted with permission from Stay et al., 1994).

known that each possesses up to 14 allatostatin peptides, with little variation between species. The deduced coding region is similar in structure in all cockroach species, with characteristic acidic "spacer" regions separating groups of allatostatin peptides in the polypeptide precursor. All the allatostatin peptides in the precursor possess a carboxyl-terminal Gly signal for α-amidation, followed by a typical dibasic cleavage site (e.g., Lys-Arg), which is a target site for prohormone convertases/endoproteases (Fig.

TABLE 1
Allatostatin Amino Acid Sequences

Cockroach family
  Diploptera/Periplaneta consensus sequence
    -Tyr-Xaa-Phe-Gly-Leu-NH₂

  Schistocerca gregaria
    Leu-Cys-Asp-Phe-Gly-Val-NH₂
    Ala-Tyr-Thr-Tyr-Val-Ser-Glu-Tyr-Lys-Arg-Leu-Pro-Val-Tyr-Asn-Phe-Gly-Leu-NH₂
    Ala-Thr-Gly-Ala-AlaSer-Leu-Tyr-Ser-Phe-Gly-Leu-NH₂
    Gly-Pro-Arg-Thr-Tyr-Ser-Phe-Gly-Leu-NH₂
    Gly-Arg-Leu-Tyr-Ser-Phe-Gly-Leu-NH₂
    Ala-Arg-Pro-Tyr-Ser-Phe-Gly-Leu-NH₂
    Ala-Gly-Pro-Ala-Pro-Ser-Arg-Leu-Tyr-Ser-Phe-Gly-Leu-NH₂
    Glu-Gly-Arg-Met-Tyr-Ser-Phe-Gly-Leu-NH₂
    Pro-Leu-Tyr-Gly-Gly-Asp-Arg-Arg-Phe-Ser-Phe-Gly-Leu-NH₂
    Ala-Pro-Ala-Glu-His-Arg-Phe-Ser-Phe-Gly-Leu-NH₂

  Calliphora vomitoria
    Asp-Pro-Leu-Asn-Glu-Glu-Arg-Arg-Ala-Asn-Arg-Tyr-Gly-Phe-Gly-Leu-NH₂
    Leu-Asn-Glu-Glu-Arg-Arg-Ala-Asn-Arg-Tyr-Gly-Phe-Gly-Leu-NH₂
    Ala-Asn-Arg-Tyr-Gly-Phe-Gly-Leu-NH₂
    Asn-Arg-Pro-Tyr-Ser-Phe-Gly-Leu-NH₂
    Gly-Pro-Pro-Tyr-Asp-Phe-Gly-Met-NH₂

  Gryllus bimaculatus
    Ala-Gln-His-Gln-Tyr-Ser-Phe-Gly-Leu-NH₂
    Ala-Gly-Gly-Arg-Gln-Tyr-Gly-Phe-Gly-Leu-NH₂

Cricket family
  Gryllus bimaculatus
    Gly-Trp-Gln-Asp-Leu-Asn-Gly-Gly-Trp-NH₂
    Gly-Trp-Arg-Asp-Leu-Asn-Gly-Gly-Trp-NH₂
    Ala-Trp-Arg-Asp-Leu-Ser-Gly-Gly-Trp-NH₂
    Ala-Trp-Glu-Arg-Phe-His-Gly-Ser-Trp-NH₂

Manduca family
  Manduca sexta/Pseudaletia unipuncta
    pGlu-Val-Arg-Phe-Arg-Gln-Cys-Tyr-Phe-Asn-Pro-Ile-Ser-Cys-Phe-OH

  Drosophila melanogaster
    pGlu-Val-Arg-Tyr-Arg-Gln-Cys-Tyr-Phe-Asn-pro-Ile-Ser-Cys-Phe-OH

2). Also, five of the peptides, which possess identical sequences across many of the species, are located in identical positions within the precursor.

The open reading frame of the cockroach allatostatin cDNA (mRNA) encodes the allatostatin prohormone, which begins with an amino-terminal signal sequence (targeting secretory proteins for translocation to the lumen of the endoplasmic reticulum). This is followed by a region of approximately 50 amino acids, immediately preceding the first allatostatin sequence. To date, all preproallatostatins in cockroaches and locusts have been found to contain the acidic spacer regions which probably serve to neutralize the basic charges associated with the allatostatins themselves and their associated processing and basic cleavage (Lys-Arg) sites. The size of the preprohormone in *Diploptera* is 370 amino acid residues, in *Periplaneta americana* (American cockroach) it is 379 residues, and in *Schistocerca gregaria* (desert locust) it is 283 residues. In *Schistocerca*, the polypeptide precursor contains only 10 allatostatin peptides, whereas the precursor in *Diploptera* contains 13 and in *Periplaneta* the precursor contains 14 allatostatins. Although there is high similarity between the allatostatin sequences across the species, there is little identity in the acid spacer regions.

The pentapeptide consensus sequence appears to be functionally important. This sequence has now been defined as the "core" sequence or region, necessary for biological activity. There appears to be a β turn in this region of the peptide and this turn is necessary for interaction with the putative allatostatin receptor since replacement of amino acid residues in the core region to prevent or alter the β turn results in a loss of biological activity. Similarly, the amide at the carboxyl terminus is necessary for biological activity and replacement with a hydroxyl function abolishes all activity.

Allatostatins in other insect orders, for example, in the Orthoptera (locusts and crickets) and Diptera (flies), show remarkable similarity to those in the cockroaches (Dictyoptera). All possess the same core pentapeptide sequence, with the exception of a Met carboxyl-terminal substitution in some blowfly (*Calliphora*) allatostatins and a Val carboxyl-terminal substitution in one desert locust (*Schistocerca*) allatostatin (Table 1). All these allatostatins show activity with respect to inhibition of juvenile hormone biosynthesis in cockroach corpora allata, although they may be ineffective in the host species; this suggests that these peptides may serve other functions in these species.

Two other families of allatostatic peptides have been isolated. One group, recently isolated from the cricket, *Gryllus bimaculatus,* comprises a family of four related peptides, with the consensus sequence Thr-Xaa-Asp-Leu-Xaa-Gly-Gly-Trp-NH$_2$, although one member of this family deviates appreciably from this sequence (Table 1). This family is less active than the cockroach allatostatins in the inhibition of juvenile hormone production by at least an order of magnitude, although it does show other types of biological activity (myo-inhibiting activity). The *Manduca sexta* allatostatin (Table 1) bears no similarity to the cockroach allatostatins nor to the cricket allatostatins. It has been isolated from both the tobacco hornworm, *Manduca sexta,* and the armyworm, *Pseudaletia unipuncta*. This peptide is not amidated at the carboxyl terminus. A related peptide, differing only at the 4 position (-Tyr- substituted for -Phe-; Table 1), has been deduced from the open reading frame of *Drosophila melanogaster* cDNA head libraries. The cDNA specifies a polypeptide of about 125 residues (in *P. unipuncta*) and contains a single allatostatin peptide, in contrast to the cockroach and locust precursor. These moth and fly allatostatins are inactive in the inhibition of juvenile hormone production in cockroaches.

Although the *Manduca* peptide is effective in inhibiting juvenile hormone biosynthesis in this species and in the related corn ear worm, it is only moderately effective in the armyworm. The *Drosophila* peptide is ineffective in inhibiting juvenile hormone production in *D. melanogaster*, the fruit fly. This suggests that the allatostatins in higher insect orders are relatively species specific and may serve functions other than inhibition of hormone production.

III. DISTRIBUTION AND EXPRESSION OF ALLATOSTATINS

The distribution of allatostatins in tissues of cockroaches, locusts, moths, and flies has been defined principally through immunocytochemistry and, to a lesser extent, using *in situ* hybridization. The allatostatins have been localized in discrete neurosecretory cell groups of the brain (Fig. 3). In general, axons project from the cells of both the medial and the lateral cell group and arborize terminally in the corpora cardiaca (neurohemal area) and the corpora allata, although some axons arborize exclusively within the tritocerebrum and the pars intercerebralis of the brain (Fig. 3). The relative contribution of the

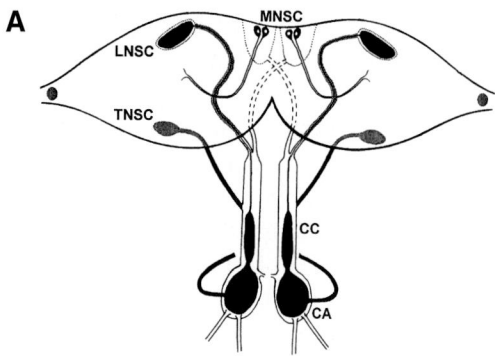

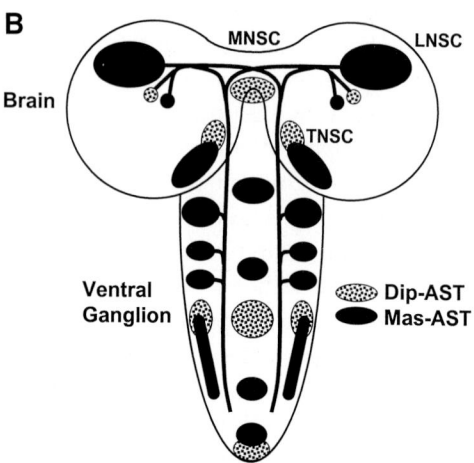

FIGURE 3 Distribution of allatostatin immunoreactivity or expression in the central nervous system (CNS) of adult female *Diploptera* (A) and larval *Drosophila* (B). In *Drosophila*, immunoreactivity to both cockroach and *Manduca* allatostatins are shown. Dorsal view. In A, shaded regions represent cell groups visible from the ventral perspective. MNSC, medial neurosecretory cells; LNSC, lateral neurosecretory cells; TNSC, tritocerebral neurosecretory cells; CC, corpora cardiaca; CA, corpora allata.

two neurosecretory cell groups to the overall quantity of material observed in the corpora cardiaca and corpora allata (retrocerebral complex) varies between the different species: In *D. punctata*, the lateral neurosecretory cells are the major contributors to the complex and the medial group contributes to the interneurons found within other regions of the brain. Nonetheless, both groups are involved in the regulation of juvenile hormone production. There are numerous other allatostatin-immunoreactive cells in the brain, including cells found adjacent to the optic lobes and to the antennal lobes in *D. punctata*. In situ hybridization studies have demonstrated allatostatin-expressing cells in both the medial and the lateral neurosecretory cell groups as well as in the cells associated with the optic lobes and the tritocerebrum of cockroaches.

Allatostatins are also widely distributed in the central nervous system of most species studied. For example, they are found in the suboesophageal ganglion and in all of the ventral ganglia of *D. punctata*. They are also found innervating many muscular tissues, including the hindgut (Fig. 4) and the pulsatile vessel associated with the antennae. *In situ* hybridization has revealed that many of these cells, including cells in the suboesophageal ganglion and in the ventral ganglia, express the allatostatin mRNA message.

Allatostatin immunoreactive cells are also distributed extensively in the midgut of both moths and cockroaches (Fig. 4). These endocrine cells are of the open type, extending across the gut wall from the hemolymph (basal) side to the lumen side, and are not neurosecretory in nature. These cells have been shown to express allatostatin mRNA in several cockroach species, although the total quantity of allatostatin message is about 20-fold lower in midgut than in brain. Allatostatin endocrine cells are not randomly distributed in the midgut but rather are concentrated at the anterior end (Fig. 4). Levels of allatostatin in the midgut fluctuate, depending on nutritional status, and can be experimentally altered by dehydration and starvation.

Allatostatins are also found in surprisingly high concentrations in the hemolymph (blood) of cockroaches. In *D. punctata*, concentrations of the allatostatin may reach >2 nM and during metamorphosis and egg production are in the low nanomolar range. It is not known if these hemolymph peptides are able to inhibit juvenile hormone production directly by way of a humoral route, but their concentration is sufficient to achieve 50% inhibition of juvenile hormone biosynthesis *in vitro*. These humoral allatostatins may also function to modulate other physiological activities. The source of these peptides in the hemolymph is not known, but about 5% of the blood

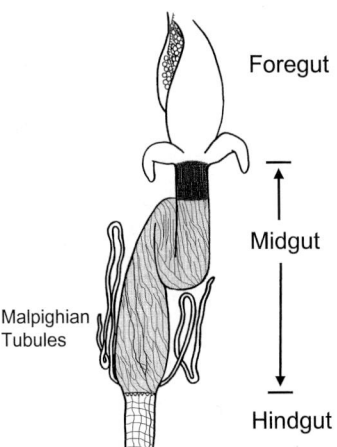

FIGURE 4 Distribution of allatostatin immunoreactivity or expression in the gut of adult *Diploptera*. Heavily shaded area indicates a concentration of endocrine cells producing/expressing allatostatin, whereas lightly shaded area indicates the presence of some allatostatin endocrine cells. Immunoreactive axons are shown as fine lines, extending over the midgut and hindgut.

cells (hemocytes) have been shown to produce allatostatins, suggesting these cells are a potential source in addition to those emanating from the neurohemal areas associated with the central nervous system.

Immunoreactive material to the cockroach allatostatin family has been reported in several other insect orders and in other phyla. For example, extensive immunoreactivity has been reported in brain, suboesophageal ganglion, and thoracic and abdominal ganglia as well in the hindgut and ovary of *Calliphora*, and similar distribution occurs in *Drosophila* and in moths. Immunoreactivity to *Manduca* allatostatin also occurs in *Drosophila* in larval brain (including optic and antennal lobes and tritocerebrum) and ventral ganglion (Fig. 3), as well as in the corpora allata.

Cockroach allatostatin immunoreactivity has also been described in the stomatogastric system of crabs, including interneurons and sensory neurons and brain; in the tick, *Dermacentor variabilis*, central nervous system; in molluscan ganglia; and in nervous systems of annelids and a range of helminths. The functions of these peptides remains to be determined in these phyla, but it is unlikely that they are involved in the regulation of juvenile hormone biosynthesis since this hormone has not been identified in any of these phyla. However, the extensive presence of these peptides in the nervous system suggests an interneuronal/transmitter role and a possible myomodulatory function.

IV. ROLES OF ALLATOSTATINS

The ability of allatostatins to inhibit the biosynthesis of juvenile hormone permitted their original isolation and characterization. However, our understanding of the mechanisms involved in this inhibition remains unclear, although it is known that a range of second messengers can be modulators of juvenile hormone production. The biosynthetic pathway for juvenile hormone is identical to that for sterol biosynthesis to the point of the production of farnesyl pyrophosphate and it is likely that mechanisms similar to those regulating the production of cholesterol in vertebrates, for example, also occur in insects. In vertebrate systems, cholesterol biosynthesis is regulated at several steps in the pathway (multivalent feedback), depending on the flow of carbon units through the pathway.

Cockroach allatostatins have been identified as inhibitors of juvenile hormone production in only a limited number of species. They are effective in those cockroach species that have been tested (e.g., *Diploptera, Periplaneta, Blattella,* as well as the cricket, *Gryllus*), but these peptides are ineffective in locusts, flies, moths, and bees. *Manduca* allatostatin similarly is effective in a few moth species but inactive in the inhibition of juvenile hormone biosynthesis in cockroaches, grasshoppers, beetles, flies, and some moths. This information indicates that the primary function of allatostatins may not be the inhibition of juvenile hormone production. The cockroach allatostatins are ubiquitous in insects and other invertebrates, based on immunocytochemical observations (see Section III), and because juvenile hormone does not appear to occur throughout the invertebrates, it is likely that this Phe-Gly-Leu-amide family originated as a family of peptides with different functionality. This family was subsequently coopted for the inhibition of juvenile hormone biosynthesis in cockroaches, crickets, and other species yet to be identified.

Based on their wide distribution within various tissues of cockroaches, it is apparent that allatostatins function not only to inhibit juvenile hormone production but also are involved in other physiological processes. The extensive innervation of muscle of the gut with allatostatin-immunoreactive neurons suggests a role in the modulation of muscle contraction, and it has been demonstrated that allatostatins do indeed decrease the frequency and amplitude of spontaneous contractions of hindgut. In other words, allatostatins are myoinhibitory peptides. Allatostatins also modulate the activity of muscle in the presence of the potent insect myotropic peptide, proctolin. Proctolin causes strong contraction of cockroach hindgut, in a dose-dependent fashion, and allatostatins are able to antagonize this effect, also in a dose-dependent manner. A similar effect has been reported on cockroach antennal heart muscle. Cricket (*Gryllus*) allatostatins of the Gly-Gly-Thr-amide family (Table 1) may also show myomodulatory activity as a consequence of their sequence similarity to locust and cockroach myo-inhibiting peptide (members of the FLRF-amide family). Few insect myoactive neuropeptides have been identified as inhibitory.

The extensive distribution of cockroach allatostatins in gut, both in fibers innervating the muscle of hind- and midgut and in endocrine cells of the midgut (Figs. 3 and 4), suggests that there may be additional functions beyond the myomodulatory effects. For example, allatostatins may contribute to the rate of movement of food through the gut, and the reduced levels of midgut allatostatins in response to nutritional status, e.g., starvation and dehydration, may similarly reflect an involvement of these peptides in digestion. In fact, allatostatins can stimulate release of carbohydrate-metabolizing enzymes (α-amylase and invertase) from the midgut. Finally, allatostatins in the gut, either from nerve fibers or from endocrine cells, may contribute to hemolymph stores of these peptides.

Although the cockroach allatostatins are widely distributed in the central nervous systems of insects, their role in neurotransmission/neuromodulation is only circumstantial. Nonetheless, allatostatin-positive cell bodies originate in medial, lateral, and tritocerebral neurosecretory cell regions and arborize extensively in protocerebral, deutocerebral, and tritocerebral regions of the brain, indicating that these cells function as interneurons. Allatostatins have been shown to function as neurotransmitters, however, in the crab stomatogastric system and the peptides have an inhibitory effect on rhythmic muscle contraction and the neurons driving this activity as well as on neuromuscular junctions.

As noted in Section III, hemolymph concentrations of allatostatins are high. The functional significance of these peptides is unclear, although the concentration is sufficient to inhibit partially the production of juvenile hormone by the corpora allata by way of a humoral route. Injection of allatostatins into the hemocoel (body cavity) of insects could therefore suppress the production of juvenile hormone. It is known that the rate of oocyte production can be slowed through repeated injection of allatostatins into adult females, and rates of juvenile hormone production and juvenile hormone titer during the same time period are slightly depressed. These peptides may also contribute to the overall excitability of muscle associated with the gut.

V. MODE OF ACTION OF ALLATOSTATINS

How is the inhibitory message of the allatostatins translated into inhibition of juvenile hormone biosynthesis? Although the precise steps in the biosynthetic pathway which are affected remain unknown, it is known that juvenile hormone production is regulated by a complex interaction between several second messengers. The corpora allata themselves are an unusual endocrine tissue since the cells comprising the glands are electrically excitable and, in fact, the cells will display action potentials in appropriate circumstances, e.g., following treatment with calcium channel blockers. This is significant because it indicates the presence of ion channels; these channels probably play a major role in regulating intracellular ion concentrations and at least some of these channels are voltage dependent, i.e., the opening of the channels is dependent on the potential difference between the inside and the outside of the cell, permitting the movement of ions across the cell membrane. In the case of the corpora allata, the cells possess voltage-dependent calcium and sodium channels which act to regulate the internal concentration of calcium. Calcium is a well-known second messenger, regulating a range of enzymatic activities within the cells, including adenylate cyclase and protein kinases.

Allatostatins may alter intracellular calcium directly, presumably through release of calcium from intracellular stores or through the modulation of calcium entry through the voltage-dependent calcium channels. Intracellular calcium stores seem to be particularly important and treatments which modulate intracellular levels of calcium have corresponding effects on juvenile hormone production—an elevation in calcium levels stimulates juvenile hormone production, whereas a reduction inhibits hormone production.

The cyclic nucleotides, cAMP and cGMP, are both able to modulate juvenile hormone production as are agents that modify intracellular calcium, including calcium ionophores. Other agents, including the adenylate cyclase activator forskolin, phorbol esters that act to activate protein kinase C, and thapsigargin, a compound that releases calcium from intracellular stores, also affect juvenile hormone biosynthesis. For example, cAMP, its analogs, or agents that stimulate its production (forskolin) strongly inhibit the production of juvenile hormone, as does cGMP and its analogs. Both of these groups of compounds probably alter ion flux in the cells of the corpora allata, but different channels are affected by the two groups. Conductance of the channels is regulated by phosphorylation/dephosphorylation of the channel proteins.

Signal transduction of the allatostatin message by cells of the corpora allata invariably involves the interaction between receptor(s) on the cells and the allatostatin. Although the receptors remain to be isolated and identified, current evidence indicates that (i) because of the large number of different cockroach allatostatins, there is likely to be more than one receptor type; (ii) the receptor is G protein linked; and (iii) the pentapeptide core sequence, including the amide, of allatostatin is necessary for binding.

VI. DEGRADATION OF ALLATOSTATINS

As regulators of important physiological processes, allatostatins, in common with other physiologically active peptides, undergo degradation to terminate their biological actions. The effects of allatostatins on both juvenile hormone biosynthesis and myotropic activity show a rapid onset and these actions are rapidly reversible. For example, *in vitro* experiments have shown that juvenile hormone production is affected within 1 hr of treatment with allatostatins, with a similar time course for recovery following withdrawal of the peptide. *In vivo*, the abolition of the biological effects of the allatostatins can occur by the cessation of release of the allatostatins, the rapid degradation of the peptides, or both.

Cockroach allatostatins are rapidly degraded by a range of proteolytic enzymes, including endopeptidases and aminopeptidases or carboxypeptidases, which cleave peptide bonds at internal basic amino acids or at the amino or carboxy termini, respectively. The rate at which peptides, including the allatostatins, are degraded is usually expressed as half-life, the time required for 50% of the peptide to be catabolized. The half-life of allatostatins differs between tissues but invariably involves a cleavage of the carboxyl-terminal leucine-amide (Fig. 5). This cleavage effectively abolishes the biological activity of the peptides since the carboxyl-terminal amide is required for inhibition of juvenile hormone biosynthesis. For those allatostatins studied, half-lives for gut or brain tissue range between 10 and 50 min at physiological concentrations, although in hemolymph, Dip-allatostatin 5 has a much longer half-life. This indicates that rates of degradation differ not only between tissues but also between the various allatostatins.

Degradative pathways, and the enzymes effecting the cleavages, also differ among tissues. As Fig. 5 shows, a specific pattern of cleavage can be associated with specific tissue types. For example, the tissues possess carboxypeptidases which invariably remove the Leu-amide function and effectively inactivate the peptides, but these cleavages occur at different sites, depending on the tissue (brain, midgut, hindgut, or corpora allata) (Fig. 5). Significantly, the hemolymph

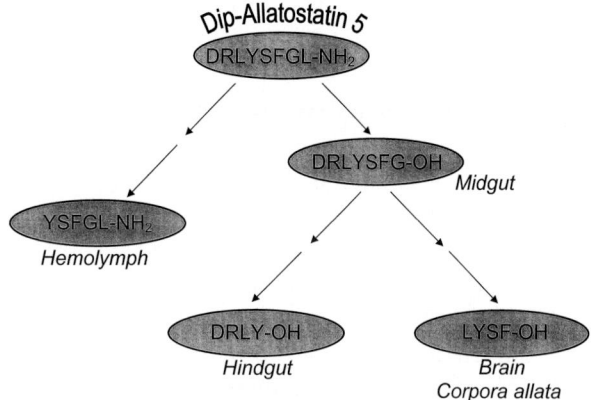

FIGURE 5 Pathways of degradation for *Diplotera* allatostatin 5 in representative tissues of adult female *Diploptera*. Each tissue shows a distinctive pattern of degradation. The final product of hemolymph degradation retains some biological activity because it retains the core region of the peptide, whereas the final products of other tissues lack biological activity.

does not appear to possess such carboxypeptidases but rather aminopeptidases, which cleave only at the N-terminal portion of the peptide (Fig. 5). The pentapeptide that remains after hemolymph cleavage is the core consensus sequence necessary for biological activity (see Section II). Thus, at least in the case of Dip-allatostatin 5, peptidase cleavage does not result in the total loss of biological activity. This core sequence shows some potency (the dose sufficient to inhibit juvenile hormone production) but full efficacy (the maximal inhibition of juvenile hormone production at saturation). These peptides may also be capable of modifying the activity of other allatostatins or may serve as primers for other physiological events.

See Also the Following Articles

CORPUS ALLATUM; *DIPLOPTERA PUNCTATA*; JUVENILE HORMONE

Bibliography

Bendena, W. G., Garside, C. S., Yu, C. G., and Tobe, S. S. (1997). Allatostatins: Diversity in structure and function of an insect neuropeptide family. *Ann. N. Y. Acad. Sci.* **814**, 53–67.

Ding, Q., Donly, B. C., Tobe, S. S., and Bendena, W. G. (1995). Comparison of the genes encoding allatostatin neuropeptides in the distantly related cockroaches *Periplaneta americana* and *Diploptera punctata*. *Eur. J. Biochem.* **234**, 737–746.

Donly, B. C., Ding, Q., Tobe, S. S., and Bendena, W. G. (1993). Molecular cloning of the gene for the allatostatin family of neuropeptides from the cockroach *Diploptera punctata*. *Proc. Natl. Acad. Sci. USA* **90**, 8807–8811.

Rachinsky, A., and Tobe, S. S. (1996). The role of second messengers in the regulation of juvenile hormone production in insects with particular emphasis on calcium and phosphoinositide signaling. *Arch. Insect Biochem. Physiol.* **33**, 259–282.

Stay, B., Tobe, S. S., and Bendena, W. G. (1994). Allatostatins: Identification, primary structures, functions and distribution. *Adv. Insect Physiol.* **25**, 267–338.

Tobe, S. S., Yu, C. G., and Bendena, W. G. (1994). Allatostatins, peptide inhibitors of juvenile hormone production in insects. In *Perspectives in Comparative Endocrinology* (K. G. Davey, R. E. Peter, and S. S. Tobe, Eds.), pp. 12–19. National Research Council of Canada, Ottawa.

Weaver, R. J., Edwards, J. P., Bendena, W. G., and Tobe, S. S. (1997). Structures, functions and occurrence of insect allatostatic peptides. *J. Exp. Biol.* (Symp. Se), 3–32.

Altitude, Effects on Humans

Lorna G. Moore

University of Colorado at Denver and University of Colorado Health Sciences Center

I. The High-Altitude Environment
II. Fecundity and Fertility
III. Pregnancy and Fetal Life
IV. Neonatal, Infant, and Maternal Mortality

GLOSSARY

abruptio placenta Premature separation of the placenta from the uterine wall.

adaptation A characteristic of structure, function, or behavior that enables an organism to live and reproduce in a given environment.

completed fertility The number of live births to a woman at age 45 or after the completion of her reproductive period.

fecundity The production of sperm or eggs and of viable embryos.

fertility The number of live births.

hypoxemia Decreased tissue oxygen delivery.

hypoxia Partial pressure of oxygen (PO_2) below values present at sea level.

IUGR Intrauterine growth retardation or less than optimal birth weight for survival.

oxygen transport system The system, consisting of the lungs, heart, and blood vessels, that actively transports oxygen from the atmosphere to the site of oxidative phosphorylation in the tissue mitochondria.

preeclampsia An elevation in blood pressure (>140/90 mm Hg, a systolic rise >30 mm Hg, or a diastolic rise >15 mm Hg) after Week 20 accompanied by significant proteinuria (>0.3 g/liter in a 24-hr collection) in a woman who is normotensive when nonpregnant.

total birth rate The number of births in a given year per 1000 persons.

The predominant characteristic of the high-altitude environment, hypoxia, results from the decreased partial pressure of oxygen in the atmosphere. The health effects of high altitude are of interest from public health, clinical, and basic science perspectives. Historically, decreased fertility and fecundity have been reported from high altitude but such effects are contested by recent findings. Current data indicate that IUGR, preeclampsia, and other complications of fetal and maternal life are more common at high altitude. Thus, studies of the effects of high altitude on reproduction have relevance for understanding fetal and maternal complications of pregnancy as well as other conditions involving the heart, lungs, blood, and metabolic pathways involved in oxygen transport or utilization. A broad range of disciplines, including anthropology, biochemistry, biology, genetics, and physiology, are actively involved in investigating the mechanisms by which human populations respond and adapt to the conditions of reduced oxygen availability at high altitude.

I. THE HIGH-ALTITUDE ENVIRONMENT

"High altitude" refers to elevations above 2500 m or ~8000 ft since, at these levels, oxygen saturation of hemoglobin in the blood usually begins to fall below sea level values. Nearly 140 million people live at altitudes above 2500 m worldwide and more than 35 million travel there yearly for recreational and other purposes. Air travel exposes even larger numbers to moderate altitudes since airplane cabins are routinely maintained at ~2500 m to conserve fuel. Populations reside permanently at high altitudes in the Rocky Mountains of North and Central America, the Andes of South America, the East African highlands, and the Asian Himalayan Plateau. The Himalayan Plateau is the largest, both in terms of geographic area and number of inhabitants (nearly 80 million). Approximately 15 million live in North and Central America, 21 million in South America, and 24 million in Africa. Tourism and other travel bring a large number of visitors annually to North America and other regions.

Oxygen availability decreases at high altitude as the result of a nonlinear fall in barometric pressure. At 3100 m (10,000 ft), the partial pressure of oxygen in the inspired air (P_IO_2) is 98 mm Hg or two-thirds the sea level value. At 4300 m (14,000 ft), P_IO_2 is 83 mm Hg. The highest permanent human habitations are at approximately 5000 m (16,500 ft), where P_IO_2 is 75 mm Hg or half the sea level value. While hypoxia is the predominant environmental stress, the high-altitude environment is also characterized by cold, dryness, increased ultraviolet radiation, and marked diurnal temperature variation.

II. FECUNDITY AND FERTILITY

A. Fecundity

Fecundity in humans is difficult to measure since accurate, acceptable ways of monitoring sperm or egg production and embryo viability are not readily available. Fertility may not be equivalent to fecundity since intrauterine mortality occurs during the long interval between conception and birth in eutherian (placental) mammals. Nonhuman animal studies can more readily distinguish between fecundity and fertility. Animals gestated at high altitude have reduced fertility as demonstrated by decreased litter size and numbers of litters due, in turn, to more reabsorptions and fetal deaths. Thus, greater intrauterine mortality rather than diminished fecundity is implicated for lowering fertility. These effects are generally greater among discontinuously exposed animals or animals brought to altitude as adults (newcomers) than those born at high altitude (natives) or descended from multigeneration natives. Thus, hypoxia's effects on fertility may lessen with increasing duration of altitude exposure.

B. Fertility

Human fertility has been reported to be reduced at high compared with low altitude but it is not clear whether the effects are attributable to decreased fecundity or increased mortality in the pre- or postnatal period. The earliest accounts stem from the Spanish conquest of the Incan Empire. According to Antonio de la Calancha in 1639 (Monge, 1948),

All children born of Spanish parents [in Potosi, Bolivia at 4000 m; 13,200 ft] died either at birth or within a fortnight thereafter, because the great cold and freezing air would kill them; the mothers used to leave in order to give birth in the neighboring valleys and until their child was more than a year old the mothers would exile themselves from this city. (pp. 36–37).

Fifty-three years was said to have elapsed before a child born to Spanish parents survived. The circumstances confronted today by the Han (ethnic Chinese) today in Lhasa and other high-altitude regions of Tibet appear analogous. The Han usually descend to near sea level in China when they become pregnant and remain there for months to years after birth. The experience of those remaining in Lhasa is instructive; their newborns weigh less than babies born to Tibetan women. The Han babies also have lower arterial oxygen saturations and greater susceptibility to a syndrome of pulmonary hypertension and right heart failure during the first months of life. Thus, IUGR, impaired neonatal transition, and other factors decreasing the chance of survival were more likely responsible for the reproductive difficulties encountered by the Spaniards than diminished fertility.

The question of whether fertility is reduced among native high-altitude populations of South America and Asia has been addressed using census data and community surveys. Increasing altitude in South America and Nepal is associated with lower total birth rates and completed fertility in some but not all studies. Where observed, the reductions are likely attributable to infrequent and incomplete enumeration of births, high neonatal (birth to 28 day) or infant (birth to 1 year) mortality, cultural practices decreasing exposure to intercourse (e.g., husband absenteeism and a high proportion of the population living as celibate nuns), or other conditions not directly related to the hypoxia of high altitudes. Despite delayed menarche and earlier menopause, higher fertility is achieved in highland Peru and Chile as the result of shorter intervals between births and more frequent conception during lactation. Sherpa males at high compared with low altitude have lower serum luteinizing hormone and a trend for lower follicle-stimulating hormone levels, suggesting less stimulation of the pituitary gland. Lower serum testosterone in males and lower maternal estradiol, estriol, and prolactin levels during pregnancy have also been reported at high compared with low altitude. Thus, while there is some alteration in reproductive function at high altitude, it appears insufficient to limit fertility.

III. PREGNANCY AND FETAL LIFE

A. General Effects on Maternal and Fetal Well-Being

At sea level, the fetus lives at oxygen tensions between ~35 and 24 mm Hg, conditions that would be extremely hypoxic for an adult (Fig. 1). However, these oxygen tensions are clearly sufficient for development and do not appear to limit growth since the postnatal increase in oxygen tensions does not raise growth rates above prenatal levels. Several adaptations maintain a stable fetal oxygen environment, including high maternal placental blood flow, the high oxygen affinity of fetal hemoglobin, and the fetus' ability to increase oxygen extraction.

Are the pregnant woman and her fetus at high altitude equally well adapted? The answer is no as judged by the reduction in fetal oxygen tensions (Fig. 1), increased fetal hematocrit, decreased birth weight, and more frequent occurrence of maternal complications of pregnancy. The data on human maternal and fetal oxygen tensions during pregnancy or gestation are extremely limited. In pregnant ewes chronically exposed to ~4000 m, fetal O_2 tensions were about 10 mm Hg below sea level values (Fig. 1). No direct measurements have been reported for human uterine or umbilical O_2 tensions *in utero* but fetal arterialized scalp vein or cord O_2 tensions at delivery in 4200 m residents or women with cyanotic congenital heart disease (whose arterial oxygen saturations were in the range of high-altitude residents) were reduced by 5–10 mm Hg (Fig. 1). Better documented for humans is a reduction in birth weight, averaging 100 g per 1000 m altitude gain, as the result of retardation of intrauterine growth which becomes apparent after 32 weeks' gestation. Gestational age is not appreciably shortened in North America although some reports suggest an increased occurrence of preterm births in South America. The effect of altitude on birth weight is greater than that and operates independently from parity, cigarette smoking, maternal size, and other known factors. Since similar birth weight reductions are seen under conditions in which fetal–placental oxygen supply is reduced by maternal smoking, hypoxic lung disease,

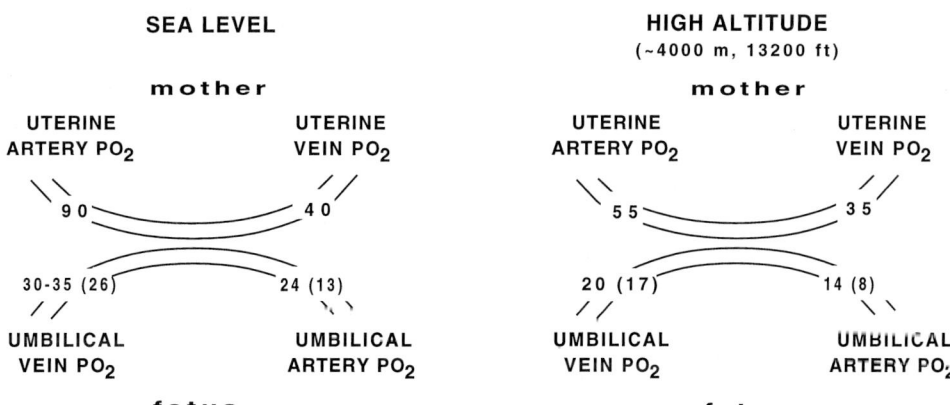

FIGURE 1 Maternal uterine and fetal umbilical oxygen tensions (mm Hg) at sea level and high altitude. Umbilical values were obtained near term *in utero* or at delivery (in parentheses) prior to the infant's first breath. Values are averaged from humans and sheep that had been at their respective altitudes throughout most of their pregnancy as reported by Dawes (*Handbook of Physiology, Respiration II*, pp. 1313–1328, 1964), FASEB (*Biological Handbooks—Blood and Other Bodily Fluids*, pp. 158–163, 1961), Jacobs *et al.* (*J. Dev. Physiol.* **10**, 97–112, 1988), Kitanaka *et al.* (*Am. J. Physiol.* R1340–R1347, 1989), Makowski *et al.* (*Am. J. Obstet. Gynecol.* **100**, 852–861, 1968), Meschia (*J. Reprod. Med.* **23**, 160–165, 1979), Novy *et al.* (*Am. J. Obstet. Gynecol.* **100**, 821–828, 1968), and Sobrevilla *et al.* (*Am. J. Obstet. Gynecol.* **111**, 1111–1118, 1971).

cyanotic congenital heart disease, and severe anemia, the hypoxia of high altitude is likely responsible for the IUGR observed. A three- or fourfold increase in the incidence of preeclampsia and abruptio placenta has been observed at high compared with low altitudes in Colorado and South America and likely contributes to the occurrence of IUGR.

B. Acute Hypoxia

There is a lack of comprehensive study detailing the effects of minutes to days or weeks of hypoxia on the pregnant woman and her fetus. Minutes of moderate hypoxia (~15% inspired O_2) produce transient, modest (<15 beats/min) increases or decreases in fetal heart rate. More severe hypoxia (~10% inspired O_2) is accompanied by prolonged fetal bradycardia which, in complicated but not normal pregnancies, is predictive of fetal distress during subsequent labor. Existing studies are without control for maternal arterial O_2 and CO_2 tensions or blood flow. While maternal ventilation and ventilatory responsiveness to hypoxia increase during pregnancy, the magnitude of change and thus the degree of hypoxia (and hypoxemia) at a given inspired O_2 concentration are likely to vary among women.

There is a reassuring lack of published case study reports of serious complications in the thousands of pregnant women who travel yearly to the Rocky Mountains and other high-altitude regions for recreational purposes. There has also been a singular lack of epidemiological study of the consequences of brief altitude sojourns. After days at moderate altitudes (1830–2225 m; 6000–7300 ft), third-trimester women showed the expected reduction in maximal exercise capacity but no change in fetal heart rate or maternal lactate, epinephrine, and norepinephrine responses to exercise. The limited information available is inconclusive as to the time course and period of pregnancy during which the maximal effects of acute hypoxia are likely to occur. In third-trimester sheep, the greatest hypoxemia occurred within the first week of exposure to 4320 m (14,260 ft); maternal and fetal oxygenation fell abruptly with ascent and then returned over a week's time toward low-altitude values as the result of elevations in maternal arterial oxygen saturation, uterine blood flow, and fetal oxygen capacity. In a small sample of women who moved to La Paz, Bolivia (3600 m; 11,880 ft), serious complications occurred in all four who arrived in their first trimester.

Given the available data, it seems prudent to recommend that women with known pregnancy complications—particularly those characterized by IUGR, preeclampsia, or placental insufficiency—avoid exposure to altitudes >2500 m. Healthy women with no known complications should limit their altitude exposure to <3660 m, or the altitude at which arterial oxygen saturation falls below 85%. Supplemental oxygen may be required to maintain arterial oxygen saturation above 85% for the first days of altitude exposure while ventilatory acclimatization is proceeding. It is important to recall that it is not the altitude but the maternal arterial oxygenation and uteroplacental blood flow that determine whether the mother and her fetus become stressed. Reductions in maternal oxygen transport stemming from smoking, lung disease, placental insufficiency, heavy exercise, or altitude illnesses such as high-altitude pulmonary edema render the pregnant patient and her fetus at the physiological equivalent of a higher altitude. Clinicians concerned about unmasking placental insufficiency at high altitude may find it useful to determine fetal response to an hypoxic gas breathing challenge. Clearly, much more clinical research is essential for the pregnant patient and her health care practitioner to make informed decisions about the risks of high-altitude exposure.

C. Chronic Hypoxia

Compensatory responses of the maternal and fetal oxygen transport systems defend oxygen supply; in their absence, maternal uterine and fetal umbilical oxygen tensions would be well below estimated values (Fig. 1). The compensatory responses to chronic hypoxia vary among and within populations in ways that relate to the occurrence of IUGR. The magnitude of birth weight reduction appears inversely related to the length of population ancestry at high altitude. This is true when carefully controlled comparisons are made between low- vs high-altitude samples from the same population (Fig. 2) or between women of different ancestry living at the same altitude. This

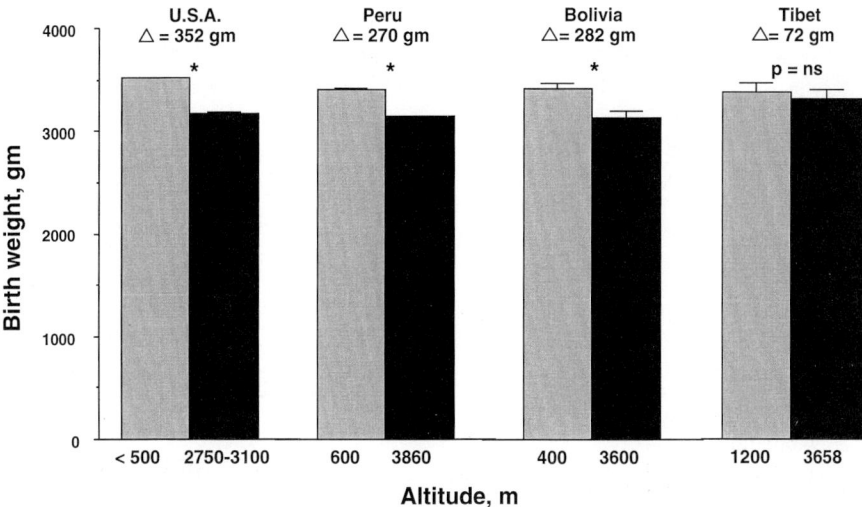

FIGURE 2 Comparing well-matched samples obtained by the same investigators at low vs high altitude; the reduction in birth weight is greatest in North Americans, intermediate in South Americans, and least in Tibetans. This gradient parallels the estimated duration of population residence at high altitudes (~50,000 years for Tibetans, ~10,000 years Aymara or Quechua residents of South America, and <150 years for European-derived residents of North America) (reprinted from S. Zamudio et al., Am. J. Physical Anthropol. 91, 215–224, 1993).

variation in IUGR probably reflects developmental as well as evolutionary influences on characteristics of maternal oxygen transport.

Pregnancy increases the ventilatory sensitivity to hypoxia and other chemosensory stimuli (Fig. 3) as the result of peripheral (carotid body) and central stimulatory effects of progesterone, estrogen, and increased metabolic rate. At high altitude, the increased ventilation raises arterial oxygen saturation. Furthermore, the magnitude of rise in a woman's hypoxic ventilatory response, ventilation, and arterial oxygen saturation relates positively to the birth weight of her infant. However, women at 4300 m in whom increased ventilation and oxygen capacity restored arterial oxygen contents to near sea level values still had smaller infants, suggesting that other factors are involved in the birth weight reductions observed. We have found lower uterine artery blood flow in normal Colorado residents of 3100 m due to a less pregnancy-associated increase in uterine artery diameter and lower total blood volume than at low altitude. Preeclamptic compared with normotensive women at high altitude demonstrate a third-trimester decline in uterine artery blood flow velocity and blood flow redistribution to favor the uterine artery, suggesting a further reduction in uterine artery blood flow. The relationship between the percentage reduction in flow estimated for normal and preeclamptic women at 3100 m and their infant's birth weight

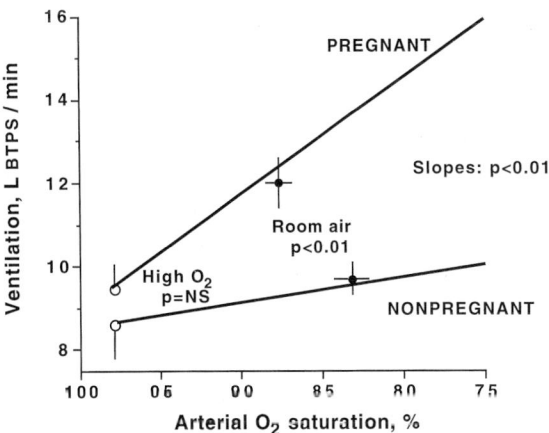

FIGURE 3 Increased maternal ventilatory sensitivity to hypoxia during pregnancy at high altitude raises resting ventilation while breathing room air (●). Straight lines show the average ventilatory response to acute isocapnic hypoxia in 21 near-term residents of 4300 m in Peru compared with values in the same women 13 ± 0 weeks postpartum (reprinted from L. G. Moore et al., J. Appl. Physiol. 60, 1401–1406, 1986).

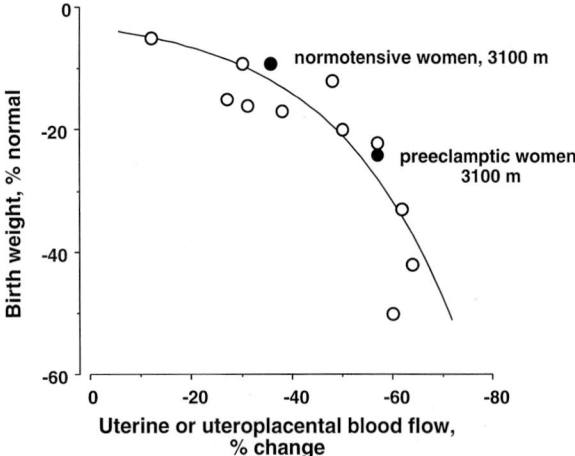

FIGURE 4 Relative to low-altitude or control values, chronic reductions in uterine or uteroplacental blood flow are associated with decreased birth weight in experimental animals (○) and normotensive and preeclamptic women residing at 3100 m (adapted from S. Zamudio et al., 1995).

agrees with that observed in a variety of experimental animal models of IUGR (Fig. 4). An attractive hypothesis is that hypoxia-related impairment of the growth and remodeling of uteroplacental vessels underlies the increased occurrence of both IUGR and preeclampsia at high altitude. Hypoxia *in vitro* inhibits the processes of cytotrophoblast differentiation and invasion involved in the uterine spiral arterial remodeling of normal pregnancy and is demonstrably deficient in preeclampsia. Differences in uterine artery blood flow may also explain the higher birth weights of babies born to Tibetan compared with Han high-altitude residents; during pregnancy, the Tibetan women distributed a greater proportion of their pelvic (common iliac) blood flow to the uterine artery than did the Han women.

IV. NEONATAL, INFANT, AND MATERNAL MORTALITY

In Colorado and other mountain states, IUGR is not associated with an increase in neonatal or infant mortality, although such an increase, particularly for infants who were both growth-retarded and premature, existed until recently. In South America, Ladakh, and other areas of the world, neonatal and infant mortality are markedly elevated at high compared with low altitudes. Maternal mortality is unrelated to altitude in the United States but increases at high altitude in South America. The more frequent occurrence of IUGR, preeclampsia, and other maternal and fetal complications at high altitude is likely to contribute to the greater mortality, but other factors, such as nutrition, cold, and access to and quality of health care, are also certainly involved.

See Also the Following Article

PREECLAMPSIA/ECLAMPSIA

Bibliography

Carter, A. (1989). Factors affecting gas transfer across the placenta and the oxygen supply to the fetus. *J. Dev. Physiol.* 12, 305–322.

Genbacev, O., Joslin, R., Damsky, C., Polliotti, B., and Fisher, S. (1996). Hypoxia alters early gestation human cytotrophoblast differentiation/invasion in vitro and models the placental defects that occur in preeclampsia. *J. Clin. Invest.* 97, 540–550.

Goldstein, M. C., Tsarong, P., and Beall, C. M. (1983). High altitude hypoxia, culture and human fecundity/fertility: A comparative study. *Am. Anthropol.* 85, 28–49.

Gonzales, G. F. (Ed.) (1993). *Reproduccion Humana en la Altura.* Consejo Nacional de Ciencia y Tecnologica, Lima, Peru.

Hackett, P. (1998). Interaction of altitude with other stressors. In *High Altitude* (T. F. Hornbein and R. B. Schoene, Eds.). Dekker, New York.

Monge, C. M. (1948). *Acclimatization in the Andes.* Johns Hopkins Univ. Press, Baltimore.

Moore, L. G., Zamudio, S., Curran-Everett, L., Torroni, A., Jorde, L. B., Shohet, R. V., and Thupten, D. T. (1994). Genetic adaptation to high altitude. In *Advances in Exercise and Sports Medicine* (S. Wood, Ed.), Vol. 76, Lung Biology and Health, pp. 225–262. Dekker, New York.

Niermeyer, S., Zamudio, S., and Moore, L. G. (1998). The people. In *High Altitude* (T. F. Hornbein and R. B. Schoene, Eds.). Dekker, New York.

Tatsumi, K., Hannhart, B., and Moore, L. G. (1995). Influences of sex steroids on ventilation and ventilatory control. In *Regulation of Breathing* (J. A. Dempsey and A. I. Pack, Eds.), 2nd ed., Vol. 79, Lung Biology and Health, pp. 829–864. Dekker, New York.

Zamudio, S., Palmer, S. K., Stamm, E., Coffin, C., and Moore, L. G. (1995). Uterine blood flow at high altitude. In *Hypoxia and the Brain* (J. R. Sutton and C. S. Houston, Eds.), pp. 112–124. Queen City Press, Burlington, VT.

Altricial and Precocial Development in Birds

F. M. Anne McNabb

Virginia Tech

I. General Patterns of Altricial and Precocial Development
II. Growth and Development
III. Development of Temperature Regulation
IV. Development of Endocrine Control Systems
V. Evolution and Adaptive Significance of Altricial and Precocial Development

GLOSSARY

altricial Referring to a pattern of development in which birds are hatched at a relatively early stage of development and remain in the nest with parental care for some time.
development All of the processes involved in the differentiation of specific tissues plus growth in body mass.
homeothermic Having a regulated, constant body temperature independent of changes in environmental temperature; warm-blooded or endothermic, i.e., using heat produced by metabolism.
poikilothermic Having body temperatures that vary with environmental temperature; cold-blooded or ectothermic, i.e., obtaining heat from outside the body.
precocial Referring to a pattern of development in which birds are hatched at a relatively mature stage of development and are capable of leaving the nest.
thermoregulation The regulation of constant body temperature by metabolic activities that produce or dissipate heat depending on the environmental temperatures to which the animal is exposed.

The terms altricial and precocial refer to patterns of development in which the young are hatched at a relatively early stage of development (altricial) or at a much more advanced stage of development (precocial). In general, altricial nestlings are nestdwelling (nidicolous) and dependent on parental care; precocial chicks are nest-fleeing (nidifugous), i.e., they leave the nest soon after hatching or immediately after hatching, and are much more self-sufficient. The terms altricial and precocial also refer to a suite of developmental characteristics at hatch as well as to behavioral and parental care patterns. Although most birds can easily be categorized as either altricial or precocial, these terms really represent the ends of a continuum and intermediate patterns can be subdivided into a more detailed classification. Examples of birds with altricial development include songbirds, pigeons, parrots, woodpeckers, and hummingbirds. Examples of precocial species include ducks, chickens, quail, grouse, and ostriches. Intermediate patterns include examples such as semiprecocial gulls and terns, which are able to leave the nest after a few days but stay in the general area, and semialtricial herons and hawks, which stay in the nest for several weeks but are capable of coordinated movements within the nest. It should be noted that mammals, like birds, also may be categorized as altricial or precocial with respect to their developmental patterns.

I. GENERAL PATTERNS OF ALTRICIAL AND PRECOCIAL DEVELOPMENT

Altricial development in birds is characterized by hatching of the young at a relatively early stage of development. There is a general trend toward altricial species laying smaller eggs with lower energy content (for embryonic nourishment) and having shorter incubation periods. Thus, it appears that these species

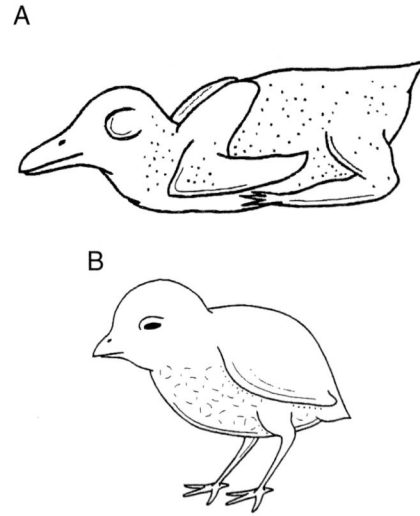

FIGURE 1 General characteristics of altricial doves (a) and precocial quail (b) at 1 or 2 days posthatching.

put less energy into egg production and incubation and more energy into posthatching care of the nestlings. Altricial hatchlings typically are unfeathered or only sparsely feathered (i.e., they lack insulation), their eyes are not open, and they are incapable of coordinated body movements (Fig. 1a). A consequence of hatching relatively early in development is that the young are dependent on parental care and confined to the nest for some time after hatch.

The parents of altricial nestlings must provide food and protection until the nestlings are capable of flight (fledging) and of leaving the nest to assume independent life. The relative helplessness and vulnerability of altricial nestlings means they require considerable protection. The nests of these species often are in locations inaccessible to most (but not all) predators, and parents may alternate with each other in spending time on the nest and foraging for food for the young. An adult bird arriving at the nest is greeted by begging nestlings with wide-gaping mouths into which the parent regurgitates food it has captured and stored in its crop. Nestlings often have their crops greatly distended with food. Characteristically, the parents of altricial young spend most of their daylight hours supplying food to the nestlings for 1–3 weeks after the young hatch.

Precocial development in birds is characterized by hatching of the young at much more advanced stages of development than is the case for altricial birds. Precocial species generally lay larger eggs with a larger proportion of yolk and have longer incubation periods than altricial species. Thus, precocial hens have relatively higher energy use in egg production and incubation but relatively less energy use in the care of chicks than do altricial birds. However, in some cases very mobile precocial young may require considerable attention on the part of parents, especially in cases in which predation is likely.

Precocial chicks typically look like miniature adults of the species; their body proportions and shape are generally like those of the adults (Fig. 1b). At hatching, precocial chicks typically are covered with down feathers, their eyes are open, their nervous systems are well developed, and they are capable of coordinated locomotion within minutes to hours after hatching. Precocial chicks typically begin to drink and eat within the first day after hatching. However, during hatching the yolk sac and its remaining nutrient stores are pulled into the body and are an important source of nutrition for the next several days.

Early learning of precocial chicks is facilitated by their "imprinting" to the mother. Imprinting is the process whereby a chick develops its sense of species identity by associating with its mother or another adult during a critical learning period in early posthatching life. In precocial young that are capable of locomotion and independent feeding, this identification with the mother results in behaviors that enhance the survival of young chicks that must learn about food sources and are vulnerable to predation or harsh environmental conditions. Thus, chicks imprinted to the mother will remain relatively close to her or other members of the same species, can follow and learn about food sources from adults, can respond to the mother's vocalizations, and will be brooded by her for protection from severe weather.

II. GROWTH AND DEVELOPMENT

Altricial nestlings typically have "distorted" body proportions and do not look like their parents. They have well-developed heads, large mouths with wide gapes, large crops, and distended abdomens due to a large visceral mass. Organ systems that must assume function during the perihatch period are the respiratory system, those of the gastrointestinal tract, and, to a lesser extent, the nervous system. The appendages of altricial nestlings are poorly developed and incapable of coordinated locomotion; their legs cannot support the body in an upright position. The wings must undergo considerable growth, maturation, and flight feather development before the young are capable of flight. Water content of the tissues is high until the tissues reach functional maturity and the proportion of body water (relative to dry solids) has been used as an index of the degree of maturation of both altricial and precocial young.

Precocial chicks have much more mature body development and the muscular, skeletal, and sensory/nervous systems are considerably more advanced at hatch than is the case for altricial nestlings. As in altricial young, the respiratory and gastrointestinal organs must have functional maturity from the perihatch period onward. In addition, the functional maturity of most other organ systems is also necessary for the relative independence of precocial chicks early in posthatching life.

Investigations of tissue differentiation and maturation in altricial vs precocial chicks have not shown the striking differences in key maturation events that might have been expected based on the organ system patterns described previously. For example, the beginning of cartilage to bone transition in many skeletal elements occurs at about the same developmental stage in altricial and precocial young. However, there may be differences in the extent of bone formation (ossification), with altricial nestlings having proportionately less ossification and precocial chicks having proportionately more at a given stage of posthatching development. Generally, these studies show that the proportion of a tissue remaining in the less mature state (e.g., cartilage) is a positive indicator of the remaining growth potential of that tissue. Similar results have been found in tissues such as muscle.

The inverse relationship between tissue functional maturity and growth potential seems to be a key factor in the differences in growth of altricial and precocial species. High growth rates are characteristic of most but not all avian species with altricial development and delayed maturation of function in many tissues. Precocial young, which have earlier maturation of many tissue functions, have slower body growth rates (Fig. 2).

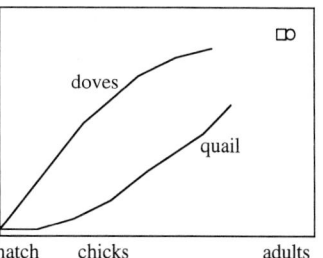

FIGURE 2 Body weight patterns in altricial doves and precocial quail from hatching to 3 weeks of age. Adult body weights of doves (□) and quail (○) are equivalent.

III. DEVELOPMENT OF TEMPERATURE REGULATION

The patterns of body temperature (thermoregulatory) development are strikingly different in altricial and precocial species of birds. Altricial young are hatched in a poikilothermic or cold-blooded state. Although altricial nestlings are capable of detecting temperature change and some may shiver in the cold, their capacity for heat production is very limited. In addition, their insulation often is poor, so their capability for heat retention is also limited. Consequently, their body temperatures change with the temperature of the environment (except when brooded by the parents) and their metabolic rates are low. Altricial young remain poikilothermic for some time (generally 1 or 2 weeks, depending on the species) then develop homeothermy relatively rapidly. During their period of poikilothermy, the nestlings increase in body size (which decreases their relative surface area) and they begin to grow an insulative feather covering. When the ability of the nestlings to produce metabolic heat (endothermy) increases, both the relative reduction in surface area due to growth and the insulation provided by the feathers are important in conserving metabolic heat for attaining homeothermy. Body temperatures begin to increase and the nestlings shiver vigorously in response to a decrease in the environmental temperature. These processes continue and typically result in the maintenance of constant body temperatures similar to those of adults by the time fledglings leave the nest.

Altricial young may develop the capability for coping with hot temperatures before they are capable of dealing with cold. Gular flutter, which refers to rapid shallow "panting" that evaporates water and withdraws heat to cool the body, appears earlier than the mechanisms for dealing with cold in some altricial nestlings. Heat may potentially be of more danger to young confined to the nest than cold.

Several altricial nestlings huddled in a group may develop "effective homeothermy or endothermy" before each is capable of homeothermy as an isolated individual. This is because huddling as a group in the nest exposes less total surface area relative to the mass of the group than would be the case for isolated individuals. For nestlings that are in the process of developing their capacity for metabolic heat production, the reduced heat loss achieved by huddling is of considerable advantage and allows them to achieve high and stable body temperatures while in a group.

Precocial chicks, which are much more mature metabolically than altricial chicks, begin to thermoregulate from hatching onward. They are capable of sufficient metabolic heat production to regulate their body temperatures well above those of the environment. Precocial hatchlings further increase their metabolic rates by shivering in response to a decrease in environmental temperature. However, precocial hatchlings are imperfect thermoregulators, their capability for metabolic heat production is limited, and the body temperatures maintained by young chicks are lower than those of adults. During the first few days to weeks posthatch, the chicks attain adult body temperatures and the range of environmental temperatures at which they can remain homeothermic is extended. Body growth (which reduces the relative surface area) and replacement of down by juvenile plumage with improved insulative qualities are both important factors in this refinement of their thermoregulatory ability. As in altricial young, huddling in groups or brooding by parents can be very important to precocial young during harsh weather conditions.

IV. DEVELOPMENT OF ENDOCRINE CONTROL SYSTEMS

Hormones play important roles in the development and growth of individual tissues and of the body as a whole. Among the most important hormones in controlling the development of individual tissues are thyroxine and triiodothyronine, the hormones produced by the thyroid gland. Of these two hormones, triiodothyronine is considered to be responsible for most thyroid hormone actions. Thyroid hormones trigger differentiation (the development of individual tissue characteristics and functions) in a number of tissues and are necessary along with other hormones for general body growth. Thyroid hormones have developmental effects in all classes of vertebrate animals. In addition, in homeothermic birds and mammals, thyroid hormones are the key controllers of the high metabolic rates that are necessary for homeothermy.

The patterns of thyroid development are distinctly different in altricial and precocial birds, as might be expected from the differences in thermoregulatory development and the importance of thyroid hormones in controlling metabolic heat production. In altricial embryos the thyroid gland shows little functional development during embryonic life and around hatching (the perihatch period). After hatching the thyroid gradually increases its capability for hormone production and release but the thyroid hormone concentrations in the blood remain low for several days. Then thyroid hormone concentrations in the blood increase and the young develop thermoregulatory responses to cooling and are able to maintain constant body temperatures (Fig. 3a).

In contrast to the relatively late thyroid development in altricial nestlings, the thyroid glands of precocial species attain a relatively high level of activity during the latter half of embryonic life. During the perihatch period, there are dramatic peaks of thyroid hormone concentrations in the blood and these high hormone concentrations are important in the initial thermoregulatory responses that appear at hatch. After the perihatch period, thyroid hormone concentrations in the blood decrease, then gradually increase to adult levels (Fig. 3b).

The other hormones that are most important in growth and development are growth hormone, the insulin-like growth factors (IGF-1 and IGF-2), insulin, and corticosteroid hormones. Growth hormone exerts its effects on body growth by stimulating the liver to produce IGFs that are released into the circulation and directly trigger growth in many tissues

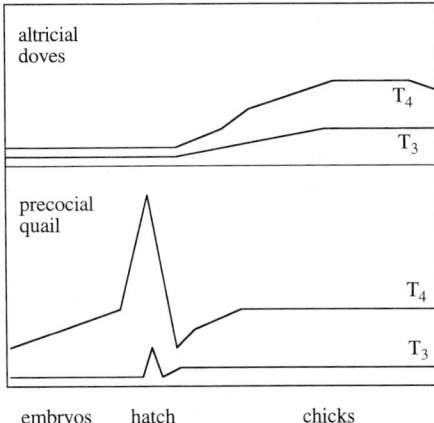

FIGURE 3 Patterns of plasma concentrations of thyroid hormones (T4, thyroxine; T3, triiodothyronine) in altricial and precocial birds from midincubation to 2 or 3 weeks posthatch. The characteristic patterns shown here are for altricial doves and precocial quail. For references to the data on individual species, see McNabb and Olson (1996) and McNabb et al. (1998).

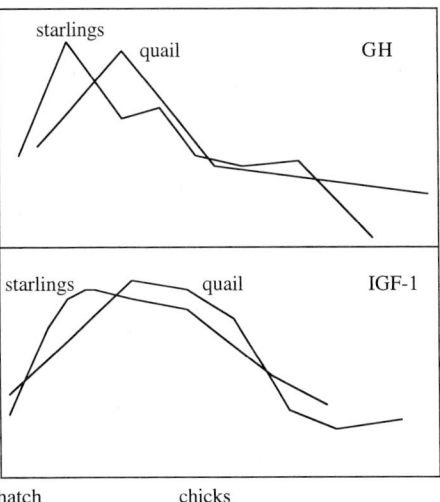

FIGURE 4 Patterns of plasma concentrations of growth hormone (GH) and insulin-like growth factor-1 (IGF-1) in altricial and precocial birds. The characteristic patterns shown here are for altricial starlings and precocial quail from hatch to 4 weeks of age. For references to data on other species see McNabb et al. (1998).

throughout the body. Circulating concentrations of growth hormone and IGFs have similar patterns, with some shifts in the time scale, in altricial and precocial birds (Fig. 4). Thus, circulating patterns of these growth-promoting hormones do not appear to be the key controlling factors in the different tissue/organ growth patterns observed in these two developmental modes. However, circulating patterns of growth factors may not provide all the relevant information because some tissues independently produce IGFs or other growth factors which have local effects within the tissue. Although insulin and corticosteroids are known to affect some aspects of development in chicken embryos, the effects of these hormones have not been studied systematically throughout development in either precocial or altricial birds. Likewise, differences in how thyroid hormones, growth hormone, or IGFs may be involved in triggering the different functional maturation rates of organs in altricial vs precocial embryos and chicks have not been investigated. Examples of known effects of these hormones on differentiation suggest such studies are likely to aid our understanding of these developmental modes.

V. EVOLUTION AND ADAPTIVE SIGNIFICANCE OF ALTRICIAL AND PRECOCIAL DEVELOPMENT

Evolutionarily, precocial development appears to have been the original developmental pattern, based on a number of birds that are considered primitive. Altricial patterns have evolved independently in a number of types of birds. Both developmental patterns appear to have adaptive significance in the environmental conditions specific to certain habitats. Thus, one or the other of these developmental modes may be particularly well suited for reproductive success in a particular habitat with its specific food sources and potential nest locations as well as its predators and characteristic environmental conditions.

See Also the Following Articles

AVIAN REPRODUCTION, OVERVIEW; IGF (INSULIN-LIKE GROWTH FACTORS); PARENTAL BEHAVIOR, BIRDS; THYROID HORMONES, IN SUBAVIAN VERTEBRATES

Bibliography

de Pablo, F. (1989). Insulin and insulin-like growth factors (IGFs) in avian development. In *Growth Factors in Mammalian Development* (I. Y. Rosenblum and S. Heyner, Eds.). CRC Press, Boca Raton, FL.

Faaborg, J., and Chaplin, S. B. (1988). The anatomy and physiology of reproduction. In *Ornithology. An Ecological Approach*. Prentice Hall, Englewood Cliffs, NJ.

Gill, F. B. (1995). Growth and development. In *Ornithology*, 2nd ed. Freeman, New York.

McNabb, F. M. A., and King, D. B. (1993). Thyroid hormone effects on growth, development and metabolism. In *The Endocrinology of Growth, Development, and Metabolism in Vertebrates* (P. K. T. Pang, C. G. Scanes, and M. P. Schreibman, Eds.). Academic Press, New York.

McNabb, F. M. A., and Olson, J. M. (1996). Development of thermoregulation and its hormonal control in precocial and altricial birds. *Poultry Avian Biol. Rev.* 7, 111–125.

McNabb, F. M. A., Scanes, C. G., and Zeman, M. (1998). The endocrine system. In *Avian Growth and Development: Evolution within the Altricial–Precocial Spectrum* (J. M. Starck and R. E. Ricklefs, Eds.). Oxford Univ. Press, New York.

Nice, M. M. (1962). Development of behavior in precocial birds. *Trans. Linnean Soc. New York* 8, 1–211.

Ricklefs, R. E. (1983). Avian postnatal development. In *Avian Biology* (D. S. Farner J. R. King, and K. C. Parkes, Eds.), Vol. VII. Academic Press, New York.

Ricklefs, R. E., and Starck, J. M. (1998). The evolution of developmental mode in birds. In *Avian Growth and Development: Evolution within the Altricial–Precocial Spectrum* (J. M. Starck and R. E. Ricklefs, Eds.). Oxford Univ. Press, New York.

Schreck, C. B. (1993). Glucocorticoids: Metabolism, growth, and development. In *The Endocrinology of Growth, Development, and Metabolism in Vertebrates* (P. K. T. Pang, C. G. Scanes, and M. P. Schreibman, Eds.). Academic Press, New York.

Starck, J. M., and Ricklefs, R. E. (1998). Patterns of development: The altricial–precocial spectrum. In *Avian Growth and Development: Evolution within the Altricial–Precocial Spectrum* (J. M. Starck and R. E. Ricklefs, Eds.). Oxford Univ. Press, New York.

Visser, G. H. (1998). Development of temperature regulation. In *Avian Growth and Development: Evolution within the Altricial–Precocial Spectrum* (J. M. Starck and R. E. Ricklefs, Eds.). Oxford Univ. Press, New York.

Altruism in Insect Reproduction

Laurence Packer
York University

I. Introduction
II. Examples of Altruistic Behavior
III. The Evolution of Altruism in Social Insects
IV. Caste Determination
V. Nestmate Recognition

Glossary

caste Any set of individuals that are specialized in behavior.

eusocial Describing a condition in which a reproductive division of labor between generations occurs, typified by adults living together in a group, cooperative brood care, and overlapping generations.

haplodiploidy The sex determining system whereby unfertilized eggs develop into males and fertilized eggs usually into females.

relatedness The proportion of genes shared among individuals that are identical by descent (rather than simply identical in state).

semisocial Describing a condition in which a reproductive division of labor occurs among individuals of the same generation.

Reproductive altruism occurs when one or more individuals in a group are morphologically and/or behaviorally modified for reduced direct reproduction while augmenting the survival and/or reproductive success of others. Such altruism may be temporary, with altruists helping in the natal nest before initiating their own, or it may be permanent, with the altruists being sterile; there are many gradations between these two extremes. Altruists may augment their parents' reproductive success in a eusocial society such as in most familiar social insects (ants, honeybees, yellowjackets, and termites), or, less commonly, they may aid the reproduction of members of their own generation in a semisocial society. Our understanding of the evolution of a reproductive division of labor has been aided by the concept of kin selection whereby the altruist increases its genetic contribution to further generations by increasing the number of copies of its own genes through aiding the reproduction of relatives. However, other factors such as the existence of a defensible resource such as a nest also seem to be important.

I. INTRODUCTION

The existence of sterile workers within social insect colonies was a difficulty for Charles Darwin to explain: How can the survival of the fittest with individuals maximizing their genetic contributions to future generations lead to a situation in which most individuals in a colony are sterile? It was not until theoretical developments in the 1960s that a detailed explanation of the evolution of reproductive altruism became possible: An individual can increase the number of copies of its own genes in future generations by augmenting the reproductive success of close relatives which share many of the same genes identically by descent. This kin selection hypothesis potentially provides a powerful explanation for phenomena such as the self-sacrificial behavior of soldier aphids, the death by partial evisceration of worker honeybees, and the efficient gathering of food for their larval siblings by army ant workers.

An individual may be selected to be altruistic if $rb > \frac{1}{2} c$, where c is the cost of the act of altruism in terms of the number of offspring lost to which the altruist would have been related by $\frac{1}{2}$, and b is the benefit in terms of the relative's gain in offspring to which the altruist is related by r (this is one of several possible formulations of Hamilton's rule, named after W. D. Hamilton, the originator of the concept of kin selection).

Students of insect sociobiology have put a great deal of effort into measuring the degrees of relatedness among nestmates in order to test the kin selection hypothesis. Use of allozyme markers provides very rough, populationwide estimates of relatedness. Recently, the application of molecular methods, such as DNA fingerprinting and microsatellites, has increased the precision of relatedness estimates and permitted estimates to be made between specific individuals, although thus far few studies have employed such precise techniques. Relatedness values between nestmates in many social insects are often very low, casting some doubt on the efficacy of kin selection. However, the other two parameters, b and c, need to be measured for an adequate test of the hypothesis, but this is extremely difficult under natural conditions. Future efforts will have to combine accurate estimates of relatedness between individuals with detailed field estimates of b and c.

Another problem with many of the empirical studies is that they concentrate on the more familiar social insects such as termites, honeybees, ants, and paper wasps. The fossil record indicates that these taxa have been eusocial since at least the Cretaceous and so their social origins are obscured by the mists of time and their have been many generations for factors other than kin selection to shape the dynamics of colony organization in these insects. Other groups, such as thrips, aphids, and sweat bees, contain many species which are solitary as well as those which have altruistic workers. There are some sweat bee species in which both solitary and social forms are found in different populations or even within the same nesting aggregation. This suggests that tests of the kin selection hypothesis might be more profitably directed at these organisms which appear to be closer to the origins of a reproductive division of labor. However, additional phylogenetic research is required so we can be sure that it is the origin of a reproductive division of labor that is being investigated rather than its loss.

II. EXAMPLES OF ALTRUISTIC BEHAVIOR

The examples given below concentrate on the lesser known instances of altruistic behavior in insects and/or those examples which seem to throw most light on the origins of altruism. Detailed accounts of the biology of the more familiar social insects can be found in the literature noted in the bibliography.

A. Partial Altruism in Australian Gall Thrips

Thrips are generally tiny, hemimetabolous insects that feed on plant juices using their asymmetrical stylet-like mouthparts. Two species of the genus *Oncothrips* are known to be eusocial: They construct galls on acacias in arid regions of Australia. Galls are initiated by single, mated, macropterous (large winged) females in the spring. The gall grows around the foundress (the female initiating the gall) and when she is completely enclosed she oviposits and the offspring feed on the inner surface tissues of the gall. Adult offspring of both sexes are dimorphic, being either macropterous or micropterous (small winged). The latter, in addition to having smaller wings, have enlarged forelegs and eclose earlier than their macropterous sibs. These more muscular individuals are disproportionately involved in gall defense against caterpillars, ants, and inquiline thrips of the genus *Koptothrips*. (Inquilines are a socially parasitic form which have no workers of their own and in which the entire life cycle is spent inside the host's nest.) The soldiers do not attack conspecifics from other galls. The soldier thrips are not sterile but appear to have reduced fecundities and are also more likely to die in gall defense than their nonaltruistic gallmates.

B. Temporary and Permanent Altruistic Soldiers in Gall Aphids

It is difficult to imagine soldier behavior being exhibited by tiny, soft-bodied aphids. Nonetheless, soldier forms have arisen at least four times among the aphids, always in species which live in galls. Soldier aphids are either first- or second-instar individuals which have more heavily sclerotized bodies and stronger legs than their normal conspecifics. Defensive behaviors involve piercing and stinging with the stylet mouthparts, scratching with enlarged tarsal claws, or butting and piercing with horns on the head. Predatory hover fly larvae, caterpillars which might consume the gall, and even vertebrates that may disturb the aphids can be repulsed by the attacks of aphid soldiers. One study found that although the proportion of individuals inside the gall that were soldiers ranged from 15 to 53%, 98% of the aphids that attacked intruders were soldiers. In some aphid species, such as *Pemphigus spyrothecae*, the soldiers are temporarily altruistic in that they are capable of developing past the soldier stage into adults but presumably are less likely to survive as a result of their defensive behavior. In taxa such as *Pseudoregma shitosanensis* the second-instar soldiers are incapable of moulting and are therefore permanently altruistic, sterile individuals.

C. Eusociality in an Ambrosia Beetle

Austroplatypus incompertus is a wood-boring beetle that makes gallery tunnels inside the heartwood of *Eucalyptus* trees. Single mated females initiate the burrows in autumn and take 7 months to tunnel through the bark and sapwood before reaching the heartwood where propagules of symbiotic fungi are deposited. The larvae and adults feed on the fungi that line the galleries. In colonies over 4 years old, there was an average of 4.7 unmated, apparently sterile, daughter beetles in addition to the gallery foundress and an average of 36 larvae. These colonies are eusocial in that the queen beetle appears to concentrate on egg laying while the workers enlarge, clean, and defend the galleries without reproducing directly. The sterile daughters had lost their tarsi (the terminal segment of the leg) suggesting that they could not gain sufficient purchase on a tree trunk to be able to initiate a new gallery system even if they were to attempt independent reproduction away from the natal burrow.

D. Sterile Defenders in a Polyembryonic Wasp

Some parasitoid wasps lay single eggs which develop into large numbers of larvae, a situation referred to as polyembryony. In several genera of the parasitoid wasp family Encyrtidae, dimorphic larvae have been found with normally developing individuals coexisting inside the flour moth larva host along with sterile defenders which have scythe-like mandibles. These defenders search out other, competing, parasitoids living in the host and dismember them. It has also been shown that in cases in which the adult wasp oviposits one female and one male egg and each develop into hundreds of larvae, the defender morphs seek out male larvae and kill them. This apparently less than altruistic behavior serves to remove excess males (only a few are needed to mate with their sisters) and provide a female-biased sex ratio (see below).

E. Temporary Altruism in a Stenogastrine Wasp

The wasp family Vespidae contains solitary potter and honey wasps in addition to the familiar yellowjackets, paper wasps, and hornets. A less well-known subgroup is the Stenogastrinae or hover wasps, which form small colonies in the Asian tropics. The species *Liostenogaster flavolineata* was studied in detail in Malaya. Nests were usually initiated by single females, although 10% were started by two individuals. Mature colony sizes average approximately two adults in that almost half have only one female and a mere 23% have more than two individuals. In multifemale nests the dominant, egg-laying females average 3% larger than the subordinate(s). Eclosing adult daughters function as temporary workers on their natal nest; approximately 25% of them replace the dominant individual later and others leave the nest to initiate their own or to join an already established nest where they later attempt to become the dominant individual.

F. Cyclical Oligogyny in Swarm Founding Wasps

Vespid wasps of the tribe Epiponini initiate new nests by swarming; the original colony fissions and the daughter colonies contain several to many queens and dozens to thousands of workers. There is a great diversity of nest architecture in the tribe, with some species making flat, camouflaged nests on tree trunks or large leaves and others making enormous, multicombed, pendulous nests hanging from branches. Some species, such as *Epipona guerini*, construct the entire nest down to the last brood cell before the first egg is laid, a most prescient form of family planning! Given that the number of queens on a nest of these species can approximate 1000, it is difficult to envisage a situation in which kin selection among close relatives can operate in these colonies. However, the colony cycle seems to permit higher relatednesses than might be expected: Cyclical oligogyny is a brood production schedule whereby the number of queens is greatly reduced over time and a new batch of queens is produced only when queen number is at its lowest. As a result, the queens in the swarm are themselves closely related.

G. Social Diversity in the Sweat Bees

The bee subfamily Halictinae contains thousands of species in several dozens of genera and subgenera at least 6 of which contain both solitary and social species. This taxonomic distribution of sociality suggests that a reproductive division of labor has originated many times independently among these bees. Nests are initiated in spring by mated females, often alone, although in some species multiple foundress associations functioning as a semisocial society may contain up to 6 individuals. Some species produce a single brood of workers, of at most 6–8 daughters in single foundress colonies, and some produce as few as an average of less than one (i.e., some foundresses remain solitary with no workers, whereas others may have one or at most two). Still other species have several worker broods during the summer and the number of adults in a nest may exceed 1000, as in *Seladonia lutescens*. One species, *Lasioglossum marginatum*, is known to have perennial colonies with one worker brood each spring for 5 or 6 years before males and potential queens are raised. An interesting intermediate between the more normal annual and the perennial colony cycles occurs in at least six species. This "delayed eusociality" involves a founding female producing a small brood, the females of which overwinter in their natal nest. If their mother survives a second winter, one or more of these daughters may become a worker and their mother the dominant egg layer. If the mother does not survive the winter, the sisters may form a semisocial society the following spring.

H. Eusociality in a Sphecid Wasp

The Sphecidae, commonly referred to as digger wasps, are a diverse family of insects which collect hexapods or spiders for their developing offspring. Most are solitary, a few are known to share nests, and only one species (*Microstigmus comes*) has been shown to be eusocial (although other related species are likely to also be eusocial). These tiny wasps (adults are approximately 2 mm long) construct nests from fibers scraped from leaves. They gather these fibers together into a mass which, with the aid of silk secretions, is fashioned into a tiny pendulous nest within which brood cells are hollowed out. From three to eight adults can be found on mature nests, although nest initiation is usually by one to several wasps. Dominant individuals are larger than the others and are responsible for most or all of the oviposition. Subordinates, whether at the semisocial or eusocial stage of colony development, behave submissively to the queen and are responsible for gathering the springtails that serve as food for the larvae.

I. Temporary Altruism in Single-Site Nesting Termites

The more primitive termites of the families Termopsidae and Kalotermitidae have colonies that live in a single piece of wood that serves as both a nest and as food. Colonies grow to a few hundred to a few thousand individuals and die out as the resource is exhausted after between 4 and 15 years or so, longer in larger pieces of wood. Colonies are initiated by one reproductive pair, which cares for the first brood of offspring until they reach the third instar, when they are old enough to care for themselves and for younger siblings. Most individual termites in these colonies develop from egg to young helpless offspring, to functional worker, and then to an alate (a winged dispersing reproductive). Most workers eventually become reproductives initiating their own nests. Exceptions are those that become (i) replacement reproductives if the original founding royals die or the piece of wood becomes fragmented, (ii) soldiers (although even here reversionary molts are possible such that a soldier can lose its fighting morphology and become a replacement reproductive), and (iii) permanent workers—a minority of workers remain as such throughout their lives. As the resources of the log are depleted, most individuals develop directly to the alate form to disperse and initiate nests elsewhere. This flexible type of altruism where the majority of individuals end up reproducing directly seems to be ancestral to the more complex and less flexible caste organizations that predominate in other termite families.

III. THE EVOLUTION OF ALTRUISM IN SOCIAL INSECTS

It is the very origin of a reproductive division of labor which requires explanation. There are two components to this: pattern and process. We need to know where in the evolutionary history of a group of organisms castes initially developed (i.e., the pattern of distribution of sociality) and then we can test the predictions of the various process models using the most appropriate test organisms.

A. The Phylogeny of Altruism

Worker altruism has arisen many times within the insects (Tables 1 and 2). By mapping the solitary or social nature of taxa onto a phylogeny for groups containing both behavioral types, we can establish the most parsimonious estimates for the number of evolutionary changes in behavior. Take, for example, the hypothetical phylogeny shown in Fig. 1a. There are three species which are solitary and five which are social. The most parsimonious explanation of evolutionary change in behavior in these organisms is for a switch from solitary behavior to sociality in the common ancestor of species 3–8 with a reversal to solitary behavior in species 8; this scenario requires two evolutionary changes (Fig. 1b). An alternative hypothesis stating that the solitary behavior of species 8 has to represent the complete lack of eusociality in this species' ancestry would require each of species 3–7 to have evolved sociality independently to give a total of five evolutionary changes

TABLE 1
The Taxonomic Distribution of Eusociality in Nonhymenopterous Insects along with the Number of Times Eusociality Is Thought to Have Arisen in Each Group, the Number of Eusocial Species, and a Comment on How Likely It Is That These Estimates Will Change Based on How Good Our Knowledge Is of Each Particular Group

Order	Distribution within the order	Number of origins	Number of species	State of knowledge
Isoptera	All	1	Thousands	Good
Thysanoptera	One genus—*Oncothrips*	≥1	≥2	Poor
Homoptera	Only in the family Pemphigidae	≥4	≥10	Poor
Coleoptera	Only in *Austroplatypus*	≥1	≥1	Poor

(Fig. 1c). The former scenario is preferred on logical grounds.

By performing formal phylogenetic analyses it has been possible to show that sociality originated once in ants, once in the hover wasp/paper wasp/yellowjacket lineage, once in termites, and probably once in the bumble bee/honeybee/stingless bee lineage. These are the groups for which the required detailed information is available. In other instances, it has been possible to show that for most of the Halictine bee species that possess both solitary and social behavior, either in the same or in different populations, it is the solitary behavior which represents the most recent evolutionary innovation; i.e., the recent evolutionary change has been a reversal to solitary behavior. These reversals seem to be associated with occupation of high-latitude or high-altitude habitats where the summer is too short to permit the development of both a worker and a reproductive brood.

Nonetheless, phylogenetic data do seem to suggest that Halictine bees, Thrips, *Austroplatypus* beetles, and aphids are good organisms for testing process theories of the evolution of worker altruism. Additionally, communal behavior (the situation in which individuals share a nest but exhibit no reproductive division of labor—basically sharing nothing more than the nest entrance and acting as solitary individuals otherwise) occurs in many different genera of sweat bee. However, phylogenetic analysis indicates that communal behavior is found in different lineages from those which have societies with altruists. This indicates that different selective processes are responsible for forming communal societies and those with reproductive altruism.

Additional insights from such phylogenetic ap-

TABLE 2
Taxonomic Distribution of a Reproductive Division of Labor in Hymnenopterous Insects

Suborder	Superfamily/family/subfamily	Number of origins	Number of species	State of knowledge
Parasitica	Encyrtidae	≥1	≥6	Poor
Aculeata	Vespoidea/Vespidae/three subfamilies	1	Hundreds	Good
	Formicidae—all subfamilies	1	Thousands	Good
	Apoidea/Sphecidae/one subfamily	≥1	Several	Poor
	Halictidae/one subfamily	≥4	Hundreds	Reasonable
	Apidae			
	Allodapini	≥2?	Dozens	Poor
	Apini	1	~700	Good

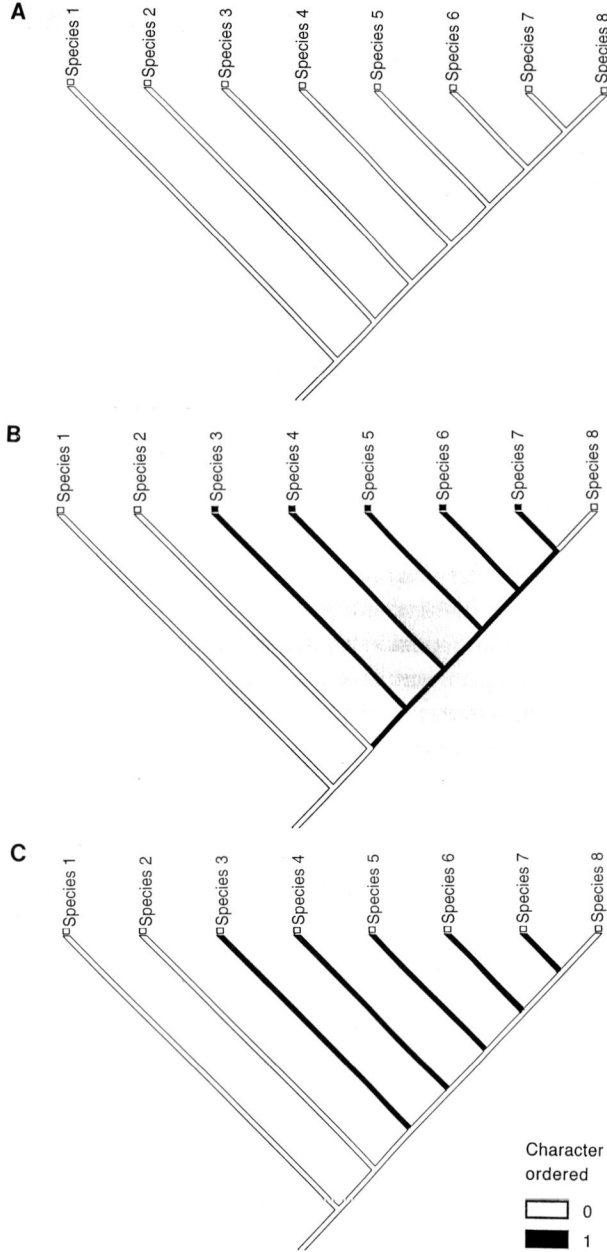

FIGURE 1 Hypothetical phylogeny.

analysis of the roaches and termites because the wood roach comes out nested within other cockroaches and does not occupy an intermediate position basal to the termites.

Similarly, because almost all examples of eusociality arise from within taxa that exhibit some form of parental care in the solitary relatives, it is likely that extended parental care is a precursor of the evolution of a reproductive division of labor in most cases.

Phylogenetic mapping of semisociality suggests that this is a much more labile evolutionary characteristic although occurring almost entirely within groups already characterized by eusociality, i.e., there is no phylogenetic evidence that semisociality preceded eusociality historically. Multiple queen societies appear to have arisen and/or been lost more frequently in ants than in vespids and in vespids more often than in bees.

B. Evolutionary Processes Leading to Reproductive Altruism

From the previous accounts, it seems clear that altruism in insects occurs disproportionately in taxa which are clonal (aphids and polyembryonic wasps), haplodiploid (Hymenoptera and Thysanoptera), and/or which live in a well-defined nest or gall (thrips, aphids, and termites).

1. Clonal Organisms

Reproductive altruism in clonal aphids and polyembryonic wasps is no more difficult to explain than the development of gonadal and somatic tissue within single multicellular organisms. Genetically, the individuals within a clone are identical and so the evolution of the sterility of some of them in order to defend the others is not difficult to explain. What may be less readily understood is the mechanism whereby different castes are produced; we have no information on this for these organisms, although it presumably involves some maternal influence.

2. The Role of Haplodiploidy

A great deal of theoretical and empirical work has been performed in attempts to explain the multiple origins of eusociality in haplodiploids because of the

proaches include establishing the probable ancestral conditions from which eusociality arose. For example, the long-held belief that the wood roach, *Cryptocercus punctulatus*, occupied an intermediate position between termites and cockroaches. This view, based on the facts that, like termites, the wood roach lives in wood and that it has extended parental care of the juveniles, has been overturned by phylogenetic unusual relatedness asymmetries that this sex-determination mechanism generates.

Female haplodiploids are related to their full sisters (those sharing the same mother and the same father) by 0.75, which gives them a 50% genetic payoff per individual raised over that which pertains if they were to raise a daughter. However, this benefit is canceled out by their lower relatedness to brothers—0.25 in comparison to their 0.5 relatedness to sons. Potential workers may still benefit if they can either raise sisters and sons or bias the sex ratio toward female sibs.

i. Worker-Produced Males A problem with the first possibility is that an individual worker should prefer to raise her own sons rather than those of a sister and so intracolony conflict is expected over male production not only between queens and workers but also among workers. However, this may be less of a problem in the incipient stages of sociality in which the number of workers is expected to be small. *Lasioglossum laevissimum* is a species in which the average number of workers per nest is less than three and dissection of ovaries indicates a substantial amount of ovarian development in workers, commensurate with them being capable of producing all of the males in the population. However, genetic data indicate that much less than a fifth of the males in the population result from worker oviposition in queenright nests. The difference between estimates based on observation of ovaries and from genealogical data results from a higher rate of ovarian resorption in workers than queens, from worker eggs being consumed by other females more often than those of the queen, or a combination of both. There is evidence for each of these processes in other Halictine species.

ii. Sex Ratio Biasing An oversimplified view is that workers should invest three times as much in female reproductives as in males because this is the point at which their increased relatedness to sisters is canceled out by increased mating success of the rarer males. Conversely, queens are equally related to sons and daughters and prefer a 1:1 investment ratio. Thus, there is a genetic conflict of interest within social hymenopteran colonies over the sex investment ratio. Numerous studies have shown that workers win this conflict, exerting considerable influence on the sex investment ratio within the colony. In ant species in which oviposition is monopolized by a single queen the average investment in males was found to be 28% in a survey of 34 species, which is in very good agreement with the theoretical expectation of 25%. In reality the situation is usually much more complex because of factors such as multiple egg layers, multiple matings by single queens, and orphaning of colonies. These factors introduce a wide range of different relatedness patterns among nestmates within colonies, complicating the predicted sex ratio patterns. Additionally, intercolony differences in social structure influence the population sex ratio in a manner that leads to greater divergence in sex ratio preferences in different colonies. Nonetheless, the basic theoretical prediction is that workers should bias the sex investment ratio facultatively to reflect their own genetic interests. Consider a population of colonies initiated by monogynous, singly mated queens in which, by the time reproductive production starts, a substantial proportion of the colonies have become orphaned. In the semisocial, orphaned colonies, the workers are equally related to their nephews as they are to their nieces, whereas in the queenright colonies workers are three times more closely related to sisters than to brothers. Given that the semisocial societies will be overproducing males in comparison to the population as a whole, this selects workers in the eusocial colonies to concentrate more of their resources in females. The net result of this is that orphaned colonies should produce mostly or entirely males, whereas the queenright ones should concentrate on female production. Experimental orphaning of colonies of the primitively eusocial bee *Augochlorella striata* indicate that the trend is in the expected direction and that orphaned colonies specialize in male production even in instances in which the replacement queen has mated and is capable of producing female offspring.

In summary, haplodiploidy would seem to confer a predisposition to the evolution of sociality, and despite the low coefficients of relatedness that are often found among nestmates in those taxa with a long history of eusociality, there is strong evidence that in situations in which it is possible, workers do manipulate the functioning of the colony to serve their own genetic interests.

iii. Relatedness in Termites Given the impressive potential of haplodiploidy in explaining sociality in Hymenoptera and Thysanoptera, it is not surprising that researchers have looked at the possibility of similar genetic mechanisms existing in termites despite their being diploid. Several ideas have been put forward. First, some termite species have a large proportion of their genome occurring in sex-linked translocation complexes and therefore males may be more closely related to their brothers and females to their sisters than is usually the case in diploid organisms. However, a major prediction of this scenario is that males should provide aid primarily to brothers and females to sisters, and this appears not to be the case. Cyclical inbreeding could also provide relatedness asymmetries favorable to kin rearing. If alates result from generations of inbreeding within their natal nest then they will be highly homozygous. If these inbred individuals then outbreed, their offspring will be highly heterozygous and more closely related to each other than to their own outbred offspring, thereby favoring the rearing of sibs over offspring. This theoretically attractive suggestion also fails empirical test because long cycles of inbreeding are required to provide the necessary homozygosity and termite colony longevities are rarely sufficient to permit this. Additionally, there is no evidence that the more basal groups of termites have intracolonial inbreeding to any great extent.

Nonetheless, the genealogical structure of termite colonies is an area of research that has received disappointingly little empirical attention to date.

3. The Nest

Clearly, the nest as a resource cannot be discounted as an incentive for offspring to remain at home and rear sibs rather than attempt independent reproduction in a new nest. There are several reasons for this. First, nest construction is often energetically expensive: Some sweat bee nests exceed 1 m in depth, paper wasp nest construction involves hours of wood fiber gathering, etc. Additionally, nest sites themselves may be limited, placing a premium on previously occupied sites. A second aspect of the importance of the nest is as a locus for individuals who are probably related to meet one another, and it is known that kin recognition in paper wasps is mediated through the nest itself (see below). Third, the presence of a nest and, in certain circumstances eggs, brood, and food stores, not only provides an emerging individual with the sensory stimuli for performing caring acts but also gives them a head start in rearing copies of its genes. In termites the royal pair and their offspring live within and feed on the structure that surrounds them, and so the nest takes on added significance as a resource in that it is itself the food, at least in the single-site nesters. A similar situation occurs in the gall-forming thrips and aphids; the living space and the food are more or less synonymous and so multiple advantages accrue from remaining where one is born—at least until such a time as the resource is used up. Annual seasonal variation places constraints on the duration of occupation of galls by eusocial aphids and thrips. For the more primitive termites it is the size of the durable lump of wood that determines the longevity of the resource and thereby the colony.

4. The Sting

Several authors have suggested that the ability to defend a living space is a prime determinant of group living and thereby sociality. Although this hypothesis may apply to Hymenoptera and perhaps even to those aphids that use their stylet mouthparts in gall defense, it is clearly of limited utility in explaining the other origins of eusociality. Furthermore, some of the more impressive social insects have lost the sting, including many ants and the stingless bees. Nonetheless, there is some utility to this idea in terms of explaining some aspects of the social organization of some taxa. Early in the evolution of the social Vespidae, there was a split between the Stenogastrinae, which persist in tiny colonies which are usually extremely cryptic, and the larger colonies of the paper wasps and yellowjackets—organisms with some of the most painful stings one can experience: A mass attack by hundreds of tropical wasps from a disturbed colony is something one never forgets if one lives through the experience! Whole nest destruction by predators is the major cause of mortality in the social wasps and it seems as if the two strategies adopted by tropical species are either to have tiny, inconspicuous colonies or massive ones which are exceptionally aggressively defended.

IV. CASTE DETERMINATION

Caste determination may take place at any stage in the development of the individual. In most semisocial societies and in many of the more primitively eusocial insects, caste is determined in the adult stage and the mechanism is often the physical suppression of individuals which become subordinates. In polistine wasps and in foundress associations and orphaned colonies of bees, the dominant individual is either the largest or the oldest. Experimental interference with isolated individuals demonstrates in at least some species that prolonged physical attack suppresses ovarian development and leads to increased rates of ovarian resorption. In paper wasps, the loss of aggressive contests early in adulthood leads to irreversible ovarian inhibition.

There is considerable evidence for the role of juvenile hormone levels in caste determination. In *Polistes dominulus,* minor differences in the size of the corpora allata among cofoundresses lead to differences in the wasp's position in the dominance hierarchy. As time progresses, these glandular size differences become enhanced as a result of dominance interactions.

In the perennially social sweat bee *L. marginatum,* mating determines caste. This bizarre situation seems to have arisen partly because males and females of this species are not active at the same time of year: Foundresses and workers are active in spring and males in autumn. Because only nests in their last year produce males it is only these nests that open to permit males to leave. Consequently, it is only these last-year nests that can permit the entry of males from other nests to mate with the females inside (females of both castes overwinter in their natal brood cells). Experimental early opening of nests permits females which would otherwise have become workers to mate and initiate nests as queens the following spring.

Nutritional differences play a major role in caste determination in most social insects in which canalization of an individual's reproductive options take place in the larval stage. Usually, there is a particular critical point during which the switch in caste takes place. In the honeybee, this is early in larval development such that neuromorphological differences are already detectable by the third instar even before juvenile hormone titer increases can be detected, although topical application of this hormone can lead to morphological changes from worker-like to queen-like characteristics. In the ant *Pheidole,* queen development is initiated by the laying of large eggs; thus, caste is at least partially determined in the egg stage in this insect.

The great diversity of tasks that require attending to in the more complex social insect societies yields a greater complexity of caste variation than that of simple queens and workers. Even in the primitively eusocial sweat bees, workers can often be differentiated into guards which remain at the nest entrance to repel intruders and foragers. Often the guard is the first individual to eclose from among the worker brood.

The need to defend the large amount of resources in social insect colonies has led to the evolution of soldier castes in many of them. As noted previously, in thrips and aphids, the altruists are the soldiers, but in ants and termites soldiers are a subgroup of altruists. Soldier production is generally controlled by a negative feedback process whereby soldiers give out pheromones that inhibit further development of individuals as soldiers. The larger the colony, the larger the soldier force required to exert this inhibition, thus ensuring a reasonable proportion of defenders in the colony.

Many ants have a very complex caste system among workers. In addition to a discrete group of soldiers, major and minor workers are commonly found. In *Eciton* army ants there is even a caste of porters, individuals with unusually long legs that are particularly efficient at carrying large prey items back to the bivouac.

Task specialization can be achieved not only by morphologically different individuals but also by temporal variation in activities. Thus, honeybee workers generally start their adult lives attending to nest duties and foraging is the last activity that they engage in. However, even in the most advanced eusocial insects there is considerable differentiation among individuals as to the propensity with which different tasks are performed, and in the honeybee at least some of this variation is due to genetic differences between workers arising from different patrilines (i.e., having the same mother but different father). This helps provide an explanation for the promiscuous mating of honeybee queens, a situation not expected on kin selection grounds but which presumably arose a long evolutionary time after the origin of eusociality in these insects.

V. NESTMATE RECOGNITION

The aggressive interactions that result when ants from different colonies are placed together is indicative of their ability to differentiate nestmates from nonnestmates. The mechanisms whereby this recognition occurs have received much attention and in some cases the chemicals involved have been identified.

In the eusocial sweat bee *L. zephyrum*, guards will keep out nonnestmates. If semisocial groups of worker-brood individuals are created in the laboratory by removing pupae from other nests, it is possible to set up colonies in which all bees except the guard are full sisters to which the guard is unrelated. In such nests the guard will let relatives of her nestmates into the nest but will exclude her own relatives. This indicates that the guard learns the odors of her nestmates but not her own smell. Similar experiments have involved a variety of degrees of inbred and outbred lines to provide a wide range of relatednesses among individuals. Using a guarded nest in which all nestmates (guard included) are full sisters, the effect of relatedness between guard and intruder on the probability of an intruder's acceptance into the nest can be investigated. The result was a very strong correlation between the probability that an individual would be allowed to enter the nest and its relatedness to the guard and her nestmates. Biochemical analysis has indicated that the chemicals that the bees use to discriminate among conspecifics are a series of cuticular hydrocarbons called macrocyclic lactones.

In paper wasps the chemicals used to mediate nestmate recognition are borne by the nest material. Individuals removed from the nest as soon as they eclose never learn to discriminate kin from nonkin, and individuals fostered into a foreign nest learn to tolerate their unrelated nestmates rather than their own genetic relatives. The effects of this imprinting can persist for months, even if the female is removed from its natal nest after a few hours and then held in isolation. This is presumably advantageous for species in which potential queens eclosing in the autumn have to leave the nest to hibernate and then associate with relatives in semisocial societies the following spring.

Kin recognition in honeybees has been demonstrated in a variety of situations but not in all instances in which it may be expected. When a colony swarms, the daughter colonies generally show a statistically significant association of workers from the same patriline as the queen with which they remain. However, when it comes to choosing among potential queens, the workers accept the first virgin queen that ecloses, irrespective of her genetic relatedness to the majority of workers in the nest.

See Also the Following Articles

INSECT REPRODUCTION, OVERVIEW; SOCIAL INSECTS

Bibliography

Choe, J. C., and Crespi, B. J. (1997). *The Evolution of Social Behavior in Insects and Arachnids.* Cambridge University Press, Cambridge, UK.

Engels, W. (Ed.) (1990). *Social Insects: An Evolutionary Approach to Castes and Reproduction.* Springer, Berlin.

Fletcher, D. J. C., and Michener, C. D. (Eds.) (1989). *Kin Recognition in Animals.* Wiley, Chichester.

Holldobler, B., and Wilson, E. O. (1990). *The Ants.* Belknap, Cambridge, UK.

Keller, L. (Ed.) (1993). *Queen Number and Sociality in Insects.* Oxford Univ. Press, Oxford, UK.

Ross, K. G., and Matthews, R. W. (1991). *The Social Biology of Wasps.* Comstock, Ithaca, NY.

Seeley, T. D. (1958). *Honeybee Ecology.* Princeton Univ. Press, Princeton, NJ.

Seger, J. (1991). Cooperation and conflict in social insects. In *Behavioural Ecology* (J. R. Krebs and N. B. Davies, Eds.), 3rd ed., pp. 338–373. Blackwell, Oxford, UK.

Turillazzi, S., and West-Eberhard, M. J. (1996). *Natural History and Evolution of Paper-Wasps.* Oxford Univ. Press, Oxford, UK.

Altruistic Behavior, Vertebrates

R. Haven Wiley

University of North Carolina, Chapel Hill

I. Introduction
II. Kin Selection
III. Reciprocal Altruism
IV. Cooperative Breeding

GLOSSARY

cooperative breeding Collaboration by the members of a stable social group in rearing the offspring produced by only some of the group's members.

Hamilton's rule A sufficient condition for the evolution of altruistic behavior toward genealogical relatives.

iterated Prisoner's Dilemma A model in game theory for repeated social interactions in which individuals can play either of two strategies: "cooperate" (accept benefits and return them) or "defect" (accept but not return benefits).

kin selection Changes in the spread of alleles in a population as a result of their influence on the survival or reproduction of genealogical relatives.

reciprocal altruism The return of benefits to the original actor by the beneficiary of an altruistic act.

tit for tat A conditional strategy which can eventually win (or spread during evolution) when played against defect in a game of iterated Prisoner's Dilemma.

Altruism occurs when an individual's actions benefit another at some cost to itself. Until the 1960s the occurrence of such behavior in animals was explained as a consequence of evolution for the overall good of the species. If an individual's actions resulted in a net benefit for its species, population, or social group, then its evolution seemed adequately explained. The appearance of W. D. Hamilton's paper, "The Genetical Evolution of Social Behavior" and G. C. Williams' book, *Adaptation and Natural Selection* abruptly reoriented thinking about the evolution of altruistic behavior. Although an allele associated with altruistic behavior might make the persistence of a species or a population more likely in the long term, it is not obvious how it could spread within a population in the short term. Because altruistic individuals incur costs in order to provide benefits for others, alleles associated with altruism do not spread within a population as rapidly as others. Because of this apparent paradox, altruistic behavior quickly became one of the central issues in the study of the evolution of behavior. In the ensuing three decades much mathematical theory and intensive fieldwork has clarified these issues and raised some new ones.

I. INTRODUCTION

To appreciate these advances in understanding the evolution of altruism, it is necessary to be clear about some basic issues. Altruism is defined in terms of benefits and costs to individuals. In evolutionary biology, these benefits and costs are increases and decreases in "fitness." This term is used in modern population genetics for the rate of spread of alleles in populations. To apply it to individuals requires some care: A change in an individual's fitness means a change in its prospects for reproduction or survival that in turn affects the spread of its alleles. Altruistic behavior is thus action that tends to increase the spread of another individual's alleles and to decrease the spread of the actor's alleles. Like many words adopted by science, fitness has acquired a more restricted meaning than in ordinary usage.

Another issue that needs attention is the association between alleles and altruistic behavior. Most comparative studies of altruistic behavior have not identified the genes involved nor attempted standard genetic analyses. Most information about the genetics of behavior comes from a relatively small number of organisms, mostly those permitting large-scale breeding. Nevertheless, diverse kinds of behavior have received attention from behavioral geneticists: aggression, mating behavior, learning, responses to stimulation, and others. The variety of behavior investigated suggests that virtually any aspect of behavior, like any other feature of organisms, has some—however small—association with particular alleles.

To reach this conclusion, we do not need to assume that any one or a few alleles "determine" a particular behavior, like a switch, nor do we need to suppose that much of the behavioral variation among individuals is associated with genetic variation. We do not need to assume that learning or culture has no influence on individuals' behavior. Indeed, alleles might be associated with a propensity to learn a particular form of behavior. Often, the development of behavior depends as much on environmental as on genetic variation. Evolutionary biologists studying altruism, however, do assume that behavioral differences in natural environments are associated, to some degree, with genetic differences. In technical terms, this position is equivalent to assuming that altruistic behavior, like other forms of behavior, often is heritable in natural environments.

To keep the evolutionary arguments clear, altruism must refer to the immediate consequences of an action, not to the eventual possibilities. The new developments in evolutionary theory have sought conditions under which the long-term advantages of alleles associated with altruism can overcome the short-term disadvantages. The fundamental problem is to explain how alleles associated with disadvantages in the short term can persist or increase in a population in the long term.

Two prominent possibilities for the evolution of altruism in the long run are kin selection and reciprocal altruism. Kin selection occurs when the behavior of individuals affects the survival or reproduction of their genealogical relatives. Reciprocal altruism occurs when an individual receiving altruism is likely to return the favor in the future. Both possibilities were first proposed in detail in the 1960s—kin selection by W. D. Hamilton and reciprocal altruism by R. Trivers. Both have subsequently received much empirical and theoretical investigation.

II. KIN SELECTION

Kin selection occurs whenever individuals affect the reproduction and survival of their genealogical relatives. Because relatives have a greater chance of sharing identical alleles than do randomly chosen individuals, an allele can spread in a population if an individual carrying it has a sufficiently large influence on the survival or reproduction of a relative, even at some cost to its own survival or reproduction. This cost (c) to the actor must be less than the benefit (b) to the relative weighted by the coefficient of genealogical relatedness (r, the probability that the relative has a copy of the actor's allele as a result of descent from a common ancestor): $c < br$. This expression is a sufficient condition for the evolution of altruism among genealogical relatives and is known as Hamilton's rule.

This rule applies strictly only to large randomly mating populations and to alleles not subject to strong selection. If there is significant inbreeding in the population, then the coefficient of relatedness needs adjustment for the background sharing of alleles identical by descent. Alternatively, the rule can be rephrased to include any background relatedness. Consider the decision an individual faces when it can promote the spread of its alleles either by direct or by collateral descent. An allele associated with altruism will spread provided that the number of additional progeny produced by a relative as a result of an altruist's actions (g_a) × the probability that those progeny will have the altruist's allele (r_a) exceeds the number of progeny a nonaltruist (egotist) individual could instead produce (g_e) × the probability that these progeny will have the egotist's allele (r_e): $g_e r_e < g_a r_a$. This corollary of Hamilton's rule incorporates any background sharing of alleles identical by descent as a result of inbreeding in the population because inbreeding increases the probability that an individual will share alleles with its own progeny.

This rephrased rule also resolves the problem of whether or not to consider parental behavior as altruism. Clearly, parental behavior often involves costs for the parents and benefits for the offspring, but in this case the recipients of benefits are also the direct bearers of the actors' genes. Application of Hamilton's original rule to this sort of cross-generational helping is problematic. The situation is clearer when we focus on alternatives for producing new copies of alleles at any one time: copies of alleles in an individual's own offspring versus copies of alleles identical by descent in the contemporaneous offspring of relatives. The rephrased rule, which applies explicitly to this case, makes the condition for the evolution of altruism by kin selection clear: Altruism evolves when alleles associated with it are more likely to be passed to new individuals in the population than are alleles associated with egotism.

In many cases of helping, several individuals assist a recipient. For instance, the cooperatively breeding species discussed later often have several helpers associated with each breeding pair. Note that Hamilton's rule (in the original or rephrased form) requires careful accounting for each individual's contribution to the breeders' reproduction. Each individual's assistance is that portion of the breeders' reproduction that would not occur if the helping individual were not present. Studies of kin selection in vertebrates have focused on measuring the effects of helpers on survival and reproduction of relatives, on establishing genealogical relationships within groups, and on determining which members of groups in fact reproduce (see Section IV).

III. RECIPROCAL ALTRUISM

If an altruist incurs an immediate cost that is more than recuperated in the future, then alleles associated with altruism will have a long-term advantage in a population. Thus, reciprocal altruism addresses the long-term advantages of altruism by considering the probability that an altruist and its beneficiary will exchange roles at some time in the future. A simple model from game theory, the iterated prisoner's dilemma, clarifies a fundamental issue for reciprocal altruism. This two-person game supposes that individuals can follow either of two strategies, "cooperate" or "defect," in repeated encounters. Because there is usually no necessary connection between the providing of assistance and its subsequent return, there is always the possibility that some individuals might defect by accepting but not returning favors. The question thus becomes, under what conditions can alleles associated with altruism spread in a population that also includes alleles associated with defection?

In their seminal analysis of the iterated prisoner's dilemma from an evolutionary point of view, Axelrod and Hamilton proposed that a combination of conditional behavior (cooperate only when your partner has cooperated in the preceding encounter) and memory (recognition) of individual partners can prevail against defection. To ensure cooperation when possible, individuals must also either try cooperation at the outset or randomly at low frequencies. This conditional behavior, called tit for tat, requires substantial cognitive abilities. Few studies have managed to demonstrate all the assumptions: conditional behavior, memory of interactors, and costs for the actors and benefits for the recipients.

Both field and laboratory studies have sought to confirm that animals are capable of the kinds of contingent behavior required by these theoretical analyses. For instance, both baboons and antelopes are more likely to groom individuals that have groomed them previously than other individuals of equal kinship. Rhesus monkeys are more likely to behave aggressively toward an individual that finds food but does not announce it with a characteristic call than toward those that do announce food. Vampire bats in Central America share their meals of blood only with others that have shared their meals in the past. A territorial male hooded warbler mostly ignores an established neighbor singing near its territorial boundary, but only when the neighbor has not recently intruded into its territory. Small fish provide a laboratory model of reciprocal altruism. Guppies and three-spined sticklebacks will approach a predatory fish visible in an adjacent aquarium, apparently for inspection. An inspecting fish will come closer to the predator when a parallel mirror makes it appear that another fish accompanies it than when a diagonal mirror makes the accompanying fish appear to lag behind. These studies of a variety of vertebrates suggest that individuals are often capable of behaving contingently toward social partners in the way required for the evolution of reciprocal altruism.

IV. COOPERATIVE BREEDING

The most prominent example of altruistic behavior in vertebrates is cooperative breeding, as practiced by many birds and mammals. In these species, individuals live in stable social groups that include more than two individuals old enough to breed when an opportunity arises. In many cases, only a subset of these individuals actually breed, yet all individuals participate in activities that promote the breeders' reproduction. These activities can include building or maintaining nests or dens, defending territories, guarding against predators, feeding the young, and feeding the parents while they in turn care for the young. Because these groups often consist of genealogical relatives, some of which breed and others help, they provide possibilities for the evolution of altruism by kin selection. In addition, the stability of membership in these groups provides opportunities for reciprocity. Several examples can illustrate both these common themes and also the diversity among them.

White-fronted bee-eaters in east Africa nest in colonies in earth banks, where each pair digs a long burrow for its nest. Offspring from previous years often assist in feeding the young. The number of progeny produced in a nest increases linearly with the number of adults feeding the young, with each additional helper adding about 0.3 additional fledged young. Complexities arise as helpers eventually acquire mates of their own. They then often concentrate on their own nests. If, however, a nest fails, the birds often return to their parents' nests to help there; mates usually separate temporarily in this case. On occasion, a parent interferes with the nesting attempts of its previous offspring so that they abandon their own nests and return to help the parents.

Florida scrub jays nest in large territories in which the birds find all of their food. Young often remain in their parents' territories for 1 or more years and help to defend the territory against other jays, to feed the parents' young (usually their siblings), and to defend the parents' nests against predators. The help from a previous years' offspring increases a pair's annual reproductive success by almost one additional offspring, but there is no additional increase as the number of helpers increases above one. Males stay longer on average than females, in some cases long enough to inherit the territory if their father dies. Females eventually leave to find a mate on another territory. Male helpers provide significantly more assistance in raising the young than do females. By helping, males thus adopt a long-term strategy for enhancing their chances for reproduction.

Stripe-backed wrens illustrate this pattern even more clearly. They live in groups averaging five or six adults, although some pairs and single birds also defend territories. Young almost always remain in their natal territories until opportunities arise to breed. A male usually stays in his natal territory for life; if he survives long enough, he succeeds to the breeding position after the deaths of his father and older brothers. A female remains until she finds a vacancy in a nearby group. DNA fingerprinting has established that only one pair breeds in most groups. Exceptions occur in 10% of groups, always those in which the female breeder has recently been replaced. In these groups, the oldest male helper sires on average 15% of offspring. Males never breed with their mothers because the oldest helping male in a group never mates with the breeding female unless she has immigrated since he was born. On average, all ages and both sexes of helpers provide equal contributions to raising the young, although individuals vary considerably in the amount of help they provide. Pairs without helpers, or with only one, produce very few young on average, and consequently few individuals attempt to breed as unassisted pairs. The assistance of two helpers, in contrast, results in a significant increase in nesting success, apparently mostly as a result of better protection of the nest from predators. Because there is such a premium on breeding in a large group, competition among young females to fill vacancies in large groups is intense. Probably for this reason, young females suffer higher mortality than any other segment of the adult population.

Acorn woodpeckers also live in groups on large territories in which the entire group works to store acorns during autumn and winter. Each territory includes several trees in which innumerable small holes are prepared in the bark or dead wood, each to receive one acorn. A group cannot breed successfully without this stored food to provide nutrients in late winter and early spring. Parentage is much more complex in this species. Often a group of males defends a territory, and these males, as revealed by DNA fingerprinting, mate jointly with the females present. Females compete to have their eggs retained in the nest by removing each other's eggs. Eventually one or more females produce the eggs raised by the entire group. Young remain in their natal territory for a year or more and during this time do not mate with their parents nor interfere with their reproduction, so some groups can eventually resemble the extended families seen in stripe-backed wrens or Florida scrub jays. Most groups, however, consist of several males, all of which mate with several females, which in turn compete with each other to have their eggs incubated.

Fairy wrens of several species represent a few of the dozens of species of Australian birds that breed cooperatively. In this case young remain in their natal territory but, unlike the species already considered, the males do not usually breed there. Instead, DNA fingerprinting shows that they routinely court and mate with females in neighboring territories. In these species, the adult sex ratio is skewed strongly toward males, apparently because of high mortality of females during nesting. Because there are so many vacancies for females, young females usually emigrate at an early age, whereas males remain in their natal territories and attempt to mate with neighboring females. The presence of these extra males does not have a significant influence on nesting success.

Pied kingfishers in African woodlands often remain with their parents after independence and help to raise subsequent broods, particularly when conditions are harsh. Sometimes individuals abandon their parents and take subordinate positions helping other adults. In this case, the advantage of taking a subordinate position in a good territory can pay off if the breeder of the same sex dies. The subordinate is then in line to succeed to the breeding position.

Gray wolves and African hunting dogs provide cases of mammals that breed cooperatively. In both species, normally one pair in a pack reproduces, although exceptions apparently occur when food is abundant. Nevertheless, all members of the pack share food with the lactating mother and later with the growing puppies. Pack members are often previous years' offspring, but at least in African hunting dogs a cohort of the same sex sometimes leaves its natal pack to evict like-sex individuals from another pack. Reports of multiple paternity in packs might apply to such cases, much as in acorn woodpeckers.

Dwarf mongooses provide a more thoroughly documented case of cooperative breeding in mammals. Social groups average about nine individuals at least 1 year old, of which two normally breed, although the extent of multiple paternity has not yet been thoroughly investigated. In this species, as in many cooperatively breeding birds, helpers have only modest effects on reproductive success in the group. As a consequence, helping to produce collateral relatives makes a relatively minor contribution to the transmission of an individual's genes.

Naked mole rats' bizarre social behavior has evoked much interest. These obligately subterranean rodents live in large social groups, each of which includes a single exceptionally large reproductive female. The analogy with social insects is inviting, except that the reproductive female mates with several males in her own colony. Almost nothing is known about acquisition of breeding status or dispersal among colonies in natural circumstances.

African lions, like wolves and hunting dogs, hunt together and rear their young communally. One female in a pride usually remains with young cubs of several pride members while the others are hunting, and the attendant female then nurses all cubs indiscriminately. Females usually remain in their natal pride, so the females in a pride are usually close relatives. Nevertheless, pride members do not all contribute equally to the group's welfare; for instance, individual females differ consistently in their roles in approaching simulated intruders in the pride's territory.

Another example of communal breeding is the groove-billed ani, one of several species of black tropical American cuckoos. Breeding often involves two or three pairs using the same nest. Most young disperse from their natal territory in their first year, with the exception of a minority of males, so that pairs sharing a nest are usually not closely related genealogically. Females compete to have their own eggs retained in the communal nest. Each female ejects others' eggs from the nest until she herself begins to lay, so one female in each group, the last to begin laying, usually produces most of the eggs incubated. The size of a group has modest effects on the survival of adults in the group and on young reared to independence.

These examples of cooperative breeding vary in important ways. In particular, there is much variation in the influence of potential helpers on reproductive success by breeders, in the relatedness of helpers to breeders, and in the distribution of reproduction among group members. Nevertheless, four important generalities emerge.

First, nonbreeding individuals usually remain with their close relatives. Although the contributions of helpers to the success of breeders are too low, in most cases, for kin selection to provide a sufficient explanation for helping, kin selection does successfully predict whom an individual helps. If nonbreeders help at all, they overwhelmingly help relatives.

Second, most long-term studies of cooperative breeders have revealed that individuals stand to gain by joining groups as nonbreeders and queuing for subsequent opportunities to breed. In this way, cooperative breeding is often a long-term strategy for maximizing reproductive success combined with some short-term advantages from kin selection. For the breeders, retention of offspring in their territory is a long-term form of parental care.

Third, the association of relatives in cooperative breeders does not preclude competition among them. Although it is often clear that individuals help relatives rather than unrelated individuals, they do not necessarily help relatives rather than promote their own interests; for example, competition for breeding opportunities within groups of bee-eaters or acorn woodpeckers.

Finally, individuals do not necessarily have to discriminate degrees of relatedness in order to associate with relatives. In many cooperatively breeding species, individuals associate with relatives simply as a consequence of remaining in their natal groups or with individuals that raised them.

Much about cooperative breeding remains to be clarified. In only a few cases has DNA fingerprinting actually established the patterns of paternity. Even so, only long-term studies of large marked populations can establish the genealogical structure of populations with such complex social structure. In addition, little is known about the behavioral mechanisms that produce the complex social interactions in cooperatively breeding species, especially individuals' capabilities for discriminating degrees of relatedness or recognizing other group members.

See Also the Following Articles

AGGRESSIVE BEHAVIOR; ALTRUISM IN INSECT REPRODUCTION

Bibliography

Axelrod, R., and Hamilton, W. D. (1981). The evolution of cooperation. *Science* **211**, 1390–1396.

Brown, J. (1987). *Helping and Communal Breeding in Birds*. Princeton Univ. Press, Princeton, NJ.

de Waal, F. (1989). *Peace-Making among Primates*. Harvard Univ. Press, Cambridge, MA.

Dugatkin, L. (1997). *The Evolution of Cooperation*. Princeton Univ. Press, Princeton, NJ.

Emlen, S. T. (1997). Predicting family dynamics in social vertebrates. In *Behavioural Ecology* (J. R. Krebs and N. B. Davies, Eds.), 4th ed., pp. 228–253. Blackwell, Oxford, UK.

Godard, R. (1993). Tit for tat among neighboring hooded warblers. *Behav. Ecol. Sociobiol.* **33**, 45–50.

Grafen, A. (1984). Natural selection, kin selection, and group selection. In *Behavioural Ecology* (J. R. Krebs and N. B. Davies, Eds.), 2nd ed., pp. 62–84. Blackwell, Oxford, UK.

Hamilton, W. D. (1964). The genetical evolution of social behaviour. I and II. *J. Theor. Biol.* **7**, 1–52. (The classic account but a difficult paper to read!)

Hamilton, W. D. (1996). *The Narrow Roads of Gene Land*. Freeman, San Francisco. (Republished by Oxford Univ. Press, Oxford, UK, 1998)

Koenig, W. D., and Mumme, R. L. (1987). *Population Ecology of the Cooperatively Breeding Acorn Woodpecker*. Princeton Univ. Press, Princeton, NJ.

Moehlman, P. D. (1986). Ecology of cooperation in canids. In *Ecological Aspects of Social Evolution* (D. I. Rubenstein and R. W. Wrangham, Eds.), pp. 64–86. Princeton Univ. Press, Princeton.

Sherman, P. W., Jarvis, J. U. M., and Alexander, R. D. (Eds.) (1991). *The Biology of the Naked Mole-Rat*. Princeton Univ. Press, Princeton, NJ.

Solomon, N. G., and French, J. A. (Eds.) (1997). *Cooperative Breeding in Mammals*. Cambridge Univ. Press, Cambridge, UK.

Stacey, P. B., and Koenig, W. D. (Eds.) (1990). *Cooperative Breeding in Birds*. Cambridge Univ. Press, Cambridge, UK.

Stevens, E. E., and Wiley, R. H. (1995). Genetic consequences of restricted dispersal and incest avoidance in a cooperatively breeding wren. *J. Theor. Biol.* **175**, 423–436.

Trivers, R. L. (1971). The evolution of reciprocal altruism. *Q. Rev. Biol.* **46**, 35–57.

Wiley, R. H., and Rabenold, K. N. (1984). The evolution of cooperative breeding by delayed reciprocity and queuing for favorable social positions. *Evolution* **38**, 609–621.

Wilkinson, G. S. (1988). Reciprocal altruism in bats and other mammals. *Ethol. Sociobiol.* **23**, 85–100.

Williams, G. C. (1966). *Adaptation and Natural Selection.* Princeton Univ. Press, Princeton, NJ.

Woolfenden, G. E., and Fitzpatrick, J. W. (1984). *The Florida Scrub Jay: Demography of a Cooperative-Breeding Bird.* Princeton Univ. Press, Princeton, NJ.

Amenorrhea

Sarah L. Berga

The University of Pittsburgh School of Medicine

I. Causes
II. Diagnostic Strategies
III. Treatment Implications

GLOSSARY

amenorrhea The absence of uterine bleeding for an extended period of time.

anovulation Lack of ovulation in a woman with an ovary or ovaries; may present as amenorrhea, oligomenorrhea, or eumenorrhea.

endometrium The glandular epithelium that lines the uterine cavity.

gonadotropin-releasing hormone (GnRH) A decapeptide releasing factor for luteinizing hormone and follicle-stimulating hormone made by GnRH neurons residing in the hypothalamus and released at the level of the median eminence into the portal vasculature in a pulsatile manner.

hypothalamus The part of the brain located around the inferior portion of the third ventricle which receives ascending neural signals from the brain stem and descending neuronal input from the limbic lobe and frontal cortex and projects primarily to the median eminence and posterior pituitary.

oligomenorrhea Uterine bleeding that occurs at infrequent and irregular intervals of time.

ovary Gonadal tissue composed of the cortex of the undifferentiated fetal gonad into which germ cells that become oocytes migrate prior to 20 weeks of fetal life.

pituitary Epithelium composed of the anterior and intermediate glandular lobes and the neurohypophysis or neural lobe which is an extension of the hypothalamus; responds to GnRH by releasing into the circulation the gonadotropins, luteinizing hormone and follicle-stimulating hormone.

Amenorrhea refers to the absence or cessation of menses for an extended interval of time in a woman of reproductive age. There are many causes of amenorrhea, including pregnancy; thus, amenorrhea does not necessarily signal a pathological condition or state. The absence of regular menstrual periods in nongravid women between the ages of 14 and 40 years, however, does warrant medical attention so that the cause can be determined and appropriate management instituted. The health consequences of amenorrhea depend on the cause. Because the health sequelae rarely are entirely benign or limited to the reproductive system, it is imperative that an amenorrheic woman receive prompt medical attention.

I. CAUSES

The causes of amenorrhea can be broadly categorized as reproductive tract anomalies; primary ovarian disorders; adrenal, thyroidal, and other glandular conditions; pituitary abnormalities; and hypothalamic conditions. These categories form the basis for developing a rational approach to the diagnosis of amenorrhea. A working differential diagnosis should consider the conditions described in the following sections.

A. Reproductive Tract Anomalies

Congenital reproductive tract anomalies that present as amenorrhea include imperforate hymen, transverse vaginal septum, partial or complete Müllerian agenesis, androgen insensitivity syndromes, 5α-reductase deficiency, and gonadal dysgenesis.

Androgen insensitivity is due to mutations in the gene for the androgen receptor that is located on the long arm of the X chromosome. Testosterone and dihydrotestosterone are absolutely required for fetal differentiation of the external genitalia. There is a spectrum of anatomic presentation depending on the extent of androgen receptor function. In the complete form, the testis secretes androgens and Müllerian-inhibiting substance (MIS). The MIS leads to regression of the Müllerian anlage, but the absence of androgen receptivity leads to a blind-ending and shortened vagina. The testes may be sequestered in the inguinal canal or pelvis. Puberty occurs when the hypothalamus matures and the pituitary release of luteinizing hormone (LH) stimulates the testes to secrete testosterone that is aromatized to estradiol. Estradiol causes the development of female secondary sexual characteristics, namely, thelarche, whereas pubic and axillary hair is scant due to the androgen receptor insensitivity. The risk of gonadoblastoma is lower in individuals with androgen insensitivity than in those with 46,XY gonadal dysgenesis; therefore, the clinical dictum is that the testes do not need to be removed until after puberty is complete. Fertility is not possible in the complete form of androgen insensitivity.

5α-Reductase is the enzyme that converts testosterone to dihydrotestosterone (DHT). DHT binds with much greater avidity to the androgen receptor and its presence in fetal life is needed for complete masculinization of the external genitalia. This syndrome occurs when the enzyme has a reduced capacity to convert testosterone to DHT. Neonates present with microphallus and often hypospadias. The testes secret MIS and regression of Müllerian anlage occurs. The testes may be located in the scrotum, which may look like labia, or in the inguinal canal. Fertility is possible. At puberty, when the testes secrete increasing amounts of testosterone, the production of DHT increases because the enzyme defect is rarely complete. Thus, at puberty, the external genitalia may further masculinize. This condition, which can be familial, is often termed "penis at 12 years" syndrome. The current standard of practice is to induce further neonatal masculinization by giving exogenous DHT.

The cause of imperforate hymen, vaginal septum, and Müllerian agenesis is a defect in the migration, fusion, or canalization of the Müllerian anlage during fetal life. A spectrum of anatomic anomalies may result, including renal anomalies such as a single pelvic kidney. Typically, ovaries form and puberty occurs on time. Anomalies of the Müllerian tract range from complete absence to partial obstruction of the outflow tract. When Müllerian remnants exist or if there is obstruction, a common presenting complaint in addition to primary amenorrhea is cyclic pelvic or abdominal pain.

An acquired cause of amenorrhea is traumatic endometrial ablation, termed Asherman's syndrome. It frequently results from a vigorous endometrial curettage in the presence of endometritis or prolonged hypoestrogenism.

B. Primary Ovarian Disorders

Premature menopause is a major cause of amenorrhea in women of reproductive age. In a study of adult-onset amenorrhea, about 35% had probable functional hypothalamic amenorrhea (FHA), 10% hyperprolactinemia, 10% premature ovarian failure, and 30% polycystic ovary syndrome. The store of primordial follicles is fixed *in utero* and thereafter the number of primordial follicles declines with age. Across each interval of time, a certain percentage of these resting follicles die or become atretic. It has been estimated that menopause occurs when the number of primordial follicles falls below 1000. The usual age for this degree of follicular depletion is 51 years. The age of menopause is not thought to be influenced by socioeconomic factors. It is estimated that the number of primordial follicles at birth is about 1 million and that the rate of atresia is relatively constant until the number of remaining follicles numbers 25,000. The customary age at which follicular atresia accelerates is estimated to be 37.5 years.

Menopause is termed premature if it occurs before age 40 years. Primary amenorrhea in conjunction with an elevated follicle-stimulating hormone (FSH) level is generally due to gonadal dysgenesis. The most common cause is Turner's syndrome, the chromosomal constitution of which is 45,XO. Deletions of the long arm of X in patients with 46,XX may cause premature follicular atresia and primary amenorrhea or premature menopause. Deletions confined to the short arm of X cause the physical stigmata characteristic of women with Turner's syndrome, but without premature follicular atresia, that is termed Noonan's syndrome. Patients with gonadal dysgenesis also may have other chromosomal derangements, including 46,XY or 47,XXX, or mosaicism, such as 45,XO/46,XX. Primary amenorrhea or premature menopause in women with these various chromosomal abnormalities is attributed to accelerated follicular atresia. Family history is a good predictor of early menopause. A history of menopause before age 46 in a mother or sister increases the probability that a woman will have her menopause before age 46 years from 5 to 25%. In one study, cases with a family history of early menopause were not more likely to have inborn errors of galactose metabolism or stigmata of Turner's syndrome, but they were less likely to have brothers. Given that the cause of ovarian failure in girls with Turner's syndrome and those with deletions of the long arm of the X chromosome is thought to be an accelerated rate of atresia, these authors suggested that the low incidence of male siblings suggested that microdeletions of the long arm of the X chromosome might account for these observations. The exact gene product encoded for at the distal end of the long arm of the X chromosome is unknown, but it presumably influences oocyte longevity and thus it would be potentially useful to identify. Inborn errors of metabolism, such as galactosemia, account for some cases of premature menopause. Another major cause other than genetic defects in the long arm of the X chromosome is autoimmune. Autoimmune premature ovarian failure is thought to be due to autoantibodies that block FSH receptors or damage oocytes. Women with autoimmune disorders, such as Hashimoto's thyroiditis, systemic lupus erythematosis, and juvenile arthritis, are at risk for autoimmune premature ovarian failure and vice versa.

Polycystic ovary syndrome (PCOS) is often viewed as a primary ovarian disorder. PCOS may present as amenorrhea or oligomenorrhea. The distinctive ovarian morphology includes thecal–stromal hyperplasia and multiple atretic follicles located in the cortex. Classically, this morphology is referred to as the "pearl necklace" sign. Increased LH stimulation generally causes the ovary to oversecrete androgens, primarily androstenedione, while the relative decrease in FSH leads to anovulation and relative hypoestrogenism. PCOS may be accompanied by insulin resistance and obesity, either of which can exacerbate ovarian androgen secretion. In women with sensitive pilosebaceous glands, the increased androgen secretion may provoke acne or hirsutism.

Androgen-producing ovarian tumors also may present as amenorrhea and masculinization or frank virilization. Typically, the history is of an acute onset of symptoms. Ovarian tumors which produce testosterone usually are Leydig cell remnants of the medullary portion of the undifferentiated fetal gonad that failed to completely regress and they may be LH responsive. Their hilar location, lack of echogenicity, and small size make them difficult to detect by ultrasound or other imaging techniques.

C. Adrenal Conditions

Congenital adrenal hyperplasia is due to absent or reduced activity of one of the adrenal enzymes necessary for the biosynthesis of cortisol. Enzyme deficiencies that permit the synthesis of androgens typically present neonatally as masculinization of females and if insufficient aldosterone is made, both males and females may develop neonatal hyponatremia. Less complete enzyme deficiencies may present as pubertal masculinization in females, with phenotypic features similar to those of women with PCOS. If the enzyme defect does not permit the synthesis of androgens, then males may present with neonatal feminization because the enzyme deficiency blocks testicular as well as adrenal androgen synthesis.

Autonomous adrenal overproduction of cortisol, Cushing's syndrome, commonly presents as amenorrhea and masculinization in women. Likewise, adrenal tumors may present as virilization and amenorrhea in women. Typically, the onset of clinical symptoms is abrupt and rapid. Addison's disease, or adrenal failure, also may present as amenorrhea accompanied by fatigue and other symptoms of adrenal insufficiency. Each of these conditions may present a significant metabolic challenge and suppress gonadotropin-releasing hormone (GnRH) drive, resulting in hypogonadism.

D. Thyroidal Conditions

Hyperthyroidism is an uncommon cause of amenorrhea. Most thyrotoxic women describe regular menses and have in-phase endometrial biopsies, suggesting that they remain ovulatory. On the other hand, menstrual irregularities are common in women with untreated hypothyroidism. The primary disturbance appears to be altered endometrial maturation in response to characteristic sex steroid excursions. Interestingly, functional hypothalamic hypogonadism is accompanied by a functional impairment of thyroidal hormone release that is either due to reduced thyrotropin-releasing hormone (TRH) drive or reduced thyroid-stimulating hormone (TSH) response to TRH. Clinically, women with functional hypothalamic amenorrhea may have an eating disorder, engage in disordered eating or excessive exercise, or experience psychogenic stress. Low-energy intake seems to potentiate the effects of psychogenic stress and improved calorie or nutritional intake may be required to reverse the associated metabolic compromise and permit recovery of GnRH drive and resumption of ovulation.

E. Pituitary Abnormalities

Pituitary adenomas commonly present in women as amenorrhea. The most frequent pituitary adenoma that causes amenorrhea is a prolactin-secreting microadenoma and its presence is generally signaled by galactorrhea. Hyperprolactinemia induces increased opioidergic tone that then suppresses GnRH release. Treatment involves the suppression of prolactin by dopamine agonists so that GnRH drive is restored. Other pituitary adenomas that may disrupt reproductive function include prolactin macroadenomas and those that secrete growth hormone, TSH, adrenocorticotropin hormone (ACTH), or FSH. Macroadenomas are pituitary adenomas >1 cm in diameter. They have the potential to grow beyond the sella turcica and may compress the optic chiasm, causing bitemporal hemianopsia. Headaches are common and are described as an unremitting intense pain behind the eyes.

Panhypopituitarism also commonly presents as amenorrhea due to reduced or absent pituitary release of LH and FSH. If the pituitary compromise is due to disrupted delivery of hypothalamic-releasing factors due to stalk damage or compression, then prolactin levels will be high and other pituitary hormone levels will be low. Elevated prolactin in this condition reflects the absence of hypothalamic dopamine delivery, which is needed to inhibit prolactin release. Pituitary hormones other than prolactin require a stimulatory signal to trigger release, whereas prolactin release occurs unless it is inhibited. If there is primary damage to the pituitary, then all the pituitary hormone levels will be low, including prolactin. There are many causes of pituitary damage, but a common cause in women is postpartum pituitary apoplexy, which is referred to as Sheehan's syndrome. The pituitary expands during pregnancy to the point that the blood supply from the arterial and portal vasculature is marginal. Sudden drops in blood pressure due to hemorrhage or dehydration then may trigger pituitary necrosis.

F. Hypothalamic Conditions

Hypothalamic causes of amenorrhea may be organic or functional. Organic causes include a variety of brain tumors and hydrocephalus. Kallmann's syndrome is due to a failure of the GnRH neurons to migrate from the olfactory placode into the hypothalamus during fetal life. In its complete form, there is primary hypogonadism with sexual infantilism and anosmia. GnRH drive also can be reduced by functional causes. In this situation, the reduction or disruption in GnRH release is a homeostatic response to psychogenic or metabolic challenge. The hypothalamic response to sustained challenge involves increased corticotropin-releasing hormone drive, leading to elevated levels of cortisol; reduced TRH or reduced TSH response to TRH, leading to reductions in triiodothyronine and thyroxine; disrupted GnRH drive, leading to slowing of LH pulse frequency and FSH levels insufficient to support ovulation; modest increases in growth hormone; marked reductions in prolactin; and elevated nocturnal secretion of melatonin. Psychogenic stress sensitizes the neuroendocrine axes to the effects of exercise and, vice versa, exercise and stress potentiate neuroendocrine responses to psychogenic challenge.

II. DIAGNOSTIC STRATEGIES

Making the correct diagnosis of the cause of amenorrhea is the cornerstone of medical management. The history and physical examination are essential. Pertinent historical information that must be elicited includes the chronology of sexual development, including adrenarche, thelarche, and menarche. The occurrence of irregular menses since menarche suggests polycystic ovary syndrome, whereas the abrupt cessation of menses following weight loss suggests hypothalamic disruption of pulsatile GnRH secretion. The physical examination should include palpation of the thyroid, careful inspection of the skin for acanthosis nigricans, striae, and hirsutism, Tanner staging, and a pelvic examination. The overall habitus should be noted.

If a congenital abnormality of the external reproductive tract is found on physical examination, then the cause must be further defined. A blind-ending vagina suggests androgen insensitivity. This can be confirmed by measuring serum testosterone and obtaining a karyotype to look for 46,XY chromosomal constitution. A karyotype is indicated also in girls with abnormal developmental chronology to look for gonadal dysgenesis. Defining anatomic anomalies often requires radiological imaging with ultrasound or magnetic resonance. An intravenous pyelogram may be needed to define renal anatomy, including that of the collecting system. Finally, an examination under anesthesia with or without laparoscopy may aid in determining anatomy.

A progestin withdrawal test is often suggested as a means to estimate the patient's estrogen status and to exclude traumatic endometrial destruction and adhesions of the uterine cavity, which is commonly termed Asherman's syndrome. This maneuver is of limited diagnostic value, however, because it is an inexact measure. If the patient's estrogen status cannot be inferred from physical examination, then a serum estradiol is a more accurate way to determine it. If the history includes past uterine infection or trauma and Asherman's syndrome is suspected, then sonohysterography, hysteroscopy, or a hysterosalpingogram allows visualization of intrauterine adhesions. In contrast, absent or scant withdrawal bleeding in response to progestin exposure may indicate Asherman's syndrome, minimal estrogen exposure, or a congenital abnormality of the uterine anatomy.

Clinical chemistries generally help to clarify the diagnosis. If there is evidence of hyperandrogenism, then the diagnostic possibilities include Cushing's disease and syndrome, congenital adrenal hyperplasia, adrenal tumors, ovarian tumors, and polycystic ovary syndrome. Screening tests include a 24-hr urinary free cortisol, serum 17-hydroxyprogesterone, dehydroepiandrosterone sulfate, testosterone, androstenedione, LH, and FSH. A high LH and a LH:FSH ratio >2 in the presence of an elevated androstenedione suggest PCOS. The higher the serum testosterone, the more likely there is to be an adrenal or ovarian tumor. If there is a tumor, then generally the history also includes the abrupt onset of masculinization or virilization and the physical examination may reveal a male habitus or clitoromegaly. Insulin resistance often accompanies PCOS. Either a fasting blood sugar or a 75-g glucose challenge to assess glucose and insulin responses is recommended. While Cushing's syndrome or disease generally is heralded by distinctive changes in the habitus, such as abdominal striae or truncal obesity, amenorrhea is an early symptom in women. Autonomous glucocorticoid secretion can be detected with a 24-hr urinary-free cortisol followed by a dexamethasone suppression test or by determining plasma ACTH levels. Like PCOS, acromegaly also causes insulin resistance and can be confused with PCOS. Insulin-like growth factor-1 is elevated in patients with excess GH secretion. Galactorrhea suggests either a prolactin-secreting pituitary adenoma or hypothyroidism. A serum prolactin and TSH should be obtained. Unlike PCOS, hyperprolactinemia, acromegaly, and Cushing's disease or syndrome are accompanied by low LH and FSH. Panhypopituitarism, which includes postpartum pituitary apoplexy or Sheehan's syndrome, is accompanied by fatigue and insufficient glandular secretion due to reduced pituitary capacity. If an organic hypothalamic or pituitary condition is suspected, a magnetic resonance image of the brain should be obtained. Provocative pituitary testing may be indicated also to discern "pituitary reserve."

The differential diagnosis of amenorrhea includes premature ovarian failure (POF), particularly if the amenorrhea is accompanied by hot flashes. An elevated FSH suggests menopause, particularly if the estradiol is low. An elevated FSH in the presence of an estradiol >150 pg/ml, however, could indicate the presence of a preovulatory follicle and a midcycle gonadotropin surge. The higher the FSH, the more likely it is that the oocyte depletion is complete. Many paradigms are now available to assess what is commonly referred to as "ovarian reserve." The simplest is determining FSH and estradiol levels on the third day following the onset of menses. A FSH ≥30 IU/liter with an estradiol ≤30 pg/ml indicates menopause, whereas FSH ≥15 IU/liter but <30 IU/liter indicates compromised reserve. If menopause occurs before age 40, it is considered to be premature. Premature menopause is often associated with autoimmune glandular compromise, particularly of the thyroid. If an autoimmune cause is suspected, TSH and thyroxine levels should be assessed along with antithyroid antibodies. It is common for women with POF to be euthyroid but positive for thyroid antibodies. Hypothyroidism may occur later, so frequent screening of thyroid function is recommended. Other clinical chemistry studies that may be warranted include calcium and parathyroid hormone to look for hypoparathyroidism; dehydroepiandrosterone sulfate or a 24-hr urinary-free cortisol to evaluate adrenal function; insulin and glucose to check pancreatic sufficiency; a complete blood count (CBC) with indices to screen for pernicious anemia; and an antinu-

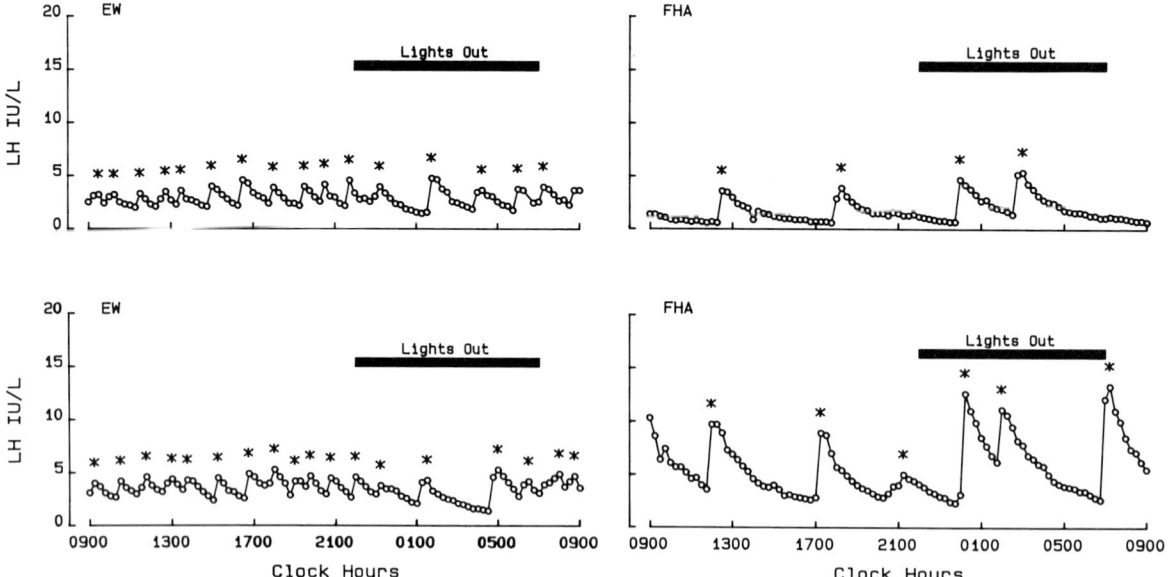

FIGURE 1 Representative LH pulse patterns in two eumenorrheic women and two women with functional hypothalamic amenorrhea. Blood samples were obtained at 15-min intervals for 24 hr.

clear antibody titer, erythrocyte sedimentation rate, and rheumatoid factor to check for other autoimmune processes.

A normal FSH with a LH:FSH ratio <1 suggests decreased hypothalamic GnRH input. In the clinical research setting, reduced GnRH drive can be documented by determining LH pulse patterns over a 24-hr duration. LH pulse patterns are a reasonably accurate representation of GnRH pulse frequency. The normal LH pulse frequency is about one pulse every 90 min or 15 pulses per day. On average, women with functional hypothalamic amenorrhea will have about 7 pulses in a day. Representative LH pulse patterns from eumenorrheic women and those with functional hypothalamic amenorrhea are shown in Fig. 1. Organic causes of the disruption of the hypothalamic release of GnRH must be considered in the differential diagnosis and magnetic resonance imaging should be undertaken unless there is a straightforward history of significant weight loss or excessive exercise. Anosmia suggests Kallmann's syndrome or isolated gonadotropin deficiency. The accompanying hypoplasia of the rhinencephalon may be visualized by magnetic resonance imaging. Functional causes of hypothalamic amenorrhea are a diagnosis of exclusion. In general, however, the history and physical examination suggest the diagnosis so that the utilization of clinical chemistry studies can remain targeted.

In summary, the laboratory evaluation of a woman with anovulation and no clear history or physical stigmata to direct the investigation should begin with a determination of LH, FSH, TSH, thyroxine, prolactin, and estradiol levels in a venous sample ideally obtained at least 2 hr after exercise, a meal, coitus, or a physical examination and between 1000 and 1200 hr to exclude menopause, thyroid conditions, and hyperprolactinemia. The possibility of PCOS can be evaluated by adding an androstenedione level. Further evaluation depends on the index of suspicion for more occult, but rare, disorders.

III. TREATMENT IMPLICATIONS

It is a tenet of medicine that diagnosis dictates therapy. In the past, however, the pathophysiology of anovulation was not well understood and it was not possible to target therapies. Also, the health consequences of anovulation other than infertility and menometrorrhagia were not generally appreciated. We are now aware that each cause of anovulation has unique clinical sequelae that must be considered when developing a management plan. Once a diagnosis is made, the patient needs to be apprised of the prognosis, risks, and benefits of various treatment options as well as prospects for fertility. Representative examples of how our ability to recognize and diagnose the specific cause of anovulation impacts on the management plan are outlined as follows:

Polycystic ovary syndrome:

1. At higher risk for cardiovascular disease, hypertension, and diabetes mellitus (Talbott): Screen for diabetes mellitus and determine lipoprotein profile; advise dietary modifications if needed.

2. At higher risk for weight gain: Dietary education a must; advise weight loss if needed.

3. Oral contraceptive use does not worsen the insulin resistance significantly, suppresses androgen secretion and action on the pilosebaceous unit, and prevents or reverses endometrial hyperplasia.

4. At low risk for osteoporosis: Advise average vitamin D and calcium intake; advise exercise primarily to control weight and reduce cardiovascular risks.

5. Ovulation induction carries a high risk of ovarian hyperstimulation and multiple gestation, particularly if exogenous gonadotropins are used: Protocols in which FSH administration is carefully adjusted so as to not exceed the FSH threshold for folliculogenesis more than necessary are recommended and follicular development must be monitored carefully with ultrasound. There is some suggestion that suppression of LH or androgen levels prior to ovulation induction will lessen the miscarriage rate. Screen for diabetes mellitus before inducing ovulation. After conception, monitor for gestational diabetes. Women with insulin resistance are at higher risk for ovarian hyperstimulation than those women with PCOS and normal insulin action. Theoretically, the use of insulin sensitizers may lessen this risk, but this has not been demonstrated.

6. Glucocorticoids should be used for adrenal suppression or supplementation only if an attenuated form of adrenal hyperplasia or other adrenal condition has been confirmed.

Functional hypothalamic amenorrhea:

1. Hormonal interventions do not ameliorate stress and its consequences. In particular, appropriate bone accretion cannot occur in the presence of ongoing stress and sustained metabolic derangements. There is no evidence that bisphosphonates will stimulate bone accretion in the face of persistent metabolic compromise.

2. This is the only truly potentially reversible cause of anovulation.

3. Encourage lifestyle changes and nonpharmacologic psychological interventions before resorting to pharmacologic options. Recommend alternative strategies other than strict dietary control and excessive exercise as ways of coping with life's inevitable stresses.

4. Resort to ovulation induction only if weight is adequate and nonpharmacologic interventions are ineffective in restoring ovulation. Ovulation induction in underweight women is associated with a higher risk of intrauterine growth retardation and preterm delivery. The exogenous administration of pulsatile GnRH mimics physiology and results in high rates of conception. Most women with functional hypothalamic hypogonadism will be unresponsive to the antiestrogen clomiphene citrate when it is employed to induce ovulation because the hypothalamus is "insensitive" to circulating estradiol levels.

5. The use of psychotropics in those without an established psychiatric diagnosis has not been studied and their use would not be recommended as a first option in those attempting to conceive.

6. If a woman with FHA has a coexisting psychiatric disorder, such as depression, anxiety, phobia, drug dependence, anorexia nervosa, or bulimia nervosa, prompt psychiatric consultation is advised.

7. Screen for organic causes such as brain tumors with magnetic resonance imaging if there are any unusual symptoms. In cases of primary amenorrhea, be sure to inquire about sense of smell. Anosmia or hyposmia indicate Kallmann's syndrome.

Premature ovarian failure:

1. Screen for other glandular failure, particularly if there is any evidence of an autoimmune disorder. Screen annually for thyroid disease, even if the cause is genetic.

2. Intermittent spontaneous ovarian activity is common, but fertility in this setting is rare, presumably because of accelerated "oocyte aging." Donor oocytes or embryos are an option that should be mentioned.

3. Hormone replacement therapy is a must to prevent the sequelae of hypoestrogenism. For those with residual ovarian activity, oral contraceptive use may prevent endometrial desynchronization and give less breakthrough bleeding.

4. Expect good bone accretion in response to exogenous sex steroids because of the lack of a concurrent metabolic disorder. Advise the usual amount of vitamin D (200–800 IU/day) and calcium. If there was a prolonged episode of anovulation, measure bone density and consider the concurrent use of estrogen and bisphosphonates.

In summary, good medical practice now dictates that the cause of anovulation be identified so that an appropriate evaluation can be conducted and so that appropriate therapeutic options can be instituted. The risks and benefits of oral contraceptive use will depend on the underlying condition. Monitoring strategies and dietary recommendations are disease specific. Fertility management is gated by the diagnosis. The era of reductionistic therapeutic approaches to anovulation has passed.

See Also the Following Articles

ADRENAL HYPERPLASIA, CONGENITAL VIRILIZING; DIHYDROTESTOSTERONE; FOLLICULAR ATRESIA; HYPERPROLACTINEMIA; MENOPAUSE; MENSTRUATION; OVARIAN FUNCTION, LACK OF; POLYCYSTIC OVARY SYNDROME; TURNER'S SYNDROME

Bibliography

Berga, S. L. (1997). Behaviorally induced reproductive compromise in women and men. *Sem. Reprod. Endocrinol.* 15, 47–53.

Berga, S. L., Mortola, J. F., Girton, B., Suh, B., Laughlin, G., Pham, P., and Yen, S. S. C. (1989). Neuroendocrine aberrations in women with functional hypothalamic amenorrhea. *J. Clin. Endocrinol. Metab.* **68**, 301–308.

Cramer, D. W., Xu, H., and Harlow, B. L. (1995). Family history as a predictor of early menopause. *Fertil. Steril.* **64**, 740–745.

Faddy, M. J., Gosden, R. G., Gougeon, A., Richardson, S. J., and Nelson, J. F. (1992). Accelerated disappearance of ovarian follicles in mid-life: Implications for forecasting menopause. *Hum. Reprod.* **7**, 1342–136.

Hurley, D. M., Brian, R., Outch, K., Stockdale, J., Fry, A., Hackman, C., Clarke, I., and Burger, H. G. (1984). Induction of ovulation and fertility in amenorrheic women by pulsatile low-dose gonadotropin-releasing hormone. *N. Engl. J. Med.* **310**, 1069–1074.

Klibanski, A., Biller, B. M. K., Schoenfeld, D. A., Herzog, D. B., and Saxe, V. C. (1995). The effects of estrogen administration on trabecular bone loss in young women with anorexia nervosa. *J. Clin. Endocrinol. Metab.* **80**, 898–904.

Klingmuller, D., Dewes, W., Krahe, T., Brecht, G., and Schweikert, H. U. (1987). Magnetic resonance imaging of the brain in patients with anosmia and hypothalamic hypogonadism (Kallmann's syndrome). *J. Clin. Endocrinol. Metab.* **65**, 581–584.

Korytkowski, M. T., Mokan, M., Horwitz, M. J., and Berga, S. L. (1995). Metabolic effects of oral contraceptives in women with polycystic ovary syndrome. *J. Clin. Endocrinol. Metab.* **80**, 3327–3334.

Koskinen, P., Penttila, T.-A., Anttila, L., Erkkola, R., and Irjala, K. (1996). Optimal use of hormone determinations in the biochemical diagnosis of the polycystic ovary syndrome. *Fertil. Steril.* **65**, 517–522.

Miller, D. S., Reid, R. R., Cetel, N. S., Rebar, R. W., and Yen, S. S. C. (1983). Pulsatile administration of low-dose gonadotropin-releasing hormone: Ovulation and pregnancy in women with hypothalamic amenorrhea. *J. Am. Med. Assoc.* **250**, 2937–2941.

Petrides, J. S., Mueller, G. P., Kalogeras, K. T., Chrousos, G. P., Gold, P. W., and Deuster, P. A. (1994). Exercise-induced activation of the hypothalamic–pituitary–adrenal axis: Marked differences in the sensitivity to glucocorticoid suppression. *J. Clin. Endocrinol. Metab.* **79**, 377–383.

Reindollar, R. H., Novak, M., Tho, S. P. T., McDonough, P. G. (1986). Adult-onset amenorrhea: A study of 262 patients. *Am. J. Obstet. Gynecol.* **155**, 531–543.

Talbott, E., Guzick, D., Clerici, A., Berga, S., Detre, K., Weimer, K., and Kuller, L. (1995). Coronary heart disease risk factors in women with polycystic ovary syndrome. *Arterioscler. Thromb. Vasc. Biol.* **15**, 821–826.

Van der Spuy, Z. M., Steer, P. J., McCusker, M., Steele, S. J., and Jacobs, H. S. (1988). Outcome of pregnancy in underweight women after spontaneous and induced ovulation. *Br. Med. J.* **296**, 962–965.

Amniocentesis

Tzazil Ayala and Nancy C. Rose

University of Pennsylvania School of Medicine

I. Introduction
II. Technique
III. Indications
IV. Safety
V. Amniocentesis in Multiple Gestations
VI. Early Amniocentesis
VII. Conclusions

GLOSSARY

amniocentesis The transabdominal aspiration of amniotic fluid.
amnion The innermost membrane that surrounds the fetus in utero.
cytogenetics The study of chromosome structure.
ultrasound The use of high-frequency sound waves that are converted from sound into electrical energy to form an image.

Prenatal diagnosis is an integral component of modern obstetrical care. Although various procedures, such as chorionic villous sampling and percutaneous umbilical blood sampling, can be used to obtain fetal tissue prenatally, midtrimester amniocentesis is the most commonly used prenatal diagnostic technique available.

I. INTRODUCTION

As originally reported in the medical literature in the nineteenth century, Liley in 1961 integrated amniocentesis into obstetrical practice by reporting the use of serial amniocentesis to indirectly detect the degree of fetal hemolysis in patients with rhesus isoimmunization during pregnancy. Currently, the most common indication for amniocentesis is midtrimester testing for fetal chromosomal aneuploidy. This technique can also be used to test for multiple disorders, such as open fetal defects, rhesus isoimmunization, inborn errors of metabolism, and the molecular analysis of many inherited disorders. The most common indication for amniocentesis is advanced maternal age. Other indications include a positive serum screening test for fetal Down syndrome, open fetal defects, or a fetal malformation which is observed during ultrasound.

II. TECHNIQUE

Amniocentesis is most often performed between 14 and 20 weeks gestation, but it can be performed from 12 weeks to term (Fig. 1). Before the procedure, an ultrasound examination should be performed. Sonography is used to evaluate the number of fetuses, assess viability, confirm gestational age, detect major fetal structural abnormalities, assess placental location and amniotic fluid volume, select an optimal location of fluid for sampling, and to rule out significant uterine or adnexal pathology.

When the ultrasound examination is completed, a site is selected for needle insertion. Location of the small bowel and bladder must be made in order to avoid injury. The needle insertion site is cleansed with iodine solution and draped with sterile towels. Some obstetricians use a local anesthetic (1 cc of 1% lidocaine) before inserting the needle. A 20- or 22-gauge spinal needle is inserted transabdominally into the amniotic cavity under continuous ultrasonographic guidance. Once satisfactory placement of the needle is obtained, the stylet is removed and in most instances free flow of the amniotic fluid will occur. Approximately 3 ml of amniotic fluid is aspirated

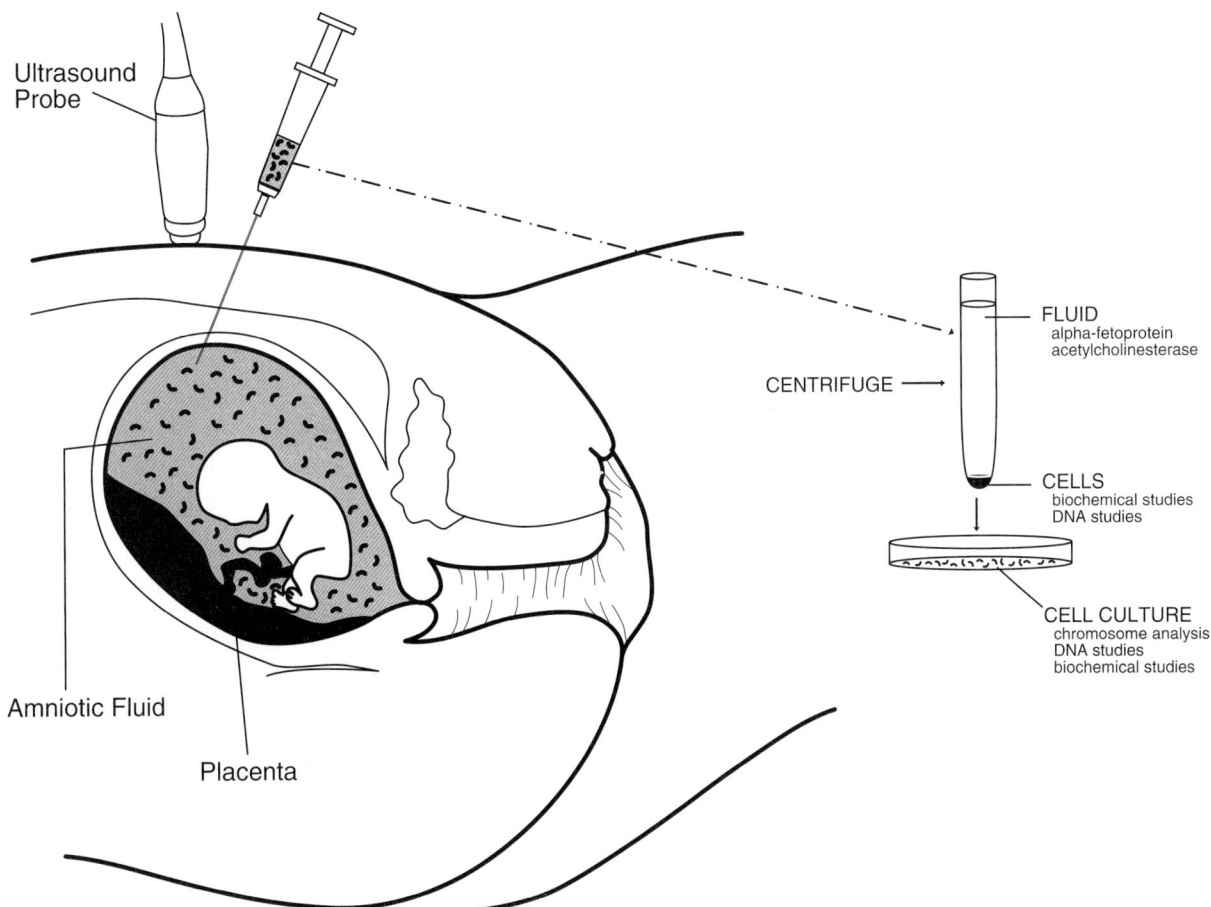

FIGURE 1 Prenatal diagnosis by amniocentesis (reproduced with permission from Greenwood Genetic Center, 1989).

initially; the syringe and fluid are discarded to reduce the risk of maternal cell contamination. For second-trimester amniocentesis obtained between 14 and 20 weeks, 20–30 ml of fluid is removed. Earlier sampling limits the amount of fluid withdrawn; the amount of fluid obtained should equal the number of weeks of gestation in milliliters.

Once fluid aspiration is completed and the needle has been removed, normal fetal cardiac activity should be confirmed. Administration of 300 μg of Rh immune globulin should be given to all unsensitized, Rh-negative patients. The patient may resume her normal activities after the procedure but is generally asked to avoid strenuous activity for 1 or 2 days. The patient is asked to inform her physician of any vaginal leaking, bleeding, or persistent cramping after the procedure. Additionally, patients are instructed to notify their physician if they develop a fever.

III. INDICATIONS

Amniocentesis may be performed for a variety of fetal or maternal indications. In the United States, it is considered standard care to offer prenatal cytogenetic analysis to all women who will be 35 years of age or older at their expected time of delivery. There is an increased risk of numerical chromosomal abnormalities as maternal age advances due to nondisjunction occurring during maternal meiosis I. In Table 1, the relationship between maternal age and the estimated risk of chromosomal abnormalities is shown.

TABLE 1
Fetal Chromosomal Aneuploidy: Estimated Term Risks

Age	Risk of trisomy 21	Risk of chromosomal aneuploidy (all types)
20	1:1667	1:526
25	1:1250	1:476
30	1:952	1:385
35	1:385	1:202
36	1:295	1:162
37	1:227	1:129
38	1:175	1:102
39	1:137	1:82
40	1:106	1:65
41	1:82	1:51
42	1:64	1:40
43	1:50	1:32
44	1:38	1:25
45	1:30	1:20
46	1:23	1:16
47	1:18	1:13
48	1:14	1:10
49	1:11	1:7

The risk for chromosomal aneuploidy in the fetus is determined by maternal age if the maternal age is >35 years or can be ascertained by maternal serum screening in younger women. The serum test is voluntary and is typically performed between 16 and 18 weeks' gestation. Three serum analytes are used to screen for aneuploidy: α-fetoprotein, human chorionic gonadotropin, and unconjugated estriol. Together, these three serum markers detect about 60% of fetal Down syndrome and fetal trisomy 18 with a 5% false-positive rate. If a patient screens positive for one of the aspects of this test, nondirective counseling for fetal diagnosis with amniocentesis is pursued. All patients who undergo amniocentesis for chromosomal aneuploidy should have an amniotic fluid α-fetoprotein performed routinely to evaluate the fetus for an occult neural tube defect. If this value is elevated, a measurement of acetylcholinesterase can be performed. The latter test is highly specific for neural tube defects with a diagnostic accuracy of >99%.

Another indication for amniocentesis is rhesus isoimmunization. This procedure can be performed for either fetal molecular diagnosis to determine the rhesus blood type of the fetus or, if felt to be affected, to determine the degree of fetal hemolysis. The latter is performed by evaluating the amniotic fluid for the spectrophotometric estimation of bilirubin, a by-product of fetal hemolysis. Liley provided a framework for the management of rhesus isoimmunized gestations. Depending on the degree of hemolysis estimated by this technique, serial amniocenteses are performed every 1–3 weeks. If the fetus is suspected to have severe hemolysis with a significant risk for fetal anemia, percutaneous umbilical blood sampling can be performed to test for fetal hematocrit and to transfuse rhesus negative blood to the fetus.

Amniocentesis is also used to evaluate the fetus at risk for preterm delivery. Aspiration of fluid for fetal lung maturity, to diagnose chorioamnionitis, and to verify ruptured membranes are possible uses for amniocentesis depending on the clinical presentation. Premature delivery is the leading cause of perinatal morbidity and mortality worldwide. The rate of preterm delivery varies but is estimated at 5–10% of all births in developed countries. Microbial invasion of the amniotic cavity has been reported in 21.6% of women with preterm labor and intact membranes who subsequently deliver a preterm neonate. The gram stain and bacterial culture of amniotic fluid is used for the diagnosis of microbial invasion of the amniotic cavity.

The prediction of respiratory distress syndrome (RDS) by amniocentesis was developed by Gluck *et al.* and first reported in 1974. RDS is caused by a deficiency of pulmonary surfactant, a substance that helps decrease the pressure needed to distend pulmonary alveoli. Phospholipids account for more than 80% of the material needed to distend the lung; two of its major proteins are lecithin and sphingomyelin. Gluck and colleagues observed a dramatic increase in amniotic fluid lecithin concentrations at 35 weeks' gestation in contrast to levels of sphingomyelin, which remained relatively constant. A lecithin/sphingomyelin (L:S) ratio of >2.0 is a common interpretation of this clinical test for fetal lung maturation. Many other types of fetal lung maturation tests are also available, although all are screening tests with associated false-positive and -negative rates. While the L:S ratio is perhaps the most common screening test for lung maturity, the finding of phosphatidal glycerol is the marker of completed pulmonary maturation. Phosphatidal glycerol (PG) is present after 35 weeks of gestation and increases until term. Most infants who lack PG but have a mature L:S profile fail to develop RDS, but infants with PG demonstrate completed pulmonary maturity. The latter is the preferred test for fetuses of diabetic mothers, who may have RDS despite mature L:S profiles.

Some inborn errors of metabolism are tested for by measuring the level of amino acids in amniotic fluid. These may include argininosuccinic aciduria, sulfite oxidase deficiency, and homocystinuria. Additionally, enzymatic study of cultured amniocytes after the 13th week of gestation can be used in the prenatal diagnosis of amino acidopathies.

IV. SAFETY

The safety of amniocentesis has been addressed in many large-scale studies. The United States National Institute of Child Health and Human Development published the first prospective study of genetic amniocentesis evaluating 1040 subjects with 992 matched controls. The amniocentesis group had a total fetal loss rate of 3.5% compared with 3.2% for controls; the difference was not statistically significant. Furthermore, newborn examination indicated no significant differences between the two groups in the incidence of congenital anomalies and no evidence of physical injury resulting from amniocentesis.

The most current randomized control trial is that of Tabor and colleagues, reported in 1986. Women (4606) between the ages of 25 and 34 years were recruited who were without risk factors for genetic disorders. Unlike several large earlier reports, all procedures were performed under ultrasound guidance. The spontaneous loss rate in the procedure group was 1.7% compared with 0.7% in the unexposed group. To our knowledge, no other studies have been performed using this truly randomized, low-risk population study design. Given the current sophistication of both ultrasonography and the practitioner, the standard quoted risk for patients is usually 1:200 or less for pregnancy loss following the procedure.

Minor maternal complications, such as transitory vaginal bleeding, persistent abdominal cramping, or amniotic fluid leakage, occur in 2 or 3% of cases. The major maternal risk is that of rhesus sensitization, which is largely preventable with the administration of anti-Rh (D) immune globulin following any prenatal diagnostic procedure. Fetal risks include needle puncture, umbilical cord hematoma and occlusion, placental separation, chorioamnionitis, and premature labor.

V. AMNIOCENTESIS IN MULTIPLE GESTATIONS

Twinning occurs in approximately 1 of 80 Caucasian women, in which one-third are monozygotic or identical twins and two-thirds are nonidentical or dizygotic twins. While the incidence of identical twins remains constant at 1 in 300 births and is unaffected by genetic factors, dizygosity is affected by race and maternal age. Between the ages of 35 and 39, the incidence increases to 1 in 65 offsprings. If a twin pregnancy is discovered at the time of an amniocentesis, gestational dating should be established. The ability to differentiate between monochorionic and dichorionic gestations is of clinical importance because of the difference in perinatal morbidity and mortality between these two groups. Traditionally, the ultrasonographic determination of dichorionicity included visualization of two sacs before 10 weeks' gestational age, discordant fetal sex, two separate placentas, and intertwin membrane characteristics, such as width and number of layers as well as the visualization of the twin peak/lambda sign.

Since the risk of fetal malformation is higher in multiple gestations, a fetal anatomic survey should be performed for each fetus. If the fetuses are believed to be identical, they should be evaluated for early signs of twin–twin transfusion syndrome, in which a vascular connection exists pumping blood from the donor to the recipient twin.

Patients need to be counseled about a somewhat higher risk of pregnancy loss from sampling fluid from each gestational sac with a separate insertion site. They also must be prepared for the possibility of discordant results. The available options are continuation of the pregnancy, termination of the pregnancy, or selective termination of the affected fetus with its associated 15% risk for miscarriage.

The needle should be introduced into the first sac under ultrasound guidance; after aspiration of the fluid, a small amount of indigo carmine, an inert dye is instilled. When the other sac is entered, aspiration of clear fluid indicates that the initial sac has not been resampled, and blue amniotic fluid signifies that the same sac has been reentered.

VI. EARLY AMNIOCENTESIS

The safety and efficacy of early amniocentesis performed between 11 and 14 weeks is currently being investigated, but there are no large randomized control trials available to evaluate this technique with either (CVS) or standard second-trimester amniocentesis. The technique differs in that the fluid volume of amniotic (after 2 or 3 ml has been aspirated for maternal cell contamination) equals that of the number of weeks of gestation using either a 20- or 22-gauge needle. The advantage of this technique is that information about the fetus can be obtained via a high-quality "G" banded amniocyte culture at an earlier gestational age. Of the studies available, one randomized control trial has been performed comparing early amniocentesis between 10 and 13 weeks of gestation with chorionic villous sampling. The rate of successful sampling was the same for both procedures (97.5%). The total number of patients was 1301, and of these 62% voluntarily chose either CVS (320) or early amniocentesis (493). Of the randomized subgroup, 238 patients were randomized to early amniocentesis, whereas 250 were randomized to CVS. The median age range across all groups was 38 years; the median gestational age for all procedures was 11 weeks. In the randomized subgroup, the fetal loss rate was 5.9% in the early amniocentesis group [confidence interval (CI), 3.3–9.7%] versus 1.2% in the CVS subgroup (CI 0.3–3.5%). In each group, 2.5% of patients had to undergo repeat sampling: The majority of CVS patients were resampled for fetal mosaicism and the early amniocentesis patients for culture failure.

Of the multiple observational studies, the spontaneous loss rate varies from 1.4 to 2.3%. In all reports, the majority of patients were of advanced maternal age. Given that the incidence of aneuploidy is higher in these patients, and that the amnion and chorion have not fused prior to 14 or 15 weeks of gestation, this may account for the higher degree of losses with this procedure.

VII. CONCLUSIONS

Prenatal diagnosis is a standard component of obstetric care. Improvement of ultrasound technology has allowed the development of safer techniques for fetal sampling, as well as a better physical assessment of the fetus. Invasive prenatal testing carries a small risk for fetal loss; therefore, this approach is generally limited to patients at significant risk for specific problems. Noninvasive techniques, such as the isolation of fetal cells from the maternal circulation or those obtained from the maternal cervix, are currently being investigated.

See Also the Following Articles

FETAL MONITORING AND TESTING; RESPIRATORY DISTRESS SYNDROME; TWINNING; ULTRASOUND

Bibliography

Canick, J. A., Palomaki, G. E., and Osathanondth, R. (1990). Prenatal screening for trisomy 18 in the second trimester. *Prenat. Diagn.* 10, 546–548.

Gluck, L., Kulovich, M. V., Borer, R. C., *et al.* (1971). Diagnosis of the respiratory distress syndrome by amniocentesis. *Am. J. Obstet. Gynecol.* 109, 440–445.

Haddow, J. E., Palomaki, G. E., Knight, G. J., *et al.* (1992). Prenatal screening for Down syndrome with use of maternal serum markers. *N. Engl. J. Med.* 327, 588–593.

Hook, E. B. (1981). Rates of chromosome abnormalities at different maternal ages. *Obstet. Gynecol.* 58, 282–285.

Hook, E. B., Cross, P. K., and Schreinemachers, D. M. (1983). Chromosomal abnormality rates at amniocentesis and in live-born infants. *J. Am. Med. Assoc.* 249, 2034–2038.

Liley, A. W. (1986). Liquior amnii analysis in the management of the pregnancy complicated by rhesus isoimmunization. *Am. J. Obstet. Gynecol.* **82**, 1359–1370.

NICHD National Registry for Amniocentesis Study Group (1976). Midtrimester amniocentesis for prenatal diagnosis. Safety and accuracy. *J. Am. Med. Assoc.* **236**, 1471–1476.

Nicolaides, K., De Lourdes Brizot, M., Patel, F., and Snijders, R. (1994). Comparison of chorionic villous sampling and amniocentesis for fetal karyotyping at 10–13 weeks gestation. *Lancet* **344**, 435–439.

Tabor, A., Madsen, A., Obel, E. B., Philip, J., Bang, J., and Nørgaard-Pedersen, B. (1986). Randomized controlled trial of genetic amniocentesis in 4,606 low-risk women. *Lancet* **1**, 1287–1292.

Wood, S. L., Onge, R. St., Connors, G., and Elliot, P. D. (1996). Evaluation of the twin peak or lambda sign in determining chorionicity in multiple pregnancy. *Obstet. Gynecol.* **88**, 6–9.

Amniotic Fluid

Robert A. Brace
University of California, San Diego

I. Introduction
II. Amniotic Fluid Volume
III. Composition of Amniotic Fluid
IV. Fluid Movements into and out of the Amniotic Sac
V. Normal Flows
VI. Regulation of Amniotic Fluid Volume
VII. Summary

GLOSSARY

amnion A membrane that surrounds the amniotic fluid.
amniotic fluid The liquid that surrounds the embryo/fetus prior to birth.
chorion A membrane extending outward from the edges of the placenta that surrounds the amnion.
fetal swallowing A primary route of clearance of amniotic fluid.
fetal urine production A primary source of amniotic fluid.
intramembranous exchange Any direct exchange of water and/or solutes between amniotic fluid and fetal blood.
lung liquid secretion A primary source of amniotic fluid.
oligohydramnios Pathologically small amounts of amniotic fluid.
polyhydramnios Pathologically large amounts of amniotic fluid.

transmembranous exchange Exchange of water and/or solutes across the amniotic and chorionic membranes between amniotic fluid and maternal blood within the wall of the uterus.

During development, the embryo and fetus are surrounded by a liquid called amniotic fluid. Amniotic fluid is surrounded by a thin, tough membrane, the amnion, which is in turn surrounded by another membrane, the chorion. The chorion extends outward from the edges of the placenta and lies adjacent to the inner wall of the uterus. Early in gestation, there are two extraembryonic/extrafetal fluids within the uterus. In addition to amniotic fluid inside the amnion, coelomic fluid fills the space between the amnion and chorion. At approximately the time of the transition from embryo to fetus at 10 weeks' gestation, the amount of coelomic fluid begins to decrease. Shortly after the end of the first trimester (13 weeks' gestation), the coelomic fluid disappears and the amnion contacts the chorion throughout the rest of pregnancy. In contrast with the disappearance of coelomic fluid, the volume of amniotic fluid increases as the fetus grows.

I. INTRODUCTION

Although there remains a great deal which is not know about the function of amniotic fluid, it is clear that the amniotic fluid provides several important benefits to the embryo/fetus and there are speculations as to other potential benefits. For example, the amniotic fluid provides a cushion for the fetus by protecting its delicate tissues from bumps and trauma to the maternal abdomen. The amniotic fluid also has antibacterial properties to help protect against infection. Another important benefit is that the amniotic fluid prevents the uterus from compressing the fetus, thereby providing space for the fetus to move as it exercises its developing muscles. The amniotic fluid is a reservoir which can provide not only water but also limited amounts of minerals and metabolic substrates in times of need when placental transfer is not adequate to meet fetal demands. Finally, growth factors in swallowed amniotic fluid enhance the growth, development, and maturation of the gastrointestinal tract. Thus, amniotic fluid provides many essential functions for the embryo and fetus.

II. AMNIOTIC FLUID VOLUME

In order for the fetus to grow and develop normally, the volume of amniotic fluid must be within its normal range. Early in gestation, amniotic fluid volume averages about 25 ml at 10 weeks' gestation, when the fetus weighs approximately 15 g. Thereafter, amniotic fluid volume and fetal weight increase roughly in parallel until about 20 weeks' gestation, when the fetus weighs 300 g and amniotic fluid volume averages 350 ml. Amniotic fluid volume increases more slowly after 20 weeks' gestation to reach its maximum at 30–35 weeks which averages approximately 800 ml. The volume of amniotic fluid often decreases as term approaches and can decrease dramatically in pregnancies which go beyond term, i.e., in postdate pregnancies. Figure 1 illustrates these changes across gestation and further shows the 95% confidence interval about the mean amniotic volume at each gestational age. Note that the variability in individual amniotic fluid volumes changes dramatically across gestation.

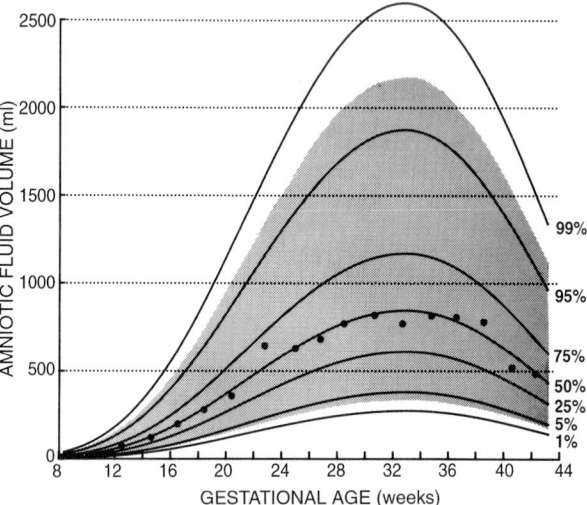

FIGURE 1 Nomogram showing amniotic fluid volume as a function of gestational age. Dots are means for each 2-week interval. Shaded area is 95% confidence interval. Percentiles are calculated from polynomial regression equation and standard deviation of residuals (reprinted with permission from Brace and Wolf, 1989).

Amniotic fluid volumes below the 95% confidence interval can be considered to be oligohydramnios (too little amniotic fluid), whereas volumes above the 95% confidence interval can be considered to be polyhydramnios (too much amniotic fluid). Both of these conditions are associated with increased rates of fetal and neonatal morbidity and mortality. Although Fig. 1 provides descriptive information about changes across gestation, amniotic fluid volume is rarely measured in clinical practice because of methodological limitations and the invasive nature of the measurement. Instead, qualitative or semiquantitative ultrasonographic indices of amniotic fluid volume are routinely determined, with the four-quadrant amniotic fluid index being the most widely used.

III. COMPOSITION OF AMNIOTIC FLUID

Osmolality is a measure of the total concentration of all dissolved solutes. In nonpregnant women, blood osmolality averages 290 mOsm/kg of water and this decreases to 280 mOsm/kg shortly after implantation. Early in gestation, amniotic fluid osmolality is the same as maternal and fetal blood osmolality (280 mOsm/kg). The major solutes, such as sodium and chloride, are present in amniotic fluid in essentially the same concentrations as in maternal and fetal plasma. Beginning at the end of the embryonic period, amniotic fluid osmolality begins a progressive decrease which continues until term, when amniotic osmolality averages 265 mOsm/kg. This fall in osmolality is paralleled by decreases in amniotic sodium and chloride concentrations, whereas maternal and fetal plasma osmolality and electrolyte concentrations are unchanged. The gestational fall in amniotic osmolality and electrolytes is believed to be due to entry into the amniotic sac of increased amounts of dilute fetal urine, which normally has an osmolality on the order of 150 mOsm/kg water and low sodium and chloride concentrations.

Essentially all other substances, including electrolytes, amino acids, metabolic substrates and by-products, hormones, and many enzymes are found in the amniotic fluid in varying concentrations. These are derived from many sources, including fetal urine and lung secretions, as well as from secretions of the amniotic and chorionic membranes and fetal skin. There have been many studies attempting to identify changes in the concentration of any amniotic solute which would provide clinically useful information. Somewhat surprisingly, there are no abnormalities in amniotic osmolality, sodium, or chloride concentrations associated with abnormal fetal conditions or disease except if the fetus dies *in utero*. The presence of pulmonary surfactants in amniotic fluid has long been used to indicate whether the fetal lungs are sufficiently mature to allow sustained ventilation after birth. The field is evolving rapidly and there are many new markers being suggested as indicators of fetal genetic anomalies, fetal infection, or impending labor.

IV. FLUID MOVEMENTS INTO AND OUT OF THE AMNIOTIC SAC

The source of amniotic fluid early in gestation is not well established but secretion by the membranes appears to be the likely source. Later in gestation, fetal urine is a major source of amniotic fluid and swallowing of amniotic fluid by the fetus is a major route of amniotic fluid removal. Fetal urine first en-

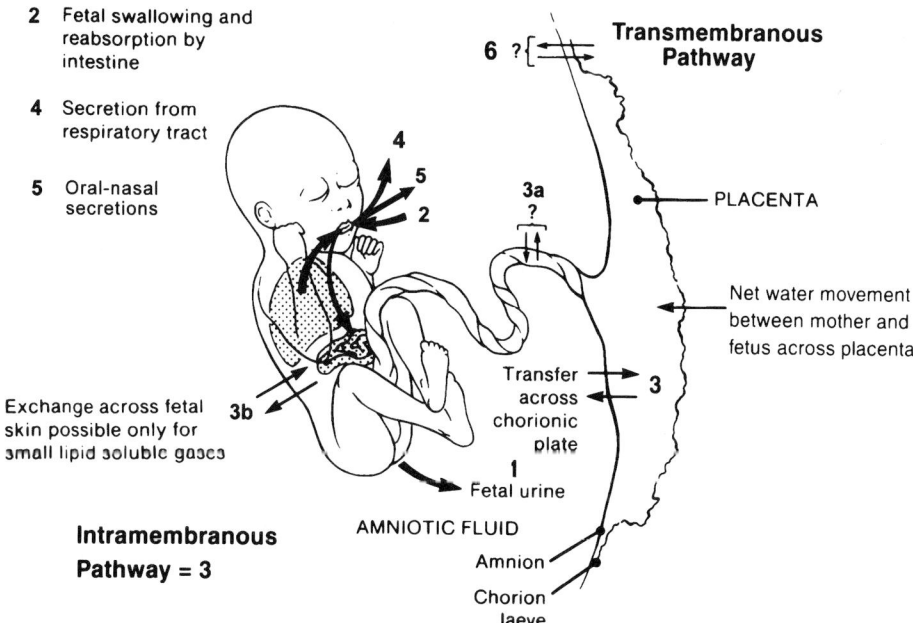

FIGURE 2 Schematic showing the six routes for fluid entry into and exit from the amniotic sac for the late-gestation fetus. The pathways are numbered in descending order relative to the magnitude of their associated flow. Multiple components of the intramembranous pathway are also shown (modified from Seeds, 1980).

ters the amniotic sac and the fetus begins to swallow amniotic fluid shortly after transition from embryo to fetus at 10 weeks' gestation. The fetal lungs begin to secrete fluid into the amniotic space shortly after this time.

These and other pathways for fluids to enter and leave the amniotic space are summarized in Fig. 2. In general, there are two primary sources of and two primary routes for clearance of amniotic fluid during the latter half of gestation. The two primary sources of amniotic fluid are fetal urine and lung liquid, and these become increasingly important as the fetus grows. Furthermore, there is a minor contribution of fetal oral–nasal secretions. The two primary routes of amniotic fluid clearance are fetal swallowing and absorption into fetal blood perfusing the fetal surface of the placenta. The latter is referred to as "intramembranous absorption," a phrase which has been generalized to include all passive exchanges which occur directly between amniotic fluid and fetal blood, e.g., across fetal skin and surface of the umbilical cord, but these are not likely to be significant during the latter half of gestation after the skin keratinizes. A final potential pathway is exchange between amniotic fluid and maternal blood within the uterine wall across the amniotic and chorionic membranes—referred to as "transmembranous" exchange. Although once thought to be significant, transmembranous fluxes now appear to be insignificantly small even for highly lipid-soluble substances such as urea and carbon dioxide. Thus, there are a total of six potential routes by which water and solutes may enter and/or leave the amniotic compartment during the latter half of gestation. It is important to recognize that, within each of these routes, water and solutes always move in the same direction (i.e., by bulk flow), except for the intramembranous and transmembranous pathways, in which water and solutes can and often do move in opposite directions (i.e., osmotic flow of water and diffusion of solutes).

V. NORMAL FLOWS

Understanding flows through the previously discussed pathways is important because, relative to amniotic volume, large volumes flow into and out of the amniotic fluid each day. For the six potential pathways described previously, quantitative estimates of the daily volume flows in human fetuses across gestation are available only for urinary output. The remaining flows have been estimated from studies in experimental animals. From ultrasound imaging of the fetal urinary bladder in three dimensions, the human fetus produces a volume of urine equal to approximately 30% of its body weight or 900 ml each day near term. The latter is in agreement with direct measurements in ovine fetuses. The volume of fluid swallowed daily by the human fetus has been difficult to estimate. Swallowing may be on the order of 17 ml/day at 20 weeks' gestation and 400–500 ml/day near term. These numbers represent the estimated volume of amniotic fluid swallowed but do not include the considerable volume of lung liquid swallowed each day. Nonetheless, it is clear that this estimate of roughly 15% of body weight per day is considerably less than the daily urine volume of 30% per day which enters the amniotic compartment. Swallowing has been measured in the fetal sheep and found to be on the order of 20–25% of body weight or 600–750 ml per day in 3-kg fetuses. Again, this is less than the volume of fetal urine which enters the amniotic compartment each day.

Less is known about lung liquid secretion rates. Although no measurements have been made in human fetuses, it is clear that all mammalian fetal lungs normally secrete fluid in quantities far in excess of that needed to expand the fetal lungs with growth. In fetal sheep, over the last third of gestation, lung secretion rates are relatively constant and average approximately 10% of fetal body weight per day. The secretion is mediated by an active transport of chloride ions into the future airways with sodium ions and water following passively. Approximately 1% of the secreted fluid is required to expand the growing lungs and the remainder exits the lungs via the trachea. Animal studies suggest that an average of half of the fluid secreted by the lungs enters the amniotic compartment, which is equivalent to 5% of body weight or 150 ml per day in late gestation. The remainder is swallowed as it exits the trachea.

Experimental estimates of intramembranous flows

are available only for the fetal sheep and suggest that, in late gestation, approximately 400 ml/day of amniotic fluid is absorbed each day into the fetal blood which perfuses the surface of the placenta and membranes.

The two remaining flows are oral–nasal secretions and transmembranous volume flow. The rate of fluid secretion from the fetal head was found to be 25 ml/day for 3-kg sheep fetuses or <1% of body weight per day. The daily volume of transmembranous flow has been confounded by the difficulty of measurement. Current best estimates suggest that the transmembranous flux of water may be only 10 ml/day near term under normal conditions. Transmembranous flow is low despite the large osmotic gradient most likely because of a high resistance within the uterine wall rather than in the membranes.

VI. REGULATION OF AMNIOTIC FLUID VOLUME

Late in gestation, when amniotic fluid volume (AFV) averages 700–800 ml, approximately 1000–1100 ml/day of fluid flows into the amniotic compartment and the same volume leaves the amniotic compartment each day. Thus, only minor to moderate aberrations in flows over a period of days to weeks could readily lead to oligohydramnios or polyhydramnios.

In order for AFV to remain relatively stable in the presence of these high flow rates, it would seem logical to conclude that AFV is well regulated. However, it is not known whether AFV is in fact regulated. That is, there are no known sensors for AFV which could be part of a control loop to return AFV toward normal whenever it becomes too high or too low. On the other hand, fetal urine flow, lung liquid secretion, and swallowing are all known to be regulated. Furthermore, intramembranous absorption is undoubtedly regulated by the factors which control intramembranous permeability and surface area. Thus, all the primary flows into and out of the amniotic compartment are regulated and it may be the interaction among these flows which provides the regulation of AFV. Furthermore, only slight changes in intramembranous permeability can have very large effects on intramembranous flux rates, raising the possibility that the intramembranous pathway may be one of the primary factors regulating AFV. Clearly, any substance, e.g., prostaglandins, excreted by the fetal kidneys, secreted by the fetal lungs, or released by the amnion or chorion which enters the amniotic fluid could potentially alter intramembranous permeability and thus lead to alterations in AFV.

Another possibility is that the ultimate controller of AFV may be the placental transfer of water and solutes to and from the fetus. Multiple animal studies have shown that, under normal conditions, the fetal kidneys are extremely capable of transferring huge amounts of exogenous water and/or salts to the amniotic compartment. Other studies have made it clear that maternal dehydration, particularly over a period of days, is associated with a reduction in AFV. Thus, maternal–fetal interactions across the placenta most likely will play an important role in regulating AFV. This concept is supported by the observation that amniotic fluid volume increases in pregnant women after they drink 2 liters of distilled water, particularly if AFV was initially below normal. Furthermore, since no change in fetal urine production occurs, the changes in AFV may be mediated by changes in intramembranous flow.

VII. SUMMARY

Although there are fairly wide variations, AFV normally undergoes characteristic changes across gestation in which it increases from 20–30 ml at 10 weeks' gestation to average 800 ml at 24 weeks. Little change occurs from then until near term, when AFV begins to decreases, and large decreases can occur in post-term pregnancies. Variations in amniotic fluid volume from its normal range are associated with increased rates of fetal and neonatal morbidity and mortality.

Relative to amniotic fluid volume, large volumes of fluid flow into and out of the amniotic compartment each day. Although there are six pathways through which fluid and solutes can enter and/or leave the amniotic sac, there are only four primary pathways which contribute to AFV during late gestation: Fetal urine and lung liquid secretion are the two primary sources of fluid, and fetal swallowing and intramembranous absorption are the two primary routes of amniotic water clearance. The intramembranous pathway also appears to be a primary source of amniotic solutes such as sodium and chloride. Variations in the concentration of certain substances in the amniotic fluid can be used to provide information about fetal conditions, including maturity of its lungs, genetic anomalies, infections, and pending labor and delivery.

See Also the Following Articles

FETAL LUNG DEVELOPMENT; FETAL MEMBRANES; FETAL-PLACENTAL UNIT

Bibliography

Brace, R. A. (1995). Progress toward understanding the regulation of amniotic fluid volume: Water and solute fluxes in and through the fetal membranes. *Placenta* 16, 1–18.

Brace, R. A., and Wolf, E. J. (1989). Normal amniotic fluid volume changes throughout pregnancy. *Am. J. Obstet. Gynecol.* 161, 382–388.

Chamberlain, P. F., Manning, F. A., Morrison, I., Harman, C. R., and Lange, I. R. (1984). Ultrasound evaluation of amniotic fluid volume. *Am. J. Obstet. Gynecol.* 150, 245–254.

Gilbert, W. M., and Brace, R. A. (1989). The missing link in amniotic fluid volume regulation: Intramembranous absorption. *Obstet. Gynecol.* 74, 748–754.

Harding, R. (1994). Development of the respiratory system. In *Textbook of Fetal Physiology* (G. D. Thorburn and R. Harding, Eds.), pp. 140–167. Oxford Univ. Press, London.

Seeds, A. E. (1980). Current concepts of amniotic fluid dynamics. *Am. J. Obstet. Gynecol.* 138, 575–586.

Amphibian Ovarian Cycles

Alberta M. Polzonetti-Magni

University of Camerino

I. Introduction
II. Structure and Shape of the Ovary
III. Seasonal Ovarian Changes
IV. Hormonal Regulatory Mechanisms
V. Ovarian Growth and Yolk Storage

GLOSSARY

autocrine A system by which the cell is controlled by substances that it produces.

endocrine A system through which hormones regulate their secretion by feedback. Endocrine systems are equipped with mechanisms for monitoring the magnitude of the biological effects controlling their secretory rate.

gonadotropin-releasing hormone A decapeptide that stimulates the pituitary gland to produce gonadotropin.

jelly Glycoprotein secreted by the oviducal gland localized above the vitelline envelope.

maturation-inducing substance A substance able to induce oocyte maturation.

maturation-promoting factor A complex of proteins that stimulate the resumption of meiosis.

oviducal glands Typical glands present in the oviduct tract.

paracrine A system by which substances produced by the cell are able to regulate short-distance cells.

secondary sexual characters Sexually distinguishing anatomical features whose development depends on hormonal control, in contrast with those determined genetically, i.e., thumb pad in anurans, nuptial crest in urodeles, and skin pigmentation.

vitellogenesis A complex process which includes vitellogenin synthesis by the liver, its release in the blood, and its selective uptake by the growing oocyte.

yolk proteins Proteins stored in the eggs of oviparous and ovoviviparous vertebrates and utilized during embryo development.

In amphibians, the reproductive processes are seasonally dependent; therefore, the ovary undergoes morphological and biochemical changes by which the oocytes become mature eggs ready for laying and fertilization. The study of the ovarian cycle involves analyses of relationships between environmental and endogenous factors in timing the breeding season; thus, attention must be paid to the regulatory mechanisms within the hypothalamic gonadotropin-releasing hormone, to gonadotropins, and to ovarian sex steroid production. The study of the ovarian cycle is intimately related to the study of oocyte growth; in fact, the most distinguishing feature of the egg is its large size, which is required for storage of reserves for development of the embryo. This process is consistent with a high rate of synthetic activities, and the reserves consist of messenger RNAs, ribosomes, enzymes, and other precursors of macromolecular synthesis in the egg prior to fertilization. The other requirement for development and growth is metabolic energy, the reserve for which is stored in the oocytes as "yolk."

I. INTRODUCTION

Amphibian ovarian cycles are closely related to patterns of reproduction which have evolved to maximize an individual's contribution of genetic information to the next generation; moreover, timing of reproduction is determined by when the most offspring will survive and when parents, mostly females, are capable of energetically supporting the production of viable young at the least cost to themselves. Temperate zone amphibians primarily breed in early spring, oogenesis occurs in summer, and the ovary is quiescent in winter, whereas tropical amphibians reproduce seasonally in environments that have a seasonal water/dry period since it is well-known that

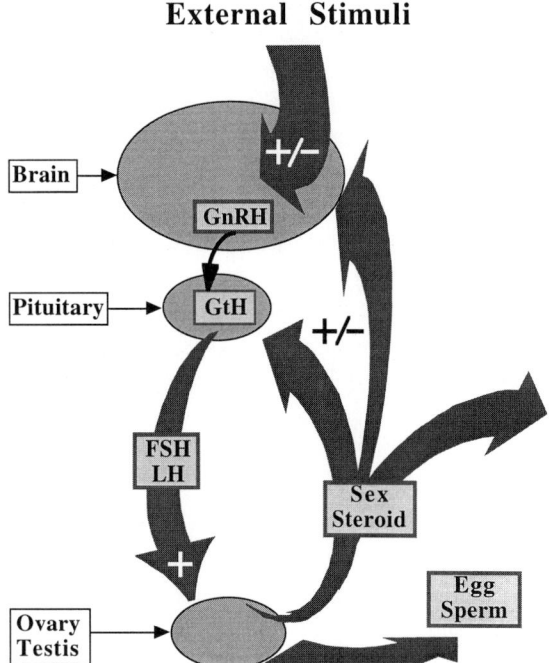

FIGURE 1 Hypothalamic–hypophysial–gonadal axis regulating the seasonal reproductive cycle through a long-loop feedback mechanism.

the factor limiting the success of amphibian reproduction is water availability. Temperature is the other cue that regulates seasonal reproduction in most amphibians; photoperiod has not been considered of primary importance. Therefore, the cyclical changes of amphibian ovary must be discussed in the frame of their seasonality, bearing in mind the role of environmental factors influencing the reproductive cascade, i.e., the hypothalamic secretion of gonadotropin-releasing hormone (GnRH) and its effects on gonadotropin (GtH) release by pituitary pars distalis. GtH, then, regulates ovarian steroidogenesis and oogenesis (Fig. 1); the latter consists not only of oocyte maturation but also of the vitellogenic processes that allow the storage of metabolic reserve as yolk (Fig. 2).

II. STRUCTURE AND SHAPE OF THE OVARY

The ovary plays two essential functions in the reproductive process: It produces the gametes (eggs)

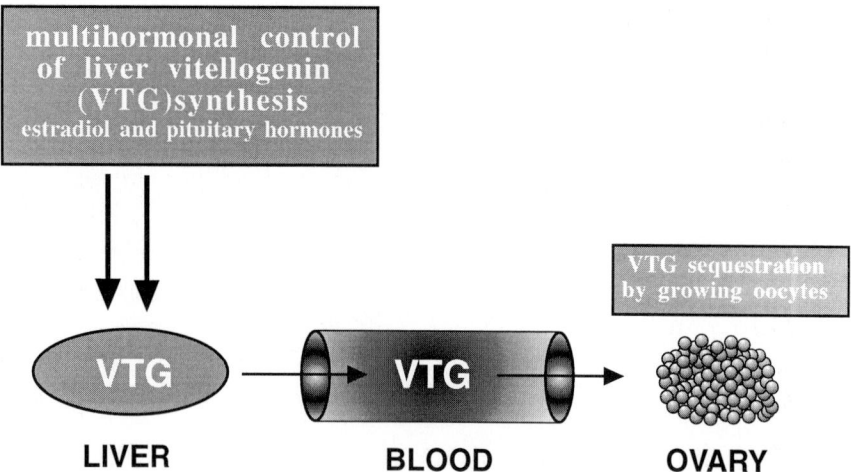

FIGURE 2 Scheme representing the vitellogenin processes in amphibians.

for the development of a new individual, and it produces steroid hormones. The functional units of the ovary are the follicles, each consisting of a single ovum surrounded by granulosa cells (the inner zone, which represents the major source of estradiol) and the theca cell (outer zone) synthesizing androgen precursors utilized by granulosa cells for estrogen synthesis.

Light microscopic study shows that follicles of different size are present in the frog ovary during the recovery phase (Fig. 3A). Oocyte yolk proteins (YPs) are present in numerous structures that have a somewhat geometric shape. During oocyte growth, the small nascent platelets increase in size by fusion with one another, and the mature egg contains a heterodisperse size range (up to 50 μm) of platelets and has a bimodal frequency distribution with peaks at 2 and 35 μm. Besides their increase in size, YPs increase in number and extend as far as the innermost areas of the oocyte cytoplasm. Oocytes sampled during the recovery phase show a thin granulosa layer containing follicle cells and a large nucleus provided with numerous nucleoli (Figs. 3B and 3C).

III. SEASONAL OVARIAN CHANGES

The ovarian changes are related to the timing of breeding season: For instance, in the wild population of the anuran *Rana esculenta* living in a mountain pond at 820 m a.s.l., changes reported as gonadosomatic index (GSI) are depicted in Fig. 4a. This frog breeds in early May, when the ambient and air temperature rise to 12 and 17°C, respectively; no more than one clutch has been observed each season, whereas in other populations of *R. esculenta* living at sea level the breeding season occurs earlier—in March and April. During the breeding season, the ovarian weight reaches peak values; after egg laying, the GSI sharply decreases since the ovary contains only postovulatory follicles together with the atretic ones. During the summer months, the ovary becomes refractory to the still favorable environmental conditions (water temperature and food availability) so that no more yolk deposition is observed; in sea-level frogs, more clutches are formed. At the beginning of autumn, the GSI significantly increases, and in 2 months it reaches the highest values; this recrudescent period is very short in the mountain population and is accomplished before early winter, when the cold temperature sets in. Once the ovary has accomplished the vitellogenic process after the winter stasis, the oocyte resumes meiotic maturation at the beginning of the spring months under hormonal control.

Regarding urodele amphibians, there are differences in the ovarian changes compared with those found in anurans. For the population of the urodele *Triturus carnifex* living in the same mountain pond at 820 m a.s.l., the ovarian changes are depicted in

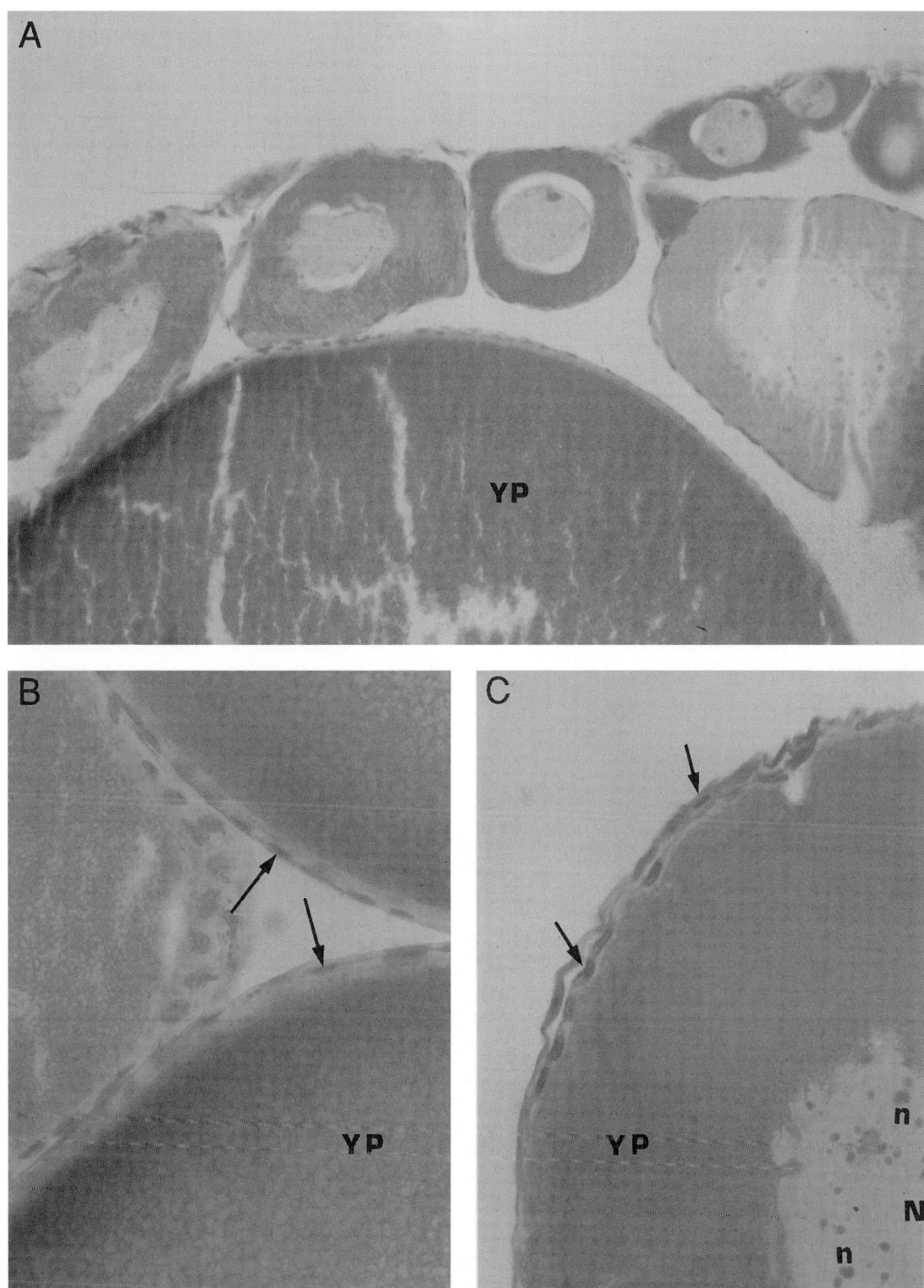

FIGURE 3 Light micrographs of sections from *Rana esculenta* ovary during the recovery phase (September). (A) The section of frog ovary shows follicles at different stages of growth. (B and C) Details of oocytes showing the follicle cells (arrows). N, nucleus; n, nucleoli; YP, yolk protein. Magnification, ×1100.

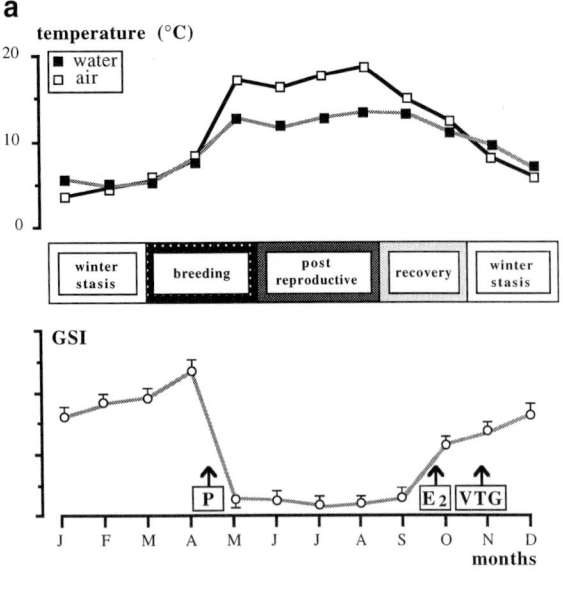

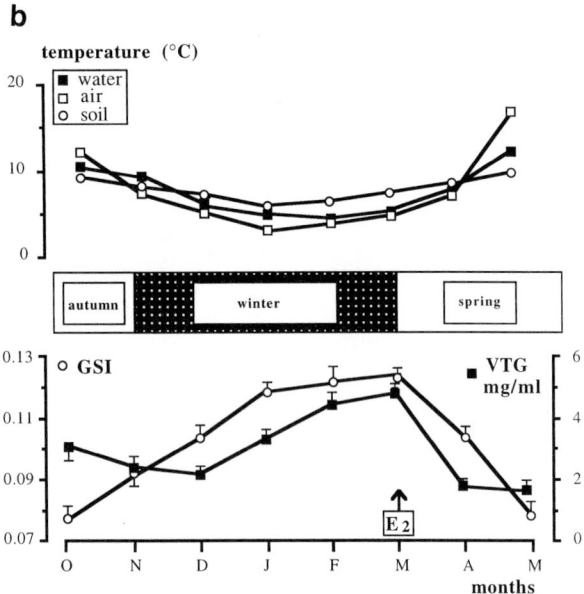

FIGURE 4 GSI, air, water, and soil temperature changes during the reproductive cycle of frog, *Rana esculenta* (a) and *Triturus carnifex* Laur (b).

Fig. 4b as a model of the reproductive pattern of an aquatic amphibian. In contrast to the frog *R. esculenta*, which can be collected year-round in both the field and the pond, this newt lives in the pond almost all year, with the exception of the breeding season. In summer, however, it can only be caught underground, where it avoids high water temperature and, at the same time, dehydration. The newt ovarian cycle, therefore, is less discontinuous in its GSI trend, and since the breeding period starts in November and lasts until March or April, several clutches can be observed; in fact, GSI is continuously high during the autumn and winter months, whereas ovarian weight decreases in the spring. As in the frog, in the newt the environmental conditions influence ovarian changes; the cold temperatures are primarily responsible for the water drive of the newts from their underground shelter to the pond in which they accomplish their mature egg laying.

The ovarian changes correlate with those of peripheral steroids, which are responsible not only for oocyte maturation but also for development of the secondary sexual characteristics and reproductive behavior. For instance, in females, ovarian steroids regulate the oviduct changes which primarily consist of jelly production from the oviductal glands, which protects ovulated eggs.

IV. HORMONAL REGULATORY MECHANISMS

Full-grown postvitellogenic oocytes in the ovary are in prophase I of meiosis and cannot yet be fertilized. For this to occur, they must complete the first meiotic division. Under appropriate hormonal stimulation, full-grown oocytes resume their first meiotic division, which involves breakdown of the germinal vesicle, chromosome condensation, assembly of the first meiotic spindle, and extrusion of the first polar body (Fig. 5). Shortly thereafter, mature fertilizable oocytes are ovulated. The meiotic process leading to extrusion of the second polar body is resumed at the time of fertilization, immediately after sperm penetration. The period between the resumption of meiosis and the second meiotic metaphase has been referred to as the period of oocyte maturation. Thus, the process of oocyte maturation is a prerequisite for successful fertilization. Oocyte maturation requires a hormonal cascade that begins with the brain decapeptide GnRH, which stimulates the synthesis and release of pituitary gonadotropins, which in turn influence the ovarian production of sex steroids, estra-

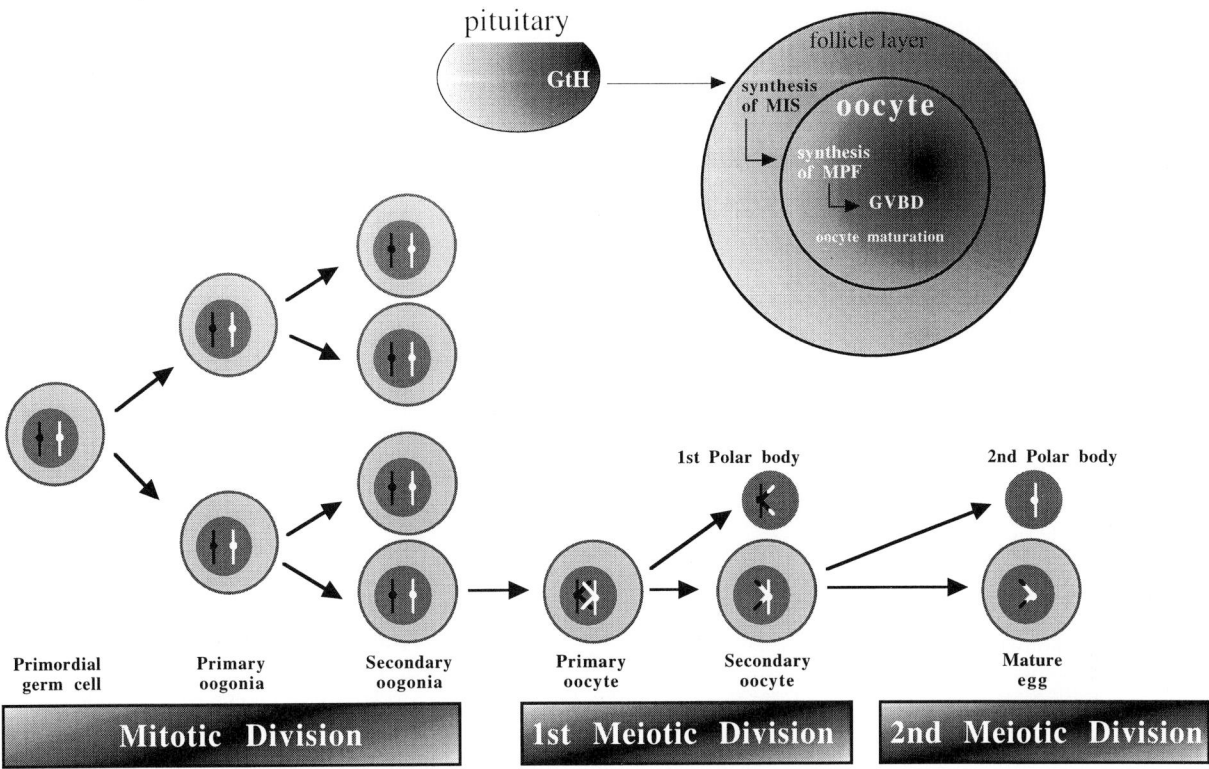

FIGURE 5 Control of oocyte maturation.

diol and progesterone, maturation-inducing substance (MIS), and maturation-promoting factor.

Two chemically distinct GtHs, which appear homologous to mammalian luteinizing hormone (LH) and follicle-stimulating hormone (FSH), have been purified from pituitaries of several amphibian species. Biochemical, immunological, and biological data indicate a marked structural homology between the two GtHs of amphibians and the FSH and LH of mammals. Both *in vivo* and *in vitro* experiments have established that LH plays a major role in the maturational process. However, the maturational action of LH is not direct but is mediated by the follicular production of MIS, and accumulating evidence suggests that progesterone is the natural MIS. Since the microinjection of maturation-inducing steroids into full-grown oocytes fails to induce maturation, it has been suggested that maturational steroids act indirectly on the germinal vesicle to induce resumption of meiosis; therefore, some cytoplasmic factor which directly brings about oocyte maturation is newly produced under the influence of MIS.

These data clearly indicate that ovarian growth, development, and secretion are ultimately controlled by a long-loop feedback mechanism; however, amphibian ovarian function is also regulated in paracrine and autocrine fashion by nonsteroidal substances synthesized locally. An example of the paracrine/endocrine role is GnRH in the regulation of ovarian activity. Moreover, opioid peptides such as β-endorphin are involved in regulating the reproductive process at endocrine (hypothalamo–pituitary) as well as at paracrine/autocrine levels (ovary).

V. OVARIAN GROWTH AND YOLK STORAGE

In amphibians, as in other oviparous vertebrates, yolk is not synthesized *in situ* but derived from a precursor protein known as vitellogenin, synthesized in the liver. A sequence of complex but well-coordinated processes is involved in the production of yolk and its accumulation in amphibian eggs. This sequence of events, collectively referred to as vitellogenesis, includes the following steps: (i) the induction of vitellogenin synthesis in the liver and its subsequent release into the blood circulation, (ii) the transport of vitellogenin in the bloodstream, (iii) the uptake of vitellogenin by the growing oocytes, and (iv) the conversion of vitellogenin into storage forms. As described in Section III, through this process the ovary acquires energy reserve and increases in size (Fig. 2). The first important stage is hepatic vitellogenin synthesis; this precursor of protein yolk components is a molecule of high molecular weight (400 kDa) encoded by a small gene family (four genes in *Xenopus. laevis*). It has been well established that estradiol-17-β is the most potent hormone regulating hepatic vitellogenin synthesis. However, in the anuran *R. esculenta*, pituitary hormones, such as growth hormone, prolactin, and GtH, may directly induce hepatic vitellogenin synthesis. Once synthesized, vitellogenin is released in the blood circulation and sequestered by growing oocytes.

Observations on selective vitellogenin incorporation have indicated that a receptor-mediated process is involved in vitellogenin uptake. Examination of the vitellogenic oocyte surface by transmission electron microscopy shows extensive numbers of coated pits and vesicles, suggesting that an endocytic mechanism is involved. However, little is known about the hormones involved in controlling protein uptake into the oocyte. During vitellogenic growth, follicle cells respond to circulating gonadotropin by producing estrogen, which in turn induces hepatic synthesis of VTG. However, once the growing oocyte acquires the competence to resume meiosis and become an egg, the follicle cells appear to restrict vitellogenin uptake and to alter steroid synthesis by producing progesterone, which in turn induces ovulation.

In conclusion, the most important factors occurring in the amphibian ovarian cycle are the following: Ovarian cycles are dependent on environmental conditions; the ovarian compartments respond to hormone stimulation since GtHs induce oocyte maturation and estradiol and progesterone secretion; estradiol in turn regulates vitellogenesis together with GtH; and progesterone is responsible for oocyte maturation.

Acknowledgments

I thank Dr. O. Carnevali and Dr. G. Mosconi for critical reading of the manuscript and for their help with the figures.

See Also the Following Articles

Amphibian Reproduction, Overview; Female Reproductive System, Amphibians; GnRH (Gonadotropin-Releasing Hormone)

Bibliography

Nagahama, Y. (1987). Endocrine control of oocyte maturation. In *Hormones and Reproduction in Fishes, Amphibians and Reptiles* (D. O. Norris and R. E. Jones, Eds.), pp. 171–202. Plenum, New York.

Polzonetti-Magni, A. M., Carnevali, O., Yamamoto, K., and Kikuyama, S. (1995). Growth hormone and prolactin in amphibian reproduction. *Zool. Sci.* **12**, 683–694.

Polzonetti-Magni, A. M., Mosconi, G., Carnevali, O., Yamamoto, K., Hanaoka, Y., and Kikuyama, S. (1997). Gonadotropins and reproductive function in the anuran amphibian, *Rana esculenta. Biol. Reprod.*, in press.

Schuetz, A. W., and Glad, R. (1985). *In vitro* production of meiosis inducing substance (MIS) by isolated amphibian (*Rana pipiens*) follicle cells. *Dev. Growth Differ.* **27**, 201–211.

Whittier, J. M., and Crews, D. (1987). Seasonal reproduction; Patterns and control. In *Hormones and Reproduction in Fishes, Amphibians and Reptiles* (D. O. Norris and R. E. Jones, Eds.), pp. 385–409. Plenum, New York.

Amphibian Reproduction, Overview

Marvalee H. Wake

University of California, Berkeley

I. Reproductive Modes
II. Parental Care
III. Reproductive Cycles

GLOSSARY

anurans Frogs and toads; they have probably the greatest diversity of reproductive modes expressed among any group of vertebrates.

caecilians Elongate, limbless amphibians that inhabit the tropics of most of the world; they, too, have a diversity of reproductive strategies.

direct development Laying eggs on land, with development through metamorphosis occurring before hatching so that the free-living larval stage is absent from the life cycle.

endogenous factors Internal regulatory mechanisms provided by the central nervous system via hypothalamic secretions and by hormones secreted by the pituitary and the gonads.

exogenous factors Environmental variables, such as humidity, light cycles, temperature, and nutritional state, that are sensed and neurally integrated to mediate the endogenous regime of an animal.

oviparity Egg-laying; it may be part of several kinds of reproductive modes, including simply laying eggs in water where a male fertilizes them, then the parents abandon them, and elaborate nest building and parental maintenance.

reproductive modes Combinations of features such as egg size, number, and oviposition, developmental biology, hatching, and parental care that characterize species or groups of species.

urodeles Salamanders and newts; like other amphibians, they have a variety of reproductive modes.

viviparity Live-bearing; the developing young are retained in or on the body of a parent. Nutrition in addition to the yolk of the egg may or may not be supplied; a placenta of the amniote sort (extraembryonic egg membranes in association with the epithelium of the oviduct) does not develop in amphibians, but they have evolved nonplacental and (possibly) pseudoplacental means of obtaining maternal nutrition. In some species, young are born as fully metamorphosed juveniles.

The reproductive biology of amphibians is generally considered to be well-known, but information about it is based on research on relatively few species. We know much more about a few species that live in temperate regions than we do about tropical species; this is of concern because the majority of amphibians are tropical, and a great diversity in biology is exhibited by tropical species. A biphasic life cycle (including a free-living, aquatic larva or tadpole that metamorphoses into an adult that is much "reorganized" and often terrestrial) characterizes many, but not all, amphibians; many species have eliminated the free-living larval period. Reproduction in amphibians is hormonally regulated (endogenous factors) and environmentally mediated (exogenous factors). More research on amphibian reproductive modes and their regulation is warranted both to understand the evolution of physiology, development, morphology, and behavior, and because amphibian species throughout the world are in decline. Better understanding of reproductive biology might facilitate ways of reducing the loss of amphibian biodiversity.

I. REPRODUCTIVE MODES

There are nearly 6000 species of amphibians in the world. Members of each of the three groups of living amphibians [order Anura (frogs and toads), order Caudata (salamanders and newts), and order Gymnophiona (caecilians)] have performed a number of "natural experiments" with their reproductive biology. The presumed ancestral condition for amphibians is aquatic reproduction in fresh water, with the male courting the female and then fertilizing the large number of small eggs that she sheds by flooding them with sperm (external fertilization); the parents abandon the clutch and development ensues. The ova develop, and the embryos hatch from the egg membranes, usually in a short time. They develop into free-living, feeding larvae; the larval period may be short (a few days) or long (more than a year). Metamorphosis then occurs, and the larva is extensively "remodeled" into the adult form through a complex, hormonally mediated series of morphological and physiological changes.

Many amphibians, however, have evolved derived modes of reproduction, often involving terrestrial sites and eliminating the free-living larval phase of the life cycle. Salamanders, caecilians, and frogs have accomplished these modifications of the ancestral condition in a variety of ways; a brief summary follows. Duellman and Trueb identify six derived modes for salamanders. All include internal fertilization [during courtship, the male deposits a spermatophore (a pedicel topped by a packet of sperm) on the substrate and lures the female over it; she grasps it with the lips of her cloaca, taking it into her reproductive tract], but the laid eggs and the larvae remain aquatic; eggs may be laid terrestrially, with the larvae aquatic; eggs may be terrestrial and the larvae terrestrial and nonfeeding; or eggs may be terrestrial, and the animals have direct development (there is no free-living larval stage). Direct development occurs in only one family of salamanders, the Plethodontidae, but because of the number of species in which it has evolved, direct development is the dominant mode of reproduction in salamanders. The evolution of direct development in plethodontids is thought to underlie the extensive adaptive radiation of the family, which is expressed in both the large number of species (more than half of all salamanders) and their geographic distribution (e.g., the only salamanders to have radiated in the tropics) and their extensive diversity of morphology and ecology. In addition, in a very few species in the family Salamandridae, eggs may be retained in the oviducts (viviparity), with development dependent either on yolk for nutrition or on maternal nutrients secreted by the oviductal epithelium after yolk is resorbed. (The alpine species in which the latter mode occurs, *Salamandra atra*, has a gestation period of 2–5 years.)

All caecilians apparently have internal fertilization (the male everts the posterior part of his cloaca; it serves as an intromittent organ that is inserted into the vent of the female to transfer sperm directly to her cloaca). Many species lay their eggs terrestrially, and the embryos upon hatching wriggle into nearby streams to spend a free-living larval period, often a year long, before metamorphosis; some species have direct development, with fully metamorphosed juveniles hatching. In contrast to salamanders and frogs, many species of caecilians retain the developing young in the female's oviducts, and nutrients are supplied to the young after yolk is exhausted, with young being born effectively fully metamorphosed. There are no known instances of oviductal retention of the developing clutch in which the embryos are yolk dependent without any additional maternal nutrition, but the evidence for many species is indirect.

Frogs are the most extensive experimenters, however; Duellman and Trueb describe 29 different modes of reproduction, and more modes have been described recently by other authors. Nearly all frogs have external fertilization, and most lay their eggs in water; the water, though, can be in tree holes or aerial plants as well as streams, ponds, or temporary pools. The tadpoles (larvae) can be highly modified for various types of feeding or not feed at all. As part of the courtship sequence, frogs in several families make foam nests by using their hindlimbs to beat the jelly secreted around the eggs with the water in which the eggs are laid; such nests can be in burrows, on the water surface, or aerial. The developing eggs are often carried by a parent, either male or female depending on the species, on the hindlegs, on the back, in pouches in the skin of the back, even in the stomach and in the vocal sacs of the male. These components of reproductive modes are discussed later in the consideration of parental care. Eggs may be laid on land in various sites and arboreally. In some of these cases, tadpoles drop into or are carried to water, where they complete development; other species are direct developers. Only seven or eight species are known to have internal fertilization; some lay their eggs and have a tadpole stage. Several of these species retain their young in their oviducts for a significant part of their development: Some species provide maternal nutrition to the embryos after yolk is exhausted, some are yolk-dependent; some give birth at any of several stages late in development, and in three species the young are born as fully metamorphosed juveniles. The ecological–physiological–endocrinological–developmental interaction for most derived reproductive modes is not known; however, it has been elegantly described for *Nectophrynoides occidentalis*, a toad that lives at high elevations on Mt. Nimba in Liberia and has a 9-month gestation period that is strongly influenced by environmental conditions. During the dry season (November to April), the pregnant female stays underground and does not feed. Her embryos grow very slowly; her oviduct is secretory, but not extensively. Corpora lutea are maintained during the 5 months of the dry season, and their secretion of progesterone inhibits oocyte maturation. When the rains begin in April, the female emerges and begins feeding. Her oviducts become hyperemic and highly secretory; her corpora lutea degenerate. Oocyte development increases. The embryos grow rapidly, and metamorphosis occurs before birth in June. Following parturition, the oviducal epithelium undergoes necrosis and exfoliation. The June–October period includes the follicular phase of the cycle, with restructuring of the oviducal epithelium, oocyte maturation and yolking, and secretion of estrogens and progesterone. Mating and internal fertilization follow, establishing the next cycle.

II. PARENTAL CARE

Care of the mass of developing eggs (the clutch) and/or the hatchlings is exhibited by some members of each of the three orders of amphibians (maintenance of developing embryos in or on the body of a parent, and nest-building, are not included as modes of parental care by most authors; parental care is restricted to active attention to eggs or larvae). Parental care after oviposition has evolved in several lineages and includes guarding the clutch, carrying tadpoles from site to site, and bringing food to the tadpoles before they leave the clutch (the latter two modes occur only in frogs). Parental care is particularly prevalent among species in all three orders that have direct development. Among caecilians, it has been known for more than 100 years that female *Ichthyophis glutinosus* coil their bodies around the clutch, which is laid in a burrow near a stream. They guard the clutch until the larvae hatch from the egg membranes and wriggle into the streams. *Idiocranium russelli*, a miniaturized African species, builds a pedicel of soil under the overlying grass mat and lays her clutch atop the pedicel. The female coils around her eggs and was reported to "spit" at a researcher who uncovered a pedicel. Guarding is also reported for other species.

Among salamanders, the majority of species practice some form of parental care. Such care occurs primarily in terrestrially egg-laying species, but a few attend clutches laid in water. In some aquatic species, males drive away conspecifics except for reproductive females; in others, parents of either sex rock the eggs or create water currents over them, which may provide for aeration or normal development, which requires egg rotation. Females of several species guard their clutches, either in streams or in terrestrial sites. Such guarding is thought to have a number of effects, depending on the species: prevention of infection of the eggs by fungi or bacteria, protection from predators, provision for aeration, or prevention of adhesion of the eggs and young.

Frogs exhibit the greatest diversity of parental care. Several species agitate aquatic eggs; one or the other parent attends the clutch in a number of species. Some species that build nests practice defense of the nest. For example, males of the gladiator frog, *Hyla rosenbergi*, aggressively protect their nests against all intruders because the eggs are laid as a surface film, and if the eggs fall to the bottom of the water because an intruder breaks the surface layer, they die from lack of oxygen. Parental attendance, often by males, characterizes a number of species whose clutches are laid in tree holes, leaf axils, the upper or undersides of leaves, or in leaf litter. Such care can prevent access by specific predators. In some species, a parent visits the clutch to moisten the eggs, especially if the clutch is in a potentially desiccating situation. Several species of arboreal frogs provide food to their hatchlings by bringing eggs for them to eat. The eggs may be unfertilized eggs laid by their own mothers or may be the eggs of conspecifics or even other species of frogs, often brought by the male of the pair. The latter is a mode that promotes the fitness of the parent's clutch at real cost to that of other frogs. Males of the obstetrical toad, *Alytes obstetricans*, transport their eggs on their hindlimbs; the eggs stick because of the jelly coats, and the father periodically wets the eggs by taking them to water. The males then sit in the water when the embryos are ready to hatch. A number of species of dendrobatids in the New World tropics carry their tadpoles on their backs, often by the male of the mated pair. Usually he carries them for a part of their developmental period, then deposits them in a stream to complete development and metamorphosis. More extreme means of "carrying" the embryos are found in several species. Males of *Rhinoderma darwini* of southern Chile and Argentina ingest their fertilized clutch and "incubate" the embryos in their vocal sacs. Fully metamorphosed young are "born" via their father's mouths. In the Australian myobatrachid frog, *Rheobatrachus silus*, the female swallows her fertilized clutch and incubates it in her stomach. She does not eat during her "pregnancy," and prostaglandins secreted by the embryos suppress her secretion of digestive enzymes. Again, fully metamorphosed froglets are born via her mouth. Unfortunately, further work on this fascinating reproductive mode is not possible because the species, discovered only in the 1960s, is now considered extinct.

III. REPRODUCTIVE CYCLES

Amphibian reproductive cycles are hormonally controlled and, in most species, environmentally mediated. The reproductive cycle includes several components: oogenesis and spermatogenesis, including cell maturation and vitellogenesis; courtship and mating; oviposition; and the timing of these events in terms of seasonality or its absence. These topics are briefly summarized.

The exogenous and proximate factors that affect reproductive cycles are best known for a few temperate species of frogs and salamanders. Temperature and rainfall are the principal factors; temperature can be indirect in that it allows animals to feed and thus improves nutritional status, essential to growth of the follicles. In the fall and during hibernation (or during spring and estivation for some species), temperatures can act directly to prevent follicles from becoming atretic. For some frogs, photoperiod is important in stimulating gonadotropin secretion and testicular activity, and the frogs react to increasing photoperiods via sensory input from both eyes and pineal organs. Some authors consider vocal communication in frogs to be an exogenous component of reproduction. In the many frogs that do call, courtship and territory establishment are important components of reproductive activity. In some species, vocalization, rather than amplexus (body grasping that facilitates copulation), induces ovulation. Furthermore, the brains of the calling males include several areas in which testosterone is aromatized to estrogen so that androgen-induced calling is mediated complexly. In frogs that live in deserts, as well as the tropics and subtropics, rainfall is usually the primary factor that times the breeding period. It may have an indirect effect by increasing the availability of food items.

For temperate salamanders, temperature, humidity, and photoperiod are the proximate causes for induction of breeding cycles; the effect of each varies among species. Olfactory signals are important in the courtship of many species of salamanders; pheromones are major mate attractants. However, there are species of tropical frogs, caecilians, and salamanders that appear to be reproductively active throughout the year so that reproduction is aseasonal. It should be noted that this is a species-level phenomenon; individuals do cycle reproductively. Males often are in active spermatogenesis for 11 of the 12 months of the year, with a very short period of regression and recrudescence; males may be asynchronous with regard to other males of the species. This pattern characterizes males of species of viviparous caecilians. The females of the species that have been studied are pregnant for 6–11 months, depending on the species, and yolk up ova during the next year, so reproduction is biennial. However, in the few species that have been studied, pregnancy is synchronous, and females of a population give birth at approximately the same time, usually the beginning of the rainy season during which prey abundance increases. It also appears that fertilization and the inception of pregnancy are synchronous, so the long-term spermatogenic cycles of males are a paradox, and reproduction overall is seasonal. Some tropical frogs have a more intuitively obvious pattern in that males and females in amplexus and tadpoles in various stages of development can be found throughout all or most of the year.

Spermatogenesis and oogenesis have been described for a number of temperate frogs and salamanders. In certain frogs, the interstitial cells of the testis are at peak secretory activity in February, when the secondary sex characters (e.g., thumb pads and fangs) become well developed, and the cells subsequently regress, as do the secondary sex characters. Many frogs have a pattern of a brief spawning period, usually in the spring, with spermatogenesis completed before amplexus. In frogs, three types of spermatogenesis occur: continuous, in which spermatogonia mature into spermatocytes throughout the year; discontinuous, in which primary spermatogonia cannot divide during part of the year, even under the administration of gonadotropins; and potentially continuous, in which primary spermatocytes divide slowly and spermatocysts continue to develop, but development arrests and many spermatocytes degenerate. However, spermatogenesis can be stimulated either by exposure to temperatures equivalent to those during the normal period of spermatogenesis or by injection of gonadotropins. Typically, there is a relatively long period of testicular regression following the breeding season which is followed by recrudescence. However, in some species recrudescence is relatively rapid (in those with types two and three spermatogenesis, as mentioned previously). A number of experiments using hypophysectomized frogs have demonstrated the complex interaction of the pituitary, follicle-stimulating hormone (FSH), luteinizing hormone (LH), and spermatogenesis, and androgen production. Spermatogenesis in salamanders is controlled similarly. Different lobules in the testis may contain sperm at different stages of maturation. Among several of the few tropical species of amphibians studied, spermatogenesis is continuous throughout the year, with a very brief period of regression. At the same time, in some tropical species spermatogenesis is mediated strongly by pronounced cycles of rain. There is abundant evidence that the interstitial cells of the testis secrete steroid hormones under the control of LH, and that spermatogenesis is mediated by FSH; spermiation may be induced by either gonadotropin, depending on the species. There is an abundant literature on the endocrinology of amphibian reproduction, again attributable to only a few species that have been studied.

Oogenesis has been extensively studied in *Bufo bufo* (the European toad); it serves as the general model for temperate amphibian species. Follicular growth is characterized by a previtellogenic growth phase in which the oocyte cell diameter increases greatly and microvilli extend toward the follicular cells. The second growth phase includes yolk accumulation, again with extensive growth of the oocyte, and its investment with epithelial, thecal, and granulosa layers. Estrogen secretion by the follicle stimulates synthesis of yolk precursors in the liver and the mobilization of lipids from the fat bodies. With maturation of the follicle and the oocyte, the later stages of meiosis occur; secretion of progesterone increases, inducing ovulation. The biochemical pathways for meiosis resumption and ovulation have been characterized for only a few species but are presumed to be generalizable. A number of amphibians are now known to have corpora lutea following ovulation; progesterone secretion has been demonstrated for some of them. The presence of corpora lutea throughout the pregnancies of live-bearing species is indirect evidence for their influence in maintenance of the pregnancy. The considerable literature on the endocrinology of amphibian reproduction has examined very few species that have any of the many derived reproductive modes.

In summary, this general discussion of the reproductive modes and reproductive cycles of amphibians illustrates that (i) a vast literature exists that deals with only a few species; (ii) great diversity of reproductive biology characterizes amphibians as a class; (iii) the interplay of environmental and physiological mechanisms both regulates reproduction and permits considerable variation and "natural experiments"; (iv) more extensive and comparative study of the reproductive biology of amphibians will facilitate our understanding of the evolution of the morphology, physiology, endocrinology, ecology, and behavior associated with reproductive patterns in vertebrate animals.

See Also the Following Articles

AMPHIBIAN OVARIAN CYCLES; FEMALE REPRODUCTIVE SYSTEM, AMPHIBIAN; MALE REPRODUCTIVE SYSTEM, AMPHIBIAN

Bibliography

Angelini, F., and Ghiara, G. (1984). Reproductive modes and strategies in vertebrate evolution. *Boll. Zool.* **51**, 121–203.

Duellman, W. E. (1985). Reproductive modes in anuran amphibians: Phylogenetic significance of adaptive strategies. *S. Afr. J. Sci.* **81**, 174–178.

Duellman, W. E. (1989). Alternative life-history styles in anuran amphibians: Evolutionary and ecological implications. In *Alternative Life-History Styles of Animals* (M. N. Bruton, Ed.), pp. 101–126. Kluwer, Dordrecht.

Duellman, W. E., and Trueb, L. (1986). *Biology of Amphibians* (see chapters on reproductive strategies, courtship and mating, vocalization, eggs and development, larvae, and metamorphosis). McGraw-Hill, New York.

Guttman, S., and Taylor, D. (1977). *Reproductive Biology of the Amphibia.* Plenum, New York.

Hödl, W. (1990). Reproductive diversity in amazonian lowland frogs. *Fortschr. Zool.* **38**, 41–60.

Lofts, B. (1974). Reproduction. In *Physiology of the Amphibia, Vol. II* (B. Lofts, Ed.), pp. 107–218. Academic Press, New York.

Salthe, S. N., and Duellman, W. E. (1973). Quantitative constraints associated with reproductive pattern in anurans. In *Evolutionary Biology of the Anura* (J. L. Vial, Ed.), pp. 229–249. Univ. Missouri Press, Columbia.

Salthe, S. N., and Mecham, J. S. (1974). Reproductive and courtship patterns. In *Physiology of the Amphibia, Vol. II* (B. Lofts, Ed.), pp. 309–321. Academic Press, New York.

Wake, M. H. (1982). Diversity within a framework of constraints. Amphibian reproductive modes. In *Environmental Adaptation and Evolution* (D. Mossakowski and G. Roth, Eds.), pp. 87–106. Gustav Fischer, Stuttgart.

Wake, M. H. (1993). Evolution of oviductal gestation in amphibians. *J. Exp. Zool.* **266**, 394–413.

Xavier, F. (1986). La réproduction des *Nectophrynoides*. In *Traité de Zoologie Amphibiens* (P. Grasse and M. Delsol, Eds.), Vol. 14, pp. 497–513. Masson, Paris.

Androgen Inhibitors/Antiandrogens

Vivian L. Fuh and Elizabeth Stoner
Merck Research Laboratories

I. Androgens and Hyperandrogenism
II. Androgen Inhibitors
III. 5α-Reductase Inhibitors
IV. Antiandrogens
V. Summary

GLOSSARY

androgens Steroid hormones that effect masculinizing activities.

androgen inhibitors Drugs that decrease androgen production or expression.

androgen receptors Cytosolic proteins that, upon binding to androgens, bind to nuclear DNA and regulate specific gene expression to effect the physiologic actions of androgens.

antiandrogens Drugs that directly inhibit androgen receptors in target tissues and block the physiological actions of both testosterone and dihydrotestosterone (DHT).

hyperandrogenism A clinical state characterized by an excessive production or expression of androgens.

5α-reductase inhibitors Androgen inhibitors which block the enzymatic conversion of testosterone to DHT in target tissues, decreasing the physiologic effect of DHT without inhibiting the effect of testosterone at the androgen receptor.

An androgen inhibitor is any drug that results in a net reduction of androgen production or expression. The term antiandrogen usually refers to direct inhibitors of androgen receptors in target tissues. In this article, we review the mechanisms of action and therapeutic use of androgen inhibitors in women with clinical manifestations of hyperandrogenism.

I. ANDROGENS AND HYPERANDROGENISM

Hyperandrogenism in women results from increased androgen production in the ovary [such as in polycystic ovarian syndrome (PCOS) or ovarian carcinoma], the adrenal gland [such as in Cushing's disease/syndrome, congenital adrenal hyperplasia, or adrenal carcinoma], or the peripheral tissues (such as in idiopathic hirsutism, androgenetic alopecia, or acne vulgaris). However, regardless of the underlying pathophysiology, hyperandrogenism may clinically manifest as anovulation, menstrual abnormality, infertility, hirsutism, androgenetic alopecia, acne, glucose intolerance, or dyslipidemia. In women, weak androgens, such as androstenedione and dehydroepiandrosterone (DHEA), are the major precursors for the more potent androgens, testosterone and its 5α-reduced metabolite, dihydrotestosterone (DHT). It is the latter two potent androgens which induce the predominant physiologic or pathophysiologic androgen activity. Testosterone and DHT have differential effects in target tissues, with DHT specifically implicated in the pathogenesis of hirsutism, acne vulgaris, and androgenetic alopecia. Studies have demonstrated increased 5α-reductase activity and DHT levels in scalp skin of balding scalp versus nonbalding scalp skin. Therefore, although the mechanism by which androgens differentially regulate hair follicle growth on the scalp (inhibit) and face (promote) remains largely unknown, local production of excess DHT appears to contribute to the pathogenesis of androgenetic alopecia and hirsutism.

As shown in Fig. 1, the biological effects of androgens are mediated through intracellular receptors that belong to a superfamily of ligand-inducible transregulators including receptors for steroids, thyroid hormones, and vitamins. The affinity of DHT

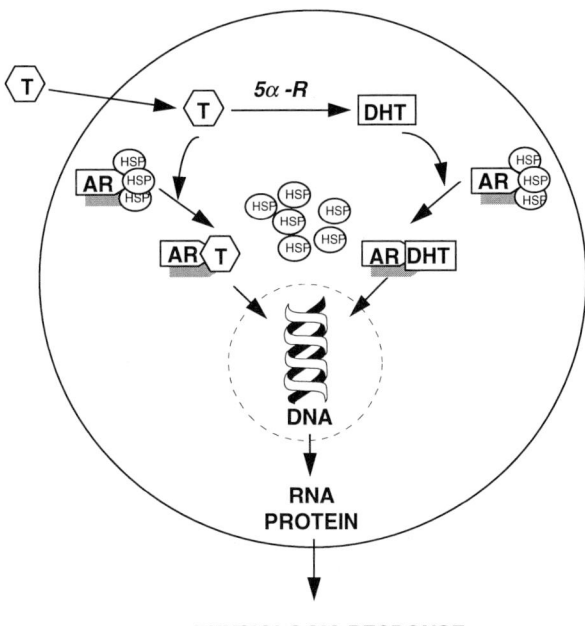

FIGURE 1 Mechanism of action of androgens. T, testosterone; DHT, dihydrotestosterone; AR, androgen receptor; 5α-R, 5α-reductase; HSP, heat shock proteins.

for the androgen receptor in human prostatic or skin cytosol is five times greater than that of testosterone. In addition, androgen effects may be mediated through interactions with sex hormone binding globulin and its membrane-bound receptor to produce G-protein-coupled cAMP generation; the clinical significance of this mechanism is not well understood.

II. ANDROGEN INHIBITORS

In Fig. 2, the mechanisms of action of antiandrogens and other androgen inhibitors used in women are illustrated. Androgen inhibitors can be grouped by the primary site of action: ovarian, adrenal, and target tissues (Table 1).

A. Ovarian Suppression

Drugs that induce ovarian androgen suppression, such as oral contraceptives and gonadotropin-releasing hormone (GnRH) agonists/antagonists, inhibit ovarian androgen production through hypothalamic–pituitary suppression. In addition, the estrogen

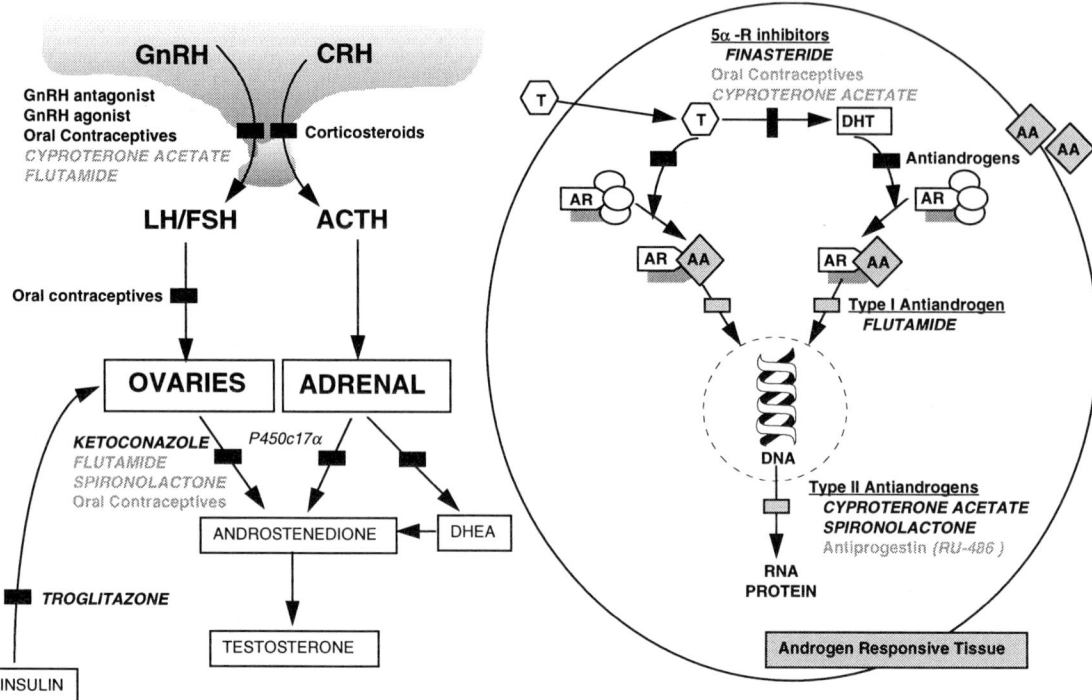

FIGURE 2 Androgen inhibitors and mechanisms of action. Androgen inhibitors are any agents which inhibit the production or expression of androgens. Antiandrogens (AA) specifically refer to those compounds which directly inhibit the androgen receptors (AR) in androgen-responsive tissues. Secondary mechanisms of androgen inhibition are highlighted in gray.

component of oral contraceptives increases sex hormone-binding globulin levels, decreasing free-androgen levels. Progestins used in oral contraceptives may also directly inhibit adrenal and gonadal steroid synthesis, directly inhibit androgen receptors, or have some 5α-reductase inhibitory activity. Some progestins (e.g., levonorgestrel or norethindrone) are androgenic and should be avoided, although lower doses of these progestins may have fewer adverse effects. Oral contraceptives containing progestins with low androgenic potential (desogestrel, norgestimate, gestodene, and cyproterone acetate) may have advantages over the more androgenic progestins. Low-dose cyproterone acetate in combination with ethinyl estradiol has been approved in many countries for the treatment of acne and moderately severe hirsutism (see Section IV). Although cyproterone acetate is not available in the United States, norgestimate in combination with ethinyl estradiol is approved for treatment of moderate acne in women.

Because GnRH agonists or antagonists suppress both ovarian estrogen and androgen secretion, prolonged treatments, unless contraindicated, would require concomitant hormone replacement therapy to prevent the effects of medical castration (e.g., osteoporosis). GnRH antagonists are currently only available for investigational use.

A new class of compounds, thiazolidinediones, may potentially offer a novel approach to inhibiting excess androgen production in some women with PCOS. In these women, insulin resistance due to a postbinding defect in insulin transduction, with consequential hyperinsulinemia, appears to be the underlying mechanism for ovarian and adrenal hyperandrogenism. Thiazolidinedione increases insulin sensitivity in peripheral tissues and reduces hepatic glucose production, thereby reducing the drive for hyperinsulinemia. In two recent small studies, treatment of women with PCOS with troglitazone led to the amelioration of insulin resistance and

TABLE 1
Androgen Inhibitors and
Mechanism of Action

Ovarian suppression
  Oral contraceptives
  GnRH agonists/antagonists
  Insulin sensitizer
    Troglitazone
Adrenal suppression
  Corticosteroids
  Ketoconazole
Peripheral tissues
  5α-reductase inhibitors
    Type 1
    Type 2
      Finasteride
  Antiandrogens
    Steroidal
      Spironolactone
      Cyproterone acetate
    Nonsteroidal
      Flutamide
      Bicalutamide
      RU 58841

---

hyperinsulinemia and concomitantly reduced testosterone and luteinizing hormone (LH) levels toward the normal range. A larger study is required to confirm these findings.

B. Adrenal Suppression

Corticosteroids inhibit adrenal androgen production through hypothalamic–pituitary suppression and are indicated when the major source of the increased androgen is adrenal. Though corticosteroids can be effective in PCOS and idiopathic hirsutism, the risk of prolonged use needs to be carefully considered. Drugs such as ketoconazole which interfere with adrenal and gonadal steroid synthesis by inhibiting cytochrome P450 enzymes may be useful for controlling severe hyperandrogenism associated with adrenal carcinoma or Cushing's syndrome. However, because these drugs have significant side effects, they are not routinely used in treating ovarian or idiopathic hyperandrogenism.

C. Inhibiting Androgen Expression in Target Tissues

Antiandrogens directly inhibit androgen receptors in target tissues and block the physiological actions of both testosterone and DHT. In contrast, 5α-reductase inhibitors, by blocking the enzymatic conversion of testosterone to DHT, specifically decrease the expression of DHT in target tissues without inhibiting the physiological actions of testosterone at the androgen receptor. Two types of 5α-reductase regulate the local production of DHT in skin. Type 1 5α-reductase is the predominant form in sebaceous glands. Type 2 5α-reductase, the primary isoenzyme in genitourinary tissues such as the prostate, was also recently immunolocalized to the inner layer of the outer root sheath of scalp hair follicles. Because of the differential distribution and expression of the two isoenzymes, selective 5α-reductase inhibitors may have specific clinical applications in the future. Further clarification of the distribution and physiologic significance of these two isoenzymes in skin is needed. Specific antiandrogens and 5α-reductase inhibitors will be discussed in Sections III and IV.

It should be noted that although antiandrogens and 5α-reductase inhibitors have been used to treat clinical manifestations of hyperandrogenism in women, most of these compounds have not been approved for these uses. Unlike oral contraceptives and GnRH agonists/antagonists which render women anovulatory, antiandrogens and 5α-reductase inhibitors are likely to maintain ovulation in women with androgenic disorders and pose a risk of feminizing the male fetus. Therefore, the use of antiandrogens and 5α-reductase inhibitors is contraindicated in women who are or may potentially be pregnant.

III. 5α REDUCTASE INHIBITORS

Finasteride is a type 2 5α-reductase inhibitor. Although structurally similar to testosterone, finasteride has no androgenic, antiandrogenic, estrogenic, progestational, or other steroid activity. Oral administration of finasteride in men produces a rapid decline (approximately 70%) in systemic DHT which is accompanied by a slight increase (approximately 10%) in serum testosterone. Gonadotropin levels in men are not significantly altered. No data are available in women. Finasteride is approved at the 5-mg dose (Proscar) for the treatment of men with benign prostatic hyperplasia. In addition, placebo-controlled clinical trials have evaluated the efficacy and safety of 1 mg finasteride in treating androgenetic alopecia in men. Efficacy data for women with androgenetic alopecia are not available. Potential uses of finasteride in women for treatment of hirsutism have been limited to small studies that were often uncontrolled. In these studies, finasteride treatment resulted in mild to moderate improvement in Ferriman–Gallwey scores. Type 1 5α-reductase inhibitors are still under investigation. The only currently available 5α-reductase inhibitor, finasteride, is not indicated for use in women. Because of the ability of type 2 5α-reductase inhibitors to inhibit conversion of testosterone to DHT in some tissues, these drugs, including finasteride, may cause abnormalities of the external genitalia of a male fetus when administered to a pregnant women. Crushed or broken tablets of finasteride should not be handled by women when they are or may potentially be pregnant because of the possibility of absorption of finasteride and the subsequent potential risk to a male fetus.

IV. ANTIANDROGENS

A. Mechanism of Action

Antiandrogens can be subdivided based on structure and function. Nonsteroidal antiandrogens, such as flutamide and bicalutamide, are considered pure antiandrogens with no inherent steroid agonist activity. Steroidal antiandrogens, such as cyproterone acetate and spironolactone, can also bind to other steroid receptors.

Antiandrogens can also be subdivided based on the mechanism of action causing specific conformational changes of the receptor. Molecular genetic analysis has identified separate domains in the receptor responsible for hormone binding, DNA binding, and transcription activation. In the absence of androgen binding, the androgen receptor remains in a latent state associated with a number of cytoplasmic proteins (e.g., heat shock protein and immunophilins). As shown in Fig. 2, upon androgen binding, the activated androgen receptor enters the nucleus and interacts with specific DNA sequences located in the regulating (flanking) regions of the target gene, resulting in activation or modulation of gene transcription and expression. Which gene is targeted depends on the individual tissue.

Type 1 antagonists, the so-called pure antagonists such as flutamide, block androgen receptor binding to DNA. Presumably, these compounds are unable to induce the necessary changes in the conformation of the androgen receptor to become activated. Type 2 antagonists, the nonpure antagonists, permit some DNA binding by the activated receptor but partially or completely block the interaction of the receptor with the transcription initiation complex. Type 2a antagonists induce an abnormal conformation of the C-terminal hormone-binding domain and almost completely inhibit transcriptional activity. The antiprogestin, RU 486, which has some antiandrogenic activity, presumably works through this mechanism. In comparison, type 2b antagonists, such as cyproterone acetate, do not cause abnormal binding at the C-terminus hormone-binding domain and permit partial agonistic activity by the transcription initiation complex in the N terminus of the receptor.

B. Spironolactone

Spironolactone is an aldosterone antagonist which is indicated in the treatment of primary hyperaldosteronism and essential hypertension. Administered orally, spironolactone reaches maximal plasma levels within 30–60 min. Spironolactone is a strong competitor for androgen receptors and therefore a potent antagonist *in vitro*. However, *in vivo* spironolactone is rapidly metabolized to canrenone and potassium canrenoate, weak competitors for the androgen receptor. In humans, the primary androgen inhibitory effect of spironolactone appears to be due to the antiandrogen effect of these metabolites. Spironolactone, at doses of 200 mg/day, also inhibits testosterone biosynthesis by inhibiting gonadal and adrenal 17α-hydroxylase and 17,20-lyase ($P450_{c17\alpha}$) activity and reducing Δ^4 androstenedione production. In addition, minor inhibition of 3β-ol dehydrogenase and

11β and 18-hydroxylase occurs. Although the effect on testosterone biosynthesis is measurable and appears to be dose related, the clinical impact of this inhibition is small. Nevertheless, because of its dual effects as an antiandrogen and inhibitor of testosterone biosynthesis, spironolactone can be useful in treating clinical manifestations of excess androgen production associated with PCOS.

Though not approved for use as an androgen inhibitor, spironolactone has been used in women primarily for treating hirsutism and acne. Doses needed for the antiandrogenic effect are higher than that prescribed for hypertension. It is usually given at a starting dose of 50 mg a day. Although higher doses (up to 400 mg daily) may improve efficacy, side effects are dose related. The most common side effect is increased frequency of menses (in about 20% of women given 100 mg daily), breast tenderness or enlargement, fatigue, and orthostatic symptoms. The cause of the menstrual abnormality is unclear, though it has been reported to be related to anovulatory menstrual cycles. It usually resolves when an oral contraceptive is added or when the dose of spironolactone is decreased. There is no evidence that giving spironolactone cyclically (3 of every 4 weeks) will prevent spotting, and androgen blockade would be less effective. There has been some concern about possible tumorigenicity, especially breast carcinoma, based on early animal studies and a case report in humans; however, on the basis of 25 years of animal studies and human epidemiologic studies, the use of spironolactone does not appear to cause malignancy. Spironolactone or its metabolites may cross the placental barrier. The safety of spironolactone in pregnant women has not been established, and any benefit must be weighed against possible hazard to the fetus. However, impaired masculinization of a male fetus has not been reported. Polyuria on spironolactone is unusual. Hyperkalemia does not occur with normal renal function but the concomitant use of agents that raise potassium levels must be done with caution.

C. Cyproterone Acetate

Cyproterone acetate has been widely used outside of the United States since 1964. Oral contraceptives containing 2 mg cyproterone acetate in combination with 35 μg ethinyl estradiol (Diane 35 or Dianette) or 50 μg ethinyl estradiol (Diane) are approved in many countries for the treatment of acne and moderately severe hirsutism. Although higher doses of cyproterone acetate (50–100 mg per day cyclically) have been used to treat excess hair growth, the 50-mg tablet (Cyprostat and Androcur) is approved only for use in adult men for treatment of prostate cancer or control of libido for severe hypersexuality.

In addition to being an androgen inhibitor, cyproterone acetate is a strong progestin with antigonadotropic effects centrally and a weak glucocorticoid. There is some evidence that cyproterone acetate may also inhibit 5α-reductase activity in the skin. Therefore, cyproterone acetate may be effective in the treatment of disorders resulting from both androgen excess and tissue hypersensitivity to androgens. Although cyproterone acetate has approximately fivefold lower affinity for the androgen receptor when compared with spironolactone, randomized trials using the Ferriman–Gallwey scoring as an endpoint found cyproterone acetate (50–100 mg per day) with ethinyl estradiol to be equally effective in treating hirsutism as 100 mg per day spironolactone with an oral contraceptive.

Absorption of cyproterone acetate is poor after oral dosing (5–30%); however, cyproterone acetate has a long biological half-life (approximately 38 hr). Because cyproterone acetate is a potent progestin with a long duration of action, it should be administered cyclically (Days 5–15 of the menstrual cycle), preferably with an estrogen, to allow for regular withdrawal bleeding.

Side effects attributable to the oral contraceptive and with high doses of cyproterone acetate include mood swings, weight gain, fluid retention, breast tenderness, loss of libido, thrombosis, and nausea. Menstrual irregularity occurs when cyproterone acetate is used alone. Cyproterone acetate has been shown to cause alteration in the DNA of rat hepatocytes and there have been rare reports of hepatocellular carcinoma with its use in women. Fulminant hepatitis has also been reported in men receiving high doses (50 mg twice a day) of cyproterone acetate for prostate cancer. Metabolic changes may occur, including a tendency toward hyperinsulinemia, glucose intolerance, and small decreases in high-density lipoprotein₂ (HDL$_2$), although the ratio of total HDL/low-density lipoprotein rises. Animal studies showed some feminization of male rat fetuses when cyproterone acetate was administered during the phase of embryogenesis at which differentiation of the external genitalia occurs. Therefore, pregnancy is an absolute contraindication for treatment with cyproterone acetate.

D. Flutamide

Flutamide (Eulexin) is approved for the treatment of prostate cancer, although in women flutamide has been used to treat idiopathic hirsutism, acne vulgaris, and androgenetic alopecia. Flutamide is a pure antiandrogen with no inherent glucocorticoid, progestational, androgenic, or estrogenic activity. There are discrepant findings regarding the central effect of flutamide on gonadotropin levels, which may be related to the differential effects of androgens on gonadotropin secretion in men, women with normal androgen levels, and hyperandrogenic women. In normal women and women with idiopathic hirsutism, flutamide (500–750 mg per day) did not alter gonadotropin levels. In contrast, in hyperandrogenic women with irregular gonadotropin secretion at baseline, flutamide normalized gonadotropin pulse frequency and caused a decrease in gonadotropin levels. This effect may be specific to flutamide and may not generally be observed with other pure antiandrogens. Though not a consistent finding, flutamide may decrease DHEAS levels through inhibition of adrenal 17,20-lyase activity.

Flutamide is well absorbed after oral administration and is rapidly metabolized to the active metabolite hydroxyflutamide. Because of the short serum half-life of hydroxyflutamide (5.5 hr), flutamide requires twice-daily dosing. Flutamide has a weak affinity for the androgen receptor compared with cyproterone acetate (approximately 150-fold lower), and high doses of 500 or 750 mg per day have been used in the treatment of hirsutism, acne, and seborrhea. However, in a nonrandomized clinical trial, lower doses of flutamide (125 mg twice daily) were found to also be effective in treating hirsutism. In one randomized trial using the Ferriman–Gallwey scoring as an endpoint, 250 mg flutamide twice a day was found to be equally effective in treating hirsutism as 100 mg spironolactone per day.

Flutamide is generally well tolerated, although clinical experience in women is little. Reported side effects in women included dry skin, mild gastrointestinal side effects, and increased appetite. Menstrual abnormalities rarely occurred. Unlike in men, women did not experience hot flashes, reflecting the differential effects of androgen withdrawal on hypothalamic–gonadotroph axis in men and women. Hepatotoxicity, though rarely reported, is a potential side effect and appears to be dose related.

Flutamide is not indicated for use in women and may cause fetal harm when administered to a pregnant woman. In animal studies, there was decreased survival rate in the offspring of rats and rabbits treated with flutamide during pregnancy. At high doses (equivalent to ≥9 times the human dose of 750 mg/day), a slight increase in minor variations in the development of the sternebrae and vertebrae was seen in fetuses of treated pregnant rats. Feminization of the male fetuses of pregnant rats treated with flutamide also occurred at these doses.

E. Other Nonsteroidal Antiandrogens

Bicalutamide (Casodex) has an affinity for the androgen receptor that is approximately four times higher than that of hydroxyflutamide and a half-life (approximately 1 week) that is compatible with once-daily administration. It does not appear to have intrinsic agonist activity. Bicalutamide, unlike flutamide, appeared to have little effect on serum LH and testosterone in the rat, possibly due to poor penetration across the blood–brain barrier. However, peripherally selectivity was not observed in humans. Nevertheless, this drug appears to be better tolerated than flutamide with less gastrointestinal and hepatic side effect. Bicalutamide is currently approved for the treatment of prostate cancer.

Bicalutamide has not been studied in women and there have been no reports of treating hyperandrogenism in women with this drug. Because bicalutamide may cause fetal harm when administered to pregnant women, bicalutamide is contraindicated in women who are or may be pregnant. The male offspring of pregnant female rats receiving doses of 10 mg/kg/day (approximately two-thirds the human therapeutic concentration based on a dose of 50 mg/day for an average 70-kg man) and higher were observed to have reduced anogenital distance and hypospadias. Affected male rat offspring were also impotent.

RU 58841 is currently being developed as a topical application for the treatment of acne vulgaris, androgenetic alopecia, and hirsutism. In animals, this compound is devoid of systemic antiandrogenic activity. Efficacy in humans have yet to be proven.

V. SUMMARY

The use of specific androgen inhibitors ideally depends on the source and magnitude of increased androgen production. Because of the multiple mechanisms of androgen inhibition, oral contraceptives, spironolactone, and cyproterone acetate have all been used as first-line treatments for hirsutism, acne vulgaris, or androgenetic alopecia in women regardless of the source of androgen production. New potential therapies on the horizon include 5α-reductase inhibitors, which specifically block DHT production, and the topical antiandrogen RU 58841. In addition, troglitazone may prove to be efficacious in decreasing hyperinsulinemia-induced hyperandrogenism.

See Also the Following Articles

Adrenal Androgens; Androgens; DHEA (Dehydroepiandrosterone); Dihydrotestosterone; SHBG (Sex Hormone-Binding Globulin); Testosterone Biosynthesis

Bibliography

Agarwal, M. K. (1995). Antihormonal steroids revisited. *Drugs Future* **20**(9), 903–910.

Carson-Jurica, M. A., Schrader, W. T., and O'Malley, B. W. (1990). Steroid receptor family: Structure and functions. *Endocrinol. Rev.* **11**(2), 201–220.

Dunaif, A., Scott, D., Finegood, D., Quintana, B., and Whitcomb, R. (1996). The insulin-sensitizing agent troglitazone improves metabolic and reproductive abnormalities in the polycystic ovary syndrome. *J. Clin. Endocrinol. Metab.* **81**(9), 3299–3306.

Ehrmann, D. A., Schneider, D. J., Sobel, B. E., Cavaghan, M. K., Imperial, J., Rosenfield, R. L., Polonsky, K. S. (1997). Troglitazone improves defects in insulin action, insulin secretion, ovarian steroidogenesis, and fibrinolysis in women with polycystic ovary syndrome. *J. Clin. Endocrinol. Metab.* **82**(7), 2108–2116.

Kaufman, K. D. (1996). Androgen metabolism as it affects hair growth in androgenetic alopecia. *Dermatol. Clin* **14**(4), 697–711.

Kuil, C. W. and Mulder, E. (1996). Deoxyribonucleic acid-binding ability of androgen receptors in whole cells: Implications for the action of androgens and antiandrogens. *Endocrinology* **137**(5), 1870–1877.

Martindale The Extra Pharmacopoeia, 31st ed. (1996). Royal Pharmaceutical Society, London.

Physicians' Desk Reference, 51st ed. (1997). Medical Economics, Montvale, NJ.

Redmond, G. P. (1995) Androgenic disorders of women: Diagnostic and therapeutic decision making. *Am. J. Med.* **98**(1A), 120S–129S.

Sawaya, M. E. (1997). Clinical updates in hair. *Dermatol. Clin.* **15**(1), 37–43.

Shaw, J. C. (1996). Antiandrogen therapy in dermatology. *Int. J. Dermatol.* **35**(11), 770–778.

Uno, H., Obana, N., Cappas, A., Bonfils, A., Battmann, T., and Philibert, D. (1996). Stimulation of follicular regrowth by androgen receptor blocker (RU58841) in macaque androgenetic alopecia. In *Hair Research for the Next Millennium* (D. J. J. Van Neste and V. A. Randall, Eds.), pp. 349–353.

Androgen Insensitivity Syndromes

James Aiman

Medical College of Wisconsin

I. History
II. Physiology of Androgen Action
III. Mechanism of Androgen Action
IV. Structure and Function of the Androgen Receptor
V. Defects in the Androgen Receptor Gene
VI. Clinical Diagnosis and Management

GLOSSARY

male pseudohermaphroditism A condition in 46,XY males who have defective virilization of the external (and possibly internal) genitalia and/or the presence of Müllerian duct structures.

Müllerian ducts The paired embryonic ducts that develop into the upper vagina, cervix, uterus, and Fallopian tubes in 46,XX females and in 46,XY males whose Sertoli cells do not excrete anti-Müllerian hormone (congenital anorchia, absent Sertoli cells, and deficient Sertoli cell secretion of anti-Müllerian hormone).

Wolffian ducts The paired embryonic ducts that develop into the epididymis, vas deferens, and seminal vesicle on each side of 46,XY embryos with normal testicular testosterone secretion and normal androgen action.

Androgen insensitivity is a collection of syndromes that all have some abnormality of the androgen receptor. The deficiency in androgen action may result in complete or partial androgen insensitivity. Phenotypic expression varies from normal-appearing female external genitalia to infertile men with defective spermatogenesis as the only evidence of androgen receptor deficiency. These syndromes occur in 46,XY individuals who have testes that secrete normal adult male levels of testosterone. The cervix, uterus, and Fallopian tubes are absent in this disorder because the testes secrete anti-Müllerian hormone.

I. HISTORY

Case reports of phenotypic females with testes and no uterus were published as early as the late nineteenth century. In 1953, John Maclean Morris, an obstetrician at Yale University, reported 2 cases of his own and included 80 cases of the complete androgen insensitivity syndrome published previously. Morris described the clinical features of the affected subjects: a female habitus including fat deposition, normal female breast development, absent or sparse pubic and axillary hair, and a female scalp hair pattern (i.e., no temporal recession). He further described female-appearing external genitalia, absent or rudimentary Müllerian duct derivatives (cervix, uterus, and Fallopian tubes), and testes with germ cell aplasia and interstitial cell (Leydig cell) hyperplasia. Laboratory results were reported in some of the patients and included increased androgen, estrogen, and gonadotropin serum concentrations.

Various forms of partial androgen insensitivity were reported as early as 1947 when Reifenstein reported an X-linked familial syndrome of hypospadias, infertility, and gynecomastia. Gilbert-Dreyfus *et al.* reported a similar family in 1957, although the degree of external genital virilization was less than that described by Reifenstein. In 1959, Lubs *et al.* described a family with affected members whose external genitalia were closer to the female end of the phenotypic spectrum. Rosewater *et al.*, in 1965, described a syndrome of familial gynecomastia and pubertal undervirilization. This was not recognized as a mild form of androgen insensitivity until 1978. In 1979, Aiman *et al.* described three phenotypically normal men with severe oligospermia or azoospermia who had androgen receptor-binding capacities in scrotal skin biopsies that were similar to those with the complete form of androgen insensitivity.

Though Morris described the clinical characteristics of the complete androgen insensitivity syndrome in detail, Lawson Wilkins seems to be the first to recognize that the condition was the result of insensitivity to androgens. In his study, the affected 46,XY subject failed to show any virilizing signs in response to daily high-dose androgens. Furthermore, he demonstrated that the gonads were actively secreting androgens and estrogens by observing a significant decrease after gonadectomy. Pituitary gonadotropins increased in his patient after removal of the gonads, indicating an intact negative feedback between the gonads and the hypothalamus–pituitary.

In the 1960s and 1970s, metabolic studies by many demonstrated that androgen production and testicular secretion rates of testosterone were normal or even increased compared to men with normal androgen receptor function. During this time, *in vivo* steroid kinetic studies and *in vitro* studies of testicular slices demonstrated that the testes of affected subjects secreted increased quantities of estradiol.

In the 1970s, Keenan *et al.* established the presence of an intracellular receptor specific for testosterone and dihydrotestosterone. Measurement of high-affinity specific androgen binding capacity in cultured genital skin fibroblasts provided a means to establish a diagnosis of androgen insensitivity with certainty. Multiple studies during this time confirmed the low receptor binding in affected subjects and demonstrated that there was variability in androgen receptor binding capacity as well as phenotypic expression. During this time studies of androgen receptor binding in cultured genital skin fibroblasts established that the syndromes of complete androgen insensitivity (testicular feminization) and those described by Lubs, Gilbert-Dreyfus, Reifenstein, and Rosewater were the result of decreased androgen receptor activity. These plus the infertile men with diminished androgen receptor binding capacity comprise a spectrum of phenotypic variants with comparable levels of androgen receptor binding activity.

II. PHYSIOLOGY OF ANDROGEN ACTION

Differentiation of undifferentiated gonads into testes begins at about 6 postmenstrual weeks in 46,XY fetuses. The gonads are histologically recognized as testes by 7 or 8 weeks' gestation. Testicular differentiation is not an androgen-dependent process. Rather, testicular development is controlled by a DNA-binding protein transcription factor encoded by a gene on the short arm of the Y chromosome. It is likely that additional gene products from autosomes or the X chromosome also contribute to testicular differentiation. Testosterone secretion from interstitial (Leydig) cells begins soon after testicular differentiation. Testosterone induces differentiation of the Wolffian ducts into the epididymata, vasa deferentia, and seminal vesicles. About this stage, the anlage of the external genitalia acquire the expression of 5α-reductase, which causes the conversion of testosterone to dihydrotestosterone. Dihydrotestosterone is responsible for masculinization of the external genitalia into normal male structures. This results in a fetus recognizable as male by 12–14 weeks' gestation. The process of internal and external genital masculinization is complete by 20 weeks' gestation. Active androgen receptor is necessary for testosterone and dihydrotestosterone to masculinize the internal and external genitalia.

Until 6 weeks' gestation all fetuses possess both Wolffian and Müllerian duct structures. While testosterone stimulates masculinization of the Wolffian ducts and external genitalia, it does not suppress formation of Müllerian ducts in 46,XY fetuses. Secretion of anti-Müllerian hormone by Sertoli cells suppresses the formation of Müllerian ducts. This is a glycoprotein hormone that is a member of the transforming growth hormone factor-α family. The presence of Sertoli cells that secrete anti-Müllerian hormone during the first trimester is an important event that helps to distinguish androgen insensitivity from other forms of male pseudohermaphroditism since a cervix, uterus, and Fallopian tubes are absent in the syndromes of androgen insensitivity.

After puberty, production rates of testosterone are 2–8 mg/day in normal adult men. Testicular secretion of estradiol in normal adult men is only 20–30 μg/day. Testosterone production rates in subjects with any of the syndromes of androgen insensitivity are equivalent to those of normal adult men or even increased. In all subjects with androgen insensitivity except those whose only manifestation is infertility, estradiol secretion rates are increased. Gynecomastia occurs in these subjects because of increased estrogen secretion plus decreased testosterone action. The role of testosterone in inhibiting the development of gynecomastia is suggested by the absence of breast development in subjects with 5α-reductase deficiency but normal androgen receptor activity.

III. MECHANISM OF ANDROGEN ACTION

Testosterone is transported in serum largely bound with high affinity to testosterone–estradiol binding protein and with lower affinity to albumin. In androgen target tissues, testosterone dissociates from its carrier proteins and enters target cells. There is no evidence of an active transport mechanism of testosterone or dihydrotestosterone into target cells. The prevailing evidence is that the androgen receptor resides in the cytosolic fraction and not within the nucleus. Within the cytoplasm, testosterone or dihydrotestosterone binds to the androgen receptor. This induces a conformational change in the androgen receptor that facilitates its dimerization, transport to the nucleus, and interaction with target DNA. This results in regulation of target gene transcription and subsequent protein synthesis leading to androgen action. These events are illustrated in Fig. 1.

IV. STRUCTURE AND FUNCTION OF THE ANDROGEN RECEPTOR

The androgen receptor (AR) is one of four related steroid receptors that include a glucocorticoid receptor, a mineralcorticoid receptor, and a progesterone

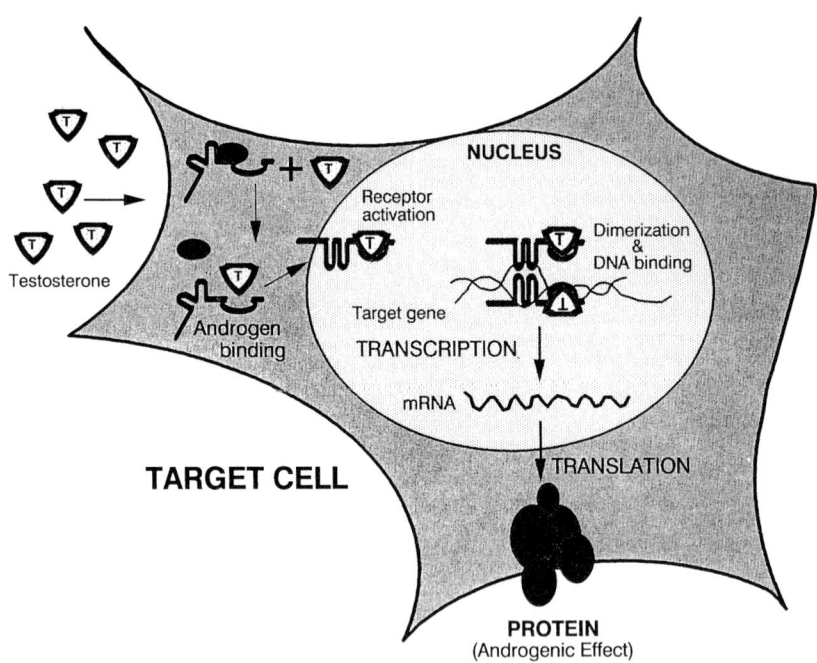

FIGURE 1 The molecular events leading to the induction of an androgen effect within target cells. The initial binding results in the removal of receptor-associated proteins, which may uncover functional domains. After nuclear translocation, the androgen–receptor complex undergoes dimerization and binding to a hormone response element, regulatory DNA sequence near a target gene. Dihydrotestosterone and testosterone induce identical changes (reproduced with permission from Quigley et al., 1995, p. 277).

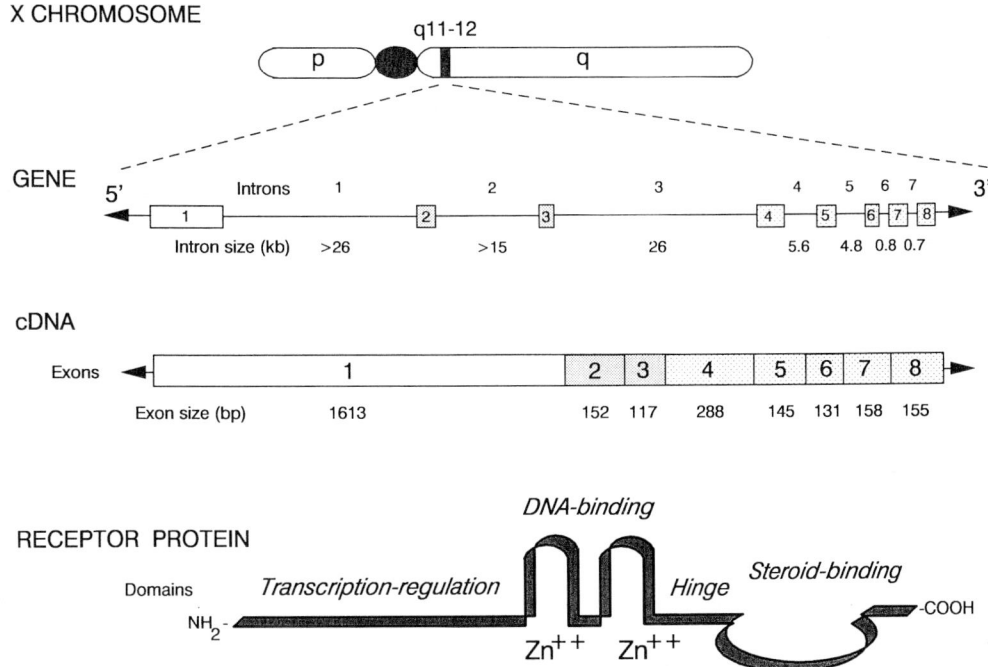

FIGURE 2 The androgen receptor gene and the structure of the androgen receptor protein (reproduced with permission from Quigley *et al.*, 1995, p. 275).

receptor. These receptors are related by their sequence homology and by their ability to activate target gene transcription via the same hormone response element. The androgen receptor, like other members of this receptor family, interacts directly with its target genes to regulate its transcription.

The AR gene is located on the long arm of the X chromosome near the centromere. This is shown in the top of Fig. 2. The structure of the AR gene is diagrammed in Fig. 2. The AR gene is approximately 75–90 kilobases of genomic DNA. There are 8 exons labeled A–H or 1–8 beginning at the 5' end, each separated by an intron, numbered above the second panel with the intron size shown below. Near the 5' end of exon 1 is a sequence of nucleic acid repeats that vary from 11 to 31 in a normal population. Approximately 20 alleles have been identified in this region, which is structurally the most heterogeneous part of the AR receptor.

The complementary DNA is shown in Fig. 2 and totals approximately 2760 base pairs. The structure of the androgen receptor is shown schematically in the bottom panel in Fig. 2. Exon 1 encodes the amino terminus of the androgen receptor. Exons 2 and 3 encode the DNA-binding region of the AR. The 5' region of exon 4 encodes the hinge region. The steroid-binding domain is encoded by the 3' region of exon 4 and exons 5–8.

The molecular weight of the androgen receptor is approximately 110–114 kDa, with 910–919 amino acids. The variation is due to the variable length of the triplet repeats in exon 1 of the complementary DNA. The three-dimensional structure of the AR protein is unknown.

V. DEFECTS IN THE ANDROGEN RECEPTOR GENE

Defective androgen action may occur as a result of a defect in the sequence of events from androgen biosynthesis to androgen binding, nuclear translocation of the AR, binding of AR to target DNA, or transcriptional activation of target genes. The most common cause of male pseudohermaphroditism is some abnormality of androgen receptor binding or abnormality of AR binding to DNA.

A variety of AR gene defects have been described. Major structural defects in the AR gene (such as complete or partial gene deletions) comprise only a small fraction of individuals with androgen insensitivity. Minor structural defects in the AR gene (1 to 4-bp insertions or deletions) also account for few instances of androgen insensitivity. The third class of AR abnormalities is a number of single base mutations that introduce a premature termination codon or alter the splicing of mRNA. These also account for a small fraction of those with androgen insensitivity. Expansion of the CAG trinucleotide repeat of exon 1 is found in Kennedy disease (spinal bulbar muscular atrophy), a neurologic abnormality in males who also develop gynecomastia and secondary infertility with testicular atrophy (many were fertile until disease onset about the third or fourth decade of life). Single base mutations that result in amino acid substitutions in the receptor protein account for the majority of AR defects in subjects with complete or partial androgen insensitivity. At least 90 single base mutations at more than 70 different amino acid codons (of a total of 919 codons) have been described in about 160 individuals with some form of androgen insensitivity. Nearly all the known substitutions occur in the steroid-binding domain or the DNA-binding domain, and most of these are in the steroid-binding domain. No mutations have been reported in the hinge region.

VI. CLINICAL DIAGNOSIS AND MANAGEMENT

A. Diagnosis and Differential Diagnosis

The diagnosis of male pseudohermaphroditism is rarely made *in utero* but may be if there is a discordance between ultrasound findings of the external genitalia (i.e., the absence of a penis) and a 46,XY karyotype from an amniocentesis. The diagnosis is made at birth only if the external genitalia are female in a child known to have a 46,XY karyotype from amniocentesis or if there is obvious genital ambiguity. Such children should have a karyotype in the neonatal period.

Most instances of male pseudohermaphroditism are discovered at or after puberty. There are only two conditions in which a normal-appearing female (with breast development) has no cervix and uterus: complete androgen insensitivity and Müllerian agenesis (the Mayer–Rokitansky–Küster–Hauser syndrome). The diagnosis of complete androgen insensitivity is suggested by the absence of pubic and axillary hair and confirmed by a 46,XY karyotype. A karyotype of peripheral lymphocytes should be obtained from anyone with primary amenorrhea or genital ambiguity. A 46,XY karyotype establishes the categorical diagnosis of male pseudohermaphroditism. This may occur as a result of deficient androgen synthesis or defective androgen action. Normal male concentrations of testosterone exclude any cause of deficient androgen synthesis, such as an enzyme defect, congenital anorchia, or the congenital absence of interstitial cells. Male pseudohermaphrodites with normal male testosterone serum concentrations must have either 5α-reductase deficiency or androgen insensitivity due to a defect in the androgen receptor. Subjects with 5α-reductase deficiency do not develop gynecomastia. All subjects with androgen receptor defects except those whose only manifestation is infertility and defective spermatogenesis develop gynecomastia. The external genitalia of subjects with 5α-reductase deficiency usually undergo some masculinization at the time of puberty. This may also occur in subjects with partial forms of androgen insensitivity. A genetic pedigree from a subject with 5α-reductase is consistent with an autosomal recessive pattern. A pedigree analysis in those with androgen insensitivity is consistent with an X-linked recessive inheritance pattern.

B. Management

Sex assignment must be made at birth. With a significant degree of undervirilization of the external genitalia it is probably best to tell the parents they have a girl because it is technically easier to create a functional vagina than a functional penis. If the undervirilization is limited to penile hypospadias (Reifenstein syndrome), assignment of a male gender and correction of the hypospadias is appropriate. Surgery to correct ambiguous genitalia should be performed, although the timing of surgery is debatable. No corrective surgery is necessary if the external genitalia are unambiguously female.

Gonadectomy is essential for those with any form of androgen insensitivity. This should be done early in puberty in subjects with partial androgen insensitivity to prevent further masculinization of the external genitalia. Gonadectomy can be deferred safely until completion of puberty in those with complete androgen insensitivity to allow the estradiol secreted by the testes to stimulate breast growth and development.

Gonadectomy is also necessary because the risk of testicular tumors increases with increasing age. The exact frequency is difficult to determine because of the small number of subjects in the series reported. The best estimate is that the overall risk of a gonadal tumor in those with any form of androgen insensitivity is 6–9%. In other studies the risk of malignant testicular tumors appears to increase from 3% at age 20 years to 30% by age 50.

Hormone replacement is essential after gonadectomy. For those raised as females, estrogen replacement equivalent to menopausal doses is adequate. In the absence of a uterus, progesterone supplements are unnecessary. For those raised as males, testosterone replacement in the form of long-acting testosterone is necessary in doses approximately 100–200 mg parenterally every 2–4 weeks. The dose and interval should be adjusted to maintain serum testosterone concentrations in the adult male range.

See Also the Following Article

ANDROGENS

Bibliography

Balducci, R., Ghirri, P., Brown, T. R., Bradford, S., Boldrini, A., Boscherini, B., Sciarra, F., and Toscano, V. (1996). A clinician looks at androgen resistance. *Steroids* 61, 205–211.

Beato, M., and Sanchez-Pacheco, A. (1996). Interaction of steroid hormone receptors with the transcription initiation complex. *Endocr. Rev.* 17, 587–609.

Brinkman, A., Jenster, G., Ris-Stalpers, C., van der Korput, H., Bruggenwirth, H., Boehmer, A., and Trapman, J. (1996). Molecular basis of androgen insensitivity. *Steroids* 61, 172–175.

Evans, B. A., Hughes, I. A., Bevan, C. L., Patterson, M. N., and Gregory, J. W. (1997). Phenotypic diversity in siblings with partial androgen insensitivity syndrome. *Arch. Dis. Child.* 76, 529–531.

LeBlanc, G. A., Bain, L. J., and Wilson, V. S. (1997). Pesticides: Multiple mechanisms of demasculinization. *Mol. Cell. Endocrinol.* 126, 1–5.

MacLean, H. E., Warne, G. L., and Zajac, J. D. (1997). Localization of functional domains in the androgen receptor. *J. Steroid Biochem. Mol. Biol.* 62, 233–242.

Purvis, K., and Christiansen, E. (1992). Male infertility: Current concepts. *Ann. Med.* 24, 259–272.

Quigley, C. A., DeBellis, A., Marschke, K. B., El-Awady, M. K., Wilson, E. M., and French, F. S. (1995). Androgen receptor defects: Historical, clinical, and molecular perspectives. *Endocr. Rev.* 16, 271–321.

Roy, A. K., and Chatterjee, B. (1995). Androgen action. *Crit. Rev. Eukaryotic Gene Expression* 5, 157–176.

Wiener, J. S., Teague, J. L., Roth, D. R., Gonzales, E. T., Jr., and Lamb, D. J. (1997). Molecular biology and function of the androgen receptor in genital development. *J. Urol.* 157, 1377–1386.

Androgens

Bernard Robaire

McGill University

I. Introduction
II. Biosynthesis and Metabolism
III. Testosterone as a Prohormone
IV. Mechanism of Action
V. Sites of Action
VI. Androgen Formulations
VII. Therapeutic Uses of Androgens
VIII. Illicit Uses of Androgens
IX. Adverse Effects of Androgens
X. Inhibitors of Androgen Action

GLOSSARY

anabolic steroids Androgens that stimulate protein synthesis and inhibit protein breakdown, having the ability to enhance the growth of muscle and bone tissue.

androgen receptors Intranuclear proteins that, when activated by binding to androgens, will act as a *trans*-acting factor and initiate the transcription of selected genes.

androgens Steroid molecules, having 19 carbon atoms, that are responsible for male characteristics.

dihydrotestosterone the 5α-reduced metabolite of testosterone that is synthesized in many androgen target tissues and is the naturally produced steroid that binds most avidly to the androgen receptor.

Leydig cells Interstitial testicular cells in which essentially all the testosterone produced in a male is synthesized.

steroid A polycyclic ring structure composed of three six-membered and one five-membered rings.

testosterone An androgen produced by the interstitial Leydig cells in the testis.

The word androgen is derived from the Greek words *andro* (male) and *gennum* (to produce) and refers to any substance that possesses masculinizing activity.

I. INTRODUCTION

The observation that the testes were the source of androgens dates back several millennia as evidenced by the knowledge in China, India, and the Middle Eastern countries that the castration of males, resulting in eunuchs, provided valuable harem attendants. This practice may well have ushered in the beginning of endocrinology. Credit for the first report that testes can secrete substances into the circulation that will mediate an action at a remote site is given to Berthold, who demonstrated that comb and wattle regression in a castrated male cockerel could be prevented by a testicular transplant.

The first demonstration that a testicular extract contained a substance that could affect seminal vesicle growth was made by Loewe and Voss. The active principle, testosterone, was isolated, the structure elucidated, and the molecule synthesized by 1935. Since that time, we have acquired a detailed understanding of the biosynthesis, metabolism, and mechanism of action of testosterone. Testosterone and several of the hundreds of analogs synthesized in the past 60 years have acquired a major place in our pharmacological armamentarium.

II. BIOSYNTHESIS AND METABOLISM

The three major sites of androgen biosynthesis in mammals are the Leydig cells of the testis, the cortical cells of the adrenal, and the thecal cells of the ovary. These cells can be induced to synthesize androgens *de novo* from acetyl CoA; however, they normally synthesize androgens from the mandatory precursor cholesterol, which is taken up from extracellular space. The transport of cholesterol to mitochondria is regulated by the steroidogenic acute regulatory protein. In mitochondria the side chain of cholesterol is cleaved to produce pregnenolone. This steroid is then transported to the smooth endoplasmic reticulum, where it is further metabolized to androgens. The primary androgen produced by Leydig cells is testosterone, whereas adrenal cells produce androstenedione, dehydroepiandrosterone, and its sulfated ester. Adrenal androgens bind weakly to the androgen receptor and seem to play a very minor role in mediating androgen action.

The production of testosterone by Leydig cells is primarily under the control of luteinizing hormone (LH). LH binds to cell surface receptors and stimulates the production of testosterone via a cAMP-dependent mechanism. The amount of androgen synthesized is directly proportional to the amount of smooth endoplasmic reticulum in Leydig cells. Testosterone is neither packaged in vesicles nor does it significantly accumulate inside Leydig cells; rather, it is secreted by facilitated passive transport as it is synthesized. The presence of extracellular binding proteins (albumin, low affinity and high-capacity; sex hormone-binding globulin, high affinity and low capacity) allows for this type of transport. In seminiferous tubules, androgen-binding protein, a protein with the same amino acid backbone as sex hormone-binding globulin, is secreted by Sertoli cells and acts as a sink to concentrate testosterone.

The synthesis of testosterone by Leydig cells is regulated at two levels. Increased intracellular testosterone accumulation results first in a rapid negative feedback inhibition by testosterone of the 17α-hydroxylase /C17,20 lyase enzyme activity and second in a more delayed inhibition, at the genomic level, of the synthesis of the complement of genes needed to convert pregnenolone to testosterone; this inhibition of gene expression is apparently mediated by androgen receptors found within Leydig cells.

The two proximate metabolites of testosterone, dihydrotestosterone and estradiol, play an essential role in mediating androgen action; further metabolites of testosterone are devoid of hormone activity.

III. TESTOSTERONE AS A PROHORMONE

Although testosterone is the hormone produced in largest quantity by the male, it is not the hormone that mediates all of its actions intracellularly. Testosterone can be converted in target tissues to dihydrotestosterone or estradiol or remain as testosterone to mediate androgen action. Thus, of all steroids, only testosterone can be considered to be both a prohormone and a hormone.

Testosterone is not the androgen that binds most avidly to the androgen receptor. The 5α-reduced metabolite, dihydrotestosterone, has a higher affinity for the androgen receptor than testosterone, by nearly one order of magnitude, and a slower dissociation rate from the receptor than testosterone. Dihydrotestosterone is synthesized in many androgen-dependent tissues, such as the prostate, epididymis, seminal vesicles, and skin, but not in others, such as muscle and testis. The enzyme that forms dihydrotestosterone from testosterone is 5α-reductase. Two genes, found on different chromosomes, have been reported for all mammalian species studied to date. Though in man type 2 5α-reductase is clearly responsible for the 5α-reductase-deficiency syndrome, the relative importance of these two genes in the regulation of androgen action has yet to be determined. Testosterone can also be aromatized to estradiol to mediate its action in tissues such as the brain and the efferent ducts via the estrogen receptor.

IV. MECHANISM OF ACTION

Steroids are highly lipophilic substances that can freely diffuse across membranes. Only those cells that can retain testosterone by either metabolizing it to a more active steroid or binding it to a receptor will exhibit a response to androgens. Intracellular retention of steroids occurs through the formation of stable complexes with specific intracellular receptor proteins. The androgen receptor is intranuclear. The binding of the steroid to its receptor occurs at high affinity, is reversible, and induces a conformational/oligomeric activation of the receptor protein followed by phosphorylation. The "activated" receptor has DNA-binding properties that allow it to recognize and to bind with high affinity to specific locations on chromatin (the nuclear acceptor site with the consensus sequence TGTTCT). The activated steroid–receptor complex can then act as a transcriptional regulator, initiating a cascade of events that ultimately lead to the synthesis, inhibition, or modification of proteins.

Androgen receptors are members of the family of proteins containing steroid, thyroid, vitamin D, and retinoic acid receptors. The steroid-specific recognition site is at the C-terminal end of the protein. The adjacent domain is the hinge domain that plays a key role in protein dimerization and binding to heat shock protein 90. This is followed by a zinc finger DNA-binding domain, whereas the N-terminal end of the protein contains the regulatory domain that allows for the selective transcription of specified genes.

In man, the androgen receptor gene is a single-copy gene that is localized on the X chromosome and is over 90 kb. A large number of mutations have been identified in this gene which result in androgen-insensitivity syndrome. These mutations range from partial or complete deletion of the androgen receptor to single base substitutions in the steroid-binding domain.

V. SITES OF ACTION

Androgens act on any cell that can retain the steroid once it has entered it and that contain a receptor that can mediate its action. Within the male reproductive system, all tissues can respond to androgens. In the testis, Leydig cells autoregulate their testosterone production and Sertoli cells regulate spermatogenesis; it remains unclear as to whether germ cells can directly respond to androgens. The epididymis, the vas deferens, the seminal vesicles, and the prostate all regress when androgens are withdrawn; this regression is mediated by apoptosis and/or cell atrophy depending on which tissue is examined and the time point posttreatment.

There are many tissues that are not part of the male reproductive system per se that also respond to androgens. These include higher brain centers, the hypothalamus, the pituitary, bone, kidney, muscle, skin, and hair follicles.

VI. ANDROGEN FORMULATIONS

A. Injectable

Androgens are highly lipid soluble and are rapidly absorbed from the gastrointestinal tract. The free unesterified form of testosterone is very rapidly metabolized and inactivated by liver enzymes (first-pass effect), rendering oral administration an impractical approach. Consequently, the method of testosterone administration that was first developed and that remains most commonly used is an intramuscular injection of testosterone esters—testosterone enanthate, propionate, or cypionate—which have an effect that lasts several days. Testosterone buciclate, a recently developed testosterone ester, exerts its androgenic effect for 3 months after a single injection. However, intramuscular injection as a means of drug administration has many obvious drawbacks and alternative modes of administering testosterone have been developed.

B. Oral

To minimize the first-pass effect, both micronized and 17β-substituted analogs of testosterone (testosterone undecanoate) have been developed for oral administration. Though requiring frequent dosing, this formulation is gaining popularity. Numerous additional efforts have been made to circumvent the first-pass effect of testosterone metabolism; testosterone and several of its esters have been tested using nasal, sublingual, or rectal administration. These have provided inconsistent results and have thus far proved to be impractical.

C. Transdermal

The lipid solubility of androgens has allowed the development of transdermal testosterone administration via skin patches. The first generation of such patches required placement on the scrotum to obtain maximal absorption of the steroid, and the amount of steroid delivered was 2 or 3 mg/day. Subsequent generations of skin patch design have allowed greater amounts of steroid to be delivered and do not require scrotal contact. Though effective for testosterone replacement in hypogonadal men, this type of delivery cannot provide sufficient daily testosterone administration for contraceptive or other uses.

D. Pellets and Implants

One of the first ways testosterone was administered after its discovery was via subdermal placement of compressed crystalline testosterone. This method was abandoned for several decades, but by using fused crystalline testosterone, stable pellets are now available and are receiving greater attention. They require a small incision to place them subdermally. These pellets give a peak serum testosterone concentration 1 month after placement; the values decrease linearly over a 6-month period. These pellets can provide effective suppression of gonadotropins for approximately 3 months.

It was discovered in the 1960s that crystalline steroids can cross polydimethylsiloxane (Silastic) membranes. This observation has been used in the development of several female contraceptives. In animal studies, testosterone-filled Silastic capsules have been used for sustained release of testosterone in the development of male contraceptives. Though such capsules are not being used in human studies, they have served as a model for the development of steroid-containing biodegradable implants that can be used either for androgen replacement or for contraception.

VII. THERAPEUTIC USES OF ANDROGENS

A. Hypogonadal States

Hypogonadal functions may be caused by lesions in the hypothalamus [decreased or absent gonadotropin-releasing hormone (GnRH) secretion], the pituitary (inability to respond to GnRH), the testis (lack of responsiveness by Leydig cells to LH), or the inability of target tissues to respond to androgens (androgen insensitivity and androgen receptor mutations). The first three types of lesions may be effectively treated by androgen replacement therapy. Therapy should be initiated as soon as the diagnosis is made peri- or postpubertally. This condition constitutes the main indication for androgen therapy. This therapy is effective in the treatment of most of the characteristic symptoms of hypogonadism (e.g., decreased libido, atrophic muscles, anemia, osteoporosis) but not adequate for the stimulation of spermatogenesis. This process requires intratesticular concentrations of testosterone that are several-fold higher than those normally found in serum and is enhanced by follicle-stimulating hormone (FSH). Consequently, for the treatment of infertility in a hypogonadal male, GnRH or human chorionic gonadotropin therapy is necessary. This therapy is used transiently, whereas androgen replacement therapy for the hypogonadal male requires life-long treatment.

B. Constitutionally Delayed Puberty

Androgen treatment should only be used when it has been clearly established that the onset of puberty has been substantially delayed. This therapy is highly effective. Since large sustained doses of androgens can result in early epiphyseal bone closure and have long-term effects on the hypothalamic–pituitary axis, treatment should be limited to short periods (usually 3 months). Such treatment will result normally in rapid bone growth, enlargement of the testis, and establishment of puberty.

C. Use as Protein Anabolic Steroid

After prolonged immobilization, surgery, or trauma, androgens have been administered, along with appropriate diet and exercise, to reverse protein loss. It is clear that androgen receptors in muscle do not differ from those found in other tissues. Hence, there are no good data to suggest that any androgen has relatively superior anabolic, as opposed to androgenic, action. The anabolic muscle cell hyperplasia induced by androgens is well established for children and women, but no convincing data have been presented to date for normal adult males since, in the adult male, the androgen receptor in muscle is fully saturated. The beneficial effects of androgen supplementation after surgery or prolonged immobilization seem to be minimal; appropriate exercise and proper diet prevent the untoward and marginal benefits of this therapy.

D. Idiopathic Oligozoospermia

Since the early 1950s, a series of studies have indicated that when testosterone is administered to oligozoospermic men, there is first a suppression of spermatogenesis followed by a rebound in the quantity and quality of spermatozoa produced after termination of drug treatment. Testosterone rebound therapy has provided highly variable results that differ only slightly from the effects of placebo in other studies. Essentially, all trials were poorly controlled and they were not double blind. In a double-blind trial in which orally administered testosterone undecanoate was tested for this condition, no drug effect was found. Although the definitive study on the value of testosterone rebound therapy for the management of idiopathic oligozoospermia is yet to be done, the preponderance of evidence indicates that this therapy has little, if any, value.

E. Aplastic and Renal Anemia

Androgens increase the proliferation of stem cells by an as yet unidentified mechanism. The increased erythropoietic cell mass of men, when compared to that of women, is due to androgenic action as demonstrated by castration and hormone replacement studies. The combined actions of androgens on increasing the synthesis of red blood cells as well as the concentration of erythropoietin has been taken advantage of in the management of anemia. The cell proliferative effect of androgens is greatest on stem cells for erythropoiesis and least on those for thrombopoiesis. Androgens are effective in the management of approximately half of the cases of hypoplastic or aplastic anemia; however, due to the large doses of androgen used, the treatment is less well tolerated by women. With the advent of human recombinant erythropoietin, the use of androgens in the management of renal anemia has declined; it is still not clear whether androgens can synergize with erythropoietin in treating this condition.

F. Gynecological Disorders

1. Postpartum Breast Engorgement

In combination with estrogen, androgens have been used to reduce the postpartum engorgement after parturition. This is an effective therapy, and due to the transient nature of the therapy, the androgenizing side effects are often well tolerated.

2. Breast Cancer

Testosterone administration to women has been used for palliative therapy of breast cancer. The effectiveness of this therapy is similar to that obtained with high doses of estrogens. This steroid seems to be the most effective of all androgens thus far tested for this condition; however, conventional chemotherapy is more effective and a preferred therapeutic approach.

3. Menopause

Hormone replacement therapy in the peri- and postmenopausal phase of a women's life is highly recommended for its beneficial effects on a number of systems, including the cardiovascular system, bone, and tissues of the reproductive system. Androgens seem to act synergistically with estrogens to increase bone mass in postmenopausal women; nandrolone decanoate causes minimal virilizing effects while increasing peripheral and axial bone mass. The beneficial effects of androgens in this condition include a sense of well-being as well as increased activity.

G. Osteoporosis

Healthy men lose bone mass with age; in hypogonadal men, this loss of bone mass is rapidly accelerated. Androgen replacement therapy in hypogonadal males clearly increases bone mass and prevents osteoporosis associated with this condition. In healthy men bone degradation decreases and bone mineral density increases after testosterone therapy; however, it is still not clear whether new bone formation ensues after testosterone administration. Based on animal studies, it appears that androgens act by stimulating osteoblast proliferation and differentiation, and that this action is via the androgen receptor since the nonaromatizable androgens are at least as effective as testosterone.

H. Aging

Beyond the third decade there is a gradual decrease in both total and free serum testosterone as well as in other androgens. The necessity for androgen replacement therapy in men as they age is still controversial. Though a number of studies report an improvement in mood and sexual behavior, decreased body fat, increased muscle mass, and increased hand grip strength, these studies have been short term. It is important to appreciate that administration of low doses of testosterone to healthy aging males results in a suppression of LH secretion and thus decreased intragonadal testosterone production; the net effect is to maintain a homeostatic serum testosterone concentration. Potential adverse effects of moderate doses of testosterone therapy on cardiovascular disease or prostate cancer have not revealed any increased risk, but longer studies with more patients are necessary before androgen replacement therapy is considered for the healthy aging male.

I. Contraception

Testosterone administration, at a dose that suppresses LH and FSH but does not raise serum testosterone, has been shown in many animal studies to act as an effective form of male contraception. The underlying basic mechanism is that the concentration of testosterone needed to sustain spermatogenesis is several-fold greater than that needed for all other actions of testosterone. Thus, by "short-circuiting" the feedback loop between the hypothalamic–pituitary complex and the testis, testosterone can suppress spermatogenesis. However, testosterone can also, on its own, partially or completely maintain spermatogenesis. A dose of testosterone that will consistently suppress spermatogenesis and result in infertility has not been found for man. Azoospermia is attained in 60–80% of subjects. The development of either improved long-lasting analogs or a biodegradable sustained-release formulation could provide an effective male contraceptive. Alternatively, combining an androgen with a progestin, an estrogen, or a GnRH analog to help suppress gonadotropin secretion may provide realistic alternatives.

VIII. ILLICIT USES OF ANDROGENS

The major illicit use of androgens is by athletes. Androgens were first used in the 1950s for body building by weight lifters; this use grew rapidly in the 1970s and 1980s so that athletes in many sports were taking anabolic steroids/androgens despite the explicit regulations most sports federations have against such drug use. It was not until the 1976 Olympics that the urine of athletes was begun to be tested routinely for the presence of anabolic steroids.

While it is clear that androgen administration will enhance muscle mass and performance in children and women, most of the data available to date indicate that androgens do not enhance performance for adult, active men. Prolonged and excessive physical exercise results in reduced GnRH output in both men and women, with the consequence that gonadal steroid production is suppressed; this effect is often transient. Small doses of steroids would be required to restore normal physiological levels of steroids. However, the doses used by athletes are often 100 times those recommended for treating hypogonadal conditions; this renders proper assessment of the effects of such treatments very difficult because of the ethics associated with conducting such trials, because of the difficulty of running a properly controlled double-blind trial, and because of the difficulty of assessing the effects of such drug use on small changes in performance, often the reason for which the drugs are taken. Some of the more important actions of androgens administered for sports competition may be on motivational or aggressive behavior; these are difficult to assess and often disregarded in assessing the action of androgens on athletes. The placebo effects of this therapy may also be quite significant. While acknowledging these limitations, it is still interesting to note that essentially all the studies done to date on normal adult males have been unable to clearly demonstrate a beneficial effect of androgen administration on strength, muscle mass, or performance.

IX. ADVERSE EFFECTS OF ANDROGENS

Most of the adverse effects of androgen administration are related to the mechanism of action and to the target tissues that can respond to exposure to androgens. Thus, it is predictable that women taking androgens could have some or all symptoms of virilization, which include decreased or arrested menses (suppression of gonadotropins), deepening of the voice, hirsutism, acne, and clitoral enlargement. If women take androgens during pregnancy, androgenization of the female fetuses is likely to result. Adverse effects of androgens in men include decreased hypothalamic and pituitary functions, diminished spermatogenesis, and edema.

A severe consequence of androgens that have a 17α-alkyl substitution, such as mesterolone, is liver toxicity. This toxicity can result in jaundice and, after prolonged exposure, hepatic adenocarcinoma. It is for this reason that many countries have removed this family of analogs from their pharmacopoeia. There is no evidence that androgens that do not have this substitution can exhibit this untoward effect.

X. INHIBITORS OF ANDROGEN ACTION

A. Inhibitors of Testosterone Biosynthesis

The most effective way to block testosterone biosynthesis is to inhibit the secretion of LH, thus arresting steroidogenesis in Leydig cells. This may be achieved effectively by administration of GnRH antagonists or long-lived GnRH agonists that cause downregulation of GnRH receptors on gonadotropes, thus resulting in what can be termed a chemical castration. Estrogens (e.g., diethylstilbestrol) have also been used to suppress the hypothalamic–pituitary testicular axis and hence induce chemical castration. However, because of the higher incidence of side effects associated with this form of chemical castration and because of the growing realization that there are estrogen receptors in several tissues of the male reproductive tract that seem to play major physiological roles, e.g., resorption of fluid in the efferent ducts, this form of therapy is disappearing rapidly.

Several compounds can partially or completely block testosterone biosynthesis by inhibiting one or several of the enzymes in the biosynthetic pathway. Spironolactone, a competitive inhibitor of aldosterone, is also an effective inhibitor of 17α-hydroxylase activity; it has proved itself effective in the treatment of hirsutism in women. Ketoconazole, an imidazole antifungal agent, inhibits several of the enzymes involved in testosterone biosynthesis, including the P450 cholesterol side chain cleavage enzyme, 17α-hydroxylase/C17,20 lyase, and 3β-hydroxysteroid dehydrogenase. Though extensively tested as a potential drug in the management of prostatic cancer, this drug is unlikely to be widely used because of its gastrointestinal side effects as well as the induction of gynecomastia.

B. Inhibitors of Testosterone Metabolism

The fact that testosterone is metabolized to a more potent metabolite, dihydrotestosterone, has been used to advantage in designing drugs that can partially or completely inhibit androgen action in those tissues that have an elevated 5α-reductase activity. The first such inhibitor to have obtained regulatory agency approval is finasteride. This drug was designed to reduce the symptoms of benign prostatic hyperplasia by reducing intraprostatic dihydrotestosterone formation. This orally active azasteroid dramatically suppresses circulating and intraprostatic dihydrotestosterone without elevating circulating testosterone, but it does cause an increase in intraprostatic testosterone. This drug is only moderately effective in the management of benign prostatic hyperplasia; this is likely due to the fact that it is a highly effective inhibitor of type 2 5α-reductase but only a moderate inhibitor of the type 1 form of the enzyme (both forms of the enzyme are present in the prostate). Either less selective inhibitors of 5α-reductases or a combination of inhibitors of each of the two forms will likely yield better therapeutic results. Effective blockade of 5α-reductase activity should not only result in decreased benign prostatic hyperplasia but also could be effective in managing other conditions, such as prostatic cancer, alopecia, and female hirsutism. Finasteride has either been approved or is currently being tested for these conditions. Effective blockade of 5α-reductase activity could also provide a highly desirable male contraceptive because it would block the maturation of spermatozoa in the epididymis without affecting the synthesis of spermatozoa or circulating testosterone concentrations.

C. Androgen Receptor Antagonists

In the search for active analogs of progesterone to be used in female contraceptive formulations, cyproterone acetate was discovered to be more than 100-fold more active than progesterone in animal bioassays. This active progestagen also inhibits the binding of dihydrotestosterone to the androgen receptor. The antiandrogen action of this agent is in part due to its blockade of LH secretion by the pituitary and the consequent suppression of testosterone production by Leydig cells and in part due to its antagonism of dihydrotestosterone at the androgen receptor. This drug has been tested in the treatment of prostate cancer, hirsutism in women, and to reduce libido in men showing deviant sexual behavior. It has been approved for the latter two uses in some countries.

Flutamide is a nonsteroidal analog (a substituted amilide) that is an effective competitive antagonist of dihydrotestosterone at the androgen receptor. By blocking the negative feedback effect of testosterone on the hypothalamic–pituitary axis, it causes a significant increase in the concentrations in serum of both LH and testosterone. Its use is therefore limited to suppression of androgen action in women and to suppression of the action of androgens in target tissues when used in combination with GnRH analogs in treating prostate cancer. Gastrointestinal side effects are the major complications. Bicalutamide is a recently developed androgen receptor antagonist which has few side effect, needs to be taken only once daily, and seems to be at least as effective as flutamide in the management of prostate cancer.

See Also the Following Articles

ADRENAL ANDROGENS; ANDROGEN INHIBITORS/ANTIANDROGENS; DIHYDROTESTOSERONE; LEYDIG CELLS; TESTOSTERONE BIOSYNTHESIS

Bibliography

Berns, J. S., Rudnick, M. R., and Cohen, R. M. (1992). A controlled trial of recombinant human erythropoietin and mandrolone decanoate in the treatment of anemia in patients on chronic hemodialysis. *Clin. Nephrol.* 37, 264–267.

Berthold, A. A. (1849). Transplantation der Hoden. *Arch. Anat. Physiol. Wiss. Med.* 16, 42–46.

Brown, T. R. (1996). Androgen receptor structure, function, regulation and dysfunction. In *Pharmacology, Biology and Clinical Applications of Androgens* (S. Bhasin, H. L. Gabelnick, J. M. Spieler, R. S. Swerdloff, C. Wang, and C. Kelly, Eds.), pp. 45–56. Wiley-Liss, New York.

Christiansen, C., and Riis, B. J. (1990). 17β-estradiol and continuous norethisterone: A unique treatment for established osteoporosis in elderly women. *J. Clin. Endocrinol. Metab.* 71, 836–841.

Clark, B. J., Wells, J., King, S. R., and Stocco, D. M. (1994). The purification, cloning and expression of a novel luteinizing hormone-induced mitochondrial protein in MA-10 mouse Leydig tumor cells: Characterization of the steroidogenic acute regulatory protein (StAR). *J. Biol. Chem.* 263, 28314–28322.

Ewing, L. L., and Zirkin, B. R. (1983). Leydig cell structure and steroidogenic function. *Recent Prog. Horm Res.* 39, 599–635.

Ewing, L. L., Chubb, C., and Robaire, B. (1976). Macromolecules, steroid binding and testosterone secretion by rabbit testes. *Nature* 264, 84–86.

Loewe, S., and Voss, H. E. (1930). Der Stand der Erfassung des männlichen Sexualhormona (Androkinins). *Klin. Wshr.* 9, 481–487.

Niesclag, E., and Behre, H. M. (1990). *Testosterone Action Deficiency Substitution*. Springer-Verlag, Berlin.

Pusch, H. H. (1989). Oral treatment of oligozoospermia with testosterone-undecanoate: Results of a double-blind-placebo-controlled trial. *Andrologia* 21, 76–82.

Rittmaster, R. S. (1994). Finasteride. *N. Engl. J. Med.* 330, 120–125.

Robaire, B., Ewing, L. L., Irby, D. C., and Desjardins, C. (1979). Interactions of testosterone and estradiol-17β on the reproductive tract of the male rat. *Biol. Reprod.* 21, 455–463.

Russell, D. W., and Wilson, J. D. (1994). Steroid 5α-reductase: Two genes/two enzymes. *Annu. Rev. Biochem.* 63, 25–61.

Ruzicka, L., and Wettstein, A. (1935). Synthetische Darstellung des Testishormons, Testosteron (Androsten-3-on-17-ol). *Helv. Chim. Acta* **18**, 1264–1275.

Santen, R. J., Manni, A., Harvey, H., and Redmond, C. (1990). Endocrine treatment of breast cancer in women. *Endocr. Rev.* **11**, 221–265.

Sherwin, B. B. (1996). Androgen use in women. In *Pharmacology, Biology and Clinical Applications of Androgens* (S. Bhasin, H. L. Gabelnick, J. M. Spieler, R. S. Swerloff, C. Wang, and C. Kelly, Eds.), pp. 319–324. Wiley-Liss, New York.

Tenover, L. (1994). Androgen administration to aging men. *Endocrinol. Metab. Clin. North Am.* **23**, 877–892.

Viger, R. S., and Robaire, B. (1996). The mRNAs for the steroid 5β-reductase isozymes, type 1 and type 2, are differentially regulated in the rat epididymis. *J. Androl.* **17**, 27–34.

Wu, F. C. W. (1997). Androgen/progestin combination in male contraception: Efficacy and safety. In *Current Advances in Andrology* (G. M. H. Waites, J. Frick, and G. W. H. Baker, Eds.), pp. 221–226. Monduzzi Editore, Bologna.

Androgens, Effects in Birds

Cheryl F. Harding
City University of New York

I. Behavior
II. Role of Androgen Metabolism in Regulating Reproduction
III. Effects on Morphology

GLOSSARY

aromatizable androgen Androgens that, like testosterone, can be metabolized (aromatized) to estrogens. They can also be metabolized to other androgens and thus provide a source of both androgenic and estrogenic metabolites. Nonaromatizable androgens, such as 5α- and 5β-dihydrotestosterone, cannot be converted to estrogens and provide only androgenic metabolites.

nuptial plumage Increases in pituitary or gonadal hormones cause many birds, particularly males, to assume brilliant or showy feathers at the onset of breeding. This is known as the prealternate molt.

pharmacological effect A hormone treatment that activates a behavior pattern, although it is not the hormone responsible for stimulating the behavior in normal circumstances.

postnuptial molt After breeding, most birds replace their plumage with more conservatively colored feathers. Timing of the postnuptial (or prebasic) molt is important so that feather replacement occurs before fall migration and/or the onset of adverse weather.

sexual dimorphism Any difference between males and females, including differences in size, metabolism, plumage, or behavior. Many sexual dimorphisms are established by hormone exposure during critical periods in development. Hormone exposure in adulthood may further increase the dimorphism between the sexes.

steroid receptor Most actions of gonadal hormones are mediated through their binding to receptors inside target cells. The hormone receptor complex then binds to DNA to regulate gene transcription and the production of specific proteins. There are specific receptors for each class of hormone (e.g., androgen and estrogen receptors).

Androgens play a central role in coordinating the physiological and behavioral changes that male birds need for a successful reproductive cycle. Most nontropical birds show marked seasonal changes in gonad size and hormone secretion correlated with profound changes in aggressive and sexual displays. The fact that the testes are larger in breeding than nonbreeding birds was even noted by Aristotle. Experiments investigating the effects of castration and androgen replacement therapy have demonstrated the importance of androgens in stimulating a variety of behavior patterns involved in reproduction in male birds. While there is great variation in specific behavior patterns shown by different species, androgens have been clearly implicated in stimulating aggressive, courtship, copulatory, and nest-building behavior. However, high androgen levels appear incompatible with parental behavior, including incubation, brooding, and feeding young, and with the postnuptial molt. If androgen levels are kept artificially high during the latter part of the breeding cycle, both breeding success and survival may be adversely affected. It is now known that many effects of androgens on avian reproductive behavior are mediated by conversion of androgens to estrogenic metabolites. Morphological male secondary sexual characters, such as brightly colored combs and wattles or male nuptial plumage, are generally maintained by androgens during the breeding season. In contrast, many effects of androgen on the brain which ultimately stimulate performance of sexually dimorphic behaviors, such as male aggressive or sexual displays, involve the metabolism of androgens to estrogens. In all avian species which have been studied, both androgens and estrogens are needed to stimulate the full repertoire of male reproductive behavior.

I. BEHAVIOR

Regardless of the type of mating system used or the behavior patterns shown, male reproductive behavior is modulated by changing androgen levels over the breeding cycle. Heightened androgen levels at the onset of breeding stimulate aggressive interactions between males, whether used to establish large all-purpose territories used for nesting and feeding, small territories on leks (communal display grounds used only for breeding), or dominance hierarchies. High androgen levels also stimulate high rates of visual and auditory displays which attract females. However, the pattern of androgen secretion over the cycle appears to be related to particular mating strategies. In species in which males attempt to attract and copulate with multiple females and provide no parental care, androgen levels typically remain high as long as females are receptive. In species in which males provide paternal care, androgen levels typically decrease during the parenting phase of the cycle and then rise when birds begin courting again prior to initiating additional clutches.

A. Aggressive Behavior

Data suggest that positive correlations between androgen levels and aggressive behavior are most clearly seen during periods of social instability, when territories or dominance hierarchies are being established. Increases in testicular size have been correlated with the transition from feeding flocks to territorial defense in birds as disparate as pheasants and hummingbirds. Measurements of plasma hormone levels over the reproductive cycle have consistently revealed strong correlations between intermale aggression and high androgen concentrations, with both typically peaking early in the cycle when males are establishing territories (Fig. 1). Testicular size or circulating androgen levels have often been found to be smaller/lower in floaters (males who do not defend territories) compared to territory holders (e.g., red grouse and red-winged blackbirds). Similarly, male satin bower birds without bowers (arches constructed of grass or twigs that are often "decorated" with brightly colored or shiny objects and used to display to females) had lower androgen levels than males which held bowers. Males have also been found to have higher androgen levels in areas with higher population densities, presumably reflecting increased aggressive interactions (e.g., cowbirds, yellow-headed blackbirds, song sparrows, and starlings).

One strategy widely used to study the relationship between hormones and aggressive behavior is to simulate a territorial intrusion by placing a decoy male in a cage on a male's territory while playing back recorded vocalizations. Males of many species (e.g., song and white-crowned sparrows and red-winged blackbirds) respond to such simulated intrusions with increased androgen levels. A second strategy is to remove territorial males and measure circulating androgens in males which take over the territories. Both replacement and neighboring males had heightened androgen levels compared to control males in stable areas. In studies of captive birds (e.g., house sparrows, Japanese quail, and red jungle fowl), higher ranking individuals or those that initiate attacks typically have higher androgen levels than lower ranking birds. In cases involving behavioral interactions, it is not clear if androgen levels increase early enough in the interaction to influence the outcome or if the increase in androgen levels is merely a response to engaging in aggressive interactions.

These correlational data suggesting high androgen levels stimulate high levels of aggression are born out by studies which have experimentally manipulated androgens. As early as 1939, Allee and co-workers demonstrated that testosterone (T) treatment increased the aggressiveness of low-ranking hens, often increasing their rank within their home flocks. In other cases, although T treatment increased aggressive interactions, it did not increase the treated bird's status with familiar birds, though it increased the probability that treated birds would dominate unfamiliar conspecifics. The fact that androgen treatment increases aggressive behavior, but elevated androgen levels alone are insufficient to overcome previously established social relationships between birds in a flock, between territory owners, or between owners and floaters, has been replicated in a variety of species, including doves, grouse, Japanese quail, and red-winged blackbirds. However, in some studies (e.g., song and white-crowned sparrows and pied flycatchers), males receiving T implants increased their aggressive displays so dramatically that their territories increased in size. The androgen levels of neighboring males increased as well, presumably be-

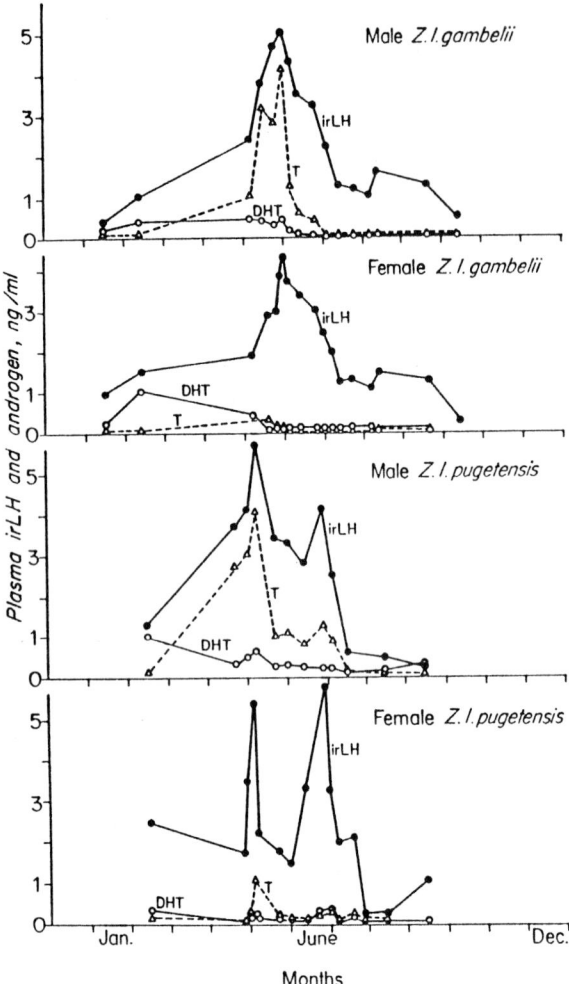

FIGURE 1 Annual cycle in plasma levels of immunoreactive luteinizing hormone (irLH), testosterone (T), and dihydrotestosterone (DHT) under natural conditions in two subspecies of the white-crowned sparrow. Spring migration occurs in April and early May, with nesting for the first clutch in early to mid-May in the southern race, Zonotrichia leucophrys pugetensis, and nesting for a second clutch in late June to early July. The northern race, Z. l. gambelii, nests only once in late May to June. Photorefractoriness sets in by late June to early July, followed closely by the onset of the postnuptial molt. Autumn migration occurs sometime between August and October. This comprehensive study of the behavioral endocrinology of the two races found that territorial encounters were most frequent when T levels peaked shortly after males arrived in their breeding areas. At the time when females were arriving and pairing with territorial males, males sang at high rates, averaging 400 songs per hour. T levels fell markedly during incubation, as did the frequency of aggressive interactions and singing, averaging 5–10 songs per hour. In the double-brooded Z. l. pugetensis, song frequency increased to 300 songs per hour during courtship preceding the second clutch, but no significant increase in T levels or aggressive behavior was detected. Males of this race maintained higher average T levels and remained in their territories longer than males of the single-brooded Z. l. gambelii [from J. C. Wingfield and D. S. Farner, Control of seasonal reproduction in temperate-zone birds, In Progress in Reproductive Biology (P. O. Houbinot, Ed.), 1980. Reproduced with permission of S. Karger, Basel].

cause of increased aggressive interactions with the T-treated birds. Decreasing androgen levels or administering antiandrogens during the breeding season almost always decreases the intensity and persistence of aggressive displays and the ability of males to achieve high rank or successfully defend territories.

Although androgen levels are clearly related to aggressive behavior and territorial defense during breeding, the contribution of androgens to aggression and territorial behavior outside the breeding season is unclear. Androgen treatment can induce birds to become aggressive and territorial outside the breeding season. However, in species which normally defend territories outside the breeding season (e.g., mockingbirds and song sparrows), this behavior occurs when androgen levels are basal, and antiandrogen treatments have no effect (e.g., European robins). Even castrated song sparrows established territories in the fall. Aggressive displays given in response to simulated territorial intrusions at this time were less intense and less persistent than those seen in the spring when androgen levels were high. Similarly, dominance status in wintering feeding flocks does not appear to be related to androgen levels (e.g., juncos and white-crowned sparrows).

Studies suggest that androgens may also play a role in modulating aggressive interactions in female birds. The avian ovary secretes fairly high levels of androgen and the incidence of female territorial defense and/or singing has been associated with increased androgen levels in a wide variety of species (e.g., gulls, robins, shrikes, mockingbirds, phalaropes, white-crowned sparrows, and wren-tits). As mentioned previously, treating females with exogenous androgen typically increases their aggressive behavior, but it is unclear to what extent this might represent a pharmacological effect. Recent studies which have measured endogenous hormone levels have suggested that high estrogen levels during egg laying and incubation contribute to females' heightened aggressive behavior.

B. Sexual Behavior

The period of highest circulating androgen levels tends to be not only the period of territory establishment but also the period when males are first interacting with females (Fig. 1). In songbirds, the high rate of singing stimulated by androgens both attracts females to the territory and deters other males from trespassing. In some species, different song types are used for these two different functions. Interacting with sexually active females increased androgen levels in some species (e.g., Gambel's, Puget Sound, and song sparrows) but not in others (e.g., eider ducks). Wingfield has hypothesized that the increase in T levels found in several species of sparrows during egg laying is important in stimulating high frequencies and intensities of aggressive displays while the female is fertile, guarding the male's paternal investment. Recently developed techniques that allow scientists to assess which male actually fertilized the eggs documented the high percentage of clutches sired by multiple males in some species and confirmed the importance of mate-guarding behavior. In barn swallows, male T levels correlated with the mean frequency of both mate guarding and singing. In other species, giving females estrogen prolonged and intensified their sexual behavior, increased aggressive behavior and mate guarding by their males, and in several species also increased the males' T levels.

Nest building occurs during this period in many birds, and increased androgen levels stimulate nest building by males in species as diverse as brush turkeys, night herons, mockingbirds, ringdoves, starlings, and zebra finches. Nest building by females often peaks around the time of ovulation when their androgen levels are high, but levels of several other hormones are also high at this time. Although androgen treatment has also been shown to stimulate nest building by females, it is unclear whether this is a normal or pharmacological effect.

In some cases, increased androgen levels have been correlated with some measure of reproductive success. In black grouse, plasma T concentrations were strongly correlated with male mating success, presumably because males with high androgen levels held the most central territories on the lek, and females prefer males possessing central territories. In satin bower birds, T levels were positively correlated with male mating success, presumably because T levels were consistently correlated with the quality of male displays, which are important to female choice.

Species with different mating strategies typically show different patterns of androgen levels over the reproductive cycle (Fig. 1). In monogamous species in which males provide parental care, high androgen levels appear to interfere with normal paternal care. The cyclic nature of androgen levels over the nesting cycle is particularly striking in house sparrows, which may incubate up to five clutches in one season. The high T levels found during each bout of egg laying correlated with high rates of intrusion by conspecifics and high rates of nest defense and mate guarding by males. T levels fell precipitously with incubation and then began to rise. Male feeding rates declined as T levels rose prior to initiating the next clutch. Androgen levels tend to remain higher for a greater proportion of the breeding season in polygynous or parasitic species in which males typically provide minimal, if any, parental care (e.g., redwinged blackbirds and cowbirds). Ultimately, androgen levels must decrease in these species as well to allow birds to enter the postnuptial molt. If this molt is not completed in a timely fashion, birds will be unable to migrate on schedule. Maintaining high androgen levels blocks the onset of molt.

In a variety of species, if T levels are kept high by implants, heightened aggressive behavior persists into later phases of the reproductive cycle but males show little or no paternal behavior, often leading to clutch failure. In two normally monogamous species (song and white-crowned sparrows) and in pied flycatchers, which are opportunistically polygynous, T treatment caused males to sing more frequently and increase territory size, and these males often attracted additional females to their territories. This, however, did not increase their reproductive success since their heightened T levels interfered with normal paternal behavior. These males did not feed fledglings, and their reproductive success suffered. T-treated male juncos visited their nests less frequently, were found at greater maximum distances from their nests, and sang more often than controls. The same pattern is seen in "sex-reversed" species in which females may mate with multiple males, and males supply most of the parental care (e.g., Wilson's phalaropes and spotted sandpipers). Male T levels fall during incubation, and T implants increase male sexual behavior and decrease incubation. Other studies have found that androgen treatment caused males to become so aggressive that they attacked females and were unable to keep a mate. One study found that T treatment increased reproductive success. T-treated red grouse gave more frequent flight song displays, had more frequent aggressive interactions, intruded into neighbors' territories more frequently, and engaged in more aerial chases of females. Implanted males expanded their territories at the expense of unimplanted neighbors, had larger broods, and successfully raised more chicks. The authors hypothesized that the T implants were exhausted by the time the chicks hatched so that androgen levels decreased fairly normally and males could show normal paternal behavior. The androgen implants did have a cost: Implanted males were less likely to survive to the following year. A variety of factors may contribute to lowered survival in T-treated birds. High androgen levels may increase metabolic rates, rates of predation, and the risk of injury in aggressive interactions, as well as depress the immune system. Even in species in which high T levels can increase reproductive success, decreased survival may counter any tendency for the evolution of higher androgen levels.

Castration has been shown to reduce or abolish courtship displays and copulatory behavior in a wide variety of species, including ruffs, bronze turkeys, pigeons, doves, chickens, quail, red-winged blackbirds, and zebra finches. In fact, Berthold is usually credited with the first experiment in behavioral endocrinology: He demonstrated in 1849 that castration reduced the sexual behavior of roosters, but their behavior was restored by replacing the testes in the body cavity. An exception to androgenic control of male sexual behavior occurs in white-crowned sparrows, in which mounting behavior appears to be sensitive to seasonal changes in day length but not hormones. However, other behavior patterns necessary to establish territories and attract females are androgen dependent, so it is unlikely that in the wild castrated males would have much opportunity to mount. Additional research is needed to determine if copulatory behavior is independent of gonadal androgens in other avian species.

II. ROLE OF ANDROGEN METABOLISM IN REGULATING REPRODUCTION

Because T treatment reverses the effects of castration, and T is one of the most common androgens secreted by the testes, it was long assumed that T was the hormone controlling male reproductive behavior in birds and other vertebrates. However, research over the past 30 years has demonstrated repeatedly that this aromatizable androgen serves primarily as a prehormone, being metabolized to other hormones to exert its effects. Hormone-dependent tissues, such as the sexual accessory tissues and specific brain areas which regulate male behaviors, selectively concentrate T, which is then metabolized by the tissues to other hormones through the use of endogenous enzymes. The identity of the metabolite(s) formed depends on the enzymes available in the target tissue. In some cases, metabolites are released from target tissues into general circulation. For example, recent studies in male zebra finches revealed that several brain areas metabolize gonadal androgens to estrogens, which are then released into general circulation. In many cases, metabolism is obligatory, and T itself does not stimulate the target tissue.

The three metabolic pathways most relevant to reproduction are (i) aromatization, which provides estrogens, and (ii) 5α and (iii) 5β reduction, which provide other classes of androgens. These three pathways compete for T; T metabolized by one pathway is not available to the other two. One of the most common metabolites of T is 5α-dihydrotestosterone (5α-DHT). 5α-DHT is the primary metabolite of T in sexual accessory tissues, such as combs, wattles, and foam glands, and 5α-DHT rather than T is the active hormone which maintains these tissues. The brain is also a target tissue for gonadal androgens, and comparative studies indicate that avian brains typically have the ability to metabolize T by all three pathways. 5β reduction appears to be an inactivation pathway important in lowering levels of behaviorally active androgens in prepubertal birds, aging males, and in males outside the breeding season. In every avian species examined, both 5α-reduced androgens and estrogens have been implicated in stimulating some portion of the repertoire of male reproductive behaviors. This research has been carried out by (i) castrating males and examining the ability of individual metabolites to restore behavior, (ii) treating males with pharmacological agents to block particular metabolic pathways, or (iii) treating males with antiandrogens and/or antiestrogens.

While both androgenic and estrogenic metabolites have been implicated in controlling male reproductive behavior in every avian species examined, there is significant variation across species in the relative abilities of different hormones to stimulate behavior. This has made it difficult to develop a general model describing the role of androgen metabolism in regulating male reproductive behavior. For example, in doves, pigeons, quail, and chickens, vocalizations which occur during sexual interactions can be stimulated by treatment with androgens alone. In contrast, in those songbird species which have been studied, including canaries, red-winged blackbirds, and zebra finches, male singing can only be stimulated by treatments which provide both androgenic and estrogenic metabolites. Hormonal control of a behavior may depend on context. In doves, cooing in the context of the bow-coo display is stimulated by androgens, but nest-coos are stimulated by estrogens. Different tissues involved in the same behavior may be stimulated by different metabolites. In songbirds, the avian vocal organ, the syrinx, is under purely androgenic control, whereas the functioning of brain areas which control the syrinx requires both androgens and estrogens. Interestingly, patterns of hormone metabolism are labile, and androgen metabolism is affected by a variety of internal and environmental factors, including availability of other hormones metabolized by the same enzymes, changes in photoperiod, social interactions, and age. Additional research is needed to clarify the role of androgen metabolism in modulating reproduction in birds.

III. EFFECTS ON MORPHOLOGY

A. Peripheral Effects

Androgens are important for sperm production and maturation. At the onset of the breeding season, the accessory sexual structures or sperm ducts develop synchronously with the testes. Castration keeps the ducts in their regressed state, and size and function can be restored by androgen therapy. Similarly, specialized glands, such as foam glands in quail and copulatory organs (e.g., penis in ducks and cloacal protuberance in some songbirds), which play an important role in fertilization, are also strongly androgen dependent.

One important effect of androgens is their maintenance of male secondary sexual characteristics—any sexually dimorphic feature not directly involved in gamete production or transport, including body weight, plumage pattern, bill, leg, and skin color, and the size and color of combs, wattles, and spurs. Sexually dimorphic behavior patterns, such as singing, are also included in this category, but were discussed previously. Androgen effects on morphological characteristics are important because they signal an individual's general hormonal status to interested individuals, and such effects have been shown to influence the outcome of aggressive interactions between males as well as female mate choice. For example, in red grouse, comb size varies with a variety of environmental and social variables, including age, season, dominance, and reproductive status. In captive birds, comb size covaries with aggressiveness; dominant males have larger combs than subordinates. Presumably, these differences in comb size reflect differences in circulating androgen levels. In many species, comb size is the best predictor of success in aggressive interactions.

For many other androgen-modulated characteristics, the direction of the effect varies from species to species. For example, androgen treatment increases body weight in some species, such as pigeons, but decreases it in others such as zebra finches. In other cases, androgens stimulate the development of morphological characteristics in both sexes which signal the onset of breeding. In starlings, increasing T levels cause the deposition of carotenoids, changing bill color from black to yellow in both sexes at the onset of breeding. Similarly, plumage patterns are often controlled by androgens, but once again there is great interspecies variation. In several species of gulls, androgens stimulate both sexes to develop adult plumage. In ruffs, androgens stimulate development of male nuptial plumage. In other species, nuptial plumage is controlled genetically or by other hormones, including luteinizing hormone or estrogens.

A recent series of studies has found a very novel way in which androgens may affect avian reproduction. Females deposit varying amounts of androgen in their eggs, affecting posthatching growth and behavior. In canaries, eggs with higher T content produced chicks which grew faster, regardless of sex. Injecting physiological levels of T into the egg yolk was sufficient to enhance chick growth within 22 hr of hatching, and these chicks showed increased levels of begging behavior compared to chicks from sham-treated eggs.

B. Effects on Brain Structure and Function

Androgens have important effects on the structure and function of brain regions controlling male reproductive behavior. Two of the best studied avian examples are the sexually dimorphic nucleus of the preoptic area in Japanese quail and the vocal control system of songbirds. The medial preoptic nucleus (POM) of quail plays an important role in coordinating male sexual behavior. It is larger in males than in females, and its size in males varies with circulating androgen levels, being larger when T levels are high. Exposure to estrogens during normal female development appears to demasculinize the POM. Treating male eggs with estrogen demasculinizes POM, and later androgen treatment does not activate male sexual behavior. POM shows a great deal of plasticity in response to androgen exposure in adulthood. T treatment increases POM's volume, alters concentrations of hormone-metabolizing enzymes, neuropeptides, and neurotransmitters, and modifies the connectivity of neurons in this area. Many of these T-induced changes are related to androgen's ability to stimulate male sexual behavior. Interestingly, POM contains high levels of aromatase, the enzyme which converts androgens to estrogens, and many effects of androgen treatment on POM are elicited through its conversion to estrogenic metabolites. When inhibitors were implanted in POM to block conversion of androgens to estrogens, T treatment did not activate male sexual behavior.

The vocal control system consists of a series of

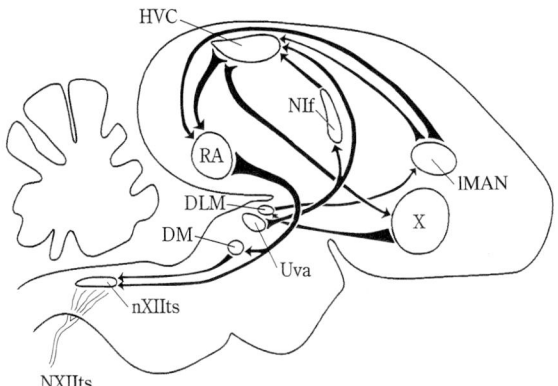

FIGURE 2 A schematic sagittal view of the songbird vocal control system. Singing is controlled by a system of eight discrete, interconnected brain nuclei. This system is divided into the motor loop responsible for song production (Uva → NIf → HVC → RA → nXIIts → syrinx) and the anterior forebrain loop involved in song learning (HVC → X → DLM → lMAN → RA). DLM, medial portion of dorsolateral thalamic nucleus; DM, dorsomedial portion of the intercollicular nucleus; HVC, high vocal center (previously called hyperstriatum ventrale pars caudalis); lMAN, lateral portion of the magnocellular nucleus of the anterior neostriatum; NIf, nucleus interfacialis; nXIIts, tracheosyringeal portion of the hypoglossal or 12th nucleus; NXIIts, tracheosyringeal portion of the 12th nerve which innervates the syrinx, the avian vocal organ; RA, nucleus robustus of the archistriatum; Uva, nucleus uvaeformis; X, area X of the parolfactory lobe.

clearly delineated nuclei which control song learning and production (Fig. 2). It is found only in those birds which learn their songs (the oscine songbirds). The ability to produce learned vocalizations has also evolved in parrots and in hummingbirds, but their vocal control systems are only beginning to be studied. The oscine songbirds account for more than half the extant bird species, and song functions in both intrasexual aggressive interactions and intersexual mate attraction and pair bonding. Not only do females choose males according to the characteristics of their songs but also male songs stimulate ovarian development, hormone secretion, and ultimately female sexual behavior.

The vocal control nuclei (VCN) differ from other recently evolved brain nuclei in their exquisite sensitivity to hormones both during development and in adulthood. In many temperate and subtropical species, the VCN are sexually dimorphic, being larger in males than females, and this dimorphism in brain structure is typically associated with a similar dimorphism in behavior. Zebra finches have one of the most sexually dimorphic vocal control systems. Female zebra finches totally lack area X and their motor pathway is incomplete: high vocal center (HVC) does not project to the nucleus robustus of the archistriatum (RA). Those VCN which they do possess are one-fifth or one-sixth of their size in males. Not surprisingly, female finches do not sing, and no hormone treatment can induce singing in adult females. In contrast, the VCN of female canaries are smaller than those of males, but the system appears complete. T treatment of adult females causes their VCN to increase in size and significantly increases the frequency and complexity of their songs. In tropical species such as bay wrens, the VCN are not sexually dimorphic, and males and females sing at similar rates, often engaging in complex duets.

The development of these dimorphisms is controversial since treating newly hatched females with estrogen induces male-like development of their VCN, and they sing if given aromatizable androgens as adults. However, researchers have not been able to block development of the VCN in males by using antiestrogens or aromatization blockers, calling the role of estrogens in normal male development into question. However, there is absolutely no doubt that the VCN are androgen sensitive in adult males. High levels of androgen receptors have been extensively documented in six VCN (nucleus interfacialis, HVC, RA, the dorsomedial portion of the intercollicular nucleus, the tracheosyringeal portion of the hypoglossal or 12th nucleus, and the lateral portion of the magnocellular nucleus of the anterior neostriatum). Only area X is reported not to contain androgen receptors; there have been no reports of receptors in the nucleus uvaeformis or medial portion of dorsolateral thalamic nucleus. Although estrogenic metabolites are necessary to induce singing in adult males, estrogen receptors are much more limited in the VCN than androgen receptors. Treatment with T not only stimulates high rates of singing but also, in seasonal breeders, it typically stimulates a variety of structural changes in the VCN, including increases in neuron size and increases in the size and complexity of connections between neurons, causing VCN volumes to increase. In canaries, the size of the VCN doubles during the breeding season when T levels are high compared to the fall. Several studies have shown that growth of the VCN depends on the interaction between photoperiod and hormone exposure. Maximal increases in VCN volume are seen when androgens are administered to photosensitive birds exposed to long days. These androgen-induced changes in neural structure are amplified by changes in neurochemistry. Hormone treatments which induce high rates of singing modulate levels of enzymes, hormone receptors, neurotransmitters, peptides, and neurotransmitter receptors in the VCN. These two avian brain systems have become important models for exploring sexually dimorphic circuits and determining how hormones affect brain structure and function.

See Also the Following Articles

AGGRESSIVE BEHAVIOR; ANDROGEN SUBAVIAN SPECIES; AVIAN REPRODUCTION, OVERVIEW; CASTRATION, EFFECTS IN NONHUMANS; MOLT AND NUPTIAL COLOR; SEXUAL IMPRINTING; SONGBIRDS AND SINGING

Bibliography

Allee, W. C., Collias, N. E., and Lutherman, C. Z. (1939). Modification of the social order in flocks of hens by the injection of testosterone propionate. *Physiol. Zool.* **12**, 413–439.

Berthold, A. A. (1849). Transplantation der Hoden. *Arch. Anat. Physiol.* **16**, 42–46.

Bottjer, S. W., and Arnold, A. P. (1997). Developmental plasticity in neural circuits for a learned behavior. *Annu. Rev. Neurosci.* **20**, 459–481.

Harding, C. F. (1983). Hormonal influences on avian aggressive behavior. In *Hormones and Aggressive Behavior* (B. B. Svare, Ed.), pp. 435–468. Plenum, New York.

Harding, C. F. (1986a). The role of androgen metabolism in the activation of male behavior. *Ann. N. Y. Acad. Sci.* **474**, 371–378.

Harding, C. F. (1986b). The importance of androgen metabolism in the regulation of reproductive behavior in the avian male. *Poultry Sci.* **65**, 2344–2351.

Kelley, D. B., and Brenowitz, E. (1992). Hormonal influences on courtship behaviors. In *Behavioral Endocrinology* (J. B. Becker, S. M. Breedlove, and D. Crews, Eds.), pp. 187–218. MIT Press: Cambridge, MA.

Panzica, G. C., Viglietti-Panzica, C., and Balthazart, J. (1996). The sexually dimorphic medial preoptic nucleus of quail: A key brain area mediating steroid action on male sexual behavior. *Frontiers Neuroendocrinol.* **17**, 51–125.

Silver, R., O'Connell, M., and Saad, R. (1979). Effect of androgens on the behavior of birds. In *Endocrine Control of Sexual Behavior* (C. Beyer, Ed.), pp. 223–278. Raven Press, New York.

Wingfield, J. C., and Moore, M. C. (1987). Hormonal, social, and environmental factors in the reproductive biology of free-living male birds. In *Psychobiology of Reproductive Behavior* (D. Crews, Ed.), pp. 149–175. Prentice Hall, Englewood Cliffs, NJ.

Androgens, Effects in Mammals

Shalender Bhasin
Charles R. Drew University of Medicine and Science

I. Testosterone Secretion, Transport, and Metabolism
II. Testosterone as a Prohormone
III. Mechanism of Androgen Action
IV. Testosterone Effects on Reproductive Organs
V. Androgen Effects on Nonreproductive Body Systems
VI. Testosterone Effects on Behavior
VII. Testosterone Replacement in Androgen-Deficient Men
VIII. Anabolic Effects of Androgens in Chronic Illnesses and Age-Associated Frailty

GLOSSARY

anabolic steroids Testosterone derivatives that are claimed to promote nitrogen retention in the body and thereby increase muscle mass.

androgenic steroids Steroid hormones such as testosterone that produce masculinizing effects, such as increased hair growth, male body habitus, deepening of voice, temporal recession of hair, and phallic enlargement.

aromatization The conversion of testosterone or androstenedione (both androgens) to an estrogen by aromatization of the A ring. Some testosterone effects, such as those on the bone and in sexual differentiation of the brain, require aromatization of testosterone to estradiol.

free testosterone The fraction of testosterone in blood that is not bound to binding proteins and is presumably biologically active at the tissue level.

Leydig cells The interstitial cells (that lie in between the seminiferous tubules) of the testis that secrete testosterone.

Müllerian structures The internal organs that develop from the Müllerian duct of the fetus. In the male fetus, Müllerian ducts undergo regression under the influence of Müllerian-inhibiting factor. In the female fetus, the uterus, Fallopian tubes, and the upper third of the vagina are derived from the Müllerian ducts.

5α-reductase An enzyme that converts testosterone to dihydrotestosterone. Two isoforms of the enzyme, types 1 and 2, have been described. 5α-Reduction of testosterone is necessary for mediating some of its effects on the prostate and skin.

STAR protein An acute steroidogenic regulatory protein that makes cholesterol available to the cholesterol side chain complex and regulates testosterone biosynthesis.

Wolffian structures The internal organs that develop from the Wolffian ducts in the fetus. In the female, the Wolffian ducts regress, but in the male, under the influence of testosterone, the epididymis, vas deferens, and seminal vesicles are derived from these ducts.

Testosterone, a 19-carbon steroid secreted by the testis, is the predominant androgen in most mammalian species. Androgens directly or indirectly affect almost all body systems during fetal and pubertal development and in adult life. The sexual differentiation of the mammalian fetus is a complex, multifactorial process that not only requires the genetic information carried on the sex chromosomes but also the hormonal secretions of the fetal testis, namely, testosterone and Müllerian-inhibiting hormone. Testosterone masculinizes the Wolffian structures and causes the external genitalia to form a scrotum and penis. The androgen effects on sexual differentiation are discussed in detail elsewhere in this encyclopedia. In addition, increasing testosterone levels during puberty promote somatic growth and virilization of boys. Testosterone plays a critical role in mammalian reproduction; it is essential for maintaining sexual function, germ cell development, and accessory sex organs. In the adult animal, testosterone has additional effects on the muscle, bone, hematopoeisis, coagulation, plasma lipids, and protein and carbohydrate metabolism.

I. TESTOSTERONE SECRETION, TRANSPORT, AND METABOLISM

A. Testosterone Secretion

In males of most mammalian species, 95% of circulating testosterone is derived from testicular secretion. Only a small amount of dihydrotestosterone ($\approx$70 μg daily) is secreted directly by the human testis; most of the circulating dihydrotestosterone is derived from peripheral conversion of testosterone. Testosterone is produced in the testis by a heterogeneous group of cells that includes the adult Leydig cells, Leydig cell precursors, and immature Leydig cells. In man, 3–10 mg of testosterone is secreted daily by the testis; direct secretion of testosterone by the adrenal, and the peripheral conversion of androstenedione that is secreted by the adrenal and converted to testosterone, collectively account for another 500 μg of testosterone daily. Leydig cells arise from poorly characterized mesenchymal precursor cells under the influence of luteinizing hormone, insulin-like growth factor-1, transforming growth factor-β, transforming growth factor-α, interleukin-1, and basic fibroblast growth factor. Leydig cells exist in two distinct generations in higher mammals, fetal and adult, that are separated in higher mammals by a prepubertal period during which the testis is devoid of Leydig cells.

Testosterone secretion by Leydig cells is under the control of luteinizing hormone (LH), a pituitary glycoprotein hormone. LH binds to specific G-protein-coupled receptors on the Leydig cells and activates the cAMP pathway. Although LH also activates phopholipase C pathway, it is unclear if this pathway is essential for LH-mediated stimulation of testosterone production. The rate-limiting step in testosterone biosynthesis is the delivery of cholesterol to the inner mitochondrial membrane, which is the site of the cholesterol side chain cleavage complex that converts cholesterol to pregnenolone. A steroidogenesis acute regulatory protein (STAR) makes cholesterol available to the cholesterol side chain complex and regulates the rate of testosterone biosynthesis. The peripheral benzodiazepam receptor that is present in a high concentration in the outer mitochondrial membrane has also been proposed as an acute regulator of Leydig cell steroidogenesis. In addition, another group of poorly characterized mitochondrial proteins has been implicated in control of steroidogenesis in Leydig cells.

Leydig cell production of testosterone is modulated by a number of paracrine factors within the seminiferous tubule and the interstitium of the testis, including insulin-like growth factor-1, insulin-like growth factor-binding proteins, inhibins, activins, transforming growth factor-β, epidermal growth factor, interleukin-1, basic fibroblast growth factor, gonadotropin-releasing hormone, and vasopressin.

B. Androgen Transport in the Body

Ninety-eight percent of circulating testosterone is bound to plasma proteins, sex hormone-binding globulin and albumin. Sex hormone-binding globulin binds testosterone with much greater affinity than albumin. Only 1–3% of testosterone is unbound and biologically active, although others have argued that albumin-bound hormone may dissociate readily in the capillaries and thus become bioavailable. The binding of testosterone to these plasma proteins is not essential for androgen action or steroid homeostasis; rats that are deficient in both sex hormone-binding globulin and albumin are fertile and have normal mating behavior.

C. Testosterone Metabolism

Testosterone is metabolized predominantly in the liver (50–70%) although some degradation also occurs in peripheral tissues, particularly the prostate and the skin. Liver takes up testosterone from the blood and, through a series of chemical reactions that involve 5α- and 5β-reductases, 3α- and 3β-hydroxysteroid dehydrogenases, and 17β-hydroxysteroid dehydrogenase, converts it into androsterone and etiocholanolone (both inactive metabolites) and dihydrottestosterone and 3α-androstanediol. These compounds undergo glucoronidation or sulfation before their excretion by the kidneys. Free and conjugated androsterone and etiocholanaolone are the predominant urinary metabolites of testosterone.

II. TESTOSTERONE AS A PROHORMONE

Testosterone can be converted in many peripheral tissues into its active metabolites. Aromatization of the A ring converts it into 17β-estradiol. In addition, reduction of the δ-4 double bond can convert testosterone into 5α-dihydrotestosterone. Testosterone actions in many tissues are mediated through these metabolites. For instance, testosterone effects on the trabecular bone and the sexual differentiation of the brain require its aromatization to estradiol. Similarly, testosterone effects on the prostate and sebaceous glands of the skin require its 5α-reduction to dihydrotestosterone. During embryonic life, testosterone controls the differentiation of the Wolffian ducts into epididymis, vasa deferentia, and seminal vesicles. The development of structures from the urogenital sinus and the genital tubercle, such as the scrotum, penis, and penile urethra, require the action of dihydrotestoserone (DHT). Although, the enzyme 5α-reductase is expressed in the muscle at low levels, it is generally believed that conversion of testosterone to dihydrotestosterone is not obligatory for mediating its anabolic effects on the muscle. Two types of 5α-reductase enzymes have been cloned. Type 1 isoenzyme is expressed in many nongenital tissues, has been mapped to chromosome region 5p15, and has a pH optimum of 8. Type 2 isoenzyme is expressed in the prostate and other genital tissues, has been mapped to 2p23, and has a pH optimum of 5.0.

III. MECHANISM OF ANDROGEN ACTION

Most androgen actions are mediated through its binding to an intracellular androgen receptor that acts as a ligand-dependent transcription factor. The androgen receptor has homology to other nuclear receptor proteins, including the receptors for glucocorticoids, progesterone, and mineralocorticoids. The predominant 919-amino acid, 110- to 114-kDa androgen receptor protein has three conserved functional domains: the steroid-binding domain, the DNA-binding domain, and the transcriptional-activational domain; of these, the central, cysteine-rich, DNA-binding domain is the most conserved. The single copy androgen receptor gene spans a 90-kb region on chromosomal region Xq11–12. In the absence of its ligand, the androgen receptor protein is distributed both in the nucleus and in the cytoplasm. However, androgen binding to the receptor causes it to translocate within the nucleus; amino acid sequences between 617 and 633 of the androgen receptor are important for its nuclear migration and *trans*-activation function. There is emerging evidence that some androgen effects may be mediated through nongenomic receptors on the cell membrane.

The mutations of the androgen receptor gene have been associated with a wide spectrum of phenotypic abnormalities. Some patients with complete androgen insensitivity present with male pseudohermaphroditism, characterized by female external genitalia, a blind vaginal pouch, and well-developed breasts. Other patients with androgen receptor mutations may have a male phenotype and milder abnormalities, such as hypospadias, gynecomastia, and infertility. An abnormal length of the polyglutamine repeat in exon 1 of the androgen receptor has been associated with spinal and bulbar muscular atrophy, also known as Kennedy's disease.

IV. TESTOSTERONE EFFECTS ON REPRODUCTIVE ORGANS

A. Testosterone and the Prostate

Androgens are required for normal prostate development and play some permissive role in the initiation of prostate neoplasms. Conversion of testosterone to dihydrotestosterone is obligatory for mediating its effects on the prostate. The prostate size increases during the peripubertal period in parallel with the rise in serum testosterone levels. Androgen imprinting prior to puberty is an important determinant of the number of stem cells within the prostate and hence the adult prostate size in rats and dogs. Castration before puberty, 5α-reductase deficiency, and inactivating mutations of the androgen receptor prevent subsequent development of benign prostatic hypertrophy. Many of the androgen effects on the prostatic epithelium are mediated through the stromal cells; thus, the interaction between the stromal and epithelial cells is important for normal prostatic development. Androgen effects on prostatic growth are mediated in part through regulation of apoptotic genes. Androgen deficiency in adult males is associated with increased expression of many apoptotic genes resulting in involution of the prostate.

Only the human and the dog develop benign prostatic hypertrophy. Androgens are essential, but not sufficient, for the development of benign prostatic hypertrophy. Abnormal stromal activation and dysregulation of several growth factors play a role in the pathogenesis of this complex disorder. Although androstanediol and dihydrotestosterone can induce prostatic enlargement in the dog, physiologic testosterone administration to hypogonadal men has not been associated with pathologic prostatic enlargement beyond levels observed in healthy, age-matched controls. Estradiol accentuates prostatic hypertrophy induced by androgen administration in the dog. The critical role of androgens in prostatic growth is supported by clinical experience that lowering serum testosterone levels by surgical orchiectomy or by pharmacologic inhibitors of testosterone production decreases prostate volume in men with benign prostatic hyperplasia. Similarly, clinical trials have established that finasteride, an inhibitor of 5α-reductase enzyme, produces clinically significant reduction in prostate size in men with benign prostatic hypertrophy.

B. Testosterone and Spermatogenesis

High intratesticular testosterone concentrations are required for initiation and maintenance of spermatogenesis. Gonadotropin deficiency resulting from hypophysectomy or administration of a gonadotropin-releasing hormone (GnRH) antagonist is associated with marked depletion of intratesticular testosterone concentration and azoospermia in the rat, monkey, and the human male. High doses of testosterone can reinitiate and maintain spermatogenesis in rats made gonadotropin deficient by hypophysectomy or GnRH antagonist treatment. In men with acquired gonadotropin deficiency, LH can reinitiate and maintain spermatogenesis by stimulating testosterone production within the testis. The mechanism by which testosterone maintains germ cell development is not well understood. Testosterone receptors are present on the Sertoli and peritubular cells, some Leydig cells, and endothelial cells of the small arterioles. However, it has not been established whether androgen receptors are also present on germ cells. It is generally believed that androgen effects on spermatogenesis are mediated indirectly through Sertoli cells; therefore, it is surprising that testosterone increases protein secretion by round spermatids but not Sertoli cells. The expression of androgen receptors is maximal in stages VI and VII of the seminiferous epithelium; testosterone effects might be exerted on the germ cells as they pass through these stages.

C. Testosterone Effects on the Epididymis

Androgens are important regulators of epididymal structure and function. A variety of epididymal biochemical processes, such as transport of small molecules across the epididymal epithelium, DNA and RNA synthesis, protein synthesis, the activity of a number of epididymal enzymes, and maturation and storage of spermatozoa, are under androgen control. The regulation of epididymal function by androgens is highly region specific. Some genes such as proenkephalin are downregulated after castration but are not restored by androgen replacement, whereas expression of other genes such as E-cadherin is decreased by castration in a region-specific manner and restored to varying degrees by androgen administration.

V. ANDROGEN EFFECTS ON NONREPRODUCTIVE BODY SYSTEMS

A. Androgen Effects on the Bone

Androgens are important regulators of bone maturation and maintenance of bone mass. Androgens promote linear bone growth and endochondral ossification in rats. In humans, the peripubertal bone accretion is androgen dependent; consequently, men with androgen deficiency during the peripubertal period have diminished bone mass. Adult men with testosterone deficiency have lower bone density than age-matched, healthy controls; testosterone replacement therapy of hypogonadal men increases but does not normalize bone density. Suppression of serum testosterone levels by administration of GnRH agonist analogs or surgical orchiectomy leads to progressive decrease in bone density. Testosterone effects on the bone are mediated in part through its conversion to estradiol. The critical role of aromatization in mediating androgen effects on the bone is supported by observations that 46,XY men with mutations of the genes that code for the estrogen receptor or the aromatase enzyme have delayed epiphyseal fusion and decreased bone density. Testosterone may have additional direct effects in stimulating cortical bone formation. Androgen receptors have been demonstrated on the osteoblasts.

B. Androgen and Body Composition-Effects on Muscle and Fat Metabolism

Testosterone is a major determinant of body composition in mammalian males. The sexual dimorphism of several androgen-responsive muscles such as masseter and levator ani in several mammalian species has been attributed to differences in androgen levels in males and females. Testosterone administration increases nitrogen retention in castrated males of several mammalian species, hypogonadal men, women, and boys before puberty. Recent studies have established that testosterone replacement of hypogonadal men increases fat-free mass. Supraphysiologic doses of androgens further increase muscle mass and strength in eugonadal men, especially when given in association with resistance exercise. Widespread abuse of androgenic steroids by athletes and recreational bodybuilders is based on the premise that androgens promote muscle hypertrophy. Testosterone effects on muscle performance have not been well studied, although the available evidence indicates that testosterone does not improve aerobic performance. The published studies do not agree on whether testosterone administration decreases fat mass in hypogonadal men. Androgen-deficient men have a greater amount of fat than eugonadal men. In middle-aged men, serum testosterone levels correlate inversely with visceral fat and cardiovascular risk. Physiologic testosterone replacement of middle-aged men with midsegment obesity decreases visceral fat, glucose, and insulin levels. Androgen effects on body composition in men are modulated by pretreatment body composition, genetic and nutritional factors, the growth hormone secretory status, existence of comorbid conditions, cytokines, and exercise status.

C. Testosterone Effects on Intermediary Metabolism

Testosterone is an anabolic hormone that increases muscle mass by stimulating fractional muscle protein synthesis. The mechanism by which testosterone increases protein synthesis is not known. It has been proposed that testosterone increases muscle protein synthesis by increasing the expression of insulin-like growth factor-1 gene within the muscle. We do not know if testosterone affects muscle protein degradation. Androgen receptors are present on adipocytes and testosterone stimulates lipolysis in some experimental models. The effects of testosterone on lipid uptake and metabolism are region specific. Cross-sectional studies of middle-aged men demonstrate a direct correlation of serum testosterone levels with insulin sensitivity and visceral fat. However, physiologic replacement or slightly supraphysiologic doses of testosterone have not been shown to affect glucose tolerance or insulin sensitivity in young men.

D. Androgen Effects on the Skin

The hair follicles, sebaceous glands, and the apocrine sweat glands in the skin are androgen-sensitive structures. Increasing testosterone levels during pubertal development stimulate the growth of terminal hair in the axillary and pubic areas in both men and women and on the face of men. Excessive androgen production or action has been implicated in the pathophysiology of clinical disorders such as acne, hirsutism, androgenic alopecia, and suppurative hidradenitis. Many investigators believe that local abnormalities in androgen metabolism within the skin in hirsute women make their skin more sensitive to the effects of circulating androgens. Testosterone effects on the skin are mediated through its conversion to dihydrotestosterone. Type 1 isoenzyme is the major form of 5α-reductase in the skin. The highest activity of the 5α-reductase enzyme has been located in the apocrine and the sebaceous glands. Inhibitors of type 1 5α-reductase enzyme in the skin, such as aliphatic unsaturated fatty acids, are being explored as therapeutic agents for the treatment of androgen-responsive skin disorders such as hirsutism and acne. Finasteride, a weak inhibitor of type 1 isoenzyme, was recently approved by the Food and Drug Administration for the treatment of androgenic alopecia.

E. Androgen Effects on Hemopoiesis, Coagulation System, and the Vascular System

Testosterone increases red cell mass; this effect is mediated primarily by increasing the production of erythroid precursors from the pluripotent stem cells in the bone marrow. Testosterone has also been reported to increase erythropoeitin production in the kidney, although other data show that testosterone can increase red cell production independent of its effects on erythropoietin. There are numerous anecdotal reports of sudden death, vascular thrombosis, myocardial infarctions, and cardiomyopathy in relatively young athletes abusing large quantities of anabolic steroids, leading to speculation that testosterone increases the thrombogenicity of plasma. However, the effects of testosterone on the coagulation system remain controversial. Fibrinogen and plasminogen activator inhibitor concentrations in plasma correlate inversely with endogenous testosterone. These changes may be viewed as favoring plasma thrombogenicity. However, androgens also increase tissue plasminogen activator, protein C, and antithrombin III levels; these factors protect the body from vascular thrombosis. Testosterone and dihydrotestosterone increase platelet aggregation at the site of endothelial trauma. Danazol, a weak synthetic androgen, has been reported to increase platelet count. Androgens may increase vascular reactivity to trauma and decrease aortic smooth muscle prostaglandin I_2. Therefore, some effects of androgens appear to favor hypercoagulability and others appear to oppose it. The net effect in a person may vary with the type and dose of androgen used, the genetic predisposition, simultaneous use of other medications, and the existence of comorbid conditions.

F. Androgens and Plasma Lipids

The effects of androgens on plasma lipids depend on the dose (physiologic or supraphysiologic), the route of administration (oral or sytemic), and the type of androgen used (aromatizable or not aromatizable). Supraphysiologic doses of testosterone and other androgenic steroids decrease plasma high-density lipoprotein (HDL) levels in men. Androgenic steroids that can be aromatized produce a smaller decrease in HDL levels than those that cannot be aromatized. Lowering serum testosterone levels experimentally by GnRH antagonist administration increases plasma HDL levels; conversely, testosterone replacement of hypogonadal men modestly lowers plasma HDL levels. Other studies using physiologic replacement doses in older men or HIV-infected men have reported either no change or only minor changes in plasma lipid profile. However, cross-sectional epidemiological studies demonstrate a direct relationship between serum testosterone levels and HDL levels and an inverse relationship between testosterone levels and visceral fat. Therefore, these data suggest that testosterone levels in the mid- to high-normal range may be optimum for cardiovascular health in men. Indeed, physiologic testosterone replacement decreases visceral fat, glucose, and insulin levels in middle-aged men with visceral obesity. We do not know what testosterone levels are consistent with optimum cardiovascular risk in men.

VI. TESTOSTERONE EFFECTS ON BEHAVIOR

A. Androgen Effects on Sexual Function

Testosterone regulates many aspects of sexual function in male mammals. Observations that eunuchs and hypogonadal men can have erections suggest that testosterone regulates libido rather than erectile function. It has been demonstrated that although overall sexual activity is reduced in hypogonadal men, these patients have normal erectile response to visual erotic stimuli. Testosterone is necessary but not sufficient for maintaining normal sexual desire. Spontaneous erections as measured by nocturnal penile tumescence are androgen dependent, but erections in response to visual erotic stimuli are less dependent on androgens. Ejaculation of seminal fluid is also sensitive to androgens. However, it is not known whether orgasmic capacity is androgen dependent.

In rats, mating behavior can be maintained at serum testosterone levels that are at the lower end of the normal male range. Similarly, in men in whom endogenous testosterone secretion has been suppressed by a GnRH agonist, low doses of testosterone that bring serum testosterone levels in the lower end of the normal range can maintain sexual desire and activity and nocturnal erections. However, supraphysiologic doses of testosterone can further increase some aspects of sexual arousability.

Although dihydrotestosterone can restore sexual behavior in castrated monkeys, patients with 5α-reductase deficiency have normal libido and erectile function. Furthermore, the relatively low frequency of sexual dysfunction in men treated with finasteride, an inhibitor of 5α-reductase enzyme, suggests that testosterone conversion to dihydrotestosterone is not required for androgen effects on sexual function.

B. Testosterone Effects on Nonreproductive Behaviors

During fetal life, testosterone is essential for sexual differentiation of the brain along male lines. There are structural sex differences in the volume and the synaptic organization of many brain nuclei or regions. Testosterone promotes aggression among males, particularly at the time of mate selection. The scent-marking behavior, a marker for territoriality, is testosterone dependent in many mammalian males. The effects of testosterone on human aggression remain controversial. There is a significant though not a strong correlation between serum testosterone levels and some aspects of aggressive behavior in men. A significant proportion of anabolic steroid users report mood disorders, manic and hypomanic syndromes characterized by aggressiveness, and other psychiatric disorders. Administration of high doses of testosterone to men is associated with increased aggressive responding compared to placebo. However, other well-controlled studies have failed to find clear increases in aggressive behaviors in androgen-treated men. Testosterone affects the development of sex-typed nonreproductive behaviors such as maze learning and juvenile play. Androgens are believed to be responsible for the higher visuospatial ability and lower verbal ability in men than in women. Exposure to testosterone through its conversion to estradiol during the neonatal period is responsible for the development of brain structures that subserve spatial functions. Alterations of androgen levels in adult animals do not affect spatial processing but may influence memory storage and affective properties.

VII. TESTOSTERONE REPLACEMENT IN ANDROGEN-DEFICIENT MEN

Testosterone, when administered by mouth, is absorbed readily from the gastrointestinal tract; however, most of it is destroyed during its first pass through the liver. Therefore, an appropriate delivery system is needed to achieve sustained levels of testosterone in the blood. Because of the potential contraceptive and anabolic applications of androgens, there has been considerable interest in developing more physiologic and long-acting testosterone delivery systems.

A. Testosterone Esters

Esterification of the 17β-hydroxyl group makes the compound more hydrophobic. Therefore, when these esters are injected in an oil suspension in the muscle, they are absorbed very slowly from the muscle depot into the bloodstream. It is the slow release rather the slow deesterification that accounts for the extended duration of testosterone esters. The longer the side chain, the more hydrophobic the compound and the longer the duration of action. Testosterone enanthate and cypionate, two commonly used esters, have identical kinetics. After injection of 200 mg testosterone enanthate or cypionate, the usual replacement dose in hypogonadal men, serum testosterone levels rise into the supraphysiologic range within 24–48 hr and then gradually decline into the hypogonadal range 10–17 days later. These fluctuations in serum testosterone levels are associated with changes in the patient's mood and energy level and are a major drawback of this formulation. Variations in serum testosterone levels during treatment with testosterone esters can be minimized by giving 100 mg every week. With physiologic doses, serum levels of estradiol and DHT are normal. Given an adequate dose, testosterone esters can produce virilization, restore sexual function, and increase muscle and bone mass and hematocrit in hypogonadal men.

B. Testosterone Transdermal Systems

Three transdermal testosterone systems are commercially available to treat hypogonadal men: a scrotal testosterone patch (Testoderm, ALZA Corp., Palo Alto, CA) and two nongenital patches (Androderm, Smithkline Beecham Pharmaceuticals, Collegeville, PA; and Testoderm TTS, ALZA Corp., Palo Alto, CA). Scrotal patches deliver either 4 or 6 mg/day of testosterone depending on their size. Serum testosterone levels in hypogonadal men peak in the midnormal range 4–8 hr after application of the patch and then decline through the day in a pattern similar to the circadian pattern of normal young men. Hypogonadal men treated with the scrotal patch have normal serum estradiol levels but much higher DHT levels compared to untreated eugonadal men. This is likely the result of the high activity level of 5α-reductase in the scrotal skin. It is not known whether long-term exposure to high serum DHT levels will have deleterious effects on the prostate and whether intraprostatic DHT levels are also elevated. The nongenital transdermal systems, Androderm and Testoderm TTS, can maintain physiologic levels of testosterone in hypogonadal men. Four to 12 hr after application of the patches, serum testosterone and estradiol levels are in the midnormal range. The nongenital patches also produces physiologic levels of serum DHT and normal DHT to testosterone ratios.

C. Oral Testosterone Formulations

17α-Alkylation of testosterone makes it less susceptible to hepatic degradation. However, these 17α-alkylated androgens should not be used because of the potential for hepatotoxicity. Testosterone undecanoate, an oral form of testosterone esterified to oleic acid, is absorbed preferentially through the intestinal lymphatics into the lymphatic duct and general circulation and is therefore spared the first-pass degradation in the liver. Its short half-life necessitates dosing two or three times per day for clinical effects. Circulating levels of testosterone vary among subjects receiving the same dose of this formulation, overnight levels are low, and clinical response is not as good as it is with other forms of testosterone. Sublingual preparation of cyclodextrin-complexed testosterone is rapidly absorbed from the sublingual mucosa into the bloodstream, leading to rapid peaks in serum testosterone levels followed by a decline within 2 hr. Dosing several times a day is necessary to maintain normal serum levels.

D. Novel Androgen Formulations under Development

Interest in developing more physiologic, sustained-release testosterone formulations has increased due to the potential application of testosterone as an anabolic agent and for male contraception. With long-acting injectable testosterone preparations (testosterone buciclate, 600 mg intramuscular every 3 or 4 months) serum levels peak at 6 weeks and can remain in the normal range for 12 weeks. Testosterone pellets (three 200-mg pellets or six 100-mg pellets) are implanted under the skin and can provide normal testosterone levels as well as physiologic levels of estradiol and DHT for up to 6 months. A single intramuscular injection of a biodegradable testosterone microsphere formulation produces normal levels of testosterone in hypogonadal men for up to 11 weeks; serum estradiol and DHT levels are maintained in the normal range.

VIII. ANABOLIC EFFECTS OF ANDROGENS IN CHRONIC ILLNESSES AND AGE-ASSOCIATED FRAILTY

Many chronic illnesses, such as cancer, HIV infection, chronic obstructive lung disease, and end-stage renal disease, are associated with substantial loss of muscle mass and function. Although we cannot cure these diseases, disease stability can often be achieved. Muscle wasting in these illnesses produces debility and is associated with poor quality of life, adverse disease outcomes, and increased utilization of health care resources. There is a high prevalence of low testosterone levels in these sarcopenic disorders. Furthermore, low testosterone levels correlate with loss of muscle mass and exercise capacity. Because testosterone replacement of hypogonadal men increases lean body mass and muscle strength, it has been speculated that testosterone supplementation of men with chronic illnesses such as HIV infection will have similar anabolic effects. Clinical trials have shown that androgen administration of HIV-infected men is associated with modest increases in fat-free mass, hematocrit, and some aspects of health-related quality of life. However, further studies are needed to determine if testosterone replacement can produce clinically meaningful changes in muscle function or disease outcomes in patients with chronic illnesses.

There is agreement that total and free testosterone levels and testosterone production rates are lower in older men compared to young healthy men. Some studies which exclusively recruited healthy, middle-income older men failed to detect significant differences in serum testosterone levels between old and young men. Recent studies with more representative population samples have reconfirmed the age-related decrease in serum testosterone levels. Interpretation of serum testosterone levels in the older men in these studies has been complicated by several confounding factors: the cross-sectional nature of these studies, the loss of diurnal variation in serum testosterone levels in the elderly so that studies that obtain samples in the afternoon underestimate the age-related changes in testosterone levels, and the higher sex hormone-binding globulin levels in older men so that total testosterone levels underestimate the greater decline in free testosterone levels (both dialysable and bioavailable). A meta-analysis of 44 studies of testosterone levels in older men demonstrated an unequivocal decrease in morning testosterone levels. Recent longitudinal studies of normal men have verified the age-related decline in serum testosterone levels. Lower testosterone levels are the result of changes at multiple levels of the hypothalamic–pituitary–gonadal axis. Testicular response to gonadotropins is diminished in older men, gonadotrope responsiveness to androgen suppression is attenuated, and the pulsatility of the hypothalamic GnRH pulse generator is altered. Coexisting diseases, malnutrition, and concomitant medications can also affect serum testosterone levels. A number of clinical problems prevalent in older men may be related to androgen deficiency, including sexual dysfunction, muscle weakness and wasting, changes in body composition, osteopenia and increased prevalence of hip and vertebral fractures, decreased body hair, decreased hematopoiesis, and memory loss. It has been speculated that androgen replacement may help prevent or reverse these disorders. A few short-term studies have provided preliminary evidence of modest gains in muscle strength and lean body mass with replacement doses of testosterone and long-term clinical studies to test this hypothesis are now in progress.

See Also the Following Articles

Androgens, Effects in Birds; Androgens, Subavian Species

Bibliography

Archer, J. (1991). The influence of testosterone on human aggression. *Br. J. Psychol.* **82**, 1–24.

Barrett-Connors, E., and Khaw, K. T. (1988). Endogenous sex hormones and cardiovascular disease in men: A prospective population-based study. *Circulation* **78**, 539–545.

Bhasin, S., and Bremner, W. J. (1997). Emerging issues in androgen replacement therapy. *J. Clin. Endocrinol. Metab.* **82**, 3–8.

Bhasin, S., Storer, T. W., Berman, N., Callegari, C., Clevenger, B. A., Phillips, J., Bunnell, T., Tricker, R., Shirazi, A., and Casaburi, R. (1996a). The effects of supraphysiologic doses of testosterone on muscle size and strength in men. *N. Engl. J. Med.* **335**, 1–7.

Bhasin, S., Gabelnick, H. L., Spieler, J. M., Swerdloff, R. S., and Wang, C. (Eds.) (1996b). *Pharmacology, Biology, and Clinical Applications of Androgens: Current Status and Future Prospects.* Wiley-Liss, New York.

De Kretser, D. M. (1987). Local regulation of testicular function. *Int. Rev. Cytol.* **10**, 89–112.

Finkelstein, J. S., Klibanski, A., Neer, R. M., Greenspan, S. L., Rosenthal, D. I., and Crowley, W. F. (1987). Osteoporosis in men with idiopathic hypogonadotropic hypogonadism. *Ann. Intern. Med.* **106**, 354–361.

Jenster, G., van der Korput, H. A. G. M., van Vroonhoven, C., van der Kwast, T. H., Trapman, J., and Brinkman, A. O. (1991). Domains of the human androgen receptor involved in steroid binding, transcriptional activation, and subcellular localization. *Mol. Endocrinol.* **5**, 1396–1404.

Kasperk, C. H., Wergedel, J. E., Farley, J. R., Linkhart, T. A., Turner, R. T., and Baylink, D. J. (1989). Androgens directly stimulate proliferation of bone cells in vitro. *Endocrinology* **124**, 1576–1578.

Kerrigan, J. R., and Rogol, A. D. (1992). The impact of gonadal steroid hormone action on growth hormone secretion during childhood and adolescence. *Endocr. Rev.* **13**, 281–298.

Kwan, M., Greenleaf, W. J., Mann, J., Crapo, L., and Davidson, J. M. (1983). The nature of androgen action on male sexuality: A combined laboratory self-report study in hypogonadal men. *J. Clin. Endocrinol. Metab.* **57**, 557–562.

Lubahn, D. R., Brown, T. R., Simental, J. A., Higgs, H. N., Migeon, C. J., Wilson, E. M., and French, F. S. (1989). Sequence of the intron/exon junctions of the coding region of the human androgen receptor gene and identification of a point mutation in a family with complete androgen insensitivity. *Proc. Natl. Acad. Sci. USA* **86**, 9534–9538.

Marin, P., Holmang, S., Jonsson, L., Sjostrom, L., Kuist, H., Holm, G., Lindstedt, G., and Bjorntorp, P. (1992). The effect of testosterone treatment on body composition and metabolism in middle aged obese men. *Int. J. Obesity* **16**, 991–997.

Mooradian, A. D., Morley, J. E., and Korenman, S. G. (1987). Biological actions of androgens. *Endocr. Rev.* **8**, 1–28.

Nieschlag, E., and Behre, H. M. (Eds.) (1990). *Testosterone: Action, Deficiency, Substitution.* Springer-Verlag, Heidelberg.

Oesterling, J. E. (1995). Benign prostatic hyperplasia. *N. Engl. J. Med.* **332**, 99–109.

Pope, H. G., and Katz, D. L. (1994). Psychiatric and medical effects of anabolic–androgenic steroid use. A controlled study of 160 athletes. *Arch. Gen. Psych.* **51**, 375–382.

Russell, D. W., and Wilson, J. D. (1994). Steroid 5-alpha reductase: Two gene/two enzymes. *Annu. Rev. Biochem.* **63**, 25–61.

Skinner, M. K. (1991). Cell–cell interactions in the testis. *Endocr. Rev.* **12**, 45–77.

Smith, E. P., Boyd, J., Frank, G. R., Takahashi, H., Cohen, R. M., Specker, B., Williams, T. C., Lubahn, D. B., and Korach, K. S. (1994). Estrogen resistance caused by a mutation in the estrogen receptor gene in a man. *N. Engl. J. Med.* **331**, 1056–1061.

Stoner, E. (1994). 5-Alpha-reductase inhibitors for the treatment of benign prostatic hyperplasia. *Rec. Prog. Horm. Res.* **49**, 285–292.

Wilson, J. D. (1988). Androgen abuse by athletes. *Endocr. Rev.* **9**, 181–200.

Zhou, Z.-X., Wong, C.-I., Sar, M., and Wilson, E. M. (1994). The androgen receptor: An overview. *Rec. Prog. Horm. Res.* **49**, 249–274.

Androgens, Subavian Species

Kyle W. Selcer and Jeffrey W. Clemens

Duquesne University

I. Introduction
II. Chemical Features of Androgens
III. Androgen Synthesis
IV. Metabolism and Degradation
V. Types of Androgens in Subavian Species
VI. Androgen Cycles in Subavian Species
VII. Androgen Effects
VIII. Mechanism of Androgen Action

GLOSSARY

aromatization A process mediated by the enzyme cytochrome P450 aromatase that converts the androgens testosterone and androstenedione to the estrogens 17β-estradiol and estrone, respectively.

Leydig cells Steroidogenic cells found in the interstitial spaces between the seminiferous tubules of the testis; also called *interstitial cells*.

nuclear receptor superfamily An evolutionarily related group of ligand-activated transcription factors that includes the steroid hormone receptors, vitamin D receptors, and retinoic acid receptors.

Sertoli cells Cells found in the seminiferous tubules of the testis that support the development of male germ cells; also called *nurse cells*, *sustentacular cells*, or *somatic cells*.

zinc finger Highly conserved amino acid sequences containing four cysteines that complex with Zn^{2+}; present in the central DNA-binding domain of nuclear receptors where they mediate the specific, high-affinity binding of nuclear receptors to target genes.

Androgens play an important role in male reproduction throughout the vertebrate classes. These steroid hormones are responsible for the growth and development of the male reproductive tract, are involved in sperm production, and are important in development of male secondary sexual characteristics and regulation of male behavior. Androgens are also the precursors to estrogens, which are sex steroid hormones involved in regulation of female reproduction. There are indications that androgens may also function directly in regulation of female reproduction in subavian vertebrates. Circulating androgens of subavian vertebrates show considerable diversity among species and vertebrate classes. For example, testosterone, dihydrotestosterone, and 11-ketotestosterone represent the major androgens present in different vertebrate groups. Androgen target tissues of subavian vertebrates include not only the reproductive tract and brain but also sexually dimorphic tissues such as the larynx and thumb pads of frogs. Androgens elicit their responses primarily through an intracellular androgen receptor that is localized to target tissues. Activation of androgen receptor by the steroid hormone causes alterations in the rate of transcription of hormone-dependent genes and ultimately changes in protein synthesis and secretion. There appears to be significant conservation of the androgen receptor structure among the vertebrate classes.

I. INTRODUCTION

Vertebrates often show striking differences in appearance between males and females. While the sexes obviously differ in their reproductive tracts and external genitalia, males and females may also be different sizes, have different coloration, and may behave in completely different manners. Male vertebrates commonly have certain physical features that are sex specific, such as the enlarged comb in roosters, the throat pouch and thumb pads of frogs, and the elongated claws of slider turtles. Males of many species also show seasonal cycles of coloration and behavior that correspond with the breeding season. The observed differences between males and females are for the most part determined by differences in steroid hormone secretions from the gonads. For males, androgens are the most important steroid hormones in determining the phenotypes and behaviors that are characteristic of their sex.

II. CHEMICAL FEATURES OF ANDROGENS

Androgens possess the four-ring chemical structure (cyclopentanoperhydrophenathrene nucleus) characteristic of cholesterol and its steroid derivatives. They are known as C19 steroids for the 19 carbon atoms that make up their skeleton. The specific properties of a given androgen are determined by the substituent groups attached to the carbon skeleton and by the number of bonds between adjacent carbon atoms.

Approximately a dozen androgens have been reported in a variety of vertebrate species, and these vary widely in their androgenic potency. Common features of the more potent androgens (e.g., testosterone, dihydrotestosterone, and 11-ketotestosterone) are a ketone group at carbon 3 and a hydroxyl group at carbon 17. Additional modifications of androgen structure include other hydroxyl or ketone side groups and conjugation with sulfate or sugar moieties. These modifications alter the properties of the molecule, including the water solubility of the steroid and the ability of the androgen to interact with the androgen receptor and other steroid-binding proteins.

III. ANDROGEN SYNTHESIS

Androgens are synthesized in steroidogenic tissues, including the adrenal glands of males and females, as well as the testis and the ovary. Testicular androgens have a primary role in regulation of male reproduction; however, the function of adrenal and ovarian androgens is less clear. In male mammals, the Leydig cells of the testis are the only significant source of gonadal androgens; however, in nonmammalian vertebrates, the Sertoli cells may also contribute to androgen production.

Androgens are derived from cholesterol, which is converted to pregnenolone through a process called side chain cleavage. Pregnenolone is then converted to various other steroid hormones through a series of enzymatic reactions. There are two alternative pathways by which androgens can be synthesized, called the 4-ene and 5-ene (or $\Delta 4$ and $\Delta 5$) pathways. These pathways lead to testosterone and androstenedione, respectively, which are interconvertible. The relative importance of the two pathways of androgen synthesis differs markedly between tissues and between species. There may also be seasonal differences in relative amounts of androgen synthesized by the two pathways.

IV. METABOLISM AND DEGRADATION

Testosterone appears to be synthesized by the testis of most, if not all, vertebrates. In some species, testosterone is the major androgen produced by the testis and is also the most potent androgen. However, in other species, testosterone is modified to androgens which may be more abundant or more potent. The conversion of testosterone to other forms may take place in either steroidogenic or peripheral tissues. One important metabolite of testosterone in many species is 5α-dihydrotestosterone (DHT), which is formed by the action of the enzyme 5α-reductase. DHT is often more potent than testosterone at eliciting androgenic responses. In birds and mammals, DHT is synthesized primarily in peripheral tissues and does not reach high levels in the circulation. In some subavian vertebrates, including frogs and toads, DHT is produced by the testis and is present in the bloodstream. Other important metabolites of testosterone in some vertebrates are 11β-hydroxytestosterone and 11β-ketotestosterone. These testicular products may be significantly more potent than testosterone.

Androgens are also metabolized to estrogens. Testosterone and androstenedione can be converted to 17β-estradiol and estrone, respectively, by the enzyme aromatase. Aromatase is commonly found in peripheral tissues but may also be present in the testis. Indeed, the testis of a few species (e.g., the mudpuppy) produce substantial quantities of estrogen. Some of the more potent androgens cannot be aromatized to estrogens, including DHT and 11-ketotestosterone.

Testosterone and other androgens may also be metabolized to less active forms in both steroidogenic and peripheral tissues. Such conversions often make the steroid more readily removed from circulation. Typical modifications involved in androgen degradation are hydroxylations at various carbons (1, 2, 6, 7, 15, and 16), reductions of the double bond, and reductions of ketones at carbon 3. These degradative reactions usually occur in the liver.

Another type of androgen modification is the conjugation of the steroid to sulfate or glucuronide moieties. These conversions render the steroid more water soluble and presumably more easily filtered by the kidneys. Steroid conjugations occur almost exclusively in the liver of mammals and are usually considered mechanisms to remove steroids from the body. However, steroid conjugation occurs in both steroidogenic and peripheral tissues of nonmammalian vertebrates, and there are indications of physiological roles for conjugated steroids in some species. For example, glucuronides (e.g., etiocholanolone glucuronide) may serve as pheromones in certain teleost fishes. Since they are water soluble, these glucuronides can be readily transmitted through the water. In other species, conjugated steroids may be involved in regulation of steroid availability. Some glucuronides are preferentially formed at elevated temperature and may serve as an inactivation mechanism, interfering with androgen action during periods when the temperature is too high for optimal breeding. Androgen sulfates may function in some organisms as a pool of circulating androgen that can be converted back to an active form by steroid sulfatases located in peripheral tissues.

V. TYPES OF ANDROGENS IN SUBAVIAN SPECIES

The major androgen synthesized by the testis varies considerably among vertebrate groups (Table 1). In birds and mammals, the major androgen released from the testis is testosterone. Testosterone is also

TABLE 1
Types of Androgens in Different Vertebrate Groups

Vertebrate group	Types of androgens
Mammals	Testosterone, 5α-dihydrotestosterone
Birds	Testosterone, 5α-dihydrotestosterone
Reptiles	Testosterone, 5α-dihydrotestosterone
Amphibians	
Urodeles	Testosterone, 11β-hydroxytestosterone, 11-ketotestosterone
Anurans	5α-Dihydrotestosterone, testosterone
Teleosts	11β-Hydroxy and 11-ketotestosterone, testosterone, 5α- and 5β-androstanes
Elasmobranchs	Testosterone
Agnathans	
Lampreys	Testosterone, 15α- and 15β-hydroxy-testosterone
Hagfishes	Testosterone, 6β- and 7α-hydroxy-testosterone

the primary androgen synthesized by the reptilian and elasmobranch (shark) testis. For other vertebrates, different androgens are secreted from the testis at levels equal to or greater than those of testosterone. The agnathans (jawless fishes) differ from other vertebrates in the primary androgens they secrete. In addition to testosterone, the lamprey testis secretes 15α- and 15β-hydroxylated androgens, whereas the hagfishes secrete 6β- and 7α-hydroxylated androgens. Teleosts (bony fishes) are a very diverse group and the information available may not be representative of the taxon as a whole. In some species, 11β-hydroxytestosterone and 11β-ketotestosterone are produced and secreted by the testis. These 11-oxygenated androgens are as much as 10 times more potent than testosterone in eliciting androgenic responses in bony fishes. Other androgens that are thought to be important in certain teleost fishes are 5β-hydroxy androstanes and androgen glucuronides. Urodele amphibians (salamanders) also produce 11-oxygenated androgens, but this is not the case in the anuran amphibians (frogs and toads) in which the testes produce substantial amounts of DHT. It should be noted that the data on distributions of androgens in the various vertebrate groups are based on a limited number of species. It is likely that other androgens will be found to be important as information on more species becomes available.

VI. ANDROGEN CYCLES IN SUBAVIAN SPECIES

Vertebrates show tremendous diversity in reproductive patterns; consequently, there are many variations in annual cycles of gonadal steroids. Nevertheless, some generalizations can be made about androgen cycles in subavian vertebrates. High levels of circulating androgens are typically associated with the breeding period, when they are responsible for the physiological and behavioral changes associated with breeding. Elevations of androgen levels may also correspond with the period of spermatogenesis. However, in many subavian vertebrates, the breeding season and the period of spermatogenesis do not correspond temporally. This is because sperm are often stored between the time of spermatogenesis and the next breeding season. Thus, there are often two distinct periods of elevated androgens in the bloodstream, one associated with breeding and one with spermatogenesis. The source and the concentrations of the androgens associated with these two reproductive events may differ. Androgens produced during the breeding season typically come from the Leydig cells (interstitial cells) and frequently reach high levels in the bloodstream during this time. Androgens associated with spermatogenesis may be produced by Leydig cells in some species but are derived from Sertoli cells (somatic cells) in others. Androgen levels in the bloodstream during spermatogenesis may be considerably lower than levels during breeding.

VII. ANDROGEN EFFECTS

Androgens are involved in regulation of many aspects of male reproduction and may have important functions in females as well. Androgen action may be mediated directly by the androgen receptor or indirectly by conversion to estrogens, with estrogen then acting through the estrogen receptor. The relative importance of these modes of action varies between species and between tissues. Interestingly, some of the major androgens cannot be aromatized to estrogens, including DHT and 11β-ketotestosterone. This may allow a physiological separation of direct androgen effects from indirect (estrogen-mediated) effects. Some species show seasonal changes in the relative amounts of aromatizable androgens versus nonaromatizable androgens.

Androgen effects are most pronounced on the brain, the reproductive tract, and the secondary sexual structures. However, other tissues throughout the body are also affected by these steroids.

Androgen effects on the brain begin at the time of sexual differentiation and continue throughout the life of the organism. In many species, metabolic conversions are important for androgen action on the brain, especially aromatization to estrogens and 5α-reduction to DHT. Both androgen receptor and aromatase are commonly found in brain tissues; however, the amount of each varies tremendously between species. For example, teleost fish brains contain 100–1000 times more aromatase and androgen receptor than mammalian brains. Also, aromatase and androgen receptor show different distributions within the brain, indicating that androgens may act indirectly in one region but may act directly in another.

Androgens are instrumental in sexual differentiation (masculinization) of the brain, often by local conversion of androgens to estrogens. Specific regions of the brain are altered by androgen actions, and these alterations result in male-specific physiological and behavioral patterns. In the adult brains, androgens are also involved in regulation of gonadotropin secretion, usually through a negative feedback relationship. There may be differences between aromatizable and nonaromatizable androgens in their effects on gonadotropin secretion.

Androgens have profound influences on behavior. In general, androgens stimulate male-associated behaviors in subavian vertebrates. Behaviors affected include aggression, migration, territoriality, courtship, and mating. Androgens may act alone in regulating a specific behavior. However, there are often interactions between androgens and other hormones, particularly those from the anterior pituitary. Hormones known to interact with androgens in regulating male behavior include gonadotropins, prolactin, and estrogens.

Androgens are instrumental in development of the male gonads and reproductive tract and in regulation of spermatogenesis. The exact role of androgens varies considerably, due to the diversity of mechanisms of sex determination and reproductive processes present in vertebrates. In most vertebrate groups, androgens are responsible for stimulation of growth and development of the derivatives of the Wolffian ducts, including the epididymis, ductus deferens, and seminal vesicles. In subavian vertebrates, androgens may also play a role in sex determination, particularly in species such as turtles that have temperature-dependent sex determination. Androgens also appear to be important in the natural sex reversals that occur in some teleost fishes. Furthermore, administration of androgens during critical periods of development have resulted in sex reversals of genotypic females in amphibians and fishes. In adult animals, androgens are necessary for spermatogenesis, although gonadotropins are largely responsible for regulation of this process.

The secondary sexual characteristics are often the most visible signs of androgen action. Androgens are responsible for many of the male-specific features found throughout the vertebrates. These include the distinctive male color patterns of teleost fishes, the enlarged thumb pad and specialized larynx of male frogs, the extension of the tail in swordtail mollies, and the elongated front claws of male slider turtles. Some androgen-dependent sexual features are developmental differences that persist throughout the life of the males, and others show seasonal patterns that correspond with the onset of the breeding season.

Although androgens are typically considered male sex hormones, they may also function in female reproduction. Serum androgen levels of female non-mammalian vertebrates often reach high levels, exceeding estrogen levels by as much as 10-fold in some species and even approaching levels found in males. The prevailing explanation for high androgen levels in females is that these steroids are serving as precursors for estrogen production. However, they also may act directly via the androgen receptor. Indeed, androgen receptor has been found in the oviducts of some reptiles.

Androgens may function in the development and regulation of the female reproductive tract of non-mammalian vertebrates. In birds, androgens are known to synergize with estrogens in stimulation of egg white protein production by the oviduct. A similar situation may exist in other oviparous vertebrates. Also, androgens have been shown to stimulate oviductal growth in reptiles and amphibians, although aromatization may be involved in this action. Aside from the reproductive tract, androgens have effects on sexual characteristics of some females. For example, androgens stimulate the changes in color pattern that occur during gravidity in some lizards. Androgens also may have modulating effects on vitellogenesis, the production of egg yolk. Androgens have been shown to retard estrogen-induced vitellogenesis in some reptiles and to stimulate vitellogenesis in certain fishes.

VIII. MECHANISM OF ANDROGEN ACTION

Androgen receptors are members of the nuclear hormone receptor superfamily that includes receptors for other steroid hormones as well as receptors for vitamin D and retinoids. Nuclear receptors are ligand-dependent transcription factors. Binding of the hormone (ligand) to these receptors results in alterations in transcription of hormone-dependent genes.

The process of androgen action is similar to that of other steroid hormones. Androgens enter cells by diffusion and bind to the androgen receptor. Androgen binding transforms the androgen receptor to a form that is capable of selective binding to DNA. The DNA binding occurs at specific locations termed androgen response elements, which are located in or near androgen-dependent genes and have a characteristic nucleotide sequence that is recognizable by the androgen receptor. The interaction of the androgen/androgen receptor complex with the response element results in changes in transcriptional rates of androgen target genes.

Androgen receptors of subavian vertebrates have not been well characterized; however, the available information indicates that there is considerable conservation of androgen receptor structure throughout the vertebrates. The androgen receptor in mammalian species is a relatively unstable protein that has an intact molecular weight of about 120 kDa. Extrapolation of molecular weight from the nucleotide sequence of the African clawed frog androgen receptor indicates that it is of similar size to the mammalian protein. Furthermore, antibodies designed to recognize mammalian androgen receptors cross-react with androgen receptors of birds, reptiles, and fishes. The biochemical properties of subavian androgen receptors also appear to be similar to those of other vertebrates. Androgen-responsive tissues of turtles, frogs, and fishes have an intracellular receptor that shows androgen-binding characteristics (affinity and specificity) much like those of the mammalian androgen receptor.

Steroid receptors have at least three functionally distinct protein domains; an amino-terminal domain involved in transcriptional activation, a central domain that is important in DNA binding, and a carboxy-terminal domain that binds to androgens. The structure of these domains appears to be largely conserved among the vertebrates, based on a comparison of mammalian (mouse), avian (canary), and amphibian (African clawed frog) androgen receptor amino acid sequences (Fig. 1).

The amino-terminal domain of androgen receptor is absolutely necessary for the process of transcriptional activation. For other steroid receptors (e.g., progesterone receptor), differences occur in the amino-terminal domain of the mature protein that alter the function of the receptor. Differences in the amino-terminal domain may also occur in the subavian vertebrate androgen receptor. For example, in male African clawed frogs there appears to be a tissue- and developmentally specific expression of different androgen receptor mRNA isoforms. Expression of one of these isoforms may be necessary for androgen-driven differentiation of the male larynx.

The central DNA-binding domains are highly conserved among members of the nuclear receptor superfamily members, including androgen receptor. Within this domain, there are common motifs, or highly similar sequences of amino acids. A pair of zinc fingers is the most striking feature of the central DNA-binding domain. These zinc fingers contact the androgen response element in target genes. Sequence specificity is mediated in part by amino acids located in the first zinc finger (the P or proximal box). A

FIGURE 1 Alignment of androgen receptor domains of the African clawed frog, the common canary, and the mouse. Amino acid sequences of the androgen receptor DNA-binding and hormone-binding domains are highly conserved among vertebrate classes, as shown in this comparison of representative frog, bird, and mammalian sequences. The P (proximal) and D (distal) boxes are associated with the first and second zinc fingers, respectively, and mediate the binding of the androgen receptor to specific DNA sequences. Dashes within the canary sequence represent a lack of available sequence data. Consensus represents the similarity of the amino acids between the androgen receptors of the three classes. +, complete conservation; . . . , only conservative changes; spaces, lack of similarity.

distal (D) box near the second zinc finger may provide additional specificity for androgen response elements. The deduced amino acid sequence for the zinc fingers of the African clawed frog is almost identical (two highly conservative changes of serine to threonine; Fig. 1) to both bird and mammalian sequences. It is likely that subavian androgen receptors recognize and bind to androgen response elements in the same manner as other vertebrate androgen receptors.

The mammalian androgen receptor carboxy-terminal domain mediates the specific, high-affinity (approximately 10^{-9} to 10^{-11} M) interaction of androgen receptor with C19 steroids. The hormone-binding domain of the African clawed frog's androgen receptor is similar in amino acid composition to the hormone-binding domains of the canary and mouse androgen receptors (Fig. 1). Changes in specific amino acids of the hormone-binding domain can alter the androgen-binding properties of the receptor and lead to clinical disease. However, the amino acids that have been shown to alter androgen specificity are not different in frog or canary compared to the mouse. In addition, a serine at amino acid position 650 of the human androgen receptor that serves as a phosphorylation site is important for transcriptional activation. This serine is conserved in mammalian, frog, and canary androgen receptors. This comparison suggests that the androgen receptor of subavian vertebrates will have similar androgen-binding specificities to those of other vertebrates.

See Also the Following Articles

ANDROGENS, EFFECTS IN BIRDS; ANDROGENS, EFFECTS IN MAMMALS; LEYDIG AND SERTOLI CELLS, NONMAMMALIAN

Bibliography

Bourne, A. (1991). Androgens. In *Vertebrate Endocrinology: Fundamentals and Biochemical Implications* (M. Schriebman and P. K. T. Pang, Eds.), Vol. 4, Part B, pp. 115–146. Academic Press, New York.

Callard, I. P., and Callard, G. V. (1987). Steroid hormone receptors and non-receptor binding proteins. In *Hormones and Reproduction in Fishes, Amphibians, and Reptiles* (D. Norris and R. E. Jones, Eds.), pp. 355–384. Plenum, New York.

Kime, D. E. (1987). The steroids. In *Fundamentals of Comparative Endocrinology* (I. Chester-Jones, P. M. Ingleton, and J. G. Phillips, Eds.), pp. 3–56. Plenum, New York.

Ozon, R. (1972). Androgens in fishes, amphibians, reptiles and birds. In *Steroids in Nonmammalian Vertebrates* (D. R. Idler, Ed.), pp. 328–389. Academic Press, New York.

Zhou, Z.-X., Wong, C.-I., Sar, M., and Wilson, E. M. (1994). The androgen receptor: An overview. *Recent Prog. Horm. Res.* **49**, 249–274.

Andrology: Origins and Scope

Philip Troen

University of Pittsburgh School of Medicine

I. Introduction
II. Early Twentieth Century
III. Late Twentieth Century
IV. Organization and Structure
V. Scope

Andrology, the study of the male, especially reproductive health and dysfunction, has been of documented interest for millennia. Not until recently, however, has it become a formal discipline. Building on important findings in the past three centuries, there has been in this century an explosion of knowledge in the field because of the work of many specialties and disciplines. Andrology now plays a major role in health and society in general.

I. INTRODUCTION

Man's interest in—and, indeed, preoccupation with—male reproductive activities undoubtedly predates recorded history. Pictorial records of 2 millennia ago such as those at Pompeii and at Luxor attest to this. Efforts to understand male reproductive health and dysfunction also go back thousands of years. As carefully documented in Rosner's translation of Preuss' major work on biblical and Talmudic medicine, the *Old Testament Bible* and the *Talmud* contain numerous references indicating awareness, for example, of the relationship of the testis to procreative function and overall well-being. The early philosophers also sought to understand the origin of semen and were aware of its relationship to procreation. Aristotle thought semen arises from the brain, Hippocrates wrote that semen was transported to the testicles via the arteries behind the ears, and Plato considered that semen originates from the spinal cord. The *Talmud* discusses the relationship of testicular injury and the absence of one testis to procreative ability and provides an accurate clinical description of eunuchoidism. During the Middle Ages, there was an unending list of remedies for impotence. Also, as recently as 1585, Pope Sixtus V decreed that all marriages in which men do not have two testicles in the scrotum should be dissolved.

The emergence of andrology as a discrete and important discipline warrants a brief historical review, not only to delineate the roots of andrology with some of its important milestones but also to demonstrate andrology's broadening impact on health and social well-being. Andrology may be considered to have its major roots in the seventeenth century (as noted by Nieschlag) with the demonstration of sperm in seminal fluid by Leeuwenhoek, in 1780 with the discovery of artificial insemination by Spallanzani, or in the mid-nineteenth century with the demonstration by Berthold in 1849 that the testis produces a substance responsible for virility and the description of the Leydig cell in 1850 and the Sertoli cell in 1865. However, it was not until the early twentieth century that the isolation and chemical synthesis of testosterone took place by Butenandt and others.

The earliest use of the term andrology, as reported by Niemi, apparently was in an editorial in the *Journal of the American Medical Association* in 1891 to announce the formation of the American Andrological Association. The subsequent activities of the new organization are not recorded and the term andrology appeared to be dormant until Schirren called attention to its reintroduction by a gynecologist, Harald Siebke, in Germany in 1951 to demonstrate that the male and female are equally important in reproduction. Andrology has been variously defined since then, ranging from "the scientific study of the masculine constitution and of diseases of the male sex; especially the study of the male organs of generation" (*Dorland's Medical Dictionary*) to Nieschlag's definition as the "science of male reproductive health and dysfunction."

II. EARLY TWENTIETH CENTURY

Until the beginning of the twentieth century, very few clinicians had described clinical problems associated with male infertility. As Jequier points out, Dr. Edward Martin in 1902 was the first clinician to apply surgical techniques successfully to the treatment of male sterility using an epididymovasotomy for the treatment of obstructive azoospermia. He also stressed the need for semen analysis in diagnosis. Jequier proposes that Martin, therefore, deserves the title of "Founding Father of Modern Clinical Andrology."

It is noteworthy that andrology as applied to animals has a history which parallels and often antedates the evolving nature of andrology as applied to humans. As summarized by Amann (personal communication), Nishikawa refers to efforts at the end of the nineteenth century to apply artificial insemination (AI) for livestock breeding in Russia. During the early twentieth century, advances in sperm handling led to progress in the practical application of AI. Studies of semen and AI techniques were widespread in the 1920s. These and the development of extenders to increase the fertility of animal semen were reviewed by Anderson.

Fawcett and Clermont have summarized and reviewed some of the events in the first half of the twentieth century that laid the scientific basis for the rapid development and growth of andrology. Following the "memorable descriptions" (in Clermont's phrase) of Regaud in 1901 on the cycle and wave of the seminiferous epithelium of the rat, over the next five decades the general architecture of the testis was described and the cyclic nature of spermatogenesis was recognized. The classic experiments of P. E. Smith in the 1920s revealed the dependence of the gonads on the pituitary. Later investigators helped develop a broad understanding of the nature of the control and feedback between the pituitary and testis. The endocrine functions of the testis, including function of the Leydig cells, and the nature, biosynthesis, and sites of action of steroid hormones were studied. In addition to the previous work by anatomists, physiologists, and biochemists, cytologists and geneticists, working with a large variety of animal species and plants, elucidated many aspects of meiosis, a key feature in the formation of haploid germ cells. During this period, other advances of importance to male reproductive health were also being made. For example, the pioneering investigation by Huggins which led to the demonstration that castration can cause regression of prostate cancer, opening the era of hormonal management of cancer of the prostate. The role of the testis in fetal sex determination was uncovered by Jost and the relationship of varicocele to infertility was proposed.

III. LATE TWENTIETH CENTURY

The discovery of the deep-freeze method for sperm preservation in 1952 and the prophylactic use of antibiotics in extended semen led to further rapid progress in AI. The monograph by Mann on the biochemistry of semen covered both mammalian and nonmammalian physiology from man to the sea urchin and even plants. Emmens provided a review of fertility in the male in farm animals and Salisbury and Van Demark reviewed the physiology of reproduction in cattle.

The introduction of electron microscopy in the 1950s and the availability of radioisotopes provided the foundation for research in many areas of andrology and during the 1960s and 1970s rapid further progress was made. As summarized by Fawcett, studies by anatomists clarified organization of the seminiferous epithelium and the blood–testis barrier; details of spermatogenesis, stem cell renewal, and spermatogenetic cycle; morphogenesis of spermatozoa; and structure and function of the peritubular and interstitial tissue of the testis. Biochemical studies using immunoassays and binding assay traced steroid biosynthetic pathways and demonstrated receptors for luteinizing hormone (LH) on Leydig cells and follicle-stimulating hormone on Sertoli cells. Additional information about biosynthesis and metabolic conversions (including 5α reduction) of androgens was developed. Physiologists obtained more information about the role of the epididymis in sperm maturation.

The isolation and synthesis of gonadotropin-releasing hormone (GnRH), the demonstration of pulsatility of LH and later GnRH, and improved assays for gonadotropins led to treatment of some forms of infertility. At the same time, the World Health Organization in 1980 published a laboratory manual for the examination of human semen, an important step for standardization.

During the past two decades, rapid progress has been made with new tools of molecular biology as well as with clinical studies. Some of the former include unraveling of the inhibin story, including purification and cloning; elaboration of paracrinology (cell to cell interaction) in the testis; cloning of the androgen and gonadotropin receptors; demonstration of GnRH neuronal migration; identification of the Kallmann, SRY, and 5α-reductase genes; and the linking of Y chromosome defects to some cases of male infertility.

Clinical advances have been achieved from widescale use of gonadotropins for treatment of infertility and GnRH analogs for contraception and prostatic cancer and trials of testosterone treatment for contraception. Other clinical studies have assessed the effectiveness of varicocele treatment in infertility, the long-term safety of vasectomy, and the use of intracavernosal injection of vasoactive substances for treatment of male sexual dysfunction. Advances have been made for both man and animals in semen examination and preservation and in the role of the andrology laboratory in assessing human male factor infertility. Placebo-controlled trials have applied more rigorous criteria to the study of effectiveness of various agents for infertility and the relationship of androgens to health in elderly men has been subjected to increased scrutiny.

IV. ORGANIZATION AND STRUCTURE

Despite the widespread activity in basic science and clinical aspects of male reproduction, it was not until the latter half of this century that there emerged an andrology journal (*Andrologie* in 1969) and an active organization (Comité Internacional de Andrologia in 1970). In the three decades since, there has been a veritable explosion of journals and publications, societies and congresses, and workshops and symposia. The explosive growth of andrology as a discipline is demonstrated by the growth of the International Society of Andrology (ISA). Founded in 1981 with 6 member national societies, ISA has grown to 37 member national and regional societies representing almost 8500 members worldwide.

With the maturation of andrology as a discipline, particularly in the past three decades, questions arose as to the nature of participants in and education for the discipline. Rosemberg pointed out that andrology should be recognized as an area of science and medicine which fosters a multidisciplinary and multifaceted approach to the study of male reproduction. By encompassing both basic and clinical sciences, andrology includes research in and applications of biochemistry, genetics, histology, immunology, molecular biology, nutrition, pathology, pharmacology, physiology, and endocrinology and also includes urology, microsurgery, gynecology, internal medicine, pediatrics, psychology, theriogenology, and animal husbandry. The broad scope of this approach led to attempts to define a sharper focus for the role of the andrologist, particularly in the clinical setting. These efforts by Chichinadze, Jequier, Nieschlag, Schirren, and Steinberger range from the suggestion that infertility is the major or sole focus for andrology to attempts to define andrology as a specialty or subspecialty of medicine. A recent survey by Forti for the European Academy of Andrology reveals the wide variation in current practice. In no European country is andrology recognized as a specialty, whereas in some countries it is recognized as a subspecialty, for example, of endocrinology, internal medicine, urology, gynecology, or dermatology. In most European countries, as well as in the United States, training in clinical andrology is incorporated into the training of the various clinical disciplines or as a basic science. In some countries, such as Egypt and Indonesia, andrology has the full status of a specialty. Basic science research in andrology is accordingly carried out under varying sponsorship depending on the circumstances.

V. SCOPE

The American Society of Andrology, founded in 1975, recently published a handbook of andrology to provide an overview of the nature and content of andrology for those interested in pursuing opportunities in andrology. An excellent indication of the current breadth and depth of andrology in clinical and basic areas is demonstrated by the program of the 1997 VIth International Congress of Andrology sponsored by the ISA. The nine plenary lectures of the congress covered the following topics: (i) imprinted genes and embryonic growth control, (ii) role of components of the sea urchin egg receptor in sperm binding, (iii) effect of androgens and 17β-estradiol on the androgen receptor, (iv) impact of andrological research on treatment of male infertility, (v) overview of the World Health Organization varicocele trial, (vi) reproductive effects of estrogen receptor gene disruption in male mice, (vii) genetic disorders of human spermatogenesis, (viii) neuroendocrinology of sexual reaction, and (ix) peptide analogs in the treatment of androgen-independent prostate cancer.

The 15 symposia of the congress surveyed the following: differentiation of male germ cell, the use of Doppler technology in andrology, advances in prostatic research and disease, molecular mechanisms of sperm–egg interaction, assisted fertilization, causes of defective spermatogenesis, hormonal aspects of male contraception, epididymal function, testicular toxicology, endocrinology of aging in males, impact of transgenic and gene studies on the understanding of male reproduction functions, nutritional and metabolic effects on male reproduction, testicular immunology, protection of fertility from cytotoxic treatment, and cultural diversity in attitudes toward intervention in reproduction.

Andrology has benefited greatly from advances in technology which have allowed the use of computerized semen analysis and ultrasound, have permitted advances in microsurgery, and have provided a wide range of approaches in assisted reproduction, notably intracytoplasmic sperm injection. Progress in hormone synthesis (steroid and peptide) and pharmacokinetics has fueled major advances toward effective male contraception and treatment of prostate cancer. With improved understanding of hormone action and enzymatic function, there is pharmacologic treatment for benign prostatic hypertrophy. Rapid advances in molecular biology, including transgenic animals and gene knockout models, have helped us understand the physiology and pathology of many andrological conditions and have pointed the way to new directions for basic and clinical studies.

The content of andrology has become broader as the interrelationship of androgens to various aspects of physiology and pathophysiology of the male has become apparent. In addition to the early areas of interest, such as infertility, contraception, impotence, the prostate, sexuality, artificial reproduction techniques, and reproductive toxicology, the andrologist has become involved in aging and other aspects of male health. Male reproductive function and dysfunction may thus affect male health and andrology may indeed be defined as the study of the male.

It should be noted that andrology has not evolved in scientific or clinical isolation. Societal pressures, as for population control, and the great cultural diversity of attitudes toward reproduction have been major influences in the speed and direction of growth of the discipline of andrology. In turn, andrology and andrologists will be important for the health and well-being of the male and society in general.

Acknowledgments

I acknowledge the helpful suggestions concerning important landmarks in andrology from my colleagues Rupert P. Amann (USA), David J. Handelsman (Australia), T. B. Hargreave (United Kingdom), Ilpo Huhtaniemi (Finland), Eberhard Nieschlag (Germany), Hiroyuki Oshima (Japan), Robert Schoysman (Belgium), Geoffrey M. H. Waites (Switzerland), Stephen J. Winters (USA), and Frederick C. W. Wu (United Kingdom).

See Also the Following Articles

Gonadogenesis, Male; Leydig Cells; Male Reproductive Disorders; Male Reproductive System, Humans; Male Reproductive System, Nonhuman Mammals; Penis; Semen; Sertoli Cells, Overview; Spermatogenesis, Overview; Spermiogenesis; Testis, Overview

Bibliography

Anderson, J. (1945). *The Semen of Animals and Its Use for Artificial Insemination.* Imperial Bureau of Animal Genetics, Cambridge, UK.

Chichinadze N. K. (1989). An attempt at creating the structure of andrological science. *Andrologie* 21, 511–515.

Clermont, Y. (1991). Four decades of research on the biology of the male reproductive system: A few landmarks. In *The Male Germ Cell: Spermatogonium to Fertilization* (B. Robaire, Ed.), pp. 17–25. New York Academy of Sciences, New York.

Editorial (1891). Andrology as a specialty. *J. Am. Med. Assoc.* 17, 691.

Eliasson, R. (1974). Comité Internacional de Andrologia (CIDA). *Andrologia* 6, 3.

Emmens, C. W. (1959) Fertility in the male. In *Progress in the Physiology of Farm Animals* (J. Hammond, Ed.), pp. 1047–1116. Butterworth, London.

Fawcett, D. W. (1976). The male reproductive system. In *Reproduction and Human Welfare: A Challenge to Research* (R. W. Greep, M. A. Koblinsky, and F. S. Jaffe Eds.), pp. 165–277. MIT Press, Cambridge, MA.

Forti, G. (1996). The status of andrology in Europe. *Int. J. Androl.* 19, 261–262.

Greep, R. O., and Koblinsky, M. A. (Eds.) (1977). *Frontiers in Reproduction and Fertility Control. A Review of Reproductive Sciences and Contraceptive Development.* MIT Press, Cambridge, MA.

Jequier A. M. (1990). Andrology: A new sub-specialty. *Br. J. Obst. Gynecol.* 97, 969–972.

Jequier A. M. (1991). Edward Martin (1859–1938). The founding father of modern clinical andrology. *Int. J. Androl.* 14, 1–10.

Mann, T. (1954). *The Biochemistry of Semen.* Methuen, London.

Niemi, M. (1987). Andrology as a specialty: Its origin. *J. Androl.* 8, 201–202.

Nieschlag, E. (1997). Scope and goals of andrology. In *Andrology: Male Reproduction Health and Dysfunction* (E. Nieschlag and H. M. Behre, Eds.), pp. 3–8. Springer, Berlin.

Nieschlag, E. (1997). Andrology at the end of the twentieth century: From spermatology to male reproductive health. Inaugural Address at the VIth International Congress of Andrology, Salzburg, May 1997. *Int. J. Androl.* 20, 129–131.

Nishikawa, Y. (1964). History and development of artificial insemination (AI) in the world. *VIth International Congress on Animal Reproduction and Artificial Fertilization (Trento)* 7, 162–257.

Preuss, J. (1978). *Biblical and Talmudic Medicine* (F. Rosner, Ed.). Sanhedrin, New York.

Rosemberg, E. (1986). American Society of Andrology: Its beginnings. *J. Androl.* 7, 72–75.

Salisbury, G., and Van Demark, N. L. (1961). *Physiology of Reproduction and Artificial Insemination of Cattle.* Freeman, San Francisco.

Schirren, C. (1985). Andrology: Origin and development of a special discipline in medicine: Reflection and view in the future. *Andrologia* 17, 117–125.

Steinberger, E. (1981). The past, the present, and the future of andrology. *Int. J. Androl. Suppl.* 5, 210–216.

Troen, P. (1992). Future prospects for andrology. In *Update on Therapy in Andrology* (G. F. Menchini-Fabris and D. Canale, Eds.), pp. 5–8. ETS Editrice, Pisa, Italy.

Troen, P. (1995). Foreword. In *American Society of Andrology Handbook of Andrology* (B. Robaire, J. L. Pryor, and J. M. Trasler, Eds.). Allen Press, Lawrence, KS.

Waites, G. M. H., Frick, J., and Baker, G. W. H. (Eds.) (1997). *Current Advances in Andrology. Proceedings of the VIth International Congress of Andrology.* Monduzzi Editore, Bologna, Italy.

Anestrus

see Seasonal Reproduction

Annelida

Damhnait McHugh

Harvard University

I. Annelida, Including Echiurans and Pogonophorans
II. Sexual Reproduction
III. Asexual Reproduction

GLOSSARY

annelid cross The cross-shaped cleavage pattern formed by blastomeres $1a^{112}-1d^{112}$ during embryological development of annelids.

architomy A form of asexual reproduction in which simple subdivision of the worm is followed by regeneration of the missing parts.

clitellum The glandular girdle involved in cocoon formation in oligochaetes and leeches.

epitoky The reproductive phenomenon seen in some polychaetes in which a benthic adult is morphologically transformed to produce a pelagic worm that swims to the surface of the water to spawn gametes.

paratomy A form of asexual reproduction in which complete individuals are produced prior to subdivision of the worm.

spermatophore A bundle of sperm enclosed in an external covering which is transferred directly or indirectly to a mating partner.

spermatozeugmata A bundle of sperm not surrounded by any sheath or covering which is transferred directly or indirectly to a mating partner.

trochophore A top-shaped, planktonic larva with an equatorial band of pre-oral cilia used in locomotion or feeding.

I. ANNELIDA, INCLUDING ECHIURANS AND POGONOPHORANS

The Annelida has been traditionally considered as a group of segmented, coelomate worms in which the nerve cord is located ventrally and paired chitinous chaetae occur segmentally; this traditional group includes the mainly marine polychaetes and the aquatic and terrestrial clitellates (oligochaetes and leeches). Recently, however, morphological and molecular evidence has been presented that supports inclusion of the deep-sea tubeworms, the pogonophorans, in the Annelida. Furthermore, molecular data indicate that echiurans represent a group of marine annelids in which segmentation has been lost. Both pogonophorans and echiurans share the following features with polychaetes and clitellates: the presence of paired chitinous chaetae, a ventral nerve cord, and spiral cleavage resulting in a characteristic annelid cross formation of blastomeres $1a^{112}-1d^{112}$. For the purposes of this article, the Annelida will be taken to include the polychaetes, oligochaetes, leeches, pogonophorans, and echiurans; the monophyly of each of these groups and the relationships among them, within the Annelida, will not be addressed here.

Annelids are important members of marine, freshwater, and terrestrial communities worldwide. They range in length from a couple of millimeters to several meters. Many annelids are deposit feeders and thus important bioturbators of aquatic and terrestrial sediments. Some polychaetes are suspension feeders, other annelids are predators, and some, including many leeches, are parasitic; pogonophorans depend on endosymbiotic, sulfur-oxidizing bacteria for their nutrition.

II. SEXUAL REPRODUCTION

A. Hermaphroditism

The majority of polychaetes and all echiurans are dioecious, and only a single pogonophoran species, *Siboglinum poseidoni*, is reported to be hermaphroditic. Simultaneous hermaphroditism is known in sabellid and serpulid polychaetes; protandry is known in some serpulids and eunicids, and some syllids are protogynous. Self-fertilization is known in some simultaneous hermaphroditic polychaetes, e.g., the freshwater nereid, *Nereis limnicola*.

All oligochaetes are simultaneous hermaphrodites and leeches are protandrous hermaphrodites. The complex reproductive system of clitellates functions in the transfer of sperm from one hermaphroditic animal to another, often reciprocally, with the sperm then stored in spermathecae until eggs are fertilized.

B. Spermatogenesis

Testes in polychaetes and echiurans are simply layers of germinal epithelium located ventrally or lateroventrally; a few polychaete families, e.g., nereids, lack a testes entirely. In most polychaete species, spermatogonia are released into the coelom, where they proliferate to form syncitial masses of spermatocytes called "morulae," "platelets," or "rosettes." In some species, the spermatogonia undergo few mitotic divisions before meiotic maturation, so four or eight spermatids develop in a cluster; in other species, many mitotic divisions occur, resulting in hundreds of cells per cluster. Such large clusters may share a central cytoplasm, called a cytophore, which may function as a cytoplasm dump for the developing sperm cells or as a means of synchronizing sperm development within each cluster.

In pogonophorans, sperm development proceeds in discrete tubular testes; spermatophores are formed and stored in sperm ducts of perviates, whereas sperm masses in which spermiogenesis is completed are released by vestimentiferans. In clitellates, clusters of spermatogonia are released from the paired testes into testis sacs; further spermatogonial divisions and sperm maturation occur there or, more often, in the seminal vesicles.

In annelids generally, there is a strong functional relationship between sperm structure and the mode of sperm transfer. Spermatozoa of species in which fertilization occurs in seawater are usually bullet shaped with a simple, cap-like acrosome and a simple midpiece of four or five mitochondria. Spermatozoa in annelid species in which copulation, spermatophore transfer, sperm storage, cocoon production, or hypodermic injection occur have elongate sperm in which the acrosome, midpiece, or both are modified. In clitellates, a prominent feature of spermiogenesis is the formation of a manchette of microtubules, which apparently function in nucleus elongation. Unlike other annelids, the nucleus and axoneme in clitellate sperm are interpolated by the midpiece mitochondria. Leeches have helicoidal, filiform spermatozoa, with corkscrew acrosomes and acrosomal filaments, and twisted or undulatory midpieces.

C. Oogenesis

Ovaries containing germ cells are small but distinct in most polychaetes and all echiurans; these paired organs are closely associated with segmental septa, nephridial blood vessels, or coelomic peritoneum. A few polychaete families, e.g., Nereidae, are atypical in lacking distinct ovaries; in these cases, oogonial development takes place in the coelom. In clitellates, oogenesis takes place in the paired ovaries, which, in oligochaetes, have an outer layer called the ovisacs in which oogenesis is completed.

Oogenesis in annelids can be extraovarian or intraovarian. In extraovarian oogenesis, which occurs in most polychaetes and all echiurans, primary oocytes are released into the coelom, where the oocytes grow and undergo vitellogenesis; coelomic oocytes may develop solitarily or in association with follicle cells, nurse cells, or coelomocytes. In intraovarian oogenesis, seen in some polychaetes, pogonophorans, and all clitellates, oocytes are contained within the ovary (or ovisacs), often in close association with blood vessels, follicle cells, or nurse cells, during development. Vitellogenesis is completed in ovisacs before oocytes are released for fertilization in clitellates, in the cocoon in oligochaetes, or internally in the leeches.

D. Fertilization

Many polychaete and all echiuran species release gametes into the water column in which fertilization and development take place. In others species, eggs are fertilized in the maternal tube or a gelatinous egg mass and brooded there during development. In some cases, sperm, spermatozeugmata, or spermatophores are released by the male and are taken up by the female. For example, in some sabellids and serpulid polychaetes, freely spawned sperm are stored in spermathecae; in some spionid polychaetes and in perviate pogonophorans, spermatophores are released by males, taken up by females, and then may be stored in spermathecae. More direct transfer of sperm during pairing of males and females is also known in some polychaetes and has been postulated for vestimentiferan pogonophorans. For example, in the terebellid polychaete, *Nicolea zostericola*, the male deposits spermatozeugmata in or near the tentacles of the female; eggs are then released by the female and passed over the tentacles for fertilization.

Hypodermic insemination is known in the parasitic polychaete group, the Histriobdellidae. Rhynchobdellid leeches also use hypodermic impregnation of a spermatophore as the mode of transfer; for example, in the rhynchobdellid leech *Placobdella parasitica*, a spermatophore is injected into the dorsal surface of the mating partner, the injected tip of the spermatophore contains chemicals that dissolve the skin of the leech, and as the spermatophore shrinks on contact with water, sperm are forced into the coelomic cavity that houses the ovaries. True copulation, in which a penis or penises are inserted into female receptacles, is known in some hesionid and dorvilleid polychaetes; for example, males in *Pisione remota* use modified ventral cirri to transfer sperm into female segmental gonopores during pairing.

Copulation is the sole mode of sperm transfer in all oligochaetes and in arhynchobdellid leeches. In these cases, sperm is transferred when mating worms pair up so that the male gonopores of one are lined up with the spermathecal opening of the other. Fertilization in oligochaetes takes place in the cocoon when a worm deposits its own eggs and stored sperm, and fertilization in leeches takes place internally, before eggs are deposited in the cocoon. The cocoon is produced within a few hours or days of copulation. Clitellar glands secrete the proteinaceous cocoon and a layer of albumin that will nourish the developing embryos. Once secreted, the cocoon and albumin layer are moved anteriorly by muscular contractions along the worm's body. Eggs are deposited in the cocoon as it passes over the female gonopores, and stored sperm are received from the spermathecae (in leeches, zygotes are received from the female gonopore). The cocoon is deposited on the benthos in aquatic species, in the soil in terrestrial species, or, in the case of some leeches, the cocoon is attached to the host, e.g., piscicolids.

E. Epitoky and Swarming

The formation of a reproductive individual, or epitoke, that leaves the benthos and becomes pelagic during spawning is a phenomenon unique to polychaete annelids of the families Nereidae, Syllidae, and Eunicidae. Modifications seen in epitokes include enlargement of the parapodia, elongation of dorsal and ventral parapodial cirri, enlargement of the eyes, histolysis of the gut and body wall, and development of swimming setae. Epitokes can arise from an atokous form either through direct transformation of an entire adult worm (epigamy), as in nereids; or by transformation and separation of the posterior end from the atoke (schizogamy), as in syllids. Syllid epitokes, often called stolons, can develop as a single body or as a chain or cluster of individuals. For example, in *Syllis ramosus* numerous epitokes bud from the side of the body and sometimes form secondary buds. In epigamous epitokes, e.g., nereids, the reproductive segments become highly modified so that the body of the worm appears to be divided into two distinct regions; these epitokes are termed heteronereids, and they die after spawning. Metamorphosis of the atoke into the epitokous individual is under the control of a hormone originating from the supraoesophageal ganglion in nereids; removal of the brain of juveniles induces precocious development of gametes and epitokal characteristics. In syllids, a hormone secreted from the proventriculus appears to be the inhibiting factor for stolonization in immature individuals.

Usually, epitokous worms swarm at the surface of the water during the spawning of gametes. Synchronous spawning and swarming of a population of epitokes occurs if an environmental stimulus provides the cue or the worms themselves provide a coordinating stimulus. Environmental cues result from changes in light intensity and lunar periodicity. For example, the eunicid, *Palola viridis* (the Samoan palolo worm), releases epitokes in October or November at the beginning of the last lunar quarter; this predictable occurrence allows local islanders to harvest the epitokes as a delicacy. In the West Indian syllid, *Odontosyllis enolpa*, epitokes swarm to the surface approximately 1 hr after sunset following the full moon. In some species, e.g., syllids of the genus *Autolytus*, simultaneous swarming of epitokes is enhanced by pheromones emitted by females or their eggs, which stimulate males to swarm; male and female worms then circle each other in the water column, releasing gametes as they swim to the surface.

F. Embryology

Cleavage in annelids, including echiurans and pogonophorans, is spiral, holoblastic, and may be equal or unequal. Annelids have determinate development, their coelom formation is by schizocoely, and the larval blastopore becomes the mouth. Although little is known about the embryology of pogonophorans, in all other annelids derivatives of the D quadrant are important in morphogenesis, and the $1a^{112}–1d^{112}$ blastomeres form a characteristic "annelid cross" pattern.

In polychaetes and echiurans, a blastocoel usually develops; however, stereoblastulae are known in nereids, capitellids, and also in boneillid echiurans. Gastrulation occurs by invagination, epiboly, or both. After gastrulation, the embryo develops as a top-shaped trochophore larva. The segmented trunk of polychaetes is formed by a pair of ectodermal (2d) and mesodermal (4d) cells that arise as the second and fourth D-quadrant micromeres. The 4d mesentoblast divides to form a pair of teloblasts, which proliferate anteriorly as a pair of mesodermal bands in the posterior growth zone; segments are produced when these mesodermal bands split and form paired coelomic spaces.

In clitellates, the D-quadrant generates a single micromere and then undergoes symmetric cleavage to produce a large animal blastomere, which will form the trunk ectoderm, and a large vegetal blastomere, which will form the trunk mesoderm.

G. Larval Development

In all clitellates, development is direct, with up to 20 zygotes developing in a cocoon and emerging as juveniles; development may take from 1 week to several months. Many polychaetes also undergo direct development, either in the maternal tube, in a gelatinous egg mass, or directly on the worms body (e.g., on the ventral surface in some syllids or on the operculum in spirorbids). With the exception of vestimentiferans, which have a planktonic lecithotrophic larval stage, pogonophorans are all thought to brood their embryos in the maternal tube.

Many polychaetes exhibit indirect development, in which the larva spends some time in the plankton. The presegmental, top-shaped larva is called the trochophore; it develops into a segmented, ciliated larva called a metatrochophore, and further development to produce chaetae and more segments leads to the nectochaete larva. Such larvae are characterized by ciliary bands used for locomotion and, in some cases, feeding. The prototrochal band encircling the larva anterior to the mouth is common to all trochophores, a metatrochal band may occur posterior to the mouth, and a telotroch encircles the pygidium.

Polychaete species that spawn many small eggs (<150 μm) to be fertilized in the water column by bullet-shaped sperm usually develop as planktotrophic (feeding) larvae that can remain in the plankton for weeks or months before settling and metamorphosing into a juvenile; other free-spawning species undergo a shorter lecithotrophic (yolk-dependent), planktonic development; still others may brood relatively few large, yolky eggs that hatch as planktonic larvae that disperse in the water column before metamorphosis.

Feeding larvae of several polychaetes families, including the Polygordiidae, Capitellidae, and Serpulidae, and echiuran larvae use a ciliary system involving both the prototroch and the metatroch to capture food particles; a special case of such "opposed-band feeding" has also been described for the mitraria larva of the oweniid polychaetes. In the feeding larvae of nephtyids, glycerids, polynoids, and phyllodocids, however, the metatroch is absent and the mechanism of particle capture in these larvae is very poorly known.

Settlement of larvae from the plankton precedes metamorphosis in some species (e.g., *Sabellaria alveolata*), and the opposite occurs in others (e.g., *Antinoella sarsi*). The changes at metamorphosis involve loss of ciliary bands and larval setae; a catastrophic loss or rearrangement of tissues is not characteristic of metamorphosis in this group. Many species are selective in their choice of settlement sites, with some choosing to settle with their conspecifics, e.g., reef-building sabellariids. Such settlement behavior is attributed to chemical cues from the substrate or from the adult worms or their tubes.

III. ASEXUAL REPRODUCTION

Asexual reproduction is not known in echiurans, pogonophorans, or in leeches. This mode of reproduction is well-developed in freshwater oligochaetes and in some polychaete species; however, it occurs by subdivision of the body into two or more pieces, followed by regeneration of the missing parts from each piece. In most polychaetes, there is no morphogenesis prior to the subdivision (architomy), whereas in naidid oligochaetes, in which asexual reproduction is the chief mode of propagation, complete individuals are produced prior to subdivision (paratomy).

Some polychaete groups are renowned for their remarkable powers of regeneration; for example, in the chaetopterid, *Chaetoperus*, and the cirratulid, *Dodecaceria*, asexual reproduction by means of regeneration of complete individuals from isolated segments occurs frequently. Spontaneous splitting of the body into two or more fragments followed by regeneration is also known in sabellids and syllids.

See Also the Following Articles

Hermaphroditism; Marine Invertebrates, Modes of Reproduction in

Bibliography

Eckelbarger, K. J. (1992). Annelida, Polychaeta, Oogenesis. In *Microscopic Anatomy of Invertebrates. Volume 7. Annelida* (F. W. Harrison and S. L. Gardiner, Eds.), pp. 109–128. Wiley-Liss, New York.

Fernandez, J., Tällez, V., and Olea, N. (1992). Annelida, Hirudinea. In *Microscopic Anatomy of Invertebrates. Volume 7. Annelida* (F. W. Harrison and S. L. Gardiner, Eds.), pp. 322–394. Wiley-Liss, New York.

Gould-Somero, M. (1975). Echiura. In *Reproduction of Marine Invertebrates. III. Annelids and Echiurans* (A. C. Giese and J. P. Pearse, Eds.), pp. 277–312. Academic Press, New York.

Jamieson, B. G. M. (1992). Annelida, Oligochaeta. In *Microscopic Anatomy of Invertebrates. Volume 7. Annelida* (F. W. Harrison and S. L. Gardiner, Eds.), pp. 217–322. Wiley-Liss, New York.

Jamieson, B. G. M., and Rouse, G. W. (1989). The spermatozoa of the Polychaeta (Annelida): An ultrastructural review. *Biol. Rev.* 64, 93–157.

Lasserre, P. (1975). Annelida:Oligochaeta. In *Reproduction of Marine Invertebrates. III. Annelids and Echiurans* (A. C. Giese and J. P. Pearse, Eds.), pp. 214–276. Academic Press, New York.

Olive, P. J. W., and Clark, R. B. (1979). Physiology of reproduction. In *Physiology of Annelida* (P. J. Mill, Eds.), pp. 271–368. Academic Press, London.

Pilger, J. F. (1997). Sipunculans and Echiurans. In *Embryology. Constructing the Organism* (S. F. Gilbert and A. M. Raunio, Eds.), pp. 189–218. Sinauer, Sunderland, MA.

Rice, S. A. (1992). Annelida, Polychaeta, spermatogenesis and spermiogenesis. In *Microscopic Anatomy of Invertebrates. Volume 7. Annelida* (F. W. Harrison and S. L. Gardiner, Eds.), pp. 129–153. Wiley-Liss, New York.

Schroeder, P. C. (1989). Annelida-Polychaeta. In *Reproductive Biology of Invertebrates. Volume IV Part A: Fertilization, Development and Parental Care* (K. G. Adiyodi and R. G. Adiyodi, Eds.), pp. 383–442. Wiley, Chichester, UK.

Schroeder, P. C., and Hermans, C. O. (1975). Annelida: Polychaeta. In *Reproduction of Marine Invertebrates. III. Annelids and Echiurans* (A. C. Giese and J. P. Pearse, Eds.), pp. 1–213. Academic Press, New York.

Shankland, M., and Savage, R. M. (1997). Annelids, the segmented worms. In *Embryology: Constructing the Organism* (S. F. Gilbert and A. M. Raunio, Eds.), pp. 219–236. Sinauer, Sunderland, MA.

Wilson, W. H. (1991). Sexual reproductive modes in polychaetes: Classification and diversity. *Bull. Mar. Sci.* 48, 500–516.

Anovulation

see Amenorrhea

Anterior Pituitary

Robert B. Page

M. S. Hershey Medical Center of the Pennsylvania State University

I. Protein Synthesis in Epithelial Cells of the Adenohypophysis
II. Epithelial Cells of the Adenohypophysis
III. Stromal Cells of the Adenohypophysis
IV. Organization of the Adenohypophysis

GLOSSARY

adenohypophysis The glandular organ made up of secretory epithelial cells that is applied to the neurohypophysis and divided into a par tuberalis, a pars intermedia, and a pars distalis.

infundibular process The caudal region of the neurohypophysis which lies within the sella turcica and is also called the *neural lobe*.

infundibular stem The neurohypophysial contribution to the pituitary stalk.

infundibulum The rostral region of the neurohypophysis; also called the median eminence.

neurohypophysis The diverticulum from the hypothalamus which contains neurosecretory axon terminals, modified astrocytes, microglia, capillaries, and pericytes but lacks a blood–brain barrier and neuronal cell bodies and is subdivided into the infundibulum (median eminence), the infundibular stem, and infundibular process (neural lobe).

pars distalis The region of the adenohypophysis which lies within the sella turcica and below the diaphragm sella and is separated from the infundibular process by the hypophysial cleft.

pars intermedia The region of the adenohypophysis which lies within the sella turcica applied to the lower infundibular stem and the infundibular process and separated from the pars distalis by the infundibular cleft.

pars tuberalis The region of the adenohypophysis which lies above the sella turcica and is applied to the median eminence.

primary capillary plexus The capillary bed of the neurohypophysis which extends throughout the infundibulum (median eminence), the infundibular stem, and infundibular process (neural lobe).

The adenohypophysis releases peptide hormones into the systemic circulation. Some of these hormones regulate secretion from other endocrine glands, e.g., thyrotropin-stimulating hormone, adrenocorticotropic hormone, follicle-stimulating hormone, and luteinizing hormone. Others have more diverse functions, e.g., growth hormone and prolactin regulate protein synthesis, melanocyte-stimulating hormone regulates the aggregation of melanophores in the squamous epithelium that comprises the skin, and endorphin acts as an endogenous opiate. The release of these hormones is controlled by feedback mechanisms and by instructions from the hypothalamus and other sites in the form of neuropeptides, biogenic amines, and cytokines. Although each of the pituitary hormones arises from a unique family (clone) of secretory epithelial cells, the origins of each family member and their relationships to each other have not been entirely clarified.

I. PROTEIN SYNTHESIS IN EPITHELIAL CELLS OF THE ADENOHYPOPHYSIS

The epithelial cells of the adenohypophysis carry out their role as the effector elements of the neuroendocrine system by synthesizing peptide hormones and releasing them into nearby capillaries. They have been the subject of intense scrutiny by electron microscopists. These epithelial cells are round to polygonal in shape. Their nucleus is usually eccentrically located and is characterized by the presence of a prominent nucleolus and marginated chromatin. Their cytoplasm contains the organelles usually associated with high rates of protein synthesis. Mitochondria are plentiful. The Golgi apparatus is prominent and composed of a half-moon-shaped system of stacked parallel cisterns. The *cis* face is the convex surface that faces the rough endoplasmic reticulum. Small lucent vesicles are associated with the *cis* face. The *trans* face is the concave surface which faces the plasmalemma. Plate-like cisterns are associated with the *trans* face of the Golgi. Larger electron-dense vesicles bud off the *trans* face of the Golgi apparatus. These electron-dense granules have been shown by immunoelectron microscopy to be the sites of storage of pituitary peptide hormones.

Pituitary hormones are assembled on the polyribosomes of the rough endoplasmic reticulum into prohormones. Nascent hormones are transported to the Golgi apparatus in small lucent vesicles and passed through the system of parallel lamellae. As prohormones pass through the Golgi they are sequentially modified by phosphorylation in the *cis* region, glycosylation in the middle region, or sialation in the *trans* region. In the most *trans* region, analogous to the GERL in dorsal root ganglia neurons, acid phosphatase has been demonstrated. The mature hormone is stored in larger electron-dense granules that bud off from the *trans* face of the Golgi. This process of sequential, compartmentalized protein processing and storage is characteristic of the manner in which adenohypophysial cells construct their final secretory product.

The release of these hormones from the cell is accomplished by transport of the electron-dense granules to the plasmalemma. This transport appears to be mediated by the constriction of actin filaments attached to the granules. At the plasmalemma the membranes of the granules and of the cell fuse to form Ω; figures and the contents of the vesicle are extruded from the cell into the extracellular space. There is ample evidence that hypothalamic-releasing hormones cause degranulation of adenohypophysial cells with consequent hormone release. In addition, pituitary hormones are released under the stimulus of negative feedback from target organs and by the actions of biogenic amines and cytokines released from distant sites.

II. EPITHELIAL CELLS OF THE ADENOHYPOPHYSIS

Light microscopists have long classified adenohypophysial epithelial cells by their staining characteristics. Acid dyes such as eosin stain the cytoplasm of some epithelial cells red. Basic dyes, or dyes that bind carbohydrate moieties, stain the cytoplasm of other cells blue. Most of the epithelial cells remain unstained. Hence, the epithelial cell have been classified as chromophobes (52%), acidophils (eosinophils) (37%), and basophils (11%). Acidophils have been found to secrete either growth hormone (GH) or prolactin (PRL), whereas basophiles secrete follicle-stimulating hormone (FSH), luteinizing hormone (LH), thyrotropin-stimulating hormone (TSH), melanocyte-stimulating hormone (MSH), or adrenocorticotropic hormone (ACTH). Because the proteins secreted by basophils are glycosylated, they stain with the periodic acid-Schiff reagent. Although this broad classification may have some usage, it should be noted that chromophobes can secrete any of the previously mentioned hormones.

Immunohistochemistry, when combined with light microscopy, has permitted classification of adenohypophysial cells along functional lines according to the protein each synthesizes, stores, and releases.

In general, gonadotropes synthesize FSH or LH, corticotropes synthesize ACTH and β-lipotropin (β-LPH), thyrotropes synthesize TSH, melanotropes synthesize α-MSH, somatotropes synthesize GH, and lactotropes synthesize PRL. Whereas immunohistochemistry has permitted localization of particular proteins within specific epithelial cells, hybrid histochemistry has located the mRNA responsible for that protein's secretion. Because pituitary adenomas consist of populations of cells with similar and unique immunohistochemical and hybrid histochemical characteristics, a concept of clones of functional pituitary cells has arisen.

Attempts to characterize cell function by the size of their secretory granules as seen in the electron microscope have not been very successful. When immunohistochemistry or hybrid histochemistry is combined with transmission electron microscopy, functional cell types in the adenohypophysis can be morphologically characterized. These studies have revealed that the picture of a specific cell type, while characteristic, is no more immutable than a picture of the author whose appearance has varied with his age and with his level of activity. In the rat pituitary gland three types of cells react with the antibody to growth hormone. Type I cells contain secretory granules that are $\sim$350 nm in diameter. Type II somatotropes contain secretory granules of $\sim$150–200 nm in diameter and type III somatotropes contain granules that are $\sim$100 nm in diameter. In immature rats, the pituitary glands are characterized by a high proportion of type III somatotropes, i.e., those containing secretory granules that are $\sim$100 nm in diameter. In juveniles, type II cells predominate and in adults type III somatotropes containing granules that are $\sim$350 nm in diameter are most frequently found. Analogous findings have been reported in studies of lactotropes and thyrotropes. Appearance also alters with function. Lactotropes in postpartum rats become degranulated and develop large laminar areas of endoplasmic reticulum when the dam is suckled. While the findings cited evoke a vision of clonal adenohypophyseal cells, each of which secretes a unique hormone, other findings somewhat cloud the picture.

FSH and LH are found in the same gonadotrope in 80% of the cells studied in adult male rats. FSH alone is seen in only 10% of the gonadotropes and LH alone in only 10%. Colocalization of GH and PRL and of ACTH and TSH in the same secretory granules in a small number of cells has been reported. Such pleuripotential cells may contribute to the adult population of secretory cells under special conditions. Proopiomelanocortin, a glycosylated protein, is synthesized by both corticotropes and melanotropes. In the rat this 31-kDa prohormone is processed into ACTH, β-LPH, and β-endorphin in corticotropes in the pars distalis. It is processed into α-MSH and corticotropin-like intermediate peptide in melanotropes in the pars intermedia. Corticotropes and melanotropes in the rat adenohypophysis both synthesize POMC but the protein is processed differently in the two cell types.

III. STROMAL CELLS OF THE ADENOHYPOPHYSIS

Folliculostellate cells are nonsecretory cells in the pars distalis. These angular cells contain microfilaments, lysosomes, and lipid bodies. They are characterized by the presence of S-100 protein and fibronectin in their cytoplasm. They are joined to each other by gap junctions and form a syncytium. Adjacent cells may unite to form a follicle which contains colloid with its lumen. The luminal surface of the follicle is characterized by the presence of microvilli and cilia.

Capillaries in the adenohypophysis are fenestrated and the organ lacks a blood–brain barrier. Passage of materials between the extracellular space and the lumen may be influenced by ionic microdomains impeding or facilitating the transfer of cations or anions.

IV. ORGANIZATION OF THE ADENOHYPOPHYSIS

The mammalian adenohypophysis is divided into a pars tuberalis, which lies above the diaphragm sella applied to the median eminence (infundibulum) and infundibular stalk, and a pars intermedia and pars distalis, which lie within the sella turcica beneath the diaphragm sella. The pars tuberalis is composed of pars tuberalis-specific cells, invasive cells, and fol-

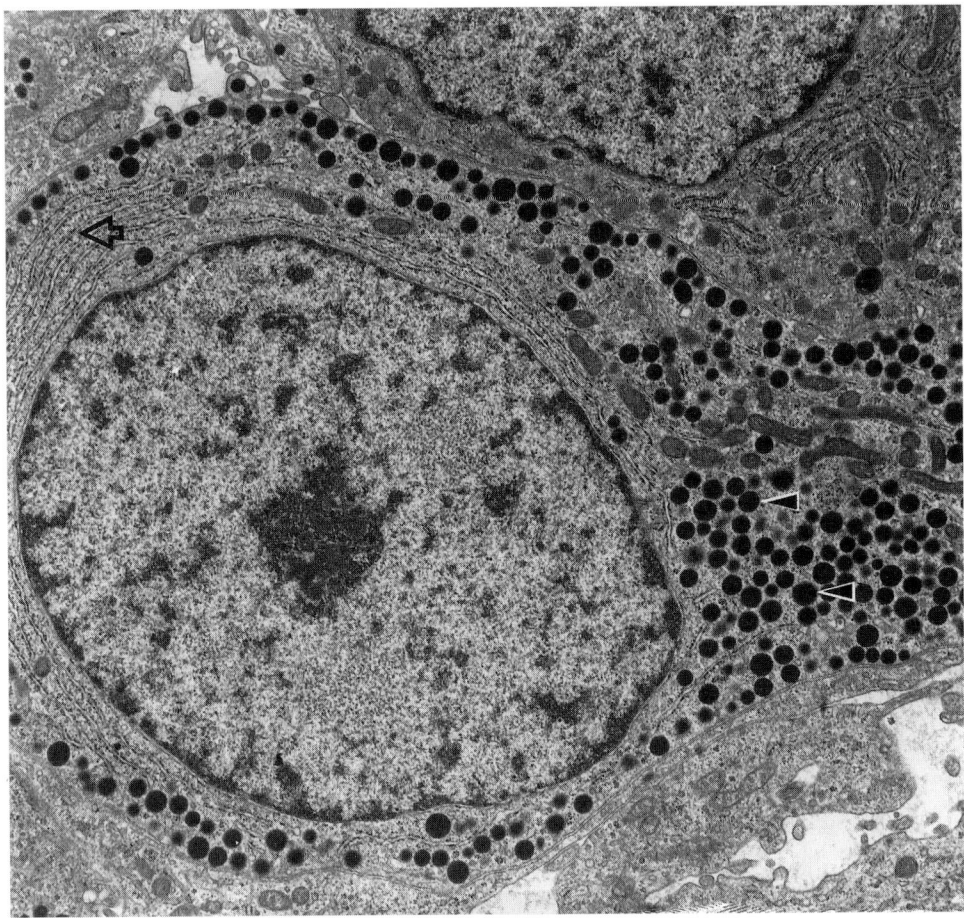

FIGURE 1 Transmission electron micrograph of a rabbit pars distalis secretory epithelial cell. The nucleus contains a central nucleolus. Dark chromatin material is scattered throughout the nucleus but also aggregated at the nuclear membrane. The secretory granules (solid arrowheads) vary somewhat in size. Mitochondria are plentiful. The rough endoplasmic reticulum (open arrowhead) is prominent (reproduced with permission; based on Figure 12 in Page, R. B., and Dovey-Hartman, B. J., Neurohemal contact in the internal zoner of the Rabbitt Median Eminence. *J. Comp. Neurol.* 226, 99286, 1984, published by Wiley Liss).

licular cells. Follicular cells are nonsecretory and resemble the folliculostellate cells in the pars distalis. Pars tuberalis cells are believed to be secretory on the basis of their transmission electron microscopic appearance and their immunohistochemical localization of the α peptide chain common to TSH, LH, and FSH. They are characterized by a spheroid shape with a few secretory granules (120–150 nm diameter in the adult rat) located at their vascular pole. A large number of characteristic lysosomes are present in the cytoplasm which consist of a dense core with a cup-like extension. The presence of glycogen granules in the cytoplasm of these cells distinguishes them from the invasive cells from the pars distalis. These are typical gonadotropes. With hypophysectomy the number of gonadotropes and the intensity of their immunohistochemical staining increased. Melatonin receptors have been identified in the pars tuberalis, and there is evidence that they are localized to pars tuberalis cells. The morphology of these cells varies with the photoperiod in some species. Several roles have been postulated for the pars tuberalis. In one, the pars tuberalis cells, under the influence of melatonin secreted from the pineal, secrete an uncharacterized protein hormone that is released into the portal system and carried to the pars distalis, where it influences the release of gonadotropins. In another, gonadotropins released from gonadotropes in the pars tuberalis form a short loop feedback mechanism to inhibit the release of gonadotropins from gonadotropes in the pars distalis. In this latter scheme, the release of gonadotropins from cells in the pars tuberalis is regulated by melatonin release from the pineal, which in turn is regulated by the photoperiod.

The pars intermedia lies applied to the infundibular stem and process (neural lobe). α-MSH has been localized in pars intermedia cells by immunohistochemistry and has long been recognized as the agent responsible for regulation of skin color in amphibians. In humans the pars intermedia is identifiable in the fetus and in the pregnant female and α-MSH has been localized in their melanotropes. In other mammals its is composed of corticomelanotropes. In the rat, light and dark melanotropes have been identified by transmission electron microscopy. Light cells contain many clear vesicles (200–300 nm diameter) and many secretory granules of the same size. Dark cells contain closely packed particles about 20 nm in diameter which give the cells their dark appearance in addition to secretory granules. Granules in both light and dark cells stain for both ACTH 17–39 and for β-MSH. β-LPH rather than β-MSH may have been identified by immunohistochemical staining for β-MSH because the entire sequence of β-MSH is found in β-LPH. The pars intermedia is less well vascularized than the pars distalis. Its extracellular space is continuous with Rathke's cleft. Portal (short) connections between the pars intermedia and the neural lobe provide a route by which secretions of the neural lobe (the side chain of oxytocin) can reach the pars intermedia to inhibit the secretion of MSH. Aminergic axons from the hypothalamus (dopaminergic, noradrenergic, serotonergic, and GABAergic) terminate near melanotropes and participate in the regulation of their hormone secretion and release.

The pars distalis lies within the sella turcica adjacent to the neural lobe. However, in the center of the gland the pars distalis is separated from the neural lobe by Rathke's cleft. Studies suggest that lactotropes and somatotropes are found more frequently in the lateral regions of the pars distalis. Somatotropes in the rat pituitary are round, oval, or polygonal in shape and contain characteristic short-clubbed mitochondria as well as secretory granules in their cytoplasm. Their morphology varies with the age of the rat. Lactotropes can be crescent shaped with an eccentric oval nucleus with mitochondria aggregated in a perinuclear location (type I), polygonal or elongated with a kidney-shaped nucleus (type II), or small and polygonal (type III). The predominant cell type depends on the developmental stage of the animal. The morphology of the endoplasmic reticulum, Golgi, nascent vesicles, and secretory granules depends on activity. Granule size can vary from 150 to 600 nm. The cells secreting glycosylated peptides (thyrotropes, gonadotropes, and corticomelanotropes) predominate more centrally in the mucoid wedge. Thyrotropes in the adult rat are small angular cells whose cytoplasmic secretory granules measure 120–180 nm in diameter. The morphology of the secretory granules varies as a function of development. Gonadotropes are usually described as large ovoid or polyhedral cells containing a spherical nucleus, prominent Golgi, and abundant rough endoplasmic reticulum. Corticotropes in the rat's pars distalis are angular in shape (as are thyrotropes). Caution should be exercised, however, in identification of cell type on the basis of transmission electron microscopy alone. Caution may also have to be exercised in the assumption of the immutability of functional cell types in the pars distalis. Perhaps some cells can change their pattern of hormone secretion depending on their hormonal environment and the demands placed on them.

Classical descriptions of the pars distalis deny the presence of nerve terminals adjacent to secretory epithelial cells. Control of secretory function in the pars distalis has been held to be solely regulated by neurosecretion in the median eminence and the transport of secreted hypothalamic peptides to the pars distalis by the portal system. Recently, there have been reports of peptidergic innervation of pars distalis epithelial cells. The function and physiological importance of such innervation have yet to be determined.

See Also the Following Articles

Hypothalamic-Hypophysial Complex; Neurohypophysial Hormones

Bibliography

Childs, G. V. (1991). Multipotential pituitary cells that contain adrenocorticotropin and other pituitary hormones. *TEM* **2**, 112–117.

Eipper, B. A., and Mains, R. E. (1980). Structure and biosynthesis of Pro-Acth/endorphin and related peptides. *Endocr. Rev.* **1**, 1–27.

Farquhar, M. G. (1985). Progress in unraveling pathways of Golgi traffic. *Annu. Rev. Cell Biol.* **1**, 447–488.

Gauer, F., Masson-Pevet, M., and Pevet, P. (1994). Seasonal regulation of melatonin receptors in rodent pars tuberalis: Correlation with reproductive state. *J. Neural Transm.* **96**, 187–195.

Paden, C. M., Moffett, C. W., and Benowitz, L. I. (1994). Innervation of the rat anterior and neurointermediate pituitary visualized by immunocytochemistry for growth-associated protein GAP-43. *Endocrinology* **134**, 503–506.

Page, R. B. (1994). The anatomy of the hypothalamo-hypophysial complex. In *The Physiology of Reproduction* (E. Knobil and J. D. Neill, Eds.), 2nd ed., pp. 1527–1619. Raven Press, New York.

Senda, T., Okabe, T., Matsuda, M., and Fujita, H. (1994). Quick-freeze, deep etch visualization of exocytosis in anterior pituitary secretory cells: Localization and possible actions of actin and annexin II. *Cell Tissue Res.* **277**, 51–60.

Soji, T., Sirasawa, N., Kurono, C., Yashiro, T., and Herbert, D. C. (1994). Immunohistochemical study of the postnatal development of the folliculo-stellate cells in the rat anterior pituitary gland. *Tissue Cell* **26**, 1–8.

Thorpe, J. R., Ray, K. P., and Wallis, M. (1990). Occurrence of rare somatomammotrophs in ovine anterior pituitary tissue studied by immunogold labeling and electron microscopy. *J. Endocrinol.* **124**, 67–73.

Wittkowski, W., Schulze-Bonhage, A. H., and Bockers, T. M. (1992). The pars tuberalis of the hypophysis: A modulator of the pars distalis. *ACTA Endocrinol. (Copenhagen)* **126**, 285–290.

Antiestrogens

Donald P. McDonnell
Duke University Medical Center

I. Antiestrogens: A Historical Perspective
II. Redefining the Term "Antiestrogen"
III. Antiestrogens: Beyond Tamoxifen
IV. Estrogens and Antiestrogens Manifest Their Activity through Two Distinct Receptor Isoforms
V. Antiestrogens and Receptor Structure
VI. Antiestrogens and Estrogens Promote High-Affinity Associations of ER with Target Gene Promoters
VII. Different ER–Antagonist Complexes Are Recognized Differently in Different Cells
VIII. Receptor-Associated Proteins as Determinants of the Cell-Selective Action of ER–Ligands
IX. A Working Model to Explain the Biological Activity of ER Agonists and Antagonists

GLOSSARY

coactivators Upon binding DNA estrogen receptor (ER) influences target gene transcription by contacting the general transcription machinery directly or indirectly through intermediary proteins called coactivators. A number of different coactivators have been identified and it has been shown that their ability to interact with ER is influenced by the overall conformation of the receptor–ligand complex; agonists facilitate whereas antagonists disrupt these interactions. It is generally considered that it is the differential expression of these cofactors in different cells which is responsible for the tissue-selective biological activity of different ER modulators.

corepressors Like coactivators whose interactions with ER are influenced by agonists, corepressors are proteins which preferentially interact with ER in the presence of antagonists. Corepressor proteins serve to inhibit the transcriptional activity of ER, though the precise mechanism remains to be determined.

inverse agonists A pharmacological term for compounds which interact with ER and reduce target gene transcription below basal level (no agonist added). Thus, these compounds do more than just inhibit estradiol activation of ER transcriptional activity; they convert ER into a transcriptional repressor.

Antiestrogens have in the past been defined as synthetic small molecules which oppose the biological activity of estrogens by competitively binding to the estrogen receptor (ER) and inhibiting agonist binding. As the molecular pharmacology of these molecules has been unraveled it is apparent that this definition must be expanded. It appears that the determination of whether a compound should be classified as an agonist or an antagonist must be made only with respect to a specific process within a specific cell and cannot be generalized to estrogen action as a whole. For instance, the pharmaceutical agent tamoxifen can function as a pure antagonist in breast by opposing the activity of estradiol, whereas in bone and the cardiovascular system, it functions as an agonist mimicking estradiol. This article provides a discussion of the current understanding of the pharmacology of ER agonists and antagonists and an explanation as to how compounds, classically defined as antagonists, can function as cell-selective estrogen receptor agonists.

I. ANTIESTROGENS: A HISTORICAL PERSPECTIVE

The steroid hormone estrogen is a key intracellular regulator of the processes involved in the establishment and maintenance of reproductive function. However, it is now well established that the nonreproductive activities of this hormone are likely to be of equal importance. Much of this appreciation has come from clinical studies, both prospective and retrospective, which have evaluated the effect of estrogen replacement therapy (ERT) on a wide range of physiological processes in postmenopausal women. Interestingly, ERT was first approved in the mid-1940s for the relief of the vasomotor symptoms associated with cessation of ovarian function. However, the subsequent millions of women-years of clinical exposure to this agent have revealed that estrogen has also had positive activities in the skeleton, the cardiovascular system, and the central nervous system. Thus, pharmaceuticals which contain estrogen have become mainline therapies for relief of the climacteric symptoms associated with menopause and also for the prevention and treatment of osteoporosis. Specific labeling for ERT as a cardioprotective agent is likely to occur in the near future.

Although the positive activities of estrogen have provided the impetus to develop drugs which activate the estrogen receptor (ER), there are circumstances in which inhibition of estrogen action is desired. By far the most important condition in this regard is breast cancer, in which several pieces of evidence have served to establish a link between estrogen and breast cancer. Estrogen is a mitogen and it can stimulate the growth of a cell(s) which has suffered a transforming mutagenic insult and which expresses a high-affinity nuclear estrogen receptor. It is not surprising, therefore, that estrogen ablation therapy is an effective way of managing tumor growth. In the past, this was accomplished by surgical ovarian ablation or by using high doses of progesterone to interrupt ovarian steroidogenesis. However, because of extraovarian production of estrogens by adipose tissue, total removal of estrogens is difficult to achieve in this manner. The appearance of compounds in the late 1970s which competed with estrogen for binding to ER but which did not display mitogenic activity in the breast or reproductive system heralded a new era in ER pharmacology. These compounds, called antiestrogens, were originally developed as contraceptives; however, it is their activity in the treatment of breast cancer which has received the most attention. The first antiestrogen approved for breast cancer, tamoxifen, has evolved to be a major weapon in the armory of the oncologist. When used as chemotherapy in the adjuvant setting, tamox-

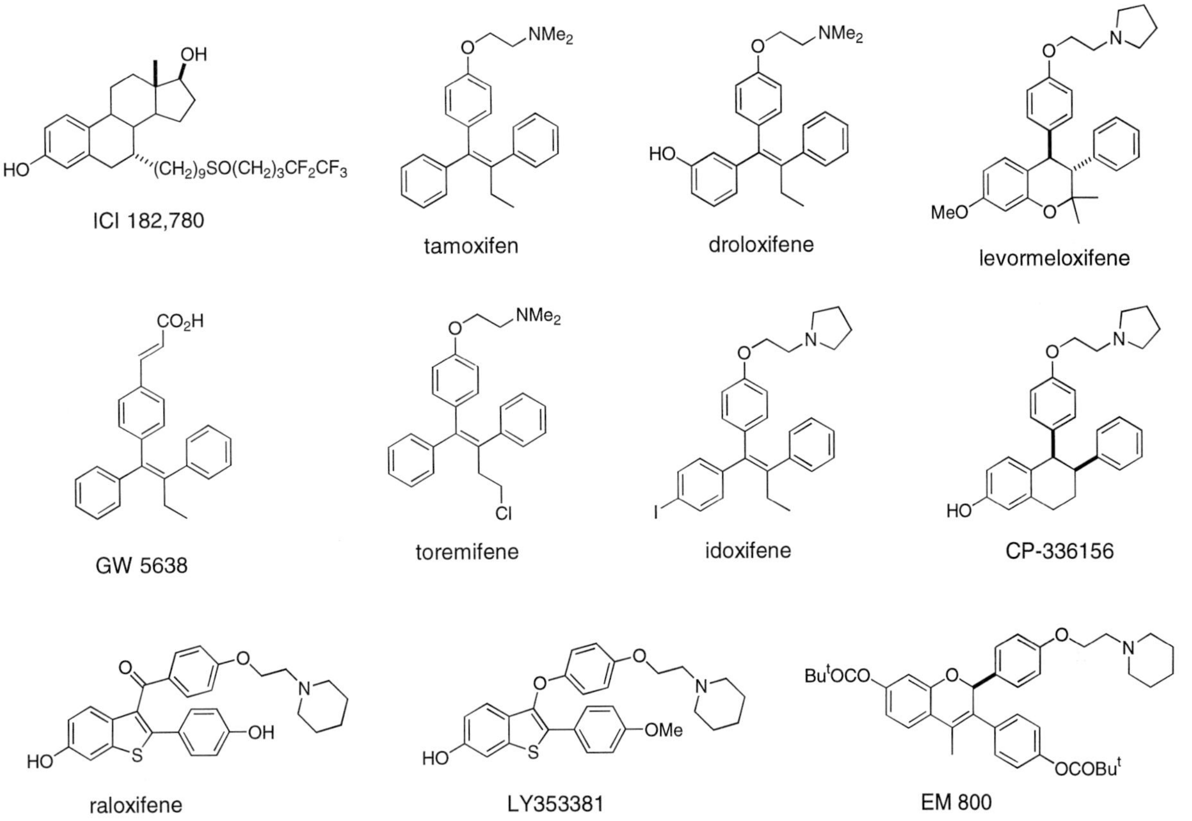

FIGURE 1 Chemical structures of the most commonly studied antiestrogens.

ifen reduces the incidence of second primary tumors by about 30% and demonstrates a similar response rate in metastatic disease, in which it inhibits cell growth and halts tumor progression. Despite its success as a chemotherapeutic, there has been a tremendous amount of interest in developing additional antiestrogens which may operate in a manner distinct from tamoxifen and which may display improved therapeutic benefit. This endeavor has now turned to reality and we have seen in the past few years the discovery and development of a plethora of new antiestrogens (Fig. 1). Although clearly interesting from a clinical perspective, these new compounds and the studies into their mechanism of action have indicated that antiestrogen action is far more complex than originally anticipated. This complexity has suggested new avenues for drug discovery but has also indicated that competitive inhibition of estradiol binding to ER is not a mechanism which adequately describes antiestrogen action.

II. REDEFINING THE TERM "ANTIESTROGEN"

The classical models of estrogen action have suggested that in the absence of hormone the nuclear ER resides in target cell nuclei in an inactive form in a complex composed of heat shock proteins and other associated proteins. Upon binding ligand, the receptor undergoes an activation event facilitating its conversion from a transcriptionally inactive to an active form, a process which permits the receptor to interact with the regulatory regions of target genes. The DNA-bound receptor then exerts a positive or a negative influence on target gene transcription. In these simple models the proposed role of estrogen was that of a switch which converts the receptor from an inactive to an active form (Fig. 2). Thus, by this definition, all estrogens are qualitatively the same and, when adjusted for affinity, are biological equivalent. Within the framework of this model antiestro-

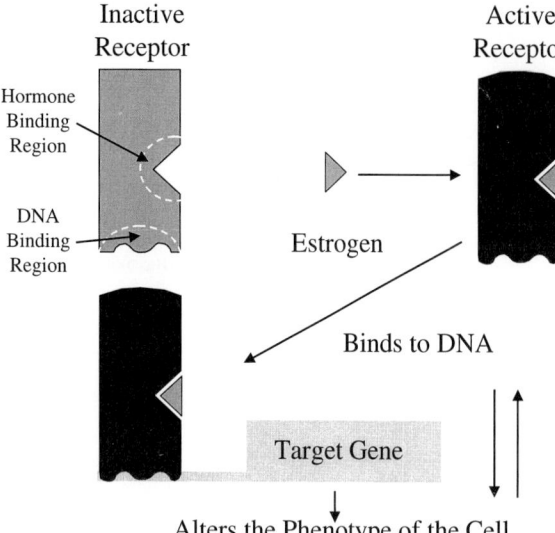

FIGURE 2 Early models of ER action do not explain the complex pharmacology of known ER–ligands. The classical models of estrogen action suggested that in the absence of ligand the ER existed in the nuclei of target cells in an inactive form. Upon binding an agonist, ER underwent an activating transformation event which facilitated the interaction of the receptor with specific DNA response elements within target gene promoters. In this model the role of estrogen was that of a "switch" which merely converted ER from an inactive to an active form. Thus, when corrected for affinity, all agonists were quantitatively the same and would evoke the same phenotypic response. By inference, antiestrogens, compounds which oppose the actions of estradiol, were considered to competitively bind to ER and freeze the receptor in an active form. As with agonists, this model predicted that all antagonists were qualitatively the same. Within the confines of this model it was hard to explain the molecular pharmacology of tamoxifen and other ligands whose agonist/antagonist activity differs from cell to cell. Thus, this model, although beautiful in its simplicity, does not adequately describe the pharmacology of the known ER–ligands.

gens are also considered to be mechanistically simple compounds which act by physically binding to ER and blocking agonist access. Thus, when corrected for affinity all antiestrogens were considered to be qualitatively similar. Initially proposed in the mid-1970s, this model in one form or another was perceived to adequately describe the actions of all the known hormones and antihormones. This picture has changed, however, in recent years. Specifically, examination of the biology of the antiestrogen tamoxifen in several systems has indicated that this compound can function as an antagonist in most environments but can actually manifest agonist activity in others. Although several studies could be used to highlight this point, it was the landmark study by Love and coworkers which examined tamoxifen action in bone that is the most illustrative. Prior to this study it had been established that estrogen was required for maintenance of bone mineral density in postmenopausal women. Therefore, it was unclear what effect tamoxifen, an antiestrogen, would have in breast cancer patients receiving this drug. However, the startling finding from these studies was that tamoxifen did not function as an antiestrogen in the bone but actually functioned as a partial ER agonist displaying 30% efficacy with respect to estrogen in the lumbar spine. This study, and similar studies, supported the hypothesis that tamoxifen could function as an agonist or an antagonist, depending on the cell context in which it was examined. These data introduced the "tamoxifen paradox," a concept which describes our inability to describe the action of tamoxifen by the molecular models which were established at the time. It suggests instead that the mechanism by which ER manifests activity is not the same in all cells and that the classification of a compound as an agonist or an antagonist could only be made with reference to a specific cell or processes. Based on these observations and other supporting studies, tamoxifen is generally considered to be the first *selective estrogen receptor modulator* (SERM), a pharmaceutical definition used to define compounds which can regulate ER function in a context-specific manner. Additionally, the realization that tamoxifen, a compound previously defined as an antiestrogen, could function as an agonist in bone suggested that it may be possible to develop compounds which could function as tissue-selective estrogens, mimicking estrogen action in some tissues and exhibiting antiestrogen action in others.

III. ANTIESTROGENS: BEYOND TAMOXIFEN

Although the discussion thus far has focused on tamoxifen, there exists several additional antiestrogens in use or under development. Although by no means a complete list, the structures of the most frequently used and studied compounds are shown in (Fig. 1). Chemically, these compounds can be separated into two groups: steroidal antiestrogens, such as ICI182,780 (and its predecessor ICI164,384), and nonsteroidal antiestrogens, such as tamoxifen, idoxifene, and raloxifene. Although there is much less information available for antiestrogens other than tamoxifen, the concept that different antiestrogens function differently in different cells is one that appears to hold for all the nonsteroidal antiestrogens. The steroidal antiestrogens, unlike tamoxifen, do not appear to have agonist activity in bone or the reproductive system, a finding which resulted in their classification as "pure" antiestrogens. However, it has been determined that ICI182,780, like estrogen, can upregulate the transcriptional activity of the retinoic acid receptor-β promoter in an ER-dependent manner. Thus, even the complex of a pure antiestrogen with ER can effectively activate transcription in some circumstances.

The emerging concept is that ER does not appear to operate in the same manner in all cells and; as a consequence, different ER–ligand complexes can have different activities in different cells. This finding begs a redefinition of the term antiestrogen and suggests that a hard and fast pharmacological classification is not possible. It indicates instead that there may be different types of antiestrogens which exert their activity by affecting ER signaling in different manners. This concept was difficult to reconcile with the simple models of ER action which proposed that the ligand was functioning merely as an all-or-nothing activating switch. However, the cloning of the ER cDNA and its reconstitution into hormone responsive transcription units in heterologous cells, combined with advances in our understanding of the biochemistry of the receptor, have resulted in the explanation of how the unique biology of different classes of ER antagonists is manifest. Although a complete description of ER action is not possible in this forum, a consideration of some of the key differences between the classical models of ER action and more updated models will help to explain the complexities of ER.

IV. ESTROGENS AND ANTIESTROGENS MANIFEST THEIR ACTIVITY THROUGH TWO DISTINCT RECEPTOR ISOFORMS

One possible explanation for the tissue-selective activity of different ER–ligands is that they interact with different receptors, each of which may be differentially expressed in target tissues. Until recently, however, this possibility was not considered likely because only a single ER gene had been identified in humans and the protein it encoded appeared to be biochemically identical in all target cells. This changed in 1995 with the cloning of a second estrogen receptor, ER-β, which displayed similar though distinct ligand-binding specificity from that of ER-α (Fig. 3). The identification of ER-β, whose tissue distribution appears to be different from that of ER-α, indicates that the overall biological activity of a compound reflects the product of its relative agonist/antagonist activities on the two different receptors. A comparison of the amino acid sequences of these two receptors reveals that they have virtually identical DNA-binding domains and most likely recognize the same DNA regulatory regions. However, within the ligand-binding domain, the amino acid similarity diverges considerably. Thus, although both receptors have an equivalent affinity for 17β-estradiol and closely similar affinities for other known estrogens and antiestrogens, it is likely that novel agonists and antagonists will be developed which bind selectively to ER-α and ER-β. In support of this hypothesis, it has been observed that the antiestrogen raloxifene is somewhat selective for ER-α, whereas genistein, a phytoestrogen, interacts more avidly with ER-β.

The discovery of ER-β has clearly provided a major leap forward in our understanding of estrogen action and may explain why phenotypic responses to estrogen were observed in some tissues in which ER-α was not detectable. However, the existence of a second receptor does not appear to be sufficient to explain the complex pharmacology of tamoxifen because the agonist/antagonist activity of this compound can vary between different cells which express ER-α alone. It appears, therefore, that different estrogens, or antiestrogens, acting through the same receptor subtype can manifest different biology in different cells.

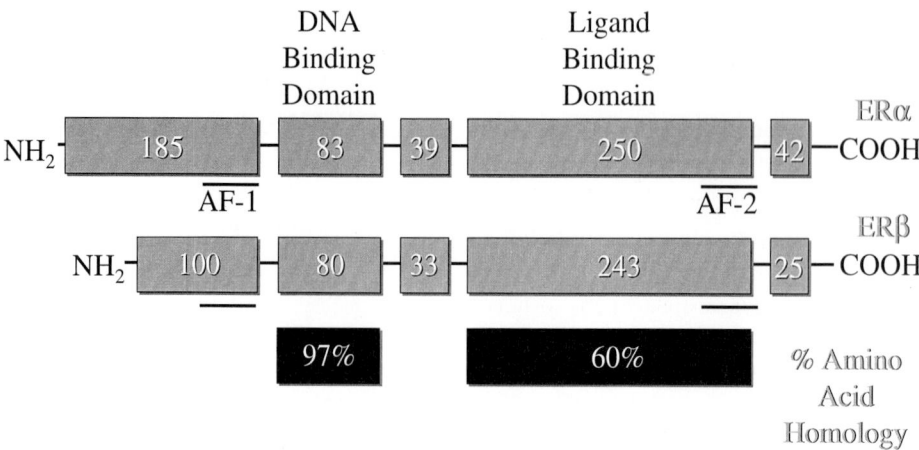

FIGURE 3 Estrogens and antiestrogens manifest their biological activity through two distinct receptors. Until recently, it was believed that a single estrogen receptor existed in target cell nuclei However, in 1995 a second high-affinity estrogen receptor was identified. Although the results are preliminary, it appears that the expression pattern of the newly discovered ER-β is distinct from that of ER-α. It is likely, therefore, that specific functions will be ascribed to this receptor and that in the future pharmacological agents will be developed which will selectively activate or inhibit ER-α or ER-β. Numbers in boxes indicate number of amino acids.

V. ANTIESTROGENS AND RECEPTOR STRUCTURE

Within the confines of established models of ER action it was difficult to explain how different ER–ligands could manifest qualitatively different biological activities. Specifically, these models suggested that ER could exist in either of two states within the cell—active or inactive. Thus, the role of estrogen was that of a switch converting receptor from an inactive to an active form. An extension of this concept was that antiestrogens, like tamoxifen, functioned by competitively blocking agonist binding and freezing ER in an inactive form. Clearly, however, this simple model was not sufficient to explain the tissue-selective agonist/antagonist activity exhibited by tamoxifen. It was not until techniques became available to study the structure of ER complexed to different antiestrogens that an explanation for the unique biology of tamoxifen became apparent. One of the most powerful techniques developed was the "protease digestion assay." In this assay the differential sensitivity of different ER–ligand complexes to the protease trypsin was evaluated. The principle of this assay is that as the conformation of the protein changes so too does the accessibility of specific tryp-

sin cleavage sites on the protein, resulting in specific fragmentation patterns. These fragmentation patterns function as surrogates for conformational changes within ER which occur upon ligand binding. As shown in Fig. 4, the protease digestion analysis revealed that the structure of ER in the presence of estradiol is different from that of an unoccupied receptor. The surprising finding, however, was that ER, when occupied by tamoxifen, raloxifene, or other SERMS, could adopt conformations which were intermediate between that induced by estrogens and that of unliganded receptor. With specific reference to tamoxifen, it appeared that this compound was not merely a competitive antagonist, freezing ER in an inactive state, but it was also an "active" antagonist activating ER and committing it down a pathway the consequence of which was not the same in all cells. Although the protease digestion assay is a relatively insensitive method of accessing receptor structure, it adequately demonstrated that different ER antagonists can induce different alterations within ER, suggesting a link between the overall structure of the ER–ligand complex and biological activity.

The fundamental observation that the structure of the ER–ligand complex was differentially affected by the nature of the bound ligand implied that com-

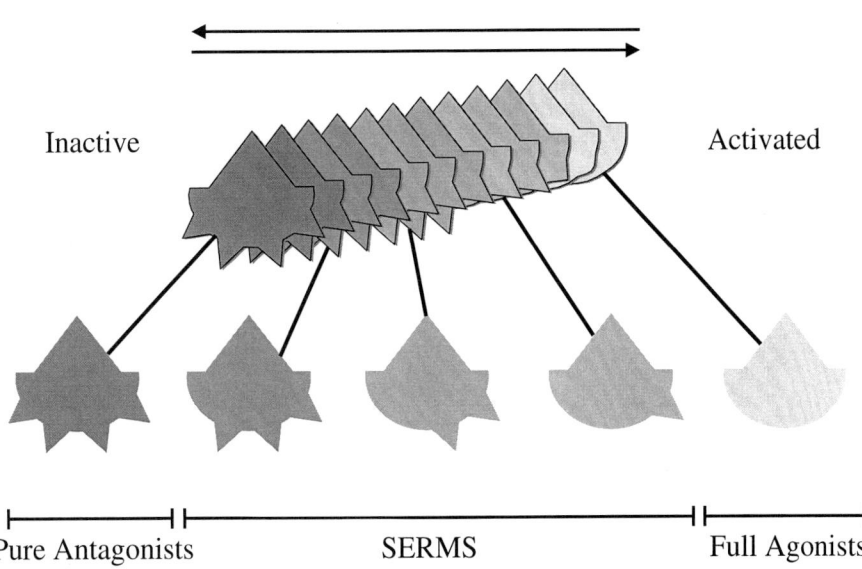

FIGURE 4 Different compounds induce different structural alterations within the estrogen receptor. The proposed role of ligand in regulating ER function has changed considerably during the past 10 years. Initially, it was believed that ER existed in the cell in either an active or an inactive form. However, as techniques have become available to study the structure of the ER–ligand complex, it has become apparent that the overall structure of the receptor is influenced by the bound ligand. Thus, ER can exist in a series of conformations ranging from the inactive receptor to the estradiol-activated receptor. This observation was particularly important with respect to the selective estrogen receptor modulators such as tamoxifen and raloxifene. These compounds can exhibit pure antagonist activity in some cells but manifest agonist activity in others. It is now believed that these compounds induce unique structural alterations within ER upon binding and that cells differ in their ability to recognize these conformations. The establishment of a relationship between the structure of the ER–ligand complex and function suggests that additional selective receptor modulators with unique biological activities will be developed.

pounds may interact in different ways within the ligand-binding domain. The recent crystallization of ER-α in the presence of the agonist estradiol and the antagonist raloxifene affirms this hypothesis.

VI. ANTIESTROGENS AND ESTROGENS PROMOTE HIGH-AFFINITY ASSOCIATIONS OF ER WITH TARGET GENE PROMOTERS

It has been shown using a variety of structurally different compounds that the binding of any ligand is sufficient to induce activation of ER. Thus, although competitive binding may be important for antiestrogenic activity, it is likely that the ability of antagonist-activated ER to interfere with agonist signaling at one or more points downstream of activation is equally important. In support of this, it has been determined that all the known classes of antiestrogens efficiently promote ER homodimerization and facilitate the interaction of ER with DNA. Thus, the ability of antagonist-activated ER to compete with agonist-activated ER for EREs within target genes contributes to antagonist efficacy. It is the ability of the cellular transcription apparatus to discriminate between the different ER–ligand complexes positioned on their response elements which determines the relative agonist/antagonist activity of the bound ligand. This finding is important from a pharmaceutical perspective because it indicates that antiestrogens can be developed which can function in either of two ways. One type of antagonist could interact with ER, facilitating its association with DNA. However, the efficacy of this class of antagonists would depend on the ability of the ligand–receptor complex to engage the transcription apparatus. All the current antiestrogens fall into this latter category. Compounds which competitively inhibit estradiol–ER interactions but which do not promote receptor–DNA interactions represent the second class of antagonist. Theoretically, at saturating doses, compounds of this class should function as pure antagonists, tying up ER, effectively preventing it from participating in estradiol signaling. Although none of the currently available antiestrogens appear to function in this manner, the steroidal antiestrogen ICI182,780 is closest because it impairs, but does not totally inhibit, the ability of ER to interact with DNA. It has been difficult to determine the degree to which this altered DNA-binding affinity contributes to antagonist efficacy because ICI182,780-activated receptor is rapidly degraded in most cells. However, it is likely that pure antiestrogens will be developed which do not affect ER content within the cells but which do not deliver ER to DNA.

VII. DIFFERENT ER–ANTAGONIST COMPLEXES ARE RECOGNIZED DIFFERENTLY IN DIFFERENT CELLS

Although all antiestrogens can activate ER, it is clear that the structures of the resulting ER–ligand complexes are differentially recognized in different cells. Specifically, the biological activity of tamoxifen, raloxifene, and ICI182,780-activated ER are quite distinct. Consequently, a great deal of attention has focused on determining how differences in receptor shape influence biological activity. Insight into this issue came following the mapping of the domains within ER required for transcriptional activation. Independently, several groups defined two regions within ER which were required for transcriptional activity. One of these domains, called activation function-1 (AF-1), was located within the amino terminus, whereas another, AF-2, was wholly contained within the receptor ligand-binding domain (Fig. 3). Importantly, it was shown that in most cell contexts both AF-1 and AF-2 were required for optimal ER transcriptional activity. However, in some cell backgrounds AF-1 or AF-2 alone was sufficient. Thus, ER did not appear to function in an identical manner in all cells. Subsequent studies revealed that all antiestrogens efficiently inhibited AF-2 activity. Thus, in cells in which AF-2 was required, all ER antagonists, regardless of which class they represented, functioned as antagonists. However, in cellular contexts in which AF-2 was not required, it was observed that tamoxifen, droloxifene, nafoxidene, clomiphene, and toremifene (Fig. 1) exhibited partial ER agonist activity, whereas ICI182,780, GW5638, and raloxifene were inactive. Consequently, this suggested that the tissue-selective agonist activity of tamoxifen in bone, the cardiovascular system, and uterus was related to its ability to function as an AF-1 agonist. However, this hypothesis did not hold up to scrutiny when additional compounds were profiled. In the first instance it was observed that raloxifene, which was shown to be as effective as estrogen and tamoxifen as a bone protective agent in ovariectomized rats, did not function as an AF-1 agonist. In fact, it was observed that this compound could function as an "inverse agonist" on some ER targets. A second compound, GW5638, gave a similar profile in that it mimicked estrogen in the bone and the cardiovascular system but was devoid of AF-1 agonist activity. Interestingly, neither GW5638 nor raloxifene function as estrogens in the uterus. It became apparent, therefore, that AF-1 activity tracked with agonist activity in the uterus, whereas the bone protective activity of these compounds must be related to another, elusive, functional activity which these compounds share. It must be cautioned, however, that very little of the *in vitro* characterization of ER–ligands has been done in bone cells. Thus, it is quite possible that a completely different classification would result if these compounds were studied in a cell line which was more relevant to bone. The recent identification of receptor-associated factors, whose expression level influences the pharmacology of ER–ligands, gives support to this hypothesis. It is possible, therefore, that in the cell systems used thus far to study ER agonists and antagonists some bone cell-specific "enabling" transcription factor is missing. Testing of this hypothesis awaits the identification of relevant ER-responsive cells from bone and a reanalysis of the current ligand classifications.

VIII. RECEPTOR-ASSOCIATED PROTEINS AS DETERMINANTS OF THE CELL-SELECTIVE ACTION OF ER–LIGANDS

The information presented thus far on the pharmacology of ER–ligands indicates strongly that the mechanism of action of ER is not the same in all cells and that the receptor may contact the general transcription apparatus in each target cell in a different manner. Thus, ER pharmacology is likely to be influenced by transcription factors which are expressed in a cell-restricted or cell-specific manner. The identification over the past 3 years of a series of proteins which interact directly with ER in a ligand-dependent manner provides evidence in support of this hypothesis. One of the first receptor-associated coactivators (also called cofactors or adaptors) identified was the protein SRC-1 (steroid receptor coactivator). When this protein was overexpressed in target cells it had the predictable effect of potentiating the transcriptional activity of estradiol activated ER. However, more important, its overexpression also permitted tamoxifen-activated ER to manifest partial agonist activity. Similar results were obtained when additional coactivators were identified and assayed. Emerging from these studies, and a host of similar studies, was the concept that differential expression of SRC-1 and other receptor-associated proteins is at the root of ER–ligand specificity. It is clear that additional coactivators will be identified. Possibly among them will be a factor which permits tamoxifen, raloxifene, and GW5638 to function as bone-selective estrogens. One of the major challenges will be to define the systems in which each of the cofactors are important. It is anticipated that this activity will enable the development of highly selective ER modulators which target specific ER–coactivator complexes.

The identification of receptor-associated coactivators and the demonstration that the agonist activity of a wide variety of ER–ligands could be enhanced by their overexpression begged the question as to whether there were receptor-associated proteins which when overexpressed suppressed ER action and improved antagonist efficacy. This issue was addressed in a round-about manner. It had been demonstrated that the thyroid hormone receptor (TR), in the absence of hormone, could function as a potent suppressor of thyroid hormone-responsive genes. Recently, it was determined that this occurred because, in the unliganded state, TR was able to recruit corepressor proteins which actively shut down transcription. Two such proteins, NCoR and SMRT, have been identified. Interestingly, however, it has been shown that these corepressors could also interact with ER in the absence of ligand or in the presence of antagonists, and in doing so they could suppress its transcriptional activity. Not surprisingly, therefore, the overexpression of SMRT, or NCoR, increased the antagonist efficacy of tamoxifen and other antiestrogens. Thus, as is the case for coactivators, it is likely that the cellular concentration of corepressors will be an important component of ER pharmacology.

The identification proteins which function as ER coactivators or corepressors and the demonstration that the nature of the bound ligand influences the interaction of the receptor with these proteins have provided an important link between the structure of the ligand–ER complex and biological activity. Although many receptor-associated proteins have been found in the past 3 years, it is clear that many more will emerge and that it will be their relative rather than their absolute expression level which will determine ER agonist and antagonist selectivity.

IX. A WORKING MODEL TO EXPLAIN THE BIOLOGICAL ACTIVITY OF ER AGONISTS AND ANTAGONISTS

Our understanding of the molecular pharmacology of ER agonists and antagonists has evolved considerably during the past few years, helping us to demystify the concept of tissue selectivity. Currently, it is considered that different compounds have different affects on ER structure and that these conformational changes influence the interaction of ER with receptor coactivators and corepressors. The overall phenotypic response to a ligand, therefore, is a product of the affinity of these proteins for the particular

ER–conformer and their expression level within the cell. In the future, as additional coactivators and corepressors are identified, it is likely that it will be possible to screen for compounds which influence one particular cofactor–receptor interaction and that, as a consequence, a new generation of tissue-selective agonists and antagonists will be developed.

See Also the Following Articles

ESTROGEN REPLACEMENT THERAPY; ESTROGENS, OVERVIEW

Bibliography

Draper, M. W., Flowers, D. E., Huster, W. J., and Neild, J. A. (1993). Effects of raloxifene (LY 139481 HCL) on biochemical markers of bone and lipid metabolism in healthy postmenopausal women. In *Proceedings 1993. Fourth International Symposium on Osteoporosis and Consensus Development Conference* (C. Christiansen and B. Riis, Eds.), pp. 119–121. Handelstrykkeriet, Aalborg.

Horwitz, K. B., Jackson, T. A., Bain, D. L., Richer, J. K., Takimoto, G. S., and Tung, L. (1996). Nuclear receptor coactivators and corepressors. *Mol. Endocrinol.* 10, 1167–1177.

Jordan, V. C. (1992). The strategic use of antiestrogens to control the development and growth of breast cancer. *Cancer* 70, 977–982.

Katzenellenbogen, J. A., O'Malley, B. W., and Katzenellenbogen, B. S. (1996). Tripartite steroid hormone receptor pharmacology: Interaction with multiple effector sites as a basis for the cell- and promoter-specific action of these hormones. *Mol. Endocrinol.* 10, 119–131.

Kuiper, G. G. J. M., Carlson, B., Grandien, K., Enmark, E., Haggblad, J., Nilsson, S., and Gustafsson, J.-A. (1997). Comparison of the ligand binding specificity and transcript distribution of estrogen receptors α and β. *Endocrinology* 138, 863–870.

McDonnell, D. P., Clemm, D. L., Herman, T., Goldman, M. E., and Pike, J. W. (1995). Analysis of estrogen receptor function in vitro reveals three distinct classes of antiestrogens. *Mol. Endocrinol.* 9, 659–669.

Smith, C. L., Nawaz, Z., and O'Malley, B. W. (1997). Coactivator and corepressor regulation of the agonist/antagonist activity of the mixed antiestrogen, 4-hydroxytamoxifen. *Mol. Endocrinol.* 11, 657–666.

Antiprogestins

Irving M. Spitz
Shaare Zedek Medical Center

I. Structure
II. Progesterone Receptor
III. Mechanism of Action
IV. Pharmacokinetics
V. Clinical Applications
VI. Untoward Effects
VII. Conclusions

GLOSSARY

abortifacient An agent producing an abortion.
abortion The expulsion from the uterus of an embryo or fetus prior to the stage of viability.
antiprogestin A substance that inhibits the synthesis of progesterone and its transport or stability in the blood or that reduces its uptake by or effects on target organs.
mifepristone (RU 486) The first synthetic antiprogestin steroid synthesized which competes with progesterone for binding to the progesterone receptor and blocks the action of progesterone.
progesterone A hormone secreted by the corpus luteum, placenta, and in minute amounts by the adrenal cortex. It prepares the uterus for the reception and development of the fertilized ovum by transforming the endometrium from the proliferative to the secretory stage and maintains an optimal intrauterine environment for sustaining pregnancy.

progestin A term used for a synthetic or naturally occurring progestational agent.

receptor A structural protein molecule on the cell surface or within the cytoplasm that binds to a specific factor, such as a hormone, antigen, or neurotransmitter.

Progesterone plays a critical role in mammalian reproduction. After the discovery of the progesterone receptor by O'Malley and coworkers, it was realized that a progesterone receptor antagonist would have a major impact on female reproductive health. The search for such an antiprogesterone lasted for more than a decade and in 1981 Philibert, Deraedt, and Teutsch from the French pharmaceutical company, Roussel Uclaf, reported a newly synthesized glucocorticoid receptor antagonist known as RU 38486 which also displayed marked antiprogestin activity. RU 38486 was subsequently abbreviated to RU 486 and is now currently known by the generic name mifepristone. This article describes the structure, mechanism of action, pharmacokinetics, clinical applications, and untoward effects of antiprogestins. The focus is on mifepristone since almost all clinical studies have utilized this antiprogestin.

I. STRUCTURE

Mifepristone is similar in structure to progesterone and glucocorticoids, but it lacks the C19 methyl group and the 2-carbon side chain at C17 and has a conjugated C9–C10 double bond (Fig. 1). It is a derivative of norethindrone with a (4-dimethylamino) phenyl group at the 11β position (which is responsible for its antagonistic activity) and a 1-propynyl chain at the 17α position [accounting for its high binding affinity to the progesterone receptor (PR)]. Following the publication of the structure of mifepristone, over 400 additional antiprogestins have been synthesized. Although a pure antiprogestin which does not bind to the glucocorticoid receptor (GR) has not been reported, a number of compounds have been described with minimum antiglucocorticoid activity. Limited clinical studies have been conducted with lilopristone and onapristone. Their structures are very similar to mifepristone (Fig. 1).

II. PROGESTERONE RECEPTOR

The antiprogestin action of mifepristone is mediated by the PR, a ligand-activated transcription factor with domains for DNA binding, hormone binding, and transactivation. Members of this nuclear receptor superfamily also include the androgen, estrogen, glucocorticoid, and mineralocorticoid receptors, as well as receptors for thyroid hormones, retinoids, and vitamin D. Philibert and coworkers originally showed that mifepristone has a binding activity five times greater than that of progesterone to the rabbit uterine PR and three times greater than that of dexamethasone to the rat thymus GR. Binding affinity for the androgen receptor was 25% of that of testosterone, and there was no binding to the estrogen or mineralocorticoid receptors.

The progesterone receptor exists as two separate isoforms, PR-A and PR-B. The structures of both isoforms are similar, although the PR-B isoform contains a fragment of 164 amino acids at the amino-terminal end which is absent from the PR-A isoform. Both forms of PR are derived from a single gene as a consequence of alternate initiation of transcription from distinct promoters.

Benhamou and coworkers demonstrated that the amino acid glycine at position 722 (Gly722) in the hormone-binding domain of the human PR and at the comparable position of the PR of most other species is critical for mifepristone binding and action. The hamster and the chicken are insensitive to mifepristone and have a cysteine rather than a glycine residue at this position. The PR of these species binds progesterone but not mifepristone. Following substitution of this cysteine with glycine in the chicken PR, binding of mifepristone and antagonistic action of this compound are observed. Since glycine is the only amino acid without a side chain, the presence of amino acid side chains in this position may sterically impede mifepristone binding to PR.

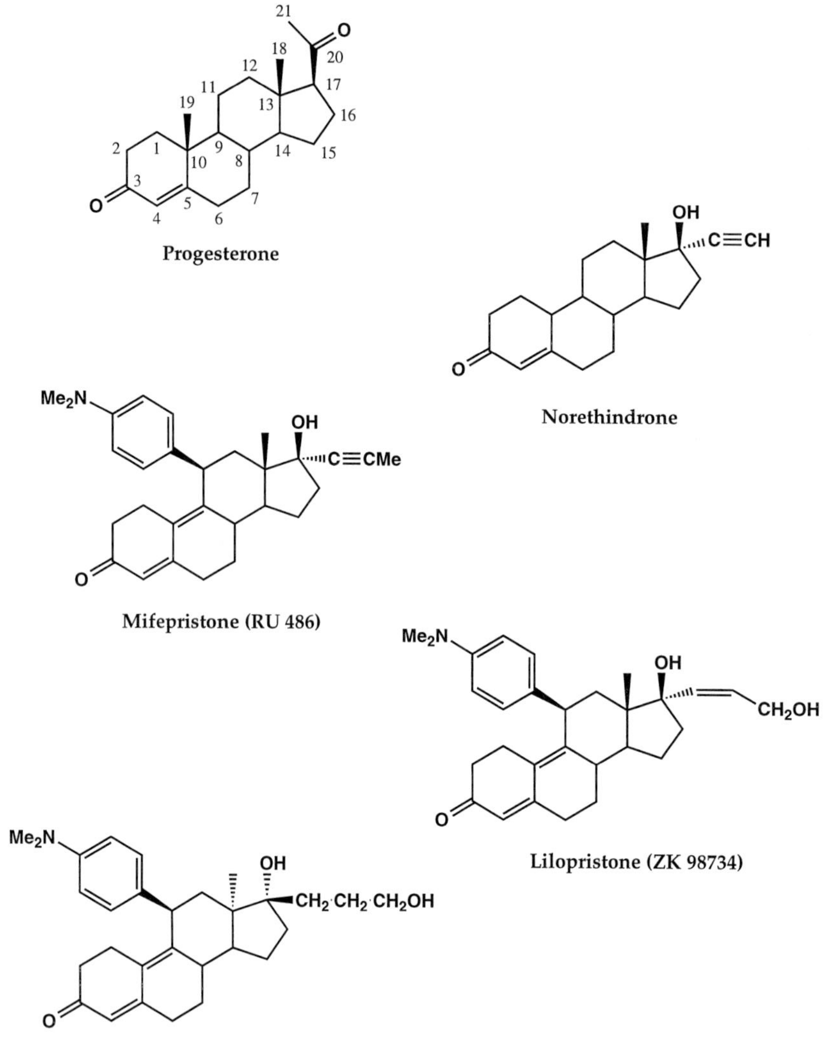

FIGURE 1 Structural formulas of progesterone, norethindrone, mifepristone, lilopristone, and onapristone.

III. MECHANISM OF ACTION

Following binding to the PR, both progestins and antiprogestins produce changes in conformation of the PR and convert it from a non-DNA-binding form to one which will bind DNA (Fig. 2). This transformation is accompanied by a loss of heat shock proteins and dimerization. The activated progesterone–PR dimer binds to progesterone response elements in the promoter region of progesterone responsive genes, and in the presence of other nuclear transcription factors it increases the rate of transcription of these genes producing agonistic effects at the cellular and tissue levels. The mifepristone–PR complex also binds to progesterone response elements but these DNA-bound receptors are transcriptionally inactive. It has been suggested that the onapristone–PR complex fails to bind to progesterone response elements. This may be due to its lower binding affinity.

In certain circumstances, antiprogestins may also function as progestin agonists. This may be related to the fact that the C-terminal end of the PR contains a repressor function which plays an important role

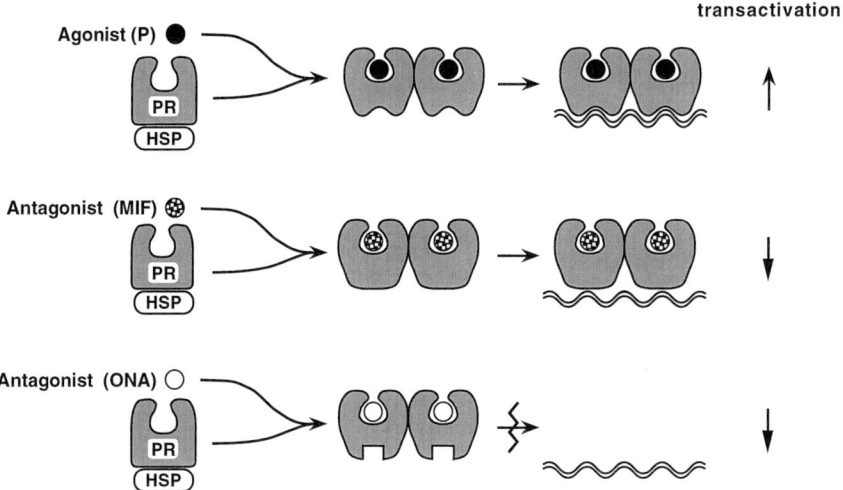

FIGURE 2 Proposed mechanisms of action of progesterone (P) and the progesterone antagonists mifepristone (MIF) and onapristone (ONA). HSP, heat shock protein; PR, progesterone receptor (reproduced with permission from Robbins and Spitz, 1996).

in determining antagonistic activity. PR mutants lacking the C-terminal 42 amino acids can be activated by mifepristone. In the presence of cGMP and cAMP, both progesterone and mifepristone (but not onapristone) activate transcription with the PR-B but not the PR-A isoform. This cross talk between two signal transduction pathways, one involving steroid receptors and the other cAMP-regulated proteins, overrides the antagonist effects of mifepristone and may also explain its agonist action (Fig. 3).

The two PR isoforms have variable effects on transcription even in the absence of cAMP. When expression vectors that encoded exclusively for PR-A or PR-B were transiently transfected into cells together with progesterone-responsive reporters in the presence of both progesterone and a progesterone antagonist, there was always activation of transcription with PR-B. Although PR-A was often inactive, it acted as a potent transdominant repressor of PR-B-mediated transcription (Fig. 4). When both isoforms were transfected into the same cell, the effect of the PR-A dominated and there was failure of transcription. These results imply that a progestin antagonist may display agonist actions depending on which PR isoform predominates.

On occasion, antiprogestins may also display antiestrogenic (antiproliferative) activity. In the presence of antiprogestins and certain progestins, the PR-A but not the PR-B isoform inhibits estrogen receptor-mediated gene transcription. This PR-A inhibition of estrogen-mediated gene transcription may explain the antiestrogenic activity of antiprogestins (Fig. 5). Thus, the PR-A isoform is crucial in determining the biological response of both progestins and antiprogestins and its modulating effect on transcription suggests how cells can generate dissimilar responses to antiprogestins. Depending on which PR isoform is predominant, antiprogestins may demonstrate pro-

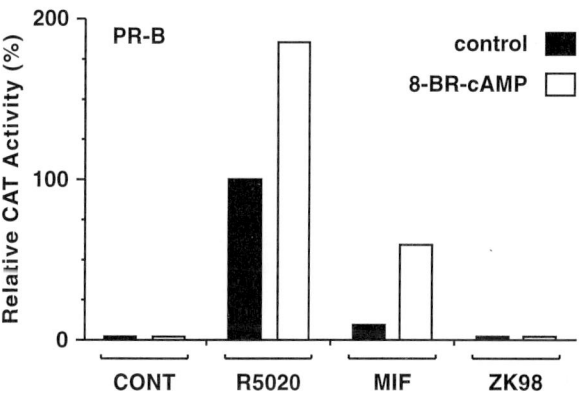

FIGURE 3 T47D-YB cells transiently transfected with the MMTV-CAT reporters were treated with progestins with or without 8-Br-cAMP. CONT, control; R5020, synthetic progestin; MIF, RU 486; ZK98, onapristone (modified with permission from Sartorius *et al.*, 1994).

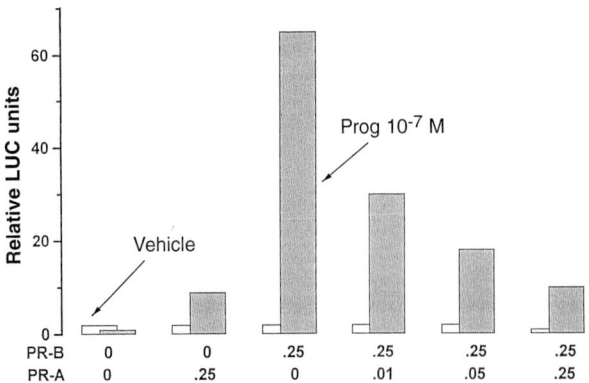

FIGURE 4 HeLa cells were transiently transfected with either 0.25 mg PR-A or 0.25 mg PR-B alone or 0.25 mg PR-B in the presence of increasing concentrations of phPR-A together with 5 mg MMTV-LUC reporter. Cells were treated with or without 10^{-7} M progesterone as indicated for 24 hr and assayed for LUC activity. PR-A is a cell-specific *trans*-dominant repressor of PR-B function (modified with permission from Vegeto et al., 1993).

gestin antagonistic, agonistic, or antiestrogenic effects (Fig. 6). Since dimerization of the PR is an important step for its binding to DNA, it is possible that PRs can form homodimers (PR-A/PR-A and PR-B/PR-B) or heterodimers (PR-A/PR-B), both of which can interact with the same DNA sequences. The type

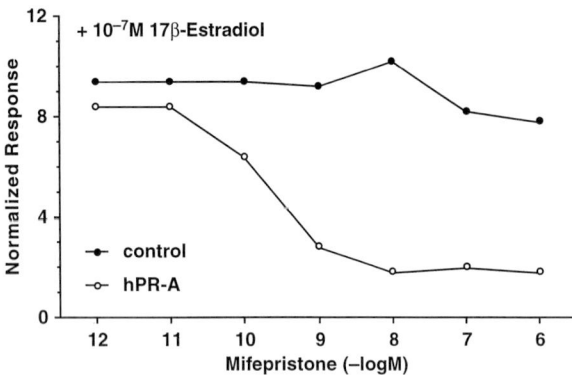

FIGURE 5 Monkey kidney CV-1 cells were transiently transfected with a MMTV-ERE-LUC reporter and vectors producing the human estrogen receptor alone or in combination with a vector producing PR-A. The transcriptional activity of the estrogen receptor was measured following the addition of 10^{-7} M estradiol alone or in combination with increasing concentrations of mifepristone. There is a dose-dependent inhibition of ER activity by mifepristone (modified with permission from McDonnell and Goldman, 1994).

PR isoforms	B B	A A	A B
Progestin agonist activity (with P or AP)	++++	0, +	0
Estrogen antagonist activity	0	++++	+++

FIGURE 6 Response to antiprogestins. The type of PR dimer formed (homodimer or heterodimer) determines whether the response will be as a progestin agonist, antagonist, or an estrogen antagonist. P, progestin; AP, antiprogestin.

of dimer formed may thus explain the biological response (Fig. 6). The expression levels of the A and B isoforms may vary in different tissues and change during the follicular and luteal phases of the menstrual cycle.

IV. PHARMACOKINETICS

Serum mifepristone levels reached a maximum 1 hr following oral administration. After single doses of 100 mg or less, the disappearance of mifepristone follows first-order kinetics with a half-life of 20–25 hr. Following higher doses (200–800 mg), there is an initial redistribution phase of 6–10 hr followed by a plateau in serum levels for 24 hr or more. With these larger doses, there is no significant dose-dependent difference in serum concentrations within the first 48 hr (Fig. 7). This is probably related to the fact that in humans, mifepristone binds to orosomucoid, an $\alpha 1$ acid glycoprotein. Orosomucoid binding sites are saturated at doses of mifepristone more than 100 mg. Different pharmacokinetic profiles are observed with other antiprogestins. For example, the half-life of onapristone is only 2 or 3 hr since it does not bind as avidly to orosomucoid.

V. CLINICAL APPLICATIONS

A. Pregnancy Termination

Since progesterone is essential for the initiation and maintenance of pregnancy, an antiprogestin would be expected to disrupt this process (Fig. 8).

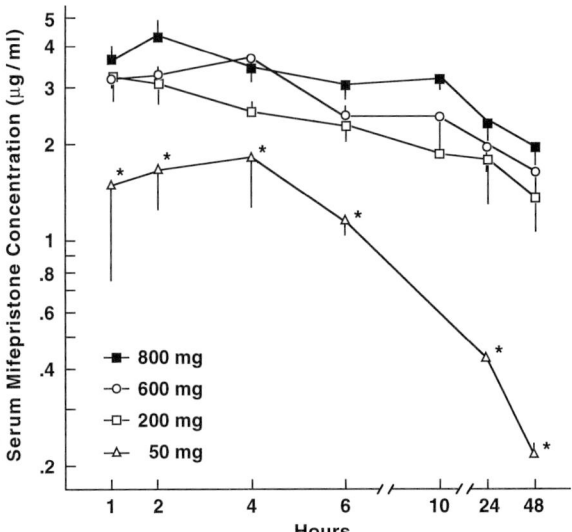

FIGURE 7 Half disappearance time of mifepristone after administration of 50–800 mg in the midluteal phase. Mean + SEM. *$p < 0.05$ compared to 200–800 mg (reproduced with permission from Robbins and Spitz, 1996).

This was first documented by Herrmann and coworkers in 1982. Subsequently, numerous studies have documented that when 200–600 mg mifepristone is administered to women with duration of gestation of 49 days or less, there is complete termination of pregnancy in 64–85%. This success rate is inadequate for general clinical use.

The addition of a prostaglandin administered 36–48 hr after mifepristone significantly improves the success rate. Prostaglandins enhance uterine contractions (Fig. 8). Antiprogestins increase endogenous endometrial prostaglandin concentrations and enhance the myometrial sensitivity to exogenous prostaglandins. Prostaglandins used are misoprostol (cytotec), which is given either orally or vaginally, and gemeprost, which is a vaginal pessary. This antiprogestin–prostaglandin combination provides an effective method of medical termination of pregnancy. In women with duration of gestation of under 49 days, the success rate reaches 95%. Failures include women with incomplete abortion or ongoing pregnancy and they are treated by surgical termination of pregnancy. Dilatation and curettage may be required for severe bleeding and rarely for other medical reasons (e.g., severe pain or vomiting). The success rate for successful pregnancy termination falls with increase in duration of gestation (Fig. 9). Gemeprost is a more potent prostaglandin than misoprostol. If it is used, the window of application can be extended to a duration of gestation of up to 63 days with similar results (Fig. 9). This method has been

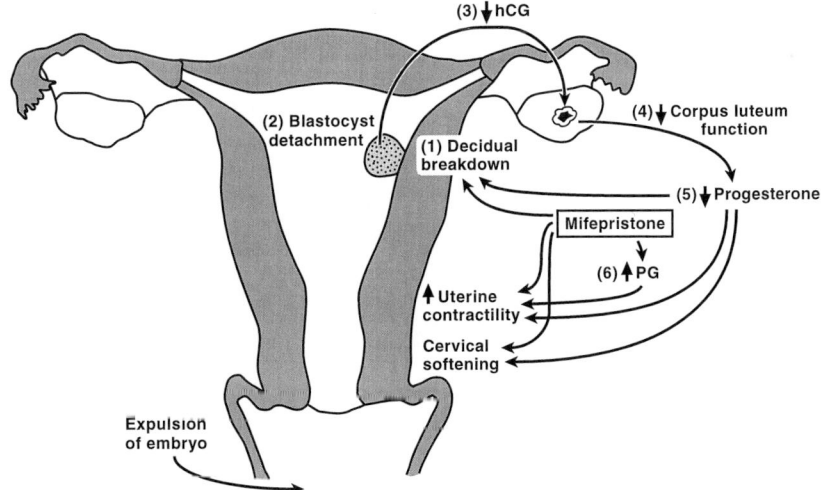

FIGURE 8 Action of mifepristone in the pregnant uterus. Initially, mifepristone promotes decidual breakdown, leading to detachment of the blastocyst. With the resulting decrease in secretion of human chorionic gonadotropin (hCG), there is a decrease in progesterone secretion by the corpus luteum. The decrease in progesterone secretion further accentuates decidual breakdown. Mifepristone also stimulates the accumulation of prostaglandin (PG) and enhances uterine sensitivity to prostaglandin, disturbing the balance between prostaglandin and progesterone and thereby increasing uterine contractility. An independent action of mifepristone results in cervical softening, which facilitates expulsion of the blastocyst (reproduced with permission from Spitz and Bardin, 1993).

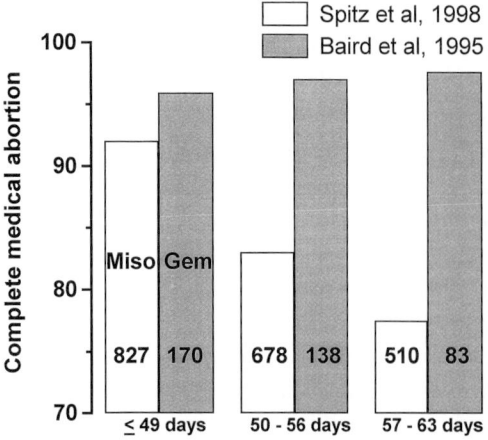

FIGURE 9 Pregnancy termination in women with duration of gestation of <49 days, 50–56, days and 57–63 days with misoprostol (Miso) or gemeprost (Gem). Doses of mifepristone used were 600 mg with the misoprostol study and 200 mg with the gemeprost study. Numbers of women studied are indicated in each column. Data on misoprostol are from Spitz et al. (1998) and those on gemeprost are from Baird et al. (1995).

approved for use in France for up to 49 days gestation and in the United Kingdom, Sweden, and China for up to 63 days.

B. Cervical Softening

In addition to their ability to enhance myometrial contractility, antiprogestins also dilate and soften the uterine cervix. Antiprogestins do not act on the cervix by stimulating endogenous prostaglandin production; rather, their effect originates from inflammatory cells and chemotactic agents such as cytokines. Because of their action on the uterine cervix, antiprogestins have many other clinical applications. They are useful in the preoperative preparation of women for first-trimester vacuum aspiration. Mifepristone is usually administered 48 hr prior to surgical abortion, is as effective as prostaglandins, and has significantly fewer side effects. In second-trimester abortions, pretreatment with mifepristone reduced the interval between prostaglandin administration and expulsion. Furthermore, the dose of prostaglandin required was reduced and the women experienced considerably less pain. Mifepristone is very effective in promoting expulsion of the uterine contents following intrauterine fetal death. This antiprogestin has also been used effectively to induce labor at the end of the third trimester; however, it does cross the fetal–placental barrier and further studies are required to ensure that no untoward effects are observed on the fetus.

Pretreatment with mifepristone also reduces the force required to dilate and soften the cervix, particularly in nonpregnant primigravid women. This could prove to be useful for planned outpatient procedures such as insertion or removal of an intrauterine device, hysterography, dilatation and curettage, or any other procedures requiring access to the uterine lumen.

C. Contraception

In addition to its critical role in pregnancy, progesterone facilitates the luteinizing hormone (LH) surge, transforms the endometrium to a secretory state, and maintains endometrial integrity. Thus, an antiprogestin may have contraceptive potential by (i) inhibiting the LH surge and ovulation, (ii) preventing secretory transformation of the endometrium and delaying the emergence of the implantation window, and (iii) shedding the endometrium, resulting in menstrual bleeding and prevention of implantation.

1. Inhibition of the LH Surge and Ovulation

Mid- or late follicular-phase mifepristone administration blocks follicular development and delays the LH surge and ovulation by 12 days. To inhibit ovulation, mifepristone must be given either continuously or intermittently. Daily mifepristone doses of 2–10 mg for 1 month blocked ovulation, although this did not consistently occur with intermittent dose schedules administered at weekly intervals. In all anovulatory cycles, estradiol levels were in the mid-follicular phase range and associated with suppressed progesterone levels, raising the possibility of unopposed estrogen action on the endometrium. For this reason and because ovulation inhibition may not be consistently observed, this strategy has been disbanded as a potential contraceptive method.

2. Prevention of Secretory Transformation of the Endometrium

Because endometrial maturation and development is critically dependent on progesterone, it is possible that this may be altered by antiprogestins. In women, a single 200-mg dose of mifepristone on the second day after the LH surge resulted in retardation of endometrial development but did not alter the length of the cycle or serum follicle-stimulating hormone or progesterone levels. When the strategy was used as the only contraceptive method in 21 unprotected women for up to 12 months, only one clinical pregnancy resulted. Similar results have been observed in monkeys.

These contraceptive effects are explained on the basis of a delay in endometrial maturation with a shift of the implantation window. Very low doses of mifepristone (0.5 mg daily or 2.5 mg once a week) can postpone endometrial maturation without altering bleeding patterns, ovulation, or hormonal profiles. These studies support the concept of endometrial contraception based on a delay in endometrial maturation rendering it nonreceptive to implantation of a blastocyst. Further studies must be conducted to identify precise endometrial markers critical for development of the implantation window and to determine which are altered by antiprogestins. Armed with this knowledge, the precise antiprogestin dose which retards the development of these markers can be selected. Thereafter, studies can be conducted in unprotected women.

3. Shedding of the Endometrium Resulting in Menstrual Bleeding and Prevention of Implantation

With the fall in estradiol and progesterone at the end of the luteal phase, endometrial bleeding occurs. Indeed, one of the major effects of progesterone is to maintain endometrial integrity. Unlike the marked sensitivity of the endometrial morphology to antiprogestins, higher doses (≥ 50 mg) are required to produce endometrial shedding and bleeding, which occurs within 3 days of mifepristone administration.

When used occasionally after missed menses, mifepristone together with a prostaglandin prevented pregnancy in women with menstrual delay of up to 11 days. Studies have also been conducted using mifepristone as a "menses regulator," i.e., mifepristone was administered every month at the end of the cycle independent as to whether the woman was pregnant or not. With this repeated monthly use, the results are not as promising. In the largest studies published to date, it was shown that the failure rate ranged from 17 to 19%. This is similar to that when mifepristone is used alone to terminate early pregnancy without a prostaglandin. This regimen is clearly not clinically acceptable and further studies using other approaches must be undertaken to improve its efficacy.

D. Postcoital Administration

Mifepristone has also been used as an emergency contraceptive. Two randomized trials have compared 600 mg of mifepristone with the classical estrogen–progestin (Yuzpe) regimen. Mifepristone given within 72 hr of unprotected intercourse was 100% effective in preventing pregnancy. Side effects such as nausea, vomiting, headache, and breast tenderness were less frequent among women given mifepristone; however, women using mifepristone often had a delay of more than 3 days in the onset of the next menstrual period. This is clearly an obvious drawback of mifepristone because the onset of menses reassures the woman who has used emergency contraception that she is not pregnant.

E. Gynecological Indications

As discussed previously, antiprogestins may also display antiestrogenic (antiproliferative) activity. In view of these properties, antiprogestins have a role in the treatment of endometriosis, an estrogen-dependent condition. In clinical studies with daily mifepristone doses of 50 mg for 6 months, there was an improvement in pelvic pain and a decrease in the extent of disease as determined by laparoscopy.

Antiprogestins also have clinical application in the treatment of uterine fibroids. Although this might also represent an antiproliferative effect of mifepristone, there is both clinical and *in vitro* evidence that progestins promote fibroid growth. Thus, under these conditions, mifepristone may be acting as a classical antiprogestin. After 3 months of daily treatment with mifepristone in doses of 25 and 50 mg, there were significant decreases in the volume of the fibroid. Unlike treatment with gonadotropin-releasing hormone agonists, there was no decrease in bone mineral density in the treatment of patients with endometriosis and fibroids with antiprogestins.

F. Tumors

Antiprogestins have also been proposed in the treatment of tumors which contain steroid receptors. Many meningiomas contain progesterone receptors and antiprogestins inhibit growth of meningioma cells in culture and reduce the size of human meningioma implanted into nude mice. In one clinical trial, patients with unresectable meningiomas received 200 mg mifepristone daily and one-third of patients demonstrated objective responses as shown by reduced tumor size on computer tomography or magnetic resonance imaging scan and improvement in visual field examination. Mifepristone has also been used in advanced breast carcinoma, but the results in the three published studies have been disappointing. Other work in animals has suggested that antiprogestins could be used in other steroid-dependent tumors, including those of the ovary, prostate, and endometrium.

G. Other Clinical Indications

Since mifepristone is also an antiglucocorticoid, it has been used to treat Cushing's syndrome due to ectopic adrenocorticotropin hormone secretion and adrenal carcinoma. Other situations in which the antiglucocorticoid properties of this class of compound may prove useful include local application in eyedrops to lower intraocular pressure in glaucoma and in the prevention of the progression of viral diseases in humans. There is also some evidence for their use in the treatment of burns, certain forms of hypertension, depression, arthritis, and cataracts.

VI. UNTOWARD EFFECTS

In view of the antiglucocorticoid properties of mifepristone, hypoadrenalism must be considered as a possible consequence of long-term treatment. Although this has been reported with doses exceeding 200 mg/day, it is an uncommon occurrence in humans with an intact pituitary–adrenal axis. Nevertheless, if the clinical picture does suggest hypoadrenalism, then glucocorticoid replacement is indicated.

No untoward adverse events have been reported during single administration for pregnancy interruption. Those which occur are invariably due to the prostaglandin component of the regimen and to the associated pregnancy and abortive process. Common side effects observed during long-term treatment include fatigue, nausea, anorexia, and vomiting. Weight loss, skin rashes, cessation of menses in premenopausal women, intermittent hot flashes, transient thinning of the hair, biochemical hypothyroidism, and occasional decrease in libido and gynecomastia in males have also been reported. The latter is presumably due to the fact that mifepristone binds with low affinity to androgen receptors. When administered chronically to women in a dose ≥ 50 mg daily, there is progesterone suppression and the endometrium is exposed to unopposed estrogen levels. The clinical consequences of this remain to be fully resolved.

Since some women fail to abort and may continue with their pregnancy following mifepristone administration, it is important to determine if there are any teratogenic effects. In rabbits, because of progesterone deficiency, skull deformities are observed due to mechanical effects secondary to uterine contractions. There are isolated case reports of normal pregnancies and offspring when women have taken mifepristone alone or in combination with a prostaglandin, have not aborted, and have elected to continue their pregnancies. It should be mentioned that the prostaglandins used in association with antiprogestins for pregnancy termination may be associated with teratogenic effects.

VII. CONCLUSIONS

Antiprogestins are among the most controversial and yet the more interesting therapeutic compounds developed in the past 20 years. It is evident that antiprogestins have numerous proven and potential clinical applications. To regard antiprogestins exclusively as abortifacients is to grossly underestimate their usefulness. In addition to providing an effective and safe means of medical abortion, these agents may be used for other obstetric indications, in various gynecologic disorders, as postcoital agents, as potential contraceptives, as well as in the treatment of patients with cancer, various forms of Cushing's syndrome, and other miscellaneous conditions. It is of interest that few of these clinical applications were envisaged when antiprogestins were first described 15 years ago.

The initial clinical trial of mifepristone for medical termination of pregnancy was conducted in 1982. However, currently antiprogestins are only available for pregnancy termination in France, the United Kingdom, Sweden, and China. This is probably related to economic, political, and legal reasons as well as opposition from antiabortion groups. It is expected that the U.S. Food and Drug Administration will give approval for the use of mifepristone for medical termination of pregnancy in the United States in the near future. It is hoped that this will be an impetus for the introduction of mifepristone and other antiprogestins into other countries. Further studies can then be conducted to define the therapeutic values and benefits of these fascinating compounds.

See Also the Following Articles

ABORTION; CERVIX; ENDOMETRIUM; PROGESTERONE ACTIONS ON REPRODUCTIVE TRACT; PROGESTINS

Bibliography

Baird, D. T., Sukcharoen, N., and Thong, K. J. (1995). Randomized trial of misoprostol and cervagem in combination with a reduced dose of mifepristone for induction of abortion. *Hum. Reprod.* **10**, 1521–1527.

Baulieu, E. E. (1989), Contragestion and other clinical applications of RU 486, an antiprogesterone at the receptor. *Science* **245**, 1351–1357.

Beier, H. M., and Spitz, I. M. (1994). Progesterone antagonists in reproductive medicine and oncology. *Hum. Reprod.* **9**(Suppl.).

Benhamou, B., Garcia, T., Lerouge, T., Vergezac, A., Gofflo, D., Bigogne, C., Chambon, P., and Gronemeyer, H. (1992). A single amino acid that determines the sensitivity of progesterone receptors to RU 486. *Science* **255**, 206–209.

Donaldson, M. S., Dorflinger, L., Brown, S. S., and Benet, L. Z. (1993). *Clinical Applications of Mifepristone (RU 496) and Other Antiprogestins*. National Academy Press, Washington, DC.

Horwitz, K. B. (1992). The molecular biology of RU 486. Is there a role for antiprogestins in the treatment of breast cancer? *Endocr. Rev.* **13**, 146–163.

McDonnell, D. P. (1995). Unraveling the human progesterone receptor signal transduction pathways (insights into antiprogestin action). *Trends Endocrinol. Metab.* **6**, 133–138.

McDonnell, D. P., and Goldman, M. E. (1994). RU 486 exerts antiestrogenic activities through a novel progesterone receptor A form-mediated mechanism. *J. Biol. Chem.* **269**, 11945–11949.

Philibert, D. (1984). RU 38486: An original multifaceted antihormone *in vivo*. In *Adrenal Steroid Antagonism* (M. K. Agarwal, Ed.), pp. 77–101. de Gruyter, Berlin.

Robbins, A., and Spitz, I. M. (1996). Mifepristone: Clinical pharmacology. *Clin. Obstet. Gynecol.* **39**, 436–450.

Sartorius, C. A., Groshong, S. D., Miller, L. A., Powell, R. L., Tung, L., Takimoto, G. S., and Horwitz, K. B. (1994). New T47D breast cancer cell lines for the independent study of progesterone B- and A-receptors: Only antiprogestin-occupied B-receptors are switched to transcriptional agonist by cAMP. *Cancer Res.* **54**, 3868–3877.

Spitz, I. M., and Bardin, C. W. (1993). Mifepristone (RU 486)—A modulator of progestin and glucocorticoid action. *N. Engl. J. Med.* **329**, 404–412.

Spitz, I. M., Croxatto, H. B., and Robbins, A. (1997). Antiprogestins: Mechanism of action and contraceptive potential. *Annu. Rev. Pharmacol. Toxicol.* **36**, 47–81.

Spitz, I. M., Bardin, C. W., Benton, L., and Robbins, A. (1998). Early pregnancy termination with mifepristone and misoprostol in the United States. *N. Engl. J. Med.* **338**, 1241–1247.

Vegeto, E., Shahbaz, M. M., Wen, D. X., *et al.* (1993). Human progesterone receptor A form is a cell- and promoter-specific repressor of human progesterone receptor B function. *Mol. Endocrinol.* **7**, 1244–1255.

Ants

see Social Insects

Apes

see Primates, Nonhuman

Apgar Score

Karen D. Hendricks-Muñoz and Joseph Dancis

New York University

I. Introduction
II. History
III. Method
IV. Interpretation

GLOSSARY

Apgar Score A clinical scoring system to evaluate the newborn condition beginning at 1 min after birth.
asphyxia Hypoxemia associated with hypercapnia and metabolic acidosis.
meconium aspiration The inhalation of meconium while *in utero*.
prematurity Birth prior to 37 weeks gestation.

I. INTRODUCTION

The Apgar Score, named after its originator Virginia Apgar, was proposed in 1952 as a rapid method for evaluating the condition of the newborn infant. It was hoped that it might also serve to predict survival, to monitor the effectiveness of resuscitation interventions, and to compare the level of perinatal care in different hospitals. An incidental but important advantage would be that it would ensure close observation of the infant during the first vital minutes after birth. Not all of these aspirations have been achieved, but the Apgar Score has gained near universal usage in the delivery room.

II. HISTORY

Virginia Apgar, an obstetric anesthesiologist, was dissatisfied with the care of the newborn infant during the perinatal period in the early 1950s. She noted that the infant was usually set aside after birth and attention was focused on the care of the mother. Dr. Apgar stressed that the period immediately after birth was critical particularly if the infant required resuscitative assistance. She felt that "unscientific" subjective conclusions about the condition of the infant should be replaced by an objective quantifiable index that would alert the physician to the need for intervention. She designed a scoring system that could be easily taught to anesthesiologists, nurses, and physicians-in-training because she recognized that the obstetrician was often distracted by concerns for the mother and was also inclined to give a higher score than a less "emotionally" involved observer.

Originally, the Apgar Score was limited to the 1-min score. In 1966, serial measurements were instituted as evidence accumulated that repeated scoring at later intervals provided additional prognostic information. Interpretation of the Apgar Score has been

TABLE 1
Apgar Score

Clinical sign	Rating		
	0	1	2
Heart rate[a]	Absent	<100 beats/min	>100 beats/min
Respiratory effort	Absent	Irregular, weak cry	Regular, strong cry
Muscle tone	Flaccid	Some flexion of upper extremities	Well flexed, active motion
Reflex irritability[b]	No response	Grimace	Cough or sneeze
Color	Central cyanosis	Acrocyanosis	Completely pink

Note. Adapted from V. Apgar, A proposal for a new method of evaluation of the newborn infant, *Curr. Res. Anesth. Analg.* 32, 260, 1953. The Apgar Score is derived by adding the ratings for the five measurements.

[a] Heart rate is estimated by auscultation or palpating the umbilical cord or radial artery for 6 sec and multiplying by 10.

[b] Response to nasal suction.

modified over the years but the major elements have been retained. Dr. Apgar has been particularly successful in stimulating interest in the immediate newborn period and in attracting the pediatrician into active participation in the delivery room.

III. METHOD

The Apgar Score is derived from five clinical signs. Each sign is rated as 0, 1, or 2, with a total score of 10 indicating the most satisfactory clinical condition (Table 1). A score is assigned at 1 and 5 min. One minute is defined as 60 sec after the head and feet of the infant are both visible. If the score is less than 7 at 5 min, the infant is reevaluated at 10, 15, and 20 min or until two scores of 8 or more are reached.

IV. INTERPRETATION

The 1-min score provides an assessment of the infant's condition at birth and of the need for intervention. If an infant is severely depressed or apneic, at birth, resuscitation should be started without delay. Of the five clinical signs, heart rate, respiratory effort, and color are the best indicators of the need for resuscitation and are recommended as such by the American Heart Association and American Academy of Pediatrics. The three signs are interdependent. Bradycardia is often caused by inadequate respiratory exchange and is commonly associated with central cyanosis. Ensuring a patent airway and assistance in breathing will often improve cardiac output and circulation, with return to a normal heart rate and improved color in the infant. The remaining two signs, tone and reflex irritability, are signs that are closely linked to the neurologic condition or physiologic maturity of the infant.

Initially, a low 1-min score was interpreted as indicating "birth asphyxia." This was an unwarranted assumption because many factors can depress the score, such as maternal sedation, newborn sepsis and immaturity. The term "asphyxia" should be abandoned as recommended by the American College of Obstetricians because it lacks precision. More accurate and informative methods are available for the diagnosis of oxygen deprivation.

The 5-min score provides an assessment of the infant after this first critical period of life and of its response to any intervention. It is of little value in predicting eventual outcome.

The later Apgar scores (10, 15, and 20 min) have some prognostic validity. This use of the Apgar Score has caused most controversy. The best data on the subject come from the Neonatal National Collaborative Perinatal Project (Fig. 1). A persistent low score has grave implications for survival and neurological damage, using cerebral palsy as an indicator. To put this in proper context, however, many infants with a persistent low score survive and do not develop

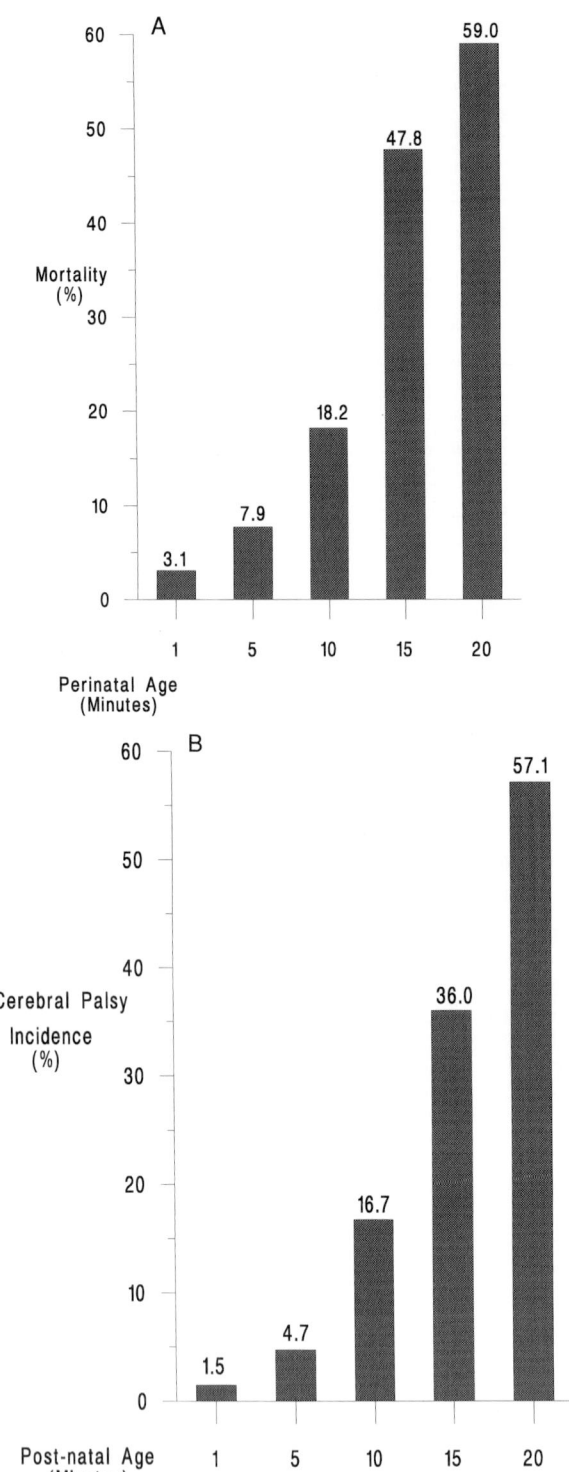

FIGURE 1 Outcome of low Apgar Score (0–3). Persistent low Apgar Scores have grave implications for mortality (A) and neurological damage (cerebral palsy) (B). Reproduced from Paneth (1993, p. 98).

cerebral palsy. Furthermore, cerebral palsy is seen in a small percentage of infants who have never had a low Apgar Score.

The premature infant often has a low Apgar Score because three of the five signs are affected by immaturity (respiratory effort, tone, and reflex irritability). The infant of less than 30 weeks gestation will commonly have a 1-min score of less than 6 and a 5-min score of less than 8. There are no data on the prognostic value of the Apgar Score in the immature infant which reflect recent modifications in neonatal care.

Cesarean section is often associated with a low 1-min score as a result of the retention of fluid in the lungs that is normally expressed during labor. In the absence of complications, the score rapidly increases to normal levels.

Meconium aspiration may severely depress the Apgar Score by compromising respiratory exchange and as a result of repeated suction to remove meconium. The score is useful in monitoring the effectiveness of treatment.

See Also the Following Articles

CESAREAN DELIVERY; FETAL MONITORING AND TESTING; PRETERM LABOR AND DELIVERY

Bibliography

Ehrenkranz, R. A. (1994). Newborn intensive care, the newborn intensive care unit. In *Principles and Practice of Pediatrics* (F. A. Oski, C. D. DeAngelis, R. D. Feigin, and J. B. Warshaw, Eds.), pp. 285–287. Lippincott, Philadelphia.

Fanaroff, A. A., Martin, R. J., and Miller, M. J. (1989). Identification and management of high-risk problems in the neonate. In *Maternal Fetal Medicine: Principles and Practice* (R. J. Creasy and R. Resnik, Eds.), pp. 1150–1152. Saunders, Philadelphia.

Freeman, J. M., and Nelson, K. B. (1988). Intrapartum asphyxia and cerebral palsy. *Pediatrics* **82**, 240–249.

Manganaro, R., Mami, C., and Gemelli, M. (1994). The validity of the Apgar Score in the assessment of asphyxia at birth. *Eur. J. Obstet. Gynecol. Reprod. Biol.* **54**, 99–102.

Paneth, N. (1993). Etiologic factors in cerebral palsy. *Clin. Invest. Med.* **16**, 98.

Stanley, F. J. (1994). Cerebral palsy trends: Implication for perinatal care. *Acta Obstet. Gynecol. Scand.* **73**, 5–9.

Aphids

Jim Hardie
Imperial College of Science, Technology, and Medicine

I. Introduction
II. Parthenogenesis
III. Sexual Reproduction
IV. Cytogenetics

GLOSSARY

anholocyclic A life cycle involving continuous parthenogenetic reproduction; sexual forms are not produced.
apomixis Oocytes produced by mitosis.
autoecious Remaining on one host throughout the year; not having an alternation between primary and secondary hosts.
cyclical parthenogenesis The alternation between parthenogenesis and sexual reproduction.
heteroecious A life cycle with a seasonal alternation between primary and secondary hosts.
holocyclic Life cycles in which parthenogenetic and sexual modes of reproduction alternate; cyclical parthenogenesis.
morph The form of the adult aphid.
parthenogenesis The development of an unfertilized egg into a new individual.
pedogenesis A process in which the reproductive organs undergo an accelerated development in relation to the rest of the body.
polyphenism The occurrence of two or more distinct phenotypes which can be induced in individuals of the same genotype by extrinsic factors.
primary host The host plant on which the sexual part of the life cycle occurs; winter host.
secondary host The host on which only parthenogenetic reproduction occurs in heteroecious aphids; summer host.
telotrophic A type of ovary in which trophocytes in the germarium provide nutrients to developing oocytes via a trophic cord.
trophocytes Polyploid cells within the germarium which produce nutrients for the developing oocyte/embryo (nurse cells).

Aphids are small insects and major pests that have a fascinating biology. Adults from a single species may be so different in morphology that some early attempts misclassified them as different species even though they might share the same genotype. Both winged and wingless adult morphs are found. In addition, aphids can reproduce both sexually and asexually, with sex being associated with egg laying and parthenogenesis with the birth of live young (vivipary). Some species alternate between the two reproductive modes (cyclical parthenogenesis), whereas others are purely asexual. Sex and egg laying are considered primitive; no extant aphids reproduce purely sexually.

I. INTRODUCTION

Aphids are plant-feeding insects. Their mouthparts pierce plants stems and leaves, move between superficial cells, and eventually penetrate the phloem system which provides nutrients. They have an incomplete metamorphosis, growing through four larval stages that closely resemble the adult form (except where the adult is winged). They belong to the order Hemiptera and are closely related to scale insects and whiteflies, but the true aphids belong to the family Aphididae, which has approximately 4500 species worldwide. Externally, the most diagnostic feature of aphids is a pair of tubular siphunculi or cornicles which are present on the posterior/dorsal surface of the abdomen (Fig. 1). These vary in shape and size and function in defense against predators. If attacked, a waxy liquid is exuded which solidifies and interferes with the attacker. It also contains a volatile alarm pheromone which causes nearby aphids to move away or drop from the plant.

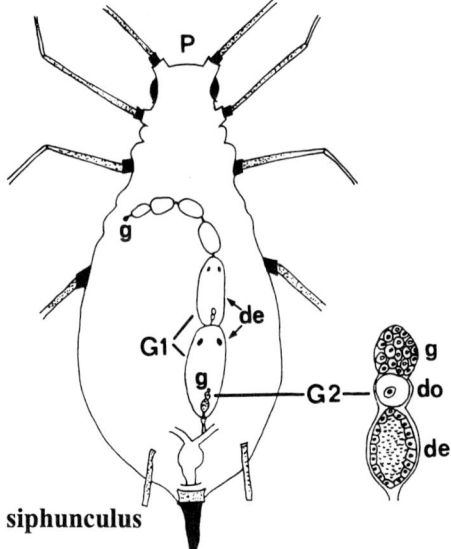

FIGURE 1 The telescoping of generations in an adult (generation P) parthenogenetic aphid. A single ovariole from one of the two ovaries is shown. It comprises the terminal germarium and a chain of developing embryos (these will form generation G_1) which become larger (more mature) toward the oviduct. The larger embryos already contain developing embryos, parthenogenetic oogenesis has taken place, and the developing embryos will form generation G_2. de, developing embryo; do, developing oocyte; g, germarium (reprinted from Hardie and Lees, 1985, with kind permission from Elsevier Science Ltd., The Boulevard, Langford Lane, Kidlington OX5 1GB, UK).

One of the main reasons for the success of aphids and their pest status is the capacity to respond to environmental cues by changing gene expression and initiating the development of different adult morphs (phenotypes) and different modes of reproduction. Such a complex polyphenic response gives rise to elaborate life cycles. The simplest are purely parthenogenetic (anholocyclic) on one host plant species and the more complex show cyclical parthenogenesis (holocycly) on one host plant and even alternation between two host plants (heteroecy). In the latter case the primary, or winter, woody host is alternated with a secondary, or summer, herbaceous host. Aphids tend to be fairly host specific, with even the most polyphagous species feeding on only about 1% of flowering plants.

II. PARTHENOGENESIS

The Swiss naturalist, Claude Bonnet, first reported parthenogenetic reproduction in animals. In 1740 he took a wingless aphid from a spindle tree (*Euonymus europaeus*) and isolated it on a cut twig under a glass cover. A single offspring was born, reared in isolation, and gave birth 12 days later, producing a total of 95 offspring over 3 weeks. The lack of mating requirement and a postnatal egg stage allows for rapid reproduction.

The ovaries of an adult parthenogenetic female aphid contain a number of ovarioles (2–>20). The number differs between species and even between different morphs (the first parthenogenetic spring generation hatching from the egg has more ovarioles than later generations and is correspondingly more fecund). Each ovariole contains a string of embryos that have developed from unfertilized ova. The embryos gradually increase in size toward the oviduct (Fig. 1), with the largest, most mature embryos being just prenatal and adjacent to the paired oviducts, which lead into the common oviduct. The ovarioles are telotrophic, with nurse cells (trophocytes) in the germarium providing nutrients to the nearest embryos via a trophic cord (Fig. 2). More mature embryos have lost this nurse cell contact and obtain nutrients directly through the ovariole sheath. At their anterior end, the germaria are joined by long, thin terminal filaments which may then attach to the body wall.

In aphids, parthenogenesis has been coupled to a form of pedogenesis. There is a telescoping of generations such that the larger embryos developing in the mother's ovaries already contain developing embryos that will form the next generation (Fig. 1). This "Russian Doll" strategy results in three generations being present (mother, daughter, and grand-daughter), minimizes the generation time, and exaggerates the reproductive rate. Comparison with a sexually reproducing insect with the same development time with each female producing 50 offspring indicates that after three generations the sexual insects would have achieved a population of 1250, whereas over the same time scale the aphid strategy results in a population of 318,750,000 individuals. Aphids are *r*

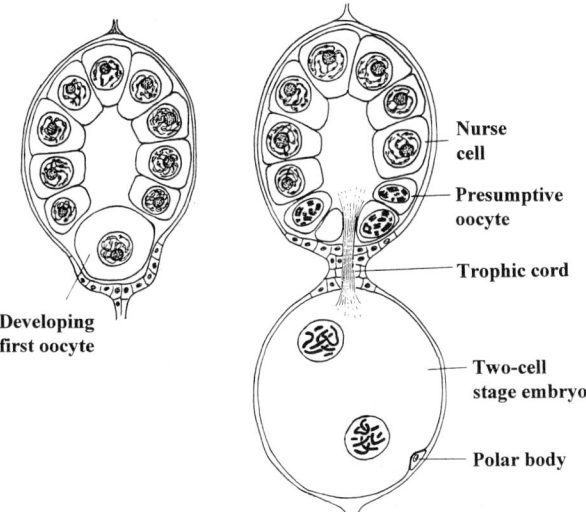

FIGURE 2 The aphid germarium is attached via a terminal filament (top) to other germaria and the body wall and via the pedicel to the oviduct. Logitudinal sections show development of the first oocyte at the base of the germarium (left), whereas other cells remain undifferentiated. Once the first oocyte has undergone mitosis, further basal cells differentiate as presumptive oocytes, whereas the others become nurse cells (tropohocytes) (reproduced from Blackman, 1978).

strategists and maximize populations on the summer, often ephemeral, host plants.

III. SEXUAL REPRODUCTION

When sexual reproduction occurs, it tends to be in the autumn and results in the laying of cold-hardy, overwintering eggs. This occurs in temperate regions and is controlled by a photoperiodic response where the short day lengths as autumn approaches result in the switch from asexual to sexual reproduction. Temperature interacts with photoperiod, with low temperatures favoring the production of distinct sexual morphs—the sexual female and the male. Indeed, the first demonstration of animal photoperiodism was in an aphid in the early 1920s.

Like the asexual females, the sexual females possess two ovaries comprising a number of ovarioles. Eight cells at the base of each germarium enter meiosis during the sexual female's embryonic development but no further growth occurs until after birth at c, the second stadium. Usually only one or two eggs are ovulated per ovariole, fecundity is much reduced compared to the asexual females, and large quantities of yolk (vitellogenin) accumulate within the egg in meiotic prophase. These gamic females also possess accessory glands and a single receptaculum seminis for sperm storage after mating. These open into the common oviduct. There are no such structures in the asexual females and very little vitellogenin is produced. In certain circumstances, under environmental conditions balanced at the point of determination of embryos as asexual or sexual females, germaria in the same ovary can produce embryos or sexual ova. In some cases, the same germarium produces both oocyte types.

Male aphids are winged in many species (always so in host-alternating aphids) but can be wingless. The male reproductive system comprises two multilobed and fused testes leading into an ejaculatory duct via two vas deferens with accessory glands. Sexual females release a volatile sex pheromone from their hindlegs to attract males and increase the chances of mate location. These pheromones appear to be species specific and can be detected by flying males at some distance away.

IV. CYTOGENETICS

A. Parthenogenesis

Parthenogenesis can often be maintained indefinitely, even in aphids that normally turn sexual in autumn, by using artificial long-day conditions and/or high temperatures. Attempts have been made to examine genetic changes over sequential asexual generations, but all indications are that truly clonal lineages are produced. One recent development is that in some aphid clones, different adult morphs appear to have banding changes after PCR of genomic DNA. The precise mechanism and understanding of this are not known. Mutations will and do undoubtedly occur, giving rise to changes in fitness such as resistance to certain pesticides. In this case, resistance can arise from gene amplification which results in increased production of esterase enzymes which break down certain insecticides.

The production of aphid female clonal lineages indicates that a simple mitosis (apomixis) gives rise to the asexual egg. There does not appear to be an initial meiosis followed later by a doubling of the chromosome number and return to the diploid state. With such a mechanism offspring could differ from their mother. Nevertheless, researchers have different interpretations of the precise mechanism. It has been suggested that chromosome homologs pair in the nucleus and undergo crossover prior to a single maturation division (so-called endomeiosis). If this occurred it would indicate an automixis and again progeny could differ from their mother. Such a mechanism seems unlikely.

The mature germarium initially contains 32 similar oogonial cells at prophase with large nuclei and nucleoli. A terminal cell enlarges to become the first oocyte (Fig. 2). At this stage the germarium is within an embryo of a last stadium larva (telescoping of generations). During initial growth the cell is in early prophase and the nucleus changes little. Then the chromosomes condense and become visible as the egg leaves the germarium. There are a diploid number of chromosomes and the nucleus enters metaphase with the chromosomes lining up along the equator (Fig. 3). Aphid chromosomes are holocentric, centromeres are not present, and the whole chromosome is drawn along its entire length to the spindle poles. A single polar body is found, again indicating a single maturation division (rather than two divisions with two polar bodies in meiosis). This is followed by mitotic division as embryo development rapidly follows. At this stage there is a differentiation of a number of oogonial cells at the base of the germarium as oocytes, whereas the remainder develop as polyploid (up to 512 n) trophocytes (nurse cells).

B. Sexual Reproduction

The autumn, short-day, low-temperature conditions bring about the appearance of sexual females and males in holocyclic aphids. These morphs are also produced parthenogenetically and the genotype is the same for both parthenogenetic and sexual fe-

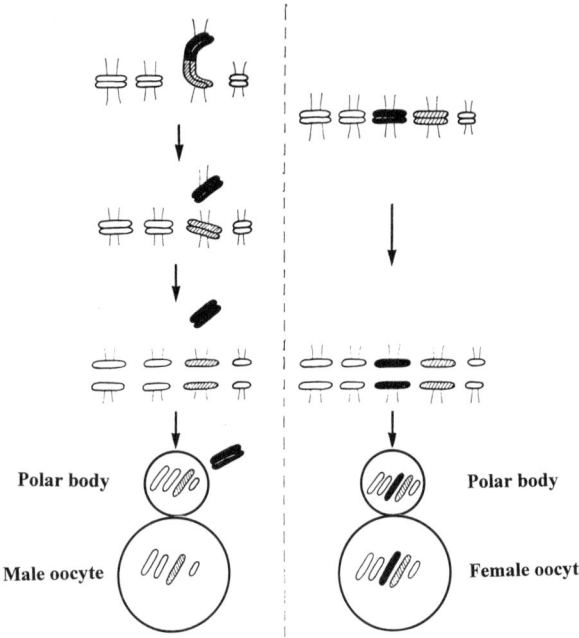

FIGURE 3 Behavior of chromosomes during cell division to produce male (left) and female (right) oocytes via parthenogenesis. Three autosomes are shown and the two X chromosomes are shown as black and shaded (reprinted with permission from Blackman, 1980).

males, with phenotypic differences appearing later during development. Males differ in lacking an X chromosome so sex determination is of the XX/XO type. Figure 3 shows the chromosomal arrangements during maturation of male and female parthenogenetic oocytes.

For males to develop, half of the sex chromatin is lost during the maturation division by a form of "minimeiosis" in which the X chromosomes pair and separate (Fig. 3). The pair that remains on the equator separates with the autosomes and gives rise to the XO male, whereas the other passes to the polar body. The X bivalent may form a distinct spherical body which condenses at the end of the growth phase prior to separation. In the female oocyte development, both X chromosomes separate with the autosomes, giving rise to an oocyte with both sex chromosomes and a single degenerate polar body.

During sexual reproduction, oogenesis involves meiotic division of the oocyte but spermatogenesis requires all viable sperm to possess an X chromosome because all viable offspring are female.

See Also the Following Article

PARTHENOGENESIS AND NATURAL CLONES

Bibliography

Blackman, R. L. (1978). Early development of the parthenogenetic egg in three species of aphids (Homoptera: Aphididae). *Int. J. Insect Morphol. Embryol.* **71**, 33–44.

Blackman, R. L. (1980). Chromosomes and parthenogenesis in aphids. In *Insect Cytogenetics* (R. L. Blackman, G. M. Hewitt, and M. Ashburner, Eds.), Symp. R. Ent. Soc. London, Vol. 10, pp. 133–148. Blackwell, London.

Dixon, A. F. G. (1985). *Aphid Ecology*. Blackie, Glasgow.

Dixon, A. F. G. (1996). Parthenogenesis in insects with particular reference to the ecological aspects of cyclical parthenogenesis in aphids. In *Insect Reproduction* (S. R. Leather and J. Hardie, Eds.). CRC Press, Boca Raton, FL.

Hardie, J., and Lees, A. D. (1985). Endocrine control of polymorphism and polyphenism. In *Comprehensive Insect Physiology, Biochemistry and Pharmacology* (G. A. Kerkut and L. A. Gilbert, Eds.). Pergamon, Oxford, UK.

Aplysia

S. Arch
Reed College

I. Introduction
II. Reproductive Life Cycle
III. Reproductive Anatomy
IV. Reproductive Neuroendocrinology
V. Reproductive Behavior
VI. Concluding Observations

GLOSSARY

afterdischarge The occurrence of repetitive and synchronous discharges of the bag cells following a brief effective stimulation; the afterdischarge is thought to ensure that a suprathreshold level of the egg-laying hormone is released into the hemolymph.

atrial gland An elaboration of the secretory epithelium of the reproductive tract that secretes several peptides related to the egg-laying hormone.

auto/allosperm Designations to distinguish sperm produced by an animal from those released into the reproductive tract by a copulatory partner.

bag cells Neuroendocrine cells associated with the central nervous system; known to synthesize and secrete the hormone responsible for initiating the egg-laying response.

egg cordon A continuous strand of adherent gelatinous material containing a substrate packaged in compartments; during egg laying, the eggs are deposited onto this substrate.

egg-laying hormone A polypeptide (36-amino acid residues) neurohormone, secreted by the bag cells, that has sites of action in the ovotestis and in the central nervous system causing egg laying and its associated behaviors.

ovotestis The gonadal structure responsible for the simultaneous production of both eggs and sperm.

The Aplysiids are widely distributed marine tectibranch mollusks with vestigial shells. They are intertidal, or nearshore subtidal, herbivores.

I. INTRODUCTION

Although Aplysiids are the subjects of attention in some extended treatments, their principal contemporary interest has come from the use of a few species as convenient preparations for neuroscientific investigation. One line of this research has been an extensive investigation of the neuroendocrine control of egg laying. In this article, I have drawn chiefly from information developed about the reproductive anatomy and physiology of Aplysia californica. Observations from studies of A. brasiliana, A. dactylomela, A. depilans, A. fasciata, and A. punctata are also included but not separately noted.

II. REPRODUCTIVE LIFE CYCLE

Aplysia have a prodigious reproductive potential. They can deposit hundreds of thousands of fertilized eggs at a time. Predation pressure on the planktonic larvae must be ferocious. After settling on a limited number of algal food sources the veligers metamorphose into crawling, macrophytic feeders. They are cryptic for several weeks after metamorphosis because of both their small size and their assumption of the coloration of their algal food. Ultimately, they begin foraging more widely and, in the few cases documented, are able to grow quite rapidly. Hence, the size of an animal is likely to be only an approximate indicator of its age. From commercial collecting records and a few systematic observations, large and reproductively active A. californica are most commonly found in the intertidal and shallow subtidal during the late spring through the middle of fall. Subsequently, animals of any size are in low abundance until the late winter, when small animals begin to appear in increasing numbers. This pattern points most directly to a life cycle of 12–18 months duration. On the other hand, since the adults produce one or more toxins and are not known to be significant prey items, it is not obvious why they should have so brief a life span. It may be that larger animals disappear from the intertidal because they move into deeper water in the winter to avoid heavy wave action and because food resources are more reliable below the intertidal. In any case, reproductive activity assessed by frequency of mating and egg laying in captivity, as well as responsiveness to hormone injection, is seasonal, with its peak in the late summer.

III. REPRODUCTIVE ANATOMY

The reproductive tract (Fig. 1) is hermaphroditic. It consists of unpaired structures that sequentially move and process both autosperm and eggs, as well as allosperm, from copulatory partners.

A. Ovotestis

The gametes originate in an ovotestis that is essentially embedded in the hepatopancreas. The color of the ovotestis ranges from a dull brown to a bright yellow-green. It seems likely that the color is derived from pigments taken up from food algae since animals collected from the same site generally show the same gonadal coloration.

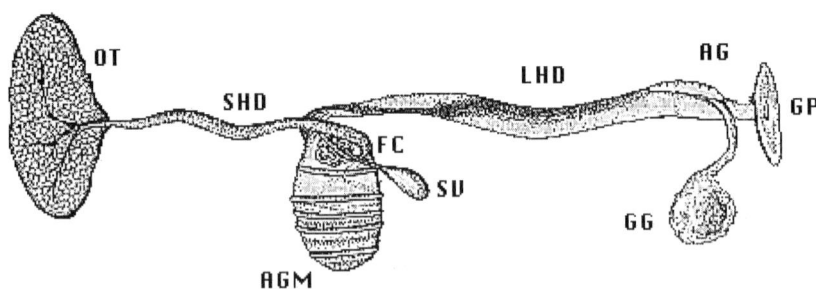

FIGURE 1 The Aplysia californica reproductive tract is shown as it would appear after removal from the animal. The main subdivisions are ovotestis (OT), small hermaphroditic duct (SHD), accessory genital mass (AGM), seminal vesicle (SV), fertilization chamber (FC; lying to the left of the SHD), large hermaphroditic duct (LHD), atrial gland (AG), gametolytic gland (GG), and the gonopore (GP). The entire length of this duct system can be 10 cm or more in a large mature animal.

The ovotestis is follicular in organization. Each follicle produces both oocytes and spermatozoa. No report describes the organization of the follicle with respect to sperm and egg production. Presumably, both originate from the loose germinal epithelium underlying the thick basal lamina that surrounds each follicle. Both show close and persistent associations with follicular nurse cells derived from the same epithelium.

Reticulate muscle cells extend over the surface of each follicle and it is likely that their contraction leads to compression of the follicle's contents and extrusion of mature oocytes into the ciliated ductules that lead to and coalesce to form the small hermaphroditic duct. While such a mechanism would logically also force free sperm into the duct, it appears that sperm movement is at least in part constitutive since the small duct in sexually mature animals is typically full of nonmotile autosperm but devoid of oocytes.

B. The Duct System

The complicated duct system that conveys the eggs and autosperm to the outside, while continuous from the gonad to the exterior, is by no means simple or monotonic enough to justify designation with any single name. Hence, the small hermaphroditic duct is the first element of this system. How the individual eggs extruded from the ovotestis in response to an ovulation stimulus move along the small duct while the sperm do not has not been determined. It may be that here, as in other components of this duct system, functionally isolated channels are formed by the advantageous placement of septa and only the portion that conveys eggs is ciliated.

At the confluence of the small hermaphroditic duct with the proximal end of the large hermaphroditic duct, it seems that the individual eggs could simply continue distally toward the gonopore, as the autosperm do, but they do not. Instead, they enter the fertilization chamber. No one has commented on the presence of sperm in this structure, despite its name. Indeed, in order for allosperm to reach the eggs they have to "backtrack" rostrally from the seminal receptacle to a point of confluence of the copulatory canal and the spermoviduct, then proceed caudally in the spermoviduct to the fertilization chamber. An alternative would be for fertilization to occur as the eggs pass the point of duct confluence. This is not considered likely, however, since by the time the eggs pass this junction, they are already ensheathed in egg cordon material.

The fertilization chamber and its adjunct winding gland are the site of the initiation of egg cordon formation. By the time eggs leave the fertilization chamber to enter the spiral canals of the mucous gland portion of the accessory genital mass, they are moving as parts of an assembled egg strand. This formation is almost certainly the product of secretions from the epithelia of the winding and albumin glands. No biochemical analysis is available to document the composition of the egg strand or the contributions of the several secretory surfaces along which it passes. The structure of the cordon is complex in that it is subdivided into a series of chambers, or capsules, each containing a number of eggs, averaging from 5 to 50 depending on the species.

After leaving the spiral tubes of the mucous gland, the egg strand proceeds into the large hermaphroditic duct (LHD). This substantial duct is surrounded by muscle bands and internally composed of several secretory epithelia. For most of its length, the lumen of the duct, though continuous, is subdivided by the extensively folded epithelium. The subdivisions thus formed have been variously named. From a functional perspective three activities must be supported. The formed egg strand has to be conveyed to the outside. Autosperm must also be guided to the penis during copulation, and intromission and routing of allosperm have to be accommodated. The least cumbersome terminology designates the spermoviduct as the portion of the LHD devoted to transport of endogenous germ cells to the outside and the copulatory duct as the element that receives the penis, and thus the allosperm, of the copulatory partner.

The LHD ends at the gonopore, or common genital aperture. Beyond this point, gametes continue to the base of the anterior tentacles in a ciliated external groove along the right side of the animal. As the egg cordon emerges from the groove it is guided by the tentacles and tamped into place on the substrate by bobbing movements of the head. During copulation, the penis everts hydrostatically from beneath the termination of the external groove and forms an extension of the sperm path into the recipient copulatory duct. It is not uncommon to find an animal that is laying eggs also serving as the female partner in copulation with another animal; there is, however, no record of an animal acting as a male in copulation and also depositing eggs.

C. Atrial Gland

Although the several species investigated differ to a greater or lesser extent, it is generally the case that the two "ducts" that comprise the LHD can be differentiated superficially. For example, in *A. californica* the spermoviduct is bounded by an epithelium with a reddish color, whereas the copulatory duct is white. The source of the color difference is not known but it is evident from electron microscopic analyses that the secretory products of the epithelial cells differ. Again, but with a notable exception, the makeup of the secretions is unknown and only the object of speculation. The exception is the atrial gland that comprises about 25% of the rostral portion of the spermoviduct component of the LHD of *A. californica*. This glandular subdivision differs in its appearance and position in various species, but it is always associated with the spermoviduct and, even when disperse, more rostrally than caudally situated.

Although the identification of this specialized secretory epithelium as the atrial gland may be in error, there appears to be little to gain from any attempt to change its name. Considerable attention has been paid to this tissue since the gland was found to contain a peptide that induces egg laying when injected into a recipient. Behavioral, physiologic, and molecular studies have confirmed that the egg-laying-inducing peptide is secreted into the lumen of the spermoviduct. It seems certain that this exocrine secretion is conveyed to the environment and does not have any endocrine effect. Since the principal products known to be secreted by the atrial gland are derived from genes closely related to that which codes for the endocrine egg-laying hormone, there has been considerable interest in the possibility of sexual pheromonal effects.

IV. REPRODUCTIVE NEUROENDOCRINOLOGY

A. The Bag Cells

The major elements of the proximate neuroendocrine regulation of egg laying in *Aplysia* have been known for about 30 years. Two clusters of neurons associated with the parietovisceral ganglion located in the "abdominal" region of the animal are known to synthesize and secrete a polypeptide hormone that initiates ovulation and assembly of the strand. These neurons are called the bag cells because they appear as two closely clustered groups of 250 or more cells, each enclosed in a "bag" of connective tissue. The unusually favorable morphology of the bag cell clusters and their hardiness in simple organ culture conditions have permitted their extensive exploitation as model preparations for the study of peptidergic neuroendocrine function. In many instances, findings from studies of the bag cells have presaged observations in members of widely disparate phyletic groups.

The extensive electrophysiologic, molecular, and cellular biology literature on bag cells will only be summarized here. The bag cell clusters, more generally the bag cell organs to reflect their neuroendocrine role, appear to be essentially homogeneous with respect to neuron type. The bag cell commitment to the regulation of the egg-laying response is most evident in the devotion of nearly 50% of their short-term protein synthesis to the production of the egg-laying hormone (ELH). More precisely, the primary translational product in the bag cells is a much larger precursor to ELH. Within the sequence of this precursor are several unique segments that may be excised and sequestered in secretory vesicles in addition to ELH. There is electron-microscopic and functional evidence that secretory vesicles in the bag cells are heterogeneous in terms of their contents.

To date, only two peptides have been found reliably and in quantity in superfusion solutions taken from stimulated bag cell organs *in vitro*: the 4.5-kDa ELH and the sequence immediately C terminal to it on the precursor. The latter polypeptide is 3.6 kDa, acidic, and has no known hormonal role. On a variety of grounds, it has been inferred that two or three additional short peptides [bag cell peptides (BCPs)] are also secreted. The strongest evidence is for the secretion of the α-BCP. This peptide may act as an autocrine agent that influences the probability of prolonged bag cell discharges. The major weakness in the case for the importance of α-BCP secretion is evidence that little of it enters axonal transport to secretory terminals and that as much as 90% is degraded in the cell somata.

B. Activation of Bag Cell Secretion

No reliable natural stimulus to bag cell secretion has been found. It is known that the cells are electrically coupled and that the sites of coupling are remote from the cell somata. Once activated, the cells will become entrained and enter a prolonged period of repetitive and synchronous discharging. These afterdischarges can be initiated by briefly stimulating certain neuron groups and commissural connections in the circumesophageal ring of ganglia. Hence, it would appear that a descending pathway is capable of controlling bag cell activation. Consistent with this inference is the fact that an afterdischarge can be activated easily by stimulation of the pleurovisceral connective nerves. Whether this reflects synaptic activation as well as antidromic stimulation of bag cell processes is not known.

Copulation is not a likely stimulus to bag cell activation since egg laying can occur without close temporal correlation with copulation—indeed, in the complete absence of copulation. There is, however, some empirical support for a temperature dependency to egg-laying activity. Naturally occurring egg laying is at its highest frequency in the late summer and early fall when intertidal temperatures are warmest. Under laboratory conditions, exposure to warmer water (20°C) for a few weeks will increase the synthesis of ELH and the spontaneous egg-laying frequency. Nonetheless, egg laying will occur spontaneously in temperatures below 15°C, albeit infrequently.

Despite the intensive examination of bag cell physiology over the past few decades, we remain only marginally informed about the regulation of their function in normal circumstances. It is likely that they are activated by presynaptic neurons in the circumesophageal ring of ganglia. Whether there is one or more pathways to their activation is not known. Ambient temperature is an important determinant of activation but so is the presence of conspecifics, especially animals laying eggs. The secretion of ELH is necessary and sufficient for egg laying, but, in view of the secretory granule heterogeneity in the somata and terminals, it is likely that other compounds are also secreted. It is unknown how the secreted stoichiometries of these additional products may vary. Finally, nothing is know about the control of autosperm propulsion through the reproductive tract during copulation. Is it hormonally activated as well, or the consequence of some sort of neuromuscular reflex?

V. REPRODUCTIVE BEHAVIOR

The few systematic behavioral studies of *Aplysia* reproduction have been done in the laboratory and on individual animals or pairs. Purified and synthetic ELH will, upon injection, cause mature animals to lay eggs and to display characteristic egg-laying behaviors. The initiation of egg laying is a consequence of the direct action of the hormone on the follicles of the ovotestis. The several characteristic behavioral activities are presumed to ensue from ELH actions on central neurons. Since the circuits responsible for organizing egg-laying behaviors are not known, only indirect indications of specific neural targeting are available. These include evidence of direct actions on an identified neuron in the parietovisceral ganglion and on neuron clusters in the head ganglia. There is also evidence of enhanced activity in nerves leading to body wall musculature from the head ganglia. Whether all stages of the behavioral sequence that can persist for 90 min are the products of ELH activation or a more complicated series of sequential neural responses and sensory feedbacks has not been established. It should be noted that activation of the bag cells by electrical stimulation in freely behaving animals leads to the same behavioral sequence that is seen after injections. There are no reports of initiation of uniquely male behavior following injections or *in vivo* stimulation.

Aplysia are commonly seen laying eggs and copulating in groups. Speculation about protandry would seem mooted by the formation of copulatory chains of from three to tens of animals. In such chains, excepting the terminal animals, each animal is playing both copulatory roles at the same time. When such chains close, as occasionally happens, the question of role preference disappears entirely. Most commonly, copulatory chains are thought to be initiated by an egg-laying animal. An interpretation of this behavior is that egg masses, egg-laying animals, or copulating animals release attraction pheromones. It is consistent with this inference that reproductive aggregations are much more common in areas where water exchange is slow than in locations, e.g., the open coast intertidal, where water exchange is frequent and thorough. The studies reported to date imply the existence of one or more aggregation pheromones associated with the egg strand but not with copulating or egg-laying animals. These studies employed preference tests in T-mazes, so the evidence must be considered inferential at best. The molecular identities of the putative active factors are not known but they are probably not derivatives of the egg-laying-inducing substance synthesized in the atrial gland. They are more likely derived from secretions of the albumin gland that are added to the egg strand in the process of its assembly.

VI. CONCLUDING OBSERVATIONS

I have tried to capture a sense of the marvelous structural and functional complexity of reproduction in *Aplysia*. It should also be apparent that there are important issues about which we know very little. For example, how long do these animals live? What is the natural trigger for egg laying? and What regulates male behavior? It might seem that the answers to these questions are unlikely to come from the current research focus on the *Aplysia* nervous system as a model preparation for the study of fundamental neuroscience. However, it was this impetus that led to the discovery of the neuroendocrine control of egg laying, the behavioral sequence during oviposition, and the widespread expression in glandular as well as neural tissues of the egg-laying hormone gene family. It bears remembering that we cannot understand the broader significance of what we find until we have understood its significance for *Aplysia*.

See Also the Following Articles

Hermaphroditism; Mollusca

Bibliography

Arch, S., and Berry, R. (1989). Molecular and cellular regulation of neuropeptide expression: The bag cell model system. *Brain Res. Rev.* **14**, 181–201.

Audesirk, T. E. (1979). A filed study of growth and reproduction in *Aplysia californica*. *Biol. Bull.* **157**, 407–421.

Beeman, R. D. (1970). The anatomy and functional morphology of the reproductive system of the opisthobranch mollusk *Phyllaplysia taylori* Dall, 1900. *Veliger* **13**, 1–31.

Coggeshall, R. E. (1970). A cytologic analysis of the bag cell control of egg laying in *Aplysia*. *J. Morphol.* **132**, 461–485.

Eales, N. B. (1921). *Aplysia*. Liverpool Marine Biology Committee, Memoir 24. *Proc. Trans. Liverpool Biol. Soc.* **35**, 183–266.

Hyman, L. H. (1967). *The Invertebrates*, Vol. VI, Mollusca I, pp. 480–494. McGraw–Hill, New York.

Kandel, E. R. (1979). *Behavioral Biology of Aplysia*. Freeman, San Francisco.

Mazzarelli, G. F. (1891). Richerche sulla morfologia e fisiologia dell'apparato riprodduttore nelle *aplysiae* nelle Golfo di Napoli. *Mem. Real. Accad. Sci. fis. mat. Napoli* (2) **4**(3),1–50.

Painter, S. D., Kalman, V. K., Nagle, G. T., Zuckerman, R. A., and Blankenship, J. E. (1985). The anatomy and functional morphology of the large hermaphroditic duct of three species of *Aplysia*, with special reference to the atrial gland. *J. Morphol.* **186**, 167–194.

Thompson, T. E., and Bebbington, A. (1969). Structure and function of the reproductive organs of three species of *Aplysia* (Gastropoda, Opisthobranchia). *Malacologia* **7**, 347–380.

Apoptosis (Cell Death)

Paul F. Terranova and Christopher C. Taylor

University of Kansas Medical Center

I. Introduction
II. Molecular and Biochemical Events
III. Apoptosis in the Ovary
IV. Apoptosis in the Uterus
V. Apoptosis in the Mammary Gland
VI. Apoptosis in the Testis
VII. Apoptosis in the Prostate

GLOSSARY

apoptosis The orderly suicide and disposal of cells in the absence of an inflammatory reaction.

Bcl-2 A family of proteins in mammals homologous to the Ced family of proteins in *Caenorhabditis elegans*. The Bcl-2 family of proteins may be apoptotic and antiapoptotic.

Caenorhabditis elegans A nematode that has been extensively studied leading to the identification of a family of genes that direct apoptosis and that have mammalian homologs.

ced 3/4 Represents *C. elegans* death-defective genes that direct and carry out apoptosis.

DNA laddering The characteristic pattern of apoptotic DNA staining in an agarose gel after electrophoresis. The DNA had been degraded into multiples of 185–200 bases within cells by endonucleases.

endonuclease An enzyme that cleaves DNA into approximately 185–200 base segments at internucleosomal sites.

Fas A cell surface receptor for Fas ligand which induces apoptosis through a series of death domain proteins linked to interleukin-1β converting enzyme-like proteases.

follicular atresia The death of an ovarian follicle with characteristic signs of apoptosis, including pyknotic nuclear and apoptotic bodies.

interleukin-1β-converting enzyme (ICE) An enzyme that converts prointerleukin-1β to interleukin-1β and induces apoptosis when overexpressed.

knockout mouse A mouse lacking a specific gene; produced using special gene targeting procedures.

luteal regression The demise of a corpus luteum, morphologically by apoptosis and biochemically by a decline in progesterone and an increase in DNA laddering.

NF-κB A nuclear transcription factor thought to participate in mediating cell survival through the tumor necrosis factor receptor(s).

pyknosis A shrunken nucleus with condensed darkened chromatin usually characteristic of apoptosis.

reactive oxygen species (ROS) Molecules produced by the transfer of an electron to molecular oxygen. ROS are associated with apoptosis by initiating DNA strand breaks.

tumor necrosis factor receptor type 1 One of two cell surface receptors for tumor necrosis factor-α which mediates apoptosis through a series of death domain proteins linked to ICE-like proteases.

Apoptosis, or programmed cell death, is the name given to the orderly suicide and disposal of cells in the absence of an inflammatory reaction. It appears that all nucleated cells contain a dedicated suicide program that, upon stimulation by the appropriate signals, will trigger apoptosis.

I. INTRODUCTION

Apoptosis may take place during development or in response to physiologic events regulating homeostasis. One of the initial reactions associated with apoptosis is activation of calcium–magnesium-dependent endonucleases resulting in cleavage of DNA into 185–200 base fragments at internucleosomal sites (Fig. 1). Agarose gel electrophoresis of DNA from apoptotic cells gives a characteristic DNA laddering caused by digestion of the linker DNA yielding nucleosomal fragments in multiples of 185–200 bases (Fig. 2). Apoptosis generally affects individual cells, causes the cells to detach from neighboring cells (and extracellular matrix *in vivo*) as well as from the culture substrate (*in vitro*). In addition, apoptosis

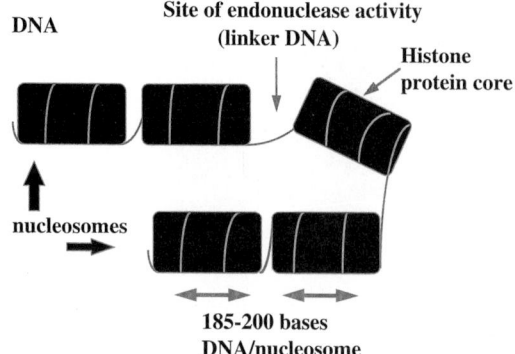

FIGURE 1 Schematic diagram demonstrating internucleosomal cleavage of DNA by calcium–magnesium-dependent endonucleases. Endonuleases cleave the DNA between nucleosomal units resulting in DNA fragments in multiples of 185–200 base pairs.

causes cells to shrink as evidenced by condensation of chromatin and reduction in the amount of cytoplasm. The nuclear lamin is destroyed by proteolysis, leading to the irreversible breakdown of the nuclear envelope. Subsequently, apoptotic bodies, vesicular membranous structures containing nuclear fragments, are formed (Fig. 3). Apoptotic cells and the apoptotic bodies are rapidly phagocytized by neighboring cells, thus avoiding an inflammatory reaction and tissue damage.

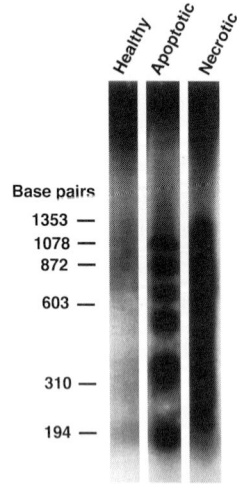

FIGURE 2 Autoradiogram of 3' end-labeled DNA demonstrating intact DNA extracted from healthy cells, internucleosomal ladder of DNA extracted from apoptotic cells, and smear of DNA extracted from cells from necrotic tissue.

TABLE 1
Morphologic and Biochemical Differences between Apoptosis and Necrosis

Apoptosis	Necrosis
Affects individually isolated cells	Affects large groups of contiguous cells
Cells shrink as exhibited by condensation of chromatin and cytoplasm; nuclear pyknosis	Cells swell uncontrollably resulting in rupture of plasma membranes
Formation of apoptotic bodies that contain membrane-bound constituents of the dying cell	Reduced ATP
Normal ATP level	Signs of inflammation and/or immune response
Lack of inflammation; phagocytosis by neighboring cells	Nonspecific DNA cleavage with random breakdown of DNA resulting in smearing of DNA on agarose gel electrophoresis
Internucleosomal DNA cleavage by endonucleases resulting in DNA ladder pattern on agarose gel electrophoresis	

Based on biochemical and morphologic criteria cells may die by either apoptosis or necrosis (Table 1). Necrosis most often results from noxious stimuli, whereas apoptosis is a physiologically regulated event. During necrosis the plasma membranes and nuclear integrities of a contiguous group of cells are disrupted due to cellular swelling soon after the traumatizing stimulus. Thus, there is uncontrolled rupture of the cells and the cellular contents are released into the extracellular spaces resulting in a classic immune response and/or inflammatory reaction at the site of injury.

II. MOLECULAR AND BIOCHEMICAL EVENTS

A. Apoptosis in *Caenorhabditis elegans*

Study of the nematode, *C. elegans*, led to the identification of a family of genes, which have mammalian homologs, that direct apoptosis. The embryonic development of *C. elegans* has been precisely mapped. The adult hermaphrodite has 1090 somatic cells, of which 131 undergo programmed cell death. By studying mutant nematodes with defects in specific

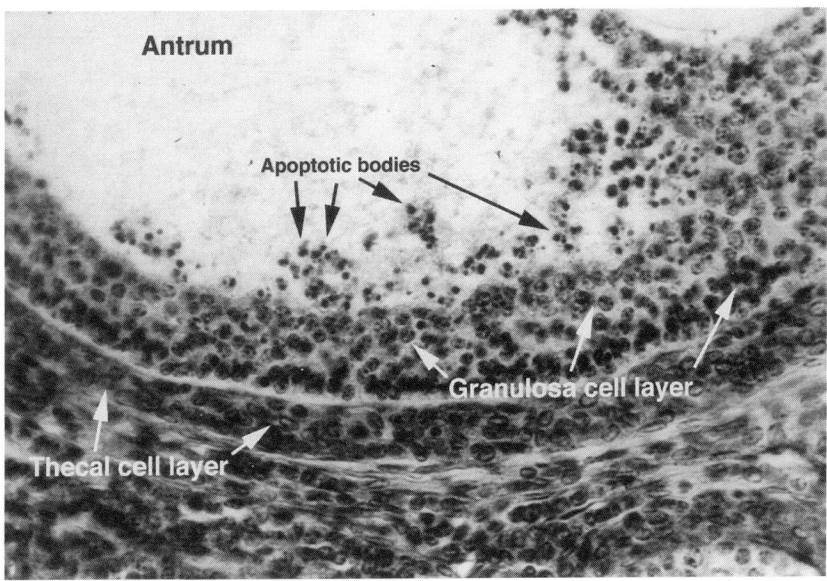

FIGURE 3 Photomicrograph of a portion of an atretic antral follicle. Note the apoptotic bodies from the antral-most layer of granulosa cells.

aspects of the cell death program, gene products that trigger and carry out the death program, and direct disposal of apoptotic cells, have been identified.

ced-3 (for *C. elegans* death defective) and *ced-4* gene products direct and carry out the death program. Loss of function of either *ced-3* or *ced-4* results in impaired apoptosis and survival of the 131 cells which would normally undergo cell death. Cloning of the *ced-3* gene revealed that it encodes a cysteine protease that is homologous to the mammalian interleukin-1β-converting enzyme (ICE; now termed caspase 1). It has recently become apparent that ICE and ICE-like proteases form a subfamily of a class of cysteine proteases with an unusual specificity, cleaving after aspartate residues. This family of proteases has been termed caspase (for cysteine–aspartate proteases or cysteine aspases) and is currently composed of three subfamilies based on their sequence homology; ICE-like proteases (caspase 1), Nedd/Ich-1 (caspase 2), and Yama (CPP32 or caspase 3). Overexpression of *ced-3*, ICE, CPP32, or Nedd2/Ich in mammalian cells results in apoptosis. The ability of *ced-3* to induce apoptosis in mammalian cells demonstrates the well-conserved nature of the cell death program throughout evolution.

Currently, no mammalian homolog for *ced-4* has been identified; however, overexpression of *ced-4* in mammalian cells does result in apoptosis. Furthermore, Ced-4 physically interacts with ICE and another ICE-like protease, FLICE (FADD-like ICE; FADD, Fas-associated death domain). These findings suggests that mammalian homologs to *ced-4* probably exist.

The *ced-9* gene encodes an inhibitor of cell death in *C. elegans* that has homology to the mammalian Bcl-2 family of proteins. The cell killing effect of Ced-4, when expressed in mammalian cells, can be inhibited by overexpression of Bcl-x_L, a mammalian inhibitor of apoptosis. This presumably occurs by blocking activation of ICE-like proteases by *ced-4*. Thus, in *C. elegans*, apoptosis is controlled by three major gene products: Ced-4 promotes the activation of the cell killing cysteine protease, Ced-3. Ced-4 is, in turn, negatively regulated by direct association with the death suppresser, Ced-9.

B. Apoptosis in Mammalian Cells

Because two of the three components in the *C. elegans* death program have mammalian homologs, and expression of *C. elegans* genes can predictably activate (*ced-3* and *ced-4*) or suppress (*ced-9*)

TABLE 2
The Bcl-2 Family and Their Role in Apoptosis

Antiapoptotic	Apoptotic
Bcl-2	Bax
Bcl-x_L	Bcl-x_S
Bclw	Bak
Mcl-1	Bad
Bag	Blk
	Bik

apoptosis in mammalian cells, it is likely that the cell death program in mammalian cells is similar, albeit more complicated.

1. Bcl-2 Family

The human *bcl-2* gene family encodes products with homology to *ced-9*. The family includes 11 members, some as promoters of apoptosis and others as negative regulators of apoptosis (Table 2). *bcl-2*, like its *C. elegans* homolog, *ced-9*, is a suppresser of apoptosis. Bcl-x_L, Bclw, and Mcl-1 also suppress apoptosis. Overexpression of either Bcl-2 or Bcl-x_L can block cell death in interleukin-3 (IL-3)-dependent cells following withdrawal of IL-3. In contrast, Bcl-x_S, a truncated form of Bcl-x_L, is a promoter of apoptosis. The pattern of expression of each of these gene products differs within different tissues. Most tissues express Bcl-x_L, whereas expression of Bcl-x_S tends to be higher in tissues with a high rate of apoptosis. Other family members include the negative regulator of apoptosis, Mcl-1, as well as the promoters of apoptosis, Bax, Bak, and Bad. Several of the family members, including Bcl-2, Bax, Bcl-x_L, Bcl-x_S, and Bad, can form heterodimers. It is thought that the ratio of apoptosis promoters to suppressors within a cell determines whether the cell will undergo apoptosis or survive.

How the Bcl-2 family regulates apoptosis is not well understood; however, there is increasing evidence that several of the family members are membrane pore-forming proteins. It has recently been determined that Bcl-2 regulates release of cytochrome c from mitochondria and this may then promote activation of caspases.

2. Caspases

The caspases include three separate subfamilies of cysteine proteases, the ICE-like proteases (currently three family members have been identified), Ich-1 (two members), and Yama/CPP32 (seven family members). ICE cleaves pro-interleukin-1β (IL-1β) from the inactive 31-kDa precursor to the active 17-kDa form. Overexpression of ICE in mouse fibroblasts can induce apoptosis. Whether ICE can induce apoptosis in a physiological context is not clear, but IL-1β does not appear to play a major role in apoptosis. Another substrate for ICE is the cysteine protease, Yama (the Hindu god of death)/CPP32. Cleavage of the 32-kDa CPP32 precursor by ICE produces the 17- and 12-kDa subunits of the active enzyme. Active CPP32 in turn cleaves and inactivates poly(ADP-ribose) polymerase (PARP). PARP is involved in DNA repair and is thought to suppress Ca^{2+}/Mg^{2+}-dependent endonucleases responsible for the internucleosomal cleavage of DNA observed during apoptosis. Thus, loss of PARP activity as a result of CPP32 activation may lead to the activation of the DNA cleaving endonucleases in the dying cell (Fig. 4).

C. Signals for Apoptosis

It is widely believed that outside signals are required to induce apoptosis in the normal physiological context. Perhaps the best characterized apoptosis-inducing signals are those represented by the cell surface receptor for Fas and tumor necrosis factor receptor 1 (TNFR1) cell surface receptors (Fig. 4). Both of these receptors contain within their cytoplasmic regions a 60- to 70-amino acid "death domain." Upon activation, receptors aggregate as trimers, and the receptor death domains interact with assembling protein–protein modules.

FADD associates with the Fas death domains. Fas activation recruits binding of FLICE(MACH) to FADD(MORT). FLICE then triggers the ICE proteolytic cascade leading to apoptosis.

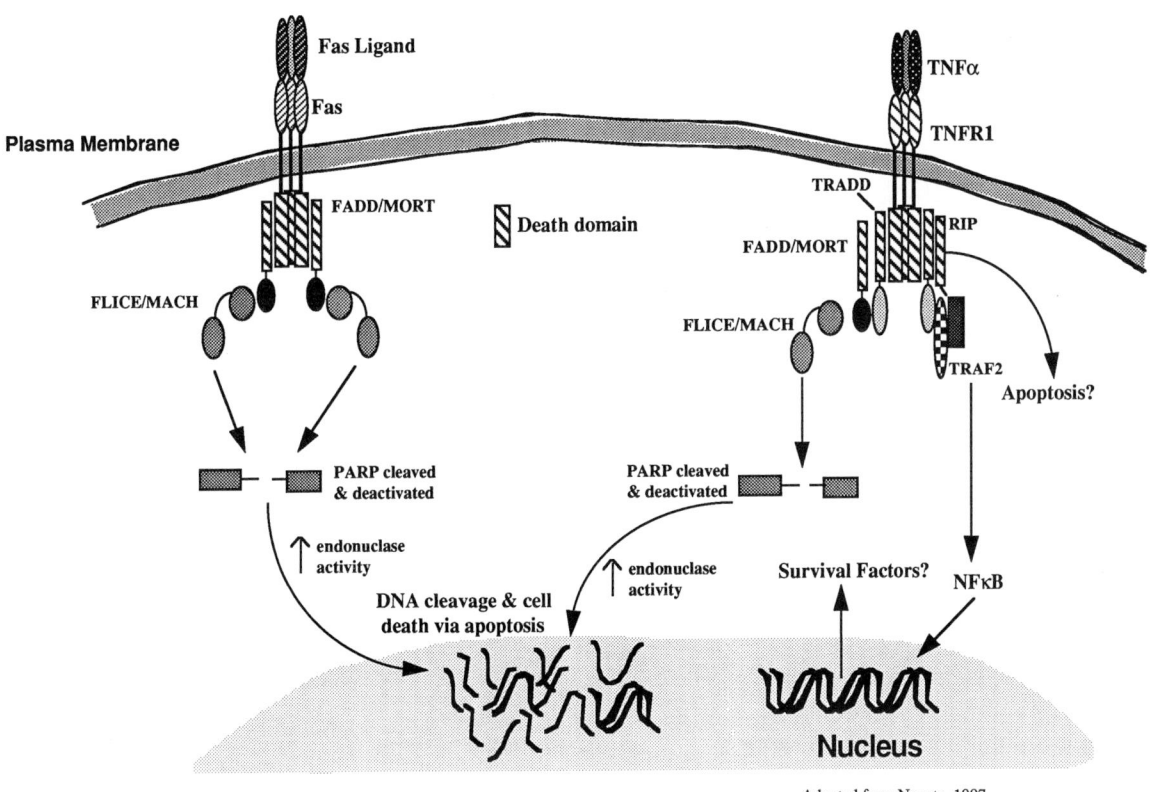

FIGURE 4 Signaling cascades stimulated by Fas ligand and TNF. Binding of ligand to their respective receptors induces receptor trimerization and formation of protein–protein complexes via interactions between death domains (■). This ultimately leads to activation of an ICE-like protease (FLICE/MACH) and DNA endonucleases, leading to apoptosis in the case of Fas and TNF and/or activation of NF-κB (via TRAF2) and induction of survival factors in the case of TNF (adapted with permission from Naganta, 1997).

TNF binding to TNFR1 also induces trimerization of the receptor and attracts binding to TRADD by the interaction with their respective death domains. TRADD then binds to FADD(MORT) and activates FLICE(MACH). ICE and CPP32 are then activated, leading to deactivation of PARP and onset of apoptosis. An alternative apoptotic pathway includes RIP binding to TRADD. However, RIP binding to TRAF2 may activate NF-κB, leading to expression of survival genes.

A major difference between Fas and TNFR1 activation is that Fas activation is associated almost exclusively with apoptosis, whereas activation of TNFR1 results in apoptosis only under certain, as yet poorly defined conditions. A key difference between the two receptors is that TNFR1 activates nuclear translocation of the transcription factor NF-κB. In contrast, Fas does not have this activity. TRADD, which associates with the TNFR1 but not Fas, appears to be the source of this difference. Overexpression of TRADD induces both activation of NF-κB and apoptosis through an association with RIP, a protein with both a death domain and kinase activity. This induced apoptosis can be blocked by expression of the ICE inhibitor encoded by the cowpox virus gene, crmA, without any effect on NF-κB activation. Thus, NF-κB activation and apoptosis can be dissociated. In fact, NF-κB activation has been found to be protective against TNF-induced apoptosis, perhaps by inducing expression of survival genes. Again, a balance between apoptotic-promoting and -suppressing signals governs whether a cell will undergo apoptosis or survive. The regulation of this balance, and how the scales may be tipped in one direction or the other, is not yet understood.

III. APOPTOSIS IN THE OVARY

Apoptosis accounts for death of three major groups of cells within the ovary: germ cells (oogonia and oocytes), granulosa cells, and luteal cells. Apoptosis in those cell types will be discussed. In the ovary, other cells that are affected by apoptosis are endothelial cells and thecal cells. However, thecal cell apoptosis appears to be highly species specific. Endonuclease activity has been detected in granulosa cells and luteal cells in many species, but there are scant reports on endonuclease activity in germ cells of the ovary or testis. Most studies have focused on granulosa cells primarily because of the easy access to large numbers of cells. It appears that apoptosis in the ovary is under endocrine and possibly paracrine regulation, with most of the studies concentrated on granulosa cell and follicular apoptosis.

A. Germ Cell Apoptosis

Reduction in the number of oogonia and oocytes through apoptosis is a normal process in ovarian development. Apoptosis plays a major role in determining the final number of primordial follicles in the ovary. The pool of primordial follicles serves as the resource from which several follicles are recruited each cycle. A reduction in the number of primordial follicles in the adult would likely coincide with a reduction in the number of oogonia during development. In addition, a reduction in the number of ova shed per cycle and premature loss of cyclicity may also be indicators of the failure of the normal number of oogonia to survive during development.

There are several reports delineating the *in vivo* morphologic changes of oogonia demonstrating that indeed apoptosis is similar to what is known in other cells types *in vivo*. In addition, there are supportive *in vitro* data indicating that apoptosis occurs in oogonia, primordial follicles, and oocytes deprived of serum, growth factors, or cytokines needed for maintenance. A major growth regulator in the early ovary appears to be stem cell (SCF) or Steel factor since early studies revealed defects in gametogenesis and subsequent fertility in mice lacking either the SCF receptor (kit) or SCF (kit ligand). These studies were further supported by *in vitro* observations that both SCF and leukemia inhibitory factor independently prevented loss of germ cells *in vitro* and appear to be major survival factors.

As indicated earlier, the ratio of various members of the Bcl-2 family of proteins as apoptotic and antiapoptotic agents may dictate the pathway taken by a cell. Similarly, in germ cells the Bcl-2 (antiapoptotic) to Bax (apoptotic) ratio may be important in apoptosis. The data to support this are from Bcl-2-deficient mice which exhibit fewer primordial follicles when compared with control mice. In contrast, Bax-deficient mice have a greater number of primordial follicles as early adults than control mice. Collectively, these results indicate that Bcl-2 and Bax are major players in determining the final number of oogonia available for recruitment into the pool of primordial ovarian follicles. Factors regulating expression of Bcl-2 and Bax in the developing ovary are unknown, but recent studies indicate that p53 may enhance apoptosis by increasing expression of Bax. In support of this concept are recent results that indicate that gonadotropins prevent apoptosis by reducing p53 expression and Bax expression. Thus, gonadotropin treatment would favor Bcl-2, an antiapoptotic factor.

B. Follicular Atresia

Follicular atresia is a major physiological event in the ovary. Approximately 99% of the follicles become atretic in the human ovary and ~80% become atretic in the mouse ovary. The primordial follicle is the earliest stage of follicular development, characterized by a small oocyte surrounded by flattened granulosa cells; no zona pellucida or theca exists at this stage of development. The establishment of the pool of primordial follicles coincides with the ability of the follicles to exhibit follicular atresia. Follicles can become atretic at any stage. However, the most prominent stage in which follicular atresia occurs is at the time of early antrum formation. Morphological changes in the oocyte and granulosa cells appear either simultaneously or independently during follicular atresia. The major morphological changes in the oocyte of small or large follicles undergoing apoptosis may be any one of the following, each occurring alone or in combination: shrinkage of the oocyte, increased argyrophilic substances and granulation in the ooplasm, breakdown of the germinal vesicle, and alignment of the chromosomes in metaphase and possibly extrusion of a polar body (extra set of chromosomes). In granulosa cells, the onset of follicular atresia is characterized by cell shrinkage, a pyknotic nucleus, the formation of apoptotic bodies, and the appearance of lipid droplets and lysosomal enzymes in the cytoplasm (Fig. 4). Hormonal regulation of apoptosis of granulosa cells has been a topic of several investigations. The initial evidence of biochemical degradation of DNA in ovarian sections and granulosa cells was from *in situ* end-labeling of DNA in ovarian sections and agarose gel electrophoresis of DNA extracts of granulosa cells, respectively. Several *in vitro* models were used to investigate follicular atresia associated with apoptosis, including culture of whole ovaries, preantral follicles, preovulatory follicles, oocytes, and granulosa cells under defined conditions. *In vivo* models have utilized gonadotropin- or estrogen-treated immature rats and knockout mice mentioned previously (SCF, Bcl-2, Bax, and p53). The following is a description of the various hormones and growth factors that exhibit antiapoptotic and apoptotic activity (Table 3) (see review by Hsueh).

1. Hormonal Regulation of Follicular Atresia and Apoptosis

i. In Vitro Apoptosis

Gonadotropin–Growth Hormone–Insulin-like Growth Factor Pathway Using an *in vitro* approach, antral follicles from immature rats were collected 48 hr after a priming dose of equine chorionic gonadotropin and were cultured in serum-free medium. A multifold increase in DNA fragmentation occurred in those follicles when compared with freshly isolated follicles. follicle-stimulating hormone (FSH) and luteinizing hormone [LH; human chorionic gonadotropin (hCG)], each alone and in combination, suppressed, by approximately 60%, DNA fragmentation in whole follicles. Interestingly,

TABLE 3
Antiapoptotic and Apoptotic Hormonally Related Factors Associated with Follicular Atresia

Antiapoptotic	Apoptotic
In vivo	*In vivo*
eCG	GnRH
Estrogens	Androgens
In vitro	*In vitro*
FSH	IL-6
LH/hCG	IGFBP-3
EGF, TGF-α	TNF-α
bFGF	
IGF-1	
GH	
IL-1β	
Nitric oxide	
cGMP, cAMP	
Superoxide dismutase	
Ascorbic acid	
Glutathione peroxidase	

growth hormone (GH), either natural or recombinant, suppressed apoptosis dose dependently and IGFBP-3 suppressed the effect of IGF-1, hCG, and GH, whereas IGFBP-3 alone had no effect. hCG increased IGF-1 mRNA in cultured preovulatory follicles indicating that the antiapoptotic action of gonadotropins may be mediated by IGF-1. Of concern with this theory is that IGF-1 cannot suppress apoptosis in isolated granulosa cells. However, an interaction among theca, granulosa, and oocyte may be necessary for this action to occur.

Gonadotropin–IL-1β–Nitric Oxide–cGMP Pathway IL-1β suppressed follicular apoptosis *in vitro* while stimulating nitric oxide (NO) and cGMP production. The antiapoptotic action of IL-1β could be suppressed by an IL-1β receptor antagonist. In addition, sodium nitroprusside, a NO generator, and cGMP, a second messenger of NO, each decreased apoptosis. In support of a gonadotropin, hCG stimulated IL-1 synthesis is the antiapoptotic action of hCG that can be blocked by the IL-receptor antagonist.

TNF-α–Ceramide Pathway TNF, a cytokine with origins within ovarian cells and macrophages within the ovary, blocks the antiapoptotic effect of FSH. Apparently, TNF acts through its type 1 receptor (p55/60) to activate a serine/threonine kinase resulting in subsequent activation of sphingomyelinase. Sphingomyelinase induces the conversion of sphingomyelin to ceramide and phosphorylcholine. Ceramide then activates ICE-related proteins. Sodium aurathiomalate, an inhibitor of ICE, suppresses ceramide-induced apoptosis in whole follicles in culture.

Oxidative Stress Pathways Reactive oxygen species (ROS) are mediators of cell death and initiate DNA strand breaks. ROS are usually produced through increased phosholipase activity, lipid peroxidation, or several other normal metabolic processes. The transfer of an electron to molecular oxygen produces the superoxide anion. The electron transfer occurs during normal reduction–oxidation reactions in mitochondria and in the endoplasmic reticulum of cells. Induction of apoptosis correlates well with elevated superoxide anions and reduction of superoxide dismutase (SOD), the enzyme that destroys superoxide anions. SOD suppressed apoptosis of granulosa cells from whole follicles cultured *in vitro*. Apoptosis was also prevented by ascorbic acid, a free radical scavenger. N-acetyl-L-cysteine, also a free radical scavenger and stimulator of glutathione peroxidase, prevented apoptosis *in vitro*. *In vivo* studies revealed that equine CG treatment of 25-day-old immature rats increased ovarian SOD.

2. In Vivo Apoptosis

Estrogen treatment of hypophysectomized rats results in the development of numerous preantral follicles, and withdrawal of estrogen treatment results in massive apoptosis of granulosa cells. In addition, testosterone and gonadotropin-releasing hormone (GnRH), both induce ovarian apoptotic DNA fragmentation in granulosa cells in hypophysectomized estrogen-treated rats.

3. Toxicological Induction of Apoptosis

Oogonia, primordial follicles, and later stages of follicular development are susceptible to a variety of external stimuli that induce apoptosis of follicular cells. For example, ionizing radiation, chemotherapeutic agents, and environmental toxicants induce apoptosis in the ovary. Two recently studied ovotoxicants, polycyclic aromatic hydrocarbons (PAH) and 4-vinylcyclohexene (VCH), have been implicated in the induction of apoptosis of ovarian cells. Their mechanisms of action may involve alterations in c-kit and Bcl-2 for PAH and Bax for VCH. VCH and its epoxide derivative(s) likely induce apoptosis in oocytes and granulosa cells through increased expression of Bax.

C. Luteal Regression

Morphological and biochemical indicators of apoptosis have been identified in the corpus luteum (CL) of several species. CL from spontaneous and prostaglandin $F_{2\alpha}$ ($PGF_{2\alpha}$)-induced luteal regression exhibit oligonucleosomal DNA fragmentation, endonucleases, and increased expression of mRNA encoding the $PGF_{2\alpha}$ receptor. There also appears to be an increase in the DNA fragmentation from early to midluteal phase and further increases in DNA fragmentation from midluteal to late luteal phases. DNA fragmentation is present in large as well as small luteal cells and in endothelial cells. Connexin 43, a component of gap junctions, has been observed immunohistochemically in healthy CL but absent in regressing CL. The lack of connexin in regressing CL may be hormonally regulated and represent an initial change in primate and human CL making the luteal cells susceptible to apoptosis. Cells are susceptible to apoptosis when cell–cell contacts have been lost.

1. Oxidative Stress

There are several hypotheses that ROS mediate regression of the CL. Presumably, those events occur after the initiation of luteal regression by $PGF_{2\alpha}$. Sulfated glycoprotein-2, ROS, and heat shock protein 70 (hsp 70) increase within functional CL either shortly after injection of $PGF_{2\alpha}$ or after the onset of spontaneous regression, and prior to a decrease in progesterone secretion and onset of DNA fragmentation. hsp 70 has been localized to large luteal cells which are $PGF_{2\alpha}$ sensitive. In addition, a metabolite of ROS, hydrogen peroxide, has also been correlated with reduced progesterone secretion associated with luteal regression. Progesterone inhibits ROS formation in mononuclear cells indicating that it may inhibit events associated with luteal cell apoptosis, i.e., a trophic action of progesterone. Enzymes associated with inhibition of ROS activity, SOD, and catalase are at high levels in functional CL and reduced in regressing CL; however, these changes may be species specific.

TNF and interferon-γ (IFN-γ) induce luteal cell apoptosis and a decline in progesterone *in vitro* and together they exhibit synergistic effects on apoptosis. Specifically, IFN-γ enhances the ability of TNF to induce apoptosis and cellular events initiating luteolysis. *In vivo* expression of TNF in CL and secretion of TNF by CL increase after progesterone secretion declines, indicating that TNF participates in the regression of CL only after the onset of luteolysis.

2. bcl-2 Family

bcl-2 cDNA has been detected in an ovine luteal cell cDNA library and Bcl-2 has been observed immunocytochemically in human granulosa–lutein cells, theca–lutein cells, and endothelial cells without changes during the menstrual cycle or after treatment with hCG. Mcl-1, an antiapoptotic member of the *bcl-2* family, has been identified immunocytochemically in granulosa lutein cells. Current studies are investigating the role of several members of the Bcl-2 family of proteins and p53 in luteal function.

IV. APOPTOSIS IN THE UTERUS

Apoptosis has important functions in the uterus, such as maintenance of the uterine cell number during the cycle, tissue remodeling during blastocyst implantation, and placental development (see Abrahamsohn *et al.*, 1993; Wride and Sanders, 1995). Apoptosis in cyclic uterine function and its control by hormones will be discussed.

It is well-known that estrogen and progesterone dictate the proliferative and secretory phases of the uterine cycle. Apoptosis is associated with the cyclic growth and regression of the endometrium and is thought to be involved in the regulation of menstruation. Bcl-2, an antiapoptotic factor, is expressed in the fetus as early as 22 weeks of gestation and in adult uterine endometrium, stroma, and myometrium. Bcl-2 is expressed in uterine glandular epithelium in the early follicular phase and reached its highest intensity in the late follicular phase and early secretory phase endometrium, but it waned during the late secretory phase and was low to nonexistent in the menstrual phase of the cycle. Staining was present in the stromal and mainly in lymphocytes (not epithelium) during the late secretory phase of the cycle. In women treated with hCG to rescue the CL, Bcl-2 expression was absent in the endometrium indicating that it is not a mediator of prolonged survival of the endometrium in the luteal phases of the cycle and thereafter. Similar changes in Bcl-2 have been observed in rat uteri. Bax, an apoptotic factor, is increased by estrogen/progesterone in uterine luminal and glandular epithelial cells, the periluminal stroma, and myometrium. The mechanism of increased apoptosis in the uterus is unknown, but TGF-β1 and -2 induce apoptosis of endometrial stromal cells *in vitro* by autocrine and paracrine mechanisms. Ovariectomy of cyclic rodents induces a wave of apoptosis in the uterine endometrium and both estrogen and progesterone are antiapoptotic, but progesterone is most effective. However, progesterone, 5α-dihydrotestosterone, and dexamethasone given singly with estradiol opposed the antiapoptotic action of estradiol on the endometrium. The actions of progesterone and 5α-dihydrotestosterone could be blocked by RU 486, an antiprogestin and flutamide, and antiandrogen, respectively. No changes in stromal cell apoptosis are observed after ovariectomy. Progestins prevent apoptosis in rat endometrial cell lines by increasing the ratio of Bcl-XL, the apoptosis-inhibiting form of Bcl-X, to Bcl-XS, the apoptosis-inducing form of Bcl-X.

Fas and Fas ligand, mediators of apoptosis, are coexpressed in the uterus. In addition, GnRH receptor sites have been detected in mouse uterus and GnRH induces endometrial regression by apoptosis in the luteal phase. Old mice (24 months old) exhibit increased apoptotic bodies in the uterus when compared with young mice (5 months old). Terminin-30, a protein identified by unique monoclonal antibodies, is found in senescent cells and has been isolated in old uteri in association with apoptosis.

V. APOPTOSIS IN THE MAMMARY GLAND

A. During Development and Reproductive Cycles

There is relatively little information in the literature regarding apoptosis during mammary development; however, Bcl-2 expression is high in the basal cell layer of developing fetal breast bud and surrounding mesenchyme. The significance of this observation is not known but suggests that apoptosis in the developing mammary gland is low.

In adult women, mammary tissues undergo continual cycles of growth and involution which coincide with the menstrual cycle. Again, little is known about regulation of apoptosis during regular reproductive cycles. However, available evidence suggests that the apoptotic index (the proportion of cells undergoing apoptosis in relation to the whole) in the mammary gland is dependent on stage of the reproductive cycle and that it is hormonally regulated. The apoptotic index in mammary tissues of adult women is cyclical, with a peak toward the end of the menstrual cycle. Expression of the antiapoptotic effector, Bcl-2, increases during the follicular phase in mammary lobules, peaks at midcycle, and then decreases during the luteal phase. A similar pattern is observed in the glandular epithelium of the endometrium, which suggests hormonal regulation of Bcl-2 expression. There is a paucity of information regarding the expression of other important regulators of apoptosis, i.e., other Bcl-2 family members and caspases.

B. During Mammary Gland Involution

During pregnancy there is proliferation and differentiation of mammary tissue resulting in functional secretory epithelium at parturition in preparation for lactation. Once suckling is terminated the mammary gland undergoes involution to its prelaction/prepregnancy state. The process of mammary gland involution in mice is at least partially the result of increased apoptosis occurring in a two-step process. The first step is apoptosis of secretory epithelial cells while specific differentiated functions, such as β-casein mRNA expression, are maintained. Up to this point mammary gland involution is reversible, despite the loss of secretory epithelial cells.

The second stage of involution is highlighted by proteolytic degradation of the extracellular matrix (ECM) which maintains the structural and functional integrity of the gland. This second phase is irreversible.

Pup removal and termination of suckling results in an increase in the expression of apoptosis-related effectors, including Bax, Bcl-x_s, ICE, and p53, a key regulator of cell cycle progression. DNA laddering is observed in isolated mammary tissues within 24 hr of pup removal. The induction of these death-promoting factors and increase in apoptosis in the involuting mammary gland appears to be under local control. Sealing of individual teats results in apoptosis within those individual glands, whereas adjacent glands with open teats remain functional, with no induction of apoptotic effectors. In addition to the induction of death-associated factors, there may be a loss of survival signals. The intracellular signal cascade stimulated by the lactogenic hormone, prolactin, is disrupted within 24 hr of pup removal.

The local stimulus inducing these effects is not clear. It is uncertain whether the buildup of milk within alveoli produces mechanical stimuli, such as disruption of cell adhesion due to increased alveolar pressure, that induce apoptosis or whether there is concentration of apoptosis-inducing stimuli. In fact, milk accumulation may have little to do with stimulating apoptosis during mammary gland involution since natural weaning leads to a decline in milk production with little increase in intraalveolar pressure. However, the pattern of involution is similar to that after pup removal, albeit at a more protracted rate.

During the first stage of mammary gland involution the extracellular matrix is maintained, probably by systemic lactogenic hormones, which maintains structural integrity of the gland and allows for the reversibility of the early stages of involution. If suckling is resumed within 48 hr, mammary involution is halted and functional integrity of the glands is restored, presumably by proliferation and differentiation of secretory epithelial stem cells. If suckling is not restored within 48–72 hr, increased expression of the matrix metalloproteinases, gelatinase A and stromolysin, as well as urokinase-type plasminogen activator by fibroblast-like cells in the stroma surrounding the collapsed alveoli result in degradation of the extracellular matrix. Systemic administration of glucocorticoids at pup removal has no effect on the induction of alveolar epithelial apoptosis but blocks the induction of the matrix metalloproteinases and breakdown of the ECM. This reinforces the idea that the first stage of apoptosis is locally regulated, whereas the second stage of involution may be regulated by systemic factors. Finally, the majority of epithelial cells are replaced by adipose tissue to restore the mammary gland to a resting state.

VI. APOPTOSIS IN THE TESTIS

The reduction in the number of germ cells that occurs during mitotic divisions of spermatogonia, during meiosis of spermatocytes, and during spermiogenesis is largely by apoptosis. Hypophysectomy or chemical castration with GnRH antagonists induces massive apoptosis in testes of immature and mature rats, with reduced serum concentrations of FSH and, to a lesser extent, LH. However, concurrent treatment with FSH, LH/hCG, and androgens suppresses apoptosis following hypophysectomy, indicating that gonadotropins and androgens are antiapoptotic.

Testicular apoptosis has been studied during development and throughout maturity in rats. Apoptosis increases in the seminiferous tubules between Days 8 and 16–28 of age and then decreases in testis of adult animals at ~70 days of age. Treatment with a GnRH antagonist increased apoptosis between Days 16 and 32, but the GnRH antagonist had no effect in younger rats or adult rats indicating age-related differences in apoptotic capability within the testis and differences in gonadotropin dependence of the testis. Most cell types undergoing apoptosis in those studies were spermatocytes. Three major factors appear to control the onset of apoptosis in the testis: (i) age of the animal, (ii) serum concentration of FSH (and probably LH), and (iii) the stage of the development of the seminiferous epithelium.

Recently, the use of gene targeting and transgenic technologies has revealed effects of specific genes on apoptosis in the testis (Table 4). In mice lacking either the bone morphogenetic protein-8B (BMP8B),

TABLE 4
Various Factors and Processes Affecting Apoptosis in the Testes

Antiapoptotic	Testicular cellular target	Apoptotic	Testicular cellular target
FSH	Germ	Hypophysectomy	Germ, Leydig, Sertoli
LH/hCG	Germ and Leydig	Cryptorchidism	Germ, Leydig, Sertoli
		Short days (species specific)	Testis
		γ-Irradiation	Sertoli
Testosterone	Germ	Anti-FSH	Spermatogonia, spermatocytes
Bcl-2 overexpression	Spermatogonia	Anti-c-kit	Spermatogonia, spermatocytes
		GnRH antagonist/agonist	Spermatogonia, spermatocytes
		c-Myc	Spermatocytes
		BMP8B	Spermatocytes
		hsp 70-2	Spermatocytes
		Xenobiotics	Cell/organ depleted
		2,5-Hexadione	Sertoli, spermatogonia
		Ethylene dimethanesulfonate	Leydig
		Mono-2 ethylhexylphthalate	Sertoli cell
		Cadmium chloride	Testis
		Methoxyacetic acid	Germ cells, spermatocytes
		2-Methoxyethanol	Spermatocytes

a member of the TGF-β family of proteins, hsp 70-2), increased apoptosis was observed in spermatocytes. BMP8B is expressed in male germ cells of the testis and its role in spermatogenesis was largely unknown until a targeted mutation revealed two distinct defects in homozygous mutant testes. During the initial 2 weeks after birth, the germ cells failed to proliferate or exhibit reduced proliferation and delayed differentiation. In adults, there was increased apoptosis in spermatocytes leading to germ cell depletion and sterility. The Sertoli and Leydig cells were unaffected in BMP8B null mice. Normally, spermatogenic cells synthesize hsp 70-2 during meiosis. Null hsp 70-2 mice lacked postmeiotic spermatids and mature sperm and were infertile (females were unaffected). Synaptonemal complexes are associated with hsp 70-2 and, in the null mice, structural abnormalities were evident by late prophase and development only rarely proceeded to meiotic divisions. In addition, analysis of nuclei and genomic DNA revealed that failure of meiosis in null mice coincided with increased apoptosis in spermatocytes.

Normally, during spermatogenesis, more than half of the germ cells undergo apoptosis. However, in mice overexpressing Bcl-2, an antiapoptotic protein, massive accumulation of spermatogonia was observed in seminiferous tubules at 4 weeks of age. Interestingly, the increased number of cells were apoptotic by 7 weeks. In older Bcl-2-overexpressing mice, abnormal accumulation of spermatogonia and degeneration of germ cells were observed. In transgenic mice with c-Myc overexpression, atrophy of the seminiferous tubules and depletion of sperm was observed. In fact, spermatogenesis was arrested at prophase of meiosis in primary spermatocytes exhibiting a high degree of apoptosis.

Reduced fertility is exhibited by male mice with alterations in gene expression. However, whether apoptosis within the testis accounts for the reduction in testicular size and infertility is unknown. For example, estrogen receptor knockout (ERKO) mice exhibit reduced fertility. At 10 weeks of age, spermatogenesis is disrupted in the caudal and cranial pole of the testis. There is degeneration and disorganization of seminiferous tubules with few spermatogenic cells, reduced epididymal sperm, dilated tubule lumens, and reduced sperm motility. Sperm from ERKO mice cannot fertilize eggs. Testes of ERKO mice are descended but the cremaster sacs were reduced and this was coupled with excessive development of the cremaster muscle. The thickened muscle was associated with retraction of the testes into the inguinal canal or abdomen; therefore, testicular volume and spermatogenesis were reduced. Other examples of null mice exhibiting a disrupted testicular axis are the steroidogenic factor-1 null mice, which lack gonads, and the apoplipoprotein B-knockout mice, which exhibit reduced fertility because the sperm do not fertilize eggs. However, once the zona pellucida is removed from the eggs of ApoB null mice, sperm could fertilize the eggs. The sperm from the apoB knockouts exhibit reduced count, motility, and survival time. Interestingly, fertility could be restored by introducing apoB into the animals. Bax-deficient mice exhibit disordered seminiferous tubules and an accumulation of atypical premeiotic germ cells but not mature haploid sperm. Inactivation of the HR6B ubiquitin-conjugating DNA repair enzyme in mice causes male sterility associated with chromatin modification. Spermatogenesis is disrupted in the HR6B knockouts during the postmeiotic condensation of chromatin in spermatids. Thus, HR6B is needed for chromatin remodeling. Lastly, TGF-β1 overexpressing transgenic mice exhibit testicular atrophy but there is no indication as to whether the reduced testicular size is due to apoptosis.

Disruption of testicular function after exposure to environmental toxicants has been localized to specific cell types (Table 4). Because of the varied effects of these compounds on testicular function, only a listing is provided.

VII. APOPTOSIS IN THE PROSTATE

In the epithelial cells of the adult human prostate, the rate of prostatic cell death is offset by an equal rate of proliferation such that neither involution nor excessive growth of the gland occurs. Castration of an adult male induces a prompt decline in serum concentrations of testosterone followed by involution of the prostate due to apoptosis of glandular epithelial cells. However, the basal epithelial and stromal cells of the prostate are unaffected by castration. Androgens stimulate glandular cell proliferation while inhibiting apoptosis. Within 10 days after castration, ~80% of the glandular epithelial cells die and are eliminated from the prostate. The earliest signs of apoptosis after androgen withdrawal are inhibition of glandular cell proliferation coupled with a generalized atrophy of the secretory cells in individual acini. Tall columnar secretory cells become cuboidal in shape and there is a concurrent decrease in protein synthesis. At this point the initiation of cell death is totally reversible by replacement therapy with exogenous androgen. In the next stage the cells become spherical, undergo changes in nuclear chromatin structure, a series of proteins increase, and histone H1 and polyamine levels decrease (maintain DNA compaction allowing for opening of the genomic DNA in the interlinking region between nucleosomes). Then a TGF-α1-directed increase in intracellular calcium occurs and a calcium–magnesium-dependent endonuclease becomes activated. At this point cell death is inevitable and irreversible. The characteristic DNA fragments are formed, but plasma and lysosomal membranes are intact and mitochondria are still functional. The ICE-like proteases are next activated and hydrolyze PARP and laminins in the nuclear membrane and in the nucleus fragment. Then, plasma membrane blebbing and cellular fragmentation occur, resulting the formation of apoptotic bodies. Calcium-dependent tissue transglutaminase activity (which cross-links membrane proteins) becomes activated and the apoptotic bodies are phagocytized by macrophages and/or neighboring epithelial cells.

Apoptosis induced by androgen withdrawal does not involve retinoblastoma protein phosphorylation, c-Myc, c-Fos, or p53. Interestingly, cyclin D1 and C (G_0 to G_1), cyclin E, cdk2, thymidine kinase, and H4 histone (G_1 to S), and cyclin A (progression through S) were reduced by castration and increased by androgen replacement.

Within the epithelial compartment of the normal prostate of the rat, Bcl-2 is expressed by basal epithelial cells, neuroendocrine cells, and intraacinar lymphocytes but not by glandular epithelial cells. The Bcl-2-negative cells are the major androgen-dependent cell type present within the gland. Castration results in a significant increase in Bcl-2 expression, which is inhibited by androgen replacement therapy. These results indicate that under normal conditions, Bcl-2 expression is held in check, preventing hyperproliferation of the prostatic epithelium.

See Also the Following Article

Caenorhabditis elegans; Mammary Gland Development; Ovary, Overview; Prostate Gland; Testis, Overview; Uterus, Human

Bibliography

Abrahamsohn, P. A., and Zorn, T. M. (1993). Implantation and decidualization in rodents. *J. Exp. Zool.* **266,** 603–628.

Denmeade, S. R., Xiaohui, S. L., and Isaacs, J. T. (1996). Role of programmed (apoptotic) cell death during the progression and therapy for prostate cancer. *Prostate* **28,** 251–265.

Hsueh, A. J. W., Billig, H., and Tsafriri, A. (1994). Ovarian follicle atresia: A hormonally controlled apoptotic process. *Endocr. Rev.* **15,** 707–724.

Hsueh, A. J. W., Eisenhauer, K., Sang-Young, C., Hsu, S.-Y., and Billig, H. (1996). Gonadal cell apoptosis. *Recent Prog. Horm. Res.* **51,** 433–456.

Li, M., Liu, X., Robinson, G., Bar-Peled, U., Wagner, K.-U., Young, W. S., Hennighausen, L., and Furth, P. A. (1997). Mammary-derived signals activate programmed cell death during the first stage of mammary gland involution. *Proc. Natl. Acad. Sci. USA* **94,** 3425–3430.

Nagata, S. (1997). Apoptosis by death factor. *Cell* **88,** 355–365.

Spencer, S. J., Cataldo, N. A., and Jaffe, R. B. (1996). Apoptosis in the human female reproductive tract. *Obstet. Gynecol. Surv.* **51,** 314–323.

Tilly, J. (1996). Apoptosis and ovarian function. *Rev. Reprod.* **1,** 162–172.

Tilly, J. L., and Perez, G. I. (1997). Mechanisms and genes of physiological cell death: A new direction for toxicological risk assessments? In *Comprehensive Toxicology* (I. G. Sipes, C. A. McQueen, and A. J. Gandolfi, Eds.), Vol. 10, pp. 389–395. Elsevier, Oxford, UK.

Tilly, J. L., and Ratts, V. S. (1996). Biological and clinical importance of ovarian cell death. *Contemp. OB/GYN* **41,** 59–86.

Wride, M. A., and Sanders, E. J. (1995). Potential roles for tumour necrosis factor alpha during embryonic development. *Anat. Embryol. (Berlin)* **191,** 1–10.

Arachnids

see Chelicerate Arthropods

Armadillo

Richard D. Peppler

University of Tennessee College of Medicine

I. Introduction
II. Female Armadillo
III. Male Armadillo
IV. Other Organs That Influence Reproduction

GLOSSARY

delayed implantation A period of time when a fertilized egg lies in a resting state in the uterine cavity without attachment.
follicle The ovum and its surrounding cells.
gestation The period of development of a fetus in the uterus.
progesterone A hormone produced by the corpus luteum, adrenal, and placenta which prepares the uterus for development of the ovum.
testosterone A hormone produced by the testis that is responsible for spermatogenesis.

In 1971, Kirchheimer and Storrs reported that lepra bacilli replicate in the armadillo producing a disease that resembles lepromatous leprosy. This finding rekindled interest to develop the armadillo as a laboratory animal, but this has been slow to materialize because of the failure to breed this species in captivity. However, its potential is unlimited not only as a model for leprosy research but also for use in genetic, transplantation, and teratology studies. Its potential is based on characteristics such as long life span (12–15 years), size (4 or 5 kg), low body temperature (32–34°C), and monozygous quadruplet offspring. Most of the knowledge about the female armadillo has originated from the investigations of Newman, Hamlett, Talmage and Buchanan, and Enders. It ovulates one egg in July and, if fertilized, the egg enters the uterine cavity and remains in diapause during obligate delayed implantation until gestation begins in mid-October. After the single blastocyst implants and before the primitive streak stage is reached, the embryo buds to form four separate areas of organization. Pregnancy lasts 4–4.5 months, with parturition occurring in late March or early April. The female armadillo is the only species which regularly produces four identical offspring from a single ovum.

I. INTRODUCTION

A. General

The armadillo is a placental mammal which differs from all others by its ossified dermal plates normally associated with reptiles or fish. It is this characteristic which leads to the name of the species, which is derived from the Spanish word *armada*—one that is armed. The armadillo belongs to the order Edentate, characterized by poorly developed enamel or the absence of teeth, which also includes the sloth, ant bears, scaly anteater, and aardvark. In the armadillo, there are 16 teeth in each jaw. All are molars and without enamel.

B. Zoogeography

There are approximately 20 species of living armadillos grouped into 9 or 10 genera. Armadillos belong to the family Dasypodidae and vary in size. The largest is the giant armadillo of eastern South America, which has a 3-ft body and weighs as much as 130 lbs. The smallest, the pygmy armadillo, lives in Argentina and is about the size of a small rat. Armadillos are located throughout the Americas from Argentina to the United States. Although populations of armadillos in general are declining, the best known and only species native to the United States, the nine-banded armadillo, appears to be increasing and extending its range. This animal migrated from South America and Mexico to the United States over 170 years ago. It was located in Texas along the U.S.–Mexico border in 1830. By 1905 it had migrated into western Texas as far north as the Colorado river. Since that time its migration has been eastward to the Carolinas, Georgia, and Florida; westward to New Mexico; and northward to Kansas and Missouri. The migration of this species has been limited by such factors as cold climates, droughts, and hard soils. It does not hibernate as some animals do and must seek food year-round. This is the reason severe weather conditions such as cold and drought have limited its migration.

C. Diet

The armadillo is an insectivore, with 75% of its diet consisting of insects such as grubs, scorpions, roaches, tarantulas, and fire ants. Fifteen percent of the diet consists of animal material such as small lizards and snake eggs. The remaining 10% comes from vegetable matter, wild berries, and dirt. It has limited vision and locates its food by sense of smell.

D. Gross Anatomy

The armadillo is nocturnal, which is the result of its response toward cooler temperatures and toward darkness. It has a low body temperature (32–34°C) and prefers coming out of its burrow when temperatures are cooler such as at dusk and in the evening.

The adult nine-banded armadillo is about the size of a large cat and weighs 3–5 kg. It has a long snout and large ears. The dorsal portion of its body is protected by a carapace as is found in turtles, lobsters, and crabs. This carapace consists of small plates of bone covered by tough skin. These plates are separated from each other by soft, hairy skin. The carapace is divided into two portions: One covers the forelimbs and the other covers the hindlimbs. These portions are linked by transverse bands of plates connected by skin. The number of these bands varies from species to species (three, six, eight, or nine); the number of bands is the basis for the name of some armadillos (i.e., nine-banded armadillo). The bands also allow the carapace to be flexed. On the ventral surface the soft skin contains a few clumps of coarse hair. The limbs are short but extremely muscular. The forelimbs have four toes and the hindlimbs have five. Each toe has a good size claw. This anatomy of the limbs assists the animal with its burrowing and food-gathering. The armadillo has a tail about the same length as its body and it is also covered with scales. In the female, there is a vestibular cleft anterior to the anus which functions as the terminal urethra and vagina. The absence of a true vagina makes it difficult to determine stages of the cycle by vaginal smears as in other species, although cyclic changes of 4 days duration occur in the urogenital smear. The uterus is simplex in form and thus is like the highly specialized uterus of primates. The male system is typical mammalian. The male has an external penis but no scrotum; the testis are intraabdominal. Because the animal's body temperature is lower than that in humans, a scrotum is not necessary for spermatogenesis. Each sex has six to eight elongated nipples and two anal glands which function in a pheromone-like fashion as in other animals. These are scent glands and account for the animal's musty odor.

E. Diseases

The armadillo has been reported to be susceptible to human diseases such as leprosy, Chagus disease, relapsing fever, African sleeping sickness, trichinoses, exanthematic and murine typhus, schistosomiaris, and *Nocardia brasiliensin* infection. Because of its characteristics and susceptibility to so many diseases, the armadillo has unlimited potential as an animal model in biomedical research. However, investigators have not established it as a model, despite its uniqueness, due to the lack of success in breeding the armadillo in captivity.

II. FEMALE ARMADILLO

Details of the growth and maturation of an individual follicle were reported by Newman. The armadillo spontaneously ovulates one egg once a year, at about the middle of July, with the subsequent formation of a single corpus luteum (CL). Accessory CLs do not appear.

A. Immature

1. Follicular Development

At birth, numerous primordial follicles, but no secondary follicles, are found in the ovary. The number of these follicles demonstrate a progressive pattern of follicular development (more larger sized follicles) from 3 to 21 months of age. This may explain why the female does not ovulate until after 15 months of age. Follicles likely to ovulate (>978 μm) are present in some ovaries at 9 months of age but not found in all ovaries until 27 months of age (Table 1). Even when these large follicles are present, nests of primordial follicles are still present at 9, 12, 15, and 18 months of age in the ovary. A CL is present in some ovaries at 15 months of age and in the ovary in all armadillos at 18 and 21 months of age (Table 1).

2. Plasma Progesterone

Correlating with the previously discussed histological observations is the fact that plasma progesterone concentrations remain below the adult concentration of 5 ng/ml until 12 months of age. A progesterone concentration >10 ng/ml, indicative of a viable CL, does not occur until 17–20 months of age.

B. Adult Ovary

1. Follicular Development

Peppler and Canale reported on the annual pattern of follicular development. The total number of follicles >358 μm in diameter remains constant from January until June but is 50% less during the remaining months of the year except in October and November. More large follicles (>978 μm in diame-

TABLE 1
Synopsis of Ovarian Histology in the Armadillo at Different Ages from 0 to 27 Months[a]

Age (months)	Primordial follicles	Secondary follicles	Total number of follicles >202 μm	Total number of follicles >978 μm	Corpus luteum
0	Yes	No	—	—	No
3	Yes	Yes	16.6 ± 1.7	0	
6	Yes	Yes	20.7 ± 1.9	0	No
9	Yes	Yes	25.2 ± 2.1	1.5 (2)[b]	No
12	Yes	Yes	23.0 ± 1.8	1.0 (2)	No
15	Less	Yes	15.5 ± 1.3*	1.0 (1)	Yes
18	Less	Yes	22.8 ± 1.7*	1.5 (1)	Yes
21	Very few	Yes	26.3 ± 1.9	2.5 (2)	Yes
27	Very few	Yes	21.7 ± 1.7	1.5 (3)	Yes

[a] Three animals killed at each time point.
[b] Numbers in parentheses signify number of animals in age group with follicles of the respective size range.
* $p < 0.05$ with value shown for previous age.

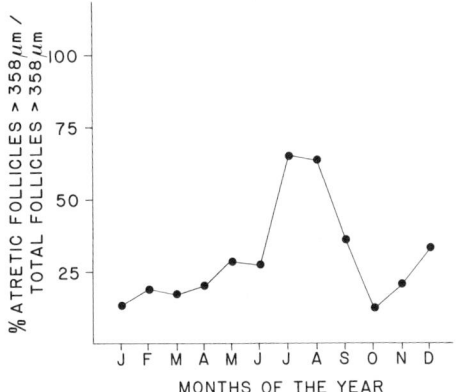

FIGURE 1 Percentage of atretic follicles >358 μm of the total number of follicles (both normal and atretic) >358 μm in the nine-banded armadillo during each month of the year.

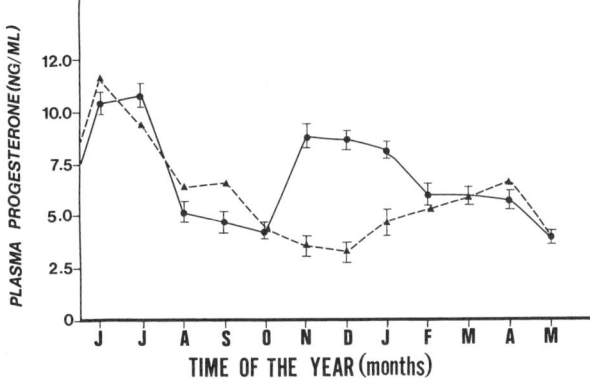

FIGURE 2 Plasma progesterone level in the nonpregnant female armadillo throughout the year. ▲, monthly value of 30 animals; ●, monthly value of 19 animals.

ter) are present in ovaries (2.7–4.1 follicles/ovary) in the months of April, May, June, and October than during any of the other months, when follicle numbers vary from 0 (December) to 1.6 (March). Maturation of all but one follicle in both ovaries ceases before ovulation and atresia of the arrested follicles occurs. Figure 1 shows the percentage of atretic follicles >358 μm of the total number of normal and atretic follicles >358 μm in the female nine-banded armadillo during each month of the year. The percentage increases dramatically during the months of July and August, during which the adult armadillo ovulates.

2. Gonadotropin Response

The armadillo ovary will respond to exogenous and endogenous gonadotropins. A dose–response relationship occurs, although the response results in more than one oocyte being shed. Animals injected with clomiphene citrate show an increase in plasma progesterone to a level indicative of ovulation within 6 days. CLs appear in the ovaries of animals 4–27 days postinjection. The urogenital smear pattern changes from a diestrous pattern preinjection to a proestrous and subsequently estrous pattern 4–8 days postinjection.

3. Plasma Progesterone

Peppler reported on seasonal progesterone concentrations in peripheral plasma from 49 nonpregnant animals (Fig. 2). An increase in the plasma progesterone level occurs in June or July and correlates with the presence of a CL. Diurnal variation in these levels is not apparent during 4-hr sampling. In most animals ($N = 30$) progesterone decreases after 3 weeks and remains at a baseline value of 5 ng/ml from August through May. In the other 40% of animals, progesterone increases during the months of November–January and then returns to a baseline level the remainder of the year (February–May). Some investigators have suggested ovulation occurs in the armadillo at other times of the year and this finding in the 19 animals discussed previously would support such a concept.

C. Delayed Implantation

Reproduction in the female armadillo is of interest because following ovulation, if mating occurs, there is a long period of arrested development of the blastocyst prior to implantation. The ovum is fertilized near the ovarian end of the oviduct and then enters the uterus in about 7–10 days as a monodermic vesicle and remains in diapause until implantation in October. During this period of obligate delayed implantation, the CL grows until it accounts for 75–90% of the ovary and reaches its maximum size when the blastocyst is in the uterus. Luteal cells are fully luteinized at this time and the appearance of the uterus does not differ from that of an animal in the late follicular phase. The progestational changes which normally occur in other species are not apparent during delayed implantation and probably account for the different opinions of whether the CL during delayed implantation is non- or hypofunctional. Hamlett considered the CL to be nonfunctional based on the lack of histological changes and the absence of progestational changes in the uterus at this time. Others found it to be functional as determined by light and electron microscopy observation and by progesterone content of this structure and that in the plasma.

Following ovulation and during delayed implantation, progesterone concentration remains elevated (13 ng/ml) during July–September. When implantation occurs, there is an increase in activity of the CL with concomitant progestational changes of the uterus. Bilateral ovariectomy during the delayed implantation period does not prevent implantation and suggests that nonovarian tissues, possibly adrenal, maintain the uterus. Accessory adrenal corticular tissue is located in the mesovarium near the cranial pole of the ovary. Implantation occurs 18–20 days following ovariectomy. In contrast, unilateral ovariectomy, whether of the ovary containing the CL or not, does not induce implantation. This suggests that the follicles in the remaining ovary inhibit the implantation and may be responsible for the delay period.

D. Pregnancy

Following the delayed implantation period, plasma levels of progesterone increase as gestation begins during October and remain at a high level (20 ng/ml) through March. Unlike many species which exhibit a decrease in plasma progesterone before or shortly after parturition, the progesterone level in the armadillo remains elevated until 55 days after delivery. Estrogen levels are thought to be low during pregnancy.

III. MALE ARMADILLO

A. Testis

The reproductive system of the male armadillo is typical of that of other mammalian species except that the testes descend only to the entrance of the inguinal canal. There is no scrotum and the intraabdominal location of the testis is not deleterious to spermatogenesis as it is in some species because of the animal's low body temperature. Unlike the female, the male reaches sexual maturity within the first year following birth. In general, the studies on the histology of the testis which have been done in recent years confirm that the structural and developmental changes of the maturing germ cells, as well as the epididymal morphology, are similar in the armadillo to those observed for most mammals. The seminiferous epithelium of the testis is organized into cellular associations of mixed composition and most cross sections of tubules show only one type of cellular association. The spermatogenic cycle is divided into 8 or 10 stages with respect to the morphology of the nuclei of the spermatids, their position and arrangement in the seminiferous epithelium, the presence of secondary spermatocytes, and the liberation of spermatozoa into the lumen of the tubule. Morphological changes in testicular histology do not occur on a seasonal basis. The duration of the cycle as determined by intratesticular injection of [^{3}H]thymidine and removal of the tissue at different times after the injection was 8.15 days. The cycle in the male armadillo is among the shortest in duration when compared with other species and the formation of spermatozoa is thought to extend over four consecutive cycles. Based on this duration, the entire spermatogenic process of armadillos lasts 32.6 days. Spermatozoa mature in the epididymis of the armadillo similar to that in scrotal animals. The interstitial tissue in the testis is composed of blood vessels, clusters of Leydig cells, connective tissue elements, and a network of lymphatic sinusoids. No major changes in the morphology of the interstitial tissue occur on a seasonal basis. In essence, testicular histology is identical in captive and free-ranging males at all times of the year.

B. Prostate

The prostate gland is a compound tubuloalveolar bilobed gland with a thin connective tissue capsule and is located anterior to the seminal vesicles. The gland is of the body type since it penetrates the urethral muscle and is encapsulated by connective tissue. There are 10–20 prostatic ducts which terminate on a tubercle on the posterior wall of the urethra much like in the human. However, unlike in the human, some of the ducts of the prostate gland empty into the vasa deferentia and the ducts of the seminal vesicle and the vasa deferentia open independently into the urethra rather than being joined. The histological characteristics of the gland do not exhibit seasonal variations.

C. Testosterone

In animals followed in captivity for 3–16 months, plasma testosterone level does not vary during the different months of the year. Diurnal variation in these levels is not apparent during 4-hr sampling. The circulating testosterone level in captive male armadillos appears to be similar to that in animals in the wild throughout the year. Although data concerning seasonal variation conflict, Czekala et al. reported some of the highest mean circulating testosterone levels for the armadillo compared with those in other species. This level ranged between 100 and 200 ng/ml (human range is 3–7 ng/ml). The lowest values were detected in January and the highest occurred in July. Peppler and Stone reported plasma testosterone levels between 9 and 14 ng/ml. Differences were not apparent between the levels measured in the different months. Most of the investigations on the histology of the testes and prostate gland in combination with the finding of constant plasma testosterone levels throughout the year indicate that the male reproductive system in the armadillo does not undergo seasonal regression.

IV. OTHER ORGANS THAT INFLUENCE REPRODUCTION

A. Placenta

Implantation in the armadillo is readily compared to that in primates rather than to members of other orders with hemochorial placentas. The trophoblast penetrates the epithelium and spreads out in the endometrium. The syncytium is much like that in the human. It is thin in older placentas and both microvilli and endoplasmic vesicles are present. Steroid aromatizing enzyme systems are present at an early stage of development and produce estrone, 17β-estradiol, and estriol. The metabolism of these estrogens is qualitatively like that in other species but not quantitatively similar. The amount of estriol is much less in the armadillo (100 times less) than in the human when compared on a weight by weight basis. The four fetuses within the uterus are enclosed in their own amnion and attached by separate umbilical cords to the placenta. There is no communication between villous circulation of one fetus and that of an adjacent fetus. Placental shape is much more uniform than it is in the human and anastomoses of the four vessels in the cord do not exist. The two umbilical arteries are branches of the internal iliac artery and the two umbilical veins, unlike in the human in which there is only one, enter the liver.

B. Pituitary

The anterior and posterior lobes of the pituitary gland are about the same size but the intermediate lobe is absent. Cells of both of these lobes bind estradiol with the effect being 10 times greater with the anterior lobe cells than with pituicytes. There is a similarity in the development of the hypothalamo–hypophyseal neurosecretory system between the armadillo and human fetus.

Bibliography

Anderson, J. M., and Benirschke, K. (1966). The armadillo, *Dasypus novemcinctus*, in experimental biology. *Lab. Anim. Care* 16, 202–216.

Czekala, N. M., Hodges, J. K., Gause, G. E., and Lasley, B. L. (1980). Annual circulating testosterone levels in captive and free ranging male armadillos (*Dasypus novemcinctus*). *J. Reprod. Fertil.* 59, 199–204.

D'Addamio, G. H., Roussel, J. D., and Storrs, E. E. (1977). Response of the nine-banded armadillo (*Dasypus novemcinctus*) to gonadotropins and steroids. *Lab. Anim. Sci.* 27, 482–489.

Enders, A. C. (1966). The reproductive cycle of the nine-banded armadillo (*Dasypus novemcinctus*). In *Comparative Biology of Reproduction in Mammals* (I. W. Rowlands, Ed.), pp. 295–310. Academic Press, New York.

Enders, A. C., and Buchanan, G. D. (1966). The reproductive

tract of the female nine-banded armadillo. *Texas Rep. Biol. Med.* **18**, 323–340.

Hamlett, G. W. D. (1932). The reproductive cycle in the armadillo. *Zeit. Wiss. Zool.* **141**, 143–157.

Humphrey, S. R. (1974). Zoogeography of the nine-banded armadillo (*Dasypus novemcinctus*) in the United States. *Bioscience* **24**, 457–462.

Kirchheimer, W. F., and Storrs, E. E. (1971). Attempts to establish the armadillo (*Dasypus novemcinctus* Linn.) as a model for the study of leprosy. I. Report of lepromatoid leprosy in an experimentally infected armadillo. *Int. J. Lepr.* **39**, 692–702.

Meritt, D. A., Jr. (1973). Edentate diets. I. Armadillos. *Lab. Anim. Sci.* **23**, 540–542.

Nagy, F., and Edmonds, R. H. (1973). Morphology of the reproductive system of the armadillo. The spermatogonia. *J. Morphol.* **140**, 307–321.

Newman, H. H. (1912). The ovum of the nine-banded armadillo: Growth of the ovocytes, maturation and fertilization. *Biol. Bull.* **23**, 100–141.

Peppler, R. D. (1979). Reproductive parameters in the nine-banded armadillo. *Anat. Rec.* **193**, 649–650.

Peppler, R. D., and Canale, J. (1980). Quantitative investigation of the annual pattern of follicular development in the nine-banded armadillo (*Dasypus novemcinctus*). *J. Reprod. Fertil.* **59**, 193–197.

Peppler, R. D., and Stone, S. C. (1980a). Plasma progesterone level during delayed implantation, gestation and the postpartum period in the female armadillo. *Lab. Anim. Sci.* **30**, 188–191.

Peppler, R. D., and Stone, S. C. (1980b). Clomiphene-induced ovulation in the 9-banded armadillo (*Dasypus novemcinctus*). *Lab. Anim.* **14**, 329–330.

Peppler, R. D., and Stone, S. C. (1981). Annual pattern in plasma testosterone in the male armadillo (*Dasypus novemcinctus*). *Anim. Reprod. Sci.* **4**, 49–53.

Peppler, R. D., Hossler, F. E., and Stone, S. C. (1986). Determination of reproductive maturity in the female armadillo (*Dasypus novemcinctus*). *J Reprod Fertil.* **76**, 141–146.

Talmage, R. V., and Buchanan, G. D. (1954). The armadillo. A review of its natural history, ecology, anatomy and reproductive physiology. *Rice Institute Pamphlet* **41**, 1–135.

Torres, C. N., Godinho, H. P., and Setchell, B. P. (1981). Frequency and duration of the stages of the cycle of the seminiferous epithelium of the nine-banded armadillo (*Dayspus novemcinctus*). *J. Reprod. Fertil.* **61**, 335–340.

Weaker, F. J. (1977a). Spermatogonia and the cycle of the seminiferous epithelium in the nine-banded armadillo. *Cell Tissue Res.* **179**, 97–109.

Weaker, F. J. (1977b). The fine structure of the interstitial tissue of the testis of the nine-banded armadillo. *Anat. Rec.* **187**, 11–28.

Weaker, F. J. (1980). Morphology of the prostate gland in the nine-banded armadillo. *Acta Anat.* **106**, 405–414.

Aromatization

Alan J. Conley and Karen W. Walters

University of California, Davis

I. The Aromatase Enzyme Complex
II. The Aromatase Complex Is Highly Conserved
III. Principal Sites of Aromatase Expression
IV. Tissue-Specific Regulation of $P450_{arom}$ Expression
V. Conclusion

GLOSSARY

aromatase The catalytically active protein complex formed by association of $P450_{arom}$ with reductase.

aromatization The enzymatic process by which estrogens are synthesized.

CYP19 The gene encoding P450 aromatase.

cytochrome P450 A superfamily of heme-containing enzymes that catalyze the oxidation of a large variety of organic substrates.

cytochrome P450 aromatase The enzyme that catalyzes the conversion of androgens to estrogens.

NADPH-cytochrome P450 reductase (reductase) nicotinamide adenine dinucleotide phosphate–cytochrome P450 reductase (reductase), the flavoprotein that forms a complex with microsomal cytochrome P450 enzymes facilitating electron transfer and substrate oxidation.

redox Pertaining to oxidation-reduction reactions.

steroid hydroxylase The subclass of cytochrome P450 enzymes involved in the synthesis and metabolism of steroid hormones.

steroidogenic Pertaining to the synthesis or metabolism of steroids.

steroids A class of hormones with a characteristic four-carbon ring structure that are synthesized from cholesterol.

Aromatase is the enzyme system responsible for the synthesis of estrogens from androgens, the sex steroid hormones essential for the endocrine control of reproduction in all mammals. Specifically, androgens and estrogens influence most, if not all, aspects of the establishment and maintenance of gender. Reproductive development fails in the absence of these steroids or their receptors, and many endocrine abnormalities affecting reproduction in mammals are accompanied by abnormalities of androgen or estrogen metabolism. Moreover, the phenotypes of individuals suffering from recently recognized genetic abnormalities affecting steroidogenic enzymes and steroid receptors add considerable weight to the historically recognized role for sex steroids in somatic growth, mineral metabolism (particularly calcium), and perhaps even cardiovascular function. Therefore, the balance between androgen and estrogen production is essential not only for normal sexual development and reproductive function but also for normal growth and the general physiological well-being of both sexes. Systemically, and at the local tissue level, expression and activity of aromatase, catalyzing the last committed step in estrogen formation, maintains this critical balance of androgen and estrogen production.

I. THE AROMATASE ENZYME COMPLEX

The catalytically active aromatase enzyme complex consists of two components: aromatase cytochrome P450 (P450$_{arom}$), which binds the substrate, and, coupled to it, a ubiquitous flavoprotein, nicotinamide adenine dinucleotide phosphate (NADPH)–cytochrome P450 reductase (reductase), which facilitates flow of the electrons necessary for substrate oxidation (Fig. 1). Cytochrome P450$_{arom}$ is a member of

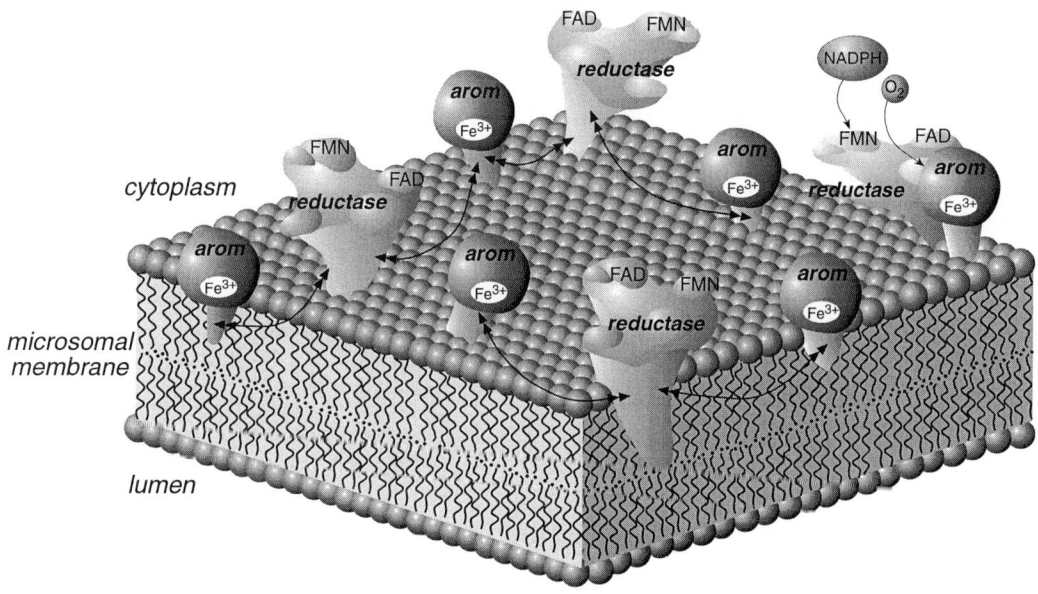

FIGURE 1 Diagrammatic representation of aromatase cytochrome P450$_{arom}$ and NADPH–cytochrome P450 reductase (reductase) anchored in the outer membrane of the microsomal membrane of the endoplasmic reticulum. Reductase is shown with flavin prosthetic groups (FMN, flavin mononucleotide; FAD, flavin adenine dinucleotide) and aromatase cytochrome P450 is depicted with Fe^{3+}, signifying the heme prosthetic group of the cytochrome P450. A catalytically active aromatase enzyme complex is formed when P450$_{arom}$ and reductase associate, as shown by utilization of NADPH (electron donor) and oxygen (O$_2$). Arrows represent molecular movement in the microsomal membrane.

the P450 superfamily of enzymes in a subgroup referred to as the steroid hydroxylases, which show greater catalytic specificity for steroid metabolism than other cytochromes P450.

A. Aromatase Is Anchored in the Endoplasmic Reticulum

All known eukaryotic cytochromes P450 are membrane bound, and $P450_{arom}$ and the redox-partner protein reductase are believed to be associated primarily with microsomal compartment. In general, both targeting of P450 enzymes to the microsomal compartment and membrane anchoring in it are processes dependent on membrane insertion of the P450 through association with signal recognition proteins. This is believed to involve a region of hydrophobic, N-terminal amino acids which constitute a conserved signal sequence. Unlike mitochondrial P450s, the N-terminal sequence is not cleaved from the microsomal P450s but is retained as a halt signal, preventing translocation across the membrane and, as expected, it is absent from soluble, prokaryotic P450s. Membrane-bound cytochromes P450 also tend to exhibit strong sequence homology in this N-terminal region which can be effectively switched from one P450 to another in some cases. Unfortunately, $P450_{arom}$ does not fall easily into this same P450 category and computer models suggest that the N-terminal sequence of human $P450_{arom}$ is uncharacteristic of other microsomal P450s. Deletion of the first 20, or even 40, amino acids seems not to preclude enzymatic activity of the human $P450_{arom}$. Several earlier subcellular fractionation studies reported that human placental $P450_{arom}$ was associated with both the mitochondrial and microsomal compartments, a result which was supported subsequently by immunocytochemistry performed on human placenta. Recently, $P450_{arom}$ and reductase were reported to be associated with the plasma membrane of cell lines expressing recombinant enzymes. Indeed, the plasma membrane is another subcellular fraction that routinely exhibits significant aromatase activity. Thus, although both components of the aromatase enzyme complex are commonly thought to reside in the endoplasmic reticulum, controversy remains regarding the membrane association of $P450_{arom}$ which seems to differ somewhat from other microsomal P450 enzymes.

B. The Catalytic Function of $P450_{arom}$

Most of what is known about the catalytic function of $P450_{arom}$ has been learned from studies conducted on the human enzyme, collected from term placenta in classic studies by Thompson and Siterii, Kellis and Vickery, Brodie, Osawa, and others. Typically, cytochromes P450 incorporate molecular oxygen into substrates by reactions that are absolutely dependent on efficient electron transfer from donor molecules. As noted previously, reductase performs this function in the case of $P450_{arom}$, and the two must associate with one another for catalytic activity to be expressed. Substrates enter the binding pocket of the P450 and the redox partner (reductase) associates, facilitated perhaps by electrostatic interactions, at an additional but distinct site (Fig. 1). Computer-generated models of $P450_{arom}$ have been constructed based on known crystal structures of soluble prokaryotic cytochromes P450. These studies have incorporated data on catalytic activities after site-directed mutagenesis of specific amino acid residues of $P450_{arom}$ and have greatly advanced our understanding of aromatase function, and recent reports by Graham-Lorence, Laughton, Chen, and others should be consulted.

1. Reductase Facilitates Three Consecutive Oxidation Events

The aromatase reaction, catalyzing the formation of an aromatic ring, is unusual if not unique in animals. The chemistry has been intensively studied by several investigators and readers should refer to articles by Robinson and Akhtar for detailed discussions. In general, however, the association of $P450_{arom}$ with reductase allows the transfer of a pair of electrons from a donor, NADPH, the first of which goes to the ferric ion of the heme prosthetic group of the P450. Oxygen then binds the iron and is activated during transfer of the second electron. The activated oxygen attacks the nearest carbon atom, hydrogen bonding and forming a hydroxyl group. In the case of some steroid hydroxylases, such as $P450_{arom}$, the hydroxylated aliphatic side chain or group is subse-

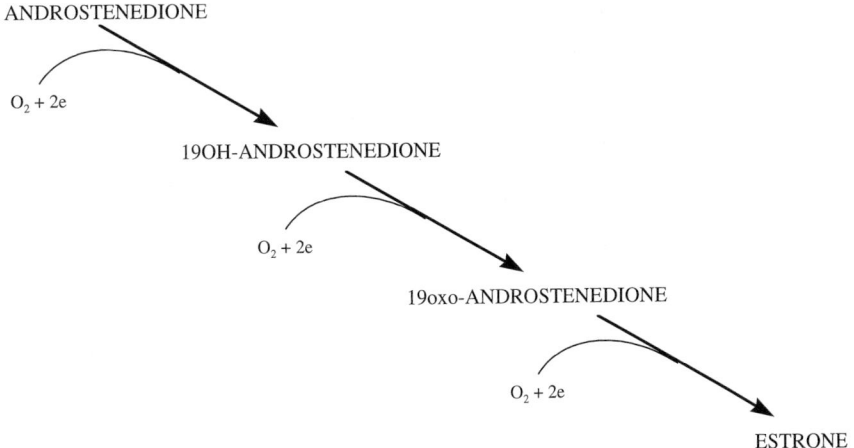

FIGURE 2 Simplification of the aromatase reaction showing the incorporation of oxygen along with electrons (from NADPH) in three sequential reactions as androstenedione is metabolized to estrone. The intermediates 19-hydroxy-androstenedione and 19-oxo-androstenendione may be released as additional reaction products in some tissues.

quently cleaved (lyase or desmolase activity). This general catalytic mechanism is uniquely complex for estrogen synthesis catalyzed by P450$_{arom}$ because it necessitates the sequential transfer of three pairs of electrons, consuming 3 mol of oxygen and 3 mol of NADPH in the synthesis of 1 mol of estrogen (Fig. 2). In fact, the reaction is thought to involve two consecutive hydroxylations at the C19 methyl group (19-hydroxylase activity) of the steroid substrate forming 19-hydroxy and eventually 19-oxo-intermediates. A third oxidative event, still debated, culminates in the cleavage of the angular C19 methyl group (19-demethylase or desmolase activity), aromatization of the steroid A ring, and estrogen formation. In fact, intermediates such as 19-OH-androstenedione and 19-oxo-androstenedione, which are known to be released as a result of androgen metabolism by purified P450$_{arom}$, become additional reaction products on occasion. Therefore, the reactions catalyzed by the aromatase complex that culminate in estrogen synthesis are uniquely dependent on reducing equivalents and reductase association.

2. Catalytic Activities of the Aromatase Complex

Historically, aromatase is the estrogen synthase but estrogen synthesis results from at least two, apparently distinct, catalytic activities, namely, 19-hydroxylase and 19-demethylase, each dependent on reductase association and electron transfer. Three consecutive oxidation events in estrogen synthesis means a greater reliance on reductase association for final estrogen production; otherwise, inefficient metabolism may result in the release of intermediates. This may occur, in fact, and 19-OH-androstenedione concentrations in women increase with advancing pregnancy at a greater rate than estrogen itself. The physiological or pathological importance of this and other intermediates is unknown, but both 19-OH- and 19-oxo-androstenedione are potent hypertensive agents in rats and are associated with hypertension in men and pregnant women. In addition, P450$_{arom}$ exhibits other catalytic activities including 2-hydroxylase, which may be responsible for catechol estrogen synthesis in human and equine placentas. Aromatases of some species may also be involved in the synthesis of norandrogens. Specifically, equine P450$_{arom}$ is believed to synthesize 19-norandrogens, which can be further aromatized to estrogens in equine placental and gonadal microsomes. Porcine testes and granulosa cells, which both express P450$_{arom}$, appear similarly capable of 19-norandrogen synthesis, which is inhibited by specific inhibitors of P450$_{arom}$. Whether aromatase catalyzes their formation directly or whether it is just one essential step among others is not entirely clear. Regardless, the 19-norandrogens may be more androgenic than androstenedione itself, suggesting the novel concept that P450$_{arom}$ function may be associated with androgen activation rather than metabolism to estrogen in some tissues. P450$_{arom}$ may be responsible directly or indirectly for the synthesis of a number of steroids that might prove to be of physiological importance in certain species.

C. Physiological and Biochemical Adaptations of Aromatase Function

Kinetic analyses have been conducted largely using human placental P450$_{arom}$ in addition to a handful of tissues from other mammals from which P450$_{arom}$ has been cloned and expressed recombinantly. Reported affinity constants (K_m) range from approximately 10 to 100 nM depending on species (human, porcine, and equine), tissue (placental and gonadal), and whether or not microsomes or whole cells are used for the determination. However, there appears to be relatively little difference in K_m between androstenedione and testosterone, although the former is generally metabolized to estrogen at a higher rate (V_{max}). 16α-Hydroxydehydroisoandrosterone from the human fetal liver is aromatized to estriol in the placenta much less efficiently. Though few species have been well characterized, a higher V_{max} for androstenedione is consistent with the lack of expression of an isozyme of 17β-hydroxy-steroid dehydrogenase that efficiently converts androstenedione to testosterone, at least in human ovarian follicles and placenta. That is, efficient estrogen synthesis may favor the metabolism of androstenedione over testosterone. Data are not available to compare turnover numbers between species, in part because of the difficulty in expressing recombinant P450$_{arom}$ in high yield. It is noteworthy that the affinity between P450$_{arom}$ and androgen substrates is considerably greater than those of other microsomal steroid hydroxylases. This suggests that P450$_{arom}$ is catalytically efficient at much lower substrate concentrations, which facilitates estrogen synthesis from P450$_{arom}$ expression in peripheral, nonclassical steroidogenic environments in which substrates are not synthesized locally in high concentrations.

Whether or not P450$_{arom}$ is "adapted" for catalytic activity in peripheral tissues, it is noteworthy that even when estrogen synthesis and P450$_{arom}$ expression is high in a tissue, androgen synthesis occurs in another tissue or organ. The "two-cell theory" of follicular estrogen synthesis is a term used to describe this phenomenon which is encountered in the ovary of most mammals. Specifically, androgens synthesized in the theca interna are metabolized to estrogen in the stratum granulosum expressing P450$_{arom}$. Another example of the compartmental separation of androgen and estrogen synthesis is seen in human pregnancy, the so-called fetoplacental unit. In this case, fetal adrenal androgens are metabolized to estrogens in the placenta. Similar examples exist among animal species in which androgens produced by the fetal gonads are aromatized in the equine placenta or placental androgens are aromatized in maternal ovaries of the rat during late gestation. In each case, androgen synthesis, catalyzed by 17α-hydroxylase/17,20-lyase cytochrome P450 (P450$_{c17}$) also in the endoplasmic reticulum, is physically separated from the cellular site of P450$_{arom}$ expression where the androgen is metabolized to estrogen. Logically, the dependence of efficient estrogen synthesis on androgen substrate supply suggests the need for coordinate regulation of P450$_{c17}$ and P450$_{arom}$ expression. Consistent with this idea, evidence from studies on ovarian follicular tissues from several species indicates that P450$_{arom}$ expression in the granulosa is tightly correlated with the expression of P450$_{c17}$ in the theca interna. However, the greatest level of coordination would be expected by expressing all the necessary enzymes in a single steroidogenic cell that responds to a single physiological cue. Separation of enzymes into different cells does provide the opportunity for differentially regulated expression of P450$_{c17}$ and P450$_{arom}$. However, the ovary does not normally secrete androgens and there seems to be no obvious need to regulate androgen synthesis independently of estrogen synthesis. Therefore, exactly why androgen and estrogen synthesis are so often partitioned into different cellular or tissue compartments is unknown, but it may represent unknown adaptive constraints on enzyme function that might be evident from studies on ancestral isozymes.

TABLE 1
Sequence Comparisons among Cloned Mammalian Aromatases (P450$_{arom}$)

Amino Acid Sequence Identity (Similarity)/Nucleotide Identity (%)

	Human	Bovine	Equine	Rabbit	Porcine blastocyst	Porcine placenta	Porcine ovary	Mus	Ratus
Human		86.3	NA	86.5	84.9	84.3	85.2	91.3	80.7
Bovine	83.7 (93.6)		NA	84.0	90.6	89.6	90.6	80.6	80.3
Equine	78.1 (90.9)	78.3 (90.5)		NA	NA	NA	NA	NA	NA
Rabbit	85.1 (93.2)	84.3 (92.3)	76.7 (89.5)		83.4	82.9	83.7	82.0	82.2
Porcine blastocyst	81.0 (89.7)	87.9 (93.4)	77.1 (88.1)	82.7 (89.7)		95.5	93.1	80.6	80.3
Porcine placenta	80.3 (90.1)	86.1 (93.4)	76.1 (88.9)	81.1 (89.8)	92.6 (96.6)		98.0	79.7	79.7
Porcine ovary	81.4 (92.2)	87.8 (94.6)	77.1 (89.0)	82.2 (90.2)	88.4 (94.0)	87.4 (94.2)		80.0	79.8
Mouse	79.1 (91.3)	78.7 (90.9)	74.0 (88.9)	80.9 (91.3)	76.7 (88.0)	76.5 (88.3)	76.6 (89.8)		92.9
Rat	76.5 (89.2)	76.5 (88.5)	71.1 (86.9)	78.1 (89.4)	74.9 (86.2)	74.3 (86.5)	73.8 (87.4)	91.8 (96.0)	

Note. Nucleotide identity (%) is shown on the right for all except equine P450$_{arom}$, which was not available (NA) for comparison. Amino acid sequence identity and similarity (in parentheses) are shown on the left. Note that the lowest, interspecies amino acid similarity is between the rat and porcine blastocyst isoform (86%) and the highest between the rat and mouse sequences (96%). Sequences were obtained largely from GenBank and comparisons (University of Wisconsin GCG program) were performed by Dr. Sandy Graham, University of Texas, Southwestern Medical Center, Dallas. For more detailed information, readers are directed to the following investigators: E. Simpson and N. Harada (human), M. Hinshelwood (bovine), J. Sirois (equine), P. Leymarie (rabbit), F. Simmen (porcine blastocyst), A. Conley (porcine gonadal and placental), Y. Shizuta (mouse), and J. Richards (rat).

II. THE AROMATASE COMPLEX IS HIGHLY CONSERVED

Consistent with the fundamental biological importance of estrogen synthesis, each component of the aromatase complex is highly conserved. P450$_{arom}$ is highly homologous among vertebrates, demonstrating 50–90% peptide sequence identity between fish and mammalian forms (Table 1). This degree of conservation is even more impressive at the genomic level wherein the exon–intron boundaries are identical for the genes encoding P450$_{arom}$ (designated CYP19) in humans and the teleost fish medaka (*Oryzia latipes*). Moreover, this level of gene organization is conserved despite immense differences in the sizes of each of the genes in these species—over 70 kb in the case of human CYP19 compared with only 2.6 kb for medaka. Few other CYP19 genes have been cloned to date, but partial sequence data suggest the porcine gene utilizes the same splice sites, even in the 5′ untranslated region of the transcript. The amino acid sequence of reductase is known for fewer vertebrate species but it appears to be even more highly conserved than P450$_{arom}$. Typically, peptide identity is over 90% among mammalian species and there is even 76% identity between the trout and mammalian reductases. Although reductase is evolutionarily ancient, there is 33% peptide identity between the mammalian protein and both yeast and bacterial species. Thus, based on sequence analysis alone, both components of the aromatase complex (P450$_{arom}$ and reductase) are highly conserved, which is consistent with the relatively recent evolution of the gene, a high degree of selection pressure, and functional constraints on the enzyme.

III. PRINCIPAL SITES OF AROMATASE EXPRESSION

Aromatase is expressed in a wide variety of tissues, depending on species. For instance, human P450$_{arom}$ is expressed in the testes and ovaries, the brain, liver, adipose tissues, and skin, as well as the placenta and a number of tissues in the fetus. A broad range of tumors, including those from breast, gonads, placenta, prostate, adrenal gland, and even myeloid leukemia cells, also express P450$_{arom}$. This diversity of tissue expression is seen similarly in primates and other mammalian species. Studies in the pig have demonstrated that P450$_{arom}$ is expressed in the theca interna and granulosa of preovulatory follicles, Leydig cells, fetal adrenal reticularis, placenta, and trophectoderm of preattachment blastocysts. The physiological significance of expression at such diverse sites is unclear in most circumstances, and it is doubtful that it is physiologically indispensable in all but the ovary and possibly the placenta.

A. Gonadal Aromatase

The pattern of expression of P450$_{arom}$ in mammalian gonads is quite tissue specific. Ovarian P450$_{arom}$ expression is restricted to the stratum granulosum of antral follicles in all mammals studied to date. The only variation to this general theme is seen in the porcine follicle, in which both the theca interna and granulosa layers express P450$_{arom}$. Estrogen secretion by the preovulatory follicle is directly correlated with the level of P450$_{arom}$: Both increase with preovulatory development. The ovulatory surge of luteinizing hormone is associated with a decrease in P450$_{arom}$ expression and estrogen synthesis, which remains minimal thereafter in luteal tissues of domestic animal species. However, in the human, primate, and rat corpus luteum, P450$_{arom}$ expression remains adequate to maintain significant estrogen synthesis. This is reflected in luteal phase estrogen concentrations in women and primates, and the rat corpus luteum is also the site of estrogen synthesis in late gestation in this species.

The regulation of P450$_{arom}$ expression in the granulosa is believed to involve follicle-stimulating hormone trophic stimulation that is mediated by intracellular cAMP. Molecular studies by Richards, Simpson, and others that characterized the ovarian promoter of human CYP19 have shown the existence of a cAMP-responsive element which is likely to be involved in the control of ovarian P450$_{arom}$ expression. In addition, it seems likely that other second messenger pathways will be shown to impinge on the regulation of expression at this level. Growth factors, such as insulin-like growth factor-1, are well-known modulators of granulosa P450$_{arom}$ expression *in vitro* and probably *in vivo*, consistent with the involvement of other cytoplasmic signaling cascades. Finally, the ovarian promoter of P450$_{arom}$ in several species contains a recognition site for the important transcriptional activator protein steroidogenic factor-1 (SF-1), which is also involved in the expression of several steroid hydroxylase enzymes. In fact, how P450$_{arom}$ expression is restricted to the stratum granulosum of most species or is regulated independently of other enzymes in the pathway is unknown, but it may involve enhancer or silencer elements that have yet to be recognized.

Aromatase is expressed in the male gonad, and at quite high levels in some species. Specifically, testicular aromatization of androgens in boars and stallions raises systemic estrogen concentrations above those seen in estrous females, although the physiological relevance of these constant, high levels remains obscure. Early studies demonstrated that P450$_{arom}$ was expressed in the Sertoli cells of the immature rat testes. The Sertoli cell might be viewed as the male homolog of the ovarian granulosa cell; this is consistent with these observations. However, subsequent studies in rodents and domestic animal species have shown that P450$_{arom}$ expression is localized to Leydig cells in the interstitial compartment of the testes of mature males. There are relatively few studies on the regulation of testicular P450$_{arom}$ expression, but it is likely to be responsive to gonadotrophic stimulated, intracellular cAMP as seen in the ovary. Even so, studies in the pig demonstrate that testicular P450$_{arom}$ expression is driven by a different promoter than that utilized for ovarian expression. Therefore, although gonadotrophic stimulation of P450$_{arom}$ expression in either the Sertoli or the Leydig cell is probably mediated by cAMP, transcription may be regulated in a fundamentally different fashion than that seen in the ovary of the pig at least.

B. Brain Aromatase

The expression of P450$_{arom}$ in the central nervous system is believed to mediate the metabolism of testosterone to estradiol in prenatal and/or neonatal animals and to induce sexual differentiation in specific regions of the brain. Sexual dimorphism established by estradiol is typified by the development of male-type mounting behavior at sexual maturity. The absence of exposure to estrogen, by preventing the rise in testosterone synthesis or subsequent aromatization, appears to lead to female neural differentiation that is required for inducing cyclic ovarian function. Studies by Callard and others have demonstrated aromatase activity in the brains of a wide variety of vertebrate species, including all mammals that have been examined to date. However, the vast majority of the literature on brain aromatase concerns rodent species, principally the rat. In general, $P450_{arom}$ levels decline progressively from late in fetal development to adulthood. Although some evidence suggests that males have larger sexually dimorphic nuclei than females, and slightly higher levels of $P450_{arom}$ expression, androgen availability rather than aromatization limits the estrogen effects on neural development. Expression occurs primarily in the preoptic area of the medial basal hypothalamus and the limbic system in which neural rather than glial cells are involved. Although discrepancies between immunocytochemical studies have been noted, these largely involve regions where $P450_{arom}$ is low relative to hypothalamic and limbic areas. Reports of $P450_{arom}$ expression in other areas such as specific regions of the frontal lobe suggest the possible involvement of estrogen synthesis and aromatization in olfactory development or function. What regulates brain $P450_{arom}$ expression remains controversial. Androgens have been found to induce $P450_{arom}$ levels in the hypothalamus but not in the amygdala. However, androgens have not been shown convincingly to induce $P450_{arom}$ expression in neonates when expression is highest. In addition, $P450_{arom}$ levels drop during puberty. Therefore, if androgens do regulate levels of $P450_{arom}$ expression, other unknown factors modulate their effects. Recent studies have also identified brain-specific transcripts, the expression of which corresponds to the use of alternative untranslated first exons. It is interesting to note that unlike other untranslated exons, those associated with $P450_{arom}$ expression in the brain exhibit a high degree of sequence homology and conservation. The promoter driving expression contains TATAA and CAAT elements, but unlike expression in other tissues such as the ovary, cAMP appears to inhibit promoter activity and $P450_{arom}$ expression. Interested readers are directed to reports by Naftolin, Roselli, and Harada and to a recent comprehensive review by Lephart.

C. Adipose Tissue

Adipose tissue is a particularly important site of $P450_{arom}$ expression in humans, although little evidence exists that it is important in any other species. Studies by MacDonald and others have shown that adipose tissue is the major site of estrogen production in postmenopausal women. Indeed, the peripheral conversion of adrenal androgens to estrogens may have a particularly important influence on the development of osteoporosis in the elderly. Estrogen has profound effects on calcium metabolism and remains a common therapy for the prevention of osteoporosis in women after menopause. In addition, recent studies have demonstrated a correlation between levels of $P450_{arom}$ expression and the incidence of neoplasia in certain quadrants of the human breast. Based on the known estrogen-dependent growth characteristics of certain mammary carcinomas, it seems reasonable that stromal expression of $P450_{arom}$ predisposes to the onset of this important cancer. Consequently, considerable interest in inhibitors of aromatase activity has evolved, as reviewed by Brodie, Henderson, Santen, and others. Therefore, while $P450_{arom}$ expression in adipose tissue may mitigate against osteoporosis in the elderly, in certain other peripheral sites of expression it may increase the likelihood of disease. The probable importance of $P450_{arom}$ and local estrogen synthesis in the development of breast cancer has stimulated much interest in the regulation of adipose $P450_{arom}$ expression. Simpson and others have demonstrated that expression of $P450_{arom}$ in adipose stromal cells utilizes different promoters through the use of alternatively spliced exons and is activated in vitro by a number of cytokines, including IL-11, IL-6, and tumor necrosis factor-α, in the presence of glucocorticoid or serum. Recent studies have suggested that prostaglandin E_2 produced by breast tumor fibroblasts may activate an alternative promoter and accentuate $P450_{arom}$ expression in adjacent stromal cells.

D. Aromatase in the Embryo and Fetus

Estrogen synthesis increases, or estrogens accumulate in fetal fluids at least, during pregnancy in many mammalian species. This is associated with $P450_{arom}$ expression in the conceptus which is seen as early as the preimplantation stage of development in some species. This was first observed to occur in porcine blastocysts and was subsequently demonstrated in a number of others, including those of the rabbit, horse, donkey, camel, hamster, and perhaps even human. The physiological role of embryonic estrogen synthesis is not fully understood but is presumed to mediate maternal–conceptus communication that may be important in implantation or otherwise for the establishment of pregnancy. For instance, the noninvasive blastocysts of domestic animal species may require estrogen to stimulate additional endometrial secretions, thus ensuring adequate nutritional support for conceptuses until placental formation and attachment is complete. In addition, a brief transient period of $P450_{arom}$ expression and estrogen secretion by the 12-day-old porcine blastocyst is important in extending luteal progesterone secretion and is a required embryonic signal for establishing pregnancy in this species. Studies in the equine suggest a slightly different role for estrogen in early pregnancy. Aromatase expression and estrogen synthesis by the horse embryo begins as early as Day 6 and continues to increase with conceptus age and size. This probably facilitates intrauterine migration, which is also thought to be necessary for preventing luteal regression in this species. Finally, others have demonstrated transient aromatase activity in rabbit blastocysts on Day 6, around the time of implantation. Therefore, it appears that the early expression of $P450_{arom}$ and estrogen secretion is an important component of various strategies adopted by several mammalian species for initiating implantation or ensuring intrauterine growth and survival. Aromatase is also expressed in fetal tissues. The human fetus expresses $P450_{arom}$ in a number of tissues, including the liver and adrenal glands. The porcine fetus expresses $P450_{arom}$ in the fetal Leydig cells and the zona reticularis of adrenal glands. Currently, these remain incidental observations. It is not known if this is associated with estrogen synthesis of any significance in these organs or what the physiological relevance might be.

E. Placental Aromatase

The expression of $P450_{arom}$ in the human term placenta is impressive, and this tissue has provided a source of the enzyme for the vast majority of studies that have been conducted on the catalytic function of this unusual cytochrome P450. Expression of $P450_{arom}$ in the human placenta is strongest in the syncytiotrophoblast and expression increases as cytotrophoblast differentiate. A strong basal promoter that binds the transcription factor C/EBP− is believed to be partially responsible for $P450_{arom}$ expression in choriocarcinoma cells and placenta. Expression may also be regulated through other trophoblast-specific enhancer elements that have yet to be clearly defined. Regardless, expression increases along with placental growth during gestation, and estrogen concentration continues to increase correspondingly. Human placental $P450_{arom}$ metabolizes the large quantities of androgens produced by the fetal adrenal glands and, consequently, genetic defects in the CYP19 gene are associated with masculinization of female fetuses *in utero*. The placentas of several other mammals (ungulates including horses, cattle, sheep, pigs, camels, and related species) are known to synthesize estrogen and therefore to express $P450_{arom}$, although direct evidence has been obtained in relatively few species. Aromatase activity in the extraembryonic membranes of horse and donkey conceptuses is highest in the chorionic girdle area but is also present in the yolk sac and allantochorionic membranes. Virtually nothing is known of how $P450_{arom}$ expression is regulated in these species, but it does involve the expression of a placenta-specific $P450_{arom}$ gene in the pig. In fact, placental estrogen production in these species is probably determined more by the supply of androgen substrate, which is modest for most of gestation in many cases. Rats and mice are notable in not exhibiting placental $P450_{arom}$ expression, even though placental androgen secretion increases during the last third of gestation. Placental estrogen secretion increases at term in several species, and placental $P450_{arom}$ expression, as well as androgen synthesis, may be enhanced under the influence of fetal glucocorticoids at this time. This increase in placental P450$_{arom}$ expression and estrogen synthesis prepares the reproductive tract for birth and is thought to actually initiate parturition in domestic animals at least.

IV. TISSUE-SPECIFIC REGULATION OF P450$_{arom}$ EXPRESSION

A. Alternative Splicing of Untranslated Exons

Despite the broad expression of P450$_{arom}$ in various tissues which might otherwise suggest a lack of specificity, the expression of P450$_{arom}$ is controlled in a very particular fashion in human tissues at least. This involves the alternative use of multiple untranslated first exons, the existence of which was first recognized in the human CYP19 gene (Fig. 3). Subsequent studies have recognized five or six such exons located upstream of the coding exons (II–X), all of which utilize a common splice junction about 20 bp 5' of the translational start site. Some of these transcripts are expressed almost exclusively in certain tissues, whereas others utilize two or more depending on physiological state. The human placental transcript utilizes an untranslated exon located more than 40 kb upstream of exon II, which contains the initiation codon. P450$_{arom}$ transcripts in adipose tissue utilize a different first exon about 20 kb downstream of that used in the placenta. However, it appears that transcripts expressed in adipose from tumor-containing breast tissue exhibit a high proportion of two other exonic sequences matching neither placental nor other adipose sequences. Therefore, it seems that human P450$_{arom}$ expression may be driven by the activation of certain gene promoters that are activated in a tissue-specific manner under certain physiological or pathological influences.

The control of P450$_{arom}$ expression is particularly notable in the ovary. Specifically, the 5' terminus of the P450$_{arom}$ transcript expressed in the ovary is contiguous in sequence with exon II, extending through and beginning beyond this splice junction. This was demonstrated several years ago for human

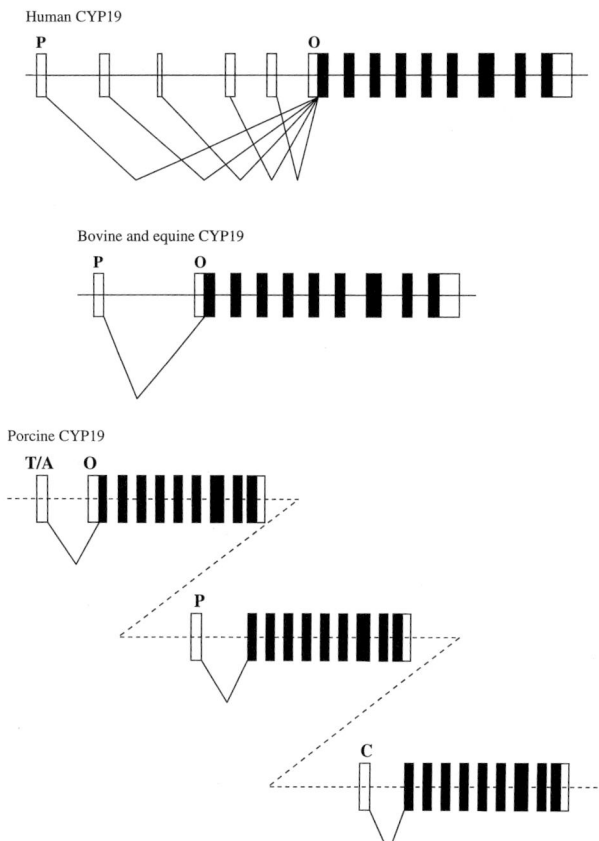

FIGURE 3 Proposed genomic structure of CYP19 encoding cytochrome P450$_{arom}$ in mammals. Human CYP19 (top) is known to be a single-copy gene which utilizes multiple untranslated exons in a tissue-specific manner. Those utilized for the placental (P) and ovarian (O) expression are indicated as the most distal and proximal (actually contiguous), respectively. Other exons shown are utilized in the brain, adipose tissue, and additional sites of expression depending on physiological state of activation. The bovine and equine CYP19 genes (yet to be cloned) are known to utilize a distal promoter and alternative first, untranslated exon when expressed in the placenta as shown (middle). The porcine CYP19 genes (bottom) have yet to be completely cloned, but it is known that multiple copies, arising by gene duplication, encode functional isozymes that are also tissue specific in expression. Shown is a proposed model based on existing evidence that suggests that expression in the testis and adrenal gland (T/A) utilizes an alternative first exon distal to the ovarian promoter (O) that is contiguous with the start site of translation as it is in human, bovine, and equine CYP19.

P450$_{arom}$ and is now known to be the case in the cow, pig, and the horse (Fig. 3). These species are also known to express P450$_{arom}$ in the placenta, and these transcripts have been shown to begin with sequence corresponding to an untranslated first exon, as seen in human placenta. The common origin of ovarian transcripts in the genomic region contiguous with the first coding exon (II) in all these species is a feature which is thought to reflect the ancestral origins of gene expression in ovarian tissues. Correspondingly, the general usage of an alternative distal untranslated first exon in the placenta is taken as evidence of the later evolution of the placenta itself and of placental P450$_{arom}$ expression in vertebrates. There is a high degree of identity among ovarian transcripts from these species in the 5' terminus but a distinct lack of homology in those from the placenta. This supports the common origins of P450$_{arom}$ expression in the ovary and further suggests that placental expression was achieved in a less conserved manner among mammals.

B. Multiple Genes Encode Tissue-Specific Isozymes of P450$_{arom}$ in the Pig

Studies conducted to date into the genes encoding P450$_{arom}$ in most mammals have suggested that a single gene encodes this enzyme. That is, although multiple first exons may exist and control tissue-specific expression, these exons are untranslated in most species, and P450$_{arom}$ expressed in whatever tissue represents only one protein species with one amino acid sequence. Moreover, it was expected that the extreme evolutionary pressure on P450$_{arom}$ as an enzyme essential for successful reproduction of the species would maintain and conserve gene organization at this level. However, this concept was challenged recently by the results of studies on P450$_{arom}$ in the pig. This species represents the only mammal investigated to date that appears to have duplicated genes encoding functional isozymes of P450$_{arom}$. These appear to be clustered in a discrete region on the same chromosome (chromosome 1), and each gene has evolved sequence differences in coding exons that result in a significant number of amino acid changes. To date, three different transcripts encoding distinct isozymes of porcine P450$_{arom}$ have been cloned and these include a gonadal, placental, and trophoblastic isozyme (Fig. 3). The gonadal or ovarian P450$_{arom}$, which some argue is ancestral, is two amino acids shorter than either the placental or the conceptus isozymes. The physiological significance of this unusual situation in the pig is unknown, but reports of a P450$_{arom}$ pseudogene in the bovine suggest that gene duplication may have occurred in other mammals. If so, perhaps these mutations encoded nonfunctional proteins or, for other reasons, these genes subsequently became redundant. Thus, the pig remains the only mammal known to expresses distinct isozymes of P450$_{arom}$, and these appear to be tissue-specific forms of the enzyme.

V. CONCLUSION

In summary, P450$_{arom}$ is a unique member of the cytochrome P450 superfamily of enzymes both biochemically and physiologically. By way of catalytic function, the activity of this enzyme can have both subtle and profound effects on reproduction and general health. Although it is now known that P450$_{arom}$ deficiency is not life threatening, this enzyme system remains essential for normal reproductive function in mammals. Estrogen formation by aromatization of androgens is characterized by catalytic specificity and efficiency as well as tissue-specific and regulated expression that must be coordinated with cellular differentiation and steroidogenesis in numerous tissues in the body. The challenge in future studies on aromatase will be to determine the biochemical and physiological basis of this balance, how it is achieved, and how it has evolved in different mammalian species.

See Also the Following Articles

ANDROGENS; ESTROGENS, OVERVIEW

Bibliography

Anonymous (1993). Proceedings of the IV International Aromatase Conference, Bologna, Italy, June 14–17, 1992. *J. Steroid Biochem. Mol. Biol.* 44, 4–6.

Anonymous (1997). Proceedings of the III International Aromatase Conference, Tahoe City, USA, June 7–11, 1996. *J. Steroid Biochem. Mol. Biol.* **61**, 3–6.

Federman, D. D. (1994). Life without estrogen. *N. Engl. J. Med.* **331**, 1088–1089.

Graham-Lorence, S., Amarneh, B., White, R. E., Peterson, J. A., and Simpson, E. R. (1995). A three-dimensional model of aromatase cytochrome P450. *Prot. Sci.* **4**, 1065–1080.

Lephart, E. D. (1996). A review of brain aromatase cytochrome P450. *Brain Res.* **22**, 1–26.

Lephart, E. D., and Simpson, E. R. (1991). Assay of aromatase activity. *Methods Enzymol.* **206**, 477–483.

Porter, T. D. (1991) An unusual yet strongly conserved flavoprotein in bacteria and mammals. *TIBS* **16**, 154–158.

Richards, J. S. Hormonal control of gene expression. *Endocr. Rev.* **15**, 725–751.

Simpson, E. R., Zhao, Y., Agarwal, V. R., Michael, M. D., Bulun, S. E., Hinshelwood, M. M., Graham-Lorence, S., Sun, T., Fisher, C. R., Qin, K., and Mendelson, C. R. (1997). Aromatase expression in health and disease. *Recent Prog. Horm. Res.* **52**, 185–214.

Artificial Insemination, in Animals

William L. Flowers
North Carolina State University

I. Introduction
II. Semen Collection
III. Semen Extension and Preservation
IV. Insemination
V. Summary

GLOSSARY

artificial insemination (AI) In animals, a process by which spermatozoa are collected, diluted with semen extenders for either short- or long-term preservation, and then manually inseminated into the reproductive tract of sexually receptive females.

cryopreservation The storage of cells or tissues at temperatures below 0°C.

ejaculation The process by which semen is emitted from the penis during mating in males.

erection A process whereby blood is trapped in the penis, causing it to become rigid in preparation for breeding.

extender A solution that provides sperm with nutrients for metabolic functions and buffers for neutralization of waste products during storage.

extension A process in which semen is mixed with extenders to increase its viability.

insemination The process by which spermatozoa are placed into the female reproductive tract.

semen A complex mixture of biological fluids, spermatozoa, and other cells emitted from the urethral opening of the penis during ejaculation.

I. INTRODUCTION

Artificial insemination (AI) in animals is a biotechnology that is at least 600 years old. Ancient Arabic writings dating to 1322 contain references about the use of AI. These texts describe procedures for collecting semen from stallions for use in breeding mares. The first documented scientific investigations concerning AI are credited to the Italian physiologist, Spallanzani, who, in the late 1700s, conducted experiments evaluating semen preservation in dogs and horses. However, it was not until the early 1900s, when it was proposed than AI could be used for propagation of superior genotypes, that major research efforts devoted to development of semen handling and insemination techniques were initiated. Major technical advancements during the mid-

1900s, such as development of artificial vagina for semen collection and use of citrate and glycerol for fresh and frozen semen preservation, respectively, facilitated the practical application of AI in animal breeding programs and led to the development of commercial AI centers. The first cooperative artificial breeding association in the United States was founded in New Jersey through the efforts of E. Perry in 1937. Currently, there are hundreds of artificial breeding centers throughout the world that provide a variety of reproductive services for many species of animals.

The primary reason AI is used in animal breeding programs is for genetic improvement. The average ejaculate contains considerably more spermatozoa than is needed to ensure adequate fertility. Consequently, because semen is collected and diluted, a greater number of females can be bred by AI than naturally. Moreover, geographic boundaries that limit the movement of animals and, hence, genetic resources can be overcome with the use of AI. Semen can be shipped to virtually any place in the world, without significant decreases in fertility, for dissemination of superior genes. While this aspect of AI may not be a major concern in the production of food animals, it is of critical importance for the continued survival of endangered species. Finally, because there are several bacterial diseases that are not transmitted via semen, AI has been used successfully in certain situations to salvage genetic information from diseased populations without significant health risks.

From a technical perspective, the ultimate goal of AI is to create a situation in which adequate numbers of fertile spermatozoa are present in the oviduct prior to ovulation leading to fertilization of ova and the birth of genetically superior live offspring. In order to accomplish this, three basic procedures must be performed successfully, regardless of the species of animal in which AI is performed. These are semen collection, semen extension and preservation, and insemination.

II. SEMEN COLLECTION

Three techniques are used to collect semen from animals: electrojaculation, digital pressure or massage, and artificial vagina. The preferred method varies among species of animals due to inherent differences in male reproductive physiology and behavior, quality of semen produced, and other practical considerations. It is generally true that the best collection procedure for a given species of animal is the one that most closely mimics social and physiological stimuli that occur during natural mating.

A. Electroejaculation

Electroejaculation involves the application of a series of short, low-voltage, electrical charges to the pelvic region via a mechanical probe inserted into the rectum. Electrical pulses emanating from the probe stimulate contractions of smooth muscles involved with erection and emission of semen. In most domesticated species of animals, electroejaculation is not the preferred method of semen collection for AI because semen obtained via this technique often contains low numbers of poor-quality spermatozoa. In addition, varying degrees of discomfort from the electrical pulses and intense muscle contractions accompany electroejaculation. However, for many large, exotic predatory mammals, electroejaculation is the only practical and effective technique for obtaining semen.

A low-voltage, bipolar electrode inserted into the rectum is the most common means of applying electrical stimulation for electroejaculation. Prior to collection, males are either secured upright in a mechanical chute or anesthetized and restrained in a prone position. It is also advisable to stimulate urination via massaging the sheath or rectal palpation prior to electroejaculation; otherwise, the risk of obtaining semen contaminated with urine is increased. The rectal probe is lubricated with obstetrical jelly and inserted into the rectum such that it lies dorsal to the secondary sex glands. The penis is then extended by either straightening the sigmoid flexure or manipulation of the shaft such that it is withdrawn from the prepuce and the glans penis is grasped with a clean, gloved hand. An insulated collection vessel with a funnel or wide mouth is held just underneath the urethral opening in order to collect semen. If possible, it is preferable to hold the glans penis and the collection vessel in one hand so that the other hand is free to massage the penis between electrical stimuli. Failure to remove the penis from the sheath prior to electroejaculation often results in emission of semen inside the sheath.

Once the collection technician is positioned, a series of 3- to 8-sec pulses are applied every 15–20 sec. The onset of ejaculation is evident when clear fluid begins to flow from the glans penis. This is referred to as the presperm fraction of semen. The presperm fraction is usually followed by a viscous and creamy-white fluid known as the sperm-rich fraction. Ejaculation in most males ends with the emission of a postsperm fraction which varies in consistency, volume, and color.

B. Digital Pressure or Massage

Digital manipulation of the penis involves applying pressure or gently massaging the glans penis of a sexually aroused male in order to stimulate ejaculation. It is the preferred method of collection for males in which pressure on the glans penis is the primary stimulus for ejaculation. A training period in which males are taught to initiate mating behaviors when appropriately stimulated is a critical aspect of semen collection by digital pressure. Pigs, dogs, and fowl are species in which digital pressure or massage techniques are used almost exclusively for semen collection.

The primary stimulus for the initiation of mating behavior in boars is an object that resembles an immobile sow or gilt. Thus, boars can be trained to mount a bench that is similar in size to a receptive female. This bench is referred to as a collection dummy or dummy sow. Most boars can be trained to mount and be collected off a dummy sow in 3 weeks provided that three criteria are met: training begins when boars are between 160 and 180 days of age, no prior experience with natural mating has occurred, and boars associate the collection dummy and process with a pleasurable experience.

Upon entry into the collection area, boars usually require 2–5 min of precoital behaviors such as nuzzling and chewing on the collection dummy. This varies considerably among individual males. Once stimulated, boars mount the dummy sow and initiate a thrusting motion by pushing their hips forward. It is common for the individual performing the collection to gently massage the sheath as the boar begins to thrust. This serves to expel any fluids trapped in the preputial diverticulum which could contaminate semen and it also sexually excites the boar. As the excitement level of the boar rises, thrusting activity increases and the penis protrudes from the sheath in a rhythmic fashion. A gloved hand is placed adjacent and anterior to the penile opening of the sheath and, as the penis is pushed out, pressure is applied on the glans penis by encircling it with two to four fingers. Pressure on the glans penis mimics cervical contractions which occur during natural mating and stimulates the boar to fully extend his penis and begin ejaculation. Semen is collected in a wide-mouth container covered with gauze or cheesecloth. It is necessary to cover the mouth of the collection vessel with gauze to separate the liquid and gel fractions of semen. In pigs, the gel fraction significantly reduces sperm motility and viability. The average ejaculation time in boars is 7 min with a normal range of 3–20 min.

In contrast to pigs, male dogs are allowed to mount a bitch that is restrained during the collection of semen. After mounting, the prepuce is gently pushed caudally behind the bulbus glandis with a gloved hand, the thumb and forefinger are manipulated such that they encircle the penis, and firm pressure is applied. Erection usually occurs spontaneously. Within 10–20 sec of erection, ejaculation begins and lasts, on average, for 10–15 min. Semen is collected into graduated cylinders or tubes with either a flared mouth or fitted with a funnel. After the sperm-rich fraction has been collected, the male dog will try and lift one hindleg over the arm of the collection technician and turn around. This is a normal mating behavior that occurs during natural service. It is common to stop collection at this point in order to prevent mixture of the sperm-rich fractions with fluids from the accessory sex glands. In dogs, secretions from the prostate gland which compose most of the postsperm fraction have been shown to have detrimental effects on viability of spermatozoa.

The reproductive anatomy of male birds is such that a massage technique is used for semen collection. The cloaca is the common exit for the bird's digestive, urinary, and reproductive tracts. Within the urodeum of the cloaca are small nipple-like projections called papilla. Manual collection of semen is achieved by massaging or "milking" the papilla. Initially, male birds are restrained and the area surrounding the vent is gently massaged. Massaging this area while simultaneously stroking the back triggers a spinal reflex arc that sexually arouses the male and causes the papilla to become erect and protrude above the floor of the cloaca. Once exposed, the papilla can be gently massaged resulting in the expression of semen. Semen is collected in an eye cup or a smooth-edged vessel that is held just underneath the papilla. In some birds, especially waterfowl, the papilla may not extend completely when the male is sexually stimulated. In these situations, semen must be collected as it flows over the surface of the partially everted vent.

C. Artificial Vagina

The artificial vagina (AV) is an imitation of the female vagina and provides thermal and mechanical stimulation that is required for ejaculation. Most artificial vagina have the same basic design which includes an outer casing made from heavy rubber or plastic, an inner lining composed of synthetic materials, and an insulated collection pouch that is normally attached to one end of the outer casing (Fig. 1). Water is added to the space between the outer casing and inner lining prior to collection. The temperature of the water is used to regulate the thermal environment inside the artificial vagina, whereas the volume controls the amount of pressure surrounding the penis during collection. Temperature and pressure requirements are species specific and intraspecies variation among males can be extreme, especially in the case of stallions, male dogs, and dairy bulls. In general, the recommended internal temperature of the artificial vagina prior to collection is between 42 and 48°C. Once the proper temperature has been achieved, the artificial vagina is lubricated in preparation for collection. A nonspermicidal obstetrical jelly commonly is placed inside the AV along the upper two-thirds of its length, with the largest amount applied in the first third where initial contact with the penis will occur.

During the collection process, the male is allowed to mount a female or a neutered male, in the case of cattle. These animals are referred to as teaser animals. When the male enters the collection pen or area, it is common for him to engage in premating courtship behaviors, such as sniffing, licking, nuzzling, and sometimes biting. These behaviors play an important role in sexually stimulating the male in preparation for collection. In cattle, it is common to allow the bull to mount the teaser animal two or three times before he is allowed to ejaculate. False mounts, as they are called, have been shown to increase sperm output.

When the male mounts the teaser animal, the penis is diverted to the side and into the AV by handling the sheath. In sheep and cattle, ejaculation is complete within a few seconds, occurs simultaneously with mounting, and involves a vigorous upward and forward thrust. The AV is not removed until the male begins to dismount and retract his penis. In dogs and cats, ejaculation occurs over a longer period of time (2–20 min). Consequently, manual stimulation of the male prior to collection is often required. For tomcats, erection and ejaculation are stimulated by application of firm pressure to the dorsal portion of the pelvic region. For dogs, the bulbus glandis is grasped and erection stimulated with a massaging motion. The penis is then directed into the AV, where additional stimulation is provided during collection

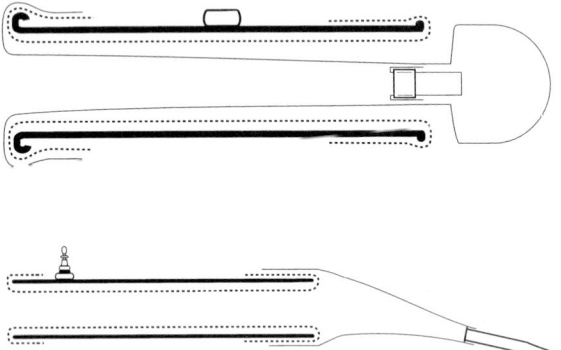

FIGURE 1 Schematic diagrams of artificial vagina used in collection of semen from stallions (top) and bulls and rams (bottom). Heavy lines represent the outline of the outer casing, and light or dotted lines illustrate the positioning of the inner lining.

by gently moving it back and forth. This simulates the pulsing action of the bitch's vagina during mating. Some AVs that have been specially designed for dogs contain a rubber bladder between the outer casing and inner lining into which air can be pumped with a syringe during collection. Pumping air into the rubber bladder also creates a pulsing sensation on the dog's penis. When the thrusting movements have stopped, the AV is moved backwards between the dog's hindlegs to mimic the position of the penis during natural mating.

Collection of semen in horses is similar to that of other animals in which the AV is used, except that the stallion mounts a collection bench or "phantom" as opposed to a sexually receptive female. Stallions tend to be the most temperamental and unpredictable males in terms of their behavioral patterns during semen collection. Thus, it is important to have an experienced individual handling the stallion as well as a qualified technician performing the collection. Many semen collection areas have a chute adjacent to the phantom. Some stallions require exposure to an estrous mare prior to and during collection for maximal sexual stimulation. As the stallion approaches the phantom, erection usually occurs before he actually mounts, especially if the stallion received previous exposure to an estrous mare. In contrast to bulls, stallions are allowed to mount immediately because false mounts are detrimental to semen volume and quality. When the stallion mounts, the technician steps forward and deflects the penis into the AV. As the stallion thrusts, the technician moves the AV toward the stallion's abdomen because this is usually more stimulating than pushing it down on the penis. Pulsations of semen enter the AV during ejaculation and usually are accompanied by a pumping movement of the tail called flagging. Flagging is the behavioral cue used to signify the termination of ejaculation and once it has occurred, the AV is gradually lowered until it is almost vertical as the stallion loses his erection. In the event that a particular male cannot be trained to collect off a phantom, then a sexually receptive mare can be used instead. However, this poses an increased risk of injury to both the horses and people involved in the collection process.

III. SEMEN EXTENSION AND PRESERVATION

After collection, semen is usually evaluated, diluted with an extender, and stored prior to use. For most species of animals, use of fresh semen is preferred over frozen semen. This is largely due to observations that reproductive performance is lower with frozen than fresh semen. Exceptions to this are dogs and cattle. In dogs, fertility is comparable between fresh and frozen semen. However, fresh semen is used more frequently than frozen semen because of technical and economic reasons. In cattle, fertility with frozen semen is equivalent to that achieved with natural mating. Consequently, use of fresh semen for AI is uncommon in cattle except in New Zealand, where most dairy cattle are bred exclusively with liquid semen. The prevalent use of fresh semen in New Zealand is due to the extensive use of forage-based production systems.

A. Semen Evaluation

Semen routinely is evaluated for sperm concentration and quality. The concentration of spermatozoa is determined directly by counting the number of cells on a specialized microscope slide called a hemocytometer or indirectly by use of a spectrophotometer which measures the amount of light that is absorbed by a suspension of semen. Assessment of semen quality includes estimates of the percentage of spermatoza exhibiting progressive forward motility and normal morphology. Procedures for semen evaluation tend to be universal in nature and will not be discussed in detail in this chapter because they are described elsewhere. In addition, acceptable values for motility, morphology, and other criteria for different species can be found in the bibliography at the end of this chapter. It is important to note that while semen evaluation is a common practice with AI, most of the quality tests are subjective and do not have a high positive correlation with fertility.

B. Liquid Semen Preservation

Extenders used in preservation of liquid semen must perform five basic functions: (i) provide nutrients for sperm metabolism, (ii) neutralize metabolic waste products, (iii) stabilize sperm membranes and prevent capacitation, (iv) maintain an osmotic equilibrium, and (v) retard bacterial growth during storage. Apart from slight modifications, the chemical composition of extenders is remarkably similar across species. Common ingredients are used in extenders to perform the five required functions. Simple sugars such as glucose and fructose are added as energy subtrates for spermatozoa. Buffering systems consist of an acid and its conjugate base to neutralize metabolic wastes and maintain pH. Examples include sodium bicarbonate, phosphate buffers, and organic zwitterionic molecules such as hydroxyethylpiperazineethanesulfonic acid. Macromolecules and chelating agents help promote membrane stability and prevent capacitation. Sodium citrate, ethylenediaminetetraacetic acid, and tris(hydroxymethyl)aminomethane are common extender ingredients used to bind free calcium, whereas egg yolk, milk proteins, and albumins, in some animals, are added because they improve semen viability, presumably due to their ability to bind to and stabilize sperm membranes. Finally, neomycin sulfate, gentamycin, pencillin, and polymixin B sulfate are effective antibiotics in preventing the growth of bacteria commonly found in semen without affecting sperm viability.

Specific formulations and preparation procedures for extenders used in a number of animals are contained in the bibliography. Many of these references are available commercially.

Maintenance of semen at reduced temperatures effectively maintains its viability during storage by reducing the metabolic activity of spermatozoa. For most species, the ideal storage temperatures for liquid semen is a topic of much debate. However, they probably are somewhere between 5 and 18°C. At temperatures higher than 18°C, metabolic activity is not reduced sufficiently to prolong viability, whereas the appearance of ice crystals is common as temperatures approach freezing. The formation of ice causes significant damage to mitochondria and other organelles, which leads to sperm death. Apart from maintenance of a stable thermal environment, few other requirements are necessary for storage of liquid semen in most animals except birds. Oxygen must be provided for avian semen during storage to maintain viability. This is accomplished by either gentle agitation or aeration. Table 1 contains current estimates of the optimal storage temperatures for liquid semen in several animal species.

An important concept associated with the use of fresh semen is that sperm viability and fertility begin to decrease as soon as the ejaculate is collected and

TABLE 1
Characteristics of Insemination Doses for a Variety of Animals

Animal	Type of semen	Number of sperm cells and volume per dose[a]	Storage temperature (°C)[b]	Maximum length of storage[c]	Theoretical number of doses per ejaculate[d]
Cats	Fresh	$10-50 \times 10^6$ in 0.1–0.5 ml	Used at time of collection	Not stored	3–5
Cattle	Frozen	$15-25 \times 10^6$ in 0.5–1 ml	−196	>1 year	200–600
Dogs	Fresh	$100-200 \times 10^6$ in 1.5–10 ml	5	4–6 days	3–7
Chickens	Fresh	$150-200 \times 10^6$ in 0.2–0.5 ml	5	24–48 hr	3–17
Horses	Fresh	$250-500 \times 10^6$ in 10–25 ml	5–20	24–36 hr	10–20
Pigs	Fresh	$2-4 \times 10^9$ in 70–100 ml	15–18	3–5 days	10–20
Sheep and Goats	Fresh	$100-300 \times 10^6$ in 0.1–0.5 ml	5	4–6 days	5–10
Turkeys	Fresh	$150-250 \times 10^6$ in 0.2–0.5 ml	5 with aeration	12–24 hr	3–17

[a] Number of motile spermatozoa for all animals except pigs, which is total number of sperm cells.
[b] Most commonly used range of storage temperatures.
[c] Storage time refers to the length of time semen can be stored from collection to insemination without significant losses in fertility.
[d] Calculated by division of the total number of spermatozoa in an ejaculate by the average number of spermatozoa in an insemination dose. In practice, ejaculates may not be extended to produce the maximum number of insemination doses.

TABLE 2
Preferred Collection and Insemination Strategies and Pregnancy Rates with AI for Different Animals

Animal	Type of collection	Insemination location	Insemination strategy[a]	Pregnancy rates (%)[b]
Cats	Artificial vagina	Cervix or deep vagina	Once 12–18 hr after hCG	60–70
Cattle	Artificial vagina	Uterus	Once 12–18 hr after detected estrus	60–70
Chickens	Manual massage	Caudal portion of oviduct	Once per week during breeding season	80–90
Dogs	Manual massage	Cervix	Once 3–5 days after first interest in male dog	70–80
Horses	Artificial vagina	Uterus or cervix	Every other day of estrus beginning at day 2 or 3	60–80
Pigs	Digital pressure	Cervix	Once each day of estrus	80–90
Sheep and Goats	Artificial vagina	Cervix	Once each day of estrus	70–80
Turkeys	Manual massage	Caudal portion of oviduct	Once per week during the breeding season	80–90

[a] Refers to the number and timing of inseminations.
[b] Pregnancy is defined as giving birth.

the addition of extenders, at best, only slows the rate at which this decline takes place. The rate at which fertility decreases is a function of many factors. Some of these have been investigated in detail, including type of extender, storage temperature, dilution ratio of the insemination dose, length of storage, and type of ejaculate. In contrast, many as yet unidentified factors probably are also involved. This is evident by the large variation in viability of semen among individual males that is stored under identical conditions. Consequently, two objectives for AI programs using liquid semen are to (i) begin the extension process as quickly as possible after collection and (ii) minimize the time period semen is stored prior to insemination. The best, current estimates for the fertile "shelf life" of fresh semen from different species are summarized in Table 2.

C. Frozen Semen Preservation

Cryopreservation of spermatozoa is possible in most mammalian species as evidenced by reports of the birth of live offspring resulting from insemination of previously frozen semen. However, as mentioned earlier, except for dogs and cattle, fertility is significantly lower with frozen than fresh semen. One of the detrimental effects that freezing has on spermatozoa is referred to as cold shock. Cold shock is a poorly understood, partially irreversible, series of changes that occur in spermatozoa when they are cooled from 25 to 5°C. Sperm cells subjected to cold shock have a decreased metabolism, an enhanced loss of intracellular molecules and ions, and reduced fertility. In addition, when spermatozoa are cooled below freezing, extracellular ice crystals form, creating higher solute concentrations outside than inside the cell. This difference in osmolarity causes water to move out of the sperm cell, leading to dehydration. The resulting high intracellular concentrations of solutes, under certain conditions, can damage cellular organelles and also reduce the fertilizing potential of spermatozoa.

The use of a two-step cooling process and cryoprotectants have proved effective in circumventing many of the problems associated with cryopreservation of semen. Penetrating cryoprotectants such as glycerol are effective because they behave like a solvent with a lower freezing point than water. Consequently, their presence increases the proportion of unfrozen solvent and decreases extracellular solute concentrations and the damage that they can cause. Nonpenetrating cryoprotectants, such as raffinose and trehalose, function by causing water to leave spermatozoa during cooling, which lowers the incidence of intracellular ice formation. Specific details of cryopreservation procedures vary according to species of animal whose semen is being frozen and the type of cryoprotectant and package that is used. However, a general outline is as follows. After collection, spermatozoa are mixed with an extender that contains either milk or egg proteins and is allowed to cool slowly from 18°C (or higher) to about 4°C at rates near 0.05°C per minute. Around 4°C, cryoprotectants are added and semen is packaged into individual doses. Finally, semen is cooled rapidly at rates between 5 and 50°C per minute from 4°C to below −100°C and stored in liquid nitrogen (about −196°C) until used. When frozen semen is prepared for insemination, a rapid thawing rate is beneficial for sperm viability and fertility. Frozen semen is usually thawed by placing the semen package directly into water at 37°C upon its removal from liquid nitrogen.

D. Insemination Dose

The insemination dose consists of three components: the number of spermatozoa, the volume of extended semen, and the semen package. Semen processing procedures, the occurrence of ovulation during estrus, and the insemination technique are several of many factors that influence the individual characteristics of the insemination dose. For example, in species in which semen is deposited into the uterus fewer numbers of spermatozoa are needed to optimize fertility compared to their counterparts in which cervical insemination is practiced. This is due to the fact that many spermatozoa are trapped in the cervical crypts after insemination and during transport to the oviduct. Consequently, a reduced volume of semen and small package are used when uterine deposition is possible. A summary of "average" insemination doses is contained in Table 1. Obviously, an inverse relationship exists between the number of spermatozoa required per insemination dose and the number of insemination doses produced from a single ejaculate. Based on the average numbers of spermatozoa in an ejaculate and the insemination dose, the number of doses per ejaculate can be determined. Estimates of these for several species of animals are summarized in Table 1.

IV. INSEMINATION

The insemination process is considered by many experts to be the most important component of AI. This is due to the observation that a positive relationship exists between the precision and accuracy with which it is performed and subsequent pregnancy rates. Thus, the breeding technician plays a significant role in the success of AI.

A. Insemination Technique

Female reproductive anatomy determines the type of insemination technique that is used. From an anatomical perspective, the goal of insemination is to deposit semen as close as possible to or in the posterior portion of the uterine body. As discussed previously, uterine deposition requires considerably fewer spermatozoa and a smaller insemination volume compared to placement in the cervix or vagina. Different strategies and equipment are used to accomplish this goal. However, before they are discussed, it is important to acknowledge two universal features of all insemination techniques.

First, procedures referred to as minimal contamination techniques are practiced. These differ slightly according to insemination technique but, in general, they involve maintenance of strict hygiene before, during, and after insemination. Minimal contamination techniques include use of disposable gloves and equipment, whenever possible; thorough cleansing of the female's perineal area prior to insemination; and lubrication of gloves and equipment with a nonspermicidal obstetrical jelly. Second, for safety reasons, estrous females are placed in a breeding chute to limit physical activity during AI. In some cases, such as with large predatory mammals, tranquilization of the female may be necessary in order to perform the insemination. Restraint of sows during insemination is not required because they normally exhibit an intense and protracted immobilization response (lordosis) during estrus.

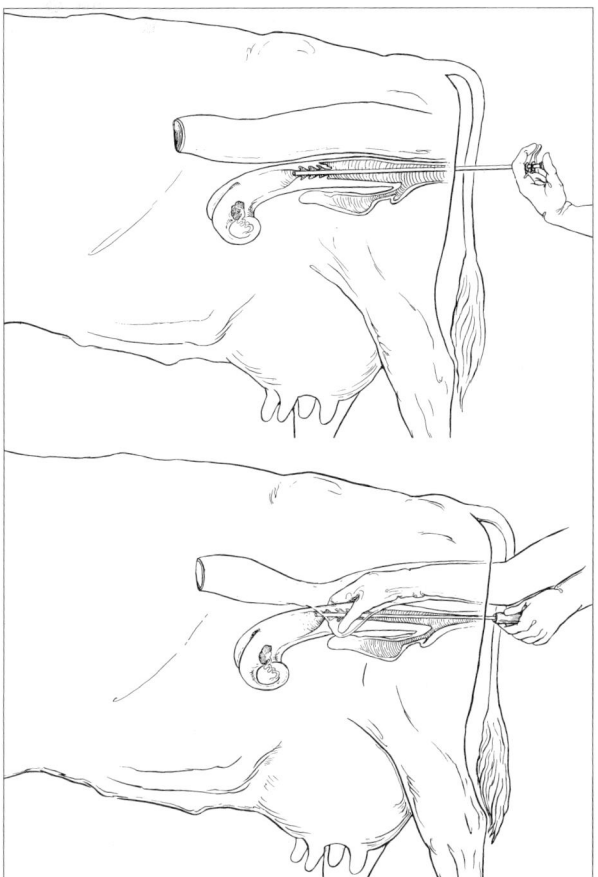

FIGURE 2 Illustrations of the rectovaginal insemination technique for cattle. In the rectovaginal procedure, one hand is placed in the rectum and grasps the cervix through the rectal wall, while the other hand guides the insemination gun through the vagina and cervix into the uterus. Correct placement of the hand in the rectum is shown (bottom) and the correct placement of the insemination rod is shown (top) (illustrations by Brenda Bunch, North Carolina State University).

A rectovaginal procedure is the most common technique used to inseminate cattle and related species (Fig. 2). With this technique, one hand is passed through the anal sphincter and grasps the cervix through the floor (ventral portion) of the rectum. The insemination gun consists of a hollow, metal rod into which a 0.5-ml straw of semen is loaded into its tip. The entire rod is covered with a disposable plastic sheath. The insemination gun is guided carefully through the vagina and cervix into the uterine body with the free hand. Gentle manipulation of the cervix through the rectum facilitates movement of the insemination gun through the cervical rings and into the uterus. Once in position, semen is expelled by pushing the plunger forward. This physically forces the extended semen out of the straw; through the tip of the gun; and into the uterus.

The insemination catheter or pipette in horses is a flexible plastic tube with a rounded tip. The insemination dose is loaded into a syringe with a plastic plunger and connected to the insemination rod with a short piece of pliable tubing. Equipment containing rubber is avoided in equine AI because of reports that it has spermicidal properties. The tip of the catheter is placed in the palm of the hand with the index finger covering the tip. The gloved hand and insemination pipette are then passed through the vulva into the cranial portion of the vagina, where the index finger is used to locate and penetrate the cervix. The insemination rod is advanced through the cervix into the uterine body and semen is slowly deposited (Fig. 3).

In small domestic mammals, such as sheep and dogs, and in some large exotic felines, cervical insemination is accomplished via a speculum (or vaginoscope) and a long, narrow insemination rod attached to a syringe or gun containing the insemination dose (Fig. 4). A speculum is basically a hollow tube with a handle that can be inserted into the vagina and positioned just caudal to the external opening of the cervix. Visualization of the cervical os through the speculum is enhanced by the use of a standard medical head lamp. Once the cervix has been located, the insemination rod is inserted into the speculum, passed through the vagina, and threaded into the cervix. In sheep and goats, the insemination rod is inserted at least 3 cm into the cervix before semen deposition. In dogs, the tip of the rod is placed just inside the cervix and upon completion of insemination the hindquarters of the bitch are gently elevated for 4–6 min to prevent backflow of semen out of the cervix.

Palpation or visualization of the cervix is not necessary for insemination in swine due to the lack of an external cervical os that projects into the vagina. Instead, the vagina becomes progressively smaller as

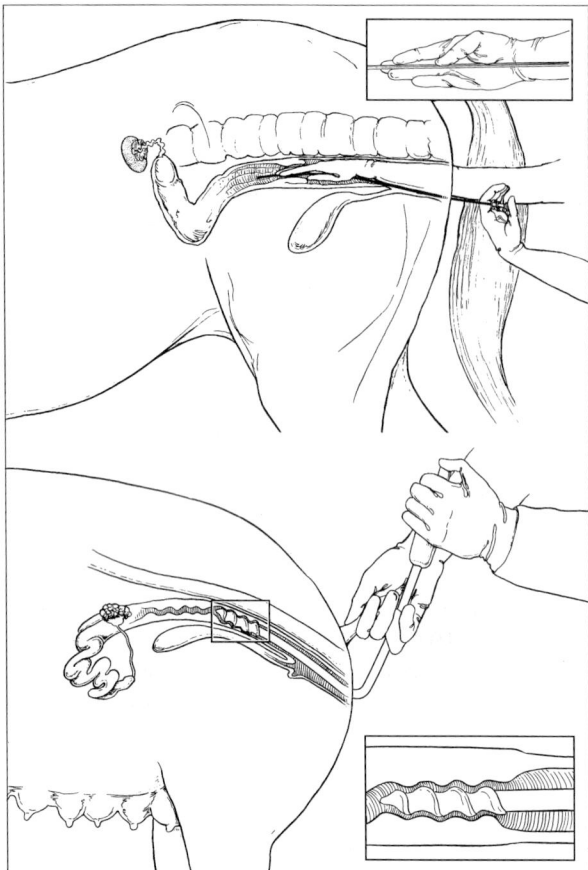

FIGURE 3 Illustrations of insemination using the vaginal approach in horses (top) and a breeding spirette in pigs (bottom). In the vaginal approach, the hand is used to guide the insemination rod through the vagina and into the cervix. The correct placement of the insemination rod in the guide hand is shown in the top insert. In swine, the breeding spirette is pushed through vulva and vagina with its tip pointed up. When resistance is encountered, the spirette is rotated counterclockwise in order to penetrate the cervix. The anatomical relationship between the cervix and breeding spirette in swine is shown in the bottom insert (illustrations by Brenda Bunch, North Carolina State University).

one moves cranially, much like the mouth of a funnel, and forms a smooth junction with the cervix. The cervix in pigs is unique in that its central canal is, anatomically, a left-handed or counterclockwise spiral. Consequently, specialized catheters with a left-handed spiral or a compressable foam tip are used for insemination. Insemination doses are packaged in small plastic bottles, tubes, or flexible bags

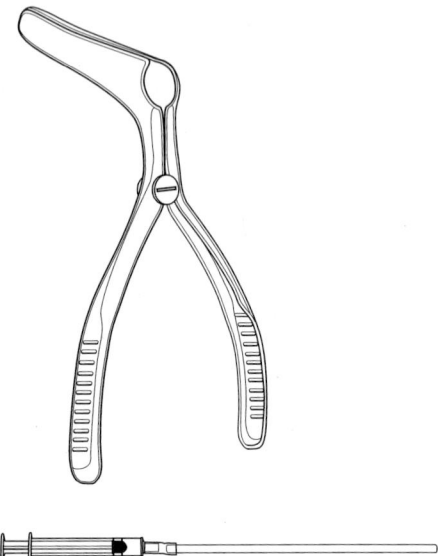

FIGURE 4 Schematic diagrams of insemination equipment for sheep and dogs: a duck-billed speculum (top) and a plastic insemination rod attached to a plastic syringe (bottom). When syringes are used for deposition of semen, it is customary for the plunger to be made from materials other than rubber.

called cochettes. With the tip pointed up, the catheter is passed through the vulva and into the vagina until resistance is encountered. This is an indication that the tip of the catheter is at the opening of the cervix. At this point, the catheter is rotated in a counter-clockwise direction. When rotation becomes difficult, insertion is stopped and the catheter is gently pulled backwards. Resistance to this backward motion indicates that the catheter is firmly locked into the cervix (Fig. 3). The insemination dose is connected to the posterior end of the cathether and semen is slowly deposited. In swine, physical stimulation of the sow by massaging her back and rear flanks and the presence of a boar during insemation facilitates deposition of semen by enhancing gamete transport and the immobilization response. Due to the large volume of semen deposited, it is not uncommon for semen deposition to take 3–7 min.

The insemination rod used for domestic cats consists of a 9-cm, 20-guage cannula with a smaller piece of polyethylene tubing fastened to the end. The cannula is passed blindly into the anterior portion of the vagina or posterior segment of the cervix and semen is deposited. For AI in birds, the insemination rod consists of a medicine dropper or small plastic rod connected by flexible tubing to a 1-ml syringe or similar device. The hen is held in much the same way that the male is handled for semen collection so that the vent is exposed. The area around the vent is stroked at the same time her back is massaged. As the hen become sexually aroused, the vent everts and an opening appears on the left side. Most of the time, this orifice resembles a round rosette, but it can look like a cleft or overfold of skin. This opening is the posterior end of the hen's reproductive tract which, anatomically, does not have a true cervix or uterus as found in mammals. The tip of the insemination rod is placed 0.5–2.5 cm deep into the opening for semen deposition.

B. Timing and Frequency of Inseminations

In addition to the technique itself, the timing and frequency of inseminations during the period of sexual receptivity significantly influence the fertility level achieved with AI. The ultimate goal of insemination strategies is to ensure that adequate numbers of spermatozoa are present in the oviduct prior to ovulation. As a result, variations among the duration of estrus, the timing of ovulation, and the viability of spermatozoa in the female reproductive tract are important considerations in developing optimal insemination strategies. In general, species in which estrus occurs for long periods of time and the timing of ovulation varies require frequent inseminations. In contrast, short periods of sexual receptivity and a consistent and predictable time of ovulation allow for single insemination with excellent reproductive success. Current, recommended insemination strategies for AI in animals are summarized in Table 2.

Due to their reproductive physiology, two species of animals have quite unique insemination strategies. Avian species have sperm storage tubes located at the uterovaginal junction from which fertile spermatozoa can be released for up to a month. Consequently, a single mating with AI can result in the production of fertilized eggs for several weeks. However, due to normal changes in female and male fertility during the laying season, inseminations are administered at least once per week in commercial situations.

Domestic cats are induced ovulators, i.e., physical stimulation of the cervix by the tom's penis stimulates ovulation. Since the AI breeding catheter commonly used in cats does not usually provide sufficient tactile stimulation of the cervix for ovulation to occur, hormonal therapy is required to control the time of ovulation. Administration of human chorionic gonadotropin (hCG) on Days 1 or 2 of estrus is used to induce ovulation, on the average, 25–26.5 hr postinjection.

V. SUMMARY

Artificial insemination in animals is an ancient assisted reproductive technology that is primarily used for selective breeding purposes. Its development is an excellent example of the development of a practical technique based on our basic understanding of reproductive biology. To date, AI techniques have been developed for many species. The preferred method of semen collection, semen preservation, and insemination along with estimations of fertility achieved with AI and frequency of use for some of these are summarized in Tables 1 and 2. Based on these results, AI in animals is a successful assisted reproductive technology. Its use will likely increase as research leads to a better understanding of semen physiology, sperm retention in the female reproductive tract, and fertilization. For example, it currently is possible to sex semen by separating spermatozoa containing the X and Y chromosomes on the basis of their DNA content. As this technology is refined, it may be possible to package insemination doses that produce either male or females. In species in which frozen semen is not a viable option for AI, characterization of plasma membrane components and changes they undergo during cooling are important areas of investigation and should yield improvements in fertility following cryopreservation of spermatozoa. Finally, the manner in which spermatozoa and noncellular components of semen interact with the oviductal environment after insemination should provide information about factors that influence the population of spermatozoa that ultimately fertilize the ova. Application of this information to AI should result in the reduction of numbers of spermatozoa required for insemination and development of single insemination strategies without compromising fertility.

See Also the Following Articles

CRYOPRESERVATION OF SPERM; HORSES (EQUIDAE); PREGNANCY IN FARM ANIMALS; REPRODUCTIVE TECHNOLOGIES, OVERVIEW; SEMEN

Bibliography

Bakst, M. R., and Whishart, G. J. (1995). *Proceedings, First International Symposium on the Artificial Insemination of Poultry.* Poultry Science Association, Savoy, IL.

Christiansen, I. J. (1984). *Reproduction in the Dog and Cat.* Bailliere Tindall, London.

Evans, G., and Maxwell, W. M. C. (1987). *Salamon's Artificial Insemination of Sheep and Goats.* Butterworth Scientific, London.

Flowers, W. L., and Esbenshade, K. L. (1993). Optimizing management of natural and artificial matings in swine. *J. Reprod. Fertil. Suppl.* 48, 217–228.

Hafez, E. S. E. (1993). *Artificial Insemination. Reproduction in Farm Animals,* 6th ed. Lea & Febiger, Philadelphia.

Johnson, L. A., and Rath, D. (1991). *Boar Semen Preservation II.* P. Parey, Berlin.

McKinnon, A. O., and Voss, J. L. (1993). *Equine Reproduction.* Lea & Febiger, Philadelphia.

Artificial Insemination, in Humans

Douglas T. Carrell, Deborah Cartmill

University of Utah School of Medicine

I. History of Artificial Insemination
II. Types of Artificial Insemination
III. Ethical and Legal Aspects of Artificial Insemination

GLOSSARY

artificial insemination The placement of spermatozoa into the female reproductive tract by any method other than intercourse.

asthenozoospermia Poor motility of sperm within the ejaculate.

capacitation A series of chemical processes, usually occurring within the female reproductive tract, and a precursor to the acrosome reaction, resulting in an influx of calcium ions into the sperm and increased membrane fluidity.

donor A provider of spermatozoa used for insemination other than the spouse or partner of the recipient of artificial insemination.

homologous insemination Use of the semen of the spouse or current partner for artificial insemination.

oligozoospermia A decreased number of sperm within the ejaculate.

ovulation induction The pharmacological induction or regulation of the production of an oocyte by the ovary.

sperm wash The removal of seminal plasma by centrifugation of semen diluted with culture medium followed by resuspension of the sperm in culture medium.

Artificial insemination is the introduction of spermatozoa into the female reproductive tract by a method other than intercourse, with the intention of causing a pregnancy. Patients may undergo artificial insemination for numerous reasons, including various aspects of both male and female infertility, sexual dysfunction including spinal cord injury, or for the use of donor spermatozoa. Numerous methodologies have evolved which have improved the success of artificial insemination. These methodologies include various techniques of ovulation induction, improvements in monitoring the growth and ovulation of oocytes and subsequent timing of artificial insemination, advances in the preparation of functionally competent spermatozoa prior to artificial insemination, and improvements in the actual techniques used for artificial insemination. Artificial insemination is often the first, and in many cases the only, type of infertility treatment couples undergo.

I. HISTORY OF ARTIFICIAL INSEMINATION

Although the underlying physiological principles were incorrect, the earliest references to artificial transfer of semen date to the second century BC. In the myth told by Apollodoros, Gaia, the goddess of earth, was inseminated when a wool rag containing sperm fell into her lap. Early Christian references to the artificial transfer of semen solely pertain to the transfer of semen by demons. In an effort to procreate, the demons stole semen from unknowing men and subsequently introduced the sperm into the uteri of unsuspecting recipients. In Islamic writings from the ninth and tenth centuries, semen transfer techniques are described and the paternity of the semen donors was discussed.

During the following several centuries, little progress was made, and there was little discussion within the literature regarding artificial insemination. While the Hippocratic theory that gametes from both the male and female were necessary for procreation was commonly known, many physicians such as Paracelsus preferred the theory of Aristotle which held forth that the semen contained homunculi and was sufficient for procreation as long as a suitable medium or growth was obtained. This incorrect theory misdirected studies on artificial insemination. Misguided research efforts were made to look for suitable replacements to the uterus as a culture medium for the semen. Studies included attempts to culture human semen in horses, blood, and cow dung.

The first true scientific studies of artificial insemination began in the eighteenth century. In 1765 Jacobi reported that he could produce salmon by mixing salmon row with the milt of fish. These studies were confirmed by others, and further studies by Spallanzani showed that artificial insemination could be successfully performed in the frog and later in dogs.

The first successful artificial insemination in humans was reported by John Hunter in the late 1770s. Hunter collected semen from a cloth merchant who had suffered from cryptorchism and, using a warm syringe, inseminated the merchant's wife. Hunter reported that a pregnancy did result. An evaluation of artificial insemination was performed in 1866 by Marion Simms. Simms performed 55 inseminations in six women with cervical abnormalities but whose husbands had normal sperm quality. Unfortunately, Simms believed that ovulation occurred near the time of menstruation; therefore, the chance of pregnancy was very low. However, one insemination did result in a pregnancy.

A major advance in the process of artificial insemination occurred in 1953, when Bungi and Sherman reported the use of cryopreserved human semen for artificial insemination. Throughout the latter half of the twentieth century, advances in sperm cryopreservation and improvements in culture medium and preparation of semen prior to artificial insemination greatly improved the success of artificial insemination, which was becoming much more accepted in society. Successful techniques for removing sperm from seminal plasma and concentrating and improving sperm quality led to the development of intrauterine insemination as the standard technique of artificial insemination. In addition, increased awareness of the risk of sexually transmitted diseases, such as HIV, has led to the exclusive use of cryopreserved semen for nonhomologous artificial insemination.

In 1884, William Penncoast performed the first insemination of donor semen. Insemination of donor semen was also reported by R. L. Dickenson in 1890; however, the use of donor semen received little attention in the following years. In the 1930s, Sophia Kleegman and Margaret Jackson reported extensively on the use of donor semen. Based largely on their writings, the use of donor semen for artificial insemination became accepted in the medical community and to some extent in the public. In 1979 the American Fertility Society issued the first guidelines for screening and testing of semen donors.

Artificial insemination of homologous and nonhomologous semen has become a cornerstone of successful infertility treatment. It is currently estimated that more than 200,000 inseminations are performed each year in the United States. More than 90,000 donor inseminations were performed in the United States in 1994, and demand has continued to grow. Along with scientific and medical advancements of improving the quality of sperm and timing of artificial insemination, regulatory agencies such as the American Society of Reproductive Medicine and American Association of Tissue Banks have set forth guidelines which have minimized some of the risks of artificial insemination of donor and homologous semen. While it is possible that improvements resulting in improved success of *in vitro* fertilization may shift some patients' therapies more quickly to *in vitro* fertilization (IVF), and while the advent of intracytoplasmic sperm injection (ICSI) has eliminated some of the prior need for donor insemination, it is evident that artificial insemination will remain a major infertility therapy.

II. TYPES OF ARTIFICIAL INSEMINATION

A. Husband versus Donor Semen

Artificial insemination of husband semen is a relatively inexpensive, low-risk procedure which in certain indications can be quite successful in improving the fecundity rate. Indications for homologous artificial insemination include sperm quality defects, cervical abnormalities, sexual dysfunction in which intercourse and ejaculation are not possible, and spinal cord injury. In addition, patients with ejaculatory disorders such as retrograde ejaculation are candidates for artificial insemination. In recent years numerous improvements in techniques to prepare sperm from both electroejaculation and retrograde ejaculation urine samples have improved and have resulted in improved pregnancy rates for patients with these disorders.

During recent years the use of homologous insemination due to low sperm counts, decreased sperm motility, or other functional defects has greatly increased. Sperm washing and concentration followed by intrauterine insemination may be beneficial in increasing the number of spermatozoa capable of reaching the oocyte. More advanced techniques of sperm preparation have been reported to improve fertilization rates in patients with teratospermia or other sperm function defects. The basis for this improvement of sperm function has not been clearly identified but may include stimulation of sperm capacitation, which normally occurs within the female reproductive tract prior to fertilization. The removal of spermatozoa from seminal fluid, which contains capacitation inhibitors, removal of nonmotile spermatozoa and cellular debris, and incubation of the spermatozoa in artificial culture medium allow sperm capacitation to occur and, in certain cases, may be of use in stimulating sperm with capacitation defects. Recently, the use of test yolk buffer (TYB) to store spermatozoa at 5°C prior to sperm preparation for artificial insemination has been reported to improve capacitation and fertilization potential in samples with capacitation defects.

Homologous artificial insemination is also commonly performed on cryopreserved semen samples. Sperm banking of semen prior to chemotherapy or radiation has greatly increased in frequency during the past decade. The increased use of these therapies, along with increased awareness of physicians and society of the potential benefit of sperm banking prior to chemotherapy or radiation, has increased the number of homologous artificial insemination patients in this category and will likely result in increased use of the treatment in the future.

More controversial uses of artificial insemination include the treatment of immunological infertility in both the male and female. Rapid processing of sperm after semen collection combined with oral steroid treatment has been reported to improve pregnancy rates in males with anti-sperm antibodies. In addition, some studies have reported improved fertility following artificial insemination in patients with hostile cervical mucus or females with anti-sperm antibodies. Homologous artificial insemination is also performed in females with age-related fertility decline, with or without ovulation induction, with varying reports of success.

While the use of donor spermatozoa for artificial insemination is more complex than the use of husband spermatozoa in terms of moral, ethical, and legal questions, donor semen is commonly used for insemination. Advances in *in vitro* fertilization and ICSI have resulted in successful treatment options for many patients that were recently candidates for nonhomologous artificial insemination. Current indications for use of donor spermatozoa include oligozoospermia or severe sperm defects in patients unable or unwilling to use more advanced reproductive techniques, azoospermia, irreversible sterilization, prior exposure to chemicals or radiation which would preclude fertilization, nontreatable ejaculatory dysfunctions, or the presence of a genetic disorder not able to be treated by preimplantation genetic diagnosis or prenatal screening. Additionally, changes in society have increased the number of women in single or nontraditional relationships requesting artificial insemination of donor spermatozoa.

Recent guidelines issued by the American Society of Reproductive Medicine and other regulatory agencies have improved the safety of donor spermatozoa provided by physicians and clinics abiding by those guidelines. Screening of semen donors includes evaluation of fertility, medical history, genetic history, lifestyle risk, psychological history, screening for sexually transmitted diseases, and genetic screening. In addition, permanent and accurate records must be maintained by all clinics or sperm banks offering donor spermatozoa. In recent years the use of fresh donor spermatozoa has been prohibited by the regulatory agencies due to the increased risk of HIV and other sexually transmitted diseases. Current guidelines indicate that donor semen samples must be cryopreserved for a minimum of 6 months prior to insemination to allow repeated screening of the donor during a possible incubation period prior to sero conversion. These changes have generally resulted in improved safety for the recipient and decreased anomalies in the offspring. A large study of 11,808 conceptions from frozen donor sperm indicated a very low rate of fetal anomalies (Tables 1 and 2).

TABLE 1
Outcomes of Pregnancies

Parameter	
Conception cycles	n = 11,808
Information not available (%)	2.3
Pregnancies with known outcomes	n = 11,535
Ectopic pregnancies (%)	0.8
Fetal losses (%)	17.7
Medical terminations (%)	0.4
Deliveries	n = 9,357
Multiple pregnancies	4.4

Note. Reprinted with permission from Thepot *et al.* (1996).

TABLE 2
Incidence of Birth Defects from 9794 Births and 35 Fetuses after Conception by Use of Frozen Donor Semen and Prenatal Diagnosis

Defect	Incidence Per thousand	n
Monomalformations		
Cardiovascular	3.1	30
Limbs[a]	2.9	28
Urinary	1.5	15
Genital	0.4	4
Gastrointestinal	1.3	13
Central nervous	1.2	12
Respiratory[b]	0.4	4
Facial	1.2	12
Associated malformations (nonchromosomal)	1.0	10
Chromosomal anomalies		
Trisomy	2.5	25
Others	1.1	11
Not documented[c]	0.1	1

Note. Reprinted with permission from Thepot et al. (1996).
[a] Excluding congenital dislocation of the hip and hip click.
[b] Including diaphragmatic hernia.
[c] Polymalformation with chromosomal anomaly not identified.

B. Sperm Preparation

Due to the presence of capacitation inhibitors and prostaglandins within the seminal fluid, sperm must be removed from seminal fluid prior to intrauterine artificial insemination and reconstituted in culture medium conducive to normal sperm function. Sperm washing, the centrifugation of semen diluted with culture medium followed by reconstitution of the sperm pellet in culture medium, is the most basic sperm preparation technique used and the precursor to several recent advanced techniques of sperm preparation. Studies indicate that in addition to concentrating the sperm into a volume small enough to be placed within the uterus, sperm washing may be beneficial for improving sperm motility, longevity, and induction of capacitation. The sperm wash is commonly used for preparation of donor semen samples and by laboratories without the capability to perform more complex sperm preparation techniques.

Recent studies have shown that sperm preparation techniques which employ sperm washing, but also attempt to remove dead spermatozoa and cellular debris, stimulate sperm capacitation, or improve motility, may result in a higher pregnancy rate than sperm washing. In addition, numerous techniques have been developed to attempt to minimize the production of reactive oxygen species due to centrifugation and/or the presence of white blood cells within the seminal fluid. These techniques include the swim-up technique, swim-down procedure, glass wool filtration, percoll gradient centrifugation, and recent commercially available gradient centrifugation solutions.

Another family of sperm preparation techniques involves the chemical stimulation of sperm motility by incubation of the spermatozoa in the medium containing phosphodiesterase inhibitors such as pentoxyfilline or caffeine. These techniques increase intracellular cAMP levels and have been shown to improve sperm motility in severe cases of asthenozoospermia, including samples collected via electroejaculation.

Lastly, TYB medium has been used to store spermatozoa at 5°C prior to processing for artificial insemination. In addition, other chemical stimulators of capacitation, such as heparin and progesterone, have been evaluated. It is believed that this technique may stimulate sperm capacitation in samples with severe capacitation or fertilization defects. The sperm penetration assay (SPA) can be used to evaluate penetration defects and the ability of these treatments to stimulate capacitation and fertilization. Reports have indicated that patients showing stimulation of fertilization ability in the SPA can use this technique prior to artificial insemination or IVF with improved fertilization ability.

Numerous studies have attempted to evaluate the optimal sperm preparation technique for all patients. Each sperm preparation technique has its own unique benefits and faults; therefore, individualization of sperm preparation, based on semen parameters and sperm functional ability, has become more common. In studies comparing preparation techniques in a general population of patients, percoll gradient centrifugation or recent commercially available gradient preparations appear to be the most beneficial techniques for sperm preparation prior to artificial insemination (Table 3). These advanced preparation techniques have been shown to result in an improved pregnancy rate compared to sperm washing or other techniques.

TABLE 3
Pregnancy Rates following 898 Inseminations with Five Sperm Preparation Techniques

Pregnancy	Technique					
	Wash	Swim-up	Percoll	Swim-down	Refrig./hep.	All
% Chemical[a]	8.9 (14/157)[b]	14.7 (29/197)	16.1 (33/204)	7.7 (15/195)[b]	11.0 (16/145)	11.9 (107/898)
% Miscarriage[a]	21.4 (3/14)	10.3 (3/29)	21.2 (7/33)	20.0 (3/15)	25.0 (4/16)	18.7 (20/107)
% Delivery[a]	7.0 (11/157)[c]	13.2 (26/197)	12.7 (26/204)	6.1 (12/195)[c]	8.3 (12/145)	9.7 (87/898)

Note. Reprinted with permission from Carrell et al. (1997). © Elsevier Science Inc.
[a] Values in parenthesis are actual data.
[b] $p < 0.05$ compared to swim-up and percoll treatments.
[c] $p < 0.10$ compared to swim-up and percoll treatments.

C. Fresh versus Frozen Semen

The difference in pregnancy rates reported when using fresh versus frozen semen has varied from only a slightly higher chance of pregnancy using a fresh sample compared to frozen to as much as a threefold difference in the pregnancy rate. The differences in these reported pregnancy rates may be due to such factors as differences in the population of patients studied and differences in the number of progressively motile sperm inseminated. Since cryopreservation results in a decreased number of progressively motile sperm within a sample, it is logical that the pregnancy rate would be lowered on that basis only. Other studies have found that cryopreservation decreases sperm longevity, membrane function, and fertilization ability. Therefore, routine use of cryopreserved husband semen is not recommended; however, it can be performed in situations in which a fresh sample cannot be obtained. Since cryopreserved sperm have also been shown to have a decreased ability to penetrate and survive in cervical mucus, it is especially important that cryopreserved sperm be used for intrauterine insemination only.

As mentioned previously, recent regulations and increased risk for HIV require that all donor semen samples be cryopreserved and stored for at least 6 months prior to insemination. Again, a wide range exists in the reported differences of pregnancy rates comparing fresh versus frozen donor sperm. No study has definitively shown the optimal number of progressively motile sperm which should be inseminated, and the number of progressively motile sperm provided by different sperm banks for each insemination varies. It is recommended that a minimum of 20 million progressively motile sperm be inseminated.

D. Timing of Artificial Insemination

Methods for predicting the optimal time of artificial insemination are diverse and partially dependent on any ovulation induction treatment that the recipient may be receiving. While still occasionally used, monitoring of base body temperature and evaluation of cervical mucus quality are generally considered nonideal methods of evaluating the optimal time of insemination. Mastorianni *et al.* reported that as many as 35% of inseminations based on base body temperatures and mucus quality are not performed in the periovulatory period. Radioimmunoassay and immunofluorescent assays of serum LH values are highly accurate in predicting ovulation; however, they are both expensive and cumbersome for the patient. Therefore, the development of simple kits for measurement of urinary LH by immunoassay techniques has greatly aided the monitoring of urinary LH values and prediction of ovulation. Currently, most natural cycles, and ovulation induction cycles using clomiphene citrate, rely on the use of home urinary LH kits. These kits allow the patient to easily monitor their urine one or two times daily to precisely predict the time of ovulation. While these assays are generally all easy and relatively inexpensive, comparative studies have shown that the accuracy in picking up the LH surge varies between the kits. Additionally, although the kits are relatively easy to use, it is not unusual for mistakes to be made in the assay procedure which may result in inaccuracies. Therefore, careful assistance needs to be given to patients using the kits, and verification of the surge by the clinician or laboratory is recommended.

Patients undergoing ovulation induction with gonadotropins, and in some cases with clomiphene citrate, undergo artificial insemination based on the time of injection of hCG. hCG is given following the monitoring of follicular growth by both ultrasonagraphy and measurement of serum estradiol levels. The minimum follicular size and serum estradiol level used to determine maturity of the follicle varies from physician to physician; however, a general guideline is a follicular size of ≥ 1.5 cm in diameter and a serum estradiol value of $\geq 200-250$ pg/ml per mature follicle.

Numerous studies have attempted to evaluate the optimal time for insemination after the LH surge or hCG injection. Additional studies have evaluated the effect of more than one insemination at various times throughout the menstrual cycle. Differences in study parameters make these data difficult to compare; however, generally it has been shown that with optimal timing of the menstrual cycle very little or no difference exists between one insemination and more than one insemination in cycles in which the LH surge is closely pinpointed by twice-daily monitoring of the LH level. The optimal time for insemination is generally considered to be 24–36 hr following the LH surge.

E. Methods of Insemination

Two types of artificial insemination are used for nearly all patients; however, slight variations and other techniques have been introduced. Cervical insemination involves insemination of semen or washed spermatozoa directly on the cervix or slightly within the external cervical os. The patient remains in the lithotomy position for 10–30 min following insemination, and a cervical cup can be inserted to minimize loss of semen from the vagina. Advantages of cervical insemination include the ease of the procedure and lack of necessity of sperm washing and preparation prior to insemination. Negative aspects of cervical insemination include the loss of a large percentage of spermatozoa within the vagina and cervix and the decreased number of spermatozoa reaching the oocyte due to low sperm concentration, decreased sperm motility, or hostile cervical mucus. In addition, it has clearly been shown that cryopreserved semen do not penetrate cervical mucus well.

Numerous studies have evaluated pregnancy rates following cervical insemination versus intrauterine insemination. Generally those studies have indicated an improved pregnancy rate following placement of processed sperm directly within the uterus compared to the cervical region (Table 4). Even more dramatic improvements in the resulting pregnancy rate are seen in patients with decreased sperm concentrations or motility or large-volume semen samples with a low concentration of spermatozoa. Therefore, intrauterine insemination has become the standard method of artificial insemination. Intrauterine insemination involves processing of the semen sample and concentration of the spermatozoa into a volume of <0.8 ml and then passing a small catheter through the cervix into the uterine cavity. Numerous types of intrauterine catheters are commercially available, and generally the technique is simple and atraumatic. However, intrauterine insemination can be difficult in patients with cervical stenosis, increased presence of blind passages within the cervical canal, or flection of the uterus. Risks include slight to severe cramping

TABLE 4
Meta-analysis of Intrauterine versus Cervical Frozen Donor Insemination[a]

	Study				
	Patton et al. (12)	Byrd et al. (11)	Peters et al. (13)	Subak et al. (5)	Hurd et al. (22)
Per cycle PR (%)[b]	23.2 versus 5.1	9.7 versus 3.9	17.5 versus 17.1	15.8 versus 7.8	18.3 versus 3.8
OR[c]	5.6 (1.69–20.82)	2.6 (1.12–6.20)	1.03 (0.28–3.74)	2.21 (0.18–11.67)	5.6 (1.08–28.26)
First cycle PR (%)[d]	28.6 versus 9.1	Not reported	38.4 versus 30.8	Not reported	17.9 versus 5.4
OR[c]	4 (0.65–29.8)		1.4 (0.21–9.66)		3.8 (0.65–27.4)
Cumulative PR	78 versus 23[e]	70.6 versus 38.2		22 versus 23	78 versus 5[e]
Maximum No. of cycles	6	12	4	3	4
Ovarian stimulation	Clomiphene citrate in ovulatory disorders	Clomiphene citrate in ovulatory disorders	No induction	Spontaneous clomiphene citrate, hCG	Clomiphene citrate in ovulatory disorders
Cervical insemination	Intracercival[f]	Intracervical	Pericervical	Pericervical	Intracervical
Remarks			Only single normal ovulatory woman		

Note. Reprinted with permission from Matorras *et al.* (1996). © Elsevier Science Inc.
[a] In all series the IUI specimen was prepared by washing, and the ovarian cycle was monitored by means of LH surge. In one series (11), one or two inseminations per cycle were performed, whereas in the remaining series only one insemination per cycle was performed.
[b] Per cycle PR: 63/439 (14.3%) versus 28/478 (5.8%); $p = 0.00005$; OR, 2.69 (CI, 1.65–4.40).
[c] Values in parentheses are 95% CIs.
[d] First cycle PR: 20/80 (25.0%) versus 8/64 (12.5%); $p = 0.046$; OR, 2.67 (CI = 1.01–7.18).
[e] Approximate values.
[f] Per woman PR: 19/28 (67.9%) versus 4/22 (18.2%); $p = 0.001$; OR, 9.5 (CI = 2.12–46.54).

due to incomplete removal of seminal plasma molecules and infection due to the passage of vaginal or seminal contaminants into the uterine cavity. One study has shown an increase of the titers of anti-sperm antibodies following intrauterine insemination in women who already possessed low to moderate titters of anti-sperm antibodies.

Recent variations of insemination techniques have included the use of fallopian tube perfusion of spermatozoa. This technique is more complex than intrauterine insemination and results in increased cost to the patient. Some studies have indicated that it is not a cost-effective improvement over intrauterine insemination; however, other studies have indicated an improved pregnancy rate. Additionally, a recent study analyzed the use of intrafollicular insemination of spermatozoa. This technique showed no improvement in the pregnancy rate.

III. Ethical and Legal Aspects of Artificial Insemination

Artificial insemination of husband sperm is a relatively safe, low-cost, proven treatment to aid infertile couples. Since the sperm of the legal husband is used for the insemination, no legal debate exists as to the paternity of the offspring. Therefore, homologous insemination has been a routine initial treatment for a large percentage of infertile couples. While some debate has existed regarding the use of artificial insemination for women over 40 years old or specific categories of infertility with a poor prognosis of pregnancy via artificial insemination, the technique is nearly unanimously accepted both within society and within the medical practice as a valuable therapy for infertile couples.

While only a very small proportion of husband inseminations involve procedures to influence gender selection, this technique is the one area of husband artificial insemination which receives significant ethical debate. This technique can be of extreme benefit in families with a sex-linked genetic defect; however, the potential for abuse and negative effect on society is obvious. In practice, this technique is used by a very small percentage of couples, generally following the birth of several children of the same sex.

Although donor insemination has been performed for several decades, ethical and legal debates regarding its usage are numerous and diverse. Due to the potential problems from donor insemination, both federal agencies and state governments have enacted regulations and laws involving the use of donor sperm. This has resulted in a situation in which the laws regulating the sperm banks and the laws regulating paternity may vary from state to state. However, due to the potential for abuse and extreme harm to patients, these regulations have generally been helpful in improving the standard of practice.

One current debate regarding donor insemination involves the question of access. As the percentage of nontraditional households increases, questions regarding the availability of donor semen for single persons or lesbian couples increases. Since the effects of being raised in a single parent or nontraditional household have not been conclusively shown, the limiting of donor semen to these persons has remained controversial. However, recent surveys have indicated that approximately 40% of the American population is opposed to the insemination of single adults, and this practice has remained at the discretion of the individual physician. Physicians opposed to donor insemination of single women generally refer the person to another physician willing to complete the therapy.

Most couples receiving artificial insemination of donor sperm do not inform the child or relatives that the child was conceived by use of donor semen. Indeed, some studies have shown that a majority of clinicians suggest that couples not divulge this information. This has led some to question the effect of this secrecy on the relationship between the child and parents. It can be argued that a foundation of parenthood based on secrecy cannot be as strong as one built on complete honesty and openness. Additionally, problems or mistrust can develop years later if for some reason the child does learn of their conception via donor sperm, which was previously hidden by the parents. Arguments for the use of secrecy include protection of the child from any negative effects of self-esteem as a result of being conceived differently than others and protection of the identity of the semen donor. Convincing arguments can be made both for and against each of these points; therefore, the decision of whether to divulge the use of donor semen to the child remains a highly personal decision that couples should make after careful consideration.

Another question, which has been thoroughly discussed by Lauritzen, involves the question of balance or symmetry within the marriage if the mother is both the social and the genetic parent of the child, whereas the father is solely the social parent. Lauritzen defines this problem as the problem of "asymmetry." Couples should be informed and questioned regarding this potential problem, and counseling should be offered if any doubt exists. If any doubt does exist in the husband as to his ability to be comfortable and complete the role of fatherhood in every sense, the use of donor insemination should not be performed. While little data exist regarding this question, clinical experience indicates that this problem is minimal in couples properly counseled prior to insemination.

Another area of ethical debate involves the screening of potential donors. Since the industry of donor sperm banking is market driven, sperm banks generally place requirements not only on the medical status of the donor but also sometimes on the educational background of potential donors. One highly publicized sperm bank advertises the sole use of "high IQ" and Nobel laureate semen donors. This has led to questions of exclusivity and the effect on the gene pool. In reality, most sperm banks continue to maintain a wide variety of semen donors with diverse backgrounds.

These and other ethical questions continue to be raised and debated and are generally beneficial for both the patients and the medical establishment. It is evident that the advancements in treatments for infertile couples will continue to provoke new questions and problems, and that the legal and legislative systems will continue to address these questions with new laws and regulations.

See Also the Following Articles

INFERTILITY; IN VITRO FERTILIZATION; REPRODUCTIVE TECHNOLOGIES, OVERVIEW; SPERM CAPACITATION; STERILITY

Bibliography

Baird, P. A. (1996). Ethical issues of fertility and reproduction. *Annu. Rev. Med.* 47, 107–116.

Byrd, W., Drobnis, E. Z., Kutteh, W. H., Marshburn, P., and Carr, B. R. (1994). Intrauterine insemination with frozen donor sperm: A prospective randomized trial comparing three different sperm preparation techniques. *Fertil. Steril.* 62, 850–856.

Carrell, D. T., Kuneck, P., Peterson, C. M., Hatasaka, H. H., Jones, K. P., and Campbell, B. (1998). A randomized, prospective analysis of five sperm preparation techniques before intrauterine insemination of husband sperm. *Fertil. Steril.* 69, 122–126.

Edwards, R. G., and Brody, S. A. (1995). *Principles and Practice of Assisted Human Reproduction.* Saunders, Philadelphia.

Fanchin, R., Hazout, A., Olivennes, F., Schwab, B., Righini, C., and Frydman, R. (1995). A new system for fallopian tube sperm perfusion leads to pregnancy rates twice as high as standard intrauterine insemination. *Fertil. Steril.* 64, 505–510.

Frederick, J. L., Denker, M. S., Rojas, A., Horta, I., Stone, S. C., Asch, R. H., and Balmaceda, J. P. (1994). Is there a role for ovarian stimulation and intrauterine insemination after age 40? *Hum. Reprod.* 9, 2284–2286.

Kahn, J. A., von During, V., Sunde, A., and Molne, K. (1992). Fallopian tube sperm perfusion used in a donor insemination programme. *Hum. Reprod.* 7, 806–812.

Karow, A. M., and Critser, J. K. (1997). *Reproductive Tissue Banking: Scientific Principles.* Academic Press, San Diego.

Lauritzen, P. (1996). Donor insemination and responsible parenting. In *Biomedical Ethics* (T. A. Mappes and D. DeGrazia, Eds.). McGraw-Hill New York.

Matorras, R., Gorostiaga, A., Diez, J., Corcostegui, B., Pijoan, J. I., Ramon, O., and Rodriguez-Escudero, F. J. (1996). Intrauterine insemination with frozen sperm increases pregnancy rates in donor insemination cycles under gonadotropin stimulation. *Fertil. Steril.* 65, 620–625.

Peterson, C. M., Hatasaka, H. H., Jones, K. P., Poulson, A. M., Carrell, D. T., and Urry R. L. (1994). Ovulation induction with gonadotropins and intrauterine insemination compared with in vitro fertilization and no therapy: A prospective, nonrandomized, cohort study and meta-analysis. *Fertil. Steril.* 62, 535–544.

Plosker, S. M., Jacobson, W., and Amato, P. (1994). Predicting and optimizing success in an intrauterine insemination programme. *Hum. Reprod.* 9, 2014–2021.

Thepot, K., Mayaux, M. J., Czyglick, F., Wack, E., Selva, J., and Jalbert, P. (1996). Incidence of birth defects after artificial insemination with frozen donor spermatozoa: A collaborative study of the French CECOS Federation on 11535 pregnancies. *Hum. Reprod.* **11**, 2319–2323.

Urry, R. L., Middleton, R. G., Jones, K. P., Poulson, M., Worley, R., and Keye, W. (1988). Artificial insemination: A comparison of pregnancy rates with intrauterine versus cervical insemination and washed sperm versus swim-up sperm preparations. *Fertil. Steril.* **49**, 1036–1038.

Vollenhoven, B., Selub, M., Davidson, O., Lefkow, H., Henault, M., Serpa, N., and Hung, T. T. (1996). Controlled ovarian hyperstimulation using human menopausal gonadotropin in combination with intrauterine insemination. *J. Reprod. Med.* **41**, 658–664.

Wolf, D. P. (1993). Sperm preparation for in vitro fertilization (IVF): Individualization or one method for all? *J. Assisted Reprod. Genet.* **10**, 246–247.

Asexual Reproduction

Kerstin Wasson

Humboldt State University

I. Types of Asexual Reproduction
II. Taxonomic Distribution and Frequency
III. Ecological and Evolutionary Consequences

GLOSSARY

asexual reproduction The production of new modules or ramets without meiosis or the fusion of gametes having taken place.

budding The formation of new modules by growth and differentiation of blebs of primordial tissue; the module which formed the bleb is not significantly diminished by the process.

clone A genet composed of multiple ramets that are physiologically separated and able to function independently.

colony A ramet composed of multiple, physiologically interconnected modules.

fission The division of a ramet into two or more fragments that become independent new ramets, sometimes after a period of regeneration.

genet A genetic individual; the entire mitotic product of one zygote.

modular animal A genet composed of multiple modules; umbrella term which includes all clonal and colonial animals.

module The fundamental functional unit of construction that is repeated in sequential iterative fashion in the development of a colony or clone.

parthenogenesis The development of a new module from an unfertilized egg.

polymorphism A discontinuous variation in the morphology and function of modules.

ramet A somatic individual; a coherent "body" able to function independently; in purely clonal animals, a single module is the ramet, whereas in colonial animals, the colony is the ramet.

unitary animal An animal that does not undergo asexual reproduction; with a single ramet per genet.

zooid A colonial module; the modular unit of a colony.

Asexual reproduction is the production of new bodies (modules or ramets) by mitotic growth. All animals that undergo asexual reproduction are modular, composed of multiple modules per genet. Animals that do not undergo asexual reproduction, and which thus only have a single body per genet, are unitary. Both modular and unitary animals generally undergo sexual reproduction. So asexual reproduction can best be understood as an alternative to unitary growth processes rather than as an alternative to sex. The purpose of somatic growth for both unitary and modular animals is to increase the size and thus the survival and fecundity of an existing genet, whereas the purpose of sex is to make new genets. Therefore, growth (including asexual reproduction) and sex should not be equated (Pearse et al., 1989). Asexual reproduction is just one way for a genet to get bigger.

I. TYPES OF ASEXUAL REPRODUCTION

A. Clones and Colonies

Modular animals can be divided into two main types: clonal and colonial animals. In a purely clonal animal, such as an aphid that undergoes parthenogenesis or an anemone that undergoes fission, the modules detach and live separately after forming. There is only a single module per ramet in such clonal animals. In a colonial animal, such as a bryozoan or a hydroid, the modules (often called zooids) remain physiologically interconnected. The whole colony is the ramet, and a colonial ramet is thus composed of multiple modules. However, all colonial animals are also potentially or actually clonal—separated portions of a colony will function independently and grow and thus can be considered multimodular clonal ramets. Some colonial animals also produce many multimodular ramets by processes such as encapsulation or polyembryony. Furthermore, in the life cycle of many hydroids, there is a colonial polypoid phase followed by a clonal medusoid phase. Despite this overlap in clonality and coloniality, it is useful to distinguish between those modular animals that are purely clonal (never form colonies) and those that are colonial (always form colonies). Table 1 summarizes the differences between these two groups.

There are many different types of asexual reproduction—many different ways of making multiple modules per genet. These have been variously classified (Hughes, 1989), but will here be treated in three categories: fission, budding, and parthenogenesis.

B. Fission

Fission is the division of a ramet to produce two or more new ramets. The original ramet is typically diminished in size as a result of the process. Following fission, the new ramets often regain the size of the original ramet by regeneration. While fission is generally not a way of producing zooids (colonial modules), both colonies and clones can undergo fission resulting in multiple ramets of the same genet. Therefore, fission can be thought of as a way of making ramets—either the multimodular ramets of colonies or the unimodular ramets of clones.

The most familiar kind of fission is binary fission, in which a ramet pulls itself into approximately equal halves. For example, many clonal anemones undergo

TABLE 1
Summary of the Differences between Purely Clonal versus Colonial Animals

Animal	Number of modules per ramet	Number of ramets per genet	Modules produced by	Common examples
Purely clonal	One	Many	Budding, fission, parthenogenesis	Aphids, rotifers, anemones
Colonial	Many	One or more[a]	Budding[b]	Corals, bryozoans, compound ascidians

[a] Multiple ramets (colonies) per genet may be produced by fission or parthenogenesis; if there are multiple ramets, the animal is clonal as well as colonial.
[b] A few anthozoan colonies produce modules by fission, but this is an extremely rare means of making modules for colonial animals.

binary fission, resembling cells undergoing mitosis. However, fission does not necessarily result in two identical ramets. For instance, other anemones shift their position on the substrate, leaving behind a small portion of the attachment disk which regenerates into a whole new anemone. Similarly, some seastars shed a single arm, which grows into a whole new seastar.

Fragmentation is a type of fission in which the force separating the original ramet into new ramets is external rather than internal. For instance, a nemertean that is damaged by a rock moving in the surf may be separated into fragments that regenerate into two independent worms. Many colonial genets, especially long-lived ones, probably consist of multiple ramets (colonies) due to fragmentation from damage done by physical factors, predators, or competitors.

The previously discussed types of fission occur during the adult phase of the life cycle. Polyembryony is fission of the zygote. Multiple ramets of the same genet are produced by division of the egg or early embryo. Identical twins in humans are a familiar example. In an armadillo species, octuplets are the norm; there are eight ramets per genet. Some invertebrate taxa (e.g., cyclostome bryozoans and encyrtid wasps) take polyembryony to an extreme, producing hundreds or thousands of genetically identical larvae by fission of a single zygote.

C. Budding

Another type of modular growth is budding, the formation of new modules by growth and cellular differentiation of a bleb of primordial tissue. Budding differs from fission in that the "parental" module is not significantly altered or diminished by the process of producing new modules, since it only donates a vesicle of cells rather than a substantial portion of its body. Budding can result in clone formation if the modules separate from the module that formed them. An example of clonal budding is the hydra, in which buds detach after an initial phase of attached development. If budded modules remain permanently interconnected after forming, the result is a colony. Budding is by far the most common way that zooids (colonial modules) are formed. While new colonial ramets may be produced by fission, size increase (by addition of zooids) of existing ramets almost always occurs by budding.

A special type of budding is encapsulation, which is the formation of dormant, nutrient-laden buds which are enclosed within resistant capsules. Encapsulation is common among freshwater sponges and bryozoans, which form gemmules and statoblasts, respectively. These dormant buds can resist seasonal desiccation and temperature stress which is lethal to normal ramets. When favorable conditions return, they germinate and produce new modules and ramets.

D. Parthenogenesis

The final type of asexual reproduction is parthenogenesis, which is the development of a new module from an unfertilized egg. Some authors regard parthenogenesis as a form of sexual reproduction. Certainly the ancestry of the process is sexual, but we consider parthenogenesis as a type of asexual reproduction which happens to exploit the sexual machinery of an organism. In many obligate parthenogens and in cyclical parthenogens such as aphids, parthenogenesis is apomictic, involving no meiosis. Apomictic parthenogenesis results in genetically identical modules of the same genet, fitting the definition of asexual reproduction. However, in other parthenogens, such as brine shrimp, parthenogenesis is automictic. Meiosis occurs together with some mechanisms for restoring the diploid genome. In automictic parthenogenesis, some genetic mixing may occur, and so the resultant modules may not be exact genetic replicates. Although these modules do not perfectly fit our definition of a genet (the mitotic product of one zygote), we include automictic parthenogenesis as a type of asexual reproduction.

II. TAXONOMIC DISTRIBUTION AND FREQUENCY

Asexual reproduction is ubiquitous in the living world. Virtually all bacteria, protists, fungi, and plants can be considered modular. Within the animal kingdom modularity is widespread (Table 2). Most

TABLE 2
Taxonomic Distribution of Unitary versus Modular Growth in the Animal Kingdom[a]

Animal taxon	Unitary	Modular Purely clonal	Colonial	Animal taxon	Unitary	Modular Purely clonal	Colonial
Porifera		XXX		Pogonophora	XX	X	
Placozoa		XXX		Vestimintifera	XXX		
Mesozoa		XXX		Sipunculida	XX	X	
Cnidaria				Echiura	XXX		
Hydrozoa	X	X	XX	Mollusca			
Scyphozoa		XXX		Aplacophora	XXX		
Cubozoa		XXX		Polyplacophora	XXX		
Anthozoa	XX	XX	XX	Monoplacophora	XXX		
Ctenophora	XX	X		Scaphopoda	XXX		
Platyhelminthes				Bivalvia	XX	X	
Turbellaria	XX	XX		Gastropoda	XX	X	
Monogenea	XX	X		Cephalopoda	XXX		
Cestoida	XX	X		Arthropoda	XX	X	
Trematoda		XXX		Onychophora	XX	X	
Gnathostomulida	XXX			Tardigrada	XX	X	
Gastrotricha	XX	XX		Pentastomida	XXX		
Kinorhyncha	XXX			Chaetognatha	XXX		
Priapulida	XXX			Echinodermata			
Loricifera	XXX			Asteroidea	XX	X	
Rotifera	X	XX		Ophiuroidea	XX	X	
Acanthocephala	XXX			Holothuroidea	XX	X	
Nematomorpha	XXX			Echinoidea	XXX		
Nematoda	XX	XX		Crinoidea	XXX		
Nemertea	XX	X		Concentricycloidea	XXX		
Bryozoa			XXX	Hemichordata			
Phoronida	XX	XX		Enteropneusta	XX	X	
Brachiopoda	XXX			Pterobranchia		XX	XX
Kamptozoa				Chordata			
Solitaria		XXX		Urochordata	XX	XX	XX
Stolonata			XXX	Cephalochordata	XXX		
Cycliophora		XXX		Vertebrata	XX	X	
Annelida							
Oligochaeta	X	XX					
Hirudinea	XXX						
Polychaeta	XX	X					

[a] Modular animals are further broken down into those that are purely clonal (never form colonies) and those that are colonial (always form colonies). Information for the table was taken from Bell (1982), Hughes (1989), and various recent invertebrate zoology textbooks. A phylum is broken down into subtaxa only when these display divergent patterns different from the pattern shown at the level of the phylum as a whole. X, some species; XX, many species; XXX, all species in the taxon.

taxa have at least some modular members, and a few taxa are entirely modular.

The vast majority of modular animals undergo sexual reproduction. Some fissiparous anemones, flatworms, and oligochaetes appear to never have sex, but there is no concrete evidence for the complete absence of sex in an animal that undergoes fission or budding. The only animals that have been convincingly shown never to undergo sexual reproduction are some parthenogens (e.g., bdelloid rotifers and some cladocerans, insects, and fish). The ubiquity of sex among modular animals suggests sexual and asexual reproduction cannot be equated; the former serves a function which cannot be accomplished by the latter (Pearse et al., 1989). Making multiple modules or ramets does not free modular animals from the selective pressures driving them, as well as their unitary counterparts, to make new genets.

While asexual reproduction has evolved repeatedly in many different lineages, it is absent from others. This absence may partly be explained by constraints: Animals with complex and specialized body plans may not be able to regenerate from fragments or rudiments, and some skeletons (e.g., bivalve shells, arthropod exoskeletons, and urchin tests) cannot readily be divided. So perhaps only simpler animals are capable of budding and fission. However, selective pressures must be invoked to explain the rarity of parthenogenesis and polyembryony, since it seems that virtually any animal could evolve the ability to undergo these processes.

Coloniality has evolved much less frequently than clonality in the animal kingdom; only five phyla (Cnidaria, Bryozoa, Kamptozoa, Hemichordata, and Chordata) have colonial members. The rarity of colonies may be due to constraints (only relatively simple body plans can undergo the budding required to form colonies) or due to selection against permanently interconnected modules. Colonies are limited to aquatic habitats, perhaps because they are all suspension feeders. Clones are found in all habitats, including terrestrial ones, in which most clones are mobile parthenogens.

Overall, asexual reproduction is very common in the animal kingdom (Table 2). However, since we ourselves are not modular, and since most of the large, conspicuous animals in our terrestrial environment are not modular either, the frequency and significance of asexual reproduction among animals is often overlooked.

III. ECOLOGICAL AND EVOLUTIONARY CONSEQUENCES

To understand any aspect of the ecology or evolution of modular animals—from demography to morphology to sexual reproduction—one must examine multiple organizational levels. Both the modules and the whole genet must be accounted for in clonal forms, whereas modules, colonies, and genets must be considered in colonial species. Their complexity complicates the study of modular animals, but also results in new insights into ecology and evolution. In recent years, researchers have explored the differences between the biology of modular animals and their unitary counterparts. There are many consequences of having the genet spread into numerous little bodies rather than contained in a single one. The consequences of asexual reproduction are more thoroughly reviewed in Jackson et al. (1985) and Hughes (1989).

A. Growth and Scaling

The only way for a unitary animal to increase in size is for its single body to grow. Modular animals have two options for size increase: Either existing modules may grow or new modules may be formed. As a general rule, most of the size increase of modular genets is due to the addition of modules. The modules of most modular animals are smaller than the bodies of their unitary counterparts and the fossil record reveals no trend toward increased module size in most modular lineages, whereas many unitary lineages do tend to show a size increase over time (Coates and Jackson, 1985).

Why do modules remain small, over both developmental and evolutionary time, while unitary bodies get bigger? Many animals rely on surface area for functions such as feeding, respiration, and locomotion. The optimal surface area to volume ratio may occur at small body size. Nevertheless, many unitary animals may grow larger because survival and fecundity often increase with body size. In modular animals, there is no such trade-off. The genet can benefit from increased survival and fecundity by making additional modules, but the modules can remain small, maintaining a favorable surface area to volume ratio. There is thus no advantage for a modular genet to increase module size equivalent to the advantage of unitary animals to increase the size of the body.

Therefore, one consequence of asexual reproduction is that biomass can be increased without changing the scaling of the bodies. For example, unitary anemones have an upper size limit above which the surface area of their tentacles (upon which they rely for food gathering) can no longer supply the metabolic energy demands of their body mass. In contrast, there is no upper size limit for a clonal anemone genet. An anemone clone can potentially increase its biomass indefinitely.

B. Phenotypic Plasticity

Asexually reproducing organisms make superb study systems for researchers who are interested in untangling environmental from genetic effects. By exposing ramets of the same genet to differing environmental conditions, in the field or laboratory, one can measure phenotypic plasticity within a single genotype—an impossibility for a unitary genet. With modular replicates, one can precisely partition genetic from environmental effects and examine interactive effects between the two.

One aspect of phenotypic plasticity and modular animals which has received much attention is the growth form of colonies. Most colonial animals are sessile, forming branching structures resembling plants and lending themselves to computer simulations. While the zooids of colonies tend to have a fairly determinate morphology, comparable to that of unitary bodies, the morphology of the colony as a whole is largely indeterminate: its final shape or size cannot be predicted. This is because colonies are typically very developmentally plastic, altering the number and placement of their modules (and in polymorphic colonies, the kinds of modules) in response to environmental variation. Many experiments (reviewed in Hughes, 1989) have demonstrated that ramets of the same colonial genet adopt different morphologies depending on factors such as nutrition or wave action. For instance, in many species, colonies placed in quiet water grow upright and extensively branched, maximizing their surface area, whereas colonies of the same genet placed in surfy water grow as flat sheets or solid mounds as protection from breakage.

C. Modular Modes of Sex

A unitary genet can either be gonochoric or hermaphroditic; the single body can be male, female, or both. The same is true of a module. If all modules of a genet synchronously display the same sexual mode, the genet and module sexual modes will be the same. For instance, if modules are always female, the genet is a female. If modules are always simultaneously hermaphroditic, the genet is a simultaneous hermaphrodite. However, the sexual mode of a genet is not always so straightforward.

In many modular animals, the sexual mode of the module is different from that of the whole genet. Consider, for instance, a cyclical parthenogen such as an aphid. Each aphid, each module, is gonochoric, either male or female. However, both male and female aphids are parthenogenetically produced by one clone. Therefore, the whole genet must be considered a hermaphrodite because it encompasses male and female functions. Another example is a colony composed of protandric zooids. Each module is a sequential hermaphrodite. However, if the modules are not synchronized in the timing of their sex change, the colony as a whole is a simultaneous hermaphrodite. In short, each organizational level (module, colony, and genet) may have its own sexual mode, and what happens at one level is not necessarily what happens at others.

According to one classification scheme (Wasson and Newberry, 1997), unitary animals have 3 sexual modes (gonochorism, sequential hermaphroditism, and simultaneous hermaphroditism), but clonal animals have 6 and colonial animals have 10. Modular animals have many more pathways to genet hermaphroditism than do unitary ones.

D. Polymorphism

The fitness of a unitary individual depends on its ability to survive and reproduce. In a modular animal, the fitness of the whole genet may actually increase if some modules do not survive or do not reproduce. In a number of modular species, some modules are specialized for other functions which contribute to the survival or reproduction of other modules. For instance, in clonal anemones, warrior anemones protect peripheries of a clone. Their aggression results in a significant reduction in size and fecundity for the warrior modules but presumably increases fitness at the level of the whole clone. Similarly, in some parthenogenetic insects, there are soldier or nurse modules which have low fitness themselves but which increase the fitness of the clone. Such variation in the morphology and function of modules within a genet is called polymorphism.

The most spectacular examples of polymorphism can be found in colonial animals. Three phyla (Cnidaria, Bryozoa, and Chordata) have colonial members which show polymorphism. While resource translocation probably occurs in most colonies, modules in these three phyla are generally more highly integrated than are those in the two phyla in which there is no colonial polymorphism (Kamptozoa and Hemichordata). The shared gastrovascular cavity of cnidarian zooids, the funicular system of bryozoans, and the interzooidal blood vessels of urochordates allow for nutrient exchange between zooids. In these taxa, some members have nonfeeding modules specialized for defense (e.g., cnidarian dactylozooids and bryozoan avicularia), reproduction (e.g., cnidarian gonozooids and bryozoan ovicells), and other functions. These polymorphs are often strikingly different in form than the feeding zooids in the colony. Despite lacking feeding organs, they may persist as long as do feeding zooids, apparently sustained by nutrients received from neighboring zooids. Polymorphism has evolved repeatedly in various lineages and is apparently not the plesiomorphic condition for any phylum (Harvell, 1994). The fossil record reveals that over evolutionary time, there has been a trend in many colonial lineages toward increased colonial integration and thus toward increased polymorphism (Coates and Jackson, 1985).

E. Life Span

Mortality from biological factors (e.g., predation, competition, and starvation) and physical factors (e.g., thermal stress and mechanical damage) may generally be lower in modular animals than in unitary ones because the risk of mortality is spread over many bodies. Partial mortality, with some modules dying but others surviving, is probably typical of many modular animals living in stable habitats. For instance, many colonial animals are only partly consumed by predators. Partial predation can select for inducible defenses; some bryozoan colonies begin to make spines only after exposure to a nudibranch predator (Harvell, 1994). Waiting until some modules have been eaten before investing in defensive structures is a good strategy if defense is costly. Like polymorphism, this strategy in effect sacrifices the fitness potential of some modules in the interest of enhancing the fitness of the whole modular genet.

In addition to the life span of modular genets surpassing that of unitary ones due to decreased risk of genet mortality, the life span of modular genets may also extend far beyond that of many unitary ones because of an absence of senescence. Senescence is a process of deterioration eventually resulting in death, apparently due to cumulative damage at the molecular level. Many unitary animals undergo senescence of the soma. When they are young, repair mechanisms keep up with molecular damage that is incurred, but as they age the repair mechanisms lag behind. The only cell lineage that does not senesce in most unitary animals is the germline. Germ cells avoid senescence by, on the one hand, minimizing molecular damage (they only divide a few times and are metabolically inactive much of the time), and, on the other hand, maximizing repair mechanisms. Therefore, healthy, nonsenescent offspring can be produced by aged, senescent parents.

The modules of most modular animals undergo a process of senescence comparable to that of unitary bodies. Nevertheless, the new modules that are produced by senescent ones are generally not themselves senescent. In the case of parthenogens, this is readily understood; parthenogenetic, like sexual offspring are a product of the nonsenescent germline. Therefore, while the modules senesce, the modular genet does not. For instance, an aphid module may live only a few weeks but may be part of a clone that is decades or even centuries old.

Animals that undergo asexual reproduction via budding do not sequester a germline. However, they apparently have populations of undifferentiated stem cells that likewise are protected from senescence. These undifferentiated stem cells form buds that grow into healthy, nonsenescent modules, even when the progenitor module is aged and deteriorating. Again, the life span of the modular genet far exceeds that of its component modules. For example, the age of some coral genets has been estimated at thousands of years, even though the polyps may only live a matter of months. Given the potential immortality of these animals, it seems certain that the earth's oldest animals are modular.

Fissiparous animals, in contrast, produce new modules built mostly of differentiated cells, since existing modules are partitioned to produce two (or more) new ones. It seems unlikely that a senescent module could divide to produce two nonsenescent modules. Thus, fission may not automatically confer immortality the way parthenogenesis and budding do. However, the growth of the new modules from the fragments resulting from fission may largely involve undifferentiated, nonsenescent cells. Therefore, in recovering to the original size, the two new modules may be substantially rejuvenated relative to the module that produced them. Some fissiparous genets may therefore gain immortality after all or at least be quite long-lived. This may explain the observed absence of senescence in some fissiparous clonal anemones. However, since some unitary anemones seem to escape senescence (or at least outlive their human observers), it is not clear whether it is the ability to undergo fission rather than the simple morphology and the absence of highly differentiated tissues which is responsible for this longevity.

F. Genetic Mosaics

In animals which sequester a germline, heritability is limited to this single cell lineage. Only mutations in the germline can be passed on to the offspring. In parthenogenetic animals, germline mutations can be passed on to asexually as well as sexually produced offspring. Thus, the odd situation in which a single genet is composed of genetically differing modules can occur. If modules can have heritable variation, selection can occur at the modular level, within genets (Boss, 1983). For instance, an aphid pest introduced to North America evolved insecticide resistance within a single clone (Blackman, 1981). In this example, mutant modules had an advantage over the wild-type ones, and selection occurred within a single genet.

Animals that reproduce asexually by fission or budding do not sequester a germline. In these animals, somatic mutations are heritable if they occur in cell lineages that give rise to germ cells or in lineages that give rise to new modules. Again, there is a potential for the genet to be a mosaic—composed of modules with genetic differences, which may compete directly with each other, with the fitter ones making more modules and/or sexually producing more offspring. Genetic mosaicism has never been demonstrated in a fissiparous or budding animal (Carvalho, 1994) but seems likely to occur, as seen in plants. Indeed, somatic mutations may be important in helping long-lived and large modular genets, such as reef-building stony corals, adapt to temporal or spatial environmental variation. The multiple levels of selection in modular organisms have only just begun to be explored.

In conclusion, asexual reproduction has remarkable consequences for ecology and evolution. Modular animals differ in many fundamental ways from their unitary counterparts. Many lessons remain to be learned from modular animals, which are abundant and diverse in terrestrial and particularly in aquatic habitats.

See Also the Following Articles

BRYOZOA; CNIDARIA; HEMICHORDATA; HERMAPHRODITISM; KAMPTOZOA; PARTHENOGENESIS AND NATURAL CLONES

Bibliography

Bell, G. (1982). *The Masterpiece of Nature: The Evolution and Genetics of Sexuality*. Univ. of California Press, Berkeley.

Blackman, R. L. (1981). Species, sex and parthenogenesis in aphids. In *The Evolving Biosphere* (P. Forey, Ed.), pp. 75–85. Cambridge Univ. Press, Cambridge, UK.

Buss, L. W. (1983). Evolution, development, and the units of selection. *Proc. Natl. Acad. Sci. USA* 80, 1387–1391.

Carvalho, G. R. (1994). Genetics of aquatic clonal organisms. In *Genetics and Evolution of Aquatic Organisms* (A. Beaumont, Ed.), pp. 291–323. Chapman & Hall, London.

Coates, A. G., and Jackson, J. B. C. (1985). Morphological themes in the evolution of clonal and aclonal marine invertebrates. In *Population Biology and Evolution of Clonal Organisms* (J. B. C. Jackson, L. W., Buss, and R. E. Cook, Eds.), pp. 67–106. Yale Univ. Press, New Haven, CT.

Harvell, C .D. (1994). The evolution of polymorphism in colonial invertebrates and social insects. *Q. Rev. Biol.* 69, 155–185.

Hughes, R. N. (1989). *A Functional Biology of Clonal Animals.* Chapman & Hall, London.

Jackson, J. B. C., Buss, L. W., and Cook, R. E. (1985). *Population Biology and Evolution of Clonal Organisms.* Yale Univ. Press, New Haven, CT.

Pearse, J. S., Pearse, V. B., and Newberry, A. T. (1989). Telling sex from growth: Dissolving Maynard Smith's paradox. *Bull. Mar. Sci.* 45, 433–446.

Wasson, K., and Newberry, A. T. (1997). Modular metazoans: Gonochoric, hermaphroditic, or both at once? *Invertebr. Rep. Dev.* 31, 159–175.

Assisted Reproduction
see Reproductive Technologies

Atresia
see Follicular Atresia

Autoimmunity
see Immunology of Reproduction

Autonomic Nervous System and Reproduction

Karen J. Berkley

Florida State University

I. Limiting the Subject Matter
II. Requirements for Successful Fertilization
III. Synopsis of Autonomic Innervation of Internal and External Reproductive Organs
IV. Effects of Denervation of Reproductive Organs on Successful Fertilization
V. Coitus as a Social Behavior
VI. Further and Future Issues

GLOSSARY

allodynia Pain due to a stimulus that does not normally provoke pain.

coitus Sexual union between a female and male involving insertion of the penis into the vagina.

external pelvic reproductive organs (external genitalia) In females, the vulva and clitoris; in males, the penis and scrotum; in both sexes, striated perineal muscles and some accessory sex glands.

fertilization The union of male and female gametes (ova and sperm) to form a single cell (zygote).

hyperalgesia An increased response to a stimulus that is normally painful.

hypoalgesia A decreased response to a stimulus that is normally painful.

internal reproductive organs In females, the ovaries, oviducts, uterus, cervix and vagina; in males, the testis, vas deferens, epididymis, seminal vesicles, prostate, and some accessory sex glands.

parasympathetic nerve fibers A division of the autonomic nervous system composed of those nerve fibers innervating internal organs whose preganglionic nerve cell bodies are located in the medulla (dorsal motor nucleus of the vagus nerve) or in the sacral (S2–S4) spinal cord segments (humans) or L6/S1 spinal cord segments (rats) and that synapse with postganglionic neurons in ganglia located adjacent to or in the walls of the target organs. The postganglionic fibers then travel only a short distance to innervate the peripheral organs, often carrying out their functions there using acetylcholine, although many other neuroactive agents, including purines, nitric oxide, and various peptides, are also involved.

pelvic ganglion (pelvic plexus) An aggregation of nerve cell bodies located adjacent to the cervix in females or the prostate in males that receives preganglionic fibers from the pelvic and hypogastric nerves and gives rise to postganglionic fibers supplying many of the internal pelvic organs. In males, it is also called the *major pelvic ganglion*, whereas in females, in which it is smaller, it is called the *paracervical ganglion, uterine cervical ganglion,* or *Frankenhäuser's ganglion.*

perineum The area between the thighs from coccyx to pubis, below the pelvic diaphragm. Thus **perineal**.

plasticity or **neural plasticity** Long-term changes in peripheral or central innervation density, synaptic efficacy, or phenotypic action of neurons (e.g., neurotransmitter and neuromodulator characteristics) that can be induced by pathophysiology and other means.

sympathetic nerve fibers A division of the autonomic nervous system composed of those nerve fibers innervating internal organs whose preganglionic nerve cell bodies are located in thoracolumbar spinal cord segments and synapse with postganglionic neurons in sympathetic chain ganglia or various para- and prevertebral ganglia. The postganglionic fibers then travel a relatively long distance to innervate the peripheral organs (long adrenergic neurons). However, exceptions to this pattern exist for some of the fibers innervating pelvic viscera; these fibers synapse on postganglionic neurons in parasympathetic ganglia located closer to the target organ (short adrenergic neurons). The sympathetic postganglionic fibers often carry out their functions on target organs using noradrenaline, although many other neuroactive agents, as for the parasympathetic nerve fibers (see above), are also involved.

I. LIMITING THE SUBJECT MATTER

The autonomic nervous system, as generally defined, comprises all afferent and efferent nerve fibers that exit the spinal cord and brain stem to supply internal organs. "Reproduction" is defined by the *American Heritage Dictionary of the English Language* (3rd ed., 1992) as "the process by which organisms generate others of the same kind." Thus, at first glance, the title of this article might seem straightforward, to wit, how do fibers that supply internal organs affect the processes by which organisms generate others of the same kind?

It takes little further reflection, however, to realize the difficulty of addressing this huge question sensibly. This frustration holds true even if the question is severely narrowed by restricting the definition of reproduction to "successful fertilization in mammals" and by focusing attention only on those afferent and efferent sympathetic, parasympathetic, and somatic nerve fibers that supply mammalian internal and external reproductive organs.

II. REQUIREMENTS FOR SUCCESSFUL FERTILIZATION

Successful fertilization has the following minimal requirements: (i) the production by males of a sufficient quantity of healthy sperm and a fully operational sperm-delivery system that protects the sperm's viability; (ii) the well-timed production by females of a sufficient quantity of healthy ova and a fully functional delivery system that protects the ova(s)' viability on route from ovary to uterus to meet incoming sperm; (iii) maintenance by the female of an effective mechanism for ensuring the viability and directional mobility of newly arriving sperm entering her reproductive tract; and (iv) a fully operational system in females for securing, nourishing, and protecting the newly formed zygote and developing embryo within the uterus. Of course, there is also the requirement for successful coitus; that is, proper sexual union between male and female that enables their gametes to form a zygote.

III. SYNOPSIS OF AUTONOMIC INNERVATION OF INTERNAL AND EXTERNAL REPRODUCTIVE ORGANS

The organization of the mixture of autonomic efferent sympathetic and parasympathetic fibers as well as afferent fibers that supply the mammalian internal reproductive organs is illustrated in Fig. 1 for the female rat and in Fig. 2 for males. Innervation of the external organs is not shown, but it is described later.

For the ovary and testis, the efferent innervation is by way of sympathetic fibers (derived from the intermesenteric and renal plexus, the superior hypogastric plexus, and the presacral hypogastric nerve) in the superior ovarian nerve (ovary), ovarian plexus nerve (ovary), and testicular nerves (testis) and possibly parasympathetic fibers in the vagus nerve (both sexes). For the other internal reproductive organs in both males and females, efferent innervation is by way of fibers in the pelvic and hypogastric nerves that pass into or through the pelvic ganglion to their target organs. Whereas the pelvic nerve is composed entirely of preganglionic parasympathetic nerve fibers, most of which synapse with postganglionic neurons mainly in the pelvic ganglion (with some coursing back up to synapse in the inferior mesenteric ganglion), the hypogastric nerve is composed of both postganglionic fibers that have synapsed with neurons in the inferior mesenteric ganglion and preganglionic fibers that continue on to synapse with postganglionic neurons in the pelvic ganglion. The external genitalia are supplied by autonomic postganglionic efferent sympathetic fibers that originate in the sympathetic chain but travel to the organs via branches of the somatic pudendal nerve as well as by sympathetic and parasympathetic fibers that travel in the hypogastric and pelvic nerves to and through the pelvic ganglia via the cavernous (or penile) nerve. Recent evidence suggests that the vagus nerve innervates, in addition to ovary and testis, most or all of the other internal reproductive organs as well as some components of the external genitalia.

The preganglionic neuronal cell bodies for of all of these efferent fibers except the vagus are located within the spinal cord; preganglionic neuronal cell bodies of the vagus are located in the caudal medulla.

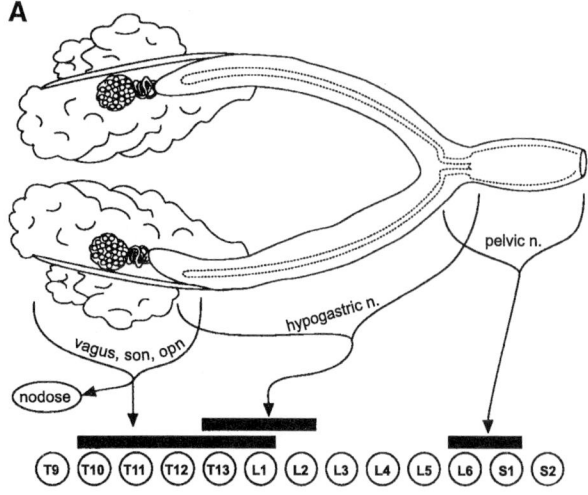

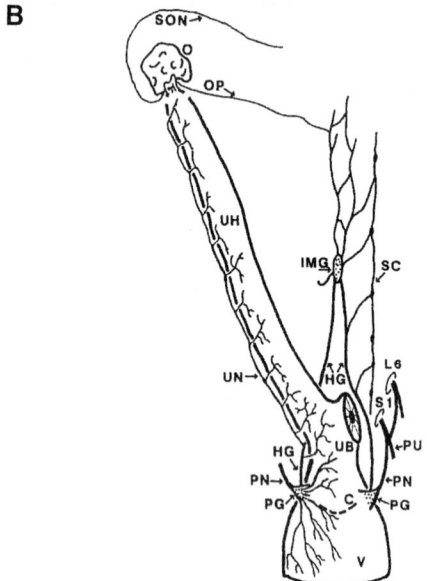

FIGURE 1 Innervation of internal reproductive organs of the female rat. (A) Pattern of afferent innervation of the internal reproductive tract. T10–S2 indicate dorsal roots and dorsal root ganglia of spinal segments T10–S2 (adapted from Berkley and Hubscher, 1995). (B) The general innervation pattern of all the nerves except the vagus. L6 and S1 indicate spinal nerves L6 and S1, respectively, emerging from the spinal cord (reproduced with permission from Papka and Traurig, 1993). Abbreviations: C, uterine cervix; HG, hypogastric nerve; IMG, inferior mesenteric ganglion; O, ovary; OP, ovarian plexus nerve; PG, pelvic ganglion; PN, pelvic nerve; PU, pudendal nerve; SC, spinal cord; SON, superior ovarian nerve; UB, uterine body; UN, uterine horn; V, vagina.

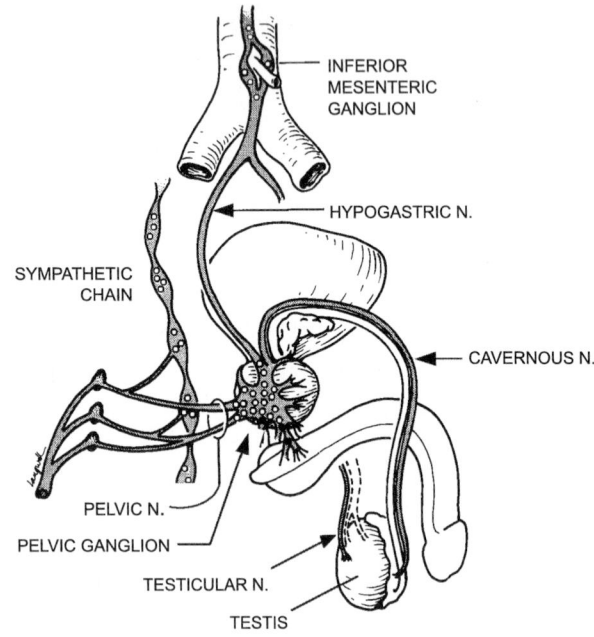

FIGURE 2 General innervation pattern of reproductive organs in the male (reproduced and modified with permission from Dail, 1993). See text for details. Of value is to compare this figure with a more detailed one showing the innervation pattern of the bladder and urethra (see Fig. 1 under The Pelvic Nerve).

All the nerves described previously also contain a large component of afferent fibers that convey information about events occurring in the various reproductive organs to the spinal cord and brain. The information is conveyed both by fast electrical action potentials and by slower axonal transport processes.

Of importance here is that substantial and significant modulatory regulation and coordination of the efferent functions of these nerves is exerted not only by intraspinal mechanisms but also by influences arising in higher brain structures, particularly the hypothalamus, acting by way of multiple descending routes to affect the activity of preganglionic neurons in both the medulla and the spinal cord. Significantly, additional coordination occurs directly within the peripheral organs themselves, in large part orchestrated by hormonal controls via hypothalamic–pituitary mechanisms.

IV. EFFECTS OF DENERVATION OF REPRODUCTIVE ORGANS ON SUCCESSFUL FERTILIZATION

The ways by which the nerve fibers act together to affect the various reproductive functions of internal and external reproductive organs in both sexes is complex and remains poorly understood. Much of what is currently known is detailed in other chapters of this encyclopedia or in the references cited in this chapter, particularly McKenna and Larson (1997). In general, the effect of denervation or of injury or disease of any of the peripheral autonomic nerves supplying the reproductive organs appears to have severe repercussions for the male, particularly in various components concerning successful delivery of viable sperm to the female (e.g., sperm transport from testis, erection, and ejaculation). Effects on sperm production are unclear. In the female, effects of denervation, injury, or disease at first appear less severe than they are in the males since coitus, conception, pregnancy, and parturition can still occur when much of the innervation is removed. However, disruption of the nerves supplying the ovary reduces follicular maturation and ovarian cyclicity; disruption of the hypogastric nerves affects secretions of the cervix important for viable sperm transport and parturition; disruption of the pelvic nerve can have profound effects on various reproductive functions associated with cervix stimulation, such as reflex controls on other pelvic organs during coitus (interfering with fertilization) and parturition (producing dystocia) and the timing of ovulation (affecting fertility); and disruption of the pelvic and/or pudendal nerves can give rise to severe pain, reduce vaginal/perineal sensations, or affect coordination of perineal striated and smooth muscles, all of which would interfere with many aspects of bodily sensation and coordination important during coitus and parturition.

V. COITUS AS A SOCIAL BEHAVIOR

Although all the previous considerations are indeed important for successful fertilization, what often goes unappreciated in such considerations is the fact that successful fertilization (unless "assisted") requires successful coitus. The act of coitus is obviously a form of social behavior, involving the initiation and consummation of active cooperation between a female and a male. Although research in this arena is currently on the increase, little is yet known about how afferent and efferent fibers in the pelvic, hypogastric, pudendal, and vagus nerves described previously cooperate with those of the somatosensory, somatomotor, olfactory, visual, auditory, and gustatory systems in the organization and control of social behaviors that constitute coitus. An important poorly understood component in this process involves controls descending from the hypothalamus and other higher brain structures to act on preganglionic neurons and how those controls may be influenced by these other systems. Another important, and perhaps even less well understood, component involves the significance of sensory afferent fibers (which in fact comprise a huge proportion in all the nerves). A factor significant to both of these realms, and until recently also largely uncharted, involves the plastic capacity of peripheral and central nervous system structures for long-term changes in their interactions in response to previous circumstances.

Examples abound that reflect each of these poorly understood components; two examples in humans—one in males and the other in females—are discussed. In men, perceived erectile or ejaculatory dysfunction can give rise to problems in the initiation of coitus that affect sexual function by positive feedback-type effects that can erode these functions still further, resulting in severe consequences for consummation of coitus in a manner necessary for successful fertilization. Testimony to this situation is the enormous demand for improved development of pharmaceutical products or mechanical devices (none of which currently are totally effective or acceptable) to provide self control for enhancing or diminishing erectile and ejaculatory processes peripherally, thereby circumventing presumably centrally induced problems in otherwise demonstrated fully functional autonomic nerves supplying the internal and external genitalia. In women, any disease of or trauma to the vagina, however major or minor, that affects activity in pelvic nerve fibers supplying it has two potential effects. One is a reduction in the generalized bodily hypoalgesia that can be induced by vaginal stimulation. Another is the development of vaginal allodynia and hyperalgesia that can continue, maintained by sensitization in the central nervous system, long after the initial injury or pathophysiology has disappeared. While sexual motivation (i.e., the desire for initiating coitus) may or may not be severely affected by either of these consequences, consummation of coitus can be severely disrupted, including poor coordination of bodily position, reduced duration of intromission and consequent disruption of the male's erectile and ejaculatory functions, and so forth.

VI. FURTHER AND FUTURE ISSUES

The previous discussion illustrates only a few of the complexities associated with considerations of how autonomic nerves supplying reproductive organs affect successful fertilization. Expanding to the full title of this chapter clearly gives rise to even further complexities. Regarding the autonomic nervous system, it is well-known that the autonomic nervous system is composed not only of those afferent and efferent fibers traveling in the pelvic, hypogastric, pudendal, and vagus nerves to supply the reproductive organs but also of (i) fibers in the same nerves traveling to supply other pelvic organs (bladder, urethra, colon, and anorectum); (ii) fibers in other autonomic nerves throughout the body that supply all other internal organs; (iii) fibers in autonomic nerves associated with the hypothalamus that supply the pituitary, which in turn has an enormous impact on reproductive and other structures throughout the body by hormonal mechanisms and (iv) fibers in nerves usually associated with the somatic division of the nervous system that supply blood vessels in skin and muscles throughout the body. Little is known about how these other fibers interact with those directly associated with the reproductive organs to affect successful fertilization, let alone reproduction in its full sense. Obviously, these fibers must be important because even successful fertilization requires coordination of the activities of other pelvic structures (e.g., changes in bladder and gastrointestinal activities during coitus and the involvement of urethral structures in orgasm, i.e., in what has been recently dubbed the "urethral–genital reflex") and other physiological functions (e.g., respiration, cardiovascular activity, and regional blood pressure regulation during coitus).

Regarding reproduction, if we consider the entire set of processes by which "organisms generate others of the same kind," some of these processes, as described previously, clearly require controls by autonomic nerve fibers supplying reproductive organs not only to effect successful fertilization and implantation but also to effect care of the developing embryo and parturition. Happily, we are beginning to know a great deal about each of these peripheral processes in the individual male or female. However, we know very little about neural processes important for "generating others of the same kind" that concern how social dynamics between males and females either affect or are affected by the actions of the autonomic nerves carrying out their effects on their particular peripheral target structures.

Thus, it is hoped that the next edition of this encyclopedia will provide some answers on the involvement in reproduction of (i) autonomic systems other than those supplying the reproductive organs; (ii) the effects of the social dynamics of reproductive behaviors on neural activity, and, reciprocally, the neural effects on those social dynamics; and (iii) mechanisms of integration between peripheral, spinal cord, brain, pituitary, and hormonal control processes in all these arenas.

See Also the Following Articles

OVARIAN INNERVATION; PELVIC NERVE; SEXUAL DYSFUNCTION

Bibliography

Berkley, K. J., and Hubscher, C. H. (1995). Visceral and somatic sensory tracks through the neuroaxis and their relation to pain: Lessons from the rat female reproductive system. In *Visceral Pain, Progress in Pain Research and Management* (G. F. Gebhart, Ed.), Vol. 5, pp. 195–216. IASP Press, Seattle.

Bonica, J. J. (1990). General considerations of pain in the pelvis and perineum. In *The Management of Pain*(J. J. Bonica, Ed.), 2nd ed., pp. 1283–1312. Lea & Febiger, Philadelphia.

Clarke, I. J. (1996). Effector mechanisms of the hypothalamus that regulate the anterior pituitary gland. In *Autonomic–Endocrine Interactions* (K. Unsicker, Ed.), pp. 45–88. Harwood Academic, Amsterdam.

Clemens, L. G., and Weaver, D. R. (1985). The role of gonadal hormones in the activation of feminine sexual behavior. In *Handbook of Behavioral Neurobiology* (N. Adler, D. Pfaff, and R. W. Goy, Eds.), Vol. 7, Reproduction, pp. 183–227. Plenum, New York.

Dail, W. G. (1993). Autonomic innervation of male reproductive genitalia. In *Nervous Control of the Urogenital System* (C. A. Maggi, Ed.), pp. 69–101. Harwood Academic, Amsterdam.

De Groat, W. C., and Booth, A. M. (1993). Neural control of penile erection. In *Nervous Control of the Urogenital System* (C. A. Maggi, Ed.), pp. 467–524. Harwood Academic, Amsterdam.

Komisaruk, B. R., and Whipple, B. (1995). The suppression of pain by genital stimulation in females. *Annu. Rev. Sex Res.* **6**, 151–186.

McKenna, K. E., and Marson, L. (1997). Spinal and brain stem control of sexual function. In *Central Nervous Control of Autonomic Function* (D. Jordan, Ed.), pp. 151–187. Harwood Academic, Amsterdam.

Newton, B. W., and Hammill, R. W. (1996). Sexual differentiation of the autonomic nervous system. In *Autonomic–Endocrine Interactions* (K. Unsicker, Ed.), pp. 425–463. Harwood Academic, Amsterdam.

Oakley, A. (1992). Measuring the effectiveness of psychosocial interventions in pregnancy. *Int. J. Technol. Assess. Health Care* **8**(Suppl. 1), 129–138.

Papka, R. E., and Traurig, H. H. (1993). Autonomic efferent and visceral sensory innervation of the female reproductive system: Special reference to neurochemical markers in nerves and ganglionic connections. In *Nervous Control of the Urogenital System* (C. A. Maggi, Ed.), pp. 423–466. Harwood Academic, Amsterdam.

Pfaff, D. W., and Schwartz-Giblin, S. (1988). Cellular mechanisms of female reproductive behaviors. In *The Physiology of Reproduction* (E. Knobil and J. Neill, Eds.), pp. 1487–1568. Raven Press, New York.

Rose, J. D. (1990). Forebrain influences on brain stem and spinal mechanisms of copulatory behavior: A current perspective on Frank Beach's contribution. *Neurosci. Biobehav. Rev.* **14**, 207–215.

Sachs, B., and Meisel, R. L. (1988). The physiology of male sexual behavior. In *The Physiology of Reproduction* (E. Knobil and J. Neill, Eds.), pp. 1393–1485. Raven Press, New York.

Stjernquist, M. (1996). Innervation of ovarian and testicular endocrine cells. In *Autonomic–Endocrine Interactions* (K. Unsicker, Ed.), pp. 231–256. Harwood Academic, Amsterdam.

Vollrath, L. (1996). Innervation of the pituitary. In *Autonomic–Endocrine Interactions* (K. Unsicker, Ed.), pp. 89–127. Harwood Academic, Amsterdam.

Avian Reproduction, Overview

Tony D. Williams

Simon Fraser University

I. Male Reproductive System
II. Female Reproductive System and Egg Formation
III. Egg and Embryo Development
IV. Endocrine Control of Reproduction
V. Seasonal Breeding
VI. Sexual Maturation
VII. Costs of Reproduction and Reproductive Variability

GLOSSARY

altricial Describing nestlings that are born blind and naked and remain in the nest for a variable period after hatching during which they are dependent on their parents for food and thermoregulation.

atresia A process by which the final stage of oocyte maturation (rapid yolk development) is halted and the yolk material in the developing follicle is reabsorbed rather than the yolk being ovulated.

costs of reproduction Trade-offs between investment in the current breeding attempt and future fecundity and survival or between investment in different stages of a single breeding attempt (e.g., egg production versus chick rearing).

deferred sexual maturation Delaying of the first breeding attempt for 1 or more years, and up to 12 years, after hatching.

intraspecific variability Marked interindividual or between-female variation in reproductive traits (egg size, clutch size, and timing of laying) typical of all or most wild birds, often repeatable within individual females.

photorefractoriness A photoperiod-dependent mechanism by which birds spontaneously regress their gonads and terminate their breeding attempt; this ensures that young are not hatched and reared at an inappropriate time late in the breeding season.

precocial Describing young that leave the nest shortly after hatching and which are capable of feeding and thermoregulating largely independently; some parental care, such as brooding and guarding, may still occur.

rapid yolk development The final stage of oocyte maturation during which large amounts of yolk material (very low-density lipoprotein and vitellogenin) are sequestered from the blood and deposited in the developing follicle.

very low-density lipoprotein One of the two main yolk precursors in birds and the main source of yolk lipids; synthesized in the liver and transported to the developing follicle in the blood.

vitellogenin The second main yolk precursor in birds and the main source of yolk proteins; also synthesized in the liver and transported to the developing follicle in the blood.

The gonads of birds consist of paired testes in the male and usually a single ovary in the female. These sex organs are responsible for production of gametes and synthesis and secretion of sex steroids which control expression of secondary sexual characteristics and reproductive behavior. Following fertilization, the female completes formation of the complex egg structure and eggs are laid, typically in discrete clutches. All birds reproduce by laying eggs which are incubated externally, usually by the bird's own body heat, in specially constructed nest structures. Eggs contain all the nutrients and energy to meet the embryo's requirements through development from blastoderm to hatchling. Reproduction is a cyclic phenomenon and the majority of avian species breed seasonally, laying eggs and rearing young during a restricted part of the year (often spring and early summer). Seasonal reproduction, as well as sexual maturation, is regulated by the hypothalamic–pituitary–gonadal axis, which in turn is integrated with and synchronized by environmental cues, in particular the annual cycle of day length (photoperiod). The reproductive system is therefore activated and "switched off" annually in most birds, with the gonads undergoing large variation in size and functional activity. Sexual maturation occurs at 1 year of age in many birds, although other species show deferred sexual maturation that delays their first breeding attempt for up to 6–12 years. Physiological mechanisms controlling sexual maturation are poorly understood, though in some cases, at least, this also seems to be a photoperiod-dependent process. Most of our knowledge of avian reproductive physiology has come from a handful of domesticated species which have been selected for near-continuous laying, reduced reproductive variability, and maximum reproductive output. Reproduction in these species is totally unlike that of the ≈9000 wild bird species which show marked interspecific and intraspecific variation in all reproductive traits, the physiological basis of which remains largely unknown.

I. MALE REPRODUCTIVE SYSTEM

The paired testes of birds are internal and are suspended from the dorsal wall of the body cavity at the cranial end of the kidneys (Fig. 1). Testis size varies with sexual activity, with testes weighing as little as 0.005% of body mass in immature or nonbreeding males but increasing 500- to 1000-fold during breeding. For example, in the chicken paired testis mass increases from 0.05 g at 1 month to 30 g at 18 months. Each testis consists of thousands of coiled seminiferous tubules which contain two main types of cells: germ cells, responsible for spermatogenesis, and Sertoli cells, which assist in sperm maturation. Lying between the seminiferous tubules, interstitial or Leydig cells secrete steroid hormones (androgens) which control sexual behavior (song and copulation) and development of some, but not

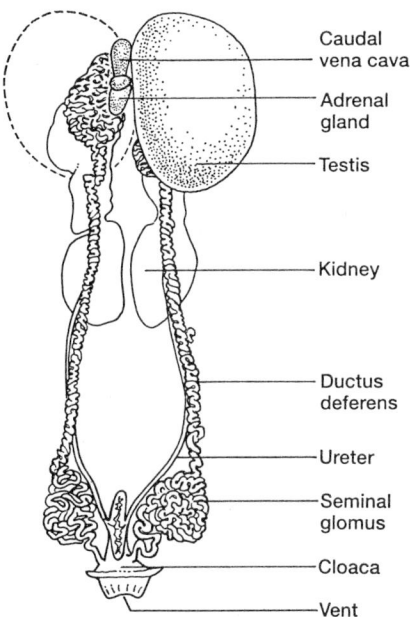

FIGURE 1 Avian male reproductive system (adapted with permission from Brooke and Birkhead, 1991; © Cambridge University Press).

all, secondary sexual characteristics (plumage and bill color). The seminiferous tubules join to form a long epididymis and ductus deferens, along which the mature sperm are carried in seminal fluid to the cloaca. In passerines, sperm can be stored in the seminal glomus, the coiled caudal end of the ductus deferens, where the temperature is ≈4°C lower than the core body temperature. The glomus is then everted into the cloaca during copulation. In most species copulation simply involves close cloacal contact between the male and female. Some ratities (e.g., ostrich), ducks, and geese have a well-developed and erectile phallus or pseudopenis and some passerines and the domestic fowl possess a nonintromittent-type phallus.

II. FEMALE REPRODUCTIVE SYSTEM AND EGG FORMATION

A. Ovary

Two bilaterally symmetrical gonads and oviducts develop in the avian embryo, and two fully developed ovaries are common in birds of prey (Falconiformes) and the kiwi (*Apteryx*). In nearly all other species only the ovary and oviduct on the left side are functional in adult life (≤5% of individuals may have two developed ovaries in these species). The ovary is suspended from the dorsal wall of the body cavity near the top of the kidneys (Fig. 2) and consists of a large number of oocytes embedded in a sparse stroma of connective tissue. Oogenesis is terminated at hatching, at which time the ovary of the chick embryo contains about 480,000 primary oocytes. In contrast to mammals and most other vertebrates, the female is the heterogametic sex in birds (ZW) and thus determines offspring sex. In domesticated species, only 200–500 oocytes will eventually mature and be ovulated, and in wild species as few as 1–20 oocytes mature, depending on life span. Oocyte de-

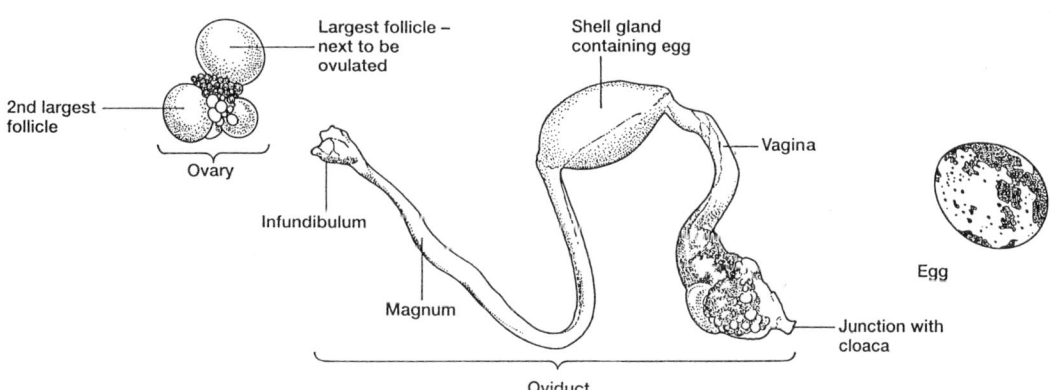

FIGURE 2 Avian female reproductive system (adapted with permission from Brooke and Birkhead, 1991; © Cambridge University Press).

velopment proceeds in two stages: an initial slow stage lasting 60 days to several years and a rapid (yolk development) stage lasting 4–16 days, depending on species. During breeding one or more oocytes sequentially enters the rapid yolk development (RYD) stage, usually at approximately 24-hr intervals, becoming surrounded by a protective follicle. What determines which follicles will be recruited from the large pool of white follicles for the final maturation stage is unknown and remains one of the major unanswered questions in avian reproductive physiology. The ovarian follicle wall comprises (i) the oocyte plasma membrane, (ii) a perivitelline membrane, (iii) a granulosa cell layer, (iv) a basal lamina, and (v) theca interna and externa layers. It is highly vascularized, except for the stigma or site of follicle rupture, and is innervated by cholinergic and adrenergic fibers. Granulosa and thecal cells synthesize and secrete hormones required for follicular maturation and control of female sexual behavior. Ovary size varies markedly with reproductive status and with the number of follicles entering the final rapid growth phase of yolk development (maturation). In the laying hen four or five developing large yellow follicles form an ovarian follicle hierarchy, with the largest 40 mm in diameter, as well as thousands of smaller undeveloped yellow and white follicles (0.5–6.0 mm). In the non-breeding female zebra finch, the ovary weighs 20 mg but increases in size over 5 or 6 days to 250 mg in breeding females. This increase is due to uptake of yolk materials from the blood: vitellogenin and very low-density lipoproteins (VLDLs), which are synthesized in the liver under estrogenic control and transported to the ovary in the blood. The duration of the yolk development phase is variable, depending on egg size: It lasts 3 or 4 days per follicle in small passerines, 6–10 days in the domestic hen, and 14–16 days in large seabirds (Procellariformes). The fully developed yolk is ovulated at maturity by rupturing of the follicle wall, and the yolk passes into the oviduct where completion of egg formation occurs. The number of ovulations in a sequence determines clutch size. Postovulatory follicles are metabolically active and persist for 8–10 days in the hen, but there is no structure in birds analogous to the corpus luteum of mammals. Follicles which start to mature but which are not ovulated become atretic and the oocyte and yolk material are reabsorbed. The physiological basis of atresia is still not well understood, though it appears to involve apoptosis (programmed cell death). The role of follicular atresia in determining clutch size and in mediating ovarian regression at the end of the breeding season is also poorly known.

B. Oviduct

The oviduct extends from the ovary to the cloaca and comprises five morphologically and functionally distinct regions (Fig. 2), each having a specific role in the final part of egg formation: the laying down of albumen, shell, and pigmentation around the yolk. Oviduct size again varies with sexual activity, increasing from <50 mg to 650 mg over 5 days at the onset of egg production in the female zebra finch. In the laying hen, the oviduct is 40–80 cm in length and weighs about 40 g. Fertilization of the oocyte occurs within an hour of ovulation in the first part of the oviduct, the infundibulum. Females of most, if not all, species can store viable sperm in storage tubules located near the junction of the shell gland and the vagina. Fertilization can therefore occur some time after mating and the interval may reach 2 months in some seabirds and the turkey. Females also appear to have some control over fertilization, in the case of multiple matings, through differential transport and selection of sperm within the reproductive tract. The fertilized ovum passes into the largest part of the oviduct, the magnum, where most of the albumen is laid down over a 2- or 3-hr period (some albumen may be laid down by the infundibulum). Soft shell membranes are added in the isthmus (1 hr) and the egg then moves to the shell gland where it is retained for 18–24 hr. Here, salts and water are taken up by the albumen (a process known as "plumping") and the calcium shell and pigmentation are added. Finally, the egg passes via the vagina to the cloaca and is eventually laid. Egg transport within the oviduct is brought about by peristaltic contractions of longitudinal circular muscles, perhaps aided by ciliary action, and is hormonally mediated by prostaglandin F2α and arginine vasotocin.

III. EGG AND EMBRYO DEVELOPMENT

Eggs consist of a germinal disc (blastoderm if fertilized), yolk, yolk membranes, albumen, and shell (Fig. 3). The relative proportions of each component vary in different taxa, in particular in relation to the mode of development of the chick. In species with altricial chicks, yolk, albumen, and shell comprise 22, 70, and 8% of total egg mass, respectively. In contrast, in precocial species, which hatch in a more advanced state of development, corresponding values are 36, 54, and 10%.

A. Yolk

Most of the yolk is yellow yolk, which is more or less heavily pigmented (depending on the bird's diet), but about 3% of the total yolk volume is paler, more fluid white yolk. White yolk occupies a small region under the blastodisc (the nucleus of Pander) and extends to the center of the yolk or latebra via a thin channel. Whole yolk contains three main components: (i) yolk granules consisting largely of two proteins, lipovitellin and phosvitin, which are synthesized as the precursor molecule vitellogenin in the liver. Vitellogenin is transported in the blood, taken up by the developing follicle by receptor-mediated endocytosis, and enzymatically cleaved into lipovitellin and phosvitin following uptake (Fig. 4); (ii) livetins, which are essentially serum blood proteins, mainly serum albumen; and (iii) low-density lipoproteins, synthesized as yolk-precursor VLDLs (mainly triglycerides) in the liver and taken up from the blood by endocytosis. In the hen's egg, granules, livetin, and VLDL form 24, 16, and 60% of the total yolk dry mass, respectively. Yolk is laid down in a stratified manner and distinct yolk rings can be discerned, probably caused by carotenoid pigments. These light and dark layers are laid down diurnally and allow determination of the duration of yolk formation (RYD). Yolk provides all of the energy and most of the nutrients for embryo development and, because many chicks retain yolk in a yolk sac at hatching, it also feeds the newly hatched chick for up to 6–8 days following hatching.

B. Albumen

Albumen consists of protein and glycoprotein, but it contains 88% water with only trace amounts of lipid and carbohydrate (<1%). Protein is mainly

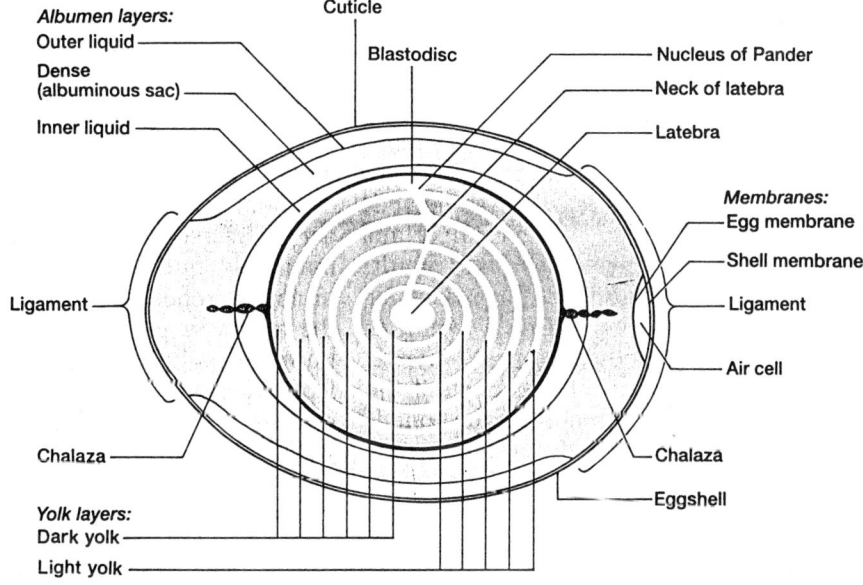

FIGURE 3 Structure of a newly laid chicken's egg (adapted with permission from Brooke and Birkhead, 1991; © Cambridge University Press).

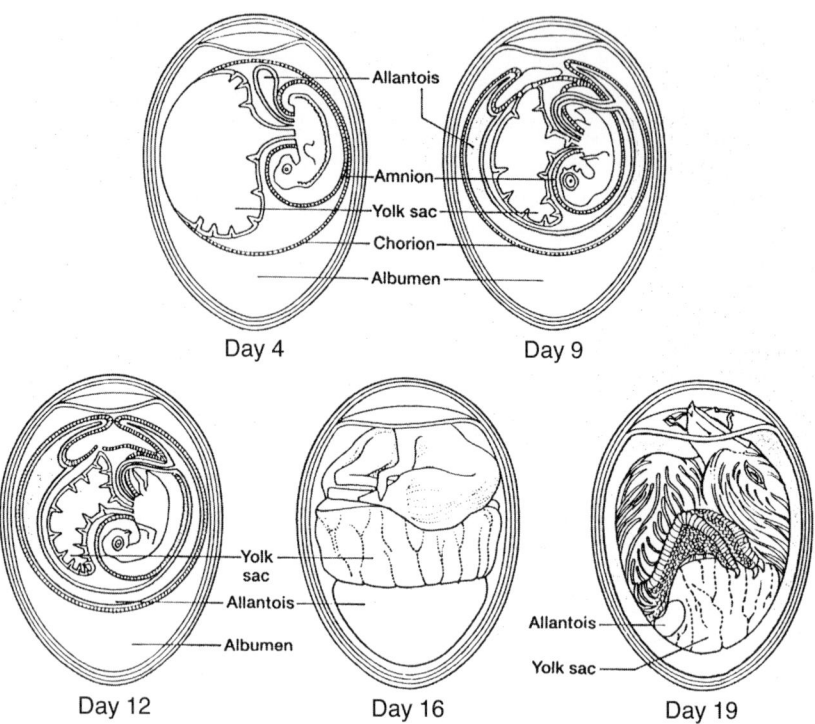

FIGURE 4 Yolk formation during the final stage of oocyte maturation. (Adapted with permission from Brooke and Birkhead, 1991. © Cambridge University Press.)

ovalbumin (54%), ovotransferrin (12%), and ovomucoid (11%). These form four distinct layers of albumen in the fully formed egg (Fig. 3): (i) the chalaziferous layer, consisting of two dense, twisted strands of protein fibers which extend to the ends of the egg and suspend the yolk in the center of the egg; (ii) an inner liquid layer; (iii) a dense or thick layer; and (iv) an outer, thin or fluid layer. Unlike yolk, which is transported to the developing ovary from the liver, albumen is synthesized *de novo* within the oviduct. Albumen is important for water balance in the developing embryo, and it provides physical protection, has antibacterial properties, and is an important source of protein for structural growth of the embryo.

C. Shell

The shell has a complex structure with four main components: an inner membrane (referred to as the egg membrane) and outer (shell) membrane, the main calcareous portion or testa, and a thin, external waxy cuticle. The testa comprises a series of layers of protein and mucopolysaccharides, the organic matrix, in which crystalline calcareous deposits are laid down. It provides the main strength of the egg with shell thickness generally being proportional to the size of the egg. The shell surface is penetrated by a large number of simple funnel-shaped pores that allow for exchange of gases and water vapor during embryo development.

D. Embryo Development

The germinal disc consists of two cell layers in the early embryo, the ectoderm and endoderm, separated by a fluid-filled space. Within a few hours ectoderm cells invade the space and form a third cell layer, the mesoderm. The ectoderm gives rise to the skin and nervous system of the adult bird, the mesoderm forms the muscles, heart, skeleton, and kidneys, and the endoderm forms the stomach, liver, and lungs. To meet the needs of nutrient utilization, gas exchange, and excretion, the growing embryo develops four membranes or auxiliary structures. First, the yolk sac membrane grows out from the blastoderm and fully encloses the yolk by Day 9 of incubation in the hen. This contains a network of capillaries that absorb nutrients from the yolk and transport them back to the embryo. The yolk sac also interacts with the yolk, mobilizing amino acids and breaking down the yolk structure through enzyme action. Before hatching, the yolk sac is absorbed into the body cavity, attached to the embryonic small intestine, and the yolk remaining is used by the chick in the first few days after hatching. Second, the amniotic membrane encloses the embryo forming a flexible, fluid-filled sac, the amnion, providing physical protection for the embryo. Third, the allantois membrane starts growing at the same time as the amnion and surrounds it. It receives and accumulates waste products from the kidney, is involved in transport of ions and water, and is eventually left behind in the egg at hatching. Finally, the allantois fuses with the chorionic membrane to form the chorionallantois, which, in hens, surrounds the contents of the egg on the inside of the inner shell membrane by Days 11 or 12 of incubation. The chorion, like the yolk sac membrane, is well supplied with blood vessels, and its outer surface is close to the shell and exposed to the atmosphere and functions in gas exchange during embryo development. It is also responsible for the absorption of calcium from the shell for skeletal growth.

In the hen, by Day 4 of incubation the embryo has formed the rudiments of all its organs, including two pairs of limb buds (Fig. 5), and by Day 10 it is clearly recognizable as a bird with beak, wings, feet, and the beginnings of feathers. The degree of development of each organ prior to hatching depends on whether the chick is altricial (less developed) or precocial (more developed). Just before hatching the embryo's metabolism is too high for carbon dioxide to be lost through the shell pores. The chick thrusts its beak into the airspace inside the shell and air breathing begins (Fig, 5, Day 19). The lungs expand and pulmonary circulation is initiated while the chorionallantois circulation stops. Shortly afterwards the chick breaks through the shell using its egg tooth on the tip of the bill and is ready to hatch.

IV. ENDOCRINE CONTROL OF REPRODUCTION

Sexual maturation and gametogenesis, during both puberty and repeated seasonal breeding, are controlled by the hypothalamic–pituitary–gonadal axis. Gonadotropin-releasing-hormone (GnRH) is released from neurosecretory cells in the median eminence of the hypothalamus, often in response to environmental stimuli. In the chicken two GnRHs have been identified—cGnRH I and cGnRH II—and the occurrence of these two forms of GnRH may be widespread in birds. Current evidence suggests that cGnRH I is the primary releasing hormone involved in regulating pituitary function. GnRH is released into the portal blood system and travels to the anterior pituitary gland (adenohypophysis), where it stimulates synthesis and secretion of two gonadotropin hormones which are released into the systemic circulation: follicle-stimulating hormone (FSH) and luteinizing hormone (LH). In males, FSH in turn stimulates testicular maturation and spermatogenesis in Sertoli cells and LH stimulates steroidogenesis and release of the androgens, testosterone and 5α-dihydrotestosterone, by Leydig cells. In females, FSH regulates oogenesis, yolk uptake, and follicular development. A three-cell model of ovarian steroidogenesis has been proposed for preovulatory follicles (cf. a two-cell model in mammals): progesterone produced by the granulosa cell layer is converted to androgens by cells of the theca interna, and these are subsequently metabolized to estrogens by theca externa cells. Intragonadal hormones (e.g., inhibin

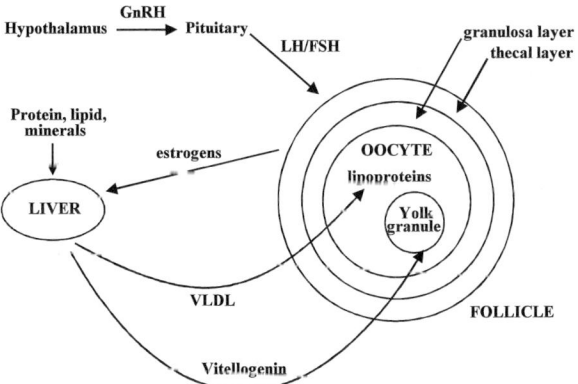

FIGURE 5 The avian embryo and its development from laying to hatching.

and follistatin) also appear to be involved in follicle maturation, as well as regulating FSH secretion in both sexes, through negative feedback on the pituitary. Gonadal steroids have many roles, affecting both physiology and behaviors such as aggression, territorial behavior, nest building, copulation behavior, and development of secondary sexual characteristics (plumage and bill color). Estradiol also stimulates the liver to produce the yolk precursors vitellogenin and VLDL and, together with progesterone, stimulates oviduct development and albumen secretion. Prolactin, which is also released by the pituitary in response to the releasing factor vasointestinal peptide, plays a major role in initiation and maintenance of parental care (incubation and chick rearing).

V. SEASONAL BREEDING

A. Timing of Breeding

In almost all nondomesticated birds breeding is restricted to a particular time of year coincident with a seasonal improvement in environmental conditions required for successful rearing of offspring: increased food availability, higher temperatures or rainfall, etc. (Fig. 6). Annual cycles of reproduction are most common, with birds in temperate and higher latitudes laying eggs in spring and rearing chicks over the summer. In contrast, in the tropics, in particular, some species can lay in any month of the year, although individual birds do not lay continuously and many still show approximate annual cycles. In other species, such as the zebra finch, breeding is more opportunistic and is related to short and unpredictable periods of improved environmental conditions. A few species have regular breeding cycles of <1 year; for example, the sooty tern has a breeding periodicity of 9.6 months on the Ascencion Islands (7.5°S). Timing of breeding is set by two types of factors. In a seasonal environment birds which can rear chicks when food is most available will breed more successfully. Food availability thus sets a selective or evolutionary value on timing of breeding, explaining why birds have evolved to breed when they do, and is called an ultimate factor. However, before young birds hatch, their parents must undergo extensive physiological and behavioral "preparation" for breeding, e.g., maturation of the reproductive system, finding a nest site and mate, and laying and incubating eggs. These events can take 6–8 weeks to complete and must be initiated well in advance of any increase in food availability (Fig. 6). Therefore, birds must be able to predict the onset of the breeding season well beforehand and they do so using information from environmental cues or proximate factors. Proximate factors therefore explain how birds time their breeding season. The main proximate factor used by birds to time their breeding season, outside of the tropics, is the annual cycle of day length or photoperiod (Fig. 7). Day length varies absolutely predictably each year and increasing day length in spring provides the initial predictive information for timing of breeding. Because there is usually small-scale, local variation in environmental conditions from year to year, other factors provide essential supplemental information allowing birds to fine-tune the onset of egg laying itself, including temperature, food availability, local weather (e.g., storms), and social factors. In females in particular, social factors such as the presence of a mate or nest

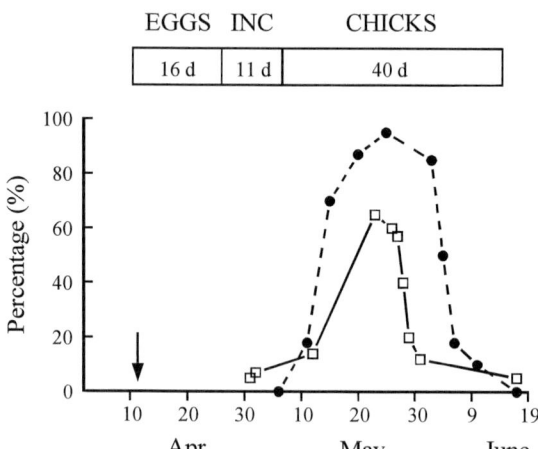

FIGURE 6 Timing of breeding in relation to seasonal variation in food availability in the great tit. ●, Number of broods in the nest; □, caterpillar abundance; the arrow indicates how far in advance of any increase in food abundance egg formation has to be initiated.

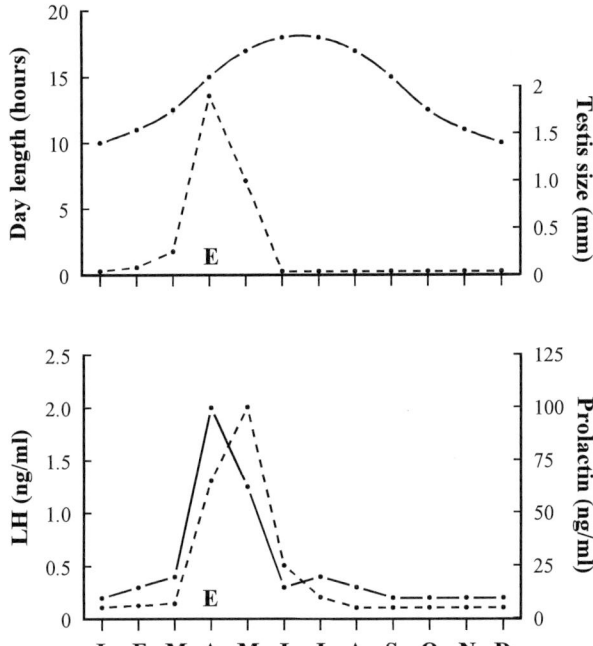

FIGURE 7 Avian reproductive cycles: (top) seasonal changes in day length (solid line) and testis size (dashed line) and (bottom) plasma levels of LH (solid line) and prolactin (dashed line). E, initiation of egg laying. Data are for European starling (*Sturnus vulgaris*) breeding in Cambridgeshire, England.

are required for yolk development and the final stages of egg formation to occur. However, none of these secondary factors is able to stimulate gonadal maturation in the absence of appropriate day length cues.

Just as it is important for birds to time onset of breeding, it is equally important that they can predict the end of favorable conditions for rearing young. The simplest way to achieve this would be to breed at any time when day length exceeds some stimulatory threshold (e.g., >12 hr of day light). However, this would generate a relatively long breeding season of about 6 months which would be symmetrical about the summer solstice. For many birds the period of increased in food availability is much less than 6 months and is skewed toward spring (Fig. 6). In the majority of species, the reproductive system regresses and breeding ends during the summer at a time when day length is still long and "stimulatory" and well before the return of short days in fall (Fig. 7). Birds time the termination of breeding through a second photoperiodic-dependent process called photorefractoriness.

B. Photoperiodic Control of Breeding

Photoperiodic control of seasonal breeding comprises four main components: (i) extraretinal photoreceptors located in the hypothalamus (cf. the eyes in mammals); (ii) integration and transduction of environmental information by the central nervous system (CNS) and an endogenous circadian clock, required to measure day length; (iii) the neurosecretory GnRH release system; and (iv) the peripheral endocrine system, including the pituitary, gonads, and thyroid gland. Birds have traditionally been considered to be classical short-day breeders, with reproduction being "switched on" by long or increasing day length in spring and "switched off" by short or decreasing day length in late summer or fall. However, it is now clear that synthesis of GnRH increases in the hypothalamus during the short days in fall, and the reproductive system is effectively switched on at this time. During the winter, because day length is short GnRH release is low and gonadal maturation proceeds only very slowly. In the spring, increasing long day lengths increase the "photoperiodic drive" on the hypothalamic system, directly causing a large increase in GnRH release, increased gonadotropin secretion, gonadal maturation, and breeding (in some species photoperiod may also temporarily decrease the negative feedback effect of gonadal steroids on the pituitary, enhancing gonadotropin release). In addition, long or increasing day lengths initiate a second process, which develops somewhat more slowly but ultimately leads to cessation of GnRH synthesis, onset of photorefractoriness, and switching off of the reproductive system. The bird then remains insensitive to the stimulatory effects of long days until refractoriness is terminated by short days in the fall. Most birds therefore show alternating cycles of photosensitivity to long or stimulatory day lengths (late fall/winter), photostimulation (spring), and photorefractoriness (late summer/fall) which are reflected in changes in hypothalamic GnRH content (cf. mammals) or hypothalamic activity. Therefore, the mechanism of photoperiodic control of breeding ultimately lies at the level of the hypothalamus or some other neural regulatory system within the CNS that controls hypothalamic function and not at the more peripheral level. The gonads and pituitary can be shown to be responsive to exogenous LH and GnRH, respectively, even during the nonbreeding period. Changes in photosensitivity and refractoriness also appear to be thyroid hormone dependent (as in mammals) although the precise mechanism(s) of thyroid involvement in seasonal reproduction is not fully understood. The way in which other environmental cues such as temperature or social factors are integrated with the hypothalamic–pituitary–gonadal axis remains poorly understood (although these most likely are mediated by sensory receptors of the visual and auditory system). Similarly, there is currently little information available on the pathways which connect the GnRH system to other areas of the brain in birds.

VI. SEXUAL MATURATION

Although domestic fowl come into lay between 5 and 7 months of age, depending on breed, most small, short-lived, seasonally breeding birds breed for the first time in the first spring after hatching, that is, when they are about 1 year old. Other, typically larger, species delay breeding for 1 or more years after hatching. Some long-lived species, such as albatrosses and penguins, show long deferred sexual maturation, breeding for the first time at 6–12 years of age. In contrast, other species can show precocial sexual maturation. For example, the zebra finch can breed at 2 or 3 months of age, and captive Japanese quail are sexually active at 6 weeks.

The mechanisms underlying timing of sexual maturation in wild birds are not well known, although nutritional status, body growth, or age per se do not appear to be important determining factors (cf. mammals). In starlings and red partridge, which first breed at 1 year of age, there is evidence that timing of the initial sexual maturation (puberty) is photoperiodically determined. Chicks hatch in a physiological state very similar, if not identical, to adult photorefractoriness which has been termed juvenile refractoriness. This ensures that nestlings and fledglings do not show an immediate response to the stimulatory effects of long day lengths with gonadal maturation in their first summer. Young birds remain photorefractory until the fall, when they become photosensitive due to the effect of short day lengths. Sexual maturation then occurs in response to long day lengths the following spring. In essence, seasonal breeding is equivalent to repeated puberty. Precocial sexual maturation can be induced in starlings as young as 12–14 weeks old using treatments which are known to terminate refractoriness in adults (thyroidectomy or exposure to short days followed by long days). Even less is known about the cause of long-deferred sexual maturation in birds. Immature penguins and albatrosses have very low circulating sex steroid and gonadotropin levels during their early prebreeding year (Fig. 8). As with control of seasonal breeding in adults, however, the pituitary and gonads appear fully functional in 1-year-old birds and will

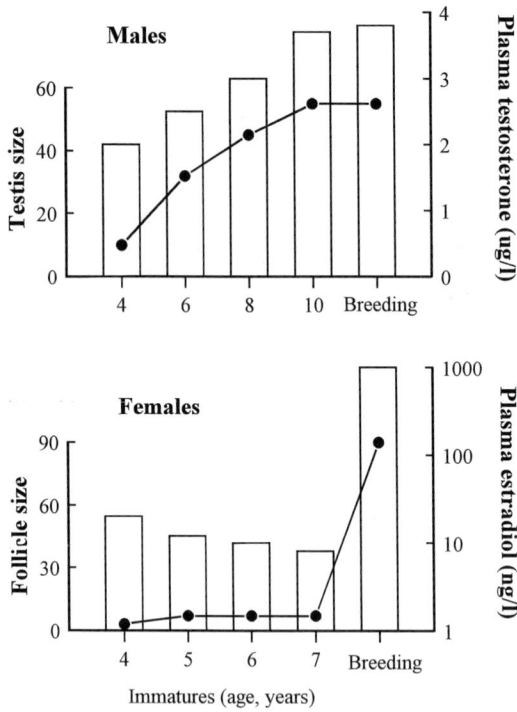

FIGURE 8 Changes in circulating steroid hormone levels (bars) and maximum gonad size (●) with age in relation to deferred sexual maturation in male and female wandering albatrosses.

respond appropriately to treatment with exogenous GnRH or LH, respectively. In males, and females of some species (e.g., macaroni penguin), plasma levels of sex steroids and gonad size increase with age and older prebreeding birds have plasma hormone profiles indistinguishable from those of adult birds 1 or more years before actual onset of breeding (Fig. 8). In other species (e.g., wandering albatross) females appear to have different plasma steroid profiles even in their final prebreeding year, compared to breeding females, with high progesterone and low estradiol levels and no evidence of follicular maturation (Fig. 8). Delayed breeding does not appear to be due to some type of deep refractory state because artificially accelerated "annual" cycles of short and long days do not advance onset of puberty. Rather, it seems likely that deferred sexual maturation is due to some chronological process at the level of the hypothalamus or CNS that takes several years to "unwind." In some species, there may also be a behavioral or social component to age of first breeding, with several years being required for birds to learn complex sexual displays or courtship rituals. In one study of herring gulls in which 75% of adult birds were removed from the breeding population, the age of first breeding among the remaining "immature" birds decreased from 6.2 to 4.3 years.

VII. COSTS OF REPRODUCTION AND REPRODUCTIVE VARIABILITY

Almost everything that is known about avian reproductive physiology comes from detailed studies of a handful of domesticated species: the chicken, Japanese quail, turkey, and Pekin duck. In many cases these species have been selected over centuries for maximum yields with continuous daily laying of large eggs. For example, hens and ducks can produce 270–350 eggs in 365 days and lay eggs 33 and 46% larger, respectively, than their wild counterparts, the red jungle fowl and mallard. Different strains of domesticated birds have also been selected for standard-sized eggs to meet marketing requirements, thereby reducing or eliminating variability in reproductive parameters. This pattern of reproductive effort is therefore totally unlike that of the ≈9000 wild bird species, most of which lay one or a few discrete clutches during a very restricted part of the year and show marked variability in all reproductive traits, such as egg size, clutch size, laying interval, and timing of breeding.

The "cost" of reproduction is self-evidently high in domesticated species from a purely economic perspective. However, in wild birds it is widely assumed that reproduction is both energetically and nutritionally expensive. Many studies have shown that during chick rearing the average daily metabolic rate of parents provisioning young is close to the theoretical maximum sustainable metabolic rate ($4 \times$ BMR). Absolute energy requirements for egg production in altricial species are generally thought to be lower than chick rearing ($0.3–2.2 \times$ BMR). However, these estimates are derived from the total energy content of eggs or clutches (assuming 100% production efficiency) and data are lacking on physiological costs of biosynthesis and egg formation for any wild bird. Egg formation often occurs early in the season before females can benefit from increased food availability (Fig. 6). As a consequence, the demands of egg production may actually be greater than those of chick rearing relative to the energy or food available to meet these demands. In precocial species, egg production probably represents the most energy-demanding phase of the breeding cycle because chicks are self-feeding and parental care is limited. Nutrient demands may also be more important than energy in determining patterns of reproductive effort; for example, different species appear to be protein-limited, lipid-limited, or calcium-limited during egg production. Specific micronutrient requirements, such as sulfur-rich amino acids, vitamins, or minerals, may also limit egg production, though virtually nothing is known about this in nondomesticated species. Finally, recent studies have shown that costs of egg production can be important in an evolutionary sense. Gulls and terns which are made to lay an extra egg (by removal of the first-laid egg) show decreased hatching success of the additional egg and reduced provisioning performance during the subsequent chick-rearing period. Therefore, in relation to reproductive effort, trade-offs occur not only between current and future breeding attempts but also between breeding stages within a single season.

Among different species egg size varies by five orders of magnitude from 0.3 g in hummingbirds to 1600 g in the ostrich. Egg weight is an allometric function of body mass (exponent = 0.67) and in general larger birds lay proportionally smaller eggs. For example, egg weight equals 4% of body mass in some penguins and cuckoos but up to 26% in some petrels and the kiwi. Total clutch mass can represent 100–120% of female body mass in some shorebirds (Charadrii) and tits (Paridae). Clutch size similarly varies greatly among species, from a single egg in many seabirds to 12–15 eggs in some ducks and game birds. Some species appear to mature a fixed number of follicles per breeding attempt. These species, called determinate layers, do not lay additional eggs in response to egg removal. In contrast, indeterminate layers will respond to egg removal by laying larger clutch sizes than normal; for example, one Northern flicker is recorded as laying 71 eggs in 73 days compared to its normal clutch of 6–8 eggs. Details of the physiological or biochemical basis of these interspecies differences in reproduction are only very poorly understood. For example, the only model available to explain laying or oviposition rates is derived from the domestic hen and is relevant to an ≈24-hr laying interval with skips (days with no egg laid). This model can not easily explain laying intervals of 28, 30, 33, or 48 hr which occur in some wild birds.

Finally, there is also marked intraspecific or interindividual variability in all reproductive traits in wild birds. For example, both mean egg size and clutch size can vary twofold among different females breeding in the same year under the same conditions. Much of this variability is repeatable within individual females; that is, individual females maintain the same level of reproductive effort between breeding attempts relative to other females. Individual variation is also maintained when birds are brought into captivity and bred in constant or controlled conditions on *ad libitum* food. Intraspecific variability in reproductive effort has clear fitness consequences.

In general, females which lay more or larger eggs and which lay earlier have the highest lifetime reproductive success. Again, however, the physiological mechanisms underlying interindividual variability within species are almost completely unknown.

See Also the Following Articles

ALTRICIAL AND PRECOCIAL DEVELOPMENT IN BIRDS; AVIAN REPRODUCTIVE SYSTEM, DEVELOPMENTAL ENDOCRINOLOGY; FEMALE REPRODUCTIVE SYSTEM, BIRDS; MALE REPRODUCTIVE SYSTEM, BIRDS; OVARIAN CYCLE, BIRDS; PARENTAL BEHAVIOR, BIRDS; PHOTOPERIODISM, VERTEBRATES; SEASONAL REPRODUCTION, BIRDS

Bibliography

Burley, R. W., and Vadehra, D. V. (1989). *The Avian Egg: Chemistry and Biology.* Wiley, New York.

Deeming, D. C., and Ferguson, M. W. J. (Eds.) (1991). *Egg Incubation: Its Effects on Embryonic Development in Birds and Reptiles.* Cambridge Univ. Press, Cambridge, UK.

Etches, R. J. (1995). *Reproduction in Poultry.* Oxford Univ. Press, Oxford, UK.

Johnson, A. L. (1986a). Reproduction in the female. In *Avian Reproduction* (P. D. Sturkie, Ed.), pp. 403–431. Springer-Verlag, New York.

Johnson, A. L. (1986b). Reproduction in the male. In *Avian Reproduction* (P. D. Sturkie, Ed.), pp. 432–451. Springer-Verlag, New York.

King, A. S., and McLelland, J. (1984). *Birds: Their Structure and Function.* Bailliere Tindall, London.

Nicholls, T. R., Goldsmith, A. R., and Dawson, A. S. (1988). Photorefractoriness in birds and comparison with mammals. *Physiol. Rev.* 68, 133–176.

Wada, M., Ishii, S., and Scanes, C. G. (1990). *Endocrinology of Birds: Molecular to Behavioural.* Japan Scientific Societies, Tokyo.

Williams, T. D. (1994). Intraspecific variation in egg size and egg composition in birds: Effects on offspring fitness. *Biol. Rev.* 68, 35–59.

Wingfield, J. C., and Kenagy, G. J. (1991). Natural regulation of reproductive cycles. In *Vertebrate Endocrinology: Fundamentals and Biomedical Implications* (M. Schreibmann and R. E. Jones, Eds.), pp. 181–241. Academic Press, New York.

Avian Reproductive System, Developmental Endocrinology

James E. Woods and Robert C. Thommes

Laboratory of Developmental Endocrinology, Sarasota, Florida

I. Embryogenesis of Gonads, Accessory Sex Ducts, Genital Tubercle, and Syrinx
II. The Hypothalamo–Adenohypophyseal–Gonadal Axes
III. Functional Development of the Hypothalamo–Adenohypophyseal–Gonadal Axes
IV. Neuroendocrine Regulation of Avian Reproductive (Sex) Behavior
V. Role of Maternal Sex Hormones Deposited in the Yolk of the Fertilized Egg

GLOSSARY

epiblast The outer layer of the blastoderm, which will later segregate into ectoderm and mesoderm.

Hamburger and Hamilton stages A classification of chick embryo developmental "stages," which utilizes a series of morphological characteristics, that are independent of the days of incubation.

hypoblast The inner layer of the blastoderm, which ultimately gives rise to the entoderm.

in situ hybridization A method for localizing a specific gene or DNA sequence within a chromosome based on the binding of a complimentary, radioactively labeled segment of RNA or DNA to it.

reverse-transcription polymerase chain reaction (RT-PCR) A technique in which repeated cycles of DNA synthesis are carried out to produce a large number of a specific DNA sequence.

Sox genes *Sry*-like *Box* genes.

sry gene A Y-linked gene central to the process of sex determination in mammals.

Embryonic development (embryogenesis) of the avian reproductive system encompasses an orchestrated series of morphological, physiological, biochemical, and molecular events, occurring over developmental time, in which the sex hormones play a key regulatory role. Most information on avian embryonic development has been obtained utilizing embryos of the domestic fowl (chick embryo), and thus, unless stated otherwise, this article will relate to the embryonic chick, which has an incubation period of 20–21 days.

I. EMBRYOGENESIS OF GONADS, ACCESSORY SEX DUCTS, GENITAL TUBERCLE, AND SYRINX

The gonads (ovary and testes) and accessory sex ducts (oviduct, uterus, and vagina in the female and epididymis and vas deferens in the male) are termed the primary sexual characters because of their essential roles in reproduction. Secondary sexual characters are external (phenotypic) features, such as the external genitalia (genital tubercle), voice box (syrinx), feathers, beak, spurs, head furnishings, bill color, and vocalizations that distinguish males from females. Another secondary sexual character is the sexually dimorphic behavioral (mating) pattern of the adult bird, which reflects the influence of hormones on the central nervous system at precise times during both the embryonic and the posthatch periods.

A. The Gonads (Ovary and Testis)

Gonadal differentiation into ovary or testis occurs according to the genotype of the zygote (fertilized egg). In birds, the heterozygote ZW is a genetic female, whereas the homozygote ZZ is a genetic male.

The initial morphological indication of gonad formation occurs at approximately 2.0–2.5 days of development [Hamburger and Hamilton (H&H) stage 19] in the form of a strip of peritoneum (germinal epithelium; genital ridge) at the base of the dorsal mesentery and on the median surface of the "primitive kidney," the mesonephros. Further growth from the undersurface of the germinal epithelium (GE), with possible contributions from cells of the "mesonephric blastema," gives rise in both ZW and ZZ embryos to the indifferent gonad, an undifferentiated structure consisting of cords of cells (medullary cords). Primordial germ cells (PGCs), which will eventually differentiate into sperm in the testes and ova in the ovary, originate, prior to the time of egg laying, in the epiblast of the area pellucida of the preprimitive streak egg and translocate to the hypoblast, from where they are carried to the germinal crescent. The PGCs subsequently migrate into the vascular system by ameboid movement and about the second day of incubation move out of the bloodstream, again by ameboid movement, and into the genital ridges (stage 17), homed-in by chemoattractants of a glycoprotein nature. However, unless they are colonized by adequate numbers of PGC, the genital ridges are unable to differentiate into gonads. By Day 5.5 in female embryos, the GE of the left indifferent gonad generates secondary sex cords (cortical cords); these cords give rise to the ovarian cortex and are the distinguishing feature of the left ovary. Such proliferation of the cortical cords (CCs) does not occur in the right female gonad or in the male gonad (testis), both of which develop from the medullary component of the indifferent gonad. By Day 6.5, gonadal sex differentiation is apparent, i.e., ovaries and testes are distinguishable. As the left ovary grows, by the proliferation of the CCs, groups of PGCs become incorporated into them. These cords increase in size over embryonic time, largely through multiplication of the enclosed germ cells, which differentiate into oogonia on Days 7 or 8 and become oocytes shortly after hatching. In the male chick embryo, the testes form as a result of continued growth of the medullary cords. By Day 13, the PGCs located within the medullary cords (MCs) transform into spermatogonia and rapidly increase in number. The MCs, which are now called seminiferous cords, are solid up to about Day 20, at which time they acquire a lumen and are referred to as seminiferous tubules. It should be noted that the corticomedullary environment of the PGCs determines whether they differentiate into ova or sperm, i.e., those that are positioned within medullary tissue differentiate into sperm and those that reside in cortical tissue become oocytes (Fig. 1).

Morphological gonadal sex differentiation in birds, with the exception of the birds of prey (falcons, vultures, and hawks) which possess two ovaries, is asymmetric in the female (only the left ovary develops) and symmetrical in the male (two testes). In the chick embryo, the left and right ovaries develop at the same rate until Days 7–9, when retardation in growth of the right gonad first becomes apparent. In the majority of cases, the right gonad of birds becomes a rudimentary medulla (there are traces of the germinal epithelium in ~21% of chick embryos), which has regressed by Day 15 to a small, barely discernible structure. However, if the left ovary (or its cortex) is removed (sinistral ovariectomy) or if testosterone (T) is injected, the right rudimentary gonad will enlarge (hypertrophy) and differentiate into a testis. Thus, the medulla of the right ovary has the potential of differentiating in the male direction, but is, in normal circumstances, prevented from transforming into a testis by the inhibitory influence of the female sex hormone 17β-estradiol (E_2), which is synthesized by the left ovarian cortex. (The genetic constitution of the embryo ZW or ZZ-determines whether the cortical component will or will not develop.) However, if the left ovary is removed or is destroyed by disease during the posthatch period, the female domestic fowl (hen chicken) will transform into an intersex, termed a poulard, which has all the external features (secondary sex characters) as well as the behavioral characteristics of a male chicken (rooster), including attempts at copulation. If the left ovary is removed within 30 days after hatching, the right gonad will develop into a testis, which in some cases contains sperm; however, since the right Müllerian duct degenerates in genetic female chick embryos, there is no duct for passage of the sperm to the exterior. If removal of the left ovary occurs after approximately 30 days, the right testis that forms is sterile, i.e., the primordial germ cells have a limited life span unless they are incorporated within a functional gonad. After 30 days posthatch, E_2 inhibition has destroyed the potential of the PGCs

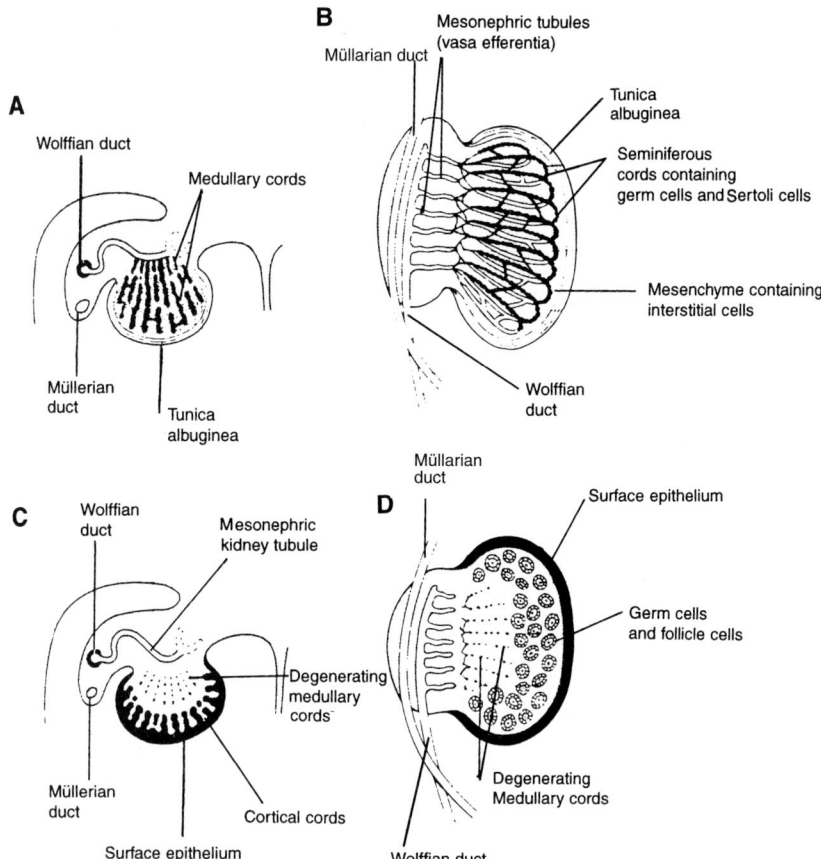

FIGURE 1 Early development of the testis and the ovary. In the male, the gonadal medullary tissue develops into the medullary cords (A), which give rise to the seminiferous tubules and the Sertoli cells (B). Mesonephric tubules give rise to the vasa efferentia (B). In the female (C), the medullary cords degenerate and the cortical cords give rise to an ovary (D). Some mesonephric elements remain in the female as well. Eventually, the Wolffian ducts (mesonephric ducts) degenerate in the female, whereas they are retained in the male, becoming the vas deferens. In both male and female, ureters develop from the caudal-most portion of the Wolffian (mesonephric) duct to drain the metanephric kidneys (from D. O. Norris, *Vertebrate Endocrinology*, 3rd ed., Academic Press, San Diego, 1996, which was modified from M. Johnson and B. Everitt, *Essential Reproduction*, 3rd ed., Blackwell, Oxford, UK, 1988.

to differentiate and thus no proliferation occurs even when the source of the E_2 is removed. At varying times, from months to years, after the transformation of the female into a poulard, the testis begins to synthesize E_2 and the poulard undergoes a sex reversion, resulting in an individual with a composite of male and female external features. This E_2 originates from a reactivated GE or remnants of cortical tissue. In muscovy ducks and ringdoves, sinistral ovariectomy produces intersexes comparable to those in the domestic fowl, but in the Japanese quail there is only hypertrophy of the right gonad. In the female zebra finch, treatment of 3- to 8-day-old embryos with an aromatase inhibitor blocks E_2 production by the left ovary and allows the right gonad to transform into a testis. This testicular tissue secretes androgens, as evidenced by plasma T levels and growth of the hypertrophied androgen-dependent syrinx, but does not bring about masculinization of courtship song or copulation. In contrast, growth of the right gonad after removal of the left ovary does not occur in turkeys, mallard ducks, bobwhite quail, or starlings. The sex-reversal phenomenon described previously produces intersexes and not hermaphrodites; hermaphrodites possess both a functional left ovary and a right testis and are quite rare.

B. Genetic Basis of Gonadal Sex Differentiation

There is a genetic "switch" that determines whether the indifferent gonad develops into an ovary or a testis; however, the details of the mechanism, whereby the gene(s) regulates the production of sex-differentiating molecules (sex morphogens) that control gonadal transformation, are unknown. The "sex" of embryos can be determined by the presence or absence of sex chromatin in the nucleus of the embryo's cells; the sex chromatin (Barr body) is found in the cells of female embryos but is not observed in males, and it can be readily observed microscopically in a variety of tissues, including blood cells and cells of the extraembryonic membranes. In recent years, molecular biology techniques that identify particular DNA sequences have been employed to determine the genotype of avian embryos. One such technique is the reverse-transcription polymerase chain reaction (RT-PCR) assay based on the presence of repeated sequences on the W chromosome of females. In mammals, a gene on the Y chromosome (*SRY* in humans and *sry* in mice) is responsible for male development; however, whether or not birds possess a gene that is homologous to *SRY/sry* remains unclear. In mice and humans, the *SRY/sry* gene exerts its action via other genes, referred to as Sox genes, one of which (*Sox9*) is of particular importance. A homolog in chick embryos (*cSox9*) has been identified in the genital ridges of both ZW and ZZ chick embryos as early as Day 4 (H&H stage 25), a time at which the genital ridges of both genotypes are morphologically identical. At Days 5.5–6 (stage 28), a stage at which differentiation into testes and ovaries occurs, the expression of *cSox9* gene in the gonads in ZZ embryos (males) is at a high level, whereas ZW embryos (females) express the gene at a low level.

C. Hormonal Regulation of Gonadal Sex Differentiation

Historically, there are two major theories regarding the mechanism of avian gonadal sex differentiation. In both, the sex hormones play pivotal roles. According to Wolff's monohormonal theory, the medulla of the indifferent (presumptive) ovary synthesizes both estrogens and androgens, whereas the medulla of presumptive testes produces only androgens. This theory states that the estrogens synthesized by the medulla of the genetic female-indifferent gonad regulate the morphological differentiation of this organ into an ovary by stimulating the growth of the cortex while simultaneously inducing regression of the medulla. Wolff's theory maintains that in the genetic male embryo, sexual differentiation of the indifferent gonad into a testis is a result of the inherent growth ability of the medulla, whereas the cortex ceases to grow because the medulla does not synthesize estrogens. Thus, androgens produced by the medulla of the ZZ-indifferent gonad are not considered as playing an active role in male gonadal sex differentiation. Willier, on the other hand, hypothesized that both androgens and estrogens regulate gonadal sex differentiation (dihormonal theory). He postulated that the germinal epithelium (incipient cortex) of the sexually indifferent gonads of both sexes synthesizes estrogens and the medulla produces androgens. According to this theory, the indifferent gonad of genetic females synthesizes estrogens preferentially, whereas in genetic males the indifferent gonad produces primarily androgens. Additionally, it postulated that ovaries are formed from the indifferent gonad of the genetic female as a result of the action of estrogens, which stimulate growth of the cortex and suppression of the medulla, and that in genetic males, the transformation of the indifferent gonad into a testis is regulated by androgens which activate medullary growth and inhibit the cortex.

A large amount of data support the key role of estrogens in avian gonadal sex differentiation. When fragments of sexually differentiated gonads from 6- to 11-day-old embryos are transplanted into the coelom of a 2-day-old host embryo in juxtaposition to the host gonad, the grafted tissue modifies the direction of sex development of the host's reproductive organs if the host is of the opposite sex. Thus, ovarian tissue grafted into genetically male (ZZ) embryos stimulates the formation of an ovarian cortex on the testis of the host, resulting in the formation of an ovary or an ovotestis. The same feminizing results are obtained with estrogens, i.e., the cortex of the embryonic testis is stimulated and undergoes differentiation and the medulla is reduced. Testis tissue grafts in female (ZW) embryos masculinize the left gonad and transform it into a testis; they also bring about a total regression of the Müllerian ducts of the female host embryos. Thus, it is apparent that Müllerian-inhibiting hormone (MIH), the hormone responsible for the regression of the MDs, is being released by the testis grafts. The addition of T to female embryos modifies ovaries in the male direction by repression of the cortex and hypertrophy of the medulla; however, the effect is not as complete as testis grafts. Together these findings implicate both MIH and T in testis formation.

The sex hormones are present early enough in the embryonic period to be available to act as gonadal sex differentiator substances. Androgens (testosterone and dihydrotestosterone) and E_2 have been identified immunohistochemically in the morphologically undifferentiated (indifferent) gonad as early as Day 3.5 (H&H stage 19). Also, from Day 5.5 until hatching the indifferent gonads of ZW embryos produce more E_2 than the indifferent gonads of ZZ embryos and the indifferent gonads of ZZ embryos produce more T than do ZW embryos. From Days 3.5 to 20.5, in both testes and ovaries androgens are synthesized by the interstitial cells (ICs), located between the MCs. From Days 3.5 to 12.5 in the female embryo, E_2 is synthesized in the medullary ICs of the left ovary, but on Day 13.5, E_2 synthesis is initiated in the cortical cords. In the testis, E_2 is present in the medullary interstitial cells throughout embryonic development (Days 3.5–20). In the biosynthesis of E_2, T is directly converted to E_2, a reaction which is catalyzed by the enzyme aromatase ($P450_{arom}$); this reaction is termed "aromatization." On Days 5 and 6, the gene that regulates the production of $P450_{arom}$, identifiable by RT-PCR and in situ hybridization, is present in much higher levels in the indifferent left ovary than in the testes; the high level of $P450_{arom}$ in the left ovary is associated with high levels of E_2 in ovarian extracts, as determined by radioimmunoassay. Treatment of female (ZW) chick embryos before the time of gonadal differentiation (Day 5.5) with an aromatase inhibitor causes the presumptive left ovary to develop into a testis (sex reversal) and the hatched bird that develops is a poulard (male phenotype). This verifies the fact that expression of the aromatase gene, via its role in the biogenesis of E_2, is one of the essential steps in the process of gonadal sex differentiation. Further evidence for a role for estrogens in ovarian differentiation includes the observations that (i) the estrogen receptor (ER), which binds E_2 and is essential for the cellular action of this estrogen, occurs as early as Day 4.5 (stage 26) in much higher levels in the presumptive left ovary than in testes; (ii) E_2 binds preferentially to the left ovary at the same time; and (iii) ER and $P450_{arom}$ are colocalized in the presumptive left ovary but not in the testes.

D. The Accessory Sex Ducts

The accessory sex ducts (the Müllerian and Wolffian ducts) of the reproductive system are laid down in both ZZ and ZW embryos and coexist in both sexes for a time during the ambisexual period of embryonic development. The Wolffian ducts (WDs) are derived from the mesonephros and its duct system. The Müllerian ducts (MDs), on the other hand, are derived from a strip of thickened peritoneum on the dorsolateral surface of the mesonephros. One type of duct system persists normally and gives rise to the definitive accessory ducts and glands, whereas the heterologous system disappears except for important vestiges. In the genetic female embryo, the left MD system survives and the right MD and both WDs regress, and in the male embryo both Wds develop and both MDs regress. In adult avian genetic females, the MD forms the oviduct, including the shell gland (uterus) and vagina, and is the tubular system in which sperm are widely distributed after copulation and the egg is transported to the cloaca from where it is extruded to the external world ("laid"). In males, the WD forms the excurrent ducts of the testis (epididymis and vas deferens), through which the sperm travel to the cloaca for later exit during copulation. The maintenance of the homologous duct and regression of the heterologous duct in each sex is under hormonal control. In the female chick embryo, growth and differentiation of the left MD is regulated by estrogens produced by the ovarian cortex, and there is a

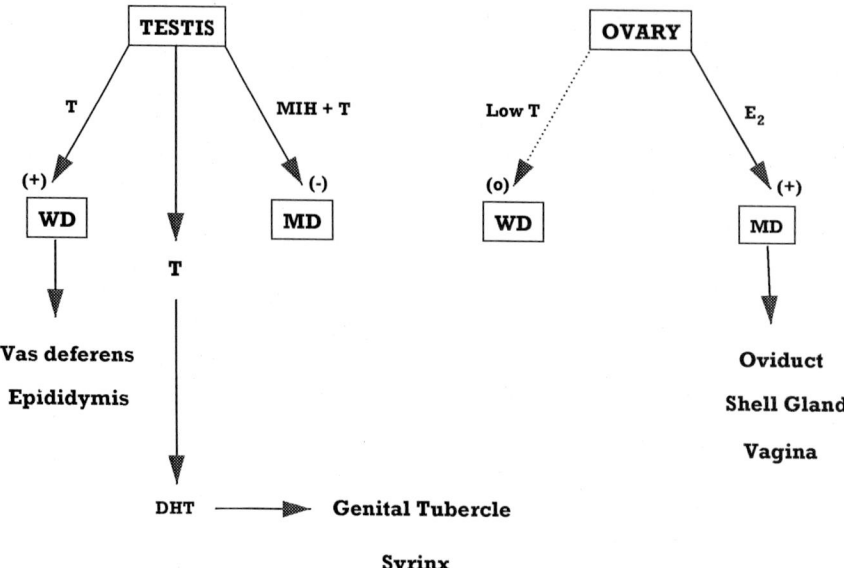

FIGURE 2 Patterns of development for primary sexual characteristics. In the male embryos the testes secrete testosterone (T), which stimulates the differentiation of the Wolffian ducts (WD), and Müllerian-inhibitory hormone (MIH), which causes regression of the Müllerian ducts (MD). Dihydrotestosterone (DHT) is either produced by the testes or converted from T in the genital tubercle causing it to differentiate in the male direction. In the female, E_2 from the ovary prevents MIH (also secreted by the ovary) from causing MD regression, whereas the absence of sufficient androgen synthesis/secretion by the ovary leads to regression of the WD.

correspondence in time between the growth of the MD, the synthesis and secretion of E_2 by the left ovary, and the binding of both E_2 and ER to the cells of this tubular structure. Regression of the WD begins about Day 7 and is a consequence of the inadequate production of T by the ovarian medulla to maintain its growth. In the male chick embryo, growth of the WD is regulated by T produced by the medullary component of the testes. Regression of the MD, which begins on Day 8 and is complete by Days 12 or 13, is a function of the cooperative action of MIH and T, both of which are synthesized by the medulla of the testis. MIH is synthesized by the Sertoli cells of the medullary cords and T is produced by the interstitial cells located between these cords (Fig. 2).

E. Genital Tubercle and Syrinx

There are other sex structures in birds, besides the sex accessory ducts, whose differentiation is also under hormonal control. The genital tubercle is the primordium of the copulatory organ of male birds (a true penis is present in only a few avian species). It is present in both species prior to sexual differentiation, and in adult males of most species it exists in a rudimentary form. In male ducks, the genital tubercle develops into a twisted organ which protrudes from the cloaca. In the male chicken, its rudimentary presence (phallus) can only be determined by everting the cloaca, a procedure that is utilized to "sex" newly hatched chicks. The syrinx of birds (a vocal organ) is located at the point where the trachea and bronchial tubes join. It is sexually dimorphic in ducks and some songbirds. In the female duck, it is small and symmetrical, whereas in the male it is large and asymmetrical. In embryonic ducks, the first indication of a sex difference occurs on Days 10–11 of incubation (28-day incubation period), at which time the genital tubercle of the female regresses, whereas the genital tubercle and syrinx grow in the male. Differentiation of the genital tubercle and syrinx is regulated by the sex hormones. This has been shown by X-irradiation of the gonadal primordium (castration) of embryonic ducks on Days 3 or 4, which is prior to gonadal sex differentiation on Day 8. When the embryos were examined between Days 17 and 22 of incubation, both sexes demonstrated a male-type genital tubercle and syrinx. Thus, the male is the neutral or ahormonal sex with regard to these secondary sex characters. If the ovary is not destroyed, male differentiation is prevented, and the female form develops. E_2 injected into incubating duck eggs causes the genital tubercle and syrinx of males to assume the female form, i.e., neither structure develops. In the female duck embryo, maximum binding of E_2 occurs on Day 12 of incubation. On the other hand, the administration of T stimulates the growth of the genital tubercle and syrinx. Similar results have been obtained using chick embryos. The action of T on the genital tubercle is via its conversion to dihydrotestosterone.

II. THE HYPOTHALAMO–ADENOHYPOPHYSEAL–GONADAL AXES

In all vertebrate organisms, including birds, there are regulatory systems that allow the organ systems to function in a coordinated fashion. One such system is the endocrine system; it consists of spatially separated tissues and organs (endocrine glands), which produce substances, termed hormones, that are carried in minute quantities in the bloodstream and exert specific physiological effects on both endocrine and nonendocrine organs. The integration of the component endocrine glands into a functional organ system is by means of the effect of the hormones on the activity of each other's glands. The most basic category of activity between the endocrine glands is termed an endocrine axis and involves the reciprocal actions of hormones produced by a trophic gland and other hormones produced by a "target gland." An endocrine axis functions by means of feedback systems, in which plasma levels of the trophic hormone regulate the rate of synthesis and release of the hormones produced by its target glands, i.e., it exhibits a feedforward regulation of the target glands. The target gland's hormones in turn accelerate or retard hormone production by the trophic gland, i.e., they demonstrate a feedback effect on the trophic gland. Thus, an endocrine axis is composed of a feedforward component and a feedback component. Negative feedback systems are the simplest and most prevalent type, in which the hormonal output of the target endocrine gland retards the activity of the trophic endocrine gland. One type of negative feedback system is the neuroendocrine feedback system. It is composed of an aggregation of nerve cells (neurons), in a part of the brain called the hypothalamus, that produce hormones (neuroendocrines), which in turn control a trophic gland (pituitary gland or adenohypophysis). One must keep in mind that higher brain centers, via neural circuitry, are capable of modulating the hormonal output of the hypothalamus and thus indirectly the adenohypophysis. In the case of the avian reproductive system, the hypothalamus produces gonadotropin-releasing hormone (GnRH), which regulates the synthesis and release of the gonadotropins [luteinizing hormone (LH) and follicle stimulating hormone (FSH)] produced by the adenohypophysis. The hypothalamus (H) and adenohypophysis (A) act as an endocrine unit, the hypothalamo–adenohypophyseal (HA) complex, which can be considered a "controller" since it regulates (controls) the synthesis of male and female sex hormones by its target glands, the gonads: the ovary and the testes. This functional interrelationship forms the feedforward component of the HA–ovarian (HAO) and HA–testicular axes (HAT), which can be referred to collectively as the HA–gonadal (HAG) axes. To modulate the amount of T and E_2 secreted by the gonads, and thus their blood levels, both of these sex hormones regulate the release of hypothalamic GnRH and adenohypophyseal LH and FSH by the HA unit and thus ultimately the degree of stimulation of synthesis and secretion of T by the testes and E_2 by the ovary. When the blood levels of T and E_2 become elevated beyond a particular programmed fixed point (set point) of the controller (HA unit), they inhibit hormone production by the HA complex. This regulation is referred to as the negative feedback component of the HAG axes since a deviation in one

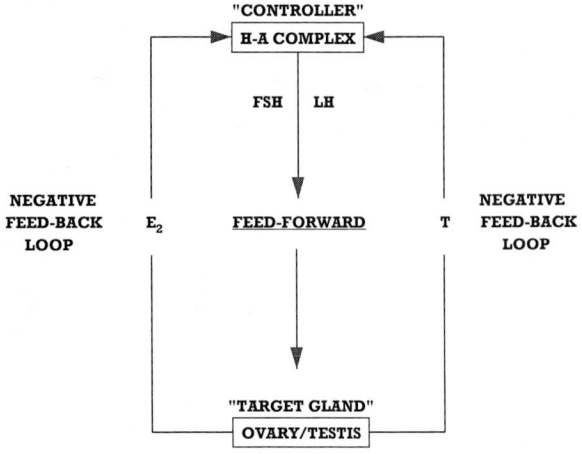

FIGURE 3 Diagrammatic representation of the HAG negative feedback system. The hypothalamic component of the HA complex ("controller") secretes LHRH into the blood supply of the adenohypophysis; LHRH stimulates the synthesis and secretion of FSH and LH by the adenohypophysis. These latter two hormones are transported in the circulatory system and stimulate the synthesis of E_2 and T by the target glands (ovaries and testes). This constitutes the feedforward component of the HAG axis. At a given blood level (set point) of E_2 and T, these hormones inhibit the production and secretion of LHRH and FSH/LH by the HA complex; this constitutes the negative component of the HAG axis. The feedforward and feedback components are referred to collectively as the negative feedback system. As the blood titers of E_2 and T decrease, the controller (HA complex) is no longer inhibited and thus reinstitutes its feedforward stimulation of the "target" glands, i.e., the ovary and testes.

direction (elevated T and E_2 levels) induces a reaction in the opposite direction (decreased production of T and E_2) in order to decrease the difference between the set point for these hormones and their blood levels. Since in adult vertebrates, whose endocrine axes are fully developed, the feedforward (stimulatory) constituent acts in concert with the negative feedback (inhibitory) component, the two components are considered as constituting together a negative feedback loop, and the term feedforward is rarely used. However, in embryonic/fetal systems, which mature gradually over developmental time, there is no reason to infer that both components of endocrine axis loops develop concurrently, and thus it is important in the discussions of such systems to make the distinction that there are two components of the HAG axis (Fig. 3).

III. FUNCTIONAL DEVELOPMENT OF THE HYPOTHALAMO–ADENOHYPOPHYSEAL–GONADAL AXES

B. H. Willier first hypothesized that in the embryonic development of vertebrates, the glandular elements of the endocrine axes are initially autonomous in their individual growth, differentiation, and function and that only at a later time in embryogenesis are endocrine feedback systems established. In the developing chick embryo, this phenomenon has been studied in detail with regard to the functional maturation of the HAG axes. Our initial understanding of hypothalamo–adenohypophyseal–gonadal interrelationships in the chick embryo utilized two experimental procedures: (i) surgical partial decapitation, at 33–38 hr of incubation, which removes the anterior portion of the brain (prosencephalon) and produces an embryo devoid of the derivatives of that cranial segment, including the HA complex, and (ii) transplantation of the adenohypophysis or addition of hormones produced by the HA complex to the chorioallantoic membrane (CAM) of these partially decapitated embryos. Fugo was the first to implicate the components of the HA unit in the growth and development of the chick embryo gonad; he demonstrated that the gonads of embryos devoid of the HA complex exhibit a normal course of development up to Day 13, after which time there is no further increase in gonad size or weight. Following this lead, other investigators confirmed and extended these observations by utilizing such experimental embryos in conjunction with techniques (biochemical, autoradiographical, histochemical, and radioimmunoassay) that are capable of demonstrating the ability of the embryonic gonads to synthesize and secrete the sex hormones. The results of these studies are consistent with Willier's hypothesis that during about the first two-thirds of chick embryo development, each endocrine component (hypothalamus–adenohypophysis–gonad) of the HAG axes initially grows, differentiates, and functions independently. The gonads initially synthesize T and E_2 on Day 3.5 and the HA unit synthesizes GnRH and FSH and LH on Day 4.5. On Day 6.5, the day of gonadal sex differentiation, there is already a preferential synthesis of E_2 by the ovaries and of T by the testes. The vascular communication between the hypothalamus and the adenohypophysis is fully formed by Day 12.5 and on Day 13.5, plasma LH and T concentrations reach peak values. Following removal of the hypothalamo–adenohypophyseal complex by decapitation, the plasma concentrations decrease significantly on Day 13.5 and transplants of the adenohypohysis to the CAM of decapitated embryos restore plasma T levels. These findings, in conjunction with the results of the administration of antibodies against LH-releasing hormone (LHRH), indicate that on Days 12.5 and 13.5 (the 13th day) of the 20- to 21-day embryonic period, the HA complex initiates, by its secretion of LHRH and LH, control of T synthesis/secretion and thus feedforward regulation of the HAT axis is established. Control of the hypothalamus, and its secretion of LHRH, by higher brain centers, via their secretion of opioid peptides, also occurs during the same time span, as shown by the effects of an antibody against LHRH and the opiate receptor antagonist, naloxone, on plasma T levels. The negative feedback component of this axis becomes operational on the same day as the feedforward component, at which time plasma LH and T concentrations reach maximal levels and there is optimal binding of LH to the testes and T to androgen receptors in both the hypothalamus and the adenohypophysis (Fig. 4).

In the chick embryo, both the feedforward and feedback components of the HAO axis also develop on Day 13. Thus, FSH, which is synthesized early in the development of the adenohypophysis (Day 4.5), is present in the circulation by Day 8 and attains maximum binding to the cortical cords of the ovary on Day 12.5, which is the same day that E_2 synthesis is initiated in these cords. Maximal binding of E_2 and ER to both the hypothalamus and the adenohypophysis (HA complex) also occurs on Day 13. On Day 13, ablation of the HA complex by surgical partial decapitation brings about both a retardation in the growth of the ovarian cortex and a statistically significant decrease in plasma E_2 levels. These observations indicate that the feedforward component of

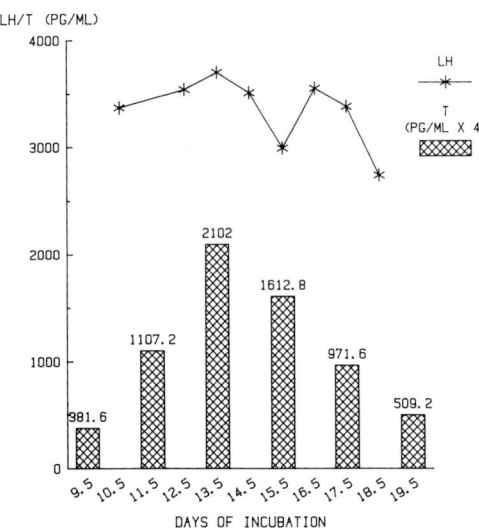

FIGURE 4 Development of the negative feedback component of the HAT axis in the male chick embryo. On Day 13.5, plasma LH levels reach a maximum level. On the same day plasma T titers are at peak levels, the result of the stimulation of T synthesis/secretion by the heightened LH concentrations. At this time (Day 13.5), T begins to exert a negative feedback effect on the production/synthesis of LH by the adenohypophysis. This results in a decrease, by Day 14.5, in plasma LH levels. The effect of the lowered plasma LH levels on T synthesis/secretion by the testes brings about a decrease in plasma T levels on Day 15.5 (from Woods et al., 1989).

the HAO axis is established on Days 12.5–13 of incubation. Also, 48 hr after the administration, on Day 12, of an aromatase inhibitor to female chick embryos (thus inhibiting ovarian E_2 synthesis and its inhibitory negative feedback effect on FSH secretion), plasma FSH concentrations were markedly increased, reaching 300% of control values. This demonstrates that the negative feedback component of this axis is operative sometime between Days 12 and 14, presumably Day 13. There is also evidence in the female Peking duck of HA regulation of ovarian estrogen secretion. Thus, in female ducks in which the HA unit has been removed by the partial decapitation method, there is a reduced secretion by the ovaries beginning on Day 14 (28-day incubation period), as indicated by the fact that the normal rudimentary condition of the genital tubercle and syrinx does not occur. Injection of E_2 into these experimental embryos restored the characteristic rudimentary development of both these structures. Although there is no direct information on the role(s) of the hypothalamus and/or the adenohypophysis on the development of the gonad in the Japanese quail, there is indirect evidence that the HA unit regulates the gonadal production of the sex hormones by Day 14 (16–17-day incubation period) as well as evidence regarding the manner in which higher brain centers regulate the HA complex—via neuropeptides and neurotransmitters.

IV. NEUROENDOCRINE REGULATION OF AVIAN REPRODUCTIVE (SEX) BEHAVIOR

Besides its role in endocrine regulation per se, the HAG axis is intimately involved in the regulation of avian posthatch behavior relating to courtship and mating. Bird species can be divided into precocial and altricial types. Precocial species have extended gestational periods and are hatched in a relatively advanced stage of development, whereas atricial species have shorter gestational periods and are hatched in an immature state. Adult avian sexual behavioral patterns reflect developmental changes that occur in the brain neural networks, particularly the hypothalamus, as a result of the action of the sex hormones. The first event in this transformation of the brain's activity is termed the organizational phase and consists of the action of the sex hormones on the proliferation of neuronal processes and the release of neurotransmitter substances. During this phase the sex hormones organize the neural substrate in a male-typical or female-typical manner. Later these sensitized neurons are acted upon by a second surge of sex hormones that activate this sexually dimorphic neuronal network so that the appropriate male or female sex behavior is played out, an event called the activational phase. Both T and E_2 are involved in the maturation of those brain centers that are involved in this embryonic imprinting of adult mating behavior. In precocial species, such as the domestic fowl and Japanese quail, the organizational phase occurs during the embryonic period and the activational phase is observed during posthatch, whereas in atricial species, including songbirds such as the Zebra finch, both the organizational and the activational phases occur posthatch. In the domestic fowl, the sex-specific effect of the sex hormones on the organization of the hypothalamus is exerted prior to Day 13 of the 21-day embryonic period, by which time there is maximum binding of both T and E_2 to the hypothalamic nuclei, whereas in the Japanese quail these same hormones bring about their effect by Day 11 of the 16- or 17-day incubation period. This "window" of embryonic time during which the sex hormones exert their organizing effect is referred to as the sensitive period. Although the pioneer investigations on avian sex behavior utilized the domestic fowl, the Japanese quail and the Zebra finch have recently become the birds of choice for the study of the endocrine regulation of mating behavior based on the shorter incubation period, the smaller size of the bird, and its more stereotyped behavioral patterns. The sex hormones bind intracellularly to specific groups of neurons in the hypothalamus, referred to as hypothalamic nuclei, the most important of which is the preoptic nucleus (POM). Female sexual behavior in genetic females results from the binding of E_2, secreted by the ovary, to the neurons of the POM. Thus, during embryonic development, genetic females are feminized (demasculinized, i.e., altered from the neutral male pattern) by ovarian estrogen. It appears, however, that with regard to the organization of male brains, there may be differences between species. Thus, in the male chick embryo, masculinization of copulatory activity requires both the high plasma T levels and the low E_2 levels, which normally exist in this sex. The major portion of E_2 that acts on the hypothalamic nuclei of the male is produced at these sites by the "aromatization" of T, i.e., the conversion of T to E_2 by the aromatase enzyme, $P450_{arom}$. In the male Japanese quail embryo, there are indications that male sexual behavior potential develops in the relative absence of gonadal steroids; however, a slight demasculinization of sexual behavior occurs transitorily in the normal male as a consequence of the E_2 formed in the hypothalamus by the aromatization of endogenous T.

It is well-known that in posthatch birds, environmental stimuli, via visual, auditory, and olfactory cues, activate the hypothalamus and thus contribute to behavioral patterns. However, the environmental stimuli that the embryo responds to are not as well documented. It is nevertheless obvious, by way of example, that in precocial species such as the domestic fowl (chicken), the embryo imprints on vocal calls in order to establish in the posthatch chick the mother–chick bond necessary for maintaining proximity to the mother and thus survival. In bobwhite quail, Japanese quail, and the domestic chicken, embryological development and the length of the incubation period can be shortened by up to 32 hr by auditory cues from adjacent newly hatched chicks or synthetic auditory stimuli.

V. ROLE OF MATERNAL SEX HORMONES DEPOSITED IN THE YOLK OF THE FERTILIZED EGG

The mother hen incorporates sex hormones into the yolk of her fertilized eggs. The possible role of these yolk-stored steroid hormones in the "physiological economy" of the developing embryo has been investigated in the canary, an altricial species, and in the precocial domestic fowl and Japanese quail. In the Japanese quail, treatment of laying hens with E_2, which increased the amount of E_2 stored in the embryonic yolk, resulted in an increased incidence of right oviducts in adult female offspring. In the canary, those chicks of a clutch that received the greatest amounts of yolk-stored T exhibited accelerated brain development, as indicated by posthatch behavioral patterns. These findings warrant further investigation.

See Also the Following Articles

AVIAN REPRODUCTION, OVERVIEW; FEMALE REPRODUCTIVE SYSTEM, BIRDS; MALE REPRODUCTIVE SYSTEM, BIRDS

Bibliography

Abinawanto, Shimada, K., Yoshida, K., and Saito, N. (1996). Effects of aromatase inhibitor on sex differentiation and levels of P45017α and P450arom messenger ribonucleic acid of gonads in chicken embryos. *Gen. Comp. Endocrinol.* **102**, 241–246.

Adkins-Regan, E. (1981). Early organizational effects of hormones: An evolutionary perspective. In *Neuroendocrinology of Reproduction* (N. Adler, Ed.), pp. 159–228. Plenum Press, New York.

Andrews, J. A., Smith, C. A., and Sinclair, A. H. (1997). Sites of estrogen receptor and aromatase expression in the chicken embryo. *Gen. Comp. Endocrinol.* **108**, 182–190.

Domm, L. V. (1939). Modifications in sex and secondary sexual characters in birds. In *Sex and Internal Secretions* (E. Allen, C. H. Danforth, and E. A. Doisy, Eds.), 2nd ed., pp. 227–327. Williams & Wilkins, Baltimore.

Fugo, N. W. (1940). Effects of hypophysectomy in the chick embryo. *J. Exp. Zool.* **85**, 271–297.

Hamburger, V., and Hamilton, H. L. (1951). A series of normal stages in the development of the chick embryo. *J. Morphol.* **88**, 49–92.

Ottinger, M. A., and Abdelnabi, M. A. (1997). Neuroendocrine systems and avian sexual differentiation. *Am. Zool.* **37**, 514–523.

Rogers, L. J. (1995). Environmental influences on development of the embryo. In *The Development of Brain and Behaviour in the Chicken*, pp. 41–71. Cab International, Wallingford, UK.

Taber, E. (1964). Intersexuality in birds. In *Intersexuality in Vertebrates Including Man* (C. N. Armstrong and A. J. Marshall, Eds.), pp. 285–310. Academic Press, London.

Thorne, M. H. (1995). Genetics of poultry reproduction. In *Poultry Production* (P. Hunton, Ed.), pp. 411–434. Elsevier, Amsterdam.

Willier, B. H. (1939). The embryonic development of sex. In *Sex and Internal Secretions* (E. Allen, C. H. Danforth, and E. A. Doisy, Eds.), 2nd ed., pp. 64–144. Williams & Wilkins, Baltimore.

Willier, B. H. (1955). Ontogeny of endocrine correlation. In *Analysis of Development* (B. H. Willier, P. A. Weiss, and V. Hamburger, Eds.), pp. 574–619. Saunders, Philadelphia.

Wolff, E. T. (1959). Endocrine function of the gonad in developing vertebrates. In *Comparative Endocrinology* (A. Gorbman, Ed.), pp. 568–581. Wiley, New York.

Woods, J. E. (1987). Maturation of the hypothalamo-adenohypophyseal–gonadal axes in the chick embryo. *J. Exp. Zool. Suppl.* **1**, 265–272.

Woods, J. E., Scanes, C. G., Seeley, M., Cozzsi, P., Onyeise, and Thommes, R. C. (1989). Plasma LH and gonadal LH-binding cells in normal and surgically decapitated chick embryos. *Gen. Comp. Endocrinol.* **74**, 1–13.

Yoshida, K., Shimada, K., and Saito, N. (1996). Expression of P45017α hydroxylase and P4540 aromatae genes in the chicken gonad before and after sexual differentiation. *Gen. Comp. Endocrinol.* **102**, 233–240.

Benign Prostatic Hyperplasia (BPH)

Kevin T. McVary and John T. Grayhack

Northwestern University Medical School

I. Incidence
II. Anatomy
III. Physiology
IV. Natural History
V. Etiology
VI. Pathophysiology

GLOSSARY

apoptosis A genetically programmed and organized cell death sequence common to many organs which undergo growth and atrophy.

benign prostatic hyperplasia (BPH) (Histologic/clinical) involutional changes that occur in specific regions of the prostate early in adult life and result in a marked increase in cell number; (histologic) when seen in conjunction with certain symptoms it is noted as clinical BPH.

hyperplasia An abnormal increase in the number of normal cells in an organ or tissue which increases its volume.

paraurethral glands The prostatic glands nearest to or surrounding the prostatic urethra which are generally thought to develop into BPH.

transurethral prostatectomy Transurethral operation performed on the prostate to alleviate obstructive tissue causing adverse symptoms or conditions in men with enlarged prostates.

verumontanum That portion of the prostatic urethra into which the ejaculatory ducts empty.

The prostate is the major accessory secretory sex gland of the male and provides fluid that constitutes about 15% of the ejaculate. Aside from producing a vehicle for sperm, it has no definitive function in reproduction. The major interest in the growth and function of the prostate has resulted from the frequency with which it is the site of benign and malignant neoplasms. The most common of these pathologic processes is benign prostatic hyperplasia. The anatomic relationship of the prostate gland at the bladder neck increases the importance of the pathologic changes which the patient so frequently undergoes.

I. INCIDENCE

The association of benign prostatic hyperplasia (BPH) and aging has been demonstrated repeatedly in autopsy studies with the use of weight, prostatic volume, and histologic criteria. Randall found the incidence of definite or probable BPH to exceed 50% in men older than 50 years of age. This occurrence rose to 75% as men entered the eighth decade. The autopsy-specific prevalence of histologic BPH is similar in several countries despite differing racial mixes. Berry *et al.* reviewed the major reports in the literature relating to the prevalence and growth rate of human BPH with age. Based on these combined data, curves were constructed for the prevalence of BPH with age (Fig. 1). This analysis implies that the development of BPH is probably initiated before the patient is 30 years old. The calculated doubling time for the weight of BPH also varies with age, being 4.5 years in the 31- to 50-year age group, 10 years in the 51- to 70-year age group, and more than 100 years in the >70-year age group.

Most reports suggest that black and white populations in North America have a similar incidence of BPH, although the symptoms of prostate enlargement

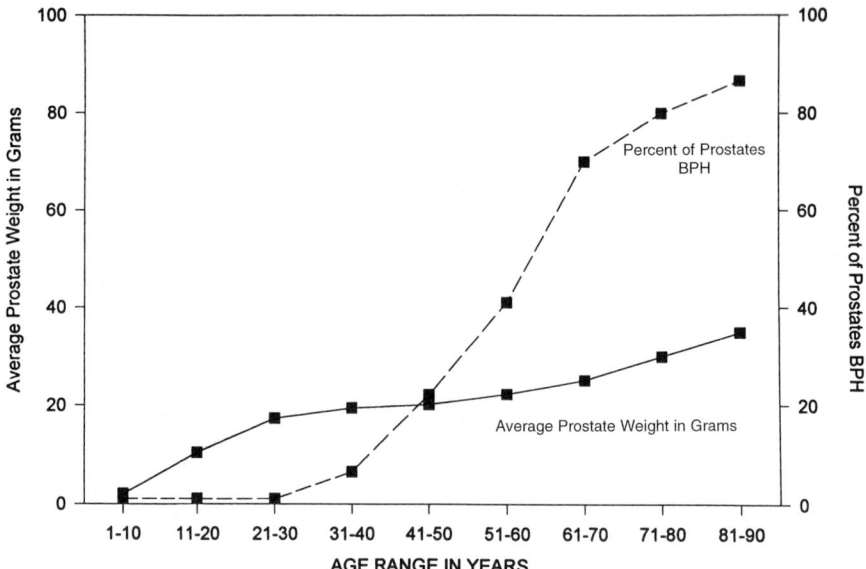

FIGURE 1 Age-related changes in histologic BPH and size of the human prostate. Autopsy prevalence of histologic BPH and changes in average prostate gland weight with advancing age. The increasing prevalence of BPH is much more striking than the increase in average weight. These concomitant observations reflect to some degree the discrepancy between the prevalence of histologic and gross BPH (modified from Berry et al., 1984).

requiring intervention probably occur earlier in blacks. The incidence of BPH is reported to be much lower in Chinese and Japanese living in Asia than in white populations. The low mortality rate for BPH from Asia reported to the World Health Organization provides confirmatory evidence for this observation. The data on the racial background of patients subjected to prostatectomy in Hawaii also provide evidence suggesting a relatively lower incidence of BPH in Chinese and Japanese than in white males. A higher than expected incidence of BPH in Welsh-speaking than in English-speaking men living in the same area of Wales and a threefold increased incidence compared with that expected in Jews requiring surgical treatment for BPH in New Haven, Connecticut, are observations cited to support a genetic factor in the development of this lesion.

II. ANATOMY

The prostate is a compound tubuloalveolar gland whose base abuts the bladder neck and whose apex merges with the membranous urethra to rest on the urogenital diaphragm. It resembles a blunted cone and weighs approximately 20 g. The urethra traverses the prostate by entering near the middle of its base and exits the gland on its anterior surface above and in front of its apical portion. The capsule of the prostate gland is an inseparable condensation of stromal elements that is incomplete at the apex; it does not represent a true capsule. Fibrous septa emanate from the capsule, pierce the underlying parenchyma, and divide it into 50 lobules. These glandular units drain into branched tubules, which lead into 20–30 prostatic ducts. Most of these ducts empty their contents into the prostatic urethra adjacent or distal to the verumontanum (Fig. 2).

A long-held belief that described discrete lobes of the prostate has fallen in disfavor. Currently the prostate is usually described as consisting of ventral (fibromuscular) and dorsal portions (glandular).

The glandular prostate thus delineated can be separated into four distinct regions: peripheral zone, central zone, transition zone, and periurethral gland region (Fig. 3). This zonal anatomy is important

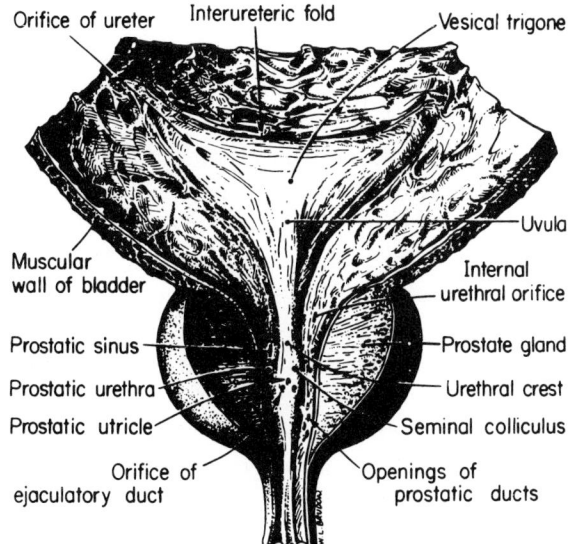

FIGURE 2 A frontal view of the bladder and prostate gland. The prostatic utricle sits atop the verumontanum (seminal colliculus) and represents one of two Müllerian duct remnants in man, the other being the appendix testis. Note that the majority of prostatic ducts drain adjacent or distal to the verumontanum. The area extending from the trigone to the termination of the prostatic urethra constitutes the internal urethral sphincter or bladder neck mechanism [From R. T. Woodburne, Pelvis, In *Essentials of Human Anatomy* (R. T. Woodburne, Ed.), Oxford Univ. Press, New York, 1978].

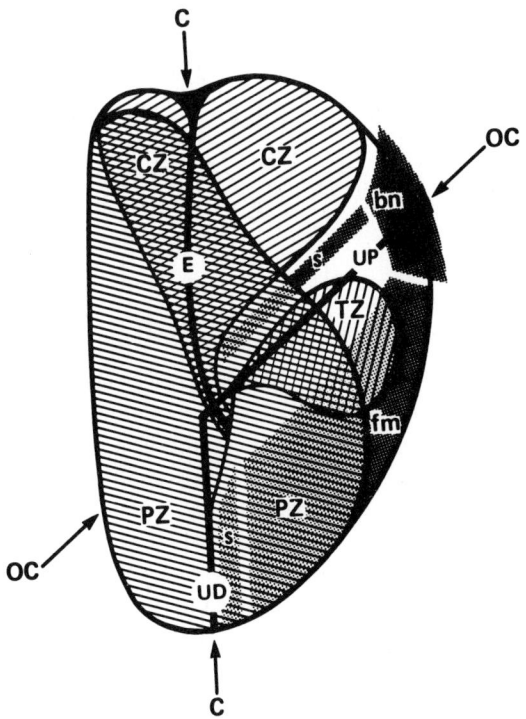

FIGURE 3 Saggital diagram of distal prostatic urethral segment (UD). Proximal urethral segment (UP), and ejaculatory ducts (E) showing their relationships to a sagittal section of the anteriomedial nonglandular tissues (bn, bladder neck; fm, anterior fibromuscular stroma; s, preprostatic sphincter; s, distal striated sphincter). These structures are shown in relation to a three-dimensional representation of the glandular prostate. CZ, central zone; PZ, peripheral zone; TZ, transitional zone (from J. E. McNeal, *Am. J. Surg. Pathol.* 12, 619, 1988).

because of the frequency with which various neoplastic diseases occur in each of these zones.

The peripheral zone constitutes approximately 75% of the glandular prostate. Its ductal system enters the urethra along the posterolateral recesses of the distal urethra extending from the prostatic apex to the verumontanum (Fig. 4).

BPH in man is a nodular, regional growth with a variegated gross appearance resulting from the inhomogeneous and irregular mixture of glandular and stromal tissue. The hyperplastic tissue is almost always located centrally in the periurethral portion of the enlarged gland.

In many instances the nodular hyperplasia is separated from the compressed peripheral prostate by a distinct cleavage plane that is smooth and resembles a capsule. The gross configuration and the weight, ranging from a few grams to more than 200 g, are highly variable. Surprisingly, no clear relationship between the size of the adenoma, the patients symptoms, or the degree of bladder neck obstruction has been established.

Observation of the hyperplastic prostate with light microscopy confirms the variable findings suggested by the gross appearance. All the glandular and stromal elements of the normal prostate are involved to a variable degree by hyperplasia. Franks identified five types of nodules based on their histologic characteristics: (i) stromal (fibrous or fibrovascular), (ii) fibromuscular, (iii) muscular ("leiomyoma"), (iv) fibroadenoma, and (v) fibromyoadenoma.

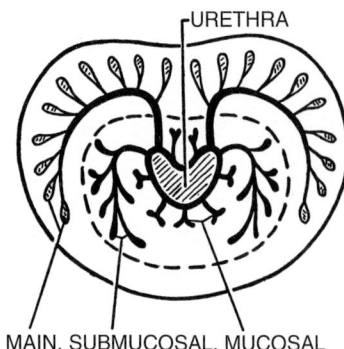

FIGURE 4 Cross section of normal prostate gland with the dorsal aspect superiorly placed demonstrating ductular architecture and three relatively concentric glandular arrangements. The mucosal component consists of delicate glands draining into short ducts, which empty circumferentially into the prostatic urethra. The submucosal group contains more extensive arborizations, which empty lateral to the urethral crest. These two groups constitute the periurethral area involved in benign prostatic hyperplasia. The acini constituting the lateral most group of glands have a distinct cystic configuration. This outer or peripheral aspect constitutes the surgical capsule of hyperplastic glands (from J. C. B. Grant and J. V. Basmajian, *Grant's Method of Anatomy*, 7th ed., Williams & Wilkins, Baltimore, 1975).

III. PHYSIOLOGY

A. Steroid Biochemistry

The growth and function of this exocrine gland are under partial endocrine control. Androgens are essential for the prostate to achieve and maintain normal tissue mass, composition, and secretory function. Estrogens, prolactin, thyroxin, and insulin influence androgen-stimulated growth but have limited or equivocal independent effect. There is an essential role of normal testicular endocrine function for the development and the maintenance of the prostate.

Normally, prostate growth is stimulated primarily by the reduced form of testosterone (T), dihydrotestosterone (DHT). DHT is produced in the prostate as a result of the activity of the type 2 isoenzyme of 5α-reductase. The observations that DHT and androstenediol were the principal metabolites of radioactive T incubated with slices of human BPH tissue has been confirmed. Although prostatic stroma has a much higher concentration of 5α-reductase, physi-

ologically significant conversion of diffused T to DHT has traditionally been thought to occur in the cytoplasm of the epithelial cells. This concept is currently being reevaluated because the DHT concentration in the prostatic tissue from younger men (age 45 years) is similar to the peripheral prostate of older men with BPH. DHT and T concentrations in BPH and normal tissue are comparable, as are the DHT concentrations reported for BPH tissue obtained by transurethral resection.

Evidence that 5α-reductase activity is increased two or three times in BPH and is predominantly localized to the stroma is substantial. These observations supported a causative relationship between the relatively increased amount of stromal tissue and the increased 5α-reductase reported in BPH tissues compared with normal tissue. The availability of an antibody to 5α-reductase type 2, the isoenzyme present in prostatic tissue, has permitted immunohistochemical evaluation of a variety of tissues for this enzyme. Observations have demonstrated perinuclear staining of the antigen in basal and stromal cells in the normal and BPH tissue. The predominant staining in BPH as contrasted to prostatic tissue designated as normal was in the stroma rather than basal cells. No antigen was detected in luminar epithelial cells.

Currently, the evidence that intracellular DHT, whatever its source, is bound to an androgen receptor in the nucleus seems substantial. T also binds to this receptor but with a significantly reduced affinity. The DHT receptor complex interacts with specific DNA sites with the ultimate production by mechanisms that are incompletely characterized of regulatory and secretory proteins. Current evidence suggests a significantly increased nuclear androgen receptor content in BPH compared with the normal prostate. Cellular proliferative, apoptotic, and secretory activities are variably affected by this interaction and by a number of other steroid/protein hormones and paracrine/autocrine growth factors. In general the reduced potency or quantity of other androgens secreted by the adrenal or testis or resulting from the peripheral metabolism of T reduces their potential effect on prostatic growth. The bioavailability of T is further altered by relatively stable binding to serum proteins such as testosterone-binding globulin. Estrogen and progestational agents are known to increase prostate size under some controlled situations, but this role *in vivo* appears less likely. Most investigators now have confirmed the presence of estrogen receptors in the prostate. Although the relative amounts and localization of this receptor remain unsettled, estrogen-binding sites are reported to be more numerous in stroma than in epithelial cells and to be decreased in concentration in BPH compared with normal tissue. Progesterone receptors are also present in appreciable concentrations in the cytosol of prostatic epithelial cells. Although the receptor–steroid complexes clearly have significant roles in the growth and function of prostatic cells, the exact mechanisms involved and the significance of receptor data with regard to the development and growth of BPH are still unclear. Despite extensive and continuing efforts, no qualitative differences in steroid content or metabolism, enzyme activity, organic compounds, or metal ions has been documented between BPH tissue and either normal prostatic tissue or the peripheral prostate in a gland with BPH. The few quantitative differences in biochemical characteristics that have been reported require further confirmation and have not helped as yet to identify a unique aspect of the metabolic activity characterizing BPH.

B. Secretory Function

The secretions of the prostate are involved in the process of insemination and also probably provide some protection against ascending bacterial infection. They are thought to be the exclusive products of the epithelial cells. The secretory process is described as both apocrine and merocrine in character. As a result, the prostatic secretion has a composition that reflects that of the intracellular milieu. This secretory capability of the epithelial cell is androgen controlled. Consequently, significant prostatic secretion is recognized to occur in the postpubertal period. Evidence from assessment of fractionally collected urine specimens suggests that the secretion of prostatic fluid is an ongoing process. Episodic expulsion of prostatic fluid as part of the ejaculatory process (emission) is readily recognizable gross evidence of prostatic secretion. This function is accomplished by a complex, integrated neurophysiologic mechanism.

The average volume of the normal human ejaculate is 3.5 ml (range, 2–6 ml), of which the prostate gland contributes only about 0.5 ml. The remainder is derived from the glands of Littre and Cowper (0.1–0.2 ml) and the seminal vesicles (2.0–2.5 ml). The secretions of the seminal vesicles are found in the terminal portion of the split ejaculate, following the secretions of the prostate, and appear to be biochemically complex, containing (i) a factor responsible for the coagulation of semen, (ii) fructose, and (iii) prostaglandins of the E, A, B, and F series.

C. Secretory Products

Prostatic secretions are a complex mixture of organic and inorganic compounds. This fluid is the sole source of seminal fluid zinc, magnesium, calcium, and citrate. Citrate, present in a mean concentration of 376 mg/dl, has been used as a biologic indicator of hormonal stimulation. The nitrogenous compounds, phosphorylcholine and the polyamines and spermine and spermidine, are secreted in prostatic fluid. Phosphorylcholine is probably a specific substrate for prostatic acid phosphatase; this enzyme liberates free choline and phosphate ion. Spermine binds to phosphate ions, nucleic acids, and phospholipids. The association between polyamine production and cellular proliferation is suggestive of a role for the former in malignant transformation and cell growth. Enzymatic degradation of spermine generates reactive aldehydes, giving semen its characteristic odor. These compounds and their polyamine precursors function as antibacterial agents.

Two-dimensional electrophoretic gel analysis indicates a very large number of distinct proteins in human prostatic fluid. Of these, only a small number have been characterized and identified, often by several differing descriptive terms. Prostate-specific antigen (PSA), prostate-secreted acid phosphatase (PAP), β-microseminoprotein, and prostate-binding protein are among these, with the first three being the predominate secretory proteins. The clinical roles for PSA and PAP in the evaluation and treatment of prostate cancer have served to focus investigative effort on these enzymes. The evidence indicates a major role for androgen in their secretion.

D. Epithelial–Stromal Interactions

Stereologic studies have demonstrated that the normal adult human prostate contains approximately 45% stromal tissue, 21% epithelial tissue, and 34% acinar lumens. Despite the recognized histologic variations in BPH this type of analysis has revealed a fourfold absolute increase of the stromal component compared with a nearly twofold increase in glandular elements in prostates with BPH. Recent studies demonstrated that the mean area density of smooth muscle, fibrous tissue, glandular epithelium, and glandular lumina in human BPH tissues is approximately 22, 54, 16, and 9%, respectively. A similar methodologic approach was applied to the analysis of prostate adenomas from men with symptomatic (after open prostatectomy, transurethral prostatectomy, or needle biopsy) and asymptomatic (after cystoprostatectomy for transitional cell cancer) BPH. Of interest, Shapiro demonstrated a statistically significant increase in the stromal compartment (62 vs 54%) and a decrease in the epithelial compartment (15 vs 21%) in men with symptomatic BPH. These stereologic and morphometric studies serve to reinforce the potential importance of stromal–epithelial interaction in the pathogenesis of BPH.

Evidence suggests that epithelial–mesenchymal interactions play a strategic role in androgen-induced regulation of prostate epithelial cell growth and differentiation via a multifaceted signal cascade that may involve cell-to-cell contact, the release of neurotransmitters, the impact of basement membrane extracellular matrix elements, and the production of soluble mediators. However, no critical differences between the cells of the normal and hyperplastic gland have been identified.

IV. NATURAL HISTORY

The first pathologic evidence of BPH occurs in <10% of the men in the 31- to 40-year-old group (Table 1). This observation may indicate that the initiating factor is present in most men of this age, with its effect being recognizable only in a few, or it may indicate that the young men with recognizable BPH have a discrepancy between physiologic and chronologic aging. Histologic evidence of BPH in-

TABLE 1
Age Prevalence of Human Benign Prostatic Hyperplasia[a]

	Autopsy studies					Combined data	
	Pradhan and Chandra (1975)	Swyer (1944)	Harbitz and Haugen (1972)	Franks (1954)	Moore (1943)	Prevalence of human BPH	
Age range (years)	No. with BPH/total no. (%)	No. with BPH/total no. (%)	No. with BPH/total no. (%)	No. with BPH/total no. (%)	No. with BPH/total no. (%)	No. with BPH/total no. (%)	% Mean ± SEM
1–10	0/11 (0)	0/16 (0)				0/27	0 ± 0
11–20	0/21 (0)	0/13 (0)		0/1 (0)		0/35	0 ± 0
21–30	0/37 (0)	0/21 (0)		0/4 (0)	0/24 (0)	0/86	0 ± 0
31–40	7/38 (18)	0/31 (0)		0/8 (0)	1/28 (4)	8/105	8 ± 8.5
41–50	6/19 (31)	2/28 (7)	4/6 (67)	3/18 (17)	7/23 (30)	22/94	23 ± 30.4
51–60	9/17 (53)	11/33 (33)	21/38 (55)	16/38 (42)	24/65 (37)	81/191	42 ± 9.7
61–70	7/12 (58)	23/33 (69)	49/66 (74)	40/54 (74)	52/77 (67)	171/242	71 ± 7.2
71–80	3/4 (75)	14/17 (82)	64/67 (96)	57/70 (81)	43/63 (68)	181/221	82 ± 11.1
81–>90	2/2 (100)		27/29 (93)	16/19 (84)	18/24 (75)	65/74	88 ± 10.9
Total	34/161	50/192	165/206	132/212	145/304	528/1075	

[a] Used with permission from Berry *et al.* (1984).

creases in each succeeding decade so that by the ninth decade it affects approximately 90% of men. The initial lesion of BPH almost always occurs in the periurethral area proximal to the verumontanum. Although the descriptions of the ductal and glandular structure of this area vary, there is general agreement that BPH arises from an inner set of prostatic ducts and glands that reside within the urethral wall or adjacent to it. McNeal indicated that this paraurethral tissue near the verumontanum, which he designates the transition zone, generally composes approximately 5% of the normal gland. Early lesions that develop within the urethral wall are usually composed of a tiny mass of loose embryonic-appearing stromal devoid of glands; however, within the nodule that develops in the paraurethral area, glandular tissue predominates. Once the process is initiated, all elements of the normal prostate—stromal, muscular, and glandular—participate to a variable degree in its progression. The glands in the hyperplastic nodules seem to have the capacity to bud and form new ducts and acini; in contrast to normal tissue, these new glandular elements grow toward each other. The variable local response to a postulated inductive agent is evident from the nodular nature of the BPH. Not all the nodules of BPH are in the same phase of development, as is clearly indicated by earlier observations that small stromal nodules were present in every enlarged prostate. This observation does not preclude the possibility that BPH results from a sequence of initiating and promoting effects that are episodic. Both the average weights of the prostate and the incidence of prostatectomy by decade suggest that once clinical BPH has developed, it is progressive in most men. The rate of growth calculated by various investigations indicates a prolongation of the doubling time with age. The important question of whether established BPH ever stabilizes or regresses spontaneously cannot be evaluated from the information available.

V. ETIOLOGY

Identification of the exact etiology of BPH has been a source of speculation and controversy since the entity was first recognized. The critical role of the testes and aging in the development of BPH has been appreciated but has not been characterized. Local factors promoting and altering growth and stromal epithelial interactions are under investigation. However, several observations continue to provide support for the possibility that BPH results from stimulation from local or systemic exposure or both to prostatic growth-promoting or growth-altering hormones or growth factors produced at extra organ sites. Among these, several deserve emphasis.

A. Role of Aging

As previously stated, about 90% of men in their ninth decade have histopathologic evidence of BPH. The increasing frequency of this process with age lends credence to the likelihood of a systemic pathophysiologic event. This almost uniform occurrence, making the absence rather than the presence of the lesion unexpected, seems to weigh heavily against an essentially spontaneous intraorgan proliferative phenomenon.

B. Role of the Testis

Development of recognizable BPH requires the presence of the testis in man. In the dog various hormonal manipulations using combinations of estrogens and androgens have produced a pathologic enlargement of the prostate that fails to replicate naturally occurring BPH completely. Although pharmacologic amounts of exogenous DHT can produce marked prostatic enlargement in the dog, BPH rarely occurs after the administration of physiologic or somewhat greater amounts of DHT or T to the castrated dog. However, administration of DHT to dogs with intact testes results in a significant incidence of induced BPH. A series of observations in rats and dogs, including radiation of the testis in dogs with BPH, support the secretion by the testis of a nonsteroidal factor that stimulates androgen-induced normal prostatic growth and BPH.

C. Age-Associated Changes in Accessory Sex Glands

Postmortem studies have revealed that male accessory sex glands other than the prostate are rarely the site of recognized pathologic changes. Assessment of age-related anatomic and functional changes in a hormone-dependent gland, the seminal vesicle, has demonstrated biologic evidence for the absence of a purely androgen-driven accessory sex gland—stimulating environment in the aging male. The seminal vesicle fluid fructose concentration, an androgen-dependent secretory product, decreased linearly with each decade. This observation corresponds to that expected from the decreased free T in the serum of aging males. On the other hand, the wet and dry weight of the seminal vesicle, an endocrine- but not totally androgen-dependent characteristic, is maintained with age. The maintenance of vesicle weight in the aging male implies the presence of a systemic sex gland-stimulating hormonal milieu. The divergent status of two endocrine-dependent biologic indicators can be explained by the fact that fructose concentration in the seminal vesicle fluid is totally androgen dependent and that seminal vesicle weight, proliferative as contrasted to a functional characteristic, is responsive to stimulation by a multiplicity of hormones usually acting with androgens. These observations employing established biologic indicators of hormonal status in the seminal vesicle support the presence of important, systemic, nonandrogenic and androgenic accessory sex gland—stimulating substances in the aging male. Other observations that suggest stimulatory mechanisms which are not primarily systemic androgen controlled as a potential etiologic factor(s) in accessory sex gland growth in man as he ages include (i) the acid phosphatase and lactic dehydrogenase concentrations in expressed prostatic fluid, which are thought to be androgen dependent, decrease progressively with age; (ii) the concentration of the citric acid, known to be increased by other steroids and protein hormones in androgen-maintained accessory sex glands of animals, increased in the 40- to 60-year-old male age group and was comparable in men older than 60 years and younger than 40 years old; and (iii) the observation that the DHT concentration in the peripheral prostate of older men with BPH is comparable to that in the prostate of young men, suggesting a paradoxic, locally mediated, but not necessarily controlled endocrine phenomenon. The limited observations suggesting that the peripheral prostate also shows an increase in weight support the concept of general accessory sex gland stimulation with differing regional effects and mechanisms. Overall, the evidence for extrinsically produced factors that play a significant role in the development of BPH seems substantial. This factor(s) may reach the prostate by a regional delivery system such as semen or urine, systemically, or all three.

The diverse histologic character of individual BPH lesions in a single gland also argues for an important role for extrinsically or intrinsically derived local mediators of cellular growth in the development of this lesion. The role of paracrine and autocrine growth factors as responders to and signal transducers of the hypothesized essential external stimuli seems likely to be a major one.

VI. PATHOPHYSIOLOGY

Concepts regarding the etiology of the urinary symptoms and obstructive sequelae resulting from BPH have traditionally focused on the development and progression of mechanical obstruction from the prostatic mass. The perception that the mass and configuration of the hyperplasia dictated the degree of outflow blockage resulted from the early experiences in treatment of patients with acute and chronic urinary retention. Renal failure, urinary tract infection, and calculi were common indications for various approaches to relieve bladder neck obstruction. Removal of the obstructing prostatic mass usually reversed these serious secondary phenomena and often restored voiding patterns. Although failures in these therapeutic goals occurred, they were overshadowed by the frequency of correction of the various problems that existed. Recently, a mass-independent, modifiable, functional component in the bladder neck obstructive phenomena in many individuals has received consideration. The demonstration of functional α-adrenergic receptors, relative and absolute increases in stromal smooth muscle in the prostate, and improved voiding patterns that have resulted from administration of α-adrenergic antagonists provided supportive evidence for a functional component in the development and progression of prostate-related voiding dysfunction. Furthermore, recent evidence supporting the presence of endothelin receptors in the prostatic stroma and epithelium and a significant contractile response of human prostatic smooth muscle to this vasoconstrictive peptide suggests that other potential mechanisms exist for functionally mediated BPH phenomena affecting voiding.

The sequence of pathophysiologic changes leading to and effects of increasing bladder outflow obstruction on bladder function are multifactorial and are incompletely characterized. The bladder reacts to increased outlet resistance in a manner very much akin to the heart operating within the Frank–Starling curve. The bladder responds to this increased workload by passing through the stages of (i) irritability, (ii) compensation, and (iii) decompensation. Investigators have found that the intravesical pressure at opening and closing of the bladder neck increased in patients with prostatic disease. Compared with the normal bladder, the detrusor muscle in the bladder with established mature bladder neck obstruction has a higher resting tonus, generates a greater force during voiding, and cannot maintain a constant increased intravesical pressure for as long a time. The anatomic changes accompanying these hemodynamic findings have been documented in experimental studies of varying degrees of lower urinary tract obstruction in several animal models and in histologic studies in humans. Histologic studies in humans with bladder neck obstruction have demonstrated hypertrophy and connective tissue infiltration of smooth muscle. Recent ultrastructural observations of myohypertrophy with wide intercellular spaces containing collagen fibrils and fibers, as well as numerous elastic fibers in the obstructed bladder, provide evidence of structural change that seems reasonably correlated with physiologic observations. The varying mixes of anatomic and physiologic alterations described in response to obstruction probably play a major role in the specific bladder and renal changes that occur in individual patients. Currently, loss of bladder compliance, thought to be primarily caused by altered collagen deposition and composition, is probably the major factor in producing upper urinary tract functional and anatomic damage. Evidence of obstruction-associated disordered bladder innervation characterized in some studies as partial denervation of smooth muscle and probably resulting in both urinary urgency and incontinence has accumulated. Fortunately, these involuntary contractions, like other phenomena associated with bladder outlet obstruction, are demonstrably reversible in animal models and, with variable frequency, in clinical practice.

See Also the Following Articles

Prostate Cancer; Prostate-Specific Antigen; Testosterone Biosynthesis

Bibliography

Berry, S. J., Coffey, D. S., Walsh, P. C., et al. (1984). The development of human benign prostatic hyperplasia with age. J. Urol. 132, 474.

Franks, L. M. (1954). Atrophy and hyperplasia in the prostate proper. J. Pathol. Bacteriol. 68, 617.

Harbitz, T. B., and Haugen, O. A. (1972). Histology of the prostate in elderly men: A study in an autopsy series. Acta. Pathol. Microbiol. Immunol. Scand. (A) 80, 756.

McNeal, J. E. (1976). Developmental and comparative anatomy of the prostate. In Benign Prostatic Hyperplasia (J. T. Grayhack, J. D. Wilson, and M. J. Scherbenske, Eds.), NIAMDD Workshop Proceedings, February 20–21, 1975, U.S. Department of Health, Education, and Welfare Publ. No. (NIH) 76-1113, pp. 1–10. National Institutes of Health, Bethesda, MD.

Moore, R. A. (1993). Benign hypertrophy of the prostate: A morphology study. J. Urol. 50, 680.

Pradhan, B. K., and Chandra, K. (1975). Morphogenesis of nodular hyperplasia-prostate. J. Urol. 113, 210.

Randall, A. (1931). Surgical Pathology of Prostatic Obstruction. Williams & Wilkins, Baltimore.

Rotkin, I. D. (1976). Epidemiology of benign prostatic hypertrophy: Review and speculations. In Benign Prostatic Hyperplasia (J. T. Grayhack, J. D. Wilson, and M. J. Scherbenske, Eds.), NIAMDD Workshop Proceedings, February 20–21, 1975, U.S. Department of Health, Education, and Welfare Publ. No. (NIH) 76-1113, pp. 105–118. National Institutes of Health, Bethesda, MD.

Shapiro, E., Becich, M. J., Hartanto, V., et al. (1992). The relative proportion of stromal and epithelial hyperplasia is related to the development of symptomatic benign prostate hyperplasia. J. Urol. 147, 1293.

Swyer, G. I. (1944). Post-natal growth changes in the human prostate. J. Anat. 78, 130.

Birds, Diversity of

Scott V. Edwards

University of Washington

I. Origin and Fossil Record of Birds
II. Phylogeny and Biogeography
III. Diversity of Passerine Birds
IV. Diversity of Nonpasserine Birds
V. Behavioral Diversity of Birds
VI. Threats to the Diversity of Birds

GLOSSARY

Cretaceous era The geological time period spanning the period roughly from 150 to 65 million years ago.

DNA–DNA hybridization A molecular biological technique for assessing the genetic divergence between two complex genomes belonging to two species. This technique was used by Charles Sibley and Jon Ahlquist to develop a phylogenetic tree for birds that spans all orders and families, the first such broad study for a vertebrate group.

keel The large breastbone to which the muscles for flight attach; also called a *carina*. Some birds, such as flightless ratites, do not have a keel.

molecular clock A condition in which the probability of nucleotide changes in a sequence of DNA is approximately constant per unit time. If this condition holds, and if the approximate number of changes per unit time is known by calibration with the fossil record, then the number of DNA differences between species can be used to estimate the time of divergence of those species.

monophyletic group or *clade* Any group of organisms including all the descendants, living and extinct, from a common ancestral species. Recent trends in taxonomy designate monophyletic groups of organisms rather than the traditional Linnaean categorical ranks of genus, family, order, etc. This trend recognizes that different groups of the same categorical rank [e.g., two avian families, such as the Old World flycatchers (Muscicapidae) and the woodpeckers (Picidae)] may nonetheless differ drastically in age and time of origin, and some traditional taxonomic groups do not include all the descendants from a particular common ancestor (i.e., are not monophyletic). In general, where older Linnaean names are assumed to designate monophyletic groups, the traditional name is now used to denote the same group that the old name denoted, but without the categorical rank (e.g., family Muscicapidae is now simply Muscicapidae).

node A common ancestor, depicted as the intersection of two branches in a phylogenetic tree.

passerine bird Any one of approximately 5300 species of generally small-bodied birds with a foot structure evolved for perching and a complex syrinx; perching bird, distributed on all continents except Antarctica.

phylogeny The history of ancestor–descendant relationships for a group of organisms; genealogy or historical descent.

ratite bird Any one of about 60 species of extant birds belonging to the clade Paleognathae, one of two primary branches in the avian tree, distributed primarily in Africa, Australia, and South and Central America. Many ratite birds, such as ostriches (*Struthio camelus*; Africa), emus (*Dromaius novaehollandiae*; Australia), or brown kiwis (*Apteryx australis*; New Zealand), are flightless, but some, such as Tinamous in the New World (Tinamidae), have limited capabilities of flight.

syrinx The avian voice box; a structure unique to birds, situated ventral to the point where the trachea splits into two bronchi and capable of producing a wide variety of sounds by air passing through a series of complex membranes.

Birds are one of the more diverse clades of vertebrates, with approximately 10,000 living species known worldwide. These supremely adapted flyers inhabit a wide variety of habitats, from polar regions to tropical rain forests, and display an unparalleled diversity of mating systems and cooperative strategies.

I. ORIGIN AND FOSSIL RECORD OF BIRDS

In recent years there has been renewed interest in the origin of birds, from both paleontological and molecular perspectives. In the past, both paleontological and molecular perspectives have suggested an origin for lineages of living birds about as long ago as that for mammals, e.g., around 100 million years ago (mya). Early studies using the technique of DNA–DNA hybridization, as well as classification of certain fossils from the Cretaceous era such as the loon-like *Hesperornis* and *Ichthyornis*, suggested such an early origin around 100 mya. However, recent reinterpretations of the available fossil data by some researchers suggest that some of these early Cretaceous fossils are not in fact related to modern forms, and that most lineages of the first round of avian evolution went extinct at the end of the Cretaceous and were replaced at the Cretaceous–Tertiary boundary (about 65 mya) by a second, rapid radiation of forms immediately ancestral to clades alive today. At the same time, however, recent analyses of molecular data suggest again an older date of origin for birds, around the mid-Cretaceous. These recent molecular studies tested for a molecular clock and, for some genes, were unable to reject a constant rate of DNA change. Using these clock-like genes, these researchers used the divergence of mammals and birds 310 mya to calibrate the clock, then used this calibration to date the divergences of various bird groups. Surprisingly, the molecular estimates of divergence were in general much older than the oldest fossils suspected to have been common ancestors of those groups, implying that the fossil record for birds was patchy and inconclusive at critical deep roots in the genealogical tree for modern bird groups. Further coordination of available molecular and fossil data undoubtedly will refine even these recent estimates for the origin of modern bird groups.

II. PHYLOGENY AND BIOGEOGRAPHY

There are currently between 9000 and 10,000 species of birds known from every continent. Although birds appear diverse in morphology and body plan, as a group they have generally been considered conservative and surprisingly restricted in morphology compared to the morphological diversity of mammals such as whales and bats. It is therefore not surprising that determining the phylogenetic relationships of this large number of species has been difficult. Indeed, the phylogenetic relationships of various clades of birds to one another are still poorly known, primarily because of a paucity of well-defined morphological and molecular characters brought to bear on the question. In addition, the likelihood of a very rapid radiation of avian lineages in the past makes reconstruction of genealogical relationships of birds particularly difficult with either molecular or morphological data. It is generally agreed that birds are a monophyletic group, i.e., all extant species (the Neornithes) have descended from a single common ancestral species (Fig. 1, node A), which presumably had the characteristics that define all birds: feathers, endothermy, a skeleton highly fused and modified for powered flight, bipedal locomotion, and a hard, keratinous beak without teeth.

It has also generally been recognized that all birds can be divided into two major groups, the Paleognathae and the Neognathae (Fig. 1, nodes B and C, respectively). Paleognathous birds, also called ratites, are defined by a distinctive, primitively shaped palate (roof of the mouth). Paleognathous birds include many well-known flightless species without keels, such as ostriches, emus, and cassowaries. These large, heavy birds are distributed on all the southern continents except Antarctica and are thought to have achieved this distribution not by flying but by drifting along with the continents as they spread apart from a former unit called Gondwanaland in the mid-Cretaceous. Another group of paleognathous birds, the Tinamous (Tinamidae), can fly short distances and are found throughout the New World tropics from Mexico to the southern tip of South America.

The Neognathae consist of all birds other than ratites (Fig. 1, node B). Amino acid sequences of proteins as well as morphological characters delineate the Neognathae as a valid clade. The basal branch of the Neognathae, again supported by molecular and morphological traits, includes many species familiar on farms and rural areas of the world—

FIGURE 1 Phylogenetic tree of major clades of birds based on molecular and morphological data (after Cracraft, 1988, and Mindell, 1997).

ducks and geese (Anseriformes) and chickens and other gallinaceous birds (Galliformes), together referred to as the Galloanseriformes (Fig. 1, node D). The Galliformes include such groups as the pheasants (Phasianidae), many with bright, metallic plumage and reaching their greatest diversity in Asia; the chachalacas (Cracidae), primitive neotropical relatives of game birds; and the megapodes (Megapodidae), which are distinguished by their habit of incubating their eggs by placing them near external heat sources (see Section V,C). The sandgrouse (Pteroclidiformes) is a little known Old World group with no true affinities to Galliformes.

The Anseriformes, generally believed to be the sister group of Galliformes, consist of a morphologically fairly uniform group of waterbirds. In addition to the more familiar kinds of ducks, geese, and swans that make up the 150 species of Anseriformes is the pied goose (*Anseranas semipalmata*), which, as its Latin name implies, is intermediate between geese (exemplified by *Anser*) and ducks (exemplified by *Anas*). With their strong wings and flight, geese have reached the far corners of the earth: Among the highest flying birds are bar-headed geese, which have been recorded migrating over the top of Mount Everest, the world's highest mountain, and the Hawaiian goose, now restricted to the island of Hawaii, having evolved from a small-bodied Canada goose that flew from the Aleutian Islands to the archipelago <3 mya.

III. DIVERSITY OF PASSERINE BIRDS

The Passeriformes, or perching birds, comprise the largest and most diverse commonly recognized clade of birds (Fig. 1). Perching birds usually cap most traditional discussions of bird diversity, revealing the tendency of systematists to portray them as the pinnacle of the bird world: These birds are very speciose, small-bodied, rapidly evolving, adaptable, and intelligent. However, recent molecular analyses of avian phylogeny suggest that perching birds may not hold the derived phylogenetic position commonly assumed: both DNA–DNA hybridization studies and DNA sequence analyses place Passeriformes if not in a basal position within Neognaths then on an equal phylogenetic level with other Neognaths. Furthermore, recent fossil specimens of the oldest perching bird, found in Australia and recently described by

Walter Boles, indicate that this group is at least 50 million years old. This raises the intriguing possibility that many of the traits considered recent and derived for perching birds may in fact be relatively primitive. In fact, the phylogenetic position of Passeriformes is still very uncertain, and they are discussed at this point simply to portend what may be a major upheaval in the genealogical position of this group in the near future.

The Passeriformes (or passerine birds) are synonymous with what are commonly known as "perching birds"; this group also contains within it a major radiation commonly known as songbirds (oscine passerines). Over half (~5300) of all extant birds are perching birds. Perching birds have a worldwide distribution, with representatives on all continents except Antarctica, and have their greatest diversity in the tropics. Body sizes of passerines vary from about 1.4 kg in northern populations of ravens (*Corvus corax*) to just a few grams. Perching birds include some of the most colorful and mysterious of all birds, such as birds of paradise from New Guinea and the bright orange cock of the rock from tropical South America. Because of their high diversity, generally small body size, and relative ease of observation, collection, and field study, perching birds have historically attracted the attention of a wide range of descriptive and experimental biologists, including systematists, behavioral ecologists, and evolutionary biologists.

Historically, it is generally agreed that the Passeriformes constitute a monophyletic group. Robert noted that Passeriformes possess a suite of distinguishing characteristics, including a unique sperm morphology, a distinctive morphology of the bony palate, a simple yet functionally diverse foot for perching with three toes forward and one (the hallux) oriented backwards (anisodactyly), and a distinctive wing and hindlimb musculature. There are few if any species which pose problems for avian systematists as to whether they are or are not passerines. Most of the controversy lies in relationships within the clade.

The basal lineages of passeriformes have for over a century included two major clades: the oscines and suboscines (Fig. 2). Suboscines are a largely tropical group of about 1000 species that reaches its greatest

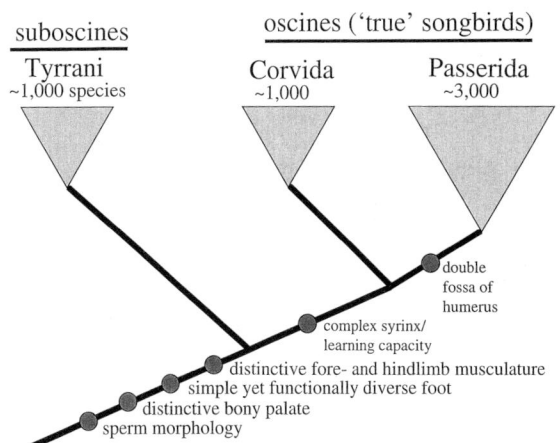

FIGURE 2 Phylogenetic tree of major clades of passerine birds, showing divisions Corvida ("crow-like" songbirds, many Australian) and Passerida (other songbirds from Africa, Eurasia, and the New World) (after Sibley and Ahlquist, 1990).

diversity in the South America; most suboscines are thought to sing "innate" songs. Oscines include about 4000 species and are what many laypersons refer to as "songbirds"; they are worldwide in distribution and are distinguished from suboscines by a complex voice box (syrinx) and song-learning capacity. However, there are a few species, such as the two lyrebirds in Australia—famous for their elaborate vocal virtuosity and mimicry as well as their lyre-shaped tail—and the diminutive New Zealand wrens (Acanthisitidae), that defy easy placement into either of these groups.

It is worth mentioning one more division within the perching birds that is within the oscine passerines (Fig. 2). Prior to the application of molecular data to questions of higher level passerine phylogeny, there was very poor if any resolution of the major clades, primarily because there were few obvious morphological and skeletal traits that united large subgroups within oscines. However, DNA–DNA hybridization comparisons have shed enormous light on this issue and delineated two major groups within the oscines: the Corvida, consisting of crow-like species and others that originated in the Australo–Papuan region, and the Passerida, consisting of sparrow- and thrush-like species and which are found mainly in Eurasia, Africa, and the Americas.

This molecular division of oscines into Passerida and Corvida is corroborated by at least one morphological character, namely, the condition of the tricipital fossa at the head of the humerus, and the division clarifies passerine biogeography by positing a single origin for the distinctive, endemic lineages of Australian songbirds—implying an extensive and diverse radiation of forms on par with the marsupials—rather than multiple independent invasions of Australia by ancestors from Eurasia.

The temporal and geographic origin of passerine birds is obscure. Traditionally, the group was thought to have originated in the Tertiary, at about the same time as extant orders of mammals. Some recent workers favor a later, Eocene origin, but the DNA–DNA hybridization data favor an earlier origin. The recent Australian fossils and other paleobiogeographical data suggest that passerines may have in fact originated in the Southern Hemisphere and spread northward. An early time for this spread leaves open the possibility that some aspects of passerine biogeography have been influenced by Gondwanan continental drift in the same way as for ratites, although an origin for passerines this early is not very likely given the current data.

IV. DIVERSITY OF NONPASSERINE BIRDS

The group of birds that are neither perching birds nor Galloanseriformes (Fig. 1, node E, not including Passeriformes) is not a natural (i.e., monophyletic) group; indeed, there are few clades (traditionally different orders) among these remaining birds whose affinities to other clades are not debated. For example, whether or not the predatory hawks and owls form a natural group is highly contentious; traditional and modern morphological analyses suggest that the traits that are similar between these groups (some of the most obvious being strong feet and talons and exceptionally keen vision) evolved only once in the common ancestor of the groups. By contrast, recent results from molecular systematics imply that hawks and owls are only distantly related, a scenario which requires two independent origins for these traits associated with predation.

One uncontested clade containing more than one major type of bird is the hummingbird/swift clade (Apodiformes; Fig. 1). Hummingbirds and swifts, which appear outwardly quite different, are nonetheless related to each other in possessing very small, weak feet in addition to a number of internal characteristics. Swifts are small to medium in body size and have a large, gaping mouth designed for catching insects on the wing, whereas hummingbirds are minute and generally have a long bill adapted for securing nectar from flowers. The smallest bird in the world is the bee hummingbird (*Mellisuga helenae*) found in Cuba; it is 2.25 in. from bill to tail tip. Molecular biology also supports this clade, one of the few examples of congruence between morphological and molecular results.

Another diverse clade but one on which there is general agreement on monophyly is the cranes and relatives, or Gruiformes, although recent molecular data question the monophyly of this group (Fig. 1). Found worldwide and frequently associated with marshes, floodplains, or other wetland habitats, Gruiformes include some of the larger land birds such as the whooping crane (*Grus americana*), an endangered North American species which stands over 1.5 m tall and has an elaborate "jumping" dance which is performed by both sexes on the mating grounds in the central United States and Canada each year. At the other extreme of body size are the rails, which have evolved flightlessness in numerous lineages but are nonetheless worldwide in distribution. Despite their small size, generally poor flying capabilities, and habit of skulking low in marsh grasses, rails have dispersed to some of the most remote of Micronesian archipelagos and atolls, repeatedly evolving flightlessness, an energetic advantage in the absence of predators. (Although no species has evolved flightlessness, megapodes share with rails the characteristic of poor flying but high dispersing capabilities, as evidenced by their having reached such isolated island chains as the Marianas in the Pacific). Some of the more enigmatic gruiformes are the limpkin, a species intermediate between cranes and rails found in the southeastern United States, and the kagu (*Rhynochetos jubatus*), a crested waterbird found only in New Caledonia.

The other major clades of birds associated with

water include birds of the open ocean, lakes, seashores, and marshes (Fig. 1, node F): Because of their resemblance to extinct toothed birds such as *Hesperornis*, the most primitive looking of this group include the loons (Gaviiformes). Although the phylogenetic affinities of loons are still uncertain, it is now known that the similarities between loons and Cretaceous birds are the result of convergence, and that loons as a group are not as ancient as previously supposed. The five species in this clade are holarctic in distribution and breed on isolated lakes in the tundra, where their undulating, mournful calls can be heard. The masters of flight over the open oceans are the Procellariformes, or tube-nosed seabirds. These birds range in size from the sparrow-sized Leach's storm petrel (*Oceanodroma leucorhoa*) of the north Pacific and Atlantic Oceans to the huge, soaring wandering albatross (*Diomedia exulans*) of the southern oceans, with a wingspan of nearly 4 m. Procellariformes spend much of their lives over the open seas in between breeding seasons on isolated atolls, islets, and archipelagos. They have a keen sense of smell by which they locate krill, crustaceans, fish, and other food kilometers away. Penguins (Spheniciformes) are the major component to the avifauna of Antarctica. The 18 species evolved from flying tube-nosed ancestors and all have flattened, highly fused wing bones that support broad flippers by which they "fly" underwater. Megellanic penguins (*Speniscus magellanicus*), a relatively small species distributed around southern South America, are known to dive up to 80 m in search of food, whereas emperor penguins (*Aptenodytes forsteri*), an Antarctic species weighing up to 40 kg (100 lbs), typically dive up to 200 m and have been recorded at depths over 450 m (1500 ft). The Pelacaniformes are a diverse clade consisting of such lineages as pelicans, frigate birds, boobies, cormorants, and tropic birds. The frigate birds, tropic birds, and anhingas are pantropical in distribution, whereas the remaining groups are more widespread. The Anhingas (*Anhinga*), or snake birds, of which there are four species, often swim through still waters and rivers with their bodies completely submerged and only their long, thin necks and streamlined head and bills breaking the surface. Frigate birds (*Fregata*) are the pirates of the open ocean, frequently harassing other birds into disgorging their food and snatching it in flight as it falls before it hits the ocean surface. Finally, the grebes (Podicipediformes) are a group of about 20 species inhabiting lakes, marshes, and estuaries around the world. Like some Anseriformes, adult grebes often carry young chicks on their backs while swimming. Grebes build floating nests of reeds and marsh grasses, and some species have elaborate courtship displays as in the case of the western grebes (*Aechmophorus occidentalis*) of western North America, in which courtship consists of pairs of birds skimming rapidly over the water surface with their heads and necks arched backward in unison.

With over 300 species, the gulls, terns, auks, puffins, and shorebirds (Charadriiformes) are the largest clade of birds generally associated with water (Fig. 1). The gulls (Laridae) are the most diverse and colonial of the group, with 88 species. Close relatives of gulls, the jaegers (Stercorariidae), breed in the polar regions and are carnivorous, frequently feeding on chicks of nearby nesting species and small rodents. Shorebirds (Charadriidae) are famous for their annual migrations in both hemispheres. The millions of shorebirds, such as dunlin, black-bellied plovers, long-billed dowitchers, and western and least sandpipers, that congregate at "staging areas" around western North American bays, estuaries, and marshes en route to their wintering areas on South America are a memorable sight to bird watchers.

The herons, night herons, egrets, bitterns, and tiger herons (Ciconiiformes) form a large, widespread clade typically found in marshes and slowly moving bodies of water. Species in this group typically have long necks and stab fish and crustaceans just under the still surface. When disturbed, bitterns sit still in the marsh reeds with their bill pointing straight up and eyes directed forward, revealing the heavily streaked underside of the neck and belly that aids in camouflage. Herons are one of the only groups in which there is thought to be a plumage polymorphism maintained by selection. Pacific reef herons (*Egretta sacra*), a species with blue and white morphs, are common throughout the south Pacific. In the Cook Islands, individuals foraging in the frothy ocean shores are white, whereas blue individuals of the same species prefer foraging in the dark, shaded waters of inland forest streams. It is thought that background matching of both the white and blue morphs offers a selective advantage in their respective microhabitats so that prey cannot easily detect them, although the larger geographical variation of this species is also consistent with other hypotheses.

Conspicuous and largely terrestrial nonpasserine bird groups include the following (Fig. 1, node E in part):

Nightjars (Caprimulgiformes): This group includes birds that forage for insects in flight, like swifts and swallows, but often at night. They are characterized by very large, dark eyes with sensitive retinae and long "whiskers" (actually modified feathers) at the base of the bill for sensing prey. The group includes frogmouths and owlet nightjars from the Australo–Papaun region, potoos from the neotropics, and the nightjars, which are found throughout the world except in New Zealand. Nearly all of these species are drably colored in mottled browns and grays and often sit motionless, camouflaged on a low branch during the day. The most bizarre caprimulgid (if it is in fact a member of this group) is the enigmatic oilbird (*Steatornis caripensis*), which with the aid of a bat-like clicking sonar device and detection system can fly accurately in the complete darkness of the South American caves that are its usual haunt.

Hornbills, rollers, and kingfishers (Coraciiformes), mousebirds (Coliiformes), cuckoos (Cuculiformes), woodpeckers (Piciformes), and trogons (Trogoniformes): These clades exhibit a wide variety of body forms and habits, most noteworthy of which is the diversity of foot types. Woodpeckers and cuckoos have a zygodactyl foot in which two toes point forward and two point back, a versatile arrangement that evidently suits the varied climbing, grasping, and perching habits of these groups quite well. Many coraciiform birds have two front toes fused, a condition known as syndactyly. This very diverse clade is largely tropical in distribution, except for the kingfishers, which have a number of representatives in the extreme Northern and Southern Hemispheres. Trogons (Trogoniformes), a pantropical group of 37, are found in Africa, Asia, and the New World. Despite their widespread distribution, trogons, like woodpeckers and other piciformes, do not cross the famous Wallace's line, a line drawn north–south between the islands of Borneo and Sulawesi demarcating the Asian and Australo–Papuan biogeographic realms that also divide the points of origin for the Corvida and Passerida within the songbirds. Trogons are distinguished by a unique foot arrangement (heterodactyly) in which toes 1 and 2 point backwards rather than toes 1 and 4 pointing backwards as in zygodactyly. African mousebirds, a strange family of 6 species named for their habit of scurrying rodent-like among leaf litter and along branches, can reverse the outer two toes to bring them backward or forward (pamprodacyly). The turacos (Musiphagiformes) are an African group of 18 brightly colored species with only the outer toe reversible. This group is distinguished by the unique deep olive greens and reds in the plumage of several species, which are produced by two pigments unique to the group, turacoverdin and turacin, respectively.

Parrots (Psittaciformes): The 340 species of parrots, lories, and cockatoos are pantropical in distribution, with major centers of diversity in South America, Africa, New Guinea and Australia, and the Pacific, suggesting an influence of the Gondwanan breakup on current diversity. Parrots, like woodpeckers, have zygodactyl feet, which they use for grasping nuts and other food and placing it in their bills. Although parrots seem ungainly and cumbersome in the wild, they are in fact highly intelligent and adaptable. The parrots of New Zealand are some of the most divergent, including the kakapo (*Strigops habroptilus*), a large, nocturnal and flightless species also called the owl parrot, and the predatory kea (*Nestor notabilis*), which has an extremely sharp, decurved bill which it uses to pry the rubber lining off the windshields of parked cars among other uses more fundamental to its survival. Many parrots are highly endangered due to habitat destruction and illegal pet trade; the Carolina parakeet, once abundant in the southeastern United States, is now extinct due primarily to hunting, whereas the thick-billed parrot, although still found in Mexico, has been locally exterminated in the mountains of southern Arizona and only recently have there been attempts to reestablish it.

Pigeons and doves (Columbiformes): A widespread

and speciose group, the pigeons and doves range in size from the diminutive ground dove of the southeastern United States and Mexico to the extinct flightless dodo (*Raphus cucullatus*), which lived on the island of Mauritius and weighted several kilograms. In the tropics, particularly in New Guinea and the Pacific, doves can be quite colorful, e.g., the elegant crowned victoria crowned pigeons (*Goura*) of New Guinea. Although in general of less conservation concern than parrots, there is no better illustration of the susceptibility of once abundant birds to extinction than the passenger pigeon (*Ectopistes migratorius*) of the eastern United States, the last of which died in the Bronx Zoo in 1914.

V. BEHAVIORAL DIVERSITY OF BIRDS

A. Migration

The diversity of sizes and shapes of birds is matched if not surpassed by the diversity of behaviors and life histories they exhibit. Most birds have a well-defined annual cycle of breeding, parental care of young, molting of feathers, dispersal, and migration. This cycle can very not only among species but also between different populations of the same species. For example, northern populations of white-crowned sparrows (*Zonotrichia leucophrys*) breeding in Alaska migrate south to California, whereas populations breeding in California are nonmigratory. Fox sparrows undergo a "leapfrog" pattern of migration in which individuals breeding in northern localities migrate the farthest south, with populations migrating less extreme distances as they breed in more latitudinally intermediate localities. Indeed, migration cycles are one of the more variable aspects of avian life history. Many tropical species do not migrate but rather remain sedentary year-round. Numerous Old and New World suboscine species maintain territories throughout the year, whereas many tiny songbirds, such as New World warblers, weighing only 10 g, undergo long-distance migration every fall from their breeding grounds in the northern United States and Canada, heading well out over the Atlantic Ocean to catch northeast trade winds that aid them on their way south to northern South America. The record-holder for long-distance migration is the Arctic tern (*Sterna*), which travels 25,000 km round-trip each year from its breeding grounds in the Arctic to where it winters at the tip of South America and back. The benefits of migration for such species, which actively exploit geographically and temporally varying food abundance on wintering grounds, must indeed be substantial. Indeed, migration in some species has evolved extremely rapidly in response to changing environments. Between the 1960s and 1990s, blackcap warblers (*Sylvia atricapilla*) from central Europe evolved a novel migration route to Britain instead of to their traditional western Mediterranean wintering grounds, quite possibly due to human supplementation of food in winter and ameliorating climate. Birds utilize a variety of cues to orient during migration, including the earth's magnetic field, the sun, moon, and stars, low-frequency sound produced by ocean waves and tides, and inborn genetic maps, which can be modified by selective breeding experiments. European blackcaps are an example of this last point. Amazingly, hybrids of birds that typically migrate southeast from Europe to North Africa and birds that typically migrating southwest to Morocco migrate in an intermediate direction. These experiments show that inborn maps, in addition to substantial learned experience and environmental cues, aid birds on the seasonal wanderings.

B. Mating Systems and Sexual Dimorphism

The mating systems of birds are among the most diverse of any vertebrate group. Until a few years ago it was believed that roughly 90% of all species were monogamous, i.e., one male mated with one female to produce a clutch each season. This seemed obvious from the pattern of associations of males and females observed in many species in the field: Two males were rarely if ever seen mating with an individual female. However, in the past 10 years, genetic studies in the form of DNA fingerprinting have changed this view dramatically. Ornithologists now find that a substantial fraction of the chicks in nests of supposedly monogamous have been sired by males other than the one attending the nest! Birds once considered classically monogamous, such as eastern bluebirds (*Sialia sialis*) and cardinals (*Cardinalis cardinalis*) in North America, are now known to have a cryptically polygynous mating system in which any given female may mate with several males and vice versa. Why males would feed and defend nestling that are the result of such "extrapair fertilizations" is not understood, although some species, such as the European dunnock (*Prunella modularis*), have supernumerary males at nests which feed nestlings at random at a rate proportional to the fraction of nestlings they have sired. Apparently, these males judge their contribution to the genetic makeup of a clutch by the time of exclusive access to the female they have enjoyed. Cooperative breeding, in which young birds delay breeding and assist other individuals (often parents) in raising young and defending the territory, is a social system that can be associated with any number of mating systems. This trait, which is frequently first identified in the field by the presence of birds in groups, is particularly prevalent in Australia and is common in several passerine groups, such as Australian fairy wrens (Maluridae) and New World jays (Corvidae).

It is generally believed that in many types of mating systems the intense selective pressures on males by females looking for a mate ("female choice") or, less frequently, between males competing for mates ("male–male competition") has produced the dramatic differences in size, shape, and color between the sexes and this is known as sexual dimorphism. In general, males are larger than females by about 5% of body weight, but in some groups, such as hawks, owls, and shorebirds, females are larger than males. For example, in shorebirds such as the pectoral sandpiper (*Calidris melanotos*) it is thought that female preference for tight maneuverability during aerial flight displays has selected for smaller body size of males. However, in most groups of birds, choosiness of females has created larger-bodied males that are in addition adorned with elaborate and striking plumage colors and shapes. The peacock's tail, the booming calls of grouse, the brilliant red inflatable air sacs of sage grouse and frigate birds are all thought to be the result of sexual selection on males by females. Many passerines, such as the wattlebirds of New Zealand (Callaeidae) and honeyeaters (Meliphagidae), have fleshy, bright blue, red, or yellow wattles on the face and neck. Other passerines, such as some Old World flycatchers (Muscicapidae) and African widowbirds (Viduinae), have extremely long tail feathers or highly modified plumes (birds of paradise: Paradisaeidae) used in courtship displays. In some groups exaggerated male traits have become so extreme that they likely hinder basic daily activities and movement, e.g., the 0.7-m long tail feathers of African long-tailed whydas (*Euplectes progne*). However, in other species, such as barn swallows, exaggerated traits favored by females also benefit males: Males with longer forked tails can turn more sharply than males with experimentally reduced tails. More symmetrical tails are also favored by females and provide better aerial rudders than less symmetrical tails. Males of many sexually dimorphic species, such as grouse, neotropical manakins (Pipridae), and some shorebirds, display in leks (spatial clusters of males displaying in groups to females). Many exaggerated male traits are thought to signal some aspect of male quality, as harbored in his superior genes or in the territory he is defending, either of which would be favored by females. For example, in North American house finches, males with brighter plumage visit their nests during feeding of young at a greater rate than drably plumaged males, indicating that in addition to being brightly plumaged, they are also superior parents. In many of these examples, parasites are thought to play an important role in the evolution of exaggerated male traits (see Section V,E).

C. Diversity of Bird Nests and Nesting

Most birds build a nest each year to provide a safe place to incubate their eggs and raise chicks. Nests often consist of cups of woven sticks, grass, and moss. In some species, such as hummingbirds and small passerines, these are held together by spiderwebs. Many species, such as ospreys (*Pandion haliaetus*), return to the same nest and maintain and augment it year after year, resulting in extremely large (10-ft across) and heavy nests for some raptors. Some swifts from southeast Asia build nests exclusively of hardened, dried saliva; these nests are a very expensive delicacy and are exported and served in the finest restaurants. Dippers (Cinclidae) build their nests under moss roofs in the banks of fast-flowing rivers. Several groups build nests of mud: swallows, swifts, and the peculiar Australian perching bird family consisting of only three species of "mud nest builders" (Corcoracidae).

Some of the most bizarre nesting habits are found in the megapodes. These birds, of which there are 22 species distributed from Australia and Micronesian Islands to Nicobar Islands west of Malaysia, incubate their eggs with geothermal or biothermal heat sources rather than by applying body heat to them as do all other birds. Some megapodes incubate their eggs in underground burrows next to volcanic heat sources, whereas others deposit the eggs in mounds of rotting vegetation that produce heat by biodegradation. The phylogenetic relationships of the megapodes indicate that the burrowing habit, although less common among contemporary species, is nonetheless the primitive condition for the group as a whole. These unique nesting habits are accompanied by complete independence of the young upon hatching, a condition known as precociality. Megapode chicks are able to walk and fly within hours of hatching, a trait shared by other Galliformes, Anseriformes, and some seabirds, although not to the extent found in megapodes.

Certain lineages of birds lay their eggs in the nests of other species, a strategy known as interspecific brood parasitism. This trait is particularly prevalent in Old World cuckoos (Cuculidae), some Old World finches (Ploceidae), African honeyguides (Indicatoridae), and cowbirds (Icteridae). Female cowbirds are induced to lay eggs when in the presence of nests made by other species, typically those of other North American passerines. They will continue to lay eggs with such visual stimulation, and some individual females have been known to lay up to 40 eggs in a season in numerous nests of other species. While some of the larger host species with large bills are able to grasp and eject or puncture the foreign cowbird eggs, some species appear not to recognize them at all or physically cannot remove them from their nest. Frequently, the adults of such species are more smaller bodied than cowbird chicks, and the parents work to an excessive degree trying to feed and maintain the fast-developing young, often to the disadvantage of their own young. In populations of some species, over 80% of the nests are parasitized by cowbirds, and such parasitism has had a severe impact on some neotropical migrants such as Bell's vireo, which breeds in North America (see Section V,E).

D. Bird Speciation

The emergence of biogeographic barriers to gene flow and other extrinsic factors causing the isolation of populations, a process known as allopatric speciation, has always been regarded as the primary diversifying force in avian populations. Ernst Mayr has argued convincingly for the power of allopatric speciation to promote speciation in birds, although some researchers suggest that geographic isolation has not played a major role in diversifying some avian gene pools, such as certain populations of Darwin's finches (*Geospiza*) and North American crossbills (*Loxia*). Unlike traditional speciation scenarios for birds, recent research using molecular markers has suggested that the most recent round of avian speciation occurred prior to the Pleistocene epoch, which began about 2 mya. These new data suggest that habitat refugia caused by the relatively recent advance of glacial ice sheets during the Pleistocene were not as important to avian speciation as previously thought.

E. Bird Diversity and Parasites

Aside from geographic isolation, there is perhaps no factor influencing the evolution of avian diversity more than parasites. The constant bombardment to individual birds of both ecto- and endoparasites has undoubtedly molded avian behavioral and morphological diversity, and regulated population numbers and speciation rates themselves. Birds are in a constant race to escape the detrimental effects of fast-changing, ubiquitous parasites.

Because parasites can directly or indirectly adversely affect the acquisition of resources for avian growth and development, perhaps the most conspicuous aspect of avian diversity altered by parasites is avian plumage color. As mentioned earlier, plumage, skin color, and other aspects of the phenotype of birds act as signals to potential mates of an individual's health and vigor. Bright colors in birds are the result of a variety of compounds, from endogenously synthesized keratin structures (blue and iridescent colors) and pigments such as melanin (browns and blacks) to pigments obtained from foods and the environment, such as carotenoids (reds, oranges, and yellows). Parasites can reduce the brightness of males plumage by impairing host vigor and reducing production or acquisition of pigments. Studies in a wide variety of bird groups, including grouse, pheasants, and swallows, suggest that birds with high numbers of parasites have on average less extreme secondary sexual traits, such as tail length or wattle brightness. Indeed, there is some indication from interspecific comparisons that highly dimorphic species with very bright males suffer generally higher rates of parasite infestation. In addition, some comparative studies have suggested that avian clades characterized by strong sexual dimorphism are also more diverse than nondimorphic clades. However, these trends are controversial and require more study.

Parasites are also thought to influence the diversity of avian mating systems. For example, it is known that in North American and European perching birds, blood parasite prevalence (the proportion of individuals infected by parasites) is lower in polygynous species than in monogamous species. Depending on which way the causal arrow points, this correlation could mean that females in polygynous species favor males resistant to parasites, resulting in a spread of resistance genes through the population. Parasites, mate choice, and mating system have been explored in a variety of species, most notably red-winged blackbirds, ring-necked pheasants, and barn swallows, but the generality of conclusions from these species must await further studies. Additionally, one study found that there was no relationship between mating system and clade diversity for a series of sister-group comparisons based on the DNA hybridization tree; thus, the effect of parasites on avian diversity via influence on mating system must await corroboration.

VI. THREATS TO THE DIVERSITY OF BIRDS

The diversity of birds is currently under siege by a variety of forces, most of them anthropogenic. Habitat destruction and change, introduced competitor species, direct predation, pollutants and their synergistic adverse effects on the endocrine and immune systems, and so on are drastically and irreversibly threatening avian diversity. The vast majority of avian extinctions in historic times have occurred on islands. For example, an estimated 2000 bird species have become extinct in the islands of Oceania since the arrival of humans, which is a 20% reduction in the global avifauna! The vast majority of these species were flightless rails which had evolved in the absence of predators and hence were powerless to flee or defend themselves when humans and their introduced mammalian predators arrived.

Human-induced habitat change has caused measurable evolutionary change in some bird populations. For example, in historic times, the endangered i'iwi (*Vestiaria coccinea*), a member of the radiation of Hawaiian finches (Drepanidinae) found only on the Hawaiian islands, once searched for nectar by probing the flowers of a native lobelioid with its strongly decurved bill. Modern populations of i'iwis have less decurved bills than their ancestors because their principal food source has now shifted from the lobelias, which are critically endangered, to plant species with straighter flowers.

Rapid deforestation and spread of agricultural lands have prompted a dramatic spread of brown-headed cowbirds (*Molothrus ater*) in North America, a species whose habit of parasitizing the other species' nests makes naive species—those that historically have not been exposed to cowbird tactics and hence had no need to evolve the ability to discriminate between their own and cowbirds eggs—particularly vulnerable. Clearly, globally wise habitat management, particularly in tropical rain forests, reduction of toxic wastes, and greater research on avian diversity must be combined to preserve both the deep roots and recent twigs of the avian genealogical tree.

See Also the Following Articles

AVIAN REPRODUCTION, OVERVIEW; GLOBAL ZONES AND REPRODUCTION; MIGRATION, BIRDS; NESTING, BIRDS; SEXUAL SELECTION

Bibliography

For a current and evolving portrayal of avian phylogenetic diversity, consult the bird sections of the Tree of Life, an interactive web page maintained by David Maddison and Wayne Maddison, University of Arizona (*http://ag.arizona.edu/tree/phylogeny.html*).

Andersson, M. (1994). *Sexual Selection*. Princeton Univ. Press, Princeton, NJ.

Austin, O. L. (1961). *Birds of the World*. Golden Press, New York.

Birkhead, T. R., and Møller, A. P. (1992). *Sperm Competition in Birds*. Academic Press, London.

Clayton, D. H., and Moore, J. (Eds.) (1997). *Host–Parasite Evolution: General Principles and Avian Models*. Oxford Univ. Press, Oxford, UK.

Cracraft, J. (1988). The major clades of birds. In *The Phylogeny and Classification of the Tetrapods, Volume 1: Amphibians, Reptiles, Birds*, (M. J. Benton, Ed.), Vol. 35A, pp. 339–361. Clarendon, Oxford, UK.

Feduccia, A. (1996). *The Origin and Evolution of Birds*. Yale Univ. Press, New Haven, CT.

Gill, F. B. (1995). *Ornithology*, 2nd ed. Freeman, New York.

Loye, J. E., and Zuk, M. (Eds.) (1991). *Bird–Parasite Interactions: Ecology, Evolution and Behaviour*. Oxford Univ. Press, Oxford, UK.

Mayr, E. (1963). *Animal Species and Evolution*. Harvard Univ. Press, Cambridge, MA.

Mindell, D. P. (1997). *Avian Molecular Evolution and Systematics*. Academic Press, San Diego.

Olson, S. L. (1985). The fossil record of birds. *Avian Biol.* 8, 79–238.

Sibley, C. G., and Ahlquist, J. E. (1990). *The Phylogeny and Classification of Birds: A Study in Molecular Evolution*. Yale Univ. Press, New Haven, CT.

Terborgh, J. (1989). *Where Have All the Birds Gone? Essays on the Biology and Conservation of Birds That Migrate to the American Tropics*. Princeton Univ. Press, Princeton, NJ.

Welty, J. C., and Baptista, L. (1990). *The Life of Birds*. Harcourt Brace Jovanovich, New York.

Birth Control

see Contraceptive Methods and Devices; Family Planning

Blastocyst

Richard M. Schultz
University of Pennsylvania

I. Origin of the Inner Cell Mass and Trophectoderm
II. Origin and Function of the Blastocoel
III. Embryonic Stem Cells
IV. Gene Expression
V. Apoptosis in the Blastocyst

GLOSSARY

blastocoel The fluid-filled cavity that occupies the center of the blastocyst and is a result of fluid transport by the trophectoderm cells.
blastocyst hatching The process by which the blastocyst escapes and emerges from the zona pellucida prior to implantation.
embryonic stem cells Totipotent cells that are derived from the inner cell mass cells and that are used in homologous recombination experiments.
inner cell mass cells The cells that are present on the inside of the blastocyst and that will become the future embryo.
trophectodermal cells The outer cells of the blastocyst that comprise a fluid-transporting epithelium that will ultimately give rise to extraembryonic tissues.

Preimplantation development is a unique aspect of mammalian development and culminates in the formation of the blastocyst (Fig. 1). The blastocyst consists of two cell types. The cells that occupy the outer portion of the blastocyst are called trophectoderm cells and are differentiated cells that constitute a fluid-transporting epithelium that is responsible for the formation of the blastocoel, which is a fluid-filled cavity. The cells of the trophectoderm give rise to extraembryonic tissue as well as provide the initial point of contact for the implanting embryo into the uterine wall. The cells that occupy the inner portion of the blastocyst are called inner cell mass (ICM) cells. The outer cells of the ICM differentiate into the primitive endoderm and the inner cells remain as primitive ectoderm. The primitive ectodermal cells give rise to the embryo proper and the germ cells. The primitive ectodermal cells also contribute to components of the extraembryonic tissues. The primitive endodermal cells contribute exclusively to extraembryonic lineages. The study of the formation of the blastocyst entails basic problems of cell–cell interaction, ion transport, and gene expression.

I. ORIGIN OF THE INNER CELL MASS AND TROPHECTODERM

The following discussion will focus primarily on the mouse embryo. Although the timing of the events that are presented differs in other mammalian species, the underlying mechanisms appear to be common. Prior to the eight-cell stage, the individual blastomeres are loosely associated with each other and are visually distinct. Associated with compaction at the eight-cell stage is the formation of gap junctions and tight junctions. In addition, the cell adhesion molecule E-cadherin, which is also called uvomorulin, is involved in compaction since antibodies to E-cadherin inhibit compaction. Following compaction the blastomeres are no longer visually distinct. The presence of tight junctions at the apical aspect generates a polarized epithelium. For example, the apical surface is rich in microvilli, whereas the basolateral portion is essentially amicrovillar. The nucleus tends to be present in the basolateral portion of the blastomere and endosomes, and cytoskeletal components such as actin are enriched in the apical region. Subsequent cleavage divisions, which are asynchronous, result in the formation of the early blastocyst that contains ~32 cells.

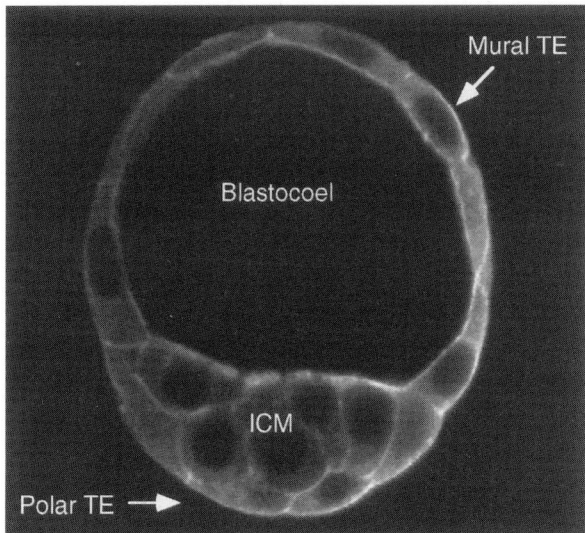

FIGURE 1 Laser-scanning confocal image of mouse blastocyst. The trophectoderm (TE) is composed of two cell types—mural TE and polar TE. Polar TE is directly apposed to the inner cell mass (ICM). Mural TE is not in contact with the ICM. The blastocoel is the fluid-filled cavity that occupies the center of the blastocyst. The embryo has been stained for actin.

These cleavage divisions result in the direct allocation of cells to the inside of the developing embryo. Results of numerous cell lineage studies indicate that the inner cells are allocated between the 8-cell and 16-cell stage and then again between the 16-cell stage and the 32-cell stage. The cleavage plane of the 8-cell blastomeres is essentially random, i.e., the daughter blastomere may be allocated to the inside or remain on the outside. This results in ~3 or 4 cells occupying the interior of 16-cell embryo. The cleavage plane of the 16-cell blastomere, however, is not random and shows a bias such that 80% of daughter cells remain on the outside and 20% are allocated directly to the inside. This results in some 10–12 cells that comprise the inner cell mass (ICM) of the early 32-cell blastocyst. Subsequent cleavage divisions of the differentiated trophectoderm cells always results in trophectoderm descendants. Lineage studies also indicate that cells of the ICM can give rise to trophectoderm descendants and thus the ICM cells can function as a stem cell population. This is consistent with the observation that when isolated ICM cells obtained from early blastocysts are cultured, trophectoderm cells form, i.e., the inner cells remain pluripotent. This potency is lost, however, by the late blastocyst stage. In contrast, culture of differentiated trophectoderm cells never gives rise to inner cell mass cell descendants. Although outer cells obtained from 16-cell embryos that are cultured preferentially give rise to trophectoderm descendants, ICM progeny can be generated. This restriction in potency is consistent with the aforementioned changes in inner cell allocation that accompany development. In addition, developing restrictions in intercellular communication that is mediated by gap junctions correlate with the differences in cell fate of the ICM cells and trophectoderm cells. Cells of the ICM and trophectoderm form separate communication compartments in that cells of the ICM communicate with each other but not with cells of the trophectoderm and vice versa. Lastly, proliferation of the polar trophectoderm cells, i.e., the trophectoderm cells in direct apposition to the ICM, requires the presence of the ICM. It should be noted that in marsupials the blastocyst is composed of a single-cell epithelium in which are embedded the cells that give rise to the embryo, i.e., the marsupial blastocyst does not contain a distinct ICM.

II. ORIGIN AND FUNCTION OF THE BLASTOCOEL

Between the 16-cell and 32-cell stages, the outer cells of the embryo start to transport fluid into the interior of the embryo. This results in a fluid-filled cavity called the blastocoel. The presence of tight junctions between the trophectoderm cells is essential for blastocoel formation since the tight junctions are required to prevent leakage of ions present in the blastocoel. The cell adhesion molecule E-cadherin is also required for the formation of a trophectoderm epithelium. Embryos that do not contain a functional E-cadherin gene fail to form a trophectodermal epithelium and blastocoel. The compaction that is observed in these embryos is likely due to residual maternal E-cadherin that persists up to the 8-cell stage and is sufficient for compaction but insufficient for trophectoderm formation.

The degree of blastocoel expansion differs among mammals. In rodents and humans, the extent of blastocoel expansion is minimal and results in a blastocyst of a couple of hundred microns in diameter. In contrast, blastocoel expansion in the rabbit is quite large and the resulting blastocyst is several millimeters in diameter. Nevertheless, the underlying molecular mechanisms that are responsible for fluid transport are likely the same.

Fluid accumulation in the blastocoel is due to an energy-dependent transepithelial sodium flux. Sodium ions enter the trophectoderm cells via numerous apically located channels that are permeant for sodium, e.g., an amiloride-sensitive sodium channel. The intracellular sodium is then actively pumped into the blastocoel via a basolaterally located sodium/potassium ATPase. (Although transcripts for the α subunit are present throughout preimplantation development, an increase in the abundance of transcripts for the β subunit starting at the morula stage correlates with cavitation, i.e., blastocoel formation.) The sodium ions are retained in the blastocoel due to the presence of the apically located tight junctions between the trophectoderm cells, and passive water transport across the trophectoderm is then osmotically driven. The observation that blastocoel expansion occurs in the presence of a sucrose gradient with the increased accumulation of sodium and chloride is consistent with the passive movement of water in response to the movement of sodium and chloride. Other ions are also transported and include chloride and bicarbonate. Both cAMP and growth factors such as TGF-α stimulate the rate of blastocoel expansion. The molecular basis for this stimulatory effect is unknown but could be a consequence of an increase in the activity of the sodium/potassium ATPase and/or a further tightening of the tight junctions, which would reduce paracellular ion efflux from the blastocoel.

Blastocoel formation and expansion are likely to have several important roles in development. The formation of the blastocoel is essential for further differentiation of the ICM. As indicated previously, the outer cells of the ICM differentiate into the primitive endoderm and the formation of a free surface on the ICM may be a key regulatory event in this differentiation. In addition, the blastocoel fluid can be viewed as a "culture medium" for the ICM. In this regard, the ion composition and concentration of ions present in blastocoel fluid may be critical for ICM cell proliferation and differentiation. In support of this proposal is that very small changes in the sodium ion concentration of the culture medium profoundly influence development of the preimplantation embryo. The blastocoel also contains proteins that are secreted by both the ICM and trophectoderm. Although the identity and function of these proteins is unknown, it is most likely that they influence ICM proliferation and differentiation.

Blastocoel expansion likely contributes to the process of blastocyst hatching. The mammalian embryo is surrounded by an extracellular glycoprotein coat called the zona pellucida. One of the functions of the zona pellucida is to mediate the initial events of sperm–egg interaction and to keep the blastomeres from becoming separated prior to compaction during embryo transit in the oviduct. The zona pellucida also prevents the developing embryo from precociously attaching the oviduct. Nevertheless, once the embryo reaches the blastocyst stage and is present in the uterus the zona pellucida forms an impediment to implantation. Blastocyst hatching results in the escape of enclosed blastocyst from the zona pellucida, and in vitro two factors contribute to hatching. The first factor is the secretion of a trophectoderm-derived protease that locally digests the zona pellucida. The second factor is blastocoel expansion that appears to extrude actively the blastocyst from the perforated zona pellucida.

Blastocoel expansion may also facilitate the initial cell–cell contacts required for implantation, which involves the interaction of two epithelia, namely, the trophectoderm and the endometrium. The continued expansion of the blastocoel following hatching, in conjunction with removal of uterine fluid, contributes to the apposition of the trophectoderm with the endometrium.

III. EMBRYONIC STEM CELLS

Embryonic stem (ES) cells are pluripotent cells that are derived from culturing ICM cells. In the presence of leukemia inhibitory factor these cells will continue to proliferate in culture and retain both their normal euploid chromosome content and pluripotency. ES cells injected into the blastocoel are incorporated into the ICM and contribute to all of the tissues of the animal as well as the germline. Their ability to contribute to the germline likely reflects the retention of a normal chromosome content. This contrasts with embryonal carcinoma cells that are karyotypically abnormal and thus poorly contribute to the germline. Embryo cells grafted to ectopic sites can give rise to malignant, transplantable tumors (teratocarcinomas) from which embryonal carcinoma cells are derived.

The ability of ES cells to contribute to all cell types, including the germline, at a high frequency has made them ideal candidates for targeted mutagensis via homologous recombination. This powerful approach has permitted researchers to approach in a rationale and systematic manner the function of any cloned gene in development. Moreover, the generation of such mutant "knockout" mice has resulted in several model systems for the study of human disease that include disorders of neural crest derivatives and vision and hearing; diseases of bone, skin, and connective tissue; neurological and neuromuscular disorders; and neoplastic, immunological, hematological, metabolic, and hormonal diseases. A problem intrinsic to this approach is that if the mutation results in an early embryonic lethal, analysis of the gene's function at later developmental stages is precluded. The recently developed strategy using the bacteriophage P1 Cre-lox system of site-specific recombination circumvents this problem since it permits the generation of the null allele at either a specific time in development or in a specific tissue. ES cells also provide model systems to study cell differentiation. In suspension, ES cells can form aggregates that are called embryoid bodies that can spontaneously differentiate to a variety of cell types.

IV. GENE EXPRESSION

Results of experiments using inhibitors of protein and RNA synthesis indicate that both translation and transcription are required for blastocyst formation and that the requirement for translation and transcription for blastocoel formation is completed only a few hours prior to the onset of cavitation. In contrast, the transcriptional events that underlie hatching are completed many hours in advance, whereas the translational needs are completed just a few hours in advance of hatching. This suggests that posttranscriptional control may be a major mechanism regulating hatching.

As described previously, by the early blastocyst stage, the trophectoderm cells are clearly differentiated and unable to generate inner cells, whereas the ICM cells can still give rise to trophectoderm cells. Accompanying these changes in cell specification and fate (i.e., inner and outer cells of the morula give rise to the ICM and trophectoderm, respectively, of the blastocyst) are changes in the spatial pattern of gene expression that were first detected by two-dimensional gel electrophoresis and that may initiate in the morula. These spatial changes in gene expression are likely critical for the establishment and differentiation of the two cell lineages in the blastocyst, i.e., they are intimately coupled to cell specification and the maintenance or loss of totipotency. The identity of these genes, however, is largely unknown. Recently, a few ICM-specific genes have been identified serendipitously. For example, *fgf-4* is preferentially expressed in the ICM. In fact, ICM cell proliferation is severely impaired in embryos homozygous for the null allele of *fgf-4*, whereas trophectoderm development is relatively unaffected. The expression of TGF-β_2 also appears restricted to the ICM. This preferential expression of growth factors within the ICM may reflect a higher growth factor requirement for ICM proliferation. It should be noted, however, that TGF-α is expressed in both the ICM and trophectoderm. Another ICM-specific gene, *etl-1*, was recently identified during an enhancer trap screen. It is related to the *Drosophila brahma* gene, which is a transcriptional regulator of homeotic genes, and to the yeast transcriptional activator SNF2/SWI2. *Etl-1* is expressed at all stages of preimplantation development, but its expression becomes restricted to the ICM in the blastocyst. The retinoic acid-regulated, zinc finger gene *rex-1* is also preferentially expressed in the ICM. The expression of *oct-4*, which is a POU domain-binding transcription factor and may play a critical role in establishment of the germline, also becomes restricted to the ICM during blastocyst formation.

A few trophectoderm-specific genes have also been identified. The cytokeratin-8, also called endo A, whose expression starts to increase around the 8-cell stage, becomes preferentially expressed in the trophectoderm. Similarly, the expression of the imprinted *H19* gene, which is expressed from the maternally derived allele, is first detected in the blastocyst and essentially only in the trophectoderm cells. Later in development, *H19* is expressed in mesodermal and endodermal derivatives in the embryo. Although the *hxt* gene, which was recently detected in a screen for bHLH transcription factors, has been proposed to direct the fate of undifferentiated blastomeres into the trophectoderm lineage, it is not known whether *hxt* expression is restricted to the trophectoderm in the blastocyst. Virtually nothing is known regarding the molecular basis for the restricted spatial pattern of gene expression in the blastocyst. Although cell–cell contact is clearly critical in cell specification, how this contact is interpreted in the context of restricted patterns of gene expression is not apparent.

In the early female preimplantation embryo both X chromosomes are active. Dosage compensation is achieved during preimplantation development by random inactivation of the X chromosome in the ICM cells, but preferential inactivation of the paternally derived X chromosome occurs in the trophectoderm cells. This preferential inactivation is due to the expression of *Xist* from the paternal allele; *Xist* encodes for an RNA that apparently is not translated.

V. APOPTOSIS IN THE BLASTOCYST

Apoptosis (programmed cell death) is observed in the blastocyst and is essentially restricted to the ICM. Apoptosis is not observed at earlier stages of preimplantation development. The cell death machinery, however, is present in these embryos since staurosporine can induce apoptosis in these earlier stage preimplantation embryos. The regulation of cell death in the blastocyst is likely to be critical for later development. For example, the ICM-specific cell death may eliminate ICM cells that retain the potential to form trophectoderm. Conversely, unregulated cell death in the ICM would also compromise later development since a critical threshold number of ICM cells is required for normal postimplantation development.

Although early blastomeres do not require survival signals, it is possible that by the time of blastocyst formation the two cell types formed (ICM and trophectoderm) do have such a requirement.

Peptide growth factors are known to regulate cell proliferation and differentiation during mammalian preimplantation development; embryos express functional growth factor receptors for ligands present in the maternal tract and those synthesized by the embryo itself. A role for growth factors serving as cell survival factors is suggested by the observation that when the epidermal growth factor receptor gene is deleted, one of the phenotypes shows periimplantation lethality on Day 6.5 due to failure of the ICM to develop. Also consistent with this proposed role is the observation that the incidence of apoptosis is dramatically increased in the ICM cells of embryos that are cultured *in vitro* and that addition of TGF-α reduces this incidence. Moreover, the incidence of apoptosis in the ICM cells is higher in mice homozygous for a null allele of TGF-α when compared to the heterozygote.

See Also the Following Articles

APOPTOSIS; IMPLANTATION

Bibliography

Adamson, E. D. (1993). Activities of growth factors in preimplantation embryos. *J. Cell Biochem.* **53**, 280–287.

Biggers, J. D., Bell, J. E., and Benos, D. J. (1988). Mammalian blastocyst: Transport functions in a developing epithelium. *Am. J. Physiol.* **255**, C419–C432.

Feldman, B., Poueymirou, W., Papaioannou, V. E., DeChiara, T. M., and Goldfarb, M. (1995). Requirement for FGF-4 for postimplantation mouse development. *Science* **267**, 246–249.

Fleming, T. P., and Hay, M. J. (1991). Tissue-specific control of expression of tight junction polypeptide ZO-1 in the mouse early embryo. *Development* **113**, 295–304.

LaRue, L., Ohsugi, M., Hirchenhain, J., and Kemler, R. (1994). E-cadherin null mutant embryos fail to form a trophectoderm epithelium. *Proc. Natl. Acad. Sci. USA* **91**, 8263–8267.

Majumder, S., and DePamphilis, M. L. (1994). TATA-dependent enhancer stimulation of promoter activity in mice is developmentally regulated. *Mol. Cell. Biol.* **14**, 4258–4268.

Pedersen, R. A.(1986). Potency, lineage, and allocation in preimplantation mouse embryos. In *Experimental Approaches to Mammalian Embryonic Development* (R. Rossant and R. A. Pedersen, Eds.), pp. 3–33. Cambridge Univ. Press, Cambridge, UK.

Pedersen, R. A. (1994). Studies of in vitro differentiation with embryonic stem cells. *Reprod. Fertil. Dev.* **6**, 543–552.

Robertson, E. J. (1991). Using embryonic stem cells to introduce mutations into the mouse germ line. *Biol. Reprod.* **44**, 238–245.

Stewart, C. L. (1994). Leukaemia inhibitory factor and the regulation of pre-implantation development of the mammalian embryo. *Mol. Reprod. Dev.* **39**, 233–238.

Threadgill, D. W., Dlugosz, A. A., Hansen, L. A., Tennenbaum, T., Lichti, U., Yee, D., LaMantia, C., Mourton, T., Herrup, K., Harris, R. C., Barnard, J. A., Yuspa, S. H., Coffey, R. J., and Magnuson, T. (1995). Targeted disruption of mouse EGF receptor: Effect of genetic background on mutant phenotype. *Science* **269**, 230–233.

Watson, A. J. (1992). The cell biology of blastocyst development. *Mol. Reprod. Dev.* **33**, 492–504.

Watson, A. J., Kidder, G. M., and Schultz, G. A. (1992). How to make a blastocyst. *Biochem. Cell Biol.* **70**, 849–855.

Weil, M., Jacobson, M. D., Coles, H. S. R., Davies, T. J., Gardner, R. L., Raff, K. D., and Raff, M. C. (1996). Constitutive expression of the machinery for programmed cell death. *J. Cell Biol.* **133**, 1053–1059.

Yeom, Y., II, Fuhrmann, G., Ovitt, C. E., Brehm, A., Ohbo, K., Gross, M., Hübner, K., and Schöler, H. R. (1996). Germline regulatory element of Oct-4 specific for the totipotent cycle of embryonal cells. *Development* **122**, 881–894.

Blood–Testis Barrier

B. P. Setchell
University of Adelaide

I. Functional Evidence for a Blood–Testis Barrier
II. Structural Evidence
III. Significance of the Blood–Testis Barrier

GLOSSARY

"difference" technique for determining seminiferous tubule fluid composition When the efferent ducts of one testis of a rat are ligated, the testes removed and decapsulated 16–24 hr later, and the cells dispersed and centrifuged, more supernatant fluid can be recovered from the ligated than from the control testis. The composition of the additional secreted fluid can be calculated by subtracting the total amount of the relevant substance in the control supernatant fluid from that in the ligated fluid and dividing the difference in amount by the difference in volume between the two fluids. It has been shown that the composition of the fluid calculated in this way gives a valid estimate of the composition of seminiferous tubule fluid.

germ cells The cells which give rise to spermatozoa, beginning with diploid stem and differentiating spermatogonia, primary and secondary spermatocytes undergoing meiosis, and haploid spermatids.

interstitial extracellular fluid The fluid between the seminiferous tubules surrounding the Leydig and other interstitial cells.

peritubular myoid cells Contractile cells situated in the tissue around the seminiferous tubule which cause the peristaltic movements of the tubules.

rete testis fluid The fluid which can be collected from a catheter in the rete testis or by puncturing the rete in rats 16–24 hr after ligation of the efferent ducts to trap the fluid secreted during that time inside the testis.

seminiferous tubule fluid The fluid in the lumina of the seminiferous tubules which can be sampled by micropuncture for analysis, or its composition can be calculated by the "difference" technique.

Sertoli cells The only somatic cells inside the seminiferous tubules, which form the main element of the blood–testis barrier and secrete the fluid environment in which the germ cells develop.

Evidence for a blood–testis barrier derives from differences in composition between the fluids outside and inside the seminiferous tubules and rete testis, from differing rates of entry of various markers from blood into these fluids, from the distribution of injected dyes as determined by light microscopy, and from the distribution of electron-opaque markers in the tissue, as determined by electron microscopy.

Functional evidence for a blood–testis barrier is provided by the differences in composition of the fluids inside the seminiferous tubules and rete testis, on the one hand, and interstitial extracellular fluid, testicular lymph, and venous blood plasma on the other hand. Substances which show such differences include ions, particularly potassium, small organic molecules such as glucose, inositol, and certain amino acids, and proteins. The rate of entry of substances from the blood into tubular or rete fluid depends largely on the lipid solubility, with a smaller influence of molecular size. While the existence of a barrier was suggested many years ago in studies involving the distribution of injected dyes, the location of the principal element of the barrier has been established using electron microscopes and electron-opaque markers as the specialized junctions between adjacent pairs of Sertoli cells. Spermatogonia and preleptotene spermatocytes are outside the Sertoli cell barrier in a basal compartment, but spermatocytes from zygotene onwards and spermatids develop inside the barrier in an adluminal compartment. However, endothelial cells in the testis share some features with endothelial cells in the brain, where they constitute the principal element of the blood–brain barrier, and therefore may have some influence on the entry of substances into the testis. Furthermore, the peritubular myoid cells do provide some restriction to movement of tracers, particularly in rodents, so the barrier may be composed of these three elements in series. The blood–testis barrier develops only at puberty, at about the time that fluid secretion from the testis begins. The barrier presumably is required for fluid secretion to occur, but it probably also provides the appropriate conditions for the germinal cells to stop dividing mitotically and to enter the meiotic prophase. The barrier also has important endocrinological significance because it excludes peptide hormones of the size of follicle-stimulating hormone (FSH) and luteinizing hormone (LH) from direct contact with developing germ cells in the adluminal compartment of the seminiferous tubules, and it also renders the interior of the tubule an immunologically privileged site by excluding circulating antibodies and preventing proteins expressed on meiotic and postmeiotic cells and spermatozoa from being "seen" by the body's immunological system. In developing germ cells a new barrier is formed outside them and the old barrier luminal to the cells then separates. The barrier persists even when spermatogenesis is disrupted, for example, by heating the testis, but does break down at least temporarily when intratubular pressure is raised after ligation of the efferent ducts, which carry the secreted fluid and the sperm it contains out of the testis and into the epididymis.

I. FUNCTIONAL EVIDENCE FOR A BLOOD–TESTIS BARRIER

A. Composition of Fluids from the Seminiferous Tubules and Rete Testis

The composition of fluid from the lumen of a seminferous tubule (STF) is quite different from that of blood plasma or of lymph collected from vessels on the surface of the testis or from the spermatic cord. This is so whether the composition of the tubular fluid is estimated directly on tiny samples obtained by micropuncture or by calculation by the "difference" technique; the two techniques in rats give virtually identical values for potassium, with the tubular fluid about 10 times higher than blood plasma (Fig. 1); sodium (which is much lower); and small organic substances such as inositol (100 times higher), glucose, and total protein (much lower), with androgen-binding protein much higher. If the walls of the tubules were freely permeable to some or all of these substances, the relevant differences would be dissipated, so there must be a barrier between blood plasma and the tubular fluid. Fluid can also be collected in rats, rams, and a number of other species from the rete testis, into which all the seminiferous tubules empty and which carries the spermatozoa and the fluid in which they are sus-

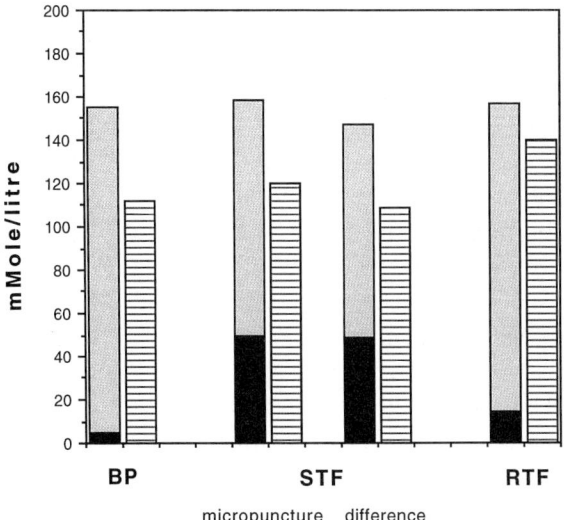

FIGURE 1 The concentrations of potassium (solid columns), sodium (dotted shading), and chloride (horizontal lines) in rat blood plasma (BP), seminiferous tubule fluid (STF)—obtained by direct analysis of micropuncture samples or by calculation using the difference technique—and rete testis fluid (RTF) collected from a catheter in the rete testis.

pended into the epididymis. This fluid (RTF) is also different in many ways from blood plasma (Figs. 1 and 2), suggesting that a barrier extends to the excurrent ducts system Interestingly, however, RTF in rats is also appreciably different from STF, with the potassium levels being intermediate between STF and blood plasma. It is assumed that this is because the walls of the rete are more permeable than those of the tubules, for which there is some direct evidence. Both STF and RTF are isosmotic with blood plasma.

B. Entry of Tracers from Blood Plasma into STF and RTF

The rate of entry of tracers from blood plasma into STF or RTF is determined mainly by the lipid solubility of the tracer and, to a much lesser extent, its molecular size. Labeled proteins are virtually excluded, as one would expect from the low concentration of protein in the STF and RTF, as are small hydrophilic molecules such as inulin, Cr-EDTA, and p-aminohippurate. The ions, sodium, potassium, rubidium, and chloride, show intermediate entry rates,

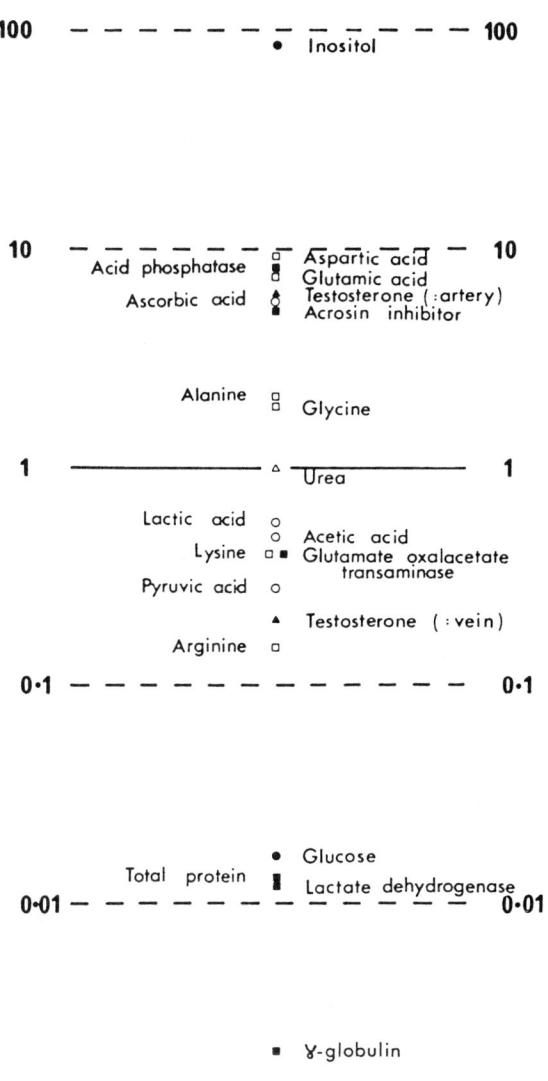

FIGURE 2 The ratios of the concentrations of various substances in ram rete testis fluid (RTF) to those in blood plasma. Note that there are two points for testosterone—one the ratio RTF/arterial blood plasma and the other RTF/testicular venous blood plasma.

whereas tritiated water, urea, and ethanol equilibrate between blood and the fluids within minutes or slightly longer. The relationship of entry rate and lipid solubility was shown most clearly with a series of barbiturates and sulfonamides with differing solubility characteristics. A nonmetabolized analog of glucose, 3-O-methyl glucose, appears to enter the fluid with the aid of a specific saturable carrier, and the entry of testosterone is much faster than that of 5α-dihydrotestosterone, which is more lipid soluble, and also appears to be reduced when the concentration of unlabeled testosterone is elevated, suggesting that a transport system may be involved. As expected, the entry of a number of tracers is faster into RTF than into STF.

II. STRUCTURAL EVIDENCE

A. Light Microscope Studies

At the turn of the century, a series of studies involving the distribution in the body of dyes injected into the circulation gave rise to the concept of the blood–brain barrier because some of the dyes which stained most tissues were excluded from the brain. In some of these experiments, it was noticed that the seminiferous tubules were also unstained, but this aroused little interest at the time. However, in the 1960s studies confirmed this finding, and some dyes were also found to penetrate quite slowly into the interstitial tissue while being excluded from the tubules. This suggested that the endothelial cells in the testis might influence the entry of some substances into the testis as well as some tubular component. The endothelial cells of the testis do share some characteristics with those of the brain, where they form the principal element of the blood–brain barrier. Testicular endothelial cells contain high concentrations of alkaline phosphatase and γ-glutamyl transpeptidase, both enzymes associated with transport process and the latter particularly with amino acid transport. The testicular endothelial cells also contain high levels of the glucose transporter GLUT-1, which is abundant in tissues with barrier properties and of P-glycoprotein, the multidrug resistance gene product, which may be involved in the exclusion of some drugs from not only the brain but also the testis.

B. Electron Microscope Studies

Specialized junctions between adjacent Sertoli cells were noticed in the 1960s, and it was postulated by Nicander that they "would impede the intercellular transport of substances to spermatocytes and spermatids and into the luminal fluid." Subsequently, it was shown that large electron-opaque markers, such as carbon particles, did not penetrate the layer of myoid cells around each semniferous tubule, whereas smaller markers such as colloidal thorium could be found between the myoid cells and the basal lamina. Even smaller markers, such as ferritin, horse radish peroxidase, and particularly lanthanum and nickel salts, penetrate between the myoid cells and between pairs of Sertoli cells or between a Sertoli cell and a spermatogonium as far as the specialized junctions between pairs of Sertoli cells. From this evidence, it was suggested that the seminiferous epithelium could be divided into a basal compartment, containing spermatogonia and early spermatocytes, and an adluminal compartment containing older spermatocytes and spermatids (Fig. 3). The adluminal compartment appears to be in communication with the luminal fluid and the spaces between the luminal surfaces of the Sertoli cells and the spermatozoa embedded in crypts.

III. SIGNIFICANCE OF THE BLOOD–TESTIS BARRIER

A. Development at Puberty and in Seasonal Breeders

The blood–testis barrier, whether demonstrated by the space of distribution of water-soluble markers such as Cr-EDTA, dye studies using light microscopy, or electron microscope studies with electron-opaque markers, is not present in animals before puberty and develops at about the time that first cells enter meiosis and fluid secretion by the Sertoli cells can first be demonstrated. In seasonal breeders such as mink, the barrier becomes less effective outside the breeding season.

B. Relation to Mitosis and Meiosis of Germ Cells

Spermatogonia, which divide by mitosis, and spermatocytes in the earliest stages of the prophase

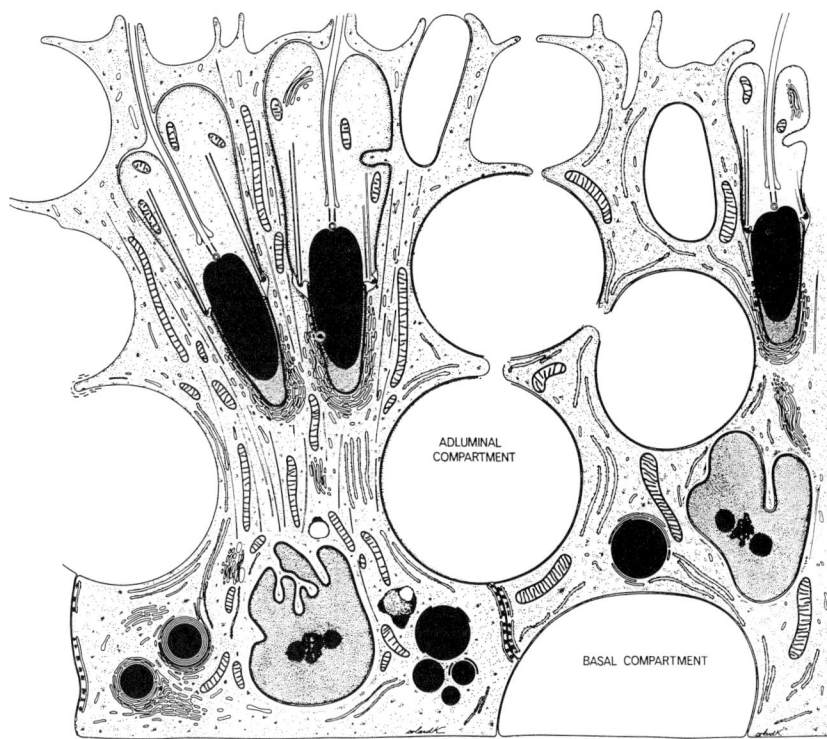

FIGURE 3 Diagram of the structure of two Sertoli cells illustrating the specialized junctions separating the intercellular space into a basal compartment, containing spermatogonia, preleptotence, and leptotene primary spermatocytes, and an adluminal compartment, containing later spermatocytes and early spermatids. Some late spermatids are illustrated in crypts in the luminal surfaces of the Sertoli cells (reproduced with permission from Fawcett, 1975).

of the first meiotic division are located outside the Sertoli cell barrier, in the basal compartment. Spermatocytes in the later stages of the first meiotic prophase and both actual meiotic divisions take place in the adluminal compartment. However, no direct link between the position of the cells in relation to the barrier and their mode of cell division has been established. It is likely that one of the functions of the barrier is to establish and maintain the conditions appropriate for meiosis to occur.

C. Passage of Developing Germ Cells through the Barrier

Because of the location of the Sertoli cell barrier in relation to the developing germ cells, it must be possible for the cells to pass through the barrier during their development. This is achieved by a new barrier forming basal to the cells in leptotene stage of the meiotic prophase so that the cells are separated both from the extratubular basal environment and from the adluminal compartment. The old junctions luminal to the developing germ cells open so that after zygotene, the cells are in an environment which is fully in communication with the adluminal compartment (Fig. 4).

D. Fluid Secretion by the Sertoli Cells

The specialized junctions between adjacent pairs of Sertoli cells are probably a vital factor in fluid secretion by these cells. The probable basis for fluid transport is the secretion of a nonpermeant substance, probably potassium but possibly also inositol, by the Sertoli cells into the intercellular spaces above the specialized junctions. This secretion creates an

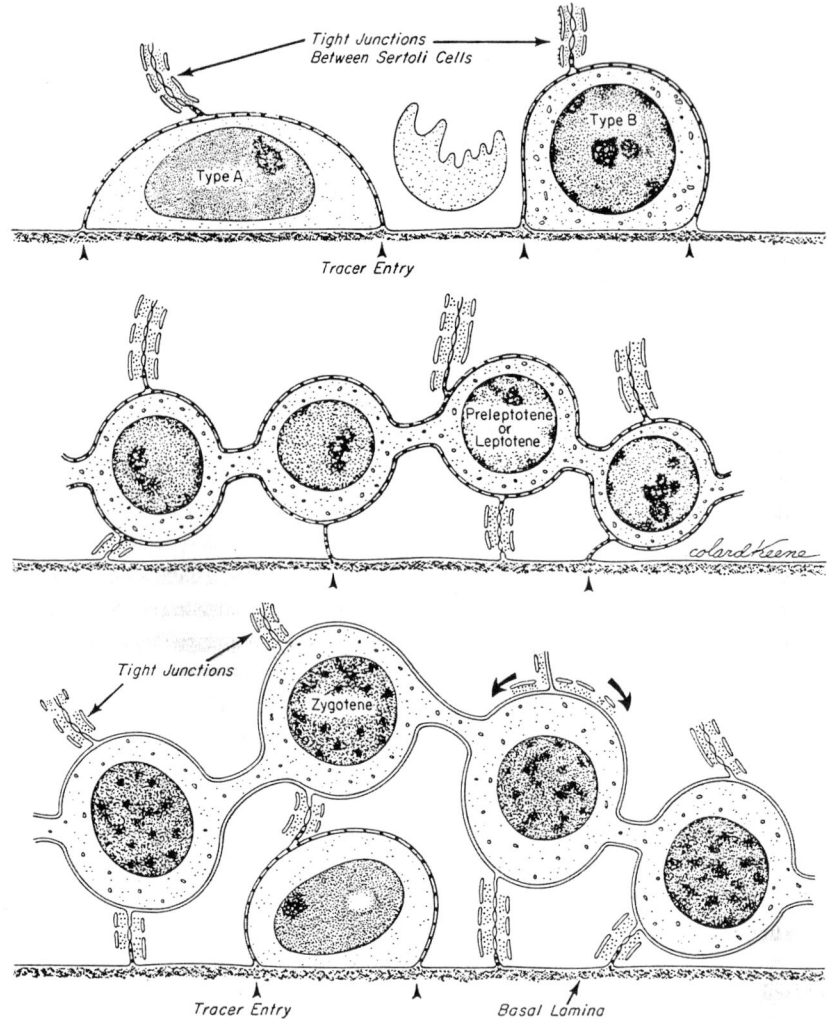

FIGURE 4 Diagram illustrating how electron-opaque marker penetrates between the Sertoli cells and the type A and type B spermatogonia (top) and between the Sertoli cells and the preleptotene and leptotene spermatocytes (middle). Then a new Sertoli cell barrier forms basal to the zygotene spermatocytes, whereas the old barrier on the luminal side of these cells separates (bottom) (reproduced with permission from Dym and Cavicchia, 1977).

osmotic gradient, which then draws fluid through the cells or along the intercellular space to maintain isomolarity. The temporal association between the beginning of fluid secretion and the formation of the barrier at puberty gives strong support to this hypothesis.

E. Endocrinological Consequences

Because of the existence of the Sertoli cell barrier, it is not possible for FSH or LH, or any other peptide hormone, to gain direct access to germ cell from zygotene onwards. Any effect of these hormones would have to be indirect via the Sertoli cells, and it is probably relevant that these cells have receptors for FSH. However, access of peptide hormones, albeit somewhat restricted to spermatogonia, would be possible, and there is some evidence that these cells also have receptors for FSH. However, because there is no evidence of how much of each pulse of LH in blood is reflected in the interstitial extracellular fluid bathing the Leydig cells, the endothelial barrier may influence the response of extratubular cells to circulating peptide hormones.

F. Immunological Consequences

There is evidence that the testis is an immunologically privileged site; that is, grafts of tissue from another animal have an increased chance of survival in the testis. The Sertoli cell junctions would prevent access of immunoglobulins into the seminiferous tubules and likewise prevent the immune system of the body from "seeing" the surface antigens expressed on the postmeiotic cells. Such cells generate an immune reaction if injected directly into the same animal, but they normally do not do so. However, many of the germ cells in the basal compartment are immunologically foreign to the body's immune system, and there is evidence that the interstitial tissue also has some immune privilege. It is not clear whether this is a function of the endothelial barrier or of the presence in the interstitial extracelullar fluid of immunosuppresive substances.

G. Breakdown of Blood–Testis Barrier

The structure of the Sertoli cell junctions is altered in cryptorchidism, and following treatment with gossypol, cis-platinum, or cytochalasin, the permeability to lanthanum and immunoglobulins was increased. However, many other treatments which lead to disruption of spermatogenesis do not apparently cause changes in barrier function, although most studies have examined the function of the barrier once changes in spermatogenesis have occurred, not during or immediately after the treatment. Following ligation of the efferent duct in rats, fluid secreted by the Sertoli cells accumulates in the tubules, distending them and causing a linear increase in testis weight. After about 36 hr, the barrier breaks down and the weight of the testis ceases to increase and then falls sharply. Subsequently, spermatogenesis is severely affected, with only Sertoli cells and some spermatogonia remaining, although by then the barrier has been reestablished.

See Also the Following Articles

LEYDIG CELLS; SERTOLI CELLS, OVERVIEW; TESTIS, OVERVIEW

Bibliography

Byers, S., Pelletier, R.-M., and Suarez-Quian, C. (1993). Sertoli cell junctions and the seminiferous epithelium barrier. In *The Sertoli Cell* (L. D. Russell and M. D. Griswold, Eds.), pp. 431–446. Cache River Press, Clearwater, FL.

Dym, M., and Cavicchia, J. C. (1977). Further observations on the blood–testis barrier in monkeys. *Biol. Reprod.* 17, 390–403.

Fawcett, D. W. (1975). Ultrastructure and function of the Sertoli cell. In *Handbook of Physiology, Section 7 Endocrinology, Vol. V: Male Reproductive System* (D. W. Hamilton and R. O. Greep, Eds.), pp. 21–55. American Physiological Society, Washington, DC.

Hinton, B. T., and Setchell, B. P. (1993). Fluid secretion and movement. In *The Sertoli Cell* (L. D. Russell and M. D. Griswold, Eds.), pp. 249–267. Cache River Press, Clearwater, FL.

Pelletier, R. M., and Byers, S. W. (1992). The blood–testis barrier and Sertoli cell junctions: Structural considerations. *Microsc. Res. Technique* 30, 3–33.

Plöen, L., and Setchell, B. P. (1992). Blood–testis barriers revisited. A homage to Lennart Nicander. *Int. J. Androl.* 15, 1–4.

Russell, L. D., and Petersen, R. N. (1985). Sertoli cell junctions: Morphological and functional correlates. *Int. Rev. Cytol.* 94, 177–211.

Setchell, B. P. (1978). *The Mammalian Testis*. Elek/Cornell Univ. Press, London/Ithaca, NY.

Setchell, B. P. (1980). The functional significance of the blood–testis barrier. *J. Androl.* 1, 3–10.

Setchell, B. P., and Waites, G. M. H. (1975). The blood–testis barrier. In *Handbook of Physiology, Section 7 Endocrinology, Vol. V: Male Reproductive System* (D. W. Hamilton and R. O. Greep, Eds.), pp. 143–172. American Physiological Society, Washington, DC.

Setchell, B. P., Uksila, J., Maddocks, S., and Pollanen, P. (1990). Testis physiology relevant to immunoregulation. *J. Reprod. Immunol.* 10, 19–32.

Setchell, B. P., Maddocks, S., and Brooks, D. E. (1994). Anatomy, vasculature, innervation and fluids of the male reproductive tract. In *The Physiology of Reproduction* (E. Knobil and J. D. Neill, Eds.), 2nd ed., pp. 1063–1175. Raven Press, New York.

Bovidae

see Cattle

Brachiopoda

Stephen A. Stricker

University of New Mexico

I. Modes of Reproduction
II. Anatomy of the Reproductive Systems
III. Gametogenesis
IV. Spawning and Fertilization
V. Embryology
VI. Larval Development and Metamorphosis
VII. Conclusions

GLOSSARY

articulate Belonging to the class Articulata in which the adult shell has a posterior articulation and is composed of calcium carbonate.

inarticulate Belonging to the class Inarticulata in which the adult shell lacks a posterior articulation and is typically composed of calcium phosphate plus chitin.

lophophore A complex of ciliated tentacles used in filter feeding.

mantle The bilayered region of body wall that secretes the calcified shell.

pedicle A posteriorly positioned anchoring organ that typically attaches to hard substrata.

protegulum The first shell secreted after larval settlement.

Brachiopods, or "lamp shells," are solitary marine invertebrates that occur in all oceans, from the intertidal realm to deep sea floor. Based simply on current species diversity, the 300–400 extant species of brachiopods constitute a relatively small phylum. However, their rich fossil record of approximately 30,000 species indicates a much more dominant presence dating back to the early Cambrian. As adults, brachiopods possess a calcified shell consisting of a dorsal and ventral valve, and the posterior end of the animal tends to be permanently attached to hard substrata by an anchoring organ called the pedicle (Fig. 1). In species belonging to the class

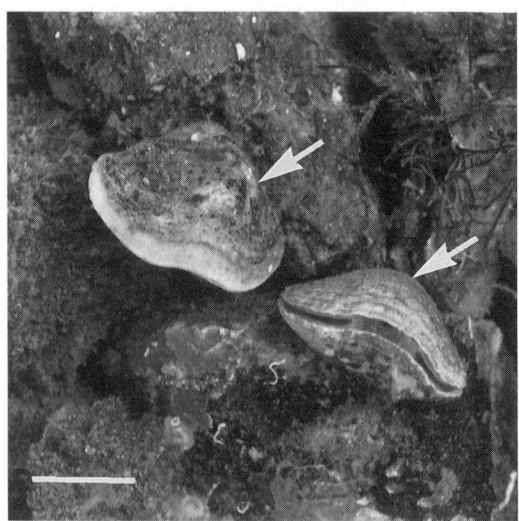

FIGURE 1 Photograph of two *Terebratalia transversa* adults (arrows) attached to the substratum. Scale bar = 10 mm.

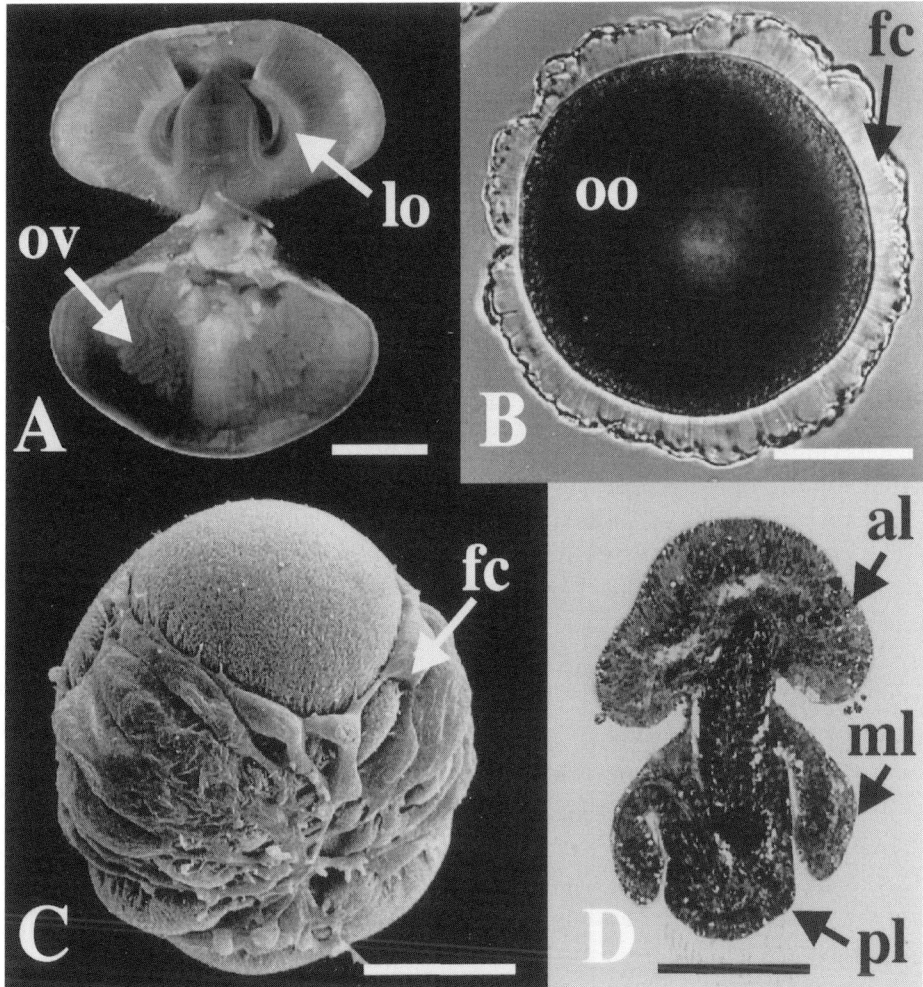

FIGURE 2 (A) An adult *Terebratalia transversa* with the two valves of the shell spread apart to show the lophophore (lo) and ovary (ov). Scale bar = 10 mm. (B) Oocyte (oo) and surrounding follicle cells (fc) of *T. transversa*. Scale bar = 50 μm. (C) Scanning electron micrograph of follicle cells (fc) in the process of forming an aggregated cap at one pole of the oocyte during ovulation in *T. transversa*. Scale bar = 50 μm. (D) Photomicrograph of a 1-μm section of the free-swimming larva of *T. transversa*, showing the apical lobe (al), mantle lobe (ml), and pedicle lobe (pl). Scale bar = 100 μm.

Articulata, the shell contains calcium carbonate and has a posteriorly positioned articulation of interlocking teeth and sockets. Shells produced by members of the class Inarticulata lack an articulation and are typically composed of calcium phosphate and chitin.

A spacious mantle cavity occurs beneath the mantle, which is the bilayered region of the body wall that secretes the shell. Within the mantle cavity, a well-developed assemblage of ciliated tentacles, the lophophore, serves as a filter-feeding organ for capturing microscopic food particles suspended in the seawater (Fig. 2A).

Brachiopods have been classified along with phoronids and bryozoans as part of a triumvirate superassemblage of lophophorate phyla that have traditionally been grouped near deuterostomes such as echinoderms and hemichordates. However, the phylogenetic affinities of brachiopods is controversial; recent molecular-based analyses indicate that inarticulate and articulate brachiopods may actually constitute separate phyla within the protostome lineage that includes molluscs and annelids.

I. MODES OF REPRODUCTION

Asexual reproduction, including polyembryony, has not been recorded in the phylum Brachiopoda. Thus, all reproduction is via the sexual mode.

Brachiopods are typically dioecious; sequential or simultaneous hermaphroditism has been documented in only a few species (e.g., *Platidia davidsoni*, *Pumilus antiquatus*, and members of the genus *Argyrotheca*). Among dioecious brachiopods, sex ratios approaching 1:1 are routinely observed, although in *Notosaria nigricans* the ratio deviates substantially from equality. The precise genetic and/or environmental factors that determine sex in brachiopods have not been identified.

II. ANATOMY OF THE REPRODUCTIVE SYSTEMS

Sexual dimorphism is generally lacking except in *Lacazella mediterranea*, in which the ventral valve of the female bulges to accommodate a brood pouch. Even the color of the gonads can be misleading, and thus microscopic examinations of gonadal preparations are typically needed for positive identifications of sex.

In adult brachiopods, the gonads constitute relatively simple accumulations of developing gametes arising from a series of folds ("genital lamellae") that are covered throughout by a germinal epithelium. In most inarticulates, the gonads are housed within the main coelomic cavity, or perivisceral coelom, that surrounds the digestive tract and other viscera. In articulate species, however, the gonads develop along coelomic channels ("vasa genitalia") that extend into the space separating the inner and outer layers of the mantle.

Adults typically have a pair, or in some cases two pairs, of relatively flattened gonads. In articulate species, the ventral half of the animal tends to contain slightly more gonadal tissue that does the dorsal half. However, in species of *Argyrotheca* the gonads are restricted to the dorsal valve, and *L. mediterranea* has only ventrally located gonads.

Secondary sexual organs are lacking except in brachiopods that generate discrete brood chambers or specialized regions of the lophophore for incubating developing embryos. There are also no copulatory organs or well-defined outgrowths of the gonads for discharging gametes because the paired nephridia serve as gonoducts during spawning.

III. GAMETOGENESIS

A. Spermatogenesis

Early stages of spermatogenesis have not been widely investigated, and the ultrastructure of the fully developed spermatozoon has been described for only a handful of species. Based on this limited database, brachiopods characteristically produce a "primitive" or "ect-aquasperm" type of sperm with a relatively compact ($\sim 2\mu 5$ μm long) head, a pericentriolar network of radiating fibers around the distal centriole, and a tail containing a $9 \times 2 + 2$ arrangement of microtubules. The sperm of inarticulates typically possess a relatively large acrosome, an anterior nuclear fossa, and several mitochondria. Alternatively, the sperm produced by articulates characteristically have a smaller acrosome, a nucleus lacking an anterior indentation, and a single fused mitochondrial component that encircles the proximal centriole.

B. Oogenesis

Little is known regarding the transition of primordial germ cells into oogonia or the mitotic divisions that oogonia undergo before meiosis. Once a primary oocyte has entered meiosis, however, most brachiopods display a "follicular" type of oogenesis in which each developing oocyte becomes surrounded by a layer of follicle cells that can attach to the oolemma via junctional complexes (Fig. 2B). Oogenesis in *Megathyris* sp. and *Terebratulina retusa*, however, is "mixed" because a few discrete accessory cells of unknown function also occur within the space separating the oocyte from its surrounding follicular sheath. Alternatively, species of *Lingula* may utilize a "nutritive" form of oogenesis in which follicle cells are lacking and neighboring nurse cells are engulfed by the oocyte during gametogenesis.

At the onset of oogenesis, previtellogenic oocytes that are in prophase I of meiosis arise from the germinal epithelium overlying the proximal side of the genital lamella. The young oocytes are encapsulated by a thin layer of follicle cells and are gradually transported toward the distal end of the lamella as they grow in size and as new oocytes are added in the basal regions. Enlargement of the oocyte is due to the hypertrophy of the prophase-arrested nucleus (the "germinal vesicle") as well as to the accumulation of cytoplasmic stores of yolk-like substances during the vitellogenic phase of oogenesis. Such stores typically consist of lipid droplets, membrane-bound proteinaceous granules, and peripherally located cortical granules that do not seem to discharge during fertilization. Based on ultrastructural analyses of *T. retusa*, vitellogenesis involves both an autosynthetic type of yolk formation, in which all the yolk is assembled within the ooplasm, and heterosynthetic vitellogenesis, in which yolk precursors formed in extraovarian locations are actively endocytosed by the vitellogenic oocyte prior to yolk assembly.

Fully grown oocytes range in diameter from about 20 to 200 μm and typically possess a well-developed array of microvilli that are either covered by a fibrous envelope or interconnected at their apices by a network of cross-linked filaments. Most brachiopod oocytes seem to be arrested at prophase I at the time of spawning and undergo germinal vesicle breakdown (GVBD) and ovulation (i.e., the emergence from the follicular sheath) following release from the genital lamellae and/or the vasa genitalia surrounding the ovaries.

In *Glottidia pyramidata*, oocyte maturation can be triggered by an aqueous extract of the lophophore that apparently acts via a cyclic AMP-based signaling system in follicle cells to elicit the reinitiation of meiosis. The fully grown oocytes of *Terebratalia transversa*, on the other hand, can be stimulated to mature simply by detaching follicles (i.e., oocytes with their surrounding sheath of follicle cells) from the genital lamellae (Fig. 2C). Conversely, GVBD can be prevented in *T. transversa* by mechanically removing follicle cells within ~30 min after follicular detachment and/or by pretreating follicles with agents that uncouple gap junctions. Collectively, such findings suggest that the detachment of follicles from the genital lamella causes follicle cells to produce a positive stimulus of meiotic maturation that reaches the oocyte via the follicle cell–oocyte junctional complexes by ~30 min after follicular detachment.

IV. SPAWNING AND FERTILIZATION

As would be expected of sessile animals that lack copulatory appendages, brachiopods spawn their gametes to undergo external fertilization. Most adults can live at least several years and have multiple spawnings during a lifetime and/or a single reproductive season. In *Lingula anatina*, reproduction has been described as occurring during summer in Japan or throughout the year near Singapore. Similarly, the reproductive seasons that have been reported for other brachiopods vary considerably depending on the species, geographic location, or method used to assess fecundity.

Factors that have been shown to induce spawning include (i) an increase in water temperature after collecting several species of winter-breeding articulates; (ii) the presence of sperm introduced near ripe specimens of *T. retusa* (but not five other species of articulates tested); and (iii) lunar cycles in the case of *G. pyramidata* in which most animals maintained in the laboratory spawn just before nighttime high tides. In *L. anatina*, females spawn in multiple bursts that last up to a few days. Such spawnings are separated by periods of inactivity so that the entire reproductive period spans at least 2 or 3 months for a given individual. Thus, in a 24-hr period, a *Lingula* female spawns about 6–2000 oocytes, and over 188 days a large specimen can release nearly 30,000 oocytes.

During spawning, gametes are discharged from the gonads into the mantle cavity via the nephridia. In male brachiopods and inarticulate females, the emitted gametes are then expelled into the seawater by means of the exhalant feeding current generated by the lophophore. However, in most articulate species, oocytes are retained external to the body proper in a part of the mantle cavity for subsequent fertilization, and the developing embryos are then brooded in a chamber or a specialized region of the lophophore. Whether or not self-fertilization can occur in hermaphroditic brachiopods has not been clearly established, although such a mode of fertilization has been postulated for species of *Argyrotheca*.

The state of oocyte maturation at the time of fertilization is not generally known for naturally spawned specimens. Reports based on laboratory cultures are variable because oocytes of *L. anatina* can reach metaphase II before being fertilized. Alternatively, oocytes of several articulate species including *T. transversa* are still in meiosis I at fertilization, and thus both polar bodies are generated after fertilization. According to laboratory studies of *G. pyramidata* and *T. transversa*, successful fertilization occurs only if the oocytes have undergone GVBD prior to insemination.

In *L. anatina*, which is the only species in which sperm penetration has been documented, sperm typically fuse at the vegetal pole of the oocyte. Following fertilization, the oocytes of a few species are reported to form a fertilization envelope.

V. EMBRYOLOGY

The following synopsis refers to free-spawning species whose embryos are easier to obtain and analyze than those produced by brooding brachiopods. Cleavage in brachiopods is radial and holoblastic. The first two cleavages are meridional and thus pass through the animal–vegetal axis. The orientations of the next cleavages, on the other hand, tend to be variable. For example in some *T. transversa* embryos, the third cell division is meridional and the fourth is equatorial. However, in others the third cleavage is equatorial and the fourth is meridional.

Based on most brachiopods examined, the cleavages are more or less equal. However, small micromeres reportedly arise at the animal pole during the early development of *L. anatina* and *T. septentrionalis* before normal blastulae with relatively equal-sized blastomeres are eventually generated.

The later cell divisions in brachiopod embryos tend to be somewhat asynchronous and eventually culminate in the formation of a hollow blastula, which develops approximately 8–14 hr postfertilization in *T. transversa*. As the wall of blastula becomes more columnar, a single cilium forms at the apex of each blastomere, and the ciliated blastula/early gastrula eventually becomes mobile. Within a few hours after the formation of the blastula, gastrulation occurs at the vegetal pole, typically by invagination. A blastopore forms along the site of invagination, and in *G. pyramidata* the blastoporal opening is retained as the larval mouth. Alternatively, in *T. transversa*, the blastopore closes as postgastrular development proceeds, and the mouth forms following larval metamorphosis at a site that probably corresponds to the former blastopore. The blastopore also closes in *Neocrania anomala*, but the mouth forms opposite to where the blastopore had developed.

The archenteron that connects to the blastopore following gastrulation eventually fills much of the interior of the developing embryo. As the archenteron enlarges, coelomic compartments are formed in postgastrular embryos and early stage larvae. Coelomogenesis in the inarticulate *L. anatina* and the articulate *T. retusa* is reported to occur by schizocoely. Alternatively, in virtually all articulate species and in the inarticulate *N. anomala*, one to four pairs of coelomic pouches bulge off the archenteron in a form of enterocoely.

Through the use of vital markers, blastomere deletions, and manipulations of cleavage planes, detailed fate maps have been constructed for the embryos of *G. pyramidata* and *T. transversa*. In both brachiopods, the vegetal half of the egg gives rise to mesoderm and endoderm. Moreover, during embryogenesis in these two species, the vegetal region of the developing larva provides an inductive signal that allows the ectoderm of the presumptive "apical lobe" to differentiate properly.

However, the apical lobe-forming region of *Glottidia* occurs in the anterolateral part of the unfertilized egg, whereas the equivalent portion in *Terebratalia* is initially in the animal half of the egg but eventually shifts to an anterolateral position during embryogenesis. In addition, the anterior–posterior axis is essentially established by the end of oocyte maturation in *Glottidia* but is not specified until the late blastula stage in *Terebratalia*.

VI. LARVAL DEVELOPMENT AND METAMORPHOSIS

In nonbrooding species, a free-swimming larva is typically formed within a few days after fertilization, and brooding brachiopods tend to release postgastrular specimens of undetermined ages. All brachiopod larvae are lobate and contain superficial grooves that separate the larval lobes without segmenting the internal structures in the larva. The larvae are also similar in containing one to four pairs of bundles of chaetae that resemble polychaete chaetae in their ultrastructure and mode of formation.

The exact morphologies of the larvae that are produced, as well as the modes of nutrition and types of metamorphosis employed during development, vary considerably within the phylum. Articulates give rise to a trilobed nonshelled larva that comprises (i) an anterior apical lobe that forms the lophophore and viscera of the adult following metamorphosis, (ii) a middle mantle lobe that gives rise to the shell-secreting mantle, and (iii) a posteriorly positioned pedicle lobe that develops into the pedicle used in anchoring postsettlement specimens to the substratum (Fig. 2D). Inarticulate species such as *Lingula*, on the other hand, generate a bilobed shelled larva that consists of (i) an anterior apical lobe that forms the lophophoral cirri used in larval feeding and (ii) a posterior mantle lobe that produces the mantle and eventually gives rise to the pedicle much later in development. The inarticulate *Neocrania* also forms a bilobed larva, but the larva lacks a shell or feeding apparatus.

The larvae of articulate brachiopods do not feed and generally spend on the order of one to several days in the plankton prior to settlement. However, development in the antarctic articulate *Liothyrella uva* takes at least 115–160 days before the larva is capable of settling.

Except for the lecithotrophic larva of *Neocrania* that settles within about 4 days after fertilization, inarticulate larvae are typically long-lived and feed in the plankton with their developing lophophore for about 3–5 weeks before assuming a benthic existence. In fact, some inarticulate specimens fail to settle even after several weeks and continue to spend a prolonged time in the plankton as "drift larvae."

In inarticulates that produce a shelled larva, overt signs of metamorphosis are lacking except for changes in the pattern of shell secretion during development. The first shell produced before hatching in lingulid inarticulates is the "embryonic shell," which becomes supplanted at its anterior and lateral edges by a "juvenile shell" in the free-swimming larval phase. After larval settlement, a third type of shell, the "protegulum," accrues in front of the juvenile shell. In disciniscid inarticulates, on the other hand, a shell is not secreted before hatching so that only a juvenile shell is produced by the free-swimming larva prior to settlement and the formation of a protegulum.

In contrast to the shelled larvae of inarticulates, the nonshelled larvae of articulate species and the inarticulate *N. anomala* exhibit a dramatic metamorphosis upon settlement. Larvae of *Neocrania* undergo a ventral flexure at settlement and become cemented to the substratum by the ventral side of the mantle. In articulates, the larva typically attaches by the distal tip of the pedicle lobe, and the mantle lobe reverses position to cover the apical lobe. Mantle lobe reversal in *T. transversa* occurs relatively rapidly, and the protegulum, which contains calcium carbonate, is secreted only after mantle reversal except in laboratory cultures immersed in high-potassium- and low-calcium-containing seawater in which a protegulum can form without the mantle being reversed. In *Waltonia inconspicua*, on the other hand, the process of mantle reversal can take more than a day, and protegulum formation typically begins before the mantle is reversed.

The exact cues that trigger settlement and metamorphosis are not well understood. Bacterial films seem to be important in inducing larvae to settle since little settlement and metamorphosis occur in sterile cultures. Conversely, in laboratory cultures of *T. transversa*, seawater with excess potassium ions can stimulate settlement and metamorphosis by depolarizing excitable cells that are apparently located in the distal tip of the pedicle lobe of fully grown larvae.

VII. CONCLUSIONS

Within the phylum Brachiopoda, the modes of reproduction and development exhibit a few unifying patterns as well as some marked differences. All brachiopods are solitary benthic organisms with simple reproductive systems. Asexual reproduction and copulation are lacking. Cleavage is holoblastic, radial, and fairly equal. Moreover, all species produce lobate larvae.

Among the various brachiopods examined, however, distinct differences occur in (i) sperm ultrastructure, (ii) fate of the blastopore, (iii) mode of coelomogenesis, (iv) larval morphology, (v) types of shell secreted, (vi) presettlement nutrition, and (vii) patterns of metamorphosis. Such differences tend to coincide with a systematic dichotomy between inarticulates and articulates (plus craniacean inarticulates), but a more robust database is needed before definite conclusions can be drawn.

Brachiopods also display supposedly deuterostome characteristics (e.g., radial cleavage, enterocoelous coelomogenesis, and failure of the blastopore to form the larval mouth) along with protostome features that are believed to be either characteristic of the entire phylum (e.g., ultrastructure of the larval chaetae and sequence of the 18S ribosomal DNA) or attributable to only a few species (e.g., schizocoelous coelomogenesis and retention of the blastopore as the larval mouth). Additional analyses should help to confirm or correct these putative trends and thereby aid in providing a better understanding of this ancient and fascinating group.

See Also the Following Articles

BRYOZOA; ECHINODERMATA; MARINE INVERTEBRATES, MODES OF REPRODUCTION IN

Bibliography

Chuang, S.-H. (1990). Brachiopoda. In *Reproductive Biology of Invertebrates. Vol. IV, Part B. Fertilization, Development, and Parental Care* (K. G. Adiyodi and R. G. Adiyodi, Eds.), pp. 211–254. Wiley, Chichester, UK.

Freeman, G. (1993). Regional specification during embryogenesis in the articulate brachiopod *Terebratalia*. *Dev. Biol.* 160, 196–213.

Freeman, G. (1995). Regional specification during embryogenesis in the inarticulate brachiopod *Glottidia pyramidata*. *Dev. Biol.* 172, 15–36.

Halanych, K. M., Bacheller, J. D., Aguinaldo, A. M. A., Liva, S. M., Hillis, D. M., and Lake, J. A. (1995). Evidence from 18S ribosomal DNA that the Lophophorates are protostome animals. *Science* 267, 1641–1643.

Hodgson, A. N., and Reunov, A. A. (1994). Ultrastructure of the spermatozoon and spermatogenesis of the brachiopods *Discinisca tenuis* (Inarticulata) and *Kraussina rubra* (Articulata). *J. Invertebr. Reprod. Dev.* 25, 23–31.

Hyman, L. H. (1959). *The Invertebrates: Smaller Coelomate Groups*, Vol. V. McGraw-Hill, New York.

James, M. A., Ansell, A. D., Collins, M. J., Curry, G. B., Peck, L. S., and Rhodes, M. C. (1992). Biology of living brachiopods. *Adv. Mar. Biol.* 28, 175–387.

Long, J. A., and Stricker, S. A. (1991). Brachiopoda. In *Reproduction of Marine Invertebrates. Vol. VI. Echinoderms and Lophophorates* (A. C. Giese, J. S. Pearse, and V. B. Pearse, Eds.), pp. 47–84. Boxwood Press, Pacific Grove.

Nielsen, C. (1991). The development of the brachiopod *Crania (Neocrania) anomala* (O. F. Muller) and its phylogenetic significance. *Acta Zool.* 72, 7–28.

Reed, C. G. (1987). Phylum Brachiopoda. In *Reproduction and Development of Marine Invertebrates of the Northern Pacific Coast* (M. F. Strathmann, Ed.), pp. 486–493. Univ. of Washington Press, Seattle.

Stricker, S. A., and Folsom, M. W. (1997). Oocyte maturation in the brachiopod *Terebratalia transversa*: The role of follicle cell–oocyte attachments during ovulation and germinal vesicle breakdown. *Biol. Bull.*, in press.

Stricker, S. A., and Reed, C. G. (1985). The ontogeny of shell secretion in *Terebratalia transversa* (Brachiopoda: Articulata) II. Formation of the protegulum and juvenile shell. *J. Morphol.* 183, 251–271.

Breast

see Mammary Gland

Breast Cancer

Laura Esserman and Hope Wallace
University of California, San Francisco

I. Introduction
II. The Epidemiology
III. Incidence Rates
IV. Risk Factors
V. Prevention
VI. Screening
VII. Diagnostic Approaches
VIII. Molecular Basis of Breast Cancer
IX. The Biology of Breast Cancer
X. Treatment Options for Breast Cancer
XI. Hormone Replacement Therapy
XII. Psychosocial Support
XIII. Breast Cancer and Pregnancy
XIV. Genetic Testing

GLOSSARY

adjuvant therapy Treatment that is given to patients to prevent or delay the spread or recurrence of cancer. Adjuvant therapy is given to patients who do not have any indication that their cancer has spread outside of the breast. It is administered after surgery and can be given before or after radiation therapy.

biopsy The removal of a sample of tissue which is microscopically examined for cancer cells.

chemotherapy A treatment for cancer that works by killing all rapidly reproducing cells.

clinical trial A research study that evaluates a new treatment, prevention, detection, or diagnosis method by comparing the new method to the standard method. Also referred to as studies.

ductal carcinomal in situ (**DCIS**) Cancer cells that develop from the lining of the milk duct but are confined to the ducts of the breast. DCIS is considered to be a precursor to invasive cancer because it does not have the ability to spread to tissues outside of the breast.

hormone therapy A treatment for cancer that works by removing, blocking, or adding hormones.

incidence rate The number of people who develop a disease (in this case, breast cancer) in a certain amount of time.

in situ A term that refers to cancers which have not grown beyond their original site.

lobular carcinoma in situ (**LCIS**) Abnormal cells that develop from the lining of the lobules in the breast. LCIS is not considered to be a precursor of cancer, but it is a marker of high risk.

local therapy Treatment that is used to remove or kill cancer cells in the breast region. Local therapy can include surgery (lumpectomy or mastectomy) and/or radiation therapy.

lumpectomy The surgical removal of the breast lump and a portion of healthy breast tissue.

lymph node Glands found throughout the body which defend the body from bacteria or other foreign invaders. A lymph node dissection is usually done in the axilla (or the underarm area) during a lumpectomy or mastectomy to determine the extent of the spread of cancer outside of the breast area.

magnetic resonance imaging A noninvasive diagnostic technique that uses an electromagnet to evaluate a tumor.

mammogram A low-dose X-ray used to identify abnormalities in the breast.

mastectomy The surgical removal of the breast.

metastasis The spread of cancer beyond the primary site. The extent of the spread of cancer is measured by the number of lymph nodes that have cancerous cells in them.

radiation therapy A local treatment consisting of short X-ray treatments to the breast. Radiation therapy is used to either shrink a tumor before surgery or to kill any cancer cells that remain after surgery.

recurrence The return or spread of breast cancer.

screening The diagnostic process of examining the breasts for abnormalities. The most common type of screening is mammography.

stage The method used to establish the extent of cancer. Staging is determined by the tumor size and whether it has spread to lymph nodes or other distant sites around the body.

systemic therapy Therapy that is given to a patient to treat the whole body. Systemic therapy is given to patients who have metastases and to patients who want to prevent or delay the development of metastases.

tumor A mass of abnormal cells that can interfere with the tissues surrounding it.

tumor grade The aggressiveness of the tumor. This is determined by a pathologist who categorizes the cells by their appearance under a microscope.

Breast cancer is the most common cancer diagnosed in women and is a major health problem in the Western world. Despite breast cancer's high incidence rate, it is important to recognize that most women do not develop the disease. Of those women who do develop breast cancer, most do not die from their disease. Breast cancer can be a devastating and lethal disease but it is not usually rapidly progressive. It grows in terms of months and years, not in terms of days and weeks. Thus, a diagnosis of breast cancer is not an emergency. Every woman diagnosed with breast cancer has the time to learn more about her disease and the time to become involved in the decision-making process. Decisions regarding therapy should be made according to each woman's values because these are important decisions that will affect the rest of her life.

I. INTRODUCTION

Cancer is the general term used to describe over 100 diseases that develop when cells divide and grow unrestrained. Over 1.3 million new cases of cancer were diagnosed in 1996. Breast cancer is the most common cancer diagnosed in women and the second largest cause of cancer deaths in women. Each year 186,000 Americans are diagnosed with breast cancer and 45,000 die of the disease. It is the leading cause of death among American women 40–55 years of age. Breast cancer has a major impact not only on the woman with the diagnosis but also on her family, friends, partner, and community. This article will provide a broad review of the epidemiology (the patterns and rates of breast cancer occurrence), incidence rates, risk factors, diagnosis, treatment, molecular basis, genetic testing, psychosocial treatments, and future therapies of breast cancer.

II. THE EPIDEMIOLOGY

Recent studies have shown a significant increase in the incidence of breast cancer. Fortunately, this increase is not as discouraging as it appears. Better and more consistent mammographic screening has led to the early detection of many small tumors, including ductal carcinoma *in situ*, and the incidence of larger tumors has actually decreased. Improved screening and greater patient and physician awareness has aided in the early diagnosis and treatment of tumors. This has led to a decline in mortality in the United States. The most pronounced decline in mortality has occurred in Caucasian women, who have the highest screening rates.

III. INCIDENCE RATES

Age is the most significant risk factor for developing breast cancer. Although the statistic that breast cancer affects one woman in eight is true, a woman's chance of developing breast cancer is a function of her age. The 1 in 8 statistic reflects the lifetime risk a woman has of developing breast cancer. A 30-year-old woman's risk of getting breast cancer is 1 in 2500. A 50-year-old woman has a risk of 1 in 50. Only if a woman lives to 85 or 90 does the 1 in 8 statistic apply (Table 1).

In addition, the 1 in 8 statistic applies only to Caucasian women. Rates for other ethnicities are lower. By age 75, Caucasian women have a 1 in 11 chance of developing breast cancer, whereas African American women have a 1 in 14 chance (incidence rates between 1989 and 1993 per 100,000 women: African American, 97.3; Caucasian, 112.8). While the lower incidence rates appear to be good news for African American women, their mortality rates are actually higher than those of Caucasian women (survival rates for women between 1986 and 1992: African American, 69.8%; Caucasian, 85.0%). Five-year survival rates for African American women from 1983 to 1988 are 62% compared to 70% for Cauca-

TABLE 1
The Chance of Developing Breast Cancer by Age[a]

Age	Chance
25	1 in 19,608
30	1 in 2,525
35	1 in 622
40	1 in 217
45	1 in 93
50	1 in 50
55	1 in 33
60	1 in 24
65	1 in 17
70	1 in 14
75	1 in 11
80	1 in 10
85	1 in 9
Ever	1 in 8

[a] Data from the National Cancer Institute Surveillance Program.

sian women. The difference in survival rates is thought to be due to later detection and lower screening rates for African American women. There has been speculation about other possible causes for the discrepant rates, including differences in estrogen metabolism, environmental exposures, and levels of dietary fat.

IV. RISK FACTORS

A. World Patterns

The incidence rates for women in the United States and Europe are significantly higher than those for women throughout the rest of the world. The factors responsible for the disparity in these rates are Western women's reproductive histories, lifestyles, and other factors associated with a higher socioeconomic class. Women in industrialized countries often have good overall health, few competing causes of illness, long exposures to estrogen, and a diet higher in fat and alcohol consumption. Western women have a greater exposure to estrogen because they are often nulliparous or delay childbearing until after age 30, have an early age of menarche, a late age at menopause, and use postmenopausal estrogen replacement for at least 5 years. These factors contribute to the incidence of over half of all breast cancer cases. The lowest incidence rates occur in developing countries, especially Asia. However, Asian women who have lived in the United States for at least 10 years develop breast cancer at rates similar to women born in the United States. This fact has led to speculation that diet is an important risk factor, although there may also be other factors not yet recognized.

B. Family History

A family history of breast cancer is also a risk factor. If a woman has a direct family member (mother, sister, or daughter) who developed breast cancer premenopausally, her chance of developing breast cancer doubles. However, the single risk factor of alcohol consumption is more significant than having a family member with premenopausal breast cancer. Although family history is important, 80% of breast cancers occur in women who have no history of breast cancer. Clearly, being a woman is the greatest risk factor. (See Section XIV.)

C. Other Potential Risk Factors

The relationship between birth control pills and breast cancer is controversial. It is thought that birth control pills increase a woman's risk of breast cancer, but studies show that once a woman has discontinued using birth control pills for over 10 years, she is no longer at a higher risk. There is also no increase in risk in women who have had spontaneous or induced abortions.

D. The Environment

Recent studies have shown that environmental causes are not as important as the combination of the other risk factors. Many people point to pesticides, dioxins, and other toxins as contributors to breast cancer. Although these may have an effect on the development of cancer, it is interesting to note that women who work in fields and who are exposed to many of these toxins often have lower incidence rates than women in affluent areas who are less exposed.

It is also important to note that unlike lung cancer, in which cigarettes are the known causative carcinogen, no carcinogen has been identified for breast cancer. More women die from lung cancer, which could be prevented by eliminating cigarettes, than from breast cancer.

V. PREVENTION

There are several lifestyle interventions which should reduce a woman's risk of developing breast cancer. Abstaining from alcohol and cigarettes, following a high-fiber, low-fat diet, and exercising are thought to reduce one's risk and certainly contribute to general health. Vigorous exercise 4 hr a week can reduce a premenopausal woman's risk of breast cancer by over 50%. This means that if a woman's lifetime risk is 1 in 8, her risk will be reduced to 1 in 16. In Norway, studies of exercise patterns suggest that premenopausal women can reduce their risk by 25% simply by walking 4 hr a week. Postmenopausal women can reduce their body fat through exercise. By reducing body fat, a woman also reduces estrone production which is a hormone thought to stimulate breast cancer cell growth. While these lifestyle changes are thought to help reduce a woman's risk, they have not been proven to prevent breast cancer.

There are other modes of prevention called chemoprevention which involve using drugs to prevent the development of breast cancer. A national prevention trial was initiated to determine if tamoxifen could be used as a chemoprevention drug. Although preliminary results suggest that tamoxifen does not prevent the development of breast cancer in a high-risk population but it has some side effects and does not prevent all cancers. There is great interest in developing drugs that will prevent the development of breast cancer in high-risk women.

VI. SCREENING

Until there is a known cause and cure for breast cancer, screening for early detection is the best weapon against death from breast cancer. While not every type of tumor can be detected by mammography, it is not known which types of tumors women will develop. For this reason, a woman's screening plan should consist of self breast exams, annual clinical exams, and mammograms beginning at age 40 or 50 depending on her risk factors and personal preference. For those women with a family history, current recommendations are to start screening 10 years prior to the age at which her first-degree relative (mother or sister) was diagnosed with breast cancer.

Mammography is the most effective method of screening for breast cancer. Mammography is an inexpensive, reliable, and safe tool for detecting lesions or abnormalities that are often not detected in a clinical or self breast exam. Mammograms can detect tumors <1 cm in diameter. Mammography does not prevent breast cancer but it is clearly the most effective tool for reducing the mortality of breast cancer.

Screening women in their 50s reduces the mortality from breast cancer and can be an effective population intervention. The benefit of routine mammographic screening has been shown in many studies, the largest of which occurred in Sweden, where mammograms performed every other year showed a 30% reduction in mortality from breast cancer in women aged 50–70. Mammography does not detect all cancers, however, and a new lump should be investigated even if a mammogram appears normal.

While mammography is effective in younger women, screening has a greater impact on women in their 50s and 60s than on women in their 40s. This is largely because only 20% of all breast cancer cases occur in women under the age of 50. Mammograms are not as effective at finding cancers in younger women partly because the breast tissue in younger women is more dense. Thus, it is important for women not to ignore a growing mass even if a mammogram appears normal. Between 5 and 15% of younger women's mammograms are considered to be "false positive." A false-positive reading occurs when a mammogram appears to have an abnormality when there is no abnormality in the breast. Women who have a false-positive reading have additional procedures (including surgery) which cause great anxiety and additional cost. Younger women also have a higher percentage of more rapidly growing tumors which are not likely to be detected by screening. Because tumors are thought to have higher growth rates in younger women, many experts recommend that women who decide to have mammograms in their 40s should do so annually. Despite mammography's decreased effectiveness in detecting cancers in women under age 50, the National Cancer Institute recommends mammographic screening begin at age 40. At a conference held in February 1997, it was determined that women who had mammograms in their 40s had a decrease in mortality.

Every woman should review information and make a decision with her provider about when she should begin screening. Women should also conduct monthly breast self-exams and have annual clinical exams. Although breast self-exams have not been proven to find cancer at an early stage, they will increase women's awareness of their bodies and empower them to take better care of their health.

VII. DIAGNOSTIC APPROACHES

In this section, we will review the interventions used to evaluate a lump or a suspicious mammogram to determine if a woman has breast cancer. The advent of new diagnostic tools, which differ according to whether or not a lump is palpable, has enabled the diagnosis of breast cancer prior to a surgical biopsy.

A. Palpable Lumps (Lumps That Can Be Felt)

1. Aspiration

Aspiration is a procedure in which a thin needle is inserted through the skin into the suspicious area. If the lump is a cyst, it is filled with fluid; the fluid can be withdrawn through the needle and the lump disappears.

2. Fine Needle Aspiration

This procedure is used when the mass is solid. Local anesthetic is usually used before a very thin needle is inserted. The needle withdraws a few cells, which are smeared onto a glass slide and then sent to a pathologist to be examined.

3. Core Needle Biopsy

A slightly wider needle is used for this procedure which removes a very small piece of tissue the size of the inside of the needle. The tissue is sent to a pathologist to be examined.

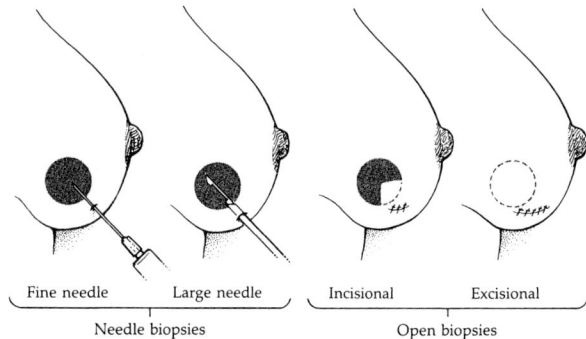

FIGURE 1 Needle and open biopsies are diagnostic approaches used to evaluate a lump or a suspicious area found on a mammogram (reprinted with permission from Love, 1995).

4. Open Biopsy

This is a surgical procedure consisting of opening the skin to remove the lump. If the entire lump is removed, it is an excisional biopsy; if only part of the lump is removed, it is an incisional biopsy. Again, the tissue is sent to a pathologist for examination (Fig. 1).

B. Nonpalpable Lumps (Lumps That Cannot Be Felt)

1. Stereotactic Needle Biopsy

This procedure uses the mammogram and a computer to identify the suspicious areas. The needle position is guided by the computer to the exact site of the abnormality on the mammogram. A piece of tissue (as in the core biopsy discussed previously) is removed and sent to the pathologist to be analyzed.

2. Wire (or Needle) Localization Biopsy

In this procedure, mammograms are used to help place a skinny wire at the site of the abnormality on the mammogram so that the surgeon can find the exact area and remove the tissue in the operating room. The tissue is sent to a pathologist to be examined.

3. Ultrasound

Ultrasound consists of sound waves that are used to characterize abnormalities seen on mammography. Ultrasound can identify if the abnormality is a cyst (fluid filled), which can be left alone. Ultrasound can also be a guide to finding a nonpalpable lump for a fine needle aspiration or core biopsy.

C. Emerging Imaging Technologies

1. Digital Mammography

Mammography is not likely to be replaced as a screening tool for the asymptomatic woman because it is sensitive, easily performed, reasonably inexpensive, and well tolerated. However, it is likely to be significantly enhanced by the introduction of digital mammography, which will allow the radiographic images to be acquired in a digital format and stored or sent electronically. This is significant because it will facilitate comparison with prior mammograms, avoid the loss of mammograms, enable access to expert mammographers, and speed the development of computer-aided detection systems.

2. Magnetic Resonance Imaging

Much research is being devoted to magnetic resonance imaging (MRI) of the breast. For the evaluation of the symptomatic breast, MRI is likely to play a significant role in the near future. MRI has the ability to distinguish benign from malignant mammographic lesions in most cases. It is highly sensitive, can help determine the stage of the disease, and can identify extensive microscopic disease. MRI potentially has a role in surgical planning because it is able to visualize the extent of tumor involvement in breast tissue.

VIII. MOLECULAR BASIS OF BREAST CANCER

Cancer is a disease that develops when cells divide and grow unrestrained. Healthy cells divide in an orderly fashion and controls inside the cell tell the cell when to stop dividing. Cancer cells lose these controls through a series of noninherited genetic changes. Genes called suppressor genes prevent normal genes or proteins from being expressed, whereas promoter genes encourage cells to divide. Occasionally, an error, or mutation, occurs in the suppressor or promoter genes which makes them function abnormally and causes cells to grow and divide without restraint. This step is critical in the transformation of a normal cell to a malignant cell. There are a number of changes that must occur for a cell or group of cells to become malignant, or cancerous. These include the group of cells' ability to

1. Escape the tumor mass and enter the bloodstream.
2. Invade the surrounding tissue outside the primary tumor.
3. Set up their own blood supply (angiogenesis).
4. Turn off the mechanisms that limit the number of cell divisions and the life span of the cell.

These mechanisms are found in every cell. The limiting of the number of cell divisions is also known as programmed cell death, or apoptosis.

IX. THE BIOLOGY OF BREAST CANCER

Tumors can be benign (noncancerous) or malignant (cancerous). A malignant tumor usually does not have a defined border or limit and invades surrounding tissues. Sometimes cancerous cells travel from the tumor into the bloodstream. These cells travel through lymphatic vessels to the lymph nodes and can be rerooted in other organs in the body. This spreading of cancer beyond the original site is called metastasis (Fig. 2).

Breast cancer is not a single disease but many heterogeneous diseases. Some breast cancers are slow growing, whereas others are very aggressive. The two most important predictors of tumor behavior are tumor size and whether or not the cancer has spread to the lymph nodes. Most women will not have a cancer that has spread to other organs, such as the liver or bones, when they are first diagnosed. When a tumor is small and it has not metastasized, it is usually a stage I cancer. Women with stage I cancers usually do very well.

Another factor that predicts the aggressiveness of a tumor is the tumor grade. The tumor grade is the characterization of cells which is determined by a pathologist who examines them under a microscope.

a patient should have additional therapy, the cell grade can sometimes aid in the decision.

A. Carcinoma *in Situ*

There are two types of carcinoma *in situ*: ductal (or intraductal) carcinoma *in situ* (DCIS) and lobular carcinoma *in situ* (LCIS). Both types are considered to be "precancer" or stage 0 cancer. Carcinoma *in situ* is contained inside ducts or lobules and predicts the possibility of progression to invasive disease.

1. DCIS

DCIS consists of a group of cancerous cells that start within the milk ducts from the cells that line the milk ducts (Fig. 3). While the cells themselves have changed to cancerous cells, they have not yet gained the ability to invade the surrounding tissue. Prior to 1972, the incidence of DCIS was less than 2 or 3%. Because of screening, it now accounts for up to 50% of mammographically detected breast cancer in women aged 40–50 and 20–25% of mammographically breast cancers detected in women aged 50–70. Although the natural history of DCIS is thought to be long and DCIS is presumed to develop into invasive cancer only 20–25% of the time, it is not closely tracked and observed because of the risk of the development of invasive cancer at the same site. Instead, DCIS is usually treated with either lumpectomy alone, lumpectomy and radiation, or mastectomy. Because it is localized in the breast, it does not need to be treated with chemotherapy. New local

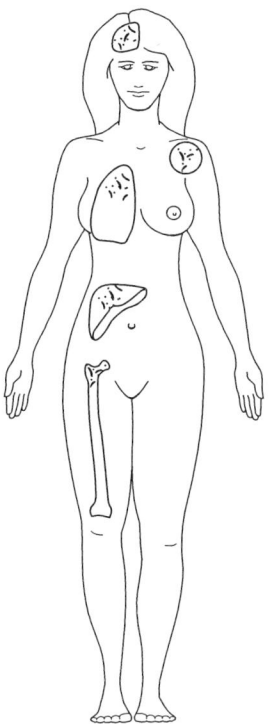

FIGURE 2 A metastasis occurs when cancer cells spread beyond the primary site. Common sites of metastasis are the bones, lung, liver, brain, and skin.

There are three tumor grades: well differentiated (grade I), moderately differentiated (grade II), and poorly differentiated (grade III). Grade I cells look most like normal cells. Higher grade tumors have larger, more variable cells that do not look like normal cells of the breast. When it is unclear whether

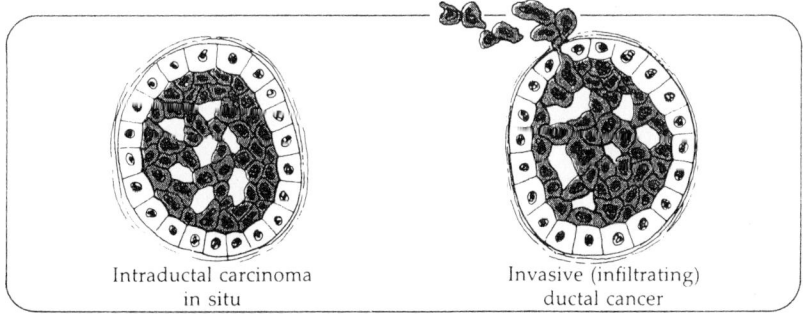

FIGURE 3 Ductal (or intraductal) carcinoma *in situ* (left) consists of cancerous cells that line the milk ducts but do not have the ability to invade surrounding tissue. Invasive or infiltrary ductal cancer (right) occurs when cancerous cells spread beyond the duct into the surrounding tissue (reproduced with permission from Love, 1995).

treatments, different from those used to treat invasive cancer, are needed to treat *in situ* cancer before it progresses to invasive cancer. DCIS is an area of intense research and is an important target for tumor prevention.

2. Lobular Carcinoma in Situ

LCIS arises from cells that line the lobules of the breast (where the milk is made; Fig. 1). LCIS is a marker that predicts approximately a 25% lifetime risk of developing breast cancer in either breast. It is not believed to be a precursor of tumor development in the site in which it is found. Most people who have this diagnosis choose close monitoring and follow-up.

B. Invasive or Infiltrating Cancer

Invasive breast cancers are those that have spread beyond the ducts or lobules of the breast into the surrounding tissue. They are not contained like *in situ* cancers and have the potential to spread to the lymph nodes and to other organs in the body. Although there is some controversy about the derivation of invasive cancer, many experts believe that the majority of invasive tumors arise from carcinoma *in situ*. Infiltrating ductal comprises 90% of invasive cancers. When palpated, infiltrating ductal cancer is usually identified as a lump, hard to the touch. Infiltrating lobular comprises only 10% of invasive cancer and is characterized by a thickening of the breast. It is sometimes more difficult to detect by clinical exam and is more often missed on mammography. However, in many cases, it can be palpated and radiographically visible.

C. Inflammatory Cancer

Inflammatory cancer is a rare type of cancer that makes up <3% of all breast cancer cases but is very aggressive, grows rapidly, and often metastasizes. It is distinctive in that the skin of the breast often swells, is warm to the touch, or has a reddish appearance. Because of the aggressive nature of this cancer, chemotherapy usually precedes surgery.

X. TREATMENT OPTIONS FOR BREAST CANCER

Breast cancer is treated both locally, to remove and control the growth of cancer cells in the breast, and systemically, to treat cancer cells that have left the breast or metastasized. Local treatment consists of surgical procedures to remove the tumor and radiation therapy to kill any remaining cancerous cells in the breast region. Systemic treatment is used to treat the rest of the body with the intent to kill or control any microscopic tumor cells that may have spread beyond the breast. Systemic treatments consist of chemotherapy and hormonal therapy which generally decrease the risk of a distant recurrence by 25%.

A. Local Therapy

1. Surgery

The first clinical procedure for breast cancer was the Halsted radical mastectomy, which included the removal of the breast, the lymph nodes, and the muscles in the chest wall. This procedure is no longer used because research has shown that other procedures are equally as effective at removing breast cancer. Today, the two most commonly used surgical procedures are the modified radical mastectomy (the complete removal of the breast tissue, including the nipple, areola, some skin, and some of the lymph nodes under the arm) and the lumpectomy (the removal of the tumor and a rim of normal breast tissue) combined with a lymph node dissection (a separate incision to remove some of the lymph nodes under the arm). Lumpectomy is followed by radiation to reduce the chance of the tumor returning in the breast. Many clinical trials comparing mastectomy and lumpectomy with radiation have shown that these medical interventions are equally beneficial and have equivalent survival rates. In other words, women with early breast cancer are now able to have a lumpectomy, which conserves the breast, with radiation and have survival rates equivalent to those seen with women who undergo mastectomies.

Lumpectomy should be the therapy of choice for women who have tumors smaller than 4 cm and who prefer breast conservation. Some women who have tumors larger than 4 cm can also be treated with breast-conserving methods if they use neoadjuvant chemotherapy (chemotherapy that precedes surgery) to shrink the tumor before surgery.

When a women has a tumor larger than 4 cm, has multifocal disease (or cancer in several areas of her breast), or a small breast in relation to the size of her tumor, a modified radical mastectomy is used to remove her breast. Regardless of whether a woman has a mastectomy or lumpectomy, use of radiation therapy is recommended if she has a very large tumor or cancerous cells in many lymph nodes.

2. Lymph Node Dissection

During a lumpectomy or mastectomy, a surgeon will usually perform a lymph node dissection from the axilla, or armpit area. Once removed, lymph nodes are sent to a pathologist who examines them for disease. Lymph node dissections help to determine the stage of the disease and if the cancer has metastasized. This helps patients make decisions about chemotherapy and/or hormonal therapy, which is used to control the disease if cancerous cells are present in the lymph nodes. If cancer cells are found in the lymph nodes, patients are usually advised to take a more aggressive approach with systemic therapy.

Complications can arise from lymph node surgery and can include some loss of arm mobility, lymphodema (swelling of the arm), and pain in the arm area. Many surgeons now perform a more limited operation which has reduced the complications. In addition, arm exercises and physical therapy can be done which should eliminate most immobility problems.

New surgical techniques are being tested to identify women who will not need a full lymph node dissection. One technique is the sentinel node dissection. In this procedure, a blue dye, or radioactive tracer, is injected into the tumor site. The dye will travel to the first lymph node along the path that the cancer cells would take. This helps surgeons identify the first lymph node and perform a less invasive procedure to examine the extent of disease. If the first lymph node is noncancerous, no further surgery is needed. In the future, two other techniques may be used to identify women who will need axillary lymph nodes removed. One is the examination of blood from the bone marrow, and the other is positive emission tomography scanning, which can detect cancer in the lymph nodes and in other parts of the body without surgery.

3. Reconstruction

Almost every woman who has a breast removed can have reconstruction if she chooses. Women can choose from a variety of options, including saline implants with expanders, tissue flaps including a latissimus dorsi muscle flap (which uses back muscle and tissue to rebuild the breast) or a transverse rectus abdominis muscle flap (which uses abdominal muscle, fat, and skin to build the breast while giving the patient a "tummy tuck"), and tattoos and skin grafts to create areolas. Women can also have changes made in their other breast to make the two breasts match. For women who are going to have a mastectomy and wish to have breast reconstruction, the best results are obtained with immediate reconstruction.

The transverse rectus abdominis muscle flap is made from an "island" of skin, muscle, and fat from a woman's abdomen (Fig. 4). The island is still attached to its original source at one end and is, therefore, nourished by its original blood supply. The island is tunneled under the skin to the place where the breast used to be and is then molded and stitched into place. These flaps give the most natural appearance. A woman's skin can be spared during a mastectomy and the breast will change size as a woman's weight changes.

The latissimus dorsi flap is formed from an island of skin and muscle from the woman's back. It does not remain attached to its original blood supply. In experienced hands, flaps are very safe and can be formed with minimal morbidity and scarring, especially if the skin is spared during the surgery.

After a mastectomy, there is often little or no sensation in the skin over the breast area. With an implant, the skin still lacks sensation. The breast with the implant feels somewhat different to the touch (firmer) than the other breast, although the skin feels the same. With the flap surgeries, the new breast has some skin sensation from the transplanted skin but does not have nipple and areola sensation.

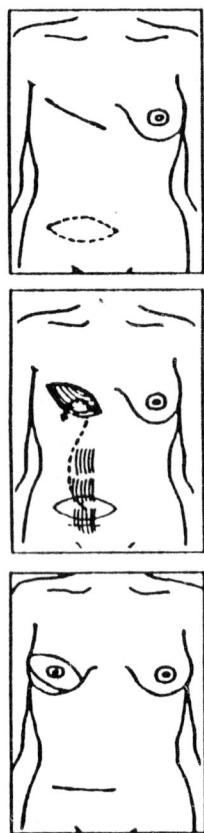

FIGURE 4 The transverse rectus abdominis muscle flap is a type of breast reconstruction that involves replacing the breast with an "island" of skin, muscle, and fat from a woman's abdomen. The island is tunneled under the skin to remain attached to the abdomen, where it is nourished by its original blood supply.

4. Radiation Therapy

Radiation therapy is an effective method of killing localized cancer cells because the rapidly dividing nature of cancer cells makes them highly susceptible to radiation. Radiation therapy usually consists of short treatments of radiation 5 days a week for 6 weeks. Radiation has been shown to benefit all patients by greatly reducing the chance of a local recurrence. It reduces the chances of a local recurrence from 30–40 to 10–15% (over 15–20 years). However, radiation is not thought to affect survival unless the tumor is large and many lymph nodes are involved.

B. Systemic Therapy

Systemic therapy is used to treat cancer cells that have spread through the body, outside of the breast. When the tumor is well established in organs outside the breast (metastasized), it is not curable. However, microscopic spread is thought to be treatable, which is why many oncologists recommend further treatment after surgery. The benefit of therapy is proportional to the risk of recurrence. The greater the risk a woman has of recurring, the greater her benefit will be from therapy. If her risk is lower, she will not receive as much benefit from therapy.

The effect of adjuvant chemotherapy or hormonal therapy (treatment that is given in addition to local treatment) for early stage breast cancer has been extensively studied in clinical trials involving over 100,000 patients. All women who choose systemic therapy will have a reduction in their mortality risk. Women with stages I and II cancer who take adjuvant therapy can reduce their mortality risk by approximately 25–35%. Although adjuvant therapy reduces mortality for patients with small tumors, most of these patients are cured by local therapy. The majority of women with tumors <1 cm in size have disease-free lymph nodes and a risk of metastatic recurrence of 10%. Adjuvant therapy reduces this risk by 2%–3%.

Women with node-positive disease may have their survival improved by 10–15% because their risk of recurrence is approximately 40%. Decisions to use adjuvant therapy are made by weighing the extent, or stage, of a patient's disease against the patient's tolerance of potential side effects and acceptance of risk. The type of therapy chosen usually depends on the menopausal status of the woman and whether the cancer cells contain estrogen and progesterone receptors (which is determined by an examination of the cancer cells).

1. Chemotherapy

Chemotherapy is the combination of anticancer drugs that target and damage rapidly dividing cells. It is usually given in an outpatient setting intravenously or in pill form and tends to be most effective in premenopausal women. The most common combinations of drugs are CMF (cyclophosphamide, methotrexate, and fluorouracil), AC (adriamycin and cyclophosphamide), and CAF (cyclophosphamide, adriamycin, and fluorouracil). Because chemotherapy targets rapidly dividing cells, it can damage the digestive tract (causing mouth sores, nausea, or vomiting), the blood producing cells in the bone marrow (causing fatigue and lower resistance to infection), and cells in the hair roots (causing temporary hair loss). Most of its side effects are reversible. Chemotherapy can also cause premature menopause or infertility in premenopausal women (40–60% risk depending on age).

2. Hormonal Therapy

Hormonal therapy is administered in pill form and interrupts cancer development by blocking estrogen receptors. Hormonal therapy includes either the drug tamoxifen or an oopherectomy, which is the surgical removal of the ovaries. Tamoxifen, one of a new class of drugs called selective estrogen receptor modifiers (SERMs), is a drug that works by blocking the estrogen receptors on cancer cells and works best on women whose tumors have estrogen receptors on the tumor cell surface. In patients who have positive estrogen and progesterone receptors, tamoxifen can reduce a woman's risk of recurrence by 35–45% (equivalent to chemotherapy). In women who have negative estrogen and progesterone receptors, it is not effective. The combination of chemotherapy and hormonal therapy can reduce risk slightly more than either treatment alone.

The benefits of hormonal therapy include the following: It is thought to reduce the risk of contralateral cancer (which is cancer that forms in the other breast), it has fewer toxic side effects than chemotherapy, and it does not drive premenopausal women into premature menopause. Women desiring children after treatment often opt for tamoxifen.

Most women have very few side effects from tamoxifen. The side effects that are reported include hot flashes, depression, and vaginal dryness, most of which can be treated. Although rare, there is a slightly increased risk of uterine cancer, blood clots, visual disturbances, and endometriosis.

When patients receive local and systemic treatment, there are concerns about the sequence of therapy. When radiation precedes chemotherapy, patients have lower local recurrence rates, but higher metastatic recurrence rates, than patients who choose chemotherapy first. Because overall survival is usually the highest priority, systemic therapy is usually initiated first.

3. Biological Therapy

Biological therapy is designed to bolster the body's immune system to fight cancer or to reduce the ability of individual cancer cells to grow. One example of biological therapy currently being used in clinical trials is HER-2/*neu* antibody. HER-2/*neu* is an oncogene, a gene that is turned on in cancer cells. It is thought to turn on the switch for cell growth. When HER-2/*neu* is switched on, it causes the overproduction of a protein on the cell surface that speeds cell growth and can lead to uncontrolled cell growth. This kind of breast cancer occurs in 20–25% of patients and it is associated with aggressive disease and a poor prognosis. The HER-2/*neu* antibody attaches to the HER-2/*neu* cells and can be used to treat women with HER-2/*neu* cancer. Studies testing this therapy are under way and early results are promising. Another class of biological drugs are the angiogenesis inhibitors, drugs to stop the development of small blood vessels that bring nourishment to developing tumors and allow them to grow. Biologic therapy is an active area of research and is potentially very exciting because scientists believe that many of these therapies are less toxic than current therapies. It is hoped that we will learn how to use them to make greater progress in the fight against breast cancer.

XI. HORMONE REPLACEMENT THERAPY

Recently, hormone replacement therapy (HRT) has generated much press and public concern. With recent studies showing a connection between HRT and breast cancer, many women fear taking estrogen at the time of menopause. Although studies show that using HRT for 5 or more years slightly increases a woman's chance of developing breast cancer to 13–17% (1.5-fold increase in rise), HRT provides many other health benefits to women. These benefits include a reduction of the risk of heart disease, the number one killer of women, from 46 to 33%; a reduction in osteoporosis and the risk of hip fracture from 15 to 13%; a decrease in colon cancer; and the potential to decrease the risk of developing Alzheimer's disease (data is controversial). Since the increased risk of developing breast cancer is very small, the benefits of HRT may often outweigh the risks of developing breast cancer. The decision to take HRT is very individual and is primarily a quality of life decision rather than one of prolonged survival.

For women at risk for bone loss who are not, or who choose not to be, candidates for HRT, there is a new class of drugs called bisphosphonates that are at least as effective as estrogen in preventing bone loss and fractures. In addition, there is a new class of synthetic estrogens (SERMs) that have been developed to avoid the stimulatory effects of estrogen on breast and uterine tissue. One of these drugs has been approved by the Food and Drug Administration for the prevention of osteoporosis. These drugs are now being studied for the treatment of breast cancer, because of the observation that women on the SERM developed less mammographically detected breast cancer.

XII. PSYCHOSOCIAL SUPPORT

Psychological factors have been correlated to both improved quality of life and survival time for breast cancer patients at all stages. Many factors influence psychological adjustment to breast cancer and quality of life, including choice of treatment type such as the election of mastectomy versus lumpectomy. Poor psychological adjustment to cancer can influence quality of life issues such as physical functioning, sexual functioning, social relationships, and psychological well-being.

Psychological interventions and one's attitude may influence medical outcomes. It is thought that emotional states may be able to cause changes in one's immune system, which in turn affects medical outcomes. Women who are able to express emotions (about their cancer and in general) and have a "fighting attitude" tend to do better clinically and have longer survival rates than women who passively accept or deny their cancer. Although the popular press usually associate expressing one's emotions with positive thinking, research shows that women who acknowledge stress and express emotions, including negative emotions and anger, have better survival rates than women who are more stoic. Women who have good interpersonal support and participate in a structured support group also have longer survival times. A controlled clinical trial of women with metastatic disease who participated in support group therapy vs those who did not participate showed an average of 36 months vs 14 months of survival, respectively.

Other specific psychosocial factors have been found to be related to distress among cancer patients. Women who accept or deny their cancer, who do not participate in support groups, and who do not have strong emotional support at home often do not attempt to engage in activities which could promote recovery. As a result, various psychosocial interventions have been developed to decrease the amount of psychological morbidity associated with the disease and to assist breast cancer patients to improve their daily functioning and quality of life. Interventions can be educational, supportive, cognitive–behavioral, or psychodynamic. Specific techniques may include supportive–expressive therapy, relaxation, imagery, hypnosis, behavioral modification, biofeedback, or meditation and can be conducted in groups or as individual treatments. The overall goal of interventions is to improve quality of life through increasing a sense of control, self-esteem, morale, and coping ability while decreasing negative side effects from surgery and adjuvant treatment.

XIII. BREAST CANCER AND PREGNANCY

It is uncommon for a pregnant women to develop breast cancer. A prenatal or postpartum breast cancer diagnosis can have a very poor prognosis and is usually diagnosed at a late stage. Little is known about the best treatment choices or optimal time to treat a pregnant woman for breast cancer. A pregnant woman can have surgery at the end of the first trimester and chemotherapy during the second, or preferably, during the third trimester. Pregnant women should not receive radiation therapy. It is also not known how long a woman can go untreated before her cancer progresses to a more advanced stage.

Many women worry about becoming pregnant after a diagnosis of breast cancer, but there is no reason why a woman should not consider having a child. Data from the Swedish Tumor Registry suggest that a woman's risk of recurrence may actually decrease with subsequent pregnancies. Decisions about future pregnancies and the timing of pregnancies involve many factors. Each woman should consider her choices carefully, discuss her situation with her physician, and know that there is a choice. Due to the sensitive time of the development of breast cancer, women in these situations need extra support.

XIV. GENETIC TESTING

Although the medical community has long known that family history plays a role in the development of breast cancer, only recently have geneticists identified some of the genes (inherited sets of DNA) that contribute to inherited breast cancer. Women with strong family histories of breast and ovarian cancer (>2 first-degree relatives) and very young women with breast cancer are at the most risk for genetically inherited breast cancer. Most breast cancer is not hereditary. Only 5–10% of all breast cancer cases are thought to result from mutation (or errors) in inherited genes. Errors that occur in the genes can be inherited from the father or mother. A person with a mutation has a 50% chance of passing the gene to his or her offspring. Breast cancer in younger women is rare and a large percentage of these cases are thought to have a hereditary component. Approximately 25% of all cases diagnosed before age 30 are thought to be due to inherited genes.

The two genes thought to account for 90% of inherited breast cancer are *BRCA1* and *BRCA2*. Special procedures to map out each gene allow scientists to identify the abnormal areas and mutations in the gene. Shortly after the cloning of *BRCA1* and *BRCA2*, tests were developed for identifying *BRCA1*, *BRCA2*, or both in individuals with breast cancer and their unaffected family members.

Mutation of *BRCA1* is thought to account for 45% of breast cancers in families with a high incidence of early onset breast cancer and for 80–100% in families with a high incidence of hereditary breast and ovarian cancers. *BRCA2* has been mapped to chromosome 13q and also appears to account for about 45% of early onset breast cancer. Unlike *BRCA1*, *BRCA2* may not influence the risk of ovarian cancer as much. The gene involved with male breast cancer is *BRCA2*. Women in families who inherit the "mutated" *BRCA1* and *BRCA2* genes were thought to have an 65–85% lifetime risk of developing breast cancer.

The identification of persons carrying breast cancer susceptibility genes is a promising approach to understanding the cause of breast cancer and developing more effective early detection and prevention strategies. At the same time, this opportunity raises numerous ethical and practical issues, both scientific and social. The implications and limitations of testing are not well appreciated by the increasing number of women who are requesting the test. Currently, there is little protection against insurance and employment discrimination. Although protective laws have been enacted, it is not yet clear how easy it will be to enforce such legislation.

There has been criticism that genetic testing for breast cancer is premature since there is no known cure. Some women with large family histories of breast or ovarian cancer have opted for prophylactic surgeries. Data are not yet available to prove the effectiveness of prophylactic mastectomies, but historical studies suggest that surgery does reduce the risk of developing cancer by up to 90%. It is possible that early detection through intensive surveillance could be equally as beneficial as prophylactic mastectomy, but this has not been confirmed.

While we cannot yet determine how to prevent the development of breast cancer, there are actions, other than prophylactic surgery, that can be taken which could reduce or even eliminate one's chances of developing breast cancer. Fenretinide, a synthetic drug similar to vitamin A, has been shown in pilot studies to reduce the chances that a woman will develop breast cancer and is currently in clinical trials in Europe. It is hoped that the discovery of *BRCA1* and *BRCA2* will generate excitement and interest in developing other chemoprevention or preventive hormonal treatments which could be used to delay or prevent the initiation of cancer cells. Another intervening step would be to participate in an intensive diet, exercise, and lifestyle modification program. Finally, a woman with a mutated *BRCA1* or *BRCA2* gene could employ an intensive surveillance program using advanced diagnostic tools such as MRI.

In addition to the planning and treatment benefits of being tested for *BRCA1* or *BRCA2*, a recent study showed that there is no increased stress or depression in those who tested positive for *BRCA1*. Instead, there was a decrease in stress among those individuals who received negative results. Thus, one of the potential benefits of genetic testing is likely to be the reduction of uncertainty about risk.

The field of genetics and breast cancer is very complex. Information changes constantly and anyone considering testing should use a program in which there are genetic counselors and physicians with a special interest in this area. It is important to learn as much as possible and to be in a setting in which counselors are available to ensure that testing is appropriate and to conduct follow-up on people who are tested. The National Cancer Institute is developing a national registry to better understand the significance of genetic mutations and cancer.

See Also the Following Articles

BREAST DISORDERS; ESTROGEN REPLACEMENT THERAPY

Bibliography

Love, S. M. (with K. Lin (1995). *Breast Book*, 2nd ed. Addison-Wesley, Menlo Park, CA.

Michnovicz, J. J. (1994). *How to Reduce Your Risk of Breast Cancer*. Warner, New York.

NCI Hotline (multiple pamphlets available upon request): 1-800-4CANCER.

Simone, C. B. (1995). *Breast Health*. Avery, Garden City, NY.

Breast Disorders

Diana O. Cua and Stefanie S. Jeffrey

Stanford University School of Medicine

I. Breast Lumps
II. Breast Pain
III. Nipple Discharge
IV. Abnormal Screening Mammogram

GLOSSARY

ANDI Aberrations of normal development and involution; a benign (noncancerous) departure from the normal tissue structure of the breast, related to normal growth or development patterns.

calcification An abnormal formation or deposit of calcium deposits in body tissue, e.g., the breast.

ecchymosis The occurrence or site of a bruise.

fibroadenoma A growth of breast tissue that is discrete, well circumscribed, movable, and firm or rubbery in consistency and of uncertain cause but associated with an imbalance of progesterone.

lipoma A benign growth of breast fatty tissue.

mammogram An X-ray examination of breast tissue.

mastalgia A general term for any recurring or persistent sensation of pain in the breast that is not associated with an immediate trauma.

Most breast disorders are benign and may be related to changes in growth and involution of the breast. This has been termed "aberrations of normal development and involution." However, it is usually a woman's concern that a breast problem is due to breast cancer that causes her to seek medical attention. There are four main problems of the breast: breast lumps, breast pain, nipple discharge, or an abnormal mammogram.

I. BREAST LUMPS

Normal breast tissue is composed of multiple structures: lobular and ductal tissue (the glands that make milk and the tubes that carry the milk to the nipple), stroma (the connective tissue framework around the breast ducts and lobules), fatty matrix, and ligaments that attach these breast structures to the skin. Because of the multiple components, normal breast tissue tends to be lumpy. Occasionally, normal architectural changes may be perceived as a breast lump. Very commonly, a normal subcutaneous fatty lobule may feel like a separate entity because it is bordered on either side by ligaments to the skin. This is termed a fibrofatty lobule and it is recognizable by its soft consistency. It may be located throughout the breast, but a fibrofatty lobule along the inframammary fold or in the subareolar region may feel particularly distinct. If the breast is examined carefully, other fibrofatty lobules of similar consistency can be identified in both breasts. Premenopausal women may also feel lumpy areas of normal fibroglandular breast tissue which become more prominent during the mid to late part of the menstrual cycle. This change is due to the proliferative impact of cyclic hormones on the breast and may create a focal or generalized change in the breast consistency, especially in the upper outer quadrants. Similar changes may also occur in postmenopausal women on hormone replacement therapy.

A breast mass may be defined as a dominant lump with measurable dimensions which is different from any other area in either breast. A breast thickening may be an area of tissue which is less compressible compared to surrounding or opposite breast tissue. Masses which are well circumscribed, mobile, soft, or multiple tend to be benign. In contrast, malignant masses generally have ill-defined perimeters and are usually hard or poorly compressible. Most palpable malignant masses are solitary at time of presentation. As malignant tumors increase in size, they invade surrounding tissue that has ligamentous attachment to overlying structures, causing skin dimpling and nipple retraction; very large tumors may invade the underlying pectoralis muscle and become fixed and unmovable over the chest wall. Breast masses associated with dermal edema (known as peau d'orange because the skin edema makes the skin pores look prominent and causes an appearance similar to the skin of an orange), areolar edema (which causes the nipple to look like a flattened pea as the areola swells up around the nipple with a change in the compressibility of the areolar skin compared to the opposite breast areola), and erythematous streaks of the skin (caused by blockage of dermal lymphatics and known as inflammatory changes) may be present in patients with very advanced breast cancer.

A hematoma may occur suddenly following breast trauma and is usually associated with ecchymosis of the overlying skin. Fat necrosis may also occur after breast trauma, in which an island of fatty tissue becomes devascularized and ultimately becomes a hard mass. However, a history of breast trauma alone does not mean that a breast mass is benign: A woman may touch her breast after it has been hit and then discover a tumor that has been growing at that site that is unrelated to the trauma and was previously unnoticed. Without a clear history of ecchymosis, needle or surgical biopsy of a posttraumatic mass is indicated.

In general, there are specific breast masses that may be associated more frequently with certain age groups: Fibroadenomas may develop during the early reproductive period, cysts are noted in the premenopausal and perimenopausal years, and cancer is more commonly diagnosed in the perimenopausal and postmenopausal years. The great majority of masses will be benign. Fibroadenoma and cyst occur with the highest frequency. A lipoma is usually clinically self-evident and may appear as a well-circumscribed lucent area on mammography.

A fibroadenoma is a discrete, well-circumscribed, movable, sometimes multilobulated mass that feels firm or rubbery in consistency. It is made of ductal and stromal tissue which organizes itself into a well-defined mass. Although its cause is not known, it seems to be a growth of ductal–lobular units and their surrounding stromal tissue, postulated by some as due to an imbalance of progesterone. Multiplicity may be another feature in 15–20% of patients with fibroadenomas and may be even more common in young Asian women. Fibroadenomas are observed most often in young females in their early reproductive state, usually <25 years of age, occurring when breast lobule formation is at its peak. They may, however, be noted in women up to age 50 and are occasionally seen in postmenopausal women, especially those on hormone replacement therapy, which can prevent the expected involution of breast hyperplastic tissue within the fibroadenoma. Cyclic breast pain may sometimes be associated with this lesion. Since a fibroadenoma is made up of breast glandular and ductal elements, cancer can develop within a fibroadenoma. This is extremely rare (estimated to occur in 2 or 3 of 1000 cases) and, when found, the cancer is frequently noninvasive. The natural history of fibroadenomas has been studied in patients undergoing needle biopsy, but long-term series are scarce since most patients ultimately opt for surgical removal. Several studies of patients undergoing fine-needle aspiration biopsy of fibroadenomas who were followed from 2 to 5 years have shown spontaneous regression or resolution of the fibroadenoma in about one-third to one-half of patients, growth in another third, and stability in the rest. Fibroadenomas also have a specific appearance on breast ultrasound, although there is some overlap with malignant solid masses. If a patient chooses clinical surveillance, a baseline breast ultrasound followed by needle biopsy, either by fine-needle aspiration biopsy or core needle biopsy, provides baseline status for follow-up. It is recommended that fibroadenomas that grow or new fibroadenomas diagnosed in women over ages 35–40 be surgically excised.

Breast cysts are quite commonly observed in women in their late 30s and 40s, when lobular involution may occur. Apocrine change, occurring when ductal epithelium develops a granular cytoplasm and apical cytoplasmic "snout-like" protrusions, is associated with increased ductal secretions. Fibrosis or epithelial proliferation may block the terminal duct leading into a lobule with resultant secretory dilatation of the lobule and cyst formation. In general, simple cysts (fluid-filled sacs with atrophic or apocrine linings) may be left alone unless significant discomfort is experienced. In this case, aspiration will relieve the tension produced by the accumulated fluid. The presence of a new cyst, especially in a postmenopausal woman, warrants further investigation. Benign intracystic papilloma or intracystic carcinoma may be masked during clinical examination of a breast cyst. A breast cyst that is aspirated contains fluid that may be golden, dark brown, green, or black. Following aspiration, the underlying breast tissue should be palpated for any residual mass in the same area that might represent a growth in the cyst wall. When no residual mass is palpable, the cyst fluid may be discarded unless it is bloody or there is particulate matter that would justify sending the specimen for cytologic analysis. If a residual mass is present despite complete collapse of the cyst, needle biopsy of this residual tissue or surgical excision is indicated. Conversely, breast ultrasound may be performed to evaluate the cyst wall and its adjacent tissue without cyst fluid aspiration. Some women have multiple breast cysts bilaterally (previously termed cystic mastitis), making follow-up difficult. Bilateral breast ultrasound has become a very useful tool in following these patients. Since women with gross cystic disease (having cysts large enough to cause a palpable mass) are at increased risk for developing future breast carcinoma, any new mass needs to be evaluated to confirm that it is indeed a cyst rather than a solid carcinoma clinically mistaken for a cyst.

Another cause of breast mass is termed fibrocystic change of the breast. Nonproliferative fibrocystic lesions include cysts, apocrine change, and mild hyperplasia (increased growth of epithelial cells lining a breast duct). Proliferative fibrocystic changes include florid hyperplasias, intraductal papillomas, and florid sclerosing adenosis. Women with nonproliferative fibrocystic change are not at increased risk for developing future breast cancer (except for women with gross cysts), whereas women with proliferative fibrocystic change have a mildly increased relative risk for developing future breast cancer. Women with atypical hyperplasia, especially if it is associated with a family history of breast cancer, have a marked increased risk for the future development of breast malignancy within the first 10 years of diagnosis, particularly if the women are premenopausal.

The incidence of breast cancer increases with increasing age: 1 in 227 (0.44%) women under age 40 develop breast cancer; 1 in 25 (4%) women 40–59 years old develop breast cancer; 1 in 15 (7%) women 60–79 years old develop breast cancer; and overall, 1 in 8 (12.5%) women from birth to death will develop breast cancer. Although cancer is less common in the young patient, all dominant masses in women of any age group require workup. When a solid breast mass is identified, it must be biopsied by fine-needle aspiration (FNA) biopsy, core needle biopsy, or surgical biopsy. Surgical biopsy may be excisional for smaller masses or incisional for larger masses (although most surgeons will perform needle biopsy of larger masses). When a patient presents with a dominant mass, she should undergo a diagnostic triple test consisting of clinical breast examination, mammography, and FNA or core needle biopsy. Breast sonography should be used in place of mammography for women under age 35 who do not have a strong family history of premenopausal breast cancer. It is preferable that no intervention be made prior to breast imaging since a needle biopsy may potentially produce slight bleeding that could obscure or change the lesion radiographically or sonographically. Series examining FNA biopsy of breast masses performed by an experienced physician and read by an experienced cytopathologist show an accuracy of 98% with a sensitivity of 97 or 98% and specificity of 99 or 100%. These results drop precipitously when less experienced persons perform or read the FNA biopsy. In this situation, core needle or surgical biopsy may be the preferred diagnostic method. If a benign result is obtained following FNA or core needle biopsy, the mass may be followed clinically. When a nonrepresentative sample is obtained, repeat FNA biopsy is indicated except if the mass clinically feels like and/or sonographically looks like a fibrofatty lobule. If atypical cells are identified on FNA or core needle biopsy, then a surgical biopsy is indicated. When there is any clinical suspicion of malignancy, a biopsy is warranted despite a negative FNA or core needle biopsy. A leading cause of medical malpractice is the diagnosis of breast cancer in women who complained to their primary care provider about a breast mass but were told that they did not need a biopsy diagnosis because their mammograms were negative. Masses in women with negative mammograms still need to be evaluated by a needle or surgical biopsy. Mammograms in premenopausal women and in postmenopausal women on exogenous hormones may be reported as negative due to adjacent dense breast tissue that can visually obscure a cancer.

Breast masses in pregnant or lactating women must also be evaluated. The vast majority will be due to preexisting lesion such as a fibroadenoma, lipoma, or papilloma, and others will be benign lesions, such as a galactocele (clogged milk duct which may be aspirated of its milky content), a lactating adenoma (hyperplastic lobule), or an area of mastitis (easily recognizable by FNA biopsy). Breast ultrasound and needle biopsy should be performed for definitive diagnosis. Surgical biopsy may be indicated in cases in which there is no experienced cytopathologist available who can distinguish the sometimes subtle differences between lactational proliferative change and carcinoma. It is recommended that women who are pregnant or lactating not undergo mammography until at least 3–6 months following cessation of nursing unless there is a strong suspicion of malignancy for the theoretical risk of exposing proliferating epithelial cells to radiation.

Breast masses in patients augmented with breast implants pose a special challenge both in clinical examination and in performing a needle biopsy without puncturing the implant. Usually, these patients are best evaluated and managed by breast specialists. A submammary or retromammary breast implant is located under the breast tissue but over the pectoral muscle; a submuscular or subpectoral breast implant in located under the pectoral muscle and is associated with a lower incidence of capsular contracture around the breast implant. Patients with breast implants may develop any of the breast masses described previously but may also develop masses related to the fibrous capsule surrounding their implant or in reaction to implant rupture with release of silicone. Some women from Southeast Asia may have a history of direct injections of liquid silicone or paraffin into their breast tissue for augmentation purposes with resulting rock-hard masses that may be clinically indistinguishable from cancer. These masses are due to foreign body reaction and usually have a characteristic mammographic and cytologic appearance.

Patients with a personal history of breast cancer who underwent breast conservation therapy have a 3–10% incidence of tumor recurrence in the ipsilateral breast. When the tumor recurs, the lesion is often located in the same quadrant as the original tumor. Lumps located close to lumpectomy or mastectomy scars may also be secondary to suture granuloma or maturing hypertrophic scar tissue or late scar contracture. Biopsy is indicated to differentiate recurrent tumor from scar. In the postmastectomy patient with autogenous breast reconstruction by transverse rectus abdominis myocutaneous (TRAM) flap, a dominant firm mass is more likely to be due to fat necrosis of ischemic adipose tissue within the TRAM flap than tumor recurrence. Again, FNA biopsy is diagnostic. Nodules on a postmastectomy chest wall usually require surgical biopsy since scar tissue fibrosis or granuloma may produce a nondiagnostic FNA biopsy and a malignant nodule would require excision in any case.

II. BREAST PAIN

Breast pain, also known as mastalgia or mastodynia, is a very common problem and affects over half of all women at some point in their lives. Only a minority of symptomatic women require treatment; most appreciate reassurance that the pain is not related to malignancy and are able to function well without treatment, often with eventual spontaneous resolution of the pain.

Breast pain can be classified as cyclic versus noncyclic in nature. Cyclic breast pain comprises over two-thirds of the cases and is seen in higher frequency among women in their third decade. It most commonly occurs at the time of ovulation or in the week prior to onset of menses. It is hormonally related and multiple causes have been suggested: excess or unbalanced estrogen, a "failed corpus luteum" with deficient progesterone in the luteal phase leading to unopposed estrogen, elevated prolactin release, unbalanced follicle-stimulating hormone (FSH) and luteinizing hormone (LH) secretion, diminished level of androgen, and even cyclic aberrations in lipid metabolism. Clinically, a patient may present with pronounced breast ache, heaviness, sensitivity, or tenderness localized in the upper outer quadrants or present in different areas throughout the breast. The pain may be unilateral or bilateral, diffuse, and can radiate to the axilla or upper arm (intercostobrachial nerve distribution). The pain is typically accentuated during the luteal phase and usually resolves or improves at the onset of menses. In some cases, the breast pain becomes long-standing and may occur throughout the menstrual cycle. Cyclic breast pain tends to disappear when ovarian function ceases at menopause, although the condition may be aggravated in women taking hormone replacement therapy. Daily pain charts are helpful to document the pattern and intensity of the pain.

Noncyclic breast pain is less common and tends to affect women in their fourth decade. In contradistinction to cyclic mastalgia, it is not regulated by hormones and occurs randomly or continuously throughout the menstrual cycle. Clinically, it has a shorter course and more often exhibits a unilateral or localized sensation and may be described as a "burning" or "drawing" type of the pain. Spontaneous resolution may be observed in as many as 50% of patients.

Most mastalgia has no identifiable cause. However, lesions that can produce breast pain are multiple: cyst under tension, ruptured cyst, lipoma, tender fibroadenoma, subareolar duct ectasia, infection, posttraumatic hematoma, fat necrosis, or malignancy. Although most breast cancers are not painful, it has been reported that as many as 7% of women with breast cancer will have pain as their sole presenting symptom, whereas 15% will have breast pain in association with other symptoms. Breast cancer pain may be characterized as persistent and fixed to a focal area, as if it were caused by compression of a nerve in a particular area or related to local pH or growth factor changes. Breast pain may also be the reason for performing a mammogram that detects a nonpalpable cancer that may or may not be located at the site of or even in the same breast as the breast pain. The workup of breast pain must include a thorough breast examination and mammography in women over age 35. Breast ultrasound is useful in looking for discrete solid or cystic lesions when the pain is focal and can be helpful in younger women in whom mammography may be contraindicated.

Certain chest wall conditions that are nonbreast in origin may radiate pain to the breast. In order to distinguish true breast pain from extramammary causes, the clinician may instruct the patient to lean forward on breast examination or examine the patient in the lateral decubitus position to set the breast apart from the chest wall. Costochondritis (also known as Tietze's syndrome) may present with medial chest wall or breast pain; Mondor's disease (thrombophlebitis of the thoracoepigastric or lateral thoracic vein of the breast and chest wall) causes more lateral pain. Other chest wall entities that may very rarely radiate to the breast are usually more clinically obvious, including rib fractures, radionecrosis of the ribs from radiation therapy, bone metastasis, fibromyositis, cervical radiculopathy, shingles, pneumonia, sickle cell disease, embolus, angina, and biliary colic.

Idiopathic breast pain may resolve spontaneously or not be as bothersome following reassurance that the pain is not related to breast cancer. Treatment is indicated in the 15% of patients who suffer severe mastalgia 7 days a month or if the pain has been present for over 6 months. Treatment recommendations, such as use of support bras (especially in women with noncyclic mastalgia) and restriction of fat intake, may be helpful to some individuals. Prospective randomized trials have shown no benefit from abstinence from methylxanthines (caffeine and chocolates), although individual response may vary. Women on hormone replacement therapy may benefit from readjustment of dosages of estrogen and progesterone; anecdotally, changes in the mode of estrogen replacement (such as switching from estrogen pills to a patch) have also helped breast pain in some women. Randomized controlled trials comparing vitamin E to placebo and trials comparing vitamin B6 to placebo have failed to show any benefit, although some patients may feel that their pain has been relieved (probably due to placebo effect or an individual response). It should be noted that large doses of vitamin E have antiplatelet properties that necessitate its discontinuation prior to any surgery. Diuretics are not indicated since studies have disproved fluctuations in total body water throughout the menstrual cycle. Exogenously administered progesterone is not more effective than placebo in controlled trials. Studies are examining the use of thyroid hormone therapy and its effect on prolactin secretion. Medical management of breast pain with evening primrose oil at 3 g/day [240 mg of γ-linolenic acid (GLA)] in two divided doses has been shown to be effective in English studies in almost 60% of patients with cyclic pain and almost 40% of patients with noncyclic breast pain. Its mechanism of action is to increase the level of GLA, an unsaturated fatty acid, and change the ratio of unsaturated to saturated fatty acids, which may affect breast sensitivity. It has low toxicity and is a recommended first therapy for patients with breast pain. If the pain is severe and continues unabated after 4 months of treatment with evening primrose oil, then danazol may be tried. Danazol is an androgen variant that binds to estrogen and progesterone receptors in the breast and brain and can affect FSH and LH release. It may be used at a dosage of 100 mg twice daily for 2 months, reduced to 100 mg/day for another 2 months, and then 100 mg every other day or only during the luteal phase of the menstrual cycle (especially Days 14–18) for a final 2 months, and it is discontinued after 6 months. If a patient does not improve within the first 2 months, the dose may be doubled to 200 mg twice a day. Danazol has a faster onset of action than evening primrose oil and provides the most effective relief of painful breast symptoms (79% in cyclic mastalgia and 40% in noncyclic mastalgia). It also may cause significant side effects at higher doses, such as amenorrhea or menstrual irregularity, weight gain, hirsutism, lower voice pitch, oily hair and skin, decreased libido, headache, and nausea. A history of thromboembolic disease contraindicates its use and contraception must be used since it is potentially teratogenic. Another androgen, gestrinone, may be more potent than danazol and appears promising in initial studies. Luteinizing hormone releasing hormone agonists, such as nafarelin and goserilin, are also effective but have been associated with menopausal side effects due to cessation of ovarian function, including hot flashes, depression or irritability, vaginal atrophy, decreased libido, and bone loss; visual changes, headaches, nausea, and hypertension may also be seen. Bromocriptine suppresses prolactin release via its dopaminergic agonist effects on the hypothalamus. Starting with an initial dose of 1.25 mg/day, with a gradual increase to a plateau dose of 2.5 mg twice a day after 2 weeks, it proved to be efficacious in 54% of patients with cyclic breast pain but only worked in 33% of patients with noncyclic pain. Bromocriptine may be used in place of danazol for patients with cyclic pain, but it is associated with a higher percentage of side effects, including nausea, dizziness, headaches, postural hypotension, seizures, and rare strokes and fatalities. Another dopamine agonist which decreases prolactin secretion, cabergoline, is also being tested in Europe since it may be better tolerated and has an easier dosing schedule than bromocriptine (twice weekly rather than twice daily). Tamoxifen is an estrogen antagonist with some estrogen agonist properties. It competitively inhibits the binding of estrogen receptors in the breast. The recommended dosage is 10 mg once or twice daily for 3 months. Side effects are rare but include increased blood coagulability that may lead to venous thrombosis or stroke and very rare uterine neoplasia, including endometrial carcinoma, with long-term usage. Of associated interest, tamoxifen has recently been shown to be effective in the chemoprevention of breast cancer in high-risk patients. As with the other medications mentioned previously, it should not be used in pregnant or lactating women. Consideration of excision of painful breast tissue should be reserved as a last resort for focal noncyclic breast pain and should be resisted in patients with cyclic mastalgia. Resulting scar tissue or surrounding tissue may again become painful. Focal pain in the breast or chest wall may instead be injected with a mixture of local anesthetic and steroid (2 ml of 1% lidocaine and 40 mg of methylprednisolone in 1 ml). Finally, although nonsteroidal antiinflammatory medications have only been anecdotally successful for mastalgia, they can be quite helpful in the treatment of chest wall pain.

III. NIPPLE DISCHARGE

Nipple discharge may be physiologic or pathologic. Physiologic nipple duct secretions may be manually expressed or aspirated by pump from both nipples in a majority of nonlactating women. Duct secretion that originates in multiple ducts in both breasts and that requires manual expression is not clinically worrisome. A spontaneous nipple discharge that may present itself by staining a woman's nightgown or brassiere and that originates from a single duct opening from one breast is considered pathologic. If the discharge is bloody or associated with a palpable mass, then it is even more suspicious for malignancy, especially if it occurs in a postmenopausal woman. Approximately 30% of postmenopausal women with a bloody nipple discharge have cancer as the etiology of their discharge. This compares to a 10% chance of malignancy for any woman with a pathologic nipple discharge. The etiology of pathologic nipple discharge varies. A central intraductal papilloma is a benign hyperplastic lesion noted in the wall of a single lactiferous duct and is the most common cause of pathologic nipple discharge. It may present in a premenopausal woman as unilateral and occasionally overtly bloody discharge arising from a single nipple duct. It is sometimes associated with a corresponding palpable subareolar mass. If the intraductal lesion is too small to palpate, the involved duct may sometimes be identified with sequential focal pressure around the areola. A ductogram may help delineate the location of the lesion within the proximal or more distal duct prior to surgical excision. Mammary duct ectasia is another nonmalignant cause of nipple discharge, usually seen in perimenopausal or postmenopausal patients and associated with cigarette smoking. It is a secretory disease that is associated with periductal inflammation and plasma cell infiltrate (hence the name periductal mastitis or plasma cell mastitis) with eventual subareolar fibrosis. There is loss of duct wall elasticity; the ducts dilate and fill with a serous, bloody, or brown discharge that may become thick and inspissated over time. Subareolar lumpiness or mass and nipple retraction may occur over time. Mammographic findings of large rod-like calcifications in a segmental distribution are diagnostic of this disease, but in the earlier stages of the disease, mammographic findings may be absent. Since the ducts may be colonized with bacteria, antibiotic coverage is important when a surgical nipple duct exploration is performed for biopsy purposes. Other benign reasons for nipple discharge are fibrocystic changes of the breast with accumulation of ductal secretion or rupture of a cyst into a breast duct. Breast cancer involving a breast duct as an intraductal (*in situ*) process or invasive breast cancer is the least common cause of nipple discharge but the main reason for surgical exploration and biopsy of the involved duct.

The nipple discharge may be tested for the presence of occult blood. The secretion may also be sent for cytologic examination, especially if it is hemoccult positive or grossly bloody. For breast cancers with nipple discharge, 70% will demonstrate a bloody discharge and about 85% will show an abnormal cytologic examination. Although it is usually negative, a mammogram should be performed to investigate any nonpalpable occult lesion. A ductogram may delineate the site of any intraductal lesion, whether proximal or peripheral. Although most surgeons perform exploration and excision of the involved nipple duct which shows the pathologic secretion, some are now following patients conservatively if there are no masses on clinical examination, the nipple discharge is hemoccult negative, cytology shows no epithelial cells, and mammography with or without ductogram is negative.

Extramammary causes of nipple discharge are numerous and typically present with bilateral, milky nipple duct secretions arising from multiple ducts. Certain medications have been implicated in the cause of ductal secretions, including cimetidine, haloperidol, reserpine, antihypertensives, phenothiazines, tranquilizers, methyldopa, tricyclic antidepressants, prolactin, and prolonged use of oral contraceptives. Tumors such as pituitary adenoma and, rarely, bronchogenic cancer may induce a state of hyperprolactinemia that may elicit copious amounts of milky discharge. Likewise, trauma and surgery to the chest have been reported to cause transient hyperprolactinemia. Other endocrine-related conditions that may cause nipple discharge include hypothyroidism and Chiari–Frommel syndrome (galactorrhea and amenorrhea without radiologic evidence of pituitary adenoma, possibly secondary to occult microadenomas).

Pregnant women in their second trimester may notice bilateral, serous, or bloody discharge. This may be related to the physiologic proliferative changes in the ducts that subsequently lead to the sloughing off of a necrotic ductal lining. Similarly, pubescent females may have bilateral serous nipple secretions that can be explained by the underlying ductal proliferation during the peak of breast development. Induced mechanical stimulation of the nipple may elicit a clear discharge in some women and is not worrisome. None of these situations require surgical treatment and reassurance should suffice.

IV. ABNORMAL SCREENING MAMMOGRAM

Screening mammograms are low-dose X rays of the breast usually taken in two views: the mediolateral oblique view, which visualizes the majority of the breast tissue, including the axillary tail, but may miss some high medial tissue or posterior tissue if positioning is suboptimal; and the craniocaudal (CC) view, which visualizes breast tissue compressed in a horizontal plane. Screening mammograms are performed on asymptomatic patients to detect malignancies that are not associated with palpable masses. The films are taken and batch processed after the patient has left the breast imaging facility. If any abnormalities are identified, the patient may be recalled for a diagnostic mammogram, a more expensive study that is directed by the mammographer and may require additional views of the breast such a lateral view, an extended CC view, and magnification and/or compression views of areas of concern in the breast. Breast ultrasound may also be performed to complement the diagnostic mammography.

Screening mammography is used for the early detection of breast cancer, with the goal of finding lesions at a smaller size with less chance of metastatic spread. Screening mammography has been shown to decrease the breast cancer mortality rate by about 30% in women age 50 and over. Although more controversial, studies also suggest an impact on premenopausal women and the American Cancer Society recommends yearly screening mammograms starting at age 40.

There are four types of abnormalities that may be mammographically identified: (i) breast masses, (ii) breast microcalcifications, (iii) areas of architectural distortion, and (iv) asymmetric densities. A mammographic breast mass is characterized by its shape, margins, and how dense (mammographically white) it appears. Dense surrounding breast tissue (also white on the mammogram) is often present in younger patients or postmenopausal women on hormone replacement therapy and may obscure a mammographic breast mass, leading to a false-negative mammogram. In cases of uncertainty, breast ultrasound may readily identify a discrete solid or cystic mass. Calcifications in the mammogram are not as affected by the density of the surrounding breast tissue. They may be present in normal tissue, benign disease, and malignant disease. Calcifications may be distinguished by their size, morphology, distribution, and associated findings. Scarring from previous breast biopsy or trauma and malignant tumors that distort the adjacent breast tissue may cause architectural distortion. New asymmetric changes in breast tissue density require further workup and possible tissue biopsy, especially if there is an associated palpable asymmetry on clinical examination.

Mammography is excellent at detecting breast abnormalities but it may not be able to distinguish between benign and malignant tissue since many mammographic features of benign and malignant lesions may overlap. In this case, tissue biopsy is indicated. When mammographic abnormalities are identified that are clearly benign (such as calcifications lining the wall of a blood vessel within the breast), no further workup is necessary. For masses or microcalcifications of low suspicion (termed "probably benign"), mammographic follow-up every 6 months for 1 year, then yearly, is adequate. Findings that are stable for 3 or more years usually, but not always, indicate benignity. For more suspicious masses or microcalcifications, FNA or core needle biopsy may be performed to sample the lesion. When needle biopsy of a nonpalpable mammographic lesion is performed under computerized mammographic guidance, it is termed stereotactic needle biopsy. FNA or core needle biopsy may also be performed under ultrasound guidance for nonpalpable lesions that are sonographically visible. Finally, hookwire localization surgical biopsy may be used to surgically excise a nonpalpable lesion. A hookwire is placed into a nonpalpable lesion using mammographic or sonographic guidance and the breast tissue surrounding the hookwire is then surgically removed. A specimen radiograph is performed to verify the excision of the abnormal lesion within the breast biopsy or lumpectomy specimen and the specimen is then sent for pathologic evaluation. Other new technologies, such as breast magnetic resonance imaging and digital mammography, are being tested to see if they can provide improved diagnostic acumen.

See Also the Following Articles

Breast Cancer; Mammary Gland Development; Mammary Gland, Overview

Bibliography

Bland, K. I., and Copeland, E. M. (Ed.) (1998). *The Breast: Comprehensive Management of Benign and Malignant Diseases.* Saunders, Philadelphia.

Carty, N. J., Carter, C., Rubin, C., *et al.* (1995). Management of fibroadenoma of the breast. *Ann. R. Coll. Surg. Eng.* 77, 127–130.

Dupont, W. D., and Page, D. L. (1985). Risk factors for breast cancer in women with proliferative breast disease. *N. Engl. J. Med.* 312, 146–151.

Harris, J. R., Lippman, M. E., Morrow, M., and Hellman, S. (Eds.) (1996). *Diseases of the Breast.* Lippincott-Raven, Philadelphia.

Holland, P. A., and Gateley, C. A. (1994). Drug therapy of mastalgia. *Drugs* 48, 709–716.

Hughes, L. E., Mansel, R. E., and Webster, D. J. (1987). Aberrations of normal development and involution (ANDI): A new perspective of pathogenesis and nomenclature of benign breast disorders. *Lancet* 2, 1316–1319.

Smallwood, J. A., and Taylor, I. (Ed.) (1990). *Benign Breast Disease.* Urban & Schwarzenberg, Baltimore.

Breastfeeding

Eloise D. Clawson and Carla D. Harris
Brigham and Womens Hospital

I. Anatomy and Physiology of Breastfeeding
II. Role of the Infant in Breastmilk Production
III. Unique Factors of Breastmilk
IV. Nutrition and Breastfeeding
V. Getting Started
VI. Pumping and Storage of Breastmilk
VII. Weaning
VIII. Sexuality and Contraception
IX. Summary

GLOSSARY

areola The pigmented area surrounding the mother's nipple which overlies the milk sinuses of the breast.

colostrum The fluid in the breast at the end of the pregnancy and during the first few days following the birth. Colostrum is the first milk and is viscous yellow, high in protein, low in fat, and contains protective immunoglobulins.

cradle/Madonna position A natural position used by many mothers while breastfeeding. The baby is cradled in the mother's arms and brought to the level of her breast for breastfeeding.

engorgement Temporary swelling of the breasts due to vascular dilation and the production of early milk.

football/clutch position A breastfeeding position often used with newborns and by mothers following a cesarean birth. The baby is held supported at the mother's side and brought up to her chest, with the baby facing the mother's breast.

latch-on This term is used when the baby is properly positioned at the breast with the gums of the mouth placed well back on the areola in order to suckle effectively.

mature milk Breastmilk that is usually present after approximately 10 days following birth. Volume and content of the mature milk adapt to the growth needs of the infant.

milk ejection reflex (letdown) A reflex initiated by the suckling of the infant at the breast causing the release of the hormone oxytocin and the ejection of milk from the collecting ductules in the breast.

side-lying position A position used for breastfeeding when both the mother and infant are lying on their sides and facing each other.

sucking or *suckling* The milking action of the baby's mouth at the breast.

transitional milk Breastmilk produced within the first few days following birth as the continuum between colostrum and mature breastmilk.

In June 1939, Alex Carrel, Nobel prize winner and author of *Man, the Unknown*, wrote a series of articles for *Reader's Digest*. In his first article, which addressed breastfeeding for babies, he wrote:

Should our baby be breastfed or bottle fed? Every year this question is discussed in thousands of homes. Under its deceptive simplicity, this question hides profound significance. It requires an answer from mothers and their physicians. Also from every man and woman in the nation. Ultimately from democratic society itself. In nursing her child, the mother fulfills her high duty with regard to the community. But the community must give her the education, moral and material help indispensable to the fulfillment of this duty. (p. 1)

Almost 60 years later, families and society are asking the same question and still the answer is not simple. Preparation for pregnancy and childbirth are fundamental experiences for each family and affect future generations. For parent, decisions about childrearing and learning are not made in isolation, and these decisions impact not only the newborn child but also society in general.

Most physicians believe that breastfeeding is best for a baby's health and that it enhances the mother's relationship to her newborn infant. Pediatricians recommend that all babies be breastfed during the first year of life unless medical conditions interfere with that process. Even breastfeeding for shorter periods of time can benefit the child. It is recognized that premature babies may not be able to breastfeed while they are very tiny, but small amounts of breastmilk provided for these fragile infants are of significant help to them as they grow and develop.

The Surgeon General's goal of Healthy People 2000 is that 75% of woman will breastfeed their infants in the early postpartum period and 50% will continue to breastfeed for 5 or 6 months. Although the United States has no national breastfeeding policy, President Clinton signed the 1994 World Health Assembly resolution (WHA 47.5) which reaffirmed the World Health Organization's code supporting breastfeeding and called for an end to free and subsidized breastmilk substitutes in all parts of the health care system. Unfortunately, President Clinton's signing went almost unnoticed in the United States.

Why did the Surgeon General and the WHA take such strong positions on breastfeeding? Research has demonstrated that babies are healthier when they have been breastfed, and mother–child bonding or attachment is enhanced through the interdependent process of lactation. Benefits to the newborn are numerous. All classes of immunoglobulins are found in human milk. As a result breastfed babies have fewer gastrointestinal, ear, and respiratory infections. It has been shown that these infants have decreased allergies and a lower incidence of sudden infant death syndrome and juvenile diabetes. In addition, breastfeeding is less expensive than the cost of bottles and formula, and breastmilk is ecological, readily available, sanitary, warm, and easily digested by the newborn.

Among the many factors associated with the choice of method for infant feeding are the women's perceptions, work or school intentions, influence of friends and family, knowledge base, and support by health care providers.

Culture also exerts a major influence on a mother's attitude toward her pregnancy, how she will feed her newborn, and the childrearing practices she will follow. The influence of the extended family and friends cannot be overemphasized. Age, education, life's experiences, traditions, and economic status all shape one's values, beliefs, and practices. Cultural attitudes toward modesty and breastfeeding are further influenced by the degree of acculturation to the dominant society. Cultural rituals, beliefs, and practices are often integral to the successful experience of breastfeeding mothers.

I. ANATOMY AND PHYSIOLOGY OF BREASTFEEDING

Breastfeeding is a symbiotic interchange between the mother and her child. In order to understand the female breast and the mechanism of milk production it is equally necessary to understand the anatomy of the infant's mouth and the mechanisms of sucking.

Culturally, the female breast is an organ of sexual attention in addition to providing nourishment and nurturing to the breastfeeding infant. Breast development begins in the human embryo and continues during childhood. At the time of puberty the hormone estrogen exerts the major influence on breast growth. As the young girl begins to menstruate, each cycle of estrogen and progesterone stimulates further development of the breast.

The role of hormones continues to be integral to the entire process of pregnancy, birth, and lactation. In response to hormones the breasts grow larger during pregnancy, and at the same time the diameter and pigmentation of the areola increases. Estrogen and progesterone influence the ductal system within the breast and promote the increase of size of the lobes, lobules, and alveoli (Fig. 1). The high levels of estrogen and progesterone prior to birth inhibit prolactin from initiating lactation until after the birth of the baby. At delivery the levels of estrogen and progesterone drop and prolactin is increased at the time of the birth of the placenta. At that time, the release of prolactin stimulates the production of milk. The suckling infant further stimulates the release of prolactin and the continued milk production.

Oxytocin is the fourth hormone of influence during pregnancy, birth, lactation, and postpartum. It stimulates contractions of the uterus during labor and causes the uterus to return to its prepregnant stage during the postpartum period. Oxytocin is released by the posterior pituitary. During breastfeeding the pituitary in the brain receives messages from the nipple and areola in response to the infant's sucking and releases oxytocin. Oxytocin stimulates

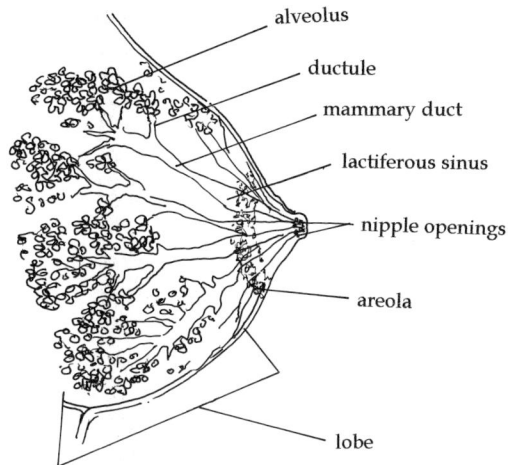

FIGURE 1 Diagram of female breast (illustrated by Mario Martins).

the myoepithelial cells of the breast, propelling the milk from the alveoli, where it is produced, into the ductal system, where it awaits the massage and suckling of the newborn.

The size of the breast is no indication of how successful a woman will be with breastfeeding, and neither is it correlated to the quantity of breastmilk which will be produced. The infant's suckling and stimulation of the breasts initiates the hormonal responses in the mother to make milk. The process becomes a supply and demand system. The more frequently the mother breastfeeds, the sooner her milk will "come in" and the more milk she will produce. As the infant becomes more efficient with suckling and its growth increases, the mother's body responds to these needs by producing more milk. The volume of breastmilk changes from 500 ml daily during the first week to 750 ml by the fourth month and to approximately 800 ml by the sixth month. At each stage of infant development, the produced milk contains precisely the right amounts of protein, fat, carbohydrates, and minerals. For example, the breastmilk of the newborn changes to meet the new demands of the infant at 3 months, 6 months, or 1 year of age. Normal growth is greatest during the newborn period when the newborn requires the high protein and low fat content of colostrum. By 1 month the total protein has decreased from 2.3 to 0.9 g/100 ml. At the same time fat has increased from 2.9 to 4.2 g/100 ml. In the later part of the year the fat content decreases in response to growth demands at that stage.

A psychogenic stimulus can also stimulate the milk ejection reflex, formerly called the "letdown reflex." Such things as hearing the infant's cry or thinking of the infant may initiate the ejection reflex. On the other hand, emotional stress can inhibit this same ejection reflex.

II. ROLE OF THE INFANT IN BREASTMILK PRODUCTION

The suckling infant is a key determinant for the release of the hormones responsible for breastmilk production and the milk ejection reflex. The newborn must master the techniques of breastfeeding in order to receive the nutrients uniquely produced by the mother's breasts.

Prior to birth the fetus is able to suck on its hands and swallow some of the surrounding fluid while *in utero*. After birth the small mouth, lips that easily close around the areola of the breast, and taste buds on the tip of the tongue of the newborn all adapt and facilitate breastfeeding. Each cheek has thick fatty tissues, known as sucking pads, which provide stability and prevent the cheeks from collapsing during suckling. Newborns are naturally nose breathers which further accommodates breastfeeding as the baby is held very close to the mother's breast.

Breastfeeding is a natural but learned process and the more frequent the contact between mother and baby, the greater likelihood of breastfeeding success. Time and experience are necessary so that each will learn more about the other. The mother will develop greater confidence in her abilities and the baby will learn the new skills of breastfeeding. Timing of feedings is most successful if initiated while the infant is in the quiet-alert state and thus more attentive and responsive to learning. Crying is a late hunger cue and it is often necessary to console the crying newborn before he or she will settle down to feed well. Some early hunger cues observed in the infant are smacking of the lips, sucking on fingers or hands, flexed arms and clenched fists, placing hands to mouth, rooting movements of the head, or opening the mouth in response to tactile stimulation.

III. UNIQUE FACTORS OF BREASTMILK

During pregnancy some of the mother's antibodies are passed to her fetus through the placenta. These proteins continue to circulate within the baby's blood for weeks to months after birth, neutralizing microbes and marking them for destruction by phagocytes which break down bacteria and viruses. Antibodies are also called immunoglobulins. The most abundant type is secretory IgA, which is found in the gut and respiratory system. As a result, maternal antibodies that pass to the fetus provide protection against respiratory, gastrointestinal, and ear infections. The mother also synthesizes specific antibodies when she comes in contact with a disease-causing agent and these are passed to the breastfeeding child. Because the mother is responding to her environment, the baby receives the protection it most needs during the first weeks of life. In addition, the immunoglobulins ignore useful bacteria normally found in the gut which provides another measure of resistance for the infant. These are important because a child's immune system does not reach full strength until around 5 years of age. The following are beneficial breastmilk components (Newman, 1995, p. 78):

B lymphocytes: Antibodies targeted against specific microbes

Macrophages: Kill microbes in the gut, produce lysozyme, and activate other components of the immune system

Neutrophils: May act as phagocytes, ingesting bacteria in digestive system

T lymphocytes: Kill infected cells directly or send out messages to mobilize other defenses; also manufacture compounds that strengthen the immune response

Antibodies of secretory IgA: Bind to microbes in digestive tract and prevent them from passing through gut into body's tissues

B_{12}-binding protein: Reduces amount of B_{12}, which bacteria need in order to grow

Bifidus factor: Promotes growth of *Lactobacillus bifidus*, a harmless bacterium in the gut which helps prevent dangerous varieties

Fatty acids: Disrupt and destroy membranes surrounding certain viruses

Fibronectin: Increases antimicrobial activity of macrophages; helps repair damaged tissues due to immune reactions in the gut

γ-Interferon: Enhances antimicrobial activity of immune cells

Hormones and growth factors: Stimulate baby's digestive tract to mature and become less vulnerable to microorganisms

Lactoferrin: Binds to iron which many bacteria need to survive; reduces available iron; interferes with growth of pathogenic bacteria

Lysozyme: Kills bacteria by disrupting their cell walls; adheres to bacteria and viruses, preventing them from attaching to mucosal surfaces

Oligosaccharides: Bind to microorganisms and prevent them from attaching to mucosal surfaces

There are three stages of breast-milk development which are significant in the adaptation of the newborn infant to extrauterine life. Colostrum is a thick, yellow fluid present in the breast at birth until approximately 48 hr. It contains relatively high levels of protein and low levels of fat. It is a lactose solution rich in IgA and other immunoglobins which provide protection against bacteria and viruses. Transitional milk is usually present by the second to fifth days after birth and may last 2 weeks. In this phase, the concentration of immunoglobulins and total protein decreases, whereas the lactose, fat, and calories increase. Mature milk is usually present by the ninth day. It contains increased water content which contributes to the temperature-regulating mechanism of the newborn. Increased lipids or fats provide the largest source of calories which are essential in rapid growth and development of the infant. The amount of fat fluctuates during each day. Breastmilk contains low fat in the morning and at the beginning of each feeding. There is a high fat content to breastmilk by mid-afternoon and at the end of each feeding. Because the fat content increases during the suckling session, it is important that infants are allowed to nurse until they are satisfied and thus receive the milk enriched with fat and calories needed for growth and development.

The newborn will breastfeed at least 8–10 times in each 24-hr period. The emptying time of the stomach of human milk is 1.5–2 hr; thus, breastfed infants who feed frequently are responding to the physiologic emptying of the stomach. Cow's milk and formulas are digested more slowly and babies fed with those products will probably eat less often. In response to the breast-milk consumed, a baby older than 5 days will have at least three bowel movements and urinate at least six time in each 24-hr period.

IV. NUTRITION AND BREASTFEEDING

Lactation is the physiological completion of the reproductive cycle and the maternal body prepares during pregnancy for lactation not only by developing the breast to produce milk but also by storing additional nutrients and energy. The Committee on Recommended Dietary Allowances of the Food and Nutrition Board recommends that the diet of a lactating mother includes slightly more of each nutrient than that recommended for the nonpregnant female. In other words, a balanced diet should be followed by selecting foods from all the basic food groups. Strict dietary rules are no longer enforced because they were hard to follow and too restrictive. Mothers on special diets should seek council of a nutritionist if they have questions while breastfeeding.

The volume of breastmilk produced by a mother varies over the duration of lactation from the first few weeks to 6 months and beyond, but it is remarkably predictable except for extreme cases of malnutrition or severe dehydration. Mothers of twins produce more breastmilk in response to the increased stimulation and suckling needs of the two infants.

Because of the "demand/supply" system of milk production the human body also adapts to the developmental growth spurts of the infant. Rapid growth periods usually occur at 10 days to 2 weeks, 6 weeks, 3 or 4 months, and 4–6 months. As the mother responds to her infant's demand for more frequent feedings, the milk supply will "catch up" within 1 or 2 days.

Considerable interest has focused on the impact of dietary fat and cholesterol on the composition of human milk. Fat is the main source of kilocalories in human milk for the infant. The caloric content of milk from well-nourished mothers does vary somewhat but averages about 65–75 kcal/100 ml. Since fat is the chief source of kilocalories, the fat content has the greatest impact on total kilocalories, with lactose and protein also contributing to the total.

A nutritious diet is recommended. Caffeine, alcohol, drugs, and certain medications should be avoided. Lactating mothers should check with their pharmacist or primary care provider when taking over-the-counter drugs or other medications while breastfeeding.

V. GETTING STARTED

Correct positioning is perhaps the most critical single measure for getting breastfeeding off to a good start. Positioning the infant at the breast in a way that facilitates good "latch-on" will enhance the baby's learning to suckle and increase the mother's sense of confidence (Fig. 2). Positioning is also key in preventing complications such as sore, cracked, or damaged nipples. It will promote satisfactory emptying of the breast at each feeding, thereby decreasing the potential of breast engorgement and providing the baby with adequate nutrition.

Engorgement is the result of congestion and increased vascularity of the breast ducts or the accumulation of milk within the breast. The best management of engorgement is prevention. Engorgement may be limited to the areolar area or may extend

LATCH - ON

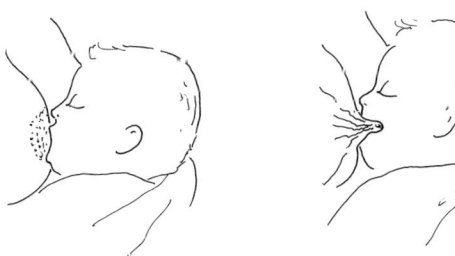

FIGURE 2   Latch-On

throughout the breast. In either situation it is painful for the mother and very difficult for the infant to latch on correctly. It is important to express the milk to relieve the pressure and soften the nipple so that the infant can grasp the areolar area into his or her mouth. Relieving the pressure may be done manually or with the use of a breast pump until the nipple becomes graspable by the baby. Gentle breast massage is also helpful, massaging in circular movement from the chest wall toward the nipple. Warm water compresses to the breasts prior to breastfeeding and will facilitate a "letdown." Some mothers find comfort in alternating hot and cold water to the breasts, warm showers, and the use of aspirin or acetaminophen. The best treatment is frequent breastfeeding.

Good positioning is key in preventing engorgement. The term positioning refers to the alignment of both the mother and her infant's body. It is how she holds the baby and how she supports her own breast. It includes the infant's mouth, lip, and tongue positions while at the breast. The latter is frequently referred to as latch-on. The baby should be positioned up onto the breast while the infant is learning to breastfeed. The mother should avoid leaning forward over the baby because this will cause the infant's mouth to grasp the nipple instead of the areolar area. Incorrect positioning of the mouth to the breast causes damage to the nipples and incomplete emptying of the breasts. Simultaneously, it is important that the mother support her breasts, facilitating good latch-on by her infant.

The following are the three most common positions (Fig. 3):

1. Cradle hold or Madonna hold is a natural position used by most mothers. In this position the baby is cradled in the mother's arms and brought to the level of her breast. The baby's chest is held toward the mother's chest.

2. Football or clutch hold is often used with the newborn, a small infant, or with the mother who has had a cesarean birth. It provides more complete control of the infant's head. The baby is held supported at the mother's side and brought up to her chest, with the baby facing the mother's breast.

3. Side-lying position is used mostly for early feedings, after a cesarean birth, or for nighttime feedings. In this position both mother and infant lie on their sides, facing each other.

FIGURE 3 a) Madonna/cradle position. b) Football/clutch position. c) Side-lying position. (Illustrated by Mario Martins.)

VI. PUMPING AND STORAGE OF BREASTMILK

Often mothers desire to express their breastmilk and store it for use with the infant at a later time. This may be done when the infant is left in the care of another reliable person, the mother returns to work, or the couple is sharing a social activity. It is also desirable to pump and store breastmilk when the infant is hospitalized or is too premature or too ill to suckle at the breast. Meticulous attention to cleanliness must be maintained in whatever method is used for expressing breastmilk. There are many varieties of breast pumps on the market which are manual or electric. They can be used on one breast or both breasts simultaneously. Some mothers prefer to hand express breastmilk without the use of artificial pumps.

In general, freshly expressed breastmilk can be stored in the refrigerator for about 48 hr. It can be frozen for 3–6 months in a freezer and later carefully defrosted and given to the infant. Once the milk has been defrosted, it should be used within 24 hr. Breastmilk should not be heated in a microwave oven because some of the properties of the milk may be altered by the microwave. In addition, the uneven heating of the milk in the microwave may result in the inadvertent burning of the infant's mouth. Lactation consultants are available for teaching and support of mothers who are breastfeeding their infants.

VII. WEANING

When a mother decides to stop breastfeeding, the weaning process is best accomplished if done gradually. In that way, it will be easier on the baby as well as on the mother. Most mothers and babies have established the milk supply to meet their needs and these cannot be shut down instantly. Weaning over a period of 2 weeks or more allows the milk production to gradually decrease. This will avoid painful engorgement of the breasts and will allow time for the baby to adjust to a new method of feeding. Usually, the mother first eliminates the breastfeeding in which the infant has demonstrated the least interest. During that feeding, an alternate method is used (bottle or cup). After several days a second feeding is switched from breastfeeding to bottle or cup, and so the process continues. It is important that the process remain consistent so that the baby will adapt to the new routine and the mother's body will adjust to the decreased demand to produce milk.

VIII. SEXUALITY AND CONTRACEPTION

Following the birth of a baby, many woman report changes in their sexual desires. The presence of a newborn into the family requires adjustments in the relationship of the parents. Caring for the infant requires time and physical energy to adapt to nonstop schedules and sleepless nights.

Some woman feel less sexual desires during this time, whereas others report heightened responsiveness and a deeper sexual bond with their partner. Masters and Johnson report that breastfeeding mothers are more comfortable with their own sexuality, whereas other researchers point out that the lower levels of estrogen following the birth result in lowered levels of interest in sex. The hormonal changes that occur during lovemaking may stimulate a milk letdown in some women. This is unpredictable; therefore, having a sense of humor is helpful during these times. Recognition of these possible changes in a couple's sex life will allow them to make modifications in their sexual relationship. The key is keeping the channels of communication open during this time of parental adjustments.

Different cultures have different attitudes and practices regarding sexual relations while breastfeeding. Some have strong feelings of abstinence, whereas others do not. Spending time together as a couple will assist in the adjustments necessary as a new family, regardless of the frequency of sexual relations.

Breastfeeding should not be used as a means of contraception. Although lactation provides some protection early in the postpartum period, pregnancy can occur once menstruation has returned. To avoid unplanned pregnancies other forms of contraception or natural family planning should be instituted.

There are very few circumstances in which a mother should not breastfeed her infant. The following is a list of usual contraindications for breastfeeding, but each mother should discuss this with her primary care provider if she desires to breastfeed her newborn:

The use of illegal drugs, such as cocaine, PCP, or marijuana
Increased alcohol consumption
Mothers who are receiving radioactive drugs and/or chemotherapy or have life-threatening illness
Mothers who have hepatitis virus
Mothers who are HIV positive, although this is controversial; the World Health Organization and Centers for Disease Control have differing opinions based on experience in industrial countries vs developing countries

IX. SUMMARY

In summary, today, as in 1939, each mother and her family are faced with the question posed by Alex Carrel: Should our baby be breastfed or bottle fed? The answer has profound significance and impacts each member of the family and society in general. Driscoll states that,

Breastmilk is the ideal food substance for human infants. But breastfeeding is much more than a feeding method; it is a complex relationship and method of communication and connection involving two people: a lactating woman and her infant. To consider breastfeeding only as a feeding method is to objectify it, to ignore the complexity and emotional aspects of the dynamic process and the vulnerability of the relationship. The breast is a living organ and part of a woman with her own personal history. That woman's life experiences, maternal self-esteem, and sense of sexuality can significantly affect the outcome of the breastfeeding experience. In addition, the infant is a critical player in the relationship, bringing a unique temperament and development sequelae. (p. 566)

Mothers and babies are members of families and extended families. Each member of society is the product of learned values and beliefs, cultural traditions, and a vision for future generations. How we feed our newborns is but one question to be answered during the formative years of childhood.

See Also the Following Articles

BREAST CANCER; LACTATION, HUMAN; MILK, COMPOSITION AND SYNTHESIS

Bibliography

American Academy of Pediatrics (1994, November). Infant feeding practices and their possible relationship to the etiology of diabetes mellitus. *Pediatrics* 94(5), 752–754.

Anderson, P. (1995). Alcohol and breastfeeding. *J Hum Lact* 11(4).

Carrel, A. (1939, June). *Reader's Digest* 34(206), 1.

Department of Health and Human Services. *Healthy People 2000: National Health Promotion and Disease Prevention Objectives*, DHHS Publ. No. (DHS) 91-50213. U.S. Government Printing Office, Washington DC.

Department of Health and Human Services (1987, April). *The Report of the Surgeon General's Workshop on "Children with HIV Infection and Their Families."* U.S. Government Printing Office, Washington, DC.

Driscoll, J. W. Breastfeeding success and failure. In *NAACOG's Clinical Issues in Perinatal and Women's Health Nursing*, Vol. 3, No. 4. Lippincott, Philadelphia.

Frederickson, D. D., Sorenson, J. R., Biddle, A. K., et al. (1993). Relationship of sudden infant death syndrome to breastfeeding duration and intensity. *Am. J. Dis. Child.* 147, 460.

Huggins, K., and Ziedrich, L. (1994). *The Nursing Mother's Guide to Weaning*. Harvard Common Press, Boston.

La Leche League International (1991). *Breastfeeding Answer Book*. Franklin Park, IL.

Lawrence, R. (1994a). Host resistance factors and immunologic significance of human milk. In *Breastfeeding: A Guide for the Medical Profession*, 4th ed. Mosby, St. Louis.

Lawrence, R. (1994b). Management of the mother–infant nursing couple. In *Breastfeeding: A Guide for the Medical Profession*, 4th ed. Mosby, St. Louis.

Lawrence, R. (1994c). Collection and storage of human milk and human milk banking. In *Breastfeeding: A Guide for the Medical Profession*, 4th ed. Mosby, St. Louis.

Masters, W. H., and Johnson, V. E. (1966). *Human Sexual Response*. Little, Brown, Boston.

National Alliance for Breastfeeding Advocacy (NABA) (1996). *National Report Card. The State of Breastfeeding United States: 1/50 Work Needs to be Done*. NABA.

Newman, J. (1995, December). How breastmilk protects newborns. *Sci. Am.*, 76–79.

Riodan, J., and Auerbach, K. (1993). *Breastfeeding and Human Lactation*. Jones & Bartlett, Boston.

Subcommittee on Nutrition during Lactation, Food and Nutrition Board, Institute of Medicine (1991). *Nutrition during Lactation*. National Academy of Sciences, Washington, DC.

World Health Organization (WHO) (1987). *Breastfeeding, Breastmilk and Human Immunodeficiency Virus*. WHO, Geneva.

Breeding Strategies for Domestic Animals

George R. Foxcroft
University of Alberta

I. Introduction
II. Theoretical Concepts
III. Regulation of Breeding Performance in Practice
IV. Future Prospects for the Regulation of Reproduction in Domestic Species

GLOSSARY

agonist A compound exerting a similar effect to another.

analogues Compounds that have similar chemical structure to natural hormones.

anestrus An infertile state in which sexual receptivity (estrus) is not observed.

antagonist A compound exerting an opposing effect to another.

endogenous Of hormones, naturally produced from within the animal.

equine chorionic gonadotropin A gonadotropin-like hormone secreted by trophectoderm/chorion, extraembryonic membrane of embryos/fetuses.

exogenous Of hormones, applied to the animal as a treatment.

farrowing The birth process in swine.

gonadotropes Pituitary cells making and releasing the gonadotropins, luteinizing hormone, and follicle-stimulating hormone.

gonadotropin releasing hormone A hormone from the hypothalamus controlling pituitary gonadotropin secretion.

gonadotropins Pituitary hormones (luteinizing hormone and follicle-stimulating hormone) that control the gonads (ovary and testis).

hormones Classically defined as blood-borne chemicals regulating organ systems.

humoral Blood-borne.

hypothalamus Part of the base of the brain that is important for controlling reproductive activity.

inhibin A hormone from the gonads that regulates follicle-stimulating hormone secretion.

luteal Related to the corpus luteum, a progesterone-releasing body formed on the ovary after ovulation; primarily responsible for maintaining early pregnancy in many mammals.

luteolytic Causing destruction (lysis) of the corpus luteum.

neuroendocrine system A complex of nervous and hormonal pathways controlling function.

opiate An endogenous secretion from neural tissue exerting morphine-like effects.

para-, *auto-*, and *juxtacrine* Hormone signaling between cells, within cells, and through cell-to-cell connections, respectively.

parturition The process of giving birth.

pheromonal signals Chemicals used to transmit information between animals.

pituitary gland The key endocrine gland directly regulated by the hypothalamus.

post-partum After parturition or birth.

Encyclopedia of Reproduction VOLUME 1
Copyright © 1999 by Academic Press. All rights of reproduction in any form reserved.

progestagens Progesterone-like compounds used to regulate reproductive activity.
progesterone The principal hormone secreted from the corpus luteum.
prostaglandins A family of hormones among which are members with luteolytic actions.
receptor The recognition component of cells that recognize specific hormones.
relaxin A hormone involved in controlling the birth process.
reproductive axis The integrated system of reproductive organs and control systems.
reproductive cycle Periodic changes in reproductive activity (can be short-term or lifetime).
suckling The act of milk removal from the mammary glands in mammals.

Breeding strategies for domestic farm species involve both direct and indirect manipulation of mechanisms controlling reproduction. Rapid advances in our understanding of the control of reproduction provide bases for improvement of techniques for managing breeding performance. A failure to appreciate even subtle species differences in physiological control mechanisms frequently limits the success of otherwise effective treatments. Changes in the social environment, acting through pheromonal signals, and changes in the physical environment, such as light and temperature, may affect reproductive function. Likewise, changes in suckling intensity and nutritional availability can affect breeding performance of farm species. Direct intervention to modify reproductive activity is practiced at all stages of the reproductive cycle. The range of exogenous compounds used to manipulate breeding activity exert their effects at all levels of the reproductive axis and may promote or inhibit the synthesis and release of hormones and other regulators of the reproductive system as well as the actions of hormones in target tissues. Regulation of the bioavailability of hormones, both endogenous and exogenous, has also proved to be an effective means of controlling fertility. In contrast to the application of fertility control in humans and in nondomesticated species, the socioeconomic climate of the modern animal husbandry industry demands precise control of the timing of reproductive activity that may be as economically important as an increase in actual fertility.

I. INTRODUCTION

Many of the techniques currently available for regulating breeding activity in domestic farm species involve either direct or indirect manipulation of the neuroendocrine system controlling reproduction. Management techniques used to control breeding performance should ideally be based on a sound understanding of the physiological mechanisms involved. Such information will indicate the most appropriate means of applying regulators of breeding activity in practice and also suggests the causes of between-animal, and between-species, variability in the response to a particular treatment. Because of the preponderance of females within domestic breeding populations, emphasis has been placed on the development of techniques to control reproductive events in the female.

In contrast to the extensive range of techniques applied on an individual basis for the regulation of fertility in humans and even in veterinary clinical practice, techniques for controlling breeding activity in domestic species are usually applied to whole herds or flocks of animals. Such techniques must therefore allow for expected between-animal variation and even for differences in the reproductive state of animals at the time of treatment.

This article describes some of the concepts that underlie the application of management strategies and regulatory substances to the control of reproductive activity. It will also illustrate the application of such techniques in different domestic species and at different stages of the reproductive life cycle. Finally, the prospects for developing further techniques will be discussed.

II. THEORETICAL CONCEPTS

Reproductive function depends on interactions between the hypothalamus, pituitary gland, and gonads that are mediated by neural and humoral pathways operating in a classic endocrine manner. The synthesis and release of hormones in this classic system nevertheless involves complex inhibitory and stimulatory pathways and diverse receptor-mediated actions of these hormones on their target tissues. It is now known that reproductive function is also mediated by important paracrine, autocrine, and juxtacrine mechanisms acting at a local level. In theory any hormone involved in, or any factor which is a mediator of, such regulatory systems could be manipulated in an attempt to control reproduction. However, because of the complexity of the regulatory process, manipulation of any single component may not necessarily produce consistent responses in different reproductive states.

A. Hypothalamic Function

At the hypothalamic level, release of gonadotropin releasing hormone (GnRH) is controlled by the feedback effects of gonadal steroids and peptides and a variety of neural pathways of intra- and extrahypothalamic origin. In principle, therefore, exogenous steroid treatment can be used to mimic the feedback effects of endogenous steroids; alternatively, the use of steroid agonists, and immunization against steroids, can effectively change steroid feedback activity. Similarly, opiate agonists and antagonists and other pharmacological agents can be used to regulate GnRH and hence gonadotropin release. The activity of the extrahypothalamic pathways, such as those mediating the effects of pheromonal and photoperiodic stimuli, can be indirectly activated by manipulating the social and physical environment of the animal. Additionally, treatment with the intermediaries of these pathways, such as pheromonally active steroids or melatonin, may be effective in promoting reproductive activity.

B. Pituitary Function

Release of the gonadotropins, luteinizing hormone (LH) and follicle-stimulating hormone (FSH), is regulated by an interaction between the pattern of GnRH stimulation emanating from the hypothalamus and those steroidal and nonsteroidal factors which affect sensitivity of the gonadotropes to GnRH stimulation. These regulatory systems affect not only the quantity but also the quality of the gondatropic stimulus to the gonad. For example, the physiological significance of episodic LH secretion is well documented. Furthermore, because the LH:FSH ratio is critical to the level of ovarian stimulation obtained, selective regulation of FSH release at the hypophysial level by inhibin is of considerable interest and may be open to manipulation in practice.

Direct stimulation of gonadotropin secretion by the pituitary can be achieved using preparations of GnRH or its analogs, whereas superagonists of GnRH are effective inhibitory agents because they decrease active GnRH receptors on target cells. A major difficulty in applying GnRH treatment to domestic animals in a commercial context has been the apparent need to mimic the pulsatile pattern of endogenous GnRH release. However, evidence suggests that continuous release of GnRH from implants at a low concentration may provide an effective stimulus to ovarian development.

C. Ovarian Function

Gonadotropic hormones are essential for the development of gonadal function and treatment with exogenous gonadotropins, such as equine chorionic gonadotropin (eCG) or human chorionic gonadotropin, continues to be an effective means of promoting follicular growth and ovulation in domestic species. Nevertheless, complex paracrine and autocrine mechanisms regulate the sensitivity of ovarian tissues to gonadotropin stimulation. Therefore, some instances of ovarian inactivity are not associated with an apparent lack of gondadotropin stimulation and local regulators of ovarian function are probably involved. Similarly, nutritional, genetic, and other influences may act at the hypothalamic–pituitary level to change LH and FSH secretion; however, changes in sensitivity to gonadotropins at the gonadal level may be of great physiological significance. The development of techniques for controlling ovarian sensitivity is therefore of considerable practical interest.

Information on the regulation of luteal function in domestic species is extensive and the use of known luteolytic agents such as the F-series prostaglandins ($PGF_{2\alpha}$) has been exploited as a means of synchronizing ovarian cycles and controlling the time of parturition. Evidence for important luteotropic and antiluteolytic factors in early pregnancy is rapidly expanding and may suggest practical ways of regulating this critical stage of the reproductive cycle. Differences in the sensitivity of the corpus luteum to luteolytic effects of $PGF_{2\alpha}$ in different species have led to very different efficacies in the control of breeding activity.

III. REGULATION OF BREEDING PERFORMANCE IN PRACTICE

A. Pubertal Animals

Although seasonal, social, and nutritional factors can affect the timing of puberty in domestic species, the endocrine mechanisms mediating these effects are poorly understood. Because an increase in gonadotropin secretion may be one essential trigger for onset of cyclic ovarian activity, use of exogenous gonadotropins to induce puberty in a practical situation has been widely evaluated. Although follicular development and ovulation may occur, there is often a failure to establish normal ovarian cyclicity, suggesting that other components of the reproductive axis lack maturity. A reliable and practicable method of inducing true precocious puberty with exogenous hormone treatment is still needed. Stimulation with male pheromones has been used as an effective trigger for precocious pubertal development in pigs; however, the inconsistency of the response to this stimulus often encountered in the long term suggests that other important factors, such as season and growth status, affect the sensitivity of the system to pheromonal stimuli.

B. Stimulation of Ovarian Activity during Anestrus

In domestic mammals ovarian activity is usually suppressed during pregnancy, during the early postpartum period, and in early lactation. Depending on species and breed, seasonal anestrus may also be evident. In some instances, such as the early postpartum period and deep seasonal anestrus, the suppression of reproductive function affects a number of different components of the reproductive axis and this inhibition is not easily broken. In late lactation and late seasonal anestrus, however, it is possible to hasten the return to reproductive activity with appropriate treatment. In sheep, an increase in LH secretion as a result of introducing the "ram effect" (Fig. 1a) or in response to multiple injections or infusions of GnRH (Fig. 1b) will trigger follicular development and an early return to fertile estrus. Pretreatment with progesterone may, however, be necessary in GnRH-treated ewes if normal luteal function is to occur. Treatment of postpartum cows with GnRH may also elicit ovarian cyclicity. Similar responses have also been obtained by treating late anestrous ewes with progesterone sponges and eCG, and this treatment may even induce mild superovulation.

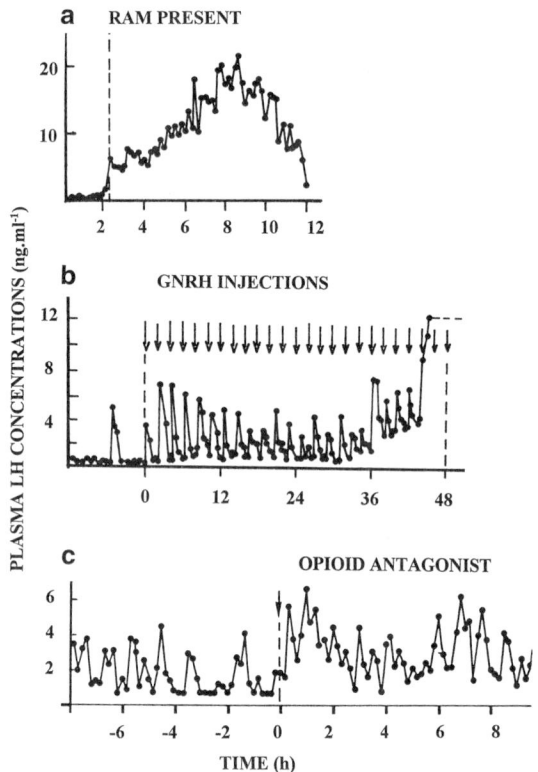

FIGURE 1 Different treatments can induce similar changes in the pattern of luteinizing hormone (LH) secretion and result in stimulation of reproductive activity. (a) LH responses in an anestrous ewe to the pheromonal stimulus of a mature ram. (b) LH response in an anestrous ewe to multiple injections of exogenous GnRH. (c) LH response in a sow during lactational anestrus to treatment with an opioid antagonist which blocks the inhibitory effect of suckling (modeled on data from (a) G. B. Martin et al., Anim. Reprod. Sci. 3, 125–132, 1980; (b) B. J. McLeod et al., J. Reprod. Fertil. 65, 223–230, 1982; and (c) M. Mattioli et al., J. Reprod. Fertil. 76, 167–173, 1986).

Evidence that inhibition of LH secretion during lactation in the sow is in part an opioid-dependent phenomenon suggests that use of opiate antagonists may provide a mechanism for promoting LH secretion and ovarian activity at this time (Fig. 1c).

C. The Synchronization of Reproductive Activity

Two approaches to synchronizing randomly cycling females can be adopted. The first is to cause premature regression of the corpus luteum and hence an early return to estrus. Because the corpus luteum is sensitive to luteolytic agents for most of the luteal phase in these species, treatment of groups of cyclic ewes or cows with luteolytic agents at 10-day intervals usually results in complete synchronization of estrus after the second injection. In practice, results may be variable, but in cattle acceptable conception rates have been obtained using $PGF_{2\alpha}$ or its analogs to induce luteolysis and then natural mating or artificial insemination (AI) at observed estrus. Conception rates in programs using fixed-time mating or AI after the second $PGF_{2\alpha}$ injection have generally been lower. Induced luteolysis is less appropriate in swine because the corpus luteum is only sensitive to the

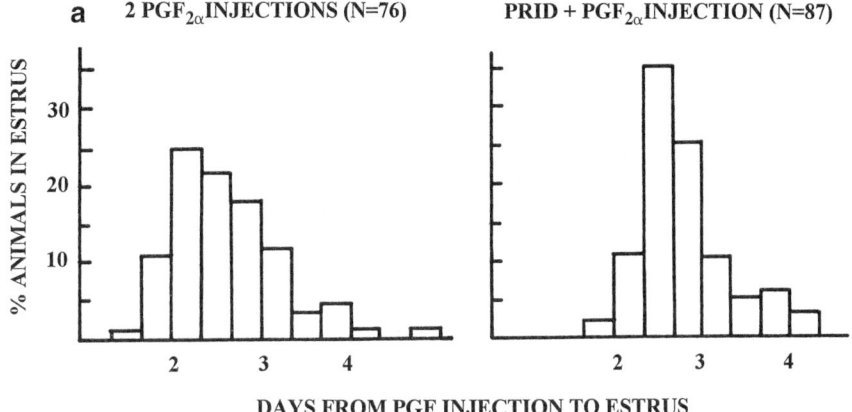

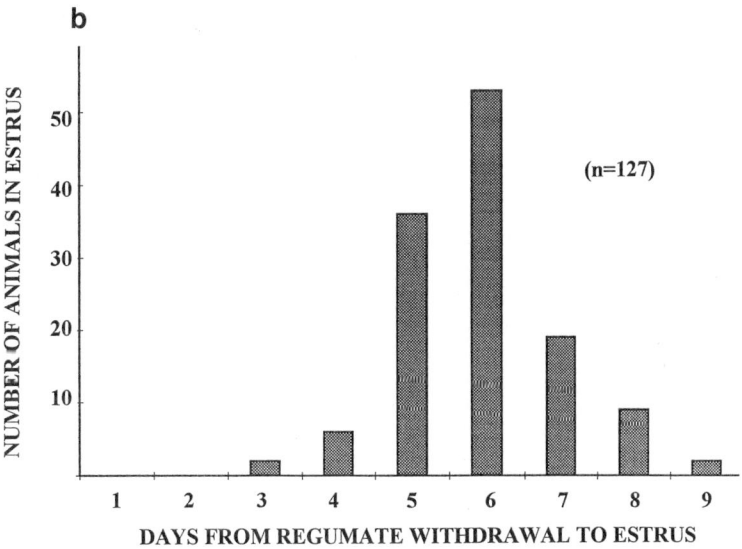

FIGURE 2 Examples of how estrus is synchronized after treatment of randomly cyclic populations of animals with synchronizing agents. (a) Synchronization of estrus in cattle or sheep after combined treatment with progestagens and luteolytic agents. PRID, progesterone-releasing intravaginal devices. (b) Synchronization of estrus in cyclic pigs after 18 days of feeding an oral progestagen (modeled on data from (a) R. D. Smith *et al.*, *J. Anim. Sci.* 58, 792–800, 1984; and (b) unpublished data from the author's laboratory).

luteolytic effects of $PGF_{2\alpha}$ for a relatively short period in the late luteal phase of the cycle.

The second approach is to prolong the luteal phase of the estrous cycle by treating cyclic animals with natural or synthetic progestagens. Treatment of groups of animals must continue for a sufficient period to allow any existing corpora lutea to undergo natural luteolysis; withdrawal of treatment should then lead to a synchronous return to estrus. In cattle and sheep, various progestagens have been delivered either from subcutaneous implants or from intravaginal sponges or other devices such as progesterone-releasing intravaginal devices. In cattle the efficiency of progestagen treatment can be increased with the combined use of luteolytic agents such as $PGF_{2\alpha}$ or estrogen (Fig. 2a). In swine the only effective treatments with progestagens have involved the use of orally active synthetic compounds such as allyl trenbolone. Feeding groups of randomly cyclic gilts or sows for 18 days, followed by withdrawal, produces effective estrus synchronization as shown in Fig. 2b.

Treatment with $PGF_{2\alpha}$ in late pregnancy has also been effectively applied in sows for synchronization of farrowing in an attempt to provide the greatest level of supervision of the animals during parturition and the critical early neonatal period. The synchrony of farrowing achieved has been further improved with combination treatments involving other hormones known to be involved in the parturient process, as shown for treatment with $PGF_{2\alpha}$ and relaxin in Fig. 3.

IV. FUTURE PROSPECTS FOR THE REGULATION OF REPRODUCTION IN DOMESTIC SPECIES

Practical techniques are available for controlling reproductive activity for much of the breeding life of domestic species. An example of such a scheme in a breeding pig herd is shown in Fig. 4. Similar programs have been applied in large commercial dairy herds. In order for such schemes to be successful there must be an inherently high fertility in the population of animals treated, and an appropriate level of stockmanship and veterinary supervision must be available. The use of the exogenous hormones to treat subfertile and totally infertile animals is also possible but is unlikely to be of long-term benefit unless the underlying causes of such problems are identified and corrected.

When considering the relevance of initiating a scheme of controlled breeding, it should be accepted that attainment of the highest levels of fertility and fecundity in individual animals may not in itself produce the highest economic returns. Development of breeding programs which allow optimal use of capital

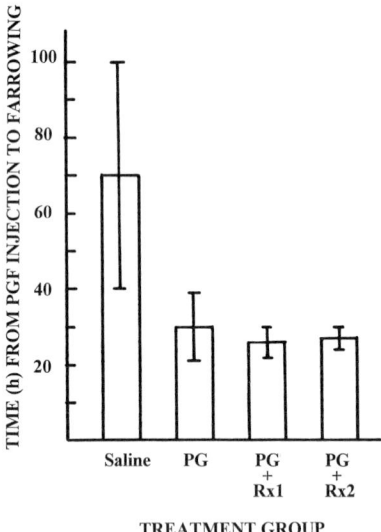

FIGURE 3 Example of how the time of birth can be synchronized in groups of late pregnant females. Compared to sows injected with saline as a control, the interval between injection of $PGF_{2\alpha}$ alone (PG) or in combination with relaxin (R) is reduced and would allow better supervision of farrowing (modeled on data from W. R. Butler and R. D. Boyd, *Biol. Reprod.* 28, 1061–1065, 1993).

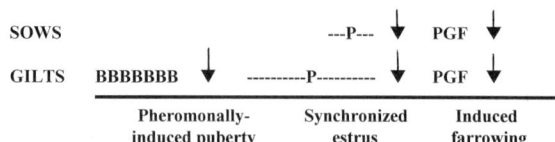

FIGURE 4 A possible integrated scheme for the management of reproduction in a sow herd. Boar pheromones (BBBBB) are used to induce onset of breeding activity, estrus of randomly cyclic females is synchronized with oral progestagen (P), and farrowing is synchronized with injection of luteolytic agents (PGF). (Illustration based on the review of F. Martinat-Botte et al., *J. Reprod. Fertil. Suppl.* 33, 211–228, 1985).

investments such as buildings, make more efficient use of labor, or reduce overall feed costs in a breeding enterprise may be effective in economic terms and yet associated with no actual improvement in the fertility of the treated population of animals. Thus, synchronization and timing of reproductive activity in domestic animal populations, or groups within populations, will continue to be of major interest to commercial livestock producers, and the reliability and consistency of the response to treatment will be of paramount importance. To this end, improved techniques to regulate reproduction are needed.

Nevertheless, as further information on the basic mechanisms controlling reproductive activity becomes available there will be opportunities to enhance the overall fertility of domestic populations in a practicable and cost-effective way. Development of procedures for immunizing against ovarian steroids and inhibin in sheep is an example of this type of development and appears to produce a more predictable physiological response than that obtained after treatment with exogenous gonadotropins. It is realistic to suggest that necessary techniques for both improving the fecundity of already fertile animals and controlling their entire breeding cycle will be available within the next decade. In populations of domestic livestock such as dairy cattle, a marked improvement in inherent fertility is needed if the application of controlled breeding programs is to be successful.

See Also the Following Articles

CATTLE; CHORIONIC GONADOTROPIN, HUMAN; EQUINE CHORIONIC GONADOTROPIN; FSH (FOLLICLE-STIMULATING HORMONE); GnRH (GONADOTROPIN-RELEASING HORMONE); LH (LUTEINIZING HORMONE); PIGS; SHEEP AND GOATS

Bibliography

Foxcroft, G. R. (1992). Female reproduction. In *Reproduction in Domestic Animals* (G. J. King, Ed.), Elsevier World Animal Science Series, Vol. 14. Elsevier, Amsterdam.

Knobil, E., and Neill, J. D. (1988). *The Physiology of Reproduction*, Vols. 1 and 2. Raven Press, New York.

Brood Parasitism in Birds

Alfred M. Dufty, Jr.
Boise State University

I. Obligate Brood Parasitism
II. Facultative Brood Parasitism

GLOSSARY

altricial young Birds that hatch naked, blind, and immobile and are completely dependent on parental care for survival.

facultative brood parasitism A reproductive strategy in which (i) females sometimes lay eggs in other birds' nests but also raise a clutch of their own or (ii) some females lay eggs in other birds' nests and other females raise their own young.

interspecific Between two species.
intraspecific Within a single species.
obligate brood parasitism A reproductive strategy in which all females of a species lay their eggs in other species' nests.
precocial young Birds that are well developed at hatch and require little or no parental care to survive.

Brood parasitism in birds is the deposition of eggs in other species' nests, with the recipients of parasitic eggs (called hosts) assuming all subsequent parental care. A distinction is made between avian species that use this breeding strategy exclusively (obligate brood parasites) and those in which some members breed parasitically on occasion but also rear their own young or some are parasitic and some are not (facultative brood parasites). The hosts usually are not genetically related to the brood parasites. Indeed, they often are not of the same species. However, hosts frequently accept parasitic eggs, providing them with the same degree of parental care that they provide their own young. Brood parasitism can have a deleterious effect on the reproductive success of hosts because of competition between host and parasitic young for limited resources (e.g., food and brooding) or because of direct attacks by adult or young brood parasites on host eggs or young. However, in some instances brood parasitism has little or no effect on host reproductive success and, in one case, may actually improve host breeding success.

I. OBLIGATE BROOD PARASITISM

Obligate brood parasites rely exclusively on other avian species to raise their young. Such birds are found within six families (Table 1) and represent approximately 1% of all avian species. These are distributed throughout the world's continents (except Antarctica). The Old World cuckoos (family Cuculidae) contain the largest number of parasitic species, including possibly the best known brood parasite in the world, the common cuckoo (*Cuculus canorus*). The remaining families are honeyguides (Indicatoridae), cowbirds (Emberizidae; including the only North American obligate brood parasites), African parasitic finches (Passeridae), American ground cuckoos (Neomorphidae), and one duck (Anatidae).

The lack of parental care in obligate brood parasites frees them to produce more eggs than do birds of similar size but which raise their own young. For example, parasitic cuckoos lay between 16 and 25 eggs per breeding season, and cowbirds produce twice that number. Despite this elevated reproductive output, many parasitic eggs fail to yield viable young because some are laid in inappropriate nests or at inappropriate times, and some hosts reject parasitic eggs.

A. Parasite–Host Relationships

1. Degree of Host Specialization

Obligate brood parasites range from those that are highly specialized, parasitizing a single host species, to extreme generalists, depositing eggs in any of hundreds of potential host species' nests. An example of a specialist is the common cuckoo. Within a given locale one or more races of cuckoos may coexist, each parasitizing members of a single host species. Typically, hosts are common species whose young have a similar diet to that of the parasite. Furthermore, hosts usually are smaller than the cuckoo and lay smaller eggs. This enhances the opportunities for successful incubation of the larger parasitic eggs because they are in closer contact with the host's incubation patch. The mechanism whereby cuckoo females develop a preference for a particular host probably relates to their own experiences as nestlings, with each female imprinting on the host species that raised them and on the breeding habitat in which this occurred.

The best known host generalist is the brown-headed cowbird (*Molothrus ater*). Cowbird eggs have

TABLE 1
The Six Avian Families That Include Obligate Brood Parasites and an Example of Each

Family	Common name (scientific name)
Emberizidae (cowbirds)	Brown-headed cowbird (*Molothrus ater*)
Indicatoridae (honeyguides)	Greater honeyguide (*Indicator indicator*)
Cuculidae (Old World cuckoos)	Common cuckoo (*Cuculus canorus*)
Neomorphidae (American ground cuckoos)	Striped cuckoo (*Tapera naevia*)
Passeridae (African parasitic finches)	Village indigobird (*Vidua chalybeata*)
Anatidae (one duck species)	Black-headed duck (*Heteronetta atricapilla*)

been found in nests of over 200 host species, although not all hosts rear young cowbirds successfully (often because the host's diet is unsuitable for cowbirds). Despite the breadth of overall host use, some individual female cowbirds show a great deal of specialization, and the factors involved in the choice of host nests remain under investigation.

2. Effects of Brood Parasitism on Hosts

i. Direct Effects

Some brood parasites have an immediate and obvious effect on the breeding success of individual hosts and can play a proximate role in the destruction of entire host populations. For example, African honeyguides parasitize cavity-nesting species. Young honeyguides completely eliminate their host's nestlings shortly after hatching by attacking them with sharp hooks on the end of the bill. These hooks fall off during nestling development and appear to function solely to eliminate host young, which compete with parasitic young for food. Nestling cuckoos of some species take a different, but no less thorough, approach to the elimination of competition from host young. A young parasite, when only a few hours old and still featherless, blind, and weighing only a few grams, will maneuver its nestmates (or the host's eggs) onto its back and push these objects up to the nest rim and out of the nest (Fig. 1). This behavior wanes after a few days. Both

FIGURE 2 An adult female common cuckoo removing (and eating) an egg of the reed warbler (reproduced with permission from Wyllie, 1981).

FIGURE 1 A nestling common cuckoo balancing an egg of its host, the reed warbler, on its back as it backs its way up the side of the nest and ejects the egg (reproduced with permission from Wyllie, 1981).

approaches result in occupancy of the nest by the brood parasite alone and in complete failure of the host's reproductive effort. In addition, these processes preclude multiple parasitism of the same nest because the first-hatched brood parasite would eliminate all subsequent conspecifics.

Adult brood parasites also can directly affect host reproduction. Adult cuckoo and cowbird females are known to remove a host egg from nests they parasitize (Fig. 2). The eggshells of some cuckoo and cowbird species are thicker than expected and can crack host eggs when deposited in the nest. Furthermore, adult cuckoo, cowbird, and honeyguide females sometimes peck host eggs, killing the embryos. This functions in two ways: It reduces competition from host young if the nest is parasitized, and it may cause hosts to renest. The latter may give the parasite a second chance to lay in a nest that had escaped brood parasitism previously. Finally, cowbirds and cuckoos have been observed removing nestlings from host nests, which may also effect renesting and afford parasites additional breeding opportunities.

ii. Indirect Effects

In contrast to the direct attacks made by some young cuckoos and honeyguides, other brood parasites may eliminate their nestmates through competition for resources. For example, nestling brown-headed cowbirds frequently are larger than their warbler, sparrow, or vireo nestmates. Host young have difficulty competing with nestling cowbirds for food and consequently may starve to death. This is most likely to occur in nests that suffer multiple cowbird parasitism. Because cowbirds do not eject their nestmates (although a recent report suggests that this deserves additional investigation), multiple cowbird parasitism is not uncommon.

iii. Neutral or Positive Effects Although a parasitic relationship, by definition, requires that one member benefits at the expense of the other, in some instances it is difficult to ascribe a significant cost of brood parasitism to the species being parasitized. An example is the black-headed duck (*Heteronetta atricapilla*), a South American species that parasitizes marsh-nesting waterbirds. Young parasites are highly precocial (i.e., require little posthatching parental care) and remain with their hosts for only a few days after hatching, at which time they leave the host and feed on their own. Thus, black-headed ducklings parasitize the body heat provided by their adult hosts during incubation and brooding. They have little or no effect on the reproductive success of their hosts, although in multiply parasitized nests the effectiveness of incubation (and, thus, hatching success) could be reduced. In addition, it is possible that some host eggs could be damaged if hosts attempt to repel female parasites discovered while egg laying. Overall, however, this species represents the most benign of the obligate brood parasites.

An example in which a brood parasite enhances the reproductive success of its host involves the giant cowbird (*Scaphidura oryzivora*) of Central and South America. Giant cowbirds primarily parasitize colonial nesting blackbirds, oropendolas (*Psarocolius* spp.), and caciques (*Cacius* spp.). Young of these blackbirds often are infested with parasitic botfly (*Philornis* spp.) larvae, which can weaken and kill the nestlings. However, nestling cowbirds remove botfly larvae from their nestmates, increasing the reproductive success of the hosts. In these circumstances, adult oropendolas and caciques tolerate brood parasitism and accept cowbird eggs. Alternatively, some blackbird nests are located near nests of stinging wasps or biting bees, which protect the blackbirds from botfly activity. In such cases the hosts have no need of the cowbird's grooming behavior and are intolerant of brood parasitism. Thus, there may be an advantage to being parasitized in some conditions, but it is tempered by the fact that if more than one nestling cowbird is present in a nest, then the increased competition for food outweighs the advantage of having botflies removed and host reproductive success suffers.

3. The Coevolutionary "Arms Race" between Hosts and Brood Parasites

The reduction in reproductive success imposed by brood parasites on hosts provides a strong selective pressure on the latter to evolve mechanisms that deter parasitism. That is, natural variation in the behavior of some individuals of a host species may produce adaptations that reduce the success of brood parasites. Through the process of natural selection, those individuals that hinder brood parasitism will produce more young than individuals that are parasitized successfully, and the behavior(s) that reduces parasitism will spread throughout the host population. However, brood parasites are also subject to natural selection, and any mechanism that circumvents host deterrents and facilitates successful brood parasitism will likewise spread throughout the population of parasites. Thus, hosts and brood parasites are engaged in a coevolutionary arms race, with the reproductive strategy of one being shaped, at least in part, by the behavior of the other.

i. Host Aggression Hosts exhibit a variety of responses to obligate brood parasitism. In principle, one of the most effective deterrents is to drive off a brood parasite before it has the opportunity to locate the host's nest and deposit an egg. Indeed, many hosts are selectively aggressive toward brood parasites. However, some brood parasites appear to use aggressive behavior on the part of the host as a cue that the host's nest is nearby. Indeed, some parasitic pairs take advantage of host aggression, with the male luring the aggressive hosts away from the nest, leaving it unguarded for the parasitic female.

ii. Recognition and Removal of Parasitic Eggs Another adaptation to brood parasitism found in some hosts involves egg recognition. That is, hosts learn

FIGURE 3 A brown-headed cowbird egg in the nest of a wood thrush.

the color, size, and shape of their own eggs and reject dissimilar eggs found in their nest. For example, eggs of the parasitic brown-headed cowbird are white to whitish-blue in ground color with brown spots or speckles, although they vary considerably among females in their specific patterning. Cowbird eggs look slightly like the eggs of a large number of their hosts, reflecting the broad range of hosts used by this brood parasite. American robins (*Turdus migratorius*) are a common species in the cowbird's geographic range, yet cowbird eggs are rarely found in robin nests. Robins, with their distinctive "robin's-egg blue" eggs, quickly identify and remove foreign eggs from their nests. However, other species with similarly distinctive eggs, such as the wood thrush (*Hylocichla mustelina*), fail to distinguish between their eggs and those of cowbirds (Fig. 3). Finally, relatively large hosts, such as robins and Baltimore orioles (*Icterus galbula*), can grasp or puncture cowbird eggs and eject them from the nest.

iii. Eggshell Thickness and Incubation Time Probably in response to the evolution of parasitic egg ejection behavior by hosts, cowbirds and other brood parasites (such as the common cuckoo) have evolved eggshells that are thicker than those of other species that produce eggs of similar size. As with other components of this arms race, any variation in the thickness of eggshells produced by female cowbirds would select for thick eggshells that are not easily punctured. The short time cowbirds and cuckoos spend on host nests to deposit parasitic eggs could also contribute to the evolution of thick eggshells. That is, brood parasites lay eggs quickly, perhaps to avoid detection by hosts, and a thick eggshell could help to prevent damage to the parasite's egg (while possibly inflicting damage to the host's).

However, there is a limit in the extent to which thick eggshells are advantageous. For example, thick eggshells could become difficult for embryos to pierce, impeding hatching. Similarly, thick eggshells provide a barrier that hinders gas exchange during embryonic development. This could extend the incubation period of cowbird eggs beyond that of their hosts, putting nestling cowbirds at a competitive disadvantage compared to their earlier hatched nestmates. However, cowbirds do not have an extended incubation period; in fact, it is significantly shorter than those of other species with eggs of similar size and also shorter than those of nonparasitic blackbirds. Cowbirds apparently resolve this problem by having more and/or larger pores in their eggshells, facilitating gas exchange. The benefit to cowbirds of rapid embryonic development is that cowbird eggs will hatch at the same time or earlier than those of their hosts. Newly hatched cowbird nestlings often are larger than host nestlings, which gives cowbird nestlings an advantage when competing for food.

iv. Nest Desertion Smaller hosts that cannot physically remove parasitic eggs sometimes abandon their nest and start a new nest elsewhere. While this avoids parasitism (unless the new nest is also parasitized), it incurs its own costs in terms of lost time and resources invested in the original nest. Still other hosts, such as yellow warblers (*Dendroica petechia*) or reed warblers (*Acrocephalus scirpaceus*), sometime build a new nest lining over parasitic brown-headed cowbird or common cuckoo eggs, respectively.

v. Egg Mimicry The removal of dissimilar eggs by hosts has driven the evolution of egg mimicry in some parasitic species. That is, given the natural variability in egg coloration within and among avian females, a parasitic egg that sufficiently resembles those of the host has a higher probability of being accepted. The resulting parasite will produce eggs that also are similar to those of the host species in whose nest it was reared. Over time, as hosts become more discriminatory and brood parasites produce more exact matches, egg mimicry results. While egg mimicry facilitates parasitism of host species that produce a given type of egg, it limits the range of hosts a brood parasite can parasitize successfully. For instance, within several cuckoo species are "gentes" or races, and each gens is specialized by the color, size, and shape of its egg to parasitize a single host species or a group of closely related species. Because they are so distinctive, cuckoo eggs laid in nests other than those of the primary host(s) often (but not always) suffer reduced hatching success.

vi. Gape and Vocal Mimicry Although some avian species reject foreign eggs, few can make the distinction between their own and parasitic young in the nest. Most adult hosts of altricial young seem to feed whichever nestling is gaping and vocalizing the strongest at the time they deliver food. A remarkable exception are five firefinch (*Lagonosticta*) species of Africa. These hosts are parasitized by five species of indigobirds (*Vidua* spp.), with each parasitic species specializing on a single host species. The coevolutionary arms race between firefinches and indigobirds includes egg mimicry by the latter on the former, with the hosts and parasites producing white eggs. However, mimicry also extends to the young. Firefinch nestlings possess distinctive mouth markings that include colored spots on the palate and tongue as well as reflective papillae at the corners of the gape. The pattern of mouth markings in young firefinches generally is similar among the different host species, although the color of the mouth and the gape papillae seems to be species specific. These patterns are important because adult firefinches will not feed young with inappropriate mouth markings. Young indigobird brood parasites are accepted by their firefinch hosts because they possess similar mouth markings, and the coloration is specific to the particular firefinch that is parasitized.

Furthermore, in addition to producing their own species-specific vocalizations, adult indigobirds mimic the vocalizations produced by their firefinch hosts. These vocalizations are learned while the young indigobird is in the host's nest. The accuracy of the mimicry is remarkable, but the mimicked songs and calls are not produced in the same context as those used by the host (e.g., mimicked alarm calls are produced at times other than when confronted with a predator). As female indigobirds search for mates, their search image may include the vocalizations of adult firefinches. Adult male indigobirds that sing mimetic firefinch songs may signal that they were raised by the same host species as the prospecting female and may stimulate her ovarian development.

4. Some Areas of Research Interest

Because brood parasites have such an unusual breeding biology, scientists are interested in the physiological and behavioral differences between parasitic and nonparasitic species. The parasitic brown-headed cowbird is abundant, easily maintained in captivity, and the subject of much current research. For example, cowbirds exhibit patterns of sex hormone secretion that are very different from those of nonparasitic species but are consistent with the extended egg-laying period typical of this brood parasite.

One intriguing phenomenon that has not been explored is the mechanism underlying the absence of a brood patch in cowbirds (and other brood parasites). Nonparasitic species lose feathers in the abdominal region when incubating and brooding young. Increased blood flow into this bare area facilitates the transfer of heat to the eggs or young, aiding their development. Production of a brood patch in many species is influenced by the pituitary hormone prolactin, and it was initially thought that brood parasites did not produce much prolactin. However, it is now known in cowbirds that both females and males show increased prolactin during the breeding season, yet no brood patch results. The tissue in the abdomen of cowbirds may no longer be sensitive to prolactin (i.e., there may be no receptors to the hormone in that area), although more definitive studies are needed.

Many behavioral questions also remain unanswered. How does a young brood parasite recognize that it is different from its host nestmates and begin to associate with its own species? How does a female brood parasite determine whether a species is a suitable host, and how does she find its nest? What is the mating system of brood parasites—monogamy, polygamy, or promiscuity? Until recently, this last question has been largely untestable because it has been difficult to determine which parasitic female produced a given egg and even more difficult to determine which male fertilized that egg. However, with the development of DNA fingerprinting techniques, answers to this question should be forthcoming rapidly.

II. FACULTATIVE BROOD PARASITISM

Not all brood parasitism is generated by obligate brood parasites. A growing number of instances of facultative brood parasitism have been reported in a wide range of species. Facultative brood parasitism involves species in which a female may lay some parasitic eggs but also incubates and rears a clutch of her own or in which some females lay parasitically and others do not. Frequently, such parasitism involves conspecific hosts (i.e., birds of the same species), but in some instances interspecific hosts are exploited. Facultative brood parasitism is seen most often among waterfowl of the family Anatidae, which includes swans, geese, and ducks (Table 2). It is especially common in ducks, with the redhead (*Aythya americana*) representing one of the most active and best studied facultative brood parasites. Although many waterfowl species remain poorly studied, at least some level of parasitism probably occurs in more than half of them.

The extent to which intraspecific facultative brood parasitism occurs has been difficult to document, largely because of problems in distinguishing parasitic eggs from those of the host. However, if the degree of within-female variability in egg appearance is determined to be less than the variability among females, then any unusual eggs can be attributed to intraspecific brood parasitism. A species in which this approach has been used to discover a high level of intraspecific brood parasitism is the northern masked weaver (*Ploceus taeniopterus*) of Africa.

Researchers recently have begun to pay closer attention to the detection of facultative brood parasitism (e.g., noting when two eggs appear in a nest in a single day or using molecular genetics to assess kinship), and the extent and breadth of its occurrence are only now becoming fully appreciated.

A. The Causes, Benefits, and Costs of Facultative Brood Parasitism

Although occasional instances of facultative brood parasitism can be attributed to mistakes on the part of the adult female (e.g., laying in the wrong nest or two females occupying the same nest), such events do not account for the vast majority of cases. It is more likely that some birds engage in brood parasitism as part of a breeding strategy that enhances their reproductive success.

Several factors contribute to the high level of occurrence of facultative brood parasitism in waterfowl. Principal among these is the fact that waterfowl produce precocial offspring, so young brood parasites require little parental care of their adult hosts. Other factors may include limited nest sites, especially for cavity-nesting birds. Also, many waterfowl species nest on the ground in high concentrations and with little or no territoriality, which facilitates the detection of potential host nests by parasitic females.

Interestingly, evidence suggests that some birds parasitize only when their prospects for reproductive success are either very good or very bad. Otherwise, they raise their own young and do not parasitize other nests. Alternatively, birds may lay parasitic eggs as a way of reducing the chances that all of their young are taken by predators, thereby ensuring at least some level of reproductive success.

1. Good Times/Bad Times

A female in good body condition, that is, one with ample energy reserves for producing eggs, may be

TABLE 2
Examples of Waterfowl Species Known to Frequently Engage in Facultative Brood Parasitism

Common name	Scientific name
Wood duck[a]	*Aix sponsa*
Redhead	*Aythya americana*
Canvasback	*Aythya valisineria*
Common goldeneye[a]	*Bucephala clangula*
Ruddy duck	*Oxyura jamaicensis*

[a] Cavity-nesting species.

able to lay more eggs than she could incubate effectively and could increase her reproductive success by depositing additional, parasitic eggs prior to completing her own clutch. To the extent that her parasitic eggs are accepted by hosts and survive, she will experience increased breeding success that year.

Conversely, females in poor body condition may be unable to incubate and care for their young without jeopardizing their own chances of survival as well as those of their offspring. Rather than refraining from breeding entirely, these birds may attempt to make the best of a bad situation by laying parasitic eggs. Young birds that cannot compete successfully for suitable nesting sites, that do not obtain a mate, or that lack the experience to raise nestlings to independence may do likewise. Similarly, females whose nests are destroyed by weather or predators during the egg-laying stage may opt to lay parasitically those eggs that are developing in their reproductive tracts, especially if the breeding season is short and offers little chance to renest successfully.

2. Avoiding Putting All Eggs in One Basket

Regardless of her physical condition, by laying at least some parasitic eggs a female can avoid the catastrophic effect on her reproductive success of the loss of her nest through predation, the death of her mate (in species with biparental care), etc. Dispersing eggs into two or more nests increases the chances that at least some young will survive. Although the average overall egg success does not differ among females that are parasitic and those that are not, the chances of at least one young surviving are better when eggs are deposited in more than one nest. However, the costs of dispersing eggs parasitically, in terms of locating suitable nests, host aggression, rejection of parasitic eggs, desertion of parasitized nests, neglect of young, etc., probably outweigh the benefits of facultative brood parasitism for many birds.

3. Benefits and Costs

Some aspects of facultative brood parasitism may actually benefit hosts. For example, the addition of parasitic young to a brood reduces the chances that a predator would select one of the host's young. In the same vein, parasitic young may look or behave differently than host young (especially if they are of different species), which may catch the attention of a predator. Finally, if the parasitic and host females breeding in a given area are related, then parasitic young may be related to the host female, and the latter would gain reproductive benefits through kin selection (i.e., enhancing the survival of relatives).

Despite these possible advantages, facultative brood parasitism can instill considerable costs to hosts. For example, if the level of parasitism is high, then a significant reduction in egg success can occur because of reduced heat transfer during incubation. Similarly, host (or parasitic) eggs can break if the parasitic and host females jostle on the nest when the parasitic female attempts to lay her egg. Furthermore, the addition of parasitic eggs could extend the incubation period, increasing the chances of the nest being detected and destroyed by a predator.

It must be emphasized that the causes, costs, and benefits of facultative brood parasitism are still poorly understood in waterfowl (and other species). Additional research is warranted to document more completely which birds are likely to engage in this behavior, the extent to which potential hosts have evolved mechanisms to deter parasitism, and the effects of brood parasitism on the lifetime reproductive success of both parasites and hosts.

B. Facultative Brood Parasitism in Altricial Species

While the majority of cases of facultative brood parasitism have been seen in waterfowl, a growing number of instances have been reported in altricial avian species. As with waterfowl, breeding in cavities or in colonies and nesting at high population densities seem to favor the occurrence of facultative brood parasitism in altricial birds. For example, European starlings (*Sturnus vulgaris*) and house sparrows (*Passer domesticus*), both cavity nesters, can experience significant levels of facultative intraspecific brood parasitism. Similarly, colonially breeding cliff swallows (*Hirundo pyrrhonota*) often parasitize their neighbors in the colony. Other birds, such as the black-billed (*Coccyzus erythropthalmus*) and yellow-billed (*C. americanus*) cuckoos, engage in facultative brood parasitism (intraspecific and/or interspecific) in years when food for egg production is abundant.

Altricial species require considerable posthatching parental care. Thus, the costs to these birds of being facultatively parasitized probably is considerably greater than that for precocial species; conversely, being parasitized is unlikely to benefit them in any way. Therefore, altricial birds likely have developed more effective deterrents to facultative brood parasitism than have precocial species. Indeed, many behaviors that altricial birds use to hinder obligate brood parasitism also impede facultative parasitism and may help to explain the relative rarity of facultative brood parasitism in altricial species compared to precocial species. However, systematic studies of many more avian species need to be conducted before such a generality can be defended with confidence.

See Also the Following Articles

AGGRESSIVE BEHAVIOR; ALTRICIAL AND PRECOCIAL DEVELOPMENT IN BIRDS; FEMALE REPRODUCTIVE SYSTEMS, BIRDS; PARENTAL BEHAVIOR, BIRDS

Bibliography

Friedmann, H. (1929). *The Cowbirds: A Study in the Biology of Social Parasitism*. Thomas, Springfield, IL.

Johnsgard, P. A. (1997). *The Avian Brood Parasites: Deception at the Nest*. Oxford Univ. Press, New York.

Lowther, P. (1993). Brown-headed cowbird. In *The Birds of North America No. 47* (A. Poole and F. Gill, Eds.). Academy of Natural Sciences, Philadelphia.

Payne, R. B. (1973). Behavior, mimetic songs and song dialects, and relationships of the parasitic indigobirds (*Vidua*) of Africa. In *Ornithological Monograph No. 11* (R. M. Mengel, Ed.). American Ornithologists' Union, Lawrence, KS.

Payne, R. B. (1977). The ecology of brood parasitism in birds. *Annu. Rev. Ecol. Syst.* **8**, 1–28.

Rothstein, S. I. (1990). A model for coevolution: Avian brood parasitism. *Annu. Rev. Ecol. Syst.* **21**, 481–508.

Sayler, R. D. (1992). Ecology and evolution of brood parasitism in waterfowl. In *Ecology and Management of Breeding Waterfowl* (B. D. J. Batt, A. D. Afton, M. G. Anderson, C. D. Ankney, D. H. Johnson, J. A. Kadlec, and G. L. Krapu, Eds.), pp. 290–322. Univ. of Minnesota Press, Minneapolis.

Weller, M. W. (1968). The breeding biology of the parasitic black-headed duck. *Living Bird* **7**, 169–207.

Wyllie, I. (1981). *The Cuckoo*. Universe, New York.

Bruce Effect

Peter Brennan
University of Cambridge

I. Introduction
II. The Nature of the Pheromonal Signal
III. Physiological Mechanisms Underlying the Bruce Effect
IV. An Ecological Significance for the Bruce Effect?

GLOSSARY

accessory olfactory bulb The first stage of processing of the pheromonal information from the vomeronasal nerves; it has a similar structure to the main olfactory bulb, but it projects subcortically.

inbred strain A strain of mouse that has been inbred for at least 20 generations of sibling matings and is thus genetically homozygous.

major urinary proteins (MUPs) Small pheromone-binding proteins that are present in large quantities in rodent urine and that are thought to convey the genetic individuality of the pheromonal signal.

pheromones Chemical signals produced by an individual that cause a relatively stereotyped behavioral or endocrine response in another individual of the same species.

TIDA neurons The tuberoinfundibular dopaminergic neurons that are found in the arcuate nucleus of the hypothalamus and release dopamine into the pituitary portal system, thereby inhibiting prolactin secretion from the anterior pituitary.

vomeronasal organ The sensory organ of the vomeronasal system containing the pheromone receptors and connected to the nasal cavity by the vomeronasal duct.

The Bruce effect is the phenomenon whereby the exposure of a recently inseminated female to the pheromones from an unfamiliar male conspecific causes the termination of her pregnancy. This effect has been most extensively studied in laboratory mice but has also been reported to occur in wild mice and a few other rodent species. The neuroendocrine mechanism underlying pregnancy failure involves an estrous-inducing pheromone present in male urine which inhibits prolactin secretion from the anterior pituitary of females. This results in a decline in progesterone release from the corpora lutea and the consequent return to estrus. Although all males produce the pregnancy-blocking pheromones, the female is able to recognize those of her mate and prevent them from eliciting the abortion of his own offspring. Despite many attempts to explain the occurrence of the Bruce effect in terms of adaptive advantages for males or females, its ecological significance remains unclear.

I. INTRODUCTION

The Bruce effect is named after Hilda Bruce, who first reported it in the late 1950s. She found that housing newly mated female mice with males which were different from the ones that had mated resulted in a high rate of failed pregnancies. Bruce and Parkes went on to show that physical contact was not necessary, and that exposure of the female to fresh male urine, or bedding soiled by the strange male, was sufficient to block pregnancy. They proposed that this pregnancy blocking effect was mediated by pheromones present in male urine.

Since these pioneering studies, male-induced pregnancy block has been shown to occur in prairie deer mice (*Peromyscus maniculatus bairdii*), collared lemmings (*Dicrostonyx groenlandicus*), and several species of vole (*Microtus ochrogaster, M. pennsylvanicus, M. agrestis, M. montanus,* and *M. pinetorum*). In mice, the Bruce effect only occurs during the first 4 days after mating, before the implantation of the developing embryos. This preimplantation pregnancy failure is typical of many species in which a Bruce effect has been observed, although in some (e.g., *M. ochrogaster*) pregnancy block can occur up to 17 days postmating, well into the gestation period. However, the male-induced pregnancy failure in *M. ochrogaster* has been reported to require direct contact between the male and female and is not induced by exposure to male urine alone. This species is an induced ovulator, with ovulation occurring 8–12 hr after mating, so it seems likely that the physical stimulation of mating with a strange male could induce a return to estrous by activating the ovulatory reflex. Therefore, not all instances of male-induced pregnancy block may involve the action of male pheromones that is characteristic of the Bruce effect.

II. THE NATURE OF THE PHEROMONAL SIGNAL

Despite extensive research, the exact nature of the pheromonal signal responsible for the Bruce effect in mice has not been fully characterized. Only the urine from sexually mature males has pregnancy-blocking activity, and this activity is lost following castration. Moreover, testosterone injection into castrated males or normal females restores activity, indicating that production of the pheromone is dependent on the level of circulating androgens.

Pheromonal activity is predominantly found in the nonvolatile fraction of male urine, although it can be extracted from this fraction using dichloromethane or prolonged dialysis against phosphate buffer. It is generally accepted that the pheromone consists of low-molecular-weight ligands that are tightly bound to proteins. Mice excrete large amounts of proteins in their urine of which the most likely candidates for the ligand binding role are the major urinary proteins (MUPs). These are members of the lipocalycin family of ligand-binding proteins and have been shown to bind putative pheromonal molecules. MUP synthesis occurs in the liver under the control of testosterone, growth hormone, and thyroxin. However, the complex of MUP and ligand has been found in various tissues, including the preputial gland, and it is likely that there are multiple sites for pheromone production.

A mouse produces a variety of MUPs in its urine, the composition of which is related to its genetic identity. Even mice that differ in only the H2 locus of the major histocompatibility complex can be distinguished by females in the context of the Bruce effect. It is unclear whether the MUP proteins have any pregnancy-blocking activity on their own or whether it is the complex of MUP and ligand that activates the pheromone receptors. Alternatively, the different MUPs might act solely as transporters for a specific cocktail of ligands. In any case, it seems reasonable to conclude that the MUPs confer the individual identity on the pheromonal signal. Individual genetic differences appear to affect the transmission and reception of the pheromonal signal. For instance, urine from CBA males has been reported to be more potent in eliciting the Bruce effect than urine from males of the C57/BJ strain. Similarly, CBA females are more sensitive to the pregnancy-blocking effects of male pheromones than those of the C57/BJ strain.

III. PHYSIOLOGICAL MECHANISMS UNDERLYING THE BRUCE EFFECT

Although the nature of the pregnancy-blocking pheromones remains elusive, intensive research has clarified the endocrinological basis for the Bruce effect and the neural pathway by which it is mediated (Fig. 1).

A. The Neural Pathway

Like most mammals, in addition to the main olfactory system, mice possess a vomeronasal system that is specialized for the detection and transmission of pheromonal information. The pheromone receptors are located in the vomeronasal organ, the selective destruction of which abolishes the ability of male pheromones to block pregnancy. In contrast, lesions

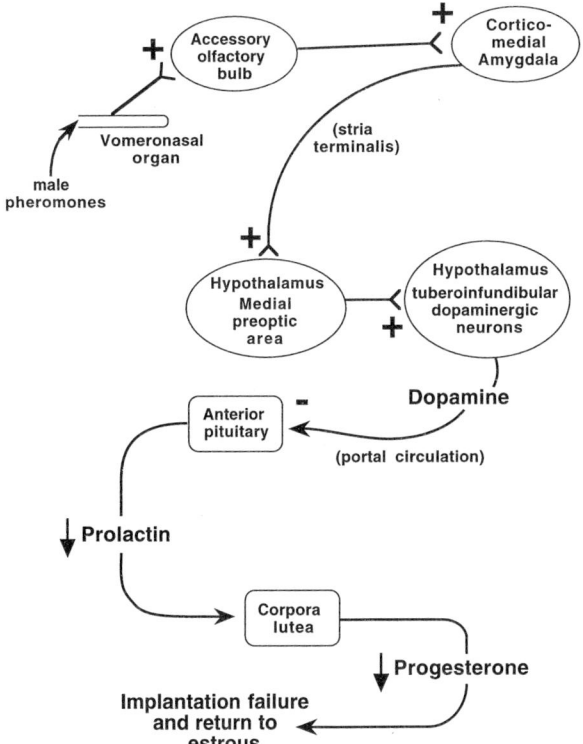

FIGURE 1 The neuroendocrine mechanism responsible for the Bruce effect in mice. Pheromonal stimulation of receptors in the vomeronasal organ initiates a neuroendocrine reflex that suppresses prolactin secretion from the anterior pituitary. The removal of luteotrophic support results in a fall in progesterone levels and a return to estrous.

of the olfactory receptors of the main olfactory system do not disrupt the estrous-inducing effects of male pheromones, implying that the Bruce effect is mediated by the vomeronasal system alone. Putative pheromonal receptor genes have been cloned from the vomeronasal epithelium. Not only do they share little homology with receptors isolated from the main olfactory epithelium but also the transduction components differ between the two systems, suggesting that they have distant evolutionary origins.

The vomeronasal organ is a closed tubular structure, entirely separate from the main olfactory epithelium and connected to the nasal cavity in mice via the narrow vomeronasal duct. Stimulus access to the vomeronasal organ is dependent on a pumping mechanism under autonomic control, which is activated in situations of novelty. This results in a large influx of mucus into the organ, carrying with it nonvolatile substances such as the pheromonal components present in dried urine marks. This sensitivity to nonvolatile components distinguishes the vomeronasal system from the main olfactory system, which is primarily concerned with the detection of volatile odor components.

The vomeronasal organ receptors project, via the vomeronasal nerves (VNNs), to mitral cells of the accessory olfactory bulb (AOB), which is located in the dorsocaudal part of the main olfactory bulb (MOB), although entirely separate from it functionally. Unlike the MOB, which has extensive cortical projections, the AOB projects subcortically to the hypothalamus, by which it mediates pheromonal influences on the reproductive endocrine state. The mitral cells in the AOB make excitatory connections onto neurons in the medial and posteromedial cortical nuclei of the amygdala. Although the MOB also projects to the amygdala, its fibers terminate in adjacent, nonoverlapping areas, maintaining the functional separation of the two systems. The vomeronasal pathway, mediating pheromonal activity, projects on from the amygdala to the medial preoptic area of the hypothalamus, which in turn projects to the tuberoinfundibular arcuate (TIDA) hypothalamic neurons. Electrical stimulation of mitral cells in the AOB has been shown to excite these TIDA neurons, causing them to release dopamine into the hypothalamic portal circulatory system, thus forming the final link in the neuroendocrine reflex linking pheromonal stimulation to endocrine state.

B. Endocrine Effects

The neuroendocrine reflex, by which male pheromones exert their estrous-inducing effect, has been extensively characterized. In mice, the successful implantation of the developing embryos depends on the endocrine environment. During the second and third days following mating, progesterone produced by the corpora lutea primes the uterine endometrium and, in combination with estrogen, induces implantation on Day 4. Injections of progesterone over the first 5 days of pregnancy prevent the Bruce effect, suggesting male pheromones exert their pregnancy-blocking effect by interfering with progesterone synthesis. Progesterone production by the corpora lutea is stimulated by prolactin released from the anterior pituitary, and hyperprolactinemia, induced by an ectopic pituitary homograft or prolactin injections, also prevents the Bruce effect. Prolactin release from the anterior pituitary is itself inhibited by dopamine released into the portal circulation from the TIDA neurons in the hypothalamus. Therefore, the pregnancy-blocking effects of male pheromones can also be mimicked by injections of the dopamine agonist bromocriptine.

Prolonged periods of pheromonal exposure of between 48 and 72 hr are typically used to elicit the Bruce effect. However, the physical stimulation at mating initiates a semicircadian pattern of prolactin release, with nocturnal and diurnal peaks. Therefore, it is not surprising that the effectiveness of male pheromones in blocking pregnancy varies with the timing of exposure (Fig. 2). More restricted exposure to soiled male bedding, for as little as two periods of 4 hr, can result in pregnancy failure provided that the exposure is coincident with the prolactin peaks. Similarly, two periods of electrical stimulation of the AOB for 2 hr are also sufficient to block pregnancy, if they are given during the prolactin peaks. In contrast, similar periods of bedding exposure or electrical stimulation given between the peaks are ineffective. These experiments provide strong evidence that the Bruce effect is the result of luteal failure induced by inhibition of prolactin release from the anterior pituitary.

C. Pheromonal Learning

In the laboratory, where inbred strains of mouse are used, the strange male has to be of a different strain from the mating male to cause pregnancy failure, and prior exposure of a female to male pheromones without mating does not reduce their pregnancy-blocking effect. Although the mating male also produces these pheromones, they do not block the pregnancy of the female with whom he mates. This is because the female recognizes the pheromones to which she was exposed at mating, preventing them from inducing the Bruce effect. This pheromonal

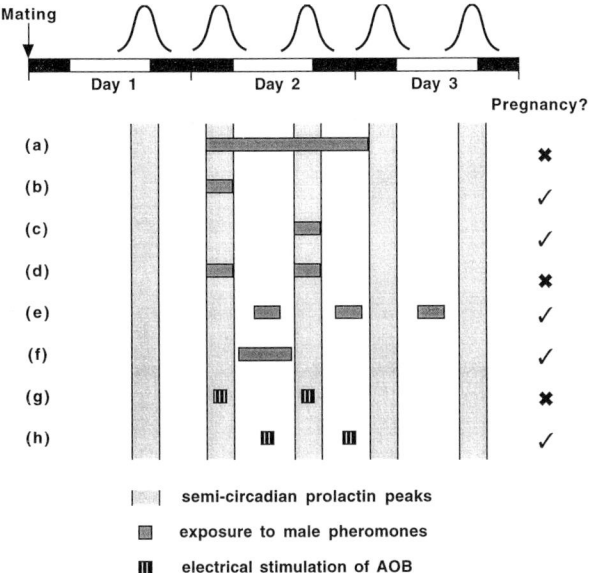

FIGURE 2 Susceptibility to the Bruce effect depends on the timing of pheromonal exposure. The timing of the semicircadian prolactin peaks initiated by mating is shown diagrammatically in relation to the light/dark cycle at the top of the figure. The timing of pheromonal exposure or accessory olfactory bulb (AOB) stimulation (horizontal bars) to the prolactin peaks (vertical bars) is shown below, along with their effects on pregnancy maintenance. A cross indicates a high level of pregnancy failures, i.e., that the Bruce effect has occurred. Exposure to strange male pheromones or electrical stimulation of the AOB during the prolactin peaks on Day 2 are effective in blocking pregnancy (a, d, and g). Exposure during only one of the peaks or exposure or stimulation between the peaks are ineffective (b, c, e, f, and h).

learning is contingent on mating and requires a prolonged pheromonal exposure, during a sensitive period of about 3–4.5 hr immediately after mating, for a robust memory to be formed. The pheromonal memory is long-lasting, being maintained for at least 30 days if the female does not become pregnant. Potentially, the memory could outlast the gestation period of 20 days in mice and affect pheromone recognition at the outset of a second pregnancy. However, the endocrine environment during pregnancy appears to curtail the duration of the memory so that it does not interfere with the Bruce effect at a subsequent mating.

Temporary disruption of neural activity at various points along the vomeronasal pathway has shown that the neural changes underlying the memory for the mating male's pheromones are localized to AOB. Although the AOB is the first stage of processing of the pheromonal information, and has a relatively simple structure, it is sufficient to account for pheromonal learning.

Learning is dependent on the association of pheromonal input from the VNNs and a mating-induced increase in noradrenaline release in the AOB. Pheromones from the mating male would be expected to activate a subpopulation of mitral cells and their associated network of granule cell inhibitory interneurons. Association of this pheromonal input with the increased levels of noradrenaline after mating is proposed to result in a long-lasting increase in the inhibitory gain of the synapses with these inhibitory interneurons. During subsequent pheromonal exposure, the mitral cells responding to the mating male's pheromones would thus be subject to increased inhibition, disrupting the pregnancy-blocking signal. Strange male pheromones would excite a different, although probably overlapping, subpopulation of mitral cells, some of which would not be subject to the enhanced feedback inhibition. In this case, the male pheromonal signal would be transmitted centrally without disruption and induce pregnancy failure. Therefore, the female's ability to recognize the mating male's pheromones can be explained quite simply as the selective disruption of the transmission of his pheromonal signal at the level of the AOB, thus preventing the abortion of his own offspring.

IV. AN ECOLOGICAL SIGNIFICANCE FOR THE BRUCE EFFECT?

Many attempts have been made to attribute an ecological role to the Bruce effect. Obviously, the Bruce effect might be advantageous for males in intermale competition by increasing their own reproductive success at the expense of their competitors. However, this is reproductively costly for the females and it is not clear if such a mechanism would evolve unless it also increased the females' overall reproductive fitness. It has been proposed that strange males will kill unrelated offspring and the Bruce effect could be a means of preventing this infanticide, with consequent savings of energy and reproductive time for the female. However, given the time restriction of preimplantation exposure, it is not clear how often such a situation would arise in natural contexts. Moreover, female mice have a postpartum estrous and would be expected to be lactating during the vulnerable preimplantation period, and the high prolactin levels during lactation have been shown to prevent the Bruce effect. Currently, there is no good evidence that the Bruce effect occurs in mice under natural conditions, and female mice do not appear to adapt their behavior on the basis of the priming effects of male or female pheromones. Some evidence has been reported that a male-induced implantation failure might be common in the wild in other species of rodent. However, considering the wide variety of social structures and reproductive strategies adopted by different species, any role for the Bruce effect in one species should not be generalized to others.

Indeed, just because the Bruce effect occurs in mice under laboratory conditions does not mean that it has evolved to play an important role in the behavioral ecology of this species. The estrous-inducing activity of male pheromones in the Bruce effect is similar to the effect of male pheromones to accelerate puberty in prepubertal females (Vandenbergh effect) and induce estrous in grouped anestrous females (Whitten effect). Seen in this context, the Bruce effect may be a laboratory artifact due to the enforced exposure to strange male pheromones and their general estrous-inducing effects. However, regardless of whether pregnancy block by strange males has any adaptive significance, it is vital that the female's ability to recognize the pheromonal signal from the mating male should have evolved to prevent the estrous-inducing effects of his pheromones from aborting his own offspring. More needs to be known about the behavioral ecology of mice and the incidence of the Bruce effect in a natural context before any definite conclusions can be drawn about its ecological significance.

See Also the Following Articles

PHEROMONES; WHITTEN EFFECT

Bibliography

Brennan, P., Kaba, H., and Keverne, E. B. (1990). Olfactory recognition: A simple memory system. *Science* 250, 1223–1226.

Dominic, C. J. (1966). Observations on the reproductive pheromones of mice; II. Neuro-endocrine mechanisms involved in the olfactory block to pregnancy. *J. Reprod. Fertil.* 11, 415–421.

Dulac, C., and Axel, R. (1995). A novel family of genes encoding putative pheromone receptors in mammals. *Cell* 83, 195–206.

Hurst, J. L. (1994). Do female house mice, *Mus domesticus*, regulate their exposure to reproductive priming pheromones?

Kaba, H., and Nakanishi, S. (1995). Synaptic mechanisms of olfactory recognition memory. *Rev. Neurosci.* 6, 125–141.

Li, C.-S., Kaba, H., Saito, H., and Seto, K. (1990). Neural mechanisms underlying the action of primer pheromones in mice. *Neuroscience* 36, 773–778.

Li, C.-S., Kaba, H., and Seto, K. (1994). Effective induction of pregnancy block by electrical stimulation of the mouse accessory olfactory bulb coincident with prolactin surges. *Neurosci. Lett.* 176, 5–8.

Marchlewska-Koj, A. (1983). Pregnancy blocking by pheromones. In *Pheromones and Reproduction in Mammals* (J. G. Vandenbergh, Ed.), pp. 151–174. Academic Press, New York.

Rosser, A. E., Remfry, C. J., and Keverne, E. B. (1989). Restricted exposure of mice to primer pheromones coincident with prolactin surges blocks pregnancy by changing hypothalamic dopamine release. *J. Reprod. Fertil.* 87, 553–559.

Storey, A. E. (1990). Pregnancy disruption by unfamiliar males in meadow voles: A comparison of chemical and behavioural cues. In *Chemical Signals in Vertebrates 5* (D. W. Macdonald, D. Müller-Schwarze, and S. E. Natynczuk, Eds.), pp. 199–208. Oxford Univ. Press, Oxford, UK.

Bryozoa (Ectoprocta)

Robert M. Woollacott

Harvard University

I. Bryozoa (Ectoprocta)
II. Sexual Reproduction
III. Asexual Reproduction

GLOSSARY

ancestrula The founding member of a colony; derived from metamorphosis of a larva.

coronate larva A nonfeeding larval type in gymnolaemates; it lacks a digestive tract.

cyphonautes larva A feeding larval type in gymnolaemates; it possesses a digestive tract.

cystid The living and nonliving parts of the body wall of individual colony members.

lophophore A coelomic-lined tentaculated feeding structure.

polypide The basic feeding unit of an individual colony member composed of the lophophore, digestive tract, and associated musculature and ganglion.

zooid An individual member of a colony. Feeding zooids are autozooids, and zooids modified for other functions are kenozooids.

Bryozoans or "moss animals" first appear in the fossil record during the Ordovician. About 15,000 extinct and 5,000 extant species are described. Bryozoans are aquatic and colonial eumetazoans. Although a small number of species live epifaunally on mud or sand, most bryozoans are sessile and occur on solid abiotic or biotic substrata. In some cases, though the bryozoan is sessile, the substrata can be highly mobile such as floating algae and plastic debris or swimming sea snake. Living species are organized into three classes: Phylactolaemata (freshwater with gelatinous exoskeletons), Stenolaemata (marine with calcareous exoskeletons), and Gymnolaemata (predominately marine with chitinous or calcareous exoskeletons). All living stenolaemates are in the order Cyclostomata. Gymnolaemates are subdivided into two orders: the Ctenostomata and the Cheilostomata. Most contemporary species are contained in the Cheilostomata, a group that began radiation in the Jurassic. Individual colony members are microscopic in size, but colonies themselves can assume massive proportions and bryozoans are often conspicuous components both in numbers and biomass of benthic communities. Some species have significant and deleterious economic impact as biofoulers, but others may be potentially rich sources of medically important bioactive compounds. At least one species is implicated in human disease causing a form of dermatitis that affects fishermen.

I. BRYOZOA (ECTOPROCTA)

Bryozoans are eucoelomates possessing a specialized food-gathering device, the lophophore, and a recurved digestive tract. Excretory and respiratory systems are absent. A cord-like plexus of mesodermally derived tissue, the funiculus, is thought to serve as a circulatory system. Colonies are formed of modular units, zooids, that are often present as a wide array of polymorphs with various specialized functions including feeding, attachment, defense, gamete formation, brooding, and communication. Feeding individuals, autozooids, possess a body wall, or cystid, containing acellular and cellular layers and a feeding device, or polypide, composed of the lophophore and digestive tract. The lophophore can be withdrawn into the case-like cystid when not actively feeding. The cystid is reinforced by a gelatinous, chitinous, or calcium carbonate exoskeleton. Polymorphic individuals, kenozooids, are modified zooids that have undergone polypide reduction and often have elaborate cystid specializations.

Most authorities consider bryozoans oligomerus metazoans. The lophophore is a tentaculated extension of the middle body compartment, or mesosome, and contains a mesocoelomic lumen. The lophophore surrounds the mouth but not the anus. A structure considered by many, but not all, authorities to be homologous to the bryozoan lophophore exists in the phyla Brachiopoda and Phoronida. On the basis of this supposedly autapomorphic trait, the three phyla are often united in one supraphyletic assemblage, the Lophophorata. The relationship of lophophorates to the deuterostome and protostome phyla is the source of ongoing debate. Recent molecular and fossil data suggest a protostome affinity, whereas anatomical and developmental data indicate a deuterostome association. Other molecular studies of branching order, however, indicate that "lophophorates" are polyphyletic and, as such, that the lophophore is a convergent feature. Finally, there is some molecular and developmental evidence that bryozoans are themselves polyphyletic, with phylactolaemates clustering in a fashion distinct from gymnolaemates and stenolaemates. Much remains unresolved about the evolutionary affinities within the Bryozoa and of this phylum to other metazoans.

II. SEXUAL REPRODUCTION

A. Hermaphroditism, Sexual Polymorphism, and Germ Cells

Bryozoans produce hermaphroditic colonies, but individual reproductive zooids can be hermaphroditic or gonochoristic depending on the species. If hermaphroditic, they can be protandrous, protogynous, or simultaneous hermaphrodites. In some species, there is sexual differentiation even in male or female phases of sequential hermaphrotidic zooids. In species with gonochoristic zooids, sexual dimorphism, when present, can range over a broad spectrum. The distinctions between male and female zooids may be restricted to variations in polypide size or tentacle number. In other cases, cystids can be strikingly dimorphic and, hence, male and female zooids are easily distinguished on the basis of superficial appearance. Germ cells are not segregated early in ontogeny and arise from totipotent cells found between the mesothelial and epithelial linings of the polypide, cystid, or polypide bud. Apparently, some of these totipotent cells can enter a somatic pathway leading either to formation of a new polypide for cyclic polypide replacement or to replication of new zooids. Alternatively, these cells may differentiate as germ cells. Germ cells are recognized by their round shape and large nucleus containing a prominent nucleolus. Nothing is known of how germ cells are specified.

The cheilostome *Celleporella hyalina* possesses gonochoristic zooids and there is elegant evidence that sex ratios within colonies can be controlled by environmental as well as genetic factors. A shift is observed in sex ratio to more male zooids in colonies grown in regimes of higher food concentrations and to more female zooids in colonies grown at higher temperatures.

B. Spermatogenesis

Sperm develop within the mesothelium usually in association with the funiculus adjacent to the polypide or in the cystid wall. In the gymnolaemate *Membranipora* spp., sperm develop in syncytial clusters containing 64 cells. At the conclusion of spermatogenesis, sperm are filiform with an elongate nucleus and they apparently lack an acrosome. The midpiece is extensive and contains two highly modified mitochondrial rods that have lost their cristae and matrices. The concave surfaces of these mitochondria face the axoneme and they develop electron-dense ridges that extend centripetally from the mitochondrion toward the axoneme. Sperm are released as clusters (spermatozeugmata) through pores at the distal tips of tentacles. Once in the seawater, a sperm packet swims as a unit until it reaches and enters the female zooid. In other gymnolaemates, the number of cells in a packet may vary and the cluster may separate into individual sperm either within the coelom or at the time of spawning, depending on the species. In stenolaemates, sperm develop in aggregates of four. Stenolaemate sperm differ from those of gymnolaemates by possessing an acrosome, having four rather than two mitochondrial rods that retain their lumena and four rods of electron-dense material that alternate between adjacent mitochondrial rods. A specialized junction exists between the midpiece and tail. Additionally, eight longitudinally arrayed peripheral fibers are found in the region of the sperm posterior to the midpiece. Such fibers are not reported in gymnolaemate sperm. Phylactolaemate sperm possess an acrosome and midpiece–tail junctions like those of stenolaemates; however, unlike other bryozoans, they have only slightly modified mitochondria that spiral around the axoneme. The tail is about the same diameter as the midpiece and contains glycogen-like material and microfilaments that are associated with the plasmalemma.

In summary, bryozoan sperm are fundamentally uniform in structure within each class and are of the "advanced" type often associated in other groups with internal fertilization.

C. Oogenesis

In gymnolaemates, the events of oogenesis are heavily dependent on differences in patterns of development between nonbrooders, nonplacental brooders, and placental brooders. Ctenostomes are nonplacental brooders that produce eggs heavily vested with yolk. The oocyte is surrounded by a follicular epithelium. This follicle layer undergoes a marked hypertrophy and the adjacent polypide begins to degenerate coincident with the onset of vitellogenesis. Cheilostomes contain nonbrooders, nonplacental brooders, and placental brooders. Nonbrooders such as *Membranipora* spp. produce numerous eggs, with each containing small amounts of yolk. These eggs are associated with follicle cells that appear not to be involved with yolk transfer but do phagocytize degenerating oogonia. Gap junctions exist between follicle cells and oocytes as well as follicle cells and coelomocytes, indicating that follicle cells may transport low-molecular-weight substances or effect communication among these cell types in some undetermined fashion. The prophase I nucleus breaks down at the time of ovulation and primary oocytes accumulate in the coelom prior to spawning. Macrolecithal eggs are produced in nonplacental brooding species. In ctenostomes with this pattern of oogenesis, the ovary occurs in association with either the caecum of the polypide or the lateral wall of the cystid. Oocytes are surrounded by follicle cells and these latter cells hypertrophy at the time of vitellogenesis. The polypide also undergoes degeneration during the period of vitellogenic growth and may contribute to egg nutrient stores via processing of breakdown products by the follicle cells. A different course of oogenesis evidently occurs in at least some cheilostomes with brooded but nonplacental development. For example, in *Chartella papyracea*, eggs develop in syncytial duplets surrounded by a layer of follicle cells. One of these cells is destined to complete oogenesis and the other to function as a nurse cell synthesizing and transporting stores of ribosomes to the future ovum. Both oocyte and hypertrophied follicle cells synthesize nutrients. A similar packing of oocyte and nurse cell is found in some placental brooders and may be widespread, but not ubiquitous, among cheilostomes. Oogenesis in two other placental brooders studied involves a single oocyte enclosed in a layer of cells whose origin is unclear and may be either follicle or nurse cells. In stenolaemates, oocytes are initially attached to the polypide, but this connection does not persist once vitellogenesis begins. At the completion of oogenesis, eggs contain little yolk. In phylactolaemates, oocytes mature in close contact with peritoneal cells that likely serve a nutritive function. The ooplasm becomes stratified into outer granular and inner homogenous layers. There are no ultrastructural studies of oogenesis in either stenolaemates or phylactolaemates.

D. Fertilization

Ovulation occurs when eggs detach from surrounding cells and appear free within the main body coelom. Eggs of noncoelomic brooders are spawned from the base of the lophophore which necessitates passage through a coelomopore formed between the main body and lophophoral coeloms. Bryozoans generally lack established gonoducts; however, in some species, regions of the polypide or adjacent peritoneum modify and assist in egg deposition. Eggs pass through a pore situated at the base of two dorsomedial tentacles. In some species, the two tentacles neighboring this pore fuse at their lateral margins to create with their exterior surfaces a ciliated tube or funnel, the intertentacular organ, through which eggs are spawned. Intertentacular organs occur in free-spawning species but also in some brooders. It is reported in certain phylactolaemates that eggs are ameboid and move through the cystid wall to the embryo sac. The means of ovulation in stenolaemates is unknown.

Sperm as well as eggs are spawned via the lophophore. Sperm are released through pores that exist at the distal tips of some or all tentacles. Such pores have been identified in all three classes. Sperm release at tentacle tips presumably reduces the possibility of sperm becoming entrapped in the feeding current of the parent lophophore. In gonochoristic males, some tentacles of the lophophore lack cilia, a feature that might further assist sperm escape. There are numerous examples in which male zooids are clustered around regions of feeding zooids where accumulated current flow is directed outward (chimneys). Finally, in some species, tentacles bend away from the lophophore during spawning, serving to further reduce rates of self-fertilization.

Thus, it appears that there are anatomical and behavioral means of reducing the likelihood of intrazooid self-fertilization, but intracolony self-fertilization remains a possibility. Indeed, several authors report the appearance of larvae from colonies raised in isolation since ancestrulae indicating that self-sterility factors or other barriers to colonywide self-fertilization do not exist. In a detailed study of reproduction of *Celleporina hyalina*, evidence was presented that colonies raised in isolation produced fewer offspring that those living together and that embryos from isolated colonies were often aborted or experienced other developmental difficulties suggesting that offspring of selfed colonies may experience reduced fitness. The limited population genetic information available supports the notion that outcrossing dominates over inbreeding as populations exist near or at Hardy–Weinberg equilibrium.

Fertilization can occur at a variety of locations and stages during egg development including at the time of spawning, after ovulation but while oocytes are still within the maternal coelom, and prior to ovulation while still residing in the ovary as oogonia or oocytes. One study reports a period of 4 days between sperm–egg fusion and eventual spawning. Sperm fusion triggers neither nuclear nor cytoplasmic activation. Egg activation occurs during spawning or in the process of transfer to an external brood chamber. The means of activation in coelomic brooders is unknown. The existence of behavioral mechanisms to concentrate sperm from water, having syngamy occur near the site of egg production thereby reducing fertilization loss as a consequence of egg dispersal, and enabling syngamy to occur at a wide range of stages in oogenesis, apparently result in high rates of successful fertilization.

E. Embryology

In gymnolaemates, early development is similar in brooders and nonbrooders. Cleavage is holoblastic and equal to subequal, sometimes producing larger cells in the vegetal hemisphere. The cleavage pattern can be radial or biradial, but, despite some reports to the contrary, it is not spiral. The spindle axes remain aligned parallel or orthogonal to the animal–vegetal axis and do not tilt 45° and rotate between cleavages as in spiral cleavage. As such, the use by some authors of spiral cleavage nomenclature for descendent blastomeres and reference to the ciliated band of the larva as a prototroch are inappropriate. A coeloblastula forms and gastrulation by primary delamination and/or ingression occurs around the 64- to 128-cell stage. Mesoderm is reported to arise from delaminated entoderm and, hence, is entomesodermal. The blastocoel becomes obliterated by proliferating mesodermal cells. No coelomic compartments exist. A ring of usually 32 cells situated immediately aboral to the equator becomes heavily ciliated and will eventually form the larval swimming organ, the corona. Depending on the species, entoderm in the larva may form a complete and functional digestive tract, may be present as an incomplete digestive tract, may exist only as small groups of cells connecting with the oral surface, or may be absent in any recognizable form. A group of epidermal cells immediately posterior to the site of gastrulation invaginate and differentiate as the metasomal (internal) sac. This sac will provide the adhesive for permanent attachment and, depending on the species, may also form components of the body wall of the ancestrula. A group of epidermal cells in the aboral hemisphere, the pallial epithelium, form the shells of the cyphonautes larva and later the frontal membrane of the twin ancestrulae at metamorphosis. In most other species, this aboral tissue is invaginated in a sulcus (the pallial sinus) that may, depending on the species, contribute at metamorphosis to varying extents the body wall of the ancestrula. This sinus may be quite shallow or extend deeply into the oral hemisphere. Other cells form the anlage of the ancestrular polypide with its splanchnic peritoneum and these cells can be identified by their possession of large masses of free ribosomes. Such anlage may exist in the aboral, oral, or both hemispheres of the larva. A sensory–glandular complex, the pyriform organ, forms in the anterior midline of the developing larva. Other sensory structures, including eyespots and intercoronal sensory cells of various functions, may develop depending on the species. A neural plate forms at the aboral pole in conjunction with an apical disc complex and it sends processes associated with muscle cells deep into the interior. These nerve–muscle cords establish connections with the roof of the metasomal sac and form a ring immediately underlying the nuclear region of coronal cells. Additional nerve and muscle tracts are associated with the pyriform complex and other epidermal tissues. The bulk of the larval interior is filled with nutrient-ladened cells.

In stenolaemates, events in early development are obscured by the occurrence of polyembryony (see Section III,C). In phylactolaemates, a coeloblastula forms, but events of gastrulation are not clear and controversy surrounds the assignment of anterior and posterior. In any event, mesodermal cells form a layer underneath the epidermis creating a two-layered embryo. Certain epidermal cells form a placental-like connection with the maternal cystid. Polypide buds arise at one end of the embryo and certain cells of the embryonic cystid form a tissue that grows to enclose the developing polypides. The outer surface becomes ciliated and is the larval locomotory organ.

F. Larval Eclosion

Light is the most common cue triggering larval release. Nothing is known of the site of photoreception in the maternal individual that might initiate the release response or of the molecular identity of the receptor itself. An action spectrum for release is not available. Release in many species follows a circadian pattern. It has long been known that it is possible, however, to manipulate the timing of release by removing colonies to the dark and then exposing them to light at times other than that of the natural release. Most species of gymnolamates that possess phototactic larvae release in conjunction with daylight. In species in which larvae are indifferent to light (e.g., most stenolaemates and some gymnolaemates) eclosion is reported to occur mainly at night.

The mechanism of release depends on the specifics of brooding in a given species. The relationship between larval and maternal activity is largely obscure. In cheilostomes with external brood chambers, such as *Bugula* spp., release involves contraction of maternal muscles that creates a pore-like opening between the lumen of the chamber and the ambient seawater through which the larva escapes. Larval shape is radically, but only transitorily, altered as it actively squeezes through this narrow opening. Evidently, eclosion involves active participation by larva and adult in these species. In contrast, release from brood chambers of another cheilostome, *Cribrilina* spp., reportedly involves only activity by the larva. In cheilostomes that brood within the vestibule (e.g., *Watersipora* spp.), the operculum of the maternal zooid is depressed by muscular activity and the tentacle sheath can be partially everted enabling the larvae to escape. In ctenostomes, brooding often occurs in the tentacle sheath and eclosion results from combined eversion of the sheath and larval ciliary activity. In stenolaemates, brooding occurs in specialized gonozooids and release is also thought to involve increased ciliary activity on the part of the larva and muscular activity in the gonozooid. In phylactolaemates, larvae develop in embryo sacs and, prior to release, they begin to rotate in these sacs and eventually escape at the site where the embryo sac is attached to the maternal body wall.

G. Larval Types

Gymnolaemates produce either feeding or nonfeeding larvae, depending on the species. Poecilogenous species are unknown. Gymnolaemate larvae are characterized by a collective suite of characters: the absence of coelomic or blastocoelic spaces, an invagination of the oral hemisphere forming a metasomal sac, an aborally situated apical disc with central neural plate, a ring of ciliated cells, the corona, and a glandular–sensory complex, the pyriform organ, situated in the anterior midline. Substantial variability exists in the relative development of these and other features among larvae of different species which makes it possible to assign gymnolaemate larvae to categories based on anatomical criteria. The primary division is into shelled versus nonshelled (coronate) larval types. Shelled larvae are provided with chitinous valves and can be either feeding or nonfeeding. Feeding larvae, the cyphonautes larva, are likely convergent because they occur in two suborders of ctenostomes and one of cheilostomes. These larvae are long-lived and are commonly encountered meroplankton in temperate waters. Nonfeeding shelled larvae occur only in the ctenostome superfamily Halcyonelloidea and very little is known of their biology. The coronate type is the most commonly encountered larval form in gymnolaemates. These larvae lack an exoskeleton and cannot feed on particulate matter. Most larvae lack even vestiges of mouth and gut, but larvae of a few species possess rudimentary digestive systems. Most coronate larvae are true lecithotrophs, whereas a small number, mostly in the cheilostome superfamily Cellularioidea, gain nutrient stores after the egg is transferred to the brood chamber via an extraembryonic nutrition system. There is evidence that larvae of at least the cellularioid *Bugula neritina* take up palmitic acid and alanine dissolved in seawater, but the fates of these molecules after incorporation and the role(s) in larval biology remain unexplored. Coronate larvae can be subdivided into a series of categories based on the relative extent of differentiation of various larval components, but there is little concordance between phylogenies based on these larval features and on those derived from adult characters. The number of species for which information on larval anatomy is available varies widely depending on the taxon. In some superfamilies, such as the Cellularioidea, larvae possess a uniform set of anatomical features across a wide range of species that have been examined. In most groups, however, either the number of examples known is so small that no comparative statement can be made or larvae differ widely in their anatomical features within a given superfamily. Despite these problems with classifications at higher taxonomic levels, it is clear that larvae often vary in distinctive ways among closely related species. The number, kind, and distribution of sensory cells, for example, are often species specific and, hence, useful diagnostic tools in systematic analyses.

Little is known about the larvae of stenolaemates compared with the extensive data available on larvae of gymnolaemates. The larvae are simple and lack a digestive system, pyriform organ, and neural plate. The surface is uniformly ciliated and there is no evidence of photoreceptors or other sensory cells, yet larvae do exhibit behavior. Larvae do have a metasomal sac. There is also a pore-like invagination at the aboral pole that opens into an aboral field. The epithelium forming this field secretes a cuticle. Other than size and shape, there is apparently little differentiation among larvae of stenolaemates.

As mentioned in Section II,E., the difficulty in assigning the anterior–posterior axis in embryology and the pronounced heterochronic acceleration of budding during ontogeny make homologizing features of phylactolaemate larvae with those of marine bryozoans difficult at best. In fact, some authorities consider phylactolaemates as direct developers and phylactolaemate "larvae" as swimming colonies. However, the existence of structures that function exclusively during the swimming phase, such as a ciliated epithelium that effects locomotion and gland cells situated at the site of attachment, suggests that phylactolaemates do possess a larva, albeit one that is highly modified from the larvae of marine bryozoans.

H. Larval Swimming Behavior

Bryozoan larvae are denser than seawater and their position in the water column results from a combination of active swimming, passive dispersal via currents, and turbulent mixing. The bryozoan literature contains a single Reynolds number: 0.5 calculated for a ctenostome larva. This value is in the range generally reported for larvae of marine invertebrates.

Light is the most frequently cited environmental cue affecting directional swimming activity. However, the existence of a response to light and sign of phototaxis vary from species to species. The most common pattern is an initial positive phototaxis followed as swimming time increases by a change to either photonegative or photoneutral. The larvae of some species, however, are indifferent to light, others are photonegative on release, and some remain photopositive throughout the larval period. Numerous studies of gymnolaemate larvae document structures assumed on anatomical criteria to be photoreceptors. Each eyespot contains a single sensory cell with a complex bundle of cilia at its apical surface thought to be the site of photopigment localization and a basal contact with an underlying nerve ring. Depending on the species, these sensory cells either contain shielding pigment in their cytoplasm or are encased in a recess or pit whose walls are formed by adjacent cells that are pigmented and provide for differential illumination of the supposed photoreceptor. The molecular identity of the photoreceptor remains unknown because an action spectrum of swimming, immunocytochemcial assay for known light-active molecules, and molecular genetical identification of prospective candidates have not been published. Some preliminary information is available addressing how the switch in sign of phototaxis from positive to negative is modulated in at least larvae of B. neritina. When bath applied, the monoamine serotonin instantly changes photopositive larvae to photonegative, whereas dopamine, another monoamine, slows or prevents the naturally occurring switch from positive to negative. These findings are supported with immunohistochemical evidence for the existence of serotonin and high-performance liquid chromatography evidence for the presence of dopamine. In a comparative study, the authors were unable, however, to document a similar response in larvae of two congeneric and three additional bryozoan species.

Finally, larvae of some species orient to geographic up and down even in the absence of light. In at least one species, centrifuge experiments document that this distribution is explained by a direct and active influence of gravity. The receptor for this response, however, remains unidentified. Interestingly, larvae of a congener failed to respond to the reoriented gravity vector in parallel companion experiments.

I. Induction of Metamorphosis

After an initial swimming phase, larvae begin substrate exploration. The duration of the swimming phase can be weeks to months in the case of feeding larvae or minutes to hours in the case of nonfeeding larvae. The factors responsible in nature for this change from free-swimming to exploratory behavior are unknown. It is documented in laboratory experiments, however, that larvae of *Bugula* spp. can be blocked from entering the substrate selection period and will remain in the free-swimming phase until eventual death when exposed to continuous bright illumination and periodic agitation.

At the conclusion of substrate exploration, a cataclysmic irreversible metamorphosis occurs. There have been a number of investigations regarding factors that promote or retard metamorphosis. Physical factors such as structural complexity of the substrate, pH, and surface wettability affect settlement patterns in controlled studies. Biological factors such as the presence of bacteria- or diatom-filmed surfaces, conspecific residents, or a dominant competitor can strongly promote or retard settlement in some species. Waterborne signals from macroalgae are also important positive and negative cues for settlement. Apparently, bryozoan larvae have evolved sophisticated means of assessing the environment and selecting or avoiding sites for attachment. In some species, adults produce larvotoxic chemicals that provide a defense against fouling of the colony surface by heterospecific larvae. Few studies, however, attempt to address the identity of these signal molecules and how this information is transduced and translated during the process of metamorphosis. There are preliminary observations that certain ligands may be implicated as receptors, but these studies have not been pursued. Elevated KCl in ambient seawater causes rates of settlement greater than those observed with microbially filmed surfaces used as controls. It is not known, however, if KCl acts by depolarizing the natural receptor, interceding in the induction process downstream from the receptor, or by activating an alternate pathway.

Duration of the larval swimming period is an important determinant of dispersal ability and for shaping the genetic structure of populations. The consequences of varying swimming duration are most dramatically encountered in nonfeeding larvae that are presumably dependent on nutrient reserves stored prior to release and on any dissolved organic matter incorporated during swimming. For example, as the swimming period increases beyond a certain point in *Bugula* spp., larvae permanently lose the ability to initiate metamorphosis. The reason for this is unclear but may relate either to depletion of energy reserves beyond some unidentified threshold or to degradation and lack of renewal of receptors. Between an initial period of hours when the success rate of metamorphosis can reach almost 100% and before the point of metamorphic cutoff lies a time zone in which metamorphosis can be initiated, but the rate of successful completion declines. In addition to this lethal effect, many individuals completing metamorphosis during this period will take longer to develop feeding ancestrulae, bud more slowly, and produce smaller ancestrulae and first budded autozooids. Thus, there exists evidence that both lethal and sublethal costs are acquired after a certain period resident in the plankton. Whether these sublethal effects ultimately translate into reduced reproductive fitness remains to be determined.

J. Patterns of Metamorphosis

Metamorphosis occurs in two phases: (i) a period of rapid morphogenetic movements that brings about a wholesale change in arrangement of tissues and (ii) a slower period of histogenesis in which adult structures differentiate. In general, the first phase requires only minutes, whereas the second may take several days. Two categories of tissues exist in bryozoan larvae. The first type are transitory tissues specific to performing activities during larval life and metamorphosis. These tissues include the swimming organ, sense organs, glandular tissues, feeding organs (in cyphonautes), certain other epidermal cell types of unknown function, and nutrient-filled mesenchymal cells of the larval interior. The second category of cells are those that will be retained to form structures within the ancestrula. The two most important of these are tissues that will form the polypide of the ancestrula and, depending on the species, portions of the metasomal sac and/or pallial epithelium that are anlage of the body wall.

The ancestrular polypide of all bryozoans is the first budded polypide of the colony. Even in the supposedly primitive cyphonautes, the digestive tract of this larva is degraded at metamorphosis and the polypide of the ancestrula arises from totipotent cells associated with the apical disc. In gymnolaemates, the totipotent precursors of the polypide can occur in the aboral, oral, or both hemispheres. In stenolaemates, the polypide rudiments are situated aborally in the larva. In phylactolaemates, the initial polypide(s) is preformed during embryogenesis and appears largely differentiated in the larva.

The body wall of the gymnolaemate ancestrula can have a composite or single origin depending on the species. In some species, the entire wall comes from the metasomal sac situated in the oral hemisphere of the larva, whereas in other species the entire wall derives from the pallial epithelium located in the aboral hemisphere of the larva. Finally, in some species, both tissues contribute to the wall. In stenolaemates, which are all heavily calcified, the body wall forms from metasomal sac and aboral epithelium. Finally, it is difficult to homologize the epidermis of phylactolaemates with those of other bryozoans.

III. ASEXUAL REPRODUCTION

A. Astogeny

In general, the first individual of a colony, the ancestrula, arises from metamorphosis of a larva. Curiously, the ancestrula is incapable of sexual reproduction and its polypide is the first budded polypide of the incipient colony (see Section II,J). The concomitant occurence of accelerated polypide replacement and delayed sexual maturity represents a combination of heterochronic events that form a recurrent pattern throughout bryozoans. In phylactolaemates, colonies can also arise from asexually produced statoblasts (see Section III,B). The colony grows by the asexual production of interconnected individuals. Daughter polypides of phylactolaemates arise from the oral side of the parental polypide, whereas daughter autozooids of stenolaemates and gymnolaemates are produced from the dorsal side of the parent zooid. This distinction between phylactolaemates and the other two classes has been advanced as one of several reasons for considering the Bryozoa polyphyletic.

In phylactolaemates, colonies may be branching or massive and globular. Polypides are produced at budding sites. In stenolaemates, the colony grows by expansion of a terminal common bud in which polypides develop and are partitioned into zooids. In gymnolaemates, a simple ancestrula is generally formed, but there are examples of heterochrony in which accelerated budding leads to the formation of complex ancestrulae. Gymnolaemates colonies have an initial cluster of zooids that change in size and shape with each generation of budding, but, later in colony growth, repeated rounds of budding produce similar zooids. Buds may develop as isolated units or in common budding zones.

Colonies can be encrusting or arborescent and take elaborate forms, including uniserial strings of zooids, multiserial plate-like colonies, and multilaminate complex masses. Most species produce colonies that are attached to some surface, but species with free-living bottom-dwelling colonies are known in the cheilostomes and, possibly, in ctenostomes. There is one recent report of a free existing planktonic ctenostome found in the Antarctic. It is unclear, however, if these colonies were not attached to ice balls that subsequently melted before the specimens were collected and observed.

B. Propagules

Many bryozoans develop resistant stages in response to changes in temperature, salinity, or nutrient availability. Phylactolaemates produce resting stages called statoblasts. These propagules are buds enclosed by a pair of chitinous valves. On germination, the valves separate and the polypide develops. A central mass of cells provides nutrients for the differentiation of the polypide and cystid. Some statoblasts float and provide dispersal, whereas others remain attached at the site of the former colony. At least one species, *Plumatella casmiana*, produces a lightly sclerotized bud, a leptoblast, that germinates soon after formation and, thus, is not a resting stage for overwintering and the like. In gymnolaemates, dormant stages involve modified stolons, rhizoids, or autozooids.

Propagation by fragmentation of existing colonies also occurs, as does regeneration of damaged zones in colonies of gymnolaemates and phylactolaemates.

C. Polyembryony

Polyembryony or embryonic fission apparently occurs in all living stenolaemates and there is some indirect evidence that this mode of reproduction may have also existed in extinct stenolaemates. Early cleavages are asynchronous and eventually a primary embryo with internal and external cell layers forms. Secondary embryos bud from this mass and secondary embryos may bud tertiary embryos in some groups. The overall increase in mass of these progeny necessitates that some form of extraembryonic nutrition exists, but the mode of nutrient acquisition has not been studied in detail.

Acknowledgment

I thank Dean E. Wendt (Harvard University) and Russel L. Zimmer (University of Southern California) for their helpful comments on the manuscript and Anya Toomre (Harvard University) for her editorial assistance. Their collective efforts greatly improved this article.

See Also the Following Articles

ASEXUAL REPRODUCTION; BRACHIOPODA; HERMAPHRODITISM

Bibliography

McKinney, F. K., and Jackson, J. C. B. (1989). *Bryozoan Evolution*. Unwin Hyman, Boston.

Mukai, H., Terakado, K., and Reed, C. G. (1997). Bryozoa. In *Microscopic Anatomy of Invertebrates. Volume 13. Lophophorates, Entoprocta, and Cycliophora* (F. W. Harrison and R. M. Woollacott, Eds.), pp. 45–206. Wiley-Liss, New York.

Reed, C. G. (1991). Bryozoa. In *Reproduction of Marine Invertebrates. Volume VI. Echinoderms and Lophophorates* (A. C. Giese, J. S. Pearse, and V. B. Pearse, Eds.), pp. 86–246. Boxwood Press, Pacific Grove, CA.

Robison, R. A. (Ed.) (1983). *Treatise on Invertebrate Paleontology. Part G Bryozoa Revised. Volume 1: Introduction, Order Cystoporata, Order Cryptostomata*. Geological Society of America/Univ. of Kansas, Boulder, CO/Lawrence, KA.

Ryland, J. S. (1970). *Bryozoans*. Hutchinson Univ. Library, London.

Woollacott, R. M., and Zimmer, R. L. (Eds.) (1977). *Biology of Bryozoans*. Academic Press, New York.

Zimmer, R. L. (1997). The lophophorate phyla: Phoronida, Brachiopoda, Bryozoa. In *Developmental Biology of Animals and Plants* (S. F. Gilbert, Ed.). Sinauer, Sunderland, MA.

Caenorhabditis elegans

Dave Pilgrim
University of Alberta

I. Introduction
II. Life Cycle
III. Anatomy
IV. Sexual System
V. Development
VI. Sex Determination and Dosage Compensation
VII. Behavior

GLOSSARY

bursa The specialized, elongated copulatory structure of the tail in male *Caenorhabditis elegans*.

cell lineage The map or description of the lineal descent of the cells in the adult from the single cell of the zygote. The somatic cell lineage of *C. elegans* is largely invariant from animal to animal.

cloaca An opening in the male tail connecting both the vas deferens and intestinal lumen to the exterior.

dimorphic Relating to the physical or anatomical differences between males and females. These can be gross (such as the difference in shape of the gonads) or biochemical (such as the synthesis of yolk proteins by the cells of the hermaphrodite, but not male, intestine).

distal and proximal gonad arms The gonads of both males and hermaphrodites are reflexed, and the region closest to the cloaca or vulva is referred to as the proximal arm. The distal arm is located dorsally to the vulva.

distal tip cell Specialized cells that cap the distal arms of the gonad in both sexes and that play a role in mitotic divisions of the germ cells.

hermaphrodite An organism that possesses male and female reproductive organs. The female of *C. elegans* makes a few sperm during a brief period in the last larval stage, and subsequent germ cells develop as oocytes. These sperm are stored and can be used for self-fertilization but not cross-fertilization. Hence, she is a functional hermaphrodite.

hypodermis The external epithelium, which surrounds the animal and secretes the cuticle. The hypodermis is composed of a sheet of cells, many of which fuse during development to form multinucleate syncytia.

larval stage After hatching from the egg, the animal proceeds through four larval stages (L1–L4) before maturing into an adult. Each larva is larger than that of the previous stage, and the progression from one to the next is punctuated by a molt in which the old cuticle is shed and a new cuticle is synthesized by the hypodermis.

ovotestis A reproductive gland that supports the development of both oocytes and sperm (albeit at different times) in the same tissue.

P-granules Cytoplasmic granules that are found only in germ-line cells and their developmental precursors. Also, *germ-line granules*.

pseudocoelom The internal compartment formed between the concentric tubes of the intestine and the hypodermis.

vulva The specialized opening in the ventral hypodermis of the hermaphrodite which serves as a passageway for male sperm during copulation and eggs following fertilization.

X/A ratio The ratio of the number of X chromosomes to sets of autosomal chromosomes. *Caenorhabditis elegans* are normally diploid, carrying two sets of autosomes, so sex is determined by the number of X chromosomes; two in hermaphrodites (X/A ratio of 1) and one in males (X/A ratio of 0.5).

Caenorhabditis elegans is a free-living soil nematode, about 1 mm in length, that is found around the world. It is currently a common laboratory model for many aspects of cellular, developmental, and molecular biology. Its popularity comes from its transparency (allowing all nuclei to be followed in living animals at all stages of development), its anatomical simplicity (1000 cells), its small genome (100 Mbp), an invariant somatic cell lineage, ease of laboratory culture, rapid generation time, and a mode of reproduction which facilitates classical genetic analysis. An interested beginner needs only a petri plate, some *Escherichia coli,* and a stereo dissecting microscope to begin study of this fascinating creature.

I. INTRODUCTION

In 1963, Sydney Brenner proposed to the British Medical Research Council that he begin a project to "tame a small metazoan" to study control mechanisms in cellular differentiation and development. *Caenorhabditis* was chosen because it is arguably the "simplest possible differentiated organism" (Wood, 1988). One of the benefits of the worm is that it exists as both a hermaphrodite (capable of both self-fertilization and cross-fertilization) and as a cross-fertilizing male, so new stocks can be created by genetic means and stably maintained by hermaphrodite self-fertilization. The power of genetics has allowed the isolation and characterization of mutants in almost every developmental process. This, coupled with the invariant and completely mapped cell lineage and the small genome (which is due to be completely sequenced by 1998), means that *C. elegans* is arguably the most well understood of all metazoans at the cellular and genomic level.

II. LIFE CYCLE

The *C. elegans* life cycle begins with the embryo, developing through four larval stages (L1–L4) to an adult. The entire cell lineage from egg to adult is largely invariant and has been mapped in its entirety (Sulston, 1988). *Caenorhabditis elegans* development time is temperature dependent. In the laboratory, *C. elegans* can be raised at temperatures between 15 and 25°C, with slower development at the lower temperatures. At 25°C, the embryo hatches into an L1 larva about 14 hr after fertilization, with 558 cells in the hermaphrodite and 560 in the male. Most of those cells are generated within the first 7 hr, with the remainder of embryogenesis dedicated to mor-

phogenesis and differentiation. Each larval stage lasts 7–11 hr, followed by a molt where the old collagenous covering or cuticle is discarded. Following the last molt (L4 to adult), the animal is fertile, and the overall cell number has increased to 959 (hermaphrodite) and 1031 (male) somatic nuclei. The entire life cycle from fertilization to fertility is about 50 hr at 25°C. The hermaphrodite produces oocytes for 4 or 5 days, whereas the male can produce sperm for somewhat longer.

III. ANATOMY

The general body plan of nematodes including *C. elegans* is simple: essentially an inner tube (intestine) and an outer tube (cuticle, hypodermis, and musculature) (Fig. 1). The space between the two tubes is the pseudocoelom, and it contains the gonad. The anterior end of the intestine has been specialized into a feeding organ, the pharynx, which functions to pump food into the intestine. In the laboratory, *C. elegans* feeds on bacteria (*E. coli*) which are crushed by the muscular pharynx as they are pumped through. The other major tissues are a syncytial external tissue layer called the hypodermis (epidermis), the nervous system (~300 cells), and the musculature (body wall, pharyngeal, and anal and sex-specific muscles).

IV. SEXUAL SYSTEM

The two sexes in *C. elegans* are easily distinguished under the dissecting microscope (Fig. 1). Hermaphrodites produce both oocytes and sperm, whereas males can produce only sperm. However, the period of spermatogenesis in the hermaphrodite is very brief and occurs early in gametogenesis, whereas mid- to late gametogenesis is exclusively devoted to the production of oocytes. There is no facility for hermaphrodite-to-hermaphrodite gamete transfer, so mating occurs only between hermaphrodites and males. For these, and several other reasons, it is convenient to think of the hermaphrodite as a female with the added ability to make a few sperm. For example, there are other *Caenorhabditis* species

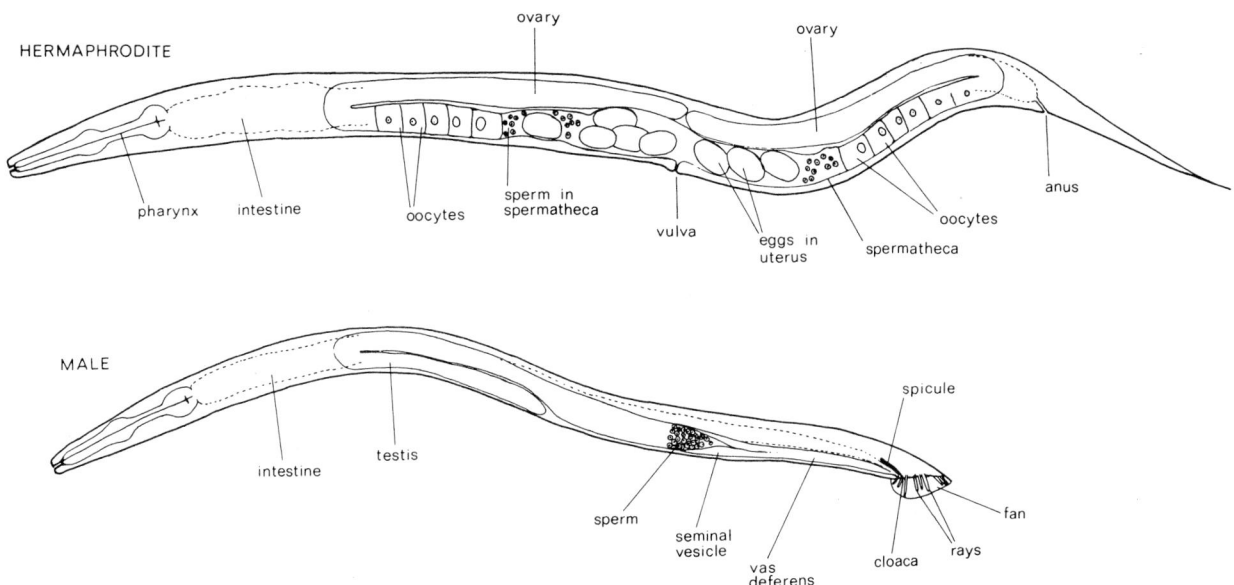

FIGURE 1 Schematic anatomy of *C. elegans* hermaphrodite and male. The actual size of the hermaphrodite is 1 mm in length, and it consists of approximately 1000 somatic cells (reproduced with permission from Wood, 1988).

which look extremely similar to *C. elegans* but exist as males and true females. Also, there are several mutations which cause the hermaphrodites to lose the ability to make sperm (see below), yet they still produce oocytes and are fertile when mated to a male.

Sex is chromosomally determined in *C. elegans*, which is normally diploid. Hermaphrodites carry two X chromosomes (XX) and males only one (XO); there is no Y chromosome. Each haploid gamete produced by a hermaphrodite carries an X chromosome; therefore, the progeny from a hermaphrodite self-fertilization are exclusively XX hermaphrodites. Rarely (~0.1%), a hermaphrodite will produce a nullo-X gamete as a result of X chromosome nondisjunction, allowing the production of a male. Certain mutations (the *him* mutants, for example) have an increased rate of spontaneous X chromosome nondisjunction during meiosis, resulting in spontaneous production of 20–35% male progeny following hermaphrodite self-fertilization. Males normally produce X-containing and nullo-X gametes in equal proportions; therefore, a male–hermaphrodite mating will produce outcross progeny consisting of 50% males. Sperm are transferred to the hermaphrodite a few at a time during mating (Ward and Carrel, 1979). Males are rarely found in the wild, so it is likely that most *C. elegans* in their natural environment arise due to self-fertilization by their mothers.

A. Reproduction

An unusual feature of *C. elegans* sexual reproduction is that male sperm is used preferentially over hermaphrodite sperm. Although sperm produced by hermaphrodites are morphologically identical to those produced by males, when a hermaphrodite mates with a male, the male sperm are used preferentially. Thus, following mating, all progeny produced result from the incoming male sperm, until such sperm have been used up. At that time, the hermaphrodite will return to using her own sperm for oocyte fertilization (Ward and Carrel, 1979). Another unusual feature is that a hermaphrodite will make only about 300 sperm but more than 1000 oocyte nuclei. During self-fertilization, sperm are used with nearly 100% efficiency, resulting in a brood size of about 300 self-progeny (Ward and Carrel, 1979). Therefore, *C. elegans* is a rather unique system in which the brood size is limited by sperm production rather than by oocyte production.

In the wild, contact of a hermaphrodite with a male is thought to occur rather rarely. This mode of reproduction seems perfectly suited to its environment. One may naively think that increasing sperm production in the hermaphrodite would increase brood size, thereby increasing the contribution of genes to the next generation and thus providing a selective advantage. However, since both sperm and oocytes are produced in the same gonad and are alternative developmental fates of the same germ cell precursors, increasing the production of sperm would delay the beginning of oogenesis. This would increase the "egg-to-egg" generation time. A mutation that increases sperm production has been isolated in the laboratory and has been shown to be less successful in competition with the wild-type strategy (Hodgkin and Barnes, 1991). Thus, the production of sperm in the hermaphrodite seems to have struck a balance between maximum number of progeny per generation and the shortest possible generation time.

B. Sexual Dimorphism

The two sexes of *C. elegans* are highly dimorphic: Not only are they different in size, shape, and cell number but also a large fraction of the cells of the adult have different fates in males and hermaphrodites. Only 60–70% of the cells in the adult have the same lineage, position, and function in the two sexes. The majority of these differences arise postembryonically because the only difference between male and hermaphrodite embryonic development (besides the differences in the rates of X chromosome transcription described below) is a pair of sex-specific cell deaths. However, larval development differs significantly in the two sexes, and most of the cells generated during this time lead to the production of sexually specialized structures such as the somatic gonad and the male tail.

The sexual differences extend through most of the animal and include all three tissue layers. The intestine is morphologically similar in both sexes but is the site of yolk protein synthesis only in hermaphrodites. The mesodermal differences include sex-specific muscles (for egg laying in hermaphrodite and tail and copulation muscles in male) and nervous system differences [12 neurons are unique to hermaphrodites, such as hermaphrodite-specific neurons (HSNs) controlling egg laying, whereas more than a fifth of the neurons in males are sexually specialized]. These nervous system differences are directly related to the sex-specific behaviors which are discussed later.

The hypodermis is specialized in the hermaphrodite to form the vulva, the opening through which copulation with the male occurs and the eggs are laid. The most conspicuous difference between the sexes is the male tail, which has been modified to aid in copulation with the hermaphrodite. The tail consists of a fan equipped with 18 sensory rays, a cloaca with its own set of sensilla, and a proctodeum containing two spikes or spicules which can be extended into the hermaphrodite vulva during copulation to aid in proper contact while sperm are transferred.

V. DEVELOPMENT

A. Gonad Development

The gonads in the adult stages of the two sexes are quite distinct in their shape. In the hermaphrodite, the gonad is symmetric and bilobed, with the proximal arms meeting at the vulva in the ventral midbody (Fig. 1). Each arm is reflexed such that the distal end is located dorsally to the vulva. The distal arm of the gonad consists of a syncytium of immature germ cell nuclei and contains at its distal end a single cell (distal tip cell) which is necessary for mitotic division of the nearby germ cells. At the anterior and posterior loops of the symmetrical arms, cell membranes begin to extend to enclose the nuclei, and oogenesis begins. The oocytes enlarge as they move down the oviduct toward the vulva and arrest in diakinesis. The proximal end of the oviduct opens into the spermatheca, where the sperm made by the hermaphrodite or transferred from the male are stored. Oocytes are fertilized as they pass through the spermatheca. The spermatheca is separated from the uterus by a valve, which regulates passage of the egg. The uterus connects the two proximal arms of the gonad with the vulva and serves to store the developing embryos until they are laid.

In the male, the gonad has only one arm, opening through the cloaca at the tail (Fig. 1). The distal end of the male gonad is partially reflexed so that it ends in the same relative position as the distal end of the hermaphrodite gonad arms. The single armed male gonad contains two distal tip cells. As in the hermaphrodite, the distal tip cells provide signals to maintain the nearby germ cells in mitosis. The germ cells enter meiosis as they progress down the gonad. The spermatocytes mature to spermatids which are stored in the seminal vesicle.

The gonad arises from the same embryonic precursor cells in the two sexes. In the L1 stage of both males and hermaphrodites, four cells (Z1–Z4) are found linearly arranged in the ventral midbody region. The two outer cells (Z1 and Z4) are the precursor cells of the somatic gonad, whereas the two inner cells (Z2 and Z3) are the germline precursors. Subsequent divisions of the Z2 and Z3 cells do not follow a fixed lineage, but they proliferate during larval growth to between 500 (males) and 1000 (hermaphrodites) germline nuclei. The somatic gonad, on the other hand, follows a precise lineage in each of the two sexes. In the hermaphrodite, the two arms of the ovotestes elongate anteriorly and posteriorly within the ventral part of the pseudocoelom up until the third larval stage. The growth is led by the distal tip cells. In L4, each gonad arm turns dorsally and then grows back toward the midbody region, forming the U-shape of the adult. In the male, only one arm is formed, and the testis elongates anteriorly in the ventral pseudocoelom, then dorsally, posteriorly, ventrally, and posteriorly to join up with the cloaca in the tail. The final shape of the male gonad is similar to a sideways question mark. In contrast to the hermaphrodite, the growth of the male gonad is led by the linker cell, whereas the distal tip cells remain at the nongrowing end of the gonad.

B. Germline Development

As in many other systems, the *C. elegans* germline is set aside early in embryonic development. The P_4 cell, which is the clonal precursor of the germline, arises from the smaller posterior daughter of each of four asymmetric cell divisions from the zygote (P_0). In each division, cytoplasmic particles called germline granules or P-granules are specifically segregated to the germline blastomere (P_1–P_4). The molecular composition of the P-granules, and their mechanism of asymmetric segregation, remains an area of speculation. The P_4 cell divides once during embryogenesis to give two cells, Z2 and Z3, both of which are germline precursors. P-granules are segregated equally to all germline progeny from the P_4 division onward. By experimental manipulation and genetic analysis, it appears that the presence of P-granules is not sufficient to specify a germline fate, but they seem to be necessary because missegregation of P-granules results in sterility.

As mentioned, the postembryonic germ cell precursors divide mitotically in a syncytium in a lineage-independent manner. The earliest meiotic nuclei are detected in the L3 and L4 stages in both sexes. The distal tip cells regulate germline proliferation and maintain nearby germ cells in a mitotic state throughout the life span (Kimble and White, 1981). As the gonad grows, germ cells in the proximal arm find themselves further and further from the distal tip cells, enter meiosis, and begin gametogenesis. Gamete maturation depends on the position of the germ nucleus in the gonad with respect to the distal tip cells, and all stages of germ cells can be found in a single gonad arm. Killing of the distal tip cell can be conveniently accomplished in living animals using a laser microbeam. When such surgery is performed, all germ cells arrest mitosis and begin meiosis (Kimble and White, 1981).

In the hermaphrodite, both sperm and oocytes are produced in the same gonad arm, from the same pool of germ nuclei, but the two processes are separated temporally. The first germ cells to mature in the L4 hermaphrodite become sperm. About 150 sperm are produced in each arm and are stored in the spermatheca. At about the time of the L4 to adult molt, the germline switches permanently to the production of oocytes: The adult XX animal is functionally a female. In normal circumstances, the hermaphrodite produces several hundred further germ cells. In the absence of a male, the hermaphrodite uses her own sperm with 100% efficiency (300 sperm give ~300 progeny). When the sperm are depleted, oogenesis is arrested. However, oogenesis can be stimulated again by mating with a male (Ward and Carrel, 1979). If a hermaphrodite is mated with several males so that sperm does not become limiting, she is capable of producing over 1400 progeny (Hodgkin, 1986). A single male, on the other hand, has been recorded siring over 2500 progeny (Hodgkin, 1983a).

Spermatogenesis processes are similar in the two sexes, despite the differences in the somatic gonads. Caenorhabditis elegans spermatozoa are pseudopodal, not flagellated. There is no apparent morphological difference between spermatozoa produced by males or hermaphrodites. However, some difference must exist because male sperm are used preferentially over hermaphrodite sperm following mating (Ward and Carrel, 1979). The hermaphrodite sperm are not lost or permanently inactivated since once the male sperm are used up (they are only transferred in small numbers to the hermaphrodite with each copulation), the hermaphrodite can revert to using her own sperm for the production of self-progeny.

Spermatids stored in the male before copulation are sessile, with no pseudopodia, and mature to spermatozoa with pseudopod formation and motility following deposition in the hermaphrodite uterus. Spermiogenesis can also be mimicked *in vitro* by a variety of chemical treatments, suggesting that activation is a chemical process due to the change in local environment. A substance in the vas deferens secretion may be responsible *in vivo*.

The spermatogenesis pathway in *C. elegans* has been well characterized both morphologically following observation under the light microscope and by the analysis of a plethora of mutants (such as the *fer* and *spe* genes) which block the production of functional sperm. These mutants block many different stages of development, from the earliest stages of spermatocyte production to spermatid and spermatozoa production, and have been recently reviewed elsewhere (L'Hernault, 1997).

VI. SEX DETERMINATION AND DOSAGE COMPENSATION

A. Sex Determination

Typically, hermaphrodites develop from embryos which inherit a single X chromosome from each gamete, whereas males develop when one of the gametes lacks an X chromosome. It is not strictly the number of X chromosomes that determines sex. Rather, it is the ratio of X chromosomes to sets of autosomes (X/A ratio) that is important (Madl and Herman, 1979). *Caenorhabditis elegans* are normally diploid, with two sets of each of the five autosomal chromosomes. However, two X chromosomes are not sufficient for female development because in tetraploid animals, males develop from embryos carrying two X chromosomes. A ratio of 0.5 directs male development, whereas a ratio of 1 results in hermaphrodites. This system is similar to that seen in *Drosophila*, although *Drosophila* males require a Y chromosome for fertility.

The molecular nature of the X/A ratio has not been completely established. There appear to be a limited number of sites on the X chromosome which act as "numerator" elements. These were initially recognized due to the partial sexual transformations or sex-specific lethality caused by duplications of different sequences of the X chromosome. Recently, specific genes have been associated with at least some of these regions (Meyer, 1997). The nature of the autosomal "denominator" element in the X/A ratio is completely mysterious.

Downstream of the X/A signal, the cellular mechanism of *C. elegans* sex determination has been largely worked out as a result of the molecular characterization of genes whose mutant phenotypes alter normal sex determination (Fig. 2). This system involves several sets of genes regulating tissue-specific and temporal dimorphism and has recently been well reviewed elsewhere (Meyer, 1997). A brief but necessarily superficial overview follows.

Among the earliest mutants isolated were those resulting in viable animals which ignored the X/A ratio. These included the *tra* mutants (transformer), where XX and XO animals develop as males; *her-1*, where XO animals transformed into hermaphrodites while XX animals are unaffected; and the *fem* mutants (feminization), where XX and XO animals are transformed into females (both types of animals have lost the ability to make sperm). Recently, several classes of mutations have been identified which lie in genes acting early in the sex determination pathway and affect both sex determination and the related process of X chromosome dosage compensation (see below). These genes include the *sdc* (sex determination and dosage compensation) and *xol-1* (XO-lethal) genes.

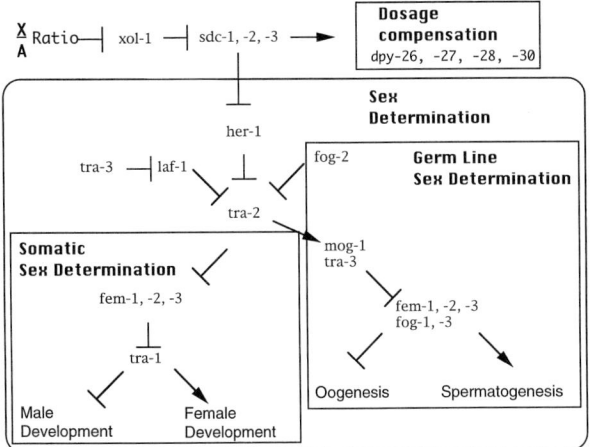

FIGURE 2 Genetic pathway for somatic and germline sex determination and X chromosome dosage compensation in C. elegans. The figure is simplified and is intended only to convey the general nature of the regulatory hierarchy, although molecular analysis of the genes involved has so far supported this order of action. Bars indicate negative regulation (repression), whereas arrows indicate that the next process is activated by the genetic activity.

A combination of genetic and molecular analysis of these genes has led to an understanding of the control of somatic and germline sex determination. From a genetic perspective, the pathway can be generalized as sequential negative interactions, where the activity of a gene at one point causes a reduction in activity at the next step. A high X/A ratio (XX) turns off the *xol-1* activity, allowing the *sdc* genes to be active. This represses *her-1*, allowing *tra-2* to repress the *fem* genes and *tra-1* to remain active. In males (low X/A ratio), *xol-1* is on, turning off the *sdc* genes. *her-1* is active, inhibiting *tra-2* and allowing the *fem* genes to repress *tra-1*. In the soma, the state of *tra-1* is the determining factor for sexual differentiation: *tra-1* activity is necessary and sufficient for female development. The *tra-1* gene encodes a transcription factor with zinc finger domains, consistent with a role as a terminal regulator of sex determination, although no direct targets of the TRA-1 protein have yet been identified.

In contrast to other systems such as *Drosophila*, sex determination in C. elegans seems not to be completely cell autonomous. Although the action of a transcription factor such as TRA-1 must be cell autonomous in the strictest sense, the regulation of *tra-1* activity can be affected in a nonautonomous manner. This has been seen with genetic mosaics in several genes, but most clearly with *her-1* (Hunter and Wood, 1992). *her-1* has been shown to encode a secreted protein, HER-1, and molecular analysis of the *tra-2* gene has predicted that TRA-2 encodes a membrane receptor for HER-1. The *fem* genes encode proteins predicted to act cell autonomously downstream of the TRA-2 receptor. The only one of the *fem* genes with a demonstrated biochemical activity is *fem-2*, which is a type 2C Ser/Thr protein phosphatase. Since no kinase has yet been found to have an effect on sex determination, it is clear that there are elements of the control mechanism which have yet to be elucidated.

In the germline, a slightly different pathway is in effect, where the *fem* genes seem to play a more direct role in directing spermatogenesis (Schedl, 1997). In addition, germline-specific activities such as the *fog* (feminization of germline) and *mog* (masculinization of germline) genes are involved. It was not unexpected to find germline-specific sexual regulatory genes, given that the hermaphrodite mode of reproduction produces both types of gametes. In XX animals genes controlling female germline development must be kept turned off while sperm are produced, then rapidly induced to promote oogenesis. This temporal control is thought to be exerted through the *tra-2* gene.

B. X Chromosome Dosage Compensation

Equalization of gene expression is a common problem among animals with differing numbers of sex chromosomes between the two sexes. In contrast to *Drosophila*, which increases transcription from the X in males, C. elegans reduces by half the X transcription in the hermaphrodite (Meyer and Casson, 1986). This control is dependent on a set of genes referred to as the "dosage compensation dumpies," named because of their mutant phenotype of short, fat, and dumpy animals. The first steps of X chromosome dosage compensation are controlled by genes common to the sex determination pathway, namely, *xol-1* and the three *sdc* genes. Genes downstream of *xol-1* and the *sdc*'s are specific to either sex determination or dosage compensation.

dpy-26 was the first dosage compensation dumpy gene to be identified. The XX self-progeny of *dpy-26* homozygous animals are dumpy and inviable, whereas the XO progeny are viable and nondumpy (Hodgkin, 1983b). A similar dumpy phenotype is seen in otherwise wild-type diploid animals which carry a third X chromosome. A fourth X chromosome in diploid animals causes lethality. Hence, it was proposed that mutations in the *dpy-26* gene cause an overexpression of X-linked genes. Mutations in *dpy-26*, as well as *dpy-21, -27, -28*, and *-30*, caused mutant phenotypes, which are most serious in XX animals. The products of the *dpy-26, -27*, and *-28* genes appear to differentially bind to the X chromosomes of XX animals, consistent with a role in X chromosome dosage compensation (reviewed in Cline and Meyer, 1996).

VII. BEHAVIOR

It may be initially surprising to learn that an animal with as simple a nervous system as *C. elegans* can have relatively sophisticated behaviors. However, in addition to feeding, chemotaxis, thermotaxis, osmotaxis, and dauer larvae formation behaviors, which are displayed by both males and hermaphrodites (and will not be discussed here), there are complicated but reproducible sex-specific behaviors (Chalfie and White, 1988).

A. Male Mating Behavior

The male seems to have the ability to detect the presence of hermaphrodites by chemosensation; however, the most distinctive aspect of male movement is the use of the tail and associated sensory structures to "sample" the surroundings. Contact of the tail with another object (hermaphrodite, other male, or itself) causes the male to move backwards with the bursa of the tail extended and pressing against the other animal. This backward movement is coupled with a coiling motion as the male searches for a hermaphrodite vulva. The male can be quite persistent, often completely circling the target animal several times. The contact of a male with another animal can be quite prolonged until a vulva is found, the other animal moves off, or the male gives up. Often several males can be found grouped in such a mutual coil even in the absence of any hermaphrodite. Once a vulva has been located by the sensory apparatus of the bursa, the spicules are extended, which locks the male in place and opens the vulva (Emmons and Sternberg, 1997). Sperm are then transferred (a process that takes less than a minute), and the male moves away. The hermaphrodite seems to take scant notice of this male behavior and does not respond in any noticeable manner throughout this episode. Hermaphrodites which have been completely paralyzed by mutation in any of a number of genes affecting movement can mate with high efficiency, indicating that no hermaphrodite responses are required for successful copulation. However, there seems to be some signal in the hermaphrodite by which the male recognizes the proper connection with the uterus. Mutations which affect chemosensation show a reduced efficiency of male mating behavior (Hodgkin, 1983a), demonstrating the necessity of both chemical and physical cues which allow the male to find the hermaphrodite vulva.

B. Hermaphrodite Egg-Laying Behavior

Egg laying is controlled by sex-specific muscles found in the hermaphrodite vulva and uterus. These muscles in turn form synapses with two HSN cells and six ventral cord motorneurons (VC) cells. Laser microbeam ablation of the HSN and VC neurons has demonstrated that only the HSNs are required for egg laying.

The rate of egg laying is controlled by food supply. In the absence of food, hermaphrodite egg laying is suppressed, and the eggs are retained (Trent, 1982). This is thought to be designed to allow the eggs to hatch in as favorable an environment as possible. The change in behavior is very rapid and is seen almost immediately when food is removed. It does not appear that the rate of fertilization of eggs within the gonad is reduced in the absence of food. If deprived of food for extended periods of time (as when a gravid hermaphrodite is placed on an agar plate in the absence of bacteria), some eggs are eventually laid and appear to be forced out to some extent by the pressure of newly fertilized eggs continuously entering the uterus. Eggs which remain in the uterus continue to develop and can hatch within the parent. This is seen most dramatically in hermaphrodites in which normal vulval development has been blocked by laser surgery or mutation. In these animals, embryos are fertilized normally using self-sperm, develop, and hatch within the uterus, where they proceed to devour the mother from within. This is the characteristic "bag-of-worms" phenotype which has been used successfully to identify and characterize mutations involved in the process of vulval, sex muscle, and sex neuron development (Trent et al., 1983).

Due to the nature of this review, only a limited bibliography is provided. It is necessarily incomplete and does not begin to give credit to the large community of *C. elegans* researchers whose contributions are mentioned without citation. For a more complete bibliography, interested readers are directed to the two monographs edited by Wood (1988) and Riddle et al. (1997). In addition, Leon Avery has established and maintains a comprehensive central web site for *C. elegans* information (*http://eatworms.swmed.edu*) which includes a searchable archive of *C. elegans* references.

Bibliography

Chalfie, M., and White, J. (1988). The nervous system. In *The Nematode Caenorhabditis elegans* (W. B. Wood, Ed.), pp. 337–391. Cold Spring Harbor Laboratory Press, Cold Spring Harbor, NY.

Cline, T. W., and Meyer, B. J. (1996). Vive la difference: Males vs females in flies vs worms. *Annu. Rev. Genet.* 30, 637–702.

Emmons, S. W., and Sternberg, P. W. (1997). Male development and mating behavior. In *C. elegans II* (D. L. Riddle, T. Blumenthal, B. J. Meyer, and J. R. Priess, Eds.). Cold Spring Harbor Laboratory Press, Cold Spring Harbor, NY.

Hodgkin, J. (1983a). Male phenotypes and mating efficiency in *Caenorhabditis elegans*. *Genetics* 103, 43–64.

Hodgkin, J. (1983b). X chromosome dosage and gene expression in *Caenorhabditis elegans*: Two unusual dumpy genes. *Mol. Gen. Genet.* 192, 452–458.

Hodgkin, J. (1986). Sex determination in the nematode *Caenorhabditis elegans*: Analysis of *tra-3* suppressors and characterization of the *fem* genes. *Genetics* 114, 15–52.

Hodgkin, J., and Barnes, T. M. (1991). More is not better: Brood size and population growth in a self-fertilizing nematode. *Proc. R. Soc. London B Biol. Sci.* 246, 19–24.

Hunter, C. P., and Wood, W. B. (1992). Evidence from mosaic analysis of the masculinizing gene *her-1* for cell interactions in *C. elegans* sex determination. *Nature* 355, 551–555.

Kimble, J., and White, J. G. (1981). On the control of germ cell development in *Caenorhabditis elegans*. *Dev. Biol.* 81, 208–219.

L'Hernault, S. W. (1997). Spermatogenesis. In *C. elegans II* (D. L. Riddle, T. Blumenthal, B. J. Meyer, and J. R. Priess, Eds.). Cold Spring Harbor Laboratory Press, Cold Spring Harbor, NY.

Madl, J. E., and Herman, R. K. (1979). Polyploids and sex determination in *Caenorhabditis elegans*. *Genetics* 93, 393–402.

Meyer, B. J. (1997). Sex determination and X chromosome dosage compensation. In *C. elegans II* (D. L. Riddle, T. Blumenthal, B. J. Meyer, and J. R. Priess, Eds.). Cold Spring Harbor Laboratory Press, Cold Spring Harbor, NY.

Meyer, B. J., and Casson, L. P. (1986). *Caenorhabditis elegans* compensates for the difference in X chromosome dosage between the sexes by regulating transcript levels. *Cell* 47, 871–881.

Riddle, D. L., Blumenthal, T., Meyer, B. J., and Priess, J. R. (1997). *C. elegans* II. Cold Spring Harbor Laboratory Press, Cold Spring Harbor, NY.

Schedl, T. (1997). Developmental genetics of the germ line. In *C. elegans II* (D. L. Riddle, T. Blumenthal, B. J. Meyer, and J. R. Priess, Eds.). Cold Spring Harbor Laboratory Press, Cold Spring Harbor, NY.

Sulston, J. (1988). Cell lineage. In *The Nematode Caenorhabditis elegans* (W. B. Wood, Ed.). Cold Spring Harbor Laboratory Press, Cold Spring Harbor, N. Y.

Trent, C. (1982). Genetic and behaviour studies of the egg-laying system of *Caenorhabditis elegans*. PhD thesis, Massachusetts Institute of Technology, Cambridge..

Trent, C., Tsung, N., and Horvitz, H. R. (1983). Egg-laying defective mutants of the nematode *Caenorhabditis elegans*. *Genetics* 104, 619–647.

Ward, S., and Carrel, J. S. (1979). Fertilization and sperm competition in the nematode *Caenorhabditis elegans*. *Dev. Biol.* 73, 304–321.

Wood, W. B. (1988). *The Nematode Caenorhabditis elegans*. Cold Spring Harbor Laboratory Press, Cold Spring Harbor, NY.

Canidae

see Dogs

Captive Breeding of Wildlife

Barbara S. Durrant
Zoological Society of San Diego

I. Historical Perspective
II. Natural Breeding
III. Evaluation of Reproductive Potential
IV. Assisted Reproduction
V. Contraception
VI. Conclusion

GLOSSARY

assisted reproductive technology Techniques developed to enhance reproduction in subfertile or geographically distant individuals.

ex situ Outside the native habitat.

founder A wild-born individual assumed to possess maximum genetic diversity.

herptile An artificial taxon composed of reptiles and amphibians.

in situ Within the native habitat.

in vitro fertilization The coincubation of capacitated sperm and mature ova to achieve fertilization outside the body.

proximate factor An immediate cause for the appearance of a phenomenon.

ultimate factor An evolutionary reason for the development of a phenomenon.

Captive breeding of wildlife comprises the propagation of endangered, threatened, or rare animal species in zoos, wildlife parks, and private collections. Once merely the occasional serendipitous result of cohabitation, captive breeding has now become the sole prospect for the continued existence of many species. Multidisciplinary investigations and the application of advanced technologies to the challenges of *ex situ* reproduction are imperative for the preservation of our faunal heritage.

I. HISTORICAL PERSPECTIVE

Ice Age hunters were probably the first to domesticate a wild species, using dogs as retrievers, herders, and guards. Since about 12,000 years ago, it has been a facet of human civilization to capture and utilize wild animals. With the advent of organized agriculture, certain wildlife species were domesticated and subjected to artificial selection to increase their usefulness to humans. Those species remaining in the wild then became objects of curiosity. Private collections of exotic species were common and Alexander the Great established the first public zoo in the third century BC.

The practice of exhibiting rare animals for their entertainment value exemplified human fascination with different life forms while simultaneously demonstrating a disregard for the quality of life of captive individuals and the survival of their species. The natural history of the species held in captivity was unknown or ignored; thus, the animals' environmental, dietary, and social needs were neglected and most individuals died without leaving progeny. Wild-caught replacements were imported, often at a very high cost to the free-ranging group from which they were taken (e.g., the practice of taking a single infant gorilla usually involved killing all or a large portion of its family).

The legitimate zoo appeared in the late 1700s, during a time of intense biological curiosity. Systematic naming of species and anatomical studies of "trophy" animals brought back from the colonies flourished until the mid-1800s. Then, for the next 100 years, scientific interest in zoo collections waned. The primary benefit of this "spectacle phase" was the education of the public.

In the past 50 years the world has witnessed a tremendous evolution of the animal exhibitry, politics, and philosophy of zoos. The breakup of European empires after World War II closed borders and restricted the exchange of animals between zoos. The end of colonial rule in many African countries saw the end of exploitive rates of animal exportation from native habitats. These factors forced a realization that zoos must become breeding facilities if only to ensure continued exhibit stock. During this time of zoo restructuring, the academic world made some startling discoveries; chief among them was the fact that the natural world was far less resilient to the by-products of the human population explosion than had been previously acknowledged.

Thus, with little intent or foresight, zoological institutions found themselves particularly well positioned by the 1960s to accept a new and interesting challenge—the preservation of endangered species in captivity. Today's zoos cooperate in a three-pronged conservation strategy comprising captive breeding of species rescued from hostile environments, preservation and/or restoration of natural habitats, and reintroduction of captive-bred animals into secure native environments.

II. NATURAL BREEDING

Stimulation of natural breeding in captivity is arguably the most important long-term goal of zoological institutions. Although assisted reproductive technology (ART) may be occasionally necessary to ensure the genetic contribution of a valuable individual, routine reliance on artificial reproduction is neither feasible nor desirable. Success rates of ART are significantly lower than those of natural breeding, and the financial burden of the specialized equipment and personnel needed for ART is beyond the means of most zoos. The most compelling argument in favor of natural breeding is the fact that species losing the ability to reproduce without human assistance will not survive reintroduction into their native habitats.

A. Environment

The failure of many species to reproduce in captivity results from our inability to re-create the integral elements of the natural environment. Detailed knowledge of the habitats in which species flourish provides essential information on the ultimate and proximate factors impinging on gametogenesis, breeding, pregnancy maintenance, and infant rearing.

Humidity, temperature, and photoperiod are among the key proximate factors affecting reproduction. Each can trigger the initiation or the cessation of reproductive efforts in diverse species. For example, humidity signals reproductive seasons in certain invertebrates, and specific humidity levels are necessary for eggs to hatch. Thus, for some species, a captive environment with constant humidity could preclude reproduction even though it may appear to be suitable for maintaining other basic life functions.

The sex of at least two herptile species is determined by the temperature at which eggs are incubated (interestingly, these two species exhibit opposite responses to high incubation temperatures). Artificial incubation conditions, therefore, must be carefully controlled but could be employed to correct sex ratio imbalances in captive populations. Certain fish and most reptiles require thermoclines for self-regulation of body temperature that is essential for timing gonadal activity. Species-specific requirements must be known and implemented for maximum reproductive efficiency.

Although light does not appear to be a critical proximate factor in herptile reproduction, the sexual seasons of many avian and mammalian species are regulated by photoperiod. Thus, long-day (some avian species) and short-day (certain deer species) breeders may require artificially extended or shortened daylight cycles, respectively, for optimum reproduction in captivity. In addition to the effects of photoperiod and photointensity, inappropriate phototransitions (i.e., sudden rather than gradual transitions between light and dark) have been shown to significantly impact the physiological state of certain fishes and may affect other species as well.

Other facets of captive environments affecting reproductive success include three-dimensional space, substrates, and furnishings. Overcrowding in herptile enclosures can cause the social system to shift from primarily territorial to more hierarchical, resulting in increased fighting, injury, and death. Adequate space replete with suitable hiding places is necessary for all potentially aggressive species as well as for those that conceal their young during the neonatal period. Without sufficient denning facilities, many mammalian species experience high rates of infant mortality due to exposure or excessive translocation by the dam. Likewise, burrowing, arboreal, oviparous, and aquatic species may experience diminished reproductive success without appropriate substrates (soil, trees, sand, and vegetation, respectively) in which to nest or lay eggs.

Loud, sudden noises have been reported to increase stress as depicted by elevated corticosteroid levels. The effect of the acoustical environment on reproduction is often noted by animal care personnel or researchers, but empirically derived data are scant. Discovery and accommodation of these and other environmental influences on reproduction will no doubt result in markedly improved rates of captive reproduction.

B. Nutrition

From invertebrates to great apes, it is tacitly and empirically understood that the nutritional state of an individual affects its ability to reproduce. However, the complex issues of how larval/prepubertal nutrition may augment or diminish an animal's reproductive potential are unexplored except for a small number of insect species. Food availability is the ultimate factor driving the evolution of nutritional state as a proximate factor in controlling seasonal reproduction. Insightful design of dietary programs aimed at optimal propagation rates will consider both ultimate and proximate nutritional factors.

Maintaining a high plane of nutrition year-round for their charges is the goal of most animal care personnel. However, there is a growing body of compelling evidence against this scheme, which may override the proximate triggers of reproductive cyclicity. For example, although well-nourished captive individuals may produce more and larger offspring than their free-ranging counterparts, this procreational burden can prove to be detrimental to both neonates and their dams. A case in point is the demonstration of higher rates of infant and maternal mortality in golden lion tamarins stimulated to overproduction by continuous feeding of highly nutritious diets. Seasonal cycles of increasing and decreasing planes of nutrition may serve to give females periods of rest between litters, limit the size and number of offspring, and stimulate the ebb and flow of sexual fitness that characterizes normally reproducing wild populations.

C. Social Needs

Other than the obvious predator/prey relationships, few noninsect species rely on interspecific interaction for survival or reproduction. However, it is not known how replacement of hunting with provisioned food may influence captive propagation of carnivores. The nutritional benefits of whole animal consumption for predators notwithstanding, the social cooperation and salubrity fostered by the act of hunting is likely to contribute substantially to the overall reproductive fitness of the group. By the same token, species which normally spend most of the day foraging may suffer socially, physically, and psychologically if allowed only discrete feeding periods.

An abundant literature describes the fundamental importance of appropriately balanced social groups to encourage natural reproduction. Age structure, number of group members, and sex ratio determine the extent of reproductive success in virtually all species. Few zoos are capable of allocating sufficient space to house socially appropriate groups, especially for megavertebrates such as elephants and gorillas. Extremely high-density populations (e.g., cave-dwelling bats) are avoided in captivity due to the difficulties in maintaining levels of cleanliness acceptable to caretakers and the public. Equally disruptive to propagation efforts is the pairing or grouping of normally solitary animals. The economic hardship of housing animals individually combined with the public's desire to see "families" in every exhibit may lead to reproductive suppression or stress levels incompatible with breeding.

To various extents, visual, auditory, and olfactory conspecific communications are integral parts of all mating strategies. While housing animals near or with each other normally allows for adequate optical and aural stimulation, the practice of removing feces, urine, and scent marks during daily enclosure cleaning reduces or even eliminates pheromonal communication. Species as disparate as rhinoceroses and lizards frequently investigate fecal piles to identify individual conspecifics and determine their hormonal status, whereas other forms rely on scent marks to signal the onset of the breeding season. With proper education the public can be taught to accept, even appreciate, the distinctive odors emanating from animal enclosures.

Proximity to humans is tolerated to various degrees by different types of species. Most herptiles, for example, do not solicit or even habituate to handling. It is not uncommon for wild-caught captive animals to injure themselves in attempts to flee from humans. Early rearing experience greatly alters flight responses and injury is much less common in captive-born animals. Nevertheless, allowance for adequate flight distances and hiding places can ameliorate the effects of interspecies proximity, thereby reducing stress, injury, or even death. Novel viewing schemes which feature visitors behind sight barriers such as vegetation or fenestrated walls may create the illusion of distance needed to allay the fears of shy species. Serious consideration must be given to housing especially vulnerable species off-exhibit to reduce the stress of public display.

The importance of adequate rearing environments must be understood for all species but especially for those in which learning is critical. For example, hand-reared primates often lack the social skills to properly interpret vocal or postural cues from conspecifics. Deprived individuals often compensate by ignoring potential mates, directing sexual energy to autostimulation, or exhibiting incompetence when attempting to mate. In addition to reproductive traits, other survival skills can be affected by deficient rearing. Hand-reared predators may retain the instinct to hunt but lack the coordination of behaviors necessary to make the kill. These examples demonstrate the importance of providing captive animals with the early learning environment that will shape their adult behaviors.

III. EVALUATION OF REPRODUCTIVE POTENTIAL

Accurate, instructive evaluation of reproductive potential comprises genetic, behavioral, and physiological analyses. Beginning with the least invasive procedures, data are compiled for the development of a comprehensive profile of the animal. Problems that can be solved without disrupting the target animal or its social group may be detected early in the evaluation, obviating the need for further, more invasive investigation.

A multidisciplinary triage scheme to evaluate reproductive potential in captive animals is outlined in Fig. 1. An individual, pair, or group is examined in a logical sequence to discover the cause of reproductive failure and its most efficacious remedy.

Noninvasive assessment of hormone levels and cycles may begin simultaneously with genetic or behavioral studies. Knowledge of the endocrine status of an animal gives depth to the interpretation of behavioral data and may illuminate the area(s) in most immediate need of physiological analysis.

A. Genetic Analysis

First, a complete karyotypic analysis is performed to uncover or rule out obvious genetic defects or incompatibilities. A single skin biopsy or blood sample is adequate to analyze an individual's chromosomal makeup. An interesting example is the dik-dik. Poor reproductive performance in North American

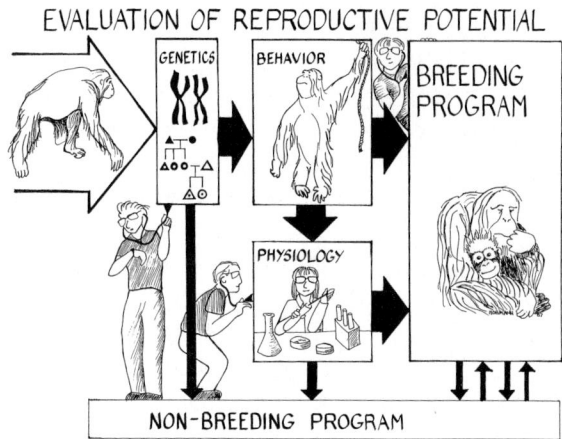

FIGURE 1 Evaluation of reproductive potential (drawn by Sue Hohmann).

captive groups of Kirk's (*Madoqua kirkii*) and Guenther's (*Madoqua guentheri*) dik-diks prompted the genetic screening of several key individuals. In addition to three distinct cytotypes (Guenther's and Kirk's A and B), a number of intercytotypical hybrids were described. Reproductive evaluation demonstrated the complete absence of mature sperm in the ejaculates and testicular tubules of hybrid males. The solution to this problem has been to pair dik-diks by cytotype, not phenotype, and to remove hybrid individuals from breeding groups. Referring to Fig. 1, proceeding no further than the first step, the problem was identified and rectified. Should genetic analysis fail to uncover the source of reproductive failure the evaluation strategy proceeds to the next stage.

B. Behavioral Assessment

Most behavioral deficits compromising natural mating in captive animals can be detected through systematic observation. The success of this part of the triage scheme depends on a thorough grasp of the normal behavior of the species. If data are not available from self-sustaining wild populations, adequate information may be obtained by the study of thriving captive groups of the same or similar species. The cheetah (*Acinonyx jubatus*) serves as a notable example. A long history of poor reproductive success in captive cheetahs has been ascribed to the species' lack of genetic diversity. However, despite poor semen quality and low copulation frequencies, captive cheetahs display normal fecundity when husbandry practices mimic free-range conditions in which females are solitary. For the cheetah, although genetic analysis has revealed extremely low rates of pleomorphism, natural breeding can be stimulated by restructuring the captive environment to accommodate natural behaviors. In the event that identified behavioral impediments to reproduction cannot be ameliorated, the last level in the triage scheme is activated.

C. Physiological Evaluation

Unproductive copulation suggests physiological disturbance. The differential diagnosis of physiological malfunction begins with the analysis of the most fundamental necessity, gamete production and transport.

1. Sperm

The ability to generate adequate numbers of normal sperm can be determined by semen collection and evaluation. A small proportion of males may be trained for artificial vagina or masturbatory semen collection. In addition to obviating the need for physical or chemical restraint, these methods provide ejaculates similar in quality to those produced during natural mating, thereby giving the most accurate data on volume, concentration, motility, and morphological abnormalities. The majority of males, however, are not sufficiently tractable for semen collection methods requiring human contact or proximity. Electroejaculation has been successfully employed in most mammalian species, but the variable nature of individual response limits its use as a single point-in-time semen evaluation. Failure of multiple collection attempts to produce sperm may necessitate the investigation of sperm production via testicular or epididymal biopsy. Normal biopsy findings indicate obstructed sperm transport which may be visualized by ultrasound or exploratory surgery. Correction of obstruction requires surgical intervention that may be outside the repertoire of many zoo veterinarians. The possibilities for restoration of normal sperm production in compromised males are limited at best.

Males incapable of producing normal sperm (e.g., the hybrid dik-dik discussed previously) necessarily become part of the nonbreeding population. Those producing poor quality sperm may contribute to the gene pool through manipulations such as "swim-ups" which separate normal from abnormal cells, increasing sperm concentration through the combination of several ejaculates, or the use of other assisted fertilization techniques.

2. Ova

Failure of mated females to produce offspring has myriad causes which may be formidably difficult to diagnose. In vertebrate species, there is no nonhormonal method analogous to semen collection that will stimulate the release of ova for evaluation. Ultrasound observation of ovarian events such as follicle growth, ovulation, and formation of a corpus luteum (CL) provides visual confirmation that ova have been released for fertilization. Unfortunately, few captive exotic females will tolerate transdermal or transrectal ultrasound without chemical restraint, which severely limits the frequency with which the technique can be utilized. Indirect evidence of follicular growth and CL formation is obtainable through measurement of circulating, urinary, fecal, or salivary levels of estrogen and progesterone metabolites, respectively. Perturbations of normal hormone patterns can often be corrected by treatment with exogenous hormones which act directly on the ovary to stimulate follicle growth [follicle-stimulating hormone (FSH)], ovulation, and CL formation [luteinizing hormone (LH)]. Alternatively, pulsatile or short-term continuous administration of gonadotropin-releasing hormone (GnRH) stimulates the pituitary to release FSH and LH which, in turn, activate the ovary.

Genital tract infection or adhesions may result in oviduct obstruction which may be diagnosed with hysterosalpingography, which involves the injection of radioopaque dye into the uterus and oviducts. As in the male, female reproductive tract obstructions can occasionally be resolved by arduous surgical procedures. If blocks to normal transport, in either sex, cannot be removed, extraction of gametes directly from the gonad offers an alternate opportunity for reproduction.

It is possible to harvest viable gametes for a proscribed period of time following the death of an animal. Epididymal extraction has been practiced for many years as a source of sperm from genetically valuable males, but studies have only recently been initiated to develop techniques to rescue ova from females postmortem. At birth a female's ovaries contain her life's supply of ova in the form of primordial follicles. Beginning prior to puberty and continuing throughout the reproductive life of the animal, cadres of primordial follicles are periodically stimulated to grow. However, the majority of ova undergo atresia (follicle death) before ovulation. Follicle isolation from whole ovaries permits the recovery of all remaining ova from a female, regardless of her age or reproductive status at the time of death. Early stage follicles can then be grown and matured *in vitro* in preparation for *in vitro* fertilization.

3. Embryonic Death

Preimplantation embryonic death is nearly impossible to detect except in species experiencing prolonged preimplantation periods such as equids, in which ultrasonography can be used to visualize the free-floating embryo. In other species, fertilization success or failure cannot be determined without removing ovulated ova from the female's reproductive tract for examination. The causes of preimplantation embryonic death are varied and include poor nutritional state, genetic anomalies of the zygote, and hormonal miscommunication between the embryo and the uterus which thwarts maternal recognition of pregnancy.

4. Pregnancy Interruption

Spontaneous abortion may occur at any stage but is most common in early gestation. Hormonal monitoring of pregnant females may reveal a common cause of early miscarriage—luteal insufficiency. This condition is characterized by failure of the CL to maintain adequate progesterone output and can be treated with exogenous progestins. Certain disease states are known to cause pregnancy interruption, as is physical or social trauma to the dam, and genetic anomalies of the fetus. Analysis of the aborted fetus or placenta may reveal chromosomal defects known to be lethal.

D. Health and Nutrition

The appearance of the veterinarian at each step in the evaluation process (Fig. 1) represents the ongoing health assessment that is vital to the success of all captive breeding efforts. The most careful analyses of genetics, behavior, and physiology lose their diagnostic impact if the animal is suffering from nutritional or other health deficits. Coordinating medical examinations with sample collection for the reproductive evaluation protocol minimizes animal handling and offers the opportunity for multidisciplinary discussion.

IV. ASSISTED REPRODUCTION

When all reasonable attempts to set the stage for natural breeding do not yield the desired sequelae, artificial reproduction becomes necessary. The least invasive procedures are generally those with the smallest negative impact on the natural breeding process and, as such, are the methods of first choice for exotic species. Techniques designed to enhance reproduction in subfertile humans and genetically valuable domestic animals are being gradually incorporated into management schemes for exotic species. Just as their use in humans and food animals is restricted (by low success rates and considerable expense) to a minute portion of those populations, ART does not offer salvation for most endangered species. Judiciously employed, however, methods of artificial reproduction have the potential to augment the genetic diversity of captive groups by infusing genes from temporally or geographically distant captive and free-ranging populations.

A. Hormonal Stimulation of Natural Breeding

Hormonal manipulation of the estrous or menstrual cycle to encourage a normal array of ovarian, uterine, and behavioral events is preferable to the highly invasive techniques described later. For animals with normal gametes, *in vivo* oocyte maturation and sperm capacitation will invariably result in higher fertilization rates than *in vitro* methods.

As discussed previously, administration of FSH or GnRH may be adequate to stimulate growth of ovarian follicles, leading to estrogen levels sufficient to induce estrous behavior and mating. Alternatively, the creation of an artificially elevated progesterone level followed by withdrawal of exogenous progestin has been shown to trigger the secretion of FSH which begins the cascade of hormonal events preceding estrus and ovulation.

B. Artificial Insemination

Artificial insemination (AI) comprises the surgical or nonsurgical deposition of fresh, refrigerated, or frozen/thawed semen into the reproductive tract of a female. This most fundamental artificial reproductive technique circumvents behavioral impediments to natural mating as well as provides the means for genetic exchange between populations without transfer of live animals. In species capable of prolonged sperm storage (e.g., birds, herptiles, and some bats), the timing of AI may be somewhat imprecise because the exact moment of ovulation need not be known. Unfortunately, in most mammalian species, sperm longevity is characteristically brief in the female reproductive tract. AI must be performed close enough to the time of ovulation to allow sperm capacitation but prevent sperm and ovum aging. Synchronization of gamete convergence, then, becomes of paramount importance and is dependent on determination of the time of ovulation. Behavioral and hormonal monitoring of the estrous cycles of multiple females will establish correlates between estrous behavior and mating and rising progesterone levels (indicative of ovulation). Subsequently, observation of estrous behavior alone may be adequate to time AI. However, the effects of chemical immobilization or the stress of handling for AI may interfere with ovulation by ablating the LH surge, necessitating more concise delineation of the hormonal events surrounding ovulation.

Pregnancy rates following AI are highest with fresh semen, but use of refrigerated or frozen semen obviates the need to inseminate immediately after collection. Refrigeration extends the functional life span of sperm to accommodate storage until optimum insemination time has been determined or for transport to distant females. Cryopreservation of semen extends the interval between collection and insemination to years, decades, or perhaps even centuries.

C. Embryo Transfer

Females incapable of maintaining a pregnancy throughout gestation are candidates for embryo transfer. This technique also provides the means to dramatically increase the number of offspring that a single female is capable of producing. Normally, superovulation is induced with exogenous hormone therapy and multiple embryos are collected following natural or artificial insemination. Successful transfer of embryos requires precise natural or artificial synchronization of the estrous cycles of the donor and recipient.

Interspecies embryo transfer is an attractive option for captive breeding of wildlife. It offers the potential for rapid expansion of a population without the physiological expense of gestation and lactation. Successful to date in a mere handful of attempts, interspecies embryo transfer presents an array of complicating nuances. Chief among them are donor/recipient incompatibilities, such as response to estrus synchronization regimes, maternal recognition of pregnancy, implantation and placentation, milk composition, maternal care, and species recognition. Selection of pairs of species which naturally hybridize can reduce or eliminate the deleterious effects of many of these differences. However, the reproductive potential of the offspring produced must be considered when planning its rearing (by the recipient dam or humans) and introduction into a conspecific group. Hand-reared offspring or those with early imprinting on another species often experience social deficits that prevent their development into reproductively competent adults.

D. *In Vitro* Fertilization

Preceding *in vitro* fertilization (IVF), hormonal stimulation of the ovaries provides multiple preovulatory follicles from which ova are aspirated. The direct application of sperm to these ova *in vitro* increases the reproductive potential of males producing poor quality semen by reducing the numbers of sperm needed for fertilization. Embryo transfer (usually back to the egg donor) is performed after confirmation of sperm penetration or when embryos reach a specific developmental stage. IVF is a complex, minimally successful technique even in species with significant market-driven research such as cattle and humans. However, extrapolation of methodologies from those species to exotic bovids and nonhuman primates, respectively, offers some hope for individuals for whom other, less invasive assisted reproduction procedures have failed.

The related technique of gamete intrafallopian transfer (GIFT) eliminates the embryo transfer step of IVF by placing aspirated oocytes and sperm directly into the oviduct for *in vivo* fertilization. Pregnancy rates are higher with GIFT than with IVF (in humans), but the former does not allow observation of the fertilization process or the cryopreservation of excess embryos.

At yet another level of reproductive assistance, the IVF process may be enhanced by means of a number of relatively new micromanipulative techniques. Zona drilling is the mechanical or chemical fenestration of the zona pellucida to allow motility-impaired sperm easier access to the oocyte. Severely compromised sperm may be placed under the zona pellucida [subzonal sperm injection (SUZI)] or directly into the ooplasm [intracytoplasmic sperm injection (ICSI)]. Because both SUZI and ICSI utilize a single sperm per ovum, they represent appealing possibilities for optimizing the reproductive impact of a single ejaculate or postmortem sperm extraction from an irreplaceable individual of an endangered species.

E. Cryopreservation

Long-term storage of germ plasm has become an integral part of captive breeding master plans for many species. Cryopreservation secures genetic resources in a minute fraction of the space required to house multiple large groups of reproducing animals. Timely infusion of genes into populations separated from donors by space and time ensures continued maintenance of maximum genetic diversity.

It has long been known that sperm from various domestic animal species exhibit different requirements for cryosurvival. While the use of freezing protocols developed for related domestic species may be adequate for some exotics, development of optimal methods for each species is desirable. Experimental designs for protocol development include comparisons of types and concentrations of cryoprotectants in various freezing media, prefreeze cooling rates and durations, freeze and thaw rates, and postthaw treatments. Because no one semen parameter is a reliable indicator of fertility, comprehensive analysis of sperm prior to freezing and subsequent to thawing should include factors such as motility and speed; integrity of plasma membranes including the acrosome; morphology; and one or more sperm functions tests, such as migration through cervical mucus, zona pellucida penetration, or heterologous *in vitro* fertilization.

V. CONTRACEPTION

Of growing importance to captive breeding efforts is the regulation of reproductive efficiency to equalize founder effect. Stimulation of reproduction in underrepresented families along with curtailment of the contribution of overrepresented lines will result in maximum genetic diversity. The simplest, and surest, method of contraception is separation of the sexes. Useful for some strictly seasonal species (e.g., the monovular giant panda), the disturbance of group dynamics and the need for redundant holding enclosures makes this scheme untenable for most captive populations.

Contraceptive choices are based on the need for reversibility, the sex to be contracepted, and the cost and labor required to establish and maintain suppression of reproduction. Reversible methods are significantly more expensive than sterilization by removal of gonads or ligation of gamete transport ducts (vas deferens or oviducts) because they require continuous administration or frequent boosters. However, permanent contraception is generally not recommended for a healthy animal until it has produced offspring with demonstrated reproductive capacity. Even then, the management of the animal must be factored into the choice of contraceptive methods. For example, castration renders a male sterile but deprives him of the testosterone-influenced behaviors that may be desirable for the social integrity of his group. If the suppression of male behavior is contraindicated, vasectomy is the logical alternative to castration.

Complete reversibility, resulting in the resumption of full fertility in all treated individuals, has not been achieved for any contraceptive method or device. Thus, the collection and storage of gametes is advised prior to the initiation of treatment as well as before sterilization.

Currently, there are more choices of effective contraception methods for females than for males. Oral contraceptives developed for women are equally effective in great apes, but their daily administration can be problematic. Progestins, in oral, injectable, or implantable forms, have been used in female felids, bovids, and cervids for many years with variable success. Increased rates of reproductive tract carcinomas in felids with prolonged use limits the duration of treatment recommended for progestins. Prevention of fertilization by immunocontraception is an active field of investigation with successes reported in equids, bovids, cervids, and elephants. Pregnancy may be safely terminated early in gestation with prostaglandin, but the use of this potent hormone should be reserved for emergency situations.

Male contraception research in animals, like in humans, has not kept pace with studies of female methods. Inhibition of sperm production by suppression of FSH and LH through GnRH agonist or antagonist administration has been investigated but results are equivocal and the cost is prohibitive. Until low-cost, long-acting forms of agonists or antagonists can be developed, this contraceptive avenue is not feasible for exotic species. The insertion of vas plugs in a small number of hoof-stock species has been discontinued due to surgical complications and inability to restore fertility upon removal. Oral or implanted progestins have provided variable levels of contraception in male deer and antelope. Physical and social side effects (mild to extreme fattening and loss of male behaviors, respectively) may make this treatment less desirable for males on exhibit and/or within a family group. The mechanisms of action for these contraceptives are discussed in other articles in this encyclopedia.

VI. CONCLUSION

It would be difficult to dispute the notion that exotic animals belong in the wild, observed and revered at a distance and allowed to live undisturbed by human habitation. However, we cannot discount the importance of captive populations as reservoirs of genetic diversity and temporary safe havens in this era of wholesale habitat destruction. It behooves conservationists to balance their efforts between *in situ* and *ex situ* propagation of endangered species and to apply the most thoughtful, creative, and noninvasive methods to the achievement of those goals.

See Also the Following Articles

BREEDING STRATEGIES FOR DOMESTIC ANIMALS; CASTRATION, EFFECTS IN NONHUMANS; CRYOPRESERVATION OF EMBRYOS; CRYOPRESERVATION OF SPERM; IN VITRO FERTILIZATION; REPRODUCTIVE TECHNOLOGIES, OVERVIEW

Bibliography

Caro, T. M. (1994). *Cheetahs of the Serengeti Plains: Group Living in an Asocial Species*. Univ. of Chicago Press, Chicago.

Carson, R. (1962). *Silent Spring*. Houghton Mifflin, Boston.

Durrant, B. S. (1990). Semen collection, evaluation and cryopreservation in exotic animal species: Maximizing reproductive potential. *Instit. Lab. Anim. Resour. News* **32**, 2–9.

Gibbons, E. J., and Durrant, B. S. (1987). Behavior and development of offspring from interspecies embryo transfer: Theoretical issues. *Appl. Anim. Behav. Sci.* **18**, 105–118.

Gibbons, E. J., Wyers, E. J., Waters, E., and Menzel, E. W. (1994). *Naturalistic Environments in Captivity for Animal Behavior Research*. State Univ. New York Press, Albany.

Gibbons, E. J., Durrant, B. S., and Demarest, J. (1995). *Conservation of Endangered Species in Captivity: An Interdisciplinary Approach*. State Univ. New York Press, Albany.

Harrison, P. F., and Rosenfield, A. (Eds.) (1996). *Contraceptive Research and Development: Looking to the Future*. National Academy Press, Washington, DC.

Lindburg, D. G., and Lasley, B. L. (1985). Strategies for optimizing the reproductive potential of lion-tailed macaque colonies in captivity. In *The Lion-Tailed Macaque: Status and Conservation* (P. G. Heltne, Ed.), pp. 343–356. A. R. Liss, New York.

Lindburg, D. G., Durrant, B. S., Millard, S. E., and Oosterhuis, J. E. (1993). Fertility assessment of cheetah males with poor quality semen. *Zoo Biol.* **12**, 97–103.

Ryder, O. A., Kumamoto, A. T., Durrant, B. S., and Benirschke, K. (1989). Chromosomal divergence and reproductive isolation in dik-diks. In *Speciation and Its Consequences* (D. Otte and J. A. Endler, Eds.), pp. 208–225. Sinauer, Sunderland, MA.

Cardiovascular Adaptation to Pregnancy

Margaret K. McLaughlin and Robin E. Gandley

University of Pittsburgh School of Medicine

I. Oxygen Consumption and Cardiovascular Function
II. Cardiac Output, Blood Volume, and Blood Pressure
III. Heart Structure
IV. Autonomic Nervous System
V. Mechanisms for the Cardiovascular Adaptation to Pregnancy
VI. Summary

GLOSSARY

arterial blood pressure The product of cardiac output and total peripheral vascular resistance.

cardiac output The amount of blood pumped by the heart every minute; it is a product of the heart rate and stroke volume (blood ejected by each heart ventricle) and is expressed in liters per minute.

endothelium The inner cell surface of the heart chambers and vessel walls; the surface in contact with blood.

glomerular filtration rate The volume of fluid filtered from the glomeruli into the Bowman's capsule of the kidney resulting in urine formation.

oxygen consumption The amount of oxygen consumed by cells to meet metabolic needs.

vascular resistance A measure of how difficult it is for blood to flow between any two points at a given blood pressure. Total peripheral vascular resistance is the sum of the resistance to flow offered by all the blood vessels of the system.

vascular smooth muscle The cells in the vascular wall that constrict and relax to control vessel diameter.

The cardiovascular adaptation to pregnancy involves a complex physiological response by the mother to the presence of the developing fetus and placenta. The hemodynamic changes include, but are not restricted to, significant increases in cardiac output and blood volume, a decrease in blood pressure, and a marked reduction in total peripheral vascular resistance. While the pattern of these changes varies, the integrated cardiovascular response to pregnancy is reasonably consistent across species. The hemodynamic alterations are normal and critical to fetal growth and development as evidenced by the high incidence of fetal and neonatal mortality and morbidity associated with the various cardiovascular disorders of pregnancy.

I. OXYGEN CONSUMPTION AND CARDIOVASCULAR FUNCTION

During human pregnancy the amount of oxygen consumed by the mother during resting conditions increases 40–50 ml/min from prepregnancy values. This 20% increase in oxygen consumption is necessary to meet the metabolic needs of the mother and those of the growing placenta and fetus. Since increases in peripheral oxygen consumption provide a potent stimulus to increase cardiac output and respiration, it is reasonable to suggest that this increased oxygen demand drives the cardiovascular response during pregnancy. The most significant increase in oxygen demand occurs during the period of greatest fetal growth (last trimester). However, the cardiovascular response to the pregnancy peaks during the first half of gestation, providing more oxygen than is required during that time period. The cardiovascular response is greater than necessary to meet metabolic needs over the entire course of pregnancy. This is evident in a narrowing of the difference between the arterial and venous oxygen contents. The arteriovenous difference narrows when the oxygen content of the venous blood increases due to an oversupply of oxygen to the tissue. Since the supply is greater than the demand, it is unlikely that changes in consumption are the cause of the increases in cardiac output and blood flow. In fact, the initial cardiovascular response to pregnancy occurs within days of implantation and precedes any major change in oxygen requirements.

Increases in function are due to a complex interplay between the nervous system, circulating hormones, and functional and structural alterations that occur within the heart, arteries, and veins. Figure 1 schematically presents this integrated response. Due to the highly dynamic nature of the cardiovascular system it has been difficult to determine the cause and effect relationships responsible for the hemodynamic profile observed during the course of pregnancy.

II. CARDIAC OUTPUT, BLOOD VOLUME, AND BLOOD PRESSURE

Cardiac output increases during pregnancy in all species studied although the pattern and extent of change differ. In the human, cardiac output is elevated about 18% by 5 weeks after the last menstrual period, which is within 2 weeks of implantation. This peaks to approximately 50% above nonpregnant levels by midgestation and either plateaus or continues to slightly rise until term. The change in output is the sum of a 17–24% increase in heart rate and a 30–35% increase in stroke volume (the volume of blood ejected from the heart with each heartbeat). The increase in heart rate precedes that of stroke volume and is the earliest hemodynamic response to pregnancy since it occurs during the luteal phase of the fertile cycle.

Alterations in cardiac function occurring during pregnancy depend on an increase in total blood volume (increased preload), a reduction in peripheral vascular resistance (reduced afterload), and changes in heart structure. Determining the total blood volume during pregnancy involves measures of plasma volume using radioactive tracers or Evans blue dye and red blood cell mass using radioisotope labels. The most significant increase in plasma volume occurs during the second trimester, with multiparous women showing greater volume expansion than nulliparous women. Depending on parity, plasma volume expansion ranges from 40 to over 100%. The increase in plasma volume is disproportionately greater than the increase in red blood cell (RBC)

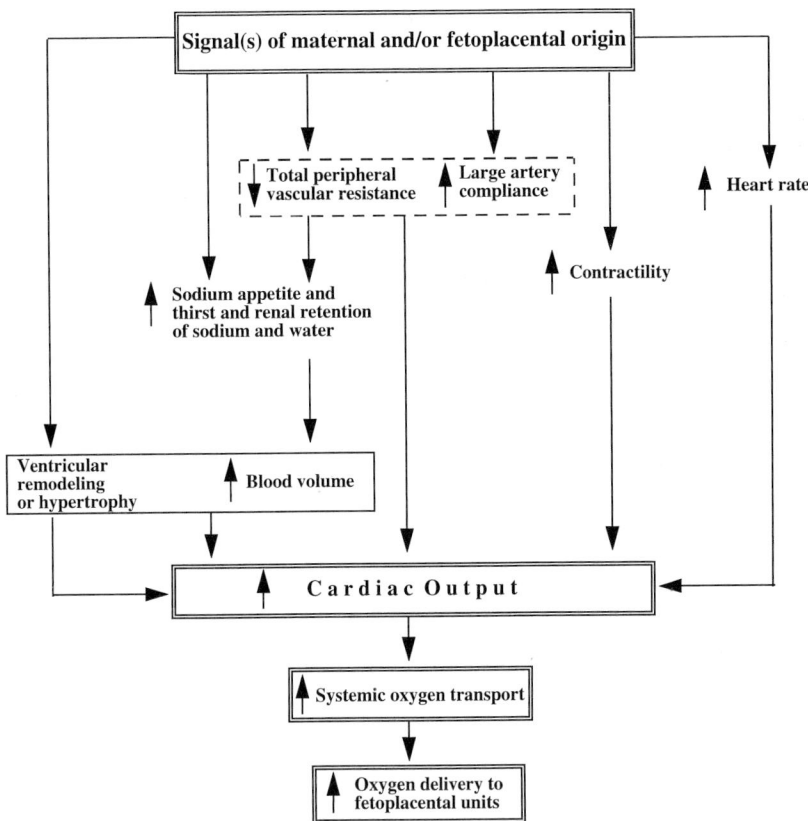

FIGURE 1 Proposed integrated scheme for the cardiovascular response to pregnancy depicting an increase in oxygen delivery to the fetoplacental unit as the common endpoint. Adaptations shown inside dashed and rectangular boxes affect afterload and preload, respectively. Factors not included that also contribute to delivery of O_2 to the fetus are reduced uterine vascular resistance, increased arterial PO_2, red blood cells mass, and decreased affinity of hemoglobin for O_2 (reprinted with permission from Gilson et al., 1992).

mass resulting in a reduced hematocrit or the "physiological anemia of pregnancy." The net effect is a 25–90% increase in RBC mass dependent on parity and degree of iron supplementation. Overall blood viscosity is decreased. This expanded blood volume increases the filling of the left ventricle when the heart is relaxed (diastole), which increases the end-diastolic left ventricular dimension and thus ventricular volume. The expanded ventricular volume increases stroke volume and thus cardiac output.

An interesting question is whether the size of the vascular compartment increases to accommodate the gestational increase in blood volume. The mean circulatory filling pressure provides an indication of the relationship between changes in blood volume compared to the circulatory volume. This is determined by the vascular capacitance. Vascular capacitance is a function of both the unstressed vascular volume (the volume of the circulation at zero pressure) and compliance (the ratio of the change in volume to a change in pressure). Any increase in mean circulatory filling pressure during pregnancy would indicate that the vascular expansion that occurs does not fully accommodate the increase in blood volume.

In general, mean circulatory filling pressure remains constant during pregnancy indicating that vascular capacitance is increased. Factors that can contribute to changes in vascular capacitance include reduced sympathetic nervous stimulation to the capacitance vessels (the veins), relaxation of the large capacitance veins in response to increased volume, and structural changes in the vessel walls. Inhibiting sympathetic nervous activity while making direct measures of vascular capacitance has shown that the autonomic nervous system does not appear to contribute to the regulation of capacitance during pregnancy. Measures of the stiffness of arteries and veins indicate that blood vessel structure changes during gestation. It is likely that the capacity of the circulation is increased during pregnancy due to increases in venous distensibility. Changes in blood volume, venous distensibility, capillary permeability, and blood pressure contribute to the tendency for pregnant women to develop edema in their extremities. In the last trimester the gravid uterus can compress femoral veins, further contributing to edema formation.

The second major contributor to the increases in cardiac output during pregnancy is the reduced afterload. Afterload is determined by the total peripheral vascular resistance, the structural characteristics of the blood vessels, and the blood viscosity. Along with heart rate, the reduction in peripheral vascular resistance is most likely the earliest cardiovascular response in the mother following fertilization. The reduction in peripheral vascular resistance is accompanied by a loss of sensitivity to stimuli which increase blood pressure. The degree of change in pressor sensitivity varies according to species, gestational age, and the vasoactive stimuli examined.

Peripheral vascular resistance decreases through a loss of vascular tone and is reflected in the fall of blood pressure during pregnancy despite an increase in cardiac output. Blood pressure measurements during pregnancy are sensitive to posture and the method of measurement. There is a small drop in systolic pressure and a marked reduction in diastolic pressure, the peak change occurring by 20 weeks of gestation. The pressures tend to rise thereafter, reaching nonpregnant levels near term. However, cardiac output remains significantly elevated such that calculated vascular resistance remains low.

The loss of vascular tone that results in the marked decrease in vascular resistance is due to active relaxation of the vascular smooth muscle in small arteries as well as to a structural remodeling of the vessel wall which affects diameter regulation. Structural changes in the resistance vasculature can affect diameter regulation by influencing signal transduction through the vascular wall, i.e., the response to changes in flow and shear stress. Structural changes that increase vascular compliance are evident by late gestation in the resistance arteries of the rat and also occur in conduit arteries such as the aorta of the human and main uterine arteries in all species studied.

Individual organ blood flows serve to identify the major contributors to the reduced peripheral resistance. Flow increases to the kidneys, uterus, heart, breasts, skin, diaphragm, and skeletal muscles, although the gestational pattern of change differs between species. The increase in uterine blood flow becomes more pronounced as gestation advances, mirroring the increase in fetal growth. The most profound circulatory change in nonreproductive tissues is that which occurs in the kidneys. There is an increase in glomerular filtration rate (GFR) due to an increase in kidney blood flow that appears to coincide with the very early rise in heart rate and cardiac output. This represents the earliest significant change in organ blood flow associated with pregnancy. There is not a sustained increase in glomerular capillary blood pressure, which explains why GFR can increase to 20–40% above prepregnant values for a prolonged period of time and not result in renal damage. The change in renal hemodynamics precedes any demonstrable alterations in plasma or blood volume. Despite the marked increases in flow and GFR, the kidney retains its ability to maintain flow in response to increases or decreases in blood pressure (autoregulation). The kidney can vasodilate further in response to such challenges as amino acid infusion, indicating that while operating in a marked state of vasodilation the renal homeostatic controls have been reset to maintain the feedback loops necessary for responding to hemodynamic challenges.

III. HEART STRUCTURE

There are a number of structural changes in the heart that are associated with the increase in cardiac output. There are marked similarities between the

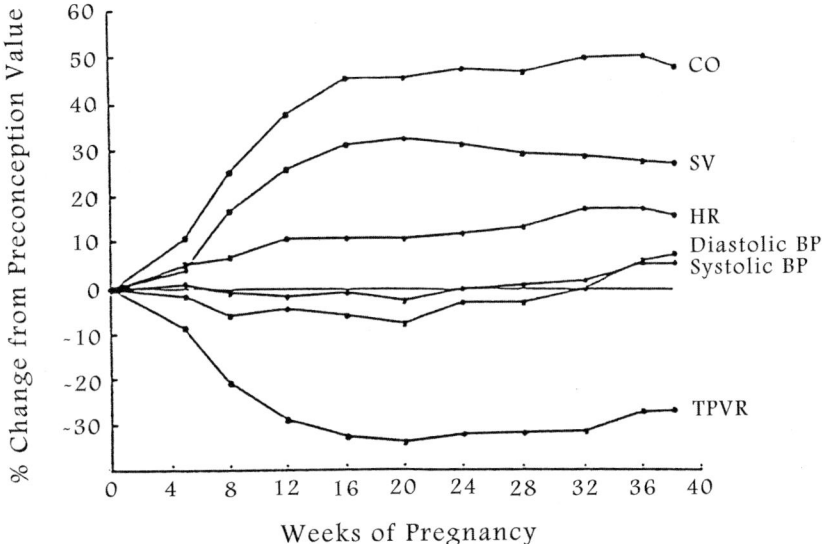

FIGURE 2 Pattern of change in blood pressure, Doppler, and cross-sectional echocardiographic results obtained in healthy women throughout pregnancy. Data are normalized to the corresponding value measured before conception. CO, cardiac output; SV, stroke volume; HR, heart rate; TPVR, total peripheral vascular resistance (reprinted with permission from Woods, 1997).

structural cardiac response to pregnancy and individuals with chronic increases in blood volume such as athletes who run. Ventricular wall mass increases nearly 50% by term. Other structural changes include an increase in the volume of the left ventricle when it is relaxed (end-diastolic dimension) and a 7–14% increase in the cross-sectional areas of the heart valves as measured in serial studies using combined Doppler and echocardiography (Fig. 2). These structural changes are significant by 12 weeks' gestation, which is 7 weeks after a measurable rise in cardiac output. Functional changes thus precede the structural adaptation. By the second trimester, there is an observable increase in ventricular wall thickness and mass which is most marked at term. The adaptive structural response of the heart and increase in cardiac output thus appear to be due first to the decrease in afterload, followed by the rise in blood volume in the second trimester. The increase in cardiac output precedes any significant change in uterine blood flow. This negates the idea that maternal cardiac enlargement is due to the uterus acting as an "arterio venous shunt," a condition that results in an increase in blood volume. The blood volume expansion of pregnancy occurs after the left ventricle has already enlarged. This increase in cardiac size is sufficient to explain an increase in stroke volume. The mechanisms responsible for the chamber enlargement are unknown, although estrogen treatment of guinea pigs and sheep can cause an effect similar to pregnancy itself.

IV. AUTONOMIC NERVOUS SYSTEM

The autonomic nervous system (ANS), which is composed of parasympathetic and sympathetic nerves, contributes significantly to the control of blood pressure and heart rate. It is a prime target for modification during pregnancy. Increasing or decreasing parasympathetic nervous activity to the heart will decrease and increase heart rate, respectively. Withdrawal of sympathetic activity will reduce peripheral vascular resistance; conversely, increases in sympathetic activity increase resistance and raise blood pressure. Changes in arterial blood pressure provide a major stimulus to the central nervous system for altering ANS activity. The study of ANS activity during pregnancy is complicated by the gestational baseline differences in heart rate, blood pressure, and blood volume. Early pregnancy studies relied on examining the heart rate response to changes in blood pressure as an index ANS activity. Recent studies entail direct measures of sympathetic nervous activity outflow in response to a variety of stimuli that influence blood pressure.

When studied in the awake state, the decrease in heart rate (bradycardia) to increasing blood pressure is enhanced, unchanged, or depressed during pregnancy. These apparent discrepancies probably reside in the time of gestation studied and the differences in the control heart rate between pregnant and nonpregnant subjects. In experiments in the rat, sheep, and human in which the normal gestational increase in resting heart rate exists prior to imposed pressure steps, the reflex bradycardia is accentuated by pregnancy. This suggests that pregnancy augments the increase in parasympathetic activity that occurs in response to pressure increases.

Decreases in blood pressure result in a withdrawal of parasympathetic activity. Pregnancy does not appear to affect this arm of the heart rate response. The baroreflex-mediated tachycardia (increase in rate) resulting from drug-induced reductions in blood pressure are unaffected by pregnancy in rats and humans. In contrast to the heart rate response, pregnant rats demonstrate an attenuated ability to increase renal sympathetic nerve outflow above baseline in response to hypotensive challenges. Increasing renal sympathetic nervous activity in the face of reduced blood pressure is a normal homeostatic mechanism evoked to restore blood pressure to normal. In women, the basal sympathetic nerve activity in skeletal muscle blood vessels does not differ between pregnant and nonpregnant women, suggesting no change in basal sympathetic tone during gestation. In addition, there is no pregnancy effect noted upon the increase in its activity that occurs in response to a cold pressor test which causes an increase in sympathetic outflow. In general, systemic vasoconstriction in response to hypotension appears diminished by pregnancy, whereas reflex tachycardia is maintained. The ability to increase cardiac output while buffering systemic vasoconstriction can serve to protect regional blood flows such as the uterine circulation during hypotensive challenges.

Remarkably little is known about mechanisms responsible for pregnancy-induced changes in ANS function. 3β-Hydroxy-dihydroprogesterone, a progesterone metabolite that is elevated during pregnancy, has been shown to reduce renal sympathetic nerve activity through the activation of certain receptors in the brain.

V. MECHANISMS FOR THE CARDIOVASCULAR ADAPTATION TO PREGNANCY

Since the early increase in cardiac output appears to be due primarily to the decrease in afterload and increase in heart rate, identifying the mechanism(s) responsible for the reduction in peripheral vascular resistance would appear key to understanding the causes of the overall cardiovascular response to pregnancy.

The mechanisms responsible for initiating a reduced vascular tone may be quite different than those that contribute to its maintenance throughout gestation. The vascular wall is very responsive to mechanical forces. Early changes in blood pressure, blood flow, and wall stress due to active vascular smooth muscle relaxation can potentially induce further changes in endothelial cell and vascular smooth muscle structure and function that serve to maintain the initial vasodilation.

While numerous mechanisms are postulated to explain the loss of vascular tone and reactivity during pregnancy, a unified theory remains to be determined. A reduction in sympathetic nervous outflow to the resistance vasculature during pregnancy could readily explain the reduction in peripheral vascular resistance. However, during pregnancy, the alterations in ANS activity appear to be more involved in the mother's ability to respond to pressure stimuli rather than changing the basal tone maintenance per se. It is more likely that there are local mechanisms within the arterial wall itself that account for the pregnancy relaxation and these in turn are initiated by endocrine signals deriving from the ovary, the placenta, or possibly the pituitary.

Remarkably little is known about vascular smooth

muscle (VSM) regulation during pregnancy. Observations have included an overall reduction in VSM contractility in arteries from certain vascular beds, hyperpolarization of VSM resting membrane potential, and numerous potential alterations in receptor function. Little is known about pregnancy effects on signal transduction within the VSM cell. Observations are limited to describing the organ or cellular response to pregnancy and the initiating mechanisms require further examination.

Steroid hormones, particularly estrogen and progesterone, are likely candidates for initiating the early vasodilation. However, the evidence for their role is strongest with regard to volume homeostasis, cardiac remodeling, and uterine blood flow regulation. Steroid hormone treatment of the nonpregnant rat does not duplicate the reduction of peripheral vascular resistance nor the changes in renal hemodynamics with pregnancy. Nonetheless, the initial rise in GFR, the increased cardiac output, and the reduction in peripheral vascular resistance in the rat are due to signals maternal in origin because these changes also occur in pseudopregnant animals in which the hormonal profile mimics early gestation in the absence of any fetoplacental contribution. Undetermined placental factors contribute to the maintenance of GFR later in pregnancy in the rat.

Endothelial cells lining the arterial wall serve to control vascular tone. Vasoactive substances produced by the endothelium are likely candidates for modifying vascular smooth muscle tone during gestation. Prostaglandins are highly vasoactive molecules that are produced in the endothelium as well as the VSM and platelets. Under normal conditions, vasodilatory prostaglandins do not appear to participate in the pregnancy-induced vasodilation. Prostaglandins can compensate when other important vascular mediators such as nitric oxide, a molecule produced by nitric oxide synthase (NOS) in the endothelium, is inhibited.

NOS inhibition in most studies during late gestation in the rat causes a greater potentiation of pressure responsiveness to angiotensin II and norepinephrine than in nonpregnant controls, suggesting that NO participates in the loss of sensitivity to vacular stimuli that occurs during pregnancy. To determine whether NO contributes significantly to the reduction in total peripheral resistance, measures of the cardiac output are required in order to calculate effects on vascular resistance.

During rat pregnancy there is an increase in the excretion of cyclic GMP, the second messenger for NO. This increase parallels that of nitrite and nitrate, which are metabolites of NO. This demonstrates that endogenous NO production increases in rat gestation, although the source of production remains to be identified. Whether it occurs in humans still awaits longitudinal analysis of 24-hr urine samples before, during, and after pregnancy while subjects are on controlled diets. In the rat renal circulation, NOS inhibition at midgestation equalizes glomerular filtration, renal plasma flow, and renal vascular resistance between pregnant and nonpregnant animals, indicating that NO contributes to the change in renal hemodynamics at midgestation.

In blood vessels from humans, a large component of the relaxation response to endothelial stimulation is mediated by non-NOS mechanisms. There is no convincing evidence that basal NOS activity has increased during pregnancy either from measures of cyclic GMP in large vessels such as the aorta or from measures of arginine to citrulline conversion in other vascular beds. The signal transduction pathway distal to NO is unchanged during pregnancy, a remarkably consistent finding among the numerous reported studies. Since the complete time course and extent of the endothelial changes occurring in pregnancy are unknown, the initiating signals for the early vasodilation remain to be determined.

VI. SUMMARY

In summary, the cardiovascular adaptation to pregnancy involves a marked change in circulatory homeostatic control mechanisms (Fig. 1). The needs of the growing fetus and placenta as measured by increased oxygen consumption are met and exceeded. It is assumed that the signals responsible for accomplishing the endpoint of increased cardiac output through changes in preload, afterload, and cardiac structure are maternal and/or fetoplacental in origin. The reduction in peripheral vascular resistance early in pregnancy is key to the sequelae of events and at this stage of pregnancy involves primarily nonreproductive organs. Further study is needed regarding the mechanisms responsible for initiating the earliest changes in vascular function and the subsequent maintenance of the elevated cardiac output and reduced peripheral vascular resistance.

See Also the Following Articles

Pregnancy in Humans, Overview; Pregnancy, Metabolic Changes in

Bibliography

Baylis, C. (1994). Glomerular filtration and volume regulation in gravid animal models. *Bailliére's Clin. Obstet. Gynaecol.* 8, 235–264.

Brooks, V. L., and Keil, L. C. (1994). Changes in the baroreflex during pregnancy in conscious dogs: Heart rate and hormonal responses. *Endocrinology* 135, 1894–1901.

Conrad, K. P., and Russ, R. D. (1992). Augmentation of baroreflex-mediated bradycardia in conscious pregnant rats. *Am. J. Physiol.* 262, R472–R477.

Ekholm, E. M. K., and Erkkola, R. U. (1996). Autonomic cardiovascular control in pregnancy. *Eur. J. Obstet. Gynecol. Reprod. Biol.* 64, 29–36.

Everett, D. C., Morris, K. G., and Moore, L. G. (1991). Regional circulatory contributions to increased systemic vascular conductance of pregnancy. *Am. J. Physiol.*, H1842–H1847.

Fan, L., Daher, S. M., Gutkowska, J., Nuwayhid, B. S., and Quillen, E. W. (1996). Enchanced natriuretic response to intrarenal infusion of atrial natriuretic factor during ovine pregnancy. *Am. J. Physiol.* 270, R1132–R1140.

Geva, T., Mauer, M. B., Striker, L., Kirshon, B., and Pivarnik, J. M. (1997). Effects of physiologic load of pregnancy on left ventricular contractility and remodeling. *Am. Heart J.* 133, 53–59.

Gilson, G. J., Mosher, M. D., and Conrad, K. P. (1992). Systemic hemodynamics and oxygen transport during pregnancy in chronically instrumented, conscious rats. *Am. J. Physiol.* 263, H1911–H1918.

Hart, M. V., Hosenpud, J. D., Hohimer, A. R., and Morton, M. J. (1985). Hemodynamics during pregnancy and sex steroid administration in guinea pigs. *Am. J. Physiol.* 249, R179–R185.

Hines, T., Lindheimer, M. D., and Barron, W. M. (1993). Total autonomic blockade eliminates the attenuated pressor response to angiotensin II in pregnant rats. *Am. J. Physiol.* 265, R1270–R1275.

Humphreys, P. W., and Joels, N. (1994). Effect of pregnancy on pressure–volume relationships in circulation of rabbits. *Am. J. Physiol.* 267, R780–R785.

Irons, D. W., Baylis, P. H., and Davison, J. M. (1996). Effect of atrial natriuretic peptide on renal hemodynamics and sodium excretion during human pregnancy. *Am. J. Physiol.* 271, F239–F242.

Mackey, K., Meyer, M. L., Stirewalt, W. S., Starcher, B. C., and McLaughlin, M. K. (1992). Composition and mechanics of mesenteric resistance arteries from pregnant rats. *Am. J. Physiol.* 263, R2–R8.

Magness, R. R., Rosenfeld, C. R., and Carr, B. R. (1991). Protein kinase C in uterine and systemic arteries during ovarian cycle and pregnancy. *Am. J. Physiol.* 260, E464–E470.

Masilamani, S., and Heesch, C. M. (1997). Effects of pregnancy and progesterone metabolites on arterial baroreflex in conscious rats. *Am. J. Physiol.* 272, R924–R934.

Nisell, H., Hjemdahl, P., and Linde, B. (1985). Cardiovascular responses to circulating catecholamines in normal pregnancy and in pregnancy-induced hypertension. *Clin. Physiol.* 5, 479–493.

Pan, Z.-R., Lindheimer, M. D., Bailin, J., and Barron, W. M. (1990). Regulation of blood pressure in pregnancy: Pressor system blockade and stimulation. *Am. J. Physiol.* 258, H1559–H1572.

Pascoal, I. F., Lindheimer, M. D., Nalbantian-Brandt, C., and Umans, J. G. (1995). Contraction and endothelium-dependent relaxation in mesenteric microvessels from pregnant rats. *Am. J. Physiol.* 269, H1899–H1904.

Poston, L., McCarthy, A. L., and Ritter, J. M. (1995). Control of vascular resistance in the maternal and feto-placental arterial beds. *Pharmacol. Ther.* 65, 215–239.

Robson, S. C., Hunter, S., Boys, R. J., and Dunlop, W. (1989). Serial study of factors influencing changes in cardiac output during human pregnancy. *Am. J. Physiol.* 256, H1060–H1065.

Sladek, S. M., Magness, R. R., and Conrad, K. P. (1997). Nitric oxide and pregnancy. *Am. J. Physiol.* 272, R441–R463.

Slangen, B. F. M., Iris, C. M., Verkeste, C. M., Smits, J. F. M., and Peeters, L. L. H. (1997). Hemodynamic changes in pseudopregnancy in chronically instrumented conscious rats. *Am. J. Physiol.* 272, H695–H700.

Stock, M. K., and Metcalfe, J. (1994). Maternal physiology during gestation. In *The Physiology of Reproduction.* 2nd ed., pp. 947–973. Raven Press, New York.

Woods, L. L. (1993). Role of angiotensin II and prostaglandins in the regulation of uteroplacental blood flow. *Am. J. Physiol.* 264, R584–R590.

Castration, Effects in Humans (Male)

Harald H. J. Hoekstra, Mels F. van Driel, and Pax H. B. Willemse

Gröningen University Medical Center

I. Introduction
II. History
III. Surgical Aspects of Orchiectomy
IV. Testicular Tumors and Their Hormonal Consequences
V. Sexual Function of Testicular Cancer Patients
VI. Castration and Body Image
VII. Semen Cryopreservation
VIII. Conclusion

GLOSSARY

castration The surgical removal of both testes.
chemical castration The induction of gonadal atrophy by prolonged treatment with female sex hormones (estradiol), antiandrogen, or gonadotropin-releasing hormone (luteinizing hormone-releasing hormone) analogs.
follicle-stimulating hormone A pituitary hormone; stimulates spermatogenesis.
gonadotropin- or luteinizing hormone-releasing hormone A hormone secreted by the hypothalamus; releases both gonadotropins (luteinizing hormone and follicle-stimulating hormone) from the pituitary.
luteinizing hormone A pituitary hormone; stimulates testosterone production by the Leydig cells.
orchiectomy The surgical removal of one testicle.
spermatogenesis The process of sperm production.
steroidogenesis The process of sex hormone (testosterone) production.
testis or testicle One of the two male reproductive glands; located in the scrotum.
testosterone A naturally occurring androgen (daily production, 7 mg); 90% is produced by the interstitial (Leydig) cells of the testis and 10% by the adrenals.

The male gonads or testes have a dual function: the production of sperm for reproduction and of male sex hormones, the most important of which is testosterone. Removal of the gonad may be performed by surgery or medically by the administration of gonadotropin-releasing hormone analogs (luteinizing hormone-releasing hormone), estrogens, or antiandrogens. The surgical removal of one or both testicles is called bilateral orchiectomy, which is performed for palliation in patients with disseminated prostate cancer and also in men with bilateral testicular cancer, but rarely in men who have benign intrascrotal tumors, genital tuberculosis, or other chronic infections of testis or epidydimis.

I. INTRODUCTION

Prostate cancer is a common disease with an incidence of 60 per 100.000, representing almost one-third of all newly diagnosed cancers in the male. In Europe, it is still true that 40–50% of all patients presenting with prostate cancer already have evidence of widespread and therefore incurable disease. In the United States, statistics show that the ratio of patients deemed curable to those not curable is 5:1. The corresponding values for African, South American, and Asian countries are unknown. During the past decade, progress has been achieved in earlier diagnosis by regular rectal examination and screening of blood for tumor markers such as prostate-specific antigen. Diagnosis at an early, curable stage has so far led only to small improvements with regard to the overall tumor-specific mortality. Radical surgery or radiation therapy are the types of treatment for nondisseminated prostate cancer. Castration has been the mainstay of palliative treatment (alleviation of complaints) in advanced prostate cancer for over 50 years.

Unilateral inguinal orchiectomy (semicastration)

is performed for the diagnosis as well as treatment of testicular cancer. The incidence of testicular cancer is approximately 5 per 100,000 men per year, but there is an increasing incidence in industrialized countries. It is one of the most common malignancy among young men in the second and third decades of life, the most sexually active years. Two types of testicular cancer can be distinguished: nonseminomatous tumors (nonseminoma) and seminomatous testicular germ cell tumors (seminoma). Today, more than 90% of patients can be cured by combined modality therapies, based on tumor type and stage. These therapies include surgery, cisplatin-based polychemotherapy for nonseminoma, and radiation for seminoma. Compared to the normal population, these patients have a 60-fold higher risk to develop cancer in their remaining testis, i.e., 2–6% overall incidence, which will finally result in complete castration. Testis-preserving surgery has been introduced in the 1990s but remains an experimental approach, with undeniable endocrine and psychological advantages.

The increased survival chance of testicular as well as prostate cancer patients has allowed attention to shift from survival toward improving quality of life. Compared to surgical (semi)castration, the psychotrauma of cytostatic-induced sterility may be no less dramatic. It is important to recognize and to deal with the psychological consequences of (semi)castration and chemotherapy which may originate in individual fantasies, beliefs, and myths about the genitals and in moral values concerning sexuality.

II. HISTORY

The relationship between testosterone and sexual behavior has remained a continuously researched subject throughout the ages. Testosterone plays a role in the pubertal growth of the genitals and sexual behavior, but it is also necessary for the development of male secondary sex features during puberty: low voice, beard growth and male pattern baldness, stimulating hemoglobin, muscle mass, and bone mineralization. All these features can be induced by giving testosterone. Aristotle (384–322 BC) wrote in his *Historia Animalium* about the effects of castration: "If the cock is castrated, he foregoes sexual passion." Later (150 BC) Araetus of Cappadocie wrote:

For it is the semen when possessed by vitality which makes us to be men, well braced in limbs, hairy, well-voiced and strong to think and act as characteristics of men prove. When the semen is not possessed, persons become shrivelled, have a sharp tone of voice, lose their beard and become effeminate as the characteristics of eunuchs prove.

It is an ancient perception that sexual potency can be restored by testicular extracts. Without any success, preparations of animal and male testes and even transplantation of testes were applied to restore impotence. In the eighteenth century Theophile de Bordue, the court doctor of the French King Louis XIV, predicted that the testes secrete a potent substance necessary for normal sexual functioning. This substance was finally discovered in 1935 by the Dutch biochemist Ernst Lacqueur. He named it "testosterone." Two years later, James Hamilton in the United States restored erection in an impotent eunuch by intramuscular testosterone injections. Adolf Butenlandt in Germany extracted testosterone from urine and received the Noble prize for chemistry at the end of the 1930s. During the presentation of this prize he said, "dynamite gentlemen, testosterone is pure dynamite!"

III. SURGICAL ASPECTS OF ORCHIECTOMY

Orchiectomy can be performed by two different surgical techniques. If there is suspicion of testicular cancer, one performs an inguinal incision for semicastration. To reduce the risk of recurrence ("tumor spill") the spermatic cord is ligated at the internal inguinal ring followed by the orchiectomy itself, which includes the removal of the testis as well as the epididymis.

Subcapsular orchiectomy—removal of the testicular contents (germinal epithelium and Leydig cells) by the transscrotal route, leaving the tunica albuginea and epidymis *in situ*—has improved the cosmetic appearance, but the somatic effects remain the same, i.e., decreased libido, impotence, tiredness and inability to concentrate, hot flashes, changed body image, and irreversibility. In patients with disseminated prostate cancer subcapsular orchiectomy is the treatment of choice. For the younger prostate cancer patient these side effects may be even less acceptable. All these side effects occur equally with medical castration, e.g., by administration of estrogen, antiandrogen, or luteinizing hormone-releasing hormone (LHRH), but the latter therapy has the advantage that the treatment can be stopped if the side effects prove too much of a burden. In the past, surgical castration was frequently used for the treatment of sexual offenders, but this has been replaced by medical castration. In case of trauma to the scrotal region, surgery is related to the extent of the trauma. With microvascular techniques, even reimplantation and vascular reconstruction are possible.

In general, the removal of a testis is a simple surgical procedure, but the loss of one testis may have major psychological effects.

IV. TESTICULAR TUMORS AND THEIR HORMONAL CONSEQUENCES

Testosterone is the most abundant androgen and is synthesized by the Leydig cells under the influence of LH from the pituitary. The testes produce about 90% of the daily production of 7 mg of testosterone, whereas the adrenal glands account for the remaining 10%. The minimal plasma testosterone level required to maintain normal sexual function is approximately 10 nmol/liter. Follicle-stimulating hormone (FSH) from the pituitary will stimulate the process of spermatogenesis in the male.

Nonseminomas as well as seminomas may produce an ectopic hormone, human chorion gonadotropin (hCG), normally produced by the placenta to maintain pregnancy. hCG is able to stimulate testosterone secretion directly and may also stimulate the Leydig cells to secrete estradiol, thereby suppressing the normal LH and FSH production. The increased estradiol:testosterone ratio may cause breast development (gynecomastia) but probably plays no role in sexual functioning. Other tumor markers, such as the embryonal albumin α-fetoprotein and lactate dehydrogenase, do not influence the pituitary–gonadal axis.

Semicastration will result in elevated plasma LH but normal testosterone levels if the function of the remaining testis is normal. Shortly after operation decreased plasma testosterone and elevated LH levels may be found, which is normal after surgical stress. This in turn may hamper spermatogenesis, which may be detrimental to the cryopreservation of semen.

Today, patients with disseminated nonseminomas are treated by orchiectomy and cisplatin-based polychemotherapy—usually four courses of cisplatin, etoposide, and bleomycin (BEP). Polychemotherapy will affect spermatogenesis, resulting in sterility in 80–90% of cases, but does not influence Leydig cell function. Experimentally, pretreatment with LHRH analogs does not protect spermatogenesis against cytotoxic polychemotherapy because there may be "endogenous stimulation" by tumor hCG, which will supervene the effect of LHRH. The BEP polychemotherapy may also have effects on the vascular and peripheral nervous system, resulting in Raynaud's phenomenon, peripheral neuropathy, hypercholesterolemia, hypertension, and even erectile dysfunction.

Adjuvant surgery, e.g., resection of a residual retroperitoneal tumor mass or even a nerve preserving retroperitoneal lymph node dissection, carries the risk of ejaculatory dysfunction ("retrograde ejaculation") due to damage to the paravertebral sympathetic nerves. Orgasm is usually preserved. Treatment with tricyclic antidepressants, which increase the sympathetic tone of the bladder neck, may partly overcome this ejaculatory dysfunction, restoring fertility to potentially fertile patients.

Patients with seminoma are mostly treated with external beam radiotherapy (25–30 Gy) to the paraaortic and ipsilateral iliac lymph nodes. Scatter radiation may compromise the Leydig cell function of the remaining testis, resulting in low-normal testosterone and compensatory increased LH levels. Scatter radiation may also compromise the feeding branches of the penile blood vessels, which may result in erectile dysfunction.

V. SEXUAL FUNCTION OF TESTICULAR CANCER PATIENTS

Scientific data on the physical, mental, and sexual consequences of castration or severe hypogonadism in young men are very scarce. It seems that sexual fantasies and erotically elicited erections are relatively testosterone independent. The effects on the penis of low testosterone levels induced by castration are still unclear and controversial, but animal studies suggest that testosterone plays an important role in erectile function by a pre- or postsynaptic action on the corpus cavernosum.

Currently, testosterone suppletion is given to patients with Kallmann's or Klinefelter's syndrome, in female-to-male transsexuals, in patients affected by panhypopituitarisme, and in men castrated for bilateral testicular cancer. Suppletion is essential for maintaining muscle mass and bone mineralization, i.e., bone density. Patients receive intramuscular testosterone depot injections of a mixture of several testosterone esters (250 mg at 2- or 3-week intervals). Despite the combination of short- and long-acting testosterone esters, there is a decline of plasma testosterone levels to below the normal limit within 2 weeks. For some patients the injection site is particularly painful for several days postinjection.

In castrated, testosterone-supplemented testicular cancer patients, none of the consecutive aspects of sexual response, i.e., libido, sexual arousal, erection, orgasm (frequency and/or intensity), and ejaculation, are related to the actual testosterone level because sexual functions are not hampered by relatively low plasma testosterone levels. However, at the end of an injection interval, libido and arousal may be negatively influenced by the physical effects (tiredness) of testosterone withdrawal. In addition to hormonal factors, other organic causes, such as vascular or neurogenic disturbances or chemotherapy-related morbidity, may be responsible for decreased sexual functioning. Psychologic factors, however, play an important role, especially in younger patients castrated for bilateral testicular cancer. The adverse side effects due to fluctuations in plasma testosterone in the course of the intramuscular testosterone injections and their inconvenience should be an incentive for improving the pharmacokinetics of this compound. Oral preparations are available but may not always reach sufficiently high levels of plasma testosterone.

VI. CASTRATION AND BODY IMAGE

The Latin word *testis* means "witness," i.e., a witness or testimony to virility. "To have balls" is a popular way of saying that someone has courage or strength of character. In many cultures masculine behavior is associated with testicular size. An impaired body image and concerns about masculinity may be psychological consequences of (semi)castration. Apart from the physical effects of (semi)castration and possible adjuvant treatment, these patients have to cope with an assault on the perception of their masculinity. The majority of patients with disseminated prostate cancer prefer medical androgen ablation instead of surgical castration in order to preserve their body image. The traumatic effect of (semi)castration may be intensified by beliefs, myths, and fantasies about the testes and sexuality. Semicastration may be viewed as punishment for sexual thoughts and acts or retribution for (excessive) sexual activity. Therefore, sexual and marital counseling may be very important, especially in younger patients. In some of patients, the insertion of testicular prosthesis may decrease the psychological sequelae of castration.

VII. SEMEN CRYOPRESERVATION

Patients treated for testicular cancer or other diseases with risk of infertility should be offered semen cryopreservation. Today, suboptimal semen parameters should not exclude for sperm banking because considerable progress has been made regarding assisted reproduction methods. Recently, the method of intracytoplasmic spermatid injection has become feasible, enabling the fertilization of one ovum by one selected spermatozoon. By this *in vitro* procedure, several embryo's can be obtained and replaced in the womb. Even in patients with severe oligopermia, it is possible to obtain immature spermatids from the epidydimis. In the near future, it will prove possible to obtain immature gonadal tissue from prepubertal gonads for cryopreservation (e.g., from children treated for pediatric cancer), to let these mature and proliferate *in vitro*, and to use the mature germ cells for *in vitro* fertilization as described previously. The use of frozen spermatogonia may even be preferable over frozen sperm because their harvesting will be more certain in view of time limits set by emergency treatment. Apart from ethical and moral issues, even the cloning of immature spermatogonia should become a realistic possibility in the near future. This way, the use of radiation or combination chemotherapy will no longer be a major obstacle to future patient reproductive capacity.

VIII. CONCLUSION

Castration, chemically or surgically, is a rather simple procedure with major consequences—psychological and physiological. Semicastration for testicular cancer followed by cisplatin-based polychemotherapy with or without retroperitoneal surgery may also be associated with treatment-related morbidity. The sexual side effects are not caused by testosterone ablation. Especially young patients castrated for bilateral testicular cancer may need counseling to cope with the psychosexual consequences of the successful treatment of this formerly lethal affliction.

See Also the Following Articles

FSH (Follicle-Stimulating Hormone); GnRH (Gonadotropin-Releasing Hormone); LH (Luteinizing Hormone); Prostate Cancer; Spermatogenesis, Overview; Steroidogenesis, Overview; Testicular Cancer; Testosterone Biosynthesis

Bibliography

Avarbock, M. R., Brinster, C. J., and Brinster, R. L. (1996). Reconstitution of spermatogenesis from frozen spermatogonial stem cells. *Nature Med.* **2**, 693–696.

Gosden, R. G., Baird, D. T., Wade, J. C., *et al.* (1994). Restoration of fertility to oophorectomized sheep by ovarian autografts stored at −196°C. *Human Reprod.* **9**, 597–603.

Hall, R., Anderson, J., Smart, G. A., *et al.* (1993). *Fundamentals of Clinical Endocrinology*. Pirman Medical, Bath, UK.

Heidenreich, A., Holtl, W., Albrecht, W., Pont, J., and Engelmann, U. H. (1997). Testis-preserving surgery in bilateral testicular germ cell tumours. *Br. J. Urol.* **79**, 253–257.

Jonker-Pool, G., van Basten, J. P., Hoekstra, H. J., *et al.* (1997). Sexual functioning after treatment for testicular cancer. *Cancer* **80**, 454–464.

Newerda, G. J. (1943). The history of the discovery and isolation of the male hormone. *N. Engl. J. Med.* **228**, 39–47.

Newling, D. W. W. (1997). The palliative therapy of advanced prostate cancer, with particular reference to the results of recent European clinical trials. *Br. J. Urol.* **79**(Suppl. 1), 72–81.

van Basten, J. P., Jonker-Pool, G., van Driel, M. F., *et al.* (1995). The sexual sequalae of testicular cancer. *Cancer Treat. Rev.* **21**, 479–495.

van Basten, J. P., Jonker-Pool, G., van Driel, M. F., *et al.* (1996). Fantasies and facts of the testis. *Br. J. Urol.* **78**, 756–762.

van Basten, J. P., van Driel, M. F., Jonker-Pool, G., *et al.* (1997a). Sexual functioning in testosterone-supplemented patients treated for bilateral testicular cancer. *Br. J. Urol.* **79**, 461–467.

van Basten, J. P., Hoekstra, H. J., van Driel, M. F., *et al.* (1997b). Cisplatin-based chemotherapy changes the incidence of bilateral testicular cancer. *Ann. Surg. Oncol.* **4**, 342–348.

Yildirim, M. K., Yildirim, S., Utkan, T., Sarioglu, Y., and Yalman, Y. (1997). Effects of castration on adrenergic, cholinergic and nonadrenergic, noncholinergic responses of isolated corpus cavernosum from rabbit. *Br. J. Urol.* **79**, 964–970.

Castration, Effects in Nonhuman Mammals (Female)

Sandra J. Legan

University of Kentucky

I. Introduction
II. General Effects of Ovariectomy
III. Hypothalamus
IV. Pituitary
V. Hormone Replacement Therapy

GLOSSARY

peptide hormone A subgroup of hydrophilic, polypeptide hormones composed of a sequence of <50 amino acids. The major peptide hormone affected by castration is the hypothalamic decapeptide, gonadotropin-releasing hormone, which stimulates the synthesis and release of pituitary luteinizing hormone and follicle-stimulating hormone.

steroid hormone A member of a subclass of lipids that contain a basic structure of four fused rings, referred to as perhydrocyclopentanophenanthrene. The steroid hormones are derived from cholesterol in the gonads, adrenal glands, and placenta. Cholesterol is synthesized from acetyl-CoA, which is converted to an activated five-carbon isoprene unit, six of which polymerize and subsequently undergo cyclization to form cholesterol. The major classes of steroids produced by the ovaries are the progestins, including progesterone and 17α-hydroxyprogesterone; the estrogens, estrone, estradiol-17β, and estriol; and the androgens, androstenedione, and testosterone.

Castration in female mammals consists of the removal of the ovaries and is usually referred to as either oophorectomy or ovariectomy. The ovaries secrete a variety of hormones, the major ones including the steroid hormones estradiol-17β, hereafter referred to as estradiol, progesterone, and androstenedione; and the peptide hormones, including inhibin, activin, and follistatin. All the consequences of ovariectomy result from the virtual absence of the physiologic actions of the ovarian hormones, the most important one of which is estradiol. Because generally "castration" refers to removal of both ovaries, and following unilateral ovariectomy the remaining ovary can compensate for the function of the missing one, this article will concentrate on the effects of bilateral ovariectomy. In addition, this discussion will be limited to the effects of ovariectomy in adults because in prepubertal females, in whom ovarian hormone secretion is very low, the major effects of ovariectomy other than preventing puberty are on the hypothalamic–pituitary axis. Furthermore, these effects are the same in prepubertal and adult females, and they are described in detail herein.

I. INTRODUCTION

Ovariectomy is mainly characterized by very low circulating levels of the major ovarian hormone, estradiol. Because estradiol has a multitude of actions in numerous target organs throughout the body, the effects of ovariectomy, although not acutely life-threatening, are numerous and far-reaching. The most notable effects of ovariectomy are observed in the regression or atrophy of the reproductive tract, namely, the oviduct, uterus, cervix, and vagina, and in accessory reproductive organs such as the mammary glands. In addition, ovariectomy results in removal of important feedback effects of the ovarian hormones on a number of brain areas, such as the hypothalamus, and on the anterior pituitary gland. Ovarian estradiol secretion is controlled by the pituitary gonadotropins, luteinizing hormone (LH) and follicle-stimulating hormone (FSH), which are in turn regulated by hypothalamic gonadotropin-releasing hormone (GnRH) secretion. GnRH is a decapeptide secreted from neurons that in non-primates are located in the diagonal band/septal/preoptic area of the rostral forebrain and project to the median eminence. Once released, GnRH diffuses into the stalk portal capillaries and is transported to the anterior pituitary gland, where it causes release of LH and FSH into the peripheral circulation. These hormones act on the ovaries to stimulate maturation of follicles and ova, ovulation, and biosynthesis and secretion of ovarian hormones, including estradiol. In addition to its trophic effects on the reproductive tract and on its other target tissues throughout the body, estradiol exerts both positive and negative feedback controls at the hypothalamus to modulate GnRH secretion and at the pituitary gland to alter the release of LH and FSH via a change in the pituitary response to GnRH. The negative feedback action of estradiol on the hypothalamic–pituitary axis constitutes an essential and integral part of the homeostatic mechanism whereby circulating estradiol concentrations are maintained within the physiologic range. Ovariectomy, by removing estradiol's chronic negative feedback action, opens the feedback loops at the level of both the hypothalamus and pituitary, thereby disconnecting the homeostatic control mechanism. As a result, levels of GnRH in pituitary portal blood and LH and FSH concentrations in peripheral blood increase. These latter increases in plasma LH and FSH concentrations have been demonstrated in numerous species, such as rats, hamsters, gerbils, rabbits, goats, sheep, pigs, cows, deer, horses, dogs, monkeys. Indeed, the well-known postovariectomy increases in circulating LH and FSH levels constitute the main neuroendocrine hallmark of ovariectomy. In addition to its effects on the reproductive tract, and the brain–pituitary gonadal axis, ovariectomy also has myriad nonreproductive effects. Thus, decreased estradiol levels are associated with increased blood cholesterol levels, decreased blood clotting rate, and decreased bone density, to name a few. Many of these sequellae cause a number of the symptoms commonly observed in postmenopausal women because after menopause circulating estradiol levels decrease to the levels observed after oophorectomy. Most of the effects of ovariectomy can be reversed by administration of ovarian steroids in such a manner as to replace physiologic release patterns. Thus, in women, steroid hormone replacement therapy is being used increasingly in the clinical management of the postmenopausal symptoms caused by estrogen withdrawal. Finally, it is important to note that the steroid-replaced ovariectomized animal has become well established as a simple but powerful model for investigating the roles of the positive and negative feedback actions of ovarian steroids in a variety of mechanisms controlling reproductive functions, such as the preovulatory LH surge, puberty, seasonal breeding, and aging.

II. GENERAL EFFECTS OF OVARIECTOMY

One of the first effects of castration is a rapid fall in circulating levels of the ovarian hormones to very low or undetectable levels. However, most ovarian hormones are not completely absent from the circulation after ovariectomy because there are extraovarian sources of most of these hormones, whether protein or steroid. For example, some of the steroid hormones may be synthesized in peripheral tissues from adrenal precursors. The best characterized example of this is estradiol, which circulates at very low levels in oophorectomized or postmenopausal women due to aromatization of adrenal androstenedione in other tissues, such as fat, liver, and kidney. The degree of this extragonadal aromatization varies among species.

The actions of the ovarian steroid hormones are mediated by their receptors that are located in their target tissues, including the components of the reproductive system, i.e., the brain, anterior pituitary, reproductive tract, and accessory sex glands, and in numerous other target tissues throughout the body, such as liver, adipose tissue, skin, and bone. Following ovariectomy, in the presence of low ligand concentrations, expression and number of estradiol receptors generally increase; however, this is a tissue-specific response. For example, one exception to this general rule is the anterior pituitary gland, in which estradiol upregulates its own receptors. Thus, ovariectomy is associated with a decrease in pituitary estradiol receptors. Progesterone receptors in many tissues are induced by priming with estradiol. Therefore, in the absence of estradiol after ovariectomy, progesterone receptors are generally decreased.

Because estradiol is essential for maintenance of the female reproductive tract, the relative absence of circulating estradiol levels following ovariectomy results in involution of the reproductive tract and atrophy of all accessory estrogen-dependent tissues. Thus, the oviduct, uterus, cervix, and vagina atrophy, become relatively avascular, and all their functions, such as motility and secretion, are decreased. In addition, other tissues dependent on ovarian hormones for their development and maintenance, such as the mammary glands, undergo involution.

The actions of estrogen on nonreproductive tissues are also decreased after ovariectomy. One such result of estrogen deficiency is an increase in rate of bone resorption, which may lead to increased incidence of fractures and osteoporosis. Finally, in nonprimate females, there is no reproductive behavior because sex behavior is completely dependent on the actions of ovarian hormones, especially estradiol and progesterone.

III. HYPOTHALAMUS

Among the many well-known effects of ovariectomy are the changes that occur in the hypothalamic–pituitary axis after the fall in circulating ovarian hormone concentrations releases it from negative feedback. The major hormone in this system is estradiol, which has feedback actions on both LH and FSH. In contrast, inhibin and activin exert their actions selectively on FSH. The remainder of this section will focus on the mechanisms controlling the interactions between estradiol and LH secretion because these are the best characterized.

Following ovariectomy, the absence of estradiol leads to a number of effects on the hypothalamus, including an increase in hypothalamic estradiol receptor mRNA (Shugrue et al., 1992) and a decrease in content of GnRH in the medial basal hypothalamus (Kobayashi et al., 1978). With regard to the effect of ovariectomy on GnRH synthesis, although proGnRH levels in the preoptic area and basal hypothalamus decrease, there is no change in GnRH mRNA levels in these same loci (Kelly et al., 1989). One interpretation of these results is that GnRH release increases after ovariectomy, and this possibility has been verified by the preponderance of evidence obtained from direct measurements of GnRH release. These studies have confirmed that GnRH is secreted in pulsatile fashion and have demonstrated that in long-term ovariectomized animals the pulse frequency varies among species. Thus, several weeks after ovariectomy, one pulse of GnRH occurs approximately every 25 min in rats (Sarkar and Fink, 1980; Levine and Ramirez, 1982), every 40 min in sheep (Levine et al., 1982; Barrell et al., 1992; Karsch et al., 1993), and every 60–90 min in monkeys (Levine et al., 1985; Van Vugt et al., 1985). How do these frequencies compare with those in intact animals? In breeding season sheep, both GnRH pulse frequency and amplitude are greater in ovariectomized ewes (frequency, 18 pulses/12 hr; amplitude, 75 pg/pulse) than in intact ewes (2–8 pulses/12 hr; 5–20 pg/pulse). In anestrous ewes, GnRH pulse frequency is so much greater in ovariectomized (15 pulses/12 hr) than in intact ewes (1 pulse/12 hr) that, even though pulse amplitude is slightly lower after ovariectomy (75 vs 100 pg/pulse; ovx vs intact), net GnRH release is increased in response to ovariectomy (compare Barrell et al., 1992, with Karsch et al., 1993). In rats, however, the results are somewhat controversial. Mean GnRH concentrations in pituitary portal blood are twofold greater 3 to 4 weeks after ovariectomy (50 pg/ml) than just before onset of the LH surge on proestrus (25 pg/ml) (Petraglia et al., 1987) and threefold greater on the 4th or 28th day after ovariectomy–adrenalectomy (15–20 pg/ml) than in intact diestrous rats (5 pg/ml) (Sherwood and Fink, 1980). A third study demonstrated that stalk portal plasma GnRH concentrations are two- or threefold greater between 1600 and 1830 hr in long-term ovariectomized rats (62 pg/ml) than at the same time on estrus (18 pg/ml), metestrus (34 pg/ml), and diestrus (20 pg/ml) (Sarkar and Fink, 1980). In contrast, there is one report (Levine and Ramirez, 1982) indicating that GnRH secretion, assessed by push–pull perfusion of the medial–basal hypothalamus in long-term (>28 days) ovariectomized rats, is not significantly different from the level observed in intact diestrous rats (0.6 pg/12 min). In general, the foregoing observations, in conjunction with the numerous reports of an increase in LH pulse frequency after ovariectomy, and the demonstration of a suppression of GnRH release by administration of estradiol to ovariectomized animals (Karsch *et al.*, 1993; Sarkar and Fink, 1980), strongly suggest that GnRH secretion increases after ovariectomy. These results further suggest that in rats, sheep, and monkeys, the increase in GnRH release is caused by removal of the negative feedback action of estradiol, at least in part, on hypothalamic GnRH secretion.

The pulsatile oscillations in GnRH release have been attributed to release of GnRH during transient, brief intervals that are separated by relatively long periods of quiescence. Such a release pattern suggests that there is a population of GnRH neurons, often referred to as the GnRH "pulse generator," that undergoes synchronous repetitive bursting. In this regard, it is interesting that episodic increases in multiunit activity in the vicinity of the GnRH neurons coincide with pulses of LH in ovariectomized rats (Kawakami *et al.*, 1982), sheep (Thiéry and Pelletier, 1981), and rhesus monkeys (Wilson *et al.*, 1984). However, it remains to be determined whether the multiunit activity is arising from GnRH neurons, and whether the GnRH neurons are the only components of the pulse generator. In ovariectomized rhesus monkeys, norepinephrine pulses are correlated with GnRH and LH pulses, suggesting that a subpopulation of ascending norepinephrine neurons may be a part of the pulse generator (Terasawa *et al.*, 1988). In rats, however, there does not appear to be such a close relationship between norepinephrine and LH pulses (Jarry *et al.*, 1990). Based on observations that each pulse of LH is preceded by a pulse of GnRH, and because direct measurements of GnRH secretion are so difficult to obtain, changes in the frequency of pulsatile LH release are often used as an indirect indication of changes in GnRH pulse frequency. However, it should be noted that modulations in LH pulse amplitude can only be interpreted to reflect changes in GnRH pulse amplitude if response of the pituitary to GnRH has not been altered.

IV. PITUITARY

At the level of the anterior pituitary gland, there are morphological changes following removal of the ovaries. These changes result from release of the gonadotrophs from the inhibitory influence of ovarian hormones that normally keep the biosynthesis and secretion of LH and FSH in check in intact animals. Within a period of several weeks, there is the appearance of "castration cells," or enlarged gonadotrophs. Microscopically, most of these cells assume the appearance of a signet ring due to the formation of a large vacuole which fills most of the cytoplasm. Their ultrastructure is characterized by swollen cisternae of the endoplasmic reticulum, large vacuoles, a large well-developed Golgi zone, and abundant mature secretory granules in the cytoplasm. In contrast to these changes, the lactotrophs, responsible for synthesis and secretion of prolactin, decrease in size and number, and prolactin synthesis and secretion decrease, resulting in low serum prolactin levels following ovariectomy.

With the advent of radioimmunoassays, one of the first observations of a change in the hypothalamic–pituitary axis following ovariectomy and the consequent removal of the negative feedback action of ovarian hormones was the dramatic increase in pituitary LH (Gay and Midgley, 1969; Yamamoto *et al.*, 1970) and FSH secretion (Zanisi and Martini, 1975). Mean plasma FSH levels rise much faster than circulating LH concentrations, increasing significantly within 24 hr, and continuing to rise at a rapid rate for about 7 days. Thereafter, FSH increases at a slower rate, attaining maximal levels about 3- or 4-fold above those observed in intact animals over a period of 3 or 4 weeks after castration. Plasma LH levels attain maximal levels that are about 20- to 50-fold greater than intact levels within about 4 weeks after ovariectomy, but increase much more gradually. Thus, although FSH attains half-maximal levels within 2 weeks after castration, plasma LH levels take about 3 weeks to increase to half-maximal levels. This differential rate of rise between the two gonadotropins is attributed in part to findings indicating that FSH secretion is controlled by the feedback actions of both estradiol and inhibin, the latter perhaps playing the major role, whereas LH secretion is mainly under the control of the negative feedback action of estradiol.

The increase in circulating LH levels after gonadectomy is slower in females than in males. Plasma LH levels undergo an approximate 40-fold increase in both sexes; however, in males the maximum level is attained in 3 weeks instead of in 4 weeks, as in females. Furthermore, although a small but significant increase in circulating LH concentrations occurs as early as 1 day after ovariectomy (Leipheimer and Gallo, 1983), LH levels in females take about 2 weeks to attain the same level that is observed within 1 day after castration in males (Blackwell and Amoss, 1970). In contrast, the increase in plasma FSH concentrations is similar in both sexes, increasing to a maximum level about 3- or 4-fold above that observed in intact animals over a period of 4 weeks (Zanisi and Martini, 1975).

The response to ovariectomy in terms of the early increase in plasma LH levels varies depending on the day of the cycle that surgery is performed. Plasma LH levels begin to increase sooner after ovariectomy on diestrus (1 or 2) than on metestrus or proestrus (Yamamoto et al., 1970; Tapper et al., 1972).

Gonadotropin secretion, like GnRH secretion, is pulsatile. Pulsatile LH secretion is much more tightly correlated with GnRH pulses than FSH, which is more heavily glycosylated than LH and therefore has a longer half-life. Because of this, the greater the frequency of GnRH input, the more likely serum LH is to increase than is serum FSH. Thus, when concomitant measurements of GnRH, LH, and FSH are made, LH pulses are always preceded by a GnRH pulse, but this is not the case for FSH pulses (Levine et al., 1991).

Both LH pulse frequency and amplitude increase between 17 and 24 hr after ovariectomy (Leipheimer and Gallo, 1983). Furthermore, by 8 days after ovariectomy the maximum LH pulse frequency of about one pulse every 20–30 min is attained; however, LH pulse amplitude continues to increase for at least 2 more weeks. In ovariectomized rhesus monkeys, maximal LH pulse frequency is approximately one per hour and is therefore referred to as "circhoral."

In the anterior pituitary gland, ovariectomy also causes increases in pituitary GnRH receptor mRNA (Kaiser et al., 1993) and GnRH receptors and LH content (Clayton and Catt, 1981), thereby increasing pituitary response to GnRH. Although serum LH concentrations increase three- or four-fold as early as 24 hr after ovariectomy in female rats, pituitary LH content and GnRH receptors do not begin to increase until 3 days postoperatively (Clayton and Catt, 1981). These parameters continue to increase gradually and attain their maximum levels sometime during the next several weeks. The increases in pituitary LH content and GnRH receptor content can be prevented by administration of estradiol with or without progesterone, indicating that these elevations are due to removal of the feedback actions of these ovarian steroids.

With regard to changes in LH synthesis, the postcastration increase in LH secretion is accompanied by changes in steady-state mRNA levels and transcription rates for the gonadotropin subunit genes. Thus, within 2 or 3 weeks after ovariectomy, increases in all three gonadotropin subunit mRNAs are maximal (α subunit, 4- to 8-fold; LHβ subunit, 8- to 14-fold; and FSHβ subunit, 2- to 4-fold) (Gharib et al., 1990). Furthermore, it has been demonstrated that the negative feedback action of estradiol on gonadotropin gene expression occurs primarily at the hypothalamus by altering GnRH pulse frequency (Shupnik, 1996).

V. HORMONE REPLACEMENT THERAPY

Administration of steroids following ovariectomy can reverse the effects of ovariectomy. Whether partial or full restoration is achieved depends on the specific effect, on the time after ovariectomy treatment is begun, and on the dose and type of steroid hormone administered. Replacement therapy that mimics physiologic conditions as closely as possible and begins as soon as possible after ovariectomy can restore circulating gonadotropins to their physiologic levels. In this regard, it should be noted that the steroid-treated ovariectomized animal has been and continues to serve as a valuable experimental model for investigating the role of steroid feedback mechanisms in virtually every aspect of reproductive endocrinology. For example, this model has been utilized in numerous experimental paradigms to determine the responsiveness of the brain and pituitary to the negative feedback action of the ovarian hormones and to determine whether this responsiveness plays a role in the mechanisms controlling transitions between cyclicity and acyclicity, such as puberty, seasonal breeding, or aging.

In addition to the withdrawal of the negative feedback actions of ovarian steroids, ovariectomy also leads to a complete absence of the positive feedback action of estradiol, and thus an absence of gonadotropin surges, which normally lead to ovulation. Thus, elucidation of the mechanism whereby the positive feedback action of ovarian steroids elicits gonadotropin surges is often conducted in the steroid-treated ovariectomized rat model.

See Also the Following Articles

FSH (Follicle-Stimulating Hormone); GnRH (Gonadotropin-Releasing Hormone); LH (Luteinizing Hormone); Ovarian Function in the Perimenopause; Ovarian Hormones, Overview

Bibliography

Barrell, G. K., Moenter, S. M., Caraty, A., and Karsch, F. J. (1992). Seasonal changes of gonadotropin-releasing hormone secretion in the ewe. *Biol. Reprod.* **46**, 1130–1135.

Blackwell, R. E., and Amoss, M. S., Jr. (1970). A sex difference in the rate of rise of plasma LH in rats following gonadectomy. *Proc. Soc. Exp. Biol. Med.* **136**, 11–14.

Clayton, R. N., and Catt, K. J. (1981). Gonadotropin-releasing hormone receptors: Characterization, physiological regulation, and relationship to reproductive function. *Endocrine Rev.* **2**, 186–209.

Gay, V. L., and Midgley, A. R., Jr. (1969). Response of the adult rat to orchidectomy and ovariectomy as determined by LH radioimmunoassay. *Endocrinology* **84**, 1359–1364.

Gharib, S. D., Wierman, M. E., Shupnik, M. A., and Chin, W. W. (1990). Molecular biology of the pituitary gonadotropins. *Endocrine Rev.* **11**, 177–199.

Jarry, H., Leonhardt, S., and Wuttke, W. (1990). A norepinephrine-dependent mechanism in the preoptic/anterior hypothalamic area but not in the mediobasal hypothalamus is involved in the regulation of the gonadotropin-releasing hormone pulse generator in ovariectomized rats. *Neuroendocrinology* **51**, 337–344.

Kaiser, U. B., Jakubowiak, A., Steinberger, A., and Chin, W. W. (1993). Regulation of rat pituitary gonadotropin-releasing hormone receptor mRNA levels in vivo and in vitro. *Endocrinology* **133**, 931–934.

Karsch, F. J., Dahl, G. E., Evans, N. P., Manning, J. M., Mayfield, K. P., Moenter, S. M., and Foster, D. L. (1993). Seasonal changes in gonadotropin-releasing hormone secretion in the ewe: Alteration in response to the negative feedback action of estradiol. *Biol. Reprod.* **49**, 1377–1383.

Kawakami, M., Uemura, T., and Hayashi, R. (1982). Electrophysiological correlates of pulsatile gonadotropin release in rats. *Neuroendocrinology* **35**, 63–67.

Kelly, M. J., Garrett, J., Bosch, M. A., Roselli, C. E., Douglass, J., Adelman, J. P., and Ronnekliev, O. K. (1989). Effects of ovariectomy on GnRH mRNA, proGnRH and GnRH levels in the preoptic hypothalamus of the female rat. *Neuroendocrinology* **49**, 88–97.

Kobayashi, R. M., Lu, K. H., Moore, R. Y., and Yen, S. S. C. (1978). Regional distribution of hypothalamic luteinizing hormone-releasing hormone in proestrous rats: Effects of ovariectomy and estrogen replacement. *Endocrinology* **102**, 98–105.

Leipheimer, R. E., and Gallo, R. V. (1983). Acute and long-term changes in central and pituitary mechanisms regulating pulsatile luteinizing hormone secretion after ovariectomy in the rat. *Neuroendocrinology* **37**, 421–426.

Levine, J. E., and Ramirez, V. D. (1982). Luteinizing hormone-releasing hormone release during the rat estrous cycle and after ovariectomy as estimated with push–pull cannulae. *Endocrinology* **111**, 1439–1448.

Levine, J. E., Pau, K.-Y. F., Ramirez, V. D., and Jackson, G. L. (1982). Simultaneous measurement of luteinizing hormone-releasing hormone and luteinizing hormone release in unanesthetized, ovariectomized sheep. *Endocrinology* **111**, 1449–1455.

Levine, J. E., Norman, R. L., Gliessman, P. M., Oyama, T. T., Bangsberg, D. R., and Spies, H. G. (1985). In vivo gonadotropin-releasing hormone release and serum luteinizing hormone measurements in ovariectomized, estrogen-treated rhesus macaques. *Endocrinology* **117**, 711–721.

Levine, J. E., Bauer-Dantoin, A. C., Besecke, L. M., Conaghan, L. A., Legan, S. J., Meredith, J. M., Strobl, F. J., Urban, J. H., Vogelsong, K. M., and Wolfe, A. M. (1991). Neuroendocrine regulation of the luteinizing hormone-releasing hormone pulse generator in the rat. *Recent Prog. Horm. Res.* **47**, 97–153.

Petraglia, F., Sutton, S., Vale, W., and Plotsky, P. (1987). Corticotropin-releasing factor decreases plasma luteinizing hormone levels in female rats by inhibiting gonadotropin-releasing hormone release into hypophysial–portal circulation. *Endocrinology* **120**, 1083–1088.

Sarkar, D. K., and Fink, G. (1980). Luteinizing hormone releasing factor in pituitary stalk plasma from long-term ovariectomized rats: Effects of steroids. *J. Endocrinol.* **86**, 511–524.

Sherwood, N. M., and Fink, G. (1980). Effect of ovariectomy and adrenalectomy on luteinizing hormone-releasing hormone in pituitary stalk blood from female rats. *Endocrinology* **106**, 363–367.

Shugrue, P. J., Bushnell, C. D., and Dorsa, D. M. (1992).

Estrogen receptor messenger ribonucleic acid in female rat brain during the estrous cycle: A comparison with ovariectomized females and intact males. *Endocrinology* **131**, 381–388.

Shupnik, M. A. (1996). Gonadal hormone feedback on pituitary gonadotropin genes. *Trends Endocrinol. Metab.* **7**, 272–276.

Tapper, C. M., Naftolin, F., and Brown-Grant, K. (1972). Influence of the reproductive state at the time of operation on the early response to ovariectomy in the rat. *J. Endocrinol.* **53**, 47–57.

Terasawa, E., Krook, C., Hei, D. L., Gearing, M., Schultz, N. J., and Davis, G. A. (1988). Norepinephrine is a possible neurotransmitter stimulating pulsatile release of luteinizing hormone-releasing hormone in the rhesus monkey. *Endocrinology* **123**, 1808–1816.

Thiery, J. C., and Pelletier, J. (1981). Multiunit activity in the anterior median eminence and adjacent areas of the hypothalamus of the ewe in relation to LH secretion. *Neuroendocrinology* **32**, 217–224.

Van Vugt, D. A., Diefenbach, W. D., Alston, E., and Ferin, M. (1985). Gonadotropin-releasing hormone pulses in third ventricular cerebrospinal fluid of ovariectomized rhesus monkeys: Correlation with luteinizing hormone pulses. *Endocrinology* **117**, 1550–1558.

Wilson, R. C., Kesner, J. S., Kaufman, J.-M., Uemura, T., Akema, T., and Knobil, E. (1984). Central electrophysiologic correlates of pulsatile luteinizing hormone secretion in the rhesus monkey. *Neuroendocrinology* **39**, 256–260.

Yamamoto, M., Diebel, N. D., and Bogdanove, E. M. (1970). Analysis of initial and delayed effects of orchidectomy and ovariectomy on pituitary and serum LH levels in adult and immature rats. *Endocrinology* **86**, 1102–1111.

Zanisi, M., and Martini, L. (1975). Differential effects of castration on LH and FSH secretion in male and female rats. *Acta Endocrinol.* **78**, 683–688.

Castration, Effects in Nonhuman Mammals (Male)

Graeme B. Martin and David R. Lindsay

University of Western Australia

I. Endocrine Consequences
II. Anatomical Consequences
III. Behavioral Consequences
IV. Practical Applications
V. Future Developments

GLOSSARY

androgen, estrogen, and progestagen The family names for the sex steroids; androgens are "male generating," estrogens are "female generating," and progestagens are "pregnancy generating."

castration, gonadectomy, ovariectomy, and caponization The removal of the gonads of either sex; castration is often reserved for males and ovariectomy for females; a capon is a castrated male bird.

cryptorchid (ism) The retention of the testes within the body cavity, either because they fail to descend or because the scrotum is removed.

endocrinology The discipline in which hormones are studied.

gonadotropin The hormones produced by the anterior pituitary gland that control the gonads; the principal two are luteinizing hormone and follicle-stimulating hormone.

gonadotropin-releasing hormone The neurohormone produced in the brain which controls the secretion of gonadotropins by the anterior pituitary gland.

gonads The primary sex organs; ovaries (female) and testes (male).

hypothalamus A small structure at the base of the brain that sits just above the pituitary gland; consists of many different clusters of cells which control functions such as sexual behavior, secretion of hormones, and the regulation of homeostasis.

neuroendocrinology A subset of endocrinology that concerns the study of neurohormones and the interactions between the nervous and hormonal systems.

neurohormone A hormone produced and secreted by a nerve cell in the same way as a neurotransmitter, but which enters the bloodstream that travels from the hypothalamus to the pituitary gland; neurohormones control pituitary activity (e.g., gonadotropin-releasing hormone).

neurotransmitter A substance produced by a nerve cell that is secreted so as to influence the activity of a neighboring nerve cell.

pheromone A substance produced and excreted by an animal that influences the physiology or behavior of another animal (usually of the same species).

pituitary gland A gland suspended below the base of the brain that is controlled by the brain and that, in turn, controls many other organs in the body by producing a range of protein hormones.

The practice of castration has historically been limited to males, especially in mammals, because the gonads are generally more easily accessible in males than in females. It has been a common practice for millennia, probably since the first realization that the scrotal contents were associated with fertility and sexual and aggressive behavior. The aim was, of course, to control these aspects of the animals' biology. Modern science has also made great use of castration as a tool for investigating the principles that underpin reproductive physiology and behavior in both sexes of a wide variety of species, including primates. This article discusses the anatomical, physiological, and behavioral consequences of gonadectomy in nonprimate males and in some females. Laboratory rodents have been an important source of information regarding the physiological mechanisms involved in responses to castration, but we are keen to limit our scope as much as possible to farm and pest animals, and particularly to mammals, because this is where the technique has been, and will be, applied to greatest benefit. In recent times, simple surgical practices have also been developed for females, as have noninvasive pharmacological approaches for both sexes, so that now we are beginning to see an expansion of the possibilities of gonadectomy as a management tool.

I. ENDOCRINE CONSEQUENCES

The effects of castration and gonadal hormone replacement have been intensively studied for decades in nonprimates. We will not cover the details of this early work here because it has been reviewed in-depth by Lincoln and Short and Martin.

A. Loss of Gonadal Hormones

The gonads produce an extremely wide variety of hormonally active substances. Apart from a large number of sex (and other) steroids, most of the major endocrine families are present, including prostaglandins, growth factors, and cytokines. Many of these only act locally within the gonad, so they do not concern us here, but many others are transported out of the gonad in the blood or lymphatic circulations. The most profound and recognizable consequences of castration on the physiological and behavioral functions of the animal are attributed to the loss of sex steroids. However, the roles of, and interactions between, most of the other gonadal products, or the consequences of their loss, are still poorly understood yet might contribute significantly to the responses of many animals to castration.

1. Male

Testosterone and the reduced derivative, 5α-dihydrotestosterone (DHT), are probably the most powerful of the naturally occurring androgens and their loss is probably responsible for most of the externally recognizable effects of castration. The Leydig cells in the testes produce both of these steroids and, in addition, testosterone is reduced to DHT in many target tissues, particularly the accessory sex glands.

The androgen family is popularly perceived to be the only endocrine product of the testis. However, the testis of most species has also been shown to produce the quintessential female sex steroids, the estrogens, particularly estradiol-17β. Estrogens are synthesized from testosterone by the enzyme aromatase, and there is some debate over where this actually happens in the testis. In the rat, it seems most likely to be in the Leydig cells, but it may take place in the Sertoli cells of other species. In addition, aromatase is expressed in several other areas of the body, including the brain and pituitary gland, so estrogens can be synthesized locally using the testicular testosterone as a substrate (aromatase cannot convert DHT to estrogen). Only small masses of estradiol are produced compared to the androgens, but it is a very potent steroid and we are only now beginning to realize the importance of the roles that it might play in the regulation of many physiological and behavioral processes in the male. Thus, we need to be cautious in our interpretation of the consequences of castration, some of which may be due to the loss of the small amounts of estradiol produced by the testis or to the loss of estradiol produced by aromatization of androgen locally within the target tissue.

2. Female

The most important female sex steroids are undoubtedly the estrogens (especially estradiol-17β), produced mostly by the ovarian follicle, and the progestagens (especially progesterone), produced mostly by the corpora lutea. The loss of these two probably accounts for most of the externally recognizable consequences of ovariectomy. The follicles in the ovaries also produce substantial masses of androgen, primarily Δ^4-androstenedione and testosterone. Androstenedione is relatively inactive and does not seem to play a major role in the animal, although it may be locally aromatized and thus exert effects in tissues that have the necessary enzyme. Testosterone is secreted in quantities that match those for estrogen, and it can also be aromatized, but whether it plays a role as an androgen, and whether the loss of testosterone contributes to consequences of gonadectomy, is not clear.

Progestagens play their most important roles in mammary tissue, the reproductive tract (particularly the uterus), and in the brain functions that control the secretion of gonadotropins and sexual and parental behavior. Moreover, many of these functions depend on synergistic interactions with estrogens. The absence of both of these "female" steroids thus leads to reductions in mammary and uterine tissue and reduces the expression of the behaviors that are associated with reproduction.

B. Loss of the Inhibin Family

The inhibin family, which includes activin and follistatin and is part of the transforming growth factor superfamily of hormones, has been extensively studied since the 1970s. Despite the fact that they are members of a growth factor family, the physiological and anatomical consequences of their complete removal by gonadectomy have not been clearly demonstrated, except for the important role of inhibin on the control of the secretion of follicle-stimulating hormone (FSH). This is currently a very active area of research.

C. Hypothalamus–Pituitary Axis

1. Neurotransmitters and Neurohormones

The central nervous system is a major target for gonadal hormones, particularly the steroids, so we expect to see neurophysiological, neuroendocrine, and perhaps anatomical changes in brain tissue following castration. The neuroendocrine consequences have been studied in great detail, particularly the systems controlling the activity of the cells that produce gonadotropin-releasing hormone (GnRH), the neurohormone that controls pituitary gonadotropin secretion and thus the reproductive system. GnRH is secreted into pituitary portal blood in pulses and there is an increase in the frequency of these pulses after castration. This leads to a decrease in the amount of this neurohormone that is stored in hypothalamic tissue.

The effects of castration and steroid replacement on the neurotransmitter systems that inhibit or stimulate GnRH secretion are an area of intensive research, but a clear picture has not yet emerged. In fact, the final picture is unlikely to be simple because of the wide variety of neurotransmitters that are involved and because the responses to castration will be affected by environmental factors, species, and sex, as well as by the contributions of the negative and positive feedback systems. The negative feedback system is responsible for maintaining an equilibrium in the reproductive axis in both sexes (Fig. 1). The positive feedback system is responsible for eliciting the surge of gonadotropins that induces ovulation, and it is present in both sexes in some, but not all, species.

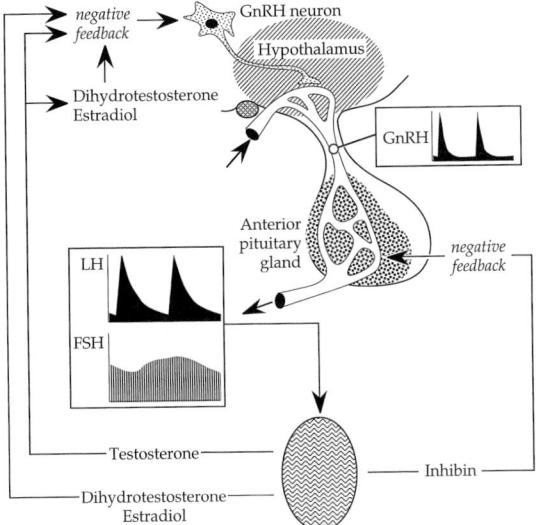

FIGURE 1 Endocrine links between the hypothalamus, pituitary gland, and testis. Note (i) the pulsatile nature of GnRH secretion; (ii) the GnRH pulses are transposed into LH pulses on a one-to-one basis; (iii) the GnRH pulses are not transposed into pulses of FSH, so this hormone is secreted in a relatively continuous manner; (iv) inhibin acts only on the pituitary gland where it selectively reduces the secretion of FSH; (v) for the sake of simplicity, we have indicated feedback directly onto the GnRH neurons, but at least one other set of neurons intervenes; and (vi) dihydrotestosterone and estradiol may be secreted by testicular tissue or they may be produced at other sites (e.g., brain) from testosterone. In any case, they will be involved in the negative feedback loop.

2. Pituitary Anatomy and Secretion of Gonadotropins

A classical histological response to castration is the appearance of "signet ring cells" in the pituitary gland due to the formation of a very large vacuole in the cytoplasm of some cells. This is apparently a pathological response of the gonadotropes to the lack of gonadal hormones (steroids and inhibin) and it seems to be estrogen that is the most important factor in both sexes.

There is an increase in the rate of secretion of both of the pituitary gonadotropins, luteinizing hormone (LH) and FSH, as a consequence of both the increase in GnRH secretion and the loss of direct inhibitory effects of the sex steroids (and inhibin for FSH) on the gonadotropes. The pattern of LH secretion is pulsatile because the release of the gonadotropin from the pituitary cells is directly controlled by the GnRH pulses (Fig. 1). Thus, withdrawal of the pituitary effects of gonadal hormones sometimes leads to an increase in the amplitude of the LH pulses (due to an increase in responsiveness to GnRH) and an increase in the basal secretion (reflecting tonic release between the GnRH-driven LH pulses). These changes vary with species and the time since castration.

In contrast to LH secretion, the secretion of FSH does not seem to be pulsatile because the gonadotropes have a different mechanism for packaging and delivering this hormone. The high rate of secretion of FSH in castrates is due to loss of inhibin as well as the sex steroids (Fig. 1). The roles of activin and follistatin in the responses to castration are not yet clear.

3. Long- vs Short-Term Castration

It has often been remarked in studies of feedback systems that it is difficult to restore completely normal patterns of gonadotropin secretion in long-term castrates with what seems to be an appropriate regime of hormonal replacement therapy. Some of these problems may be due to the simplistic nature of the hormonal replacement, because some components of the gonadal secretions have been omitted, or it may be due to the need for specific combinations. For example, the synergism between estrogen and progesterone in the inhibition of GnRH pulse frequency is probably due to induction of progesterone receptors by estrogen. It is also possible that, after long periods of very low concentrations of gonadal hormones, some permanent damage is done to the hypothalamic and pituitary tissue. Certainly, the signet ring cells are an indication that this is the case for the gonadotropes. In addition, a lack of gonadal hormones, generally assumed to be steroids but probably also including inhibin, alters the intracellular processing of the gonadotropins so that the hormone secreted has a longer half-life. This makes it more difficult to restore normal concentrations of LH and FSH in the circulation, especially with short-term regimes for replacement of gonadal hormones.

At the hypothalamic level, there seems to be a loss in responsiveness to seasonal cues with successive years after ovariectomy.

4. Responses to Environmental Cues and Factors

In addition to the effects on expression of responses to photoperiod mentioned previously, gonadectomy may also alter the way that an animal responds to other environmental factors. In small ruminants, for example, it appears that the pathways through which nutrition and stress affect gonadotropin secretion depend greatly on the presence of gonadal steroids. On the other hand, ovariectomized females still show an increase in pulsatile LH secretion when presented with a sociosexual stimulus, in much the same way that they show such responses to photoperiodic stimuli. Sex steroids serve to enhance these responses.

D. Calcium Balance

The relationships between the gonadal steroids, particularly estrogen, and osteoporosis in postmenopausal women have led to the development of several animal models for studying the mechanisms involved. The ovariectomized rat is perhaps the most common and has been used to assess the roles of phytoestrogens in the control of bone loss, for example, but bigger mammals such as sheep are also proving useful and are arguably more appropriate as a model for the human condition.

II. ANATOMICAL CONSEQUENCES

In most mammalian species, the fetal testis plays an important role in the differentiation of the internal and external genitalia and on sexual behavior. Without the two key hormones produced in the fetal testis, Mullerian-inhibiting hormone (MIH) and androgen, the primordial ducts differentiate to a female form regardless of the genetic sex of the animal. On the other hand, in the presence of a functional testis, the primordial wolffian ducts are stimulated by the androgen to develop male internal genitalia and further growth of the primordial Mullerian ducts is prevented by MIH. Thus, castration of the fetal male results in its developing female internal genitalia, whereas castration of the fetal female makes no difference to the normal development of the female form.

In domestic species, it is routine to castrate males soon after birth, by which stage the type of genitalia is already established and the effects are confined to modifying development of the secondary sexual characteristics and controlling aspects of behavior. The appearance of castrated males may be highly variable depending on the age at which castration was performed. In sexually dimorphic species, the earlier the time of castration in males, the more their appearance resembles that of females. Secondary sexual characteristics do not develop further after castration and some, such as deepness of voice, may even regress.

A. Secondary Sex Organs

The degree of departure from normality following castration also depends largely on the timing of the gonadectomy relative to the pubertal development of the animal. Thus, gonadectomy before puberty has more profound consequences than gonadectomy after puberty, by which time many of the accessory sexual organs will have developed. Basically, initial development is easy to prevent and involution is a slow process that often does not go to completion so that a return to prepubertal structures is rare. Apart from the obvious loss of the testes (and often the scrotum), there is usually a diminution in the size of the prepuce and penis and loss of erectile function. In general, these changes are accompanied by reductions in the size of the various accessory glands (the ampulla, the seminal vesicles, the prostate gland, and the bulbourethral glands). The extent and specific consequences of these reductions depend on the species, which, of course, vary greatly in the combinations of glands with which the males are endowed.

In the female, the accessory glands tend to be less numerous and less obvious. However, loss of mammary gland tissue, a poorly developed vulva, and glandular tissue lining the tract (and thus poor production of mucus) are clear consequences of withdrawal of ovarian steroids.

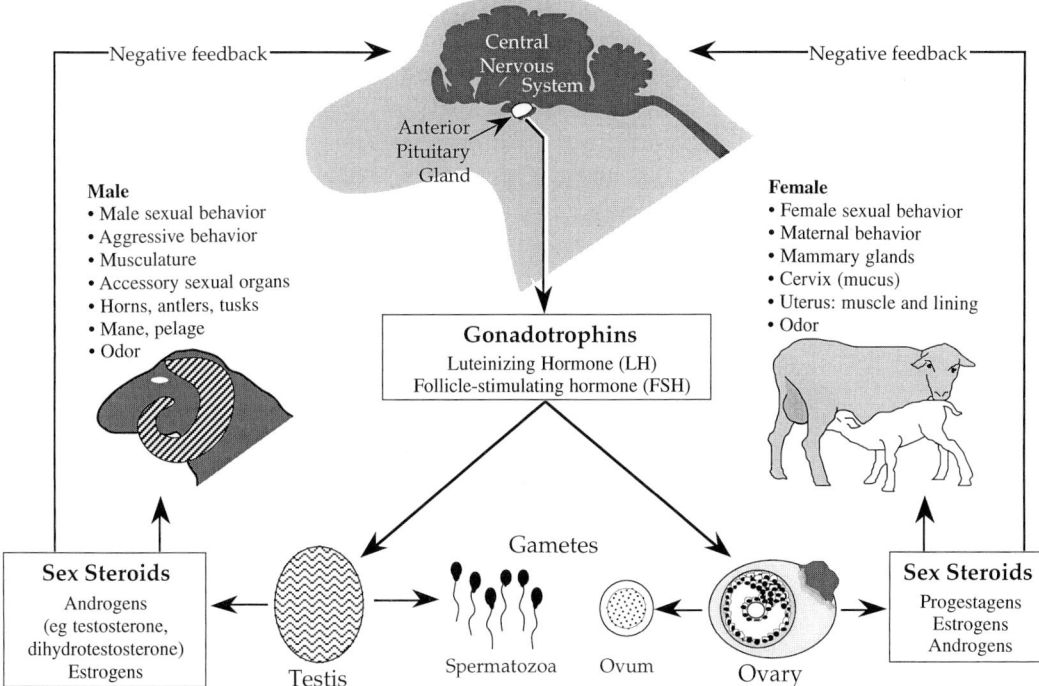

FIGURE 2 In the female, the gonadotropins stimulate the follicles and the corpus luteum in the ovary to produce steroids, and they also stimulate the growth and development of the ovum within the follicles. In the male, they stimulate the production of steroids and the activity of the seminiferous tubules where the spermatozoa are produced. In contrast to the gonadotropins, the sex steroids affect the activity of a much wider range of tissues. The broadest possible overview suggests that they are responsible for bringing together the gametes and nurturing the subsequently produced conceptus and offspring. All of these activities will be lost, some to a greater degree than others, following castration.

B. Secondary Sex Characters

Because sexual differentiation is driven by the sex steroids, the secondary sexual characters are among the first aspects of an animal that are affected (Fig. 2).

1. Sexual Display Characters

In many species, castrate males are clearly identified by the lack of overt male characteristics, particularly the lack of horns, antlers, or tusks in species and breeds that carry them and the lack of display features, such as extra pelage, colored pelage or skin, and skin folds, often from around the neck and shoulders. These are characteristics of so-called "sexually differentiated" species, an anthropomorphic convenience that belies the ability of members of nondifferentiated species to tell their sexes apart. The loss of these masculine features is usually attributed entirely to the loss of androgens because testosterone replacement can restore them. In some cases, however, the obligatory nature of this relationship is doubtful because of the potential for local aromatization of testosterone to estrogen in the responsive tissues.

2. Odoriferous Tissues

Generally, humans tend to notice the odors of male animals more than those of females, and this is often the reason that male meat animals are castrated (Fig. 3). Apart from the question of environmental "pollution," these odors also become taints in the meat.

The ethological aspects of these odors, due to their pheromonal qualities, are essential for sociosexual interactions in many species. They play critical roles in the marking of territories, defense of harems, and control of sexual functions in mates. In small ruminants, the pheromones from both sexes can induce gonadotropin secretion in the opposite sex within

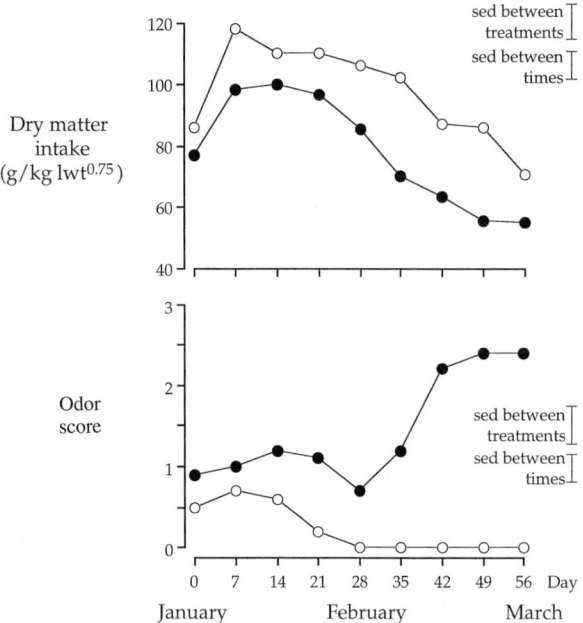

FIGURE 3 Dry matter intake and odor score of entire (●; $n = 5$) and castrated (○; $n = 5$) male goats as the autumn breeding season commences in the Southern Hemisphere (redrawn from Godfrey et al., 1996)

minutes of exposure, and the overall response of the female to the male can be sufficiently powerful to induce ovulation in females that are seasonally anovulatory. In fact, for centuries, farmers have been unknowingly taking advantage of this phenomenon to breed their flocks out of season.

Here again, there is great diversity between the species with regard to the target tissues for the sex steroids that are responsible for pheromone production. The ovulation-inducing pheromone of sheep and goats seems to be produced by the skin all over the body, although there has been some speculation that the blood supply to the skin inside the hindlegs plays a role. In cattle, the female seems to produce male-attracting pheromones from the skin around the vulva. Male pigs produce their pheromone in saliva, whereas the pheromones produced by rodents are produced by the preputial glands and are thus carried in the urine.

3. Body Conformation

If mature bulls and steers are placed together, the differences in body conformation are most evident. Bulls are usually much larger at the same age and have much heavier musculature around the shoulders and neck. This secondary sexual characteristic has evolved as a consequence of the drive to capture and defend a harem at mating time. To reach the larger size, intact males grow at a faster rate than castrates, and they deposit muscle and fat at strategically important parts of the body.

C. Longevity

Castrated males of some species appear to live longer than noncastrates. For example, in an undisturbed flock of sheep in the Western Isles of Scotland, the entire males live only about a third as long as the females, whereas castrated males achieve the same life span as the females. This might be a direct effect of testicular hormones on longevity, a consequence of sexual activity and competition for females, or the seasonal depression of appetite by testosterone during autumn (as shown for goats in Fig. 3). The result is that ewes and castrate rams are probably better buffered than intact rams against the nutritional hardships of the winter that follows the mating season. Longevity is of minor interest in domestic situations in which animals, especially castrates, seldom complete their natural life span but, as we have seen, the impact on body condition is most important.

III. BEHAVIORAL CONSEQUENCES

As can be seen in Fig. 2, the sex steroids are directly responsible for a wide range of behaviors, so the consequences of castration include effects on establishment and marking of territory, aggressive behavior, sexual behavior, and parental behavior.

A. Sexual Behavior

1. Early Castration

The timing of castration affects behavioral activity in the same way as it affects anatomical development. In most species, the central nervous system is sexually differentiated and the identity of the hormones that might trigger sexual behavior is of secondary importance. Early experiments with rats showed that there is a critical period within the few days after birth when the brain appears to be programmed for the rest of its life to elicit sexual behavior under the appropriate stimuli. Young male rats castrated at this time show a much greater propensity to display female behavior patterns if given ovarian hormones later in life. Young females injected with testosterone or estrogen at this critical stage will show, at best, only aberrant female behavior in later life and generally are totally acyclic. In species other than rodents, the critical period for the feminizing effect of castration of males is in fetal rather than postnatal life, a difference that reflects variation in physiological maturity at birth.

2. Late Castration

In most species, both sexes show incomplete sexual behavior patterns after castration, and the degree of sexual quiescence is often related to the time since castration. This is not surprising since sexual behavior is a complex characteristic with neural, psychological, and hormonal determinants. Removal of the hormonal component by castration does not automatically eliminate the others and some of the more advanced mammals may display at least some aspects of sexual activity for a considerable time after castration (e.g., mounting but not intromission). In the castrated male sheep and, no doubt, in other species, a few days of replacement therapy with testosterone restores male sexual activity, including the capacity to stimulate anestrous females to begin cyclical sexual activity.

Ovariectomy in the female of cyclically breeding species at any stage of life completely eliminates both sexual proceptivity (male-seeking behavior) and receptivity (acceptance of mounting by males). Both of these can be rapidly restored by the administration of an appropriate regimen of female hormones.

B. Aggressive Behavior

The influence of gonadal hormones on aggressive behavior is generally poorly understood. This is probably because of the very strong interactions between the social environment of animals and the type and amount of aggression that they display. Some species are stimulated more by particular social interactions than by their gonadal hormones. For example, in some species only the strongest males participate in the reproductive activity of the population and the weaker animals are totally excluded. Their strength is physically demonstrated through aggressiveness which may persist throughout the year or be confined to periods of seasonal sexual activity. Nonetheless, the primary or secondary consequence of castration, and therefore a lack of sexual hormones in both males and females, is diminished aggressiveness particularly at the time of sexual activity in noncastrated flock or herd mates.

C. Parental Behavior

The level of postnatal care is most intensive in species in which only a few offspring are born or hatched. The nesting and nurturing behaviors associated with this care are induced primarily by the sex steroids, with some assistance from other hormones, such as oxytocin and prolactin. This applies to all species studied thus far, including mammals and birds. Thus, castration blocks the expression of parental behavior, and this behavior can be induced in castrates by treatments with estrogen and progesterone. In mammals, the induction process requires other stimuli, such as vaginal dilation, to trigger the acute onset of the behavior. In birds, the behavior is triggered by prolactin.

Clearly, this work has mostly been done in females (sheep, rats, chickens, and turkeys). How parental behavior is induced in males, particularly in species such as the emu, in which only the male incubates the eggs and nurtures the offspring, remains to be resolved.

Finally, in mammals, the growth, development, and differentiation of the mammary glands also depend on the female sex steroids. In the absence of the ovaries, lactation is not feasible without treatment with estrogen and progestagen.

IV. PRACTICAL APPLICATIONS

It is a very common practice in animal husbandry to castrate males before puberty because sexual and aggressive behavior, not to mention unwanted pregnancies, can lead to unacceptable levels of interference with growth and productivity. In addition, castration can prevent some undesirable side effects, such as meat taints.

A. Control of Sexual Behavior and Fertility

In enterprises in which males reach puberty before the normal age of sale, the risks and inconvenience of separating the sexes to avoid unwanted pregnancies are high. In these cases, the solution for centuries has been to castrate all males at an early age except those being saved for later breeding.

B. Aggressive Behavior

Association by humans with large male animals, domestic or otherwise, often involves physical risk. In the domestic species, males are usually castrated to lessen this risk but, as with sexual behavior, removal of hormonal stimulation alone does not guarantee complete docility.

C. Appetite

In many species, such as deer, sheep, goats, and a variety of birds, the males show very pronounced changes in appetite with season, usually with a depression around the onset of the breeding season (Fig. 3). In the more seasonal breeds of sheep and goats, for example, the males eat voraciously over summer then reduce their feed intake in autumn, independently of the presence of females or the amount of feed supplied. There is a commercial advantage in avoiding the appetite loss and the phenomenon is partly dependent on the sex steroids. Thus, the appetite depression is strongly reduced by castration or by immunization against GnRH. On the other hand, some of the seasonal loss of appetite is retained by castrates (Fig. 3) so nongonadal factors might also be involved in appetite control.

D. Meat Quality

Meat from castrates is usually considered to be more tender, brighter in color, juicier (due to the extra fat, usually called "marbling"), and taint-free compared with meat from intact males. In many circumstances, these views do not stand up to scientific scrutiny. For example, taste panel tests for sheep meat have shown that consumers cannot detect differences between ram, wether, or ewe in tenderness, taste, or odor.

On the other hand, in the goat and the boar, odors in the carcass may make the meat totally or partially unacceptable for some consumers. In the boar, this taint is due to levels of the pheromone 5α-androstenone in the fatty component in excess of 0.5–1.0 μg/g, or to levels of skatole, a metabolite of the amino acid tryptophan, >0.22 μg/g. The level of 5α-androstenone is related to the weight of the testes in boars and becomes a problem when the animals surpass about 80 kg live weight. Castration controls these odors.

E. Testosterone-Treated Castrates

The potential for male-induced ovulation for breeding for sheep and goats out of season has been understood since the 1940s and the phenomenon is widely exploited in many areas of the world. However, one of the managerial costs of this system was the need to retain infertile males, usually vasectomized rams or bucks. Apart from the cost of the surgery, care had to be taken with these animals so that they could be readily identified and not confused with the fertile males of the flock. The problem was solved when it was shown that androgen-treated castrates were as effective as intact males at inducing ovulation. This is now common practice.

V. FUTURE DEVELOPMENTS

The advantages of prepubertal castration of all males not wanted for breeding are clear. The question to address is whether this very common practice can be improved. In the extreme, does it even have a future?

Animal welfare, costs of castration, a worldwide drift toward extensification of animal industries, and changes in the nature of meat markets (toward less fat) are affecting the usage of castration as a management tool. It is probably only a matter of time before we will see campaigns against surgical castration of mammals, in much the same way as we have seen prohibition of the caponizing of birds and tail docking of dogs. Male taints are not regarded favorably in many Western countries so most of the male meat that is consumed is from castrated mature males or from very young intact males. On the other hand, in some countries, there are specialist markets in which products from gonad-intact males are preferred.

Animal breeders and keepers will thus ultimately be driven to develop alternative technologies in the mainstream farm animals, such as cattle, sheep, and goats. This also applies to any new species that are being considered for commercial development, particularly species with internal gonads such as birds.

Some Uses for Alternative Technologies

In terms of productivity, there is a constant tension between the beneficial effects of testicular hormones on growth and carcass quality or leanness and their detrimental effects on ease of handling and management of breeding stock. As higher carcass weights become more normal in commercial animals, this conflict will become more serious.

1. Meat Production

One of the major ancillary effects of testicular hormones is their anabolic activity. In sheep, for example, intact ram lambs grow 10–20% faster than castrated lambs ("wethers"). In addition to size, androgens also affect body conformation and composition, reducing synthesis and deposition of fat and promoting deposition of muscle, leading to a leaner carcass. It was once possible to replace this anabolic activity in castrates by treatment with synthetic and other hormones that mimicked the anabolic rather than the sexual activity of testicular hormones. This approach is now largely banned for reasons of welfare, productivity, and perceptions of danger to consumers.

In some cases, such as in the pig meat and chicken meat industries, the alternative has been to improve the rate of growth of the animals so that they can be sold from the enterprise before they reach puberty. In others, such as the lamb meat industry, there is interest in converting young males to cryptorchids.

2. Cryptorchidism

Cryptorchid testes do not produce viable spermatozoa but retain some of their ability to produce androgen. Thus, rams that have been made cryptorchid, by removal of the scrotum to prevent the natural descent of the testes, do not reach the same maximum body size or remain as lean as normal rams with descended testes, but they do grow faster (by about 10%) and have a leaner carcass than rams that are castrated before puberty.

3. Blocking the Action of GnRH

Suppression of the reproductive axis, particularly by immunization against GnRH or the use of implants of GnRH analogs, is the most promising alternative to surgical castration for reducing male sexual and aggressive behavior in adult mammals. This approach is often termed "hormonal castration" or "chemical castration."

i. Birds An excellent example regarding birds is the rapidly expanding industry based on ratite birds (ostrich, emu, and rhea), in which the major sources of economic loss are damage to skins and carcass caused by fighting and seasonal loss of body fat in birds destined for the abattoir. Most likely, these problems are caused by the sex steroids and thus could be solved by gonadectomy. Independently of the issue of animal welfare, a surgical approach is impractical in birds of this size. The obvious answer is hormonal castration, an approach that has been shown to be effective in chickens.

ii. Goats At the start of the breeding season, there are marked increases in the circulating concentrations of testosterone, and this is associated with a decline in voluntary feed intake and growth, increased male odor, and increased aggressive and sexual behaviors. In some circumstances, such as the retention of male kids to allow selection of superior producers of cashmere, a reversible method to control these behaviors may be desirable. In other circumstances, such as with feral goats "harvested" from the wild, nonsurgical castration could also be used to control the behavior of mature bucks. Immunization against GnRH, using commercially available antigens, has proven successful in achieving these goals.

iii. Pigs In the boar, odors in the carcass can also be controlled by castration but this practice is widely considered unacceptable. Importantly, there appears to be little difference in efficacy between surgical, immunological, or chemical castration. Anti-GnRH vaccines can reduce the levels of the two tainting compounds, 5α-androstenone and skatole, to those of surgically castrated boars ("barrows").

iv. Tropical Rangeland Cattle In the management of tropical rangeland cattle, human interference more than once or twice a year is uneconomical so the males are left with the herd for very long periods, often throughout the year. Two types of loss are caused by this practice. First, in response to a flush of feed in the wet season, heifers enter puberty and readily conceive before being fully grown, then find it difficult to survive the following dry season when the problems of feed shortage are compounded by the drains of pregnancy and lactation. Second, older cows that are past their best years and being prepared for market also conceive, thus losing condition and value. Until recently, there has been no solution to the loss of the heifers and their offspring. The problem with the old cows has been partially overcome by new techniques of surgical castration in the field without anesthetic. This practice is susceptible to ethical pressure.

As with goats, there is now a commercial product available for immunizing rangeland beef cattle against GnRH and it provides an elegant method for blocking ovulation for up to 12 months. Such products should be successful because they replace the surgical practice, their effects are generally reversible, and the labor input is low.

v. Pest Animals In recent years, there has been a considerable effort to find ways to reduce feral and native animal populations through reduction of their reproductive competence. Among the major targets are the possum in New Zealand, and the house mouse, the rabbit, and the fox in Australia. The pressure on fox hunting in Britain might lead to a similar need there. Some of the possibilities include inducing forms of immunity against gonadal hormones or against hormones that facilitate production of gonadal hormones.

See Also the Following Articles

CASTRATION, EFFECTS IN HUMANS; CASTRATION, EFFECTS IN NONHUMANS (FEMALE)

Bibliography

D'Occhio, M. J. (1993). Immunological suppression of reproductive functions in male and female mammals. *Anim. Reprod. Sci.* **33**, 345–372.

Findlay, J. K., Robertson, D. M., Clarke, I. J., Klein, R., Doughton, B. W., Xiao, S., Russell, D. L., and Shukovski, L. (1992). Hormonal regulation of reproduction—General concepts. *Anim. Reprod. Sci.* **28**, 319–328.

Godfrey, S. I., Walkden-Brown, S. W., Martin, G. B., and Speijers, E. J. (1996). Immunisation of goat bucks against GnRH to prevent seasonal reproductive and agonistic behaviour. *Anim. Reprod. Sci.* **44**, 41–54.

Kordon, C., and Drouva, S. V. (1992). Gonadotropin regulation, oestrogens and the immune system. *Horm. Res.* **37**(Suppl. 3), 11–15.

Lincoln, G. A., and Short, R. V. (1980). Seasonal breeding: Nature's contraceptive. *Recent Prog. Horm. Res.* **36**, 1–52.

Lindsay, D. R. (1996). Environment and reproductive behaviour. *Anim. Reprod. Sci.* **42**, 1–12.

Martin, G. B. (1984). Factors affecting the secretion of luteinizing hormone in the ewe. *Biol. Rev.* **59**, 1–87.

Martin, G. B. (1995). Reproductive research on farm animals for Australia—Some long-distance goals. *Reprod. Fertil. Dev.* **7**, 967–982.

Martin, G. B., and Walkden-Brown, S. W. (1995). Nutritional influences on reproduction in mature male sheep and goats. *J. Reprod. Fertil. Suppl.* **49**, 437–449.

Martin, G. B., Oldham, C. M., Cognie, Y., and Pearce, D. T. (1986). The physiological responses of anovulatory ewes to the introduction of rams—A review. *Livestock Production Sci.* **15**, 219–247.

Martin, G. B., Walkden-Brown, S. W., Boukhliq, R., Tjondronegoro, S., Miller, D. W., Fisher, J. S., Hotzel, M. J., Restall, B. J., and Adams, N. R. (1994). Non-photoperiodic inputs into seasonal breeding in male ruminants. In *Perspectives in Comparative Endocrinology* (K. G. Davey, R. E. Peter, and S. Tobe, Eds.), pp. 574–585. National Research Council of Canada, Ottawa.

Thiery, J.-C., and Martin, G. B. (1991). Neurophysiological control of the secretion of gonadotrophin-releasing hormone and luteinising hormone in the sheep—A review. *Reprod. Fertil. Dev.* **3**, 137–173.

Catecholamines

see Autonomic Nervous System

Cats

David E. Wildt, Janine L. Brown, and William F. Swanson

Smithsonian Institution

I. Domestic Cat
II. Wild Felids
III. Future Priorities

GLOSSARY

Asiatic lion A subspecies of lion (*Panthera leo persica*) indigenous to India and restricted to a single wildlife sanctuary in the Gujarat region.

felid Of or belonging to the cat family Felidae.

Florida panther A subspecies of puma (*Puma concolor coryi*) indigenous to southern Florida and composed of fewer than 50 total individuals.

laparoscopy A method of examining the abdominal cavity content by means of a fiberoptic telescope inserted through a small incision.

normospermia A condition in which most ejaculated sperm are structurally normal.

pleiomorphic Having multiple structural forms.

teratospermia A condition in which most ejaculated sperm are structurally abnormal.

Felids have fascinated and inspired humans since the beginning of recorded history. Recent molecular studies support the concept that there are 37 distinctive cat species inhabiting Earth. All share the instinct to hunt, survive, and master their environment. Twenty-three cat species (or subspecies) are threatened or endangered with extinction. Because reproduction is the essence of species survival, an understanding of how cats reproduce may potentially contribute to their management and conservation.

I. DOMESTIC CAT

The domestic cat is one of the world's most popular companion animals and serves an important role in biomedical research, contributing to the fields of genetics, fertility, developmental biology, neurophysiology, cancer, immunology, and infectious disease.

A. Reproductive Anatomy

The ovaries are caudal to the kidneys and are attached proximally by the suspensory ligament and dorsally by the mesovaria. The mesosalphinx envelops the oviducts and covers the lateral aspect of the ovaries to form an ovarian bursa that can be removed to reveal all ovarian aspects. Each oviduct courses cranially in the medial aspect of the ovarian bursa, then caudally to the lateral aspect before terminating at the uterotubal junction. The bicornuate uterine horns are suspended dorsally by the mesometrium. The uterine body is partially divided internally by a septum, and the cervix is short, opening obliquely at the vaginal orifice.

Cat testes are descended at birth and are located in the scrotum caudodorsal to the penis. The penis is directed caudally and is covered with 100–200 cornified papillae (spines) that are testosterone dependent, regressing in size after castration. Accessory reproductive organs in the male consist of paired bulbourethral glands and a prostate gland.

B. Sexual Maturity and Seasonality

The female reaches puberty at 6–9 months of age (2.3–2.5 kg body weight) with some dependence on season of birth and growth rate. Queens are seasonally polyestrous, with onset of estrous activity commencing by February or March and continuing until September. Because seasonality is affected by photoperiod, geographic latitude influences degree of seasonality. At least 12 hr of artificial lighting/day maintains circannual cyclicity.

Males usually become sexually mature before 1 year of age (at >3.5 kg). Spermatogenesis begins within 8 months of birth, and sperm production appears unaffected by season.

C. Phases of the Estrous Cycle

There are four recognized phases: proestrus, estrus, diestrus, and anestrus (or interestrus).

1. Proestrus

Proestrus (<1 day) is associated with the presence of ovarian follicles (flat, clear areas that are 1 or 2 mm diameter), circulating estradiol-17β <20 pg/ml, and, occasionally, interest in males excluding copulation.

2. Estrus

Estrous behavior rapidly follows increased circulating estradiol-17β associated with advanced follicular development. Estrus is characterized by permitting male mounting and coitus, and usually vocalization, lordosis, and foot-treading. Estradiol-17β commonly fluctuates (Fig. 1) but usually remains above 20 pg/ml (mean range, 25–100 pg/ml).

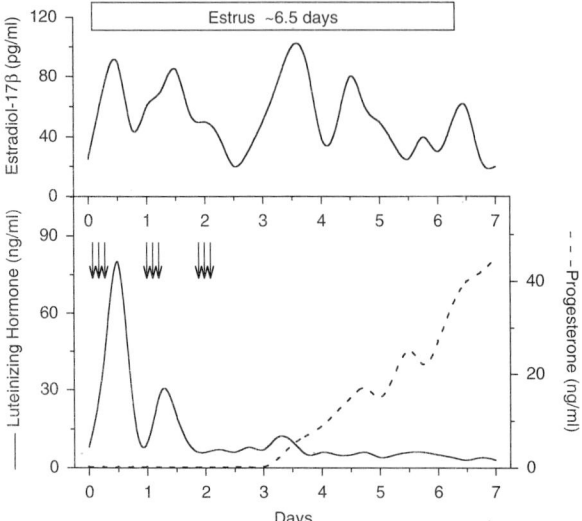

FIGURE 1 Endocrine profiles during the periovulatory period in the domestic cat. Arrows designate mating activity.

Distinct vesicular follicles (>2 mm diameter; usually one to four follicles/ovary) protrude slightly above the ovarian surface. Neither coitus nor ovulation affects the duration of elevated estradiol-17β or estrus (mean, 6 or 7 days). The mated queen can ovulate and enter diestrus, which can proceed to gestation (if conception occurs) or to a sterile luteal phase if the mating is infertile.

i. Ovulation The cat is considered an induced ovulator, but significant spontaneous ovulation occurs in some populations. Ovulation depends on coitus-induced neuronal stimulation of the medial basal hypothalamus, a reflex release of gonadotropin-releasing hormone (GnRH), and a subsequent luteinizing hormone (LH) surge from the pituitary. Cats copulate frequently during estrus (as often as 36 times within 36 hr), and early matings serve more to finalize folliculogenic events (oocyte maturation and release) than to ensure the presence of many sperm. With frequent mating, blood LH surges from <4 ng/ml (basal) to >70 ng/ml within 8 hr of coital onset and then gradually declines during the next 20–24 hr (Fig. 1). Such LH surges are sufficient to cause final follicular maturation, with ovulation then occurring at variable time points after first breeding. Variation is related to number and time of onset of matings and differences in follicular maturity. Queens mated three times/day beginning on Day 1 of estrus require ~48 hr to ovulate, whereas cats mated initially on Day 2 of estrus ovulate within 30–36 hr. Queens permitted *ad libitum* mating starting on Day 3 or 4 complete ovulation within 32 hr.

Vesicular follicles (2.5–3.5 mm diameter) become vascularized, develop a central hyperemic stigma, and release follicular contents onto the ovarian surface, with a slight inward collapse of the follicular apex. Ovulation is all-or-none because follicle number before mating corresponds to corpora lutea (CL) number after coitus (approx one to four CL/ovary). If mating is limited to one copulation, neither an LH surge nor ovulation usually occurs. This neuroendocrine pathway can become refractory because copulations on consecutive days result in fewer LH surges of lower amplitude (Fig. 1). The last copulations on a given day of unrestricted matings fail to elicit increased LH, and suppression may persist for 48–72 hr. However, injecting GnRH can induce a pronounced LH spike, indicating that decreased responsiveness is caused by depleted hypothalamic GnRH or feedback inhibition of GnRH on further hypothalamic GnRH release.

3. Anestrus

Unmated estrual queens gradually enter an anestrus (or interestrus) characterized behaviorally by indifference or active resistance to sexual advances. Mature ovarian follicles undergo atresia, and peripheral estradiol-17β is at nadir. The interestrous phase in nonovulating females varies from 7 to 21 days.

4. Diestrus and Pregnancy

Ovulating queens enter diestrus, a phase in which the queen no longer displays estrual postures and is unresponsive to the male. There is vacuolation (luteinization) of granulosa and theca interna cell cytoplasm in corpora hemorrhagica as early as 64 hr after first mating. By 100 hr (48–72 hr after ovulation), CL develop as raised, reddish-orange structures with polygonal cellular morphology (Fig. 2), each reaching maximal size within 10–16 days. Circulating progesterone increases above baseline (<1 ng/ml) ~76 hr after first mating (~40 hr after ovulation), rising to 25–90 ng/ml within 14–22 days (Figs. 1 and 2). In response to rising progesterone, uterine glandular epithelia show increased cell height and cytoplasmic vacuolation (Fig. 2). Luteal progesterone and LH receptor content are static for the first 100 hr and then increase in parallel with peripheral progesterone (Fig. 2). Rapid progesterone rise beginning at 76 hr reflects accelerated synthesis within CL. LH appears to play a luteotropic role. Progesterone is elevated for 36–38 days in nonpregnant cats, with CL gradually regressing to form persistent luteal scars. Many females then reinitiate estrous cyclicity within 7–14 days. Premature CL regression cannot be induced with exogenous prostaglandin $F_{2\alpha}$.

The interval of elevated progesterone during a sterile luteal phase (~36 days) is shorter than that during pregnancy (63–67 days). Temporal progesterone profiles after ovulation and through the demise of the CL vary among individuals, so progesterone concentrations cannot be used as a reliable indicator of pregnancy during the first 36 days after ovulation. Progesterone declines slowly, beginning ~30 days into pregnancy, returning to nadir on, or shortly after, the day of parturition. Progesterone is produced primarily by CL during the first 40–45 days of pregnancy. Ovariectomy does not induce abortion, so the placenta has been credited with producing progesterone during the last 3 weeks of pregnancy.

During the second half of pregnancy, estradiol-17β periodically fluctuates above basal levels, indicating that there are continuous waves of follicular development. A distinct estradiol-17β surge occurs 8 or 9 days before parturition. Queens occasionally mate during pregnancy, with superfetation a possible but infrequent occurrence.

Peripheral prolactin increases about Day 35 of pregnancy, plateaus 2 weeks later, is maintained through parturition and the first 4 weeks of lactation, and is at nadir shortly after weaning. Prolactin may be luteotrophic in the pregnant cat because administering the antiprolactin cabergoline at 30–40 days reduces progesterone to baseline. Peripheral relaxin increases beginning about Day 25 of gestation, reaching a plateau 5–10 days later and then declining to baseline by parturition. Circulating prostaglandin $F_{2\alpha}$ increases by approximately Day 30, plateaus at Day 45, and spikes just before parturition.

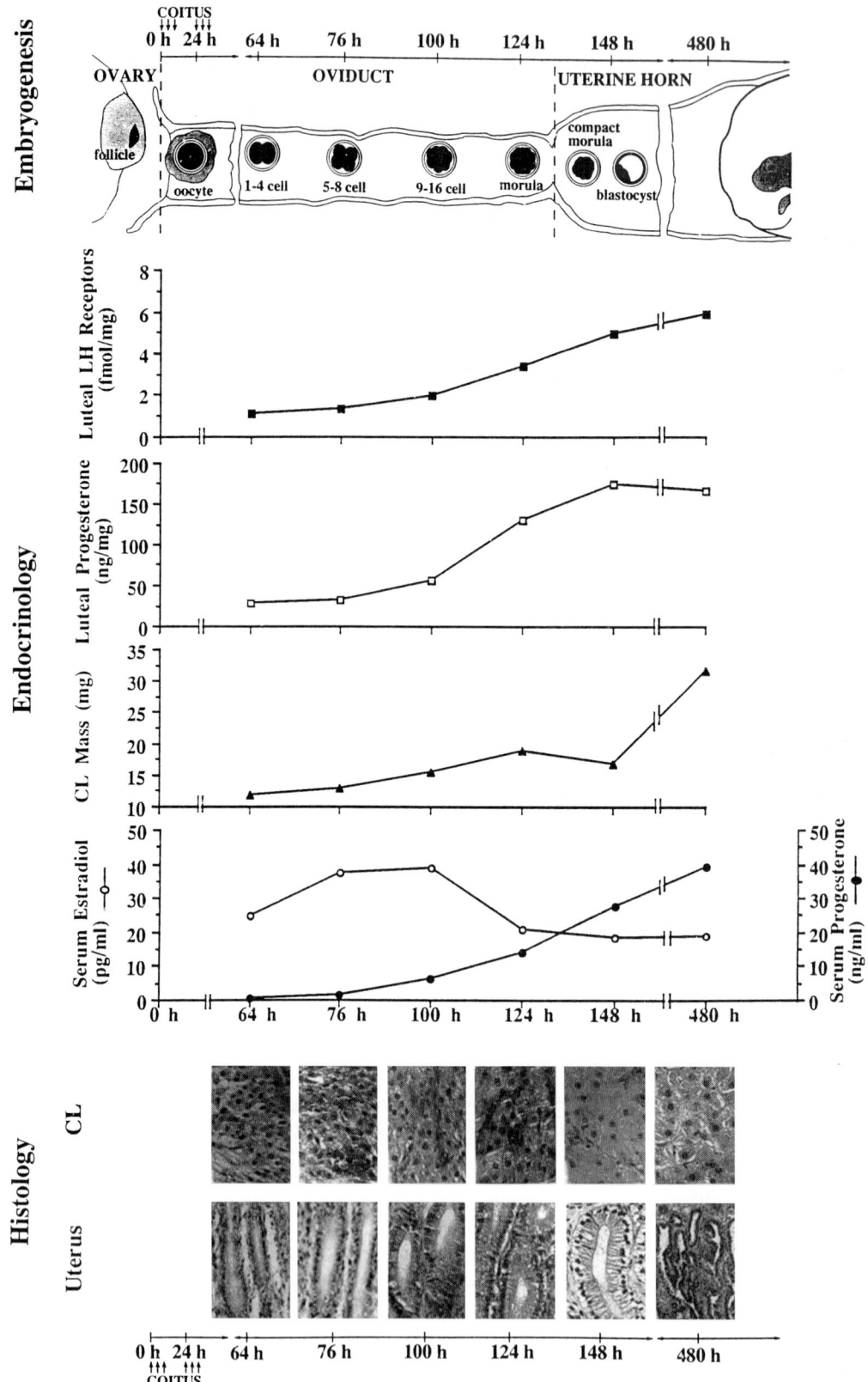

FIGURE 2 Composite diagram of *in vivo* embryogenesis, endocrine traits, and histological traits in naturally estrual, naturally mated domestic cats.

D. Oocytes and Embryo Development *in Vivo*

Number and quality of follicle/oocyte populations within the ovary vary markedly at any given time. Most follicle/oocyte complexes (>60%) are in various stages of atresia. The oocyte is in metaphase II of meiosis at ovulation and is ~160 μm in diameter (~125 μm without the zona pellucida). The cat oocyte is dark because of a high cytoplasmic concentration of lipid, which makes it difficult to identify intracellular structures and to assess meiotic status. The germinal vesicle and pronuclei can be distinguished after centrifugation. The former comprises ~15% of the intracellular volume, and the pronuclei are ~10 μm in diameter.

About 90% of mated and ovulating queens conceive (produce fertilized oocytes/embryos). Oocytes are fertilized in the proximal oviduct, and most cleave once by 64 hr after first breeding. Blastomeres cleave every 12 hr through eight-cells and then every 24 hr through the morula stage (Fig. 2). Embryos undergo compaction at the morula stage and transverse the uterotubal junction as compact morulae or early blastocysts 124–148 hr after first mating. This timing correlates with declining circulating estradiol-17β and rapidly increasing progesterone which promotes the transition. This period corresponds with pronounced morphological changes in uterine histology (Fig. 2). Cat embryos migrate between uterine horns to evenly distribute fetuses. Zona lysis occurs ~12 days after mating, with the zona thinning and dissolving over one pole, which is followed by escape of the expanded blastocyst. Implantation and formation of a zonary, endotheliochorial-type placenta begins on Days 13 or 14. About 30% of ovulated oocytes never fertilize, implant, or survive gestation, and most embryonic mortality occurs 64–100 hr after first mating (early preimplantation). Queens with high versus low fertility have negligible differences in blood hormone profiles, CL weight, and LH receptor concentrations. Cats producing poor quality embryos have consistently elevated luteal progesterone concentrations, perhaps indicating premature luteinization causing an inhospitable oviductal environment. There is no relationship between uterine or CL histology and embryo quality. The gestation interval is 63–67 days, and litter size averages four or five kittens. Pregnancy diagnosis and fetal growth can be monitored effectively by ultrasonography.

E. Gametes, Embryos, and Hormonal Stimulation

1. Semen

Semen is collected by an artificial vagina (AV), electroejaculation, or after death or castration. Approximately 60% of adult males can be trained to an AV and collected three times weekly, harvesting ~30–40 μl of semen and ~60 million total sperm/ejaculate. Electroejaculation is suitable for all healthy males and can be performed once weekly. After anesthesia, a sine wave electrostimulator and three-electrode rectal probe are used to deliver low-voltage stimuli to produce a 100- to 200-μl ejaculate containing ~30–50 million sperm of comparable quality to the AV approach. Sperm motility using both techniques usually ranges from 50 to 95%, and computer-assisted semen analysis (CASA) values are available for percentage motility, curvilinear velocity, linearity, straight line velocity, and amplitude of lateral head displacement. Electroejaculated seminal fluid has an osmolarity of ~323 mOsm/liter and a pH of ~8.0–8.5. Sperm collected from the ductus deferens or epididymis postmortem or castration are capable of fertilizing ova *in vitro*.

The normal cat ejaculate contains 20–30% abnormally shaped sperm. Certain males consistently produce >60% sperm pleiomorphisms and are commonly referred to as teratospermic. The most prevalent anomalies include a bent midpiece with a laterally displaced cytoplasmic droplet, a bent flagellum, and a tightly coiled flagellum. Transmission electron microscopy reveals that 30% of sperm from teratospermic donors have malformed acrosomes. These males often have abnormally low circulating testosterone but normal LH. Low-speed centrifugation (to produce a sperm pellet) followed by supernatant removal and sperm swim-up into fresh medium allows recovering mostly normal-appearing sperm from teratospermic ejaculates.

2. Hormonal Stimulation of Ovarian Activity

The exogenous hormones porcine follicle-stimulating hormone (pFSH) or equine chorionic gonadotropin (eCG) are used to stimulate follicular development in anestrual queens. Daily treatment of queens with 2.0 mg/day pFSH (im) for 5–7 days and natural mating has resulted in 70% pregnancy success and five or six kittens/litter. Ovarian hyperstimulation and poor embryo quality are risks. An alternative is a single im injection of 100 IU eCG which stimulates estrus and mating in most females and results in four times as many good quality embryos as produced by mated queens in natural estrus. eCG is given to individual queens no more than once every 4 months to avoid an immunologically mediated refractory response. Human chorionic gonadotropin (hCG) can be used to induce ovulation in naturally estrous (250 IU, im) or eCG-treated (75 IU, im 80 hr after eCG) cats. Compared to naturally estrual, mated, and ovulating queens, cats given hCG have some irregularities in circulating hormones, endometrial histology, and luteal progesterone content. Higher hCG dosages are associated with more degenerate or unfertilized oocytes because this hormone is folliculogenic. Pharmacokinetic studies reveal that hCG persists in circulation for ~96 hr, and eCG persists for ~120 hr. This explains why single injections of either hormone can cause ancillary follicular development several days after ovulation. Follicular oocytes on Day 1 of estrus are unable to respond to hCG, and >80 of the oocytes degenerate. Thus, hCG is given on Day 2 of estrus in naturally estrual queens.

The eCG + hCG treatment described here produces luteal progesterone and embryo development *in vivo* that mimics that of naturally estrual/mated queens while increasing total number of high-quality oocytes. Circulating progesterone may be higher because of more total CL produced. This regimen has been used with intrauterine sperm deposition or natural breeding to produce high-quality embryos capable of developing into blastocysts *in vitro* and living young. GnRH (two 25-μg im injections at a 12-hr interval) efficiently stimulates ovulation in cats in natural estrus but not in eCG-treated queens.

3. Oocytes Matured in Vitro

Oocytes can be harvested from freshly excised ovaries by mincing or follicular aspiration. Some oocytes mature *in vitro* in Eagle's minimum essential medium containing FSH, LH, estradiol-17β, and bovine serum albumin. A few oocytes (<5%) achieve nuclear maturation spontaneously without gonadotropins or supplemental protein. Significant meiotic maturation begins as early as 12 hr after culture onset, with >60% of oocytes being in telophase I or metaphase II at 24 hr and there is no increase thereafter. Fetal calf serum as a medium supplement inhibits meiotic nuclear maturation due to dialysable factors of low molecular mass. Intraovarian oocytes stored at 4°C for up to 72 hr, are capable of undergoing nuclear maturation.

4. Oocytes and in Vivo Maturation

Oocytes are recoverable by transabdominal laparoscopic aspiration of ovarian follicles in eCG/hCG-treated queens. Aspirated follicles convert to normal luteal function. From 10 to 20 oocytes can be collected per cat with a recovery rate of 90–100%. Follicle and oocyte number is unaffected by hCG dose or the eCG to hCG interval, but a lower hCG dose (100 versus 200 IU) reduces degenerate oocyte number and enhances *in vitro* fertilization (IVF) success. Eighty to 84 hr is the optimal interval between eCG and hCG injection to maximize the number of mature oocytes recovered and to produce temporal progesterone profiles in the range of naturally estrual-mated queens. eCG-primed follicles remain available for the hCG trigger for at least 92 hr but thereafter rapidly become atretic. Even by 92 hr there is a marked decline in the number of oocytes categorized as mature, and by 96 hr most are degenerate. More than 60% of oocytes aspirated at 84 hr after hCG are not at metaphase II, but >90% mature within 8 hr of culture. The ability of these oocytes to eventually fertilize *in vitro* is independent of stage of nuclear maturation at recovery. Unlike some species, cortical granule localization cannot be used as a marker for oocyte maturation. The cat lacks a cortical granule-free domain over the metaphase spindle, and premature cortical granule release does not occur.

5. Sperm Capacitation

Sperm in fresh ejaculate are short-lived (<60 min) *in vitro*. Dilution in culture medium (Ham's F10) and swim-up processing allows sperm to remain motile *in vitro* at 37°C for several days. Conditions for inducing sperm capacitation are not complex. Removing seminal plasma is unnecessary, and capacitation proceeds at room temperature in the presence of a protein source (fetal bovine serum) with a low frequency (<20%) of spontaneous acrosomal loss. Sperm capacitation is quantitated by assessing the ability to acrosome react after exposure to calcium ionophore (A23187) or solubilized zonae pellucidae. Acrosomal status can be measured using the fluorescent probe *Arachis hypogaea* (peanut) agglutinin (PNA). Transmission electron microscopy reveals that PNA lectin binding sites are localized on the outer acrosomal membrane. Under capacitating conditions, about 50% of cat sperm first acrosome react within 2 hr, with maximum activity (~75%) by 2.5 hr. The ionophore is a more potent inducer of the acrosome reaction than solubilized zonae pellucidae.

Sperm express at least two tyrosine phosphorylated proteins of 160 and 95 kDa. The level of phosphorylation increases after capacitation and homologous zona pellucida exposure. The 95-kDa protein binds with ZP glycoproteins. Adding the tyrosine kinase inhibitors, tyrphostin and genistein, inhibits the acrosome reaction and zona penetration, indicating that the 95-kDa protein participates in a critical step leading to sperm–oocyte interaction.

6. In Vitro Fertilization of Oocytes Matured in Vitro

Parthenogenetic activation (cleavage) occurs in ~6% of cultured oocytes. Fertilization success is highest when insemination (with swim-up processed sperm) is performed 32 hr after oocyte recovery, which includes 24 hr of maturation culture. Successful nuclear maturation is uncoupled from fertilization and developmental competence. Oocytes exposed to 24 hr of intraovarian cold (4°C) storage remain capable of maturing, fertilizing, and growing into morulae and blastocysts *in vitro*. After 24 hr of cold storage, cat oocytes rapidly lose the ability to fertilize and are senescent by 40 hr.

If oocytes are harvested randomly from excised cat ovarian tissue, up to 60% can be induced to mature, and ~25–35% of the total can fertilize. Rates can be enhanced if an oocyte grading system is used. More than 35% of grade I (best quality) oocytes cleave, and 24% grow into blastocysts. However, only 13% of cumulus oocyte complexes meet grade I criteria, likely related to the naturally high incidence of follicular atresia. The presence of FSH, LH, estradiol, and fetal bovine serum assist in fertilization and cleavage of *in vivo*-matured oocytes. Developmental kinetics of these embryos are no different from those after IVF of *in vitro*-matured oocytes or those recovered from mated cats.

7. IVF of Oocytes Matured in Vivo

Sperm readily interact *in vitro* with *in vivo*-matured oocytes, with 60–80% fertilization success using high-quality sperm (2×10^5 cells) and coculture for ~12–18 hr. Varying culture medium ("simple" versus "complex"), temperature, and gas atmosphere or using oviductal cell coculture have no significant effect on gamete interaction. Oocyte quality is enhanced by an eCG/hCG interval of no shorter or longer than 80–84 hr. Follicular oocytes not reaching nuclear maturation by the time of aspiration do so rapidly in culture with no effect on IVF success. The presence of luteal tissue on the ovaries at the time of oocyte aspiration reduces IVF success by ~20% perhaps because of a negative progesterone effect. Coculturing sperm and oocytes together for only 2 hr reduces the number of fertilized oocytes. The block to polyspermy appears to occur at the level of the vitelline membrane.

Cleaved embryos are first observed ~30 hr after insemination, and 70–95% readily develop into morulae in culture. IVF embryos complete the first three cell cycles quickly, with embryos reaching the 5- to 8-cell stage by 54 hr after insemination. Thereafter, IVF embryos slow to approximately one cell division every 24 hr. Cleaved embryos prefer a complex medium such as Ham's F10 containing fetal bovine serum. Only 22–30% of IVF-derived morulae become blastocysts. This morula-to-blastocyst block cannot be ameliorated by altering temperature, gas phase, culture medium, or protein supplementation. The block appears to result from a problem with the oocyte (or with the effect of the culture system on the oocyte) because embryos developing *in vivo* can grow into blastocysts *in vitro*. Furthermore, if IVF embryos are placed in culture at the 9- to 16-cell stage, >90% form blastocysts compared to only ~40% of 5- to 8-cell embryos. The developmental block is dissociated from the timing of the maternal-to-zygotic transition (onset of embryonic genomic transcription) which occurs at the 5- to 8-cell embryo stage. IVF embryos also fail to undergo compaction. The reason for the difference in blastocyst production between oocytes matured *in vitro* and *in vivo* and then fertilized *in vitro* is unclear. The two groups vary in exposure to supplemental gonadotropins and onset of insemination.

8. Importance of Normo- versus Teratospermia

Malformed sperm are not successful at IVF; some cells bind the zona but none are able to penetrate its inner layer. Even structurally normal-appearing sperm from teratospermic donors are compromised in ability to penetrate the zona pellucida and fertilize *in vivo*- or *in vitro*-matured oocytes. Injecting these sperm directly into oocytes does not improve fertilization. During cooling, cryopreservation, and thawing, sperm from teratospermic ejaculates are more sensitive to acrosome damage than sperm from normospermic counterparts. These sperm from teratospermic donors have a functional deficit. Compared to sperm from normospermic donors, normal-appearing sperm from teratospermic males are delayed in ability to undergo capacitation, and the acrosome reaction is inferior (after chemical ionophore) or nonexistent (after solubilized zona pellucida exposure). The mechanism whereby teratospermia causes acrosomal dysfunction likely occurs at the cellular level because tyrosine phosphorylation of the 95-kDa sperm receptor is compromised. Inability to acrosome react is likely related to diminished phosphorylation efficiency.

F. Assisted Breeding

Cat sperm are highly sensitive to acrosomal damage during cooling, freezing, and thawing. However, AI has been used to produce kittens using fresh or frozen–thawed sperm and after ovulation induction using various hormonal regimens. Vaginal or intracervical sperm deposition is ineffective. Pregnancy success increases markedly when sperm are deposited into the proximal half of the uterine horn (via transabdominal, laparoscopic insemination). Conventional anesthetics inhibit sperm transport and ovulation, so AI must be performed ~35–38 hr post-hCG after ovulation has commenced at ~25–30 hr post-hCG. Pregnancy has resulted after inseminating as few as 1 million fresh, motile sperm. Pregnancy success is ~50%, with an average of two kittens/litter. The lentivirus feline immunodeficiency virus can be present in cat seminal plasma and washed sperm and can be transmitted by AI.

Embryo transfer has been successful for producing kittens using embryos (i) flushed from the tract of mated females, (ii) produced by IVF (two-cell to morula) from oocytes matured *in vivo* or *in vitro*, (iii) produced after subzonal or intracellular sperm injection, or (iv) after freezing and thawing. Embryo transfer requires anesthesia and surgical placement of embryos. The biological competence of IVF-derived embryos has been proven in independent laboratories, but pregnancy success is low.

II. WILD FELIDS

There are less specific data on the reproductive physiology of wild felid species compared to the domestic cat. The physiological mechanisms regulating reproduction among these species are as diverse as the phenotypes within the taxon, in part related to differences occurring during evolution and worldwide distribution (Table 1).

A. Reproductive Anatomy

The general anatomy of the female reproductive tract of all felid species is similar to that of the domestic queen. Basic male reproductive anatomy is also similar among species with the exception that there are few or no penile spines in the margay. Testicular volumes have been measured for most wild species but appear unrelated to the number of sperm col-

TABLE 1
Life History and Reproductive Characteristics for Wild Felid Species

Common name	Genus/species	Geographic range	Female sexual maturity (months)	Male sexual maturity (months)	Estrous cycle (days)	Estrus (days)	Gestation (days)	Litter size (range)
Black-footed cat	Felis nigripes	Sub-Saharan Africa	12–21	12	nd	1–5	63–68	1–2
African golden cat	Profelis aurata	Sub-Saharan Africa	nd[a]	nd	nd	nd	nd	1–2
Cheetah	Acinonyx jubatus	Sub-Saharan Africa/North Africa and Southwest Asia	24–36	30–36	7–21	2–6	90–98	1–8
African lion	Panthera leo	Sub-Saharan Africa	24	30	18–21	4	110	1–6
Serval	Leptailurus serval	Sub-Saharan Africa	18–24	18–24	nd	4	73	1–5
Leopard	Panthera pardus	Sub-Saharan and North Africa/Southwest and Tropical Asia	24–36	24–36	46	7	90–105	2–3
Caracal	Caracal caracal	Sub-Saharan Africa/North Africa and Southwest Asia	14–16	12–15	14	1–3	78–81	1–3
African wildcat	Felis silvestris, lybica grp.	Sub-Saharan Africa	11	11	nd	nd	56–63	2–5
Asiatic lion	Panthera leo persica	India	36–48	36–96	nd	nd	nd	1–5
Sand cat	Felis margarita	North Africa and Asia	9–14	9–14	28–35	5	59–67	1–5
Tiger	Panthera tigris	Tropical Asia	40	36–72	10–39	7	103	2–5
Bornean bay cat	Catopuma badia	Tropical Asia	nd	nd	nd	nd	nd	nd
Clouded leopard	Neofelis nebulosa	Tropical Asia	26	26	25–30	3–6	93	1–5
Asiatic golden cat	Catopuma temmincki	Tropical Asia	18–24	24	39	6	80	1–3
Flat-headed cat	Prionailurus planiceps	Tropical Asia	nd	nd	nd	nd	56	nd
Rusty-spotted cat	Prionailurus rubiginosus	Tropical Asia	nd	nd	nd	5	68	1–2
Fishing cat	Prionailurus viverrinus	Tropical Asia	nd	nd	nd	nd	63–70	1–4
Marbled cat	Pardofelis marmorata	Tropical Asia	21	21	nd	nd	81	1–4
Jungle cat	Felis chaus	Tropical Asia	11–18	11–18	nd	5	63–68	1–6
Leopard cat	Prionailurus bengalensis	Tropical Asia	8	8	nd	nd	56–70	1–4
Iriomote cat	Prionailurus bengalensis iriomotensis/Incertae sedis	Tropical Asia	nd	nd	nd	nd	60–70	1–4
Snow leopard	Uncia uncia	Eurasia	24–36	24–36	15–39	2–12	98–104	1–5
Chinese mountain cat	Felis bieti	Eurasia	nd	nd	nd	nd	nd	2–4
Pallas' cat	Otocolobus manul	Eurasia	12	nd	20–46	1–5	66–75	6–8
Asiatic wildcat	Felis silvestris oranata grp.	Eurasia	10	10	nd	nd	58–62	2–8
Eurasian lynx	Lynx lynx	Eurasia	20–24	30	nd	nd	69	1–4
Iberian lynx	Lynx pardinus	Eurasia	8–10	nd	nd	nd	61	2–3
European wildcat	Felis silvestris, silvestris grp.	Eurasia	10–12	9–10	nd	2–8	63–68	1–8
Kodkod	Oncifelis guigna	Americas	nd	nd	nd	nd	72–78	1–3
Andean mountain cat	Oreailurus jacobitus	Americas	nd	nd	nd	nd	nd	nd
Jaguar	Panthera onca	Americas	24–36	36–48	37	6–17	91–111	1–4
Oncilla	Leopardus tigrinus	Americas	10–12	10–12	11–20	3–7	78–86	1–3
Margay	Leopardus wiedi	Americas	18–24	18–24	11–18	2–7	76–84	1–2
Canada lynx	Lynx canadensis	Americas	10–23	24	nd	nd	63–70	2–5
Geoffroy's cat	Oncifelis geoffroyi	Americas	18	24	20	2–3	72–78	2–3
Puma	Puma concolor	Americas	24	24	23	8	92	1–6
Ocelot	Leopardus pardalis	Americas	18–22	30	11–18	2–5	79–85	1–3
Bobcat	Lynx rufus	Americas	9–12	18	44	5–10	50–70	1–8
Pampas cat	Oncifelis colocolo	Americas	24	nd	nd	nd	nd	1–3
Jaguarundi	Herpailurus yaguarondi	Americas	24–36	24–36	nd	3	70–75	1–4

[a] nd, no data.

lected by standardized electroejaculation. The genetically depauperate Florida panther has a high incidence (>80%) of unilateral or bilateral cryptorchidism.

B. Fundamental Reproductive Traits

Known information on sexual maturity, duration of the estrous cycle and estrus, gestation length, and litter size are summarized in Table 1.

C. Reproductive Activity Based on Endocrine Studies

Ovarian steroid profiles in circulation are available for the lion, leopard, puma, Siberian tiger, and snow leopard. However, information on female reproductive patterns is based largely on longitudinal monitoring of fecal hormonal metabolite excretion. Most endogenous steroids in felids are excreted in feces, not urine. Estradiol is voided in equal amounts as unconjugated estradiol and non-enzyme-hydrolyzable estrogen conjugates. None of the progesterone is excreted in its native form, with most being non-enzyme-hydrolyzable conjugates. Steroid metabolite profiles have been quantified in the tiger, cheetah, clouded leopard, leopard cat, ocelot, margay, oncilla, Pallas' cat, serval, caracal, bobcat, lion, and snow leopard for the purposes of identifying duration of the reproductive cycle, influence of season, type of ovulation (induced versus spontaneous), or causes of infertility. This approach is entirely noninvasive, and data can be collected without animal sedation or restraint.

There is considerable variation among species in fecal estradiol and progesterone metabolite concentrations throughout the cycle and pregnancy (Figs. 3 and 4). Although steroid metabolism appears conserved within the taxon, absolute production of steroids from different ovarian compartments is fairly species specific. Increases in fecal estradiol are associated with behavioral estrus or ovarian follicular development. Distinct increases in fecal progestogens occur after observed matings or hormonal ovulation induction. Temporal excretory patterns mimic the few data available from wild felids in which serial blood samples have been analyzed.

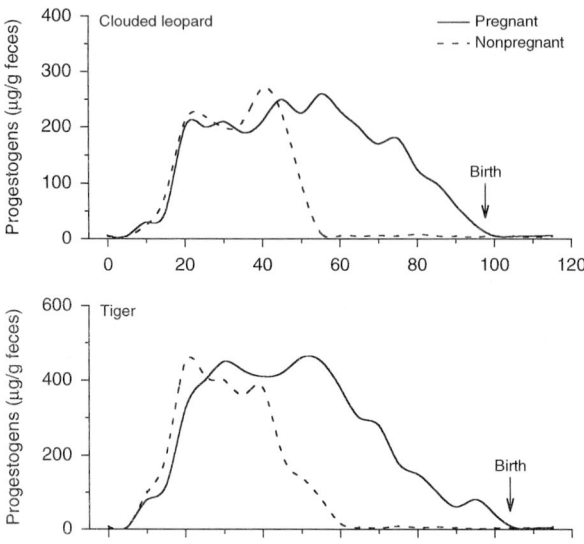

FIGURE 3 Fecal progestogen metabolite excretory patterns comparing pregnant versus nonpregnant luteal activity in the clouded leopard and tiger.

1. Seasonality

Blood hormone analysis reveals that zoo-maintained female Siberian tigers and snow leopards cycle more during late winter and spring months, in contrast to lions and pumas that cycle year-round. Fecal monitoring indicates that the clouded leopard is less reproductively active in late summer and early autumn but remains active throughout the year when maintained under at least 12 hr of artificial light. Cheetahs can cycle for long periods but usually not continuously throughout the year, a finding unrelated to season or other obvious environmental factors. Cheetahs held in the same institution often alternate periods of estrous cyclicity, whereas 25% of zoo-held females never express cyclic ovarian activity. The ocelot, margay, and oncilla cycle throughout the year, in contrast to marked seasonality in both female and male Pallas' cats. Estrous cyclicity based on fecal estradiol and maximum testosterone excretion in the male occur in January and February (Fig. 5), a time coinciding with peak semen quality and body weight. Most other felid species produce uniform amounts and quality of sperm throughout the year; an exception is the snow leopard, in which more sperm and higher circulating testosterone are

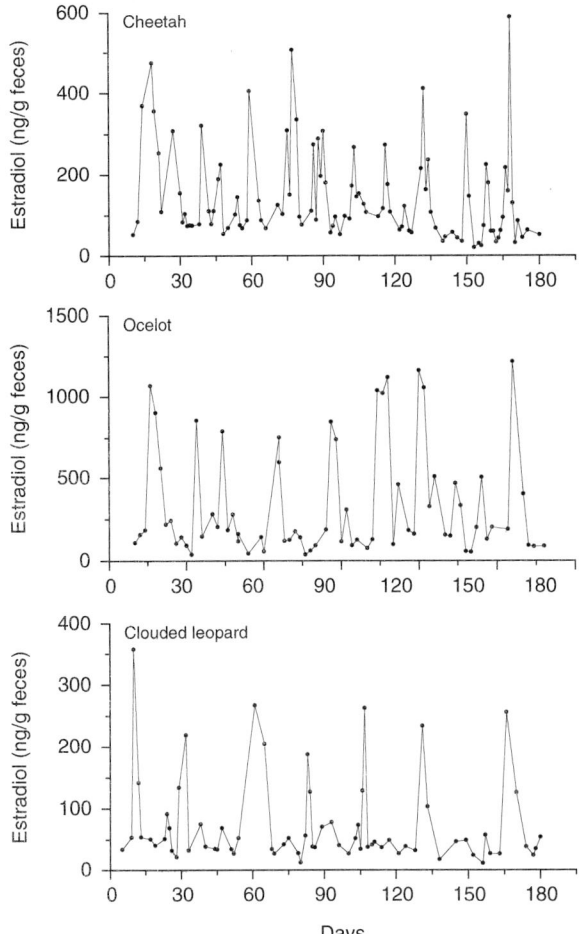

FIGURE 4 Individual excretory profiles of fecal estradiol concentrations in a cycling female cheetah, ocelot, and clouded leopard.

measured in the winter compared to summer and autumn. Normal circulating testosterone concentrations vary among species but generally not circannually within species.

2. Ovulation

No spontaneous increases in circulating progesterone occur after estradiol-17 surges in serial blood samples from tigers, pumas, or snow leopards, but random ovulations occasionally occur in lions and leopards. Spontaneous ovulation also occurs in as many as half the reproductive cycles of singleton clouded leopards based on increases in fecal progestogen metabolites. Using the same monitoring approach, only 1% of cheetahs ovulate spontaneously. Thus, ovulatory mechanisms within the Felidae are regulated by species- and/or individual-specific responses to physical and/or psychological stimuli.

3. Pregnancy

During pregnancy, progestogen metabolite concentrations increase several hundred-fold and peak at midterm (Fig. 3). Concentrations return to baseline at, or shortly after, parturition. In all species studied to date (lion, puma, cheetah, clouded leopard, tiger, and snow leopard), length of the nonpregnant luteal phase is about half that of pregnancy (Fig. 3). Except for duration, there are no other qualitative or quantitative differences in progestogen metabolite excretion between the pregnant and nonpregnant state.

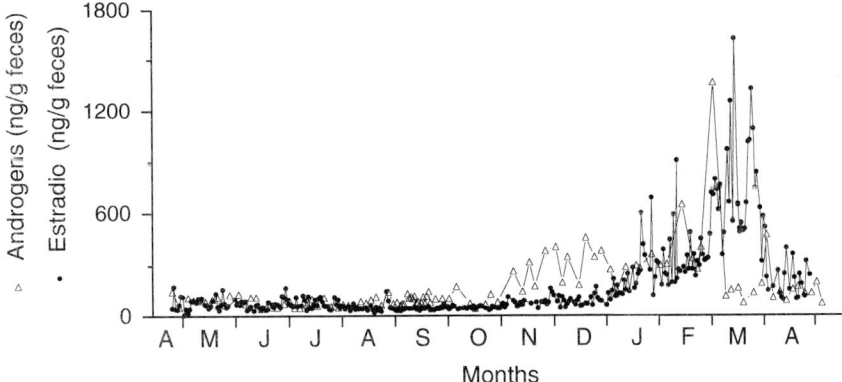

FIGURE 5 Individual excretory profiles of fecal androgen metabolite concentrations in a male and fecal estradiol concentrations in a female Pallas' cat throughout the year.

D. Gametes, Embryos, and Hormonal Stimulation

1. Semen and Sperm

Semen can be collected safely and routinely by electroejaculation in anesthetized males. Males breed naturally and sire young after electroejaculation. Sperm characteristics are available for 28 of the 37 species. There are wide ranges in reported sperm concentrations and percentage motilities, but sperm densities usually are <150 million cells/ml of electroejaculate. The taxon as a whole is composed of species or subspecies in which teratospermia is the norm. Fifteen species routinely produce >50% pleiomorphic sperm per ejaculate. Defects range from simple bending of the midpiece or flagellum to extensive derangements of the mitochondrial sheath and acrosome. Etiology is not well understood, although malformed sperm are prevalent in species or populations that have undergone genetic bottlenecks and have diminished genetic diversity. Examples include the Asiatic lion (>50% sperm malformations), cheetah (>70%), and Florida panther (>90%). These males have normal pituitary function but often low-circulating testosterone and compromised spermatogenesis.

Inadequate nutrition affects sperm production. Providing all-meat diets with a lack of appropriate vitamins and minerals reduces sperm concentration, an effect that can be partially rectified by dietary supplementation.

Freshly ejaculated sperm from all wild felids rapidly lose motility *in vitro* unless diluted immediately in medium. Swim-up processing usually allows sperm to be maintained at ambient or culture temperature for >6 hr.

2. Sperm–Oocyte Interaction and Embryos

In vivo-matured oocytes are recoverable from ovarian follicles by laparoscopic aspiration and are similar in appearance to domestic cat counterparts. High-quality oocytes are dark and uniform in shape, with an expanded, light-colored cumulus mass, whereas poor quality oocytes have inconsistent coloring of the cytoplasm and few or granular cumulus cells. Oocytes are also recoverable from freshly excised and minced ovarian tissue, and ~10–40% have the capacity to achieve metaphase II *in vitro*.

Wild felid species vary in the functional ability of sperm to undergo *in vitro* capacitation and oocyte penetration. In the presence of fetal bovine serum, leopard cat, clouded leopard, and cheetah sperm fully capacitate within 3 hr, but more total sperm capacitate in the normospermic leopard cat than the teratospermic clouded leopard and cheetah. The use of homologous serum rather than bovine serum enhances capacitation in the cheetah but not in the leopard cat, clouded leopard, or tiger. Tiger and snow leopard sperm are inexplicably difficult to capacitate *in vitro*.

The 95-kDa tyrosine phosphorylated protein identified in domestic cat sperm is present in leopard cat, clouded leopard, cheetah, and tiger sperm. Capacitation conditions enhance phosphorylation of this protein in all these species but the cheetah. Adding a tyrosine kinase inhibitor to leopard cat sperm reduces the ability to fertilize oocytes *in vitro*.

Domestic cat oocytes, either collected fresh or stored in a hypertonic salt solution, are useful for assessing functionality of wild felid sperm. Sperm from the leopard cat, clouded leopard, cheetah, and tiger bind and penetrate the outer layer of the domestic cat zona pellucida *in vitro*. Snow leopard sperm, even in the presence of pentoxifylline, fail to penetrate the outer layer of the cat zona pellucida; perhaps this is related to the unique capacitation requirements associated with the need for buffered medium (physiological pH) to maintain motility *in vitro*. Leopard cat sperm easily penetrate the inner zona pellucida and the perivitelline space of domestic cat oocytes, an ability likely related to the level of normospermia and ease of achieving capacitation.

IVF procedures developed for the domestic cat are adaptable to other felid species. Species specificities occur in the number and quality of oocytes recovered after gonadotropin treatment. Fertilization success is influenced by sperm quality. About 65% of oocytes fertilize in the normospermic tiger compared to ~30% from the teratospermic Florida panther, and all of the latter fail to cleave. Clouded leopard sperm and oocytes fail to interact *in vitro* due to an inherently high (40%) incidence of acrosomal defects. Sperm from these pleiomorphic species are compromised in ability to undergo the acrosome reaction. IVF embryos readily develop in culture to morulae, and in some species such as the tiger, they occasionally advance to blastocysts.

IVF success of oocytes matured *in vitro* is highly variable and difficult in some species such as the cheetah. Although as many as ~15–45% of oocytes fertilize, few undergo cleavage and growth *in vitro*, but success is enhanced when the donor is young and healthy. Domestic cat sperm successfully fertilize *in vitro*-matured oocytes from the tiger, leopard, jaguar, puma, serval, and Geoffroy's cat, and in the puma these embryos have grown to 16 cells. Parthenogenetic cleavage in the oocytes of wild felid species is <7%.

3. *Hormonal Stimulation*

Gonadotropins eCG and hCG are often used to induce follicle development and ovulation, respectively, because single doses can be effective and minimize animal stress by avoiding serial injections. There is no relationship between body mass of a given felid species and eCG and hCG dose required to induce ovarian activity (Table 2). Hormonal dose to elicit an ovarian response is species specific, and excessive eCG or hCG causes ovarian superstimulation that can be measured by hyper amounts of fecal estrogen metabolites. Gonadotropin treatment also affects CL morphology, with eCG doses that are too low or too high causing small-sized, underdeveloped CL. The interval between hCG administration and ovulation onset is species specific and can occur as late as 40 hr after injection. In seasonally breeding species such as the snow leopard, response to hormone treatment is affected by time of year, with females being more sensitive to a given dose in the breeding season. A 6-month interval between subsequent eCG/hCG treatments is sufficient to limit immunoglobulin production in the ocelot, cheetah, and tiger.

E. Assisted Breeding

Pregnancy will not occur in anesthetized cheetahs and tigers inseminated vaginally or intracervically near the time of ovulation because anesthesia inhibits sperm transport and ovulation. Offspring have been produced by intrauterine AI in the hormonally treated and postovulatory puma, leopard cat, cheetah, tiger, ocelot, clouded leopard, oncilla, and snow leopard, but only after surgical deposition of sperm via laparotomy or laparoscopy. As few as 3.4, 7.5, and 9 million motile sperm have been used to produce pregnancy in the cheetah, ocelot, and snow leopard, respectively, and for the latter species, AI can result in conception during the nonbreeding season. Young have been produced in the leopard cat, cheetah, and ocelot using sperm frozen as pellets on dry ice in an egg yolk–lactose–glycerol diluent. Sperm cryopreserved by this method have a high incidence of damaged acrosomes. Cheetah offspring have been produced from sperm collected from wild cheetahs in Africa, cryopreserved, and transported to the United States. AI is an effective management tool in the cheetah, a species in which ovulation can be induced consistently with exogenous gonadotropins. In most AI successes, litter size is small (one or two kittens or cubs/litter).

Embryos produced by IVF of *in vivo*-matured oocytes have been used to produce young in the Asiatic wild cat and tiger after transfer to surrogate dams. Attempts in other species, including the lion, leopard cat, cheetah, snow leopard, fishing cat, and jungle cat, have resulted in viable-appearing embryos but no confirmed pregnancies.

III. FUTURE PRIORITIES

Felids express subtle to markedly different variations in reproductive mechanisms. A better understanding of fundamental reproductive physiology could facilitate breeding, management, and conservation. There is a need to understand the cause and

TABLE 2
Gonadotropin Dosages Required to Achieve Similar Ovulation Induction Response

Species	Body weight (kg)	eCG (IU)	hCG (IU)
Domestic cat	4	100	75
Oncilla	2	200	150
Leopard cat	4	100	75
Ocelot	9	500	225
Clouded leopard	15	75	50
Snow leopard	30	600	300
Puma	35	200	100
Cheetah	35	200	100

significance of teratospermia on reproductive fitness. Priorities for females include clearly defining normative reproductive activities, especially using noninvasive fecal hormone metabolite monitoring. For both males and females, there is a need to identify environmental (seasonality) and husbandry/management (social/behavioral, nutrition, and space enrichment) factors that influence reproductive physiology and behavior. For assisted reproduction, gamete processing needs to be improved to ensure viability and functionality. Techniques for inducing ovarian activity require modification to enhance consistency, avoid hyperstimulation, and provide an optimal maternal environment for fertilization and embryo development. A high priority is understanding how to control the reproductive cycles of all cat species, especially downregulating endogenous ovarian activity and synchronizing estrus. Emphasis should be placed on sperm and embryo cryobiology, especially understanding membrane stability and developing optimal protocols for establishing genome resource banks and repositories for protecting valuable lineages and for helping conserve and manage endangered populations.

See Also the Following Articles

CAPTIVE BREEDING OF WILDLIFE; DOGS (CANIDAE); PREGNANCY IN DOGS AND CATS

Bibliography

Brown, J. L., and Wildt, D. E. (1997). Assessing reproductive status of wild felids by noninvasive faecal steroid monitoring. *Int. Zoo Yearbook* 35, 173–191.

Concannon, P. W. (1991). Reproduction in dogs and cats. In *Reproduction in Domestic Animals, 4th Edition* (P. T. Cupps, Ed.), pp. 517–554. Academic Press, New York.

Howard, J. G. (1998). Assisted reproduction techniques in carnivores. In *Zoo and Wild Animal Medicine: Current Therapy IV* (M. E. Fowler, Ed.). Saunders, Philadelphia.

Nowell, K., and Jackson, P. *Wild Cats: Status Survey and Conservation Action Plan.* IUCN Cat Specialist Group.

Swanson, W. F., and Wildt, D. E. (1997). Strategies and progress in reproductive research involving small cat species. *Int. Zoo Yearbook* 35, 152–159.

Wildt, D. E., Pukazhenthi, B., Brown, J. L., Monfort, S. L., Howard, J. G., and Roth, T. L. (1995). Spermatology for understanding, managing and conserving rare species. *Reprod. Fertil. Dev.* 7, 811–824.

Cattle

Roy Fogwell
Michigan State University

I. Introduction
II. Reproductive Biology in Bovine Females
III. Reproduction in Beef Cattle
IV. Reproduction in Dairy Cattle
V. Bulls
VI. Reproductive Technology

GLOSSARY

anovulation No ovulations before puberty, during pregnancy, or at least 21 days after parturition; also called *anestrus*.

bull A male bovine.
calf A bovine offspring between birth and weaning.
cattle The gender- and age-neutral term for all bovines.
conception The fertilization of an oocyte by a spermatozoon. Defined practically by palpation of the conceptus in uterus at 35–45 days after insemination.
cow A female bovine after at least one parturition.
estrous cycle Sexual receptivity with ovulation (estrus) followed by sexual quiescence (diestrus) terminated by the next estrus in postpubertal nonpregnant females. Estrus occurs every 17–24 days.
heifer A female bovine between birth and her first parturition.

infertility Complete lack of conception (sterility) or conception that occurs later than goals.

parturition The expulsion of fetus and placental membranes from the uterus. *Prepartum* is before and *postpartum* is after parturition.

puberty The first ovulation and first opportunity for conception in females.

weaning Change of diet for calves to all solid feed (no milk).

Cattle are members of the family Bovidae and the genus *Bos*. The two major species groups are *Bos taurus*, in and from Europe, and *Bos indicus*, the cattle humped at the shoulder. Cattle are present on six continents and the sum of *B. indicus* plus *B. taurus* is 473 distinct breeds worldwide. Cattle are domesticated animals, so breeds of cattle are categorized according to primary use. Some breeds such as Holstein are used primarily to produce milk for humans. There are only five major dairy breeds in the United States. Other breeds, such as Angus and Simmental, are used primarily to produce beef. Many breeds worldwide are considered dual purpose (milk and beef). In some areas of the world, cattle are still used to provide power for draft purposes. Finally, a central doctrine of Hinduism is reverence for cattle so cattle are a point of worship. Therefore, there are highly diverse uses of cattle around the world.

I. INTRODUCTION

Independent of the varied uses for cattle, reproduction and consequent generation of offspring is necessary. For commercial herds of cattle it is important to maintain or increase the number of animals. Calves are necessary to replace older animals that for various reasons leave a herd. In addition, reproduction is key for selection of future generations that are superior genetically to their parents. In mammals, lactation is critical for completion of reproduction. During late gestation, mammary development is abundant. Gestation ends with parturition to deliver offspring and to initiate lactation. Therefore, reproduction and lactation are interdependent in all cattle. For dairy cattle, parturition is necessary to initiate production of the primary product, milk. For beef cattle, lactation is critical for new calves to suckle until they achieve nutritional independence at about 7 months of age. Beef calves that will not be included in the herd are sold as the primary product from the herd.

Reproduction is a key determinant of long-term productivity and genetic improvement of cattle. Successful reproduction of cattle depends on many biological events, careful management of the animals by people, and attention to the environment.

II. REPRODUCTIVE BIOLOGY IN BOVINE FEMALES

As mammals, generation of bovine offspring depends on sexual reproduction. At fertilization a haploid spermatozoa combines with a haploid oocyte to establish a diploid zygote (one-cell embryo) with 60 chromosomes. Like all mammals, sexual genotype or gender of bovine offspring is determined at fertilization by the sex chromosomal complement of the spermatozoa that fertilized the oocyte. Sexual differentiation or development of sexual phenotype is directed by the genotype. If genotype is XX, phenotype is female and if genotype is XY phenotype will be male. In cattle, the incidence of twins is 3–5%, which means that about 2 or 3% of all pregnancies involve male and female cotwins. If both embryos/fetuses are alive 25–40 days after conception, at least 90% of females cogestated with males are masculinized and will be sterile (Freemartin). There are no detected consequences to sexual differentiation of males that are cotwin to females.

In females, puberty is the first ovulation or first spontaneous release of an oocyte from an ovarian follicle. Therefore, puberty is the first opportunity for conception. Age at puberty varies widely and the major variables are breed, dietary energy, and health. Larger breeds or heifers fed inadequate dietary energy will attain puberty later than smaller breeds or heifers that are fed adequately. However, most heifers should achieve puberty by 10–12 months of age.

After puberty, nonpregnant females experience recurrent estrous cycles. The duration of an estrous cycle is the time between two successive periods of estrus. Average duration of bovine estrous cycles is

21 days but the duration ranges from 17 to 24 days. For 3–5 days before estrus a single ovarian follicle grows rapidly and produces abundant estradiol to cause the behavioral and physical signs of estrus. During estrus ("heat"), females are receptive to males for an average of 12 hr but may be receptive for as short as 2 hr or as long as 28 hr. In addition, bovine females will mount other females and will stand to be mounted by another female. That bovine females display courtship and mounting (mating) behavior with other female cattle is unique among species of livestock. Additional signs of estrus are physical and include swollen vulvae and vaginal discharge of clear mucus secreted by the cervix. Ovulation occurs 24–30 hr after the onset of estrus and most cattle have a single ovulation. Ovulation is spontaneous (mating is not required) and is stimulated by increased secretion of luteinizing hormone (LH). After ovulation, the granulosal and thecal cells of the follicular wall differentiate into luteal cells. The luteal cells fill the cavity of the recently ovulated follicle to form a new endocrine tissue, the corpus luteum (CL), that secretes progesterone. If conception does not occur, the CL will be present and secrete progesterone until 17 or 18 days after estrus and then will regress. Regression of the CL and the consequent decrease of progesterone in blood are critical to allow the next estrus and ovulation to occur.

In nonpregnant females, the period when a CL is present and functional is diestrus. During diestrus, estrus and ovulation are blocked by progesterone from the CL. However, growth and development of follicles continues throughout this period. In fact, every 6–10 days a new cohort of follicles initiates growth. From each cohort a single follicle is largest and secretes more estradiol than other follicles and is considered dominant. This process that selects a single dominant follicle from a cohort is described as a wave of follicles. During an estrous cycle, most cows have three waves of follicles, some cows have two waves, and a few cows have a single wave. Waves of follicles that occur during diestrus do not lead to ovulation and all follicles in those waves experience atresia. The only follicles that ovulate are those few that are dominant after the end of diestrus when progesterone secretion is low and there is increased secretion of LH.

If conception occurs, diestrus is continuous with gestation. The CL of diestrus persists throughout gestation, which lasts for an average of 283 days. Duration of gestation is affected by breed of the parents and gender of the fetus. Large breeds and male fetuses have the longest gestations. A key to establish gestation is to avoid death of the embryo (embryo mortality) before the start of fetal life at about 35–45 days after conception.

Among fertilized oocytes, 15–20% will not survive to the fetal stage. In fact, 70% of embryo mortality occurs by 18 days after conception. To establish pregnancy, maternal recognition of pregnancy must prevent regression of CL, sustain diestrus, and thus prevent estrus. Gestation ends successfully with birth of a live offspring at parturition. Dystocia is a difficult or prolonged delivery and can reduce vitality of calves and cows. Parturition is a major stimulus to initiate lactation in beef and dairy cattle. Resumption of ovulation after calving is not immediate in most cows. This period of anovulation in beef cows that are suckled by a calf (45–60 days) is longer compared to that of dairy cows (21–35 days) that are milked by machine but not suckled. After ovulation resumes, conception should occur by 90 days postpartum and the reproductive cycle is repeated.

This overview of reproductive biology is intended to establish the sequence and some description of major events in the reproductive cycle of female cattle. This sequence is illustrated in Fig. 1. Furthermore, it is intentional that this description does not include details of biological events or models to illustrate endogenous regulation. Readers should refer to separate sections of this encyclopedia that address in-depth virtually all the major biological and regulatory processes that affect reproduction in cattle.

III. REPRODUCTION IN BEEF CATTLE

A. Sequence of Events

Major reproductive events for beef females are presented in Fig. 2 to illustrate the sequence and the duration of specific periods. The straight line segment at the left of the figure represents the interval

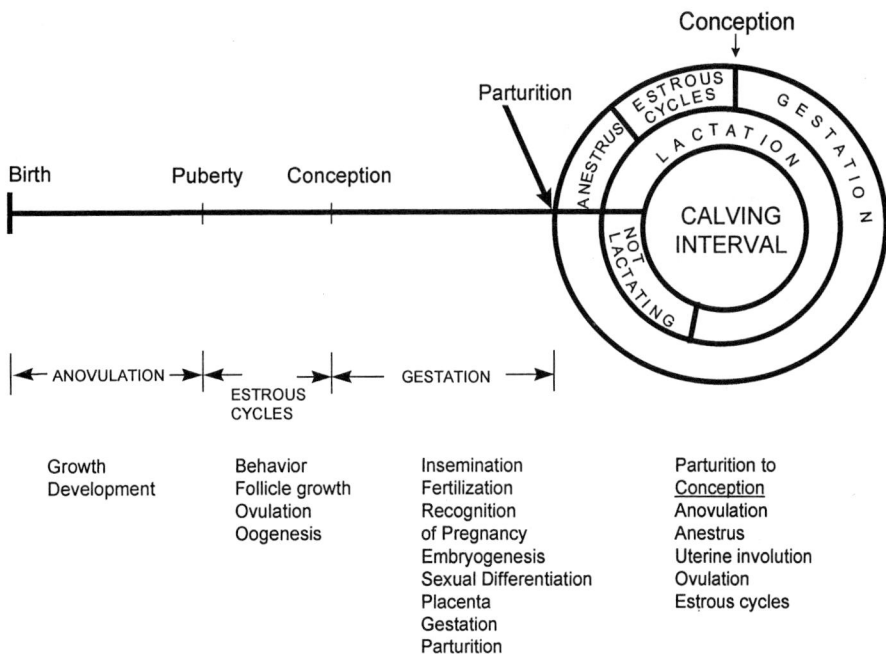

FIGURE 1 Sequence of reproductive events in an individual bovine female. The overall schedule is determined by a combination of biological limitations and managerial strategy. Specific biological events that affect reproduction are listed at age-appropriate times. Each of these specific biological events is discussed fully in other chapters of this encyclopedia.

from birth to first parturition. Females that have never experienced parturition are nulliparous and are called heifers. The circular portion of the figure represents events between successive parturitions (calving interval). Females in any phase of this cycle have experienced parturition at least once so they are parous and are called cows. There are two major goals for reproductive performance of beef cattle (Fig. 2). First, a beef heifer should experience first parturition at 2 years of age. Second, a beef cow should experience parturition every 12 months (annual calving interval). These goals are important be-

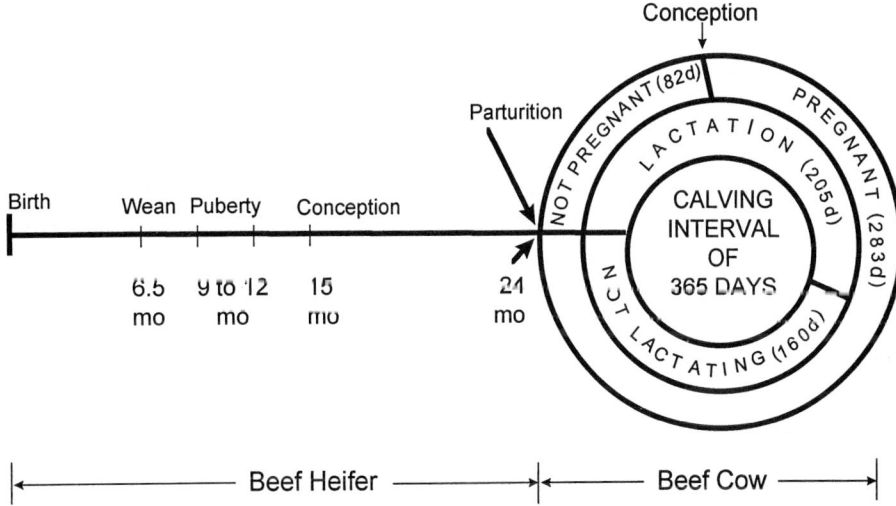

FIGURE 2 Sequence of major reproductive events in an individual beef female.

cause performance toward these goals affects economics of production. In addition, these goals are strategic because there is sufficient biological variation that performance can be influenced by management. In contrast, duration of gestation ($\approx$280 days) has minimal variation and is not a managerial goal. The start of lactation varies among cows according to date of parturition. However, the end of lactation is the same for all beef cows within a herd because in most herds all calves are separated from their mother (dam) on the same calendar date. Average duration of lactation is about 6.5 months. Cessation of suckling terminates secretion of milk. Many factors can influence when beef calves are weaned and consequently when cows stop lactating.

However, a major issue is that at 6.5–7 months after calving, most beef cows do not produce enough milk to satisfy the nutritional needs of a calf that weighs at least 180 kg and is still growing rapidly.

In the sequence of reproductive events (Fig. 2), note that concurrent with lactation and suckling by calves, beef cows must resume ovulation and conceive. In fact, to experience annual calving, a cow must be pregnant by $\approx$85 days after calving. However, suckling by calves extends anovulation and increases the difficulty to achieve conception on time. Attention to adequate dietary energy and reserves of body fat before calving are important management practices to reduce the adverse effects of suckling on reproductive performance of beef cows after calving.

For beef cattle, reserve calories stored as body fat are estimated by body condition scoring (BCS) with a scale from 1 to 9 (9 = obese). By visual evaluation of the exterior, this scoring system is a noninvasive subjective appraisal of subcutaneous fat that allows managers to judge the merit of nutritional management by comparison of BCS among cows and changes in BCS of individuals over time. An idealized profile for BCS is presented from birth through one calving interval (Fig. 3). Adequate BCS is necessary for timely occurrence of puberty in heifers and timely resumption of ovulation in cows after calving. Evaluations of BCS before puberty and before parturition are important managerial procedures so that BCS is above the minimum required for reproduction. However, if BCS is excessive, this can reduce reproductive success. For example, heifers that are obese before puberty produce less milk as cows and obese cows will have increased difficulty calving (dystocia). Therefore, nutritional management of postpubertal beef cattle should maintain moderate to high-moderate BCS (5.5–7), should avoid major acute decline of BCS in early lactation, and should ensure that any BCS lost during lactation is recovered ideally by the time the current calf is weaned (Fig. 3) and absolutely before the next parturition.

B. Seasonality

In postpubertal nonpregnant beef heifers and cows, estrous cycles will occur in all seasons of a year. Therefore, in contrast to horses and sheep, reproduction in beef cattle is not limited biologically to any season. However, in North America most beef cattle are managed for breeding and conception in the late spring to early summer. With the breeding period confined to a season, it follows that calving will occur in the subsequent winter to spring. Therefore, every year, a high percentage of the annual crop of calves is weaned in the fall. A partial explanation for this pattern is that calves were suckling during seasons that dams consumed forage at pasture. Then grains from the fall harvest would be abundant for newly weaned calves to finish their growth in feedlots.

To achieve this seasonal pattern of production from beef cows, breeding must be limited to a distinct period. In herds that do 100% natural breeding, bulls are introduced with the cows on a specific date and then bulls are separated from cows at a later specific date. The duration of this breeding season or breeding period ranges from 45 to >120 days among herds and averages $\approx$90 days. For an individual herd, the duration of the breeding period is usually consistent over time. Shorter breeding periods produce more homogeneous calves at weaning, but the risk is that some cows will fail to conceive because of too few periods of estrus while the bulls were present.

Animals that just experienced their first parturition (primiparous) at 2 years of age are physically immature and continued body growth is necessary until age is 4 or 5 years. Because these primiparous cows

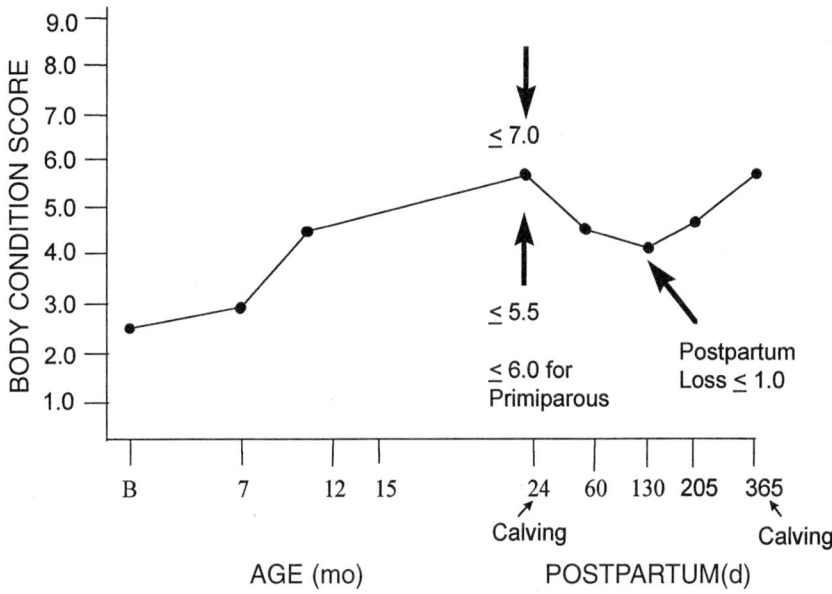

FIGURE 3 Idealized profile for body condition score (BCS) of female beef cattle. In beef cattle, BCS is an estimate of the amount of subcutaneous fat based on external visual appraisal of the rump, loin, and ribs of live animals. For beef cattle, BCS ranges from 1 to 9 (9 = obese). Puberty, parturition, postpartum ovulation, and lactation are affected adversely by extreme, low or high, BCS. Nutritional management of beef females to achieve good reproduction and lactation can be guided by BCS. Determining BCS at strategic times in the growth phase and reproductive cycle is useful to evaluate and adjust nutritional management.

are lactating, suckled, and growing, resumption of estrous cycles after calving is delayed more than the delay due to suckling and lactation of mature cows. If this situation is not addressed, a very low percentage of young cows will conceive by the end of the breeding period. Managers are encouraged to breed heifers 3 weeks before cows so that these heifers deliver calves before the mature cows and will have more time to recover between calving and the beginning of the breeding period.

Reproduction in beef cattle is managed to be seasonal. This means that at a particular time of year, all the cows in a herd have a similar reproductive status. For example, in early summer most cows will be lactating and not pregnant but in winter most cows will be in late gestation. This allows managers to focus on one aspect of reproductive management at a time. However, this approach requires intense attention and diligence to be successful during those narrow windows of opportunity.

C. Reproduction and Production of Beef

Body weight of calves at weaning is affected by birth weight plus genetic and nutritional effects on growth. Recall that within a herd, beef calves are weaned on a common date independent of age. Therefore, compared to calves conceived late in the breeding period, calves conceived early in the breeding period will be the oldest at weaning. Independent of genetics and nutrition, older calves have more time for growth, so age will be associated positively with body weight at weaning. Thus, conception early in the breeding period will increase productivity of beef cows. The number of calves weaned is affected by the number of cows that conceive during the breeding period, mortality during gestation (15–25%), and mortality between birth and weaning (<5%). To maximize the percentage of cows that conceive is fundamental to maximize the number of

TABLE 1
Average and Potential Reproductive Performance of Beef Cattle

	Reproductive performance	
	Average	Potential
Heifers		
Pregnancy rate = number pregnant ÷ number in breeding group (%)	90	98
Calving rate = number calving ÷ number in breeding group (%)	85	95
Age at calving (months)	26	24
Cows		
Pregnancy rate = number pregnant ÷ number in breeding group (%)	85	95
Calving rate = number calving ÷ number in breeding group (%)	80	90
Calf survival = number weaned ÷ number born (%)	90	97
Weaning rate = pregnancy rate × calf survival (%)	76	92

resources to maximize calving rate and to achieve timely conception. However, the priority must be for calving rate because a small calf as a result of late conception is certainly better than no calf because there was no conception.

Reproduction in beef cattle is important because of the profound effects on replacements for genetic selection and annual productivity of a herd. Whether and when beef cows conceive are major reproductive issues because there are independent and additive effects on production (Fig. 4) and thus on gross income.

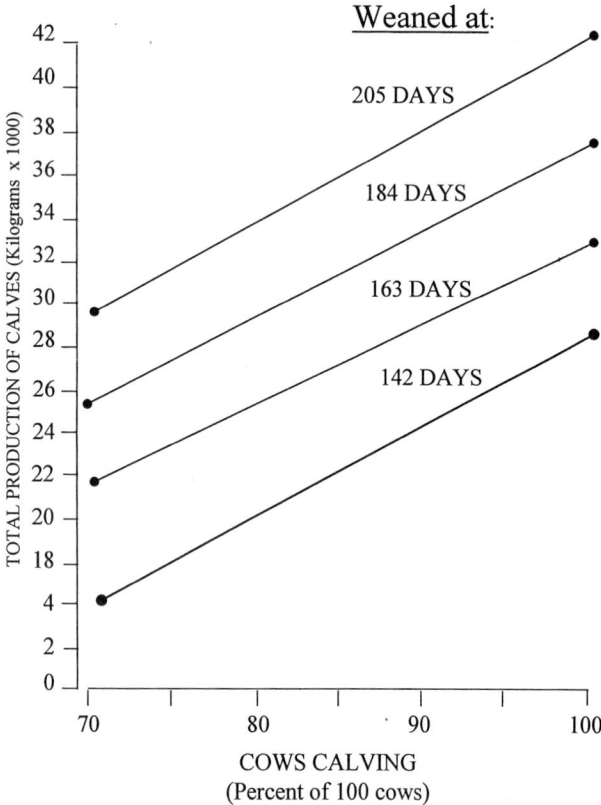

FIGURE 4 Variation in reproductive performance and consequences to productivity of 100 beef cows. Productivity of cows is mass (kg) of calf at weaning. Reproductive performance of cows is considered as (i) calving rate on the x-axis and (ii) age of calves at weaning. All calves are weaned on the same calendar date so variation in age is really variation in time of conception during the breeding period. Calves conceived late in the breeding period are younger and weigh less at weaning.

calves born and ultimately weaned. Pregnancy rate varies among herds and among years, but average pregnancy rate is about 80% (Table 1). Therefore, there is certainly opportunity to improve pregnancy rate as a means to increase the number of calves weaned.

Weaned calves are the major product available for sale from beef cows. Therefore, total production of calves is a major determinant of gross income. Reproductive performance has a major effect on number of calves, age of calves, and thus total pounds of calf available at weaning (Fig. 4). Calving rate and time of conception each have independent effects on productivity of cows. Note that conception early in the breeding period will increase age at weaning. In addition, calving rate and time of conception interact positively so that increased productivity is more than additive. Clearly, managers should direct effort and

D. Measures of Success

There are many measures of reproductive performance that are used to monitor performance toward managerial goals or to identify the cause of a problem. The measures of reproductive success presented below will be limited to those that meet three criteria: (i) The reproductive performance that is measured has substantial economic impact, (ii) the reproductive performance can be affected by management, and (iii) all females in a breeding group are included in the evaluation.

The measures of reproductive success that satisfy these criteria are defined and average performances are presented in Table 1. The potential reproductive performance, that performance that is biological feasible, is also presented (Table 1) for perspective.

E. Infertility

Failure to conceive or conception late in the breeding period represent infertility in beef cows. There are many known and some unknown causes of infertility. Certainly, anovulation that persists through the breeding period is a problem because these females have no opportunity to conceive. Other animals may have limited opportunity for conception because estrus and ovulation did not resume until after the start of breeding. Extended anovulation would be associated with suckling, low reserves of body fat, and primiparous females. At any period of estrus when a cow is mated, the probability of conception is at most 80% and averages 60–70%. In postpartum cows, the probability of conception can be reduced by dystocia, metabolic disorders, and uterine problems after calving, low body condition score before calving (Fig. 3), or low dietary energy during the breeding period. Independent of the reason for failure to conceive after mating, the consequences range from no calf to a calf that is up to 12 weeks younger at weaning. Therefore, infertility will increase the number of nonproductive days and reduce productivity of a cow.

Infertility in heifers has much in common with infertility in cows. In addition, heifers may experience delayed puberty (extended anovulation) due to inadequate dietary energy to support growth and accumulation of body fat. Adequate skeletal growth and BCS are special challenges for breeds that have large mature size (e.g., Chianina, Charolais, and Simmental). However, delayed puberty is an issue only when management of nutrition or health are not adequate. Another major source of infertility in beef heifers is difficult parturition (dystocia). Dystocia has consequences to the calf and to the dam. Incidence and severity of dystocia are variable but may occur in 50% of heifers at first parturition. In beef heifers, dystocia is usually because the pelvic area is too small for the fetus. A breeding soundness exam for heifers before the start of the breeding period should determine pubertal status and measure pelvic area. Heifers that are prepubertal should be isolated for supplemental nutrition or removed from the herd. Heifers with a small pelvic area (<200 cm^2) should be mated to bulls of proven calving ease or removed from the breeding group and marketed for beef.

F. Reproductive Performance

The purpose of this section is to provide information on average reproductive performance of beef cattle in the United States (Table 1). Readers should be aware that there is substantial variation regarding these averages. This variation is due to geographical region and associated differences in climate and managerial systems, herd, economic status of the industry, and breed. However, these averages for reproductive performance (Table 1) will provide some indication of the proximity between actual and potential reproductive performance. Because actual performance is less than goals (Table 1 and Fig. 2), there is opportunity to improve all aspects of reproductive performance in beef cattle. However, whether these incremental increases in reproduction will be cost-effective depends on the economic and managerial status of each herd.

IV. REPRODUCTION IN DAIRY COWS

A. Sequence of Events

Major reproductive events for dairy cattle are presented to illustrate the sequence and the duration of specific periods (Fig. 5). As illustrated for beef cattle

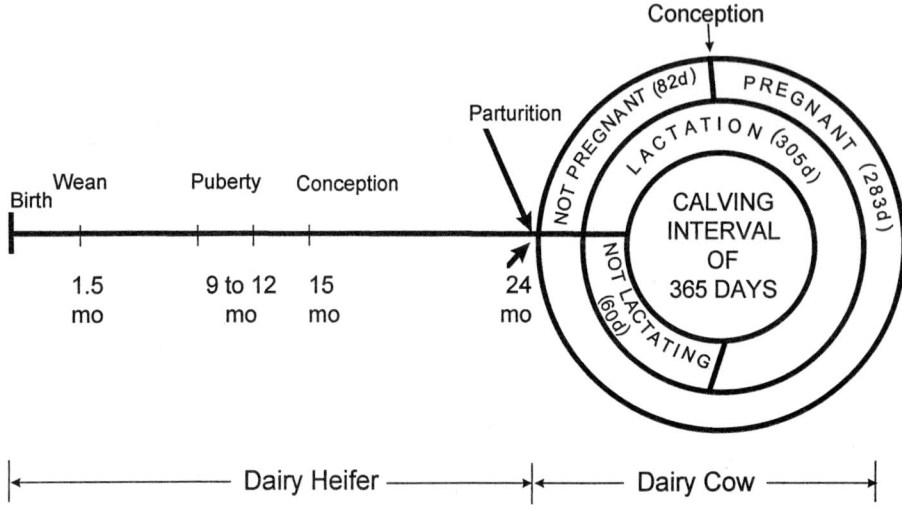

FIGURE 5 Sequence of major reproductive events in an individual dairy female.

(Fig. 2), the linear segment at the left of the figure represents the interval from birth to first parturition in dairy heifers. The circular portion of the figure represents events between successive parturitions for dairy cows. There are two major goals for reproductive performance of dairy cattle (Fig. 5). First, a dairy heifer should deliver her calf and start her first lactation at 2 years of age. Second, a dairy cow should experience parturition every 12 or 13 months (annual calving). These goals are important because performance affects the economics of milk production especially over the lifetime of a cow and because performance can be influenced by management.

After initiation by parturition, lactation in dairy cattle has an average duration of 305 days. Between successive lactations there should be a period of no lactation ("dry period") primarily for regeneration of secretory cells in the mammary glands. The lactational cycle (lactation plus dry period) is dictated by the concurrent reproductive cycle (Fig. 6). Therefore, dairy cows must be pregnant by ≈82 days postpartum to be 7 months pregnant at the end of lactation and ready to calve at the end of the 60-day dry period (Fig. 6).

Whether insemination is artificial or by bulls, onset of the breeding period for dairy cows should be delayed until at least 60 days postpartum. If conception occurs too early (before 82 days postpartum), the duration of lactation or the dry period will be shortened. In contrast to beef cows, there is no programmed or uniform end to the breeding period for dairy cows. However, every cow not pregnant by 82 days postpartum lactation will be extended into a time that daily yield of milk is usually low (Fig. 6).

B. Seasonality

As with beef heifers and cows, in dairy females reproduction can occur in all seasons of a year. Because there is no biological or economic constraints, dairy calves are born throughout a year. However, in areas that experience high ambient temperature and humidity for prolonged periods, managers may elect to avoid breeding cows or have cows in early lactation during times of high environmental challenge. Furthermore, those herds that feed cows exclusively or largely by grazing would manage for seasonal reproduction as with beef cows. However, grazing is not a common method to feed lactating dairy cows in the United States.

C. Reproduction and Production of Milk

Total yield of milk over the life of a dairy cow is affected by interdependence between reproduction

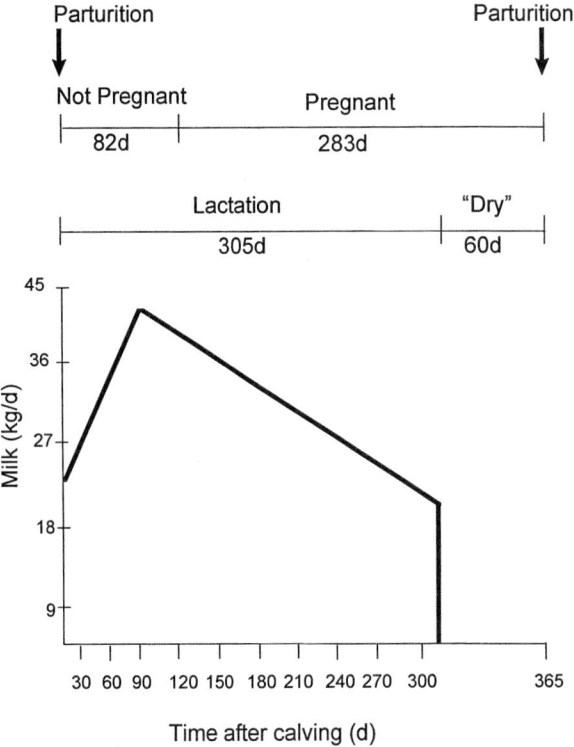

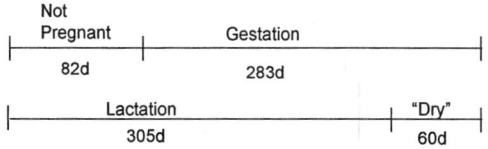

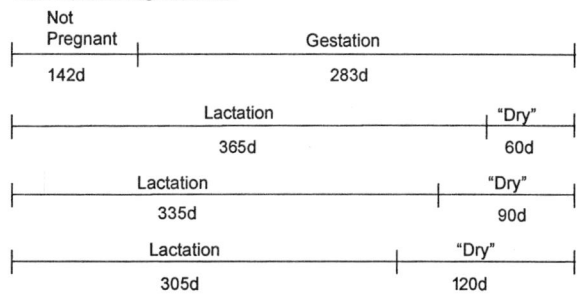

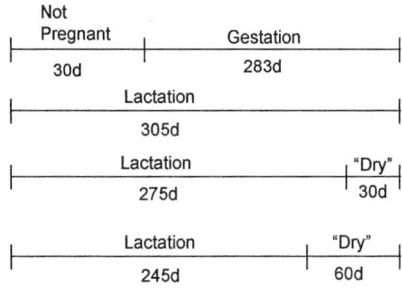

FIGURE 6 Reproduction and the lactational cycle for dairy cows. The profile for yield of milk is based on a monthly decline of 7%. The period of nonlactation (dry period) before parturition should be 60 days to allow for restructuring of mammary glands to enhance lactation after the next parturition.

postpartum the CI will be 14 months and lactation will be longer than 305 days (Fig. 7) and for some cows will extend to a time when daily yield of milk is typically very low. Therefore, timing of conception after parturition dictates the lactational cycle in dairy cattle. Conception at 75–125 days postpartum (12

and lactation (Fig. 6). This relationship between reproduction and lactation is established by and based on the profile for the yield of milk ("lactation curve"). From the onset of lactation, daily yield of milk increases to a peak at 6–8 weeks after calving. After the peak, yield of milk declines 3–8% monthly until the end of lactation. Lactation ends when yield of milk is not profitable or at 60 days before calving, whichever occurs first. For most cows, the duration of lactation is determined by the date of next expected calving. If conception occurs at 85 days postpartum, the next calving will be 365 days after the previous calf was born and the calving interval (CI) will be 12 months. With this sequence and timing of reproductive events, lactation will end at 305 days postpartum and 60 days before the next calving (Figs. 6 and 7). However, if conception occurs at 142 days

FIGURE 7 Variation in reproductive performance and consequences to the lactational cycle of dairy cows. Reproductive performance can be measured by the interval from calving to conception. With a calving interval (CI) of 12 months, conception at 82 days postpartum is considered timely because lactation of 305 days and "dry period" of 60 days occur before the next parturition. However, when conception is early (10-month CI) or late (14 month CI) the lactational cycle must be adjusted. If conception is too early and there is a CI shorter than 12 months, lactation or the "dry period" must be shortened. If conception is late and CI is >14 months, lactation is extended to when daily yield of milk is low and marginally profitable or dry period is extended which will increase nonproductive days. The principle is that reproductive performance dictates the lactational cycle.

or 13 month CI) is deemed currently to be ideal to maximize yield of milk per day of life in a herd. This schedule benefits from high yield of milk in early lactation with minimal risk that yield of milk will decline below profitable levels in late lactation. Later conception (e.g., 142 days postpartum) and consequent longer calving intervals will extend lactation and increase yield of milk per lactation but will reduce yield of milk per day of life. Although not common, if conception occurs before 75 days postpartum (less than annual calving) yield of milk per lactation will be decreased and yield of milk per day may be increased because of high yield in early lactation and fewer days of low yield in late lactation. However, when CI is <12 months, increased frequency of parturition will increase the risk that a cow will experience metabolic or infectious disease at or near calving. Such challenges to health are costly because of treatments required, reduced yield of milk, reduced fertility, and possible removal of cows from the herd (culling). Therefore, likely consequences to health of cows limit the extent that the reproductive cycle can or should be shortened. Independent to the duration of lactation, the period of nonlactation before calving must be 45–60 days. Therefore, shortening the period of lactational rest is not an acceptable practice to compensate for shortened lactations attendant to early conception.

With variation in duration of calving intervals, dairy managers are forced to change the duration of lactation, the dry period, or both. Timing of conception after calving dictates the length of the CI and duration of the lactational cycle. That reproduction dictates lactational cycles makes reproduction important. However, this relationship is a problem because the effect of reproduction on lactational cycles is independent of milk yield. The principle is that a 12- or 13-month CI will maximize time that cows are in early lactation when yield of milk is high and will minimize time that cows are in late lactation with low yield of milk. Lactational rest should always be >45 days to allow for regeneration of mammary secretory cells. Variation in CI affects only duration of lactation. However, the area of a profile for milk yield is affected by yield at peak, decline after peak, and duration of lactation. Therefore, the impact of CI on area of lactational profiles is variable. In fact,

from most economic analyses of variation in reproduction, CI must be >14 months before there is detectable loss of income. It is possible that some cows with high peak or slow decline would be more profitable with a longer CI. However, there are no reliable methods to predict total lactational yield based on very limited data from early lactation. Until these methods are developed, reproductive performance will continue to dictate the duration of lactational cycles and managers should continue to strive for annual calving in most cows.

D. Measures of Success

Compared to beef cows, in dairy cows there are more measures to monitor reproductive performance or to identify the cause of problems. More measurement of reproductive performance in dairy cows is due largely to extensive use of artificial insemination and generally more intense management compared to beef cows. However, for this discussion, the measures of reproductive success will be limited to those that satisfy the same three criteria discussed for beef cattle. The measures of reproductive success that satisfy these criteria are defined and average performance is presented in Table 2. For perspective, the reproductive performance that is biologically feasible (potential) is also presented (Table 2). To determine these and other measures of performance, extensive, detailed, and current records are required. There are many systems to record performance of dairy cows, including various software for use on farms.

E. Infertility

In dairy cows, infertility is failure to conceive or delayed conception. In healthy dairy cows, there are many known and some unknown causes of infertility. In contrast to beef cows, anovulation is not a significant cause of infertility in healthy dairy cows. This is because dairy cows are not suckled by calves. However, primiparous dairy cows that experience severe negative energy balance may have anovulation that persists to the start of the breeding period. After insemination, about 50% of cows conceive. Therefore, failure to conceive and embryonic mortality are certainly important sources of infertility. In fact, with

TABLE 2
Average and Potential Reproductive Performance of Dairy Cattle

	Reproductive performance	
	Average	Potential
Heifers		
Pregnancy rate = number pregnant ÷ number in breeding group (%)	95	98
Calving rate = number calving ÷ number pregnant (%)	90	95
AI calves = number AI calves ÷ all calves (%)	60	95
Age at calving (months)	27	24
Cows		
Pregnancy rate = number pregnant ÷ number in breeding group (%)	85	95
Calving to conception (days)	122	85
Calving interval (days)	405	365
AI calves = number AI calves ÷ all calves (%)	75	100

no definite end to the breeding period and low success of insemination, it is not unusual for some dairy cows to receive three or more inseminations ("repeat breeders"). In general, extension of the breeding period due to repeat breeding is not desirable because this extends the calving interval, delays the start of the next lactation, and increases nonproductive days by at least 21 days for each additional insemination.

About 70% of dairy calves are conceived by artificial insemination (AI). To have the opportunity and to schedule AI, people must observe cows to detect estrus. For a variety of reasons, detection of estrus is <50% successful. Therefore, for cows in the breeding period, failure to detect estrus is a major cause of infertility because these missed opportunities for AI delay conception and extend the calving interval. Failure of dairy cattle to conceive after insemination is associated with dystocia, retained fetal membranes, uterine problems, and metabolic disorders. For the 80% of dairy cows that experience negative energy balance because they cannot satisfy their caloric requirement from their diet, they mobilize fatty acids from adipose tissue to supply calories that support lactation. During this catabolic state of negative energy balance there is less atresia, increased selection, and reduced dominance of ovarian follicles within each wave. Coincident with this change in follicles is reduced function of the CL and less progesterone in blood. However, despite considerable discussion there is currently no evidence that negative energy balance affects fertility of dairy cows. Alternatively, when negative energy balance leads to clinical metabolic disease (e.g., ketosis), fertility will decline.

Body fat is especially important to dairy cows because most cows experience negative energy balance and must use calories from body reserves to sustain lactation. For dairy cattle, body fat is estimated by BCS with a scale from 1 to 5 (5 = obese). An idealized profile for BCS is presented for the period from birth through one calving interval (Fig. 8). Adequate BCS is necessary for timely occurrence of puberty and timely resumption of ovulation after calving. Fur-

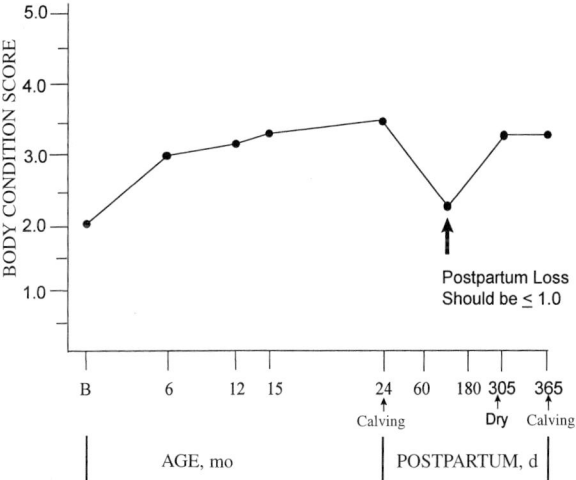

FIGURE 8 Idealized profile of body condition score (BCS) of female dairy cattle (Holsteins). To determine BCS in dairy cattle, the rump and loin are palpated to estimate the amount of subcutaneous fat. For dairy cattle, BCS ranges from 1 to 5 in increments of 0.5. A cow with BCS of 5.0 is obese. Puberty, parturition, postpartum health and ovulation, and lactation are affected adversely by extreme BCS (low or high) or by acute major decline of BCS. Nutritional management of dairy females to achieve good reproduction and lactation can be guided by BCS. Determining BCS at strategic times during growth and in the reproductive cycle is useful to evaluate and adjust management of nutrition and health.

thermore, BCS to sustain lactation during negative energy balance may be more important to dairy than to beef cows. In fact, the metabolic priority to use body fat to sustain lactation in dairy cows lowers the priority to resume ovulation after calving. High BCS (>4.0) should be avoided so that appetite is maximal and that the duration of declining energy balance is minimal. To reduce dystocia and metabolic disorders and to sustain lactation, prepartum BCS of dairy cows should be moderate to high-moderate (3.5–4.0). In addition, the nutritional management soon after calving should avoid a major or acute decline of BCS and should ensure that BCS lost during early lactation is replenished by the end of lactation (Fig. 8).

There are many infectious conditions that can affect fertility and reproductive performance of dairy cows. However, consistent use of available vaccines, rigorous sanitation of maternity areas, cautious introduction of animals into a herd, effective ventilation of facilities, and adequate nutrition will prevent most of these problems.

Cystic ovarian follicles are large or persistent follicles that fail to ovulate. The behavioral signs that a cow has a follicular cyst range from anestrus to nymphomania. Incidence of follicular cysts varies widely among herds but averages 10% of postpartum cows. Ovarian cysts receive considerable attention by veterinarians and dairy managers because this problem is associated with cows proven to have high yield of milk and the condition may persist for 30–60 days. However, it is doubtful that ovarian cysts are a source of infertility in a significant portion of the population.

Especially in the southern United States, high ambient temperature and high humidity challenge thermal regulation in lactating cows. Severe and prolonged heat stress is devastating to reproduction because of low detection of estrus and low success of conception. In those regions that experience high temperature and humidity, managers may establish seasonal breeding to avoid hyperthermic challenges to reproduction and to lactation.

F. Reproductive Performance

What is the actual reproductive performance of dairy cattle in the United States? Performance is reported as averages (Table 2). However, these averages mask the substantial variation due to climate, managerial systems, herd, and economic status of the industry. Note that average actual performance (Table 2) is consistently below goals and potential performance (Table 2 and Fig. 5) so there is opportunity to improve reproductive performance of dairy cattle. Whether reproductive performance improves to achieve goals will be determined by efforts of managers. Whether improvements are cost-effective is based on the economic status of a herd or the industry.

It is also possible that evolving technology and management in the future will warrant review of the reproductive goals for dairy cows. For example, if after peak of lactation milk yield decreased only 1 or 2% monthly, not 5%, the traditional rationale for annual calving should be reviewed. Under these conditions, there would be less justification to have cows conceive by 82 days postpartum and there could be economic benefit to calving intervals of 14 months or longer. Importantly, this issue has not been tested and traditional goals are not yet rejected by objective analysis.

V. BULLS

A. General Role

In cattle, reproduction is sexual so bulls are critical for successful reproduction. Because $\approx 70\%$ of dairy calves are conceived after artificial insemination, most bulls that sire dairy calves do not copulate with cows, are not members of dairy herds, and are simply sources of spermatozoa. In contrast, most beef calves ($>95\%$) are conceived after copulation (natural mating). Therefore, to breed cows, beef bulls must be an integral part of a herd.

B. Biology of Reproduction

Puberty in bulls is when spermatids first appear in the seminiferous tubules and occurs when bulls are 12–15 months of age. Cellular and regulatory aspects of spermatogenesis are discussed in-depth in other sections of this encyclopedia. Spermatogenesis has two phases: (i) a proliferative phase (spermatogonia to primary spermatocytes) controlled by follicle-stimulating hormone (FSH) and estradiol and (ii) a differentiation phase that includes meiosis (secondary spermatocytes to spermatozoa) and is regulated by LH and testosterone. In addition to hormonal control, successful spermatogenesis requires that testicular temperature is 1–3°C cooler than core body temperature. In bulls, testicular temperature is regulated primarily by involuntary adjustment of the distance between the scrotum and the body and secondarily by the pampiniform plexus. For example, as ambient temperature increases the scrotum is lowered. However, severe and extended heat stress or fever may overwhelm this regulatory system and will impair the differentiation phase (meiosis) of spermatogenesis and may cause bulls to be infertile at least temporarily.

In contrast to oogonia in females, spermatogonia retain mitotic activity for the fertile life of an adult bull. In healthy bulls, proliferation creates 64 gametes per parent cell and 3–9 billion sperm per ejaculate. Production of billions of structurally normal and viable spermatozoa is critical to reproductive performance of bulls.

C. Breeding Capacity

Although spermatogenesis is a prolific process, it does have a limit. If the capacity to produce spermatozoa is exceeded by too much breeding activity fertility will be reduced. Therefore, it is important to know the number of cows that can be bred or served by a bull. This capacity varies among bulls due largely to age. Within a 60-day period of natural breeding, yearling bulls can service 1–15 cows, 2-year-old bulls can service 25–30 cows, and mature bulls (>3 years) can service 40–50 cows. Note that these serving capacities ignore variation among individual bulls (see Section V,C). With these numbers of cows per bull, it is assumed that occurrence of estrus is distributed evenly among cows so that only 1–3 cows experience estrus daily. In addition to adequate numbers of spermatozoa, behavior of bulls also affects breeding capacity. First, bulls must have an adequate libido to court and mate cows. Second, if multiple bulls are together with the appropriate number of cows, a dominant bull can prevent breeding by subordinate bulls. Thus, some bulls do not perform at the level of their breeding capacity. This means that the number of cows available for breeding will exceed the breeding capacity of the dominant bull. Consequently, some cows may not be bred or the number of spermatozoa ejaculated may be too low to achieve conception. There is no easy or reliable method to determine if a bull will dominate other bulls to the extent that breeding capacity will be affected. Therefore, the ideal situation is that each bull has a harem of cows sized for his breeding capacity and there is physical separation between each group of cows and the bull assigned to them.

D. Breeding Soundness

Most postpubertal bulls are capable of acceptable reproductive performance. However, fertility varies among individuals so that 10–20% of bulls do not have acceptable fertility and an additional 10–20% are marginal. It is not feasible to conduct direct tests of fertility for even a few selected individual bulls each year. A compromise between no evaluation and fertility tests is the breeding soundness exam (BSE), which is designed to estimate potential fertility of individual bulls. During the BSE, the following primary factors can be evaluated:

1. Physical condition: ability to detect estrus and to breed cows
2. Reproductive tract: structural status of the penis and scrotal circumference as an estimate of sperm production
3. Sperm: morphology and motility
4. Genital health: infectious disease
5. Libido: sex drive

Physical condition and genital health are evaluated as pass or fail. Libido of bulls is important but is not examined routinely in the United States because during the test welfare of animals is not acceptable. Morphology and motility of sperm and scrotal circumference are scored numerically. Each numeric score is based on objective criteria for the factor evaluated. For sperm and scrotum the cumulative score must be at least 60% of maximum for a bull to pass the exam. However, failure in physical condition, health, or libido supersedes a passing score for scrotum and sperm so the bull will fail the entire exam. A bull that fails the BSE may be examined again if therapy, management, or additional sexual development can remove the original limitation to fertility.

VI. REPRODUCTIVE TECHNOLOGY

A. Artificial Insemination

This technology, including frozen semen, has been available for over 50 years. Through AI, the genetic and geographic impact of a bull are expanded several thousand-fold compared to natural breeding. A typical ejaculate will produce enough spermatozoa to inseminate up to 1000 cows. Frozen semen can be transported anywhere. Because with AI the presence of the whole bull is not required, the genetic impact of a bull can occur in multiple females at many locations simultaneously. Among dairy cattle there is an excellent system to identify superior bulls, AI is used extensively, and genetically based milk yield improves 45–135 kg per year. With AI, people must detect those cattle that are in estrus and this process is not very successful. When dairy cows fail to conceive by 82 days after calving, most of this infertility is associated with inefficiencies to detect estrus and consequent failure to perform AI. When AI occurs at the proper time, success is not 100%. However, failure to conceive is a secondary problem. In many herds genetic progress achieved by AI justifies the managerial challenges and potential limitations to reproductive performance that are consequential to AI.

In beef cattle, AI is used primarily by seed-stock producers and a few commercial producers. Total beef calves conceived by AI is ≈5%. Why is use of AI in beef cattle so low? During the breeding period, 60–120 days postpartum, most beef cows and heifers are on pasture or extensive range lands. In these circumstances, it is extremely time-consuming to detect estrus and to restrain cows for insemination. In addition, genetic selection in beef cattle involves multiple traits so genetic progress in any one trait is slow. Therefore, high costs for labor to use AI may cancel increased revenue from genetic progress and erode narrow profit margins that exist for beef cattle.

B. Synchronization of Estrus and Ovulation

By use of exogenous hormones, such as prostaglandin $F_{2\alpha}$ or progesterone, the occurrence of estrus for a group of females can be condensed from a period of 21 days to 3–7 days. Occurrence of synchronized estrus is predictable and will improve success and reduce labor required to detect estrus. A major goal is to control ovulation with sufficient precision that AI at predetermined times will be successful without need to detect estrus ("AI by appointment"). Based on products and procedures available currently, AI by appointment is 10–20% less successful than AI with detected estrus. Therefore, more development is necessary. However, synchronization currently available has great merit as a management tool to facilitate detection of estrus.

Synchronization can be integrated into routine management programs. For example, synchronization of groups of dairy females every 2 weeks is a major asset to manage time and labor. Furthermore, 100% of females in the breeding group can be synchronized and inseminated within 1–5 days of the start of the breeding period and thus will enhance timely conception. Routine scheduled use of synchronization to facilitate AI is labeled "target breeding" and is common in dairy herds. In beef herds, synchronization is primarily used to facilitate use of AI in most cows at the onset of the breeding period so that some calves are sired by genetically superior bulls and to maximize age of calves at weaning (Fig. 4).

C. Superovulation and Embryo Transfer

Most cows have one offspring after a gestation of approximately 283 days. Also, there is currently no control over gender of offspring so only 50% of calves are female. Part of the rationale for superovulation and embryo transfer is to increase the number of offspring from a cow per generation. Cows selected to be donors of embryos are deemed superior genetically by one or more criteria, so superovulation and embryo transfer is a method to intensify maternal line breeding. This procedure has generated some excellent females. However, the major impact on genetics of the bovine population is that more superior bulls are available for use in AI.

The basic procedure is to use exogenous FSH to develop multiple large follicles, which are then ovulated by endogenous LH. The number of ovulations is variable (0–40) but the average is about 10. Superovulated cows receive multiple inseminations and embryos are collected nonsurgically from the uterus at 7 days postestrus. After evaluation of quality and developmental status, embryos may be transferred fresh into nonpregnant recipients or stored frozen. Embryos that are frozen must be of very good to excellent quality before freezing. Best results are achieved when the developmental stage of an embryo matches the stage of an estrous cycle of the recipient. In general, the success of embryo transfer with fresh or frozen embryos is equal or superior to success of AI. Costs per offspring from embryo transfer (>$250) are variable due to practitioner and success but in general will be at least sixfold greater than the costs for an offspring from AI (>$25).

D. Sexing Semen

In mammals, the male parent determines gender of the offspring because each spermatozoa has an X or Y chromosome. Separation of X- from Y-bearing spermatozoa to establish X- or Y-enriched semen is a long-standing interest for livestock. For example, dairy managers want mostly female offspring and beef producers will benefit from mostly male offspring. Technology has been developed to enrich porcine semen for X or for Y spermatozoa. Despite periodic commercial claims for bovine spermatozoa, this technology has not been established to reliably affect the gender of offspring and to achieve satisfactory fertility.

See Also the Following Articles

ARTIFICIAL INSEMINATION, IN ANIMALS; EMBRYO TRANSFER; PIGS (SUIDAE); RUMINANTS; SHEEP AND GOATS

Bibliography

Bauman, D. E., and Currie, W. B. (1980). Partitioning of nutrients during pregnancy and lactation: A review of mechanisms involving homeostasis and homeorhesis. *J. Dairy Sci.* 63, 1514–1529.

Bazer, F. W., Thatcher, W. W., Hansen, P. J., Mirando, M. A., and Plante, C. (1991). Physiological mechanisms of pregnancy recognition in ruminants. *J. Reprod. Fertil.* 43(Suppl.), 39–47.

Hughes, H. (1991). Economics of reproductive efficiency of beef cow herds. *Proc. Am. Assoc. Bovine Practitioners* 23, 65–74.

Lucy, M. C., Savio, J. D., Badinga, L., De La Sota, R. L., and Thatcher, W. W. (1992). Factors that affect ovarian follicular dynamics in cattle. *J. Anim. Sci.* 70, 3615–3626.

Odde, K. (1990). A review of synchronization of estrus in postpartum cattle. *J. Anim. Sci.* 68, 817.

Schillo, K. (1992). Effects of dietary energy on control of luteinizing hormone secretion in cattle and sheep. *J. Anim. Sci.* 70, 1271–1282.

Seidel, G. E. (1981). Superovulation and embryo transfer in cattle. *Science* 211(23), 351–357.

Short, R. E., Bellows, R. A., Staigmiller, R. B., Berardinelli, J. G., and Custer, E. E. (1990). Physiological mechanisms controlling anestrus and infertility in postpartum beef cattle. *J. Anim. Sci.* 68, 799–816.

Twagiramungu, H., Guilbault, L. A., and Dufour, J. J. (1995). Synchronization of ovarian follicular waves with a gonadotropin-releasing hormone agonist to increase the precision of estrus in cattle: A review. *J. Anim. Sci.* 73, 3141–3151.

VanHorn, H. H., and Wilcox, C. J. (1992). *Reproduction: Large Dairy Herd Management,* pp. 88–209. American Dairy Science Association, Champaign, IL.

CBG (Corticosteroid-Binding Globulin)

Geoffrey L. Hammond

University of Western Ontario

I. Structure and Function of Plasma Corticosteroid-Binding Globulin
II. Influence of CBG on Sperm during Fertilization
III. Localization and Actions of CBG at the Fetal/Maternal Interface
IV. CBG in the Mother and Fetus during Gestation and Parturition

GLOSSARY

glucocorticoids Corticosteroid hormones produced by the adrenal cortex which have a wide range of homoeostatic (e.g., gluconeogenic), antistress, and antiinflammatory activities.

serine proteinase inhibitors (serpins) An extensive superfamily of protein substrates for several classes of proteinases, including serine proteinases and metalloproteinases. Many serpins act as proteinase inhibitors, but some function primarily as plasma transport proteins for hormones or cytokines.

I. STRUCTURE AND FUNCTION OF PLASMA CORTICOSTERIOD-BINDING GLOBULIN

Corticosteroid-binding globulin (CBG) was initially characterized as the protein that binds and transports glucocorticoids in the blood. It is now clear that CBG has additional functions. CBG shares remarkable structural similarity with several members of the serpin superfamily of plasma proteins that control the activities of proteinases at sites of inflammation or tissue damage and repair. This has led to the hypothesis that CBG is a substrate for proteinases at sites of inflammation or tissue remodeling, and that proteolytic cleavage of CBG causes a disruption of its steroid-binding site and affects the local release of its steroid ligands. Since most of the glucocorticoids in the blood are bound to CBG, this has obvious implications with respect to their targeted release at sites of inflammation. However, it should be appreciated that other steroids, including progesterone, bind with relatively high affinity to CBG in humans and several other mammalian species, and a similar mechanism may be involved in the delivery of these steroids to their sites of action. It is also important to realize that CBG has only a single steroid-binding site for which glucocorticoids and other steroid ligands compete. Thus, the type of steroid that occupies the CBG steroid-binding site may vary depending on its anatomical location. For instance, while CBG in the blood is largely occupied by glucocorticoids, it is probably occupied primarily by progesterone in the female reproductive tract during the luteal phase of the menstrual cycle. Similarly, in most eutherian mammals, the placenta produces large amounts of progesterone which probably saturate CBG in the placental vasculature.

In addition to representing the major transport protein for glucocorticoids in the blood, CBG functions in extravasculature compartments. For example, CBG is produced in large amounts by the immature rodent kidney and is secreted into the developing renal tubules where it presumably regulates the bioavailability of glucocorticoids. Plasma CBG also interacts with specific binding sites on the surface of some cells and activates an adenylcyclase-dependent signal transduction pathway, but the membrane components responsible for this have not been identified and the biological significance of these activities is not known. Despite the fact that CBG is structurally most closely related to several serpins with proteinase inhibitory functions, there is no evidence that CBG inhibits proteinases in any

II. INFLUENCE OF CBG ON SPERM DURING FERTILIZATION

Progesterone exerts some very rapid effects on sperm through nongenomic mechanisms that enhance the acrosome reaction and fertilizing potential of sperm. These include an increase in sperm intracellular calcium and phosphatidylinositide hydrolysis and the tyrosine phosphorylation of sperm proteins. A follicular fluid protein responsible for accentuating these actions of progesterone has recently been shown to be a progesterone-binding protein that is recognized by antisera against CBG. Since luteinized granulosa cells surround the human oocyte as part of the cumulus oopherous, it is likely that CBG trapped within the perivitaline space is saturated by progesterone. In this context, the sperm head is rich in proteinases that may sequester and cleave CBG on its surface, and these interactions could enhance the delivery of progesterone to its membrane binding sites on sperm just prior to their fertilization of oocytes.

III. LOCALIZATION AND ACTIONS OF CBG AT THE FETAL/MATERNAL INTERFACE

In the mouse placenta, the matrix between maternal and fetal blood vessels accumulates a considerable amount of immunoreactive CBG, which certainly originates from the maternal compartment because it appears at the trophoblast/decidual interface prior to the initiation of CBG production in the fetus. It is not known if the CBG in this location is intact or modified by proteinases in the extracellular matrix, but the placenta is particularly rich in proteinases and their inhibitors which act in concert to control the invasive properties of trophoblasts. These include the matrix metalloproteinases (MMPs) and their inhibitors, and this may be relevant because MMPs interact with and cleave the serpins most closely related in structure to CBG. Thus, CBG is probably a target for proteinases at the fetal maternal interface and may thereby influence the way progesterone functions to maintain the integrity of the fetal placental unit.

In some species, CBG from the maternal circulation crosses the placenta and enters the fetal compartment, but this may depend on the type of placentation in different species because it does not occur in sheep. The placental transfer of CBG has been examined in rabbits and the small proportion of CBG in the maternal circulation that crosses the placenta is rapidly excreted by the fetal kidney into the amniotic fluid with its steroid-binding properties intact. Moreover, the fetal kidney appears to discriminate between fetal and maternal CBG, and the CBG that accumulates in the rabbit amniotic fluid is therefore largely maternal in origin. The physiological significance of this remains to be determined, but the ability of the fetal kidneys to discriminate between fetal and maternal CBG is remarkable because they only differ in the way they are glycosylated. In this regard, the type of carbohydrates associated with fetal CBG change during development in a glucocorticoid-dependent manner, and different glycoforms of CBG may determine how the protein partitions between mother and fetus, as well as within the fetus itself.

IV. CBG IN THE MOTHER AND FETUS DURING GESTATION AND PARTURITION

Plasma CBG levels increase in the maternal plasma during pregnancy in humans and several other mammalian species. This increase is probably estrogen dependent because the administration of exogenous estrogen also causes an increase in plasma CBG in nonpregnant women. In pregnant women, plasma cortisol and CBG concentrations increase in parallel, and this maintains free cortisol concentrations at levels similar to those in nonpregnant women. At term, maternal plasma CBG levels fall precipitously and this is immediately preceded by an abrupt reduction in hepatic CBG mRNA levels. Whether this change in maternal CBG production at term has any impact on the final stages of parturition is not clear, but again this may be species specific and related to placental type. In fetal sheep, a progressive increase in fetal plasma CBG levels occurs slightly later in gestation when compared to most other species. This helps maintain a positive drive from the pituitary on adrenal cortisol production and sustains the increase in fetal plasma cortisol levels that is considered to play a central role in promoting the cascade of events which culminate in parturition in sheep.

Changes in CBG gene expression in the fetal and maternal livers appear to be controlled independently and quite differently. In all laboratory animals, including primates, there is a peak in fetal hepatic production of CBG between mid to late gestation, which is followed by a reduction to relatively low levels prior to or at term. The resultant changes in plasma CBG undoubtedly influence glucocorticoid bioavailability in fetal and neonatal blood at critical stages of development, and this is probably important because it will influence the way glucocorticoids serve to mature several organ systems in preparation for extrauterine life.

The mechanisms responsible for changing CBG production in the fetal liver throughout gestation are not known but are likely to be quite distinct from those that are operative at later stages of development. This is illustrated by the fact that, whereas glucocorticoids enhance CBG production by the fetal liver, this effect is lost in the neonate and is even reversed in adult mammals.

The liver is clearly the major site of plasma CBG production in the fetus, but the CBG gene is also expressed transiently in several other fetal or early postnatal tissues. The production of CBG by these tissues may serve to provide a means of regulating glucocorticoid action at a local level during stages of development when changes in plasma CBG levels influence the actions of glucocorticoids in a more global context.

See Also the Following Articles

PLACENTAL AND DECIDUAL PROTEIN HORMONES, HUMAN; PROGESTERONE ACTIONS ON REPRODUCTIVE TRACT; PROTEIN HORMONES OF PRIMATE PREGNANCY; STEROID HORMONE RECEPTORS

Bibliography

Revelli, A., Massobrio, M., and Tesarik, J. (1998). Nongenomic actions of steroid hormones in reproductive tissues. *Endocr. Revs.* 19, 3–17.

Challis, J. R. G., Berdusco, E. T. M., Jeffray, T. M., Yang, K., and Hammond, G. L. (1995). Corticosteroid-binding globulin (CBG) in fetal development. *J. Steroid Biochem. Mol. Biol.* 53, 523–527.

Scrocchi, L. A., Hearn, S. A., Han, V. K. M., and Hammond, G. L. (1993). Corticosteroid-binding globulin biosynthesis in the mouse liver and kidney during postnatal development. *Endocrinology* 132, 910–916.

Hammond, G. L. (1993). Extracellular steroid-binding proteins. In *Steroid Hormone Action: Frontiers in Molecular Biology* (M. G. Parker, Ed.), pp. 1–25. IRL Press, New York.

Hammond, G. L. (1995). Potential functions of plasma steroid-binding proteins. *Trends Endocrinol. Metab.* 6, 298–304.

Cell Cycle

see Meiosis

Cell Death

see Apoptosis

Cephalochordata

M. Dale Stokes

Stanford University

I. Description of Taxon
II. Modes of Reproduction
III. Modes of Sex
IV. Anatomy of the Reproductive System
V. Reproductive Physiology and Endocrine Control
VI. Modes of Fertilization
VII. Modes of Development
VIII. Larvae and Metamorphosis

GLOSSARY

adenohypophysis The anterior lobe of the vertebrate pituitary gland, consisting of a mixed population of cells that secrete a cell-specific complement of hormones, such as luteinizing hormone.

benthic Relating to or found in the benthos, the region at the bottom of an ocean or at the soil–water interface of an ocean.

chordate One of the Chordata; an animal having at some stage in its development a dorsal nerve cord, a notochord, and pharyngeal gill slits, e.g., mammals, birds, reptiles, amphibians, fish, and certain marine invertebrates.

Lancelets of the Chordate subphylum Cephalochordata (also commonly known as amphioxus) have historically occupied an important place in debates of the origin of the vertebrates from the invertebrates. Considered by many to be the closest living invertebrate relatives of the vertebrates, lancelets possess typical chordate features (segmented musculature, pharyngeal gill slits, notochord, and tubular dorsal nerve cord) yet seem to lack some neural crest-derived tissues that are apparently unique to the vertebrates. Recent research has shown advances in reproductive ecology, endocrinology, and the fine structure of tissues using electron microscopy. Because of their phylogenetic significance, lancelet anatomy and embryology have been extensively studied for the past century and recent advances in molecular techniques have reawakened interest in using lancelets to answer questions pertinent to developmental evolution.

I. DESCRIPTION OF TAXON

There have been 29 species of lancelets described from shallow waters worldwide. The lancelets belong to a single family, Branchiostomatidae, and are separated into two rather similar genera, *Branchiostoma* (with gonads along both sides) and *Epigonichthys* (with gonads limited to the right side). The most current treatment of lancelet systematics can be found in Poss and Boschung.

Adult lancelets have a benthic habit and typically burrow into sandy sediments intertidally to farther out on the continental shelf (<200 m depth). Although most species have a wide geographic distribution, they are often relatively rare throughout most of their range. In a few locations high densities can be found (e.g., Qinqdao, China; Kingston Harbor, Jamaica; Lagos Lagoon, Nigeria; and Tampa Bay, Florida) and consequently these populations have received the most study. Several populations (at Naples and Faro in Italy and at Xiamen, China), formerly much studied, have dwindled or vanished, largely due to pollution.

The lancelet has a simple, elongated body (Fig. 1) which is laterally compressed, colorless, and vaguely resembles a small, headless fish. At the anterior end, a series of buccal cirri surround the mouth, which leads into the extensive buccal cavity. The body has no lateral fins but does have a low median ridge that extends along the dorsal side as a dorsal fin, posteriorly as a caudal fin, and as far anteriorly as the anus on the ventral side as the ventral fin. Lancelets are relatively small creatures: Adults are typically <5 cm long.

In keeping with its chordate affinities the lancelet has pharyngeal gill slits. However, these gill slits are very numerous (up to 200) and do not open directly to the outside. Instead, they are covered by metaplural folds of the body which together enclose an atrial cavity. Water moving into the pharynx flows out through the gill slits into the atrial cavity and then posteriorly to exit the body via the atriopore. A ciliated groove, the endostyle, runs along the midventral floor of the pharynx. The endostyle secretes mucus to aid food particle capture. Ciliated tracks sweep sheets of this mucus dorsally up the inner walls of the pharynx and gathered middorsally for transport anteriorly. The postpharyngeal gut has an anteriorly projecting diverticulum believed to be the main site of digestive enzyme secretion. Digestion takes place in the lumen of the mid- and hindguts. After digestion, food remnants continue to move through the gut for expulsion out the anus, which opens posterior to the atriopore.

The adult amphioxus has a tough outer skin for protection. This epidermis is only one cell layer thick, ciliated when young and strengthened by a fibrous intracellular cytoskeleton in the adult. The epidermis lies above a dermis comprising an amorphous ground substance in which are embedded collagen fibers, nerve cell processes, hemal channels, and a few scattered fibroblasts.

Muscle myotomes are visible in the shape of rearward opening chevrons and composed of blocks of striated muscle fibers arranged in flat plates and separated by myosepta. The muscles are not connected to the nervous system by motor nerves but instead by invertebrate-like extensions of the muscle cells which form a synapse with the nerve cord. In contrast, the muscles in the ventral atrial wall have a neuromuscular anatomy more similar to that of vertebrates and are innervated by motor neurons that exit the dorsal nerve cord suggestive of a dorsal spinal root.

The notochord extends the entire length of the body, from the tail past the main myotomal musculature to the anterior end. The notochord is not directly attached to body muscles but functions instead as an incompressible elastic rod that prevents the shortening of the body when the myotomes contract and allows the body to bend. The notochord is formed of a stacked series of highly modified muscle cells

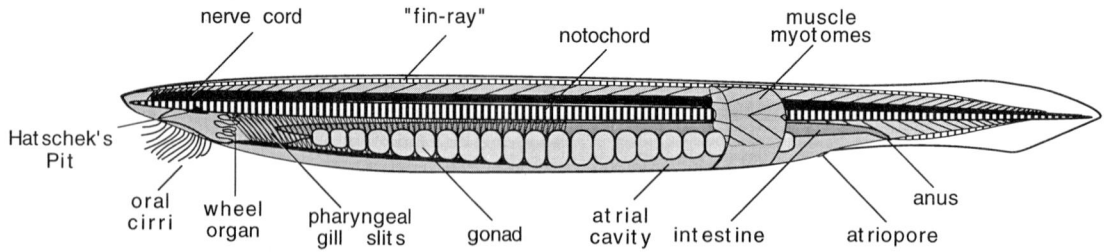

FIGURE 1 Key anatomical features of an adult cephalochordate shown in sagital section. The anterior is oriented toward the left and maximum length of typical adult shown is approximately 50 mm. With the exception of a short anterior section, the body musculature has been removed to illustrate the internal organs.

surrounded by a fibrous sheath and parts of the notochordal musculature synapse with the nervous system in a manner unlike that for vertebrates.

The circulatory system of the lancelet is very simple, lacking an endothelium. The walls of the hemal vessels are really mesothelia or myomesothelia lining adjacent coelomic spaces. There are very few blood cells, and the hemal plasma and blood is pumped by slow, irregular contraction of the vessels since no true heart is present; movement is generally posterior in dorsal vessels and toward the anterior in the ventral vessels (although transient reversals of flow occur). The blood is without any respiratory pigment (oxygen carried in solution is apparently adequate for respiratory needs) and most oxygenation may take place in vessels lying closest to the skin rather than being restricted to vessels within the gill bars.

Excretion is via the unpaired Hatschek's nephridium and the more posterior, paired branchial nephridia. These organs are intermediate in structure between an invertebrate protonephridium and the vertebrate-like metanephridial system and gather wastes via close diffusive contact with blood vessels and ultimately empty into the atrium.

The lancelet central nervous system is a dorsal hollow tube with an anterior dilation (the cerebral vesicle) which is homologous to the vertebrate diencephalic forebrain; the presence of a midbrain is controversial, although an extensive hindbrain is thought to be present. The dorsal nerve cord lies dorsal to the notochord and, as previously mentioned, is connected with the myotomal musculature via muscle fiber "roots" that directly synapse with the nerve cord. Other peripheral nerves are also present which innervate various structures and, interestingly, none of these peripheral nerves have any myelin sheathing. A few giant Rhode cells can be found which possess multiple dendrites and an axon that passes the entire length of the nerve cord. Epidermal sensory cells (thought to be mechanosensory or chemosensory) are scattered on the body. No image-forming eyes are present. However, there is an unpaired photoreceptive organ at the anterior end of the cerebral vesicle, and rows of unicellular photoreceptors (each enclosed by a cup-shaped pigment cell) occur on either side of the nerve cord. The lancelet has a very simple behavioral repertoire, best described as "phobotactic," which is a trial-and-error mode of lifestyle that is almost entirely reactionary in nature.

II. MODES OF REPRODUCTION

Asexual reproduction by Cephalochordates has never been reported. Lancelets are capable of a limited amount of tissue regeneration, particularly in caudal tissues posterior to the atriopore. However, injured lancelets are easily infected and reproduction by fragmentation is impossible.

III. MODES OF SEX

Lancelets have separate sexes; females and males have a series of approximately two or three dozen ductless, sac-like gonads (depending on the species). The gonads are arranged serially in the body alongside the myotomes and empty their gametes into the atrium during spawning. Apart from a difference in gonad coloration (ripe testis appear whitish through the translucent body wall and ovaries appear yellowish) there are no other obvious sexually dimorphic traits.

Hermaphroditism is exceedingly rare in lancelets. In an examination of over 50,000 individual *Branchiostoma floridae,* no hermaphrodites were found. In other species, an occasional hermaphrodite has been found. These individuals have either both testis and ovaries present or bisexual gonads producing a mixture of both spermatozoa and oocytes.

IV. ANATOMY OF THE REPRODUCTIVE SYSTEM

The use of electron microscopy to study the fine structure of lancelet gonads has clarified those anatomical details first examined by classical anatomists and more modern light microscopical studies. A detailed treatment of lancelet gonad ultrastructure can be found in Holland and Holland (1989, 1991) and Welsch and Fang (1996).

Lancelet gonads are composed of both germinal and nongerminal cells. The visceral peritoneum of the gonads is composed of a single layer of coelomic epithelial cells and has a hemal layer of thick blood vessels and narrow sinuses which lies at the base of the germinal epithelium. The thickness of the hemal layer varies with gonad ripeness and is squashed thin when the gonads are most full. The morphology of the female gonad is complicated by a secondary ovarian cavity which encloses the germinal epithelium. No specialized endocrine cells or nerve fibers are present in the lancelet gonads that might influence gametogenesis. Control of the gonad is presumably via hormone sources connected to the blood vascular system.

In the ovaries, the nongerminal cells are phagocytic, interconnected by cell junctions, and produce secretory granules. The oocytes contain some specialized structures in addition to the usual cell organelles. Nuage is present throughout oocyte growth near the nucleus as well as throughout the cytoplasm. When oocytes (*B. lanceolatum*) have reached approximately 20 μm in diameter, cortical granules and closely associated Golgi complexes are first seen. When oocytes reach 30 μm in diameter, vacuoles containing the presumed precursors of the vitelline layer are present. Yolk granules are found when oocytes reach 35 μm in diameter and at 50 μm a central vacuole is present within the nucleus. Small oocytes have smooth plasma membranes except in areas with abundant endocytotic pits. The oocytes continue to enlarge to their maximum diameter of approximately 120 μm and contain more numerous and larger yolk granules and a nucleolus containing a nucleolar vacuole. In addition, the larger oocytes have at their animal pole a few unusual striated fibers of unknown functional significance that resemble ciliary rootlets. The outer surfaces of large oocytes have abundant microvilli and are covered by a dense vitelline layer. During the course of oogenesis, contact between oocytes and nongerminal cells gradually shrinks to the degree that, in the largest oocytes, contact is made only at the animal pole. In this region, contact is made at adherens-type junctions between nongerminal cells and the oocyte plasma membrane.

In the testis, nongerminal cells (composed of single secretory and phagocytic epithelial cells without a blood–testis barrier) are found in low numbers scattered throughout the germinal epithelium and the testis lumen. All types of spermatogenic cells may have a simple 9 + 2 flagellum and are joined in small groups by intercellular bridges.

The Golgi complex in spermatids forms the bell-shaped acrosomal vesicle near the posterior pole of the spermatid during differentiation. The subacrosomal material is formed by the merging of two material components, one from each end of the spermatid, after the acrosomal vesicle has migrated to the anterior end of the mature spermatozoan.

V. REPRODUCTIVE PHYSIOLOGY AND ENDOCRINE CONTROL

Recent work provides compelling evidence for vertebrate-like hormonal regulation of reproduction in lancelets and has led to theories of the origin of the vertebrate adenohypophysis. Vertebrate gonadotropins or gonadotropin-like substances (sex steroids: testosterone, progesterone, estradiol, and estrone) are found within the lancelet body and in greatest concentration at the peak of the breeding season. Additionally, the injection of exogenous leutinizing hormone-releasing hormone (LHRH), gonadotropin-releasing hormone (GnRH), or mammalian gonadotropin can stimulate gametogenesis and increase the concentration of sex steroids.

Most important, immunopositive reactions to human gonadotropin, GnRH, LH, LHRH, folliclestimulating hormone, as well as thyrotropin-releasing hormone are found in cells of Hatschek's pit. This organ is a blind evagination from the roof of the buccal cavity, which is juxtaposed to but not directly contacting the dorsal nerve cord. It has been speculated that seasonal pheromones may be able to directly stimulate the epithelia of the Hatschek's pit cells to secrete hormones to link cyclical reproduction with external environmental cues. Functioning in this manner, Hatschek's pit would combine both chemosensory and endocrine roles. Despite having a proposed mechanism for reproductive regulation, the exact environmental cues that trigger lancelet spawning and/or gametogenesis have not been established.

The immunoreactive endocrine activity of Hatschek's pit has led to theories on the origin of the vertebrate adenohypophysis. It suggests a change from a protochordate to vertebrate type of reproductive control through the use of separate sensory organs and the nervous system to sample the environment while linking the adenohypophysial functions to the control of the central nervous system. What became the vertebrate adenohypophysis was then isolated from the mouth and direct contact with the environment. The transfer of the sensory receptive function to the adjacent nervous system allows the vertebrate adenohypophysis to respond to a broader range of sensory stimuli. If such speculation is correct, the primitive chordates evolved a form of vertebrate-like hormonal reproductive control before the evolution of the first vertebrates.

VI. MODES OF FERTILIZATION

Lancelets spawn after sundown during the warm summer season in temperate and subtropical regions (some tropical species may have a year-round breeding season). Males and females shed gametes out of their atriopore directly into the water column. Most of the population may spawn synchronously on the same night. In laboratory aquaria, the Chinese lancelet has been observed to leave the substrate and swim in the water column (first the males and then the females) while releasing gametes on the night of spawning. In the laboratory, lancelets have been induced to spawn when ripe using a mild electric shock on nights when they were also spawning in the field. Females are capable of shedding between 500 and 11000 eggs, depending on their body size.

When conditions allow, some lancelet species (i.e., *B. floridae*) may be capable of repeated spawning throughout the breeding season at approximately 2- or 3-week intervals. Other species breed only once a season (i.e., temperate living *B. lanceolatum*). During the breeding season, glycoproteins and lipids stored extracellulary as retroperitoneal accumulations (the so-called "fin rays," which are not homologous with the fin rays of fishes) function as nutritional reserves to supplement regular feeding for gametogenesis as well as somatic growth. These reserves are replenished after the breeding season is complete. Other aspects of lancelet reproductive ecology are reviewed by Stokes and Holland (1996).

VII. MODES OF DEVELOPMENT

The early ontogeny of lancelets has long been studied by classical embryologists and has been reviewed in detail numerous times. The first observations showing cleavage, gastrulation, and mesoderm formation similar to that found in echinoderms, and later development similar to the vertebrates, occupied evolutionary discussion by noted zoologists for many years. Recently, modern molecular techniques involving gene cloning, *in situ* hybridization, and DNA sequencing of embryonic material support the view that lancelets retain an archetypal body plan and are not merely degenerate vertebrate forms.

After fertilization, all development occurs within the confines of the fertilization membrane, a three-layer membrane formed via corticle granule exocytosis. Lancelet embryos show holoblastic (total) cleavage and early cleavage is approximately equal (blastomeres are approximately the same size). The first two cleavage stages are in the vertical plane and the third cleavage stage is in the horizontal plane. A hollow blastula with internal blastocoel is fully formed by the eighth cleavage cycle and marks the end of synchronous cleavages. Gastrulation of the embryo is via invagination whereby the vegetal hemisphere of the blastula folds inward to underlie the animal hemisphere, forming an internal cavity (the archenteron) in open communication with the environment at the blastopore. After gastrulation, the embryo comprises an inner layer (the hypoblast) and an outer layer (the epiblast). With the formation of the gastrula, the lancelet embryo begins to elongate.

The first eight mesodermal somites are formed from epithelial "outpockets" which are pinched off from opposite sides of the archenteron (enterocoely) similar to mesodermal formation in invertebrate deuterostomes. The more posterior somites are formed in the vertebrate fashion as mesenchymal cell blocks that further subdivide (schizocoely). The recent evidence of *engrailed* gene expression in the anterior eight somites of the lancelet which is also expressed in invertebrate metameres, i.e., in *Droso-*

phila, lends credence to the notion that segmentation in vertebrates and invertebrates is homologous.

The dorsal side of the gastrular epiblast becomes the neural plate which becomes overgrown by cells of the ectoderm during the first phase of neurulation. During the second phase of neurulation, the neural plate elongates, narrows, and then rolls up to form the dorsal nerve cord. The notochord, lying below the dorsal nerve chord, is formed of cells from the dorsal surface of the archenteron. The neurula continues to elongate and its outer epidermis is uniformly covered with motile cilia which cause the embryo to rotate within its fertilization membrane.

The speed at which the embryo develops is dependent on the temperature. *Branchiostoma floridae* hatches about 8 hr after fertilization at 30°C, whereas *B. lanceolatum* hatches in about 18 hr at 18°C. In *B. floridae* (at 30°C) the mouth and first gill slit appear about 21 hr after fertilization. The anus opens approximately 32 hr after fertilization near the site of the blastopore remnant, and the larvae begin feeding.

VIII. LARVAE AND METAMORPHOSIS

Lancelet larvae are strongly asymmetric; the mouth and preoral pit (a densely ciliated pit anterior to the mouth) dominate the left anterior side of the body, whereas the first gill slits appear on the right side (Fig. 2). This curious asymmetry, long a source of debate, may be an adaptation for feeding. The anus, club-shaped gland, portions of nervous system, and other organs all show additional asymmetries (e.g., the anus first appears on the right side of the body and later migrates across the midline to open on the left side). The primary behavior of the larvae is to hover almost motionless in midwater using epidermal cilia. They continuously filter feed while hovering and the larvae slowly elongate through the addition of myotomes at the posterior end of their body. Additional gill slits open posterior to the first until about a dozen are formed depending on the species (the primary gill slits). A secondary series of gill slits forms above the primary gill slits and will remain the definitive right-side gill slits when the primary gill slits move ventrally and then to the left

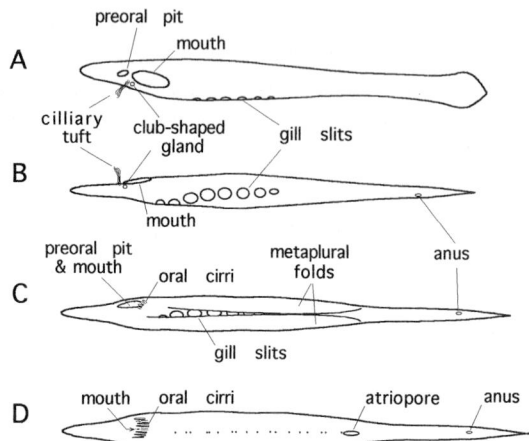

FIGURE 2 Diagram illustrating major morphological changes during lancelet larval metamorphosis. (A) Premetamorphic larva in left side view. (B) Premetatmorphic larva in ventral view. (C) Midmetamorphic larva, ventral view. (D) Postmetamophic juvenile, ventral view. In each diagram the anterior is toward the left. Length of postmetamorphic juvenile is approximately 4 mm and anatomy is nearly identical to that in an adult (Fig. 1).

side prior to metamorphosis. The gill slits are further subdivided by the downward growth of a tongue bar into two ciliated clefts.

Metamorphosis in lancelet larvae involves many morphological changes, the completion of which removes most of their external asymmetry. The length of time until metamorphosis begins increases in those species living in colder waters (e.g., the subtropical *B. floridae* begins metamorphosis after about 1 month, and the temperate *B. lanceolatum* begins metamorphosis after many months). Metamorphosis involves enclosure of the gill slits by the metapleural folds to form the atrium, rotation of the mouth to a medial position, and concurrent movement of the preoral pit into the roof of the oral cavity forming the wheel organ and Hatschek's pit. The mouth opening becomes surrounded by oral cirri and several new cell types (presumably sensory) appear on the epidermis. The epidermal cilia are gradually lost, as are the external opening to the club-shaped gland and the ciliary tuft. These changes in posthatching morphology are reviewed in detail by Stokes and Holland (1995).

After metamorphosis, the lancelets, now juveniles, settle out of the plankton and remain primarily benthic, usually buried just beneath the surface of sandy substrates with their anterior end tilted up for filter feeding. They continue to grow and reach maturation at lengths between approximately 10 and 30 mm depending on the species.

See Also the Following Articles

MARINE INVERTEBRATE LARVAE; PITUITARY GLAND, IN FISH

Bibliography

Conklin, E. G. (1932). The embryology of *Amphioxus*. *J. Morphol.* **54**, 69–151.

Gorbman, A. (1995). Olfactory origins and evolution of the brain–pituitary endocrine system: Facts and speculation. *Gen. Comp. Endocrinol.* **97**, 171–178.

Guraya, S. S. (1983). Cephalochordata. In *Reproductive Biology of the Invertebrates. Oogenesis, Oviposition and Oosorption* (K. G. Adiyodi and R. G. Adiyodi, Eds.), pp. 735–752. Wiley, New York.

Holland, L. Z., Kene, M., Williams, N. A, and Holland, N. D. (1997). Sequence and embryonic expression of the amphioxus engrailed gene (AmphiEn): The metameric pattern of transcription resembles that of its segment-polarity homolog in *Drosophila*. *Development* **124**, 1723–1732.

Holland, N. D., and Holland, L. Z. (1989). The fine structure of the testis of a lancelet (= amphioxus), *Branchiostoma floridae* (phylum Chordata: subphylum Cephalochordata = Acrania). *Acta Zool.* **70**, 211–219.

Holland, N. D., and Holland, L. Z. (1991). The fine structure of the growth stage oocytes of a lancelet (= amphioxus), *Branchiostoma lanceolatum*. *Invertebr. Reprod. Dev.* **19**, 107–122.

Holland, P. W. H. (1996). Molecular biology of lancelets: Insights into development and evolution. *Israel J. Zool.* **42**, S247–S272.

Jefferies, R. P. S. (1986). *The Ancestry of the Vertebrates*. Cambridge Univ. Press, London.

Lacalli, T. C. (1996). Frontal eye circuitry, rostral sensory pathways, and brain organization in amphioxus larvae: Evidence from 3D reconstructions. *Philos. Trans. R. Soc. Ser. B.* **351**, 243–263.

Poss, S. G., and Boschung, H. T. (1996). Lancelets (Cephalochordata: Branchiostomatidae): How many species are valid? *Israel J. Zool.* **42**, S13–S66.

Stokes, M. D., and Holland, N. D. (1995). Embryos and larvae of a lancelet, *Branchiostoma floridae*, from hatching through metamorphosis: Growth in the laboratory and external morphology. *Acta Zool.* **76**, 105–120.

Stokes, M. D., and Holland, N. D. (1996). Reproduction of the Florida lancelet (*Branchiostoma floridae*): Spawning patterns and fluctuations in gonad indexes and nutritional reserves. *Invertebr. Biol.* **115**, 349–359.

Welsch, U., and Fang, Y.-Q. (1996). The reproductive organs of *Branchiostoma*. *Israel J. Zool.* **42**, S183–S212.

Wickstead, J. H. (1975). Chordata: Acrania (Cephalochordata). In *Reproduction of Marine Invertebrates, Vol. II, Entoprocts and Lesser Coelomates* (A. C. Giese and J. S. Pearse, Eds.), pp. 282–319. Academic Press, New York.

Cervical Cancer

B. Hannah Ortiz and James A. Roberts

Stanford University

I. Epidemiology
II. Embryology and Anatomy
III. Risk Factors
IV. Screening: The Pap Smear
V. Management of an Abnormal Pap Smear
VI. Invasive Cancer
VII. Treatment of Invasive Cervical Cancer
VIII. Chemotherapy
IX. Survival

GLOSSARY

atypical squamous cells of undetermined significance A frequent interpretation of the Pap smear in which abnormal cells are observed but the degree of abnormality is uncertain.

colposcopy The process of evaluating the cervix with magnification to identify abnormal cells.

high-grade squamous intraepithelial lesion The Pap smear interpretation given when a premalignant lesion is present.

human papilloma virus A double-stranded DNA virus of approximately 8000 base pairs which forms part of a family of papillomaviruses that include common skin or genital wart viruses as well as the virus thought to produce cervical cancer.

LEEP (loop electrosurgery excision procedure) A procedure employing an electrified wire to excise a portion of the external cervix.

low-grade squamous intraepithelial lesion The Pap smear interpretation given when only an human papilloma virus or low-grade abnormality is detected.

Pap smear The collection of cells from the cervix which are reviewed microscopically for abnormal cells

squamocolumnar junction A portion of the cervix at which the squamous cells meet the glandulae cells; this is the site at which most cervical cancers begin.

transformation zone The portion of the cervix at which the squamocolumnar junction began and is now located; the change in position results from squamous metaplasia.

Cervical cancer has been a major public health problem that women have faced since the dawn of time. For much of that time it has been a somewhat silent killer of middle-aged mothers and wives throughout the world. It is without symptoms until its characteristic irregular bright red painless vaginal bleeding begins. Now in its advanced stages, it is a terminal disease. Several advances in women's health care have begun to change this dim picture into a more hopeful future. These advances will be outlined in this article. As will be shown, there is still much work to be done regarding cervical cancer.

I. EPIDEMIOLOGY

Worldwide, cervical cancer incidence is second only to that of breast cancer (excluding nonmelanotic skin cancers) among women. Over 437,000 women are diagnosed with cervical cancer each year. This constitutes 12% of all new cancer cases in women. Cervical cancer is the fifth most common cancer worldwide (cancer statistics on men and women inclusive). It is the most common cancer irrespective of gender in the developing countries of Asia, South America, and Africa. On the other hand, it is relatively uncommon in the United States, with an annual incidence of 15,800 new cases and 4800 deaths. The usual age of presentation with this disease is between the fourth and fifth decades, with a mean age at presentation of 54 years. Worldwide over 200,000 women die of cervical cancer each year. It has been estimated that each of these deaths represents a loss of between 14 and 20 years of pre-70s life. This calculates to a worldwide loss of 3.4 million woman-years of pre-70s life each year. Tragically, this loss of life may well be preventable.

II. EMBRYOLOGY AND ANATOMY

The cervix is a functional organ that makes up the lowest portion of the uterus and is positioned at the top of the vagina. It is a firm, 2- or 3-cm fibrous cylinder with a centrally located endocervical canal. The opening facing the uterine lining is the internal os; the opening facing the vagina is the external os, which is visible on physical examination of the vagina. Embryologically, the cervix is formed when the right and left Müllerian ducts begin to fuse at approximately 10 weeks of gestational age. Embryonic mesothelium gives rise to endocervical glands which line the cervical canal. Ectoderm forms the lower one-third of the vagina and gives rise to vaginal squamous cells. It is at the junction of the endocervical columnar cells and squamous cells that most cervical cancers arise. This interface of two cell types is called the squamocolumnar junction. There is a lifelong migration of the vaginal squamous cells toward the endocervical canal. Histologically, this area of migration is characterized by squamous metaplasia, or migration of mature squamous cells over the endocervical glands. This movement of the squamocolumnar junction produces the transformation zone (TZ). As the metaplastic squamous cells cover the endocervical tissues, the gland openings are obstructed, resulting in the accumulation of a mucinous discharge and the formation of variously sized Nabothian cysts. Because the TZ is the site of cell differentiation and turnover, it is thought that this area is more susceptible to agents which cause cervical cancer.

The uterus is lined by cuboidal endometrial cells which are sloughed cyclically during menses. Unlike the TZ at the external os, the endocervical os, in which the change from cuboidal cells to columnar cells occurs, is not a site of metaplasia. The uterus is a muscular organ situated between the urinary bladder anteriorly and the rectum posteriorly. The uterus receives its blood supply primarily from the uterine artery, a branch of the internal iliac, or hypogastric, artery's anterior division. Other vessels that supply the uterus are located in the uterosacral and cardinal ligaments, which are connective tissue structures. These structures together with the broad ligament (a layer of peritoneum draping over the uterus) are called its parametria. The uterosacral ligaments support the uterus through their connection to the sacrum posteriorly. The vesicouterine ligament provides a weak support anteriorly to the symphysis pubis and the cardinal ligaments extend laterally to the pelvic sidewall. The ureter courses from the kidney toward the bladder, crossing over the common iliac artery as it bifurcates into the internal iliac (hypogastric) and external iliac arteries. It then crosses under the uterine artery <1 cm from the cervix. It can be injured easily during dissection or can be obstructed by a growing cervical cancer. These relationships are important in the surgical treatment of cervical cancer as well as its staging.

Lymphatics also course along the routes of these vessels and the presence of lymphatic or vascular space involvement with cervical cancer is an important prognostic factor. Lymph node bundles are nestled among vessels of the pelvis and are sites of metastatic spread of this disease. In addition to this spread via lymphatics, cervical cancer spreads by local extension, meaning that involvement of pelvic structures, such as the vagina, uterus, and parametrium, usually precedes involvement of higher lymphatics along the aorta or intravascular dissemination to more distant sites, such as the lungs or brain. This is very useful in planning radiation treatment and surgical management.

III. RISK FACTORS

Cervical cancer has been associated with early age at first sexual intercourse, sexual promiscuity, and smoking. Celibacy results in a near zero incidence of cervical cancer. Immunosuppression, as with HIV infection or medical immunosuppression after organ transplant, also increases the risk of cervical cancer. These findings have traditionally pointed to a sexually transmitted agent as the etiologic factor of cervical cancer and its precancerous lesions. While correlations were made between such infectious agents as syphilis, gonorrhea, or herpes, molecular genetic testing confirms that the human papilloma virus (HPV) is the most likely cause of squamous cell cervical cancer. The National Institutes of Health formally acknowledged this association between cervical cancer and infection with HPV in 1996.

The HPV is a site-specific DNA virus. Infection with HPV is highly prevalent and has been detected in approximately one-third of American female college students and in 8% of men between ages 15 and 49. Different types of HPV infection exist, some of which resolve spontaneously, some result in latent genital infections, and others are associated with clinical disease ranging from benign anogenital or cervical warts to precancerous or cancerous cervical lesions. The more benign forms of HPV are types 6 and 11 and are usually associated with either condylomata (warts) or mild dysplastic changes of the cervical epithelium. Infection with these viruses rarely progresses to cervical carcinoma. The higher risk viruses include types 16, 18, and 31. These are associated with a high rate of malignant transformation. Intermediate risk viruses are associated with the low-grade changes [squamous intraepithelial lesion (LGSIL) or cervical intraepithelial neoplasia (CIN I)] and are less frequently associated with malignant change (types 33, 35, 39, 51, 56). There are currently no programs for HPV screening, although some sources estimate that HPV is present in 50–90% of dysplastic cervical lesions. Screening programs in the future may well include HPV typing and testing as part of a routine examination. Vaccination for the treatment and prevention of HPV infections are being studied and may become a way of eliminating cervical cancer.

Other forms of cervical cancer occur less frequently. The most common of these is adenocarcinoma. Historically this has been reported to occur in 5–10% of women with cervical cancer; but most referral centers are now reporting incidences in the range of 20–40%. Many studies have been published supporting HPV as a cause of this malignancy. *In utero* exposure to diethylstilbestrol has been associated with an increased risk of a woman developing clear cell adenocarcinoma of the cervix. In very rare cases, women have been diagnosed with melanoma or sarcoma of the cervix. Little information is available about the risk factors or causes of these two malignancies.

IV. SCREENING: THE PAP SMEAR

As early as 1886, Sir John Williams proposed that cervical cancer was the final step in a continuum of cellular change. Cervical lesions were noted to have variable degrees of atypia suggesting various stages of change from benign to malignant disease. In his 1943 monograph, *Diagnosis of Uterine Cancer by the Vaginal Smear*, George Papanicolaou presented a summary of years of investigation into the cytology of exfoliated vaginal cells and its usefulness for everything from the diagnosis of malignancy to the timing of the menstrual cycle. He described a technique for obtaining cells from the vagina and introduced a classification system to characterize various premalignant and malignant conditions, a variant of which was used until very recently. His technique involved the suctioning of cells from the posterior vaginal fornix using a bulb aspirator. These cells were then sprayed onto a slide and placed in small jars containing ethanol fixative. With time the bulb aspirator was replaced by a cytobrush and a wooden spatula (Fig. 1). Cells are obtained by introducing the brush into the cervical canal and turning it at least 180°. This maximizes the endocervical sampling. The wooden spatula is then used to scrape cells from the external os.

The Papanicolaou system of reporting smear results was based on observation and involved a classification based on the relative appearance or suggestion of malignancy in the cells. It was greatly affected by interobserver variability. There were limited criteria or specific terminology to characterize the appearance of cells and in this respect the classification may have been more an art form than a science. Nevertheless, this system resulted in a remarkable decline in morbidity and mortality from cervical cancer. In the 1950s and 1960s, screening programs were first conducted in northern Europe and Canada. In some of these countries the incidence of cervical cancer decreased by as much as 50%; in British Columbia it decreased by 85%. Between 1947 and 1984, the mortality from cervical cancer in these countries decreased by over 50%.

Even though the original classification system contributed considerably to a reduction in cervical can-

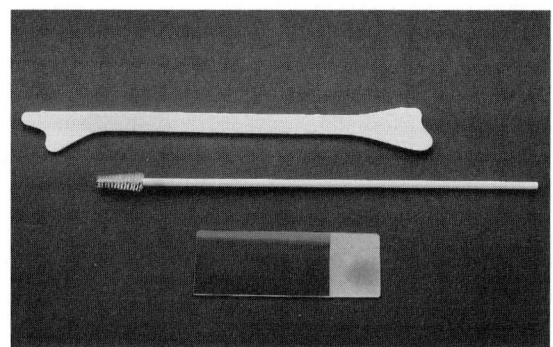

FIGURE 1 A Pap smear is obtained by using the spatula and endocervical brush to scrap the cervix. The specimen is then applied to a glass slide.

cer, its classes did not have precise uniform standards for histological diagnosis, and, therefore, standardized treatment options could not be established. The system originally described three grades of cell types but it gradually evolved into a five-class (I–V) system. Class I smears were defined as normal. Class II smears were atypical or reactive but did not represent malignancy. Class III smears had atypical cells suggestive but not diagnostic of malignancy. Class IV had cells strongly suggestive of malignancy. Class V smears were consistent with malignancy. These vague descriptions begged further clarification. The term carcinoma *in situ* was defined in 1961 and described epithelial lesions that were not frankly invasive through the cellular basement membrane but were poorly differentiated throughout the epithelial layer. Invasive lesions were those which invaded the cervical stroma by breaking through the basement membrane. These histologic descriptions, used to describe biopsy specimens, elucidated the need for more precise grading of precursor lesions based on the degree of cytologic atypia.

In 1973, an attempt to provide cytologic criteria for the interpretation of Pap smears was made by the World Health Organization. It proposed a new system with more cytologically descriptive terminology. Rather then categorize cells based on overall appearance, the new system proposed evaluation by cellular characteristics, such as nuclear chromatin content, nuclear to cytoplasmic ratio, and loss of normal polarity. Pap smears were either normal or atypical, corresponding to the benign cells or inflammatory changes of classes I or II in the Papanicolaou system. The term dysplasia was introduced to identified cells suggestive of premalignancy, or class III. This category was further subdivided into mild, moderate, or severe dysplasia. Class IV was translated into carcinoma *in situ*. Class V was defined as invasive squamous cell carcinoma or adenocarcinoma. Reichert described a more histologically precise terminology for mild, moderate, and severe dysplasia using the term cervical intraepithelial neoplasia. This terminology was intended to imply more clearly the presence of neoplastic or preneoplastic lesions. The terms mild, moderate, and severe dysplasias were used interchangeably with the terms CIN I, CIN II, and CIN III, respectively; however, CIN III often included carcinoma *in situ*. The purpose of these classification changes was to establish guidelines for treatment of these noninvasive dysplasias.

The identification of HPV as the etiologic agent in most cervical cancers has led to the reassessment and redefinition of the Pap smear. Until recently a gradual progression from mild to severe dysplasia and then to cancer, as proposed by Williams, seemed logical. With increasing understanding of HPV, it has become clear that certain HPV types are associated with benign changes and others with malignant transformation. The Bethesda classification, introduced in 1989 by the National Cancer Institute, proposed a two-tiered system. This allowed cytologists to provide an even more accurate description of the squamous cell changes and to classify lesions as high-grade squamous intraepithelial lesions (HGSIL) (associated with more malignant types of HPV), requiring aggressive intervention, or LGSIL (associated with the more benign HPV types). For the first time, it also established criteria for assessing the adequacy of the Pap smear and provided room for cytologists to describe findings such as inflammation or other benign changes that might contribute to an abnormal Pap smear but that might not necessarily mean a premalignant change. The current Pap smear report, therefore, should include a statement about the adequacy of the specimen, the cytologic diagnosis, and a comment about other changes or processes noted on the Pap smear.

The terminology of the Bethesda system describes squamous cell abnormalities. The term, atypical squamous cells of undetermined significance, correlates with squamous atypia describe in the CIN system or class II of the Papanicolaou system The problem with this category is that high- or low-grade lesions may be hiding within this group. Current opinions on the management of this classification are numerous and conflicting. Atypical cells may not be a result of infection with HPV but simply a result of hypoestrogenized epithelium in a postmenopausal woman. In addition, they may be atypical as a result of inflammation from various sexually transmitted diseases or from changes in the vaginal pH or environment. Management may therefore vary. This category was introduced to help management but has resulted in overtreatment in a medicolegal climate that leaves no room for error. The other categories include the LGSIL, closely related to the CIN system's HPV atypia and CIN I or mild dysplasia and the Papanicolaou class III classification. The term HGSIL correlates most closely with the CIN III, moderate to severe dysplasia or carcinoma *in situ*, and the Papanicolaou class IV. Squamous cell carcinoma is invasive cancer, or Papanicolaou class V. While the Bethesda system was introduced to simplify while creating an international standard for Pap smear reporting, it is unclear if it has been accepted on a worldwide scale. The evolution of the Bethesda system is the result of continued efforts to provide more precise cytological and histological information about cervical cells. In this respect, reporting of cellular abnormalities has become more specific. Previous assumptions about cervical neoplasia as a continuum of disease as proposed by Sir John William may have been challenged by our knowledge about HPV. The current system reflects that knowledge. It will necessarily continue to evolve as our research further characterizes HPV and screening for HPV becomes more widespread. Table 1 demonstrates the comparison between these reporting systems.

The original technique of specimen collection described by Papanicolaou has been modified to obtain a higher quality specimen from the uterine cervix. The major drawback to either of these techniques has been the presence of debris on the smear. This debris can make interpretation difficult and increase the chance of a misdiagnosis. Methods for obtaining and interpreting specimens to eliminate this problem have been reported. The newest Food and Drug Administration-approved attempt to eliminate this problem is the Thinprep system, which uses the technique of immersing the sampling device (e.g., cytobrush) into a fixative solution which breaks up blood, mucus, and debris. Cells are then collected by a filter and mounted onto a slide. Because these cells have been "washed," interpretation is less likely to be confounded by debris and, therefore, should be more representative and accurate. This may become a screening tool for the future, but it is currently being used to rescreen abnormal smears. Its major drawbacks are its expense and the potential environmental hazard of disposing of the liquid fixative (methanol).

Another advance in the interpretation of the Pap smear is in automated screening. These methods attempt to eliminate the interobserver variability which has plagued the earlier screening systems. Even with stricter cytologic criteria, it is argued that human error in reading Pap smears may still result in misdiagnoses since only a fraction of the approximately 300,000 cells on a Pap smear slide are actually evaluated by the cytologist. This new technology utilizes automated image analyzer technology coupled to a neural network computer technology to create a system that can review every cell on the Pap smear slide and capture images of abnormal cells or cell groups and analyze how far these images diverge from the normal. This system selects the 128 most abnormal-appearing areas on the slide and creates a video image of them for review by the cytologist. Another system

TABLE 1
Different Classification Systems for the Pap Smear

Pap	World Health Organization	Bethesda
Class I	Negative	Negative
Class II	Atypical	ASCUS
Class III	CIN I	LGSIL
Class III	CIN II–III	HGSIL
Class IV	CIN III/CIS	HGSIL
Class V	Cancer	Cancer

produces a point score for the Pap smear which dose not require image review by a cytologist. The specimen for these two techniques is prepared in the traditional manner of spreading cells onto the slide and fixing them immediately, resulting in no added expense from obtaining the specimen. This system is currently being used as a rescreening tool for questionable Pap smears and laboratory quality control.

V. MANAGEMENT OF AN ABNORMAL PAP SMEAR

With the Bethesda system's frequent reporting of atypia, usually described as atypical squamous cells of undetermined significance (ASCUS), the current practice is to repeat the Pap smear in these women after treating the inflammations or treating the postmenopausal women with estrogen. If this ASCUS persists, or if the lesions are reported as LGSIL or HGSIL, the next step is to further evaluate the cervix using a colposcope to obtain a directed biopsy specimen from the worst-appearing section of the cervix for histologic study.

The colposcope is a lighted low-power magnifying device used to inspect the cervix. Colposcopy makes use of the fact that a solution of 3% acetic acid will cause changes in the epithelial cell membrane proteins resulting in abnormal cells taking on a white color. This acetowhite epithelium (AWE) is examined first with white light and then using a green filter to better visualize the blood vessels. Specific colposcopic criteria have been developed to characterize the changes in AWE, such as the margins of the lesion, its surface contour, and the presence of and appearance of any vessels within the lesion. These are best assessed by an experienced colposcopist because changes are sometimes subtle and may represent anything from normal squamous metaplasia to invasive carcinoma. Application of a 25% solution of Lugol's iodine, which stains higher grade lesions yellow, is another technique for assessing abnormal epithelium. Directed biopsies are obtained from what appears to be the most suspicious lesions. First, an endocervical curettage (ECC), a biopsy of the endocervical canal, must be performed by scraping the area extending from the internal to external os. This ECC is performed to sample columnar cells that are not visible with colposcopy and to assess the presence of possible skip lesions. A colposcopy report includes a statement of the adequacy of the colposcopy and a drawing of the cervix delineating areas of AWE. Biopsies are sent for histological review by the pathologist. Colposcopy can be described as satisfactory only if the TZ has been completely visualized. Colposcopy is considered inadequate if (i) the TZ is not visualized fully, (ii) a visible lesion extends into the endocervical canal, (iii) the endocervical curetting reveal dysplastic fragments, or (iv) there is a discordance of more than one grade between the Pap smear and the directed biopsy. If colposcopy is unsatisfactory, other diagnostic measures should be undertaken, such as cold knife cone biopsy or a LEEP (loop electrosurgery excision procedure). Conization or LEEP should also be performed if the colposcopic biopsy indicates any degree of invasion. Another reason to perform a conization or a LEEP is the absence of a visible lesion. These procedures can at once be diagnostic and therapeutic.

Both these procedures involve the removal of tissue surrounding the external os, including the entire TZ and a portion of the endocervical canal. The purpose is to obtain a wedge of tissue for diagnosis and treatment. A cold knife cone biopsy requires general anesthesia in an operating room setting, thereby raising its costs. A cone biopsy can also be performed using a laser as the cutting instrument, yielding similar results. On the other hand, a LEEP procedure is an office-based procedure. Because it uses cautery to fulgurate as it cuts, there has been concern that the margins of a LEEP specimen cannot be evaluated adequately, although this has not been borne out in clinical practice. If the margins of these excisional biopsies are free of disease and there are no signs of invasion, then definitive treatment has been performed and the patient is followed with quarterly Pap smears for 1 year. Positive margins require more definitive treatment, such as repeat excisional biopsy or, in rare cases, a hysterectomy. Complications of these excisional biopsies include hemorrhage, sepsis, cervical stenosis, infertility, and cervical incompetence. They are reported to occur in 2–12% of patients. In the absence of invasive carcinoma, treatment of LGISL, HGSIL, or CIS can also be performed using cryotherapy, laser ablation, or cautery of TZ. Care must be taken to (i) include the TZ, the visibility of which is essential for these noninvasive treatments, and (ii) be certain that no invasion is present.

VI. INVASIVE CANCER

Screening for cervical cancer involves persistence in the search for subtle changes and the identification of preinvasive, precancerous lesions. Lesions can be elusive and there are numerous gray zones, the management of which remains open to interpretation. Despite stalwart attempts to provide ever stricter criteria for determining which lesions should be excised and which should be followed conservatively, management of the abnormal Pap is as much based on judgment and experience as on histopathologic criteria.

The most common presenting symptom in patients with cervical cancer is painless bright red postcoital bleeding. Some patients present with spontaneous vaginal bleeding or postmenopausal spotting. In more advanced cases, there may be a complaint of weight loss or malodorous vaginal discharge, suggesting tumor necrosis and superinfection with vaginal flora. Patients who present with unilateral leg edema, lower extremity pain, or low back pain may have extensive tumor involving the pelvic sidewall with sciatic nerve compression. Patients may also have recurrent urinary tract infections or pyelonephritis resulting from urinary stasis related to hydronephrosis. Back or flank pain may also represent this. This is caused by compression of the ureters at the pelvic brim by tumor-laden lymph nodes. Rarely, cervical cancer may present as severe hemorrhage requiring transfusion, vaginal packing, and emergent radiation therapy.

The management of a patient presenting with such symptoms begins with a complete physical examination and evaluation. The clinical staging of cervical cancer is the most important factor in determining prognosis. The International Federation of Gynecology and Obstetrics has proposed that clinical rather than surgical or radiographic staging of cervical cancer be performed because imaging modalities such as magnetic resonance imaging, computer tomography scanning, and lymphangiography may not be available worldwide (Table 2). The staging, therefore, places great emphasis on initial evaluation with physical examination. Radiographic studies included in the staging of cervical cancer are intravenous pyelogram for assessment of hydronephrosis and chest radiography to exclude the presence of disseminated disease. Procedures performed for staging purposes include cervical or cone biopsy, cystoscopy, and sigmoidoscopy.

Routine physical examination should place particular emphasis on palpation of lymph nodes (cervical, supraclavicular, axillary, and inguinal). Biopsy of enlarged nodes is indicated. An enlarged supraclavicular node implies widespread metastatic disease involving the pelvic and paraaortic lymph nodes. Pelvic examination includes careful tactile assessment of the perimetrium. Speculum examination should be thorough, involving inspection of each vaginal fornix, the vaginal mucosa from the fornix to the hymen, and external vulva. The speculum should be moved and repositioned as needed for thorough inspection. Localization of lesions on the vagina will affect stag-

TABLE 2
The International Federation of Gynecology and Obstetrics Staging System for Cervical Cancer

Stage	Description
0	Carcinomas *in situ* (CIS)
I	Carcinoma confined to cervix
IA1	<3 mm invasion
IA2	<5 mm invasion and <7 mm lesion
IB1	<IA2 and cervix <6 cm
IB2	cervix >6 cm
IIA	Vaginal invasion
IIB	Parametrial invasion
IIIA	Lower one-third vagina invasion
IIIB	Invasion of pelvic sidewall or hydroureter
IVA	Bladder or rectal invasion
IVB	Distant disease

ing, treatment, and cure. In the presence of a large mass, an accurate assessment of size should be made and biopsy performed to confirm its histology. While most lesions that arise on the cervix are squamous cell cancers, the increasing incidence of adenocarcinomas supports the importance of a biopsy. Colposcopy is performed to evaluate small preclinical lesions for contiguous or nearby clusters of lesions and to further inspect the vagina. Note should be made about tumor margins, their appearances, the presence of abnormal vasculature such as punctate or serpiginous vessels, mosaicism, or reticular patterns of vessels. Irregular contours with erosion by tumor through the epithelium indicate carcinoma. Because irregular contours of a cervical mass could also represent a benign papilloma, biopsy should be performed.

Bimanual examination is performed and should always include examination of the rectum as well as vagina. The rectovaginal examination reveals crucial information that vaginal examination alone cannot. The examination begins with a simple vaginal exam. One hand is used to examine the vagina and the other is placed above the symphysis pubis. This should be done with an empty bladder because a full urinary bladder may limit palpation of the uterus and adnexa. The size, shape/texture, and position of the uterus are assessed. A normal uterus is smooth, mobile, nontender, and ranges in size from 5 to 9 cm in a woman with no other coexisting pathology (e.g., fibroids). The cervix is palpated from below as the uterus is stabilized from above. This provides another estimate of tumor size. Attention is next turned to the adnexa, which are also assessed for smoothness and contour. The presence and size of an adnexal mass are noted. Adnexal masses should be characterized as soft, firm, nodular, woody, mobile, or immobile. Contiguity with the uterus should be noted. It should be noted that adnexal structures are normally tender when palpated.

The rectovaginal exam proceeds systematically and begins with reassessment of the position, texture, and mobility of the uterus. The cervix is palpated with the abdominal hand stabilizing the uterus, the vaginal finger, and the rectal finger, giving another measurement of cervical tumor volume. The adnexa are assessed again and any masses previously palpated are again assessed for size, texture, and relationship to the uterus. The rectovaginal exam gives excellent access to the uterosacral ligaments, which should be palpated for involvement with tumor or nodularity. Pelvic sidewalls are palpated, and extension of tumor from the uterus or vagina to the pelvic sidewalls should be assessed. Determination of whether or not the tumor is fixed to or free from the sidewall is critical for staging purposes. The rectovaginal septum should be palpated and normally is smooth; if there is nodularity or if the rectum is fixed to the vagina in this space, tumor involvement should be suspected. The paravaginal and pararectal spaces are palpated for the presence of tumor or nodularity. The rectum is palpated for the presence of masses. At the end of the exam, a test for occult rectal blood should be performed.

VII. TREATMENT OF INVASIVE CERVICAL CANCER

The treatment of cervical cancer is somewhat stage dependent. The uterus is an intraabdominal organ covered by peritoneum. While it can be removed without dissection into the retroperitoneal space, dissection in this area is required in more radical procedures for cervical cancer. The blood supply to the uterus and ovaries is retroperitoneal, as are the ureters and lymph nodes draining the cervix and uterus. The type of surgery will be dictated by the stage of the disease. Because the disease spreads by local extension, generally it is safe to treat noninvasive lesions conservatively. This may be especially important for patients wishing to preserve fertility. Stage 0 through IA (CIS or microinvasive disease) may be treated with conization or a LEEP procedure as described previously. Lesions with <3 mm of invasion will have nearly a 0% chance of nodal involvement and can therefore be treated with conization; however, extrafascial or vaginal hysterectomy are more widely accepted for treatment of these lesions. More invasive surgery involving modifications of a radical hysterectomy and/or radiation are used for more extensive disease. Radical hysterectomies are classified according to the amount of pelvic tissue removed. Oophorectomy is not necessary because metastases to the ovaries are rare. The ovaries can be suspended in the peritoneal cavity out of a future radiation field in premenopausal women.

Radical hysterectomies have been divided into four classes. Class I is an extrafascial abdominal hysterectomy and includes removal of the cervix and uterus, while sparing the ovaries and not requiring entry into the retroperitoneum. This type of hysterectomy is recommended for stages 0–IA1 disease. Stages IA2–IIA, and in very rare cases stage IIB, can be treated by any of the following: Class II or a modified radical (extended) hysterectomy involves removal of the cervix and upper vagina and includes paracervical tissues located inside of the ureters. While the dissection in a class I hysterectomy remains intraperitoneal, a class II hysterectomy requires dissection of the ureters to the point of entry into the urinary bladder. A class II hysterectomy is recommended for patients with 3–5 mm depth of invasion or stage IA2 disease. Class III or radical abdominal hysterectomy with bilateral pelvic lymphadenectomy is the most commonly performed procedure in the United States for cervical cancer. First described in 1897, it is also known as the Wertheim *or* Meigs procedure. In addition to the structures removed with a class II hysterectomy, this procedure extends the dissection to include the parametrial tissues dissected out to the pelvic sidewall. The cardinal, uterosacral, and uterovesical ligaments are isolated and clamped close to the pelvic sidewall. The uterine artery is clamped and cut as close to its origin at the hypogastric artery as possible. The procedure also involves the complete dissection of ureters from their beds as well as mobilization of the bladder and rectum. At least 2 or 3 cm of the vaginal cuff is removed. The class IV or extended radical hysterectomy involves removal of more vaginal cuff and necessitates sacrifice of the superior vesical artery. In a class V procedure, a portion of the rectum, bladder, and/or distal ureter is also removed.

Comparisons of surgery versus radiation have yielded comparable 5-year survival statistics in stages IB and IIA disease. Either radiation or surgery can, therefore, be offered as treatment options. Radiation may be administered to patients who either do not desire surgery or those who are too ill or debilitated to undergo a surgical procedure. Patients with positive tumor margins after definitive surgery or lymph node involvement also undergo radiation treatment. Radiation can be intracavitary, meaning that it is administered by locally placed applicators to deliver higher doses to the cervix and avoid radiation effects to other pelvic tissues. External beam radiation therapy is used for more extensive nodal involvement and may be used to treat the pelvis or the whole abdomen and pelvis, depending on the presence of and extent of nodal disease.

Stage IIB (except in rare instances) and more advanced stages are treated with radiation alone. Stage IVA has one surgical treatment option: pelvic exenteration, which can also be used in cases of localized recurrent disease. In 1948, Brushwchwig described the first successful total pelvic exenteration (performed in 1946) involving complete excision of the pelvic viscera for advanced recurrent cervical cancer. Variants of this operation are performed today and involve en bloc resection of the uterus, vagina, bladder, and rectum. This procedure requires urinary diversion, construction of a reservoir for urine from bowel, colostomy and/or rectal reconstruction, and pelvic and periaortic lymph node sampling. The operation is performed for proven stage IVA disease or documented centrally recurrent pelvic disease. The absence of pelvic sidewall, extrapelvic, or nodal disease must be well documented prior to this surgery. Patients must be otherwise healthy because the surgery is physiologically challenging and requires careful patient selection and education.

VIII. CHEMOTHERAPY

The role of chemotherapy is limited. This is in part due to the fact that good results have been achieved with surgery and radiation, and chemotherapy seldom is a first line treatment. Patients who have had recurrences have frequently undergone previous irradiation, limiting access of the drug to the tumor because of a decreased blood supply to the tumor as a result of this prior treatment. Tumor cells may also have become resistant to treatment as a result of irradiation. Patients who have received radiation may also have decreased bone marrow reserve, limiting the doses of chemotherapy they can safely receive. Because recurrent disease may involve the pelvic sidewall with compression of the ureters, compromised renal function can also limit the feasibility of chemotherapy. Despite these limitations, attempts to use chemotherapy began in the mid-1970s with mixed result. Chemotherapy has been used to treat recurrent disease no longer amenable to radiation and concomitantly with radiation as a sensitizing agent to potentiate the effects of radiation.

The most active agent studied is cisplatin. Cisplatin is thought to work by binding to DNA, producing cross-links and formation of DNA adducts, thus blocking DNA replication. Its major side effect is nephrotoxicity. Cisplatin has been used as a single agent and in combination with other cytotoxic drugs and as a radiation sensitizer. Response rates are approximately 23%. Some improvement in survival has been seen in some studies when it is used in combination with radiation as opposed to chemotherapy alone. Ifosfamide is an alkylating agent similar to cyclophosphamide. Overall response rates to ifosfamide range from 14 to 30%. Mitolactol is a halogenated sugar with responses of about 29%. Other drugs studied include cyclophosphamide, chlorambucil, melphalan, galactitol, carboplatin, iproplatin, doxorubicin, porfiromycin, piperazinedione, 5-flourouracil, methotrexate, Baker's antifol, VM-26, vincristine, vinblastine, vindesine, ICRF-159, hexamethylmelamine, and CPT-11. Most have given response rates of <20%. Combination treatment is the most likely to have a major impact on survival but surgery and radiation remain the mainstay of treatment.

Recently, many investigators have begun to explore the role of neoadjuvant chemotherapy. In this treatment plan, chemotherapy is administered as the primary therapy, which is followed by radical surgery. The initial results from several studies suggest that this approach may be more effective than the standard treatment: radiation therapy. It remains to be determined if this form of treatment will assume the primary role of therapy for stage IIB and greater cervical cancer.

IX. SURVIVAL

The morbidity free 5-year survival statistics is very good for early stage disease but declines rapidly with advanced stages. These statistics are a testament to the importance and value of screening and diagnosis of less advanced disease. For stage IA1 treated with extrafascial hysterectomy or cone biopsy, the disease-free 5-year survival rates approaches 100%. For medically inoperable patients who choose radiation, 5-year survival is in the same range. Stage IA2 or IB disease treated with radical hysterectomy and lymph node dissection or brachytherapy with external beam radiation has survival rates of 90% at 5 years. These rates drop with stage IIA, treated either with surgery or radiation, to about 75%. Stage IIB disease treated with surgery results in a survival of around 44%. The standard of treatment for stage IIB disease is radiation with a 5-year morbidity-free survival ranging from 60–65%. Stage IIIB carcinoma has a 5-year survival rate ranging from 25 to 48%. Stage IVA disease treated with radiation carries survival rates 18–34%. Stage IVB disease is treated with chemotherapy and has a 5-year survival rate of 5%.

See Also the Following Articles

BREAST CANCER; CERVIX; OVARIAN CANCER; TUMORS OF THE FEMALE REPRODUCTIVE SYSTEM

Bibliography

Bauer, et al. (1991). Genital human papillomavirus infection of female university students as determined by a PCR-based method. *J. Am. Med. Assoc.* **265**, 472–477.

Berek, J. S., and Hacker, N. F. (Eds.) (1994). *Practical Gynecologic Oncology*, 2nd ed. Williams & Wilkins, Baltimore.

Brunschwig, A. (1948). Complete excision of pelvic viscera for advanced carcinoma. *Cancer* **1**, 177–183.

Cannistra, S. A., and Niloff, J. M. (1996). Cancer of the uterine cervix. *N. Engl. J. Med.* **334**, 1030–1038.

Grussendorf-Conen, et al. (1986). Human papillomavirus genomes in penile smears of healthy men. *Lancet* **2**, 1092.

Hoskins, W. J., Perez, C. A., and Young, R. C. (Eds.) (1996). *Principles and Practice of Gynecologic Oncology.* Lippincott-Raven, Philadelphia.

Mitchell, R. R., Stastny, J. F., *et al.* (1996). PAPNET-directed rescreening of cervicovaginal smears: A study of 101 cases of atypical squamous cells of undetermined significance. *Am. J. Clin. Pathol* **105**, 711–718.

Papanicolaou, G. N., and Traut, H. F. (1943). *Diagnosis of Uterine Cancer by the Vaginal Smear.* Commonwealth Fund, New York.

Schneider, A. (1988). HPV infection in women and their male partners. *Contemp. OB/GYN* **32**(5), 131–144.

Cervidae

see Deer

Cervix

Kamran S. Moghissi

Wayne State University/Hutzel Hospital

I. Cervical Secretion
II. Cervical Contraception
III. Conclusion

GLOSSARY

cervicitis Inflammation of the cervix.

cervix A neck-like part that forms the lower and narrower end of the uterus, separating the vagina from the uterine cavity. It is the site of numerous crypts secreting cervical mucus.

endocervix The membrane lining the canal of the cervix; the region of the opening of the cervix into the uterine cavity.

exocervix The portion of the cervix lined with squamous epithelial cells.

Müllerian ducts The pair of embryonic ducts that develop into the Fallopian tubes, vagina, and uterus in females and degenerate into a vestigial structure (appendix testis) in males.

sperm migration The movement of sperm from the point of insemination to the site of fertilization.

The cervix represents the terminal portion of the uterus and separates the vagina from the uterine cavity. The cervix is a thick-walled, cylindrical structure that tapers at its inferior extremity (Fig. 1). The canal of the cervix measures 2.5–3 cm in length, is fusiform in shape, and is slightly dilated in its middle third. The average transverse diameter at its widest

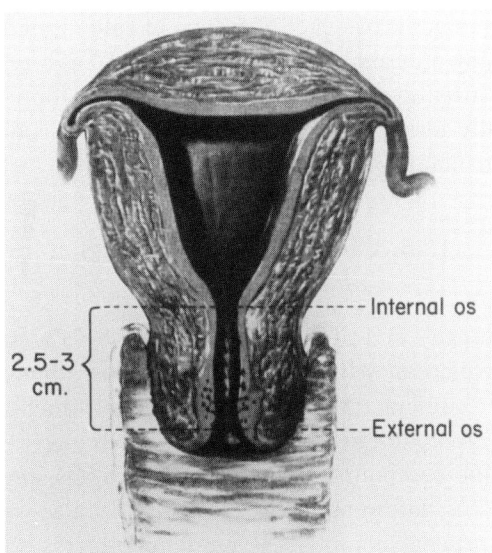

FIGURE 1 Human uterine cervix.

point is 7 mm. The external os is the opening in the portio vaginalis that connects the cervical canal with the vagina. The cervix is lined with stratified squamous epithelium similar to the lining of the vagina and it normally shows no cornification. The epithelial cells of the endocervix comprise different types of nonciliated secretory and ciliated cells that are tall, columnar, and rest on a thin basement membrane. There are no true glands in the cervix. Instead, there is an intricate system of crypts or grooves which grouped together give an illusory impression of glands. These crypts may run in an oblique, transverse, or longitudinal direction. They never cross one another, although they may bifurcate or extend downward. The junction of columnar epithelium of the endocervix and squamous epithelium of the exocervix is known as the squamocolumnar junction. Unless active metaplasia is present, the junction is very sharp.

The cervix, like the rest of the uterus, is derived from the female or Müllerian ducts. The blood supply of the cervix is derived from the uterine artery and the nerve supply is derived from three plexuses of the pelvic autonomic system: the superior, middle, and inferior hypogastric plexuses.

Nerve fibers positive for cholinesterase-specific staining have been observed in the vicinity of cervical crypts. An extensive network of cholinergic and adrenergic nerves has also been described at the level of internal os and throughout the cervix in relation to blood vessel walls.

From the functional point of view, the cervix acts as a biological valve, which at certain periods during the reproductive cycle allows the entry of sperm into the uterus and at other times bars their admission. Other properties of the cervix and its secretion include (i) protecting sperm from the hostile environment of the vagina and from being phagocytized; (ii) supplementing the energy requirement of the sperm; (iii) filtering and selection process, that is, discarding abnormal or unfit sperm and allowing only the entry of vigorous normal sperm into the uterus; and (iv) providing a site for sperm capacitation. The structure of the cervix and secretion of cervical mucus from endocervical epithelia contribute to these functions.

I. CERVICAL SECRETION

Cervical mucus is a complex secretion produced by the secretory cells of the endocervix. A small amount of endometrial, tubal, and possibly follicular fluid may also contribute to the cervical mucus pool. The average mucus production amounts to 20–60 mg/day in normal women of reproductive age. During midcycle this rate increases 10- to 20-fold and may increase up to 700 mg/day. The amount, physical properties, and chemical constituents of the cervical mucus also undergo cyclic variations.

A. Physical Properties of Cervical Mucus

Cervical mucus is a heterogeneous secretion with a number of rheological properties including consistency, flow elasticity, spinnbarkheit, thixotropy, and tack or stickiness. Spinnbarkheit is the capacity of liquids to be drawn into threads. To demonstrate this property a sample of cervical mucus is stretched between a glass slide and a coverslip. An estimate of spinnbarkheit is made by measuring the length (in centimeters) of the thread before it breaks (Fig. 2). Increasing estrogen levels during the preovulatory period result in increasing spinnbarkheit.

Another clinically used property of cervical mucus

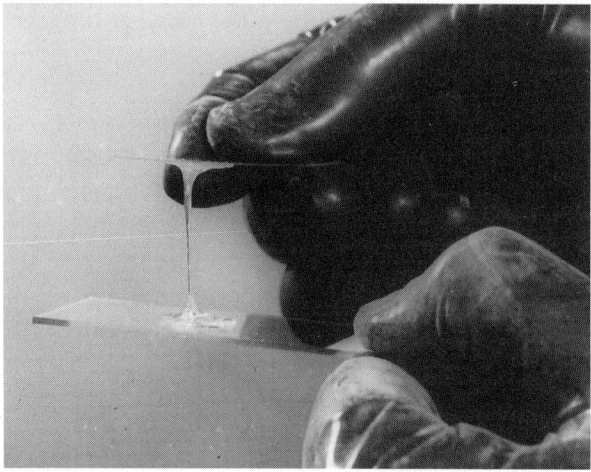

FIGURE 2 Technique for determining spinnbarkheit of cervical mucus.

is ferning or crystallization. Ferning occurs during the ovulatory period and is used for detection or prediction of ovulation time and as an indirect index of circulating estrogen level. Ferning appears when cervical mucus is spread on a glass slide and allowed to dry (Fig. 3). An intriguing pattern of arborization develops which is the result of crystallization of electrolytes in the presence of protein or colloid solution. Ferning appears between Days 8 and 10 of a typical menstrual cycle, reaches its peak at ovulation, and usually disappears after ovulation. Cervical mucus of pregnancy does not show ferning.

B. Chemical Constituents of Cervical Mucus

Human cervical mucus contains 92–94% water. At ovulation, when the mucus is most abundant, water content rises to 98%. Other constituents include inorganic salts (1%) and low-molecular-weight organic compounds such as sugars, lipids, prostaglandins, and amino acids. High-molecular-weight constituents comprise serum-type proteins (albumins and globulins), polysaccharides, and enzymes. The major constituent of cervical mucus is a high-molecular-weight glycoprotein of mucin type. Mucin constitutes 45% of proteins in cervical mucus and is responsible for the distinct viscoelastic property of cervical mucus which facilitates sperm penetration into cervical mucus. Detailed structural analyses of the cervical mucin is lacking. Early proposals of a relatively ordered micellar structure have been challenged by an alternative model suggesting a random network of entangled mucin molecules, a structure similar to that of raw rubber.

C. Cervical Infections

Inflammatory conditions of the cervix are of specific interest to clinicians interested in infertility since they have the potential of interfering with sperm transport through the cervix and impairing fertility. Acute cervicitis may result from direct infection by several microorganisms including streptococci, staphylococci, and enterococci. During the purpurium the cervix is particularly susceptible to these bacteria. The most common secondary invader of the cervix is *Neisseria gonorrhoeae*.

Clinically, acute cervicitis is characterized by purulent, malodorous vaginal discharge and turbid, thick, and purulent cervical mucus.

Cervical mucus has bacteriostatic properties against staphylococci *and Streptococcus hemolyticus* but appears to enhance the growth of certain strains

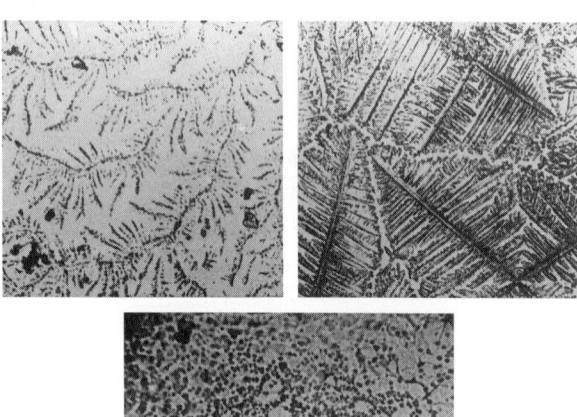

FIGURE 3 Ferning (crystallization) of cervical mucus. Preovulatory ferning (top right); atypical ferning from low estrogen level (top left); and absent ferning (bottom).

of *N. gonorrhoeae in vitro*. Muramidase (lysozyme), an enzyme that is capable of hydrolyzing the structurally important β 1–4 linkage between *N*-acetyl muramic acid and *N*-acetyl glucosamin in the cell wall of certain bacteria, is present in the cervical mucus. Another antimicrobial peptide, defensin-5, is expressed in the endocervical tissue and is believed to be active against a variety of yeasts, gram-negative and gram-positive bacteria, and enveloped viruses. Bacteriocidal activity of the human cervical mucus is present in all phases of the menstrual cycle but is least pronounced at ovulation. Based on *in vitro* studies it has been speculated that most aerobic and anaerobic bacteria attach to spermatozoa and migrate through the cervical mucus to the upper reproductive tract. Since sperm traverse cervical mucus only in the periovulatory period, it is assumed that antibacterial defense of the cervix is least effective during this period.

D. The Cyclic Changes of Cervical Mucus and Their Relations to Ovulation

The secretion of cervical mucus is regulated by ovarian hormones. Estrogen stimulates the production of copious amounts of watery mucus, whereas progesterone inhibits the secretory activity of cervical epithelia. The physical properties and most chemical constituents of cervical mucus show cyclic variations and their determination may be used to evaluate indirectly the ovulation time. Additionally, cyclic changes of cervical mucus properties and its constituents influence sperm penetrability, nutrition, and survival. Figure 4 shows serial determination of important properties of human cervical mucus tested during a normal ovulatory menstrual cycle in 10 women in relation to pituitary and ovarian hormones and *in vitro* sperm penetration. These data clearly demonstrate that optimal changes of cervical mucus properties, such as greatest increase in quantity, spinnbarkheit, ferning, pH, and decrease in constancy (viscosity) and cell content, occur immediately prior to ovulation and are reversed after ovulation. Preovulatory mucus is most receptive to sperm penetration.

E. Sperm Migration through Cervical Mucus

During coitus 50–500 million spermatozoa are deposited on the cervix and posterior vaginal fornix. Human semen coagulates immediately after ejaculation and traps most sperm cells until seminal proteolytic enzymes, which are activated in the presence of acid vaginal pH, liquefy the coagulum. The first portion of the ejaculate contains the highest concentration of spermatozoa (75%), which under favorable conditions promptly penetrate cervical mucus. In addition to favorable quality of cervical mucus, sperm motility is essential for cervical mucus penetration and subsequent fertilization.

Two phases of sperm migration through the cervix are recognized: (i) a rapid phase during which the leading spermatozoa penetrate the central portion of the cervical canal and advance in a line parallel to mucin fibrils originating in the vicinity of the internal os and (ii) a delayed phase. During the delayed phase, sperm enter the cervical mucus around the periphery of the central core and are oriented by mucin fibrils originating from the crypts and colonize them (Fig. 5). Thus, the crypts acting as a reservoir are responsible for the storage of sperm and their gradual release over an extended time of several days into the uterus and tubes. Once in the uterine cavity, further migration of sperm is aided by rhythmic uterine contractions which are modulated by estrogen and progesterone, a process that has been elegantly demonstrated by several investigators.

F. Clinical Evaluation of Cervical Function

The condition of cervical mucus greatly influences sperm receptivity; therefore, it should be evaluated in patients complaining of infertility. Traditionally, two procedures have been used to test cervical factor in infertility: postcoital test and *in vitro* sperm cervical mucus penetration tests.

Preovulatory mucus receptive to sperm is profuse, thin, clear, acellular, and alkaline. It exhibits intense ferning (4+) and high spinnbarkheit.

The postcoital test (PCT), known also as the Sims–Huhner test, has been used since 1866 when first

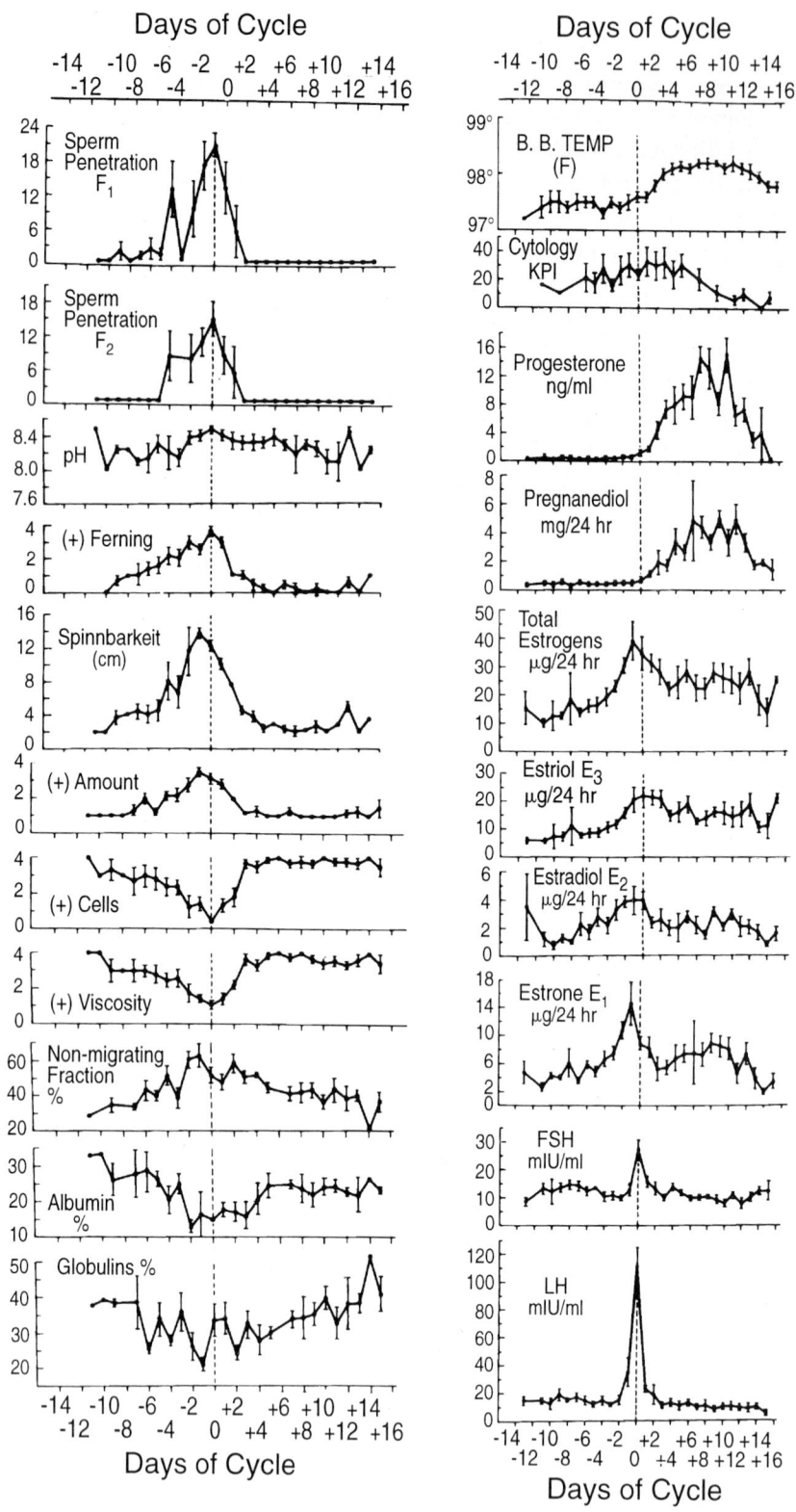

FIGURE 4 Composite profile of serum gonadotropin and progesterone levels; urinary estrogens and pregnanediol levels; basal body temperature (BBT), karyopyknotic index (KPI) of vaginal cells; and cervical mucus properties throughout the menstrual cycle in 10 normal women. Day 0 is the day of Luteinizing hormone (LH) peak (dotted line). Vertical bars represent 1 SEM. F1 and F2 indicate the number of sperm in the first and second microscopic fields (magnification, ×200) from interface, 15 min after the start of *in vitro* sperm–cervical mucus penetration test (reproduced with permission from Moghissi, 1984).

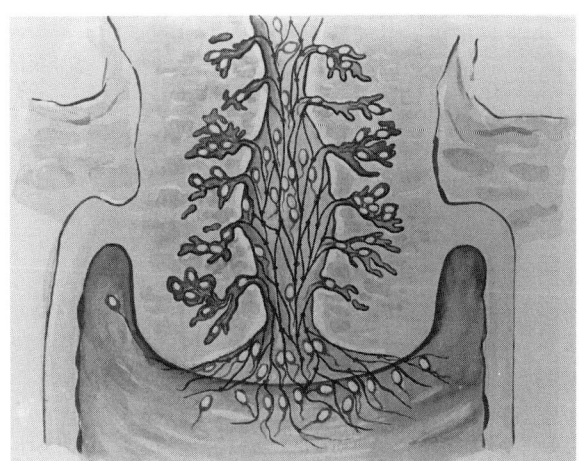

FIGURE 5 Schematic representation of current concept of sperm transport through the cervix. Note the long mucus filaments, sperm penetration along the molecular line of strains, and aggregation of spermatozoa in the crypts and clefts of the cervical canal (reproduced with permission from Moghissi, 1984).

described by Sims. It is performed immediately before ovulation. After an abstention of 2 days the couples are instructed to have intercourse approximately 2–8 hr prior to being tested. The test is performed by removing samples of cervical mucus from the exocervix and endocervical canal using a tuberculin syringe or pipette and counting the number of sperm per high power field (HPF) and evaluating their motility status on microscopic examination. At the same time the quantity and important characteristics of cervical mucus are assessed using a scoring system approved by the World Health Organization which scores the amount, consistency, ferning, spinnbarkheit, and cellularity of the mucus.

In vitro sperm cervical mucus penetration tests are performed using a capillary tube system or a slide. The capillary tube system measures the ability of spermatozoa to penetrate a column of cervical mucus in a flat capillary tube.

For the slide test a small sample of midcycle cervical mucus is placed on a glass slide next to a drop of semen and covered with a coverslip. Sperm penetration into cervical mucus is then observed under microscope. In normal circumstances large numbers of motile (>25) sperm/HPF are observed in cervical mucus 6–10 hr after coitus. Ten or more motile sperm per HPF is considered by most investigators to be consistent with normal PCT, whereas <5, particularly when sperm motion is sluggish, is deemed to be abnormal.

Major causes of abnormal PCT include inappropriate timing of the test, ovulatory disorders, male ejaculatory disorders, sperm and semen abnormalities, antisperm antibodies in male or female reproductive tract, anatomical anomalies of the cervix (cervical amputation, conization, and stenosis), and hostile cervical mucus. Abnormal PCTs are usually repeated and may be confirmed by in vitro tests. There is considerable controversy regarding the validity, sensitivity, specificity, and diagnostic value of the PCT.

PCT is a simple noninvasive clinical procedure which provides valuable information relative to sperm deposition in the female reproductive tract, the quality of sperm, and its interaction with and survival in cervical mucus. On the other hand, it is clearly a subjective test and requires some training and experience for its performance and interpretation. The test should not be used as a prognostic tool to assess the chance of pregnancy since it measures only one factor among many responsible for infertility, i.e., sperm–cervical mucus interaction.

In the past, cervical factor infertility was characterized by abnormal PCT and/or in vitro sperm–cervical mucus interaction and was treated with medication. In the past decade, however, this condition was almost universally treated by intrauterine insemination using washed and concentrated sperm preparation. Several retrospective and prospective studies have established the efficacy of this technique.

II. CERVICAL CONTRACEPTION

Because of its important function in sperm transport, the uterine cervix is to be considered a major site of action for some existing contraceptives and presents several possibilities for the development of newer fertility-controlling modalities.

A. Sex Steroids

Secretory activities of cervical epithelium are controlled by estrogen and progesterone. An increase in the amount of estrogen during the preovulatory phase of the cycle or the administration of synthetic estrogens produces copious amount of thin, watery, alkaline cervical mucus with intense ferning and spinnbarkheit which is highly receptive to sperm penetration.

Endogenous progesterone during the luteal phase of the cycle, or in pregnancy, or exogenous administration of pregestational preparations produces scanty, viscous, and cellular mucus with absent ferning and spinnbarkheit which inhibits sperm penetration in the cervix. Other constituents of cervical mucus, such as proteins, enzymes, and electrolytes, are also sensitive to hormonal changes. Thus, a major function of progestin-only contraceptives whether administered orally (minipills) or parenterally (Norplant, DepoProvera, and vaginal ring) appears to be the ability of these formulations to inhibit sperm migration through cervical mucus. Similarly the effectiveness of combination-type oral contraceptives appears to be in part due to the property of progestin component to induce cervical mucus hostility and impair sperm migration.

B. Ovulation Prediction and Periodic Abstinence

Changes of cervical mucus during menstrual cycle have been used to determine the time of ovulation for the purpose of the sexual abstinence method or so-called natural family planning. Natural family planning is devised to accommodate those couples that, because of their religious or moral beliefs, are opposed to artificial methods of birth control.

The cervical mucus or ovulation method relies on self-observation and perception of midcycle mucorrhea for ovulation timing. Five phases of cervical mucus pattern are recognized: phase 1, dry days; phase 2, early preovulatory days; phase 3, wet days; phase 4, postovulatory days; and phase 5, late postovulatory days. The fertile or unsafe period is presumed to start on the first day during which postmenstrual mucus is observed (phase 2) and to continue until the fourth day after the clear lubricative cervical mucus (peak day) appears—a period that can last 7–14 days. All subsequent days are considered infertile and safe for intercourse. Prospective studies have reported a pregnancy rate of 22.5 per 100 women-years. However, almost all failures could be attributed to a conscious departure from the rules.

C. Cervical Changes in Pregnancy

The function of the cervix during pregnancy is to retain the conceptus until the uterus is prepared to evacuate its content and to dilate sufficiently to allow delivery of the fetus. Physical changes caused by pregnancy include softening of the cervix, proliferation of endocervical epithelia and crypts, and progressive squamous metaplasia of endocervical columnar epethelium resulting from the eversion of the cervix and exposure of endocervix to acid pH of the vagina. Cervical mucus becomes thick and tenacious, producing the so-called "cervical plug" which protects the contents of the uterus and remains in place until labor. Several weeks before the onset of labor the cervix begins to unfold downward from above and becomes effaced. When labor starts, the cervix is soft and mushy and about half of the canal has been taken up. As the contractions of labor become established, further shortening and dilatation occurs. Much of these changes are probably due to a loss of collagen concentration of the connective tissue framework resulting from increased activity of collagenase during pregnancy.

III. CONCLUSION

The cervix plays a unique role in human reproduction. In addition to its important role during parturition, the cervix acts as a biologic valve separating and protecting the upper reproductive tract from the lower genital tract. By virtue of its structure and secretion, it is actively involved in the process of sperm migration within the female reproductive system.

See Also the Following Articles

CERVICAL CANCER; CONTRACEPTIVE METHODS AND DEVICES, FEMALE; FEMALE REPRODUCTIVE SYSTEM, HUMANS; VAGINA

Bibliography

Eimers, D. M., TeVelde, E. R., Gerritse, R., *et al.* (1994). The validity of the postcoital test for estimating the probability of conceiving. *Am. J. Obstet. Gynecol.* **171**, 65–70.

Moghissi, K. S. (1984). The function of the cervix in human reproduction. *Curr. Problems Obstet. Gynecol.* **7**, 1–58.

Moghissi, K. S. (1986). Evaluation and management of cervical hostility. *Sem. Reprod. Endocr.* **4**, 343–355.

Sims, J. A., and Gibbons, W. E. (1996). Treatment of human infertility: The cervical and uterine factors. In *Reproductive Endocrinology, Surgery and Technology* (E. Y. Adashi, J. A. Rock, and Z. Rosenwaks, Eds.), Vol. 2, pp. 2141–2169. Lippincott-Raven, Philadelphia.

Svinarick, D. M., Wolf, N. A., Gomez, R., Gonik, B., and Romero, R. (1997). Detection of human defensin-5 in reproductive tissues. *Am. J. Obstet. Gynecol.* **176**, 470–475.

Trussel, J., and Grummer-Strawn, L. (1990). Contraceptive failure of the ovulation method of periodic abstinence. *Family Planning Perspect.* **22**, 65.

Cesarean Delivery

Linda J. Heffner
Harvard Medical School

I. Introduction
II. Indications for the Use of Cesarean Delivery
III. Available Techniques for Cesarean Delivery
IV. Complications of Cesarean Delivery
V. Vaginal Birth after Cesarean
VI. Use of Cesarean Delivery in Veterinary Practice

GLOSSARY

breech Buttocks, used to refer to the presenting fetal part.

cephalopelvic disproportion A disparity during labor between the dimensions of the fetal head and those of the maternal pelvis.

dehiscence Occult rupture of a surgical incision.

disposition A veterinary term relating orientation of the fetal axis and extremities in addition to presentation and position.

dystocia Abnormal labor or childbirth.

fascia Strong connective tissue which lies deep to the skin and forms a protective covering for muscles and body organs.

fetal position The direction that a given fetal landmark is facing in the maternal pelvis, i.e., left occipitoanterior and right sacroposterior.

fetal presentation The relationship of the long axis of the fetus to that of the mother; also called *lie*. The presenting part of the fetus is that which enters the maternal pelvis first.

fundus A thick contractile portion at the top of the human uterus.

hysterotomy Uterine incision.

laparotomy Incision through the abdominal wall.

nonreassuring fetal status Changes in the fetal heart rate that cause doubt about fetal well-being.

peritoneum The strong colorless membrane which covers the abdominal organs and the walls of the abdomen and pelvis.

theriogenology Diagnosis, treatment, and prevention of reproductive diseases in animals.

vaginal birth after cesarean Successful trial of labor after prior cesarean delivery.

vertex Top; used to refer to the fetal head when presenting in labor.

A cesarean section is the delivery of a fetus through incisions in the abdominal and uterine walls (laparotomy and hysterotomy). It is usually performed for maternal or fetal conditions in which vaginal birth will result in significant injury to, or death of, either mother or fetus. Techniques for cesarean delivery exist for many mammalian species, including but not limited to humans and other primates, large animals such as cows and horses, smaller farm animals such as pigs, sheep, and goats, pets such as dogs and cats, and laboratory animals.

I. INTRODUCTION

The origin of the term cesarean is unknown. It is commonly thought to have originated from the delivery of Julius Caesar, who was reportedly cut from the womb of his mother, Aurelia, in 100 BC. Given that she lived to hear of her son's conquest of Britain, such a derivation is unlikely. The procedure was included in the codification of Roman law in 715 BC as a means of salvaging a living fetus from a dead or dying mother or providing for the infant's separate burial in the event of the mother's death. A third possible origin may be from the Latin word *caedare*, to cut. The term cesarean section is redundant because section also means a cut. Many authorities prefer the term cesarean delivery.

The first written record of cesarean delivery of a surviving woman and infant is that of Jacob Nufer's wife in early sixteenth-century Switzerland. Cesarean delivery remained a desperate surgical procedure, designed to save only the soul of infants of dead women and an occasional lucky woman until well into the nineteenth century. As late as the midnineteenth century, maternal mortality rates following cesarean delivery remained as high as 50%. The introduction of anesthesia in 1846, antisepsis in the 1860s, and uterine suture techniques in 1882 made the procedure much safer. The introduction of antibiotics into medical practice in the midtwentieth century reduced the last significant obstacle to the operation as a safe alternative to vaginal delivery. Fetal salvage became a significant reason for cesarean delivery and remains the single largest indication today. It is estimated that 1 million women undergo cesarean delivery annually in the United States, making it the most common major operation performed.

II. INDICATIONS FOR THE USE OF CESAREAN DELIVERY

The four most frequent indications for cesarean delivery in humans are (i) history of a prior cesarean; (ii) dystocia or failure to progress in labor; (iii) breech presentation, and (iv) nonreassuring fetal status.

The dictum "Once a cesarean, always a cesarean" first appeared in a 1916 article in *The New York Medical Journal*. The practice of recommending cesarean delivery once a first had been performed originated from the high rate of uterine rupture in subsequent pregnancies. The risk of uterine rupture is related to the type of hysterotomy and is highest (6–18%) with a large vertical incision extending to the fundus of the uterus. The introduction of the low transverse hysterotomy by the British surgeon Kerr in 1926 dramatically improved the safety of the procedure by reducing the risk of uterine rupture to <1%. Despite the low risk of uterine rupture with a subsequent trial of labor, repeat cesarean delivery accounts for 8.5% of all births in the United States and about 3% of all births in Europe.

Labor dystocia is the most frequent indication for cesarean delivery in the United States. The definition of obstructed or difficult labor is imprecise, but generally dystocia is the consequence of four abnormalities that may occur singly or in combination. Dystocia may result from (i) abnormalities of the expulsive forces, (ii) abnormalities of the bony pelvis, (iii) abnormalities of the birth canal other than those involving the bony pelvis, and (iv) abnormalities of the fetus. The most frequently encountered type of dystocia is the first category and results from abnormalities of either uterine contractile forces or inadequate voluntary muscle efforts once the fetus passes through the fully dilated cervix. Often abnormalities of the contractile forces can be overcome with judicious use of the uterine stimulant, pitocin, a synthetic oxytocin. Cesarean delivery is performed when oxytocin (pitocin) augmentation of labor fails to dilate the maternal cervix completely or when the infant cannot be pushed down far enough into the vagina to make vaginal delivery assisted by forceps or a vacuum extractor a safe alternative. Abnormalities of the bony pelvis are uncommon in the developed world but include rickets and healed pelvic fractures in the Third World. Maternal obesity leading to excessive soft tissues in the vagina is an example of an abnormality of the birth canal not involving the maternal bony pelvis. Fetal abnormalities that can result in a need for cesarean delivery include abnormal presentations and congenital anomalies.

Abnormal fetal presentation accounts for at least 20% of all cesarean deliveries with breech presentation the most common. Breech presentation occurs in 3 or 4% of term fetuses; current estimates are that 90% of all breeches are delivered abdominally because of the risk of birth injury to the infant. Shoulder and transverse presentations at term necessitate abdominal delivery. Infants in the vertex presentation are at greater risk for cesarean delivery if they are in the occipitoposterior (facing up) or transverse (facing sideways) position because the leading diameter of the fetal head is greater than in the occipitoanterior (facing down) position. Such patients are said to have relative cephalopelvic disproportion.

Nonreassuring fetal status or "fetal distress" accounts for about 15% of cesarean deliveries performed. Like dystocia, the definition of fetal distress is broad and vague. The introduction of electronic fetal heart rate monitoring into obstetric practice in the late 1960s was intended to predict fetuses at risk for asphyxial birth injury from labor. Despite more than 30 years of use and research into electronic fetal monitoring, there is little hard evidence linking specific fetal heart rate patterns and subsequent brain damage. To some degree, there is an element of fetal distress or danger inherent during normal birth. Thus, uncertainty about the meaning and prognostic ability of the fetal heart rate tracing has led to use of the terms "reassuring" or "nonreassuring" fetal heart rate patterns. Most cesarean deliveries occur because the obstetrician is no longer reassured enough about the well-being of the fetus to allow the labor to proceed.

III. AVAILABLE TECHNIQUES FOR CESAREAN DELIVERY

The major components of cesarean delivery include the abdominal incision, the uterine incision, removal of the fetus followed by the placenta and fetal membranes, repair of the uterine incision, and abdominal closure.

The abdominal incision for a cesarean delivery must traverse the maternal skin, abdominal wall, supporting musculature, and parietal peritoneum which covers the inside of the abdominal cavity. Two general approaches are an infraumbilical vertical incision and a low transverse abdominal incision, known as a modified Pfannenstiel technique. The incision type is chosen after considering the skill of the surgeon, the time available to reach the fetus in cases of fetal distress, the mother's habitus, and any history of prior abdominal surgery. The vertical incision is the quickest and is used when speed of delivery is critical. It involves less dissection and less potential blood loss because it is made directly in the midline where the layers of the abdominal wall do not include any muscle but only the sheath of the rectus muscles. It has the advantage of being immediately extendable above the umbilicus should additional operative field be necessary. Large midline vertical incisions are more prone to dehiscence in the postoperative period.

The Pfannenstiel incision, which is both more cosmetic and stronger, involves a curvilinear incision through the skin and subcutaneous tissue, followed by a larger curvilinear incision of the exposed fascia overlying the rectus muscles. Once the fascia is dissected off the rectus muscles, the muscles are split bluntly in the midline and the underlying peritoneum entered vertically. Additional room in the operative field may be obtained by securing the blood vessels to the rectus muscles bilaterally and cutting laterally through them, as described by Maylard. The Pfannenstiel incision is preferred for most cases because of its cosmetic acceptability and tensile strength during healing. Pfannenstiel incisions are more prone to concealed bleeding.

In both types of incisions, care must be taken when entering the peritoneal cavity to ensure that tissues, such as the omentum, bowel, or uterus, are not adherent to the undersurface of the peritoneum following previous abdominal surgery.

Currently, the low transverse or Kerr hysterotomy is the most widely used uterine incision for cesarean delivery. A lower uterine segment incision has several technical advantages. It requires only modest dissection of the bladder from the underlying myometrium and makes the incision in the thinnest and most easily repaired part of the uterus. In turn, this leads to less blood loss, a lower risk of postoperative adhesion of bowel or omentum to the uterine incision, and the lowest risk of uterine rupture in subsequent pregnancies. It is important when making the uterine incision to consider the size of the infant's head and trunk so as to minimize the risk of tearing into the uterine vessels that course along the lateral margins of the uterus. When it is likely that the uterine incision may extend laterally into the region of the uterine vessels, a vertical lower segment incision first described by Kronig in 1912 is advocated. A Kronig incision may also be used when there are multiple fetuses, a breech presentation, or a very premature infant in which the thickness of the lower segment is questionable. The locations of the uterine incisions used in women are shown in Fig. 1.

Critical to the performance of a cesarean delivery in humans is avoidance of injury to the maternal bladder, which lies upon the lower pole of the uterus. This is accomplished by incising the loose layer of peritoneum above the bladder and overlying the anterior surface of the uterus, followed by dissection to separate the bladder from the underlying myometrium. The bladder may then safely be retracted downward and away from the lower uterine segment.

The so-called classical hysterotomy, which is a large vertical incision of the uterus starting in the lower segment and extending high into the uterine fundus, is used far less frequently than the lower segment hysterotomies. Blood loss is greater and the operative repair more difficult. Typical indications for classical hysterotomy include an inability to separate the bladder and safely expose the lower segment, a transverse fetal lie at term, placenta previa with anterior implantation, and extreme prematurity with a very thick and undeveloped lower uterine segment.

Following the uterine incision, the membranes surrounding the fetus are ruptured if this has not occurred and the fetus is gently and expeditiously delivered. This typically involves lifting the fetal head or buttocks through the incision and applying some pressure to the maternal abdomen at the top of the uterus. The umbilical cord is doubly clamped and cut, and the infant is handed off the operating field to an attendant. An umbilical blood sample is removed from the cord and the placenta delivered.

Repair of the uterus is performed in one or two layers of suture depending on the position, size, and hemostasis of the original incision. The peritoneum overlying the bladder and abdominal wall are usually not sutured. The fascia is closed using standard surgical techniques followed by loose approximation of the subcutaneous tissue. The skin may either be sutured or closed with removable stainless-steel clips.

IV. COMPLICATIONS OF CESAREAN DELIVERY

Complications of cesarean delivery are both immediate and delayed. Cesarean delivery typically results in a greater blood loss than vaginal delivery because of bleeding from the hysterotomy until repair is complete. Blood loss during cesarean delivery averages

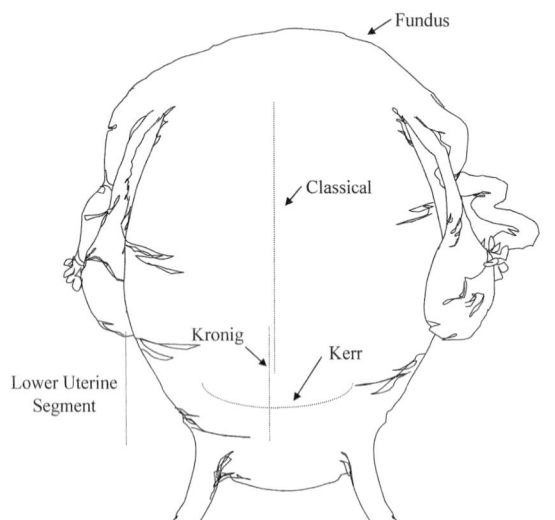

FIGURE 1 Types of hysterotomies performed in women.

1000 ml, but due to the increase in circulating blood volume encountered in pregnancy women normally tolerate blood losses of up to 1500 ml without difficulty. A need for transfusion is usually limited to patients with inadequate uterine contraction after delivery, which results in continued bleeding, and those requiring additional surgery such as hysterectomy following cesarean delivery. Operative injuries to the adjacent maternal organs or to the fetus are rare but do occur in the settings of prior abdominal surgery, unusual fetal presentations, and highly emergent deliveries. Postpartum endomyometritis, a bacterial infection of the lining and musculature of the uterus, complicates as many as 85% of cases performed with the mother in labor. Its incidence can be decreased to 20% by intraoperative administration of prophylactic antibiotics following clamping of the umbilical cord. Venous thrombosis is a rare but serious complication of cesarean delivery. Diagnosis of thromboses in the lower extremities can usually be made based on symptoms, but pelvic thromboses often are accompanied by fever unresponsive to adequate antibiotic therapy. Delayed complications of cesarean delivery can occur in subsequent pregnancies. Uterine dehiscence or rupture occurs with a frequency directly related to the type of uterine incision, with classical incisions carrying the greatest risk. Thirty percent of uterine ruptures with classical incisions occur prior to the onset of labor. Women with a prior cesarean delivery are at increased risk for development of placenta previa, a condition in which the placenta implants on the cervix. The risk of placenta previa in women with a prior cesarean increases with the number of prior cesarean deliveries from 1.9% with one to 4.1% with three or more.

V. VAGINAL BIRTH AFTER CESAREAN

Prior to the mid-1980s, fewer than 10% of women in the United States with a prior cesarean delivered a subsequent infant pelvically. This was generally true throughout Europe, with the exception of Sweden, where for two decades approximately 50% of women with a prior cesarean delivery delivered pelvically. Repeat cesarean section had been recommended because of fear of uterine rupture. In the past 15 years, vaginal birth after cesarean (VBAC) has emerged as a safe and effective way to lower the escalating cesarean delivery rate. Recent statistics indicate that about half of American women eligible for VBAC now choose this option with a success rate of 60–80%.

Critical to the decision to allow a woman to labor after a prior cesarean delivery is the type of hysterotomy performed. Women with a classical incision should not undergo a trial of labor because 6–12% of these women will have a catastrophic rupture. Low segment transverse incisions, by contrast, have <1% risk of rupture and this appears not to be increased by a trial of labor. Rupture or dehiscence of the hysterotomy is usually not catastrophic after a low transverse incision. More than one low transverse cesarean delivery is not considered a contraindication to VBAC. Little information is available about the safety of VBAC following a Kronig low vertical hysterotomy because of its infrequent use.

Success rates for VBAC vary according to the reason for the previous cesarean deliveries. Women with indications other than dystocia appear to have the highest success rates for VBAC. Women with a prior cesarean delivery for dystocia have a 50–60% success rate, often delivering a comparable or larger size infant than previously.

VI. USE OF CESAREAN DELIVERY IN VETERINARY PRACTICE

For many of the same reasons, cesarean delivery has become a safe alternative to vaginal delivery in animals as well as people. It has assumed an important role in safe fetal delivery in the dairy and cattle industries where loss of a valuable dam or fetus has significant financial implications. Similar circumstances have developed among highly valued breeds of dogs.

Cesarean delivery of cows is most commonly performed for fetal oversize due to relative immaturity of the dam or fetal conditions due to genetic mismatching or prolonged pregnancy. Calves that present posteriorly are at risk for birth injury and asphyxia similar to the human in breech presentation. The components of a cesarean delivery in cows are generally the same as in humans with three exceptions. First, the procedure is most commonly done in the standing position through a flank incision. Second, separation of the bladder from the uterus is not necessary in cows because the flank incision dictates a longitudinal uterine incision. Finally, in cows, the placenta and fetal membranes are not removed at surgery. This is because the bovine placenta is a different type than that of the human and is diffuse and densely adherent to the uterine wall. Antibiotics placed within the uterine cavity prior to closure of the hysterotomy reduce the risk of postoperative infection. Closure of the uterine and abdominal incisions follows general surgical procedures in all animal species.

In horses, cesarean delivery is indicated when the foal lies transversely in the maternal abdomen during labor and in cases in the which the ends of the fetus occupy both horns of the uterus. Occasionally, a cesarean may be necessary when there is a maternal pelvic deformity or life-threatening injury to the mare. Again, placement of the abdominal incision along the flank results in a longitudinal uterine incision with no risk of injury to the maternal bladder. Equine placentae are not removed because they are diffuse and densely adherent like bovine placentae.

Use of cesarean delivery in sheep and goats is more limited in veterinary practice because of marginal economic benefits to farmers and breeders. Indications for cesarean delivery in sheep include a failure of the cervix to adequately dilate (ringwomb), fetal oversize, and maternal malnutrition. Prevention of dystocia is an important principle in avoiding cesarean delivery in sheep. The recurrence risk for dystocia is 30% and hence dams experiencing dystocia are not typically returned to the breeding flock. Large, well-managed herds will have dystocia rates of 1 or 2%. Cesarean delivery as part of research protocols using sheep is common. Sheep are remarkably tolerant of fetal manipulation and surgery *in utero* and rarely abort, making them a valuable subject for fetal research. Like other large animals, the abdominal incision is usually through the flank. A longitudinal uterine incision is made, followed by delivery of the lamb. Unless readily removable, the fetal membranes are left in the uterus.

Among domestic animals, dogs require cesarean delivery far more frequently than cats. Most cesareans are performed in both species because of breeding techniques favoring specific physical attributes. Breeds such as bulldogs, which have disproportionately large heads, and small domestic breeds with extreme variations in fetal size often require cesarean delivery. Other spontaneous causes of dystocia in dogs include small litter size which is associated with fetal overgrowth, prolonged gestation, and uncorrectable fetal maldispositions. In both dogs and cats, the abdominal incision is ventral. In dogs, the hysterotomy is typically in the uterine body, which joins the two horns. Care is taken to avoid extending the incision into the horn(s) because of possible interference with placentation in a subsequent pregnancy. In cats, the uterine incision is begun in the midline but may be extended slightly into one uterine horn. In both species, it is usually sufficient to make a single hysterotomy because the pups or kittens in both horns can be "milked" toward the incision for delivery. The placentae are typically removed at delivery in both dogs and cats.

There is a high incidence of Q fever in pregnant sheep, goats, and cows, making cesarean delivery an occupational hazard for the surgeon. The placenta of an infected ewe can contain up to 10^9 organisms per gram of tissue. The organism is resistant to dessication and persists in the environment for long periods of time, thereby contributing to significant risk of transmission in research facilities handling pregnant sheep. Failure to observe rigorous recommendations for the control of Q fever in such facilities can result in infection of personnel and their contacts. Cats, dogs, and small rodents used in research also can be infected, but the risk of transmission is significantly lower in these species.

See Also the Following Articles

FETAL MONITORING AND TESTING; FETAL SURGERY; PREGNANCY IN DOGS AND CATS

Bibliography

Committee on Occupational Safety and Health in Research Animal Facilities, Institute of Laboratory Animal Resources, Commission on Life Sciences, National Research Council (1997). *Occupational Health and Safety in the Care and Use of Research Animals.* National Academy Press, Washington, DC.

Cragin, E. B. (1916). Conservatism in obstetrics. *N. Y. Med. J.* **104**, 1–3.

Cunningham, F. G., MacDonald, P. C., Gant, N. F., Leveno, K. J., Gilstrap, L. C., III, Hankins, G. D. V., and Clark, S. L. (1997). *Williams Obstetrics,* 20th ed. Appleton & Lange, Stamford, CT.

Dorland, N. (Ed.) (1994). *Dorland's Illustrated Medical Dictionary.* Saunders, Philadelphia.

Morrow, D. A. (1986). *Current Therapy in Theriogenology: Diagnosis, Treatment and Prevention of Reproductive Diseases in Small and Large Animals,* 2nd ed. Saunders, Philadelphia.

Sewell, J. S. (1993). *Cesarean Section—A Brief History.* American College of Obstetricians and Gynecologists, Washington, DC.

Cetacea

see Whales and Porpoises

Chaetognatha

George L. Shinn
Truman State University

I. Modes of Reproduction
II. Anatomy of the Reproductive System
III. Reproductive Physiology and Endocrine Control
IV. Mating, Fertilization, and Ovulation
V. Modes of Development

GLOSSARY

accessory fertilization cells In chaetognaths, paired somatic cells in the ovary to which eggs are attached through which sperm must pass in order to fertilize the eggs.

chemotaxis The process locating the source of a chemical by detecting and moving toward ever-increasing concentrations of the chemical.

coelom In some animals, a type of fluid-containing body cavity that is completely lined by a mesodermally derived epithelium.

deuterostome Animals in which the mouth forms at the opposite end of the body from the blastopore, which is the first opening to the embryonic digestive tract.

enterocoely In animals, a method of coelom formation in which the epithelial lining of the coelom develops as an elaboration of the embryonic digestive tract; this involves folding of the epithelium comprising the embryonic gut.

seminal vesicle Part of the male reproductive tract; the function varies between animal groups. In invertebrates, it is commonly a site of sperm storage prior to mating; in vertebrates, it is the source of secretions that are added to sperm to make semen.

spermatophore In animals, a packet of sperm held together by a secreted coating. These are commonly transferred to mates through complex behaviors and enable internal fertilization without copulation.

syncytium A multinucleate "super cell" formed by the fusion of uninucleate cells.

vitellogenesis During the formation of eggs, the process of yolk formation.

The Chaetognatha is a phylum of highly cephalized predatory marine worms. About 130 species are known to exist and most of them are planktonic. Chaetognaths are among the most abundant and ecologically important animals in the sea. Chaetognaths are routinely hermaphroditic and their reproduction is exclusively sexual. Studies of chaetognath reproductive biology have involved only a few species but these have revealed unusual complexities. A mating "dance" involving species-specific movements and posturing leads to deposition of sperm onto the mate. This may be at the gonopore or elsewhere on the skin, depending on the species. Sperm find their own way to and into the female gonopores. This is thought to involve chemotaxis. Internal fertilization is mediated by somatic "accessory fertilization cells." Chaetognaths produce and release numerous clutches of eggs and, if food is abundant, this can occur daily. Most arrowworms presumably spawn fertilized eggs directly into the sea, but developing eggs adhere to the sides of the body in some species and benthic arrowworms attach eggs to the substrate on which adults live. Chaetognaths have radial cleavage and deuterostomous, enterocoelic development. Precocious determination of primordial germ cells is thought to be controlled by cytoplasmic "germinal determinants." Development is rapid and direct; tiny juveniles hatch in one or a few days.

I. MODES OF REPRODUCTION

Chaetognaths are hermaphroditic and have only sexual reproduction. The male system commonly matures first but adults typically produce eggs and sperm simultaneously. Well-fed individuals produce successive batches of gametes. Gametogenesis, mating, and release of fertilized eggs occur on a synchronous daily cycle in some species (e.g., *Adhesisagitta hispida* and *Aidanosagitta crassa*) but not in others (e.g., *Paraspadella gotoi*).

II. ANATOMY OF THE REPRODUCTIVE SYSTEM

The streamlined body has three segments—head, trunk, and tail—divided internally by transverse septa (Fig. 1). The ovaries are discrete cylindrical organs located in the trunk, on either side of the intestine. The female gonopores open laterally, just in front of the posterior septum. The male reproductive system occupies the entire tail. The male organs, including a pair of testes, two sperm ducts, and two seminal vesicles, are embedded in the lateral body walls of the tail. Most of spermatogenesis is intracoelomic; circulating masses of spermatogenic cells fill the tail coelom of adults.

A. Male Reproductive System

The two inconspicuous testes are situated anterolaterally in the trunk. Each consists of a thin column of mitotically active spermatogonia ensheathed by specialized cells of the coelomic lining (Fig. 2). Most or all spermatogonia in the testes are parts of small syncytial clusters. The testis sheath appears to function mainly as a compartment boundary. Its cells contain few synthetic organelles, no elaborate contacts with spermatogonia, and no ultrastructural features indicative of a transport function. The testis sheath is incomplete anteriorly, where small groups of spermatogonia are released to the tail coeloms.

Spermatogenic masses in the tail coelom may be at several stages of differentiation, suggesting that spermatogonia are released from the testes in pulses. Spermatogenic cells remain connected in syncytia as they divide by mitosis and meiosis and then differentiate into spermatozoa. Mature sperm are filiform and highly derived compared to those of animals such as cnidarians and echinoderms, which free spawn and have external fertilization. A centriolar derivative at the narrow anterior tip gives rise to a

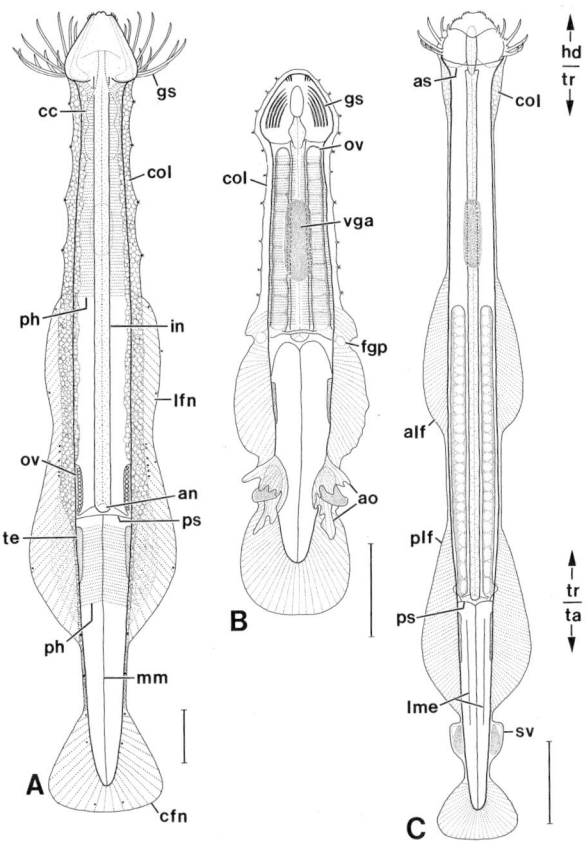

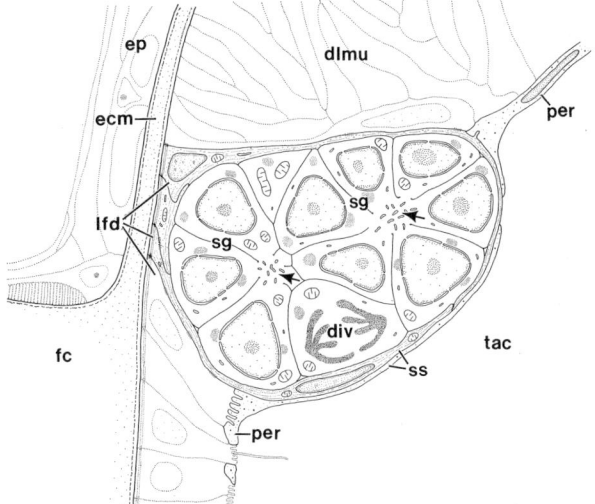

FIGURE 1 Diagrams showing general morphology of chaetognaths. (A) *Heterokrohnia involucrum*, dorsal view; (B) *Paraspadella gotoi*, ventral view; (C) *Adhesisagitta hispida*, dorsal view. alf, anterolateral fin; an, anus; ao, adhesive organs; as, anterior septum; cc, corona ciliata; cfn, caudal fin; col, collerette; fgp, female gonopore; gs, grasping spines; hd, head; in, intestine; lfn, lateral fin; lme, lateral mesentery; mm, medial mesentery; ov, ovary; ph, phragma muscles; plf, posterolateral fin; ps, posterior septum; sv, seminal vesicle; ta, tail; te, testis; tr, trunk; vga, ventral ganglion. Scale bars = 1 mm (reproduced with permission from Shinn, 1997).

FIGURE 2 *Adhesisagitta hispida*. Diagram showing transverse section through testis. div, dividing spermatogonium; dlmu, longitudinal muscles of dorsal body wall; ecm, extracellular matrix of body wall; ep, epidermis; fc, fin core; lfd, lateral field cells; per, peritoneocytes; sg, syncytial spermatogonial mass; ss, somatic sheath of testis; tac, tail coelom; arrows indicate central cytoplasmic masses of syncytia. Scale bar = 2.5 μm (reproduced with permission from Shinn, 1997).

conventional axoneme that extends the length of the sperm. The axoneme is paralleled by a thin, highly modified mitochondrion. Numerous Golgi-derived vesicles of several types and of unknown function are arranged in series in the anterior half of the sperm. The thin elongate nucleus lies posteriorly in the sperm.

Spermatogenic masses break apart upon completion of spermatogenesis. In species having daily reproductive cycles, the oldest spermatogenic masses disintegrate synchronously and the mature sperm are quickly drawn into the expanded ciliated anterior ends of the sperm ducts. The latter are short, straight ciliated tubes situated on either side of the tail, just anterior to the seminal vesicles. As sperm move down the sperm ducts, fluids among them are resorbed by the ductal epithelium so that sperm in the seminal vesicles are densely packed.

The seminal vesicles are blind-ended, double-walled sacs in which sperm are stored until mating. They are lined by a simple glandular epithelium and covered by stratified epidermis. The glandular lining produces adhesive secretions that enable sperm masses to adhere to mates. In some species, the secretions form a definitive spermatophore coat. The seminal vesicles of some species have chitinous teeth, papillae, or other ornamentations of uncertain function. Although the seminal vesicles have no permanent openings, specialized epithelial cells form a suture line, along which the seminal vesicles rupture during mating. The suture is normally kept closed by apical adhesion of opposing suture cells. The suture seems to open mechanically when a seminal vesicle contacts the mate's body.

B. Female Reproductive System

The two cylindrical ovaries are complex composite organs mediating a variety of functions, including oogenesis, storage of sperm, internal fertilization, ovulation, and egg laying (Fig. 3). A thin myoepithelial ovarian wall encloses a fluid-containing ovarian space within which oocytes differentiate and, laterally, an oviducal complex.

The ovaries contain several size classes of oocytes, each representing a separate clutch of eggs. Oocytes arise from clusters of germinal cells embedded in the medial side of the oviducal complex. As growing oocytes separate from the germinal clusters, they become attached to pairs of specialized oviducal cells that differentiate into "accessory fertilization cells." Only the largest oocytes are involved in vitellogenesis and will be fertilized and released during the next egg-laying event. Within the ovarian space, oocytes are ensheathed by a delicate network of mesenchymal tissue called the follicular reticulum. Secretory vesicles and synthetic organelles dominate its cytoplasm. Suggested functions include secretion of the jelly coat, bulk transport of yolk precursors to oocytes from the ovarian space or oviducal complex, and contraction during ovulation.

The oviducal complex opens out at the female gonopore but is blind-ended internally. It is a double-walled structure, having of an outer "cellular sheath" and a central "syncytium." The cellular sheath consists of secretory cells that exocytose Golgi-derived vesicles, thought to contain yolk precursors, into the ovarian space. The oviducal syncytium is a multinucleate cytoplasmic mass extending the length of the oviducal complex. Sperm entering at the female gonopores after mating occupy an elongate vacuole that extends the length of the syncytium. The syncytium is capable of resorbing sperm that do not participate in fertilization.

III. REPRODUCTIVE PHYSIOLOGY AND ENDOCRINE CONTROL

Nothing is known about the chemical control of reproduction in arrowworms. The existence of a daily reproductive cycle in certain species suggests that reproductive hormones exist. Physiological integration of reproductive processes may involve a recently discovered hemal system. This consists of a narrow blood space in the intestinal wall and paired "posterior sinuses" in the posterior septum. Specialized tissues bordering the posterior sinuses are thought to allow passage of nutrient-laden hemal fluids into the tail coelom and into the ovarian spaces.

IV. MATING, FERTILIZATION, AND OVULATION

Laboratory observations reveal that cross-fertilization via mating is probably typical but that self-insemination and/or parthenogenesis may also occur.

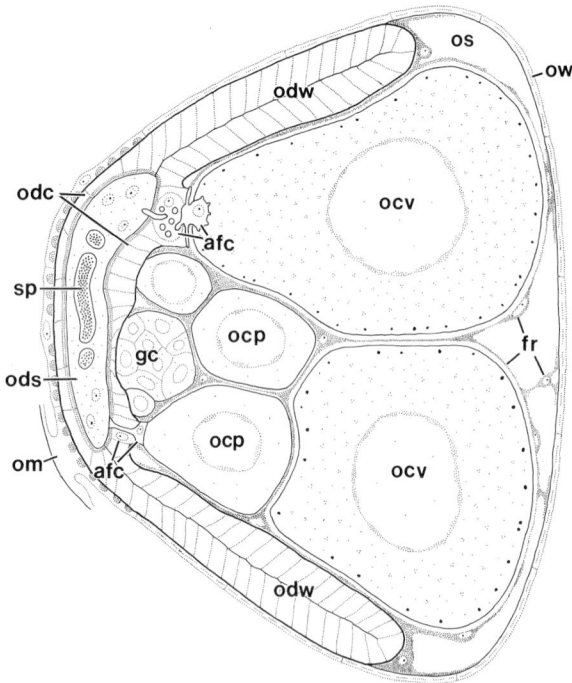

FIGURE 3 *Adhesisagitta hispida*. Diagram showing transverse section through left ovary. afc, accessory fertilization cells; fr, follicular reticulum; gc, germinal cluster; ocp, previtellogenic oocyte; ocv, vitellogenic oocyte; odc, cellular sheath of oviducal complex; ods, syncytium of oviducal complex; odw, wing of oviducal complex; om, ovarian mesentery; os, ovarian space; ow, ovarian wall; sp, spermatozoa (reproduced with permission from Shinn, 1997).

Mating activities are brief and frenetic. Pairing involves rapid flexures and undulations of the body. These are thought to create distinctive waterborne disturbances that signal prospective mates. The worms align either head-to-tail or head-to-head, depending on the species. In some, the grasping spines normally used in prey capture are locked onto the head of the mate. Sperm transfer is routinely reciprocal in some species (*Spadella cephaloptera*) but rare in others (*A. hispida* and *P. gotoi*). The seminal vesicle of a sperm donor is hit against the body of the mate, the seminal vesicle ruptures, and either a naked mass of sperm or a spermatophore is left attached to the mate. In some species, the sperm are deposited precisely at the opening of a female gonopore, but in others it is deposited elsewhere and sperm migrate in coherent columns over the body surface to the female gonopores. Observations on *S. cephaloptera* suggest that sperm follow a trail of secretions produced by the corona ciliata. The latter is a ring of ciliated secretory cells located dorsally behind the head in all chaetognaths.

Sperm received during mating are temporarily stored in the syncytium of the oviducal complex. Fertilization occurs before ovulation by passage of sperm through a pair of specialized oviducal cells called accessory fertilization cells. These differentiate in coordination with an attached oocyte, participate in fertilization, and then, during ovulation, are sloughed into the oviducal syncytium and resorbed. Ultrastructural studies reveal that one of the accessory fertilization cells contains a complexly shaped "fertilization canal" through which sperm reach the egg. The canal is normally occluded by a close-fitting cytoplasmic extension of the other accessory fertilization cell. At fertilization, by an unknown mechanism, the cytoplasmic process disappears from the fertilization canal, opening the pathway for the sperm. Details of sperm–egg fusion remain unknown.

Ovulation occurs immediately after fertilization. In sagittid chaetognaths, zygotes pass individually through the medial wall of the oviducal complex through pores that form when the accessory fertilization cells enter the syncytium. In benthic chaetognaths, whose eggs are relatively large, zygotes become free in the ovarian space and then enter the oviducal complex posteriorly, through a single pore.

V. MODES OF DEVELOPMENT

Pelagic chaetognaths typically release zygotes into the water column. Species of *Eukrohnia*, however, retain the eggs throughout development and for some time afterwards in gelatinous brood pouches surrounded by the curled lateral fins. *Pterosagitta draco* releases its eggs in a pelagic jelly mass. Benthic species (e.g., *S. cephaloptera*) and some neritic species (e.g., *A. hispida*) attach eggs to a substratum.

Development of chaetognaths is direct and rapid. Juveniles typically hatch in 1–4 days depending largely on temperature and, thus, season. General features of development appear to be stereotypic for the phylum, despite the fact that eggs vary in size from about 300 μm (*A. crassa*) to about 900 μm (*Eukrohnia fowleri*). Cleavage is equal, total, and radial. The blastula has a small blastocoel, and gastrulation by invagination results in obliteration of the blastocoel. Two to four primordial germ cells are evident among the endodermal cells at the inner end of the archenteron. They are relatively large and contain a so-called germ cell determinant. The latter is an RNA-rich cytoplasmic inclusion that forms after fertilization and becomes segregated to prospective germ cells during cleavage.

The mesoderm and embryonic coeloms form by enterocoely. Two vertical folds arise at the closed inner ends of the archenteron and grow posteriorly toward the blastopore, gradually dividing the archenteron into a central gut and two lateral coelomic spaces. A head coelom pinches off the anterior end of each developing lateral coelom. The primordial germ cells move posteriorly in association with the enterocoelic folds so that the primordial germ cells are ultimately associated with the mesoderm and coeloms of the trunk and tail. The blastopore closes before enterocoely is complete. Anteriorly, the endodermal epithelium becomes confluent with an invaginating stomodeum. The embryo elongates and coils within the egg membrane. This coincides with differentiation of longitudinal muscles in the mesodermal epithelia. The gut and coelomic cavities collapse during this stage of development, reopening only after hatching.

Newly hatched young have the same body regions as adults, although hatchlings are much simpler than adults in histological construction. Hatchlings typically lack grasping spines and teeth and have a nonfunctional intestine. Feeding structures differentiate within several days after hatching. In brooding species, hatchlings adhere to the parent as differentiation continues. In *E. fowleri*, the young are 3.0–3.5 mm long, have five or six pairs of grasping spines, and are probably capable of feeding by the time they leave the parent.

See Also the Following Articles

Marine Invertebrate Larvae; Marine Invertebrates, Modes of Reproduction in

Bibliography

Ghirardelli, E. (1968). Some aspects of the biology of chaetognaths. *Adv. Mar. Biol.* 6, 271–375.

Reeve, M. R., and Cosper, T. C. (1975). Chaetognatha. In *Reproduction in Marine Invertebrates* (A. C. Giese and J. S. Pearse, Eds.), Vol. 2. Academic Press, New York.

Shinn, G. L. (1997). Chaetognatha. In *Microscopic Anatomy of Invertebrates, Vol. 15: Hemichordata, Chaetognatha, and the Invertebrate Chordates* (F. W. Harrison and E. E. Ruppert, Eds.), pp. 103–220. Wiley-Liss, New York.

Chalaza

see Egg, Avian

Chelicerate Arthropods

W. Reuben Kaufman

University of Alberta

I. What Are Chelicerates?
II. Aspects of Reproduction Shared among the Chelicerates
III. Merostomata and Pycnogonida
IV. Arachnida
V. Concluding Remarks

GLOSSARY

dioecious Describing species in which the individuals are either male or female. The opposite is hermaphrodite.

20E 20-Hydroxyecdysone; the ecdysteroid which is most commonly the true molting hormone.

ecdysteroid The generic name for the family of steroid hormones based on the arthropod molting hormone precursor, ecdysone.

Juvenile hormone (JH) A family of hormones characteristic of arthropods. A high titer of JH during molting prevents metamorphosis to the adult. In the adult, JH serves as a vitellogenic hormone in many insects.

opisthosoma The posterior body segment of chelicerates corresponding to an abdomen. In scorpions, the opisthosoma comprises two regions, a proximal mesosoma and a distal metasoma.

pedipalps Sensory and protective appendages associated with the mouthparts of chelicerates.

prosoma The anterior body segment of chelicerates, roughly corresponding to a combined head and thorax (cephalothorax).

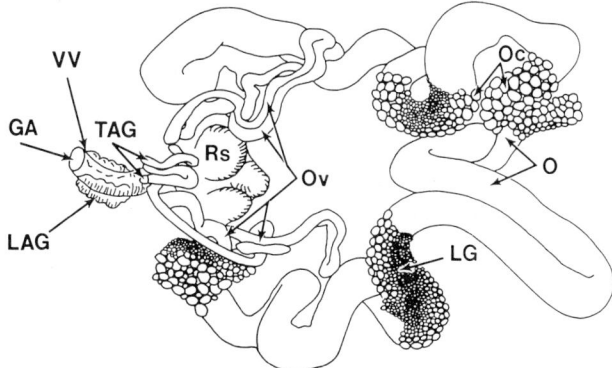

FIGURE 1 Diagram of the female reproductive system of a chelicerate: the ixodid tick, *Dermacentor andersoni*, dorsal view. GA, genital aperture; LAG, lobular accessory gland; LG, longitudinal groove; O, ovary; Oc, oocyte; Ov, oviducts; Rs, seminal receptacle; TAG, tubular accessory gland; VV, vestibular vagina. Modified by Sonenshine (1991) after Brinton and Oliver (1971), *J. Parasitol.* 57, 708–719. Reproduced with permission of the *Journal of Parasitology*.

I. WHAT ARE CHELICERATES?

There are four arthropod subphyla: Trilobita (about 4000 extinct species), Chelicerata (about 65,000 extant species), Crustacea (about 40,000 species), and Uniramia (about 1 million species, most of which are insects). Although all arthropods develop from a fundamentally similar segmented body plan, the specific pattern of segmentation and appendage structure varies among the subphyla. Thus, the chelicerates all lack antennae and have mouthparts comprising a pair of pedipalps and chelicerae. The body is divided into an anterior prosoma, bearing the mouthparts and walking legs, and a posterior opisthosoma. The major classes are Merostomata (horseshoe crabs), Pycnogonida (sea spiders), and Arachnida. The arachnids are the most prominent chelicerates and comprise the following orders: Scorpiones, Pseudoscorpiones, Araneae (spiders and "harvestmen"), and Acari (mites and ticks).

II. ASPECTS OF REPRODUCTION SHARED AMONG THE CHELICERATES

Space does not permit a review of the morphology/histology of the reproductive system and of egg development for all the groups discussed in this article. I have chosen instead to focus mostly on a few physiological and behavioral aspects of reproduction.

With rare exceptions chelicerates are dioecious. The gonads are concentrated in the prosoma, although expansion of the ovary throughout the body occurs during the reproductive phase. The ovary forms a loop which empties into both sides of the oviduct (Fig. 1). Immature oocytes lie in the ovarian wall; as they grow, they protrude outward on short stalks. The eggs ovulate into the lumen through the cavities of their stalks (Fig. 2). Once eggs are ovulated into the central lumen of the ovary, they can migrate in either direction to reach the oviduct.

During the asynchronous gonotropic cycle, oocytes in all stages of development are present simultaneously. Chelicerates are also unusual in lacking a follicular cell layer around the oocytes. Exchange of nutrient and uptake of yolk occur directly from the hemolymph.

III. MEROSTOMATA AND PYCNOGONIDA

Horseshoe crabs are the closest living relatives of the trilobites. We know very little about their physiology of reproduction and nothing about hormonal regulation. This is likely because they require about 10–15 years to reach sexual maturity (Sekiguchi, 1988).

The "sea spiders" are small, elusive creatures. Very little is known about their reproduction beyond the morphology of the reproductive system and a histological description of vitellogenesis (King, 1973).

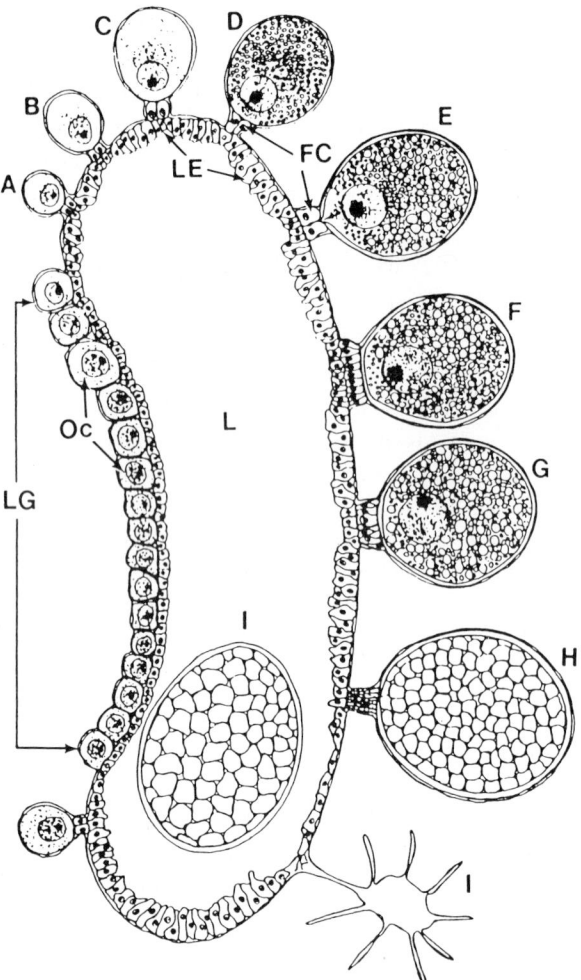

FIGURE 2 Developing oocytes (stages I–V) of the tick, *Dermacentor andersoni*, as seen in cross section of an ovary. FC, funicle cells; I, ovulated oocyte; L, lumen of ovary; LE, luminal epithelial lining; LG, longitudinal groove; Oc, early stage oocytes. In the tick, oocyte development is stimulated by feeding on a blood meal. In other chelicerates, the natural stimuli of egg development are poorly understood. The oocytes in the longitudinal groove represent the stage characteristic of the unfed tick. (A–C) Previtellogenic growth (stages I and II) which occurs during the feeding period. (D–G; stage III) Uptake of yolk and the major growth of the shell which occur following mating and engorgement. Stage IV oocytes (H) are ready for ovulation. Stage V (I) is the ovulated egg and the remnant of the tunica propria in the ovarian epithelium. Fertilization usually occurs any time after ovulation or occasionally at stage V After Brinton and Oliver (1971), *J. Parasitol.* 57, 708–719, as reproduced by Sonenshine (1991). Reproduced with permission of the *Journal of Parasitology* and Oxford University Press.

IV. ARACHNIDA

A. Scorpions and Pseudoscorpions

Scorpions are unusual among the arthropods in being viviparous (Fig. 3) and even more unusual for devoting maternal care to the young for up to several months. Courtship behavior continues to be the major research emphasis in scorpion biology. There are about 12 components to the behavior (Polis, 1990). After recognition of a potential mate, male juddering occurs, which is a rapid rocking/shaking motion. Clubbing involves striking the other partner with the metasoma and is believed to inhibit aggression. Then follows the promenade-à-deux in which male and female clasp pedipalps and promenade together, usually for 30–60 min. Often, this dance continues for hours or even a few days, but this is now believed to be a laboratory artifact associated with observation chambers which lack a suitable substrate for depositing the spermatophore. During the promenade, the male punctures the female's body near the membrane adjacent to the tibial joint of the pedipalps (sexual sting); the sting remains within the female for up to 20 min, but it is not known to what extent venom is released; it is possible that this serves to reduce female aggression. Toward the end of the promenade, the male grasps and kneads the female chelicerae with his own (cheliceral massage). When a suitable substrate is eventually located, the male sweeps the substrate with his pectines, clears away the sand, and deposits the spermatophore. [The pectines (singular =pecten), unique to scorpions, are comb-like sensory structures on the ventral surface of the mesosoma. They are probably homologous to the gill lamellae of Merostomata. They are richly endowed with sensilla, which probably function as mechanoreceptors and contact chemoreceptors.] He then induces the female to place her genital aperture over it, and she assumes a headstand-like posture during the time that the spermatophore breaks open and releases the sperm. There then follows a rapid post-mating escape which has been interpreted as an action to avoid cannibalism by the (generally larger) female. The remains of the spermatophore are usually eaten by one of the partners.

Nothing has been reported on hormonal control

they develop very slowly, with a pregnancy cycle requiring about 8 months.

In pseudoscorpions, the surprisingly complex spermatophores consist of a stalk bearing an apical sperm packet, and sometimes a liquid droplet (the latter supposedly containing a pheromone). Mating patterns range from no body contact ("indirect spermatophore transfer") to full body contact including a "courtship" dance. The developing embryos are incubated in a brood sac from which, aided by a "pumping organ," they take up a nutritive fluid. This pumping organ is incorporated into the mouth apparatus at metamorphosis. During the incubation period, the female may spin a silken brood nest in which she retreats until the protonymphs hatch. Nothing has yet been published on the hormonal regulation of reproduction in pseudoscorpions.

B. Araneae

Courtship/copulatory behavior and guarding of the eggs are two conspicuous aspects of spider reproductive biology. It is commonly believed that the courtship rituals of the male spider evolved because the female would otherwise kill and eat him. The male ".was dramatically pictured as occupying a terrifying position between fear and desire, seeking, be it only temporarily, the appeasement of his formidable partner" (Savory, 1977, p. 310). The latter view derives largely from observations on a single species held in captivity under conditions which prevent ready escape. Such extreme aggression rarely occurs in the wild.

There are at least three types of courtship behavior. In the first, represented by the harvestmen (Opiliones), when male and female encounter each other, insemination occurs with little formality. In the second, represented by most of the web-spinning spiders, the male is attracted by pheromones to the web, which he then vibrates or drums on in such a way as to inhibit the normal aggression of the female toward intruders. Insemination occurs soon after physical contact. In the third group [the large-eyed jumping spiders (Salticidae) of the tropics and the huntsmen spiders (Oxyopidae)] the male performs an elaborate dance to inhibit the female's aggressive tendency. Such rituals may, of course, also arouse

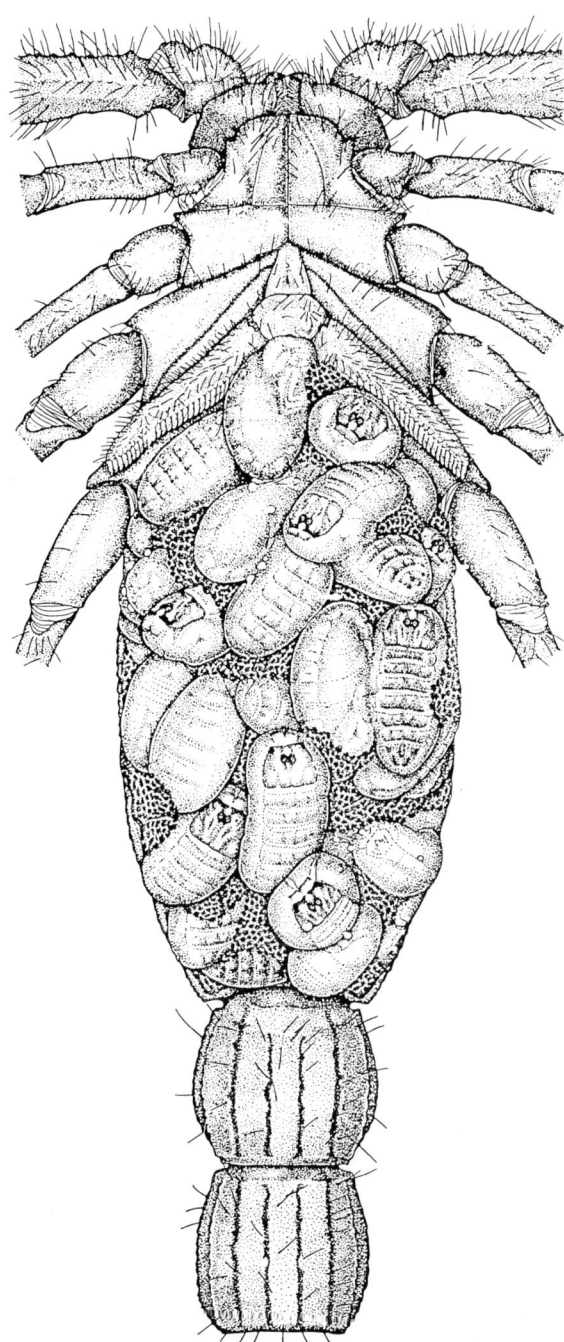

FIGURE 3 A gravid female of the scorpion, Buthotus tamulus, showing the embryos developing in the ovariuterus. Reproduced from Keegan (1980) with permission of University Press of Mississippi.

of reproduction in scorpions, despite their being relatively large, cosmopolitan organisms of considerable medical importance. However, like horseshoe crabs, the female. In all groups, the sperm or spermatophore is transferred by the pedipalps.

Ovarian development in spiders has been correlated to fluctuations in hemolymph ecdysteroid titers. The highest ecdysteroid levels occurred at the transition between the previtellogenic and the vitellogenic stages. A potential role for an ecdysteroid as the vitellogenic hormone reminds one of the dipteran insects (Hagedorn, 1994). A material similar to JH diol has been detected in the hemolymph of 8th-instar *Pisaura*. There is thus at least preliminary indication for a JH-like compound in a spider but no evidence that it plays any role in reproduction.

C. Acari

The male gonad (Fig. 4) consists of a paired or a loop-like testis which leads to vasa deferentia, a seminal vesicle, and an ejaculatory duct, the terminal segment of which may be an intromittent organ or a spermatophoric organ. The accessory glands are usually prominent in the male. Spermatogenesis has been very well described in ticks (Sonenshine, 1991).

The spermatophore is usually delivered to the female gonopore by the palps. In the case of some ticks, the male secretes a special saliva which prevents the spermatophore from sticking to either partner. Indirect sperm transfer occurs in some mites (Evans, 1992).

The female reproductive tract comprises a single (looped) or paired ovary leading to oviducts, median uterus, vagina, and seminal receptacle (Fig. 1). Accessory glands of the female are less elaborate than those of the male. Although some yolk may be synthesized in the oocyte, most of it is produced in the fat body (and sometimes the gut) and released into the hemolymph, from which it is transported to the oocyte. Some Acari are parthenogenetic, but there is much variability among the groups.

1. Mites

Since most mites weigh only a few micrograms, endocrinological studies have relied solely on testing the effects of topically applied hormones and drugs (Kaufman, 1997). In one study on the chicken mite, *Dermanyssus gallinae*, the anti-JH drug precocene-2

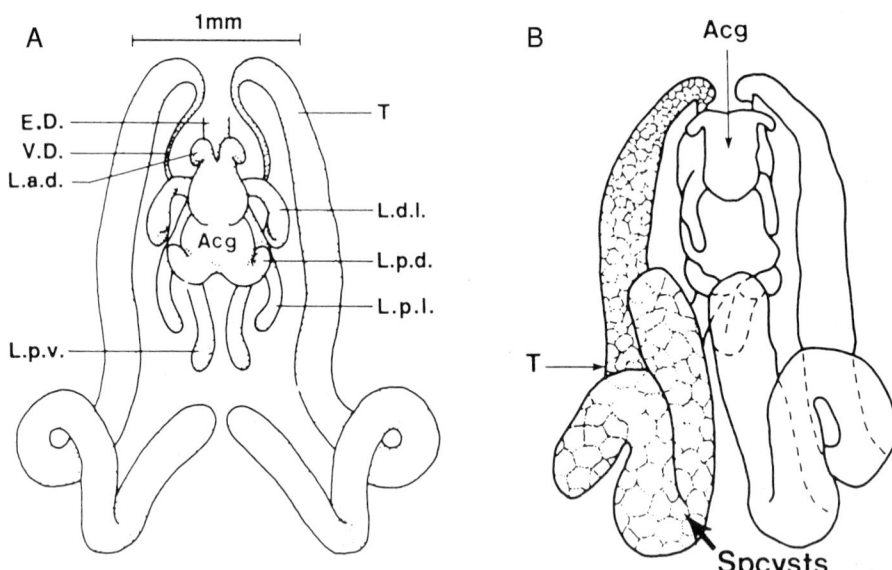

FIGURE 4 Diagram of the male reproductive tract of an unfed (left) and fed (right) *Dermacentor andersoni* tick. Acg, accessory gland; E.D., ejaculatory duct; L.a.d., anterodorsal lobe; L.d.l., dorsolateral lobe; L.p.d., posterodorsal lobe; L.p.v., posteroventral lobe; Spcysts, spermatocysts; T, testis; V.D., vas deferens. From Sonenshine (1991). (A) After Douglas (1943), The internal anatomy of *Dermacentor andersoni* Stiles, *Univ. Calif. Public. Entomol.* 7, 207–272; reproduced with permission of Univ. California Press. (B) After Oliver and Brinton (1972), *J. Parasitol.* 58, 365–379; reproduced with the permission of *Journal of Parasitology* and Oxford University Press.

reduced the number of progeny, an effect which could be partially reversed by treatment with JHIII. In another study, the JH analog, fenoxycarb, stimulated egg production in the grain mite *Acarus siro*.

Varroa mites are well-known parasites which feed on the hemolymph of bees. Female mites can produce eggs only when parasitizing the brood, and there is a higher reproductive success among mites reared on drone vs worker broods. Mites produce eggs only following contact with a spinning larva. These and other observations suggest that reproduction in the mite is attuned to the prevailing hormonal milieu of its host. Several authors have tested ecdysteroids and JH as potential hormonal signals exploited by the mite. There are significant differences between queen, drone, and worker larvae with respect to hemolymph titers of JH and ecdysteroids, and these differences can sometimes be correlated with mite fertility. Fertility is maximal when parasitizing drone cells. Such observations suggest that JH in the host stimulates the mite gonocycle, although Kaufman (1997) cites some conflicting evidence.

2. Ticks

More is known about reproduction in ticks than in all the other chelicerates probably because of their medicoveterinary importance, relatively large size, and ease of rearing large numbers quickly in a laboratory setting.

Pheromones have been particularly well studied in ticks (Sonenshine, 1991). The elaborate system, only hinted at here, serves to avoid wasteful, interspecific matings among sympatric genera which share the same hosts. The two major types are (i) aggregation/attachment pheromones (AAPs) and (ii) sex pheromones.

The AAPs are volatile substances, produced only by feeding males, that attract unfed males and females. They are a cocktail of phenols, so far detected only in *Amblyomma* species. Together, they excite searching, aggregation, attachment, and clasping behaviors. Unfed *A. hebraeum* females can detect steers which harbor feeding males from as far as 3 or more meters away. Because of this ability, females tend to choose hosts which are infested with males.

The sex pheromones are of three types: (i) attractant sex pheromone (ASP), (ii) mounting sex pheromone (MSP), and (iii) genital sex pheromone (GSP). The ASP (e.g., 2,6-dichlorophenol), released from the foveal glands of feeding females, attracts fed males. The MSP (cholesteryl oleate in *Dermacentor variabilis*), probably secreted by the dermal glands of the female, excites the male to mount in the correct orientation for copulation. The GSP, so far identified only in the genus *Dermacentor,* is the most species specific of the three. It is a mixture of numerous long-chain fatty acids and includes an ecdysteroid. It is probably secreted by the gonadal accessory gland of the female and is detected by sensilla on the cheliceral digits of the male.

Laid eggs are coated with a waxy material secreted by Gené's organ, an unusual structure found exclusively in ticks (Fig. 5). As each egg is released by

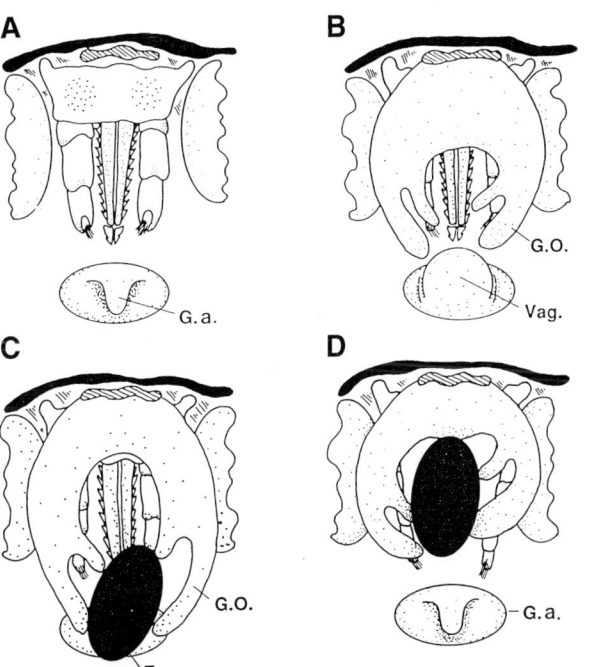

FIGURE 5 The function of Gené's organ, the egg-waxing organ of ticks. (A) The capitulum flexes ventrally toward the genital aperture (G.a.) in preparation to receive an egg. (B) The glandular portion of Gené's organ (G.O.) everts through the camero-stomal fold and vagina (Vag.) prolapses through genital aperture. (C) The egg (E.) emerges and is taken by Gené's organ aided by the mouthparts. (D) The egg is coated with the lipid secretion of the glandular tissue and is passed dorsally. From Sonenshine (1991), reproduced with permission of Oxford University Press.

prolapse of the vestibular vagina, it is taken by the mouthparts and transferred to lobes of Gené's organ which have protruded through the camerostomal cavity. Eggs deprived of this waterproofing secretion dry out very quickly (Sonenshine, 1991). Resistance to desiccation is probably the result of (i) the direct waterproofing property of the secretion and (ii) the eggs sticking together so as to minimize surface-to-volume ratio of the egg mass as a whole.

Chemical messengers other than hormones play a major role in reproduction. In insects, the spermatophore contains "fecundity-enhancing substances" and "receptivity-inhibiting substances," quite possibly the same material in a given species. These are all proteins (Gillott, 1988). Likewise, injection of spermiophores into the hemocoel of fed virgin *Ornithodoros moubata* stimulates both vitellogenesis and oviposition. The active substance is a very large protein. It acts either (i) directly on the gut to stimulate digestion and transport of protein, (ii) on the fat body cells to stimulate yolk synthesis, or (iii) on the endocrine system (Connat *et al.*, 1986). At least three other "mating factors" have been identified in ticks. Both ixodid and argasid ticks use a peptide, produced by the male accessory gland, which promotes sperm maturation. An "engorgement factor" from the male gonad, transferred to the female at copulation, induces her to complete the meal rapidly. A "male factor," likewise transferred to the female at copulation, hastens the release of the ecdysteroid hormone which induces salivary gland degeneration and possibly vitellogenesis; the latter two may be the same protein, but they are distinct from the sperm maturation peptides. We are generally ignorant about the modes of action of these mating factors.

i. Family Argasidae Adult female argasids can feed numerous times and lay a clutch of eggs following each meal. Clutch size is about 100–150 eggs in early gonotropic cycles but falls to about 50 eggs in later cycles. There is evidence that JH stimulates oocyte development in fed argasid ticks. Topical application of JH or JH analogs break reproductive diapause and induce oviposition in fed virgins. On the other hand, JH and related analogs are totally ineffective in stimulating yolk synthesis and ovarian development in unfed *O. moubata*, even though unfed specimens are capable of responding to other exogenous stimuli. Likewise, precocenes 1 and 2 did not reduce yolk titers in the hemolymph nor inhibit the oocyte development that occurs under a variety of experimental regimes that stimulate egg development and oviposition. How feeding modulates sensitivity to JH has not yet been determined.

It is possible to ligate argasid ticks after feeding in such a way as to isolate a hemocoel compartment bathing the synganglion from one bathing the ovary. From such experiments, one can show that blood feeding triggers the release of a factor in the anterior region of the hemocoel which stimulates ovarian development. This factor, a peptide, comes from the synganglion and has been named vitellogenesis-inducing factor (VIF). VIF acts somewhere in the posterior part of the tick to produce a fat body stimulating factor. The latter stimulates yolk synthesis. Its chemical nature and source in the posterior of the tick is unknown. It might be an ecdysteroid based on preliminary experiments which suggest that ecdysone and 20E increase Vg titer in the hemolymph of unfed mated females. This is another instance in which chelicerates may display at least a superficial similarity to the dipteran reproductive model (Hagedorn, 1994).

ii. Family Ixodidae Female ixodid ticks are the gluttons of the animal kingdom. They imbibe well over 200 times their unfed weight in host blood during a 7- to 10-day feeding period. With the exception of the genus *Ixodes*, mating occurs only while on the host, and copulation is required to stimulate full engorgement. Oviposition begins within 2 weeks of engorgement. There is only a single gonotropic cycle, after which the female dies. However, clutch size is enormous; in large genera (*Hyalomma* and *Amblyomma*) 15,000–20,000 eggs are laid.

JH and ecdysteroids are probably involved in reproduction, but what they actually do is poorly understood. For example, in one study, hexane extracts of vitellogenic *Boophilus microplus* possessed juvenilizing activity in two bioassays, and hemolymph extracts from such females reacted with antisera created against JHI, JHII, and JHIII. However, such extracts did not unambiguously share the spectra of JH0, JHI, JHII, JHIII, or methylfarnesoate derived from analysis by gas chromatography followed by mass spectrometry (reviewed by Kaufman, 1997). JH may be a natural hormone in *D. variabilis* and *H. dromedarii*, but its role in reproduction has not yet been defined (Sonenshine, 1991).

The idea that an ecdysteroid controls vitellogenesis in ixodid ticks arises from the following observations. Hemolymph ecdysteroid titer rises 10- to 100-fold during the first week postengorgement, i.e., during the period when yolk synthesis is most rapid, and 20E stimulates isolated fat body to synthesize yolk and release it into the culture medium; the JH mimic methoprene has no such effect. However, concerted attempts to elicit oocyte development in partially fed female *A. hebraeum* with a variety of hormone treatments involving JH, JH mimics, or ecdysteroids have so far failed (Kaufman, 1997).

Ecdysteroids may stimulate spermatogenesis. One can correlate spermatocyte differentiation with ecdysteroid titers reported in nymphs, and germ cell DNA synthesis is stimulated by injecting 20E into unfed males. There is also some indication for the involvement of 20E in spermatogenesis for argasids (Sonenshine, 1991).

V. CONCLUDING REMARKS

Relative to what is known about some insects and crustaceans, our understanding of the reproductive physiology of chelicerates is still rather meager. However, it seems likely that this ancient group of arthropods will teach us important things about the evolution of reproduction in the phylum as a whole. First, however, more zoologists must be willing to take notice of this major, but sadly neglected, group of organisms.

Bibliography

Connat, J.-L., Ducommun, J., Diehl, P. A., and Aeschlimann, A. A. (1986). Some aspects of the control of the gonotrophic cycle in the tick *Ornithodoros moubata* (Ixodoidea, Argasidae). In *Morphology, Physiology and Behavioural Biology of Ticks* (J. R. Sauer and J. A. Hair, Eds.), pp. 194–216. Ellis-Horwood, Chichester, UK

Diehl, P. A., Aeschlimann, A. A., and Obenchain, F. D. (1982). Tick reproduction, oogenesis and oviposition. In *The Physiology of Ticks* (F. D. Obenchain and R. Galun, Eds.), pp. 277–350. Pergammon, Oxford, UK

Evans, G. O. (1992). *Principles of Acarology*. CAB International, Wallingford, UK.

Foelix, R. F. (1982). *Biology of Spiders*. Harvard Univ. Press, Cambridge, MA.

Gillott, C. (1988). Arthropoda—Insecta. In *Reproductive Biology of Invertebrates, Volume III, Accessory Sex Glands* (K. G. Adiyodi and R. G. Adiyodi, Eds.), pp. 319–471. Wiley, Chichester, UK.

Hagedorn, H. H. (1994). The endocrinology of the adult female mosquito. In *Advances in Disease Vector Research* (K. F. Harris, Ed.), Vol 10, pp. 109–148. Springer-Verlag, Berlin/New York.

Kaufman, W. R. (1997). Arthropoda—Chelicerata. In *Reproductive Biology of Invertebrates* (K. G. Adiyodi and R. G. Adiyodi, series Eds.), *Volume VIII, Progress in Reproductive Endocrinology* (T. S. Adams, Ed.). Oxford & IBH.

Keegan, H. L. (1980). *Scorpions of Medical Importance*. Univ. Press of Mississippi, Jackson.

King, P. E. (1973). *Pycnogonids*. Hutchinson, London.

Legg, G., and Jones, R. E. (1988). *Pseudoscorpions*. Brill/Backhuys, Leiden.

Polis, G. A. (1990). *The Biology of Scorpions*. Stanford Univ. Press, Stanford, CA.

Savory, T. H. (1977). *Arachnida*, 2nd ed. Academic Press, London.

Sekiguchi, K. (1988). *Biology of Horseshoe Crabs*. Science House, Tokyo.

Sonenshine, D. E. (1991). *Biology of Ticks*, Vol. 1. Oxford Univ. Press, New York.

Weygoldt, P. (1969). *The Biology of Pseudoscorpions*. Harvard Univ. Press, Cambridge, MA.

Chemical Communication

see Pheromones

Chickens, Control of Reproduction in

Peter J. Sharp

Roslin Institute

I. Introduction
II. The Female Reproductive System
III. The Male Reproductive System
IV. Fertilization and Embryogenesis
V. Sexual Maturation and Reproductive Behavior
VI. Nutrition and Photoperiodism
VII. Commercial Reproductive Problems

GLOSSARY

albumen Egg white, a mixture of more than 40 proteins secreted by the oviduct.
altricial Hatched well developed and able to feed independently.
broiler A chicken selected for rapid growth and meat production.
broiler breeder Parents of chickens bred for meat production.
broodiness Female maternal behavior including incubating eggs and caring for young.
cloaca The terminal part of the gut into which the kidney and reproductive ducts open.
clutch Eggs accumulated in a nest prior to incubation; it may comprise one or several sequences.
commercial egg layer Light body-weight chickens bred to lay eggs for consumption.
ductus deferens The male reproductive tract.
egg A calciferous shelled body containing a yolky ovum and albumen.
incubation The act of sitting on eggs to provide the correct environment for embryo development; provision of an artificial environment for eggs to hatch.
oviduct The female reproductive tract; lays down albumen and shell around the yolky ovum.
oviposition Expulsion of a hard-shelled egg from the female reproductive tract.
sequences Series of eggs laid by a hen on successive days separated by "pause days" on which no egg is laid.
table egg An egg produced for human consumption.
yolk A hen's ovum filled with yellow lipoprotein; lipoprotein is synthesized in the liver under the influence of estrogen and is taken up from the bloodstream to be deposited in the maturing ovum.

The domestic chicken (*Gallus domesticus*) is a form of wild jungle fowl, originating from Southeast Asia. Reproduction in the chicken involves the production and fertilization of large yolk-filled ova which are packaged within the oviduct into hard-shelled eggs. After oviposition, eggs are incubated for 21 days until the chicks hatch. Newly hatched chicks are altricial but require brooding until they become fully self-sufficient. Reproductive activity is influenced by genotype, nutrition, and changes in day length. Commercial problems center on optimizing nutrition and lighting patterns for maximum egg production and quality. Selection for growth rate in broilers creates problems of poor egg production and fertility.

I. INTRODUCTION

Chickens (*Gallus domesticus*) are medium-sized Gallinaceous birds in the same family (*Phasianidae*) as pheasants and quails. They are domesticated forms of wild jungle fowl from Southeast Asia, weighing 1–5 kg, with males being larger than females. Domestication has been for recreational purposes (e.g., plumage color and fighting prowess) and for egg and meat consumption. Early records of domestication are from 6000 BC in southeast China. Cockerels (males) have larger combs, longer neck and tail feathers, and better developed spurs than hens (females). In nonwhite breeds cockerels have brightly colored plumage, whereas the female plumage is drab. The sex differences in plumage and spur size are controlled by estrogen, originating from the ovary, whereas the sex difference in comb size is controlled by testosterone, originating from the testes.

Approximately 24% of the world's meat consumption is derived from poultry, and table egg production is about 280×10^9/year. During the late nineteenth century, a large number of chicken breeds were developed for exhibition purposes. In the twentieth century, new techniques in quantitative genetics exploited these breeds to produce synthetic commercial lines for table egg or meat production (Fig. 1). A commercial hen kept for table eggs produces about 310 eggs, whereas a broiler breeder (meat-type bird) produces about 160 eggs in a laying year. The breeding of broilers typically involves two male and two female grandparent lines selected for commercial traits. These lines are crossed to produce hybrid male and female parent lines. Similar breeding strategies are used to produce synthetic lines of birds for table egg production.

The domestic chicken has 78 chromosomes (i.e., 39 pairs). Of these, the largest 6 pairs are in the same size range as human chromosomes. The remainder are much smaller. The karyotypes of cockerels and hens differ by the presence of a heteromorphic pair of chromosomes. The hen is the heterogenetic sex with ZW sex chromosomes, whereas the male is the homogenetic sex with ZZ sex chromosomes. The Z chromosome is the fifth largest in size and many genes have been mapped onto it. In contrast, the W chromosome is very small and contains a repeated sequence of noncoding DNA.

II. THE FEMALE REPRODUCTIVE SYSTEM

The structure and function of the hen's reproductive system reflects the requirement to produce hard-shelled eggs which are then incubated until the chicks hatch. In commercial chickens, eggs are incubated artificially. In contrast with mammals, a corpus luteum does not form from the chicken postovulatory follicle since there is no requirement to maintain a pregnancy. The laying hen is therefore always in a reproductive state equivalent to the follicular phase of the mammalian estrous cycle. Since the chick embryo develops outside the hen's body, nutrients for embryogenesis are provided in the form of yolk deposited within the ovum. The hen's ovary, when fully developed, contains large yolky follicles and occupies a substantial volume of the body cavity. To accommodate the large size of the developed ovary, the development of the right ovary is suppressed, together with the development of the right oviduct.

A. The Hypothalamic–Pituitary–Ovarian Axis

Reproductive activity in the hen is ultimately controlled by about 1000 neurons in the preoptic–

FIGURE 1 The two main types of commercial chicken (top) and their ovaries (bottom). The smaller commercial egg layer (top left), bred to produce table eggs, has an ovary with a well-organized yellow-yolky follicular hierarchy (bottom left) with successive follicles due to ovulate on successive days. The larger broiler breeder (top right), bred to produce meat-type chickens (broilers), has an overdeveloped ovary with a disrupted yellow-yolky follicular hierarchy (bottom right). The broiler breeder must be fed a restricted diet to reduce the excessive numbers of yellow-yolky follicular hierarchy and to allow an acceptable rate of egg production (courtesy of Dr. P. M. Hocking, Roslin Institute, Edinburgh).

anterior hypothalamus producing the decapeptide, gonadotropin-releasing hormone-I (GnRH-I). The activity of these neurons is influenced by age, environmental factors such as the photoperiod and nutrients, and by the inhibitory and stimulatory actions of ovarian steroids. GnRH-I is released from the median eminence into a hypophysial–portal vascular system to stimulate the release of luteinizing hormone (LH) and follicle-stimulating hormone (FSH) from the anterior pituitary gland. Luteinizing hormone stimulates ovarian steroidogenesis, whereas FSH stimulates ovarian follicular growth. The secretion of LH is controlled by the inhibitory and stimulatory effects of ovarian steroids, whereas the secretion of FSH is probably controlled, at least in part, by ovarian inhibin.

B. The Ovary

1. Gross Anatomy

The fully developed hen's ovary weighs about 45–55 g and the most prominent feature is five to seven yellow-yolky preovulatory follicles, arranged in a hierarchy ranging between 10 and 35 mm in diameter (Fig. 1). These are attached to the ovary by follicular stalks and are destined to ovulate on successive days separated by occasional days on which ovulation does not occur. The resulting eggs are therefore laid in a series of "sequences" in which eggs are laid daily, interrupted by pause days on which no egg is laid. Yellow-yolky preovulatory follicles enter a rapid growth phase about 9 days before ovulation. The major components of yolk are synthesized in the liver in response to increased plasma estrogen and are transported to the ovary in the blood. Yolk is actively transported from the blood into the growing follicle by a specific receptor-mediated mechanism. If the preovulatory follicles are removed, the remaining ovary contains up to about 3000 visible follicles divided into three classes according to size and color: small yellow follicles (4–10 mm diameter), large white follicles (2–4 mm diameter), and small white follicles (<2 mm in diameter). The latter are the most abundant. Atresia is common among small yellow and large white follicles and is probably part of the mechanism which selects follicles destined to enter the large yellow-yolky preovulatory phase of development. The ovary of a laying hen contains several postovulatory follicles which gradually shrink back into the ovarian stroma over a few days. The postovulatory follicle does not develop luteal cells as in mammals.

2. Endocrine Secretions

The developing ovum is surrounded by several tissue layers, including steroidogenic granulosa and thecal tissues. About 50% of the estrogens secreted by the ovary are produced by thecal tissue surrounding small and large white follicles using 17α-hydroxypregnenolone, produced by the adjacent granulosa cells, as substrate. Thecal cells also produce androgens but not progesterone. During the final rapid yellow-yolky follicular growth phase, thecal tissue progressively loses its capacity to produce estrogen but continues to produce androgens, whereas the granulosa layer progressively acquires the ability to produce progesterone. The ability of the granulosa cells to produce progesterone is held in check by either androgens or estrogen produced by the theca or because progesterone produced by the granulosa is metabolized by the theca. This change in the pattern of steroidogenesis in the granulosa layer is associated with a decrease in FSH responsiveness and an increase in LH responsiveness.

Inhibin is produced by granulosa cells in yellow-yolky follicles and the amount produced per milligram protein decreases during follicular maturation. Its production is controlled by FSH and it is believed to act in a paracrine manner to regulate ovarian function.

3. The Ovulatory Cycle

Ovulation in the hen is preceded by a surge of plasma LH and progesterone but not of FSH. This preovulatory surge lasts about 8 hr and is generated by a stimulatory action of progesterone, originating from the preovulatory follicles, on gonadotropin-releasing hormone (GnRH)-I release. Unlike mammals, in the hen estrogen does not induce preovulatory release of LH. The preovulatory release of LH and progesterone is responsible for the rupture of the wall of the largest preovulatory follicle. Rupture occurs along an avascular strip in the follicle wall known as the "stigma." Immediately before ovulation, proteolytic enzyme and collagenase activity increases in the stigma, resulting in protein and collagen degradation. This causes the follicle wall to weaken and rupture and to release the ovum.

The pause day separating sequences of eggs is caused by a neuroendocrine mechanism which permits the preovulatory release of LH and progesterone to occur only during a 9- or 10-hr "open period" each day. This open period is controlled by a circadian mechanism and in hens entrained to a 24-hr lighting cycle it occurs every 24 hr. In hens producing mature ovulable follicles at intervals of more than 24 hr, some follicles occasionally mature after the end of an open period. Consequently, ovulation is delayed until the arrival of the next open period.

C. The Oviduct

The oviduct transports the newly ovulated ovum to the external world, and on the way surrounds it with albumen and a hard shell. The oviduct in a laying hen weighs about 48 g and is about 74 cm long. Its development and maintenance depend on increased blood estrogen and progesterone. Removal of these steroids results in the complete regression of the oviduct to a structure weighing about 0.2 g. The oviduct has five segments, each with a different function. The infundibulum is a funnel-shaped structure at the end of the oviduct nearest the ovary. The infundibulum actively engulfs the ovulating follicle and is the site of fertilization. The infundibulum leads to the magnum, which is the longest section of the oviduct (about 37 cm). The magnum secretes egg albumen which coats the ovum. The most abundant proteins in egg albumen are ovalbumin (54%), which contains all essential amino acids, ovotransferrin (12%), and ovomucoid (11%), a protease inhibitor. The isthmus forms the next section of the oviduct and secretes shell membranes around the albumen layer. The isthmus leads to the shell gland, in which water is taken up by the albumen in a process called "plumping" and the shell, composed of calcium carbonate, is deposited on the egg membrane. The formation and deposition of calcium carbonate in the shell gland is controlled by estrogens, vitamin D, and parathyroid hormone. About 2 or 3 g of calcium carbonate is required to form an eggshell. The calcium ions in the deposited calcium carbonate are derived from food in the gut and from the bones. The carbonate anions are catalyzed by carbonic anhydrase in the shell gland from carbon dioxide, obtained through respiration, and water. The final part of the oviduct is a short vagina. Its junction with the shell gland contains approximately 25,000 sperm storage tubules.

After ovulation, the ovum spends about 15 min in the infundibulum, 3 hr in the magnum, 1.5 hr in the isthmus, 19–29 hr in the shell gland, and 1–10 min in the vagina.

D. Oviposition

Oviposition is induced by contractions of the shell gland, vagina, and abdominal muscles. It is associated with increases in plasma prostaglandin E_2 and prostaglandin $F_{2\alpha}$, followed by an increase in plasma arginine vasotocin. Prostaglandin $F_{2\alpha}$ and arginine vasotocin are potent stimulators of smooth muscle contraction, and prostaglandin E_2 causes relaxation of the vaginal sphincter. Prostaglandin $F_{2\alpha}$ originates from the largest pre- and postovulatory follicles and its release is stimulated during the preovulatory surge of LH and progesterone. Arginine vasotocin is believed to be released from the posterior pituitary gland.

III. THE MALE REPRODUCTIVE SYSTEM

The structure of the cockerel's reproductive system differs in several ways from that in most mammals. The paired testes are internal and spermatogenesis occurs at the normal avian body temperature of 41°C. There are no accessory reproductive glands, such as the prostate or seminal vesicles, and there is no intromittent organ. Semen is stored at the lower end of the ductus deferens rather than in the epididymis.

A. The Hypothalamic–Pituitary–Testicular Axis

Reproductive activity in the cockerel is controlled by GnRH-I neurons in the preoptic anterior hypothalamus. The activity of these neurons is influenced by age, photoperiod, nutrition, and the inhibitory actions of testicular steroids. GnRH-I released from the median eminence stimulates LH and FSH secretion from the anterior pituitary gland. LH stimulates androgen production by the Leydig cells, but the function of FSH in the testis is less well established. It is likely to act on Sertoli cells to stimulate estrogen and inhibin production and it is assumed to stimulate the growth of the seminiferous tubules.

B. The Testes and Reproductive Tracts

In the sexually mature cockerel, the combined testes weight is about 1% of body weight (25–35 g). Each testis contains a network of seminiferous tubules with Leydig cells interspersed between them. The tubes have a large diameter and high fluid content and are surrounded by blood vessels. Spermatogenesis and spermatozoa maturation is completed in 13 or 14 days and daily spermatozoa production is about $8–120 \times 10^6$ per gram of testis. Spermatozoa are shed into the lumen of the seminiferous tubules where they are suspended in a fluid (seminiferous tubule fluid) secreted by Sertoli cells. The suspended spermatozoa are immotile and are transported out of the testis by hydrostatic pressure generated by myoepithelial cells surrounding the seminiferous tubules. The spermatozoa suspended in seminiferous tubule fluid collect in cavernous channels, the rete testes, before entering a small epididymis. This structure includes efferent, connecting, and epididymal ducts. A substantial portion of the epididymis contains efferent ducts, which absorb 90% of the seminiferous tubule fluid flowing into them. As a consequence, the concentration of spermatozoa leaving the testis is increased about 60-fold during passage through the epididymis.

The ducts in the epididymis secrete proteins which are believed to contribute to the potential of spermatozoa to become motile. In mammals, epididymal proteins are also responsible for capacitation, the process whereby spermatozoa acquire fertilizing capacity. Capacitation is not required for chicken spermatozoa to acquire fertilizing capacity.

Spermatozoa pass through the epididymis in about 1.5 hr and enter the ductus deferens suspended in seminal plasma. This semen reaches the lower end of the ductus deferens after a further 22–24 hr, propelled by peristalsis. About 90% of extratesticular spermatozoa are stored in the ductus deferens for up to 20–30 days in cockerels not allowed to copulate.

C. Ejaculation

Each ductus deferens opens into the cloaca as a laterally positioned papillus. The ventromedial lip of the cloaca is composed of a median phallic body with a groove, bounded by lateral phallic folds. In a sexually excited cockerel, these phallic structures become engorged by lymph flowing from the adjacent paracloal vascular body and protrude from the cloaca. At ejaculation, the semen flowing from the reproductive tracts is guided through the medial groove of the phallic body to the exterior world. A lymph-like fluid, known as "transparent fluid," from the phallic structures is added to the ejaculated semen. This transparent fluid is thought to play a role in stimulating spermatozoa into motility, thus ensuring that they travel through the hen's vagina into the oviduct.

A sexually active cockerel may mate up to 20–30 times/day but not all copulations end in ejaculation. A cockerel kept in isolation for 2 days and then allowed to copulate produces a first ejaculate of 0.3 or 0.4 ml containing 2 or 3×10^9 spermatozoa/ml. Subsequent ejaculates on the same day are smaller with lower concentrations of spermatozoa.

D. Spermatozoa Storage in the Oviduct

After a single natural mating, spermatozoa deposited in the oviduct will result in fertile egg production for about 14 days. After deposition in the vagina, spermatozoa are actively attracted to enter sperm storage tubules located at the junction of the shell gland and vagina. The mechanism responsible for this attraction is unknown. A sphincter at the junction of the shell gland and vagina prevents the transport of nonmotile spermatozoa into the oviduct. Spermatozoa stored in the sperm storage tubules are progressively released and subsequently travel approximately 60 cm up the oviduct to the site of fertilization in the infundibulum. Spermatozoa survive for long periods in clefts and shallow tubule glands in the infundibulum, thus increasing the chances of successful fertilization.

IV. FERTILIZATION AND EMBRYOGENESIS

The newly ovulated ovum is fertilized in the infundibulum. During its passage down the oviduct, embryogenesis is initiated, transforming the blastodisc of the unfertilized ovum into the blastoderm of the fertilized ovum. In a new-laid fertilized egg, the blastoderm contains 40,000–60,000 cells.

A. Fertilization

The maternal chromosomes of the unfertilized ovum are located on the surface of the yolk in a white spot, 3.5 mm in diameter, called the germinal disc. About 1 hr before ovulation, the first meiotic division begins and the first polar body is extruded from the pronucleus. The female pronucleus then progresses to the second meiotic division and at ovulation is in anaphase. Within 15 min of ovulation, several spermatozoa penetrate the perivitelline layer and are transformed into spheroidal male pronuclei. At about the same time, the female pronucleus extrudes the second polar body and after about 4 hr fuses with one of the male pronuclei. This is followed by the first meiotic division. The remaining male pronuclei move to the periphery of the germinal disc.

Within 15 min after ovulation, after several spermatozoa have penetrated the perivitelline layer above and around the germinal disc, an outer perivitelline layer is deposited by the infundibulum. This prevents further spermatozoa from reaching the ovum.

B. Embryogenesis and Germline Formation

A fluid-filled cavity develops under the blastoderm (the developing embryo) as cell division proceeds. This makes the central area of the blastoderm appear translucent in the newly laid fertile egg and is referred to as the "area pellucida." After the egg is laid, further embryo development is arrested at temperatures below 10–15°C. An increase in temperature, to that under an incubating hen, stimulates the subsequent stages of embryogenesis.

About 18 hr after an egg begins to be incubated, germline cells, called "primordial germ cells," appear in the germinal crescent, anterior to the head of the embryo. During the next 18 hr the primordial germ cells migrate laterally into the extraembryonic membranes where they enter blood vessels as the vascular system forms. After about 33 hr, the primordial germ cells begin to migrate to the bilateral genital ridges via ameboid movement. This migration of primordial germ cells continues for up to 12 or 13 days of incubation in males and up to 8 days of incubation in females. Sexual differentiation occurs between 5 and 7 days of incubation when the primordial germ cells become incorporated into the developing primary sex cords.

V. SEXUAL MATURATION AND REPRODUCTIVE BEHAVIOR

Both cockerels and hens become sexually mature at 5–7 months of age depending on genotype, nutrition, and lighting pattern. The onset of egg laying in hens is not influenced by the presence or absence of sexually active cockerels. Conversely, the onset of semen production in males is not influenced by the presence of hens.

Reproductive behavior in the chicken is directed to successful mating, nesting, incubation, and brooding of young.

A. Mating Behavior

The ancestors of domestic chickens were probably polygamous, and the process of domestication has resulted in promiscuous sexual behavior in both cockerels and hens. In commercial practice, breeding flocks are maintained with a ratio of about 1 cockerel: 10–15 hens.

The onset of sexual maturation is preceded by gonadal development stimulated by increased gonadotrophin secretion and by increased secretion of gonadal steroids. In males increased testicular steroid secretion stimulates comb growth, crowing, aggressive behavior toward other males, and courtship behavior. Aggressive behavior is particularly expressed if sexually mature or maturing males are introduced to males they have not previously encountered. Courtship is initiated by males with a "waltz," in which one wing is dropped and the hen is approached with shuffling side steps. Associated behavior includes "tidbitting," in which a cockerel attracts the attention of a hen by pecking or scratching on the ground while emitting a characteristic food call.

In females, at the onset of sexual maturation, increased secretion of gonadal steroids stimulates comb growth and "squatting" behavior which allows the male to mount and copulate successfully. This behavior occurs several days before the first egg is laid and is induced by estrogen.

B. Nesting, Incubation, and Brooding Behavior

Hens show nesting behavior before each oviposition. This involves a reduction in eating and drinking and an increased interest in potential, or established, nesting sites. If given the opportunity, a hen will enter a nest box and sit for 1 or 2 hr before each oviposition. After the egg is laid, the hen cackles for several minutes. When kept in battery cages nesting behavior is expressed as stereotyped pacing for about 1 hr after oviposition. Nesting behavior is controlled by increased plasma estrogen and progesterone and probably by the increase in plasma prostaglandins and arginine vasotocin which precede oviposition. After a clutch of eggs has accumulated, tactile information from these eggs is thought to trigger the release of prolactin from the anterior pituitary gland. Increased plasma prolactin secretion transforms nesting behavior into incubation behavior, which involves almost continuous sitting in the nest.

After the onset of egg production, there is a progressive increase in a tendency to express incubation behavior. This is enhanced by the presence of a nest site and a quiet, secure environment. If hens are allowed to incubate eggs, the chicks hatch after 21 days incubation. Hens brood their newly hatched chicks by providing warmth and shelter, and they aggressively defend them against attack by intruders.

VI. NUTRITION AND PHOTOPERIODISM

Reproductive function depends on the birds being in good health with adequate nutrition. The timing of the onset of reproductive activity and its overall intensity is influenced by day length.

A. Nutrition

The onset of sexual activity involves an interaction between age and a minimum mature body composition and weight. Chickens fed to grow as fast as possible may exceed the minimum mature body weight required to support reproductive activity before they reach an age which allows them to become sexually mature. Hens can be kept permanently out of breeding condition by maintaining them below the minimum mature body weight.

The nutritional requirements of the hen change dramatically 10–14 days before the onset of egg production. During this period the ovary and oviduct develop rapidly, the liver increases in size to accommodate yolk production, and calcium stores are laid down in medullary bone for eggshell formation. The hen's appetite increases during this period to ensure a greater intake of the nutrients required to support egg production.

B. Photoperiodism

Domestic chickens are photoperiodic, and when exposed to natural seasonal changes in day length they stop breeding in response to the shortening day lengths of the fall. During the winter months this inhibitory effect of short day lengths weakens and egg laying and semen production starts again in early spring while day lengths are still short. The photoperiodic mechanism involved has been extensively explored in nondomesticated seasonally breeding photoperiodic birds. Prolonged exposure to long days induces an inhibitory input to the reproductive neuroendocrine system, which in wild birds may result in the development of reproductive photorefractoriness. In the chicken, the development of this long-day-induced inhibitory input to the reproductive neuroendocrine system only becomes apparent when the photoperiod is reduced after a prolonged period of photostimulation. Subsequent prolonged exposure of chickens to short days dissipates the inhibitory effects of long days and reproductive activity resumes.

Domestic chickens become sexually mature at 5–7 months when reared on either constant short or long days. However, from as early as 4 weeks of age, they are photoresponsive. Transfer of chicks reared on short days to long days at 4 weeks of age or older stimulates increased LH secretion. An increase in photoperiod from short days between about 8 and 14 weeks of age results in a significant advance in the onset of sexual maturation. The effects of photostimulation before or after these ages are less effective in advancing sexual maturation. In chickens reared on long days, transfer to short days after about 8 weeks of age results in a significant delay in the onset of sexual maturation. Before this age the effects of decreasing day length on the rate of sexual maturation are less pronounced.

In chickens reared on short days, and not yet sexually mature, the minimum or "critical day length" required to stimulate increased LH secretion is 10 or 11 hr/day. Photoperiods in excess of 12 hr/day do not increase LH secretion further.

In husbandry systems using light-controlled housing, it is customary to rear chickens on short days and to photostimulate them at about 18 weeks of age. The objective of this procedure is to advance and synchronize the onset of lay and to ensure maximum total egg output during a laying year. After prolonged exposure to long photoperiods, egg production begins to decrease. This is due to several factors, including age and metabolic exhaustion, but it can be ameliorated to some extent by further increases in photoperiod up to a maximum of about 17 or 18 hr. It is for this reason that when chickens are first brought into breeding condition by photostimulation it is preferable not to increase the photoperiod beyond 12 hr. This allows for increases in photoperiod during the laying year in order to further stimulate egg production.

VII. COMMERCIAL REPRODUCTIVE PROBLEMS

Major objectives of commercial poultry breeders are to improve the efficiency of food conversion into eggs, improve persistency of laying and total output of eggs, maintain fertility, and prevent the deterioration of eggshell quality toward the end of the laying year. These problems are approached through selective breeding and manipulation of dietary composition and food intake and lighting patterns.

Broiler breeders present management problems which are a consequence of selection for rapid growth rate. A 6-week-old broiler, of marketable age, is the same weight (about 2.3 kg) as a sexually mature hen bred for table egg production (Fig. 1). If growth is left unchecked, broilers are massively obese at 5–7 months of age when they become sexually mature. This obesity results in a loss of fertile egg production. The females lay only small numbers of eggs and the weight of the males is so great that successful natural mating becomes difficult. These problems are solved by dietary restriction which must be applied in a carefully programmed manner during juvenile growth and after the onset of puberty.

The main effect of selection for growth rate in broiler breeder hens is to increase massively the number of yellow-yolky ovarian follicles at the onset of lay (Fig. 1). Instead of developing in a well-ordered hierarchy, with one follicle developing each day, follicles develop as pairs or triplets. This results in multiple ovulation with two or three ova entering the oviduct after each preovulatory LH surge. If two or three newly ovulated ova enter the oviduct around the same time, they may be processed into a double- or triple-yolked egg. Such eggs are rejected by the hatcheries because chicks rarely hatch out of them. More commonly, multiple ovulations result in two or three ova entering the oviduct at intervals of 3 or 4 hr. In this case, the oviduct expels membraneous soft-shelled eggs or a mixture of soft-shelled and membraneous eggs. As laying broiler breeder hens become older, the incidence of multiple ovulation decreases in association with a rapid loss of the capacity to develop yellow-yolky follicles and the development of ovarian atresia. This loss of reproductive function in broiler breeder females parallels a similar decrease in the production of spermatozoa in aging broiler breeder males.

Bibliography

Bakst, M., Wishart, G., and Brillard, J. P. (1994). Oviducal sperm selection, transport and storage in poultry. *Poultry Sci. Rev.* **5**, 117–143.

Crawford, R. D. (Ed.) (1990). *Poultry Breeding and Genetics*. Elsevier, Amsterdam.

Etches, R. J. (1996). *Reproduction in Poultry*. CAB International, Wallingford, UK.

Gilbert, A. B. (1979). Female genital organs. In *Form and Function in Birds* (A. S. King and J. McLelland, Eds.), Vol. 1, pp. 237–360. Academic Press, New York.

Howarth, B. (1995). Physiology of reproduction. The male. In *Poultry Production* (P. Hunter, Ed.), World Animal Science C9, pp. 243–270. Elsevier, Amsterdam.

Johnson, A. L. (1996). The avian ovarian hierarchy: A balance between follicle differentiation and atresia. *Poultry Avian Biol. Rev.* **7**, 9–119.

Johnson, P. (1997). Avian inhibin: *Poultry Avian Biol. Rev.* **8**, 21–31.

Lake, P. E. (1980). Male genital organs. In *Form and Function in Birds* (A. S. King and J. McLelland, Eds.), Vol. 2, pp. 1–62. Academic Press, New York.

Lewis, P. D., and Perry, G. C. (1995). Effects of lighting on reproduction in poultry. In *Poultry Production* (P. Hunter, Ed.), World Animal Science C9, pp. 359–389. Elsevier, Amsterdam.

Norgren, R. B., Jr. (1996). Development of LHRH neurons in the avian embryo: *Poultry Avian Biol. Rev.* **7**, 163–182.

Sharp, P. J. (1996). Strategies in avian breeding cycles: *Animal Reprod. Sci.* **42**, 505–513.

Chimpanzees

see Primates, Nonhuman

Choriocarcinoma

Anthony C. Evans, Jr.

University of Utah

I. Disease Characteristics
II. Epidemiologic Characteristics
III. Clinical Manifestations
IV. Persistent GTD Following Molar Evacuation
V. Metastatic GTD
VI. Diagnostic Aids
VII. Atypical Presentation
VIII. Staging and Treatment
IX. Prognosis
X. Posttreatment

GLOSSARY

gestational choriocarcinoma A malignancy that arises from trophoblastic tissue of term pregnancies, ectopic gestations, spontaneous or induced abortions, or molar elements and contains malignant cytotrophoblasts and syncytiotrophoblasts.

gestational trophoblastic disease A spectrum of disease characterized by abnormal trophoblastic proliferation and including hydatidiform moles, invasive mole, and choriocarcinoma.

human chorionic gonadotropin A glycoprotein hormone liberated by trophoblastic tissue that serves as a sensitive marker of tumor burden among patients with gestational trophoblastic disease.

hydatidiform mole Abnormal placental tissue exhibiting trophoblastic hyperplasia and hydropic chorionic villi; complete hydatidiform moles are androgenetic, diploid, and lack fetal elements; partial hydatidiform moles are triploid, contain both maternal and paternal contributions to the genome, and may have fetal elements present.

invasive mole Molar placental tissue that invades into the wall of the uterus and has the potential to metastasize.

Choriocarcinoma is a trophoblastic malignancy that is the most virulent form of the spectrum of ailments known as gestational trophoblastic disease. This article outlines the most common clinical manifestations, epidemiologic characteristics, diagnostic aids, treatment strategies, and pathology of choriocarcinoma.

I. DISEASE CHARACTERISTICS

Gestational choriocarcinoma (CCA) is a malignancy that arises from trophoblastic tissue of term pregnancies, ectopic gestations, spontaneous or induced abortions, or molar elements. CCA invades and metastasizes early and is often widespread at the time of diagnosis. Lung metastases are present in almost all patients with extrauterine disease; other sites often involved include brain, liver, vagina, kidney, and intestines. Fifty percent of cases arise from molar gestations (2% of all complete hydatidiform moles will degenerate into choriocarcinoma), 25% develop from aborted pregnancies, and the remaining cases follow term deliveries.

CCA differs from invasive mole (IM) in that IM is a histologically benign condition that evolves from hydatidiform moles. Although benign appearing, IM results from the invasion of abnormal trophoblasts into uterine myometrium or embolization of molar tissue through pelvic veins. IM metastasizes in about 15% of cases; the lungs and vagina are the most commonly involved sites.

Because many cases of CCA are identified by clinical criteria and no pathologic diagnosis exists, treatment of this tumor is based on clinical parameters. Thus, there is little distinction between IM and CCA in terms of the way that treatment strategy is planned. More important than the type of gestational trophoblastic disease (GTD) is the accurate and rapid diagnosis so that treatment may be administered promptly.

II. EPIDEMIOLOGIC CHARACTERISTICS

In the United States, approximately 3000 cases of hydatidiform mole and 500–750 cases of IM or CCA are diagnosed each year. The incidence of complete hydatidiform mole is between 50 and 150 per 100,000 pregnancies. Choriocarcinoma occurs in approximately 2.2 per 100,000 pregnancies in the United States. Risk of developing CCA is increased 1000–2000 times for women with hydatidiform moles. Epidemiologic risk modifiers for CHM probably also apply to the one-half of CCA cases that follow molar pregnancies. Therefore, Asian heritage, age <20 or >40, and prior molar gestation would be expected to increase risk. It is much more difficult to determine risk factors for gestations for which CCA arises from apparently normal placental tissue.

It is well established that an increased incidence of CCA occurs at the upper and lower extremes of the reproductive years. It has been suggested that nonwhite populations are at increased risk. Certain parental ABO blood groups are linked to increased risk of CCA development but do not affect risk for hydatidiform mole. No HLA locus antigens have been found to be associated with development of CCA. No other epidemiologic variables have been consistently linked with CCA, probably owing to the rarity of the disease.

III. CLINICAL MANIFESTATIONS

The most common symptom of invasive mole is irregular uterine bleeding following evacuation of a molar gestation. If metastasis has occurred, bleeding from vaginal or pulmonary sources may be present if these sites are affected. Diagnosis of persistent GTD after molar evacuation is usually made by observation of persistently elevated or rising human chorionic gonadotropin (hCG) levels.

Presenting signs and symptoms of gestational choriocarcinoma are highly variable. The invasive nature of this tumor often leads to early metastasis. Additionally, some tumors are diagnosed long after the antecedent gestation and are widely disseminated. Vaginal bleeding is the most common symptom.

Symptoms and signs associated with metastatic sites of CCA are often present owing to extensive tissue invasion with secondary necrosis and hemorrhage. Vaginal metastases may bleed extensively, especially when biopsied. Uterine enlargement and irregularity may be noted. Metastases to the lung are present in more than 80% of women with metastatic GTD. Pulmonary involvement may cause cough, dyspnea, hemoptysis, and/or pleuritic chest pain due to pulmonary infarction or pleural space invasion. Patients may experience abdominal pain if visceral involvement or hemoperitoneum is present. Neurologic manifestations may include headaches, seizures, loss of consciousness, and paralysis. Profound anemia may be another feature of metastatic CCA.

IV. PERSISTENT GTD FOLLOWING MOLAR EVACUATION

About 20% of patients with complete hydatidiform mole develop malignant sequelae. Eighty-five percent of these individuals have disease localized to the uterus. Fewer than 4% of women with partial hydatidiform mole contract persistent GTD. Postmolar GTD arising from partial hydatidiform mole is almost always invasive mole (as opposed to CCA), tends not to metastasize, and readily responds to chemotherapy.

Clinical features suggesting persistent GTD include plateauing or rising hCG levels, elevated hCG at 4 months postevacuation (>20,000 mIU/ml), hemorrhage from uterine or other metastatic foci, demonstrable metastases, and histologic verification of choriocarcinoma. Often no histologic diagnosis exists and treatment is initiated based on hCG criteria; therefore, most investigators refer to categories of invasive and metastatic disease generically as "GTD." However, some researchers have advocated clinicopathologic diagnosis of invasive mole or choriocarcinoma based on type of antecedent pregnancy and histologic criteria . Such classification has been demonstrated to affect prognosis for cure.

Several clinical findings are associated with increased risk for development of postmolar GTD. About half of women with uterine size >20 weeks gestational equivalent or with theca–lutein cysts develop persistent GTD after molar pregnancy. Thirty-five to 40% of women with a second molar gestation have progressive disease. Postmolar GTD ensues in one-quarter to one-third of patients with uterine size greater than dates or over age 40. Postevacuation hemorrhage, trophoblastic embolization to lungs, and eclampsia have also been identified as high-risk factors. The pattern of hCG regression and ratio of free hCG β subunit to total β subunit have been reported to predict persistent disease. Early studies suggested that extent of trophoblastic proliferation and degree of anaplasia were associated with increased risk of postmolar GTD, but this has not been confirmed. About 35% of patients with molar pregnancies have one or more of the cited "high-risk" factors; therefore, vigilant follow-up is warranted since those individuals destined to require further therapy for postmolar GTD cannot be predicted with certainty.

V. METASTATIC GTD

CCA metastases have been reported in almost all conceivable sites. Organs most commonly involved are lung, vagina, central nervous system, liver, kidneys, and gastrointestinal tract. Symptoms and physical findings reflect the degree of involvement of the affected organ(s) and whether hemorrhage or necrosis is the prominent feature.

Causes of death among patients with metastatic GTD are variable depending on the organ systems affected. Hemorrhage from metastatic foci is the most common cause of death (42%). Sites involved (in descending order) are lungs, liver, gastrointestinal tract, and peritoneal cavity. Intraabdominal bleeds are more prevalent among patients with advanced disease. Pulmonary insufficiency secondary to alveolar destruction, tumor burden, and hemorrhage is the cause of death in about one-third of patients. Death results from chemotherapeutic toxicity in about 10% of cases, but toxicity is frequently a complicating factor. Less common causes of death (about 2%) include sepsis, uremia secondary to obstructive uropathy, acute tubular necrosis, and pulmonary embolism. Five to 8% of patients (with extensive metastases) die away from the hospital and the fatal events cannot be assessed.

VI. DIAGNOSTIC AIDS

Several laboratory and radiologic tools are available to secure the clinical diagnosis of GTD and determine extent of disease. In general, hCG levels correspond to tumor burden. Radioimmunoassay, fluoroimmunoassay, and enzyme immunoassay are sensitive to below 5 mIU/ml and provide reproducible results.

Measurement of hCG in cerebrospinal fluid has been used to monitor therapeutic response in those with proven central nervous system involvement. Detection of hCG in symptomatic older, postmenopausal women strongly suggests the diagnosis of latent CCA. Serial hCG monitoring of women with disease in prolonged remission (>1 year) often provides early evidence of recurrence. CCA may rarely be associated with undetectable levels of hCG. Significant amounts of hCG may be liberated by a number of tumors other than GTD; this should be borne in mind when working up any woman with elevated hCG levels who is suspected of harboring a malignancy.

A hematocrit should be obtained in all women with GTD to assess for anemia. The presence of hyperemesis, preeclampsia, hyperthyroid symptoms, excessive hemorrhage, infection, hematuria, oliguria, melena, or dyspnea may suggest the need for further studies, such as white blood count, platelets, electrolytes, BUN, creatinine, liver enzymes, thyroid profile, coagulation studies, urinalysis, stool guaiac, blood cultures, or arterial blood gases. Central hemodynamic monitoring may be required to monitor cardiopulmonary status in unstable patients.

Radiologic evaluation is often helpful for diagnosing molar gestation and in assessing patients for metastases of GTD. A chest X ray is useful in evaluating women experiencing respiratory distress with suspected pulmonary edema, adult respiratory distress syndrome, or extensive lung metastases. Chest roentgenology has utility as a screening tool for trophoblastic emboli in women with complete hydatidiform mole. Chest X-ray findings in women with pulmonary metastases of GTD are variable. Early findings include generalized increased lung markings. As lesions progress, patchy infiltrates and small nodules are seen. More advanced disease results in medium (3–5 cm) to large (>5 cm) round, low-density shadows with hazy or well-defined margins. Atelectasis, pneumothorax, and effusion (usually hemothorax) are seen in many patients. Extensive pulmonary disease leads in some cases to cardiac enlargement.

Ultrasound is a useful noninvasive imaging method, especially when combined with hCG data. CHM has the appearance of a vesicular mass in association with uterine enlargement and focal areas of hemorrhage. Theca–lutein cysts may be demonstrated. Location and size of invasive mole or CCA lesions can be determined. Transvaginal sonography is especially helpful for diagnosing and subsequently following GTD lesions during chemotherapy. Color flow Doppler can define regions of increased vascularity representative of invasive disease and enhanced uterine perfusion distinct from that seen in conditions related to pregnancy or abortion. Ultrasound and color Doppler imaging of extrauterine disease may demonstrate tumor vascularity and blood flow patterns and can be employed to target lesions for percutaneous sampling.

Computed tomography (CT) scan is an effective means of screening for central nervous system and abdominopelvic metastases. Chest CT is more sensitive than chest X ray for detecting pulmonary metastases; about 40% of patients with presumed nonmetastatic GTD have pulmonary lesions when screened with CT scan. Magnetic resonance imaging has been shown to accurately define the degree of uterine invasion by GTD and is useful for monitoring tumor responsiveness to therapy.

VII. ATYPICAL PRESENTATION

Several reports of complete hydatidiform mole and choriocarcinoma in association with intrauterine pregnancy exist. Fetal compromise is often found in such cases. Grossly, CCA in a term placenta may be misinterpreted as a placental infarct. Sometimes pathologic exam of the placenta yields the first sign of CCA and has led to the diagnosis of metastases in otherwise asymptomatic patients.

Metastases of GTD may occur almost anywhere in the body. Unusual sites, such as bone, breast, and others, must be diagnosed since their presence may affect therapeutic decisions and prognosis. Rare instances of metastases to the fetus have been documented.

Postmenopausal women may harbor latent GTD. Serum hCG levels may be undetectable in some patients. Choriocarcinoma may rarely arise from totipotent cells in extragonadal sites and may be misdiagnosed as GTD.

VIII. STAGING AND TREATMENT

Malignant GTD is classified according to the extent of its spread to other sites (Table 1). Treatment is administered on the basis of this classification. Nonmetastatic and good prognosis metastatic GTD are treated with single agent chemotherapy using either methotrexate or actinomycin D. Poor prognosis metastatic GTD requires therapy with multiagent chemotherapy. Typically a regimen including etoposide, methotrexate, actinomycin, cyclophosphamide, and vincristine (EMA-CO) is employed. Surgery is usually reserved for removal of chemotherapy-resistant

TABLE 1
Classification of Malignant Gestational Trophoblastic Disease

Nonmetastatic
Metastatic
  Good prognosis
    hCG <40,000 mIU/ml serum
    Disease present <4 months
    No brain or liver metastases
    No prior chemotherapy
    Antecedent pregnancy molar, ectopic, abortion
  Poor prognosis
    hCG >40,000 mIU/ml serum
    Disease present >4 months
    Brain or liver metastases
    Prior chemotherapeutic failure
    Antecedent term pregnancy

foci of tumor. However, hysterectomy is often adequate therapy for nonmetastatic malignant GTD in women who have completed childbearing.

Salvage chemotherapy with a number of agents, such as bleomycin, cisplatin, and paclitaxel, has been employed for failures of first-line multiagent chemotherapy, but these agents are rarely effective. Radiation therapy is often used as an adjuvant treatment for patients with brain, liver, or extensive pelvic involvement with tumor.

Treatment is continued until hCG falls below detectable levels. Disease resistant to single-agent therapy (with methotrexate or actinomycin D) may be treated by switching to the other agent or, if high-risk features are present, administering EMA-CO. Nonmetastatic or good prognosis metastatic malignant GTD patients should receive one additional chemotherapy dose after hCG remission. Poor prognosis metastatic malignant GTD patients should be given three additional cycles after tumors are treated to remission.

IX. PROGNOSIS

Nonmetastatic and good prognosis metastatic malignant GTD is virtually 100% curable with current treatment strategy. However, poor prognosis metastatic malignant GTD is curable in only about 70% of cases. The highest risk factor is failure of prior chemotherapy treatment. Thus, it should be emphasized that the patients in the poor prognosis metastatic category should be treated aggressively up front with multiagent chemotherapy. Administration of single-agent chemotherapy will likely result in resistant disease foci among these patients, significantly diminishing their chance of cure.

X. POSTTREATMENT

The diagnosis of mole or malignant neoplasia in a woman anticipating routine pregnancy has obvious psychological consequences. GTD patients have been shown to have clinically significant fatigue, anger, anxiety, confusion, and sexual dysfunction. These concerns should be addressed and psychiatric referral obtained if indicated.

Current chemotherapeutic regimens allow preservation of fertility in most GTD patients. Normal reproductive outcome should be anticipated in women with prior molar gestations; there is no increased risk of stillbirth, abortion, preterm delivery, or congenital anomalies. The risk of recurrent mole is about 2%. Similarly, patients with postmolar GTD and CCA cured by chemotherapy can expect therapeutic outcomes no different from that of the general population.

Patients should employ reliable contraception to prevent confusion caused by intercurrent pregnancy and the attendant rise in hCG. Hormonal contraception has been proven to be safe and very effective.

Serial serum hCG levels should be determined for at least 1 year after remission is accomplished. Patients should also be queried on each visit regarding any symptoms suspicious for recurrence (especially irregular vaginal bleeding). Most recurrences of GTD occur within 1 year of achieving remission. However, late recurrences have been reported years after primary remission; the incidence of such cases is <1%.

See Also the Following Article

CHORIONIC GONADOTROPIN, HUMAN

Bibliography

Bagshawe, K. D. (1992). Choriocarcinoma. A model for tumour markers. *Acta Oncol.* 31, 99–106.

Bandy, L. C., Clarke-Pearson, D. L., and Hammond, C. B. (1984). Malignant potential of gestational trophoblastic disease at the extreme ages of reproductive life. *Obstet. Gynecol.* 64, 395–399.

Hammond, C. B. (1988). Gestational trophoblastic neoplasms: History of the current understanding. *Obstet. Gynecol. Clin. North Am.* 15, 435–441.

Hammond, C. B., and Soper, J. T. (1984). Poor-prognosis metastatic gestational trophoblastic neoplasia. *Clin. Obstet. Gynecol.* 27, 228–239.

Hammond, C. B., Borchert, L. G., Tyrey, L., Creasman, W. T., and Parker, R. T. (1973). Treatment of metastatic trophoblastic disease: Good and poor prognosis. *Am. J. Obstet. Gynecol.* 115, 451–457.

Hammond, C. B., Clarke-Pearson, D. L., and Soper, J. T. (1984). Management of patients with gestational trophoblastic neoplasia: Experience of the Southeastern Regional Center. *Adv. Exp. Med. Biol.* 176, 369–381.

Hunter, V., Raymond, E., Christensen, C., Olt, G., Soper, J., and Hammond, C. (1990). Efficacy of the metastatic survey in the staging of gestational trophoblastic disease. *Cancer* 65, 1647–1650.

Lurain, J. R., and Sciarra, J. J. (1991). Study and treatment of gestational trophoblastic diseases at the John I. Brewer Trophoblastic Disease Center, 1962–1990. *Eur. J. Gynaecol. Oncol.* 12, 425–428. [Review]

Miller, J. M., Jr., Surwit, E. A., and Hammond, C. B. (1979). Choriocarcinoma following term pregnancy. *Obstet. Gynecol.* 53, 207–212.

Mutch, D. G., Soper, J. T., Babcock, C. J., Clarke-Pearson, D. L., and Hammond, C. B. (1990). Recurrent gestational trophoblastic disease. Experience of the Southeastern Regional Trophoblastic Disease Center. *Cancer* 66, 978–982.

Newlands, E. S., Bagshawe, K. D., Begent, R. H., Rustin, G. J., and Holden, L. (1991). Results with the EMA/CO (etoposide, methotrexate, actinomycin D, cyclophosphamide, vincristine) regimen in high risk gestational trophoblastic tumours, 1979 to 1989. *Br. J. Obstet. Gynaecol.* 98, 550–557.

Soper, J. T., and Hammond, C. B. (1987). Role of surgical therapy and radiotherapy in gestational trophoblastic disease. *J. Reprod. Med.* 32, 663–668.

Soper, J. T., Clarke-Pearson, D., and Hammond, C. B. (1988). Metastatic gestational trophoblastic disease: Prognostic factors in previously untreated patients. *Obstet. Gynecol.* 71, 338–343.

Soper, J. T., Hammond, C. B., and Lewis, J. L., Jr. (1992). Gestational trophoblastic disease. In *Principles and Practice of Gynecologic Oncology* (W. J. Hoskins, C. A. Perez, and R. C. Young, Eds.), pp. 795–825. .Lippincott, Philadelphia.

Soper, J. T., Evans, A. C., Clarke-Pearson, D. L., Berchuck, A., Rodriguez, G., and Hammond, C. B. (1994a). Alternating weekly chemotherapy with etoposide–methotrexate–dactinomycin/cyclophosphamide–vincristine for high-risk gestational trophoblastic disease. *Obstet. Gynecol.* 83, 113–117.

Soper, J. T., Evans, A. C., Conaway, M. R., Clarke-Pearson, D. L., Berchuck, A., and Hammond, C. B. (1994b). Evaluation of prognostic factors and staging in gestational trophoblastic tumor. *Obstet. Gynecol* 84, 969–973.

Chorion

see Fetal Membranes

Chorionic Fluid

see Fetal Membranes

Chorionic Gonadotropin, Human

Steven Birken, John O'Connor, Leslie Lobel, and Robert Canfield

Columbia University

I. Chemistry and Function
II. Molecular Biology
III. Measurement of hCG
IV. Clinical Aspects

GLOSSARY

β core fragments Proteolytically produced products of the β subunits of either human chorionic gonadotropin or human luteinizing hormone which are composed of approximately half of the mass of the β subunit. They are end products of a natural digestion process and are found at high concentrations in urine. These fragments have proven useful for some diagnostic measurements.

exon The sequences of the primary RNA transcript (or the DNA that encodes them) that are spliced together to form a messenger RNA molecule. In the primary transcript neighboring exons are separated by introns.

glycoprotein hormone family Four related hormones containing protein and sugar components. The species usually precedes the abbreviation in lowercase: hCG, human chorionic gonadotropin; hLH, human luteinizing hormone; hFSH, human follicle stimulating hormone; hTSH, human thyroid-stimulating hormone.

glycoprotein hormones Protein hormone molecules composed of two nonidentical subunits which also contain sugar groups covalently linked to the polypeptide chain(s).

gonadotropins Protein hormones that stimulate the gonads to secrete sex steroids and promote maturation of the oocyte in the female and of spermatozoa in the male. All gonadotropins can be produced by the pituitary, but hCG is primarily produced by the conceptus, and its detection is commonly used as a marker of pregnancy.

nicked hCG Nicked forms of protein hormones refer to hormones that display breaks in the peptide bond structure, apparently caused by protease activity. All of the amino acids remain bound together because other peptide bonds in the polypeptide chain are linked by disulfide bridges, thus preventing the release of any peptides.

palindromic sequence A nucleic acid sequence that is identical to its complementary strand when each is read in the 5′ to 3′ direction (e.g., TGGCCA). Palindromic sequences are often the recognition sites for restriction enzymes.

promoter A region of DNA where RNA polymerase binds before initiating the transcription of DNA into RNA. Most factors that regulate gene transcription bind at or near the promoter and thereby affect the initiation of transcription. Most eukaryote promoters have the Goldberg–Hogness or TATA box that is centered around position −25 (from the start of transcription) and has the consensus sequence 5′-TATAAAA-3′. Several promoters have a CAAT box around −90 with the consensus sequence 5′-GGCCAATCT-3′.

pseudogene A nonfunctional DNA sequence that is very similar to the sequences of a known genes. Some probably arise from gene duplications that become nonfunctional because of the loss of regulatory elements or the accumulation of mutations.

β sheet A protein conformation in which the polypeptide chain is nearly fully extended and individual peptide strands aggregate side by side by forming hydrogen bonds between carbonyl and amino groups of the backbone structure. The strands appear planar and can be parallel or antiparallel in orientation to each other.

transcription The synthesis of RNA from a DNA template by RNA polymerase.

translation The process that is mediated by the ribosome whereby information in mRNA is used to specify the sequence of amino acids in a growing polypeptide chain.

triple screen A blood assay measuring molecules associated with α-fetoprotein, unconjugated estriol, and hCG which has proven useful for diagnosis of likely Down's pregnancies as well as other problem pregnancies.

I. CHEMISTRY AND FUNCTION

A. Introduction

Chorionic gonadotropin (CG) is known as the pregnancy hormone and its measurement is the basis of all pregnancy tests in humans. Although it was discovered nearly 80 years ago, recent advances in understanding of its structure and the capability to measure it at low concentrations in blood and urine have enabled implementation of new diagnostic tests for problem pregnancies and cancer.

Chorionic gonadotropin is a glycoprotein hormone composed of two nonidentical subunits, α and β. Early in pregnancy the CG molecule is produced by the trophoblast and later by the placenta, an organ derived from trophoblastic tissue. The function of CG is to stimulate the corpus luteum to produce steroid hormones that help to maintain pregnancy. Later in pregnancy, the placenta itself can secrete the necessary steroid hormones. Chorionic gonadotropin occurs only in primates. Other animals rely on different signals to the mother that are necessary to maintain pregnancy. These communication chemicals stimulate maternal tissues to produce a variety of other molecules, such as steroids, in order to maintain the proper environment for growth of the embryo. For example, the horse uses a hyperglycosylated form (extra sugar groups added to the protein) of its equine luteinizing hormone (LH), whereas the rodents use forms of prolactin.

B. Structure

1. Protein Portion

Human CG (hCG) is one of a family of four homologous glycoprotein hormones. These hormones all have a common α subunit that is encoded by a single gene. The α subunit is composed of 92 amino acids in the human. The β subunit of each hormone has a distinctive structure and amino acid composition (the β subunits vary in size from approximately 102 to 145 amino acids). Recent studies support the earlier notion that the hormone-defining β subunits confer conformational alterations on the common α subunits as part of the structural endowment for hormone specificity. The four human hormones are hCG, hLH, follicle-stimulating hormone (hFSH), and thyroid-stimulating hormone (hTSH). hCG and hLH are both directed to the same receptor in the ovary and in the testis, and they induce secretion of sex steroids in males (testosterone) and females (estrogen and progesterone). The β subunits of hLH and hCG are highly homologous in primary structure (amino acid sequence). The amino acid sequence of the α subunit is shown in Fig. 1, whereas that of the hCGβ subunit is shown in Fig. 2. Receptors for hCG/hLH have recently been reported to be expressed in low densities in nonreproductive tissues such as brain, uterus, and kidney, suggesting that these hormones possibly have other functions in addition to inducing secretion of sex steroid hormones.

Of the other hormones in this family, FSH promotes growth of the ovarian follicles from which eggs are derived in the female and sperm formation in the male. The fourth hormone of the family, TSH, has a nonreproductive function—that of promoting synthesis and release of the thyroid hormone, thyroxine, which controls the rate of metabolism.

2. Carbohydrate Portion

The carbohydrate groups on hCG, and on the other homologous members of its family, perform several

```
        1                              10
Ala-Pro-Asp-Val-Gln-Asp-Cys-Pro-Glu-Cys-Thr-Leu-Gln-Glu-Asn-Pro-Phe-
          20                              30
Phe-Ser-Gln-Pro-Gly-Ala-Pro-Ile-Leu-Glx-Cys-Met-Gly-Cys-Cys-Phe-Ser-
                          40                              50
Arg-Ala-Tyr-Pro-Thr-Pro-Leu-Arg-Ser-Lys-Lys-Thr-Met-Leu-Val-Gln-Lys-
CHO                              60
Asn-Val-Thr-Ser-Glu-Ser-Thr-Cys-Cys-Val-Ala-Lys-Ser-Tyr-Asn-Arg-Val-
          70                      CHO       80
Thr-Val-Met-Gly-Gly-Phe-Lys-Val-Glu-Asn-His-Thr-Ala-Cys-His-Cys-Ser-
              90
Thr-Cys-Tyr-Tyr-His-Lys-Ser
```

FIGURE 1 The amino acid sequence of the α subunit. It is identical in all the glycoprotein hormones since it is encoded by a single gene in humans. Carbohydrate moieties are the N-Asn biantennary type (see Fig. 3) and are indicated by CHO above the appropriate Asn residue.

```
                                          CHO
  1                          10
Ser-Lys-Glu-Pro-Leu-Arg-Pro-Arg-Cys-Arg-Pro-Ile-Asn-Ala-Thr-Leu-Ala-
             20                                    CHO
Val-Glu-Lys-Glu-Gly-Cys-Pro-Val-Cys-Ile-Thr-Val-Asn-Thr-Thr-Ile-Cys-
                         40                                     50
Ala-Gly-Tyr-Cys-Pro-Thr-Met-Thr-Arg-Val-Leu-Gln-Gly-Val-Leu-Pro-Ala-
                                    60
Leu-Pro-Gln-Val-Val-Cys-Asn-Tyr-Arg-Asp-Val-Arg-Phe-Glu-Ser-Ile-Arg-
         70                                80
Leu-Pro-Gly-Cys-Pro-Arg-Gly-Val-Asn-Pro-Val-Val-Ser-Tyr-Ala-Val-Ala-
                     90                                    100
Leu-Ser-Cys-Gln-Cys-Ala-Leu-Cys-Arg-Arg-Ser-Thr-Thr-Asp-Cys-Gly-Gly-
                             110
Pro-Lys-Asp-His-Pro-Leu-Thr-Cys-Asp-Asp-Pro-Arg-Phe-Gln-Asp-Ser-Ser-
 120 CHO                        CHO          130     CHO
Ser-Ser-Lys-Ala-Pro-Pro-Pro-Ser-Leu-Pro-Ser-Pro-Ser-Arg-Leu-Pro-Gly-
 CHO       140                       145
Pro-Ser-Asp-Thr-Pro-Ile-Leu-Pro-Gln
```

FIGURE 2 The amino acid sequence of the hCGβ subunit. Carbohydrate groups are indicated by CHO above the appropriate residue. Those attached to Asn differ from those attached to serines (see Fig. 3). Light shaded area comprises the structure of the hCGβ core fragment, composed of residues 6–40 disulfide linked to residues 55–93. The dark shaded area is the CTP portion of hCGβ, which is an additional peptide portion absent in highly homologous hLH.

important functions, some of which probably remain to be elucidated. Among their known functions are aiding in the folding of the proteins during biosynthesis within their secreting cells, controlling their circulating half-lives, and playing an important role in their capability to induce a response signal upon binding to their receptors. Multiple investigators have contributed information concerning the role played by carbohydrate during proper folding of the glycoprotein hormones at the time of biosynthesis or during *in vitro* efforts to refold hormones outside of the cell. Ruddon and colleagues performed pioneering work in studies of the biosynthesis and folding of the β subunit during synthesis within the cell. Several distinct intermediates during synthesis were discovered, and the order of disulfide bond formation was determined, as well as the likely association of particular chaperone proteins that aid in intracellular folding of the hormone prior to heterodimer formation and secretion.

The terminal sugars, such as sialic acid or galactose, as well as the inorganic sulfate at the end of the glycoprotein sugar moieties, are crucial in determining the circulating lifetimes of the molecules. The liver contains a receptor that recognizes glycoproteins with sulfate/N-acetyl galactosamine (Fig. 3) on a receptor that very rapidly removes glycoproteins with a galactose terminus. Such proteins are formed when sialic acid-terminating sugar groups lose their sialic acid to expose the underlying galactose residues. The carbohydrate moieties attached to hCG are summarized in Fig. 3. hCG has the longest circulating half-life of all the glycoprotein hormones, in part because it has the most sialic acid-terminating sugar groups, including the presence of four O-linked carbohydrate groups within the COOH-terminal peptide portion (CTP; composed of residues 112–145) on its β subunit. This extra peptide, it has been suggested, developed from a mutation of a termination codon into a reading codon from its ancestral hLH parent. Horse LH and CG, which have identical amino acid sequences, also contain a CTP extension similar to that of hCG, but with even more carbohydrate than is present in hCG. The CTP on hCGβ greatly adds to the hormone's hydrated radius in the bloodstream and decreases its clearance rate through the glomerulus of the kidney. Horse CG contains a glycosylated CTP with so much more carbohydrate that the hormone remains a serum protein (pregnant mare's serum gonadotropin) with very little of it

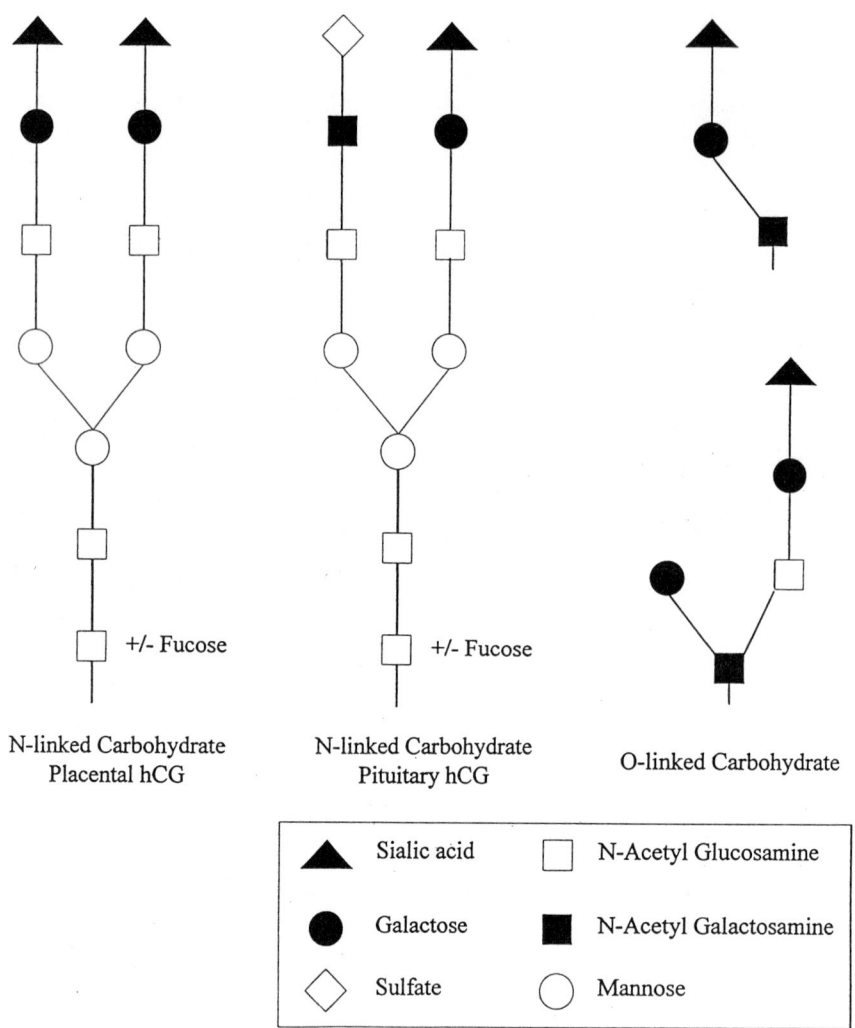

FIGURE 3 The carbohydrate groups commonly found in forms of hCG. The structure of the most commonly found N-linked (to Asn) carbohydrate is shown for natural placentally produced hCG of pregnancy. Some variation is found. The N-linked form found in pituitary hCG is similar to that of pituitary hLH. Two common O-linked structures attached to the serines of hCGβ are also shown. (There are at least six different O-linked moieties.)

appearing in horse urine. That is, its hydrated radius becomes so large due to the high sugar content that it cannot pass through the glomerulus of the kidney to enter the urine. This highly glycosylated peptide can prolong the circulating half-life of hCG by incorporating it onto the FSH structure and showing a corresponding increase in the circulating lifetime of FSH with this addition. If hCG is desialyated (sialic acid can be cleaved by enzymes in the circulation), it is very rapidly removed (within minutes) from the circulation by the liver receptor for asialo glycoproteins compared to its usual 29- or 30-hr circulating half-life.

3. Nonpregnancy Forms of hCG

A form of hCG was recently isolated from an extract of human pituitaries. Although there have been reports of a likely pituitary source of minute quantities of hCG in nonpregnant individuals (on the basis of immunochemical measurements) during the past two decades, none had been isolated until recently. The presence of hCG in the blood of nonpregnant individuals in pulsatile pattern to homologous hLH has been reported. The pituitary form of hCG contains sulfate as well as sialic acid, similar to its homologous hLH. The placental form of hCG does not contain sulfate.

hCG is also produced by a variety of gonadal and nonreproductive tissue tumors. The tumor forms of hCG, which are measured to monitor certain cancers, are usually sialylated but are not sulfated. Despite the wide use of hCG testing in monitoring of therapy for certain cancers, the only diagnostic tests for hCG approved by the FDA are for determination of pregnancy in urine or blood.

4. Three-Dimensional Structure

More than two decades after publication of the amino acid sequences of both of the subunits of hCG, the three-dimensional structure was determined by X-ray crystallographic analysis of chemically deglycosylated hCG by two different methods, each of which yielded the identical structure. This represents a major milestone in the understanding of the structure of hCG. Structural determination was delayed by inability to grow crystals of quality suitable for X-ray diffraction studies. This difficulty was eventually overcome by the use of partially deglycosylated hormone. To date, this technique has only worked for hCG, which is the most stable of the glycoprotein hormones. The structural solution of hCG, which is presumably closely applicable to those of the other members of the glycoprotein hormone family, demonstrated that both subunits have similar structures exhibiting three loops composed of β sheets. The two subunits are held together noncovalently by a loop of the β subunit which wraps itself around the α subunit like a seat belt and interacts by forming hydrogen bonds (Fig. 4). The β subunit has 12 half-cystine residues that form six disulfide bridges. The region where the β loop wraps around α comprises residues between half-cystines 10 and 12. The portion of the seat belt termed "the determinant loop" by Ward and Moore is located between the 10th and 11th half-cystines of the β subunit and plays an important role in receptor specificity by regulating charge distribution. This was an insightful early observation which has been confirmed by recent studies. Molecular biology studies by multiple investiga-

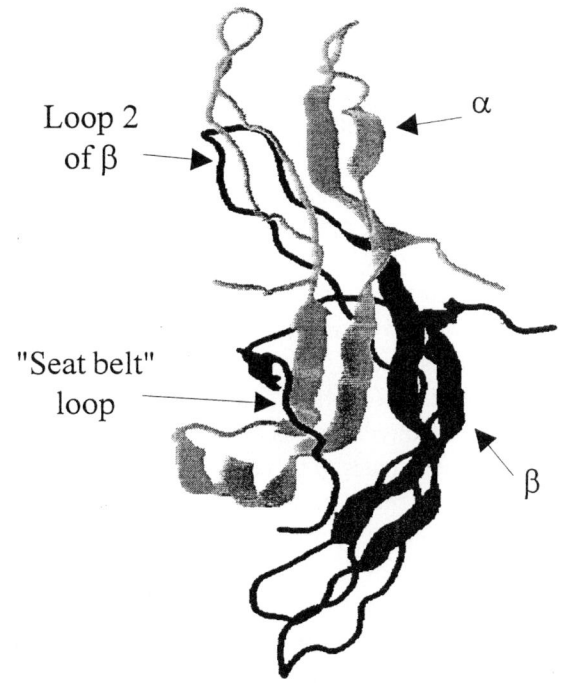

FIGURE 4 The three-dimensional structure of hCG. The α subunit is lightly shaded, whereas the β subunit is black. The area of β which wraps around α is termed the "seat belt" as labeled. Loop 2 of β is exposed to solvent and is easily cleaved by proteases.

tors indicated the importance of this seat belt region in receptor specificity. Swapping regions among the glycoprotein hormones (called chimeric construction since parts of two distinct molecules are used to create an artificial hybrid) can be used to define function. Swapping the hCG or hLH seat belt region into hTSH produces hLH specificity and vice versa. Swapping the hFSH seat belt into hCG produces hFSH specificity. FSH seems to require more than the seat belt region to alter its specificity, as does TSH, since the FSH seat belt did not produce FSH binding in TSH, nor did the CG seat belt convert FSH to CG type binding. However, the mechanism of signal transduction, when hormone has bound to receptor, remains undefined, although many models have been proposed such as ligand interactions with protease-like receptor regions, ligand binding to a lectin-like region of the receptor, dimer formation of receptors on the membrane surface similar to growth hormone, or steric interactions of ligand carbohydrates with receptor structures. It has been proposed that the simple bulk of the carbohydrate moieties (which are actually very large in three-dimensional space) induces structural change in portions of the receptor such that elements of the intracellular portions of the receptor move, in turn inducing signal generation.

Nuclear magnetic resonance (NMR) methodology based on new techniques for uniform NMR labeling with stable isotopes of proteins and carbohydrates has been used to determine the three-dimensional structure in solution of the carbohydrate moiety attached on Asn52 of the α subunit in solution. This particular carbohydrate was demonstrated earlier by Boime and colleagues to be critical for bioactivity by participation in the induction of signal after hormone binding to receptor. Homans showed that the carbohydrate group does not interact with the hormone's protein backbone (i.e., fold back upon the peptide and bond to its side chains) but is in the same conformation as it would be as free carbohydrate in solution. This suggests that the protein does not direct the carbohydrate into a unique structure to induce a signal upon binding to its receptor.

5. Relating Structure and Function

A molecule with TSH-binding properties has been produced from a combination of segments of hCG and FSH. The abilities of such hCG/hFSH chimeric molecules with no TSH-specific residues to induce signal when binding at all three types of receptors, i.e., LH, TSH, and FSH, support the hypothesis that signal transduction occurs between structures in common among the glycoprotein hormone family. Although the regions swapped in these studies were mainly in the seat belt region, these residues were not primary contact regions with the receptor. This is consistent with the hypothesis that specificity of the glycoprotein hormones is derived from adjustment of structural elements that prevent binding to nontarget receptors rather than from regions of the molecule that react with particular portions of the target receptors. Essentially, this theory proposes that gonadotropins would tend to bind similarly to all gonadotropin receptors were it not that individual gonadotropins evolved along with their receptors to acquire deviations in structure that prevent binding to receptors other than the one which defines specificity.

Using high concentrations of peptides produced by solid-phase synthesis, it has been demonstrated that the 38–57 region of the β subunit of hCG (loop 2; Fig. 4) binds to receptor and can stimulate signal transduction within the LH/CG receptor but not the FSH receptor. These studies appeared to be quite specific since alterations of the sequence produced inactive peptides. However, molecular biology methods have been used to show that replacement of the hCG loop 2 (Fig. 4; essentially the 36–58 region of the β subunit) with the hFSH loop 2 region did not affect the specificity of hCG for the human LH/CG receptor and did not enable hCG to bind to the hFSH receptor. The role of this loop 2 region in binding to the LH/CG receptor remains unclear, but it does have an influence on receptor binding since peptide bond cleavages in this loop (namely, the formation of proteolytically-nicked hCG) greatly reduce receptor binding and signal transduction. In addition, it has been reported that both in $vitro$ and in $vivo$ activities are displayed by synthetic peptides from this region for various glycopeptide hormones, including FSH. Extensive site-directed mutagenesis has been used to identify residues crucial to subunit combination and secretion as well as receptor interaction. Residues conserved among homologous hormones usually prove to be important for subunit and hormone secretion.

A "tandem," covalently linked form of hCG with β, linked via its COOH-terminal peptide to α, has been produced. This recombinant construct endows hCG with enhanced stability since subunits can no longer dissociate, and it has enabled a series of structure–function studies that were previously impossible to conduct. The tandem structure avoided trapping of subunits within the secreting cell caused by the inability of subunits to associate and was especially useful for LH studies because LHβ is not secreted well from cells without the α subunit. This tandem construct led to the unexpected observation that, while lack of some proper disulfide bridges (cystines) in both α and β subunits leads to drastic conformational alterations, these altered molecules nevertheless bind receptor and induce signal transduction. It was found that, even when some of the disulfide bridges which form early in biosynthetic folding of the subunits were deleted, the tandem construct was still able to bind receptor and induce signal after receptor binding. Such a discovery was not possible without the covalently linked construction because the subunits would not associate properly into the heterodimer when disulfide bridges were missing and only the heterodimer can bind to the receptor. It is possible to make major alterations in the primary and the three-dimensional structure of hCG and still maintain binding and signal transduction as long as the two subunits can be held together. The receptor is much more forgiving of shape alterations of the hCG ligand, but both secretion of hCG during biosynthesis and binding to the receptor itself require that the two subunits associate and retain the capability to induce folding changes in each other after receptor binding. These findings explain why such extensive mutations and segment splicing of hCG and the other gonadotropins frequently yielded ligands that would bind to the receptor and were bioactive.

6. Summary of Structure and Function Studies

Major advances in understanding the relationship of structure to function have taken place during the past decade of studies of hCG and the other glycoprotein hormones. This has been made possible by the molecular biology approach, not only by site-directed mutagenesis but also by the technique of swapping of loops and domains among the homologous family of glycoprotein hormones. These advances have accelerated with the solution of the three-dimensional structure of hCG 3 years ago. The next major advance awaits further understanding of the interaction of hCG when it binds to receptor. The seat belt region has proven to contain important determinants of receptor specificity, although its three-dimensional orientation does not seem important in the actual induction of signal, but it is important in the noncovalent assembly with its complementary subunit. The precise atomic interactions between ligand and receptor and the nature of the signal transduction process after hormone binding remains to be elucidated after the three-dimensional structure of the hormone within the receptor is determined.

II. MOLECULAR BIOLOGY

A. The Chorionic Gonadotropin Genes

As noted, chorionic gonadotropin (hCG) is a heterodimer consisting of an α and β subunit and the primary structure of the α subunit of hCG is identical to that of the α subunits of hLH, hFSH, and hTSH. The α subunit is encoded at a single genetic locus on chromosome 6 and is expressed in cells that produce any of these hormones. As a result, the gene encoding the α subunit is regulated by a number of different hormonal pathways that are involved in the biosynthesis of these hormones. In contrast, the β subunit of hCG is unique to that hormone. Nonetheless, it displays a high degree of homology with the β subunit of hLH. This high degree of homology is likely a result of the evolution of the hCGβ subunit from the hLHβ subunit.

Molecular cloning of the loci encoding the hCG and hLHβ subunits indicates that they both map to the same location on chromosome 19 and form a gene cluster. This cluster spans approximately 52 kilobases and is composed of six CGβ genes, one LHβ gene, and one CGβ pseudogene. The hCGβ genes are numbered 1, 2, 3, 5, 7, and 8 in the cluster (Fig. 5). hLHβ corresponds to gene No. 4 in the cluster, whereas the hCGβ pseudogene is No. 6.

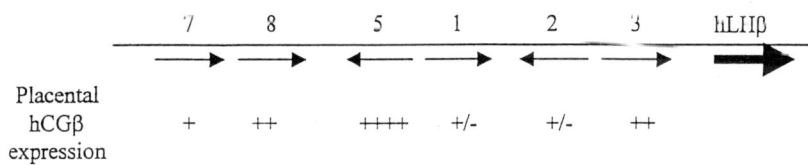

FIGURE 5 The organization of the hLH/CGβ gene cluster and placental expression of the different CGβ genes: The hLH/CGβ subunit gene cluster is outlined with the relative positions of each of the hCG genes and the hLH gene and their direction of transcription with respect to the cluster. Beneath the hCG gene cluster is a series of "+" symbols that denote the relative placental transcriptional activity of each of the hCGβ subunit loci (++++, the highest transcriptional activity; ±, a gene that is marginally transcribed).

The hCGβ genes appear to have evolved from the hLHβ gene by deletion of a single nucleotide (22 base pairs upstream from the hLH translational stop codon) that causes a read-through mutation of the translational stop codon in the hLH gene. As a result, hCG has a 24-amino acid extension at its carboxy terminus that corresponds to the peptide encoded in the 3' untranslated region of the hLH gene beyond its translational termination codon. This carboxy-terminal extension has been termed the CTP (as previously noted) of hCG and is the major distinguishing feature between hCG and hLH.

Although the hCG and hLHβ subunit genes have a common evolutionary ancestry, a single locus encodes the hLHβ subunit and six genes encode the β hCG subunit. The hCG gene cluster likely arose through a process of gene duplication. The sequences of these genes display identities of 96% in both coding and noncoding domains. Although this high degree of conservation may be due to the process of "gene conversion," it more likely indicates that the evolution of this cluster was a recent evolutionary event. In support of this is the finding that CG is produced primarily by primates. Most nonprimate mammals have developed alternative physiological mechanisms to maintain early pregnancy.

Whereas hCGβ and hLHβ subunits display striking genetic conservation in both coding and noncoding domains, hCG evolved a unique promoter sequence that accounts for its placental-specific expression. The LHβ gene 5' untranslated region consists of only nine nucleotides with a consensus TATA box sequence located at an appropriate distance upstream of the transcription initiation site. In contrast, the hCGβ gene has a long 5' untranslated region with a transcription initiation site occurring 365 base pairs upstream of the homologous promoter region in the hLHβ gene. By constructing a series of chimeras between the promoter regions of the hLHβ and hCGβ genes, it has been determined that the sequences surrounding the hCGβ promoter account for its placenta-specific expression. Furthermore, the vestigial hLHβ promoter, found in the hCGβ gene 5' untranslated region, is not required for expression of the hCGβ subunit.

B. Human Chorionic Gonadotropin Gene Expression

All of the hCGβ subunit loci are transcribed *in vivo* in first-trimester placenta but with highly variable levels of expression. The relative levels of expression of these genes have been reported to be as follows: hCGβ 5 ≫ 3 = 8 > 7 > 1, 2. The presence of many active β subunit loci may be indicative of the critical role that hCG plays in development. This may serve to ensure high levels of β CG production during the first trimester of pregnancy. Furthermore, the presence of multiple β subunit genes might also protect the species during evolution of the CGβ gene cluster.

In contrast to the β CG loci, the singular α subunit gene is expressed in a number of different cell types in parallel with its coexpression with the β subunits of hCG, hFSH, hLH, and hTSH. Although this indicates that it is regulated by a number of different hormone pathways, experiments suggest that one promoter initiates transcription at the same site in all the cell types in which the α subunit is expressed. Nonetheless, distinct regulatory elements might exist within the promoter or elsewhere in the gene that mediate cell-specific expression of the α subunit locus.

C. Regulation of Chorionic Gonadotropin Gene Expression

The biochemical pathways that regulate the expression of the hCGα and -β subunit genes are poorly understood. Nonetheless, a few details of the regulation of these genes have been determined. First, cAMP appears to regulate expression of the α and β subunit genes at the transcriptional level. Second, cAMP also plays a role in stabilizing the messenger RNA from these genes following transcription. Finally, the kinetics of transcription of the α and β loci following stimulation with cAMP are different. Transcription at the α locus appears to initiate and peak earlier than that of the β loci. This suggests that the regulatory pathways for these loci are different and that the β loci might require additional regulatory steps.

Regulation of the α and β subunit loci of hCG (at

the transcriptional level) by cAMP suggests that the promoters of these genes contain a cAMP response element (CRE). Studies of the α promoter and the promoter of CG gene 5 have revealed that both genes are regulated by a CRE. In addition, the CRE of the α gene appears to bind a different transcription factor than the CRE for β CG gene 5. Therefore, although both subunits are regulated by cAMP, there are probably two distinct pathways that modulate their activities.

To date, the promoter of the α locus has been examined more extensively. Studies have revealed that the CRE of the α promoter consists of tandem repeats of an 18-base pair (bp) sequence. Each copy of the 18-bp repeat contains an 8-bp palindrome, TGACGTCA. This imparts a twofold symmetry to the CRE, which is important for its recognition by transcription factors. The CRE element itself also exhibits all the characteristics of a classical enhancer element in that it can impart cAMP responsiveness to heterologous promoters, and its function is independent of orientation. In addition, it can also function in heterologous systems when it is placed upstream or downstream from the promoters. Nonetheless, the activity of the core CRE element is influenced by the surrounding "environment" of sequences that flank it. This may serve to influence interactions with cell-specific transcription factors or promoter elements that are responsible for cell-specific expression.

Further studies of the α promoter CRE revealed that it was the site for binding of a specific protein known as CREB. CREB is a protein of approximately 35 kDa and is a member of the "leucine zipper" family of transcription factors. It forms homodimers (unlike most other members of the leucine zipper family) that recognize the palindromic CRE element. The transcriptional activation domain is in the amino terminus of CREB, which also contains targets for a number of different kinases.

In addition to the CRE, there are other elements in the α promoter that are involved in either enhancing transcription (like the CRE) or in basal level activity of the promoter. One such enhancer element which is noteworthy is the trophoblast-specific element (TSE), which is likely involved in placental-specific expression of the α subunit gene. The TSE cannot stimulate promoter activity by itself but, instead, operates in conjunction with the CRE elements that lie downstream in the promoter. All the regulatory elements of the α gene promoter operate in a combinatorial fashion such that the combined activities are greater than the activities of the individual elements.

Whereas the α promoter has been studied extensively, the CGβ promoter has not been well-defined. Noteworthy is the absence of a TATA box even though a discrete transcriptional start site exists. The CRE of the β promoter differs from that of the α gene and does not bind CREB, although two protein binding sites have been mapped to one region of the β CRE. Studies have also suggested that a second region in the first exon of the β gene is responsible for stimulation of transcription by cAMP. Therefore, the mechanism for activation of the α subunit gene is likely distinct from that of the β subunit gene, and these processes are mediated by unique transcription factors.

III. MEASUREMENT OF hCG

The discovery of hCG nearly 80 years ago by Hirose and by Aschem and Zondek preceded the development of modern immunoassays by many years. The first means of hCG measurement were biological activity assays utilizing gross changes in animal organs (e.g., mouse uterine weight and rat ventral prostate weight) or steroid production in tissue or cell isolates (e.g., testosterone production in rat or mouse Leydig cells). These early assays were based on receptor binding and biological response and hence were restricted to detecting only biologically active forms of hCG. These assays did not distinguish between hLH and hCG since both molecules, as noted earlier, bind to the same receptor.

Immunological assays for hCG were developed in the 1950s and 1960s. These early tests generally involved hemagglutination or latex agglutination inhibition or complement fixation assays of quite limited sensitivity and specificity, especially with respect to hCG, because of the extensive structural homology between hLH and hCG. As has been shown by several studies over the years, the ratio of biological activity to immunological activity can vary substantially, depending in part on the forms of hCG detected in the immunoassay. Modern research bioassays are increasingly based on an *in vitro* system in which the functional human LH/CG receptor is genetically engineered into a cell line (e.g., Chinese hamster ovary cells). This approach has the advantages of using the more specific human receptor in comparison to the rodent receptor and providing a consistently reproducible biological reagent.

One important caveat applicable to all *in vitro* bioassays for hCG is that the net charge of the hCG influences the response in the *in vitro* bioassay opposite to that of an *in vivo* bioassay. That is, more basic isoforms of hCG (missing some sialic acid) are more potent within the *in vitro* bioassays but less potent *in vivo* because of reduced circulating half-life.

The advent of radioimmunoassay technology was soon followed by the production of a polyclonal hCG (SB-6) antisera. The problem in developing antisera specific to hCG is the great similarity of structure between hCG and hLH. The α subunits are identical in primary structure and offer no possibility of hormone differentiation but the β subunits have sufficient differences to permit development of distinguishing antibodies. Because the structural differences between the β subunits of hCG and hLH are modest, mainly within the CTP region, development of a rabbit polyclonal antiserum that was relatively specific to hCG in the presence of hLH was unusual. SB-6 was one such rare antiserum and was used for several years until the advent of monoclonal antibody technology. SB-6 exhibited some cross-reaction with hLH but had sufficient specificity to measure hCG in the blood without significant interference from hLH. However, urinary hCG assays remained a challenge because the urinary form of hLH interfered extensively with the SB-6 assay. In addition, Schroeder reported that a major urinary metabolite of hCG is a molecule about half the size of intact hCGβ subunit. This molecule, later identified as hCGβ core fragment, was detected in many of the intact hCG assays using polyclonal antisera. The structure of the hCGβ core fragment is shown in Fig. 2. One approach to the continuing specificity problem was the development of antibodies to the CTP region of the hCGβ subunit. Such assays employing anti-CTP antibodies had absolute specificity for hCG with no cross-reaction to hLH (the β CTP extension is not present on the hLH molecule). The CTP-based assays could be applied to both blood and urine but were not particularly sensitive for hCG measurement because of the modest affinity of most CTP antisera when compared to antisera elicited to more potent antigens such as the entire β subunit. Thus, the assay of hCG has developed over the years so that, with modern assay technology, a sensitivity has been achieved which will permit the deletion of pregnancy-associated hCG in maternal urine before the time of the anticipated menses.

Investigations of the metabolic processing of the intact hCG molecule increasingly revealed, however, that multiple forms of hCG were being detected in the blood and urine not only of pregnant subjects or of those with gestational trophoblastic disease but also in subjects with a variety of nontrophoblastic malignancies. In addition, hCG was detected in nonpregnant women and men.

An appreciation has emerged that the individual measurement of hCG molecular variants and associated metabolites can provide information in substantial excess of that provided by determinations of just the intact hCG molecule. The most explicit example of this is the excretion of hCGβ core fragment in malignancy, which is detected with a much greater frequency than is hCG itself. The characterization of the various types of hCG has been made feasible by the development of both monoclonal antibody technology and the immunometric assay format. Monoclonal antibodies provide the ability to detect a single molecular epitope and immunometric assays offer enhanced specificity of measurement. The specificity of the immunometric assay system is a composite of the individual specificities of the two antibodies used in its construction. The most widely used format for an immunometric assay is the "two-site" assay in which one monoclonal antibody, immobilized on a solid phase, commonly a plastic tube or microtiter plate well, is used to extract the analyte from solution and a second monoclonal antibody, labeled with a probe (^{125}I, enzyme, europium, or chemiluminescent) which binds to another epitope on the same molecule, is used for detection. This design generally affords the high specificity needed to measure the spectrum of hCG molecular forms in blood and urine. Thus, it has been possible to construct quite specific assays for intact hCG, hCG free α and β

subunits, nicked hCG, hCGβ core fragment, and, recently, for hCG forms which are sensitive to glycosylation differences. An idealized representation of the profile of hCG and related molecular forms in blood and urine throughout pregnancy is presented in Fig. 6. Individual assay of these forms of hCG has been shown to have clinical utility, as outlined in Section IV.

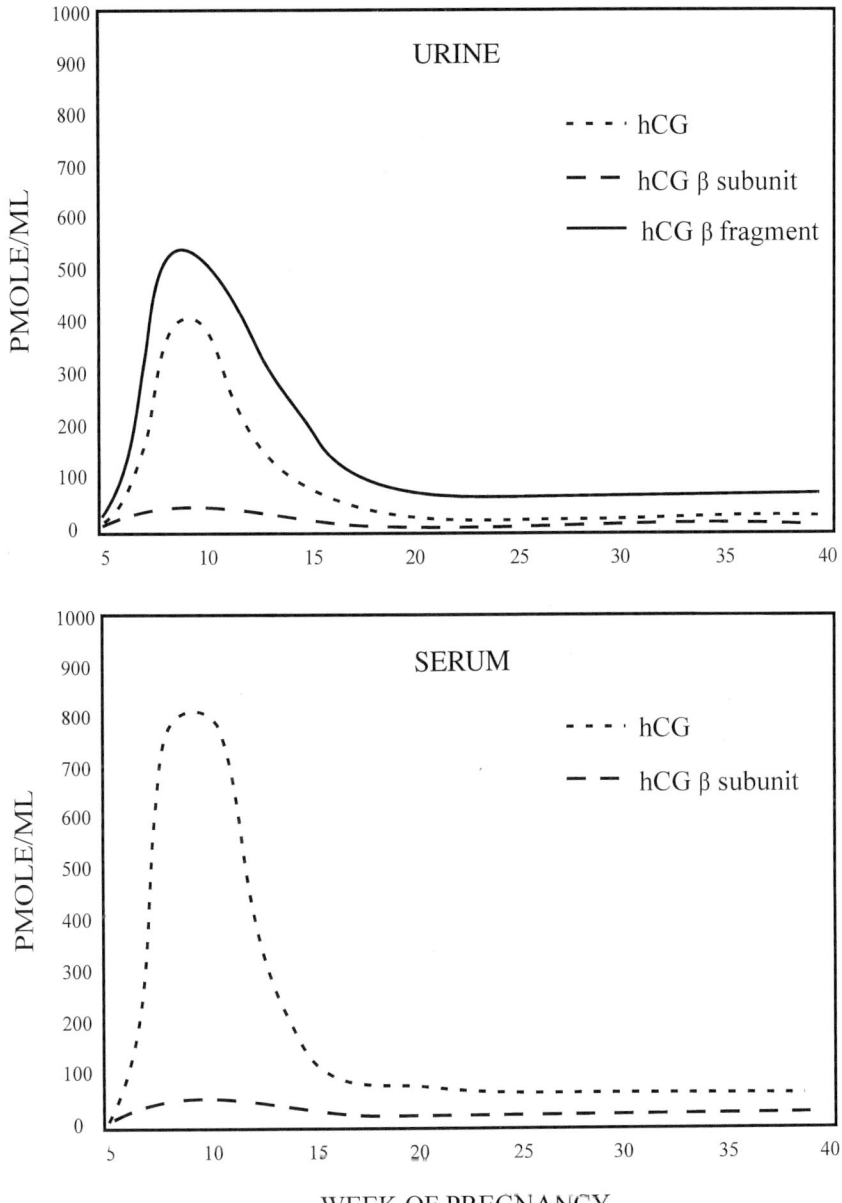

FIGURE 6 An idealized representation of the profiles of hCG and its associated analytes in blood (bottom) and urine (top) throughout pregnancy. hCG values rise exponentially and peak in the first trimester; they then fall to lower levels that are maintained with relatively little variation until the third trimester, when a slight elevation in values occurs. Elevated levels that persist into the second trimester have been associated with various pregnancy disorders. Because there are several isoforms of both intact hCG and, to a lesser degree, its associated metabolites, different assays can provide somewhat different profiles. There is consensus, however, that hCG free β subunit is present in an amount much less than that of the intact molecule in serum. Its presence in urine, however, can be more variable. hCGβ core fragment is detected only in vanishingly small amounts in blood but generally constitutes a major mole fraction of urinary hCG immunoreactivity.

IV. CLINICAL ASPECTS

The clinical utility of hCG measurements has undergone considerable expansion since its initial use in the diagnosis of pregnancy, trophoblast disease, and other forms of germ cell malignancy. Modern hCG testing, in addition to its use in both clinical pregnancy testing and pregnancy detection with home test kits, has found increasing utility in the diagnosis of pregnancy disorders, in the detection of early pregnancy loss, and as a marker of nontrophoblastic malignancy, especially gynecological malignancy. In these applications, measurements of hCG free β and β subunits in blood and urine and of hCGβ core fragment in urine alone have demonstrated utility, often exceeding that derived from the measurement of the intact hCG molecule itself.

Two recent reports provide illustrations. In studies of hCG elevations in Asian women who screened positive for Down's syndrome but delivered a chromosomally normal child, it was determined that elevations of α fetoprotein (AFP) and hCGβ subunit (genetic analysis showed that the fetuses of these women did not have trisomy 21 but the screening assays for AFP and hCGβ were at levels suggestive of trisomy) were associated with low birth weight, abnormal placentation, fetal structural abnormalities other than neural tube defect, prematurity, fetal death, or other perinatal problems. In women who had higher than normal hCG levels in the second trimester, there was a significantly increased risk for preeclampsia, interuterine growth retardation, or premature delivery. Therefore, abnormal expression of hCG alone can be sufficient to predict pregnancy outcome, that is, to identify troubled pregnancies that were not associated with genetic abnormalities. Additional diagnostic uses of hCG and hCG-associated molecules are described in the following sections.

A. Screening for Down's Syndrome

An association between high levels of hCG and Down's syndrome has been documented. The determination of serum-intact hCG, coupled with serum AFP and unconjugated estriol, forms the basis of the so-called "triple screen" for Down's syndrome. Implementation of this screen in a larger population indicated a detection rate of 48% with a false-positive rate of 4.1%. Pregnancies conceived through the aid of assisted reproduction techniques such as *in vitro* fertilization (IVF) appear to display higher incidences of positives in such triple-screen assays, although the reasons are not clear. Additionally, a recent study has shown that there are variations in the pattern of the triple-screen assay among racial groups. Therefore, the results of the triple-screen assay for Down's syndrome or other problem pregnancies may be affected both by the patient's racial background and by the type of conception. These findings make it necessary to develop separate triple-screen databases for women of different racial backgrounds as well as for women in assisted reproduction programs.

Other variations in the hCG portion of the triple-screen assays have led to an examination of the potential role of serum hCG free β subunit and hCG β core fragment instead of the intact hCG molecule as a marker for Down's syndrome. Some investigators claim that free β subunit is superior to hCG itself as such a marker. The most recent use of an hCG-associated molecule for the detection of Down's syndrome involves measurement of urinary hCGβ core fragment. The hCGβ core fragment may prove to be the most valuable component of a triple-screen assay for Down's syndrome as well as for other problems of pregnancy.

B. hCG Molecular Forms as Cancer Markers

hCG determinations have long played a role in the detection of both gestational trophoblastic disease and some testicular cancers. Both choriocarcinoma and hydatidiform mole produced large quantities of hCG, generally related to tumor burden. Nonseminomatous germ cell tumors of the testes produce a detectable serum level of hCG in about 80% of cases, whereas only about 40% of seminomatous tumors produce hCG. Serial measurements of hCG have proven invaluable in monitoring therapy and early recurrence of disease in these cases. However, with other forms of cancer, early studies concerning the expression of hCG in the blood revealed that only approximately 20% of all cancers, especially those of gynecological, lung, melanoma, or head and neck tumors, produced measurable levels of hCG. The levels expressed are most often low, near the threshold of detection, and do not bear any clear relationship to tumor burden or stage of disease. Recently, circulating levels of hCG free α and β subunits have been measured in a variety of malignancies, with sometimes conflicting findings. Few of these patients had measurable serum levels for either of the free subunits. Measurement of intact hCG and free subunits in patients with cancers compared to controls revealed low levels of intact hCG (<1 ng/ml) in nearly all subjects. Serum levels were diagnostic only in cases of testicular cancer. Free α subunit concentrations showed no significant differences between cancer and controls. However, free β subunit levels in excess of 100 pg/ml were found in 70% of the cases of nonseminomatous testicular cancer, 47% of bladder cancers, and approximately 30% of both cervical cancers and pancreatic cancers. The previous examples reinforce the initial findings regarding the circulating hCG molecular forms in malignancy; namely, that they are present in low levels and can be found in both cancer patients and normal controls, originating most probably from the pituitary in the latter group.

The measurement of hCG-associated molecular forms in urine has been more productive, however. Shortly after the development of the first specific immunoassay for hCG, it was noted that the urinary form of hCG found in cancer patients was smaller than the intact hCG molecule. Subsequently, this molecule was identified as hCGβ core fragment (Fig. 2). It has since been demonstrated by several investigators that this molecule is the predominant urinary form of hCG produced in cancer. Initially, it was reported that in a cohort of 87 women with active gynecological cancer, only 18% had detectable levels of intact hCG in their blood. The figure increased to 32% for urinary intact hCG. However, 74% had detectable levels of urinary gonadotropin fragments (primarily hCGβ core fragment) in their urine. These initial findings have been generally confirmed by other investigators, i.e., that urinary β core fragment is the most highly expressed urinary form of hCG in gynecological malignancy as well as other forms of malignancy.

The utility of this marker is compromised by the documented existence of the structurally analogous hLHβ core fragment. This molecule, which can be expected to react in many assays for hCGβ core fragment and has been shown to be present in the urine of pre- and postmenopausal women and men, has the potential to reduce the specificity of assays in cases of malignancy. However, hCGβ core fragment measurements do appear to correlate with disease progression and have found some utility in disease staging, even though their utility as a screening vehicle remains moderate.

C. hCG in Early Pregnancy Loss Detection

Early pregnancy loss (EPL), defined as the loss of a pregnancy before either the woman or her physician are aware of its existence, is used as an endpoint in epidemiological studies of exposures to putative reproductive toxins in the environment, the workplace, and by personal use. Early investigations of EPL gave widely divergent estimates of its incidence in the general population, probably because of differences in assay specificity and sample collection protocol. Measurement of urinary hCG has become the most widely used technique for EPL. A large-scale epidemiological study of EPL in a population of women free of known reproductive dysfunction has been performed using an immunometric assay for urinary hCG, specific for the intact molecule with the β CTP intact, and with a sensitivity of 0.01 ng/ml hCG. The incidence of EPL was found to be 22%. Combined with the observed incidence of clinical loss, the total rate of pregnancy wastage was 32%.

The many technical problems associated with this assay format included labor intensity, requirement for large volumes of urine, and an extensive collection protocol. A concern was raised that measurement of just the intact hCG molecule might not be sufficiently sensitive to detect all EPL events. These considerations led to investigations of more abbreviated collection protocols and improvement in assay design. The latter consideration gave rise to the microtiter plate-based assay format which could be modified to measure simultaneously the three major urinary forms of hCG, i.e., the intact molecule, the free β subunit, and hCGβ core fragment. A slightly different format allowed for the individual quantitation of these analytes. Although there is currently no firm consensus as to the optimum assay for EPL studies, most of the studies have employed the simultaneous assay of hCG, hCG free β subunit and hCGβ core fragment, or the intact hCG format described previously.

A recent study using the intact hCG version of the assay found in a presumably reproductively normal population that 31% of all conceptions ended in loss. Using the triple-analyte version of the assay, and collecting three specimens per cycle, it has been found that the overall loss rate in the population was 17–20.5% depending on the urine sampling technique employed (i.e., which urine samples were analyzed). Current EPL studies include exposure to electromagnetic fields and occupational exposure to industrial toxicants. It is likely that future studies of EPL will be extended to populations exposed to the large number of identified environmental endocrine disrupting agents.

D. Ectopic Pregnancy

The incidence of ectopic pregnancy has been established to be in the range of 9–14 per 1000 pregnancies. Ectopic pregnancy is the leading cause of maternal death in the first trimester. hCG determinations can be useful for the diagnosis of ectopic pregnancy and also for monitoring response to treatment, especially when the intervention is nonsurgical, e.g., administration of methotrexate. The immunological methods used in the diagnosis of ectopic pregnancy have included a comparison of absolute levels of hCG or serial sampling of hCG. Levels of hCG are usually lower in ectopic pregnancies than in normal intrauterine gestations, presumably because of abnormal placentation. Additionally, the normal doubling time of hCG in intrauterine pregnancy is generally not achieved in ectopic pregnancy. Only 15% of ectopic pregnancies achieve the normal rate. Although both absolute levels of hCG and its rate of increase can be used to differentiate ectopic pregnancy, usually in conjunction with ultrasound, the specificity of this approach is less than perfect, leading some to investigate alternative combinations of markers.

The utility of hCG free α and β subunit measurements has been evaluated in ectopic pregnancy in a population of 49 IVF pregnancies. There were 11 ectopic pregnancies in this group. Low intact hCG values could distinguish most ectopic pregnancies by Day 12. The addition of hCG free α and/or β subunit measurements provided only marginal improvement in detection. However, later in the pregnancy hCG free α subunit measurements proved to be the most sensitive marker of a normal pregnancy.

The addition of a serum progesterone measurement to that of hCG materially improves the efficiency of ectopic pregnancy diagnosis. An hCG value below 3000 IU/liter, coupled with a serum progesterone of <40 nmol/liter, had a positive predictive value of 91% for an abnormal pregnancy outcome and a negative predictive value of 95%. The utility of urinary hCGβ core fragment as a marker in ectopic pregnancy measurements has been assessed in a recent report. A series of 12 women, at 38–80 days of gestation, with surgically proven ectopic pregnancy were compared to 36 women of the same gestational age who had normal intrauterine pregnancy. Measurements of urinary intact hCG, free β subunit, and β core fragment were compared between the two groups. The greatest quantitative distinction between the two groups was afforded by the β core fragment assay, which displayed a much greater reduction in concentration than did either of the other analytes and provided a positive predictive value in excess of 98%.

See Also the Following Articles

Ectopic Pregnancy; FSH (Follicle-Stimulating Hormone); Gonadotropin Receptors; Gonadotropin Secretion; LH (Luteinizing Hormone); Pregnancy in Humans, Overview

Bibliography

Albanese, C., Colin, I. M., Crowley, W. F., Ito, M., Pestell, R. G., Weiss, J., and Jameson, J. L. (1996). The gonadotropin genes: Evolution of distinct mechanisms for hormonal control. *Recent Prog. Horm. Res.* **51**, 23–61.

Baenziger, J. U. (1994). Protein-specific glycosyltransferases: How and why they do it. *FASEB J.* **8**, 1019–1025.

Ben-Menahem, D., and Boime, I. (1996). Converting heterodimeric gonadotropins to genetically linked single chains. *TEM* **7**, 100–105.

Campbell, R. K., Bergert, E. R., Wang, Y., Morris, J. C., and Moyle, W. R. (1997). Chimeric proteins can exceed the sum of their parts: Implications for evolution and protein design. *Nat. Biotechnol.* **15**, 439–443.

Hussa, R. O. (1987). *The Clinical Marker HCG.* Praeger, New York.

Jameson, J. L., and Hollenberg, A. N. (1993). Regulation of chorionic gonadotropin gene expression. *Endocr. Rev.* **93**, 203–221.

Lapthorn, A. J., Harris, D. C., Littlejohn, A., Lustbader, J. W., Canfield, R. E., Machin, K. J., Morgan, F. J., and Isaacs, N. W. (1994). Crystal structure of human chorionic gonadotropin [see comments]. *Nature* **369**, 455–461.

Muyan, M., and Boime, I. (1997). Secretion of chorionic gonadotropin from human trophoblasts. *Placenta* **18**, 237–241.

O'Connor, J. F., Birken, S., Lustbader, J. W., Krichevsky, A., Chen, Y., and Canfield, R. E. (1994). Recent advances in the chemistry and immunochemistry of human chorionic gonadotropin: Impact on clinical measurements. *Endocr. Rev.* **15**, 650–683.

Ruddon, R. W., Sherman, S. A., and Bedows, E. (1996). Protein folding in the endoplasmic reticulum: Lessons from the human chorionic gonadotropin beta subunit. *Protein Sci.* **5**, 1443–1452.

Ryu, K. S., Ji, I., Chang, L., and Ji, T. H. (1996). Molecular mechanism of LH/CG receptor activation. *Mol. Cell Endocrinol.* **125**, 93–100.

Wu, H., Lustbader, J. W., Liu, Y., Canfield, R. E., and Hendrickson, W. A. (1994). Structure of human chorionic gonadotropin at 2.6 A resolution from MAD analysis of the selenomethionyl protein. *Structure* **2**, 545–558.

Chorionic Gonadotropins, Nonhuman Mammals

Marylynn Barkley and Mary B. Zelinski-Wooten

University of California, Davis, and Oregon Regional Primate Research Center

I. Functions of Chorionic Gonadotropins
II. The Structure of Chorionic Gonadotropins
III. Mechanism of Action
IV. Secretion
V. Regulation
VI. CG-like Hormones in Nonprimates
VI. Summary

GLOSSARY

blastocyst The stage of embryonic development at which implantation occurs; consists of an inner cell mass that will develop into the fetus and an outer covering of cells called the trophoblast which will differentiate into the placenta.

conceptus The sum of derivatives from a fertilized ovum at any stage of development from fertilization until birth, including extraembryonic membranes as well as the embryo or fetus.

decidua The endometrium of the pregnant uterus, all of which except the deepest layer is shed at parturition.

deciduoma (plural, ***deciduomata***) An intrauterine mass containing decidual cells.

hypophysectomy Surgical removal of the pituitary gland.

luteotropic The ability to stimulate the corpus luteum, including cellular growth and hormone synthesis.

parturition The act or process of giving birth.

progesterone The steroid hormone produced initially by the corpus luteum of the ovary during early pregnancy and later by the placenta of primates, essential for the maintenance of pregnancy, that prepares the uterine endometrium for implantation of the embryo and prevents contraction of the uterine myometrium during gestation.

pseudopregnancy False pregnancy; can be induced in some mammals by cervical stimulation.

receptor A protein that recognizes and binds with other specific molecules.

signal transduction The biochemical pathway by which molecules send information from the cell surface to the nucleus.

trophoblast A layer of extraembryonic tissue on the outside of the blastocyst; from it are derived the chorion and the amnion.

Chorionic gonadotropins are glycoprotein hormones that provide embryonic communication essential for maternal recognition of early pregnancy. Chorionic gonadotropins are produced by the placentae of primates and certain other mammals. These hormones circulate in the blood and bind to specific receptors in target tissues to induce cellular responses, the most important of which is progesterone production by the corpus luteum. Rescue of the corpus luteum by chorionic gonadotropin (CG) is dependent on maternal signal transduction. The fetal portion of the placenta secretes CG which provides separate and synergistic regulation of luteal activity early in pregnancy. CG has other actions, including immunomodulation and neurotropic effects on the developing brain. Several mammalian species produce chorionic gonadotropins. These include women, primates, and equines, but CGs seem to be absent in bovines, ovines, canines, and felines. Controversy surrounds the existence of CGs in guinea pigs, hamsters, rats, and mice. Their placentae contain factors that have CG-like immunoreactivity, but as yet there is no convincing evidence that CG genes or gene products are present in these species. The study of CG production, function, and bioactivity continues to stimulate critical thinking about normal and abnormal ovarian and fetal–placental function in diverse mammals, particularly those on the endangered species list.

I. FUNCTIONS OF CHORIONIC GONADOTROPINS

Three to four days after ovulation and fertilization of the primate ovum, the early embryo is propelled from the oviduct into the uterus. At the blastocyst stage of development, the embryo invades the uterine endometrium between 7 and 9 days postovulation in women and macaques and on Days 11 or 12 in marmosets. All species of nonhuman primates studied thus far synthesize and secrete chorionic gonadotropin (CG) from the trophoblast cells of the blastocyst during early pregnancy. Chorionic gonadotropin is the essential signal that announces the intrauterine presence of the conceptus to the mother. In a fertile menstrual cycle, embryonic CG acts as a luteotropin which "rescues" the corpus luteum from regressing in the maternal ovary as it normally would during a nonfertile cycle, allowing continued luteal production of progesterone during early pregnancy. Unlike many nonprimates, extended progesterone synthesis by the corpus luteum is subsequently replaced by placental production of progesterone during an interval designated as the "luteal–placental" shift. In women and rhesus monkeys, the luteal–placental shift occurs at approximately the sixth week and after the third week of pregnancy, respectively. A somewhat analogous period, referred to as the "luteotropic" shift, occurs in certain other mammals. During the luteotropic shift in rodents, the regulation of ovarian progesterone synthesis changes at midpregnancy from maternal to fetal–placental control, but the site of progesterone production remains luteal throughout the remainder of gestation.

The crucial need for CG to extend luteal function during early pregnancy has been established in nonhuman primates. Passive immunization with anti-CG antiserum or active immunization with modified forms of CG prior to the luteal–placental shift causes a decline in serum progesterone levels, pregnancy termination, or infertility which can be overcome by progesterone administration. Using a different approach, treatment of Old and New World monkeys and women with exogenous CG extends the life span of the corpus luteum and progesterone production during the menstrual cycle. Limited evidence suggests that CG transiently stimulates progesterone production by enhancing the uptake and availability of cholesterol sources rather than by stimulating the activity of key steroidogenic enzymes. In addition to progesterone, CG promotes luteal production of additional steroids including estrogens and androgens. In the macaque corpus luteum, CG increases the expression of aromatase, the enzyme that converts androgen substrates such as testosterone and androstenedione into estradiol and estrone, respectively. CG may also enhance androgen synthesis in theca luteal cells in addition to stimulating aromatase activity in the granulosa luteal cells in the central portion of the corpus luteum. This effect of CG on androgen production provides a model for CG-like activity in nonprimates, particularly during the luteotropic shift.

Luteal synthesis of the peptide hormones relaxin and inhibin also increases in women and nonhuman primates following exposure to CG. The production of significant amounts of inhibin is apparently unique to the primate corpus luteum. Whether these hormones are important for the maintenance of early pregnancy in primates is unknown. Other functions of CG include effects on the immune system and possible stimulation of Leydig cells in the fetal gonad thereby facilitating sexual differentiation. This may explain the presence of human CG (hCG) in amniotic fluid and fetal serum. hCG also has neurotropic effects in fetal rat brain neurons which express functional luteinizing hormone (LH)/CG receptors.

II. THE STRUCTURE OF CHORIONIC GONADOTROPINS

A. Biochemical Aspects

Chorionic gonadotropins are members of the glycoprotein hormone family including thyrotropin (TSH), follicle-stimulating hormone (FSH), and LH, to which CGs bear the most functional similarity. The glycoprotein hormones are heterodimers composed of a conserved α subunit and a hormone-specific β subunit which discriminates between similar glycoprotein hormone receptors in target tissues. Neither free α nor β subunit has significant biological activity. Limited similarity between α subunits and the individual β subunits within a mammalian species suggests that the α and β subunits have a common evolutionary origin but diverged very early in the evolution of the glycoprotein hormones. A common evolutionary ancestor for all the β subunits is also likely given the varying degrees of β subunit identity and the observation that CGs are present only in some mammals. For example, the first 114 amino acids of human LHβ and CGβ are highly homologous in sequence which suggests that the β subunit of CG evolved from β-LH or a β-LH-like gene. The unique presence of a 24- to 31-amino acid extension at the C terminus strengthens this idea and confers upon CG "last-in-line" status for emergence as a functional glycoprotein hormone.

The primate CGβ gene probably evolved through a single base pair deletion in exon III and a two-base pair insertion that extended the translated region through the C terminus of a LHβ-like sequence. Marmoset CGβ contains the C-terminus extension suggesting that the CGβ gene evolved around 40 million years ago before the divergence of Old and New World monkeys. This serine-rich C-terminus extension accommodates serine-O-glycosylation of CGβ which markedly prolongs its half-life in the circulation and increases its biological potency. The recent discovery that heterodimer stabilization of CG is accomplished by wrapping part of the β subunit around the α subunit and fastening this by a "seat belt" covalent linkage of Csy26–Cys110 led to the surprising realization that the glycoprotein hormones are part of a superfamily of endocrine/paracrine factors that contain a cysteine knot. This group includes nerve growth factor, transforming growth factor-β, and platelet-derived growth factor.

B. The Chorionic Gonadotropin Genes

The α and individual β subunits of the known CGs are encoded by separate genes. Only one single-copy gene for the α subunit of hCG in humans was detected and localized to chromosome 6. As could be predicted by the species variation and hormonal specificity conferred by individual β subunits, the organization of the LHβ and CGβ genes is complex. Multiple copies of structurally related human β genes are clustered together and form an array of tandem and inverted copies suggestive of gene duplication on chromosome 19. All these genes are actively transcribed but with different levels of expression. The hLHβ gene is inactive in the placenta. The different patterns of expression of LHβ and CGβ genes are partly explained by variation in the locations of their transcriptional start sites and promoter regulatory regions. The human α gene is expressed in different cells, including thyrotropes, gonadotropes, and syncytiotrophoblasts, and is regulated by a variety of hormonal pathways that control the biosynthesis of TSH, FSH, LH, and CG. Embryonic expression of the α gene precedes expression of all other classic pituitary hormone genes and CGβ, presumably controlled by transcription factors that appear early in development. Different transcription factors are involved in the subsequent extinction of α gene expression in other pituitary cell lineages. In the immature primate trophoblast, it appears that β subunits are produced in excess of α subunits and only as the cytotrophoblast begins to differentiate does production of α subunit equilibrate with β subunit availability for heterodimer formation. This provides indirect evidence for independent regulation of subunit biosynthesis in the placenta in keeping with what occurs in pituitary cell lines. Positive and negative regulators of CG gene transcription have been sought and several factors have been identified including gonadotropin-releasing hormone (GnRH); cAMP; activin/inhibin; CG itself; steroids including progesterone, glucocorticoids, and local growth factors; and cytokines acting in autocrine, paracrine, or endocrine fashion. It appears that the placenta contains the same factors or analogs that the hypothalamic–pituitary–gonadal axis produces, and, although speculative, placental CG expression may involve essentially the same physiological controls as pituitary gonadotropin production.

C. CG and the LH/CG Receptor

The proposition that CGβ evolved from a LHβ-like sequence is supported by the binding of both LH and CG to a common LH/CG receptor. The LH/CG receptor is a member of the seven-transmembrane (TM) receptor family encoded by 11 exons and composed of two functional domains consisting of an extracellular N-terminal half and a membrane-associated C-terminal half. The extracellular N-terminal domain of glycoprotein hormone receptors is by itself responsible for high-affinity hormone binding. Exon 11 of the LH/CG receptor encodes the membrane-associated domain which can bind hormone with low affinity and is capable of receptor activation. When compared to CG, LH has a lower receptor binding affinity and steroidogenic potency due to faster rates of dissociation, internalization, and sulfated terminal carbohydrate moieties which shorten its half-life. Ultimately, the structural characteristics of the peptide chains confer functional differences in these glycoprotein hormones despite their close sequence homology. Species variation in the ability of LH and CG to bind to nonspecies LH/CG receptor has long been recognized, e.g., hCG and rat LH will bind the rat LH/CG receptor and activate signal transduction, but rat LH will not bind the human LH/CG receptor. The greater ligand specificity of the human LH/CG receptor is further demonstrated by the inability of ovine or equine LH to displace hCG binding. Surprisingly few amino acids confer the ability of primate CGs to recognize domestic animal and rodent LH/CG receptors. Small portions of the LH molecule likewise exclude recognition of rodent LHs by primate LH/CG receptors.

The crystal structure of hCG reveals extensive regions of contact between the α and β subunits and holohormone surface sites likely to be available for receptor binding. Residues throughout both hormone subunits may participate in essential, high-affinity hormone receptor contacts, with species differences in the affinity of ligands for the LH/CG receptor probably caused by a few key residues in the primary receptor contact site. Using chimeras of hCG and bovine LH (bLH) to bind human and rat LH/CG receptors, the low affinity of bLH for human LH/CG receptor seems to involve an interaction between bLH subunits that either distorts the region of the high-affinity contact and/or creates a steric interaction between the hormone and the receptor. Specifically, differences in the abilities of hCG and bLH to bind human LH/CG receptor are due primarily to an influence of the seat belt on subunit interaction. Thus, subtle changes in hormone conformation have a dramatic effect on ligand binding. This is best illustrated by an experiment of nature in which a point mutation encodes an arginine in place of glutamine at position 54 in the β subunit of human LH resulting in an immunologically active hormone virtually devoid of LH/CG binding activity. A glutamine residue at this sequence position is conserved among all gonadotropins, emphasizing its important influence on both subunit association and receptor binding, a feature further demonstrated by site-directed mutagenesis.

III. MECHANISM OF ACTION

In women and monkeys, pituitary secretion of LH is episodic and LH pulse frequency slows during mid- to late luteal phase. Since CG has a much longer half-life than LH and is produced in increasing amounts by the developing conceptus, any vestiges of LH pulses are obscured by continuously rising CG levels. The duration of CG action on steroidogenesis is longer than that of LH, and this correlates with differences in the rate of LH/CG receptor movement or turnover in luteal membranes when the binding of CG is contrasted to that of LH. How the LH/CG receptor carries out its function of relaying hormone activation to the target cell's response system is actively under study. The LH/CG receptor is coupled to the adenylate cyclase and phospholipase C (PLC) signal pathways, presumably through Gs and Gq, respectively, which induce the production of cAMP and inositol phosphates (IPs) as intracellular signal molecules. Apparently, the cAMP signal and the IP signal originate and diverge upstream of G protein coupling. Generation of these two signals independently occurs at the cell surface near or at the hormone binding site. The two signals diverge and transduce through lysine 583 in exoloop 3 of the rat LH/CG receptor, the cAMP signal presumably toward a Gs coupling site and the IP signal to a Gq site in cytoloops.

In mammals other than primates, exposure of target cells to elevated concentrations of CG results in a progressive loss of specific receptors (downregulation) and a reduction of the cell's biological responses to renewed hormonal stimulation. However, rising titers of CG do not have this effect in primates and women. When early pregnancy is simulated in the macaque, elevated levels of CG do not induce a loss of LH/CG receptors per se. In this situation, the total LH/CG receptor population remains constant and is characterized by a decrease in numbers of unoccupied (free) receptors and an increase in CG-occupied receptors. A similar phenomenon occurs in women, i.e., receptor downregulation does not explain declining progesterone production in the face of rising CG levels. How, then, does the primate ovary limit its response to CG stimulation? In the macaque corpus luteum, the affinity of the LH/CG receptor for CG decreases significantly (fivefold) after 3–6 days of CG exposure. Furthermore, the ability of the adenylate cyclase system to respond to CG is severely impaired after 3 days of CG exposure *in vivo* and, by Day 6, stimulation of cAMP production by gonadotropin is lost. This is a specific impairment which is expressed even though the adenylate cyclase system remains functional and capable of transducing signals from prostaglandin receptors but not from gonadotropin receptors. If factors other than CG are present, including prostaglandins of the E and I series and nonhormonal activators such as nucleotides and forskolin, continued stimulation of cAMP production takes place. This specific reduction in luteal responsiveness to CG stimulation suggests that homologous desensitization occurs in the macaque corpus luteum during early pregnancy which may contribute to a transient decline in progesterone production. The uncoupling of LH/CG receptors from cAMP production in early pregnancy may be an important step in the progestational response of the corpus luteum to CG. The role of gonadotropin-independent stimulation of cAMP-mediated events during this gestational period is yet to be defined.

IV. SECRETION

A. Nonhuman Primates

Assays have been developed for the detection of all primate CGs during pregnancy using various methods that are either nonspecific, i.e., able to measure CGs from a variety of nonhuman primates (bioassays or radioreceptor assays), or specific for a particular primate (radioimmunoassay). Although human and nonhuman primate CGs share some structural and functional similarity, the reagents developed for detecting hCG in the urine of pregnant women can measure hCG and the CGs of great apes but not the CGs of other nonhuman primates. Levels of CG in these other species are typically measured in peripheral blood using bio- or radioimmunoassays (RIA) that do not distinguish among species. For example, a bioassay developed to measure CG in rhesus monkeys has also been used to detect CGs in other Old World as well as New World primates (Table 1). Measurement of CG at distinct intervals throughout pregnancy has been accomplished to varying degrees in New World primates (chimpanzees, gorillas, orangutans, baboons, and macaques including rhesus, cynomolgus, and bonnet monkeys) and Old World primates (marmosets, tamarins, and owl and squirrel monkeys). Although the patterns of CG secretion differ among nonhuman primates, three basic examples have generally emerged from the available data (Table 1). First, in women and the great apes, CG in blood or urine is detectable at the time of implantation, attains maximal levels during the first trimester of pregnancy, and can be measured continuously throughout the remainder of gestation. Second, in baboons and macaques, CG is secreted coincident with implantation, but in contrast to women and the great apes, a transient increase in this hormone occurs during the first trimester. Thereafter, CG declines to very low or nondetectable levels during mid- and late gestation in these nonhuman primate species. The third pattern of CG secretion is found in New World monkeys (marmosets, tamarins, and squirrel and owl monkeys) and consists of detectable levels at midgestation that decline continuously until parturition. In relative terms, peak levels of CG are highest in women, 10-fold less in the great apes, and 100- to 1000-fold less in baboons, macaques, and New World primates.

TABLE 1
Comparative Aspects of CG Secretion during Pregnancy among Various Primate Species

Species	Duration of gestation (days)	First detection of CG, day postovulation	Peak levels of CG, days of gestation	Duration (days) of CG production
Human (*Homo sapiens*)	270	8–10	56–82	8–270
Great apes				
Chimpanzee (*Pan troglodytes*)	229	11	30–50	9–229
Orangutan (*Pongo pygmaeus*)	230	10	30–50	30–230
Gorilla (*Gorilla gorilla*)	265	10	21–60	21–245
Old World primates				
Baboon (*Papio anubis*)	180	11–12	27	11–51
Macaques	165–170	10	20	10–40
Rhesus monkey (*Macaca mulatta*)				
Cynomolgus monkey (*Macaca fasicularis*)				
Bonnet monkey (*Macaca radiata*)				
New World Primates				
Marmoset (*Callithrix jacchus*)	135–145	14–17	50–70	14–134
Squirrel monkey (*Saimiri sciuresus*)	146	25	50–70	25–146
Owl monkey (*Aotus trivigatus*)	145	15	50–90	15–145
Golden lion tamarian (*Leontopithecus rosalia*)	126–129	25	50–60	25–60
Prosimians				
Ruffled lemur (*Lemur variegatus*)	95	nd	nd	nd

Note. Data represents estimates from determinations of urinary or serum CG concentrations. nd, nondetectable.

Among the prosimians, lemurs apparently do not produce CG during pregnancy.

In vitro culture of nonhuman primate embryos obtained by surgical flushing of the uterus revealed that CG production precedes its detection in the peripheral circulation. The rhesus monkey blastocyst can produce CG prior to attachment, whereas CG secretion by the marmoset blastocyst commences at the time of attachment; in both cases, CG production increases exponentially for several days. Using human embryos, it was determined that the structural forms of secreted hCG vary throughout pregnancy. Prior to implantation, the β subunit of CG is preferentially secreted, with subsequent production of dimer hCG around implantation, followed by excess α subunit secretion at the end of the first trimester. Independent regulation of CG subunit production by nonhuman primate embryos and its physiological relevance during early pregnancy have not been examined.

V. REGULATION

The mechanisms controlling the temporal patterns of CG synthesis and secretion during pregnancy in primates are not well understood. *In vitro* studies on human placental cells and CG-producing cell lines have yielded information indicating there are several hormonal and local modulators of CG production. Collectively, the evidence suggests that hypothalamic (i.e., GnRH) and ovarian hormones (steroids, inhibin, activin, and growth factors) that regulate gonadotropin secretion by the pituitary gland are also produced by the placenta and can be considered potential modulators of CG production possibly via direct, GnRH receptor-mediated actions. The presence of the inner cell mass in the marmoset blastocyst is required prior to CG gene expression in the trophoblast in order for timely CG secretion to occur. Once expression of the CG genes is initiated, removal of the inner cell mass does not affect subsequent CG production. In rhesus monkeys and marmosets, GnRH was localized to and secreted by preimplantation embryos. Levels of GnRH secreted by blastocysts *in vitro* increased over time as the embryos attained peri- and postimplantation stages of development. Secretion of CG by blastocysts *in vitro* was suppressed by a GnRH antagonist and stimulated by a GnRH agonist, implicating GnRH as a modulator of CG production in nonhuman primate embryos. However, embryo attachment was reduced following exposure to either the antagonist or the agonist. Therefore, GnRH may affect CG synthesis directly or indirectly by influencing early embryonic development in nonhuman primates. CG may also autoregulate its own placental biosynthesis, particularly during late gestation.

For several decades it has been clear that CG is the major embryonic factor required to maintain early pregnancy in primates. The functions of CG are shared among human and nonhuman primates, and information continues to emerge about the regulation of CG subunit gene expression and synthesis in these species. The divergence in CG structure and secretory patterns during pregnancy among Old and New World primates is providing valuable insight into the embryo–maternal dialogue in mammals. Studies in nonprimates are also providing a better understanding of the actions of CG as well as its role in the regulation of luteal function.

VI. CG-LIKE HORMONES IN NONPRIMATES

For over 50 years reproductive biologists have questioned how the embryo communicates with the mother to maintain early pregnancy if not through the actions of a CG or CG-like hormone. A variety of experimental approaches were used to search for a CG in species other than equines and primates. A trophoblastic protein with luteotropic properties was found in bovines and ovines, but there is no apparent CG in these species, nor has a CG been found in canines and felines. This is the case in rodents as well, despite a large body of physiological evidence that indirectly supports the existence of a placental product with gonadotropic properties. The validity of a rodent CG has subsequently become a subject of debate because the absence of molecular proof has created controversy about whether a placental gonadotropic hormone is encoded in murine species.

A. Historical Perspectives

Early experiments with rodents provided fundamental information about hormonal communication between the conceptus and the maternal ovaries. It appeared that the fetus and its associated membranes elaborated chemical messengers essential for the prolongation of luteal life span. By 1930 it was known that bilateral removal of the ovaries terminated rat gestation. Thereafter, the maternal pituitary was found to control luteal function during the first half of pregnancy by a mechanism that also maintained pseudopregnancy. Although deciduomata (stimulated endometrium) could not prolong pseudopregnancy, placental membranes alone were sufficient to maintain rodent gestation. The 1938 observation that crude extracts of rat placentae induced deciduomata after uterine trauma suggested that the rodent placenta secreted a product that resembled chorionic substances found in the woman and the mare. This information led to the proposal that a similar hormonal mechanism maintained the luteal function of pregnancy in many mammals.

For several years the assumption that the hormone produced by the rodent placenta was similar in structure and function to primate CGs went unchallenged. The 1950 report that, unlike human CG, the ability of rat placental luteotropin to stimulate pigeon crop sac activity was weak or nonexistent called into question the similarity between rodent and other mammalian CGs. By the 1960s it appeared that rat trophoblastic cells produced a substance with both luteotropic and mammotropic actions. The subsequent study of a rat choriocarcinoma revealed that trophoblastic giant cells were the likely origin of the placental luteotropin(s). Improved biochemical technology stimulated active pursuit of the molecular identification of these rodent placental hormones. Initial studies supported the production of more than one placental molecule, consistent with the physiological information that both luteotropic and mammotropic activity were provided by rat chorionic products. The rapid expansion of placental hormone isolation and characterization in the late 1970s and early 1980s documented the existence of several prolactin-like molecules with distinct temporal secretory patterns and inherent luteotropic and mammotropic bioactivity. In both the rat and the mouse, a placental lactogen with luteotropic and mammotropic activity was isolated, identified, and sequenced from placental material obtained at midpregnancy. This hormone was designated placental lactogen I (PL-I). A second hormone, placental lactogen II, was found in greater abundance during the second half of rodent gestation and is mammotropic. Additional hormones with prolactin-like activity were also identified. The hormones of the prolactin-like group originate first from the decidua and then from the trophoblast.

Prior to the molecular conformation of rodent placental lactogens, the reported immunoreactivity of guinea pig, rat, mouse, and hamster placental extracts in hCG-β and α-subunit RIAs supported the idea that these mammals produce a CG structurally similar to hCG. Although later attempts to identify these compounds would fail, in chromatographic studies of the murine placental material, CG-like α-subunit and β-subunit immunoreactivity eluted in the same positions as those of hCG, hCG-α, and hCG-β subunits. The pursuit of a rodent CG also produced evidence that placental CG-like secretion occurred with a biphasic pattern. The first detectable increase in mouse CG-like activity occurred at midpregnancy, followed by a second rise during late gestation. Placental extracts from all three species also had significant gonadotropic bioactivity in a mouse uterine test. Extracts of rat and hamster placentae had gonadotropic activity similar to that found in normal human placentae at term. Stimulation of fetal mouse testes with placental extracts induced testosterone production, providing even more support for the existence of a rodent hormone with CG-like bioactivity. In a more specific study, androgen values were calibrated against an hCG standard and expressed in terms of gonadotropin-like activity following placental stimulation of Leydig cell testosterone production. Androgen synthesis in response to placental material obtained throughout pregnancy occurred with the biphasic periodicity established for mouse CG-like immunoreactivity. Even the gestational profile of maternal testosterone levels showed a remarkable resemblance to the secretory pattern and placental content of bioactive, placental gonadotropin (Fig. 1). Support for a rat placental hormone apart from the placental lactogens also emerged. The prolactin-like activity of rPL-I could not account for the ability of rat serum or placental extracts to stimulate Leydig cell production of testosterone. In the absence of the maternal pituitary, the persistence of gonadotropin-like stimulation of androgen production by placentae and sera from pregnant rats supported the rodent CG concept. The same study showed that both the secretory pattern and response to hypophysectomy of the putative rat CG were distinct from those observed for rPL, including rPL radioreceptor activity.

B. Attempts to Isolate and Identify Rodent CG

Scientific interest and the successful identification of placental lactogens led to several attempts to purify "rodent CG," but none were successful. Despite this problem, many reproductive biologists remained convinced that rodents rescued and maintained luteal function as did primates—through the production and activity of a CG. The ability of primate CGs to recognize LH/CG binding sites in nonprimate species strengthened this notion. It seemed just a matter of time before the rodent CGs would be isolated, identified, and sequenced. However, as early as the 1950s, some attempts to detect gonadotropin activity in the rat placenta had failed. Other studies reporting a similar lack of success led to disagreement regarding the existence of a rodent CG. This dissension became more pronounced following the inability to detect any mRNA encoding the α subunit or a LH or CGβ-like species in rat placenta using a rat pituitary α-subunit cDNA. Nor could anyone find a mRNA encoding the β subunit of a LH-like molecule in the rat placenta. Rat genomic DNA fragments obtained by restriction enzyme digestion were assayed for LHβ and LHβ-like genes, and only a single subunit gene encoding pituitary LHβ subunit mRNA was detected using hybridization conditions of high stringency. Even when cloned cDNAs coding for both the α and β subunits of rat LH were used to look for gonadotropin-like subunit mRNAs in rat placenta under conditions of low stringency, neither α-subunit nor β-subunit mRNAs were detected. It became widely accepted that only a single LHβ or CGβ-like subunit gene is present in the rodent which

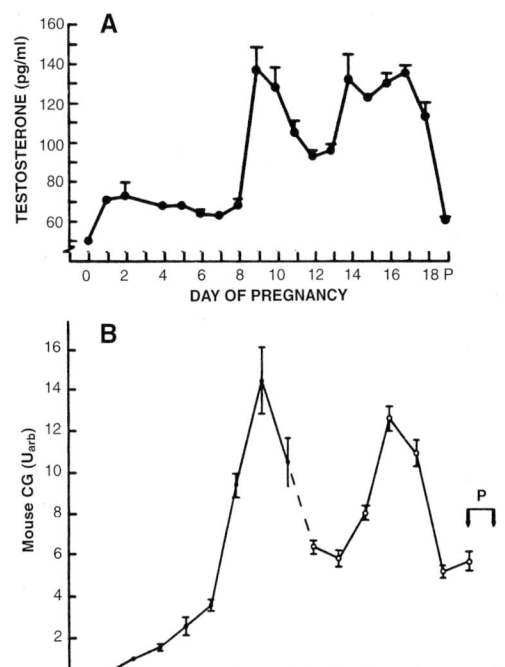

FIGURE 1 (A) The pattern of testosterone levels in maternal blood throughout pregnancy and at parturition (P) in mice. Each point represents the mean + SE (reproduced with permission from Barkley et al., 1977). (B) Average content of mouse CG per implantation site (●) or placenta (○) from early pregnancy to term in the mouse. Estimations were made with a RIA on extracts of pooled implantation sites (Days 5–12) and of placentae (Days 13–19). The average content of a Day 6 implantation site was designated 1 U$_{arb}$. P, parturition (reproduced with permission from Wide and Wide, 1979).

corresponds to the β subunit of pituitary LH. This is also the case for equines, in which only one gene for LHβ is found. Although it was proposed that rodent CG-like activity could reside in a placental molecule sufficiently unlike LH to allow detection by LH subunit cDNA probes, or in a molecule of extraplacental origin, these hypotheses mustered little scientific investigation.

A CG-like hormone has not been detected in mouse trophoblast, but the human gonadotropin α promoter coupled to a chloramphenicol acetyltransferase expression vector is expressed in both mouse placenta and pituitary cells. This indicates that the transcriptional environments of human and mouse trophoblast are similar enough to support expression of the α subunit. This observation led to study of placental gene transcription in transgenic mice constructed to express the human CGβ gene cluster. Mouse placenta did not just express human CGβ transcripts—it did so in a developmentally regulated manner. The ratios of CGβ genes were comparable with human placenta, and the promoter of the human gonadotropin α-subunit gene was also transcriptionally active. CGβ transcripts were also expressed in the cerebral cortex and pituitary of these transgenic mice.

CG gene expression by the mouse placenta stimulated another attempt to find a CGβ-like molecule in murine species. If rat CG exists in rat placenta, it was argued that the amino acid sequence and nucleotide sequence should be highly homologous or identical to that of rat LH, consistent with the report that the rat has only one LHβ and no CGβ gene. Mixed primers were designed for the CGβ subunit based on the amino acid sequences common to rat LHβ, human LHβ, equine LHβ, and human CGβ subunits and for the α subunit designed from the rat LHα subunit. The information produced by this approach is equivocal but nonetheless interesting. PCR with mRNA extracted from rat placentae amplified several bands, including cDNA identical to part of rat LHα and LHβ. Expression of these genes was confirmed in the placenta on several days of gestation from midpregnancy until shortly before parturition. The specific bands for these subunit mRNAs were also amplified in rat pituitary but not in other tissues, including decidua, liver, spleen, brain, and kidney. Expression of the mRNAs was observed in both the junctional and the labyrinth zones of placental tissue. The authors of this study claim that rLHα- and β-mRNA expression during rat gestation is so small that rat CG may act in a paracrine fashion in the fetal–placental unit. Physiological studies of the luteotropic shift indirectly support this idea.

C. The Luteotropic Shift as a Model for the Study of Rodent CG

In both the rat and the mouse, LH stimulates luteal cell androgen production, providing substrate for estradiol which acts as an intracrine signal to maintain progesterone production throughout pregnancy. This hormonal cascade operates until midgestation, when the rodent placenta begins to produce androgen during the luteotropic shift in response to an as yet unidentified signal. Maternal LH appears to inhibit placental androgen production by a specific effect on $P450_{17\alpha}$, the key rate-limiting enzyme for androgen synthesis. An abrupt increase in the placental activity of this enzyme takes place once LH secretion declines or is actively suppressed at midgestation. A direct stimulus for placental androgen production remains elusive, but in the rat it is thought to reflect conversion of luteal progesterone provided as substrate for placental androgen synthesis. In this respect, the luteal–placental relationship in the rodent resembles the primate fetal–placental unit in which fetal adrenal androgen serves as substrate for placental estrogen synthesis.

In the rat maternal levels of androgen increase steadily from midgestation until shortly before parturition. A different pattern is seen in the mouse, in which maternal androgen concentration mirrors the biphasic bioactivity speculated to reflect placental production of a CG-like molecule (Fig. 1). The initial rise in circulating androgen is confined to a fixed stage of fetoplacental development 5 days after implantation. This androgen surge on Day 9 allows fetal viability to be monitored during the period of endocrine transition from pituitary to fetal–placental regulation of ovarian progesterone synthesis. In rodents, the maternal ovaries and conceptus both contribute to elevated androgen secretion during the luteotropic shift at midpregnancy (Fig. 2). A fetal testes contribution to the day 9 androgen surge is unlikely since testicular differentiation does not occur until Days 12–13, and appreciable androgen production by the mouse fetus does not take place until late gestation. This is consistent with the results of *in situ* hybridization analysis of pituitary hormone gene expression in fetal mice. Although mRNA for the alpha gonadotropin subunit appears in the anterior wall of Rathke's pouch on embryonic day 9.5, β subunits for LH, FSH, and TSH are first detected on gestational day 16.5.

Removal of mouse fetus(es) dramatically reduces both placental and ovarian androgen secretion. This

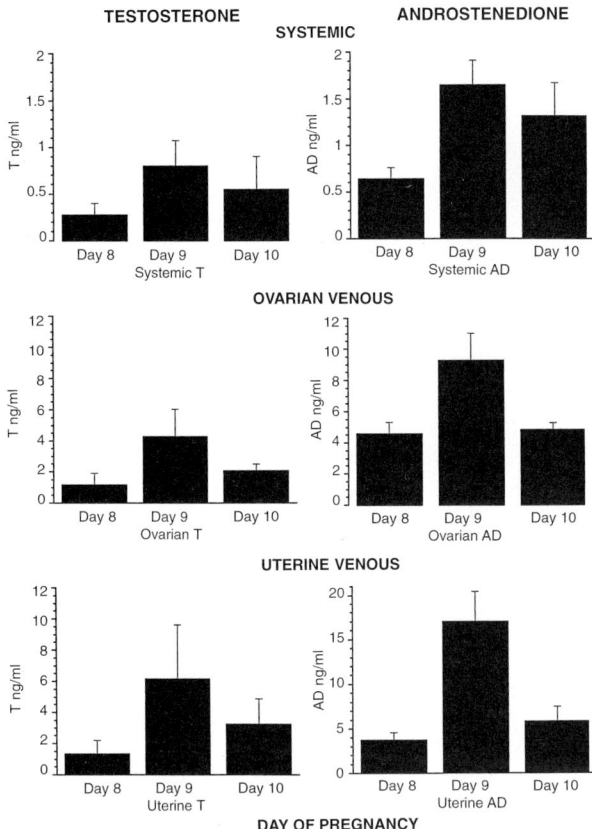

FIGURE 2 Plasma testosterone (left) and androstenedione (right) levels in systemic (top), ovarian (middle), and uterine (bottom) venous blood on Days 8–10 of mouse gestation. Each bar represents the mean + SE of five to eight samples. Scales for systemic values differ from those for ovarian and uterine venous samples (reproduced with permission from Murphy-Hackley, 1997).

indicates that viable fetus(es) (or some associated membranes) must be present to support the midpregnancy androgen surge. Histological examination of placentae removed from fetectomized mice revealed normal development of the decidua, trophoblast, and labyrinth with the presence of giant cells and yolk sac. The normal appearance of this placental tissue, particularly the giant cells which produce androgen in both the rat and the mouse, suggests that placental androgen secretion may be regulated by the fetus during the luteotropic shift. Possibly the fetus provides some factor(s) to the placenta which, in response, produces a hormone with activity that stimulates androgen synthesis. The source of this molecule is not the maternal pituitary because ovarian and placental components of the Day 9 androgen surge are preserved in mice hypophysectomized on Day 7 if a nonconvertible progestin (medroxyprogesterone acetate) is provided to maintain fetal viability (Fig. 3). This demonstrates that the tropic stimulation produced by the conceptus is sufficient to maintain ovarian as well as placental androgen production in the absence of maternal LH stimulation, i.e., the conceptus is the source of a putative regulatory signal that can stimulate maternal and placental androgen production. Other observations indicate that this signal is probably of trophoblastic origin. A previously reported rescue of androgen production in rats fetectomized at midpregnancy may reflect placental conversion of exogenous progesterone which was provided daily.

If a rodent placental gonadotropin does exist, available information suggests its structure resembles pituitary LH. The reportedly low production of this molecule may reflect a paracrine function with minimal systemic activity. What purpose might this serve? During rodent pregnancy, the luteotropic actions of estrogen are essential, but estrogen in high amounts can induce an inappropriate LH surge and is embryotoxic. Precise regulation of luteal estrogen production is particularly critical during the luteotropic shift. Perhaps the fetus is the primary sensor operating at the time of endocrine transition to determine the setpoint for placental gonadotropic activity. As in other mammals, progesterone and other steroids may, together with the gonadotropin itself, regulate the production and secretion of the placental LH or CG-like hormone. If progesterone production wanes sufficiently, the putative placental gonadotropin may act as an endocrine signal. An interesting transient decline in progesterone occurs around the time of the initial rise in CG-like immunoreactivity and bioactivity found in crude extracts of mouse placentae. Another characteristic of the luteotropic shift in rodents is an increase in fetal death. A similar phenomenon occurs in primates during the luteal–placental shift. Switching from maternal to fetal–placental control of pregnancy is clearly imprecise and may offer a potential escape if environmental

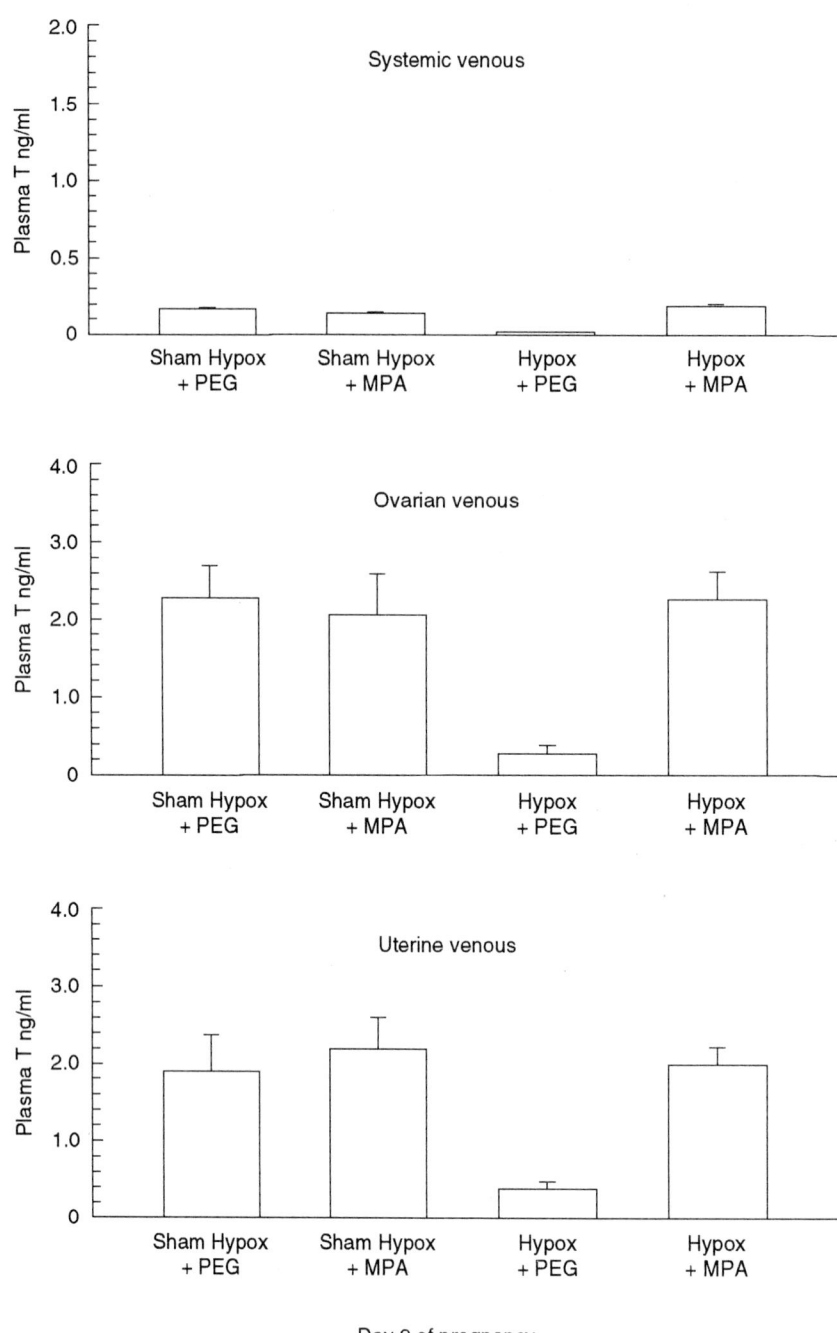

FIGURE 3 Plasma testosterone in systemic (top), ovarian (middle), and uterine (bottom) venous blood on Day 9 of mouse gestation. Mice were hypophysectomized (Hypox) on Day 7. Medroxyprogesterone acetate (MPA) was provided as a single 0.05 ml IM dose of a 100 mg/ml solution or 0.05 ml of polyethylene glycol (PEG) was delivered IM as a control to Hypox or sham Hypox females. Testosterone was measured by RIA using different antisera than for sample values depicted in Fig. 2. Testosterone values are shown as the mean ± SE of four to six samples for each treatment group (reproduced with permission from Murphy-Hackley, 1997).

conditions are detrimental for the continuation of gestation.

VII. SUMMARY

The synthesis and secretion of CG by the periimplantation embryo in nonhuman primates is essential for establishing pregnancy by extending the production of progesterone by the corpus luteum until the time of the luteal–placental shift. A similar requirement for CG may exist in other mammals during the luteotropic shift when hormonal control of luteal progesterone synthesis resides within the conceptus. Considerable debate about the existence of a rodent CG-like hormone continues to thrive. It was recently claimed that low levels of LHα and LHβ gene expression occur in the rat placenta from midpregnancy until shortly before parturition. The possibility that these gene products act in a paracrine fashion and have previously escaped detection due to technical problems is speculative but intriguing. Validation of the idea that luteal function in the pregnant rodent is sustained by a hormonal mechanism like the mare's will depend on conclusive identity of a rodent placental gonadotropin. If this is forthcoming, the search for other mammalian CGs may flourish. Despite the pivotal role of CG during early pregnancy, only a few studies have addressed the factors that regulate its production in the periimplantation period. Continued research on CG production using nonhuman primates as animal models will facilitate studies on early embryonic development and implantation for alleviating infertility or controlling fertility in women. For example, knowledge of the factors necessary for timely CG production could be applied to improve pregnancy rates of human embryos obtained by assisted reproductive technologies cultured *in vitro* from the zygote to the blastocyst stage prior to uterine transfer. Further elucidation of CG–maternal interactions could also lead to clinical therapies for preventing recurrent miscarriage, and to novel, more effective methods of contraception.

See Also the Following Articles

EQUINE CHORIONIC GONADOTROPIN; FSH (FOLLICLE-STIMULATING HORMONE); LH (LUTEINIZING HORMONE)

Bibliography

Albanese, C., Ides, C. M., Crowley, W. F., Ito, M., Pestell, R. G., Weiss, J., and Jameson, J. L. (1996). The gonadotropin genes: Evolution of distinct mechanisms for hormonal control. *Recent Prog. Horm. Res.* **51**, 23–59.

Astwood, E. B., and Greep, R. O. (1938). A corpus luteum-stimulating substance in the rat placenta. *Proc. Soc. Exp. Biol. Med.* **38**, 713–716.

Barkley, M. S., Michael, S. D., Geschwind, I. I., and Bradford, G. E. (1977). Plasma testosterone during pregnancy in the mouse. *Endocrinology* **100**, 1472–1475.

Gibori, G., Khan, I., Warshaw, M. L., McLean, M. P., Puryear, T. K., Nelson, S., Durkee, T. J., Azhar, S., Steinschneider, A., and Rao, M. C. (1988). Placental-derived regulators and the complex control of luteal cell function. *Recent Prog. Horm. Res.* **44**, 377–429.

Hodgen, G. D., and Iskowitz, J. (1988). Recognition and maintenance of pregnancy. In *The Physiology of Reproduction* (E. Knobil and J. D. Neill, Eds.). Raven Press, New York.

Jameson, J. L., and Hollenberg, A. N. (1993). Regulation of chorionic gonadotropin gene expression. *Endocr. Rev.* **14**, 203–220.

Murphy-Hackley, P. (1997). Maternal–fetal–placental interaction at midgestation in the mouse. MS thesis, University of California at Davis.

Soares, M. J., Faria, T. N., Roby, K. F., and Deb, S. (1991). Pregnancy and the prolactin family of hormones: Coordination of anterior pituitary, uterine, and placental expression. *Endocr. Rev.* **12**, 402–423.

Shinozaki, M., Uchida, H., Ikeda, S., Min, K., Shiota, K., and Ogawa, T. T. (1997). Expression of LH-alpha and -beta subunit mRNAs in the rat placenta. *Endocr. J.* **44**, 79–87.

Spies, H. G., and Chappel, S. C. (1984). Mammals: Nonhuman primates. In *Marshall's Physiology of Reproduction* (G. E. Lamming, Ed.), 4th ed., Vol. 1. Churchill Livingstone, New York.

Stouffer, R. L., and Hearn, J. P. (1998). Endocrinology of the transition from menstrual cyclicity to establishment of pregnancy in primates. In *The Endocrinology of Pregnancy* (F. W. Bazer, Ed.). Humana, Totowa, NJ.

Tullner, W. W. (1974). Comparative aspects of primate chorionic gonadotropins. *Contrib. Primat.* **3**, 235–257.

Webley, G. E., and Hearn, J. P. (1994). Embryo–maternal interactions during the establishment of pregnancy in primates. In *Oxford Reviews of Reproduction Biology* (H. M. Charlton, Ed.). Oxford Univ. Press, Oxford, UK.

Wide, L., and Hobson, B. (1978). Chromatographic studies on a chorionic gonadotropic activity in the placenta of the rat, mouse, and hamster. *Uppasala J. Med. Sci.* **83**, 1–6.

Wide, L., and Wide, M. (1979). Chorionic gonadotropin in the mouse from implantation to term. *J. Reprod. Fertil.* 57, 5–9.

Zeleznik, A. J., and Fairchild Benyo, D. (1994). Control of follicular development, corpus luteum function, and the recognition of pregnancy in higher primates. In *The Physiology of Reproduction* (E. Knobil and J. D. Neill, Eds.), 2nd ed. Raven Press, New York.

Chorionic Villi

see Placenta: Implantation and Development

Chromosomal Abnormalities

see Sex Chromosomes

Circadian Rhythms

Fred W. Turek
Northwestern University

I. Introduction
II. Generation and Entrainment of Circadian Rhythms
III. Location of Internal Circadian Clock System
IV. Genetic Control of Circadian Rhythms

GLOSSARY

circadian clock An internal timing device that regulates the expression of circadian rhythms. In mammals, a master circadian clock is located in the hypothalamic suprachiasmatic nucleus.

circadian rhythms Biochemical, physiological, and behavioral events that reoccur in a cyclic fashion with a period of about 24 hr.

entrainment The process by which an internal circadian clock is entrained or synchronized to the period of an external stimulus that usually has a period of about 24 hr. In nature, the light–dark cycle is the primary external environmental signal that entrains the circadian clock to the 24-hr period of the day that is due to the rotation of the earth on its axis.

free-running Describing circadian rhythms persisting in the absence of any 24-hr synchronizing information from the external environment. The expression of free-running circadian rhythms with a period close to (e.g., 23–25 hr) but rarely exactly 24 hr demonstrates the endogenous nature of the circadian clock underlying the expression of circadian rhythms.

Circadian rhythms refer to the processes in living organisms that fluctuate with a period of about 24 hr. A remarkable feature of the daily rhythms that are observed in organisms as diverse as algae, fruit flies, and humans is that they are not simply a response to the 24-hr changes in the physical environment imposed by the principles of celestial mechanics, but instead arise from an internal timekeeping system. This timekeeping system, often referred to as the "biological clock" or "circadian clock," allows the organism to predict and prepare in advance for the changes in the physical environment that are associated with night and day. Thus, the organism adapts, both behaviorally and physiologically, to meet the challenges associated with the daily changes in the external environment, and there is temporal synchronization between the organism and the external environment. The most obvious example of such an adaptation to the physical environment is the finding that many animals are only active during the light period (diurnal species) or the dark period (nocturnal species) and are inactive during the other part of the day. Such "external synchronization" is of obvious importance for the survival of the species and ensures that the organism does the "right thing" at the right time of the day. Of equal, but perhaps less appreciated, importance is the fact that this biological clock, like a conductor of a symphony orchestra, provides internal temporal organization and ensures that internal changes take place in coordination with one another. Just as living organisms are organized spatially, they are also organized temporally to ensure that there is "internal synchronization" between the myriad of biochemical and physiological systems in the body.

I. INTRODUCTION

Successful reproduction involves a variety of precisely timed events. From the initiation of gonadal development through pregnancy, parturition, and the raising of the young, numerous endocrine and neural events occur within a well-defined temporal program. Therefore, it is not surprising to find a great deal of interaction between the reproductive system and the major system responsible for supplying temporal organization in animals: the circadian system. The importance of the circadian system in the overall temporal organization of reproductive phenomena is emphasized by the central role it plays in many noncircadian reproductive rhythms, including ovarian and seasonal cycles.

II. GENERATION AND ENTRAINMENT OF CIRCADIAN RHYTHMS

Under laboratory conditions devoid of any external time-giving cues from the physical environment, it has been found that just about all diurnal rhythms that are present under natural conditions continue to be expressed in the laboratory. However, under constant environmental conditions, the period of the rhythm rarely remains exactly 24 hr but instead is "about" 24 hr. Because the period of diurnal rhythms is close to but not exactly 24 hr in duration, they are referred to as "circadian rhythms," from the Latin *circa diem*, meaning about a day. When a circadian rhythm is expressed in the absence of any 24-hr signals in the external environment, it is said to be "free-running," i.e., the rhythm is not synchronized or entrained by any cyclic change in the physical environment. For a population of animals within a given species, the period of a given rhythm (e.g., drinking, body temperature, or locomotor activity) will be different between animals, but in general all will lie in close proximity to 24 hr (e.g., from 23–24 hr in duration). Strictly speaking, a diurnal rhythm should not be referred to as "circadian" until it has been demonstrated that such a rhythm persists under constant environmental conditions. The purpose of this distinction is to separate out those rhythms which are simply a response to 24-hr changes in the physical environment from those which are driven by some internal time-giving system. However, for practical purposes, there is little reason to make a distinction between "diurnal" and circadian rhythms since almost all diurnal rhythms expressed under natural conditions are found to persist under constant environmental conditions in the laboratory. Consequently, the term circadian is often used to refer to diurnal rhythms that are observed under either natural or laboratory conditions.

Genetic, physiological, and behavioral experiments have established that the timing system which underlies the generation of circadian rhythms is endogenous to the organism itself. That is, there is a circadian clock or clocks within the organism which somehow regulates the 24-hr fluctuations in diverse physiological and behavioral systems. While literally thousands of rhythms have been monitored in plants and animals, it has not been possible to assay the state of a circadian clock directly in any experimental model to date. Thus, attempts to understand the properties of circadian clocks focus on the "hands" of the clock, i.e., the expression of overt rhythms regulated by the clock. While the list of biochemical and physiological processes that show circadian fluctuations is enormous, a few select behavioral rhythms (e.g., locomotor activity and drinking) are most often utilized to characterize the basic features of the clock system in mammals. Behavioral rhythms are utilized because of their ease of measurement for many cycles without disturbing the animal. Figure 1 provides a schematic representation of the circadian rhythm of locomotor activity in a male golden hamster held under both entrained and free-running conditions. Not only can this rhythm be monitored for essentially the lifetime of the animal without any interference of the sampling procedure on the rhythm itself, but also automated sampling systems allow one to access the state of this rhythm continuously on essentially a minute-to-minute basis. For practical and economic reasons, such long-term and frequent sampling of biochemical and physiological rhythms is often not possible. Nevertheless, as discussed below, data obtained in mammals indicate that behavioral rhythms represent the hands of the same circadian clock system that underlies most if not all biochemical and physiological rhythms.

The fact that the period of circadian rhythms is not exactly equal to that of the period of the rotation of the earth on its axis demands that 24-hr changes in the physical environment must somehow synchronize or entrain the internal clock system regulating circadian rhythms. Otherwise, even a clock with a period only a few minutes shorter or longer than 24 hr would soon be totally out of synchrony with the environmental day. An endogenous circadian clock that could not be reset by environmental signals would be of little use to organisms that need to time specific activities to particular times of the day.

Except in a few exotic species living in unusual environments (e.g., blind cave fish or eyeless mole rats), the light–dark cycle appears to be the primary environmental agent which synchronizes circadian rhythms to the 24-hr environmental cycle in the physical environment. Thus, in the presence of a 24-hr light–dark cycle the period of circadian rhythms exactly matches the period of the light–dark cycle (Fig. 1). From one day to the next, the time between successive recurrences of specific phase points of a rhythm (e.g., the onset of locomotor activity or the minimum in body temperature) is the same as the period or duration of the light–dark cycle. In addition to establishing "period control," an entraining light–dark (LD) cycle establishes "phase control" such that specific phases of the circadian rhythm occur at the same times in each cycle relative to the entraining agent. For example, in a hamster entrained to a LD 14:10 cycle (i.e., 14 hr of light followed by 10 hr of darkness every 24 hr), the onset of the main bout of daily activity always occurs within a few minutes after lights off, day after day. Following a phase shift in the LD cycle, the rhythm reentrains (Fig. 1), although the development of a steady-state phase relationship between the circadian rhythm and the entraining light–dark cycle often takes many days due to the limitation on the number of hours per day that the internal circadian clock can be phase shifted by light.

Although circadian rhythms can be entrained to light–dark cycles that are not exactly 24 hr in duration, entrainment is restricted to cycles with periods

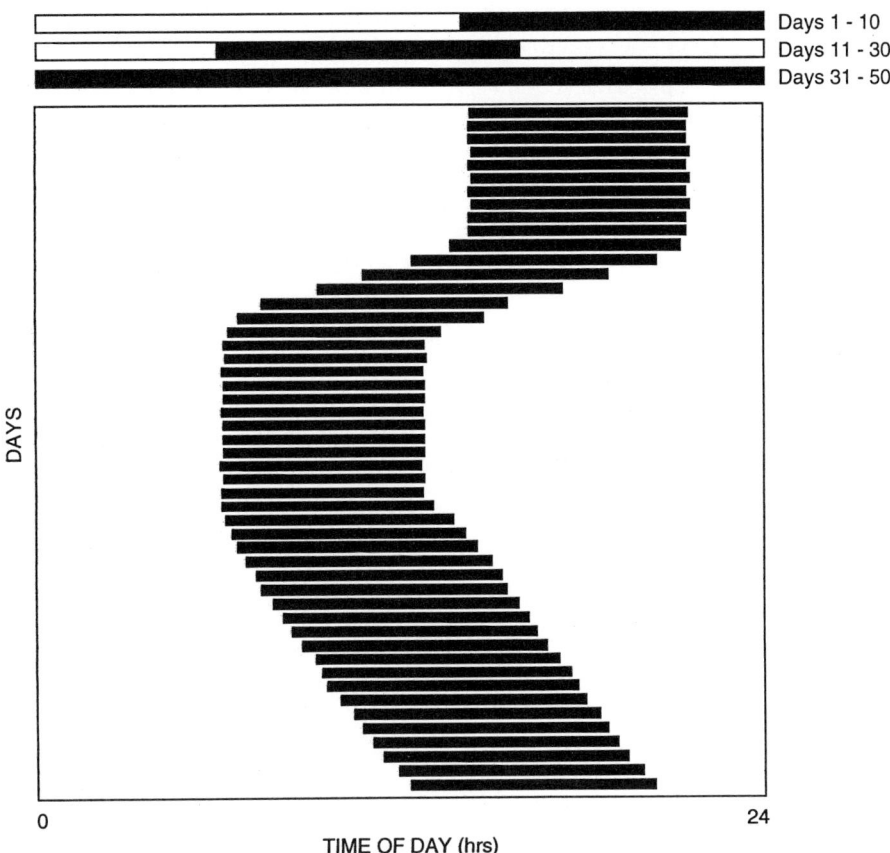

FIGURE 1 Schematic representation of the daily rhythm of locomotor activity of a nocturnal rodent exposed to an LD 14:10 light cycle or to constant darkness. Heavy black bars on each day represent the time of intense activity, and the absence of the black bar represents time of little or no activity. The activity data are plotted on the horizontal axis for each 24-hr period, and the vertical axis represents consecutive days from 1 to 50. The first 10 days of the record depict a schematic representation of activity when an animal is exposed to the LD 14:10 cycle shown on the top bar of the figure (open bar, light; dark bar, dark). On Day 11 the LD cycle is phase advanced by 8 hr (second bar at the top of record) and over the next 10 days the activity rhythm reentrains to the new LD cycle such that by Day 20 the activity period once again has the same phase relationship to the LD cycle as before the shift. After Day 30, the activity rhythm is shown to free run with a period slightly greater than 24 hr because after this time there is no light (represented by solid black bar at top of record) to synchronize the endogenous clock.

that are close to 24 hr in duration. The "range of entrainment" can vary from species to species and is dependent on the experimental conditions (e.g., intensity of light–dark cycle and whether period of light–dark cycle is changed gradually or rapidly), but in general animals do not entrain readily to LD cycles that are more than a few hours shorter or longer than the period of the endogenous free-running circadian rhythm. If the period of the LD cycle is too short or long for entrainment to occur, the circadian rhythm will free run with a period close to 24 hr.

III. LOCATION OF INTERNAL CIRCADIAN CLOCK SYSTEM

A primary objective of circadian rhythm researchers over the past two decades has been to identify and localize biological clocks in living tissue. It has

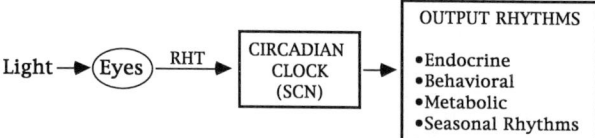

FIGURE 2 Simplified schematic overview of the circadian organization in mammals. RHT, retinohypothalamic tract; SCN, suprachiasmatic nucleus.

now been established in mammals that a bilaterally paired structure, the suprachiasmatic nucleus (SCN), located in the anterior hypothalamus, contains a master circadian clock that regulates most, if not all, endogenously generated circadian rhythms (Fig. 2). This structure receives a direct neural input from the retina via the retinohypothalamic tract, which is involved in relaying synchronizing light–dark information to the biological clock(s) in the SCN region (Fig. 2). Although the vast majority of experiments examining the role of the SCN in the circadian organization in mammals have been carried out in rodents, there are experimental data from species as diverse as sparrows and monkeys to indicate that the SCN represents a master circadian clock not only in other mammalian species but also in other classes of vertebrates.

A central role for the SCN in the regulation of circadian rhythms associated with reproduction is indicated from studies which found that lesions of the SCN disrupt a number of reproductive rhythms, including those of circulating prolactin and gonadotropin levels, pineal melatonin, sexual behavior in males, sexual receptivity in females, and the timing of ovulation. In addition, the SCN is involved in measuring the annual change in day length, information that is used to time seasonal reproductive cycles in many animals.

While there is some evidence indicating that both free-running and entrained rhythms persist even after total destruction of SCN tissue, there is no convincing consistent evidence in mammals that any 24-hr rhythm can persist for a prolonged period of time under constant environmental conditions following total destruction of the SCN. Nevertheless, following large SCN lesions, unstable ultradian (i.e., on the order of 1–8 hr) rhythms are often observed in the activity patterns of rodents, suggesting that oscillators outside of the SCN, which are normally synchronized by the master pacemaker in the SCN, fail to couple or achieve only weak or unstable coupling after SCN lesions.

There are a number of situations in which rhythms persist in SCN lesioned animals exposed to an external environment with periodic fluctuations Abnormal but persistent entrainment of the locomotor activity rhythm to a LD cycle in hamsters with complete SCN lesions is often observed. The presence of direct retinal projections to hypothalamic regions outside of the SCN is consistent with the hypothesis that LD information may influence other hypothalamic areas which can provide some temporal information to the animal. Over the years a number of investigators have demonstrated that periodic food availability can entrain circadian rhythms in mammals. For example, if food is only provided to rats for a few hours per day at a fixed time, an increase in locomotor activity and body temperature anticipates the daily mealtime. The anticipatory activity associated with mealtime appears to involve a circadian oscillatory mechanism since such a response does not occur if food is presented at intervals that are much less or greater than 24 hr. Furthermore, this food-entrainable oscillator appears to lie outside of the SCN region since periodic food presentation can entrain rhythms in SCN lesioned rats.

To date, there is no convincing data to indicate that any structure outside of the SCN acts as a master circadian pacemaker. There is occasional confusion in the literature on this, particularly with respect to the role of the mammalian pineal gland as a possible circadian pacemaker in mammals. In mammals, there is no evidence that the pineal gland itself is capable of generating circadian rhythms. Although the pineal gland expresses pronounced circadian rhythms, particularly in the synthesis and release of the hormone, melatonin, the regulation of this rhythm appears to be totally under the control of circadian neural signals from the SCN. The pineal melatonin rhythm may regulate other rhythms, and it is possible that melatonin may have some feedback effects on the circadian pacemaker in the SCN since injections of melatonin under certain experimental conditions can phase shift the circadian clock of rats, and melatonin receptors have been localized to the SCN in diverse mammalian species, including humans. Some of the confusion over the role of the pineal gland as a central circadian oscillator in mammals arises from the fact that in lower vertebrates the pineal gland can function as a self-sustained circadian pacemaker regulating other circadian rhythms. Indeed, treatment with melatonin can have dramatic effects on the circadian rhythm of locomotor activity in birds. Furthermore, the pineal gland of birds and reptiles expresses clear circadian rhythms even in culture. No convincing evidence is available to indicate that the mammalian pineal gland can sustain rhythmicity in culture, and the pineal gland is clearly not necessary for the expression of most circadian rhythms in rodents.

IV. GENETIC CONTROL OF CIRCADIAN RHYTHMS

Early studies in invertebrates demonstrated that the period of the circadian clock was clearly under genetic control since 24-hr rhythms developed in organisms never experiencing any 24-hr changes in the physical environment. Furthermore, rearing mice on non-24-hr cycles as extreme as 20 or 28 hr has only small (i.e., on the order of a few minutes) and transient effects on the free-running period of the circadian clock when animals are subsequently housed under conditions of constant darkness. Until recently, the importance of the genome in defining the characteristics of the circadian clock system in mammals has been determined primarily by comparing properties of various circadian rhythms between different inbred strains of the same species. Natural genetic differences in various circadian rhythm parameters have been observed between various strains indicating that the genetic background influences the expression of circadian rhythms in mammals. In addition, there have been a few attempts to breed animals selectively with different circadian phenotypes to determine the inheritance patterns of these phenotypes.

Identification of circadian clock mutations *Per* and *Tim* in *Drosophila* and *Frq* in *Neurospora* has effectively led to identification of molecular elements of the clock in these organisms. It is now becoming clear that an autoregulatory program of periodic gene expression forms the core of the circadian pacemaker. Efforts to identify mammalian orthologs of these invertebrate clock genes have proven unsuccessful. While a spontaneous circadian rhythm mutation, *tau*, had been identified in the hamster, sufficient information about the hamster genome to identify and study this gene at the molecular level simply does not exist. However, using a systematic chemical mutagenesis screen, we identified a mouse gene, *Clock*, which is essential for circadian behavior.

A semidominant mutant allele of *Clock* was identified which has two primary phenotypic effects on circadian behavior. First, the mutation lengthens circadian period by about 1 hr in *Clock*/+ heterozygotes and by about 4 hr in *Clock* homozygotes. Second, the mutation abolishes the persistence of circadian rhythmicity in constant darkness in *Clock* homozygotes. In the majority of *Clock* homozygous mice transferred from LD 12:12 to constant darkness, the circadian activity rhythm free runs with a 27- or 28-hr period for about 3–20 cycles before damping out. The damped rhythm is "sloppy" compared to wild type, and the disappearance of the circadian rhythm is accompanied by the emergence of a 6- to 9-hr ultradian periodicity.

Through a series of positional cloning techniques, *Clock* has recently been identified and found to be a distinctly new gene containing about 100,000 DNA base pairs. The gene contains 24 separate exons, or regions that code for segments of the protein product. The mutation was identified as being a single A to T nucleotide transversion which results in the skipping of a single exon and the deletion of 51 amino acids from the transcribed CLOCK protein. The protein is clearly a transcription factor, meaning that it can bind to DNA and regulate other genes. In addition to being the first gene with known circadian function to be identified in mammals, it is the first circadian *Clock* gene that encodes a protein with features predicting DNA binding, protein dimerization, and activation domains. Proof that the candidate *Clock* gene was indeed the gene was demonstrated by the finding that the insertion of a bacterial artificial chromosome containing the *Clock* gene into the embryos of animals carrying a mutation in the gene restored normal circadian function in both heterozygous and homozygous *Clock* mutant animals.

Examination of the DNA from other vertebrate species indicates that the gene is highly conserved among the vertebrates, including humans. The cloning and molecular characterization of the first clock gene in mammals is expected to lead to new insights into the genetic and molecular mechanisms underlying the entrainment, generation, and expression of circadian rhythms in higher organisms.

See Also the Following Articles

PHOTOPERIODISM, VERTEBRATES; OVARIAN CYCLE; PINEAL GLAND, BIRDS AND OTHERS; PINEAL GLAND, MAMMALS; SEASONAL REPRODUCTION; SUPRACHIASMATIC NUCLEUS

Bibliography

Arendt, J. (1995). *Melatonin and the Mammalian Pineal Gland.* Chapman & Hall, London.

Binkley, S. (1990). *The Clockwork Sparrow: Time, Clocks, and Calendars in Biological Organisms.* Prentice Hall, New York.

King, D. P., Zhao, Y., Sangoram, A. M., Wilsbacher, L. D., Tanaka, M., Antoch, M. P., Steeves, T. D. L., Vitaterna, M. H., Kornhauser, J. M., Lowrey, P. L., Turek, F. W., and Takahashi, F. W. (1997). Positional cloning of the mouse circadian Clock gene. *Cell* 89, 641–653.

Klein, D. C., Moore, R. Y., and Reppert, S. M. (1991). *Suprachiasmatic Nucleus—The Mind's Clock.* Oxford Univ. Press, New York.

Kryger, M. H., Roth, T., and Carskadon, M. (1994). Circadian rhythms in humans: An overview. In *Principles and Practice of Sleep Medicine* (M. H. Kryger, T. Roth, and W. C. Dement, Eds.), pp. 301–308. Saunders, Philadelphia.

Miller, J. D., Morin, L. P., Schwartz, W. J., and Moore, Y. R. (1996). New insights into the mammalian circadian clock. *Sleep* 19, 641–667.

Turek, F. W., and Van Cauter, E. (1994). Rhythms in reproduction. In *Physiology of Reproduction* (E. Knobil, J. Neill, et al., Eds.), pp. 487–540. Raven Press, New York.

Vitaterna, M. H., King, D. P., Chang, A.-M., Kornhauser, J. M., Lowrey, P. L., McDonald, J. D., Dove, W. F., Pinto, L. H., Turek, F. W., and Takahashi, J. S. (1994). Mutagenesis and mapping of a mouse gene, *Clock*, essential for circadian behavior. *Science* 264, 719–725.

Circannual Rhythms

Irving Zucker and Brian J. Prendergast
University of California, Berkeley

I. Introduction
II. Circannual Rhythms of Reproduction
III. Circannual Rhythms of LH Secretion
IV. Entrainment of CARs to the Light–Dark Cycle
V. Neural Substrates for Generation of CARs
VI. Influence of Pineal and Thyroid Hormones on CARs
VII. Conclusions

GLOSSARY

circadian rhythm An endogenous self-sustained oscillation with a period of approximately 24 hr.

entrainment The process by which an external cycle (e.g., daylight) controls the phase and period of an endogenous self-sustained biological oscillation.

melatonin An indoleamine hormone secreted by the pineal gland.

phase–response curve Graphic summary of phase shifts produced in a biological rhythm by external or internal stimuli (e.g., light, temperature, drugs, or hormones).

retinohypothalamic tract A direct projection from the retina that terminates in the suprachiasmatic nucleus of the hypothalamus.

suprachiasmatic nucleus A hypothalamic structure that contains a clock essential for generation of circadian rhythms.

Circannual rhythms (CARs) are endogenous oscillations that persist for two or more cycles, usually with a period shorter than 12 months, in animals held under constant environmental conditions. A circannual mechanism restricts breeding activities of long-lived mammals to a particular phase of each calendar year. Changes in day length, by altering nightly patterns of melatonin secretion controlled by a circadian oscillator, entrain the CAR of reproduction to a period of 12 months. A major goal of circannual research is the specification of neuroendocrine substrates that respectively generate and entrain CARs of reproduction.

I. INTRODUCTION

April is not the cruelest month for mammals in temperate and boreal regions. More offspring are issued during spring months than at any other time of year. The reproductive cycle may have been the primary target of natural selection in the evolution of seasonality. Because the cost of lactation is great, dams that nurse their young when food is most abundant enjoy a considerable advantage; this is particularly true of small short-lived species in which body mass of the litter relative to that of the mother is higher, and the costs of lactation are concentrated over a shorter time span than in larger species. Ultimately, the disproportionate number of spring births is attributable to availability of greater quantities of higher quality food acquired with relatively less energy expenditure by dams that give birth during this season. Orientation of reproduction in time increases fitness and ensures greater representation of spring breeders in the population.

Two distinct mechanisms mediate seasonal reproduction in mammals. In smaller species with short life spans, gonadal involution commences in late summer, typically in response to decreasing day lengths; reproduction ceases during autumn and winter. An interval timer limits reproductive quiescence to ~5 months; gonadal recrudescence initiated in midwinter is completed before the advent of favorable breeding conditions in the spring. Species that manifest these type I rhythms lack circannual clocks; their reproductive cycles do not persist under constant environmental conditions.

By contrast, circannual rhythms (CARs), or type II rhythms, are common among long-lived mammals, including representative rodents, carnivores, ungulates, bats, and primates, and arose independently several times in the course of mammalian evolution. Species that manifest these rhythms often have prolonged gestations, typically produce a single litter each year, and may be isolated from environmental perturbations for long intervals (e.g., animals that hibernate). CARs in reproduction presumably are initiated by circannual clocks in the central nervous system, analogous to but not isomorphic with circadian oscillators in the suprachiasmatic nucleus (SCN) of the hypothalamus.

II. CIRCANNUAL RHYTHMS OF REPRODUCTION

A typical seasonal rhythm of reproduction among sheep held outdoors is illustrated in Fig. 1. Note that lactation in ewes coincides with annual peaks of high-quality food availability. Male sheep (rams) show corresponding testicular cycles; the period of this rhythm for individuals held indoors in unvarying photoperiods is less than 12 months long. Marmots and squirrels kept in fixed day lengths also generate reproductive rhythms that free-run with a period of 10 or 11 months. A simulated natural annual photocycle entrains the reproductive cycle to a period of 12 months.

In the field, several primate species restrict copulation and births to a specific season, whereas others breed year-round but still exhibit seasonal birth peaks. Laboratory data bearing on the endogenous nature of these rhythms are sparse. Annual rhythms in testicular volume and serum concentrations of testosterone and luteinizing hormone (LH) have been described in male rhesus monkeys maintained under constant day length and temperatures. The CAR in plasma testosterone persists for several cycles in the absence of periodic environmental input. Most human populations, including those in developed urban societies, exhibit seasonal variations in the frequency of conceptions and births. According to

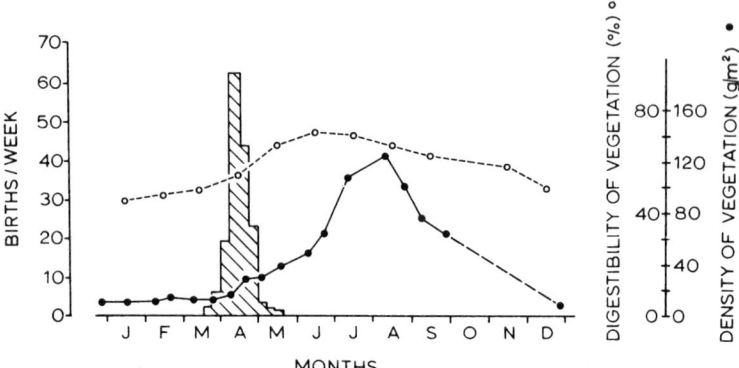

FIGURE 1 Timing of births among Soay sheep on the island of St. Kilda in relation to availability and digestibility of food. Births are restricted to a period early enough for the ewe and lambs to capitalize on abundant food of the spring pasture (from Lincoln and Short, 1980).

Bronson, there is no support for the hypothesis that human reproduction is under the control of a CAR and evidence for photoresponsiveness is mostly negative. Food availability and cultural factors probably are the major environmental factors that contribute to the worldwide seasonal variation in human births, although the contribution of circannual clocks and day length has not been put to experimental test.

III. CIRCANNUAL RHYTHMS OF LH SECRETION

A. Steroid Dependent

Ovariectomized ewes maintained outdoors at a latitude of 42°N and implanted with constant release capsules that provide physiological concentrations of estradiol manifested elevated blood concentrations of LH at intervals of exactly 12 months (Fig. 2). Ewes treated similarly but kept indoors in a fixed short photoperiod exhibited circannual LH rhythms with a period of 11 months. Intra- and interindividual variability in timing of LH secretion in the latter animals over the course of several years, and deviation from a period of 12 months in peak LH concentrations, attest to the endogenous nature of this rhythm. Seasonal anestrus in ewes reflects differential responsiveness of the neuroendocrine axis to estrogen at different phases of the circannual cycle, perhaps mediated by afferent neurons that regulate gonadotropin-releasing hormone (GnRH) secretion. The transition from seasonal anestrus to breeding condition is accompanied by a substantial increase in the number of synaptic inputs onto GnRH-immunoreactive cells in the preoptic area.

B. Steroid Independent

Blood LH concentrations were elevated at intervals of 9 or 10 months in ovariectomized squirrels held in a constant photoperiod and temperature (Fig. 3). Regardless of the time of year that squirrels were ovariectomized, they first manifested elevated plasma LH concentrations during the normal reproductive season (Fig. 4). A circannual oscillator that does not require input from ovarian hormones apparently restricts GnRH secretion to a fraction of each annual cycle. Ovarian hormones feedback to influence LH secretion during the spring breeding season but not at other times of year. In sharp contrast, the testicular cycle in this species is also controlled by a circannual mechanism but is subject to feedback inhibition of LH secretion by gonadal hormones during most of the annual cycle; gonadectomized males, unlike females, sustain chronically elevated blood LH concentrations all year. There are several distinct types of circannual control of gonadotropin secretion, but too few species have been studied to identify the most common patterns.

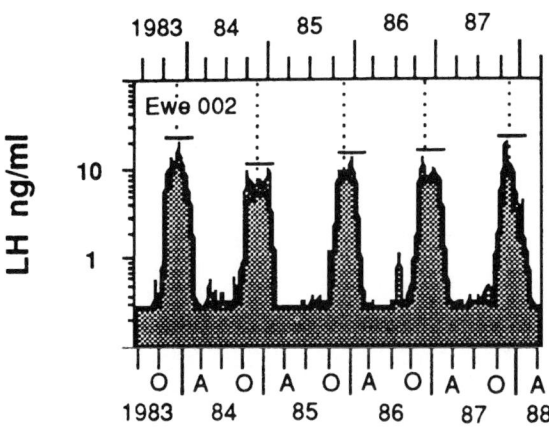
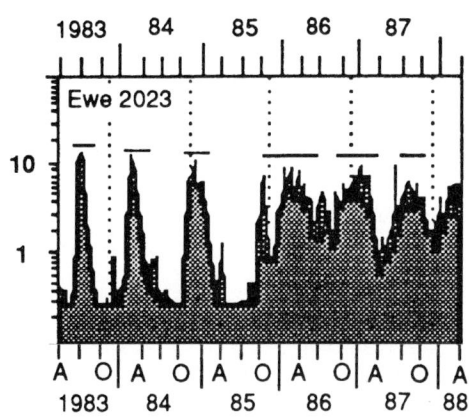

FIGURE 2 Serum LH concentrations on logarithmic scale for representative ewes maintained outdoors and exposed to natural variations in day length (left) or kept indoors and exposed to a fixed short day (8 hr light:16 hr dark; right) for 5 years. Dotted vertical lines indicate average midpoint of the high stage of the LH cycle for the entire study. Horizontal lines above elevated LH values indicate duration of elevated LH as determined by cluster analysis. The period between successive increases in LH was significantly shorter among ewes kept in a fixed short day than in ewes housed outdoors (329 ± 13 vs 364 ± 3 days) (modified from Karsch et al., 1989).

IV. ENTRAINMENT OF CARs TO THE LIGHT–DARK CYCLE

A. Actions of Light and Melatonin

Annual variations in day length are sufficient to entrain CARs of reproduction to a period of 365 days (Fig. 5, top). The nightly pattern of melatonin (Mel) secretion, which transduces the effects of light on the reproductive axis, is the principal signal for entrainment of CARs of reproduction. Thus, photic entrainment of CARs in reproduction is disrupted or absent in pinealectomized sheep and squirrels. Mel signals of differing durations are discriminated by neural tissues with high concentrations of Mel receptors. Ewes exposed to natural day lengths or corresponding Mel signals for only one-quarter of each annual cycle entrain their LH secretory rhythms to a period of 12 months. The long days of spring and summer entrain correspondingly short Mel signals that are sufficient to synchronize the reproductive rhythm; day lengths at other times of year are less important for entrainment of CARs.

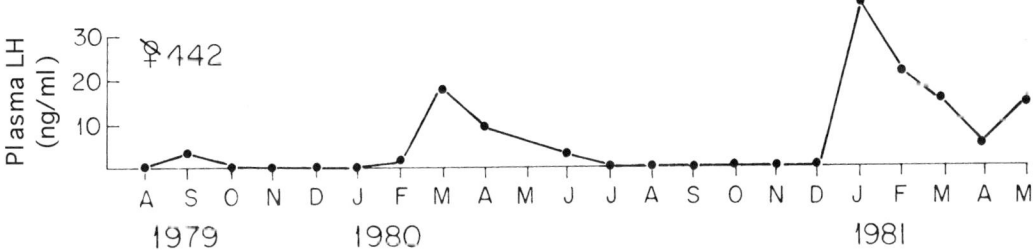

FIGURE 3 Postoperative plasma LH concentrations of an individual ovariectomized golden-mantled ground squirrel kept in a 14 hr light:10 hr dark photocycle over the course of 22 consecutive months (modified from Zucker, 1988).

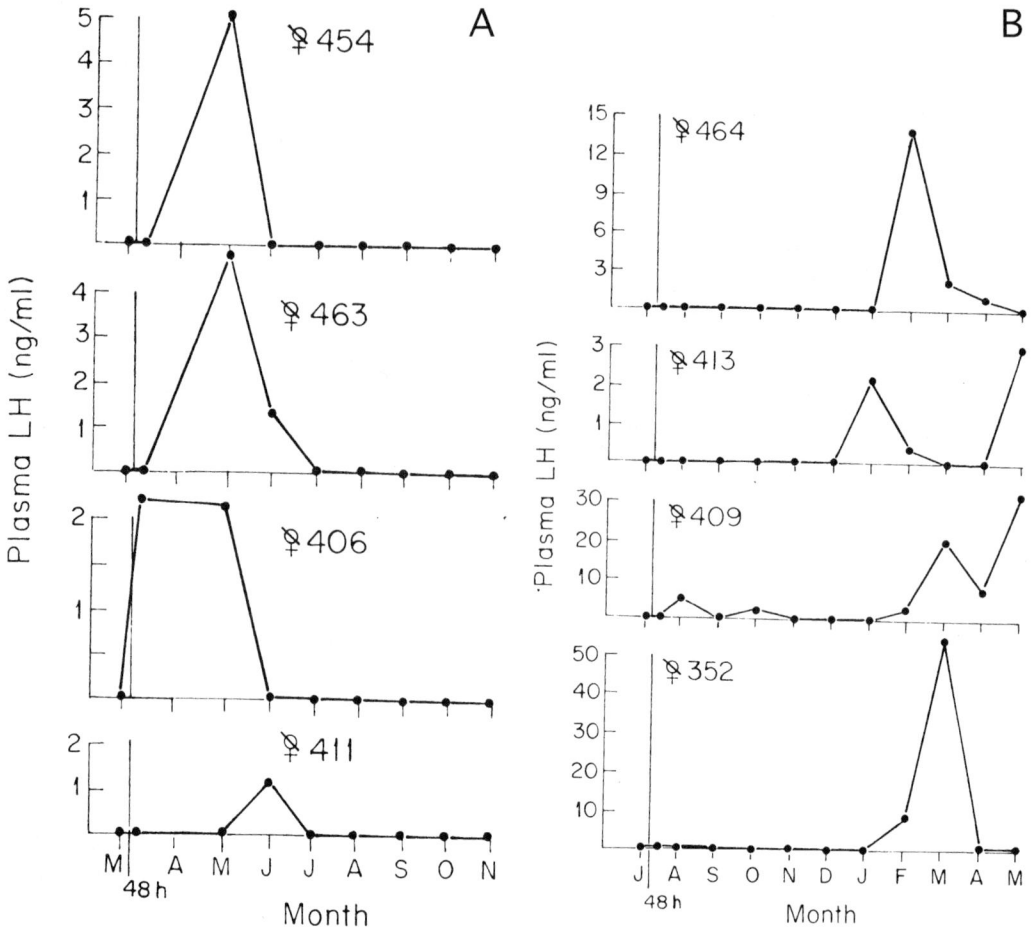

FIGURE 4 Plasma LH concentrations of individual squirrels ovariectomized in March (A) and July (B). The vertical line separates LH measurements taken at the time of ovariectomy and 48 hr postsurgically. Elevations in LH were restricted to the spring after surgery (data from Zucker, 1988).

B. Neural Pathways for Entrainment

The precise projections by which light entrains CARs in reproduction have not been specified. The retinohypothalamic tract (RHT) that terminates in the SCN is essential for photic entrainment of circadian as well as type I reproductive rhythms. Interruption of the RHT, produced by ablating the SCN, also compromises photic entrainment of the CAR of estrus in squirrels. It is not known whether the RHT, in the absence of visual input from other projections, is sufficient to mediate photic entrainment of reproduction.

C. Circannual Phase–Response Curves to Hormones

Responsiveness of reproductive rhythms to hormones varies seasonally. The onset of seasonal estrus can be delayed or advanced several weeks with appropriately timed Mel treatments restricted to separate quadrants of the circannual cycle of squirrels and sheep. In pinealectomized ewes, Mel treatments restricted to midsummer (Fig. 5, bottom) synchronized the annual LH rhythm with period and phase similar to those of pineal-intact controls. Spring Mel treatment likewise entrained the LH rhythm to a period

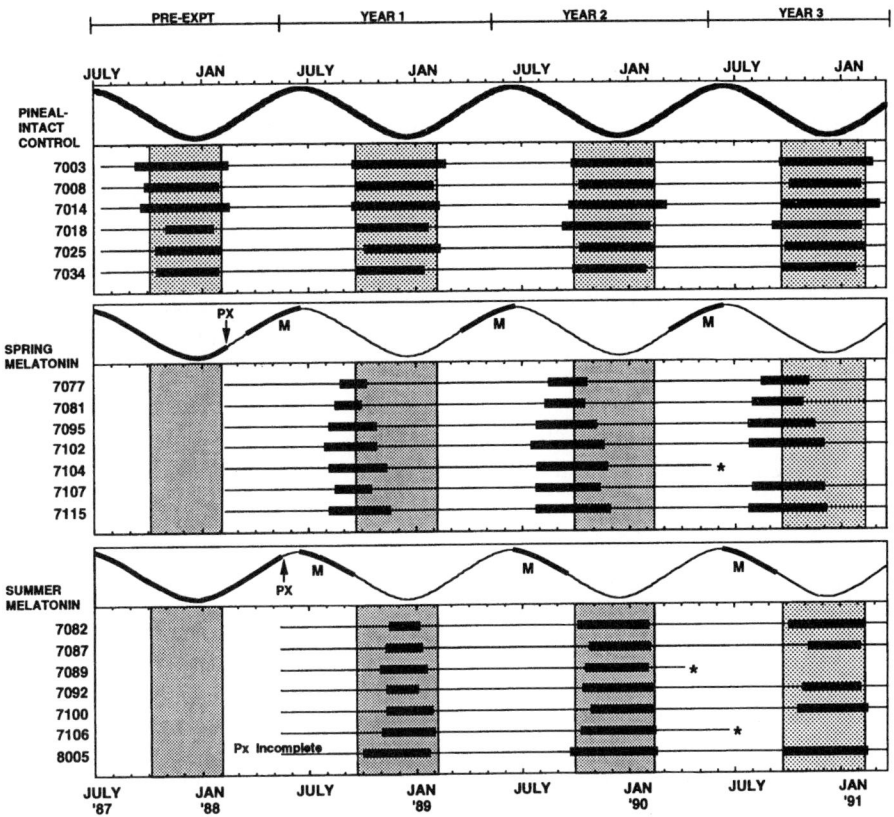

FIGURE 5 Summary of timing of high and low LH stages in pineal-intact control ewes (top) and in pinealectomized ewes treated with melatonin (M) for 3 consecutive months each year for 4 years. Thick curved lines indicate when melatonin was present and its changing pattern. The timing of the high LH stage of each cycle relative to the calendar year (abscissa) is plotted as a thick horizontal bar, with the high LH stage of the first observed cycle depicted by the left-most horizontal bar for each animal. Asterisks indicate that an animal died before observations were completed (modified from Woodfill et al., 1994).

of 12 months, but onset of the high LH stage was phase advanced relative to that of controls (Fig. 5, middle). Winter treatments were largely ineffective. The efficacy of Mel as a zeitgeber in sheep corresponds to times of year during which day length produces its greatest effects on entrainment of CARs.

In golden-mantled ground squirrels 3 months of Mel treatment beginning in late summer phase advanced the onset of estrus by an average of 25 days; the advance persisted and grew larger in subsequent cycles. Similar duration Mel treatment beginning in early spring phase delayed estrus the following cycle, but this effect waned over the next two cycles and may reflect masking rather than a permanent phase shift of the circannual oscillator(s). Circannual clocks in both sheep and squirrels respond to Mel in a phase-dependent manner over the circannual cycle.

V. NEURAL SUBSTRATES FOR GENERATION OF CARs

No single localized clock has been identified as essential for generation of CARs in reproduction. Many neural tissues can be ablated without compromising reproductive rhythms of male and female mammals. Specification of the neural networks that mediate CARs of GnRH secretion remains a major challenge.

Specific neural substrates are implicated in photoperiodic (melatonin) control of LH and prolactin secretion. Melatonin microimplants in medial basal hypothalamic structures (perhaps most critically the ventromedial nucleus) influence secretion of LH; the pars tuberalis of the pituitary is the locus for control of prolactin secretion by Mel. Although ablation of anterior hypothalamic structures extends the reproductive phase of ewes and removal of the olfactory bulbs lengthens the period of the circannual rhythm of testosterone secretion of squirrels, these brain regions are not suspected of harboring clocks critical for generation of CARs of reproduction. Rather, these neural substrates appear to affect expression of the rhythm, perhaps by modifying the several effector systems interposed between the clock and hormonal events that mediate reproduction.

VI. INFLUENCE OF PINEAL AND THYROID HORMONES ON CARs

The period of the CAR of reproduction is substantially shorter in pinealectomized than in intact female squirrels but most animals deprived of pineal melatonin secretion nevertheless still generate CARs of estrus for several years postoperatively. CARs of LH secretion also persist in pinealectomized ewes but the rhythm damps more rapidly than in squirrels.

The duration of the breeding season is extended in ewes and rams that have been thyroidectomized; the transition from reproductive to quiescent status is facilitated by thyroid hormone secretion during the latter part of the breeding season. Again, neither melatonin nor the several thyroid hormones appear to be part of the mechanism that generates the underlying circannual rhythm but are instead endocrine "enablers" that permit rhythm expression.

VII. CONCLUSIONS

Some long-lived mammals employ a circannual clock to generate repeated reproductive cycles over several years; unlike noncircannual seasonal oscillations, circannual rhythms do not require environmental resetting for the rhythm to be expressed on successive years. However, both types of rhythms must be adjusted to local environmental conditions, specifically food availability, to maximize fitness consequences of seasonal reproduction. In both cases this "fine-tuning" involves animals attending to annual variations in day length, transduced by changes in Mel secretion. There are several distinct types of circannual organization of reproduction and sex differences in circannual control of gonadotropin secretion. Future research will focus on neural mechanisms that underlie generation of CARs of GnRH secretion and pathways by which melatonin entrains this rhythm to the day length cycle.

See Also the Following Articles

CIRCADIAN RHYTHMS; NUTRITIONAL FACTORS AND REPRODUCTION; PHOTOPERIODISM; SEASONAL REPRODUCTION

Bibliography

Bronson, F. H. (1989). *Mammalian Reproductive Biology*. Univ. of Chicago Press, Chicago.

Bronson, F. H. (1995). Seasonal variation in human reproduction: Environmental factors. *Q. Rev. Biol.* 70, 141–164.

Gwinner, E. (1986). *Circannual Rhythms*. Springer-Verlag, Berlin.

Gwinner, E., Ball, G. F., Goldman, B. D., Karsch, F. J., Saunders, D. S., and Zucker, I. Circannual rhythms and photoperiodism. Submitted.

Karsch, F. J., Robinson, J. E., Woodfill, C. J. I., and Brown, M. B. (1989). Circannual cycles of luteinizing hormone and prolactin secretion in ewes during prolonged exposure to a fixed photoperiod: Evidence for an endogenous reproductive rhythm. *Biol. Reprod.* 41, 1034–1046

Lincoln, G. A., and Clarke, I. J. (1994). Photoperiodically-induced cycles in the secretion of prolactin in hypothalamo-pituitary disconnected rams: Evidence for translation of the melatonin signal in the pituitary gland. *J. Neuroendocrinol.* 6, 251–260.

Lincoln, G. A., and Short, R. V. (1980). Seasonal breeding: Nature's contraceptive. *Recent Prog. Horm. Res.* 36, 1–52.

Malpaux, B., Daveau, A., Maurice, F., Gayrard, V., and Thiery, J.-C. (1993). Short-day effects of melatonin on luteinizing hormone secretion in the ewe: Evidence for central sites of action in the mediobasal hypothalamus. *Biol. Reprod.* 48, 752–760.

Michael, R. P., and Bonsall, R. W. (1977). A 3-year study of an annual rhythm in plasma androgen levels in male rhesus monkeys (*Macaca mulatta*) in a constant laboratory environment. *J. Reprod. Fertil.* 49, 129–131.

Wickings, E. J., and Nieschlag, E. (1980). Seasonality in endocrine and exocrine testicular function of the adult rhesus monkey (*Macaca mulatta*) maintained in a controlled laboratory environment. *Int. J. Androl.* **3**, 87–104.

Woodfill, C. J. I., Wayne, N. L., Moenter, S. M., and Karsch, F. J. (1994). Photoperiodic synchronization of a circannual reproductive rhythm in sheep: Identification of season-specific time cues. *Biol. Reprod.* **50**, 965–976.

Xiong, J.-J., Karsch, F. J., and Lehman, M. N. (1997). Evidence for seasonal plasticity in the gonadotropin-releasing hormone (GnRH) system of the ewe: Changes in synaptic inputs onto GnRH neurons. *Endocrinology* **138**, 1240–1250.

Zucker, I. (1988). Neuroendocrine substrates of circannual rhythms. In *Biological Rhythms and Mental Disorders* (D. J. Kupfer, T. H. Monk, and J. D. Barchas, Eds.). Guilford, New York.

Zucker, I. (1999). Circannual rhythms: Mammals. In *Handbook of Behavioral Neurobiology* (J. Takahashi *et al.*, Eds.).

Zucker, I., Lee, T. M., and Dark, J. (1991). The suprachiasmatic nucleus and annual rhythms of mammals. In *The Suprachiasmatic Nucleus: The Mind's Clock* (D. C. Klein, R. Y. Moore, and S. M. Reppert, Eds.), pp. 246–259.

Circumcision

M. Sean Esplin
University of Utah

I. History and Background
II. Indications and Reported Benefits of Circumcision
III. Complications and Disadvantages of Circumcision
IV. Female Circumcision
V. Conclusion

GLOSSARY

balanitis Inflammation of the glans penis.
circumcision Excision of the foreskin, or prepuce, that covers the tip, or glans, of the penis.
paraphimosis Retention of the preputial ring proximal to the coronal sulcus causing swelling of the foreskin and glans penis.
phimosis Scarring or stenosis of the preputial ring with resultant inability to retract the fully differentiated foreskin back to expose the glans penis.
posthitis Inflammation of the prepuce, or foreskin, alone.
prepuce The free fold of skin that covers the glans penis, also called the *foreskin*.
preputial ring The ring-like opening formed by the prepuce at the tip of the penis.
preputial space The space between the glans penis and the prepuce.

Circumcision (Fig. 1) refers to the surgical removal of the foreskin or prepuce which covers the glans of the penis in males or the clitoris in the female. Although it originated as a part of religious ceremony or tradition, some argue that medical benefits exist to support its continued practice. The modern practice of circumcision has attracted passionate advocates and critics and has become the source of considerable debate.

I. HISTORY AND BACKGROUND

Circumcision is one of the oldest operations known dating back to the time of the ancient Egyptians and is believed to have been part of a religious ceremony or a mark of slavery or punishment. Today, circumcision is rarely performed throughout the world except in North America, among Jews and some African and aboriginal tribes. Only one in seven males worldwide are circumcised in the neonatal period. However, in the United States, more than 1 million newborn males (60%) are circumcised yearly.

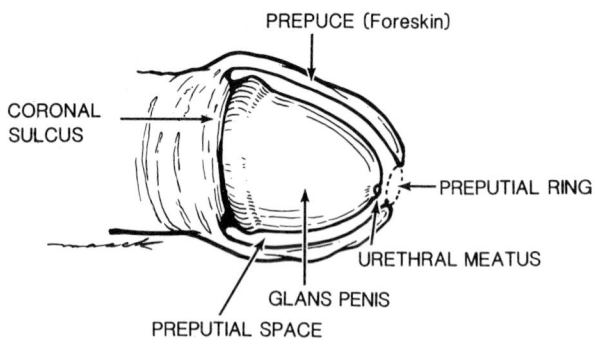

FIGURE 1 Anatomy of the penis.

The rate of circumcision varies greatly by region and is lowest on the West Coast (50%) and highest in the Midwest (70%).

Advocates of male circumcision argue that recently reported medical benefits among circumcised males far outweigh the potential risks of the procedure. Others contend that these benefits are unproven and argue that the risks caused by the procedure are too great. No consensus has been reached within the medical community and no strong recommendation has been made. This debate has also led to the formation of nonmedical groups, such as the National Organization of Circumcision Resource Centers and the International Organization Against Circumcision Trauma, which attempt to educate and raise public awareness about the problems associated with circumcision.

Female circumcision is a traditional practice that consists of the removal of the clitoral hood or removal of the clitoris itself with part or all of the labia minora in its most severe form. It is a common practice in many countries in Africa, the Middle East, and Southeast Asia. In Somalia, more the 99% of women have been subjected to some form of female circumcision.

II. INDICATIONS AND REPORTED BENEFITS OF CIRCUMCISION

Circumcision is performed for medical, social, and religious reasons. For the purpose of this chapter, only the medical indications of the procedure will be discussed. The reported medical benefits of circumcision include decreased incidence of urinary tract infections, penile cancer, sexually transmitted disease, and inflammatory conditions. Although abundant literature exists regarding these possible benefits, none has been conclusively proven.

A. Urinary Tract Infections

The overall incidence of urinary tract infections (UTI) is low for males. However, it is highest during the first 12 months of life when it is even greater than the incidence in females. Some believe this is due to inadequate hygiene of the foreskin. One study has shown that circumcision prevents bacterial colonization of the prepuce and subsequent periurethral colonization and ascending infection. Ascending infections and acute pyelonephritis (kidney infections) early in life may increase the risk of renal failure later in life.

Studies comparing the incidence of UTI in circumcised and uncircumcised males have noted the incidence is 6–10 times higher in uncircumcised children. However, these studies are limited by their retrospective design and possible bias. No study has proven an association between circumcision and a protection against renal failure. Although the incidence of UTI is higher in uncircumcised males, it is still very low at 0.6–0.8% and may not pose a significant risk.

B. Cancer of the Penis

Cancer of the penis is a rare problem with an annual incidence of 1/100,000 in the United States. Circumcision may protect against penile cancer by improving hygiene and decreasing the amount of smegma, a cheese-like buildup of dead cells that collects inside the foreskin. Smegma has been shown to produce tumors when applied to the skin of mice but the actual carcinogen has not been identified.

Early reports from the 1930s found no cases of penile cancer in circumcised men and recent studies have confirmed these findings. Of 1600 cases of penile cancer reported over a 20-year period, there were no cases in circumcised men. A 1993 review by the National Cancer institute found the risk for penile cancer to be 3.2 times higher for uncircumcised males. However, the rates of penile cancer in developed countries, such as Denmark and Japan, where circumcision is not performed, are very similar to rates in the United States suggesting that stringent standards of hygiene may be sufficient to reduce the risk of cancer in uncircumcised populations.

C. Sexually Transmitted Disease and Inflammation

Uncircumcised men appear to be more susceptible to sexually transmitted diseases that disrupt epithelial surfaces such as genital herpes, syphilis, human papilloma virus, and chancroid. Some believe that diseases causing genital ulceration are important risk factors in the transmission of HIV. These risk factors can be decreased by appropriate penile hygiene and the practice of safe sex.

Inflammation of the foreskin (posthitis) and paraphimosis (constriction of the glans penis by retraction and swelling of the foreskin) cannot occur in circumcised males. However, balanitis does occur in both circumcised and uncircumcised men. Circumcision may reduce the incidence of penile inflammatory disorders but, again, a similar benefit may be obtained by proper hygiene alone.

III. COMPLICATIONS AND DISADVANTAGES OF CIRCUMCISION

A. Technical Errors

The complications of circumcision usually arise from surgical technique. The most common complication is bleeding, which occurs in 0.2–0.6% of all procedures but is usually mild and can be controlled with pressure alone. More severe cases will require suturing and surgical repair. Infection is also a common problem that usually responds to proper cleaning but may require antibiotic treatment. Phimosis, which can occur when too little foreskin is removed, can be severe and cause urinary obstruction. Disfigurement occurs when too much of the foreskin is removed, resulting in skinning of the penile shaft. This can often be treated by allowing reepithelialization of the shaft but if extensive, a skin graft may be required to produce an acceptable cosmetic result. Other problems include the accidental total amputation of the glans penis or necrosis and total loss of the penis following the use of cautery or epinephrine to control bleeding.

Technical errors are preventable by following basic principles. Experienced operators should be involved in all procedures. The penis should be examined prior to the procedure to identify any abnormality, such as hypospadias, that may prevent the safe performance of the procedure. The coronal sulcus should be marked on the foreskin prior to beginning the procedure to ensure that the appropriate amount of skin will be removed. When an abnormality or problem is first noted the procedure should be stopped and prompt consultation with a urologist should be obtained. Any viable foreskin that remains may be useful if a repair is required.

Many different instruments have been used over time to perform circumcisions. There are currently three popular techniques which include the Plastibell, the Gomco, and Mogan clamps. All three techniques give uniformly good results without significantly different rates of complication that would warrant recommendation of one over the other.

B. Pain and Anesthesia

Although anesthesia is not routinely used for circumcision, anyone who has witnessed the procedure would agree that it is painful to the infant. The anatomic and functional neural pathways necessary to perceive pain are present in the newborn. Behavioral changes, such as crying, irritability, and varied sleep patterns, as well as physiologic changes, including increased pulse and blood pressure, have been used to document infants' responses to circumcision and seem to indicate that this is a painful procedure. These changes usually disappear within minutes to hours following the circumcision.

The most common form of anesthesia used during circumcision is the dorsal penile nerve block. This is relatively safe although the usual complications of local anesthetic injection, including hematoma formation and toxic effects of intravascular injection, have been reported. The dorsal penile nerve block has been shown to decrease the behavioral and physiologic changes associated with circumcision. Other types of anesthesia have been suggested, including the application of topical anesthetics which have been effective and safe.

IV. FEMALE CIRCUMCISION

Female circumcision is practiced in over 30 countries throughout the world and affects as many as 80 million women. It is usually performed prior to puberty by untrained practitioners under nonsterile conditions. The extent of the circumcision ranges from removal of the prepuce alone to removal of the prepuce and the labia with complete obliteration of the vaginal introitus. Complications include dysmenorrhea and dyspareunia as well as obstruction during the second stage of labor and an increase in vesicovaginal fistulas. Women with circumcision will often require both anterior and posterior episiotomies at the time of delivery.

Female circumcision is considered by many to be a form of mutilation and the practice has been outlawed in many countries. However, in some places the practice is entrenched in tradition and religious belief and it remains commonplace. One study in 1992 surveyed women in Somalia who had undergone female circumcision themselves and found that all planned to subject their daughters to the custom as well.

V. CONCLUSION

Circumcision is a relatively simple procedure that is safe when performed by an experienced operator. It does decrease the risk of UTI and penile cancer and may decrease the incidence of sexually transmitted diseases including HIV. However, these benefits may also be obtained with proper hygiene in the uncircumcised male. In 1989, the American Academy of Pediatrics appointed a task force to review the indications and complications of routine circumcision. They concluded that the potential benefits do not warrant the routine circumcision of all males. However, the benefits and risks of circumcision should be explained to parents who are considering the procedure so that an informed decision can be made.

See Also the Following Articles

CLITORIS; PENIS

Bibliography

American Academy of Pediatrics (1989). Report of the task force on circumcision. *Pediatrics* **84**(2), 388–390.

Anderson, G. (1989, March). Circumcision. *Pediatr. Ann.* **18**(3), 205–213.

Benini, F., Johnston, C., Faucher, D., et al. (1993). Topical anesthesia during circumcision in newborn infants. *J. Am. Med. Assoc.* **270**(7), 850–853.

Dirie, M., and Lindmark, G. (1992). The risk of medical complications after female circumcision. *East Afr. Med. J.* **69**, 479–482.

Erian, M., and Goh, T. (1995). Female circumcision. *Aust. N. Z. J. Obstet. Gynecol.* **35**(1), 83–85.

Fontaine, P., and Toffler, W. (1991). Dorsal penile nerve block for newborn circumcision. *Am. Family Phys.* **43**, 1327–1333.

Harkavy, K. (1987). The circumcision debate. *Pediatrics* **79** 649–650.

Holman, J., Lewis, E., and Ringler, R. (1995). Neonatal circumcision techniques. *Am. Family Phys.* **52**(2), 511–518.

Kochen, M., and McCurdy, S. (1980). Circumcision and the risk of penile cancer: A life-table analysis. *Am. J. Dis. Child* **134**(5), 484–486.

Moses, S., Plummer, F., Bradley, J., et al. (1994). The association between lack of male circumcision and risk for HIV infection: A review of the epidemiologic data. *Sex Trans. Dis.* **21**, 201–210.

Niku, S., Stock, J., and Kaplan, G. (1995). Neonatal circumcision. *Urol. Clin. North Am.* **22**(1), 57–65.

Parker, S., Stewart, A., Wren, M., et al. (1983). Circumcision and sexually transmissible disease. *Med. L. Aust.* **2**, 288–290.

Roberts, J. (1986). Does circumcision prevent urinary tract infection? *J. Urol.* **135**, 991.

Robson, W., and Leung, A. (1992). The circumcision question. *Postgrad. Med.* **91**(6), 237–244.

Schoen, E. (1991). The relationship between circumcision and cancer of the penis. *CA Cancer J. Clin.* **41**, 306–309.

Smith, G., Greenup, R., and Takafujii, E. (1987). Circumcision as a risk factor for urethritis in racial groups. *Am. J. Public Health* **77**, 452–454.

Stang, H., Cunnar, M., Snellman, L., *et al.* (1988). Local anesthesia for neonatal circumcision: Effect on distress and cortisol response. *J. Am. Med. Assoc.* **10**, 1507–1511.

Tyndall, M., Ronald, A., Agoki, E., *et al.* (1996). Increased risk of infection with human immunodeficiency virus type 1 among uncircumcised men presenting with genital ulcer disease in Kenya. *Clin. Infect. Dis.* **23**(3), 449–453.

Wiswell, T. (1996). Circumcision circumcspection. *NEJM* **336**(17), 1244–1245.

Wiswell, T., Smith, F., and Buss, J. (1985). Decreased incidence of urinary tract infections in circumcised male infants. *Pediatrics* **75**, 901–903.

Wiswell, T., Miller, G., Gelston, H., *et al.* (1988). Effect of circumcision status on periurethral bacterial flora during the first year of life. *J. Pediatr.* **113**(3), 442–446.

Cleavage

see Embryogenesis

Clitoris

Ursula Kuhnle
Universitätskinderklinik München

I. Anatomy of the External and Internal Female Genitalia
II. Function of the Clitoris
III. Normal Development of Male and Female Genitalia
IV. Abnormal Development of the Clitoris and the External Genitalia
V. Therapy
VI. Female Circumcision

GLOSSARY

androgens Male hormones synthesized and secreted by the testis causing the masculinization of the body.
cliteromegaly Abnormal enlargement of the clitoris.
ectoderm The outer of three layers of cells formed during early embryonic development.
endoderm The inner of three layers of cells formed during early embryonic development.
estrogens Female hormones synthesized and secreted by the ovary causing the feminization of the body during puberty.
gestation or *pregnancy* The time period from conception to birth.
hirsutism Excessive hair growth.
mesoderm The middle of three layers of cells formed during early embryonic development.
Müllerian ducts Female fetal ducts of the internal genitalia, which have the capacity to differentiate into the fallopian tubes, uterus, and upper two-thirds of the vagina.
ovary The female gonad, which produces estrogens and ova.
postnatal Referring to the time period after birth.
prenatal Referring to the time period before birth, during pregnancy.
primordial Describing or referring to the very first stages of development.

SRY gene A gene within the sex-determining region on the Y chromosome that is necessary to initiate testicular development.

testis The male gonad, which produces androgens and sperms.

Wolffian ducts Male fetal ducts which have the capacity to differentiate into the epididymis, vas deferens, and the ejaculatory ducts.

The clitoris is the erectile organ of the female external genitalia and is equivalent to the phallus of the male genitalia. Androgens play a decisive role in the differentiation of not only the phallus but also the entire external genitalia. Therefore, abnormal development of the clitoris usually includes abnormal development of the entire external genitalia.

I. ANATOMY OF THE EXTERNAL AND INTERNAL FEMALE GENITALIA

The female external genitalia consists of labia minora, labia majora, the clitoris, and the perineum and is shown schematically in Fig. 1H. The clitoris is located at the anterior angle of the labia majora. The size, color, and the visibility of the clitoris varies widely from one woman to another, but in general the length of the clitoris does not exceed 0.5 cm and it is usually fully covered by the labia majora.

Developmentally, the male and female genitalia originate from the same indifferent anlagen of the fetus and the parallel development in both sexes is illustrated in Fig. 1. The clitoris as well as the penis develop from the genital tubercle, and the clitoris consists of the glans (head) clitoris, corpora cavernosa clitoridis, and the bulb of the vestibule (cavity) corresponding in the male to the glans penis, corpora cavernosa penis, and corpus spongiosum penis, respectively (Fig. 1). The male phallus develops and grows under the influence of testicular androgens; thus, abnormal development of the clitoris is usually caused by androgen excess. The severity depends on the timing and duration of exposure to male hormones. Early prenatal androgen excess causes not only marked clitoromegaly but also complex abnormalities of the female external genitalia, whereas postnatal androgen excess will result in cliteromegaly only.

The female internal genitalia include the ovaries, which originate from the primordial germ cells, and the vagina, uterus, and fallopian tubes, which differentiate from the fetal müllerian ducts.

II. FUNCTION OF THE CLITORIS

The clitoris is a highly sensitive organ and possesses, in contrast to the vagina, numerous free nerve endings. The clitoris is stimulated by sexual activity, but it is not known to play a direct role in any reproductive function. Its role during sexual intercourse and its function for reaching orgasm have been discussed by Kinsey and co-workers (1953). This group has put forward the idea of a difference between clitoral and vaginal orgasm. While clitoral orgasm should be achieved mainly through masturbation, vaginal orgasm is reached through penetration and thought to be more satisfying. The practical significance of such differentiations seems doubtful.

In this context it may be of interest to note that there is apparently an operative procedure to treat frigidity or nonorgasmy by which adhesions from the ventral parts of the clitoris are removed. So far a therapeutically beneficial effect has not been established.

III. NORMAL DEVELOPMENT OF MALE AND FEMALE GENITALIA

The initial step of gonadal differentiation occurs during the first 4 or 5 weeks of fetal development when primordial cells migrate into the genital ridge, which is close to the kidneys and the adrenals. Until about 42 days of gestation, the gonads in both sexes are morphologically indistinguishable. The differentiation into either an ovary or a testis is determined by the sex chromosomes and depends on whether the ovum has been fertilized by either an X- or a Y-bearing sperm. From studies of animal and human disorders it was concluded that in the human the

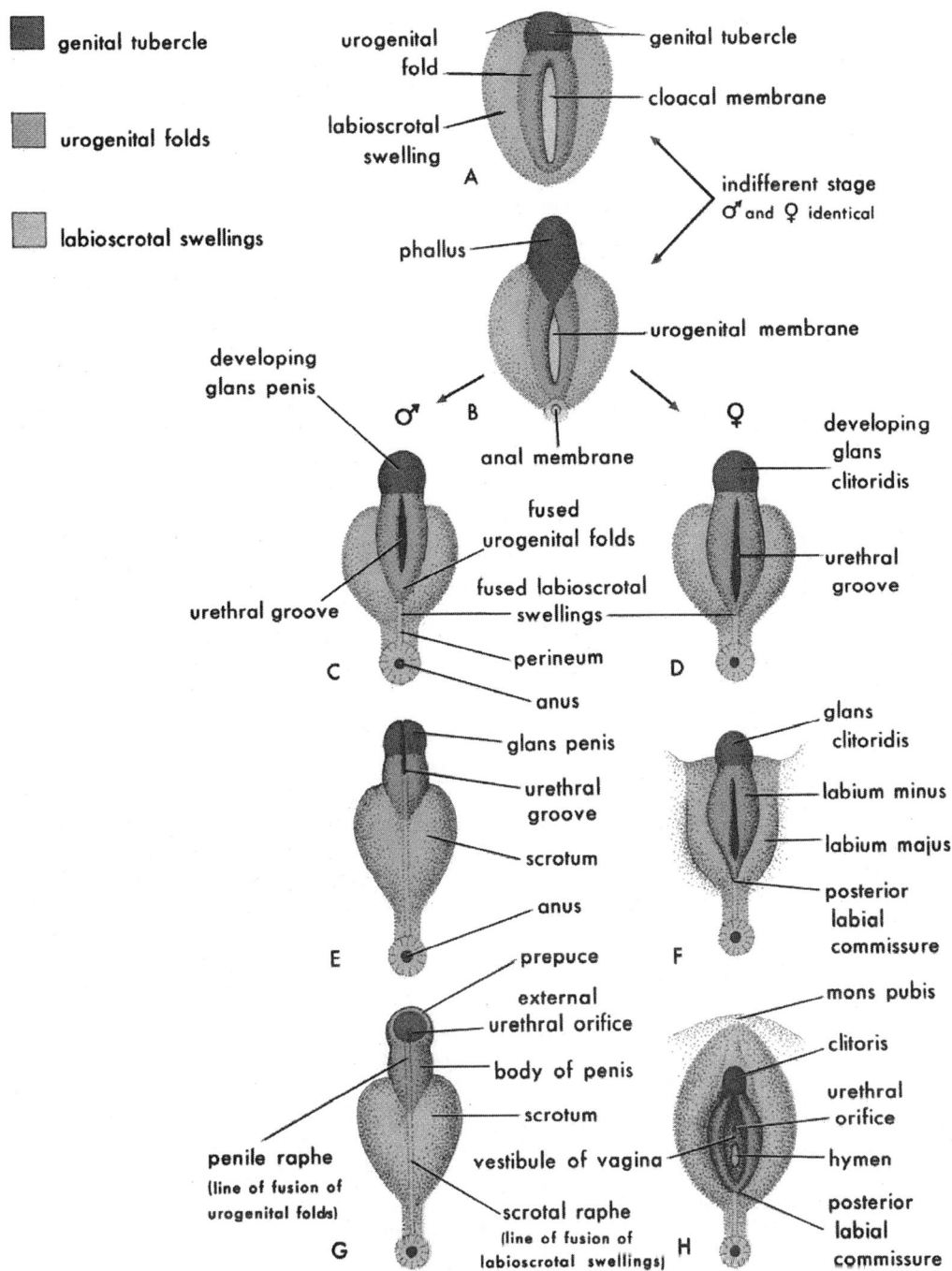

FIGURE 1 Schematic illustration of the parallel development of the male and female external genitalia originating from the same indifferent anlage. (Reproduced with permission from Moore and Persaud (1993).

contribution of two intact X chromosomes is necessary for the development of normal fertile ovaries, whereas a gene located on the short arm of the Y chromosome, the SRY gene, is required for the development of the testis.

A. Development of the Ovary and Female Genitalia

In the presence of two X chromosomes, ovarian development commences slowly and primordial follicles containing oogonia can be seen only at 16 weeks of gestation. Many oogonia degenerate before birth but 1 or 2 million continue to develop into oocytes before birth. The process of degeneration is increased when only one X chromosome is present, as is the case in patients with Turner's syndrome. Even though the ovary seems to be capable of estrogen synthesis, it contributes little or nothing to the high circulating hormone levels present during pregnancy.

The feminization of the indifferent external genitalia does not need the presence of gonadal hormones. The growth of the genital tubercle decreases and forms the clitoris, the urogenital folds and labioscrotal folds form the labia minora and labia majora, respectively. Thus, while the clitoris of the female corresponds to the phallus in the male, the labia majora correspond to the scrotum.

The wolffian ducts regress and, during the absence of müllerian-inhibiting factor, the paramesonephric or müllerian ducts develop into an uterus, whereas the vagina develops from the urogenital sinus. As in the male, the development of the internal and external genitalia in the female fetus is completed by 12 weeks of gestation.

B. Development of the Testis and Male Genitalia

To fully understand abnormal development of the female genitalia, it is necessary to understand the development of the normal male genitalia. Development of the testis is initiated by the SRY gene on the Y chromosome, which triggers the activation of a coordinated sequence of genes resulting in the development of testes, which consist of a dense tunica albuguinea, the rete testis, and the seminiferous tubules. The seminiferous tubules give rise to the interstitial cells of Leydig, which secrete testosterone, and the Sertoli cells derived from the surface epithelium, which secrete müllerian-inhibiting factor. Further development of male genitalia depends on the secretion of these two hormones.

The androgen, testosterone, stimulates the mesonephric or wolffian ducts to differentiate into male internal genitalia—that is, the epididymis, vas deferens, and ejaculatory ducts—and stimulates the external genitalia to masculinize. For the development of the external genitalia, testosterone is reduced by the target tissue to dihydrotestosterone. Stimulated by androgens, the phallus enlarges and the genital tubercle elongates to form the glans penis, whereas the labioscrotal folds grow and fuse to form the scrotal sac. The line of fusion is visible as scrotal raphe. In complete masculinized males the urethra opens at the tip of the penis, whereas hypospadia describes a condition in which, depending on the severity, the urethra opens somewhere along the shaft of the phallus, along the scrotal raphe or the perineum.

The müllerian-inhibiting factor secreted by the Sertoli cells inhibits the development of the female internal ducts and, hence, the development of a uterus and the upper two-thirds of the vagina. The entire process of differentiation is completed during the 12th week of fetal development. A defect or a deficiency of any of the two testicular hormones prior to the 12th week of gestation leads to incomplete male sexual development.

IV. ABNORMAL DEVELOPMENT OF THE CLITORIS AND THE EXTERNAL GENITALIA

It is of importance to remember that the difference between male and female external genitalia is the result of androgenic hormones secreted by the fetal testis (Fig. 1). Ovarian hormones or estrogens do not play a role during fetal life. In normal sexual differentiation the chromosomal sex corresponds to the appearance of the internal and external sex organs. In abnormal sexual development there is a discrepancy between the chromosomal sex and the appearance of the external genitalia, and various degrees of intersexuality, ambiguity, or "hermaphroditism" will develop. The degree of intersexuality of the external genitalia is sometimes so severe that the aspect per se does not allow differentialization between male or female. It might well be that an incompletely virilized male or a masculinized female may look identical. In these cases, the decision of which sex a child should be brought up is difficult and complex, and the medical, psychological, social, and legal aspects have to be carefully considered.

Clitoromegaly Associated with Ambiguity of the Genitalia in Genetic Females

1. Prenatal Androgen Excess

Female fetuses exposed to high androgens during the sensitive first 12 weeks of gestation develop complex abnormalities of the external genitalia. Androgens during this time period not only cause hypertrophy of the clitoris but also induce various degrees of masculinization of the external genitalia. The most common disorder in a female fetus is congenital adrenal hyperplasia in which, due to an enzyme defect of adrenal steroid synthesis, high amounts of androgens are secreted by the fetal adrenal. The degree of masculinization varies widely from girls with mild clitoromegaly only to girls with severely ambiguous genitalia and, in the extreme, to girls with a phallus-like enlarged clitoris (Fig. 2). Rarely is androgen excess in the female fetus due to other causes such as androgen-producing tumors in the mother or maternal ingestion of androgens. Since only the testes secrete müllerian-inhibiting factor, even severely virilized girls have normal internal female genitalia—that is,

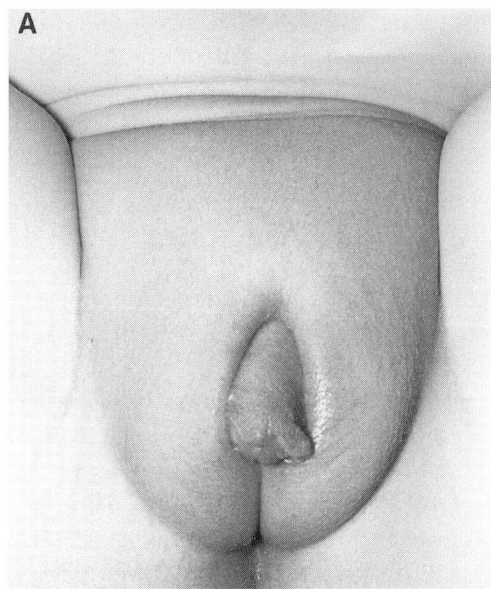

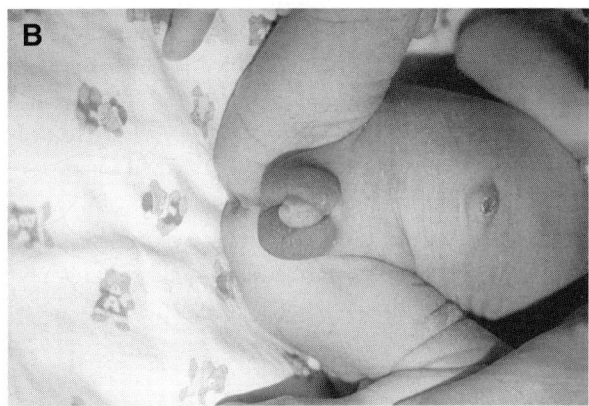

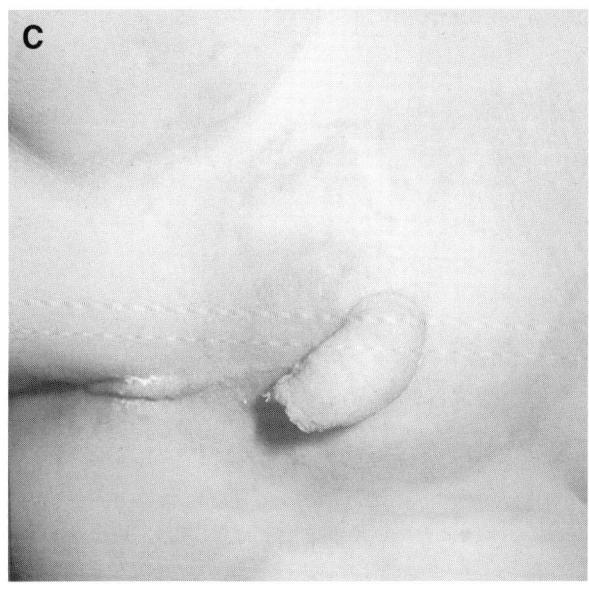

FIGURE 2 The wide variability of masculinization of the female genitalia in female pseudohermaphroditism. The picture shows the virilization of the external genitalia in three girls with congenital adrenal hyperplasia ranging from mild clitoromegaly (A), to severely ambiguous genitalia (B), to completely masculinized external genitalia (C) with the urethra opening at the tip of the phallus-like enlarged clitoris. The labia majora in B and C are fused to give a scrotum-like appearance. In these children the mere aspect per se does not allow to decide whether the infant is a girl or a boy.

ovaries, uterus, and fallopian tubes. This condition is commonly referred to as female pseudohermaphroditism.

True hermaphroditism describes a situation in which ovaries as well as testis develop in the same individual. This disorder most commonly develops in genetic females and is due to a mutation in a sex determining gene. Male hormones will be secreted and, depending on the amount of hormones secreted, the internal as well as external genitalia will develop male and female features.

In male pseudohermaphroditism the internal genitalia are male but the external genitalia are incompletely masculinized. In the most severe form, the "testicular feminization syndrome," genetic males develop female external genitalia despite high levels of androgens. Their body cells lack the receptors for androgens and thus do not masculinize. The external genitalia are indistinguishable from normal females. In less severe cases of male pseudohermaphroditism, the genitalia are ambiguous, with a small phallus and various degrees of hypospadia (Fig. 3).

2. Postnatal Androgen Excess

Isolated cliteromegaly develops when a female is exposed to androgens after the external genitalia have been fully developed—that is, after the 12th week of gestation.

Besides ingestion of androgen-containing drugs, excessive androgens in women originate either in the adrenals or in the ovaries. Androgen-producing adrenal tumors can arise at any age. More common, though, are mild enzymatic defects of adrenal steroid synthesis which cause so-called nonclassical or late-onset adrenal hyperplasia. Androgen excess in these cases increase with age.

Ovarian overproduction of androgens can be a sign of often malignant ovarian tumors or is present in a disorder called polycystic ovarian disease, usually associated with increased body weight. In addition to cliteromegaly, acne and increased growth of body hair are often present which are distributed in the classical male pattern.

V. THERAPY

In most cases of intersexuality, corrective surgery of the genitalia is needed. In the case of the virilized female infant, it is important to shorten the enlarged clitoris and reconstruct the labia majora to allow the parents to identify the child as a girl. Current operative procedures maintain the innervation of the sensitive glans clitoridis and remove only the enlarged corpus spongiosum or the erectile portion of the clitoris. In most cases, the vagina does not develop adequately and has to be reconstructed. The cliteromegaly which develops postnatally rarely requires surgical correction.

Cliteromegaly Unrelated to Androgen Excess

An enlarged clitoris at birth can be associated with a variety of tumors, such as lipomas, lymphangiomas, and gliomas, the latter of which might be the first manifestation of neurofibromatosis (Fig. 4). It should not be difficult for an experienced physician to recognize and differentiate cliteromegaly due to androgen excess or another cause.

Congenital absence of the clitoris (or phallus) is extremely rare and probably due to inadequate ectodermal and mesodermal interactions of the genital tubercle.

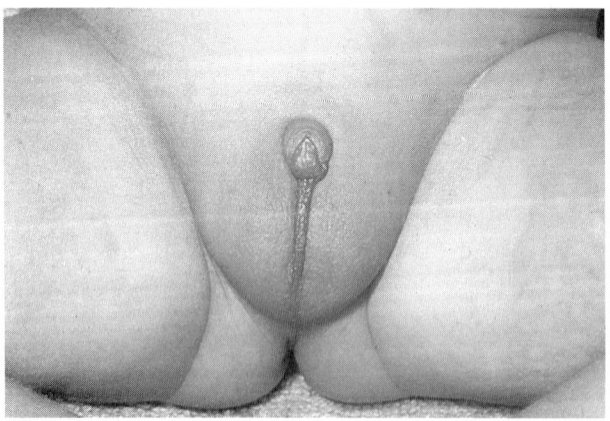

FIGURE 3 The genitalia in male pseudohermaphroditism with an extremely small phallus and severe hypospadia.

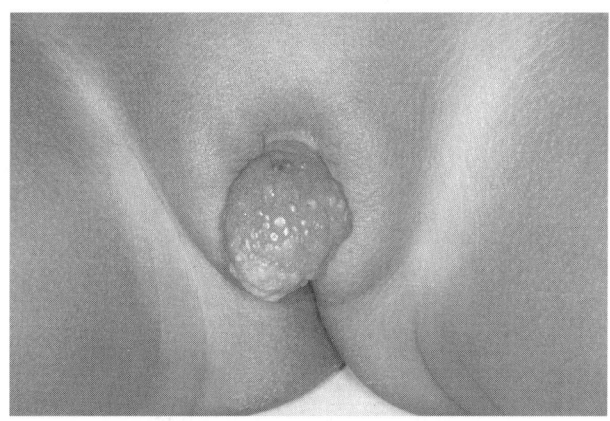

FIGURE 4 Clitoromegaly due to a lymphangioma.

VI. FEMALE CIRCUMCISION

In the nineteenth century in Europe, female circumcision or rather clitoridectomy was performed for a variety of reasons, including insanity, epilepsy, hysteria, masturbation, and a disorder called nymphomania. Female circumcision as a therapeutic procedure is obviously obsolete and no longer performed.

Female circumcision is still performed in some countries and cultures, mainly of northern African origin, where it is known among differing ethnic and cultural groups such as Christians, Muslims, Jews, and among the followers of indigenous religions. It is practiced among all socioeconomic classes and it is estimated that at least 100 million women are affected.

During male circumcision, the prepuce or foreskin of the penis are removed, whereas the glans and the erectile tissue of the phallus remain unharmed. This skin flap develops during the 12th week of gestation from surface ectoderm and has no equivalent in the female. In contrast, female circumcision is performed by removing portions or by totally excising the clitoris and, in some instances, also the labia majora, which correspond to the scrotal sac. Thus, this procedure corresponds to an amputation of the phallus and scrotum. Since neither the clitoris nor the labia majora are essential for reproduction, clitoral resection or amputation does not interfere with fertility but merely with pleasure and satisfaction during intercourse.

Female circumcision is similar to male circumcision in terms of having deep cultural roots and profound social meanings. How it enhances religious and ethnic identity is poorly understood.

See Also the Following Articles

ADRENAL HYPERPLASIA, CONGENITAL VIRILIZING; CIRCUMCISION; FEMALE REPRODUCTIVE DISORDERS; GENITALIA

Bibliography

Griffin, J. E. (1992). Androgen resistance—The clinical and molecular spectrum. *N. Engl. J. Med.* **326**, 611–620.

Grumbach, M. M., and Conte, F. A. (1992). Disorders of sex differentiation. In *William's Textbook of Endocrinology* (Wilson and Forster, Eds.), 8th ed. Saunders, Philadelphia.

Kinsey, A. C. (1953). *Sexual Behaviour in the Human Female.* Saunders, Philadelphia.

Moore, K. L., and Persaud, T. V. N. (1993). *The Developing Human*, 5th ed. Saunders, Philadelphia.

New, M. I., White, P., Pang, S., Dupont, B., and Speiser, P. (1993). The adrenal hyperplasia. In *The Metabolic Basis of Inherited Disease* (C. Scriver, A. Beaudet, S. Sly, and D. Valle, Eds.). McGraw-Hill, New York.

Ploss, A., and Bartels, R. (1935). *Women.* Heinemann, London.

Toubia, N. (1994). Female circumcision as a public health issue. *N. Engl. J. Med.* **11**, 712.

Cloning Mammals by Nuclear Transfer

Randall S. Prather

University of Missouri at Columbia

I. Theory of Cloning by Nuclear Transfer
II. Procedures for Cloning by Nuclear Transfer
III. Limitations of Nuclear Transfer
IV. Applications and Implications

GLOSSARY

clone A genetic copy or genetically identical individual.
differentiation The process of increased tissue specialization; specialized structure and function from a precursor cell(s).
DNA Deoxyribonucleic acid; the molecular basis of inheritance.
homologous recombination The process by which similar sequences of DNA pair and crossover with resulting changes in the DNA sequence.
mitochondria The organelles within cells that directs ATP production and has its own genome separate from the nucleus.
RNA Ribonucleic acid; transfers the DNA code to the cytoplasm to direct protein synthesis.
totipotent Having the potential to direct complete development from the one-cell stage to a fertile adult.

Nuclear transfer is a procedure by which genetically identical individuals can be created. The recent birth of a cloned lamb resulted from nuclear transfer and focused a great deal of attention to the field of reproductive biology. The ramifications of this report will be far-reaching. The applications of these techniques for nuclear transfer will be in agricultural, biomedical, and basic research. This article will cover the theory, procedures, limitations, applications, and implications of technologies for cloning mammals.

I. THEORY OF CLONING BY NUCLEAR TRANSFER

An understanding of the theory of nuclear transfer procedures requires knowledge of the differentiation of cells. Cell differentiation may be thought of as dominos falling over. Fertilization is one point at which the first domino falls over and initiates the developmental program of a zygote or 1-cell embryo. The next major point is when the embryo first begins producing significant amounts of RNA that will direct the synthesis of its proteins and play a major role in regulation of differentiation. The stage at which RNA synthesis is initiated after fertilization is species specific. In mouse embryos, RNA synthesis begins during the 2-cell stage, whereas RNA synthesis begins at the 4-cell stage in humans, pigs, and rats and at the 8- to 16-cell stage in sheep and cattle. Thus, at this species-specific stage another domino falls.

The differentiation process in embryonic development can be thought of in two distinct ways. One is to consider that differentiation results in cells becoming different from their predecessors. For example, the eight-cell stage pig embryo is different from the two-cell stage pig embryo because the eight-cell stage embryo is producing significant amounts of RNA and the two-cell stage is producing negligible amounts of RNA. Another type of differentiation is that of tissue differentiation. This is first observed at the blastocyst stage (Fig. 1). When the blastocoel cavity forms in the developing blastocyst, two tissues can be isolated, the inner cell mass cells and the trophectoderm. These two tissues produce a different repertoire of RNA and are destined to form different tissues. Thus, more dominos fall over as cells diverge

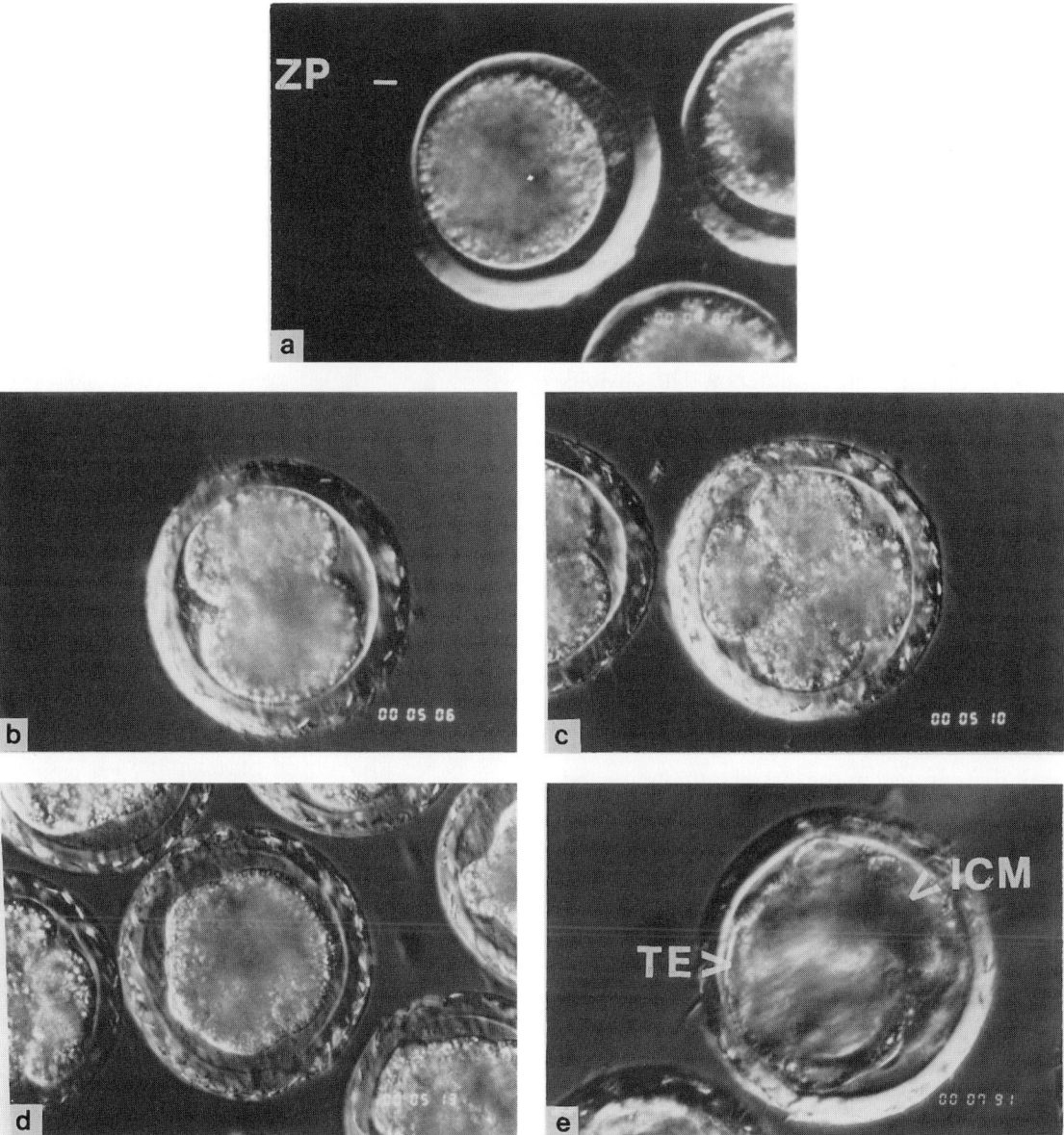

FIGURE 1 Development of the pig early embryo. Fertilization initiates a series of events within the oocyte that start the differentiation process. The fertilized oocyte (a) then cleaves to the two-cell (b), four-cell (c), and eight-cell stage. At or beyond the eight-cell stage the embryo begins to compact as the cells flatten against one another (d) and individual cells are no longer visible. A cavity is then formed by the accumulation of fluid within the ball of cells (e). As the cavity grows, two cell types become identifiable, the inner cell mass cells (ICM) and the trophectodermal cells (TE). The outer diameter of the zona pellucida (ZP) is about 150 μm. Note the accumulation of sperm in the zona pellucida between the one-, two-, and four-cell stage.

into two tissue-specific pathways. This process continues throughout development of the embryo, fetus, adolescent, mature adult, and aging adult into senescence. As tissues become more differentiated, it was believed by most that the genetic material in nuclei of the cells became increasingly difficult, if not impossible, to reprogram.

One of the basic questions in biology early in this century was whether nuclei in early embryos have unequal inheritance during development and if this results in tissue differentiation. In 1938, Hans Spemann proposed a set of experiments to determine when the first irreversible differentiation events, or unequal inheritance, occurred. He proposed to take nuclei from embryos that were progressively more advanced in development and transfer them to unfertilized eggs that had their chromosomes removed. He suggested that unequal inheritance would be evident if nuclei progressed to a point that they could longer recapitulate development. This would indicate the stage at which the nuclei began to lose their ability to direct development, or lose their totipotency. The obvious outcome of such experiments is to determine the potential to produce many animals that are derived from nuclei from the same donor embryo, that is, a method of cloning embryos.

Spemann's experiments, conducted with frogs, were not reported until after his death. However, results of his experiments suggested that development of nuclear transfer embryos was dramatically reduced when the donor cells were beyond the stage at which the embryo initially began producing RNA. This suggested that there was a significant differentiation event that resulted in the loss of totipotency and that another domino had fallen over.

II. PROCEDURES FOR CLONING BY NUCLEAR TRANSFER

Procedures for nuclear transfer are technically difficult and require specialized equipment. First, an unfertilized egg is treated with cytoskeletal inhibitors to depolymerize the microtubules and/or microfilaments. Such treatments impart elasticity to the plasma membrane. The egg is then held in place with a fire-polished holding pipette (Fig. 2) and another beveled micropipette is inserted through the outer membrane (zona pellucida) of the ovum and thrust into the cytoplasm. Since the plasma membrane is now very elastic it invaginates around the pipette. The chromosomes are located, either directly or after treatment with a DNA-specific dye and ultraviolet light illumination, and aspirated into the pipette. The pipette is then withdrawn and the cell membrane of the ovum closes the wound. This results in a membrane-bound nucleoplast in the pipette and an

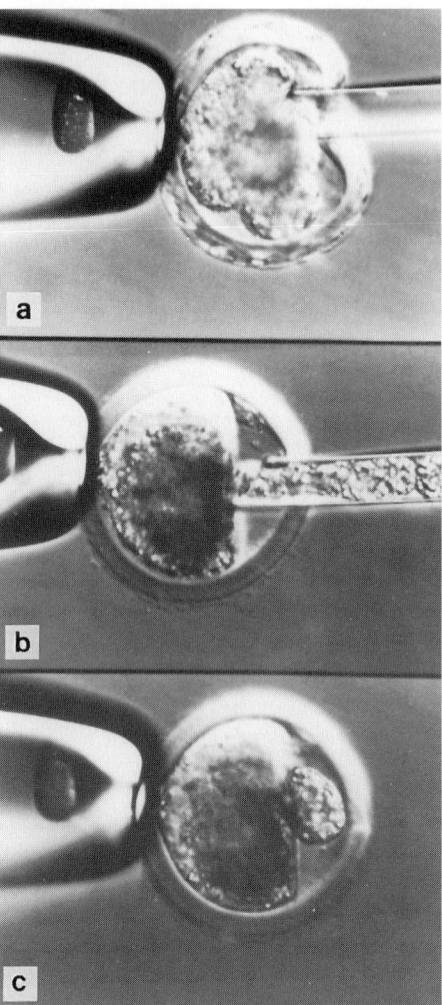

FIGURE 2 A procedure for nuclear transfer. An unfertilized oocyte is held in place and another pipette is used to aspirate the chromosomes from the oocyte. A donor cell, in this case from an eight-cell stage embryo (a), is aspirated into the pipette and inserted under the zona pellucida of the unfertilized oocyte (b and c). The two cells are then fused together and allowed to develop (from M. A. Mayes and R. S. Prather, unpublished results).

ovum with no nucleus/chromosomes that can serves as a nuclear transfer recipient. A nuclear donor cell is then located and, using a micropipette, is placed within the zona pellucida next to the recipient ovum. The two cells are then induced to fuse using viral, chemical, or electrical stimuli to begin the fusion process. This results in a deposition of the nucleus into the cytoplasm of the recipient ovum and a mixing of the cytoplasmic contents of the two cells. There can be many technical variations on the previous description without loss of subsequent development.

A. Nuclear Reprogramming

After the nucleus is transferred into the cytoplasm of the recipient ovum, there are proteins within the nucleus that migrate into the cytoplasm as well as proteins in the cytoplasm that enter the nucleus. This nuclear–cytoplasmic exchange of proteins results in reconfiguration of the structure of the nucleus such that RNA synthesis is altered. If the nucleus is truly reprogrammed, it will reinitiate its developmental pathway and recapitulate early developmental events of the embryo; that is, the dominos, no matter which pathway was followed, will be stood up again.

An example from studies of early cattle embryos illustrates this point. The nuclei from a normally developing embryo at the 32-cell stage will, after 24 hr, direct the development of the blastoceole cavity and formation of a blastocyst. If nuclear transfer does not reprogram the donor nuclei to recapitulate development, then, 24 hr after the nuclear transfer, these nuclei will attempt to direct the development of a blastoceole cavity while the nuclear transfer embryo is only at the 2-cell stage. If this were to happen, the resulting embryo would likely form just the trophectoderm and not the inner cell mass, i.e., just a placenta without a fetus. On the other hand, if the nucleus is reprogrammed, then 24 hr after nuclear transfer the nucleus should produce very little RNA and participate in cleavage to the 2-cell stage. After retraversing the early cleavage divisions, followed by compaction and blastoceole formation a few days later, the nuclear transfer embryo would then be in the blastocyst stage and be suitable for transfer to the uterus of a cow. The latter example of the recapitulation of development is what occurs after nuclear transfer.

A more specific example of nuclear reprogramming is illustrated in Fig. 3. Ribosomal RNA (rRNA) synthesis can be indirectly evaluated based on nucleolar morphology. In the early pig embryo nucleoli are tight and compact (Fig. 3c) as expected due to the absence of rRNA synthesis. After the four-cell stage the pig embryo is actively producing rRNA, which changes the morphology of nucleoli within the nucleus from the tight appearance to a vacuolated and granular appearance (Fig. 3a); however, an intermediate type nucleolus can be observed sometimes (Fig. 3b). After nuclear transfer the previously reticulated (active) nucleoli again becomes compact and agranular (inactive) (Fig. 3d). These indicators sug-

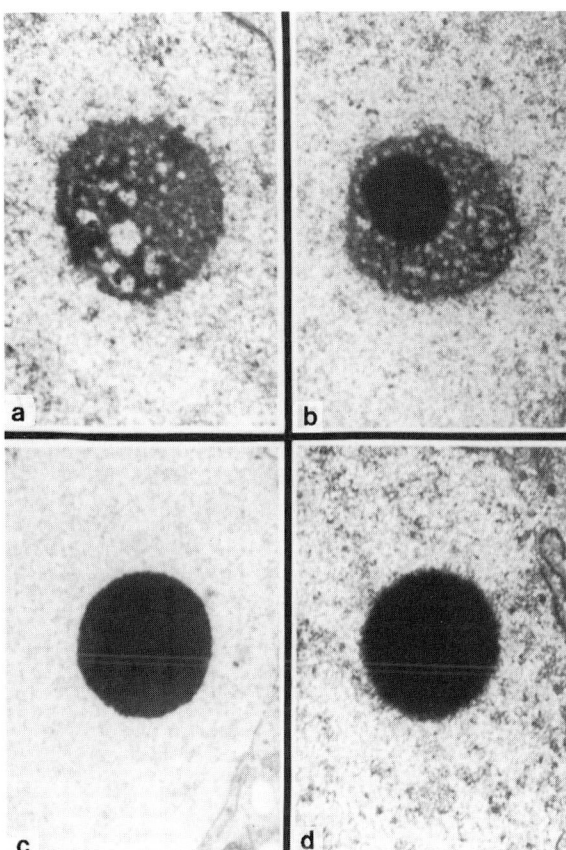

FIGURE 3 Nucleolar morphology before and after nuclear transfer. rRNA) synthesis can be indirectly evaluated based on nucleolar morphology. In the early pig embryo nucleoli are tight and compact (c, ×28,800), consistent with the absence of rRNA synthesis. After the four-cell stage the embryo is actively producing rRNA and this can observed by a change in the morphology of the nucleoli within the nucleus from the tight appearance to a vacuolated and granular appearance (a, ×36,000); however, sometimes an intermediate type nucleolus can be observed (b, ×24,800). After nuclear transfer the previously reticulated nucleoli again becomes compact and agranular (d, ×35,000). This suggests that the program of rRNA synthesis starts over after the nuclear transfer (from M. A. Mayes and R. S. Prather, unpublished results).

gest that the nuclei are reprogrammed when the nuclear transfer is synchronous with activation of the oocyte, and that the fidelity of the reprogramming is remarkable. In frogs, genes for muscle-specific actin are turned on to begin RNA synthesis at a specific stage in development in a specific tissue, but these genes can be turned off (RNA synthesis stops) after nuclear transfer and reprogramming. These genes are not only turned on again when the embryo reaches the correct stage of development but also they are turned on only in the correct tissues! These results illustrate the regulatory control of cytoplasm of the oocyte on gene expression.

B. Ovum Activation

Another event that must occur for subsequent development of nuclear transfer embryos is activation of the ovum. The mammalian ovum is generally arrested at metaphase II of meiosis. Fertilization breaks this arrest and activates development of the oocyte. When performing nuclear transfer it is necessary to artificially break this meiotic arrest, otherwise the nuclear transfer recipient embryo will remain arrested at the one-cell stage with condensed chromosomes. Artificial activation of meiosis can be accomplished with the electrical pulse used for fusion or chemically. Studies have shown that it is better to activate the ovum before or after cell fusion, depending on the stage of the cell cycle of the donor cell, i.e., before DNA synthesis, during DNA synthesis, or after DNA synthesis.

III. LIMITATIONS OF NUCLEAR TRANSFER

Until recently, it was thought that only nuclei from early embryos could recapitulate the early developmental process. In fact, some believed that it was "biologically impossible" to reprogram a mammalian nucleus. Recent results, however, have been from scientists who have pushed the envelope by using nuclei from progressively more differentiated cells as donors. The most differentiated nuclear donor cells used to date was from the udder of an aged pregnant ewe. Since the ewe was late in gestation, the mammary gland was likely proliferating rapidly. Although uncertain, this was likely a good source of nuclei that were competent to divide rapidly and to be compatible with early embryo development.

There are at least two limitations to the procedures of cloning mammals by nuclear transfer. The first is the possibility that resulting offspring are not true clones of the donor cells since one cannot know the extent of or document the source of mitochondria in the recipient ovum. While the nucleus contains the bulk of the DNA within a cell, mitochondria have their own DNA or genome. Since there are differences in the sequence of mitochondrial DNA among animals, one must confirm that the donor and recipient cells have identical mitochondrial DNA to ensure a genetic clone. This is important because mitochondria are the powerhouse of the cell where most of the energy (adenosine triphosphate) is generated. The mitochondrial DNA directs synthesis of RNAs and specific proteins, some of which must interact with RNAs and proteins whose synthesis is directed by nuclear DNA. These proteins must interact with a lock and key specificity. If these interactions are not specific, that is, nuclear and mitochondrial proteins do not interact exactly, then the cell may fail to thrive.

A second limitation is that the environment affects phenotype. For example, offspring derived from nuclear transfer have exceptionally high birth weights in cattle and sheep. While the large birth weight animal syndrome has been associated with nuclear transfer procedures, it may result from *in vitro* culture conditions and not from nuclear transfer. It is of considerable interest that simply culturing an embryo *in vitro* for a few days during the first week of development results in the "large calf" and "large lamb" phenotype at the end of gestation. All calves from a "clutch" of nuclear transfer embryos are not large at birth, and even the large ones are of normal size as adults and produce offspring of normal size. Thus, the large offspring phenotype is only observed in the first generation.

Another method to produce genetically identically animals, which illustrates the environmental effect on phenotype, is embryo splitting in which blastocysts are cut in half and placed in empty zona pellucidae. Each can give rise to an offspring. When the offspring are born, however, they may not look alike! For example, in calves, one may have a red patch over the left eye and the "identical twin" may also have a red patch but not in the identical location. This difference in phenotype results from differential migration of the melanocytes, the pigment-producing cells. The two halves of the blastocyst were exposed to a different environment from the moment of separation. Even if they were placed in the same uterus, they had a different uterine environment. This may be due to crowded conditions in the uterus and/or lack of equal blood flow to the different conceptuses. It should also be noted that antibody-producing cells of the body undergo DNA rearrangements to create the diversity of antibodies. Thus, the types of antibodies produced by two clones would be different between two animals that were exposed to different pathogens. While there are many more examples, these two examples illustrate the importance of environmental conditions on phenotype of animals that one would expect to be "identical." While both nuclear and mitochondrial DNA genetics play an important role in how an animal looks and behaves, the environment also contributes significantly to the phenotype.

IV. APPLICATIONS AND IMPLICATIONS

The application of cloning technology will likely be wide and far-ranging. Genetically identical animals will be used in agriculture, biomedicine, and basic science. Their uses will be as numerous as the imagination of scientists. Some of these applications are outlined.

A. Biomedical

The first use of cloned animals may be for biomedical uses. However, this will require the adaptation of another technology called transgenesis. Currently, it is possible to add function to a domestic animal by injecting the DNA encoding a gene into the pronucleus of a one-cell embryo. These procedures have severe limitations because it is not possible to direct the insertion of the gene into a specific region of a chromosome, much less a specific chromosome. However, a procedure termed homologous recombination does allow one to direct this insertional event to a specific chromosome and to a specific site on that chromosome. If this step were conducted immediately before the nuclear transfer, then all the offspring would have the gene integrated correctly and functioning in an appropriate manner. Homologous gene recombination permits both the removal ("gene knockout") and the addition ("gene knock in") of a specific gene. The importance of this is illustrated in the following example. When organ transplants are performed there are cell surface molecules that allow the body to recognize it as self (accepted) or foreign (rejected). If one can "knockout" genes for pig cell surface molecules from pig organs and "knock in" genes for the equivalent human molecules, then pig organs could be transferred to humans and not be rejected. While most of the examples of cloning deal with adult tissues, isolation of cell lines that have not undergone any *in vitro*-induced DNA alterations may prove more useful for making genetically identical animals. There will likely be many uses for cloned animals for biomedical research and for use in biomedicine.

B. Basic Science

Basic science has already benefited from cloning technology due to contributions to the regulation of the differentiation process. It is now possible to study many new aspects of the regulation of transcription. Factors present in cytoplasm of oocytes that globally downregulate transcription may be identified and they may have a use in treating cancers. In addition, it is now possible to study nuclear–cytoplasmic interactions and, through transfer of mitochondria of one genotype to the cell of another nuclear genotype, interactions between products of those two genomes.

Genetically identical animals are extremely useful to researchers because they have no genetic variation, which reduces the number of animals needed for many experiments to allow valid statistical comparisons, thus resulting in researchers being more efficient and experimentation being more cost-effective. Cloned animals will be most useful in species in which there is a large degree of genetic diversity, i.e., domestic animals. Laboratory mice are already very inbred and there is little genetic variation; therefore, less will be gained from the use of cloned mice.

C. Agricultural

Genetic progress in domestic animals is most rapid when the generation interval can be reduced to a minimum and maximum selection pressure can be applied. Currently, the greatest selection pressure is placed on males because they can sire many offspring which can be evaluated to estimate genetic value with some degree of confidence. Embryo transfer has allowed more selection pressure to be placed on the females and hence increased the genetic progress. By incorporating cloning technology into a system of superovulation of genetically superior cows bred artificially to the best bulls, it has been estimated that genetic progress will be even more rapid and result in genetically superior breeding stock.

Agricultural applications of cloning for food production will likely be the last to be implemented because the immediate profit margins will be low. However, one might expect improved efficiency in cloning procedures to result in them being applied to domestic animals for food production. One such scenario may be a cloned line of females with known desirable traits for calving ease, milk production, and mothering ability. These females could become recipients through embryo transfer of a cloned line of embryos with genetics for superior preweaning gains, excellent feedlot performance, and desirable carcass quality. The cows should calve easily without dystocia and provide high yields of milk for the calves, whereas the calves should exhibit highly desirable rates of growth, feedlot performance, grow well, and have carcass traits that are consistent with consumer demand. Furthermore, it is possible that clonal lines may be developed that are suitable to different regions of the country where management and environmental factors place different demands on animal production practices. While the scenario will not likely happen in the near future, it is a definite possibility for the future. Final applications of cloning technologies will be species specific, but goals similar to those just described for cattle would be of general interest, regardless of species. Our perceptions of mammalian cloning technologies will be limited only by our imagination, and its final applications and implications will become visible only as time marches on and the technology is implemented and its value proven.

See Also the Following Articles

Blastocyst; Cattle (Bovidae); Embryo Transfer; Meiosis; Pregnancy, in Farm Animals; Sheep and Goats

Bibliography

First, N. L., and Prather, R. S. (1991). Genomic potential in mammals. *Differentiation* 48, 1–8.

Prather, R. S. (1996). Progress in cloning mammalian embryos. *Proc. Soc. Exp. Biol. Med.* 212, 38–43.

Prather, R. S., and Robl, J. M. (1991). Cloning by nuclear transfer and splitting in laboratory and domestic animals. In *Animal Applications of Research in Mammalian Development* (R. A. Pedersen, A. McLaren, and N. L. First, Eds.), pp. 205–232. Cold Spring Harbor Laboratory Press, Cold Spring Harbor, NY.

Smith, C. (1989). Cloning and genetic improvement in beef cattle. *Anim. Prod.* 49, 49–62.

Spemann, H. (1938) *Embryonic Induction.* Hafner, New York.

Wilmut, I., Schnieke, A. E., McWhir, J., Kind, A. J., and Campbell, K. H. S. (1997). Viable offspring derived from fetal and adult mammalian cells. *Nature* 385, 810–813.

Cnidaria

Daphne Gail Fautin
University of Kansas

I. Life Cycle and Systematics
II. Sexual Reproduction
III. Asexual Reproduction

GLOSSARY

Anthozoa A class of Cnidaria comprising 5000 extant species that includes sea anemones, reef-forming corals, and sea fans; these animals are strictly polypoid (i.e., they lack a medusa phase).

clone A group of genetically identical units derived asexually (i.e., vegetatively) from a single progenitor; in the case of cnidarians, a clone typically refers to polyps that are not physically connected but that are derived from a single polyp, or to a group of medusae derived from a single polyp or polyp colony.

cnida (plural, *cnidae*) The microscopic "stinging capsule" that is the sine qua non and the source of the name of phylum Cnidaria; all cnidarians and only cnidarians produce cnidae, each of which can be fired only once; used in offense and defense, they were considered "independent effectors" but are now known to be under nervous control.

Coelenterata The name of a taxon that is, in modern parlance, usually used as a synonym of phylum Cnidaria, but that previously included the phylum Ctenophora and originally may have included phylum Porifera.

colony A clonal group of organisms that are physically connected to one another; cnidarian colonies are polypoid (e.g., most hydroids and reef-forming corals) or, rarely, include both polypoid and medusoid individuals (e.g., *Physalia*, the Portuguese man-o'-war).

Cubozoa A class of Cnidaria comprising about 50 species called the "box jellies"; the polyps are small and inconspicuous, but the medusae of some species are capable of killing humans with their sting.

Hydrozoa The most diverse class of Cnidaria comprising nearly 3000 extant species, including the only freshwater members of the phylum; in some taxa the polypoid phase is minor or absent, whereas in others the medusoid phase is minor or absent.

medusa The pelagic (free-swimming) phase in the typical cnidarian life cycle; these animals reproduce primarily or exclusively sexually.

mesoglea The central body layer of a cnidarian, which varies from being an acellular adhesive holding together the two cellular layers (ectoderm and endoderm) in Hydrozoa, to a substantial layer containing many cells in Anthozoa, to being the thick, mostly acellular central layer of Scyphozoa that gives "jellyfish" their name.

nematocyst One of three types of cnidae, and the only one to occur in members of all four classes.

planula (plural, *planulae*) The typical larva of cnidarians which, in most species, is planktonic; pear-shaped and entirely ciliated; planulae of some species feed, those of other species live entirely off yolk, and those of some species have zooxanthellae.

polyp The sessile (attached) phase in the typical cnidarian life cycle; these animals reproduce primarily or exclusively sexually.

Scyphozoa A class of Cnidaria comprising 200 extant species in which the medusa ("true jellyfish") predominates and the polyps are small and inconspicuous or even absent.

zooid The term for an individual organism, especially one that is a member of a colony.

zooxanthellae (singular, *zooxanthella*) Intracellular symbiotic dinoflagellates (golden-brown algae) possessed by some cnidarians.

Cnidaria (jellyfish, sea anemones, corals, and their relatives) is a phylum of simple animals placed in four classes. Cnidarians are considered primitive because they are at the tissue level of organization and lack a head and central nervous system. However, their distinguishing feature—the cnida (Fig. 1)—is the most complex secretory product known. Cnidarians have a fossil record extending back to pre-Cambrian times, more than 600 million years ago. They are found throughout the world's oceans

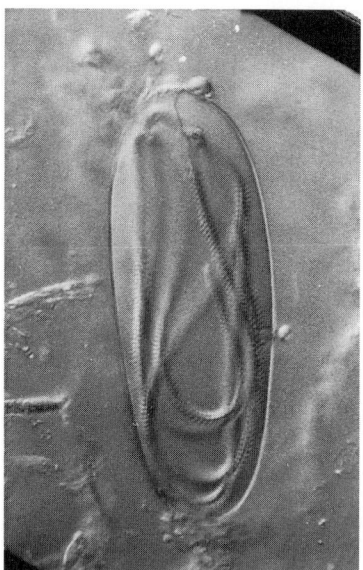

FIGURE 1 An unfired nematocyst from an anthozoan cnidarian. Coiled inside the 100-μm-long capsule is a tubule; when the nematocyst is fired, the tubule is shot out and penetrates, sticks to, or wraps around predator or prey. In some types of nematocysts, it delivers a toxin.

to the greatest depth of the sea; few species live in fresh water and none is terrestrial. Virtually all facets of cnidarian systematics and biology are poorly known, presumably because these animals are not of direct economic importance. The best known species are those that inhabit shallow waters of North America and Europe, but increasing attention is being paid to tropical species, particularly those inhabiting coral reefs.

I. LIFE CYCLE AND SYSTEMATICS

The typical cnidarian life cycle (Fig. 2) consists of alternating medusa and polyp phases characterized by sexual and asexual reproduction, respectively. The paradigm is that the medusa, being motile, constitutes the dispersal phase of the life cycle in contrast with the sessile polyp.

In fact, only a minority of cnidarian species exhibit a "typical" life cycle, with alternation of polypoid and medusoid phases. Although it is comparatively rare, medusae of some species can propagate asexually. Polyps of species in which there is no medusa phase, such as some hydrozoans and the entire class Anthozoa, reproduce sexually (many also reproduce asexually). Ironically, members of the scyphozoan order termed Stauromedusae lack a medusa phase: gametes produced by the solitary, sessile individuals join, and the resulting larvae metamorphose into individuals of the same phase. Because they are gametogenic, these individuals have been referred to as medusae, but they are now considered to be polyps. Taxa lacking a medusa phase seem, on average, no less widely distributed than do those with one; and although sexual reproduction in most species involves a planula larva, which can effect dispersal, even species in which the larva is benthic, brooded, or lacking do not appear to have unusually small ranges or localized genotypes. These observations reinforce the theory that the primary function of larvae is not dispersal.

Higher taxa of cnidarians are defined in part morphologically and in part on life-cycle characters, which, to a great extent, covary. Asexual reproduction occurs in nearly bewildering variety, with some variants defining taxa. Sexual reproduction occurs, as far as is known, in every species of Cnidaria; it is relatively uniform within the phylum and similar to that in other animals.

The ecological and evolutionary importance of various modes of reproduction manifest among cnidarians is poorly understood. Most models that purport to explain the adaptive significance of repro-

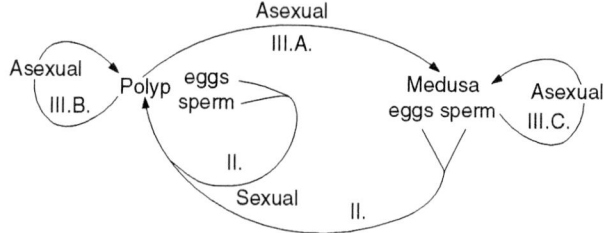

FIGURE 2 Idealized life cycle of members of phylum Cnidaria. The Roman numerals and letters refer to the sections of this article in which that portion of the life cycle is discussed. The medusa—the dispersal phase—forms gametes. An initial polyp, which develops from a zygote, asexually (vegetatively) produces additional polyps and, ultimately, also medusae, which swim off to undergo sexual reproduction (reproduced with permission from Fautin and Romano, 1997).

ductive strategies of animals were developed for obligately sexual, solitary organisms, and thus may not be appropriate to apply to animals with other sorts of life histories. Cnidarians appear to have some ecological and reproductive attributes of plants (see Section II,A), but knowledge is too limited to be certain how precisely botanical models apply to cnidarians and how far such analogies should appropriately be carried. Presumably a major difference is that both phases of the cnidarian life cycle are diploid, but this has not been verified.

A. Coloniality

Asexual reproduction results in a colony if the progeny remain physically attached to one another. Among the most familiar cnidarians are colonial ones such as stony corals (anthozoan order Scleractinia), sea fans and whips (anthozoan order Gorgonacea), and hydroids (hydrozoan order Hydroida). Some taxa, such as sea pens (anthozoan order Pennatulacea), siphonophores (hydrozoan order Siphonophora), and black corals (anthozoan order Antipatharia), consist entirely of colonial species. Only sea anemones (anthozoan order Actiniaria) and some closely related taxa are exclusively solitary; most higher taxa of cnidarians constitute mostly colonial species with a small number of solitary ones.

B. Polymorphism

An individual cnidarian consists of a sac-like body (the space of which is termed the gastrovascular cavity or coelenteron) that connects to the outside through a single opening, the mouth, which is surrounded by tentacles. This basic morphology is subject to considerable modification, most fundamental of which is the dimorphism between polyp and medusa. A polyp (Fig. 3) is typically anchored to the substratum at its aboral end, whereas a medusa (Fig. 4) is planktonic, drifting freely in the water. Although usually rendered with the mouth facing the sea or lake bottom and the tentacles hanging down, many medusae spend only a small proportion of their time in this position. In most species, a medusa is derived asexually from a polyp; thus, despite their morpho-

FIGURE 3 A polypoid cnidarian, as represented by a sea anemone of the species *Urticina lofotensis*. The animal is attached to the substratum at its aboral end; the single body opening, the mouth, is located at the center of the oral disc and is surrounded by tentacles (photograph by Steve Wedi).

logical differences, a polyp and its daughter medusae are genetically identical.

Members of a colony are ultimately derived asexually from a single founding individual. In stony corals of nearly all species and in scyphozoans, the polyps of a colony are monomorphic; that is, they are morphologically identical. In other cnidarians, despite their genetic identity, members of a colony may differ in structure and associated function. The original, sexually produced primary polyp of a sea pen grows large and gives rise to smaller secondary polyps; the primary polyp serves to anchor, as well as to create, the colony, whereas the secondary polyps feed, form gametes, etc. Most colonial hydrozoan species exhibit polymorphism to some degree. Gastrozooids are individuals specialized morphologically and functionally for capturing food, which is then distributed to other, nonfeeding members of the colony. Dactylozooid, nematophore, and tentaculozooid are terms that have been applied to polyps specialized to defend the colony. Only some members of polymorphic colonies are gametogenic; typically these gonozooids lack feeding and protective structures such as tentacles.

In contrast to the taxa discussed previously, a siphonophore colony consists of both polyps and medusae, linked in a pelagic structure that may be many

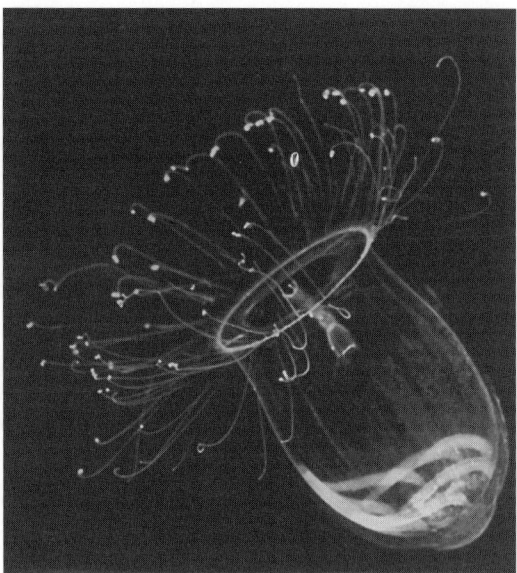

FIGURE 4 A medusa (jellyfish) of the hydrozoan species *Aglantha digitale*. Its diameter is about 5 mm. The gonads are suspended from the roof of the animal's coelenteron (photograph by Claudia Mills).

meters long; the best known species is the Portuguese man-o'-war, *Physalia*. Siphonophores exhibit the greatest polymorphism in the phylum, for which a complex terminology has been created.

II. SEXUAL REPRODUCTION

A. Sexuality

An individual cnidarian may be gonochoric (producing gametes of only one sex) or, if it produces gametes of both sexes, may be a simultaneous or a serial hermaphrodite. At least two species of sea anemones are gynodioecious, with populations consisting of females and hermaphrodites (but no males), a pattern once known only in plants. Gonochorism is the most common sexual condition of individual cnidarians.

The basis of sex determination is unknown for most species. It appears to be affected by environmental conditions in some cnidarians, but whether that is mediated genetically is unknown. Genetic determination is strongly implied in species in which all individuals of a clone or colony are of the same sex. Multiple genes are thought to influence sex in *Hydra*, with the degree of manifestation of either sex being dose-dependent.

Sexuality is complicated in colonies. A zooid may be gonochoric, but a colony may have both sperm- and egg-producing zooids and thereby functionally be hermaphroditic. A colony can also be hermaphroditic by virtue of consisting of hermaphroditic zooids.

B. Gonads

Because cnidarians lack organs, they do not truly possess gonads; structures termed gonads actually are nests of gametogenic tissue. This tissue hypertrophies during gametogenesis but regresses entirely when gametogenesis is not occurring. Thus, no attribute distinguishes the sexes of most cnidarians other than the form of gametes (i.e., cnidarians lack secondary sex characteristics), so there is no way of determining the sex of a sterile individual.

Hydroids of some species do not release free medusae; rather, their gametes are formed in structures evolutionarily derived from medusae. A complete sequence can be traced: (i) a typical medusa that is released from the polyp less than fully grown, in which gametes mature after release; (ii) a medusa that is released fully grown and with mature gametes; (iii) a medusa-like structure that is not released from the hydroid but from which gametes are released; (iv) a gonophore, which is a gametogenic structure that does not resemble a medusa.

C. Gametogenesis

Gametogenesis appears to begin in most cnidarians only upon attainment of a particular size or age. Sexual maturity in some stony corals and hydrocorals occurs once the colony constitutes a minimal number of polyps. The scyphomedusa *Aurelia aurita* begins producing gametes when it achieves a particular diameter, and its gametogenic tissue will regress if it shrinks below that critical size (like many cnidarians, this jellyfish will degrow when starved).

Germ cells arise from the endoderm in Anthozoa,

Scyphozoa, and Cubozoa. The gametogenic tissue of an anthozoan polyp lies in some or all of its mesenteries—radial partitions of the body space that grow inward from the body wall, some as far as the actinopharynx (the tubular "gullet" that extends from the mouth into the gastrovascular cavity and that is a distinguishing feature of Anthozoa). Scyphomedusa gametes form on the floor of the gastrovascular cavity; some polyps of order Coronata are gametogenic. Cubomedusae are presumably like scyphomedusae in regard to gametogenesis. In part due to their low diversity and in part due to their dangerous sting, cubozoans are the least known cnidarians. Hydrozoan gametes are commonly stated to be of ectodermal origin but the gametes of some taxa arise from the endoderm. Thus, in this attribute as in so many, Hydrozoa is the most variable class.

1. Spermatogenesis

The process of spermatogenesis appears to be like that of many other animals. Spermatozoa mature in "packets," "follicles," or "nests," with the least ripe cells peripheral and the ripest ones at the center. The packet ruptures to release tailed spermatozoa. Although documented in some species, an acrosome appears to be lacking in most.

2. Oogenesis

Mature cnidarian eggs range in diameter from about 100 to 1000 μm. Oogenesis is unremarkable. Oogonia of scyphozoans and anthozoans arise in the endoderm but move into the mesoglea to mature. Maturing oocytes remain near the gastrovascular cavity, where digestion occurs. Nutritive material is thought to derive from or be passed across endodermal epithelial cells (trophocytes) in some scyphozoans. In stauromedusans, oocytes develop in follicles that project into the gastrovascular cavity; accessory cells around each oocyte are presumed to mediate passage of material from the gastrovascular cavity. Nutritive material passes through an endodermal channel (trophonema) between the growing oocyte and the gastrovascular cavity of some anthozoans.

Polar body formation has been documented in few species. In some it seems to occur prior to fertilization, whereas in others it follows fertilization.

D. Spawning

Like many marine animals, a typical cnidarian releases its gametes into the water, where fertilization and embryogenesis occur ("broadcast spawning"). Spawning, whether by a polyp or a medusa, may be triggered by exogenous or endogenous events. In most species, release of gametes is simultaneous among many individuals in an area ("mass spawning" or "epidemic spawning"). The frequency of spawning ranges from annual to daily.

More than 100 species of anthozoans, as well as other invertebrates, spawn on a single night in early summer on the Great Barrier Reef. This predictable annual event is clearly related to time of year and phase of moon, although the proximate cues to the animals are uncertain and may not be the same for all species. (Far less spawning may occur during the preceding and following nights, and a month earlier and later.) As in other marine organisms, some cnidarians spawn monthly during a particular phase of the moon or tides. Some scyphomedusae and hydromedusae spawn daily during part of the year (commonly summer), typically at dawn or dusk.

For species in which gametes are of endodermal origin (the majority), release of gametes is through the single opening of the body, the mouth. This can occur only after gametes have broken from their mesogleal nests into the gastrovascular cavity.

E. Fertilization

External fertilization is typical in cnidarians, but internal fertilization occurs as well. Species-specific sperm attractants have been identified in a few species of anthozoans and hydrozoans. They have been sought especially where many species spawn simultaneously. Sperm entry is at the site of polar body emission in some hydrozoans studied but is unknown in most species.

F. Brooding

In some species, eggs are fertilized in the gastrovascular cavity; rarely, fertilization occurs before the egg is released into the gastrovascular cavity. Fertilization is typically by sperm from other individuals carried into the gastrovascular cavity with the water that also carries oxygen and food, although self-fertilization is at least theoretically possible in simultaneously hermaphroditic individuals. Embryos may be brooded internally only a short time, being released to complete development in the plankton. However, in many species, brooding extends through the entire developmental process, so juveniles are ultimately released.

In most hydroids with gonophores, sperm that are released into the water fertilize eggs that are retained in the gonophores, where embryogenesis also occurs. Development occurs in pockets on the oral arms of some scyphomedusae. In externally brooding anthozoan species, eggs are fertilized either before or after being spawned and immediately attach to the exterior surface of the female parent, where development occurs. Continuous, or "dribble," spawning is particularly prevalent in organisms that brood their young.

III. ASEXUAL REPRODUCTION

Typical cubozoan, hydrozoan, and scyphozoan polyps reproduce only vegetatively, forming medusae or polyps. Anthozoan polyps of many but not all species produce additional polyps asexually, and some medusae reproduce asexually. Animals of species that are capable of asexual reproduction also appear to be mostly those that can regenerate, a linkage that is not unique to cnidarians. Cnidarians exhibit a bewildering variety of asexual reproductive modes. Some types are no doubt evolutionarily related but detailed discussion of that aspect is beyond the scope of this article.

A. Formation of Medusae by Polyps

The most obvious manifestation of asexual reproduction in cnidarians is the formation of medusae, which, in the "typical" life cycle, swim free of the polyps that form them. Mode of medusa formation differs among the three classes of cnidarians that possess a medusa and is one basis for their systematic distinction.

1. *Cubozoa*

The life cycle of the poorly known Cubozoa is characterized by a polyp metamorphosing completely into a medusa. Therefore, unlike in the other two classes in which a medusa exists, medusa production does not directly involve a process that can be construed as reproduction (because it does not result in an increased number of individuals).

2. *Hydrozoa*

Medusae develop on special branches of a hydroid colony or in specialized pits of hydrocoral colonies; normally such structures develop only in particular circumstances, such as when the colony achieves a minimal size or at certain times of year. The primordium of a typical medusa begins as a hollow bud containing both cell layers. It bulges at the distal end, where mouth and tentacles ultimately form, and constricts at the attachment end; pulsations at the tentacle end break the medusa free, and it swims off. Reduction in elaboration and duration of the medusa phase is discussed in Section II,B.

3. *Scyphozoa*

The typical scyphozoan polyp is a small organism, whereas a scyphomedusa may attain a diameter of 1 m or more. A constriction develops at the distal end of a polyp that is on the verge of forming a medusa; in the tissue that has thereby been segmented off, eight broad tentacles form as the thin tentacles of the polyp regress (thereby depriving the polyp, now termed a strobila, of the ability to feed). Within a few days, the young medusa, called an ephyra, has developed and swims free. If 1 ephyra is produced at a time, the process is referred to as monodisk strobilation. As many as 20 ephyrae may be produced successively by a strobila through polydisk strobilation; they resemble a stack of plates atop the strobila. Strobilation, as scyphomedusa formation is called, is initiated by factors such as temperature, salinity, dissolved oxygen, light, zooxanthellae, and pH, depending on the species and the time of year. Number of strobilae produced may differ with time of year in a particular species. After strobilating, a strobila may resume life as a scyphopolyp and may strobilate again later.

B. Formation of Polyps by Polyps

The term "budding" is commonly used as a synonym for vegetative propagation in cnidarians. Its use should be restricted to production of hydromedusae and to one of the many processes by which polyps produce polyps.

1. Budding

A bud is a hollow outgrowth of part of an existing organism containing both cell layers (see Section III,A,2). The bud differentiates into a polyp, forming a mouth at the distal end and tentacles around it. In *Hydra*, for example, as in the case of medusa buds, the connection between parent and bud constricts and the new progeny drops off. In others, a daughter polyp remains connected to its parent, ultimately resulting in a colony. In hydroid and soft coral colonies, the gastrovascular cavities retain continuity as well, so fluids and food can be moved among polyps; in stony corals, there is no such continuity, which is probably related to the absence of polymorphism in such animals.

A bud arises from the polyp in cnidarians such as some stony corals, some scyphozoans, and some hydrozoans. More commonly, a bud arises from a stolon or other living tissue that connects polyps of a colony but is not itself part of any polyp. Colonies of many soft corals (anthozoan order Alcyonacea) and most hydroids develop in this manner. The spacing and position of buds are responsible for the shape of a colony.

An entire animal can regenerate from a tentacle of the sea anemone *Boloceroides mcmurrichi*. Indeed, regeneration may begin before a tentacle detaches from the parent organism so that the tentacle crown is a mass of animals in various stages of development. Although highly unusual, this process does meet the definition of "budding" in that the primordium, a tentacle, is a hollow outgrowth of an existing organism.

2. Fission

Division of an existing polyp into two (typically) or more large pieces is defined as fission. Because cnidarians rely on hydrostatic pressure for locomotion, food capture, and other vital functions, rapid healing is essential following fission.

i. Longitudinal Colonies of some stony corals are formed in this manner; fission is recorded in the skeleton as a bifurcation of the trace of a polyp's lineage. In a sea anemone undergoing longitudinal fission, the rent edges of each daughter polyp draw together and heal so it is about half the diameter of the mother polyp. Longitudinal fission has been especially well studied in the clonal sea anemone *Anthopleura elegantissima*. A polyp of the scyphozoan *Aurelia aurita* adds tentacles along each side before dividing into two.

ii. Transverse This rare form of division has two variants in sea anemones. In those such as *Nematostella vectensis*, the animal divides (occasionally into more than two pieces) and then forms tentacles at the new anterior end, whereas in those such as *Gonactinia prolifera*, a circlet of tentacles grows from the body of the progenitor before fission occurs so that both resulting progeny are tentaculate immediately. Transverse fission is also known in a hydrozoan, the tentacleless *Protohydra*, and in some stony corals.

3. Fragmentation

Small multicellular pieces of tissue each can reorganize and develop into an entire animal of some species. This can result, under favorable conditions, in rapid production of large populations. As a sea anemone such as *Haliplanella luciae* and *Metridium senile* glides along the substratum, small bits of its base are left behind; each can regenerate. This phenomenon is referred to a "pedal laceration."

Planula-like fragments break from the polyps of some members of hydrozoan order Trachymedusae. Several types of fragment from the stolon of hydrozoan or scyphozoan colonies have been distinguished on the basis of morphological origin, fate, and conditions of formation. Most such multicellular propagules that have been referred to as frustules, cysts, gemmae, and gemmules are believed to be or have been demonstrated to be able to persist through conditions that might kill the animal. Because many cysts can be derived from a single individual, their formation may legitimately be considered reproductive.

C. Formation of Medusae by Medusae

Hydromedusae of few species are able to form medusae asexually. Fission is only longitudinal. The freshwater hydromedusa *Limnocnida* buds medusae from its stomach region.

D. Other Vegetative Processes

1. Colony Propagation

Every cnidarian colony was previously assumed to have originated with a sexually produced founder. The assumption of each colony's genetic uniqueness is now known to be flawed, although the extent to which colonies propagate is unknown. These processes have the potential to increase local population density rapidly. They are considered reproductive because they increase the number of colonies (and thereby probably the number of component polyps compared to what would have been produced had there been only a single colony).

i. Breakage A piece of a stony coral or sea fan that broke off was once thought to be doomed; it is now known that it may reattach to the substratum and continue to live. Indeed, a fragile skeleton may be an adaptation to local dispersal. Aquarium specimens of stony corals are being propagated in this manner, which relieves natural populations of collecting pressure. In some sea whips and soft corals (members of anthozoan subclass Alcyonaria), the terminal portion of a colony can autotomize. This mode of reproduction is distinguished from fragmentation (see Section III,B,3) because a piece of a colony typically consists of tens to hundreds of intact polyps (even though polyps at the break points may be damaged or killed).

ii. Transmigration Small groups or individual polyps of stony corals can leave their skeleton to settle on the substratum and initiate a new colony. At least three variants on this poorly understood process have been documented, one of which—sometimes referred to as "polyp bail-out"—seems to be a response of the organism to stress.

2. Parthenogenesis and "Parasexual" Reproduction

Development of an egg without benefit of fertilization appears to occur in some cnidarians. A well-documented instance is the sea anemone *Sagartia ornata*, in which the egg seems not to have completed reduction division before it begins to develop.

What appear to be larval or postlarval life history stages that are not derived from conventional embryogenetic events have been reported in a variety of cnidarians. Some scyphistomae form buds that resemble planulae but that are larger and contain a greater diversity of cells. Otherwise, they follow a course much like that of typical planula larvae: swimming, attaching, and metamorphosing into polyps. Some sea anemones and stony corals brood juveniles that appear not to be sexually derived: the brooding individual may be a male, or the genotype of the juveniles is identical to that of the parent. The source of such progeny remains a mystery; efforts to produce them artificially by a variety of means have not succeeded. Adult males of the deep-sea scyphomedusa *Stygiomedusa gigantea* have never been collected, so the young medusae being brooded by females are suspected of being asexual in origin.

See Also the Following Articles

ASEXUAL REPRODUCTION; HYDRA; MARINE INVERTEBRATES, MODES OF REPRODUCTION IN; PARTHENOGENESIS AND NATURAL CLONES; PORIFERA

Bibliography

Arai, M. N. (1997). *A Functional Biology of Scyphozoa*. Chapman & Hall, London.

Campbell, R. D. (1974). Cnidaria. In *Reproduction of Marine Invertebrates, Volume I* (A. C. Giese and J. S. Pearse, Eds.), pp. 133–199. Academic Press, New York.

Cornelius, P. F. S. (1995a). *North-West European Thecate Hydroids and Their Medusae. Part 1*. Linnean Society of London, London.

Cornelius, P. F. S. (1995b). *North-West European Thecate Hydroids and Their Medusae. Part 2*. Linnean Society of London, London.

Dunn, D. F. (1982). Cnidaria. In *Synopsis and Classification of Living Organisms, Volume 1* (S. P. Parker, Ed.), pp. 669–706. McGraw-Hill, New York.

Fautin, D. G. (1992). 2. Cnidaria. In *Reproductive Biology of Invertebrates* (K. G. Adiyodi and R. G. Adiyodi, Eds.), Vol. V, pp. 31–52. Oxford & IBH, New Delhi.

Fautin, D., and Romano, S. (1997). Cnidaria (Coelenterata). In *Tree of Life* (D. Maddison, Ed.). http://phylogeny.arizona.edu/tree/eukaryotes/animals/cnidaria/cnidaria.html.

Fautin, D. G., Spaulding, J. S., and Chia, F.-S. (1989). 2. Cnidaria. In *Reproductive Biology of Invertebrates* (K. G. Adiyodi and R. G. Adiyodi, Eds.), Vol. IV, pp. 43–62. Oxford & IBH, New Delhi.

Hyman, L. H. (1940). *The Invertebrates Volume I: Protozoa through Ctenophora.* McGraw-Hill, New York.

Kirkpatrick, P. A., and Pugh, P. R. (1984). *Siphonophores and Velellids.* Linnean Society of London, London.

Richmond, R. H., and Hunter, C. L. (1990). Reproduction and recruitment of corals: Comparisons among the Caribbean, the tropical Pacific, and the Red Sea. *Mar. Ecol. Prog. Ser.* **60**, 185–203.

Shick, J. M. (1991). *A Functional Biology of Sea Anemones.* Chapman & Hall, London.

Columbidae

see Pigeons

Conjugation in Ciliates

Peter J. Bruns

Cornell University

I. Introduction
II. Nuclear Dimorphism and Vegetative Growth
III. Conjugation in the Life Cycle
IV. Pair Formation and Induction of Conjugation
V. Nuclear Events in Normal Conjugation
VI. Patterns of Inheritance Following Conjugation
VII. Variations in Conjugation
VIII. Transformation during Conjugation

GLOSSARY

autogamy A self-mating in individual cells that yields whole genome homozygotes.

cytogamy A self-mating in paired cells that yields whole genome homozygotes.

exconjugant clones Clones derived by isolating the two members of a mating pair. They are useful for studying cytoplasmic inheritance.

genomic exclusion Mating between normal cells and sterile (so-called star) cells. It induces genetic events that yield germinal nuclei with whole genome homozygotes and parental somatic nuclei.

karyonides The clones derived by isolating the four cells that result after the first cell division following pair separation; they are clones of the four independently developed macronuclei.

macronucleus The somatic nucleus, which is usually highly polyploid but the amount of polyploidy and size of the chromosomes are species specific. It is actively transcribed.

micronucleus The germinal nucleus, usually diploid; it is not expressed during vegetative growth.

synclone The clones derived from isolated mating pairs.

I. INTRODUCTION

Conjugation in the ciliated protozoa is composed of a series of developmental events typically involving pair formation, meiosis, and fertilization in the germline (the genetic phase) followed by the differentiation of new germinal and somatic nuclei (the development phase). Figure 1 is a scanning electron micrograph of mating *Tetrahymena thermophila*. Mating is not for growth but rather for the production of new, recombinant genomes. The entire process proceeds in starved cells; there are no specialized gametic cells or spores. Conjugation has been most intensely studied in species of the class Oligohymenophorea, including species in the genera *Paramecium* and *Tetrahymena*, as well as in several species of the class Spirotrichea, including *Euplotes*, *Oxytricha*, and *Stylonychia*. These last three species are often referred to as hypotrichs.

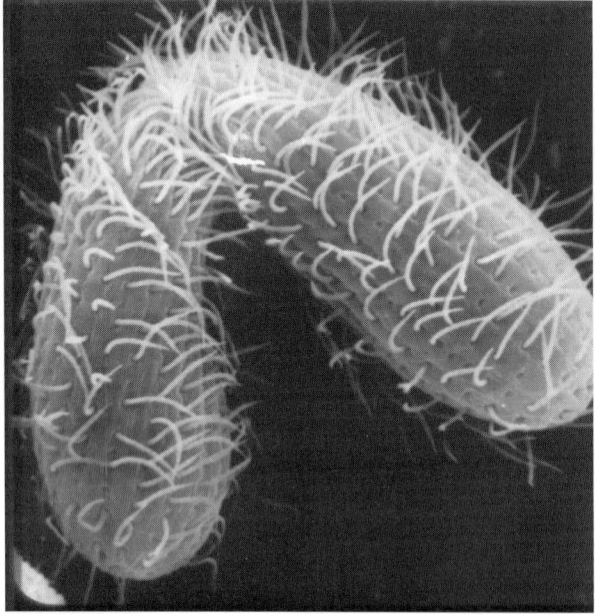

FIGURE 1 Scanning electron micrograph of mating *Tetrahymena thermophila*.

II. NUCLEAR DIMORPHISM AND VEGETATIVE GROWTH

Ciliates in which conjugation occurs have one, several, or even many copies of two very different nuclei: the germinal micronucleus and the somatic macronucleus. These two nuclei have very different structure and function, replicate at separate times, and have been viewed as model systems in which to study aspects of nuclear differentiation, chromatin remodeling, and apoptosis. Much recent work has been directed at understanding the molecular basis of the differences between micro- and macronuclei.

The micronucleus is in many ways a typical eukaryotic nucleus, undergoing mitosis during vegetative growth and meiosis in conjugation. Where it has been most studied, the micronucleus is diploid: *Tetrahymena thermophila*, for example, has five metacentric micronuclear chromosomes and a haploid genome size of 2×10^8 base pairs. It is an unusual nucleus because it is transcriptionally silent, providing little or no contribution to cellular phenotype. Because it does not direct cellular function, some species have sterile amicronucleate forms; high degrees of aneuploidy have been found in others. In *T. thermophila*, for example, this lack of somatic function of the micronucleus has made it possible to experimentally produce and exploit for genetic studies nullisomic strains missing both copies of one or more than one of the chromosomes in the micronucleus.

In contrast, the macronucleus consists of many copies of a subset of the genome, contained on much smaller chromosomes than those in the micronucleus. In all cases studied, the macronucleus is amitotic and never undergoes meiosis. During vegetative growth, it elongates and constricts, delivering a macronucleus to each daughter cell. There are no mechanisms to ensure delivery of equal amounts of DNA to daughter cells; some ciliate species regularly yield nonequivalent daughter nuclei at cytokinesis. The highly polyploid content of the macronucleus apparently allows a certain degree of variation in gene dosage. Ultimately, the number of macronuclear chromosomes is probably maintained in a reasonable range by a DNA replication mechanism. Some ciliates maintain the same approximate number of macronuclear chromosomes throughout the lifetime of the clone. Others (notably hypotrichs) show clonal variation in specific macronuclear chromosomes during continued vegetative growth, with single, apparently random, macronuclear chromosomes becoming highly amplified. The very large number of chromosomes in each macronucleus (thousands) has recently made this nucleus a very attractive system to study replication and function of the specialized structures found at the ends of eukaryotic chromosomes (telomeres).

III. CONJUGATION IN THE LIFE CYCLE

Figure 2 relates the separate pathways of conjugation and the vegetative cell cycle. For conjugation to occur, three conditions must be met: starvation, diverse mating types, and sexual maturity. The starvation phase has been termed initiation and, where studied, requires protein synthesis. The role of mating type is somewhat variable, depending on the mechanism by which cells recognize each other (see below). Maturity plays a role since, as a general rule, most ciliates cannot remate immediately after successfully completing conjugation, as defined by the production of both new micro- and macronuclei. Mating unresponsive progeny are said to be immature and require a clonally specified number of cell fissions before they are once again mating reactive. The need for diverse mating type and maturity for entry into the conjugation pathway ensures dispersal before remating, promoting outbreeding. Notable exceptions occur when a new outbred macronucleus is not produced, as in autogamy or genomic exclusion (see below). In addition, it has been observed in *T. thermophila* that vegetatively compromised progeny become mature after only a few divisions, suggesting that cells abandon outbreeding to mate again and produce a new macronucleus if somatic function appears to be impaired. D. L. Nanney has referred to this as a case in which incest is preferable to extinction.

At the other end of the life cycle, cultures typically show varying degrees of senility. Some ciliates lose mating-type differentiation, allowing intraclonal mating. Many eventually lose the ability to form pairs, and some lose vegetative vigor, eventually dying out altogether. In some species, the onset of senility occurs after a predictable number of cell fissions, suggesting a regulatory mechanism; in others, senility appears to be the result of the accumulation of damage in the macro- or micronuclei.

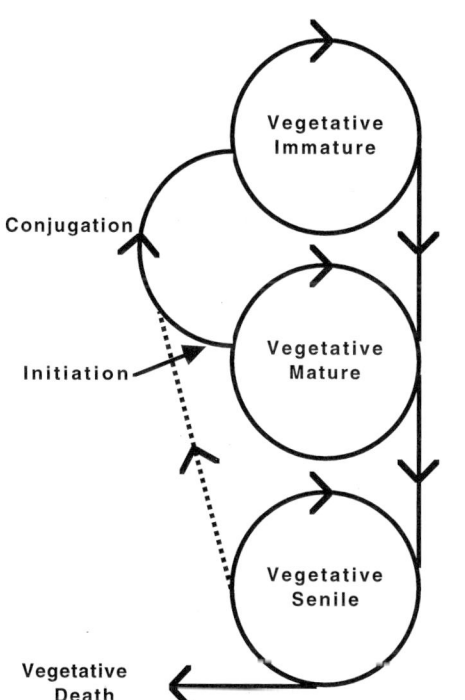

FIGURE 2 Generalized diagram of the life cycle of ciliates. The three circles represent vegetative growth cycles. Vegetative mature and immature cycles are connected by conjugation, a separate pathway, which is entered by a process termed initiation. Vegetative senility and death follow extended growth in the absence of mating.

IV. PAIR FORMATION AND INDUCTION OF CONJUGATION

When all the conditions described in the previous section have been met, potential mates interact, either at a distance through pheromones or by direct contacts. In some species, pheromones act as chemical attractants, allowing potential mates to find each other. Pheromones can induce loose pairing both between cells of the same mating type (homotypic pairing) and between cells of different mating types (heterotypic pairing). The nuclear events of conjugation only proceed in the heterotypic pairs. In some species, homotypic pairing of cells of one mating type can be induced by adding cell-free pheromone from the other. If mating-reactive cells of the second mating type are then added, heterotypic pairs quickly form and meiosis is induced. The action of a diffusible substance for the induction of meiosis is suggested if homopolar doublets (cells with a duplication of surface features at the same pole of the cell) are used as the cells to form the homotypic pairs. Chains of mating cells result since each cell has two sites for pair formation. When cells of the second mating type are added, a heterotypic pair forms at one end of the chains and meiosis is seen to occur down the chains, starting at the heterotypic pair and spreading to the rest of the paired cells in the chain.

Other species form firm heterotypic pairs only after they have physically interacted with each other in a process termed costimulation. During this process, the membranes of the two conjugants are cleared of membrane proteins at the site where the cells will pair.

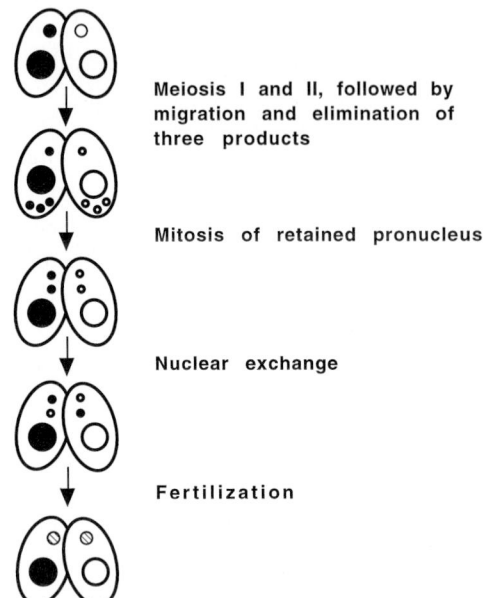

FIGURE 3 Genetic phase of conjugation. General illustration of the events in conjugation in *T. thermophila*. The small circle in the anterior end of the starting cells is the diploid micronucleus; the large circle in all is the polyploid macronucleus; the smallest circles, in the center cells, are haploid pronuclei. Some ciliates have more micro- and/or macronuclei, but the events are essentially the same since only one product of meiosis is retained for mitosis, exchange, and fertilization.

V. NUCLEAR EVENTS IN NORMAL CONJUGATION

Conjugation can be divided into two parts: the genetic phase, which yields a new genome in the germline, and the development phase, which results in the development of new somatic nuclei with the new genome and therefore the expression of new phenotypes.

A. Genetic Phase: Micronuclear Meiosis and Cross-Fertilization

Figure 3 diagrams the general events in the genetic phase. Three prezygotic nuclear divisions (meiosis I, meiosis II, and a third prezygotic mitosis) yield at least two haploid germinal pronuclei in each conjugant. Some species begin with more than one micronucleus, but usually the two pronuclei in each conjugant that will be involved in subsequent exchange and fertilization are mitotic daughter nuclei of the same haploid meiotic product and are therefore genetically identical. Nuclear division is followed by an exchange of one of the pronuclei from each conjugant and fertilization (karyogamy), resulting in new germinal genotypes in the resulting zygote nuclei. Note that the cells still contain only the parental macronucleus and, thus, up to this point still express the parental phenotype. In some instances [natural (e.g., see Genomic Exclusion) or experimental by the addition of drugs such as cycloheximide], the process can be halted at this point and the resulting cells are genetic heterokaryons, with different genotypes in micro- and macronuclei.

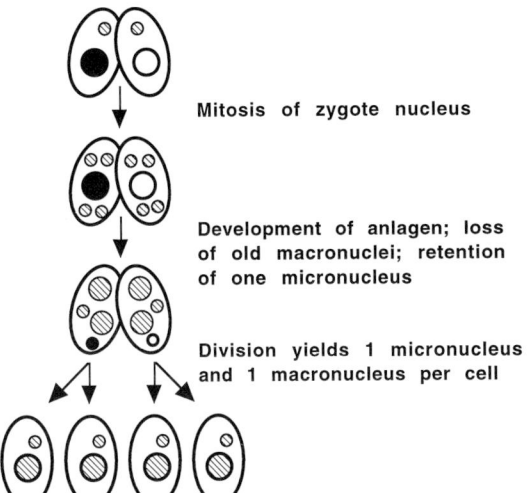

FIGURE 4 Development phase of conjugation. Following fertilization, the diploid zygote nuclei (striped circles in starting pairs) divide twice, and the resulting three different classes of nuclei (macronuclear anlagen, old macronuclei, and new micronucleus) all suffer different fates. At the first cell division following conjugation, the new micronuclei divide mitotically and the new macronuclei are segregated, each to a different cell.

B. Development Phase: Reestablishing Nuclear Dimorphism

Figure 4 diagrams the general features of the development phase. Two or more postzygotic mitoses occur, depending on the species. The figure sketches the process in *T. thermophila*, in which two divisions yield four new nuclei. The products of these divisions are positioned in specific locations in the cells and suffer very different fates, reestablishing the nuclear dimorphism. The new nuclei become either macronuclear anlagen or new micronuclei.

1. Macronuclear Anlagen Development

The development of macronuclear anlagen follows and involves significant changes in chromosome composition, including the following.

i. Change in Chromosome Size The large micronuclear chromosomes are broken down into much smaller macronuclear chromosomes. In some species (notably, the hypotrichs) the resulting macronuclear chromosomes mostly contain single genes. In other species, macronuclear chromosomes contain multiple genes; for example, macronuclear chromosomes in *T. thermophila* range in size from 21 kilobase pairs for the chromosome containing two copies of the ribosomal locus, organized as a palindrome, to large chromosomes in the size range of 3.5 megabase pairs, carrying many genes. The resulting somatic chromosomes have telomeres added at this time.

ii. Loss of Germinal DNA Sequences In all species studied, a loss of moderately repetitive DNA is seen. The amount of DNA eliminated in this phase varies dramatically from 90% in some hypotrichs to 10–15% in *T. thermophila*. Micronuclear-specific sequences that are lost include sequences located between the ends of the newly liberated macronuclear chromosomes and the so-called internal eliminated sequences (IESs) in which there is both loss of DNA and splicing of retained sequences to form the macronuclear chromosomes. In some hypotrich species this process is associated with a rearrangement of parts of the chromosomes, bringing novel sequences together. IESs have been studied extensively in hypotrichs and have been found in strikingly large numbers in the genome. For example, it has been estimated that about 60,000 short IESs (10–500 base pairs) are removed from a species of *Oxytricha* during macronuclear development. In addition, larger transposon-related micronuclear-specific sequences have been identified that contain several open reading frames, including genes that show marked similarity to transposase genes in other organisms. These larger elements (4000–6000 base pairs) are also found in many sites. *Euplotes crassus* has 5000–7000 copies of this type of transposon-related sequence in its haploid genome; all are precisely eliminated during macronuclear anlagen development.

There is considerable variation in the cytology during this phase of development. For example, polytene chromosomes are formed in hypotrichs before the eliminated sequences are lost. In these species, fragmentation of the micronuclear chromosomes into the much smaller macronuclear chromosomes is accompanied by the formation of vesicles, which partition the germinal chromosomes into many fragments. Recently, heterochromatin-associated proteins (Pdd1 and Pdd2) have been cloned from *T. thermophila* which seem to be involved with the removal of micronuclear-specific sequences during macronuclear anlagen development. This work is considerably advancing the molecular analysis of this process.

iii. Change in Somatic Chromatin The developing anlagen chromosomes undergo a change in histone content and modification, chromatin structure, and DNA methylation. Transcription begins, changing the phenotype of the cells from parental to progeny. Much current work is focused on the molecular and structural changes that accompany the transition from silent germinal to active somatic genome.

iv. DNA Replication The resulting macronuclear chromosomes undergo repeated rounds of DNA replication, to increase the ploidy of the macronucleus. Different species vary widely in the amount of polyploidy, from 48 in *T. thermophila* to several thousand in some species of hypotrichs. In general there is coordinate replication of all the chromosomes, with the notable exception of the macronuclear chromosome carrying the genes for ribosomal RNA in *T. thermophila*, which gets replicated to about 10,000 copies.

2. Loss of Parental Macronucleus

At the same time that the macronuclear anlagen are undergoing terminal differentiation, the parental macronuclei are discarded. There is variation in the details of the programmed loss of this nucleus. Some ciliates retain parental macronuclear function for some time, even after progeny macronuclei have developed. For example, in *Paramecium* species the parental macronucleus is fragmented early in conjugation, but the fragments continue to be transcribed, even after a new macronucleus has developed; since they are normally not replicated, they are soon diluted out as the progeny grow and are lost. If a new macronucleus fails to form, the fragments aggregate and are replicated in a process called macronuclear retention, yielding conjugants with progeny micronuclei but parental macronuclei.

In contrast, in *T. thermophila* the parental macronucleus functions until the anlagen begins transcription. It then moves to the posterior end of the mating cell, ceases transcription, becomes pyknotic, and is broken down in a process reminiscent of apoptosis in higher forms. Some of the machinery that is associated with the loss of micronuclear-specific sequences in the developing anlagen has been observed to be involved in this DNA degradation step. There is a clear and very distinct transition from parental to progeny macronuclei, and therefore phenotype, in this species.

3. Retention of Progeny Micronucleus

The nuclei that are destined to become new micronuclei are retained, remaining transcriptionally silent. Different species produce different numbers of micronuclei, although, as mentioned previously, most only retain one pronucleus, derived from one of the micronuclei following meiosis in a subsequent conjugation.

VI. PATTERNS OF INHERITANCE FOLLOWING CONJUGATION

Three different levels of clones can be easily established following conjugation (Fig. 5). The clone derived by isolating mating pairs is called the synclone. The clones established by isolating the two cells that result when pairs separate are called exconjugant clones, and the clones established by isolating cells after the first division of the exconjugants are called karyonides.

A trait that is uniformly expressed in all the clones derived from a mating pair (the synclone) must be determined by nuclear genes since cross-fertilization in any pair should create identical zygote nuclei in both partners. On the other hand, any trait that is inherited in the exconjugant clones must be cytoplasmic. Mitochondrially encoded drug-resistance genes have been shown to follow this pattern in several species of ciliates. Examples of mating-type determination and surface antigen gene differentiation in *Paramecium* have also followed this pattern,

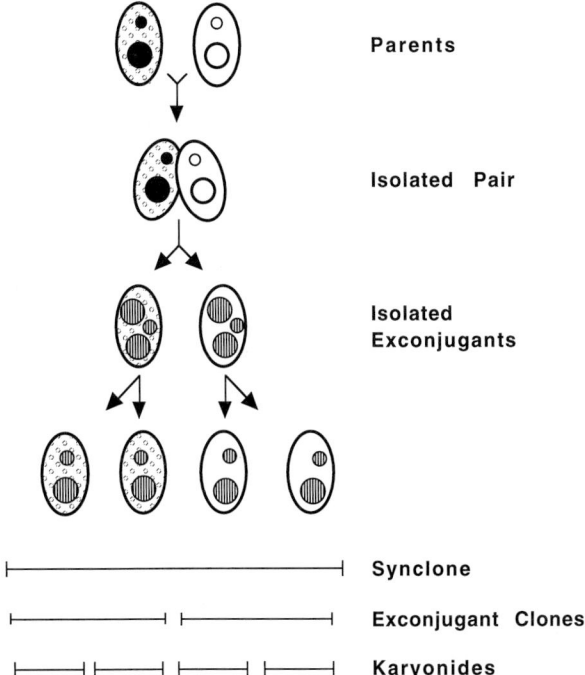

FIGURE 5 Levels of inheritance associated with different clones following conjugation. Isolating pairs, exconjugants, or karyonides (the first products of division following pair separation) facilitate identification of different patterns of inheritance.

A. Cytogamy

Cytogamy has been observed to occur naturally in a variety of species. It can be induced experimentally in *T. thermophila* by the introduction of an osmotic shock during conjugation. Cytogamy is simply conjugation in which all functions except nuclear exchange occur. Thus, following the prezygotic nuclear divisions described previously (meiosis and mitosis), instead of cross-fertilization the two pronuclei produced in each member of the pair fuse, resulting in a self-fertilization by nuclei with identical genotypes, yielding whole genome homozygotes. Normal development ensues, creating two cells which are homozygous for some set of their parent's genome and completely independent of each other. In this instance, traits encoded by nuclear genes show an exconjugant pattern of inheritance.

B. Autogamy

Autogamy is essentially cytogamy but in the absence of pairing. Some species of *Paramecium* have been shown to undergo autogamy if they have not paired within a characteristic period of time. Interestingly, exautogamous cells do not express the extended period of immaturity that is characteristic of exconjugants in these species.

C. Genomic Exclusion

Genomic exclusion occurs in *T. thermophila* when diploid cells mate with certain defective lines called star strains. In these matings, the diploid cell essentially undergoes the normal events in the genetic phase of conjugation, producing two identical haploid pronuclei and donating one to its mate. The star strain eliminates its micronucleus before producing any pronuclei, so a unidirectional exchange results instead of cross-fertilization. The pronuclei from the diploid duplicate the DNA without nuclear division (endoreduplication), yielding fully homozygous, identical diploid zygote nuclei in both cells. The cells abort further development and separate, retaining parental macronuclei (and therefore phenotypes). Thus, the exconjugants are heterokaryons, with different genotypes in micro- and macronuclei. These

suggesting a cytoplasmic influence on determining mating-type locus expression in developing macronuclei as a consequence of parental macronuclear expression. A feedback loop is formed, which maintains a constant phenotype in progeny even though there may be a genetic potential for other phenotypes. Finally, karyonidal inheritance patterns, as found in mating-type determination in several species of *Paramecium* and *Tetrahymena*, suggest independent differentiation for mating-type expression in developing macronuclei since the karyonides represent clones of independent macronuclear anlagen.

VII. VARIATIONS IN CONJUGATION

A number of variations of the process described previously occur in nature or can be induced experimentally. The best described include cytogamy, autogamy, genomic exclusion, and triplet matings.

cells are still mature and express their parental phenotypes and are therefore able to remate, this time with fully competent micronuclei. A second round of mating yields homozygous cells, which now develop new macronuclei.

D. Triplets

Mating in triplets has been documented in *T. thermophila*. Two of the three partners undergo essentially normal mating, with all three prezygotic divisions and a mutual exchange of pronuclei. The third cell also performs the three prezygotic divisions, and donates a migratory pronucleus to one of the other two cells but receives none. Thus, one cell becomes diploid, one becomes haploid, and one becomes triploid. All three cells carry out the development phase, making new macronuclei.

VIII. TRANSFORMATION DURING CONJUGATION

It has recently been found that DNA can be added to mating *T. thermophila* by electroporation or biolistic bombardment with DNA-coated particles. DNA introduced during the anlagen development phase results in incorporation into macronuclei. Of special note here is that DNA incorporated in the early phase of anlagen development undergoes the normal amount of amplification, ensuring sufficient copy number of the introduced sequences in recipient cells. Introduced DNA which contain the sequences necessary for processing during macronuclear anlagen formation undergo appropriate changes, including DNA splicing and amplification. Bombardment during the first prezygotic division (meiosis 1) yields sequences incorporated into the germline. Thus, it is now possible to introduce new sequences into either the germinal or the somatic nuclei in this species. All instances of incorporation into the genome of either nucleus have been by homologous recombination. Both gene replacement and elimination (so-called genetic knockouts) have been produced by these methods and are being actively used in a variety of studies.

See Also the Following Article

PROTOZOA

Bibliography

Beale, G. H. (1954). *The Genetics of Paramecium aurelia*. Cambridge Univ. Press, London.

Coyne, R. S., Chalker, D. L., and Yao, M.-C. (1996). Genome downsizing during ciliate development: Nuclear division of labor through chromosome restructuring. *Annu. Rev. Genet.* 30, 557–578.

Gall, J. G. (Ed.) (1986). *The Molecular Biology of Ciliated Protozoa*. Academic Press, Orlando.

Hausmann, K., and Bradbury, P. C. (Eds.) (1996). *Ciliates: Cells as Organisms*. Fischer-Verlag, Stuttgart.

Klobutcher, L. A., and Herrick, G. (1997). Developmental genome reorganization in ciliated protozoa: The transposon link. *Prog. Nucleic Acid Res. Mol. Biol.* 56, 1–62.

Sonneborn, T. M. (1975). Paramecium aurelia. In *Handbook of Genetics, 2* (R. C. King, Ed.). Plenum, New York.

Contraceptive Methods and Devices, Female

Malcolm Potts and Diana S. Wolfe

University of California, Berkeley

I. History of Contraceptive Devices
II. Efficacy and Side Effects
III. Barrier Methods and Protection against Sexually Transmitted Diseases and HIV
IV. Intrauterine Devices
V. Intravaginal Devices
VI. New Methods

GLOSSARY

barrier method Any technique of contraception that is based on providing a physical or chemical barrier to the passage of sperm and thus preventing fertilization.

diaphragm A form of barrier contraceptive consisting of a flexible rubber dome-shaped device with a rubber-covered metal or plastic rim.

intrauterine Located or occurring within the uterus.

spermicide A substance or agent that acts to kill sperm.

Contraceptives are devices that diminish the likelihood of conception. All reversible methods have a failure rate. The same person may switch from one method to another, take risks, or resort to abortion at different times in their life. Female barrier methods of contraception can be categorized as those acting in the vagina and those acting in the uterus [intrauterine devices (IUDs)]. These contraceptive methods traditionally are under her control. IUDs have low failure rates but occasionally cause serious side affects. Vaginal barriers have moderate failure rates and have no systemic effects, although they can be responsible for temporary and local problems. Some vaginal barriers offer protection against sexually transmitted diseases and HIV infection.

I. HISTORY OF CONTRACEPTIVE DEVICES

The possibility of using the vagina to hold a chemical or a physical barrier designed to separate sex from pregnancy is obvious and ancient. Spermicides are described in Egyptian texts and classical writings. Caps to occlude the cervix, and diaphragms that cover the cervix and part of the vagina wall, date back to the early nineteenth century. Spermicides and female barrier methods were commercially produced and advertised in the United States and Europe during the nineteenth century, although quality was uneven and sales outlets were limited by the law and conservative conventions. Intrauterine devices (IUDs) were also developed in the nineteenth century and their primary drawbacks and advantages had been well studied by Richter as early as in 1909.

Female barrier methods became the engine driving the early twentieth-century family planning movement in Europe and the United States. Margaret Sanger's New York clinic (closed by the police in 1916) used the female diaphragm, as did the clinic opened by Marie Stopes in England in 1921 and that of Elise Ottesen-Jensen in Sweden. In 1920 the histologist J. R. Baker was expelled from his laboratory in Oxford for daring to study spermicides and was only rescued by Lord Florey (who later received the Noble prize for his work on penicillin). Early IUDs were made of metal and they fell into disrepute between the two World Wars because of intrauterine infections. The introduction of a second generation of flexible plastic devices led to a renaissance in use in the 1960s and helped launch several nationwide family planning programs in Asia.

TABLE 1
Contraceptive Efficacy and Side Effects

Method	Failure rate/100 woman-years of use	Number of women in study	Side effects reported
Diaphragm	7.1	670	Urinary infection
Female condom	2.6	377	
Cervical cap	17.0	1394	
Lea's shield	5.6	146	
Copper T	2.2	4127	Pelvic inflammatory disease
LNG-IUD (hormone releasing)	0.5	1821	

II. EFFICACY AND SIDE EFFECTS

In the case of barrier methods, the theoretical effectiveness is often much lower than the failure rate observed in everyday use. The systematic study of female barrier methods is relatively recent and even today is not pursued as intensively as these methods would justify. It is often difficult to recruit a representative sample of volunteers to test a method that is already widely used, and clinical trial data on the efficacy of caps, diaphragms, and spermicides are limited. Study methodologies have evolved and it is now common to use pregnancy tests to detect early pregnancy, whereas when data were collected on the use of many barrier methods this sophistication was not available. Even more important, abortion was illegal when early studies were conducted; therefore, unless a pregnancy ended in a term delivery, a pregnancy was not necessarily counted as a failure. As a result, recent inventions such as the female condom are sometimes judged by stricter criteria than barriers that have been on the market for a long time. Care should be taken when comparing failure rates collected for different methods at different times (Table 1). The cost of various methods varies. Costs are often highest in the United States (Table 2) and always lowest when devices or spermicides are purchased in large quantities for use in developing countries.

A large and increasingly sophisticated literature surrounds IUD failures, expulsions, and removals. The plastic IUDs of the 1960s (Lippes loop, Margulies spiral, Birnberg bow, etc.) represented a major step forward. It was observed that the larger the device the lower the failure rate, but the more the side effects of bleeding and pain. The addition of copper to a small plastic frame, pioneered by Jaime Zipper in Chile, solved this dilemma and produced the widely used and satisfactory devices of today (e.g., Copper TCu380 and Gravigard). A third generation of IUDs, which release steroid hormones into the uterus, promise another leap forward, and perhaps may be even more important than the introduction of copper-bearing devices (Table 1).

IUDs often increase the volume and duration of menstrual loss, and they may give rise to cramping pains. Even with copper devices, a 50–60% increase in menstrual loss has been measured in the first 3 months after insertion, although the new steroid-releasing IUDs reduce blood loss significantly. Heavy bleeding is the most common reason for a woman to request removal of a device. More serious, however, is the possibility of intrauterine and pelvic infec-

TABLE 2
Cost of Female Barrier Methods and IUDs

Method	Cost
IUD (device + insertion by physician)	$200–$300 in public-sector family planning
Diaphragm/cap	$22 for device; $50–100 for medical visit
Sponge	$1.50 each
Female condom	$2.50
Spermicide	$0.25 per sexual act

tions occurring when an IUD is in place. Infections were the prime reason for the condemnation of the method between the two World Wars. In the 1970s, the Dalkon Shield, which was an aggressively promoted, badly designed IUD, was associated with illness and deaths from infection and undermined the use of second-generation devices, especially in the United States.

III. BARRIER METHODS AND PROTECTION AGAINST SEXUALLY TRANSMITTED DISEASES AND HIV

While IUDs have been implicated in the spread of sexually transmitted diseases (STDs), female barrier methods deter infection. Condoms protect both sex partners against the transmission of gonorrhea, syphilis, other STDs, and HIV. Female barriers may be less effective than condoms, but they may have a relatively large potential in controlling the spread of diseases because some women may use barriers more consistently than some men use condoms. In one study, couples using female barrier contraceptives had lower prevalence rates of STDs than did couples using condoms. In certain circumstances, a woman may need to use one method of contraception for birth control and another as a barrier against STD transmission.

It was long thought possible that physical and chemical barriers to sperm viability reduce the transmission and acquisition of sexually transmitted diseases, but the clinical demonstration that spermicides deterred the spread of gonorrhea only came in the 1980s. However, since present-day spermicides kill sperm by destroying their cell membranes, they can also damage the vaginal epithelium in some women, causing irritation. It is also possible that if the vaginal epithelium is damaged (even perhaps without the woman having any symptoms) the user could be more susceptible to HIV. Finally, spermicides may also destroy a woman's natural bacterial flora, which aid in protecting against infection. The balance of risks and benefits is complex and requires careful research.

IV. INTRAUTERINE DEVICES

There are advantages and disadvantages to the IUD that the user needs to understand. For the woman who wants to use an IUD there are few medical constraints. The most important relate to actual or potential risk of infection with a sexual disease or HIV. A couple does not have to interrupt their lovemaking when using the IUD to protect the woman from pregnancy, but side effects can be tiresome. The biological action of IUDs, like that of the pill, probably depends on interrupting a number of processes essential for conception. The device greatly reduces the number of sperm reaching the Fallopian tubes, where the egg is fertilized. If fertilization takes place, an IUD may also act by preventing implantation or attachment of the fertilized egg to the uterine wall.

There are relatively few conditions, such as an active pelvic infection, that preclude the use of an IUD (Table 3). IUDs are inserted through the cervix and into the uterine cavity using a sterile insertor about the size of a drinking straw. IUDs can be inserted in a nonpregnant woman immediately after delivery of a baby or at the time of surgical abortion, or they used as an effective emergency postcoital contraceptive. Insertion within minutes of delivering the placenta is painless, but it is associated with an

TABLE 3
Women Who Should Not Use Intrauterine Devices

Risks of use usually outweigh benefits	*Method should not be used*
	Pregnancy
	Uterine abnormality that makes it difficult to fit an IUD
HIV positive or AIDS	Sexually transmitted disease with purulent cervical discharge
	Uterine infection after delivery or abortion
	Untreated cervical cancer
	Uterine or ovarian cancer
	Pelvic tuberculosis

increased expulsion rate. Insertion 6 weeks or more after delivery, when a woman is breast-feeding, has a low expulsion rate. In a woman still experiencing menstrual cycles, insertion can take place on any day of the menstrual cycle, although many practitioners still limit insertion to the time near menstruation.

Complete or partial spontaneous expulsion of IUDs is always possible. Modern devices can have monofilaments or strings that protrude through the cervix. Women can be taught to feel the strings to check that the device is in place. The strings can also be palpated or visualized by health professionals, and they are used to pull devices out when no longer needed. Plastic devices can remain in place as long as the woman wishes; modern copper devices have a life of at least 5–7 years. Once a device is in place and properly checked after 3 months of use, a woman need not see a health professional unless she is concerned about side effects.

The vertical stem of a plastic IUD can be used as a carrier for releasing a synthetic progesterone analog. The levonorgestrel-releasing IUD (LNG-IUD) releases 20 μg a day from a silastic reservoir, which is enough to suppress endometrial growth and control the increased uterine bleeding found with other IUDs. Progestogens also make the cervical mucus more resistant to the susceptibility of infection. The Pearl index for LNG-IUDs is 0.0–0.2 per 100 women-years, the lowest recorded for any reusable method of contraception. The device lasts for 5 years before the hormone runs out and the IUD needs replacing. It is also possible that LNG-IUD could replace hysterectomy for some women suffering from heavy uterine bleeding in the menopausal years. LNG-IUDs, developed by T. Lukkainen in Finland, are used in Scandinavia and the United Kingdom but are unlikely to be sold in the United States because of litigation problems surrounding the use of silicone.

Pregnancy, while unusual, is a possibility with any IUD. IUDs reduce the possibility of intrauterine and extrauterine (or ectopic) pregnancies. However, the reduction in intrauterine pregnancies is greater than that for ectopic implantation; hence, the ratio of ectopic to intrauterine pregnancies is raised in IUD users. If pregnancy occurs with an IUD in place, the device should be removed because there is a small but life-threatening possibility of severe intrauterine infection in the case of a pregnancy in which an IUD remains in place.

V. INTRAVAGINAL DEVICES

A. Spermicides

A variety of chemicals, put together in a number of formulations, are capable of killing sperm in the vagina. Homespun methods have a long history. During World War II, Marie Stopes in Britain recommended using half a lemon, both as a barrier and because lemon juice is spermicidal. Coca-Cola, which is moderately acidic, has been used as a postcoital douche. Some women believe douches can prevent pregnancy, although there is no evidence of this and douches can disrupt bacterial flora inside the vagina, making a woman more susceptible to certain vaginal infections. Compounds containing mercury were used as spermicides until the 1970s, when they were withdrawn in all Western countries, although no clinical data exist on whether they were or were not harmful. Most commercially available spermicides use a detergent which destroys the sperm cell membrane, such as nonoxonyl-9 or menfegol.

There is practically no one who cannot use an intravaginal method of contraception. Until more is known about the relationship between spermicides and vaginal damage, a woman who suspects she is at high risk of HIV infection might be cautious of spermicides. (She should, of course, be using a condom if at all possible.) Spermicides can be formulated in a variety of vehicles, including suppositories, that melt at body temperature; spermicide impregnated water-soluble plastic films (C-film); as an aerosol, like a spray paint; and as gels stored in tubes, like toothpaste. A variety of disposable and reusable plungers are marketed for use with some gels.

B. The Diaphragm (Dutch Cap)

The diaphragm is a flexible rubber dome-shaped barrier with a rubber-covered metal or plastic rim. A diaphragm both makes a partial physical barrier to the ascent of sperm and holds a spermicide in the pathway of sperm before they reach the cervix.

Diaphragms range in sizes from 50 to 105 mm in diameter and must be fitted by a trained person. The possibility that diaphragms might work without a spermicide has been explored, but the evidence is not strong enough to recommend this use.

Diaphragms depend on the vaginal muscles to hold them in place. The thin, flat spring rim has a gentle spring that is comfortable for women with good vaginal muscle tone and the sturdy coil spring rim has a firm spring strength suitable for a woman with less muscle tone and an average pubic arch depth. The arcing spring folds into an arc shape that facilitates insertion but maintains correct position even if a woman has lax vaginal muscles. The woman needs to insert the diaphragm up to 2 hr prior to intercourse, and it must remain in place for at least 6 hr afterwards in order that the spermicide can kill all sperm.

Diaphragms and caps are designed for reuse. The method requires counseling and training from a health professional, as well as a willingness by the woman to handle her genitalia. Diaphragms have typically been used by middle-class Western women, although small clusters of users are found in developing countries. At its peak popularity, just before the introduction of the pill, the diaphragm was used by about 1 in 10 women in the United States and Britain. Today, diaphragms are used by 2 or 3% of women in these countries.

The spermicides used with diaphragms can cause irritation and diaphragms can cause mechanical discomfort by pressing on the bladder.

C. Cervical Cap

Cervical caps are small latex domes with a firm, round rim. They are filled one-third full of spermicide before use and seal tightly on the cervix. There are several designs from 20 to 50 mm in diameter. Caps can be made out of latex, plastic, or even metal (in the past). The Prentif cavity rim cervical cap provides protection for 48 hr and does not require additional spermicide applications for repeated intercourse. It is not recommended to wear a cap longer than 48 hr because of the possible risk of toxic shock syndrome. A trained person must teach the woman to fit the caps and they can be moderately difficult for the user to insert and remove; therefore, their use is not widespread. When the U.S. Food and Drug Administration (FDA) began to control contraceptive devices, methods including spermicides and diaphragms were "grandfathered" into the approval process, but the cervical cap was omitted. It was not until 1988 that the FDA approved caps for general use in the United States, and then only after lobbying by some women's' groups. Additional choices are always welcome in contraceptives, but caps have never been used by more than 1% of women.

D. Sponge

In 1983, the FDA approved for marketing a vaginal contraceptive device called the Today Sponge. It is a pillow-shaped device made of polyurethane and impregnated with 1 g of nonoxynol-9. Like the diaphragm and cervical cap, the sponge is a physical barrier to shield the cervix and, when used with spermicide, provides a chemical to kill sperm. It allows up to 24 hr of contraceptive protection but must be left in for 6 hr after intercourse. One size fits all, and unlike the diaphragm or cap, sponges are sold over-the-counter and require no special training to use. The sponge has a concave dimple on one side that is intended to fit over the cervix and decrease the chance of dislodgment during intercourse. A polyester loop located on the opposite side of the sponge facilitates removal. Before use, it must be moistened with tap water.

The sponge can cause infections if left in for more than 24 hr. Cases of toxic shock have been reported. The etiology of this phenomena is thought to be that normal vaginal flora are sensitive to nonoxynol-9, resulting in the overgrowth of anaerobic organisms. The sponge should not be reused and has been found to be less effective in parous women.

E. Female Condom

The first female condom was invented in Denmark, manufactured in England, and first marketed in Switzerland, Austria, The Netherlands, and the United Kingdom as Femidom. The FDA approved its use in the United States in 1993, where it is marketed under the brand name Reality.

The female condom is a heat-sealed polyurethane pouch 7.8 cm in diameter and 17 cm long that shields the perineum and lines the vagina completely. A silicone-based lubricant is provided with the device, but no spermicide is used. The female condom has a soft flexible rim at the proximal end that remains outside the introitus and a loose, firmer rim that holds the sac over the cervix and reaches into the posterior fornix, rather like the rim in a diaphragm. The female condom can be inserted up to 8 hr before intercourse. The polyurethane used in the sheath is a soft, impermeable material with good heat-transfer characteristics. It is stronger than latex and therefore less likely to break. The device should not be used with a male condom.

Currently, the female condom is the only female-controlled method offering protection against HIV (and STD) transmission. It can be seen during use, but it has proved acceptable to some commercial sex workers. It is sold as a disposable device in the West, but it is being studied as a potentially reusable device in developing countries, where cost currently prevents widespread use.

VI. NEW METHODS

A. IUDs

Any new design of IUD needs to be tested against a well-studied benchmark device, such as a CuT 380, to see if it offers women any real advantages. An interesting innovation, developed by D. Wildermeerch in Belgium, consists of a thread anchored in the muscle at the top of the uterus and carries the same copper sleeves as a CuT 380. There is no rigid plastic device. The anchor is placed with a special insertor. This device is performing well in comparative trials with older existing devices.

B. Femcap

This is a silicone rubber device that can be worn for up to 48 hr. Safety and efficacy studies are being conducted. It appears to be not quite as effective as the diaphragm used with spermicide.

C. Lea's Shield

Lea's Shield is a silicone barrier designed to be held in place by its volume rather than its diameter. One size fits all users, so it has the advantages of diaphragm without the need for clinician fitting. A loop, which is an integral part of the device, is used for removal and also helps orient the device in the vagina. Effectiveness increases with the addition of spermicides. Clients are instructed to insert the device any time before coitus, to leave it in place at least 8 hr after the last coital act, and not to leave the shield in place for more than 48 hr at a time.

D. New Spermicides

Many chemicals kill sperm and also act as microbicides. One new spermicide, Praneem, based on an extract from the seeds of *Azadirachta indica*, is being tested in India. Perhaps the single greatest need in the field of barrier methods is the development of a female-controlled, vaginal, chemical method that prevents HIV transmission. The empowerment of women is crucial for the prevention of HIV transmission to women. Many women are in situations in which they suspect their partners may not be monogamous but they cannot persuade the partners to use a condom. In such a situation, an effective chemical virucide, if it could be used without the partners knowledge before or after sexual intercourse, might be widely accepted and could save many lives. Unfortunately, research has moved slowly, and the most widely used spermicide (nonoxonyl-9) has not been shown to prevent HIV acquisition in women.

See Also the Following Articles

CONTRACEPTIVE METHODS AND DEVICES, MALE; HIV (AIDS); HORMONAL CONTRACEPTION; SEXUALLY-TRANSMITTED DISEASES

Bibliography

Andersson, K., *et al.* (1994). Levonorgestrel-releasing and copper-releasing (Nova T) IUDs during five years of use: A randomized comparative trial. *Contraception* 49, 56–72.

Cambell, P. (1993). Efficacy of female condom. *Lancet* 341, 1155.

Diaz, T., *et al.* (1995). Relationship between use of condoms

and other forms of contraception among human immunodeficiency virus-infected women. *Obstet. Gynecol.* 86, 277–282.
Farr, G., *et al.* (1994). Contraceptive efficacy and acceptability of the female condom. *Am. J. Public Health* 84, 1960–1964.
Faundes, A., *et al.* (1994). Spermicides and barrier contraception. *Curr. Opin. Obstet. Gynecol.* 6, 552–558.
Ferreira, A., *et al.* (1993). Effectiveness of the diaphragm, used continuously, without spermicide. *Contraception* 48, 29–35.
Gabby, M., and Gibbs, A. (1996). Does additional lubrication reduce condom failure? *Contraception* 53, 155–158.
Kawachi, I., *et al.* (1994). Long-term benefits and risks of alternative methods of fertility control in the United States. *Contraception* 50, 1–16.
Lukkainen, T., and Toivonen, J. (1995). Levonorgestrel-releasing IUD as a method of contraception with therapeutic properties. *Contraception* 52, 269–276.
Mauck, C., *et al.* (1996). Lea's Shield: A study of the safety and efficacy of a new vaginal barrier contraceptive used with and without spermicide. *Contraception* 53, 329–335.
Milsom, I., *e. al.* (1995). The influence of the Gyne-T 380S IUD on menstrual blood loss and iron status. *Contraception* 52, 175–179.
Rosenberg, M. J., *et al.* (1992). Barrier contraceptives and sexually transmitted diseases in women: A comparison of female-dependent methods and condoms. *Am. J. Public Health* 82, 669–674.
Service, R. F. (1996). Panel wants to break R&D barrier. *Science* 272, 1258.
Smith, C., *et al.* (1995). Effectiveness of the non-spermicidal fit-free diaphragm. *Contraception* 51, 289–291.
Trussel, J., *et al.* (1993). Contraceptive efficacy of the diaphragm, the sponge and the cervical cap. *Family Planning Perspect.* 25, 100–105.
Trussell, J., *et al.* (1994). Comparative contraceptive efficacy of the female condom and other barrier methods. *Family Planning Perspect.* 26, 66–72.
Xu, J.-X., *et al.* (1996). A comparative study of two techniques used in immediate postplacental insertion (IPPI) of the copper T-380A IUD in Shanghai, People's Republic of China. *Contraception* 54, 33–38.
Zhu, P., *et al.* (1995). The effect of levonorgestrel-releasing intrauterine devices (20 µg/day) (LNG-IUD-20) on the morphological structure of human endometrium: A study of the endometrial factor VIII activity in the woman before and after insertion of LNG-IUD-20 by the digital image analysis. *Contraception* 52, 63–68.

Contraceptive Methods and Devices, Male

M. Cristina Meriggiola
University of Bologna

William J. Bremner
University of Washington

I. Condoms
II. Withdrawal
III. Hormonal Contraception

GLOSSARY

azoospermia Absence of spermatozoa in the ejaculate.
Pearl index Pregnancy rate per 100 men per year.
severe oligospermia The presence of less than 3 million/ml spermatozoa in the ejaculate.
spermatogenesis A complex process in which primitive germinal stem cells (spermatogonia) differentiate into highly specialized cells (spermatozoa). This process takes place in the seminiferous tubules of the testis.
spermatozoa Motile cells produced in the testis that are capable of rising through the female genital tract, recognizing the egg, penetrating the zona pellucida, and fusing with the ovum. A normal human ejaculate may contain 40–300 million spermatozoa.

I. CONDOMS

Condoms are a type of barrier method that when placed over the penis prevent sperm from entering the vagina. They are effective not only as contraceptives but also in preventing sexually transmitted diseases (STDs).

A. History

Although sparse and scanty, evidence for condom use dates back to the Egyptians, who used penis sheaths for protection against diseases and as decorations to demonstrate social status. Tales from the Greek and Roman times report that King Minos of Crete used goat bladder sheaths to mitigate his peculiar disease (described as semen containing serpents and scorpions). The first indisputable reference to a sheath comes from the Italian anatomist Fallopio in 1564 who described them "for use against venereal disease and numerous bastard offspring." The word condom was coined later, and its etymological origin remains uncertain. It first appeared in print in 1717, and some historians reported that it may derive from the Latin word *condus* (a receptacle); others say the word may derive from the Persian word *kendu* or *kondu*. It may also derive from Dr. Condom, a court physician of King Charles II, who prescribed the use of this device to his king. In the eighteenth century, the contraceptive benefits of condoms were already appreciated and it is reported that Casanova not only used them regularly but also was aware of the varying levels of condom quality. Early condoms were made of animal intestines. They were very expensive and therefore not available to everyone. In 1870, production of rubber condoms began. Since then, these devices have become much cheaper and their use is widespread. Latex condoms appeared in the mid-1930s. About 45 million couples worldwide use condoms, with the highest percentage use in Japan (50% of couples), Hong Kong, and Taiwan. The lowest percentage use is in sub-Saharan Africa (<1%) with a 13% average use in the developed world overall. However, these surveys were performed in the 1980s, before the beginning of the major effort to promote condoms for human immunodeficiency virus (HIV) prevention. Since then, a large campaign to promote condom use as both a contraceptive and a prophylactic has been carried out. As a result of this campaign, it is believed that condom use has increased dramatically.

B. Mechanism of Action

Condoms work as barriers to prevent sperm and microbes from entering the vagina. In addition, condoms may have spermicide and microbicide additives. Condoms lubricated with silicone oil, wet jellies, dry powders, or spermicides such as nonoxynol-9 are more effective than unlubricated condoms, both as contraceptives and in preventing STDs. Spermicidal condoms have been available since the early 1980s. The spermicides are placed on the inner and outer surface of the condom. Sperm ejaculated into the condom are quickly inactivated, although the effectiveness of spermicides in the case of condom breakage is unknown. Although natural latex rubber condoms are the most widely commercialized types of condoms, natural condoms made of lamb cecum are still commercial products in many countries and, in fact, they retain up to 5% of the market, despite being very expensive. Some men prefer them because they provide more comfort and increased sensation. These characteristics of natural condoms derive from their ability to allow heat transmission and exchange of fluids. A major shortcoming of natural skin condoms is that they are permeable to HIV and do not provide a good protection against STDs.

In the past few years, the search for new materials for condoms has received new impetus because of the "life-and-death consequences" of condom failure. New polymers such as Kratos, copolymers, and polyurethanes are being evaluated in experimental condoms. The advantages of these polymers include a higher resistance to storage, the possibility of manufacturing thinner condoms, increased sensitivity, and the absence of protein impurities that may raise allergic reactions.

C. Effectiveness

There is a paucity of well-controlled studies on the clinical effectiveness of condoms, both against pregnancy and sexually transmitted disease. Breakage and slippage are the most studied indicators of the protection offered by condoms. Recent reports indicate that condom breakage rates during vaginal intercourse vary between 0 and 12%. First-year contraceptive failure rate is estimated at about 12% during typical use and 3% during perfect use. The percentage of slippage is also about the same. Although there is no good research to compare perfect users and inconsistent users, it seems that breakage and slippage happen very rarely in frequent users, whereas these are more common among infrequent users. A recent survey conducted by Family Health International reported an 8.7% failure rate for condoms among 170 participants that used 1947 condoms. Less than 10% of the couples were responsible for half of all condom failures, whereas 37% of the couples did not experience any failure. Factors that more commonly influence condom breakage are quality of manufacture; inconsistent use; bad storage conditions; age of condoms; and users' behavior, such as incorrect methods of putting on condoms, use of oil-based lubricants, reuse, and duration and intensity of intercourse.

Epidemiological studies have shown that condoms are effective mechanical barriers to HIV, herpes simplex virus, cytomegalovirus, hepatitis B virus, chlamydia trachomatis, and *Neisseria gonorrhea*. Latex condoms provide greater protection against viral STDs than do natural-membrane condoms. The actual effectiveness of condoms in preventing STDs is very difficult to assess. Cross-sectional and case-control studies have shown that condom users and their partners have a lower frequency of STDs than do people who do not use condoms. Failure of condoms in prevention of STDs is more commonly due to a user's failure to use a condom for each episode of intercourse, to put the condom on before any genital contact, to completely unroll the condom, and to use water-based lubricants. The use of spermicide-containing condoms may provide additional protection against STDs in the case of condom slippage or seepage. Also, the actual protection provided by condoms against HIV is still debated. Recent studies report that the HIV 120-nm particle does not pass through the latex condoms. However, permeability varies according to condom quality. Laboratory studies have shown that various spermicides are able to inactivate the HIV. Nonoxynol-9 inactivates the HIV within 30 sec at a concentration as low as 0.05%. Commercially available preparations contain 2–20% nonoxynol-9. Two recent meta-analysis studies report a failure rate of 31 and 50% respectively in the prevention of HIV transmission by condoms. However, the weakness of individual studies, e.g., the lack of control for a number of variables such as compliance with condom use, number of partners, etc., might have led to an underestimation of the effectiveness of condoms when they are consistently and correctly used.

II. WITHDRAWAL

Withdrawal or coitus interruptus is a male contraceptive method based on the withdrawal of the penis from the vagina before ejaculation occurs.

A. History

This method has been used for centuries and it is primarily responsible for the historical transition from high to low fertility rate. Various types of coitus interruptus have been reported according to different cultures. The "reserved coitus" was popular in a religious American community in which men were taught to pray during sexual intercourse to avoid achievement of orgasm and ejaculation. The "sassonic coitus" is based on the prevention of ejaculation achieved by exerting a pressure on the base of the penis. In this way, the excretion of the fluid can be prevented and the reflux of semen in the bladder is facilitated. Whatever means of interrupting coitus are used, this method is still very commonly used in many countries throughout the world. National surveys have reported use up to 35% in Rumania, 27% in Turkey, 24% in the Czech Republic, 16% in Mauritius, and 8% in Sri Lanka. Withdrawal is frequently used in conjunction with other methods of contraception or only during certain parts of the woman's cycle.

B. Mechanism of Action

The withdrawal contraceptive method is based on the timely removal of the penis from the vagina during sexual intercourse before ejaculation occurs. Ejaculation must occur completely away from the vagina and the external genitalia of the female partner. Because of its mechanism of action, this method should be practiced by men who have no trouble predicting when they will ejaculate and have no seepage of preejaculatory fluid. The advantage of this method is that it does not involve medications. However, the interruption of intercourse during the apex of excitement may markedly reduce the pleasure of sexual intercourse and for this reason many men dislike it.

C. Effectiveness

Very little research had been performed on the effectiveness of this method. It has been reported that it has a first-year pregnancy failure rate among typical users of 19%, but this number is based on only four studies performed since 1960. In these studies there was a wide variation in the effectiveness with a Pearl index ranging from 6.7 to 21.9%. Indeed, it is very difficult to predict effectiveness of this method because it is very user dependent. If it is used with commitment, discipline, and motivation this method can be very effective. For those men who fulfill these requirements, this can very well be the method of choice. It has also been suggested, although not all studies agree, that preejaculatory fluids can contain semen and thus be responsible for some failures of this contraceptive method. Especially in the case of a recent ejaculation, there may be semen stored in the prostate, in the penile urethra, or in the Cowper's gland that can be released before the actual ejaculation occurs. Therefore, the likelihood of failure due to preejaculatory fluid containing semen may increase if this method is used after a recent ejaculation. Couples that use this method successfully need not be counseled to switch away from it. This method requires men's involvement and commitment. In this sense, it may reinforce men's responsibility and participation in the relationship, and it may induce the man to have more dialog with the woman and respect for her. On the other hand, couples have to be aware that they are not protected from STDs or HIV transmission and they should be counseled to switch to condoms if they are at risk.

III. HORMONAL CONTRACEPTION

Since the advent of hormonal female contraception, attempts have been made to develop hormonal regimens to suppress fertility in men. However, while the female cycle is finely regulated and even small hormonal variations may disrupt ovulation, the male reproductive process, based on the continuous production of millions of spermatozoa, seems to be very resistant and hard to disrupt. In females, even a partial suppression of ovarian activity can provide an excellent contraceptive protection, whereas in males, a very profound suppression of spermatogenesis has to be achieved to obtain good contraceptive effectiveness. These differences have slowed the development of hormonal contraceptives for men. Compared to the current means of contraception such as condoms, withdrawal, and vasectomy, hormonal contraception would have the advantages of being more effective, reversible, and more acceptable because it does not interfere with the spontaneity of intercourse. As in the female, hormonal contraception in the male is also not intended to protect against STDs or HIV.

A. Mechanism of Action

The contraceptive action of hormone administration in men is based on the suppression of gamete maturation from stem cells to mature haploid cells apt to fertilization. To achieve a complete and profound suppression of sperm production, both gonadotropins must be profoundly suppressed and the testis must be depleted of testosterone (T). Since either luteinizing hormone (LH) or follicle-stimulating hormone (FSH) alone can maintain some level of sperm production, a selective suppression of only one of the two gonadotropins is not sufficient to achieve a good contraceptive effect. Also, because T alone, in the absence of both gonadotropins, can maintain spermatogenesis, testicular T suppression must be quite profound or intratesticular T action must be blocked. Sperm production and steroid production by the testis are very closely correlated and are both driven by gonadotropins. As a consequence, the suppression of LH and FSH by hormonal means not only decreases in sperm production but also decreases in T production. In this setting, exogenous T administration is necessary to ensure maintenance of androgen-dependent physiological functions. The mechanism by which hormones suppress spermatogenesis allows the preservation of original stem cells, thus ensuring complete reversibility of this contraceptive method.

B. Androgens

The administration of androgens alone has been the most extensively studied hormonal approach. This regimen has the advantage of combining in one compound the theoretical possibility of suppressing gonadotropins and replacing peripheral T concentrations. The World Health Organization (WHO) supported two multicenter trials conducted in various countries in which weekly injections of testosterone enanthate (TE) at the dosage of 200 mg were administered to normal men. In these trials, the hormonal regimen induced azoospermia in only about 70% of the men at Month 6 of treatment, and in about 28% of the subjects sperm count was suppressed below 3 million/ml, whereas a few subjects (about 2%) either did not suppress below 3 million/ml or did not maintain this suppression. Approximately 4 months were necessary to achieve azoospermia. Results of these studies showed that there is an inverse relationship between suppression of spermatogenesis and pregnancy rate. One pregnancy (Pearl index, 0.8 per 100 year/men; CI, 0.02–4.5) and zero pregnancy (Pearl index, 0 per 100 year/men; CI, 0.0–1.6) were reported among azoospermic subjects in the first and second WHO trial, respectively. Pregnancy rate was 8.1 (CI, 2.2–20.7) in subjects with sperm counts >3 million/ml. Although the contraceptive protection achieved by severe oligospermia is not as good as that provided by azoospermia, it is still better than that offered by condoms. In subjects whose sperm count remained above the 3 million/ml threshold, pregnancy rate was worse than that reported with condoms. Interestingly, in the WHO-supported study a marked difference in the ability of TE to suppress spermatogenesis was reported among Caucasian and Asian centers (Chinese, Indonesian, and Thai). In the Asian centers, 91% of the subjects in the first WHO study and 85.7% in the second study became azoospermic compared to 65 and 67% of Caucasian the population, respectively. The reason for this ethnic difference is unknown. Ethnic difference in the hormonal milieu, receptor concentration, or nutritional intake or other environmental differences may influence the susceptibility to steroids, thus resulting in a different spermatogenetic suppression and different incidence of side effects. Following weekly injections of 200 mg TE, serum testosterone levels were above the physiological range for 3 days. These peaks caused adverse effects, such as decreased high-density lipoprotein (HDL) cholesterol, increased body weight, and acne. While acne and increased body weight can be annoying effects that can reduce the acceptability of this regimen, the decrease of HDL cholesterol has the potential to increase the risk for cardiovascular disease.

In conclusion, these trials proved the validity of the basic concept of hormonal male contraception, demonstrating the high efficacy of the hormonally induced azoospermia. They also suggest that the achievement of complete azoospermia in all subjects would be a desirable goal for an optimal contraceptive regimen. In addition, they demonstrated that a regimen based on testosterone-only administration does not seem to be promising as a contraceptive, at least among Caucasian subjects, because of the failure to induce azoospermia in all subjects and because of the previously mentioned occurrence of adverse effects. Also, the need for weekly injections would make this hormonal formulation unacceptable to most men. This study also indicated a different sensitivity to the steroid-suppressive effects among individuals of the same as well as of different backgrounds.

The achievement of a more profound gonadotropin suppression could result in a greater reduction of spermatogenesis in a higher number of men. Therefore, other steroid or protein hormones were added to testosterone with the purpose of increasing the efficacy of this regimen and of reducing the dose of the androgen to minimize side effects. Two classes of hormones have been used in combination with testosterone: progestins and gonadotropin-releasing hormone (GnRH) analogs.

C. GnRH Analogs

GnRH analogs have been synthesized by substituting one or more amino acids of the GnRH protein with a natural or synthetic molecule. With this technique, two classes of GnRH analogs have been developed: GnRH agonists and GnRH antagonists. Agonist analogs of GnRH have more affinity than GnRH for the receptors and are more resistant to degradation. After GnRH agonist administration, there is a prompt release of gonadotropins followed by a desensitization of the hypophysis that results in a suppression of gonadotropin secretion and in a state of pharmacological castration. In contrast, GnRH antagonists compete with native GnRH for pituitary receptor occupancy and therefore have an immediate suppressive inhibitory action that lasts as long as antagonist molecules are present in sufficient concentration. Because of their ability to suppress gonadotropins and therefore to inhibit gonadal activity, these compounds have been tested in male contraceptive trials administered in combination with T.

Twelve human trials for male fertility regulation have been published in which GnRH agonists have been administered alone or with androgens. Results of these studies showed that in about 30% of the men sperm count decreased below 5 million/ml, in another 30% it was lower than 30 million/ml, whereas in the remaining subjects spermatogenesis was either unaffected or only slightly reduced. Only a few men achieved azoospermia. In studies conducted in monkeys, the addition of testosterone seemed to attenuate the suppressive effects of GnRH agonists on spermatogenesis. Therefore, from these studies it was concluded that GnRH agonists do not have a future in male contraception.

GnRH antagonists have also been tested in male contraceptive trials and results from these studies seem to be more promising. Like GnRH antagonists, GnRH antagonists also have to be administered in combination with testosterone to maintain androgen-dependent physiological functions. Nal-Glu, a second-generation antagonist, is the compound most tested in humans. This compound causes local irritation at the site of injection (itching, burning, and subcutaneous nodules) due to histamine release. Recently, a new GnRH antagonist with reduced histamine-releasing releasing activity named Cetrorelix has been tested. Both Nal-Glu and Cetrorelix are rapidly broken down in the stomach and have a relatively low potency. Therefore, they have to be administered by daily subcutaneous injections. In the studies performed to date, a profound suppression of spermatogenesis was achieved, but only 35 of the 40 men treated (88%) became azoospermic. The mean time to azoospermia was 12 weeks (range, 4–24 weeks). It became apparent from studies in monkeys and men that dose and timing of testosterone replacement might be crucial for the achievement of a profound sperm suppression. Higher T doses seemed to be associated with lower sperm suppression and also with side effects such as a 10–15% HDL decrease. The use of testosterone dosages that maintain testosterone peripheral concentrations in the lower normal range might have a favorable effect not only on the suppression of spermatogenesis but also on androgen-dependent physiological functions. Initial studies in monkeys suggested that by delaying the beginning of testosterone replacement, a more profound and complete suppression of spermatogenesis can be achieved.

In conclusion, this regimen seems to be, at least conceptually, a very promising approach. However, antagonists available to date for clinical trials are very expensive, have problems with local side effects, and have to be administered by daily subcutaneous injections. These characteristics make these compounds unsuitable for long-term use by the population at-large. New generations of more potent and long-acting GnRH antagonists are needed before this regimen can qualify as a real option in male contraception.

D. Progestins Plus Androgens

Since the advent of hormonal female contraception based on estrogens + progestin administration, progestins in combination with testosterone (T+P) have been tested to suppress fertility in men. In early studies, the combined administration of T+P was reported to induce azoospermia in only about 50% of the men, therefore not providing any advantages over T-only regimens. However, none of these trials had a T-only control group. Therefore, whether adding a progestin to T could provide some advantage, either in terms of gonadotropin suppression or in terms of sperm suppression compared to T-only regimens, was never tested. Only recently, the combined administration of 500 µg/day levonorgestrel

(LNG) + 100 mg/week TE was compared to 100 mg/week TE + placebo. LNG + TE induced a rapid and profound suppression of spermatogenesis in 67% of the men, whereas only 33% of the men treated with TE + placebo achieved azoospermia. Severe oligospermia or azoospermia was almost universally achieved in men treated with TE + LNG administration (94%), whereas only 61% of the men treated with TE + placebo became azoospermic or severely oligospermic. This study demonstrates that the addition of a progestin can increase the efficacy of T and suggests that the combined administration of these steroids may indeed provide a more effective contraceptive regimen than an androgen alone. The dosage of TE used in this study (100 mg/week) was one-half of the dosage used in the WHO multicenter trial; the effects of combining progestins with higher dosages of TE have never been reported.

Although LNG is very potent in terms of gonadotropin suppression, it retains some androgenic properties. When used at high doses in men in combination with testosterone, LNG causes androgen-related side effects. LNG at a dose of 500 mg/day + 100 mg/week TE caused side effects such as decrease of HDL cholesterol, weight gain, and acne. Whether lower doses of LNG can induce the same profound suppression of spermatogenesis and, at the same time, reduce the incidence of side effects remains to be determined. Progestins derived from 17-OH progesterone are free of androgenic effects but are less potent in terms of gonadotropin suppression. The best known progestin in this group is medroxyprogesterone acetate. Administered alone or in combination with testosterone to suppress fertility in men, this progestin did not induce a consistent suppression of spermatogenesis. Moreover, there was a delay in the restoration of spermatogenesis, probably due to the accumulation of this compound in the adipose tissue.

In conclusion, results of these studies demonstrate that the combined administration of T plus a progestin is indeed more effective than T alone as a contraceptive. The development of new progestins with more favorable biological properties constitutes a fundamental point in the progress of this contraceptive regimen.

E. Progestins/Antiandrogens Plus Androgens

Results from previous studies clearly show that, in addition to gonadotropin suppression, another factor(s) might be important for the achievement of a complete and consistent suppression of spermatogenesis. Even in the presence of very low gonadotropin concentrations, some level of testosterone production by Leydig cells can persist. This residual testosterone might be involved either directly or through its metabolite DHT in the maintenance of some level of sperm production in a few subjects. This effect can be blocked by achieving a very profound suppression of gonadotropins and therefore of T production and/or by blocking directly the T effect within the testis. Cyproterone acetate (CPA) is a synthetic progestin with antiandrogenic and progestin activity. We hypothesized that adding a progestin with antiandrogenic properties to T may have two advantages: (i) Because of its progestin activity, CPA can act synergistically with testosterone in the suppression of gonadotropins, and (ii) because of its antiandrogenic activity CPA can act at the gonadal level, blocking the stimulatory effect of exogenous and/or endogenous androgens on spermatogenesis. The threshold for androgen action may vary among different tissues; therefore, the dose of the antiandrogen and of testosterone could be adjusted to maintain androgen-dependent physiological functions while creating testosterone levels that do not maintain spermatogenesis and do not cause undesirable side effects. In a pilot study, two doses of CPA (100 and 50 mg/day; 5 men in each group, CPA-100 and CPA-50) were administered in combination with 100 mg/week TE to normal subjects. The effects of this regimen were compared to those of 100 mg/week TE administered alone (5 men). Regardless of the dose, all 10 men receiving CPA became azoospermic, whereas only 3 men of those receiving TE alone achieved azoospermia. Men treated with CPA achieved azoospermia in 6.8 + 0.5 and 8.4 + 1.0 weeks in the CPA-100 and CPA-50 groups, respectively, whereas 14.0 + 1.2 weeks were necessary to achieve azoospermia in men receiving TE alone. Therefore, the possibility of shortening the latency period for the onset of contraceptive action also constitutes a great advantage of the CPA + TE contraceptive regimen compared to previously tested regimens. Indeed, this is the first hormonal regimen that induces suppression of spermatogenesis in less than 72 days, which is the time required for a spermatogonia to mature into a fully mature spermatozoa. The rapidity of action of this regimen might be explained by a double effect on the process of sperm maturation: While the inhibition of FSH probably blocked spermatogonia maturation, the inhibition of androgen action might have interfered with the process of spermiogenesis. Also, a decrease in the concentration of androgen-dependent adhesion molecules might have caused a premature sloughing of spermatids, thereby contributing to the early disappearance of sperms from the ejaculate. This regimen did not cause any major side effects. A slight decrease of hematological parameters was reported which was dependent on the dose of CPA.

In conclusion these trials showed the importance of blocking intratesticular T action in addition to gonadotropin suppression. They also suggest the possibility of developing compounds with differential effects on different systems to avoid adverse effects and at the same time provide good spermatogenetic suppression.

See Also the Following Articles

CONTRACEPTIVE METHODS AND DEVICES, FEMALE; FAMILY PLANNING; SPERMATOGENESIS, OVERVIEW; VASECTOMY

Bibliography

Bebb, R. A., Anawalt, B. D., Christensen, R. B., Paulsen, C. A., Bremner, W. J., and Matsumoto, A. M. (1996). A promising male contraceptive approach: Combined administration of testosterone and levonorgestrel. *J. Clin. Endocrinol. Metab.* **81**, 757–762.

Cummings, D. E., and Bremner, W. J. (1994). Prospects for new hormonal male contraceptives. *Clin. Endocrinol. Metab. North Am.* **23**, 893–922.

Feldblum, P., and Joanis, C. (1994). Modern barrier methods: Effective contraception and disease prevention (N. Herndon and E. T. Robinson, Eds.) pp. 9–11.

Meriggiola, M. C., and Bremner, W. J. (1997). Progestin–androgen combination regimens for male contraception. *J. Androl.* **18**(3), 240–244.

Meriggiola, M. C., Bremner, W. J., Paulsen, C. A., Valdiserri, A., Incorvaia, L., Motta, R., and Flamigni, C. (1996). A combined regimen of cyproterone acetate and testosterone enanthate as a potentially highly effective male contraceptive. *J. Clin. Endocrinol. Metab.* **81**, 3018–3023.

Potts, M., and Diggory, P. (1983). *Textbook of Contraceptive Practice*, 2nd ed., pp. 106–118. Cambridge Univ. Press, Cambridge, UK.

Santow, G. (1993). Coitus interruptus in the twentieth century. *Population Dev. Rev.* **19**(4), 773.

World Health Organization Task Force on Methods for the Regulation of Male Fertility (1990). Contraceptive efficacy of testosterone-induced azoospermia in normal men. *Lancet* **336**, 955–959.

World Health Organization Task Force on Methods for the Regulation of Male Fertility (1996). Contraceptive efficacy of testosterone-induced azoospermia and oligozoospermia in normal men. *Fertil. Steril.* **65**, 821–829.

Cooperative Breeding

see Altruistic Behavior, Mammals

Copulation

see Mating Behaviors

Copulation, Mammals

Robert L. Meisel

Purdue University

I. Behavioral Description
II. Hormonal Regulation
III. Neural Regulation
IV. Synchronization of Copulation

GLOSSARY

androgen A class of steroid hormones secreted from the testes. Testosterone is the primary testicular androgen in male mammals.

castration Surgical removal of the testes.

estrogen A class of steroid hormones secreted from the ovaries. Estradiol is the primary ovarian estrogen in mammals.

medial preoptic area A forebrain region critical for the expression of copulatory behavior in males.

ovary The female reproductive organ from which steroid hormones are secreted and eggs are released.

progestin A class of steroid hormones secreted from the ovaries. Progesterone is the primary ovarian progestin in mammals.

testis The male reproductive organ from which steroid hormones are secreted and sperm are released.

ventromedial nucleus of the hypothalamus A forebrain region critical for the expression of copulation in females.

The fertilization of eggs and the production of offspring are closely tied to copulation in mammals. When rare or endangered species of animals are bred in captivity, there is disappointment surrounding those mating attempts that do not result in pregnancy or the birth of viable young. Though we cannot overestimate the role of the animal's reproductive physiology in the success or failure of such matings, quite often the problem is behavioral rather than physiological. For reproduction to occur, an intricate behavioral coordination between females and males is implied involving first sexual attraction of mating partners, then sexual arousal, and culminating in copulatory interactions permitting the delivery of sperm to the female's reproductive tract.

I. BEHAVIORAL DESCRIPTION

For the male, copulation involves motor patterns directed toward penile insertion into the female's vagina (intromission). Males of different species have copulatory patterns that may involve single or multiple intromissions to ejaculation, single or multiple intravaginal thrusts, and single or multiple ejaculations within a particular copulatory interaction. Females also have special copulatory motor patterns (i.e., receptive behaviors) involving postural changes designed to support the weight of the mounting male and to make the vagina accessible for intromission to occur.

II. HORMONAL REGULATION

Long before any formal experiments were conducted, the effects of removal of the gonads on copulation of male and female mammals were well-known. Research conducted during the past century has confirmed the conventional wisdom of the consequences of removal of the testes or ovaries and has offered new insights into the ways in which copulation depends on the biochemical consequences of steroid hormone actions. One insight was the structural similarity of the testicular and ovarian steroids and how a secreted hormone that has physiological actions of its own can also be a biosynthetic precursor for another steroid hormone. The relationships

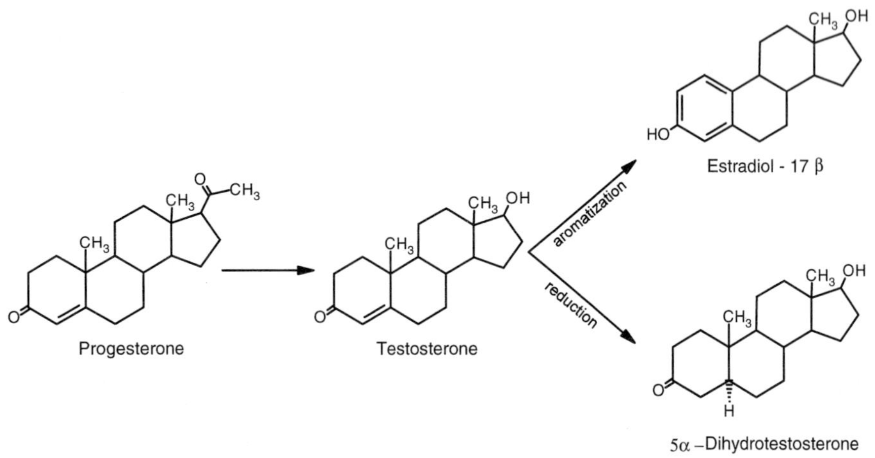

FIGURE 1 Representation of the chemical structures and biosynthetic relationships of the primary mammalian reproductive steroids.

among the steroid hormones that are primary regulators of copulation in males and females are depicted in Fig. 1.

A. Hormonal Regulation in Females

Removal of the ovaries is typically associated with the loss of copulatory function in female mammals. Which ovarian hormones are necessary to restore copulation differs among species, however. Because the expectation is that copulation in females should coincide with ovulation, it is reasonable to assume that the pattern of steroid hormone release leading up to ovulation triggers the display of sexual behavior in females. However, for spontaneously ovulating females there are two general patterns (estrous and menstrual cycles) of secretion of estrogen (e.g., estradiol) and progesterone (Fig. 2).

For females with estrous cycles there is a sustained release of estradiol followed by a brief pulse of progesterone leading to ovulation. Consistent with our expectations, this sequence of estradiol and progesterone secretion activates the expression of copulation in females. For females of some species (e.g., rats and hamsters) estradiol treatment alone can stimulate low levels of copulatory activity, though the combination of estradiol and progesterone is optimum for maximal expression of copulation. In contrast, females with menstrual cycles rely on the sustained secretion of estradiol (estrone in some species)

for the activation of mating, and in these species progesterone either has no effect on copulation or in fact will inhibit the expression of mating behavior.

Females of some mammalian species (reflex ovulators) do not ovulate spontaneously, but rather in these females ovulation is triggered by vaginal stimulation from the penis during the male's copulatory

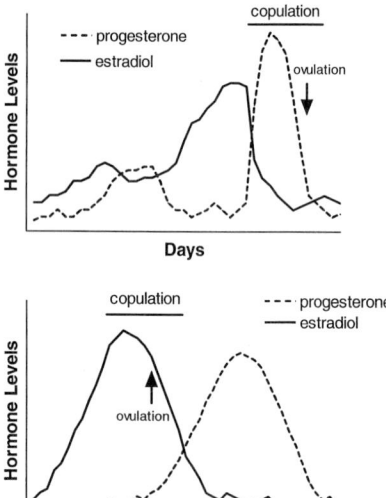

FIGURE 2 Relative blood levels of estradiol and progesterone during model estrous (top) and menstrual (bottom) cycles and the relation of these hormone fluctuations to the timing of ovulation and copulation. In this depiction the estrous cycle lasts about 4 days and the menstrual cycle about 3 or 4 weeks.

attempts. In this case estradiol increases the female's acceptance of the male's copulatory attempts, permitting the coincident delivery of sperm with ovulation.

B. Hormonal Regulation in Males

For all mammalian species castration reduces copulatory ability. Ejaculatory patterns disappear first, with sexual arousal and mounting patterns persisting for longer time periods after castration. The great deal of interspecies variability in the time course of this reduction in copulatory potential is interesting. Rats may take a few months for copulation to disappear, whereas other species, notably dogs and goats, may maintain copulatory activity for years after castration. Individuals within a species also demonstrate considerable variability in their sensitivity to castration, with individual rats maintaining considerable copulatory activity from 1 to 6 months after castration. Similarly, some rhesus monkeys may have severe copulatory impairments a few months after castration, with other monkeys maintaining copulation for a year or longer.

Testosterone replacement will maintain copulation at normal levels if administered soon after castration and will restore copulation to normal levels even if treatments are initiated at a time after castration when copulatory activity has been lost. Though we think of testosterone as activating copulation, testosterone acts primarily as a prohormone for more potent steroids. In many cases the conversion of testosterone to estradiol (termed "aromatization") is a requirement for copulatory activity to be maintained, and neurons in the preoptic area and hypothalamus have the capacity to enzymatically convert testosterone to estradiol. In other instances testosterone is reduced to dihydrotestosterone, which is the active steroid for the maintenance of copulation. Dihydrotestosterone can be formed in peripheral tissues, but again neural regions have the capacity to biotransform testosterone as well.

C. Location of Steroid Responsive Neurons

Gonadal steroids can effect actions on nerve cells through intracellular receptors that interact with the genome or through membrane receptors that affect signal transduction pathways. The distribution of intracellular receptors and their role in copulation have been characterized to a much greater extent than similar properties of membrane receptors. There are few differences between males and females in the location of neuronal regions that contain cells with intracellular androgen or estrogen receptors. Generally, these neural regions form a medial band in the basal forebrain. There is consistency among mammals in brain regions containing androgen receptors. Areas of special interest in the regulation of copulation include the medial preoptic area, the bed nucleus of the stria terminalis, and the ventromedial nucleus of the hypothalamus. The anterior hypothalamus of primates also contains androgen receptors. There is also evidence of cells containing androgen receptors outside this band, especially within limbic structures. For example, the lateral septum (though not in primates) and amygdaloid areas, including the medial and cortical nuclei, show evidence of androgen receptors. In primates, the medial nucleus of the amygdala does not necessarily contain androgen receptors, though an accessory nucleus medial to the cortical nucleus does and may represent a homolog of the medial amygdala receptor population in rodents. In the spinal cord, androgen receptors are restricted to certain pools of motor neurons.

The distribution of brain regions containing estrogen receptors is similar to that of androgen receptors though somewhat more widespread. Again with respect to regions relevant for copulation, the lateral septum, medial preoptic area, ventromedial nucleus of the hypothalamus, cortical nucleus of the amygdala, and medial amygdala (accessory amygdaloid nucleus in primates) are primary regions containing estrogen receptors. In the spinal cord estrogen receptors are mainly found in the dorsal horn, suggesting a sensory role for these receptors in copulation.

Progestin receptors have the most restricted distribution in the brain of all the steroid receptors. An interesting characteristic of intracellular progestin receptors is their regulation by estrogen. In the absence of estradiol, either during the female's cycle or following surgical removal of the ovary, progestin receptors are present at nearly undetectable levels. With the rise in estrogen levels either naturally or by estrogen replacement, there is a dramatic induction of progestin receptors, particularly in the medial preoptic area and ventromedial nucleus of the hypothalamus. This induction of progestin receptors is apparent in both males and females, though to a somewhat greater degree in females. Progestin receptors are not found in the spinal cord.

III. NEURAL REGULATION

The dependence of copulation in both male and female mammals on stimulation by gonadal hormones provided a convenient starting point for examining functional neural pathways subserving the expression of sexual behavior. Implants of steroid hormones directly into the brain of gonadectomized animals indicated that the ventromedial nucleus of the hypothalamus was the most responsive brain site for the hormonal activation of sexual behavior in females, whereas the medial preoptic area was the most sensitive site for the hormonal activation of sexual behavior in males. By following neural interconnections between these sites of hormone action and other brain and spinal cord neuronal regions, researchers were able to increase our understanding of comprehensive pathways controlling copulation in mammals. A schematic overview of the pathways mediating copulation in males and females is presented in Fig. 3.

A. Male Sexual Behavior Pathways

The importance of the medial preoptic area for the expression of male sexual behavior has been confirmed for all mammals studied to date. The medial preoptic area has the ability to integrate sensory, hormonal, and autonomic information, permitting the male to copulate. The medial and cortical nuclei of the amygdala project directly to the medial preoptic area as well as indirectly via the bed nucleus of the stria terminalis. The amygdala is well placed to receive auditory, olfactory, and tactile information relevant to the arousal and performance of copulation. Olfactory information is the most accessible with monosynaptic inputs from the olfactory bulbs. It appears that these inputs to the medial preoptic area are one pathway through which males are aroused sexually. Efferents of the medial preoptic area permit the male to act on this sexual arousal through interactions with ventral midbrain regions as well as with brain stem areas, such as the nucleus paragigantocellularis. These regions in turn send information to the spinal cord to integrate thrusting patterns with penile erectile responses.

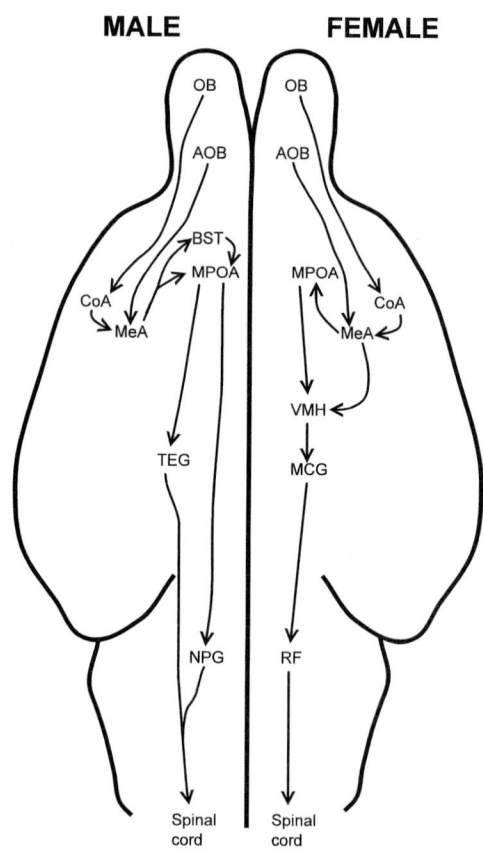

FIGURE 3 Schematic illustration of some of the neural pathways controlling copulation in male and female mammals. Pathways regulating male copulatory behavior are represented on the left side of the brain and pathways for female copulatory behavior on the right side. The brain structures are positioned within a horizontal (overhead) view of a rodent brain. AOB, accessory olfactory bulb; BST, bed nucleus of the stria terminalis; CoA, cortical nucleus of the amygdala; MCG, midbrain central gray; MeA, medial nucleus of the amygdala; MPOA, medial preoptic area; NPG, nucleus paragigantocellularis; OB, olfactory bulb; RF, reticular formation; TEG, midbrain tegmentum; VMH, ventromedial nucleus of the hypothalamus.

TABLE 1
Overview of the Effects of the Major Neurotransmitter Systems on Copulation in Males and Females

Neurotransmitter system	Male copulation	Female copulation
Norepinephrine	α_1 activation	α_1 activation
	α_2 inhibition	
Dopamine	Postsynaptic activation	Postsynaptic activation
	Presynaptic inhibition	Presynaptic inhibition
Serotonin (5-HT)	5-HT_{1A} activation	5-HT_{1A} inhibition
	5-HT_2 inhibition	5-HT_2 activation
Acetylcholine	Muscarinic activation	Muscarinic activation
γ-Aminobutyric acid (GABA)	GABA_A inhibition	GABA_A activation
Endogenous opioids	Inhibition	Inhibition

B. Female Sexual Behavior Pathways

In all mammalian species studied, damage to the ventromedial nucleus of the hypothalamus eliminates the expression of copulation. Forebrain structures, such as the cerebral cortex, septum, medial preoptic area, and olfactory bulbs, typically inhibit the expression of copulation in females. Sensory information can serve to counteract the inhibitory actions of these forebrain areas. In addition, gonadal hormones both disinhibit and activate the ventromedial hypothalamus, which conveys efferent information to midbrain structures such as the central gray. Further activation of brain stem reticular formation and spinal pathways activate skeletal muscles necessary for postural and copulatory movements.

C. Neurotransmitter Control of Copulation

Alterations in neurotransmitter systems are one means through which gonadal hormones activate copulation in males and females. Androgens, estrogens, and progestins may either increase or decrease activity in neurotransmitter systems consistent with the excitatory or inhibitory actions of these neurotransmitters on copulation. Not surprisingly, there are differences in the ways in which individual neurotransmitter systems affect copulation in males and females. This issue is further complicated by observations of neurotransmitter receptor subtypes that may have antagonistic actions and by the fact that a single neurotransmitter may function differently depending on the location within the nervous system. Still, it is possible to make some generalizations about the actions of some neurotransmitter systems on copulation, and these findings are summarized in Table 1.

IV. SYNCHRONIZATION OF COPULATION

Because there is an energetic cost of reproduction for both males and females, the necessity of integration of copulatory activity between male and female mammals is widely appreciated, though rarely studied. Within some species there is coincident developmental onset of copulatory behavior in males and females. Also, for those species with defined annual breeding seasons, males and females have coordinated fluctuations in reproductive capacity including the expression of copulation. How a behavioral system such as copulation, which requires the orchestration of environmental events with social structure and individual associations, and in turn with the individual's physiology (including molecular events within individual cells), is regulated is a question that continues to generate research relevant to a wide range of disciplines within the field of reproduction.

See Also the Following Article

MATING BEHAVIORS, MAMMALS

Bibliography

Meisel, R. L., and Sachs, B. D. (1994). The physiology of male sexual behavior. In *The Physiology of Reproduction* (E. Knobil and J. D. Neill, Eds.), 2nd ed. Raven Press, New York.

Pfaff, D. W., Schwartz-Giblin, S., McCarthy, M. M., and Kow, L.-M. (1994). Cellular and molecular mechanisms of female reproductive behaviors. In *The Physiology of Reproduction* (E. Knobil and J. D. Neill, Eds.), 2nd ed. Raven Press, New York.

Corpora Lutea of Nonmammalian Species

Giovanni Chieffi and Gabriella Chieffi Baccari
The University of Naples

I. Introduction
II. Cyclostomes
III. Elasmobranchs
IV. Teleosts
V. Amphibians
VI. Reptiles
VII. Birds

GLOSSARY

corpus albicans The white fibrous scar in an ovary produced by involution of the corpus luteum.

corpus atreticum (plural, *corpora atretica*; synonyms, *preovulatory corpus luteum, atretic follicle*) Degenerating structure occurring in all vertebrates. It develops from the ovarian follicles which do not undergo ovulation.

corpus luteum (plural, *corpora lutea*; synonym, *postovulatory follicle*) Ephemeral gland occurring in all vertebrates. This structure develops in the ovary after ovulation from the membranes of postovulatory follicle.

oviparous species Animals producing eggs that develop and hatch outside the body.

ovoviviparous species Animals producing yolky (telolecithal) eggs that develop internally.

viviparous species (of most mammals) Animals producing oligolecithal eggs that develop within the mother's body.

Corpora lutea of nonmammalian vertebrates are steroidogenic and secrete mainly progesterone as in mammals. Unfortunately, lack of experimental data (i.e., oophorectomy or extirpation of corpora lutea), despite many attempts, does not allow us to precisely define their role in many subgroups. However, the numerous histological, ultrastructural, histochemical, and biochemical observations strongly suggest an endocrine steroidogenic role of the corpus luteum in the nonmammalian vertebrates. It is likely that corpus luteum and progesterone are primarily involved in the retention of eggs and in the downregulation of hepatic vitellogenin synthesis as a corollary of viviparity and placental development.

I. INTRODUCTION

Nonmammalian vertebrates develop structures comparable to the luteal tissue of mammals. Such structures originate from the transformation of the ovarian follicles either before or after ovulation. They have been called, respectively, preovulatory corpora lutea, atretic follicles or corpora atretica, and corpora lutea, postovulatory corpora lutea, or postovulatory follicles. The former develop from the follicular epithelium (granulosa) following the atresia of the oocyte, and the latter develop through the invasion of the empty follicle by the granulosa and theca cells. In this article, the term "corpus luteum" will be used aside from its origin. Although the formation of luteal tissue is a regular phenomenon in the ovary of nonmammalian species, the endocrine nature as well as the function of these structures are still a matter of controversy.

II. CYCLOSTOMES

The question of the formation of luteal-like structures in cyclostomes is still unsettled. Processes of follicular atresia accompanied by hypertrophy of the follicular cells have been described in *Lampetra planeri*. In the smaller oocytes phagocytosis is carried out by the inwardly migrated theca cells. Once the process of yolk resorption is over, the hypertrophy of follicular cells declines rapidly. The postovulatory follicles also follow the same course of events. In *Lampetra*, the occurrence of atresia seems to be responsible exclusively for the resorption of oocytes, and the atresia coincides with the stages of previtellogenesis and the beginning of vitellogenesis. Pre- and postovulatory corpora lutea showing signs of secretory activity have been observed in *Myxine*. Postovulatory follicles are of two kinds: Some are described as solid masses of tissues derived from the follicle cells, whereas others are fluid-filled cysts.

3β-hydroxysteroid dehydrogenase (3β-HDS) activity has been reported in the follicular epithelium of *Lampetra fluviatilis*. This activity reaches a maximum in February and early March when granulosa cells reach their maximal size. No steroidogenic activity has been described in the thecal cells. However, ultrastructural observations in *L. planeri* ovary revealed the presence of organelles characteristic of steroidogenic tissues in the thecal cells but not in the follicular epithelium. In the postovulatory follicles of *Lampetra* only small quantities of cholesterol are detected by the Schultz test.

III. ELASMOBRANCHS

The formation of corpora lutea, following ovulation and as a result of atresia, has been described in numerous representative species of this class. The preovulatory corpora lutea develop from the ovarian follicles often in the stage of advanced vitellogenesis, following the phagocytosis of yolk. At full development they are characterized by hypertrophy of the granulosa cells, whereas the theca interna forms the connective stromal tissue. The formation of preovulatory corpora lutea has been studied at the electron microscopic level in the ovoviviparous species, *Torpedo marmorata*. The course of atresia may be divided into four stages. The first two comprise the dissolution of the oocyte and its phagocytosis by the small cells of the granulosa epithelium (Figs. 1A and 1B). The third stage consists of the transformation of the granulosa epithelium into an active glandular structure and is accompanied by the development of smooth endoplasmic reticulum (Fig. 1C). The fourth stage is marked by sclerosis and pigmentary degeneration of the atretic follicle (Fig. 1D). In the preovulatory corpora lutea of *T. marmorata* and *Torpedo ocellata* (ovoviviparous spp.), the Schultz reaction gives positive results; moreover, *in vitro* studies have demonstrated that these structures of *T. marmorata* synthesize steroid hormones. Furthermore, increased plasma progesterone levels in pregnant *T. marmorata* may account for both the lengthening of the uterine folds and the inhibition of oogenesis. Together, these observations suggest an endocrine steroidogenic role for the corpora atretica in *T. marmorata*.

In the majority of elasmobranchs, the postovulatory follicle gives rise to a corpus luteum in oviparous, ovoviviparous, and viviparous species but not in the viviparous species showing follicular gestation. The corpora lutea are short-lived in oviparous species but persist longer in live-bearing forms. Despite the fact that the formation of corpora lutea is a regular phenomenon in the ovary of selachians, these structures exhibit differences both in structure (cavitary or compact corpora lutea) and in cellular architecture. This variability is not associated with either phylogenetic affinities or mode of reproduction. At their full development, immediately after ovulation, the corpora lutea are characterized by hypertrophy of the granulosa cells and by their transformation in luteal cells characterized by a well-developed smooth endoplasmic reticulum (SER) and a prominent Golgi apparatus. Granulosa cells may or may not fill up the follicular cavity. In some ovoviviparous species (*T. marmorata* and *Squalus acanthias*), the thecal tissue forms a distinct sheath surrounding the central granulosa luteal cells. In some oviparous and vivipa-

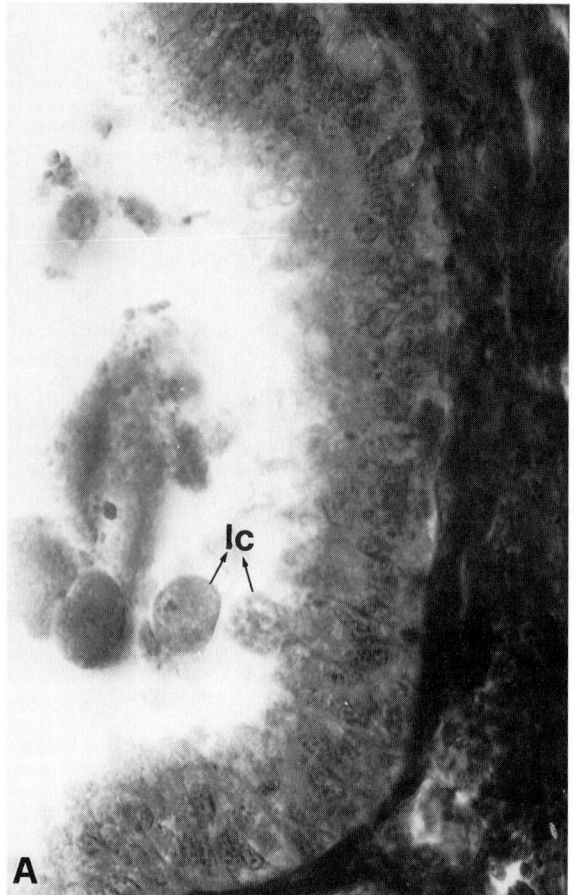 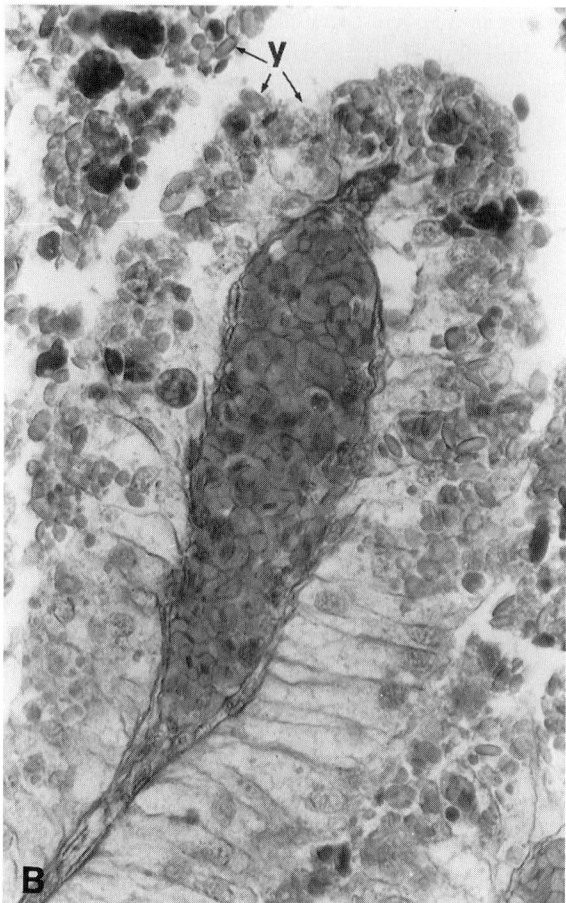

FIGURE 1 Torpedo marmorata. (A) Photomicrograph of the stage I atretic follicle showing the folding of the follicular wall and the degeneration of the large granulosa cells (lc). (B) Stage II atretic follicle showing the phagocytosis of the yolk globules by the elongated small granulosa cells. y, yolk. (C) Stage III atretic follicle (preovulatory corpus luteum) showing the richly vascularized glandular structure. Note the change in polarity of the glandular cells by the migration of the nucleus to the apex of the cell. c, capillaries. (D) Stage IV atretic follicle showing the sclerosis of the glandular vascular net. Magnification, ×250 (from Chieffi Baccari et al., 1992).

rous species (Rhinobatus granulatus, Scyliorhinus stellaris, and several species of genus Raja), both granulosa and thecal cells hypertrophy after ovulation and form the luteal cell mass. Also in the viviparous species, such as Mustelus canis, both granulosa and thecal cells hypertrophy but the latter remain located in the thecal tissue. Because the blood vessels are always located only in the thecal layer, the vascularization of the granulosa luteal cell mass is dependent on the behavior of the thecal tissue, which may or may not invade the follicular cavity along with the granulosa cells after ovulation.

When the involution of the corpora lutea is advanced, it becomes impossible to distinguish between atretic follicles and postovulatory follicles. Both structures show many similarities, such as a theca-derived outer layer, lipid-rich cytoplasm of granulosa-derived cells, minimal contribution of the theca to the central part of the structure, cellular debris, and an abundance of invading blood cells in the central lumen.

Glucose-6-phosphate dehydrogenase (G-6-PDH) and 3β-HSD activities have been demonstrated in the granulosa cells of some species of elasmobranchs.

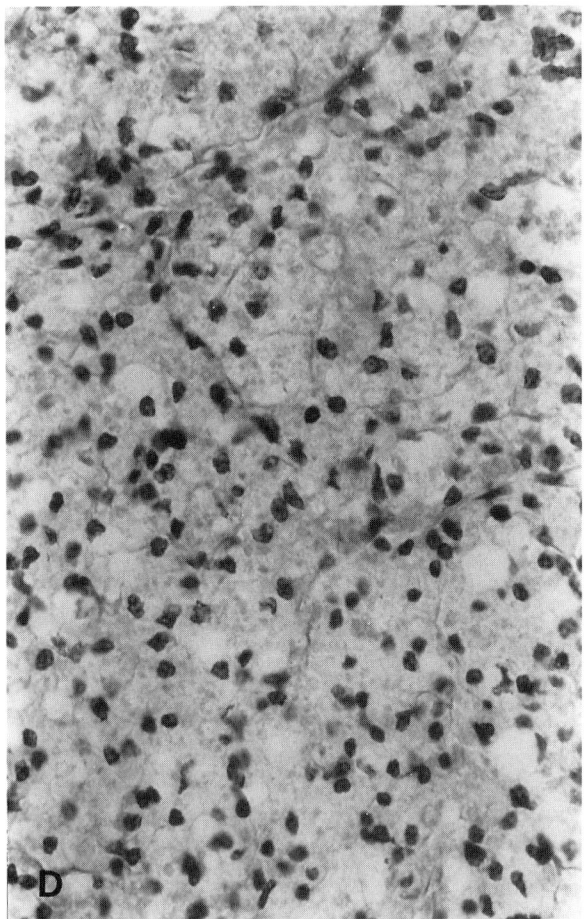

FIGURE 1 (continued)

The increased activity of both enzymes with the growth of the oocytes suggests that the follicular epithelium could be the site of estrogen biosynthesis. 3β-HSD activity has been found in the atretic follicles of *T. marmorata* but not in those of *S. stellaris*. In vitro studies have definitely demonstrated that the luteal tissue from the shark, *S. acanthias*, and from an oviparous species, *Raja erinacea*, synthesizes progesterone (Fig. 2). Interestingly, the corpus luteum of *S. achantias* is the longest-lived corpus luteum of pregnancy on record (2 years), including all known mammals.

There is little information concerning the existence of luteotropic factors in elasmobranchs. In *M. canis* (viviparous species), the corpus luteum could be formed and maintained independently of the pituitary, whereas in *S. acanthias* the corpus luteum secretes progesterone in response to pituitary hormones.

IV. TELEOSTS

The formation of pre- and postovulatory corpora lutea has been described in numerous oviparous and viviparous teleostean species. While in the majority of oviparous teleosts the postovulatory follicles degenerate and are rapidly resorbed, in some species they transform into well-organized, although in certain cases transitory, corpora lutea. No glandular structures similar to corpora lutea are formed in the viviparous species that show follicular gestation.

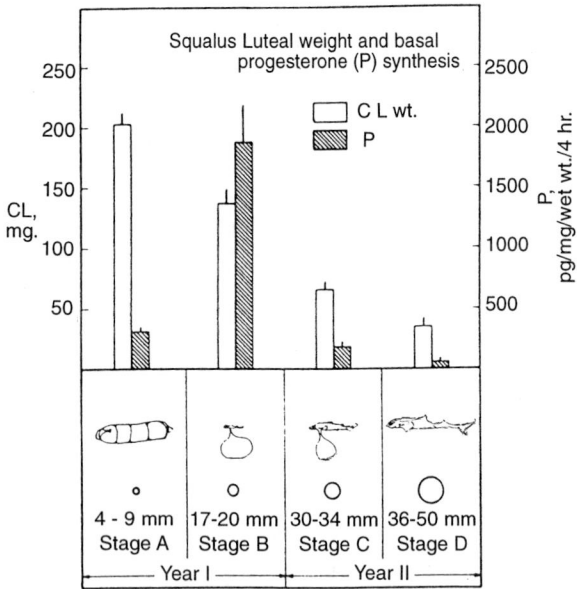

FIGURE 2 Luteal weights and basal progesterone production in *Squalus acanthias* during the reproductive cycle. Individual corpora lutea taken from pregnant animals at different stages of pregnancy were weighed and tissues from similar stages were scissors minced. Equal aliquots were dispensed into culture wells containing 1.0 ml Eagle's basal medium supplemented with urea and glutamine and incubated for 4 hr at 18°C. The amount of progesterone (P) in the medium was determined by radioimmunoassay and expressed as pg/mg corpus luteum (CL)/4 hr ± SE (Bottom) Correlated stages of follicular size in mm and embryonic–fetal development during pregnancy (from Tsang and Callard, 1987).

Two basic types of postovulatory corpora lutea can be distinguished in teleosts. In the first type the luteal cells derive only from hypertrophic granulosa cells. The thecal tissue either forms a distinct sheath surrounding the central granulosa luteal cells (*Notopterus notopterus, Merluccius merluccius, Oryzias latipes*, and *Perca fluviatilis*) or invades the follicular cavity with the hypertrophied granulosa cells and supplies the connective tissue [*Tor* (= *Barbus*) *tor*]. In the second type, both theca and granulosa hypertrophy and form the luteal cell mass (*Clarias batrachus*, an oviparous species). In a large number of oviparous teleosts (*Scomber scomber, Cichlosoma nigrofasciatum, Haplochromis multicolor, Cyprinus carpio, Brachydanio rerio, Carassius auratus, Oncorhynchus kisutch, O. gorbuscha, Salmo gairdneri*, and *Salvelinus leucomaenis*) the hypertrophic thecal cells remain separated from the granulosa luteal cells.

In teleosts, "special" thecal cells have been described in proximity of blood capillaries. These cells show ultrastructural characteristics of steroid-producing cells such as mitochondria with tubular cristae and an extensive SER.

Although atretic follicles have been observed in a number of teleostean fishes (*S. scomber, Mystus seenghala, Salvelinus fontinalis, Heteropneustes fossilis, Xenentodon cancila, Gobius giuris*, and *Sebastodes paucispinis*), they do not evolve into glandular structures. During the atretic process the granulosa cells become hypertrophic and phagocyte the yolk. Theca cells and invading blood cells rarely participate in the resorption of yolk and/or the debris of atretic follicles. Studies carried out in *Poecilia reticulata*, a viviparous species, indicated that, prior to follicular atresia, the granulosa cells are a possible source of synthesis of steroid hormones. Postovulatory corpora lutea are not formed in this species. The atretic follicles appear to have no influence on a normal pregnancy.

The localization of steroid biosynthesis varies considerably among species of teleosts studied to date. A strong reaction for 3β-HSD, 17β-HSD, and G-6-PDH has been found in the granulosa cells of the guppy, *P. reticulata*. In the preovulatory follicles of *S. leucomaenis*, neither granulosa cells nor special thecal cells showed 3β-HSD activity. Weak 3β-HSD activity has been demonstrated in the granulosa cells of *S. scomber*; the activity is strongest in some thecal cells at the beginning of vitellogenesis. In *Mystus cavasius*, the activity of some steroidehydrogenases (3β-, 17β-, and 11β-HSD) and of G-6-PDH has been demonstrated in hypertrophied granulosa cells of early atretic follicles. In later stages only thecal cells remain in the ovarian stroma, and they show a strong enzymatic activity. 3β-HSD activity has been found in both the granulosa and the thecal cells of *Acanthobrama terraesanctae* and *Sarotherodon niloticus*. In vitro studies with thecal and granulosa cells from the amago salmon, *Onchorhynchus rhodurus*, suggested that thecal cells contribute to estradiol production by synthesizing testosterone, which is transferred to the granulosa layer and aromatized to estradiol.

Histochemical studies have demonstrated 3β-HSD activity in the granulosa cells of the postovulatory follicles of *O. latipes* and *B. rerio*, in the thecal cells of *S. leucomaenis*, and in both granulosa and thecal cells of *S. gairdneri*, *S. scomber*, *Trachurus mediterraneus*, and *C. auratus*. In addition to 3β-HSD, 11β-HSD and 20α-HSD activities have been recorded in *T. mediterraneus*. Activity of 3β-HSD usually parallels the activity of G-6-PDH in *T. mediterraneus*, *S. gairdneri*, and *C. batrachus*. Ultrastructural studies have confirmed the steroid-synthesizing capacity of granulosa and thecal cells in *C. auratus*, *B. rerio*, *S. gairdneri*, *C. nigrofasciatum*, and *H. multicolor*. In these species, postovulatory follicles are equipped with a smooth endoplasmic reticulum and mitochondria with tubular cristae.

There are very few indications of hormonal control of luteal activity in the teleosts. In salmonids, pituitary gonadotropin seems to be responsible for the production of progesterone and testosterone, but not of estradiol, by corpora lutea. In the common carp, *C. carpio*, the "switch off" of aromatase activity seems to be regulated by an ovarian rather than a pituitary factor but the mechanism by which the switch operates remains unclear. In teleosts, the functional luteal phase lasts up to a few days. Involution of the luteal cells (both granulosa and special thecal cells) is characterized by cell organelle degradation and an increase of lysosome-like bodies.

V. AMPHIBIANS

The development of postovulatory corpora lutea has been observed in the ovary of both oviparous and viviparous species of anuran and urodele amphibians. In oviparous species, postovulatory follicles are rapidly resorbed, but they are more persistent in ovoviviparous and viviparous species.

In oviparous amphibians, ovulation is immediately followed by the hypertrophy of granulosa cells and thecal cells (in that order). In some oviparous (*R. esculenta*, *Triturus cristatus*, *R. catesbiana*, *R. cyanophlyctis*, and *R. verrucosa*) and viviparous species (*N. occidentalis*), the granulosa becomes multilayered and fills the cavity of the follicle. The resulting corpus luteum shows enlarged thecal cells on the outside and a granulosa cell mass in the center. Other species, such as the oviparous marsupial frog, *Gastrotheca riobambae*, and the viviparous salamander, *Salamandra salamandra terrestris*, develop cavitary corpora lutea since the granulosa cells never fill the follicular cavity. In the viviparous apodan, *Typhlonectes compressicaudes*, the granulosa cells proliferate and form the luteal cell mass which is invaded by thecal cells, accompanied by capillary blood vessels.

3β-HSD has been demonstrated in granulosa as well as thecal cells of numerous species of anurans and urodeles. In *Xenopus laevis*, 17α-, 17β-, 3α-, and 3β-HSD activities are confined to the granulosa cells. Ultrastructural features of steroid-secreting cells have been demonstrated in the granulosa but not in the thecal cells in *Necturus maculosus*, *S. salamandra*, and *Bufo bufo*.

After ovulation, the granulosa cells of the postovulatory follicle show a strong histochemical reaction for 3β-HSD activity and a positive Schultz test for cholesterol. Weak 3β-HSD activity has been demonstrated in the thecal cells of the postovulatory follicles of *S. salamandra*. Ultrastructural studies have been conducted on the postovulatory follicles of only two species. The granulosa cells of *N. maculosus* and *S. salamandra* contain a SER, mitochondria with tubular cristae, a well-developed Golgi apparatus, and lipid droplets—all features characteristic of steroidogenic cells.

The postovulatory follicles are transitory in oviparous amphibians and generally have been considered to have no functional significance. However, corpora lutea of the skipper frog, *R. cyanophlyctis*, are steroidogenic and last as long as corpora lutea of mice. In the case of ovoviviparous and viviparous urodeles, the possible functional role of the postovulatory corpora lutea is still debated. It has been demonstrated that ovariectomy performed in the viviparous *S. salamandra* at the beginning of pregnancy does not interrupt the intrauterine development of the larvae. It is also reported that ovariectomy of *N. occidentalis* at the end of gestation does not cause abortion, whereas ovariectomy performed at the beginning of gestation in females experiencing their first gestation leads to abortion. This finding has been interpreted in terms of low levels of progesterone secreted by the ovaries in females undergoing first gestation.

VI. REPTILES

In contrast with other vertebrates, all oviparous and viviparous reptiles studied exhibit postovulatory corpora lutea. Luteal morphology is relatively consistent among the squamates (lizards and snakes) and chelonians, showing morphological characters very similar to those observed in mammals. There is a strong correlation between luteal life and the period of ovulation or, in viviparous species, of pregnancy. At the end of the ovulatory period or of pregnancy the corpora lutea regress rapidly.

After ovulation, the granulosa cells of the collapsed follicle hypertrophy, filling the empty follicle, and form a central luteal mass surrounded by a relatively thin theca interna and theca externa. As observed in other lower vertebrates, in reptiles the theca interna and vascularization contribute in different ways to corpus luteum development. This variability cannot be associated with either phylogenetic affinities or reproductive mode of the species. In most reptilian species only the granulosa cells are responsible for the formation of luteal tissue, which possesses a more or less well-defined vascular system. The thecal tissue remains distinct from luteal cell mass. In other species, cells of the theca interna contribute to the composition of the luteal cell mass. The hypertrophied theca interna cells either remain separate from the granulosa luteal cells, as in *Naja naja*, or combine with the granulosa luteal cells mass, as in *Uromastix hardwicki* and *Lacerta vivipara*, in which thecal luteal cells are organized in superficial small clusters. In these species, the corpus luteum is well vascularized and closely resembles those of mammals.

Unique among reptiles is the luteal morphology of the American alligator (*Alligator mississippiensis*). As in birds, the luteal cells of alligators exhibit little hypertrophy and hyperplasia so that the granulosa never fills the postovulatory follicle. Unlike those of most other reptiles, the central luteal cell mass is composed of cells derived either from the granulosa or from theca interna. Both cell types are present throughout gestation, but only one cell type is seen during mid- to late luteolysis.

Among reptiles, phenomena of follicular atresia are less frequent. The granulosa cells, which are generally hypertrophic, are involved in yolk resorption; the thecal cells participate in the process of follicular atresia in very advanced stage of vitellogenesis. In *Zootoca vivipara*, after resorption of the yolk the granulosa cells are transformed into lutein cells.

Activity of 3β-HSD has been detected in both the granulosa and thecal cells of all oviparous and viviparous reptiles studied. 17β-HSD activity has been found in postovulatory follicles of *Sceloporus cyanogenys* and *Dipsosaurus dorsalis* but not in those of *Lacerta sicula*. In some cases the follicular components show other cytological features of steroidogenesis.

In reptiles, the physiological significance of postovulatory corpora lutea is unclear. However, numerous reports on plasma progesterone levels in oviparous and viviparous species suggest that corpora lutea secrete progesterone. In some species the corpus luteum is able to produce androgens and estrogens as well. Regarding the patterns of progesterone production, it is possible to distinguish two basic types of corpus luteum. The first type occurs in chelonia, in which a preovulatory or an ovulatory progesterone surge is observed; the corpora lutea, in comparison to preovulatory follicles, are relatively poorly steroidogenic in the ovigerous phase. The second type is exhibited in squamata, in which plasma progesterone peaks either shortly after ovulation (oviparous species) or at mid- or even late pregnancy (viviparous species). In squamata, the persistence of elevated progesterone levels suggests that the corpora lutea remain active up to near the time of oviposition or parturition; afterwards the corpus luteum undergoes rapid regression, and plasma progesterone levels dramatically decline. A positive correlation exists between plasma progesterone concentrations and histological activity of corpora lutea in all species of reptiles, except *Sceloporus jarrovi*.

The American alligator exhibits a different pattern of plasma progesterone concentration that is unique for a nonsquamate reptile. Plasma progesterone levels are elevated throughout gravidity but decline sharply and significantly after oviposition.

Little is known about the regulatory mechanisms of corpus luteum secretory activity. In oviparous reptiles, oviposition coincides with a dramatic increase of prostaglandin F (PGF) and prostaglandin E_2 (PGE_2). In *Podarcis sicula sicula* $PGF_{2\alpha}$ seems to induce luteolysis, whereas PGE$_2$ maintains the corpus luteum. Recent data indicate that salmon gonadotropin-releasing hormone and substance P favor the conversion of PGE$_2$ to PGF$_{2\alpha}$ in *P. sicula sicula* corpus luteum.

VII. BIRDS

The development of a corpus luteum from the postovulatory follicle in birds, even if transitory, is a common feature. The histological changes occurring during the life span of the postovulatory follicle have been described in the pigeon. They may be classified into four stages. Initially, the newly formed postovulatory follicles shrink in size due to contraction of the theca. The granulosa cells hyperthrophy. The membrana propria, located between the theca interna and granulosa layers, is thin. The theca interna has an inner cellular and an outer fibrous layer. Blood capillaries are present in both thecal layers. Stage two consists of the further hypertrophy of granulosa cells which become multilayered and take a syncytial appearance. The membrana propria now appears thick but broken at many places. Vacuoles appear in granulosa and theca interna cells. The theca externa becomes more fibrous. Erythrocytes invade the granulosa layers as well as the follicular lumen due to hemorrhage of thecal capillaries. The third stage is marked by a further shrinkage of the postovulatory follicle narrowing the lumen, which is now filled with the granulosa lutein cells. These often have pigment in the cytoplasm. The membrana propria begins to fragment and gradually disappears. The fibroblasts invade the theca interna and the granulosa lutein mass. In the fourth stage, the postovulatory follicle is greatly reduced in size. The granulosa cells have pycnotic nuclei and an extremely vacuolated cytoplasm. The thecal layers are not distinguishable as separate structures. When the regression is complete, the remnants of the postovulatory follicles become part of the ovarian stroma.

In the hen, the postovulatory corpus luteum derives mainly from the hypertrophic thecal cells. Similarly, in the sparrow and in fowl both thecal and granulosa cells contribute to the formation of the corpus luteum. Furthermore, granulosa luteal and thecal cells show a complete intermixing in later stages.

In the cells of granulosa and thecal layers, the presence of cholesterol and an intense activity of 3β-HDS have been observed, whereas 17β-HSD, 11β-HSD, and G-6-PD enzyme activities occur principally in granulosa cells. Ultrastructural studies carried out on the hen postovulatory follicles have shown that the granulosa cells possess structural characteristics of steroid-secreting cells, i.e., a SER and numerous lipid droplets. It has been demonstrated that the postovulatory follicles in the hen contain progesterone and estrogens.

Follicular atresia is very frequent in the ovary of birds and may affect ovarian follicles at all developmental stages. Atresia may take place according to a variety of patterns. In the larger follicles, in general, burst atresia occurs. The yolk is extruded through the granulosa into the ovarian stroma, where it is phagocyted. In the final stage of atresia, once the yolk is resorbed, only a few vacuolized cells, some capillaries, and connective tissue bands occupy the site of the resorbed oocytes. The more frequent form of atresia in *Corvus frugilevus* is the "lipoidal atresia" in which steatosis of the egg is followed by the proliferation of the granulosa and thecal layers, which subsequently remove the yolk. In other cases, the oocyte is resorbed by a mass of cells derived from the hyperplasia of the granulosa; these cells migrate into the oocytes which are not subjected to steatosis. Eventually, the massive degeneration of the atretic follicle results in the formation of hyaline areas resembling the corpus albicans, which are invaded by cells of unknown origin. They are cholesterol positive for a short period.

Cholesterol is abundant in the thecal tissue of atretic follicles and during the initial stages of atresia in the granulosa cells. In both the granulosa and thecal layers intense activity of steroid 3β-HSD is demonstrable, whereas 17β-HSD activity is shown only in the granulosa layer. A weak reaction for 3β-HSDH, 17β-HSDH, 11β-HSDH, and G-6-PDH has been demonstrated in the granulosa cells of the preovulatory follicle in the pigeon. The presence of both cholesterol and some steroidogenic enzymes in the atretic follicles indicates a potential synthesis of steroid hormones in these structures. However, conclusive evidence that the atretic follicles of birds represent preovulatory corpora lutea is not available.

See Also the Following Article

CORPUS LUTEUM

Bibliography

Callard, I. P., Klosterman, L. L., Sorbera, L. A., Fileti, L. A., and Reese, J. C. (1989). Endocrine regulation of reproduction in elasmobranchs: Archetype for terrestrial vertebrates. *J. Exp. Zool. Suppl.* **2**, 12–22.

Chieffi, G., and Botte, V. (1970). The problem of "luteogenesis" in non-mammalian vertebrates. *Boll. Zool.* **37**, 85–102.

Chieffi, G., and Pierantoni, R. (1987). Regulation of ovarian steroidogenesis. In *Hormones and Reproduction in Fishes, Amphibians, and Reptiles* (D. O. Norris and R. E. Jones, Eds.), pp. 117–144. Plenum, New York.

Chieffi Baccari, G., Minucci, S., Di Matteo, L., and Chieffi, G. (1992). Ultrastructural investigation of Corpora atretica of the electric ray, *Torpedo marmorata*. *Gen. Comp. Endocrinol.* **86**, 72–80.

Guillette, L. J., Woodward, A. R., You-Xiang, Q., Cathy Cox, M., Matter, J. M., and Gross, T. S. (1995). Formation and regression of the corpus luteum of the American alligator (*Alligator mississippiensis*). *J. Morphol.* **224**, 97–110.

Jones, R. E., and Baxter, D. C. (1991). Gestation, with emphasis on corpus luteum biology, placentation, and parturition. In *Vertebrate Endocrinology: Fundamental and Biomedical Implications* (P. K. T. Pang, M. P. Schreibman, and R. E. Jones, Eds.), pp. 205–302. Academic Press, San Diego.

Saidapur, S. K. (1982). Structure and function of postovulatory follicles (corpora lutea) in the ovaries of nonmammalian vertebrates. *Int. Rev. Cytol.* **75**, 243–285.

Tsang, P., and Callard, I. P. (1987). Luteal progesterone production and regulation in the viviparous dogfish, *Squalus acanthias*. *J. Exp. Zool.* **241**, 377–382.

Xavier, F. (1987). Functional morphology and regulation of the corpus luteum. In *Hormones and Reproduction in Fishes, Amphibians, and Reptiles* (D. O. Norris and R. E. Jones, Eds.), pp. 241–282. Plenum, New York.

Corpus Allatum

Stephen S. Tobe
University of Toronto

I. Introduction
II. Structure and Innervation of Corpora Allata
III. Biosynthesis of Juvenile Hormone
IV. Regulators of Corpora Allata
V. Evolution of the Juvenile Hormone Biosynthetic Pathway

GLOSSARY

enantiomer Mirror images of two molecules of the same chemical composition, such as one of a pair of optical isomers containing chiral carbon atoms, showing optical activity. Juvenile hormones show optical isomerism because of their chiral carbon atoms.

gap junctions Intercellular junctions between cells through which small water-soluble molecules, including ions, pass. They have a defined structure and likely function in intercellular communication.

immunoreactivity Pertaining to the ability of an antigen (usually a protein or peptide) to react with a specific antibody (usually an immunoglobulin); this interaction is usually visualized by tagging the antibody with an enzyme or dye.

methyl ester An ester having the general formula $RCOOCH_3$, in which the hydrogen of the carboxylic acid has been replaced by a methyl group; in juvenile hormones, R = carbon skeleton—C_{15}, C_{16}, C_{17}, or C_{18}.

neurosecretory cells Neuroendocrine cells, which possess characteristics of both endocrine cells and neurons, located in or associated with the central nervous system, containing peptides or amines which are released upon depolarization (e.g., following an action potential).

ring gland (Weismannn's ring) An endocrine ring-like structure, found in larvae and pupae of higher Diptera, which comes to lie behind the brain and around the aorta; contains three different types of endocrine cells corresponding to the corpora allata, corpora cardiaca, and the prothoracic glands.

stomatogastric nervous system That portion of the nervous system of arthropods that pertains to the esophagus and gut, particularly the anterior portion of the gut (foregut and midgut).

titer The concentration of a hormone in the circulatory system (hemolymph in insects); this is the effective concentration of a hormone to which target tissues are exposed.

The corpus allatum is the defined site of synthesis and release of insect juvenile hormone. Juvenile hormone regulates the processes of both metamorphosis and reproduction in most insect species. These endocrine glands have their embryonic origin from the mandibular pouches and are composed principally of ectodermal cells. The glands come to lie posterior to the brain and, depending on the species, may be located either ventrally or more dorsally. There are large species variations in the number of cells comprising the corpus allatum, ranging from tens to many thousand. The cells of individual glands are electrically coupled and, in some instances, are electrically excitable. The corpus allatum undergoes large changes in volume and in cell number and size, in association with the gonadotrophic cycle. These changes mirror, to some degree, the changes in the biosynthesis of juvenile hormone. The biosynthesis of juvenile hormone in the corpus allatum follows the conventional sterol biosynthetic pathway to the production of farnesyl pyrophosphate; four unusual biosynthetic steps subsequently convert farnesyl pyrophosphate or its higher homologs to juvenile hormone. The unique chemical nature of juvenile hormone and its exclusive occurrence in the Insecta within the animals has resulted in unusual modifications in the corpus allatum at both the structural and biochemical levels.

I. INTRODUCTION

Corpora allata (plural) are endocrine glands found in the posterior head/neck region of insects. They are located posterior, and ventral through dorsal, to the brain, depending on the species, and are usually associated, through nervous connectives, with the brain/central nervous system (CNS). They originate early in embryonic development in an ectodermal region anterior to the maxillary segment.

Corpora allata are the sites of biosynthesis and release of the juvenile hormones, unique sesquiterpenoid compounds which regulate in part the processes of growth, development, metamorphosis, and reproduction in most insect species. This group of compounds has not been identified definitively in any other group of animals, although one juvenile hormone (JH) has been isolated and identified in a small number of plant species, particularly the sedges. The presence or absence of JH during the life cycle of an insect will define the outcome of the subsequent molt and the degree of metamorphosis in the case of larval insects, and it will determine the ability to produce eggs in the case of adult females. Thus, the biosynthesis of JH by the corpora allata is a tightly regulated process because the presence or absence of the hormone determines the occurrence of both metamorphosis and reproduction. The innervation of this endocrine gland is extensive, both from the CNS and indirectly, from the stomatogastric nervous system, associated with the guts of insects (Figs. 1 and 2). This indicates a high degree of coordination of activity of the corpora allata by the nervous system, both from external signals (such as photoperiod and temperature) and from internal signals (such as quality of food, size/weight of the individual, copulation, and reproductive state), to permit hormone production only at appropriate times.

II. STRUCTURE AND INNERVATION OF CORPORA ALLATA

A. Structure

Corpora allata have been identified in all stages of the insect life cycle, including the embryo. In the locusts, beetles, and moths) or fuse (many flies and bugs) to form a single gland associated with the aorta.

The cells comprising the corpora allata are usually of a single type, although often some cells appear to be more active than others on the basis of their ultrastructure (extensive endoplasmic reticulum,

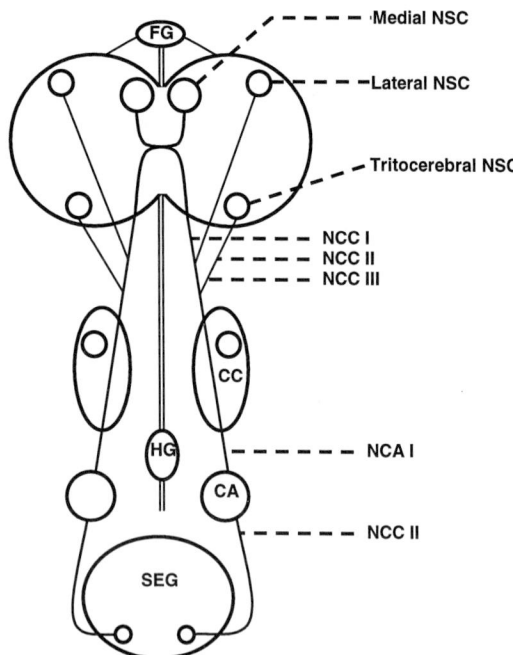

FIGURE 1 Schematic representation of the principal features of the innervation of the corpora allata (CA) and the corpora cardiaca (CC). Three nerves exit the brain and enter the CC [nervi corporis cardiaci (NCC) I–III]. Medial neurosecretory cells (NSC) contribute axons to the contralateral NCC I, lateral NSC contribute axons to the ipsilateral NCC II, and tritocerebral NSC contribute to the ipsilateral NCC III. NCA I and II are the nervi corporis allati (NCA) which extend from the CC to the CA and the subesophageal ganglion (SEG) to the CA, respectively. FG, frontal ganglion; HG, hypocerebral ganglion: both are associated with the stomatogastric nervous system (reproduced with permission from Tobe and Stay, 1985).

embryo, they arise as ectodermal pouches anterior to the maxillary segment; they form two coherent masses of cells that then migrate in dorsal and posterior directions. In more primitive insects, including the silverfish, this migration is minimal and the glands remain more ventral and anterior, whereas in the higher insects, the corpora allata undergo a marked migration. In many species, the glands either flank the esophagus (lower insects, hemimetabolous) or fuse with the ventral wall of the aorta or lie above the aorta (higher insects, holometabolous) (see Fig. 2 for examples of the two types). The corpora allata may remain as two distinct glands (e.g., cockroaches,

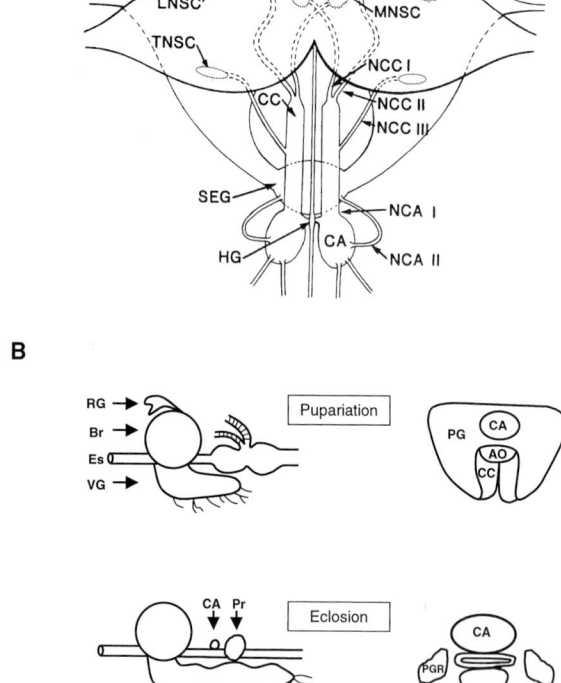

FIGURE 2 (A) Diagrammatic representation of brain–retrocerebral complex of the cockroach, *Diploptera punctata*, from a dorsal perspective. See the legend for Fig. 1 for abbreviations. MNSC, medial NSC; LNSC, lateral NSC; TNSC, tritocerebral NSC (modified with permission from Thompson and Tobe, 1990). (B) Diagrammatic representation of the changes in the fly, *Drosophila melanogaster*, brain–ring gland complex [which contains the corpora allata (CA)] during pupal–adult metamorphosis. (Left) The morphological changes in location and structure of the ring gland. (Right) A higher magnification of changes in the ring gland or corpora allata. Age of animals: top, formation of puparium; bottom, adult emergence. AO, aorta; Br, brain; CA, corpus allatum; CC, corpus cardiacum; Es, esophagus; PG, prothoracic gland; PGR, prothoracic gland remnant; Pr, proventriculus; RG, ring gland; VG, ventral ganglion (adapted with permission from Dai and Gilbert, 1991, and from Gilbert et al., 1996).

large and extensive mitochondria, swollen Golgi complexes, and nuclei). Glands may thus appear to be composed of more than one cell type. In many lower insects, the cells are relatively small (nuclei ~10 μm in diameter) and each gland comprises many cells (up to 10,000), whereas in many holometabolous insects, the cells are large and few in number (as few as 10–50). Some cells may be as large as 200 μm (in the leafworm). In some Lepidoptera, there appear to be two distinct types of corpora allata—the capsular and the isolated cell type—and it appears that the isolated type develop from the capsular type. Although in most species, the cells are relatively closely packed, the corpora allata of some species show a vesicular structure, with "epithelial" cells arranged around a central cavity, at specific times during the life cycle (e.g., after periods of intense biosynthetic activity) and this "secondary space" may represent a consequence of the completion of a cycle of activity, at which time the cells of the glands decrease in size. In all species, corpora allata are surrounded by a thin basal lamina (noncellular layer).

B. Nervous Connections

Corpora allata receive innervation from two principal nervous centers—the brain and the subesophageal ganglion (Fig. 1). It is these nervous connections that contribute predominantly to the regulation of JH production by the glands. In some primitive insects, the predominant source of innervation is from the subesophageal ganglion (e.g., spring tails and mayflies), whereas for the majority of insect species, corpora allata are innervated by axons originating in both the brain and the subesophageal ganglion. However, the glands of some beetles, flies, and bugs receive innervation exclusively from the brain.

The cerebral cells that innervate corpora allata are neurosecretory in nature and originate in three distinct regions in the brain: the medial, the lateral, and the tritocerebral regions (Fig. 1). All three areas do not always innervate corpora allata in different species, but this appears to be the general situation for those species studied in detail. The nerves arising in the brain terminate in both corpora allata and the corpora cardiaca, a neurohemal area associated with the CNS and the aorta. Neurosecretory cells in all three regions of the brain show immunoreactivity to allatostatins (peptide inhibitors of JH biosynthesis) and/or allatotropins (peptide stimulators of JH biosynthesis), depending on the species, and many of these project to the corpora cardiaca and the corpora allata.

Based on nerve transection experiments, innervation of the corpora allata by nerves arising in the subesophageal ganglion is probably less important in terms of regulation of the glands than cerebral innervation. Nonetheless, subesophageal nerve cells do innervate the corpora allata and at least in cockroaches and the moth, *Manduca*, some of these show allatostatin immunoreactivity.

The importance of nervous connections to the regulation of corpus allatum activity has been established for several species, particularly cockroaches, locusts, moths, and beetles. Experimental evidence involving either the interruption of the nervous pathways by surgical transection or the cautery of selected regions of the brain containing the neurosecretory cells demonstrated that both stimulatory and inhibitory messages are transmitted by the nervous tracts innervating the corpora allata. For example, in cockroaches, transection of the NCA I or cautery of the lateral or medial neurosecretory cell region in the brain (Fig. 1) results in a rapid and sustained increase in JH production. Such evidence indicates that the nerves in this case carry an allatostatic (inhibitory) signal to the corpora allata, restraining the production of the hormone.

The fine structure of the cells of the corpora allata is typical of steroid-producing cells, with extensive endoplasmic reticulum, well-developed Golgi complexes, and large numbers of mitochondria. These ultrastructural characteristics do change with changes in the activity of the cells: Glands producing JH at high rates have less conspicuous endoplasmic reticulum, more obvious mitochondrial substructure, and swollen Golgi complexes, whereas cells from glands producing JH at low rates show clear endoplasmic reticulum, less obvious mitochondria, and flattened Golgi complexes. Cell-to-cell junctions in the form of membrane gap junctions are obvious in some species, including cockroaches. These gap junctions are usually associated with intercellular communication, and in cockroaches the cells of the glands are electrically coupled. The corpora allata act as an electrical syncitium, and depolarizing current injected into one cell is rapidly propagated throughout the entire gland. In this fashion, all cells can respond to a neurosecretory signal at the same time; also, it is not necessary for all cells of the gland to receive the neurosecretory signal for coordination to occur. In addition to the ability to propagate electrical signals, the cells of the corpora allata are electrically excitable, and injection of depolarizing current will result in the generation of an action potential, as is observed in typical nerve cells, and in the propagation of the action potential throughout the gland. It should be remembered that the cells of the corpora allata are not neural derived but rather are of conventional ectodermal origin, as are other endocrine cells.

III. BIOSYNTHESIS OF JUVENILE HORMONE

The corpora allata produce JHs, six of which have been identified definitively (Fig. 3). JH III is the most ubiquitous of the JHs and has been found in most insect orders. JH 0, I, II, and iso-JH 0 (4-methyl-JH 1) have been found only in moths, and JH III bisepoxide has been identified as the principal JH of higher (cyclorraphan) flies. The situation in moths is further complicated because late larvae, pupae, and adult male corpora allata produce not the juvenile hormones but rather the carboxylic acids of juvenile hormones (rather than the methyl esters), which can be subsequently converted to the hormones in the accessory glands. The juvenile hormone acids lack significant biological activity.

The epoxide group common to all juvenile hormones shows optical activity; that is, there are stereoisomers/enantiomers as a result of the chiral centers at the 10 position of JH III and at the 10 and 11 positions of the other juvenile hormones. The bisepoxide of JH III also possesses two chiral centers as a consequence of the two epoxide groups. The occurrence of enantiomers of the juvenile hormones is significant because the naturally occurring isomer

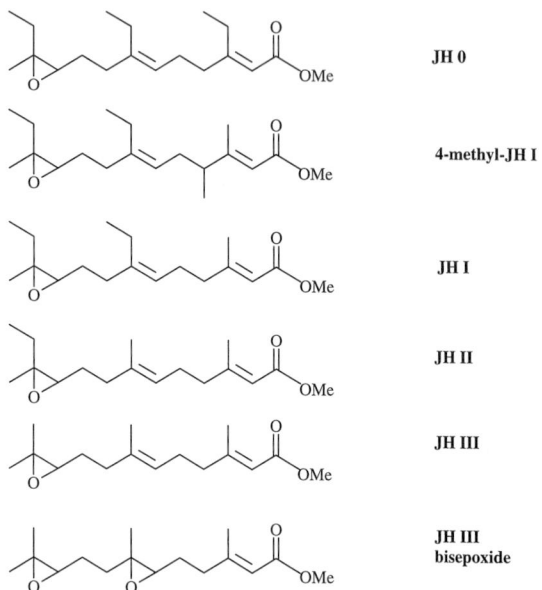

FIGURE 3 Chemical structures of juvenile hormones. JH 0, I, and II are found exclusively in Lepidoptera, whereas JH III is found throughout the insect orders and can be regarded as more primitive and ubiquitous. JH III bisepoxide is restricted to the cyclorraphan (higher) flies. JH I and 4-methyl-JH I have been isolated from moth embryos only. Asymmetric carbons at C_{10} and C_{11} (JH III, only C_{10}) give rise to enantiomers, only one of which occurs naturally.

shows higher biological activity and different rates of degradation than the unnatural enantiomers.

A. Pathway

The juvenile hormones are biosynthesized from simple precursors; acetate in the case of JH III and acetate and propionate in the case of the higher homologs (Fig. 4). For JH III, the pathway of biosynthesis is identical to that for vertebrate sterol biosynthesis to the point of production of farnesyl pyrophosphate (Fig. 4). Condensation of three C_2 units give rise to one C_6 unit, which, following decarboxylation and phosphorylation, produces a C_5 isoprenoid unit. Condensation of three C_5 units ultimately gives rise to farnesyl pyrophosphate (C_{15}). From this point, farnesyl pyrophosphate is modified through four enzymatic steps to juvenile hormone III. The biosynthesis of the higher homologs, particularly JH I and II, is somewhat more complicated, involving the utiliza-

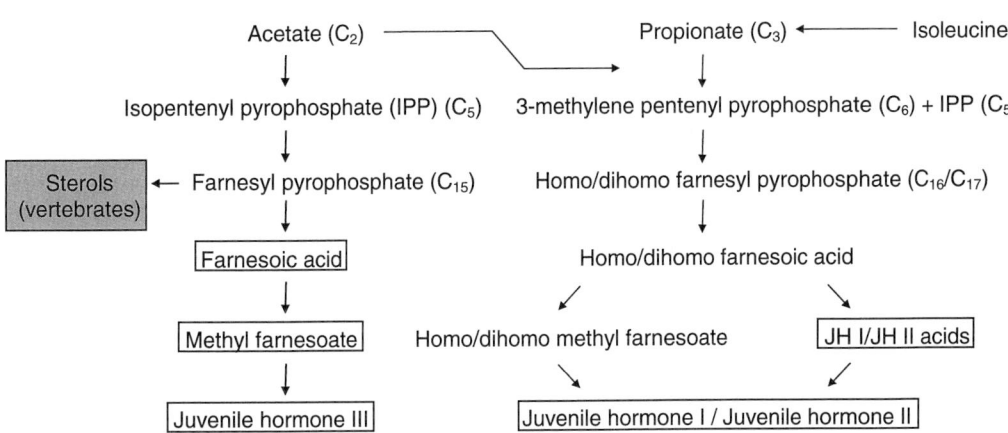

FIGURE 4 Biosynthetic pathways for juvenile hormones. JH III is derived exclusively from acetate (C_2 units), whereas JH I and II are derived from both acetate and propionate (C_3). Known release products are shown in boxes. The insect pathway for farnesyl pyrophosphate biosynthesis is probably identical to that of vertebrates. Insects convert farnesyl pyrophosphate to JH III, whereas in vertebrates, it is converted to sterols (redrawn with permission from Gilbert et al., 1996).

tion of both acetate and propionate as substrates. The formation of the side chains of these juvenile hormones involves differential utilization of these substrates to give rise to both C_5 and C_6 pyrophosphate intermediates. Subsequent condensation of two C_6 units and one C_5 unit results in the formation of the JH I skeleton (C_{17}), whereas one C_6 unit plus two C_5 units gives rise to the JH II skeleton (C_{16}) (Fig. 4).

B. Release of Juvenile Hormone from Corpora Allata

Production of the starting substrates from glucose or branched-chain amino acids occurs in the mitochondria of corpus allatum cells, whereas the production of the C_6 precursor, mevalonate, occurs on the endoplasmic reticulum; the epoxidation of methyl farnesoate or the acids of the higher homologs to the juvenile hormones occurs within the endoplasmic reticulum. The localization of the epoxidase (the enzyme responsible for epoxidation of methyl farnesoate) within the cisternae of the endoplasmic reticulum of corpus allatum cells suggests that this enzyme may be involved in the exit of JH from the cells. The movement of the substrate, methyl farnesoate, into the cisternae of the endoplasmic reticulum may facilitate the movement of newly synthesized JH out of the cell. Juvenile hormone is released from the cells as soon as it is synthesized (i.e., there is no storage of the hormone within the cells); the close association of the endoplasmic reticulum with the extensive microvilli-like projections of the plasma membranes of active corpus allatum cells and with the extracellular spaces suggests that this physical relationship is important for efficient hormone release. Rates of JH biosynthesis can thus be considered approximately equivalent to rates of hormone release, i.e., measurement of rates of juvenile hormone release provide an accurate estimate of rates of hormone biosynthesis because rates of release increase proportionally to rates of biosynthesis.

C. Measurement of Rates of Juvenile Hormone Production

Although simple C_2 precursors such as acetate are rapidly incorporated into the juvenile hormone carbon skeleton, such precursors are also incorporated into many other compounds within the cells as a consequence of their central role in energy production (e.g., in the synthesis of lipids and complex carbohydrates). Accordingly, the incorporation of radiolabeled C_2 precursors into the carbon skeleton of JH would not be an appropriate choice for monitoring rates of hormone production. Incorporation of radiolabeled substrates into a final product is a common method for determining the rate of biosynthesis of a hormone, but because of the extensive nonhormonal incorporation of C_2 units into other cellular components, time-consuming and expensive methods for the separation of the hormone of interest from other radiolabeled products are necessary. In the early 1970s, the unique incorporation of another substrate, L-methionine, into the methyl ester function of JH was demonstrated. This methyl donor is used almost exclusively in the conversion of farnesoic acid to methyl farnesoate (Fig. 4). For this reason, inclusion of radiolabeled methylmethionine as methyl donor in incubations of corpora allata *in vitro* results in the incorporation of the radiolabel into methyl farnesoate, which is subsequently converted to JH III (Fig. 4). Because virtually no other cellular products incorporate the methyl group of methionine in this system, measurement of the incorporation of this group (i.e., measurement of radioactivity) is an accurate reflection of the production of JH. This procedure, known as the "radiochemical assay for juvenile hormone," is the most common method for the determination of rates of JH production by isolated corpora allata in most insect systems. The method is rapid, accurate, efficient, and does not require extensive procedures to separate the hormonal product from other radiolabeled compounds.

IV. REGULATORS OF CORPORA ALLATA

As noted earlier, JH is not stored within glands; accordingly, regulation of hormone release is not an important component in the overall regulation of the corpora allata. It is therefore necessary to look elsewhere for regulators of corpus allatum function. Rates of JH biosynthesis change dramatically during the life cycle, both during development/metamorphosis and during reproduction, indicating that the glands are in fact precisely regulated. However, in only a few species has a convincing correlation between JH production and the titer of JH (concentration of the hormone in the hemolymph/blood) affecting the target tissues been demonstrated. In the cockroach, *Diploptera punctata*, there is a clear correlation between hormone production and titer and, at least in this species, rates of JH production provide an accurate reflection of the hormone titer and hence the concentrations of hormone to which target tissues are exposed.

Figure 5 shows the likely pathways for the regulation of the corpora allata in insects. The CNS (brain) receives input and integrates signals from both internal and external environments. This information is transmitted to endocrine centers, particularly neurosecretory cells, within the brain (and other centers associated with the CNS) which respond either with release or cessation of release of their endocrine product. These endocrine products can be classified into two general groups: the stimulatory factors (allatotropins) and the inhibitory factors (allatostatins). Allatotropins stimulate the production of JH, whereas allatostatins inhibit the production of the hormone.

A. Allatotropins

Allatotropins are widely distributed in insect species, including locusts, grasshoppers, moths, bugs, and cockroaches, but to date only one allatotropin, from the moth *Manduca sexta* (tobacco hornworm), has been isolated and characterized. It is a tridecapeptide (13 amino acid residues) that is active only on adult female corpora allata. The primary sequence of this peptide is Gly-Phe-Lys-Asn-Val-Glu-Met-Met-Thr-Ala-Arg-Gly-Phe-amide and it bears strong similarity to an accessory gland myotropin (which stimulates muscle contraction) in the locust. Despite this single identification, immunoreactive material to *Manduca* allatotropin has been described in many other moth species as well as in locusts.

B. Allatostatins

Allatostatins are also widely distributed in insects and they have been definitively isolated and characterized in several orders, including the Dictyoptera (cockroaches), Orthoptera (crickets and locusts), and Lepidoptera (moths). Allatostatins show a high degree of species specificity and it is likely that the peptides act only on a limited number of closely related species within the order, i.e., cockroach allatostatins do not act to inhibit JH production in moths

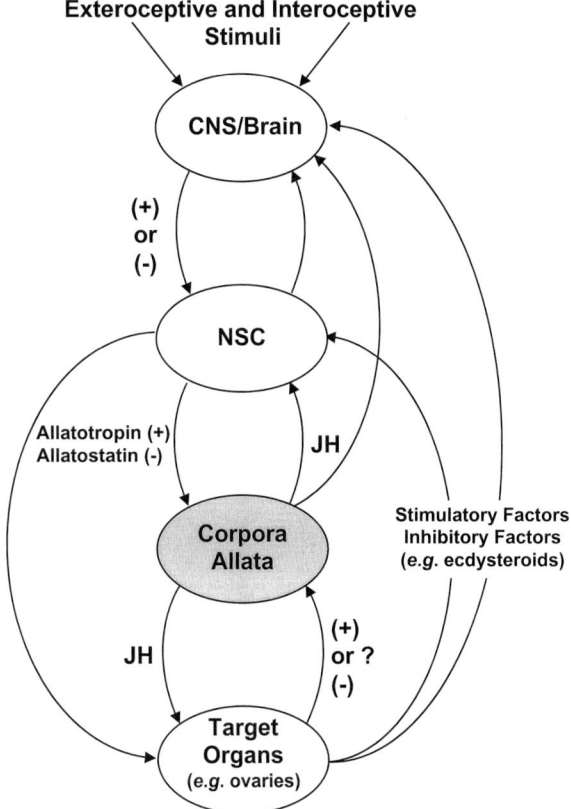

FIGURE 5 Feedback loops regulating production of JH by corpora allata in adult female cockroaches. Signals reaching the brain from external or internal sources are processed and transmitted to appropriate neurosecretory cells (NSC), most of which are also located in the brain. Stimulatory (allatotropic, +) or inhibitory (allatostatic, −) peptides are released from the terminals of the neurosecretory cells at the corpora allata, and JH production is appropriately modified. Juvenile hormone titer can modulate release of these regulatory peptides through the neurosecretory cells directly or through the CNS. Juvenile hormone stimulates ovarian growth and vitellogenesis; vitellogenic ovaries stimulate production of JH, either through the CNS or through the modulation of neuropeptide release from the neurosecretory cells. Ecdysteroid release from the mature ovary can subsequently inhibit production of JH either directly or indirectly.

or flies and moth allatostatins do not affect cockroach or cricket JH production.

The allatostatins comprise three distinct families: the cockroach family, with a common carboxy terminus -Tyr/Phe-Xaa-Phe-Gly-Leu-amide; the cricket family, with a common carboxy terminus -Trp-Xaa-Asp-Leu-Xaa-Gly-Gly-Trp-amide; and the moth family, of which only one member is known to function as an allatostatin (16-amino acid nonamidated peptide: pGlu-Val-Arg-Phe/Tyr-Arg-Gln-Cys-Tyr-Phe-Asn-Pro-Ile-Ser-Cys-Phe-OH) and is found only in moths and flies.

The cockroach allatostatins occur in moths and flies, as well as in locusts and crickets, but only in cockroaches and crickets are they effective in inhibiting JH production. Moth allatostatin has been deduced to occur in *Drosophila* on the basis of cloning and sequencing of cDNA, but its functions in the fly are uncertain. The moth allatostatin is ineffective in inhibiting JH biosynthesis in species outside the order Lepidoptera and may not even be effective in all Lepidoptera.

C. Amines

Not all neurosecretory cells are peptidergic (peptide containing). Many neurosecretory cells in the insect CNS contain another group of compounds, the biogenic amines, and these cells are termed aminergic. One such amine is octopamine, and appreciable quantities of this biogenic compound have been reported in corpora allata of several species. It is likely that the octopamine is localized within some of the many neurosecretory endings that are found in the corpora allata. Treatment of glands with this biogenic amine inhibits the production of JH in some species but stimulates hormone production in others. This difference probably reflects the differences in the underlying mechanisms of regulation of the corpora allata, i.e., glands are either normally "inhibited" (off) or "stimulated" (on), depending on the species. Octopamine can affect corpora allata by exerting a direct effect on ion channels of the cells or by effecting the release of the neuropeptides allatostatin or allatotropin from the neurosecretory cell endings within the glands.

D. Feedback Loops

The occurrence of allatotropins and allatostatins in many insect species suggests that these are fundamental regulators of JH production, and although there is evidence to suggest that both types of regulators occur in many species, only in Manduca have both been characterized. Even in this latter case, the allatotropin appears to be active only in adults, whereas the allatostatin is active particularly in larval forms; hence, there is a temporal separation of the regulators. Clearly, only one form of regulator is necessary to effect the "switching" on and off of the corpora allata, although the simultaneous use of both regulators would probably permit a more precise control of hormone production. For example, in glands that are normally switched off, the release of an allatotropin would stimulate JH production and its withdrawal would end the stimulation. Conversely, in glands that are normally switched on, the release of an allatostatin would inhibit JH production and its withdrawal would terminate the inhibition (Fig. 5).

Feedback loops involving the target tissues for JH and the corpora allata are important in regulating the circulating titer of JH, and allatostatins and allatotropins are probably instrumental in these loops. For example, artificial elevation of the titer of JH (e.g., through administration of hormone by topical application or injection) results in the suppression of hormone production by the corpora allata. On the other hand, growth of the ovaries of female cockroaches, particularly early in the reproductive cycle, stimulates hormone production by the corpora allata. In both these instances, signals from target tissues feed back, by way of the CNS/brain, to effect the change in JH production. In the first example, hormone production is reduced, probably as a result of release of allatostatin from neurosecretory cell endings (originating in the CNS) onto the cells of the corpora allata, whereas in the second example, cessation of release of allatostatin, as a consequence of a stimulatory signal from the ovary, would result in an increase in JH production. Implicit in this model is the existence of centers in the CNS that are capable of sampling the titer of JH and other signals emanating from the target tissues. This information is then integrated in the CNS and relayed to the allatostatin-containing neurosecretory cells (Fig. 5).

V. EVOLUTION OF THE JUVENILE HORMONE BIOSYNTHETIC PATHWAY

The pathway for the biosynthesis of farnesyl pyrophosphate from simple C_2 precursors seems to have changed little, both in vertebrates and invertebrates, during evolutionary time. In fact, examination of the primary structures of one of the enzymes involved in this biosynthesis, hydroxyl-methyl-glutaryl CoA (HMG-CoA) reductase—the enzyme responsible for the conversion of HMGCoA to mevalonate—shows significant similarity/homology between vertebrates and insects (cockroaches). Insects and other Arthropods are incapable of cholesterol biosynthesis *de novo* as a consequence of metabolic blocks between acetate and cholesterol; hence, sterols must be acquired from the diet.

Plants and most other animals (eukaryotes) are capable of the biosynthesis of either cholesterol or related sterols; cholesterol is an important cellular constituent not only for cell membranes but also as a substrate for a range of other products, including hormones and biological detergents. For this reason, the biosynthetic pathway for cholesterol is probably ancient. If a portion of this pathway was subsequently subsumed by insects for the production of JH, with the consequent loss of cholesterol biosynthesis, cholesterol must have been readily available in the diet of most ancient insects. For some period, both pathways must have coexisted in insects, and because of the dietary availability of cholesterol, selection pressure for those organisms able to produce JH and not cholesterol must have been operative. Cholesterol biosynthesis is an energy-intensive pathway and some selective advantage must have been conferred on those insects which produced only JH. Juvenile hormone must be an ancient metabolic, developmental, reproductive, and metamorphic regulator and the corpora allata must be organs of ancient origin. Their embryology suggests such an early appearance. Functionally, however, it is unclear which process initially became associated with JH and how the other processes were subsequently coopted. However, in higher insect orders, e.g., flies and moths, new forms of JH have appeared (JH I, II, and

III *bis*-epoxide) and the role that the hormone plays in reproduction has been reduced in some species.

Juvenile hormone occurs only in the class Insecta and a few plant species. Interestingly, its immediate precursor, methyl farnesoate (and the biosynthetic pathway), occurs in Crustacea and may act as an endocrine regulator of metamorphosis and reproduction in this group. The glands which produce methyl farnesoate in Crustacea (the mandibular organs) have a similar embryonic origin as the corpora allata. How widely JH and its precursors are distributed in the eukaryotes remains to be determined. However, the search for similar organs, the isolation, characterization, and cloning of the unique enzymes involved in JH production, and the development of antibodies to these enzymes should prove instructive to our understanding of the regulation of development and reproduction in invertebrates.

See Also the Following Articles

ALLOSTATINS; *DIPLOPTERA PUNCTATA*; INSECT REPRODUCTION, OVERVIEW

Bibliography

Cassier, P. (1990). Morphology, histology and ultrastructure of JH-producing glands in insects. In *Morphogenetic Hormones of Arthropods. II. Embryonic and Postembryonic Sources* (A. P. Gupta, Ed.), pp. 83–194. Rutgers Univ. Press, New Brunswick, NJ.

Dai and Gilbert. (1991). *Devel. Biol.* 144, 309–326.

Gilbert, L. I., Rybczynski, R., and Tobe, S. S. (1996). Endocrine cascade in insect metamorphosis. In *Metamorphosis: Postembryonic Reprogramming of Gene Expression in Amphibian and Insect Cells* (L. I. Gilbert, B. Atkinson, and J. R. Tata, Eds.), pp. 59–107. Academic Press, New York.

Johnson, G. D., Stay, B., and Rankin, S. M. (1985). Ultrastructure of corpora allata of known activity during the vitellogenic cycle in the cockroach *Diploptera punctata*. *Cell Tissue Res.* 239, 317–327.

Johnson, G. D., Stay, B., and Chan, K. K. (1993). Structure–activity relationships in corpora allata of the cockroach *Diploptera punctata*: Roles of mating and the ovary. *Cell Tissue Res.* 274, 279–293.

Thompson, C. S., and Tobe, S. S. (1990). Innervation and electrophysiology of the corpus allatum. In *Cockroaches as Models for Neurobiology: Applications in Biomedical Research* (I. Huber, E. P. Masler, and B. R. Rao, Eds.), pp. 89–101. CRC Press, Boca Raton, FL.

Tobe, S. S., and Stay, B. (1985). Structure and regulation of the corpus allatum. *Adv. Insect Physiol.* 18, 305–432.

Tobe, S. S., Yu, C. G., and Bendena, W. G. (1994). Allatostatins, peptide inhibitors of juvenile hormone production in insects. In *Perspectives in Comparative Endocrinology* (K. G. Davey, R. E. Peter, and S. S. Tobe, Eds.), pp. 12–19. National Research Council of Canada, Ottawa.

Corpus Cardiacum, Insects

Barry G. Loughton
York University

I. Introduction
II. Structure of the Corpus Cardiacum
III. Reproductive Hormones Released from the Corpus Cardiacum
IV. Conclusion

GLOSSARY

neurohemal organ A specialized site containing the axon terminals of neurosecretory cells where neurohormones are liberated into the hemolymph.

neurosecretory granules Intracellular vesicles, 100–300 nm in diameter, that are most commonly electron dense, reflecting the protein nature of their contents.

retrocerebral complex Those structures immediately posterior to the brain including the corpus cardiacum, hypocerebral ganglion, and corpus allatum.

The corpus cardiacum is a part of the nervous system of the insect. It is a paired structure lying immediately behind the brain and is applied to the surface of the aorta in the head.

I. INTRODUCTION

The corpus cardiacum is composed of glial and neurosecretory cells. One category of neurosecretory cells has its cell body elsewhere in the central nervous system and its axons project to the corpus cardiacum. These are "extrinsic" cells. It is the distended axon termini of these axons that constitute the bulk of the corpus cardiacum. The second type is confined to the corpus cardiacum and characteristically has very short axons. These are the so-called intrinsic cells of the corpus cardiacum.

Other terrestrial arthropods possess neurohemal organs derived from the distended axon termini of protocerebral neurosecretory cells. In spiders the neurohemal organ can be separated into two parts: the Tropfenkomplex, which is made up of protocerebral cell axon termini, and Schneiders organ, which is composed of intrinsic cells. These organs appear to be homologous with the corpora cardiaca of insects. In other arachnids neurohemal organs derived from protocerebral neurosecretory cell axon termini have been identified, but no evidence of intrinsic neurosecretory cells has been adduced. The axons of neurosecretory cells in the Myriapod protocerebrum terminate in cerebral glands, paired structures lying behind the brain. No evidence of intrinsic neurosecretory cells has been obtained. In all these organisms the neurohemal organ is closely associated with elements of the blood vascular system. Thus, a neurohemal organ derived from the axon termini of protocerebral neurosecretory cells appears to be a common feature of terrestrial arthropods.

II. STRUCTURE OF THE CORPUS CARDIACUM

The corpus cardiacum is the major neurohemal organ of the insect and is the release site for most but not all of the neurohormones elaborated in the brain, including hormones controlling the processes involved in growth and molting, water balance, and excretion and reproduction. The cell bodies of those cells whose axons terminate in the corpus cardiacum originate in the protocerebrum and the tritocerebrum of the brain. In some insects there are also neurosecretory cells in the suboesophageal ganglion with axons terminating in the corpus cardiacum. Figure 1 shows the disposition in the brain of the neurose-

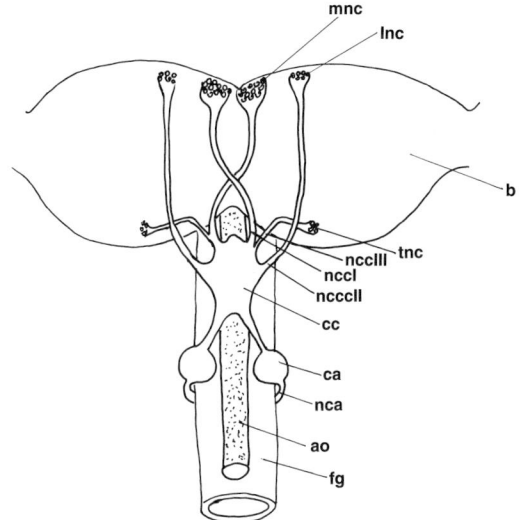

FIGURE 1 Diagram of the retrocerebral complex showing the neurosecretory cells of the brain that contribute to the corpus cardiacum. ao, Aorta; b, brain; ca, corpus allatum; cc, corpus cardiacum; fg, foregut; lnc, lateral neurosecretory cells; mnc, median neurosecretory cells; nccI, nervus corporis cardiacum I; nccII, nervus corporis cardiacum II; nccIII, nervus corporis cardiacum III; tnc, tritocerebral neurosecretory cells (redrawn after Goldsworthy et al., Endocrinology, Wiley, New York, 1981).

cretory cells contributing to the corpus cardiacum. It can be seen that in the protocerebrum there are two distinguishable groups of cells. The median neurosecretory cells (MNC) of the pars intercerebralis lie in the midline close to the anterior surface of the brain, and their axons pass posteriorly through the brain decussating en route to leave by the contralateral nervus corporis cardiacum I (NCC I) and enter the corpus cardiacum. The lateral neurosecretory cells (LNC) of the protocerebrum lie closer to the optic lobes of the brain and their axons leave the brain in nervus corporis cardiacum II (NCC II) which enters the corpus cardiacum from a dorsal aspect. In many orders of insects NCC II is absent and the LNC axons enter the corpus cardiacum via the ipsilateral NCC I. The axons of the neurosecretory cells of the tritocerebrum enter the corpus cardiaca via NCC III or the ipsilateral NCC I. In higher Diptera, such as *Drosophila* and *Musca*, the corpus cardiacum is incorporated into the ring gland, which also contains the corpus allatum and the prothoracic gland. This structure encircles the aorta. The corpus cardiacum is then ideally situated to serve as a neurohemal organ. However, radiolabeled compounds injected to the hemocoel of locusts are evenly distributed throughout the blood within 2 min of injection. Direct entry of the hormone into the aorta can only improve an already efficient distribution system.

The corpus cardiacum is easily identified under the dissecting microscope and manifests a characteristic blue-white color due to the diffraction of light by the neurosecretory granules in the axon termini. In the Acrididae the blue-white appearance of the neurosecretory axons of the "extrinsic" cells allows the recognition of two anatomically separate lobes of the corpus cardiacum. Whereas the "storage" lobe derived from extrinsic cells shows the characteristic Tyndall blue effect, the intrinsic cells of the so-called glandular lobe appear as a grayish white color. The neurosecretory granules in these particular cells are unusually large (600–800 nm). The intrinsic and extrinsic cells are less easily distinguished in other insects. The intrinsic cells are not anatomically distinct, nor are their neurosecretory granules consistently different in size from those in the extrinsic cells (generally 100–300 nm).

III. REPRODUCTIVE HORMONES RELEASED FROM THE CORPUS CARDIACUM

There are several hormones that appear to be released from the corpus cardiacum which are involved in the control of reproductive processes. Table 1 lists these hormones in the approximate order in which they affect the reproductive process and indicates their action and organism(s) in which they have been identified. This list of organisms is by no means complete and serves only to give examples of the experimental insects used. It should be noted, however, that seldom has there been a direct observation of release of hormones from the cardiacum. More frequently the presence of the hormone in the cardiacum and its apparent absence elsewhere in the retrocerebral complex has been taken to indicate that the release site is the corpus cardiacum. On occasion this evidence is strengthened by the observation that

TABLE 1
The Reproductive Hormones and Their Anatomical Source(s)

Hormone	Insect[a]	Source	Nature	Molecular weight (kDa)	Action
Ovarian ecdysteroidogenic hormone = egg development neurosecretory hormone	♀ *Aedes aegypti*	MNC	Peptide	6	Promotes follicle cell synthesis of ecdysone
Locust ovary maturing parsin	♀ *Locusta migratoria*	MNC	Peptide	6.9	Promotes vitellogenesis
Neuroparsins A and B	♀ ♂ *Locusta migratoria*	MNC	Peptide	A, 8.8; B, 8.2	Decreases titer of JH or antagonizes JH action
Adipokinetic hormone I	♀ ♂ *Locusta migratoria*	Intrinsic cells of CC	Peptide	1.1	Inhibits vitellogenesis synthesis
Ovulation hormone (myotropin)	♀ *Rhodnius prolixus*	MNC	Peptide	8	Causes ovulation (and oviposition)
Pheromone biosynthesis-activating neuropeptide	♀ *Helicoverpa zea, Pseudaletia separata*, and others	NC subesophageal ganglion	Peptide	3.9	Stimulates biosynthesis of pheromone by the pheromone gland
Accessory reproductive gland (ARG)-stimulating hormone	♂ *Rhodnius prolixus*	MNC?	Peptide	?	Stimulates growth of ARG in conjunction with JH
Vitellogenin synthesis-stimulating factor	♀ *Locusta migratoria*	Brain	?	?	Stimulates vitellogenin synthesis by fat body in the presence of JH

[a] The insect category gives the organism on which most studies have been conducted and is not intended to be comprehensive.

depolarization of the isolated corpus cardiacum results in the release of neurohormone. It is assumed that since release can occur from this organ it normally does occur there. One reason for caution lies in the fact that although it had originally been assumed that the prothoracicotropic hormone (PTTH) was liberated from axons terminating in the corpus cardiacum, it was eventually demonstrated that in some insects the axons in question terminated in the corpus allatum and PTTH release occurred there. Nevertheless, the anatomy of the corpus cardiacum and the demonstration that it is the release site for at least some neurosecretory hormones make it most likely that most claims of neurosecretory hormone release from this structure are valid.

A. Ovarian Ecdysteroidogenic Hormone

Ovarian ecdysteroidogenic hormone (OEH) is manufactured in the MNC and transported to the corpus cardiacum in female mosquitoes via the NCC I. The extrinsic cells of the corpus cardiacum in these insects terminate in a rather diffuse neurohemal area which extends along the aorta for some distance before terminating at the junction of the neck and thorax (i.e., behind the head). At this point, immediately prior to the corpus allatum are the "cardiaca" neurosecretory cells which correspond to the intrinsic cells of the corpus cardiacum. OEH is released from the corpus cardiacum as a result of a signal from the ovaries after the insect receives a blood meal. It acts on the follicle cells of the ovary to stimulate the production of edcysone and initiate the process of vitellogenesis.

B. Locust Ovary Maturing Parsin

Locust ovary maturing parsin (LOMP) is a neuropeptide of molecular weight 6.9 kDa lacking cystein, which is manufactured in the type-B MNC. The presence of LOMP in this type of cell is consistent with its staining properties since it does not stain with fuchsin, which preferentially stains cysteine-rich proteins. The hormone promotes vitellogenesis perhaps by stimulating ecdysone synthesis by follicle cells. It is present in the protocerebrum at all stages of development in both males and females. Though it has not been demonstrated in the hemolymph during vitellogenesis, the delayed vitellogenesis following electrocautery of these cells and the presence of LOMP in the corpus cardiacum is persuasive evidence that LOMP I is indeed a reproductive hormone. The relationship between LOMP and OEH in mosquitoes is not clear. No other OEH has been demonstrated in the locust.

C. Adipokinetic Hormone

There are three adipokinetic hormones (AKHs) known from locusts and all three have been shown to inhibit protein synthesis. However, the shut down of vitellogenin synthesis in the fat body of vitellogenic locusts at the time of chorionation is attributed primarily to AKH I. AKH I is present in the hemolymph of female locusts in sufficient concentration to inhibit vitellogenin synthesis in fat body. *In vitro* experiments reveal that the inhibiting effect of AKH I is primarily on vitellogenin and not on other hemolymph proteins. The function of AKH during reproduction would appear to be to limit vitellogenin synthesis between vitellogenic cycles, thus conserving energy and protein.

Adipokinetic hormone I:
Q L N F T P N W G T-amide

The isolation and characterization of AKH was facilitated as a result of the special organization of the locust corpus cardiacum wherein the intrinsic cells are organized as a structure separate from the extrinsic cell axon termini. The first recognized function of AKH was to bring about the mobilization of lipid from the fat body to provide energy for prolonged flight. The hormone has been shown to have pleiotropic effects, including the activation of glycogen phosphorylase, inhibition of lipid, RNA and protein synthesis, enhancement of lipid/lipophorin aggregation, and cardiostimulatory activity.

D. Locust Neuroparsins A and B

The neuropeptides are present in the type-A MNC and in the axon termini of these cells in the corpus cardiacum. The B neuroparsin is a truncated form of neuroparsin A, lacking the N-terminal five amino acids. They show some sequence homology with OEH but have no ecdysiotropic activity. On the contrary, they stimulate water uptake from the rectum (antidiuretic effect), mobilize lipid and carbohydrate, and exert an antigonadotropic effect. The two molecules have equal biological activity in all systems so far examined. The mode of action of the neuroparsins is not known. Injections of neuroparsin A into adult female locusts delayed vitellogenesis. Injection of antineuroparsin A serum produced results consistent with high levels of juvenile hormone (JH) in the insect, but incubation of neuroparsin with corpora allata did not result in lowered JH synthesis. Thus, the mode or site of action of the neuroparsin in inhibiting the normal effects of JH is unknown.

E. Myotropin

The corpus cardiacum of *Rhodnius prolixus* has been shown to contain a 8.5-kDa myotropic peptide which originates in 10 prominent MNCs. The myotropin is immunoreactive with antiserum directed against FMRF-amide, but is much larger than most FMRF-amide-related peptides (FaRPs) so far discovered. In female *Rhodnius* the peptide is released from the corpus cardiacum of mated insects at the time of oviposition in response to a humeral matedness signal from the spermatheca and the presence of ecdysteroid which emanates from the ovary. These two factors cause bursts of action potentials in the myotropin-containing neurosecretory cells resulting in release from the corpus cardiacum. The myotropin acts on the ovarian follicle to bring about contractions of the ovarian muscular sheath and hence ovulation of the mature oocytes and presumably on the oviducts to cause oviposition. This is one of the few cases in which the identity and activity of the neuropeptide, its cellular origin, the nature of the factors controlling its release, and its release from the corpus cardiacum into the hemolymph at the appropriate time have been documented. A peptide of similar size with similar immunological and myotropic properties has been found in the female locust. It is not known whether the myotropic activity of these peptides is specific to the ovary and oviducts.

F. Pheromone Biosynthesis-Activating Neuropeptide

The synthesis of sex pheromones in female moths is controlled by pheromone biosynthesis-activating neuropeptide (PBAN), which is itself synthesized by neurosecretory cells in the ventral midline of the suboesophageal ganglion of female Lepidoptera. The axons from the most posterior of these groups project to the corpus cardiacum and aorta via NCC III. In many moths the release of PBAN from the corpus cardiacum can be induced by providing the gravid female with either host plants or volatiles extracted from the host plant. Though the molecular weight of PBAN may vary from species to species, the five C-terminal amino acid sequence is conserved.

Helicoverpa zea PBAN

$$\underset{}{\text{L S D D M P A T P A D Q E M Y T Q D P E Q I D S R T K Y F S P R L}}\text{-amide}$$
(positions 10, 20, 30)

Pseudaletia separata PBAN

$$\underset{}{\text{K L S T D D K V F E N V E F T P R L}}\text{-amide}$$
(positions 10, 15)

G. Protein Synthesis-Stimulating Factor(s)

Several reports indicate that normal vitellogenesis requires the action of JH and the contribution of neurosecretory factors from the brain. In some cases isolated corpora cardiaca or corpus cardiacum extracts can restore normal vitellogenesis in brainless insects. The mode of action of these neurosecretory factors is not known. It is possible that they act on the ovary to stimulate ecdysteroid synthesis and thereby stimulate protein synthesis. However, in the locust, extracts of brain applied to fat body *in vitro* in the presence of the JH analog methoprene stimulated vitellogenin synthesis.

Most of the research on the endocrinology of insect reproduction has concentrated on the female. However, the growth of the transparent accessory reproductive gland (TARG) of *Rhodnius* is controlled by JH and by a peptide factor from the neurosecretory cells of the brain and corpus cardiacum which stimulates protein synthesis in the TARG *in vitro*. The TARG-stimulating peptide and the vitellogenesis synthesis-stimulating factor may well be the same molecule. Both emanate from the brain and both enhance protein synthesis in the presence of JH, though they are documented from separate insects. Thus, there is persuasive evidence that neurosecretory material from the brain can directly stimulate protein synthesis for reproductive purposes. In addition, in "whole" insects normal levels of protein synthesis appear to require both JH and the presence of an intact brain and corpus cardiacum.

IV. CONCLUSION

The known reproductive hormones which have their release site in the corpus cardiacum are all peptidergic in nature. Their physiological effects range from stimulation of some aspect of the reproductive process to its inhibition. Some of the hormones involved have pleiotropic effects and at different times in the life cycle of the insect are important in other physiological processes. The recognition of hormonal activity has often stemmed from the convenience of reproductive processes as indicators of biological activity. Thus, although hormones such as AKH and neuroparsin are known to exert biological effects on other systems, it is entirely possible that many if not all the hormones documented here have important biological functions elsewhere in the organism or at earlier stages of the insect life cycle. Finally, it must be remembered that the investigation of insect reproduction has centered on only a very few of the known insects. These insects are often those most easily reared in the laboratory or those of economic importance. Insects demonstrate an extraordinary diversity and have applied an extraordinarily imaginative range of solutions to biological problems. It is to be expected that new processes and new hormones will be discovered as the scope of entomological research widens and our appreciation of insects grows.

See Also the Following Articles

Drosophila; Juvenile Hormone; Locusts; Rhodnius Prolixus

Bibliography

Girardie, J., Richard, O., and Girardie, A. (1992). Time-dependent variations in the activity of a novel ovary maturating neurohormone from the nervous corpora cardiaca during oogenesis in the locust, Locusta migratoria migratorioides. *J. Insect Physiol.* **38**, 215–221.

Glinka, A. V., Kleinman, A. M., and Wyatt, G. R. (1995). Roles of juvenile hormone, a brain factor and adipokinetic hormone in regulation of vitellogenin biosynthesis in Locusta migratoria. *Biochem. Mol. Biol. Int.* **35**, 323–328.

Goldsworthy, G. (1994). Adipokinetic hormones of insects: Are they the insect glucagons? In *Perspectives in Comparative Endocrinology* (K. G. Davey, R. E. Peter, and S. S. Tobe, Eds.), pp. 486–492. National Research Council, Ottawa, Canada.

Gupta, A. P. (1983). *Neurohemal Organs of Arthropods.* Thomas, Springfield, IL.

Nijhout, F. H. (1994). *Insect Hormones.* Princeton Univ. Press, Princeton, NJ.

Orchard, I., and Loughton, B. G. (1985). Neurosecretion. In *Comprehensive Insect Physiology, Biochemistry & Pharmacology* (G. A. Kerkut and L. I. Gilbert, Eds.), pp. 61–107. Pergamon, New York.

Raina, A. (1993). Neuroendocrine control of sex pheromone biosynthesis in Lepidoptera. *Annu. Rev. Entomol.* **38**, 329–349.

Sedlack, B. J. (1985). Structure of endocrine glands. In *Comprehensive Insect Physiology Biochemistry & Pharmacology* (G. A. Kerkut and L. I. Gilbert, Eds.), pp. 25–60. Pergamon, New York.

Willey, R. B. (1961). The morphology of the stomodeal nervous system in Periplaneta americana (L.) and other Blatteria. *J. Morphol.* **108**, 219–262.

Corpus Luteum

Jennifer L. Juengel, Eric W. McIntush, and Gordon D. Niswender

Colorado State University

I. Introduction
II. Background
III. Development of the Corpus Luteum
IV. Control of Luteal Function during the Reproductive Cycle in the Absence of Pregnancy
V. Luteolysis
VI. Maternal Recognition of Pregnancy
VII. Future Research
VIII. Summary

GLOSSARY

luteinization The differentiation of thecal and/or granulosal cells into luteal cells.

luteolysis A process that the corpus luteum undergoes at the end of a reproductive cycle. Luteolysis is characterized by decreased secretion of progesterone from the corpus luteum and luteal cell death.

luteolytic hormone A hormone that causes decreased secretion of progesterone from the corpus luteum and luteal cell death.

luteotropic hormone A hormone that is necessary for maintenance of the corpus luteum and that increases secretion of progesterone.

maternal recognition of pregnancy The process by which the conceptus signals the mother to maintain secretion of progesterone from the corpus luteum for maintenance of pregnancy.

progesterone A steroid hormone that is essential for maintenance of pregnancy.

pseudopregnancy A condition that occurs in animals such as rats and rabbits in which an infertile mating causes secretion of progesterone from the corpus luteum to be extended, with the mother undergoing endocrinological changes similar to those seen in pregnancy. Pseudopregnancy typically is of shorter duration than pregnancy.

I. INTRODUCTION

The corpus luteum is a transient endocrine gland that forms in the ovary from residual follicular tissues following ovulation. Normal function of the corpus luteum is critical to the reproductive process since the corpus luteum produces the steroid hormone progesterone (Fig. 1), which is required for pregnancy. In general, the number of corpora lutea formed is the same as the number of follicles that ovulate. In litter-bearing species, multiple follicles ovulate resulting in multiple corpora lutea; whereas species that usually have single ovulations, such as humans, usually have a single corpus luteum.

The progesterone produced by the corpus luteum acts on the uterus to prepare it for pregnancy and acts on higher brain centers to prevent additional ovulations and suppress sexual receptivity in most mammals. The length of reproductive cycles (period of time from one ovulation to the next) in females is dictated in part by life span of the corpus luteum. If pregnancy does not occur, corpora lutea must stop producing progesterone and physically regress so that subsequent ovulation and mating can occur. This allows the opportunity for pregnancy during the next reproductive cycle.

Characteristics of luteal function that are common among most species are emphasized in this article, with important differences between corpora lutea of humans, sheep, rats, and rabbits discussed where appropriate. These species were selected because they are representative of mammals in which a variety of mechanisms are used to regulate luteal function (Table 1).

II. BACKGROUND

The mechanisms that regulate development and life span of the corpus luteum are different between species, and these differences have an important influence on the frequency with which opportunities for pregnancy occur. Reproductive efficiency for a given species is determined, in part, by the time from one opportunity for pregnancy to the next. Some species, such as humans and sheep, form corpora

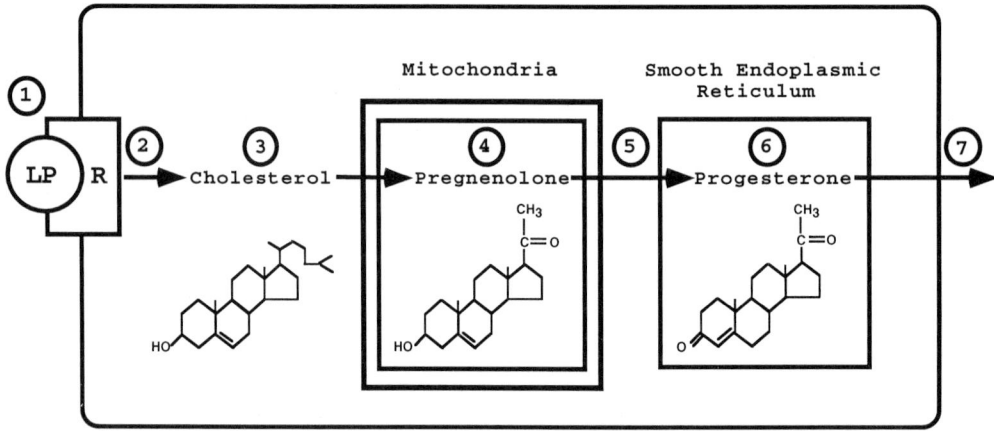

FIGURE 1 Synthesis of progesterone in luteal cells. Circulating lipoproteins (LP) containing cholesterol bind to receptors (1). Cholesterol is then internalized (2) and transported through the cell and across the outer and inner mitochondrial membranes (3). Enzymes located in the mitochondria convert cholesterol to pregnenolone (4), which then moves out of the mitochondria to the smooth endoplasmic reticulum (5), where it is enzymatically converted to progesterone (6), which then exits the cell (7). Luteotropic and luteolytic hormones can regulate synthesis of progesterone by regulating the ability of the cell to obtain lipoproteins from the bloodstream, the transport of cholesterol through the cell and into the mitochondria, or the enzymatic steps involved in conversion of cholesterol to progesterone. A major site of regulation appears to be transport of cholesterol into the mitochondria.

TABLE I
Various Parameters of Luteal Function in Humans, Sheep, Rats, and Rabbits

Parameter	Luteal function			
	Human	Sheep	Rat	Rabbit
Hormone-inducing ovulation and luteinization	LH	LH	LH	LH
Ovulation induced by mating	No	No	No	Yes
Pseudopregnancy	No	No	Yes	Yes
Length of cycle (days)[a]	28	17	4	NA
Length of gestation (days)[a]	265	145	21	32
Weight of the mature corpus luteum (mg)[a]	2000	550	4.5	17
Primary luteotropic hormone(s)	LH	LH	Estradiol and PRL	Estradiol
Primary luteolytic hormone	$PGF_{2\alpha}$	$PGF_{2\alpha}$	$PGF_{2\alpha}$	$PGF_{2\alpha}$
Mechanism(s) for maternal recognition of pregnancy	Secretion of hCG	Decreased secretion of $PGF_{2\alpha}$ due to INF-τ	Secretion of PL and androgens	Decreased secretion of $PGF_{2\alpha}$ and secretion of luteotropic(?) hormone

Note. Abbreviations used: NA, not applicable because the rabbit is an induced ovulator; LH, luteinizing hormone; PRL, prolactin; $PGF_{2\alpha}$, prostaglandin $F_{2\alpha}$; hCG, human horionic gonadotropin; INF-τ, interferon-τ; PL, placental lactogen(s).

[a] Average values; considerable variation normally occurs among individuals.

lutea that secrete progesterone for a prolonged period of time (10–16 days) during each reproductive cycle irrespective of whether mating has occurred. The extended period of secretion of progesterone results in uterine changes necessary for survival of developing embryos, but if the female is not pregnant it also delays ovulation and another chance for pregnancy. The result is reproductive cycles that are longer than those of species which do not have an extended luteal phase unless mating has occurred (such as rats and rabbits). In unmated rats progesterone is secreted by the corpus luteum for only 2 or 3 days, resulting in relatively brief reproductive cycles (4 days). Rabbits do not experience cyclic ovulation but rather mating initiates a neuroendocrine reflex arc resulting in ovulation and subsequent corpus luteum development. Corpora lutea of mated rats and rabbits produce progesterone for 10–15 days if the mating did not result in pregnancy. This condition is known as pseudopregnancy. Alternatively, in rats and rabbits that become pregnant as a result of mating, the corpus luteum produces progesterone for the duration of pregnancy (Table 1).

III. DEVELOPMENT OF THE CORPUS LUTEUM

The endocrine signal for ovulation, luteinizing hormone secreted from the anterior pituitary gland, induces breakdown of the follicular wall for release of the oocyte and triggers the cells of the follicle to undergo luteinization, i.e., differentiate into cells of the corpus luteum (Fig. 2). The transition of follicular tissue into luteal tissue is a dynamic process that includes differentiation, migration, and proliferation of cells. To appreciate the dynamic nature of development of corpora lutea, one must first appreciate the structure of the follicle. Follicles consist of four distinct layers; a theca externa, a highly vascular theca interna, a basement membrane, and an avascular granulosal layer (Fig. 2). Following ovulation, the basement membrane is degraded as cells of the theca interna and its associated capillary wreath invade the avascular granulosal layer. The previously compartmentalized follicular tissue becomes the corpus luteum, which consists of a heterogenous population of cells (Fig. 2) that differ morphologically, physio-

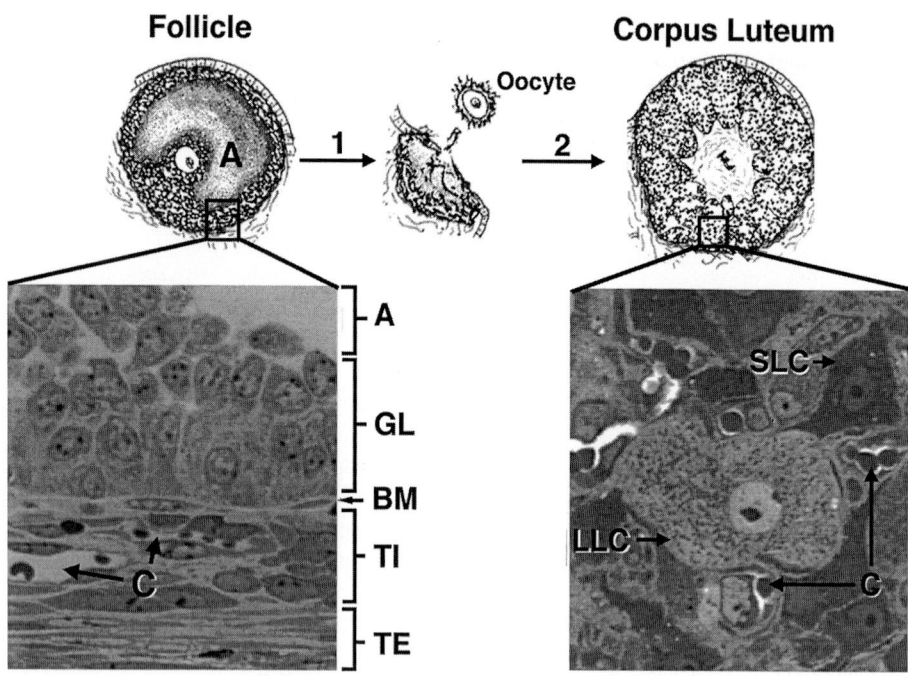

FIGURE 2 Development of a corpus luteum from a follicle. Prior to ovulation, the follicle is organized in distinct layers (see enlargement of the follicular wall). The antrum (A) of the follicle is a fluid-filled lumen that is surrounded by the granulosal layer (GL). The granulosal layer and oocyte are separated from the rest of the follicle by the basement membrane (BM). Outside the basement membrane are the theca interna (TI) and theca externa (TE) layers of the follicle. Cells of the granulosal and theca interna layers synthesize and secrete steroid hormones; however, the cells of the theca externa are not steroidogenic. Capillaries (C) of the vascular wreath surrounding follicles are present in the theca interna and externa but are absent in the granulosal layer because the basement membrane acts as a barrier to vascularization. Erythrocytes can often be observed in the lumen of capillaries. Luteinizing hormone causes breakdown of the follicular wall and release of the oocyte at ovulation (1). Following ovulation, cells of the theca interna and its associated capillaries cross the degraded basement membrane and invade the granulosal layer as the follicular tissue develops into the corpus luteum (2). The corpus luteum contains a heterogeneous population of cells that includes large steroidogenic luteal cells (LLC) and small steroidogenic luteal cells (SLC), which are proposed to be luteinized granulosal cells and thecal cells, respectively. Note the abundance of capillaries, which is indicative of the high degree of vascularization of the corpus luteum.

logically, and biochemically. In corpora lutea of sheep, rats, and rabbits, follicular-derived cells become extensively intermixed, whereas in humans intermixing of cells is less extensive.

Thecal and granulosal cells of the follicle, which previously worked in concert to produce estradiol as the predominant steroid, differentiate into separate types of luteal cells (small and large luteal cells—thecal lutein and granulosal lutein cells, respectively; Fig. 2), both of which produce progesterone. The corpus luteum also contains fibroblasts and endothelial cells. The function of these nonsteroidogenic cells is to provide the structural organization and contribute to the microenvironment of the corpus luteum. A significant characteristic of early luteal development is the rapid rate of cell proliferation, primarily of endothelial cells, associated with the growth of follicular-derived tissue into a corpus luteum that weighs 15–20 times more than the original follicular tissue. In fact, the rate of cellular proliferation during luteal development is similar to that seen in rapidly growing tumors. Many of the proliferating cells contribute to the extensive capillary network of the corpus luteum. Once established, the capillary network supports blood flow to the corpus luteum at a rate that exceeds that in other tissues. A high rate of blood flow is required to provide the substrates and nutrients necessary to sustain the high metabolic rate of fully developed corpora lutea.

IV. CONTROL OF LUTEAL FUNCTION DURING THE REPRODUCTIVE CYCLE IN THE ABSENCE OF PREGNANCY

Corpora lutea of most species have two steroidogenic cell types referred to as large luteal cells and small luteal cells. In general, large luteal cells have a diameter ≥ 20 μm and small luteal cells have a diameter <20 μm (Fig. 2). Under basal conditions, large luteal cells secrete more progesterone than small luteal cells; however, small luteal cells are more responsive to the luteotropic hormone, luteinizing hormone. A luteotropic hormone is a hormone that is necessary for maintenance of the corpus luteum and increases secretion of progesterone. Steroidogenic cells account for only 30% of the cells in the mature corpus luteum; however, because of their large size, they occupy approximately 60% of the volume of the corpus luteum. The other cells of the corpus luteum (endothelial cells, fibroblast, and resident immune cells) provide the structural framework of the corpus luteum as well as synthesize products that regulate secretion of progesterone.

Luteinizing hormone is luteotropic in most species, but the relative importance of luteinizing hormone in maintaining progesterone production differs among species. In women, removal of luteinizing hormone causes a precipitous decline in progesterone secretion. Ablation of luteinizing hormone in the ewe causes progesterone secretion to decrease; however, some progesterone is still secreted indicating that the ovine corpus luteum is less dependent on luteinizing hormone than the human corpus luteum. In the rat, during pseudopregnancy or pregnancy, both prolactin and estradiol are necessary for normal luteal function. Prolactin is essential for maintenance of luteal receptors for both estradiol and luteinizing hormone. Luteinizing hormone stimulates secretion of estradiol, which in turn enhances secretion of progesterone. Estradiol is the primary luteotropic hormone in the mated rabbit. In this species, luteinizing hormone and follicle-stimulating hormone stimulate secretion of estradiol from follicles that in turn support secretion of luteal progesterone. Growth hormone also supports luteal function in some species. The corpus luteum also secretes other products including peptides and protein hormones such as relaxin, oxytocin, insulin-like growth factor-1, and basic fibroblastic growth factor. Most of these products influence, directly or indirectly, secretion of progesterone from the corpus luteum.

V. LUTEOLYSIS

If the oocyte was not fertilized following ovulation, then progesterone secreted by the corpus luteum is not needed to maintain pregnancy and secretion of progesterone must be stopped before a new ovulation can occur. Hormones that cause luteolysis, defined as decreased secretion of progesterone and luteal cell death, are termed luteolytic hormones. In many species, including the ewe, rat, and rabbit, prostaglandin $F_{2\alpha}$ is secreted into the bloodstream from the uterine endometrium and acts on luteal cells to cause luteolysis. In humans, prostaglandin $F_{2\alpha}$ is also thought to cause luteolysis; however, the source of prostaglandin $F_{2\alpha}$ is most likely the corpus luteum itself. Prostaglandin $F_{2\alpha}$ decreases blood flow to the corpus luteum and disrupts the intracellular mechanisms that produce progesterone, thus decreasing the concentration of progesterone in the blood. Prostaglandin $F_{2\alpha}$ also causes luteal cell death and an influx of immune cells into the corpus luteum that results in decreased luteal weight.

VI. MATERNAL RECOGNITION OF PREGNANCY

If fertilization of the oocyte occurs, progesterone secretion from the corpus luteum must be maintained for pregnancy to continue. Thus, the conceptus must send a signal to maintain secretion of progesterone from the corpus luteum. This process involves secretion of a luteotropic substance(s) and/or preventing the secretion of the luteolytic hormone. Maternal recognition of pregnancy must occur prior to the time that normal luteal regression would occur. The human conceptus secretes human chorionic gonadotropin that is similar to luteinizing hormone. Human chorionic gonadotropin binds to the same receptor as luteinizing hormone and maintains secretion of progesterone from the corpus luteum. The conceptus of the ewe secretes a protein, interferon-τ, that decreases secretion of prostaglandin $F_{2\alpha}$ by the uterus. In the rat, mating initiates surges of prolactin that sustain the corpus luteum for the first half of pregnancy or the duration of pseudopregnancy if fertilization did not occur. The conceptus must secrete placental lactogen to maintain secretion of progesterone from the corpus luteum in the second half of pregnancy. In addition, the corpus luteum becomes dependent on androgens produced by the placenta to serve as substrate for the synthesis of estradiol. Thus, at about the midpoint of gestation the rat placenta secretes both placental lactogen (which binds to prolactin receptors) and androgens to maintain secretion of progesterone from the corpus luteum. In the rabbit, ovulation is induced by a neuroendocrine reflex arc that is triggered by mating. If fertilization did not occur the uterus releases prostaglandin $F_{2\alpha}$ and the corpus luteum regresses about 16 days after mating. The presence of a rabbit conceptus suppresses uterine prostaglandin $F_{2\alpha}$ secretion. In addition, the rabbit placenta secretes a luteotropic substance that enhances the ability of estradiol to increase progesterone synthesis. Thus, the conceptus maintains secretion of progesterone from the corpus luteum by secreting a luteotropic hormone(s) and/or preventing the secretion of a luteolytic hormone. In most primates and ruminants, the placenta eventually secretes adequate quantities of progesterone to ensure continued maintenance of pregnancy. The switch from dependence on luteal progesterone to placental progesterone occurs at species-specific times during pregnancy (60–70 days into gestation in humans and approximately 50 days in sheep). However, in rats and rabbits progesterone secretion from the corpus luteum is required for the duration of the pregnancy.

VII. FUTURE RESEARCH

A complete understanding of the regulation of luteal function in mammals is critical from a number of perspectives. First, inadequate secretion of progesterone and/or faulty signals for maternal recognition of pregnancy result in the loss of 20–50% of embryos early in pregnancy in all mammals including humans. This has important implications for the efficient production of food and in human infertility. The actions of luteal steroids provide a major component of current contraceptive methods and controlling the function of this gland is a potential target for future contraceptives. A number of important questions regarding luteal function remain to be answered. For example, it is clear that the different cell types in the corpus luteum communicate with each other but most of the details of the communication pathways remain to be elucidated. Second, while it is thought that $PGF_{2\alpha}$ produced by the human corpus luteum causes the demise of this organ, the mechanisms involved in this process are not well understood. The role played by nonsteroidogenic luteal cells and the extracellular microenvironment in regulating luteal function also remains to be determined. Additional research is needed to obtain the necessary information to resolve these and other issues.

VIII. SUMMARY

The primary function of the corpus luteum is synthesis and secretion of progesterone. The actions of progesterone include preparation of the uterus for pregnancy and suppression of ovulation and sexual receptivity. Because pregnancy cannot be maintained without progesterone, proper function of the corpus luteum is required for propagation of mammalian species. Recurrent opportunities to establish pregnancy are presented by cyclic ovulation followed by luteal development, maintenance, and regression. These events are highly regulated in an endocrine manner by pituitary- and uterine-derived factors. Once pregnancy has been initiated, developing embryos must interrupt the reproductive cycle by preventing regression of the corpus luteum so that production of progesterone will continue and thereby maintain a uterine environment conducive to survival and development of embryos.

See Also the Following Articles

FOLLICULAR DEVELOPMENT, CONTROL OF; LUTEOLYSIS; LUTEOTROPIC HORMONES; PSEUDOPREGNANCY

Bibliography

Gibori, G., Khan, I., Warshaw, M. L., McLean, M. P., Puryear, T. K., Nelson, S., Derkee, T. J., Azhar, S., Steinschneider, A., and Rao, M. C. (1988). Placental-derived regulators and the complex control of luteal cell function. *Recent Progress Hormone Res.* 44, 377–429.

Hillard, J. (1973). Corpus luteum function in guinea pigs, hamsters, rats, mice and rabbits. *Biol. Reprod.* 8, 203–221.

Holt, J. A. (1989). Regulation of progesterone production in the rabbit corpus luteum. *Biol. Reprod.* 40, 201–208.

Nappi, C., Gargiulo, A. R., and DiCarlo, C. (1994). The human luteal paracrine system: Current concepts. *J. Endocrinol. Invest.* 17, 825–836.

Niswender, G. D., and Nett, T. M. (1994). Corpus luteum function and its control in infraprimate species. In *The Physiology of Reproduction* (Knobil and Neill, Eds.), 2nd ed., Vol. 1, pp. 781–816. Raven Press, New York.

Stouffer, R. L. (1996). Corpus luteum formation and demise. In *Reproductive Endocrinology, Surgery, and Technology* (Adashi, Rock, and Rosenwaks, Eds.), Vol. 1, pp. 251–269. Lippincott-Raven, Philadelphia.

van Tienhoven, A. (1983). *Reproductive Physiology of Vertebrates*, 2nd ed. Cornell Univ. Press, Ithaca, NY.

Zeleznik, A. J., and Benyo, D. F. (1994). Control of follicular development, corpus luteum function, and the recognition of pregnancy in higher primates. In *The Physiology of Reproduction* (Knobil and Neill, Eds.), 2nd ed., Vol. 2, pp. 751–782. Raven Press, New York.

Corpus Luteum of Pregnancy

Richard L. Stouffer
Oregon Health Sciences University

I. Introduction
II. Long-Lived Corpora Lutea
III. Short-Lived Corpora Lutea
IV. Ultrashort-Lived Corpora Lutea

GLOSSARY

luteal–placental shift An interval in pregnancy when the essential activities (i.e., progesterone production) of the corpus luteum are assumed by the placenta.

maternal recognition of pregnancy A mechanism whereby the developing conceptus signals its presence to maternal tissues, particularly the corpus luteum, and extends luteal function in early pregnancy.

viviparity The ability to produce living offspring instead of eggs from within the body, in the manner of mammals and some reptiles and fishes.

The corpus luteum (yellow body) is an ephemeral endocrine gland that forms in the ovary of mammals (and some reptiles) from the wall of the follicle after ovulation. The organ acquired its name from the accumulation of lipid and lipid-soluble pigments in cytoplasmic droplets of the luteal cells. The differentiation of the corpus luteum from the ovulatory follicle leads to a shift in the predominant form of steroid hormone secreted by the ovary—from estrogen (estradiol-17β) by the maturing follicle to progestin (progesterone) by the functional corpus luteum. The evolution of the corpus luteum is associated with the achievement of viviparity in mammalian species; the secretion of progesterone is the critical event in preparing the uterus for implantation and in providing an intrauterine environment that maintains gestation until timely delivery of the fetus at parturition. Reinier de Graaf (1672) was the first person to describe the conversion of the mature ovarian follicle (i.e., Graafian follicle) into the corpus luteum. More than 200 years passed before Frankel (1901) demonstrated the essential role of the corpus luteum in pregnancy by terminating gestations in rabbits via removal of corpora lutea. It took another 30 years before Corner and colleagues (1929) isolated the lipoid substance from the corpus luteum, formed crystals of the purified substance, and called the steroid hormone they discovered progesterone. Since then, investigators have spent considerable effort characterizing the activities of the corpus luteum in various species and attempting to unravel the factors and processes that control the functional life span of the corpus luteum. By understanding these events, methods might be developed to improve fertility of some animals (e.g., domesticated animals that serve as food sources) as well as to treat specific types of infertility in women. Also, treatment modalities could be sought to limit fertility (e.g., in populations of rodents or pet animals) and to serve as contraceptives that prevent unwanted pregnancy in women.

I. INTRODUCTION

The corpus luteum differs from many other endocrine glands in the remarkable differences that have evolved between species for its regulation during the ovulatory ovarian cycle and in pregnancy. As adapted from Rothchild (1981), mammals (except marsupials and monotremes) can be divided into three major categories based on how long the corpus luteum functions after ovulation in the nonfertile ovarian cycle and how the luteal life span is affected by pregnancy onset and gestation. Species with long-lived corpora lutea include monestrous animals, e.g., carnivores such as the dog. In these animals, the corpus luteum functions for many days/weeks after ovulation, irregardless of whether fertilization and implantation occurs, and the luteal life span is not affected by pregnancy. Species with *short-lived corpora lutea* include polyestrous animals such as many domesticated farm animals (e.g., sheep, cow, and horse) and primates (monkeys, great apes, and humans). Herein, the corpus luteum develops and functions for a finite period during the ovarian cycle, but the luteal life span is markedly increased if pregnancy ensues. Species with ultrashort-lived corpora lutea include the polyestrous rodents (e.g., mouse and rat) and the rabbit. These animals either do not ovulate spontaneously or do not form a fully developed corpus luteum in the absence of mating. Mating results in ovulation (rabbit) and development of functional corpora lutea (rabbits and rodents) whose life spans are extended further in pregnancy.

Thus, species have developed diverse means to ensure that ovarian cyclicity returns in the event of an infertile cycle, yet adequate progesterone support is provided if pregnancy ensues. In the simplest case (e.g., marsupials and many carnivores), gestation lasts as long as or is shorter than the normal life span of the functional corpus luteum in the ovarian cycle. In this scenario, there is no need for the conceptus or mother to alter luteal function because the corpus luteum of the cycle is designed to provide the necessary progestational support. However, most other mammals have developed a means for keeping the functional life span of the corpus luteum "short" (i.e., less than necessary for successful pregnancy) in the nonfertile cycle to increase reproductive efficiency. Therefore, processes evolved to maintain adequate progestational support should pregnancy ensue, typically by circumventing luteolysis at the end of the fertile cycle and/or prolonging the functional life span of the corpus luteum. The mechanism whereby the developing conceptus extends luteal function is the critical event in "maternal recognition of pregnancy." In some species, the corpus luteum of pregnancy is a critical source of progesterone throughout pregnancy. In others, fetal tissue (i.e., the trophoblast) within the placenta ultimately replaces the corpus luteum as the primary source of progesterone. In such species, including primates, the corpus luteum becomes unnecessary after the luteal–placental shift as luteal activity is replaced by placental functions.

II. LONG-LIVED CORPORA LUTEA

There are some monestrous species in which a fully functional corpus luteum develops after ovulation and its functional life span does not differ significantly regardless of whether the corpus luteum originates during a nonfertile or fertile cycle. In the dog, for example, the length of the luteal phase in pregnant and nonpregnant animals in 64–66 days. Progesterone levels peak between Days 10 and 40 after the ovulatory luteinizing hormone (LH) surge and then gradually decline to term. The pituitary hormone, prolactin, is considered luteotropic during pregnancy and pharmacologic suppression of prolactin support will cause premature luteolysis and pregnancy loss. The only known pregnancy-specific hormone in dogs is relaxin, which is secreted by the corpus luteum (and uterus) during the latter third of pregnancy. At the end of pregnancy, there is rapid luteolysis which is thought to result from lost luteotropic (prolactin) support. In the nonpregnant animal, luteolysis is very protracted; the mechanisms are unclear. Although prostaglandin (PG) $F_{2\alpha}$ is luteolytic in dogs, hysterectomy does not alter the life span of the corpus luteum, suggesting that luteolysis is not dependent on a uterine luteolytic factor (see Section III,B).

Other animals, including the marsupials except for bandicoots, may conceptually be included in this category. Even though the life span of the corpus luteum only lasts for ~14 days, it is not altered by the occurrence or lack of pregnancy or by the uterus.

III. SHORT-LIVED CORPORA LUTEA

A. Primates

In primate species, such as Old World monkeys, great apes, and humans, a functional corpus luteum forms after ovulation and exists for a sufficient interval (~2 weeks) to permit timely movement of the early embryo through the oviduct and to prepare the uterus for implantation. However, luteolysis near the end of the menstrual cycle occurs before a developing placenta can initiate local functions (e.g., progesterone production) to keep the uterus in a supportive, yet quiescent, state throughout gestation. The mechanism(s) causing luteolysis in primates is not known; but it is clear that the demise of the corpus luteum is not due to a luteolytic factor secreted from the uterus since hysterectomy does not alter cyclic ovarian function in women or monkeys. "Self-destruct" mechanisms have been proposed, whereby substances such as prostaglandin $F_{2\alpha}$ or estrogen synthesized within the ovary or corpus luteum initiate luteolysis near the end of the nonfertile cycle.

Although the processes limiting the functional life span of the primate corpus luteum in the nonfertile cycle are unclear, it is well established that the "rescue" of the corpus luteum from its impending demise in the fertile cycle is due to an additional luteotropic signal, chorionic gonadotropin (CG), first secreted by the developing conceptus around the time of implantation. CG is a glycoprotein hormone that is structurally related to the pituitary gonadotropin, LH, which promotes luteal structure and function during the menstrual cycle. However, there are significant differences in the numbers, types, and sulfation of carbohydrate residues, particularly on the amino acid extension of the C-terminus of CG, that markedly increase the hormone's circulating half-life and hence its bioactivity. The gonadotropic hormones are members of a superfamily of endocrine/paracrine factors (including nerve growth factor, transforming growth factor-β, and platelet-derived growth factor) that contain a cysteine knot within the dimer (e.g., α and β CG subunits) structure.

Although CG secretion appears characteristic of pregnancy in all primates, the patterns and peak levels vary among species. Peak CG concentrations are highest in women, about 10-fold lower in apes, and >100-fold lower in baboons and Old World or New World monkeys. The exact time when CG is first produced by the conceptus remains unresolved, but evidence suggests that early embryos produce little CG until the time of uterine attachment (implantation). Thereafter, CG levels rise exponentially as the placenta develops; thus, it is not the fetus per se that provides the signal for maternal recognition of pregnancy but the syncytiotrophoblast of the developing placenta. The complexity and dynamics of the placenta, as well as the difficulty in obtaining normal tissues from early pregnancy, hinder efforts to understand the mechanisms controlling CG synthesis and secretion. However, an intriguing hypothesis arises from evidence that hormones originally observed in the hypothalamus [i.e., gonadotropin-releasing hormone (GnRH)] and ovary (i.e., inhibin, activin, and progesterone) which are positive or negative regulators of pituitary gonadotropin secretion are also present in the placenta and act as local modulators of CG production. Further studies are required to establish the physiologic importance of these factors and what role, if any, they have in the different patterns of CG secretion in primates. For example, in humans and apes, CG levels peak in the first third of pregnancy, then decline but remain at substantial levels throughout gestation. However, in baboons and macaques, CG levels become nondetectable by midpregnancy.

In every primate species studied, data support the concept that placental CG shares a common receptor on the luteal cell membrane with pituitary LH; occupation of the receptor by either CG or LH stimulates adenylate cyclase and cyclic AMP production in luteal cells, with subsequent enhancement of cyclic AMP-dependent protein kinase activity and progesterone synthesis. To date, most studies on CG action have focused on acute responses; little is known about subsequent effects on cellular parameters promoting luteal structure or differentiation. Limited evidence suggests that CG exposure results in divergent changes in the expression of mRNAs and proteins involved in steroid and hormone production by the corpus luteum. In monkeys, CG appears to stimulate progesterone production by enhancing the processes making cholesterol sources available for steroidogenic pathways (e.g., the low-density lipoprotein receptor), not by chronically stimulating key enzymes (e.g., P450 side chain cleavage or 3β-hydroxysteroid dehydrogenase) in progesterone synthesis. However, CG also markedly stimulates estrogen (via P450 aromatase expression) synthesis as well as the production of two protein hormones, relaxin and inhibin. The cellular mechanisms whereby CG stimulates relaxin or inhibin expression are obscure, especially since acute exposure to CG or cyclic AMP has no effect on relaxin production by luteal cells *in vitro*. The involvement of cAMP-independent effector systems (e.g., the phosphoinositol–protein kinase C pathway) in CG action, or indirect action via other local factors, awaits investigation.

It remains unclear why the production of another LH-like hormone, i.e., CG, rescues the corpus luteum in early pregnancy. LH is secreted by the pituitary in an episodic manner and the frequency of LH pulses slows during the mid- to late luteal phase of the menstrual cycle. However, maintenance of high-frequency LH pulses (via GnRH treatment or pulsatile administration of LH) does not prevent timely regression of the corpus luteum in monkeys or women. These observations suggest that a decrease in pituitary LH support is not the primary cause of luteolysis at the end of the nonfertile cycle. Due to differences in secretion patterns and circulating half-life between LH and CG, there are qualitative and then quantitative differences in the gonadotropin milieu in early pregnancy. The intermittent pulses of LH translate into intervals of gonadotropin support and withdrawal which correlate with luteal steroidogenic activity late in the nonfertile cycle. Since CG has a much longer half-life than LH in the bloodstream and is produced in increasing amounts by the developing trophoblast, CG circulates continuously and in rising levels until obscuring any remaining LH pulses. Thus, the conceptus offers uninterrupted, increasing levels of luteotrophic support for the corpus luteum. Whether the changing pattern of LH-like support is a primary factor in further differentiation of the corpus luteum in early pregnancy awaits investigation. Alternatively, there may be important differences in the cellular actions of CG vs LH that facilitate primate luteal function. Finally, it remains to be determined if CG prevents a local signal(s) for luteolysis from occurring in the fertile cycle or overcomes the actions of such luteolysins.

Thus, CG not only rescues the corpus luteum from impending luteolysis and extends its progestational function, it also promotes (e.g., estrogen and inhibin) or begins (e.g., relaxin) other luteal activities even when progesterone secretion is declining near the

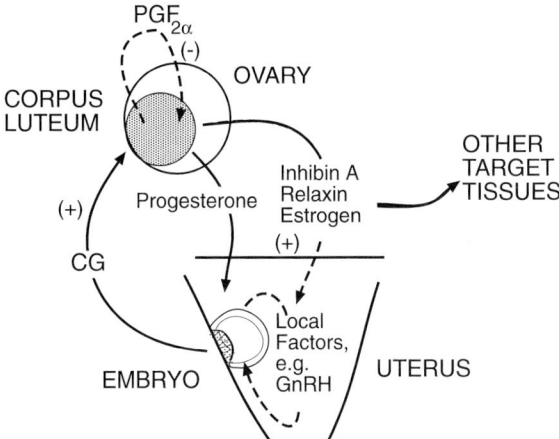

FIGURE 1 Model of the endocrine and local mechanisms active during extension of corpus luteum function at the start of pregnancy in primate species. Solid lines indicate established pathways, whereas dotted lines depict proposed pathways. Following implantation, embryo-derived CG, whose synthesis may be promoted locally by GnRH or other factors, prevents the typical regression of the corpus luteum at the end of the nonfertile menstrual cycle. CG, an LH-like luteotropic hormone, may prevent the synthesis or action of a putative local luteolytic factor in the corpus luteum, e.g., $PGF_{2\alpha}$. Hence, the corpus luteum continues to produce progesterone which sustains intrauterine pregnancy until the luteal–placental shift. CG also stimulates luteal production of other steroid (i.e., estrogen) and protein (i.e., relaxin and inhibin A) hormones that may facilitate intrauterine gestation (adapted from Stouffer and Hearn, 1998).

time of the luteal–placental shift. Early pregnancy is the final interval of luteal differentiation in primates—a process that starts with the LH surge at ovulation and ends with CG secretion postimplantation. While progesterone of luteal origin is indispensable, the roles of other products, such as estrogen, inhibin, and relaxin, are unknown. Figure 1 summarizes the current perspective on the endocrine and local mechanisms activated during rescue of the corpus luteum in early pregnancy of primate species. The luteal–placental shift occurs by about the sixth week of pregnancy in women and the third week of gestation in rhesus macaques. The corpus luteum of pregnancy then degenerates by midgestation (it is not clear if the process at this stage is comparable to the luteolytic events at the end of the cycle). However, there is intriguing evidence that the corpus luteum "rejuvenates" by term pregnancy and is again a progesterone-secreting gland, but its importance is unclear in view of the massive steroidogenic activity of the placenta.

B. Domesticated Farm Animals

In sheep and cattle, the two species most extensively used for studies on estrous cycles and pregnancy in ruminants, the corpus luteum forms after ovulation and functions until Day 17 or 21 postestrus, respectively. Research, particularly in the ewe, demonstrates the primary control of the uterus over the life span of the corpus luteum in ruminants. In sheep, luteolysis in the nonfertile cycle occurs in response to the pulsatile release of $PGF_{2\alpha}$ by the uterus. Hysterectomy during the functional life of the corpus luteum prevents timely luteolysis and its life span is prolonged to about 5 months, the duration of normal pregnancy. Moreover, injection of $PGF_{2\alpha}$ causes premature luteolysis and immunization against PGF blocks luteolysis in ewes. It is less clear what mechanisms activate endometrial production of luteolytic $PGF_{2\alpha}$ in the event that pregnancy does not occur during the estrous cycle.

It appears that a complex scenario involving actions of progesterone from the corpus luteum, estrogen from follicles maturing in the late luteal phase, and oxytocin from the posterior pituitary and/or the corpus luteum regulates $PGF_{2\alpha}$ synthesis and secretion from the uterine epithelium. It is clear that uterine exposure to progesterone for 10–12 days not only prepares the uterus for establishment of pregnancy but also primes the mechanisms for $PGF_{2\alpha}$ production in the event that pregnancy does not occur. Results from *in vivo* studies support the model that as chronic progesterone exposure downregulates its own receptor in the uterine luminal epithelium, progesterone loses its ability to suppress the expression of endometrial receptors for estrogen (ER) and oxytocin (OTR). Circulating estrogen further stimulates uterine ER and OTR expression. Oxytocin from the corpus luteum and posterior pituitary stimulates pulsatile release of uterine $PGF_{2\alpha}$ into the uterine vein where, due to their intimate anatomical relationship, it passes by countercurrent exchange into the ovarian artery and causes luteolysis.

It is well established that extension of the functional life span of the corpus luteum in the fertile cycle is due to the production of an antiluteolytic substance by the developing conceptus. Embryos transferred to synchronized recipients as late as Day 12 postestrus will prevent timely luteolysis; hence, the critical period for maternal recognition of pregnancy in the sheep is Days 12 or 13 of the fertile cycle. Evidence indicates that the conceptus secretes antiluteolytic activity, which is not an LH- or prolactin-like substance, beginning about Day 10 and increasing through Day 16, but it is nondetectable by Days 21 or 22. The antiluteolytic substance in ewes, and presumably other ruminants, is interferon-τ (IFN-τ, initially termed ovine trophoblast protein or TP-1) produced by mononuclear cells of the embryonic trophectoderm.

Ovine IFN-τ was originally called trophoblastin because it is the first major protein secreted by the trophectoderm of the peri-implantation conceptus. Following cDNA cloning and amino acid sequencing, these proteins were identified as type 1 interferons and confirmed as functional interferons with potent antiviral, antiproliferative, and immunomodulatory activities. There is high sequence homology of IFN-τs across ruminant species, in which they apparently have a unique role in the establishment of pregnancy. There are multiple IFN-τ isoforms which arise from multiple genes; but their functional significance remains unclear since individual ovine IFN-τ isoforms all extend luteal function when injected into the uterine lumen of cyclic ewes, cows, and goats.

The mechanism of action of IFN-τ to prolong the function of the corpus luteum is under active investigation. Concentrations of circulating progesterone in pregnant and nonpregnant ewes are not different through the first 14 days postestrus but are maintained at this level through Day 50 in gestation. Thereafter, the placenta begins to secrete progesterone and provides adequate progestational support until parturition. It is well established that the high-amplitude pulses of luteolytic $PGF_{2\alpha}$ coming from the uterus around luteolysis in nonpregnant ewes are absent in fertile cycles. IFN-τ appears to act locally on the uterine epithelium to prevent and/or alter the pattern of $PGF_{2\alpha}$ secretion. Although basal secretion of $PGF_{2\alpha}$ is higher in early pregnancy, the large-amplitude, episodic pulses of $PGF_{2\alpha}$ do not occur; also, secretion of $PGF_{2\alpha}$ may be directed into the uterine lumen rather than the bloodstream in some species. IFN-τ appears to interact with type 1 IFN receptors in the endometrium to block expression of estrogen and oxytocin receptors, thereby preventing estrogen and oxytocin action and consequently luteolytic pulses of $PGF_{2\alpha}$. Whether IFN-τ action to suppress estrogen receptor expression is the primary event in preventing the luteolytic process awaits further study. Also, the mechanisms whereby IFN-τ suppresses receptor expression in the endometrium have not been defined, but they likely involve the activation of latent tyrosine kinases (JAK1 and tyk2) and production of several members of the type 1 IFN-induced transcription factor family. A yin–yang interaction between IFN regulatory factor-1 (IRF-1) and IRF-2 may be responsible for regulation of IFN-τ-responsive genes in the ruminant uterus.

Since intrauterine infusions of IFN-τ alone can extend the life span of the corpus luteum, some investigators assume that this is the only pregnancy recognition factor produced by ruminant conceptuses. However, others noted that the sensitivity of the corpus luteum to the luteolytic action of $PGF_{2\alpha}$ is reduced in early pregnancy. Moreover, luteal resistance to $PGF_{2\alpha}$ was transient and coincided with the time of maternal recognition of pregnancy. The uterine endometrium secretes a different PG, PGE_2, in greater amounts during early pregnancy; this PG promotes luteal progesterone production and suppresses the luteolytic action of estrogen and $PGF_{2\alpha}$. The physiologic role of PGE_2 or an embryonic protein as a uterine luteotropin which reduces the luteolytic effectiveness of $PGF_{2\alpha}$ remains to be established.

Figure 2 summarizes the current perspective of the endocrine and local mechanisms activated during rescue of the corpus luteum in ruminant species, such as sheep, goats, and cows. The mechanisms are considerably different in other farm animals, such as pigs and horses. The corpus luteum in the pig is required throughout gestation as a source of progesterone to maintain pregnancy. It is proposed that estrogens secreted by the porcine conceptus, rather than IFN-τ, prevent the luteolytic actions of uterine

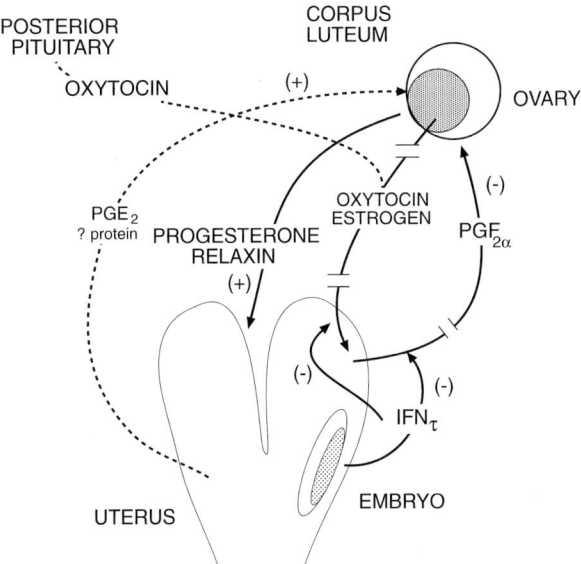

FIGURE 2 Model of the neuroendocrine, endocrine, and local mechanisms active during extension of corpus luteum function at the start of pregnancy in ruminants such as sheep. Solid lines depict established pathways, whereas dotted lines indicate proposed pathways. Following implantation, embryo-derived IFN-τ prevents the pulsatile release of the uterine luteolytic factor PGF$_{2\alpha}$, thereby preventing the timely regression of the corpus luteum at the end of the nonfertile estrous cycle. Secretion of PGE$_2$ or unknown (?) proteins from the pregnant uterus may also promote luteal function. Hence, the corpus luteum continues to produce progesterone which sustains intrauterine pregnancy until the luteal–placental shift.

PGF$_{2\alpha}$, but this awaits rigorous testing. A similar mechanism may occur in the mare; however, in addition, the fetal trophoblast begins secreting massive quantities of equine chorionic gonadotropin (eCG; also known as pregnant mare serum gonadotropin) around Day 35 of gestation. This LH-like gonadotropin induces ovulation of large antral follicles growing in early pregnancy and formation of secondary corpora lutea. The function of the primary corpus luteum of the cycle is also stimulated by eCG. The corpora lutea secrete progesterone until about Day 150 of gestation, when eCG disappears from the circulation. At this time, the luteal–placental shift occurs and progesterone support for the remainder of pregnancy in the mare is provided by the fetal–placental unit.

IV. ULTRASHORT-LIVED CORPORA LUTEA

Rodents such as the laboratory rat are spontaneously ovulating, polyestrous animals that do not form fully developed, functional corpora lutea unless mating occurs. In the absence of mating, the estrous cycle lasts 4 or 5 days and lacks a true luteal phase, thus ensuring rapidly occurring cycles with frequent opportunities for mating and establishment of pregnancy. After ovulation, the ruptured follicle takes on some of the characteristics of the corpus luteum, but the luteal gland never attains maximal size and is considered nonfunctional since it does not secrete sufficient progesterone to permit uterine decidualization (e.g., in response to implantation). Any progesterone produced by the corpus luteum is rapidly metabolized in the luteal cell to 20α-hydroxyprogesterone (20α-OHP), which is an inactive progestin that will not support pregnancy.

Mating or cervical stimulation during estrus results in development of functional corpora lutea that secrete progesterone for 12–14 days. If conception does not occur (e.g., as induced experimentally by mating with a sterile male), this interval of luteal function is known as pseudopregnancy. The event promoting luteal development is a neuroendocrine reflex arc in response to cervical stimulation that elicits a surge of prolactin secretion from the pituitary. In rats, mating causes diurnal and nocturnal surges of prolactin. Prolactin treatment will promote luteal function during the estrous cycle in rats, whereas an inhibitor of prolactin secretion prematurely terminates pseudopregnancy. Thus, in many rodents, the luteotropic actions of prolactin are essential for conversion of the undeveloped, inactive corpus luteum of the estrous cycle into the functional corpus luteum of pseudo- or early pregnancy.

The mechanism(s) of action of prolactin on the rodent corpus luteum has been only partially elucidated. It appears that prolactin increases the receptors for other luteotropic factors, i.e., pituitary LH and follicular estrogen, within luteal cells. Although LH acutely enhances progesterone synthesis, the crucial effect of LH appears to be to promote luteal formation of estradiol, which acts intraluteally to sustain progesterone production. Estradiol (not LH per se) is essential to sustain the corpus luteum; but prolactin is necessary for its action. Prolactin also abolishes the activity of the enzyme, 20α-hydroxysteroid dehydrogenase, which metabolizes progesterone to 20α-OHP. Thus, prolactin promotes the secretion of progesterone by the corpus luteum not by stimulating its synthesis but by preventing its metabolism. Prolactin also promotes the expression of elongation factor-2 (EF_2), an important intracellular protein for peptide elongation and an essential component of protein synthetic machinery. The intracellular pathways of prolactin action and the role of EF_2 in luteal development or function await further study.

Around Day 12 postmating, the diurnal surges of prolactin cease and luteolysis occurs unless implantation and embryo development occurs in the fertile cycle. Successful pregnancy in rodents requires progesterone secretion by the corpus luteum throughout gestation. To sustain luteal activity, first the maternal decidua and then the embryonic trophoblast secrete prolactin-like hormones. Implantation in rodents involves transformation of uterine stromal cells into decidual cells; a compartment of the uterine decidua secretes numerous hormones, including prolactin-like protein B (PLP-B) and decidual prolactin-like protein (dPLP). Although pituitary prolactin is secreted until Days 11 or 12 of pregnancy, its presence is not absolutely required after Day 6, i.e., 1 day after implantation. By this time, the decidual tissue is secreting a luteotropic factor which is believed to be primarily dPRP. Decidual PRP binds to prolactin receptors on luteal cells and has prolactin-like activity, e.g., it suppresses 20α-HSD activity. By midpregnancy, the decidual tissue regresses and dPRP secretion ceases around the time that maternal pituitary prolactin surges wane. These events coincide with the appearance of placental lactogens secreted by the trophoblast beginning on Day 10 of gestation.

Placental lactogens are found in a variety of mammals, but physiologic roles for these proteins are only established in rodents. In the rat, seven members of the prolactin gene family are expressed by the

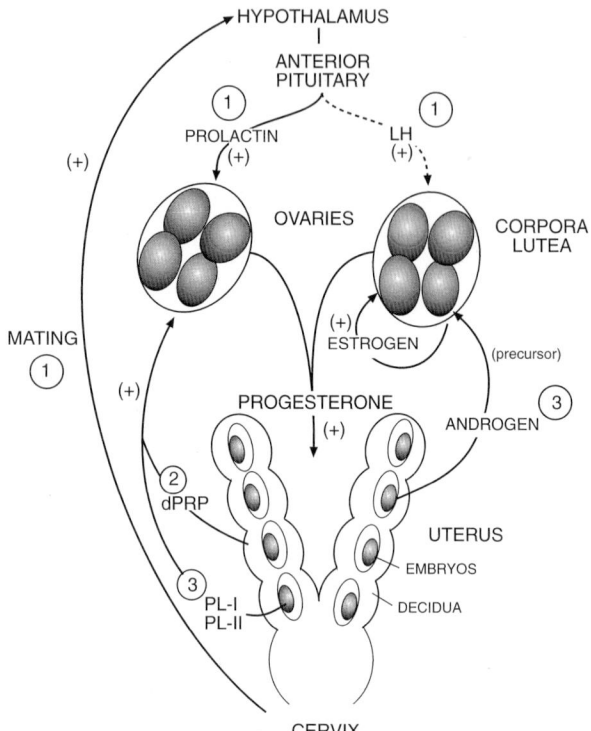

FIGURE 3 Schematic of the neuroendocrine, endocrine, and local mechanisms responsible for extension of corpus luteum function during pregnancy in rodents such as the rat. Three different sources/types of prolactin-like hormones provide luteotropic support as gestation proceeds: (1) A neuroendocrine reflex arc activated by cervical stimulation at mating results in prolactin release from the anterior pituitary; (2) by Day 6 of pregnancy, decidual PRP also provides support; and (3) by mid- to late pregnancy, placental lactogens (PL-I then PL-II) are the primary luteotropins. However, intraluteal estrogen is also a critical luteotropin. In early pregnancy, LH promotes estrogen production, but by mid- to late pregnancy androgens provided by the placenta are the primary precursors for estrogen production by the corpus luteum. Prolactin-like hormones and estrogen combine to maintain luteal structure and function throughout pregnancy, which is essential since the rodent placenta is not a significant source of progesterone.

trophoblastic cells in cell-, positional-, and temporal-specific patterns during mid- to late pregnancy. Four members, placental lactogen I (PL-I), PL-I variant, PL-II mosaic, and PL-II, bind to prolactin receptors. There is now direct evidence for luteotropic actions of PL-I and -II during Days 10–12 and after Day 12 of pregnancy, respectively. PL-I and -II, as well as other family members, may have additional roles during pregnancy, e.g., to promote mammary gland development.

Although the rodent placenta does not secrete significant amounts of progesterone or estrogen, fetal trophoblast cells acquire the capacity to secrete androgens at midpregnancy. This source of estrogen precursor appears critical for continued luteotropic support of the corpus luteum because maternal pituitary LH (as well as prolactin) secretion declines at midgestation and, hence, luteal production of androgen drops. Placental androgens secreted into the maternal circulation serve as precursors for luteal estrogen production, so the level of estrogen does not decline and it continues to act locally to promote luteal structure and function. Thus, by secreting what is generally considered a male sex steroid, androgen, it allows the formation of the intraluteal tropic factor, estrogen, and continued luteal progesterone secretion after midpregnancy.

The regression of the corpus luteum at the end of pseudopregnancy or pregnancy is believed to be controlled by the uterine luteolytic factor $PGF_{2\alpha}$. Exogenous $PGF_{2\alpha}$ is luteolytic in rodents. Thus, in rodents, the functional life span of the corpus luteum is extended by luteotropic factors (prolactin-like proteins and estrogen) whose tissue origins vary as gestation proceeds (Fig. 3). In response to mating and then intrauterine events, the structure and function of the corpus luteum are promoted. Ultimately, its demise is controlled by the uterus, in which either the absence of concepti following mating or the timely delivery of fetuses generates a luteolytic signal.

See Also the Following Articles

Chorionic Gonadotropins; Corpora Lutea of Nonmammalian Species; Graafian Follicle; LH (Luteinizing Hormone); Luteinization; Luteolysis; Pregnancy, Maintenance of; Progesterone Actions on Reproductive Tract

Bibliography

Bazer, F. W., Ott, T. L., and Spencer, T. E. (1998). Endocrinology of the transition from recurring estrous cycles to establishment of pregnancy in subprimate mammals. In *Endocrinology of Pregnancy* (F. W. Bazer, Ed.), Contemporary Endocrinology Series, pp. 1–34. Humana Press, Clifton, NJ.

Gibori, G., Khan, I., Warshaw, M. L., McLean, M. P., Puryear, T. K., Nelson, S., Durkee, T. J., Azhar, S., Steinschneider, A., and Rao, M. C. (1988). Placental-derived regulators and the complex control of luteal cell function. *Recent Prog. Horm. Res.* 44, 377–429.

Hodgen, G. D., and Itskovitz, J. (1988). Recognition and maintenance of pregnancy. In *The Physiology of Reproduction* (E. Knobil and J. D. Neill, Eds.), 1st ed., pp. 1995–2021. Raven Press, New York.

Niswender, G. D., and Nett, T. M. (1994). Corpus luteum and its control in infraprimate species. In *The Physiology of Reproduction* (E. Knobil and J. D. Neill, Eds.), 2nd ed., pp. 781–816. Raven Press: New York.

Rothchild, I. (1981). The regulation of the mammalian corpus luteum. *Recent Prog. Horm. Res.* 17, 183–299.

Stouffer, R. L., and Hearn, J. P. (1998). Endocrinology of the transition from menstrual cycles to establishment of pregnancy in primates. In *Endocrinology of Pregnancy* (F. W. Bazer, Ed.), Contemporary Endocrinology Series, pp. 35–37. Humana Press, Clifton, NJ.

Corpus Luteum Peptides

O. David Sherwood
University of Illinois

Phillip A. Fields
University of South Alabama

I. Luteal Oxytocin of the Estrous Cycle
II. Luteal Relaxin of the Estrous Cycle
III. Luteal Oxytocin during Pregnancy
IV. Luteal Relaxin during Pregnancy
V. Summary and Directions for Future Research

GLOSSARY

corpus luteum The yellow endocrine body that is formed in the ovary in the site of a ruptured graafian follicle and secretes the steroid hormone progesterone.
luteolysis The demise of the corpus luteum that is associated with reduced production and release of progesterone.
oxytocin A nonapeptide hormone that is secreted by the neural lobe of the pituitary in all mammalian species and is also secreted by the corpus luteum in some mammals.
relaxin A peptide hormone with a structure similar to that of insulin that is produced by the corpus luteum and/or other portions of the reproductive tract during pregnancy in mammals.

The corpus luteum of mammalian species secretes progesterone, and this steroid hormone acts on the uterus to enable both implantation and maintenance of pregnancy. Two peptide hormones, namely oxytocin and/or relaxin, are also produced and secreted by the corpus luteum in some mammalian species. Moreover, there is evidence that the luteal oxytocin and/or relaxin have physiological roles. This article describes the synthesis, secretion, and physiological roles of luteal oxytocin and relaxin during both the estrous cycle and pregnancy in those few species in which information on this topic is available.

I. LUTEAL OXYTOCIN OF THE ESTROUS CYCLE

Oxytocin is a nonapeptide that is synthesized in the hypothalamus as a precursor polyprotein of approximately 128 kDa. The polyprotein consists of a N-terminal signal sequence, oxytocin, and its carrier neurophysin I. Following synthesis, the hormone is transported via axons to the posterior pituitary for release into the bloodstream. Pituitary oxytocin stimulates uterine contractions during parturition and is involved in the milk ejection reflex during nursing.

In 1980, a "contractin factor" that stimulated mouse uterine contractions was identified in extracts of cow corpora lutea. Shortly thereafter, oxytocin was identified in extracts of corpora lutea from the sheep, goat, cow, and human being. The corpus luteum consists of both large and small luteal cells, and it has been demonstrated in the sheep and the cow that only the large luteal cells produce oxytocin. Electron microscopic immunochemical analysis demonstrated that both oxytocin and neurophysin are stored in secretory granules within the corpus luteum (Fig. 1).

A. Synthesis and Secretion

Luteal oxytocin synthesis and secretion has been studied most extensively throughout the approximately 21-day estrous cycles in the cow and ewe.

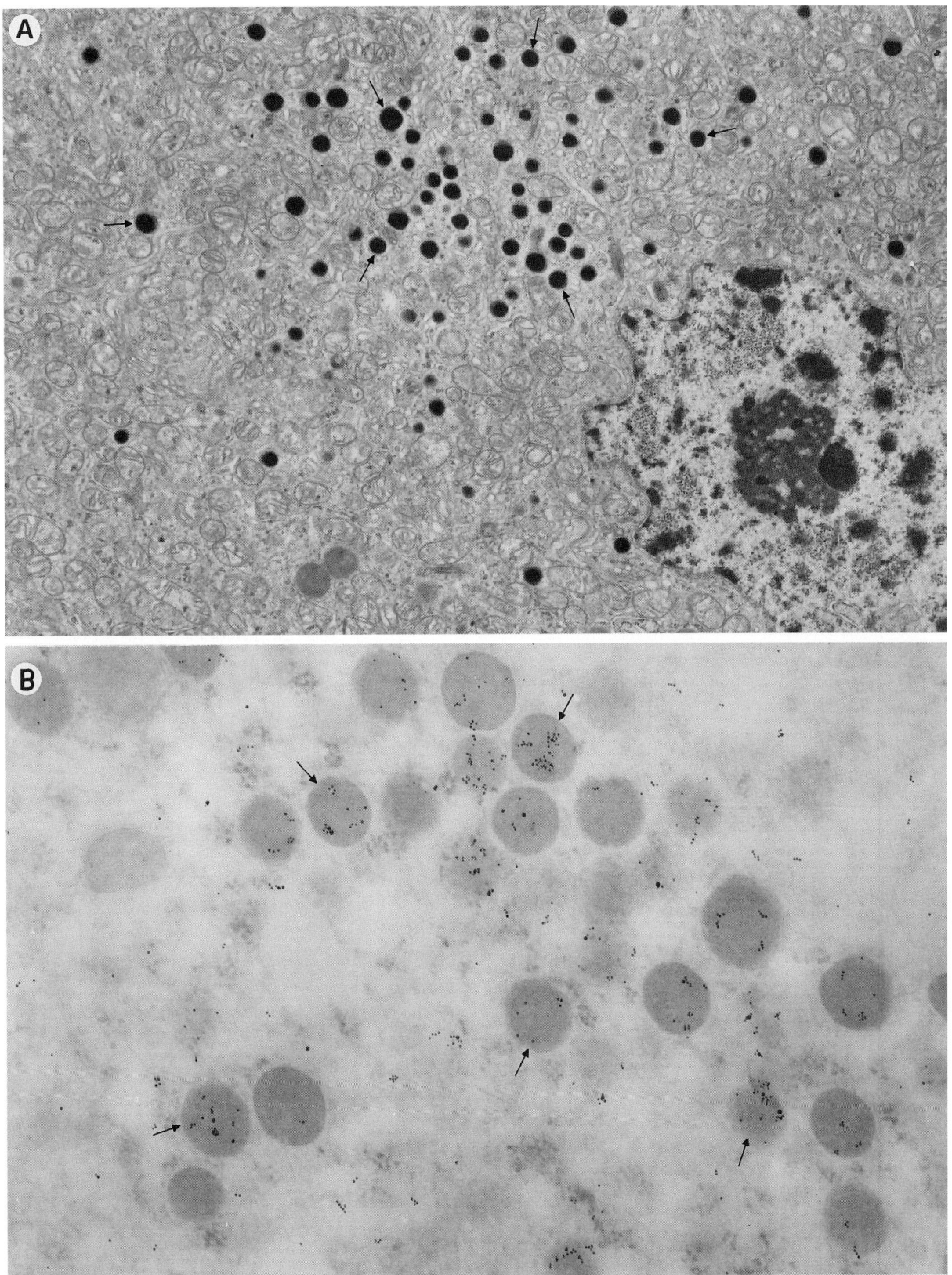

FIGURE 1 Cow corpus luteum from Day 11 of the estrous cycle. (A) Large luteal cell with a cluster of secretory granules (arrows). (B) Tissue section incubated with antioxytocin serum and gold-labeled IgG. Gold labeling of the secretory granules within the large luteal cell (arrows) indicates the presence of oxytocin. A, ×16,000; B, ×52,000.

Oxytocin–neurophysin mRNA is nearly undetectable in granulosa cells during the follicular phase of the cycle. Following ovulation (Day 0 of the cycle) oxytocin mRNA content surges to maximal levels within 3 days and then declines rapidly up to midcycle. The factors that promote luteal oxytocin mRNA production are not known, but the ovulatory surge of LH/FSH may prime the oxytocin gene. Bovine and/or ovine granulosa cells obtained from follicles collected after the ovulatory surge of LH/FSH have a greatly increased capacity for oxytocin synthesis. Experiments employing primary cultures of granulosa cells collected at about the time of the LH surge provided limited evidence that FSH, LH, IGF-1, estradiol, and catecholamines promote oxytocin synthesis. However, most of these factors then switch to an inhibitory influence. Efforts to identify factors that promote oxytocin secretion in primary cultures of luteal cells have not proved successful.

The production of luteal oxytocin lags well behind the translation of luteal oxytocin–neurophysin mRNA. Peak levels of oxytocin immunoactivity do not occur until the midluteal phase at about Days 8–12, and then luteal oxytocin levels decline rapidly. Following translation of the oxytocin–neurophysin mRNA, the prohormone is packaged into secretory granules where it undergoes a series of enzymatic steps culminating in amidation to form oxytocin. The delay in posttranslational processing of oxytocin precursor appears to be attributable to the delayed availability of the enzyme peptidyl glycine α-amidating monooxygenase, which is the terminal enzyme required for oxytocin synthesis.

In cows and ewes a portion of the oxytocin is released as it is processed so that the profile of blood levels of oxytocin is similar to that of luteal levels. The factors that regulate the release of oxytocin during this early to midluteal phase, when most of the oxytocin is released, are unknown. A gradual decrease in oxytocin secretion occurs from midluteal stage until luteolysis begins on about Day 18 when oxytocin is released in relatively small but distinct surges that last about 2 or 3 hr and occur at intervals of about 8 hr. In vivo, $PGF_{2\alpha}$ is a potent stimulator of luteal oxytocin release at the time of luteolysis. It is thought that the release of oxytocin during luteolysis is attributable, at least in part, to the direct effects of $PGF_{2\alpha}$ that emanates from the uterus. Studies with luteal cells in vitro have failed to provide evidence that prostaglandins act directly to release oxytocin, and it may be that $PGF_{2\alpha}$ requires an intermediary step. There is evidence that 5-lipoxygenase products of arachidonic acid metabolism act directly on luteal cells to release oxytocin and that this action is calcium dependent. Calcium may play a role in initiating the release of oxytocin-laden secretory granules from the cytoplasmic actin network that holds the secretory granules at the plasma membrane.

B. Physiological Role

There is evidence in cows, ewes, goats, and pigs that luteal oxytocin plays a role in bringing about luteolysis during the estrous cycle if fertilization of the ovum does not occur. The working hypothesis is that luteal oxytocin is secreted into the bloodstream, binds to receptors in the uterus, and stimulates the uterus to secrete the luteolysin $PGF_{2\alpha}$. $PGF_{2\alpha}$ then enters the bloodstream, binds to large luteal cells, and induces both luteolysis and release of more oxytocin that acts on the uterus to bring about additional release of $PGF_{2\alpha}$. Several lines of evidence support the existence of this positive feedback between the corpus luteum and uterine endometrium. In the cow and sheep, endometrial oxytocin receptors rise sharply during the late luteal phase when luteolysis occurs. The administration of oxytocin not only stimulates the release of $PGF_{2\alpha}$ from the endometrium in cows and goats but also induces luteolysis in cows and pigs. Additionally, neutralization of oxytocin extends the estrous cycle in the goat.

The impact of the midcycle release of oxytocin on luteolysis, if any, has not been explained. However, two populations of large luteal cells (α and β) with different sensitivities to $PGF_{2\alpha}$ were recently identified (Fields and Fields, 1996). The β cells release their oxytocin-laden secretory granules in response to a subluteolytic dose of $PGF_{2\alpha}$, whereas the α cells are unaffected. Thus, it is possible that the β cells are responsible for the early release of oxytocin (prior to day 14). This oxytocin does not cause luteal regression but may upregulate its own endometrial receptors. Then, under the influence of increased oxytocin-stimulated release of uterine $PGF_{2\alpha}$, the α cells release their oxytocin at a time when oxytocin receptors are plentiful. This secondary release by the α cells may be responsible for initiating the final luteolytic pulses of $PGF_{2\alpha}$.

Luteal oxytocin has also been identified in the human, cynomolgus monkey, and baboon. However, luteal oxytocin's role in humans and other primates, if any, has not been identified.

C. Estrous Cycle to Pregnancy Transition

The luteal oxytocin content during early pregnancy in the cow is identical to that observed during the first 11 days of the estrous cycle. Its release, however, does not cause luteolysis as is the case in nonpregnant animals. The failure of luteolysis to occur in pregnant cows and sheep appears to be attributable to the presence of the conceptus. The conceptus secretes INF-τ, which inhibits expression of the endometrial oxytocin receptors and oxytocin-induced secretion of endometrial $PGF_{2\alpha}$. Thus, the positive feedback loop between luteal oxytocin and uterine $PGF_{2\alpha}$ is interrupted. Consistent with this hypothesis, the depletion of luteal oxytocin takes place over a longer period in pregnant cows than in cycling cows.

II. LUTEAL RELAXIN OF THE ESTROUS CYCLE

Relaxin has a molecular weight of about 6000. This protein hormone consists of two chains with three disulfide linkages whose locations are homologous to those in insulin. Although luteal relaxin is most abundant during pregnancy, it has also been demonstrated to be present in the corpus luteum of the nonpregnant pig and human being.

A. Synthesis and Secretion

In the pig, relaxin is a product of the theca interna cells in developing follicles during the preovulatory phase of the estrous cycle. Luteinization of the granulosa and theca cells occurs at ovulation and both luteal cell types appear to produce relaxin. Relaxin immunostaining and mRNA levels are elevated during the midluteal phase (Days 11–15 postovulation) and then they decline as progesterone levels decline at luteolysis. As during pregnancy, relaxin is packaged in secretory granules of the large luteal cells during the estrous cycle (Fig. 2). In the human being and rhesus monkey, low levels of relaxin are produced by the corpus luteum of the menstrual cycle. In pigs, humans, and rhesus monkeys relaxin is secreted during the cycle, and maximal blood levels occur during the midluteal phase.

B. Physiological Role

Although relaxin is produced during the estrous cycle, no clear physiological role for the hormone in the nonpregnant animal has emerged. As a product of the theca interna layer in the developing follicle, relaxin may have a paracrine/autocrine role in modulating ovarian function. Relaxin, insulin, and insulin-like growth factor-1 (IGF-1) promote DNA synthesis by granulosa cells of small, medium, and large porcine follicles. Immunoneutralization of IGF-1 inhibits relaxin-induced DNA synthesis; therefore, relaxin-induced granulosa cell DNA synthesis may be mediated by IGF-1. Consistent with the previous findings, relaxin promotes proliferation of granulosa cells from small follicles.

In the rat, there is limited evidence that relaxin may act via a paracrine mechanism to promote ovulation. The effects of relaxin on ovulation may be mediated by enhanced secretion of metalloproteinases from ovarian interstitial as well as granulosa cells. These enzymes may be involved in digesting the connective tissue in the area of the stigma and thus weaken the ovarian wall to allow ovulation to occur.

III. LUTEAL OXYTOCIN DURING PREGNANCY

A. Production and Secretion of Oxytocin

With the exception of sheep and cattle, little is known concerning luteal oxytocin during pregnancy.

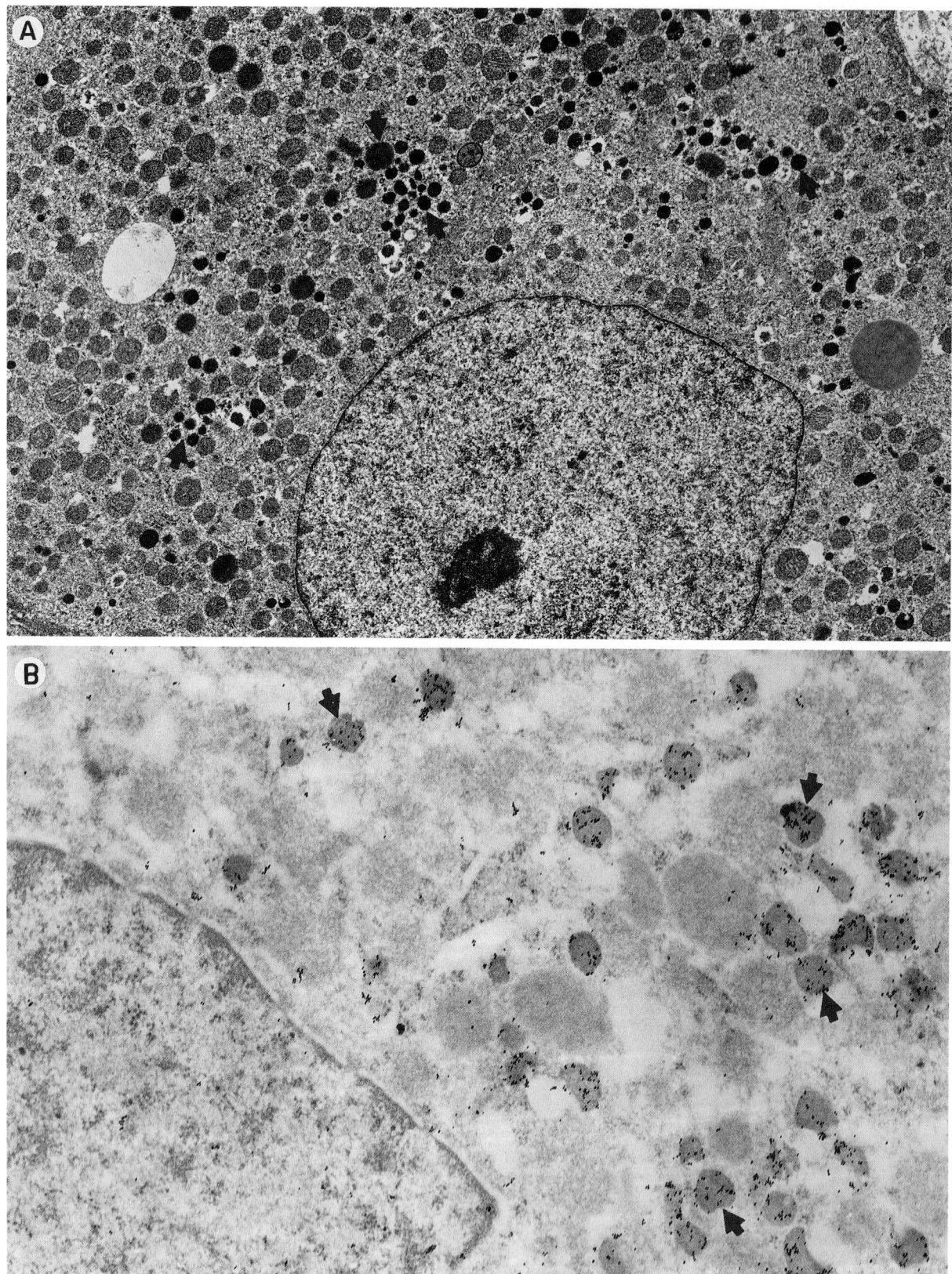

FIGURE 2 Pig corpus luteum from Day 14 of the estrous cycle. (A) Large luteal cell with secretory granules (arrows). (B) Tissue section incubated with antirelaxin serum and gold-labeled IgG. Gold labeling of the secretory granules within the large luteal cell (arrows) indicates the presence of relaxin. A, ×10,200; B, ×30,000.

In these two species, both luteal oxytocin and oxytocin mRNA are low during early pregnancy. In cattle, luteal oxytocin and oxytocin mRNA levels increase at term and there is limited evidence that luteal oxytocin may be secreted during late pregnancy in cattle.

B. Physiological Roles of Oxytocin

There is currently no evidence that luteal oxytocin has a physiological role during pregnancy in any species.

IV. LUTEAL RELAXIN DURING PREGNANCY

A. Production and Secretion of Relaxin

Relaxin is produced in highest levels during pregnancy. The tissue that is the primary source of the hormone varies among species—it may be the corpus luteum, placenta, or uterus. This chapter describes the production and secretion of relaxin from the corpus luteum of the pig, rat, and human being—the three species in which the regulation of the production, secretion, and physiological roles of relaxin are best understood.

1. Pig

In the pig the corpora lutea produce and secrete both relaxin and progesterone throughout the approximately 114-day gestation period. Luteal levels of both relaxin mRNA and relaxin increase markedly between Days 12 and 16. Relaxin then accumulates progressively in dense membrane-limited cytoplasmic granules (200–600 nm in diameter) until approximately 2 days before delivery when degranulation occurs and luteal relaxin levels decline rapidly. Luteal production of relaxin requires the pituitary but does not require the presence of the conceptuses. Luteal relaxin levels increase steadily as the age of the corpora lutea increases.

The profile of relaxin levels in the peripheral blood throughout pregnancy reflects the fact that a portion of the relaxin accumulates within the corpora lutea and is released with the rapid degranulation that occurs during the 2 days before delivery. Plasma relaxin levels increase gradually until late pregnancy. During the 2 or 3 days before birth, relaxin levels surge to attain maximal levels of 50–250 ng/ml 24–14 hr before delivery (Fig. 3A). Essentially nothing is known concerning the factors that regulate the secretion of relaxin from the corpora lutea during the first 110 days of gestation. It is known that the antepartum surge in relaxin is associated with functional luteolysis, but the factors that regulate the process are not well understood. There is limited evidence that the fetal pituitary–adrenal–placental system is involved in the initiation of the antepartum surge in serum relaxin levels and precipitous decline in progesterone levels that occurs at luteolysis. Although the nature of the factor(s) that initiates luteolysis remains to be identified, there is good evidence that prostaglandins are involved. It is currently not known whether the prostaglandins that contribute to the release of relaxin from the corpus luteum are secreted by the uterus or are produced within the ovary.

2. Rat

As in the pig, the corpora lutea are the source of both the relaxin and the progesterone released into the peripheral blood during the approximately 23-day gestation period. Ovarian levels of both relaxin mRNA and relaxin increase from about Day 10 of gestation until 2 or 3 days before birth—the so-called antepartum period. Some relaxin accumulates in cytoplasmic membrane-bound granules (100–270 nm diameter) until the antepartum period, when the luteal cells degranulate and ovarian relaxin content declines rapidly to low levels. Unlike the pig, maintenance of the corpus luteum and luteal production of high levels of relaxin requires the presence of conceptuses.

The profile of relaxin levels in the peripheral blood throughout rat pregnancy is shown in Fig. 3B. Regulation of the release of relaxin from Day 10 to Day 20 differs from its regulation during the antepartum period. During the first period, relaxin becomes detectable in the serum by Day 10, rises rapidly to 40–80 ng/ml by Day 14, and remains relatively constant until Day 20. During this period relaxin synthesis, relaxin secretion, and progesterone secretion are promoted by at least one placental luteotrophic fac-

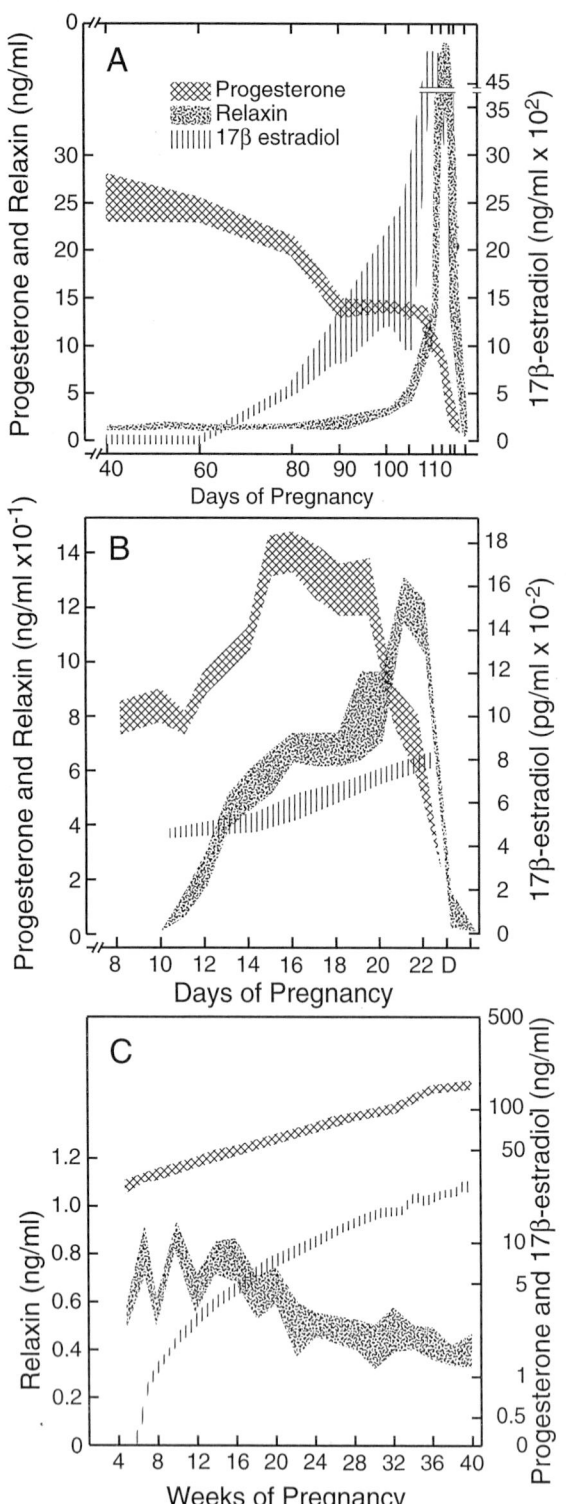

FIGURE 3 Mean peripheral serum levels (±SE) of relaxin, progesterone, and 17β-estradiol during pregnancy in (A) pigs, (B) rats, and (C) human beings. Hormone levels in A are adapted from Eldridge-White *et al.* (1989). Hormone levels in B are adapted from Sherwood *et al.* (1980), Pepe and Rothchild (1974), and Taya and Greenwald (1981). Hormone levels in C are adapted from Bell *et al.* (1987) and Buster *et al.* (1979).

tor(s) and inhibited by a pituitary factor(s). Neither the placental luteotrophic factor(s) nor the pituitary inhibitory factor(s) have been identified.

A surge in serum relaxin levels occurs during the 3-day antepartum period. Maximal serum relaxin levels generally range from 120 to 220 ng/ml and occur about 24 hr before birth. As with the pig, the antepartum relaxin surge is associated with functional luteolysis. There is evidence that the luteolytic process with its coincident antepartum surge in serum relaxin and decline in serum progesterone levels occurs at 24-hr intervals and that its timing is linked to the photoperiod. Moreover, the timing of the luteolytic process is influenced by the number of conceptuses. In rats with litters of three or fewer pups, luteolysis and birth are delayed 24 hr relative to rats with litters of five or more pups.

3. Human

In the human, the corpus luteum is the apparent source of all relaxin in the peripheral circulation. Relaxin is not detected in the peripheral serum of women with premature ovarian failure who become pregnant with egg donation and do not have a corpus luteum. Levels of relaxin within the corpus luteum of pregnancy in women are much lower than those in the corpora lutea of pregnant pigs and rats. There is no evidence that relaxin accumulates in membrane-bound granules in humans as it does in pigs and rats.

Relaxin levels are detectable in the peripheral blood of women throughout nearly all of pregnancy (Fig. 3C). Blood levels are higher during the first than during the second or third trimester. Available evidence indicates that human chorionic gonadotropin secreted by the trophoblast not only rescues the corpus luteum of pregnancy but also promotes relaxin secretion. Circulating relaxin levels in the pregnant woman remain one or two orders of magnitude lower than those attained in pigs and rats during late pregnancy. Also unlike the pig or the rat, there is neither functional luteolysis nor an antepartum surge in serum relaxin levels.

B. Physiological Roles of Relaxin

Research aimed at gaining a better understanding of the physiological roles of relaxin during pregnancy has been conducted most extensively with pigs and rats for two primary reasons. First, because the corpora lutea produce the relaxin that circulates within the peripheral blood, the source of the hormone can be readily removed for studies of relaxin's physiological roles during pregnancy. Second, the corpora lutea in pigs and rats contain relatively high levels of relaxin; therefore, sufficient relaxin to enable physiological studies of the hormone can be obtained more readily from these two species than from other domestic and laboratory species. The diverse and vital physiological roles of circulating luteal relaxin on reproductive tissues are emphasized in this section, whereas biological effects of relaxin on nonreproductive tissues are simply mentioned.

Some of relaxin's biological effects are known to be estrogen dependent in pigs and rats. Estrogen levels are elevated during the last third of pregnancy in both species (Figs. 3A and 3B). The placentas secrete increasing quantities of estrogen from Day 80 until term in pigs, and the developing ovarian follicles secrete increasing quantities of estrogen from Day 15 until term in rats.

1. Pig

During pregnancy, circulating luteal relaxin promotes growth of the entire lower reproductive tract (uterine portion of the cervix, vaginal portion of the cervix, and vagina; Figs. 4A–4C) and also increases the extensibility (softening) of both the uterine and vaginal portions of the cervix (Figs. 4D and 4E).

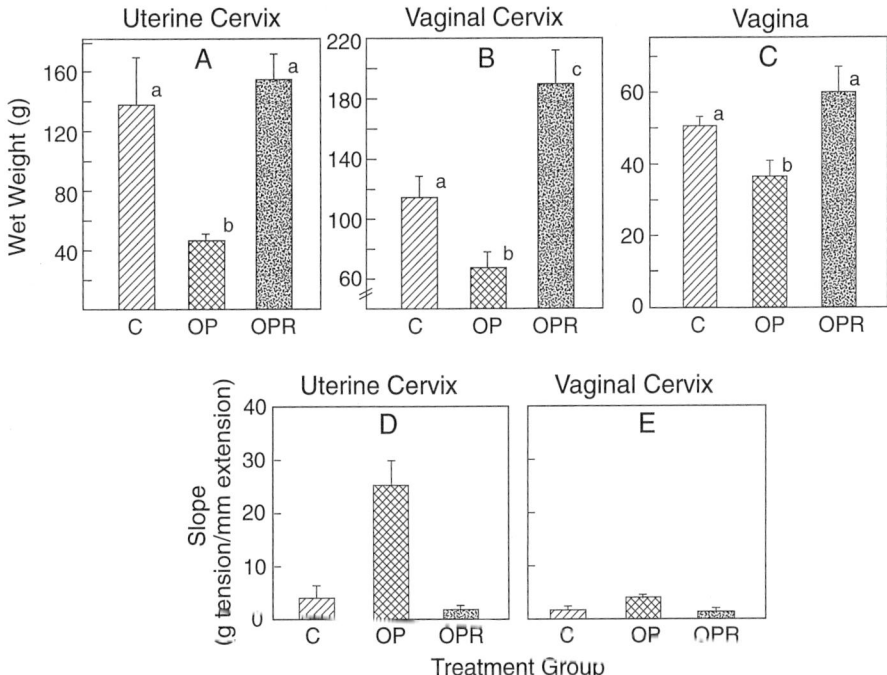

FIGURE 4 Influence of relaxin on growth and softening of the lower reproductive tract in the pregnant pig. (A–C) Mean wet weight (+SE) of the (A) uterine cervix, (B) vaginal cervix, and (C) vagina on Day 110 of pregnancy in control gilts (group C), ovariectomized gilts treated with progesterone (group OP), and ovariectomized gilts treated with progesterone plus relaxin (group OPR). Six to eight gilts were used in each group. Bars with different superscripts differ significantly ($p > 0.05$). Adapted from Min et al. (1997). (D and E) Mean slopes (+SE) of the linear regressions (grams of tension per millimeters of extension) of 1-cm segments from the uterine (D) and vaginal (E) portions of the cervix for each treatment group on Day 110 of pregnancy. Five or six gilts were used per group. Adapted from O'Day et al. (1989).

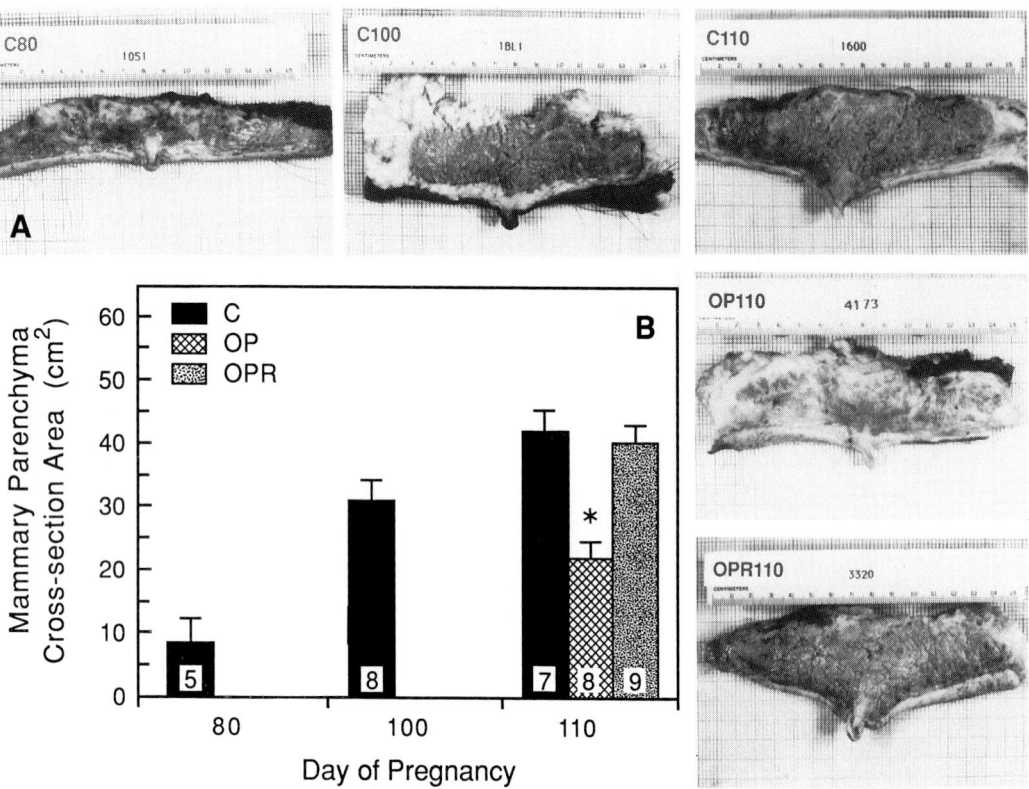

FIGURE 5 Influence of relaxin on mammary gland development in the pregnant pig. (A) Photographs of transverse cross sections of mammary glands from gilts during late gestation. Representative mammary glands are from control groups (C80, C100, and C110), ovariectomized progesterone-treated groups (OP110), and ovariectomized progesterone plus relaxin-treated groups (OPR110). (B) Mean (+SE) cross-section areas of mammary parenchymal tissue from gilts during late pregnancy. The number of animals is given at the base of each bar. *Significant difference from adjacent group ($p < 0.05$). Adapted from Hurley et al. (1991).

Relaxin's effects on the lower reproductive tract are vital at delivery. Abolishment of luteal relaxin by bilateral ovariectomy during late pregnancy prolongs the duration of delivery several fold and markedly reduces the incidence of live piglets at birth. Relaxin's effects on growth and softening of the uterine portion of the cervix are more profound than its effects on growth of the vaginal portion; therefore, it seems likely that its actions on the uterine portion of the cervix play the most important role in facilitating rapid and safe delivery.

Circulating relaxin contributes to the preparation of the mother for postpartum lactation. Relaxin plays a major role in promoting the marked development of mammary gland lobule–alveolar tissue that occurs during the last third of gestation in gilts (Fig. 5). Relaxin's effects on mammary lobule–alveolar development are estrogen dependent.

There are additional effects of relaxin in pregnant pigs whose physiological significance, if any, remains poorly understood. Circulating luteal relaxin promotes growth of the uterus and this effect may assist in accommodating the rapidly growing fetuses during late pregnancy. There is also limited evidence that relaxin acts directly on both the myometrium to inhibit uterine contractility and the skin to facilitate tissue expansion.

2. Rat

In rats, circulating relaxin promotes growth and extensibility of both the cervix and the vagina (Fig. 6). As with gilts, relaxin's effects on the lower repro-

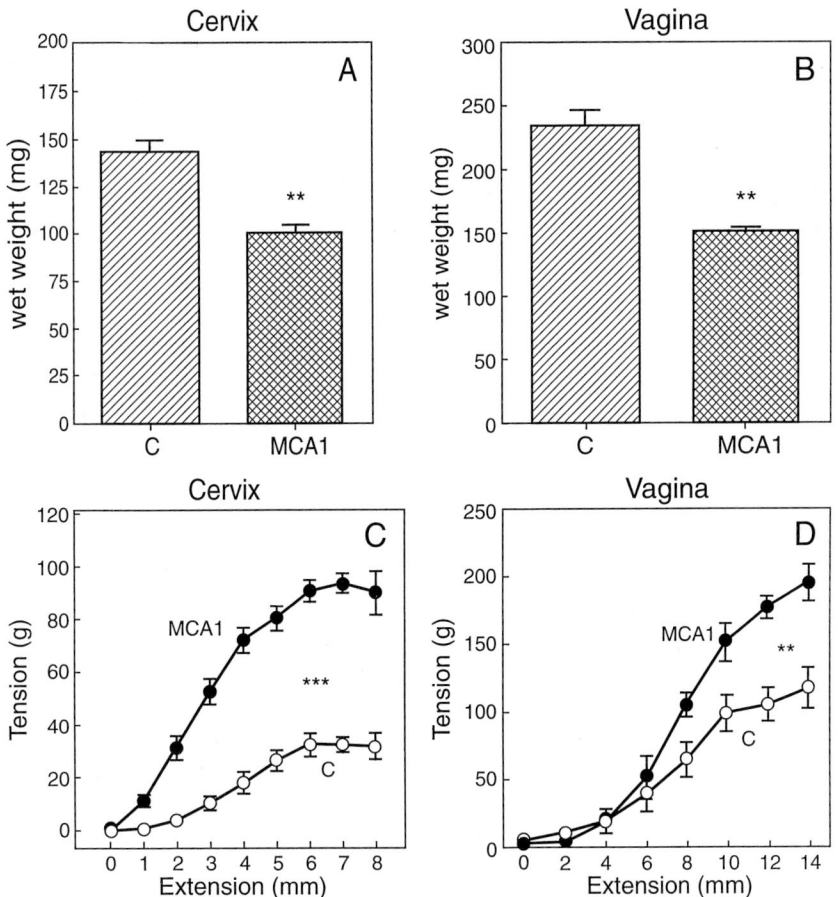

FIGURE 6 Influence of relaxin on growth and softening of the lower reproductive tract in the pregnant rat. Mean (+SE) wet weight of the (A) cervix and (B) vagina and extensibility of the (C) cervix and (D) vagina on Day 22 of pregnancy in control rats (group C) and rats in which circulating relaxin was neutralized throughout the second half of pregnancy with monoclonal antibody for rat relaxin (group MCA1). Eight rats were used in each group. Significantly different from controls (**$p < 0.01$; ***$p < 0.001$). Adapted from Hwang and Sherwood (1988) and Zhao et al. (1996).

ductive tract facilitate rapid and safe delivery. Abolishment of circulating relaxin by bilateral ovariectomy or passive immunization of relaxin with a monoclonal antibody for rat relaxin during the second half of pregnancy prolongs the duration of delivery several fold and reduces the incidence of live pups at birth. Unlike gilts, relaxin does not appear to promote growth of the rat uterus during pregnancy.

Also unlike gilts, circulating relaxin has little, if any, influence on the size of the mammary glands. Relaxin does, however, play a major role in promoting the growth of the mammary nipples that occurs during the second half of pregnancy (Fig. 7). This effect is vital. When rats are deprived of relaxin throughout the second half of pregnancy, the nipples are so small that pups cannot grasp them to obtain milk.

There are additional effects of circulating relaxin in pregnant rats, but their physiological significance remains poorly understood. Consistent with findings in gilts, circulating relaxin inhibits uterine contractility. Relaxin also promotes modest water consumption during the second half of pregnancy in rats, and it appears to do so through direct effects on the brain.

3. Human

It remains to be demonstrated that circulating relaxin has an important physiological role during pregnancy in women. There is reason to suspect that it may not. Women with premature ovarian failure

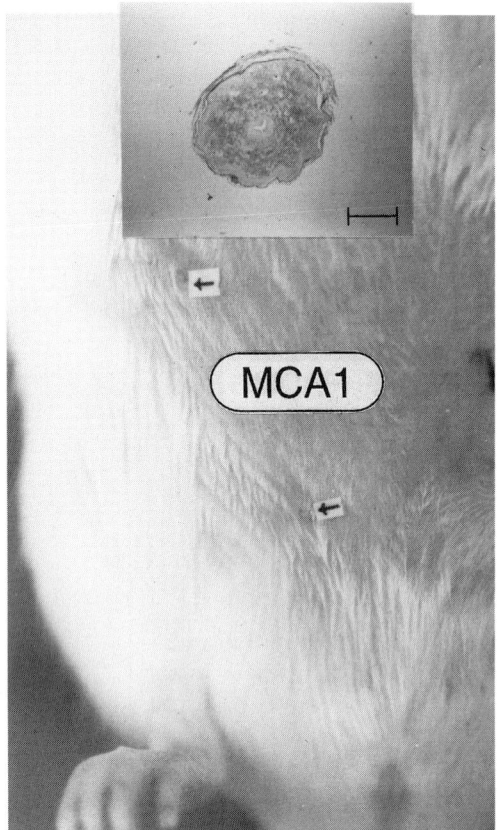 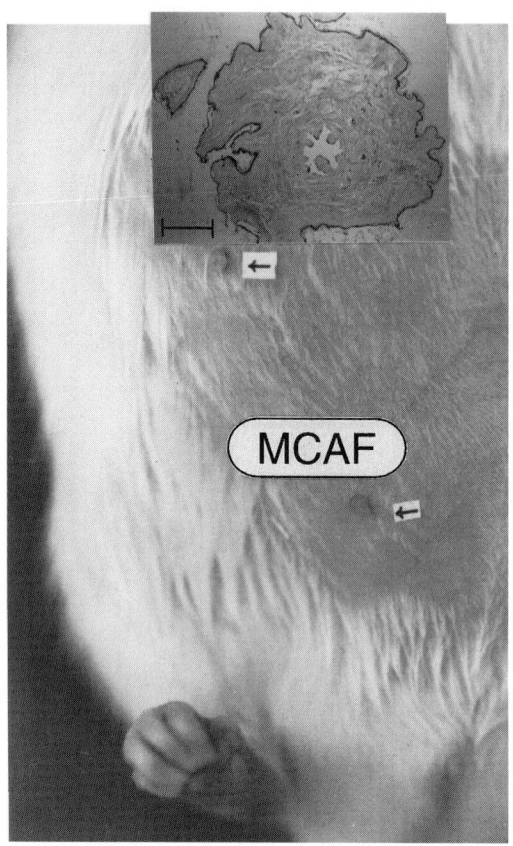

FIGURE 7 Representative photographs of nipples (arrows) of Day 22 pregnant rats following the administration of monoclonal antibody for rat relaxin (MCA1) or monoclonal antibody for fluorescein (MCAF, monoclonal antibody control). Insets show cross sections of the midpoint of nipples. Bar = 300 mm. Adapted from Kuenzi and Sherwood (1992).

who become pregnant with egg donation and therefore have no detectable relaxin in the peripheral circulation have not been reported to have either abnormal delivery or poor lactational performance.

V. SUMMARY AND DIRECTIONS FOR FUTURE RESEARCH

In summary, the two peptide hormones oxytocin and relaxin are produced within the corpus luteum in some mammalian species. Luteal oxytocin, which is produced and secreted in cows, sheep, goats, and pigs during the estrous cycle, appears to play a role in promoting luteolysis by means of promoting release of the luteolysin $PGF_{2\alpha}$ from the uterine endometrium. Little luteal oxytocin is produced during pregnancy. Small amounts of relaxin are produced by the corpus luteum during the estrous or menstrual cycle in species such as pigs, rats, and humans, but its function, if any, is unknown. The corpus luteum produces and secretes far more relaxin during pregnancy than during the estrous cycle in pigs, rats, and human beings. In pigs and rats, relaxin promotes growth of the reproductive tract and thereby facilitates delivery. Relaxin also promotes growth and development of the mammary glands in pigs and mammary nipples in rats.

Our understanding of luteal peptide hormones is in its infancy, and many fundamental areas of research require investigation. Future research is needed to identify additional species in which the corpus luteum produces oxytocin and/or relaxin. The factors that promote luteal oxytocin and relaxin synthesis and secretion remain poorly understood for all species. In only a few species have physiological roles of luteal oxytocin and relaxin been identified. Finally, it is possible that the corpus luteum produce additional nonsteroid hormones that remain to be discovered.

See Also the Following Articles

Estrous Cycle; Luteolysis; Oxytocin; Relaxin, Mammalian

Bibliography

Bagnell, C. A. (1991). Production and biological action of relaxin within the ovarian follicle: An overview. *Steroids* 56, 242–246.

Bagnell, C. A., Zhang, Q., Downey, B., and Ainsworth, L. (1993). Sources and biological actions of relaxin in pigs. *J. Reprod. Fertil. Suppl.* 48, 127–138.

Bell, R. J., Eddie, L. W., Lester, A. R., Wood, E. C., Johnston, P. D., and Niall, H. D. (1987). Relaxin in human pregnancy serum measured with a homologous radioimmunoassay. *Obstet. Gynecol.* 69, 585–589.

Buster, J. E., Chang, R. J., Preston, D. L., Elashoff, R. M., Cousins, L. M., Abraham, G. E., Hobel, C. J., and Marshall, J. R. (1979). Interrelationships of circulating maternal steroid concentrations in third trimester pregnancies. I. C_{21} steroids: progesterone, 16 alpha-hydroxyprogesterone, 17 alpha-hydroxyprogesterone, 20 alpha-dihydroprogesterone, delta 5-pregnenolone, delta 5-pregnenolone sulfate, and 17-hydroxy delta 5 pregnenolone. *J. Clin. Endocrinol. Metab.* 48, 133–138.

Eldridge-White, R., Easter, R. A., Heaton, D. M., O'Day, M. B., Petersen, G. C., Shanks, R. D., Tarbell, M. K., and Sherwood, O. D. (1989). Hormonal control of the cervix in pregnant gilts. I. Changes in the physical properties of the cervix correlate temporally with elevated serum levels of estrogen and relaxin. *Endocrinology* 125, 2996–3003.

Fields, P. A. (1991). Relaxin and other luteal secretary peptides: Cell localization and function in the ovary. In *Ultrastructure of the Ovary* (G. Familiari, S. Makabe, and P. M. Motta, Eds.), pp. 177–198. Kluwer, Boston.

Fields, M. J., and Fields, P. A. (1996). Morphological characteristics of the bovine corpus luteum during the estrous cycle and pregnancy. *Theriogenology* 45, 1295–1325.

Fields, M. J., Barros, C. M., Watkins, W. B., and Fields, P. A. (1992). Characterization of large luteal cells and their secretary granules during the estrous cycle of the cow. *Biol. Reprod.* 46, 535–545.

Fields, M. J., Ndikum-Moffor, F. M., Simmen, R. C. M., Buhi, W. C., Rollyson, K., Kowalski, A. A., Chang, S. M. T., and Fields, P. A. (1996). Bovine luteal secretory proteins of the oestrous cycle and pregnancy. *Reprod. Domestic Anim.* 31, 407–425.

Hansel, W., and Blair, R. M. (1996). Bovine corpus luteum: A historic overview and implications for future research. *Theriogenology* 45, 1267–1294.

Hurley, W. L., Doane, R. M., O'Day-Bowman, M. B., Winn, R. J., Mojonnier, L. E., and Sherwood, O. D. (1991). Effect of relaxin on mammary development in ovariectomized pregnant gilts. *Endocrinology* 128, 1285–1290.

Hwang, J-J., and Sherwood, O. D. (1988). Monoclonal antibodies specific for rat relaxin. III. Passive immunization with monoclonal antibodies throughout the second half of pregnancy reduces cervical growth and extensibility in intact rats. *Endocrinology* 123, 2486–2490.

Kuenzi, M. J., and Sherwood, O. D. (1992). Monoclonal antibodies specific for rat relaxin. VII. Passive immunization with monoclonal antibodies throughout the second half of pregnancy prevents development of normal mammary nipple morphology and function in rats. *Endocrinology* 131, 1841–1847.

Min, G., Hartzog, M. G., Jennings, R. L., Winn, R. J., and Sherwood, O. D. (1997). Evidence that endogenous relaxin promotes growth of the vagina and uterus during pregnancy in gilts. *Endocrinology* 138, 560–565.

O'Day, M. B., Winn, R. J., Easter, R. A., Dziuk, P. J., and Sherwood, O. D. (1989). Hormonal control of the cervix in pregnant gilts. II. Relaxin promotes changes in the physical properties of the cervix in ovariectomized hormone-treated pregnant gilts. *Endocrinology* 125, 3004–3010.

Pepe, G. J., and Rothchild, I. (1974). A comparative study of serum progesterone levels in pregnancy and in various types of pseudopregnancy in the rat. *Endocrinology* 95, 275–279.

Sherwood, O. D. (1994). Relaxin. In *The Physiology of Reproduction* (E. Knobil and J. D. Neill, Eds.), 2nd ed., pp. 861–1009. Raven Press, New York.

Sherwood, O. D., Crnekovic, V. E., Gordon, W. L., and Rutherford, J. E. (1980). Radioimmunoassay of relaxin throughout pregnancy and during parturition in the rat. *Endocrinology* 107, 691–698.

Sherwood, O. D., Downing, S. J., Lao Guico-Lamm, M., Hwang, J.-J., O'Day-Bowman, M. B., and Fields, P. A. (1993). The physiological effects of relaxin during pregnancy: Studies in rats and pigs. In *Oxford Reviews of Reproductive Biology* (S. R. Milligan, Ed.), Vol. 15, pp. 143–189. Oxford Univ. Press, Oxford, UK.

Spencer, T. E., Ott, T. L., and Bazer, F. W. (1996). Tau-interferon pregnancy recognition signal in ruminants. *Proc. Soc. Exp. Biol. Med.* **213**, 215–229.

Taya, K., and Greenwald, G. S. (1981). In vivo and in vitro steroidogenesis in the pregnant rat. *Biol. Reprod.* **25**, 683–691.

Wathes, D. C. (1989). Oxytocin and vasopressin in the gonads. In *Oxford Reviews of Reproductive Biology* (S. R. Milligan, Ed.), Vol. 11, pp. 226–283. Oxford Univ. Press, Oxford, UK.

Wathes, D. C., and Denning-Kendall, P. A. (1992). Control of synthesis and secretion of ovarian oxytocin in ruminants. *J. Reprod. Fertil.* **45**, 39–52.

Wathes, D. C., Gilbert, C. L., and Ayad, V. J. (1993). Interactions between oxytocin, the ovaries, and the reproductive tract in the regulation of fertility in the ewe. *Ann. N. Y. Acad. Sci.* **689**, 396–410.

Zhao, S., Kuenzi, M. J., and Sherwood, O. D. (1996). Monoclonal antibodies specific for rat relaxin. IX. Evidence that endogenous relaxin promotes growth of the vagina during the second half of pregnancy in rats. *Endocrinology* **137**, 425–430.

Corticosteroid-Binding Globulin

see CBG

Corticosteroids

see Steroid Hormones, Overview

Cotyledonary Placenta

Stephen P. Ford

Iowa State University

I. Preimplantation Embryonic Development
II. Placental Attachment
III. Placental Classification
IV. Development of the Placentomes throughout Gestation
V. Factors Affecting Fetal Nutrition
VI. Uteroplacental Angiogenesis
VII. Control of Uterine Blood Flow during Gestation

GLOSSARY

allograft A graft of tissue obtained from a donor genetically different from though the same species as the recipient.

angiogenesis The formation of new blood vessels, or neovascularization, which begins with capillary proliferation and culminates in the formation of a new microcirculatory bed composed of arterioles, capillaries, and venules.

blastocyst An early embryonic form produced by cleavage of a fertilized ovum and consisting of two distinct cell populations, the outer trophoblast cells and the embryoblast (inner cell mass) residing at one pole beneath the trophoblast.

decidua A mucous membrane lining the uterus that is modified during pregnancy and shed at parturition or during menstruation.

ruminant Any of various hoofed, even-toed, usually horned mammals of the suborder Ruminantia, characteristically having a stomach divided into four compartments and chewing a cud consisting of regurgitated, partially digested food.

The cotyledonary placenta is found in ruminants (cow, sheep, goat, deer, antelope, and giraffe) and is characterized by separate tufts of chorionic villi called cotyledons, which are widely scattered over the surface of the chorion. These localized dense areas of villi develop only on those parts of the chorion that are adjacent to aglandular proliferations of connective tissue on the uterine luminal mucosa called caruncles. Each cotyledon develops highly branched villi, which interdigitate with corresponding crypts that develop in the proliferating caruncle. The combined cotyledonary–caruncular unit is referred to as a placentome and is the site of fetal–maternal exchanges of nutrients and waste in these species.

I. PREIMPLANTATION EMBRYONIC DEVELOPMENT

A. Blastulation

Except for timing and slight structural differences, the development of the fetal membranes of all domesticated ruminants is similar. The period of the ovum extends from fertilization until a change in shape and cellular makeup occurs (Fig. 1). During this period the embryo undergoes a series of cell divisions without changing drastically in shape or size. Specifically, the zygote divides to form two blastomeres, then four, and so on until it forms a solid mass of cells called a morula. At this point, the central blastomeres begin to separate and form the blastocoelic cavity, while a single layer of cells called the trophoblast surrounds the developing embryo now referred to as a blastocyst. As this process proceeds, the cells at one pole congregate to form the inner cell mass or blastoderm and the entire embryo continues to be surrounded by the zona pellucida. During the first week or two of gestation, the spherical blastocyst floats free in the uterine lumen, hatches from the zona pellucida, and, as gastrulation commences, begins to elongate. This elongation process is so rapid that within 48–72 hr, the filamentous blastocyst has extended the entire length of the uterine horn. Elongation of the ovine blastocyst begins on about Day 12, whereas elongation of the bovine conceptus begins on about Day 16 of gestation.

B. Gastrulation

The blastocyst then enters the period of the embryo, which is marked by the specialization of previously indifferent cell types to specialized cells which will form the extraembryonic membranes (Fig. 2). This process, called gastrulation, is marked by a thickening of the trophoblast to form the ectoderm and the migration of entoderm and mesoderm cell layers out of the inner cell mass, which will form the embryo proper. The extraembryonic membranes, which form from combinations of these cell layers, consist of the yolk sac, chorion, amnion, and allantois. Entoderm cells migrate out of the inner cell mass to line the cavity of the gastrula and then fuse with splanchnic mesodermal cells to form a rudimentary yolk sac in the vicinity of the embryonic hindgut. Although the yolk sac differs from that found in birds and reptiles due to a lack of stored yolk, it plays an essential role in the initial stages of hematopoiesis (blood vessel formation) and is the original location of the primordial germ cells, which will eventually migrate into the gonads of the developing embryo. It may also function in embryonic absorption of histotrophe, a fluid secreted by the uterine luminal glands, during the preimplantation period. This material, called uterine milk, is crucial for supplying the preimplantation embryo with essential metabolic substrates and growth factors. Next, there is a folding of the outer ectoderm and the underlying somatic

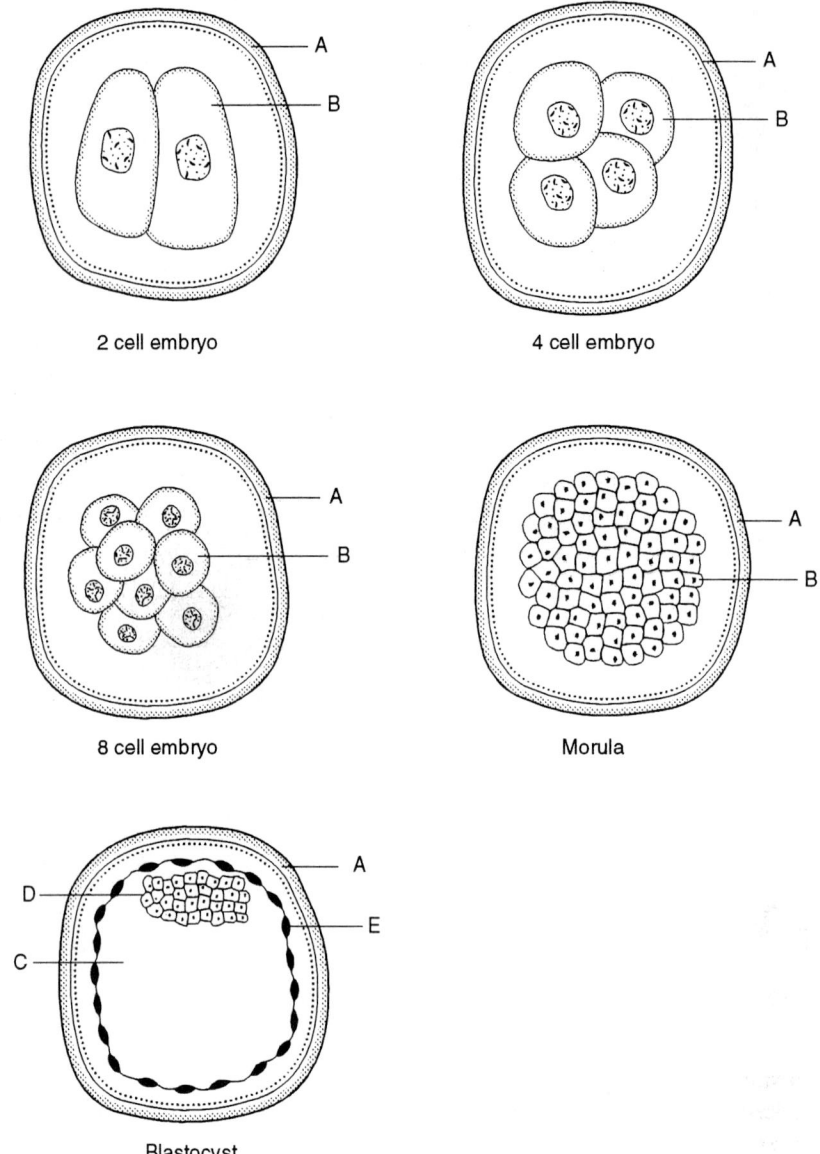

FIGURE 1 The period of the ovum which extends from fertilization to the development of a blastocyst. A, zona pellucida; B, blastomere; C, blastocoele; D, inner cell mass; E, trophoblast.

mesoderm so that an envelope forms around the embryo, which is called the amnion, and an outer envelope exposed to the uterine environment is formed called the chorion. The function of the amnion (also called embryonic vesicle) and its fluid is to protect the embryo from mechanical damage and in the later stages of gestation it collects excretions from the urinary and gastrointestinal systems. The third membrane, the allantois, forms as an outpocketing of the hindgut near the yolk sac along the umbilical cord. The allantois is entoderm covered with splanchnic mesoderm and it grows sac-like in all directions until it contacts and fuses with the outside of the amnion and the inside of the chorion to form the amnioallantoic membrane and chorioallantoic membrane, respectively. The allantois even-

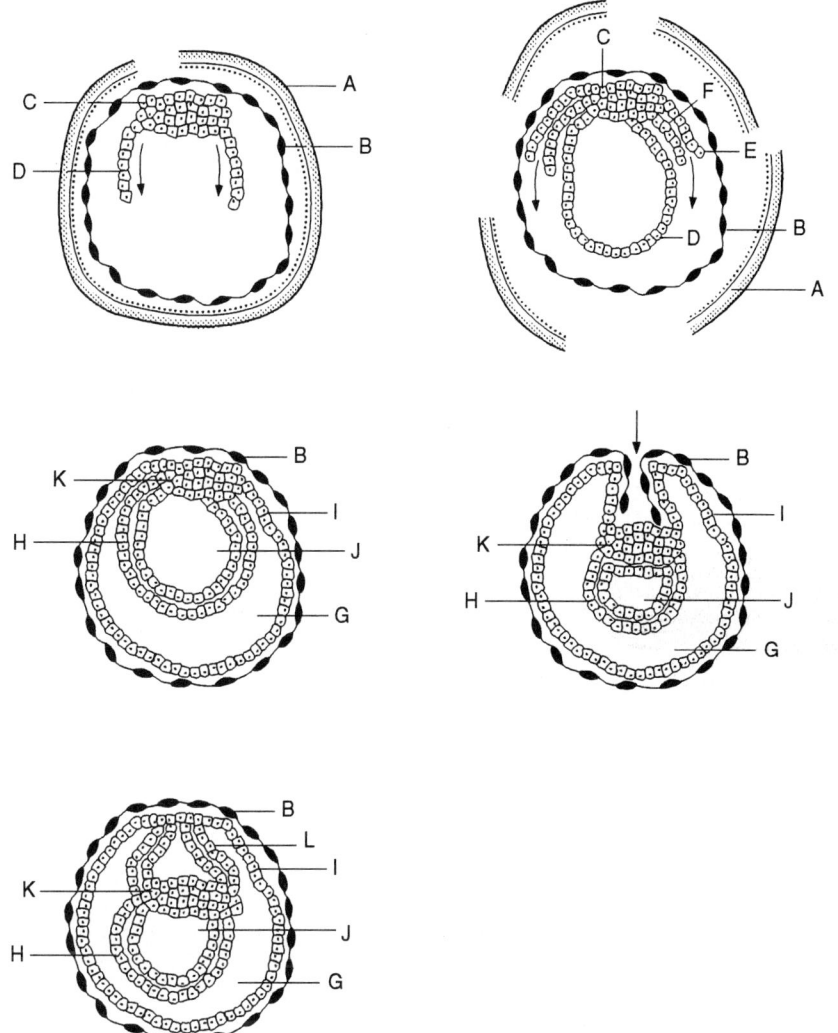

FIGURE 2 The period of the embryo which extends from the hatching of the blastocyst through the development of the extraembryonic membranes. A, zona pellucida; B, trophoblast; C, inner cell mass; D, entoderm; E, somatic mesoderm; F, splanchnic mesoderm; G, coelom; H, yolk sac; I, chorion; J, primitive gut; K, embryonic disc; L, amnion.

tually fills the much elongated chorionic cavity of the filamentous embryo so that only the very ends of the chorion are free of its attachment (Fig. 3). The allantois is responsible for vascularizing the chorion, which then develops an organized vascular system in the chorioallantoic membrane connected to the embryo (now designated a fetus) through the umbilical cord. Small ischemic zones occur at the ends of the fetal membranes and are referred to as necrotic tips and function to prevent interplacental blood exchange. The allantois collects urine through the urachus in the umbilical cord in early gestation, but this duct closes in later gestation resulting in these excretions accumulating in the amnion. The true placenta can be defined as a functional union between fetal membranes and the uterus which occurs during gestation in eutherian mammals. This structure has a primary function in the selective exchange of nutrients, oxygen, and waste products between the mother and fetus. It also serves as an endocrine

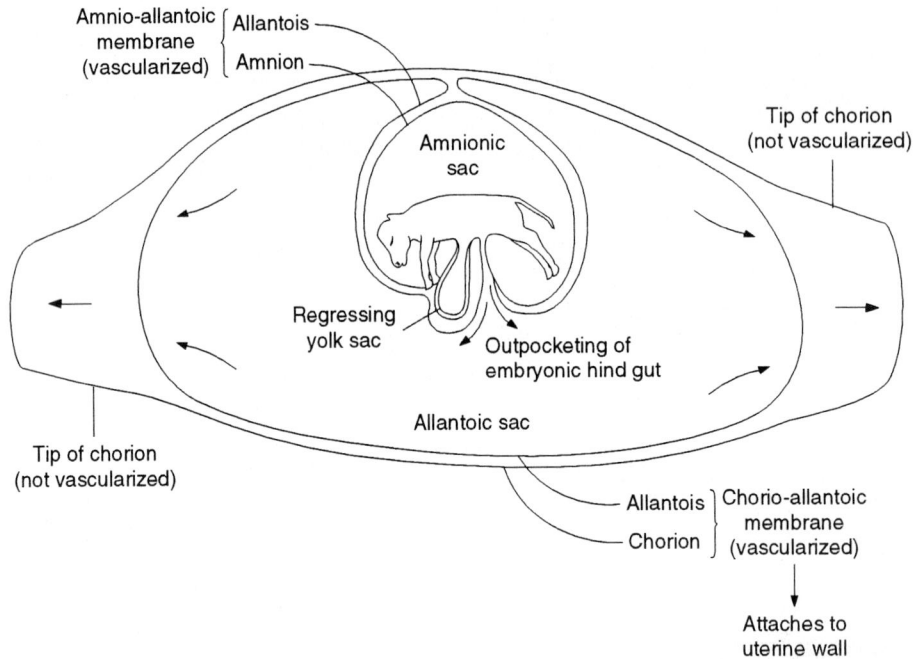

FIGURE 3 The period of the fetus extends from the vascularization and attachment of the chorioallantoic membrane to the uterine luminal surface through parturition.

organ throughout gestation to impact maternal metabolism, growth, and development of the mammary glands and to time parturition with fetal maturity at term. The placenta also functions as a barrier to prevent the mixing of fetal and maternal blood, which if allowed to occur would lead to the rejection of the fetal allograft.

II. PLACENTAL ATTACHMENT

The placentas of all eutherian mammals share common structural origins and essential functions, but in their fully developed state they present contrasting appearances both in their areas of contact with the uterine luminal surface and in the number of tissue layers separating the fetal and maternal bloodstreams. In ruminants, the placenta attaches to discrete areas on the uterine mucosa called caruncles. These caruncles are aglandular proliferations of connective tissue which appear as knobs along the uterine luminal surface. These caruncles are arranged in two dorsal and two ventral rows throughout the length of the uterine horns. The ruminant placenta membranes attach at these sites via chorionic villi at areas called cotyledons. Figure 4 depicts a Day 120 bovine conceptus in which the fetus and umbilical cord have been exteriorized through a slit in the amnion and chorioallantoic membrane. The placental vasculature can be seen radiating out from the umbilicus to the individual cotyledons. Also note the decreasing size of the cotyledons as their distance from the fetus increases, as well as the markedly reduced size of all cotyledons in the contralateral uterine horn (right). There are between 70 and 120 caruncles in the cow and between 80 and 100 in sheep. The caruncular–cotyledonary unit is called a placentome and is the functional area of physiological exchanges between mother and fetus. It is in reference to this unique type of attachment between the ruminant placental membranes and the uterine wall that results in its designation as a cotyledonary placenta. When these caruncles are contacted by fetal membranes they enlarge and form swellings, with a

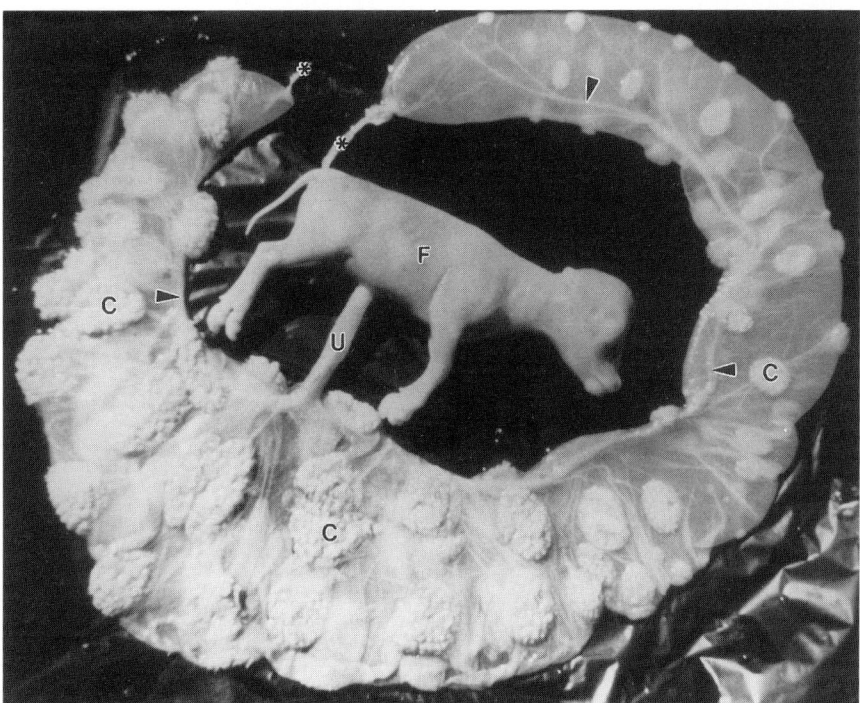

FIGURE 4 Day 120 bovine conceptus. F, fetus; U, umbilical cord; C, cotyledon; arrows, placental vasculature; asterisks, necrotic tips.

convex surface in cattle and in sheep. In concert, the fetal membranes at these sites of contact develop highly branched villous processes that interdigitate with corresponding crypts, which develop in the proliferating caruncle. In the areas between the placentomes, there is only a loose fetal–maternal apposition associated with a relative lack of villous development of the fetal membranes. Uterine endometrial glands continue to secrete histotroph in these intercaruncular areas throughout gestation.

III. PLACENTAL CLASSIFICATION

The areas of fetal–maternal exchange across placental types are classified as to structural type by the number of tissue layers separating the fetal and maternal bloodstreams (Fig. 5a). There are a maximum of six tissue layers, which include (i) endothelial cells lining blood vessels in the endometrium; (ii) connective tissue components, which form a matrix or network and function to hold other tissues together; (iii) uterine luminal epithelial cells; (iv) chorionic epithelial cells; (v) connective tissue of chorioallantoic membrane; and (vi) endothelial cells lining placental blood vessels. Figure 5b provides a listing of the placental types and shapes of different eutherian mammals for comparison.

Although some debate has occurred in the literature, it is now generally accepted that the cow has an epitheliochorial placenta. The first part of the word relates to the dam and the second part to the fetus. Thus, in the cow, there are no tissue layers missing in the placentome because maternal epithelium is present, and the chorion is present and intact. In other ruminants, such as the sheep and goat, however, the maternal epithelium is lost in the placentome and their placental type is referred to as syndesmochorial. Generally speaking, as the number of tissue layers separating the maternal and fetal bloodstreams decrease from six to one, the size of the placenta required per unit weight fetus declines in

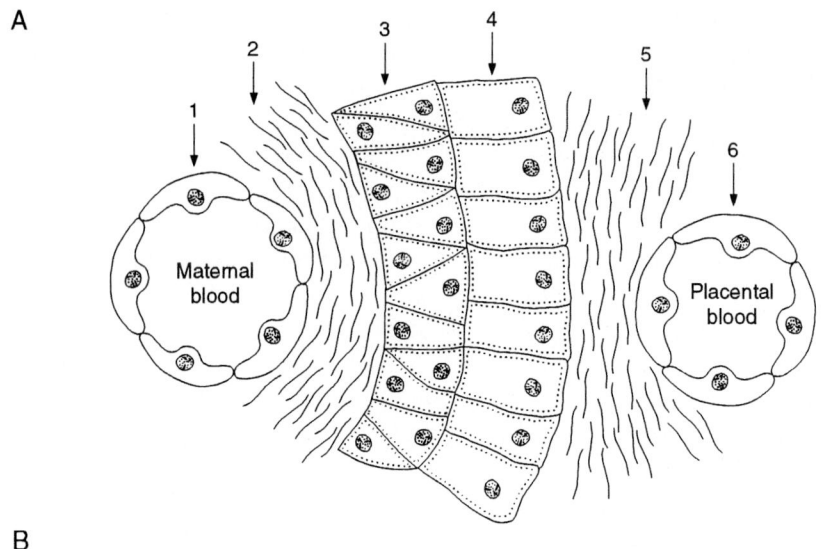

FIGURE 5 (a) Depiction of the tissue layers separating the maternal and fetal blood streams in a species with an epitheliochorial placental type. (b) Species comparisons in the number of tissue layers separating the fetal and maternal bloodstreams as well as their placental classification as to shape and type.

association with an increase in placental extraction efficiency. Thus, ruminants with five or six tissue layers between the maternal and fetal bloodstreams require a relatively large placental surface area for adequate nutrient, oxygen, and waste product exchange. The role of maximizing placental surface area is fulfilled by the cotyledonary villi. Other components which must be considered when evaluating placental extraction efficiency include any progressive thinning of maternal–fetal intercapillary distances, the relative direction of caruncular and cotyledonary capillary blood flow (i.e., concurrent, cross-current, or countercurrent), as well as increases in placental vascularity.

IV. DEVELOPMENT OF THE PLACENTOMES THROUGHOUT GESTATION

Björkman (1954) described the formation of crypts and septa on the surface of the bovine caruncle. Outgrowths of the caruncular tissue formed short and thick septa perpendicular to the surface of the caruncle. The resulting crypts were filled by villi from the fetal cotyledon. The septa continued to grow longer and thinner as the villi grew deeper into the crypts. Processes of connective tissue and blood vessels grew into these septa. The placentome continued to increase in thickness throughout most of pregnancy as the cotyledonary villi grew deeper into the deepening caruncular crypts. The shape of the placentome also changes throughout gestation. Initially, the placentome is thin and lies flat on the surface of the uterine wall. By 100 days, the placentome has developed a definite outer shape with the free surface being markedly convex. Cotyledonary tissue surrounds almost the entire caruncle so the attachment of the caruncle to the uterine wall is no more than a stalk of highly vascularized endometrial tissue. The placentome increases markedly in size during the first half of gestation, then its growth slows markedly.

About 7–10 days postpartum in the cow, the crypts within each caruncle undergo necroses due to ischemia, and maternal epithelial cells are sloughed. The uterine epithelium later heals over the regressing caruncle. Thus, the ruminant placenta can be considered as partially deciduate. Failure of the cotyledon to separate from the caruncles following birth results in a retained placenta.

V. FACTORS AFFECTING FETAL NUTRITION

During the last half of gestation, fetal growth is exponential, whereas placentome growth slows or ceases. It has been verified over the past two decades, however, that the placental transport capacity is able to keep pace with the ever-increasing fetal demands. In ruminant species with relatively noninvasive placentas (i.e., epitheliochorial and syndesmochorial placental types), these increases in fetal demands are met predominantly by increases in blood flow through the fetal–maternal interface rather than by marked alterations in nutrient uptake per milliliter of caruncular blood. This is evidenced by the fact that any prolonged decrease in blood flow to the gravid uterus of the cow or ewe during late gestation will result in a corresponding decrease in fetal weight. Uterine and umbilical blood flows increase dramatically throughout gestation in all mammalian species including ruminants. This results in increased placental vascularity. Not only do umbilical and uterine blood flows increase throughout gestation but also the proportion of the total uterine and umbilical blood flows received by the caruncular and cotyledonary tissues, respectively, also increases.

VI. UTEROPLACENTAL ANGIOGENESIS

During the first half of gestation, placental growth far outpaces fetal growth. This preferential placental growth during early gestation is required so that the accelerating demands for fetal growth during late gestation can be met. In association with the requirement for placental growth, the density of blood vessels in caruncular tissues increases substantially from early to midgestation, then increases more slowly thereafter. In contrast, the density of cotyledonary tissues (i.e., villus capillaries) remains relatively constant until midgestation and then increases dramatically thereafter. Thus, the early rapid angiogenesis in the caruncular tissues is required to supply nutrients and oxygen for placental growth and development, whereas the late acceleration in vascular development of the cotyledonary tissues is necessary for the exponential fetal growth rate required to produce a viable offspring at the end of gestation.

VII. CONTROL OF UTERINE BLOOD FLOW DURING GESTATION

As previously discussed, nutrient uptake by the gravid uterus in ruminant species is primarily dependent on the rate of uterine blood flow. Of interest is the fact that while uterine blood flow increases progressively throughout gestation via increases in uterine arterial diameter, the arteries retain their ability to contract to the point of almost total occlusion of gravid uterine blood flow for short intervals (<10 min). This adaptation is required during times of acute maternal stress and functions to shunt blood to the musculature for "fight or flight" responses necessary for maternal survival. Estrogens produced and secreted by the placenta in increasing amounts throughout gestation are known to progressively decrease uterine arterial tone (i.e., increase arterial diameter and blood flow) via their conversion to "catechol" estrogens via a unique soluble placental peroxidase enzyme. Catechol estrogens directly inhibit calcium uptake by uterine arterial smooth muscle cells, resulting in vasodilation. At parturition, catechol estrogens are no longer synthesized and delivered to arterial smooth muscle (caruncular and uterine), resulting in marked and progressive decreases in uterine arterial diameter and blood flow.

See Also the Following Articles

BLASTOCYST; PREGNANCY IN FARM ANIMALS; RUMINANTS (RUMINANTIA)

Bibliography

Björkman, N. (1954). Morphological and histochemical studies on the bovine placenta. *Acta Anat.* **22**(Suppl. 22), 5–91.

Ferrell, C. L. (1989). Placental regulation of fetal growth. In *Animal Growth Regulation* (D. R. Campion, G. J. Hausman, and R. J. Martin, Eds.), pp. 1–19. Plenum, New York.

Ford, S. P. (1995). Control of blood flow to the gravid uterus of domestic livestock species. *J. Anim. Sci.* **73**, 1852–1860.

Mossman, H. W. (1937). Comparative morphogenesis of the fetal membranes and accessory uterine structures. *Carnegie Contrib. Embryol.* **26**, 129–246.

Reynolds, L. P., and Redmer, D. A. (1995). Utero-placental vascular development and placental function. *J. Anim. Sci.* **73**, 1839–1851.

Courtship Behavior

see Mating Behaviors

Cowper's Gland

see Male Reproductive System

Cricetidae (Hamsters and Lemmings)

David A. Freeman
University of California, Berkeley

Bruce D. Goldman
University of Connecticut

I. Ovulatory Cycles: Syrian Hamsters
II. Sexual Differentiation: Syrian Hamsters
III. Photoperiodism and Annual Reproductive Rhythms: Syrian Hamsters
IV. Photoperiodism and Pineal Melatonin: Siberian Hamsters
V. Role of Photoperiod History Effects in Shaping the Annual Reproductive Cycle: Siberian Hamsters
VI. Photoperiodism: Turkish Hamsters
VII. Coordination between Seasons of Hibernation and Reproduction: Turkish and European Hamsters
VIII. Reproductive Strategies in Hamsters
IX. Seasonal Reproductive Patterns in Lemmings
X. Photoperiodism in Lemmings
XI. Population Cycles in Lemmings

GLOSSARY

estrous cycle Cyclic events associated with the maturation of ovarian follicles and characterized by a distinct period of female sexual receptivity (estrus).
estrus The period of sexual receptivity in females.
lactation Milk production and its release during suckling of offspring.
lordosis A sexual posture characterized by immobility with a species-typical posture that facilitates vaginal penetration by the penis of the male.
melatonin The primary hormone of the pineal gland; it is involved in reproductive and other physiological responses to changes in day length.
ovarian follicle Ovarian compartment housing an individual ovum. The follicle produces the hormonal changes necessary for ovulation and sexual behavior.
ovulation The release of a mature egg from the ovary.
photoperiod The length of the light phase in a light/dark cycle.
photoperiodism The ability of an organism to measure and respond to different day lengths.
pineal gland A gland lying outside of, but attached to, the brain. The pineal secretes melatonin under the rhythmic control of the circadian clock.
postpartum estrus The phenomenon in which a female ovulates and becomes sexually receptive immediately after parturition.

Of the several species of hamsters, two in particular have been widely used in laboratory studies of the physiology of reproduction in mammals. The Syrian, or golden, hamster (*Mesocricetus auratus*) has been used extensively for investigations of the neuroendocrine basis for estrous cycles and for studies of the role of the endocrine system in early sexual differentiation. This species was also the first to be widely employed in the United States for studies of the neuroendocrine mechanisms in the photoperiodic regulation of seasonal reproductive rhythms. The Siberian hamster has also been used in several laboratories in studies of photoperiodism. The lemmings constitute another group of rodents that have received significant attention regarding their reproductive patterns. In this case, most interest has focused on the population cycles and associated migrations occurring in these animals. There has been far less laboratory research in lemming reproduction compared to that for hamsters.

I. OVULATORY CYCLES: SYRIAN HAMSTERS

In the early twentieth century, the rat became widely used in studies of the neuroendocrine basis for ovulatory cycles. The rat ovulatory, or estrous, cycle generally recurs every 4 or 5 days, and an early description of the rat cycle was published in a monograph by Evans and Swezy in 1931. The use of Syrian hamsters in biological research can be traced to 1930, when Professor I. Aharoni of the Hebrew University captured a hamster with her litter of 11 pups in Syria and brought the animals to Israel. These animals were used to establish a laboratory breeding colony, and most or perhaps all Syrian hamsters that are commercially available today trace their ancestry to the animals trapped by Dr. Aharoni. Though the laboratory rat continues to be a widely used model for studies of the physiology of estrous cycles, the Syrian hamster offers two distinct advantages: Female Syrian hamsters exhibit almost exclusively 4-day estrous cycles, eliminating the element of uncertainty in predicting the time of ovulation. Also, the Syrian hamster produces a distinct vaginal discharge on the day of estrus. This discharge can be seen without use of a microscope, obviating the need to perform daily vaginal smears to monitor the cycle.

Like rats, Syrian hamsters release a preovulatory surge of luteinizing hormone (LH) during the afternoon of proestrus. This surge of LH triggers the processes that lead to follicular rupture, or ovulation, approximately 10 hr later. Thus, ovulation occurs at night, a time when this nocturnal species might be most likely to encounter potential mates. The precise timing of release of LH from the anterior pituitary is regulated by a circadian clock located in the suprachiasmatic nucleus (SCN) of the hypothalamus. The SCN has an intrinsic rhythm with a period length close to 24 hr. The SCN rhythm is entrained by photic cues received via the retinohypothalamic tract, which is a portion of the optic nerve that branches to innervate the SCN. As a result of the entraining action of light, the SCN signals LH release at a time that is determined by the light:dark cycle; for example, in a photoperiod of 16 hr light:8 hr dark, preovulatory LH release occurs about 5 or 6 hr before lights-off.

The preovulatory surge of LH is accompanied by a smaller elevation in blood concentrations of follicle-stimulating hormone (FSH). A second elevation of FSH occurs several hours later, beginning at about the time of ovulation and lasting for several hours. This second FSH increase is thought to have a major role in initiating the development of a new set of ovarian follicles that will mature during the next cycle, if the female fails to mate around the time of ovulation.

Syrian hamsters normally ovulate approximately 8–12 ova at the end of each cycle and consequently give birth to about that number of young. If one ovary is removed early in the estrous cycle, the remaining ovary shows increased follicular development and, at the end of the same cycle, ovulates a full complement of 8–12 ova. This experimental paradigm has been used to study the endocrine regulation of follicle maturation. Following hemiovariectomy, there is a rapid increase in serum FSH concentration; FSH declines to baseline levels several hours later. In the hemiovariectomy paradigm, it appears that the reduction in ovarian feedback signals—estradiol, progesterone, and inhibin—results in activation of the hypothalamic–pituitary axis to increase FSH secretion. The increase in FSH secretion persists until a sufficient number of new follicles begin development and restore the appropriate feedback signals.

During the estrous cycle, the maturing ovarian follicles begin to secrete increased amounts of estradiol on the day (metestrus) preceding the day (proestrus) of the preovulatory LH surge. This increase in estradiol primes the female for sexual behavior but does not by itself elicit the behavior. The LH surge stimulates a rapid increase in ovarian secretion of progesterone and a decrease in estradiol production. It is this rise in circulating concentrations of progesterone that provides the final endocrine stimulus for female sexual behavior. A suitably estrogen-primed female hamster begins to exhibit all the typical sexual responses within 2–4 hr following exposure to sufficient amounts of progesterone. In this way, the same set of endocrine events that are associated with development of mature follicles and ovulation also serve to stimulate female sexual receptivity.

Syrian hamsters are solitary living animals, and during most phases of the estrous cycle female hamsters are aggressive toward males. In laboratories, general husbandry methods include housing males and females separately except during the period of the female's behavioral estrus. Pairs are housed together only from late afternoon of proestrus until the morning of estrus. Males left with females at other times of the estrous cycle are liable to be attacked and are sometimes killed. One aspect of sexual behavior in Syrian hamsters is somewhat unusual among laboratory rodents: Receptive females exhibit a prolonged lordosis response, often remaining in the lordosis posture for a matter of minutes while being courted and during mating.

Social interactions are quite different for Siberian hamsters; the sexes are not generally aggressive toward each other. Breeding pairs of Siberian hamsters can be left together at all times, including during the periods of pregnancy and lactation. Unlike Syrian hamsters, female Siberian hamsters have a postpartum ovulation and frequently mate following parturition. These females give birth to a second litter shortly before the first litter has been weaned.

II. SEXUAL DIFFERENTIATION: SYRIAN HAMSTERS

A large number of studies, mostly in laboratory rats and mice, have revealed that testosterone, secreted by the testes during a critical developmental period, results in masculinization and defeminization of the brain. In rats, the critical period for this action of testicular androgens begins shortly prior to birth and extends into the first few days of postnatal life. Syrian hamsters have also been used in studies of the endocrine basis of sexual differentiation of the brain. Hamsters have a gestation period of 16 days—about 5 days less that the length of pregnancy in rats. Nevertheless, the critical period for early androgen action in hamsters also encompasses the perinatal period, which is similar to the case for rats. When genetic female hamsters are treated with androgen at the time of birth, they become permanently anovulatory as a result of failure to develop the normal female mechanism for cyclic release of preovulatory LH surges. The androgen-treated females also fail to exhibit typical female sexual behavior in adulthood, even after treatment with a regimen of estrogen and progesterone that consistently evokes lordosis behavior in normal females.

The action of androgen during a critical developmental stage to inhibit development of the feminine mechanism for cyclic LH release has been reported for several other species. In Syrian hamsters, there is also a sexual differentiation of the ability to respond to the negative feedback effects of androgen on the secretion of pituitary LH and FSH. Castrated adult male hamsters exhibit a greater inhibition of gonadotropin secretion in response to testosterone compared to ovariectomized females. This trait was also shown to become sexually differentiated as a result of the action of testicular androgen during the perinatal period. There has been no systematic study of sex differences in the negative feedback mechanism in other species.

III. PHOTOPERIODISM AND ANNUAL REPRODUCTIVE RHYTHMS: SYRIAN HAMSTERS

Syrian hamsters were studied in several laboratories that explored mechanisms of photoperiodism in mammals. Like many other small rodents, Syrian hamsters breed during the summer and reproduction ceases during the autumn and winter months. Laboratory studies showed that the major environmental cue used to time the annual reproductive cycle is day length, or photoperiod. Thus, reproduction can be inhibited at any time of year by exposing the animals to a short photoperiod. The "critical photoperiod" in this species is 12.5 hr of light; animals exposed to day lengths of 12.5 of hr light or more maintain reproductive activity, whereas the reproductive axis is markedly inhibited in photoperiods of 12 hr of light or less.

Studies in Syrian hamsters have helped provide considerable information regarding the mechanisms of photoperiodism in mammals: These animals are able to distinguish quite accurately between day lengths of 12.5 and 12 hr of light, for example. Day length measurement—also termed photoperiodic time measurement—is accomplished through a mechanism that involves a circadian clock in the SCN. This clock receives photic information from the retina via the retinohypothalamic tract. The clock, acting via a multisynaptic neural pathway, regulates a circadian rhythm of melatonin secretion by the pineal gland. Melatonin is a hormone that is derived from serotonin via the action of two enzymes—NAT and HIOMT—that are present in the pineal. Studies performed in Syrian hamsters in the 1970s yielded strong evidence that melatonin is centrally involved in the photoperiodic mechanism. Continuous-release implants of melatonin or daily injections of the hormone frequently evoked responses of the reproductive system that were remarkably similar to the responses evoked by manipulation of the photoperiod. Most notably, certain types of melatonin treatment mimicked the inhibitory effects of short day lengths on the reproductive axis. The inhibition of reproduction was not apparent until after several weeks of melatonin treatment; this was similar to the time lag observed for reproductive effects following transfer of hamsters from long to short photoperiod.

IV. PHOTOPERIODISM AND PINEAL MELATONIN: SIBERIAN HAMSTERS

In virtually all mammals studied to date, pineal and serum melatonin concentrations increase markedly at night, and the duration of the nocturnal elevation of melatonin secretion is directly proportional to night length; that is, the duration of the circadian pineal melatonin elevation is inversely proportional to day length. Though studies in Syrian hamsters implicated melatonin as a part of the photoperiodic mechanism, they did not elucidate what feature of the pineal melatonin rhythm was most important for mediating the actions of day length. Experiments with Siberian hamsters (*Phodopus sungorus*) revealed that it is the duration of the nocturnal elevation of melatonin secretion that conveys a photoperiodic message. These experiments involved daily timed infusions of melatonin in pinealectomized juvenile male Siberian hamsters and took advantage of the fact that reproductive responses to day length or to melatonin treatment occur within 2 weeks in juveniles of this species compared to the several weeks required in most species. The importance of the duration of the nightly rise in melatonin has been confirmed for Syrian hamsters and for sheep. Less thorough evidence supporting this mechanism exists for a few additional mammals.

A further benefit accrued from the use of Siberian hamsters in these studies of the photoperiodic mechanism. This species exhibits several robust seasonal changes that are cued by day length, including seasonal changes of pelage, changes in body weight and fat stores, and thermoregulatory adaptations that include the ability to enter bouts of daily torpor in the winter. Like the seasonal changes in reproductive activity, these other various seasonal changes are all dependent on actions of pineal melatonin.

V. ROLE OF PHOTOPERIOD HISTORY EFFECTS IN SHAPING THE ANNUAL REPRODUCTIVE CYCLE: SIBERIAN HAMSTERS

Photoperiodic animals are not only capable of measuring current day lengths and responding directly to them but also able to modulate their responses based on past photoperiod experience. Three instances of the importance of photoperiod history are evident in Siberian hamsters. The first observation of photoperiod history effects in this species derived from an experiment that involved transferring adult hamsters to longer or shorter photoperiods. When male hamsters were raised in 16 hr light and then transferred to 14 hr light, testis regression ensued; for this paradigm 14 hr light was an inhibitory photoperiod with respect to reproduction. In contrast, when hamsters were first transferred from 16 to 8 hr light for several weeks to induce gonadal regression, subsequent transfer to 14 hr light led to complete recrudescence of the testes. For the 8 to 14 hr light transfer, 14 hr light was a stimulatory photoperiod.

A second, and especially interesting, discovery of a photoperiod history effect in Siberian hamsters is related to the ability of pregnant female hamsters to provide photoperiod information to their fetuses that may modulate their reproductive responses to the photoperiod they experience after birth. When male hamsters were exposed to 14 hr light from the day of birth until Day 32 of life, testis growth during that time was considerably more rapid in males whose mothers had been housed under 12 hr light during gestation compared to males whose mothers had been housed under 16 hr light. In this type of paradigm, males that experience an increase in day length between the time of late fetal life and postnatal life show more rapid sexual maturation compared to males that experience a decrease in photoperiod over the same time. This mechanism could be adaptive in allowing juvenile hamsters to more quickly develop appropriate responses to an "intermediate" photoperiod such as 14 hr light that occurs in both early summer and late summer. Hamsters born in early summer, when day length is increasing, evidently employ a strategy of rapid sexual maturation so as to breed before the end of the summer season; hamsters born in late summer delay puberty until the following spring when conditions will be more favorable for reproduction. There is evidence that the maternal cue may modify the postnatal photoperiodic responses of male pups via an effect on the pattern of melatonin secretion during early juvenile life.

A third type of photoperiod history effect was uncovered in studies that examined individual differences in photoperiod responsiveness within a laboratory breeding population of hamsters. Not all Siberian hamsters cease reproductive activity during the winter. Animals have been found breeding in the field during the winter, and approximately 20–25% of the individuals bred in laboratories fail to exhibit reproductive inhibition when exposed to short days. This is true only if the hamsters are first exposed to short days in adulthood; juvenile hamsters are almost uniformly responsive to short days with respect to inhibition of reproductive development. It has recently been reported that for hamsters exposed to a very long photoperiod (18 hr light) for 10 weeks, about 90% of the animals failed to show testis regression when subsequently exposed to short photoperiod. Most laboratories raise their animals under a 16 hr light day length. It is possible that exposure to this photoperiod for several weeks is what induces a resistance to short day effects in 20–25% of the adult individuals. This observation suggests that a period of exposure to an especially long photoperiod may cause these hamsters to become unresponsive to short days. Indeed, when hamsters were raised from birth in 14 hr light rather than 16 hr light, virtually all the animals exhibited the typical short-day responses when transferred to 8 hr light.

VI. PHOTOPERIODISM: TURKISH HAMSTERS

The Turkish hamster (*Mesocricetus brandti*) is the closest living relative to the Syrian hamster. However, these species have intriguing differences with respect to their responses to day length. Male Syrian hamsters exhibit a stimulated reproductive axis in long days and in continuous illumination and show inhibition of reproduction only in short days (<12.5 hr light). Male Turkish hamsters show testis regression not only in short days (<14 hr light) but also in continuous illumination and even in very long days (>18 hr light). Thus, for this species, reproductive activity is maintained only over a fairly narrow range of photoperiods (14–18 hr light). Another difference between Syrian and Turkish hamsters is related to the effects of pinealectomy on reproduction: In Syrian hamsters, pinealectomy leads to a maintenance of reproductive activity, even if the animals are exposed to short photoperiod. In Turkish hamsters, pinealectomy leads to inhibition of the reproductive axis, even when the animals are maintained under what would otherwise be a stimulatory photoperiod. Thus, for Syrian hamsters, the total absence of pineal melatonin leads to reproductive activity, whereas for Turkish hamsters the absence of melatonin is inhibitory to reproduction. It is not clear whether this species difference has any bearing on the normal photoperiodic mechanism in these species since both species have similar pineal melatonin patterns under a variety of photoperiods and these patterns would normally be expressed on a daily basis throughout the breeding season. Turkish hamsters are hibernators though, and the pineal does not appear to synthesize melatonin during torpor.

VII. COORDINATION BETWEEN SEASONS OF HIBERNATION AND REPRODUCTION: TURKISH AND EUROPEAN HAMSTERS

Many hibernating mammals, like their nonhibernating counterparts, undergo a regression of the gonads during the late summer or early autumn and resume reproductive activity the following spring. In some hibernators, the gonads are fully regressed at the time of entry into hibernation, yet breeding may commence within 1 or 2 weeks after emergence from hibernation. For males in particular, the ability to achieve reproductive competence so soon after terminating hibernation may require an impressive feat of timing since spermatogenesis requires several weeks in mammals. Field studies in ground squirrel species that hibernate illustrate the importance of timing. Virtually all the females have mated and become pregnant soon after they emerge from hibernation; any males that were not able to mate successfully at the time of female emergence would be unlikely to find a mate later and would thus "waste" the breeding season.

Studies in Turkish and European hamsters have yielded clues to how appropriate reproductive timing might be obtained in these hibernators. Like other mammalian hibernators, the hibernation season in these species consists of alternating bouts of torpor and arousal. Bouts of torpor generally last 4–8 days, whereas arousals generally last for a few hours to 2 days. If testosterone is administered to either species during a bout of arousal the animal does not return to torpor; rather, the animal fails to return to hibernation for as long as exposure to testosterone is continued. Estrogen sometimes prevents these hamsters from entering torpor but is much less effective than testosterone in this regard. In Turkish hamsters, there is evidence of somewhat increased testicular activity shortly prior to emergence from hibernation. Stimulation of testis endocrine activity and early stages of spermatogenesis may occur during the bouts of arousal that occur during the hibernation season. Increased secretion of testosterone may lead to the termination of hibernation so that the animals would become active at a time when the testes are on the verge of achieving functional status. This hypothesis is supported by the observation that castrated male Turkish hamsters remained in hibernation for, on average, about 1 month longer compared to gonad-intact controls. Indeed, a few of the castrates failed to terminate hibernation at all, continuing to show alternating bouts of torpor and arousal for more than 2 years in some instances. Ovariectomy of female hamsters did not lead to a similar prolongation of the hibernation season in that sex. Since ovarian follicular maturation requires only a few days, there may be less need for females to evolve special mechanisms to fine-tune the coordination between the hibernation and reproductive seasons, respectively; even if ovarian stimulation does not begin until after the end of hibernation, females presumably would be ready to breed by the end of the first week of activity.

VIII. REPRODUCTIVE STRATEGIES IN HAMSTERS

The various species of hamsters, like most small rodents, have relatively short periods of gestation and lactation and are able to produce more than one litter during a single breeding season. Syrian and Turkish hamsters have gestations lengths of 16 and 15.5 days, respectively; these are the shortest pregnancies known for placental mammals. Neither of these species exhibits a postpartum ovulation. The Siberian hamster, with a somewhat longer gestation period of 19 days, does have a postpartum ovulation and females sometimes mate and carry fetuses while they are nursing the first litter. The second litter is then born shortly before the first group of pups are weaned. Short gestation lengths and postpartum ovulations are characteristics that probably evolved as mechanisms to allow a female to produce a larger number of pups during the breeding season.

Like several other rodents that experience postpartum ovulation, Siberian hamsters exhibit delayed implantation when pregnancy occurs during lactation. This appears to be an evolutionary strategy to delay birth of the second litter until a time when the preceding litter is almost independent of the mother. In addition, there is evidence that lactating Siberian hamsters are capable of a postimplantation delay of parturition, i.e., an embryonic diapause. Embryonic diapause is also known in some bats but was previously unknown in other placental mammals. In Siberian hamsters, embryonic diapause seems to occur predominantly in instances in which the nursing pups require more maternal nourishment, as when litter size is large or when sources of food other than the mother's milk are scarce. Thus, it appears that embryonic diapause may be a mechanism that evolved to permit adjustments to be made during late pregnancy that would tend to avoid extensive overlap in timing between the end of dependency of the first litter and the birth of the second.

Laboratory rats and certain other species require estrogen, secreted by the mother, to initiate the process of implantation of the blastocysts in the uterine wall. This is not the case for Syrian hamsters; in this species it appears that a signal coming from the blastocyst itself triggers implantation. It is possible that there is a relationship between the mechanism of implantation and the phenomenon of delayed implantation. Rats have a postpartum ovulation, and if the female becomes pregnant as a result of postpartum mating, implantation of the blastocysts is delayed by a few days and the length of the delay is related to the size of the first litter—greater delays with large litters. The delay of implantation in this situation staggers the litters to some extent so that there will not be much overlap between the nursing of the first litter and the birth and nursing of the second. Perhaps the need for a maternal signal (i.e., estrogen) for implantation in this species is related to the need to regulate the timing of implantation relative to the size and developmental state of the preceding litter. In Syrian hamsters, in which there is no postpartum ovulation, there would be no requirement for this type of coordination and timing of implantation may depend entirely on a signal from the blastocyst that presumably reflects its own state of development.

Syrian and Siberian hamsters have been by far the most widely studied hamster species with respect to reproductive characters, but some information is available for several additional species. *Phodopus campbelli* is a close relative of the Siberian hamster, and some authors considered the two to be subspecies. Reproductive characteristics are similar to those described for Siberian hamsters. The European hamster (*Cricetus cricetus*) is the largest hamster, with adults ranging from 250 to 500 g. Females are very aggressive toward males except during estrus, and a considerable amount of courtship behavior occurs prior to mating. These animals breed in the wild from April until August. As with many other hamsters, photoperiod is a major cue used to shape the annual reproductive cycle. The Rumanian hamster (*Mesocricetus newtoni*) is a close relative of *M. auratus* and *M. brandti*. It has been possible to cross *M. newtoni* with each of the other two species, despite the fact that each species has a different chromosome number. All offspring were sterile, however. The Chinese hamster (*Cricetulus griseus*) is a small rodent, weighing 30–35 g. The gestation length is 21 days, and litter size averages four pups. The Armenian hamster (*Cricetulus migratorius*) has an 18- or 19-day gestation and litter size is generally six or seven. Adults weigh 40–80 g.

IX. SEASONAL REPRODUCTIVE PATTERNS IN LEMMINGS

Lemmings inhabit arctic and boreal regions of the earth, which are among the most energetically challenging environments inhabited by small mammals. Lemmings exhibit reproductive strategies which differ from those of similar-sized rodents inhabiting more temperate regions. Lemmings and voles have long been known to exhibit striking cycles of population density, sometimes resulting in population explosions; population densities can range from 0.5 to 125 animals/hectare. In some populations, cycles of population density occur with peaks every 3 or 4 years, whereas other populations cycle irregularly or not at all.

Lemmings exhibit periods of intense breeding during the summer months similar to what has been observed in other small mammals inhabiting temperate regions (e.g., the Siberian hamster). During the summer, females reach sexual maturity at around 3 or 4 weeks of age (or when they attain a body weight of approximately 30 g). At high latitudes (e.g., Barrow, Alaska), the summer breeding season for brown lemmings (*Lemmus trimucronatus*) begins in June or

July and lasts until mid- to late August. During the summer breeding season, female brown lemmings are either pregnant or in estrus. Some evidence indicates that lemmings are induced ovulators; females remain in estrus for long periods of time but do not ovulate until after mating. Mating-induced ovulation is observed in several other species of mammals, including cats and rabbits. Female brown lemmings also exhibit postpartum estrus, and pregnant, lactating females have been observed in the field and in the laboratory. Concurrent lactation does not lengthen gestation in brown (*L. trimucronatus*) or Norwegian lemmings (*Lemmus lemmus*), but in collared lemmings (*Dicrostonyx groendlandicus*) the length of gestation is dependent on the number of suckling offspring, similar to what is observed in laboratory rats. Thus, female brown and Norwegian lemmings are capable of giving birth to litters of offspring approximately every 20 or 21 days, which is the length of gestation. Lactation in these species lasts for 12–16 days. This relatively short period of lactation may ensure that late pregnancy, the most energetically demanding portion of gestation, does not occur during the period of lactation of the previous litter.

The timing of the onset of the breeding season can vary by as much as 30 days in brown lemmings, whereas the termination of the summer breeding season is more tightly regulated, always ending between August 12 and 20 at Barrow, Alaska. The onset of the breeding season may be temperature sensitive and has been correlated with a 2-week period of relatively warm temperatures about 1 or 2 weeks before the initiation of breeding. The termination of the breeding season appears to be independent of temperature and may be regulated by a different environmental cue such as photoperiod.

Lemmings born early in the summer breeding season undergo rapid reproductive maturation and puberty is attained by around 3 or 4 weeks of age. If young females have not attained adult body size (100 mm in length) by August they fail to breed that season, though no data are available concerning the possibility of winter breeding in these individuals. Thus, reproduction in the season of birth is related to the length of the breeding season. Some male lemmings appeared to be reproductively competent all year and all males are competent before females in the spring. Similar to females, if young males do not reach adult body size by late July, they fail to undergo development of their reproductive systems that season. The summer breeding season described previously is very similar to what is observed in other small rodent species inhabiting cyclic environments.

In contrast to what is observed in many other species of small rodents, lemmings often reproduce during the winter months; in fact, some studies conclude that in the absence of winter breeding, some populations would become extinct. The winter breeding season lasts from October through April and may be associated with periods of heavy snow cover, which affords insulation from cold temperatures as well as cover from predators. Female brown lemmings remain in diestrus for much of the winter season, occasionally undergoing estrus and possibly pregnancy. Some males remain fecund during the winter months. The mean litter size is less for winter-born litters (winter litters averaged 2.9 pups) compared to summer-born litters (mean litter size of 7 pups). Though the environmental control of litter size remains unknown, cold temperatures may contribute to the smaller litter size.

X. PHOTOPERIODISM IN LEMMINGS

Lemmings exhibit many seasonal changes in morphology and physiology. Collared lemmings exhibit a seasonal cycle of pelage coloration including a molt to white pelage in winter. These lemmings also gain body mass during winter and develop a bifid digging claw. Exposure to short photoperiods elicits these winter responses, whereas exposure to long photoperiods results in summer-type responses, indicating that lemmings are photoperiod responsive. Melatonin administration can mimic photoperiod responses, depending on the duration of the melatonin signal. When collared lemmings housed in constant light, which eliminates endogenous melatonin synthesis, are infused with melatonin for 14 hr/day they weigh more than animals infused for 6 hr/day, suggesting a short-day response to winter-type melatonin infusions.

The role of photoperiod and melatonin in reproduction of lemmings is less clear. Several early studies in collared lemmings suggest that while photoperiod determines pelage type and body growth it does not determine reproductive condition. However, recent studies in the same species concluded that transfer of juvenile animals from short to long photoperiods stimulated reproduction in both males and females and that chronic exposure to long days resulted in inhibition of male gonadal development. Long-duration infusions of melatonin for 8 weeks in lemmings housed in constant light resulted in decreased uterine weight, whereas short-duration infusions led to stimulated uterine growth. Experiments which involve switching photoperiods at weaning suggest that lemmings may be responsive to the direction of photoperiod change rather than to day length per se. Furthermore, male lemmings chronically exposed to long day lengths (22 hr of light/day) exhibit inhibition of reproductive maturation. If these animals were transferred from long days to short days, then a subsequent shift back to long days resulted in stimulation of reproductive maturation. This unique pattern of photoperiod responsiveness may be related to the fact that in the field, periods of unchanging long days (i.e., early summer) are associated with periods of low food availability and flooding of the lemmings arctic habitat. Therefore, these are not optimal times for attempting to reproduce. In the field, lemmings born during early summer may not undergo rapid testicular development until they experience decreasing day lengths in late summer/fall or until they are exposed to short days followed by increasing day lengths after the winter solstice.

In lemmings, short-day exposure leads to an increase in body weight, a molt to more insulative winter pelage, and an increase in the length of the large intestine. All these responses may help to lower the energy intake required to maintain body temperature (i.e., increase in body mass leads to a decrease in surface area/volume ratio, thereby decreasing the tendency for heat loss, and the winter pelage may provide better insulation, whereas the increase in the length of the large intestine appears to increase the efficiency with which available food is utilized).

These adaptations to the winter environment serve to increase the energy available for reproduction during winter; thus, though photoperiod may not directly determine the status of the reproductive system, nonreproductive photoperiod responses may facilitate short-day breeding.

XI. POPULATION CYCLES IN LEMMINGS

Many studies have been directed at illuminating the causal factors underlying population cycles in lemmings. In Alaska, peak densities are reached during winter under snow cover and animal numbers decline in spring after the snowmelt. In Fennnoscandia, winter breeding contributes to the population peak during the subsequent summer. In some winters, females from the first winter litters breed that same winter; furthermore, preceding peak densities, most females contribute to winter breeding. In some populations winter breeding is always associated with the increase phase of the population cycle and the lack of winter breeding is always associated with the decline phase. The increase phase is rapid and is accomplished in 1 year, with overwinter population growth being most critical. These observations suggest that the extent of winter breeding may be important in determining population cycles in lemmings.

Many other hypotheses have been examined in relation to population cycles in lemmings and voles. Among the most prevalent are those concerning food availability or quality, predation, disease, and social stress, as well as phenotypically or genotypically driven cycles of behavior. Many of these hypotheses have been examined and to date no consensus has been reached on the causative factor(s) driving the striking population density cycles of lemmings.

In general, the various species of hamsters and lemmings have evolved several mechanisms to enable them to take advantage of optimal environmental opportunities for breeding. These include the use of photoperiod cues to anticipate seasonal changes in the environment, short gestation lengths, postpartum ovulation with delayed implantation or embryonic diapause, and endocrine mechanisms to tightly coordinate the end of the hibernation season with the onset of reproductive activity.

See Also the Following Articles

CIRCADIAN RHYTHMS; CIRCANNUAL RHYTHMS; LORDOSIS; MICROTINAE (VOLES); PHOTOPERIODISM; RODENTIA (RATS, MICE, ETC.); SEASONAL REPRODUCTION

Bibliography

Goldman, B. D. Parameters of the circadian rhythm of pineal melatonin secretion affecting reproductive responses in Siberian hamsters. *Steroids* **56**, 218–225.

Goldman, B. D., and Elliott, J. A. (1988). Photoperiodism and seasonality in hamsters: Role of the pineal gland. In *Processing of Environmental Information in Vertebrates* (M. H. Stetson, Ed.), pp. 203–218. Springer-Verlag, New York.

Goldman, B. D., Darrow, J. M., Duncan M. J., and Yogev, L. (1986). Photoperiod, reproductive hormones and winter torpor in three hamster species. In *Living in the Cold* (H. C. Heller, X. J. Musacchia, and L. C. H. Lang, Eds.), pp. 341–350. Elsevier, New York.

Gower, B. A., Nagy, T. R., and Stetson, M. H. (1997). Alteration of testicular response to long photoperiod by transient exposure to short photoperiod in collared lemmings (*Dicrostonyx groenlandicus*). *J. Reprod. Fertil.* **109**, 257–262.

Henttonen, H., and Kaikusalo, A. (1993). Lemming movements. In *The Biology of Lemmings* (N. C Stenseth and R. A. Ims, Eds.), pp. 157–186. Academic Press, New York.

Krebs, C. J. (1993). Are lemmings large Microtus or small reindeer? A review of lemming cycles after 25 years and recommendations for future work. In *The Biology of Lemmings* (N. C Stenseth and R. A. Ims, Eds.), pp. 247–260. Academic Press, New York.

Maier, H. A., and Feist, D. D. (1991). Thermoregulation, growth, and reproduction in Alaskan collared lemmings: Role of short days and cold. *Am. J. Physiol.* **261**, R522–R530.

Mullen, D. A. (1968). Reproduction in brown lemmings (*Lemmus trimucronatus*) and its relevance to their cycle of abundance. *Univ. Calif. Publ. Zool.* **85**, 1–24.

Newkirk, K. D., McMillan, H. J., and Wynne-Edwards, K. E. (1997). Length of delay to birth of a second litter in dwarf hamsters (Phodopus): Evidence for post-implantation embryonic diapause. *J. Exp. Zool.* **278**, 106–114.

Van Hoosier, G. L., Jr., and McPherson, C. W. (1987). *Laboratory Hamsters*. Academic Press, New York.

Yellon, S. M., Hutchison, J. S., and Goldman, B. D. (1989). Sexual differentiation of the steroid feedback mechanism regulating follicle-stimulating hormone secretion in the Syrian hamster. *Biol. Reprod.* **40**, 7–14.

Critical Period, Estrous Cycle

Lewis C. Krey

New York University School of Medicine

I. Introduction
II. The Brain as a Timekeeper for Periovulatory Events in Spontaneously Ovulating Species
III. Circadian Periodicity of the Critical Period: Steroid Feedback Effects and Determination of the Estrous Cycle
IV. The Critical Period: A Prominent Role in the Design of Neuroendocrine Research
V. The Critical Period: Evolutionary Aspects of the Neuroendocrine Control of Preovulatory Gonadotropin Release

GLOSSARY

corpus luteum The cellular mass formed by the luteinized Graafian follicle after ovulation.

gonadotropin-releasing hormone A neuropeptide secreted by the brain into the pituitary portal circulation to stimulate the synthesis and secretion of the gonadotropic hormones.

gonadotropins Glycoprotein hormones secreted by the anterior pituitary gland responsible for ovarian follicle development (follicle-stimulating hormone) and for ovulation and corpus luteum formation (luteinizing hormone) during the mammalian reproductive cycle.

proestrus The day of the rodent estrous cycle on which preovulatory gonadotropin secretion occurs.

The "critical period" hypothesis was one of the most important formative concepts in the study of the neuroendocrinology of reproduction. Based on experimental data collected during the 1940s and early 1950s by John Everett, Charles Sawyer, and Joseph Markee on an inbred strain of rat, this hypothesis provided an initial explanation of how the central nervous system regulates preovulatory gonadotropin secretion during the reproductive cycle of a spontaneously ovulating mammalian species. This information provided important conceptual groundwork for the design of studies to investigate how the brain integrates external (i.e., environmental) and internal (i.e., ovarian steroid) cues to time ovulation.

I. INTRODUCTION

In 1932 Hohlweg and Junkman suggested that the central nervous system plays a major role in the feedback relationships between gonadal and pituitary hormones. Based on observations that the formation of "castration cells" in the rat anterior pituitary gland could be blocked either by chronic atropine treatment or by transplantation to the kidney capsule, these investigators proposed the existence of a neural "sex center" within the hypothalamus that regulates pituitary gonadotropin secretion and is sensitive to either the direct or indirect feedback actions of the ovarian hormones. In separate studies on guinea pigs and rats during the next two decades, Dey, Hetherington, Ransom, and Hillarp each examined the effects of brain lesions on reproduction. These studies indicated that at least two specific hypothalamic areas are essential for normal reproductive cycles. Whereas lesions in the medial basal hypothalamus in close proximity to the pituitary stalk produced ovarian atrophy, lesions in the anterior hypothalamus at the caudal aspect of the optic chiasm did not influence ovarian follicle development but did block ovulation.

Such lesioned animals displayed a "persistent estrous" pattern of cornified vaginal smears previously noted by others in aged rats or in rats maintained in constant light. Subsequent research by Harris and coworkers establishing the importance of the pituitary portal circulation for normal pituitary function provided an anatomically appropriate explanation for the medial basal hypothalamic lesion effects. However, the ability of the anterior lesions to block ovulation required more information about the role of the central nervous system in this important event in the mammalian reproductive cycle.

In 1949, Sawyer, Markee, and Townsend reported that reflex ovulation in the rabbit could be blocked with large doses of the neurally acting drugs Dibenamine and atropine sulfate. However, these drugs were only effective when injected within 1 min after coitus. The rapidity required for these injections was interesting considering the observation of Fee and Parkes 20 years earlier that at least 1 hr was required for the anterior pituitary gland to secrete an ovulatory quota of gonadotropins in this species. Such a delay implied that the immediate postcoital drug injection blocked the neural reflex stimulus that initiated gonadotropin release by the pituitary gland. Utilizing a similar pharmacologic approach, Everett, Sawyer, and Markee then tested the hypothesis that these drugs could also block preovulatory gonadotropin secretion in a spontaneously ovulating species, the laboratory rat.

II. THE BRAIN AS A TIMEKEEPER FOR PERIOVULATORY EVENTS IN SPONTANEOUSLY OVULATING SPECIES

In a series of four controlled studies, Everett, Sawyer, and Markee provided data suggesting that the anterior pituitary gland secretes gonadotropins in response to a neurogenic stimulus that occurs during a "critical period" in the afternoon hours. Each study utilized the same inbred strain of rats [Osborne–Mendel (O–M)]. This was an important consideration because the O–M strain was studied extensively by Everett during the previous years and its reproductive capabilities and responses to ovarian steroid hormone treatments and external lighting conditions were well characterized. It is also important to note that the strain was maintained under controlled lighting conditions (14 hr light:10 hr dark; lights on 5:00 AM–7:00 PM) since previous studies had illustrated the importance of environmental lighting to the development of persistent vaginal estrus. Under this lighting schedule individual rats displayed regular reproductive cycles, either 4 or 5 days in duration, and cycle stage could be easily determined by daily monitoring of vaginal smears. In addition, ovulation was known to occur during the early morning hours and could be routinely assessed by recovering ova or identifying viable corpora lutea as judged by cholesterol accumulation following terminal laparotomy and microscopic evaluation of the ovaries and Fallopian tubes the next morning.

Initially, studies were conducted in rats in which ovulation was induced by steroid administration, experimental models previously characterized in detail by Everett. First, estrogen injections were used to induce ovulation in newly impregnated rats. When injected on the day after estrogen administration, both Dibenamine and atropine sulfate effectively blocked the ovulation and corpus luteum cholesterol deposition expected the following day. However, when injected concurrently with luteinizing hormone (LH), these drugs did not affect these parameters, illustrating that they prevented the release of this gonadotropin rather than its actions on the ovary.

In a second study, progesterone was administered to rats on Day 3 of a 5-day estrous cycle; this treatment advanced ovulation by 24 hr. Similar results were noted; the neurally acting drugs blocked progesterone's induction of premature ovulation even when injected after a 4-hr delay. Moreover, the interval between progesterone and atropine was unimportant as long as atropine was administered during the afternoon hours. Significantly, progesterone advancement of ovulation was also related to the time of day; the rats did not respond to the progesterone injections late in the afternoon. Thus, it appeared that progesterone stimulation of gonadotropin secretion was confined to certain midafternoon hours on Day 3. Additional evidence suggested that preovulatory gonadotropin secretion might occur at a similar time on the day of proestrus. In particular, the morphologic age of the progesterone-induced corpora lutea on the morning of Day 4 appeared closely similar to that of spontaneously ovulating rats on the morning of estrus.

In the third study Dibenamine or atropine sulfate was used to block spontaneous ovulation in proestrous cycling rats. The results of this study were definitive: Dibenamine or atropine selectively blocked spontaneous ovulation when injected on or before 2:00 PM on proestrus. Injection of these neurally acting drugs after 4:00 PM was generally ineffective. These results indicated clearly that the normal spontaneous discharge of gonadotropins responsible for ovulation occurred during proestrous afternoon and that this discharge was stimulated by a neuronal mechanism. The authors concluded that, in rats, a spontaneous stimulus passes from the hypothalamus to the anterior pituitary gland at some time during a critical period on proestrous afternoon, resulting in the discharge of gonadotropin to eventuate in ovulation 10–12 hr later. Subsequently, other investigators reported that similarly timed injections of barbiturates, morphine, reserpine, and monoaminergic blocking drugs also effectively blocked ovulation in rats; these reports provided further support for the neurogenic nature of the stimulus triggering gonadotropin release.

In a final study Everett compared the effects of timed atropine injections and parapharyngeal hypophysectomy in proestrous O–M rats. Ovulation response was assessed in terminal laparotomy the next morning. The results are summarized in Fig. 1. Both atropine and hypophysectomy blocked progressively fewer animals when performed later in the day, suggesting that the neurogenic signal proceeded in parallel with the resultant release of gonadotropins.

The major tenets of the critical period hypothesis have been confirmed by recent radioimmunoassay studies. Closely timed monitoring of circulating gonadotropin levels has revealed pronounced discharges of LH on proestrous afternoon; such patterns are depicted in Fig. 2 for CD rats. Moreover, similarly timed discharges in gonadotropin-releasing hormone (GnRH) secretion have been reported in pituitary

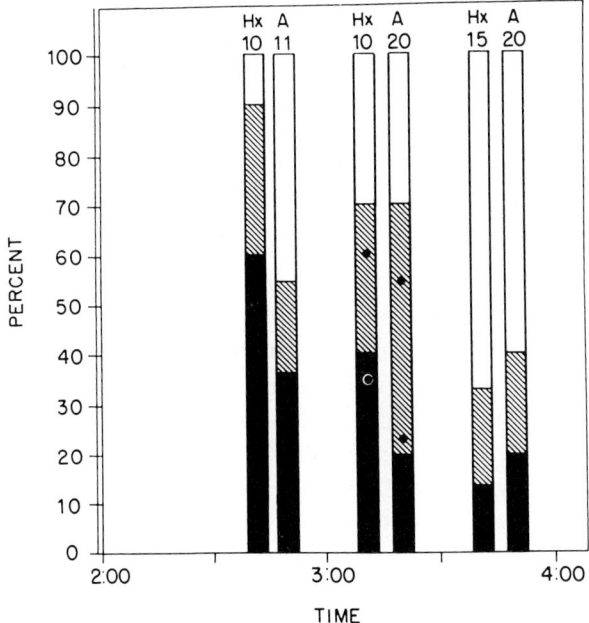

FIGURE 1 Influences of hypophysectomy (Hx) or atropine (A) injection at different times during the proestrus critical period in 4-day cyclic O–M rats. Black bars, complete blockade; cross-hatched bars, partial blockade; white bars, complete ovulation. The number of rats in each group is listed at the top of each column. The vertical pairs of dots mark the averages for all time groups for Hx and A. The lower dot in each pair indicates the percentage totally blocked; the vertical spacing between dots indicates percentage partially blocked. (From Everett, J. W., *Endocrinology* **58**, 786, 1956. © The Endocrine Society.)

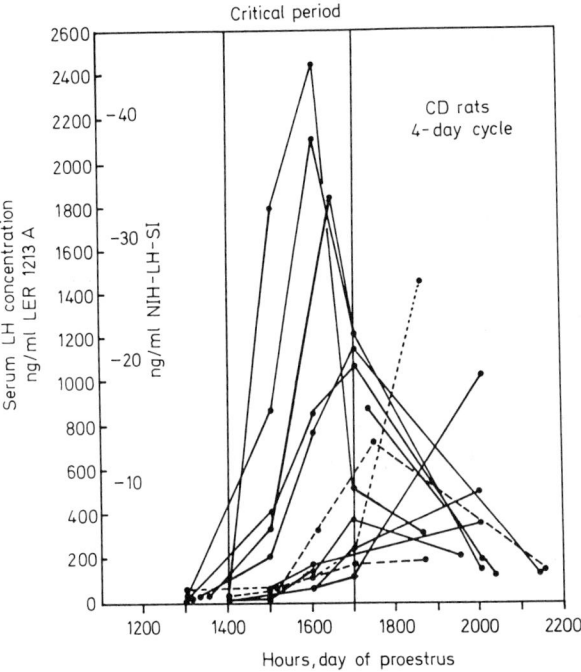

FIGURE 2 Animal to animal variation in the time of onset of spontaneous preovulatory LH surges in proestrous 4-day cyclic CD rats. The temporal parameters of the critical period presented for these rats are longer than those defined for O–M rats. (From Everett, J. W., Krey, L. C., and Tyrey, L., *Endocrinology* **93**, 947, 1973. © The Endocrine Society.

portal blood samples collected on proestrous afternoon and these patterns are described in more detail elsewhere in this encyclopedia.

It should be emphasized that the absolute parameters of a critical period are defined by many considerations, including the species and strain of the animal under study, its estrous cycle history, and, as discussed previously, the environmental conditions under which it is maintained. Thus, discussion of a critical period is only appropriate when referring to a population of animals maintained under a particular set of experimental conditions. Even then there may be individual to individual variation in the absolute timing of gonadotropin release; this is evident in the LH patterns of CD rats presented in Fig. 2. Not only does the critical period for rats of the CD strain appear to be longer than that for O–M rats but also there are individual animals in whom LH release begins extremely late relative to the majority of the population.

III. CIRCADIAN PERIODICITY OF THE CRITICAL PERIOD: STEROID FEEDBACK EFFECTS AND DETERMINATION OF THE ESTROUS CYCLE

In addition to defining the temporal borders of the critical period, the studies of Everett and Sawyer also suggested that the neurogenic mechanisms stimulating ovulatory gonadotropin secretion in the rat are characterized by a 24-hr periodicity and could occur on days other than proestrus. For example, when injected on Day 3 of a 5-day estrous cycle, progester-

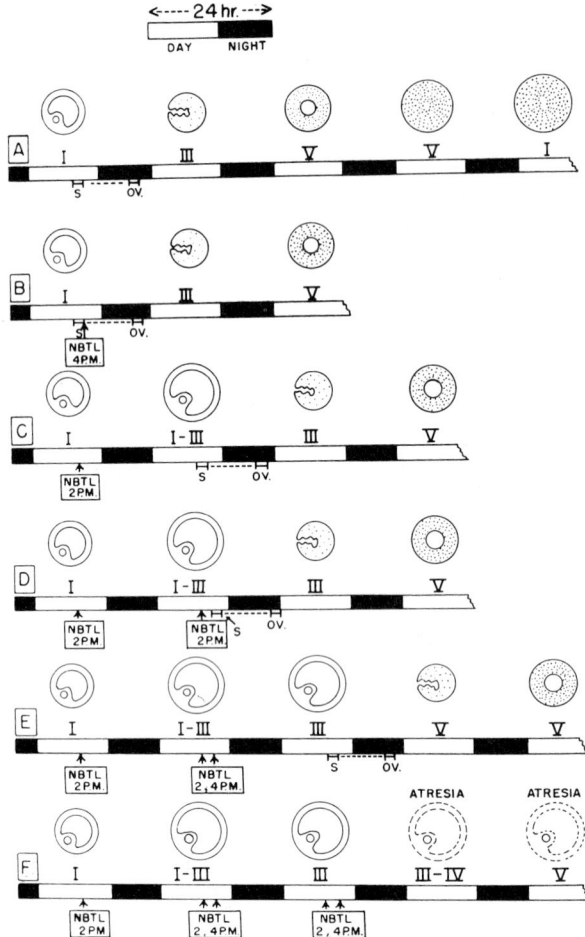

FIGURE 3 Representations of ovarian events during the normal 4-day estrous cycle in O–M rats (A) and the characteristic results of different pentobarbital regimens (B–F). Vaginal stages (I, proestrus; III, estrus; V, diestrus) are shown over each time scale and the symbols above these show the corresponding Graafian follicle and corpus luteum stages. S, the normal and experimental critical periods; OV, the normal ovulation time in A and the estimated time elsewhere; NBTL, an intraperitoneal injection of pentobarbital. Note the progressive 24-hr delays in ovulation NBTL treatment blockade is given on proestrus and on subsequent days (C and E). (From Everett, J. W., and Sawyer, C., *Endocrinology* 47, 198, 1950. © The Endocrine Society.)

one advanced ovulation 24 hr and this could be blocked by appropriately timed atropine or Dibenamine injections. More definitive illustrations of circadian periodicity were provided in additional studies in which rats were monitored for an additional 24, 48, or 72 hr following pharmacologic blockade of spontaneous ovulation (Fig. 3). Significantly, when ovulation was blocked on proestrus, the rats ovulated spontaneously during the second night. Atropine administration on the second afternoon revealed a critical period for gonadotropin release at this time, and blockade of this gonadotropin release resulted in a second 24-hr postponement and another critical period for gonadotropin release 24 hr later.

The circadian periodicity of periovulatory gonadotropin release in rats has been confirmed in studies in which ovariectomized animals are chronically treated with estradiol; these animals display daily surges in LH secretion. However, Freeman and colleagues demonstrated that these daily surges are abruptly terminated by a brief exposure to progesterone. Such steroid-induced patterns of LH secretion suggest that the single periovulatory discharge of LH observed during rat estrous cycle is the direct result of the sequential pattern of estrogen and progesterone secretion by the ovarian follicles prior to and during the periovulatory period.

Other studies from Everett's laboratory have suggested that the critical period on proestrus may delineate a time of day when important neuroendocrine events can occur on different days of the reproductive cycle. For example, the ability of estrogen to advance ovulation in rats with 4- or 5-day estrous cycles depends on the time of day of injection. Injection before noon on diestrus Days 1 or 2 consistently advances ovulation by 24 hr; however, injection at times after 4:00 PM is much less effective. Other neuroendocrine events occur daily during the afternoon hours in rats maintained on similar lighting conditions. Adrenocortical hormone secretion routinely begins at this time, as does an increase in general activity. Prolactin also is preferentially secreted in the afternoon hours particularly during pregnancy and pseudopregnancy.

IV. THE CRITICAL PERIOD: A PROMINENT ROLE IN THE DESIGN OF NEUROENDOCRINE RESEARCH

The identification and characterization of a critical period for GnRH and gonadotropin release has provided investigators with a research model that can been utilized to gain insights into the underlying neuronal mechanisms.

Extensive series of pharmacologic studies have tested the ability of numerous neurally acting drugs to block periovulatory gonadotropin secretion in rats and other rodent species. As mentioned earlier, barbiturates, morphine, and anticholinergic and antimonoaminergic drugs have been shown to block periovulatory release. These studies have provided insights into the neurochemical pathways involved. In follow-up studies the appropriate antagonists or agonists have been administered in an attempt to reinitiate gonadotropin secretion. These studies have been extended to neuropeptides, which have been demonstrated to have potent stimulatory effects on gonadotropin release. In some instances, these have been microinjected into the brain or pituitary gland to obtain morphologic evidence for site of the relevant neuronal pathways. These studies have been summarized elsewhere in this encyclopedia.

Neurophysiologic approaches to study neuroendocrine function have also been influenced by the critical period hypothesis. As mentioned previously, early studies by Dey and Hillarp reported that lesions in the rostral hypothalamus produced a "persistent estrous" condition in which follicles developed but ovulation did not occur. Subsequently, Everett and Sawyer postulated that there was a neuronal apparatus in the rostral hypothalamus that had intrinsic circadian rhythmicity and controlled ovulation. Subsequent studies by these investigators and colleagues established that electrochemical stimulation of this brain region could overcome the blocking actions of pentobarbital and stimulate the secretion of an ovulatory quota of gonadotropins. These studies led to the identification in the rat of a funnel-shaped septal–preotico–tuberal pathway that carries the relevant neuronal information to the median eminence to trigger preovulatory gonadotropin secretion. Using similarly blocked animals, electrical stimulation studies have also implicated other brain regions that can control ovulation and these have provided experimental rationales to trace and characterize the relevant neuronal pathways regulating GnRH release using lesioning or neuron-mapping techniques. Neuronal recording techniques have also been applied to study the moment-to-moment changes in brain activity important for GnRH gonadotropin release; clearly, these are most relevant in the rat when performed during the critical period. The outcomes of these approaches have also been described elsewhere in this encyclopedia.

V. THE CRITICAL PERIOD: EVOLUTIONARY ASPECTS OF THE NEUROENDOCRINE CONTROL OF PREOVULATORY GONADOTROPIN RELEASE

With the advent of radioimmunoassays, temporal patterns of LH secretion during estrous or menstrual cycles or following steroid-induced gonadotropin secretion have been examined in detail for several animal species, including man. Such studies have reported some interspecies similarities, but more often differences, in the timing of periovulatory LH secretion. Significantly, in many instances different patterns of LH secretion are also observed which suggest that different mechanisms have evolved to control this aspect of gonadotropin release. As discussed previously, spontaneous or steroid-induced periovulatory gonadotropin release in small rodent species is brief and is restricted to a critical period in the daily light:dark cycle. When exposure to an estrogen stimulus is sufficient, an LH discharge occurs; if the stimulus is insufficient, secretion is postponed for 24 hr. When estrogen levels remain elevated in the absence of progesterone, daily LH surges occur. In each case, the LH discharge can be blocked by appropriately timed treatments of neuroactive drugs. However, in other species that display estrous cycles (e.g., the guinea pig and ewe), there is no defined critical period for periovulatory gonadotropin release. The LH surge can occur at any time of day and is determined by the parameters of the estrogen stimulus, i.e., time of onset, magnitude, and duration. Importantly, only a single LH surge occurs and in some instances it cannot be blocked by treatment with the neuroactive drugs. The situation is even more complicated in primates, species in which preovulatory gonadotropin secretion can persist for days. Again, only a single LH surge occurs in these species and its timing appears to be determined primarily by the time of onset and strength-duration characteristics of the estrogen stimulus. Taken together, these considerations point to major interspecies changes in the neuroendocrine mechanisms controlling ovulation and this conclusion is supported by many other observations using other experimental approaches. These considerations also suggest that, despite its postulation almost 50 years ago, the critical period hypothesis still exerts an active influence on the design and interpretation of experiments that examine the neuroendocrine control of ovulation in mammals.

See Also the Following Articles

Circadian Rhythms; Corpus Luteum; Estrous Cycle; GnRH (Gonadotropin-Releasing Hormone); Hypophysectomy; LH (Luteinizing Hormone)

Bibliography

Everett, J. W. (1961). The mammalian female reproductive cycle and its controlling mechanisms. In *Sex and Internal Secretions* (W. C. Young, Ed.), 3rd ed. Williams & Wilkins, Baltimore.

Everett, J. W. (1989). *Neurobiology of Reproduction in the Female Rat*, Monographs on Endocrinology, Vol. 32. Springer-Verlag, Berlin.

Freeman, M. E. (1994). The neuroendocrine control of the ovarian cycle of the rat. In *The Physiology of Reproduction* (E. Knobil and J. D. Neill, Eds.), 2nd ed., Vol. 2. Raven Press, New York.

Kordon, C., Drouva, S. V., de la Escalera, G. M., and Weiner, R. I. (1994). Role of classic and peptide neuromediators in the neuroendocrine regulation of luteinizing hormone and prolactin. In *The Physiology of Reproduction* (E. Knobil and J. D. Neill, Eds.), 2nd ed., Vol. 1. Raven Press, New York.

Crowding

see Stress and Reproduction

Crustacea

Matthew Landau
Richard Stockton College

Hans Laufer
University of Connecticut

I. Introduction
II. Growth of Crustaceans
III. Modes of Reproduction
IV. Modes of Sexual Reproduction
V. Anatomy of the Reproductive System
VI. Endocrine Control of Crustacean Reproductive Physiology
VII. Crustacean Embryology
VIII. Larval Development and Metamorphosis

GLOSSARY

androgenic gland A ductless gland, found only in malacostracan crustaceans, that is associated with the sperm duct and that produces proteins, androgenic hormones, as well as some terpenes.

chitin A carbohydrate polymer that makes up the crustacean exoskeleton, which is often strengthened with proteins and mineral deposits, especially calcium.

ecdysone A steroid hormone released from the Y-organ that is converted into its active form, 20-hydroxyecdysone, which regulates molting.

eyestalk The movable stalk (or peduncle) that supports the compound eyes; eyestalks are often important because they not only aid the animal in sensing its environment but also house complex endocrine tissues.

germogen The germinal area of the ovary.

hepatopancreas The site of production of many crustacean digestive enzymes; it is also likely to be the source of extraovarian yolk lipoproteins in at least some crustaceans.

mandibular organ A small endocrine gland associated with the mandibular tendon, which is the source of methyl farnesoate.

megalopa A larval form that appears after the zoea, which swims with its abdominal appendages, and is restricted to the malacostracans since they are the only group with true pleoplods.

methyl farnesoate The hormone produced by the mandibular organ that is very similar to a juvenile hormone produced by insects and is associated with reproduction in both males and females.

nauplius The least developed of the larval stages; it has only three metameres and moves using the first three pairs of head appendages.

spermatheca An enclosed site for long-term storage of the spermatophore or sperm.

thelycum A structure often used when defining the systematics of penaeid shrimp which may be involved in long-term storage, may be the deposition site of nonsperm substances by the males, or may be involved in the transfer of the spermatophore or sperm to the spermatheca.

X-organ and *sinus gland* Neuroendocrine structures found in close association in the eyestalk of higher crustaceans which are responsible for synthesis and release, respectively, of many of their most important hormones, including those that regulate reproduction.

zoea The larval form, which follows the naupliar stage, that swims with its thoracic appendages.

I. INTRODUCTION

The crustaceans are a monophyletic group of arthropods consisting of almost 40,000 identified extant species. They are characterized by a generalized body plan (although modifications abound) consisting of (i) a head which has five pairs of appendages; (ii) a segmented trunk, often consisting of an anterior thorax and a posterior abdomen; and (iii) a telson which is devoid of appendages. Segmentation, however, is sometimes difficult to see. The appendages are biramous; the appendages of the head include two pairs of antennae, one pair of mandibles, and two pair of maxillae, although accessory mouthparts, the maxillipeds, may arise from thoracic segments. The carapace is composed of chitin but is often strengthened with proteins and mineral deposits, especially calcium. The blood typically contains hemocyanin but rarely other pigments, and gas exchange can take place via the body surface or with the aid of gills. The classification of the recent Crustacea is as follows:

1. Class Remipedia: Small group with elongate bodies and many primitive characteristics; live in marine caves; have no eyes

2. Class Cephalocardia: Small group with elongate bodies and many primitive characteristics; benthic detritus feeders; have no eyes

3. Class Branchiopoda: A group of mostly small freshwater organisms with gill-bearing limbs; included are the anostraca (fairy shrimps), cladocerans (water fleas), notostracan (tadpole shrimps), and the conchostracan (clam shrimps)

4. Class Maxillopoda: A large and diverse group; there is a somewhat abbreviated trunk; class is derived from ancestors with a common body plan of six thoracic somites and five abdominal somites and this "6-5 plan" is still present in some of the maxillopods; class includes two small subclasses, the minute mystacocarids and the branchiurids (fish lice); also included are the cirripeds (barnacles), with >1000 known species, and the copepods, with >8500 known species

5. Class Tantulocardia: A small group of tiny copepod-like ectoparasites of deep-water crustaceans; have no recognizable cephalic appendages

6. Class Ostracoda: Probably a sister group to the Maxillopoda; mostly small animals with oval bodies surrounded by a bivalved carapace; no visible signs of segmentation

7. Class Malacostraca: The largest class of the Crustacea (>23,000 known species); body is composed of a head, an eight-segmented thorax, and an abdomen of six or (rarely) seven segments; there can be extensive head and thorax fusion (= cephalothorax); among the many groups in this class are the stomatopods, cumaceans, isopods, amphipods, euphausids (krill), and the familiar decapods (shrimp, lobsters, crabs, and crayfish)

II. GROWTH OF CRUSTACEANS

Like the other arthropod groups, crustaceans are supported and protected by an exoskeleton of chitin, a carbohydrate polymer. Often the chitin is impregnated with minerals, especially calcium. Since the exoskeleton confines the animal, it must be shed for the animal to grow and develop. A typical growth cycle is shown in Fig. 1.

The physiological events associated with molting, which are largely known from the malacostracans, have been reviewed by Skinner and others. The molting hormone, ecdysone, is released from the Y-organ and is converted into its active form, 20-hydroxyecdysone, in the epidermal tissues by a monooxygenase. The level of circulating ecdysteroids in the hemolymph of crustaceans is generally low; however, just before the molt there is a dramatic rise followed by

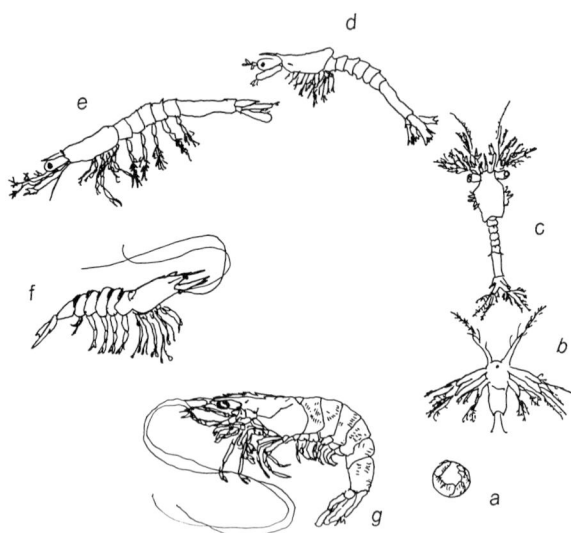

FIGURE 1 The life cycle of a penaeid shrimp (not drawn to scale). a, egg; b, nauplius; c, protozoea; d, mysis; e, postlarva; f, juvenile; g, adult (redrawn from Hanson and Goodwin, 1977, and Mock and Murphy, 1971).

a fall just before the actual molt. The production of ecdysone by the Y-organ is apparently under the control of the eyestalk. When the eyestalk is removed (from animals that are not reproductively active or have not already passed through a terminal molt) the animal will molt; the sinus gland releases a peptide, the molt-inhibiting hormone (MIH), that halts ecdysteroid synthesis by the Y-organ. Molting can be positively affected by loss of limbs and negatively affected by some specific parasitizations.

The events that make up the "molt" are well studied in the crustaceans. After the exoskeleton is shed and before the new one has had time to harden, there is a critical period during which time (i) the animal is subject to predation without its hard outer covering, and (ii) it will increase its volume by rapidly taking up water so that after the skeleton hardens it can release the water from its tissues and have "room to grow." The molt cycle is divided into stages A–E, which may be divided into substages 1, 2, etc:

1. Stage D, proecdysis (= "premolt"): The epidermis withdraws from the old exoskeleton. This is followed by a new cuticle being secreted that can be seen if part of the old cuticle is removed, whereas most of the old postecdysial cuticle is digested from within and reabsorbed. Finally, the old cuticle splits as the animal absorbs water.

2. Stage E: Ecdysis occurs when the old cuticle is pulled off by the animal.

3. Stages A–C_3, metecdysis (= "postmolt"): during these stages, the soft cuticle becomes paper-like, due to secretion of the principle layer, and calcification starts as calcium as $CaCO_3$ and $Ca(PO_4)_2$ are deposited in spaces between chitin; rigidity progresses.

4. C_4 is the intermolt stage.

III. MODES OF REPRODUCTION

Reproduction can be sexual or asexual in the Crustacea. While some invertebrates are able to reproduce asexually by budding or fission, asexual crustaceans are limited to parthenogenesis, the production of a diploid egg.

There are probably no crustacean genera that are obligate parthenogens, but many can reproduce for a number of successive generations parthenogenically. Some of the filter-feeding cladocerans have been shown to exhibit heterogony, the alternation of parthenogenic and sexual generations. This has been summarized by Kaestner. The "summer eggs" are produced parthenogenically and develop in the brood sac into young cladocerans, whereas "winter eggs" are haploid and require fertilization. Day length and temperature seem to play major roles in determination of the mode of reproduction in certain cladocerans such as *Daphnia magna*. At moderate temperatures and 12-hr days, females produce other females parthenogenically, and they can do this for a number of generations. However, increasing the day length, or lowering the temperatures to 8°C or increasing them to 30°C brings parthenogenesis to a temporary halt. Other factors, such as change in the diet or overcrowding, may also exert an effect on reproductive strategies. The brine shrimp, *Artemia*, is also able to produce offspring parthenogenically when under stressful conditions, in particular when levels of dissolved oxygen decrease in the water.

IV. MODES OF SEXUAL REPRODUCTION

Many crustacean genera are characterized by having separate sexes (i.e., they are "gonochoristic" = "dioecious"). However, a significant number are hermaphroditic. Some of these are simultaneous hermaphrodites, able to produce eggs and sperm at once, whereas others are sequential hermaphrodites, changing from one sex to another.

A. Gonochoristic Crustaceans

While gonochorism is the most common mode of sexual reproduction, there is a great deal of variation on this theme. In the copepods, the sexes look different, especially in parasitic forms, which typically have dwarf males. In some cases, the sexes are so different in appearance that it is difficult to recognize them as members of the same species. Dwarf males are also known from parasitic cirripeds, such as *Ascothorax* and *Trypetesa*.

The decapod malacostracans are the best studied of the crustaceans; while a few are protandric hermaphrodites, most are dioecious, and the sexes are discernible from secondary characteristics (such as the shape of a crab's apron). In some cases the males are larger than the females, whereas in other cases the females are larger. Copulation is the general rule, with males often having modified appendages for this purpose. Fertilization can be internal but usually takes place as the oocytes are extruded by the female while they pass a spermatophore she received during copulation. The female commonly has special seminal receptacle and usually receives the spermatophore soon after she molts; for some animals, however, this may not always be the case. Other malacostracans may have different strategies; for example, the male mysid *Mesopodopsis* sheds his sperm into the water near the female, then uses his pleopods to push the sperm through the water toward his mate.

B. Hermaphroditic Crustaceans

Sequential hermaphrodites can be protandric (males change into females) or protogynous (females change into males). There are many examples of protandry in the Crustacea; in these cases the androgenic gland, the site of a factor that controls development of male characteristics, will degenerate. Often in crustaceans, sequential protandry is partial, such as in the protandrous shrimp *Athanas*, whose populations always include some males which do not change sex, and in the crayfish *Parastacus*. Other examples of protandry include the parasitic isopods of the Cymothoidea and Cryptoniscina. In the lysianassoid amphipod *Acontiostoma*, the large females retain the penial process. There are fewer examples of crustacean protogyny. The males of the tanaidid malacostracan, *Heterotanais*, can either develop from "neuter" larvae or develop from females who have molted after brooding and whose ovaries degenerate.

True simultaneous hermaphrodites are also rare in the Crustacea, with the best examples being the cephalocarids, which have common genital openings, and most of the barnacles of the order Thoracica (although some, such as *Scalpellum*, have females and dwarf males). The male gonads of the well-studied barnacle genus *Balanus* lie along the thorax, whereas the ovaries are preoral with oviducts that open into the mantel cavity. The long, flexible penis of one individual will fertilize the oocytes of its neighbor when it enters the mantle aperture. In certain circumstances, such as isolation, these barnacles can self-fertilize.

V. ANATOMY OF THE REPRODUCTIVE SYSTEM

Gonads are typically paired, although there may be fusion, and are derived from either coelomic pouches or primary germ cells. In most Crustacea, as in the insects, the gonads are dorsal or dorsolateral in relation to the gut. The location of the gonads will vary with the group, extending through most of the body or being limited to the abdomen or thorax. Reproduction is typically seasonal, with the size, and often the color, of the gonads changing appreciably with gamete maturation.

A. Female System

There has been debate concerning the site of yolk protein synthesis (vitellogenesis) in crustaceans. The primary yolk protein, vitellin, is a high-density lipoprotein composed of at least two polypeptide subunits (apoproteins). The polypeptides of vitellogenin are similar to, or indistinguishable from, those of vitellin, yet vitellogenin is found circulating in the hemolymph of the female, suggesting an extraovarian source of the yolk protein. While the polypeptides in the ovaries and hemolymph are identical, or at least very much alike, the lipids of these lipoproteins are clearly distinct. The most likely site of extraovarian vitellogenesis is the hepatopancreas, although there is some question concerning the importance of the hepatopancreas compared to the ovaries; some researchers have found no evidence for vitellogenesis in the hepatopancreas. Along with the hepatopancreas, the subepidermal adipose tissue has been suggested as a site of nonovarian vitellogenesis.

The ovary, the site of egg production, will often fill a large portion of the body of a mature female. Ovaries are primitively simple, tube-like, paired organs but can be fused. In the higher crustacean groups the ovaries are morphologically more complex, having finger- or pouch-like projections, and in the true crabs they are H-shaped. The ovary is typically composed of inner and outer layers of epithelial cells, separated by one or more midlayers of connective tissue. The germinal area of the ovary, the "germogen," is normally distinct during ovarian development when the oocytes are accumulating yolk; however, it lacks the distinct compartmentalization typical of the insect ovary. The placement of the germogen in the ovary is variable and largely unrelated to crustacean taxonomy.

Oviducts are short tubes that usually project laterally from the ovaries, connecting with the external openings, the gonopores; the position of the gonopores varies with different crustacean groups. The oviducts are composed of an outer epithelium that is continuous with the ovary and have a distinct inner lining. In the area of the gonopores a sac or pouch, the spermatheca (= seminal receptacle), is present; these may be connected to the oviduct or not, as in the case of the Macrura; the spermatheca is lined with secretory epithelial cells. Interestingly, the snow crab *Chionoecetes* may maintain populations of symbiotic bacteria in the spermatheca to help maintain a specific pH balance. The penaeid shrimp "thelycum" may or may not be involved in long-term storage and may have other related functions.

B. Male System

The gonads are usually paired, but the testes occupy less space in the males than do the ovaries in the females. In some of the barnacles the testes are diffuse lobules found in the connective tissue of the prosoma. From the testes extend ducts, the vas deferens, that are typically long, narrow, and often highly coiled. A "seminal vesicle," i.e., an enlargement of the vas deferens near the terminal end, is seen in some crustaceans. From this simple plan, crustaceans have developed many variations.

In male anostracans there is a pair of penes. In copepods the testis is unpaired, as is the ovary, and lies dorsal to the gut; the vas deferens, seminal vesicle, and ejaculatory duct are unpaired in the calanoids but paired in the cyclopoids. The vas deferens of the thoracian barnacles join to make a single ejaculatory duct in the penis, whereas in the small male, ascothoracian and acrothoracian barnacles have simplified systems. In the malacostracans the testes are found in the thorax or cephalothorax in most groups, although there are exceptions.

During mating, males may use special devices to hold the females. These are somewhat varied and include the well-developed second antennae of anostracans such as *Artemia*, the clawed first limbs of the conchostracans, the last limbs of the fish louse *Argulus*, and the antennae, maxillipeds, and the sixth pair of thoracopods in the copepods. During mating, the transfer of the sperm to the female varies between the groups of crustaceans; the malacostracans usually rely on the first two pair of pleopods, the copepods use the last limb, and the anostracans, ostracods, and thoracid barnacles all use a penis for transfer.

VI. ENDOCRINE CONTROL OF CRUSTACEAN REPRODUCTIVE PHYSIOLOGY

It appears that hormones, as in most other organisms, play a central role in the regulation of sexual differentiation, reproductive maturation, prespawning physiology, mating, and postembryonic development of crustaceans. Several of the crustacean groups have received almost no attention from researchers, whereas others have been the subject of a modest amount of interest, such as the isopods and amphipods. Only the decapods have been well studied because they are economically important and because their large size allows easy manipulation in the laboratory. Most of the results described here refer to decapods, whose eyestalks contain the X-organs and sinus glands which are responsible for synthesis and release, respectively, of many of their most important hormones, including those that regulate reproduction. A summary of the major endocrine effectors in female decapods is shown in Fig. 2.

A. Female Endocrine Regulators

The eyestalk is the site of production of the gonad-inhibiting hormone (GIH), shown to be a small protein in the lobster. Its role is to suppress development of the gonads, and perhaps other tissues associated with reproduction, during the nonspawning seasons.

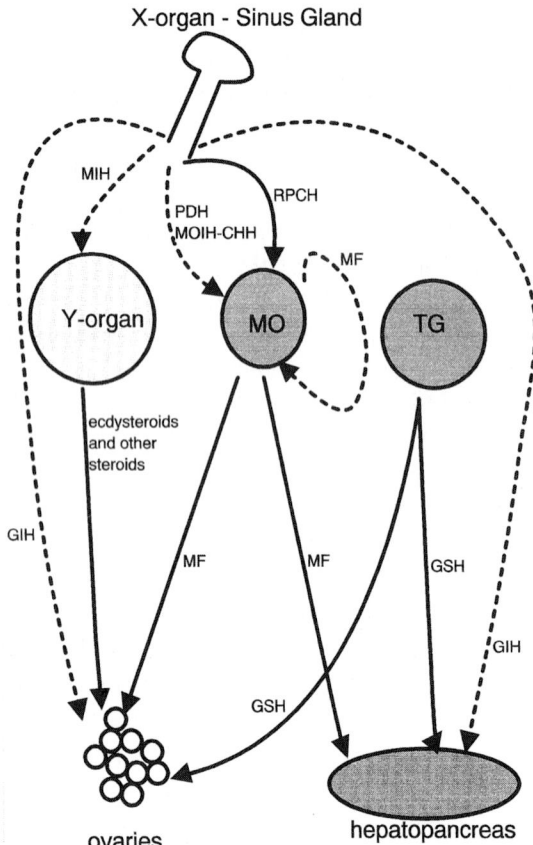

FIGURE 2 The endocrine regulation of reproduction in female decapod crustaceans. Stimulation is indicated by a solid line, whereas a broken line indicates an inhibition. CHH-MOIH, crustacean hyperglycemic hormone–mandibular organ-inhibiting hormone; GIH, gonad-inhibiting hormone; GSH, gonad-stimulating hormone; MF, methyl farnesoate; MIH, molt-inhibiting hormone; MO, mandibular organ; PDH, pigment-dispersing hormone; RPCH, red pigment concentrating hormone; TG, thoracic ganglion. (modified from Laufer et al., 1992, 1993).

GIH therefore works directly at the level of the ovary as well as at extraovarian sites of vitellogenesis, perhaps through the secondary messenger cAMP. When the eyestalks are removed during the reproductive season, ovarian growth follows.

Steroids are thought to also play a role in crustacean reproduction, although this role has not been well-defined. Some of the steroids that may be involved are the ecdysteroids from the Y-organ, whose synthesis and release are under the control of the eyestalk's MIH (see Section II). The removal of the Y-organ seems to halt vitellogenesis in isopods and amphipods. Ecdysteroids are suspected of also playing a role in the decapods, along with several other steroids.

There is also a gonad-stimulating hormone (GSH), although it is not as well studied as the GIH. The existence of the GSH was first established in the thoracic ganglion in the crab *Potamon*. A GSH appears to also be present in the brain of some decapods, whereas in the amphipod *Orchestia*, subepidermal adipose tissue production of yolk protein seems to be stimulated by protocerebral and follicular tissue products. Release of GSH is probably under the control of modulating biogenic amines.

The importance of the hormone methyl farnesoate (MF), which is a product of the mandibular organ (MO) and is very similar to juvenile hormones (JH) produced by insects, has been demonstrated. MF, like JH in insects, is known to be essential for egg production and other aspects of female reproduction in Crustacea. While MF may, as does JH in insects, act at several sites, the mode of action is not clear. However, the fact that there is high MF ester hydrolysis in the hepatopancreas, which has been suggested to be the site of vitellogenin synthesis, may indicate that this organ is a primary MF target. The release and/or synthesis of MF by the MO appears to be under the influence of an eyestalk factor(s) that works through the secondary messenger cGMP produced by a membrane-bound cyclase rather than a G protein; the eyestalk factor(s) could be pigmentary regulating peptides, such as red pigment concentrating hormone and pigment-dispersing hormone in *Procambarus*, or a member of the crustacean hyperglycemic hormone–mandibular organ inhibiting hormone family of peptides. MF itself may also regulate the MO by a feedback mechanism.

B. Male Endocrine Regulators

The most important hormonal factors of male crustaceans are probably (i) MF and its eyestalk regulators and (ii) the products of the androgenic gland (AG) and its eyestalk regulators. Steroids probably play a role, but this is not as well-defined in the male as it is in the females. Some unique peptides made by the seminal vesicles could also be involved.

MF in male decapods has been studied using the spider crab, *Libinia*, as a model because (i) the males can be divided easily into several distinct morphotypes as a function of claw size, carapace size, and the abrasion of the exoskeleton which indicates the elapsed time since the last molt, and (ii) reproductive *Libinia* appear to produce high levels of MF. Morphotype is clearly related to the development of the reproductive system, circulating levels of MF, and *in vitro* activity of the MO, as is mating behavior. An MF-binding protein has been identified in the hemolymph of male *Cancer* and *Procambarus*.

The AG is a ductless gland, found only in malacostracan crustaceans, that is found in association with the sperm duct. The AG produces proteins, androgenic hormones (AHs), as well as some terpenes. In a series of publications over two decades, using implantation studies Charniaux-Cotton showed that the AG, not the testes, is responsible for male differentiation in the amphipod *Orchestia*: (i) When the AG is removed from mature males spermatogenesis stops and oocytes develop (this has also been shown in the decapod *Cherax*, and this effect is mimicked in crabs by the parasitic barnacle *Sacculina*); (ii) when the AG is implanted into females the ovaries develop into testes, and after molting they begin to take on external male characteristics; and (iii) when ovaries are implanted into males the ovaries develop into testes as long as the AG is present but without regard to the presence or absence of the original testes. Similar results were obtained with decapods; male *Macrobrachium* can be sex-reversed in this way and then the neofemales could be mated with true males. There is an eyestalk factor that apparently regulates the AG by inhibiting the synthesis and/or release of its active factors as well as a stimulating factor in the ventral ganglion.

VII. CRUSTACEAN EMBRYOLOGY

The embryology of the crustaceans has been reviewed by both Kaestner and Anderson. The biochemical composition of crustacean embryos varies seasonally and with the stage of embryogenesis. Embryo development may be considered to take place during the period between fertilization of the oocyte and hatching; some species in the crustaceans, often those that have fewer somites, hatch into a nauplius larval stage, whereas others, often those with larger eggs and many somites, have embryonized the nauplius and begin their larval development in a postnauplial form. Some cladocerans display this condition even though the eggs are relatively small. Two schemes of development have been used to describe the degree of prehatching development: (i) Epimorphic development is the term applied to species that hatch with all the metameres (body segments) they will have in their lifetime, whereas (ii) species that go through anamorphic development are missing some of the metameres and appendages that will appear later. Anamorphic development is seen in at least some members of the Branchiopoda, Copepoda, Ostracoda, Cirripedia, Euphausiacea, and Decapoda. While there is often a free-swimming naupliar stage, some anamorphic crustaceans, especially the malacostracans, complete naupliar development before hatching, although not all the metameres are present before hatching. In invertebrates in general, epimorphy is often linked to the freshwater rather than the marine environment, but this distinction is less clear in the Crustacea.

Anderson suggests that all crustacean embryology has as its basis a total cleavage that has evolved from modifications of the spiral pattern observed in such phyla as Annelida and Mollusca. The spiral nature of cleavage is most obvious in the thoracian barnacles but is also seen in some higher groups. Radial cleavage, seen in some other crustacean groups, is probably secondarily derived from the ancestral condition. Other crustacean groups, including most malacostracans, display either total cleavage and direct blastoderm formation or intralecithal cleavage and direct blastoderm formation with no signs of a spiral total cleavage.

In the blastula, the majority of the cells constitute the presumptive ectoderm; these cells show little gastrulation movement. The presumptive mesoderm cells form an arc in front of the presumptive midgut on the blastula and move beneath the ventral ectoderm during gastrulation. The presumptive stomodaeum cells of the blastula form a circular patch in the ventral midline in front of the presumptive midgut and will thicken and fold inward in the gastrula.

Blastular proctodaeum cells are found in the posteroventral midline of the presumptive ectoderm. The location of the presumptive midgut of the blastula is more variable. The presumptive midgut is sometimes recognizable early: It is derived from the yolky 4d blastomere of the barnacles, the 2d blastomere of the cladocerans, and the 3d blastomere of the copepods; in other groups it is composed of numerous cells that make up part of the hollow blastula. During gastrulation of species with a hollow blastula, the presumptive midgut cells immigrate to the interior or invaginate, whereas the presumptive midgut cells of species that develop a blastoderm directly show no gastrulation movement.

The barnacles, mystacocarids, ostracods, copepods, many of the branchiopods, and some of the malacostracans develop from small eggs into naupliar larvae. A nauplius is typically small, round or oval, and contains three segments, the metameres. Embryonic development of external features begins with the outgrowth of antennulary, antennal, and mandibular limb buds, the elongation of the buds, and their eventual development into biramous appendages. At the anterior preantennulary region an eyespot develops and becomes pigmented; this pigmentation is often used as an easy guide to the stage of development. The labrum is formed in front of the mouth.

The fertilized eggs are often carried by the female during embryogenesis until hatching; exceptions to this include some of the penaeid shrimp and some planktonic types of crustaceans such as euphausiids, most of the calanoid copepods, some of the ostracods, and some of the planktonic decapods, which release the fertilized eggs to allow them to develop on the bottom or in the water column. Hatching is ultimately related to environmental signals. In the barnacle *Balanus*, a hatching factor is produced only by well-fed adults, thus hatching initiates only when there is enough phytoplankton in the water for the nauplii to feed on; the hatching factor is an eicosanoid fatty acid. Some crustacean eggs, including those of some copepods and branchiopods such as the well-known brine shrimp *Artemia*, have a dormant resting stage (diapause). The diapause eggs will hatch in response to particular environmental cues, including temperature, resuspension in the water column, or other stimuli. The resting egg (cyst) of *Artemia* is an extreme example of an egg diapause and will hatch after exposure to alcohols or after being kept at reduced pressure for 6 days at $-270.8°C$. Other crustaceans may overwinter as adults, undergoing a reproductive diapause during which time reproduction is suspended; there are many examples of seasonal reproduction in the Crustacea.

VIII. LARVAL DEVELOPMENT AND METAMORPHOSIS

The stage of larval development at hatching, and the subsequent metamorphoses to other larval stages and eventually to a juvenile stage, varies among crustacean groups. Development from the nauplius to the adult may be gradual as in the cephalocaridians, some branchiopods, mystacocarids, and ostracods. Crustacean development, however, is more often marked by one or more abrupt changes in morphology before the larval stages are over rather than a gradual unfolding to the adult form. The term metamorphosis is used here to designate such a sudden change, as discussed by Costlow. However, metamorphosis has been defined in various ways that do not incorporate all such shifts in morphology. Metamorphosis appears to be under the hormonal control of methyl farnesoate, whose presence or absence during the critical period before the molt affects morphogenesis (see Section VI).

Some crustaceans go through direct development, that is, at hatching they are almost identical to the adults, except they are smaller and have not developed the sexual characteristics; this is seen in many of the cladoceran branchiopods and is also typical of several of the subclasses of malacostracans. While decapod malacostracans do not often have direct development, a few do, such as freshwater crabs of the family Potamonidae and the well-studied freshwater astacoid crayfish. The newly emerged young may stay with the mother for extended periods after hatching. The majority of crustaceans do have larval stages (instars), so their development can be classified as indirect. The most widely recognized types of crustacean larvae are nauplius, zoea, and megalops.

The nauplius (Figs. 1b, 3a, and 3b), the least developed of the larval stages, has only three metameres.

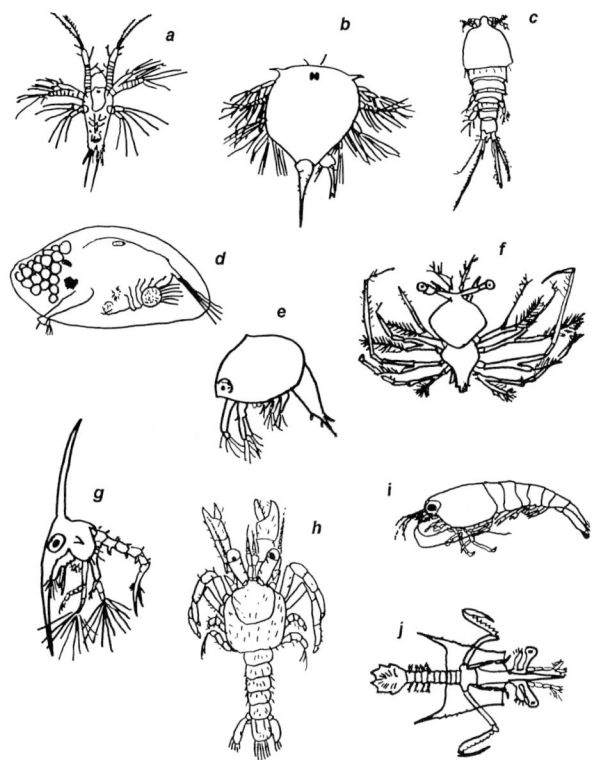

FIGURE 3 Some larval crustacean forms (not drawn to scale). a, sergestid decapod *Acetes* nauplius (redrawn from Oshiro and Omori, 1996); b, barnacle *Balanus* nauplius (redrawn from Newell and Newell, 1963); c, *Longipedia* copepodid (redrawn from Onbé, 1984); d, barnacle *Balanus* cyprid (redrawn from Walley, 1969); e, *Meganyctiphanes* calyptopis (redrawn from Newell and Newell, 1963); f, *Palinurus* phyllosoma (redrawn from Newell and Newell, 1963); g, crab *Portunus* late zoea (redrawn from Newell and Newell, 1963); h, hermit crab *Pagurus* megalopa (redrawn from Bidle and McLaughlin, 1992); i, caridean shrimp *Caridina* megalopa (redrawn from Benzie and de Silva, 1983); j, mantis shrimp *Squilla* alima (redrawn from Newell and Newell, 1963).

The nauplius moves using the first three pair of head appendages and often has a singular median eye. In the ostracods there is already a carapace, and the cirriped's nauplius can usually be recognized by its frontal horns (Fig. 3b). There is sometimes a "metanauplius" substage, which still uses only the three original swimming appendages but has additional metameres.

The typical zoea is the larval form which swims with its thoracic appendages. Often specific terms are used to describe special free-living zoeal types and phases; a further subset of terms is used for postnaupliar parasitic forms, but these specialized forms (e.g., the "trichogon" of the rhizocephalan barnacles or the "chalimus" stages of parasitic copepods) will not be discussed. The following are some of the more commonly used terms: (i) the "copepodid" (Fig. 3c) is the copepod larva that arises from the metanauplius, (ii) the "cypris" (Fig. 3d) is the nonfeeding barnacle larva that searches for a site to metamorphose into an adult. This stage was named because of its morphological resemblance to certain ostracods; (iii) the "protozoea" (Fig. 1c) are the early zoeal stages when the organism continues to use both pairs of antennae for swimming along with the thoracic appendages; (iv) the "mysis" stage (Fig. 1d) follows the protozoea and designates the transfer of the swimming function to the thoracopods. This stage is named because of its superficial resemblance to adult mysids; (e) the terms "calyptopis" (Fig. 3e) and "furcilia" are the specialized euphausid stages of the early and late zoeal stages, respectively; (f) the "antizoea" are the stomatopod zoeal stages and often have five pairs of biramous thoracic legs for locomotion, whereas in its "pseudozoea" the second pair have changed to raptorial limbs; (g) the sergesid shrimp, such as *Lucifer* and *Sergestes*, have early zoeal "elaphocaris" stages, which later change to the zoeal "acanthosoma" forms, (h) the spiny lobsters have either "eryoneicus" or "phyllosoma" (Fig. 3f) zoea; and (i) "metazoea" (Fig. 3g) is the term sometimes used for the late zoeal stage of crabs.

The megalopa (Figs. 3h and 3i), which swims with its abdominal appendages, is restricted to the malacostracans since they are the only group with true pleoplods. Some of the specialized terms include (i) the "cyrtopia" of many of the euphausids, (ii) the "erichthus" and "alima" (Fig. 3j) of the stomatopods, (iii) the "postlarvae" (Fig. 1e) of the penaeid shrimps, (iv) the sergesid shrimp "mastigopus" stage, and (v) the "puerulus" of the spiny lobster.

Recent trends in research indicate that many of the genes that function during the early development of the fruit fly, *Drosophila melanogaster*, also function during the early development of the crustacean eggs. Of particular interest is the work on regulatory genes and homeobox domains in regulating body segmentation and limb differentiation.

Acknowledgment

The authors thank Dr. Kenneth Davey for helpful comments during revisions of the manuscript.

See Also the Following Articles

ECDYSTEROIDS; HERMAPHRODITISM; JUVENILE HORMONE; MARINE INVERTEBRATE LARVAE; MARINE INVERTEBRATES, MODES OF REPRODUCTION IN

Bibliography

Adiyodi, R. G., and Subramoniam, T. (1983). Arthropods—Crustacea. In *Reproductive Biology of Invertebrates* (K. G. Adiyodi and R. G. Adiyodi, Eds.), pp. 443–495. Wiley, New York.

Anderson, D. T. (1982). Embryology. In *The Biology of the Crustacea* (D. E. Bliss, Ed.), Vol. 2, pp. 1–41. Academic Press, New York.

Benzie, J. A. H., and de Silva, P. K. (1983). The abbreviated larval development of *Caridina singhalensis* Ortmann, 1894 (Decapoda, Atyidae). *J. Crust. Biol.* **3**, 117–126.

Bidle, K. D., and McLaughlin, P. A. (1992). Development in the hermit crab *Pagurus caurinus* Hart (Decapoda: Anomura: Paguridae) reared in the laboratory. Part 1. Zoeal and megalopal stages. *J. Crust. Biol.* **12**, 224–238.

Charniaux-Cotton, H., and Payen, G. (1988). Crustacean reproduction. In *Endocrinology of Selected Invertebrate Types* (H. Laufer and R. G. H. Downer, Eds.), pp. 279–303. A. R. Liss, New York.

Costlow, J. D., Jr. (1968). Metamorphosis in crustaceans. In *Metamorphosis: A Problem in Developmental Biology* (W. Etkin and L. I. Gilbert, Eds.), pp. 3–41. North-Holland, Amsterdam.

Hanson, J. A., and Goodwin, H. L. (1977). *Shrimp and Prawn Farming in the Western Hemisphere*. Dowden, Hutchinson & Ross, Stroudsburg, PA.

Hessler, R. R., Elofsson, R., and Hessler, A. Y. (1995). Reproductive system of *Hutchinsoniella macracantha* (Cephalocardia). *J. Crust. Biol.* **15**, 493–522.

Kaestner, A. (1970). *Invertebrate Zoology. Volume 3, Crustacea*. Krieger, Huntington, NY.

Landau, M., Biggers, W. J., and Laufer, H. (1997). Invertebrate endocrinology. In *Handbook of Physiology, Section 13* (W. H. Dantzler, Ed.), pp. 1291–1390. Oxford Univ. Press, New York.

Laufer, H., Homola, E., and Landau, M. (1992). Hormonal regulation of reproduction in female Crustacea. In *Proceedings of the U. S.–Japan Joint Aquaculture Meeting, Seattle, Washington* (R. Svrjcek, Ed.), NOAA Technical Report NMFS 106, pp. 89–98. NOAA, Seattle, Washington.

Laufer, H., Ahl, J. S. B., and Sagi, A. (1993). The role of juvenile hormone in crustacean reproduction. *Am. Zool.* **33**, 365–374.

McLaughlin, P. A. (1980). *Comparative Morphology of Recent Crustacea*. Freeman, San Francisco.

McLaughlin, P. A. (1983). Internal anatomy. In *The Biology of the Crustacea, Vol. 5* (D. E. Bliss, Ed.), pp. 1–52. Academic Press, New York.

Mock, C. R., and Murphy, M. A. (1971). Techniques for raising penaeid shrimp from egg to post larvae. *Proc. First Annu. Workshop World Maricult. Soc.* **2**, 55–65.

Newell, G. E., and Newell, R. C. (1963). *Marine Plankton: A Practical Guide*. Hutchinson, London.

Onbé, T. (1984). The developmental stages of *Longipedia americana* (Copepoda: Harpacticoida) reared in the laboratory. *J. Crust. Biol.* **4**, 615–631.

Oshiro, M. L. Y., and Omori, M. (1996). Larval development of Acetes americanus (Decapoda: Sergestidae) at Paranaguá and Laranjeiras Bays, Brazil. *J. Crust. Biol.* **16**, 709–729.

Sastry, A. N. (1983a). Ecological aspects of reproduction. In *The Biology of the Crustacea, Vol. 8* (F. J. Vernberg and W. B. Vernberg, Eds.), pp. 179–270. Academic Press, New York.

Sastry, A. N. (1983b). Pelagic larval ecology and development. In *The Biology of the Crustacea, Vol. 7* (F. J. Vernberg and W. B. Vernberg, Eds.), pp. 213–282. Academic Press, New York.

Skinner, D. M. (1985). Molting and regeneration. In *The Biology of the Crustacea, Vol. 9* (D. E. Bliss and L. H. Mantel, Eds.), pp. 43–146. Academic Press, New York.

Walley, L. J. (1969). Studies on the larval structure and metamorphosis of *Balanus balanoides* (L.). *Philos. Trans. R. Soc. London Ser. B* **256**, 237–280.

Williamson, D. I. (1982). Larval morphology and diversity. In *The Biology of the Crustacea, Vol. 2* (D. E. Bliss, Ed.), pp. 43–110. Academic Press, New York.

Cryopreservation of Embryos

Akiyasu Mizukami, Douglas T. Carrell, and C. Matthew Peterson

University of Utah

I. General Principles of Cryopreservation
II. Cryoprotectants
III. Cryopreservation Techniques
IV. Quality Control
V. Clinical Applications
VI. New Developments: Blastocyst Cryopreservation
VII. Safety
VIII. Take-Home Points

GLOSSARY

blastocyst A preimplantation embryo containing typically 30–150 cells and an expanded blastocoele comprising much of the embryo volume. Blastomeres have differentiated into trophoectoderm, which produces the chorion and amnion, and an inner cell mass, which produces the developing embryo.

cryoprotectant, nonpermeating Higher-molecular-weight substances that do not penetrate the cell membrane and that assist in cellular dehydration while cooling and additionally prevent osmotic shock during thawing.

cryoprotectant, permeating Low-molecular-weight molecules (glycerol, dimethylsulfoxide, and 1,2-propanediol) that permeate the cell and allow intracellular contents to be supercooled to temperatures between -5 and $-15°C$ without forming ice crystals.

exosmosis Intracellular-to-extracellular water flow; intracellular dehydration through exosmosis during cooling decreases cryopreservation damage resulting from ice crystal formation.

morula A preimplantation embryo containing 8–16 loosely associated, spherical cells of nearly equal cytoplasmic volume.

osmotic shock Cellular damage resulting from the rapid flow of extracellular fluid into the cell; osmotic shock and recrystallization are the major impediments to successful thawing.

preembryo The pluripotent dividing cells of the fertilized ovum prior to the beginning of cellular differentiation.

recrystallization The formation of ice crystals during the thawing of frozen cells; recrystallization is promoted by an improper thawing rate which is determined by the freezing rate.

seeding The initiation of ice crystal formation by removal of latent energy by a supercooled external source; introduction of a seed crystal during cooling causes a rapid formation of ice crystals in the medium which results in exosmosis.

vitrification The process of freezing into a glassy, vitreous state; this technique depends on the exposure of cells to high concentrations of cryoprotectants.

Cryopreservation of embryos, developed as an adjunct for *in vitro* fertilization (IVF), has reduced or eliminated several complications of IVF (high-order multiple pregnancy and ovarian hyperstimulation risks), and results in a higher cumulative pregnancy rate per stimulation cycle. In addition, cryopreservation can provide practical benefits such as allowing time to perform complicated prenatal genetic analysis. Technical advances continue to improve this technology.

I. GENERAL PRINCIPLES OF CRYOPRESERVATION

Successful cryopreservation programs depend on successful cooling of the embryo to subzero temperatures ($-196°C$) followed by subsequent thawing of the cryopreserved embryo to room temperature prior to transfer. First, the cells are cooled to approximately $-7°C$ in cryoprotectant medium, a temperature at which both cells and medium are not frozen.

Next, extracellular ice formation is initiated by seeding the solution, bathing the embryo with a needle or forcep cooled to −196°C. As the temperature continues to decrease, exosmosis continues as a result of the osmotic gradient created by the nonpermeating cryoprotectants and the rising sodium concentration in the extracellular media. A sufficiently slow cooling velocity is required for intracellular exosmosis to occur. Exosmosis minimizes intracellular ice crystal formation and membrane damage. In the mouse embryo, 90% of the original intracellular water concentration must move to the extracellular space to avoid potentially lethal intracellular ice formation. Once the temperature reaches −130°C, all chemical reactions cease and the cryopreserved embryos are then kept in long-term storage at −196°C. The pregnancy rates for cryopreserved embryos stored in liquid nitrogen do not appear to diminish even after up to 60 months of storage.

Recrystallization and osmotic shock during thawing are the major impediments to successful thawing procedures. Recrystallization and osmotic shock are highly dependent on the thawing rate, which is predetermined by the cooling velocity: rapid cooling–rapid thawing or slow cooling–slow thawing. The size of crystals formed during cooling is directly related to the cooling rate and amount of intracellular hydration. For this reason, when cooling of the embryos is carried to −30 to −40°C followed by a plunge into liquid nitrogen, a rapid thawing protocol (200–500°C per minute) is required to prevent the fusion of small ice crystals that may be present into large crystals (rapid cooling plunge–rapid thawing). When the slow cooling is continued to −60°C or lower, slow thawing rates (<25°C per minute) are necessary in order to prevent osmotic shock to the extensively dehydrated cells (slow cooling–slow thawing).

II. CRYOPROTECTANTS

Permeating cryoprotectants are small molecules which allow intracellular fluid to be supercooled to temperatures between −5 and −15°C without forming ice crystals. In this temperature range ice crystals begin to form in the external media, but the intracellular fluid remains unfrozen. Cooling below −15°C uniformly results in intracellular ice formation. The process of seeding involves a controlled cessation in the freezing process between −5 and −8°C, at which time the introduction of a seed crystal causes a rapid formation of ice crystals in the medium. The cessation of cooling during seeding allows additional intracellular water to equilibrate with extracellular water through exosmosis. The intracellular dehydration during cooling lessens ice crystal formation and potential freeze injuries. Nonpermeating cryoprotectants are used to osmotically remove intracellular water for replacement with permeating cryoprotectants during cooling and additionally to prevent osmotic shock by controlling the intracellular rehydration of the cells during thawing.

A. Permeating Agents

Permeating cryoprotectants include glycerol, dimethyl sulfoxide, and 1,2-propanediol. Permeating cryoprotectants enter the cells of the embryo with a developmental stage specificity and vary in their diffusion properties. 1,2-Propanediol is most effective for the freezing of pronucleate oocytes and two- to four-cell stage preembryos. Dimethyl sulfoxide appears to be best suited for later stages of cleavage. Glycerol is appropriate for blastocyst cryopreservation. A stepwise declining dilution of cryoprotectants in the thawing protocol helps reduce the osmotic shock of the thawed embryos.

B. Nonpermeating Agents

Nonpermeating cryoprotectants are typically larger molecules such as sucrose and are used to osmotically remove water from the blastomeres, thus allowing cryoprotectants to replace the water inside the cells. They are also useful in the stepwise rehydration of cells to avoid osmotic shock from rapid rehydration. The nonpermeating molecule also aids in removal of the permeating cryoprotectant by creating a hyperosmotic medium.

III. CRYOPRESERVATION TECHNIQUES

Currently characterized cryopreservation techniques for embryos are generally classified as follows: (i) slow cooling–slow thawing, (ii) rapid cooling–rapid thawing, (iii) vitrification, and (iv) ultrarapid freezing.

A. Slow Cooling–Slow Thawing

This technique requires a programmable biological freezer. With this method, embryos are loaded into vials after equilibration with a cryoprotectant solution and then cooled at a rate of 0.5–2°C per minute down to −7°C. Seeding is then induced and a holding period of 5–15 min allows equilibration of the temperature. There is a slight increase in temperature during seeding due to the release of latent heat during ice formation. Intracellular and extracellular water also equilibrate during this time period through exosmosis. Thereafter, embryos are cooled to −60°C or lower at a rate of 0.3–0.5°C per minute before being transferred to liquid nitrogen. Frozen embryos must be slowly thawed at a rate of less than 25°C per minute to prevent osmotic shock.

B. Rapid Cooling–Rapid Thawing

This technique involves the equilibration of embryos with a cryoprotectant solution, seeding at −7°C, and a holding period of 5–15 min as in the slow cooling–slow thawing protocol. In this technique, however, cooling is terminated at −30 to −40°C and the embryos are then plunged into liquid nitrogen for rapid cooling to −196°C. Thawing is therefore performed rapidly (200–500°C per minute) to prevent recrystallization.

C. Vitrification

Vitrification eliminates the need for a controlled biologic freezer. In this technique, cryoprotectants are added at 0°C in a high concentration (40% weight/volume). These concentrations are cytotoxic at room temperature. This technique capitalizes on partial dehydration of cells in the high concentrations of cryoprotectants (6–8 M) before rapid freezing into a glassy, vitreous state. Thawing is done rapidly in ice-water temperature at a rate of 200–500°C per minute to avoid devitrification of the glassy solution, at which time serial dilutions remove the cryoprotectants. Within certain limits, the faster the temperature is changed, the lower the concentration of viscosity required and vice versa. This method avoids ice crystal formation but retains the complications of toxicity by cryoprotectants, osmotic shock, and fracture damage associated with all cryopreservation techniques.

D. Ultrarapid Freezing

This method, originally developed for oocyte cryopreservation, is thought to result in less damage to the zona pellucida than that seen with the slow cooling–slow thawing protocol. In this technique, serial equilibrations of embryos in high concentrations of DMSO (3–5 M) supplemented with sucrose (0.3–0.5 M) are required. The embryos are then plunged into liquid nitrogen. Thawing is done in a warm water bath (approximately 500°C per minute). Both this technique and vitrification are currently research protocols.

IV. QUALITY CONTROL

Quality control is a major key to success in any cryopreservation program. Feedback on the cryopreservation procedure is delayed until thawing occurs; therefore, any errors in processing must be corrected immediately to avoid continued suboptimal outcomes. It is recommended that all aspects of the cryopreservation program be double-checked by a second member of the laboratory staff. All equipment must be checked routinely and have appropriate monitors and alarms. All aspects of the cryopreservation program should be documented in log books. Seeding temperatures are logged daily for each run. Additionally, it is recommended that freezing and thawing media and procedures be tested each week in a mouse two-cell embryo toxicity bioassay.

Backup systems for the cryopreservation program include emergency power sources, a secondary liquid nitrogen source, and a backup freezing unit. Alarms must be monitored for refrigerant leakage, temperature variations, and fire and water damage. A laboratory contact for security personnel must be established, a disaster management plan should be developed, and an accessible duplicate inventory storage list should be maintained off the laboratory premises. Manufacturer's technical support and maintenance numbers and a secondary source of refrigerant must be readily available and should be included with the backup inventory storage list.

V. CLINICAL APPLICATIONS

The first delivery resulting from cryopreservation of human embryos was reported in 1985. In the 3-year period 1987–1990, more than a 600% increase in the number of cryopreserved embryos was recorded in the United States. For ethical, financial, and medical reasons, the use of embryo cryopreservation in assisted reproduction continues to increase rapidly throughout the world.

A. As an Adjunct to in Vitro Fertilization

The ability to store excess embryos increases the efficiency and minimizes the risks attendant with *in vitro* fertilization (IVF). Multiple series document the improved overall IVF pregnancy rates per retrieval provided through the transfer of cryopreserved embryos. Cryopreserved embryo transfer not only adds a 6% or greater improvement in the ongoing pregnancy rate per retrieval but also this is accomplished at a cost between 25 and 45% of the cost of any other assisted reproductive technologies. By limiting the number of fresh embryos transferred, the multiple gestation rate and its attendant obstetrical risks can also be reduced. Restriction in the number of fresh embryos transferred in some countries has made cryopreservation an integral component of successful IVF programs.

B. Avoidance of Ovarian Hyperstimulation Syndrome

The most significant complications of therapies that use controlled ovarian hyperstimulation are multiple pregnancies and ovarian hyperstimulation syndrome (OHSS). Severe OHSS may be associated with massive ovarian enlargement, intravascular volume depletion, ascites, pleural effusions, and electrolyte and coagulation disturbances. Improved methods and expertise in monitoring follicular development and estradiol response in women treated with gonadotropins have resulted in the virtual elimination of the risk of mortality; however, the risk of morbidity due to OHSS remains. This risk is particularly noted in patients with polycystic ovarian syndrome who conceive in the treatment cycle. It has been recognized that the incidence of this complication has not been eliminated with the use of GnRHa protocols for ovarian stimulation. A number of therapeutic approaches have been used in an attempt to reduce the incidence and severity of OHSS in patients considered to be at risk. Continuation with the inherent risks, cancellation of the cycle, "coasting," and cryopreservation have become the options available to these patients. Cryopreservation and a delayed primary transfer of embryos results in an acceptable pregnancy rate which rivals fresh transfer rates in published reports and eliminates the risk of OHSS. (Table 1).

In contrast, the transfer of cryopreserved embryos remaining after fresh transfer of the best quality em-

TABLE 1
Pregnancy Rates after Delayed Primary Transfer of Cryopreserved Embryos to Avoid Ovarian Hyperstimulation Syndrome

Reference	Delayed primary transfer pregnancy rate (%)
Pattinson *et al.* (1994)	31/123 (25.2)
Tiitinen *et al.* (1995)	15/46 (32.6)
Frederick *et al.* (1995)	21/63 (33.3)
Shaker *et al.* (1996)	5/13 (38.5)
Total	72/245 (29.4)

bryos generally yields a 6–15% pregnancy rate per transfer. These secondary embryos are often of poorer quality and may be cryopreserved at vulnerable developmental stages. Therefore, embryo quality and stage at freezing are critical to the outcome.

C. Oocyte and Embryo Cryopreservation for Women with Cancer

Recently, the application of cryopreservation technologies has extended to women who face the loss of gonadal function. Extirpative gonadal surgery, chemotherapy, and radiation therapy for patients with various cancers have become increasingly successful, and sustained remissions are common. The price for these successes has been acute toxicity and, in the long-term, infertility due to gonadal failure. Oocyte banking and embryo cryopreservation for women with various cancers has been the natural consequence of the technological advances in cryopreservation. As sperm banking through cryopreservation has been successfully implemented in men for many years, oocyte banking and maturation techniques will eventually be commonplace.

D. An Adjuvant to Preimplantation Genetic Diagnosis

The number of diseases capable of being diagnosed through molecular biological techniques is expanding. While some of these diagnoses can be made rapidly with precision, the ability to remove time constraints through embryo biopsy and cryopreservation until all diagnostic work is completed will expand the utilization of preimplantation genetic diagnosis.

E. Predicting the Ability to Cryopreserve Successfully

The ovarian response to various stimulation protocols, patient age, endometrial receptivity, and number of embryos transferred affect the success of cryopreservation.

1. Ovarian Response

Poor responders to ovarian stimulation (1–5 oocytes per retrieval) exhibit a lower fresh pregnancy rate than both moderate (6–10 oocytes) and high responders (16+ oocytes) in recent studies. High responder patients obviously provide a larger pool of embryos from which a number may be good candidates for successful cryopreservation and thawing. Approximately half of the patients with 6–10 preovulatory oocytes at retrieval have embryos for cryopreservation after fertilization, development in culture, and fresh transfer of three or four preembryos. High responders (16+ oocytes), however, are able to cryopreserve in over 90% of stimulation cycles and achieve a clinical pregnancy (combined fresh transfer plus thawed embryo transfer pregnancy rates) in well over half of the cases. Numerous assisted reproductive technology programs have concluded that the total number of embryos available for freezing is a critical clinical parameter of success.

2. Patient Age and Endometrial Receptivity

Age-related declines in pregnancy rates using cryopreserved embryos are apparent after the age of 35, similar to those seen in fresh embryo transfers. The percentage that survive thawing is constant (66–75%). However, compared to controls <35 years old, pregnancy rates decrease by over 33 and 66% for women 35–39 and >40 years old, respectively (Table 2). In addition to the age at cryopreservation, endometrial receptivity may also be a critical factor associated with age because some researchers have noted an age-related decline in implantation and pregnancy rates in natural cycle transfers. However, in patients 35–39 years old transferred in a GnRHa downregulation/hormone replacement cycle, no decline in pregnancy or implantation rates is seen. For women over 40 years of age most clinicians advise transfer in a GnRHa downregulation/hormone replacement cycle. Amenorrheic women on GnRHa downregulation/hormone replacement cycles versus eugonadal women in natural cycles appear to have higher pregnancy rates per transfer. For ovulatory women <35 years of age multiple studies demonstrate no particular advantage to GnRHa downregulation/hormone replacement transfer cycles compared to natural

TABLE 2
Effect of Age at the Time of Freezing on
Pregnancy Rate Following Thawed Embryo Transfer

	Age (years)		
	<35	35–39	>40
Number of survived/thaw	1352/2058	919/1397	147/197
% Survived/thawed	66%	66%	75%
Number pregnant/cycle w/transfer	126/374	70/300[a]	6/50[a]
% Pregnant/transfer	34%	23%	12%
Ongoing pregnancy/cycle w/transfer	90/374	45/300[a]	3/50[a]
% Ongoing per transfer	24%	15%	6%

[a] $p < 0.05$ compared to <35 years. All IVF cycles with subsequent thawed embryo transfer at The Jones Institute, January 1987 to December 1993 ($n = 724$). Reprinted with permission from Queenan et al. (1995).

transfer cycles, but some clinics find the advantages of lower cancellation rates, less vigilance in monitoring to predict ovulation, and the ability to decide in advance which day to transfer outweighs the additional expense and intervention. The ability to store preembryos indefinitely and the success of GnRHa downregulation/hormone replacement transfer cycles have combined to sustain high pregnancy rates for women of advancing age.

3. Ovarian Stimulation Protocols

Clomiphene citrate/human menopausal gonadotropin cycles result in lower pregnancy rates compared to GnRHa/human menopausal gonadotropin cycles. No differences in cryosurvival, fresh pregnancy rates, or cryothaw pregnancy rates among stimulation cycles have been detected using hMG, FSH, or hMG/FSH with or without GnRHa pretreatment (Table 3). Any of the currently used gonadotro-

TABLE 3
The Effect of Ovarian Stimulation and GnRHa Use on Cryopreservation and Thawing

	Protocol				
	Gonadotropins only	Gonadotropins + GnRHa	hMG + FSH	FSH	hMG
No. cycles w/thaw	199	597	507	233	56
No. survived/thawed (%)	437/645 (68)	1981/3007 (66)	1498/2304 (65)	757/1102 (69)	163/246 (66)
Average no. transferred per cycle	2.18	3.29[a]	2.93	3.23	2.89
Fresh pregnancy rate per cycle (%)	69/248 (28)	332/845[a] (39)	269/713 (38)	102/308 (33)	30/72 (42)
Current augmented pregnancy rate (%)	69 + 42/248 (45)	332 + 160/845 (58)	269 + 126/713 (55)	102 + 59/308 (52)	30 + 17/72 (65)
No. cycles w/specimens still in storage (%)	58/199 (29)	340/597[a] (57)	269/507 (53)	106/233 (45)	23/56 (41)
	Thaw results by stimulation with or without GnRHa				

Note. Reprinted with permission from Queenan et al. (1995).
[a] $p < 0.05$. All IVF cycles with cryopreservation at The Jones Institute, January 1987 to December 1993 ($n = 796$).
[b] $p < 0.05$ compared to swim-up and percoll treatments.
[c] $p < 0.10$ compared to swim-up and percoll treatments.

pin stimulation regimens with or without GnRHa result in augmentation of the cumulative pregnancy rate by over 40%. These cumulative rates demonstrate the need to report both fresh and cryothaw rates, including when describing the success from a given stimulation cycle.

4. Number of Oocytes Transferred

The number of thawed embryos transferred directly affects the pregnancy rate. A twofold increase in the pregnancy rate between the transfer of one and two preembryos has been reported. Pregnancy rates continue to improve with the transfer of three or four embryos compared to two preembryos. However, as has been found with fresh preembryo transfers, there appears to be no statistical improvement in pregnancy rates after transferring four compared to three preembryos.

VI. NEW DEVELOPMENTS: BLASTOCYST CRYOPRESERVATION

Recently, cryopreservation of cocultured human blastocysts has been shown to produce acceptable pregnancy rates. The use of cryopreserved blastocysts for patients undergoing IVF-ET yields cryosurvival rates of over 80%, the ability to transfer embryos in over 90%, and ongoing pregnancy rates of nearly 20%. It appears that GnRHa downregulation/hormone replacement transfer cycles result in higher pregnancy rates. Although there may be fewer preembryos to cryopreserve by continuing culture until the blastocyst stage, those remaining appear to be of good quality for cryopreservation and can be successfully thawed and transferred (Table 4).

VII. SAFETY

The incidence of birth defects and perinatal and obstetrical risks associated with pregnancies resulting from the transfer of frozen embryos are no different than those of normally conceived pregnancies. Programs are encouraged to monitor the outcomes of these pregnancies.

VIII. TAKE-HOME POINTS

Permeating and nonpermeating cryoprotectants utilized in cooling protocols designed to provide adequate time to prevent ice formation and facilitate exosmosis and thawing protocols which minimize recrystallization and osmotic shock have resulted in numerous applications for preembryo cryopreservation in assisted reproduction. Cryopreservation is used in the following applications:

As an adjuvant to IVF
To reduce the number of high-order multiple pregnancies
To avoid OHSS
For the banking preembryos and oocytes of women undergoing extirpative ovarian surgery or radiation or chemotherapy for cancer
As an aid to successful preimplantation genetic diagnosis

Critical clinical parameters of successful preembryo cryopreservation resulting in pregnancy are the following:

The number of oocytes collected
Patient age at the time of cryopreservation
Endometrial receptivity control through GnRHa/hormone replacement cycle transfers in women with amenorrhea, anovulatory cycles or age >35 years
The number of preembryos transferred

TABLE 4
Cryopreservation at the Blastocyst Stage after Coculture on Vero Cells

No. of cycles thawed	563
No. of transfer cycles[a]	516 (92%)
No. of thawed blastocysts[a]	1239
No. of transferred blastocysts[a]	1033 (83%)
No. of implanting embryos	138
Implantation rate	138/1033 (13.4%)
No. of clinical pregnancies	112
Pregnancies per transfer	21.7%
Ongoing pregnancies per transfer	19%

Note. Reprinted with permission from Kaufmann *et al.* (1995).

The transfer of cryopreserved embryos in GnRHa/hormone replacement cycles lowers the cancellation rates, requires less vigilance in monitoring to predict ovulation, and provides the ability to decide in advance which day to transfer. Blastocyst and pronuclear stage preembryos appear best suited for the rigors of cryopreservation and thawing.

See Also the Following Articles

Blastocyst; In Vitro Fertilization; Reproductive Technologies

Bibliography

Al-Shawaf, T., Yang, D., Al-Magid, Y., Seaton, A., Iketubosin, F., and Craft, I. (1993). Ultrasonic monitoring during replacement of frozen/thawed embryos in natural and hormone replacement cycles. *Hum. Reprod.* 8, 2068–2074.

Cohen, J., Simons, R. F., Edwards, R. G., Fehilly, C. B., and Fishel, S. B. (1985). Pregnancies following the frozen stage of expanding human blastocysts; *J. in Vitro Fertil. Embryo Transfer* 2, 59–64.

Demoulin, A., Jonan, C., Gerday, C., and Dubois, M. (1991). Pregnancy rates after transfer of embryos obtained from different stimulation protocol and frozen at either pronucleate or multicellular stages. *Hum. Reprod.* 6, 799–804.

Frederick, J. L., Ord, T., Kettel, L. M., Stone, S. C., Balmaceda, J. P., and Asch, R. H. (1995). Successful pregnancy outcome after cryopreservation of all fresh embryos with subsequent transfer into an unstimulated cycle. *Fertil. Steril.* 64, 987–990.

Fugger, E. F., Bustillo, M., Dorfmann, A. D., and Schulman, J. D. (1991). Human preimplantation embryo cryopreservation: Selected aspects. *Hum. Reprod.* 6, 131–135.

Kahn, J. A., von During, V., Sunde, A., Sordal, T., and Molne, K. (1993). The efficacy and efficiency of an in-vitro fertilization programe including embryo cryopreservation: A cohort study. *Hum. Reprod.* 8, 247–252.

Kaufmann, R. A., Nicollet, B., Menezo, Y., DuMont, M., Hazout, A., and Servy, E. J. (1995). Cocultured blastocyst cryopreservation: Experience of more than 500 transfer cycles. *Fertil. Steril.* 64, 1125–1129.

Menzo, Y., Nicollet, B., Herbaut, N., and Andre, D. (1992). Freezing cocultured human blastocysts. *Fertil. Steril.* 58, 977–980.

Pattinson, H. A., Hignettt, M., Dunphy, B. C., and Fleetham, J. A. (1994). Outcome of thaw embryo transfer after cryopreservation of all embryos in patients at risk of ovarian hyperstimulation syndrome. *Fertil. Steril.* 62, 1192–1196.

Queenan, J. T., Jr., Veeck, L. L., and Muasher, S. J. (1994). Transfer of cryopreservation-thawed preembryos in a natural cycle or a programmed cycle with exogenous hormonal replacement yield similar pregnancy results. *Fertil. Steril.* 62, 545–550.

Queenan, J. T., Veeck, L. L., and Muasher, S. J. (1995). Clinical and laboratory aspects of cryopreservation. *Sem. Reprod. Endocrinol.* 13, 69–71.

Santhanadan, M., MacNamee, M. C., Rainsbury, P., Wick, K., Brinsden, P., and Edwards, R. G. (1991). Replacement of frozen/thawed embryos in artificial and natural cycles: A prospective semirandomized study. *Hum. Reprod.* 6, 685–687.

Shaker, A. G., Zosmer, A., Dean, N., Bekir, J. S., Jacobs, H. S., and Tan, S.-L. (1996). Comparison of intravenous albumin and transfer of fresh embryos with cryopreservation of all embryos for subsequent transfer in prevention of ovarian hyperstimulation syndrome. *Fertil. Steril.* 65, 992–996.

Tiitinen, A., Husa, L. M., Tulppala, M., Simberg, N., and Seppala, M. (1995). The effect of cryopreservation in prevention of ovarian hyperstimulation syndrome. *Br. J. Obstet. Gynaecol.* 102, 326–329.

Trounson, A. O. (1990). Cryopreservation. *Br. Med. Bull.* 46, 695–708.

Van Voorhies, B. J., Syrop, C. H., Allen, B. D., Sparks, A. E. T., and Stovall, D. W. (1995). The efficacy and cost effectiveness of embryo cryopreservation compared with other assisted reproductive techniques. *Fertil. Steril.* 64, 647–650.

Veeck, L. L., Amundson, C. H., Brothman, L. J., *et al.* (1991). Significantly enhanced pregnancy rates per cycle through cryopreservation and thaw of pronuclear stage oocytes. *Fertil. Steril.* 59, 1202–1207.

Wang, X. J., Ledger, W., Payne, D., Jeffrey, R., and Matthews, C. D. (1994): The contribution of embryo cryopreservation to in-vitro fertilization/gamete intrafallopian transfer: 8 years experience. *Hum. Reprod.* 9, 102–109.

Cryopreservation of Sperm

Rupert P. Amann

BioPore, Incorporated

I. Introduction
II. Overview of Procedures Involved in Cryopreservation
III. Changes in Metabolism and Membranes Associated with Cooling and Warming
IV. Cryoprotectants
V. Movement of Solutes and Solvents
VI. Consequences of Ice Formation
VII. Changes in Sperm Volume Associated with Cryopreservation
VIII. Cold Shock
IX. Consequences of Cooling Sperm to $-196°C$ and Warming to $37°C$
X. Removal of Cryoprotectant
XI. Evaluation of Success
XII. Thoughts to Ponder

GLOSSARY

cryopreservation A procedure for preparation of a suspension of cells, or group of cells (e.g., embryo), for storage; cooling, long-term storage of the suspension of cells at a temperature colder than $-80°C$, and warming to body temperature; and removal of cryoprotectant from within the cells. All steps should be optimized to provide survival of sufficient cells to achieve a specific task, such as impregnation of a female with cryopreserved sperm.

cryoprotectant A solvent or solute incorporated into a suspension of cells to alter the nature or extent of changes occurring in both the cells and the suspending medium and, thereby, increasing the percentage of cells surviving cryopreservation. Glycerol is the most commonly used cryoprotectant.

extender A medium selected to prolong the interval over which sperm can be stored with maintenance of fertilizing potential; also used to dilute "neat semen" (as ejaculated) to a number of sperm per milliliter appropriate for storage or artificial insemination.

liquid nitrogen A liquid which boils at approximately $-196°C$ and provides an excellent coolant for storage of cells in either the liquid phase ($-196°C$) or vapor phase (typically $-180°C$). Prepared by high-pressure liquefaction of air followed by low-temperature distillation.

spermatozoa The male gametes (singular, *spermatozoon*), also termed **sperm** (both singular and plural).

Cryopreservation of sperm from common mammals is a deceptively simple-appearing process that works despite limitations. It is likely that >0.5 billion calves, >0.3 million children, and fewer young of other species have been born as a result of artificial insemination using frozen–thawed sperm. The spermatozoal genome is unaltered by cryopreservation, and sperm remain usable after storage at $-196°C$ for decades. Cryopreservation is a procedure that involves multiple steps: (i) preparation of a suspension of sperm for low-temperature storage by incorporation of cryoprotectants and placement of individual insemination doses into vials or "straws"; (ii) cooling at an appropriate rate; (iii) long-term storage of the suspension of cells at a temperature between $-180°C$ and $-196°C$; (iv) distribution at $-196°C$ to potential users; (v) warming an insemination dose at an appropriate rate to body temperature; and (vi) controlled removal of cryoprotectant so that the cells are in an environment appropriate for subsequent use. Ideally, all steps would be optimized to provide retention of all spermatozoal attributes essential for fertilizing potential by the maximum percentage of sperm in most samples from that male or species. Mere retention of "viability" or "life" is insufficient. Regardless of species or the exact cryopreservation protocol, a substantial percentage (>30%) of sperm which were motile before cooling below $5°C$ will be rendered immotile (and presumably killed) before warming back to above $5°C$. Subsequent removal of intracellular cryoprotectant can damage additional sperm. With the exception of a few species (i.e., cattle and chicken), for both biological and ethical reasons it is difficult (e.g., horse, mouse, pig, rabbit, and rat) or impossible (e.g., human and endangered species) to compare precise fertilization percentages (or pregnancy, birth, or hatching rates) obtained with alternative cryopreservation protocols. Hence, results of *in vitro* tests of sperm quality (e.g., percentage of motile sperm) after freeze–thawing conventionally are used to identify useful procedures. It is well documented, however, that attributes other than motility are required for a spermatozoon to have fertilizing potential. Hence, there is no assurance that a sample of thawed semen containing a relatively high percentage of motile sperm (>25 or 30%) will result in fertilization of even one egg. This poses a problem when cryopreserving sperm from most species, in that few stored cells may retain fertilizing capacity.

I. INTRODUCTION

The discovery of a practical procedure to cryopreserve sperm generally is attributed to Polge and colleagues (1949). Others had little success during the preceding decades, although Spallanzani had reported, in 1789, that frog sperm frozen in snow could be revived and fertilize frog eggs. The breakthrough in cryopreservation resulted from accidental usage of a mislabeled solution. The solution later was found to contain glycerol and, hence, it was found that glycerol gave protection against damage associated with freeze–thawing sperm suspensions. As a cryoprotectant, glycerol was amazingly successful and 50 years later remains the most used penetrating cryoprotectant for sperm (cryoprotectants and their mechanism of action are discussed in Section IV). Initially, sperm were stored in glass ampules immersed in alcohol cooled to −79°C with solid carbon dioxide (dry ice).

A second important step was recognition that −79°C was a marginally cold temperature, and that storage in liquid nitrogen at −196°C would be better from perspectives of biophysical changes and ease of use. J. Rockefeller Prentice, who owned a business offering semen for artificial insemination of cattle, encouraged the Linde Company to develop appropriate cryogenic containers. "Superinsulated containers" are more efficient than conventional "thermos bottles" because they have ≈1000 alternating layers of thin aluminum foil and plastic within the vacuum space. They now are used for countless applications.

The third element in the cryopreservation story is that, for reasons which remain uncertain, among common species (chicken, horse, human, mouse, pig, rabbit, and sheep) sperm from bulls are uniquely resistant to damages associated with cryopreservation. Furthermore, there are sufficient sperm in one ejaculate of bull semen to inseminate 250–800 cows (vs <20 females per ejaculate for most other species). These features enabled extensive usage of genetically superior bulls whose sperm survived freeze–thawing by standard procedures and rapid validation of important biological principles and practical results. It evolved that (i) pregnancy rate for dairy cattle inseminated with cryopreserved semen was equivalent to that obtained with fresh semen, although more sperm usually were required to offset those killed during freezing and thawing; (ii) offspring were completely normal; (iii) valuable germ plasm could be disseminated rapidly from one superior sire to thousands of offspring at low cost (>50,000 calves per year for one bull), resulting in the need for fewer dairy cows (60% fewer, 1995 vs 1960) and savings to consumers of dairy products; and (iv) selection for bulls whose sperm suffered minimal damage by common cryopreservation procedures over time gave a population of dairy bulls most of whose sperm were of good fertility after freeze–thawing, and success achieved was improved.

The success with artificial insemination of dairy cattle facilitated introduction of the general procedure with sperm from many species, including human, but also was an impediment. It commonly was assumed that what worked well with bull semen should serve as the starting (and often final) procedure for cryopreservation of sperm from other species. Only during the past 5–10 years has this flawed concept been fully discredited. Species differences in physiological and biophysical characteristics of sperm are being clarified and such information now should be used to tailor cryopreservation protocols to fit the sperm, considering both species and individual male. This will be an active area of research to improve pregnancy rates obtained with most males of species of economic or emotional importance.

II. OVERVIEW OF PROCEDURES INVOLVED IN CRYOPRESERVATION

An ejaculate is obtained by masturbation, use of an artificial vagina, or electrostimulation of the male reproductive tract, and if necessary (i.e., human) the semen is allowed to liquefy. The volume of the semen is measured and the number of sperm per milliliter (concentration) determined so that total number of sperm in the ejaculate can be calculated. As soon as practical, the semen is mixed with a volume of extender calculated to provide the desired number of sperm in each insemination dose and cooled to 4°C over 2 or 3 hr (slow cooling is unnecessary with rooster sperm and often is ignored with sperm from some species). Depending on the species and processing laboratory, a seminal extender could be a simple buffered salts solution containing sugar (e.g., 0.5–2.0% fructose, glucose, or lactose), heated skim milk or reconstituted powdered skim milk, or a simple buffered salts solution containing 4–20% yolk from a hen's egg. The sugars, protein in milk, or lipoproteins of egg yolk in the extender serve as nonpenetrating cryoprotectants (cannot enter cells), but alone are insufficient for freezing sperm. A penetrating cryoprotectant (enters cells) such as glycerol is essential. With bull sperm, the semen is partially extended and cooled to 4°C before an equal volume of glycerol-containing extender is added slowly to result in final extension. Slow addition of glycerol would be desirable with other species (e.g., human and mouse), but often a complete extender containing the desired final concentration of glycerol is added (at room temperature) to the neat semen in one step; however, this labor-saving step damages the sperm.

Ideally at 4°C, but often at room temperature, the extended semen is dispensed into a series of prelabeled containers. For human sperm, use of 1.5-ml screw-cap plastic vials is now conventional. For other species, plastic straws (0.5 or 0.25 ml for bull, dog, ram, stallion; 4 ml for boar or stallion), or special containers (1 ml for rooster) are used. Semen then is frozen by positioning the containers on a rack 4–8 cm above the surface of liquid nitrogen in a foam container (static vapor freezing), with the cooling rate determined by the thermal load and height of the rack above the liquid. Alternatively, a microprocessor-controlled freezer is used to provide a prescribed cooling rate by automatic injection of liquid nitrogen into the freezing chamber. For certain species, fewer sperm might be killed if small drops of extended semen are placed in depressions in a block of solid carbon dioxide and the resulting "pellets" then transferred into precooled vials. In any case, after temperature of the semen is below −100°C the containers of semen are immersed in liquid nitrogen. The containers then are placed in inventory-control holders (goblets on canes within canisters; boxes) and stored immersed in liquid nitrogen at −196°C.

Superinsulated storage tanks are available in many sizes and might contain 100 to >75,000 doses of semen. Semen can be shipped immersed in liquid in "field tanks" holding possibly 35 liters of liquid nitrogen and 100–5000 doses of semen. Alternatively, hazard can be reduced by use of a "dry shipper" that contains a matrix saturated with liquid nitrogen (just prior to shipment) around a chamber sized to accommodate 50–1000 doses but no liquid that can spill. Transfer of semen from one superinsulated container to another must be rapid so that containers of semen are not exposed to room temperature for more than a few seconds; warming of semen above −150°C must be avoided.

Frozen semen must be thawed at the site of final use so that it can be inseminated into a female within 5–10 min after warming above 0°C. This is because the fertile life of sperm that have been frozen and thawed is much shorter than that of fresh sperm or sperm held at 4°C for 6–72 hr without freezing. Typically semen is thawed by immersion of the container in 37 or 0°C water or (probably less desirable) by simple exposure in room temperature air. Once semen has been thawed it should never be refrozen (unless intended for intracytoplasmic sperm injection of an oocyte). For most species, it is conventional to inseminate thawed semen directly into a receptive female, with the goal of insemination 1–12 hr prior to ovulation. With boar (and sometimes stallion) semen, the suspension of thawed sperm is mixed with additional extender prior to insemination. With human semen, the thawed suspension sometimes is gently centrifuged and the sperm are resuspended in a simple salts solution prior to intrauterine insemination. With chickens, glycerol is contraceptive and must be removed by dilution, centrifugation, and resuspension or by novel technology which involves a primary cryopreservation container fabricated using a membrane with pores, which are sealed until after thawing but then opened to allow removal of cryoprotectant without dilution of the suspension. As discussed in Section X, fertility of sperm from most species probably would be improved by gradual, rather than relatively abrupt, change in the glycerol concentration around the sperm (and hence movement of water into and glycerol out from the cells) prior to insemination.

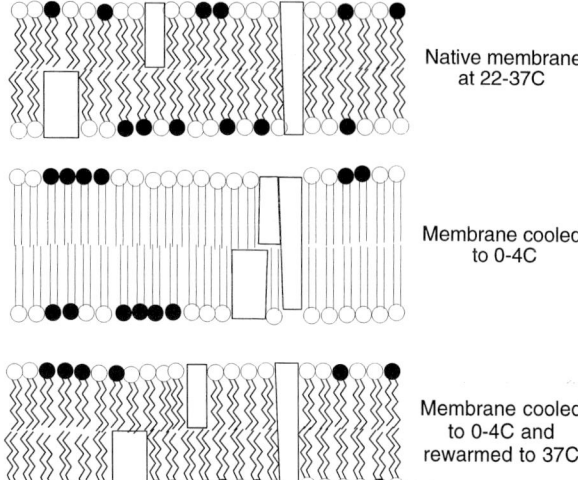

FIGURE 1 Schematic representations of a sperm plasma membrane and changes induced therein by cooling and warming. See discussion in text. From Amann *et al.* (1997) with permission.

III. CHANGES IN METABOLISM AND MEMBRANES ASSOCIATED WITH COOLING AND WARMING

It is a well-established principle of biology that metabolic rate of a cell is dependent on temperature. Each 10°C decrease in temperature below body temperature depresses metabolic activity by 50%. Hence, reducing temperature from 37 to 2°C should, in theory, reduce metabolic activity to ≈5% of the value at 37°C. Cryopreservation carries this concept further. A temperature colder than −80°C ensures that metabolic processes are essentially stopped and also that damaging aggregation of ice crystals is minimal during long-term storage (see Section IX).

A spermatozoon has a plasma membrane plus numerous internal membranous structures. Membranes basically are a lipid bilayer, with hydrophilic (water-loving) portions of the lipids and proteins oriented externally and hydrophobic (water-hating) portions of these molecules oriented toward the center of the bilayer (Fig. 1, top). There are many different lipids in each bilayer, and compositions of the inner and outer bilayers are different; limited exchange of lipids between the two bilayers can occur. Each species of lipid in the membrane has a phase-transition temperature, at which it changes from a liquid-crystalline state (liquid) to a crystalline-array state (gel), with fatty acid chains then in a parallel and rigid form. Proper function of the membrane requires a fluid membrane, and at body temperature all of the lipid molecules in the bilayer are in a liquid-crystalline state; the membrane functions "well."

When temperature of a membrane is decreased below the phase-transition temperature of an individual lipid species, all molecules of that lipid species (but not other species of lipids) aggregate in microdomains of lipid gel (a cluster of molecules in a group crystalline array) within an otherwise fluid membrane (Fig. 1). This causes (i) removal of molecules of this lipid species from normal associations with specific proteins or random associations with other lipids, and (ii) formation of gaps, or leaky and unstable borders, between microdomains of lipid gel and fluid portions of the membrane, which facilitate membrane rupture or fusion with other membranes. Each species of lipid undergoes similar phase changes as temperature is further reduced. There is a range of >10°C over which these phase transitions occur. The end results (Fig. 1, middle) are clusters of lipids by species; redistribution of proteins away from normal locations, and aggregation in fewer sites; a rather rigid membrane; gaps in membrane integrity and resultant leaks; and hampered function of enzymes or receptors. Unfortunately, when the membrane is rewarmed to body temperature and lipids return to a liquid-crystalline state, all lipid and protein molecules are not returned to their initial locations (Fig. 1, bottom). These changes occur even if cells are not cooled below 0°C.

Damage can be minimized by (i) cooling (usually slow) and warming (usually rapid) at carefully selected rates through the temperature zone where the phase transitions of the lipids occur and (ii) including additives or cryoprotectants in the extender (egg yolk to modify lipids; EGTA to alter ion movement). Nevertheless, membranes in cells cooled below 18–20°C are irreversible altered. Repeated cycles of cooling and warming usually exacerbates these changes.

IV. CRYOPROTECTANTS

A cryoprotectant is a molecule that allows a substantial percentage of sperm to survive a freeze–thaw cycle and retain fertilizing capacity. Cryoprotectants that pass through the sperm plasma membrane, and act both intracellularly and extracellularly, are termed penetrating cryoprotectants. Nonpenetrating cryoprotectants act only extracellularly. Glycerol is the most effective penetrating cryoprotectant for sperm from most species. Glycerol is of low toxicity relative to the alternative penetrating cryoprotectants ethylene glycol or dimethylsulfoxide. However, glycerol induces more osmotic flow of water because it passes through membranes slower than ethylene glycol or dimethylsulfoxide. All penetrating cryoprotectants pass through a membrane much slower than water, and this difference is temperature dependent. Nonpenetrating cryoprotectants include proteins (e.g., milk or egg yolk used with mammalian sperm); sugars such as lactose, fructose, raffinose, or trehalose; synthetic polymers such as methyl cellulose; and amides. Although glucose is slowly transported into sperm for metabolism at $\geq 0°C$, it acts as a nonpenetrating cryoprotectant.

Most penetrating cryoprotectants, such as glycerol, serve as a solute (and cause osmotic flow of water) and a solvent (to dissolve salts and sugars) miscible with water. All nonpenetrating cryoprotectants are solutes or colloids and cannot serve as a solvent. Both water and glycerol, as well as other solvents, pass through the membranes of sperm and eventually equilibrate at the same concentration in all internal structures so that the intracellular and extracellular concentrations are the same.

The solute role of a penetrating cryoprotectant likely causes damage because of induced osmotic flow of water. The solvent role is beneficial because penetrating cryoprotectants (e.g., glycerol) have a freezing point much lower than that of water. As extended semen is frozen, crystals of pure water freeze and form "blocks of ice," among which are channels of unfrozen solvent containing all of the solutes plus the sperm. In the presence of glycerol, the portion of the solvent mixture remaining unfrozen at any given temperature is greater than that if water were the only solvent. Hence, at any given temperature there is more "space" for the sperm in channels of unfrozen solvent and a lower concentration of solutes (the same amount of solutes is contained in more liquid). This phenomenon occurs both inside and outside the sperm. Furthermore, the presence of glycerol probably reduces formation of microfractures in the ice and this, in turn, minimizes damage to sperm (fewer small sharp edges).

Nonpenetrating cryoprotectants, such as sugars or lipoproteins, typically are present in high concentration. They likely act by (i) modifying the plasma membrane so it is more resistant to temperature-induced damage or (ii) simple action as a solute and lowering of the freezing point of the solvent–solute mixture.

V. MOVEMENT OF SOLUTES AND SOLVENTS

If a solute or solvent can pass through a semipermeable membrane, such as the plasma membrane of a spermatozoon, concentration of that molecule will increase or decrease as necessary, on the appropriate side of the membrane, to bring the concentrations into equilibrium. If this transmembrane movement is due only to the difference in solute concentration, it is termed simple diffusion. Placing sperm into a glycerol-containing extender results in movement of glycerol through the plasma membrane for a few minutes, until the concentration of glycerol inside the cell has risen to equal that outside. Concurrently, water diffuses out from the cell, down its concentration gradient, to the extender. However, water equilibrates across the plasma membrane in <1 sec, whereas equilibration of glycerol requires >20 sec. Rate of movement of water or glycerol into or out of sperm is reduced at 4°C or lower, but the extent of this lowering from the rate at >20°C depends on the solvent. Movement of water by simple diffusion is augmented by movement of water via osmotic flow (see below). Osmotic action can move a solvent but not a solute.

Active transport is an alternative to simple diffusion. Specific machinery within the cell "pumps" a given molecule or ion either into or out of the cell. Active transport can be up or down a concentration gradient. Sperm have active transport systems for glucose, and in some species glucose transporters in the sperm membrane apparently have a secondary water-channel function. However, glycerol and most solutes are not moved across the sperm plasma membrane by active transport.

Osmotic flow is movement of solvent (typically water, but also glycerol) across a semipermeable membrane in an attempt to maintain an equal concentration of solvent on both sides of the membrane. This biophysical process cannot be avoided, has a profound effect on the well-being of sperm, affects formulation of seminal extenders to minimize such movement of water, and affects procedures used to freeze–thaw or deglycerolate cryopreserved sperm. The more solute molecules present in a given volume, the fewer solvent molecules can be present. Hence, osmotic flow is movement of solvent but is driven by the concentration of solutes in the solution.

With sperm, only solutes that cannot permeate the plasma membrane (or intracellular membranes) affect osmotic flow of solvent. If total concentration of nonpermeable solutes (or permeable solutes not equilibrated across the membrane) and colloidal particles within a spermatozoon is less than that in the extracellular extender, water moves from inside the cell, through the plasma membrane, into the extender. This happens as extracellular water forms ice (see Consequences of Ice Formation). Conversely, if total concentration of nonpermeable solutes and colloidal particles within a spermatozoon is greater than in the extender, water will move into the cell causing it to swell; this swelling can rupture the plasma membrane. This happens when frozen–thawed sperm are deglycerolated too rapidly.

Glycerol serves as a solute within the water of a seminal extender and also penetrates into sperm. Because glycerol diffuses slowly through sperm membranes, it temporarily contributes to the total concentration of nonpermeable solutes and colloidal particles in the extender or within sperm. During an interval of nonequal concentrations, glycerol causes osmotic flow of water out from or into sperm; the latter is especially deleterious. Similarly, a high concentration of a nonpenetrating cryoprotectant (e.g., fructose and lactose) in the extender also causes osmotic flow of water out from sperm, for example, after mixture of semen with a sugar-rich extender. Osmotic flow of water out of sperm causes shrinkage (limited by the structure of sperm) and dehydration.

VI. CONSEQUENCES OF ICE FORMATION

Ice is formed from "pure water" and, consequently, formation of ice increases solute concentration. As temperature is lowered below 0°C, a portion of the water in a seminal extender starts to freeze and, as temperature is further reduced, the amount of frozen water increases. This causes the concentration of solutes (glycerol, sugars, and ions) in the extender to increase because there is less solvent in which they are dissolved. Consequently, the osmotic strength of the unfrozen portion of the extender increases, water moves out of the cell by osmotic flow, and the cell increasingly shrinks (occupies less volume) and becomes further dehydrated (the concentration of solutes increases within the cell). Nevertheless, the concentration of solutes in unfrozen extender still exceeds that within the cells, so water continues to move out of each spermatozoon until concentrations of solvents and solutes reach final equilibrium at some temperature below −40°C. At this point, sperm have a minimum volume and a maximum solute concentration; solute concentration both within and outside the cell is four to seven times normal.

The decrease in cell volume is probably good because it is a direct result of water loss by the sperm—resulting in reduced probability for ice crystals within a cell and "better fit" for cells within small channels of unfrozen solvent. However, the accompanying increase of solute concentrations (both within and outside the cells) is bad—it causes irreversible shifts in membrane structure, denaturation of proteins, disassembly of internal structures within the axoneme (a microtubular structure in the sperm tail that transduces energy into flagellar movement), and "dislodges" proteins from the external surface of the plasma membrane.

VII. CHANGES IN SPERM VOLUME ASSOCIATED WITH CRYOPRESERVATION

It is impossible to avoid changes in sperm volume during cryopreservation because they are caused by the biophysical forces discussed previously. Volume changes depicted in Fig. 2 are caused by combined effects of ice formation, diffusion rates, and osmotic flow. As shown in Fig. 2, for many species (especially cattle) a substantial proportion of sperm survive these volume and other temperature-induced changes and retain a high fertilizing potential. The down side, however, is that inevitably some sperm are rendered nonfertile without special handling after thawing to eliminate the very large increase in volume. Ideally, the concentration of glycerol in the medium surrounding the thawed sperm would be reduced in a controlled manner so that the combined actions of osmotic flow and diffusion bringing water into the cells and outward diffusion of glycerol across the plasma membrane are slow enough to eliminate potential damage from volume changes. The alternative to especially slow deglycerolation of thawed ram or rooster sperm is to inseminate them deep into the female reproductive tract rather than at the conventional site used for insemination of fresh sperm; it is not clear why deep insemination works.

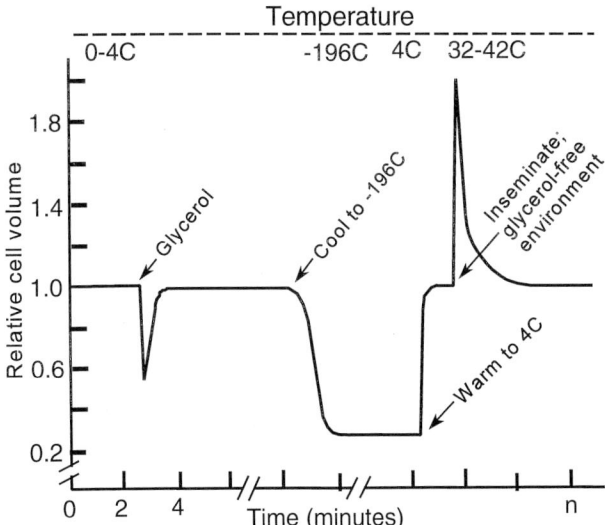

FIGURE 2 Volume changes induced in sperm by addition of extender containing glycerol, dehydration resulting from freezing of solvent during cooling to $-196°C$, rehydration with melting of solvent during warming, and placement of the sample in a glycerol-free environment. See discussion in text. From Amann *et al.* (1997) with permission.

VIII. COLD SHOCK

Sperm from many species are sensitive to cold shock. There are two types of cooling damage: direct damage and latent damage. Direct damage is a function of cooling rate, "history" of thermal changes from body temperature previously imposed on the cells, and composition of the extender. Cold shock affects cellular structures (e.g., ruptures membranes) or alters cellular functions (e.g., slows metabolic processes). This cold shock damages sperm from many mammalian species when they are cooled too rapidly from approximately 20°C to just above 0°C. Bull sperm are sensitive to cold shock, human sperm are relatively insensitive, and rooster sperm are not affected. Direct damage is evidenced by changes in pattern of sperm motion or permeability of the plasma membrane. Cooling from 20 to 0–5°C over 2 or 3 hr eliminates this damage. Latent damage is dependent on how low a temperature is reached, not cooling rate. Latent damage to sperm is difficult to discern because it might not be evident for hours or days after cooling cells to 0–4°C or after freeze–thawing. The most obvious evidence of latent damage is with frozen–thawed sperm. During incubation at 37°C, the subpopulation of sperm remaining motile shortly after thawing dies more quickly than sperm that had not been frozen.

IX. CONSEQUENCES OF COOLING SPERM TO −196°C AND WARMING TO 37°C

Building on the background presented previously, this section presents an overview of biophysical changes and responses when a suspension of sperm (0.25–2.0 ml in a glycerol-containing extender within a typical container) is cooled below 0°C. Initially, the suspension will cool a few degrees below the freezing point of the extender before formation of ice starts. This is termed supercooling. When the suspension reaches some point between −5 and −15°C, a few ice crystals will be formed spontaneously, accompanied by a rapid rise in temperature (Fig. 3, insert). Formation of ice continues for 1 or 2 min until most of the extracellular water is frozen; then the temperature of the sperm suspension again declines. Although supercooling can be almost eliminated by "seeding" to induce ice formation, this technique usually is not used with sperm. However, the time required to freeze most of the water (removal of heat of fusion) can be minimized by careful programming of the temperature around the containers of semen (Fig. 3), and this probably is desirable.

As cooling continues, more ice forms outside the sperm, the extracellular concentration of solutes rises, water moves from inside the sperm to the extracellular environment, and the sperm become progressively dehydrated. At typical cooling rates for sperm, these processes continue until the sample is below −40 or −50°C. Cooling rate determines the extent of water movement out of sperm and their dehydration, with greater dehydration at slower cooling rates. Loss of intracellular water minimizes the probability of large ice crystals forming within the cells (Fig. 4) but increases the probability of damage from high intracellular concentrations of solutes. Ultimately, most of the sperm are found in small channels of "very salty solvent" among large blocks of ice.

Appropriate cooling rate and an extender containing certain nonpenetrating cryoprotectants result in formation of relatively large channels among ice blocks and minimize the probability of damage to sperm from extracellular ice crystals.

The actual cooling (freezing) rate must balance slowness, to allow sufficient dehydration of cells, and rapidity, to minimize size of extracellular ice crystals (center path in Fig. 4). Furthermore, the ideal warming (thawing) rate is dependent in part on the cooling rate used for those samples as well as on the extender, thermal properties and volume of the package containing the sample, and unknown factors. Cooling too slowly causes excessively large extracellular ice crystals and possibly intracellular damage from a very high concentration of intracellular solutes. Cooling

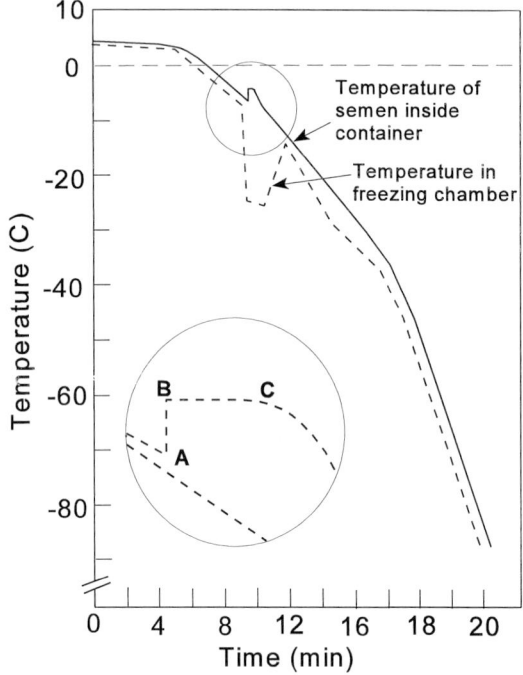

FIGURE 3 Typical cooling curve used for cryopreservation of sperm, showing temperature of the sample (solid line) and nitrogen vapor (dashed line). A microprocessor-controlled freezer, cooled with liquid nitrogen, was programed to rapidly dissipate the latent heat of fusion. The insert shows changes occurring in semen in the absence of seeding or special programing of vapor temperature. The sample supercools to point A, most of the water freezes between points B and C as the latent heat of fusion is given off, and then cooling continues with additional formation of ice at lower temperatures. Modified from Amann et al. (1997) with permission.

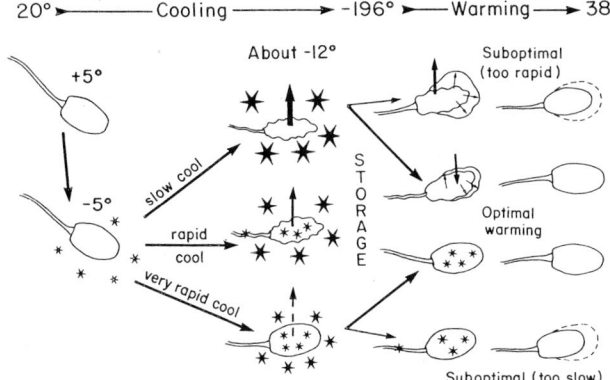

FIGURE 4 Schematic representation of probable changes in sperm, and surrounding extender, during cooling and warming at ideal rates (center path) and also at too slow or too rapid rates. See discussion in text. Stars represent the size of ice crystals. From Amann and Pickett (1987) with permission.

too rapidly does not allow sufficient dehydration of the sperm, and formation of intracellular ice is likely.

Cryopreservation of sperm (or any cell) requires storage at temperatures far below −80°C because at −75°C or warmer crystals of ice gradually become larger, resulting in damage to cells. The simplest approach is to use a cryogenic refrigerator containing the samples and liquid nitrogen. Liquid nitrogen has a temperature of −196°C, and the vapor-phase above the liquid is only slightly warmer.

Damage to sperm also occurs during warming from −196 to above 0°C. With an optimal warming rate, movement of newly formed extracellular water into the sperm (due to osmotic flow) is appropriately gradual so that intracellular microcrystals of ice melt, rather than form large ice crystals, and the cells are gradually rehydrated. Once a sample has been thawed it should not be refrozen because each freeze–thaw cycle kills cells. It should be prepared for immediate use in artificial insemination because most sperm will die within a few hours after thawing.

X. REMOVAL OF CRYOPROTECTANT

After thawing, the sperm remain suspended in extender containing 4–18% glycerol (depending on the species). For bull sperm, direct insemination of the thawed suspension is conventional and sperm move from the glycerol-rich medium to that of a cow's uterus without damage. With sperm from other species, however, it probably is desirable (e.g., for horse and human) or essential (e.g., for rooster) to carefully remove glycerol (or ethylene glycol) before artificial insemination because (i) rapid reduction of the glycerol concentration outside sperm in a suspension will cause sufficient water to move into the cells to cause irreparable damage and (ii) glycerol is a contraceptive in chickens if placed into a hen at >0.7% of the inseminate—even with nonfrozen sperm. The concentration of glycerol in the extender can be reduced gradually in six to eight steps while mixing the suspension by adding small amounts of nonglycerolated extender to provide equal molar changes in concentration of the cryoprotectant. Centrifugation is then necessary to provide a concentrated suspension appropriate for AI.

XI. EVALUATION OF SUCCESS

The obvious criterion of success for cryopreservation is birth or hatching of young. Simply quantifying the percentage of pregnant females, fertilized eggs, or young is insufficient without careful control of major factors affecting outcome and an understanding of probability. Success obtained with frozen semen (or fresh semen) is dependent on many factors including true fertility of the male, parity and true fertility of the population of females inseminated with his sperm, correctness in timing of insemination relative to ovulation, number of inseminations per ovulatory cycle, number of sperm inseminated, site of insemination, and inseminator. Results are also affected by details of protocols followed during freezing, thawing, and postthaw removal of cryoprotectant. Many published reports comparing "success" with fresh vs frozen–thawed sperm or alternative cryopreservation procedures are of little or no value because the authors failed to consider or control important variables.

Laws of probability impose limitations on precision of most data sets. With mammals, pregnancy is essentially a binominial event, with a given female becoming pregnant, or remaining nonpregnant, after timely insemination during one estral or menstrual

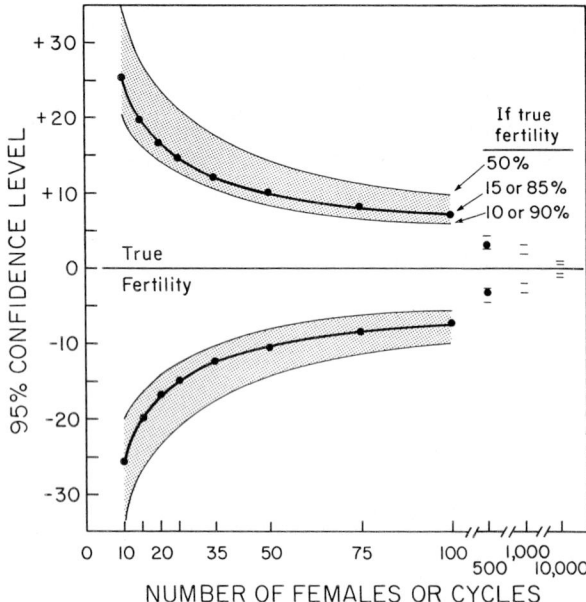

FIGURE 5 Approximate 95% confidence interval for a binomial as influenced by sample size (number of females or cycles of exposure) and "true fertility." Because biological variation also affects precision, the 95% confidence intervals for "fertility" would be >1.2 times the values shown. From Amann (1989) with permission.

cycle. Uncertainty inherent in binomial variation (Fig. 5) requires data on a substantial number of females per seminal sample, per male, or for a given treatment to obtain meaningful results concerning possible differences. Biological variation is at least 20% of that resulting from binomial variation so that actual confidence intervals for a biological data set would be >1.2 times those depicted in Fig. 5. Data based on 20–30 females typically have a 95% confidence interval of greater than ±20 percentage units, with imprecision greatest as pregnancy or birth rate approaches 50%. Data from a commercial human semen bank likely would be limited to <100 females per semen donor for ethical reasons, and uncertainty about a reported pregnancy or birth rate (95% confidence interval) will exceed ±10 percentage units.

The method used to establish that fertilization of ≧1 egg occurred, that embryo development proceeded, that implantation or attachment occurred, or that young were born will affect both the absolute percentage value for success in a population of females and the magnitude of error in measurement. It is important to consider success of cryopreservation on the basis of pregnancy rate per cycle of exposure rather than per year of exposure or per breeding season. With cattle or chickens, data could be precise with variables carefully controlled and based on sufficient inseminations. Unfortunately, for most other species data reporting success rate are flawed.

With the previous caveats implicit, and despite the paucity of comparative data for many species, the following statements are probably correct. (i) For <20% of dairy (Holstein) bulls, fertility of frozen–thawed sperm is substantially lower than that of fresh sperm when used under comparable conditions, even with more sperm per insemination; (ii) for <40% of beef bulls (many breeds), fertility of frozen–thawed sperm is substantially lower than that of fresh sperm; (iii) for >50–60% of boars, men, or stallions, fertility of frozen–thawed sperm is substantially lower than that of fresh sperm regardless of number of sperm per insemination; (iv) for the majority (or all) of males in many inbred lines of chickens or mice, fertility of frozen–thawed sperm is very low relative to that of fresh sperm; (v) for many species, including endangered species, appropriate comparative data are unavailable; and (vi) with careful selection of the males, birth rates achieved with frozen semen from beef bulls, men, stallions, and most other species can approach those achieved with fresh sperm. Setting appropriate expectations (not too high) usually will result in satisfaction when cryopreserved semen is used.

XII. THOUGHTS TO PONDER

In contrast to somatic cells, spermatozoa cannot reproduce themselves and have virtually no capability to repair damage induced by their environment, including damage associated with cryopreservation. Past and current research focuses on minimizing damage to sperm, but the track record for this approach to improve retention of fertilization potential for most sperm for a species is disappointing. Could postthaw "treatment" with exogenous molecules repair certain damaged features of sperm function or restore bioactive molecules eluted from the sperm surface?

There are two general goals for cryopreservation of sperm: (i) retention for later use of unique germ plasm from an endangered species or rare genetic strain (or animal model—cat, chicken, dog, or mouse) when recovery of a few young using artificial insemination is sufficient and cost is not a driving force; and (ii) propagation of a species (cattle, chicken, goat, horse, human, pig, and sheep) when percentage of artificial inseminations resulting in offspring is an important factor and cost might be a dominant feature. Different success rates are required for the two classes of use. To achieve the "banking" goal, current approaches for cryopreservation of sperm are probably adequate for the next decade. To improve success in achieving the propagation goal, research on cryopreservation is needed, should utilize the species of interest, and should include studies on the heritability of resistance of sperm to cryodamage or improvement of postthaw sperm quality by retention or exclusion of selected genes.

Achievement of pregnancy requires timely availability of a population of sperm in which a high percentage of the individual sperm have retained all attributes essential for fertilization and embryo development and express these attributes in the appropriate amount and correct temporal sequence. What is biochemically different about most sperm in samples that retain high fertilizing potential vs most sperm that lose one or more essential attributes in a sample with low potential for fertilization and development of a viable embryo? Individual cells must be studied. Optimum use of cryopreserved sperm might require a shorter interval between thawing–insemination and ovulation or a different site of insemination than those used for fresh sperm.

There are ample challenges in the area of cryopreservation for an individual entering a research career. Modern knowledge of cell biology opens many opportunities not available 5–50 years ago. Furthermore, observations detailed over the past 50 years should not be ignored but should be reinterpreted with a broad and current perspective. Careful study of these primary publications (not reviews thereof) will provide clues to causes of success or failure. Unfortunately, funding for research necessary to improve success rates with cryopreserved sperm from most males of most species will be limited. The early and continuing success of frozen semen with dairy cattle, and paucity of precise fertility data for most other species, masks the limited success that can be anticipated with a "typical male" (unselected) of the human or other species. When general recognition of the limited success achieved with cryopreserved sperm from unselected males occurs, funding of research to improve success with important species may become available.

See Also the Following Articles

ARTIFICIAL INSEMINATION; REPRODUCTIVE TECHNOLOGIES; SPERMATOZOA

Bibliography

Amann, R. P. (1989). Can the fertility potential of a seminal sample be predicted accurately? *J. Androl.* **10**, 89–98.

Amann, R. P., and Hammerstedt, R. H. (1993). In vitro evaluation of sperm quality: an opinion. *J. Androl.* **14**, 397–406.

Amann, R. P., and Pickett, B. W. (1987). Principals of cryopreservation and a review of cryopreservation of stallion spermatozoa. *J. Equine Vet. Sci.* **7**, 145–173.

Amann, R. P., Gill, S. P. S., and Hammerstedt, R. H. (1997). *Maximizing Genetic Impact of Roosters: Effective Handling, Extension and Utilization of Rooster Semen and Artificial Insemination.* BioPore, State College, PA.

Buss, E. G. (1993). Cryopreservation of rooster sperm. *Poultry Sci.* **72**, 944–954.

Gao, D. Y., Liu, J., Liu, C., McGann, L. E., Watson, P. F., Kleinhans, F. W., Mazur, P., Critser, E. S., and Critser, J. K. (1995). Prevention of osmotic injury to human spermatozoa during addition and removal of glycerol. *Hum. Reprod.* **10**, 1109–1122.

Hammerstedt, R. H., Graham, J. K., and Nolan, J. P. (1990). Cryopreservation of mammalian sperm: What we ask them to survive. *J. Androl.* **11**, 73–88.

Karow, A. M., and Critser, J. K. (1997). *Reproductive Tissue Banking: Scientific Principals.* Academic Press, New York.

Mazur, P. (1984). Freezing of living cells: Mechanisms and implications. *Am. J. Physiol.* **247**, C125–C142.

Parkes, A. S. (1966). *Sex, Science, and Society.* Oriel, Newcastle-upon-Tyne, UK.

Pickett, B. W., and Berndtson, W. E. (1978). Principals and techniques of freezing spermatozoa. In *Physiology of Reproduction and Artificial Insemination of Cattle* (G. W. Salisbury, N. L. VanDemark, and J. R. Lodge, Eds.), 2nd ed., pp. 494–554. Freeman, San Francisco.

Polge, C., Smith, A. U., and Parkes, A. S. (1949). Revival of spermatozoa after vitrification and dehydration at low temperatures. *Nature* **164**, 666.

Watson, P. F. (1995). Recent developments and concepts in the cryopreservation of spermatozoa and the assessment of their post-thawing function. *Reprod. Fertil. Dev.* **7**, 871–891.

Cryptorchidism

Irene M. McAleer and George W. Kaplan

University of California, San Diego and Children's Hospital and Health Center

I. Introduction
II. Embryology
III. Testicular Descent
IV. Evaluation
V. Management
VI. Sequelae

GLOSSARY

cryptorchidism Undescended or incompletely descended testis.

gubernaculum A mesodermal derivative located at the lower end of the testis and epididymis.

human chorionic gonadotropin A gonad or testis-stimulating hormone which is produced by placental trophoblasts and may be helpful in promoting testicular descent and proper germ cell maturation.

orchiopexy A surgical procedure that frees the undescended testis from the structures preventing its reaching a dependent scrotal position.

retractile testis Testis that lies in the normal pathway of descent which, historically, was located in the base of the scrotum that is located in an elevated position probably due to cremasteric muscle contraction.

Cryptorchidism or maldescent of the testis is a common condition in children. The causative factors for testicular descent and failure of complete descent, which were first described by John Hunter in 1786, are still not well understood.

I. INTRODUCTION

The incidence of cryptorchidism is between 3 and 5% of term male infants. Undescended testes are more common in preterm infants; between 17 and 33% of preterm infants may have cryptorchidism noted at birth. Many of these testes descend into the proper scrotal position during the first year of life. Most of these will descend spontaneously within the first 3 months of life so that the eventual incidence at 1 year of life is 0.8%. This incidence is identical to that in adolescence.

Patients with cryptorchidism are at increased risk for developing testicular tumors later in life and have a higher incidence of infertility as well. It has yet to be determined if timing or type of management of undescended testes can improve fertility or decrease the risk of developing testicular cancer.

Operative management of undescended testes has been reported in the literature since 1871 and has been the mainstay of therapy for this condition. Success rate as defined as obtaining a normal anatomical position has been reviewed by Docimo and approximates 90%. Varying methods have been used to relocate the testes, with some newer methods attempting to improve positioning and function of high-lying testes. Laparoscopic and microvascular approaches have been used to this end.

Medical management with hormonal manipulation has also been used to attempt to get the testis into the scrotum. Success rates have been varied. It appears that there may be benefit to using low-dose luteinizing hormone-releasing hormone analogs for improved germ cell production and, ultimately, improved fertility.

II. EMBRYOLOGY

The undifferentiated gonadal ridges are first apparent at 3–5 weeks of gestation. The sexual differentiation of the primitive gonad occurs around the sixth week of gestation. Testicular development occurs if the *SRY* gene, which is located on the short arm of the Y chromosome, is present. The gonadal ridge, which will form the testis, is located between body segments T6 through S2. The upper region of the ridge involutes so that the final germinal tissue arises from the lower aspect of the ridge. This explains why the testis is located fairly low in the abdomen before the initial phase of its descent and, in cases of cryptorchidism, is generally found in the lower abdomen and not in the upper abdomen or retroperitoneum.

The Sertoli cells develop by Weeks 6 or 7 and, shortly afterwards, begin to produce Müllerian-inhibiting substance which produces regression of Müllerian ductal structures.

Leydig cells develop by the ninth week and produce testosterone, which promotes development of the Wolffian duct structures. The Wolffian duct structures arise from the mesonephros and the mesonephric duct. The head and body of the epididymis are derived from the mesonephros and will ultimately be found in a posterolateral location on the testis. Initially, many mesonephric tubules branch and pass into the primitive testis to become the rete testis. The upper portion of the mesonephric duct will become the tail of the epididymis and the caudalmost portion ultimately becomes the vas deferens.

The urogenital ridge is formed by the testis and the Wolffian duct structures and is attached above and below. The upper attachments are the testicular artery and vein. Inferiorly, a mesodermal derivative forms an elongated attachment from the testis and epididymis to the groin called the gubernaculum. The gubernaculum forms early in gestation, well before the abdominal muscles grow forward from the lateral cell mass, and consequently produces a gap in the forming musculature allowing an obliquely oriented tunnel (the inguinal canal) to be formed in the inguinal region. The cremasteric muscles are derived from the medial and lateral aspects of the mesenchyme of the gubernaculum. The lower end of the gubernaculum travels downward and medially to terminate in the subcutaneous tissue of the developing scrotum.

The testis and its cord structures are located in the lower abdomen in the region of the internal inguinal ring by the 21st or 22nd weeks of gestation. The testis is fairly mobile except for its superior and inferior attachments and, although intraabdominal, is covered with peritoneum except in its posterior aspect.

III. TESTICULAR DESCENT

The testis must ultimately become located in the scrotum so that the cooler temperature necessary for formation of normal spermatozoa is achievable. True descent of the testis begins about the 28th week of gestation with most of the descent occurring into the eighth and ninth months of gestation. The actual cause and mechanism of testicular descent has not been determined and is a topic of much debate. Adequate animal models are not available since there is significant species variability in methods of descent and timing of descent. Study of descent in larger mammals is expensive and fetal surgery is difficult due to the size of the animals which are studied. Several authors have proposed that there are two or more stages of descent which may occur. These stages include one or two intraabdominal or transabdominal stages prior to the true descent of the testis through the inguinal canal into the scrotum.

There have been many theories regarding the initiation and mode of testicular descent. Hunter was the first to describe the descent of the testis as being directed by the gubernaculum; he named this band of tissue from the Latin word *gubenator* (helmsman or pilot). Hunter theorized that the gubernaculum was like the rudder on a ship and it directed the testis into the scrotum. Another theory states that the gubernaculum pulls the testis into the scrotum and is an attractive force; however, the cremasteric muscles which are supposed to be pulling the testis into the scrotum do not extend beyond the inguinal area until shortly before the testis descends. The force produced by contraction of the cremaster would cause the testis to retract up into the groin rather than descend into the scrotum. Scorer theorizes that the gubernaculum does not pull the testis into the scrotum but preserves the length and diameter of the tunnel of descent so that the testis can travel into the scrotum through the abdominal musculature. The gubernaculum is large and globular at the time of inguinal descent and may, therefore, allow the scrotum to rapidly enlarge to accommodate the testis. Once the testis reaches the bottom of the scrotum, the gubernaculum begins to involute and eventually atrophy over several weeks or months after the testis has descended into the scrotum (Fig. 1).

Intraabdominal pressure has also been theorized to cause descent of the testis. It has been thought that the testis may descend into the scrotum by increases in intraabdominal pressure produced by fetal respiration, development of the liver and bowel, or pressure exerted by accumulation of meconium. It is thought that the intraabdominal pressure exerts its influence on the testis in the presence of a patent processus vaginalis which allows the pressure to push the testis into the scrotum. This theory does not explain why the ovary does not also descend since intraabdominal pressure changes should also be present in females, unless descent is determined by male sex differences in the development of the gubernaculum, processus vaginalis, or the scrotum. Hadziselimovic also observed that testicular descent occurs in boys with omphaloceles (in which intraabdominal pressure is theoretically reduced) if there are no brain malformations.

Epididymal factors have also been presented as causing testicular descent by pushing the testis or secreting fluid allowing descent. In support of this theory, epididymal anomalies have been noted in 36–79% of undescended testes. Elder also found that in boys who underwent surgical correction of a patent processus vaginalis (hernia or hydrocele repair), 50% of these descended testes had an abnormal epididymal attachment. These epididymal abnormalities may have more to due with a persistent patent processus vaginalis than with the cryptorchid testis itself. In any case, testicular descent can occur despite the complete absence of the epididymis.

Endocrine factors have been associated with some control of testicular descent. An intact hypothalamic–pituitary–gonadal axis has been shown to be necessary for normal testicular descent. Patients with Kallmann syndrome, in which luteinizing hormone-releasing hormone (LHRH) is deficient, commonly have cryptorchidism. Gonadotropins have been shown to induce production of testicular androgens in the fetal testis. Blocking androgen effect with antiandrogens in laboratory animal studies has been shown to prevent testicular descent. Androgen production is initiated by placental human chorionic gonadotropin (hCG) and eventually sustained by fetal pituitary luteinizing hormone after 20–24 weeks' gestation. Leydig cells increase in number before descent of the testis but disappear during descent with decreasing testosterone levels after 17–28 weeks' gestation. Children with absent or abnormal androgen receptors will also have failure of testicular descent past the internal inguinal ring (97%) and a persistence of a patent processus vaginalis. Administration of testosterone for cryptorchidism has not been very successful, however, in obtaining testicular descent.

Ultimately, testicular descent most likely occurs in the presence of a normally developed gubernaculum, a normal hormonal axis, and a patent processus vaginalis. Animal studies are helpful but cannot directly correlate with human development.

IV. EVALUATION

The best time for determining testicular location is at birth. The scrotum is relatively larger than at any other time in life and is more often relaxed than contracted at this period, subcutaneous fatty tissue is scant in the newborn, and the cremasteric reflex is virtually absent at birth. Premature infants are more likely to have undescended testes at birth, however. The incidence of cryptorchidism at birth is 3–3.68% of cryptorchidism. Low birth infants (<2500 g) and premature infants (<37 weeks) have a higher rate of nondescent (19.83 and 17.24%, respectively). Most of these undescended testes descend typically within the first 3 months of life. Spontaneous descent is more common in low birth weight and premature infants than in term infants. The prevalence of cryptorchidism at 1 year is between 0.8 and 1.06%.

Distribution of undescended testes varied with the study group: Scorer, Cendron, and Hadziselimovic found that failure of descent was more common on the right side than the left; Berkowitz found the left side was more likely to be undescended. Bilaterally undescended testes were found in 15.5–30% of patients. Between 15 and 20% of patients will have a nonpalpable testis. Certain medical conditions have

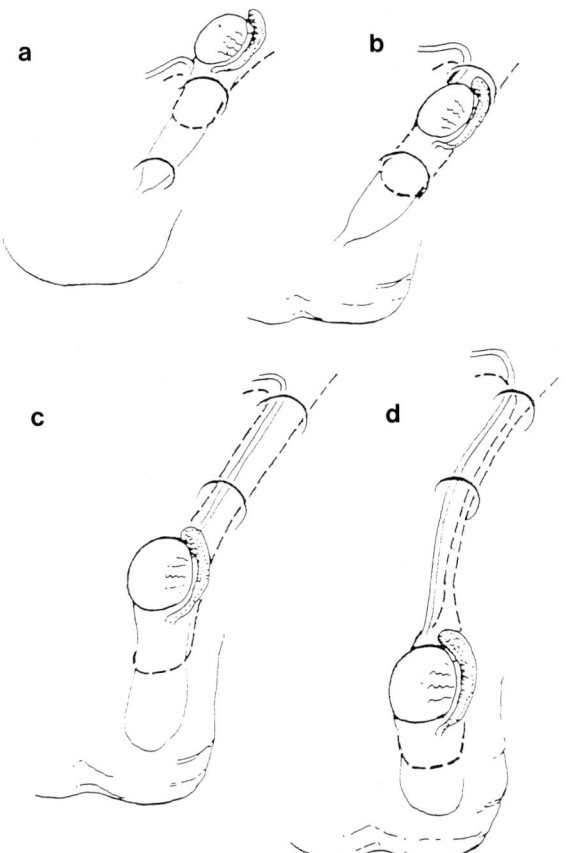

FIGURE 1 Stages in the descent of the testis. (a) Testis, epididymis, and gubernaculum before movement has begun. The scrotum is smooth and undeveloped. (b) All three parts together move down the inguinal canal, drawing a sleeve of peritoneum after them. (c) The testis, after emerging from the external inguinal ring, becomes larger. The gubernaculum changes in appearance and begins to disappear. (d) The scrotum enlarges to accommodate the testis. The gubernaculum continues to atrophy. The sleeve of peritoneum becomes narrower and begins to be absorbed (reproduced from Scorer and Farrington, 1971).

an increased incidence of undescended testes. Prune belly syndrome, meningomyelocele patients, patients with mental retardation (e.g., Prader–Willi syndrome), and patients with central nervous system abnormalities or abnormalities of their hypothalamic–pituitary–gonadal axis have a higher likelihood of having cryptorchidism.

Nomenclature for location of cryptorchid testes is confusing and conflicting. Several authors have attempted to classify undescended testes based on testicular location at palpation after manipulation or at the time of operation. These different systems make review of incidence of cryptorchidism, success of treatment, or incidence of infertility or cancer difficult to correlate.

To better define the undescended testis, it is important to define the normal testicular location (Fig. 2). Assuming ideal examination conditions, a normal testicular exam should demonstrate that the testis is well descended into the base of the scrotum by both visual inspection and manual examination. The ipsilateral scrotum should be well developed and the testicular size should be comparable to the size of its fellow. The testis should not require manipulation to get it in this dependent position and, historically, the testis should have been known to have been descended at birth or shortly thereafter.

Truly undescended testes should be palpated along the normal route of descent. The ipsilateral scrotum is generally hypoplastic or poorly developed. Generally, these boys have never had a testis noted in the scrotum. Occasionally, however, these children may have had a hydrocele in the scrotal sac which filled

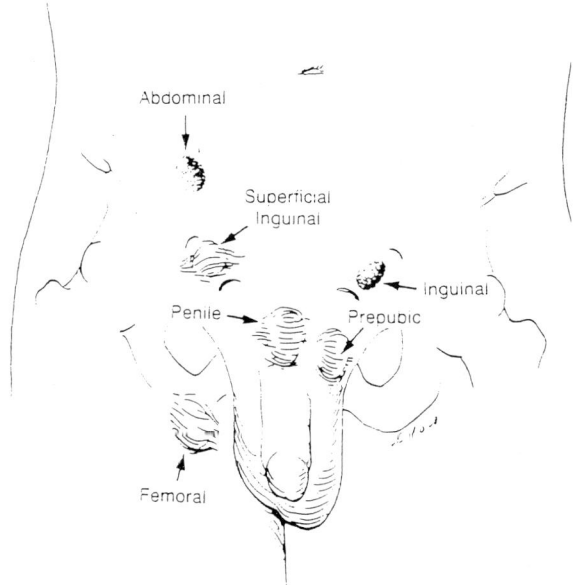

FIGURE 2 Testicular locations in cryptorchidism (reproduced from W. J. Cromie, *Decision Making in Urology*, 2nd ed., Decker, Philadelphia).

the scrotum and gave an appearance of a descended testis when in fact the scrotal fullness was due to hydrocele fluid and not a normally descended testis. Endocrine abnormalities may present with bilaterally undescended testes. These boys are typically undervirilized and have either a central endocrine or gonadotropin deficiency or androgen receptor abnormalities.

Canalicular or gliding or "peeping" testes are located along the normal path of descent and may be palpable or nonpalpable on any given examination. These testes reside near the internal inguinal ring and, because of a patent processus vaginalis, are either palpable within the inguinal canal or nonpalpable in an intraabdominal location at any given time.

Ectopic testes are no longer in the normal path of descent. The majority of the ectopically located testes may be in the superficial inguinal pouch, a potential space between the subcutaneous tissue between Scarpa's fascia and the external oblique muscle as described by Denis Browne. They may also be found in the perineum, femoral canal, penopubic area, or transversely located. This occurs in approximately 10% of unilateral undescended testes.

Intraabdominal testes are also present with either formation of patent processus vaginalis (canalicular or peeping or gliding) or closed inguinal ring indicating an absent patent processus vaginalis. These testes are nonpalpable. The scrotum is poorly developed on the side of the nonpalpable testis. Between 8 and 34% of the testes were intraabdominal. Some may have been canalicular to account for the wide variability of occurrence.

Approximately 10% of the nonpalpable testes are intraabdominal vanishing testes and between 47.5 and 54.5% of the nonpalpable testes will be atrophic or infarcted nonfunctional remnants of testes. Koff and others have speculated that patients with atrophic or dysgenetic testes will have significant contralateral compensatory testicular enlargement.

Retractile testes are often confused for truly undescended testes. These testes are not located in the normal dependent scrotal position with visual inspection but can be located in the inguinal area and manipulated into the base of the scrotum where they will remain for several minutes without traction or trapping the testis into the scrotum. The ipsilateral scrotum is typically well developed, as is the contralateral side. Historically, these boys had both testes located in the base of the scrotum at birth or shortly thereafter. Hormonal manipulation with hCG will cause the testis to descend into the scrotum. Testicular retraction occurs once the activity of the cremasteric muscle is strong enough to pull the testis up into the inguinal canal or superficial pouch. This most commonly occurs in boys between the ages of 6 and 12 years. Retractility of the testis disappears once the cremasteric activity begins to decrease, typically by the age of 13 years. Disappearance of retraction was found to occur approximately 2 years before the onset of puberty. To differentiate between truly undescended testes and retractile ones, undescended testes will be found to be smaller and softer than descended ones and the scrotum is not as developed in truly undescended testes.

Examinations should be performed as soon after birth as possible to best determine if the testis is cryptorchid or normally placed. After that time, examinations should be performed in a warm room, with warm hands, and the patient should be given time to relax. Visual inspection of the scrotum should be carried out before manual examination. The patient should be examined in either a cross-legged sitting position or with the patient supine with his hips externally rotated and abducted and the knees flexed. The testis may be manipulated into the scrotum gently using the thumb and forefinger to "trap" the testis. Testis volume and size should be determined; comparison with volume measurements on the descended mate should also be performed. If a testis is palpably undescended or thought to be so, the examiner may improve localization of the testis by applying liquid soap to the fingertips on the examining hand. This will allow the fingers to smoothly glide over the groin to isolate and manipulate a mobile testis.

V. MANAGEMENT

Treatment for cryptorchidism should be initiated once it has be determined that the testis is truly undescended or nonpalpable. Early treatment, as early as 9 or 10 months, is recommended to increase fertility potential, decrease risk for torsion, and possibly decrease malignancy rate. It has been observed that tubular scarring and loss of germ cells increases markedly in undescended testes after 18 months of life. Also, decreased fertility index measurements are noted before age 18 months. Surgery may be more successful in younger ages as well, especially in the intraabdominal testis group.

The mainstay in therapy in the United States is surgical mobilization of the undescended testis and ultimate tension-free placement of the testis into the scrotum in a dependent position. Hormonal therapy has been tried with some success in Europe.

Hormonal treatment using hCG has been effective in causing testicular descent in 30–40% of patients. The treatment consists of a series of intramuscular injections and can cause growth of the genitalia and some mood changes; consequently, it is not generally recommended except for differentiating retractile from true undescended testes or to make a nonpalpable testis more readily discernible. LHRH analogs (Kryptocur) or gonadotropin-releasing hormone analogs (buserelin) have been used for treatment of undescended testes in Europe. A recent trial of buserelin, placebo, and surgical therapies showed that buserelin was capable of inducing descent of the testis in 28% of the patients (all boys received hCG in addition to buserelin or placebo, however). It was also demonstrated that the germ cell number increased and the epididymis was found to be further developed over the placebo or surgical orchiopexy arms of the study.

Despite these results, surgical orchiopexy has the best success rate in obtaining descent of the testis. In the case of the palpably undescended testis, orchiopexy, or fixation of the testis, is performed via an inguinal incision with mobilization of the testis and cord structures, dissection and ligation of the processus vaginalis, and stripping of the cremasteric muscles so that the testis is placed in a tension-free location in the scrotum.

Recent practice has been to place the palpably undescended testis into the scrotum with the least amount of manipulation of the cord structures and testis as possible. This can be performed using a laparoscopically assisted procedure or it can be performed with an improved fixation technique in the scrotum; whether this will improve ultimate testicular growth or function is unknown.

A meta-analysis undertaken by Docimo demonstrated overall success rates for orchiopexy in the inguinal region of about 89%. The success of orchiopexy dropped off significantly when dealing with intraabdominal or canalicular testes (74 and 82–89%, respectively). Fowler–Stephens orchiopexy, in which the testicular vessels are divided and the testis blood supply is through the vasal and peritoneal pedicle vessels, had an overall success rate of only 67%, increasing to 77% for staged Fowler–Stephens orchiopexy. Microvascular autotransplantation has the best success rate (84%) for the intraabdominal testis but is a very complicated and demanding procedure that is not generally performed.

Evaluation and management of the nonpalpable undescended testis was facilitated with the advent of laparoscopic inspection and localization of the intraabdominal testes; inspection of the intraabdominal vessels and vas allow identification of the intraabdominal vanishing testis, an intraabdominal testis, or a canalicular testis. Techniques and instrumentation have improved so that a small incision for abdominal access and clear visualization of the internal ring should be accomplished in the majority of the patients with nonpalpable testes.

There is controversy regarding whether the inguinal region should be explored in the case of the nonpalpable testis with no intraabdominal testis, a closed inguinal ring, and normal appearing vas and vessels entering the ring. Koff, Belman, and others would forego laparoscopy if a patient with a nonpalpable testis was found to have compensatory hypertrophy of the normally descended contralateral testis and proceed directly to inguinal exploration alone and remove any atrophic tissue found on exploration. Currently, if laparoscopy demonstrates a closed inguinal ring with normal vas and vessels entering the ring but no apparent intraabdominal testicular tissue, an inguinal exploration is needed, in our opinion, to remove any possible atrophic testicular tissue found as germ cells, which may have oncogenic potential and can be found in the testicular remnant.

VI. SEQUELAE

Ultimately, the fate of the undescended testis concerns eventual male factor fertility or infertility and malignancy potential. It is difficult to determine if treatment, either through surgical or hormonal means, will ultimately improve or change fertility in patients with undescended testes. Typically, patients with unilateral cryptorchidism have reasonable fertility, with 50–75% of these men being able to father children. Unfortunately, most men with a history of bilateral undescended testes will have an infertility rate as high as 59%. Puri found that all patients in his series with bilateral impalpable testes who had orchiopexies performed previously were all azoospermic. Lipshultz reported that sperm counts were significantly lower in a group of men with a history of unilateral cryptorchidism compared to a similar group of age-matched controls subjects. Even with normal or near normal fertility potential in the cryptorchid testis itself, abnormalities of the vas or epididymis, which are seen in up to 70% of patients with cryptorchid testes, may preclude adequate maturation or transport of the sperm necessary for fathering children. Many of the patients involved in reported studies on fertility had orchiopexies performed in late childhood and adolescence which has been correlated with decreased fertility potential and increased tubular scarring on testicular biopsy done at the time of orchiopexy. It has yet to be determined if early orchiopexy will improve fertility parameters.

Hadziselimovic has recently been reported using buserelin after orchiopexy in a pilot group of nine males. These males were found to have decreased germ cells in the biopsies of the cryptorchid testes. Semen analyses obtained in adulthood in these patients had an increased number of spermatozoa, increased numbers of normal sperm, and improved motility compared to cryptorchid controls not given buserelin.

Undescended testes have a greater risk of developing testicular cancer than do descended testes. The risk for developing testicular cancer in the United States is about 2 or 3 per 100,000 per year. A cryptorchid patient's risk of developing a testicular malignancy is between 7.3 and 9.7 times more than that of the general male population. In Scandanavia, approximately 20% of the testis tumors that occur in cryptorchid patients will occur in the normally descended testis. Studies of Scandinavian men with a history of undescended testes found intratubular germ cell neoplasia in 2 or 3% of the testicular biopsies. This neoplasia is known to be a precursor to testicular germ cell tumors. In a recent study in Denmark, boys were biopsied at the time of orchiopexy (average age, 12.7 years); only 0.2% of those boys went on to develop germ cell testis cancer and another 0.2% went on to develop intratubular germ cell neoplasia. (Lifetime risk for developing testis cancer in the population is about 0.6%, which is much higher than that in the United States.) Only 9% of the patients were younger than 10 years old at the time of their orchiopexies. The patients who developed intratubular germ cell neoplasia or testis cancer were older than 10 years at the time of their orchiopexies. The study group also found that the patients with intratubular neoplasia had 45X/46 XY mosaic karyotypes. The recommendations from this study were to exercise more vigilance over those patients with this mosaic karyotype for possible higher risk of developing testicular malignancy. Another study from Sweden reviewed the incidence of testis cancer developing in patients who underwent orchiopexies for cryptorchid testes compared to patients who underwent hernia operations. Three of four patients who developed testis cancer had orchiopexies for cryptorchidism. There was a sevenfold increased risk for developing testis cancer in the undescended testis group and no increased risk for the hernia group. All the boys with cryptorchidism who developed testis cancer had intraabdominal testes. This may be another risk factor, although it was not found to be excessive in the previous study.

To date, most of the men with a history of cryptorchid testes who later developed testis cancer had their orchiopexies after age 10 years. It may be that earlier orchiopexy as is currently being performed may change the relative risk of developing testis cancer; one testis cancer patient in the recent study from Sweden had his orchiopexies performed at age 2 years. More investigation and surveillance are necessary to determine if there is a change in risk of cancer development.

Cryptorchidism is one of the more common birth anomalies. Investigation into improved surgical techniques, hormonal supplementation or treatment, and patient screening for testis cancer development are ongoing and, it is hoped, will improve patient fertility and survival.

See Also the Following Articles

CHORIONIC GONADOTROPIN, HUMAN; TESTICULAR CANCER; TESTICULAR DEVELOPMENTAL ANOMALIES

Bibliography

Berkowitz, G. S., Lapinski, R. H., Dolgin, S. E., et al. (1993). Prevalence and natural history of cryptorchidism. *Pediatrics* **92**, 44–49.

Bica, D. T. G., and Hadziselimovic, F. (1992). Buserelin treatment of cryptorchidism: A randomized double-blind, placebo-controlled study. *J. Urol.* **148**, 617–621.

Bloom, D. A., Duckett, J. W., Kaplan, G. W., et al. (1992, May). What is the best approach to the nonpalpable testis? *Contemp. Urol.*, 39–61.

Browne, D. (1938). The diagnosis of undescended testicle. *Br. Med. J.* **2**, 168.

Cendron, M., Huff, D. S., Keating, M. A., et al. (1993). Anatomical, morphological and volumetric analysis: A review of 759 cases of testicular maldescent. *J. Urol.* **149**, 570–573.

Cortes, D., Thorup, J., Frisch, M., et al. (1994). Examination for intratubular germ cell neoplasia at operation for undescended testis in boys. *J. Urol.* **151**, 722–725.

Docimo, S. G. (1995). Results of surgical therapy for cryptorchidism: A literature review and analysis. *J. Urol.* **154**, 1148–1152.

Docimo, S. G., Moore, R. G., Adams, J., et al. (1995). Laparoscopic orchiopexy for the high palpable undescended testis: Preliminary experience. *J. Urol.* **154**, 1513–1515.

Elder, J. S. (1992). Epididymal anomalies associated with hydrocele/hernia and cryptorchidism: Implications regarding testicular descent. *J. Urol.* **148**, 624–626.

Farmington, G. H. (1968). The position and retractibility of the normal testis in childhood with reference to the diagnosis and treatment of cryptorchidism. *J. Pediatr. Surg.* **3**, 53.

Felix, W. (1912). The development of the urogenital organs. In *Manual of Human Embryology*, pp. 752–975. Lippincott, Philadelphia.

Fowler, R., and Stephens, F. D. (1959). The role of testicular vascular anatomy in the salvage of high undescended testes. *Aust. N. Z. J. Surg.* **29**, 92.

Gier, H. T., and Marion, G. B. (1920). Development of the mammalian testis. In *The Testis*, pp. 1–45. Academic Press, New York.

Hadziselimovic, F., and Herzog, B. (1997). Treatment with LH-RH analog following successful orchidopexy improves the chance of fertility later in life. Submitted for publication.

Heyns, C. F., and Hutson, J. M. (1995). Historical review of theories on testicular descent. *J. Urol.* **153**, 754–767.

Kaplan, G. W. (1993). Nomenclature of cryptorchidism. *Eur. J. Pediatr.* **152**(Suppl. 2), S17–S19.

Koff, S. A. (1991). Does compensatory testicular enlargement predict monorchism? *J. Urol.* **146**, 632–633.

Kogan, S., Hadziselimovic, F., Howards, S., et al. (1997). Pediatric andrology. In *Adult and Pediatric Urology* (J. Gillenwater, J. Grayhack, S. Howards, et al., Eds.), 3rd ed., Vol. 3, pp. 2623–2674. Mosby-Yearbook., St. Louis, MO.

McAleer, I. M., Packer, M. G., Kaplan, G. W., et al. (1995). Fertility index in cryptorchidism. *J. Urol.* **153**, 1255–1258.

McMahon, D. R., Kramer, S. A., and Husman, D. A. (1995). Antiandrogen induced cryptorchidism in the pig is associated with failed gubernacular regression and epididmal malformations. *J. Urol.* **154**, 553–557.

Peppas, D. S., Matthews, R., Gearhart, J. P., et al. (1997, March). The undescended testis: An update. *Dialogues Pediatr. Urol.* **20**(3).

Pinczowski, D., McLaughlin, J. K., Lackren, G., et al. (1991). Occurrence of testicular cancer in patients operated on for cryptorchidism and inguinal hernias. *J. Urol.* **146**, 1291–1294.

Puri, P., and O'Donnell, B. (1990). Semen analysis in patients operated on for impalpable testes. *Br. J. Urol.* **66**, 646–647.

Rozanski, T. A., and Bloom, D. A. (1995). The undescended testis: Theory and management. *Urol. Clin. North Am.* **22**(1), 107–118.

Scorer, C. G., and Farrington, G. H. (1971). *Congenital Deformities of the Testis and Epididymis*. Appleton-Century-Crofts, Butterworth, London.

Tennenbaum, S. Y., Lerner, S. E., McAleer, I. M., et al. (1994). Preoperative laparoscopic localization of the nonpalpable testis: A critical analysis of a 10-year experience. *J. Urol.* **151**, 732–734.

Ctenophora

George I. Matsumoto
Monterey Bay Aquarium Research Institute

I. Description of the Phylum Ctenophora
II. Modes of Reproduction: Sexual versus Asexual
III. Modes of Sex: Hermaphroditic versus Gonochoric
IV. Anatomy of the Reproductive Systems
V. Reproductive Regulation
VI. Modes of Fertilization
VII. Larvae and Metamorphosis

GLOSSARY

colloblast Specialized adhesive cells used for prey capture; unique to the phylum.

ctene plate A structure consisting of thousands of cilia linked (via the three and eight doublets) together in a plane parallel to the plate. These cilia are stacked in rows. These ctenes are arranged in eight rows and beat in a synchronized fashion. The cilia are up to 2 mm long and 50 μm thick.

degrowth The ability to shrink or reduce in size when nutrients are scarce.

dioecious Having separate sexes.

dissogony The ability to exhibit two periods of reproductive activity (larval and adult) separated by a period of gonadal regression.

gonopore A ciliated, discrete invagination of the epidermis where the eggs and sperm are released.

hermaphrodite An organism possessing both male and female gonads; most ctenophores are simultaneous hermaphrodites with both gonad types becoming ripe at the same time.

protandry Possession of both male and female gonads but with the male gonads occurring first and the female gonads appearing later.

statocyst An aboral sensory organ consisting of a complex statolith balanced on four sets of cilia.

stomodeum Another term for pharynx.

substomodeal Adjacent to the stomodeal plane.

subtentacular Adjacent to the tentacular plane.

The phylum Ctenophora is often overlooked in invertebrate biology courses because its members are difficult to collect and even harder to preserve. Much has changed in the past 24 years since the last inclusion of the phylum in a volume such as this. There are two additional orders to the phylum; technology has enabled the observations and collections of both open-ocean and deep-sea ctenophores (which has generated many new species descriptions) and the importance of ctenophores in the oceanic ecosystem has been recognized. As often abundant members of the ocean, ctenophores are voracious predators and in turn are often prey items for a variety of other organisms. This article will provide a general background to the phylum and provide a summary of what little is known about this phylum in terms of reproduction. Most of the knowledge presented here is material gathered from nearshore ctenophores. There is much work left to be done: The discovery of dioecious ctenophores is a result of shallow open-ocean research and many of the deeper water species that we are aware of still do not have scientific names.

I. DESCRIPTION OF THE PHYLUM CTENOPHORA

A. Overview

Due to difficulties in making observations and in collecting specimens, the phylum Ctenophora is an often overlooked group of marine organisms. The phylum has been tentatively separated into seven orders and 21 families. The taxonomy of ctenophores is based primarily on their external morphology and over 200 species of ctenophores have been described, of which some have been synonymized or modified and others are rarely seen and may or may not be valid descriptions. Taxonomic uncertainty stems from incomplete descriptions in older work, from the uncertain evolutionary framework within the phylum, and from the recent and anticipated discovery of new species that clarify old descriptions and/or introduce new taxonomic groups. The technique of netting planktonic organisms for analysis is not suitable for ctenophores because it often results in ctenophores becoming fragmented or turning into unrecognizable mucous clumps in the cod end of the net. Net tows through areas of visibly high ctenophore density (18–305 individuals per cubic meter) have resulted in counts of zero ctenophores. More information on the distribution of ctenophores and their natural history is needed in order to determine the importance of ctenophores in oceanic ecosystems.

B. Gross Morphology

Ctenophores or "comb-bearers" are named for the characteristic eight rows of macrociliary plates that they all possess at some point in their life. These ctene rows are named either substomodeal ctene rows or subtentacular ctene rows, depending on their location relative to the stomodeal or the tentacular plane (Fig. 1).

The body consists of two tissue layers, the endodermis and the ectodermis, which enclose a poorly differentiated gelatinous layer of acellular mesoglea. This mesoglea is of ectodermal origin and the organization of the mesoglea suggests that the Ctenophora are closer to Turbellaria than to Cnidaria. Determination of phylogenetic relationships within the phylum Ctenophora and with other phyla is a difficult task. Although the Ctenophora have been closely aligned with Cnidaria, numerous alternative schemes exist.

C. Bioluminescence

Most but not all ctenophores are bioluminescent. Ctenophore photoproteins (e.g., mnemiopsin and berovin) have an oxygen-bound enzyme substrate complex ("precharged") that only requires Ca^{2+} ions to trigger an *in vitro* reaction.

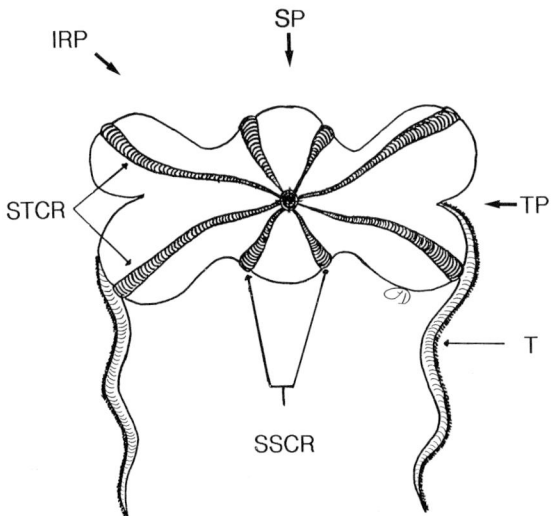

FIGURE 1 Oral view of *Mertensia ovum* shows the ctene rows, both the substomodeal ctene rows (SSCR) and the subtentacular ctene rows (STCR). The oral end is in the center, and the tentacles (T) originate from tentacle sheaths within the body and are covered with a mass of fine tentillae. The tentacular plane (TP) bisects the ctenophore, passing through the tentacle sheaths. The stomodeal plane (SP) bisects the body perpendicular to the tentacular plane and is in the same axis as the stomodeum. The ctene rows are named for their proximity to the two planes. Male gonads are generally found facing the interradial plane (IRP) which bisects the body at a 45° angle to both the tentacular plane and the stomodeal plane (modified from Matsumoto, 1991).

D. Gastrovascular System

The ctenophore gastrovascular system can be divided into an axial portion and a peripheral portion. The axial portion includes the mouth, stomadeum, infundibulum, infundibular canals, anal canals, and anal pores. The peripheral portion is composed of the following canals: perradial, interradial, adradial, meridional, tentacular, and paragastric. The degree of continuity of the gastrovascular system varies between orders as well as between genera or species so that any description of a ctenophore should include a detailed schematic diagram of the entire system (Fig. 2). Any of these canals may be missing; however, the system is essentially the same in all ctenophores. Near the statocyst are structures called pole plates for which a function is still unknown. The anal pores open through the pole plates. The defecation of undi-

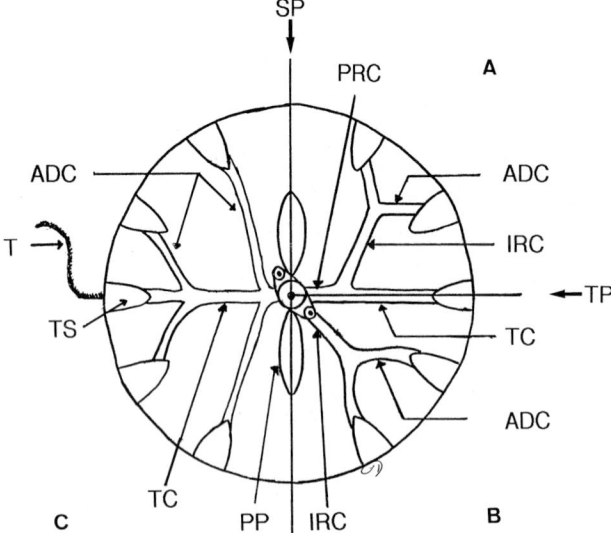

FIGURE 2 A generalized cross section through the body of (A) a pleurobrachiid, (B) a pleurobrachiid or a lobate, and (C) a mertensiid, a bathyctenid, or a platyctene. The canal structure varies within the phylum. The pole plates (PP) are on either side of the statocyst and the various canals are labeled. ADC, adradial canal; IRC, interradial canal; PRC, perradial canal; SP, stomodeal plane; T, tentacle; TC, tentacular canal; TP, tentacular plane; TS, tentacle sheath (modified from Harbison, 1985).

gested material is primarily through the mouth, although defecation through anal pores can occur in *Pleurobrachia* when the ctenophore is stressed.

E. Physiology

Ctenophores have both smooth and striated muscle but lack the epithelial–muscle cells that are characteristic of cnidarians. Metabolic rates vary among ctenophores (ranging from 1.5 mg C day^{-1} for *Beroe* to 2600 mg C day^{-1} for *Mnemiopsis*), with respiratory and excretory rates a direct linear function of animal weight (ash-free dry weight). These rates are very temperature sensitive. There is also a linear relationship between metabolism and size.

F. Nutritional Ecology

Research on the nutritional ecology of ctenophores has generally focused on the more commonly studied nearshore ctenophores, although much work has been completed in the past 20 years on the more delicate and difficult to observe open-ocean and deep-sea species.

The phylum Ctenophora is unique in the possession of specialized adhesive structures called colloblasts. These organelles are utilized to capture prey in an analogous fashion to the nematocysts in the Cnidaria, but the colloblast and nematocyst morphologies are very different. The atentaculate ctenophores and some species of tentaculate ctenophores lack colloblasts. Species in the genus *Haeckelia* possess kleptocnidae that are incorporated into the tentilla from ingested prey items.

G. Interspecific Interactions

Interspecific interactions are poorly understood due to the difficulty in sampling ctenophore abundance or even presence. Recent ctenophore population explosions have occurred in the Black Sea and the Sea of Azov, in which *Mnemiopsis mccradyi* (brought in with ballast water) increased from 10 g m^{-2} in May 1988 to 15000 g m^{-2} by April 1990. This burst of ctenophores has decimated the fisheries in the area (by depleting the larval fish and the food source of the adult fish) and the populations of *Aurelia aurita*, *Pleurobrachia pileus*, *Calanus helgolandicus*, and *Sagitta setosa* have virtually disappeared.

II. MODES OF REPRODUCTION: SEXUAL VERSUS ASEXUAL

Ctenophore reproduction is primarily external, with the majority of ctenophores being simultaneous self-fertilizing hermaphrodites. Sperm and eggs are broadcast into the water either with the sperm encased in mucus or free. Protandry has been documented for some platyctenids; platyctene ctenophores are also unusual in that internal fertilization, brooding of embryos, and asexual reproduction by laceration are common.

Some ctenophores can regenerate (*Bolinopsis infundibulum* and *Mnemiopsis leidyi* have been shown to regenerate if cut in halves and quarters but not very well if cut into eighths) and many of the ctenophores in the order Platyctenida are capable of reproducing using asexual laceration and budding; this process is accelerated if the animals are fed (see Section V,B).

III. MODES OF SEX: HERMAPHRODITIC VERSUS GONOCHORIC

Hermaphroditism is postulated as the primitive state for the phylum, with the majority of ctenophores being hermaphroditic. Diisogamy or pedogenesis (see Section V,B), in which reproductive ability as a larva and as an adult are separated by a period of gonadal regression, has been noted for *Pleurobrachia, Bolinopsis, Dryodora,* and *Mnemiopsis.*

Only one genus (*Ocyropsis*) and four subspecies have been found to be dioecious. These dioecious ctenophores exhibit no external sexual dimorphism but do display marked differences in reproductive behavior. Protandry has been suggested for the platyctenid *Coeloplana gonoctena,* which may become a simultaneous hermaphrodite as it matures.

IV. ANATOMY OF THE REPRODUCTIVE SYSTEMS

A. Gonad Location

The ctenophores have been categorized at the tissue level of construction and so their reproductive system consists of gametogenic tissues, although they have been considered as organs by some authors. The number of gonopores ranges from 8 (1 per gonad) to over 600 (one pair of pores between adjacent ctene plates). The oviducts are narrow tubes consisting of six to eight ciliated elongate cells that bridge the mesoglea. A connection between the oviduct and the ovary has not yet been found. The sperm duct is connected to the testicular sinus and is also ciliated.

The location of the gonads varies between and within the orders of the phylum Ctenophora (Fig. 3). The gonads differentiate from the meridional canals and both male and female tissue develop simul-

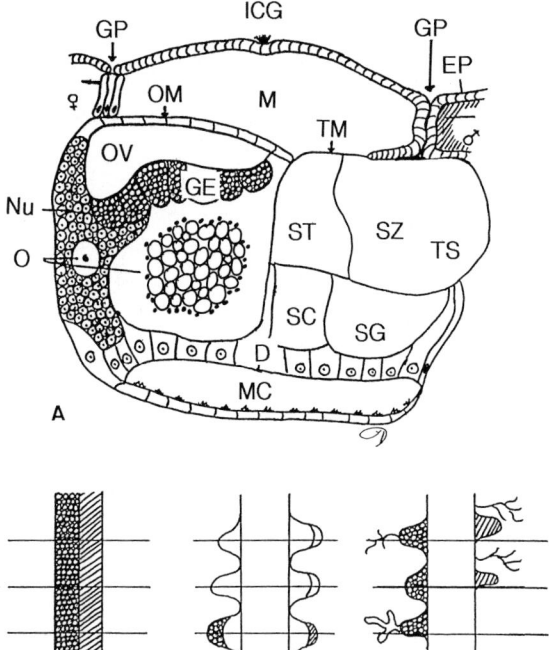

FIGURE 3 (A) A generalized cross section (modified from Hernandez-Nicaise, Fig. 32, 1991) through a meridional canal (MC). Connecting the ctene plates is the interplate ciliated groove (ICG) which arises from the epidermis (EP) and is over the mesoglea (M). The gonopores (GP) are paired (typical of *Bolinopsis*) and are connected to the female and male reproductive structures. OM, ovarial membrane; OV, ovary; GE, glandular epithelium; Nu, nurse cells; O, oocytes; TM, testicular membrane; ST, spermatids; SZ, mature spermatozooids; TS, testicular sinus; SC, spermatocytes; SG, spermatogonia. (B) Gonad arrangement typical of *Bolinopsis.* (C) Gonad arrangement typical of *Leucothea.* (D) Gonad arrangement found in Beroids.

taneously, usually with the male tissue facing the interradial plane. Gonad location in some orders is as follows:

Cydippida: The postulated ancestral group to the phylum has unbroken bands of gonadal tissues running under the ctene rows. The male gonads are located interradially, whereas the female gonads are located perradially.

Cestida: The gonadal tissues are only found under the substomodeal canals and the morphology of the gonadal tissues is a generic difference, with

Velamen having a number of genital swellings alternating with areas of no gonadal tissue. Continuous gonadal tissue is found in *Cestum*.

Lobata: Larval lobates resemble cydippids and have exhibited diisogamy (see Section V,B). There are several different locations within this order. Some lobates (e.g., *Leucothea*) have blind pockets next to the ctene plates, with the female tissue on the perradial side and the male tissue on the interradial. In other lobates (e.g., *Bolinopsis*), the gonadal tissue is only found under the ctene rows between the ctene plates.

Beroida: Gonadal tissue forms under the meridional canals or into some of the anastomosing branches. In some species, the gonads are entirely within these branches.

Platyctenida: Location and morphology vary widely within this order. The gonadal tissue can be in unbroken lines or within the branches off of the meridional canals. Some species appear to possess specialized reproductive structures distinct from the meridional canals and can even develop seminal receptacles and brood chambers in the papillae.

Thalassocalycida: Eggs have been observed in the branching canals associated with the subtentacular ctene rows. Male gonadal tissue has not been observed yet.

Ganeshida: Unknown. It is probable that this order is invalid. There is only one described species in this order and the description resembles a stage in the lobate life cycle.

B. Origin of Gonads and Germ Cells

The development of gonads from germ cells has not been well studied. The germ cells are in clusters and are associated with the endoderm, in which they are grouped into a triplet. The middle band of the triplet elongates and separates the two other clusters; the cluster facing the tentacular plane or the stomodeal plane develops into the ovaries and the cluster facing the interradial plane develops into the testes. This separation may also be due to digestive cell processes and has been depicted as such in Fig. 3.

C. Spermatogenesis

Spermatogenesis has been closely studied in the beroid *Beroe ovata* by Franc, who found that the spermatozoon were of a primitive type (spherical nucleus, few mitochondria, and the absence of a highly symmetrical organization). The spermatogonia originate from the testis, multiply, and form synchronized clusters. These cells are small and possess a dense nucleus and a double centriole. Spermiogenesis is completed when a long flagellum has differentiated from the distal centriole and the organelles inside coalesce. The mitochondria become either one or a few larger mitochondria, and dense granules (originally associated with endoplasmic reticulum) merge into a paranuclear body whose function is currently unknown.

D. Oogenesis

Ctenophores have a highly organized oocyte–nurse cell system and oogenesis appears to be similar in all ctenophores studied thus far. During development, each oocyte has ≈100 nurse cells (in three clusters) associated with it. These nurse cells contribute yolk and cytoplasm to the oocyte via a permanent syncytium. Each nurse cell–oocyte complex originates from a single oogonium. Dunlap has separated the development process into four discrete stages: previtellogenesis, early vitellogenesis, late vitellogenesis, and corticogenesis. The mature oocyte has a central endoplasm (filled with electron-lucent, PAS-positive yolk bodies), a subcortical layer (with small, electron-dense, PAS-negative yolk bodies), and a cortex (yolk-free, containing mitochondria and endoplasmic reticulum).

E. Ovulation and Spawning

During the spawning process, the mature oocyte becomes ameboid and reaches the oviduct, where it is coated with a jelly coat from oviductal gland cells. No connection between the ovary and the oviduct has been observed and the mature oocytes are thought to move through the glandular epithelium. The oviduct is a narrow tube that consists of six to eight elongate ciliated cells. The sperm accumulates in bands or clusters, often turning opaque. As the oocytes are ovulated, the first polar body forms and the second polar body forms shortly afterward. The vitelline membrane appears elevated once the egg reaches seawater (with or without fertilization having occurred). Sperm are shed first, with the eggs being shed a few minutes later and being fertilized as they exit the gonopore (see Section VI,A). Although most ctenophores spawn single eggs, *Dryodora* eggs emerge in strings of up to 20–30 eggs which stay together until hatched, *Beroe* eggs are often encased in a mass of mucus, and *Bolinopsis* eggs have also been reported to be released with mucus. Once the eggs are released, sperm entry can take place anywhere along the egg surface. The mechanism for fertilization among the platyctenes is unclear. The presence of seminal receptacles indicates that sperm may be stored for periods of time. Both eggs and sperm have been observed to be released from the mouth which would facilitate the brooding of eggs and larvae. Internal fertilization may also occur since internal brood chambers (in the papillae) have also been described.

V. REPRODUCTIVE REGULATION

A. Reproductive Physiology

The breeding season for many ctenophores is postulated to be year-round, with spawning peaks in the spring and the summer. These observed peaks may be artificial because during the fall and the winter the ctenophores descend deeper in the water column, thus making them more difficult to collect.

Ctenophores appear to be annual organisms, but this is primarily anecdotal information and studies on other organisms once thought to be annuals have clearly shown that they are long-lived [e.g., *Aurelia aurita* (Cnidaria: Scyphomedusae) can live at least 7 years]. Size cannot be used as an indication of age because ctenophores exhibit degrowth.

B. Factors Influencing Gametogenesis

1. Light

Ctenophores possess two well-known light-induced responses: (i) that of triggering spawning behavior and (ii) a reversal of geotaxis. Ctenophores may be utilizing photoreceptor cells for spawning cues. Species on the west coast of North America spawn after dawn or after exposure to a brief light period followed by a long dark period. Species on the east coast of North America spawn after midnight or after a medium-length dark period followed by a long light period.

The apical organ or statocyst consists of a pit of modified ciliated epithelium containing a calcareous ($MgCaPO_4$) statolith. The statocyst is mechanically coupled to the ctene rows by four pairs of ciliary tracts and four sets of balancer cilia (multiciliated epithelial cells). The statocyst provides directional movement in response to gravitational stimuli. The apical organ is also postulated to be a polyvalent receptor similar to the vertebrate acoustolateralis system in being able to detect gravitational and hydrostatic change. In addition, the apical organ contains four crescent-shaped groups of ciliary membranes. The membranes of 10–15 cilia (9 : 0) are compressed and are lined with granules. These granules are termed pigment cells because they move in response to light. Thus, the apical organ may also serve as a photoreceptor.

2. Temperature

There is only anecdotal evidence for the effects of temperature shifts on reproduction. Spawning has been induced in *Pleurobrachia* by moving specimens from cold water (6°C) to warm water (12°C).

3. Nutrients

Gametogenesis in both lobates and cydippids as a result of both underfeeding and overfeeding has been recorded. Spawning may occur just prior to death and 2 or 3 days after excessive feeding.

4. Endocrines

Currently, there is only indirect evidence for endocrine control of gametogenesis. Support for the presence of endocrine secretions includes experimental data (the removal of apical organ results in greater numbers of gametes) and indirect data (synchronous development). Harbison and Miller demonstrated that *Ocyropsis* spawning became more synchronous when males and females were placed together.

5. Age

Pedogenesis or dissogeny has been documented in several ctenophores (lobates and cydippids) and may be more prevalent throughout the phylum. Development of gonads occurs only under the substomodeal ctene rows, the number of eggs produced by juveniles is smaller, and the eggs are smaller in size as well. Following reproduction, the gonads regress back to germ cells and develop into adult gonads.

VI. MODES OF FERTILIZATION

A. Mechanisms of Spawning

Gametes are usually released through pores or ducts above the gonads, although some benthic ctenophores release the gametes through either the mouth or the ventral epidermis. The platyctenids are also capable of brooding embryos in the canal system with release through papillae on the aboral surface. The majority of ctenophores use ducts for release of gametes, with sperm release occurring a few minutes prior to oocyte release. Both eggs and sperm can be released as free swimming or within a mucus clump or string.

B. Fertilization

Fertilization will occur when the oocytes are released, although the release of fertilized eggs from ducts in the cydippid *Pleurobrachia* has been noted and the platyctenids brood (internally and externally) fertilized eggs. Fertilization can occur anywhere on the egg surface, with the egg pronucleus migrating toward the sperm pronucleus. In cases of polyspermy, which is common, the egg pronucleus appears to select one sperm pronucleus after probing several.

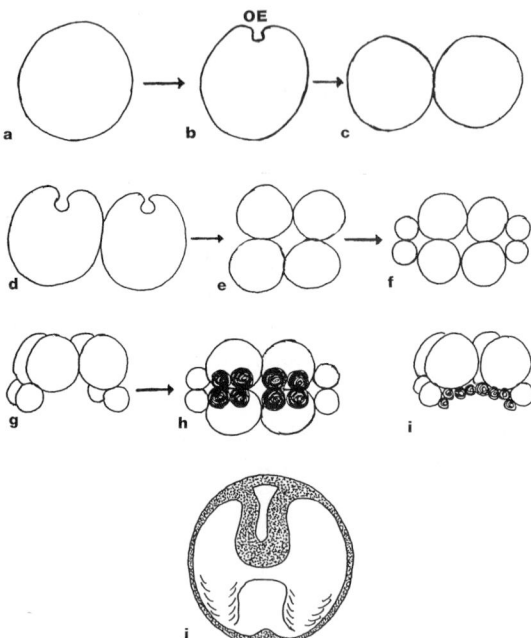

FIGURE 4 Developmental sequence of a fertilized ctenophore egg. The egg has an elevated membrane that appears when the egg is released (not pictured); upon the joining of the egg and sperm pronucleus, an invagination develops (b) that is also the oral end (OE) of the ctenophore. The second division (d, e) generates four equally sized blastomeres, whereas the third division produces unequally sized blastomeres distinguished as either the middle or the end blastomeres (f). The next division (to the 16-cell stage) generates eight macromeres (still labeled as either middle or end) and eight micromeres (smaller and shaded). The side views of the 8-cell stage (g) and the 16-cell stage (i) are provided as are the aboral views of the 8-cell stage (f) and the 16-cell stage (h). The gastrulation stage (j) shows the developing ctene rows and the stomodeum.

VII. LARVAE AND METAMORPHOSIS

A. Embryonic Development

Cleavage is unilateral, holoblastic, unequal, highly determinate, and, after the first division, biradial (Fig. 4). Sperm fusion can occur anywhere along the egg surface and the first cleavage will begin at the sperm fusion site. The first two meridional cleavages generate four blastomeres, and the third cleavage is nearly vertical and generates a boomerang shape of eight cells arranged in two rows of four. The inner four cells are larger than the outer four cells. The tentacular plane lies along this long axis. These eight macromeres then divide, forming eight macromeres and eight micromeres. Most of the ectoderm is derived from both cell types, whereas the endoderm is derived from the macromeres only.

Development is highly determinate and the cellular lineage of ctenophore eggs has been well studied. The oral–aboral axis is established at the time of first cleavage, with the site of sperm entry being the site of first cleavage and the future oral end. In cases of polyspermy, the site of first cleavage depends on the entry site of the sperm pronucleus that is fertilized. This cleavage also plays a causal role in the development of this axis. During this cleavage, the ctene plate-forming potential is localized in the aboral region of the embryo.

As gastrulation proceeds, the stomodeum becomes very evident and the interradial ctene plates form and begin to beat. Tentacle rudiments develop and either stop development (class Nuda) or invaginate to form into tentacle sheaths and tentacles (class Tentaculata). The colloblasts, statocyst, and photocytes (if present) develop at this time as well. Hatching usually occurs within 2–4 days after fertilization and is accompanied by a swelling of the embryo (due to hydration of the mesoglea).

The characteristic larva for the class Tentaculata is a cydippid larva which resembles the adult ctenophores in the order Cydippida. The larvae for the class Nuda lacks tentacles and tentacle sheaths. All larvae have eight rows of ctene (secondarily lost in some platyctene species) and a simple gastrovascular system. Feeding starts immediately, with the primary food being microzooplankton (ciliates, rotifers, and copepod nauplii). The change to the adult form for the latter larva is primarily just an increase in size and an expansion of the stomodeum. The change to the adult form for the cydippid larva is varied (depending on the order) and can involve the migration and/or reduction of the tentacles, a change in lateral compression (order Cestida), or an enlargement of the oral region into lobes (orders Lobata and Thalassocalycida). The ctene rows can be completely lost as the cydippid larva settles and changes into one of the ctenophores in the order Platyctenida.

See Also the Following Articles

Hermaphroditism; Marine Invertebrate Larvae; Marine Invertebrates, Modes of Reproduction in

Bibliography

Carre, D., and Sardet, C. (1984). Fertilization and early development in *Beroe ovata*. *Dev. Biol.* **105**, 188–195.

Cormier, M. J., Hori, K., and Anderson, J. M. (1974). Bioluminescence in coelenterates. *Biochim. Biophys. Acta* **346**, 137–164.

Dunlap, H. L. (1966). Oogenesis in the Ctenophora. PhD thesis, University of Washington, Seattle.

Franc, J. M. (1973). Etude ultrastructurale de la spermatogenese du ctenaire *Beroe ovata*. *J. Ultrastruct. Res.* **42**, 255–267.

Franc, J. M. (1978). Organization and function of ctenophore colloblasts: An ultrastructural study. *Biol. Bull.* **155**, 527–541.

Freeman, G. (1977). The establishment of oral–aboral axis in the ctenophore embryo. *J. Embryol. Exp. Morphol.* **42**, 237–260.

Harbison, G. R. (1985). On the classification and evolution of the Ctenophora. In *The Origins and Relationships of Lower Invertebrates* (S. C. Morris, J. D. George, R. Gibson, and H. M. Platt, Eds.), pp. 78–100. Oxford Univ. Press, Oxford, UK.

Harbison, G. R., and Miller, R. L. (1986). Not all ctenophores are hermaphrodites. Studies on the systematics, distribution, sexuality and development of two species of *Ocyropsis*. *Mar. Biol.* **90**, 413–424.

Harbison, G. R., Madin, L. P., and Swanberg, N. R. (1978). On the natural history and distribution of oceanic ctenophores. *Deep-Sea Res.* **25**, 233–256.

Hernandez-Nicaise, M. L. (1984). Ctenophora. In *Biology of the Integument, Invertebrates* (J. Bereiter-Hahn, A. G. Matoltsy, and K. S. Richards, Eds.), Vol. 1, pp. 96–111. Springer-Verlag, New York.

Hernandez-Nicaise, M. L. (1991). Ctenophora. In *Microscopic Anatomy of Invertebrates: Volume 2—Placozoa, Porifera, Cnidaria, and Ctenophora* (R. W. Harrison and J. A. Westfall, Eds.). Wiley-Liss, New York.

Horridge, G. A. (1974). Recent studies on the Ctenophora. In *Coelenterate Biology Reviews and New Perspectives* (L. Muscatine and H. M. Lenhoff, Eds,), pp. 439–498. Academic Press, London.

Kremer, P., Canino, M. F., and Gilmer, R. W. (1986). Metabolism of epipelagic tropical ctenophores. *Mar. Biol.* **90**, 403–412.

Martindale, M. Q. (1987). Larval reproduction in the ctenophore *Mnemiopsis mccradyi* (order Lobata). *Mar. Biol.* 94, 409–414.

Matsumoto, G. I. (1991). Functional morphology and locomotion of the arctic ctenophore *Mertensia ovum* (Fabricius) (Tentaculata: Cydippida). *Sarsia* 76, 177–185.

Pianka, H. D. (1974). Ctenophora. In *Reproduction of Marine Invertebrates Vol. I* (A. C. Giese and J. S. Pearse, Eds.). Academic Press, New York.

Reeve, M. R., and Walter, M. A. (1978). Nutritional ecology of ctenophores: A review of recent research. In *Advances in Marine Biology* (R. S. Russell and C. M. Yonge, Eds.), Vol. 15, pp. 178–199. Academic Press, New York.

Stoecker, D. L., Verity, P. G., Michaels, A. E., and Davis, L. H. (1987). Feeding by larval and post-larval ctenophores on microzooplankton. *J. Plankton Res.* 9, 667–683.

Tamm, S. L. (1982). Ctenophora. In *Electrical Conduction and Behavior in "Simple" Invertebrates* (G. A. B. Shelton, Ed.), pp. 266–358. Oxford Univ. Press, Oxford.

Cyclophora

Peter Funch
University of Aarhus

Reinhardt Møbjerg Kristensen
University of Copenhagen

I. Description of Cycliophora
II. Proposed Life Cycle
III. Modes of Reproduction
IV. Modes of Sex
V. Anatomy of the Reproductive Systems
VI. Reproductive Physiology and Endocrine Control
VII. Modes of Fertilization
VIII. Modes of Development
IX. Larvae and Metamorphosis

GLOSSARY

buccal funnel A bell-shaped organ with a ciliated mouth ring used in filter feeding in some stages of Cycliophora.
chordoid larva The dispersal stage of cycliophorans.
commensal Of an organism, living in close association with another.
hypodermic insemination Copulation and sperm transfer through the skin; male penis penetrates the female integument; known from some rotifers.
inner bud A cluster of cells with the ability to regenerate new feeding structures and new stages in Cycliophora.
notochord A supportive, longitudinal rod dorsal to the nerve cord in chordates.
Pandora larva An asexually formed larva in Cycliophora involved in the colonization of the host mouthparts.

Cycliophora is a phylum consisting of microscopic commensals having a complicated life history with several attached and several free stages. All stages have the capacity of asexual reproduction, except the female and the secondary male. Otherwise, the knowledge of reproduction in cycliophorans is rather limited since the discovery of these marine animals was very recent. Their reproductive physiology, sex determination, mode of fertilization, and early development are unknown.

I. DESCRIPTION OF CYCLIOPHORA

Cycliophora was described as a phylum in 1995. It contains only one described species, *Symbion pandora*, living in a close association with the Norway lobster *Nephrops norvegicus*. What seems to be another undescribed species lives on the mouthparts of *Homarus gammarus* and *H. americanus*, and an undescribed genus of cycliophorans might live on harpacticoid copepods. Feeding stages of *S. pandora* are microscopic commensals (221–916 μm long) firmly attached to the mouthparts of the crustacean host. Free stages includes Pandora larva, primary males, females, and a chordoid larva. The primary male attaches to the trunk of the feeding stage, and one to three smaller secondary males develop inside the primary male.

II. PROPOSED LIFE CYCLE

A proposed life cycle is shown in Fig. 1, and the numbers in the following text refer to those used in

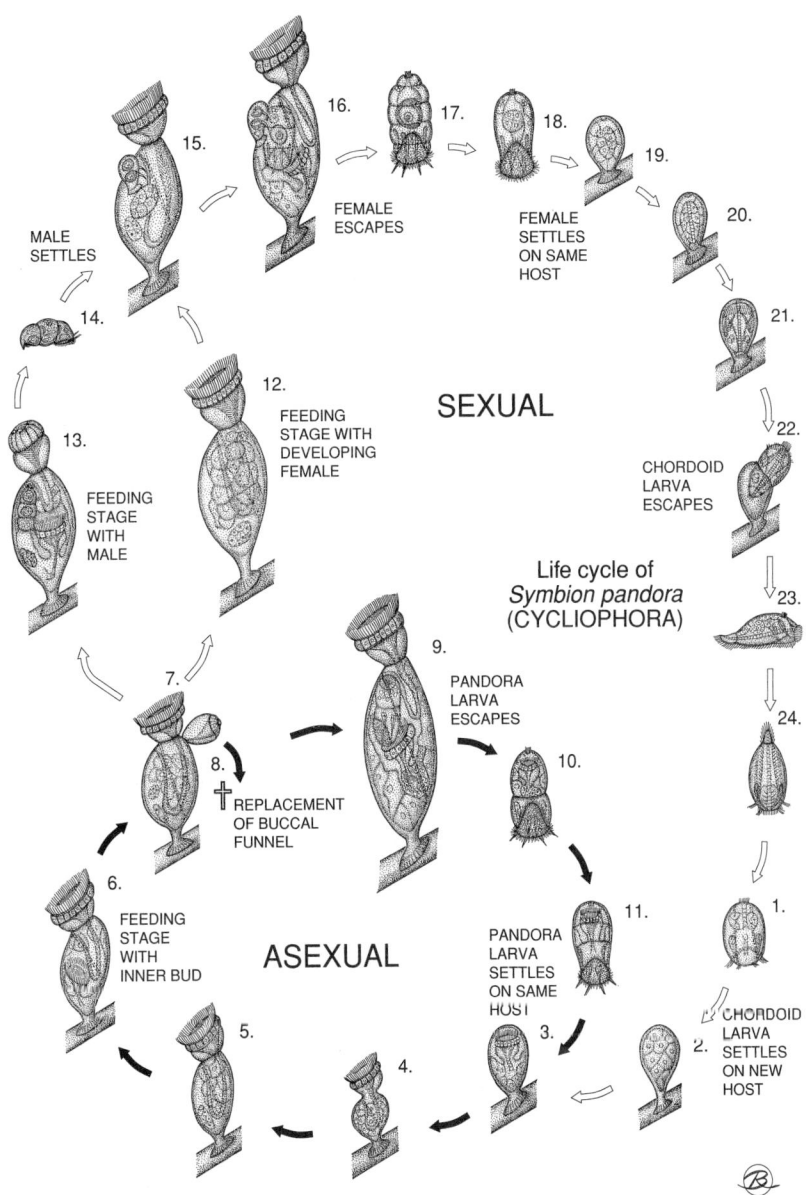

FIGURE 1 The proposed life cycle of *Symbion pandora* living on the Norway lobster, *Nephrops norvegicus*. For explanation, see Section II (reproduced from Funch and Kristensen, 1995).

the figure. The dominant stages in the life cycle are the feeding stages (4–9, 12 and 13, and 15 and 16), named so because they are the only stages with a digestive tract. A feeding stage is attached to the mouthparts of the host with an acellular stalk and adhesive disc. The trunk contains a U-shaped gut, and basal to this undifferentiated cells continually produce internal buds. These buds develop into new feeding structures and other organs (5 and 6). Eventually, a feeding stage also produces a Pandora larva asexually (9). Later, feeding stages switch to production of a primary male (13) or a sexually mature female (16). The Pandora larva (9), primary male (13), and female (16) are developed inside the feeding stages in a brood chamber. They are all short-lived and do not possess a gut. While inside the feeding stage, the Pandora larva develops a juvenile feeding stage from budding cells (9). A fully developed Pandora escapes from its maternal feeding stage (10), settles somewhere on the mouthparts of the lobster (11), and a new feeding stage develops from budding cells (4). When the primary male escapes (14), it seeks a feeding stage and attaches permanently (15). Some primary male organs then degenerate, but budding cells inside develop into several new individual, tiny, sexually mature males. Eventually, a female develops inside the feeding stage (16). The female develops a single oocyte while inside the feeding stage. This oocyte is fertilized by hypodermic insemination, during release of the female, or immediately after the release. Free females have a zygote inside (17). The female then settles on the mouthparts (19), degenerates, and an embryo develops to a chordoid larva (19–21). The chordoid larva escapes from the old cuticle of the female (22), seeks an appropriate site on the mouthparts of a lobster, settles (1), and degenerates (2), except for budding cells posteriorly, which develop into a new feeding stage (2–4).

III. MODES OF REPRODUCTION

Cycliophorans alternate between asexual and sexual reproduction (Fig. 1). Cycliophorans also regenerate several organs, producing new structures by inner budding. Such a regeneration could be regarded as an incomplete asexual reproduction since the regeneration involves the development of a new nervous system, but it does not result in a complete new individual. This is not the case in the production of a Pandora larva, a primary male, a secondary male, and a female. In all these cases new individuals arise from budding cells. The new feeding stages produced inside both the Pandora larva and the chordoid larva also develop by budding. In fact, the only stage in the life cycle which develops from the zygote is the chordoid larva, the product of sexual reproduction.

A. Asexual Reproduction

The Pandora larva, primary male, and female are produced inside one sessile feeding stage by asexual reproduction one at a time. Pandora larva formation seems to occur in young feeding stages, whereas primary male and female occur in older feeding stages. The development of all three stages is always accompanied by the development of new feeding structures (a new inner bud), which means that a feeding stage developing a free stage always produces two buds. One develops into the free stage and another into a new inner bud, which develops a new buccal funnel ("head"). The Pandora larva develops inside young feeding stages and facilitates the colonization of unexploited resources on the mouthparts of the host. One feeding stage is probably capable of producing several Pandora larvae over time, although only one Pandora larva develops at a time. In this way, large populations of *S. pandora* are established on the mouthparts of the host. At a certain point, feeding stages switch to production of primary males and females. It is unknown whether different feeding stages produce different free stages or if a single feeding stage is capable of producing all the free stages during time. One possible explanation is that the Pandora larva is produced inside a type A feeding stage originating from the chordoid larva. Type B feeding stages would then develop from feeding stages originating from Pandora larva and would be capable of producing sexually stages. However, this explanation does not seem to be correct. It is more likely that the same feeding stage is capable of producing three different stages over time, and that a succession occurs with Pandora larvae being developed first, then males, and finally females.

1. Pandora Larva

The Pandora larva is developed from undifferentiated budding cells located basally in the feeding stage near the stalk. In the early development of a Pandora larva, these budding cells divide and proliferate toward the anal region. The budding cells organize in two layers of cells, in between which develops a cuticle in small individual plates which later are fused and thereby delimit a fluid-filled brood chamber.

The undifferentiated cells closest to the anal region develop into a Pandora larva (Fig. 2), whereas other undifferentiated cells simultaneously differentiate into a new buccal funnel with digestive tract and other structures. This resembles the typical budding in a feeding stage, when no free stage develops inside.

In this brood chamber the anterior end of the Pandora larva and the mouth ring of the internal bud are recognized early. The brood chamber wall close to the outer body wall develops two rows of cilia of unknown function. A placenta-like structure with nutritive cells surrounds the posterior end of the Pandora larva. The posterior of the Pandora larva is rather undifferentiated, consisting of cells with a large, round nucleus with heterochromatin in the center. The anterior end hangs free in the lumen of the brood chamber, and early it is almost fully differentiated with a ventral ciliated disc, a dorsal brain, and numerous glands inside. The orientation of the Pandora larva inside the brood chamber is fixed, and the ventral ciliated disc is next to the two ciliary tufts of the brood chamber wall. As the Pandora larva grows, the feeding stage transports nutrients from the cells lining the gut and the large mesenchyme cells. These cells degenerate during development and disappear totally when the old buccal funnel is replaced by a new one. In the late development of the Pandora larva, a new juvenile feeding stage forms from posterior buds inside the Pandora larva. In this way, a free-swimming Pandora larva already carries a tiny new feeding stage inside (Figs. 2 and 3). A fully grown Pandora larva escapes with the posterior end first; it is squeezed out through the large slit-like anus. It has a length of 150–185 μm and uses the ventral ciliated disc to crawl. The anterior end is equipped with long sensilla. It has a brief free period, and then it settles on the mouthparts of the same host. Glue from adhesive glands is formed and is smeared over the cilia of the ventral disc, which becomes the adhesive disc of a new feeding stage. Most cells in the Pandora larva die, including those of the brain, and are used as nutrition for

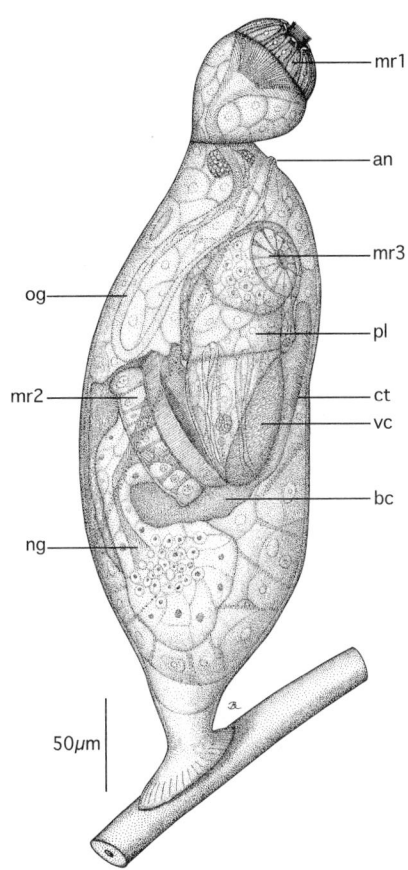

FIGURE 2 A young feeding stage of *Symbion pandora* which could be regarded as four individuals in one. The first is a feeding stage with a free mouth ring (mr1) containing the second, a Pandora larva (pl). The old gut (og) in the first feeding stage is degenerating in order to provide space for the developing Pandora larva and the third individual, developing from an inner bud. This inner bud has differentiated into a mouth ring (mr2), a ciliated buccal funnel, and an immature new gut (ng). The inner bud and the Pandora larva develop inside a common brood chamber (bc). Anteriorly, the Pandora larva has a ventral ciliated disc (vc), and inside the posterior end a bud is developing into the fourth individual, a juvenile feeding stage. At this early stage the mouth ring (mr3) can be recognized. an, anus; ct, cilia tufts (reproduced from Funch and Kristensen, 1997).

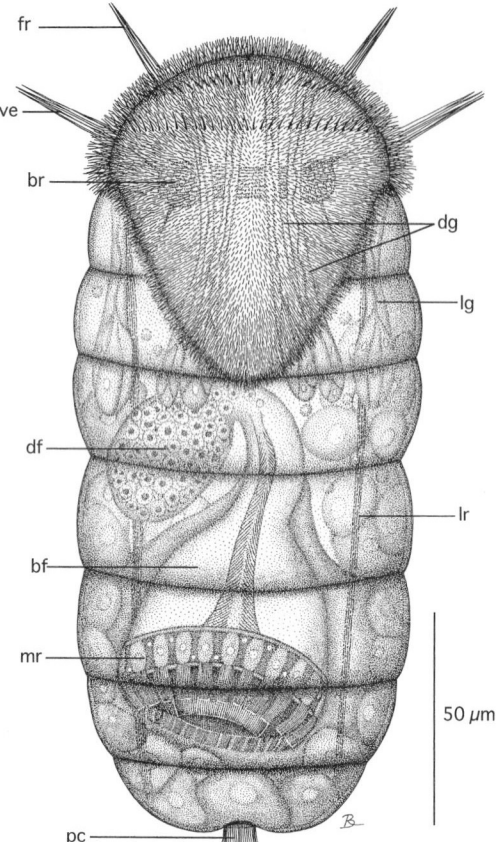

FIGURE 3 Free Pandora larva with a juvenile feeding stage inside. The buccal funnel (bf) and mouth ring (mr) of the developing feeding stage (df) are clearly recognized, but the gut is not fully developed. br, brain; dg, dorsal glands; fr, frontalia, frontal sensoria; lg, lateral glands; lr, lateral retractor muscle; pc, posterior cilia tuft; ve, ventralia, ventral sensoria (reproduced from Funch and Kristensen, 1997).

a manner similar to that of the Pandora larva. Males developing inside feeding stages have not been observed frequently, probably due to limited observations. In one particular population of brooding males and females, only about 2% of the feeding stages had males, probably because males are produced first. The development of a female resembles the development of the Pandora larva. Escaping males have not yet been observed.

3. Secondary Male

A special case of asexual reproduction is the production of sexually mature males. The dwarf primary

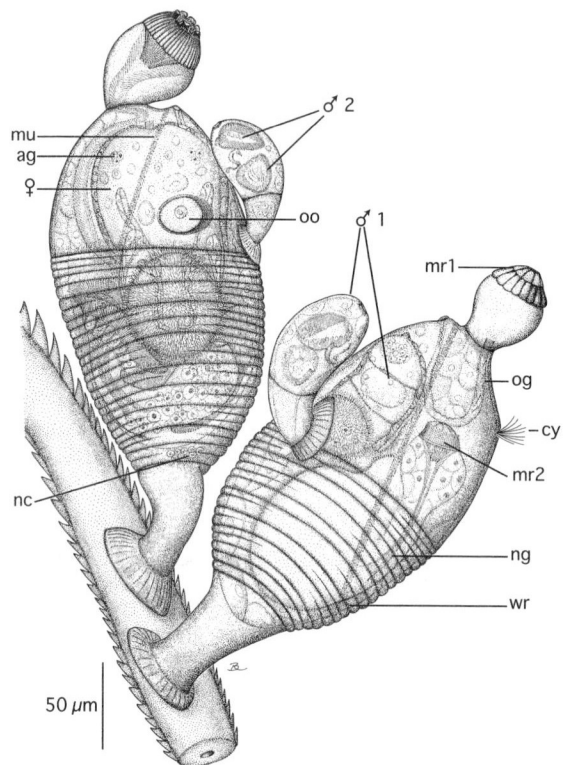

FIGURE 4 Two old feeding stages of *Symbion pandora* with wrinkles (wr) in their cuticle. One has a small primary male (♂1) in the brood chamber, and another has a larger female (♀) with an oocyte (oo). Both feeding stages have a primary male attached to the trunk. Secondary males (♂2) are developed inside the primary males. ag, accessory genital glands; cy, cyanobacteria; mr1, old mouth ring; mr2, new mouth ring; mu, muscle; nc, necrotic cells; ng, new gut; og, old gut (reproduced from Funch and Kristensen, 1997).

the developing new feeding stage inside. The mouth ring and buccal funnel are well differentiated, whereas the gut is still embryonic. The buccal funnel approaches the posterior end and an anal pore forms next to the existing posterior pore which has lost its cilia. This is where the buccal funnel is pushed out, and when this occurs a new feeding stage is complete.

2. Development of Female and Primary Male

The female and primary male (Fig. 4) participate in the sexual cycle, but they are formed asexually in male situated on a feeding stage contains one or two secondary males (Fig. 4, ♂2), and in some cases three tiny males. These tiny secondary males develop asexually from buds in the attached male. Each secondary male develops a brain, a testis, and a cuticular S-shaped penis.

B. Sexual Reproduction and the Chordoid Larva

Little is known about the sexual part of the life cycle in Cycliophora. At some point sperm from the secondary male fertilize the egg inside the female. The female then escapes from the maternal feeding stages, and after a short free existence it settles on the mouthparts of the lobster. Following settlement, the female degenerates. However, its adhesive disc and cuticle persist and form a cyst, inside which the chordoid larva develops, nourished by the dying maternal cells. The chordoid larva (156–206 μm long) is named after a characteristic notochord-like structure composed of 40–50 stacked muscle cells. The chordoid larva probably develops sexually from the zygote formed inside the female. It has no gut and nothing is known about the uptake of nutrients. A free chordoid larva is a much better swimmer than the Pandora larva and the female and is thought to be a dispersal stage for the cycliophorans.

IV. MODES OF SEX

Cycliophorans are dioecious since males and females are separated individuals, but both males and females develop from buds inside feeding stages of unknown sex. Some feeding stages contain no other stages inside, and such individuals are probably juveniles. Some feeding stages contains one Pandora larva, others one primary male, and still others one female. It is not known if a sessile feeding stage which releases a Pandora larva has the capacity to later develop a male or female. Furthermore, the fate of a feeding stage which released a female is unknown. In other words, it is unknown if such a sessile stage could switch back to asexual production of a Pandora larva.

V. ANATOMY OF THE REPRODUCTIVE SYSTEMS

Both the primary male and the female of *S. pandora* develop inside a feeding stage in a manner similar to that of the Pandora larva.

A. Male Reproductive System

The male (58–138 μm long) found in the brood chamber of the feeding stage and the newly settled male lack penises and other differentiated reproductive organs. It seems that these males never obtain sexual maturity and should be considered as the first generation of males or primary males. Primary males escape from the feeding stage, are free for a moment, and then settle on the trunk of a feeding stage. Most attached males contain a group of undifferentiated cells close to the persisting attachment disc. These cells were first interpreted as a testis, but it is more likely that these cells are homologous to the undifferentiated cells present in the asexual stages with capacity of budding. These cells might have the capacity of budding which results in independent secondary males. Some males are almost empty of living cells, whereas others contain one, two, or three tiny secondary males inside. In fact, what appears to be one individual is really several. These secondary tiny males clearly contain spermatids, spermatozoa, and penis-like structures. The tiny sexually mature males (about 30 μm long) are a second male generation, developing inside the primary male by budding, and several organs of the primary male degenerate shortly after settlement on a feeding stage. The secondary male is developed in a fluid-filled cavity delimited by a very thin cuticle, resembling the budding in a brood chamber in other stages of the life cycle. Early formation of internal cuticular plates as observed in the formation of a Pandora larva has not been observed in attached males, which could indicate that the interpretation of independent secondary males is wrong. On the other hand, ultrastructural studies of several attached males clearly demonstrate that each mature secondary male is situated freely in the chamber and has its own brain, penis, and ciliary structures. At this point, there is no cell contact with the primary male. This supports the interpretation of two generations of males.

Spermiogenesis takes place in the secondary male, but the details have not been studied. Spermatids develop in tetrads, and narrow cytoplasmic bridges connect four daughter spermatids to each other. The fully formed spermatozoon seems to be filiform with an elongated nucleus (about 4 μm in length), a middle piece with a osmiophobic vesicle and a totally transformed mitochondrial sheath, and a long flagellum with an axoneme. The axoneme exhibits the standard 9 + 2 arrangement of microtubules and seems to spiral around the nucleus, at least in the early stages of spermiogenesis. Anterior to the nucleus a small invagination is present which might be an acrosome. The dimorphic spermatozoa shown in rotiferan males are probably absent in *S. pandora*, but a more detailed study of the spermiogenesis in *S. pandora* is certainly needed.

Each mature secondary male has one penial structure. It is an S-shaped tube of epicuticle. The diameter of the lumen is as wide as 2 μm at the base but <0.4 μm distally. The base of the penis is located in a cuticular sheath connecting the penis with the general integument. A complete system of cross-striated muscles attaches to the base of the penial structure via intervening epithelial cells. It is likely that the muscles protrude the penial structure, and some kind of copulation with the female occurs. However, it is unknown when or how this occurs. The secondary males contain several glands with osmiophilic granules that have gland ducts terminating close to the penial structure.

B. Female Reproductive System

The oogenesis in Cycliophora is not known. An ovary has not been observed, and all females (155–190 μm long) observed develop a single oocyte with a diameter between 22 and 34 μm while they are still inside the brood chamber of the feeding stage. The oocyte is not surrounded by a chorion. In fact, the female reproductive system is rather simple. A well-defined oviduct is absent. A pair of accessory gland cells might be associated with the female reproductive system. These glands consist of three cells and are observed early in the development of the female (Fig. 4, ag). A ciliated pore is situated posteriorly and might serve as the site for sperm entrance during copulation, if the oocyte is not fertilized, while the female is inside the feeding stage. In free-swimming females the egg is situated in a tiny egg cavity.

VI. REPRODUCTIVE PHYSIOLOGY AND ENDOCRINE CONTROL

The control of reproduction in cyliophorans is totally unknown. It seems that the age of the feeding stage determines which stage is developed inside the brood chamber. Pandora larvae are only encountered in feeding stages attached to younger and smaller lobsters. Young lobsters often shed their cuticle, which reduces the age of the epizoans living on them. These feeding stages have a clean, smooth, and relatively thin cuticle (Fig. 2). Feeding stages with Pandora larvae have active buccal funnels just before release of the Pandora. In contrast, feeding stages with a male or a female inside are found on older, larger lobsters, and the cuticle of the feeding stages is rather thick, wrinkled, and overgrown with epizoic bacteria and algae (Fig. 4). In such feeding stages the buccal funnel is often degenerated or missing. In feeding stages containing a male or female, brownish cell debris are often present in the basal trunk or stalk. Such necrotic cells are not seen in feeding stages with a Pandora larva inside. It is possible that males develop earlier than females; if so, this could explain the low abundance of males inside feeding stages in one particular population in which only 2% of the feeding stages with sexual stages had males. It is possible that the presence of primary males on a feeding stage induces it to switch from production of a male to production of a female. The fact that males and females are common on older lobsters indicates a coordinated control with the condition of the host.

The control mechanisms of switching to a production of the sexual stages are unknown, but males and females are encountered on hosts with larger populations of *S. pandora*. Perhaps crowding triggers this change or maybe the host molting hormones are involved, as is known for the bryozoan

Triticella koreni, living as epibiont on the crustecean *Calocaris*.

VII. MODES OF FERTILIZATION

The female is fertilized by males probably by some kind of copulation. This has not been observed, however, but the complexity of the male reproductive organs, such as the presence of a cuticular tube, indicates the presence of a copulation. This tube has been interpreted as a penis; due to the facts that the tube has several muscles attached to it and that it communicates between the inside containing the sperm and the outside. It appears that the egg is already fertilized in free-swimming females. This leaves three opportunities for copulation: (i) while the female is inside the brood chamber in the feeding stage, (ii) during birth, or (iii) right after the release. If copulation takes place while the female is still inside, the penises must be capable of penetrating a rather long distance through two layers of primary male and feeding stage cuticles (the feeding stage cuticle is actually two layers since the wall of the brooding chamber has a cuticle of its own). When the tip of the penis has reached the brood chamber, it is possible that the sperm is released. Then the function of the cilia in the brood chamber wall might be aiding the sperm to reach a pore in the posterior end of the female. This pore is probably the last part of the female to be formed, ensuring the sperm reach a fully mature egg. Another possibility is that the penis also penetrates the integument of the female and releases sperm inside the female. A similar kind of hypodermic insemination is well-known from many rotifers. Primary males prefer to settle nearby the anus (cloaca) of the feeding stage, which is where the female escapes during birth. At the onset of female release the posterior end emerges first. The posterior end has a pore which seems to communicate with a duct leading to the egg. The posterior part of an escaping female always comes in close contact with the posterior end of the settled primary male, and this end of the male contains the most developed sexually mature dwarf male inside as well as a pore. It is therefore possible that copulation could happen during escape of the female when the posterior end of the female comes into close contact with a mature male. Another possibility is that the escape of the female triggers a simultaneously release of a mature dwarf male, and that copulation takes place between two sexually mature free stages.

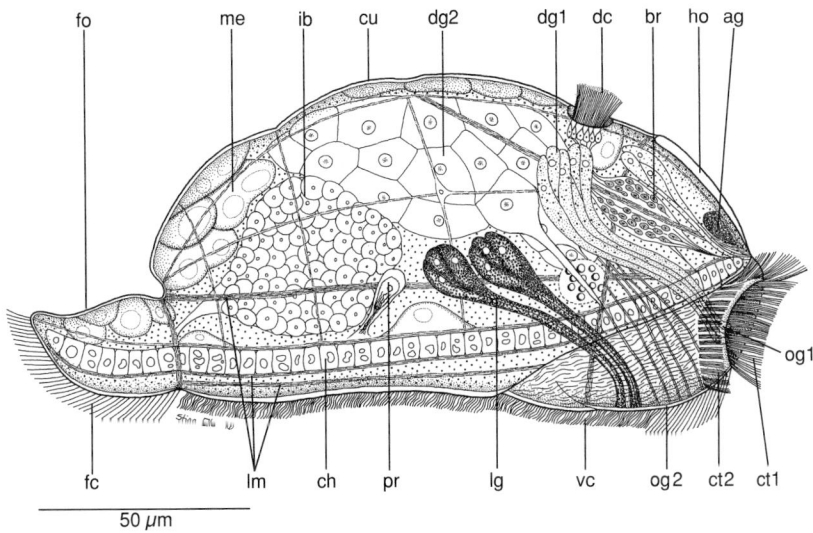

FIGURE 5 Lateral view of a free chordoid larva of *Symbion pandora*. The ventral ciliation consists of two anterior bands (ct1 and ct2), a large ventral ciliated field (vc), and foot cilia (fc). The characteristic chordoid organ (ch) is situated ventrally. ag, anterior glands; br, brain; cu, cuticle; dc, dorsal ciliated organ; dg1 and dg2, dorsal glands; fo, foot; ho, hood of rigid cuticle; ib, inner bud; lg, lateral gland; lm, longitudinal muscles; me, mesenchyme; og1 and og2, outlet complexes of glands; pr, protonephridia (reproduced from Funch, 1996).

VIII. MODES OF DEVELOPMENT

The early embryology of Cycliophora is almost unknown. The first cleavage was observed in one instance in a free-swimming female of *S. pandora*. The single zygote inside divided once, and then the female died. It is therefore not possible to determine which cleavage type occurs in Cycliophora. The female attaches to the mouthparts of the host with the anterior ventral ciliated disc by secreting an adhesive substance. Then the female itself degenerates, while the embryo inside develops at the expense of the female cells. This results in a chordoid cyst which is a fully developed chordoid larva encased in the female body cuticle. This stage rests until a sudden change in the condition of the host, such as death or molting.

IX. LARVAE AND METAMORPHOSIS

The chordoid larva develops from the zygote and it may be homologous to protostomian larvae, such as a trochophora. This is reflected in the ciliation of the chordoid larva (Fig. 5). It possesses two anterior ciliated bands, a ventral ciliated field, and a ciliated foot. This pattern of cilia could correspond to the prototroch and metatroch, gastrotroch, and telotroch, respectively. The ventral ciliated field consists of compound cilia in contrast to the gastrotroch of trochophores. The chordoid larva is only ciliated ventrally; the rest of the integument has an apical cuticle. It is acoel and contains many bottle-shaped gland cells. It has a pair of protonephridia and does not have a digestive tract. A unique feature of the chordoid larva is the chordoid organ. The chordoid organ is a longitudinal rod of similar muscles cells. Each muscle cell has a large central vacuole and contractile fibers in the periphery. It functions as support and counteracts the contraction of the longitudinal muscles. Although it functions as the notochord of cephalochordates, it has a different structure and is not homologous with a notochord. The release of the chordoid larva from the cyst is coordinated with molting or death of the host. Metamorphosis of the chordoid larva has not been studied.

Bibliography

Funch, P. (1996). The chordoid larva of *Symbion pandora* (Cycliophora) is a modified trochophora. *J. Morphol.* 230, 231–263.

Funch, P., and Kristensen, R. M. (1995). Cycliophora is a new phylum with affinities to Entoprocta and Ectoprocta. *Nature* 378, 711–714.

Funch, P., and Kristensen, R. M. (1997). Cycliophora. In *Microscopic Anatomy of Invertebrates, Vol. 13, Lophophorates, Entoprocta, and Cycliophora* (F. W. Harrison and R. M. Woollacott, Eds.), pp. 409–474. Wiley-Liss, New York.

Winnepenninckx, B. M. H., Backeljau, T., and Kristensen, R. M. (1998). Relations of the new phylum Cycliophora. *Nature* 393, 636–638.

Cyclostomes

see Agnatha

Cytochrome P450

see Aromatization

Cytokines

Sarah A. Robertson

The University of Adelaide

I. The Molecular Nature of Cytokines
II. The Biological Actions of Cytokines
III. Cytokine Receptors and Signal Transduction
IV. Regulation of Cytokine Production and Activity
V. Cytokines and Reproduction
VI. The Role of Cytokines in the Ovary
VII. The Role of Cytokines in the Endometrium
VIII. The Role of Cytokines in Pregnancy and Parturition
IX. Reproductive Success in Genetically Cytokine-Deficient Mice

GLOSSARY

angiogenesis The process of capillary formation, composed of the proliferation, differentiation, and migration of endothelial cells.

antigen-presenting cells Cells including macrophages and dendritic cells which can initiate protective or destructive immune responses, depending on their cytokine environment.

chemokine A cytokine with chemotactic properties, e.g., activity in the regulation of target cell motility.

csfm^{op}/csfm^{op} mice A strain of mice with severely depleted tissue macrophages due to a genetic deficiency in colony-stimulating factor-1.

cytokine A soluble protein or glycoprotein which acts through specific cell surface receptors to regulate the survival, proliferation, differentiation, or function of a target cell.

cytokine antagonist Soluble cytokine receptors or non-signal-transducing cytokines that block the interaction between a cytokine and its specific cell surface receptor.

cytokine receptor A membrane-bound protein or glycoprotein which binds a specific cytokine and triggers intracellular signal transduction.

immune deviation The process by which the immune system is biased toward a TH1 or TH2 cytokine pattern or a protective or destructive response.

large granular lymphocytes Phenotypically unique lymphocytes which have roles in cytokine secretion and immunoregulation in the decidua.

macrophages Lymphohematopoietic cells derived from blood monocytes which can differentiate and activate into diverse phenotypes within tissues, depending on their cytokine microenvironment.

signal transduction The biochemical pathway by which cells send signals from cytokine receptors on the cell surface to the nucleus.

soluble cytokine receptor An isoform of a cytokine receptor that is secreted from the cell surface.

TH1/TH2 cytokines Mutually antagonistic groups of cytokines classified on the basis of their association with or capacity to favor immune deviation toward cell-mediated and inflammatory immunity (TH1) or antibody-mediated humoral immunity (TH2).

Cytokines are small glycoproteins which mediate communication between cells. They are distinguished from other growth factors by having been originally identified, described, and named for their actions on lymphohemopoietic cells, but are now known to act on, and to be produced by, cells of many diverse lineages. Cytokines regulate cell growth, differentiation, activation, and movement by binding to specific receptors on the surface of the target cell. Multiple, often overlapping, functions can be attributed to individual cytokines, with the precise response induced depending on the nature of the target cell and the local concentration of other cytokines and regulatory molecules, such as cytokine antagonists. In reproductive tract tissues, where many of the events are reminiscent of inflammatory and reparative processes, the cytokine axis has emerged as a fundamental and inextricable constituent of local intercellular communication networks. Under the dominion of sex steroid hormones, cytokines produced by resident and infiltrating cells in the endometrium, ovary, and testes have key roles in orchestrating the tissue remodeling and immunological events central to ovulation, spermatogenesis, embryo implantation and development of the placenta, parturition, and menstruation.

I. THE MOLECULAR NATURE OF CYTOKINES

A. Structure

Cytokines are typically low in molecular weight (<80 kDa) but otherwise comprise an extremely heterogeneous array of glycoproteins that are highly diverse in structure and amino acid sequence. The carbohydrates associated with the protein core are not necessary for functional activity but can be influential in determining the half-life of the molecule *in vivo*. Most cytokines are secreted into the extracellular fluid, but some can be sequestered into reservoirs within the extracellular matrix through binding with glycosaminoglycans or produced in alternative isoforms which can be anchored to the cell membrane.

B. Nomenclature

Cytokines are usually named after the biological function for which they were first discovered, and so their nomenclature can be misleading and even anomalous. On the basis of structural similarities, gene organization, chromosomal location, and receptor usage they have now been classified into six families: the hematopoietins [including the colony-stimulating factors (CSFs), interferons (IFNs), and most of the interleukins]; epidermal growth factors; β-trefoils [fibroblast growth factors and interleukin (IL)-1]; the tumor necrosis factors (TNFs); the cysteine knot cytokines [including the transforming growth factor (TGF)-β family]; and the chemokines (Table 1).

II. THE BIOLOGICAL ACTIONS OF CYTOKINES

A. Cytokines and Cell–Cell Communication

The cytokines were originally identified, described, and named for their roles in mediating communication between cells of the lymphohemopoietic system, but they are now recognized to have a much wider range of activities since many diverse cell lineages produce and respond to these molecules. Thus, cytokines comprise a subdivision of the larger growth factor family, which together with neurotransmitters and hormones comprise the body's armory of soluble intercellular signaling agents. Despite their structural diversity, this family of molecules shares a number of common properties that contrast with those of the

TABLE 1
Structural Families of Cytokines and Receptors Implicated in Reproductive Tract Tissues

Family	Members	Receptor type
Hematopoietins	IL-2, IL-3, IL-4, IL-5, IL-6, IL-11	
	GM-CSF, LIF, G-CSF	Hematopoietin receptor type I
	IL-10, IFN-α, IFN-β, IFN-γ, IFN-τ	Hematopoietin receptor type II
	CSF-1	Tyrosine kinase
β-Trefoil	IL-1α, IL-1β, IL-1Rα	IL-1 receptor
β-Jelly roll	TNF-α, TNF-β, fas ligand	TNF receptor
Cysteine knot	TGF-β1, TGF-β2, TGF-β3	Serine/threonine kinase
	PDGF	Tyrosine kinase
Chemokines	IL-8, MIP-1α, MIP-1β, etc.	G-protein-coupled superfamily

[a] Abbreviations used: PDGF, platelet-derived growth factor; MIP, macrophage inhibitory protein. Other abbreviations are given in the text.

TABLE 2
Properties and Functions of Cytokines versus Endocrine Hormones

Property	Sex steroid hormones	Cytokines
Cellular origins	Few	Many
Biological redundancy	Low	High
Biological pleiotrophy	Low	High
Presence in the circulation	Yes	Rarely
Sphere of influence	Endocrine (distant)	Autocrine, paracrine, juxtacrine (local)
Inducers	Gonadotrophins, cytokines	Sex steroid hormones, cytokines
Influence of microenvironment on biological effect	Low	High
Effector action in target cells	Often indirect	Usually direct
Receptor location	Nuclear	Cell surface

endocrine hormones (Table 2). Generally, cytokines are produced in relatively small quantities and exert their actions at nanomolar or picomolar concentrations. Together with their transient existence conferred by their short half-life, this usually restricts their sphere of influence to autocrine, paracrine, or juxtacrine actions within the immediate neighborhood of production. In this way, cytokines provide the signals which promote or inhibit multiple aspects of cell behavior, including survival, proliferation, and reversible and irreversible transitions in phenotype, secretory profile, and motility. Such events are an integral component of normal physiological processes in all tissues but are particularly evident during tissue remodeling, be it associated with developmental programming, homeostasis, or unscheduled events such as inflammation, infection, injury, and healing. During acute pathophysiological states, local measures may be supplemented by endocrine effects; the release of cytokines such as IL-1, TNF-α, and IL-6 into the circulation contributes to fever through inducing production of acute phase proteins and temperature elevation.

B. Pleiotrophy and Redundancy

Cytokines are notable for the considerable degree of overlap in activity between individual family members. This functional redundancy is clearly born out by the surprising lack of severe phenotypic effects caused by disruption of many cytokine genes in mice. Also, cytokines are remarkably promiscuous and pleiotrophic in their actions; that is, they elicit extraordinarily diverse and sometimes apparently opposite responses in a range of target cells.

C. The Importance of "Context"

The response of a cell to a given cytokine is dependent not only on its lineage and differentiation state but also on the local microenvironment, most notably the concentration of other cytokines and growth factors, and the extracellular matrix. Thus, cytokines do not work in isolation but rather interact within a network to amplify, modulate, or antagonize each other's activities. Appreciation of the importance of context in cytokine signaling has led to the view that they are best regarded as individual elements of a code or alphabet, with the ultimate reaction of a cell depending on the sum total of signals converging at the cell surface. This notion limits the extent to which one can extrapolate from data concerning cytokine activity generated from *in vitro* experiments since it is virtually impossible to replicate the full range of environmental influences to which a cell is exposed *in vivo*.

III. CYTOKINE RECEPTORS AND SIGNAL TRANSDUCTION

Cytokines do not act as effector molecules directly, but rather bind to specific, high-affinity receptors in the cytoplasmic membrane on the surface of the target cell. Binding triggers a cascade of intracellular events which cause changes in the pattern of messenger RNA expression and protein synthesis, in turn ultimately leading to altered cell behavior. Cytokine receptors are expressed at low numbers (10–10,000 per cell) and the fact that occupancy rates of <10% can elicit a response accounts for the extreme potency of cytokines as biological mediators.

A. Receptor Superfamilies

Although the amino acid sequence homology between cytokine receptors is low, similarities in their secondary and tertiary structure and biochemical function provide evidence for a common ancestry between disparate cytokine–cytokine receptor pairs and allow their classification into superfamilies including the hematopoietin receptor family, the TNF receptor family, the receptor kinase families, and the G-protein-coupled receptor family (Table 1). Similarities are common even across these boundaries due to sharing of similar elements drawn from a wide range of modular structures or domains. For example, the receptors for IL-1, IL-6, and CSF-1 all contain one or more immunoglobulin superfamily domains. The largest receptor superfamily is the hematopoietin superfamily, which includes the majority of the receptors for the interleukins, interferons, and other hematopoietic cytokines.

B. Intracellular Signaling and Functional Redundancy

The initial event in signal transduction for all cytokine receptors other than the G-protein-coupled receptors is ligand-mediated interaction between receptor subunits to form multimeric complexes. This facilitates passage of information from the extracellular domain to the cytoplasmic environment without the necessity for conformational change being transmitted directly through the membrane and triggers kinase activity to initiate one of a number of possible "second messenger" cascades which amplify and redirect the binding signal to eventually culminate in altered gene expression within the cell nucleus. Cytokine receptors belonging to the receptor kinase family by definition have intrinsic tyrosine or serine/threonine kinase activity. Cytoplasmic tyrosine kinases of the Janus kinase (JAK) family catalyze events initiated by members of the hematopoietin receptor superfamily, where the intracellular regions have no intrinsic kinase activity or other obvious catalytic function.

The functional cross-reactivity and redundancy characteristic of cytokines is accounted for by their capacity to induce or activate common signaling pathways within the target cell. This is achieved through the capacity of subunits shared by different cytokine receptors to engage common intracellular signal-transducing moieties, such as the Stat family of proteins. Alternatively, the capacity of disparate cytokines to induce common outcomes in target cells can, in some cases, be explained by the sharing of common signal-transducing accessory molecules in the formation of heterodimeric complexes which act to bind ligand with higher affinity than the receptor alone (Fig. 1).

IV. REGULATION OF CYTOKINE PRODUCTION AND ACTIVITY

A. Inducibility of Cytokine Synthesis

Cytokines are not constitutively expressed, and their production depends on appropriate stimulation of the cell, usually through binding of one of a variety of agents such as immunoglobulin, cytokines, bacterial products, or, in the case of reproductive tract tissues, ovarian steroid hormones. The cytokines induced depend on the nature both of the cell and of the stimulus, such that different cell lineages can produce different combinations of cytokines in response to a given stimulus, presumably through the use of specific intracellular signaling pathways. Even within a given cell type, distinct patterns of cytokine expression are evoked after activation of different intracellular pathways specific for various stimuli. Upregulation of production of cytokine proteins is

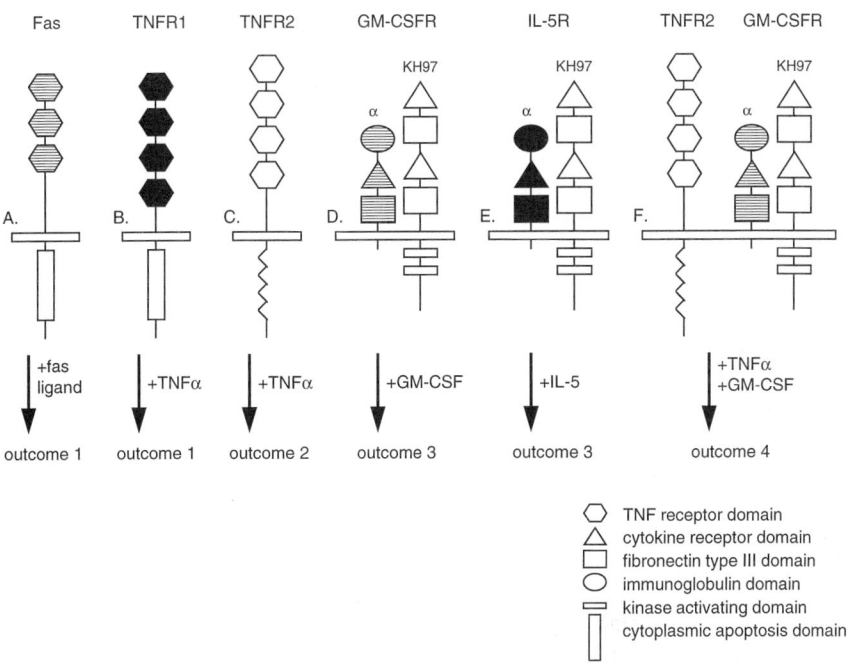

FIGURE 1 The significance of cytokine receptors in cytokine redundancy and pleiotrophy. (A and B) The capacity of two different cytokines (fas ligand and TNF-α) to evoke the same outcome (e.g., apoptosis) in a target cell by signal transduction through a cytoplasmic domain common to each receptor. (B and C) The capacity of one cytokine (TNF-α) to achieve each of two alternative outcomes (e.g., apoptosis or activation), depending on which of two receptors (TNFR1 or TNFR2) are activated. (D and E) The capacity of two different cytokines (GM-CSF and IL-5) to evoke the same outcome (e.g., proliferation) through shared usage of a common, signal-transducing β chain (KH97). (D and F) The modulation of the cellular response to a given cytokine (e.g., GM-CSF) by other cytokines in the microenvironment (e.g., TNF-α).

in most instances achieved by increased *de novo* synthesis of cytokine mRNA and sometimes also by an increase in the stability of the mRNA. In unusual cases, more rapid induction of cytokine release may occur by upregulated translation of already abundant cytoplasmic mRNA (e.g., with induction of TNF-α in macrophages) or by release of protein stored within cytoplasmic granules (such as occurs in eosinophils and basophils).

B. Quenching the Cytokine Response

A variety of mechanisms exist to limit the duration and spread of a cytokine response. This ensures that the potent effects of cytokine release are confined to the immediate vicinity of the producing cell. When cell stimulation diminishes, a reduction in transcription rate and rapid degradation of mature transcripts rapidly follows. The number and affinity of specific receptors for a given cytokine in target cells are often sensitive to cytokine concentration, with downregulation achieved through a reduction in the rate of synthesis or by internalization and subsequent degradation of receptor–ligand complexes. Targeted secretion of cytokine toward the direction of the eliciting stimulus has been demonstrated in leukocytes as well as naturally polarized epithelial and endothelial cells, and it limits the inadvertent activation of bystander cells.

C. Cytokine Antagonists and Inhibitors

The activities of cytokines, particularly those involved in provoking inflammatory reactions, are counterbalanced and modulated both by other cytokines that oppose their effects and by naturally occurring cytokine antagonists. These factors act to restore homeostasis and prevent extensive tissue destruction that would result if an inflammatory response was left unchecked. IL-4 and IL-10 are good examples of inhibitory cytokines that downregulate expression of proinflammatory cytokines, including IL-1, IL-6, IL-8, and TNF-α. Cytokine inhibitors act by interfering with binding of a cytokine to its cell surface receptor. They may take the form of either soluble receptors which compete with surface receptors by binding cytokine, for example, IL-4 or TNF-α-soluble receptors, or may be non-signal-transducing isoforms of the cytokine, such as the IL-1 receptor antagonist. Cytokine antagonists and soluble receptors are usually products of the same gene that encodes the active molecule but arise from alternatively spliced mRNA transcripts. The existence of these factors highlights the importance of the contingency of the microenvironmental context on cytokine signaling.

V. CYTOKINES AND REPRODUCTION

A. Cytokines and Reproductive Function

The normal functioning of the female reproductive tract tissues necessitates cycles of tissue growth, restructuring, and breakdown that are unparalleled in other tissues. Since many aspects of these events resemble the destructive and reparative events associated with immune responses and wound healing, it was hypothesized and has been proven that many of the same cytokines are involved. Indeed, the intrinsic nature of cytokines, particularly their capacity to be produced transiently, act locally, and to be rapidly quenched, means that they are ideally placed to act as local mediators of reproductive processes.

The specific roles of cytokines in various reproductive events are discussed in subsequent sections of this article. The intention is not to provide a comprehensive review (for which the reader is directed to the bibliography) but rather to give selected examples which illustrate common principles of cytokine function. Indeed, our knowledge of the cytokine biology of the reproductive tract is growing at a rapid rate and it is certain that many new cytokines and new actions of existing cytokines will be discovered in the coming years.

B. Interlinking the Reproductive–Immune–Neuroendocrine Axes

It is becoming increasingly clear that cytokines are a fundamental component of a ubiquitous cell–cell communication language employed by every cell in all tissues and organs of the body, including the brain. This common language provides the means for interaction between the reproductive, immune, and neuroendocrine systems at the cellular level and explains the plethora of experimental evidence showing that manipulation of one of these axes can have dramatic repercussions in each of the others. Examples include the steroid hormone-driven dimorphism of the immune response in the two genders and the effects of immunological insults or stress on pregnancy outcome.

C. The Immunology of Reproduction

One common feature shared by the ovary, testes, and uterus is that each of these tissues is charged with nurturing the growth and development of cells (gametes or the conceptus) that are essentially "foreign" to the body. This requires that the immune system servicing each of these tissues is endowed with the capacity to discriminate between these and other potentially pathogenic or otherwise dangerous agents and is able to initiate immune responses of a character and extent appropriate to each stimulus. Interestingly, an important role for the rich complexity of cytokines found in reproductive tissues may be to support this special immunological attribute. Cytokines including GM-CSF, CSF-1, TGF-β, IL-10, IL-4, and IL-12 are all implicated in the phenotypic regulation of antigen presenting cells, which are of paramount importance in the selective induction of permissive or antagonistic immune responses through their discretionary capacity to take up and process antigen, migrate to draining lymph nodes, and express surface molecules involved in the differential activation of various regulatory T lymphocyte populations. All these cytokines are expressed at various times and sites in reproductive tract tissues, and further studies will determine the extent to which their changing patterns of expression are related to immune function.

D. Angiogenesis

Reproductive events are also characterized by a requirement for profound changes in the local vascular architecture that is unsurpassed in other tissues in the body. Angiogenesis is an obligatory aspect of follicular growth and rupture, formation of the corpus luteum, endometrial growth and menstruation, and development of the placenta. Although principally regulated by specific growth factors, the activities of endothelial cells during angiogenesis can be influenced by a variety of cytokines that are expressed at appropriate times and locations within reproductive tract tissues, including TNF-α, TGF-β, IL-6, GM-CSF, and G-CSF. Other aspects of endothelial cell function are also responsive to these cytokines. These include alterations in vascular tone, which lead to endometrial edema at estrus, and the upregulation of endothelial cell surface molecules that assist the cyclic recruitment of leukocytes into the endometrium and ovary. Furthermore, although the complex changes in maternal vessels that preempt trophoblast invasion during placental development are not well understood, it is clear that maternal leukocytes contribute to these changes and that they may also be mediated through cytokine release.

E. Steroid Hormone Regulation of Cytokine Expression

Another special aspect of the biology of cytokines in reproductive tract tissues is their fundamental role, together with other growth factors, in mediating local effects of steroid hormones. This is by virtue of the remarkable capacity of epithelial and stromal cells within the endometrium, ovaries, and testes to respond to ovarian or testicular sex steroid stimulation with the production of a diverse array of cytokines. Regulation is at the transcriptional level and is specific for individual cytokines; for example, endometrial epithelial cell expression of CSF-1 and leukemia inhibitory factor (LIF) requires the synergistic action of estrogen and progesterone, whereas GM-CSF expression is dependent on estrogen but inhibited by progesterone. The means by which steroid hormones control cytokine synthesis is not yet clear, and the potential importance of steroid response elements in cytokine genes as well as indirect pathways, involving autocrine or paracrine mediators, is currently being explored. Factors other than steroid hormones, including cytokines, growth factors, prostaglandins and other agents derived from leukocytes, semen, or the conceptus, can further modulate cytokine expression. In this way it is possible to achieve a multitude of diverse and dynamic cytokine microenvironments within the reproductive tract.

VI. THE ROLE OF CYTOKINES IN THE OVARY

The actions of gonadotrophins and sex steroids in the ovary are mediated or attenuated at the local level by peptide growth factors including cytokines. The selective growth and development of follicles, release of the oocyte at ovulation, and the genesis and demise of the corpus luteum are all events that involve complex cooperative interactions between somatic cells, various leukocyte lineages, and the oocyte to achieve considerable remodeling of the tissue architecture over short time periods. Several cytokines, including IL-1, IL-2, IL-6, CSF-1, GM-CSF, TNF-α, and IFN-γ, have been detected in ovarian tissue and fluids from various species. Although reasonably comprehensive analyses of the physiological roles of IL-1 and TNF-α have emerged in recent years, in many cases a full understanding of the actions of these factors awaits detailed descriptions of receptor expression and biological activities in the relevant target cells. One notable aspect of ovarian cytokine production is the clear spatial and temporal association between leukocyte infiltration and activation within compartments of the tissue undergoing change and fluctuations in local cytokine synthesis. This has led to the notion of a cytokine loop, in which cytokines originating from somatic cells (theca and granulosa cells) act to recruit and regulate the behavior of leukocytes within the ovary, which in turn release other cytokines and enzymes that can influence the differentiation, structural association, and secretory behavior of the somatic cells.

A. Cytokines and Ovulation

During follicular growth and development, IL-1 and TNF-α appear to act on theca and granulosa cells to stimulate cell division, prevent inappropriate release of progesterone, and inhibit premature follicular rupture through regulation of plasminogen activator activity. A role in the growth of new blood vessels supplying the developing follicle is also likely. The expression of IL-1 and TNF-α is further increased at ovulation, in which a contribution to breakdown of the follicle wall is suggested by their capacity to promote prostaglandin synthesis. Additional roles in the release of collagenase and reactive oxygen intermediates have been postulated. Synthesis of these cytokines has been localized to the developing oocyte itself but more significantly to the theca–interstitial area where infiltrating macrophages reside, steadily increasing in number as ovulation draws closer. Indeed, an important role for these cells, particularly in the breakdown of the follicle wall at ovulation, is indicated by the finding of dysregulated ovulation in $csfm^{op}/csfm^{op}$ mice. Apart from CSF-1, the precise factors involved in the functional regulation of ovarian leukocytes have not been defined, but various chemokines known to recruit macrophages and neutrophils into tissues, including IL-8 and macrophage chemotactic protein (MCP)-1, are temporally associated with follicular development and ovulation, as are cytokines such as GM-CSF and CSF-1, which are implicated as regulators of differentiation and activation of leukocytes within the tissue. A diagrammatic illustration of some roles of cytokines in the ovary is given in Fig. 2.

B. Cytokines and Corpus Luteum Function and Regression

Ovulation is followed by rapid differentiation of granulosa–luteal cells, intense local angiogenic activity, and further infiltration of macrophages within the ruptured follicle to culminate in formation of the corpus luteum. TNF-α and IL-1, presumed to be of

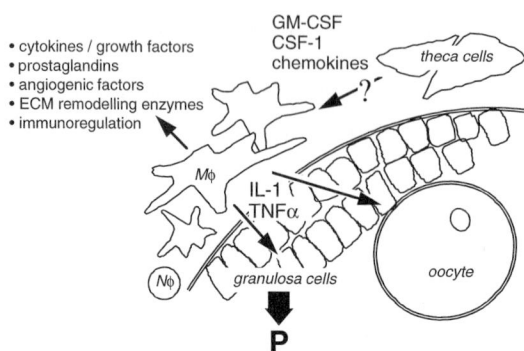

FIGURE 2 Origins of cytokines and targets for their action in the periovulatory ovary. mφ, macrophage; Nφ, neutrophils; P, progesterone.

macrophage origin, can act to promote progesterone secretion and enhance luteal cell proliferation. Both of these cytokines, together with IL-6 and possibly GM-CSF, are also implicated in the dramatic proliferation of endothelial cells early in corpus luteum development. Further changes in the local cytokine–leukocyte axis accompany the regression phase. T lymphocytes may have a role in the functional and structural demise of luteal cells through their release of antisteroidogenic and possibly cytotoxic cytokines, including IL-2 and IFN-γ, and by mediating apoptosis. Macrophages and/or dendritic cells expressing MHC class II molecules steadily increase in number during the later phases of luteal life and may act in an immunoregulatory capacity as well as to phagocytose cellular debris.

C. Cytokines and the Testes

Cytokines have integral roles in the processes of spermatogenesis and testosterone synthesis in the testes. Most notably, Sertoli cells are a potent source of IL-1α, and the finding of IL-1 receptors in spermatogonia indicates a likely role in the regulation of spermatogenesis. Macrophages comprise a substantial proportion of the cells in the interstium and produce an array of cytokines, including TNF-α, IL-1, GM-CSF, and IL-6. A likely role for these cells in regulating testosterone production, presumably mediated at least in part through cytokine release, is suggested by their close spatial and developmental association with Leydig cells and recent findings of dysregulated steroidogenesis in macrophage-deficient $csfm^{op}/csfm^{op}$ mice.

VII. THE ROLE OF CYTOKINES IN THE ENDOMETRIUM

A. Proliferation and Differentiation of Endometrial Cells

The mammalian endometrium undergoes cycles of dramatic growth and regression driven by fluctuations in steroid hormones released by the ovary. During the growth or proliferative phase, estrogen produced by the developing ovarian follicle promotes extensive proliferation of endometrial epithelial, mesenchymal, and endothelial cells. After ovulation the corpus luteum secretes progesterone, which acts to inhibit proliferation and transform the endometrium into a secretory tissue in anticipation of embryo implantation.

Of the various cells in the uterus, the luminal and glandular epithelial cells have been identified as an especially potent source of a variety of cytokines, with the precise temporal patterns of cytokine release being regulated primarily by ovarian steroid hormones. Leukocytes recruited into the endometrial tissues, particularly macrophages and to a lesser extent lymphocytes, are also a major source of a variety of cytokines, the precise combinations of which are dependent on activation phenotype of the cell which, in turn, is also related to cycle stage. Potential functions for various cytokines are implied by findings of expression of the relevant receptors, with *in vivo* and/or *in vitro* evidence for biological effects on endometrial cell behavior. A diagrammatic summary of some of cytokine networks operating in the endometrium is presented in Fig. 3.

Although epidermal growth factor and insulin-like growth factor-1 (described elsewhere) appear to be the principal stimulators of epithelial cell proliferation, cytokines including IFN-γ and TGF-β are implicated in the postovulatory switch to secretory epithelium. Both of these cytokines are induced by progesterone, bind to specific receptors on epithelial cells, and inhibit epithelial cell proliferation *in vitro*. Spatial differences in cytokine secretion may be related to microenvironmental differences in the tissue architecture; for example, in the human endometrium, IFN-γ produced by scattered foci of lymphoid aggregates acts to induce MHC class II expression in adjacent glandular epithelial cells.

Stromal cells including fibroblasts and endothelial cells also provide potential targets for cytokine action throughout the menstrual cycle. IL-1, TNF-α, CSF-1, and GM-CSF have been suggested to influence the secretion of prostaglandins and cytokines such as IL-6 and SCF as well as the differentiation state of stromal fibroblasts. Recent experiments in IL-11 receptor knockout mice suggest a critical role for this cytokine in the decidual response. Angiogenesis as well as leukocyte chemoattraction and the increase in vascular permeability which leads to edema can each be influenced by cytokines known to be secreted by uterine epithelial cells and leukocytes.

B. Menstruation

In the absence of a pregnancy, endometrial remodeling culminates in a process of regression involving atrophy and death of endometrial cells, with degradation of the extracellular matrix. In menstruating primates, including humans, the final stages of the secretory phase are marked by partial degradation and shedding of the superficial endometrium. The bio-

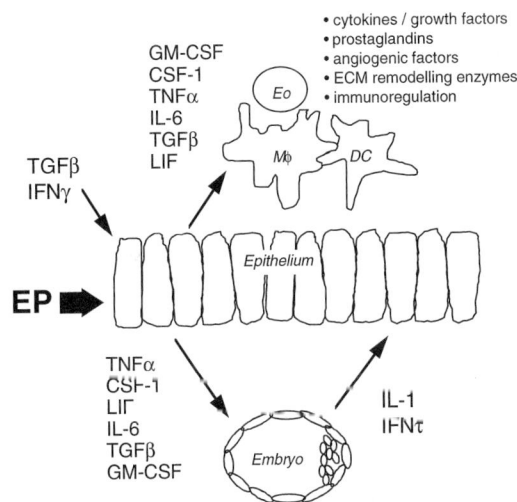

FIGURE 3 Origins of cytokines and targets for their action in the endometrium during early pregnancy. mφ, macrophage; E_o, eosinophils; DC, dendritic cells; EP, estrogen + progesterone.

chemical mechanisms leading up to menstruation are not well understood but clearly involve extensive degradation of the stromal extracellular matrix and a steady increase in apoptosis of the epithelial lining. Cytokines, particularly TNF-α and IL-1, have been implicated in the regulation of production of the matrix metalloproteinase enzymes which catalyze matrix breakdown. TNF-α, the expression of which increases during the secretory phase, is likely to be an important mediator of epithelial cell decline through its actions as an inhibitor of growth and as a potent mediator of apoptosis.

C. Leukocyte Recruitment and Phenotypic Regulation

Leukocytes recruited from the blood are abundant in the endometrium of both humans and rodents and are particularly important target cells for the actions of endometrial cytokines. Indeed, each of the cytokines known to be made by uterine epithelial cells was originally described as a regulator of the recruitment and/or behavior of myeloid leukocytes. The dramatic fluctuations in their composition, distribution, and activation phenotype which occur in concert with structural changes through the menstrual cycle are clearly orchestrated by the changing patterns of cytokines emanating from uterine epithelial cells.

During the proliferative phase, infiltrating leukocytes are predominantly of the macrophage and dendritic cell lineages which accumulate within the endometrial stroma forming a network immediately subjacent to the epithelial surface. Studies in mice have been particularly illuminating in unraveling the roles of individual factors in recruiting and regulating the behavior of these cells. Most notably, a clear role for CSF-1 in maintaining macrophages within the tissue has been established using genetically CSF-1-deficient $csfm^{op}/csfm^{op}$ mice. Further activities for a variety of chemokines, including RANTES and MCP-1 as chemotactic agents, as well as for GM-CSF in promoting the differentiation and activation of macrophages and dendritic cells have been established.

Estrogen also causes the recruitment of large numbers of eosinophils into the endometrium, where they become distributed throughout the stromal tissue, and chemokines including RANTES and eotaxin, as well as GM-CSF, are potential regulators of these cells. T and B lymphocytes are also present, but they are fewer in number and their precise lineages and phenotype vary between species. In women, progesterone is associated with an increase in the infiltration of phenotypically unique large granular lymphocytes (LGLs) which accumulate near glands and spiral arteries, but the factors associated with the infiltration and regulation of these cells remain to be identified.

The physiological functions of each of these leukocyte populations are not clear, but they are likely to participate in immune surveillance of the mucosal surface and, importantly, in immunological adjustments required to accommodate the conceptus at implantation. Nonimmunological functions are also possible; for example, eosinophils contain abundant TGF-α and could have a role in epithelial cell growth. A role for macrophages in endometrial tissue remodeling, particularly at menstruation, is suggested by the extraordinary capacity of these cells to produce an array of secretory products including proinflammatory cytokines, prostaglandins, histamine, and extracellular matrix-degrading enzymes. These secretory functions are all controlled by microenvironment signals, particularly cytokines.

D. The Postmating Inflammatory Response

Studies in mice have led to the novel finding that the introduction of semen into the reproductive tract elicits an inflammation-like response in the estrogen-primed uterine endometrium. Epithelial cytokines are intimately associated with this response. The initiating agent appears to be TGF-β and possibly other components of the seminal plasma which act to upregulate the synthesis of GM-CSF and IL-6 by uterine epithelial cells. Epithelial GM-CSF as well as various chemokines which are also induced during the response are key factors in the subsequent dramatic infiltration of the underlying uterine stroma and lumen with inflammatory leukocytes, predominantly neutrophils and macrophages. These cells become activated to release IL-1 and TNF-α and to increase their expression of molecules such as MHC class II involved in antigen presentation. Seminal plasma is already identified as having an immunomodulatory role in reproductive processes, and these observations indicate a further role for this complex fluid in providing a "priming" stimulus to the preimplantation uterus. A comparable response occurs at the cervix in women and other species, in which semen does not pass into the uterine cavity. Studies in rodents and other species have shown that while not obligatory, this response can contribute to pregnancy success, perhaps through initiating maternal immune responses that act to protect the conceptus from rejection at implantation.

VIII. THE ROLE OF CYTOKINES IN PREGNANCY AND PARTURITION

A. Preimplantation Embryo Development

Cytokines originating from the oviduct and uterine epithelium under the influence of ovarian steroid hormones are thought to exert a regulatory influence on development of the preimplantation embryo as it traverses the reproductive tract. Through both positive and negative effects on the timing and extent of proliferation and differentiation, these factors have an important role in synchronizing embryo growth with the maternal changes that lead to uterine receptivity. This concept has emerged from studies in a number of species, including human, showing that the preimplantation embryo produces and responds to a range of growth factors including cytokines. Their physiological role is illustrated by findings that embryo development is retarded or arrested *in vitro* and that to some extent this can be offset by culturing embryos with exogenous growth factors or in the presence of other cells. Furthermore, culture of embryos *ex vivo* for even a short period can have detrimental effects on pregnancy outcome in some species. Many cytokines which can exert regulatory effects on embryo development have been identified, and in most cases cognate receptors are synthesized by the embryo. Notably, these factors include all those cytokines known to be produced by the uterine and oviductal epithelia during early pregnancy. The actions of cytokines are dependent on the developmental stage of the embryo and may also vary somewhat between species and even among strains of mice. LIF, CSF-1, GM-CSF, and TGF-β improve secretory activity and/or the proportion of embryos that develop to the blastocyst stage and beyond in rodent, livestock, and human species. Attainment of an "adhesive" phenotype in trophectoderm cells at implantation is a further differentiation event likely to be influenced by cytokines. Conversely, TNF-α and IL-6 have inhibitory influences on inner cell mass proliferation and embryo attachment and outgrowth, respectively.

The embryo itself synthesizes cytokines, including IL-1, LIF, TGF-β, and IL-6, which potentially have roles in autocrine pathways or provide signals to maternal tissues. An excellent example is IFN-τ, produced by the preimplantation embryo in sheep and cattle, which acts as a mediator of maternal recognition of pregnancy through preventing regression of the corpus luteum. Embryo-derived IL-1 may also have a key role in signaling ensuing implantation since blocking its activity with antagonist can lead to implantation failure in mice.

B. Implantation and Decidualization

Embryo implantation is a particularly fragile event that is highly susceptible to disruption. The importance of the cytokine milieu at implantation is highlighted by the dramatic consequences of even minor perturbations to local or systemic cytokine levels. For example, mice bearing CSF-1-secreting tumors or animals injected with very small amounts of CSF-1 on the day of implantation have high rates of implantation failure and fetal resorption. Although the precise roles of CSF-1 and others among the array of cytokines known to be synthesized in the endometrium at implantation are not well understood, the events that comprise the earliest stages of pregnancy involve cytokine-responsive cells and are potentially influenced by cytokine levels.

The capacity of the embryo to attach and invade at implantation depends largely on the "receptivity" of the endometrial tissue. This develops during the luteal phase of the menstrual cycle under the paramount regulation of ovarian steroid hormones, mediated at the local level by growth factors and cytokines synthesized predominantly by epithelial cells and infiltrating leukocytes. LIF and IL-11 are two cytokines, produced in epithelial cells and stromal cells, respectively, which have essential roles in implantation, through regulating endometrial stromal cell differentiation into decidual cells. Other cytokines including CSF-1, TNF-α, and TGF-β are induced just prior to implantation and, together with IL-6 and GM-CSF, are implicated but not proven participants in embryo attachment, the decidual response, and the subsequent increase in vascular permeability and proliferation of trophoblast cells.

C. Placental Growth and Development

As pregnancy proceeds, the potential roles for cytokines expand in parallel with the increasing structural complexity of the interacting uterine and placental tissues. In the endometrium, stromal fibroblast cells at the placental–maternal interface continue to proliferate and large granular lymphocytes (LGLs) accumulate and differentiate into cells capable of producing an array of cytokines, including LIF, CSF-1, and GM-CSF. The complex and multilayered placenta and fetal membranes develop spatially ordered compartments with differing invasive, adhesive, secretory, and antigenic properties. There is abundant evidence indicating that both placental and decidual tissues actively participate in local cytokine networks by secreting copious quantities of various cytokines and responding to many of the cytokines that are released by uterine cells.

Many cytokines have emerged as potential regulators of proliferation, differentiation, and secretory function of placental trophoblast cells. In most instances the source of these cytokines has been localized to cells in both the maternal decidua, particularly LGLs and other leukocytes, and the placenta itself, including mesenchymal macrophages and sometimes trophoblast cells. Spatial and temporal patterns of cytokine and receptor expression on adjacent cell layers suggest precise roles for individual cytokines in the sequential differentiation and growth of the different cell types that make up the placenta. For example, the temporal pattern of CSF-1 receptor (c-*fms*) expression in trophoblast cells is clearly consistent with CSF-1 having a significant role in the regulation of early placental development, particularly in promoting trophoblast invasion and differentiation. Proliferating cytotrophoblast cells also express receptors for GM-CSF, which together with CSF-1 has been implicated by *in vitro* studies in the regulation of placental hormone secretion and may promote differentiation into hormonally active villous syncytiotrophoblast cells. In contrast, TGF-β and LIF appear to have roles as important effectors of the alternate pathway for differentiation of cytotrophoblasts into extravillous anchoring cells, which are less invasive and have a diminished capacity to secrete placental hormones. Placental tissues contain a plethora of other cytokines, including TNF-α, IFN-γ, IL-1, IL-6, G-CSF, and stem cell factor, all of which are likely to contribute in a stage-specific manner to the commitment and proliferation of the various trophoblast cell lineages. Further complexity is afforded by the finding of abundant and diverse cytokine regulatory molecules, including soluble receptors for TNF-α, GM-CSF, and unusual isoforms of cytokines such as CSF-1 and IL-2. Comprehensive studies of cytokine receptor expression, together with better methods for *in vitro* culture of and discrimination between various populations of trophoblast cells, will clarify the roles of cytokines in regulating the different trophoblast lineages that comprise the developing and mature placenta.

D. Maternal Immune Deviation during Pregnancy

One critical aspect of mammalian pregnancy is a transition in the maternal immune system which allows it to tolerate the foreign conceptus. The mechanisms underlying this are not fully understood but clearly involve complex interactions between cytokines of placental origin, maternal immune cells, and pregnancy hormones. Importantly, the placenta and placental membranes are now recognized to synthesize certain cytokines in abundance, including TGF-β, IL-4, and IL-10, which appear to have local and systemic inhibitory effects on the capacity of the mother to initiate certain types of cell-mediated immune responses. Further restrictions on the capacity of any aberrant effector cells to harm the placenta are blocked by the abundant secretion of these same cytokines by decidual leukocytes, particularly LGLs. This may be a means by which the fetal–placental unit blocks potentially harmful (TH1) responses and thereby promotes its own survival. Cytokines which oppose this immune deviation (IL-2, IFN-γ, TNF-α, and TNF-β) are not usually abundant in uterine or placental tissues and their exogenous administration can be detrimental to pregnancy outcome. Further reinforcement of this immune deviation is mediated through progesterone, which acts, through the induction of a factor called progesterone-induced blocking factor, to influence the cytokine output of any regulatory T lymphocytes activated during pregnancy.

E. Fetal Growth and Development

A largely uninvestigated role for cytokines of maternal or placental origin is in the exogenous regulation of various aspects of fetal development. An increasing number of studies implicate a role for amniotic fluid cytokines, emanating from the placental membranes, in various aspects of gut and lung development in the fetus. Cytokines of maternal origin can also cross into the fetal circulation where they presumably could influence fetal hemopoiesis and development of the immune system. This aspect of cytokine function in pregnancy is certain to become an area for increased research activity.

F. Parturition

Increased expression of certain cytokines in the placental membranes and cervix has been implicated in the initiation and propagation of labor. In contrast to earlier stages of gestation, those that feature most prominently are cytokines usually associated with inflammatory responses, including IL-6, TNF-α, IL-1, IFN-γ, and IL-8. This inversion in cytokine profile suggests that parturition reflects the natural termination of the protective maternal immune response which has nurtured the fetus *in utero*. The mechanism of action of these molecules remains obscure, but roles for IL-6 in the elevation of prostaglandin E$_2$ secretion and for IL-8 in the recruitment of neutrophils associated with cervical ripening seem likely. Inappropriate expression of these cytokines is also associated with preterm labor resulting from intrauterine bacterial infection.

IX. REPRODUCTIVE SUCCESS IN GENETICALLY CYTOKINE-DEFICIENT MICE

Studies employing transgenic mice lacking individual growth factors clearly indicate that embryonic development to implantation stage and through to birth can proceed in the absence of most cytokines, including CSF-1, GM-CSF, IFN-γ, IL-2, IL-4, IL-6, IL-10, and TNF receptor-1 and embryo-derived TGF-β and SCF. The most severe effects on reproductive performance appear in the absence of maternal LIF, which appears to be essential for implantation of the embryo (described elsewhere) and CSF-1, the absence of which causes multiple lesions in the functioning of the ovaries, testes, and uterus, presumably because of the critical role for CSF-1 in regulating macrophage recruitment and survival in these tissues. IL-11 has also been identified as a cytokine necessary for successful implantation and placental development: in this case, deficiency leads to dysregulated decidual cell transformation and overgrowth of placental giant cells. However, the degree to which individual deficiencies cause sublethal aberrations in development is unknown in most cases and will be difficult to measure, particularly when the additional impact of environmental circumstances is considered.

This somewhat surprising lack of phenotypic consequences of cytokine deletion concurs with the general conclusion that few cytokines are absolutely essential for individual cellular functions in any tissue of the body, presumably because there are very few biological responses that cannot be induced by several cytokines. Rather than suggesting that these cytokines are unimportant, experiments in cytokine knockout mice indicate that critical cellular functions in the reproductive tract are usually backed up in a "fail-safe" mechanism by which the loss of one cytokine can be compensated for by others with similar activities.

See Also the Following Articles

Corpus Luteum; Immunology of Reproduction; Implantation; Trophoblasts to Human Placenta

Bibliography

Brannstrom, M., and Norman, R. J. (1993). Involvement of leukocytes and cytokines in the ovulatory process and corpus luteum function. *Hum. Reprod.* **8**, 1762–1775.

Burger, D., and Dayer, J.-M. (1995). Inhibitory cytokines and cytokine inhibitors. *Neurology* **45**, S39–S43.

Callard, R. E., and Gearing A. J. H. (1994). *The Cytokines Facts Book*. Academic Press, London.

Daiter, E., and Pollard, J. W. (1992). Colony stimulating factor-1 (CSF-1) in pregnancy. *Reprod. Med. Rev.* **1**, 83–97.

Jägou, B., Cudicini, C., Gomez, E., and Stäphan, J. P. (1995). Interleukin-1, interleukin-6 and the germ cell–Sertoli cell cross-talk. *Reprod. Fertil. Dev.* **7**, 723–730.

Hunt, J. S., and Robertson, S. A. (1996). Uterine macrophages and environmental programming for pregnancy success. *J. Reprod. Immunol.* **32**, 1–25.

Nathan, C., and Sporn, M. (1991). Cytokines in context. *J. Cell Biol.* **113**, 981–986.

Nicola, N. A. (1994). *Guidebook to Cytokines and Their Receptors*. Oxford Univ. Press, Oxford, UK.

Pampfer, S., Arceci, R. J., and Pollard, J. W. (1991). Role of colony stimulating factor-1 (CSF-1) and other lymphohematopoietic growth factors in mouse pre-implantation development. *Bioessays* **13**, 535–540.

Pollard, J. W. (1991). Regulation of polypeptide growth factor synthesis and growth factor-related gene expression in the rat and mouse uterus before and after implantation. *J. Reprod. Fertil.* **88**, 721–731.

Robertson, S. A., Seamark, R. F., Guilbert, L. J., and Wegmann, T. G. (1994). The role of cytokines in gestation. *Crit. Rev. Immunol.* **14**, 239–292.

Sharkey, A. (1995). Cytokines and embryo/endometrial interactions. *Reprod. Med. Rev.* **4**, 87–100.

Tabibzadeh, S. (1991). Human endometrium: An active site of cytokine production and action. *Endocr. Rev.* **12**, 272–290.

Tabibzadeh, S. (1994). Cytokines and the hypothalamic–pituitary–ovarian–endometrial axis. *Hum. Reprod. Update* **9**, 947–967.

Terranova, P. F., and Montgomery Rice, V. (1997). Review: Cytokine involvement in ovarian processes. *Am. J. Reprod. Immunol.* **37**, 50–63.

Wegmann, T. G., Lin, H., Guilbert, L. J., and Mossman, T. G. (1993). Bidirectional cytokine interactions in the maternal–fetal relationship: Is successful pregnancy a TH2 phenomenon? *Immunol. Today* **14**, 353–356.

Decidua

Juan C. Irwin and Linda C. Giudice

Stanford University

I. Introduction
II. Decidualization
III. Anatomical and Cellular Aspects
IV. Autocrine/Paracrine Interactions
V. Summary

GLOSSARY

decidua The endometrium of pregnancy.

decidual cell A differentiated endometrial stromal cell of the decidua.

decidualization The process of transformation of the endometrium into the morphologically and functionally distinct decidual tissue, involving the differentiation of endometrial cells and infiltration by large numbers of lymphoid cells.

endometrium The tissue lining the cavity of the uterus composed of columnar epithelial cells, forming a simple glandular epithelium, and an underlying stroma, containing the stromal cells and extracellular matrix, as well as blood vessels and lymphoid cells.

implantation The process by which the embryo establishes an intimate connection with the maternal tissues, involving the penetration of the embryo through the endometrial epithelium and its invasion into the underlying stroma.

placenta An embryonically derived organ that connects the fetus to the uterine wall and its blood supply; it separates maternal and fetal blood mediating the exchange of molecules between them and serves as a major source of hormones during pregnancy.

trophoblast A specialized placental cell that comes in direct contact with the maternal blood and uterine tissues and produces steroid and peptide hormones throughout gestation; its active invasion of maternal tissues anchors the placenta to the uterine wall and creates the modified blood vessels of the placental bed.

The decidua is a maternally derived intrauterine tissue of pregnancy that originates as the result of endometrial differentiation and derives its name from the fact that part of it is shed at parturition. The extent of histological transformation that takes place in the endometrium as it differentiates into decidua indicates a high degree and diversity of cellular specialization in this tissue and suggests an important role of the decidua in pregnancy. A variety of functions have been attributed to the decidua, including controlling trophoblast invasion, serving as a cleavage zone for placental separation, providing nutrition to the embryo, being a source of hormones, and serving an immunoregulatory role during pregnancy. With the emergence of new molecular and cellular technological tools, and of new concepts in other areas of biology, advances in decidual biology have led to the expansion of our knowledge and concepts of the nature and physiological role of the decidual tissue and its various cellular components. Because reproductive strategies can differ widely among mammals, even between closely related species, this article focuses primarily on available information from human studies, with occasional reference to relevant animal data.

I. INTRODUCTION

During pregnancy the human endometrium undergoes a unique process of differentiation to constitute the decidua, a morphologically and functionally distinct tissue that persists throughout gestation, representing the maternal aspect of the maternal–fetal interface. Because of its strategic anatomical location the decidua lies in intimate contact with the developing embryo during the early stages of gestation and with the placenta and fetal membranes later in pregnancy. Consequently, the decidua is the tissue wherein maternal and embryonic cells actually meet and, as such, a site where information is exchanged between them. This cross-talk, believed to be critical for the successful establishment/maintenance of pregnancy, is established largely by means of secreted products. These products include signaling molecules, such as hormones, growth factors, and cytokines, and structural and effector molecules, such as extracellular matrix components, growth factor-binding proteins, proteases, and protease inhibitors. The maternal–fetal dialogue is also mediated by means of cellular receptor molecules which enable maternal and fetal cells to sense incoming signals and to interact with the surrounding cellular and extracellular environment. The decidua, therefore, is the tissue in which this intricate network unfolds throughout pregnancy to maintain the delicate balance that ensures the successful completion of gestation.

II. DECIDUALIZATION

The process of transformation of endometrium into decidua affects all the cell types present in the uterine mucosa. The end result is the formation of a morphologically and functionally distinct tissue as a consequence of endometrial cell differentiation and the infiltration by large numbers of lymphoid cells. The morphological changes that characterize decidualization are initiated independently of conception during the late secretory phase of normal ovulatory cycles. The process of cyclic endometrial development that culminates with decidualization, begins immediately following menses and the ensuing endometrial regeneration. Initially, there is intense proliferation of both epithelial and stromal cells under the influence of rising levels of circulating estrogen during the first half of the cycle or proliferative phase. At midcycle, ovulation takes place and ovarian progesterone production is initiated. During the second half of the cycle or secretory phase, progesterone action on the endometrium will suppress proliferation and induce endometrial differentiation. Epithelial differentiation, characterized by glandular secretion, is rapidly apparent, being maximal 5–7 days following ovulation. However, stromal cell differentiation, foreshadowing the cellular changes associated with the gestational decidua, is only evident 9 or 10 days after ovulation. Morphological decidualization is accompanied by the acquisition of a distinct biosynthetic and secretory phenotype by the endometrial stromal cells. Decidualized stromal cells synthesize and secrete prolactin and insulin-like growth factor binding protein-1 (IGFBP-1), express desmin intermediate size cytoskeletal filaments, and produce extracellular matrix components typical of gestational decidual cells. Cyclic development of the endometrial vasculature also peaks at the time of endometrial decidualization. The growth of the endometrial spiral arteries that had started in the proliferative phase continues throughout the secretory phase, leading to their maximal development toward the end of the cycle. In nonconception cycles, a monthlong process of endometrial development comes to an abrupt end with the onset of endometrial breakdown and bleeding at menstruation. In the event of pregnancy, decidual changes become more extensive throughout the functionalis, giving rise to the three layers of the decidual tissue, and further regional differentiation, depending on the localization of the implantation site, leads to the establishment of the distinct anatomical regions of the decidua.

III. ANATOMICAL AND CELLULAR ASPECTS

A. Regional Differentiation

The maternal decidua is the endometrium of the pregnant uterus and thus constitutes an anatomical interface between the placenta and fetal membranes and the adjacent uterine myometrium (Fig. 1). It comprises three distinct anatomical regions: (i) the decidua basalis, which forms at the site of implantation to lie directly beneath the placenta, being the maternal tissue invaded by placental trophoblast

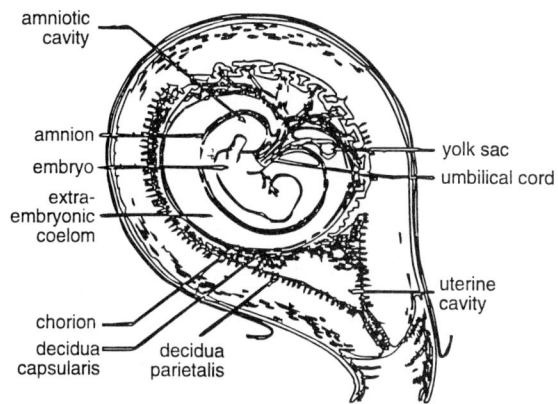

FIGURE 1 Schematic diagram of intrauterine tissues in early human gestation (from Martina et al., J. Clin. Endocrinol. Metab. 82, 1894–1898, 1997).

cells; (ii) the decidua capsularis, which overlies the gestational sac as it projects into the uterine lumen; and (iii) the decidua parietalis, which lines the remainder of the uterine cavity (Fig. 1). With the expansion of the conceptus, by the end of the fourth month of pregnancy, the decidua capsularis atrophies and the fetal membrane surrounding the conceptus (chorion leave) eventually makes contact and fuses with the decidua parietalis, with degeneration of the uterine surface epithelium and obliteration of the uterine lumen. Three layers can be discerned in the decidual tissue: the superficial compact zone (zona compacta), the middle spongy zone (zona spongiosa), and the basal zone (zona basalis) adjacent to the myometrium (Fig. 2). The compacta and spongiosa constitute the functional zone (zona functionalis) which is shed at parturition, whereas the undifferentiated basalis remains after delivery and serves as the seed for endometrial regeneration.

B. Decidual Cell Populations

The spongy layer consists of distended glands with abundant secretion, separated by a narrow rim of nondecidualized stroma. As gestation advances decidual glands atrophy and are scarce at term. The compact layer, containing decidualized stroma with attenuated nonsecretory glands, is composed of closely packed large differentiated endometrial stromal cells, with characteristic epithelioid polygonal

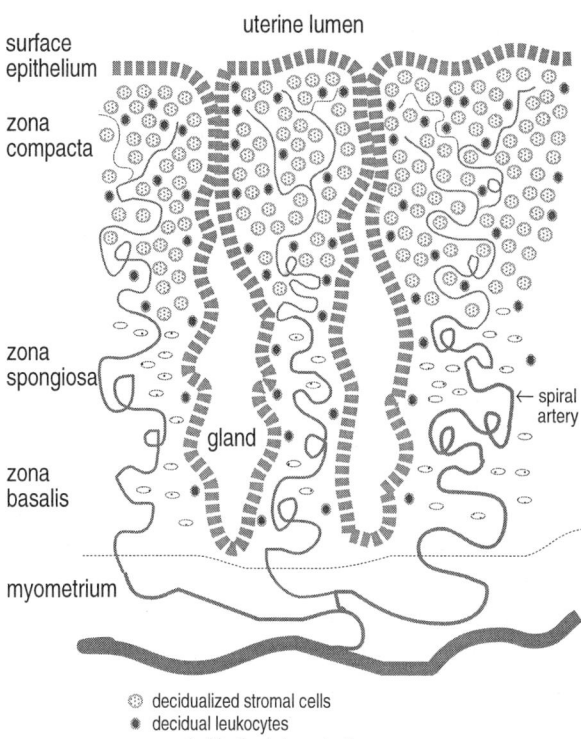

FIGURE 2 Schematic diagram of the histological structure of the decidual tissue.

morphology. Scattered among these typical decidual cells are smaller fibroblastic-like elements considered to be the precursors of the mature decidual cells. The ultrastructure of decidual cells reveals the presence of gap junctions and an extensive development of the organelles involved in protein synthesis (rough endoplasmic reticulum) and secretion (Golgi apparatus). This development of the cellular biosynthetic and secretory apparatus is more prominent in the decidua basalis. Decidual cells produce, and are surrounded by, a characteristic extracellular matrix containing laminin, type IV collagen, fibronectin, and heparan sulfate proteoglycan, all of which are typical constituents of epithelial basement membranes.

The decidua hosts a variety of bone marrow-derived immunocompetent cells, mainly large granulated lymphocytes, T cells, and macrophages. The predominant population, constituting 60–70% of stromal leukocytes in normal first-trimester decidua, is represented by natural killer (NK)-like cells of large granulated lymphocyte morphology. This vast population of decidual NK cells is unique in that the majority of them have a distinct antigenic phenotype, expressing intensely the CD56 NK cell marker but not two other antigens (CD57 and CD16) typically expressed by the majority of circulating NK cells. In early pregnancy NK cells account for at least 30% of all stromal cells in the decidua parietalis, being even more abundant in decidua basalis in which they associate closely with the invading trophoblast. Decidual NK cells decline in numbers later in pregnancy and are very scarce at term, when they represent <5% of decidual leukocytes. This distinct temporal pattern suggests the involvement of decidual NK cells in early events in gestation, possibly implantation and/or placental development. T lymphocytes are found scattered throughout the stroma, in intraepithelial location, and in basal lymphoid aggregates. They comprise 10–20% of the decidual leukocyte population, and their numbers remain constant throughout pregnancy. Almost equal numbers of helper ($CD4^+$) and cytotoxic ($CD8^+$) T cell subsets are found in decidua, with the exception of the intraepithelial T cells, which have all the $CD8^+$ phenotype. Macrophages represent about 20% of decidual leukocytes in early pregnancy and up to 80–90% at term, which suggests they may be important in late pregnancy/parturition. Like the NK cells, macrophages are more abundant in decidua basalis where they associate closely with the invading trophoblast.

Decidual blood vessels are profoundly modified by trophoblast invasion, acquiring a unique structure which leads to the establishment of the low-pressure, high-capacity vasculature of the placental bed. As the result of this process, actual lakes of arterial blood are formed at the base of the placenta which can accommodate the vast volume of maternal blood required by the placenta of the developing conceptus. During the first trimester of pregnancy, trophoblast cells invade spiral arteries of the decidua basalis, replacing the muscular media and the endothelium. Following destruction by the trophoblast, the muscular and elastic elements of the arterial media are replaced by a noncellular amorphous material termed fibrinoid. By the eighth week of pregnancy, vascular invasion by the trophoblast has affected the full thickness of the decidua basalis up to the decidual–myometrial border. However, the characteristic disruption of vascular walls is not found in decidual veins or in arteries away from the implantation site. During the second trimester, another wave of trophoblastic invasion extends to the intramyometrial segments of the spiral arteries, with similar subsequent fibrinoid change of the media.

IV. AUTOCRINE/ PARACRINE INTERACTIONS

A growing body of evidence indicates that the decidua is, together with the placenta and fetal membranes, the site of production of a variety of hormones, neuropeptides, growth factors, and cytokines that may act as autocrine/paracrine regulators of placental/fetal growth and differentiation and myometrial activity.

A. Hormones/Neuropeptides

1. Prolactin

Prolactin is a peptide hormone secreted by the anterior pituitary lobe, which has a wide variety of actions on reproduction, growth, development, and immune regulation. During pregnancy, prolactin is synthesized and secreted by the maternal decidua but not by the chorion or placenta. The transcription of the prolactin gene in decidual cells utilizes an alternative first exon, different from the one utilized in pituitary mammotropes, resulting in a decidual transcript that differs from the pituitary mRNA in the 5' untranslated region but encodes an identical protein. The use of the different promoter associated with the alternative first exon accounts for the distinct regulation of the prolactin gene in decidual cells. Known regulators of pituitary prolactin secretion (dopamine, bromocriptine, and TRH) have no effect on decidual prolactin production which, instead, depends initially on progesterone to induce the decidualization of endometrial stromal cells and stimulate prolactin secretion. However, later in gestation prolactin secretion is found to be progesterone independent but can be stimulated by a variety of agents, including relaxin, endothelin, insulin, insulin-like growth factor-I (IGF-I), free glycoprotein hormone α subunit, and a protein isolated from human placental conditioned medium which is a decidual prolactin-releasing factor. Prolactin receptors are expressed primarily by chorion and amnion cells, suggesting important paracrine actions of decidual-derived prolactin on these cells. Receptors are also detected in decidual cells and to a lesser degree in villous syncytiotrophoblast, indicating they are also possible autocrine/paracrine targets for decidual prolactin. The physiological role of prolactin in the fetal–placental–uterine compartment is not known. High levels of prolactin accumulate in the amniotic fluid during pregnancy, and an osmoregulatory function in amniotic and fetal extracellular fluid has been proposed based on studies with nonhuman primates. *In vitro* studies with human fetal membranes have shown decreased diffusion of water into the maternal side when prolactin is present on the fetal side of the membranes. Such an effect would prevent dehydration of the fetal compartment when the amniotic fluid is hypotonic. Prolactin also has known immunoregulatory actions as an important factor for lymphocyte proliferation during the course of specific immune responses. Therefore, decidual prolactin may have actions on the vast population of immunocompetent cells present in the maternal–fetal interface.

2. Relaxin

Relaxin is a member of the insulin family of peptides which, despite structural homology, has neither sequence homology nor biologic cross-reactivity with insulin. In humans there are two relaxin peptides (H1 and H2) encoded by separate genes, with H2 being the major form which circulates in women during pregnancy. The relaxin receptor has not been structurally identified, but specific binding has been demonstrated in human fetal membranes. In women, the corpus luteum is the major source of relaxin. However, relaxin is also produced by gestational intrauterine tissues, primarily the decidua. The better known biological actions of relaxin in mammals are the inhibition of myometrial activity and the remodeling of connective tissue of the reproductive tract to allow accommodation of pregnancy and successful parturition. The inhibition of myometrial contractility by relaxin has been well documented in experimental animals, but similar studies with human tissues have yielded conflicting results. Relaxin can significantly increase collagen turnover both by stimulating collagenase production and by reducing collagen synthesis in human dermal fibroblasts. When incubated with human fetal membrane explants, relaxin stimulates the expression of the extracellular matrix-degrading proteases matrix metalloproteinase-1 (MMP-1), MMP-3, and MMP-9, which suggests that local relaxin may be involved in the degradation of extracellular matrix in fetal membranes. That such mechanisms may also come into play during cervical ripening is suggested by findings that topical administration of relaxin to women at term can enhance cervical softening, effacement, and dilation before labor. The potential importance of local relaxin in the remodeling of cervical connective tissues is further suggested by observations that there is no prelabor systemic relaxin surge in women and that women without ovaries can undergo normal labor and delivery. The fact that women without ovaries, who have no detectable circulating relaxin, can carry a successful pregnancy to term further indicates that local relaxin production can adequately provide for the needs of the uteroplacental unit. *In vitro* studies also suggest potential autocrine/paracrine actions of decidual relaxin. Relaxin has been found to stimulate the production of three major secreted products—prolactin, IGFBP-1, and renin—by isolated decidual cells or *in vitro* decidualized endometrial stromal cells. Therefore, local relaxin may be involved in the maintenance of decidual secretory activity during pregnancy. In addition, the ability to secrete prolactin and IGFBP-1 is closely associated with the acquisition of the decidual phenotype by endometrial stromal cells, and endometrial relaxin is progesterone dependent, raising the possibility that local relaxin may mediate or amplify progesterone effects on decidual differentiation and function.

3. Oxytocin

Oxytocin is a nonapeptide that stimulates myometrial contractility and is produced by the pituitary and also by gestational intrauterine tissues. Because the placental content of immunoreactive oxytocin is about five times higher than that of the posterior pituitary lobe, it has been suggested that intrauterine tissues might be the main gestational source of this hormone. The potential importance of local uteroplacental–fetal oxytocin sources for myometrial activity is underscored by the fact that serum oxytocin levels are not increased with the onset, and during the first stage, of human labor. Oxytocin expression is higher in decidua than in amnion, chorion, or trophoblast, whereas myometrial oxytocin receptor expression is markedly increased at parturition. These observations suggest a paracrine action of decidual-derived oxytocin stimulating uterine contractions during labor. In addition, oxytocin receptors are also expressed by decidual cells, and oxytocin stimulates prostaglandin production by decidual but not by myometrial tissue *in vitro*. Therefore, decidual-derived oxytocin may further act via autocrine stimulation of prostaglandin production to stimulate uterine contractions during labor.

B. Growth Factors and Cytokines

The growth factors and cytokines encompass a broad heterogeneous group of secreted polypeptides that display pleiotropic effects on a variety of cells. The decidua has been found to be the site of production of many growth factors and cytokines that may serve as autocrine/paracrine regulatory agents acting in concert at the maternal–fetal interface (Table 1). Frequently, individual molecules will be produced by the decidua and also expressed by the placenta and/or fetal membranes. The precise physiological significance of this redundancy in uteroplacental tissues is not known, but is believed to serve as a safety net that would ensure the availability of critically important molecules throughout gestation. The ubiquity of some of these molecules may also reflect their specific functions in different compartments at the maternal–fetal interface. Important insights into these possibilities come from studies on the expression of the growth factors cytokines and their specific cellular receptors and on the biological effects they display on individual cellular targets *in vitro*.

1. Insulin-like Growth Factors

The IGFs (IGF-I and IGF-II) are peptides structurally related to insulin which have mitogenic and metabolic effects. They promote the growth and differentiation of a variety of cell types and are also antiapoptotic factors. The type I IGF receptor primarily mediates signal transduction and cellular actions of the IGFs. The IGF system also comprises a family of soluble IGFBPs that serve as carriers and modulators of the biological actions of IGFs. In cycling endometrium, IGF expression is restricted to the stromal cells, in which IGF-I is primarily expressed in the proliferative phase, whereas IGF-II predominates in secretory endometrium. In first-trimester intrauterine pregnancies IGF-II continues to be the main peptide expressed at the maternal–fetal interface. However, IGF-II is not detected in decidual cells, whereas high levels are expressed by the invading extravillous trophoblast. IGFBPs are, in turn, primarily expressed by the decidual tissue, with IGFBP-3 being the only one showing significant expression by trophoblastic cells at the maternal–fetal interface. IGFBP-1, by far the most abundant in late secretory endometrium, continues to be expressed at very high levels during pregnancy. The type I IGF receptor is uniformly expressed by placental and decidual cells, all of which are potential targets for IGFs and the modulatory actions of IGFBPs. The distinct spatial pattern of expression of the IGFs (mainly IGF-II) and IGFBPs (mainly IGFBP-1) provides important clues to the potential role(s) of the IGF system at the maternal–fetal interface. High levels of IGF-II expression in the trophoblastic cell columns of anchoring villi suggest that autocrine stimulation by IGF-II may be important to support the intense proliferative activity displayed by this cytotrophoblast subpopulation. Cytotrophoblasts at the distal end of the columns disperse and actively invade the decidual tissue, in which their proliferative activity abruptly stops, while they continue to express high levels of IGF-II. Immunoreactive IGFBP-1 is localized in the stromal cells and extracellular matrix of first-trimester decidua, in which it can interact with the invading trophoblast. Therefore, abundant IGFBP-1 encountered by cytotrophoblasts as they enter the decidua may act to limit IGF-II autocrine activity, promoting the transition from a proliferative to an invasive extravillous trophoblast phenotype. In addition to its regulation of IGF bioavailability, IGFBP-1 has direct effects on cells by virtue of its binding through its RGD sequence to the $\alpha_5\beta_1$ integrin (a cell surface receptor that recognizes the RGD sequence of fibronectin). Since the $\alpha_5\beta_1$ integrin is upregulated in cytotrophoblasts as they differentiate into the invasive phenotype, it has been suggested that decidual IGFBP-1

TABLE 1
Growth Factors and Cytokines at the Decidual–Trophoblast Interface, Their Cells of Origin, and Potential Functions

Growth factor/cytokine	Cells of origin	Target cells	Proposed functions
IGF-II	Extravillous trophoblast	Trophoblast	↑ Proliferation, invasion
		Decidual stromal cells	↓ IGFBP-1
EGF/TGF-α	Villous trophoblast	Trophoblast	↑ Proliferation, invasion
	Extravillous trophoblast		↑ hCG/hPL secretion
	Decidual stromal cells		↓ Apoptosis
	Decidual epithelium	Decidual stromal cells	↑ Decidualization, secretion
TGF-β	Villous trophoblast	Trophoblast	↓ Proliferation, invasion
	Extravillous trophoblast		↑ Oncofetal fibronectin immunosuppression
	Decidual stromal cells	Decidual immune cells	
	Decidual NK and T cells		
PDGF	Extravillous trophoblast	Trophoblast	↑ Proliferation
		Decidual stromal cells	↑ Proliferation, ↓ prolactin
		Decidual vascular cells	↑ Angiogenesis
VEGF	Decidual macrophages	Decidual vascular cells	↑ Angiogenesis
	Decidual epithelium		
CSF-1	Decidual stromal cells	Decidual macrophages	↑ Survival, differentiation
	Decidual NK cells	Trophoblast	↑ Syncytial differentiation
GM-CSF	Decidual stromal cells	Decidual macrophages	↑ Survival, differentiation
	Decidual NK cells	Trophoblast	↑ Syncytial differentiation
LIF	Decidual tissue	Trophoblast	↑ Implantation, ↑ hCG secretion
	Decidual NK and T cells		↑ Extravillous differentiation
TNF-α	Villous trophoblast	Trophoblast	↑ Apoptosis?
	Extravillous trophoblast		↑ Class I MHC molecules?
	Decidual stromal cells	Decidual stromal cells	↓ Prolactin, renin
	Decidual NK and T cells		
IL-1	Decidual stromal and NK cells	Trophoblast	↑ Invasion, ↑ hCG secretion
	Decidual macrophages		
	Extravillous trophoblast	Decidual stromal cells	↓ Prolactin, IGFBP-1, renin
IL-6	Villous trophoblast	Trophoblast	↑ hCG/hPL secretion
	Chorion		
	Decidual cells	Decidual stromal cells	↓ Prolactin, renin
IFN-γ	Villous trophoblast	Trophoblast	↑ Apoptosis?
	Extravillous trophoblast		↑ Class I MHC molecules?
	Decidual macrophages	Decidual macrophages	↓ Renin
		Decidual stromal cells	↓ Renin

may directly interact with the trophoblast $\alpha_5\beta_1$ integrin to promote or restrain its invasiveness. Collectively, available data suggest significant interactions between uterine and placental cells may involve the IGF system at the maternal–fetal interface.

2. Epidermal Growth Factor/Transforming Growth Factor-α

Epidermal growth factor (EGF) and transforming growth factor-α (TGF-α) belong to a family of peptides which act through the EGF receptor (the product of the c-*erb*B1 protooncogene). EGF/TGF-α and its receptor are widely distributed in uteroplacental tissues, including decidual cells and epithelium. EGF is upregulated during the *in vitro* decidualization of endometrial stromal cells, and it acts in synergy with progesterone to stimulate the secretion of decidual-specific products (prolactin, IGFBP-1, laminin, and fibronectin) by decidualized endometrial stromal cells. These observations suggest the involvement of autocrine actions of EGF in decidual differentiation and function. However, higher levels of expression of EGF/TGF-α and its receptor are found in placental tissues than in decidua, suggesting a more prominent placental EGF system may prevail at the maternal–fetal interface. Colocalization of EGF/TGF-α and its receptor in uteroplacental tissues is suggestive of an autocrine mode of action of these growth factors in early pregnancy. The pattern of expression among the various trophoblast populations further suggests the involvement of EGF/TGF-α in supporting cell proliferation in the villous cytotrophoblast and the cytotrophoblastic columns, and also the differentiated functions of the invasive extravillous trophoblast and the secretory villous syncytiotrophoblast. The proposed role(s) of EGF/TGF-α on trophoblast function is supported by *in vitro* studies showing EGF/TGF-α stimulation of trophoblast proliferation, invasion, and secretion of human chorionic gonadotrophin (hCG) and/or placental lactogen (hPL). Overall, the existing evidence suggests a wide involvement of the EGF system in normal decidual and placental development and function at all stages of gestation, but does not indicate a significant role in regulating specific trophoblast phenotypic transitions.

3. *Transforming Growth Factor-β*

Transforming growth factor-β (TGF-β) is a multifunctional peptide belonging to a family which includes, in addition to the three TGF-β isoforms, inhibin, activin, and Müllerian-inhibiting substance. TGF-β can stimulate or inhibit the growth and differentiation of a variety of cells, and its effects are known to be modulated by the presence of other growth factors. Cellular signaling by TGF-β is mediated through two receptors types (I and II), which are widely distributed among the different cell types at the maternal–fetal interface. TGF-β expression is higher in intrauterine gestational tissues than in cycling endometrium, and a variety of cells at the maternal–fetal interface produce TGF-β, including the syncytiotrophoblast, the extravillous trophoblast, and decidual stromal, NK, and T cells. The potential implications of such rich and varied sources of TGF-β in uteroplacental tissues are many-fold. TGF-β has suppressive effects on lymphocytes and other cells associated with immune responses, largely counteracting the effects of proinflammatory cytokines. At the maternal–fetal interface, large numbers of immunocompetent cells could be exposed to proinflammatory cytokines [such as tumor necrosis factor-α (TNF-α) and interleukin-1β (IL-1β)], which may be otherwise locally required during implantation and/or placental development. Therefore, TGF-β may influence the function of decidual immunocompetent cells and thereby contribute to maintain maternal tolerance to the semiallogenic trophoblast. A number of studies also suggest that interactions of decidual TGF-β with the extravillous trophoblast may play an important role in restraining trophoblast invasion. For example, TGF-β1 inhibits the proliferation and invasion of isolated trophoblasts *in vitro* and promotes the formation of multinucleated cells, similar to the noninvasive giant cells of the placental bed. The invasive capacity of the trophoblast is dependent on the secretion of proteases, particularly MMPs, whose activities are regulated by specific tissue inhibitors of the MMPs (TIMPs). TGF-β released by first-trimester decidual cells stimulates the expression of TIMP-1 by isolated trophoblasts and thereby inhibits their invasion *in vitro*. TGF-β1 also stimulates the production of oncofetal fibronectin by isolated human cytotrophoblasts. Since oncofetal fibronectin is localized *in vivo* to a specific region in the extracellular matrix where the anchoring villi connect to the maternal decidua, this finding implies that local action of TGF-β may promote adhesive interactions that contribute to anchor the placenta to the uterine wall. Therefore, the potential actions of TGF-β at the maternal–fetal interface would appear to serve diverse protective functions that prevent uncontrolled invasion of maternal tissues by the placental trophoblast and also, in turn, protect the trophoblast from potentially harmful activity of maternal immunocompetent cells, while contributing to maintain the physical integrity of the uteroplacental connection.

4. Platelet-Derived Growth Factor

Platelet-derived growth factor (PDGF) (a product of the c-*sis* protooncogene) is a protein with known mitogenic and angiogenic properties. It consists of two polypeptide subunits (A and B), encoded by separate genes, that combine to form homo- or heterodimers. The PDGF receptor is composed of two subunits (α and β) that form homo- or heterodimers in which each receptor subunit interacts with one of the PDGF ligand subunits. The PDGF-B chain binds to both α and β subunits, and the A chain only binds to the α subunit. Therefore, the cellular responses to PDGF will depend both on the receptor subunits expressed and on the PDGF ligand isoforms present in the local environment. PDGF-B is primarily expressed by the extravillous trophoblast that forms the cytotrophoblastic shell at the interface with the maternal decidua. PDGF receptor subunits are localized not only in some cells of the cytotrophoblastic shell but also in the stromal cells and spiral artery media in first-trimester decidua. PDGF stimulates cell proliferation and inhibits prolactin secretion in first-trimester decidual stromal cells. The available information therefore suggests that PDGF may play specific roles during placental development. It may have autocrine effects to stimulate the proliferation of cells within the cytotrophoblastic shell and also promote decidual cell proliferation and angiogenesis by paracrine actions.

5. Vascular Endothelial Growth Factor

Vascular endothelial growth factor (VEGF) is a homodimeric angiogenic factor known to induce endothelial cell proliferation and increase vascular permeability. Two VEGF receptors have been identified: the *fms*-like tyrosine kinase (*flt*) and a kinase domain-containing receptor (KDR). In nongestational endometrium VEGF is expressed by both epithelial and stromal cells. In first-trimester gestations VEGF is strongly expressed in decidual glandular epithelium and also very intensely by the macrophages in the decidua basalis, suggesting a contribution of VEGF to angiogenesis during placentation. KDR is only expressed by endothelial cells, whereas *flt* is additionally expressed by decidual macrophages and stromal cells in first-trimester gestations and by all extravillous trophoblast populations throughout pregnancy. Decidual macrophages are found closely associated with the invading extravillous trophoblast, suggesting potential paracrine interactions involving VEGF. It has been suggested that VEGF may be a trophoblast mitogen since it stimulates thymidine incorporation by choriocarcinoma cells *in vitro*. In addition, VEGF has been found to stimulate the release of parathyroid hormone-related protein (a known vasorelaxant) by immortalized first-trimester trophoblast cells in culture. VEGF is very likely an important angiogenic factor during placentation and may also have regulatory actions on trophoblast proliferation/invasion and decidual vascular tone.

6. Colony-Stimulating Factor-1

Colony-stimulating factor-1 (CSF-1) is a homodimeric glycoprotein, initially described as a monocyte/macrophage growth factor. CSF-1 can be secreted or produced in a membrane-bound form which can only be released after proteolytic cleavage. CSF-1 actions are mediated by a membrane receptor encoded by the protooncogene c-*fms*. CSF-1 expression is higher in first-trimester decidua than in nongestational endometrium, with decidual stromal and NK cells being the main source of secreted CSF-1. The CSF-1 receptor is detected primarily in the cytotrophoblast cell columns, the invasive interstitial trophoblast, and decidual macrophages, all of which are potential paracrine targets for decidual-derived CSF-1. This pattern of expression suggests that decidual-derived CSF-1 may regulate the activity of the invasive trophoblast. *In vitro* studies have not specifically examined the effect of CSF-1 on trophoblast invasiveness but have shown that CSF-1 induces the formation of syncytia by first-trimester cytotrophoblasts, with an increase in hCG and hPL production. These findings are consistent with a role of CSF-1 in stimulating trophoblast differentiation into a villous syncytial secretory phenotype. Stromal fibroblast from the villous core have been found to produce CSF-1 *in vitro* and may be a source of paracrine CSF-1 that acts on the adjacent villous trophoblast to promote syncytial differentiation. In the extravillous trophoblast, decidual-derived CSF-1 may act to induce the formation of noninvasive multinucleated placental bed giant cells. Existing data, therefore, are consistent with a role of CSF-1 in supporting the decidual macrophage population and promoting trophoblast syncytial differentiation. It remains to be determined if CSF-1 can also act as an inhibitor of extravillous trophoblast differentiation into an invasive phenotype.

7. Granulocyte-Macrophage Colony-Stimulating Factor

Granulocyte-macrophage colony-stimulating factor (GM-CSF) is a known growth factor for bone marrow progenitor cells and activator of leukocytes in peripheral tissues. GM-CSF binds to a heterodimeric receptor formed by a specific α subunit and a common β subunit shared by the receptors for IL-3 and IL-5. In first-trimester gestations, isolated decidual NK cells and trophoblasts secrete GM-CSF, and the receptor α subunit is detected in villous cytotrophoblast, all extravillous trophoblast populations, and decidual macrophages. *In vitro* studies suggest potential autocrine and paracrine actions of uteroplacental GM-CSF may be involved in promoting syncytial differentiation and hCG and hPL secretion. GM-CSF receptor α subunit is detected in term syncytiotrophoblast, and GM-CSF is produced by the villous core fibroblasts, suggesting that this cytokine may contribute to syncytial differentiation via an intravillous paracrine mechanism similar to CSF-1. Therefore, current data are consistent with GM-CSF cooperating with CSF-1 to regulate decidual macrophage function and promote trophoblast syncytial differentiation.

8. Leukemia Inhibitory Factor

Leukemia inhibitory factor (LIF), initially identified by its ability to inhibit myeloid leukemia cell differentiation, is a monomeric cytokine with pleiotropic actions on a variety of cells and tissues. The LIF receptor consists of a specific subunit which dimerizes with a common signal transduction subunit (gp130) shared by other cytokine receptor systems, such as IL-6 and IL-11. Targeted disruption ("knockout") of the LIF gene has conclusively shown that maternal LIF production is essential for embryo implantation in mice. Blastocysts are unable to implant in homozygous females lacking a functional LIF gene, but their transfer to wild-type pseudopregnant females results in implantation and pregnancy. In humans, available evidence is consistent with a similar role because LIF expression peaks in human endometrium at the time of implantation. LIF is also expressed in first-trimester decidua, and both the cytokine and its receptor are expressed by decidual NK and T cells, suggesting the potential autocrine regulation of these decidual leukocyte populations by LIF. In contrast, LIF is not expressed by isolated first-trimester trophoblasts which express abundant LIF receptor, suggesting maternally derived LIF may also have important paracrine actions during postimplantation development. LIF stimulates hCG production by isolated first-trimester trophoblasts, suggesting it may contribute to the maintenance of placental secretory function. LIF has been shown to inhibit the differentiation of pluripotent embryonic stem cells and may be involved in the maintenance of a placental trophoblast stem cell population during pregnancy. Alternatively, LIF action at the maternal–fetal interface may play a role in trophoblast differentiation switching to the villous or extravillous phenotypes, similar to its function in the neural system, in which it induces a neuronal differentiation switch from an adrenergic to a cholinergic phenotype. *In vitro* studies are consistent with the latter possibility. In isolated human cytotrophoblasts, LIF induces specific biochemical changes which are characteristic of the extravillous phenotype, inhibiting hCG secretion but stimulating the production of oncofetal fibronectin. These findings suggest that the paracrine action of decidual-derived LIF may induce a developmental switch that drives trophoblast differentiation toward an extravillous anchoring phenotype. Therefore, LIF appears to be a crucial mediator of maternal–embryonic interactions, which highlights the importance of the local maternal environment for normal implantation and placental development.

9. Tumor Necrosis Factor-α

TNF-α is a pleiotropic cytokine associated with inflammatory processes and originally identified as a product of activated macrophages with antitumor activity. In its functional form as a homotrimer, TNF-α can bind to two distinct types of cellular receptors (p55 and p75) which have similar extracellular ligand-binding domains but different cytoplasmic domains. In human cells, p55 seems to mediate most of the biological actions of TNF-α. Soluble forms of the receptors have been identified which are believed to regulate TNF-α bioavailability. TNF-α is expressed by the syncytiotrophoblast, endovascular trophoblast, and decidual stromal, NK, and T cells. The p55 receptor is expressed in term syncytiotrophoblast, decidua, and chorion. TNF-α inhibits the production of prolactin and renin by term decidual cells *in vitro*, suggesting autocrine/paracrine TNF-α may influence decidual cell secretory activity. In first-trimester gestations p55 receptor expression is variable in villous trophoblast but particularly high in the cytotrophoblastic cell columns emerging from anchoring villi. Since this extravillous trophoblast population does not express TNF-α, it is a likely target for paracrine actions of decidua-derived TNF-α. The expression of high levels of its receptors in the highly proliferating cells of the cytotrophoblastic columns would seem consistent with TNF-α stimulating extravillous trophoblast proliferation *in vivo*. However, TNF-α has been found to induce apoptotic cell death of term villous cytotrophoblasts *in vitro*. These conflicting data are reconciled by the finding that EGF completely blocks the induction of apoptosis by TNF-α *in vitro*. Therefore, the local action of an EGF receptor ligand serves to protect this trophoblast population from the proapoptotic actions of TNF-α. What is then the function of the high levels of p55 expression in the cytotrophoblastic columns? One possibility is the production of secreted soluble receptors; this would provide additional protection to this trophoblast population from the potentially harmful effects of TNF-α. Alternatively, TNF-α may be required for extravillous trophoblast differentiation/function. Invasive extravillous cytotrophoblasts express the nonclassical class I major histocompatibility complex (MHC) molecule HLA-G, which may be important for their recognition by decidual NK cells to prevent NK cell-mediated cytotoxicity. TNF-α is known to stimulate the expression of class I MHC molecules and may be involved in upregulating the expression of HLA-G by invading cytotrophoblasts.

10. Interleukin-1

The interleukins comprise a large group of multifunctional immunoregulatory cytokines. The major forms present in uteroplacental tissues are IL-1 and IL-6, which are known mediators in inflammatory processes. There are two forms of IL-1 (α and β) which are encoded by separate genes but bind to the same receptors and have identical biological activities. Of the two IL-1 cellular receptor types (I and II), type I is believed to be involved in signaling. The IL-1 system also includes the IL-1 receptor antagonist (IL-1ra), which is structurally homologous to IL-1 and binds to the IL-1 receptors but is biologically inactive, so it functions as a competitive inhibitor of IL-1. The expression of IL-1 increases dramatically in first-trimester decidua compared to cycling endometrium, implying a prominent role of IL-1 in early pregnancy. Placental tissues express little, if any, IL-1 in either early or late gestations, which suggests a primary decidual origin for this cytokine. There is no clear consensus on the localization of IL-1 proteins in uteroplacental tissues. However, overall, the existing evidence indicates that decidual stromal cells, NK cells, macrophages, as well as the extravillous trophoblast are potential sources of IL-1 at the maternal–fetal interface. IL-1 type I receptors are present in early pregnancy villous syncytiotrophoblast, extravillous trophoblast, and decidual glandular epithelium, suggesting the autocrine/paracrine actions of this cytokine may regulate trophoblast and decidual function. *In vitro* studies suggest that IL-1β may act on the extravillous trophoblast to promote invasion by stimulating the synthesis and secretion of the extracellular matrix-degrading protease MMP-9 and on the villous syncytiotrophoblast to stimulate hCG secretion. The latter *in vitro* effect of IL-1β on first-trimester placental trophoblast is mediated via an autocrine loop involving the induction of IL-6, similar to the classical cytokine cascade in inflammatory reactions. In contrast, IL-1β has inhibitory effects on decidual cells, suppressing secretion of prolactin, IGFBP-1, and renin. However, the actions of IL-1 at the maternal–fetal interface may be modulated by IL-1ra. Term decidual cells produce high levels of IL-1ra *in vitro*, and its secretion is stimulated by IL-1β. Therefore, local production of IL-1ra by the maternal decidua may provide a protective mechanism that limits inhibitory autocrine/paracrine effects of IL-1β on the decidual cells. Also, and perhaps more important, intradecidual IL-1ra may protect against the local proinflammatory effects of IL-1β on decidual immunocompetent cells and its potential adverse effects on pregnancy.

11. Interleukin-6

IL-6 is a homodimer synthesized by a variety of cells in response to IL-1 in the course of inflammatory responses. The IL-6 receptor is a heterodimer composed of a ligand-specific α subunit and a common signal-transducing β subunit (gp130) shared with other cytokine receptors such as LIF and IL-11. Placental syncytiotrophoblast, chorion, and decidual cells appear to be the primary sources of IL-6 in uteroplacental tissues. TNF-α and/or IL-1 stimulate IL-6 secretion by chorion, decidual, and villous core mesenchymal cells, suggesting IL-6 may also mediate or amplify some of the local effects of the other two cytokines via autocrine/paracrine mechanisms. Cells expressing the IL-6 receptor α subunit have not been detected in first-trimester decidua and placenta by immunohistochemistry or flow cytometry. However, the ligand-binding α subunit of the IL-6 receptor can occur in membrane-bound or soluble forms, and soluble ligand receptor complexes can effectively dimerize with gp130 at the cell surface and activate signal transduction. This mechanism may mediate the reported actions of IL-6 on trophoblast and decidual cells. *In vitro*, IL-6 stimulates hCG and hPL secretion by term placental trophoblasts but inhibits prolactin and renin production by decidual cells. Therefore, like IL-1, known effects of IL-6 are mainly stimulatory for trophoblast and inhibitory for decidual cells.

12. Interferons

The interferons (IFNs) comprise a group of proteins which stimulate natural defense mechanisms against viral infections and are produced by a variety of cells. There are three distinct antigenic types of IFNs: α (leukocyte), β (fibroblast), and γ (immune). IFN actions are mediated by two different but homologous monomeric cellular receptors. IFN-α and IFN-β molecules (type I INF) bind to the same (α/β) cell surface receptor and elicit similar cellular responses, whereas IFN-γ (type II IFN) binds to a different (γ) IFN receptor and displays distinct immunoregulatory actions. Villous cytotrophoblast, syncytiotrophoblast, extravillous interstitial trophoblast, and decidual macrophages appear to be the primary sources of IFNs in uteroplacental tissues. Overall, IFN-α and -β proteins are most abundant, but all three IFNs are found to be highest in early pregnancy. In contrast, the levels of both types of IFN receptors remain constant throughout gestation. IFN receptors are localized in the same cell populations as their IFN ligand, with the exception of the cytotrophoblastic columns, which have receptors but do not appear to express IFN ligands. The physiological role(s) played by INFs at the maternal–fetal interface is not known. The two types of IFNs have overlapping, but clearly distinct, sets of biological actions. However, studies with isolated human uteroplacental cells have primarily examined the effects of IFN-γ. IFN-γ inhibits the production of renin by isolated term decidual stromal cells and macrophages *in vitro*. The absence of demonstrable IFN receptor protein in decidual stromal cells *in vivo* would seem to imply that they may not be major targets of IFN-γ in normal pregnancy. However, autocrine/paracrine IFN-γ actions seem likely in receptor-expressing decidual macrophages. Decidual-derived IFN-γ may be also required for extravillous trophoblast differentiation/function. Pretreatment of trophoblasts with IFN-γ protects them from killing by lymphokine-activated decidual NK cells, probably by enhancing the expression of class I MHC molecules. However, IFN-γ can also induce apoptotic cell death of term villous cytotrophoblasts *in vitro*, but this effect is completely abrogated by EGF. These findings reveal a mechanism by which IFN-γ receptor-expressing cells of the cytotrophoblastic columns may evade the proapoptotic effects of this cytokine and underscore the need for a better understanding of how the effects of multiple interacting signaling molecules are integrated within the gestational uterine microenvironment.

V. SUMMARY

The gestational decidua is situated at the crossroads of maternal and fetal domains. In this context, the physiological role of the decidua can be envisioned as fulfilling two basic types of functions—protective and regulatory—which must serve both maternal and fetal economies during the course of gestation. Protective functions may be directed on one side at protecting the mother from uncontrolled invasion by the placental trophoblast and on the other at preventing rejection of the semiallogenic conceptus by the maternal immune system. It can also be inferred that regulatory decidual functions may influence placenta/fetal growth and development and the activity of the adjacent myometrium. The emerging picture suggests that the mechanisms underlying such vital processes in human pregnancy involve complex temporal and spatial interactions between the various maternal and placental cells. Our understanding of these processes progressively advances as the specific roles of individual cells and signaling and effector molecules involved are elucidated. However, much of the fundamental information is yet to be gathered. The availability of sensitive analytical methods and new molecular reagents, as well as refined methods for the identification, isolation, and *in vitro* study of individual cell types, will continue to contribute to these advances. The use of animal models with transgenic overexpression or gene knockout have and will continue to provide important information on decidual function. A major challenge is to unravel the patterns in which such complex signaling events are dynamically integrated within the gestational uterine microenvironment. To this end, the use and development of experimental systems to study cellular interactions should prove particularly fruitful in future research on human decidual biology.

See Also the Following Articles

ENDOMETRIUM; IGF (INSULIN LIKE GROWTH FACTORS); INTERFERONS; OXYTOCIN; PROLACTIN, OVERVIEW; RELAXIN, MAMMALIAN; TROPHOBLASTS TO HUMAN PLACENTA

Bibliography

Aplin, J. D. (1991). Implantation, trophoblast differentiation and haemochorial placentation: Mechanistic evidence in vivo and in vitro. *J. Cell Sci.* **99**, 681–692.

Bryant-Greenwood, G. D. (1994). Relaxin. In *The Uterus* (T. Chard and J. G. Gudzinskas, Eds.), pp. 252–267. Cambridge Univ. Press, Cambridge, UK.

Cross, J. C., Werb, Z., and Fisher, S. J. (1994). Implantation and the placenta: Key pieces of the of the development puzzle. *Science* **266**, 1508–1518.

Giudice, L. C. (1997). Biochemical events in the human endometrium during the menstrual cycle and implantation. In *Infertility, Contraception, and Reproductive Endocrinology* (R. A. Lobo, D. R. Mishell, and R. D. Paulson, Eds.), pp. 141–158. Blackwell Scientific, Cambridge, UK.

Giudice, L. C., and Ferenczy, A. (1996). The endometrial cycle. In *Reproductive Endocrinology, Surgery, and Technology* (E. Y. Adashi, J. A. Rock, and Z. Rosenwacks, Eds.), pp. 271–300. Lippincott-Raven, Philadelphia.

Hofmann, G. E., and Scott, R. T. (1995). Epidermal growth factor transforming growth factor-α and their common receptor: Their potential relevance in normal human endometrial development and pregnancy. *Sem. Reprod. Endocrinol.* **13**, 109–119.

Klinman, H. J. (1994). Placental hormones. *Infertil. Reprod. Med. Clin. North Am.* **5**, 591–610.

Lala, P. K., and Lysiak, J. J. (1994). Role of locally produced growth factors in human placental growth and invasion with special reference to transforming growth factors. In *Immunobiology of Reproduction* (J. S. Hunt, Ed.), pp. 57–81. Springer-Verlag, New York.

Loke, Y. W., and King, A. (1995). *Human Implantation: Cell Biology and Immunology*. Cambridge Univ. Press, Cambridge, UK.

Loke, Y. W., King, A., and Burrows, T. D. (1995). Decidua in human implantation. *Hum. Reprod.* **10**(Suppl. 2), 14–21.

Mitchell, M. D., Trautman, M. S., and Dudley, D. J. (1993). Cytokine networking in the placenta. *Placenta* **14**, 249–275.

Petraglia, F., Florio, P., Nappi, C., and Genazzani, G. R. (1996). Peptide signaling in human placenta and membranes: Autocrine, paracrine, and endocrine mechanisms. *Endocr. Rev.* **17**, 156–186.

Ramsey, E. M. (1994). Anatomy of the human uterus. In *The Uterus* (T. Chard and J. G. Gudzinskas, Eds.), pp. 18–40. Cambridge Univ. Press, Cambridge, UK.

Salzman, A., and Cooke, N. E. (1996). Prolactin. In *Reproductive Endocrinology, Surgery, and Technology* (E. Y. Adashi, J. A. Rock, and Z. Rosenwacks, Eds.), pp. 747–768. Lippincott-Raven, Philadelphia.

Starkey, P. M. (1993). The decidua and factors controlling placentation. In *The Human Placenta* (C. W. Redman, I. L. Sargent, and P. M. Starkey, Eds.), pp. 362–413. Blackwell, Oxford, UK.

Deciduoma

Yan Gu and Geula Gibori
University of Illinois at Chicago

I. Induction of Decidualization
II. Morphology of the Deciduoma
III. Regression of the Deciduoma
IV. Major Roles of Antimesometrial and Mesometrial Decidual Cells
V. Decidual Growth Factors and Cytokines

GLOSSARY

decidualization A transformation process of the endometrial stromal cells that takes place in response to a deciduogenic stimulus. It can be induced by either the implanting embryo in pregnancy or artificial stimuli to the uterus in pseudopregnant animals. The uterine endometrial cells undergo proliferation and differentiation which results in a gross increase in size and weight of the uterus.

mesometrial region An anatomical region in the rodent uterus which includes the half portion of the uterine horn located at the side near the conjunction of the mesometrium containing the blood vessels gaining access to the uterus and divided longitudinally along the uterine lumen from the other half portion of the uterus. The decidual tissue in the mesometrial region is termed *mesometrial decidua* and is, in pregnant animals, the site where trophoblasts invade.

antimesometrial region A region containing the portion of the uterine horn located at the opposite side of the mesometrial region divided longitudinally by the uterine lumen. (Compare with *mesometrial region*.)

mesometrial triangle An anatomical area in the uterine horn above the mesometrial decidua. The borders of the mesometrial triangle are limited on two sides by the outer longitudinal layer of the myometrium and on the third side by the inner circular layer of smooth muscle which is eventually disrupted by the development of the metrial glands in the mesometrial triangle.

stromal cells A major cell population in the uterine endometrium. After hormonal sensitization and deciduogenic stimulation, stromal cells undergo a remarkable transformation termed *decidualization*.

Deciduoma is the decidual tissue artificially induced in pseudopregnant animals in the absence of embryo. This tissue is comparable to the decidua of pregnant rats in its formation, regression, and secretory capacity. Deciduoma can be induced by a traumatic stimulus, such as scratching the luminal surface of uterine horns, or the stimulus can be achieved by intraluminal injection of numerous chemicals in pseudopregnant animals that are either mated with infertile males or ovariectomized and treated with a sequential progesterone and estradiol regimen to sensitize the uterus.

I. INDUCTION OF DECIDUALIZATION

In all mammalian species, pregnancy is accompanied by remarkable changes in the uterine environment to allow for contact between mother and fetus without damage to the mother or rejection of the fetus. This requires profound changes and severe reorganization of the different tissues forming the uterus. The most striking event that takes place in the uterine milieu in primates and rodents is the rapid growth and differentiation of the endometrial stroma giving rise to unique cells, termed decidual cells, that differ totally from the original fibroblast cells. Decidualization results in a significant increase in the size and the weight of the uterus. In humans, decidualization normally occurs with each menstrual cycle and the formation of the decidual tissue depends principally on levels of progesterone and estradiol in the circulation. In other species, including the rat, decidualization requires, in addition to adequate levels of hormones, an exogenous trigger which may be either the contact of the blastocyst with the endometrium or artificial (either chemical or physical) stimulation of the uterus. The reason for the physical stimulus required to induce the formation of the decidual tissue in hormonally sensitized uterus is not known. The decidual tissue triggered by the implanting blastocyst is termed decidua, whereas that due to artificial stimuli is termed deciduoma. Despite their different names, the decidua and deciduoma are similar in their formation, regression, and secretory capacities. The deciduoma has been extensively used as a model to study the induction, development, and role of this organ in the absence of contaminating trophoblast cells. Hormonal preparation of the uterus by progesterone is an absolute requirement for the induction of decidualization. However, one of the most important observations in the induction of the deciduoma in rodents is that the decidual response can only occur during a narrow period of time and always starts on the antimesometrial side. In the rat, maximal decidualization is obtained on Day 5 of pseudopregnancy (equivalent to the day of nidation in pregnant rat) and depends on priming of the uterus with progesterone throughout the 5 days and with low levels of estradiol on Day 5. Decidualization takes place first at the subepithelial area in the antimesometrial region independently of whether this process is triggered by the blastocyst during pregnancy or by artificial stimuli in pseudopregnancy. The cellular and molecular mechanisms of the induction of endometrial stromal cell transformation and the nature of the signal transduction between uterine luminal epithelial cells and stromal cells remain unclear. However, the luminal epithelial cells located at the antimesometrial side seem to be primarily, if not exclusively, responsible for signaling to the neighboring stromal cells and for the early initiation of decidualization. However, they die by apoptosis early during the development of the decidual tissue in rodent.

It has been suggested that decidualization represents an evolutionary adaption to an inflammatory reaction. This contention was evidenced by the presence of most, if not all, of the cellular elements of the normal inflammatory process during decidualization, such as cell damage, increased vascular permeability, cellular proliferation, and an increase of prostaglandins and histamine contents. However, the massive infiltration of leukocytes and macrophages seen in a real inflammatory reaction are not observed during decidual differentiation of uterine endometrium. The induction of decidualization has been subject to extensive investigation and the following factors have been shown to be involved in this phenomenon:

1. Histamine, released from uterine mast cells, has been hypothesized to be responsible for the initiation of decidual reaction. This is based on the observation that histamine receptors are present in the uterine endometrium, that intraluminal injection of histamine can induce decidualization in pseudopregnant rats, and that antihistamines prevent decidualization. Although disputing observations have been reported, it seems probable that histamine indeed plays an important role in decidualization either directly or indirectly.

2. Uterine prostaglandins (PGs) have been proposed to play an obligatory role in the induction of decidual transformation of stromal cells and the increased vascular permeability, the first observable reaction in the site of implantation in the uterus. This proposition is based on several lines of observation. An increased prostaglandin concentration in the uterus at the beginning of decidualization occurs in both pregnant and pseudopregnant rodents. Receptors for PGs are expressed in the endometrial stroma, and intraluminal injections of PGs can induce stromal cell decidualization. In addition, treatment with PG synthesis inhibitor, indomethacin, either inhibits or delays implantation in pregnant rats and prevents or greatly reduces the extent of artificially induced decidual reaction in pseudopregnant animals. The expected increase in vascular permeability is also blocked by the PG synthesis inhibitor.

3. Increased cyclic AMP concentration in the decidual tissue is thought to be due to the effect of PGs in rodents since indomethacin treatment can inhibit this cAMP increase. cAMP-induced decidualization in human endometrial stromal cells *in vitro* has been reported. However, intrauterine application of cAMP does not induce decidualization in pseudopregnant mice but instead induces implantation in pregnant animals.

4. Recently, several homeobox genes (Hoxa

genes) have been suggested to play an important role during pregnancy since implantation in mice with defective *Hoxa-10* gene, a member of the Hoxa gene family, was compromised. Hoxa genes encode transcription factors that can switch regulators to initiate genetic cascades in several cells and control the development of the uterus and uterine receptivity during pregnancy. A deficient stromal decidualization induced by artificial stimuli has been reported in mice with defected *Hoxa-10*.

5. Leukemia inhibitory factor (LIF), a pleiotropic cytokine expressed in decidual glandular epithelium in both pregnant or pseudopregnant mice, appears to be involved in the induction of decidualization and implantation. LIF expression in the uterus is regulated by sex steroids and its spatial expression in the decidual tissue coincides with the window of implantation. LIF has been considered an obligatory factor for successful implantation of blastocyst into the decidua in pregnancy since female mice lacking a functional LIF gene are fertile but their embryos fail to implant.

In summary, the induction of decidualization appears to depend on steroid hormones secreted by the ovaries and by locally expressed factors in the uterus.

II. MORPHOLOGY OF THE DECIDUOMA

Decidualization involves marked cellular transformation and reorganization of the endometrial stroma. Anatomically and morphologically the fully formed deciduoma is divided into three major regions: the antimesometrial decidua, which forms the decidua capsularis in the pregnant rodent, the mesometrial decidua, (or decidua basalis of pregnancy) and the mesometrial triangle (Fig. 1). Each region consists of different cell populations with distinct morphological characteristics in the fully developed decidual tissue.

The antimesometrial decidua primarily consists of extensively decidualized stromal cells, termed antimesometrial decidual cells. These cells represent the most extreme form of differentiated uterine endometrial stromal cells. They are large, tightly packed, polynucleated, and show high levels of ploidy (up to 32n). Most cells have large nuclei with irregular-

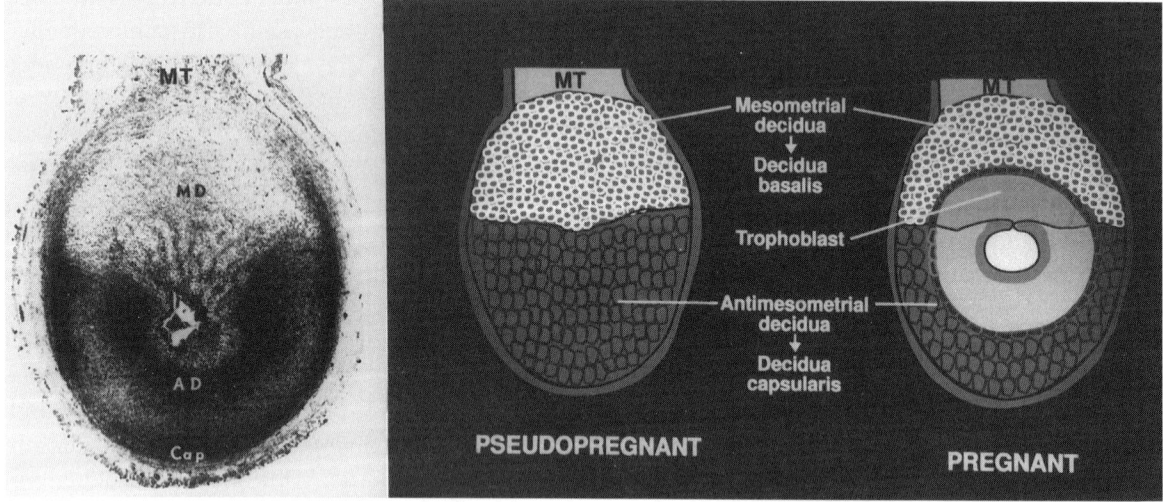

FIGURE 1 Morphology of the deciduoma. (Left) A sagittal cut in the uterus of a pseudopregnant rat 4 days after the induction of decidualization (from O'Shea *et al.*, 1983). (Right) The two types of decidual cells in the mesometrial (MD) and antimesometrial (AD) decidua. The mesometrial cells are small and loosely packed, whereas the antimesometrial cells are large and tightly packed allowing for a visual separation of the darker antimesometrial decidua from the mesometrial decidua. In the pregnant rat, the mesometrial decidua forms the decidua basalis, which together with the trophoblast become the placenta. The antimesometrial cells form the decidua capsularis in pregnancy.

shaped membranes, numerous free ribosomes, elongated mitochondria with lamellar cristae, well-developed Golgi elements, endoplasmic reticulum, and cytoplasmic filament. The reduction of intercellular space and the development of adhesion-type junctions and gap junctions are observed in the antimesometrial decidua. They form a tissue of tightly packed cells suggesting protective function and cell-to-cell communication.

The mesometrial decidua is formed by cells which undergo far less pronounced cytodifferentiation than cells of the antimesometrial region. Those differentiated cells are small, loosely packed, often binucleated, and termed mesometrial decidual cells. Polyploidy occurs to a limited extent (most 4n). The presence of large stores of glycogen and relatively few collagen fibrils in the intercellular areas suggests a role of nutrition and supervision of trophoblast invasion. As in antimesometrial decidual cells, gap and tight junctions are also present in the mesometrial decidual cells.

Another group of cells found in the decidual tissue, but that are not decidual cells per se, are termed endometrial granulocytes in humans and granulated metrial gland (GMG) cells in rodents. These cells are found principally in the metrial gland and in the mesometrial decidua. Some studies suggest that those decidual GMG cells may originate from bone marrow. Immunologically, GMG cells are estrogen-receptor and interleukin-2-receptor positive and express similar membrane antigens as those found on natural killer (NK) cell membrane. The GMG cells in rodents were recently named uterine NK (uNK) cells. uNK cells appear roughly spherical in shape and are identifiable by their cytoplasmic granules. The function of uNK cells is not clear, but they have been suggested to play a role in fetal survival and trophoblast migration.

III. REGRESSION OF THE DECIDUOMA

For pregnancy to proceed normally, great numbers of decidual cells must die to accommodate and protect the developing conceptuses. Regression and extensive reorganization of the decidual tissue do not take place at random but appear well orchestrated. The first cells to degenerate are the epithelial cells, which disappear together with the uterine lumen 5 days after the induction of decidualization. The antimesometrial decidual cells decidualize first and also degenerate first. In pregnant animals they transform into a thin layer of tissue surrounding the amnios. Cell death takes place later in the mesometrial decidua, which forms together with the trophoblast the placenta in the pregnant animal. As decidual tissues induced by the implanting blastocyst or by artificial stimuli regress with similar morphology and kinetics, decidual cell regression appears to be controlled by an inherent cell program rather than being under the control of the embryo. Decidual cell death occurs by apoptosis. Pyknotic nuclei and DNA breakdown typify cell regression. In contrast to necrosis, apoptosis applies to an active process of self-destruction, requiring new gene expression and enzyme activation. Nuclear changes and DNA cleavage are the first manifestations of this type of cell death. Although apoptotic cells do not induce an inflammatory reaction, they are rapidly phagocytized by either macrophage or adjacent viable cells. What causes this cell-specific and spatially controlled apoptosis in the deciduoma is not clear. Interestingly, progesterone withdrawal causes immediate involution of the decidual tissue suggesting that progesterone may prevent decidual cell death and that in its absence apoptosis occurs. However, cell death takes place in the deciduoma despite high levels of progesterone in the circulation and of progesterone receptor mRNA in the decidual cells. It is possible, however, that levels of progesterone receptor mRNA do not directly reflect levels of functional progesterone receptor. Progesterone receptor protein may be rapidly degraded first in the antimesometrial cells and thereafter in the mesometrial cells, leading to differential cell death at different times in the two decidual cell populations. However, other data suggest that apoptosis in the decidual tissue may be induced by locally produced signals. Transforming growth factor-β, a known inducer of apoptosis in uterine, prostate, and liver epithelial cells, was shown to induce apoptosis in primary decidual cell culture, whereas activin, which causes apoptosis in liver, is highly expressed in the decidual tissue, and its temporal and cell-specific expression appears to be well related to cell death during decidual development.

IV. MAJOR ROLES OF ANTIMESOMETRIAL AND MESOMETRIAL DECIDUAL CELLS

The physiological functions of the decidual tissues remain largely unresolved but appear to include protection of the maternal uterus against trophoblast invasion, nutrition of the implanting embryo, intrauterine immune protection, provision of a cleavage matrix for placental separation at parturition, and endocrine activity. Whereas the immune protection may be due in part to the uNK cells, the other functions of the decidua are related to the decidual cells themselves. These cells can be divided into two major different subpopulations not only according to their size, morphological appearance, and residential locations but also to their differential function. Antimesometrial and mesometrial cells differ by the genes they express and the role they play in pregnancy.

A. Mesometrial Decidual Cells

The mesometrial decidual cells appear to have the ability to limit trophoblast invasion and protect the uterus from uncontrolled damage due to trophoblast cell invasion. The mesometrial cells can restrict and direct trophoblast invasion for at least two distinct reasons. First, the mesometrial cells are loosely connected, allowing for trophoblast invasion between cells without massive cellular destruction. Second, these cells secrete potent protease inhibitors that appear to limit the destructive effect of proteases released by trophoblast cells.

The invasive behavior of the trophoblast is thought to be related to the secretion of proteolytic enzymes. In the rat, α2-macroglobulin, a potent proteinase inhibitor, is the most abundantly synthesized and secreted protein by the mesometrial decidual cells. Very little if any α2-macroglobulin is produced by the antimesometrial decidual cells in vivo. Among all proteinase inhibitors produced by the decidual tissue, α2-macroglobulin inhibits proteinase of all four classes and is exclusively synthesized and secreted in vivo by the decidual cells of the mesometrial decidua in which trophoblasts invade. The abundant secretion of potent protease inhibitors may be of critical importance for the limitation of tissue damage during placentation. The abundant expression of proteinase inhibitor in the mesometrial decidua supports the protective function of the decidua. Uncontrolled and destructive trophoblastic invasion in ectopic pregnancy can be attributed in part to the lack of mesometrial decidual tissue and the local deficiency of decidua-derived proteinase inhibitors.

The profound difference in cell transformation that occurs in the two opposite regions of the uterus has been known for a long time. The reason why the mesometrial cells differentiate much less than the antimesometrial cells is not well-known. However, mesometrial α2-macroglobulin appears to be, at least in part, responsible for the limited differentiation of the mesometrial cells in relation to the antimesometrial cells. Besides its ability to inactivate a wide range of proteinase, α2-macroglobulin also binds to and prevents the activity of a variety of growth factors. Growth factors are involved in cellular proliferation and differentiation in numerous tissues and many of them are indeed expressed in uterine endometrium and decidual tissue. The abundant secretion of α2-macroglobulin by the mesometrial decidual cells and little or none by the antimesometrial cells may be the cause of the profound differences between antimesometrial and mesometrial decidual cell differentiation.

B. Antimesometrial Decidual Cells

The antimesometrial cells differ from the mesometrial decidual cells not only by their size but also by the genes they express and the role they play in pregnancy. The antimesometrial cells are endocrine cells that secrete hormones which appear to play a role in the maintenance of pregnancy. The major proteins secreted by these cells in the rat belong to the prolactin (PRL) growth hormone family. Each family member has a molecular weight range from 27 to 29 kDa. Three different proteins have been shown to be synthesized by the rat decidua; prolactin-like protein B (PLP-B), decidual prolactin-related protein (dPRP), and decidual luteotropin (DLt). PLP-

B and dPRP have been cloned but their role has not been defined. Decidual luteotropin, the only decidual hormone that binds to PRL receptor, has an endocrine role and can maintain luteal progesterone production by the ovary. In addition to its luteotropic function, DLt plays a local role in the decidua. The decidual cells express PRL receptor and DLt acts on mesometrial decidual cells to stimulate the expression of α2-macroglobulin. By doing so the decidual PRL-like hormone appears to allow normal placentation to occur.

V. DECIDUAL GROWTH FACTORS AND CYTOKINES

Decidualization involves temporally and spatially coordinated endometrial cell proliferation, differentiation, and subsequent cell death. During this process, a variety of growth factors and cytokines are produced by decidual cells correlating to this remarkable tissue reorganization. The following are only a few examples with some speculated functions.

In the early development of decidual tissue, in addition to the increased permeability of blood vessels, increased mitosis in decidual endothelial cells is also observed suggesting an angiogenesis paralleling decidual formation. Messenger RNA of vascular endothelial growth factor (VEGF), a mitogen for endothelial cells which induces angiogenesis and increases vascular permeability, has been detected in the decidual cells in rodents. Its specific temporal and spatial expression in luminal epithelial cells and decidualizing stromal cells is closely correlated to the implantation site, vascular permeability, and angiogenesis in pregnant animals. Thus, decidual VEGF may be the signal or one of the signals responsible for the increased vascular permeability and angiogenesis at the early stage of decidualization.

Insulin-like growth factor-I (IGF-I) in the uterus is considered to be the mediator of estrogens' mitotic action and is involved in the adult uterine proliferation and differentiation. Both IGF-I and IGF-II are expressed in decidual tissue, but they may be involved more in the fetal development than in decidualization. IGF-binding protein-1 (IGFBP-1) is also a product secreted by decidual cells. It can be detected on Day 5 of pregnancy and pseudopregnancy and increases thereafter. Uterine IGFBP-1 has been proposed to play a role in regulating decidual tissue formation, embryo implantation, and trophoblast invasion since IGFBP-1 contains the RGD motif (Arg-Gly-Asp) which can directly interact with membrane-bound integrin receptors in fetal and maternal tissue membrane.

Epidermal growth factor family (including EGFα, HB-EGF, amphiregulin, β-cellulin, epiregulin, and heregulins), interleukins, tumor necrosis factors (TNFs), and many other cytokines/growth factors have been reported to be expressed in the decidual tissue with spatiotemporal patterns together with the expression of receptors corresponding to those growth factors and cytokines. Interleukins and TNFs are thought to regulate maternal homeopoiesis and immune responses. EGF and amphiregulin may play a role in the decidual induction and implantation regulation. Although the precise biological functions of each decidua-derived cytokine and growth factor are not clearly defined, it is likely that they all participate in a finely orchestrated and balanced complex network in order to provide a proper milieu for the maintenance of pregnancy.

See Also the Following Articles

Apoptosis (Cell Death); Decidua; Endometrium; IGF (Insulin-Like Growth Factors); Pseudopregnancy

Bibliography

Barbai, V., and Kraicer, P. P. (1996). Intrauterine signals and embryonic implantation. *Biological Signals* 5, 111–121.

Bell S. C. (1983). Decidualization: Regional differentiation and associated function. *Oxford Rev. Reprod. Biol.* 5, 220–271.

Gibori, G. (1994). The decidual hormones and their role in pregnancy recognition. In *Endocrinology of Embryo Endometrium Interaction* (S. R. Glasser, J. Mulholland, and A. Psychoyo, Eds.), pp. 217–222. Plenum, New York.

Gibori, G., Gu, Y., and Srivastava, R. K. (1995). Differential gene expression and programmed cell death in the two cell population forming the rat decidua. In *Molecular and Cellular Aspects of Periimplantation Processes* (S. K. Dey, Ed.), pp. 67–83. Springer-Verlag, New York.

Glasser S. R. (1990). Biochemical and structural changes in

uterine endometrial cell types following natural or artificial deciduogenic stimuli. In *Trophoblast Research* (H.-W. Denker and J. D. Aplin, Eds.), Vol. 4, pp. 377–416. Plenum, New York.,

Krehbiel R. H. (1937). Cytological studies of the decidual reaction in the rat during early pregnancy and in the production of deciduomata. *Physiol. Zool.* **10**, 212–233.

Lala, P. K., and Graham, C. H. (1990). Mechanisms of trophoblast invasiveness and their control: The role of proteases and protease inhibitors. *Cancer Metastasis Rev.* **9**, 369–379.

O'Shea, J. D., Kleinfeld, R. G., and Morrow, H. A. (1983). Ultrastructure of decidualization in the pseudopregnant rat. *Am. J. Anay.* **166**, 271–298.

Peel, S. (1989). Granulated metrial gland cells. In *Advances in Anatomy, Embryology and Cell Biology* (F. Beck, W. Hild, W. Kriz, R. Ortmann, J. E. Pauly, and T. H. Schiebler, Eds.), pp. 1–112. Springer-Verlag, New York.

Robertson, S. A., Seamark, R. F., Guilbert, L. J., and Wegmann, T. G. (1994). The role of cytokines in gestation. *Crit. Rev. Immunol.* **14**, 239–292.

Soares, M. J., Faria, T. N., Roby, K. F., and Deb, S. (1991). Pregnancy and the prolactin family of hormones: Coordination of anterior pituitary, uterine, and placental expression. *Endocr. Rev.* **12**, 402–423.

Welsh, A. O., and Enders, A. C. (1985). Light and electron microscopic examination of the mature decidual cells of the rat with emphasis on the antimesometrial decidua and its degeneration. *Am. J. Anat.* **172**, 1–29.

Deer

Edward D. Plotka
Marshfield, Wisconsin

I. Anatomy
II. Puberty
III. Estrus and Ovulation
IV. Annual Patterns of Reproduction
V. Pregnancy
VI. Concluding Remarks

GLOSSARY

antlers The solid deciduous horns of the deer family.
antlerogenesis The history and development of individual antler pairs.
buck A term applied to the adult males of the genera *Capreolus, Dama, Elaphodus, Hippocamelus, Hydropotes, Mazama, Muntiacus, Odocoileus, Ozotoceros,* and *Pudu.*
doe The current name for the female of a number of deer genera, including *Capreolus, Dama, Elaphodus, Hydropotes, Mazama, Mochus, Muntiacus, Odocoileus,* and *Pudu.*
hind The correct name for the female of most members of the genus *Cervus* except for *Cervus canadensis* (wapiti).
pedicle The bone protuberances on the top of a skull from which the antlers grow annually.
rut The mating season of the deer.
stag The adult male of certain species of deer, especially members of the genus *Cervus.*

The family Cervidae consists of approximately 40 species that are members of indigenous mammalian fauna inhabiting four continents. They range throughout North and South America, Europe, and Asia from the Artic islands to the equatorial rain forests. Most of the species still exist in their native habitat, which extends from cold polar regions to the tropics. The most northern of the species are *Alces alces* (moose) and *Rangifer tarandus* (reindeer/

caribou), with their range extending up to 80°N latitude. The greatest number of species inhabit the north temperate regions and include *Capreolus capreolus* (roe deer), *Cervus elaphus* (red deer and wapiti/elk), *Cervus nippon* (sika deer), *Dama dama* (fallow deer), *Odocoileus virginianus* (white-tailed), and *Odocoileus hemionus* (mule/black-tailed deer). South American deer include *Blastocerus* (marsh deer), *Hippocamelus* (huemul), *Mazama* (brocket), *Ozotoceros* (pampas), and *Pudu* (pudu). The remaining species of Cervidae are distributed throughout Eurasia with *Cervus elaphus* and *Rangifer tarandus* inhabiting both Eurasia and North America. *Cervus elaphus* are called red deer in Europe and Asia and wapiti/elk in the United States. Some references classify wapiti/elk as a separate species, *Cervus canidensis*. Within this presentation the specific animals used in experiments on *C. elaphus* will be referenced in parentheses behind the genus/species, e.g., *C. elaphus* (red) for Eurasian animals and *C. elaphus* (wapiti) for North American animals. Tropical species living within 20° of the equator include *Mazama*, *Ozotoceros*, *Axis*, *Muntiacus*, and *Cervus*. Approximately 15 species live primarily in the temperate and colder zones from 30° to 80° latitude, and 19 live in the tropical zones between 20°N and 20°S. The deer species with the widest distribution by latitude is *Odocoileus virginianus*, which includes many subspecies and has a range extending from approximately 50°N in North America to 15°S in South America. Other deer overlapping both temperate and tropical zones include *Hippocamelus*, *Mazama*, and *Pudu*.

I. ANATOMY

The anatomy of male and female Cervidae has been described for only a few species and, in general, is similar to that of other ungulates. The female anatomy consists of paired ovaries and oviducts, a bicornuate uterus, cervix, and vagina. *Elaphurus davidianus* and *Muntiacus reevesi* appear to be unique among the cervidae in that it has been reported that only the right horn of the uterus is developed.

The male anatomy consists of a penis, paired testes, epididymides, and accessory glands. The accessory sex glands of *Cervus elaphus* (red deer) reportedly consist of paired seminal vesicles, bulbourethral (Cowper's) glands and ampullae, as well as a prostate gland composed of a continuous body and disseminate parts. The accessory glands of *Capreolus* bucks are composed of paired ampullae and seminal vesicles, a prostate, also consisting of a continuous body and disseminate parts, and pelvic urethra. *Odocoileus hemionus*, *Muntiacus muntjak*, and *Rangifer tarandus* reportedly have a compact bulbourethral gland, whereas *Capreolus capreolus*, *Odocoileus virginianus*, and *C. elaphus* do not.

In general, spermatozoa from *Alces alces*, *C. elaphus*, *O. virginianus*, *Odocoileus hemionus*, *R. tarandus*, *Cervus nippon*, and *Dama dama* are similar in shape, size, and abnormalities to sperm of domestic ungulates. Cervid spermatozoa have been described as having a flat paddle-shaped head, with an apical ridge, that can be subdivided into a head cap, an equatorial segment, and a postnuclear sheath. A short neck connects the head with the tail, which is subdivided into a midpiece, a principal piece, and an endpiece and has a junction between principal and midpiece. Various defects have been noted, among which cytoplasmic droplets and midpiece defects appear most frequently. Representative ejaculate characteristics of some cervid species are presented in Table 1.

Normal constituents of seminal plasma include fructose, citric acid, glutamic-oxaloacetic transaminase, alkaline phosphatase, hyaluronidase, and protein. These constituents change with reproductive status. For example, total protein in seminal plasma from *C. elaphus* stags has been shown to be very high during the rutting period.

II. PUBERTY

Male *O. virginianus*, *O. hemionus*, *R. tarandus*, and *C. capreolus* in good physical condition are capable of breeding before one year of age. For *Capreolus*, *Rangifer*, and *O. virginianus*, sperm have been detected in the reproductive tract at 6, 4, and 6 months of age, respectively. In contrast, sperm are not detectable in *C. elaphus* stags until around 14 months of age. Puberty in male cervids is accompanied by phenotypic changes. The increase in secretion of testosterone by the testes at puberty stimulates the initial

TABLE 1
Representative Sperm Concentrations and Motility and Ejaculate Volume for Various Cervidae

Genus/species	Sperm concentration	Sperm motility (%)	Ejaculate volume (ml)	Reference
Alces alces	$0.85–2.85 \times 10^6$	50–60	1.0–2.5	Krzywinski et al. (1987)
Axis axis	$44–419 \times 10^7$	80–100	0.2–0.8	Haigh et al. (1993)
Cervus elaphus	$0.24–1.18 \times 10^9$	—	1.0–2.0	Strzezek et al. (1985)
Cervus eldi	$2.67–16.32 \times 10^8$	43–80	0.7–3.49	Monfort et al. (1993a)
Dama dama	$1.7–4.4 \times 10^9$	95–100	0.6–1.3	Asher et al. (1990)
Elaphurus davidianus	$420–950 \times 10^6$	80–95	3.0–8.0	Asher et al. (1988b)
Odocoileus virginianus	$60–2085 \times 10^3$	—	0.5–1.75	Bierschwal et al. (1970)
Odocoileus hemionus	$100–700 \times 10^6$	—	—	West and Nordan (1976)

TABLE 2
Timing of Puberty in Various Species of Cervidae

Genus/species	Onset of ovarian function (months)	Reference
Alces alces	24	Sheng and Ohtaishi (1993)[a]
Axis axis	14–17	Nowak and Paradiso (1983)[b]
Capreolus capreolus		Short and Mann (1966)
	14	Sheng and Ohtaishi (1993)
Cervus albirostris	18–30	Sheng and Ohtaishi (1993)
Cervus elaphus (red)	16–17	Fisher et al. (1992)
	24–28	Nowak and Paradiso (1983)
Cervus eldi	18–24	Sheng and Ohtaishi (1993)
Cervus nippon	16–18	Nowak and Paradiso (1983)
Cervus porcinus	13–18	Sheng and Ohtaishi (1993)
Cervus unicolor	18–24	Sheng and Ohtaishi (1993)
Dama dama	16	Nowak and Paradiso (1983)
Elaphurus davidianus	30	Grzimek (1975) as cited by Nowak and Paradiso (1983)
Hydropotinae inermis	9–10	Sheng and Ohtaishi (1993)
Mazama americana	13–14	Branan and Marchington (1987)
Muntiacus muntjac	10	Chapman (1992), Pei and Liu (1994)
Mochus moschiferus	18	Sheng and Ohtaishi (1993)
	12	Nowak and Paradiso (1983)
Odocoileus virginianus	6–8	Haugen (1975)
Rangifer tarandus	4–8	Leader-Williams (1979)

[a] Data from this reference are presented for species residing in China.
[b] Presented without reference to original work or citation.

development of the antler pedicles, and the subsequent cycle in secretion of testosterone dictates the seasonal casting and regeneration of the antlers.

Females of most deer species ovulate for the first time either when under 12 months of age or between 12 and 24 months. Only two species, *E. davidianus* and *C. elaphus*, have been reported to be over 24 months of age at first ovulation (Table 2). Generally, the smaller species reach puberty earlier, but this pattern is confused because seasonal breeding affects the time to the first breeding opportunity. Age at first ovulation varies considerably between populations and years and appears to be related to a critical body weight. Thus, a high percentage of deer residing in areas with more and higher quality vegetation may ovulate and become pregnant at 7–9 months of age, whereas animals of the same species residing in areas with poorer nutrition will not breed until the following year. Additionally, earlier puberty has been reported for several species maintained in pens or with access to higher quality forage, presumably due to an early achievement of necessary body weight. For *R. tarandus*, it was reported that pregnancy rates in yearlings were related to fat reserves. Although photoperiod may play a role in controlling age at first estrus, this influence appears less important than the role played by nutrition.

After puberty almost all adult female cervids ovulate every year. However, in some polytocous species, single ovulations may occur during first estrous season and pregnancy will typically result in delivery of only one young. Peak reproductive performance is generally achieved during the second or third year of life. A decline in pregnancy rate or number of fawns per doe with advancing age has been reported for *C. elephas* and *O. virginianus*. Reproductive pauses have also been reported to occur with some degree of regularity in *R. tarandus* and may occur in other cervid species as well.

III. ESTRUS AND OVULATION

Based on parturition data, various reports have demonstrated that from 69 to 95% of conceptions occur within a period of 10–21 days of the first observed estrus for *A. alces*, *D. dama*, *C. elaphus*, *E. davidianus*, *O. virginianus*, *O. hemionus*, and *R. tarandus*, indicating that in these species a very high percentage of females get pregnant during the first behavioral estrus of the season. Table 3 presents the length of breeding season and other characteristics of the estrous cycle for unmated cervidae where data have been reported. Unfortunately, in some cases, the length of the breeding season has been assessed by the spread of births or by observed rutting behavior. This method underestimates the length of the breeding season because of the high efficiency of the male at detecting estrus and fertilizing the female.

The majority of cervid species are polyestrus, with only *C. duvauceli* and *C. capreolus* stated to be monestrus. In some cervids there is considerable ovarian activity before the first overt estrus. Thus, follicular cycles have been reported in both *O. virginianus* and *O. hemionus* and ovulation without overt behavioral estrus ("silent heat") before the first observable estrus has been documented for *A. alces*, *O. virginianus*, *O. hemionus*, and *R tarandus*. Several studies suggest that *C. elaphus* also has ovulations before first estrus. It appears that in cervids, a preovulatory fall in progesterone prior to the rise in estradiol and luteinizing hormone (LH) is necessary to initiate behavioral estrus.

Hormonal changes during the estrous cycle have been reported for *O. virginianus*, *O. hemionus*, *C. elaphus*, *C. eldi*, *C. capreolus*, and *D. dama*. Basically there is an increase in estradiol prior to the onset of behavioral estrus. A preovulatory LH surge occurs just prior to or coincident with the onset of estrus (Fig. 1). Ovulation occurs during or shortly after the end of estrus and progesterone rises during the luteal phase. After a luteal phase of variable length from 13 to 30 days, depending on species, progesterone levels decline and the series is repeated in all but the two monestrous species. The fundamental changes that occur are similar to those reported for domestic species. However, the length of the luteal phase differs between species and thus minor differences occur in the length of the estrous cycle (Table 3). Progesterone levels in peripheral blood of *C. capreolus* remain elevated for about 6 months, suggesting that an obligate luteal phase similar to the length of

TABLE 3
Length of Breeding Season, Estrous Cycle, Heat, Time of LH Surge, Age at First Estrus, and Ovulation

Species	Length of breeding season (months)	Length of estrous cycle (days)	Duration of heat (hours)	Onset of LH surge relative to onset of estrus (hr)	Ovulation (hours after onset of estrus)	Reference
Alces alces		25–30				Schwartz cited by Monfort et al. (1993b)
		20–22	24			Nowak and Paradiso (1983)
	3					Shen and Ohtaishi (1993)
Axis axis		17–21	~24–36			Chapple et al. (1993)
		27–30				Asdell (1964)
Capreolus capreolus	1	Monestrus				Flint et al. (1994)
				−24–48		Sempéré et al. (1993)
Cervus albirostris	1					Sheng and Ohtaishi (1993)
Cervus elaphus (red deer)	4–5	18–21	25; 2 days			Kelly et al. (1985), Krzywinski and Jaczewiski (1978), Morrison (1960b) cited by Sadlier (1987)
				−8 to +8	20–26	Asher (1992)
		17–19	12–20		2.5–9.5 (observations by ultrasound)	Gizejewski et al. (1993)
Cervus duvauceli		Monestrus				Schaller (1987)
Cervus eldi	8–9	14–23	12–24			Monfort et al. (1990)
	5					Sheng and Ohtaishi (1993)
Dama dama	3–4	19–25, 24–26				Asher et al. (1990), Sadlier (1987)
				0	~24	Asher et al. (1986, 1992)
Elaphurus davidianus	5	18–21				Brinklow and Loudon (1993), Wemmer et al. (1989), Grzimek (1975) cited by Nowak and Paradiso (1983)
		17–25				
				−10		Curlewis et al. (1988), Loudon et al. (1990)
Hydropotinae Inermis	3					Sheng and Ohtaishi (1993)
Mochus moschiferus	3					Sheng and Ohtaishi (1993)
Muntiacus muntiacus	2–3; possibly nonseasonal	14–21	Approx 2 days			Barrette (1977) cited by Sadlier (1987)
						Medway (1978) cited by Nowak and Paradiso (1983)
Muntiacus reevesi	12	14–15				Yahner (1979) cited by Sadlier (1987)
	10–12					Sheng and Ohtaishi (1993)
						Chapmen (1992)
Odocoilues hemionus	4–5	22–25				West (1968)
		22–29				Thomas and Cowan (1975) cited by Sadlier (1987)
Odocoileus virginianus	4 (in temperate climates)	22–29	24			Severinghaus (1955)
		28	39.9 ± 2.4			Cowan (1956), Cheatum and Morton (1946)
						Thomas and Cowan (1975), Verme and Ozoga (1981)
Rangifer tarandus		13–33	50			Ropstad et al. (1995)
						Bergerud (1974) cited by Sadlier (1987)

the embryonic diapause in pregnant animals occurs in this species. Peak concentration of progesterone in peripheral blood serum during the estrous cycle is correlated with the number of ovulations and corpora lutea formed.

Studies utilizing estrous synchronization procedures have determined that the preovulatory estradiol peak occurs 36–48 hr before the onset of estrus in *C. elaphus* (red deer) and 42–54 hr before the onset of behavioral estrus in *E. davidianus*. Luteiniz-

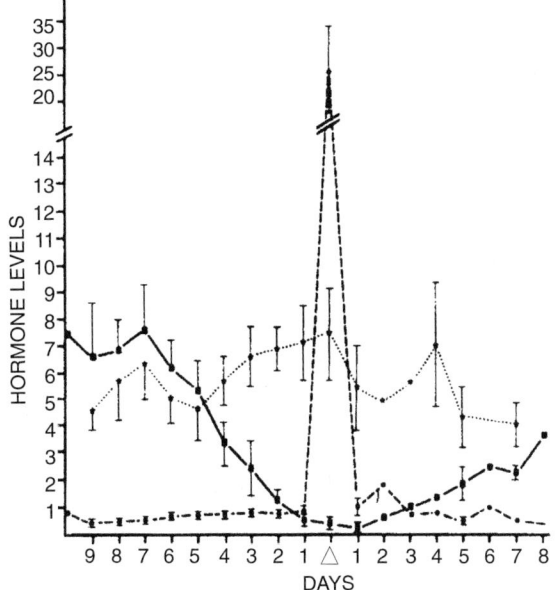

FIGURE 1 Periovulatory changes in serum LH (- - -), estradiol (· · ·), and progesterone (—) concentrations (mean ± SEM) from 10 days before estrus through 8 days after estrus. Estradiol is expressed in pg/ml, LH in ng NIH-LH-S7/ml, and progesterone in ng/ml (data from Plotka et al., 1980).

ing hormone peaks coincident with the onset of estrus in the *C. elaphus* hinds. Although a direct comparison of the timing of these events in synchronized and natural cycles has not been performed, the timing of the estradiol rise in synchronized cycles is consistent with estradiol changes reported in natural cycles in other cervidae.

Although proportional changes in estradiol around the time of estrus and ovulation appear similar to those recorded for other polyestrous ungulates, absolute levels may differ considerably. Thus, estradiol levels have been reported to be 57–107 pg/ml in nonpregnant *C. capreolus*, 5–30 pg/ml in nonpregnant *O. virginianus*, 11–25 pg/ml in *D. dama*, 5–10 pg/ml for *C. eldi*, and up to 124 pg/ml in *C. elaphus*. The variability of these results suggest that there may be considerable differences in estradiol dynamics among cervid species.

Luteinizing hormone surges have been recorded around estrus but the timing of blood sampling has been sufficiently infrequent that absolute conclusions about the time of the beginning and peak of the LH surge cannot be derived for most species. However, in all cervids studied, an LH surge has been detected within 24 hr of estrus. The duration of the LH surge has been reported to be approximately 12 hr for *E. davidianus* hinds and *D. dama* does and 15.5 + 1.2 hr for *C. elaphus* (red deer) hinds.

The pulsatile pattern of LH secretion during the estrous cycle has been reported in *E. davidianus*. The mean LH pulse frequency during the follicular phase was reported at 0.58 pulses/hr, similar to that reported for cattle and sheep. LH pulse frequency in the luteal phase was 0.24/hr. Pulse amplitude in the two phases of the cycle were similar.

The pulsatile pattern of LH secretion during anestrus has also been described for *E. davidianus*. LH frequency is low early in anestrus (3.3 pulses/18 hr), rises during midanestrus to 8.4 pulses/18 hr, and declines slightly to 6.4/hr late in anestrus. These changes were also associated with significant changes in the response to a fixed dose of gonadotropin-releasing hormone (GnRH) and have been interpreted to indicate that *Elaphurus* have an early period of deep anestrus. Ovulation was more difficult to induce during the period of early (deep) anestrus than during late anestrus with similar GnRH regimens, confirming the differences between the two periods.

IV. ANNUAL PATTERNS OF REPRODUCTION

Reproduction is the last of essential life processes to be satisfied by the body's energy requirements. Thus, for reproduction to occur, energy and nutrients are required above those needed for normal body maintenance. Consequently, those species living in temperate and colder climates have synchronized the high demands of late pregnancy and lactation to the season of greatest food abundance. Some tropical cervid species have similar synchrony with rainfall and lush vegetation. Because deer are large animals with long gestation periods, the timing of conception occurs under quite different environmental conditions than those conditions favoring survival of offspring. In temperate species, young are produced in late spring or early summer when food of high nutritional value is most abundant. Estrus occurs the previous late summer or fall. In general, there is a close relationship between latitude and the timing of the breeding season in cervidae. Thus, cervids living in temperate regions exhibit seasonal reproduction, whereas most tropical species experience little seasonality when living in their natural habitats.

A. Seasonal Cycles in the Male

Males adapt their breeding cycle to the availability of females in estrus. In temperate zone cervids this cycle is highly synchronized among individuals of the same species. Males are fertile only during the hard antler phase after there has been significant development of the testes. A seasonal cycle of testicular and accessory sex gland activity has been found in *C. capreolus*, *C. elaphus*, *C. nippon*, *D. dama*, *O. hemionus*, and *O. virginianus*. The period of maximal activity occurs during the rut, and the organs are least active during the sexually quiescent period after the breeding season. As much as a sixfold increase in testis weight has been reported for *C. capreolus*, *R. tarandus*, *C. elephus*, *D. dama*, and *O. virginianus* during the period of testicular recrudescence. These, as well as endocrine changes, are reflected in the secondary sexual characteristics, many of which are easily discernible, phenotypic changes. For example, in most cervidae only males develop antlers. Antler growth is initiated during the period of low testosterone secretion but continues while testosterone secretion is increasing. When testosterone secretion reaches an appropriate level, antler hardening and polishing occurs. Removal of the testes after initiation of antler growth prevents antler hardening and polishing. Thus, it appears that both the initial growth of the antlers at puberty and the seasonal replacement of the antlers in the adult are events dependent on the secretion of testosterone from the testes. The only species of deer in which both sexes normally produce antlers is *R. tarandus*. In the female of this species, a gonadal hormone from the ovaries, suggested to be estradiol, appears to induce and maintain the hard antler state.

The highly synchronized testicular cycles have maximum activity that corresponds with the time when females are in estrus and ovulate. For the northern temperate species, *C. capreolus* breed in July and August, *C. elaphus* in September (wapiti) or October (red deer), *C. nippon* and *D. dama* in October, and *O. virginianus* in November. Using north temperate *O. virginianus* as an example, Fig. 2 shows the relationship between the seasonal changes in LH, testicular size, testosterone secretion, and the antler and neck circumference cycles. The seasonal increase in testosterone secretion from the testes induces calcification of the antlers, the hard antler phase, and the dramatic increase in neck circumference as well as overt sexual and aggressive behavior characteristic of the rut. *Odocoileus virginianus* bucks remain fertile over the fall and early winter during the time that the nonpregnant does are in estrus. When testosterone secretion decreases below a critical level in late winter or early spring, neck circumference regresses and fertility of the bucks decline. The withdrawal of testosterone when the testes become fully regressed allows the old antlers to be cast. Differences in the timing of these events do occur with species and *C. elaphus* (red deer) stags remain fertile and retain their antlers throughout the winter during the period that females may show estrus. Generally, the larger species with the longer gestations rut and breed earlier and the smaller species breed later in the season. *Capreolus capreolus* are an exception, however, because they breed very early during the summer. In this species, implantation of the blastocyst is delayed for 5 months so that births occur in the late spring.

Despite phase differences between the various species, the interactive dynamics of the pituitary–gonadal axis appears similar. Luteinizing hormone concentrations peak approximately 3 months before follicle-stimulating hormone (FSH) and testosterone concentrations. Follicle-stimulating hormone concentrations more closely parallel the testosterone cycle and sperm production than LH levels. These data support the concept that LH is critical for initiating testicular steroidogenic activity and FSH and testosterone are important for supporting spermatogenesis during the peak of the circannual cycle. The role of inhibin in this scheme of events has only been studied in *C. eldi*. In that study, immunoreactive inhibin concentrations paralleled FSH secretion during testicular recrudescence and regression. However, in-

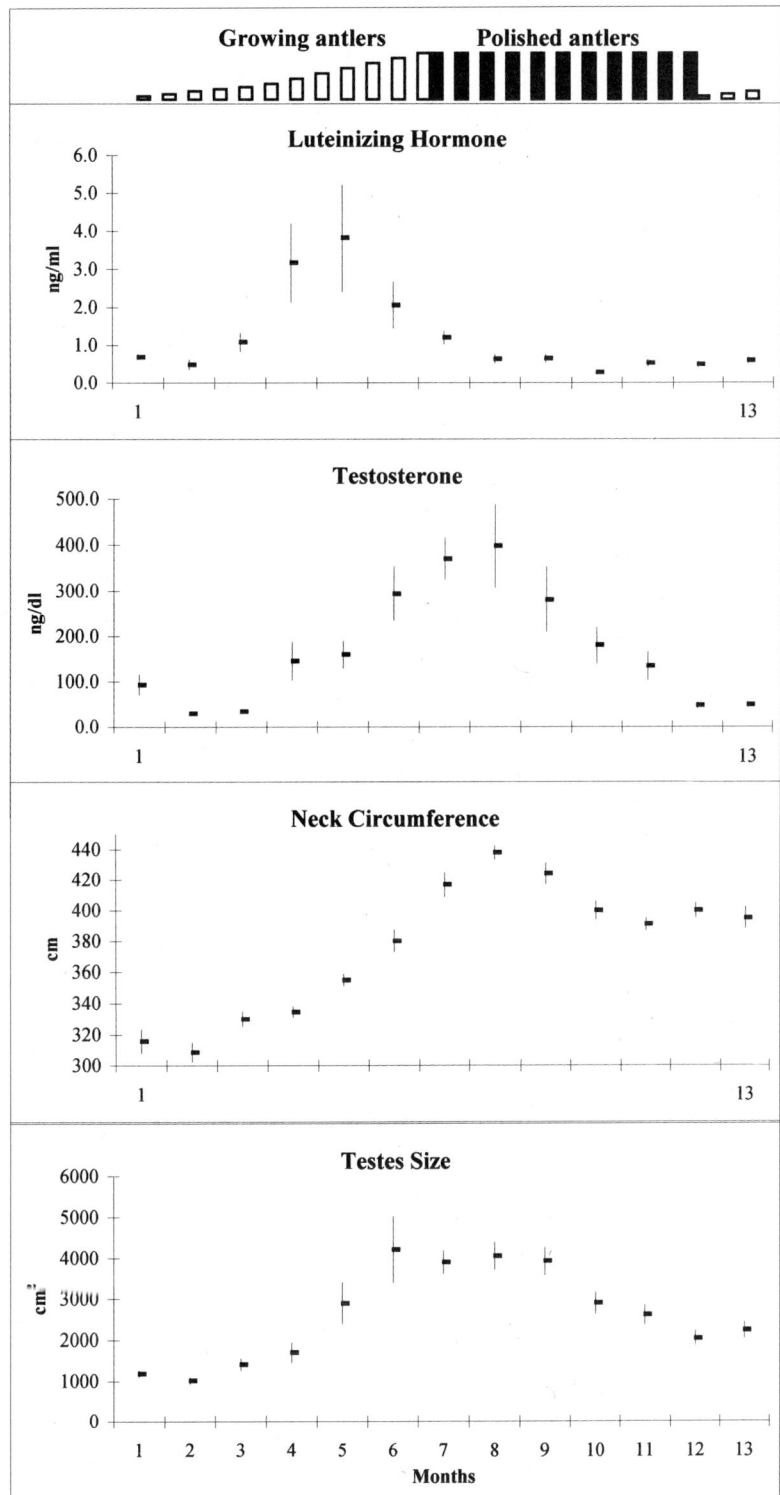

FIGURE 2 Circannual changes in serum hormone and physical parameters for seven *O. virginianus* bucks over 13 months. Summer solstice occurred during Month 3 and the vernal equinox during Month 6. Months 1 and 13 are the same but for sequential years.

hibin secretion continued to rise for 3 weeks after FSH began to decline, suggesting that changes in hormonal interrelationships are necessary for stimulating and suppressing reproductive processes.

B. Seasonal Cycles in the Female

In the species studied in detail over several sequential years, there appears to be very little difference in the range and median dates of conception between years in the same population. There is an occasional animal within each species that does not get bred or conceive during the first or second estrus and may deliver out of the normal season. Although this appears to be fairly uncommon, these events are sufficiently frequent in cervidae to suggest that modifying factors influence the ability of individuals to ovulate and get pregnant.

Tropical species of deer exhibit less synchrony and some males may be in hard antler and fertile at any given time of the year. These cervidae typically remain aseasonal breeders even after being transferred to temperate environments. For example, *A. axis* held in temperate Europe maintain a distinct 12-month cycle of antler growth, testis size, and neck circumference but with little synchrony between individuals. Such characteristics have been reported for *O. virginianus* living in the tropics, where females are potentially is estrus throughout the year. *Muntiacus muntiacus*, on the other hand, have been reported to exhibit a high degree of synchrony in antler cycles but have no marked seasonal variation in testis size or neck circumference.

C. Role of the Pineal Gland

The breeding season of temperate-zone *C. elaphus*, *O. hemionus*, and *O. virginianus* has been shown to be controlled by photoperiod. The annual change in photoperiod is a reliable clue from the environment and responding to this cue is prime importance for survival in temperate and cold climates. Support for this hypothesis has been provided by the extensive studies of Richard Goss demonstrating that antler growth and hardening can be manipulated with artificial light. Further studies on *O. virginianus*, *C. elaphus* (red deer), and *D. dama* have implicated the pineal gland and its primary hormone, melatonin, as interpreting the changing photoperiod signal in the control of the timing of seasonal reproduction. Evidence for this role has been provided by demonstrating that pineal removal (*O. virginianus*) or severing the superior cervical ganglia (SGX), the predominant route for transfer of photosensory inputs from the retina of the eye to the pineal gland (*C. elaphus*, red deer), disrupts the timing of seasonal reproduction. The effect of SGX in *C. elpahus* is similar to the effect of pinealectomy in *O. virginianus*. Interestingly, pinealectomized animals appear to have a circannual rhythm but the rhythm is delayed by 2 or 3 months and appears disassociated from photoperiodic regulation (Fig. 3). Ganglionectomized *C. elaphus* stags, however, have been reported to revert to a normal circannual rhythm of antler growth and casting. These stags also respond to changes in artificial light, suggesting that ganglionectomized animals can still interpret changes in photoperiod.

Melatonin, secreted by the pineal gland, translates the changing photoperiod effects and influences the timing of the seasonal reproductive cycle in photoperiodic species. Because melatonin is secreted at night, the duration of the increased melatonin secretion varies with the length of the dark period and provides an index of the prevailing photoperiod, i.e., long or short days. Support for the hypothesis that the change from long to short days during increasing

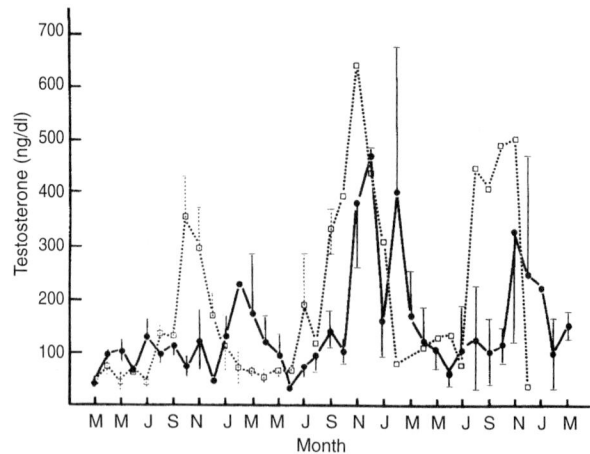

FIGURE 3 Circannual changes in baseline serum testosterone levels in pinealectomized (●) and control (□) *O. virginianus* bucks over 3 years.

LH secretion activates the reproductive axis has been provided by relating patterns of melatonin secretion and associated changes in pulsatile LH and testosterone secretion in *D. dama* and demonstrating that administration of melatonin to deer maintained in temperate climates and exposed to long days induces premature gonadal activity in both males and females (*O. virginianus*, *C. elaphus*, and *D. dama*). Thus, artificially extending the duration of elevated melatonin by administration appears to simulate short days.

Melatonin administration during long days also suppresses prolactin secretion in *D. dama* and *C. elaphus*. Prolactin has been implicated in influencing seasonal cycles in antler growth, summer coat, and appetite. Seasonal changes in prolactin secretory patterns have been documented in several cervid species to be directly related to day length, with maximum concentrations occurring during long days and minimum concentrations during short days. Schulte *et al.*, however, although demonstrating alterations in prolactin secretion in pinealectomized animals, did not observe a change in the circannual rhythm of the hormone. Pinealectomized animals had higher prolactin levels in summer under long days than in winter during short days despite a delay in the antler cycle. These data suggest that other mechanisms play a greater role in controlling antler cycles.

Although photoperiod appears to be the major mechanism by which seasonal rhythms are driven or synchronized, other factors that influence nutrition, e.g., rainfall, have been implicated in controlling seasonal reproduction. For example, 94% of births in *Hippocamelus antisensis* (North Andean deer, Taruca) occur toward the end of the rainy season, suggesting that the reproductive cycle is markedly associated with seasonal climatic patterns.

V. PREGNANCY

Within the cervidae there is considerable variation in reproductive strategies and physiology, manifest as differences in parameters such as seasonality, litter size, and gestation length (Table 4). Even the timing of embryonic implantation varies considerably, with *C. capreolus* exhibiting delayed implantation for several months. *Capreolus capreolus* is unique among the artiodactyl because it is the only species in the cervid family that exhibits the phenomenon of delayed implantation. The entire gestation period of the *Capreolus* doe lasts approximately 10 months, with a 5-month embryonic diapause. During the first 2 weeks in January (in the Northern Hemisphere), a normal rate of embryonic growth is resumed and the blastocyst rapidly elongates. Embryonic elongation is associated with a significant increase in the concentration of total plasma estrogens and is followed by placental attachment. Activated *Capreolus* blastocysts contain aromatase activity and the ability to convert androstenedione to estrone.

A high percentage of adult cervid females become pregnant. As with ovulation rate, the proportion which actually become pregnant is influenced by nutrition. Nutrition may also influence the number of ovulations in polytocous species. However, two species, *E. davidianus* and *M. reevesi*, may be physically incapable of carrying more than one fetus because it has been reported that only the right horn of the uterus is developed.

Endocrine changes in most cervid species are characterized by a rise in progesterone within 8 hr following ovulation to maximum luteal concentrations which are related to the number of ovulations and amount of luteal tissue formed. Concentrations of progesterone in peripheral blood remain relatively constant during pregnancy in several cervid species, with fluctuations related to secretion and probably sampling. Thus, studies in *O. virginianus* and *C. elaphus* (red deer) have demonstrated a close correlation between the number of corpora lutea present on the ovary and the peripheral concentration of progesterone (Fig. 4). These data suggest that the ovary is the principal, if not the sole, source of progesterone during pregnancy. The effect of ovariectomy during pregnancy, however, has been variable. Asher *et al.* demonstrated that *C. elaphus* (red deer) rely on luteal support of pregnancy for the first 75 days of pregnancy. Because this period coincides with extensive placental development in *C. elaphus*, the authors concluded that the fetoplacental unit is not a significant source of progesterone synthesis through Day 75 of gestation. Additionally, treatment of *R. tarandus* with cloprostenol during the first 3 months of pregnancy results in a marked fall in plasma progesterone con-

TABLE 4
Characteristics of Gestation in Cervidae

Species	Gestation length (days)	Ovarian dependent	Usual no. of young	Reference
Alces alces	226–244		2 polytocous	Skuncke (1949) cited by Schaller (1987)
	226–26		1–3	Nowak and Paradiso (1983)
Axis axis	231–237		Monotocous	Chapple (1993)
	238–242			Mylea et al. (1992)
Axis porcinus kuhli	243			Whitehall (1972)
	220–230		1 (or 2)	Sheng and Ohtaishi (1993)
Blastocerus dichotomass	~270			Nowak and Paradiso (1983)
Capreolus capreolus	294–300	Yes	2 polytocous	Sempéré et al. (1992)
	270–280		1–3	Sheng and Ohtaishi (1993)
Cervus albirostris	220–230		1 monotocous	Sheng and Ohtaishi (1993)
	246			Yu et al. (1993)
Cervus elaphus (red deer)	227–234	At least first 75 days	1 monotocous	Lincoln et al. (1970)
	223–227			Arman (1974)
Cervus elaphus (wapiti)	240–256		Monotocous	Flook (1970) cited by Hudson et al. (1991)
Cervus duvauceli	240–250			Schaller (1967)
Cervus eldi	236–244		Monotocous	Desai and Malhotra (1978), Blakeslee et al. (1979)
Cervus nippon	214–234		1 monotocous	Haensel (1980), Miura (1980), Sheng and Ohtaishi (1993)
Cervus timorensis	249		Monotocous	Van Mourik et al. (1986)
Cervus unicolor	240–270		1	Sheng and Ohtaishi (1993)
Dama dama	225–234	Yes	Monotocous	Prell (1938) as cited by Schaller (1987)
Elaphurus davidianus	283–284		Monotocous	Brinklow and Loudon (1973), Shaller and Hamer (1978)
	285–300			Nowak and Paradiso (1983)
	250–270		1–2	
Hippocamelus	240–270			Merkt (1987) (based on difference between fawn sightings and breeding)
Hydropotinae Inermis	180–210		2–3 polytocous	Chapman (1974)
	168–170		1–5	Sheng and Ohtaishi (1993)
Mazama americana	222–228		1 monotocous	Macnamara and Eldridge (1987)
Moschus moschiferus	160		Polytocous	Gupta and Jain (1980) cited by Schaller (1987)
	178–185		1–3	Sheng and Ohtaishi (1993)
Muntiacus muntiacus	222–238		Monotocous	Dubost (1971) cited by Schaller (1987)
	210–214			Sheng and Ohtaishi (1993)
Muntiacus reevesi	209–220			Chaplin and Dangerfield (1973)
	210–214		1–2	Sheng and Ohtaishi (1993)
Pudu pudu	202–233		Monotocous	Hershkovitz (1982) cited by Macnamara and Eldridge (1987)
Odocoileus hemionus	199–207		2 polytocous	Golly (1957) cited by Schaller (1987)
Odocoileus virginianus	196–213	Yes	2+ polytocous	Plotka et al. (1980)
Rangifer tarandus	208–230	Inconclusive data	Monotocous	McEwan and Whitehead (1972) and Bergerud (1975) cited by Schaller (1987)
	227–229		1–2	Nowak and Paradiso (1983)

centrations and abortion in most animals. Progesterone concentrations in uterine vein blood of this species have been reported to be similar to concentrations in peripheral blood during late pregnancy, further supporting the hypothesis that the placenta is not a significant source of progesterone. However, the fact that ovariectomy of pregnant *C. elaphus* and *R. tarandus* is not always followed by abortion suggests that other sources of hormones may be influencing pregnancy, particularly in the latter stages.

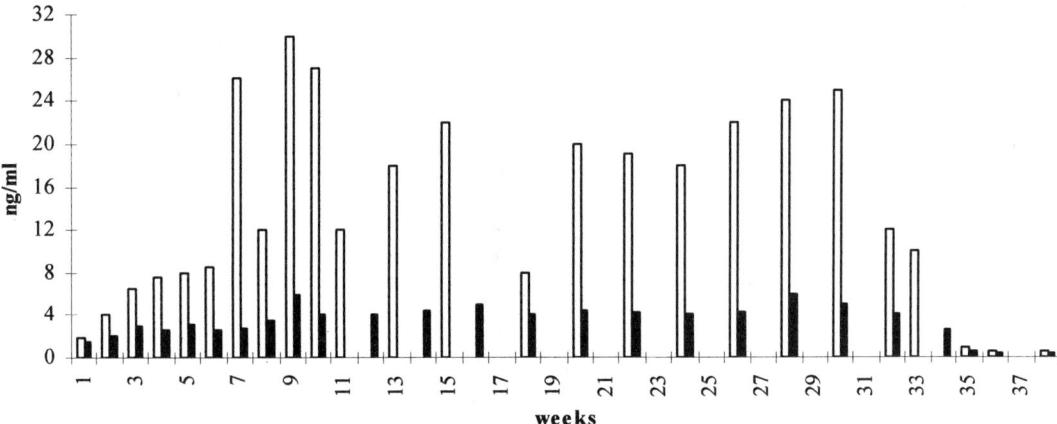

FIGURE 4 Serum progesterone concentrations in pregnant C. elaphus hinds having 1–2 or >3 corpora lutea. Parturition occurred around Weeks 34 or 35 (data from Kelly et al., 1982).

Contrary to the results in *Cervus* and *Rangifer*, ovariectomy of pregnant *O. virginianus* does results in abortion regardless of stage of pregnancy, suggesting that the ovaries are essential for the maintenance of pregnancy in this species.

In contrast to the relatively constant progesterone concentrations during pregnancy in *O. virginianus* and *C. elaphus*, *C. eldi* exhibit progressively rising pregnanediol glucuronide (the principal metabolite of progesterone) excretory patterns throughout pregnancy. The patterns exhibited in that species are similar to circulating progesterone patterns in *C. capreolus* which plateau during early pregnancy but increase following implantation. Rising progesterone concentrations during pregnancy suggest that the growing placenta is secreting more progesterone as it increases in size coincident with the growing fetus. These data, taken collectively, suggest that the corpus luteum is essential for maintenance of pregnancy in many but not all cervid species and demonstrate the variation in reproductive strategies and physiology between species of cervidae.

One explanation for the variability of results after ovariectomy may come from the observations of unexplained elevations in plasma progesterone concentrations that have been reported in *C. elaphus*, *D. dama*, and *O. virginianus*. Convincing evidence that the adrenal can be a significant source of progesterone in both males and females of *O. virginianus* and *D. dama* has been provided by adrenal suppression and stimulation studies. An adrenal source of progesterone secretion may affect various aspects of the reproductive process by helping to maintain pregnancy during periods of stress by neutralizing the pregnancy-terminating action of cortisol and/or by inhibiting estrus and ovulation during the breeding season and preventing pregnancy under highly stressful conditions.

Estrogen levels, on the other hand, are low during most of pregnancy and rise 5 or 6 weeks before parturition. This has been demonstrated in *O. virginianus*, *C. elaphus* (red deer), *C. eldi*, *E. davidianus*, and *A. alces*. Thus, estrogen secretion during pregnancy may be more consistent among cervid species.

VI. CONCLUDING REMARKS

One can see that most of the information regarding reproduction in cervidae has come from less than one-third of the species. It is clear that information from one cervid species cannot be extrapolated to other cervid species. This is exemplified by the differences in estrous cycle length, luteal and placental function during pregnancy, and dependence on environmental factors. Our knowledge of cervidae has increased dramatically over the past 20 years as a result of a large amount of research on a few species. Cervid species from temperate and colder climates have adapted to environmental changes by conforming to a rigid circannual rhythm entrained by changes in photoperiod that reflect seasonal patterns in vegetation and food availability. Tropical species are much less restricted in their breeding season but appear to operate under similar physiological mechanisms although tuned to different environmental cues. Assisted reproduction techniques in the form of artificial insemination, in vitro fertilization, and embryo transfer have been tried in some of the more commercial cervid species with variable results. The difficulty of obtaining consistent results demonstrates the paucity of knowledge regarding cervid reproduction and heralds the need for further research and understanding. A more complete understanding of cervid reproductive physiology would be helpful for applying assisted reproduction techniques for trying to preserve the 18 cervid species classified by the World Conservation Union as threatened or endangered.

Bibliography

Adam, C. L., and Atkinson, T. (1984). Effect of feeding melatonin to red deer (*Cervus elaphus*) on the onset of the breeding season. *J. Reprod. Fertil.* **72**, 463–466.

Adam C. L., McDonald, I., Moir, C. E., and Pennie, K. (1988). Foetal development in red deer (*Cervus elaphus*). *Anim. Prod.* **46**, 131–138.

Adam, C. L., Moir, C. E., and Shiach, P. (1989). Plasma prolactin concentrations in barren, pregnant and lactating red deer (*Cervus elaphus*) given melatonin to advance the next breeding season. *Anim. Reprod. Sci.* **18**, 77–86.

Aitken, R. J. (1974). Delayed implantation in roe deer (*Capreolus capreolus*). *J. Reprod. Fertil.* **39**, 225–233.

Aitken, R. J. (1981). Aspects of delayed implantation in the roe deer (*Capreolus capreolus*). *J. Reprod. Fertil. Suppl.* **29**, 83–95.

Arendt, J. (1986). Role of the pineal gland and melatonin in seasonal reproductive function in mammals. *Oxford Rev. Reprod. Biol.* **8**, 266–320.

Argo, C. M., and Loudon A. S. I. (1992). Effect of age and time of day on the timing of the surge in luteinizing hormone, behavioural oestrus and mating in red deer hinds (*Cervus elaphus*). *J. Reprod. Fertil.* **96**, 667–672.

Asdell, S. A. (1964). *Patterns of Mammalian Reproduction*, 2nd ed. Cornell Univ. Press. Ithaca, NY.

Asher, G. W., Barrell, G. K., and Peterson, A. J. (1986). Hormonal changes around oestrus of farmed fallow deer, *Dama dama. J. Reprod. Fertil.* **78**, 487–496.

Asher, G. W., Day, A. M., and Barrel, G. K. (1987). Annual cycle of liveweight and reproductive changes of farmed male fallow deer (*Dama dama*) and the effect of daily oral administration of melatonin in summer on the attainment of seasonal fertility. *J. Reprod. Fertil.* **79**, 353–362.

Asher, G. W., Barrell, G. K., Adam, J. L., and Staples, L. D. (1988a). Effects of subcutaneous melatonin implants on reproductive seasonality of farmed fallow deer (*Dama dama*). *J. Reprod. Fertil.* **84**, 679–691.

Asher, G. W., Adam, J. L., Otway, W., Bowmar, P., Van Reenan, G., MacKintosh, C. G., and Dratch, P. (1988b). Hybridization of Pere David's deer (*Elaphurus davidianus*) and red deer (*Cervus elaphus*) by artificial insemination. *J. Zool. (London)* **215**, 197–204.

Asher, G. W., Peterson, A. J., and Duganzich, D. (1989). Adrenal and ovarian sources of progesterone secretion in young female fallow deer, *Dama dama. J. Reprod. Fertil.* **85**, 667–675.

Asher, G. W., Kraemer, D. C., Magyar S. J., Brunner, M., Moerbe, R., and Giaquinto, M. (1990). Intrauterine insemination of farmed fallow deer (*Dama dama*) with frozen-thawed semen via laparoscopy. *Theriogenology* **34**, 569–577.

Asher, G. W., Fisher, M. W., Jabbour, H. N., Smith, J. F., Mulley, R. C., Morrow, C. J., Veldhuizen, F. A., and Langridge, M. (1992). Relationship between the onset of oestrus, the preovulatory surge in luteinizing hormone and ovulation following oestrous synchronization and superovulation of farmed red deer (*Cervus elaphus*). *J. Reprod. Fertil.* **96**, 261–273.

Asher, G. W., Veldhuizen, F. A., Morrow, C. J., and Duganzich, D. M. (1994). Effects of exogenous melatonin on prolactin secretion, lactogenesis and reproductive seasonality of female red deer (*Cervus elaphus*). *J. Reprod. Fertil.* **100**, 11–19.

Asher, G. W., Fisher, M. W., Berg, D. K., Waldrup, K. A., and Pearse, A. J. (1996). Luteal support of pregnancy in red deer (*Cervus elaphus*): Effect of cloprostenol, ovariectomy and lutectomy on the viability of the post-implantation embryo. *Anim. Reprod. Sci.* **41**, 141–153.

Bierschwal, C. J., Mather, E. C., Martin, C. E., Murphy, D. A., and Korschgen, L. J. (1970). Some characteristics of deer semen collected by electroejaculation. *J. Am. Vet. Med. Assoc.* **5**, 627–632.

Branan, W. V., and Marchington, R. L. (1987). Reproductive ecology of white-tailed and red brocket deer in Suriname. In *Biology and Management of the Cervidae* (C. M. Wemmer, Ed.). Smithsonian Inst. Press, Washington, DC.

Brinklow, B. R., and Loudon, A. S. I. (1993). Gestation periods in the Père David's deer (*Elaphurus davidianus*): Evidence for embryonic diapause or delayed development. *Reprod. Fertil. Dev.* **5**, 567–575.

Brinklow, B. R., McLeod, B. J., Loudon, S. I., and Curlewis, J. D. (1992). Induction of ovulation in Pere David's hinds at two stages of seasonal anoestrus. The Biology of Deer: Proceedings of the International Symposium on the Biology of Deer, May 28–June 1, 1990, Mississippi State University, Mississippi State.

Brown, R. D., Cowan, R. L., and Kavanaugh, E. J. F. (1978). Effect of pinealectomy on seasonal androgen titres, antler growth and feed intake in white-tailed deer. *J. Anim. Sci.* **47**, 435–440.

Bubenik, G. A. (1983). Shift of seasonal cycle in white-tailed deer by oral administration of melatonin. *J. Exp. Zool.* **225**, 155–156.

Bunnell, F. L. (1987). Reproductive tactics of cervidae and their relationships to habitat. In *Biology and Management of the Cervidae* (C. M. Wemmer, Ed.). Smithsonian Inst Press, Washington, DC.

Chapman, N. G. (1992). Breeding performance of female Chinese muntjac deer in England. In *The Biology of Deer* (R. D. Brown, Ed.). Springer-Verlag, New York.

Chapman, N. G., and Chapman, D. I. (1979). Seasonal changes in the male accessory glands of reproduction in adult fallow deer (*Dama dama*). *J. Zool. (London)* **189**, 259–274.

Chapman, N. G., and Harris, S. (1988). Evidence that the seasonal antler cycle of adult Reeves' muntjac (*Muntiacus reevesi*) is not associated with reproductive quiescence. *J. Reprod. Fertil.* **92**, 361–369.

Chapple, R. S., English, A. W., and Mulley, R. C. (1993). Characteristics of the oestrous cycle and duration of gestation in chital hinds (*Axis axis*). *J. Reprod. Fertil.* **98**, 23–26.

Cheatum, E. L., and Morton, G. H. (1946). Breeding seasons of white-tailed deer in New York. *J. Wildlife Manage.* **10**, 249–263.

Cowan, I. M. (1956). Life and times of the coast black-tailed deer. In *The deer of North America* (W. P. Taylor, Ed.). Stackpole, Harrisburg, PA.

Curlewis, J. D., Loudon, A. S. I., and Coleman, A. P. M. (1988). Oestrous cycles and the breeding season of the Pere David's deer hind (*Elaphurus davidianus*). *J. Reprod. Fertil.* **82**, 119–126.

Curlewis, J. D., McLeod, B. J., and Loudon, A. S. I. (1991). LH secretion and response to GnRH during seasonal anoestrus of the Pere David's deer hind (*Elaphurus davidianus*). *J. Reprod. Fertil.* **91**, 131–138.

Fisher, M. W., Meikle, L. M., and Fennessy, P. F. (1992). Duration of melatonin treatment affects the timing of puberty in young red deer hinds. In *The Biology of Deer* (R. D. Brown, Ed.). Springer-Verlag, New York.

Flint, A. P. F., Krzywinski, A., Sempere, A. J., Mauget, R., and Lacroix, A. (1994). Luteal oxytocin and monoestry in the roe deer *Capreolus capreolus*. *J. Reprod. Fertil.* **101**, 651–656.

Gadsby, J. E., Heap, R. B., and Burton, R. D. (1980). Oestrogen production by blastocyst and early embryonic tissue of various species. *J. Reprod. Fertil.* **60**, 409–417.

Gizejewski, A., Pedich, M., Jaczewski, Z., and Bartecki, R. (1993). Ultrasonic control of ovarian function in red deer females during the oestrus cycle. In *Deer of China* (N. Ohtaishi and H.-I. Sheng, Eds.). Elsevier, Tokyo.

Gosch, B., and Fischer, K. (1989). Seasonal changes of testis volume and sperm quality in adult fallow deer (*Dama dama*) and their relationship to the antler cycle. *J. Reprod. Fertil.* **85**, 7–17

Gosch, B., Bartolomaeus, T., and Fischer, K. (1989). Light and scanning electron microscopy of fallow deer (*Dama dama*) spermatozoa. *J. Reprod. Fertil.* **87**, 187–192.

Goss, R. J. (1969a). Photoperiodic control of antler cycles in deer. I. Phase shift and frequency changes. *J. Exp. Zool.* **170**, 311–324.

Goss, R. J. (1969b). Photoperiodic control of antler cycles in deer. II. Alterations in amplitude. *J. Exp. Zool.* **171**, 223–234.

Goss, R. J. (1976). Photoperiodic control of antler cycles in deer. III. Decreasing versus increasing daylengths. *J. Exp. Zool.* **197**, 307–320.

Goss, R. J., and Rosen, J. K. (1973). The effect of latitude and photoperiod on the growth of antlers. *J. Reprod. Fertil. Suppl.* **19**, 111–118.

Goss, R. J., Dinsmore, C. E., Grimes, L. N., and Rosen, J. K. (1974). Expression and suppression of the circannual antler cycle in deer. In *Circannual Clocks* (E. T. Pengellery, Ed.). Academic Press, New York.

Haigh, J. C., Cates, W. F., Glover, G. J., and Rawlings, N. C. (1984). Relationships between seasonal changes in serum testosterone concentrations, scrotal circumference and sperm morphology of male wapiti (*Cervus elaphus*). *J. Reprod. Fertil.* **70**, 413–418.

Haigh, J. C., Dradjat, A. S., and English, A. W. (1993). Comparison of two extenders for the cryopreservation of Chital (*Axis axis*) semen. *J. Zoo Wildlife Med.* **24**, 454–458.

Hamilton, W. J., and Baxter, K. L. (1980). Reproduction in farmed red deer. 1. Hind and stag fertility. *J. Agric. Sci.* **95**, 261–273.

Hamilton, W. J., Harrison, R. J., and Young, B. A. (1960). Aspects of placentation in certain cervidae. *J. Anat.* **94**, 1–33.

Haugen, A. O. (1975). Reproductive performance of white-tailed deer in Iowa. *J. Mammal.* **56**, 151–159.

Hoffman, B., Barth, D., and Karg, H. (1978). Progesterone and estrogen levels in peripheral plasma of the pregnant and nonpregnant roe deer (*Capreolus capreolus*). *Biol. Reprod.* **19**, 931–935.

Kelly, R. W., McNatty, K. P., Moore, G. H., Ross, D., and Gibb, M. (1982). Plasma concentrations of LH, prolactin, oestradiol and progesterone in female red deer (*Cervus elaphus*) during pregnancy. *J. Reprod. Fertil.* **64**, 475–483.

Kelly, R. W., McNatty, K. P., and Moore, G. H. (1985). Hormonal changes about oestrus in female red deer. In *Biology of Deer Production* (P. F. Fennessy and K. R. Drew, Eds.), Bulletin 22. Royal Society of New Zealand, Wellington.

Krzywinski, A., Niedbalska, A., and Krzywinska, K. (1987). Collection and freezing semen of the moose bull. *Sweden Wildlife Res. Suppl.* **1**, 761–765.

Leader-Williams, N. (1979). Age-related changes in the testicular and antler cycles of reindeer, *Rangifer tarandus*. *J. Reprod. Fertil.* **57**(1), 117–126.

Lincoln, G. A. (1971). Puberty in a seasonally breeding male, the red deer stag (*Cervus elaphus* L.). *J. Reprod. Fertil.* **25**, 41–54.

Lincoln, G. A. (1985). Seasonal breeding in deer. In *Biology of Deer Production* (P. F. Fennessy and K. R. Drew, Eds.), Bulletin 22. Royal Society of New Zealand, Wellington.

Lincoln, G. A. (1992). Biology of seasonal breeding in deer. In *The Biology of Deer* (R. D. Brown, Ed.). Springer-Verlag, New York.

Lincoln, G. A., and Tyler, N. J. C. (1994). Role of gonadal hormones in the regulation of the seasonal antler cycle in female reindeer, *Rangifer tarandus*. *J. Reprod. Fertil.* **101**(1), 129–138.

Loudon, A. S. I., and Curlewis, J. D. (1988). Cycles of antler and testicular growth in an aseasonal tropical deer (*Axis axis*). *J. Reprod. Fertil.* **83**(2), 729–738.

Loudon, A. S. I., McLeod, B. J., and Curlewis, J. D. (1990). Pulsatile secretion of LH during the periovulatory and luteal phases of the estrous cycle in the Pere David's deer hind (*Elaphurus davidianus*). *J. Reprod. Fertil.* **89**(2), 663–670.

Macnamara, M., and Eldridge, W. (1987). Behavior and reproduction in captive pudu (*Pudu puda*) and red brocket (*Mazama americana*), a descriptive and comparative analysis. In *Biology and Management of the Cervidae* (C. M. Wemmer, Ed.). Smithsonian Inst. Press. Washington, DC.

McLeod, B. J., Brinklow, B. R., Curlewis, J. D., and Loudon, A. S. I. (1991). Efficacy of intermittent or continuous administration of GnRH in inducing ovulation in early and late seasonal anestrus in the Pere David's deer hind (*Elaphurus davidianus*). *J. Reprod. Fertil.* **91**, 229–238.

Merkt, J. R. (1987). Reproductive seasonality and grouping patterns of the North Andean deer or taruca (*Hippocamelus antisensis*) in southern Peru. In *Biology and Management of the Cervidae* (C. M. Wemmer, Ed.). Smithsonian Inst. Press. Washington, DC.

Mirarchi, R. E., Scanlon, P. F., and Kirkpatrick, R. L. (1977). Annual changes in spermatozoan production and associated organs of white-tailed deer. *J. Wildlife Manage.* **41**, 92–99.

Mirarchi, R. E., Howland, B. E., Scanlon, P. F., Kirkpatrick, R. L., and Sanford, L. M. (1978). Seasonal variation in plasma LH, FSH, prolactin and testosterone concentrations in adult male white-tailed deer. *Can. J. Zool.* **56**, 121–127.

Monfort, S. L., Wemmer, C., Kepler, T. H., Bush, M., Brown, J. L., and Wildt, D. E. (1990). Monitoring ovarian function and pregnancy in Eld's deer (*Cervus eldi thamin*) by evaluating urinary steroid metabolite excretion. *J. Reprod. Fertil.* **88**(1), 271–282.

Monfort, S. L., Martinet, C., and Wildt, D. E. (1991). Urinary steroid metabolite profiles in female Père David's deer (*Elaphurus davidianus*). *J. Zoo Wildlife Med.* **22**, 78–85.

Monfort, S. L., Brown, J. L., Bush, M., Wood, T. C., Wemmer, C., Vargas, A., Williamson, L. R., Montali, R. J., and Wildt, D. E. (1993a). Circannual inter-relationships among reproductive hormones, gross morphometry, behaviour, ejaculate characteristics and testicular histology in Eld's deer stags (*Cervus eldi thamin*). *J. Reprod. Fertil.* **98**, 471–480.

Monfort, S. L., Schwartz, C. C., and Wasser, S. K. (1993b). Monitoring reproduction in captive moose using urinary and fecal steroid metaboilites. *J. Wildlife Manage.* **57**, 400–407.

Mueller, C. C., and Sadlier, R. M. FF. S. (1979). Age at first conception in black-tailed deer. *Biol. Reprod.* **21**, 1099–1104.

Nowak, R. M., and Paradiso, J. L. (1983). *Walker's Mammals of the World,* 4th ed. Johns Hopkins Univ. Press. Baltimore.

Pei, K., and Liu, H. W. (1994). Reproductive biology of male Formosan Reeves' muntjac (*Muntiacus reevesi micrurus*). *J. Zool. (London)* **233**, 293–306.

Plotka, E. D., Seal, U. S., Verme, L. J., and Ozoga, J. J. (1977). Reproductive steroids in the white-tailed deer (*Odocoileus virginianus borealis*). II. Progesterone and estrogen levels in peripheral plasma during pregnancy. *Biol. Reprod.* **17**, 78–83.

Plotka E. D., Seal U. S., Letellier M. A., Verme L. J., and Ozoga J. J. (1979). Endocrine and morphological effects of pinealectomy in white-tailed deer. In *Animal Models for Research on Contraception and Fertility* (N. J. Alexander, Ed.). Harper & Row, Hagerstown, MD.

Plotka, E. D., Seal, U. S., Verme, L. J., and Ozoga, J. J. (1980). Reproductive steroids in deer. III. Luteinizing hormone, estradiol and progesterone around estrus. *Biol. Reprod.* **22**, 576–581.

Plotka, E. D., Seal, U. S., Verme, L. J., and Ozoga, J. J. (1983). The adrenal gland in white-tailed deer—A significant source of progesterone. *J. Wildlife Manage.* 47, 38–44.

Ropstad, E., Forsberg, M., Sire, J. E., Kindahl, H., Nilsen, T., Pederson, O., and Edqvist, L. E. (1995). Plasma concentrations of progesterone, oestradiol, LH and 15-ketodihydro-PGF2alpha in Norwegian semi-domestic reindeer (*Rangifer tarandus tarandus*) during their first reproductive season. *J. Reprod. Fertil.* 105, 307–314.

Ropstad, E., Kindahl, H., Nilsen, T. A. B., Forsberg, M., Sire, J. E., Pedersen, O., and Edqvist, L. E. (1996). The effect of cloprostenol in non-pregnant and pregnant Norwegian semi-domestic reindeer (*Rangifer tarandus tarandus* L). *Anim. Reprod. Sci.* 43, 205–219.

Sadlier, R. M. F. S. (1987). Reproduction in female cervids. In *Biology and Management of the Cervidae* (C. M. Wemmer, Ed.). Smithsonian Inst. Press. Washington, DC.

Schams, D., and Barth, D. (1982). Annual profiles of reproductive hormones in peripheral plasma of the male roe deer (*Capreolus capreolus*). *J. Reprod. Fertil.* 66, 463–468.

Schulte, B. A., Seal, U. S., Plotka, E. D., Letellier, M. A., Verme, L. J., Ozoga, J. J., and Parsons, J. A. (1981). The effect of pinealectomy on seasonal changes in prolactin secretion in the white-tailed deer (*Odocoileus virginianus borealis*). *Endocrinology* 108, 173–178.

Sempéré, A. J., Mauget, R., Blanvillain, C., and Chemineau, P. (1993). The role of the photoperiod in the sexual cycle in female roe deer. In *Deer of China* (N. Ohtaishi and H.-I. Sheng, Eds.). Elsevier, Tokyo.

Severinghaus, C. W. (1955). *NY Fish Game J.* 2, 239.

Sheng, H., and Ohtaishi, N. (1993). The status of deer in China. In *Deer of China* (N. Ohtaishi and H.-I. Sheng, Eds.). Elsevier, Tokyo.

Short, R. V., and Hay, M. F. (1966). Delayed implantation in the roe deer (*Capreolus capreolus*). *Symp. Zool. Soc. London* 15, 173.

Short, R. V., and Mann. T. (1966). The sexual cycle of a seasonally breeding mammal, The Roebuck (*Capreolus capreolus*). *J. Reprod. Fertil.* 12, 337–351.

Sjaastad, O. V., Blom, A. K., Anstad, R., and Oen, E. O. (1990). Plasma progesterone in reindeer in relation to ovariectomy and hysterectomy. *Acta Vet. Scand.* 31, 45–52.

Strzezek, J., Krzywinski, A., and Swidowicz, K. (1985). Seasonal changes in the chemical composition of red deer (*Cervus elaphus*) semen. *Anim. Reprod. Sci.* 9, 195–204.

Suttie, J. M., Lincoln, G. A., and Kay, R. N. B. (1984). The endocrine control of antler growth in red deer stags. *J. Reprod. Fertil.* 71, 7–15.

Thomas, D. C. (1982). The relationship between fertility and fat reserves of Peary caribou. *Can. J. Zool.* 60, 597–602.

Thomas, D. C., and Cowan, I. M. (1975). The pattern of reproduction in female Columbian black-tailed deer, *Odocoileus hemionus columbianus*. *J. Reprod. Fertil.* 44, 261–272.

Verme, L. J., and Ozoga, J. J. (1981). Sex ratio of white-tailed deer and the estrus cycle. *J. Wildlife Manage.* 45, 710–715.

Wemmer, C., Halverson, T., Rodden, M., and Portillo, T. (1989). The reproductive biology of female Père David's Deer (*Elaphurus davidianus*). *Zoo Biol.* 8, 49–55.

Wesson, J. A., III, Scanlon, P. F., Kirkpatrick, R. L., Mosby, H. S., and Butcher, R. L. (1979). Influence of chemical immobilization and physical restraint on steroid hormone levels in blood of white-tailed deer. *Can. J. Zool.* 57, 768–776.

West, N. O. (1968). The length of the estrous cycle in the Columbian black-tailed deer or coast deer (*Odocoileus hemionus columbianus*). BSc thesis, University of British Columbia, Vancouver, BC, Canada.

West, N. O., and Nordan, H. C. (1976). Hormonal regulation of reproduction and the antler cycle in the male Columbian black-tailed deer (*Odocoileus hemionus columbianus*). Part I. Seasonal changes in the histology of the reproductive organs, serum testosterone, sperm production, and the antler cycle. *Can. J. Zool.* 54, 1617–1636.

Whitehead, G. K. (1972). *The Whitehead Encyclopedia of Deer*. Swan Hill Press, Shrewsbury, UK.

Whitehead, P. E., and McEwan, E. H. (1973). Seasonal variation in the plasma testosterone concentration of reindeer and caribou. *Can. J. Zool.* 51, 651–658.

Wislocki, G. B. (1949). Seasonal changes in the testes, epididymes and seminal vesicles of deer investigated by histochemical methods. *Endocrinology* 44, 167–189.

Yu, Y., Miura, S., Pen, J., and Ohtaishi, N. (1993). Parturition and neonatal behavior of white-lipped deer. In *Deer of China* (N. Ohtaishi and H.-I. Sheng, Eds.). Elsevier, Tokyo.

Zuckerman, S. (1953). The breeding season of mammals in captivity. *Proc. Zool. Soc. London* 122, 827–850.

DHEA (Dehydroepiandrosterone)

Nancy Pahle and John E. Nestler

Virginia Commonwealth University

I. Dehydroepiandrosterone in Pregnancy
II. Adrenarche
III. DHEA/DHEAS Changes with Aging
IV. DHEA and the Ovary
V. DHEA and Menopause
VI. DHEA Administration to Women
VII. Summary

GLOSSARY

adrenal Describing or relating to the pair of flattened glands located above each kidney; the adrenal cortex (outer wall) secretes certain important steroid hormones.

adrenarche A condition of intensified secretion by the adrenal cortex, especially androgens, occurring in humans at about the age of 8 in both sexes.

metabolic clearance rate The rate at which a given chemical product is removed from the body or converted to another substance.

menopause The permanent termination of the menstrual cycle in the human female, typically occurring after age 45.

preeclampsia An abnormal condition of pregnancy characterized by high blood pressure, edema, and excessive protein in the urine; progression from this condition to eclampsia poses significant risk of fetal or maternal death.

During the past decade there has been renewed interest in the adrenal steroids dehydroepiandrosterone (DHEA) and its sulfate ester DHEA sulfate (DHEAS). This stems primarily from the combined observations that circulating levels of these steroids exhibit marked aged-related declines, and that DHEA appears to exert beneficial biological actions. However, often overlooked is the fact that DHEA and DHEAS also play important roles in human reproduction.

I. DEHYDROEPIANDROSTERONE IN PREGNANCY

A. Fetal Adrenal Production of Dehydroepiandrosterone Sulfate

Dehydroepiandrosterone sulfate (DHEAS) (Figs. 1 and 2) plays an important physiologic role even before birth. Fetal adrenal androgen production increases during pregnancy due to a large fetal adrenal zone, and the simultaneous increase in circulating estrogens in the maternal circulation is primarily due to placental conversion of fetal DHEAS to estrogens. Specifically, pregnenolone derived from the placenta is sulfated by the fetal adrenal gland and then metabolized by the Δ^5 pathway to DHEAS. This pathway does not require the enzyme 3β-hydroxysteroid dehydrogenase/isomerase, which is relatively deficient in the fetal adrenal. In addition, the fetal adrenal cortex synthesizes DHEAS from lipoprotein-carried cholesterol. Fetal adrenal-derived DHEAS is then metabolized by the placenta into estrone and estradiol.

The fetal adrenal glands produce more than 200 mg of DHEAS daily. This is 10- to 15-fold greater than the daily secretion of DHEAS by the adrenal glands in an adult, which totals 20–30 mg. In a study of fetal adrenal DHEAS production during pregnancy, investigators found that the mean umbilical cord concentration was relatively constant from 18 to 34 weeks of pregnancy and then rose steadily to peak at 39 or 40 weeks. This increase correlated well with the increase in fetal adrenal weight. An intact brain–pituitary axis is required for normal fetal adrenal function since the adrenal glands are usually atrophied and maternal estrogen levels are low in pregnancies in which the fetus is anencephalic.

Adrenal production of DHEAS by the fetus has been hypothesized to influence labor and delivery.

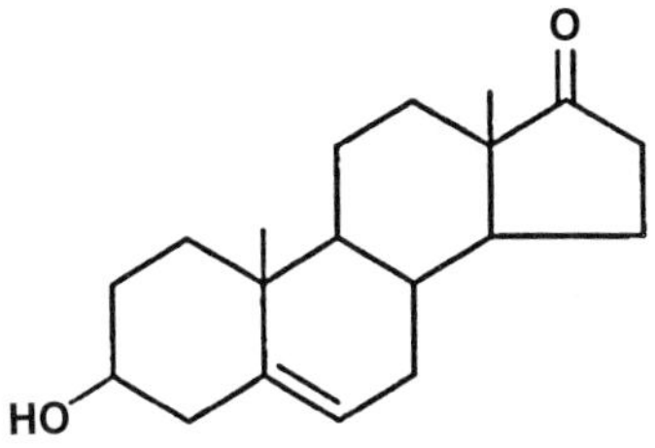

FIGURE 1 DHEA.

Several studies have reported increased Bishop scores (a measure of cervical maturity) with higher concentrations of DHEAS, and in Japan parenteral DHEAS has been administered at the onset of labor in order to produce "cervical ripening". DHEA has been found to inhibit fetal membrane production of progesterone *in vitro* and therefore may play a role in initiating labor since progestins inhibit myometrial activity.

B. Placental Metabolism of DHEAS

The metabolic clearance rate (MCR) of DHEAS increases in women during pregnancy and is 5- to 20-fold greater in the pregnant woman than in the nonpregnant woman. The increased metabolism is primarily accounted for by two major pathways not present in the nonpregnant female: namely, ~35% of DHEAS is cleared through placental aromatization to estradiol, and approximately 32% is cleared by 16α-hydroxylation in the maternal compartment. 16α-Hydroxy-DHEAS is desulfated by placental steroid sulfatase to produce estriol.

Even though the production rate of DHEAS increases during pregnancy as a result of fetal adrenal biosynthesis, maternal serum DHEAS concentrations decline due to the even greater increase in the MCR of DHEAS. Interestingly, the decrease in serum DHEAS levels persists in the mother for some time after delivery. In contrast, serum levels of unconjugated DHEA remain relatively stable in the mother during pregnancy, despite a 2.5-fold increase in the MCR of DHEA, presumably because of a concurrent and equal increase in production rate.

Placental clearance of DHEAS (the concentration of DHEAS converted to estradiol) should reflect uteroplacental perfusion and provides a means to assess uteroplacental function in different clinical situations. For example, in a prospective study in young pregnant women, the MCR of DHEAS in early pregnancy was found to be greater in those women who eventually became preeclampic than in normal pregnant women. The increased MCR of DHEAS suggested a role of placental size in the etiology of preeclampsia. Conversely, late in the course of those pregnancies at risk, the MCR of DHEAS began to decrease even before the women developed hypertension, suggesting compromised blood flow weeks before preeclampsia appeared.

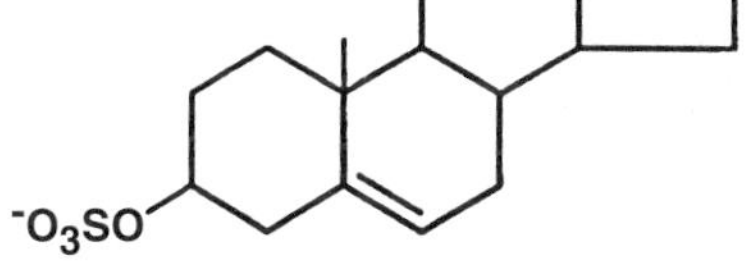

II. ADRENARCHE

In newborn babies, serum concentrations of DHEA and DHEAS decline markedly in the first month of life, as the fetal zone of the adrenal glands involutes, and remain low until 6 or 7 years of age. From age 7 or 8 years to age 13–15 years, serum DHEAS concentrations increase progressively by ~20-fold. This hormonal change parallels the development of axillary and pubic hair as well as a transient growth spurt and bone maturation. The increase in serum DHEAS levels occurs approximately 2 years prior to the increase in gonadal steroids and comprises "adrenarche."

Serum adrenocorticotropin hormone levels are unchanged during adrenarche, which suggests that adrenarche is due to a primary intraadrenal alteration in androgen biosynthesis, and there appears to be a selective increase in adrenal 17,20 desmolase and 17α-hydroxylase enzymic activities. The mechanism subserving the sudden intraadrenal increase in androgen production remains enigmatic.

FIGURE 2 DHEAS.

III. DHEA/DHEAS CHANGES WITH AGING

Serum DHEA and DHEAS concentrations continue to increase after adrenarche and peak at around 20–30 years of age. Cross-sectional studies have indicated that thereafter serum DHEA and DHEAS levels decrease progressively with aging, and this observation has been confirmed by a longitudinal study. The age-related reduction in serum DHEA and DHEAS is unique since serum levels of glucocorticoids and mineralocorticoids remain relatively stable with aging. It is unknown whether the reduction in circulating DHEA and DHEAS with aging is due to decreased production rates, increased metabolic clearance rates, or a combination of these processes. Moreover, what factor(s) regulates the production of these steroids by the adrenals remains to be clarified. A few years ago it was reported in preliminary form that a cortical androgen-stimulating hormone had been isolated from the pituitary, but subsequent reports have failed to confirm the existence of this putative factor. Nonpituitary factors may also regulate DHEA and DHEAS metabolism. For example, insulin reduces circulating DHEA and DHEAS in men. However, this action of insulin may be sex specific and germane to men only.

IV. DHEA AND THE OVARY

The importance of DHEA to female fertility remains unclear. Haning and colleagues performed infusions of [^{3}H]DHEAS and [^{14}C]testosterone in four women and found that circulating DHEAS acts as a prohormone for the production of ovarian steroids, including intrafollicular testosterone. Multiple linear regression analysis of data collected in a study of women undergoing *in vitro* fertilization and treated with gonadotropins supported the hypothesis that intraovarian testosterone derived from circulating DHEAS can influence follicular development by stimulating ovarian aromatase activity and inhibin production. Hence, circulating DHEAS may affect folliculogenesis and/or ovulation.

Not only may circulating DHEAS serve as a prohormone for ovarian testosterone biosynthesis but also ovarian steroids may influence adrenal DHEA and DHEAS production. Although there is a significant variation among subjects, a small midmenstrual cycle increase in serum DHEA has been reported as well as a trend toward higher serum DHEA and DHEAS levels during the luteal phase. A recent study performed in transsexuals also supports the idea that gonadal steroids may influence adrenal androgen production. In this study, serum DHEAS levels declined in male-to-female transsexuals when estradiol was administered, whereas they rose in female-to-male transsexuals during testosterone administration. In contrast, several studies have failed to show a change in serum DHEA or DHEAS levels during hormone replacement therapy in postmenopausal women.

V. DHEA AND MENOPAUSE

Secretion of DHEAS declines with age and is accelerated after menopause. Cumming and colleagues studied 223 women and found that premature ovarian failure and ovariectomy in young as well as postmenopausal women caused an earlier decrease in serum DHEAS concentrations. Long-term estrogen replacement showed no improvement in the age-related decrease; therefore, ovarian factors apart from estrogen are thought to be responsible for the decline in serum DHEAS related specifically to menopause (rather than to aging).

VI. DHEA ADMINISTRATION TO WOMEN

Animal studies suggest that DHEA exerts beneficial biological effects, including prolongation of life span, prevention of obesity and diabetes mellitus, enhancement of the immune system, prevention of atherosclerosis, and inhibition of carcinogenesis. As a result, there has been recent interest in the effects of DHEA administration to human beings. However, clinical studies in human beings specifically addressing this question have been limited in number, and only a handful of studies have been conducted in women.

Casson and colleagues administered micronized DHEA to postmenopausal women and reported upregulation of the immune system. Other studies, not conducted in women, support the idea that DHEA may have salutary immunomodulatory actions. In this regard, it is notable that oral DHEA administration has been reported in two separate studies (from the same investigators) to benefit women with systemic lupus erythematosus.

Another laboratory also reported on the metabolic and endocrinologic effects of oral DHEA administration to postmenopausal women, who were administered doses that ranged from 50 to 1600 mg daily. The primary benefit appeared to be an increase in physical and psychological well-being, but this was often counterbalanced by deteriorations in glucose tolerance and lipid profiles. To what degree the latter may have been secondary to the rise in more potent androgens that accompanied DHEA administration, rather than to the DHEA itself, cannot be ascertained.

VII. SUMMARY

The adrenal steroids DHEA and DHEAS play an important role during pregnancy since fetal adrenal DHEAS serves at the major substrate for placental estrogen biosynthesis. The mother's MCR of DHEA and MCR of DHEAS increase during pregnancy. After birth, serum DHEA and DHEAS levels rapidly decline and remain low until adrenarche. At this time, serum DHEA and DHEAS concentrations rise. This appears to be due to a primary intraadrenal shift in steroid biosynthesis, the mechanism of which is unexplained.

During reproductive years, circulating DHEAS serves as substrate for ovarian testosterone biosynthesis and is the major precursor of intrafollicular testosterone. Serum DHEA and DHEAS levels peak around 20–30 years of age and progressively decline thereafter. An additional decrease may occur during menopause which may be due to nonsteroidal ovarian factors. Studies examining DHEA replacement therapy in postmenopausal women are limited, and a confirmed beneficial effect has not yet been demonstrated.

See Also the Following Articles

ADRENAL ANDROGENS; ADRENARCHE; FETAL ADRENALS; MENARCHE; MENOPAUSE; PREECLAMPSIA/ECLAMPSIA

Bibliography

Beer, Jakubowicz, Beer, and Nestler (1993). The calcium channel blocker amlodipine raises serum dehydroepiandrosterone-sulfate and androstenedione, but lowers serum cortisol, in insulin-resistant obese and hypertensive men. *J. Clin. Endocrinol. Metab.* **76**, 1464–1469.

Casson and Carson (1996). Androgen replacement therapy in women: Myths and realities. *Int. J. Fertil. Menopausal Stud.* **41**, 412–422.

Casson, Andersen, Herrod, et al. (1993). Oral dehydroepiandrosterone in physiologic doses modulates immune function in postmenopausal women. *Am. J. Obstet. Gynecol.* **169**, 1536–1539.

Cumming, Rebar, Hopper, and Yen (1982). Evidence for an influence of the ovary on circulating dehydroepiandrosterone sulfate levels. *J. Clin. Endocrinol. Metab.* **54**, 1069–1071.

Everett, Porter, MacDonald, and Gant (1980). Relationship of maternal placental blood flow to the placental clearance of maternal plasma dehydroepiandrosterone sulfate through placental estradiol formation. *Am. J. Obstet. Gynecol.* **136**, 435–439.

Haning, Jr., Hua, Hackett, et al. (1994). Dehydroepiandrosterone sulfate and anovulation increase serum inhibin and affect follicular function during administration of gonadotropins. *J. Clin. Endocrinol. Metab.* **78**, 145–149.

Haning, Jr., Flood, Hackett, Loughlin, McClure, and Longcope (1991). Metabolic clearance rate of dehydroepiandrosterone sulfate, its metabolism to testosterone, and its intrafollicular metabolism to dehydroepiandrosterone, androstenedione, testosterone, and dihydrotestosterone in vivo. *J. Clin. Endocrinol. Metab.* **72**, 1088–1095.

Haning, Jr., Hackett, Flood, Loughlin, Zhao, and Longcope (1993a). Plasma dehydroepiandrosterone sulfate serves as a prehormone for 48% of follicular fluid testosterone during treatment with menotropins. *J. Clin. Endocrinol. Metab.* **76**, 1301–1307.

Haning, Jr., Hackett, Flood, Loughlin, Zhao, and Longcope (1993b). Testosterone, a follicular regulator: Key to anovulation. *J. Clin. Endocrinol. Metab.* **77**, 710–715.

Ibañez, Potau, and Carrascosa (1997). Androgens in adrenarche and pubarche. In *Androgen Excess Disorders in Women* (R. Azziz, J. E. Nestler, and D. Dewailly, Eds), pp. 73–84. Lippincott-Raven, New York.

Madden, Siitieri, MacDonald, and Gant (1976). The pattern and rates of metabolism of maternal plasma dehydroepiandrosterone sulfate in human pregnancy. *Am. J. Obstet. Gynecol.* **125**, 915–920.

Mellon, Shively, and Miller (1991). Human proopiomelanocortin-(79-96), a proposed androgen stimulatory hormone, does not affect steroidogenesis in cultured human fetal adrenal cells. *J. Clin. Endocrinol. Metab.* **72**, 19–22.

Morales, Nolan, Nelson, and Yen (1994). Effects of replacement dose of dehydroepiandrosterone in men and women of advancing age. *J. Clin. Endocrinol. Metab.* **78**, 1360–1367.

Mortola and Yen (1990). The effects of oral dehydroepiandrosterone on endocrine-metabolic parameters in postmenopausal women. *J. Clin. Endocrinol. Metab.* **71**, 696–704.

Nestler (1995). Are there sex-specific effects of insulin on human dehydroepiandrosterone metabolism? *Sem. Reprod. Endocrinol.* **13**, 282–287.

Nestler (1996). Advances in understanding the regulation and biologic actions of dehydroepiandrosterone. *Curr. Opin. Endocrinol. Diabetes* **3**, 202–211.

Nestler and Kahwash (1994). Sex-specific action of insulin to acutely increase the metabolic clearance rate of dehydroepiandrosterone in humans. *J. Clin. Invest.* **94**, 1484–1489.

Nestler, Usiskin, Barlascini, Welty, Clore, and Blackard (1989). Suppression of serum dehydroepiandrosterone sulfate levels by insulin: An evaluation of possible mechanisms. *J. Clin. Endocrinol. Metab.* **69**, 1040–1046.

Nestler, Beer, Jakubowicz, and Beer (1994). Effects of a reduction in circulating insulin by metformin on serum dehydroepiandrosterone sulfate in nondiabetic men. *J. Clin. Endocrinol. Metab.* **78**, 549–554.

Orentreich, Brind, Vogelman, Andres, and Baldwin (1992). Long-term longitudinal measurements of plasma dehydroepiandrosterone sulfate in normal men. *J. Clin. Endocrinol. Metab.* **75**, 1002–1004.

Parker (1989). *Adrenal Androgens in Clinical Medicine*. Academic Press, San Diego.

Parker, Jr., Leveno, Carr, Hauth, and MacDonald (1982). Umbilical cord plasma levels of dehydroepiandrosterone sulfate during human gestation. *J. Clin. Endocrinol. Metab.* **54**, 1216–1220.

Parker, Lifrak, Shively, *et al.* (1989). Human adrenal gland cortical androgen-stimulating hormone (CASH) is identical with a portion of the joining peptide of pituitary proopiomelanocortin (POMC). *Program Abstr. 71st Annu. Meeting Endocr. Soc.* 97. [Abstract 299]

Penhoat, Sanchez, Jaillard, Langlois, Begeot, and Saez (1991). Human proopiomelanocortin-(79–96), a proposed cortical androgen-stimulating hormone, does not affect steroidogenesis in cultured human adult adrenal cells. *J. Clin. Endocrinol. Metab.* **72**, 23–26.

Polderman, Gooren, and van der Veen (1995). Effects of gonadal androgens and oestrogens on adrenal androgen levels. *Clin. Endocrinol. (Oxford)* **43**, 415–421.

Regelson, Kalimi, and Loria (1990). DHEA: Some thoughts as to its biologic and clinical action. In *The Biologic Role of Dehydroepiandrosterone (DHEA)* (M. Kalimi and W. Regelson, Eds.), pp. 405–445. de Gruyter, New York.

Van Vollenhoven, Engleman, and McGuire (1994). An open study of dehydroepiandrosterone in systemic lupus erythematosus. *Arthritis Rheum.* **37**, 1305–1310.

Van Vollenhoven, Engleman, and McGuire (1995). Dehydroepiandrosterone in systemic lupus erythematosus. Results of a double-blind, placebo-controlled, randomized clinical trial. *Arthritis Rheum.* **38**, 1826–1831.

Worley, Everett, Madden, MacDonald, and Gant (1978). Fetal considerations. Metabolic clearance rate of maternal plasma dehydroepiandrosterone sulfate. *Sem. Perinatol.* **2**, 15–28.

Yen, Morales, and Khorram (1995). Replacement of DHEA in aging men and women. *Ann. N. Y. Acad. Sci.* **774**, 128–142.

Diapause

David L. Denlinger
Ohio State University

Seiji Tanaka
National Institute of Sericultural and Entomological Science

I. Introduction
II. Characteristics of Reproductive Diapause
III. Environmental Control
IV. Hormonal Regulation
V. Molecular Mechanisms

GLOSSARY

critical day length The day length marking the transition between a day length eliciting diapause and one that promotes nondiapause development.
diapause A stage-specific arrest in insect development or reproduction that is used to circumvent inimical seasons.
ecdysteroids Steroid-molting hormones of insects that are synthesized in prothoracic glands of larvae and in the reproductive organs of adults.
juvenile hormones Isoprenoid hormones synthesized by the insect's corpora allata and that act to maintain the status quo during larval development and serve as regulators of several reproductive processes in adults.

I. INTRODUCTION

Diapause is a form of developmental arrest that plays a critical role in enabling insects and related arthropods to circumvent adverse seasons of the year. Winter is the season most commonly avoided in temperate zones, but diapause can be exploited as well to avoid hot dry summers and periods of food shortage in the tropics. Diapause is not an immediate reaction to the adverse situation but a response that is programmed well in advance of the seasonal adversity. The arrest that occurs is analogous to mammalian hibernation in many ways. Energy reserves are sequestered, metabolism is suppressed, feeding activity halts or is greatly reduced, and the insect seeks a protected habitat. Diapause can occur at nearly any stage of embryonic development, during any of the larval instars, in the pupal stage, or in the adult, but for most species the capacity for diapause is limited to a single stage characteristic of that species. For example, the gypsy moth always diapauses as a late embryo, the European corn borer as a fifth-instar larva, the Cecropia moth as a pupa, and the Colorado potato beetle as an adult. Only a few insects have the genetic capacity to enter diapause in two or more developmental stages. Though diapause in any developmental stage has consequences for the seasonal reproductive cycle, this chapter focuses primarily on adult diapause—the seasonal cessation of reproductive activity that is a direct consequence of diapause occurring in the adult stage. Diapause in the adult stage (reproductive diapause) is especially common in beetles (Coleoptera), but it is also well documented in butterflies and moths (Lepidoptera), flies (Diptera), bees and wasps (Hymenoptera), bugs (Hemiptera and Homoptera), grasshoppers (Orthoptera), lacewings (Neuroptera), and mites (Acarina).

II. CHARACTERISTICS OF REPRODUCTIVE DIAPAUSE

The central feature of an adult diapause is that the insect does not reproduce. Ovaries of the female remain small, and the oocytes within the ovarioles contain little or no yolk. In the Colorado potato beetle, for example, no visible oocyte development is evident during diapause, whereas in the mosquito, *Culex pipiens*, small oocytes at the previtellogenic stage can be found. In males of some species, the testes are reduced in size during diapause, but in others the testes are well developed and contain sperm. Diapause in males is better characterized by a reduction in size of the accessory glands, which are responsible for activation of sperm, spermatophore production, and other functions related to reproduction. Mating behavior is strongly suppressed in most diapausing adults. Mating activity is sometimes totally dependent on the male so that females, even those in diapause, may be inseminated by sexually active males under experimental conditions. Whether insects mate before or after diapause depends on the species. Wasps mate in the autumn, the males die, and only the females have the capacity for overwintering in diapause. In many bugs, butterflies, and beetles, both sexes enter diapause and mating takes place in the spring, after diapause has been terminated. Some species, such as lacewings and weevils, mate both before and after diapause.

Diapausing adults are characterized not only by the cessation of reproductive activity but also by other morphological, physiological, and behavioral adaptations that are collectively referred to as the diapause syndrome. Newly emerged adults may first consume a large amount of food and build up their fat reserves before entering diapause, but feeding is then suppressed on entry into diapause and may not be reinitiated until diapause ends. Some species display a conspicuous change in body color associated with the induction of diapause. Reproductively active adults of a lacewing, *Chrysopa carnea*, are green but turn brown in autumn when diapause is induced. They become green again in early spring when reproduction is initiated. Such changes in body color occur in several species of lacewings and bugs and presumably serve as a camouflage against predators. Flight muscles in many beetles and bugs degenerate at the onset of diapause. This saves energy that would otherwise be expended for maintenance of the huge muscle mass. At the end of diapause the flight muscle again regenerate, enabling the adults to return to their food source for feeding and egg laying.

Diapausing insects, especially beetles, bugs, and

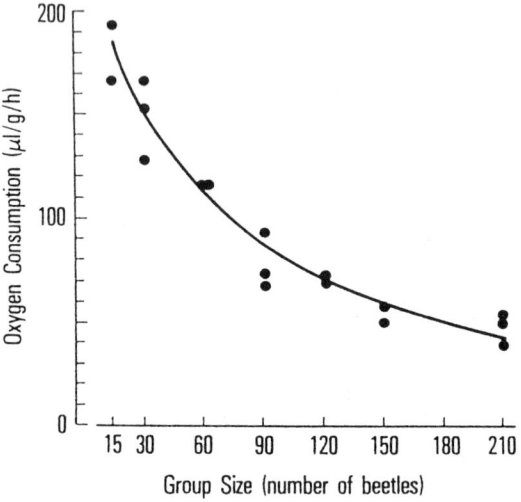

FIGURE 1 Oxygen consumption rates in diapausing adults of the tropical fungus beetle, *Stenotarsus rotundus*, when the beetles were aggregated in groups of different sizes at 25°C and 5% relative humidity. Metabolic rate decreases as group size increases (from S. Tanaka, H. Wolda, and D. L. Denlinger, *Physiol. Ent.* **13**, 239–241, 1988).

butterflies, are often found in aggregations, and on occasion different species can be found together at the same site. For species that are distasteful, aggregations are likely to provide protection from predators. However, another important function of an aggregation is that it may create a fairly stable microenvironment. An aggregation of the tropical fungus beetle, *Stenotarsus rotundus*, may consist of over 70,000 individuals. In this case, the beetle's metabolic rate is inversely related to group size and relative humidity (Fig. 1). By forming an aggregation, they create a stable, high humidity in their environment—a feature that, in this species, serves to reduce the metabolic rate.

A. Association with Migration

Adults frequently leave the site where they have spent their immature development and migrate to another place before entering an overwintering diapause. The monarch butterfly, *Danaus plexippus*, is perhaps one of the best known examples. The monarchs are reproductively active during summer in the northern regions of the United States and Canada. With the advent of autumn, the monarch begins a long southward migration to specific overwintering sites along the Pacific coast in California or to highland areas in Mexico, Guatemala, and Honduras. Huge aggregations are clustered in small geographic areas during the winter months and during this time the monarchs fly very little, metabolic rate is suppressed, and they are reproductively inactive. In February, as the days begin to lengthen, the butterflies mate and the females begin the return trip northward.

Tropical fungus beetles (*S. rotundus*) gather at the base of a palm tree in Panama and remain in reproductive diapause for up to 10 months, including 6 months of a wet season and 4 months of the subsequent dry season. Soon after the beetles arrive at the dormancy site, their flight muscles degenerate and the beetles become inactive. In response to slight increases in day length and humidity during the second half of the dry season (February and March), the gonads and flight muscles start to develop (Fig. 2). By the end of the dry season, the beetles are fully competent to fly and reproduce but they fail to do so until the first heavy rains arrive. When the rains

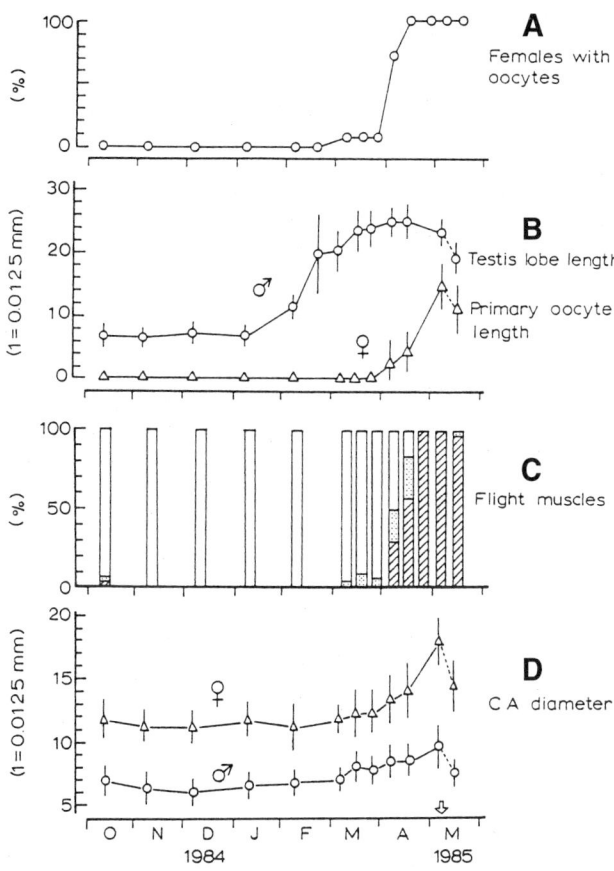

FIGURE 2 Seasonal development of a tropical fungus beetle from Panama, *Stenotarsus rotundus*. The adult beetles fly to their aggregation site in August, flight muscles degenerate, and the beetles remain in diapause throughout the rainy season and during the dry season that begins in January. At the beginning of the next rainy season, usually in late April, the beetles mate and depart the aggregation site. As day length increases following the winter solstice male reproductive organs are the first to respond by enlarging in size (B). A bit later, female reproductive organs enlarge (B) and by early April most females contain primary oocytes (A). An endocrine organ, the corpora allata (D), and flight muscles (C) also increase in size prior to the termination of diapause. Open bars indicate no development of the flight muscles, stippled area indicates slight development, and hatched area represents well-developed flight muscles. In this particular year, the first heavy rain fell on April 30 and the beetles dispersed from their aggregation site on May 6 and 7 (arrow) (from S. Tanaka, D. L. Denlinger, and H. Wolda, *Physiol. Ent.* 12; 213–224, 1987).

arrive the beetles initiate a flurry of mating activity and then quickly disperse from the tree.

Many temperate species regularly migrate to overwintering sites, though the distances covered may not be as exaggerated as in the monarch butterfly. Such short-range migratory behavior is usually a component of the diapause syndrome. The migration typically begins shortly before the insect enters diapause. Adult ladybird beetles, noctuid moths, bugs, and other insects commonly migrate to estivating or overwintering sites. The sites to which they migrate usually provide significant protection from physical extremes or from predators and parasites. After diapause ends, the insects again leave the dormancy site and return to their food source for egg laying. Thus, for most insects, migration is not an alternative to diapause but rather a component of the same syndrome.

B. Association with Cold Hardiness

Becoming dormant, by itself, does not ensure winter survival. Their small size implies that insects quickly assume a body temperature close to that of their environment. Body water is thus quite vulnerable to freezing, and the high surface to volume ratio of insects makes them particularly susceptible to desiccation during the long months of winter. Diapausing insects that overwinter in temperate and polar regions have a host of behavioral, physiological, and biochemical adaptations that enable them to survive the onslaught of low temperature.

A few insect species can actually survive body freezing, not merely the freezing of extracellular fluids but also freezing of the tissues. This strategy, freeze tolerance, is especially well studied in the goldenrod gall fly, *Eurosta solidaginis*. In this case it is the mature larva that overwinters in diapause within a gall in the stem of the goldenrod. The body becomes completely frozen in the winter months, and when spring arrives the larva thaws and proceeds with metamorphosis. Insects such as *E. solidaginis* that are freeze tolerant characteristically have supercooling points just a few degrees below 0°C. Their bodies thus freeze at relatively high subzero temperatures.

However, the majority of insects cannot tolerate body freezing. Such freeze-intolerant or freeze-avoiding insects prevent body freezing by several mechanisms. The choice of a thermally buffered microenvironment is a first line of defense. Second, freeze-intolerant species will normally void their gut and thus eliminate ice nucleators present in the digestive system. In addition, most freeze-intolerant insects synthesize polyols such as glycerol or sorbitol, which are classic antifreezes that serve to suppress the supercooling point. The polyols thus prevent the body from freezing until it reaches a much lower temperature. For example, the supercooling point of diapausing flesh fly pupae is approximately $-23°C$. The diapausing pupae can survive to temperatures down to the supercooling point but not lower temperatures. Supercooling points, however, are not a perfect indicator of the lower limit of low-temperature survival. Many insects cannot survive temperatures as low as their supercooling points. Though polyols were the first cryoprotective agents recognized, it is now clear that other components also contribute to cold hardiness. Several proteins appear to play key roles in cold hardiness: thermal hysteresis proteins, ice nucleating proteins, and stress proteins (heat shock proteins).

Problems of water management are an integral component of cold hardiness, and contributions of polyols and the various proteins ultimately all relate to managing water resources. However, several additional adaptations also affect water management. Some insects simply reduce their body water content during diapause and thus decrease the pool of free water susceptible to freezing. Such reductions usually result in increases of hemolymph osmolality, which in turn decrease the body's supercooling point. A waterproofing wax layer that coats the insect cuticle plays a critical role in preventing water loss. Insects destined for diapause frequently deposit an extra-thick coating of hydrocarbon on their cuticle, a feature that conserves water loss by reducing the rate of transpiration. Free water may be a limiting resource in the overwintering habitat, and some insects that are unable to drink have the capacity to absorb water vapor from the air.

The many dimensions of cold hardiness may or may not be directly linked to diapause. For some insects, such as the flesh fly, cold hardiness is a component of the diapause syndrome. When the fly enters diapause it is also cold hardy. In this case these two features cannot be separated. Thus, the same cues that dictate diapause entry also dictate cold hardiness. This, however, is not the case for the European corn borer. Larvae of the corn borer may enter diapause, but they do not attain cold hardiness until they have experienced low temperatures in the field. In this case diapause and cold hardiness are not triggered by the same cues. Though diapause and cold hardiness frequently coincide in time, some insects can be in diapause without being cold hardy, while others may be cold hardy without being in diapause.

III. ENVIRONMENTAL CONTROL

For a few insect species diapause is obligatory, i.e., the insect is genetically programmed to enter diapause at a certain developmental stage and will do so regardless of the environmental conditions. The gypsy moth is a good example. In each generation the embryo enters a diapause at the completion of embryonic development, and the mature embryo remains in diapause throughout the winter months. One generation is completed each year. Only an occasional aberrant individual fails to enter diapause and such individuals do not survive the winter.

However, for most insects diapause is facultative. A facultative diapause, which is by far the most common, is one that is expressed in response to an environmental cue. If a certain environmental cue is received during a sensitive period the insect will enter diapause; however, if this cue is not received or not received at the correct time, development will proceed without interruption. This design feature enables an insect to track seasonal changes and regulate its development accordingly. Many insects can produce multiple generations each year, and insects with a facultative diapause will frequently produce spring and summer generations without a diapause and then produce a generation in late summer or autumn that enters an overwintering diapause. The environmental cue most widely used to signal diapause induction is photoperiod, but temperature, food quality, and other factors can contribute to the decision.

A. Photoperiod

Because of its mathematical accuracy as a seasonal cue, photoperiod or day length is the most prevalent signal used to program diapause in insects. Unlike plants, insects normally do not require light itself for growth and development. Instead, they use photoperiod as a token stimulus to foretell the advent of winter. Quite frequently the developmental period that is sensitive to these photoperiodic cues occurs well in advance of the actual diapause stage. Thus, diapause is not usually an immediate reaction to photoperiod but occurs in response to photoperiodic signals received at an earlier stage of development. Such early programming enables the insect to prepare for diapause by sequestering additional food reserves and making other preparatory adjustments prior to the actual onset of the developmental arrest.

Reproductive diapause is frequently programmed during larval development, but adults often remain sensitive to photoperiod as well. One can observe the full range of diapause incidences by exposing insects to various photoperiods. Diapause is induced in all individuals of the linden bug, *Pyrrhocoris apterus*, at short day lengths between 8 and 15 hr per day, whereas diapause is prevented at day lengths of 17 hr or longer (Fig. 3). Day lengths shorter than 8 hr, day lengths that do not occur naturally within the range of this species, give rise to a lower incidence of diapause. The change in percentage diapause is rather abrupt at intermediate day lengths (15 or 16 hr), indicating that this is a threshold response. The inflection point of the curve, representing the day length that induces 50% of the maximum response, is called the critical day length. In temperate areas, most insects reproduce during the warm season when day lengths are long and become dormant during the cold seasons. As seen in the example of *P. apterus*, reproduction persists as long as long day lengths prevail, but diapause is induced when day lengths become shorter than the critical day length in autumn. Some species undergo a summer estivation and instead reproduce in the autumn. In such

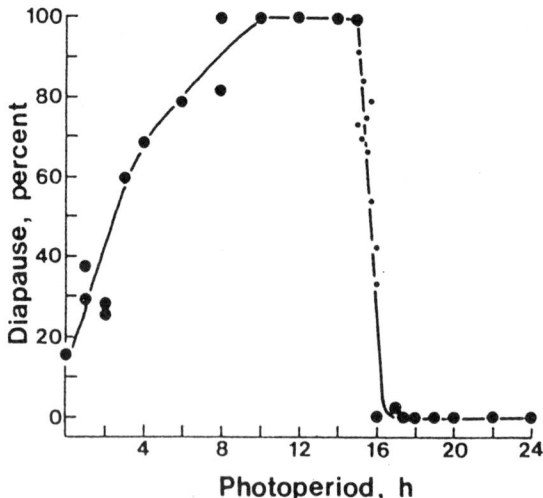

FIGURE 3 The photoperiodic response curve for diapause induction in adults of the linden bug, Pyrrhocoris apterus. The bugs were reared under the range of day lengths indicated and the incidence of reproductive diapause in the adults was recorded. The critical day length in this case is between 15 and 16 hr of light/day (from D. S. Saunders, J. Insect Physiol. 29, 399–405, 1983).

insects, diapause may be induced by long day lengths and prevented or terminated under short-day conditions.

Even close to the equator, subtle changes in day length may be utilized to extract seasonal information. Both the tropical fungus beetle, Stenotarsus rotundus (9°N), and the locust, Nomadacris septemfasiata (7–9°S) can detect slight changes in day length to regulate their reproductive diapause. However, there is no evidence that species living within 5° of the equator can utilize the changes in day length that prevail at those latitudes. Diapause still exists at those equatorial latitudes, but cues derived from temperature, rainfall, and food quality take precedence over photoperiod in those circumstances.

The duration of diapause is called diapause depth, and it too may depend on photoperiod. Diapause may be induced with different depths at different photoperiods. In such cases, the response is graded rather than all-or-none. In the lacewing, C. carnea, diapause depth is controlled by photoperiod in such a way that diapause is deeper when it is induced earlier in autumn, thus preventing an untimely termination of diapause before the onset of winter.

Photoperiodic information is perceived through a receptor, integrated and stored in the brain, and translated into the endocrine events that control the induction and maintenance of diapause. The location of the photoreceptor responsible for measurement of day length has been studied in relatively few species, but in most cases the compound eyes and ocelli of the insect are not the conduit for this information. Surgical destruction of these visual centers or coating the eyes with an opaque paint usually does not interfere with the photoreception involved in the programming of diapause. In such cases of extraretinal reception the photoperiodic signal appears to impinge directly on the brain. However, this is not always so. In a ground beetle, Pterostichus nigrita, and a bug, Riptortus clavatus, cauterization or surgical removal of the compound eyes does eliminate the sensitivity to photoperiod. The exact location of the extraretinal photoreceptors and the time-measuring mechanisms have not yet been elucidated.

B. Temperature

Temperature provides another important seasonal cue. For many temperate zone species temperature modifies the photoperiodic response. In general, the higher the temperature the lower the incidence of diapause. Induction of diapause in the blowfly, Protophormia terraenovae, is photoperiodically controlled, but no diapause occurs at high temperature under any photoperiodic conditions. In aestivating species, high temperature may act in just the opposite manner: High temperature maintains diapause, whereas low temperature serves to terminate diapause.

A period of chilling may be essential for diapause termination. Diapausing insects often do not resume reproduction immediately upon transfer to diapause-preventing conditions, such as long day lengths and high temperature, but if they are first chilled for a few months, they respond to high temperature more quickly. During the chilling period, many species lose their responsiveness to photoperiod and postdiapause development becomes totally dependent on temperature. Some species that survive for more than

1 year may overwinter twice, and in such cases they regain their responsiveness to photoperiod after the first winter.

C. Food and Moisture

Food and moisture may also influence the induction and termination of diapause. Because these factors can directly influence development of reproductive organs, it is sometimes difficult to determine whether the absence of reproduction is due to a direct influence of these factors or to the induction of diapause. However, there is good evidence that a few species use these factors as token stimuli. Adult bugs (*Eurydema rugosa*) enter estival diapause if they are grown on seeds of *Brassica* plants under long-day conditions; those grown on leaves undergo continuous reproduction. Food modifies the primary response to photoperiod in the Colorado potato beetle. This species enters diapause even at a diapause-preventing photoperiod when reared on senescent potato leaves. Diapausing beetles (*S. rotundus*) become sensitive to moisture during the second half of the dormant period. Diapause is terminated more rapidly at higher humidities, and the increasing humidity late in the tropical dry season prepares the beetles to leave the dormancy site as soon as the wet season arrives.

D. Geographic Variation

The photoperiodic response controlling diapause varies among geographic strains of the same species. Populations from lower latitudes characteristically respond to shorter critical day lengths. An increase in latitude of 5° results in an increase in critical day length of approximately 30 min. This pattern of variation is closely related to the latitudinal temperature gradient and is well documented in species of *Drosophila* that inhabit the Japanese archipelago. The species that occur in the subtropical zone exhibit only a weak diapause or no diapause at all. As one moves northward in the archipelago, the diapause response becomes more pronounced and the flies utilize longer critical day lengths for diapause induction. In the tropics many species become dormant during either the wet or dry season. While carabid beetles living along shores of permanent lakes and swamps lack diapause, species exposed to periodic inundations along rivers show a wet-season diapause and species living in areas of periodic drought enter a dry-season diapause.

IV. HORMONAL REGULATION

The key hormonal regulators of insect reproduction include the juvenile hormones (JH), ecdysteroids, and the neuropeptides that regulate synthesis of the JHs and ecdysteroids. In addition, numerous recently discovered neuropeptides and other regulatory compounds control diverse aspects of insect reproductive physiology, such as pheromone biosynthesis, mating behavior, ovulation, parturition, and sperm maturation. The classic approaches to understanding the roles of various hormones in control of reproductive diapause have included gland transplantation and ablation; histological examination of glandular tissue; nerve transection; application of hormones, hormone mimics, and hormone antagonists to diapausing adults; and measurements of hormone titers. Such approaches have yielded a wealth of data supporting a major role for JH in regulation of adult diapause, some data supporting a possible role for ecdysteroids in some species, and an important role for the brain neuropeptides that preside over the corpora allata (source of JH).

A. Juvenile Hormones

The JHs are isoprenoids synthesized by the corpora allata, a pair of glands located posterior to the brain and attached by nervous conduits to the brain. Several closely related active compounds have been identified in different insects, and in some cases more than one type of JH is present in a single species. While the JHs play a consistent role in maintaining the status quo during immature phases of insect development, in adults JHs assume diverse roles associated with reproduction. In the monarch butterfly and many other species, JH stimulates the fat body to synthesize vitellogenins; in the tobacco hawkmoth

JH stimulates the final phase of water uptake by the developing oocyte, but in the Cecropia moth JH plays no discernible role in egg maturation. Consistently, in both males and females, JH promotes maturation of the accessory glands. From these observations one might predict that a shutdown of JH synthesis could be an effective mechanism for bringing about a halt to reproductive processes. This indeed has proven to be the essence of reproductive diapause in most species.

Corpora allata are usually much smaller during adult diapause than during periods of reproductive activity (Fig. 2). Surgical extirpation of the corpora allata from diapausing adults and implantation of the glands into other individuals usually demonstrates that the corpora allata from diapausing adults are inactive, although in some instances the surgical manipulation can reactivate the gland. Conversely, transplantation of active glands into diapausing hosts usually prompts the termination of diapause. In addition, topical applications of JH or one of its many analogs have proven highly effective in terminating reproductive diapause. The opposite can be demonstrated by application of precocene, a chromene compound that chemically ablates the corpora allata. Nondiapausing females of the Colorado potato beetle treated with precocene stop laying eggs, crawl down into the soil, and enter a diapause-like state. Surgical removal of the corpora allata from nondiapausing beetles elicits this same effect. Measurements of the JH titer in the insect's hemolymph (blood) points to the same conclusion. The JH titer is normally high in nondiapausing adults that are actively reproducing, whereas the titer is quite low and sometimes undetectable in diapausing adults (Fig. 4). At the termination of diapause the JH titer quickly rises prior to the onset of reproduction. This marked difference between diapausing and nondiapausing adults can also be seen when the corpora allata are cultured *in vitro*: The rate of JH synthesis is high in glands dissected from reproducing adults but is quite low in glands dissected from diapausing adults.

Despite a solid body of evidence indicating a key role for JH in regulating reproductive diapause, it is also clear that JH does not, by itself, control the diapause fate of the insect. This fact is most evident by examining the response of diapausing adults to exogenous JH. While an application of JH definitely will stimulate signs of diapause termination, such as flight muscle development and some oviposition, the quantity needed to elicit the response is usually rather high and reproduction is usually not sustained. After an initial burst of egg laying the female will frequently revert to a diapause-like state. This most likely implies that additional events must take place to promote continuous JH production by the adult's own corpora allata. In a diapause that is broken naturally, this sustainment of activity is a role most certainly played by the brain.

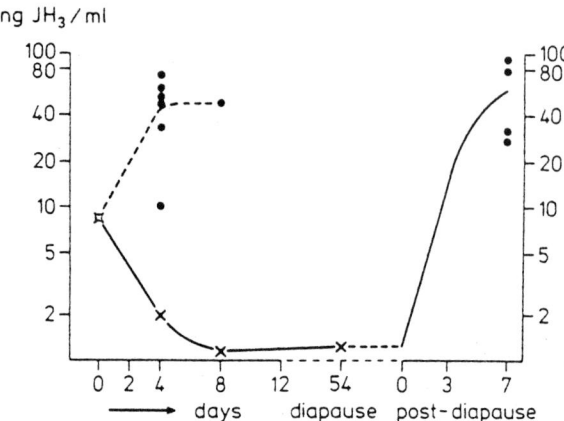

FIGURE 4 Juvenile hormone titer in the hemolymph of the Colorado potato beetle reared under diapause-inducing, short-day conditions (solid line) or nondiapause-inducing, long-day conditions (dotted line). Characteristically the JH titer drops as the beetles enter diapause, remains low throughout diapause, and then rises at the termination of diapause. A drop in the JH titer is not observed in beetles that reproduce without entering diapause (from C. A. D. de Kort, *Ent. Gen.* 7, 261–271, 1981).

B. Brain

The corpora allata does not act autonomously in regulating diapause. The brain plays the critical role of regulating the rate of JH synthesis by the corpora allata, and it does so by both humoral and nervous pathways. The pars intercerebralis (central region) is the essential area of the brain, and destruction of this region in reproductively active Colorado potato beetles causes the beetles to begin digging in the soil and assume the characteristics of diapause. When this region of the brain has been destroyed diapausing beetles cannot respond to long day lengths by terminating diapause. In this species severance of the nerve that connects the corpora allata to the brain in no way influences the activation of the corpora allata, thus suggesting that the gland is normally activated by a humoral route. In other species, such as the linden bug *Pyrrhocoris apterus,* the corpora allata appear to be regulated both humorally and by nervous connections. Severance of nervous connections to the corpora allata in diapausing adults of *P. apterus* causes the bugs to promptly become reproductively active. This suggests that diapause in this case may involve a nervous inhibition from the brain that is removed upon severance of the nerve.

The hormonal signals from the brain that regulate the corpora allata during diapause are most likely two neuropeptides, allatotropin (stimulation of gland activity) and allatostatin (inhibition of a gland activity). While little work has been done on these two neuropeptides in direct relationship to reproductive diapause, a solid body of evidence supports a role for these two neuropeptides in regulating the corpora allata in nondiapausing insects.

Though the details of how the corpora allata are regulated in different species may differ, the brain is consistently implicated in the control mechanism. The brain receives the environmental signals that program the diapause, stores this information, and then releases the hormonal messages that regulates the diapause. While the brain–corpora allata axis is central to most regulatory schemes for adult diapause, other factors are also quite possibly involved. The other major group of insect hormones, the ecdysteroids, may be involved in some instances. In the Colorado potato beetle, the ecdysteroid titer in young, diapause-destined beetles is nearly twice as high as that in beetles programmed for reproduction, but the function of ecdysteroids in adult diapause remains unknown.

Diapauses in other stages of development show an interesting diversity of regulatory schemes. Consistently, the brain is central to diapause regulation but regulation involves other sets of hormones in other developmental stages. The embryonic diapause of the silkworm is regulated by diapause hormone, a neuropeptide released by the mother's subesophageal ganglion, and the late embryonic diapause of the gypsy moth is induced and maintained by a high-ecdysteroid titer. Larval and pupal diapauses are usually caused by a shutdown in the brain–prothoracic gland axis. Ecdysteroids are not released and hence the insect does not proceed to the next developmental stage. In some cases a high JH titer also contributes to maintenance of larval diapause. Some adult females also control the diapause destiny of their offspring. Such maternal effects are especially common among flies and parasitic wasps, but little is know about how such information is transferred from the mother to her progeny.

V. MOLECULAR MECHANISMS

The environmental cues that regulate diapause have been well defined, and there is also a fairly good understanding of the downstream hormonal signals that serve to coordinate diapause. However, many of the intermediate steps remain undefined. Information about the molecular basis for diapause is especially poorly known, but diapause offers a rich potential for exploring the molecular mechanisms that can direct an organism toward two such very different developmental pathways. Is diapause simply a shutdown in gene expression or does it represent the expression of a unique set of genes? This question has not yet been addressed for reproductive diapause but has been a recent focus of attention for pupal diapause in flesh flies. In flesh flies the brains of diapausing pupae synthesize far fewer proteins than do brains from nondiapausing pupae, thus suggesting that fewer genes are being expressed during diapause. However, a small cluster of proteins are also uniquely expressed in the brains of diapausing flies. Recently, several diapause upregulated clones have been isolated from flies, a result suggesting that diapause is not simply a shutdown in gene expression but is also the consequence of the expression of novel genes. The isolation of such genes should now make it possible to determine whether there is some commonality to the molecular mechanisms involved in regulating diapause in different species and in different developmental stages.

See Also the Following Articles

Circadian Rhythms; Circannual Rhythms; Ecdysteroids; Juvenile Hormone; Migration, Insects

Bibliography

Danks, H. V. (1987). *Insect Dormancy: An Ecological Perspective.* Biological Survey of Canada, Ottawa.

Denlinger, D. L. (1985). Hormonal control of diapause. In *Comprehensive Insect Physiology Biochemistry and Pharmacology* (G. A. Kerkut and L. I. Gilbert, Eds.), Vol. 8, pp. 353–412. Pergamon, Oxford, UK.

Denlinger, D. L. (1986). Dormancy in tropical insects. *Annu. Rev. Entomol.* **31**, 239–264.

Lee, R. E., Jr., and Denlinger, D. L. (Eds.) (1991). *Insects at Low Temperature.* Chapman & Hall, New York.

Masaki, S. (1980). Summer diapause. *Annu. Rev. Entomol.* **25**, 1–25.

Saunders, D. S. (1982). *Insect Clocks,* 2nd ed. Pergamon, Oxford, UK.

Suzuki, A., Kataoka, H., and Matsumoto, S. (1995). *Molecular Mechanisms of Insect Metamorphosis and Diapause.* Industrial Publishing, Tokyo.

Tauber, M. J., Tauber, C. A., and Masaki, S. (1986). *Seasonal Adaptations of Insects.* Oxford Univ. Press, Oxford, UK.

Zaslavski, V. A. (1988). *Insect Development.* Springer-Verlag, Berlin.

Dicyemida

see Rhombozoa

Dihydrotestosterone

Richard A. Hiipakka and Shutsung Liao

University of Chicago

I. Synthesis of Dihydrotestosterone
II. Mechanism of Androgen Action
III. Effects of Mutations in 5α-Reductase and Androgen Receptor on Androgen Action
IV. Inhibitors of 5α-Reductase

GLOSSARY

alopecia Loss of hair; baldness.

androgens Steroid hormones derived from cholesterol that are responsible for inducing and maintaining a male phenotype but which have physiological and pharmacological effects in both sexes.

androstanes 19-Carbon steroids with axial methyl groups between rings A/B and C/D.

erythropoiesis The development of erythrocytes from multipotential stem cells.

erythropoietin A polypeptide hormone that stimulates erythropoiesis.

hirsutism Excessive hair growth in normal or abnormal locations.

isoenzymes Enzymes that catalyze the same reaction but are encoded by different genes.

missense mutation Single base DNA mutations that change the codon for one amino acid to another.
NADPH Enzyme cofactor; a reduced form of nicotinamide adenine dinucleotide phosphate.
nonsense mutation Single base DNA mutations that change the codon for an amino acid to a stop codon.
Proscar, finasteride A 4-aza steroidal inhibitor of steroid 5α-reductase that is prescribed for treatment of benign prostate hyperplasia and alopecia.
Wolffian duct An embryonic duct that differentiates into the epididymis, vas deferens, and seminal vesicle in males.

Testosterone is the major androgen circulating in the blood of males and is metabolized by the enzyme, steroid 5α-reductase, in many androgen target tissues to yield 5α-dihydrotestosterone, the active androgen in these tissues. Certain androgen-responsive tissues lacking steroid 5α-reductase respond to testosterone directly. The effect of androgens on cells is mediated through the androgen receptor, a ligand-activated transcription factor. Males with gene mutations that alter the activity of steroid 5α-reductase type 2 or the androgen receptor have abnormal male sexual development. Pharmacological inhibition of steroid 5α-reductase or the androgen receptor may be appropriate for treatment of various androgen-dependent disorders, such as benign prostatic hyperplasia, prostate cancer, alopecia, hirsutism, and acne. 5β-Dihydrotestosterone and certain other 5β steroids synthesized by 5β-reductase may have a role in erythropoiesis.

I. SYNTHESIS OF DIHYDROTESTOSTERONE

Testosterone is the immediate precursor for the synthesis of dihydrotestosterone. The testis is the major source of testosterone in males, but other steroid-synthesizing organs in males and females, such as the adrenals and ovaries, also produce small amounts of testosterone. Testosterone is also produced by peripheral enzymatic conversion of suitable precursors, such as dehydroepiandrosterone and androstenedione, secreted by organs such as the adrenals and ovaries. Testosterone is converted to dihydrotestosterone by the enzymes steroid 5α- and 5β-reductase. These enzymes catalyze the reduction of the double bond between carbons 4 and 5 in testosterone to produce isomeric androstanes that differ in their molecular configuration about carbon 5 (Fig. 1). Most active androgens are derivatives of 5α-androstane with a *trans* A/B ring junction, whereas 5β-androstanes with a *cis* A/B ring junction are not androgenic.

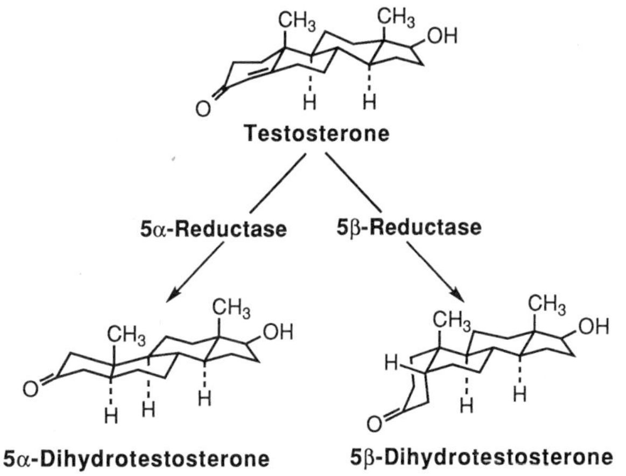

FIGURE 1 Testosterone is converted to dihydrotestosterone by 5α- and 5β-reductases. 5α-Dihydrotestosterone is the active androgen in many androgen target tissues and elicits its effects by binding to the androgen receptor. 5α-Dihydrotestosterone has a *trans* configuration at the A/B ring junction and an overall planar molecular configuration. 5β-Dihydrotestosterone is not an active androgen and does not bind to the androgen receptor. 5β-Dihydrotestosterone has a *cis* configuration at the A/B ring junction and an angular molecular configuration.

A. Steroid 5α-Reductase

Steroid 5α-reductase utilizes the cofactor NADPH to catalyze the irreversible reduction of testosterone to 5α-dihydrotestosterone (Fig. 2). A variety of other Δ⁴-3-keto steroids are also substrates for 5α-reductase, including progesterone, androstenedione, and 11-deoxycorticosterone. Two isoenzymes of 5α-reductase, types 1 and 2, are found in humans, monkeys, rats, and mice and are expressed from separate genes. The genes code for hydrophobic proteins of approximately 30 kDa. The isoenzymes share 50%

FIGURE 2 5α-Reductase catalyzes the NADPH-dependent conversion of testosterone to 5α-dihydrotestosterone, which is a critical step in androgen action. Natural (GLA and EGCG) and synthetic (Proscar) inhibitors of this enzyme may be useful in the treatment of certain androgen-dependent disorders.

identity in their amino acid sequences and differ in various biochemical parameters. The type 1 isoenzyme has a broad neutral to basic pH optima compared to a sharp acidic pH optima for the type 2 isoenzyme. At physiological pH, the type 2 5α-reductase has higher affinity for testosterone and the 4-azasteroid 5α-reductase inhibitor, finasteride (Proscar), than the type 1 isoenzyme. In humans, the type 1 5α-reductase is expressed in liver and nongenital skin, whereas the type 2 isoenzyme is found in tissues dependent on androgens for their development and function, such as the prostate, epididymis, and seminal vesicle. The type 2 isoenzyme is also expressed in human liver and genital skin. In the rat, the type 1 5α-reductase is expressed in a wide variety of tissues with highest levels found in liver, whereas the type 2 5α-reductase has a more limited pattern of expression, with highest levels found in the epididymis, as well as in other accessory sex glands. Steroid 5α-reductases are integral membrane proteins and have been localized to the endoplasmic reticulum and outer nuclear membranes. It is not clear why two different isoenzyme of 5α-reductase exist. However, in certain tissues such as the liver, the type 1 5α-reductase has a catabolic role in the metabolism of testosterone, whereas in other tissues such as the prostate the type 2 5α-reductase has an anabolic role. 5α-Reduced products are better substrates for 3α- and 3β-hydroxysteroid dehydrogenases, and their products, androstanediols, are substrates for sulfation and glucuronylation, a process that enhances excretion of steroids from the body. 5α-Reductase type 2 has been localized on or near the cell nucleus in some tissues. This location may facilitate interaction of the product, 5α-dihydrotestosterone, with the nuclear androgen receptor before it can be metabolized to androstanediols, which bind poorly to the androgen receptor.

B. Steroid 5β-Reductase

Steroid 5β-reductase is a NADPH-dependent enzyme that, in contrast to 5α-reductases, localizes to the soluble fraction of cells and not to subcellular membranes. Currently, only a single type of gene product for 5β-reductase has been cloned and sequenced from rat liver. The gene codes for a protein of 37 kDa. 5β-Reductase shows no significant amino acid sequence homology to 5α-reductases. 5β-Reductase activity has been detected in liver, brain, and adrenals. Liver 5β-reductase is required for the synthesis of bile acids that are secreted in bile and, in the small intestine, facilitate lipid absorption. Some 5β-androstanes (also called etiocholanes) are active as hematopoietic agents. These steroids increase the production of heme by inducing the rate-limiting enzyme of heme biosynthesis, Δ-aminolevulinate synthetase. Increased heme synthesis enhances hemoglobin synthesis and erythropoiesis. Liver 5β-reductase may have a role in fetal erythropoiesis since blood cell production in some embryos takes place in the liver. Testosterone and 5α-dihydrotestosterone also enhance erythropoiesis by stimulating the synthesis of erythropoietin by the kidney.

II. MECHANISM OF ANDROGEN ACTION

Although the effects of steroid hormones on animals were studied for many years after their isolation in the early part of this century, the mechanism by which these small molecules elicit their striking effects on animals remained unknown until the discovery of proteins, within cells that respond to steroid hormones, that act as specific, high-affinity receptors for the different kinds of steroid hormones. Androgen receptors were first identified in extracts of rat prostatic tissue by virtue of their ability to bind to radioactive androgens. The rat prostate responds dramatically to changes in androgen levels, losing more than 90% of its weight upon castration but rebounding to normal size again when exogenous androgens are administered. Androgen receptors have a high affinity for androgens but a low affinity for other classes of steroids (e.g., estrogens, progestins, and glucocorticoids). Androgen receptors also have a higher affinity for 5α-dihydrotestosterone than testosterone. Shortly before the discovery of androgen receptors it was observed that radioactive testosterone incubated with prostatic tissue was converted to 5α-dihydrotestosterone that was selectively retained in prostate nuclei. From these results and others, it was determined that prostatic steroid 5α-reductase converts labeled testosterone to 5α-dihydrotestosterone, which then binds to androgen receptors that are associated with the nucleus.

Androgen receptors are members of a family of ligand-activated transcription factors termed nuclear receptors that enhance the transcription of specific genes and in so doing control the development and function of different tissues containing these receptors. Some of the other members of this family include receptors for estrogens, progestins, glucocorticoids, mineralocorticoids, vitamin D3, thyroid hormones, and retinoic acid. After binding 5α-dihydrotestosterone, the androgen receptor interacts as a homodimer with a specific palindromic sequence of DNA in those genes that it regulates in the cell nucleus. These DNA sequences are called hormone response elements and have the consensus DNA sequence 5'-AGNACANNNTGTNCT-3' The same sequence is recognized by receptors for progestins, glucocorticoids, and mineralocorticoids, so specificity of gene induction by these different steroid hormones must rely on additional factors, such as the context of neighboring DNA sequences around the hormone response element, availability of ligand and receptor within the target cell, and availability of receptor-specific coactivators for gene activation. How steroid and other nuclear receptors activate gene transcription is currently under study. In some cases the receptor bound to DNA may help recruit other factors, such as histone acetylases. Normally, DNA in the cell is wrapped around an octamer of histones to form nucleosomes. Acetylation of histones reduces their overall positive charge and weakens their interaction with the negatively charged DNA backbone, and this may allow the DNA to unwind from the histones, which helps give the transcription complex access to the gene to be transcribed.

Androgen receptors are composed of about 920 amino acids and have a molecular weight of about 98,000. There is some variation in the size of the androgen receptor due to individual variation in the length of polyglycine and polyglutamine stretches in the amino-terminal portion of the protein. Androgen receptors, like other steroid receptors, are composed of three functional domains: an amino-terminal domain making up more than half of the receptor (560 amino acids) that is important for transcriptional activation, a central DNA-binding domain (70 amino acids) and a carboxyl-terminal steroid-binding domain (250 amino acids). The DNA-binding domains of steroid receptors are highly conserved; different steroid receptors have 60–80% amino acid identity in this region. The DNA-binding domain contains two zinc atoms coordinated to the sulfurs of eight cysteines in this domain forming motifs called zinc fingers. This domain is responsible for sequence-specific interactions with the DNA hormone response element. The ligand-binding domain is the next most highly conserved region with 15–50% identity among other steroid receptors. Human and rat androgen receptors have identical amino acid sequences in this domain. The amino-terminal domain is poorly conserved among all members of the steroid receptor family in both sequence and length. Only 20% of the amino acid in this domain is conserved between human and rat androgen receptors.

Androgens affect many different cell types both during development and in the adult. This is obvious given the phenotypic differences of males and females. The effects of androgens have been documented in a variety of tissues in laboratory studies, including the prostate, seminal vesicle, vas deferens, epididymis, scrotum, penis, testis, muscle, brain, liver, kidney, hematopoietic system, and skin and its appendages. Although the gross effects of androgens on various tissues are well documented, many of the underlying changes in gene expression responsible for the gross effects are not known.

All cells containing androgen receptors have the potential to respond to androgens. The presence or absence of androgen receptors, however, is not a reliable test for predicting how androgens will affect cell growth. Historically, androgens have been viewed as positive regulators of cell growth, but in some cases androgens also inhibit cell proliferation. Androgens suppress the growth of a human prostate cancer cell line both *in vitro* and as a tumor xenograft in mice. Inhibition requires conversion of testosterone to 5α-dihydrotestosterone and is blocked by antiandrogens. 5α-Reductase inhibitors that block androgen action, such as those used to treat benign prostatic hyperplasia, may stimulate prostate cancer cell growth in some cases. Androgens also have a role in inhibition of breast development. In mice, androgens block differentiation of the breast anlage in males; therefore, adult male mice do not respond to estrogens with growth of mammary tissue. In humans, the case is somewhat different since breast development is possible in both adult males and females and is dependent on the ratio of estrogen to androgens. Males exposed to excessive levels of estrogens will develop gynecomastia.

Cells lacking androgen receptors can respond to androgens indirectly by interacting with trophic factors that are synthesized and secreted by other cells that respond directly to androgens. For instance, the pattern of growth hormone secretion, pulsatile or tonic, is regulated by androgen acting at the level of the hypothalamus and pituitary. In male rats, androgens produce a pulsatile pattern of growth hormone secretion that acts on the liver to induce male-specific gene products. Certain tissues and organs, such as muscle or kidney, may contain androgen receptors and little or no 5α-reductase but respond to androgens. It is unclear whether testosterone is interacting directly with the androgen receptor in these tissues or whether dihydrotestosterone produced at other sites and released into the blood is sufficient to activate the androgen receptor in these tissues.

III. EFFECTS OF MUTATIONS IN 5α-REDUCTASE AND ANDROGEN RECEPTOR ON ANDROGEN ACTION

The higher affinity of 5α-dihydrotestosterone for the androgen receptor and its selective retention by nuclear androgen receptors in androgen-responsive tissues indicated that perhaps testosterone, in contrast to estradiol that interacts directly with the estrogen receptor, was a prohormone, requiring metabolic conversion to 5α-dihydrotestosterone before binding to and activating the androgen receptor. Definitive proof of the role of both testosterone and 5α-dihydrotestosterone in androgen action came with the characterization of individuals with mutations in the genes for 5α-reductase and androgen receptor.

A. Mutations in 5α-Reductase

More than 30 different mutations in the 5α-reductase type 2 gene have been described in individuals with a diagnosis of 5α-reductase deficiency. These mutations lead to a form of male pseudohermaphroditism, in which the level of testosterone is adequate for development of the epididymis, vas deferens, and seminal vesicle from the Wolffian ducts but is inadequate for virilization of the urogenital sinus and genital tubercle, swellings and folds from which the prostate and external genitalia develop. These mutations are rare but come to the attention of medical staff and parents due to the presentation of ambiguous external genitalia and the consequent difficulty in assigning sex at birth. The proximity of the developing testes to the Wolffian ducts may provide a high intracellular concentration of testosterone in the Wolffian duct and allow activation of the androgen receptor in this tissue. Even in normal individuals 5α-reductase activity is low or absent during Wolffian duct differentiation; therefore, testosterone must be the active androgen in this tissue at this time. The inability of testosterone to support differentiation of the urogenital sinus may reflect an inadequate supply of testosterone to this tissue or extensive metabolism of testosterone to inactive products in this tissue. At puberty, when blood levels of testosterone rise, individuals with 5α-reductase deficiency show some growth of the urogenital tract and external genitalia. A male pattern of musculoskeletal development also takes place during puberty, but growth of beard and other body hair, development of acne, and male pattern baldness are often lacking in these individuals.

Individuals with mutations in the 5α-reductase type 2 gene come from a variety of ethnic backgrounds. Mutations have been described in each of the five exons of the 5α-reductase type 2 gene and include gene deletions, nonsense and missense mutations, and a mutation at a splice junction. Some of these mutations result in a total lack of steroid 5α-reductase activity, whereas others decrease the efficiency of the enzyme by lowering the enzyme's affinity for either testosterone or NADPH. The carrier frequency of these mutations may be relatively high since 35% of the individuals with 5α-reductase deficiency are compound heterozygotes (i.e., different mutations in the two alleles of the gene).

No mutations in the steroid 5α-reductase type 1 gene have been linked to the syndrome of 5α-reductase deficiency. To gain insight into the role of this enzyme in androgen action, researchers created mice in which the steroid 5α-reductase type 1 gene was disrupted by homologous recombination in embryonic stem cells. These stem cells were then used to generate mice lacking steroid 5α-reductase type 1. Male mice lacking the type 1 steroid 5α-reductase had no obvious gross phenotypic differences from their normal littermates. However, knockout of this gene in female mice had two important consequences, both related to the ability of these mice to reproduce. Female mice lacking the type 1 steroid 5α-reductases have a defect in parturition, in which about 70% of gravid females do not deliver their pups at term, leading to either death of the mother or resorption/expulsion of the dead fetuses. This parturition defect could be overcome by administering the steroid 3α,17β-androstanediol to pregnant mice. The synthesis of this steroid from testosterone requires the enzymes steroid 5α-reductase and 3α-hydroxysteroid dehydrogenase. The activity of both these enzymes increases dramatically in the preterm uteri of pregnant mice. How androstanediol functions in parturition is unknown, but it is unlikely to work through the androgen receptor because of its low affinity for this receptor. The second consequence of a deficiency in type 1 steroid 5α-reductase in pregnant mice is a decreased litter size due to fetal death during midgestation. Fetal death appears to be due to exposure to high levels of estradiol *in utero* since antiestrogens block fetal death in 5α-reductase-deficient mice and administering estrogens to normal pregnant females causes fetal death. Testosterone is the direct precursor for the enzymatic synthesis of both estradiol and 5α-dihydrotestosterone. Competition between these two pathways may prevent the synthesis of toxic levels of estradiol. However, if steroid 5α-reductase is absent, all available testosterone may be converted to estradiol, which causes the buildup of toxic levels of estradiol leading to fetal death.

B. Mutations in the Androgen Receptor

More than 200 mutations in the androgen receptor gene have been documented. Individuals with mutations that impair the activity of the androgen receptor have complete or partial insensitivity to androgens. Since the gene for the androgen receptor is present on the X chromosome, males, with a 46,XY chromosome complement, have a single copy of the androgen receptor gene, whereas females, with a 46,XX chromosome complement, have two copies. The hemizygosity of the androgen receptor gene in males increases penetrance of mutant androgen receptor alleles in men; therefore, all documented cases of androgen insensitivity have been in males. Androgen receptor mutations should follow a recessive pattern of inheritance and, accordingly, female carriers of mutant alleles show little or no phenotypic effects. However, the random inactivation of the X chromosome that occurs in females during development may lead to penetrance of mutant androgen receptor alleles in certain females. Thus, delayed puberty or asymmetric development of pubic and axillary hair in some females carriers may be indicative of androgen insensitivity. Individuals with a 46,XY genotype and androgen receptor mutations vary from phenotypic females to undervirilized and infertile men depending on the severity of the mutation. Individuals with complete androgen sensitivity are outwardly female in appearance and have little or no pubic, axillary, or facial hair. Testes are present intraabdominally or along the inguinal canal, but spermatogenesis is absent. Normal to high circulating levels of testosterone are present, indicating normal Leydig cell function. Internal male reproductive organs, such as the epididymis, vas deferens, seminal vesicle, and prostate, that develop from the Wolffian duct are absent. A vagina is present but ends blindly, and there is no uterus, oviducts, or ovaries present.

The androgen receptor gene is composed of eight coding exons. The first exon codes for the amino-terminal domain, exons 2 and 3 code for the DNA-binding domain, and exons 4–8 code for the steroid-binding domain. The majority of documented androgen receptor mutations occur in the DNA- and steroid-binding domains of the receptor protein. The scarcity of amino-terminal domain mutations may reflect an insensitivity to structural changes in this domain. However, one mutation in this region that produces an increase in the number of CAG repeats coding for a polyglutamine repeat causes Kennedy's disease, also known as spinal and bulbar muscular atrophy. Affected males have progressive muscular weakness secondary to neural degeneration, reduced fertility, and gynecomastia. Expansion of the polyglutamine stretch from 17 to 23 residues to the 40–50 residues seen in Kennedy's disease may produce a gain-in-function mutation in which mutant receptors acquire a new property resulting in the observed phenotype. Structural mutations in the androgen receptor gene include complete and partial gene deletion as well as small deletions and insertions. These structural mutations comprise about 10% of the documented mutations in the androgen receptor. The remaining 90% of the mutations are due to single base mutations, producing premature termination codons, errors in splice junction recognition sequences, and amino acid substitutions. Approximately half of the reported androgen receptor mutations producing amino acid substitutions lead to complete androgen insensitivity.

Androgen receptor mutations have also been detected in some cases of prostate cancer, although the frequency appears to be low, and the relationship of the mutation to development or progression of prostatic disease remains unclear. However, in a few cases of prostate cancer it has been shown that mutations change the ligand specificity of the androgen receptor such that other steroids and even antiandrogens bind to and activate the mutant receptor. Prostate cancer is initially androgen dependent and first-line therapy for metastatic prostate cancer is chemical or surgical castration to remove the androgen stimulus. However, after a period of months to years the cancer usually returns and is no longer androgen dependent. In some cases, discontinuation of antiandrogen therapy will lead to temporary remission of relapsed tumors. Given the genetic instability of cancers, it is possible that a fraction of prostate cancers acquire mutant androgen receptors that may be activated by ligands other than androgens.

IV. INHIBITORS OF 5α-REDUCTASE

A variety of medical conditions are dependent on or due to excessive levels of androgens and should be responsive to antiandrogen therapies that block interaction of the androgen receptor with activating ligands. However, systemic treatment with antiandrogens has side effects, including a decrease in libido, inhibition of spermatogenesis, loss of muscle mass, and development of gynecomastia. In some cases, these side effects outweigh the usefulness of the treatment. A number of natural and synthetic compounds that are inhibitors of 5α-reductase have been identified (Fig. 2). Given the importance of 5α-reductase in a cell's response to testosterone, these inhibitors may be useful for the selective treatment of conditions dependent on 5α-dihydrotestosterone, including benign prostatic hyperplasia, prostate cancer, hirsutism, baldness, and acne. These inhibitors should not affect the availability of testosterone and physiological processes believed to be controlled by testosterone, such as libido, spermatogenesis, and muscle growth. In some cases testosterone is converted to estradiol and elicits effects through the estrogen receptor. Hormonal control of sex drive may occur in this manner.

A class of synthetic steroid 5α-reductase inhibitors that has been extensively studied is that based on 4-

aza steroids. A large number of 4-aza steroids have been synthesized and found to be potent inhibitors of steroid 5α-reductase, with varying selectivity for either the type 1 or 2 isoenzyme. One member of this class, finasteride (17β-(N-tert-butylcarbamoyl)-4-aza-5α-androst-1-en-3-one), is prescribed under the trademark Proscar for the treatment of benign prostatic hyperplasia. This steroid reacts with the cosubstrate NADPH when bound to 5α-reductase to form a bisubstrate adduct with very high affinity (half-life of 1 month at 37°C) for the type 2 isoenzyme. Inhibition of prostatic 5α-reductase lowers prostatic 5α-dihydrotestosterone levels, which decreases the activity of the prostatic androgen receptor. Finasteride appears to be effective in shrinking prostates of large size and those composed of a large percentage of epithelial cells. Prostates composed mostly of stromal cells or of small to moderate size are less responsive to finasteride treatment. Finasteride is also being tested in clinical trials as a chemopreventative for prostate cancer.

Androgens, and 5α-reduced androgens in particular, appear to control the growth and development of certain forms of body hair. Male pattern baldness does not occur in castrated men or in men with a deficiency in 5α-reductase due to mutations in the type 2 5α-reductase gene. Women with excessive blood levels of androgens or enhanced sensitivity to androgens can have significant scalp hair loss. Oral finasteride is effective treatment for some forms of both male and female alopecia.

A variety of compounds found in the diet are also potent inhibitors of 5α-reductase. Some of these phtyochemicals include unsaturated fatty acids, flavanoids, anthraquinones, and curcuminoids (Fig. 2). It is unknown if these natural inhibitors have a role in modulating 5α-reductase activity. However, diet is an important component of cancer risk and geographical and racial differences in cancer rates have been attributed, in part, to differences in diet. Asians, who have a diet rich in fruits, whole grains, and vegetables, have a lower incidence of prostate and other cancers. Components of their diet—for example, green tea, which contains high levels of the flavanoid, epigallocatechin gallate, a potent inhibitor of 5α-reductase—may modulate androgen activity in tissues such as the prostate and in so doing decrease the incidence of cancer. Ingestion of large amounts of these inhibitors through the diet may also have unwanted effects, especially on male sexual development *in utero*. Some cases of idiopathic hypospadias may be due to inhibition of 5α-reductase in the developing fetus by dietary components. Synthetic 5α-reductase inhibitors, such as finasteride, should also be avoided during pregnancy due to their potential to feminize male fetuses.

See Also the Following Articles

Androgens, Effects in Mammals; Benign Prostatic Hyperplasia; Prostate Cancer; Wolffian Ducts

Bibliography

Anderson, K. M., and Liao, S. (1968). Selective retention of dihydrotestosterone by prostatic nuclei. *Nature* **219**, 277–279.

Chang, C., Kokontis, J. K., and Liao, S. (1988). Structural analysis of complementary DNA and amino acid sequences of human and rat androgen receptors. *Proc. Natl. Acad. Sci. USA* **85**, 7211–7215.

Gardner, F. H., and Besa, E. C. (1983). Physiological mechanisms and the hematopoietic effects of the androstanes and their derivatives. *Curr. Topics Hematol.* **4**, 123–195.

Hiipakka, R. A., and Liao, S. (1995). Androgen receptors and action. In *Endocrinology* (L. I. DeGroot, Ed.), 3rd ed., Vol. 3, pp. 2336–2351. Saunders, Philadelphia.

Imperato-McGinley, J., Guerro, L., Gautier, T., and Peterson, R. E. (1974). Steroid 5α-reductase deficiency in man: An inherited form of male pseudohermaphroditism. *Science* **186**, 1213–1215.

Metcalf, B. W., Levy, M. A., and Holt, D. A. (1989). Inhibitors of steroid 5α-reductase in benign prostatic hyperplasia, male pattern baldness and acne. *Trends Pharmacol.* **10**, 491–495.

Quigley, C. A., De Bellis, A., Marschke, K. B., El-Awady, M. K., Wilson, E. M., and French, F. S. (1995). Androgen receptor defects: Historical, clinical and molecular perspectives. *Endocr. Rev.* **16**, 271–321.

Randall, V. A. (1994). Role of 5α-reductase in health and disease. *Bailliere's Clin. Endocrinol. Metab.* **8**, 405–431.

Russel, D. W., and Wilson, J. D. (1994). Steroid 5α-reductase: Two genes/two enzymes. *Annu. Rev. Biochem.* **63**, 25–61.

Umekita, Y., Hiipakka, R. A., Kokontis, J. M., and Liao, S. (1996). Human prostate tumor growth in athymic mice: Inhibition by androgens and stimulation by finasteride. *Proc. Natl. Acad. Sci. USA* **93**, 11802–11807.

Diploptera punctata

Barbara Stay

University of Iowa

I. How *Diploptera* Became a Model Insect
II. Life History
III. Regulation of Egg Development Cycles
IV. Allatostatins: Neuropeptides That Inhibit Juvenile Hormone Synthesis
V. Regulation of Milk Synthesis
VI. Male Physiology and Reproductive Behavior Studies

GLOSSARY

allatostatin A neuropeptide named for its ability to inhibit production of juvenile hormone by corpora allata; it is also found in nerve cells other than those innervating the corpora allata and appears to function as a neuromodulator and myomodulator.

brood sac An infolding of the integument in the genital region of the female that contains a brood of embryos and produces secretion to nourish them.

cockroach milk A nutritive secretion of gland cells of the brood sac that contains glycoprotein, carbohydrate, and lipid.

corpora allata Producers of juvenile hormone; a pair of spherical endocrine glands lying above the gut, behind the brain, and innervated by nerves from the cerebral nervous system.

juvenile hormone A major insect hormone necessary for maintaining juvenile characteristics during larval development and in most insects for production of vitellogenin and its uptake by maturing oocytes.

vitellogenesis The process of producing a large protein, vitellogenin (usually by the fat body), and its pinocytotic uptake by oocytes to form yolk spheres containing vitellin, the protein derived from vitellogenin.

viviparous As applied to reproduction of *Diploptera*: indicating that the mother nourishes the embryos during development and gives birth to first-stage larvae.

Diploptera punctata is a viviparous species of cockroach native to tropical Pacific regions and introduced into Hawaii. It has been identified as a notable experimental model.

I. HOW *DIPLOPTERA* BECAME A MODEL INSECT

This unique cockroach became an experimental insect when a colony sent from Hawaii was established in the laboratory of Louis M. Roth and Edwin R. Willis, who were studying the evolution of cockroach reproductive behavior. Harold R. Hagan had described the embryology of this species (then called *D. dytiscoides*) which he included in his 1951 book, *Embryology of Viviparous Insects*. From dissections of fixed females collected in the field and from histological sections of the embryos, he observed a huge increase in size of the embryos carried in the brood sac of the mother without a decrease in the initial egg yolk. From this he concluded that *Diploptera* must be a truly viviparous cockroach, i.e., the embryos were nourished by the mother during development. Roth and Willis verified his conclusion by culturing *Diploptera* in the laboratory and demonstrating that embryos increased 50-fold in dry weight during development. Many studies on the reproductive cycle and its regulation have since been performed.

II. LIFE HISTORY

A. Laboratory Culture

Diploptera is dark brown in color and intermediate in size between the domestic pest species, the American cockroach (*Periplaneta*) and the German cockroach (*Blattella*). It moves quickly but not as quickly as the previously mentioned species and is therefore easily handled. Its odor is not strong and although it has defensive glands that secrete quinones (from the first abdominal spiracles), it seldom discharges these if handled gently.

In the field, *Diploptera* lives in leaf litter, eats dried vegetation, and damages living trees and shrubs by feeding on their bark. However, they are easily reared in the laboratory on dry dog food or laboratory chow and water delivered through absorbent cotton or some system that prevents them from drowning in open water. Stock cultures may be kept in plastic boxes (e.g., mouse cages) with a thin layer of Vaseline around the top of the box to prevent escape; adults will not fly out when the cover is removed. The cover should be sufficient to exclude stray cockroaches of pest species but permeable enough to moisture so that the food does not mold. Moldy chow will not be eaten. Small paper boxes should be included to increase the surface area of the cage and provide hiding places during the light cycle. Cockroaches are most active during the dark cycle. Smaller groups of experimental animals may be housed in smaller containers with similar provisions. The description of the life cycle that follows is for insects reared at 27°C in a 12-hr light:12-hr dark cycle and about 70% relative humidity.

Care should be taken not to allow escape into a greenhouse because escapees will live happily in potted plants, feeding on favored leaves at night and hiding in the pots and under leaves during the day. Otherwise, *Diploptera* does not appear to be a household pest in the temperate region.

B. Adults

1. Body Size, Sexual Maturity, and Longevity

Males are the smaller and slimmer of the sexes (Fig. 1). Their body length is about 15 mm and at adult emergence they weight about 130 mg. Females have a slightly longer body length (about 18 mm) with a distinctly broader abdomen than that of males. At adult emergence, females weigh about 220 mg. Males are not sexually mature until 8 days after adult emergence; although some 4-day males mate, 70% of these matings did not produce offspring. In contrast, females are sexually mature at adult emergence. Adult males live about 15 months and females about 12–14 months.

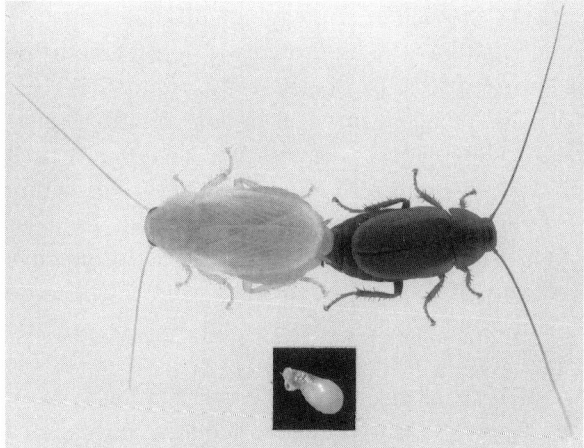

FIGURE 1 *Diploptera punctata* mating. The female (left) has recently molted from the last larval stage and is not yet tanned; the male (right) is about 8 days older. Magnification, ×1.6. Inset shows sperm-containing spermatophore that the male forms and transfers to the female during copulation. Magnification, ×6.3.

2. Mating

Initiation of courting requires that the male make contact with the female. When a sexually mature male encounters a receptive female, he rapidly moves his antennae over the female and then orients beside her, but in the opposite direction. Courtship consists of wing flutters directed to the head of the female, followed by a raised wing position. The female responds by climbing up on the male's back, and the male rapidly engages the female's genitalia. Alternatively, males sometimes back under the female to grasp her genitalia. In either case, he then swings around to face in the opposite direction from the female. Within the next 30 min the male forms, from accessory gland secretions, a packet (spermatophore) containing sperm (Fig. 1, inset). He transfers it and "glues" it (also with accessory gland secretion) into the copulatory bursa of the female. This accomplished, the pair uncouples. By 5 hr the sperm have migrated into the female's sperm storage organ, the spermatheca.

Females mate at adult emergence, sometimes before they release the swallowed air which facilitates emergence from the larval cuticle and almost always before their new cuticle darkens and hardens (Fig. 1). Females with access to sexually mature males virtually all mate on the day of adult emergence. A female that has already mated will reject a courting male by a distinctive waggle of her abdomen. A female that does not mate on the day of adult emergence gradually loses the ability to elicit male courting (by Day 4 only 60% of females mate). The nature of the female sex attractant is not known. Females that have mated and proceed through a reproductive cycle will mate again just after giving birth to a brood of embryos. Females that do not remate can use sperm stored from the first mating to fertilize the second batch of eggs, but there is a slight decrease in the number of embryos that develop. As the embryos emerge from the brood sac they molt, leaving their embryonic cuticles protruding from the female. The attractiveness of a female to males after the adult molt and after her embryos molt suggests that the sex attractant is a component of molting fluid. However, it could be some other factor produced at appropriate times. The nature of the female sex attractant is worthy of further investigation.

3. Egg Development Cycle, Ovulation, and Oviposition

The paired ovaries of *Diploptera* each contain six ovarioles, rarely seven. Each ovariole has, in linear sequence, numerous oocytes each surrounded by a layer of follicle cells. (No nurse cells contribute to the developing oocytes in cockroaches.) In the egg maturation cycle that follows mating, only the basal oocyte in each ovariole (that nearest the oviduct) becomes vitellogenic (forms yolk spheres). At adult emergence these oocytes are 0.6 mm in length (Fig. 2). By 3 days after mating the oocytes have grown

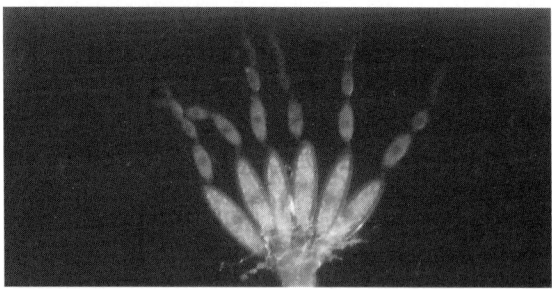

FIGURE 2 One ovary from a 0-day female showing the six ovarioles, each with a series of progressively larger oocytes from anterior (top) to posterior where the ovary joins the oviduct (bottom). Magnification, ×16.

FIGURE 3 One ovary from a 7-day mated female showing the yolk-filled mature basal oocytes. Magnification, ×16.

to 0.8 mm in length, spaces occur between the follicle cells, and light yellow yolk spheres containing vitellin (yolk protein) are visible in the oocytes. These oocytes grow rapidly in volume and yolk content until Days 5 or 6, when the process of vitellogenesis ceases and each oocyte contains about 18 μg of vitellin. There is only a 10-fold increase in protein in oocytes during maturation (Fig. 3). At that time, the oocytes are about 1.6 mm in length, the spaces between the follicles are closed, and the follicle cells secrete a porous multilayered egg shell (chorion). On Days 4 or 5 the glue holding the spermatophore in place is solubilized by secretions from the accessory spermathecal glands and the spermatophore is dropped. At ovulation the eggs leave the ovarioles and are released one at a time from the common oviduct. They are held in the genital segments with anterior ends dorsal facing the opening of the spermathecal ducts. The sperm enter the eggs through the hole in the chorion (micropyle). The second egg is aligned beside the first and subsequent eggs are aligned in pairs so that six pairs of fertilized eggs move posteriorly and receive secretions from the left and right colleterial glands (accessory glands). The secretions flow down over the eggs and form a greatly diminished egg case (ootheca) compared to that of other cockroaches. It is a thin, elastic boat-shaped membrane over the base (posterior end) of the eggs. This small batch of eggs protrudes only slightly be-

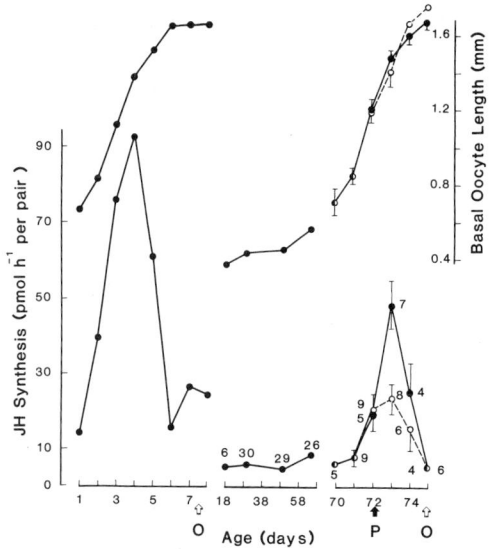

FIGURE 4 Graph showing two reproductive cycles. The change in basal oocyte lengths is shown above. This growth is supported by cycles of juvenile hormone (JH) synthesis by the corpora allata (below). After oviposition (O) of the basal oocytes on Days 7 or 8, embryos develop in the brood sac while oocytes grow slowly and JH production is low until just before parturition (P) when a second cycle begins. The oocyte growth and JH rates for females that have not mated a second time are also shown (--O--) (reprinted from *J. Insect Physiol.* 31, S. M. Rankin and B. Stay, pp. 145–157, © 1985, with kind permission from Elsevier Science Ltd, The Boulevard, Langford Lane, Kidlington OX5 1GB, UK).

yond the genital segments of the female. Once formed, it is rotated 90° to the left and within a few minutes retracted into the brood sac (this is equivalent to oviposition). The posterior end of the batch of eggs can be seen when the genital segments are separated.

At the time of oviposition, the next basal oocytes in the ovaries are about 0.4 mm in length. They grow very slowly during the gestation period of the embryos until several days before parturition, when they begin the next vitellogenic cycle and are oviposited 4 days after the embryos are born (Fig. 4). Females, on average, have three broods, but four or five have been recorded.

A female that remains unmated does not mature eggs promptly as mated females do but will sometimes oviposit infertile eggs after about 30 days and may do this repeatedly. The regularity and speed with which females mate and oviposit, as well as the fact that mating, but not feeding, is required for the egg cycle to ensue promptly, are elements of *Diploptera* reproduction that make this species a good model insect.

4. Gestation (Pregnancy)

i. Brood Sac Pregnancy takes place in a membranous infolding of the body wall at the ventral posterior end of the abdomen (Fig. 5). Thus, the brood sac is lined with cuticle under which are cell types typical of insect integumentary glands: gland cells, duct cells, and cuticle cells. The gland cell secretions are carried to the surface through cuticular ducts of the duct cells. The ducts open in clusters into depressions in the numerous papillae of the lining. Small sensory hairs protrude from the ventral surface of the sac. When the embryos are small, the sac is contracted by the surrounding muscle so that the embryos are held tightly and some of the sensory hairs are hidden in the folds of the sac. As the embryos grow, the sac unfolds and stretches to accommodate the occupants. In the fully stretched state, more sensory hairs are exposed and the papillae are flattened, leaving only a shallow depression into which the pores open. The embryos remain surrounded by air in the brood sac.

The gland cells undergo a cycle of secretion during gestation and progress from an upright position (columnar) to being almost horizontal as the brood sac stretches. The organelles of protein synthesis, rough endoplasmic reticulum (RER) and Golgi bodies, become more conspicuous from oviposition (Day 8)

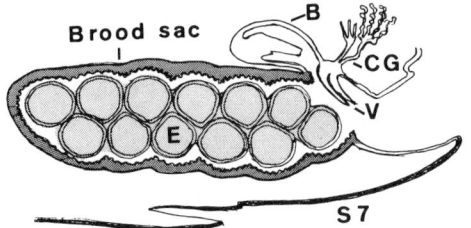

FIGURE 5 Diagram of mid-longitudinal section of a brood sac containing 12 young embryos (E) in cross section. The brood sac is continuous with sternite 7 (S7). The oviduct and spermathecal ducts open into the copulatory bursa (B). Colleterial glands (CG) secrete the oothecal materials. Ovipositors (V) are small (reproduced with permission from B. Stay and A. Coop, *Tissue Cell* 6, 669–693, 1974).

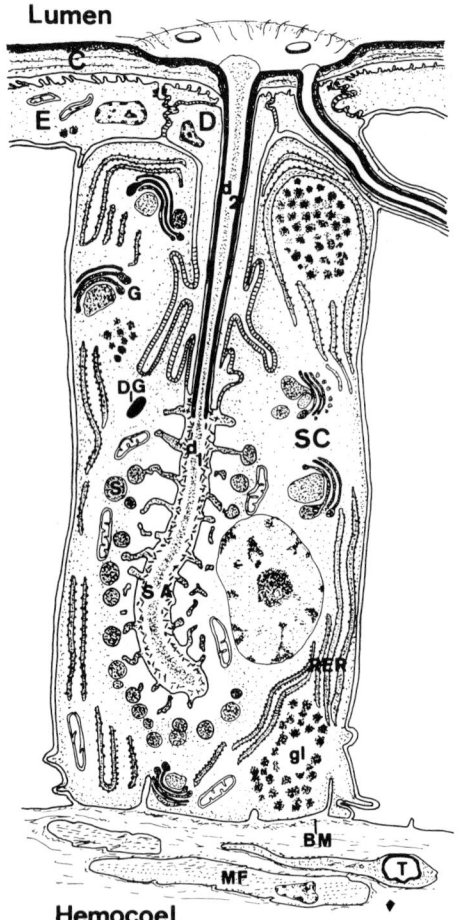

FIGURE 6 Diagram of a portion of brood sac wall. Cell (E) produces the cuticular lining (C); duct cell (D) produces the ducts (d2) carrying the secretion from the duct (d1) of the secretory apparatus (SA) of the secretory cell (SC) to the lumen. The gland cell is filled with rough endoplasmic reticulum (RER), Golgi (G), secretory granules (S), glycogen (gl), and a few dense granules (DG). BM, basement membrane; T, trachea; MF, muscle fiber (reproduced with permission from B. Stay and A. Coop, *Tissue Cell* 6, 669–693, 1974).

until milk secretion begins on Day 20. Thereafter, until shortly before the end of gestation (about Day 70), the cisternae of the RER are filled with secretion, and large vesicles are present near the Golgi bodies and along the infolded secretory surfaces of the cells (Fig. 6). The change in morphology of the gland cells is reflected in an increase in total protein content of the brood sac and an increase in its secretory product from Day 20 to the end of gestation. At the end of gestation, the RER, Golgi, and mitochondria are degraded in autophagic vacuoles, and the cells are diminished in size until the next cycle.

ii. Composition of Cockroach Milk The nutritive secretion from the brood sac, called cockroach milk, can be collected by substituting rolled filter papers for embryos and eluting the absorbed milk. From a female a few days before parturition about 400 μg of protein secretion can be obtained in 24 hr. The secretion is largely glycoprotein (51% of dry weight). The carbohydrate is 25% of dry weight and consists mostly of mannose and smaller amounts of unidentified sugars. Twenty-five percent of this carbohydrate is associated with the protein. The lipid content is about 10% of the dry weight of the secretion. There is also a wax that loosely coats the embryos. The wax is not a nutritive component and is probably produced by the cuticular cells of the brood sac. Only a small portion of the dry weight of secretion is free amino acids (about 1 or 2%). The protein contains all of the common amino acids. After removal of carbohydrate, the protein was resolved by sodium dodecyl sulfate electrophoresis into two bands, 16 and 13 kDa; the former was more abundant and sharper. It should be mentioned that milk protein in the midgut of the embryos appears to have most of the carbohydrate removed and forms a crystalline precipitate, presumably a storage reserve of nutrient.

iii. Embryonic Development The pairs of embryos develop in the brood sac with their heads to the left, dorsal surfaces against the brood sac lining and ventral surfaces facing one another. About 62 days are required for embryogenesis in *Diploptera*, i.e., the mother gives birth to her first batch of young 70 days after adult emergence. However, as Hagan surmised from his studies, the early stages of development occur quickly compared to those of other cockroaches. The gestation time elapsing before the dorsal body wall is closed (dorsal closure) and the gut is formed is only 19% compared to 40–50% in other cockroaches that carry embryos in a brood sac during embryogenesis. Dorsal closure corresponds precisely with the stage when milk secretion begins in the glands cells of the brood sac. The embryos are thus

TABLE 1
Developmental Timetable for *Diploptera* Embryos[a]

Age of female (days, 27°C)	Length of embryo (mm ± SEM)[b]	Wet weight of embryo batch (mg ± SEM)	Protein content of embryo batch (mg)	Stage characteristic
8	1.32	3.05 ± 0.10	0.3	Oviposition
16	1.38 ± 0.03	3.27 ± 0.11		Embryo moves to posterior in egg
18	1.55 ± 0.03	3.14 ± 0.30	0.3	Embryos easily separable from each other
20	1.61 ± 0.02	3.78 ± 0.15		Dorsal closure of body wall
21	1.65 ± 0.03	3.58 ± 0.21	0.3	Yellow fluid (milk) in midgut
25	1.77 ± 0.03	6.48 ± 0.01		Midgut half milk, half yolk
35	2.50 ± 0.07	10.55 ± 0.76	0.8	± Eye color; brain fills head
45	3.70 ± 0.12	26.52 ± 0.55	2.5	Eyes orange-brown; midgut in abdomen
55	4.15 ± 0.07	52.15 ± 1.9	7.0	"Bumps" on abdominal segments
65	5.30 ± 0.10	106.3 ± 2.9	17.0	Eyes and mandibles black
67				Body hairs pigmented
70	5.7 ± 0.28	140.9 ± 11.0	18.0	
71				Birth

[a] Modified from B. Stay and A. Coop, *J. Insect. Physiol.* **19**, 147–171, 1973.
[b] Dorsal view of embryo within the ootheca.

able to ingest the secretion and utilize it to grow. They increase 60 times in protein content during the remainder of development. It appears that the milk is absorbed by the hydrophilic chorion, passes through it to the hydrophobic cuticle of the embryo, and then proceeds to the ventral surface of the embryos under the appendages to the mouth where pumping motion of the pharynx sucks it into the gut. Salient features of development are given in Table 1 and some embryos are shown in Fig. 7.

Another extraordinary feature of these embryos is their pleuropodia, appendages of the abdominal segments. They have two pairs, an evaginated type on the first abdominal segment and a semiinvaginated type on the ninth abdominal segment. The latter were called adenopodia by Hagan. The pleuropodium of the first abdominal segment is very conspicuous because it has a trilobed bulb near the body and a thin extension that runs posteriorly to contact the surface of the adenopodium and then continues along the dorsal surface of the embryo and over the top of the head. Both of these organs have vast plasma membrane–mitochondrial complexes but on opposite surfaces of their cells. Those of the pleuropodia are at the external surface and those of the adenopodia are at the internal, hemolymph surface of the organ. The function of these organs has not been demonstrated but the huge mitochondria associated with infolded membrane suggest that adenopodia and pleuropodia are involved in ion and fluid regula-

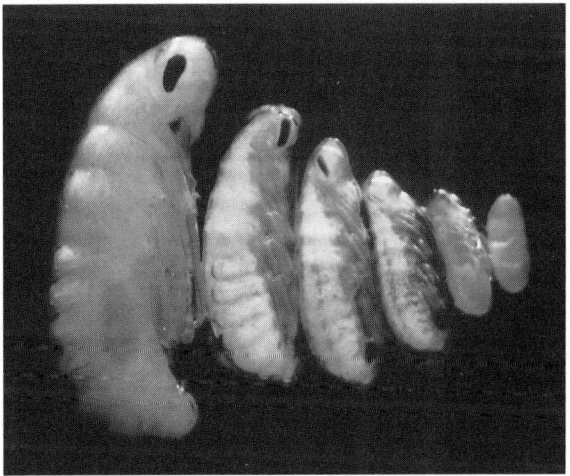

FIGURE 7 Representative stages of developing embryos removed from brood sacs of pregnant females at 18, 25, 35, 45, 55, and 67 days of adult age (from right to left). Magnification, ×8.

tion and that these organs probably function in concert. The extensions of the pleuropodia are sufficiently long so that they could be studied *in vitro* as has been done in other insects with the malpighian tubules (organs of nitrogenous excretion and salt and water balance associated with the gut).

C. Larval Stages

First-instar larvae, also called nymphs, are large, as would be expected from the extended period (about 48 days) of feeding while embryos. The duration of larval development is not much longer than that of embryogenesis. Given ample food and uncrowded conditions, most males have three larval instars and females have four. Under less favorable conditions, both sexes undergo an extra instar. Under favorable conditions the three stadia (the period of time for a stage or instar) for males are 12, 13, and 22 days. The four stadia for females are 12, 14, 17, and 23 days. Thus, larval life for males is about 47 days and that of females about 66 days. The external genitalia distinguish the sexes in all instars. It is possible to identify individuals at molts into the penultimate and last instars in the stock cultures with accuracy (70% for penultimate and almost 100% for last instars) so that animals of known ages in these stages can be used experimentally without special rearing.

III. REGULATION OF EGG DEVELOPMENT CYCLES

The predictability of the time course of the female reproductive cycle in *Diploptera* makes it an excellent insect for experimentation to determine factors that regulate this cycle. Healthy animals need only a little antibiotic applied to wounds for almost 100% survival following surgical operations such as gonadectomy and nerve transections. However, perhaps the most important reason that *Diploptera* was used in so many studies on the regulation of reproduction is because the hormone required for egg development, juvenile hormone (JH), is produced abundantly by the corpora allata (CA), glands located posterior to the brain and innvervated by nerves from the brain.

They produce JH III (C_{16}JH) and, as for other CA, the hormone is released as it is synthesized. Identification of the structure of JH by Röller and colleagues in 1967 and the development of a radiochemical method for measuring rates of JH synthesis by CA *in vitro* by Pratt and Tobe in 1974 made it possible to demonstrate directly the production of JH. Electron microscopical observation of CA, in which JH synthesis had been measured, showed a cycle of morphological changes, especially in mitochondria and smooth endoplasmic reticulum, with increasing hormone production. At the end of a cycle of JH production these organelles were degraded by autophagy. Also, the high activity of adult female CA makes them well suited for studies of the pathways of JH biosynthesis.

A. Role of Mating

One of the first questions asked was how mating starts the cycle of egg development. Is sperm necessary or is a spermless spermatophore sufficient? Does the ventral nerve cord need to be intact for mating to have an effect? Experiments done by Franz Engelmann and by Louis Roth showed that an empty spermatophore or a glass bead in the copulatory bursa was sufficient and an intact nerve cord was necessary for an egg development cycle. It was also demonstrated that mating was not required for an egg development cycle to ensue if a pair of CA were implanted into a virgin female or the nerves from the brain to the CA were transected. Thus, mating in *Diploptera* appears to release inhibition of JH production by the CA.

B. Role of the Ovary

Egg development in *Diploptera* is clearly correlated with the rate of JH synthesis by CA *in vitro* (Fig. 4). Because the quantity of JH in the hemolymph (insect blood), measured by gas chromatography–mass spectrometry, was found to parallel JH production by the CA during the reproductive cycle, it could be inferred that the *in vitro* measurement of JH production reflects the concentration of hormone that the tissues experience *in vivo*. Thus, it is possible to perform various surgical interventions on individuals and subsequently remove the CA to measure their

JH production *in vitro* to investigate factors involved in the regulation of the reproductive cycle. Also, measuring the vitellogenic growth of the oocytes in such females serves as a cumulative measure of JH titers over the interval following the surgical interventions.

Such experiments have shown that the ovary has a regulatory role in its own development in that it influences the rate of JH synthesis. In the absence of the ovaries the CA do not show a cycle of synthesis. However, after implantation of a vitellogenic ovary into an ovariectomized female, JH synthesis is stimulated, whereas after implantation of an ovary near the termination of vitellogenesis, JH synthesis is actively inhibited. Only the basal oocytes of a few ovarioles are required for these effects. Neither the stimulatory nor the inhibitory factors from the ovary have been determined. It was thought that ecdysteroids (insect growth hormones) produced by the ovary might be the inhibitory factor because application of exogenous ecdysteroid to the females inhibits JH synthesis. However, the ovarian ecdysteroids are produced at the time of chorionation of the eggs and thus after JH synthesis begins to decline.

C. "Feedback" of JH on JH Production

Exogenously applied JH analog in very low doses stimulates JH synthesis but only in the presence of the ovary. This suggests that low doses of JH start vitellogenesis in the ovary and the ovary in turn stimulates JH production. Large doses of exogenous JH analog inhibit JH synthesis but the ovary is not required for this effect.

D. Pregnancy and Inhibition of Oocyte Development

During pregnancy the basal oocytes in the ovaries do not become vitellogenic until about 2 days before parturition. JH production must be maintained at a sufficiently low rate during pregnancy so that ovarian eggs do not become vitellogenic while the embryos are completing development. Presumably, sensory information traveling along the ventral nerve cord to the brain and on through nerves to the CA prevents JH synthesis by the CA. Evidence for this comes from premature onset of JH synthesis during pregnancy following transection of the nerve cord or the nerves between the brain and the CA. Also, removal of embryos in the course of pregnancy starts a premature cycle of oocyte development. Removal of embryos and severance of the nerves to the CA have an additive effect, suggesting several pathways of inhibition of the CA. The stimulation of increasing numbers of sensory hairs in the brood sac, as the enlarging embryos stretch the brood sac and expose more sensory hairs, could be the source of information traveling to the brain to inhibit JH synthesis. Adaptation of the brain to this inhibitory stimulation late in pregnancy could result in the renewal of JH production and a renewed vitellogenic cycle. This speculation should be investigated experimentally.

IV. ALLATOSTATINS: NEUROPEPTIDES THAT INHIBIT JUVENILE HORMONE SYNTHESIS

A. Isolation of the Peptides, Cloning of the Gene, and Sensitivity of the Corpora Allata

About 20 years before JH synthesis could be measured directly, experiments done by Berta Scharrer led her to predict that in cockroaches neurosecretions from the brain inhibited JH synthesis by the CA. Because the synthesis of JH by the CA of *Diploptera* could be measured so easily *in vitro*, it was possible to use this method to test extracts of brain for their ability to inhibit JH synthesis. Indeed, saline extracts heated to 100°C did inhibit the production of JH by CA *in vitro*. Purification of the inhibitory material revealed seven small peptides that were named allatostatins. All have the same C-terminal sequence and are amidated at the C-terminal end. These peptides and six others are coded for by a single gene. All these peptides inhibit JH synthesis but with different potencies. The five C-terminal amino acids and amidation are required for activity and the variable N terminals determine the degree of activity. Allatostatins appear to act very early in the pathway of JH synthesis, probably at the transfer of two carbon units, derived from mitochondrial contribution to the processing of glucose and amino acids, into the cytoplasm.

CA in different physiological states show different sensitivities to these allatostatins. The CA are least sensitive when they are synthesizing JH at high rates and most sensitive when JH synthesis begins to fall at the end of the vitellogenic cycle of JH synthesis. However, sensitivity is also low during pregnancy. There may be multiple reasons for these changes in sensitivity. It could be that the receptors, which have yet to be isolated, change in quantity or that allatostatin acts on one pathway of JH synthesis and that the pathways change during the cycle of synthesis.

B. Regulation of Allatostatin Release

Allatostatin is transported from nerve cell bodies in the lateral part of the forebrain by axons that extend out of the same side of the brain to the CA just behind the brain. Studies have begun to determine what triggers these cells to release allatostatins. Some complex experiments, in which male *Diploptera* were treated with large doses of JH analog and nerves from the brain to the CA were severed, showed unusually high levels of allatostatin in the hemolymph and some inhibition of JH synthesis by the CA. If the nerves were left intact this elevation in allatostatin did not occur and the CA were strongly inhibited. These results suggest that when the nerves are intact, inhibition of JH synthesis by JH analog acts through the brain causing nerve cells to release allatostatin within the CA. If the nerves are transected, the allatostatin that would have been released within the CA is released into the hemolymph, where it is less effective in inhibiting JH synthesis. When an ovary was also implanted into a male along with JH analog treatment and transection of the nerves to the CA, this ovary became vitellogenic and the quantity of allatostatin released into the hemolymph was decreased and JH synthesis was elevated compared to controls without ovaries. Thus, it appears that JH feedback to inhibit JH production and ovarian stimulation of JH production may act through the brain by influencing the amount of allatostatin released at the CA.

V. REGULATION OF MILK SYNTHESIS

A. The Onset and Stimulation of Milk Synthesis

Two kinds of experiments have shown that initiation of milk synthesis is regulated humorally. Severance of nerves to the brood sac did not prevent the onset of milk synthesis on Day 20. Also, milk synthesis in an empty brood sac from a 12- to 18-day-old female could be initiated by implanting such brood sacs into the body cavity of a female that was actively producing milk (e.g., Day 50) but not into the body cavity of a female that was not producing milk. Brood sacs from younger females could not be stimulated to produce milk. Thus, brood sacs become competent to produce milk about Day 12 but normally do not produce milk until Day 20. One factor found to be associated with the onset of competence to produce milk was the breakdown of glycogen specifically in the brood sac tissues. Empty brood sacs started to produce milk after transplantation into milk-producing females and normal brood sacs containing infertile eggs started to produce milk (although slightly later than those with fertile eggs); however, in the absence of growing embryos the quantity of milk produced did not increase. The following are unanswered questions with respect to synthesis of milk: What signals the brood sac to break down glycogen and become competent to produce milk? What is the nature of the humoral factor that initiates and stimulates milk synthesis? and How do growing embryos potentiate milk synthesis? It is known that if only 6 embryos are present in the brood sac (accomplished by removing one ovary from a previtellogenic female), these embryos gain the same dry weight as a normal batch of 12, no doubt by drinking twice the normal amount of milk.

B. Inhibition of Milk Synthesis by Juvenile Hormone

A few days before parturition milk synthesis declines and ovarian oocytes become vitellogenic as JH synthesis increases. Thus, it is possible that decline in milk synthesis results from elevation in JH concentration in the hemolymph. This was shown to be true by applying JH analog or implanting active CA into pregnant females actively producing milk. Such treatment resulted in a decline in the amount of milk within brood sac tissue and in synthesis of milk by the brood sac as well as autophagy of the protein synthesis organelles within the gland cells of the brood sac. It has not been determined whether JH acts directly on the brood sac to inhibit milk synthesis or whether this is an indirect effect. Studies on the brood sac *in vitro* should answer this question.

VI. MALE PHYSIOLOGY AND REPRODUCTIVE BEHAVIOR STUDIES

The CA of males are smaller than those of females and produce less JH, but they are able to respond to the influence of the ovary when implanted into females lacking their CA. For an unknown reason, female CA implanted into males produce large quantities of JH without ovarian stimulation. This is a phenomenon worthy of further investigation.

Male reproductive behavior does not require JH. In one male:one female behavioral tests of mating in *Diploptera*, more than half of 8-, 14-, and 28-day males (previously isolated from females) initiate courtship within 10 sec of encountering a receptive female and copulation ensued within 10 sec or less of courtship. However, a greater proportion of older males exhibited this pattern of behavior than did younger ones. In competitive mating tests with one female and three males, 8, 14, and 28 days of age, the 28-day males were more successful in mating because they were better able to follow females and were more persistent in their courtship.

See Also the Following Articles

Allatostatins; Corpus Allatum; Juvenile Hormone

Bibliography

Evans, L. D., and Stay, B. (1989). Humoral induction of milk synthesis in the viviparous cockroach *Diploptera punctata*. *Invertebr. Reprod. Dev.* **15**, 171–176.

Johnson, G. D., Stay, B., and Rankin, S. M. (1985). Ultrastructure of corpora allata of known activity during the vitellogenic cycle in the cockroach *Diploptera punctata*. *Cell Tissue Res.* **239**, 317–327.

Rankin, S. M., and Stay, B. (1985). Ovarian inhibition of juvenile hormone synthesis in the viviparous cockroach, *Diploptera punctata*. *Gen. Comp. Endocrinol.* **59**, 230–237.

Roth, L. M., and Willis, E. R. (1955). Intra-uterine nutrition of the "beetle-roach" *Diploptera dytiscoides* (Serv.) during embryogenesis, with notes on its biology in the laboratory (Blattaria: Diplopteridae). *Psyche* **62**, 55–68.

Stay, B., Sereg Bachmann, J. A., Stoltzman, C. A., Fairbairn, S. E., Yu, C. G., and Tobe, S. S. (1994a). Factors affecting allatostatin release in a cockroach (*Diploptera punctata*): Nerve section, juvenile hormone analog and ovary. *J. Insect Physiol.* **40**, 365–372.

Stay, B., Tobe, S. S., and Bendena, W. G. (1994b). Allatostatins: Identification, primary structures, functions and distribution. *Adv. Insect Physiol.* **25**, 267–337.

Sutherland, T. D., and Feyereisen, R. (1996). Target of cockroach allatostatin in the pathway of juvenile hormone biosynthesis. *Mol. Cell. Endocrinol.* **120**, 115–123.

Tobe, S. S., Ruegg, R. P., Stay, B. A., Baker, F. C., Miller, C. A., and Schooley, D. A. (1985). Juvenile hormone titre and regulation in the cockroach *Diploptera punctata*. *Experientia* **41**, 1028–1034.

Discoidal Placenta

John J. Rasweiler IV and Nilima K. Badwaik

Cornell University Medical College

I. Location of the Placenta
II. Vascularization and Finer Structure
III. Extraplacental Trophoblast
IV. Development

GLOSSARY

allantois An extraembryonic membrane that arises as a highly vascular, sac-like outgrowth of the embryonic hindgut.

chorion The outermost extraembryonic membrane composed of trophoblast and extraembryonic mesoderm.

chorioallantoic placenta An organ for physiological exchange formed by the complex union of two fetal membranes, the allantois and the chorion, with the uterus of the mother.

cytotrophoblast A cellular form of trophoblast that divides mitotically.

discoidal placenta A type of chorioallantoic placenta in which the surface that is amplified (usually in the form of villi or as a labyrinth) to promote physiological exchange between the maternal and fetal circulations occupies a disc-shaped area.

endotheliochorial placental barrier A placental barrier separating maternal and fetal circulations that includes at least a layer of maternal endothelial cells, one or two layers of trophoblast, and a layer of fetal endothelial cells.

fetal membrane One of the extraembryonic membranes that usually participate in protection of the conceptus and/or the formation of placental structures that are sites of physiological exchange between conceptus and mother; these include the allantois, amnion, chorion, and yolk sac.

hemochorial placental barrier A placental barrier separating maternal and fetal circulations that includes at least one layer of trophoblast and a layer of fetal endothelial cells.

interstitial membrane A prominent basal lamina that forms between the maternal endothelium and trophoblast of the interhemal barrier in many endotheliochorial placentae.

intervillous space A sinusoidal space surrounding the chorionic villi in some hemochorial placentae through which maternal blood circulates.

intrasyncytial lamina An interrupted extracellular layer in the interhemal barrier of many hemochorial placentae that is derived initially from the basal lamina of maternal endothelial cells and engulfed by processes of the syncytiotrophoblast.

placental labyrinth A system of tubular channels found in many chorioallantoic placentae through which maternal blood flows.

syncytiotrophoblast A type of trophoblast that is formed by the fusion of cytotrophoblast cells, incapable of proliferating mitotically, and consists of a multinucleated cytoplasmic mass.

Placentae are organs that form during gestation in many vertebrates and some invertebrates to facilitate physiological exchange between mother and conceptus. In the case of mammals, this usually involves the development of intimate anatomical associations between the extraembryonic membranes of the conceptus and the uterus of the mother. In human ectopic pregnancies, however, placentae can form without the participation of uterine tissues. The placenta delivered at the end of a normal human pregnancy is created by intimate linking of two such membranes of the conceptus (the allantois and the chorion) with the uterus of the mother and is discoidal in shape. While this is often viewed as the usual form of a placenta, many important placental structures in mammals do not involve both of these membranes, and the final (definitive) chorioallantoic placentae of many species do not have a discoidal shape, being instead diffuse, cotyledonary, or zonary. Mammalian orders in which discoidal chorioallantoic placentae have been observed include the Carnivora (some bears, ferrets, and mink), Chiroptera (bats), Dermoptera (flying lemurs), Insectivora (hedgehogs, moles, shrews, solenodons, tenrecs, and water shrews), Lagomorpha (hares, pikas, and rabbits), Macroscelidea (elephant shrews), Marsupialia (bandicoots only), Primates (except Strepsirhini), Rodentia, Scandentia (tree shrews), and Xenarthra (anteaters, armadillos, and sloths). In some cases (miniopterine and natalid bats, many primates, tree shrews, and some three-toed sloths), two discoidal placentae serve each conceptus.

I. LOCATION OF THE PLACENTA

In most mammals the discoidal placentae develop in constant, species-specific locations relative to the attachment of the uterine mesenteries. This means that the placentae for each species usually have positions that are consistently mesometrial, antimesometrial, lateral (i.e., between mesometrial and antimesometrial), or intermediate between these locations. Furthermore, this positioning of the placenta is usually similar in mammals belonging to the same taxonomic group (e.g., a family or order), although the bats are a notable exception in this regard. This characteristic is therefore of interest to biologists attempting to work out systematic relationships between mammals.

II. VASCULARIZATION AND FINER STRUCTURE

During the formation of discoidal chorioallantoic placentae, the chorion (which consists of trophoblast and avascular mesoderm) becomes intimately associated with the endometrial lining of the uterus, which provides the placenta with a supply of maternal blood. Mesodermal villi of the allantois then fuse with the chorion and provide a vascular supply for the embryonic side of the placenta. The anatomical relationship between the resultant chorioallantois and the maternal blood supply varies considerably with placental type. In villous placentae (e.g., those which develop in humans), the chorionic surface within the placenta is covered with many villi that end freely or are only occasionally interconnected. These project into maternal blood spaces that are large and sinus-like. In trabecular placentae, adjacent villous branches are fused in numerous places to form an anastomosing network through which maternal blood circulates. In labyrinthine placentae, maternal blood flows through narrow channels lined either by maternal endothelial cells or by trophoblast.

The interrelationship of maternal and fetal blood flow patterns in the regions of physiological exchange is of importance in determining placental efficiency. Anatomical studies have revealed three general types of vascular arrangements. In countercurrent flow systems (which occur in the placentae of rodents and rabbits), maternal and fetal microvascular beds are arranged in parallel, but blood flow in the two beds is generally in opposite directions. This is the most efficient system for promoting exchange between the two circulations by passive diffusion. In crosscurrent flow systems (which occur in the placentae of tree shrews and some bats), exchange occurs between small maternal vascular channels and fetal capillary beds that, for the most part, pass at right angles to each other. Finally, in the multivillous flow system (which occurs in catarrhine primates), exchange takes place between fetal capillary beds within chorionic villi that are bathed by maternal blood.

Chorioallantoic placentae are also classified according to the finer structure of the barrier that separates their fetal and maternal circulations. This relies on a nomenclature system that was originally devised many years ago by Otto Grosser on the basis of light microscopic observations. Because important components of the barrier in many species are too fine to be resolved by histological analysis alone, the system has been further refined by ultrastructural studies, most notably by Allen Enders (1965). Representative types of interhemal barriers found in discoidal placentae are included in Table 1. While the diversity is remarkable, major common themes followed in the evolution of these organs have been to bring the fetal and maternal circulations into close proximity, while still maintaining their separation, and to maximize the surface area available for physiological exchange between them. In the villous placenta of humans, this involves lengthening and increased branching of the trophoblast-covered villi that are bathed by maternal blood circulating

TABLE 1
Minimal Interhemal Barrier in Selected Discoidal Chorioallantoic Placentae

Type of barrier in definitive placenta	Species	Structural components
Endotheliomonochorial	Tree shrew (*Tupaia glis*)	Hypertrophied maternal endothelium lining placental labyrinth; mesh-like basal lamina frequently penetrated by processes of maternal endothelial cells and trophoblast; multinucleated trophoblast; basal lamina of trophoblast; fetal endothelial cells lacking a basal lamina
Endotheliomonochorial	Lesser mouse-tailed bat (*Rhinopoma hardwickei*)	Hypertrophied maternal endothelium lining placental labyrinth; thick basal lamina (interstitial membrane); cytotrophoblast; basal lamina of trophoblast and fetal endothelium; fetal endothelium
Endotheliodichorial	European mole (*Talpa europaea*)	Hypertrophied maternal endothelium lining placental labyrinth; thick basal lamina (interstitial membrane); syncytiotrophoblast; cytotrophoblast; single or double basal lamina; fetal endothelium
Hemomonochorial	Human	Syncytiotrophoblast lining intervillous space; scattered cytotrophoblast cells; single or double basal lamina; fetal endothelium
Hemomonochorial	Guinea pig (*Cavia porcellus*)	Syncytiotrophoblast (often highly attenuated) lining placental labyrinth; trophoblast basal lamina; basal lamina absent or poorly developed around fetal capillaries; fetal endothelium
Hemomonochorial	Black mastiff bat (*Molossus ater*)	Highly interdigitated cytotrophoblast giant cells lining placental labyrinth; single or double basal lamina; fetal endothelium
Hemodichorial	Short-tailed fruit bat (*Carollia perspicillata*)	Syncytiotrophoblast (containing a discontinuous intrasyncytial lamina) lining placental labyrinth; cytotrophoblast (often highly attenuated with many interdigitating microvilli); single or double basal lamina; fetal endothelium
Hemodichorial	Rabbit (*Orcytolagus cuniculus*)	Outer layer of syncytiotrophoblast, with pores and sometimes highly attenuated, lining placental labyrinth; inner layer of trophoblast varying from cellular to multinucleate or syncytial; basal laminae of trophoblast and fetal endothelium; fetal endothelium
Hemotrichorial	Rat (*Rattus norvegicus*), mouse (*Mus musculus*), hamster (*Cricetus auratus*)	Attenuated layer of cytotrophoblast with pores lining placental labyrinth; middle layer of syncytiotrophoblast; inner layer of trophoblast (probably syncytial); basal laminae of trophoblast and fetal endothelium; fetal endothelium
Bandicoot chorioallantoic placental barrier	Australian bandicoots (*Isodon macrourus*, *Perameles nasuta*)	Maternal endothelium; basal lamina of maternal endothelium; thin endometrial stroma; attenuated portions of syncytial cells ("heterokaryons") apparently formed by the fusion of uterine epithelial and trophoblast cells; fetal endothelium

through the intervillous space. In species with labyrinthine hemochorial or endotheliochorial placentae, there is an expansion in the network of trophoblast- or maternal endothelium-lined channels carrying maternal blood. As gestation progresses in both types of placenta, the microvascular beds on the embryonic/fetal side of the placenta also proliferate, and the finest of these vessels frequently move closer to the maternal vascular spaces (e.g., by occupying indentations in the trophoblast of the interhemal barrier), thereby reducing the diffusion distance between the two circulations.

In order to facilitate physiological exchange, some elements of the interhemal barrier often become more attenuated, sieve-like, or are completely eliminated as gestation progresses. For example, during early pregnancy in humans, cytotrophoblast cells form a nearly complete layer underneath the syncytiotrophoblast of the placental villi (i.e., the early interhemal barrier is hemodichorial). A major function of the cytotrophoblast cells is to divide mitotically and fuse to augment the quantity of syncytiotrophoblast in the placenta. However, recent studies have also localized growth factors that might be influencing mesodermal invasion and angiogenesis in the developing placenta, or their mRNAs, in the villous cytotrophoblast cells (Ferriani et al., 1994; Mühlhauser et al., 1996; Ziche et al., 1997). Later in pregnancy, the cytotrophoblast layer becomes incomplete so that in the mature human placenta cytotrophoblast cells can be found underneath only about 20% of the villous syncytiotrophoblast. Thus, the minimal interhemal barrier by that stage is hemomonochorial. In some areas the fetal connective tissue between the fetal endothelial cells and the syncytiotrophoblast is also eliminated, their basal laminae fuse, and the overlying syncytiotrophoblast becomes very thin.

A reduction of elements in the interhemal barrier is also commonly seen in many other species. In the short-tailed fruit bat (*Carollia perspicillata*) a complete layer of cytotrophoblast persists in the hemodichorial placental barrier; however, in many regions this layer becomes highly attenuated (Figs. 1 and 2). In laboratory rats and mice (which have hemotrichorial placentae) and rabbits (which have hemodichorial placentae), pores are present in the outer layer of trophoblast lining the maternal vascular

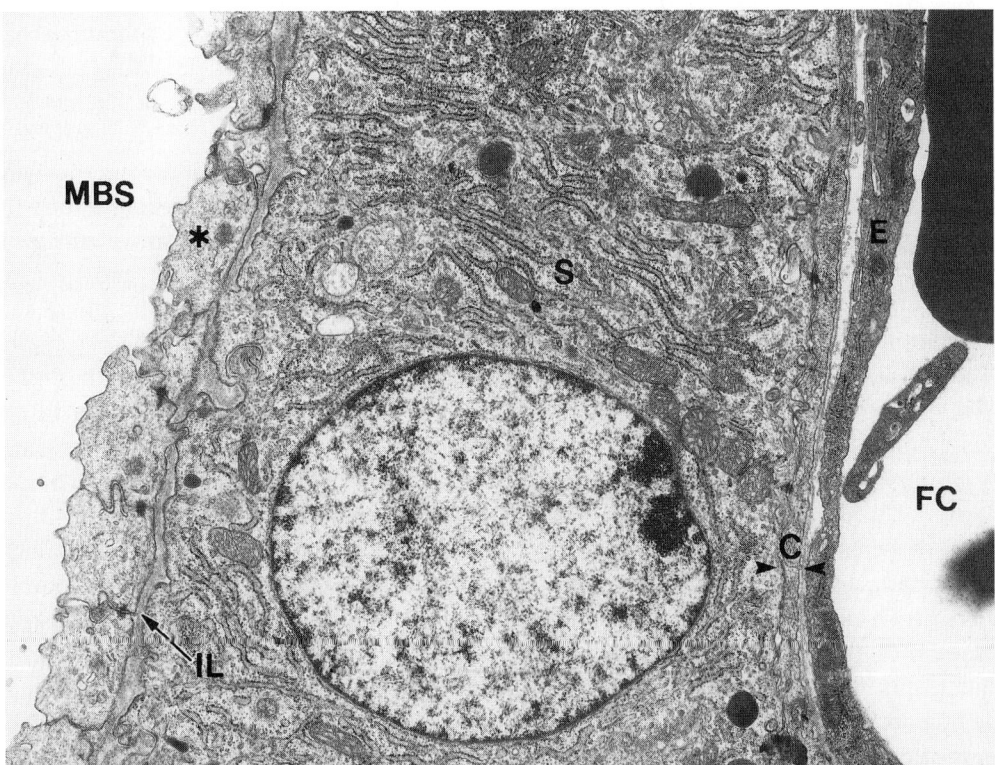

FIGURE 1 Hemodichorial barrier in the placenta of a short-tailed fruit bat, Day 70 postcoitum (18.2 mm fetus). Cellular components include the endothelium (E) of a fetal capillary (FC), an attenuated layer of cytotrophoblast (C), and a layer of syncytiotrophoblast (S). Ectoplasmic processes of the syncytiotrophoblast (*) have penetrated the intrasyncytial lamina (IL) and provide a lining for the maternal blood space (MBS). Magnification, ×11153.

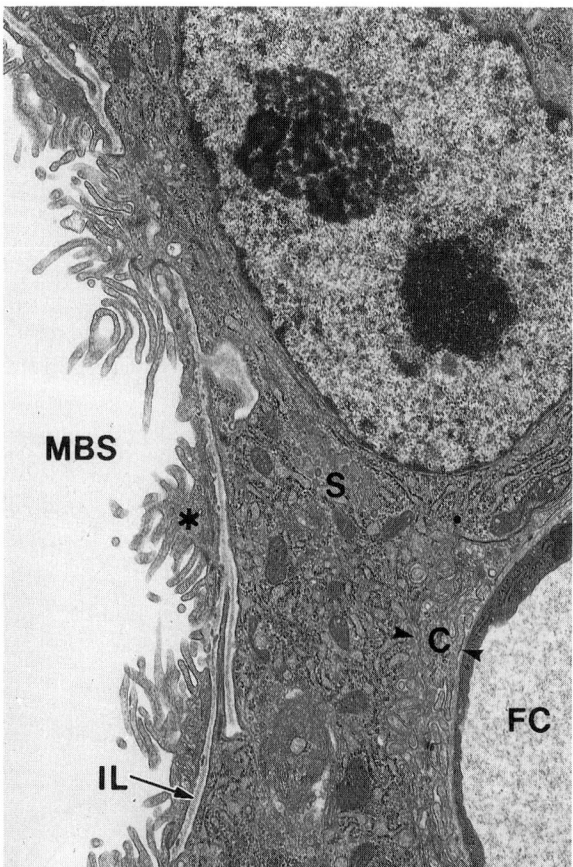

FIGURE 2 Hemodichorial barrier in the near-term placenta of a short-tailed fruit bat, Day 118 postcoitum (32.5 mm fetus). Cellular components include the endothelium of a fetal capillary (FC), a thin layer of cytotrophoblast (C), and a layer of syncytiotrophoblast (S). Ectoplasmic processes of the syncytiotrophoblast (*) have penetrated the intrasyncytial lamina (IL). These possess a profusion of microvilli which project into the maternal blood space (MBS). Magnification, ×10809.

transtrophoblastic channel system (Kertschanska et al., 1997). In the endotheliochorial placenta of the short-tailed shrew (Blarina brevicauda) the only layer of trophoblast is thin, syncytial, and composed primarily of attenuated, branching cytoplasmic processes honeycombed with extracellular spaces. These are penetrated by processes of the maternal endothelial cells and appear to provide channels through which substances might traverse the syncytiotrophoblast by an extracellular route. The fetal endothelium of the shrew placental barrier is unusual in possessing an abundance of branching cytoplasmic processes and lacking a basal lamina. The syncytiotrophoblast in the endotheliochorial placenta of the three-toed sloth (Bradypus tridactylus) is similar in having a lacy, sponge-like appearance. In the musk shrew (Suncus murinus) the only layer of trophoblast in the interhemal barrier of the definitive placenta is syncytial and often thin or discontinuous. The basal laminae of the syncytiotrophoblast and the fetal endothelium are discontinuous as well. This arrangement permits projections of the maternal and fetal endothelia to make contact with each other, thereby establishing a minimal interhemal barrier that is endothelio–endothelial.

Cellular components of the interhemal barrier frequently possess surface specializations that expand the area available for absorptive and transepithelial transport functions and/or provide extracellular channels that facilitate the diffusion of substances at least part of the way across the placental barrier. In many species with hemochorial placentae microvilli are abundant on the surface of the trophoblast exposed to maternal blood (Fig. 2), and caveolae (coated pits) or absorption canaliculi may be found between the bases of the microvilli. In humans the microvillous surface has also been found to be rich in some enzymes and to possess receptors for a number of factors (e.g., insulin and transferrin, which binds and transports iron in the blood). Microvilli, other types of surface projections, and/or plasmalemmal infoldings are commonly seen on the basal (fetal) side of the trophoblast as well in many species with hemochorial or endotheliochorial placentae.

Depending on the type of discoidal placenta (hemochorial or endotheliochorial), the spaces traversed by maternal blood are lined by trophoblast or mater-

spaces. Studies of hemotrichorial placentae using macromolecular tracers have generally indicated that this layer offers little resistance to the movement of macromolecules into the intercellular spaces between the outer and middle layers. Ultrastructural studies of the human, guinea pig, and degu (Octodon degus) placenta after the introduction of lanthanum hydroxide as an extracellular tracer have led to the discovery of narrow, membrane-lined invaginations of the syncytiotrophoblast plasmalemma. It has not yet been determined if these belong to a continuous,

nal endothelial cells. Although its functional significance has not been established for all species, these cells generally are rich in granular endoplasmic reticulum. This suggests that they are probably engaged in the synthesis of substantial amounts of protein for export. In the case of humans, the hormones human chorionic gonadotropin and human placental lactogen have been localized in secretory granules of the syncytiotrophoblast by means of ultrastructural immunocytochemistry. Caveolae, vesicles, and vacuoles associated with the uptake of material by endocytosis are frequently seen in the syncytiotrophoblast as well.

In most mammals with hemochorial placentae at least one continuous layer of syncytiotrophoblast is present in the interhemal barrier. Although this would seem to ensure that large molecules have to traverse the syncytiotrophoblast, which could act selectively on them, there are species (noted previously) in which extracellular pathways through this layer may, or do, exist. There are also some mammals with hemochorial placentae in which the only layer of trophoblast is of the cellular variety. These include members of three mammalian orders: several bats (Figs. 3 and 4), two rodents (the jumping mouse, *Zapus hudsonius*, and the jerboa, *Jaculus jaculus*), and the rock hyrax (*Heterohyrax brucei*).

In many mammals possessing endotheliochorial or hemochorial placentae, prominent extracellular layers are present in the interhemal barrier (Fig. 5). In the endotheliochorial placentae of some bats, insectivores, tree shrews, carnivores, and elephants, this takes the form of a thickened basal lamina, frequently referred to as an "interstitial membrane" that intervenes between the maternal endothelium and the trophoblast. In many other bats, several sciurid rodents, the hyena (a carnivore with a zonary placenta), and possibly the elephant shrew (a macroscelid), the maternal endothelium is lost, thereby creating a hemochorial condition. Components of the endothelial basal lamina appear to persist, however, as at least part of an interrupted layer (the

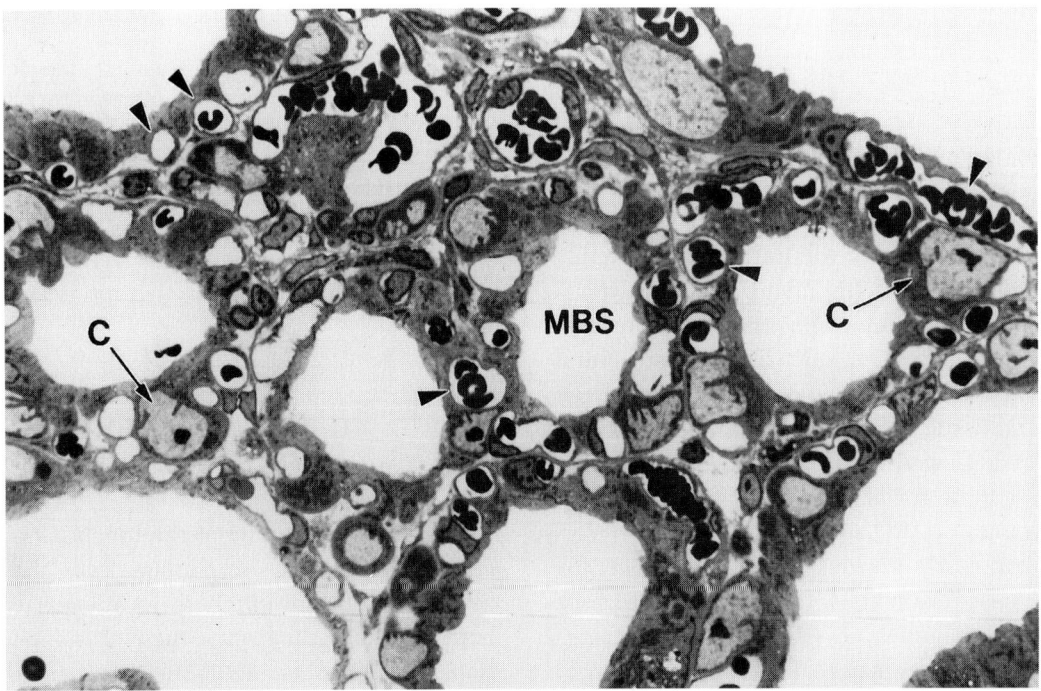

FIGURE 3 Plastic section of the near-term placenta from a black mastiff bat carrying a 32.0-mm fetus. This is a labyrinthine hemomonochorial placenta with cytotrophoblast giant cells (C) being the only trophoblast in the interhemal barrier. Many of the fetal capillaries (arrowheads) occupy indentations in the cytotrophoblast layer, thereby reducing the diffusion distance between fetal and maternal circulations. Blood was flushed from the maternal blood spaces (MBS) when the specimen was perfused with fixative. Magnification, ×752.

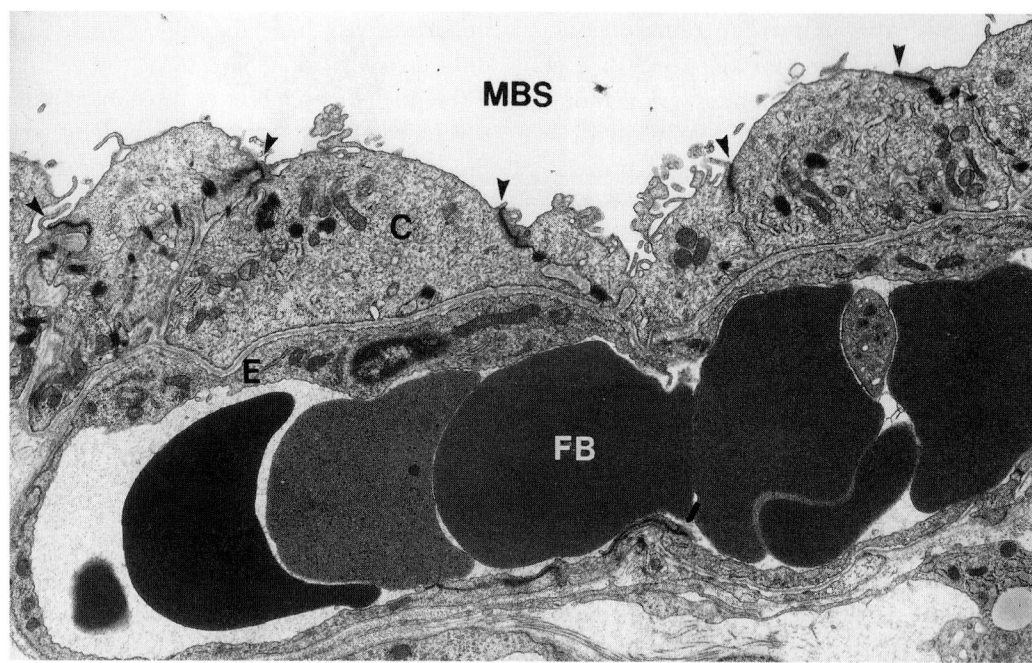

FIGURE 4 Electron micrograph of the interhemal barrier in the same near-term placenta of a black mastiff bat shown in Fig. 3. The minimal barrier consists of the endothelial lining (E) of a maternal capillary, interdigitating processes of cytotrophoblast cells (C), and the fused basal laminae of these two cellular layers. The arrowheads denote intercellular junctions. FB, fetal blood; MBS, maternal blood space. Magnification, ×7300.

"intrasyncytial lamina") within the interhemal barrier that is completely engulfed by trophoblast (Figs. 1 and 2).

Suggested functions for the intrasyncytial lamina have included roles in amplifying the apical plasmalemma of the syncytiotrophoblast for absorptive and secretory purposes, selective filtration, structural support, the establishment and maintenance of cell polarity, and tissue differentiation (Cukierski, 1987; Badwaik and Rasweiler, 1998). With the exception of the first, these might also apply to the interstitial membrane.

Our recent work provides additional support for several of these possibilities. In the short-tailed fruit bat (*Carollia*), the intrasyncytial lamina clearly thickens around the larger maternal vascular channels in the discoidal placenta as pregnancy progresses (Fig. 6). Because the degree of thickening appears to be directly related to the size of the channel, its purpose would seem to be to provide additional mechanical support for the channel wall. Studies of placentation in the black mastiff bat (*Molossus ater*) suggest that endometrial endothelial cells, interstitial membranes, and intrasyncytial laminae may also play important morphogenetic roles in controlling the growth and differentiation of trophoblast (see Section IV.

III. EXTRAPLACENTAL TROPHOBLAST

In some species, one of the functions of the placenta is to produce trophoblast cells that migrate away in significant numbers. In humans, macaques, hamsters, and guinea pigs, these cells appear to play important roles in modifying the structure and/or function of the uteroplacental arteries, thereby permitting increased blood flow as pregnancy progresses (Nanaev et al., 1995; Pijnenborg, 1996). Highly invasive trophoblast cells are also produced by the discoidal placentae of some bats (Figs. 7 and 8) (Rasweiler

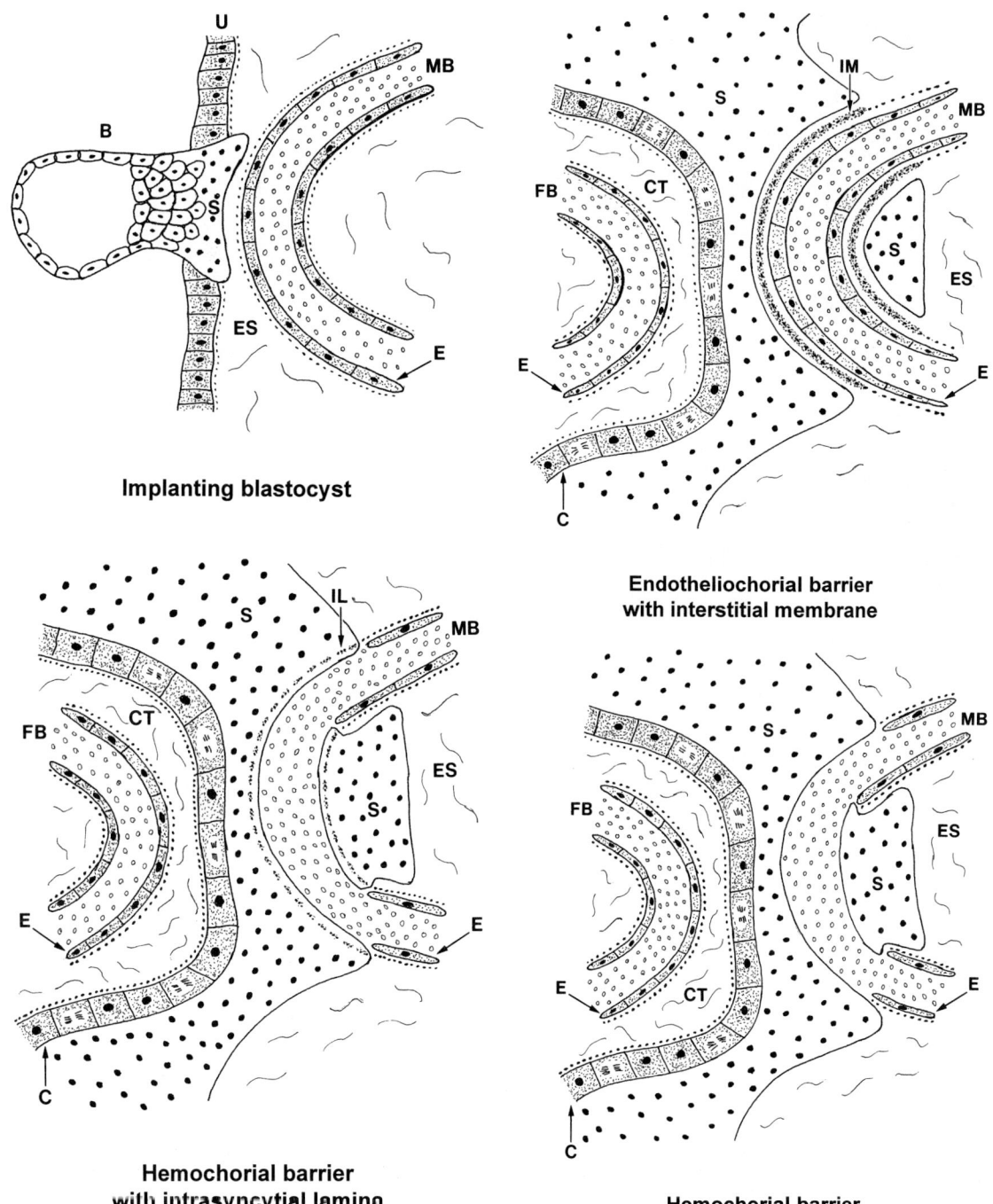

FIGURE 5 Diagrams depicting the anatomical relationships between an implanting blastocyst and the uterine endometrium and showing the interhemal barriers that develop in several types of discoidal placentae. In many endotheliochorial placentae, the maternal endothelial cells appear hypertrophied and a prominent amorphous extracellular layer, the interstitial membrane, is present between that endothelium and the trophoblast. In some hemochorial placentae, the maternal endothelium is lost, but its basal lamina is retained as part of a fragmented layer (the intrasyncytial lamina) engulfed by processes of the syncytiotrophoblast. In other hemochorial placentae, no maternal vascular components are retained as part of the interhemal barrier. B, blastocyst; C, cytotrophoblast; CT, fetal connective tissue; E, endothelium; ES, endometrial stroma; FB, fetal blood; IL, intrasyncytial lamina; IM, interstitial membrane; MB, maternal blood; S, syncytiotrophoblast; U, uterine epithelium.

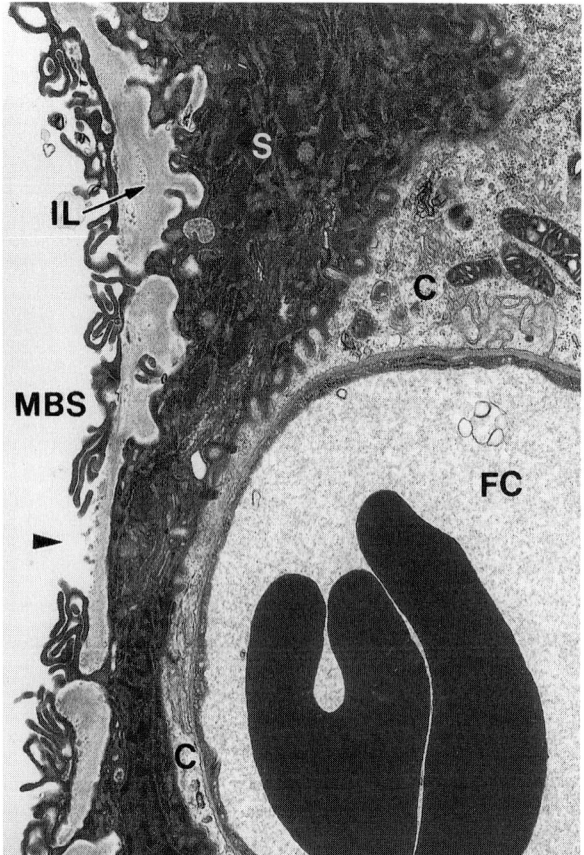

FIGURE 6 Thickened intrasyncytial lamina (IL) in the interhemal barrier surrounding a larger maternal blood space (MBS) in the same near-term placenta of a short-tailed fruit bat shown in Fig. 2. Gaps (arrowhead) in the syncytial covering over the intrasyncytial lamina are not unusual at this stage. C, cytotrophoblast; FC, fetal capillary; S, syncytiotrophoblast. Magnification, ×10545.

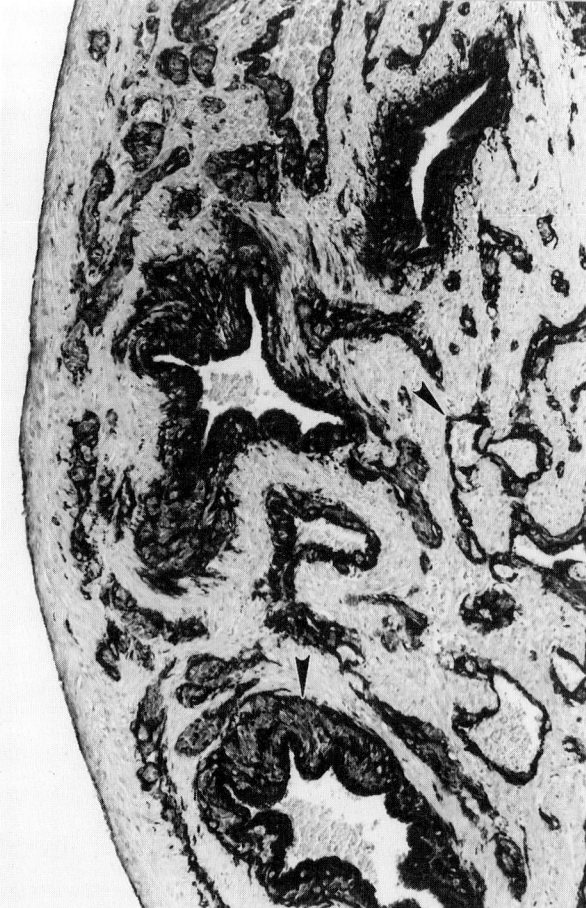

FIGURE 7 Section of myometrium beneath the developing discoidal placenta in a white-winged vampire bat, *Diaemus youngi*, carrying a primitive streak-stage embryo. This was stained immunocytochemically with monoclonal antibodies (AE1/AE3) directed against a variety of cytokeratins (intermediate filament proteins usually found in cells of epithelial origin) and lightly counterstained with hematoxylin. Large numbers of cytokeratin-positive trophoblast cells have migrated away from the placenta via an interstitial route, within the walls of the uterine blood vessels (e.g., at arrowhead). These vessels were all still lined by an intact endothelium. No uterine glands are present in this field. Magnification, ×93.

et al., 1998), but their functional significance remains to be elucidated.

IV. DEVELOPMENT

Placentae are unusual organs in that they are formed *de novo* during each pregnancy through the close interaction and association of tissues belonging to genetically distinct individuals, and they undergo substantial structural modification as the pregnancy progresses. Some knowledge of their development is therefore necessary to fully understand placental organization, function, and terminology. Furthermore, while many mammals possess discoidal placentae, these can be formed in quite different ways. It is important to recognize that such differences exist, as they undoubtedly contribute to the variation observed in placental structure.

At the end of implantation in the human, the blastocyst is completely embedded (i.e., interstitially implanted) within the endometrium. It consists, on the

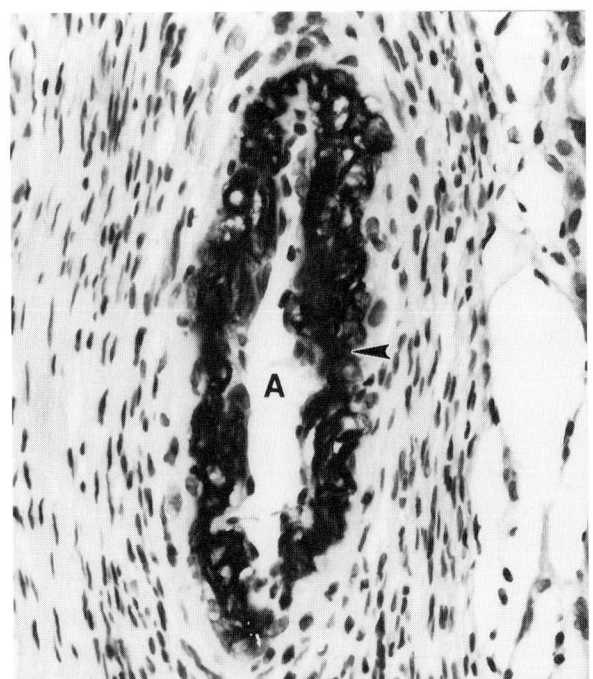

FIGURE 8 Section of myometrium from a black mastiff bat obtained in late pregnancy and carrying a 25.5-mm fetus. This was stained immunocytochemically with monoclonal antibodies (AE1/AE3) directed against a variety of cytokeratins and lightly counterstained with hematoxylin. Large numbers of cytokeratin-positive cytotrophoblast cells (e.g., at arrowhead) have left the discoidal placenta and migrated along an artery (A) via an interstitial route. This vessel was still lined by an intact endothelium. Magnification, ×238.

periphery, of cytotrophoblast (which divides mitotically) and an outer layer of syncytiotrophoblast (which is formed by the fusion of cytotrophoblast cells). At that stage the syncytiotrophoblast contains a labyrinth of lacunar spaces through which some maternal blood circulates. Much of this early labyrinth seems to be formed by the fusion of lacunae and clefts that develop spontaneously within the syncytiotrophoblast and then become connected with maternal vessels in the adjacent endometrium.

As the early conceptus grows, the mass of syncytiotrophoblast increases, cytotrophoblast begins to proliferate out into the syncytiotrophoblast forming primary villi, and mesoderm grows along the interior of the cytotrophoblast. The combination of trophoblast and its investment of mesoderm constitutes the chorion and transforms the implanted blastocyst into a chorionic sac and its contents. The primary villi are converted into secondary villi upon their invasion by mesoderm and then into tertiary villi when blood vessels that will convey embryonic blood form within their mesenchymal cores. Finally, cytotrophoblastic cells in anchoring villi on the periphery of the developing placenta break through the syncytiotrophoblast and proliferate around the chorionic vesicle to form an outer shell of cytotrophoblast. The continued proliferation of cytotrophoblastic cells within the villi and around the periphery of the placenta permit further lengthening and ramification of the villi as well as circumferential expansion of the developing placenta. Other cytotrophoblast cells that leave the anchoring villi and migrate into the uterine wall play important roles in modifying the maternal vascular supply to the placenta.

All discoidal placentae are formed by the vascularization of the chorion by vessels of the allantois—a highly vascular diverticulum of the hindgut that forms in amniote embryos. Although humans and most other anthropoid primates lack an allantoic sac, allantoic vessels and a rudimentary allantoic duct develop in the connecting stalk—a region of mesoderm that suspends the amnion, embryonic disc, and yolk sac from the interior of the chorionic sac. The connecting stalk is eventually transformed into the umbilical cord. As the allantoic mesoderm spreads over the inner surface of the chorion, a chorioallantois is created, and the allantoic vessels become linked with embryonic vessels developing in the chorion.

In humans, chorionic villi initially form over the entire surface of the chorionic sac. The villi directed toward the decidua basalis (the modified endometrium beneath the developing placenta) then become much more elaborate, whereas those facing the decidua capsularis (the modified endometrium which separates the chorionic sac from the uterine lumen) atrophy. This presumably occurs because growth of the conceptus stretches the decidua capsularis and progressively interferes with the maternal blood supply to that side of the sac. The final, discoidal shape of the placenta in humans is determined by the restricted distribution of the persisting chorionic villi.

In anthropoid primates with superficial implantation allantoic mesoderm also spreads over the entire inner surface of the chorionic sac; however, villi develop only where the chorioallantois attaches to the uterine wall. This results in the formation of two placental discs instead of just one.

The nine-banded armadillo, *Dasypus novemcinctus*, also has discoidal villous, hemochorial placentae, but these form in a much different fashion from the placentae of humans. The single armadillo blastocyst (which will give rise to identical quadruplets) always implants within a restricted zone at the fundic end of the mother's simplex uterus. In this region, highly anastomotic venous sinuses are present in the superficial endometrium, and these are continuous with other venous sinuses deep to the endometrial glands in the body of the uterus. Trophoblast penetrates into the sinuses shortly after implantation, and all subsequent growth of the placental villi occurs within the sinuses. Initially, a placental disc forms in the fundus of the uterus but, as the conceptuses develop, it separates into four discs located in the upper corpus. The villi of the armadillo placenta are also unusual in having clusters of cytotrophoblast cells at their tips that are centers of proliferation. Many of the new cytotrophoblast cells then fuse to augment the quantity of villous syncytiotrophoblast (Mossman, 1987; Enders and Welsh, 1993).

Placentation in the black mastiff bat, *Molossus ater*, is in a number of respects unusual and has provided some insights into the possible functional significance of the hypertrophied maternal endothelial cells and prominent basal laminae or amorphous extracellular layers present in the placentae of many other species (Rasweiler, 1993). *Molossus* develops three different vascular placentae: a transient choriovitelline placenta that exists only early in gestation, a diffuse endotheliodichorial placenta that is prominent in midgestation, and a discoidal hemochorial placenta that becomes the principal site of fetomaternal exchange during late gestation. The choriovitelline and diffuse placentae are extensive, surrounding most of the chorionic sac, and include both cyto- and syncytiotrophoblast. The discoidal placenta, on the other hand, always develops at the cranial end of the right uterine horn, and with possible minor exceptions, contains trophoblast only of the cellular variety.

This positioning of the discoidal placenta in *Molossus* was found to be related to the presence of an unusual vascular tuft which forms in the endometrium at that site during early pregnancy. Development of the tuft does not depend on the presence of an embryo because similar structures were observed in the nongravid left horn of pregnant bats and in both horns of nonpregnant bats during the luteal phase of their cycles. As the decidual reaction spreads through the endometrium during implantation, the endothelial cells of the tuft vessels hypertrophy, and the glycoprotein content of their basal laminae increases. These changes precede engulfment of the tuft vessels by trophoblast. Some endothelial hypertrophy was also noted in the endometrium of the nontuft decidua in both pregnant and nonpregnant bats. However, obvious basal lamina thickening occurred around these vessels only following their engulfment by trophoblast and to a variable extent later in pregnancy in the decidualized, nongravid left horn. Differences in the degree of maternal endothelial hypertrophy and basal lamina thickening between the tuft and nontuft endometrium were also observed following spontaneous decidualization of the endometrium during nonpregnant cycles.

When a blastocyst was present, trophoblast invaded the decidua all around the embryo. Much of the trophoblast that engulfed maternal vessels within the choriovitelline and diffuse placentae was syncytial, whereas that proliferating around vessels of the vascular tuft remained cellular. As allantoic mesoderm began to infiltrate the developing discoidal placenta, it became even more obvious that much of the proliferating cytotrophoblast was using the thickened basal laminae of the tuft vessels as scaffolding on which to grow.

These observations suggest that the endometrial endothelial cells of *Molossus* may be secreting factors which play important roles in controlling early trophoblastic growth and are incorporated, at least in part, into the endothelial cells' basal laminae. They also indicate that differences in the secretory activities of the maternal endothelial cells of the vascular tuft and the nontuft decidua may be partially responsible for the distinct patterns of trophoblastic growth and differentiation that occur during development of the discoidal and diffuse placentae in those regions.

Finally, it is of interest that when the embryo of *Molossus* reaches the limb bud stage, the morphogenetic relationship between the trophoblast and vessels of the vascular tuft begins to change. Solid trophoblastic tubules devoid of any maternal vascular components then start to sprout from the trophoblastic cuffs surrounding these vessels. These tubules eventually become interconnected, develop lumina, and begin to carry maternal blood. Although mitotic activity is commonly seen in the cytotrophoblast of the solid tubules, this ceases abruptly when the tubules become patent. From that point on, further growth of the tubules is due to hypertrophy of the cytotrophoblast cells, which are gradually transformed into giant cells (Fig. 3). Development of the discoidal placenta in *Molossus* provides a vivid example of how the behavior of trophoblast can change markedly as gestation progresses. It also points to the danger of using only term placentae as a source of trophoblast cells for *in vitro* studies.

The observations made on *Molossus* suggest that endometrial endothelial cells may have a similar morphogenetic role in many other mammals with invasive trophoblast. Hypertrophied endothelial cells and thickened basal laminae (interstitial membranes) are present in the interhemal barrier of the endotheliochorial placenta in a number of species. During the development of hemochorial placentae in some mammals (particularly many bats), the maternal endothelial cells are instead lost. Their basal laminae are engulfed by the processes of adjacent trophoblast cells, however, and contribute to the formation of intrasyncytial laminae (Figs. 1, 2, and 5). This peculiar morphological arrangement may have evolved in part as a biological compromise that retains tissue of maternal origin in the interhemal barrier for morphogenetic reasons (i.e., to influence trophoblastic growth and/or differentiation) but isolates potentially thrombogenic subendothelial components from exposure to the maternal blood (Rasweiler, 1993). Interestingly, late in pregnancy in the short-tailed fruit bat, gaps frequently develop in the syncytial covering over the intrasyncytial lamina (Fig. 6), but no signs of platelet adherence and clot formation have been noted. Possibly the material added to the intrasyncytial lamina by the syncytiotrophoblast is nonthrombogenic.

The possible morphogenetic role of maternal endothelial cells in other species with hemochorial placentae is still unclear. Hypertrophied endothelial cells develop in endometrial blood vessels prior to their engulfment by trophoblast in a number of anthropoid primates, including rhesus monkeys and humans. In marmosets these cells persist for some time after first being engulfed and could be significantly influencing trophoblastic growth or differentiation. In rhesus monkeys, maternal endothelial cell hypertrophy is pronounced in some of the venous capillaries and venules at the implantation site, but its functional significance is still unknown.

In many endotheliochorial placentae, the placental labyrinth appears much more extensive and exhibits a markedly different morphological arrangement than the capillary beds of the uterine endometrium prior to trophoblastic invasion. This indicates that maternal endothelial cell proliferation and cytoskeletal changes also probably play important roles in the formation of these placentae.

Acknowledgment

Previously unpublished work and preparation of the manuscript was supported in part by NIH Grant HD 28592.

See Also the Following Articles

Blastocyst; Ectopic Pregnancy; Endometrium; Epitheliochorial Placentation; Fetal Membranes; Hemochorial Placentation; Trophoblast to Human Placenta

Bibliography

Badwaik, N. K., and Rasweiler, J. J., IV (1998). The interhemal barrier in the chorioallantoic placenta of the greater mustache bat, *Pteronotus parnellii*, with observations on amplification of its intrasyncytial lamina. *Placenta* **19**, in press.

Benirschke, K., and Kaufmann, P. (1995). *Pathology of the Human Placenta, Third Edition.* Springer-Verlag, New York.

Cukierski, M. A. (1987). Synthesis and transport studies of the intrasyncytial lamina: An unusual placental basement membrane in the little brown bat, *Myotis lucifugus. Am. J. Anat.* **178**, 387–409.

Enders, A. C. (1965). A comparative study of the fine structure of the trophoblast in several hemochorial placentas. *Am. J. Anat.* **116**, 29–68.

Enders, A. C., and Welsh, A. O. (1993). Structural interactions of trophoblast and uterus during hemochorial placenta formation. *J. Exp. Zool.* **266**, 578–587.

Ferriani, R. A., Ahmed, A., Sharkey, A., and Smith, S. K. (1994). Colocalization of acidic and basic fibroblast growth factor (FCF) in human placenta and the cellular effects of bFGF in trophoblast cell line JEG-3. *Growth Factors* 10, 259–268.

Hamlett, W. B., and Rasweiler, J. J., IV (1993). Special issues: Comparative gestation and placentation in vertebrates—Part I and Part II. *J. Exp. Zool.* 266, 341–656.

Kertschanska, S., Schroder, H., and Kaufmann, P. (1997). The ultrastructure of the trophoblastic layer of the degu (*Octodon degus*) placenta: a re-evaluation of the "channel problem." *Placenta* 18, 219–225.

Loke, Y. W., and King, A. (1996). *Human Implantation: Cell Biology and Immunology*. Cambridge Univ. Press, Cambridge, UK.

Mossman, H. W. (1987). *Vertebrate Fetal Membranes*. Rutgers Univ. Press, New Brunswick, NJ.

Muhlhauser, J., Marzioni, D., Morroni, M., Vuckovic, M., Crescimanno, C., and Castellucci, M. (1996). Codistribution of basic fibroblast growth factor and heparan sulfate proteoglycan in the growth zones of the human placenta. *Cell Tissue Res.* 285, 101–107.

Nanaev, A., Chwalisz, K., Frank, H.-G., Kohnen, G., Hegele-Hartung, C., and Kaufmann, P. (1995). Physiological dilation of uteroplacental arteries in the guinea pig depends on nitric oxide synthase activity of extravillous trophoblast. *Cell Tissue Res.* 282, 407–421.

Pijnenborg, R. (1996). The placental bed. *Hypertension Pregnancy* 15, 7–23.

Rasweiler, J. J., IV (1993). Pregnancy in Chiroptera. *J. Exp. Zool.* 266, 495–513.

Rasweiler, J. J., IV, Badwaik, N. K., and Muradali, F. (1998). Coexpression of cytokeratins and vimentin by highly invasive trophoblast in the white-winged vampire bat, *Diaemus youngi,* and the black mastiff bat, *Molossus ater,* with observations on intermediate filament proteins in the decidua and intraplacental trophoblast. Submitted for publication.

Steven, D. H. (1975). *Comparative Placentation. Essays in Structure and Function*. Academic Press, London.

Ziche, M., Maglione, D., Ribatti, D., Morbidelli, L., Lago, C. T., Battisti, M., Paoletti, I., Barra, A., Tucci, M., Parise, G., Vincenti, V., Granger, H. J., Viglietto, G., and Persico, M. G. (1997). Placenta growth factor-1 is chemotactic, mitogenic, and angiogenic. *Lab. Invest.* 76, 517–531.

Dogs

Cheryl S. Asa
St. Louis Zoological Park

I. Social System and Behavior
II. Social Aspects of Reproduction
III. Seasonal Reproduction
IV. Puberty
V. Phases of the Ovulatory Cycle
VI. The Endocrine Basis of Parental Behavior
VII. The Interplay of Physiology and Social Organization

GLOSSARY

anestrus The period of the female reproductive cycle when the ovaries are quiescent.

estrus The period of the reproductive cycle around the time of ovulation when the female is receptive to mating.

luteal phase The phase of the female reproductive cycle following ovulation when the corpus luteum produces and secretes progesterone (sometimes called *diestrus* in polyestrous mammals).

monestrus (monestrum) The ovulatory pattern in which there is only one ovulation per breeding period.

monogamy The mating system in which a male and female mate only with each other.

proestrus The transition period between anestrus or diestrus (the luteal phase) and estrus.

Canidae are mammals that include dogs, wolves, coyotes, jackals, and foxes. The family Canidae is generally considered to have 16 living genera with 36 species, although there is controversy about both these numbers, depending on whether some forms are considered full species or merely subspecies or whether some species are placed in their own genera. Feral forms of the domestic dog (*Canis familiaris*) are found in Australia (dingo) and New Guinea (singing dog). The natural distribution of canids includes most land areas of the world, in climates that extend from hot deserts (fennec fox, *Fennecus zerda*, and Sechuran fox, *Dusicyon sechurae*) to arctic ice fields (arctic fox, *Alopex lagopus*). Weights range from 1 kg (fennec fox) to as much as 80 kg (gray wolf, *Canis lupus*). Although placed in the order Carnivora, all species are known to ingest some plant matter or insects as well. Because reproduction involves a dynamic interplay of behavior and physiology, social context can provide insight into the adaptive value of reproductive processes. Canids are a particularly interesting family for such study since there are many features of their reproductive physiology and behavior that are not typical of most mammalian species. These include monogamy, monestrum, obligate pseudopregnancy in adult females that fail to conceive, incorporation of postpubertal offspring into the social unit, behavioral suppression of mating in these offspring so that only the dominant pair reproduces, and parental care by other group members, including adult males. Although atypical of mammals, these characteristics can best be understood as components of a general reproductive strategy.

I. SOCIAL SYSTEM AND BEHAVIOR

An apparent continuum in sociality ranges from the maned wolf (*Chrysocyon brachyurus*), which is solitary outside the breeding season, to the highly gregarious gray wolf and African wild dog with large, complex packs that remain together year-round, a trend that has been attributed to body size. However, the basis for sociality may be more closely tied to the strategy for food acquisition. That is, advantages related to group hunting or foraging may foster sociality. Solitary foraging may be best suited to hunting rodents or searching for invertebrates or fruits (foxes and maned wolves), whereas groups of gray wolves or hunting dogs may more successfully capture large ungulates. The flexible social system of the coyote (*Canus latrans*) supports this view since larger groups have been observed hunting deer cooperatively, whereas pairs or lone individuals hunted rodents or other small game. Bat-eared foxes (*Otocyon megalotis*), however, are unusual; they forage in small family groups for invertebrates.

II. SOCIAL ASPECTS OF REPRODUCTION

A. Monogamy

Canids are a frequently cited example of monogamy, which is otherwise rare for mammals. The primary social unit of canids is the mated pair, with a strong tendency toward long-term fidelity, often for life. Even for more solitary species, such as many of the foxes, pairs separate during the nonbreeding season but the same animals are likely to be found together in successive breeding seasons. Larger social groups typically are extended families composed of the mated pair and young of that or of previous years. Exceptions to the apparent rule of monogamy [e.g., red (*Vulpes vulpes*) and kit (*V. macrotis*) foxes], although not common, are probably related to the local pattern of food distribution.

B. Paternal Care

Some degree of direct or indirect paternal care has been observed in every canid species studied. Direct care includes retrieval, carrying, grooming, providing food, baby-sitting, playing, and active defense of the young. Indirect care refers to behaviors such as the acquisition and maintenance of territory or other resources. The method of carrying food back to the den, used by larger canids, is also unusual. Stomach contents are regurgitated in response to muzzle licking by the waiting mother or pups. Smaller species, however, are more likely to carry food items in their mouths.

C. Incorporation of Postpubertal Young

Enlargement of the social unit beyond the male/female pair typically involves young that remain at least temporarily with their parents rather than disperse. With rare exceptions, postpubertal offspring forego mating while they remain in the group. Although the young of the maned wolf and some fox species (bat-eared and kit foxes) disperse before the next breeding season, young red and corsac (*V. corsac*) foxes may remain through winter. The adaptive value to the parents of incorporation of postpubertal young in the social group is believed to be their participation in the care of their younger siblings by guarding, bringing food and/or regurgitating, grooming, and playing. The benefit to postpubertal nondispersers may be not only the opportunity to practice parenting but also continued opportunity to practice hunting or foraging skills plus the safety of group living.

D. Reproductive Suppression

The mechanism for reproductive suppression has only been demonstrated for the gray wolf, in which offspring attain physiological puberty at about 22 months but are prevented from mating by their parents. This behavioral suppression may be accomplished by direct stare, if the dominance of the parents is firmly maintained, or, in cases in which subordinates challenge the parents, fighting may ensue.

Data are not available for all canid species, but results from gray wolves show that subordinate, nonbreeding males and females do not differ from the dominant, breeding pair in any aspect of reproductive physiology. That is, starting at about 22 months of age subordinate females come into estrus and ovulate, and subordinate males produce sperm. Thus, the subordinate wolves are physiologically able to reproduce.

III. SEASONAL REPRODUCTION

Reproduction is seasonal for most canids. Those living in temperate latitudes usually mate in late winter or early spring, with the season more sharply defined at increasingly higher latitudes. Although domestic dogs are not thought to be seasonal breeders, there is a trend for more breedings in late winter/early spring. Even for canid species in tropical or equatorial zones, breeding periods may relate to seasonal changes in rainfall and food availability.

Photoperiod has been assumed to mediate the timing of reproduction in temperate zone species. Translocation across the equator results in a 6-month shift in breeding season in red foxes, maned wolves, and dholes (*Cuon alpinus*). However, in male and female gray wolves, removal of the pineal glands or of the suprachiasmatic nucleus, components of the pathway involved in seasonal response to photoperiod in other mammalian species, does not alter the timing of reproduction. This suggests that, if photoperiod does affect time of breeding, other mechanisms must be operating.

Canid males living at temperate latitudes also exhibit seasonal reproductive parameters, e.g., testosterone production, testis size, and spermatogenesis. In males of seasonally breeding species for which there are data, testosterone concentrations and sperm production vary annually, declining outside the breeding season. A notable exception is the male domestic dog, which remains reproductively competent year-round, as do male dingoes in temperate regions of Australia. This continuous breeding condition of the males is likely the result of the nonseasonality of the females that can come into estrus at any time of the year. Thus, the male is ready to mate with any female as she comes into estrus. For species or conditions in which females exhibit strictly seasonal estrus, we expect male reproductive physiology to follow the same seasonal pattern. For the smaller foxes in which females sometimes exhibit more than one estrus per year, male reproductive ability might be expected to be similar to that of the domestic dog. Testosterone data are not available for crab-eating fox males, but testosterone in fennec males is not constant but varies with the estrous cycles of their partners.

A. Monestrum and the Female Cycle

Canids have only one ovulatory cycle per breeding season, a condition known as monestrum (Table 1). The only other mammalian species for which monestrum is reported are some bats and the European roe deer, *Capreolus capreolus*. Polyestrum, typical for other mammals, is characterized by successive cycles

TABLE 1
Monestrous Species

Red fox, *Vulpes vulpes*	Pampas fox, *D. gymnocercus*
Swift fox, *V. velox*	Gray wolf, *Canis lupus*
Kit Fox, *V. macrotis*	Coyote, *C. latrans*
Artic fox. *Alopex lagopus*	Domestic dog, *C. familiaris*
Andean fox, *Dusicyon culpaeus*	Maned wolf, *Chrysocyon brachyurus*

of estrus and ovulation, either seasonal or continuing year-round. Thus, if a polyestrous female fails to conceive at one ovulation, she has more chances. In contrast, female canids have only one estrus and so only one opportunity per season to conceive. A possible exception comes from observations of captive bush dogs that showed repeated periods of estrous behavior suggestive of polyestrum, but field and physiological data are lacking.

Canids produce one litter per year, except for the fennec fox (*Fennecus zerda*), crab-eating fox (*Cerdocyon thous*), and domestic dog (*Canis lupus familiaris*). The fennec and crab-eating foxes can have two litters per year in captivity, but again there are no data from the field. Most domestic dogs also have two litters per year, but this can vary by breed. The African hunting dog can produce a second litter, but only if the first is lost. However, even in these cases, there is only one estrus and ovulation during each breeding period.

Although systematic data have not been reported, breeders of domestic dogs have long been aware that females living in proximity to each other have relatively synchronous (within a few weeks) estrous periods, even though they are not seasonal breeders. These observations suggest that the estrous synchrony exhibited by wild canids may be socially facilitated and not be entrained solely by environmental factors such as photoperiod or rainfall.

B. Prolactin

Although only measured in gray wolves and in red and arctic foxes, prolactin varies seasonally in both males and females, with a spring peak that occurs after the mating period, coinciding with the birth of pups. This also is a period in males of declining testosterone concentrations, another hormonal change that may favor paternal care by reducing aggression.

IV. PUBERTY

Puberty for many species has been calculated by subtracting gestation length from birth of first young. Documentation of first ovulation and sperm production in the few species tested has verified these calculations. In general, physiological puberty occurs at 9 or 10 months in smaller species and at about 22 months in larger ones, although not all individuals mate at this age. The relationship between body size and onset of reproductive capacity also seems to apply to breeds of dogs, in which smaller breeds tend to reach puberty earlier than large breeds.

V. PHASES OF THE OVULATORY CYCLE

A. Anestrous Phase

The anestrous phase, characterized by a quiescent reproductive system, constitutes the time between breeding periods. It varies in length, depending primarily on (i) the number of reproductive periods per year (species typical) and (ii) lactation (i.e., whether conception and live births occurred). Thus, anestrus is longer in species that breed only once per year. Within a species it is shorter for individuals that give birth and lactate.

B. Proestrous Phase

During the transition preceding estrus, the female is attractive to males but not receptive to mating. In females of the genus *Canis*, proestrus is accompanied by a sanguinous discharge from the vulva. A slight bloody discharge also occurs in some female maned wolves. Although the source of blood, the uterus, is similar, canid proestrous bleeding differs from the menstrual phase of primates in the associated hormonal milieu. Menstruation occurs during a nadir of both estrogen and progesterone, following the luteal phase. Canids, in contrast, show proestrous bleeding in response to increasing concentrations of estrogen. The proestrus phase can be as long as 2 or 3 months for coyotes, 6 weeks for gray wolves, and 2 or 3 weeks for dingoes. The domestic dog typically has a proestrus of about 1 week. Greyhounds may be an exception, though, with proestrus sometimes lasting up to 3 weeks. However, a proestrous phase even of 1 week is unusually long for mammals and is most comparable hormonally to the follicular phase of many primates, also a time when consortships may begin.

Proestrus can also be detected by increased courtship activity, vulval swelling, nonsanguinous vaginal discharge, and vaginal cytology, which at this stage includes polymorphonuclear leukocytes, some mucus, progressive cornification of epithelial cells, and red blood cells in those species with a sanguinous discharge.

C. Estrous Phase

The estrous phase, when the female is receptive and mating occurs, is briefest in the fennec fox (1 or 2 days) and longest in the African hunting dog (up to 20 days). However, for most canids, estrus lasts about 1 week, longer than for most mammalian species for which 1–3 days is common. Exceptions include many primates and the equids, which, like canids, have an estrous period of about 1 week. These also are species in which males contribute to the well-being of infants, albeit usually indirectly by territory maintenance and guarding.

Although estrogen alone maintains the proestrous phase, progesterone appears necessary for full stimulation of estrous behavior, which includes receptiveness to copulation. In addition, female dogs show a preference for particular males suggesting a role for female mate choice in this and probably other canid species. This phase is distinguished by a maximally swollen vulva and vaginal cytology typified by maximum cornification of epithelial cells and a decrease in leukocytes.

Ovulation and corpus luteum formation are spontaneous in canids. Although data are not available for the other canids, in the domestic dog, primary oocytes are ovulated and sperm penetration occurs during the first meiotic division, in contrast to other mammals in which secondary oocytes are ovulated. Ovulation occurs 1 day before to 5 days after the onset of estrous behavior and 38–44 hr after the luteinizing hormone (LH) surge. The estradiol peak occurs about 24 hr before this surge, and progesterone starts to increase at or just before the LH peak. Typically, mating first occurs about 18 hr after the increase in progesterone is detected. Prostaglandin F2α may be luteolytic at the end of pregnancy and pseudopregnancy.

Copulation in canids is distinguished by the copulatory lock or tie, when the swollen bulbus glandis at the base of the penis is held by the vaginal sphincter for varying lengths of time ranging from a few minutes up to 2 or more hours (fennec fox). The African hunting dog may be an exception, with either no or very brief ties. The function of the tie is likely a combination of mate guarding, i.e., the prevention of another copulation during the period of the tie, and of stimulation of sperm transport more directly into and through the uterus.

D. Luteal Phase or Pseudopregnancy

The approximately 2-month luteal phase (the period of elevated progesterone following ovulation) of the nonpregnant canid is significantly longer (Table 2) than that of most other mammals, in which it averages 2 weeks. Because this prolonged luteal phase is approximately equivalent in length and in hormonal profile to pregnancy, it often is referred to as pseudopregnancy. (In fact, in some studies, the luteal phase has been reported to last 2 or 3 weeks longer than the species-typical gestation.) Thus, if a female canid ovulates, she either becomes pregnant or pseudopregnant.

TABLE 2
Species with Obligate Pseudopregnancy

Red fox, *Vulpes vulpes*	Coyote, *C. latrans*
Arctic fox, *Alopex lagopus*	Domestic dog, *C. familiaris*
Andean fox, *Dusicyon culpaeus*	Raccoon dog, *Nyctereutes procyonoides*
Gray wolf, *Canis lupus*	

Pseudopregnancy typically is only diagnosed in dogs by veterinarians when there are overt symptoms such as a distended abdomen, swollen mammary glands that may secrete milk, or maternal behavior. Some females may even behave maternally toward, and try to care for, a toy or phantom or surrogate pups. However, although the expression of symptoms is variable by individual, the underlying endocrine basis for pseudopregnancy is present in all nonpregnant, postovulatory females. This conclusion is supported by endocrine results from red and arctic foxes, gray wolves, coyotes, and raccoon dogs. In fact, pregnancy can only be reliably distinguished from pseudopregnancy in canids by palpation, ultrasound, or X-ray of fetuses.

E. Pregnancy and Parturition

Pregnancy duration varies from about 50 (fennec fox) to 80 (African hunting dog) days, correlated primarily with body mass. Two exceptions are the bat-eared fox and the bush dog with gestations of 65 and 70 days, respectively, despite being two of the smaller species.

Both estradiol and progesterone are elevated during pregnancy and are apparently of ovarian origin. Both hormones peak during the first trimester and gradually decline throughout gestation. A drop in basal body temperature of about 1° occurs approximately 24 hr before parturition. Female canids typically dig dens for whelping, and young remain in the natal dens for about 3 weeks before emerging. Maximum litter size occurs in arctic foxes (up to 25), although 4–6 are common in most species. Fennec foxes, however, typically have only 2. The number of pairs of teats range from two (side-striped jackal) to seven (African hunting dog and arctic fox). Prolactin is highest during early lactation.

VI. THE ENDOCRINE BASIS OF PARENTAL BEHAVIOR

As in other mammals, the hormones of pregnancy (estrogens and progestins) prepare the female to behave maternally. This is made even more clear in canids by the behavior of females that have experienced hormonal pseudopregnancy without giving birth. These females can show the entire repertoire of maternal care if presented with pups, sometimes including lactation. In the absence of appropriate stimuli, such females may attempt to care for inanimate objects or even phantom pups. Prolactin, which supports milk production and is elevated in pseudopregnant as well as pregnant females, may also influence maternal behavior.

The case for hormonal support of paternal care is much less clear. However, as with females, prolactin may be important because it has been implicated in the paternal behavior of some other mammals. In fact, although only studied in the gray wolf, it may stimulate parental behavior in other group members as well since it increases seasonally with peak levels coinciding with the birth of pups in all adults.

VII. THE INTERPLAY OF PHYSIOLOGY AND SOCIAL ORGANIZATION

Although it is not possible to determine whether reproductive physiology has influenced the evolution of canid social organization, the unique combination of social and physiological features does favor pair or pack cohesion.

First, the prolonged proestrous phase affords the male/female pair extended time to establish or reaffirm the pair bond. The increasing attractiveness of the female encourages the male to remain in close proximity. In fact, during the breeding season, the alpha pair of wolves can be distinguished from the other pack members by their closer sleeping or resting distances.

During proestrus (best documented for the gray wolf and coyote), the rate of scent marking with urine commonly increases, with the male and female marking over one another's urine. The observation that newly formed pairs of gray wolves and bush dogs tandem mark more than established pairs suggests a role for this behavior in pair formation. In addition, in a study of male and female anosmic gray wolves, tandem marking decreased and pair bonding did not occur.

It also is likely that the proestrous period is important for the female to evaluate the male's commitment since the extent of his parental contribution may determine the survival of their pups. Although the data set is not complete, there is a suggestion that, among canids, species in which the male makes the greatest contribution are those with the longest period of proestrus.

The relatively long proestrus also may have been selected for in response to monestrum. With only one estrus, and so only one opportunity for conception per year for the vast majority of canid females, ensuring access to a mate at that time is crucial. The relatively long period of estrus might be adaptive for the same reason. In species with shorter proestrus and estrus, such as the fennec fox, although the extent of paternal care has not been documented, the female fennec can conceive twice per year, possibly resulting in less pressure to conceive at each estrus.

From an evolutionary perspective, it is difficult to understand how monestrum is adaptive. Polyestrum, with its backup system of repeated cycles of estrus and ovulation, seems a much safer strategy. Two other aspects of canid reproductive systems may help explain how monestrum might be advantageous: the role of postpubertal offspring in the social group and the typical limitation of breeding to only the dominant pair.

The observation that the modal pattern for canids, with the exception of the maned wolf and perhaps some foxes, is the incorporation of adult young into the social group (Table 3) but with reproduction limited to their parents suggests a common function or advantage. A possible benefit to the youngsters of delaying reproduction comes from studies of jackals, particularly silverbacks, *Canis mesomelas*, and African hunting dogs. The presence of helpers, or alloparents, is associated with an increase in the number of young that survive. An alternative hypothesis is that helping may be incidental and adult subordinates may stay with the group for more selfish reasons. In fact, benefits of staying and helping may be related to ecological factors such as relative prey abundance which, of course, varies. The incorporation of additional members also would be advantageous for species that benefit from group hunting strategies.

Even though alloparenting may not be the sole motivation of young canids that opt to remain in their natal groups, the long-term bonds of the parents make it very likely that a female's pups of successive years all share the same sire. Thus, a subordinate that is deferring reproduction shares, on average, as many genes with its siblings as it would with its own offspring, which is an example of kin selection. This strategy apparently succeeds both because direct benefits accrue to the reproductive pair and because indirect benefits accrue to the subordinates, e.g., increased survival of siblings, practice at parenting, and biding time until conditions improve for dispersal.

Monestrum may also contribute to social cohesion since subordinate females are reproductively competent and do come into estrus and ovulate. If canids were polyestrous, nonpregnant females would continue to cycle and experience periods of estrus, which would undoubtedly be disruptive. With monestrum, aided by the relative synchrony of estrus, intrapack aggression to suppress subordinate sexual behavior can be limited to a brief period.

Still, it would seem more parsimonious for subordinate animals to be physiologically suppressed so that they would have no inclination to mate, eliminating this source of competition and aggression entirely. That monestrum has survived suggests that it has been more adaptive than either polyestrus or delayed puberty. Monestrum appears to be accomplished by the extended luteal phase, which suppresses further ovarian activity. This extended luteal activity also primes nonpregnant females to behave maternally, suggesting that the value of this behavioral contribution may be greater than the brief period of social dissention that accompanies the one period of estrus.

TABLE 3
Species with Nonbreeding, Subordinate Helpers

Red fox, *Vulpes vulpes*	Dingo, *C. familiaris*
Corsac fox, *V. corsac*	Dhole, *Cuon alpinus*
Arctic fox, *Alopex lagopus*	Golden jackal, *C. aureus*
Gray wolf, *Canis lupus*	Black-backed jackal,
Red wolf, *C. rufus*	*C. mesomelas*
Coyote, *C. latrans*	African hunting dog, *Lycaon pictus*

Maternal behavior by subordinate females, induced by the hormones of pseudopregnancy and possibly enhanced by the seasonal rise in prolactin, is expressed in guarding and bringing food to their parent's pups. If the hormonal stimulation of pseudopregnancy is necessary to support the behavior, and the behavior does indeed increase the survival rate of pups, then the selective advantage of both monestrum and pseudopregnancy is more apparent.

See Also the Following Articles

Cats (Felidae); Pseudopregnancy; Olfaction and Reproduction; Seasonal Breeding, Mammals

Bibliography

Asa, C. S. (1997). Hormonal and experiential factors in the expression of social and parental behavior in canids. In *Cooperative Breeding in Mammals* (J. A. French and N. G. Solomon, Eds.), pp. 129–149. Cambridge Univ. Press, Cambridge, UK.

Concannon, P. W., England, G. C. W., Rijnberk, A., Verstegen, J. P., and Doberska, C. (Eds.) (1997). Reproduction in dogs, cats, and exotic carnivores. *J. Reprod. Fertil. Suppl.* **51**.

Concannon, P. W., Hansel, W., and McEntee, K. (1977). Changes in LH, progesterone and sexual behavior associated with preovulatory luteinization in the bitch. *Biol. Reprod.* **13**, 112–121.

Concannon, P. W., Morton, D. B., and Weir, B. J. (Eds.) (1989). Dog and cat reproduction, contraception and artificial insemination. *J. Reprod. Fertil. Suppl.* **39**.

Concannon, P. W., England, G. C. W., Verstegen, J. P., and Russell, H. A. (Eds.) (1993). Fertility and infertility in dogs, cats and other carnivores. *J. Reprod. Fertil. Suppl.* **47**.

Ewer, R. F. (1973). *The Carnivores*. Cornell Univ. Press, Ithaca, NY.

Geffen, E., Gompper, M. E., Gittleman, J. L., Luh, H.-K., Macdonald, D. W., and Wayne, R. K. (1996). Size, life-history traits, and social organization in the Canidae: A reevaluation. *Am. Nat.* **147**, 140–160.

Kleiman, D. G. (1977). Monogamy in mammals. *Q. Rev. Biol.* **52**, 39–69.

Kleiman, D. G., and Eisenberg, J. F. (1973). Comparisons of canid and felid social systems from an evolutionary perspective. *Anim. Behav.* **21**, 637–659.

Kleiman, D. G., and Malcolm, J. R. (1981). The evolution of male parental investment in mammals. In *Parental Care in Mammals* (D. J. Gubernick and P. H. Klopfer, Eds.), pp. 347–387. Plenum, New York.

Lamming, G. E. (Ed.) (1984). *Marshall's Physiology of Reproduction, Vol. 1, Reproductive Cycles of Vertebrates*. Churchill Livingstone, Edinburgh, UK.

Mech, L. D. (1970). *The Wolf: The Ecology and Behavior of an Endangered Species*. Natural History Press, New York.

Moehlman, P. D. (1986). Ecology of cooperation in canids. In *Ecological Aspects of Social Evolution* (D. I. Rubenstein and R. W. Wrangham, Eds.), pp. 282–302. Princeton Univ. Press, Princeton, NJ.

Nowak, R. M. (1991). *Walker's Mammals of the World*, 5th ed. Johns Hopkins Univ. Press, Baltimore.

Packard, J. M., Mech, L. D., and Seal, U. S. (1983). Social influences on reproduction in wolves. In *Wolves in Canada and Alaska: Their Status, Biology, and Management* (L. N. Carbyn, Ed.), Canadian Wildlife Service Report Series No. 45, pp. 78–86.

Dorsal Bodies in Mollusca

A. Saber M. Saleuddin

York University

I. Introduction
II. Structure of the Dorsal Body
III. Function of the Dorsal Body
IV. Environmental and Neuronal Inputs of the Dorsal Body
V. Chemical Nature of the Dorsal Body Hormone

GLOSSARY

albumen gland A female sex gland concerned with the secretion of perivitelline fluid.
ecdysteroids The most widespread polyhydroxyl steroid hormones in invertebrates.
lateral lobes Two lateral extensions of the cerebral ganglia found only in basommatophoran pulmonates.
neurosecretory cell Neurons which produce peptidergic hormones/modulators.
pulmonates Air-breathing molluscs.

Next to arthropods, molluscs are the largest and perhaps the most diverse groups of animals. They occupy every possible ecological "niche," many of them are good to eat, several are hosts to debilitating parasites, and some cause damage to crops.

I. INTRODUCTION

The Mollusca has four minor (Monoplacophora, Polyplacophora, Aplacophora, and Scaphopoda) and three major (Gastropoda, Bivalvia, and Cephalopoda) classes. Of these groups, vigorous endocrinological research has been largely restricted to gastropod molluscs, particularly the aquatic and terrestrial air-breathing molluscs. The presence of the putative endocrine organs called "dorsal bodies" (DBs) was first reported in a freshwater aquatic gastropod *Lymnaea stagnalis*, and, since then, DBs have been observed in all pulmonates studied thus far. The DBs are concentrated in two separate groups just dorsal to the cerebral ganglia, as in members of Basommatophora, or are scattered in the connective tissue sheath of the central nervous system (CNS), as in the members of Stylommatophora. It is now well established that DBs are involved in female reproduction but their role in male reproduction is not well documented. Their analogous structures in prosobranchs and opisthobranchs are called juxtaganglionar cells and in cephalopods they are known as optic glands. Like the DBs in pulmonates, these analogous structures are closely associated with the CNS and are involved in the control of female reproduction.

II. STRUCTURE OF THE DORSAL BODY

A. Basommatophora

The DB cell structure was first described by light and then by electron microscopy. In basommatophoran snails, the DBs rest on the dorsal surface of the cerebral ganglia and consist of two distinct groups of cells which lie very close together and appear as one group, as in *Helisoma* and *Planobarius*, or which occur as two very distinct tissue masses, as in *Ancylus*, *Lymnaea*, *Siphonaria*. Within each group, two separate zones can be recognized: the cell body region (cortex), containing the nucleus and other major organelles, and the cell processes region (medulla), containing secretory granules and very few other organelles. The latter region lies in very close proximity to the cerebral ganglion. Only in *Lymnaea* are

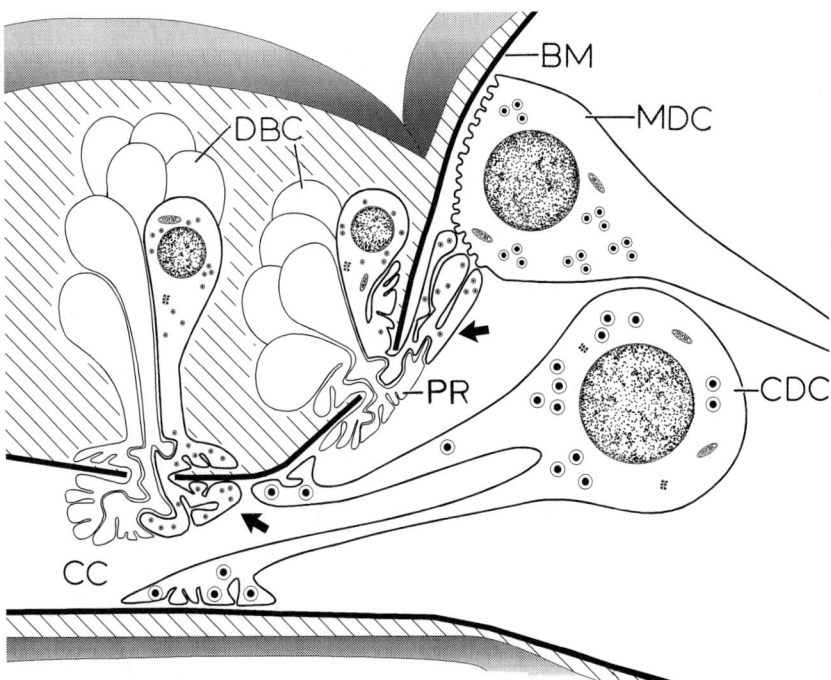

FIGURE 1 Diagrammatic representation of the organization of DB cells (DBC) and their relationship to neurosecretory mediodorsal cells (MDC) and the cerebral commissure (CC) in *Helisoma duryi*. The DB cells proceed through the connective tissue and diverge into fine processes (PR) near the basement membrane (BM) that separates the nervous tissue from the DB. Arrows indicate the presence of DB processes within the central nervous system. Neurites of neurosecretory caudodorsal cells (CDC) terminate in the CC. Not to scale.

two additional DB cell masses called lateral dorsal bodies found.

In *Helisoma duryi*, the DB cells in the cortex are grouped into distinct lobules, each separated from the other by a thin layer of collagen, smooth muscle fibers, fibroblasts, and occasional pore cells. Each lobule consists of 6–12 cells each measuring 10–15 μm in diameter. The organization of DBs in relation to the cerebral ganglia and the cerebral commissure is shown in Fig. 1. The DB cell processes are tortuous, many subdivide into finer branches which may interdigitate and end either very close to the neurosecretory mediodorsal cells (MDCs) or on the dorsal surface of the cerebral commissure where many axons of the neurosecretory caudodorsal cells (CDCs) are found. The cerebral commissure is the neurohemal area of the CDCs. In *Siphonaria*, the DB cells occur as groups but are not organized into lobules as in other basommatophoran pulmonates. Furthermore, their processes run in a random fashion and do not end in specified regions. Consequently, in this snail, the designations of cortex, where cell bodies are found, and medulla, where cell processes are known to occur, are not appropriate.

Structurally, the DB cells of the basommatophoran pulmonates are fairly uniform. The cells from reproductively active snails have numerous mitochondria, some of which contain lipid droplets. Also within the cytoplasm are many lipid droplets, prominent Golgi bodies, and both rough and smooth endoplasmic reticulum. Lysosomes are commonly seen in most cells (Fig. 2). Among all basommatophoran snails studied, only in *Siphonaria* are the DB cells ciliated, where the number of cilia per cell varies from one to three. The role of cilia is not clear but their presence suggests that DB cells are ectodermal rather than mesodermal in origin, as has been proposed earlier. The secretory granules are Golgi derived, membrane bound, and measure 60–90 nm in diameter. Various morphologies of the secretory

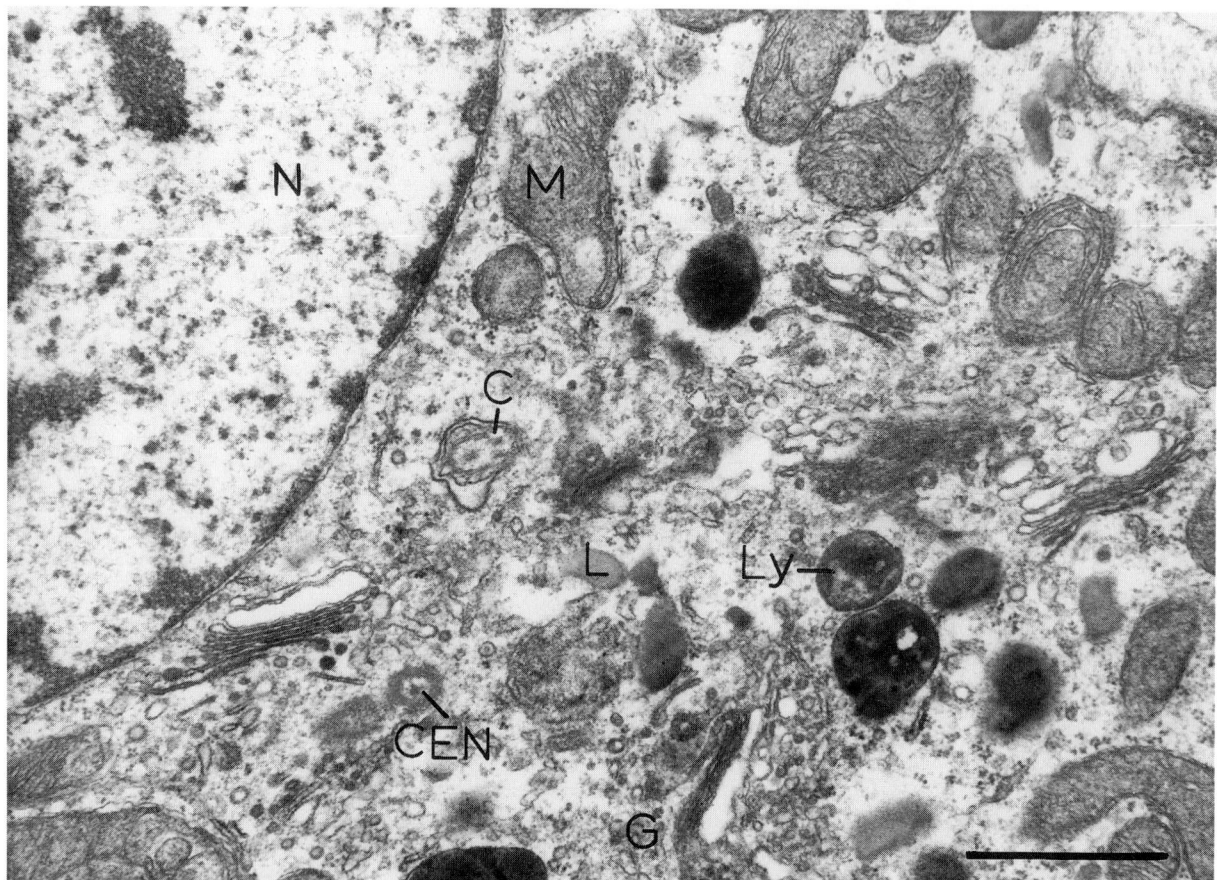

FIGURE 2 An electron micrograph of a DB cell in *Siphonaria* showing a cilium (C), centriole (CEN), Golgi (G), lipid droplet (L), lysosome (Ly), mitochondria (M), and nucleus (N). Scale bar =1 μM (from Saleuddin et al., 1997).

granules have been reported. The majority of the granules are seen in the cell processes rather than in the cell bodies and they are released outside as revealed by TARI technique. The DB cells in *Helisoma* and *Siphonaria* are coupled by gap junctions. In virgin snails, which are reproductively inactive, profiles of exocytotic release of secretory granules and the occurrence of gap junctions are fewer than in snails which are reproductively active, suggesting that DB cells are functionally active during egg production.

As previously mentioned, the cellular organization of DBs in *Siphonaria* is different from that of the rest of basommatophoran pulmonates. Two separate groups of DBs occur on the dorsal surface of each cerebral ganglion. In this species, three groups of DB cells, each consisting of about five cells, are found within each cerebral ganglion of the CNS. Like the outside cells, these cells within a group are coupled by gap junctions, are ciliated, and at least some of these cells have processes which leave the CNS in close contact with some of the outside cells. However, it is not certain if the inside and outside cells are coupled. Recently, it has been shown that cell processes of the outside cells enter the CNS and come in close proximity to the inside cells. This is the first report of the presence of DB cells within the CNS. It is not clear at what stage of life the DB cells migrate into the CNS or the reason for this migration. Perhaps the DB cells, through their product(s), control the nearby MDCs which may produce insulin-like peptide (growth hormone). It has been shown that MDCs in *Helisoma* produce growth hormone.

B. Stylommatophora

Except in *Succinea putris,* in which the DBs occur in precise locations, in all other stylommatophoran pulmonates, DB cells are found diffusely distributed either singly or in groups of four to six cells in the connective tissue covering of the CNS. The cells are 15–18 μm in diameter with irregular outlines and very often with several cell processes. No gap junctions have been found between the cells. Lipid droplets, numerous mitochondria, prominent Golgi, and glycogen granules are commonly found in the DB cells. The secretory granules, like those in *Helisoma,* are membrane bound and they measure 70–90 nm in diameter and are released outside the cell by exocytosis as revealed by the TARI technique. Associated with the DB cells are glycogen-rich support cells. The function of these cells in relation to DB cells is not known. Interspersed with the DB cells are pore cells, fibroblasts, muscle fibers, and collagen fibers.

III. FUNCTION OF THE DORSAL BODY

All pulmonate molluscs are concurrent hermaphrodites but many reproduce by cross-fertilization. It is now well accepted that the DBs are involved in several aspects of female reproduction in those pulmonates which have been studied. For example, the DB controls oocyte maturation, differentiation, and growth of the female accessory structures, secretion of sex organs, and ovulation in a number of pulmonate molluscs such as *L. stagnalis, H. duryi, Agriolimax reticulatus, Helix aspersa, Limax maximus,* and *Arion rufus.*

In *L. stagnalis,* surgical removal of the DB from juvenile animals retarded the growth of the female accessory structures and reduced ovipository activities because of the reduction in vitellogenesis. Reimplantation of the DB restored these activities. Electrocautery of the DB in adult *H. duryi* caused reduction in egg production. *Helisoma duryi* and *H. aspersa* snails raised individually in isolation to adulthood, called virgins, laid very few viable eggs, and when these virgins were allowed to mate, egg production ensued. Electron microscopic studies showed that the DBs in virgin snails were synthetically inactive and that mating brings about their activation. The criteria used to assess the activation were increased numbers of mitochondria, Golgi bodies, and exocytotic profiles of the secretory granules. When the frequency of immature and mature oocytes, based on size, was studied in DB-ablated (non-egg-laying), virgins (non-egg-laying), and mated (egg-laying) *H. duryi,* the number of immature oocytes was significantly higher in non-egg-laying than egg-laying snails. Thus, it was concluded that the DBs in non-egg-laying snails were synthetically inactive. The albumen gland is the female accessory gland in pulmonate molluscs and it secretes albumen (perivitelline fluid) which coats fertilized eggs in the carrefour. Miksys and Saleuddin (1985), using *in vitro* experiments, showed that the DBs control the synthesis of albumen. Ablation of DBs from reproducing snails reduced the volume of the female accessory structures. In the stylommatophoran pulmonate *A. reticulatus,* removal of the DBs caused a reduction in oocyte maturation and the growth of the female accessory structures. Using the organ culture system it was shown that in *H. aspersa,* only the DBs stimulated oocyte maturation. When the DBs were tested in conjunction with the cerebral ganglia, oocyte maturation was inhibited suggesting the presence of an inhibitory factor in the cerebral ganglia. Also using the organ culture system, increased incorporation of [14C]leucine and [3H]fucose in the gonadal tissue in the presence of the DBs in *H. aspersa* was found. In this species, synthetic activity of the DBs correlates with the reproductive state. It reaches maximum just prior to egg laying and decreases following mating.

The DB has been implicated in ovulation in *H. aspersa, H. duryi,* and *L. stagnalis.* In the presence of the DB extracts, mature oocytes of *Helisoma* exhibit cytoskeleton-generated ameboid movements which may assist in ovulation. However, because the DB occurs in close proximity to the neurosecretory CDC axons in the cerebral commissure, it is possible that the DB extracts contained caudodorsal cell hormone, which in *L. stagnalis* is claimed to be the ovulation hormone as well as the egg-laying hormone.

IV. ENVIRONMENTAL AND NEURONAL INPUTS OF THE DORSAL BODY

It has been well documented that photoperiod influences reproduction in pulmonate molluscs. Long-day (16 hr light:8 hr dark) photoperiod stimulates increased egg laying in *L. stagnalis* and in *H. duryi*. Furthermore, when *L. stagnalis* was exposed to long-day photoperiod the volume of the DBs increased with a concomitant increase in the number of mature oocytes. In *H. aspersa*, the DBs exhibited a diurnal cycle of synthesis-release activities; synthesis was maximum in the scotophase, whereas the release reached its peak in the photophase. In *Lymnaea, Helisoma*, and perhaps in other pulmonate molluscs, body/shell growth and reproduction are influenced by photoperiod, temperature, mating opportunity, and food availability; an inverse relationship exists between these two processes. The neurosecretory MDCs, the growth hormone-producing cells, and the DBs occur in close proximity (Fig. 1). In basommatophoran pulmonates, the cerebral ganglia have two lateral lobe (LL) extensions which are involved in reproduction and body/shell growth. In *Lymnaea*, removal of both LLs caused increased body/shell growth and thus it was suggested that the LL has a stimulatory effect on the DBs. In *Helisoma*, ablation of both LLs caused an increased incorporation of labeled amino acids in the periostracum (organic shell layer) *in vitro* which suggested an inhibitory role of LL on shell growth. In *Lymnaea*, hormonal control by LL on the DBs was postulated because no innervation from the LL to the DBs was found. In *Helisoma*, each LL has a large number of neurons, including three large neurosecretory neurons, one canopy cell (CC), and two lateral lobe cells (LLCs), and these three cells from the right LL only innervate the DBs (Fig. 3). Neurites from the CC innervate the cell bodies, whereas LLC neurites innervate the cell processes of the DBs. The neurosecretory granules from both CC and LLC neurites are released at the sites of innervation. Saleuddin and Ashton (1996) postulated that CC and LLC stimulated the synthesis and release of DB hormone, respectively. Two major branches of the optic nerve from each side enter the LL and some of the axons form synapses with LLC1 and possibly LLC2. Furthermore, an axon from each CC enters into the optic nerve (Fig. 3). Thus, both CC and LLC receive direct input from the optic nerve, which may explain the cellular basis of the stipulatory effect of long days on the DBs and thus increased egg production. The DB cells in *Siphonaria*, which occur both inside and outside the CNS, are

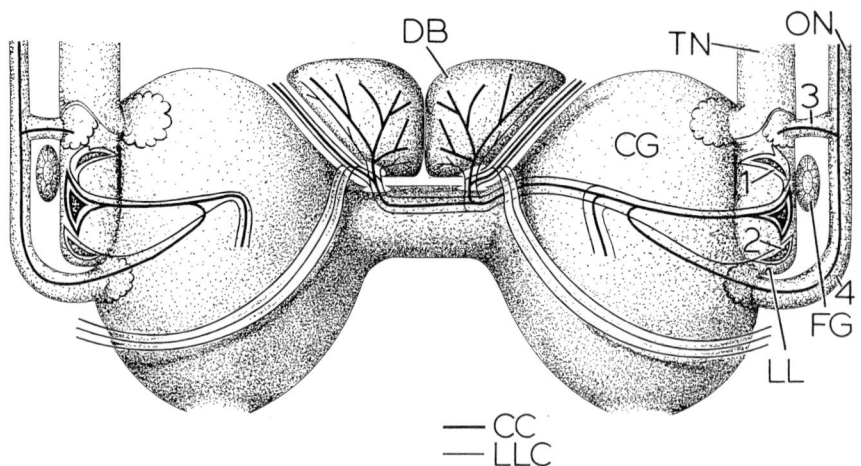

FIGURE 3 Diagrammatic representation of the cerebral ganglion (CG) showing the positions of the dorsal body (DB), tentacular nerve (TN), follicle gland (FG), optic nerve (ON), lateral lobe (LL), canopy cell (thick line, CC), and two lateral lobe cells [thin line; LLC1 (1) and LLC2 (2)]. The neurite distribution and the two major branches of the optic nerve (3 and 4) are shown. For clarity only one side has been labeled. Not to scale (from Saleuddin and Ashton, 1996).

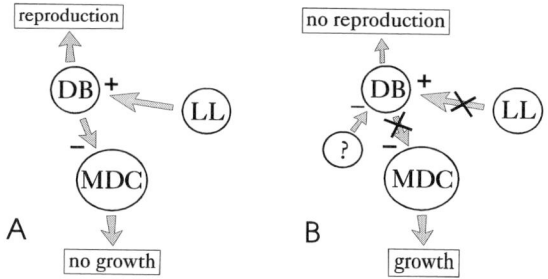

FIGURE 4 Schematic representation of two feasible models showing an inverse relationship in the endocrine and/or neuroendocrine control of reproduction and body growth in a pulmonate mollusc. (A) Lateral lobe (LL) stimulates DB, which induces reproduction but it inhibits the neurosecretory mediodorsal cells (MDC) and thus no growth occurs. (B) The stipulatory control from LL to DB is withdrawn causing inactivity of DB. This results in the loss of the inhibitory control of DB on MDC resulting body growth. This model suggests the presence of an inhibitory center controlling DB. Currently, it is unknown exactly how the stimulatory and inhibitory influences are exerted on DB.

innervated by one type of neurosecretory axon originating in the cerebral ganglia. We are in the process of tracing the location of their neurons in this species by serial sections. Based on accumulated data on the interrelationship between the LL, DBs, and MDCs on reproduction and growth, two feasible models are proposed (Fig. 4).

In a number of stylommatophoran pulmonates, such as *Helix* and *Theba*, the DB receives direct innervation from the cerebral ganglia. In *H. aspersa*, the DB cells are innervated by inhibitory neurosecretory axons, some of which belong to the cerebral green cells located in the cerebral ganglia. These cells are known to produce an insulin-like substance. Griffond and Mounzih (1990) identified four types of neurosecretory axons which innervate the DBs of *H. aspersa* and, based on immunohistochemistry, one of these four types contains FMRFamide-like material. In this snail, FMRFamide inhibited the incorporation of methionine into the protein of the DBs. Based on granule morphology and size, Saleuddin *et al.* reported two kinds of neurosecretory axons which innervate the DB cells in *H. aspersa* at sites called synapse-like-structures (SLSs). Neurosecretory granules in type I axons are 90–110 nm in diameter and conform to the type reported earlier by Griffond and Mounzih in the same species. Type II axons, on the other hand, contain larger granules (170–190 nm) and correspond to type 3 axons described by Griffond and Mounzih. Observations on limited serial sections suggest that all DB cells are individually innervated by both type I and type II axons; thus, it was speculated that these two types are involved in stimulation or inhibition of the DBs. Unlike *Helix*, the DB cells in *Helisoma* and *Siphonaria* have gap junctions and thus innervations of a group rather than individual cells can exert neuronal control over the whole organ.

V. CHEMICAL NATURE OF THE DORSAL BODY HORMONE

The exact chemical nature of the DB secretion (hormone) remains largely unknown. The small size of the DB in the pulmonate molluscs and its very close proximity to the CNS make it extremely difficult to isolate pure DB cells. In most cases, they are contaminated with neurons from the CNS during the process of isolation. Fine structural features of the DBs in pulmonate molluscs, such as membrane-bound granules, prominent Golgi, rough endoplasmic reticulum, many lipid droplets, smooth endoplasmic reticulum, and numerous mitochondria, seem to suggest that the major secretory products (not necessarily hormones) of the DBs could be both protein and steroid. Indeed, several lines of evidence to support this conclusion. The evidence for the peptidergic nature of the DB hormone (DBH) has come from *in vitro* bioassay measuring the stimulation of albumen synthesis by the albumen gland. Using this bioassay, the occurrence of DB peptides of various molecular weights has been reported from a number of pulmonate molluscs. In *Helix* and *Limax*, the putative DB hormone has a molecular weight between 4 and 7 kDa, whereas in *Lymnaea* it was proposed to be 30 kDa. This large discrepancy could be due to impure composition of the DB extracts used in these studies. In other words, the source(s) of this hormone cannot be precisely determined. Using similar bioassays, Miksys and Saleuddin (1985) suggested that the DBH in *Helisoma* is a steroid. Based on the presence of steroid-synthesizing enzymes and on electron microscopy, it was proposed that the DBs in stylommatophoran pulmonates secrete steroid(s). Later, Nolte et al. (1986) tentatively identified ecdysone in the DBs of *H. pomatia.* Saleuddin et al. (1994) reported that the gonadotrophic factor of the DB extract in *Helisoma* is insensitive to proteases but stable to heat. Using radio immunoassay, these authors showed that several immunoreactive ecdysteroids are present in the methanolic extract of medium from cultured DB cells. Furthermore, 20-hydroxyecdysone stimulated albumen synthesis in the albumen gland *in vitro*, and when injected into virgin snails it caused laying of viable eggs. High-performance liquid chromatography (HPLC) separation suggests that the DBH in *H. duryi* is hydrophobic and fine structural studies and preliminary HPLC separations indicate that it lipophilic.

See Also the Following Articles

MARINE INVERTEBRATE LARVAE; MOLLUSCA; OVIPOSITION IN MOLLUSCS

Bibliography

Boer, H. H., Slot, J. W., and Van Andel, J. (1968). Electron microscopic and histochemical observations on the relation between the mediodorsal dorsal bodies and neurosecretory cells in the basommatophoran snails *Lymnaea stagnalis, Ancylus fluviatilis, Australorbis glabratus* and *Planobarius corneus. Z. Zellforsch. Microsk. Anat.* 87, 435–450.

Gomot, A., Gomot, L., Marchand, C., Colard, C., and Bride, J. (1992). Immunocytochemical localization of insulin-related peptide(s) in the central nervous system of the snail *Helix aspersa* Muller: Involvement in growth control. *Cell. Mol. Neurobiol.* 12 21–32.

Griffond, B., and Mounzih, K. (1990). Innervation of the dorsal body cells of *Helix aspersa*: Immunocytochemical evidence for the presence of FMRFamide-like substances in nerves and synapselike structures. *Tissue Cell* 22, 741–748.

Joosse, J. (1988). The hormones of molluscs. In *Endocrinology of Selected Invertebrate Types* (H. Laufer and R. G. H. Downer, Eds.), pp. 89–140. A. R. Liss, New York.

Joosse, J., and Geraerts, W. P. M. (1983). Endocrinology. In *The Mollusca* (K. M. Wilbur and A. S. M. Saleuddin, Eds.), Vol. 4, Part 1, pp. 317–406. Academic Press, New York.

Khan, H. R., Ashton, M. L., and Saleuddin, A. S. M. (1990). Changes in the fine structure of the endocrine dorsal body cells of *Helisoma duryi* (Mollusca) induced by mating. *J. Morphol.* 203, 41–53.

Lever, J. (1958). On the relation between the mediodorsal–dorsal bodies and the cerebral ganglia in some pulmonates. *Arch. Neerl. Zool.* 13, 194–201.

Lundelius, J. W., and Freeman, G. (1986). A photoperiod gene regulates vitellogenesis in *Lymnaea peregra* (Mollusca: Gastropoda: Pulmonata). *Int. J. Invertebr. Reprod. Dev.* 10, 201–226.

Miksys, S. L., and Saleuddin, A. S. M. (1985). The effect of the brain and dorsal bodies of *Helisoma duryi* (Mollusca: Pulmonata) on albumen gland synthetic activity in vitro. *Gen. Comp. Endocrinol.* 60, 374–380.

Nolte, A. (1983). Investigations on the dorsal bodies of stylommatophoran snails. In *Molluscan Neuroendocrinology* (J. Lever and H. H. Boer, Eds.), pp. 142–146. North-Holland, Amsterdam.

Nolte, A., Koolman, J., Dorlchter, M., and Straub, H. (1986). Ecdysteroids in the dorsal bodies of pulmonates (Gastropoda): Synthesis and release of ecdysone. *Comp. Biochem. Physiol. A* 87, 777–782.

Saleuddin, A. S. M., and Ashton, M.-L. (1996). Neuronal pathways of three neurosecretory cells from the lateral lobes in *Helisoma* (Mollusca): Innervation of the dorsal body. *Tissue Cell* 28, 53–63.

Saleuddin, A. S. M., Kunigelis, S. C., Schollen, L. M., Breckenridge, W. R., and Miksys, S. L. (1983). Studies on endocrine control of reproduction in *Helisoma* and *Helix.* In *Molluscan Neuroendocrinology* (J. Lever and H. H. Boer, Eds.), pp. 138–142. North-Holland, Amsterdam.

Saleuddin, A. S. M., Mukai, S. T., and Khan, H. R. (1990). Hormonal control of reproduction and growth in the freshwater snail *Helisoma* (Mollusca: Pulmonata). In *Neurobiology and Endocrinology of Selected Invertebrates* (B. G. Loughton and A. S. M. Saleuddin, Eds.), pp. 163–182. Captus Press, Toronto.

Saleuddin, A. S. M., Griffond, B., and Ashton, M.-L. (1991). An ultrastructural study of the activation of the endocrine dorsal bodies in the snail *Helix aspersa* by mating. *Can. J. Zool.* 69, 1203–1215.

Saleuddin, A. S. M., Mukai, S. T., and Khan, H. R. (1994). Molluscan endocrine structures associated with the central nervous system. In *Perspectives in Comparative Endocrinology* (K. G. Davey, R. E. Peter, and S. S. Tobe, Eds.), pp. 256–263. National Research Council of Canada, Ottawa.

Saleuddin, A. S. M., Ashton, M.-L., and Khan, H. R. (1997). An electron microscopic study of the endocrine dorsal bodies in reproductively active and inactive *Siphonaria pectinata* (Pulmonata: Mollusca). *Tissue Cell* 29.

Drosophila

John Ewer and Marla B. Sokolowski

York University

I. Introduction
II. Oogenesis and Spermatogenesis
III. Sex Determination
IV. Courtship
V. Summary

GLOSSARY

autosome Chromosomes other than sex chromosomes.
courtship Behaviors exchanged between a male and a female of the same species prior to mating.
diploid Number of chromosomes in a fertilized egg (2n).
germline Cells from which gametes are derived.
gynandromorph Individual mosaic for male and female tissues.
haploid Number of chromosomes in a gamete.
maternal Derived from the female parent.
mosaic (or chimera) Individual with cells of different genotypes.
oviposition The process of egg-laying.
pheromone A chemical that is secreted by an organism and is used in communication.
somatic Cells other than germ cells.

The biology of reproduction in the fruit fly, *Drosophila melanogaster*, shares many features with that of other insect species. The reason for the interest in this particular species lies in its genetics. Work over the past 80+ years has resulted in the generation of an immense wealth of tools and resources which include a large number of mutations as well as genetic and molecular tricks. These resources have made it possible to address genetically and molecularly many questions about the control of this organism's reproduction. Questions that have been very fruitfully investigated using this approach include how the oocyte and egg develop, how the fly's sex is determined and maintained, and how the adult convinces its mate to start the cycle all over again. Conversely, some aspects of insect reproduction, such as its hormonal control, have been studied more successfully in other insect species. In this article, we mainly focus on areas in which *Drosophila* has provided unique contributions to the field and results from the genetic analysis of reproduction.

I. INTRODUCTION

A genetic analysis uses mutants to uncover and define biochemical pathways involved in the production of normal phenotypes. In some cases, such as most of those discussed in this article, a genetic approach is the only—or at least only practical—way to systematically identify genes involved in generating a particular phenotype. The advantages of *Drosophila melanogaster* for isolating, mutating, and cloning such genes are numerous (see Box 1).

In addition to the use of mutants to identify components of a pathway, genetic variants can be used as tools. Thus, for instance, one might use a mutation that eliminates photoreceptor function to question the role of vision in courtship, much the same way as painting the eye with black wax might be used to ask the same question in a bigger insect.

BOX 1

A Primer in *Drosophila melanogaster* Tools

It has been said that the fruit fly *D. melanogaster* is the most genetically manipulable metazoan. In addition to a relatively short generation time (about 2 weeks), only four sets of chromosomes, and a genome 1/20 the size of that of a typical mammalian genome, its superior status in modern genetics derives from the work of geneticists that have been using this organism since its genetic analysis was started in earnest by T. H. Morgan at Columbia University in 1909. Today, the number of tools available for genetic, molecular, and cellular analysis of *Drosophila* biology is overwhelming. This field has, per force, also become more technical. As a result, investigators who do not use *Drosophila* in their work often feel that "breaking into" this field represents an insurmountable challenge. However, the essentials are quite simple, and sophisticated experiments can be done even by those who last thought about genetics in terms of "big A and little a." We describe some of the tools that are available and well within the reach of any investigator. This review is not exhaustive but is merely intended to break the ice.

A. Identifying Genes Involved in a Given Process

In order to identify genes involved in, for instance, oogenesis, flies are mutagenized and their progeny screened for flies with abnormal oogenesis. The mutagen can be a chemical, such as ethylmethanesulfonate (EMS), ionizing radiation (e.g., X rays), or the mobilization of mobile elements (such as P-elements) which create a mutation when they insert into the genome. Each agent has advantages and disadvantages. For instance, EMS usually produces small lesions such as single base pair substitutions and small deletions; however, these can be difficult to identify molecularly. Mobile elements make it relatively easy to clone the DNA flanking the insertion (which is presumably in or close to the gene of interest). However, they show some selectively in their insertion sites and some genes are refractory to such an approach.

Mutations can be induced in any organism. The strength of *Drosophila* is that mutations can be maintained and manipulated with relative ease. Thus, mutations that are fully recessive and/or lethal when homozygous, or even ones that are deleterious if heterozygous with a normal copy of the gene, can be investigated. The ability to maintain and manipulate mutations is largely due to the existence of "balancer chromosomes." These chromosomes contain multiple and many times nested inversions that prevent recombination with its homolog. In addition, they carry a dominant marker and are usually homozygous lethal or sterile. If a fly bearing the mutation of interest (m) is crossed to a fly bearing a balancer chromosome (Bal), progeny bearing the mutation and the balancer (m/Bal) will produce a line that can be maintained indefinitely without selection. Indeed, in such a stock, flies homozygous for the balancer chromosome (Bal/Bal) die, and the mutation cannot be lost through recombination with the homologous (balancer) chromosome. Finally, note that the stock can be maintained even if the mutation of interest is lethal or causes sterility when homozygous. Indeed, the flies that are heterozygous for the mutation and the balancer chromosome (m/Bal) will be wild type and will produce m/Bal progeny (in addition to m/m and Bal/Bal, which will die or be sterile). While these details may be regarded as being merely practical, balancer chromosomes are probably the single most important resource of all of *Drosophila* genetics, and they set it apart from other genetically tractable metazoans.

In a balanced stock the characteristics of the mutant phenotype can be studied. Indeed, progeny that are homozygous for the mutation (m/m) can be identified by the lack of the dominant marker of the balancer chromosome and will express the mutant phenotype of interest (e.g., abnormal oogenesis); the heterozygote (m/Bal) can serve as a control.

In general it is very important to generate several alleles of each gene before one can infer the function of the wild-type product. In the case of oogenesis, for instance, it might be possible for one allele (m^1) to produce a very specific defect in oogenesis, yet the wild-type (m^+) gene might encode a protein with a general "housekeeping" function (e.g., a ribosomal protein). Oogenesis is a complicated process during which a large number of RNAs and proteins must be pumped into the oocyte over a short period of time. Thus, mild mutations in a housekeeping gene might produce defects in oogenesis only because this process requires that the cell's synthetic machinery work perfectly. A more severe allele of such a gene (m^2) might be lethal in all cells. In contrast, a severe allele of a gene involved specifically in oogenesis would be sterile.

Additional alleles can be recovered in a screen in which mutagenized chromosomes are "tested" heterozygous with the initial mutant chromosome (m^1). A new allele (m^2) in combination with m^1 (m^2/m^1) might also produce a defect in oogenesis. Alternatively, it might be lethal, suggesting that the wild-type gene has a more widespread function.

It is also possible that the product of the wild-type (m^+) gene plays a specific role in other functions in addition to oogenesis. The times and tissues in which a gene function are required can be determined using temperature-sensitive alleles and mosaic animals, respectively. Temperature-sensitive alleles produce defects at one (restrictive) temperature and not at others (permissive temperature). Thus, the role of a gene at one particular time of development can be assessed by raising the fly at restrictive temperature during that period while maintaining it at permissive temperature at other times.

Another great resource available in *Drosophila* is the ability to localize genes on the chromosome cytologically. *Drosophila* salivary glands contain "giant" chromosomes which result from endoreplication. As a result, several thousand copies of each homolog are apposed to each other, producing a giant chromosome with a pattern of banding that is characteristic for each region of every chromosome. Thus, mutations that do not complement a cytologically visible deletion can be mapped onto the chromosome.

Genes can also be mapped genetically relative to other genes, which is accomplished by measuring the frequency of recombination between the mutant of interest and existing markers. Mapping mutations is important because it might allow for the identification of other alleles previously identified in other screens. It also might reveal that the gene of interest is close to a cloned gene or to a site of a mobile element insertion, both of which allow for the cloning of the gene. An ongoing project is the construction of a database containing the position of P-elements present in available stocks. If such a database contained, for instance, 1000 entries, any gene would be expected to lie within 100 kilobases of a known P-element. The stock containing the P-element closest to the gene of interest could be determined by genetic mapping, and the gene could subsequently be cloned by "chromosomal walking" starting from the tagged DNA that flanks the P-element.

Once the gene is cloned its sequence can be revealing of its function. In addition, a wild-type copy of the gene can be stably reintroduced into the genome by "transforming" a fly with a P-element into which the gene has been molecularly engineered. Restoration of wild-type function in a mutant by such transformation is conclusive evidence that the mutant phenotype is due to a defect in the cloned gene. The ability to reintroduce cloned genes into the genome can also be used to produce flies in which expression of the gene is restricted in time and/or space, allowing for the further analysis of the gene's wild-type function. Finally, once a gene in this species has been sequenced, homologous genes can be searched for in other animals. Often conservation of sequence and function can be found.

B. Identification of Genes Encoding Particular Products

Previously, we discussed some of the screens and methods that can be used to identify genes involved in a particular process, such as oogenesis. Such screens do not make any assumption on the identity of these genes other than the fact that they are involved in oogenesis. (However, the details of the screen will determine which types of genes will be identified. Thus, for instance, a screen for mutations that produce female sterility will probably not identify genes that are required for both oogenesis and embryonic or larval development. Indeed, mutant alleles of such genes might kill the organism before it reaches the adult stage.) In other cases one might be interested in isolating the genes that encode specific products, such as an enzyme, or that contain particular molecular motifs, such as a particular DNA-binding sequence. In such cases a library (genomic, cDNA, or protein expression) is screened and the resulting clones are identified. To isolate mutations in the gene encoding this product, the clones are first localized on the cytological map. Mutations in that region can then be screened for. In Drosophila it is not generally possible to target mutations to specific sequences in the genome as it is in the mouse ("targeted mutagenesis").

II. OOGENESIS AND SPERMATOGENESIS

A. Cellular Aspects

The two ovaries of Drosophila are arranged into a series of about 16 separate units called ovarioles. At the tip of each ovariole, in the region called the germarium, a stem cell divides to produce a cystoblast and regenerate a stem cell. The cystoblast then divides four times to produce 16 germline cells. Of these, one becomes the oocyte while the remaining 15 become nurse cells that synthesize maternal components that are later transported to the oocyte. Interestingly, cytokinesis is incomplete at each of the cystoblast divisions. Thus, its progeny is interconnected by large cytoplasmic bridges or ring canals. The oocyte invariably differentiates from 1 of the 2 cells that are connected to 4 future nurse cells. As the egg chamber differentiates it moves away from the apical region and is eventually surrounded by the somatically derived follicle cells. As a result, each ovariole is composed of a developmentally ordered series of egg chambers. In the apical regions, oocyte growth is due to the transport of nutrients from the nurse cells. At later, vitellogenic stages, the oocyte endocytoses yolk proteins synthesized by the fat body and the follicle cells. During the final stages, nurse cells collapse as their cytoplasm rapidly enters the oocyte. The oocyte remains in the metaphase of the first meiotic division until it is released into the oviduct and fertilization initiates embryonic development.

Early spermatogenesis bears some resemblance to oogenesis and is initiated when a germline stem cell divides to produce a primary spermatogonial cell and regenerate a stem cell. Four gonial mitotic divisions follow producing 16 primary spermatocytes, which remain interconnected by cytoplasmic bridges, as are the progeny of the cystoblast in the female. However, spermatocytes then enter meiosis, ultimately producing the spermatids which develop into the mature sperm.

While our understanding of oogenesis and spermatogenesis has been derived mostly from cell biological experiments, genetic approaches are becoming increasingly useful. This is especially true for oogenesis, in which genetic analysis is more advanced. Indeed, the study of genes involved in oogenesis, most of which were defined by female sterile mutations, has led to a further understanding of the sequence of events that occur during oocyte differentiation and growth. In addition, the cloning of many of these genes and the subsequent localization of the corresponding gene products has provided a number of useful labels for specific cyst components (e.g., the ring canals) and has led to a more detailed view of the elements and dynamic changes involved in this complex cytological event. The same situation is likely to obtain for male germ differentiation and the control of meiosis.

B. Hormonal Control of Reproduction

Insect reproduction is to a greater or lesser extent under the control of juvenile hormones and ecdysone. These hormones transduce the multiple and complex internal and external stimuli, such as the quality of food (e.g., protein vs carbohydrate), the presence of a fertilized egg in the brood sac (as occurs in viviparous cockroaches), or the length of the day, and regulate the animal's reproductive status.

In *Drosophila*, vitellogenesis involves proteins that differ from the vitellogenins present in noncylorraphous Diptera, and are referred to as yolk proteins (YP1–YP3). They are most similar to mammalian lipases, are synthesized by both the fat body and the follicle cells, and are taken up by the oocyte.

The details of the hormonal control of oogenesis in *Drosophila* are far from clear. However, it has been proposed that it might usefully be considered to be a variant of the situation observed in the mosquito. In this dipteran species, juvenile hormone (JH) is first released shortly after eclosion and in some way prepares the oocyte and fat bodies to respond subsequently to other hormones. Following a blood meal (i.e., a protein-containing meal), a brain-derived hormone is released and causes the ovaries to produce and release ecdysone. This steroid then acts on the fat bodies to promote vitellogenesis. Vitellogenin is released into the circulation and taken up by the oocyte, a process that requires JH.

The extent to which this model applies to *Drosophila* is still under investigation. Several features of vitellogenesis in this fly have made its control difficult to study. In this organism, previtellogenic differentiation begins during the latter part of adult development. Thus, inferences derived from hormonal and surgical manipulations done after eclosion might provide an incomplete picture. In addition, the sources of relevant hormones have not all been identified. For instance, females ovariectomized at eclosion show normal fat body vitellogenesis, suggesting that there is a nonovarian source of ecdysteroids. Finally, even the exact chemical nature of the hormones involved has had to be revised, such as occurred following the discovery that the primary juvenile hormone may be JH III bisepoxide and not JHIII as had previously been believed.

Nevertheless, the features of the hormonal control of oogenesis in *Drosophila* are at least broadly in accord with the picture that has emerged from the work on mosquitoes. In addition to careful "classical" extirpation/replacement experiments, genetic experiments have helped unravel this complex process. Thus, for instance, mutations in the gene *ecdysone* (*ecd*), which severely reduces the levels of ecdysone starting in the first-instar larva, have been of use. In particular, the temperature-sensitive allele, ecd^1, has confirmed that there is a requirement for this steroid during vitellogenesis. Likewise, mutations in *apterous*[4], which causes defects in the levels of JH, cause female sterility. Most oocytes of *apterous*[4] females are underdeveloped and "escaper" vitellogenic oocytes lack microvilli and pinocytotic vesicles. Both these defects can be restored to wild type by application of a JH analog, indicating that JH (or a molecule similar to it) is required at least for the uptake of yolk proteins.

In males, spermatogenesis begins at pupariation. In contrast, the synthesis of accessory gland proteins appears to be under hormonal control in the adult, their synthesis being stimulated by low levels of JH. This effect can be demonstrated *in vitro* and is being used to investigate the elusive molecular mechanism of action of this hormone. In particular, results obtained using tissues from flies mutant for the *Resistance(1) Juvenile Hormone* (also known as *Methoprene tolerant*) suggest that this gene may encode a nuclear receptor for JH.

III. SEX DETERMINATION

In *Drosophila* the fly's sex is determined by the sex chromosome (X) to autosome (A) ratio (X:A), with a ratio of 1 (e.g., two X chromosomes:two sets of autosomes) specifying female differentiation and a ratio of 1:2 (e.g., one X chromosome:two sets of autosomes) resulting in male differentiation. The Y chromosome is required for the male's fertility but not for its somatic or behavioral sex.

Much of the mechanism by which a fly's sex is determined has been elucidated and results from the identification and analysis of genes which when mutant alter specific aspects of a fly's sexual phenotype.

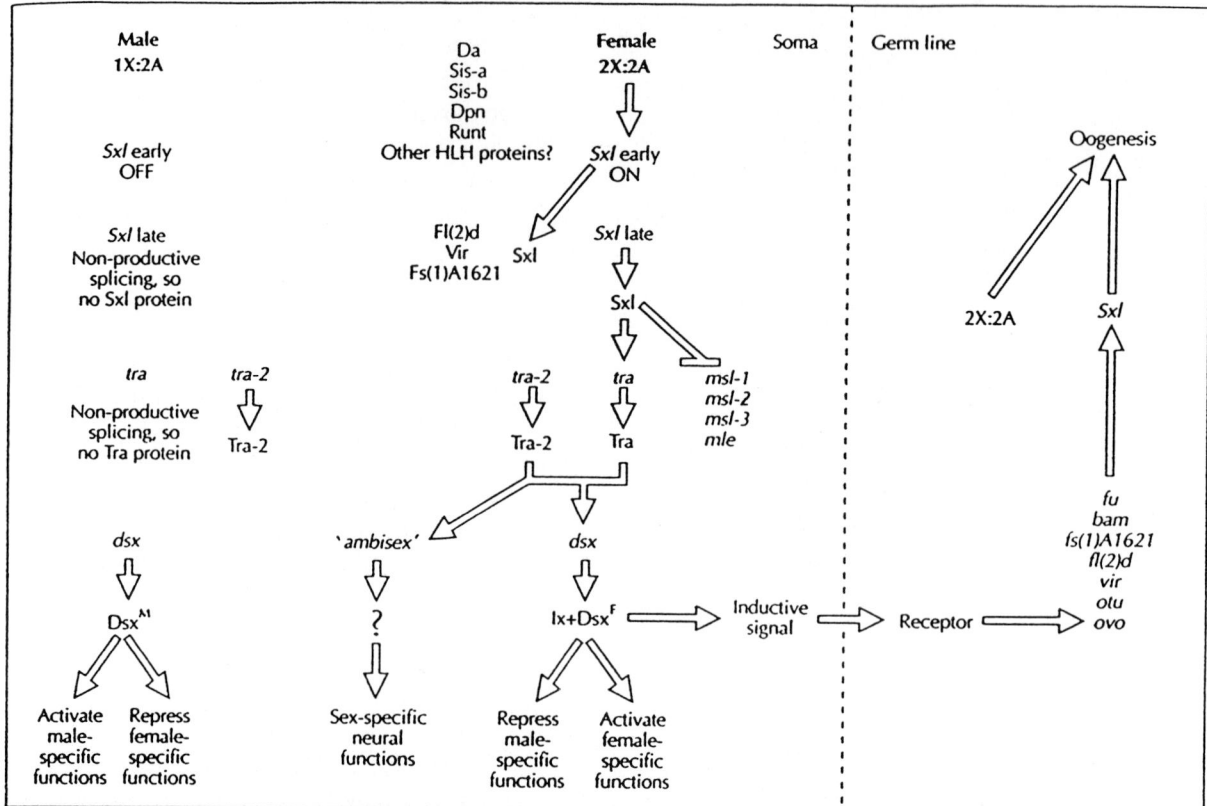

FIGURE 1 Hierarchical relationships among genes involved in sex determination of Drosophila melanogaster. The key event is whether the early *Sxl* transcript is (female) or is not (male) produced. This switch is dependent on the X:A ratio, which is "calculated" using genes such as *da, sis-a, sis-b, dpn,* and *runt*. The Sxl protein prevents the expression of genes that mediate dosage compensation in the male (*msl-1–3* and *mle*). It also causes the splicing of the *tra* mRNA into a transcript that encodes a functional protein which, in conjunction with *tra-2*, sets the pathway in the female mode. The presence of a branch specific for sex determination in the nervous system has been indicated (and includes the hypothetical gene "ambisex"). The gene *fru* is likely to act in this branch. Sex determination in the germline is well less understood but in the female requires an inductive signal from female somatic tissue and some or all of the genes *fu, bam, fs(1)A1621, vir, otu,* and *ovo*. Arrows indicate hierarchical, not necessarily direct, iterations (reproduced from Burtis, 1993, © Current Biology Ltd.).

Genes of interest are, for instance, those which when mutant produce either no males or no females or produce animals whose germline, somatic, or behavioral sex does not correspond to that appropriate to their X:A ratio. The genes involved in sex determination and our current understanding of their regulatory relationships are shown in Fig. 1.

Our understanding of the mechanism of sex determination in this species is most advanced for the somatic tissues. Here, the first and critical gene in the decision to become female or male is the *Sex-lethal* (*Sxl*) gene, which is first expressed in the early embryo and is required for female somatic differentiation. This X-chromosome gene can produce two transcripts. Of these, the first one to be transcribed is only expressed in females and its protein product regulates the splicing of the second, later transcribed, *Sxl* RNA. In females the resulting mature second transcript encodes a functional Sxl protein that is expressed throughout the female's life and puts the sex-determination pathway in a female-specific mode. In contrast, the corresponding *Sxl* male transcript is not spliced correctly and does not produce a functional protein; this lack of *Sxl* function results in the pathway being set in a male mode. Thus, the key sex-determining switch is set depending on whether the first *Sxl* transcript is (females) or is not (males) transcribed. The mechanism by which it is set is not entirely elucidated but involves "computing" the X:A ratio using specific X-chromosome and autosomal genes which include the genes *daughterless, sisterless-a* and *sisterless-b, runt,* and *deadpan*.

Sxl mutant female embryos die because the mechanism that regulates the levels of expression of genes on the X chromosome goes awry. In *Drosophila*, X-chromosome genes are transcribed at lower levels in females (XX) than they are in males (XY) in order to compensate for their difference in number in the two sexes. As might be expected, the level of expression for genes on the X chromosome depends on *Sxl*. In *Sxl* mutant females death occurs because the embryo produces male levels of transcripts from each of its two X chromosomes. Likewise, *Sxl* mutants that inappropriately produce functional (female) Sxl proteins in males are also lethal due to the underexpression of the genes from the male's single X chromosome. This aspect of sex determination, referred to as dosage compensation, involves the products of at least four genes (*male-specific lethals 1–3* and *mle*). These bind to X-chromosome genes to increase their level of expression in males.

Beyond the branch on the pathway that sets the level of dosage compensation lie a number of genes that determine the fly's somatic, germline, and nervous system/behavioral sex (Fig. 1). References in the Bibliography discuss these genes, their relationships, and how sex is regulated along the pathway that determines these various aspects of a fly's sexual phenotype. We will point out a few interesting features of sex determination in *Drosophila*.

The first point of interest is that the main mechanism by which genes upstream in the sex-determination pathway control those downstream appears to be via the regulation of splicing alternatives, such that the expression of a functional sex-specific gene product allows for the sex-specific splicing of the transcript from the next gene downstream. We have already mentioned that the protein encoded by the first *Sxl* transcript is necessary for the splicing of the second *Sxl* mRNA into a transcript that encodes a functional and female-determining protein. Likewise, the other genes in the cascade exert their control by regulating the splicing of downstream genes. Thus, for instance, the second, later-expressed Sxl protein regulates the splicing of *transformer* (*tra*), which in turn regulates that of *doublesex*. This control of splicing occurs through the direct binding of these proteins to specific sequences present on the downstream transcripts. This direct, sequence-specific interaction has been exploited to molecularly identify other members of this pathway such as the gene *fruitless*.

The second feature of interest is that the extent to which a cell's sex is dependent on that of it's neighbors is different for somatic vs germline tissues. In the former, this decision is cell autonomous (meaning that it is independent of the genotype of the rest of the organism), whereas in the germline it is to some extent dependent on influences from the surrounding somatic tissue. In addition, the sex-determination pathway for the germline involves components that are exclusive to this tissue (Fig. 1).

The following question arises from these studies: How general are the mechanisms of sex determination used by this little fly? A perusal of sex determination in different animals reveals profound differences in the mechanisms by which the female vs male decision is made both at the superficial and the molecular levels. Thus, for instance, alternative splicing mechanisms do not seem to be involved in sex determination in the nematode. Furthermore, in the housefly, it is the presence of a Y chromosome that determines a male fate. While this mechanism of sex determination appears to be very different from that of *Drosophila*, it would be surprising if sex determination in these comparatively closely related dipteran species did not share fundamental similarities. The extent to which similar mechanisms may underlie apparent differences in sex-determination mechanisms in other organisms must await further investigation.

IV. COURTSHIP

In addition to obvious somatic sex dimorphism, females and males show clear sex-specific differences during courtship behavior. During this "dance" the male performs a relatively fixed sequence of distinct behaviors, as illustrated in Fig. 2. In this sequence the male first orients and follows the female and taps her abdomen with his foreleg. He then extends a

FIGURE 2 Courtship in *Drosophila melanogaster*. In the presence of a receptive female a mature male performs a stereotyped behavioral sequence. He orients toward the female, taps her abdomen with his foreleg, and then "sings" by extending and vibrating one of his wings. He will then lick the female's genitalia, curl his abdomen, and copulate. Each step within this sequence can vary in duration depending on the female's level of interest, and the whole sequence can break off. The female plays a comparatively passive role, although she must accept the male and slow down in order for copulation to occur (from Greenspan, 1995, with permission from Suzanne Barnes, artist).

wing and vibrates it to produce a species-specific "love song." If all goes well he will then lick her genitals, mount her, and copulate. If the female is unreceptive the sequence may break off and restart, usually with the male reorienting and repeating the cycle. During courtship the female appears to play a relatively passive role. Her most obvious behavior is slowing down once she accepts to mate with her suitor. If she is unreceptive, which occurs if she has mated recently, she will continuously fend off her suitor and extrude her ovipositor.

A. "Mapping" of Male Courtship

The following is an interesting question that can be addressed with any precision only in *Drosophila*: What areas of the nervous system are involved in generating this sex-specific courtship behavior? Given that a fly's sex is determined by its number of X chromosomes, this question can be addressed by asking which tissues must be haplo- (or diplo-) X in order for a diploid fly to display male (or female) courtship behavior.

One classic way of answering this question is by analyzing the behavior of gynandromorph mosaic flies. In these organisms an initially XX embryo loses one of its X chromosomes in one of the daughter cells during one of the first cell divisions. As a consequence, the resulting fly is a mixture of XX, or female, and XO, or male, tissues. While such mosaics have been studied in a number of insect species, only in *Drosophila* is it possible to determine the genotype of a mosaic fly's internal tissues and thus establish with some precision which cells must be male in order for the fly to behave as a male. These studies showed that this behavior could be accurately mapped to the nervous system and revealed that the dorsal posterior region of the brain had to be male for the early steps of male courtship, whereas performance of the later steps required, in addition, the presence of male tissue in the ventral nervous system. One prominent structure in the dorsal posterior brain of insects is the "mushroom bodies," a pair of densely packed neuronal nuclei and their axons that are believed to be one of the fly's association areas involved in olfactory learning. The results of these mosaic studies pointed to the sex of this structure as being important for the generation of sex-specific behavior in males.

A second way of making gynandromorphs is to express Tra, the female determining product of the *tra* gene, in patches of tissue within a male. Such a fly would be male except in those tissues that express Tra, which would be female despite their XY genotype. In *Drosophila* it is possible to produce such partially "feminized" males and target the expression of Tra to restricted patches of tissue. As expected, most of these animals behave as normal males, courting females and not males. In striking contrast, some feminized males court both females and males. Interestingly, such males expressed Tra in the mushroom bodies, the antennal lobe (which receives efferents from the olfactory organs), or both. While the exact explanation of this bisexual courtship is currently under investigation, trivial explanations have been dismissed, such as it being caused by autostimulation of the male by the female pheromones produced by its own, partially feminized, body. It should be noted that male–male courtship is very rare and usually results from mutations in genes involved in the determination of the fly's behavioral sex (Fig. 1). For instance, males bearing the *fruitless* (*fru*) mutation are male in soma and germline, yet they court both females and males vigorously. Interestingly, the *fru* gene contains a Tra-binding sequence, placing it squarely in the sex-determining pathway. *fru* appears to be involved in sex determination of the central nervous system and, consequently, of sex-specific behaviors.

B. Role of Learning in Courtship

Under laboratory conditions male courtship follows a relatively stereotyped and invariant sequence. However, substantial experience-dependent modification of the male's behavior has been exposed in two contexts. In the first ("mated female conditioning"), a male is paired with a mated female, which actively and repeatedly rejects his courtship attempts. When this male is then paired with a virgin female he displays significantly less courtship than he would if he had not previously been exposed to the mated female. The second context in which behavioral plasticity is observed involves sexually immature males ("immature male conditioning"). Mature males vigorously court sexually immature males. However, this response decreases, and such a "conditioned" male will not court an immature male for several hours. These two experience-dependent decrements in courtship do not appear to show any "cross talk." Thus, mature males "conditioned" with a mated female show robust courtship toward an immature male and vice versa. Furthermore, the decrease in courtship due to prior exposure to the mated female requires both the presence of the female (which can be replaced by an inert courtship object, such as the carcass of a solvent-extracted male) and her smell, suggesting that it represents a form of associative learning. Indeed, the decrease in courtship occurs only when the courtship object is paired with the smell of a mated female and not when either of these cues are presented alone. In contrast, the immature male conditioning can be elicited by the young male's smell alone. Thus, this response may be due to sensory adaptation or to a simpler form of learning called habituation. That these two experience-dependent modifications of behavior represent some form of learning was also demonstrated using mutants that show abnormal learning and memory by unrelated criteria. Thus, for instance, a male from such mutant strains showed vigorous courtship toward a virgin female regardless of whether he had previously been paired with an unreceptive female. This "courtship conditioning" assay is now being exploited to genetically dissect learning and memory in *Drosophila*.

C. Role of Olfaction in Courtship

Courtship involves visual, auditory, olfactory, and gustatory cues, none of which is essential for mating to occur under laboratory conditions. Thus, for instance, flies kept in the dark or made genetically blind can mate; however, they may take longer to do so than when they can see. It is only when all cues are removed that flies are virtually incapable of mating. Under these conditions the sterility is behavioral due to the failure of courtship.

Flies mutant for a particular sensory modality have

been used to investigate the specific role that this modality may play during courtship. Mutants with defective olfactory responses have been especially useful since manipulating olfactory sensitivity and cues is hard to achieve experimentally (unlike, for instance, the role of vision, which can be examined by placing the flies under dim red light). Experiments using mutant flies that cannot smell revealed that this sense is used as a cue for both partners. Thus, mutant males that are forced to rely on olfaction show defective courtship when paired with wild-type females. Likewise, a wild-type male will take longer to mate if the female cannot smell him. These types of experiments have been extended by studying the courtship elicited in wild-type and mutant males exposed to "inert" courtship objects in the presence of female or young male cuticular pheromones and they have led to an understanding of the relative role that smell, taste, and contact play during courtship.

The pheromones involved in courtship consist of 15–20 cuticular hydrocarbons with chain length ranging from 21 to 31 carbons. These compounds have low volatility, consistent with the fact that olfaction plays a role only over short distances. Some of these pheromones are mostly found in either males or females and could thus be used by a fly to determine if its partner is of the opposite sex. There are also pheromonal differences between virgin and mated females, as well as between immature and older males, which contribute to the different courtship behaviors elicited by these various partners. Finally, there are also variations between natural wild-type strains as well as between different *Drosophila* species, suggesting that these pheromones could contribute to a group's reproductive isolation. Mutants with abnormal pheromonal profiles are currently being used to dissect the role played by specific compounds or pheromonal bouquets; similarly, the partially "feminized" males discussed previously are being used to determine where these pheromones are produced.

D. Female Receptivity

Under laboratory conditions, the *D. melanogaster* female apparently plays a relatively minor role during courtship. Her only active behaviors seems to be her initial running away from and/or rejecting the male followed by her gradual slowing down upon accepting him, which allows for copulation to occur. This slowing down is at least partly mediated by olfaction of the male by the female because anosmic mutant females do not slow down, resulting in an increased latency to mating. Normal females paired with males that have low levels of the male cuticular pheromone cis7-tricosene do not readily slow down during courtship, suggesting that this male pheromone is used by the female to sense the presence of her mate.

Another cue that contributes to the female's receptivity is the male's love song. Males with clipped wings take longer to mate because the female does not slow down, a situation that can be partly remedied by broadcasting this song to the courting pair. The latency to mate can also be reduced by "priming" the virgin females with this song.

Recently mated females vigorously reject males. They also show enhanced oviposition compared to that of the virgin. Interestingly, this change in behavior elicited by mating is caused by the male and is due to a 36-amino acid peptide present in his seminal fluid. This "sex peptide" (SP) is transferred during copulation and effects these behavioral and physiological changes. The cloning of the gene encoding SP has allowed for the initiation of detailed analyses of this peptide's mode of action. Of particular interest is the finding that the induction of widespread expression of SP in the female produces the same response as mating: male rejection and enhanced oviposition.

V. SUMMARY

Genetic mutants and variants of the little fruit fly *D. melanogaster* provide a unique opportunity to investigate genetically two areas important to reproductive biology: sex determination and courtship. This approach has resulted in the identification of elements and genes involved in these aspects of this fly's reproduction as well as an understanding of their mechanisms and sites of action.

Sex determination is controlled by a largely hierarchical gene pathway in which the control of genes "downstream" occurs via the sex-specific control of splicing alternatives. While the sex of the soma, germline, and sex-specific neural functions share some components, others are unique to each particular pathway. A detailed analysis of courtship has revealed that this is a relatively complex and plastic behavior that requires the coordination between the male and female partners. The genetic analysis of this behavior and the use of mutants with particular defects have allowed us to localize regions of the nervous system involved in courtship as well as identify the cues used and provided by each partner in order for courtship to result in mating.

The hormonal control of reproduction in *Drosophila* is complex and only partly understood. However, it appears to be similar to what obtains in the mosquito, but this is still under investigation. However, the mosquito model is useful at the very least because it provides a concrete framework and suggests experiments that should be performed.

See Also the Following Articles

FEMALE REPRODUCTIVE SYSTEM, INSECTS; JUVENILE HORMONE; MALE REPRODUCTIVE SYSTEM, INSECTS; MATING BEHAVIORS, INSECTS

Bibliography

Ashburner, M. (1989). *Drosophila: A Laboratory Manual*. Cold Spring Harbor Laboratory Press, Cold Spring Harbor, NY.

Bate, M., and Martinez Arias, A. (1995). *The Development of Drosophila melanogaster*. Cold Spring Harbor Laboratory Press, Cold Spring Harbor, NY.

Blackman, R. L. (1995). Sex determination in insects. In *Insect Reproduction* (S. R. Leather and J. Hardie, Eds.), pp. 57–94. CRC Press, Boca Raton, FL.

Burtis, K. C. (1993). The regulation of sex determination and sexually dimorphic differentiation in *Drosophila*. *Curr. Opin. Cell Biol.* **5**, 1006–1014.

Ferveur, J.-F. (1997). The pheromonal role of cuticular hydrocarbons in *Drosophila melanogaster*. *Bioassays* **19**, 353–358.

Gillot, C. (1995). Insect male mating systems. In *Insect Reproduction* (S. R. Leather and J. Hardie, Eds.), pp. 33–55. CRC Press, Boca Raton, FL.

Greenspan, R. J. (1995). Understanding the genetic construction of behavior. *Sci. Am.* **272**, 74–79.

Hall, J. C. (1994). The mating of a fly. *Science* **264**, 1702–1714.

Hardie, J. (1995). Hormones and reproduction. In *Insect Reproduction* (S. R. Leather and J. Hardie, Eds.), pp. 95–108. CRC Press, Boca Raton, FL

Hoffmann, K. H. (1995). Oogenesis and the female reproductive system. In *Insect Reproduction* (S. R. Leather and J. Hardie, Eds.), pp. 1–32. CRC Press, Boca Raton, FL

Jallon, J.-M. (1984). A few chemical words exchanged by *Drosophila* during courtship and mating. *Behav. Genet.* **14**, 441–478.

Kelly, T. J. (1994). Endocrinology and vitellogenesis in *Drosophila melanogaster*. In *Perspectives in Comparative Endocrinology* (K. Davey, R. E. Peter, and S. S. Tobe, Eds.), pp. 282–290. National Research Council of Canada, Ottawa.

Kubli, E. (1992). The sex peptide. *BioEssays* **14**, 779–784.

Rubin, G. M. (1988) *Drosophila melanogaster* as an experimental organism. *Science* **240**, 1453–1459.

Wyatt, G. R., and Davey, K. G. (1996). Cellular and molecular actions of juvenile hormone. II. Roles of juvenile hormone in adult insects. *Adv. Insect Physiol.* **26**, 1–155.

Drug Abuse

see Substance Abuse and Pregnancy

Dysmenorrhea

see Menstrual Disorders

Ecdysiotropins and Ecdysiostatins

Henry H. Hagedorn
University of Arizona

I. Introduction
II. Hormones That Stimulate the Prothoracic Gland
III. Hormones That Stimulate the Ovary
V. Hormones That Inhibit Ecdysteroid Production

GLOSSARY

ecdysiostatins Hormones that inhibit the production of ecdysteroids by endocrine glands.

ecdysiotropins Hormones that stimulate the production of ecdysteroids by endocrine glands.

eye stalk A structure in crustacea containing a complex group of endocrine glands and neurosecretory cells.

neurohemal organ The site of release of neurosecretory hormones.

prothoracic gland An endocrine gland in the prothorax of insects that produces ecdysteroids.

ring gland A compound gland consisting of the prothoracic gland, corpora allata, and corpora cardiaca found in higher Diptera, such as *Drosophila*.

Y-organs An endocrine gland in the thorax of crustacea that produces ecdysteroids.

I. INTRODUCTION

Hormones that regulate the production of ecdysteroids are known as ecdysiotropins. Several of these hormones have been isolated and sequenced from insects and crustaceans. In addition to ecdysiotropins there is evidence for inhibitory factors in both the insects and crustacea. These have been termed ecdysiostatins and molt-inhibiting hormones, respectively. The ecdysones were first identified because of their effects on the ecdysis (i.e., molting) of insects. In fact, they have widespread effects on many aspects of the physiology of animals, including reproduction, so the term "ecdysiotropins" with its emphasis on ecdysis is somewhat of a misnomer.

The endocrine glands stimulated by ecdysiotropins include the Y-organs of Crustacea and the prothoracic glands, ovaries, and testes of Insecta. There is experimental evidence for the existence of ecdysiotropins in many more animals. Much of this evidence is the result of the removal of the brain or other parts of the nervous system, simple decapitations, and perhaps replacement therapy. The emphasis here will be on those factors that have been isolated.

The ecdysiotropins are neurosecretory peptides that are produced by specialized nerve cells that synthesize peptides and secrete them into the blood. The release site is called the neurohemal organ. Some peptides produced by nerve cells are released in the nervous system where they can act as neuromodulators or neurotransmitters. In some cases a single neurosecretory peptide has been shown to function as a neurohormone, neuromodulator, and neurotransmitter. Thus, the roles of neurosecretory peptides can be very diverse. There is limited evidence for such diversity of function in the ecdysiotropins, but it may occur.

II. HORMONES THAT STIMULATE THE PROTHORACIC GLAND

A. Prothoracicotropic Hormone

The prothoracicotropic hormone was isolated from the lepidopteran, *Bombyx mori*, the commercial silkworm. This hormone stimulates the development of brainless *B. mori* pupae and also stimulates the production of ecdysone by incubated prothoracic glands. The hormone causes an intracellular increase in the second messenger, cAMP, which initiates a cascade of events that lead to ecdysone synthesis. The gene for the hormone codes for a 224-amino acid preprohormone that is processed to yield a 109-amino acid subunit. The mature protein is a homodimer. The sequence of the 109-amino acid subunit is as follows:

GNIQVENQAIPDPPCTCKYKKEIEDLGENSVPRFI
ETRNCNKTQQPTCRPPYICKESLYSITILKRRETKS
QESLEIPNELKYRWVAESHPVSVACLCTRDYQLRY
NNN

No sequence similarities to other known peptides were found. However, the three-dimensional structure of the prothoracicotropic hormone is determined by the position of the cysteines which were found to be very similar to those in a family of growth factors of vertebrates, the β-nerve growth factor, the platelet-derived growth factor, and the transforming growth factor-β2. All of these proteins have a cysteine-knot motif. Figure 1 shows two of the members of this family of proteins. Not shown are hydrophobic residues which are also found in conserved positions in the members of the family. The three-dimensional structures of the growth factors have been determined, but the structure of the prothoracicotropic hormone is partly hypothetical.

Prothoracicotropic hormones have been isolated, but not sequenced, from several other insects including the lepidopterans *Manduca sexta* and *Lymantria dispar* and the dipteran, *Drosophila melanogaster*. These hormones appear to be somewhat species specific since the prothoracic hormone from *B. mori* will activate its own prothoracic glands but not those of *L. dispar*. An apparently homologous pair of neurosecretory cells in the brains of *B. mori* and *M. sexta* produce the hormone. Molecules with prothoracicotropic hormone activity have also been identified in the insect gut and the embryo.

The main function of the prothoracicotropic hormone is to regulate molting and metamorphosis of insects. However, the development of the gonads is part of metamorphosis. In immature insects the gonads remain undeveloped until metamorphosis occurs. 20-Hydroxyecdysone titers are very high throughout the development of the adult. The testes in many insects are fully developed in the newly eclosed adult. In some insects—for example, the lepidopterans, *B. mori* and *M. sexta*—oocytes are nearly fully formed when the adult ecloses from the pupa. In other insects, the ovaries do not complete development until after adult eclosion. In these insects steroidogenic hormones that act on the ovaries are found (see below).

B. Bombyxin

Bombyxin is a small, 5-kDa peptide related to insulin, and the insulin-like growth factors, that was initially isolated from the lepidopteran, *B. mori*, on the basis of its ability to stimulate the prothoracic gland of another lepidopteran, *Samia cynthia*. Paradoxically, bombyxin is inactive in *B. mori*. Its physiological role is unknown, although it may function in regulating egg development because hemolymph titers are high in the adult female *B. mori*. It is produced by four neurosecretory cells in the brain with a neurohemal site near the corpora allata. Genes coding for bombyxin have been isolated. *Bombyx. mori* may have as many as 29 bombyxin genes.

Bombyxin is unusual in having a high degree of similarity to insulin and the insulin-like growth factors as shown below. Identical amino acids are shown in capital letters. The conserved position of the cysteines results in a three-dimensional structure that is very similar to that of insulin. Shown here is a comparison between three regions (chains A, B, and C)

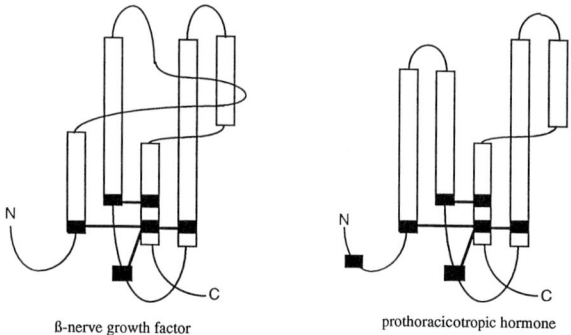

FIGURE 1 β-Nerve growth factor (left) and prothoracicotropic hormone (right). The black boxes indicate the position of the cysteines, with a thick line indicating disulfide bonds between them.

of insulin-like peptides, bombyxin (Bom), human insulin (HI), and insulin growth-related factor-1 (GRF). Capitol letters indicate identical amino acids. Gaps were inserted to improve the alignment.

B-chain
```
Bom  eqpqavhTyCGrHLartLadlCwEaGvd
HI   aa..fvnqhlCGsHLveaLylvCgErGffytpkt
GRF  ta..gpeTlCGaelvdaLqfvCgdrGfyfnkpt
```

C-peptide
```
Bom  kR......sGaqfasyGs.aw.LmP.ysEG.rgKR
HI   rReaedlqvGqvelggGpgagsLqPlalEGslqKR
         G.....ygSssrra. Pqt
```

A-chain
```
Bom  GIVDECClRpCSvdvLlsYC
HI   GIVeqCCtsiCSlyqLenYCn
GRF  GIVDECCfRsCdlrrLemYCa
```

In *M. sexta*, a bombyxin-like peptide is produced by eight cells in the brain that send axons to the neurohemal site on the corpora allata. Immunohistochemical staining patterns indicated that the peptide is released in the last larval instar just prior to the molt to the pupa. Numerous neurosecretory cells were identified in the brain of the locust, *Locusta migratoria*, using an antibody against *B. mori* bombyxin. The corresponding peptide, and its gene, have been isolated; however, the peptide has not been found to stimulate ecdysone production. There is some evidence for a bombyxin-like peptide in mosquitoes.

Thus, both these insect hormones, bombyxin and prothoracicotropic hormone, are related to known families of growth factors in other animals. The occurrence of two size classes of steroidogenic neurohormones may be a general phenomenon in insects. Two ecdysiotropins, 7 and 30 kDa, have been identified in *M. sexta*. The dipteran, *D. melanogaster*, also has two forms of steroidogenic molecules, 3–5 and 14–17 kDa.

III. HORMONES THAT STIMULATE THE OVARY

A. The Neuroparsins

Neuroparsins are a family of peptides initially identified in the corpora cardiaca, a major neurohemal site, of the locust *L. migratoria*. Three of these peptides have been isolated and sequenced. Neuroparsins A and B are closely related peptides. Both are produced by the same neurosecretory cells of the brain (type A cells) with a neurohemal site in the corpora cardiaca. A third, unrelated parsin, the ovary maturing parsin of *L. migratoria* (Lom OMP), was also sequenced. This peptide is produced in different neurosecretory cells in the brain (type B cells) with a neurohemal site in the corpora cardiaca. Below is a comparison between neuroparsin A, B (Neu A, Neu B) of the locust, and mosquito ecdysteriodogenic hormone (OEH). Capitol letters indicate identical amino acids. Gaps were inserted to improve alignment. The sequence of the ovary maturing parsin of the locust (Locusta OMP) is shown separately.

```
Neu A    NpisrSCE..GANCVVDLTRCEY
Neu B         SCE..GANCVVDLTRCEY
OEH      ptNvleirCklysgpaVqntgeCvh

GDVTdFFGRKVCAKGPGDKCG...GPYELHGK
GDVTaFFGRKVCAKGPGDKCG...GPYELHGK
GaelnpcGklsClkGvGDKCGestagiimsGK

CGVGMDCRCG.LCSGCSLH............
CGVGMDCRCG.LCSGCSLHgkcgvgmdCrcgL
CasGlmC.CGgqCvGCkng.......iCdhrL

.......NLQCFFFEGGLPSSC
CsgcslhNLQCFFFEGGLPSSC
Cpprl

Locusta OMP
YYEAPPDGRHLLLQPAPAAPAVAPA(S/A)PA
SWPHQQRRQALDEFAAAAAAAADAQFQDEEED
GGRRV
```

Several functions have been attributed to neuroparsins A and B, but they are not well understood. However, Lom OMP has been shown to have effects on the vitellogenic growth of the oocytes. It has been suggested that the effect of Lom OMP is to induce ecdysone production by the ovary, but direct evidence is lacking. The role of ecdysone in egg development in the locust is not clear because juvenile hormone has also been shown to be necessary for normal egg development in this insect.

B. The Ovarian Ecdysteroidogenic Hormone of the Mosquito

The first gonadotropin to be identified in insects stimulates ecdysone production by the ovary of the mosquito, *Aedes aegypti*. This hormone, the ovarian ecdysteroidogenic hormone (OEH, also known as EDNH) has been isolated from mosquito heads and sequenced. It was found to be about 30% similar to neuroparsins A and B as shown previously. The conservation of 11 of the 13 cysteines may be significant. The significance of the similarity to the neuroparsins is unknown because neuroparsins A and B have no known effect on ecdysone production. OEH is unrelated to Lom OMP, which is believed to function in locust egg development and may affect ecdysone production by the ovary.

OEH stimulates ecdysone production after the mosquito takes a blood meal. The ecdysone is converted to 20-hydroxyecdysone, which has important functions in egg development, particularly in stimulating the expression of genes coding for yolk proteins, especially vitellogenin.

C. The Testis Ecdysiotropin

A peptide has been isolated from the brains of pupae of the gypsy moth, *Lymantria dispar*, that stimulates the sheath of the testis to produce ecdysteroids. The sequence of this 21-amino acid peptide is shown below. No similarities to other peptides has been found.

ISDPDEYEPLNDADNNEVLDP

This peptide does not stimulate the prothoracic gland, and the prothoracicotropic hormone does not stimulate the testis sheath. The testis ecdysiotropin is most effective in the presence of low hemolymph titers of 20-hydroxyecdysone produced by the prothoracic glands.

V. HORMONES THAT INHIBIT ECDYSTEROID PRODUCTION

A. Ecdysiostatins

A hexapeptide (NPTNLH) isolated from a dipteran, the gray flesh fly, *Neobellieria (Sarcophaga) bullata*, inhibits ecdysone production by the larval prothoracic glands (the ring glands). It was initially isolated from the ovary. Its cellular source and function remain uncertain.

A larger peptide (11 kDa) with ecdysiostatic activity has been detected in the blue blowfly, *Calliphora vicina*. Two ecdysiotropins were also found that correlated in size to those seen in other insects. The rate of production of ecdysteroids may result from the integration of the effects of these positive and negative effectors.

B. Molt-Inhibiting Hormone

In contrast to the insects, ecdysteroid production in crustacea is under negative control. A molt-inhibiting hormone inhibits the production of ecdysteroids by the Y-organ, thus preventing a molt and, presumably, reproduction. Changing titers of ecdysteroids and methyl farnesoate, a terpenoid related to the juvenile hormones of insects, have been demonstrated in several crustaceans. Both of these hormones have been implicated in regulating reproduction, but the precise roles of each are not clear.

Several of the molt-inhibiting hormones have been isolated and sequenced. The molt-inhibiting hormones are neurosecretory peptides with 78 amino acids that are produced by neurosecretory cells in the X-organ of the eyestalk. The neurohemal site is in the sinus gland, also in the eyestalk. The molt-inhibiting hormones of several crustacea have distinct sequence similarities. These hormones are also related to peptides, also produced in the eyestalk, that inhibit the production of methyl farnesoate by the mandibular organ. An example of these sequences is given below in which identical amino acids are capitalized.

```
1-  RVINDeCPNLIGNRDLYKKVEWICeDCsN
2-  RVINDDCPNLIGNRDLYKKVEWICdDCaN
3-  RVINDDCPNLIGNRDLYKKVEWICeDCsN
4-  RrINnDCqNfIGNRamYeKVdWICkDCaN

IFRkTGMAsLCRrNCFFNEDFLWCVHATERSE
IyRsTGMAsLCRkdCFFNEDFLWCVrATERSE
IFRnTGMAtLCRkNCFFNEDFLWCVyATERTE
IFRkdGllnnCRsNCFyNtEFLWCidATEnTr

elrdLeeWVgILGAGRd
dlaQLkQWVtILGAGRi
```

emsQLrQWVgILGAGRe
nkeQLeQWaaILGAGwn

where Nos. 1–3 are molt-inhibiting hormones of *Carcinus maenas,* *Callinectes sapidus,* and *Cancer pagurus,* respectively, and No. 4 is the mandibular organ-inhibiting hormone of *C. pagurus.*

It is evident from these sequences that a family of related neurosecretory peptides are produced by cells in the eyestalk that regulate the production of hormones that appear during molting and reproduction. However, experimental evidence linking the neurosecretory hormones to reproduction is not available.

Bibliography

Brown, M. R., Graf, R., and Swiderek, K. M., Fendley, D., Stracker, T. H., Champagne, D. E., and Lea, A. O. (1998). Identification of a steroidogenic neurohormone in female mosquitoes. *J. Biol. Chem.* **273,** 3967–3971.

Chang, E. S. (1993). Comparative endocrinology of molting and reproduction: insects and crustaceans. *Annu. Rev. Entomol.* **38,** 161–180.

Chung, J. S., Wilkinson, M. C., and Webster, S. G. (1996). Determination of the amino acid sequence of the moult-inhibiting hormone from the edible crab, *Cancer pagurus. Neuropeptides* **30,** 95–101.

Hagedorn, H. H. (1994). The endocrinology of the adult female mosquito. In *Advances in Disease Vector Research* (K. F. Harris, Ed.), pp. 109–148. Springer-Verlag, Berlin/New York.

Hua, Y.-J., and Koolman, J. (1995). An ecdysiostatin from flies. *Regul. Peptides* **57,** 263–271.

Hua, Y.-J., Jiang, R.-J., and Koolman, J. (1997). Multiple control of ecdysone biosynthesis in blowfly larvae: Interaction of ecdysiotropins and ecdysiostatins. *Arch. Insect Biochem. Physiol.* **35,** 125–134.

Kelly, T. J., Thyagaraja, B. S., Bell, R. A., and Masler, E. P. (1995). A novel low molecular weight ecdysiotropin in post-diapause, pre hatch eggs of the gypsy moth, *Lymantria dispar. Regul. Peptides* **57,** 253–261.

Koolman, J. (1989). *Ecdysone, from Chemistry to Mode of Action.* Thieme-Verlag, Stuttgart.

Noguti, T., Adachi-Yamada, T., Katagiri, T., Kawakami, A., Iwami, M., Ishibashi, J., Kataoka, H., Suzuki, A., Go, M., and Ishizaki, H. (1995). Insect prothoracicotropic hormone: A new member of the vertebrate growth factor superfamily. *FEBS Lett.* **376,** 251–256.

Wainwright, G., Webster, S. G., Wilkinson, M. C., Chung, J. S., and Rees, H. H. (1996). Structure and significance of mandibular organ-inhibiting hormone in the crab, *Cancer pagurus. J. Biol. Chem.* **271,** 12749–12754.

Wagner, R. M., Loeb, M. J., Kochansky, D. B., Gelman, D. B., Lusby, W. R., and Bell, R. A. (1997). Identification and characterization of an ecdysiotropic peptide from brain extracts of the Gypsy moth, *Lymantria dispar. Arch. Insect Biochem. Physiol.* **34,** 175–189.

Ecdysteroids

Henry H. Hagedorn
University of Arizona

I. Introduction
II. Chemistry
III. Functional Roles

GLOSSARY

corpora cardiaca A pair of glands that are the source of some peptide hormones in insects.

ecdysone The ecdysteroid produced by an endocrine gland in many animals as a precursor to the active form, 20-hydroxyecdysone.

ecdysteroids A generic term for 27-carbon steroids with an extended side chain and numerous hydroxyl groups.

20-hydroxyecdysone The active form of ecdysteroid in most animals.

juvenile hormone A terpenoid hormone produced by the corpora allata of insects.

prothoracic glands One source of ecdysteroids in insects.

steroid A 17-carbon lipid with four rings.

I. INTRODUCTION

Ecdysteroids are steroids found in a variety of plants and invertebrates. In plants the roles of these steroids are debated. In invertebrates they function as hormones regulating a wide range of biological functions. The best known function of ecdysteroids in insects and Crustacea is the regulation of molting. In some animals ecdysteroids are also involved in the regulation of reproduction, which is the subject here.

II. CHEMISTRY

The ecdysteroids belong to the class of unsaponifiable lipids called steroids. Steroids are isoprene derivatives, as are the terpenes and carotenoids. The most common steroid is cholesterol, which is a major constituent of biological membranes. Other important steroids are the bile acids and numerous hormones, such as the ecdysteroids of invertebrates and corticosteroids and sex hormones of vertebrates.

The ecdysteroids are derived from cholesterol, which arthropods cannot synthesize. Some other invertebrates can synthesize cholesterol. Arthropods must therefore obtain the requisite substrate from their diet. Carnivorous insects can obtain cholesterol from their diet, but phytophagous insects must convert plant steroids that are alkylated at C24 to animal steroids that are alkylated at C27. However, some plant-feeding insects that obtain sitosterol from their diet, including some Hemiptera and Hymenoptera, cannot dealkylate it. They therefore use makisterone A instead of ecdysone as their active hormone. The pupa of an ant, *Acromyrmex octospinosus*, has been shown to contain 24-epi-makisterone A rather than makisterone A (Fig. 1).

The synthetic pathway of ecdysone from cholesterol in insects occurs by several alternative pathways that involve the sequential addition of hydroxyl groups to the molecule. The ecdysteroids have a large number of hydroxyl groups compared to the steroid hormones of vertebrates. This results in the molecule being water soluble, which is a unique feature of the ecdysteroids (Fig. 2).

The ecdysone molecule differs from estradiol in having a 27- rather than a 18-carbon skeleton, an unsaturated keto function in the B ring, and a bent shape caused by the hydrogen at C5 being in a β position. Makisterone A and epi-makisterone are the only ecdysteroids used as hormones in insects that have 28 carbons (Fig. 3).

Ecdysone is a precursor to the active hormone, 20-hydroxyecdysone, which has an hydroxy group in the 20 position (Fig. 4).

Relatively few of the 130,000 possible ecdysteroid structures have been discovered in nature. In animals 61 ecdysteroids have been identified, whereas 118 phytoecdysteroids have been found in plants, some of which are identical to those found in animals, including ecdysone, 20-hydroxyecdysone, and ponasterone A.

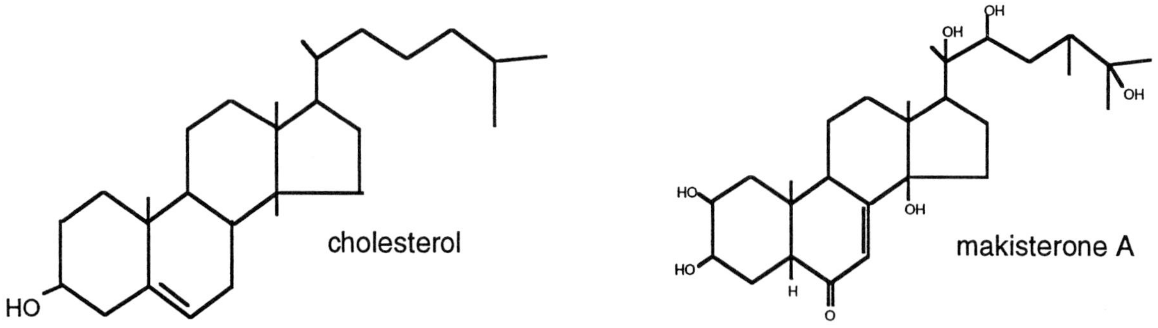

FIGURE 1

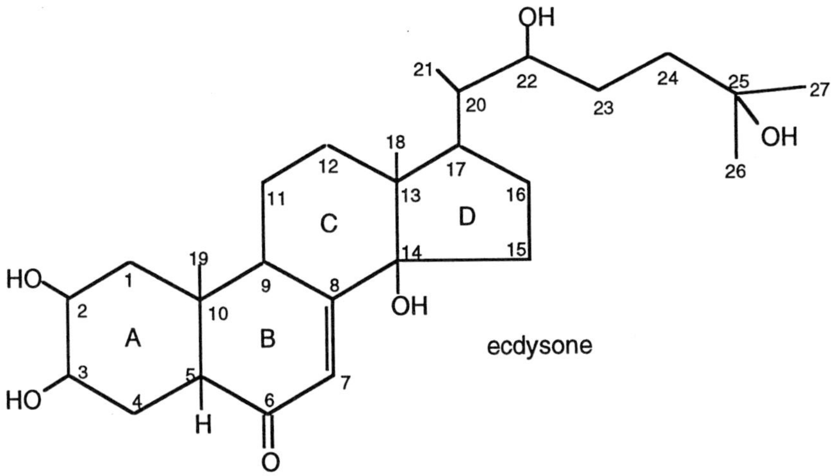

FIGURE 2

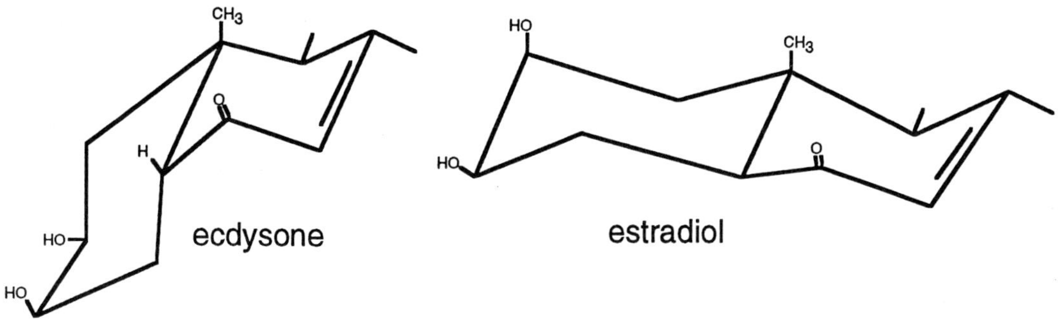

FIGURE 3

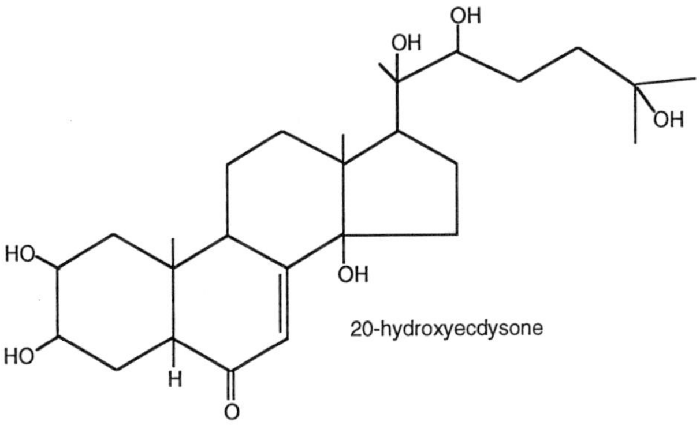

FIGURE 4

III. FUNCTIONAL ROLES

A. Sources of Ecdysteroids

The ecdysteroids of insects are made by several tissues depending on the life stage. The prothoracic glands, or ventral glands, produce ecdysteroids that regulate molting in the larvae. The ventral glands are distinct organs in primitive insects such as the silverfish, cockroach, and locust. The prothoracic glands are diffuse tissues that lie in the anterior end of the thorax which degenerate during metamorphosis to the adult. They are found in the Hemiptera and all the holometabolous insects. In most insects the prothoracic glands synthesize ecdysone, but in Lepidoptera they synthesize both ecdysone and 3-dehydroecdysone, both of which are converted to 20-hydroxyecdysone. The follicle cells of the ovary synthesize ecdysone in most adult female insects. There is evidence that the testis sheath produces ecdysteroids in the male. Other tissues may also make ecdysteroids in sufficient quantities to regulate development. For example, in the beetle, *Tenebrio molitor*, the epidermis is a major source of ecdysteroids, as it is in a tick, *Ornithodoros parkeri*.

In the crustacean, *Carcinus maenas*, the Y-organs secrete ecdysone and 25-deoxyecdysone. Ecdysone is converted to 20-hydroxyecdysone by various tissues. The testis converts 25-deoxyecdysone into ponasterone A. The major hemolymph forms are 20-hydroxyecdysone and ponasterone A. The ovaries contain large amounts of ecdysteroids, most of which is ponasterone A (650 ng/g in the lobster, *Homarus gammarus*). Ponasterone A has high biological activity. The source of ecdysteroids in other arthropods is not known (Fig. 5).

The hormones produced by the insect ovary and testis have reproductive functions. In the female insect the ecdysteroids have two distinct fates. Some enter the hemolymph and have roles in regulating gene expression, but in the majority of insects the ecdysteroids produced by the follicle cells of the ovary enter the oocyte, where they may effect embryonic development.

B. Effects on Oocyte Development of Insects

Ecdysteroids may stimulate early, previtellogenic, stages of egg development. The firebrat, *Thermobia domestica*, is a primitive insect in which molting continues in the adult and, in the female, molting is interspersed between regular cycles of egg development. The peak of 20-hydroxyecdysone that stimulates a molt also causes the previtellogenic growth of the ovarian follicles. After the molt, vitellogenic growth of the oocyte, which includes the deposition of protein yolk, is regulated by juvenile hormone. Cycles of molting and reproduction also occur in adult Crustacea.

In contrast, in the Diptera, ecdysteroids stimulate the vitellogenic stage of follicle growth. In mosquitoes, 20-hydroxyecdysone titers rise after a blood meal and stimulate the synthesis of the yolk protein, vitellogenin, by the fat body. Ecdysteroids also regulate vitellogenin synthesis in many Cyclorrhaphid Diptera, such as *Drosophila* and related flies.

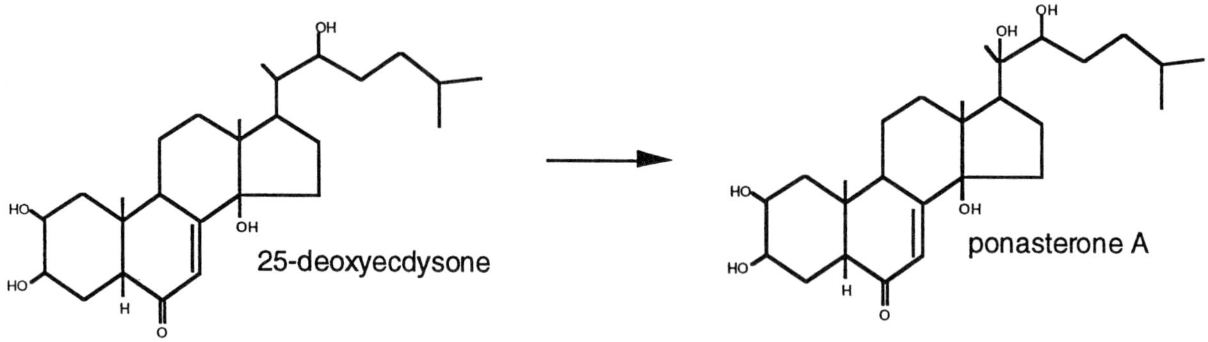

FIGURE 5

One of the roles of 20-hydroxyecdysone in the Diptera is to regulate the expression and translation of genes that code for components of the yolk. The most predominant of these proteins is vitellogenin, which can make up over 75% of the total protein in the yolk. 20-Hydroxyecdysone also stimulates the synthesis of several other proteins, such as the vitellogenin-degrading carboxypeptidase, that become part of the yolk. These proteins are synthesized by the fat body, travel through the hemolymph, and are taken into the oocyte by receptor-mediated endocytosis. This process has been intensively studied in the mosquito, *Aedes aegypti*, and the fruitfly, *Drosophila melanogaster*. Some of the genes regulated by 20-hydroxyecdysone in these insects have been isolated and sequenced. Transcription factors that mediate the effects of 20-hydroxyecdysone have been isolated and are discussed in more detail.

In some Diptera, such as the mosquito and the blowfly, the production of ecdysteroids and vitellogenin is intimately tied to the taking of a protein-rich meal which provides the substrates necessary for the large-scale production of yolk. In both of these animals the production of eggs is controlled by both juvenile hormone and 20-hydroxyecdysone. In the mosquito, juvenile hormone appears prior to the blood meal and then falls rapidly after the meal while 20-hydroxyecdysone titers rise. 20-Hydroxyecdysone titers are high for only 1 day after the meal, and as they decline, juvenile hormone titers rise again. The role of juvenile hormone in egg development of the mosquito is complex and appears to involve preparing the animal for events after a blood meal, including sexual behavior and development of the ovary and fat body that prepare them to respond to 20-hydroxyecdysone. In the blowfly and *Drosophila*, juvenile hormone stimulates the uptake of vitellogenin by the oocytes. In some Diptera, including *Drosophila*, both 20-hydroxyecdysone and juvenile hormone stimulate vitellogenin synthesis.

In some Lepidoptera, such as the silkmoth, *Bombyx mori*, vitellogenic growth of the oocyte occurs during pharate adult development in the latter part of the pupal stage. The ecdysteroid titers that are high during this stage control both metamorphosis and oocyte development.

C. Effects of Ecdysteroids on Meiotic Reinitiation

In the locust, *Locusta migratoria*, the chromosomes of the oocyte during oogenesis are arrested in prophase I of meiosis. After the chorion is laid down, and shortly before oviposition, meiosis is reinitiated and a metaphase plate develops. The oocyte then enters a metaphase arrest which is broken on oviposition. Increases in ecdysone titers occur during both meiotic reinitiations. The effect of ecdysteroids on meiotic reinitiation has been confirmed in the cockroach, *Periplaneta americana*, a cricket, *Gryllus bimaculatus* and the decapod crustacean, *Palaemon serratus*.

D. Effects on Embryonic Development of Insects

The fact that the ovary produced ecdysteroids in insects was discovered in the 1970s. The initial evidence for the function of this hormone in Diptera centered around the development of the oocyte as described previously. It soon became clear that ovaries of many other insects made rather large amounts of ecdysteroids, but most of the ecdysteroids were destined to be stored in the oocyte itself. The amounts found in the ovary ranges from over 70 ng/mg in the locust, *Shistocerca gregaria*, and the moth, *Galleria mellonella*, to 0.007 ng/mg in the fly, *Calliphora vicina*.

In contrast to the case in the mosquito and other Diptera, the hormone in the locust and cockroach appears late in oocyte development. For example, in the locust it appears between the time of chorion formation and oviposition, long after the vitellogenic phase of oocyte development is complete. Most of the ecdysteroids stored in the oocyte are present as inactive conjugates, including acetyl phosphates, ecdysteroid acids, and ecdysteroids phosphorylated at C3. In several insects, including the cockroaches, *Blaberus craniifer* and *Nauphoeta cinerea*, and the locust, *L. migratoria*, pulses of free ecdysteroids are found in the embryo. The appearance of this free, active hormone is correlated with the appearance of a phosphatase that can cleave the ecdysteroid conjugates. In these insects, and in some others, the embryo molts during embryogenesis. Most strikingly, in the locust the kind of free ecdysteroid present changes as the embryo molts so that the deposition of the serosal cuticle and the first embryonic molt are correlated with an increase in ecdysone, the second molt is correlated with the appearance of 20,26-dihydroxyecdysone and ecdysone, and the third molt is correlated with the appearance of these two molecules plus 20-hydroxyecdysone. The types of cuticles made during each of these molts are unique. The prothoracic gland of the locust embryo appears before the second molt so the ecdysteroids that appear during the later molts may be embryonic, rather than maternal, in origin.

In the moth, *Manduca sexta*, there is evidence that maternal ecdysteroids stimulate gastrulation and segmentation of the gastrula during embryogenesis.

E. Effects on Other Endocrine Organs

There is some evidence that ecdysteroids produced by the ovary can affect other endocrine organs. In the bug, *Rhodnius prolixus*, a peptide hormone (myotropin) from the corpora cardiaca has been identified that stimulates ovulation. The release of this hormone has been shown to be due to a peak of 20-hydroxyecdysone produced by the ovary prior to ovulation. 20-Hydroxyecdysone directly affects the brain. The hormone stimulates action potentials in the neurosecretory cells that produce myotropin.

The case of the cockroach, *Diploptera punctata*, is especially interesting. This is a viviparous cockroach that nurtures its embryos and has a clear pregnancy cycle. Juvenile hormone titers are high during early stages of oocyte development when yolk is being deposited. As oocyte development is completed the juvenile hormone titers rise as large amounts of ecdysteroids are being produced for deposition in the oocyte as discussed previously. It has been suggested that some of the ecdysteroid enters the hemolymph and influences the decline of juvenile hormone, perhaps via an effect on the brain.

F. Effects of Ecdysteroids on Spermatogenesis of Insects

Spermatogenesis in many insects occurs in the late larval, or nymphal, life or during the pupal stage. There is evidence that the testis sheath of the testes produces ecdysteroids in the lepidopterans, *Heliothis virescens, Lymantria dispar,* and *Spodoptera littoralis*, and the orthopteran, *Melanopus sanguinipes*. The testis ecdysteroids are believed to stimulate the growth of the genital tract via their effect on growth factors. Ecdysteroids have been implicated in the development of the testis and the regulation of spermatogenesis. For example, in the lepidopteran, *Ephestia kuhniella*, 20-hydroxyecdysone stimulates the migration and fusion of the testes. In the bug, *R. prolixus*, 20-hydroxyecdysone stimulates cell divisions in spermatogonial cells. In the moth, *Hyalophora cecropia*, 20-hydroxyecdysone stimulates spermatogenesis by promoting the penetration of factors from the hemolymph into the spermatocysts. Ecdysteroids stimulate the production of proteins in the accessory glands of the testes of some Lepidoptera.

G. Effects of Ecdysteroids on Reproductive Behavior

There is good evidence that ovarian ecdysteroids control the release of sex pheromones by the female housefly, *Musca domestica*. (Z)-9-tricosene, also known as muscalure, is produced by the female during oocyte development in response to rising titers of 20-hydroxyecdysone. The pheromone is a cuticular lipid that is attractive to males. Sex pheromone production may also be under the control of ecdysteroids in *D. melanogaster*.

H. Roles of Ecdysteroids in Other Animals

Several crustacea have been shown to excrete ecdysteroids during molting. Suggestions that these steroids were sex pheromones, attracting males to molting females, have not been confirmed. Although the existence of sex pheromones in crustacea is well established, their identity is unknown.

Increasing titers of ecdysteroids have been correlated with reproduction in some crustaceans, including the crabs, *Acanthonyx lunulatus* and *Carcinus maenas*. Both ecdysone and 20-hydroxyecdysone are the major ecdysteroids in the isopod *Armadillidium vulgare*. Hemolymph levels are five times higher in the female and fluctuate during the reproductive cycle. 20-Hydroxyecdysone levels are high during oocyte development of the shrimp, *Penaeus vannamei*, and there is some evidence that it regulates expression of ecdysteroid-responsive genes important during vitellogenesis. However, titers of methyl farnesoate, a hormone related to the juvenile hormones of insects, also fluctuate during reproduction in Crustacea. Specific roles for ecdysteroids and methyl farnesoate during reproduction have not been established.

The oocytes of crustaceans contain considerable amounts of ecdysteroids, mainly ponasterone A. Whether this hormone is involved in development of the embryo, as has been suggested in insects, is not known, although changes in titer of ecdysteroids during embryogenesis have been shown. Changes in ecdysone conjugates during embryogenesis have also been seen in several ticks.

Ecdysteroids have been found in eggs of a parasitic Platyhelminth *Hymenolepis diminuta*. There is some evidence that ecdysteroids may be involved in asexual reproduction and embryonic development in Cestodes. The Trematode, *Schistosoma mansoni*, one of the causative organisms of schistosomaisis in humans, shows fluctuations in 20-hydroxyecdysone titers during sexual maturation, vitellogenesis, and egg laying. Ecdysteroids have been detected in many nematodes. Fluctuations in 20-hydroxyecdysone levels during egg development were seen in *Ascaris suum*, *Paranemertes peregrina*, and *Pantinonemertes californiensis*. High titers of ecdysteroids have been found in the oocytes of a Polychaete, *Perineris cultifera*. Ecdysone was identified in the leech, *Nephelopsis obscura*, and found to stimulate spermatogenesis and the development of second stage oogonia. Ecdysteroids have been detected in molluscs and gastropods, but their source and biological roles are debated.

I. Ecdysteroid Nuclear Receptors/Transcription Factors

All the known steroid hormones, including the ecdysteroids, estrogens, retinoids, thyroid hormones, and vitamin D, act on target tissues via a nuclear receptor or transcription factor. These receptors bind ligands, in this case steroids, and also have binding sites for specific DNA sequences, also known as hormone response elements. Binding of the ligand usually activates the receptor so that it can bind to DNA. The receptors usually act as dimers, usually with other transcription factors, in binding to the hormone response element. It is significant that the ecdysteroid nuclear receptors are part of the superfamily of steroid hormone receptors, suggesting their common evolutionary origin.

Ecdysteroid receptors have been cloned from several insects, a tick, *Amblyomma americanum*, the nematode *Onchocerca volvulus*, and the lobster *Homarus americanus*. Three isoforms of the ecdysone receptor have been cloned in insects which are differentially expressed during development.

J. Evolution of Ecdysteroids

The presence of identical ecdysteroids in plants and animals, and the complexity of the biosynthetic pathways, suggests that these pathways were present before the separation of plants and animals. Ecdysteroids have been found in many phyla within the protostomes but have not been found in the deuterostomes (Echinodermata, Chordata, and related groups). The exception is that some ecdysteroids have been found in mammals, but they are thought to be of dietary origin or produced by internal parasites.

It has been hypothesized that the steroidal hormones appeared as a result of hydroxylation of cholesterol by monooxygenases, making them both water soluble and able to bind to lipid surfaces—admirable characteristics for a hormone. These hydroxylated steroids could have been the precursors of the ecdysteroids, which became the main steroid hormones of the prostomians. The deuterostomes developed enzymatic pathways to cleave the side chain resulting in the estrogen and corticosteroid type of steroid hormones.

Within the insects it has been speculated that the ovary was the original source of the ecdysteroids. The initial function of these hormones could have been in oocyte and sperm development and in regulating meiotic reinitiation. The large amounts of ecdysteroids stored in the oocyte functioned during meiotic reinitiation and embryonic development. Thus, they secondarily became the major regulators of molt induction during development of the immatures, with the prothoracic gland as the main source of the hormone.

See Also the Following Articles

Accessory Glands, Insects; Corpus Cardiacum; Juvenile Hormone

Bibliography

Davey, K. G. (1983). Hormonal integration governing the ovary. In *Endocrinology of Insects* (R. G. H. Downer and H. Laufer, Eds.), pp. 251–258.

Gillott, C. (1996). Male insect accessory glands: Functions and control of secretory activity. *Invertebr. Reprod. Dev.* **28**, 199–205.

Hagedorn, H. H. (1994). The endocrinology of the adult female mosquito. In *Advances in Disease Vector Research* (K. F. Harris, Ed.), pp. 109–148. Springer-Verlag, Berlin/New York.

Henrich, V. C., and Brown, N. E. (1995). Insect nuclear receptors: A developmental and comparative perspective. *Insect Biochem. Mol. Biol.* **25**, 881–897.

Horn, D. H. S. (1971). The ecdysones. In *Naturally Occurring Insecticides* (M. Jacobson and D. G. Grosby, Ed.), pp. 333–459. Deyker, New York.

Karlson, P. (1985). Evolution der chemischen Kommunikation im Tierreich. In *Information und Kommunikation* (E. Markl, Ed.), pp. 23–43. Wissenschaftliche Verlagsgesellschaft, Stuttgart.

Koolman, J. (1989). *Ecdysone, from Chemistry to Mode of Action.* Thieme-Verlag, Stutgart.

Lanot, R., and Cledon, P. (1989). Ecdysteroids and meiotic reinitiation in *Palaemon serratus* (Crustacea Decapoda Natantia) and in *Locusta migratoria* (Insecta Orthoptera). A comparative study. *Invertebr. Reprod. Dev.* **16**, 169–175.

Loeb, M. J., Bell, R. A., Gelman, D. B., Kochansky, J., Lusby, W., and Wagner, R. M. (1996). Action cascade of an insect gonadotropin, testis ecdysiotropin, in male Lepidoptera, *Invertebr. Reprod. Dev.* **30**, 181–190.

Raikhel, A. S. (1992). Vitellogenesis in mosquitoes. In *Advances in Disease Vector Research* (K. F. Harris, Ed.), pp. 1–39. Springer-Verlag, Berlin/New York.

Rees, H. H. (1985). Biosynthesis of ecdysone. In *Comprehensive Insect Physiology, Biochemistry and Pharmacology* (G. A. Kerkut, and L. I. Gilbert, Eds.), Vol. 7, pp. 249–293. Pergamon, Oxford, UK.

Yin, C.-M., Zou, B.-Z., Yi, S.-X., and Stoffolano, J. G. (1990). Ecdysteroid activity during oogenesis in the black blowfly, *Phormia regina*. *J. Insect Physiol.* **6**, 375–382.

Echinodermata

Maria Byrne
University of Sydney

I. Introduction
II. Sexuality
III. Asexual Reproduction
IV. Sexual Reproduction
V. Development

GLOSSARY

asexual reproduction Reproduction without union of gametes.
dioecious Individuals are either male or female.
hermaphroditic Individuals are both male and female either simultaneously or in sequence.
lecithotrophic development Development through a larva which utilizes endogenous reserves in the egg.
planktotrophic development Development through a larva which feeds on phytoplankton.
sexual reproduction Reproduction involving union of gametes derived from two genomes.

Echinoderms form a conspicuous and colorful component of the invertebrate fauna of the world's oceans. There are six echinoderm classes: the Asteroidea (sea stars), the Ophiuroidea (brittle stars), the Echinoidea (sea urchins, heart urchins, and sand dollars), the Holothuroidea (sea cucumbers), the Crinoidea (feather stars and sea lilies), and the recently discovered Concentricycloidea (sea daisies). The status of this new class is unclear; some systematists contend that they are highly derived asteroids.

I. INTRODUCTION

Although the shape of echinoderms varies a great deal, the overall body plan is based on pentamerous radial symmetry. This unique feature of the phylum is most easily seen in the star-shaped profile of five-armed sea stars (Fig. 1a). In contrast to the adult body plan, echinoderm larvae are bilaterally symmetrical. In adults, the surface bearing the mouth is called the oral surface and the surface bearing the anus is called the aboral surface.

Echinoderms show a diversity of reproductive and developmental patterns with among species differences in a range of life history traits, including sexuality, reproductive periodicity, breeding systems, and larval morphology.

II. SEXUALITY

Echinoderms reproduce by sexual or asexual reproduction and some species propagate by both methods. Sexual reproduction is the most common mode of propagation. Most echinoderms have separate sexes and are said to be dioecious or gonochoristic (Figs. 2a and 2b). This is particularly true of echinoderms with a large body size. In general, there are no external morphological differences between the sexes. Some sea cucumbers have genital papillae which assist in discerning the sexes.

FIGURE 1 Echinoderm diversity. (a) Class Asteroidea, *Astropecten duplicatus*; (b) class Ophiuroidea, *Ophionereis olivacea*; (c) class Echinoidea, *Strongylocentrotus franciscanus*; (d) class Crinoidea, *Florometra serratissima*; (e) class Holothuroidea, *Stichopus variegatus* (Courtesy M. Kingsford); (f) asexual reproduction in *Linckia multiflora* results in individuals with a varying number of arms. The comet-shaped individual (top) is regenerating three arms from the base of a single arm.

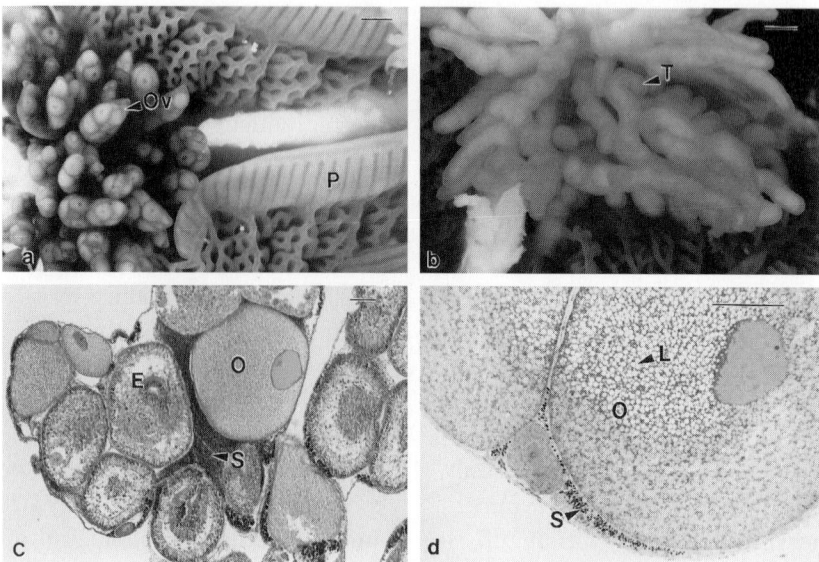

FIGURE 2 Most echinoderms such as the sea stars *Patiriella calcar* (a) and *P. gunnii* (b) are dioecious. Detection of hermaphrodites may require histology as shown here for the ovotestis of *P. pseudoexigua* (c, d). This species incubates its embryos in the gonad. E, embryo; L, lipid; O, oocyte; Ov, ovary; P, pyloric caecum; S, sperm; T, testis. Scale bars: a, b = 500 μm; c, d = 100 μm.

Echinoderms with a small body size are often hermaphroditic and may be simultaneous or sequential hermaphrodites (Figs. 2c and 2d). Simultaneous hermaphrodites have ovaries and testes or ovotestes. Sequential hermaphrodites are either male or female but change sex at some stage in their life. Echinoderms that change sex are usually protandric, maturing first as a male and becoming female later. Some species become increasingly female as they grow and so sex change is gradual and does not occur at a specific size.

Correlating with the differences in sexuality between small and large echinoderms is the tendency for large species to spawn their gametes into the water column while small species often retain their eggs and incubate the developing young (Figs. 2c and 3e). Many hypotheses have been proposed to explain the divergent trend in the life history traits of small and large echinoderms, but as yet there is no satisfactory explanation.

III. ASEXUAL REPRODUCTION

Asexual reproduction is common in asteroids, ophiuroids, and holothuroids. It does not occur in echinoids and crinoids. Some asexual species rarely if ever contain recognizable gonads. In asteroids and ophiuroids asexual reproduction is often exhibited by multiarmed species which reproduce by division or fission of the body (Fig. 1f). Transverse fission is also common in tropical sea cucumbers. Division of the body is followed by regeneration of each half into a new complete individual. The tropical asteroid *Linkia multiflora* has remarkable regenerative capabilities and can form a complete individual from a single arm, forming comet-like profiles (Fig. 1f). Asexual propagation by asteroid larvae is a recently discovered phenomenon. The larvae of some *Luidia* species bud off part of their bodies which grow to form a complete larva. The remarkable ease by which echinoderms cast off body parts is based on the mutable properties of their connective tissue. During fission or arm autotomy connective tissue along the breakage plane softens irreversibly resulting in separation of body parts.

In some habitats asexual echinoderms are the numerically dominant invertebrates. Many populations of *Holothuria atra*, a common holothuroid in the Indo-Pacific, appear to be maintained by asexual proliferation. Asexual echinoderms have a highly clonal

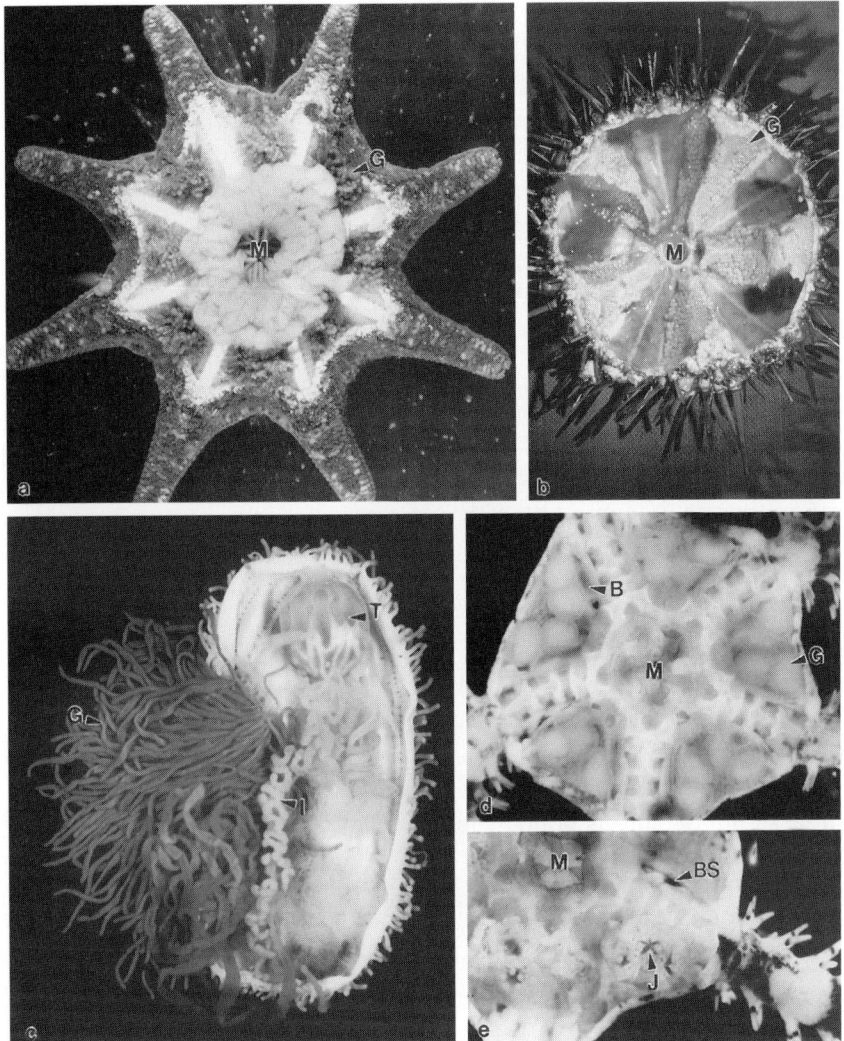

FIGURE 3 Reproductive anatomy. (a) Asteroidea, *Patiriella calcar*; (b) Echinoidea, *Heliocidaris erythrogramma*; (c) Holothuroidea, *Eupentacta quinquesemita*; (d) Ophiuroidea, *Ophionereis olivacea*; (e) *O. olivacea* with developing juveniles in bursa. (d and e reproduced with permission from M. Byrne, *Mar. Biol.* 111, 387–399, 1991. © Springer-Verlag). B, Bursa; BS, bursal slit; G, gonad, I, intestine; J, juvenile; M, mouth; T, retracted tentacular bulb.

population structure with most individuals being genetically identical.

IV. SEXUAL REPRODUCTION

A. Reproductive Anatomy

The anatomy of the reproductive system of asteroids, ophiuroids, echinoids, and, to some extent, crinoids reflects the pentamerous body plan (Figs. 3a, 3b, 3d, and 3e). Holothuroids differ from other echinoderms in having a single gonad (Fig. 3c).

1. Asteroidea

Although most asteroids have 5 arms, multiarmed species are common and some of these have up to 40 arms (Fig. 3a). In all asteroids there are two gonads in each arm and in mature specimens the gonads dominate the arm coelom. Each gonad can have one or several gonopores opening to the outside.

2. Echinoidea

Externally, the echinoid body is composed of a ridged test of closely fitting skeletal plates. In sea urchins a large elongate gonad is positioned along the inside of the test in each interradius for a total of five gonads (Fig. 3b). The gonopores are located in the genital plates around the anus. Acquisition of a secondary bilateral shape in heart urchins and sand dollars is associated with a decrease in the number of gonads. These echinoids usually have three or four gonads.

3. Ophiuroidea

The reproductive anatomy of ophiuroids is associated with 10 respiratory bursae, one on either side of the base of each arm (Figs. 3d and 3e). These are sac-like invaginations of the body wall which project into the coelom. Depending on the species, each bursa may have one or several gonads associated with it. The gametes are released into the bursae and exit through the bursal slit at the base of the arms. In some species the bursae function as a marsupium-like chamber for the developing young (Fig. 3e). *Hemipholis elongata* lacks bursae and its gonoducts connect with genital papillae at the base of the arms, through which the gametes are released. Some phrynophiurids are unusual in having gonads in the arm coelom, similar to asteroids. In this case there are either one or several gonoducts opening to the outside.

4. Holothuroidea

Holothuroids are elongated along their anterior–posterior axis and have a single gonad attached to the anterior region of the body wall (Fig. 3c). This gonad consists of elongate tubules which dominate the coelom in gravid specimens. The gonoduct opens on the anterior surface near the tentacles.

5. Crinoidea

The most striking feature of crinoids is their array of feather-like arms which accounts for them being known as feather stars (Fig. 1d). The arms range in number from five to several hundred and have delicate side branches called pinnules. The gonads are associated with the pinnules and so crinoids have a multitude of small gonads. In mature specimens the pinnules are noticeably swollen. Gonoducts are lacking and spawning occurs by rupture of the gonad wall.

6. Concentricycloidea

Concentricycloids are minute, medusa-like echinoderms with a maximum diameter of 1 cm. They have 10 gonads located in the body coelom. Concentricycloids are unusual in being sexually dimorphic, with the males having a copulatory organ. The gonoduct leading from the testes extends beyond the body margin where it is supported by spines. This structure appears to function as a copulatory organ for internal fertilization. Only two species are known so it is not possible to say what reproductive characteristics are typical of this group.

B. Gonad Structure

Echinoderms have simple sac-like gonads composed of several tissue layers (Figs. 4a and 5b). Asteroid, echinoid, ophiuroid, and concentricycloid gonads consist of two sacs, one enclosing the other, which are separated by the genital coelom (Fig. 4a). Each sac is composed of three tissue layers. The outer sac consists of the peritoneum, a central connective tissue layer, and the genital coelomic epithelium. The inner sac consists of the internal coelomic epithelium, a connective tissue layer which contains the hemal sinus, and the germinal epithelium. The lining of the inner sac includes myoepithelial cells which contract during spawning to expel the gametes. The gonads of holothuroids have a single sac structure similar to the inner sac of the gonads of the other classes.

The gametes develop in the innermost layer the germinal epithelium. The somatic cells of this epithelium include follicle cells, which surround the developing oocytes, and interstitial cells, which play a supportive role in spermatogenesis. Echinoids have specialized somatic cells called nutritive phagocytes which function in nutrient storage and which are important in gamete nutrition (Fig. 7).

The genital hemal sinus appears to serve as a nutrient store and immunocytochemical studies indicate

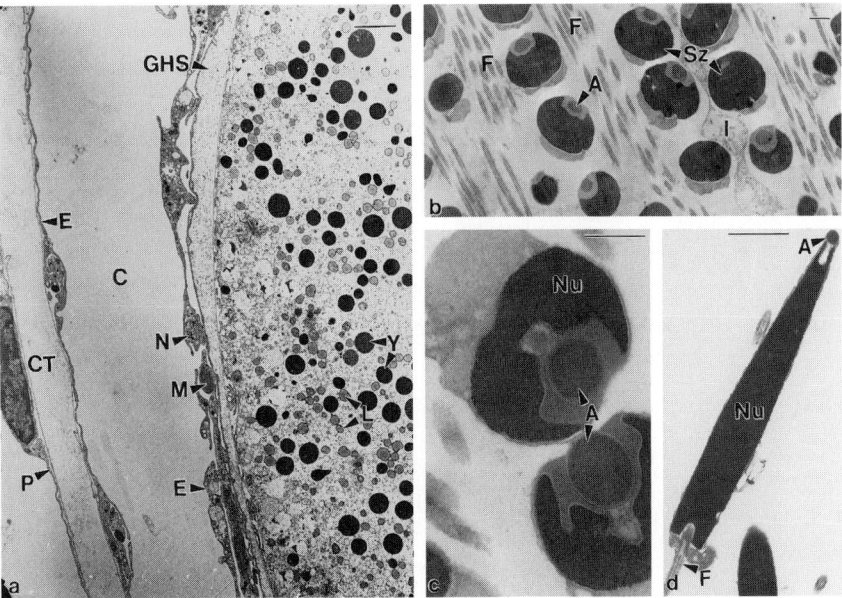

FIGURE 4 Ultrastructure of the gonads. (a) Ovary of the ophiuroid *Ophiolepis paucispina* showing the two sac structure of the gonad. (b) Spermatocyte columns in the testis of the ophiuroid *Ophionereis schayeri* [a and b reproduced with permission from F. W. Harrison and F. S. Chia (Eds.), *Microanatomy of the Invertebrates Volume 14 Echinodermata*, Wiley-Liss, New York, 1994. © Wiley-Liss, Inc.]. (c) Sperm of the asteroid *Asterias amurensis*. (d) Sperm of the echinoid *Heliocidaris erythrogramma* (courtesy J. Healy). A, acrosome, C, genital coelom; CT, outer connective tissue; E, epithelium; F, flagellum; GHS, genital hemal sinus; I, interstitial cell; M, myoepithelial cell; L, lipid; N, nerve; Nu, nucleus; P, peritoneum; Sz, Spermatozoa; Y, yolk. Scale bars: a = 3 μm; b–d = 0.5 μm.

that vitellogenin is a component of the fluid. Vitellogenin is a precursor molecule required for yolk synthesis. Hemal fluid is identified in tissue sections by its granular appearance and because it stains brilliantly with the periodic acid-Schiff's (PAS) staining method. In both ovaries and testes the size of the hemal sinus often cycles in association with the developmental state of the gametes (Figs. 5 and 6). The amount of fluid in the hemal sinus typically increases prior to gametic growth followed by its virtual disappearance as the gonads become gravid. The mechanisms underlying the dynamic properties of hemal fluid, an extracellular tissue, remain to be elucidated.

C. Gametogenesis

The gametogenic cycle of echinoderms is usually described by the staging method, which classifies gonad development according to defined histological criteria (Figs. 5–7). Gonad development is typically divided into five stages: (i) the spent stage, which occurs after spawning when the gonads are at their minimal state of development; (ii) the recovery stage, which is signaled by the renewal of gametogenesis; (iii) the growing stage, which is characterized by intense gametogenesis as the gametes increase in size and number in preparation for spawning; (iv) the mature stage, which is reached when the gonads reach the fully gravid condition; and (v) the spawning stage, which occurs when gamete release is under way.

Seasonal spawners with short breeding periods usually have synchronous gametogenesis, producing a large cohort of gametes which are at a similar stage of development. In contrast, echinoderms with long breeding periods or nonphasic reproduction have gametes at different stages of development in the gonads. The histology of gametogenesis is most readily illustrated by species with annual reproduction, such as the asteroid *Asterias amurensis* and the echinoid *Centrostephanus rodgersii* (Figs. 5–7).

In most echinoderms gonad growth is reflected by

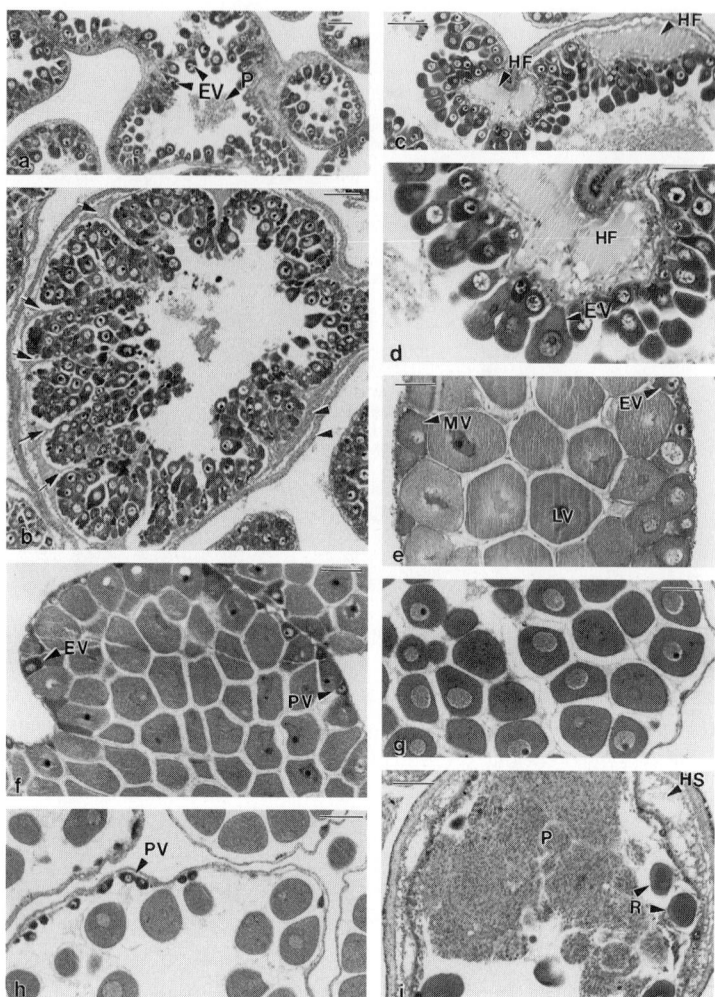

FIGURE 5 Histology of ovary development in the sea star *Asterias amurensis*. (a) Recovering ovary lined with small oocytes. (b) Growing ovary showing the hemal invaginations (arrows). Arrowheads, inner and outer sacs of gonad. (c, d) Association between hemal fluid and early oocytes which are attached by their basal stalk. (e) Growing ovary with oocytes at different stages of development. (f) Mature ovary filled with fully grown oocytes and containing a few pre- and early vitellogenic oocytes. (g) Partly spawned ovary early in the breeding season. (h) Partly spawned ovary late in breeding. (i) Spent ovary. Scale bars: a–c = 50 μm; d, e, i = 40 μm; f–h = 80 μm. EV, early vitellogenic oocyte; HF, hemal fluid; HS, hemal sinus; LV, late vitellogenic oocyte; MV, midvitellogenic oocyte; P, phagocyte; PV, previtellogenic oocyte, R, relict oocyte (reproduced with permission from M. Byrne *et al.*, *Mar. Biol.* 127, 673–685, 1997. © Springer-Verlag).

the cyclic pattern of gametogenesis. Gonadal growth in echinoids differs from that of other echinoderms because it involves two temporally separated cell cycles. As detailed later, nutritive phagocyte development precedes gametogenesis in echinoid gonads.

1. Ovary Development

When the ovaries start to recover after spawning the oogonia undergo a series of mitotic divisions to give rise to a new cohort of primary oocytes. At this stage small nests of early oocytes are present in the germinal layer. These early oocytes grow and subsequently line the germinal epithelium as seen in recovery stage ovaries (Figs. 5a, 5b, and 7a). The hemal sinus is evident as a PAS$^+$ layer under the germinal epithelium (Figs. 5b–5d).

In growing stage ovaries, the oocytes start to accumulate yolk. Vitellogenesis involves the accumula-

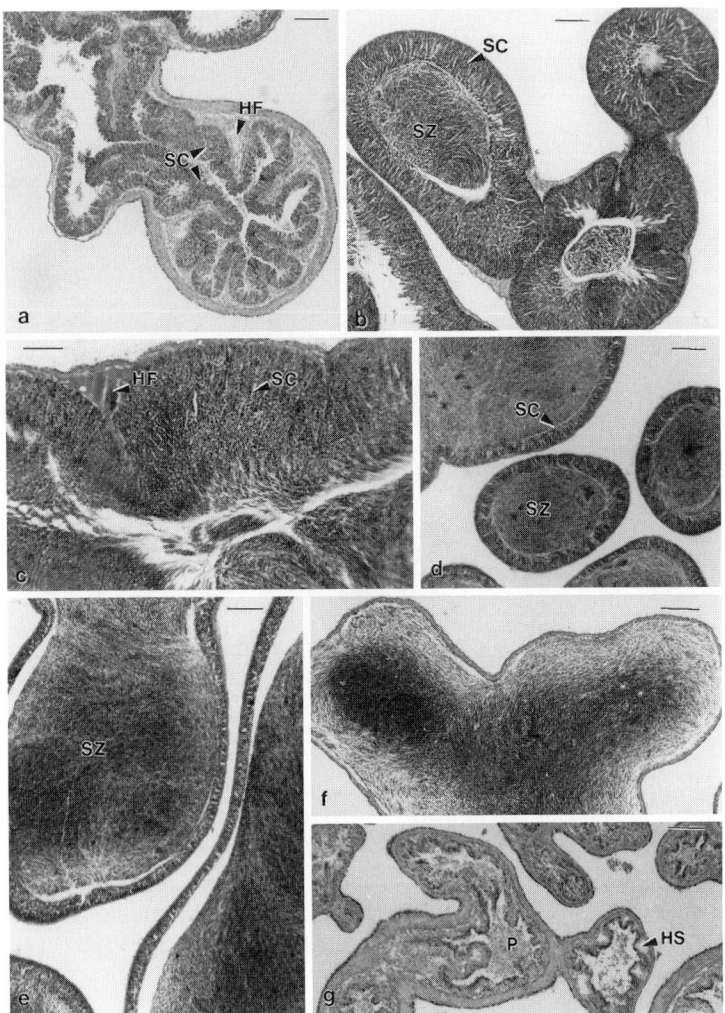

FIGURE 6 Histology of testis development in *Asterias amurensis*. (a) Recovering testis showing hemal invaginations. (b) Growing testis with spermatocyte columns and spermatozoa accumulating in the lumen. (c) Hemal fluid in a growing testis. (d) Mature testis filled with spermatozoa. (e) Partly spawned testis early in the breeding season. (f) Partly spawned testis late in breeding. (g) Spent testis. HF, hemal fluid; HS, hemal sinus; P, phagocyte; SC, spermatocyte columns; SZ, spermatozoa. Scale bars: a, b, d = 90 μm; c = 35 μm; e, g = 75 μm; f = 175 μm (reproduced with permission from M. Byrne *et al.*, *Mar. Biol.* **127**, 673–685, 1997. © Springer-Verlag).

tion of PAS$^+$ yolk granules in the oocytes. This PAS$^+$ response intensifies as the oocytes increase in size. In asteroids, ophiuroids, holothuroids, and crinoids, vitellogenic oocytes maintain a close association with the hemal sinus which provides nutrients for gamete growth. The vitellogenic eggs of asteroids are pear shaped and remain attached to the germinal epithelium with their basal stalk-like region (Fig. 5d). Nutrient transfer from the hemal sinus to the oocytes occurs across the stalk. Ophiuroid and crinoid eggs are surrounded by hemal fluid (Fig. 4a). Echinoid oocytes also avail of hemal nutrients, but the relationship between the hemal sinus and the oocytes is not understood. Ultrastructural examination indicates that echinoderm oocytes sequester material from the hemal sinus by endocytosis.

Vitellogenic oocytes are characterized by an abundance of yolk granules dispersed through the cyto-

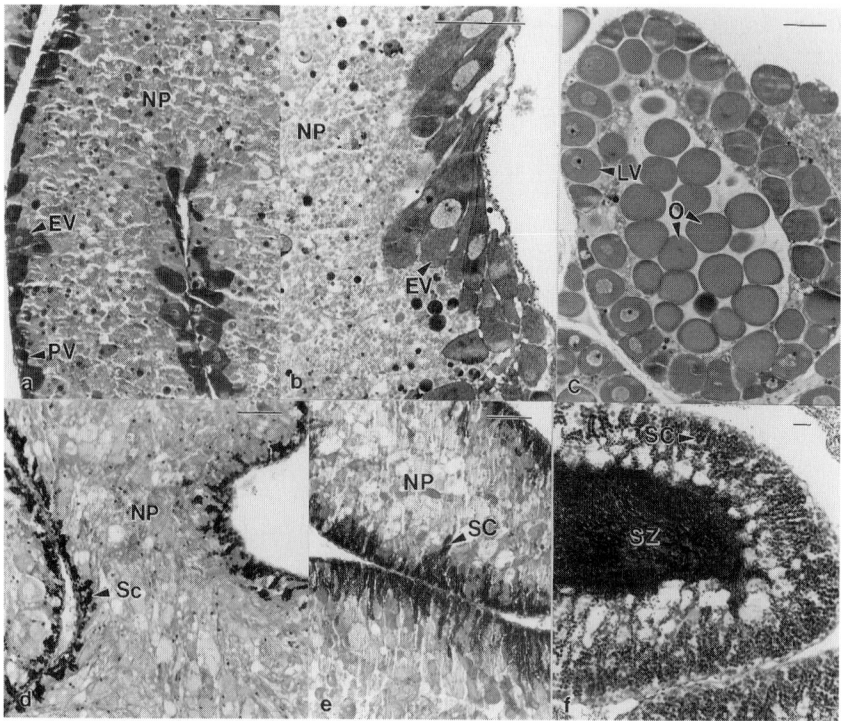

FIGURE 7 Development of the ovaries (a–c) and testes (d–f) of the sea urchin *Centrostephanus rodgersii*. (a) Recovering ovary at the beginning of the gametic phase with early oocytes lining the epithelium. The lumen is filled with nutritive phagocytes. (b) As the oocytes grow the stored nutrients are utilized. (c) Ovary approaching the mature stage containing ova and late vitellogenic oocytes and with reduced nutritive tissue. (d) Recovering testis at the beginning of the gametic phase with spermatocytes in the epithelium and nutritive phagocytes filling the lumen. (e) As the spermatocyte columns increase in length the stored nutrients are utilized. (f) Growing stage testis with spermatocyte columns and spermatozoa. The nutritive tissue is reduced. EV, early vitellogenic oocyte; LV, late vitellogenic oocyte; NP, nutritive phagocytes; O, ova; PV, previtellogenic oocyte; Sc, spermatocytes; SC, spermatocyte columns; SZ, spermatozoa. Scale bars = 100 μm.

plasm (Fig. 4a). These are formed by the endoplasmic reticulum and Golgi complex. The number of lipid droplets also increases during vitellogenesis, but the mechanism of lipid deposition is not understood.

As the ovaries become mature the oocytes detach from the germinal layer and accumulate in the lumen in preparation for spawning (Figs. 5e and 7c). Mature ovaries are packed with late vitellogenic and fully grown oocytes (Figs. 5f and 7c). In asteroids, ophiuroids, holothuroids, and crinoids final maturation of the eggs occurs in association with spawning and fertilization. In echinoids oocyte maturation occurs in the ovary and ova may be stored for some time before release (Fig. 7c).

Spawning in echinoderms results in the complete evacuation of the gonads or, more commonly, a partial release of the gametes (Figs. 5g and 5h). In species with an extended breeding season, primary oocytes continue to develop and replace oocytes lost through spawning. Spent ovaries have a shrunken appearance and the lumen is invaded by somatic cells which phagocytose relict oocytes (Fig. 5i). Unspawned eggs usually lyse, releasing yolk granules, lipid droplets, and other material into the lumen. Echinoderms often produce more eggs than they spawn and atretic oocytes are phagocytosed by somatic cells. The rationale underlying excess oocyte production in echinoderms is not known. It is thought that vitellogenic material derived from atretic oocytes may be a source of nutrients for subsequent oogenesis.

Echinoderm ova range in diameter from 60 to 3000

μm. Species that have planktotrophic development through a feeding larva have small (60–160 μm diameter) eggs (Figs. 5f, 7c, and 10). By contrast, species that have lecithotrophic development through a nonfeeding larva have large (200–3000 μm diameter) eggs (Figs. 2c, 2 d, and 10). Echinoderms with small eggs are highly fecund and are also large species. The crown of thorns starfish, for instance, releases up to 100 million eggs each year. In contrast, species with large eggs are often small echinoderms with a reduced fecundity. The reduction in fecundity with increasing egg size is illustrated by brooding echinoderms which often produce a few eggs at any one time.

2. Testis Development

Testis development follows a similar pattern to that described for the ovaries. In recovering testes the spermatogonia in the germinal layer give rise to spermatocytes (Figs. 6a and 7d). Echinoderm sperm develop in spermatocyte columns that extend toward the lumen of the testis (Figs. 4b, 6b, 6c, 7e, and 7f). In these columns the sperm are accompanied by the interstitial cells which send processes up the center of the columns to provide structural support (Fig. 4b). Interstitial cells serve a Sertoli-like function, forming a scaffold of filipodial processes within which sperm develop until their release as mature spermatozoa. When the spermatogonia divide, filipodia of the interstitial cells are extended between the daughter cells. The filipodia remain with these cells and their descendants throughout spermatogenesis and are located in the axis of the columns (Fig. 4b). In addition to their structural role, the interstitial cells may function in nutrition by providing a link between the genital hemal sinus and developing sperm. The PAS$^+$ response of the axis of the columns provides evidence for a link with the hemal sinus.

In growing stage testes, the spermatocyte columns form in distinct layers of increasing maturity from the germinal epithelium to the testis lumen (Figs. 6b, 6c, 7e, and 7f). The columns are generated at the base with dividing spermatogonia and early primary spermatocytes near the epithelium. Toward the tip of the columns the sperm develop to maturation, resulting in the presence of spermatozoa at the luminal end (Figs. 6c, 6d, and 7f). This simultaneous proliferation and differentiation of germ cells is a general feature of testis development in echinoderms.

In mature testes the spermatozoa form a dense mass in the lumen (Fig. 6d). Spawning stage testes are identified by the dispersed appearance of the spermatozoal mass and spent testes have a shrunken appearance (Figs. 6e–6g).

Echinoderm sperm are 2–10 μm in diameter or length and are of the simple type typical of species with external fertilization in seawater (Figs. 4b–4d). All echinoderms except echinoids have round-headed sperm (Figs. 4b–4c). Echinoid sperm has a conically shaped head (Fig. 4d).

3. Agametic and Gametic Phases in Echinoid Gonads

The germinal epithelium of echinoids is characterized by two distinct and inversely related cell cycles, the nutritive phagocyte cycle (agametic phase) and the gametogenic cycle (gametic phase). After spawning the increase in weight by the gonads results from expansion of the nutritive phagocytes (Figs. 7a and 7d). During this period the gonads are dominated by these cells, which increase in size as they sequester nutrients derived from digestion. In many sea urchins the gonads reach their maximum weight during the agametic period. In association with the ability of the nutritive cells to sequester nutrients, the gonads function as the main nutrient storage organ in echinoids.

Growth of sea urchin gonads is highly sensitive to food quantity and quality and this greatly influences the gametic phase. Successful gametogenesis depends on the nutrients stored in the phagocytes, and in food-poor conditions reproduction is compromised, with a reduction in the amount of gametes produced. During the gametic phase nutrients are mobilized to support gametogenesis, resulting in a reduction in the nutritive tissue (Figs. 7b, 7c, 7e, and 7f). The nutritive phagocytes shrink in size and are difficult to identify in histological sections (Fig. 7f).

Immunocytochemical evidence indicates that the products stored by nutritive phagocytes include vitellogenin, the source of which appears to be the hemal sinus. It is not known how the nutritive material is transferred to the developing gametes. Echinoid gametogenesis follows the general pattern described previously for the ovaries and testes.

D. Reproductive Cycles and Control of Reproduction

Like many marine invertebrates most echinoderms reproduce with a discrete and predictable annual cycle of gametogenesis and spawning. The gonads of seasonal spawners increase in weight as the nutritive cells or gametes increase in number and size, followed by a sharp reduction in weight due to gamete release (Fig. 8). The reproductive cycle of most echinoderms can be followed by calculating the gonad index (100 × gonad wt/body wt).

Echinoderm reproduction is controlled by a suite of endogenous and environmental factors which appear to interact in a complex synergistic fashion. Depending on the species, various seasonally variable environmental factors, including photoperiod, temperature, and food availability, may be important in regulating the different stages of gametogenesis (proliferation, growth, and maturation) and breeding. The sea urchin *Strongylocentrotus purpuratus* has been well studied with respect to the control of reproduction. Widely separated populations of this species along North America's west coast have relatively synchronous reproduction in a region where sea temperature changes are ill-defined and relatively minor. Detailed laboratory studies demonstrate that gametogenesis in *S. purpuratus* is controlled by photoperiod with little or no influence by temperature. In contrast, gametogenesis in some Japanese echinoids is regulated by an endogenous program and critical changes in sea temperature and is independent of photoperiod. These contrasting results obtained for echinoids on either side of the north Pacific may reflect the contrast between the absence of a marked sea temperature change along the American coast and the presence of annually predictable seasonal temperature changes in Japanese waters.

Spawning in some diadematoid urchins is entrained by the lunar cycle and so gametogenesis occurs with a monthly pattern. Lunar control of gametogenesis is also seen in diadematoids with annual breeding periods. In *Centrostephanus coronatus* gametogenesis proceeds with a monthly sequence culminating in monthly gamete release over a breeding period lasting several months. Spawning usually occurs over a short time period and for many echinoderms is a synchronous populationwide event. Synchronous spawning appears to be cued by a number of environmental factors, including water temperature, photoperiod, the state of the tidal currents, and the presence of phytoplankton.

Reproductive synchrony is probably also modulated by endogenous factors such as hormonal control and the release of pheromones. The influence of interindividual chemical communication on gametogenic entrainment and the influence of metabolites released with spawned gametes on gamete release by conspecifics has been described for a holothuroid and an echinoid.

In situations in which population density is high and spawning is synchronous, fertilization success would be expected to be high. Many echinoderms, however, do not live close to conspecifics and, in this case, fertilization appears to be a haphazard event. Field studies have shown a dramatic decrease in fertilization levels with a small increase in the distance between spawning males and females. One exception is the high fertilization rates recorded for the crown of thorns starfish.

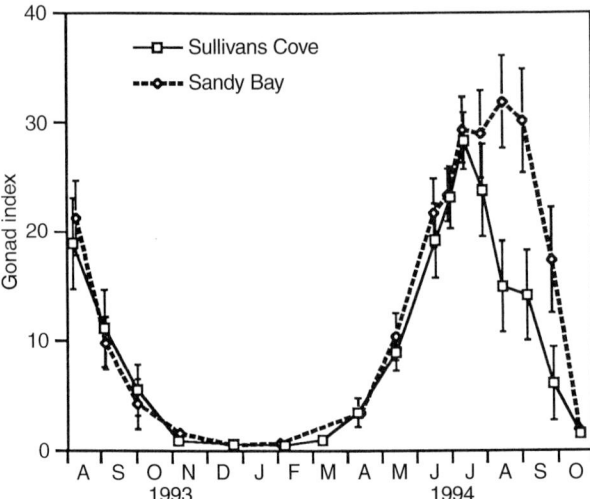

FIGURE 8 Annual cycle of the gonad index of *Asterias amurensis* at two sites in Tasmania. Gonad index = 100 × gonad weight/total weight (reproduced from M. Byrne *et al.*, *Mar. Biol.* **127**, 673–685, 1997. © Springer-Verlag).

V. DEVELOPMENT

A. Early Embryos

Free-spawning echinoderms release their gametes into the water column and this is where fertilization and development occur. Echinoderms that care for their embryos deposit their eggs in masses attached to the substratum or release them into a brood chamber located on or in the body. Sperm derived from the same or a different parent fertilizes the eggs.

Echinoderm development follows the deuterostome pattern, with radial cleavage followed by the blastula and gastrula embryonic stages. The blastula is a hollow ball of cells surrounding a central blastocoel. In some echinoderms the blastula develops deep folds in the epithelium, giving them a contorted appearance. The significance of this is not known. Hatching occurs at the blastula or gastrula stages. Gastrulation involves invagination of the vegetal pole to form the larval gut or archenteron. The external opening of the archenteron is called the blastopore. Coelomogenesis ensues, with formation of two pouches which bud off the anterior end of the archenteron.

B. Larvae

The gastrula develops into a bilaterally symmetrical larva, the structure of which differs in each echinoderm class (Fig. 9). Larval structure also depends on whether development is of the feeding or nonfeeding type. Echinoderms that have small eggs develop through free-swimming planktotrophic larvae with complete digestive tracts (Figs. 9h and 10). These larvae feed on phytoplankton and have an intricate array of ciliated bands for feeding and locomotion (Figs. 9a, 9b, and 9e). In contrast, species with large eggs develop through planktonic or benthic lecithotrophic larvae that lack a functional gut and which have a simple pattern of ciliation (Figs. 9c, 9d, 9f, and 9i). Development of these larvae is supported by nutritive reserves present in the egg.

Most asteroids have bipinnaria and/or brachiolaria larvae (Figs. 9a–9d). The bipinnaria is a feeding larva and has two ciliary bands which loop around the body. The archenteron grows forward to form the mouth, whereas the blastopore functions as the larval anus. The rest of the digestive tract is divided into an esophagus, stomach, intestine, and anus. The bipinnaria gives rise to the brachiolaria larva with formation of the brachiolar apparatus at the anterior end (Fig. 9b). This structure consists of three arms and a central adhesive disc. It is used for benthic attachment by larvae ready to metamorphose. Asteroids with nonfeeding development lack a bipinnaria and the gastrula gives rise directly to a lecithotrophic brachiolaria which is uniformly ciliated (Figs. 9c and 9d). The blastopore closes and so the gut is a closed structure.

Echinoids and ophiuroids have pluteal larvae with a ciliary band which follows the contour of the larval arms (Figs. 9e and 9h). Echinoplutei have up to six pairs of arms and ophioplutei have up to four pairs of arms. Each arm is supported by a skeletal rod. The similarity of echinoplutei and ophioplutei was taken to suggest a close relationship between echinoids and ophiuroids, but phylogenetic analyses show this is not the case. The similarity of these larvae is a striking example of parallel evolution. Echinoids with lecithotrophic development may have a reduced pluteus or have a simple spherical larva with no pluteal structures (Fig. 9i). Ophiuroids with nonfeeding development have elongate vitellaria larvae which are distinguished by their barrel-shape and transverse bands of cilia.

The planktotrophic larva of holothuroids is the auricularia. It has a single ciliary band that loops around the body and looks similar to the asteroid bipinnaria. The auricularia gives rise to a doliolaria larva by reorganization of the ciliated band into transverse rings. Holothuroids with nonfeeding development have a vitellaria larva. All crinoids have nonfeeding development through a vitellaria.

C. Metamorphosis

Echinoderm metamorphosis involves a dramatic change from the bilateral symmetry of the larva to the pentamerous radial symmetry of the juvenile (Figs. 9g and 9i). The juvenile body develops at the posterior end of the larva (Figs. 9b, 9g, and 9i). Metamorphosis involves resorption or loss of the larval body and may occur while the larva is in the

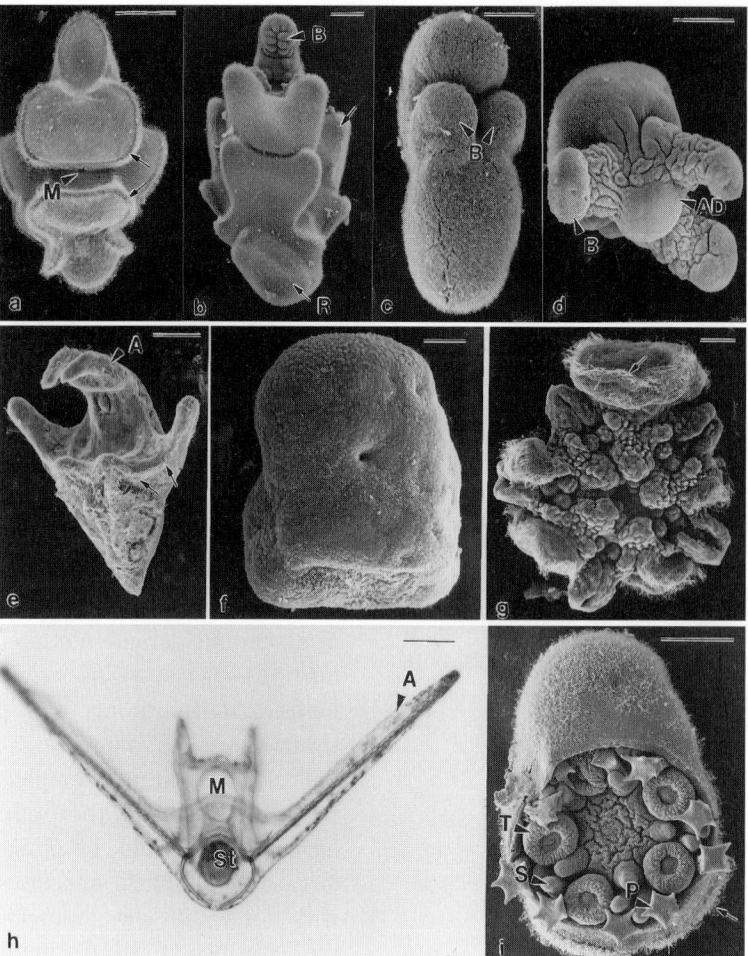

FIGURE 9 Echinoderm larvae. (a, b) Bipinnaria and brachiolaria larvae of the sea star *Patiriella regularis* (a and b, reproduced with permission from M. Byrne and M. Barker, *Biol. Bull.* **180**, 332–345, 1991). (c) Brachiolaria of *Patiriella gunnii*. (d) Brachiolaria of *Patiriella exigua* (reproduced with permission from M. Byrne, *Biol. Bull.* **188**, 293–305, 1995). (e) Ophiopluteus of *Ophiothrix spongicola* [reproduced with permission from F. W. Harrison and F. S. Chia (Eds.), *Microanatomy of the Invertebrates Volume 14 Echinodermata*, Wiley-Liss, New York, 1994. © Wiley-Liss, Inc.]. (f) Intrabursal reduced vitellaria of *Ophionereis olivacea*. (f, reproduced with permission from M. Byrne, *Mar. Biol.* **111**, 387–399, 1991). (g) Metamorphosing vitellaria of *Ophionereis schayeri* (courtesy P. Selvakumaraswamy). (h) Echinopluteus of *Centrostephanus rodgersii* (Courtesy C. King). (i) Metamorphosing larva of *Heliocidaris erythrogramma*. A, larval arms; AD, adhesive disc; B, brachiolar arms; M, mouth; P, pedicellaria; R, rudiment; S, primary spines; St, stomach; T, primary tube feet; arrows, ciliated band. Scale bars: a–d, h, i = 100 μm; e–g = 40 μm.

plankton or after settlement. In species that care for their young, metamorphosis occurs in the brood chamber and the juveniles walk away from the parent (Figs. 2c and 3e).

D. Evolution of Development

The feeding larvae of echinoderms are very ancient and the free-spawning planktotrophic life history is considered to be the ancestral pattern. The presence of nonfunctional feeding structures, such as a closed gut and reduced pluteal arms in some nonfeeding larvae, provides evidence that lecithotrophy evolved through an ancestor with planktotrophic development. The acquisition of a large egg was an essential change that freed the larvae from the need to feed, resulting in the reduction and loss of larval feeding structures. Modified development through nonfeed-

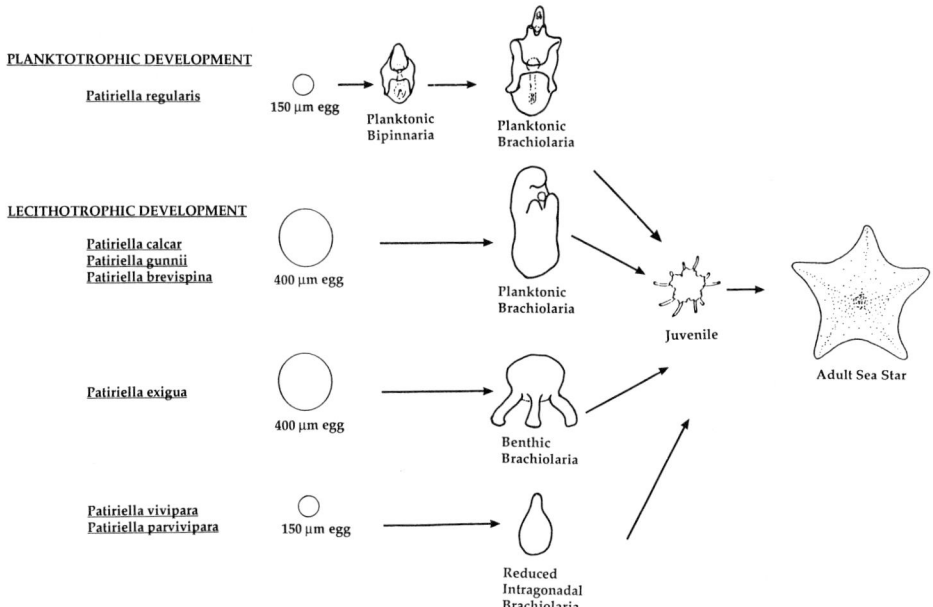

FIGURE 10 Developmental patterns in *Patiriella* (reproduced with permission from M. Byrne and A. Cerra, *Biol. Bull.* **191**, 17–26, 1996).

ing larvae is often associated with a suite of life history traits, including decreased adult size, a large egg, hermaphroditism, internal fertilization, and parental care of the young. The evolutionary significance of the linkage of these traits has been a subject of extensive discussion with many hypotheses proposed, none of which provide a widely applicable explanation.

The evolutionary changes associated with the switch from planktotrophy to lecithotrophy are exemplified by the sea star genus *Patiriella* (Fig. 10). These starfish exhibit the full range of life histories with planktotrophic development at one end of the spectrum and intragonadal development at the other end. In contrast to the range of larval forms in *Patiriella*, the adults are quite similar. The close relationship between the *Patiriella* species facilitates comparison of homologous cells and structures in development, providing a model system to investigate the evolution of development.

Patiriella regularis represents the ancestral state in spawning small eggs and in having feeding bipinnaria and brachiolaria larvae (Figs. 9a and 9b). It is also a relatively large species and is dioecious. The intermediately sized dioecious species *P. calcar*, *P. gunnii*, and *P. brevispina* spawn large eggs and have planktonic lecithotrophic brachiolaria (Fig. 9c). The bipinnaria stage has been deleted, but the brachiolar complex is retained due to its importance in benthic settlement. Most asteroids with lecithotrophic development have the type of brachiolaria seen in these species. *Patiriella exigua* is a small hermaphrodite that lays large eggs on to the substratum (Fig. 9d). The lecithotrophic brachiolaria of *P. exigua* has a tripod shape due to hypertrophy of the brachiolar complex which serves as a tenacious and permanent attachment device. At the extreme end of the developmental spectrum is the viviparous life history of *P. vivipara* and *P. parvivipara*. These are the smallest known sea stars. They are hermaphrodites, have internal fertilization, and have intragonadal development through a minute brachiolaria. Without the need for settlement the brachiolar complex is reduced. The eggs are secondarily decreased in size and so the larvae metamorphose as tiny juveniles. These juveniles prey on their intragonadal siblings and are born about a year later as large juveniles which emerge through the gonopore. Viviparity in *Patiriella* is the most derived life history pattern seen in the Echinodermata.

See Also the Following Articles

Asexual Reproduction; Marine Invertebrates, Modes of Reproduction in

Bibliography

Byrne, M. (1989). Ultrastructure of the ovary and oogenesis in the ovoviviparous ophiuroid *Ophiolepis paucispina* (Echinodermata). *Biol. Bull.* **176**, 79–95.

Hamel, J.-F., and Mercier, A. (1996). Evidence of chemical communication during the gametogenesis of holothuroids. *Ecology* **77**, 1600–1616.

Harrison, F. W., and Chia, F. S. (1994). *Microanatomy of the Invertebrates Volume 14 Echinodermata*. Wiley-Liss, New York.

Giese, A. C., Pearse, J. S., and Pearse, V. B. (Eds.) (1991). *Reproduction of Marine Invertebrates Volume VI Echinoderms and Lophophorates*. Boxwood Press, Pacific Grove, CA.

Levitan, D. R. (1995). The ecology of fertilization in free-spawning invertebrates. In *Ecology of Invertebrate Larvae* (L. McEdward, Ed.), pp. 123–156. CRC Press, Boca Raton, FL.

Pearse, J. S., and Cameron, R. A. (1991). Echinoidea. In *Reproduction of Marine Invertebrates Volume VI Echinoderms and Lophophorates* (A. C. Giese, J. S. Pearse, and V. B. Pearse,, Eds.). Boxwood Press, Pacific Grove, CA.

Pearse, J. S., Pearse, V. B., and Davis, K. K. (1986). Photoperiodic regulation of gametogenesis and growth in the sea urchin *Strongylocentrotus purpuratus*. *J. Exp. Zool.* **237**, 107–118.

Sakairi, K., Yamamoto, M., Ohtsu, M., and Yoshida, M. (1976). Environmental control of gonadal maturation in laboratory-reared sea urchins, *Anthocidaris crassispina* and *Hemicentrotus pulcherrimus*. *Zool. Sci.* **6**, 721–730.

Starr, M., Himmelman, J. H., and Therriault, J.-C. (1990). Direct coupling of marine invertebrate spawning with phytoplankton blooms. *Science* **247**, 1071–1074.

Strathmann, M. F. (1987). *Reproduction and Development of Marine Invertebrates of the Northern Pacific Coast*. Univ. of Washington Press, Seattle.

Walker, C. W. (1982). Nutrition of gametes. In *Echinoderm Nutrition* (J. M. Lawrence and M. Jangoux, Eds.), pp. 449–468. Balkema, Rotterdam.

Eclampsia

see Preeclampsia/Eclampsia

Ectoderm

see Germ Layers

Ectopic Pregnancy

Christos Coutifaris

University of Pennsylvania

I. Introduction
II. Overview of Early Embryonic Development
III. The hCG Discriminatory Zone
IV. Diagnosis
V. Management
VI. Follow-Up
VII. Conclusions and Future Directions for Research

GLOSSARY

culdocentesis Aspiration of fluid from the posterior cul-de-sac (pouch of Douglas) with a needle inserted transvaginally through the posterior vaginal fornix.

ectopic pregnancy Pregnancy established at any extrauterine site, most commonly within the Fallopian tube (oviduct).

human chorionic gonadotropin (hCG) discriminatory zone The serum hCG concentration above which an intrauterine pregnancy can be visualized reliably by an imaging modality such as ultrasound.

laparoscopy Surgical procedure performed under general anesthesia for the visual evaluation of the abdominal and pelvic cavities after the introduction of an endoscope (7–12 mm in diameter) through the umbilicus.

salpingectomy The surgical excision of the Fallopian tube.

salpingostomy Surgical incision through the full thickness of the Fallopian tube wall. Usually performed to extract an unruptured tubal pregnancy or to surgically repair (open) a distally occluded Fallopian tube.

In an ectopic pregnancy embryo implantation and placentation take place outside the endometrial/uterine cavity. Over 95% of ectopic pregnancies occur in the Fallopian tubes, including a small proportion which are found in the interstitial portion of the oviduct traversing the myometrium at the cornual regions of the uterus. Other anatomic sites at which ectopic implantations occur are the uterine cervix and the ovary. In addition, a variety of abdominal ectopic pregnancies have been described. In these cases implantation has occurred on the omentum, the intestine, or the pelvic sidewall. Curiously, some abdominal pregnancies have been reported to have progressed into the third trimester of pregnancy and have reached the stage of viability. Ectopic pregnancy is a major public health problem and, if not diagnosed in a timely fashion, can lead to significant morbidity and even death. In fact, ectopic pregnancy is the second leading cause of maternal mortality and the most common cause of maternal death in the first half of pregnancy. Successful treatment, which avoids or minimizes complications, requires early diagnosis and prompt medical intervention. Diagnosis of an ectopic pregnancy with the classic medical tools of history and physical examination alone has been notoriously difficult. Nevertheless, advances in imaging techniques have allowed the reliable definition of the absence of an intrauterine pregnancy or even the detection of an ectopic gestation. These advances, have led to the successful very early diagnosis of an ectopic pregnancy and have now opened the door for the use of conservative treatments, which only a few years ago were either rarely utilized or would not have been considered.

I. INTRODUCTION

Ectopic pregnancy has reached epidemic proportions. Although the incidence in the United States has doubled over the past 15–20 years to approximately 1 in 66 pregnancies, the incidence in pregnant patients presenting to an emergency room is much higher. A study carried out in 1986 found that 1 in

10 patients discharged with the clinical diagnosis of threatened abortion subsequently proved to have an ectopic pregnancy. This staggering statistic, along with the information provided by Dorfman et al. on the mortality secondary to ectopic pregnancies in the United States, should put the obstetrician gynecologist on alert. This is especially true considering the advances that have been made in recent years, which allow the diagnosis of pregnancy within 1 week from conception and its visualization by ultrasound within 2 or 3 weeks. Although preexisting risk factors and the classic signs and symptoms of ectopic pregnancy have guided clinicians over the years, increasing numbers of completely asymptomatic patients, with unremarkable physical examinations and no risk factors, are found to have early ectopic pregnancies. A methodical evaluation of the patient just diagnosed as being pregnant should distinguish the one who may carry an abnormal gestation and thus lead to the prompt diagnosis of an ectopic pregnancy. Such an approach provides the opportunity for early management, which can be undertaken in a controlled fashion with optimal results for the patient with regard to future fertility. Furthermore, early diagnosis allows for choices of management that the physician can review with the patient and involve the patient in the decision-making process.

II. OVERVIEW OF EARLY EMBRYONIC DEVELOPMENT

In an idealized 28-day cycle, conception occurs at approximately the 15th day of the cycle. For the subsequent 2 or 3 days the zygote develops through successive cell divisions within the Fallopian tube and reaches the morula stage of development. At approximately 4 days of life, the embryo enters the uterine cavity, continues its development to the blastocyst stage, and then hatches through the zona pellucida. The actual process of implantation, which involves adhesion of the trophectoderm to the uterine epithelium and subsequent penetration of the trophoblast, is initiated at approximately Days 6 or 7 from the time of conception (approximately 3 weeks from the last menstrual period). The process of penetration of the endometrium by the embryo continues, and by Day 14 [4 weeks from last menstrual period (LMP); date of the expected menstrual period] the embryo is completely embedded in the maternal decidua and covered by the endometrial epithelium. At this point, chorionic villi can be distinguished all around the chorionic sac and the embryo proper has also started to develop from the inner cell mass of the blastocyst. At this stage of development the diameter of the chorionic sac is approximately 2 mm. Embryonic and chorionic development continues, with the chorionic sac reaching approximately 6 mm in diameter at 21 days of postconceptual life (5 weeks from LMP) and approximately 12 mm at 28 days from conception (6 weeks from LMP).

It is now well established that a preimplantation human embryo synthesizes and secretes human chorionic gonadotropin (hCG). Nevertheless, hCG is not readily detectable in maternal blood until embryonic–maternal communication is established after implantation. Furthermore, this becomes clinically relevant when the patient appreciates that she is pregnant after missing a menstrual period. It is well documented in the literature that hCG concentrations, during the first 8 weeks of a normal singleton pregnancy, rise in the maternal serum in a predictable fashion with the concentrations doubling approximately every 2 days (lower limit of normal: 66% increase every 2 days or 114% every 3 days). Although serial hCG measurements can identify a large proportion of abnormal gestations (spontaneous abortions or ectopics), a single hCG determination cannot be used to predict outcome due to a high false-positive rate.

Appreciation of the size and functional parameters of the early embryo can help the clinician approach the pregnant patient rationally and thus make an early diagnosis of an abnormal gestation. Close examination of Dorfman et al.'s study shows that most fatal events in patients with ectopic pregnancies occurred after the sixth week following the LMP. This is expected since even in a normally growing pregnancy, the chorionic sac would have a maximal diameter of approximately 1.5 cm at 6 weeks from LMP. With the possible exception of direct trophoblastic invasion into a major vessel, a quite unlikely occurrence, a major catastrophic event secondary to an ectopic gestation would be improbable this early in gestation.

Even some of the earlier deaths reported in the study probably represented miscalculation of a true LMP. Thus, diagnosis of pregnancy soon after the missed menses allows the clinician enough time (1.5–2 weeks) for careful evaluation and prompt management of an ectopic pregnancy.

III. THE hCG DISCRIMINATORY ZONE

The concept of the hCG discriminatory zone was introduced by Kadar and coworkers and refers to the serum hCG concentration above which a normal intrauterine singleton pregnancy can be visualized by transabdominal ultrasound. This concept has also been confirmed by others and has been proposed for use in management protocols. Advances in ultrasound technology, and specifically the development of transvaginal transducers, have permitted the use of high-frequency/high-resolution equipment and have allowed the visualization of the growing chorionic sac during the very early period of development. It is now well established that the chorionic sac ("gestational sac") of a normal singleton intrauterine pregnancy can be visualized during the first week after a missed menstrual period at lower serum hCG concentrations than was previously possible (Fig. 1). Given an hCG doubling time of 2 or 3 days, accurate diagnosis can be made 4–7 days earlier than was previously possible. In addition, 3- to 5-mm ectopic gestational sacs can now be frequently visualized by transvaginal ultrasound in selected cases (Fig. 2). It should be appreciated that there is variation in

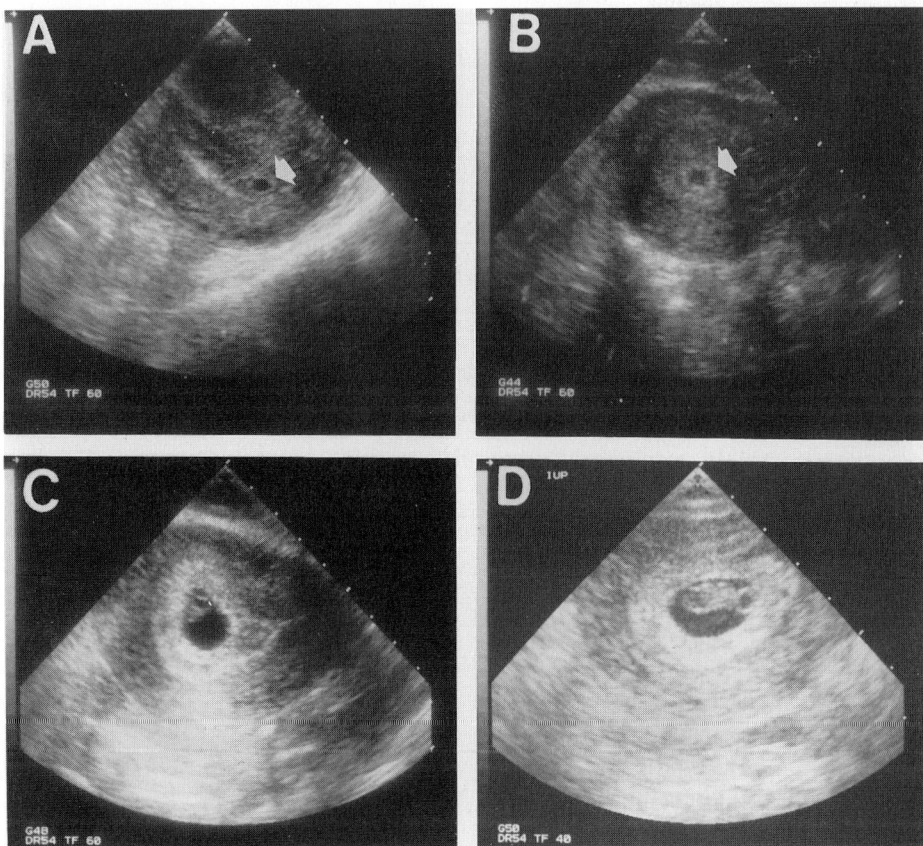

FIGURE 1 Visualizations of an intrauterine gestational sac in early pregnancy through the use of transvaginal ultrasound. (A, B) Intrauterine gestational sacs (arrow) of 3–7 mm in diameter at quantitative hCG of 2000–3500 mIU 1st IRP/ml. (C, D) Visualization of an intrauterine pregnancy at 6–9 weeks post-LMP with clear identification of the yolk sac and fetal pole. At this gestational age, hCG is > 20,000 mIU 1st IRP/ml and fetal heart activity can be seen on real-time ultrasound.

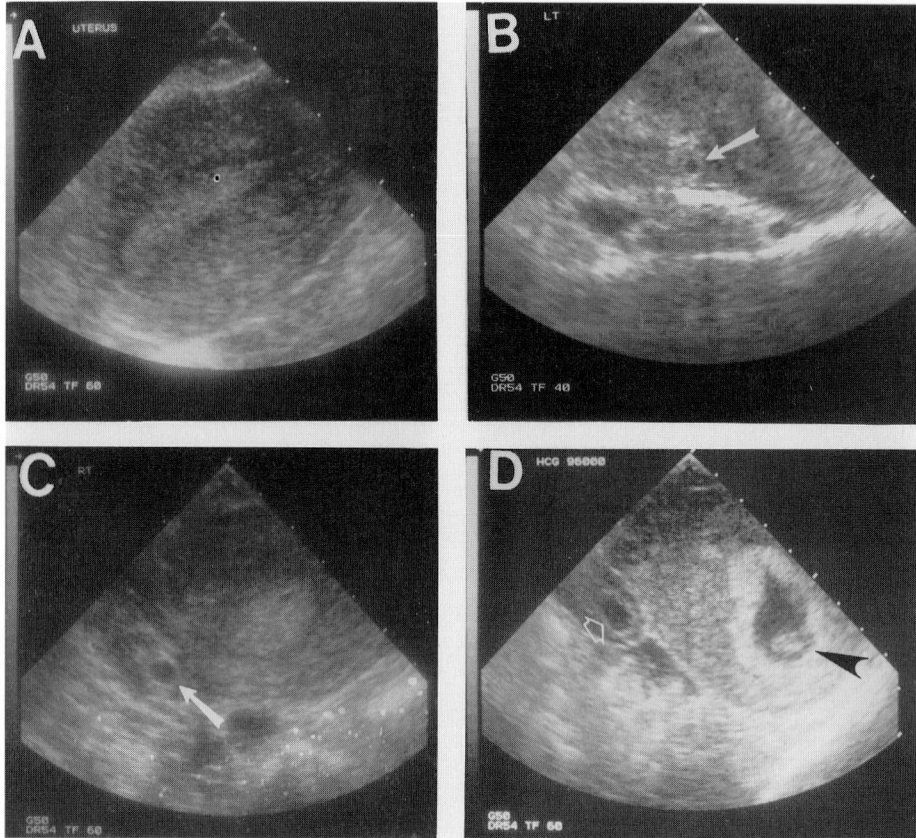

FIGURE 2 Demonstration of ultrasonographic findings in three patients suspected of having an ectopic pregnancy. (A, C) Ultrasonographic examination of the uterus disclosed no intrauterine gestational sac (A) and examination of the adnexa revealed a 7-mm gestational sac on the right (C, arrow) at an hCG concentration of 1920 mIU 1st IRP/ml. (B) Example of a gestational sac in the left adnexa measuring 4 mm at a quantitative hCG of 2030 mIU 1st IRP/ml. (D) A "sac-like" structure in the adnexa (arrow) proved to be a corpus luteum while the patient clearly had an intrauterine gestation (arrowhead) with fetal heart activity at an hCG concentration of 96,000 mIU 1st IRP/ml. Note the similarity of the sac-like structure in D and the true ectopic gestational sac in C.

equipment and expertise at different institutions and, hence, each hospital should establish its "own" discriminatory hCG zone and use that in the evaluation and management of patients carrying early gestations. At the Hospital of the University of Pennsylvania this hCG discriminatory level is currently set at 1500 mIU 1st International Reference Preparation / ml. Despite the true lowering of the discriminatory zone secondary to the improved ultrasound technology, review of the published reports uncovers some confusion because of the existence of two different hCG reference preparations used by investigators in the measurement of this hormone. Some assays use the 1st IRP and others the 2nd International Standard (2nd IS) as the hCG standard for calibration of the assay. In both, the concentrations are calculated in milli-International Units per milliliter (mIU/ml), but there is a two or threefold difference in the numerical result (1 mIU of the 1st IRP is equivalent to approximately 2 or 3 mIU of the 2nd IS). Currently, many commercially available assays utilize the 3rd IRP as their reference standard, which is equivalent to the 1st IRP. The clinician should appreciate these points in interpreting the reports in the literature and in comparing the published data to the ones at his/her own institution.

IV. DIAGNOSIS

When a patient presents pregnant, symptomatic, and hemodynamically unstable, she should be approached as an emergency and expeditiously managed given the clinical parameters. This would involve fluid resuscitation, cross-match for blood, immediate hemoglobin and hematocrit measurements, and urgent transport to the operating room. Whether a culdocentesis or laparoscopy is to be performed prior to laparotomy depends on the specific clinical situation.

In contrast, the pregnant patient who is hemodynamically stable and often asymptomatic (possibly with the exception of vaginal bleeding) most often can be managed expectantly, using the decision-making tree shown in Fig. 3.

V. MANAGEMENT

Currently, the standard management of ectopic pregnancy is surgical. In recent years, several reports of medical therapy of ectopic pregnancies have appeared in the literature and these are very promising. Methotrexate is the agent most widely evaluated with excellent success with respect to resolution of the ectopic gestation and minimal side effects. This has been achieved when the ectopic pregnancies are small (<2 or 3 cm), when they have shown no fetal heart activity on ultrasound, and when the hCG serum concentration is low (<3000–5000 mIU 1st IRP/ml). Two protocols have been proposed for use: Both involve the intramuscular administration of methotrexate, but the dosing and schedule of administration of the drug differs. In one protocol, metho-

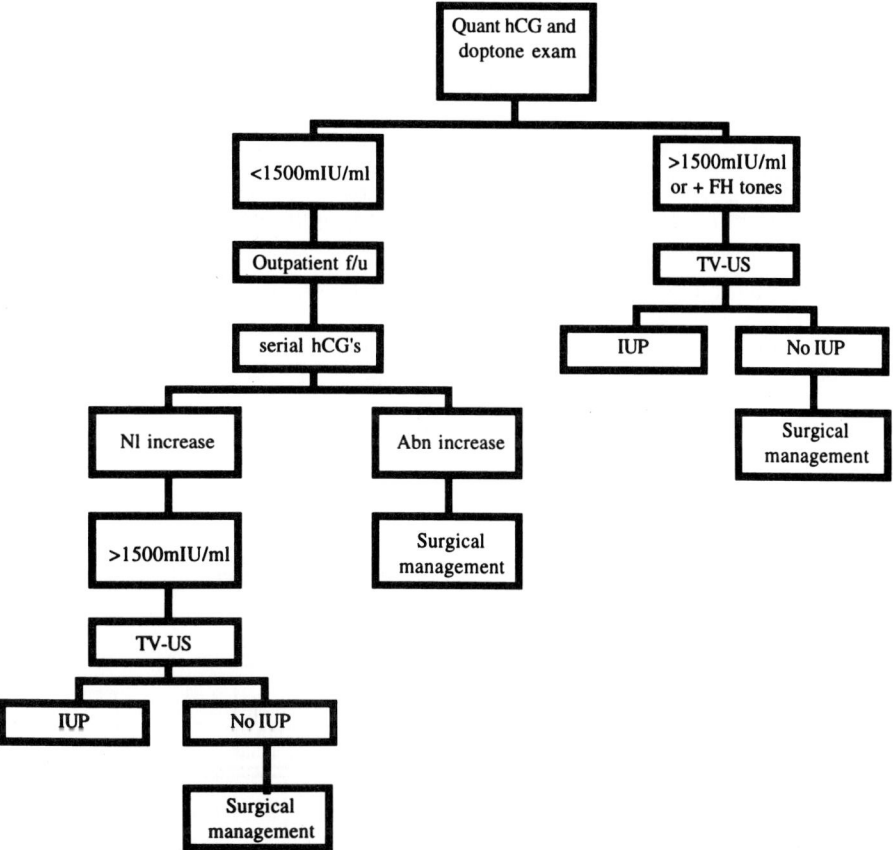

FIGURE 3 Algorithm for the early diagnosis of ectopic pregnancy utilizing quantitative hCGs and ultrasound examination in an emergency department setting. hCG, human chorionic gonadotropin; TV-US, transvaginal ultrasound; FH, fetal heart tones; IUP, intrauterine pregnancy.

trexate is administered at a dose of 1 mg/kg on Days 1, 3, 5, and 7 of treatment, alternating with leukovorum rescue on Days 2, 4, 6, and 8 at a dose of 0.1 mg/kg. This protocol is the most widely used worldwide and multiple studies have shown excellent results in the 80–90% range. A second protocol, which requires fewer visits but which may have a lower success rate, calls for the administration of 50 mg/m2 methotrexate as a single dose without leukovorum rescue. Due to a high failure rate (approximately 40%) of single-dose therapy, administration of a second equal dose a week later has been proposed. With the administration of a second methotrexate injection, results appear to be equivalent to those obtained with repeated injections. Given the excellent postoperative results obtained from laparoscopic operative management of ectopic gestations, it has been difficult to justify primary treatment of the majority of ectopics (possibly with the exception of interstitial pregnancies) with a chemotherapeutic agent. However, with improvements in the accuracy of preoperative diagnosis and visualization of very early ectopic gestations with transvaginal ultrasound, medical therapy of early ectopic pregnancies is rapidly becoming the standard of care, thus sparing the patient a surgical procedure. In addition, it has recently been suggested that direct instillation of the pharmacologic agent (i.e., methotrexate) into the ectopic chorionic sac under ultrasound guidance may provide adequate therapy. It should not be forgotten that even though such approaches may be effective in the treatment of ectopic pregnancies, the long-term effects of this therapy with respect to tubal patency, recurrence of ectopic pregnancy, and overall fertility remain to be evaluated.

Currently, operative management is the primary approach for patients with ectopic pregnancies (Fig. 4). If a ruptured ectopic pregnancy is present, a salpingectomy should be performed. This can be done laparoscopically, although, in the majority of the cases in which active bleeding is observed, a laparotomy may be required. If the ectopic gestation is not ruptured, then either "radical" or "conservational" management can be followed. The term conservational rather than "conservative" is used because, with the continued development and improvement of in vitro fertilization technology, conservational may not necessarily be conservative. Currently, the statistics suggest that this is certainly true if a particular patient has already suffered from two ectopic pregnancies. Given the current capacity for earlier diagnosis of an ectopic pregnancy, there is enough time to review treatment options with the patient and to arrive at a mutually agreeable plan of action.

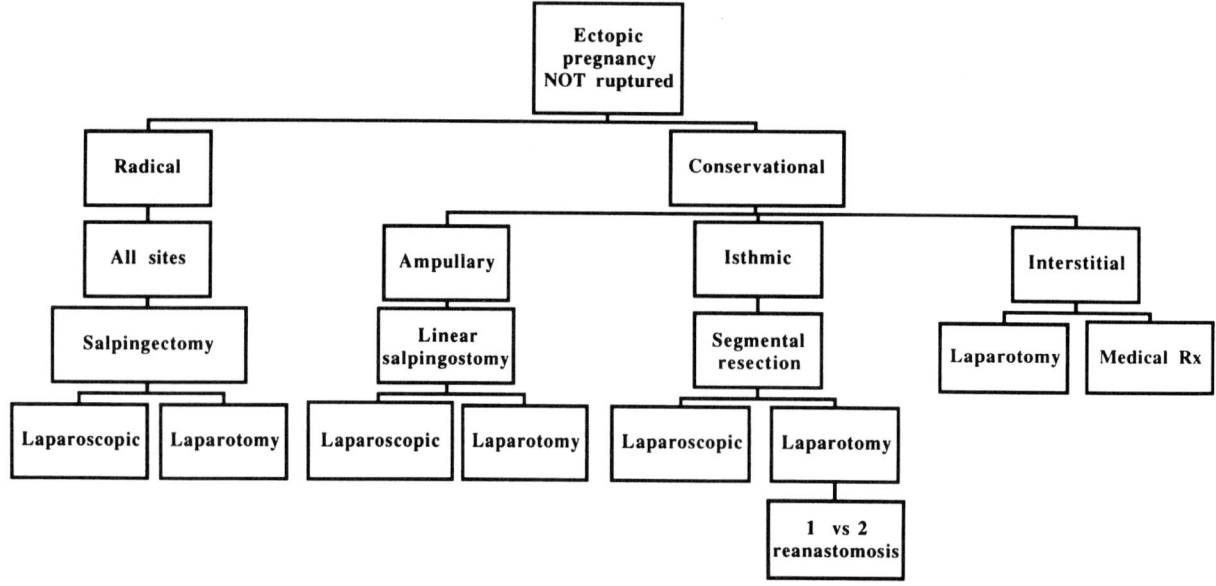

FIGURE 4 Proposed algorithm for the surgical management of an unruptured ectopic pregnancy.

It should be remembered that whether a procedure is performed through the laparoscope or by laparotomy, the principles of microsurgery should be followed: meticulous hemostasis, gentle handling of tissues, irrigation with heparinized saline or lactated Ringer's solution, and use of absorbable, fine sutures. Whether magnification is necessary is debatable. Nevertheless, most microsurgeons find magnification to be of use, and if the procedure is performed laparoscopically, some degree of magnification and close-up observation of the operative field is achieved. Regarding the operative procedure to be performed, the decision tree illustrated in Fig. 4 can be followed.

VI. FOLLOW-UP

Success of therapy of ectopic pregnancy depends on appropriate follow-up. Although the incidence of persistent ectopic pregnancies following conservative management is low, the impact of such an event on the patient's health and future fertility can be quite devastating. Thus, close follow-up and monitoring is warranted for all patients having undergone fertility preserving operative procedures for ectopic pregnancies. This allows for the prompt discovery of persistent functional trophoblastic tissue and initiation of further therapeutic procedures in a timely fashion. Although the traditional approach has involved a repeat diagnostic laparoscopy and subsequent surgical excision of the persistent ectopic tissue, use of medical therapy is proving to be an excellent adjunct in the management of these patients. As with the situation of primary medical therapy of ectopic pregnancy, the long-term effects of such an approach need to be evaluated. The decision-making tree illustrated in Fig. 5 can be followed as a guide for follow-up monitoring of patients after the surgical treatment of ectopic pregnancy without the use of salpingec-

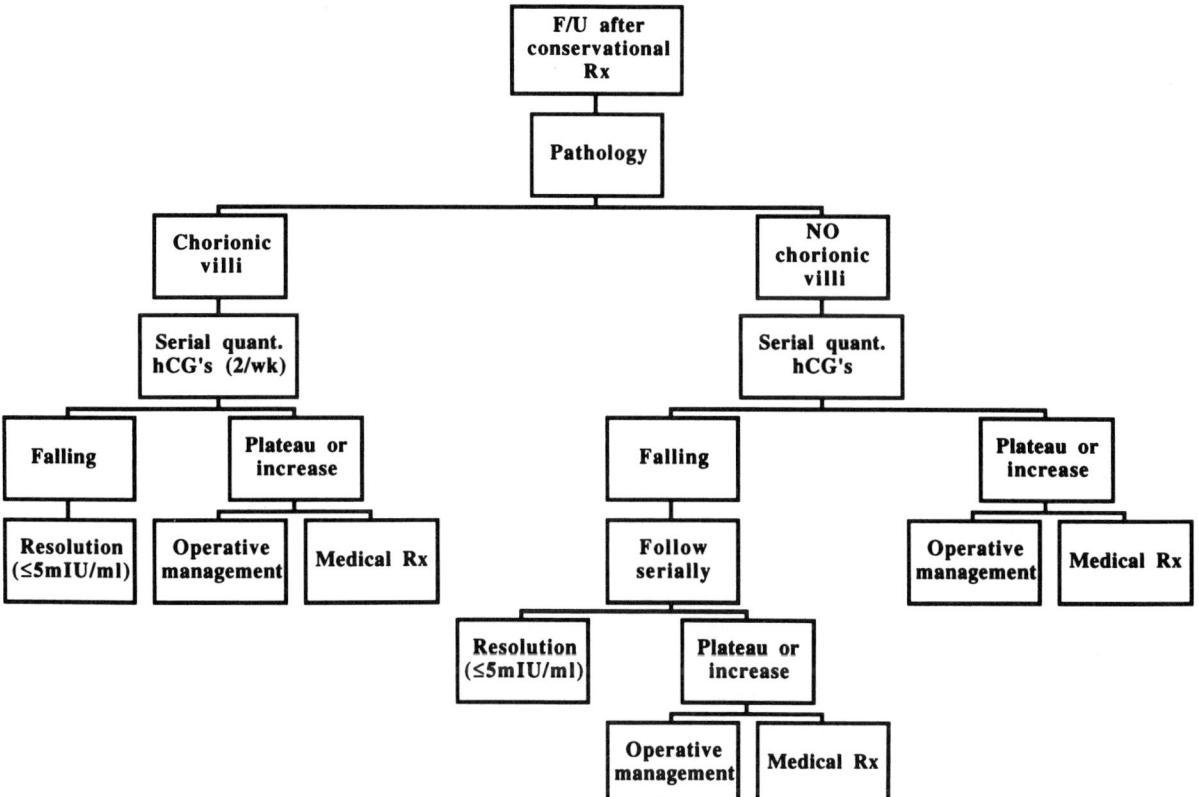

FIGURE 5 Proposed algorithm for follow-up of patients after surgical treatment for ectopic pregnancy which did not include the complete removal of the Fallopian tube. Rx, therapy.

tomy. Recently, in order to minimize the incidence of persistent ectopic pregnancy after linear salpingostomy, the suggestion has been made to treat these patients with a prophylactic, single-dose administration of methotrexate immediately following the surgical procedure. The utility of this approach remains to be determined.

VII. CONCLUSIONS AND FUTURE DIRECTIONS FOR RESEARCH

There is an evolving "standard of care" for the treatment of ectopic pregnancy. What is new today may become antiquated in a matter of a few years as technology improves diagnostic capabilities. Furthermore, advances and better designed clinical trials will open the way for the widespread use of medical therapy for ectopic pregnancies. These therapeutic modalities will have to withstand the test of time, as clinical trials prove their efficacy and their short- and long-term effects with respect to toxicity and future fertility. Nevertheless, we are currently in an era of new developments that markedly improve patient care and decrease the mortality and morbidity traditionally associated with ectopic pregnancy. With the dissemination of this new knowledge the number of adverse outcomes should be reduced dramatically.

See Also the Following Article

CHORIONIC GONADOTROPIN, HUMAN

Bibliography

Bangham, D. R., and Storring, P. L. (1982, February 13). Standardization of human chorionic gonadotropin, HCG subunits and pregnancy tests. *Lancet,* 390. [Letter]

Barnhart, K., Mennuti, M. T., Benjamin, I., Jacobson, S., Goodman, D., and Coutifaris, C. (1994). Prompt diagnosis of ectopic pregnancy in an emergency department setting. *Obstet. Gynecol.* **84,** 1010–1015.

Bernaschek, G., Rudelstorfer, R., and Csaicsich, P. (1988). Vaginal sonography versus serum human chorionic gonadotropin in early detection of pregnancy. *Am. J. Obstet. Gynecol.* **158,** 608–612.

Diamond, M. P., and DeCherney, A. H. (1987). Surgical techniques in the management of ectopic pregnancy. *Clin. Obstet. Gynecol.* **30,** 200–209.

Dorfman, S. F., Grimes, D. A., and Cates, W., Jr. (1984). Ectopic pregnancy mortality, United States, 1979 to 1980: Clinical aspects. *Obstet. Gynecol.* **64,** 386–392.

Fossum, G. T., Davajan, V., and Kletzky, O. A. (1988). Early detection of pregnancy with transvaginal ultrasound. *Fertil. Steril.* **49,** 788–791.

Goldstein, S. R., Snyder, J. R., Watson, C., and Danon, M. (1988). Very early pregnancy detection with endovaginal ultrasound. *Obstet. Gynecol.* **72,** 200–204.

Hay, D. L., and Lopata, A. (1988). Chorionic gonadotropin secretion by human embryos in vitro. *J. Clin. Endocrinol. Metab.* **67,** 1322–1324.

Ichinoe, K., Wake, N., Shinkai, N., Shiina, Y., Miyazaki, Y., and Tanaka, T. (1987). Non-surgical therapy to preserve oviduct function in patients with tubal pregnancies. *Am. J. Obstet. Gynecol.* **156,** 484–487.

Kadar, N., DeVore, G., and Romero, R. (1981). Discriminatory hCG zone; Its use in the sonographic evaluation for ectopic pregnancy. *Obstet. Gynecol.* **58,** 156–161.

Kadar, N., Taylor, K. J. W., Rosenfield, A. T., and Romero, R. (1983). Combined use of serum HCG and sonography in the diagnosis of ectopic pregnancy. *Am. J. Radiol.* **141,** 609–615.

Oelsner, G. (1987). Ectopic pregnancy in the sole remaining tube and the management of the patient with multiple ectopic pregnancies. *Clin. Obstet. Gynecol.* **30,** 225–229.

Ory, S., Villanueva, A. L., Sand, P. K., and Tamura, R. K. (1986). Conservative treatment of ectopic pregnancy with methotrexate. *Am. J. Obstet. Gynecol.* **154,** 1299–1303.

Pittaway, D. E. (1987). §-hCG dynamics in ectopic pregnancy. *Clin. Obstet. Gynecol.* **30,** 129–135.

Rodi, J. A., Sauer, M. V., Goriu, M. J., Bustillo, M., Gunning, J. E., Marshall, J. R., and Buster, J. E. (1986). The medical treatment of unruptured ectopic pregnancy with methotrexate and citrovorum rescue: Preliminary experience. *Fertil. Steril.* **46,** 811–815.

Storring, P. L., Gaines-Das, R. E., and Bangham, D. R. (1980). International reference preparation of human chorionic gonadotrophin for immunoassay: Potency estimates in various bioassay and protein binding assay systems; and international reference preparations of the α and § subunits of human chorionic gonadotropin for immunoassay. *J. Endocrinol.* **84,** 295–310.

Egg, Avian

Carol Masters Vleck
Iowa State University

I. Components
II. Egg Appearance and Number
III. Other Ecological Considerations
IV. Gas Exchange and Metabolism

GLOSSARY

air cell The gas-filled space that forms at the blunt end of the egg between the inner and outer shell membrane as water evaporates from the egg.

brood parasite A bird that does not incubate its own eggs but rather lays them in the nest of another bird, the host, which then incubates the eggs and cares for the young.

brood patch An area on the breast or abdomen of an incubating bird that is applied to the eggs and is defeathered, edematous, and highly vascularized to facilitate heat transfer to the eggs.

clutch The number of eggs a bird lays over a period of several days which are incubated together and usually hatch within a few days of each other.

external pipping The first break of the eggshell by the hatchling when it begins to emerge from the egg.

indeterminate layer Species that lay additional eggs if some eggs are removed from the nest before the clutch is complete.

internal pipping The breaking into the air cell several hours to days before hatching by the embryo which then begins to breath convectively from the air cell.

shell conductance A physical property of the eggshell determined by shell thickness and the number and geometry of pores in the eggshell that, along with the gradient in gas pressures, determines the rate at which gases move by diffusion between the egg and environment.

An avian egg is a cleidoic (closed) structure consisting of a calcified, gas-permeable eggshell, the albumen, and yolk. It contains all the organic nutrients, minerals, and water needed for the embryo and extraembryonic membranes to develop. Simultaneously, it provides a unit, protected from microbial invasion and mechanical damage, that the adult bird(s) can maintain at a temperature appropriate for development. Egg size, color, and usual number laid per clutch are characteristic for each species. Birds constitute the only vertebrate class with a significant number of species that are exclusively oviparous. Given the possible selective advantages of viviparity (e.g., close maternal regulation of temperature, nutrients, gases, water, and nitrogenous wastes and protection from predation and microbial infection), lack of viviparity in any bird may seem puzzling. Birds, however, have achieved most of the advantages that could be accrued from egg retention and viviparity by such specializations as nest construction and incubation behavior, yolk and albumen provisioning, and eggshell modification for strength, porosity, and camouflage.

I. COMPONENTS

The classic work on the composition of avian eggs was carried out by Alexis Romanoff. Much of what we know is based on poultry species of economic importance, but in the past 25 years considerable comparative research with eggs from free-living birds has been spearheaded by the work of Hermann Rahn and colleagues. The nutritive value and the warm temperature at which eggs are maintained for long periods would seem to make them very susceptible to microbial invasion, yet such invasion is relatively rare in the intact egg in nature due to the properties of the shell and egg constituents.

A. Eggshell and Membranes

The rigid, avian eggshell typically consists of an outermost organic cuticle that impedes entry of microorganisms and a much thicker layer of calcium carbonate crystals arranged in columns (Fig. 1). It constitutes 11–15% of the egg's mass. Bird embryos rely on the eggshell for most of the calcium needed for development. The crystal layer is traversed by pores that provide a gas conductance pathway between the environment and the egg contents. These pores are usually simple, funnel-shaped structures but may be elaborately branched as in swan eggs. The shell is laid down upon an underlying fibrous membrane, the outer shell membrane. An inner shell membrane becomes lined with the chorioallantoic membrane produced by the embryo as development proceeds. During incubation water evaporates from the egg, leaving a gas-filled space between the two shell membranes that usually opens at the blunt end of the egg, forming an air cell. Just before external pipping most embryos pierce the inner shell membrane (internal pipping) and begin lung ventilation from the contents of the air cell.

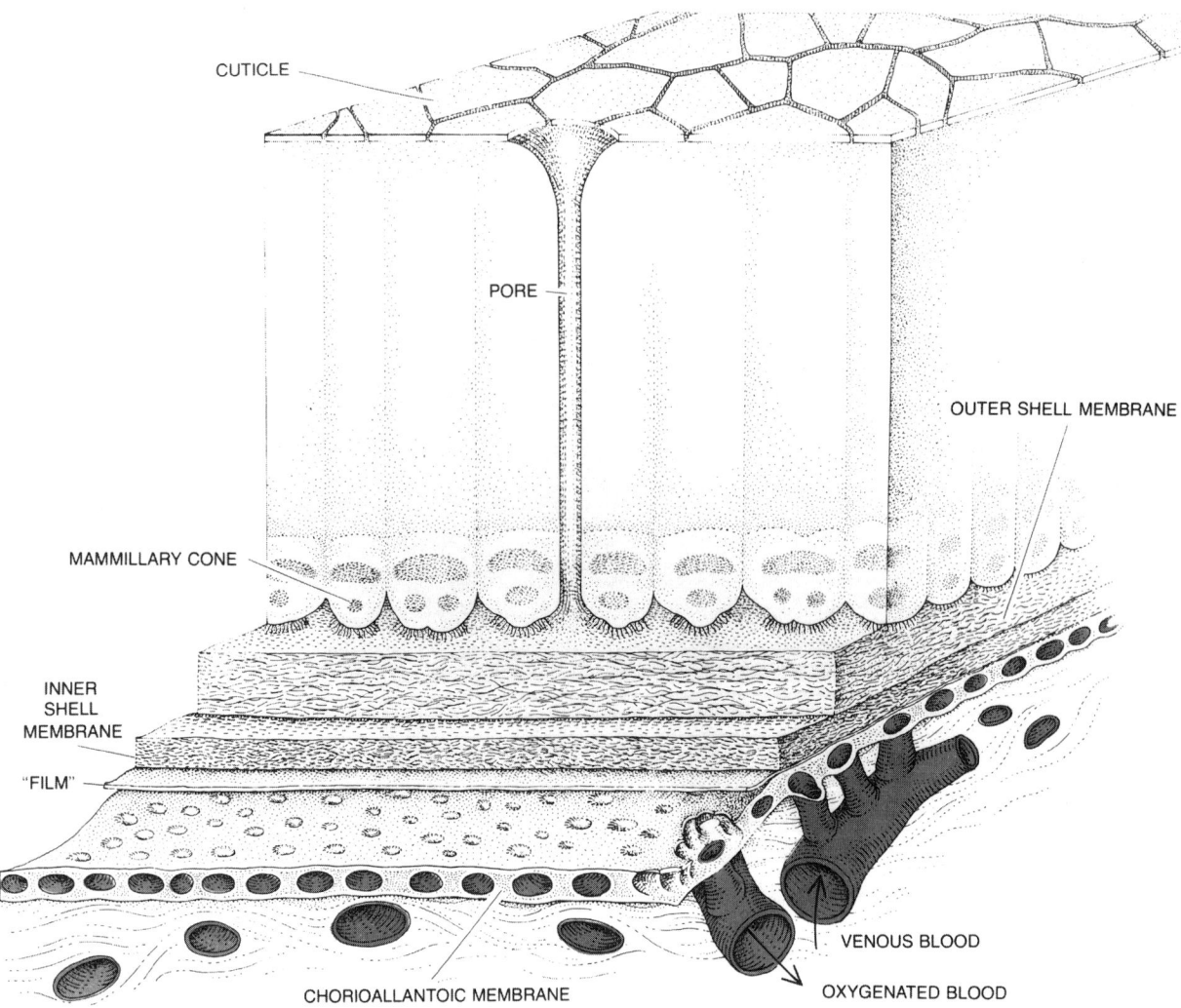

FIGURE 1 Structure of an avian eggshell with underlying shell membranes and chorioallantoic membrane. The cuticle is the outermost layer, under which are columns of calcite crystal and air-filled pores. These terminate on the outer shell membrane, which is loosely attached to the inner shell membrane. The chorioallantoic membrane contains blood capillaries that permit gas exchange between the blood and the gas within the pores (reproduced with permission from Rahn et al., 1979).

The thickness of the eggshell increases with the size of the egg. The shell must be thick enough to prevent mechanical damage during incubation, desiccation, and invasion of microbes. It must not be so thick as to prevent the entry of oxygen for embryonic metabolism and escape of carbon dioxide or to prevent the chick from breaking out at the end of development. The tensile and compressive strengths of eggshells are very high. Most bird embryos have a specialized knob on the upper bill called an egg tooth and a hypertrophied complexus muscle in the neck to facilitate breaking out of the shell. These are lost or reduced in size soon after hatching. The thickness of the eggshell and the diameter of the pores determine the conductance of the shell to gases. In general, shell conductance among species increases directly with egg mass and inversely with length of the incubation period.

B. Membranes

The extraembryonic membranes are temporary living extensions of the embryo that function as organs of exchange, transport, and excretion (Fig. 2). The yolk sac grows to enclose the yolk with projections that penetrate into the yolk and functions to absorb nutrients for transport to the embryo by way of the vitelline circulation. The chorioallantois, formed from the fusion of the chorion and allantois, contains blood vessels which lie close to the inner shell membrane and transport calcium and respiratory gases between the embryo and the egg surface. This vascularized structure thus serves as the respiratory and excretory system for the embryo. The allantoic fluid accumulates nitrogenous waste products (mostly water-insoluble urates) which are left behind in a crystalline form at hatching. The amniotic membrane encloses a sac that surrounds the embryo and bathes it in amniotic fluid which protects the embryo from infection, mechanical injury, desiccation, and adhesion. Although these membranes contain living tissue, the energy they consume is small relative to that used by the embryo proper.

C. Albumin

The largest portion of most eggs is the egg white or albumin. The albumin is an important source of water, structural proteins, and amino acids for embryonic growth. The albumen consists of a mixture

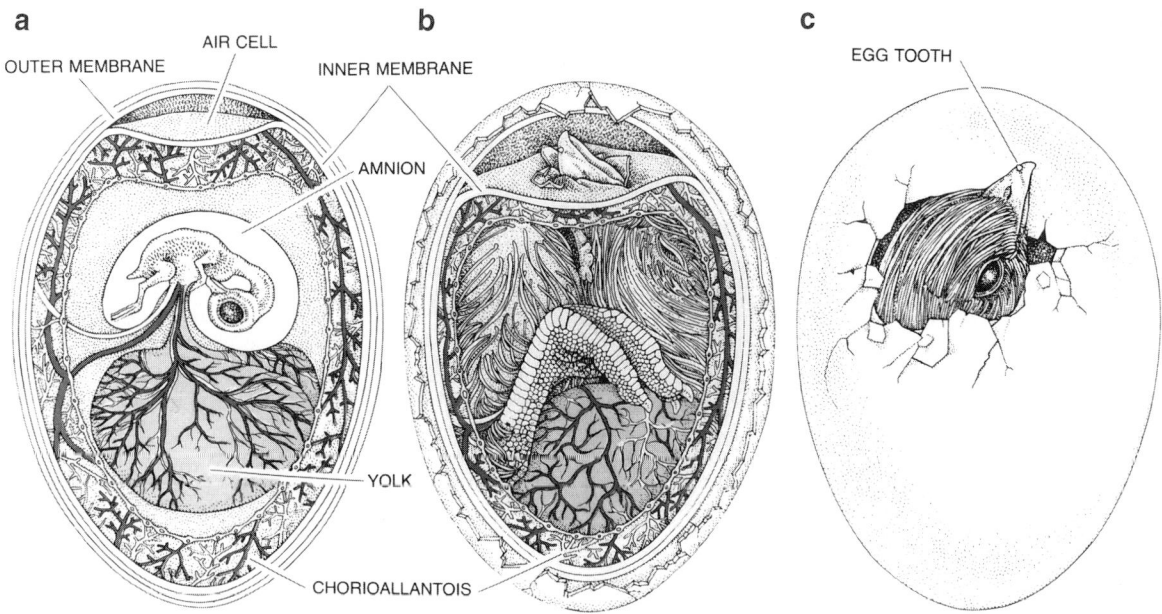

FIGURE 2 Diagram of the typical chicken egg during development. (a) Early development showing the embryo and extraembryonic membranes. (b) Internal pipping of the embryo into the air cell on approximately Day 19 of incubation. (c) External pipping of the eggshell and hatching on approximately Day 21 of incubation (reproduced with permission from Rahn et al., 1979).

of water (~90%) and proteins (~10%), principally ovalbumin and ovotransferrin. Many of these proteins bind mineral ions and/or have bactericidal and antiviral activity. The albumin is organized into a thin, watery layer and a gel-like, thick layer. At each end of the yolk strands of protein form the rope-like chalazae which suspend and center the yolk as the incubating bird turns the egg. This turning seems to keep the embryo and membranes from sticking to the inside of the eggshell.

D. Yolk

Prior to embryogenesis the egg yolk is bounded by the proteinaceous vitelline membrane that prevents mixing with the albumen. Yolk has a relatively low water content (about 50%), the remainder consisting of protein and lipid at a ratio of about 1:2. The lipid is present primarily as noncovalently bonded lipoprotein. This yolk lipid supplies the major portion (~90%) of energy used during development. Lipovitellin in the yolk complexes with calcium and iron. Special binding proteins for vitamins are also found in the yolk. The yolk is laid down in concentric bands probably related to the diurnal patterns in food availability to the laying female. The yellow color of yolk is due to carotenoid pigments. Maternal steroids can also be present in yolk that can influence the development and behavior of the hatchling. The blastoderm which develops into the embryo proper floats at the top of the yolk. At hatching some residual yolk remains and is incorporated into the embryo's abdominal cavity to serve as a source of nutrients for early postnatal growth and maintenance.

E. Comparative Composition

The composition of eggs varies among bird species that produce hatchlings of different maturities (Fig. 3). Species such as passerines and woodpeckers, in which the hatchlings emerges in a relatively undeveloped, altricial state, lay eggs that have a small yolk (<25% of contents), a high water content (>80% of contents) because albumen content is high, and low overall energy density (~4.7 kJ/g wet mass). Species such as ducks and chickens, in which the hatchling emerges in a relatively independent, precocial state, produce eggs that have a large yolk (>40% of contents), a low water content (<73% on contents), and high energy density (>7.9 kJ/g wet mass). An altricial chick at hatching also has a higher water content and less residual yolk than a precocial chick. The residual yolk comprises about 9% of hatchling mass in altricial chicks and 18% of hatchling mass in precocial chicks. Species that produce chicks of intermediate maturity have intermediate values for relative yolk, albumen, and water content.

FIGURE 3 An overview of the differences in egg composition including water content, energy density, and relative yolk content in hatchlings of different maturities from altricial to precocial (reproduced with permission from Sotherland and Rahn, 1987).

% Water	kJ·g⁻¹	% Yolk
82	4.7	20
78	6.3	30
73	7.9	40
67	9.5	50
61	12.3	70

II. EGG APPEARANCE AND NUMBER

A. Size

Eggs from living birds vary in size from approximately 0.25 g (bee hummingbird) to 1.5 kg (ostrich) (Fig. 4). The egg of the extinct elephant bird *Aepyornis* of Madagascar probably weighed about 9 kg.

FIGURE 4 Eggs of birds ranging in size from approximately 1 g to 1.5 kg. Note the variation in shape and coloration. Species included from top left to bottom right: zebra finch (*Poephila guttata*), black-throated sparrow (*Amphispiza bilineata*), brown-headed cowbird (*Molothrus ater*), red-winged blackbird (*Agelaius phoeniceus*), western scrub-jay (*Aphelocoma californica*), American robin (*Turdus migratorius*), Gambel's quial (*Callipepla gambelii*), common murre (*Uria aalge*), redhead duck (*Aythya americana*), wild turkey (*Meleagris gallopavo*), great horned owl (*Bubo virginianus*), ring-necked pheasant (*Phasianus colchicus*), ostrich (*Struthio camelus*), emu (*Dromiceius novaehollandiae*), and greater rhea (*Rhea americana*).

Egg mass in general scales with the laying female's body mass raised to about the 0.77 power, which means that large birds lay proportionally smaller eggs than small birds. Small passerine birds typically lay eggs that are more than 10% of their body mass, whereas the egg of a female ostrich is only about 1.5% of her body mass. Birds that produce altricial young tend to lay smaller eggs than similar sized species that produce precocial young. Some taxa of birds lay eggs that vary from the expected in a predictable way. For example, all Procellariiformes (tube-nosed seabirds) lay one egg that is exceptionally large for their body size. Brood parasites, such as cuckoos and cowbirds, lay small eggs for their body size. The kiwi lays an egg that can be up to 25% of its own body mass. Intraspecific variation in egg size can be considerable and is often related to age or condition of the mother, laying order, and genetic differences between females.

B. Shape

Egg shape is determined by the muscular movements of the oviduct as the eggshell is laid down. Bird eggs vary in shape from almost spherical to long and pointed. The functional significance of egg shape relates to closest packing arrangement within a nest and to the danger of the egg rolling. Cavity nesters, such as owls and kingfishers, tend to lay rounded eggs. Pointed ovoid eggs are laid by many shorebirds, with three or four eggs fitting tightly into a nest with rounded ends oriented out. Cliff-nesting species, such as murres and guillemots, lay long, pointed eggs that are likely to roll in a tight circle rather than off the cliff when disturbed.

C. Color and Texture

Pigments that give the eggshell its color are derived from pyrrole products produced in the red blood cells or in the oviduct. Pigments are confined to the cuticle and outer part of the calcified layer of the eggshell. The ground colors of eggs range from white to dark green or red (Fig. 4). Patterns on the egg surface (speckles, splotches, and elaborate scrawls) are determined by the movement of the egg within the female's uterus as pigments are deposited. These color patterns are under genetic control, but the mechanism that controls this process is not well understood. These color patterns provide for camouflage and parent recognition. The dark pigments of eggs tend to be very reflective in the near-infrared spectral region which helps prevent overheating of exposed eggs. Cavity nesters, such as woodpeckers, kingfishers, and swifts, tend to lay white eggs. Birds that either cover their eggs with plant material on leaving the nest (ducks and grebes) or begin to incubate after the first egg is laid (pigeons and penguins) also tend to have white eggs. Experimental evidence has shown that some birds can recognize their own eggs or discriminate between a foreign egg and their own based on individual differences in the color patterns. This recognition seems to be best developed in birds that are exposed to brood parasites. Egg textures range from the very glossy eggs produced by South American tinamous to the dull, chalky eggs of penguins, ibises, and megapodes. Some ratite eggs appear heavily pitted. The function of these textures is not known.

D. Clutch Size

Clutch size, the number of eggs laid over a period of several days into the nest, varies from 1 to more than 20. Clutch size is an evolutionary adaptation determined by natural selection, but is also affected by environmental conditions. Many birds lay 1 egg per day until the clutch is complete, but other species require 3–5 days or more between eggs. Inter-egg-laying interval increases during periods of low resource availability. Clutch size may be fixed, as in procellariiform birds (1 egg), hummingbirds (2 eggs), doves (2 eggs), and plovers (4 eggs), or it may vary with latitude, habitat, date in season, and female age or quality. Purple gallinules tend to have 3 or 4 eggs per clutch in the tropics and 5–10 eggs per clutch in the northern part of their range. Many hawks and owls lay larger clutches in years of higher rodent abundance. Within a taxonomic group, clutch size tends to decrease with body mass and with egg mass. Hole-nesting species lay larger clutches than those nesting in the open, possibly because of reduced predation pressure. It has been suggested that altricial species lay the maximum number of young that parents can feed at one time and that precocial species lay the maximum number of eggs which the female can produce and incubate effectively; however, experimental evidence for these generalities is only partly supportive and other factors, such as bird longevity, predation pressure, and seasonality, also affect the evolution of clutch size.

From one to more than five clutches may be laid in a year. Most large or precocial species and species in which care of young is extensive (penguins and other seabirds, hawks, and eagles) have only one nesting cycle per year. Small altricial species often attempt more than one cycle per year, particularly in years of abundant resources. At low latitudes, some species do not have strong seasonal breeding cycles and may produce clutches almost continuously as long as resources are available, although the time to raise one clutch to independence is often longer than that for birds at higher latitudes.

III. OTHER ECOLOGICAL CONSIDERATIONS

A. Egg Predation

Many vertebrates, including snakes, other birds, and mammals, use bird eggs as a major food source during the nesting season. The avian egg has been described as the most complete, natural human food, lacking only vitamin C, which the avian embryo can synthesize for itself. Many of the characteristics of nests, egg color, and parental behavior are designed to foil egg predators, yet predation is the primary cause of nest failure. Many birds will compensate for the loss of the clutch to predators by renesting, even if this means that the eggs hatch late in the regular nesting season. Most small passerines will renest within 4–7 days after nest loss. If eggs are removed during the laying process, birds known as indeterminate layers will continue to lay eggs to replace those removed from the nest. Domestic hens are indeterminate layers, as are some woodpeckers. One famous flicker laid 71 eggs in 73 days when the new egg was removed each day by a curious egg collector.

B. Brood Parasitism

About 80 species of birds, including European cuckoos, North and South American cowbirds, and African honey guides and widow weaverbirds, are obligate brood parasites. They do not incubate their own eggs but rather lay their eggs in the nests of another species. Brood parasites typically lay more eggs (up to 30 or more in some cases) than does a typical bird that produces and cares for a clutch of its own. The victims of this parasitism are usually smaller passerine birds that lay open-cup nests. Brood parasites can deposit their egg quickly into the host's nest. Their eggs are usually small for the body size and have short incubation periods. Consequently, they usually hatch before those of the host species. The brood parasite often removes one of the host's eggs at the time it lays its own. In addition, the hatchling of many of these parasitic species often destroys other eggs in the nest or kills other young so that only it remains to be fed by the host parents. The brood parasites often have quite variable egg color between individuals and birds tend to lay in the nests of species whose eggs match the ones that they produce. Some species of birds recognize a parasite's egg and will throw out the egg, abandon the nest entirely, or build a new bottom over the eggs and re-lay, but most host species incubate the parasite's egg along with their own and raise the nestling as if it were their own.

Other species (including some cuckoos, quail, ducks, pheasants, rails, and many passerine birds) occasionally lay eggs in the nest of another bird which may be another member of the same species or a different species. These same birds may also raise a clutch of their own. Such facultative brood parasitism may be a step in the evolution toward obligate brood parasitism.

C. Male/Female Role in Egg Care

In birds, postlaying care of eggs or chicks can be taken on by the male, in contrast to viviparous mammals. This is often cited as the evolutionary explanation for the fact that over 90% of bird species are categorized as monogamous. The male may provide food for the laying female, incubate the eggs, feed the incubating female on the nest, or provide care (thermal, food, or protection) for the hatchlings. In over 50% of avian families, both sexes incubate the eggs. The male alone incubates the eggs and cares for young in about 6% of avian families. In other cases the male's only contribution may be to defend the territory in which the female resides, to defend the female from other males, or just to fertilize the eggs.

IV. GAS EXCHANGE AND METABOLISM

A. Water and Respiratory Gases

The developing embryo requires oxygen for cellular metabolism and produces carbon dioxide. These respiratory gases travel through the pores of the eggshell by the process of diffusion. The rate of metabolism is very low early in development but becomes substantial late in development when the embryo nearly fills the eggs (Fig. 2). In many species the partial pressure of oxygen within the egg falls and carbon dioxide increases to levels that are remarkably similar to those found within the adult lung. Eggs that develop at high altitudes or have shell conductances that are unusually low can develop, however, even though gas composition within the egg is quite extreme. Elevation in carbon dioxide seems to serve as a cue for internal pipping, the time when the embryo breaks into the air cell and begins to aerate the lungs.

The porosity of the shell to respiratory gases means that the shell is also permeable to water, and water is lost from the saturated contents of the egg to the environment at a nearly constant rate. Water loss is replaced in part by metabolic water produced by the metabolism of lipids so that the proportion of water in the egg contents remains nearly constant. Water loss usually totals about 10–15% of the egg mass. This mandatory loss provides for the increase in the air cell space in which the embryo aerates the lung after internal pipping, but prior to external pipping. Tolerance to changes from the mean water loss is quite high in some species, such as red-winged blackbirds, which show no change in hatchability when as little as 7% or as much as 33% of the original mass is lost as water. In general, avian embryos deal with changes in water economy by differentially adjusting the volume and composition of compartments within the egg.

B. Shell Conductance

The shell conductance, determined by the eggshell thickness and the number and geometry of the pores, is usually constant through incubation although it may increase slightly due to calcium removal from the eggshell, abrasion of the shell cuticle, or decrease in shell water content. Shell conductance is under genetic control and presumably is set by natural selection to optimize gas exchange. It represents an evolutionary compromise between necessity to take up oxygen, eliminate carbon dioxide, and regulate water loss. Shell conductance tends to be relatively low in eggs laid in dry environments (penguins and desert birds) and high in eggs laid in wet environments (cavity nesters, grebes, and megapodes). Shell conductance is often low in birds with unusually long incubation periods, but these birds often pip the shell relatively early in development, which permits greater gas exchange late in development. At high altitudes conductance is often reduced, possibly to limit rates of water loss at altitudes at which diffusion coefficients increase. However, in eggs laid at very high elevations (>~2800 m), conductance of the shell is increased, possibly as partial compensation for the low partial pressure of oxygen.

C. Metabolism and Energetics

If one compares eggs of the same mass, eggs of precocial species have longer incubation periods and higher energy costs of development (~30%). They also hatch a chick with a higher energy density and more residual yolk than altricial species. Precocial species incur higher total energy costs than altricial species for the same egg mass because the embryo is larger for a greater proportion of the incubation period than is the altricial embryo. Thus, the precocial embryo's total maintenance costs to support the larger body size are greater than those of the altricial species.

If one compares eggs with same energy content, however, the differences between altricial and precocial species virtually disappear. The energy costs of development (~17 kJ/g dry mass of hatchling) and hatchling energy content (~50% of the original energy stored in the egg) do not differ between altricial and precocial species. These results suggest that natural selection has adjusted the egg energy content for each species to correspond closely to the energy needs of the embryo. If energy needs are higher because of the precocial mode of development or because of a long incubation period, then the egg energy content is also higher and vice versa. Eggs with unusually long incubation periods for their size have a higher energy content and use more energy during development for maintenance costs than those with shorter incubation periods. A 10% change in the incubation period produces about a 5% change in the energy consumed in the production of the chick.

D. Thermal Properties

The eggshell is not a great barrier to heat transfer. In nature eggs are heated to a temperature between about 32 and 35°C when the incubating adult applies the brood patch to the eggs in a nest. An exception to this rule are the mound-building megapode birds of Oceania and Australia in which the heat source for incubation can be sunlight, geothermal heat, or decomposition of vegetable material. The embryo floats on the surface of the yolk in early development, which means that it is in close apposition with the brood patch and thus warmer than the rest of the egg. Heat is usually lost across other surfaces of the eggshell by conduction and convection; heat loss due to evaporation from the shell also occurs. The heat capacity of the egg will increase with relative water content, so eggs of altricial species will have a higher heat capacity and slower rate of cooling for their size than eggs of precocial species. During development the embryo's metabolic rate increases, which could elevate egg temperature, but simultaneous increase in embryonic blood flow may facilitate increasing heat loss from the egg surface as well.

See Also the Following Articles

ALTRICIAL AND PRECOCIAL DEVELOPMENT IN BIRDS; AVIAN REPRODUCTION, OVERVIEW; BROOD PARASITISM IN BIRDS; PARENTAL BEHAVIOR, BIRDS

Bibliography

Ar, A., Arieli, B., Belinsky, A., and Yom-tov, Y. (1897). Energy in avian eggs and hatchlings: Utilization and transfer. *J. Exp. Zool. Suppl.* **1**, 151–164.

Bakken, G. S., Vanderbilt, V. S., Buttemer, W. A., and Dawson, W. R. (1978). Avian eggs: Thermoregulatory value of very high near infrared reflectance. *Science* **200**, 321–324.

Blackburn, D. G., and Evans, H. E. (1986). Why are there no viviparous birds? *Am. Nat* **128**, 165–190.

Board, R. G. (1982). Properties of avian egg shells and their adaptive value. *Biol. Rev. Cambridge Philos. Soc.* **57**, 1–28.

Burley, R. W., and Vadehra, D. V. (1989). *The Avian Egg: Chemistry and Biology*. Wiley, New York.

Carey, C. (1983). Structure and function of avian eggs. In *Current Ornithology Vol. 1*. (R. F. Johnson, Ed.), pp. 69–103. Plenum, New York.

Carey, C. (1986). Tolerance of variation in eggshell conductance, water loss, and water content by red-winged blackbird embryos. *Physiol. Zool.* 59, 109–122.

Frith, H. J. (1956). Breeding habits in the family Megapodiidae. *Ibis* 98, 620–640.

Lack, D. (1968). *Ecological Adaptations for Breeding in Birds.* Methuen, London.

Rahn, H. (1991). Why birds lay eggs. In *Egg Incubation: Its Effects on Embryonic Development in Birds and Reptiles* (D. C. Deeming and M. W. J. Ferguson, Eds.), pp. 345–360. Cambridge Univ. Press, Cambridge, UK.

Rahn, H., and Ar, A. (1974). The avian egg: Incubation time and water loss. *Condor* 76, 147–152.

Rahn, H., Ar, A., and Paganelli, C. V. (1979). How bird eggs breathe. *Sci. Am.* 240(3), 46–55.

Romanoff, A. L. (1967). *Biochemistry of the Avian Egg.* Wiley, New York.

Rothstein, S. I. (1974). Mechanism of avian egg recognition: Possible learned and innate factors. *Auk* 91, 796–807.

Schwabl, H. (1993). Yolk is a source of maternal testosterone for developing birds. *Proc. Natl. Acad. Sci. USA* 90, 11446–11450.

Sotherland, P. R., and Rahn, H. (1987). On the composition of bird eggs. *Condor* 89, 48–65.

Turner, J. S. (1994). Time and energy in the intermittent incubation of birds' eggs. *Israel J. Zool.* 40, 519–540.

Tyler, C. (1969). The snapping strength of eggshells of various orders of birds. *J. Zool. London* 159, 65–77.

Vleck, C. M., and Vleck, D. (1996). Embryonic energetics. In *Avian Energetics and Nutritional Ecology* (C. Carey, Ed.), pp. 417–460. Chapman & Hall, New York.

Egg Coverings, Insects

Michael P. Kambysellis
New York University

Lukas Margaritis
University of Athens

Elysse M. Craddock
Purchase College, State University of New York

I. Introduction
II. Egg Surface Morphology
III. Radial Complexity of Eggshell Architecture
IV. Development of the Egg Coverings
V. Functional Adaptations of Insect Egg Coverings
VI. Conclusion

GLOSSARY

aeropyle Derived from the Greek words *aeras* (air) and *pyli* (passage); refers to the channels through the insect eggshell that facilitate gas exchange.

chorion The multiple, mostly proteinaceous, layers that collectively form the outer covering or shell surrounding the insect egg.

dorsal ridge A specialized chorionic structure extending longitudinally on the dorsal face of the eggs of Hawaiian *Drosophila* (Diptera).

follicle imprints The polygonal (mostly hexagonal) pattern on the surface of many insect eggs corresponding to the arrangement of follicle cells that constructed the multilayered chorion. The ridges or imprint borders mark the positions of cell boundaries.

micropyle (micropylar apparatus) Derived from the Greek words *micro* (small) and *pyli* (passage); denotes the localized structure on the egg surface that permits sperm to traverse the chorion to fertilize the oocyte.

oocyte The female gamete cell prior to fertilization; in insects this cell grows to a large size during oogenesis as a result of the uptake of yolk and other materials necessary to support embryogenesis.

operculum The specialized structure in many insect eggs that facilitates the escape of the larva or nymph from the eggshell at the time of hatching. This region of the egg surface may also function in respiration.
pillars The columnar structures supporting the intrachorionic meshwork within the insect eggshell that form part of the respiratory system of the egg.
plastron A superficial gas film of constant volume and an extensive air/water interface that facilitates underwater respiration.
vitelline envelope (vitelline membrane) The innermost layer of the insect egg coverings directly in contact with the plasma membrane of the oocyte.

Insect egg coverings are the multiple proteinaceous and waxy layers that surround the plasma membrane of the female gamete. Generally quite thick and tough, the complex layers outside the vitelline envelope are collectively referred to as the eggshell or chorion.

I. INTRODUCTION

Eggshell construction takes place in the insect ovaries during the final phases of oogenesis when the ovarian follicle cells secrete a successive series of layers from the innermost vitelline envelope to the outermost chorionic layer. These egg coverings serve to protect the fertilized zygote and developing embryo from desiccation, an essential requirement since insects are primarily terrestrial organisms and for the most part undergo external development. In some insects, particularly those in the orthopteroid orders, the eggs are packaged in groups within a further envelope composed of secretions of the female accessory glands. These coverings are described elsewhere in the encyclopedia.

While needing a tough eggshell to protect it from desiccation, the insect embryo must still be able to undertake respiratory exchange with its environment in order to support its metabolism and growth. Furthermore, when embryogenesis is complete, the first-instar larva or nymph must be able to escape from the egg coverings in order to seek food and continue its development. An additional constraint on eggshell architecture is imposed by the fact that although insects have adopted internal fertilization, in most cases formation of the eggshell is complete before the egg has left the ovaries. Some provision for sperm entrance through the eggshell is thus a necessity to ensure successful fertilization—a prerequisite for offspring production and perpetuation of the species in all but a few parthenogenetic insects.

The evolutionary success and predominance of insects on earth is due in large measure to the fact that they have managed to invade and successfully reproduce in a wide diversity of ecological habitats. Insect eggs are laid or oviposited in the open, in dry sand or moist soil, in or on various parts of living or decaying plants, in other animals, in decaying animal flesh, and even in still or running water, with females of each species selecting a very specific habitat for oviposition. As a consequence of the plethora of substrates utilized for insect oviposition, the diversity of physical environments in which insect eggs are found, and the numerous other constraints on egg architecture, insect eggshells display a wide range of structural modifications, each egg being adapted to particular features of the species' ecological niche. Coupled with the extraordinary size range and taxonomic diversity of insects, the remarkable adaptations of insect eggs that have evolved to meet multiple functional and environmental constraints constitute a broad array of morphological solutions to covering the insect egg.

In the following sections we describe the major layers that comprise insect egg coverings and specific regional attributes of the chorion, outline their developmental patterns, and discuss similarities and differences in particular egg surface structures as they relate to functional and adaptive constraints.

II. EGG SURFACE MORPHOLOGY

A. General Features

Adult insects vary tremendously in size from minute organisms smaller than the largest protozoa to gigantic specimens of some Australian stick insects, which exceed 25 cm in length. Their eggs show a corresponding variation in size as well as in shape. Insect eggs range from the tiny eggs of the ephemeropteran *Electrogena zebrata* (0.14 μm in diameter) to the large eggs, often more than 10 mm long, of many stick insect species. Egg shape is also quite variable, with most eggs being ovoid, oblong, or elongate with distinct anterior and posterior poles (Figs. 1–3), but some are lenticular (Fig. 3c) or even spherical (Fig. 3d) or spheroidal (Fig. 1f). In the more cylindrical eggs the poles may be pointed (Fig. 1b) or rounded, and the eggs may bear various appendages,

FIGURE 1 Sample of the diversity in external morphology of insect eggs. (a) Scanning electron micrograph (SEM) of egg (×35) of *Drosophila claytonae* (Diptera) showing the long respiratory filaments (rf) (courtesy of K. Dellas). (b) SEM of egg of *Corythucha arcuata* (Hemiptera) with smooth chorion surface and anteriorly a cap (C), rim (R), and filament (F) of the operculum (reproduced with permission from G. T. Baker and R. L. Brown, *Proc. Entomol. Soc. Wash.* 96, 70–73, Fig. 2, 1994). (c) SEM of egg of *Epitedia faceta* (Siphonaptera) with a honeycomb-patterned chorion and at the posterior end (uppermost) a disk with micropylar ring (arrow). Scale bar = 100 μm (reproduced with permission from J. R. Linley, A. H. Benton, and J. F. Day, *J. Med. Entomol.* 31, 813–827, Fig. 11, 1994). (d) SEM of egg of *Baculum extradentatum* (Phasmatodea) showing operculum at the anterior pole (arrow) with a pseudocapitulum and on the dorsal face the micropylar plate bearing the micropylar cup (arrowhead). Scale bar = 50 μm (reprinted from *Int. J. Insect Morphol. Embryol.* 22, M. Mazzini, M. Carcupino, and A. M. Fausto, p. 407, Fig. 54, copyright 1993, with kind permission from Elsevier Science Ltd, The Boulevard, Langford Lane, Kidlington 0X5 1GB, UK). (e) SEM of egg of *Philonthus flavolimbatus* (Coleoptera) with longitudinally ridged chorion and anteriorly (top) a club-shaped, tubular process. Scale bar = 0.3 mm (reproduced with permission from G. Y. Hu and J. H. Frank, *Proc. Entomol. Soc. Wash.* 97, 582–589, Fig. 1.B, 1995). (f) SEM of egg of *Mamestra brassicae* (Lepidoptera) showing the distinct ribs extending from the base of the egg to the micropylar area (arrow) at the anterior end (reprinted from *J. Insect Physiol.* 36, G. A. Pak *et al.*, pp. 869–875, Fig. 3A, copyright 1990, with kind permission from Elsevier Science Ltd, The Boulevard, Langford Lane, Kidlington 0X5 1GB, UK).

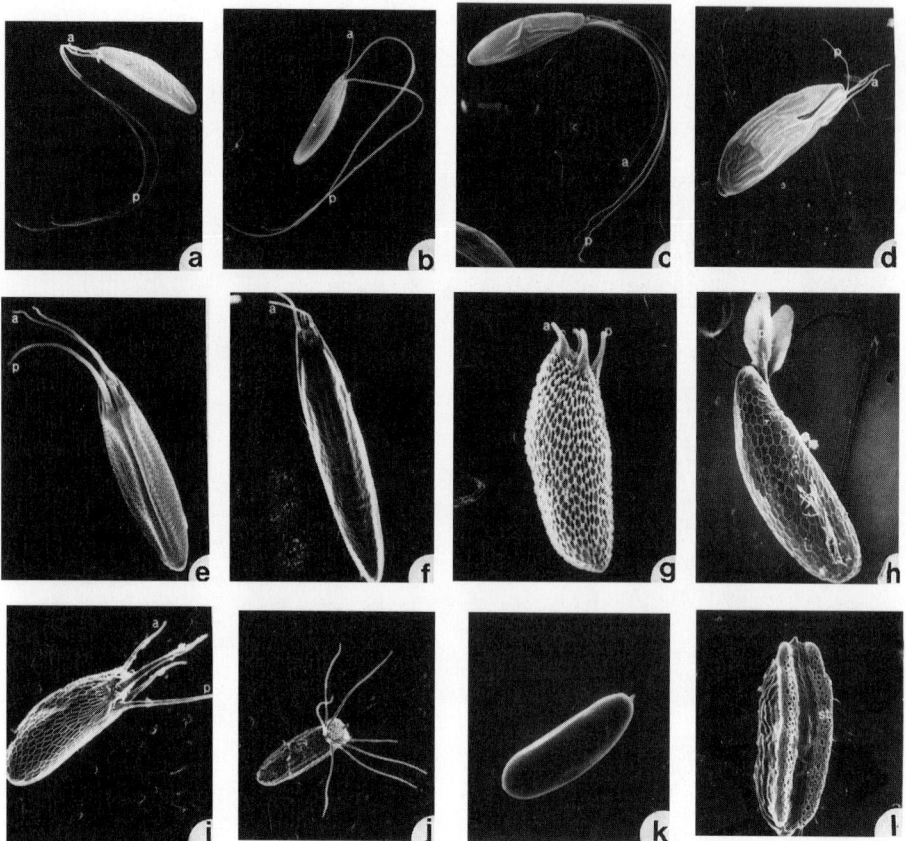

FIGURE 2 Variation among members of the family Drosophilidae [both Hawaiian (a–g, k, l) and non-Hawaiian (h–j)] in egg shape, and respiratory appendage number, length, and form, shown by scanning electron micrographs. (a) *D. murphyi*, ×56; (b) *D. claytonae*, ×35; (c) *D. heteroneura*, ×52; (d) *D. formella*, ×78; (e) *D. truncipenna*, ×71; (f) *D. mulli*, ×89; (g) *D. longiperda*, ×128 (note the prominent follicle cell imprints); (h) *D. willistoni*, ×156 (note two oar-shaped posterior filaments); (i) *D. virilis*, ×102; (j) *D. pattersoni*, ×73; (k) *Scaptomyza (Exalloscaptomyza) oahuensis*, ×71 (note smooth chorion surface and the absence of filaments; (l) *Scaptomyza (Tantalia) albovittata*, ×109 (respiratory filaments appear to have been replaced by two longitudinal respiratory flanges) (reprinted from *Int. J. Insect Morphol. Embryol.* 22, M. P. Kambysellis, pp. 417–446, Fig. 4, copyright 1993, with kind permission from Elsevier Science Ltd, The Boulevard, Langford Lane, Kidlington OX5 1GB, UK).

most often situated at the anterior pole. Whereas there are typically characteristic differences in egg morphology among insect orders (Fig. 1), even within an order or family there may be striking variation, as among species of the family Drosophilidae (Fig. 2). In these eggs the most obvious variation concerns the presence or absence, number, length, and form (filamentous or oar-shaped) of the respiratory appendages at the anterior pole.

Although gross morphology and differences among insect eggs can be seen with the naked eye or under the light microscope, superficial details of the egg coverings and regional differences in external morphology can be better appreciated via high-resolution scanning electron microscopy. Using this tool the egg surface can be seen to be smooth (Fig. 1b), variously sculptured (Figs. 1c, 1d, and 3), or even ridged (Figs. 1e, 1f, and 3c), and differences in the anterior and posterior poles and ventral and dorsal surfaces (Figs. 3a and 3b) of the egg are apparent.

B. Regional Complexity of the Egg Surface

The external covering of the insect egg is not uniform but displays many specialized regions of particular functional significance. All eggs except those of a very few insect species have a micropylar apparatus, most often located anteriorly, to permit sperm entry and fertilization. Also at the anterior pole are found the respiratory appendages, if present, and in the majority of eggs a distinct hatching region that facilitates escape of the nymph or larva from the eggshell. This hatching region may involve a collar structure surrounding the operculum, as in *Drosophila* (Figs. 3a and 3b), or a flat apical operculum as in phasmatids, which may bear a capitulum (Fig. 1d). In the posterior pole region the chorion surface is usually distinctly different from that of the main body of the egg. Even the latter is not uniform but may be interrupted by features such as the micropylar region, a dorsal ridge (Fig. 3b), or some structure for attachment to the substrate.

1. Surface Features of the Main Body of the Egg

In most cases the surface is rough, often revealing the imprints of the follicular cells responsible for eggshell protein production and secretion. These imprints are usually hexagonal (Figs. 3a, 4a–4d, and 4g), although exceptions do occur (Figs. 4e and 4f). There are also many insect eggs in which the physical egg surface is relatively smooth, indicating that the last layer formed has no apparent substructure (Fig. 1b). It is this layer that is normally responsible for the attachment of the egg to the substrate upon laying, although in several cases special attachment devices do exist.

Another surface feature of the main body of the egg is the presence of aeropyles which may be distributed on the borders of the follicle cell imprints (Fig. 4c) or within the borders on the floor of the imprints (Figs. 4c, 4d, 4l, and 4m). Since aeropyles are also located in other regions of the egg surface besides the main body, these structures will be described separately.

2. Micropylar Apparatus

A prominent feature of the surface of virtually all insect eggs is the micropyle region, which provides the point of sperm entry. In addition to the externally visible portion, this eggshell complex includes inner components, specifically a micropylar canal which passes through the eggshell to conduct the sperm to the oocyte plasma membrane. The physical dimensions of the pore in *Drosophila melanogaster* (0.8 μm)

FIGURE 3 Regional complexity of insect eggshells revealed by scanning electron micrographs. (a) Lateral view of egg of *D. mimica*. (b) Dorsal view of egg of related species, *D. infuscata*, showing typical features of Dipteran eggshells; rf, respiratory filaments; c, collar; fi, follicle imprints; m, micropyle; o, operculum; r, dorsal ridge; p, posterior pole. (a and b, reprinted from *Int. J. Insect Morphol. Embryol.* **22**, M. P. Kambysellis, pp. 417–446, Fig. 2, copyright 1993, with kind permission from Elsevier Science Ltd, The Boulevard, Langford Lane, Kidlington OX5 1GB, UK). (c) Lenticular-shaped egg of *Neoperla falayah* (Plecoptera) with ridged chorionic surface and follicle cell imprints restricted to the posterior pole. The arrows identify the micropyles located in a ring around the main body of the egg (reproduced with permission from B. P. Stark and D. L. Lentz, *Ann. Entomol. Soc. Am.* **81**, 372–376, Fig. 13, 1988). (d) Spherical egg of *Antheraea pernyi* (×15) (Lepidoptera) in which the entire chorionic surface is covered with aeropyles (see Figs. 6a–6c) [reprinted from L. H. Margaritis, *Comprehensive Insect Physiology Biochemistry and Pharmacology*, Vol. 1 (G. A. Kerkut and L. I. Gilbert, Eds.), pp. 153–230, Fig. 28.f, copyright 1985, with kind permission from Elsevier Science Ltd, The Boulevard, Langford Lane, Kidlington OX5 1GB, UK].

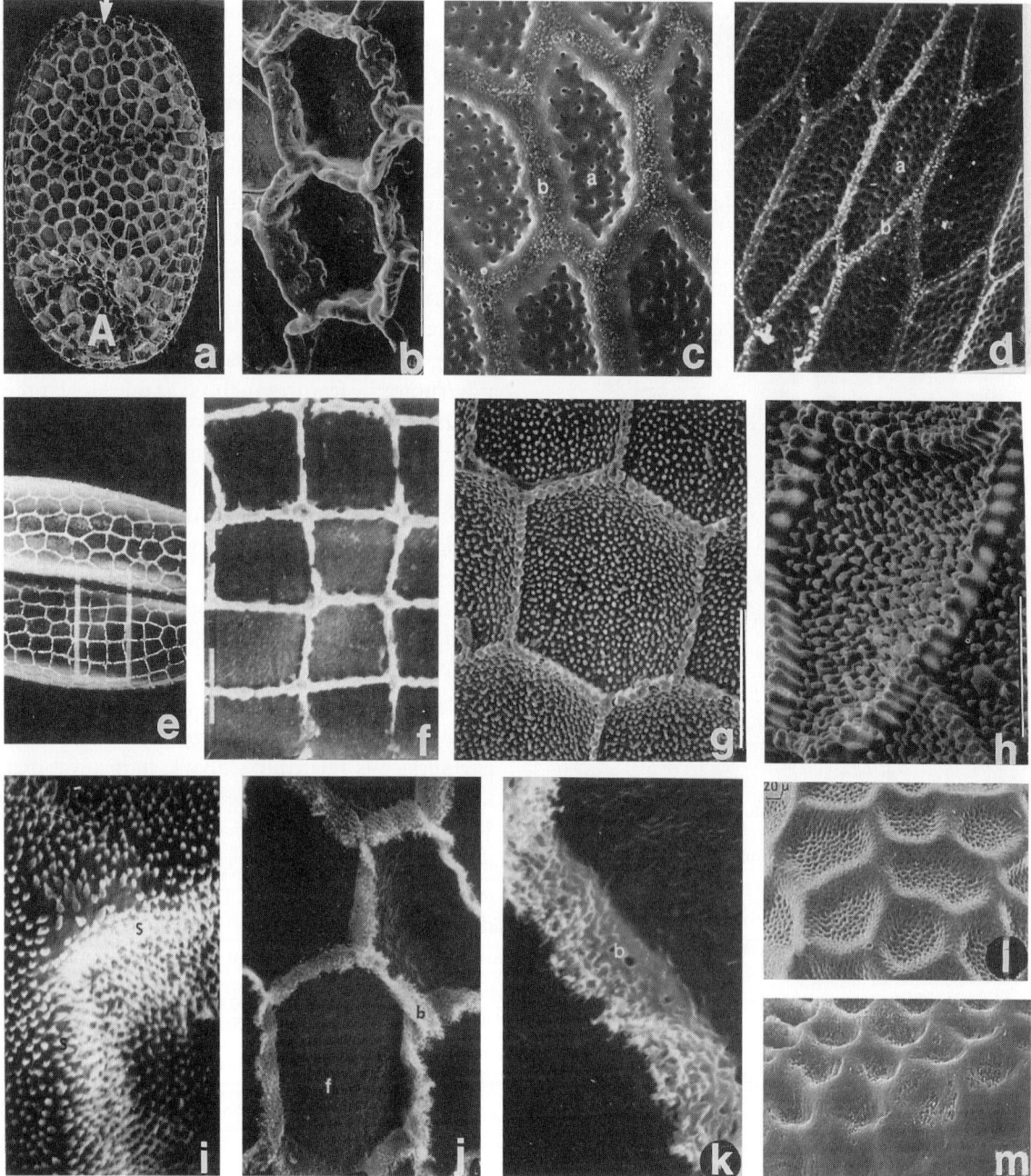

FIGURE 4 Follicle cell imprints on the chorionic surface of insect eggs. (a) Flea egg of *Conorhinopsylla stanfordi* (Siphonaptera) showing the hexagonal follicle cell imprints covering the main body of the egg, the aeropylar disk (A) anteriorly, and the micropylar disk (arrow) at the posterior pole. Scale bar = 200 μm (reproduced with permission from J. R. Linley, A. H. Benton, and J. F. Day, *J. Med. Entomol.* 31, 813–827, Fig. 10a, 1994). (b) Higher magnification of (a) (scale bar = 50 μm) to show the form of the follicle imprint borders. (c and d) Hexagonal follicle cell imprints from *D. primaeva* (×1200) and *D. pattersoni* (×1140), respectively, showing differences in border morphology (b) and density of aeropyles (a) on the imprint floor (reprinted from *Int. J. Insect Morphol. Embryol.* 22, M. P. Kambysellis, pp. 417–446, Figs. 14F and 14H, copyright 1993, with kind permission from Elsevier Science Ltd, The Boulevard, Langford Lane, Kidlington OX5 1GB, UK). (e) Portion of egg of *Lutzomyia lichyi* (Diptera) showing regional diversity in follicle imprint pattern (×230). (f) Higher magnification of chorionic area outlined in

are just sufficient to allow passage of the 0.6-μm wide sperm, thus restricting polyspermy. In the larger egg of Manduca sexta the micropylar orifice has a diameter of 4 or 5 μm. The structural organization of the micropylar apparatus (MPA) varies among different insect orders. It forms a protrusion, as in some Diptera (Figs. 3b and 5f) and Megaloptera (Fig. 5c), a depression, as in Lepidoptera (Fig. 5b), Heteroptera, and Coleoptera, or less often a cup-like structure as in some Diptera (Fig. 5e). Furthermore, the location and number of micropyles vary widely among insects. Normally positioned at or near the anterior pole of the egg, the MPA may be located on the dorsal or ventral sides, in an equatorial band (Fig. 3c), or even at the posterior pole, as found in eggs of scorpion flies and fleas (Fig. 1c). There is often just one MPA, as in *Drosophila*, but even with a single MPA, there may be up to six micropylar channels radiating from a common external orifice. In some insect eggs, there are up to 40 micropyles, either clustered together in the same region of the egg or variously distributed on the egg surface. The presence of multiple micropyles may correlate with the phenomenon of polyspermy, which is common in insect eggs. Interestingly, no micropyles have been detected in the eggs of some Apterygote insects.

In some Diptera and some Megaloptera the single micropyle is a cone-like protrusion ~20–100 μm long at the anterior end of the egg. The sides of the cone may be covered with follicle cell imprints which differ from those on the main body of the egg. At the apex there is often a tufted structure (Figs. 5c and 5f) of modified chorionic material which may play a role in chemotaxis or chemical attraction of sperm toward the oocyte. Also, there is some evidence of molecular sperm "receptors" on the micropylar tip; these molecules in *D. melanogaster* serve to bind glycoconjugates on the sperm surface. The specific localization of these molecules in this region of the egg surface is thus critical to the fertilization process.

In Lepidoptera one or more micropyles are found in a central depression at the anterior end of the egg (Fig. 1f) bordered by a rosette of petal-shaped follicle cell imprints which are surrounded by additional elongate imprints that gradually merge into the surrounding polygonal pattern (Fig. 5b). In phasmatids (Fig. 1d) the micropyle is situated on the dorsal face of the egg capsule in a cup-shaped prominence at the posterior end of the micropylar plate, a structure unique to this order.

3. Posterior Pole

The posterior pole region is, in most cases, structurally different from the main body of the egg surface. In some insects, this region bears the micropyle (Fig. 1c) or serves for egg attachment onto the substrate (Fig. 1f), but in many insects it participates in gas exchange. Often there is a high density of aeropyles at the posterior pole (possibly to provide enhanced respiratory exchange for the germline cells which form initially at the posterior pole of the embryo within). In some stick insects the posterior pole bears a peculiar raised structure, the polar mound.

(e) depicting tetragonal rather than hexagonal imprints (×1150). (e and f, reproduced with permission from M. D. Feliciangeli, O. C. Castejon, and J. Limongi, *J. Med. Entomol.* 30, 651–656, Fig. 2A, 1993). (g and h) Eggshell of *Epitedia faceta* (Siphonaptera) showing hexagonal follicle imprints with tall bilobed tubercles on the imprint borders as well as tubercles on the imprint floor. Scale bars = 10 μm (g) and 5 μm (h) (reproduced with permission from J. R. Linley, A. H. Benton, and J. F. Day, *J. Med. Entomol.* 31, 813–827, 1994). (i) High magnification (×6600) of the eggshell of *D. macrothrix* (Diptera) showing similar tubercles to those in *E. faceta* (g and h) (reprinted from *Int. J. Insect Morphol. Embryol.* 22, M. P. Kambysellis, pp. 417–446, Fig. 14D, copyright 1993, with kind permission from Elsevier Science Ltd, The Boulevard, Langford Lane, Kidlington OX5 1GB, UK). (j, ×1728; and k, ×6000) Follicle cell imprints of *D. silvarentis* (Diptera) showing tall, thin, lacy imprint borders (reprinted from *Int. J. Insect Morphol. Embryol.* 22, M. P. Kambysellis, pp. 417–446, Figs. 14A and 14B, copyright 1993, with kind permission from Elsevier Science Ltd, The Boulevard, Langford Lane, Kidlington OX5 1GB, UK). (l) Follicle imprints on eggshell of *Semiothisa signaria dispuncta* (Lepidoptera) showing broad imprint borders and numerous aeropyles (reproduced with permission from Hinton, p. 388, Plate 70.c, 1981). (m) Eggshell of *D. dolichotarsus* (Diptera) showing follicle imprints with similar morphology to those in (l) (courtesy of F. Piano).

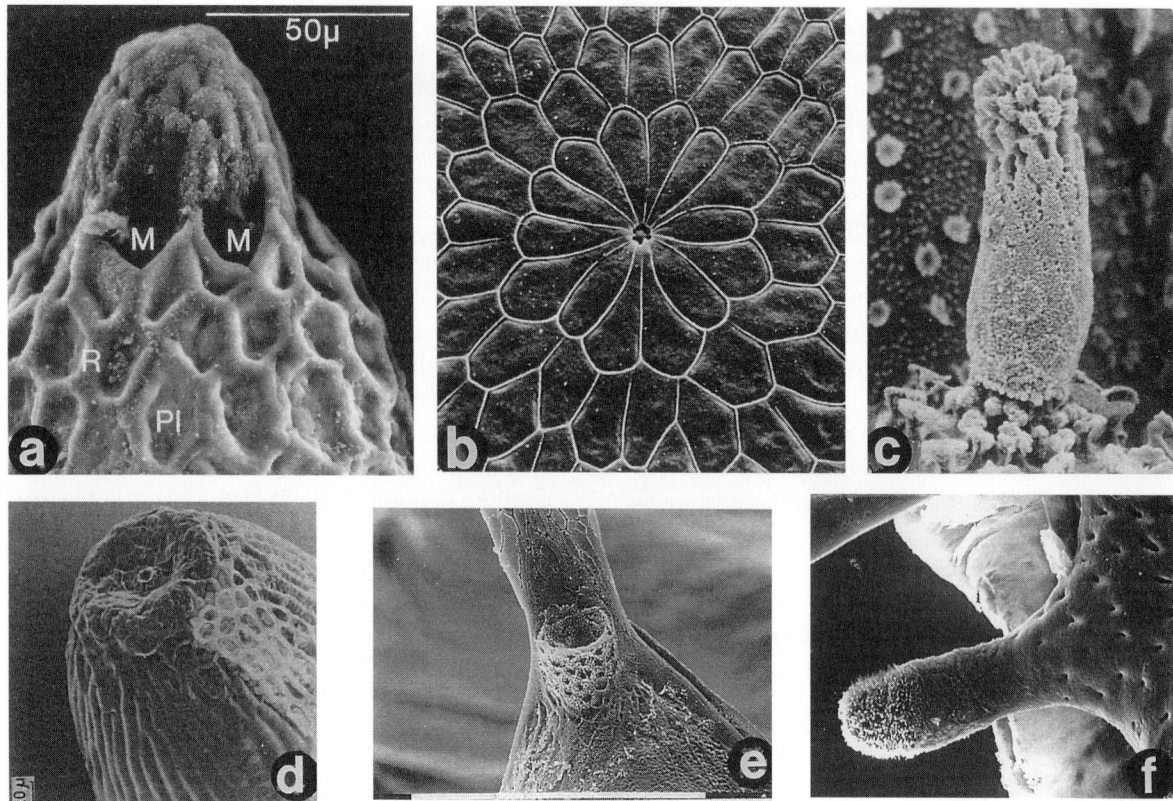

FIGURE 5 Variation in the micropylar apparatus (MPA) of insect eggshells. (a) Anterior end of *Aeshna juncea* (Odonata) egg showing the micropyles (M) located within depressions, hexagonal follicle imprints (PI), and broad follicle imprint borders (R) (reprinted from *Int. J. Insect Morphol. Embryol.* 23, G. Sahléen, pp. 345–354, Fig. 2, copyright 1994, with kind permission from Elsevier Science Ltd, The Boulevard, Langford Lane, Kidlington OX5 1GB, UK). (b) Eggshell of the silkmoth *Antheraea polyphemus* showing the numerous micropyles located in a depression (×500) [reprinted from L. H. Margaritis, *Comprehensive Insect Physiology Biochemistry and Pharmacology*, Vol. 1 (G. A. Kerkut and L. I. Gilbert, Eds.), pp. 153–230, Fig. 61b, copyright 1985, with kind permission from Elsevier Science Ltd, The Boulevard, Langford Lane, Kidlington OX5 1GB, UK]. (c) MPA of *Sialis hasta* (Megaloptera) with terminal protrusions (×600) (reproduced with permission from L. Canterbury and S. Neff, *Can. Entomol.* 112, 409–419, Fig. 19, 1980). (d–f) Dramatic variation in micropylar structure among Diptera. (d) MPA of *Leptohylemyia coarctata* (Diptera, Anthomyiidae) located in a depression at the anterior end of the egg (reproduced with permission from Hinton, Plate 152A, 1981). (e) MPA of *Nemopoda nitidula* (Diptera, Sepsidae) forms a cup-like structure dorsal to the base of the single respiratory filament. Scale bar = 0.1 mm (reproduced with permission from R. Meier, *Ent. Scand.* 26, 425–438, Fig. 19, 1995). (f) MPA of *Drosophila tanythrix* (Diptera, Drosophilidae) on a protrusion located between the four respiratory filaments. Note the aeropyles around the base and the fibrils at the tip of the MPA (courtesy of F. Piano).

4. Aeropyles

Aeropyles are respiratory devices of the eggshell that occur widely among insect orders (Fig. 6). Strictly, the term "aeropyle" refers to the tiny channels that traverse the relatively thick chorion (Fig. 6g) and conduct the ambient air through the egg coverings to the oocyte. The diameter of these pores varies from 0.2 μm to a few micrometers, a dimension several times the mean free path of the respiratory gases. In the eggs of *Bombyx mori* that undergo a long diapause, they are very narrow (presumably to limit water loss), whereas in nondiapausing eggs they can be quite wide. They are distributed in many regions of the egg surface including the main body, where they may be evenly scattered or arrayed in bands. Quite often the aeropyle openings are located

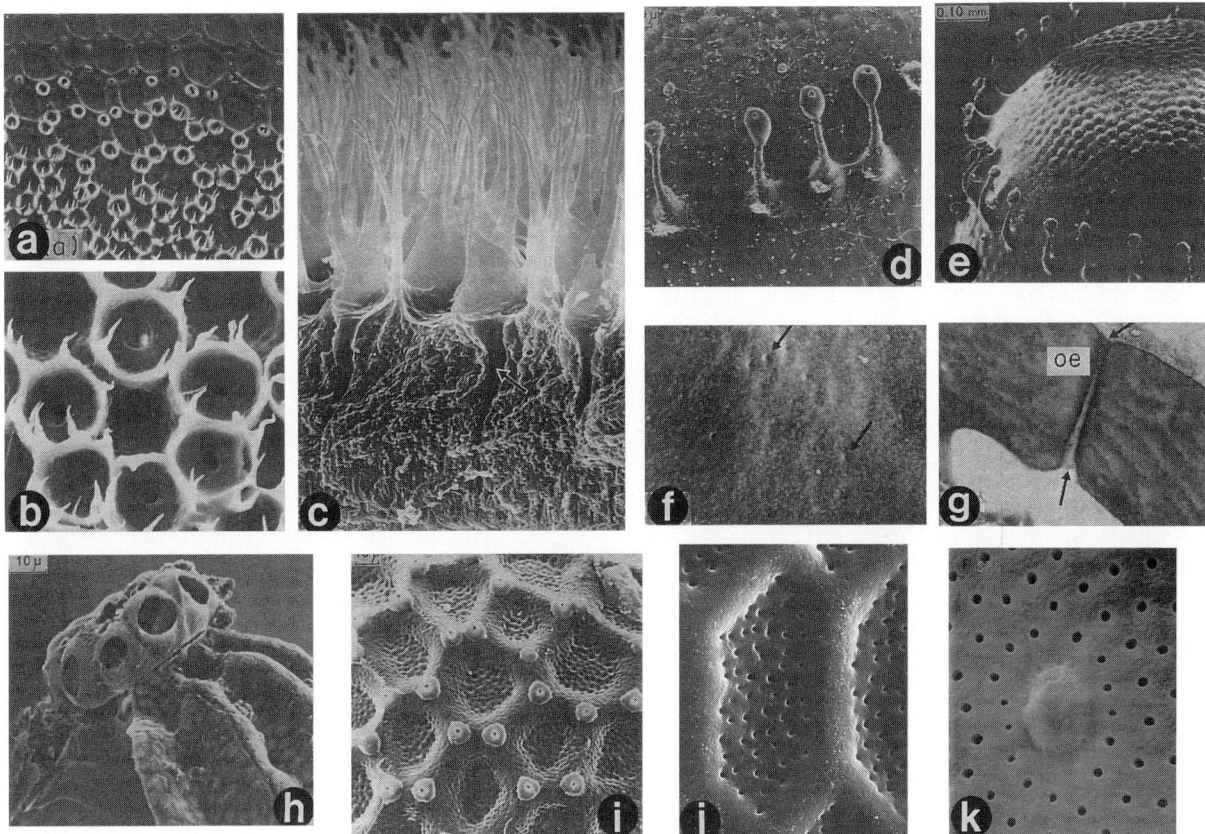

FIGURE 6 Variation in aeropyles on insect eggshells. (a–c) SEM views of egg surface of *Antheraea polyphemus* (Lepidoptera) [reprinted from L. H. Margaritis, *Comprehensive Insect Physiology Biochemistry and Pharmacology*, Vol. 1 (G. A. Kerkut and L. I. Gilbert, Eds.), p. 217, Figs. 72a–72c, copyright 1985, with kind permission from Elsevier Science Ltd, The Boulevard, Langford Lane, Kidlington OX5 1GB, UK]. (a) The striped aeropyle region showing the graded development of the crown surrounding the aeropyles which are located at junctions of three imprints (×250). (b) Higher magnification (×1100) showing form and arrangement of the crowns. (c) Torn eggshell showing in SEM side view (×1600) the small trabecular layer below, covered by multiple lamellae which are traversed by aeropyles (arrow). Around the opening of each aeropyle is a very tall crown. (d and e) Stalked aeropyles on chorion of *Picromerus bidens* (Heteroptera) egg arranged in a ring (e) around the anterior end of the egg (reproduced with permission from Hinton, Plates 32E and 32F, 1981). (f and g) Chorion of *Drosophila grimshawi* [reprinted from L. H. Margaritis, *Comprehensive Insect Physiology Biochemistry and Pharmacology*, Vol. 1 (G. A. Kerkut and L. I. Gilbert, Eds.), p. 217, Figs. 72d–72e, copyright 1985, with kind permission from Elsevier Science Ltd, The Boulevard, Langford Lane, Kidlington OX5 1GB, UK]. (f) Surface view (×5500) of the endochorion (exochorion removed by sonication) revealing the tiny aeropyles (arrows) 0.05 μm in diameter. (g) Thin TEM cross section (×25,000) through a laid egg showing a 500 Å wide aeropyle (arrows) traversing the whole outer endochorion (oe). (h) Ring of aeropyles on the anterior cap of the egg of *Psila rosae* (Diptera, Psilidae) surrounding the micropyle (reproduced with permission from Hinton, Plate 123C, 1981). (i) A single aeropyle located at each junction of three follicle cell imprints, and numerous smaller aeropyles on the floor of the imprints on the egg of *Semiothisa signaria dispuncta* (Lepidoptera) (reproduced with permission from Hinton, Plate 70B, 1981). (j) Aeropyles on the follicle imprint floor of *Drosophila primaeva* (Diptera) egg (courtesy of K. Dellas). (k) Aeropyles surrounding a tubercle on the chorion of *Philonthus ventralis* (Coleoptera) (reproduced with permission from G. Y. Hu and J. H. Frank, *Proc. Entomol. Soc. Wash.* 97, 582–589, Fig. 2.F, 1995).

at the junctions of three follicle cell imprints (Fig. 6i), but they may also be found on the floor of the follicle imprints (Fig. 6j) or along the crests of longitudinal ridges such as those on butterfly eggs.

Superficially, aeropyles vary in structure from a simple orifice on the egg surface (Fig. 6k) to an opening at the tip of a stalked body (Fig. 6d), with the aeropyles in this species being arranged in a ring around the anterior end of the egg (Fig. 6e). This arrangement of stalked aeropyles is the general rule in certain superfamilies of Hemiptera. Sometimes the stalks are very long so that even when the egg is submerged under water, the aeropyle openings are exposed to atmospheric oxygen. The aeropyles thus serve to conduct air to the embryo in both terrestrial and aquatic situations. As shown in Figs. 6a–6c and 6i, aeropyle openings can also be fringed by various types of borders.

5. Respiratory Filaments or Appendages

Another respiratory device extending from the outer layer of some insect egg coverings is provided by the single or multiple appendages variously called respiratory filaments, respiratory horns, or dorsal appendages (Figs. 1a, 2a–2j, and 7c–7f). Located usually at the anterior end of the egg, these tubular chorionic extensions are hollow (Fig. 7f), presenting very low resistance to the flow of oxygen and carbon dioxide along their axes. Respiratory filaments are found in eggs that are oviposited deeply into the substrate or in eggs laid in very humid, moist (semiaquatic) environments. In these situations the filaments dramatically increase the surface area for plastron respiration, serving as physical gills when submerged under water or in decaying plant tissue.

The number, length, and surface texture of these appendages vary widely. There may be 1 filament, a pair, two pairs (Figs. 3a and 3b), or even up to 26 filaments, as in a Hemipteran species, and filament number may even vary within a species, as in *Drosophila pattersoni* (Fig. 2j). Filaments may be rudimentary or short, intermediate, or very long, up to several millimeters in length (Figs. 1a and 2a–2j). In *D. melanogaster* there are two 300-μm long appendages, the dorsal surface having 1- or 2-μm wide plaques and the ventral surface having a tight filamentous network. Among the Hawaiian *Drosophila* there are commonly four respiratory appendages arranged as a shorter anterior pair and a longer posterior pair which may extend as much as 4 mm. In these species the filament surface may be smooth (Fig. 7d), variously porous, or covered with plaques (Figs. 7e and 7f); the texture and length are presumably adaptations to the particular environment and oviposition substrate of each species.

Egg filaments may serve purposes other than respiration. For example, in the almond wasp, *Eurytoma amygdali* (Hymenoptera), the egg has two filaments, the shorter one of which contains the micropyle, whereas the very long filament, which extends out of the almond fruit after oviposition, presumably facilitates embryonic respiration.

6. Operculum and Collar

Another of the specialized regions on the outer covering of some insect eggs is the operculum (Figs. 7g and 7h), which may be bordered around the ventral edge by a distinct ridge called the collar or the ventral ridge (Fig. 7i). This hatching region differs substantially in structure from the rest of the eggshell. In *Drosophila* the operculum functions as a flap door that allows larval escape when embryogenesis is complete (Fig. 10d). As the larva emerges, the dorsal and ventral halves are forced apart.

In phasmatid eggs the operculum is a flattened lid over the anterior pole that is lifted up by the hatching nymph. It is sometimes capped by a distinct bowl-shaped structure, the capitulum, that is easily detached. The operculum of some phasmatid eggs bears a pseudocapitulum, a central protuberance that cannot be detached (Fig. 1d).

7. Dorsal Ridge

On the dorsal side of the eggs of Hawaiian *Drosophila* species below the operculum is a dorsal ridge, extending longitudinally from anterior to posterior (Fig. 3b). The aeropyles and microplastron surface in this region suggest that this structure may facilitate respiration in the wet rain forest environments occupied by these species.

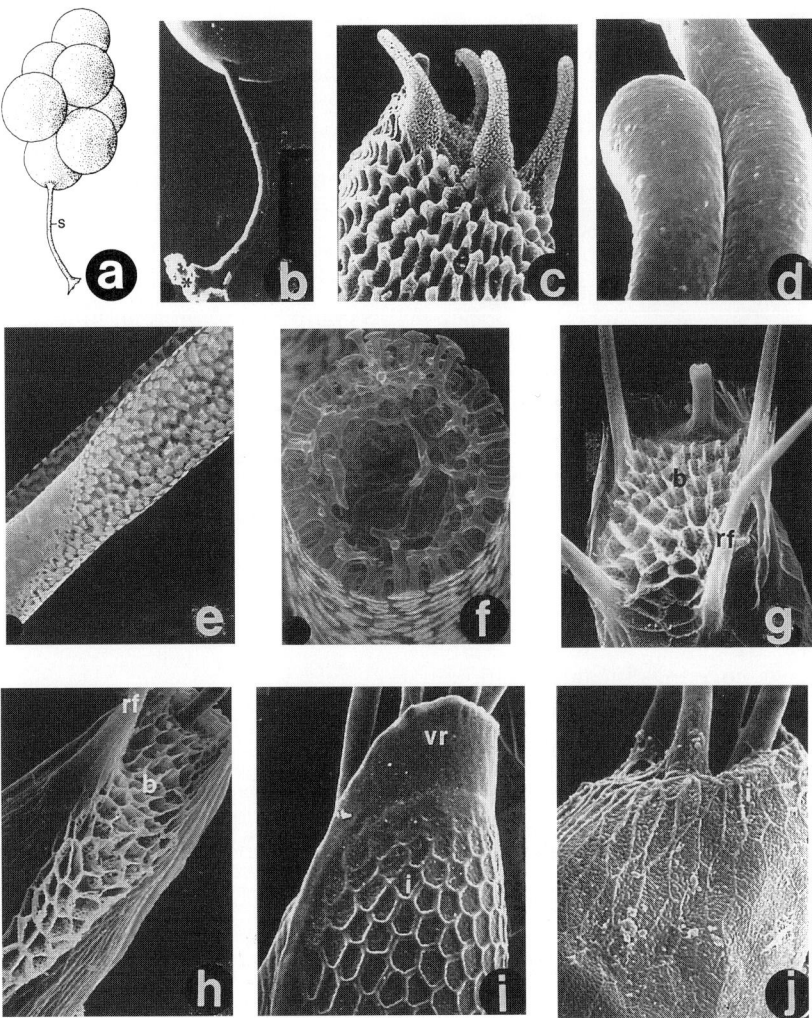

FIGURE 7 Specialized structures of insect eggshells. (a) Diagram of batch of oviposited eggs of *Campodea* sp. (Apterygota, Campodeidae) showing the stalk used to anchor the egg mass to the substrate. (b) SEM of supporting stalk structure (×160) (a and b, reprinted from *Int. J. Insect Morphol. Embryol.* 18(4), S. M. Biliński and O. Larink, pp. 199–204, Figs. 1 and 3, copyright 1989, with kind permission from Elsevier Science Ltd, The Boulevard, Langford Lane, Kidlington OX5 1GB, UK). (c) Anterior end of eggshell of *D. longiperda* (Diptera, Drosophilidae) showing the four short respiratory filaments with plastron surface and the exochorion pattern of the main body of the eggshell (courtesy of F. Piano). (d–f) Respiratory filaments of Hawaiian *Drosophila* species (Diptera, Drosophilidae) showing variation in morphology. In some species, such as *D. disjuncta* (d), the surface is smooth and solid; in others, such as *D. fasciculisetae* (e), a portion of the filament surface is covered with plaques. (f) A cross section of a broken filament showing the surface plaques and hollow interior of the filament. (d–f, courtesy of K. Dellas). (g and h) Variability in the extent of the operculum. (g) In *D. mojavensis* (×568) the operculum is restricted to the region between the bases of the four respiratory filaments (rf). (h) In *D. grimshawi* (×355) the operculum extends posteriorly down the dorsal side of the egg, connecting to the dorsal ridge. (g and h, reprinted from *Int. J. Insect Morphol. Embryol.* 22, M. P. Kambysellis, pp. 417–446, Figs. 7A and 7F, copyright 1993, with kind permission from Elsevier Science Ltd, The Boulevard, Langford Lane, Kidlington OX5 1GB, UK). (i and j) Anterior end of *Drosophila* eggs showing in (i) *D. hanaulae* (×528) a prominent collar or ventral rim (vr); this structure is essentially absent in *D. pictiventris* (j, ×540) (reprinted from *Int. J. Insect Morphol. Embryol.* 22, M. P. Kambysellis, pp. 417–446, Figs. 15B and 15C, copyright 1993, with kind permission from Elsevier Science Ltd, The Boulevard, Langford Lane, Kidlington OX5 1GB, UK).

III. RADIAL COMPLEXITY OF EGGSHELL ARCHITECTURE

Besides the regional complexity of the outer surface of the insect egg, the eggshell itself is radially complex, consisting of many layers. Most often, the eggshell consists of an outer layer, the chorion, and an inner layer, the vitelline envelope (or vitelline membrane), adjacent to the oocyte plasma membrane. The chorion layer can be further subdivided on morphological grounds. Other layers may also exist in between the vitelline envelope and chorion. High-resolution electron microscopy has revealed enormous variation and complexity in the eggshell layers (Fig. 8). The more primitive Apterygotes, however, have a simpler eggshell—in *Campodea* sp. (family Campodeidae) it consists of only a single homogeneous layer, the chorion, abutting the oocyte membrane. In the more typical case as in the dipteran *D. melanogaster*, the egg coverings include multiple layers above the oocyte membrane—the vitelline envelope, waxy layer, innermost chorionic layer, endochorion (further subdivided into inner part, pillars, outer part, and network), and the exochorion (Fig. 8). Each layer and region is responsible for specific structural functions.

A. Vitelline Envelope

The vitelline envelope is the first extracellular layer to be secreted onto the oocyte surface by the surrounding follicular epithelium. Although commonly referred to as the vitelline membrane, this erroneous terminology should be avoided since this layer is not a biological membrane. A homogeneous layer of variable thickness (from 0.1 to >1 μm), the envelope is composed of protein, lipid, and carbohydrate and has elastic properties which accommodate oocyte growth during oogenesis. At the same time, this innermost layer of the egg coverings must also allow gas exchange. It has been reported that upon fertilization, the vitelline envelope undergoes condensation, providing some additional protection against water loss beyond that provided by the subsequently deposited wax layer.

B. Intermediate Layers

Interposed between the vitelline envelope and the complex chorion may be one or more additional layers which primarily protect the egg against water loss.

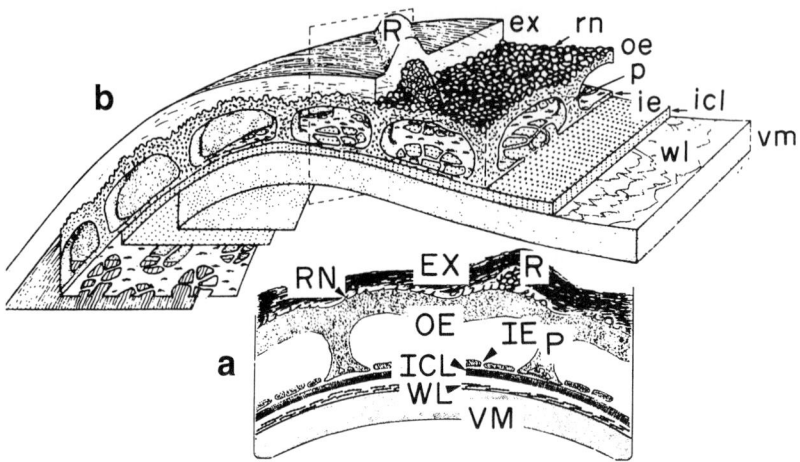

FIGURE 8 Two-dimensional (a) and three-dimensional (b) representation of the eggshell in *D. melanogaster* (Diptera) showing the vitelline envelope or membrane (VM), the wax layer (WL), the crystalline innermost chorionic layer (ICL), the inner endochorion (IE), the pillars (P), the outer endochorion (OE), the outer endochorionic network (RN), the exochorion (EX), and ridges of the follicle cell imprints (R) [reprinted from L. H. Margaritis, *Comprehensive Insect Physiology Biochemistry and Pharmacology*, Vol. 1 (G. A. Kerkut and L. I. Gilbert, Eds.), Fig. 31, copyright 1985, with kind permission from Elsevier Science Ltd, The Boulevard, Langford Lane, Kidlington OX5 1GB, UK].

1. Wax Layer

This very thin (<0.1 μm) hydrophobic waxy layer has been reported in most terrestrial insect eggs examined ultrastructurally and serves to waterproof the egg and embryo against water loss, thus promoting survival. In the parasitic Hymenoptera, a wax layer has not been detected, presumably because these embryos do not develop in a dry substrate. Freeze-fracture electron microscopy reveals that the wax layer in *D. melanogaster* is multilayered, composed of three or four planes of overlapping plaques 0.5–1 μm in diameter. In another dipteran, *Phormia regina*, the wax layer is composed of up to seven layers of hexagonally shaped lamellae. The hydrophobic nature of the wax layer confers water impermeability on the insect eggshell, yet still permits oxygen uptake by the embryo via diffusion down a partial pressure gradient generated by the embryo's metabolic activity.

2. Crystalline Layer

In several insect orders including Diptera and Coleoptera, the eggshell has been reported to include below the chorion an additional intermediate layer with periodically arranged components that construct a three-dimensional (3-D) crystal. This rigid layer is mainly proteinaceous and of variable thickness. In *D. melanogaster* this 40- to 50-nm layer is referred to as the innermost chorionic layer (ICL), although it is not strictly a component of the chorion. It is composed of four or five bilaminar sublayers with adjacent crystallites differing from one another in the direction of their lattice vectors. Fourier transform analysis and 3-D reconstruction have suggested that the basic repeating unit in this crystalline structure could plausibly be an octameric arrangement of spherical protein molecules. The holes of this crystalline lattice are large enough (about 2–4 nm) to allow passage of the respiratory gases through this layer. An additional function postulated for the ICL is to guide wax layer morphogenesis by pressing its components against the elastic vitelline envelope during oocyte expansion.

C. The Chorion

The most peripheral layer of the multilaminar egg coverings, the chorion, is the thickest and most complex (Fig. 8). This layer is largely proteinaceous (96% of dry weight in silkmoths and 94% in *Drosophila*) and forms a rigid sheath or eggshell around the egg, the main function of which is mechanical protection. In general, larger eggs have thicker chorions, but thickness varies dramatically, even within a family or order. For example, in the Coleoptera, one of the thickest known eggshells is 150 μm, whereas another species has a very thin eggshell of only 0.5 μm. A very thick chorion is typical of the large eggs of Orthoptera and Phasmatodea. Chorion structure is very elaborate and highly variable among and within insect orders, but generally consists of a complex endochorion and an exochorion.

1. The Endochorion

The endochorion or "trabecular layer" is usually the thickest and most variable portion of the chorion. It normally consists of an inner part, vertically oriented pillars, and an outer part (Fig. 8). Interestingly, in the Muscidae and Tephritidae (order Diptera) there are two trabecular layers that are inverted with respect to one another. The inner endochorion is generally perforated and this portion, together with the system of pillars above, forms an intrachorionic meshwork that connects with the aeropyles bringing oxygen from the surface. The cavities in the endochorion vary widely among and within insect orders in size, shape, and overall configuration. In bumblebees (order Hymenoptera) and Sialidae (order Neuroptera), in which eggs are laid in masses, the endochorion consists solely of pillar-like protrusions, permitting extensive air circulation. The endochorion may also consist of a sheath with small protuberances on its outer surface (order Phasmatodea) or a close network of band-like structures (order Orthoptera).

This matrix, like the other layers covering the egg, is maternal in origin and formed by the protein secretions of the ovarian follicle cells. The component molecules may form an amorphous matrix, as in Drosophilidae, a multilamellar supramolecular structure with the basic units consisting of helicoidally arranged fibrils, as in silkmoths, or even an entirely crystalline layer. Following oviposition, the eggshell undergoes a hardening process, and the chorion proteins undergo secondary stabilization that seems to involve the formation of covalent bonds.

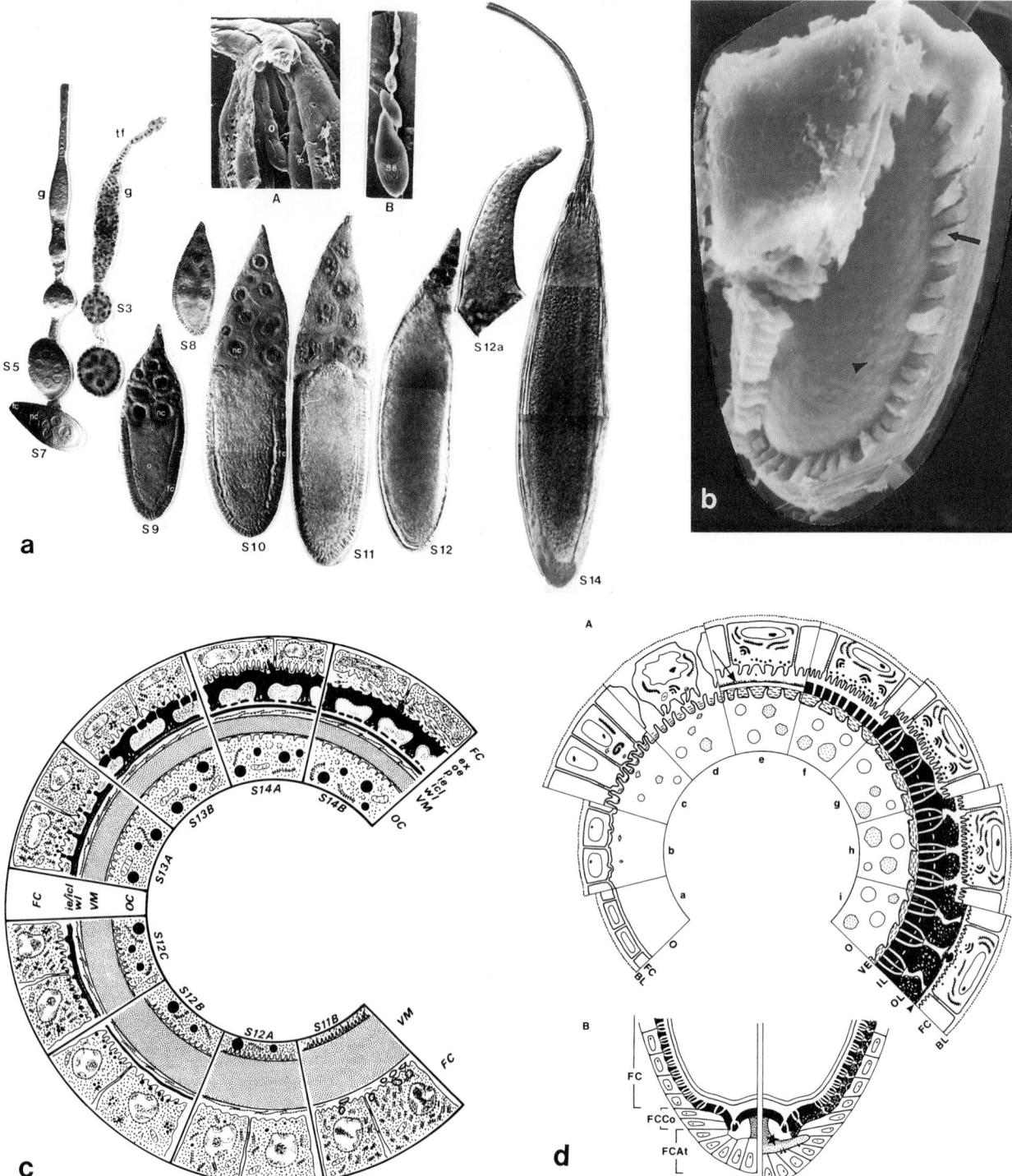

FIGURE 9 Egg and chorion development in insects. (a) Stages of egg chamber development in the Hawaiian species *Drosophila grimshawi* up to the mature egg (S14) complete with chorion and respiratory filaments. Insets (SEM) show the anterior portion of an ovary (A) and a single isolated ovariole (B) (modified from *J. Insect Behav.* 4, M. P. Kambysellis and E. M. Craddock, pp. 83–100, Fig. 1, copyright 1991, with kind permission from Elsevier Science Ltd, The Boulevard, Langford Lane, Kidlington OX5 1GB, UK). (b) SEM of a stage 10b follicle of *D. grimshawi* showing the sheet of columnar follicle cells (arrow) surrounding

2. The Exochorion

The proteinaceous or glycoprotein exochorion is again structurally variable among insects. It may be very thin and simple or thick with a complex multilayered structure. For example, the phasmatid *Carausius morosus* has an outer exochorion characterized by a columnar structure, an intermediate exochorion that is thicker than the other layers, and an inner exochorion consisting of two laminar layers of identical homogeneous structure but of different thicknesses. This outermost layer of the egg coverings may, in some cases, be synthesized by cells of the female accessory glands. This is the layer that is exposed to the environment and, as described earlier, shows a great deal of variation in various regions of the main body of the egg. In many cases it is perforated by aeropyles that conduct oxygen to the meshlike spaces in the endochorion below, and ultimately to the embryo within the eggshell, while limiting the evaporative loss of water.

IV. DEVELOPMENT OF THE EGG COVERINGS

In contrast to the situation in birds in which eggshell construction takes place after the egg has left the ovaries and after it has been fertilized, in insects the eggshell is deposited around the developing oocyte during the final phases of oogenesis, prior to the time of fertilization. (An exception to this typical order of events occurs in cimicoids, in which fertilization takes place in the ovaries following an unusual pattern of hemocoelic insemination, and another possible exception is some Apterygota which lack a micropyle). Insect egg coverings are secondary egg coats synthesized not by the oocyte itself, but by a monolayer of follicle cells surrounding the oocyte (Fig. 9b) according to a strictly developmentally regulated program of stage-specific protein synthesis. Thus, immediately following secretion of the proteins, successive layers are formed either by apposition, as in the case of *D. melanogaster*, or by intercalation, as in the case of the silkmoth *B. mori*.

Eggshell morphogenesis is an integral part of insect oogenesis, occurring mainly in the postvitellogenic phase of egg development, following the oocyte's uptake of yolk proteins from the hemolymph. Generally referred to as choriogenesis, it also includes synthesis of the vitelline envelope and any other intermediate layers below the chorion. The radial and regional complexity of insect egg coverings results from both temporal and spatial changes in the pattern of follicle cell synthesis, with the various specialized regions of the insect egg coat being synthesized by distinct subpopulations of follicle cells which often migrate to new positions in the developing egg chamber. The complex process of egg coat synthesis is best known in *Drosophila* (Figs. 9b and 9c) and in the silkmoth. Both of these have polytrophic meroistic ovarioles, in which nurse cells are included in the egg chamber, but they serve as models for chorion formation in other types of ovarioles (Fig. 9d).

In silkmoths the process of chorion protein synthesis and secretion is under the control of more than 100 genes that are switched on and off in rapid succession by exquisitely controlled mechanisms. Moreover, the demand for specific protein produc-

---

the growing oocyte (arrowhead) the egg chamber being enclosed within the ovarial sheath. (c) Diagrammatic representation of the sequential stages of choriogenesis (counterclockwise from the bottom, S11B–S14B) in the meroistic polytrophic ovaries of *D. melanogaster*. The follicle cells (FC) surrounding the oocyte (OC) progressively secrete the vitelline envelope or membrane (VM) (beginning earlier than stage 11B), then the wax layer (wl), the innermost chorionic layer (icl), the inner endochorion (ie), the pillars (p), the outer endochorion (oe), and the exochorion (ex) [reprinted from L. H. Margaritis, *Comprehensive Insect Physiology Biochemistry and Pharmacology*, Vol. 1 (G. A. Kerkut and L. I. Gilbert, Eds.), Fig. 13, copyright 1985, with kind permission from Elsevier Science Ltd, The Boulevard, Langford Lane, Kidlington OX5 1GB, UK). (d) (A) Diagrammatic presentation of the sequential secretory activity (a–i) of the follicle cells (FC) during oogenesis in the panoistic ovaries of *Perla marginata* (Plecoptera, stoneflies). Notice the progressive changes in the oocyte (O), the vitelline envelope (VE), and the inner (IL) and outer layers (OL) of the chorion. (B) Posterior pole region of the oocyte showing the two additional subpopulations of follicle cells that synthesize the specialized collar (FCCo) and the attachment structure (FCAt) (reprinted from *Int. J. Insect Morphol. Embryol.* 24, E. Rościszewska, pp. 253–271, Fig. 26, copyright 1995, with kind permission from Elsevier Science Ltd, The Boulevard, Langford Lane, Kidlington OX5 1GB, UK).

tion is so high that some chorion proteins are encoded by multigene families, and in some species single-copy genes are selectively amplified in the follicle cells earlier in oogenesis to provide additional templates for transcription. The massive synthesis of chorion and other egg coat proteins occurs very rapidly over a brief interval of only 6.5 hr in *D. melanogaster*, which has a 0.8-μm thick chorion, and over a 51-hr period in the silkmoth *Antheraea polyphemus*, in which the chorion is 60 μm thick. Obviously, any individual protein is synthesized for a much shorter period; the chorion proteins are produced asynchronously in overlapping temporal sets. Although complex, this developmental process provides a superb model system for molecular investigations of the control of gene expression and the morphogenesis of a supramolecular system.

The sequence of events in *Drosophila* oogenesis has been subdivided into 14 stages (Fig. 9a). The terminal events of eggshell production or choriogenesis occur during the postvitellogenic stages S11B–S14. Initially during previtellogenesis (S1–S7) the follicle cells are tightly packed into a columnar epithelium and in extremely close contact with the oocyte surface. As vitellogenesis ensues (S8–S11), junctions between the follicle cells become localized, opening up large spaces between the cells that allow yolk passage through the follicular layer to the oocyte surface, where yolk uptake takes place. In the postvitellogenic period the follicle cells once again resume close junctional contacts with one another. At this time the follicle cell cytoplasm becomes loaded with secretory granules, the contents of which are released from the apical surface into the perivitelline space to construct the consecutive layers of the egg coverings.

The vitelline envelope material is the first to be deposited at the follicle cell/oocyte interface; this process begins during the latter phases of vitellogenesis. As shown in Fig. 9c, the vitelline envelope or membrane becomes progressively thinner throughout choriogenesis in part due to stretching as a result of the terminal growth of the oocyte and in part due to pressure from the radially expanding chorion. Following synthesis of the vitelline envelope, the follicle cells progressively secrete the subsequent layers (the wax layer, the innermost chorionic layer, the inner endochorion, the pillars, the outer endochorion, and the exochorion) in a defined temporal sequence as shown diagrammatically in Fig. 9c.

Specific subpopulations of follicle cells construct localized portions of the egg coverings. For example, in *D. melanogaster* a group of 9 border cells cooperate with about 36 peripheral cells to form the micropylar apparatus. Other follicle cells elongate to construct the respiratory filaments. Although there are about 1000 follicle cells in the *D. melanogaster* egg chamber, in the larger silkmoth *A. polyphemus* there are about 10,000 follicle cells. Although there are regional and timing differences in the structural patterns of eggshell development among insects with various types of ovaries (Figs. 9c and 9d), a common principle applies—sequential protein synthesis by the follicle cells and secretion of these follicle cell products onto the oocyte surface.

V. FUNCTIONAL ADAPTATIONS OF INSECT EGG COVERINGS

Reproduction in the majority of insects is oviparous, and once oviposited the fertilized egg develops quite independently of the female parent. The self-sufficiency of the insect egg is due in large measure to the complex architecture of the coverings surrounding the egg cell. These coverings constitute a remarkable device, the insect eggshell, that has evolved to protect the embryo within from desiccation, flooding, predation, and parasitism and at the same time facilitates respiratory exchange of the embryo with its environment. The conflicting demands of respiration and protection against water loss impose a particularly severe problem for insect eggs because of their small size (hence, a high surface to volume ratio), and because the oxygen molecule is larger than the water molecule. Any surface layer that admits oxygen to the egg will also permit the exit of water molecules. The structural modifications of the insect egg represent a variety of solutions to this central functional dilemma as well as adaptations to the particular oviposition substrate and microenvironment of each species and adaptations that facilitate all the various aspects of egg function from fertilization through oviposition to hatching. Clearly, natural selection has played a pivotal role in driving the development of an array of egg structures that ensure reproductive success and species survival in the great diversity of aquatic, semiaquatic, and terrestrial habitats in which insects thrive.

A. Respiratory Adaptations

Successful insect embryogenesis depends on passage of the respiratory gases through the eggshell, with oxygen from the ambient environment being exchanged for the carbon dioxide and water vapor produced as a result of metabolism. For an actively developing embryo gas exchange is a passive process based on diffusion and governed by the coefficients and relative partial gas pressures on either side of the eggshell. Although the insect chorion is often quite thick, the majority of terrestrial eggs have some form of air-containing meshwork within the chorion, together with aeropyles extending through the shell to connect the chorionic layer of gas with the ambient atmosphere. The intrachorionic oxygen pressure is maintained below that of the atmosphere so that oxygen diffuses through the aeropyles into the intrachorionic meshwork and from there through the inner layers and vitelline envelope to the embryo. The relatively small size of insect eggs and thus high surface to volume ratio (typically 50 cm^2/cc) is an advantage in terms of access to oxygen but a distinct disadvantage in terms of evaporative water loss. This respiratory/desiccation problem is especially acute for the smallest eggs, but insects have evolved a variety of adaptive solutions, explaining why insect eggshells are often much more complex than those of larger animals. The various adaptations depend largely on whether the eggs are exposed to a dry terrestrial or an aquatic environment, or a dry environment with intermittent periods of flooding.

1. Adaptations to Dry Environments

Insect eggs laid in dry environments are generally shielded below the chorion by a hydrophobic wax layer that waterproofs the egg (Fig. 8). In addition, the evaporative surface area is greatly reduced by the presence of a thick, tough, and largely impermeable chorion, with restriction of water loss (and oxygen intake) to the aeropyle openings. The number and distribution of aeropyles and size of the aeropyle openings are highly variable (Fig. 6) and presumably are optimized in each case to the needs of the embryo in that particular environment. As already noted, the aeropyles connect to an extensive intrachorionic network of air-spaces, a major but highly variable feature of insect chorions.

2. Adaptations to Aquatic and Semiaquatic Environments

Eggs that are oviposited and develop in water must avoid drowning while maintaining adequate oxygen intake to sustain embryogenesis. The latter problem is significant because the rate of diffusion of oxygen in water is about a million times less than that in air. Even in terrestrial situations, many insect eggs are temporarily flooded by heavy rains or waterlogging of the substrate in which they are embedded, alternately facing wet and dry conditions. In rain forest insect species that insert their eggs deeply into decaying plant or fungal material (such as the majority of Hawaiian drosophilids), the chorionic surface of the body of the egg may not be directly exposed to the ambient air, in which case the egg's environment is essentially aquatic. Various adaptations have evolved to facilitate embryogenesis under these aquatic or semiaquatic conditions. Notably, in fully aquatic eggs it is less common to find a broad distribution of aeropyles and an extensive intrachorionic meshwork.

i. Stalked Aeropyles One solution to the respiratory problem under occasional aquatic conditions is to restrict the aeropyle openings to the apices of stalks that extend above the egg surface. Even when the egg is immersed, the stalks or tubercles may project above the water surface, thereby enabling the embryo to continue to utilize atmospheric oxygen. Stalked aeropyles may be distributed all over the chorionic surface as in some Psocoptera, some Staphylinidae, and some Geometridae, or they may be localized in a ring at the anterior end of the egg, as found in some Hemiptera (Figs. 6d and 6e).

ii. Respiratory Horns or Appendages Another similar solution that has evolved numerous times among insects is to have one or a few long appendages projecting from the anterior end of the egg that are specialized for respiration. Aeropyles are clustered on the apex or on the apex and sides of the respiratory horn or appendage which usually arises from the operculum or from the anterior margin of the eggshell (Figs. 3a, 3b, 7c, and 7g). Frequently, the egg is embedded in the host plant up to the base of the appendage, which extends above the substrate surface, thereby facilitating atmospheric respiration. In some insect eggs the respiratory appendages are also apparently adapted for underwater respiration as indicated by the plastron organization on part or all of the appendage surface.

iii. Plastron Respiration In eggs that develop in aquatic or semiaquatic situations, a common structural modification is the presence of one or more hydrophobic plastron regions on the egg surface that effectively permit underwater respiration. When an egg is submerged, chorionic surface irregularities, such as an aggregation of aeropyles or a localized superficial network of plaques (Figs. 7e and 7f), can function to trap a film of air that acts as a physical gill. Because these regions resist wetting, an extensive air–water interface can be created. Respiration of the embryo can continue underwater by virtue of extraction of oxygen from that dissolved in the surrounding water into the plastron and passage from there through the eggshell to the embryo. For effective embryonic respiration, it has been estimated that the water–air interface should be of the order of 10^5 or 10^6 μm^2/mg of egg weight.

Commonly found in the adults of aquatic insects that typically live in well-aerated waters such as rapidly flowing streams, plastron respiration is just as common in the eggs of terrestrial insects as in the eggs of aquatic insects. Eggs of terrestrial insects are frequently flooded when it rains, but the advantage of a plastron region is that it provides a large area for the extraction of dissolved oxygen when underwater yet also allows direct passage of atmospheric oxygen into the egg under dry conditions. Plastrons are often localized to the operculum region of the egg between the hatching lines (Figs. 7g and 7h) or to the respiratory appendages (Figs. 7e and 7f) when present. The great increase in surface area provided by these appendages greatly facilitates plastron respiration, as can be demonstrated by comparison of an egg with such appendages versus one lacking appendages. In *D. melanogaster* plastron respiration is restricted to the two respiratory appendages (each 300 μm long and 20 μm thick) that comprise an area of about 4×10^4 μm^2, and have an effective plastron surface of 4×10^6 μm^2/mg egg weight. In the larger egg of the silkmoth *B. mori*, which lacks respiratory appendages, the plastron surface is an order of magnitude less (estimated as $<4 \times 10^5$ μm^2/mg egg weight).

B. Adaptations to the Oviposition Substrate

Chorionic structure and function can only be meaningfully evaluated in the context of the particular environment and oviposition substrate utilized by each species. Whether the egg is laid in the open, in fresh or decaying plant or animal tissue, in dung, in soil, or in water will dramatically affect its chance of desiccation and its access to atmospheric oxygen. Hence, adaptations to the oviposition substrate are intimately related to respiratory/dessication/flooding adaptations, some of which have been outlined previously. Specifically, the number and distribution of aeropyles, the nature and extent of the intrachorionic meshwork, the overall thickness and structure of the chorion, and the presence or absence of plastron regions on the egg surface vary widely among insect eggs. The particular combination of egg traits in each individual case must have evolved in response to the forces of natural selection and is presumably more or less adaptive with respect to the substrate and ambient conditions typically encountered by eggs of that species.

It should also be noted that during oviposition, insect eggs are subject to substantial mechanical pressure as they pass through the ovipositor; even when maximally distended, the ovipositor is often narrower than the diameter of the egg. The egg coverings are initially quite pliable, allowing for considerable deformation during oviposition, but shortly thereafter, secondary stabilization of the chorion proteins leads to permanent hardening of the eggshell; this molecular change is adaptively significant. Where eggs are inserted into woody or otherwise firm substrate material, this hardening gives them

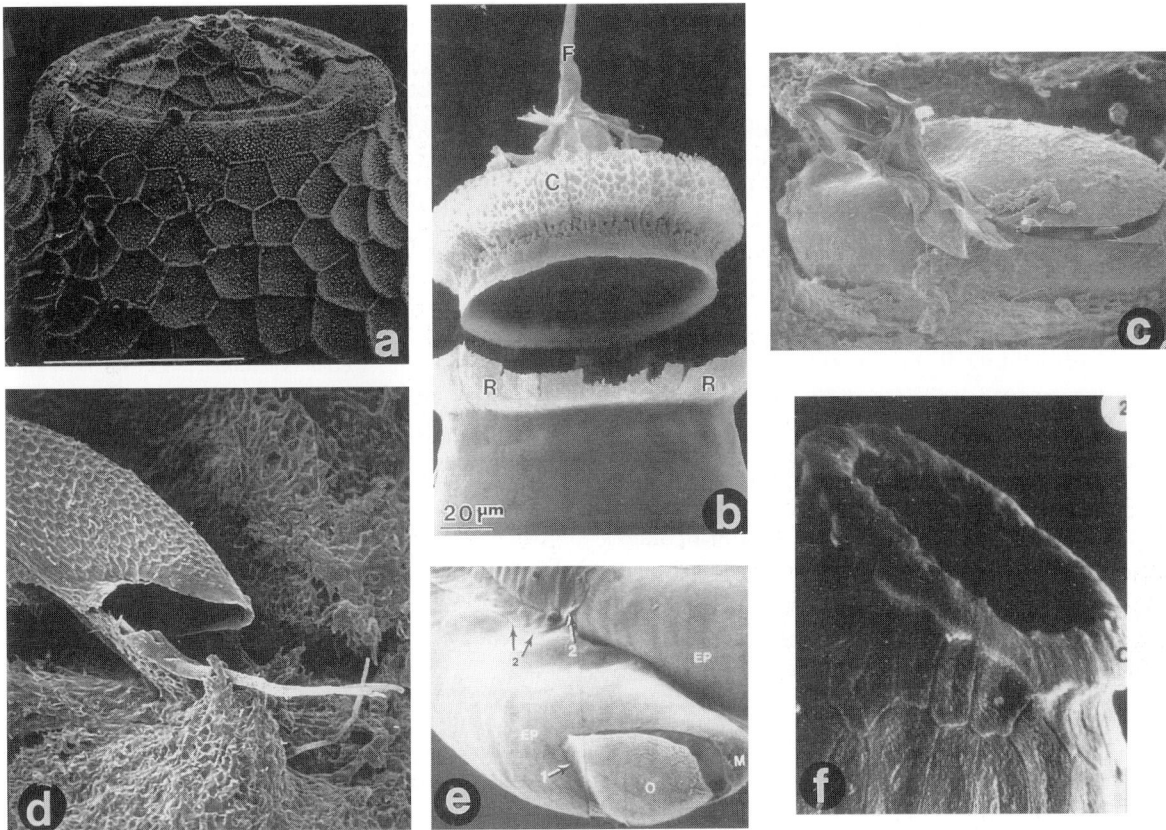

FIGURE 10 Devices for larval escape from the egg coverings. (a) Posterior end of egg of *Epitedia faceta* (Siphonaptera) showing disk for larval escape, micropylar ring, and mound bearing the micropylar orifice. Scale bar = 50 μm (reproduced with permission from J. R. Linley, A. H. Benton, and J. F. Day, *J. Med. Entomol.* 31, 813–827, Fig. 11, 1994). (b) Anterior end of egg of *Corythucha arcuata* (Hemiptera, lacebugs) (see also Fig. 1b) showing rim (R), cap (C), and filament (F) of the operculum used for larval escape (reproduced with permission from G. T. Baker and R. L. Brown, *Proc. Entomol. Soc. Wash.* 96, 70–73, Fig. 4, 1994). (c) An egg (×120) of *Metcalfa pruinosa* (Homoptera) shown immediately following hatching (reproduced with permission from A. Lucchi, *Proc. Entomol. Soc. Wash.* 96, 548–582, Fig. 2c, 1994). (d) An egg (×60) of *Drosophila mimica* (Diptera) found on the natural oviposition substrate just after hatching (reprinted from *Int. J. Insect Morphol. Embryol.* 22, M. P. Kambysellis, pp. 417–446, Fig. 5H, copyright 1993, with kind permission from Elsevier Science Ltd, The Boulevard, Langford Lane, Kidlington OX5 1GB, UK). (e) Egg of *Dermatobia hominis* (Diptera, Cuterebridae) following hatching of larva showing the posteriorly hinged (arrow 1) operculum (O) below the micropylar region (M) (reprinted from *Int. J. Insect Morphol. Embryol.* 18, T. P. Cogley and M. C. Cogley, p. 243, Fig. 4, copyright 1989, with kind permission from Elsevier Science Ltd, The Boulevard, Langford Lane, Kidlington OX5 1GB, UK). (f) Anchoring pole of egg of *Perla marginata* (Plecoptera) showing collar (c) into which the attachment disk fits to anchor the egg in aqueous environments. This region may also allow for larval escape (reprinted from *Int. J. Insect Morphol. Embryol.* 20, E. Rościszewska, pp. 189–203, Fig. 2, copyright 1991, with kind permission from Elsevier Science Ltd, The Boulevard, Langford Lane, Kidlington OX5 1GB, UK).

the mechanical strength to withstand the pressure of the substrate.

One example of morphological divergence in the egg coverings which is directly allied to the group's adaptive radiation concerns the endemic Hawaiian *Drosophila*. This large assemblage of hundreds of species of flies has radiated to occupy a diversity of ecological habitats, and members use a wide array of oviposition substrates, including fungi, decaying flowers, leaves, fruits, stems, bark, slime fluxes, and even spiders' eggs. Chorionic structure varies widely as expected; one particularly prominent feature is the length (Fig. 2) and surface texture of the respiratory filaments or appendages (Figs. 7c–7f). At one extreme, in species that simply drop their eggs on leaf or flower surfaces, these appendages are missing (Fig. 2k). At the other extreme, in the bark-breeding species that insert their eggs very deeply in cracks in decaying bark, the filaments are several millimeters long and several times the length of the egg (Figs. 1a, 2b, and 2c). Although the body of the egg may be completely embedded in the substrate, the long filaments protruding from the substrate provide a large surface area for respiration, using either atmospheric oxygen or dissolved oxygen via the plastron regions (Figs. 7c and 7e) when flooded—a frequent condition for eggs of these rain forest species. Species using other substrates have filaments of intermediate lengths (Figs. 2d and 2e); the variations in length and surface texture of the respiratory filaments correlate with the oviposition substrate.

C. Egg Regions to Facilitate Hatching

A final feature of the egg coverings functions at the end of embryogenesis. All insect eggs have regions of weakness, the so-called hatching lines generally bordering the operculum, that allow the operculum to be raised like a flap (Figs. 10d and 10e) or removed completely (Fig. 10b) at the time of hatching. In some instances, this region of the eggshell may serve other functions as well (Fig. 10f).

VI. CONCLUSION

The previous examples provide just a small sample of the many functional adaptations of insect egg coverings. It is impossible to be comprehensive given the size of the class Insecta and its morphological and ecological diversity. In addition, only relatively few insect eggs have been studied at the ultrastructural level, so it is certain that we have a great deal more to learn about the fascinating variations in the structure and function of the coverings of the insect egg.

See Also the Following Articles

DROSOPHILA; EGG, AVIAN; FEMALE REPRODUCTIVE SYSTEM, INSECTS; INSECT ACCESSORY GLANDS; INSECT REPRODUCTION, OVERVIEW

Bibliography

Hinton, H. E. (1981). *Biology of Insect Eggs*, Vols. I–III. Pergamon, New York.

Kambysellis, M. P. (1993). Ultrastructural diversity in the egg chorion of Hawaiian *Drosophila* and *Scaptomyza*: Ecological and phylogenetic considerations. *Int. J. Insect Morphol. Embryol.* 22, 417–446.

Kambysellis, M. P., and Craddock, E. M. (1997). Ecological and reproductive shifts in the diversification of the endemic Hawaiian *Drosophila*. In *Molecular Evolution and Adaptive Radiation* (T. J. Givnish and K. J. Sytsma, Eds.), pp. 475–509. Cambridge Univ. Press, Cambridge, UK.

Margaritis, L. H. (1985). Structure and physiology of the eggshell. In *Comprehensive Insect Physiology Biochemistry and Pharmacology*, Vol. 1 (G. A. Kerkut and L. I. Gilbert, Eds.), pp. 153–230. Pergamon, New York.

Margaritis, L. H., and Mazzini, M. (1998). Structure of the egg. In *Microscopic Anatomy of Invertebrates*, Vol. 11c, Insecta, pp. 995–1037. Wiley, New York.

Regier, J. C., and Kafatos, F. C. (1985). Molecular aspects of chorion formation. In *Comprehensive Insect Physiology Biochemistry and Pharmacology*, Vol. 1 (G. A. Kerkut and L. I. Gilbert, Eds.), pp. 113–151. Pergamon, New York.

Zeh, D. W., Zeh, J. A., and Smith, R. L. (1989). Ovipositors, amnions and eggshell architecture in the diversification of terrestrial arthropods. *Q. Rev. Biol.* 64, 147–168.

Eicosanoids

William J. Silvia

University of Kentucky

I. History of Eicosanoids
II. Synthesis of Eicosanoids
III. Mechanism of Eicosanoid Action
IV. Metabolism of Eicosanoids
V. Eicosanoids in Male Reproduction
VI. Eicosanoids in Female Reproduction

GLOSSARY

arachidonic acid The most common substrate for eiconsanoid biosynthesis. It is a 20-carbon polyunsaturated fatty acid with double bonds at positions 5, 8, 11, and 14.

lipoxygenase An enzyme that catalyzes the dioxygenation of an unsaturated fatty acid.

Eicosanoids are oxygenated metabolites of long chain, polyunsaturated fatty acids, most commonly arachidonic acid. Included among the eicosanoids are prostaglandins, thromboxanes, leukotrienes, and lipoxins that serve as local intercellular communication signals secreted by specific cell types within most tissues of the body and play roles in the reproductive physiology of both males and females.

I. HISTORY OF EICOSANOIDS

In the 1930s, three research groups reported that extracts of human seminal plasma were able to reduce blood pressure, relax uterine smooth muscle, and contract intestinal smooth muscle. von Euler extracted these myokinetic factors from ovine seminal vesicles and prostate glands. He was the first to refer to these factors as prostaglandins, due to their presence in the prostate extracts. Subsequent studies conclusively demonstrated that these factors were primarily from the seminal vesicles, not the prostate; however, the name prostaglandin was retained. In the early 1960s, Bergstrom and colleagues isolated the first individual prostaglandins from seminal vesicle extracts and described their chemical structures. The first two prostaglandins were designated PGE and PGF because of the way they partitioned during extraction, either in the ether (E) phase or in the aqueous, phosphate (F: fosfat in Swedish)-buffered saline phase. Since that time, many bioactive metabolites of arachidonic acid have been identified and new ones continue to be discovered.

II. SYNTHESIS OF EICOSANOIDS

A. Mobilization of Substrate for Eicosanoid Synthesis

The first step in synthesis of eicosanoids is the release of the substrate, primarily arachidonic acid, from phospholipid storage pools (Fig. 1). This is accomplished by activation of an arachidonic acid-specific, cytoplasmic phospholipase A_2 ($cPLA_2$). When cells are not synthesizing eicosanoids, $cPLA_2$ is found in the cytoplasm in an inactive form. Activation of $cPLA_2$ is accomplished through a complex cascade of protein phosphorylations. Agonists that stimulate eicosanoid synthesis bind to specific receptors on the cell surface. Most of these receptors belong to the seven-transmembrane domain family of receptors. Some of these receptors interact with heterotrimeric GTP-binding proteins that, in turn, activate phosphatidylinositol-specific phospholipase C_β (PLC). This results in the production of diacylglycerol (DAG) and inositol-1,4,5-trisphosphate (IP_3). IP_3 induces release of Ca^{2+} from the endoplasmic

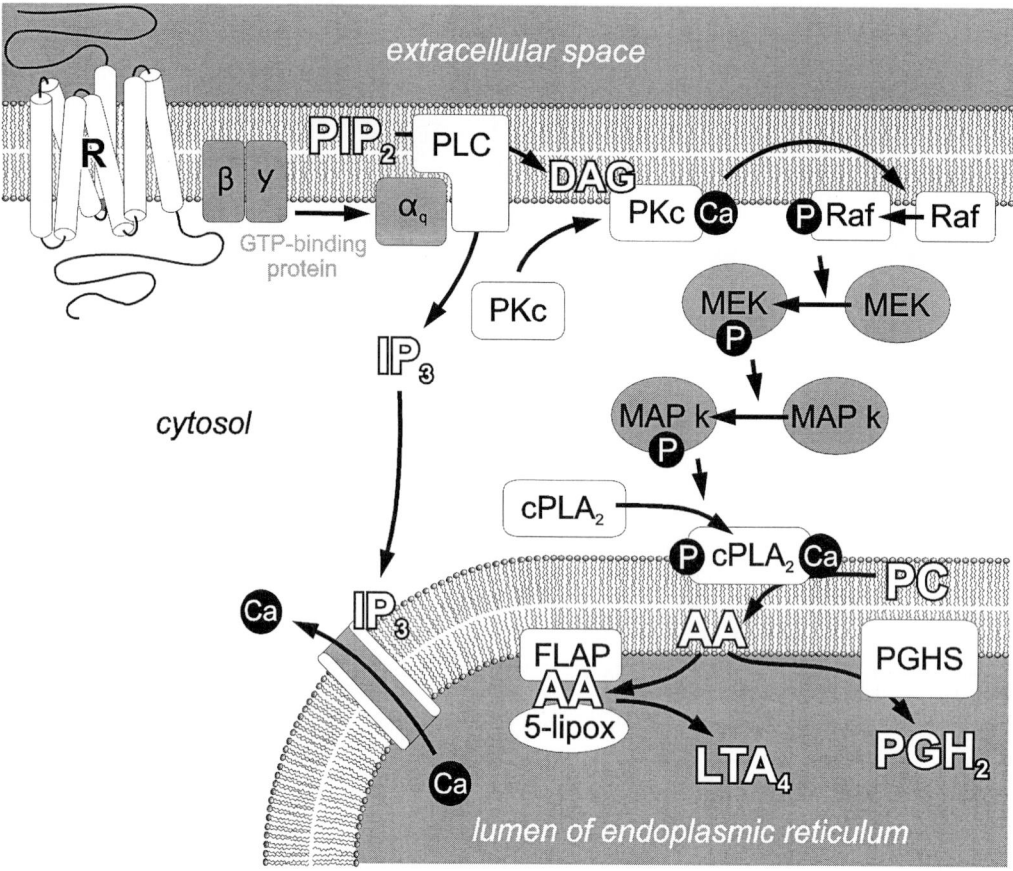

FIGURE 1 Intracellular signaling cascade that may regulate arachidonic acid availability for eicosanoid biosynthesis. Details are described in the text. R, seven-transmembrane domain receptor; $G\alpha_q, \beta, \gamma$: α_q, β, and γ subunits of the heterotrimeric GTP-binding protein; PLC, phosphatidylinositol-specific phospholipase C_β; PIP_2, phospatidylinositol bisphosphate; DAG, diacylglycerol; IP_3, inositol-1,4,5-trisphosphate; Pkc, protein kinase C; RAF, raf protein; MEK, MAP kinase/extracellular signal-regulated kinase; MAPk, mitogen-activated protein kinase; $cPLA_2$, cytoplasmic phosphatidylcholine-specific phospholipase A_2; PC, phosphatidylcholine; AA, arachiodonic acid; P, phosphate; Ca, calcium; PGHS, prostaglandin H endoperoxide synthase; 5-lipox, 5-lipoxygenase; FLAP, 5-lipoxygenase activating protein.

reticulum. DAG binds to protein kinase C (PKc). This promotes binding of Ca^{2+} to PKc, its translocation to the plasma membrane, and its activation. The active PKc can phosphorylate, and thereby activate, Raf proteins that are serine/threonine kinases themselves. Raf proteins can phosphorylate MEK [also known as mitogen-activated protein (MAP) kinase kinase or extracellular-signal regulated kinase; MEK is a hybrid acronym). MEK can phosphorylate MAP kinase. One of the protein substrates for MAP kinase is $cPLA_2$. Once phosphorylated, $cPLA_2$ binds Ca^{2+} and becomes associated with intracellular membranes. $cPLA_2$ specifically cleaves arachidonic acid from phosphatidylcholine, making it available to eicosanoid synthesizing enzymes. The second step in the synthesis of eicosanoids is the oxygenation of arachidonic acid by lipoxygenases. There are several lipoxygenases in mammalian cells. The most significant of these are prostaglandin H (PGH) endoperoxide synthase and the 5-, 12-, and 15-lipoxygenases.

B. Prostaglandin H Endoperoxide Synthase

Prostaglandin H endoperoxide synthase (PGHS) catalyzes the cyclooxygenation of arachidonic acid

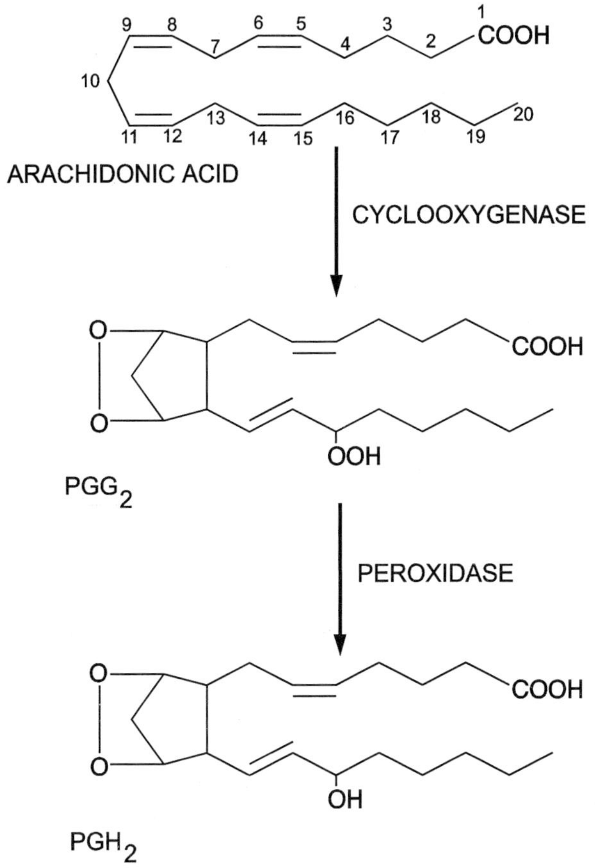

FIGURE 2 Biosynthetic pathway for conversion of arachidonic acid to PGH$_2$ by prostaglandin H endoperoxide synthase. The conventional carbon numbering scheme for arachidonic acid and its metabolites is indicated for arachidonic acid.

between carbons 9 and 11 to form PGG$_2$ and the immediate peroxidation/reduction to form PGH$_2$ (Fig. 2). These two activities occur at distinct sites in the enzyme. The cyclooxygenation reaction is also associated with the "suicide" inactivation of PGHS. Each PGHS molecule loses enzymatic activity after the synthesis of approximately 1300 molecules of PGG$_2$. Therefore, in the presence of large quantities of arachidonic acid, the ability to synthesize new PGHS enzyme can limit the amount of arachidonate metabolized.

PGHS exists in two forms, referred to as PGHS-1 and PGHS-2. They are products of separate genes and differ slightly in apparent molecular weights on SDS–PAGE. PGHS-1 has an M_r of 72,000. PGHS-2 appears as a doublet with M_r bands at 72,000 and 74,000. The amino acid sequences are 75% homologous within species examined. Both forms appear to be localized to the endoplasmic reticulum and nuclear membrane. PGHS-1 is referred to as the "constitutive" form because its synthesis does not appear to be regulated acutely. PGHS-2 is referred to as the "inducible" form because dramatic changes in its synthesis have been observed in response to stimuli over relatively short time periods (2–4 hr).

Once PGH$_2$ is synthesized, it is rapidly converted to one of the biologically active prostanoids through the action of specific enzymes. The prostanoids derived from PGH$_2$ include PGD$_2$, PGE$_2$, PGF$_{2\alpha}$, thromboxane A$_2$, and prostacyclin (Fig. 3). The specific prostanoids synthesized from PGH$_2$ in a particular cell type are determined by the relative amounts and activities of each of these enzymes. For example, tissues that possess more PGH$_2$–PGE$_2$ isomerase than PGH$_2$–PGF$_{2\alpha}$ reductase will produce more PGE$_2$ than PGF$_{2\alpha}$. The prostanoid products derived through the action of PGHS have been shown to play important roles in many reproductive processes as will be described.

C. 5-Lipoxygenase

5-Lipoxygenase catalyzes the oxidation of arachidonic acid at carbon 5 to form 5-hydroperoxy eicosatetraenoic acid (5-HPETE; Fig. 4). This enzyme can also convert 5-HPETE to leukotriene (LT) A$_4$. The synthesis of LTA$_4$ by 5-lipoxygenase is facilitated by 5-lipoxygenase activating protein (FLAP). FLAP is an integral membrane protein that binds free arachidonic acid and "presents" it to the soluble 5-lipoxygenase. The regulation of 5-lipoxygenase and FLAP synthesis has just begun to be studied in cells of the immune system and nothing is known about their regulation in the reproductive system.

LTA$_4$ is extremely unstable and will undergo rapid, nonenzymatic hydrolysis to 6-trans-LTB$_4$, a biologically inactive metabolite. Alternatively, LTA$_4$ may be converted to one of three possible products enzymatically (Fig. 4). First, it can be hydrolyzed to LTB$_4$, a potent chemotactic signal in the immune system. Second, it can have glutathione added to carbon 6 to form LTC$_4$. LTC$_4$ can have the glutamyl and glycine residues from glutathione removed sequentially to

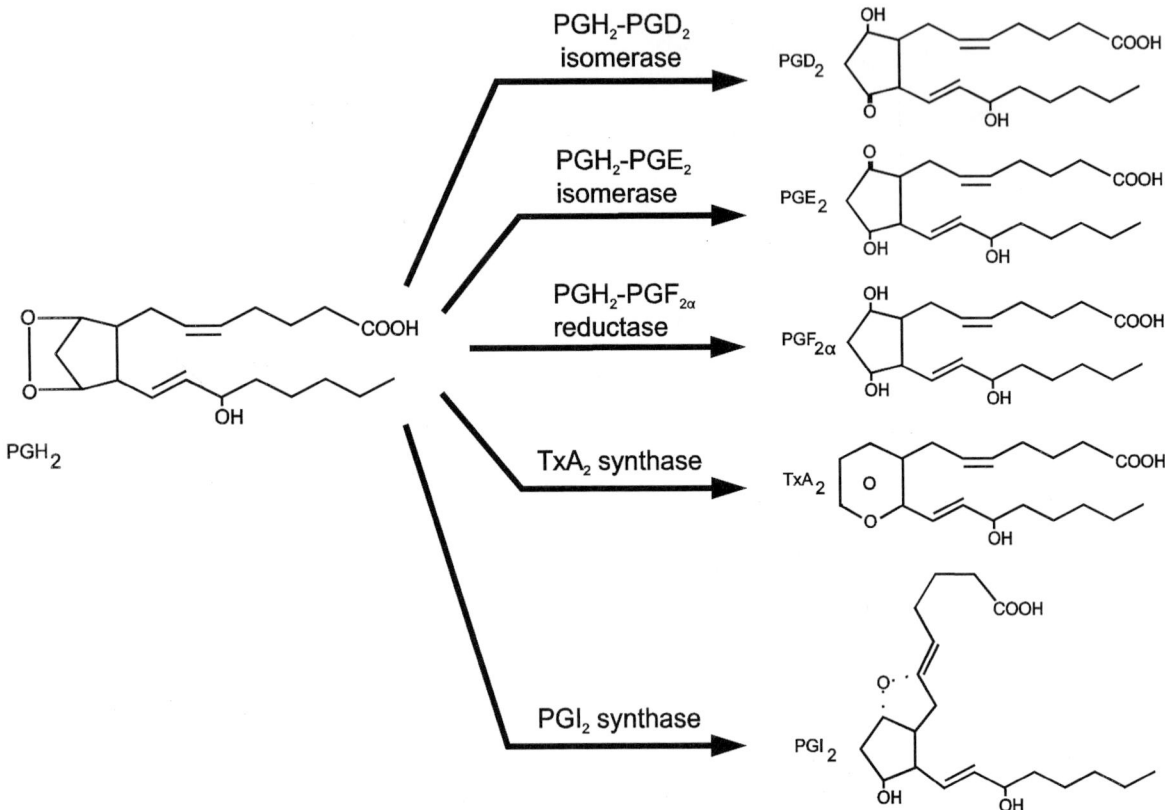

FIGURE 3 Biosynthetic pathways and enzymes involved in the conversion of PGH_2 to bioactive prostanoids.

form LTD_4 and LTE_4, respectively. These cysteinyl leukotrienes are referred to as the slow-reacting substances of anaphylaxis. Finally, LTA_4 can undergo a second oxidation at carbon 15 to form lipoxins A and B. This reaction is catalyzed by one form of 12-lipoxygenase (platelet or p-form). This enzyme has both 12- and 15-lipoxygenase activity. Just like PGHS, this enzyme is prone to suicide inactivation. While evidence will be presented for a role of 5-lipoxygenase products in luteal regression and parturition, 5-lipoxygenase-deficient mice are fertile.

D. Other Lipoxygenases

In addition to PGHS and 5-lipoxygenase, several other lipoxygenases are present in various mammalian cell types. The 12- and 15-lipoxygenases appear to have the most biological significance. 12-Lipoxygenase exists in two forms derived from two genes. The "platelet-derived" or p-form is responsible for enzymatic conversion of LTA_4 to the lipoxins as previously described. The "leukocyte-derived" or l-form is directly involved in 12-HPETE and 12-HETE formation. This ultimately leads to the synthesis of hepoxilins, a recently characterized group of eicosanoids. 15-Lipoxygenase catalyzes the oxidation of arachidonic acid to 15-HPETE and on to 15-HETE. 15-HETE may contribute to completion of the acrosome reaction in sperm. 15-HETE can be further oxidized by 5-lipoxygenase as an alternate route for lipoxin formation.

In addition to the lipoxygenases that catalyze dioxygenation of fatty acid substrates, renal cytochrome P450 type 4s can catalyze the monooxidation of fatty acids to form a variety of products including 20-HETE. 20-HETE is involved in regulation of water and ion reabsorption in the kidney. It remains to be determined if eicosanoids derived from 12-lipoxy-

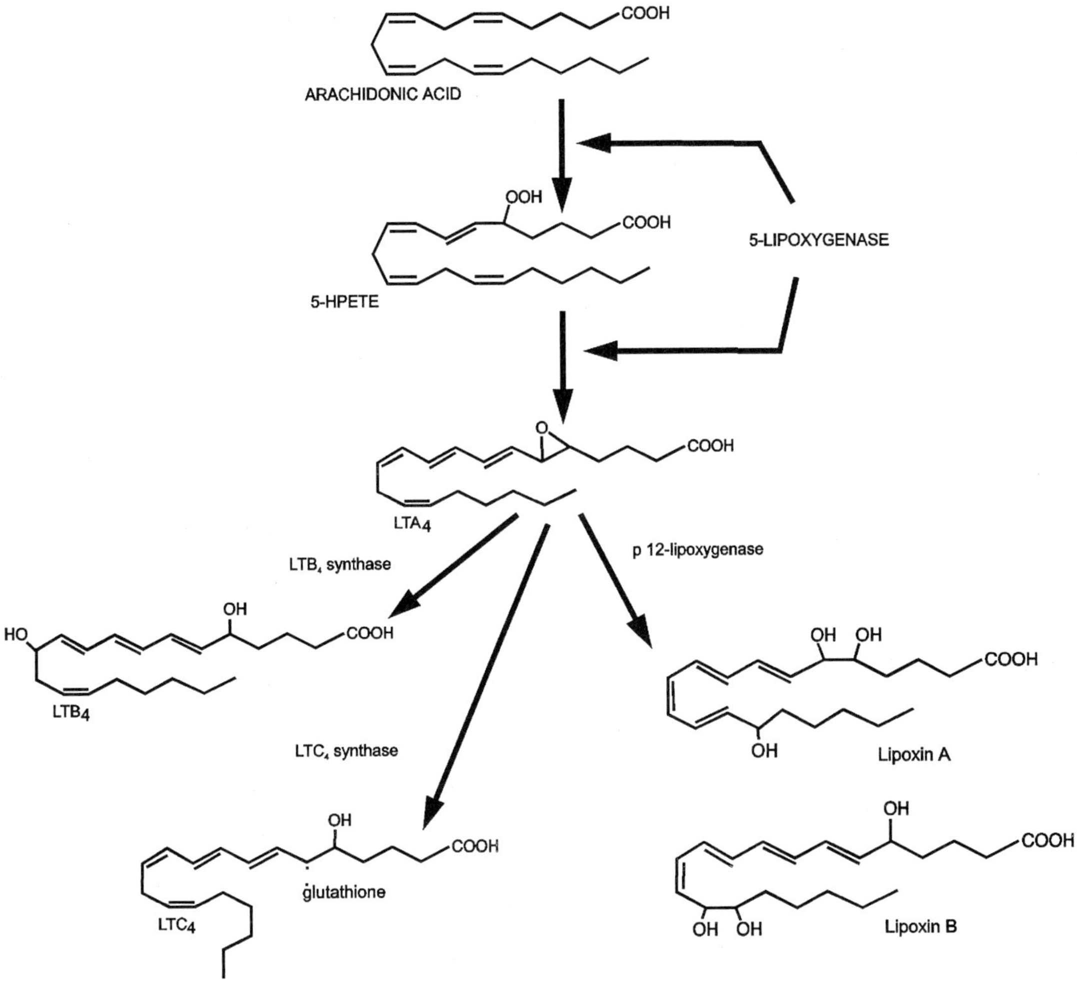

FIGURE 4 Biosynthetic pathways for the conversion of arachidonic acid to bioactive eicosanoids initiated by 5-lipoxygenase.

genase or cytochrome P450 metabolism are involved in any reproductive processes. Rats deficient in the leukocyte form of 12-lipoxygenase are fertile.

III. MECHANISM OF EICOSANOID ACTION

Although eicosanoids are synthesized from relatively hydrophobic fatty acid precursors, the oxygenation and other modifications that occur in their synthesis reduce their hydrophobicity so that they do not easily penetrate the cell membrane. Their transport out of cells that synthesize them appears to be regulated by specific transport proteins. Once released, they exert their effects on target cells by binding to receptor sites on the cell surface. The mRNA and deduced amino acid sequences of receptors for PGD_2, PGE_2 (four different isoforms designated EP_1–EP_4), $PGF_{2\alpha}$, PGI_2, TxA_2, LTB_4, and LTC_4 have been determined and all belong to the seven-transmembrane domain family of receptors. These receptors, upon binding ligand, interact with one of several heterotrimeric, GTP-binding proteins that, in turn, interact with second messenger generating enzymes. In most situations, PGD_2, PGE_2, and PGI_2 stimulate activity of adenylate cyclase. The EP_1 type of PGE_2 receptor appears to be linked to a ligand-gated Ca^{2+} channel in the plasma membrane. $PGF_{2\alpha}$, TxA_2, LTB_4, and LTC_4 receptors interact with G-proteins that activate PLC. The lipoxins also stimulate activity of PLC upon binding, implying that they bind to a seven-transmembrane domain receptor as well.

IV. METABOLISM OF EICOSANOIDS

Most tissues synthesize and/or respond to eicosanoids. Eicosanoids are typically used as local signaling molecules to convey information in a paracrine or autocrine manner. Because of their potency and diverse actions on many cell types, it is extremely important to prevent the errant transfer of eicosanoids from one tissue to another. For prostaglandins, this is accomplished by two enzymes, prostaglandin 15-dehydrogenase and Δ^{13} reductase. These enzymes convert the active prostaglandins to their corresponding inactive 13,14-dihydro, 15-keto metabolites. These enzymes are found in the kidney, liver, and lung as well as in many tissues that make prostaglandins. Levels in the lung are very high and result in the efficient breakdown of prostaglandins in a single pass through the pulmonary circulation. Carbons 1–4 are then cleaved through β and ω oxidation to produce primary urinary metabolites. PGI_2 and TxA_2 are inherently unstable at physiological temperature and pH and rapidly break down to the inactive metabolites 6-keto-$PGF_{1\alpha}$ and TxB_2. Metabolism of the other eicosanoids is poorly understood but it is likely that similar spontaneous and enzymatic mechanisms contribute to their rapid inactivation at sites of release and in the kidney, liver, and lung.

V. EICOSANOIDS IN MALE REPRODUCTION

A. Seminal Vesicles and Eicosanoids in Semen

Early research on prostaglandins and PGHS was done using seminal vesicles as a source of the enzyme. Seminal vesicles of rams, men, chimpanzees, and rhesus monkeys secrete large quantities of prostaglandins into semen (>10 μg/ml of semen). Ram seminal vesicles are the richest source of PGHS of any tissue examined. PGHS-1 is the predominant form of the enzyme in this tissue. In rams, PGE_1 is the principal prostaglandin secreted into semen. In contrast, 19-OH PGE_1 is the predominant form in primates. This is the only reproductive tissue in which prostaglandins of the 1 series are the major products. This is because eicosatrienoic acid is the primary substrate made available to PGHS, not arachidonic acid. The size, structural differentiation, and secretory activity of the seminal vesicles are regulated by testosterone. Castration results in atrophy of the seminal vesicles and the disappearance of PGHS-1 protein and mRNA. Both can be restored by androgen replacement therapy.

Several roles have been proposed for seminal prostaglandins; however, none has been rigorously tested. Suggested roles include effects on the male and female reproductive tracts to facilitate sperm transport and direct effects on spermatozoa. There is indirect evidence to support all of these actions. The concentrations of prostaglandin are much less (<1 μg/ml) in semen from bull, boar, stallion, and rabbit. This leads one to question if the seminal prostaglandins play an important role in fertility.

B. Spermatozoal Eicosanoids and the Acrosome Reaction

Spermatozoa can synthesize PGE_2 and 15-HETE from arachidonic acid. Administration of arachidonic acid or either of these eicosanoids to sperm cells will induce the acrosome reaction. PGE_2 appears to exert its effect through the EP_1 class of receptor and the influx of extracellular Ca^{2+}. This leads to activation of PKc and subsequent mobilization of arachidonic acid from phospholipid stores by $cPLA_2$. The arachidonic acid is used as substrate for 15-HETE synthesis by 15-lipoxygenase. Pretreatment of spermatozoa with nordihydroguaiaretic acid (NDGA, an inhibitor of lipoxygenase activity) blocks the acrosome reaction induced by either arachidonic acid or PGE_2 but not by 15-HETE. Therefore, 15-HETE appears to induce the acrosome reaction by an unknown mechanism.

VI. EICOSANOIDS IN FEMALE REPRODUCTION

A. Ovulation

The role of eicosanoids in the ovulatory process has been studied extensively in mammals. The biochemical and intercellular signaling processes activated in the follicle during ovulation have been likened to an inflammatory response. Nonsteroidal, antiinflammatory drugs (NSAIDs), which inhibit activity of PGHS, block ovulation in all mammalian species studied. There is an increase in follicular tissue concentrations of $PGF_{2\alpha}$ and PGE_2 within 4 hr of the preovulatory surge in luteinizing hormone (LH) or administration of an ovulatory dose of human chorionic gonadotropin (hCG). This increase in PG synthesis is associated with a concurrent increase in PGHS-2 mRNA induced by gonadotropins. PGE_2 stimulates adenylate cyclase activity, particularly in granulosa cells, which may reinforce the steroidogenic action of LH within the follicle. It may also contribute to the increase in follicular blood flow (hyperemia) that occurs during the early phase of the ovulatory process. Prostaglandin $F_{2\alpha}$ stimulates activity of phospholipase C and has been implicated in the reduction of follicular blood flow, particularly at the rupture point, during the latter portion of the ovulatory process. This contention is based primarily on the well-described vasoconstrictive properties of $PGF_{2\alpha}$ and the failure to observe the follicular ischemic response when ovulation is blocked with NSAIDs. Prostaglandins may also activate proteolytic enzymes that contribute to breakdown of the follicular wall, particularly collagenase. While specific receptors for prostaglandins have not been identified in follicular tissues, biological effects on granulosa and vascular cells imply that they exist in these cell types.

Leukotrienes, 12- and 15-lipoxygenase metabolites of arachidonic acid, also accumulate in follicular tissue during the ovulatory process. Some NSAIDs inhibit activity of lipoxygenases as well as PGH synthase. Indomethacin, a NSAID commonly used to establish the role of PGs in ovulation, is among this group. Interestingly, the dose–response curve for indomethacin blockade of ovulation is much more comparable to its dose–response inhibition of 15-HETE accumulation than of PGE_2. NDGA, BW755c, and FPL55712 (inhibitors of lipoxygenase activity) also block ovulation in some species. Therefore, lipoxygenase products may contribute to the ovulatory process as well.

Increases in follicular or ovarian prostaglandin levels around the time of ovulation have been reported in many nonmammalian species, including chickens, reptiles, fish, and some invertebrates such as snails and scallops. Ovulation was blocked in some of these species with NSAIDs. Therefore, the involvement of eicosanoids in the rupture of ovarian follicles to release ova appears to be a common phenomenon in the animal kingdom.

B. Luteal Function

1. Luteotropic Effects

Both PGE_2 and PGI_2 stimulate progesterone secretion from luteal cells *in vitro* through their ability to activate adenylate cyclase and induce accumulation of cAMP. PGE_2 also protects luteal tissue from luteolytic effects of $PGF_{2\alpha}$ when both of the prostaglandins are administered together.

2. Luteolysis

i. Prostaglandin $F_{2\alpha}$ from the Uterus Prostaglandin $F_{2\alpha}$ induces premature luteal regression in many mammalian species, including cattle, sheep, goats, pigs, horses, and guinea pigs and in pseudopregnant rats, mice, hamsters, and rabbits. Luteolysis is blocked in these species by NSAIDs, or immunization against $PGF_{2\alpha}$, implying that endogenous eicosanoids, presumably $PGF_{2\alpha}$, play an essential role in luteal regression. The source of $PGF_{2\alpha}$ in these species is the uterus. Two lines of evidence support this conclusion. First, in most subprimate species, luteolysis can be prevented by hysterectomy (surgical removal of the uterus). Second, the uterus synthesizes and secretes $PGF_{2\alpha}$ at the appropriate time of the estrous cycle to cause luteal regression. This appears to be the only situation in which a blood-born eicosanoid acts as an endocrine signal. The efficient metabolism of $PGF_{2\alpha}$ in the pulmonary circulation makes it impossible for luteolytic quantities of active $PGF_{2\alpha}$ to reach the corpus luteum by a conventional systemic circulatory route. A small fraction (<10%) of the PGF$_{2\alpha}$ leaving the uterus passes locally from the uteroovarian vein to the adjacent uteroovarian artery, by a poorly understood countercurrent transfer system, to reach the ovary. The uteroovarian vascular architecture in this region is similar in structure to the vascular pampiniform plexus that supplies the testes in the male and facilitates the countercurrent exchange of heat.

Several luteolytic responses occur shortly after exposure to PGF$_{2\alpha}$. The concentration of progesterone in the peripheral circulation and in luteal tissue declines within 3 or 4 hr. This is associated with a decline in luteal blood flow, infiltration of the corpus luteum by eosinophils, and exocytosis of content of secretory granules from large luteal cells. More extensive structural breakdown of luteal cells is evident after 12–24 hr when oligonucleosomes, evidence of apoptotic cell death, can be detected.

PGF$_{2\alpha}$ exerts its effects by binding to receptors on the surface of large luteal cells and activation of phospholipase C. This results in the production of two intercellular second messengers, DAG and IP$_3$. The increase in IP$_3$ increases cytosolic concentration of Ca^{2+} responsible for structural breakdown of luteal cells. The Ca^{2+} influx has been linked to programmed cell death in other tissues and may be responsible for triggering exocytosis of the contents of secretory granules of large luteal cells.

The accumulation of DAG increases activity of PKc, which is responsible for the antisteroidogenic action of PGF$_{2\alpha}$. The precise mechanism by which PKc exerts this effect is not known. PKc may initiate the Raf, MEK, MAP kinase phosphorylation cascade that activates cPLA$_2$ and promote local eicosanoid biosynthesis by the luteal cells. Luteal cells possess PGHS and synthesize prostaglandins *in vitro*. An increase in LTB$_4$ can be detected in luteal tissue within 2 hr of exposure to PGF$_{2\alpha}$. LTB$_4$ is a potent chemoattractant for eosinophils and could be responsible for the influx of eosinophils observed during luteolysis. An essential role for lipoxygenase products is further supported by the fact that NDGA delays luteal regression in cattle. Thus, one eicosanoid (PGF$_{2\alpha}$) may exert some of its luteolytic effects by inducing synthesis of another eicosanoid (LTB$_4$) in luteal tissue.

ii. Regulation of Uterine PGF$_{2\alpha}$ Secretion Ovarian steroids contribute to the regulation of uterine PGF$_{2\alpha}$ secretion. Progesterone induces the accumulation of lipid droplets which serve as a reservoir of arachidonic acid esterified in phospholipids and triglycerides in epithelial cells. Progesterone is also required for expression of PGHS-2 in epithelial cells.

In the large domestic species, PGF$_{2\alpha}$ is secreted by the uterus at luteolysis as a series of pulses associated with synchronous pulses of oxytocin. Oxytocin is a potent stimulus for PGF$_{2\alpha}$ secretion from the uterus. The corpus luteum in ruminants is a rich source of oxytocin, which is stored in secretory granules within large luteal cells. As noted previously, PGF$_{2\alpha}$ induces the exocytosis of these granules and secretion of luteal oxytocin. Therefore, uterine PGF$_{2\alpha}$ and luteal oxytocin constitute a positive feedback loop. The high concentrations of both hormones associated with each luteolytic pulse of PGF$_{2\alpha}$ are believed to be due to the activation of this positive feedback loop.

The expression of oxytocin receptors on epithelial cells is regulated by progesterone and estradiol; therefore, these steroids contribute to the regulation of PGF$_{2\alpha}$, indirectly, through this mechanism. The epithelial cells of the endometrium appear to be the source of luteolytic PGF$_{2\alpha}$ because they possess much higher concentrations of both PGHS-2 and oxytocin receptors than stromal cells. The oxytocin receptor is a member of the seven-transmembrane domain receptor family. Binding of ligand to this receptor results in an increase in activity of PLC. This is assumed to result in initiation of the Raf, MEK, MAP kinase phosphorylation cascade and activation of cPLA$_2$. Inhibition of PLA$_2$ activity will completely eliminate PGF$_{2\alpha}$ synthesis in response to oxytocin. The active cPLA$_2$ provides free arachidonic acid to PGHS-2 for conversion to PGH$_2$. While uterine tissue can synthesize PGE$_2$ and PGI$_2$ from PGH$_2$, the majority of PGH$_2$ is converted to PGF$_{2\alpha}$. PGH$_2$–PGF$_{2\alpha}$ isomerase (also referred to as PGF synthase) has been detected in uterine tissue from mice. During a luteolytic pulse of PGF$_{2\alpha}$, the uterus secretes 1 nmol PGF$_{2\alpha}$/min. Due to the suicidal nature of the PGHS enzyme, this would require 1 pmol (63 ng) of PGHS enzyme to be synthesized every minute during each pulse. To meet this enormous need for PGHS enzyme, oxytocin induces a rapid and transient increase in PGHS-2 gene expression.

iii. Prostaglandin $F_{2\alpha}$ from the Corpus Luteum

The uterus does not participate in luteolysis in most or all primates, including humans, but also dogs and cats. The precise mechanisms regulating luteal regression in these species are not known. The corpus luteum appears to be much less sensitive to the luteolytic action of $PGF_{2\alpha}$ in these species. In rhesus monkeys, exogenous $PGF_{2\alpha}$ must be administered directly into luteal tissue by continuous infusion for several days to induce complete luteal regression. Luteal cells from primates synthesize a variety of eicosanoids including $PGF_{2\alpha}$. *In vivo*, an increase in $PGF_{2\alpha}$ secretion from the corpus luteum has been detected around the time of luteolysis. As in ruminants, the human corpus luteum contains oxytocin. Exogenous oxytocin, administered directly into human corpora lutea, stimulates luteal secretion of $PGF_{2\alpha}$. Thus, an intraluteal oxytocin : $PGF_{2\alpha}$ positive feedback loop may contribute to regulation of $PGF_{2\alpha}$ secretion and luteolysis in primates. It may be analogous to the uterine:luteal feedback loop described for ruminants.

C. Maternal Recognition of Pregnancy

In order for pregnancy to proceed normally, luteal regression must be prevented. This process is termed maternal recognition of pregnancy. The conceptus can suppress secretion of luteolytic $PGF_{2\alpha}$ from the uterus and/or the conceptus can secrete a substance that protects the corpus luteum from the luteolytic effect of $PGF_{2\alpha}$.

The conceptus secretes a variety of substances that can suppress uterine secretion of luteolytic $PGF_{2\alpha}$ into the uteroovarian vasculature. These include interferon-τ in ruminants and estrogens in pigs. Interferon-τ inhibits synthesis of estrogen and oxytocin receptors by uterine luminal epithelial cells and may induce the synthesis of intracellular factors that reduce activities of PGHS and PLA_2. In the pig, estrogens alter the direction of $PGF_{2\alpha}$ secretion from basal surface (toward the vasculature) to apical surface (toward the uterine lumen). The result is that $PGF_{2\alpha}$ is sequestered in the lumen of the uterus rather than being secreted into the vasculature to induce luteolysis. This implies that estrogens exert an effect on an unidentified directionally specific, intracellular PG transport system.

Large amounts of PGE_2 are secreted from the gravid uterus in some large domestic species at the expected time of luteolysis in nonpregnant animals. PGE_2 is a possible mediator of maternal recognition of pregnancy due to its luteotropic and luteoprotective properties. Both the gravid uterus and conceptus synthesize PGE_2.

D. Menstruation

In 1957, Pickles described myokinetic properties of extracts from human menstrual fluid which could be partitioned into three fractions. One of these fractions contains very high concentrations of $PGF_{2\alpha}$ and smaller amounts of PGE_2. High concentrations of TxB_2 and smaller amounts of 6-keto-$PGF_{1\alpha}$ and cysteinyl leukotrienes have been detected subsequently. Maximal concentrations of both $PGF_{2\alpha}$ and PGE_2 in endometrial tissue were detected during the late secretory (luteal) phase of the menstrual cycle, around the time of luteolysis and menstruation. This period of eicosanoid synthesis is characterized by an increase in amounts of $PGF_{2\alpha}$ relative to other eicosanoids.

$PGF_{2\alpha}$ has been proposed to serve as a stimulus for constriction of spiral arterioles in the endometrium. Their constriction is required to reduce blood loss during menstruation. Women who experience excessive blood loss at menstruation (menorrhagia) typically produce abnormally high concentrations of the vasodilatory prostaglandins, PGE_2 and PGI_2. Leukotrienes may also contribute to the influx of leukocytes into the endometrium at this time.

As in species that employ uterine $PGF_{2\alpha}$ as a luteolytic agent, the production of prostaglandins by human endometrial tissue is regulated by ovarian steroids. Maximal prostaglandin synthesis is achieved by administration of estradiol after priming with progesterone which maximizes PLA_2 and PGHS expression in endometrial tissue. The acute stimulus for prostaglandin synthesis at menstruation has not been identified. Several substances associated with inflammatory responses will trigger prostaglandin synthesis by human endometrial tissue *in vitro*, including endothelin-1, interleukin-1 (IL-1), tumor necrosis factor (TNF)-α, bradykinin, and histamine. Both endothelin-1 and IL-1 are synthesized by endometrial tissue. Concentrations of TNF-α increase in epithelial cells around the time of menstruation.

E. Implantation

In species that have a true invasive implantation, eicosanoids appear to play an essential role in that process. This has been studied extensively in rodents. An embryonic factor (possibly IL-1) stimulates production of prostaglandins and leukotrienes from endometrial epithelial cells in close proximity to the implanting blastocyst. This is associated with a local increase in vascular permeability. Inhibition of eicosanoid biosynthesis by NSAIDs prevents this increase in vascular permeability and will delay or prevent implantation. A similar vascular response can be induced by PGE_2 which stimulates an increase in adenylate cyclase activity in uterine stromal cells. Stromal cells release platelet-activating factor (PAF) in response to PGE_2 and PAF stimulates production of PGE_2 by epithelial cells. Therefore, a positive feedback loop system is established between epithelial PGE_2 and stromal PAF that leads to the synthesis of very large amounts of each.

The uterine response to the implanting blastocyst only occurs if the uterus has been properly preconditioned with progesterone and estradiol. This is due to the ability of steroids to regulate endometrial expression of receptors for embryonic ligands that initiate this response. There is a very dramatic increase in the concentration of IL-1 receptors on epithelial cells around the time of implantation. The steroids regulate endometrial concentrations of PGHS, PLA_2, and the accumulation of lipid precursors for eicosanoid biosynthesis. The concentration of receptors for PGE_2 in stromal cells is also regulated by ovarian steroids.

F. Parturition

Local production of eicosanoids plays an important role in parturition. In many species, inhibition of eicosanoid synthesis with NSAIDs delays the onset or progress of labor. Amnion, chorion, and decidual endometrium synthesize a variety of eicosanoids *in vitro*, including PGE_2, $PGF_{2\alpha}$, PGI_2, TxA_2, LTB_4, LTC_4, 5-HETE, 12-HETE, and 15-HETE. Concentrations of many eicosanoids or their stable metabolites (including PGE_2, $PGF_{2\alpha}$, TxA_2, PGI_2, and 5-HETE) increase in the peripheral circulation, amniotic fluid, endometrium, decidual tissue, myometrium, and cervix during labor. The increase in prostanoids may be due to the increase in PGHS-2 gene expression in human amnion and decidua and in ovine placental tissues at parturition. The signal responsible for inducing expression of PGHS-2 has not been identified; however, IL-1β, epidermal growth factor, and renin stimulate PGHS-2 gene expression by human amnion cells *in vitro*.

Another factor that may contribute to eicosanoid synthesis at parturition is oxytocin. In many species, concentrations of oxytocin in the peripheral circulation increase during labor. This is due to the reflex release of oxytocin from the posterior pituitary gland. In other species (rat and human), the uterus and/or placental tissues synthesize and release oxytocin locally. In some species, concentrations of receptors for oxytocin increase in the endometrium and myometrium during labor. Changes in PGHS-2 and oxytocin receptor gene expression may be induced by a shift from progestational to estrogenic dominance of the uterus during parturition. Increases in the level of oxytocin and its receptor in uterine and placental tissues may induce eicosanoid biosynthesis, much as it does in endometrial tissue during luteolysis. Oxytocin stimulates activity of PLC and prostaglandin synthesis by human amnion tissue collected near term.

Eicosanoids exert several effects that contribute to successful parturition. In species that rely on the corpus luteum as the major source of progesterone during gestation (marsupials, dog, goat, and pig), uterine $PGF_{2\alpha}$ contributes to luteolysis to remove the inhibitory effect of progesterone on myometrial contractions. Several eicosanoids ($PGF_{2\alpha}$, thromboxane A_2, and 5-HETE) stimulate myometrial contractions and some prostaglandins (PGE_2 and PGI_2) tend to relax the myometrium. As in other tissues, binding of $PGF_{2\alpha}$ to its myometrial receptors induces an increase in PLC activity and myometrial contractions occur in response to an increase in cytosolic Ca^{2+} concentrations induced by IP_3. Prostaglandins, particularly PGE_2, promote ripening of the cervix but increase the activity of collagenolytic enzymes, similar to the response in ovarian follicles at ovulation. PGE_2 is used clinically to promote cervical relaxation during difficult births. Prostaglandins probably contribute to the release and expulsion of the placenta and continue to increase during human parturition until the placenta is shed. In cattle, the incidence of retained placentas following induced parturition can be dramatically reduced by administration of $PGF_{2\alpha}$ during labor. This is also likely to be due to the ability of the prostaglandins to promote collagenolytic enzyme activity.

G. Oviposition

Many nonmammalian species produce a nutrient-rich oocyte encased in a protective shell shortly after fertilization. Other species release nutrient-rich oocytes into the environment for external fertilization. The process of passing these large, nutrient-rich oocytes or recently fertilized embryos through the reproductive tract and out into the environment is termed oviposition. In many ways, it is analogous to parturition in mammals including the essential involvement of eicosanoids. In both the domestic hen and the lizard, oviposition can be induced by vasotocin. This is associated with an increase in oviductal prostaglandin concentrations. Oviposition can be delayed in those species by administration of NSAIDs.

See Also the Following Articles

CORPUS LUTEUM OF PREGNANCY; LUTEOLYSIS; OXYTOCIN; PARTURITION, NONHUMAN MAMMALS

Bibliography

Clark, J. D., Schievella, A. R., Nalefski, E. A., and Lin, L. L. (1995). Cytosolic phospholipase A2. *J. Lipid Med. Cell Signaling* 12, 83–117.

Funk, C. D. (1996). The molecular biology of mammalian lipoxygenases and the quest for eicosanoid functions using lipoxygenase-deficient mice. *Biochem. Biophys. Acta* 1304, 65–84.

Kanai, N., Lu, R., Satriano, J. A., Bao, Y., Wolkoff, A. W., and Schuster, V. L. (1995). Identification and characterization of a prostaglandin transporter. *Science* 268, 866–869.

Mitchell, M. D., Romero, R. J., Edwin, S. S., and Trautman, M. S. (1995). Prostaglandins and parturition. *Reprod. Fertil. Dev.* 7, 623–632.

Murdoch, W. J., Hansen, T. R., and McPherson, L. A. (1993). A review—Role of eicosanoids in vertebrate ovulation. *Prostaglandins* 46, 85–115.

Otto, J. C., and Smith, W. L. (1995). Prostaglandin endoperoxide synthases-1 and -2. *J. Lipid Med. Cell Signaling* 12, 139–156.

Poyser, N. L. (1995). Review: The control of prostaglandin production by the endometrium in relation to luteolysis and menstruation. *Prostagl. Leukocyte Essen. Fatty Acids* 53, 147–195.

Psychoyos, A., Nika, G., and Gravanis, A. (1995). The role of prostaglandins in blastocyst implantation. *Hum. Reprod.* 10(Suppl. 2), 30–42.

Silvia, W. J., Brockman, J. A., Kaminski, M. A., DeWitt, D. L., and Smith, W. L. (1994). Prostaglandin endoperoxide synthase in seminal vesicles. *Mol. Androl.* 6, 197–207.

Ushikubi, F., Hirata, M., and Narumiya, S. (1995). Molecular biology of prostanoid receptors; An overview. *J. Lipid Med. Cell Signaling* 12, 343–359.

Ejaculation

Kevin E. McKenna

Northwestern University School of Medicine

I. Peripheral Innervation of the Pelvic Organs
II. Peripheral Neurophysiology and Pharmacology
III. Neural Control
IV. Spinal Afferents
V. Spinal Efferents
VI. Spinal Interneurons
VII. Spinal Reflexes
VIII. Supraspinal Control

GLOSSARY

afferent Moving toward a certain region. Often used in neurobiology to refer to sensory nerve fibers projecting from peripheral structures to the spinal cord or brain.

bladder neck The smooth muscle at the base of the bladder. When contracted, it blocks the flow of urine into the urethra or prevents the flow of semen into the bladder.

efferent Moving away from a region; often used for nerve fibers which leave the brain or spinal cord to innervate an organ.

interneurons Neurons in a brain region or the spinal cord which are interposed between sensory fibers and the efferent neurons.

paracrine cells Cells that release messenger substances which modulate the function of cells in the immediate vicinity.

perineal muscles The skeletal muscles of the floor of the pelvis, including the ischiocavernosus, bulbospongiosus, external urethral and anal sphincters, and the levator ani.

Ejaculation is defined as the expulsion of seminal fluid from the urethral meatus. It occurs at sexual climax. However, it should be noted that ejaculation is not exactly synonymous with sexual climax or orgasm. Orgasm refers to the entire response of sexual climax: ejaculation, extragenital responses, and subjective pleasurable experience. Ejaculation consists of two distinct successive phases: emission and expulsion. Both are largely mediated by spinal reflexes. Pudendal sensory fibers mediate most of the ejaculatory responses and supraspinal control exerts both inhibitory and excitatory influence. Emission involves the secretion of seminal fluids from the accessory sex glands, contraction of the ductus deferens containing spermatozoa, and closure of the bladder neck and the external urethral sphincter. Expulsion is produced by the rhythmic contractions of smooth muscle of the urethra and the striated perineal muscles, primarily the bulbospongiosus muscle, which act to expel the semen. Following ejaculation is a refractory period during which sexual arousal is inhibited.

I. PERIPHERAL INNERVATION OF THE PELVIC ORGANS

The innervation of the reproductive organs has been the subject of extensive investigation in many species, including man. The general conclusion is that the visceral organs involved in ejaculation receive both sympathetic and parasympathetic innervation. These derive largely from the pelvic plexus (formerly referred to as the inferior hypogastric plexus) and the caudal sympathetic chain, respectively. There is a common misconception that emission is largely a sympathetic phenomenon, but in most pelvic organs, a synergism of sympathetic and parasympathetic control is commonly seen. Parasympathetic mechanisms have a strong role in stimulating secretion of seminal fluids by epithelial cells of the accessory sex glands and sympathetic mechanisms control contraction of glandular smooth muscle to expel seminal fluids from the glands. Adrenergic and cholinergic mechanisms are involved in the neural control of ejaculation as well as nonadrenergic/noncholinergic (NANC) mechanisms served by autonomic innervation using neuropeptides, purines, and nitric oxide. Some afferent innervation has been identified but its physiological role is unclear.

A. Epididymis

The major part of the innervation of the epididymis is the inferior spermatic plexus from the pelvic plexus. A lesser contribution is provided by the superior spermatic plexus which follows the spermatic artery. The innervation is most dense in the cauda epididymis, with fewer fibers in the caput epididymis and the fibers are quite sparse in the efferent ducts. Smooth muscle of the epididymis receives innervation from both adrenergic and cholinergic terminals. Many adrenergic fibers also contain neuropeptide Y (NPY) and cholinergic fibers may contain vasoactive intestinal polypeptide (VIP). Calcitonin gene-related peptide (CGRP) has been identified in sensory fibers.

B. Ductus Deferens

The ductus deferens receives dense adrenergic and cholinergic innervation. The sympathetic and parasympathetic preganglionic fibers travel in the hypogastric and pelvic nerves, respectively. Both classes of postganglionic neurons are located within the pelvic plexus. The vasculature of the ductus is innervated by fibers originating in the caudal sympathetic chain. Adrenergic fibers predominate and form an extensive plexus throughout the muscle layers. The cholinergic fibers are primarily seen innervating the epithelium, but cholinergic innervation of the inner smooth muscle layer has also been identified. In addition to the classical neurotransmitters, fibers containing VIP and NPY are common, probably colocalized in cholinergic and adrenergic fibers, respectively. CGRP, substance P (SP), and enkephalin fibers are numerous, primarily in afferent fibers. These may provide modulatory control of efferent function.

C. Seminal Vesicle

The seminal vesicles also receive a dual sympathetic and parasympathetic innervation from the pelvic plexus with preganglionic fibers provided by hypogastric and pelvic nerves. The adrenergic innervation is distributed to both inner and outer smooth muscle layers, without an innervation of the epithelium. In contrast, the cholinergic innervation has an epithelial distribution, with sparse, if any, innervation of the smooth muscle layers. Thus, it appears that the sympathetic innervation provides contractile control and parasympathetic stimulates secretion. As in the other organs, a colocalization of NPY and VIP has been identified.

D. Prostate

The prostate receives sympathetic and parasympathetic innervation from the hypogastric and pelvic nerve through the pelvic plexus. The smooth muscle of the stroma receives a very dense adrenergic innervation. A less dense cholinergic innervation is found on both smooth muscle and the epithelium. VIP is the most common peptidergic neurotransmitter in the prostate. As in other urogenital organs, a colocalization with acetylcholine is likely. A strong NPY innervation is found in the prostate, possibly colocalized with norepinephrine. An afferent innervation of the prostate has been identified, but its physiological significance is unclear.

E. Bladder Neck and Urethra

In the bladder neck, a dual sympathetic and parasympathetic innervation is prominent. The sympathetic innervation has its origin in both the hypogastric nerve and the paravertebral sympathetic chain. Sympathetic stimulation elicits a closure of the bladder neck and proximal urethra. A cholinergic parasympathetic innervation controls active relaxation of the smooth muscle of the bladder neck and prostatic urethra. NANC agents released from parasympathetic fibers, such as VIP and especially nitric oxide, probably play a vital role in relaxation of urogenital smooth muscle. Sensory fibers within the bladder neck and urethra contain SP and CGRP.

F. Paracrine Cells

The epithelium of the urethra and prostatic ducts contain a dense network of paracrine cells. These cells contain serotonin and some peptides such as somatostatin. Many of these paracrine cells possess microvilli which extend into the lumen of the urethra. This feature suggests that these cells may monitor the chemical composition of urethral fluids. Most of the paracrine cells possess a single large process extending into the subepithelial layer. These processes are often associated with sensory nerve terminals, suggesting that serotonin modulates urethral sensory processes.

G. Perineal Muscles

The striated perineal muscles, including the ischiocavernosus, bulbospongiosus, and levator ani, are innervated by the pudendal nerve.

II. PERIPHERAL NEUROPHYSIOLOGY AND PHARMACOLOGY

Activation of both sympathetic and parasympathetic nerves produces seminal emission. Emission involves both epithelial secretion and smooth muscle contraction throughout the seminal tract. Adrenergic mechanisms tend to predominate in contraction of smooth muscle in the seminal tract. Epithelial secretion often is mediated by cholinergic and peptidergic stimulation. The effects of adrenergic stimulation tend to be more dramatic due to the rapid expulsion of fluids caused by smooth muscle contraction and this has led to an overemphasis on sympathetic mechanisms in the control of emission. The role of parasympathetic control in the prior secretion of fluids must be recognized.

Stimulation of sympathetic nerves elicits a strong contractile response in the ductus deferens which promotes a transport of sperm and epididymal fluid toward the urethra. Contraction of the ductus is also produced by stimulation of the pelvic nerve. The sympathetic stimulation of contraction is mediated by the α1-adrenergic receptor and the parasympathetic in part by the muscarinic cholinergic receptor. Purinergic mechanisms may also contribute to the sympathetic effect. Epithelial secretion in the ductus may be mediated by cholinergic mechanisms.

Both adrenergic and cholinergic agents elicit contraction of the seminal vesicle. Contractions induced by sympathetic stimulation are only partly due to adrenergic mechanisms, indicating another contractile neurotransmitter, possibly purines. The neural control of seminal vesicle secretion is not fully understood. A cholinergic parasympathetic control of epithelial secretion is likely.

Prostatic secretion is stimulated by both sympathetic adrenergic and parasympathetic cholinergic mechanisms. Stimulation of sympathetic nerves causes the expulsion of prostatic fluid into the urethra, whereas pelvic nerve stimulation elicits considerably less fluid. Pharmacological experiments indicate that both adrenergic and cholinergic stimulation induce prostatic secretion. Prostatic epithelial secretion is stimulated by cholinergic muscarinic mechanisms and smooth muscle contraction is mediated by α1-adrenergic mechanisms.

Bladder neck contraction is elicited by sympathetic nerve stimulation mediated by α1-adrenergic receptor. Relaxation is produced by nitric oxide. Sympathetic stimulation in the human produces a coordinated emission pattern: closure of the bladder neck, contraction of the seminal vesicles, and contraction of prostate and ductus deferens. The fluid expelled by this smooth muscle contraction is likely the result of prior parasympathetic stimulation of epithelial secretion. Thus, a general conclusion about the peripheral autonomic control of emission is that parasympathetic cholinergic mechanisms predominate in the control of epithelial secretion and that sympathetic adrenergic stimulation produces smooth muscle contraction.

III. NEURAL CONTROL

While a detailed wiring diagram of the neural control of ejaculation is not available, the general organizational plan can be offered. Ejaculation is produced by a spinal pattern generator. This system produces a coordination of sympathetic, parasympathetic, and somatic outflow to induce emission by secretion from accessory sex glands and contraction of gland smooth muscle, contraction of the bladder neck, and rhythmic contractions of pelvic smooth and striated muscles. This response is produced in response to continued sexual arousal mediated by genital afferents primarily carried in the pudendal nerve. The spinal ejaculatory system is under descending control, both excitatory and inhibitory, from brain stem and hypothalamic sites. These supraspinal sites are themselves influenced by genital stimulation.

The exact neural mechanisms which trigger ejaculation are not known, but ejaculation involves a summation of segmental sensory stimulation and descending excitatory drive. Genital sensory stimulation serves to activate the ejaculatory system as well as to facilitate descending excitatory drive and possibly reduce descending inhibitory control. Descending control is also affected by higher sensory stimuli (olfactory, visual, etc.) and internally generated arousal. Ejaculation can be induced purely by spinal sensory stimuli (as in patients with spinal cord injury) or purely by descending excitation (as in extreme cases of premature ejaculation or sleep-related ejaculation). The neural basis for the postejaculatory refractory period is unknown.

IV. SPINAL AFFERENTS

Pudendal nerve afferents are the most important in mediating emission and expulsion reflexes. The majority of these travel in the dorsal nerve of the penis. The role of the visceral sensory fibers innervating the seminal tract in the ejaculatory process is unknown. Pudendal nerve afferents terminate in the lower lumbar and upper sacral spinal segments. These sensory fibers terminate bilaterally in the medial portions of the dorsal horn and the dorsal gray commissure. There is a smaller projection to the lateral portions of the dorsal horn. The pudendal sensory fibers probably do not make direct synaptic connections with somatic or autonomic efferents.

Contrary to popular perception, tactile (light touch) sensitivity of the penis is very low compared to that of other skin despite the fact that it has one of the highest densities of sensory fibers in the body. However, painful sensations are activated at low thresholds. The majority of penile afferents in man are small, thinly myelinated or unmyelinated fiber of the $A\delta$ or C fiber category. The majority of penile afferents are free nerve endings which probably mediate low force, slowly adapting afferent responses as well as noxious stimuli. Noxious stimuli have an inhibitory effect on reflex contractions of the penile muscles. The glans penis has an encapsulated nerve ending, which is unique to the penis and is referred to as the genital end bulb or lamellated corpuscles. These sensory structures have received considerable attention for their possible role in mediating sexual arousal. Penile sensitivity is greatly enhanced during erection. This may be due to mechanical changes in the tissue, increases in tissue temperature, as well as modulation by autonomic activity. Despite considerable investigation, no definitive conclusions can be drawn regarding the exact sensory mechanisms responsible for erection, emission, and expulsion.

V. SPINAL EFFERENTS

The sympathetic preganglionic neurons innervating the pelvis are located in the intermediolateral cell column and in the central autonomic region of the lower thoracic and upper lumbar segments. There is evidence that preganglionics in the intermediolateral cell column preferentially project to ganglion cells in the sympathetic chain and are more likely to be involved in control of the vasculature. Conversely, the medial neurons preferentially project to the prevertebral ganglia and are mainly involved in visceral control, such as smooth muscle contraction. Parasympathetic preganglionic neurons are located in the intermediolateral cell column of the upper sacral segments, referred to as the sacral parasympathetic nucleus.

Pudendal motoneurons innervating the striated perineal muscles are located in Onuf's nucleus in the lumbosacral spinal cord. The dendrites of these neurons form distinctive bundles, which have been suggested to be involved in synchronizing the contractions of these muscles.

VI. SPINAL INTERNEURONS

All of the known sexual reflexes are polysynaptic; that is, interneurons are required for converting sensory signals into motor or secretory outputs. However, currently little is known about the spinal interneuronal networks involved in sexual function. Spinal neurons which are presynaptic to pudendal motoneurons or autonomic preganglionic neurons have been identified in anatomical studies in animals using substances that are retrogradely transported from a peripheral injection site and then transferred across synapses to label interneurons. Interneurons innervating the penis, urethra, and striated penile muscles are located bilaterally in the medial gray matter, including the dorsal gray commissure and surrounding the central canal, as well as in the ventral horn of the spinal cord. These interneurons extend for several segments, from lower thoracic to sacral segments. The location of interneurons overlaps the distribution of pudendal sensory fibers. The central gray matter of the spinal cord is also a prominent target for fibers from supraspinal sites. Thus, a multisegmental network of neurons in the central core of the lower spinal cord provides an anatomical substrate for linking genital sensory stimuli and descending control with sympathetic, parasympathetic, and somatic outflow.

VII. SEGMENTAL REFLEXES

A. Bulbocavernosus Reflex

The bulbocavernosus reflex is a spinal segmental reflex. It is activated by light pressure or stroking of the penis and consists of a contraction of the perineal muscles. There is a simultaneous contraction of all of the muscles innervated by the pudendal nerve. The afferent limb is mediated by pudendal nerve sensory fibers and the efferents are pudendal nerve motoneuron axons. It has been demonstrated that this reflex involves spinal interneurons interposed between the sensory fibers and the motoneurons. The purpose of the bulbocavernosus is not definitely known. However, during sexual arousal, the contraction of the bulbocavernosus and ischiocavernosus muscles would cause an increased erection of the glans and shaft of the penis, respectively. The contraction of the external urethral sphincter may promote the buildup of seminal fluid in the posterior urethra.

B. Bladder Neck Closure

The contraction of the bladder neck is primarily mediated by sympathetic innervation. Stimulation of pudendal afferents gives rise to an inhibition of the bladder muscle and contraction of the bladder neck. This reflex also requires the participation of spinal interneurons. The bladder neck closure is necessary to prevent semen from being ejaculated into the bladder (retrograde ejaculation) rather than expelled from the meatus.

C. Emission

Stimulation of penile afferents induces seminal emission and contraction of the vas deferens and seminal vesicles. Both sympathetic and parasympathetic efferents are probably involved in this response. The sensory fibers mediating emission travel in the pudendal nerve.

D. Ejaculatory Reflex

The final phase of ejaculation consists of several rhythmic contractions of the perineal muscles and smooth muscles of the urethra. The response is characterized by synchronous activation of ischiocavernosus, bulbospongiosus, anal and urethral external sphincters, and levator ani muscles. The contractions are extremely regular, with an interval between contractions starting at approximately 0.6 sec and increasing by approximately 100 msec for each subsequent interval. The ejaculatory response in men typically is about 10–15 contractions and tends to be rather consistent in individual subjects. An example of the ejaculatory reflex in men is shown in Fig. 1. It should be noted that the striated pelvic contractions seen at sexual climax are extremely similar in men and women. These findings, as well as similarities in subjective experiences and extragenital responses, have led to the hypothesis that the physiological events of orgasm in the two sexes are produced by homologous neural mechanisms.

The sensory trigger which elicits the ejaculatory motor pattern is unknown. One possible theory is that the ejaculatory motor pattern is elicited by the buildup of seminal fluid in the urethra. However,

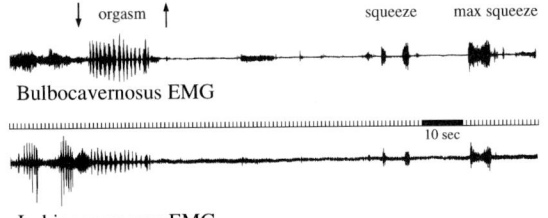

FIGURE 1 The electromyographic (EMG) recording from the bulbocavernosus (bulbospongiosus) and ischiocavernosus muscles during ejaculation induced by masturbation in healthy volunteers. Note that the activity in the two muscles is synchronized and consists of a series of highly regular rhythmic contraction. The arrows indicate the beginning and end of subjectively recorded orgasm (reproduced with permission from T. C. Gerstenberg, R. J. Levin, and G. Wagner, Erection and ejaculation in man. Assessment of the electromyography activity of the bulbocavernosus and ischiocavernosus muscles, Br. J. Urol. 65, 395–402, 1990).

ejaculatory motor patterns were still observed in cancer patients after surgical removal of the bladder, prostate, and seminal vesicles and in volunteers given drugs to inhibit seminal emission. Clearly, sensory fibers in the penis are crucially involved in transmitting the sensory trigger, but the exact nature or location of the afferents are unknown. Stimulation of pudendal nerve afferents in animals can elicit ejaculatory motor patterns and anesthesia of the glans penis prolongs the onset to ejaculation in animals and human. Presumably, sexual stimulation causes activation of spinal systems and when this activation reaches a certain level, the ejaculatory reflex is evoked. However, the mechanisms of this process are not known.

The ejaculatory pattern is a spinal reflex since it is present even after complete spinal transection in both humans and experimental animals. This indicates that the generation of the characteristic rhythmic contractions of the pelvic muscles occurs within the spinal cord. Likewise, emission reflexes, closure of the bladder neck, and the bulbocavernosus reflex are all spinal reflexes elicited by pudendal sensory stimulation. Thus, all components of ejaculation are contained within the lumbosacral spinal cord. These spinal reflexes are under complex inhibitory and excitatory control from brain stem and forebrain sites.

VIII. SUPRASPINAL CONTROL

The control of ejaculation by supraspinal sites is still poorly understood. Some brain stem and forebrain areas have been implicated in the control of sexual function, but a complete map of the pathways controlling ejaculation is not currently available. A crucial brain site for male sexual behavior is the medial preoptic region of the hypothalamus (MPOA). The MPOA contains a large concentration of neurons which have receptors for steroid hormones and implantation of testosterone restores sexual behavior in castrated animals. Lesions of the MPOA have been shown to cause massive deficits in sexual behavior in every vertebrate species tested, including primates. Electrical or chemical stimulation of this area activates male sexual behavior. For example, stimulation reduces the number of intromissions prior to ejaculation and shortens the postejaculatory refractory period in copulating animals and it can produce ejaculation without any genital stimulation.

Clearly, such studies indicate an important role for the MPOA in ejaculation and sexual behavior in general. However, studies of animals following lesions of the MPOA demonstrate that, while copulatory behavior may be almost completely suppressed, animals are still sexually motivated and capable of erection and ejaculation. Thus, MPOA lesions may abolish copulatory behavior by disrupting some aspect of the male's ability to engage in the appropriate sociosexual interactions, but the neural substrates for sexual arousal and ejaculation remain intact.

Another area that is important for the expression of sexual behavior is the amygdala. Using methods which label neurons which have been activated, neurons in the amygdala (especially in the medial portion) are labeled following copulatory behavior. Discrete amygdala lesions cause disruption of sexual behavior. Motivational components of sexual behavior are much more impaired than the copulatory motor performance. Specific control of the ejaculatory process by the amygdala is not clear.

An area which is receiving increasing attention for its role in sexual function is the paraventricular nucleus of the hypothalamus (PVN). The PVN has been identified as a key site for neuroendocrine and autonomic integration. With respect to sexual function, the parvocellular portion of the PVN projects directly to the lumbosacral spinal cord to interneuronal areas as well as pudendal motoneurons. This hypothalamic–spinal pathway uses oxytocin as a neurotransmitter. Oxytocin levels in the circulation and in the cerebrospinal fluid are greatly increased by sexual arousal, especially ejaculation/orgasm, indicating that there is a widespread activation of this nucleus. The PVN receives an extensive input from the MPOA and neurons within the PVN are activated by stimulation of penile sensory nerve fibers. Stimulation of the PVN elicits penile erection and ejaculation. Since the PVN is both excited by penile afferents and directly stimulates erection and ejaculation, it may serve to act in a positive feedback manner to intensify the sexually arousing effects of genital stimulation.

Lesion and stimulation studies in animals have demonstrated that midbrain structures, especially the periaqueductal gray (PAG), have important roles in ejaculatory function. The PAG has reciprocal connections with the MPOA and may serve as both a descending and ascending relay for hypothalamic communication with brain stem and spinal sites controlling sexual responses.

It has long been known that spinal sexual reflexes are under an inhibitory control from supraspinal sites. One site providing such inhibitory control has recently been identified. A small cluster of neurons in the rostral portion of the nucleus paragigantocellularis (nPGi) mediate a powerful inhibitory influence over spinal ejaculatory-like responses. These neurons project directly to pelvic efferents and interneurons in the lumbosacral spinal cord. The vast majority of the neurons in this area which project to the lumbosacral spinal cord are serotonergic, and application of serotonin to the spinal cord suppresses spinal ejaculatory reflexes. These findings provide a basis for the inhibitory effects on ejaculation of drugs which increase serotonin levels such as clomipramine, which has been used as a treatment for premature ejaculation, and serotonin selective reuptake inhibitor antidepressants, which are associated with a high incidence of anorgasmia and ejaculatory disruption.

See Also the Following Articles

Epididymis; Erection; Orgasm; Penis; Prostate Gland; Semen; Seminal Vesicles

Bibliography

Andersson, K.-E., and Wagner, G. (1995). The physiology of penile erection. *Physiol. Rev.* 75, 191–236.

Dail, W. G. (1993). Autonomic innervation of male reproductive organs. In *Nervous Control of the Urogenital System* (C. A. Maggi, Ed.). Harwood Academic, London.

De Groat, W. C., and Booth, A. M. (1993). Neural control of penile erection. In *Nervous Control of the Urogenital System* (C. A. Maggi, Ed.). Harwood Academic, London.

Kinsey, A. C., Pomeroy, W. B., and Martin, C. E. (1948). *Sexual Behavior in the Human Male*. Saunders, Philadelphia.

Marberger, H. (1974). Mechanisms of ejaculation. In *Physiology and Genetics of Reproduction* (E. M. Coutinho and F. Fuchs, Eds.). Plenum, New York.

Masters, W. H., and Johnson, V. E. (1966). *Human Sexual Response*. Little, Brown, Boston.

McKenna, K. E., and Marson, L. (1997). Spinal and brainstem control of sexual function. In *Central Control of Autonomic Function* (D. Jordan, Ed.). Harwood Academic, London.

Meisel, R. L., and Sachs, B. D. (1994). The physiology of male sexual behavior. In *The Physiology of Reproduction* (E. Knobil and J. D. Neill, Eds.). Raven Press, New York.

Rose, J. D. (1990). Brainstem influences on sexual behavior. In *Brainstem Influences on Sexual Behavior* (W. R. Klemm and R. P. Vertes, Eds.). Wiley, New York.

Elasmobranch Reproduction

Thomas J. Koob

Shriners Hospital for Children and The Mount Desert Island Biological Laboratory

I. Introduction
II. Reproductive Modes
III. Reproductive Cycles
IV. Physiological Mechanisms
V. Conclusion

GLOSSARY

lecithotrophic viviparity A live-bearing reproductive strategy in which the ovulated eggs contain all the essential organic nutrients for embryonic development but development to term takes place in the uterus.

matrotrophic viviparity A viviparous reproductive strategy in which embryonic development occurs within the uterus and essential nutrients are provided during gestation *in utero* over and above that incorporated in the ovulated egg.

oviparity A reproductive mode in which eggs are laid in protective capsules and the yolk provides all the organic nutrients for embryonic development to hatching.

Contemporary elasmobranchs utilize complex and diverse reproductive modes operating over protracted cycles, all of which result in a small number of relatively large progeny. Internal fertilization and an adaptable reproductive system, coupled with neural and ovarian control mechanisms, have allowed the evolution of a specialized form of oviparity and divergent viviparous strategies, including lecithotrophy and three forms of matrotrophy.

I. INTRODUCTION

Of contemporary vertebrate taxa, the elasmobranchs, or the sharks, skates, and rays, exhibit the most widespread and diverse reproductive strategies (Fig. 1). Oviparity, viviparous lecithotrophy or ovoviviparity, and distinct forms of matrotrophic viviparity operate quite successfully in extant species. Approximately 42% of all known elasmobranch species are oviparous, including several shark species and the entire skate taxon comprising over 200 species. Viviparous lecithotrophy occurs primarily in sharks but also in a few batoid species and occurs in 27% of the species. Placental viviparity occurs in only 9% of elasmobranchs and all of these are sharks. Uterotrophic mechanisms operate in 19% of elasmobranchs, and most of these are rays. Oophagy, while very intriguing, occurs in only 2% of the species. The diversity in the physiological mechanisms responsible for the sustained reproductive success within this group of fish attests to their long phylogenetic history.

II. REPRODUCTIVE MODES

A. Oviparity

Oviparity in elasmobranchs as well as chimaeroids differs from oviparity in most other fish in that few eggs of a large size relative to the mother are laid over protracted spawning periods and development to hatching requires months to years. Prolonged incubation periods are due in part to the substantial yolk invested in the ovulated egg and the water temperature wherein incubation occurs. There is no larval stage as in teleosts but rather the term embryo resembles the adult morphologically and is able to forage soon after hatching. Following a single mating, pairs of eggs are ovulated, fertilized, encapsulated, and oviposited at intervals of several days or more; the exact interim, at least in captivity, is species specific. Spawning periods last for months. Once the

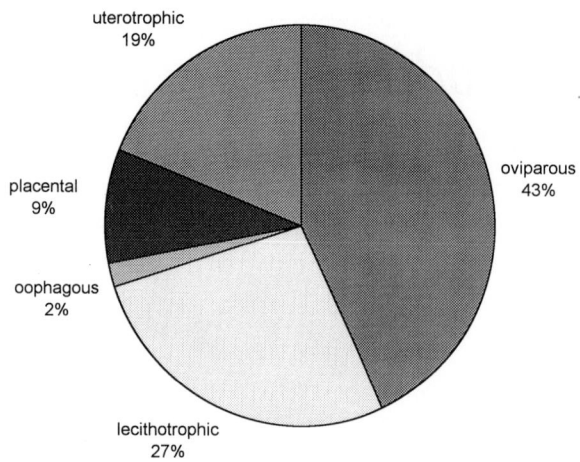

FIGURE 1 Relative number of species utilizing distinct reproductive modes in contemporary elasmobranchs (data adapted from Compagno, 1990).

around submerged objects such as corals and sponges during the process of oviposition. The heterodontid sharks produce unusually shaped capsules with two spiral flanges. It is believed, although it has never been observed, that females pick up the oviposited capsules and wedge them into rock or coral crevices since they are normally found in such situations. Skate egg capsules have horns emanating from the four corners, and it is speculated that these may function as anchoring devices. However, since naturally oviposited skate eggs have never been observed *in situ*, it is unclear whether the horns serve as mooring devices.

After approximately one-third of embryonic development is completed, the gelatinous albumen has disappeared and the embryo completes the last two-thirds of development in seawater which gains access

eggs are laid and positioned, the female takes no further notice of her progeny. Enough evidence exists to suggest that females of many species select traditional oviposition sites and repeatedly return to these same nursery grounds each spawning season.

All the organic nutrients required for embryonic development to hatching are present in the oviposited egg, although water, minerals, and possibly other solutes may be sequestered from the environment. Once the digestive organs differentiate in the embryo, yolk platelets are transferred via the yolk stalk by ciliary action from the external yolk sac to the internal yolk sac and then to the intestine. At hatching, the external yolk sac has disappeared, but the internal yolk sac remains full of yolk and is able to nourish the hatchling for the first few weeks of life outside the capsule.

The fertilized eggs are packaged by the shell gland in the upper oviduct, together with a gelatinous material often referred to as albumen, in morphologically complex, mechanically and chemically stable capsules (Fig. 2). Encapsulated eggs must endure the lengthy incubation in the corrosive and unpredictable marine environment. Shark egg capsules are equipped with devices for attaching eggs to objects above the sea floor. Cat shark egg capsules have extraordinarily long, coiled tendrils arising from each of the four corners. Females seek out and actively attach the eggs by winding the lengthy filaments

FIGURE 2 Representative egg capsules of oviparous elasmobranchs. (Top) *Cephaloscyllium ventriosum*; (middle) *Heterodontus mexicanus*; (bottom) *Raja eglanteria*.

to the capsule lumen through slits opened by liquefaction of the dense gel. Skate embryos actively flush water through the capsule with a specialized and transitory tail appendage. The fully developed offspring emerge through a preformed hatching seam in the capsule located at one of the narrow ends between the filaments or horns. The exact means by which the term embryo liberates itself from the confines of the capsule is not known. Once freed, the hatchling behaves remarkably like the adults.

B. Viviparity

A live-bearing form intermediate between oviparity and matrotrophic viviparity with respect to the source of nutrition for the developing embryo has evolved in many species of sharks. This reproductive mode has been traditionally called ovoviviparity, but recently it has been relegated to a subgroup within the viviparous forms and called lecithotrophy, a descriptor indicating that the ovulated egg contains most if not all of the organic materials required for embryonic development. This successful reproductive strategy occurs in live-bearing shark species, including the nurse shark, whale shark, guitarfish, angel sharks, and spiny dogfish. The relatively large fertilized eggs are initially encapsulated by the shell gland and enter the uterine portion of the reproductive tract, wherein embryonic development proceeds in a manner similar to that in oviparous species. The size of the yolk-rich ovulated egg is very large in these species since embryonic development relies exclusively on the yolk for organic nutrients. Embryonic development is similar to that in oviparous species. Yolk is transferred from the external yolk sac to the embryonic intestine and is digested in the yolk syncitial–endoderm complex. The capsules in ovoviviparous forms vary from substantial structures, such as those produced in the whale shark and nurse shark, to extremely thin membranes, as in the spiny dogfish. The developing embryos eventually emerge from these capsules to complete development free in the uterine lumen. In these forms, embryonic development occurs entirely within the uterus, but no additional organic nutrients appear to be provided by the uterus. However, the uterus becomes morphologically specialized and thereby regulates the intrauterine milieu, especially by providing oxygen and removing wastes. At birth, when the external yolk sac is entirely internalized, the offspring of ovoviviparous species are equally as advanced as those of other viviparous species.

Three forms of matrotrophic viviparity have evolved in the sharks and rays: a yolk sac placental form, a type utilizing uterotrophic mechanisms, and oophagous viviparity. Certain important features are common to these viviparous reproductive modes. Ovulated eggs are encapsulated in the shell gland and remain in the capsule for at least a portion of pregnancy. The initial stages of embryonic development rely almost exclusively on nutritional elements present in the ovulated and encapsulated eggs, primarily the yolk. Once the yolk is consumed, further maternal provisions are supplied via the uterus either through vascular transfer mechanisms (placenta) or by production of organic nutrients by the uterus (uterotrophic). In many cases a combination of placental and uterotrophic mechanisms is employed. The embryos themselves develop additional specializations for efficient acquisition of nutritional elements.

Placental viviparity occurs in 9% of contemporary shark species, including hammerhead sharks, sharpnose sharks, sandbar sharks, blacknose sharks, and several species of dogfish. The initial stages of embryonic development rely almost exclusively on yolk provisions in the ovulated egg; however, uterine secretions augment the declining yolk stores during the middle gestational stages in some species. Placenta develop just prior to the period when the yolk reserves are depleted. Further sustenance for development is then acquired through the morphogenesis of the yolk sac to form the embryonic portion of the placenta and the corresponding modifications in the maternal uterine wall. While the embryonic portion of the placenta is analogous to that in mammals, it differs in that it is a yolk sac placenta. Transfer of nutrients from mother to offspring is achieved via hemotrophic routes. Uterine specializations occur but are principally composed of increased vascularity, an increase in surface area, and a decrease in the diffusion distance from the maternal circulation at the site of apposition of the embryonic yolk sac. Based on morphological data, the maternal placental site functions in respiratory gas exchange, osmotic and ionic regulation, and nutrient transfer. Typical of viviparous sharks with placental structures is the formation of individual uterine compartments for each embryo.

Rays utilize an altogether different form of viviparity. After the yolk is depleted, the morphologically specialized uterus produces an organically rich material which is secreted directly into the uterine lumen. The embryos develop within this liquid and they use the organic material as the source of nutrients for the remainder of gestation. Uteri in viviparous rays develop extensive villi and in some species these villi grow to remarkable size with extensive secretory inclusions and an elaborate vascular circulation, structures called trophonemata. Acquisition of the histotroph by the developing embryo can be through a variety in adsorptive surfaces and directly through the gut.

Another rather unique form of viviparity occurs in some species of lamnoid sharks, for example, the porbeagle shark, great white shark, and sand tiger shark. Many eggs are ovulated but only two embryos survive to birth. Following depletion of the relatively small amount of yolk reserves in the ovulated egg, one embryo in each uterus devours its mates and then is "fed" eggs that are continually ovulated for the first part of pregnancy. The successful embryo emerges at birth at an extraordinarily large size. Only two offspring are produced for each reproductive cycle, but they enter the sea with an advantage in size and are well equipped to begin hunting.

The result of each of these reproductive strategies is the production of relatively few, morphologically, physiologically, and biomechanically advanced offspring. At birth or hatching, elasmobranch offspring are on their own and begin foraging soon after liberation from the confines of capsules or uterus. In oviparous species, fecundity from one breeding season is estimated to range between 30 and 60 offspring. In viviparous species, from 1 neonate in the cownose ray to 130 progeny in the blue shark are produced, with the number being entirely species specific rather than specific to reproductive mode. In the oophagous species and some species of rays, females concentrate all their nutritional efforts into producing 2 vary large progeny at birth.

III. REPRODUCTIVE CYCLES

Mating occurs at specific times of the year in nearly all elasmobranch species, and breeding is generally synchronized within the population. Exactly what factors mediate mating activity is unknown for any species. Whether the female signals receptivity by behavioral or chemical cues is not known. Internal fertilization is ubiquitous among elasmobranch species. Specialized copulatory organs in the male, called claspers, are derived and modified from the pelvic fins during sexual development. Claspers are used for delivering sperm during insemination of the female. Based on direct observation, males inseminate females by inserting one clasper into the urogenital sinus while grasping the female. Sperm bundles are delivered via a groove in the clasper. Sperm are then stored in the upper reaches of the oviduct, in or near the shell gland. Sperm storage allows continued production of fertile eggs over periods of months in oviparous species and effectively decouples mating with ovulation in viviparous species.

In general, elasmobranch reproductive cycles are notably protracted, requiring several months to as much as 2 years (Fig. 3). Oviparous skates breed either seasonally, as does the clearnose skate in subtropical waters where egg laying occurs only in the winter months, or year-round, as exemplified by little skates in northern seas where the population lays eggs throughout the year. Most oviparous shark populations inhabiting cold waters lay eggs year-round. The common dogfish in the northeastern Atlantic lays eggs all year, but a peak of egg-laying activity occurs in late winter–early spring. In warmer waters, some oviparous shark species have more limited laying cycles. The Port Jackson shark in Australian waters is believed to lay eggs primarily in August and September. Whether the seasonal cycles of certain oviparous species are tied to environmental factors such as temperature is not known.

Viviparous species exhibit well-circumscribed annual breeding cycles. Each population is synchronized with respect to mating, gestation, and birth. Among shark species, the prevalent pattern is approximately a year-long pregnancy followed immediately by ovulation, mating, and the subsequent pregnancy. However, there are important exceptions,

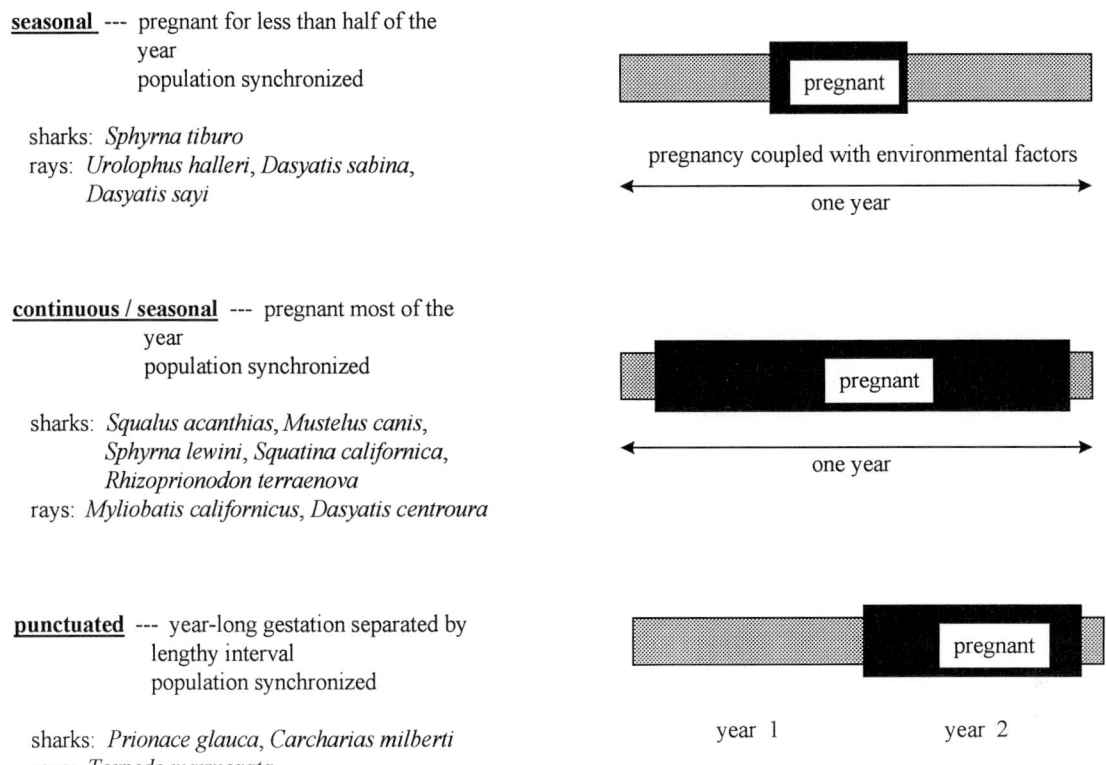

FIGURE 3 Reproductive cycles in viviparous elasmobranchs.

such as the blue shark (2-year cycle), spiny dogfish (2-year pregnancy), and bonnethead shark (5-month pregnancy). Most rays produce young every year, but the length of pregnancy is species specific. Because few young are produced for each pregnancy and pregnancies are annual cycles, fecundity in these fish is very low compared to that of bony fish and, for some species, low in absolute numbers as well.

Rays utilize yearly cycles wherein pregnancy is either short, lasting approximately one-fourth of the year, or long, lasting most of the year. Stingrays in the warm coastal waters of the subtropical eastern United States reproduce with short gestations taking place in late spring to early summer. A cycle typical for this group occurs in the Atlantic stingray; ovulation takes place in March or April and pregnancy ensues immediately and lasts approximately 4 months, with parturition occurring in midsummer. The round stingray of the west coast of the United States is pregnant for only 3 months of the year, and the thorny stingray in Tunisian waters is pregnant for about 4 months. Other closely related species of rays utilize pregnancies approaching a year. Of these, some species undergo a complete cycle in 1 year, i.e., producing offspring annually, whereas in others, consecutive year-long pregnancies are interrupted by 1 or 2 years of reproductive quiescence.

Annual cycles with pregnancies lasting nearly 1 year predominate among shark species, e.g., hammerhead sharks, angel sharks, guitarfish, sharpnose sharks, and sand tiger sharks. Mating, ovulation, and pregnancy ensue almost immediately after parturition. Females of these species are pregnant most of their adult life. Other species of sharks have adopted variations of this pattern. Blue shark and sandbar shark females produce offspring every 2 years. Pregnancy in these species lasts only 1 year, but an intervening year is spent nonpregnant. Pregnancy in spiny dogfish lasts nearly 2 years; at any one time there are two groups of females—the stage of pregnancy in one group is separated by 1 year from that in the other group. A few species of sharks are pregnant for only a minor portion of the yearly cycle, similar to the cycle most stingrays. The bonnethead shark in Florida waters mates in spring and gives birth in late summer.

Little is known about what factors regulate these reproductive cycles. Nearly all elasmobranch species mate, are pregnant, and give birth at specific times of the year, and these cycles, whether they are short seasonal cycles such as those in the rays or year-long pregnancies as occur in many sharks, are repeated year after year. Environmental factors, such as water temperature, light cycles, and food resources, are likely involved, but the primary determinants remain a mystery.

IV. PHYSIOLOGICAL MECHANISMS

Follicular cycles are central to reproductive cycles in elasmobranchs. Follicle development involves accumulation of yolk in the developing oocyte and differentiation of steroidogenic pathways. Accumulation of yolk is essential regardless of reproductive mode since all elasmobranchs incorporate yolk in the oocyte before ovulation. In oviparous, lecithotrophic, and oophagous species, yolk is the principal source of organic nutrients for embryonic development. In viviparous species, yolk fuels the initial stages of development until morphogenesis of the placental site (sharks) or initiation of histotroph production (rays). While the identity and source of every yolk constituent has not been delineated, it is clear that an authentic vitellogenin, homologous to that in reptiles and birds, is produced by the liver in at least two elasmobranch species. Vitellogenin is secreted by the liver, transported to the ovary, and finally incorporated in the developing oocyte. Regulation of vitellogenin synthesis involves both estradiol and progesterone. Based on the few experimental studies completed to date and correlations between circulating steroids and folliculogenesis, estradiol induces *de novo* synthesis of vitellogenin in the liver, whereas progesterone inhibits its production.

Unequivocal evidence that follicles and corpora lutea in elasmobranchs produce steroids derives from studies on the oviparous little skate and lecithotrophic viviparous spiny dogfish. *In vitro* studies have shown that follicle cells produce estradiol and testosterone during follicle development and corpora lutea are the source of progesterone in both species. The dynamics of steroidogenesis is species specific as might be expected from the differences in reproductive modes. While estradiol and testosterone are the predominant steroids circulating during the follicular phases of the cycles in both species as well as in a third species, the viviparous bonnethead shark, elevations in progesterone levels differ. In the little skate, which ovulates and oviposits pairs of eggs every 7 days on average, progesterone titers peak 1 day before ovulation and remain elevated for 1 or 2 days only. In contrast, progesterone titers in spiny dogfish are elevated in the periovulatory period but remain elevated for several months during the first half of pregnancy. In the bonnethead, progesterone titers are elevated for brief period after ovulation and during the initial stages of pregnancy.

Endocrine regulation of specific physiological mechanisms during reproductive cycles has likewise been examined in only a few species. As mentioned previously, estradiol regulates liver vitellogenin synthesis. Estradiol also appears to be involved in regulating the growth and activity of the shell gland which is responsible for synthesis, secretion, and assembly of the egg capsules in all elasmobranchs. Estradiol is involved in potentiating the effects of another ovarian hormone, the peptide relaxin. In experimental studies, relaxin has been shown to inhibit myometrial contractions in late pregnant spiny dogfish and to increase the compliance of the cervix in both the spiny dogfish and the little skate. Both these actions are similar to the role of relaxin in mammalian species during late pregnancy.

Egg encapsulation is nearly universal in elasmobranchs. Oviparous species produce robust capsules, the function of which is to protect the egg and developing embryo from mechanical damage and from predatory attacks. The walls of these capsules are laminated structures assembled by an extrusion process of several distinct proteins in the shell or nidamental gland. While it was once believed that capsules of oviparous species were composed primarily of collagen, recent analyses have established that, if collagen is present, it represents a relatively minor proportion of the proteins involved. Stabilization of the assembled proteins relies in large part on a quinone tanning mechanism in which catechols are introduced in the uterus and oxidized to quinones.

Egg capsules in viviparous species are generally much thinner than those of oviparous species, particularly in species utilizing placental and uterotrophic viviparity. The composition of these capsules and the chemical stabilization process are unexplored. The fate of these capsules is species specific. In placental species, it persists throughout pregnancy, whereas in uterotrophic species as well as some lecithotrophic species, it disappears around the time when the yolk reserves are depleted.

The uterus must regulate the intrauterine milieu in lecithotrophic viviparous species. In early pregnancy in the spiny dogfish, when the embryos are developing in the egg capsule, the ionic composition of the uterine fluids is between that of maternal plasma and seawater, but the urea concentration approaches that in maternal tissues. After the embryos have broken out of the capsules and thereafter develop within a fluid-filled uterine lumen, the fluids are essentially identical to seawater and urea concentrations are low. Thus, the uterus actively regulates the intrauterine milieu in early pregnancy but does little metabolically in late pregnancy except to acidify the fluid, thereby detoxifying the principal waste product, ammonia. The uterus itself becomes highly vascular during the latter half of pregnancy in order to meet the increasing respiratory demands of the large embryos and for the acidification process. Embryonic osmoregulatory mechanisms develop concomitant with the transition to seawater *in utero*.

In matrotrophic species, the uterus, in addition to mediating respiratory, ionic, and osmotic requirements, provides nutrients via hemotrophic transfer of metabolites from the maternal circulation or *de novo* synthesis and secretion of histotroph by the endometrium (Fig. 4). Uteri in species employing histotroph production to supplement yolk reserves undergo extensive proliferation and differentiation. The differentiated uterine endometrium appears to

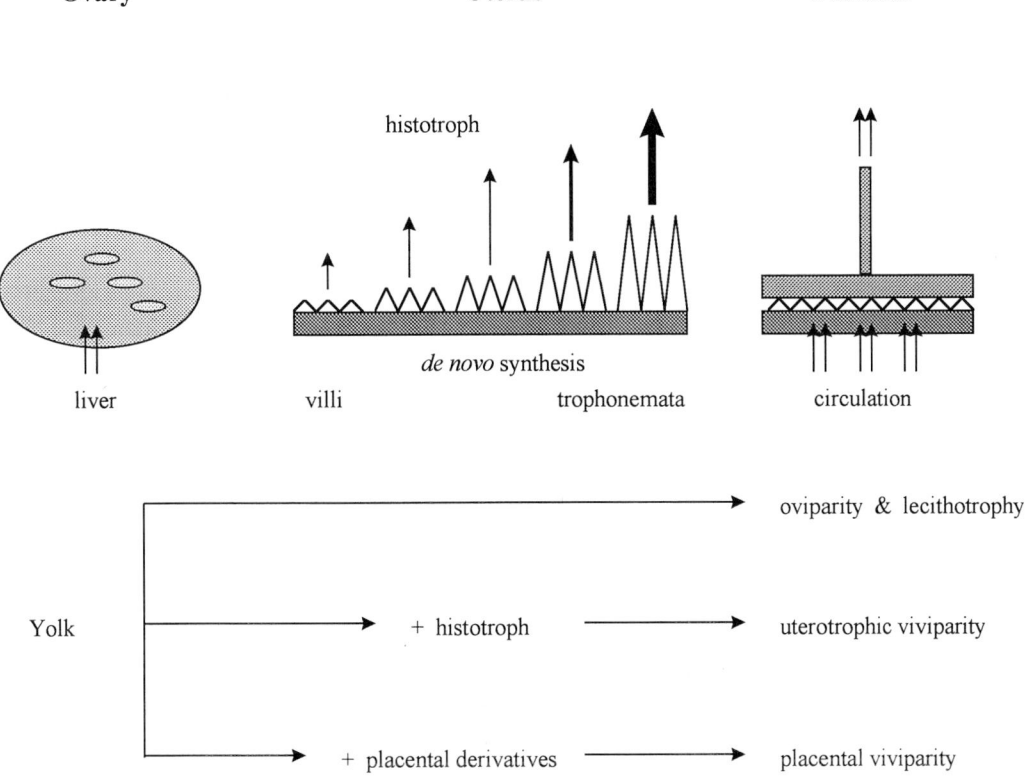

FIGURE 4 Source of essential nutrients for embryonic development in oviparous and viviparous elasmobranchs.

be the principal source of the histotroph. The endometrium develops expansive villiform appendages supplied with elaborate vascular beds. The epithelial cells join to form secretory assemblages. In late pregnancy, these cells are extremely metabolically active, exhibiting extensive rough endoplasmic reticulum that appears to be involved in producing abundant materials destined for secretion into the uterine lumen. A number of morphologically distinct cytoplasmic granules that may likewise be secretory vesicles populate these cells. While the chemical nature of the histotroph has not been examined with contemporary analytical methods, earlier analyses indicated that it contains significant amounts of proteins, carbohydrates, and lipids. The relative concentration or amount of organic materials in the histotroph varies among rays, with the highest levels correlating with the lowest yolk investitures.

The uterine contribution to the placental complex in sharks involves localized modifications of the endometrium and subjacent vascular bed. In general, placental attachment sites are highly vascular, rugose elevations that interdigitate with the yolk sac placenta. The uterine epithelium is not eroded but persists throughout the functional life of the placenta. The extremely thin egg capsule, which remains intact throughout pregnancy in placental species, separates the uterine epithelium from the yolk sac epithelium. All metabolites reaching the embryo from the maternal circulation must permeate the capsule wall. The size and chemical nature of the nutrients normally supplied from the maternal circulation through the placenta have not been characterized in any species to date, but it is clear that low-molecular-weight compounds (<1 kDa) are able to cross the capsule membrane.

Embryos in matrotrophic species either elaborate specialized organs for acquiring nutrients or accelerate functional development of the gastrointestinal tract. In placental species, the yolk sac, after being depleted of yolk, is transformed into a yolk sac placenta that functions until delivery of the fully developed embryo. These yolk sac placenta are noninvasive and nondeciduate. The yolk sac differentiates into two structurally distinct sections. The portion in apposition with the uterine epithelium is a highly vascularized and extensively ridged tissue composed of a surface epithelial bilayer with an underlying endothelium with fenestrated vasculature. The proximal, saccular portion of the placenta is made up of metabolically active cells containing massive smooth endoplasmic reticulum, suggesting dynamic secretory activity. The yolk stalk emanating from the placenta, sometimes termed an umbilical chord, is composed of umbilical artery, umbilical vein, extraembryonic coelom, and the ductus vitellointestinalis. In some species it is decorated with appendiculae. These protruding structures can be highly developed including multiform branching as in the sharpnose shark or minimally developed as in certain hammerhead species. While the exact function of these structures is unclear, based on their morphological characteristics, they appear to both absorb materials and secrete materials. Embryonic rays obtain much of their nutrients by ingesting the uterine histotroph. Both the stomach and spiral valve begin functioning rather early in development. This is also true for embryos of oophagous species in which the ingested eggs need to be processed and absorbed.

V. CONCLUSION

The mechanism for provisioning the developing offspring is the defining feature for the diverse reproductive modes utilized by contemporary elasmobranchs. The nutrients for supporting development in oviparous species accumulate in the yolk of the oocyte during folliculogenesis and entirely before ovulation and encapsulation. A significant loss of mass occurs during development as some of the nutrients are utilized as metabolic fuel. Hatchlings of oviparous species are fundamentally smaller than the ovulated egg (Fig. 5). Nonetheless, the oviparous strategy allows sustained production of relatively large eggs over protracted cycles and thereby high fecundity compared to most viviparous species. The nutritional provisions for embryonic development in lecithotrophic viviparous species are essentially similar to those in oviparous species. A similar loss of mass from metabolic costs occurs during development (Fig. 5) and the offspring are constitutively smaller than the ovulated egg. It should be remembered, however, that the ovulated eggs in both ovipa-

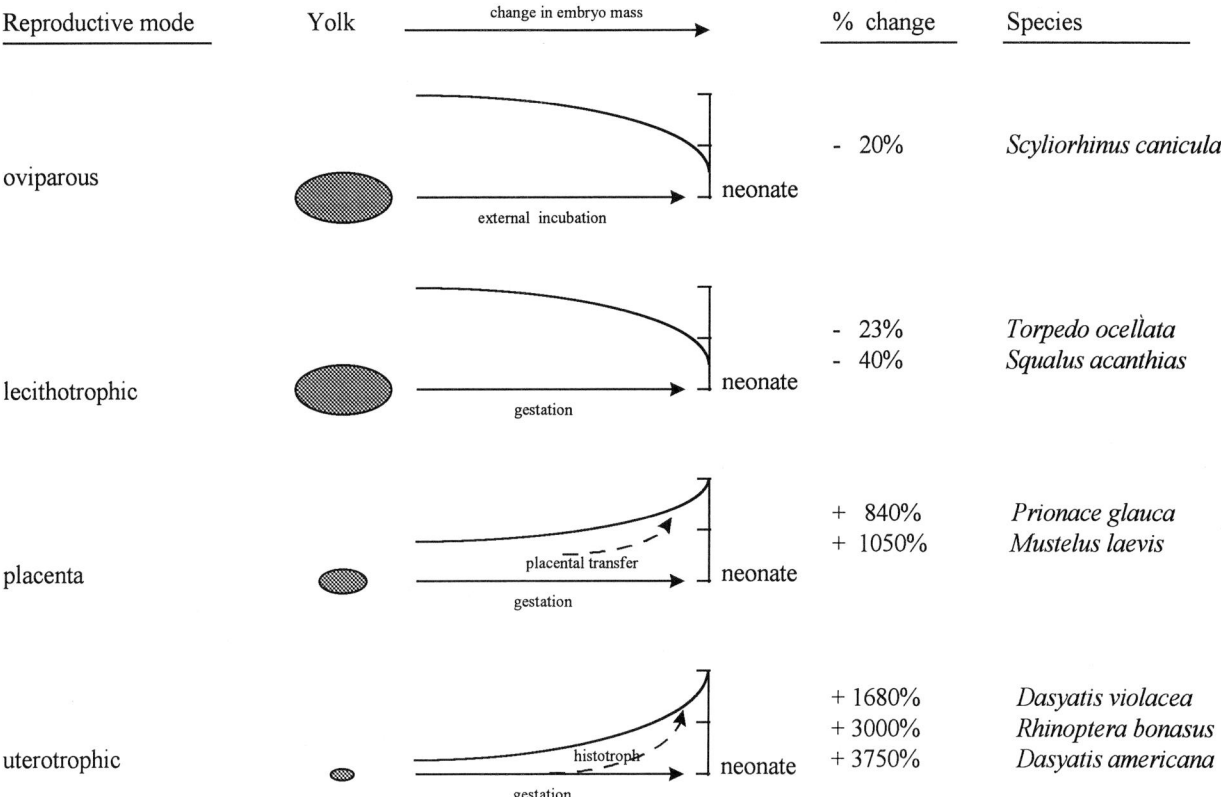

FIGURE 5 Mass relations of ovulated egg and offspring in distinct reproductive modes of representative elasmobranch species.

rous and lecithotrophic viviparous species are relatively large so that the mass loss does not significantly sacrifice relative size at hatching or birth.

In matrotrophic species the yolk contains only a portion of the nutrients essential for the completion of development. Yolk reserves are supplemented by one of several mechanisms during gestation. The yolk sac placenta in sharks is fairly efficient at acquiring additional nutrients from the maternal circulation resulting in significant increases in mass. The mass of the progeny ranges from 5- to 10-fold higher, with the exact amount being species specific. In uterotrophic species, wherein the uterine endometrium synthesizes and secretes a nutrient-rich histotroph, the increase in mass over the ovulated egg occurring during gestation can be enormous, especially in rays (Fig. 5). Fifteen- to 30-fold increases in mass occur in some species.

The physiological mechanisms that have been recruited in elasmobranchs for reproduction result in precocial progeny able to fend for themselves soon after hatching or birth. Their large size, both in relation to the parents and in absolute terms, allows them to enter the environment at a trophic level higher than that of other fish. This advantage has been instrumental in their remarkable success in the marine environment during the past 350 million years.

See Also the Following Article

Hormonal Control of the Reproductive Tract, Subavian Vertebrates

Bibliography

Callard, G. V. (1991). Reproduction in male elasmobranch fishes. In *Oogenesis, Spermatogenesis and Reproduction* (R. K. H. Kinne, Ed.). Karger, Basel.

Callard, I. P., Klosterman, L. L., Sorbera, L. A., Fileti, L. A., and Reese, J. C. (1989). Endocrine regulation in elasmobranchs: Archetype for terrestrial vertebrates. *J. Exp. Zool. Suppl.* 2, 12–22.

Compagno, L. J. V. (1990). Alternative life-history styles of cartilaginous fishes in time and space. *Environ. Biol. Fishes* 28, 33–75.

Demski, L. S., and Wourms, J. P. (Eds.) (1993). *The Reproduction and Development of Sharks, Skates, Rays and Ratfishes*. Kluwer, Dordrecht.

Hamlett, W. C., (Ed.) (1999). *Sharks, skates, and rays:* The Biology of Elasmobranch Fishes. Johns Hopkins Univ. Press, Baltimore.

Koob, T. J., and Callard, I. P. (1991). Reproduction in female elasmobranchs. In *Oogenesis, Spermatogenesis and Reproduction* (R. K. H. Kinne, Ed.). Karger, Basel.

Pratt, H. L., Jr., Gruber, S. H., and Taniuchi, T. (Eds.) (1990). Elasmobranchs as living resources: Advances in the biology, ecology, systematics, and the status of the fisheries, NOAA Tech. Rep. No. NMFS 90. U. S. Department of Commerce, Washington DC.

Wourms, J. P. (1977). Reproduction and development in Chondrichthyan fishes. *Am. Zool.* 21, 379–410.

Wourms, J. P., Grove, B., and Lombardi, J. (1988). The maternal–embryonic relationship in viviparous fishes. In *Fish Physiology* (W. S. Hoar and D. J. Randall, Eds.), Vol. XIB. Academic Press, San Diego.

Elephants

Cheryl Niemuller
Stouffville, Canada

Janine Brown
Smithsonian Institution

Keith Hodges
German Primate Centre

I. Introduction
II. Female Reproduction
III. Male Reproduction
IV. Conservation and Management

GLOSSARY

estrous cycle The ovarian cycle from one period of estrus to the beginning of the next period of estrus.
estrus Physical and behavioral signs of sexual receptivity.
flehmen Pheromonal behavior typically carried out by the male elephant on the urine of the female to determine if a female is in estrus.
follicular phase The period of recruitment and development of follicles during a reproductive cycle.
luteal phase The period during the estrous cycle which is characterized by elevated progesterone levels and during which the uterus is receptive to implantation.
musth A physiological and psychological condition in the male elephant characterized by elevated androgens, increased aggression, and mating behavior.

Both the Asian and African species of elephants occupy a unique position among the living mammals of today. They have no close living relatives and are themselves at the end of a once great evolutionary branch which was so diversified that various forms could be found throughout the world. Their future, however, is not secure and the survival of indigenous, wild populations of both species will remain endangered until humans and elephants are able to coexist peacefully in the same environment.

I. INTRODUCTION

Studies of elephant reproductive physiology have been conducted during the past 40 years in order to better understand the mechanisms controlling reproductive function for both the Asian and African elephant. Furthermore, zoos and private wildlife reserves are becoming increasingly important as genetic harbors for ensuring the means of preserving species-specific gene diversity. However, for both the

Asian and African elephant, breeding success in captivity remains inconsistent and our knowledge of basic reproductive physiology is limited. The information presented in this article represents a compilation of knowledge gathered from two separate genera: Asian (*Elephas maximus*) deriving predominantly from studies of captive individuals, and African (*Loxodonta africana*) from work carried out mostly on culled animals during population control efforts.

II. FEMALE REPRODUCTION

A. General Reproductive Biology

Observations on mating behavior of wild and captive animals indicate that the onset of puberty (or ovarian cyclicity) usually occurs between 7 and 15 years of age for African elephants and slightly earlier, at 6–9 years, for the Asian species, with captive animals generally reaching puberty at the earlier age. The range for wild animals is considerable and varies according to prevailing environmental conditions and social factors. Both species are polyestrus and, if unmated, will cycle continuously throughout the year.

Gestation lengths based on mating observations and endocrine evaluations are on average 21 and 22 months for Asian and African elephants, respectively (range: 18–24 months). Both species have an allantochorionic placenta. Single offspring are usually produced; the incidence of twinning is low (1%) for both species. Birth weights for singleton offspring are approximately 120 kg for males and 90 kg for female calves.

In the wild, lactational anestrus typically lasts 12–24 months for African elephants, although a calf may suckle for 3 or 4 years or until the birth of the next offspring. The sex of the calf appears to play a role since male calves suckle more frequently and tend to delay resumption of estrus by about 6 months. Based on hormonal data, shorter periods of lactational anestrus of 40–45 weeks have been reported in captive Asian elephants and both species can resume reproductive cyclicity earlier (within 2 or 3 months) if the calf does not survive.

The interbirth interval ranges from 3 to 9 years in African and 2 to 9 years in Asian elephants and is profoundly affected by the nature and quality of the food supply. Since food availability is usually related to rainfall, drought years tend to be associated with reduced incidence of estrous cows and conceptions. There is also evidence suggesting that under conditions of poor nutrition dominant females come into estrus sooner (and thus give birth sooner) than subordinate females, irrespective of any effects of lactation. In regions with distinct seasons, the highest incidence of conceptions usually occurs during or shortly after the peak in rainfall. Typically conception rate in the wild is between 75 and 85%, whereas in captivity it is <20%.

Delayed, suppressed or infertile ovulations, unsuccessful implantations, or spontaneous abortions all impede reproduction in females. High population density, as well as physiological and social stresses, also can inhibit reproductive activity. Dystocias are not uncommon in zoo elephants and may be caused by nutritional imbalances or anatomical abnormalities. There is a high prevalence of genital tract leiomyomas (especially in the uterus) in captive Asian but not African elephants. The presence of leiomyomas does not appear to interfere with reproductive cyclicity; however, their cause and effect on subsequent fertility are unknown.

Asian and African elephants remain reproductively active into their early 50s although the conception rate is typically very low (8.3%) and the interbirth interval increases significantly. The majority of animals show a decline in reproductive activity by age 40.

B. Reproductive Anatomy

1. Urogenital Canal

The most prominent feature of the reproductive tract is the long urogenital canal (85–100 cm) which extends from the external orifice (vulva) to a hymen leading to the vagina near the level of the urethral opening. The vulva is located on the ventral and caudal aspect of the abdomen between the rear legs. Great variation in the configuration of the vaginal orifice (hymen) has been reported in nulliparous animals. The most common description is the occurrence of three small openings, with the central opening being patent with short, blind ending orifices on either side. Other variations include two orifices divided by a septum with one opening only and three orifices of which two are patent. In a naturally reproducing population the hymenous membrane is present in nulliparous and primigravid elephants but is torn in multiparous animals, leaving one wide vaginal orifice surrounded by ragged folds of mucous membrane. The site of ejaculation during natural mating is the urogenital canal.

2. Ovary

Elephant ovaries are well innervated and vascularized, paired organs composed of stroma, developing follicles, and corpora lutea (CL) surrounded by a single layer of surface epithelium. An ovary weighs approximately 60 g, with an overall dimension of $7 \times 5 \times 3$ cm. Follicles of up to 20 mm in size have been reported in ovaries removed from culled African elephants, although size at ovulation is not known. Ovulation rate in both species is almost certainly variable; single ovulation has been described, but multiple ovulations are probably more common.

The CL of the elephant have been classified into three categories: those that are formed by luteinization of granulosa/thecal cells of either ovulated or unovulated follicles and those that result from luteinization of thecal and stromal cells which invade atretic follicles. Generally, classification of CL has been based on gross morphological appearance, with an "active" CL being large (18 mm) and yellow in color, and degenerate CL, termed corpora rubra or corpora nigra, being smaller (7 mm) and brown in color. Ovaries almost always contain more than one luteal body (range: 2->20)—some with and some without visible ovulation points. Thus, in addition to CL arising from ovulation, CL-like structures lacking ovulation stigmata (often termed accessory CL) are also formed during normal ovarian cycles.

The CL of pregnancy have large, binucleate, and rounded cells and together can comprise up to 200 g of luteal tissue. Various researchers have determined that CL endocrine activity is greatest during early to midpregnancy. The structure of the corpus luteum has been shown to persist 2 months past an entire pregnancy and continue on in a degenerate condition for at least 4 years postpartum although it is no longer steroidogenically active.

The steroidogenic function of the elephant CL has been the subject of considerable debate. Although initial studies in the 1950s and 1960s reported low or nondetectable concentrations of progesterone in the CL, electron microscopic examination clearly demonstrated that the CL contained structures characteristic of steroidogenic cells (e.g., predominance of smooth endoplasmic reticulum along with a close association with mitochondria, small amounts of rough endoplasmic reticulum, and large stores of lipid). Thus, it is generally accepted that the CL are steroidogenic tissue but with limited ability to secrete progesterone.

C. Reproductive Behavior

1. Estrous Behavior

It is well-known that in captivity, female elephants in the absence of a male do not exhibit any outward physiological or behavioral changes signifying the onset of estrus, a phenomenon that has led most researchers to believe that estrus can only be detected when a female is seen mating. Sexual activity during estrus in both captive and wild Asian elephants includes contact-promoting behaviors such as the flehmen response by the male and the smelling of the male's temporal gland by the female. These behaviors are followed by precopulatory behavior involving head to head wrestling with intertwined trunks, head resting by the male on the back of the female followed sometimes by biting, and particularly herding the female before actual mounting attempts. Studies of behavioral estrus in free-ranging African elephants indicate that the behavior typically includes an exaggerated walk, chasing by males, mounting, consort behavior, and loud yet low-frequency calls which appear to advertise their receptive estrous state. Because these infrasonic calls can be heard over distances of several kilometers, it has been hypothesized that females call during midestrus to attract distant high-ranking males. In captivity, estrous synchrony is sometimes, though not always, observed in females housed at the same facility. Continuous reproductive cyclicity typically seen in captive elephants cannot be considered the norm in the wild because it is unlikely that a female would go unmated during estrus.

2. Chemical Signaling in Female Elephants

The most reliable indication of impending estrus is the male flehmen response, whereby the male examines the urine of the female by placing the tip of his trunk into his mouth and blowing the urine into the vomeronasal organ. This activity is believed to be triggered by the excretion of chemical pheromones mainly in urine but possibly also in feces, mucus secretions from the genital tract, saliva, mucus secretions of the eye, and wax compounds produced by the ear and interdigital glands.

Chemical signal mediation of sexual interactions may have evolved, in part, because of the reproductive limitations imposed by the female and the fact that females are sexually active for a relatively short portion of their life. Chemical cues that announce a female is in estrus might serve to attract males, even those from a long distance, to ensure mating. Asian elephants appear to advertise their receptivity by excreting (Z)-7-dodecenlyl acetate in urine. This is the same pheromonal compound used by females of many insect species to attract mates. In elephants, concentrations increase markedly during the luteal phase to a peak just before ovulation. Asian bulls respond to this compound by eliciting multiple flehmen responses, erections, and mounting behaviors.

African elephants probably also produce sex pheromones. African females regularly contact the anogenital region of other females and exhibit heightened chemosensory responsiveness to urine during the follicular phase of the cycle. In the wild, African elephant bulls entering a herd will flehmen test most if not all females for estrus. Females remain still while the male touches or blows on the urogenital area; some will urinate prior to or just following this contact.

D. Reproductive Physiology

1. Endocrinology of the Ovarian Cycle

The ovarian cycle of the Asian and African elephant was not characterized until 1983 and 1988, respectively, when various progesterone assays were identified that could monitor luteal activity. Based on the profile of immunoreactive progesterone (iP; measurements of progestogens made with antisera of varying specificity), the ovarian cycle of the elephant has a duration of 13–16 weeks, comprising an 8- to 11-week luteal phase and a shorter interluteal (follicular) phase of 4–6 weeks. In both species, estrogen-related events (e.g., estrous behavior, vaginal cytology, and urinary estrogen excretion) suggested that multiple waves of follicular growth occurred in 3-week cycles, although upon analysis estradiol concentrations were found to be low (generally <15 pg/ml) and fluctuated without showing any clear pattern. Multiple discrete luteinizing hormone (LH) peaks are observed during the cycle. A characteristic feature of the interluteal period is the presence of two LH surges that occur 3 weeks apart. As measured by immunoassay, the two surges are qualitatively and quantitatively indistinguishable, but only the second is associated with a detectable increase in iP. Cyclic patterns of follicle-stimulating hormone (FSH) lasting 12–14 weeks reach peak concentrations during the late luteal phase and may be involved with stimulating the 3-week waves of follicular growth (Fig. 1).

Luteal phase levels of circulating iP are variable (Asian, 0.2–0.9 ng/ml; African, 0.1–0.9 ng/ml) and low in relation to most other mammalian species, primarily because progesterone is a relatively minor secretory product of the corpus luteum. In the Asian elephant, 17α-hydroxyprogesterone (17α-OHP) is more abundant than iP during the cycle, although in both species the predominant circulating gestagens are the 5α-reduced compounds 5α-dihydroprogesterone (5α DHP) and 3α-hydroxy-5-pregnan-20-one. Studies in the African elephant have shown that the biosynthetic potential and luteal content of 5α-reduced compounds exceeds that for progesterone by as much as 100- to 1000-fold. Thus, the African elephant is unique among mammals studied to date in that 5α-reduced steroids are the major progestins of ovarian origin. Furthermore, since 5α-DHP demonstrates high-affinity binding to the elephant endometrial progesterone receptor equal to progesterone itself, it may represent a more biologically relevant progestin in this species. The significance of 5α-reduced progestins in Asian elephant

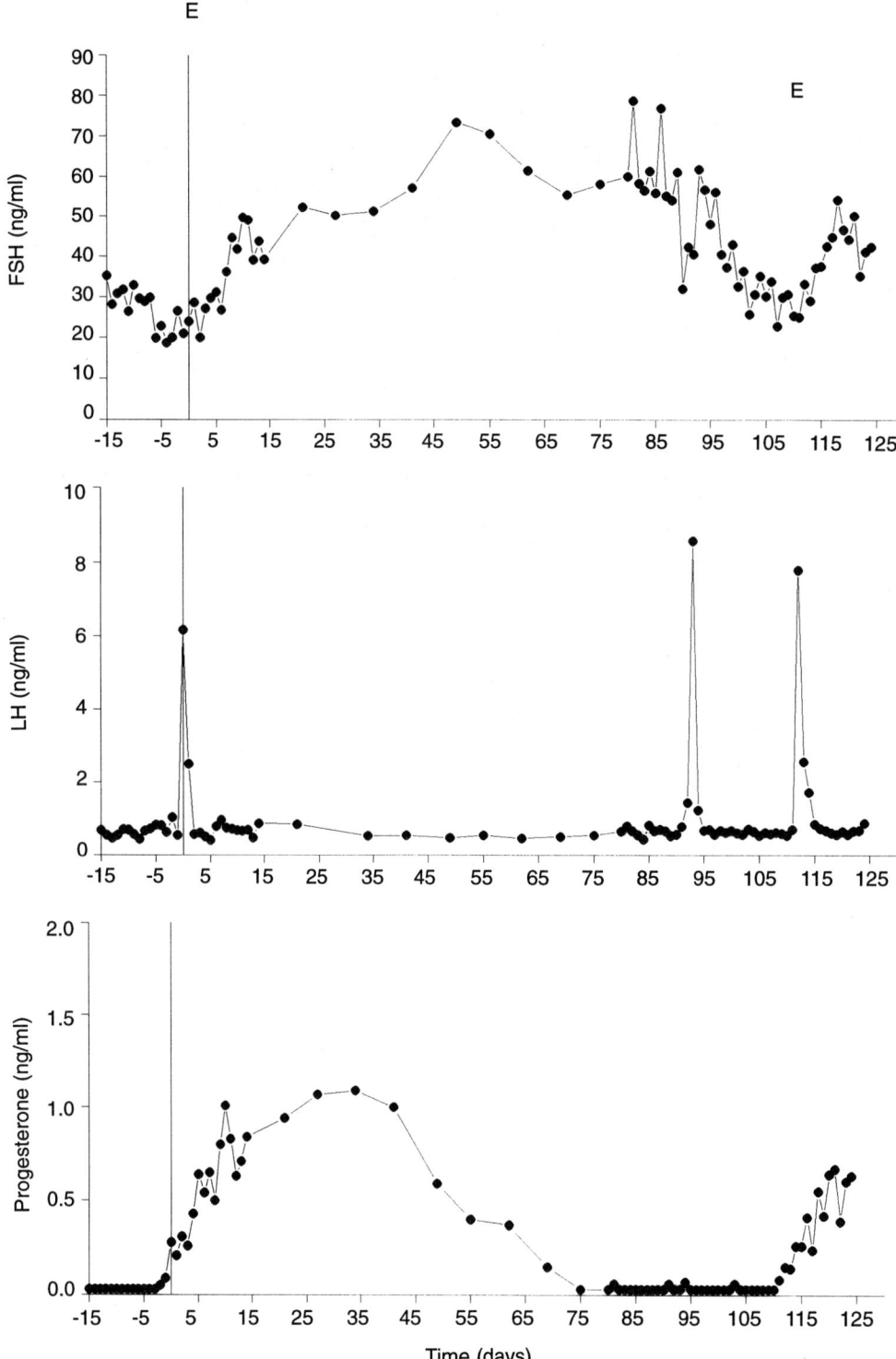

FIGURE 1 Daily plasma concentrations of progesterone, LH, and FSH throughout one estrous cycle in a female Asian elephant (*Elephas maximus*). E, time of presumed estrus and ovulation.

ovarian physiology is less clear, although they have also been shown to predominate in circulation in this species.

Analysis of circulating iP (provided a nonspecific antiserum that cross-reacts with 5α-reduced compounds is used) is commonly employed for routine monitoring of ovarian cycles in both Asian and African elephants. In both species, the 5α pregnanes predominate in urine and feces and their measurement provides a useful noninvasive method for monitoring ovarian function. In the Asian elephant, the 17α-OHP metabolite, 5β-pregnanetriol, is an abundant urinary progestin successfully used for noninvasive monitoring of reproductive cycles in this species.

2. Endocrinology of Pregnancy

Elevated iP beyond the normal luteal phase indicates that conception prolongs the steroid secretory action of the CL. Histological and endocrine findings indicate that the CL is most steroidogenically active between 3 and 15 months of pregnancy. Thereafter, luteal content (5α-reduced progestins) and peripheral iP decline gradually toward birth. Circulating levels of iP during pregnancy in both species are generally low (compared with most other mammalian species), fluctuating between 0.5 and 2 ng/ml. In the Asian elephant ratios of iP : 17α-OHP are significantly lower 3–8 weeks after ovulation in conception vs nonconception cycles. The physiological significance of this observation is not known but it provides a potentially useful method of early pregnancy diagnosis.

The factors responsible for prolonging luteal function during early pregnancy are not known and a placental gonadotropin has not been conclusively demonstrated. In the African elephant immunoreactive LH and FSH concentrations during pregnancy do not appear to differ from values obtained from nonpregnant animals. Levels of immunoreactive relaxin in the Asian elephant rise significantly by Week 20 of gestation and remain elevated throughout most of the pregnancy (Fig. 2), suggesting a role in the establishment and maintenance of pregnancy possibly by assisting progesterone in the growth and maintenance of a quiescent uterus.

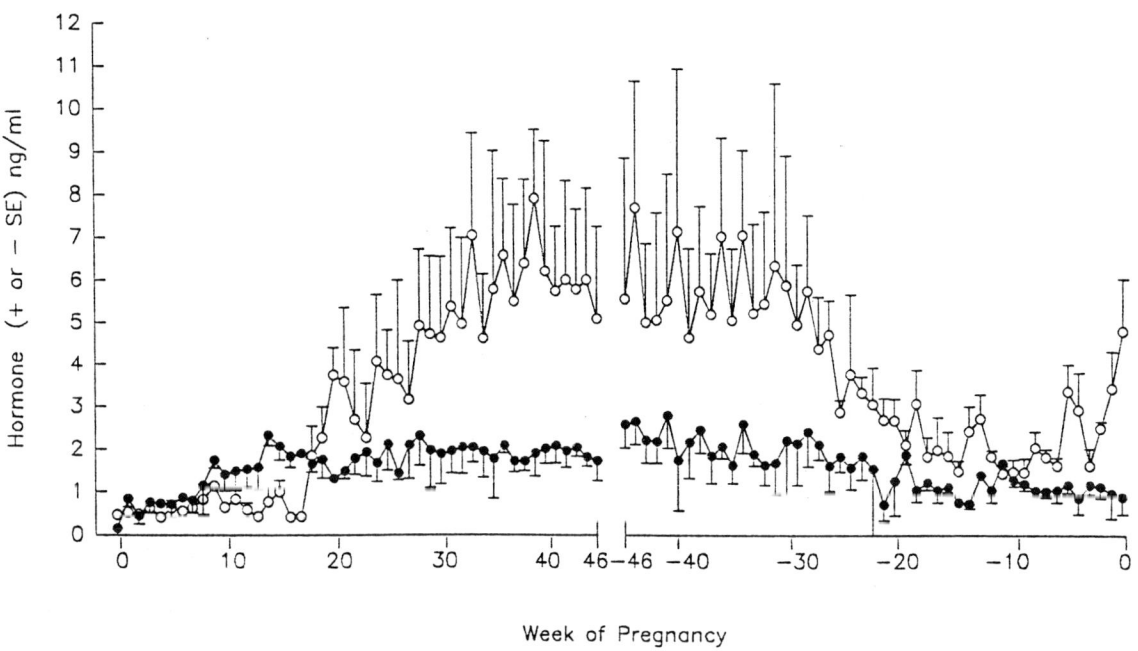

FIGURE 2 Mean (± SEM) concentrations of immunoreactive relaxin (○) and progesterone (●) in plasma of four Asian elephants (*Elephas maximus*) throughout five pregnancies.

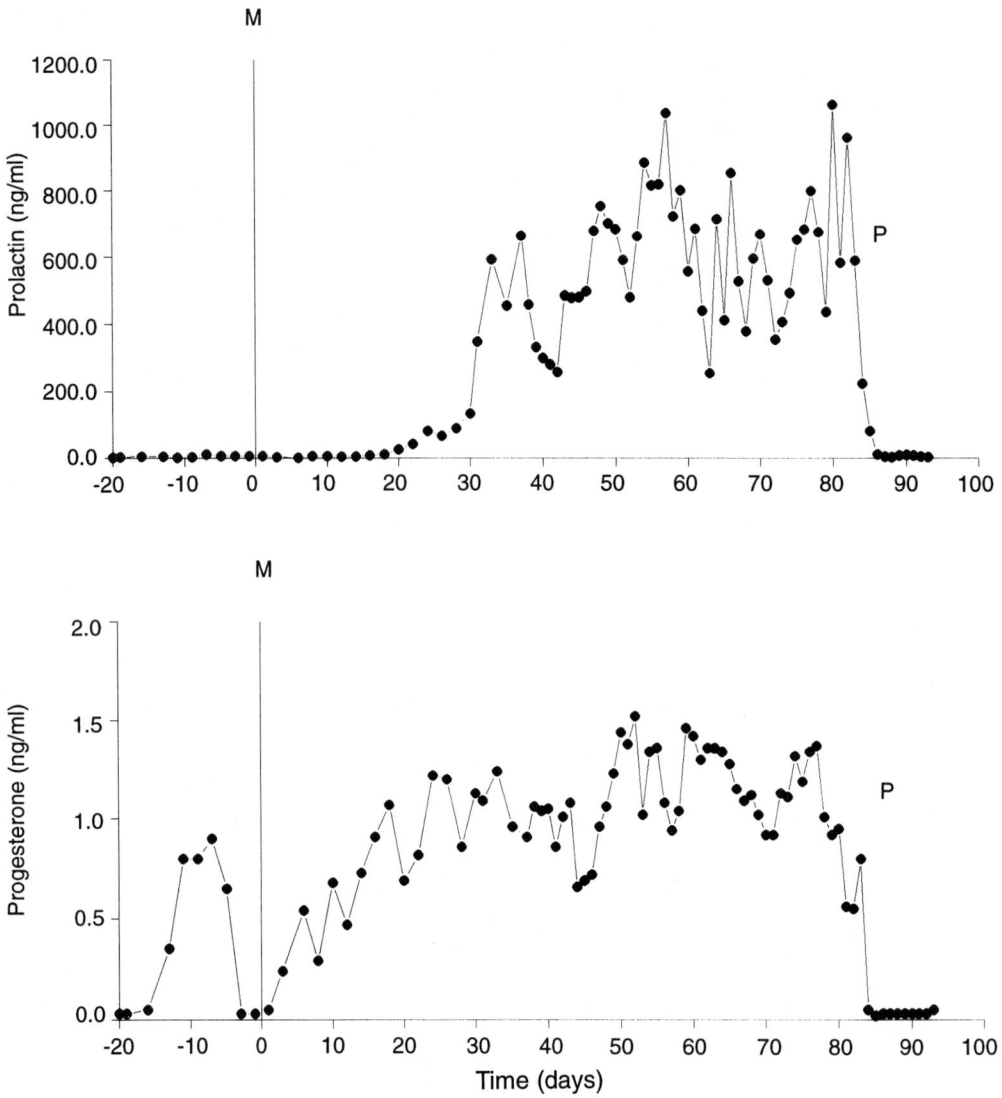

FIGURE 3 Weekly plasma concentrations of progesterone and prolactin throughout pregnancy in a female Asian elephants (*Elephas maximus*). M, mating and conception; P, parturition.

One conspicuous endocrine event during pregnancy in both species is a several hundred-fold increase in circulating immunoreactive prolactin, first detectable at about 4–6 months of gestation (Fig. 3). Although a luteotrophic function for prolactin is plausible, there is no evidence for this, and the timing of the increase is too late to account for the initial extension of CL function beyond the normal luteal phase length. A relationship between prolactin and estrogen secretion may exist since in both species circulating and/or urinary estrogen concentrations show a gradual but progressive increase beginning between 6 and 8 months of pregnancy.

In African elephants, a poor correlation between iP and luteal volume, declining levels of luteal 5α-DHP content during the second half of pregnancy, and a positive relationship between fetal iP and gestational age all suggest a role for the placenta in progestin production. Direct evidence is lacking, however, and placental tissue was not found to be active in metabolizing pregnenelone in *in vitro* incubations. Considerable enlargement of the fetal gonads of both sexes during the second half of pregnancy, due mainly to hypertrophy of interstitial tissue, suggests a possible steroidogenic function.

3. Parturition

A cow in stage 2 labor (undergoing contractions) usually shows little behavioral change. Stage 3 labor involving active bearing down is typically over within 4 hr. During this time the cervical plug is expelled; bloody discharge is produced from the cervix, and the female vocalizes. The majority of calves are born hindlegs first. Breech births and high-birth-weight calves are not uncommon but are a major cause of higher neonatal mortality (40% in one study).

The endocrine events triggering parturition are under investigation. Detailed studies carried out on the Asian elephant show that circulating iP concentrations drop precipitously 2–5 days before parturition. Urinary estrogen levels also decline at the time of birth and in one animal studied a peri- and immediately postpartum elevation in cortisol was detected. A secondary, significant rise in relaxin also occurs at the end of gestation.

III. MALE REPRODUCTION

A. General Reproductive Biology

Male elephants generally reach puberty (capable of sexual activity) between 11 and 20 years of age. In wild populations of African elephants, studies have shown that reproductive success is unlikely under 20 years due to social/dominance factors. In the wild, once male elephants reach puberty, they are forced out of the matriarchal herd and associate in loosely knit bull coalitions or remain solitary. They are often seen in the vicinity of a cow herd, although they are not part of the immediate herd. Male Asian and African elephants display no detectable seasonality with respect to breeding, although peak mating periods tend to occur coincident with increased rainfall. There appears to be a definite social hierarchy based on size, age, and experience with respect to dominance; however, this order changes with the onset of specific behavioral and physiological changes known as "musth."

B. Anatomy

In gross morphological appearance, the male reproductive tract of the Asian and African elephants are similar. The testes are located internally and hang from the dorsal wall of the abdominal cavity medial to the kidneys. The mesorchium folds around the efferent ducts and anterior aspect of the epididymis which is medial to the dorsal pole of the testis. The efferent ducts are highly convoluted and radiate dorsally from the extragonadal rete testis located on the superior aspect of the testis about one-third caudal from the anterior end. Although structurally similar to other species, unusual characteristics of the elephant efferent duct system include a reddish-brown pigmentation and the presence of anastomoses between separate regions. The epididymis is comparatively primitive and courses posteriorly along the testis within the mesenteric fold. Structural studies indicate that the duct is differentiated and can be classified into proximal, isthmus, and distal regions which appear to be homologous to those described for scrotal mammals, albeit not as distinct. The epithelium lining the efferent ducts is a low pseudostratified columnar type composed of ciliated and stereociliated cells and some basal and halo cells. The epididymis is lined by pseudostratified stereociliated, columnar epithelium composed mainly of principal and basal cells. Live spermatozoa in high concentration are found in the distal epididymis. The ductus deferens contains few spermatozoa and is made of principal cuboidal and basal cells surrounded by a thick muscular wall. The ampullae are large glandular structures that open into the ducts of the corresponding seminal vesicles. The seminal vesicles are large, thick-walled sacs containing a fructose-rich watery secretion. The prostate glands are lobulated and situated on the dorsal wall of the urethra, immediately posterior to the seminal vesicles. The large bulbourethral glands are tubuloalveolar in nature and covered by a thick muscular capsule containing sialic and citric acid. The bull elephant has no inguinal canal, cremaster muscle, or pampiniform plexus.

The penis (weighing up to 27 kg) has an extremely well-defined corpus cavernosum and large paired levator muscles on its dorsal surface, allowing for considerable voluntary control during mating. It is sigmoid in shape during erection.

C. Testicular Physiology

1. Gametes

Studies on sperm physiology and cryopreservation in the male Asian and African elephant are few, most likely due to the difficulty of semen collection and low number of accessible males. Spermatozoa mature during passage through the epididymis, developing mature motility, a condensed acrosome, and biochemical changes in the plasma membrane. Migration of the cytoplasmic droplet also occurs along the middle region. The typical male Asian elephant ejaculate collected by artificial vagina (AV) or masturbation is approximately 200–300 ml with a concentration of $0.5–2 \times 10^9$ sperm/ml. Season, temporal gland activity, and circulating testosterone concentrations do not appear to affect sperm quality in African or Asian elephants. Recent work carried out on Asian elephants noted that fresh sperm collected using an AV, washed and extended in egg yolk extender, had normal morphology (>95%), high motility (>90%), and good viability (>90%) but exhibited decreasing acrosomal integrity over time, falling to <20% by 5 hr after collection. Although few reports on cryopreservation of spermatozoa for both species exist, they document low post-thaw viability (<40%), motility, and acrosomal integrity compared with fresh sperm.

D. Reproductive Behavior

1. Musth

The phenomenon of musth has been recognized for centuries to occur in the Asian elephant and recently (within the past 20 years) was recognized in the African elephant. Musth may occur annually or semiannually in the mature male and although not absolutely necessary for breeding, it is regarded as an important reproductive strategy. There is no obligate seasonality associated with musth and males are not in synchrony, although high-ranking bulls will be in musth during those times of the year when there are the most estrous females (e.g., prior to the onset of the rainy season). In the wild musth promotes outbreeding because the males will travel more widely to reach more herds, thus ensuring genetic diversity.

A male in musth has elevated androgen levels and shows increased aggression toward other males and as such is generally avoided by nonmusth males regardless of their dominance rank. Musth males may also exhibit continuous urine dribbling, loss of appetite, changes in skin coloration (white patches may become yellow), and have swollen temporal glands (tubuloalveolar apocrine glands) which eventually drain, producing a pungent, oily, viscous liquid. Comparison of standard hematological and biochemical parameters between the musth and nonmusth states indicate significant increases in the level of two liver enzymes, alkaline phosphatase and γ-glutamyltransferase, during musth as well as an increase in creatinine levels.

In the wild, a musth male can outcompete a normally more dominant male for access to estrous females. In captivity, musth presents a major problem because the males become extremely aggressive and dangerous to handle. It is likely that the improved plane of nutrition can result in an exaggerated form of musth in which the animals become so agitated that they have little interest in breeding. Musth can last from several days to several months, with the intensity ranging from mild to very strong. Generally, the longer an animal is in musth, the more severe are the manifestations. In captive Asian elephants the intensity of musth appears to be unrelated to age, with some males exhibiting a strong musth from the beginning, others showing increasing intensities with increasing age, and still others exhibiting a relatively mild musth throughout their life. In African elephants musth is age related. There are no documented musth in wild bulls below the age of 25 and it is typically only seen in older bulls in peak physical condition.

E. Reproductive Endocrinology

In general, the limited information available regarding male elephant reproductive endocrinology follows the classical description of the hypothalamic–pituitary–gonadal axis. Frequent sampling of Asian bulls both in and out of musth demonstrated that LH is secreted in pulses (four every 12 hr) under the influence of gonadotropin-releasing hormone (GnRH). Pulses of testosterone closely followed those of LH. The limited data available show little change between LH and testosterone pulse frequency during musth compared to nonmusth. However, both pulse amplitude and pulse area increased significantly during musth. Additionally, testes of musth bulls appear to be hyperresponsive to GnRH-induced LH secretion. The length of time the testes are exposed to an increase in LH secretion may be an important factor in determining the extent of increased testosterone production and, thus, the severity of associated musth behavior.

Throughout most of the year plasma androgens, including testosterone, dihydrotestosterone, and androstenedione, remain at basal levels (<1 ng/ml), with androstenedione concentrations greater than those of testosterone. During the time of musth, however, androgenic steroid secretion increases dramatically in both species (>15 ng/ml) and a switch in the androstenedione/testosterone ratio (A/T) in favor of testosterone occurs. It has also been noted that brief shifts (up to 3 days) in A/T can occur during nonmusth in male Asian elephants that resemble miniature musth profiles which appear to be responses to specific behavioral events such as contact with an estrous female (Fig. 4). It appears that "anticipatory" endocrine responses triggered by external cues may play a substantial role in the mating behavior by inducing a general or sexual arousal.

Due to the many problems associated with managing captive male elephants in musth, there is considerable interest in finding a treatment to suppress the aggressive behavior. Current methods for controlling musth include isolation and reducing food and water intake, actions that could elicit animal welfare concerns. Although castration offers a permanent solution, the surgery is both difficult and often unaccept-

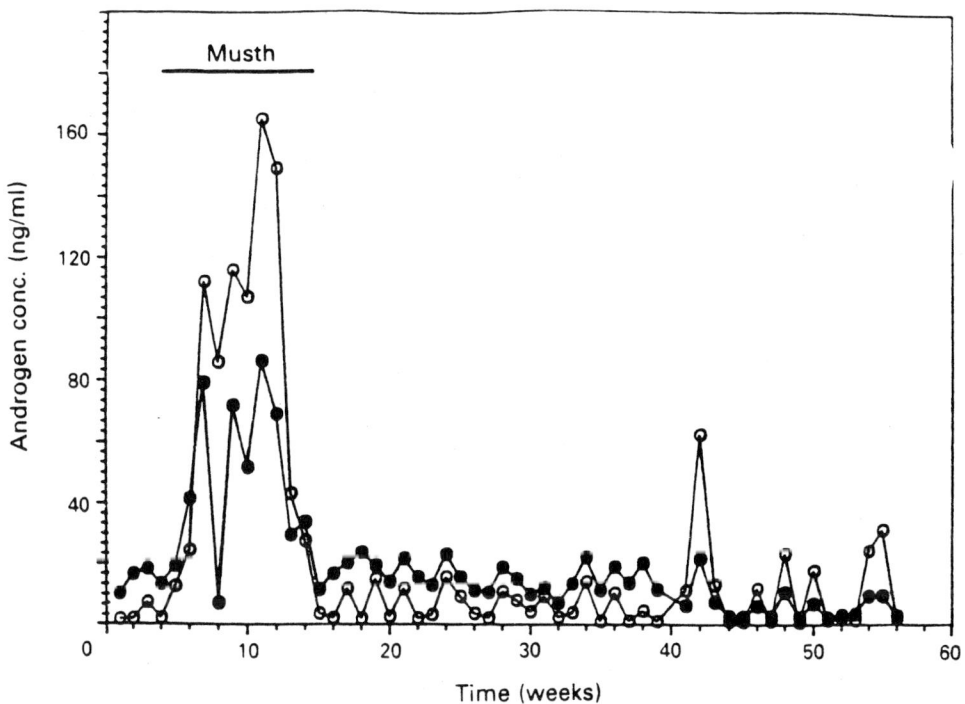

FIGURE 4 Weekly plasma concentrations of androstenedione (●) and testosterone (○) throughout 1 year in a mature male Asian elephant (*Elephas maximus*). Behavioral signs of musth were observed from Weeks 4 to 16. The switch in the A/T ratio during nonmusth as a result of specific behavioral events such as mating occurs during Weeks 41–56.

able since some bulls may be needed for breeding. Because musth appears to be related to elevated circulating testosterone concentrations, any therapy that temporarily suppresses pituitary LH release and subsequent testosterone secretion might attenuate or even alleviate behavior problems until the musth cycle ends.

The use of GnRH analogs has been shown to disrupt normal pituitary gonadal function in African and Asian elephants, although dosages, length of treatment, and number of injections remain to be standardized in order to definitively determine the effect on testosterone secretion and musth behavior. Preliminary trials with cyproterone acetate (an antiandrogen) given during musth to Asian bulls reduced testosterone to baseline levels but musth behavior was not altered.

IV. CONSERVATION AND MANAGEMENT

Asian elephants were once widely distributed over the Indian subcontinent and Southeast Asia, but now they are restricted largely to the few uninhabited, forested areas. The total wild population is estimated to range between 35,000 and 56,000, with continuing gradual population declines due primarily to habitat loss and capture. Asian elephants today are still domesticated and used extensively in the timber industries of Myanmar, Thailand, Nepal, and India as well as for ceremonial and religious purposes and ecotourism. Because of poor breeding success in captivity, these domestic stocks are generally replenished by capturing wild elephants.

In Africa, heavy poaching for ivory reduced elephant numbers from 1.3 million in 1979 to less than 500,000 by 1989. Since that time, the African elephant population has remained relatively stable due in part to the ban on ivory trading adopted by the CITES in 1989 and to increased antipoaching activities. In some southern African countries (e.g., South Africa, Namibia, Malawi, Botswana, and Zimbabwe) elephants are becoming overpopulated in protected areas and the potential negative impact on biodiversity is now a major concern. Until recently, routine culling of elephants in some regions and national parks was used to control animal numbers; however, public anguish over this practice has forced the consideration of other methods such as translocation and contraception. Trials are now under way in Africa to test the efficacy of various contraceptive agents to suppress reproductive activity in females, including immunocontraception with porcine zona pellucida antigen and subcutaneous implantation of steroid-containing devices.

Mortality rate in free-ranging elephants ranges from 5 to 40% in calves and from 2 to 10% annually among adults and is influenced by many factors, both natural and man-made (e.g., disease, injury, drought, and poaching). Calf mortality is high in captivity (30% in the first year) especially for African elephants, in which a recently discovered herpes-type virus is now know to be a major cause of death among subadults (25% of animals born between 1983 and 1996).

In captivity, fewer than 20% of Asian and 10% of African elephants have produced offspring due in large part to the physical, economical, and logistical problems associated with transporting elephants between institutions for breeding and the danger and difficulty of maintaining bulls in captivity. In the future, acquisition of replacement elephants from wild populations likely will be restricted and in some Asian regions it is already illegal. This situation emphasizes the need to improve natural breeding as well as to develop assisted reproduction techniques, including artificial insemination (AI) and genome resource banking, to maintain stable, genetically healthy populations. Recent developments of ultrasound-guided insemination procedures, endocrine assays identifying the presumed ovulatory LH surge, and semen collection via rectal palpation from unanesthetized bulls are increasing the feasibility of AI, which may prove useful in helping to promote genetic diversity among captive populations.

See Also the Following Articles

CAPTIVE BREEDING OF WILDLIFE; ESTROUS CYCLE; ESTRUS; PHEROMONES

Bibliography

Brown, J. L., Citino, S. B., Bush, M., Lehnhardt, J., and Phillips, L. G. (1991). Cyclic patterns of luteinizing hormone, follicle-stimulating hormone, inhibin and progesterone in the Asian elephant (Elephas maximus). *J. Zool. Wildlife Med.* **22**, 49–57.

Brown, J. L., Bush, M., Wildt, D. E., Raath, J. R., de Vos, V., and Howard, J. G. (1993). Effects of GnRH analogues on pituitary–testicular function in free-ranging African elephants (Loxodonta africana). *J. Reprod. Fertil.* **99**, 627–634.

Hildebrandt, T., Goritz, F., Pratt, N. C., Schmitt, D. L., Quandt, S., Raath, J., and Hofmann, R. R. (1998). Reproductive assessment by ultrasonography in elephants (Loxodonta africana and Elephas maximus). I. Sonomorphology of the male urogenital tract. *J. Zool. Wildlife Med.*, in press.

Hodges, J. K. (1998). Endocrinology of the ovarian cycle and pregnancy in the Asian and African elephant. *Anim. Sci. Reprod.*, in press.

Mikota, S. K., Sargent, E. L., and Ranglack, G. S. (1994). The reproductive system. In *Medical Management of the Elephant* (S. K. Mikota, E. L. Sargent, and G. S. Ranglack, Eds.), pp. 159–186. Indira, Oak Park, MI.

Niemuller, C. A., and Liptrap, R. M. (1991). Altered androstenedione to testosterone ratios and LH concentrations during musth in the captive male Asian elephant. *J. Reprod. Fertil.* **91**, 139–146.

Niemuller, C. A., Gray, C., Cummings, E., and Liptrap, R. M. (1998). Plasma concentrations of immunoreactive relaxin and progesterone in the pregnant Asian elephant (Elephas maximus). *Anim. Sci. Reprod.*, in press.

Rasmussen, L. E. I. (1997). Chemical communication: An integral part of functional Asian elephant (Elephas maximus) society. *Ecoscience*, in press.

Schmidt, M. J. (1993). Breeding elephants in captivity. In *Zoo and Wild Animal Medicine, Current Therapy 3* (M. Fowler, Ed.), pp. 445–448. Saunders, Philadelphia

Sikes, S. K. (1971). *The Natural History of the African Elephant.* Am. Elsevier, New York.

Sukumar, R. (Ed.) (1989). *The Asian Elephant: Ecology and Management.* Cambridge Univ. Press, Cambridge, UK.

Embryogenesis, Mammalian

Carol A. Burdsal
Tulane University

I. Introduction
II. Early Cleavage
III. The Blastocyst
IV. Implantation
V. Postimplantation Development

GLOSSARY

blastocyst The embryonic stage in mammals that implants into the wall of the uterus.

cleavage A specialized series of mitotic cell divisions that occur following fertilization.

compaction A process that occurs during embryogenesis in placental mammals in which cells flatten upon one another and display increased cell to cell adhesion.

gastrulation The process that generates the three primary germ layers of the embryo: the ectoderm, the endoderm, and the mesoderm.

inner cell mass The group of cells present on the inside of the blastocyst that will give rise to the embryo.

primitive streak The region in the embryo from which prospective mesoderm and endoderm cells migrate and differentiate during gastrulation.

I. INTRODUCTION

Following the fertilization of the mammalian oocyte, a complex program of gene activity directs the process of embryogenesis. The genetic program in mammalian embryos coordinates an elaborate series of cell divisions, cell migrations, and cellular differentiations. These embryonic processes change a single cell, the fertilized oocyte, into the complex multilayered mature embryo. Early in mammalian embryogenesis, cell lineages which will give rise solely to extraembryonic support structures are separated from those that will give rise to the embryo proper. This segregation of embryonic and extraembryonic lineages at the blastocyst stage of development is unique to placental mammals. Early development in placental mammals culminates in another singular event, implantation (the invasion of the embryo into the wall of the uterus). In contrast to early embryogenesis, many steps of mammalian postimplantation development occur similarly to those of other vertebrates such as reptiles and birds. These events include the formation of the primitive streak and the differentiation of the three primary germ layers of the embryo—the ectoderm, the endoderm, and the mesoderm—during gastrulation. Cells in the three primary germ layers interact and eventually give rise to all the tissues and structures found in the mature embryo. This article will focus on embryogenesis in two placental mammals: the human and the mouse, the most extensively studied mammal.

II. EARLY CLEAVAGE

Fusion of the sperm and oocyte in mammals triggers the completion of meiosis in the oocyte and initiates the developmental program of the embryo (Fig. 1A). The first phase of this program is known as cleavage. Cleavage in mammals consists of a series of mitotic cell divisions spaced at approximately 12-hr intervals. The cell divisions that occur during cleavage are unusual in that no cellular growth occurs between each division. Therefore, as cleavage continues, the large volume of cytoplasm present in the fertilized oocyte is partitioned into more numerous and smaller cells. Early cleavage in mammals occurs while the embryo is present in the oviduct where it was fertilized and while the embryo remains inside the zona pellucida (Figs. 1A–1F). The zona pellucida is a hard extracellular layer which protects the cleaving embryo and prevents the embryo from interacting with the oviduct as it travels toward the uterus.

A. Development to the Eight-Cell Stage

The first cleavage in the mammalian zygote produces two identical cells called blastomeres (Fig. 1B). An important event that occurs at the two-cell stage in the mouse embryo is the initiation of gene transcription from the zygotic genome. In many other organisms, early embryonic development is controlled by stored maternal factors (mRNAs and proteins). In those embryos, maternal products regulate early embryogenesis because the onset of zygotic transcription is delayed until the embryo is composed of hundreds to thousands of cells (i.e., until the midblastula transition). Therefore, a unique aspect of mammalian embryogenesis is that beginning at the two-cell stage, the process is controlled by zygotic gene products.

Another feature that distinguishes mammalian cleavage is that cell divisions following the two-cell stage are asynchronous. This means that en route to producing the four-cell stage embryo (Fig. 1C), the blastomeres of the two-cell stage embryo do not undergo the second round of cleavage at the same time. This asynchrony often results in the formation of a three-cell stage embryo. Embryos containing odd numbers of blastomeres are often observed as mammalian development proceeds to the eight-cell stage.

B. Compaction

A crucial event that distinguishes mammalian cleavage from the process in all other organisms occurs at the 8-cell stage when the embryo undergoes compaction. Before the process of compaction all the blastomeres are packed loosely about one another and display relatively minimal cell to cell contacts (Fig. 1C). However, at compaction, the blastomeres dramatically increase their level of cell to cell adhesion and flatten upon one another. As shown in Fig. 1D, the increased cell adhesion in a compacted 8- to 16-cell stage embryo makes it difficult to identify

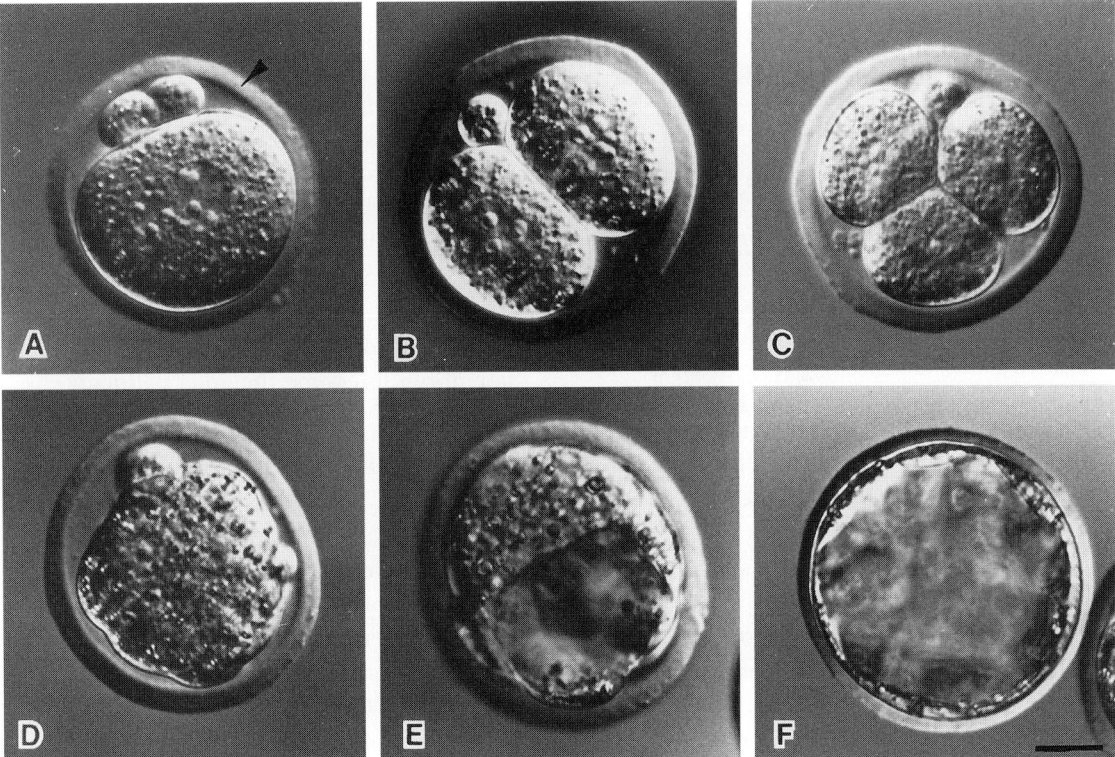

FIGURE 1 Preimplantation stages of embryonic development in the mouse. A, Fertilized oocyte with two polar bodies (which were generated during meiosis) and surrounding zona pellucida (arrowhead); B, two-cell stage embryo; C, four-cell stage embryo (one cell is out of view behind the other blastomeres); D, compacted 8- to 16-cell stage embryo; E, early blastocyst (32-cell stage embryo); F, late or expanded blastocyst (64-cell stage embryo), the inner cell mass is out of the focal plane in this photograph. Scale bar = 25 μm (modified from Pedersen and Burdsal, 1994).

individual blastomeres. In the compacted embryo, the outer flattened blastomeres stabilize their association by forming adhesive complexes called tight junctions between one another.

C. The Morula (the 16-Cell Stage Embryo)

The cells of the compacted 8-cell embryo proceed through the fourth cleavage and produce the 16-cell stage embryo, also called the morula (Fig. 2). The two distinct cell types found in the morula, the small group of inner cells and the larger number of external or outer cells, were brought about by the events of compaction. During this phase of embryogenesis, outer cells become organized as an epithelium, whereas inner cells become connected by gap junctions.

III. THE BLASTOCYST

A. The Inner Cell Mass and the Trophoblast

Cleavage continues in the morula and, at approximately the 32-cell stage, the embryo enters the uterus. Between the 32- and 64-cell stage another unique phase of mammalian development, the blastocyst, is reached (Fig. 1E). The differentiation of the blastocyst begins when the outer epithelial layer of the embryo, now called the trophoblast (or troph-

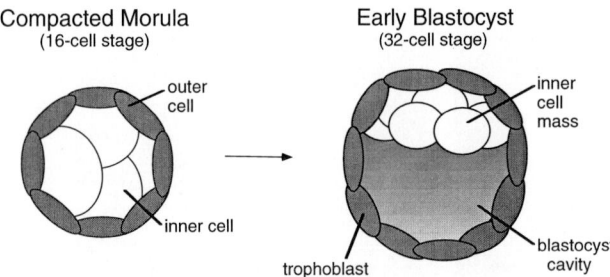

FIGURE 2 The differentiation of the blastocyst. During the differentiation of the blastocyst, cells present on the outside of the compacted morula (outer cells) give rise to the trophoblast. The trophoblast cells transport fluids that fill and expand the blastocyst cavity. Cells found on the inside of the compacted morula give rise to the inner cell mass at the blastocyst stage of development in the mouse.

ectoderm), transports fluid from the outside to the inside of the embryo. This activity generates a fluid-filled cavity called the blastocyst cavity or blastocoel (Fig. 2). During cavitation, the descendants of the inner cells found in the morula remain separate from the trophoblast. This group of cells assumes a position on the inner wall of the blastocyst cavity and is now called the inner cell mass.

During 64- to 128-cell stage of development, the trophoblast cells continue to transport fluid causing the blastocyst cavity to expand (Fig. 1F). The distinction between the trophoblast and the inner cell mass in the blastocyst is not just morphological. Fate mapping has demonstrated that only the inner cell mass will contribute cells to the embryo itself. Trophoblast cells are not fated to form embryonic structures but rather to contribute to the placenta, the major extraembryonic support structure of the human and mouse embryo.

B. The Hypoblast

As the blastocyst matures, a new epithelium forms over the inner cell mass (Fig. 3A). This layer is termed the hypoblast in the human embryo and the primitive endoderm in the mouse. The fate of hypoblast cells, like that of cells in the trophoblast, is to contribute only to an extraembryonic structure—in this case, the yolk sac.

C. Hatching

Because the early blastocyst has entered the uterus, the zona pellucida is no longer needed to prevent the premature attachment of the embryo. The final step of preimplantation development in placental mammals is the hatching of the blastocyst. During hatching, the blastocyst secretes proteolytic enzymes which weaken the zona pellucida. This enzymatic activity, along with pressure exerted by the expansion and contraction of the blastocyst cavity, permits the embryo to escape the zona pellucida.

IV. IMPLANTATION

Following hatching, the mammalian embryo settles onto the epithelium that lines the uterus, triggering the process of implantation. Implantation begins with the differentiation of two cell types from the trophoblast, the cytotrophoblast and the syncytiotrophoblast. Cytotrophoblast cells continue to surround the inner cell mass and the blastocyst cavity (Fig. 3A). However, near the site of blastocyst contact with the uterine wall, proliferating cytotrophoblast cells fuse with one another and form a multinucleate mass of tissue called the syncytiotrophoblast. During implantation, the syncytiotrophoblast secretes proteolytic enzymes which disrupt the uterine epithelium and which allow the blastocyst to embed itself in the wall of the uterus (Fig. 3B).

As implantation continues, cytotrophoblast and syncytiotrophoblast cells further differentiate and produce the chorion, the embryo's contribution to the placenta. The first step in the development of the chorion is the formation of small holes or lacunae in the syncytiotrophoblast (Fig. 3C). In the mature placenta, finger-like projections of trophoblast cells called chorionic villi extend into the lacunae and maternal blood fills these spaces now called intervillus spaces. Because embryonic blood flows through vessels that develop in the chorionic villi, gas and nutrient exchange can occur between the embryonic and the maternal circulation. Implantation and the establishment of communication between the embryonic and maternal circulation are exceptional events which distinguish the development of placental mammals from the process in all other vertebrates.

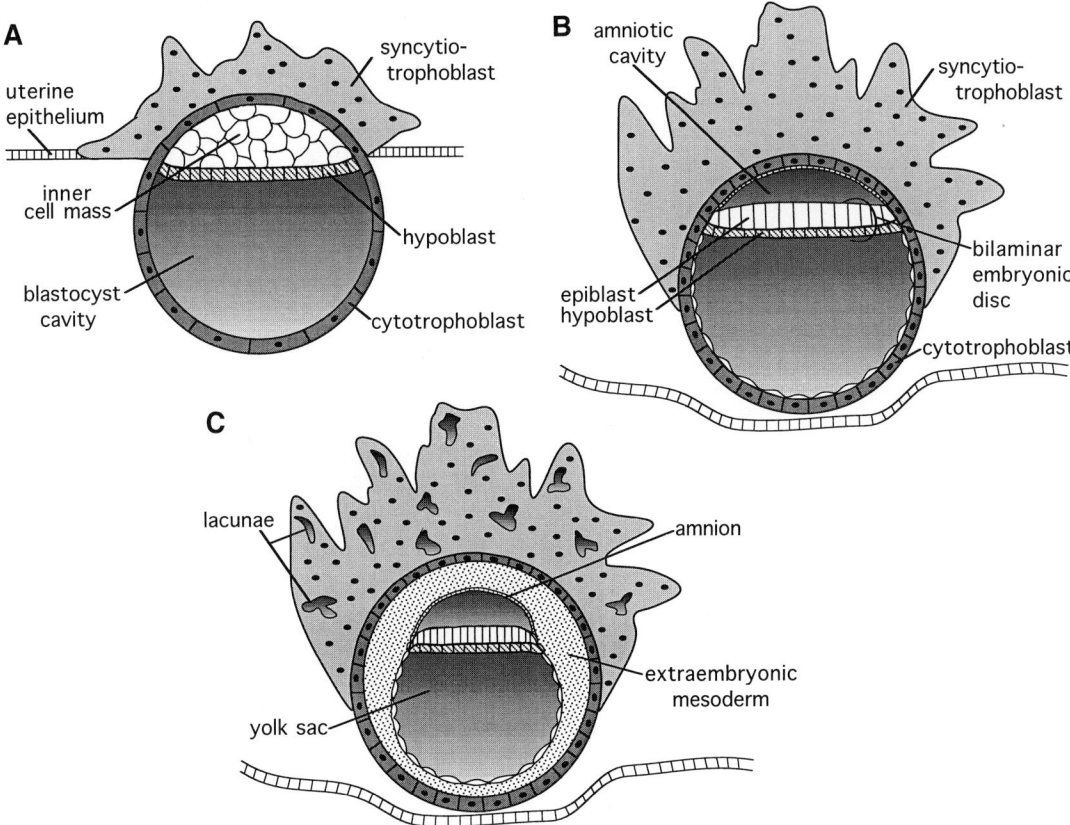

FIGURE 3 Implantation during mammalian embryogenesis. A, Trophoblast cells at the site of uterine contact form a syncytium, the syncytiotrophoblast, which secretes proteolytic enzymes that disrupt the uterine epithelium; B, the embryo migrates into or invades the wall of the uterus until it is completely embedded under the uterine epithelium; C, once implanted, the formation of the chorion, the embryo's contribution to the placenta, begins with the formation of lacunae in the syncytiotrophoblast.

V. POSTIMPLANTATION DEVELOPMENT

A. The Bilaminar Embryonic Disc

As the mammalian embryo invades the uterus, the inner cell mass reorganizes and forms a simple epithelial sheet called the epiblast (Fig. 3B). The sheet of epiblast directly overlies the hypoblast (Figs. 3B and 3C). This stage of development in the human is called the bilaminar embryonic disc. In the mouse, the sheet of epiblast is not flat but rather curved into the shape of a cylinder. Despite the different shape of the embryos, the majority of the next steps of embryogenesis occur similarly in the human and the mouse.

As diagrammed in Fig. 3C, the bilaminar embryonic disc comes to lie between two fluid-filled spaces, the amniotic cavity and the yolk sac. The amniotic cavity forms between the bilaminar disc and the overlying trophoblast and is lined with a layer of cells called the amnion. The amniotic cavity will provide a space for fetal growth and act as a cushion against mechanical shock or injury to the fetus. The yolk sac, which lies below the bilaminar embryonic disc, becomes lined with cells derived from the hypoblast. The yolk sac plays a vital role in postimplantation development because it is the site where blood cells are produced until the embryonic liver matures and becomes hematopoietic. At this stage of development in the human, the embryo is also surrounded by a layer of loosely associated cells called the extraem-

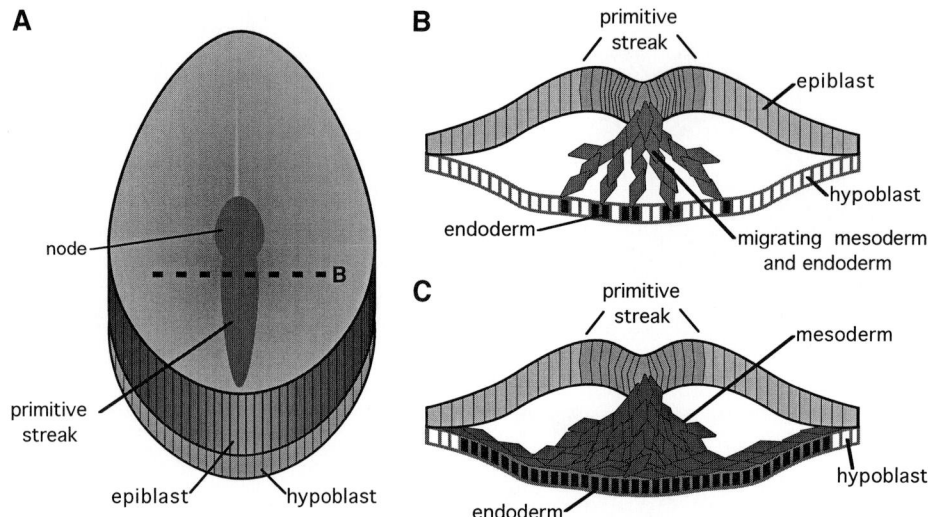

FIGURE 4 Gastrulation in the mammalian embryo. A, Dorsal view of the mammalian bilaminar embryonic disc shown diagrammatically. The future anterior of the embryo lies at the top of the disc, whereas the primitive streak marks the future posterior of the embryo; B, diagram of a cross section taken at the level of the dotted line in A showing the morphology of tissues in the primitive streak region of a gastrulating mammalian embryo; C, organization and morphology of mesoderm and endoderm as gastrulation continues in the mammal. Definitive endoderm cells displace hypoblast cells laterally and mesoderm cells lie between the endoderm and the epiblast (the future ectoderm of the embryo).

bryonic mesoderm (Fig. 3C). In the mammalian bilaminar embryonic disc, only the epiblast will give rise to the embryo proper.

B. Gastrulation

The differentiation of true embryonic tissues in the mammal begins with the process of gastrulation. During gastrulation (Fig. 4), cell migration transforms the embryo from a bilayered to a trilayered structure composed of three tissues: the ectoderm, the endoderm, and the mesoderm. These three primary germ layers will give rise to all the tissues and organs found in the mature mammal.

Mammalian gastrulation is similar to the process in avian embryos. In both instances, the process begins with formation of the primitive streak (Fig. 4). The primitive streak appears when epiblast cells proliferate and accumulate forming an elongated mound or streak in the prospective posterior of the epiblast (Fig. 4A). The primitive streak cells also secrete proteolytic enzymes, which digest underlying extracellular matrix. In mammals, the anterior limit of the primitive streak is marked by a small knob of cells called the node.

Cell migration during gastrulation begins when cells in the primitive streak loose their cell to cell contacts and move into the space between the epiblast and the hypoblast (Fig. 4B). Prospective endoderm cells migrate from the primitive streak and infiltrate into the hypoblast layer. These embryonic endoderm cells eventually displace hypoblast cells into extraembryonic regions (Fig. 4C). Prospective mesoderm cells also migrate through the primitive streak to the inside of the embryo (Figs. 4B and 4C). Mesoderm and endoderm cells migrate laterally and to the future anterior of the embryo during gastrulation. Epiblast cells that remain on the surface of the embryo constitute the third germ layer, the ectoderm. Thus, after gastrulation, a layer of mesoderm is present between the ectoderm and the endoderm. This stage of development in the human embryo is called the trilaminar embryonic disc.

The combination of cell proliferation and cell migration during gastrulation elongates the embryo along the anterior–posterior axis. The cell migrations occurring during gastrulation also bring new embryonic tissues into contact. These contacts lead to interactions between the three germ layers which induce organ systems during the next phase of embryonic development—organogenesis. Therefore, gastrulation is one of the most critical periods of postimplantation development.

C. Differentiation in the Embryonic Mesoderm

While gastrulation continues in the posterior of the embryo, two tissues, the notochord and the somites, differentiate in more anterior embryonic mesoderm (Fig. 5). These tissues are present transiently but are important because they help to organize the differentiation of other tissues along the anterior–posterior axis of the embryo. The notochord is a small dense rod that extends anteriorly from the node of the primitive streak and is thought to provide structural support for the embryo. The cells that form the notochord are also important because they induce the differentiation of neural tissue in the anterior of the embryo (see Section V,D). To either side of the notochord, a subset of mesodermal cells join with one another and form small balls of tissue called somites (Fig. 5A). Somites are repeating structures that form as the embryo elongates, and during this phase of

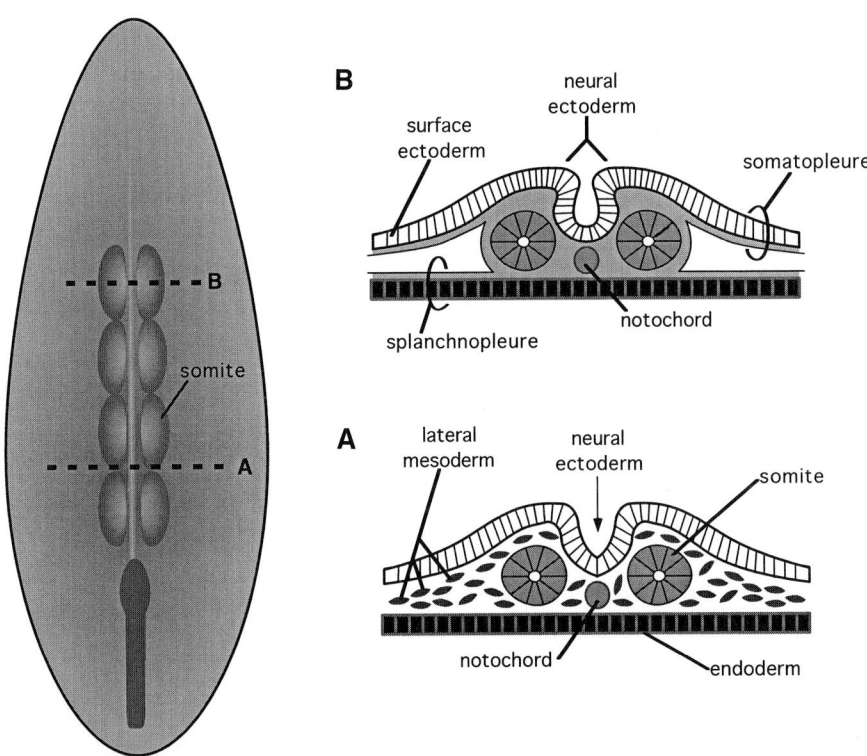

FIGURE 5 Differentiation in the mammalian embryonic mesoderm. Diagrammatic dorsal view of a mammalian embryo which has elongated along the anterior–posterior axis during gastrulation. Dotted lines indicate the level of the cross sections shown in A and B. A, Two tissues, the notochord and the somites, have differentiated in the mesoderm in regions anterior to the primitive streak. In the surface layer, at the dorsal midline of the embryo, prospective neural tissue invaginates; B, more anterior cross section in which the lateral mesoderm has split into two populations, the somatic and the splanchnic mesoderm. The somatic mesoderm and its associated dorsal ectoderm constitute the somatopleure, whereas the splanchnic mesoderm and its associated endoderm constitute the splanchnopleure. The neural ectoderm continues to invaginate and will form a closed tube that will eventually give rise to the brain and spinal chord.

development they divide the embryo into segments. With continued development somites will give rise to the axial skeleton (the vertebrae and ribs), the skeletal muscles of the body wall, back and limbs, and the dermis of the back.

Differentiation proceeds from the anterior to the posterior of the embryo during this phase of postimplantation development. The more anterior cross section diagrammed in Fig. 5B indicates how the lateral mesoderm shown in Fig. 5A splits into two populations as it differentiates. Part of this mesoderm associates dorsally with ectoderm forming the somatopleure. The other part of the mesoderm associates ventrally with the endoderm forming the splanchnopleure. The splanchnic mesoderm is important because it will give rise to the first fully functioning organ in the embryo, the heart.

D. Differentiation in the Embryonic Ectoderm

Two basic tissues differentiate in the ectoderm: epidermis and nervous tissue. Neural differentiation begins when the prechordal mesoderm (the mesoderm that will give rise to the notochord) migrates under the anterior ectoderm. This mesoderm signals or induces the overlying ectoderm to differentiate into neural tissue (Fig. 5). The neural ectoderm invaginates and eventually forms a tube which drops to the dorsal midline of the embryo just under the surface ectoderm. The neural tube is the embryonic precursor of the central nervous system—the brain and spinal chord. Later in development, a small population of cells, called the neural crest, migrate from the neural tube and establish a number of tissues, including the peripheral nervous system. The remainder of the embryonic ectoderm—that is, tissue not induced to differentiate into neural ectoderm—is fated to give rise to the outer covering of the embryo—the epidermis.

E. Differentiation in the Embryonic Endoderm

As organogenesis continues, the embryonic endoderm also assumes the shape of a tube. In mammals, a complex series of folds allows the ectoderm and mesoderm to surround the tube of endoderm. In this way, the basic body plan of vertebrates, in which an outer tube of ectoderm and an inner tube of endoderm are separated by a layer of mesoderm, is achieved. In anterior regions the endoderm contributes to the bronchi and tracheae of the lungs. The endoderm also gives rise to the organs and glands which constitute the digestive tract.

A summary of the developmental fate of the mammalian embryonic and extraembryonic cell lineages discussed in this chapter is presented in Fig. 6. Fol-

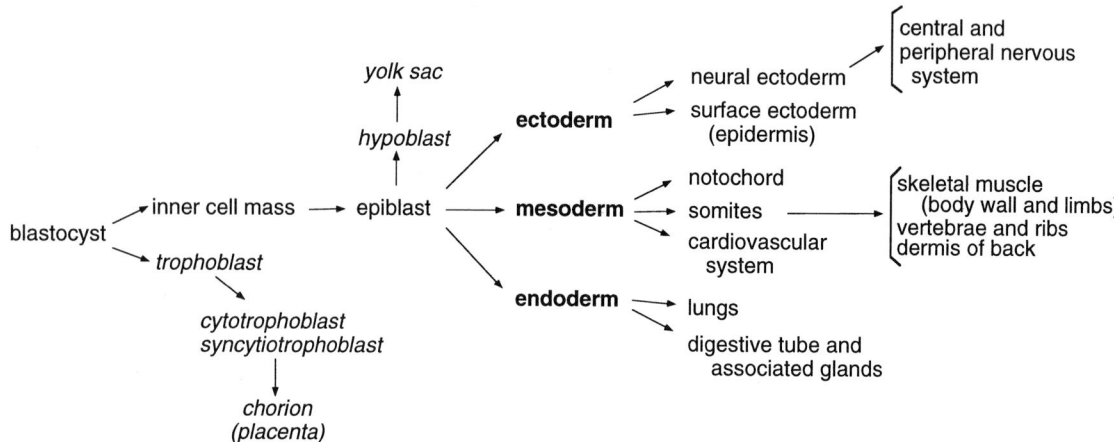

FIGURE 6 Developmental fate of cell lineages in the mammalian embryo. Schematic diagram showing the developmental fate of the major embryonic and extraembryonic cell lineages in the mammalian embryo. Extraembryonic lineages are shown in italics and the three primary germ layers which differentiate from the epiblast during gastrulation are shown in bold.

lowing organogenesis, a period of extensive growth precedes birth in placental mammals.

Acknowledgments

The author gratefully acknowledges R. W. Burdsal for the preparation of Figs. 4 and 5 and thanks Dr. Roger A. Pedersen for contributing Fig. 1 and for his mentorship.

See Also the Following Articles

BLASTOCYST; OOCYTE, MAMMALIAN; TROPHOBLAST

Bibliography

Cruz, Y. P. (1997). Mammals. In *Embryology. Constructing the Organism* (S. F. Gilbert and A. M. Raunio, Eds.), pp. 459–489. Sinauer, Sutherland, MA.

England, M. A. (1994). The human. In *Embryos. Color Atlas of Development* (J. B. L. Bard, Ed.), pp. 207–220. Wolfe Publishing, London.

Moore, K. L., and Persaud, T. V. N. (1993). *Before We Are Born*, 4th ed. Saunders, Philadelphia.

Pedersen, R. A., and Burdsal, C. A. (1994). Mammalian embryogenesis. In *The Physiology of Reproduction* (E. Knobil and J. D. Neill, Eds.), 2nd ed., pp. 319–390. Raven Press, New York.

Snell, G. D., and Stevens, L. C. (1966). Early embryology. In *Biology of the Laboratory Mouse* (E. L. Green, Ed.), 2nd ed., pp. 205–245. McGraw-Hill, New York.

Embryo Transfer

George E. Seidel, Jr.
Colorado State University

I. Superovulation
II. Recovery of Embryos
III. Maintenance of Embryos *in Vitro*
IV. Embryo Transfer
V. Applications of Embryo Transfer and Related Biotechnologies

GLOSSARY

blastocyst An embryo with a cavity (blastocoele). This stage of development usually begins about Days 5–7 after fertilization in most species.

cervix The organ connecting the vagina and uterus.

cryoprotectant A chemical that protects cells during cooling.

donor A female from which embryos are recovered.

embryo The fertilized egg or ovum; also includes stages of development until major organs form.

embryo transfer A general term encompassing superovulation, recovery of ova from donors, and transfer of embryos to the reproductive tract of recipients.

estrus The period of sexual receptivity in the female; usually lasts a few hours to a few days, depending on the species.

fetus The stage of prenatal development from major organ formation to birth.

fimbria The funnel-shaped structures by which ovulated ova gain access to the oviducts.

follicle A fluid-filled sac in the ovary containing the immature ovum (oocyte) plus somatic cells.

gonadotropin A hormone that stimulates the ovaries and testes.

hormones Products of living cells that circulate in body fluids and produce a specific effect(s) on the activity of cells other than those that produced the hormone.

morula Latin for mulberry. The stage of embryonic development between the 8-cell and blastocyst stages, when cells are too numerous to count easily. Morulae usually contain from 16 to 60 cells.

oocyte The unfertilized ovum.

oviduct A tube-like organ that conveys the ovum to the uterus; it is the site of fertilization. Also known as *Fallopian tube* or *uterine tube*.

ovulation Release of the ovum from the ovarian follicle.
ovum (plural *ova*) A Latin term meaning egg. Some researchers limit usage to indicate unfertilized egg.
recipient A female that receives a transferred embryo.
superovulation Stimulation of a larger than normal complement of follicles to mature into large follicles ready for ovulation.
uterus The organ in which the embryo implants and develops into a fetus.
zona pellucida An acellular gelatin-like covering of the oocyte and embryo.

Embryo transfer has several levels of meaning. The term literally means transferring an embryo from the female reproductive tract of one mammal to the reproductive tract of another. Prior to the 1950s, the only source of mammalian embryos was the female reproductive tract. Recently, it has been possible to create embryos by fertilization *in vitro*; this is becoming commonplace for human reproduction for infertile couples. Perhaps if the term were coined today, one would describe the process as "embryo placement" rather than "embryo transfer."

For historical reasons, embryo transfer also refers to a multistep process that includes recovery of embryos from donor animals, holding them *in vitro* for a period of hours to decades (if they are cryopreserved), as well as the actual transfer to the reproductive tract of a recipient female. Associated techniques, such as superovulation, cryopreservation, biopsy of embryos to diagnose genetic defects or sex, and cloning by nuclear transplantation, often fall under the heading of embryo transfer technology or the embryo transfer industry, at least in the case of nonhuman species. Indeed, there are numerous national and international scientific societies and trade associations with "embryo transfer" in their names. Curiously, the umbrella term that has evolved for describing these technologies for human applications is "assisted reproductive technologies," often abbreviated ART.

Although embryo transfer procedures can be used in virtually any mammalian species, there are species-specific differences in procedures. For example, nonsurgical procedures appropriate for cattle and horses are impossible for rats and rabbits. Also, species characteristics such as litter size, seasonal reproduction, and nature of the reproductive cycle affect procedures. However, embryo transfer techniques can be used to modify intrinsic species characteristics, for example, to circumvent the seasonal nature of reproduction, or to make species that bear single young have litters. This occurs increasingly frequently, though unintentionally, in women treated for infertility and in cattle to cause deliberate twinning. In the following sections, individual steps in the broad process of embryo transfer will be explained.

I. SUPEROVULATION

Superovulation refers to treating females with hormones to increase the number of oocytes (unfertilized eggs) ovulated. Superovulation is a low-cost, logical step and is almost always used in embryo transfer programs. Superovulation is imposed on the process of normal follicular growth. The usual reason for superovulation is to increase the reproductive rate of valuable females. Even in human females, in which case the objective usually is to obtain one baby, superovulation is used to compensate for inefficiencies of *in vitro* procedures and the natural reproductive process.

Oocytes (unfertilized ova) develop in fluid-filled structures in the ovary termed "follicles." Follicles with their oocyte grow from tiny structures to mature, preovulatory follicles of 2–30 or more millimeters in diameter over a period of 1.5–5 months (both parameters depend on the species). The final stages of follicular growth are stimulated by a gonadotropic hormone secreted by the anterior pituitary gland, follicle-stimulating hormone (FSH). To effect superovulation, excess FSH or other hormones that mimic it are injected into the female with the result that excess follicles develop and ovulate.

Superovulation typically increases the number of oocytes (and hence embryos) produced by a factor of 5- to 10-fold (2- or 3-fold in the case of litter-bearing species). However, there is wide variability in response. For example, in cattle, a typical response to superovulation might result in 8–10 oocytes (and 5–7 normal embryos); however, in 30% of cases, 0 or 1 embryo is produced. On the other hand, the same superovulatory treatment results in 30 or more normal embryos in about 2% of superovulations.

II. RECOVERY OF EMBRYOS

Except when embryos are to be fertilized *in vitro*, the donor female is bred naturally, or more frequently, inseminated artificially with semen from a valuable male. For the first few days after fertilization, the embryonic cells divide every 12–24 hr, and thus are at 2-, 4-, 8-, etc., cell stages. These early cell divisions take place in a tubular structure, the oviduct (Fig. 1). About 3 or 4 days after fertilization, the embryos pass from the oviduct into the uterus, where they normally continue developing and eventually implant.

To recover embryos from the oviduct efficaciously requires surgery in most species. This is sometimes done for research purposes, but in most cases one waits until embryos have migrated to the uterus because recovery procedures are less invasive. Recovery from the uterus is done by flushing it with fluid to rinse out the embryos. The fluids for maintaining the embryo have many of the same ingredients as blood and reproductive tract fluids, mainly water and sodium chloride.

In cattle, horses, goats, humans, and several other species, tubing to recover embryos can be inserted nonsurgically through the vagina and cervix into the uterus (Fig. 1). In pigs, sheep, small laboratory animals, and most zoo animals, embryos usually are recovered surgically because of the difficulty in entering the uterus through the cervix. For both surgical and nonsurgical methods, the routine rate of recovery is 70–80% of embryos present. Embryos may be recovered any time prior to implantation, although they are easily damaged if too large (e.g., after Day 14 in cattle).

Oocytes can be recovered before they are fertilized by two methods: (i) from the oviduct after ovulation or (ii) from the preovulatory ovarian follicles just prior to ovulation. The latter procedure is used almost exclusively and is done by inserting a long needle into the body cavity either through the abdominal wall or, more commonly, through the vaginal wall. In most cases, ultrasonography is used to visualize the process so that the needle can be inserted into the follicle correctly. Oocytes are then recovered by aspiration of the follicular contents and

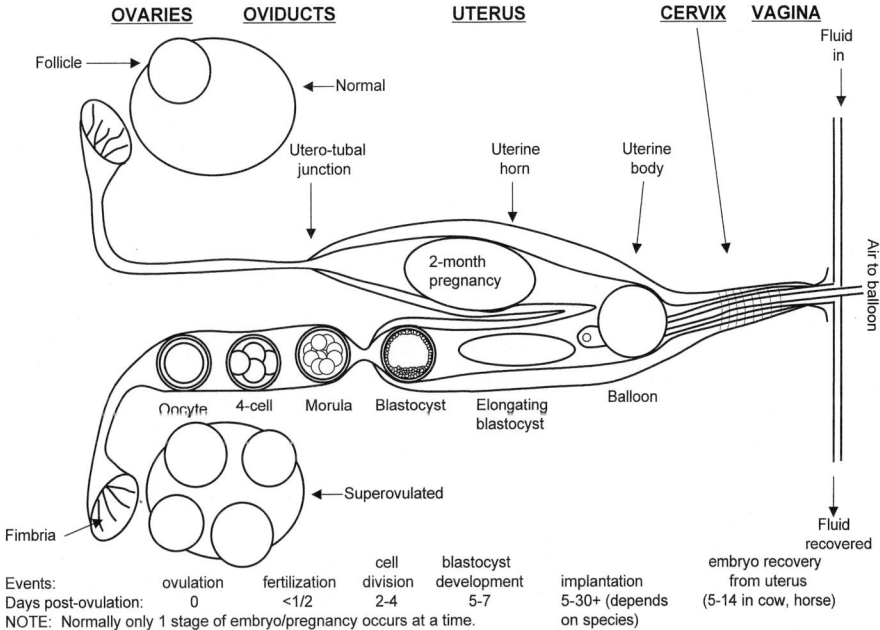

FIGURE 1 Anatomy (not to scale) and selected physiology and embryology of the bovine reproductive tract.

fertilized *in vitro*. *In vitro* fertilization is discussed in detail elsewhere in this encyclopedia.

III. MAINTENANCE OF EMBRYOS *IN VITRO*

Embryos frequently are maintained *in vitro* for several hours to half a day between collection and cryopreservation or transfer. However, in the case of *in vitro* fertilization in cattle, embryos usually are maintained for an entire week in a body-temperature incubator so that they develop to late morulae or blastocysts; these stages result in best pregnancy rates with nonsurgical transfer to the uterus. Unless embryos are cryopreserved, the longer that embryos are kept *in vitro*, the more important it becomes to simulate conditions in the female reproductive tract. Ideal conditions for embryo development include maintaining body temperature of the species, a 5% CO_2 atmosphere, a 5–8% O_2 atmosphere, pH slightly above 7.0, and the absence of microorganisms. In addition, the fluid culture should have the appropriate concentration of cations (e.g., sodium, potassium, calcium, and magnesium), anions (e.g., chloride and phosphate), energy sources (e.g., glucose and lactate), amino acids (e.g., glycine), and protein (e.g., serum albumin). For prolonged culture, attention also needs to be paid to vitamins, minerals, and antioxidants. Often developing embryos are cocultured with somatic cells from the oviduct, ovary, or cell lines maintained *in vitro*.

Other chemicals, such as glycerol or ethylene glycol, are required if embryos are to be cryopreserved. Cryopreservation of embryos is discussed elsewhere in this encyclopedia, so only a few points will be made. Cryopreserved embryos are stored in liquid nitrogen at −196°C. It is unclear how long embryos can be stored in liquid nitrogen and still be viable after thawing; some have been maintained successfully for longer than 20 years. Storage for the equivalent of two millennia has been simulated by bombarding liquid-nitrogen canisters containing cryopreserved embryos with 400 times natural background radiation; normal young resulted when the embryos were thawed and transferred. It appears that considerable damage would occur to embryos or any other cells stored for 10,000 years or longer due to radiation damage from naturally occurring ^{40}K.

IV. EMBRYO TRANSFER

Like embryo recovery, transfer can be performed surgically or nonsurgically, depending on the species. As a rule, embryos can be transferred successfully nonsurgically in species from which embryos can be recovered nonsurgically. The nonsurgical transfer is done very similarly to artificial insemination and basically involves inserting a tube through the vagina and cervix to deposit the embryo(s) into the uterus. Surgical transfer, of course, involves anesthesia, making an incision, and either making a puncture wound in the uterine wall or threading tubing into the opening of the oviduct to deposit embryos (Fig. 1).

Success of embryo transfer depends on several principles. First, early stage embryos should be transferred to the oviduct and later stage embryos to the uterus. Second, the reproductive cycle of the recipient should be at the same stage as that of the donor at the time of embryo collection. In most cases, asynchrony of 1 day is tolerated well, but asynchrony of 2 or more days results in low pregnancy rates. Because of this, any successful embryo transfer program requires huge efforts in dating and/or hormonally synchronizing reproductive cycles. For most species that normally ovulate a single ovum, a third principle is that embryos must be transferred to the uterine horn adjacent to the ovary that ovulated to obtain high pregnancy rates.

How successful are embryo transfer procedures? In cattle that are naturally mated or are on well-managed artificial insemination programs, the pregnancy rate averages 70% (50% in lactating dairy cattle). In well-managed embryo transfer programs with cattle, pregnancy rates are 70% (65% with cryopreserved embryos). In young human couples having regular, unprotected intercourse, pregnancy rates are approximately 30% per reproductive cycle; pregnancy rates from transfer of single, high-quality embryos to fertile, young women are also about 30%. Similar data also are available from other species. That is, pregnancy rates with well-managed embryo transfer programs are similar to pregnancy rates with conventional reproduction.

V. APPLICATIONS OF EMBRYO TRANSFER AND RELATED BIOTECHNOLOGIES

Currently, more embryo transfer is done in cattle worldwide than in all other species combined, with nearly half a million embryos transferred annually. The main application in this species is to amplify the reproductive rates of valuable cows. Normally, a cow only has one calf a year. By combining superovulation and other embryo transfer procedures, it is common for a donor cow to have 10–20 or more genetic offspring annually, without giving birth to any of them herself. Similarly, increasing reproductive rates of genetically or commercially valuable donors is practiced in many other species of farm animals as well as in some zoo animals and endangered species.

A second major application of embryo transfer is to circumvent infertility. For example, due to disease or injury, there may be blockages in the reproductive system that can be circumvented. Another common cause of infertility is due to age. In many species, the uterus becomes senescent before the oocytes. One can transfer embryos from such females to younger females for gestation to term. Although both the ovum and the uterus age in all species, in the human it appears that the ovum ages more rapidly than the uterus. Circumventing infertility is the major use of embryo technology in human beings. The annual number of embryo transfers for this purpose, usually with embryos derived from *in vitro* fertilization, currently exceeds 100,000 worldwide. Interestingly, up to 20% of human embryo transfer is done to circumvent infertility in the male partner; fewer normal sperm are required for fertilization *in vitro* than with natural mating. Use of embryo transfer to circumvent infertility also is important for aging horses and cattle.

Another extremely important application of embryo transfer is to move germ plasm from one area of the world to another. Transportation of large animals over long distances is complicated, expensive, and sometimes debilitating to the animals. An even more important consideration is that it is easy to introduce new diseases when transporting animals from one area to another. Because the early embryo is surrounded by a gelatin-like capsule, the zona pellucida, it is impermeable to most viruses. Additionally, embryos can be treated with antibiotics and antiviral agents, effectively sanitizing them in a way that is impossible with postnatal animals. Internationally accepted protocols have been developed for safe import and export of embryos. Not a single case of disease transmission has been reported in relation to international commercial trading of more than 100,000 embryos. Thus, introducing new genetic material via embryos, usually cryopreserved, is relatively inexpensive and avoids many problems.

Embryo transfer is an essential step for dozens of other applications, including some forms of prenatal genetic testing, producing clones, producing transgenic animals, and many experimental techniques, including use of prepubertal animals as donors. People are fascinated by embryo transfer. It is a simple elegant approach to tricking Mother Nature in a way that is easy to explain. Since the first transfer of an embryo in a rabbit in 1890, millions of offspring have been born in dozens of species. Except for a few special situations, the incidence of abnormalities in these offspring has been no higher than that of natural reproduction. Undoubtedly, embryo transfer will continue to be used in agriculture, medicine, and conservation programs.

See Also the Following Articles

Artificial Insemination, in Animals; Cattle (Bovidae); In Vitro Fertilization; Reproductive Technologies, Overview; Sperm Transport

Bibliography

Adams, C. E. (Ed.) (1982). *Mammalian Egg Transfer*. CRC Press, Boca Raton, FL.

Betteridge, K. J. (1981). An historical look at embryo transfer. *J. Reprod. Fertil.* **62**, 1–13.

Betteridge, K. J. (1986). Increasing productivity in farm animals. In *Reproduction in Mammals: 5. Manipulating Reproduction* (C. R. Austin and R. V. Short, Eds.), pp. 1–47. Cambridge Univ. Press, Cambridge, UK.

Edwards, R. G., and O'Brody, S. A. (1995). *Principles and Practice of Assisted Human Reproduction.* Saunders, Philadelphia.

Evans, J. W., and Hollaender, A. (Eds.) (1986). *Genetic Engineering of Animals.* Plenum, New York.

International Embryo Transfer Society (1978–present). Proceedings of annual conferences. (January issues of Theriogenology; Available from Elsevier or IETS, 1111 N. Dunlap Avenue, Savoy, IL 61874)

Kuzan, F. B., and Seidel, G. E., Jr. (1986). Embryo transfer in animals. In *Manipulation of Mammalian Development* (R. B. L. Gwatkin, Ed.), pp. 249–278. Plenum, New York.

Mastroianni, L., Jr., and Biggers, J. D. (Eds.) (1981). *Fertilization and Embryonic Development in Vitro.* Plenum, New York.

Seidel, G. E., Jr. (1981). Superovulation and embryo transfer in cattle. *Science* **211**, 351–358.

Seidel, G. E., Jr., and Seidel, S. M. (1981). The embryo transfer industry. In *New Technologies in Animal Breeding* (B. G. Brackett, G. E. Seidel, Jr., and S. M. Seidel, Eds.), pp. 41–80. Academic Press, Orlando, FL. (Also available in Spanish)

Seidel, G. E., Jr., and Seidel, S. M. (1989). Analysis of application of embryo transfer in developing countries. *Theriogenology* **31**, 3–16.

Willadsen, S. M. (1986). Nuclear transplantation in sheep embryos. *Nature* **320**, 63–65.

Endocrine Disruptors

see Environmental Estrogens

Endoderm

see Germ Layers

Endogenous Opioids

J. A. Russell, C. H. Brown, and R. W. Carón

University of Edinburgh

I. Exogenous Opiates
II. Endogenous Opioid Peptides
III. Endogenous Opioids and Reproduction
IV. Conclusion

GLOSSARY

affinity A measure of the ability of a moiety, or ligand, to bind reversibly to a receptor, and indicating potency or effectiveness. The dissociation constant, K_D, is the concentration of the ligand at which 50% of the receptors are occupied and ideally at which the effect is also 50% of maximum. The lower the K_D, the more effective or potent is the ligand: Opioids and their antagonists bind to the opioid receptors with K_D in the subnanomolar range.

arcuate nucleus A group of several different types of neuron in the ventral hypothalamus, adjacent to the median eminence, including dopamine (tuberoinfundibular dopamine) neurons projecting to the hypothalamo–hypophysial portal vessels to regulate prolactin secretion, and β-endorphin neurons projecting to other brain regions.

dynorphins Octa- and heptadecapeptide endogenous opioids derived from the proenkephalin B precursor (preprodynorphin).

β-endorphin A 31-amino acid endogenous opioid peptide derived from the proopiomelanocortin precursor.

enkephalins Pentapeptide endogenous opioids (Met5- or Leu5-enkephalin) derived from the proenkephalin A precursor (also called preproenkephalin).

opiate A drug such as morphine; one of the many similar natural alkaloids present in extracts of the opium poppy (*Papaver somniferum*), having a range of actions through opioid receptors, reversed or prevented by naloxone.

opioid A peptide produced by neurons, and other cell types, having actions such as those of morphine (which is also an opioid), belonging to one of three groups: enkephalins, endorphins, or dynorphins.

opioid receptor Seven-transmembrane domain protein in the plasma membrane of opioid-sensitive cells to which opioid agonists or antagonists bind; agonist binding activates a G protein link to intracellular regulating mechanisms. There are three types (δ, κ, and μ) derived from distinct genes and with different pharmacological properties.

preoptic area Anterior to the hypothalamus, with sex steroid-sensitive and opioid neurons; receives inputs from β-endorphin neurons; locus of gonadotropin-releasing hormone neuron cell bodies; involved in organizing expression of male and female sexual behavior, and maternal behavior.

receptor downregulation Reduction in the number of functioning receptors on a cell, reducing the maximal effect of an opioid agonist (desensitization); may result from inactivation of the receptor gene, phosphorylation of receptor, binding of an irreversible antagonist, or agonist-induced internalization.

selective opioid agonist A moiety, or ligand, that binds selectively to δ, κ, or μ receptors with high affinity, activating postreceptor mechanisms so that only cells expressing the target receptor are affected by the opioid.

selective opioid antagonist A moiety that binds selectively to δ, κ, or μ receptors without activating postreceptor mechanisms and blocks the binding and action of an agonist. Naloxone is not selective because it acts as an antagonist on all three types of opioid receptor. A competitive antagonist, such as naloxone, can be overcome by greatly increasing the concentration of opioid agonist.

selectivity Property of an opioid receptor agonist or antagonist with a much greater affinity for one of the types of opioid receptor. Typically, the K_D is >1000-fold less than that for the other two receptor types.

supraoptic nucleus Adjacent to the optic tracts in the ventral hypothalamus, containing magnocellular oxytocin and vasopressin neurons projecting to the posterior pituitary gland; these neurons also produce opioid peptides and express opioid receptors.

ventromedial nucleus A major group of neurons in the hypothalamus, including estrogen-sensitive enkephalin neurons; a site of oxytocin receptors; involved in expression of female receptive behavior.

Endogenous opioids, which are peptides produced in the body and especially in the brain and which have actions like those of opiate drugs (e.g., morphine), are now known to have important roles in regulating several neuroendocrine mechanisms that are pivotal in reproduction. Much of the information about the actions of endogenous opioids on these mechanisms comes from studies of the effects of opiate drugs on reproductive processes. This article includes an account of the properties of opiate drugs and the classification of three classes of opioid receptors upon which they and the endogenous opioids act as well as an outline of the nature of the three classes of endogenous opioids and their sites of production. All three classes of endogenous opioids are involved in regulating reproductive processes, acting in particular through κ and μ opioid receptors. The actions of opiates and the roles of endogenous opioid peptides are particularly well described for three reproductive neuroendocrine systems: (i) the regulation of gonadotropin-releasing hormone (GnRH) neurons and thus of luteinizing hormone and follicle-stimulating hormone secretion, gametogenesis, sex steroid production, and sexual behavior (ii) the control of the secretion of oxytocin from the posterior pituitary gland, with a particular role for endogenous opioids in moderating oxytocin secretion in pregnancy and parturition; and (iii) the regulation of prolactin secretion from the anterior pituitary gland by actions on the neurons producing hypothalamic factors controlling prolactin secretion, which is especially important in lactation. The actions of opioids on GnRH and oxytocin secretion are almost uniformly inhibitory, although the endogenous opioid mechanisms are not always active since they are sensitive to the levels of sex steroid hormones. In contrast, opioids generally stimulate prolactin secretion, and thus milk synthesis and secretion, yet opioids can have inhibitory and disruptive effects on maternal behavior.

I. EXOGENOUS OPIATES

A. Agonists: Receptors

The antinociceptive, soporific, and euphoric effects of extracts of the opium poppy *Papaver somniferum* have been known to man for many hundreds of years. Depressant effects on reproductive functions are well recognized. The active natural compounds, the opiates, are alkaloids and include morphine. Many synthetic derivatives have been produced in a search for compounds with specific, selective effects and without the potential to lead to addiction (heroin, which is widely abused and more addictive than morphine, was originally advertised as free from addictive properties). Careful comparisons of the effects of these compounds in pharmacological studies on different tissues, including the brain, led to the concept that there are three principal opiate receptor types (μ, morphine; δ, detected in vas deferens; and κ, binding *k*etocyclazocine), whereas subtypes ($\mu 1$, $\mu 2$, etc.) were defined pharmacologically within these types. A further receptor type (ε) was proposed to explain the actions of β-endorphin, whereas some opiates act atypically, with effects not reversed by naloxone, and the σ receptor was thus defined. Thus, the agonists are classified according to their affinities at the different receptors (morphine and, after metabolism, heroin are μ agonists; U50,488 and U69,593 are κ agonists).

B. Antagonists

Opiate antagonists were developed to treat patients suffering from morphine overdose (i.e., respiratory depression), of which naloxone remains the most effective. Naloxone antagonizes opiate actions at μ, δ, and κ receptors, but it has greatest potency on μ receptors. Naloxone freely crosses the blood–brain barrier and begins to act within a few seconds of systemic injection to reverse central morphine actions; naltrexone is similar to naloxone but longer lasting. There are antagonists with high affinity and selectivity for the κ receptor (*nor*-binaltorphimine, of which the first available, MR2266, has poorer selectivity for κ over μ receptors) and δ receptors (including naltrindole, 7-benzylidene naltrexone, and naltriben, but H-Dmt-Tic-DH is the most effective, with very high subnanomolar affinity, selectivity, and antagonist potency); β-funaltrexamine is an irreversible μ antagonist.

II. ENDOGENOUS OPIOID PEPTIDES

The knowledge about high-affinity selective binding sites in the central nervous system and elsewhere led to the postulate that there are endogenous ligands for these receptors, and the endogenous opioid peptides were in due course isolated and characterized in the mid-1970s. Subsequent analysis of their precursor peptides revealed three families, with considerable homology to each other, so that the same active peptide sequences are present in more than one precursor. These endogenous peptides are called opioids to distinguish them from the alkaloid or synthetic opiates (which are also generically opioids). The classical opioid peptides are (i) the pentapeptides Met5-enkephalin and Leu5-enkephalin, products of the proenkephalin A precursor, but these sequences are also in the proopiomelanocortin (POMC) or proenkephalin B (prodynorphin) precursors, so their production depends on the selective processing of the precursor by endopeptidases in the cells expressing the proenkephalin A gene; (ii) β-endorphin (31 amino acids), one of the active products [along with adrenocorticotrophic hormone (ACTH), αMSH, LPH, and CLIP] of the POMC precursor; and (iii) dynorphins, principal products of the proenkephalin B (or prodynorphin) precursor, processed to yield dynorphin A_{1-8} or A_{1-17}. The gene sequences and chromosomal locations for these three precursors are known. Of relevance to reproductive processes is the presence of an estrogen response element in the upstream regulatory sequences of the proenkephalin A gene.

Recently, novel tetrapeptides with opioid actions have been identified through a combinatorial chemistry approach; these peptides, named endomorphins 1 and 2, are present in the brain, including the hypothalamus. The precursor and gene for these peptides are unknown.

Another heptadecapeptide, with homology to dynorphin A_{1-17}, has been identified. This peptide, called nociceptin or orphanin FQ (OFQ), is the product of a precursor and gene distinct from proenkephalin B but with high homology. However, this peptide does not act on opioid receptors (it is not antagonized by naloxone), so it is "opioid-like" but not an opioid.

Each endogenous opioid peptide generally has poor selectivity for the different types of opioid receptor, although they are potent. An exception are the endomorphins, which have marked selectivity for the μ receptor, whereas dynorphin A_{1-17} shows preference for the κ receptor, the enkephalins prefer μ or δ receptors, and the extended enkephalins (Met-enkephalin-Arg-Phe and Met-enkephalin-Arg-Gly-Leu), products of the proenkephalin A precursor, have higher affinity for the κ receptor.

Synthesis of peptides modified from the natural ligands has yielded a range of more receptor type-selective agonists and antagonists. CTAP (D-Phe-Cys-Tyr-D-Trp-Arg-Thr-Pen-Thr-NH$_2$) is a selective μ antagonist, [D-Ala^2, N-methyl-Phe^4, Gly^5-ol]-enkephalin is a selective μ agonist, and [D-Penicillamine^{2,5}]-enkephalin is a selective δ agonist.

The endogenous opioids are subject to degradation by peptidases, and this is one mechanism to limit their duration of action. Drugs that inhibit these peptidases have been developed, principally to promote antinociceptive actions of endogenous opioids.

A. Opioid Receptors

The conclusion from pharmacological studies that there are three types of opioid receptor has been validated by the identification of the amino acid sequences of the receptors, following characterization of their mRNAs in 1993. The μ, κ, and δ opioid receptors are members of the seven-transmembrane domain G protein-coupled receptor family. A distinct ε receptor has not been identified. The opioid receptor sequences are highly conserved across vertebrates but are not found in invertebrates. There are single copies of each gene, and their chromosomal locations are known (the gene for the μ receptor is on chromosome 10 in the mouse, that for the κ receptor is on chromosome 1, and that for the δ receptor is on chromosome 4). Each opioid receptor has ~400 amino acids, with ~60% identity between types and ~90% identity between species for each type.

Different subtypes of receptor (revealed by earlier pharmacological studies) can arise by alternative splicing of the mRNA. The μ receptor has two splice variants (A and B): B is translated from a mRNA with an additional exon. Like other G protein-coupled receptors, following agonist binding, μ receptors are internalized, and this involves an internal C-terminus domain on the receptor. Expression of the μ opioid receptor gene can be reduced by chronic agonist (morphine) treatment. Regulatory elements of the receptor genes include sites sensitive to protein kinase A (PKA) and PKC pathway activation. The cytoplasmic domains have phosphorylation sites that are probably important in regulated desensitization (contributing to tolerance).

A fourth member of this family which has low affinity for opioids or naloxone but has homology to the κ receptor has been identified (from the cDNA studies which led to characterization of the classical μ, δ, and κ receptors) and named opioid receptor-like1 receptor (ORL$_1$). The high-affinity endogenous decaheptapeptide ligand, nociceptin or orphanin FQ, was then isolated from brain. A selective antagonist is now available for ORL$_1$.

B. Mechanisms of Opioid Actions on Neurons

Opioids inhibit the firing of action potentials and thus reduce the secretion of neurohormone or transmitter by the neurons acted on by opioid. There are two kinds of action: presynaptic, or preterminal, and postsynaptic actions. Presynaptic action involves reduced release of transmitter with depolarization by action potentials, either through hyperpolarization or reduced Ca^{2+} entry or mobilization. Thus, the amplitude of excitatory postsynaptic potentials is decreased, reducing the probability of firing. Postsynaptic inhibition can result through the same mechanisms in the postsynaptic cell, directly reducing excitability; in addition, inhibition of cAMP generation may be involved. μ Receptors may be directly coupled via $G_{i/o}$ protein to activate an outward K^+ conductance, leading to hyperpolarization or reduced excitability; κ receptors may be coupled by a different mediator, with similar effects, and both may inhibit a voltage-activated Ca^{2+} channel.

In the posterior pituitary, κ opioids inhibit oxytocin secretion by opening K^+ channels in neurosecretory nerve terminals, making action potentials less effective; attenuation of the increase in cytoplasmic Ca^{2+} caused by the depolarization as action potentials arrive and trigger exocytosis appears not to be involved. However, κ agonists (but not μ or δ agonists) decrease the peak amplitude of Ca^{2+} currents (especially L type but also N- and P-type currents) in neurosecretory nerve terminals, with decreased exocytosis, measured by capacitance change.

C. Endogenous Antiopioids

It is appropriate to consider whether the predominantly inhibitory actions of endogenous opioids may have counterbalancing endogenous antiopioid partners.

Several endogenous neuropeptides have been proposed to have antiopioid actions. These include melanocyte-inhibiting factor (MIF) and ACTH (a product of POMC) which may act via opioid receptors. In studies on nociception, centrally administered cholecystokinin$_{1-8}$ (CCK) has antiopioid actions, but via CCK and not opioid receptors; CCK prevents the inhibitory effect of β-endorphin on maternal behavior. FMRFamide (Phe-Met-Arg-Phe-NH$_2$), related to neuropeptide FF (an amidated octapeptide, F8Famide) is found in the brain and has antiopioid effects on nociception but acts via a distinct, nonopioid receptor. Immediately following its discovery, behavioral tests indicated that OFQ (nociceptin) has pronociceptive, antiopioid actions but via ORL$_1$ not opioid receptors. To date, effects of OFQ on defined neurons relevant to reproduction (oxytocin and β-endorphin neurons) show inhibitory effects; in the case of β-endorphin neurons such an action is functionally "antiopioid."

D. Tolerance and Dependence

A consequence of chronic administration of the μ agonist morphine is the development of tolerance and dependence; these are features of addiction. Tolerance is a requirement for increased dose of opiate to achieve the same effect or substantial loss of effect of the same dose. Dependence is the development of an adaptation such that acute withdrawal, usually by administration of naloxone, leads to an amplified mirror image of the acute effect. Assuming an initial inhibitory action of the opioid, withdrawal leads to a large excitation of activity. Mechanisms underlying tolerance and dependence are not clear. In tolerance, downregulation of opioid receptor expression is not a constant feature, and inactivation or uncoupling of receptors (e.g., by phosphorylation) may be involved. To explain dependence, some compensatory overactivity of excitatory mechanisms opposing the inhibitory effect of the opioid has been proposed which is revealed when the opioid occupation of receptors is antagonized by naloxone; the nature of the compensation is not clear. Whether tolerance and dependence occur during prolonged exposure to endogenous opioid is uncertain, but this could produce altered excitability states of opioid-sensitive neurons.

In contrast, chronic antagonist administration (naloxone and naltrexone) can lead to agonist supersensitivity through upregulation of opioid receptors; however, this does not seem to involve increased gene expression.

III. ENDOGENOUS OPIOIDS AND REPRODUCTION

As a general principle, while actions of exogenous opioids, appropriately disposed opioid peptide neurons and processes, and evident opioid receptors in relevant brain regions can all indicate a capacity for endogenous opioids to influence neural networks regulating reproductive processes, the critical test is to study the consequences of blocking actions of endogenous opioid. In principle, this can be achieved with opioid receptor antagonists, immunoneutralization of opioid peptides, antisense oligonucleotide knockdown of opioid peptide or receptor gene expression, or deletion or inactivation of the respective genes.

The principal roles of endogenous opioids in relation to reproduction are in the brain, with the opioids being synthesized by specific sets of neurons, and generally only one of the genes is expressed per opioid neuron. The direct effects of opioids on other neurons are inhibitory; however, since the neurons affected may also be inhibitory, it is possible for opioids to have a net stimulating effect on a neural circuit, although this is unusual. Some of the neurons affected by opioids are themselves the source of the opioid, and these then have autoregulatory inhibitory actions.

There are cells expressing the POMC gene in both the anterior and intermediate lobes of the pituitary gland, but ACTH is the principal product (β-endorphin and αMSH are also secreted) of the former, and β-endorphin (although it is acetylated and thus not active as an opioid) of the latter. The anterior pituitary essentially lacks opioid receptors, so it is unlikely to be a site of action of opioids.

One of the characteristics of brain opioid systems is that they show adaptation or plasticity. One underlying mechanism involves the development of tolerance and dependence with continual exposure, whereas responsiveness of opioid peptide genes to sex steroids leads to changes in expression with reproductive state, and altered expression, or properties, of opioid receptors on target neurons is another mechanism.

A. Opioids and Gonadotropin-Releasing Hormone Neurons: Luteinizing Hormone Secretion

The involvement of endogenous opioid peptides in the regulation of luteinizing hormone (LH) secretion can be postulated on the basis that heroin addicts are hypogonadal and/or amenorrhoeic. While opioid inhibition of LH secretion has been demonstrated in the human, the most extensive investigations of opioid–LH interactions have been carried out using rats as a model for spontaneous ovulation.

The synthesis and secretion of both LH and follicle-stimulating hormone (FSH) from anterior pituitary gonadotrophs is stimulated by gonadotropin-releasing hormone (GnRH). The neurons which secrete this decapeptide are located in the hypothalamic preoptic area and project to the median eminence where they release GnRH into the hypophysial portal blood system.

In both males and females, GnRH secretion, and

consequently LH secretion, is pulsatile. Pulsatile activity may arise from rhythmic activity at synaptic inputs to GnRH cells, generated by reciprocal interactions among the GnRH neurons and their inhibitory and excitatory inputs. However, since immortalized GnRH-producing neurons release GnRH in an episodic pattern, pulsatility may be an intrinsic property of GnRH neurons.

In intact male rats, the timing and amplitude of LH pulses are irregular. Castration causes an acute increase in LH secretion by increases in pulse frequency and amplitude which are reversed by testosterone administration. Castration-induced LH hypersecretion is associated with an increased frequency of GnRH pulses. However, the mean level of GnRH secretion is unchanged because the GnRH pulses are of smaller amplitude. The acute LH rise which follows castration may be attributed in large part to the removal of testosterone suppression of pituitary responsiveness to GnRH.

Ovariectomy also results in an increase in LH pulse frequency and amplitude. The basic mechanisms underlying acute gonadectomy-induced rises in LH secretion appear to be similar in males and females; release of pituitary gonadotrophs from gonadal steroid suppression of responsiveness to GnRH is the primary mediator. Nonetheless, increases in GnRH pulse frequency will underlie the increases in LH pulse frequency, indicating negative feedback effects of sex steroids on the mechanisms regulating GnRH pulses. These effects are indirect since GnRH neurons lack sex steroid receptors.

Endogenous opioid peptides inhibit LH secretion by reducing GnRH release, and the opioid receptor antagonist, naloxone, stimulates GnRH release *in vivo*, especially in the luteal phase, and *in vitro*, indicating that GnRH neurons are under tonic endogenous opioid inhibition. Since opioid antagonists can reverse steroid-induced suppression of LH secretion and gonadectomy reduces endogenous opioid inhibition of LH secretion, endogenous opioids may mediate the negative feedback effects of gonadal steroids. However, unlike the effects of ovariectomy, naloxone administration increases GnRH pulse amplitude rather than frequency in intact rats and its ability to do so is not impaired in ovariectomized rats. Thus, endogenous opioids may simply reduce the amount of GnRH released during each pulse rather than reducing the intrinsic activity of the GnRH pulse-generator mechanism. However, GnRH neurons do not themselves express δ, κ, or μ receptors, but they are surrounded by neurons with these receptors so the actions of opioids will be mediated by such neurons close to the cell bodies of GnRH neurons or presynaptically, or they will be exerted indirectly at the level of the terminals of GnRH neurons in the median eminence. There are endogenous opioid (enkephalin) neurons in the anteroventral periventricular nucleus of the preoptic region, but few enkephalin, dynorphin, or β-endorphin terminals contact GnRH neuron cell bodies. A subset of β-endorphin neurons projects to the medial preoptic area. Endogenous opioid-containing nerve fibers are sparse in the vicinity of GnRH fibers in the median eminence. One established mechanism for opioids to influence GnRH neurons is via preterminal inhibitory action on noradrenergic synapses on GnRH neurons, and another involves inhibition of NO generation.

1. Arcuate Nucleus POMC Neurons

POMC neurons in the brain are almost all, apart from a few in the nucleus tractus solitarius in the brain stem, aggregated in the arcuate nucleus in the ventral part of the hypothalamus. These neurons thus have long axonal projections from the arcuate nucleus to the multiple sites in the brain where the axonal terminals release β-endorphin. They project to the preoptic area to influence maternal behavior and GnRH neurons and are sensitive to estrogen and progesterone. POMC mRNA expression in the rostral arcuate nucleus varies in the rat estrous cycle, with a low level at proestrus, rising after the LH surge; this pattern is consistent with an inhibitory action of β-endorphin on GnRH neurons after the surge and is mimicked in ovariectomized rats by estrogen and progesterone priming leading to an LH surge. Many POMC neurons in the arcuate nucleus have estrogen receptors in a range of mammals, and estrogen treatment in ovariectomized animals induces progesterone receptors. The initial effect of estrogen treatment in ovariectomized animals is to stimulate POMC gene expression. Similarly, in castrated males the decline in rostral arcuate nucleus POMC expression is reversed by testosterone treatment.

POMC expression in arcuate nucleus neurons is also regulated by activity of their inputs, for example, being stimulated by their tachykinin input and inhibited by local opioid action. μ Opioids inhibit the electrical activity of arcuate nucleus neurons, and this action is attenuated by estrogen. Such an effect of estrogen would release β-endorphin neurons from autoinhibition with high estrogen level, potentially increasing β-endorphin release, for example, onto GnRH neurons.

There are few POMC neurons in the arcuate nucleus of old female rats, and chronic estrogen treatment is neurotoxic selectively to this set of arcuate nucleus neurons. This effect, due to generation of free radicals, is followed by increased μ opioid receptor binding in the medial preoptic area, perhaps with increased inhibition of GnRH neurons, and could contribute to reproductive senescence. In pregnant, pseudopregnant, and lactating rats more arcuate nucleus neurons express the proenkephalin A gene; this could contribute to increased prolactin or suppressed GnRH secretion.

2. The Preovulatory LH Surge

In the female rat, every 4 days on proestrus the amplitude of the GnRH secretory pulses increases to such an extent that a surge of GnRH release is generated. This GnRH surge is the final direct stimulus for the LH surge on proestrus which stimulates ovulation and it is dependent on an increase in circulating gonadal steroids, principally estrogen. The magnitude of this preovulatory LH surge is amplified by an increase in responsiveness of the gonadotrophs to GnRH, primarily as a result of GnRH "self-priming" whereby preexposure of gonadotrophs to GnRH increases their responsiveness to subsequent GnRH exposure.

The LH surge is entrained to the light–dark cycle, occurring only on the afternoon of proestrus. While spontaneous surges occur every 4 days, it is possible to induce preovulatory-like surges in ovariectomized rats on a daily basis by administration of estrogen, indicating that the neural signal which results in the surge is generated in a circadian pattern but is only successful in producing an LH surge in the presence of the appropriate steroid milieu.

i. Endogenous Opioids Play a Pivotal Role in the Generation of the LH Surge Agonists active at μ opioid receptors decrease LH secretion by reducing GnRH secretion and block the preovulatory LH surge on proestrus, but convincing evidence of the involvement of endogenous opioids in the generation of the LH surge is the ability of naloxone to advance the timing of the surge when given on the morning of proestrus. Thus, it seems that a reduction in opioid tone may free the GnRH neurons from tonic inhibition at a time when the pituitary gonadotrophs are most sensitive to the effects of GnRH. However, blockade of GABAergic neurotransmission can also advance the timing of the LH surge in rats. Indeed, it has been proposed that there are separate LH pulse- and surge-generating mechanisms and that GABAergic systems may be involved in the LH surge-generating apparatus, whereas the ability of naloxone to advance the surge may simply reflect actions on the pulse mechanism rather than surge generation.

Nevertheless, the preovulatory LH surge depends critically on the interaction of circulating steroids with GnRH neurons. However, GnRH neurons do not possess steroid receptors so the effects of steroids on the activity of GnRH neurons must be indirect; one important set of neurons expressing estrogen receptors are nearby GABA neurons. β-Endorphin neurons of the arcuate nucleus also possess estrogen receptors. β-Endorphin levels in the medial preoptic area and arcuate nucleus fluctuate over the estrous cycle in a steroid-dependent manner, reaching a nadir on proestrus, whereas in ovariectomized rats the expression of μ opioid receptors and the density of β-endorphin fibers in the medial preoptic area are increased a day after estrogen and progesterone treatment, with parallel stimulation of the POMC gene in the arcuate nucleus neurons producing the β-endorphin. Thus, it appears that endogenous opioid peptides may provide a functional link between circulating steroids and their stimulatory influence on GnRH neurons; the elevated sex steroid concentrations of proestrus reduce the activity of β-endorphin neurons to free GnRH neurons from tonic opioid inhibition, after which opioid inhibition is reimposed.

While endogenous opioids interact directly with GnRH neurons in the generation of the LH surge, a major influence of opioids on GnRH secretion may be indirect by presynaptic inhibition of other excitatory afferent inputs. μ Agonists reduce noradrenaline efflux from the medial preoptic area *in vitro* and the naloxone-induced increase in LH secretion can be prevented by prior administration of drugs which eliminate noradrenergic activity. Thus, with low levels of estrogen, opioid activity is high, suppressing activity in a stimulatory noradrenergic input. Under the influence of circulating sex steroids, opioid activity is reduced and this frees the stimulatory input from tonic presynaptic inhibition, causing an increase in the excitatory synaptic input to GnRH neurons and consequently increasing GnRH secretion.

The endogenous opioid restraint of excitatory inputs to GnRH neurons may not be limited to noradrenergic systems. Recently, nitric oxide synthase (NOS) neurons, generating NO as a neuromodulator, have been visualized throughout the hypothalamus and are especially abundant in the preoptic area and median eminence. While GnRH neurons do not contain NOS, a significant proportion are surrounded by NOS-containing cells. These NOS cells express NMDA receptors (a class of glutamate receptor) and a NOS inhibitor suppresses NMDA-induced LH release and reduces the magnitude of the LH surge; β-endorphin neurons synapse on NOS neurons in the vicinity of GnRH neurons and naloxone increases NO efflux in the medial preoptic area and causes a concomitant rise in LH secretion. Thus, a reduction of endogenous opioid tone on preoptic NOS neurons may also contribute to the generation of the preovulatory LH surge.

Recently, rapid effects of steroids have been reported in several neuronal systems and it appears that GnRH neurons may also be sensitive to the nongenomic actions of estrogen since the time course of the acute inhibition of LH (GnRH) release by estrogen is too rapid to be explained by a purely transcriptional mechanism of action. These actions may also contribute to the generation of the preovulatory LH surge. In the guinea pig, both μ agonists and estrogen inhibit GnRH neurons by opening K^+ channels to hyperpolarize these neurons. These effects of estrogen and opioids may be synergistic; removal of opioid tone as occurs on proestrus may cause a marked reduction in the rapid inhibitory effect of estrogen on GnRH neurons and so result in an increase in excitability of GnRH neurons at a time when excitatory synaptic inputs are freed from opioid inhibition.

Endogenous opioids are intricately involved in the generation of the preovulatory LH surge at several levels; removal of the tonic opioid inhibition of the excitatory synaptic inputs to GnRH neurons as a result of the inhibitory action of the high proestrus level of estrogen on opioid neurons, when other mechanisms driving the GnRH neurons are stimulated by estrogen, will facilitate the GnRH/LH surge.

ii. Pregnancy, Seasonal Breeding, and Aging In pregnant and in seasonally breeding animals, pituitary gonadotropin secretion is reversibly suppressed. Endogenous opioids have a role in this suppression in the former but not the latter. In pregnancy, κ opioid mechanisms acting in the medial preoptic area and medial basal hypothalamus are involved in suppressing the activity of GnRH neurons. In seasonal breeders, such as the sheep, endogenous opioids are involved in regulating LH secretion in the active phase or in the regressing phase of the cycle but are not involved in suppression of LH secretion in the inhibited phase.

Similarly, while naloxone reliably increases LH secretion in adult humans with normal gonadal function, it is ineffective before puberty or after the menopause, unless estrogen and or progesterone replacement is given; naloxone is without effect on LH secretion in the amenorrhoea of anorexia nervosa.

B. Reproductive Behavior

1. Female Receptive Behavior: Estrogen Stimulation of Proenkephalin A

Although morphine reduces sexual behavior in female rats, there are inconsistent results, generally negative, with intracerebroventricular administration of endogenous opioid peptides. However, the transient postejaculatory inhibition of receptivity involves a central opioid action since it is reversed by naloxone. Otherwise, naloxone has little acute effect, except in lactating rats when it restores sexual receptivity. Nonetheless, there is other evidence for a role of Met5-enkephalin neurons in the ventromedial nucleus in the estrogen induction of lordosis behavior.

Proenkephalin A neurons are found in many regions of the brain, so their axons generally project short distances from their cell bodies, functioning as 'local' opioid neurons. The ventromedial nucleus of the hypothalamus is a key site for the initiation of female receptive (lordosis) behavior. In appropriately estrogen/progesterone-primed female rats, oxytocin facilitates lordosis by action at this site and has similar actions, along with other peptides, in the medial preoptic area and periaqueductal gray in the midbrain. In contrast with the inhibitory interactions between opioids and oxytocin neurons, central administration of Met5-enkephalin, or β-endorphin, facilitates the priming of lordosis by estrogen. Proenkephalin A is expressed in ventromedial nucleus neurons, and this is correlated with estrogen-dependent sexual behavior. An effective site for Met5-enkephalin in this behavior may be the periaqueductal gray. In ovariectomized rats, 17β-estradiol stimulates expression of the proenkephalin A gene in neurons of the ventrolateral part of the ventromedial nucleus. It is relatively ineffective in males and acts more slowly in female mice. Progesterone prolongs the stimulation by estrogen. Thus, it is not surprising that the proenkephalin A gene contains an estrogen response element to which the receptor binds.

Other factors are likely to be involved; thus, in the anteroventral periventricular nucleus 17β-estradiol rapidly induces phosphorylation of cAMP response element binding protein (pCREB) in proenkephalin A and B neurons, and genes for both precursors contain cAMP response elements. Enkephalin innervation of this region is sexually dimorphic, with a much denser Met-enkephalin-containing nerve fiber plexus in females that is regulated by estrogen.

The presence of estrogen receptor in up to 70% of proenkephalin A-expressing neurons in the rat spinal cord provides a possible means for estrogen to influence nociceptive or sensory processing, perhaps relevant to analgesia of pregnancy.

2. Male Copulatory Behavior: Penile Erection

Copulatory activity of male rats with a receptive female is acutely depressed by morphine administration, without depressing general locomotor activity. Intracerebroventricular injection of β-endorphin has similar effects, which are exerted specifically in the medial preoptic area evidently via δ opioid receptors, and without altering ingestive behavior, for example. Dynorphin A_{1-17} or Met5-enkephalin are less effective than β-endorphin. In otherwise normal males showing reduced sexual activity, naloxone intensifies copulatory activity, and a δ opioid receptor antagonist infused into the preoptic area has similar effects. This indicates that β-endorphin terminals (from arcuate nucleus POMC neurons) have the capacity to restrain male copulatory behavior by action in the preoptic area, but they are normally quiescent; activation of these neurons in adverse circumstances would inhibit male sexual activity.

Depression of sexual activity by opioids includes inhibitory effects on penile erection. In the rat, morphine (but not a κ agonist) acts in, or via, the paraventricular nucleus to prevent penile erection induced by centrally administered oxytocin seemingly by decreasing NO generation. Naloxone or naltrexone has been reported to facilitate (e.g., apomorphine-induced) or precipitate penile erection (cat and rat), including in men with idiopathic impotence, and interact with α_2 antagonists in normal men.

C. Opioids and Oxytocin

There are well-defined inhibitory actions of opioids on oxytocin secretion, and endogenous opioids have an important role in regulating the responsiveness of oxytocin neurons, although in particular reproductive states. The established roles of oxytocin are in female mammals: Oxytocin secreted from the posterior pituitary gland has a major role in promoting uterine contractions at parturition and an essential role in stimulating myoepithelial cells in the mammary gland to contract to cause milk ejection during suckling; in marsupials mesotocin substitutes for oxytocin. Oxytocin is synthesized in the cell bodies of the magnocellular neurons in the supraoptic and paraventricular nuclei in the hypothalamus and transported within their axons to be stored in the axon terminals in the posterior pituitary. Secretion of oxytocin into the systemic circulation follows the depolarization of the nerve terminals by the arrival of action potentials conducted along the axons from the cell bodies, causing Ca^{2+} influx and exocytosis of oxytocin from the neurosecretory vesicles in which it is stored. When it matters most, oxytocin is secreted from the posterior pituitary in pulses, generally several minutes apart, with each pulse being the consequence of a coordinated high-frequency burst of action potentials in all the oxytocin neurons; this is the most effective and efficient pattern of secretion of oxytocin for inducing the intermittent contractile responses of the target organs that respectively expel the fetuses and eject milk. However, during parturition the basal secretion of oxytocin is also increased. In parturition, the triggering of the bursts depends on distension of the uterine cervix or vagina by the presenting fetal part, and in lactation it depends on the suckling stimulus applied to the nipples. The afferent neural signals are processed in the spinal cord, brain stem, and hypothalamus. Clearly, opioids could modulate the secretion of oxytocin during parturition or in lactation by action at any of the steps described. In addition, opioids could act at the target organ to modify the response to oxytocin.

The cell bodies of centrally projecting parvocellular oxytocin neurons, distinct from the magnocellular neurons, are located in the paraventricular nucleus. These neurons have axon terminals in inter alia, the brain stem, ventromedial nucleus, bed nucleus of stria terminalis, and other limbic system components, in which through oxytocin receptors the peptide is involved in facilitation of female sexual receptivity or affiliative behavior, the initiation of maternal behavior, and lactation-related adaptations (modulating the milk-ejection reflex and modifying responses to stressors). Action of opioids on the parvocellular oxytocin neurons would be expected to alter expression of these central mechanisms.

1. Coexistence of Oxytocin and Endogenous Opioids

Involvement of endogenous opioids in autoregulatory mechanisms in oxytocin neurons is indicated by the coproduction of endogenous opioids of the proenkephalin A and proenkephalin B families in magnocellular oxytocin neurons; the opioid peptides are contained in the same secretory vesicles as those for oxytocin. Evidence comes from *in situ* hybridization studies of gene expression and immunocytochemistry and radioimmunoassay of posterior pituitary peptides. Direct measurement of release is difficult because of the small amounts of opioid peptides (the stores are ~1000-fold less than the oxytocin content), although depletion studies demonstrate release. The particular opioid peptides produced by the oxytocin neurons from proenkephalin A are Met5-enkephalin and the extended forms Met-enkephalin-Arg-Phe and Met-enkephalin-Arg-Gly-Leu and from proenkephalin B the dynorphins (principally processed to dynorphin A_{1-8}, but some dynorphin A_{1-17} is present).

Another important source of local endogenous opioid peptide is the magnocellular vasopressin neurons, whose cell bodies are close to those of the magnocellular oxytocin neurons and whose axon terminals are intermingled with those of oxytocin neurons in the posterior pituitary gland. The vasopressin neurons coproduce prodynorphin B, processed to dynorphin A_{1-8}. Thus, when secretion of vasopressin is stimulated, the coreleased dynorphins could act on adjacent oxytocin neurons to cross-inhibit oxytocin secretion; such a mechanism may have a survival advantage since vasopressin is secreted when homeostasis is threatened by dehydration, hypovolemia, or hypotension.

The posterior pituitary contains dynorphins (principally A_{1-8}, but A_{1-17} in some species) in all mammals studied (guinea pig, pig, and rhesus monkey), as does that of the axolotl but not the chicken, which contains only Met5-enkephalin. Within the posterior pituitary, some nerve terminals of nonmagnocellular neurons contain Met5-enkephalin, as do the modified glial cells (pituicytes), so they could contribute to a pool of opioids in the environment of oxytocin terminals.

2. Opioid Receptors and Oxytocin Neurons

Information about the presence and distribution of opioid receptors on magnocellular oxytocin neurons is derived from radioligand or biotinylated ligand binding studies, *in situ* hybridization for detection of the respective mRNAs, and immunocytochemistry using antibodies against synthesized receptor peptide fragments; functional studies (discussed later) confirm activity.

The posterior pituitary contains a high density of κ opioid receptors in all species examined (human, pig, and rat), and only the human shows μ receptors. The κ receptors may be on both oxytocin and vasopressin nerve terminals, but those on oxytocin terminals are functionally important. As would be expected, the cell bodies of magnocellular oxytocin neurons also have κ receptors and a lower density of μ receptors, which are also on presynaptic terminals of certain inputs on the neurons, especially of noradrenergic neurons projecting from the brain stem. The cell bodies and posterior pituitary lack δ receptors.

3. Opioid Peptide Neurons and Oxytocin Neurons

i. POMC Neurons POMC is expressed in two loci which may influence oxytocin neurons. First, the pars intermedia of the pituitary produces opioid peptides from POMC, including β-endorphin; the intimate relations of the pars intermedia and posterior pituitary suggest that β-endorphin from the pars intermedia may act on oxytocin terminals in the posterior pituitary, although this is not supported by *in vitro* studies.

At the level of the cell bodies of oxytocin neurons, β-endorphin-containing fibers from the arcuate nucleus POMC neurons pass close to the supraoptic nucleus, and a few fibers enter the nucleus.

ii. Enkephalin Neurons Met-enkephalin-containing parvocellular neurons in or close to the paraventricular, or supraoptic, nucleus may contribute to local endogenous opioid circuitry regulating magnocellular oxytocin neurons. Enkephalins are also coproducts of brain stem neurons in the nucleus tractus solitarii projecting to magnocellular oxytocin neurons and could thus modulate the actions of noradrenaline (or other cotransmitters, e.g., neuropeptide Y, purines, somatostatin, and inhibin β) on oxytocin neurons by pre- or postsynaptic actions.

iii. Endomorphins The precise topography of endomorphin neurons and terminals in relation to oxytocin neurons is still to be described. Endomorphins are present in the hypothalamus, and because they are selective μ-opioid receptor agonists they may be important in endogenous opioid regulation of oxytocin neurons.

iv. Endogenous Opioid Antagonist? From studies of antinociceptive actions of opioids, both CCK8S and OFQ (nociceptin) have been proposed to have antiopioid actions. Interestingly, CCK8S is a coproduct of oxytocin neurons, but although CCK8S has direct excitatory actions on magnocellular oxytocin neurons, it has not been demonstrated that release of CCK8S by oxytocin neurons functions to regulate the inhibitory actions of opioids released in the vicinity of oxytocin neurons.

The supraoptic nucleus has ORL_1 receptors and OFQ inhibits the spontaneous firing of oxytocin neurons *in vitro* (or *in vivo* if given centrally). OFQ thus acts similarly to opioids on these neurons but without antagonism by naloxone, and currently there is no evidence for any antiopioid actions of OFQ on oxytocin neurons.

4. Opioid Actions on Oxytocin Neurons

Exogenous opiates, given systemically or by direct central administration, inhibit oxytocin secretion regardless of the stimulus. Similarly, endogenous opioids given by an appropriate route to circumvent the blood–brain barrier also inhibit oxytocin secretion. The important actions are mediated via μ or κ receptors.

i. Opiates in Parturition Although opiates have been used for many years in obstetrics to reduce pain in childbirth, it is remarkable that effects on oxytocin secretion were not considered or investigated until the late 1970s and that the first studies were on lactating mice. Prior to this, a substantial literature on the actions of opiate analgesics [e.g., morphine and pethidine (Demerol and meperidine)] on the progress of human labor was established—with conflicting conclusions. Some studies concluded that opiates hastened labor (supposedly as a consequence of reducing pain and reducing anxiety), whereas others indicated the converse, with clear reduction in uterine contractile activity. Detailed studies in the rat have shown that μ (morphine and pethidine) or κ (U50,488) agonists slow established parturition by a central inhibitory action on oxytocin neurons so that oxytocin secretion or other indicators of oxytocin neuron activation (e.g., expression of the immediate early gene product Fos) are inhibited, whereas the effects of the opiate are reversed by intravenous oxytocin infusion, most effectively if given in pulses. It is unlikely that opiates have any significant action on the myometrium, except that pethidine inhibits uterine contractions directly but not through opioid receptors (pethidine was originally developed as a cholinergic antagonist). The few studies in humans show that opiates decrease oxytocin secretion in the first stage of labor. Thus, it can be expected that the use of systemic opiate analgesia in parturition will reduce oxytocin secretion and potentially decrease uterine contractile activity; however, the common use of exogenous oxytocin administration in obstetric practice will offset this action of the opiate.

ii. Opiates in Lactation Exogenous opiates inhibit the milk ejection reflex, and the fact that the reflex persists in the anesthetized rat has allowed detailed investigation of the site(s) of action. The μ agonist morphine probably does not inhibit oxytocin release by acting at the axon terminals of oxytocin neurons in the posterior pituitary, whereas this is a site of action of the κ agonist, U50,488.

Morphine acts first by inhibiting the basal, continuous firing of the cell bodies of oxytocin neurons and second by variably reducing the intensity of the intermittent burst firing during suckling. The first action reduces basal oxytocin secretion, and the second reduces the amount of oxytocin secreted in a pulse; since the reduced secretion of oxytocin per pulse is on a background of reduced basal secretion, the mammary gland myoepithelial cells fail to respond to the attenuated pulse and milk ejection is inhibited. U50,488 inhibits the firing of oxytocin neurons and also acts at the posterior pituitary. In addition to acting on the oxytocin neurons directly, or preterminally on excitatory inputs in the magnocellular nuclei, μ and κ opiates have been shown by intrathecal injection to act at the spinal cord, presumably inhibiting transmission at synapses in the dorsal horns of the barrage of excitatory input from the sensory nerve endings in the suckled nipples. Similar multiple sites of action of μ and κ opioids will explain the inhibition of oxytocin secretion by opiates in parturition.

iii. Endogenous Opioid Actions The actions of exogenous opiates on oxytocin neurons clearly demonstrate the potential effects of endogenous opioid peptides. The critical test of actions of endogenously released opioid peptides is examination of the consequences of blockade of endogenous opioid action. To date, only opioid receptor antagonists have been used to analyze endogenous opioid actions on oxytocin neurons, with most studies carried out on rats. Given *in vivo*, naloxone generally does not alter basal oxytocin secretion, but it does increase stimulated oxytocin secretion, indicating activation of endogenous opioid restraint when oxytocin neurons are excited by the range of stimuli that impinge on them. Studies in anesthetized rats, or in hypothalamic slices, show no increase in the firing rate of oxytocin neurons after naloxone, except during pregnancy.

Therefore, the predominant site of endogenous opioid action is normally the posterior pituitary. This is confirmed by studying actions of opioid antagonists on oxytocin secretion evoked by electrical stimulation of the pituitary stalk *in vivo* or *in vitro*. In either case, naloxone enhances stimulated oxytocin secretion, as does the selective κ antagonist, nor-binaltorphimine. Thus, stimulated oxytocin secretion is usually restrained by κ-selective endogenous opioid released from the oxytocin neurons themselves or possibly from adjacent vasopressin terminals.

iv. Changes in Pregnancy The posterior pituitary κ opioid mechanism is less active toward the end of pregnancy, with downregulation of κ receptors and decreased dynorphin A_{1-8} and enkephalin content in the posterior pituitary. These changes may be a result of increasing secretion in the last week of pregnancy but have the consequence of effectively increasing the efficiency of excitation–secretion coupling in the oxytocin terminals. This will aid the secretion of oxytocin in parturition.

Naloxone increases basal oxytocin secretion toward the end of pregnancy, and if the opioid mechanism at the posterior pituitary is less effective, it follows that now there must be a central endogenous opioid inhibition of oxytocin neurons. This is clearly confirmed by several lines of evidence which show that naloxone activates the cell bodies of oxytocin neurons in late pregnancy. μ Receptors are involved since the κ-selective antagonist *nor*-binaltorphimine is ineffective, whereas the endogenous opioid may act presynaptically on the noradrenergic input to the oxytocin neurons from the brain stem. The identity of the responsible endogenous opioid is not clear, although there is increased activity of arcuate nucleus β-endorphin neurons in pregnancy, and some β-endorphin fibers enter the supraoptic nucleus.

The dual changes in the endogenous opioid mechanisms impinging on oxytocin neurons—downregulation of the κ mechanisms inhibiting release at the level of the terminals in the posterior pituitary and activation of μ mechanisms inhibiting excitation of the neurons centrally—will have the consequence of enhancing stimulus-secretion coupling in the posterior pituitary and enabling central opioid braking of the responses of oxytocin neurons to their excitatory inputs. There are two functional consequences. First release of this central opioid inhibition would result in strong excitation of the oxytocin neurons, ensuring the provision of adequate oxytocin for expulsive uterine contractions. In pregnancy, blockade of the opioid restraint of oxytocin neurons by naloxone greatly increases the oxytocin secretory response to a stressor or to systemic CCK compared with virgin rats after naloxone. During parturition, naloxone increases oxytocin secretion (rat, pig, and human) and speeds up the process, indicating that some opioid restraint on oxytocin neurons persists in parturition It is possible that as a result of continuous exposure to endogenous opioid the oxytocin neurons develop dependence, as they do when exposed chronically to morphine; in this dependent state withdrawal of the opioid would contribute to increased oxytocin secretion at parturition.

Second, the central opioid inhibition of oxytocin neurons provides a new mechanism to suppress the secretion of oxytocin in response to circumstances in which delay or suspension of parturition is advantageous. Thus, environmental stress during parturition, for example, removal from the nest to an unfamiliar environment, slows the delivery of fetuses in mice, rats, and pigs, and this is correlated with reduced oxytocin secretion. Both the slowing of delivery and the reduced secretion of oxytocin caused by this stress are reversed by naloxone. Since oxytocin is a stress hormone in the rat, the inhibition of oxytocin secretion by a stressor in parturition seems paradoxical; however, the stimulatory effect on oxytocin neurons of the stressor alone may be weaker than that of the afferent positive feedback stimuli from the reproductive tract in parturition, and it is the latter that are inhibited by the stressor through opioid mechanisms.

In lactation, there is no evidence from electrophysiological studies of any persisting central endogenous opioid inhibitory action on oxytocin neurons. The milk-ejection bursts of oxytocin neurons are not affected by naloxone. At the posterior pituitary, the pulsatile release of oxytocin during suckling is variably and only transiently enhanced by naloxone in the rat, but there is emerging evidence for a facilitation by naloxone of the milk-ejection reflex in the cow.

v. *Causes of Changes in Pregnancy* There is conflicting evidence concerning whether the production of endogenous opioids by oxytocin neurons themselves is increased in pregnancy. How expression in these neurons is regulated is not known. The recent demonstration of mRNA for estrogen receptor (ER) β in oxytocin neurons suggests the possibility of direct genomic actions of estrogens on these neurons. To simulate pregnancy levels, estrogen and progesterone treatment induces enhanced opioid restraint of oxytocin secretory responses to a stressor. The reduced posterior pituitary contents of Met5-enkephalin and dynorphin$_{1-8}$ at the end of pregnancy, which probably reflects increased release, would cause the observed downregulation of the posterior pituitary κ opioid mechanisms. Prior to this, the release of κ

opioids from the oxytocin terminals will, by restraining oxytocin secretion, have contributed to the accumulation of the large store of oxytocin in the posterior pituitary by the end of pregnancy. This implies a continual drive to oxytocin neurons in pregnancy which may be provided by relaxin from the corpora lutea. Relaxin excites oxytocin neurons and is secreted maximally in the last week of pregnancy in the rat. Chronic relaxin infusion to nonpregnant rats induces endogenous opioid tone since naloxone increases oxytocin secretion after this treatment.

The increased activity of arcuate nucleus β-endorphin neurons in pregnancy is likely to be driven by the greatly increased secretion of sex steroids since some of these neurons have both ERα and progesterone receptors.

D. Other Opioid Mechanisms Emerging in Pregnancy

1. Analgesia of Pregnancy

The threshold for nociceptive stimuli is increased in pregnancy (mouse, rat, pig, and human) or pseudopregnancy (rat), and reversal by naltrexone or *nor*-binaltorphimine demonstrates involvement of endogenous opioid, in particular in acting on δ and κ receptors after local release at synapses in the dorsal horns of the spinal cord. A selective μ antagonist (CTAP) is ineffective; nonetheless a selective μ agonist (e.g., fentanyl) is used as an effective analgesic, given epidurally in human obstetric practice. Interestingly, the analgesic action of the related sufentanil in nonpregnant rats is potentiated by intrathecal progesterone, which is abundant in pregnancy, or the metabolite 5α-pregnane-3-α-ol-20-one. The opioids involved may be enkephalins or dynorphins since dynorphin antibody applied intrathecally blocks the analgesia, whereas inhibitors of enkephalin metabolism (i.e., of enkephalinase) enhance the analgesia; OFQ (nociceptin) reverses the analgesia, perhaps by inhibiting enkephalin neurons. This opioid-mediated analgesia of pregnancy, which is maximal at term, is likely to be important in restricting noxious input from the birth canal at parturition.

Pseudopregnancy (rat) also leads to opioid-mediated analgesia, whereas this can be replicated in nonpregnant rats by estrogen/progesterone treatment, indicating that sustained action of the female sex steroids induces the changes in spinal cord opioid mechanisms partly via increasing dynorphin A$_{1-17}$ production, with increased sensitivity to κ agonist as well. Although circulating β-endorphin concentration is increased in pregnancy and further increased in parturition, it is unlikely that this is important in the hypoalgesia or analgesia of pregnancy.

2. Placental Opioids

The placenta is a large, ephemeral endocrine gland producing sex steroids and a range of peptide or glycoprotein hormones as well as functioning as an exchange organ for the fetus. The products include opioid peptides, which have actions within the placenta rather than on the mother. The placenta produces dynorphin$_{1-8}$, β-endorphin, and, at least in sheep, Met5-enkephalin. β-Endorphin, but not dynorphin, release can evidently be stimulated by dopamine (via D1 receptors). However, most of the β-endorphin is acetylated and is thus inactive, whereas little of the circulating β-endorphin in pregnancy is acetylated; consequently, the maternal anterior pituitary is the probable source of the increased circulating β-endorphin in pregnancy.

The placenta expresses the κ opioid receptor, and κ agonist (including dynorphin) stimulates pulsatile chorionic gonadotropin (hCG) release from human trophoblast, probably via stimulation of trophoblast GnRH release. Endogenous dynorphin seems to be important since κ antagonist decreases hCG secretion; β-endorphin may act similarly. κ Agonists also stimulate placental lactogen release.

The placenta (dolphin, rat, and human) or amniotic fluid contain a factor that, when given orally, enhances opioid-mediated analgesia (placental opioid enhancing factor) without disrupting maternal behavior. This is clearly of interest in the assessment of the beneficial effects of placentophagia.

3. Opioids and Maternal Behavior

In many species, maternal behavior emerges only close to, or immediately after, birth of the young. Endogenous opioids have an important role, as does centrally acting oxytocin, although there are evident species differences in direction of action. Morphine or other μ agonists can disrupt maternal behavior in the parturient or lactating rat by a central mechanism, which involves interference with the suppression of aversive responses to olfactory signals from the pups. Centrally acting oxytocin facilitates maternal behavior, but it is not clear whether inhibitory actions of opioids involve inhibition of central oxytocin release or actions. The medial preoptic area is important in the organization of maternal behavior, and both μ receptor density and the density of β-endorphin fibers are maximal in pregnancy; these changes can be replicated by appropriate estrogen and progesterone treatment. Neurons in this region show inhibitory effects of morphine and show inhibition by arcuate nucleus stimulation that is prevented by naloxone. In lactation, μ opioid receptor density and β-endorphin content are reduced, so the concept has been developed that the expression of maternal behavior requires a withdrawal of endogenous opioid action in the medial preoptic area. However, naloxone (or naltrexone) can reduce maternal performance (e.g., interfere with pup cleaning and placentophagia) but not motivation, indicating that central endogenous opioid mechanisms are involved in rewarding the performance of maternal behavior. μ Opioid actions in the ventral tegmental area in rats evidently facilitate the onset of maternal behavior. Furthermore, in nonpregnant ewes primed with estrogen and progesterone and given vaginocervical stimulation to induce maternal behavior, opioids (morphine) enhance central oxytocin release and intensify maternal behavior; opioid antagonist (naltrexone) has opposite effects.

The apparent species differences may well reflect multiple sites of actions of opioids and different actions among the sites.

E. Opioids and Prolactin Secretion

Prolactin is a protein hormone secreted by the anterior pituitary gland which is essential in mammals for stimulating mammary gland maturation and for initiating and maintaining lactation. In lactation its secretion is maintained by the suckling stimulus. Its synthesis and secretion are regulated by the action on the pituitary lactotrophs of hypothalamic prolactin inhibitory factors (dopamine) and releasing factors [vasoactive intestinal peptide (VIP), oxytocin, thyrotropin-releasing hormone, and 17β-estradiol]. The regulation of prolactin secretion is unusual in that it is tonically inhibited by the hypothalamus in mammals and stimulated in all species, including humans, by a range of opioids including morphine. Studies with opioid antagonists in most species (except humans) show that endogenous opioids have an important role in the physiological stimulation of prolactin secretion. Since opioids do not have direct actions on the secretion of prolactin at the level of the anterior pituitary, opioids will act by modifying the release of the hypothalamic-releasing or release-inhibiting factors.

In rodents, prolactin is secreted at proestrus, and its secretion is further stimulated reflexly at copulation, with increased secretion continuing in early pregnancy to act as an essential luteotrophin, thus sustaining progesterone secretion and pregnancy (or pseudopregnancy). Stressors stimulate prolactin secretion, probably reflecting its cytokine-like action on the immune system.

1. Opioids and Tuberoinfundibular Dopamine Neurons

The stimulation of prolactin secretion involves inhibition of the tuberoinfundibular dopaminergic (TIDA) neurons in the mediobasal hypothalamus and/or stimulation of prolactin-releasing factor neurons, with serotonin [5-hydroxytryptamine (5-HT)] and endogenous opioids as important neurotransmitters acting on these neurons.

The principal mechanism involves inhibition of TIDA neurons. Both κ and μ opioid agonists but not δ; agonists stimulate prolactin secretion, acting via opioid receptors since selective opioid antagonists prevent their actions; treatment with an antisense oligonucleotide complementary to the μ receptor mRNA also blocks the actions of morphine. Central administration of β-endorphin can evoke secretion of prolactin quantitatively similar to that seen during suckling, and Met5-enkephalin is similarly effective. The effects of β-endorphin seem to be complex since its stimulation of prolactin secretion is prevented by δ-, κ-, or μ-selective antagonists. In male rats, restraint stress stimulates prolactin secretion, and morphine or bremazocine (κ agonist) administration produces a similar response. Since an antagonist that does not readily cross the blood–brain barrier (namelfene methyl iodate, a quaternary compound) prevents the stimulation of prolactin secretion by morphine, one site of morphine's action is probably the median eminence. Morphine or other opioids decrease the synthesis and turnover of dopamine at the median eminence and decrease dopamine concentration in hypothalamo–hypophysial portal blood; these opioid actions can be exerted through κ or μ opioid receptors. This inhibition of TIDA neurons will lead to increased prolactin secretion. The use of pharmacological blockers of 5HT action has shown that the stimulation of prolactin secretion by a selective μ agonist (sufentanil) is prevented by a $5HT_{1A}$ receptor antagonist, but the stimulation by a κ agonist (U50,488) is not. Thus, μ agonists may act indirectly via 5HT neurons, although other evidence disputes this, and κ agonists more directly. Developmental studies also indicate independent stimulatory actions of κ and μ opioids on prolactin secretion, with μ agonist action via a 5HT input to TIDA neurons, and another perhaps direct action of κ agonists. In male rats, μ agonist (but not β-endorphin) stimulation of prolactin secretion is shown by pharmacological studies to involve histamine (H_2, not H_1) receptors.

The persisting stimulation of prolactin secretion by morphine with blocked dopamine receptors on prolactin cells indicates that morphine does not just inhibit TIDA neurons but may stimulate prolactin-releasing factor neurons; indeed, Met5-enkephalin or β-endorphin stimulate VIP (which acts as a prolactin releasing factor) secretion from the hypothalamus.

2. Endogenous Opioids

Studies with opioid receptor antagonists or intracerebroventricular injection of β-endorphin antibody show that endogenous opioids stimulate basal prolactin secretion; but there are sex differences since the κ-selective antagonist *nor*-binaltorphimine is effective only in males. Studies on stimulated prolactin secretion show clearly the importance of endogenous opioids.

Thus, the secretion of prolactin at proestrus or stimulation by suckling in lactation is prevented by naloxone. Similarly, prolactin secretion in response to a stressor is prevented by prior naloxone administration, and this involves opioid actions within the blood–brain barrier. Prevention of the increase in prolactin secretion during exercise in trained athletes by naloxone provides one of the few instances of endogenous opioid regulation of prolactin secretion in humans. There is a reduced prolactin response to morphine after a stressor, which indicates desensitization of μ receptors (the response to a κ agonist is not decreased) and hence indicates that stress activation involves μ receptors.

3. POMC (β-Endorphin) Neurons and Prolactin

β-Endorphin is likely to be an endogenous opioid inhibiting the activity of TIDA neurons and thus stimulating prolactin secretion. The actions may not be direct since few TIDA neurons are contacted by β-endorphin terminals; more are contacted by enkephalin terminals and most by dynorphin.

i. Preovulatory Prolactin Surge The content of β-endorphin in the arcuate nucleus decreases at proestrus, whereas the content in the median eminence increases, coincident with the rise in prolactin secretion; as reviewed previously, a proportion of POMC neurons have estrogen receptors through which their activity can be coordinated with the ovarian steroid signals regulating LH secretion. The increase in β-endorphin content in the median eminence at proestrus correlates with a reciprocal decrease in median eminence dopamine content, consistent with inhibition of TIDA neurons by β-endorphin, which would be followed by an increase in prolactin secretion. Conversely, induced hyperprolactinemia is accompanied by reduced POMC mRNA expression in the arcuate nucleus, which indicates a short-loop negative feedback regulation by prolactin.

ii. Lactation Endogenous opioids mediate the stimulation of prolactin secretion in response to suckling (in the rat or pig) since prior administration of μ or κ opioid antagonist inhibits suckling-stimulated prolactin secretion. Activation of μ opioid mechanisms is inferred from the failure, through cross-tolerance, of morphine to stimulate prolactin secretion in lactation. β-Endorphin is thought to be involved since immunoneutralization by intracerebroventricular injection of β-endorphin antiserum also prevents suckling stimulation of prolactin secretion. Intracerebroventricular injection of β-endorphin increases prolactin secretion in lactating rats, and this action is inhibited by a selective κ or μ antagonist. The increased release of β-endorphin into hypothalamo–hypophysial portal blood is likely to be from the pars intermedia (or pars tuberalis) of the pituitary and not relevant to the regulation of prolactin secretion by arcuate nucleus POMC (β-endorphin) neurons. β-Endorphin neurons in the rostral arcuate nucleus are stimulated by suckling since their expression of Fos, as an indicator of neuronal activation, is increased. In lactating sheep, the level of POMC mRNA in the arcuate nucleus is increased. The neuropeptide galanin, given by intracerebroventricular injection, stimulates prolactin secretion, whereas 90% of β-endorphin neurons are contacted by galanin-containing terminals. β-Endorphin neurons mediate at least part of the action of galanin (part is via noradrenergic mechanisms) since it is less effective on prolactin secretion after naloxone treatment. This mechanism could be involved in activating β-endorphin neurons in lactation. Similarly, substance P stimulates prolactin secretion, and both β-endorphin and TIDA neurons are contacted by axon terminals containing the related peptide tachykinin. β-Endorphin neurons are also contacted by enkephalin-containing terminals, although this input would be expected to be inhibitory. Proenkephalin A gene expression in rostral arcuate nucleus neurons, and not in other hypothalamic neurons, is increased in lactation.

The stimulation of prolactin secretion by suckling involves inhibition of the activity of TIDA neurons, measured as reduced turnover of dopamine in the median eminence. μ Opioid receptors are involved since this inhibition of dopamine secretion by suckling is prevented by a selective μ opioid antagonist (β-funaltrexamine) but not by a selective κ antagonist (nor-binaltorphimine). It has been proposed that β-endorphin neurons are activated by suckling and then inhibit TIDA neurons. However, few TIDA neurons are contacted by β-endorphin terminals, and β-endorphin antiserum does not block the inhibition of TIDA neurons by suckling. Since this prevents the stimulation of prolactin secretion by suckling, β-endorphin may act to increase the secretion of prolactin releasing factor(s), perhaps via κ receptors; another opioid(s), acting on μ receptors, may inhibit TIDA neurons in response to suckling since this is prevented by a μ antagonist. Suckling induces analgesia (in the pig), and this is mediated by opioid mechanisms since it is prevented by naloxone.

In sum, suckling inhibits TIDA neurons by a μ opioid mechanism and thus increases prolactin secretion. β-Endorphin release from POMC neurons is also involved but may act through another, possibly κ, opioid mechanism activating prolactin releasing factor neurons.

4. Reversal of Endogenous Opioid Action on Prolactin Secretion: Pregnancy

In some circumstances opioids inhibit prolactin secretion. In the rat, the surge of prolactin secretion accompanying the LH surge induced by estrogen priming is inhibited by morphine; this probably involves stimulation of TIDA neurons. In human females, naloxone increases prolactin secretion as the follicular phase of the cycle progresses, indicating emergence of an inhibitory endogenous opioid influence on prolactin secretion.

In early pregnancy in the rat the nocturnal surges of prolactin are blocked by naloxone treatment or advanced by exogenous opioid (e.g., β-endorphin injected into a cerebral ventricle); these effects are mediated by inhibition of TIDA neurons. However, β-endorphin is less effective after midpregnancy, and following progesterone withdrawal at the end of pregnancy naloxone increases prolactin secretion. The inferred emergence of an inhibitory action of endogenous opioid on prolactin secretion seems to be a consequence of chronic estrogen action in the arcuate nucleus. These findings have not yet been replicated in another species; in the pregnant pig naloxone decreases prolactin secretion until quite late in pregnancy.

Naloxone has no effect on prolactin secretion in sheep during the short-day phase of the seasonal cycle when prolactin secretion is low, indicating lack of endogenous opioid activity on mechanisms regulating prolactin. Similarly, in postmenopausal women naloxone has no effect on prolactin secretion, but after sex steroid replacement is given, naloxone reduces prolactin secretion, indicating that the sex steroids have reactivated endogenous opioid neurons.

IV. CONCLUSION

Activation of endogenous opioid mechanisms in the brain will inhibit reproductive processes leading to conception by restraining gonadotropin secretion in both sexes and at least male reproductive behavior, perhaps in adverse external or internal environmental circumstances. However, endogenous opioids, through the sex steroid sensitivity of opioid neurons, have a physiological role in sex steroid feedback regulation of gonadotropin secretion, including the preovulatory LH surge, and in priming females for their intermittent display of sexual receptivity so that it is timed to coincide with ovulation. In pregnancy, the endogenous opioid mechanisms are activated to suppress pituitary gonadotropin secretion, preventing ovulation; to restrain oxytocin secretion, conserving oxytocin stores for parturition and providing a means to interrupt the birth process in the face of external threats; to inhibit prolactin secretion (a reversal of the normal direction of opioid action on prolactin secretion); to reduce pain perception, lessening the discomfort of giving birth; and to hold back the expression of maternal behavior until the appropriate time. In lactation, endogenous opioids mediate the rewarding, or satisfying, feelings of motherhood and the stimulation by the suckling infant of prolactin secretion for milk production. The endogenous opioid mechanisms affecting reproductive processes are not active during periods of reproductive quiescence in the life cycle, i.e., before puberty, in the repressed phase in seasonal breeders, and in female reproductive senescence. While the use of selective opioid antagonists can provide convincing information about the type of opioid receptor involved in endogenous opioid action, the activity of any endogenous opioid peptide generally at more than one type of opioid receptor makes more problematic the definition of the nature of the endogenous opioid peptide involved in regulating any particular reproductive process.

See Also the Following Articles

FSH (Follicle-Stimulating Hormone; GnRH (Gonadotropin-Releasing Hormone); Lactation, Human; LH (Luteinizing Hormone); Milk-Ejection Reflex; Oxytocin

Bibliography

Akil, H., Watson, S. J., Young, E., Lewis, M. E., Khachaturian, H., and Walker, J. M. (1984). Endogenous opioids: Biology and function. *Annu. Rev. Neurosci.* 7, 223–255.

Bicknell, R. J. (1993). Opioids in the neurohypophysial system. In *Handbook of Experimental Pharmacology, Vol. 104/II. Opioids II* (A. Herz, Ed.), pp. 525–550. Springer-Verlag, Berlin.

Borsook, D., and Hyman, S. E. (1995). Proenkephalin gene regulation in the neuroendocrine hypothalamus: A model of gene regulation in the CNS. *Am. J. Physiol.* 269, E393–E408.

Cameron, J. L. (1997). Stress and behaviorally induced reproductive dysfunction in primates. *Sem. Reprod. Endocrinol.* 15, 37–45.

Cesselin, F. (1995). Opioid and anti-opioid peptides. *Fundam. Clin. Pharmacol.* 9, 409–433.

Desjardins, G. C., Beaudet, A., Meaney, M. J., and Brawer, J. R. (1995). Estrogen-induced hypothalamic beta-endorphin neuron loss: A possible model of hypothalamic aging. *Exp. Gerontol.* 30, 253–267.

Dhawan, B. N., Cesselin, F., Raghubir, R., Reisine, T., Bradley, P. B., Portoghese, P. S., and Hamon, M. (1996). International Union of Pharmacology. XII. Classification of opioid receptors. *Pharmacol. Rev.* 48, 567–592.

Douglas, A. J., Bicknell, R. J., and Russell, J. A. (1995). Pathways to parturition. *Adv. Exp. Med. Biol.* 395, 381–394.

Genazzani, A. R., Gastaldi, M., Bidzinska, B., Mercuri, N., Genazzani, A. D., Nappi, R. E., Segre, A., and Petraglia, F. (1992). The brain as a target organ of gonadal steroids. *Psychoneuroendocrinology* 17, 385–390.

Gu, G., and Simerly, R. B. (1994). Hormonal regulation of opioid peptide neurons in the anteroventral periventricular nucleus. *Horm. Behav.* 28, 503–511.

Hammer, R. P. J., Zhou, L., and Cheung, S. (1994). Gonadal steroid hormones and hypothalamic opioid circuitry. *Horm. Behav.* 28, 431–437.

Keverne, E. B., and Kendrick, K. M. (1994). Maternal behavior in sheep and its neuroendocrine regulation. *Acta Paediatr. Suppl.* 397, 47–56.

Knapp, R. J., Malatynska, E., Collins, N., Fang, L., Wang, J. Y., Hruby, V. J., Roeske, W. R., and Yamamura, H. I.

(1995). Molecular biology and pharmacology of cloned opioid receptors. *FASEB J.* **9**, 516–525.

Kristal, M. B. (1991). Enhancement of opioid-mediated analgesia: A solution to the enigma of placentophagia. *Neurosci. Biobehav. Rev.* **15**, 425–435.

Li, X., Keith, D. E. J., and Evans, C. J. (1996). μ Opioid receptor-like sequences are present throughout vertebrate evolution. *J. Mol. Evol.* **43**, 179–184.

Panksepp, J., Nelson, E., and Siviy, S. (1994). Brain opioids and mother–infant social motivation. *Acta Paediatr. Suppl.* **397**, 40–46.

Raynor, K., Kong, H., Chen, Y., Yasuda, K., Yu, L., Bell, G. I., and Reisine, T. (1994). Pharmacological characterization of the cloned kappa-, delta-, and mu-opioid receptors. *Mol. Pharmacol.* **45**, 330–334.

Rosenblatt, J. S., Mayer, A. D., and Giordano, A. L. (1988). Hormonal basis during pregnancy for the onset of maternal behavior in the rat. *Psychoneuroendocrinology* **13**, 29–46.

Russell, J. A., Leng, G., and Bicknell, R. J. (1995). Opioid tolerance and dependence in the magnocellular oxytocin system: A physiological mechanism? *Exp. Physiol.* **80**, 307–340.

van Furth, W. R., Wolterink, G., and van Ree, J. M. (1995). Regulation of masculine sexual behavior: Involvement of brain opioids and dopamine. *Brain Res. Brain Res. Rev.* **21**, 162–184.

Endometriosis

Camran Nezhat, Farr Nezhat, and Ceana Nezhat
Stanford University

I. Symptoms and Diagnosis of Endometriosis
II. Endometriosis and Infertility
III. Nonsurgical Treatment Options
IV. Recognition and Diagnosis of Typical and Atypical Endometrial Implants
V. Laparoscopic Surgical Treatment of Endometriosis

GLOSSARY

adhesions Bands of inflammatory scar tissue.

GnRH agonist A drug that lowers estrogen to menopausal levels for the purpose of controlling endometriosis. As estrogen thickens the endometrial lining and catalyzes the growth of ectopic endometrial implants, a decrease in estrogen atrophies the endometrium, thus giving relief to sufferers of endometriosis.

hydrodissection A technique used in laser treatment of endometriosis that involves using a fluid backstop [the carbon dioxide (CO_2) laser does not penetrate water], which allows the surgeon to work on selected tissue with a greater safety margin than would otherwise be available. Fluid is injected between the lesion to be excised and the underlying ureter or blood vessels, and the fluid under the implant absorbs the CO_2 laser energy, buffering the underlying tissue.

theory of coelomic metaplasia A theory proposing that endometriosis is caused by multipotential cells, under the proper stimuli, changing from peritoneal epithelium to endometrium. Possible stimuli are cyclic hormonal changes and irritation from menstrual debris.

videolaseroscopy A variation on the standard laparoscopic technique that incorporates the use of a CO_2 laser and miniature video camera attached to the scope. The camera projects the laparoscopic image on television screens above the operating table.

Endometriosis, often called "the career woman's disease," afflicts women of reproductive age and is one of the most common reasons for pelvic pain and infertility. There are three theories as to the cause of this elusive disease, but the exact etiology remains a mystery. The most common and scientifically valid theory is a combination of immunological alteration and retrograde menstruation—direct transportation of viable endometrial cells through the fallopian tubes into the pelvis which then implant in ectopic sites. Other suggestions are (i) coelomic metaplasia, which proposes that multipotential cells may change from peritoneal epithelium to endometrium, and (ii) lymphatic spread/mechanical transport, where the cells are picked up by the lymphatic or circulatory and ferried about the body.

I. SYMPTOMS AND DIAGNOSIS OF ENDOMETRIOSIS

Unfortunately, there is no current noninvasive test for diagnosing endometriosis. However, the typical symptoms include infertility, ovarian cysts, or chronic pelvic pain such as generalized, vague, and sometimes sharp pain; painful periods (dysmenorrhea); painful intercourse (dyspareunia); back pain; and painful ovulation. The symptoms are often related to the menstrual cycle, generally resulting from functioning endometrial tissue or scarring in the affected site, but the symptoms are not always cyclic, which can be very puzzling to both the patient and the clinician. A pelvic examination may confirm suspicion, but in mild to moderate stages of endometriosis, the physical and pelvic examinations may be completely normal. A common finding on pelvic examination is pelvic tenderness and tender nodularity of the posterior cul-de-sac and uterosacral ligaments. With more advanced disease, in addition to thickening and nodularity of these structures, fixed tender retroversion of the uterus and thickening or masses of the rectovaginal septum may be palpable. Adnexal involvement is characterized by ovarian enlargement, with or without tenderness. Pain or bleeding from any site coinciding with menses should raise the index of suspicion and lead to careful evaluation of the specific anatomic areas involved, such as lungs, inguinal canal, umbilicus, or previous incisions.

Symptoms of extragenital endometriosis, such as that involving the bowel, bladder, ureters, and diaphragm, may include cyclic organ dysfunction, painful evacuation, or chest pain. Laparoscopy remains the most reliable procedure for detecting pelvic or abdominal endometriosis.

II. ENDOMETRIOSIS AND INFERTILITY

The relationship between endometriosis and infertility is complex and poorly understood. However, the most logical link involves adhesion formation and subsequent anatomic distortion of pelvic organs. Prostaglandin-induced tubal and ovulatory dysfunction, spontaneous abortion, luteinized unruptured follicle syndrome, alterations in the immune system, and intraperitoneal inflammation could also be connected to endometriosis because they all rely on the presence of ectopic endometrial implants as the basis for the infertility. Medical or hormonal therapy was used in the past as a means of enhancing fertility. However, the efficacy of the method has not been proven in controlled studies.

III. NONSURGICAL TREATMENT OPTIONS

Different hormonal therapies, such as birth control pills, progesterone, danazol, GnRH analog, and, recently, RU 486, have been used with some success to decrease the size of endometriotic lesions. This treatment can be used alone or in conjunction with surgery. Preoperative hormonal therapy using danazol has been shown to create a pelvic environment different from the pseudopregnant or normal ovulatory state. There is much less blood flow and hyperemia in a hypoestrogenic state. Capillaries are less abundant and less dilated. The reduced inflammatory reaction makes identification and removal of the endometrial implants and endometriomas easier. GnRH agonists seem to have a similar effect on the pelvic environment.

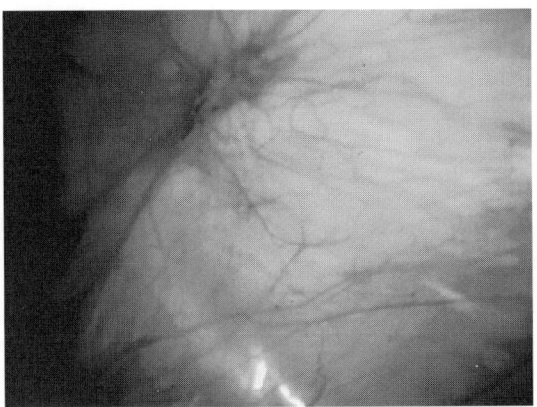

FIGURE 1 Typical lesion.

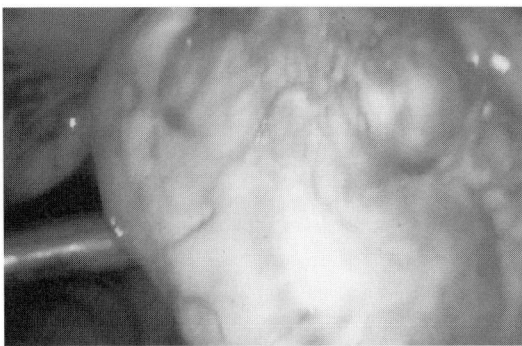

FIGURE 3 Endometrioma (right ovary).

IV. RECOGNITION AND DIAGNOSIS OF TYPICAL AND ATYPICAL ENDOMETRIAL IMPLANTS

Endometriotic implants can take on various forms (Figs. 1–7). They may be apparent as a smattering of cysts, as bubble-like lesions, or as deep and invasive lesions. They can appear as typical, black puckered pigmented lesions (resulting from tissue bleeding and retention of blood pigment) or as nonpigmented lesions classified as red (flame-like, glandular excrescences, petechial peritoneum, and hypervascularization areas) and white (white opacification, subovarian adhesions, and yellow-brown patches).

Red flame-like lesions of the peritoneum or red vesicular excrescences most often affect the broad ligament and the uterosacral ligaments. Histologically, these are due to the presence of active endometriosis surrounded by stroma. In color, translucency, and consistency, glandular excrescences closely resemble the mucosal surface of the endometrium seen at hysteroscopy. Biopsy reveals the presence of numerous endometrial glands. Areas of petechial peritoneum or areas with hypervascularization resemble the petechial lesions that result from manipulation of the peritoneum or from hypervascularization of the peritoneum. Generally, they affect the bladder flap and the broad ligament; histologically, red blood cells are numerous and endometrial glands very rare.

White opacification of the peritoneum, which appears as peritoneal scarring or as circumscribed patches, is often thickened and sometimes raised. Histologically, white opacified peritoneum results from the presence of an occasional retroperitoneal

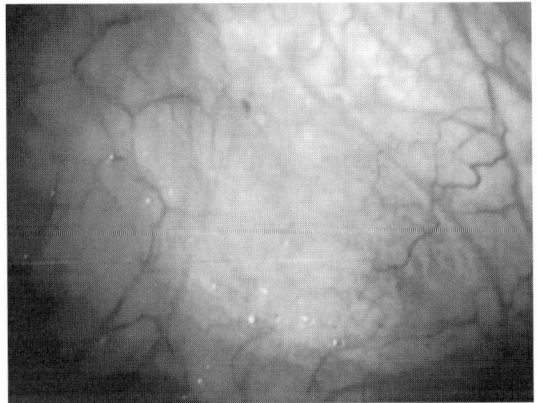

FIGURE 2 Atypical spots of endometriosis.

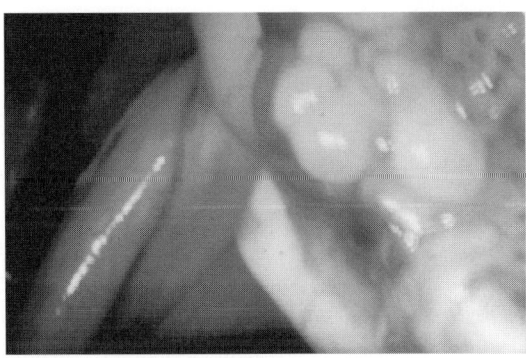

FIGURE 4 Lesion of endometrioma of the right ovary.

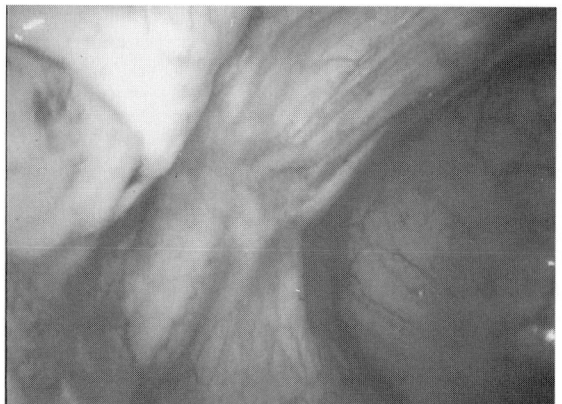

FIGURE 5 Infiltrative endometriosis with partial left ureter obstruction and hydro ureter.

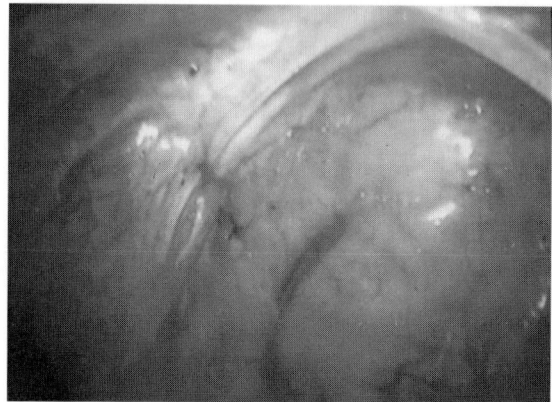

FIGURE 7 Different appearing endometriosis in posterior cul-de-sac on birth control.

glandular structure and scanty stroma surrounded by fibrotic tissue or connective tissue.

Subovarian adhesions or those between the ovary and peritoneum of the ovarian fossa differ from adhesions characteristic of previous salpingitis or peritonitis. Histologically, connective tissue with sparse endometrial glands is found.

Yellow-brown peritoneal patches resemble "cafe au lait" patches. The histological characteristics are similar to those observed in white opacification, but blood pigment (hemosiderin) among the stroma cells produces the cafe au lait color.

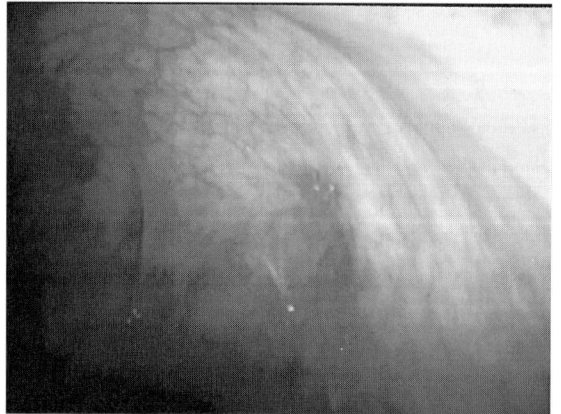

FIGURE 6 Vesicle-type lesion of posterior cul-de-sac on birth control.

V. LAPAROSCOPIC SURGICAL TREATMENT OF ENDOMETRIOSIS

In the past, surgical treatment of endometriosis included laparotomy and imminent hysterectomy. This was mainly due to the fact that surgeons were unfamiliar with endometriosis and how to treat it successfully. The presumption was that, if the patient was no longer menstruating, the endometriosis pain would cease. However, it has been found that this theory no longer applies. A fairly recent improvement in technology, called operative videolaparoscopy, has revolutionized the way patients are being treated for endometriosis. The advantages over laparotomy include smaller incisional sites ($\frac{1}{4}-\frac{1}{2}$ in. versus 4–6 in.), shorter hospital stay (1 day versus 3–5 days), and a shorter recovery period (1 or 2 weeks versus 4–6 weeks).

A. Peritoneal Implants

Peritoneal implants can be coagulated (desiccated), excised, or ablated *in situ*. They must be completely eradicated. Superficial lesions up to 5 mm in diameter can be excised, electrodesiccated, or vaporized with the laser. It is preferable to excise larger and infiltrating lesions. Care should be taken not to damage any surrounding structures. Cutting in close proximity to vital structures should be done using scissors or the CO_2 laser.

B. Ovarian Implants

Endometriosis of the ovary can exist as small superficial lesions or endometriomas of less than 2 cm in diameter, cysts of 2–5 cm in diameter, and endometriomas that are greater than 5 cm. Like peritoneal lesions, endometrial implants on the ovary can be coagulated, excised, or vaporized. The goal should be to achieve complete treatment while causing the least amount of trauma to ovarian function. Hemorrhagic ovarian cysts clinically resembling endometriomas can be classified into two major types. Type I endometriomas, or pure endometriomas, are small (1 or 2 cm), contain dark fluid, develop from surface endometrial implants, and are difficult to remove surgically. Histologic analysis reveals endometrial tissue in all of them. Although small type I endometriomas are difficult to remove intact because of associated fibrosis and adhesions, they can be biopsied, drained, and vaporized using laser or electrosurgery or removed in pieces. The larger type I lesions (2 or 3 cm) must be removed completely.

Type II endometriomas, or secondary endometriomas, are caused when follicular or luteal ovarian cysts have been involved or invaded by cortical endometriotic implants or by primary endometrioma. Type II endometriomas are classified into three subgroups based on the relationship of cortical endometriosis with the cyst wall.

Type IIA endometriomas are large, the cyst wall is separated easily from the ovarian tissue, and, if endometrial implants are seen, they do not penetrate the cyst wall. These hemorrhagic cysts are either follicular or luteal in origin. The surgical technique for this type of lesion includes lysis of periovarian adhesions, evaluation of the ovarian cortex, and aspiration of the cyst.

Types IIB and IIC are endometriomas with features of functional cysts involved deeply with surface endometriosis, with histologic findings of endometriosis in the cyst wall. They usually are large and associated with periovarian adhesions attaching them to the pelvic sidewall and back of the uterus and tend to rupture during separation. In type IIB, the lining is separated easily from the ovarian capsule and stroma except where adjacent to the areas of endometriosis. In Type IIC, surface endometrial implants penetrate more deeply into the cyst wall, making excision more difficult. The degree of endometrial invasion of the cyst wall forms the basis for differentiating between these two subtypes and is characterized by the progressive difficulty in removing the cyst wall. Surgical treatment of types IIB and IIC includes mobilizing the ovary and removing the contents of the cyst and irrigating the cavity using the suction–irrigator probe.

C. Genitourinary Endometriosis

Ureteral and bladder involvement has been reported in 1–11% of women in whom a diagnosis of endometriosis has been confirmed. It tends to be superficial but can be invasive on occasion and may even result in complete obstruction of the ureter. In most cases the diagnosis is made at the time of diagnostic laparoscopy for complaints of infertility or pain. Bladder instability and decreased capacity that is unresponsive to conventional medical therapy with oral contraceptives, progesterones, or GnRH agonists may be due to endometriosis. Invasive bladder involvement may be suspected clinically if there is a complaint of hematuria or dysuria, particularly if it is temporally related to menstruation.

If there is reason to suspect genitourinary involvement preoperatively, urinalysis; measurement of serum electrolytes, creatinine, and urea levels; ultrasonography of the kidneys; and intravenous pyelography should be performed. Cystoscopy is usually reserved for cases in which there is evidence of recurrent hematuria.

1. Ureteral Endometriosis

Hydrodissection has greatly simplified the laparoscopic management of superficial peritoneal implants that are sited over or in close proximity to the ureter. Implants that are firmly embedded, with scarring to the subperitoneal connective tissues, can be more difficult to remove but, once again, hydrodissection can often be used to tunnel beneath the lesion and thus protect the ureter when vaporizing or excising the endometriosis. Fortunately, cases of partial or complete ureteral obstruction due to endometriosis are rare and can be managed by ureterolysis and occasionally by resection of a segment of ureter and reanastomosis of the obstructed segment.

2. Bladder Endometriosis

Deeply infiltrative endometriosis of the bladder wall is uncommon. The lesions are usually superficial. When invasive, they usually penetrate the muscularis but spare the mucosa. Rarely, they may infiltrate the full thickness of the bladder wall. They can be dealt with by laparoscopic means.

Superficial lesions can be treated by excision, electrodesiccation, or laser vaporization as for other peritoneal implants. This approach can also be used when the muscularis is involved, but it will be necessary to excise the affected muscularis and close the defect. If the entire thickness of the bladder wall has been infiltrated by endometriosis it is necessary to excise the affected section of the bladder wall and reconstruct the defect.

D. Cul-de-Sac Obliteration

The cul-de-sac may be partly or totally obliterated, which is often a source of pain and may hinder oocyte pickup by the fimbria. There usually is a great degree of anatomic disorganization involving the rectosigmoid, vaginal fornix, posterior aspects of the cervix and uterus, ureter, and major blood vessels. There may be endometriosis of the rectovaginal septum. The bowel should be prepared preoperatively in such cases.

Laparoscopic restoration of the cul-de-sac is not simple and should only be undertaken if general surgical and/or urologic assistance is readily available or by a surgeon who is familiar with the performance of surgery on the gastrointestinal and urinary tracts.

E. Intestinal Endometriosis

The intestinal tract is involved in 3–37% of women with endometriosis. The rectosigmoid is most commonly affected. Clinical presentations vary from the asymptomatic, in which the disease is noted at the time of laparoscopy or laparotomy, to one of complete intestinal obstruction. The presence of intestinal endometriosis should be suspected in any woman of child-bearing age who presents with gastrointestinal symptoms and a prior history of known endometriosis or intestinal stricture in the absence of an intraluminal mass. Cyclical rectal bleeding is a rare presentation and occurs when the lesion infiltrates the bowel mucosa. Because, in addition to the rectosigmoid, the uterosacral ligaments and rectovaginal septum are often involved in the disease process, the patient may complain of low abdominal pain, back pain, dysmenorrhea, and dyspareunia. She may experience diarrhea, constipation, and tenesmus. Radiographic studies that suggest constriction of the bowel, proctoscopy, and colonoscopy may provide helpful clues but are rarely diagnostic. Occasionally, microscopic examination of a biopsy of the bowel mucosa may reveal the presence of endometrial tissue. Laparoscopy is the best method of diagnosis.

The implants can occur at any location from the small bowel to the anal canal. The rectum and sigmoid colon are most commonly affected and are involved in 76% of such cases. The appendix and cecum are involved in 18 and 5% of cases, respectively.

In cases of severe disease of the bowel wall, resection may be necessary. The technique involves laparoscopic mobilization of the lower colon; transanal, transvaginal, or transabdominal prolapse; resection; and anastomosis. Four major techniques are disk excision (shaving the lesion off the bowel), full-thickness resection of the affected area of the colon wall with reconstruction of the defect, bowel resection and end-to-end anastomosis (for circumferential lesions), and intraabdominal segmental resection.

F. Endometriosis of the Diaphragm

Endometriosis of the diaphragm is rare. In centers that deal with a very large number of patients with endometriosis, such lesions are identified in less than 1% of cases. Since most of these cases are asymptomatic, identification of such lesions requires a thorough examination of the undersurface of the diaphragm, which is more easily effected by laparoscopy than by laparotomy. The symptoms included pleuritic, shoulder, or upper abdominal pain which may occur in conjunction with menses. Rarely, it can cause catamenial pneumothorax.

Laparoscopic removal of such lesions is hazardous because the phrenic nerves, lungs, and heart may all be at risk for injury. All other treatment options should be fully explored before recourse to surgery. A course of medical suppressive therapy may be all that is required, and cure may be effected if radical treatment of pelvic endometriosis is undertaken. If fertility is still an issue and symptoms have not responded to medical therapy, laparoscopic intervention may be justified.

See Also the Following Articles

ENDOMETRIUM; MENSTRUAL DISORDERS

Bibliography

Barbieri, R. L., and Gordon, A.-M. C. (1991). Hormonal therapy of endometriosis: The estradiol target. *Fertil. Steril.* 56, 820–822.

Buttram, V. C., Belue, J. B., and Reiter, R. (1982). Interim report of a study of danazol for the treatment of endometriosis. *Fertil. Steril.* 37, 478–483.

Gomel, C., and Taylor, P. J. (eds.) (1995). *Diagnostic and Operative Gynecologic Laparoscopy*. Mosby-Year Book, St. Louis, MO.

Nezhat, C. R., Nezhat, F. R., Luciano, A. A., Siegler, A. M., Metzger, D. A., and Nezhat, C. H. (Eds.) (1995a). *Operative Gynecologic Laparoscopy: Principles and Techniques*. McGraw-Hill, New York.

Nezhat, C. R., Berger, G. S., Nezhat, F. R., Buttram, V. C., and Nezhat, C. H. (Eds.) (1995a). *Endometriosis: Advanced Management and Surgical Techniques*. Springer-Verlag, New York.

Endometrium

Linda C. Giudice

Stanford University

I. Steroid Hormone Receptors
II. Growth Factors
III. Implantation
IV. Secretory Phase Proteins
V. Cell Adhesion Molecules
VI. Menstruation/Remodeling
VII. Clinical Disorders

GLOSSARY

decidua The endometrium of pregnancy.

endometrium The lining of the uterus, composed of epithelial cells, stromal cells, underlying stroma, and lymphoid cells.

growth factors Peptides or proteins that affect cellular growth and differentiation by interacting with specific target cell membrane receptors.

hemostatic Preventing hemorrhage.

implantation In humans, the process whereby an embryo apposes itself next to the endometrial epithelial lining, attaches to it, and then invades into the endometrial stroma.

proliferative phase The first half of the menstrual cycle in which proliferation and increase in extracellular matrix occur in the endometrium under the influence of increasing circulating estradiol secreted by the growing dominant ovarian follicle (also known as the *follicular phase*).

secretory phase The second half of the menstrual cycle in which glandular secretion and stromal decidualization occur under the influence of progesterone secreted by the corpus luteum (also known as the *luteal phase*).

steroid hormones Hormones that have four ring structures that interact with nuclear receptors to effect cellular differentiation or proliferation.

vasoactive Affecting vascular tone.

Uterine endometrium is a dynamic tissue that displays predictable changes in response to cyclic variations in circulating ovarian-derived steroid hormones. In humans the tissue is composed of two distinct layers: the basalis layer and the functionalis layer. It is primarily the latter that undergoes dramatic changes observed cyclically, although regeneration of this layer derives from a functioning basalis. During the reproductive years the functionalis layer undergoes a series of histologic and biochemical changes that culminate in a remarkable 10-fold increase in endometrial thickness during the aptly named proliferative phase of the menstrual cycle. In the secretory phase, it undergoes primarily differentiation in preparation for an implanting blastocyst. In the absence of implantation, unique events occur resulting in uniform and efficient shedding of the functionalis layer and its regeneration in the subsequent cycle. Numerous biochemical principles are expressed and play important roles in endometrium, including steroid hormone receptors, growth factors and their receptors, enzymes and their inhibitors, angiogenic and vasoactive factors and their receptors, prostanoids, hemostatic factors, cell adhesion molecules, and a group of proteins uniquely synthesized in this tissue in the secretory phase. There are also several cell types in endometrium that contribute to the changes observed and functions of this tissue. These include glandular epithelium, luminal epithelium, stromal cells, endothelial cells, vascular smooth muscle cells, and resident and transient cells of the immune system. Abnormal development of the cellular components or biochemical principles that they synthesize in the functionalis and basalis layers of the endometrium can result in infertility, repetitive miscarriage, abnormal placentation, and dysfunctional uterine bleeding. The events occurring in human endometrium during the menstrual cycle and related clinical disorders are described herein. Discussion is primarily limited to the functionalis layer, which has been widely studied, due primarily to availability of this tissue.

I. STEROID HORMONE RECEPTORS

Key to endometrial development is its responsiveness to steroid hormones. In human endometrium, estrogen receptor (ER) and progesterone (PR) expression vary throughout the cycle and are expressed differently in the glands and stroma (Fig. 1). ER levels are highest primarily in glandular epithelium during the proliferative phase in response to rising circulating levels of estradiol (E_2), and then decrease after ovulation, reflecting progesterone's (P) suppressive effects. PR levels peak at the time of ovulation, reflecting induction by E_2. PRs are prominent in glands in the proliferative phase but are undetectable by the midsecretory phase, in contrast to stroma, in which moderately high levels persist throughout the secretory phase (Fig. 1).

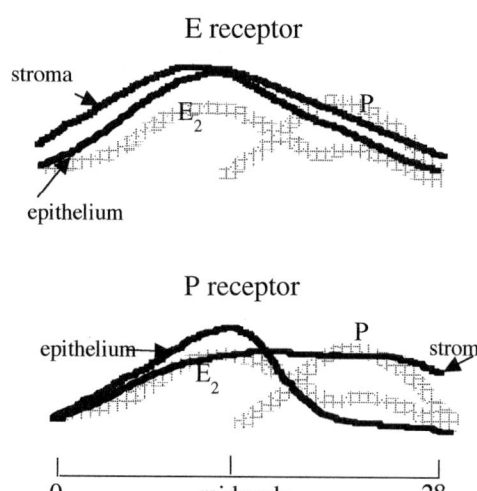

FIGURE 1 Estrogen (E) and progesterone (P) receptor levels in human endometrial stroma and glandular epithelium during the menstrual cycle. Shaded lines represent circulating levels of estradiol (E_2) and P throughout the cycle (from Lessey et al., 1986).

II. GROWTH FACTORS

There are many growth factors in the endometrium that are expressed primarily or exclusively in the estradiol-dominant proliferative phase or the progesterone-dominant secretory phase (Fig. 2). Their functions are believed to be related to the phase of the cycle in which they are expressed (i.e., proliferation or differentiation). In addition, they are usually expressed in cell-specific fashions, underscoring

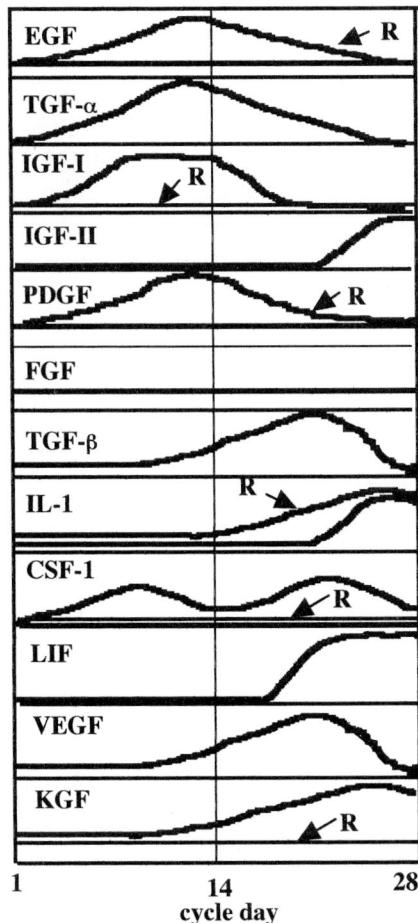

FIGURE 2 Cyclic variation of growth factors and their receptors in human endometrium. Growth factors include the following: EGF, epidermal growth factor; TGF-α, transforming growth factor-α; IGF-I, insulin-like growth factor-I; IGF-II, insulin-like growth factor-II; PDGF, platelet-derived growth factor; FGF, fibroblast growth factor; TGF-β, transforming growth factor-β; IL-1, interleukin-1β; CSF-1, colony-stimulating factor-1; LIF, leukemia inhibitory factor; VEGF, vascular endothelial growth factor; KGF, keratinocyte growth factor; R, signal transducing receptor for particular growth factor.

their roles in the changes that occur in endometrium, including glandular proliferation, stromal decidualization, angiogenesis, implantation, menstruation, and repair. They act by autocrine, paracrine, or juxtacrine mechanisms. Figure 2 shows schematically the changes in several growth factor families in human endometrium across the menstrual cycle. The two growth factors suggested to play a major role in endometrial proliferation and endometrial thickness are epidermal growth factor/transforming growth factor-α (EGF/TGF-α) and insulin-like growth factor-I (IGF-I). Both EGF and IGF-I are potent mitogens of endometrial cells *in vitro*. Thus, their actions *in vitro* and their cycle-specific expression suggest mitogenic roles for them *in vivo* in this tissue. IGF-II is believed to play a role in stromal decidualization. All the other growth factors and cytokines shown, including TGF-β, interleukin-1α and -β (IL-1α and IL-1β), colony-stimulating factor (CSF), and leukemia inhibitory factor (LIF), likely play roles in either implantation or initiation of endometrial shedding and endometrial cellular apoptosis in the absence of implantation. Vascular endothelial growth factor (VEGF) and fibroblast growth factor (FGF) are believed to play a role in angiogenesis, and VEGF also probably plays a role in water permeability in the endometrium. Keratinocyte growth factor is believed to be a differentiation factor for glandular epithelium, derived from stroma.

III. IMPLANTATION

Implantation is a complex interplay between the conceptus and the maternal endometrium ("decidua") and involves growth factors, cytokines, and a variety of proteins and peptides. In humans there are three phases of implantation (Fig. 3): apposition of the embryo with the maternal endometrial epithelium, attachment and passage through the epithelium, and invasion into the stroma. In the invasive phase, the major goal of the invading conceptus is to reach the maternal blood supply, thereby ensuring its access to nutrients for its own sustenance and continued growth. The maternal host is selective in that it is "receptive" to an implanting blastocyst only within a "window of implantation," which is temporally and spatially restricted. In addition, recent studies support a proactive role for the endometrium in nourishing the conceptus, protecting it from the maternal immune system, and preventing its overzealous invasion. Several growth factors, binding proteins, and cytokines are differentially expressed in endometrium during the secretory phase of the cycle (Fig. 2). For the most part, their expression

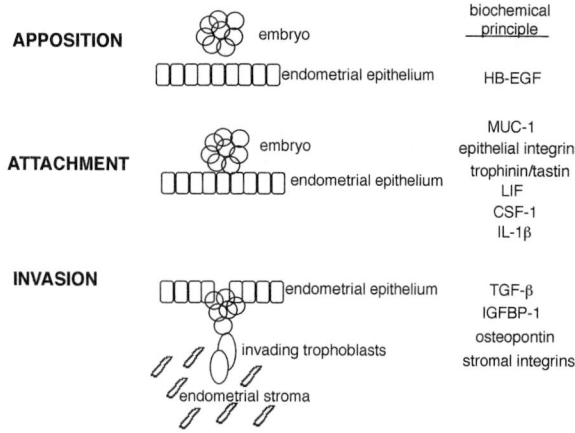

FIGURE 3 Schematic of phases of implantation in humans. Biochemical principles listed on the right are believed to be involved in specified phases (see text).

continues during early pregnancy, suggesting that they play a major role in the cross-talk between mother and conceptus. These effector molecules include members of the EGF family, CSF-1, LIF, IL-1β, TGF-β, and insulin-like growth factor binding protein-1 (IGFBP-1).

A. Early Uterine:Embryonic Interactions

Recent evidence indicates that a member of the EGF family, namely, heparin-binding EGF-like growth factor (HB-EGF), may play a significant role in embryo–uterine interactions. The HB-EGF gene is expressed in mouse uterine luminal epithelium surrounding the blastocyst just prior to implantation. In addition, HB-EGF also promotes blastocyst growth, zona hatching, and trophoblast outgrowth *in vitro*. These findings suggest an important role for HB-EGF in the very early stages of implantation. The expression and role in implantation of HB-EGF and other newly discovered members of the EGF family in the female reproductive tract of the mouse and human await further investigation.

B. Apposition/Attachment

1. Colony-Stimulating Factor-1

A role for CSF-1 in implantation was first suggested by the observation that CSF-1 levels in mouse uterus increase about 1000-fold during pregnancy. Furthermore, the role of this cytokine in implantation has been underscored in the osteopetrotic (op/op) mouse model. Animals that are homozygous for a naturally occurring null mutation in the CSF gene are toothless and have multiple skeletal defects, decreased numbers of macrophages, and infertility (low implantation rates and decreased fetal viability). When homozygotes are treated with exogenous CSF-1, their fertility is restored. Thus, it has been proposed that CSF-1, produced by uterine epithelium, interacts with the CSF-1 receptor on the trophectoderm and that this interaction promotes blastocyst attachment.

2. Leukemia Inhibitory Factor

LIF is another protein whose role in implantation comes from studies using animal models. LIF is expressed in mouse endometrium prior to ovulation and also on the fourth day of pregnancy, which is the day on which implantation is initiated (where Day 1 is the day of the vaginal plug). Similar LIF expression has been observed in psuedopregnant mice, suggesting that it is not induced by the presence of an embryo and is under maternal control. The importance of LIF in implantation has been shown conclusively in a mouse model lacking a functional LIF gene; implantation was achieved by gene targeting and homologous recombination. When homozygous females are mated with normal males, they produce normal blastocysts that are recoverable from uteri on Day 4 of pregnancy. In addition, transfer of these embryos to wild-type pseudopregnant females results in implantation and pregnancy. Thus, endometrial epithelial LIF expression appears to be very important for attachment of embryos to the epithelium. Mechanisms underlying regulation of maternal endometrial LIF expression and embryonic implantation await further investigation, as does a role for this cytokine in human implantation.

3. Interleukin-1

Human embryos secrete IL-1α and IL-1β *in vitro*, and high concentrations of these cytokines in the conditioned media correlate with successful implantation after embryo transfer in humans. Using a mouse model, it has been demonstrated that intraperitoneal injection of high levels of IL-1 receptor antagonist into pregnant mice on the third day of pregnancy results in a significant decrease in the number of implantation sites (6.7%) compared to noninjected and buffer-injected animals (59 and 74%, respectively). These studies suggest that IL-1 may be necessary for implantation and that elevated antagonist levels are detrimental to a successful pregnancy. However, similar experiments in other strains of mice do not demonstrate the same impairment of fertility, and IL-1 receptor knockout mice have normal fertility. Therefore, a role for IL-1 in implantation awaits further investigation.

C. Invasion

1. Transforming Growth Factor-β

TGF-β isoforms are expressed in specific cell types and at distinct times within the uterus, suggesting specific roles for them. In humans, TGF-β1 mRNA is equally distributed in endometrial glands and stroma, and levels are lowest in the proliferative phase. Levels increase twofold in the early secretory phase, decrease in the late secretory phase, and increase approximately fivefold in early pregnancy decidua. The cycle dependence of TGF-β mRNA expression suggests that it may contribute to inhibition of cellular proliferation during periods when endometrial cellular differentiation, induced by TGF-β or other growth factors, is dominant. In addition, the abundance of TGF-β mRNA expression in decidua also suggests a possible role for this growth factor in implantation. TGF-β promotes differentiation of human cytotrophoblasts into noninvasive syncytiotrophoblasts. The invading trophoblast secretes a variety of proteases, including plasminogen activators and matrix metalloproteinases (MMPs). Since TGF-β induces plasminogen activator inhibitor (PAI) mRNAs and induces tissue inhibitor of metalloproteinase-1 (TIMP-1), it has been postulated that TGF-β may act to control trophoblast invasion by regulating proteases and their inhibitors in the embryo and maternal decidua. Also, since the latent form of TGF-β is activated by plasmin, it is possible that maternal decidua stores latent TGF-β in the extracellular matrix, waiting for the invading trophoblast and its production of plasmin. TGF-β could then increase PAIs and inhibit trophoblast-derived MMPs (by increasing TIMP-1). The net result would be limiting trophoblast invasion. In support of the latter hypothesis is the observation that TGF-β produced by first-trimester decidual cells in culture inhibits the in vitro invasiveness of first-trimester human trophoblasts due to the induction of synthesis and secretion of TIMP-1.

2. Insulin-like Growth Factor Binding Protein-1

IGFBP-1 is a major protein product of human endometrium. It has dual functions: as a high-affinity binding protein of the IGFs (mostly inhibiting their actions) and as a ligand for cellular adhesion molecules, in particular the $\alpha_5\beta_1$ and $\alpha_v\beta_3$ integrins, resulting in altered cellular motility. The anchoring villus of the invading fetal placenta expresses primarily $\alpha_5\beta_1$ integrin. Recent studies show that IGFBP-1, which contains the Arg-Gly-Asp (RGD) sequence, a ligand for $\alpha_5\beta_1$, specifically binds to human cytotrophoblasts, binds to the $\alpha_5\beta_1$ integrin in cytotrophoblasts, and inhibits trophoblast invasion into maternal decidua in vitro. Thus, IGFBP-1 in maternal decidua may have a dual role as a modulator of IGF action (trophoblasts produce high levels of IGF-II) and as one of likely several "maternal restraints" on trophoblast invasiveness.

IV. SECRETORY PHASE PROTEINS

During the secretory phase of the menstrual cycle, there is intense secretory activity of the glands and protein secretion from the stroma, accompanying glandular development and stromal decidualization. Many proteins produced by these cells have autocrine actions, others may act by paracrine mechanisms on neighboring myometrium, and others are secreted into the circulation, acting as classical endocrine factors. Proteins secreted apically appear in uterine luminal fluid, and those transported across fetal membranes appear in amniotic fluid. Major secretory proteins in endometrium include placental protein-14 (PP-14), IGFBP-1, extracellular matrix proteins, prolactin, relaxin, and an ever-growing list of other

TABLE I
Secretory Phase Proteins, Their Cells of Origin, and Their Potential Functions

Protein	Cells of origin	Proposed functions
PP-14 (PEP, α_2-PEG)	Glandular epithelium	Immunosuppressant
IGFBP-1 (α_1-PEG)	Decidualized stromal cells	Regulating IGF action; modulating trophoblast invasiveness
Extracellular matrix proteins (laminin, fibronectin, tenascin)	Decidualized stromal cells	Extracellular matrix–trophoblast and –endometrial cell interactions
Integrins and other cell adhesion molecules	Glandular epithelium and decidualized stromal cells	Extracellular matrix–cell and cell–cell interactions
Prolactin	Decidualized stromal cells	Osmoregulation of amniotic fluid; pulmonary surfactant synthesis regulation; inhibition of uterine contractility; immunosuppresant
Relaxin	Decidualized stromal cells	Stimulates stromal aromatase activity; collagen breakdown
24-kDa protein	Glandular epithelium	Unknown
CA-125	Glandular epithelium	Unknown
Uteroglobin-like protein	Decidualized stromal cells	Prostaglandin metabolism?
PAPP-A	Glandular epithelium	Unknown
Mucins	Glandular epithelium	Unknown
Renin	Decidualized stromal cells	Unknown
Lactoferrin	Glandular epithelium	Unknown

Serum proteins
  Albumin
  Ceruloplasmin
  β-Lipoprotein
  α_2-Macroglobulin
  α_1-antitrypsin
  Fibronectin
  Complement components
	Undefined	Transporting growth factors, steroids, ions, lipids; protease inhibition; controlling free radical oxidation; structural components; immune response

Note: From Giudice, 1997.

proteins whose function(s) awaits further definition (Table 1).

A. Placental Protein-14

Placental protein-14 (also known as pregnancy associated endometrial protein and α_2-pregnancy associated endometrial globulin) is not a placental protein at all; rather, it is a major secretory product of glandular epithelium during the secretory phase. It is also known as glycodelin. It is a glycoprotein that shares sequence homology with β-lactoglobulin (59%), bilin-binding protein (27%), and retinol-binding protein (23%), although it does not bind retinoic acid. Serum levels increase during the luteal phase, although they are not predictive of luteal phase adequacy of endometrial maturation, and no difference in serum levels has been found in conception versus nonconception cycles. Recently, PP-14, also known as glycodelin, has been described as interacting with sperm, affecting motility.

B. Extracellular Matrix Proteins

A variety of extracellular matrix proteins are produced in the scretory phase, changing from a collagen-like matrix to a basement membrane-like matrix. For example, laminin production by endometrial stroma is more abundant in secretory compared to proliferative endometrium, and it increases with implantation and in early pregnancy. Fibronectin, a major secretory product of decidualized stromal cells, has the RGD sequence and may interact with the invading trophoblast. It is likely that stromally derived laminin and fibronectin (as well as IGFBP-1) are among several proteins to which the trophoblast attaches during the invasive phase of implantation and which play a role in enhancing or limiting trophoblast invasiveness.

C. Prolactin

Prolactin is produced by decidualized endometrial stromal cells in the late secretory phase, as well as in the decidua throughout pregnancy, and is a progesterone-dependent protein. Endometrial stromal cells decidualized *in vitro* assume the decidual phenotype and secrete prolactin. A protein has been isolated from human placental conditioned medium which is a decidual prolactin-releasing factor. The precise regulation of decidual prolactin and interactions with placentally derived prolactin-releasing factor *in vivo* have yet to be determined. During pregnancy, prolactin accumulates in amniotic fluid by traversing the fetal membranes by an unknown mechanism. Receptors for prolactin are in fetal membranes, and one of the proposed functions of prolactin is as an osmoregulator of amniotic fluid. In addition, decidual prolactin is believed to regulate pulmonary surfactant synthesis, inhibit uterine contractility, and may be an immunosuppressant.

D. Relaxin

Relaxin is structurally similar to proinsulin and the insulin-like growth factors. It stimulates IGFBP-1 production by endometrial stromal cells decidualized *in vitro* as well as IGFBP-1 production by term decidual monolayers in culture. This is in contrast to the action of its structural homologs, which have the opposite effect. Relaxin may participate in endometrial estradiol economy since it stimulates aromatase activity in cultured endometrial stromal cells. Another potential role for relaxin in endometrium is in the breakdown of collagen. It activates MMPs in some systems and may do so in endometrium as well. Decidualized stromal cells in the late secretory phase and particularly early gestational decidua parietalis immunostain heavily for relaxin, and relaxin genes are transcribed in both the endometrial glands and stroma. Relaxin may be involved in the preparation and maintenance of early pregnancy by increasing collagenase and tissue plasminogen activator activities to break down collagen.

E. Other Proteins

A major product of glandular epithelium during the secretory phase is 24-kDa protein. Maximal immunohistochemical staining is seen in glandular cells at ovulation, in surface epithelium at the time of implantation, and in decidualizing cells later in the secretory phase. Although its function is not understood, it may prove useful as an adjunct in endometrial dating for predicting endometrial luteal "adequacy." CA-125 is an ovarian epithelial glycoprotein that is also found in endometrium. As in many tissues of similar embryological origin, its function remains undefined. A uteroglobin-like protein has been identified in human endometrium. Uteroglobin is a phospholipase A2 inhibitor, and the uteroglobin-like protein in human endometrium may be involved in endometrial prostaglandin metabolism. Pregnancy-associated plasma protein-A (PAPP-A) is produced by endometrial stromal cells. It is a high-molecular-weight glycoprotein whose function is unknown. Mucins are high-molecular-weight glycoproteins that are synthesized and secreted by endometrial epithelial cells into the glandular lumen during the middle of the secretory phase. They are mucopolysaccharides, and they function in preventing bacterial adhesion to the endometrial epithelium. While a decrease in a mucin named MUC-1 during the window of implantation in mice suggests that regulation of mucin synthesis is important in implantation, in humans levels of mucins do not appear to change. Rather, changes in their carbohydrate moieties are believed to play a role in uterine receptivity during the window of implantation in humans. Renin and lactoferrin are other proteins in secretory endometrium whose functions in this tissue remain to be determined. Several plasma proteins are synthesized *de novo* by glandular epithelium of gestational endometrium monolayer cultures *in vitro*. These include albumin, ceruloplasmin, β-lipoprotein, fibronectin, α_2-macroglobulin, α_1-antitrypsin, and complement factors C3 and C4. These proteins may participate in transport mechanisms (albumin, ceruloplasmin, and β-lipoprotein), inhibition of protease activity (α_2-macroglobulin, α_1-anti-trypsin), control of free radical oxidation (ceruloplasmin), tissue structure (fibronectin), and the immune response (complement components). The list of additional proteins in endometrium continues to grow, although we have little information regarding their functions.

V. CELL ADHESION MOLECULES

The complex and changing nature of endometrial cellular constituents during the menstrual cycle and the invasiveness of the trophoblast in early pregnancy underscore the importance of cell–cell and cell–matrix interactions. These interactions are mediated through different families of cell adhesion molecules, including integrins, adhesive molecules of the immunoglobulin superfamily, cadherins, and those with lectin-like domains. The most intensely studied group of cell adhesion molecules in endometrium are the integrins, a family of heterodimeric glycoproteins composed of 1 of 11 α and 1 of 6 β subunits. They are receptors for extracellular matrix proteins, including collagen, fibronectin, and laminin. Glandular epithelium expresses all integrin molecules, as do surface epithelium and endothelium, except for $\alpha_4\beta_1$ and $\alpha_1\beta_1$, respectively. Lymphoid cells express $\alpha_1\beta_1$, $\alpha_4\beta_1$, and $\alpha_6\beta_1$. Regional differences have also been observed. Basalis epithelium exhibits higher expression of $\alpha_4\beta_1$, $\alpha_1\beta_1$, and $\alpha_6\beta_1$ compared to the functionalis epithelium, and $\alpha_1\beta_1$ and $\alpha_4\beta_1$ expression is cycle dependent. Glandular epithelium expresses collagen/laminin receptors, whereas stromal cells express mainly the fibronectin receptor. $\alpha_v\beta_3$ has recently been suggested to be a marker of receptivity of the endometrium to implantation in humans. In subsets of women with infertility, while the endometrium appears normal histologically during the window of implantation (between Postovulatory Days 5 and 9), it lacks β_3 expression. In limited studies, therapies to increase β_3 expression have resulted in restored fertility. These findings are a dramatic demonstration that assessment of endometrial markers of uterine receptivity are of more utility than histology in assessing normalcy of endometrium for the purposes of infertility evaluation. The challenges are to elucidate mechanisms underlying embryo–endometrial interactions during implantation and to conduct large scale clinical trials on predicting uterine receptivity and efficacious therapies to convert a nonreceptive to a receptive endometrium. Alternatively, for contraceptive purposes, maintaining the endometrium in a nonreceptive state is an appealing concept.

VI. MENSTRUATION/REMODELING

In the absence of implantation, there is no chorionic gonadotropin secreted by an implanting conceptus to maintain the corpus luteum, and thus progesterone production by this tissue begins to decline. This occurs in the mid- to late secretory phase and is one of several signals that begin the orderly process of shedding the endometrial functionalis layer culminating in menstruation, a controlled and self-limited process of endometrial bleeding and tissue desquamation. During menstruation, there are focal areas of blood vessels that begin to be broken down such that blood escapes beyond the basement membrane and into the endometrial interstitial space. Vasospasm occurs and growth factors and proteins are produced to prevent hemorrhage. There is a combination of tissue involution by apoptosis initiated by cytokines and extracellular matrix degradation which is the result of enzymatic activities triggered by falling progesterone levels, cytokines, and growth factors. The enzyme/enzyme inhibitor systems involved in the complex process of menstruation include the MMPs and the TIMPs, lysosomal enzymes, and plasminogen/plasminogen activator/inhibitor/plasmin cascade. Figure 4 illustrates potential mechanisms for the initiation of MMP activation, as well as the involvement of cytokines in apoptosis, at the time of menstruation. These are explained in more detail below.

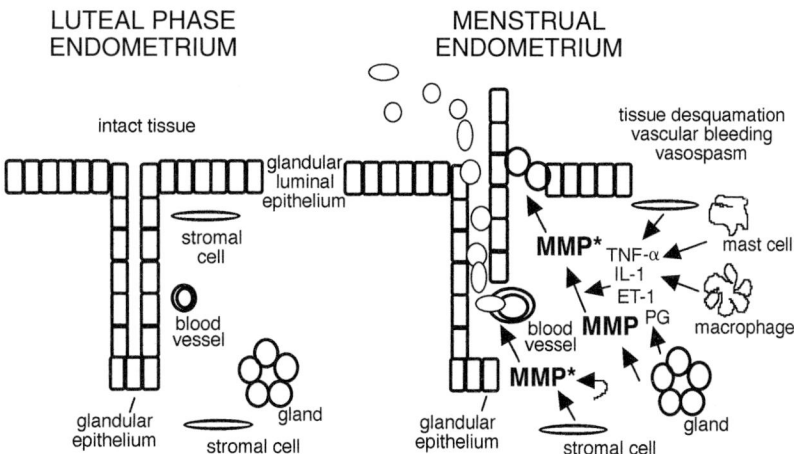

FIGURE 4 Model of intact luteal phase endometrium (left) and menstrual endometrium (right). Progesterone withdrawal results in stromal production of MMPs, and cytokines from resident and transient cell populations in the endometrium in the late secretory phase and in the menstrual phase (see text).

A. Enzymes and Their Inhibitors

1. Matrix Metalloproteinases and Tissue Inhibitors of Metalloproteinases

MMPs comprise a multigene family of metal ion-requiring enzymes that degrade components of the extracellular matrix. They are subdivided into three classes: collagenases (substrates: Types I, II, and III collagen), the gelatinases (substrates: basement membrane collagens), and the stromelysins (substrates: proteoglycans, fibronectin, and laminin). These enzymes are important in embryogenesis, tissue remodeling, wound healing, and tumor invasion. Most MMPs in human endometrium are expressed in stroma, although MMP-7 is expressed in glandular epithelium. MMP mRNAs are expressed in the proliferative phase and, except for MMP-1, are conspicuously absent in the midsecretory phase during the window of implantation. TIMPs are specific inhibitors of MMPs. TIMP-1 and TIMP-2 are constiuitively expressed throughout the cycle, although TIMP-3 is differentially expressed in the secretory phase and especially in early pregnancy decidua. The expressions of the MMPs and of the TIMPs likely reflect different functions—as inhibitors of endogenous MMPs within the endometrium at the time of menstrual shedding and tissue regeneration and also as part of the maternal "defense" to trophoblast invasion into the stroma by trophoblast-derived MMP production.

In vitro systems have demonstrated that progesterone withdrawal is an effective inducer of several MMPs, including MMP-1, -2, and -3. TIMP-1, -2, and -3 are produced by endometrial stromal but not epithelial cells. Regulation of MMPs by progesterone withdrawal has been demonstrated with cultured endometrial stromal cells or explants. However, it is likely that other factors contribute to MMP production, including the proinflammatory cytokines, TNF-α and IL-1, which are produced by epithelial and stromal cells in the mid- to late secretory phase. In addition, they are produced by migratory cells that are transient residents of the endometrium during the late secretory phase. Other induces of MMPs include endothelin-1 (ET-1), a potent vasoconstrictor and present abundantly in menstrual endometrium, and prostaglandins. The MMPs and other enzymes activate the precursor forms of the MMPs (pro-MMPs), and migratory cells and resident mast cells may secrete enzymes that activate the pro-MMPs at the time of menstruation. In addition to progesterone, TGF-β_1 has a suppressive effect on MMP production. Thus, the temporal and spatial expression and control of MMPs and TIMPs must be tightly regulated to ensure lack of MMP production or activation, e.g., during the window of implantation when extracellular matrix degradation in the stroma would be incompatible with pregnancy, and controlled activation at the time of menstrual shedding.

2. Lysosomal Enzymes

Estradiol stimulates development of Golgi-derived primary lysosomes in glands, stroma, and vascular endothelium of the functionalis which contain potent proteolytic enzymes. During the first half of the postovulatory period these enzymes are confined to the lysosomes, and they are not released because progesterone stabilizes lysosomal membranes. However, when progesterone (and estradiol) levels decrease in the latter part of the cycle, the lysosomes are no longer maintained, and their enzymes are released, leading to local tissue proteolysis.

3. Plasminogen Activators/Plasmin

The plasminogen activators (PAs), urokinase (uPA) and tissue PA (tPA), are specific serine proteases which cleave the inactive zymogen plasminogen to the potent protease, plasmin. tPA is the main PA in fibrinolysis. For nonfibrinolytic actions of extracellular proteolysis (e.g., tissue remodeling, cell proliferation, migration, and invasion), uPA is primarily involved. Human endometrial and uterine luminal fluid PA activity increases during the proliferative phase, is maximal at midcycle, decreases during the luteal phase, and increases again prior to menstruation. In addition, PA activity in proliferative phase human endometrium organ culture is increased by E_2 and decreased by P. There are two types of PA inhibitors (PAI-1 and -2), and PAIs are also present in endometrium. The PA/PAI system may play a role in tissue remodeling during the proliferative phase of the cycle as well as in endometrial shedding and in nonclotting of menstrual blood. TGF-β stimulates PAI activity, thereby inhibiting the PA activation and thus plasmin activation, likely to be important to maintain endometrial structural integrity as well as in the invasive process of implantation.

B. Vasoactive Factors

Vascular epithelium produces a plethora of paracrine substances that maintain a delicate balance between inhibition of vascular growth and stimulation of vascular growth, vasodilation and vasoconstriction, and antithrombotic and hemostatic mechanisms. These processes are probably mediated by a variety of factors, including prostanoids, endothelins, endothelial cell relaxing factor (commonly known as nitric oxide), and tissue factor.

1. Prostanoids

Prostanoids, depending on their structure, are either potent vasoconstrictors or vasodilators. Prostaglandin E_2 and $F_{2\alpha}$ are in the former group, whereas prostacyclin is in the latter group. These likely play a major role in endometrial hemostasis, as do vasoactive peptides.

2. Endothelins

Endothelins (ET) are potent vasoconstrictors derived from endothelial cells and synthesized from larger precursors, preproendothelins. PreproET-1 mRNA is expressed in endometrium, with highest abundance in menstrual, compared to proliferative or secretory, endometrium. It has been found in cultured glandular epithelium and endometrial stromal cells, which also secrete immunoreactive ET into the conditioned medium. Specific binding sites for ET-1, ET-2, and ET-3 have been detected on endometrial vascular endothelium and on endometrial glandular epithelium. These data suggest that ET-1 derived from endometrial stromal and/or epithelial cells may act on the adventitial surface of neighboring spiral arterioles in the endometrium promoting vasoconstriction, a requisite for menstrual shedding. ET-1 is degraded by enkephalinase [membrane metalloendopeptidase (MMEP)] which proteolyzes a number of peptides. MMEP is stimulated by progesterone and levels are highest in midsecretory endometrium. It is likely that MMEP activity is highest during the window of implantation when endometrial vasoconstriction must be minimized to ensure adequate oxygenated blood for an implanting conceptus. In addition, low levels of MMEP during menstruation would allow for ET action for vasoconstriction, which is important during this phase of the cycle.

3. Fibroblast Growth Factor

Basic fibroblast growth factor is a heparin-binding angiogenic protein. It is highly mitogenic for capillary endothelial cells *in vitro* and can induce angiogenesis *in vivo*. By immunohistochemistry, it has been localized to the basement membranes of medium-sized blood vessels and to smooth muscle cells of the tunica media and to capillary endothelial cells and basal lamina in most tissues examined, including endometrium. While not cycle dependent, its receptors may be, and thus it likely plays a role in endometrial angiogenesis.

4. Nitric Oxide

Endothelium-derived relaxing factor, released by vascular endothelial cells, has been identified as nitrous oxide. It is a potent vasodilator, perhaps more important than prostacyclin in maintaining vascular tone. It may modulate endometrial bleeding, and preliminary studies have shown the presence of a vascular relaxant in rabbit uterine perfusate which is believed to be nitric oxide. Precise identification, cellular localization, and regulation in rabbit endometrium and the possible presence and role in human endometrium await further investigation.

5. Tissue Factor

Tissue factor is a membrane-bound glycoprotein in cells of nonendothelial origin that initiates coagulation by direct contact with blood. In early secretory and early gestational endometrium, tissue factor is present in decidualized endometrial stromal cells and is stimulated by progestins. It has been postulated that perivascular decidual cell tissue factor may serve to promote endometrial hemostasis during trophoblast invasion of the maternal endometrial vasculature.

C. Apoptosis

Apoptosis or programmed cell death occurs in endometrium of nonmenstruating species as a way to effect tissue resorption in the absence of implantation. In menstruating species, however, endometrial apoptosis has been observed, primarily in the basalis layer. TNF-α and IL-1 are believed to play a role in this process, involving only endometrial basalis glandular epithelium. Whether there is a teleological reason for apoptosis in humans and nonhuman primates or if this is an evolutionary "remnant" remains to be determined.

VII. CLINICAL DISORDERS

There are several clinical disorders that involve the endometrium, including Asherman's syndrome (scarring of the endometrium), submucous leiomyomas (benign tumors of the myometrium that grow adjacent to the endometrium), luteal phase deficiency and nonreceptivity of the endometrium to implantation, nongenetic causes of repetitive miscarriages, breakthrough bleeding associated with contraceptive steroid use and with hormonal replacement therapy, and endometriosis (a disorder of the endometrium infiltrating into the myometrium or found in the abdominal/pelvic cavity and growing and invading into pelvic structures). The causes of most of these disorders are still unknown, and mechanisms underlying their pathogenesis remain to be determined. This is most vexing to patients and clinicians because therapies associated with these abnormalities are limited or are empiric because our fundamental knowledge of mechanisms underlying normal physiologic events in endometrium is, at best, incomplete. It is envisioned that with additional understanding of normal endometrial physiology, abnormalities associated with the endometrium can be treated in the future with rationally designed and efficacious therapies. While major advances have been made over the past decade in this field using animal models and human endometrial cell and explant cultures, much additional research is needed to fulfill the goals and expectations of patients and clinicians who treat women with these disorders.

See Also the Following Articles

APOPTOSIS (CELL DEATH); DECIDUA; ENDOMETRIOSIS; GROWTH FACTORS; IGF (INSULIN-LIKE GROWTH FACTORS); LEIOMYOMA; STEROID HORMONE RECEPTORS

Bibliography

Giudice, L. C. (1997). Biochemical events in the endometrium during the menstrual cycle and implantation. In *Infertility,*

Contraception, and Reproductive Endocrinology (R. A. Lobo, D. R. Mishell, and R. J. Paulson, Eds.), pp. 141–158. Blackwell, Cambridge, UK.

Giudice, L. C., and Ferenczy, A. (1995). The endometrial cycle: Morphologic and biochemical events. In *Reproductive Endocrinology, Surgery, and Technology* (E. Y. Adashi, J. A. Rock, and Z. Rosenwaks, Eds.), pp. 171–194. Raven Press, New York.

Lessey, B. A., Killam, A. P., Metzger, D. A., Haney, A. F., Greene, G. L., and McCarty, K. S. (1988). Immunohistochemical analysis of human uterine estrogen and progesterone receptors throughout the menstrual cycle. *J. Clin. Endocrinol. Metab.* 67, 334–340.

Lessey, B. A., Damjanovich, L., Coutifaris, C., Castelbaum, A., Albelda, S. M., and Buck, C. A. (1992). Integrin adhesion molecules in the human endometrium. *J. Clin. Invest.* 90, 188–195.

Lockwood, C. J., and Schatz, F. (1996). A biological model for the regulation of peri-implantational hemostasis and menstruation. *J. Soc. Gynecol. Invest.* 3, 159–165.

Lopata, A. (1996). Blastocyst–endometrial interaction: An appraisal of some old and new ideas. *Mol. Hum. Reprod.* 2, 519–525.

Salamonsen, L. A., and Woolley, D. E. (1997). Matrix metalloproteinases and their tissue inhibitors in endometrial remodeling and menstruation. *Reprod. Med. Rev.* 5, 185–203.

Seppala, M., Julkunen, M., Riittinen, L., *et al*: (1992). Endometrial proteins: A reappraisal. *Hum. Reprod.* 7, 31–40.

Stewart, C. L. (1994). The role of leukemia inhibitory factor (LIF) and other cytokines in regulating implantation in mammals. *Ann. N. Y. Acad. Sci.* 184, 157–165.

Strickland, S., and Richards, W. G. (1992). Invasion of the trophoblasts. *Cell* 71, 355–357.

Endosalpinx

see Fallopian Tube

Endotheliochorial Placentation

Vibeke Dantzer

Royal Veterinary and Agricultural University

I. Introduction
II. Morphological Events: Cat and Mink
III. Vascular Interrelationship
IV. Placental Subunit, Hemophageous Zone, or Organ

GLOSSARY

angiogenesis The modulation and formation of new generations of capillaries from existing ones.

chorioallantois The fused vascular allantois and nonvascular chorion.

chorion frondosum Different types of projections of the chorion that create a substantial surface area.

chorion laeve The smooth part of chorion.

conceptus The fertilized ovum, which through subsequent development comprises the embryo/fetus and its membranes including the fetal placenta.

decidua The transformed part of the uterine endometrium, inclusive of decidual cells, which is shed together with the fetal membranes at birth.

hemotroph Nutrients transferred from the maternal blood across the interhemal barrier of the placenta to the fetal circulation.
histotroph Uterine glandular epithelial secretions and products of cell degeneration.
placenta The well-vascularized embryonic/fetal membranes and endometrium in intimate contact.
trophoblast The outer cell mass of the blastocyst and later the outer epithelium of the chorion. It has to express paternal genes in order to participate in establishment of the placenta.
vasculogenesis The first or *de novo* development of blood vessels.

Endotheliochorial placentation is the establishment and development of a placenta in which the chorion, vascularized by the allantois of the conceptus, comes into intimate contact and erodes the maternal or uterine mucous membrane. In such placentae, the interhemal barrier consists of three cell layers and two compressed or modified connective tissue layers. On the fetal side, the layers consist of endothelium, mesenchyme, and trophoblast. On the maternal side the layers consist of an interstitial layer and the endothelium of the maternal vascular system. The contact area is increased by the development of folds or lamellae in order to maximize the area available for the exchange of oxygen, carbon dioxide, nutrients, and metabolic products across this interhemal barrier. This increase is related to changes in the shape and extension of the surface area of chorion frondosum and complementary uterine changes. This develops into a complete or incomplete belt or zonary placenta. Due to the invasiveness of the endotheliochorial placenta, the transformed maternal tissue intimately interconnected to the fetal membranes is lost with the afterbirth. This type of placenta is designated as deciduate. Studies of placentation are closely related to processes involved in the establishment of primary contact, implantation, and subsequent placental development. While this process is conserved within a species, it displays marked variations among species. This process between two genetically different tissues is governed by the interaction between a wide range of signal transmitter substances such as hormones, growth factors, and cytokines. The morphology of the different development stages and biochemical data on possible factors involved in the regulation of these dramatic changes is the fundamental basis for further studies. A correlation between morphology, physiological, and biochemical factors can be shown by immunohistochemistry and *in situ* hybridization. This will contribute to the understanding of the regulatory mechanisms inclusive of gene activation, and their interplay to accomplish an efficient exchange system, for complementary feto–maternal epithelial and vascular growth and cellular differentiation during initial placentation as well as subsequent growth and specialization of the placenta throughout gestation.

I. INTRODUCTION

Endotheliochorial placenta is characteristic of carnivores but is also found among a variety of other species, i.e., Proboscidae (elephant), Phocidae (seal), Tubulidentata (aardvarks), Rodentia (*Dipodomys*, the kangaroo rat), Insectivora (*Talpa*, the European mole), and Bradypodidae (the three-toed sloths). The cat and mink are used as examples of this placental type.

II. MORPHOLOGICAL EVENTS: CAT AND MINK

A. Yolk Sac

The yolk sac in the cat forms a transient lamellar choriovitelline placenta, with trophoblast villi invading and eroding the uterine mucous membrane. It is initially extensive but is substituted by the definitive, well-vascularized, chorioallantoic placenta. Although the yolk sac persists to term, its function remains to be elucidated (Fig. 1).

B. Placentation

Implantation begins on Day 12.5 postcoitum (pc) in the cat, whereas the mink has delayed implantation. It can be subdivided into apposition, adhesion, and intrusion phases, preceding development of the definitive placental zones.

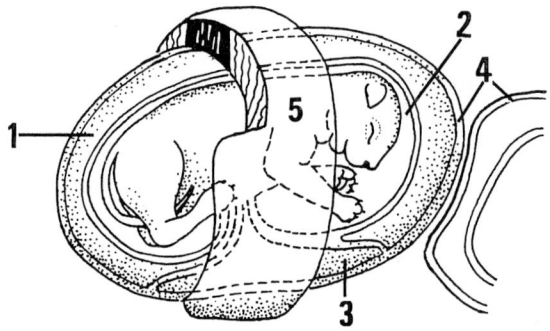

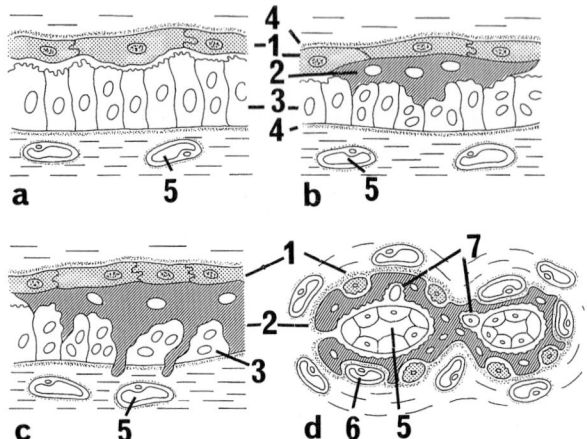

FIGURE 1 Schematic drawing of a cat fetus in its fetal membranes to show the placental shape in carnivores with a zonary (girdle) lamellar or tightly folded placenta. 1, Allantoic cavity; 2, amnionic cavity; 3, yolk sac; 4, allantochorion; 5, zonary placenta. The uterine compartment is not included except for the transformed part, which is part of the zonary placenta.

1. Implantation

It is characteristic that a syncytial trophoblast layer develops from the cytotrophoblast and becomes apposed to the uterine epithelium (Figs. 2, 3a, and 3b). In the apposition phase, seen antimesometrially along the periembryonic ring zone of the conceptus, the trophoblast establishes the first points of contacts, although there is a 20-nm gap separating the still nonvascularized chorioallantois. Shortly after, the chorion develops extensions into the lumen of

FIGURE 3 Diagrammatic sketches showing the penetration through the uterine epithelium by the trophoblast during invasive implantation. (a) Apposition, (b) intrusion, (c) penetration, and (d) established placenta, late gestational stage (carnivores, mink). The trophoblast is shown in two gray shades. 1, Cytotrophoblast; 2, syncytiotrophoblast; 3, uterine epithelium; 4, basal lamina; 5, maternal capillary; 6, fetal capillary; 7, decidua-like cells (a–c modified from Denker, 1990).

the uterine glands and the process of contact between syncytiotrophoblast and uterine epithelium by microvillous interdigitations and embryo–endometrial junctional complexes are established (Figs. 3 and

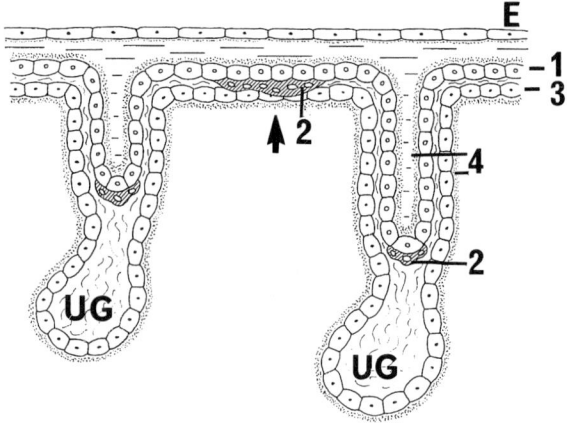

FIGURE 2 Schematic drawing from early implantation showing the initial anchoring by chorionic extensions into the uterine glands (UG) and the beginning of trophoblast intrusion into the uterine epithelium (arrow) during endotheliochorial placentation. 1, Cytotrophoblast; 2, syncytiotrophoblast, with a gray shade; 3, uterine epithelium, continuous with glandular epithelium; 4, basal lamina; E, endoderm.

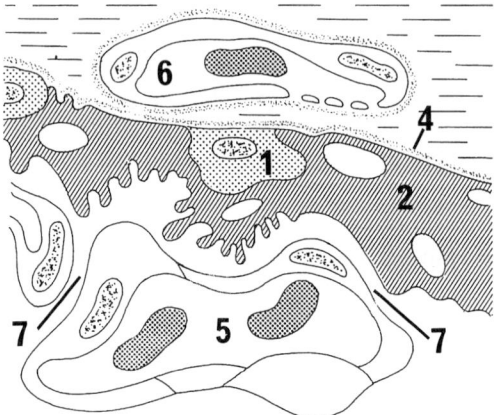

FIGURE 4 The tissue layers of the feto–maternal interhemal barrier of the endotheliochorial placenta. 1, Cytotrophoblast; 2, syncytiotrophoblast; 4, basal lamina; 5, maternal capillary; 6, fetal capillary; 7, interstitial layer, seen as a clear space (modified from R. Leiser and P. Kaufmann, P., Placental structure: In a comparative aspect, *Exp. Clin. Endocrinol.* **102**, 122–134, 1994).

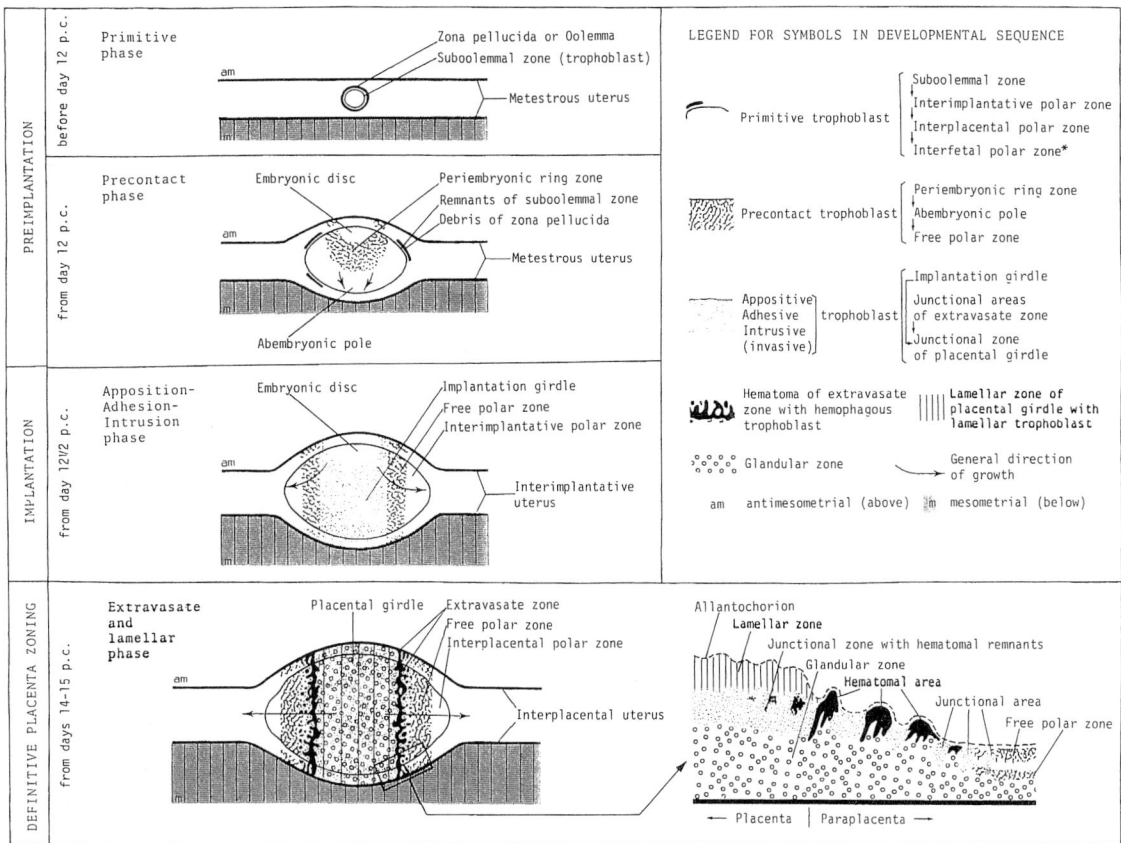

FIGURE 5 Synopsis of changes in the trophoblast and opposing endometrial tissue in the pregnant cat. The size relationships of the different developmental stages are not drawn to scale (from Leiser, 1982, reproduced with permission of S. Karger AG, Basel).

6). This is followed by the intrusion phase. Here a degeneration of the uterine epithelium and progressive intrusion of the trophoblast into the degenerating maternal stroma takes place without affecting the maternal capillaries, which become almost enclosed by the trophoblast (Figs. 3c and 3d). Thus, the first step in an endotheliochorial placenta with tightly arranged lamellae (cat) is established and can be observed from Day 14 pc. It is characteristic that some of the extracellular matrix persists and a basal lamina-like layer, the interstitial layer, is formed between the maternal endothelial cells and the syncytiotrophoblast (Fig. 4). During vascularization of the chorioallantois, a network of subtrophoblastic fetal capillaries develops. It follows the trophoblast and envelopes the much wider maternal capillaries, thereby giving the appearance of a labyrinth (Figs. 3 and 7). This process proceeds in a mesometrial direction to establish the placental girdle, localized around the equator of the chorionic sac (Fig. 1). The chorioallantois then becomes subdivided into the zonary placenta, the paraplacenta, and the interplacental polar zone (Fig. 5).

2. Zonary Placenta

The zonary placenta is subdivided into the lamellar/labyrinth (cat/mink) and junctional zones which are connected to the uterine glandular layer or zone of the endometrium. The lamellar zone is located on the fetal side and beneath it is the junctional zone (Fig. 6). The lamellae are almost regular in the cat; however, they remain villous, but twisted and highly irregular in the mink. In the lamellar/labyrinth zone, the placenta is composed of lamellae/villi of the well-vascularized allantochorion with a thin-walled capillary network beneath the trophoblast layer of syncytio- and cytotrophoblast. The complementary maternal side is composed of wide

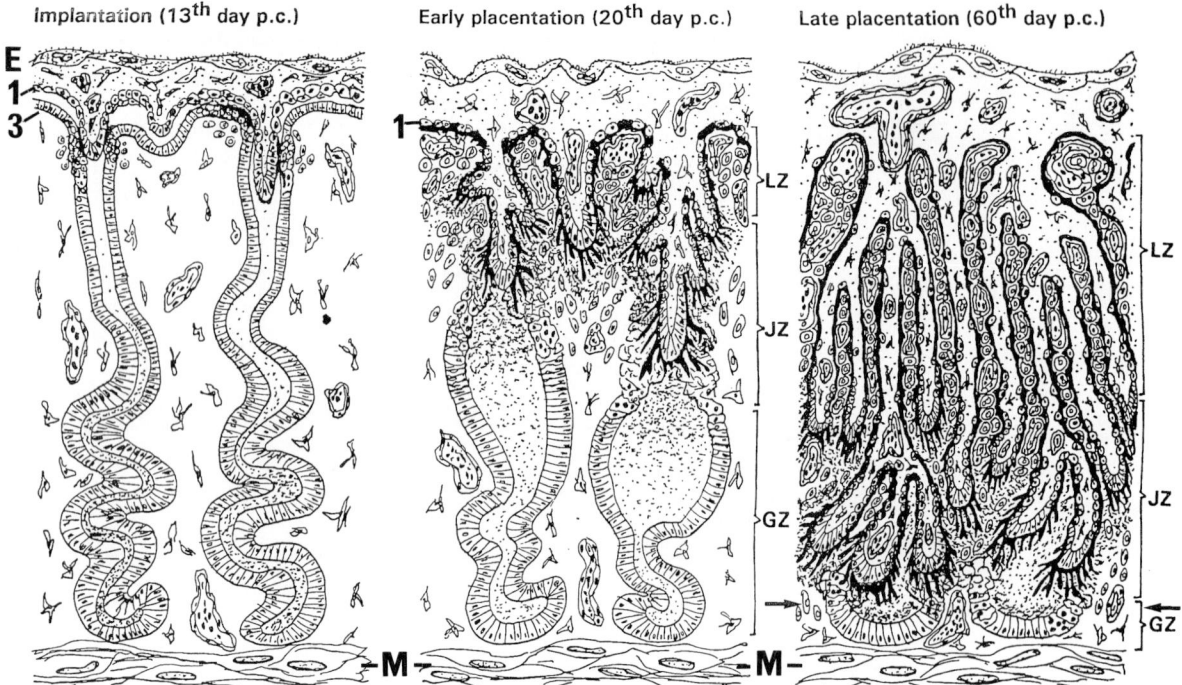

FIGURE 6 Diagrammatic scheme of the different feto–maternal relationships during placentation in the cat as seen from left to right. The typical zonar division and lamellar structure in the late stage of gestation is seen to the right. LZ, lameller zone; JZ, junctional zone; GZ, glandular zone; M, myometrium; E, allantoic endoderm; 1, cytotrophoblast (syncytiotrophoblast is indicated in black); 3, uterine epithelium, continuous with uterine glandular epithelium. The arrows above the glandular zone to the right indicates the prospective area for separation of the placental afterbirth (from Development and characteristics of placentation in a carnivore, the domestic cat, Leiser and Koob *J. Exp. Zool.*, copyright © 1993 John Wiley & Sons, Inc. Reprinted by permission of Wiley-Liss, Inc., a subsidiary of John Wiley & Sons, Inc.).

capillaries and giant decidual-like cells, which are typical for the cat but not clearly identified in the mink (Fig. 3d). The wide maternal capillaries with prominent endothelial cells, most pronounced in the mink, are delineated by an irregular interstitial layer enclosed by a layer of syncytiotrophoblast, a discontinuous layer of cytotrophoblast, and at the mesenchymal side of the lamellae a fine fetal capillary network. The interstitial layer is breached in places by either syncytiotrophoblast or maternal endothelial cell extensions.

The maternal endothelial cells have a well-developed rough endoplasmic reticulum and in the mink many lipid droplets as well as basally located cytoskeletal elements indicative of contractility. The syncytiotrophoblast has a well-developed endoplasmic reticulum, many mitochondria, some dense bodies (lysosomes), and lipid droplets, most prominent in the cat.

The junctional zone of the placenta is composed of terminal parts of fetal lamellae, maternal vessels, uterine glandular secretion, and cell debris. In this area, the trophoblast cells phagocytose histotroph. In the late stages of gestation, the invasion of the uterine glands has progressed almost to the deep end of the uterine glands, leaving only a small shell-like area of glandular epithelium at parturition (Fig. 6). This glandular epithelium then multiplies to restore the uterine epithelium after parturition.

III. VASCULAR INTERRELATIONSHIP

Because the vascular system is so closely related to placental structure, the description was included in the previous paragraph. The three-dimensional interrelationships of maternal and fetal capillaries represent a cross-current type of blood flow (Fig. 7).

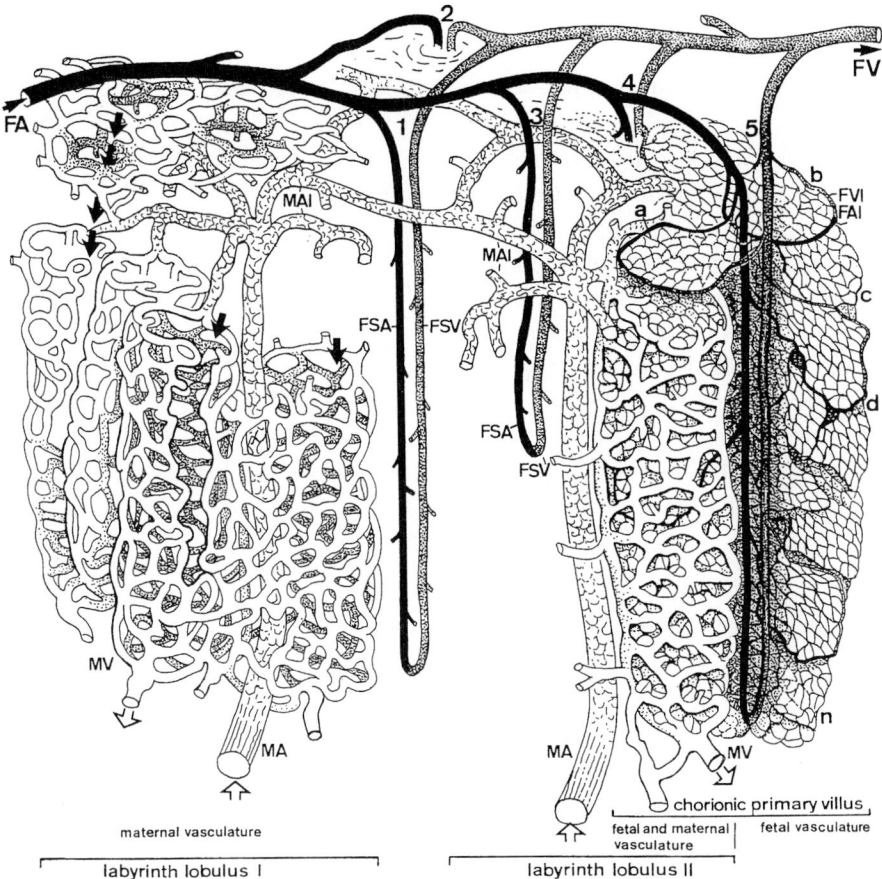

FIGURE 7 Schematic drawing of the mink zonary placenta including the vascular architecture of the placental labyrinth-like zone and the geometric aspect of the materno–fetal capillary interrelationship. Two lobules of the labyrinth (I and II) illustrate a rough hexagonal pattern, each with a maternal stem artery (MA) as a central axis peripherally delimited by five pairs (1–5) of fetal stem arteries (FSA) and stem veins (FSV), each pair marking the central axis of a chorionic primary villus. Lobule I shows the maternal vasculature of the labyrinth with a star-like shape formed by radially oriented arteriolar branches of the stem artery. Complexes of sinusoidal capillaries form hollow laterally split columns. They arise from the radial arterioles, forming a layer at the fetal side of the labyrinth (top), going to the venous outlets (MV) on the maternal side of the labyrinth (bottom). Lobule II shows the same situation only drawn on its left side and with the fetal vasculature included. A maternal column or crypt wall of sinusoid capillary complex is penetrated over its full length by fetal capillary complexes of secondary villi, which are tributary branches of a chorionic primary villus (5). On the upper right side of this chorionic primary villus (5), secondary villi with tributary capillary complexes of terminal villi (a–d and n) are irrigated by blood from a pair of vessels (arteriole and venule) from the stem vessels situated in the axis of this primary villus. The chorionic villous system is supplied by the allantochorionic artery (FA) and vein (VF). The three-dimensional interrelationship of maternal and fetal capillaries represents a cross-current type of blood flow.

IV. PLACENTAL SUBUNIT, HEMOPHAGEOUS ZONE, OR ORGAN

At the paraplacental zone, extravasation of maternal blood becomes more distinct in late gestation. It is visible as a brownish border on either side of the placental girdle in cat (Fig. 5) or as one large hematoma at the antimesometrial side in mink. The hematoma provide, among other elements, iron for the fetus.

A simple proliferative cytotrophoblast in the junctional zone contributes to development of the syncytiotrophoblast and to the hemophagous trophoblast, which grows in height concomitantly with increasing contact with maternal blood, followed by phagocytosis and digestion. The syncytial trophoblast invades the intact endometrium and penetrates into the maternal vessels with consequent blood extravasation, which accumulates in hematomal pouches. These are thus lined by uterine epithelium, hemophagous trophoblast, and tall columnar trophoblast cells. The latter differentiation takes place in a functional cycle of three steps: (i) development of proliferative cytotrophoblast at the junctional area; (ii) followed by developmental changes into hemophagous trophoblast with phagocytotic activity; and (iii) thereafter a digestive phase in which the hemophageous trophoblast is integrated into the junctional area (Fig. 5).

See Also the Following Articles

CATS (FELIDAE); DOGS (CANIDAE); EPITHELIOCHORIAL PLACENTATION; PLACENTA: IMPLANTATION AND DEVELOPMENT

Bibliography

Anderson, J. W. (1969). Ultrastructure of the placenta and fetal membranes of the dog. 1. The placental labyrinth. *Anat. Rec.* 165, 15–36.

Denker, H.-W. (1990). Trophoblast–endometrial interactions at embryo implantation: A cell biological paradox. In *Trophoblast Invasion and Endometrial Receptivity. Novel Aspects of the Cell Biology of Embryo Implantation* (H.-W. Denker and J. D. Aplin, Eds.), Trophoblast Research Vol. 4, pp. 3–29. Plenum, New York.

Denker, H.-W. (1993). Implantation: A cell biological paradox. *J. Exp. Zool.* 266, 541–588.

Enders, A. C., Blankenship, T. N., Lantz, K. C., and Enders, S. S. (1998). Morphological variation in the interhemal areas of chorioallantoic placentas. *Trophoblast Res.* 13, in press.

Krebs, C., Winther, H., Dantzer, V., and Leiser, R. (1997). Vascular interrelationship of near term mink placenta: Light microscopy combined with scanning electron microscopy of corrosion casts. *Microsc. Res. Technique* 38, 125–136.

Leiser, R. (1982). Development of the trophoblast in the early carnivore placenta of cat. *Bibl. Anat.* 22, 93–107.

Leiser, R., and Koob, B. (1993). Development and characteristics of placentation in a carnivore, the domestic cat. *J. Exp. Zool.* 266, 642–656.

Mossman, H. W. (1987). *Vertebrate Fetal Membranes: Comparative Ontogeny and Morphology, Evolution, Phylogenetic Significance, Basic Functions, Research Opportunities.* Macmillan, New York.

Pfarrer, C., Winther, H., Leiser, R., and Dantzer, V. (1998). The development of the endotheliochorial mink placenta: Light microscopy and scanning electron microscopical morphometry of maternal vascular casts. *Anat. Embryol.* (accepted 2/7/98)

Wooding, F. B. P., and Flint, A. P. F. (1994). Placentation. In *Marshall's Physiology of Reproduction* (G. E. Lamming, Ed.), 4th ed., Vol. II, Chap. 4. Chapman & Hall, London.

Energetics of Reproduction

Cynthia Carey

University of Colorado

I. General Concepts and Methods
II. Energetics of Reproduction
III. General Conclusions

GLOSSARY

allometry A study of body size and its consequences. Allometric relations are usually described by a power function: $X = aM^b$, where M is body mass in grams, X is some size-related function or anatomical property, and a and b are constants denoting the intercept of the line on the abscissa and the slope of the line, respectively. Alternatively, the function can be written as $\log X = \log a \pm b (\log M)$.

ectotherms Animals such as all invertebrates and vertebrates except for birds and mammals that have variable body temperatures, usually very close to air/substrate temperature.

endotherms Animals such as birds or mammals that have a high and relatively constant body temperature produced by internal heat production.

metabolic rate The sum total of all metabolic or biochemical processes ongoing in an animal per unit time. The lowest metabolic rate necessary to keep an animal alive is termed the "basal" or "standard" metabolic rate; it is usually estimated by measuring oxygen consumption, heat production, or carbon dioxide production. Energy expenditures for reproduction, activity, thermoregulation (in birds and mammals), and growth elevate metabolic rate above standard or basal levels. The "field metabolic rate" is calculated using water labeled with isotopes and represents the total metabolic rate of a free-ranging animal over the duration of the test, frequently a few days.

oviparity The mode of reproduction in which eggs are laid by the female. Embryos complete most of their development to hatching outside the female's body. The fertilized eggs may be retained in the oviduct for some time before laying in certain species.

reproductive effort The proportion of the total energy budget of an animal allocated to reproductive processes.

viviparity The mode of reproduction in which fertilized eggs are retained in the body of the female. Embryos develop to a particular stage and then are "born." Viviparous species are separated by some workers from "ovoviviparous" species on the basis of how much nutrition the embryo receives from the female during gestation. Many intermediate stages between viviparity and ovoviviparity exist in fish, amphibians, and reptiles.

Energetics of reproduction is defined as the amount of energy that an animal expends to reproduce. The energetics of reproduction can include the costs of gamete manufacture, synthesis of secondary sexual characteristics and sex-attractant chemicals (pheromones), and reproductive behavior including territorial defense, nest building, courtship rituals, and parental care. Reproductive costs are usually evaluated as a proportion of the amount of energy that the animal expends per unit time—per day, breeding season, year, or lifetime. Because reproduction can be one of the most energetically demanding events in the life of an animal, energy availability and utilization influence the ability of a species to produce a succeeding generation. If energy or essential nutrients are limited just prior to or at reproduction time, reproduction may be reduced or may not even occur.

I. GENERAL CONCEPTS AND METHODS

The chemical energy contained in the food that an animal absorbs from the digestive tract is allocated among the competing demands of maintenance (usually defined as basal, routine, or standard metabolic rate), growth, storage for future use, activity, and reproduction. Allocation of energy among these necessities varies at different times of the annual cycle. Furthermore, energetically costly activities may be separated in time so that energy intake is not divided among too many requirements simultaneously. For instance, some mammals avoid the high costs of thermoregulation during winter when food supplies may be diminished or unavailable by hibernating, i.e., undergoing a regulated decrease in body temperature and activity. Reproduction occurs in the spring when food supplies increase and thermoregulatory costs decrease. Fat deposition for the next hibernation period begins only after the young have been weaned.

Darwin was probably the first to speculate about the significance of reproductive energetics: "On the whole, the expenditure of matter and force by the two sexes is nearly equal, though effected in very different ways and at different rates" (C. Darwin, 1871, p. 63). Two general approaches have been used to determine the cost of reproduction. Evolutionary biologists view the costs of reproduction in terms of how reproduction at one time affects the survival and future reproductive fecundity of the adult. In other words, "costs" are not quantified energetically but are viewed in the context of demography, probability of survival to the next breeding season, age-specific mortality, environmental predictability, and fitness. The theoretical interrelations among these parameters are complex and beyond the scope of this review. Briefly, each species is thought to have evolved a set of traits that govern how energy is allocated for current reproduction in relation to the probability of survival for future reproduction. In certain environments, long-lived animals may have late maturity, multiple clutches, fewer but bigger eggs, and parental care, whereas selection in other environments may favor short life spans, early maturity, large clutches with small eggs, one or few reproductive attempts, and no parental care. Because of the physiological emphasis of this encyclopedia, this approach will receive limited attention in the subsequent discussion.

Other approaches have evaluated the costs of reproduction by estimating how much energy is expended in reproductive activities, including gamete production, sexual displays, synthesis of pheromones or anatomical secondary sexual characteristics, territorial defense (related to reproduction), movement to breeding area, parental care, nest building, and courtship behavior. While quantifying all these costs, both in absolute terms and as a proportion of the total annual or lifetime energy budget of an animal, is a most important goal, the techniques necessary to quantify accurately either reproductive costs or total energy budgets do not exist. A number of methods have been attempted: Costs of gamete production in oviparous females have been estimated by measuring the caloric content of an egg and multiplying by the number of eggs produced in one reproductive event (clutch or litter size). This method underestimates the cost of gamete production since egg caloric content does not include the costs of ovarian recrudescence (growth of the ovary, which may be substantial in a seasonally breeding animal, in preparation for egg production), accumulation of nutrients through foraging, mobilization of nutrients from storage depots, and transport of the nutrients from the blood into the egg. Studies that estimate clutch size from ovary mass or volume may overestimate the number of eggs to be laid if not every follicle in the ovary produces an egg during a given reproductive event. Estimates of costs of reproductive behavior have also been made by measuring oxygen consumption in closed systems, but errors can occur if the animals alter their normal behavior in response to testing conditions.

After the absolute values of the energetic costs of reproduction have been estimated by these methods, they have generally been evaluated in the contexts of how reproductive costs vary in relation to other factors, such as the total annual or lifetime energy budget, food intake, or body mass.

1. In some studies, oxygen consumption of animals exhibiting various types of behaviors (sitting, sleeping, foraging, etc.) has been used in conjunction with time budgets (how much time spent in each activity) to estimate how much energy is used per day. Since behavior of most animals is impacted by captivity and by confinement in closed containers during measurements of oxygen consumption, errors can easily be introduced into the calculation of these

energy budgets. Although this type of model cannot be used to estimate the costs of manufacture of gametes, it could be useful for predicting the costs of parental care or other reproductive behaviors.

2. Another method of estimating metabolic expenditures during a 24-hr period has used biophysical models of energy exchange between the animal and its environment through radiation, conduction, convection, and evaporation. These models suffer in their overall utility for estimating reproductive costs because these costs cannot be separated from overall metabolic heat exchange.

3. Injection of water, labeled with either deuterium (^{2}D) and oxygen 18 (^{18}O) or tritium (^{3}H) and ^{18}O, has led to fairly accurate predictions of metabolic rate of field-active animals as long as a number of assumptions and criteria are met. However, this method calculates an estimate of the total energy expenditure during the sampling time period, from which it is not possible to separate out metabolic costs of reproduction.

4. Another approach uses the caloric content of the food, measured with bomb calorimetry, in conjunction with caloric value of tissue and eggs to estimate the production costs of these materials as a proportion of food intake. A variant of this method estimates caloric content of body mass, eggs, and stored lipid as a measure of the total calories in storage for future reproduction or maintenance needs. This method assumes that all stored energy is in the form of extractable lipid, an assumption which ignores glycogen as a storage form of glucose and structural protein as a mobilizable form of amino acids for egg manufacture. Furthermore, it does not account for the energetic costs of acquisition and transport of substrate into these materials.

5. In general, body size is one of the most important factors that determine fecundity. Allometric equations can be calculated to describe the relation of energetic costs of reproduction to body mass. Since measurement of body mass is relatively less complicated than other measurements noted previously, the accuracy of this approach is limited largely by the methods used to estimate energetic costs of reproduction. The utility of this approach is, of course, also limited in that its predictive value is restricted to effects of body mass alone and it ignores other factors that could affect allocation of energy to reproduction.

The attempts to define how energy is allocated among the competing demands of maintenance, activity, storage, growth, and reproduction and how food limitations and environmental factors alter energy allocation are important. However, errors and omissions in quantifying the amount of energy devoted to reproduction in absolute terms are compounded by problems associated with accurately calculating daily, monthly, and yearly energy budgets. Therefore, conclusions based on reproductive costs as a proportion of the total amount of energy expended per unit time should be viewed with caution.

The following overview reviews the approaches taken to understand the energetic costs of reproduction of vertebrates, excluding Chondrichthyes (sharks, rays, etc.).

II. ENERGETICS OF REPRODUCTION

A. Fish

The economic value of aquaculture has led to the calculation of energy budgets for a number of fish species, but most of these estimates emphasize energy allocation to growth and have no reproductive component. No studies exist that provide estimates of annual costs of reproduction including all reproductive activities (gamete production and reproductive behaviors). Energetic costs of egg production have been estimated on the basis of ovarian mass just prior to spawning relative to body mass, ovarian caloric content relative to body caloric content, or energy in eggs produced per year to amount of food consumed per year. Ovarian mass may vary from <8% of body mass during nonspawning seasons to over 30% just before spawning. Annual fecundity has proven difficult to calculate for many species because some species can breed multiple times during a breeding season. The average energy content of eggs or ripe ovaries in teleost fish is 23.48 kJ.g-1 ($n = 50$ species). Costs of annual egg production estimated from the energy contained in eggs produced each year relative to the energy of food consumed varied between 1.3 and 11 or 12% in 2- and 8- to 25-year-old cod (*Gadus morhua*), respectively. Costs of egg production for several other temperate saltwater teleosts varied from 1 to 10% of the annual energy budget, irrespective of whether fish are short-lived and breed once or long-lived, multiple spawners. The efficiency of gamete production is in fish appears to be lower than that for converting food to body tissue during growth (10–25%). However, if all costs of reproduction were included in the calculations, efficiencies of growth and reproductive energy expenditures might prove to be similar.

While most species are oviparous, some species are viviparous. In these species, many different anatomical arrangements exist for transfer of nutrients to the embryo. In some cases, yolk stores adjacent to the embryo are mobilized to provide all the nutrients, or nutrients may be transferred to the embryo through a placenta or placenta-like organ. In some species, embryos obtain nutrients by feeding on sibling eggs. Estimates of annual energy expenditures for reproduction based on ovarian mass or energy content in viviparous fish do not include the costs of nutrient supply to embryos.

Fish lay several to millions of eggs per spawning. Tropical and subtropical fish may reproduce several times a year, whereas temperate and arctic fish generally have one breeding season per year. The most critical factor determining how many eggs are laid is body mass. Most fish appear to grow throughout their lifetime. Since larger fish usually produce more eggs than smaller fish, energy devoted to growth ultimately may contribute positively to reproductive effort. However, at any point in time, fecundity in fish is generally negatively correlated with growth rates. When food is limited, some species reabsorb some of their eggs or do not develop all the oocytes, whereas others support near-normal levels of ovarian growth during food restriction by depleting body stores. In experiments in which ration size is varied, females on higher rations produce slightly more eggs than those provided less food.

B. Amphibians

Caecilians reproduce biennially and salamanders annually or biennially. Tropical anurans can produce up to several clutches a year, whereas temperate ones typically reproduce once annually. While the anuran populations at high latitudes or altitudes usually breed yearly, individual females may lay eggs only once every 2 or 3 years. Annual fecundity of amphibians can vary between 1 and 2 to more than 80,000 offspring. While eggs of most species are typically laid in water and develop into aquatic larvae that later undergo metamorphosis to the adult stage, amphibians exhibit a variety of other breeding modes: Some species lay eggs in overhanging vegetation, in burrows, or on land, and the tadpoles wriggle into the water; eggs of some species are carried on the legs, in the stomach, or in dorsal pouches on the backs of adults; some eggs develop directly without a larval stage; and some viviparous species retain the eggs in the oviduct until hatching.

The testes increase in size and mass during spermatogenesis in temperate anurans just prior to breeding but the organs of tropical males remain functional throughout the year, with sperm in various stages of maturation. Egg production involves deposition of yolk (45% phosphoprotein, 25% lipids, and 8% glycogen), during which time the size of an oocyte can increase 27,000-fold or more.

Estimates of the physiological costs of reproduction have been made by measuring clutch size (1–48,000 eggs) and the size and caloric content of an eggs. In general, a positive correlation exists between clutch size and female snout–vent length. For instance, the relation between "ovarian complement" (OC, no units given) and snout–vent length (SVL, mm) for tropical plethodontid salamanders having direct development of terrestrial eggs is as follows:

$$OC = -1.58 + 0.377\ SVL,\ r = 62.4,\ p < 0.001$$

In general, larger females tend to lay larger eggs, clutch size and egg size are negatively correlated, and larger eggs tend to produce larger hatchlings. Few attempts to estimate the cost of reproduction relative to an annual energy budget exist, but about 70% of the annual secondary production of a female terrestrial salamander *Desmognathus ochrophaeus* is thought to be allocated to egg production. While the age at first reproduction is known for many species, information on life span and total reproductive effort of females over their lifetime is available for only a handful of species. Considerable effort has been made to identify the energetic costs of reproductive behavior in amphibians. Oxygen consumption of terrestrial plethodontid salamanders, which are typically inactive in the field except for short bouts of foraging, aggressive behavior, or courtship, was only 38% higher during courtship than that of noncourting animals. Amphibians differ interspecifically in relation to the types of metabolic pathways which support activities. Some amphibians support almost all vigorous locomotor activity anaerobically, whereas others are almost exclusively aerobic. However, the most active male *Bufo americanus* during breeding choruses are not the ones in the population with the highest aerobic ability to maintain prolonged muscle contraction during locomotion, nor is the ability of males of a number of species to call loudly frequently correlated with their locomotory aerobic capacity. These results indicate that the large body of information that has accumulated on aerobic and anaerobic capacities for locomotion will not prove particularly useful in predicting the costs of various reproductive behaviors or which males may prove to be the most successful reproductively.

C. Reptiles

Reptiles utilize both oviparity and viviparity in the production of young. Viviparity appears to have evolved independently in several different families in the squamate reptiles but not in crocodiles or turtles. Some snakes or lizards may be oviparous in some populations but viviparous in others. Almost all reptiles, even those in tropical climates, breed cyclically. Both testes and ovaries regress during nonbreeding periods and grow to functional size prior to the breeding season. Clutch sizes may range upwards of several hundred eggs in turtles.

Total annual energy budgets have been estimated for several species of oviparous lizards and turtles using calorimetry of lean dry body mass, eggs, lipid extraction and growth rates, and field metabolic rates. A similar trade-off between growth rate and reproductive energetics exists in reptiles as that noted previously in amphibians. Estimates based on these measurements indicate that reproduction, in terms of egg production only, accounts for 6–40% of the total energy budget per year, with most values falling below 25%. These estimates are based on populational averages of lipid storage, growth rates, and caloric content of body and reproductive tissues. Therefore, this type of model does not allow determination of an energy budget for a specific individual. No information is available on the costs of reproduction of viviparous reptiles.

D. Birds

Birds are strictly oviparous. Temperate and high-latitude species have one breeding season per year, although they may lay more than one clutch or renest more than once in that season. Tropical species may have more than one breeding season per year. The gonads of both males and females regress between breeding events, except in some desert species that maintain gonads in a near state of readiness so that breeding can commence rapidly after a rain.

Egg size varies from about 0.2 g in hummingbirds to 1500 g in ostriches. Egg mass (E_m, in grams) is directly related to female body mass (M, in grams) as described for passerine birds by the equation:

$$E_m = 0.258 \, M^{0.729}$$

This equation indicates that egg mass becomes a proportionally smaller fraction of female body mass as birds get larger. For a 4-g bird, egg mass is 18% of body mass, whereas it is only 4% of the mass of a 1200-g bird. The slope of this line is statistically indistinguishable from that relating metabolic rate to body mass in birds, indicating that the relative cost of producing a egg is similar regardless of body mass.

Egg contents at laying vary according to the developmental maturity of the chick. Eggs producing precocial young (relatively mature at hatching, feathered, may be able to feed self, mobile, and can thermoregulate to some extent) contain an average of 46% more yolk than similarly sized altricial eggs (young are relatively helpless at hatching, have few feathers, and cannot thermoregulate). The caloric content of eggs increases with the yolk content: Altricial and precocial eggs contain 4.7 and 12.3 kJ g^{-1} wet mass, respectively. The costs of producing more energetically inexpensive altricial eggs are presumably more than balanced by the costs of parental care for these immature hatchlings but the overall total expenditures for reproduction have not been quantified for either precocial or altricial eggs.

The cost of ovarian growth prior to egg production is estimated to be about 2–9% of the basal metabolic rate of birds. Costs of egg production have been estimated by the equation:

$$E = 1.3 \times 4.23 \times W$$

where E is the cost of producing an egg (kJ), 1.3 is the inverse of 0.77, the net efficiency of egg production (kJ kJ^{-1}), 4.23 is the average energy content of passerine eggs (kJ g^{-1} egg), and W is the egg mass (grams). Production of an egg takes about 41% of the daily basal metabolic rate of a passerine bird.

The relation between clutch size (between 1 and 20 eggs) and body mass has been analyzed in several families of birds. In ducks (Anatidae), body mass varies over a 50-fold range, but egg mass increases only 14-fold within this range and clutch mass increases only 7.6-fold. Clutch mass is almost 100% of body mass in the smallest duck, whereas it is only 16% of the body mass of the largest duck.

E. Mammals

Monotremes lay eggs, whereas marsupial and eutherian mammals are viviparous. Marsupials have a short gestational period and a long lactation period, whereas the opposite is true of eutherians. The gestational period of elephants is about 20 months. Despite theories that the marsupial pattern of reproduction might be cheaper energetically than that of eutherians, marsupial and eutherian expenditures are similar when energetic expenditures are compared as a function of relative time from conception to weaning.

Litter size ranges from 1 offspring in shrews and some bats, primates, whales, elephants, and deer to about 13 in some canids and rabbits. The total energetic expenditures during pregnancy (kcal) in eutherian mammals as a function of litter size (W_L, in kilograms) are predicted by the following equation:

$$\text{kcal} = 4090 \, W_L^{1.24}$$

This cost was calculated by measuring oxygen consumption throughout pregnancy and subtracting maintenance costs.

Since lactation can be more energetically expensive than gestation, some authors have argued that costs of reproduction in mammals are better described by equations relating weaning mass of the offspring to body mass of the female. The slopes of the regression lines are not significantly different ($b = 0.77$ for litter weight and 0.73 for weaning weight), indicating that the larger the mammal, the less energy invested per young. Since the slope of the allometric equation relating basal metabolic rate of mammals to body mass is 0.75, the relative caloric investment of the mother in producing an offspring is independent of body mass. Estimates for reproductive effort for several species of rodents as a proportion of total energy budget ranged from 27 to 32%.

III. GENERAL CONCLUSIONS

The differences in methods used to compare absolute and relative values for reproductive costs of various taxa make comparisons difficult. However, certain generalities have emerged that may or may not hold up as more about reproductive energetics is learned. The clutch and litter sizes of birds and mammals are significantly smaller than those of many ectotherms. Parental care, while present in some lower vertebrates, is typical in birds and mammals. The overall cost of production of 1 g of offspring appears to be relatively constant across invertebrate and vertebrate taxa, but endotherms (birds and mammals) devote a smaller proportion of their total energy budget to reproduction than do ectotherms. This phenomenon may be best explained by the fact that thermoregulatory costs comprise such a large proportion of the total energy expended per year in birds and mammals. Within homeotherms, larger organisms expend less energy per offspring than smaller ones.

Bibliography

Brett, J. R., and Groves, T. D. D. (1979). Physiological energetics. In *Fish Physiology*, (W. S. Hoar, D. J. Randall, and J. R. Brett, Eds.), Vol. VIII, pp. 279–352. Academic Press, Orlando, FL.

Carey, C. (Ed.) (1996). *Avian Energetics and Nutritional Ecology*. Chapman & Hall, New York.

Congdon, J. D., Dunham, A. E., and Tinkle, D. W. (1982). Energy budgets and life histories of reptiles. In *Biology of the Reptilia* (G. Gans and F. H. Pough, Eds.), Vol. 13, pp. 233–271. Academic Press, New York.

Darwin, C. (1871). *The Descent of Man, and Selection in Relation to Sex.* John Murray, London.

Duellman, W. E., and Trueb, L. (1986). *Biology of Amphibians.* McGraw-Hill, New York.

Duvall, D., Guillette, L. J., and Jones, R. J. (1982). Environmental control of reptilian reproductive cycles. In *Biology of the Reptilia* (G. Gans and F. H. Pough, Eds.), Vol. 13, pp. 233–271. Academic Press, New York.

Hayssen, V., van Tienhoven, A., and van Tienhoven, A. (1993). *Asdell's Patterns of Mammalian Reproduction: A Compendium of Species-Specific Data.* Cornell Univ. Press, Ithaca, NY.

Loudon, A. S. I., and Racey, P. A. (Eds.) (1987). *Reproductive Energetics in Mammals.* Clarendon, Oxford, UK.

Packard, G. C., Tracy, C. R., and Roth, J. J. (1977). The physiological ecology of reptilian eggs and embryos, and the evolution of viviparity within the class Reptilia. *Biol. Rev.* 52, 71–105.

Rahn, H. (1982). Comparison of embryonic development in birds and mammals: Birth weight, time and cost. In *A Companion to Animal Physiology* (C. R. Taylor, K. Johansen, and L. Bolis, Eds.), pp. 124–137. Cambridge Univ. Press, Cambridge, UK.

Rahn, H., Sotherland, C. V., and Paganelli, C. V. (1985). Interrelationships between egg mass and adult body mass and metabolism among passerine birds. *J. Ornithol.* 126, 263–271.

Reiss, M. J. (1989). *The Allometry of Growth and Reproduction.* Cambridge Univ. Press, Cambridge, UK.

Ryan, M. J. (1992). Costs of reproduction. In *Environmental Physiology of the Amphibians* (M. E. Feder and W. W. Burggren, Eds.), pp. 426–434. Univ. of Chicago Press, Chicago.

Stearns, S. C. (1976). Life history tactics: A review of the ideas. *Q. Rev. Biol.* 51, 3–47.

Wooton, R. J. (1979). Energy costs of egg production and environmental determinants of fecundity in teleost fish. *Symp. Zool. Soc. London* 44, 133–159.

Energy Balance, Effects on Reproduction

George N. Wade
University of Massachusetts at Amherst

I. Energy Balance and Fertility
II. Reproductive Physiology
III. Reproductive Behaviors
IV. Metabolic Cues and Sensory Mechanisms
V. Summary

GLOSSARY

area postrema A neural structure located in the floor of the fourth ventricle in the caudal hindbrain; situated so that it is able to sample the composition of both cerebrospinal fluid and blood.

2-deoxy-D-glucose A glucose analog that inhibits cellular glucose oxidation and induces glucoprivation.

glucoprivation Reduced availability of glucose for intracellular oxidation and ATP production.

leptin A 166-amino acid polypeptide secreted by adipose tissue; animals deficient in leptin or its receptors are obese and infertile.

lipoprivation Reduced availability of fatty acids for intracellular oxidation and ATP production.

methyl palmoxirate A compound that inhibits fatty acid transport into mitochondria, inhibits fatty acid oxidation, and induces lipoprivation.

nutritional infertility Reduced reproductive capacity resulting from insufficient availability of metabolic fuels.

The food supply is the single most important environmental factor controlling reproduction. Energy availability has immediate, direct effects on reproduction in opportunistic breeders, and it shapes responses to long-term cues for seasonal breeding, such as photoperiod. That is, food serves as both a proximate factor and an ultimate factor in the regulation of fertility.

I. ENERGY BALANCE AND FERTILITY

A. Adaptive Significance

In times of abundance, when animals are in positive energy balance, they are able to reproduce at will, maintain other physiological processes at optimal levels, and still have calories left over to sequester as triglycerides in adipose tissue storage depots. However, in nature both food availability and energetic demands can fluctuate dramatically and sometimes unpredictably. When energy expenditure exceeds intake, as during winter or periods of famine, an obvious physiological response is to mobilize adipose tissue triglycerides, accumulated when food was abundant, so that the constituent glycerol and fatty acids can be oxidized to provide energy. Animals in negative energy balance partition their remaining calories among the various energy-consuming processes according to a set of priorities that maximizes chances of individual survival and optimizes long-term reproductive success (Fig. 1).

Some physiological processes, such as basic cellular maintenance and cardiorespiratory function, cannot be compromised and must be maintained at all costs. Other activities may or may not change, depending on the circumstances. For example, some mammals maintain elevated body temperatures during negative energy balance, whereas others may save energy by hibernating or entering a shallow torpor. Similarly, locomotor activity may be reduced during food shortage, or it may increase as animals forage for food. Other processes, not essential for individual survival, are uniformly diminished during food shortage. One example is the growth of juvenile animals; another is reproduction.

That reproduction is inhibited during famine or excessive energy expenditure is not unexpected. A complete cycle of mating, conception, pregnancy, and lactation is one of the most energetically expensive activities that female mammals can undertake, particularly in species in which females bear multiple offspring concurrently. If a reproductive attempt is made when energy is not sufficiently abundant, it is unlikely that the offspring will survive. In addition, an inappropriately timed reproductive attempt may cost the mother her life. Thus, most species have adopted the strategy of temporarily deferring reproductive attempts during periods of energy shortage,

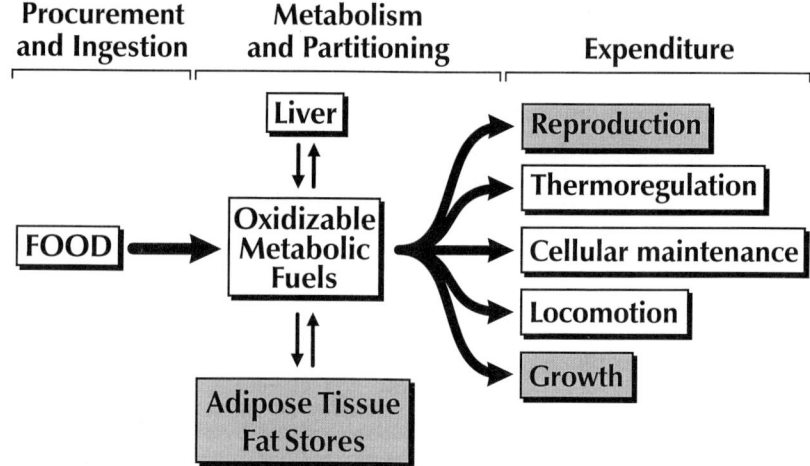

FIGURE 1 Partitioning of metabolic fuels. When energy availability is insufficient to meet demands, processes such as reproduction and growth are deferred, and adipose tissue fat stores are mobilized. Essential processes such as cellular maintenance, thermoregulation, and foraging for food are typically maintained (modified from Wade et al., 1996).

resuming breeding when energy is once again abundant.

Nutritional infertility has been observed in all species in which it has been examined, ranging from shrews to whales and including human beings. This can manifest itself in a number of ways, including a delay in the onset of puberty, suppression of ovulatory cycles and estrous behavior in adult animals, a reduction in lactational performance, and an inhibition of parental behavior.

Although nutritional infertility is a perfectly normal adaptive response to prevailing environmental conditions, it is often regarded as a pathology. Indeed, it has a significant biomedical impact during famine or even when food is plentiful and women restrict their intakes for cosmetic and other reasons, engage in prolonged strenuous exercise, or both. Furthermore, the health consequences of reduced estrogen levels (a characteristic of nutritional infertility) extend beyond fertility and include secondary changes such as osteoporosis, cardiovascular disease, and cognitive impairments. In the animal production industry, nutritional infertility has a substantial economic impact. Diversion of energy to competing processes, such as meat or milk production, can seriously impair fertility and limit the reproduction of breeds with otherwise economically desirable traits.

B. Causes of Nutritional Infertility

A number of types of metabolic perturbations can result in nutritional infertility, but they all share a common feature—a decreased availability of metabolic fuels for oxidation. Not all these disturbances are associated with a negative energy balance.

The most obvious cause of nutritional infertility is undereating, whether it is involuntary (e.g., during a famine) or voluntary (e.g., in the case of eating disorders). Another cause is elevated levels of energy expenditure unaccompanied by compensatory increases in food intake. (In fact, this could be regarded simply as undereating.) Examples include participation in endurance sports, such as distance running, or living in cold environments. The latter is particularly demanding of energy in small animals due to their low surface-to-volume ratios. Individuals whose food intake is limited and who expend a great deal of energy (e.g., ballet dancers, gymnasts, and wild animals in winter) are particularly susceptible to nutritional infertility.

Note that in all these examples nutritional infertility is associated with negative energy balance, i.e., eating too little given the level of energy expenditure. In these cases, infertility can be prevented or reversed simply by eating more. Thus, endurance athletes can prevent amenorrhea by increasing their intakes to compensate for the calories used in training. Small animals, such as mice, remain fertile when living in cold, energy-demanding environments as long as ample food is available. Therefore, maintenance of fertility depends on the balance between energy intake and expenditure rather than on the absolute levels of either one.

Some forms of nutritional infertility are seen in individuals in positive energy balance. In these cases metabolic fuels may not be generally available because they are being used for processes such as growth or fattening. Indeed, some types of obesity are due to excessive uptake and storage of metabolic fuels by adipose tissue rather than to overeating. In these cases animals overeat because they are becoming obese rather than vice versa. This disorder of energy partitioning deprives other physiological processes of energy and can result in nutritional infertility. The phenomenon can be duplicated experimentally by treating animals with pharmacological doses of insulin, which stimulates adipocyte uptake and storage of glucose and fatty acids and inhibits mobilization of metabolic fuels. If these insulin-treated animals are not permitted to compensate for the excess storage of metabolic fuels by increasing their food intake, fertility is suppressed.

Untreated or poorly controlled diabetes mellitus is another metabolic disorder in which positive energy balance is often accompanied by nutritional infertility. In diabetics, blood glucose levels are greatly elevated, but the glucose is not available for oxidation in most tissues due to the fact that insulin is required for glucose utilization by many types of cells. Correction of disordered glucose metabolism by insulin treatment can restore fertility in diabetics. Unlike glucose, fatty acid utilization is not impaired in diabetes, and in some experimental animal models diabetic infertility can be reversed or ameliorated by feeding high-fat diets.

Another example of nutritional infertility without

negative energy balance, discussed at some length later, is that caused by metabolic inhibitors. Pharmacological agents that inhibit cellular oxidation of glucose and fatty acids mimic the effects of food deprivation and suppress various indices of fertility. These drugs do not result in a negative energy balance and, in some species, actually cause a positive energy balance by increasing food intake.

Therefore, nutritional infertility is not invariably associated with absolute levels of energy intake, absolute levels of energy expenditure, negative energy balance, positive energy balance, or any aspect of body size or composition. Rather, it is consistently associated with a decrease in the availability of readily oxidizable metabolic fuels, regardless of the status of these other conditions.

C. Physiological Controls

In attempting to decipher the control of fertility by metabolic fuel availability, a number of questions arise. These are illustrated in Fig. 2, and they include the following: What are the changes in reproductive physiology (Fig. 2, No. 1) and behavior (Fig. 2, No. 2) that are seen in nutritional infertility? How are the neural circuits that control reproductive physiology and behaviors affected by the availability of metabolic fuels (Fig. 2, No. 3)? What are the metabolic cues that signal these neural circuits about the availability of metabolic fuels (Fig. 2, No. 4)? Where are these signals detected (Fig. 2, No. 5)? and How is this information conveyed to the forebrain circuits that control reproduction (Fig. 2, No. 6)? Each of these questions is considered in turn.

II. REPRODUCTIVE PHYSIOLOGY

All levels of the hypothalamo–pituitary–gonadal axis are affected by the availability of metabolic fuels, but the primary dysfunction appears to lie at the level of hypothalamic gonadotropin-releasing hormone (GnRH) neurons (Fig. 2, No. 1). In food-deprived animals both gonadal steroidogenesis and gametogenesis are reduced. However, if deprived animals are treated with gonadotropins, the ovaries and testes appear to respond normally. Similarly, pulsatile secretion of luteinizing hormone (LH) is suppressed when metabolic fuel availability is diminished (Figs. 3–5), but the pituitaries of food-deprived animals respond with normal levels of LH secretion when stimulated with GnRH. Therefore, although both pituitary and gonadal function are altered in nutritional infertility, they are able to respond normally if provided with the appropriate hormonal signals.

Two other lines of evidence are consistent with the hypothesis that decreased GnRH secretion is the cause of reduced pituitary and gonadal function during caloric deprivation. First, direct measurements of hypophyseal portal hormone levels indicate that GnRH secretion is reduced in food-deprived sheep. Second, in hamsters, treatments that suppress ovulatory cycles (i.e., food deprivation, insulin treatment, pharmacological inhibition of glucose, and fatty acid oxidation) also decrease the activation of GnRH-producing neurons.

Reproductive physiologists typically use pulsatile LH release as an index of reproductive function during manipulations of metabolic fuel availability. There are a number of principles that emerge from this work. First, the suppression of pulsatile LH release in fuel-deprived animals is readily reversible

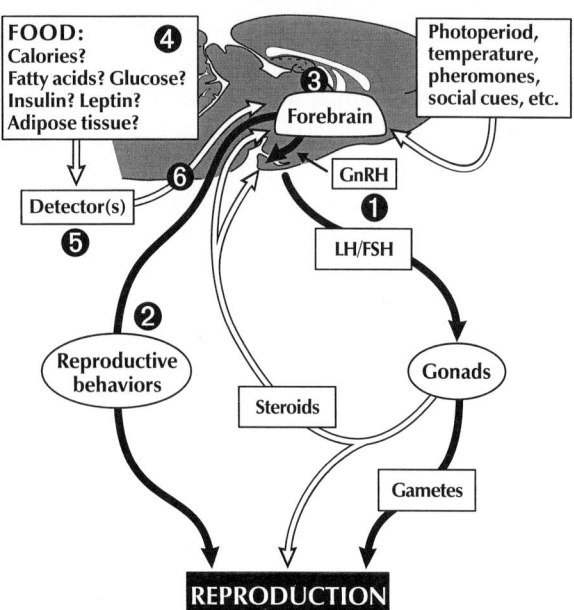

FIGURE 2 Diagram identifying areas of investigation in nutritional controls of reproductive physiology and behavior. GnRH, gonadotropin-releasing hormone; LH, luteinizing hormone; FSH, follicle-stimulating hormone.

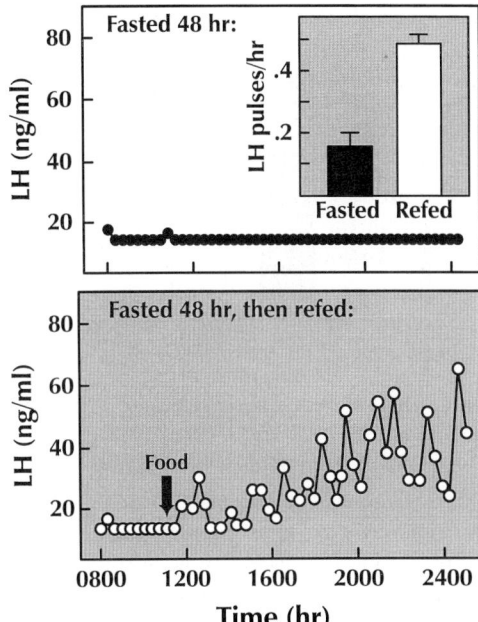

FIGURE 3 Immediate resumption of pulsatile LH release in male rhesus monkeys fed a meal following 48 hr of food deprivation. Inset: Number of LH pulses/hour in the 13 hr following refeeding (modified from D. B. Parfitt, K. R. Church, and J. L. Cameron, Restoration of pulsatile luteinizing hormone secretion after fasting in rhesus monkeys (*Macaca mulatta*): Dependence on size of the refeed meal, *Endocrinology* 129, 749–756, 1991).

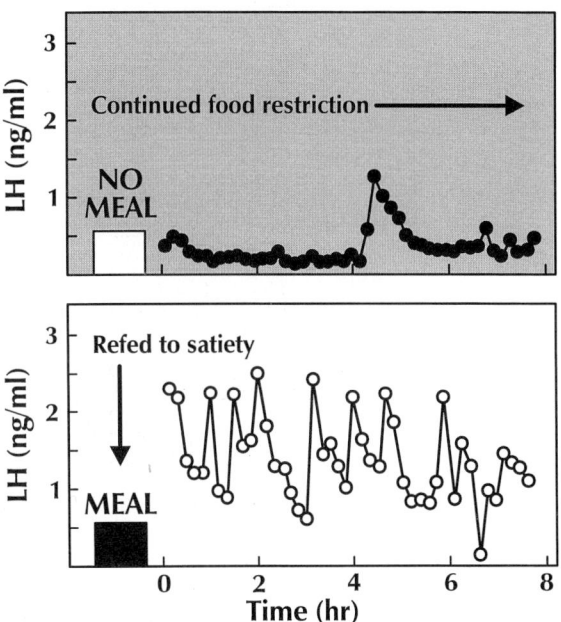

FIGURE 4 Immediate resumption of pulsatile LH release following a single meal to satiety in pigs that had been food-restricted for 7 days (modified from J. R. Cosgrove, P. J. Booth, and G. R. Foxcroft, Opioidergic control of gonadotrophin secretion in the prepubertal gilt during restricted feeding and realimentation, *J. Reprod. Fertil.* 91, 277–284, 1991).

(Figs. 3 and 4). This is consistent with the fact that nutritional infertility ceases once animals are refed. Second, metabolic fuel deprivation inhibits pulsatile LH release by both steroid-dependent and -independent processes. In gonadally intact animals, food deprivation enhances the negative feedback effects of gonadal steroids on LH release. Food deprivation also inhibits pulsatile LH release in gonadectomized animals (i.e., in the absence of gonadal steroids). Third, pulsatile LH release responds rapidly, within minutes or hours, when there are abrupt changes in metabolic fuel availability. Refeeding food-deprived animals causes an immediate resumption of pulsatile LH release (Figs. 3 and 4). Similarly, when metabolic fuel oxidation is inhibited by treatments with metabolic inhibitors, there is an immediate inhibition of pulsatile LH release (Fig. 5). This means that the neural mechanisms that control GnRH and LH release respond to minute-to-minute or hour-to-hour changes in metabolic fuel availability.

Currently, relatively little is known about the neural circuits that convey metabolic cues to forebrain GnRH-secreting neurons (Fig. 2, No. 3). There are a number of neuropeptide systems that respond to changes in metabolic fuel availability and also have been shown to affect eating behavior and various aspects of reproduction. These peptides include neuropeptide Y, galanin, cholecystokinin, corticotropin-releasing hormone, and various opioids, but none has yet been shown conclusively to mediate the effects of nutrition on LH secretion. It is likely that several of them participate in nutritional controls of reproduction.

III. REPRODUCTIVE BEHAVIORS

Reproductive behaviors, including both copulatory and parental behaviors, are inhibited when metabolic fuel availability is reduced (Fig. 2, No. 2). These behavioral changes are due in part to the disruption of secretion of the gonadal hormones that

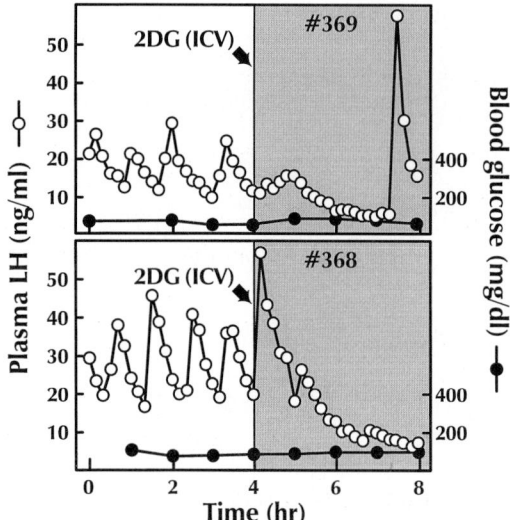

FIGURE 5 Immediate suppression of pulsatile LH release in prepubertal lambs following an intracerebroventricular (ICV) infusion of 2-deoxy-D-glucose (2DG), a drug that inhibits cellular oxidation of glucose (modified from D. C. Bucholtz, N. M. Vidwans, C. G. Herbosa, K. K. Schillo, and D. L. Foster, Metabolic interfaces between growth and reproduction. V. Pulsatile luteinizing hormone secretion is dependent on glucose availability, Endocrinology 137, 601–607, 1996).

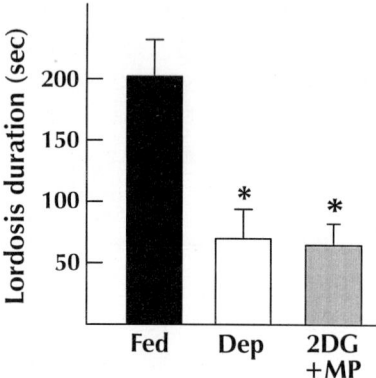

FIGURE 6 Inhibition of sexual receptivity in hamsters following 48 hr of food deprivation (Dep) or treatment with 2-deoxy-D-glucose (2DG) and methyl palmoxirate (MP), drugs that inhibit oxidation of glucose and fatty acids, respectively. Animals were ovariectomized and brought into heat with injections of estradiol and progesterone (modified from R. W. Dickerman, H.-Y. Li, and G. N. Wade, Decreased availability of metabolic fuels suppresses estrous behavior in Syrian hamsters, Am. J. Physiol. 264, R568–R572, 1993).

act on the brain to induce these behaviors. However, they are also due to a reduced neural responsiveness to gonadal hormones. That is, metabolic fuel deprivation inhibits reproductive behaviors in gonadectomized animals that have been treated with controlled amounts of hormones. For example, when ovariectomized hamsters are food deprived or treated with metabolic inhibitors, estrous behavior is diminished following treatment with estradiol and progesterone when compared with *ad libitum*-fed controls (Fig. 6).

These decreases in behavioral responsiveness to ovarian steroids are consistently associated with alterations in levels of neural steroid receptors across numerous nutritional manipulations. Again, in ovari-

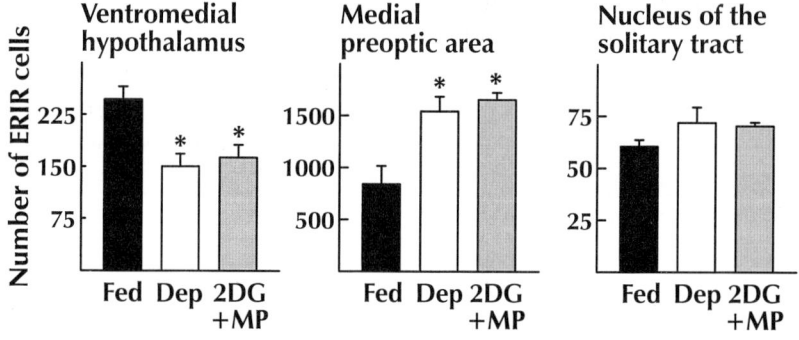

FIGURE 7 Changes in the number of cells containing estrogen-receptor immunoreactivity (ERIR) in three brain regions of hamsters following 48 hr of food deprivation (Dep) or treatment with 2-deoxy-D-glucose (2DG) and methyl palmoxirate (MP), drugs that inhibit oxidation of glucose and fatty acids, respectively (modified from H.-Y. Li, G. N. Wade, and J. D. Blaustein, Manipulations of metabolic fuel availability alter estrous behavior and neural estrogen receptor immunoreactivity in Syrian hamsters, Endocrinology 135, 240–247, 1994).

ectomized hamsters, food deprivation or treatment with metabolic inhibitors decreases estrogen-receptor immunoreactivity (ERIR) in the ventromedial hypothalamus, the primary area where ovarian steroids are thought to act to facilitate estrous behavior (Fig. 7). Thus, it has been hypothesized that the reduced behavioral responsiveness in metabolic fuel-deprived animals may be due in part to reduced neural binding of estradiol (Fig. 2, No. 3).

In other brain areas (e.g., paraventricular nucleus of the hypothalamus, arcuate nucleus, and medial preoptic area) metabolic fuel deprivation increases the number of ERIR-positive cells in hamsters. These are sites at which estradiol can act to affect LH (via GnRH) secretion, raising the possibility that the enhanced negative feedback potency of estradiol on LH release could be due in part to increases in neural estrogen receptor in some loci.

Thus, changes in neural binding could underlie both increased (negative feedback on LH release) and decreased (estrous behavior) responsiveness to estradiol in food-deprived animals. To date, fuel deprivation-induced changes in neural ERIR have been demonstrated in rats, mice, sheep, and hamsters.

IV. METABOLIC CUES AND SENSORY MECHANISMS

A major question in the field of nutritional infertility concerns the nature of the metabolic cues that inform the neural circuits controlling reproductive physiology and behavior as to the availability of usable energy (Fig. 2, No. 4). Three kinds of hypotheses have been advanced, and of course they are not mutually exclusive.

A. Body Fat Content

At one time it was thought that fertility was controlled by body fat stores. That is, if an individual fell below a certain body fat content, he/she became infertile until the percentage body fat again exceeded some hypothetical threshold for reproduction. This hypothesis was based solely on correlative data in human beings. However, when subjected to experimental scrutiny, this hypothesis was found wanting.

Indeed, there is now a large and growing experimental literature that is inconsistent with a critical body fat hypothesis. Although this notion is still widely accepted clinically, it is no longer taken seriously by most investigators who study nutrition–reproduction interactions. Of course, there is a consistently high correlation between body fat levels and reproductive competence, but this simply reflects the fact that fat storage and reproduction both have very low priorities in nutritionally challenged animals.

B. Short-Term Availability of Oxidizable Metabolic Fuels

1. Metabolic Inhibitors and Eating Behavior

Contemporary research in nutritional infertility has drawn heavily on studies of the physiological controls of eating behavior for both concepts and experimental tools. This literature shows that eating behavior is responsive to the minute-to-minute availability of oxidizable metabolic fuels. This has been shown most clearly using pharmacological inhibitors of metabolic fuel oxidation (Fig. 8), including glucose oxidation [e.g., 2-deoxy-D-glucose (2DG) and 5-thio-D-glucose] and fatty acid oxidation [e.g., methyl palmoxirate (MP) and mercaptoacetate]. When satiated rats are treated with any of these drugs, they exhibit a rapid increase in food intake, and a combination of 2DG (glucoprivation) and MP (lipoprivation) is particularly effective at increasing intake.

2. Metabolic Inhibitors and Reproduction

Treatment with these drugs mimics the effects of food deprivation on reproductive physiology. Pulsatile LH secretion is particularly sensitive to glucoprivation, and systemic or intracerebroventricular administration of 2DG rapidly suppresses LH pulses in rats and sheep (Fig. 5). In hamsters, too, treatment with 2DG inhibits ovulatory cycles, and this effect is greatly exacerbated by concurrent MP treatment (glucoprivation + lipoprivation).

Treatment with 2DG + MP mimics the effects of food deprivation on estrous behavior and on neural estrogen receptors (Figs. 6 and 7). Unlike LH secretion, the effects on estrous behavior and neural estrogen receptors require both glucoprivation and lipo-

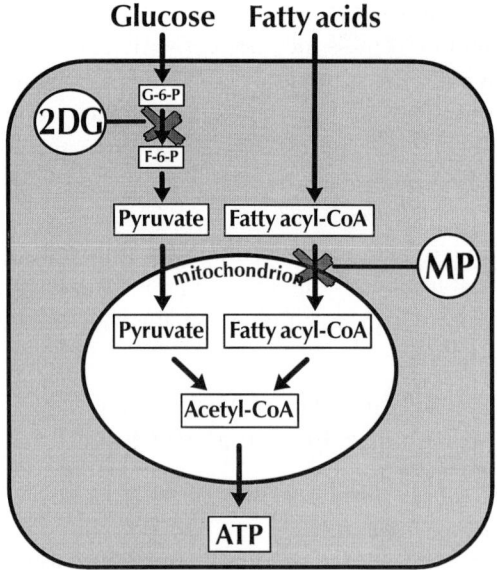

FIGURE 8 The drug 2-deoxy-D-glucose (2DG) inhibits intracellular glucose metabolism and induces glucoprivation. After phosphorylation to 2DG-6-phosphate, 2DG competitively inhibits glycolysis at the phosphohexoseisomerase step, an action that may also inhibit cellular glucose uptake. Methyl palmoxirate (MP) inhibits cellular fatty acid oxidation and induces lipoprivation by binding to carnitine palmitoyltransferase I and blocking transport of long-chain fatty acids into mitochondria. CoA, coenzyme A (modified from Wade et al., 1996).

privation (2DG + MP). Neither drug alone is effective, even when given in very high doses that are sufficient to have other physiological effects. These, and other, data indicate that nutritional effects on reproductive physiology and behaviors are separate and independent responses, although they normally occur concurrently. Thus, reproductive function, like eating behavior, is responsive to short-term (minute-to-minute or hour-to-hour) changes in the availability of oxidizable metabolic fuels.

3. Detection of Metabolic Cues

Where are these metabolic cues detected (Fig. 2, No. 5)? Work on the physiology of food intake indicates that glucoprivic and lipoprivic cues are detected in the viscera (liver and upper GI tract) and in the area postrema (AP), a circumventricular structure in the caudal hindbrain. Lesions of the AP or subdiaphragmatic vagotomy, which prevents transmission of visceral sensory information to the brain, abolish metabolic inhibitor-induced increases in food intake in rats.

The AP and vagus nerves appear to play a similar role in metabolic controls of reproduction. AP lesions prevent the inhibition of estrous behavior (Fig. 6) and the decrease in ventromedial hypothalamic estrogen receptors (Fig. 7) in hamsters deprived of metabolic fuels. Lesions of the AP also prevent glucoprivation (2DG)-induced anestrus in hamsters. Vagotomy prevents the effects of food deprivation on pulsatile LH release in rats and the effects of metabolic inhibitors on preoptic area estrogen receptors (Fig. 7) in hamsters. The fact that AP lesions and/or vagotomy abolish most reproductive responses to metabolic fuel deprivation suggests that the critical metabolic signals are detected in the viscera and hindbrain and then relayed to the forebrain circuits that control GnRH secretion and reproductive behaviors (Fig. 9). There is no evidence that the forebrain effector cells detect these metabolic cues directly.

4. Transmission of Metabolic Cues from Detectors to Effectors

Relatively little is known about the neuroanatomical pathways and neurotransmitters/neuropeptides that convey metabolic information from the visceral and hindbrain detectors to the forebrain effectors (Fig. 2, No. 6). As noted previously, there are a number of neuropeptides that respond to changes in metabolic fuel availability and also affect eating behavior and various aspects of reproduction. It is likely that one or more of these is involved in nutritional controls of reproduction. Similarly, the neuroanatomical pathways connecting the visceral hindbrain and the forebrain have been described in some detail, but precisely which ones mediate nutrition–reproduction interactions remains to be determined.

C. Hormones, Peptides, and Amino Acids

Another possibility is that metabolic fuel availability could be cued by circulating levels of metabolic hormones, such as insulin, growth hormone, glucagon, or cholecystokinin, or by amino acids (which

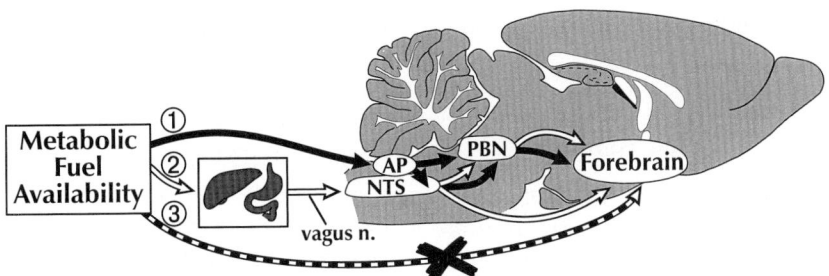

FIGURE 9 Putative pathways by which metabolic fuel availability affects reproductive physiology and behaviors. 1, Some metabolic cues are detected in the caudal hindbrain, probably in the area postrema (AP), and then relayed to the forebrain via the nucleus of the solitary tract (NTS) and the lateral parabrachial nucleus (PBN). 2, Some metabolic cues are detected in the viscera, most likely the liver and upper GI tract, and then relayed to the forebrain via the vagus nerves, the NTS, and the PBN. 3, There is no evidence to support the possibility that the metabolic cues are detected directly by forebrain effector cells (modified from Wade et al., 1996).

may serve as precursors for synthesis of neurotransmitters). These are interesting and viable hypotheses, although the participation of some hormones in some species seems to have been ruled out. Clearly, much more work is needed to test these ideas adequately.

There has been a great deal of interest in the possible role of leptin in nutrition–reproduction interactions. Leptin, the product of the *Ob* gene, is a peptide hormone that is produced by adipocytes, and circulating levels are directly related to body fat content. Animals lacking leptin or its receptors are obese and infertile, and most of the reproductive deficits of leptin-deficient strains (e.g., *ob/ob* mice) are corrected by treatment with the hormone. In addition, leptin treatment can prevent the effects of food deprivation on a number of indices of fertility in wild-type animals. Leptin receptors are found in numerous sites throughout the body, including hypothalamus, hindbrain, pituitary, and the gonads.

There are a number of ways that leptin could act to affect fertility. One possibility is that circulating leptin levels could signal the availability of oxidizable metabolic fuels. If this is the case, then leptin levels should change immediately following abrupt changes in metabolic fuel availability (Figs. 3–5). A second possibility is that circulating leptin levels could fluctuate more slowly but modulate neural responsiveness to more rapidly occurring metabolic cues. Finally, leptin could act centrally and/or peripherally to alter metabolic fuel metabolism and, thus, influence reproduction indirectly. Leptin has significant effects on the partitioning and utilization of metabolic fuels in a number of tissues. None of these possibilities has been examined in detail.

V. SUMMARY

Reproduction is critically dependent on an abundant supply of oxidizable metabolic fuels. Nutritional infertility results when metabolic fuels are unavailable due to reduced food intake (e.g., famine and eating disorders), excessive energy expenditure by other physiological processes (e.g., exercise and thermoregulation), or disturbances of fuel partitioning (e.g., obesity and poorly controlled diabetes mellitus). Nutritional effects on reproductive physiology are mediated by GnRH-secreting neurons in the forebrain, whereas suppression of reproductive behaviors seems to be due, at least in part, to changes in steroid receptors in neural tissues. Reproductive physiology and behaviors respond to short-term (minute-to-minute or hour-to-hour) changes in metabolic fuel oxidation rather than to any aspect of body size or composition. These metabolic cues seem to be detected in the viscera and in the hindbrain and then transmitted to forebrain GnRH-secreting or steroid-binding effector neurons. There is no evidence for direct detection of metabolic cues by the forebrain effector cells. Recent work suggests that the peptide hormone, leptin, could participate in nutritional controls of reproduction, but the mechanisms by which leptin might act have yet to be elucidated.

See Also the Following Articles

GnRH (GONADOTROPIN-RELEASING HORMONE); LH (LUTEINIZING HORMONE); NUTRITIONAL FACTORS AND LACTATION; NUTRITIONAL FACTORS AND REPRODUCTION; PHOTOPERIODISM, VERTEBRATES; SEASONAL REPRODUCTION

Bibliography

Bronson, F. H. (1989). *Mammalian Reproductive Biology*. Univ. of Chicago Press, Chicago.

Cameron, J. L. (1996). Nutritional determinants of puberty. *Nutr. Rev.* **54**, S17–S22.

Flier, J. S. (1997). Leptin expression and action: New experimental paradigms. *Proc. Natl. Acad. Sci. USA* **94**, 4242–4245.

Friedman, M. I. (1990). Making sense out of calories. In *Handbook of Behavioral Neurobiology 10. Neurobiology of Food and Fluid Intake* (E. M. Stricker, Ed.), pp. 513–529. Plenum, New York.

I'Anson, H., Foster, D. L., Foxcroft, G. R., and Booth, P. J. (1991). Nutrition and reproduction. *Oxford Rev. Reprod. Biol.* **13**, 239–311.

Kalra, S. P., and Kalra, P. S. (1996). Nutritional infertility: The role of the interconnected hypothalamic neuropeptide Y-galanin-opioid network. *Front. Neuroendocrinol.* **17**, 371–401.

Loewy, A. D. (1990). Central autonomic pathways. In *Central Regulation of Autonomic Functions* (A. D. Loewy and K. M. Spyer, Eds.), pp. 88–103. Oxford Univ. Press, New York.

Loucks, A. B., Vaitukaitis, J., Cameron, J. L., Rogol, A. D., Skrinar, G., Warren, M. P., Kendrick, J., and Limacher, M. C. (1992). The reproductive system and exercise in women. *Med. Sci. Sports Exercise* **24**, S288–S293.

Maeda, K.-I., and Tsukamura, H. (1996). Neuroendocrine mechanism mediating fasting-induced suppression of luteinizing hormone secretion in female rats. *Acta Neurobiol. Exp.* **56**, 787–796.

Ritter, S., Calingasan, N. Y., Hutton, B., and Dinh, T. T. (1992). Cooperation of vagal and central neural systems in monitoring metabolic events controlling feeding behavior. In *Neuroanatomy and Physiology of Abdominal Vagal Afferents* (S. Ritter, R. C. Ritter, and C. D. Barnes, Eds.), pp. 249–277. CRC Press, Boca Raton, FL.

Wade, G. N., and Schneider, J. E. (1992). Metabolic fuels and reproduction in female mammals. *Neurosci. Biobehav. Rev.* **16**, 235–272.

Wade, G. N., Schneider, J. E., and Li, H.-Y. (1996). Control of fertility by metabolic cues. *Am. J. Physiol.* **270**, E1–E19.

Environmental Estrogens

Stephen H. Safe
Texas A&M University

I. Introduction
II. Different Structural Classes of Xenoestrogens
III. Bioassays for Characterizing Xenoestrogens
IV. Xenoestrogens as Estrogen Receptor Agonists and Antagonists
V. Xenoestrogen Interactions and Human Health Effects

GLOSSARY

endocrine disruptors Anthropogenic or natural chemicals that modulate an endocrine response pathway.

ER_α and ER_β Two forms of the estrogen receptor.

xenoestrogens Anthropogenic compounds that bind to the ER and elicit ER agonist or antagonist responses.

I. INTRODUCTION

During the 1990s, there has been considerable public, regulatory, and scientific concern regarding the potential adverse effects of environmental and human exposure to a class of anthropogenic compounds designated as endocrine disruptors. Several reports have suggested that environmental estrogenic compounds (xenoestrogens) may contribute to decreased wildlife reproduction in contaminated areas, a global decrease in male reproductive capacity, and the increased incidence of breast cancer women. In 1992, results of a meta-analysis of 61 sperm-count studies showed a nearly 50% worldwide decrease in sperm counts between 1940 and 1990. It was then hypothesized that industrial-derived and naturally occurring estrogenic compounds and other endocrine-active chemicals may be responsible for a global decrease in sperm counts and male reproductive capacity. This hypothesis is based, in part, on human and laboratory animal effects associated with *in utero* exposure to the potent estrogenic drug diethylstilbestrol which causes a diverse spectrum of reproductive problems in male offspring. However, it has recently been reported that sperm counts have decreased in some studies but not others. Moreover, it is now known that male sperm counts can be variable within the same country suggesting that a global decrease is unlikely. The reasons for temporal and region-specific differences in male sperm counts are unknown and require further investigation.

Although lifetime exposure to endogenous estrogens is a risk for breast cancer in women, the role of xenoestrogens as preventable causes of breast cancer has been questioned. This chapter briefly outlines the structures and hormonal activity of the major structural classes of xenoestrogens.

II. DIFFERENT STRUCTURAL CLASSES OF XENOESTROGENS

Synthetic estrogenic compounds were initially developed for use as therapeutic agents; however, the estrogenic activity of diverse industrial products, including 1,1,1-trichloro-2-(*p*-chlorophenyl)-2-(*o*-chlorophenyl)ethane, kepone, phenolics, and polychlorinated biphenyl (PCB) mixtures, has been known for several decades. In 1988, it was also reported that hydroxy PCBs exhibited estrogenic activity and, with recent development of estrogen-responsive short-term *in vitro* bioassays, the number of different structural classes of compounds (Fig. 1) that exhibit estrogenic activity has significantly expanded and includes phthalates, other phenolics (e.g., *p*-alkylphenols and bisphenol A), PCB congeners, polynuclear aromatic hydrocarbons, several organochlorine insecticides (toxaphene, endosulfan, chlordane, and dieldrin), and *t*-butylhydroxyanisole. These results clearly demonstrate that the estrogen receptor exhibits a surprising lack of binding specificity for various chemical classes, some of which exhibit minimal structural similarity to the natural hormone, 17β-estradiol (E2).

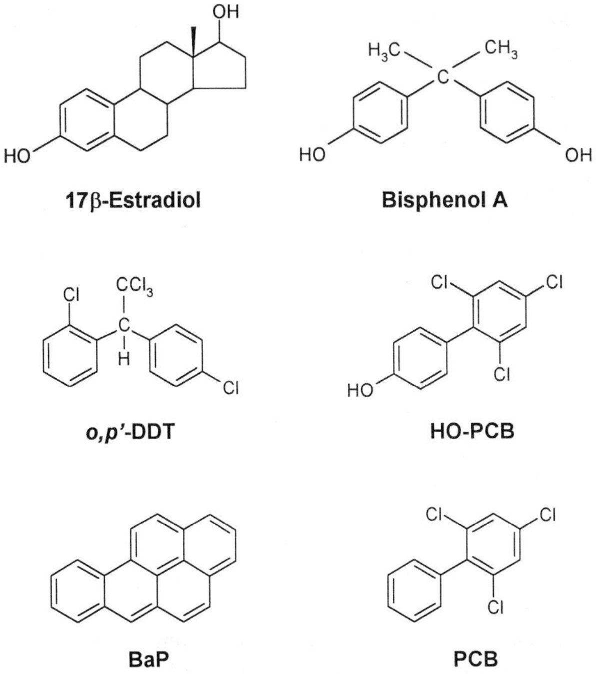

FIGURE 1 Structures of different classes of compounds which bind the estrogen receptor.

III. BIOASSAYS FOR CHARACTERIZING XENOESTROGENS

The recent concern regarding the hypothesized adverse health impacts of xenoestrogens has resulted in development of a host of short-term *in vitro* bioassays for detecting these compounds. Some of these assays include estrogen receptor (ER) binding using recombinant ER, MCF-7 cell proliferation ("E-screen"), expression of E2-induced genes and related proteins/activities, yeast-based assays transformed with mammalian ER and promoter–reporter constructs, and assays using chimeric ER promoter–reporter constructs. These assays, coupled with more traditional *in vivo* uterine/vaginal responses, have been extensively used to identify the increasing number of xenoestrogens.

The structure-dependent ER binding and ER agonist activities of E2 and related steroidal estrogens has been extensively investigated. For most assays, the different structural classes of xenoestrogens exhibit one common property, namely, these compounds elicit weak estrogenic activity as either full or partial agonists (assay dependent) and are >100 times less potent than E2. Despite the lower ER agonist activities of most classes of xenoestrogens, this does not preclude the importance of various structural determinants that govern estrogenic potency. For example, based on an E2-responsive yeast-based assay for a series of alkylphenols, it was found that alkyl substitution (*para* > *meta* > *ortho*), degree of alkyl branching (tertiary > secondary = normal), and number of carbon atoms (six to eight were maximal) markedly influenced estrogenic potencies of these analogs. Recent studies have identified a second form of the ER (i.e., ER_β) and structure–ER_β binding studies for a series of steroidal estrogens and related compounds were similar for both forms of the ER. The tissue-specific role of ER_β is unknown and, therefore, the importance of this new intracellular receptor as a target for xenoestrogens requires further study.

IV. XENOESTROGENS AS ESTROGEN RECEPTOR AGONISTS AND ANTAGONISTS

The estrogenic activity and potency of various xenoestrogens will depend on a number of factors, including pharmacokinetics and metabolism, serum levels, binding to serum proteins, binding to intracellular proteins, and target organ/tissue. For example, bisphenol A and nonyl/octyl-phenols exhibit weakly estrogenic activity in most *in vitro* bioassays with a potency of 0.001 compared to E2. Adult male mice exposed *in utero* to bisphenol A exhibited increased prostate weights, whereas comparable doses of octylphenol were inactive. In contrast, our studies showed that nonylphenol was a potent estrogen in the immature female rat uterus, whereas bisphenol A was only weakly estrogenic and, in combination with E2, exhibited antiestrogenic activity. In a battery of *in vitro* assays in MCF-7 and other cancer cell lines, seven hydroxy-PCB congeners identified in human serum also exhibited antiestrogenic activity. These studies indicate that ER-mediated activities of xenoestrogens are response/species-specific and both ER agonist and antagonist effects can be observed.

V. XENOESTROGEN INTERACTIONS AND HUMAN HEALTH EFFECTS

The potential impacts of xenoestrogens on human health are unknown; however, their effects will be dependent on levels of exposure, potency, and target cell/organ uptake during critical periods of development. It has previously been pointed out that humans are exposed to relatively high levels of natural estrogenic compounds in foods as well as other natural and xenoendocrine disruptors that inhibit E2-induced responses. Thus, adverse health impacts of relatively trace levels of xenoestrogens in the diet will have to be induced over the high background of natural estrogens/antiestrogens. A recent study reported that binary mixtures of weakly estrogenic pesticides (toxaphene, chlordane, endosulfan, and dieldrin) exhibited up to 1600-fold higher estrogenic activities in yeast-based assays than predicted from the effects of the compounds alone. These synergistic interactions heightened concern regarding the potential human health impacts of xenoestrogens. However, recent studies using a battery of other *in vivo* and *in vitro* assays did not observe any synergistic interactions among the same set of organochlorine pesticides, and the original paper was recently withdrawn. The potential hazards and risks of xenoestrogens and other endocrine disruptors on human health are controversial and various aspects of this problem are being extensively investigated worldwide.

See Also the Following Articles

BREAST CANCER; ESTROGENS, OVERVIEW

Bibliography

Ahlborg, U. G., Lipworth, L., Titusernstoff, L., Hsieh, C. C., Hanberg, A., Baron, J., Trichopoulos, D., and Adami, H. O. (1995). Organochlorine compounds in relation to breast cancer, endometrial cancer, and endometriosis: An assessment of the biological and epidemiological evidence. *Crit. Rev. Toxicol.* **25**, 463–531.

Carlsen, E., Giwercman, A., Keiding, N., and Skakkebaek, N. E. (1992). Evidence for the decreasing quality of semen during the past 50 years. *Br. Med. J.* **305**, 609–612.

Colborn, T., Vom Saal, F. S., and Soto, A. M. (1993). Developmental effects of endocrine-disrupting chemicals in wildlife and humans. *Environ. Health Perspect.* **101**, 378–384.

Davis, D. L., Bradlow, H. L., Wolff, M., Woodruff, T., Hoel, D. G., and Anton-Culver, H. (1993). Medical hypothesis: Xenoestrogens as preventable causes of breast cancer. *Environ. Health Perspect.* **101**, 372–377.

Safe, S. (1995). Environmental and dietary estrogens and human health—Is there a problem? *Environ. Health Perspect.* **103**, 346–351.

Sharpe, R. M., and Skakkebaek, N. F. (1993). Are oestrogens involved in falling sperm counts and disorders of the male reproductive tract. *Lancet* **341**, 1392–1395.

Soto, A. M., Sonnenschein, C., Chung, K. L., Fernandez, M. F., Olea, N., and Serrano, F. O. (1995). The E-screen assay as a tool to identify estrogens—An update on estrogenic environmental pollutants. *Environ. Health Perspect.* **103**, 113–122.

Zacharewski, T., Berhane, K., and Gillesby, B. (1995). Detection of estrogen- and dioxin-like activity in pulp and paper mill black liquor effluent using in vitro recombinant receptor/reporter gene assays. *Environ. Sci. Technol.* **29**, 2140–2146.

ISBN 0-12-227021-5

90038